2025 IEEE Applied Power Electronics Conference and Exposition (APEC 2025)

Atlanta, Georgia, USA
16-20 March 2025

Pages 1959-2653

IEEE Catalog Number: CFP25APE-POD
ISBN: 979-8-3315-1612-3

**Copyright © 2025 by the Institute of Electrical and Electronics Engineers, Inc.
All Rights Reserved**

Copyright and Reprint Permissions: Abstracting is permitted with credit to the source. Libraries are permitted to photocopy beyond the limit of U.S. copyright law for private use of patrons those articles in this volume that carry a code at the bottom of the first page, provided the per-copy fee indicated in the code is paid through Copyright Clearance Center, 222 Rosewood Drive, Danvers, MA 01923.

For other copying, reprint or republication permission, write to IEEE Copyrights Manager, IEEE Service Center, 445 Hoes Lane, Piscataway, NJ 08854. All rights reserved.

*** This is a print representation of what appears in the IEEE Digital Library. Some format issues inherent in the e-media version may also appear in this print version.

IEEE Catalog Number:	CFP25APE-POD
ISBN (Print-On-Demand):	979-8-3315-1612-3
ISBN (Online):	979-8-3315-1611-6
ISSN:	1048-2334

Additional Copies of This Publication Are Available From:

Curran Associates, Inc
57 Morehouse Lane
Red Hook, NY 12571 USA
Phone: (845) 758-0400
Fax: (845) 758-2633
E-mail: curran@proceedings.com
Web: www.proceedings.com

TABLE OF CONTENTS

Versatile Controller Architecture for a Universal DC Fast Charging Front-End 1
    Anurag Singh, Sayan Paul, Tejas Bhuse, Trent Martin, Hien Nguyen, Inder Vedula, Nikola Milivojeviæ, Dragan Maksimoviæ, Luca Corradini

A 10 kV SiC MOSFET based Three-Phase Single-Stage Isolated MVAC/LVDC Converter for Solid State Transformer Applications 9
    Anup Anurag, Chi Zhang, Rudy Wang, Peter Barbosa

Direct Digital Control Applied to T-Type Vienna Rectifiers for Power Factor Correction 16
    Jun-Yang Chang, Tsai-Fu Wu, Chien-Chih Hung, Jui-Yang Chiu

Active Power Decoupling Method based on Dual Active Bridge Converter without Additional Components 21
    Kosuke Takeuchi, Takashi Ohno, Hiroki Watanabe, Yuki Nakata, Jun-Ichi Itoh

An ANPC-Based Building Block for Medium-Voltage Applications 27
    Ahmed Rahouma, Hui Cao, David A. Porras, Zhuxuan Ma, Yue Zhao, Juan C. Balda

Analog Control of a 2.5 kW GaN based CRM PFC with Input Filter Optimization 34
    Naveed Ishraq, Ayan Mallik

An iTHD and Efficiency Optimized Control Method for Triangular Conduction Mode Totem Pole Bridgeless PFC with Zero Current Detection 42
    Brent McDonald, Sheng-Yang Yu

Resonance Current Suppression for AC-DC Active-Clamp Flyback Converter by Triangular Current Mode 48
    Yasuo Uchida, Hiroki Watanabe, Jun-Ichi Itoh

A Universal DC Fast Charging Front-End with Optimized Film Capacitor Design 54
    Sayan Paul, Anurag Singh, Tejas Bhuse, Trent Martin, Hien Nguyen, Inder Vedula, Nikola Milivojeviæ, Dragan Maksimoviæ, Luca Corradini

Power Characterization of a 1200-V/800-V 22-kVA 30-kHz Unity-Gain Dual-Active-Bridge Converter Prototype 62
    Radhika Sarda, Abishek Sethupandi, Madasamy Palavesha Thevar, Howe Li Yeo, Praveenkumar Palani, Vaisambhayana B. Sriram, Anshuman Tripathi

Design of Fully Soft-Switched Semi-Dual Active DC-DC Converter for Battery Charging Application 69
    Siva Prabhakar, Shiladri Chakraborty, Sandeep Anand

A ZCS-ZVS Strategy for Low Impedance Dual Active Bridges in MHz Range 77
    Pushkar Saraf, Michael Solomentsev, Alex Hanson

A 6.6 kW Highly Efficient Reconfigurable Dual Active Bridge Converter Designed using Planar Transformer, SiC-Fets and Monolithic Bidirectional Devices 90
    Reza Barzegarkhoo, Fabian Groon, Arkadeb Sengupta, Marco Liserre

Interleaved Switched-Inductor-Based SIPO Partial Power Converter Module for Battery Management Systems 98
    Fengwang Lu, Henry Shu-Hung Chung

Single Sensor-Based Fault Localization and Detection in GaN Three-Phase Dual Active Bridge Converters 103
  Satyam Sa, Yi Han, Cheng Feng Wang, Olivier Trescases

Enhanced Cocharge Operation Scheme in Bidirectional PhaseShift Full-Bridge Converters with Eliminated Voltage Overshoot and Reduced Freewheeling Current..111
  Tien-Sheng Li, Minh Ngo, Rolando Burgos, Dong Dong

DC Bias Elimination in Isolated DC-DC Converters using Fundamental-Frequency Ripple 118
  Arkadeb Sengupta, Thiago Antonio Pereira, Marco Liserre

Tunable Matching Network with Dual Phase-Switched Impedance Modulation Actuators............ 124
  Alexander Jurkov, David Perreault

Soft-Switched Pulsed Bias Plasma Supply System 132
  Julia Estrin, Alexander Jurkov, David J. Perreault

Analysis and Design of a Cyclo-Active-Bridge Inverter for Single-Stage Three-Phase Grid Interface.. 139
  Mian Liao, Tanuj Sen, Yang Wu, Minjie Chen

Modular Nanosecond Pulse Generator Leveraging GaN and SiC for Versatility and Performance 147
  P. Briz, H. Sarnago, O. Lucía

A Variable Frequency Technique for EMI and Efficiency Improvements in High-Level Count Flying Capacitor Multilevel Converters 151
  Francesca Giardine, Sahana Krishnan, Logan Horowitz, Robert C. N. Pilawa-Podgurski

Analysis and Implementation of Minimum-Sensor Capacitor Voltage Estimators for Flying Capacitor Multilevel Converters 157
  S. Tahmid Mahbub, Rahul K. Iyer, Ivan Z. Petriœ, Robert C. N. Pilawa-Podgurski

Single-Stage Bidirectional High-Frequency Link DC to Three-Phase AC (4-Wire) Grid-Tied Microinverter.. 164
  Aniruddh Marellapudi, Satish Belkhode, Joseph Benzaquen Sune, Deepak Divan

Analysis and Design of a Constant Current LCC Class-E Inverter 171
  Ju Gao, Ziheng Liu, Jiayin He, Hongjie Peng, Chengkang Ao, Jinyan Wang

Series Connected Class-E Push-Pull Converters using GaN HEMT for High-Efficiency RF Generators in Float Zone Silicon Production 175
  Faheem Ahmad, Thore Stig Aunsborg, Jannick Kjær Jørgensen, Stig Munk-Nielsen

State of the Art 1.7kV Lateral GaN HEMTs, an Alternative to SiC 180
  Karthick Murukesan, Robert Yang, Kamal Varadarajan, Sorin Georgescu, Doug Kang

Modeling and Characterization of Current and Future 1.2 kV Wide Bandgap Semiconductor-Based MOSFETs.. 185
  Sushanta Gautam, Austin M. Szczublewski, Samuel K. Atwimah, Aidan P. Fox, William M. Collings, Tolen Nelson, Daniel G. Georgiev, Raghav Khanna, Andrew D. Koehler, Karl D. Hobart

2.5-kV 6.4-ns 100-kHz Repetitive GaN Marx Generator.. 192
  Ruize Sun, Ci Pan, Wanjun Chen, Bo Zhang

Novel Dual Output LDO Architecture in 650-V GaN Technology for Power ICs 195
  Plinio Bau, Thanh Hai Phung, Deniz Aygun, Bart Coomans, Mike Wens

Impact of Substrate Bias on the Stability of Bidirectional GaN HEMT in Hard- and Soft-Switching 202
Qihao Song, Hongchang Cui, Qiang Li, Yuhao Zhang

Characterization of LED Driven GaN-Based Photoconductive Switches 207
Samuel K. Atwimah, Tolen M. Nelson, Geoffrey M. Foster, Daniel G. Georgiev, Andrew D. Koehler, Alan G. Jacobs, Karl D. Hobart, Micheal R. Hontz, Raghav Khanna

Development and Validation of Repetitive Transient Gate Overvoltage Rating for GaN HEMTs 214
Ricardo Garcia, Angel Espinoza, Siddhesh Gajare, Shengke Zhang

Junction Temperature Monitoring of GaN HEMT by using On-Resistance with Voltage Clamp and
Current Shunt 219
Xiao Wang, Mingrui Zou, Jiakun Gong, Yulei Wang, Zheng Zeng

False Turn-On Failure and Protection of p-Gate GaN HEMT in MHz Class-E Resonant Inverter 225
Ziheng Liu, Ju Gao, Hongjie Peng, Jiayin He, Jinyan Wang, Maojun Wang

Heat Extraction from Ferrite Cores using Metallic Laminations 231
Alyssa Brown, Duy T. Nguyen, Alex J. Hanson

Folded Flex-PCB Winding Planar Transformer for High-Frequency Isolated DC-DC Converters 238
Soundhariya G. Soundararajan, Hans Wouters, Wout Vanderwegen, Wilmar Martinez

Winding Strategy Analysis and Optimization for High-Current Matrix Transformer 246
Bima Nugraha Sanusi, Pinhe Wang, Michael A. E. Andersen, Ziwei Ouyang

Investigation on Impact of Transformer Parasitic Capacitance on Standby Power Consumption in
Power Converters 252
Kamran Kamran, Andrea Russo, Federica Cammarata, Claudia Malannino, S. Yuri Ciardo, Ziwei Ouyang

PCB-Winding Integrated Transformer for 800-V Dual Active Bridge Converter using 1.2-kV GaN
Devices 258
Hans Wouters, Wei-Ren Lin, Nicolas Pirson, Thomas Jochmans, Yu Zuo, Wilmar Martinez

Comparative Assessment of Inductance Modeling for PCB-Based Circular Spiral Coils in Inductive
Power Transfer Systems 266
Gaia Petrillo, Drazen Dujic

Compact Air-Core Inductors for Variable Frequency Soft-Switching in 3 Phase Inverters 272
Youssef A. Fahmy, Matthias Preindl

Simulation and Experimental Research on Cooling Performance of Fully-Immersed Evaporative
Cooling High-Frequency Transformer 278
Zhanlei Liu, Lingyu Zhu, Yuntian Gao, Yongliang Dang, Cao Zhan, Shengchang Ji

High-Efficiency PCB-Embeddable Inductor for Vertical Power IVR Applications 285
Youssef Kandeel, Liang Ye, John Flannery, Cian Ó Mathúna, Ranajit Sai, Seamus O'Driscoll, Takayuki Tsuchida, Naoya Terauchi, Sumiaki Kishimoto, Toshio Hiraoka, Masanori Nagano

An Adaptive Zero Current Switching Control Technique for Multi-Resonant Switched-Capacitor
Converters 291
Haifah B. Sambo, Rose A. Abramson, Sahana Krishnan, Robert C. N. Pilawa-Podgurski

Small-Signal Analysis and External Ramp Design for Multiphase Current-Mode Constant On-Time
Control with Phase Overlapping 299
Sundaramoorthy Sridhar, Qiang Li

Multiphase Constant-on-Time Minimum-Deviation Controller for Modern Processors 307
Duo Li, Gianluca Roberts, Aleksandar Prodić, Alan Wu

Closed-Loop Control of a Dual-Side Series/Parallel Piezoelectric-Resonator-Based DC-DC
Converter 315
Wen-Chin B. Liu, Gaël Pillonnet, Patrick P. Mercier

High-Bandwidth Embedded Rogowski Coil on Multilayer Substrate with Minimal Contribution to
Power Loop Inductance 321
Takahiro Okamoto, Masataka Ishihara, Kazuhiro Umetani, Eiji Hiraki

Operating and Switching Frequency Circulating Current Control in Paralleled High Power
Adjustable Speed Drives with Common DC Link 327
Kevin Lee, Zhihao Song, Wenxi Yao, Bo Wei

Mixed-Signal Sliding Mode Controller for Non-Inverting Buck-Boost Photovoltaic DC Optimizers 334
Anurag Singh, Sayan Paul, Dragan Maksimović, Luca Corradini

A Current Sensorless Output Voltage Tracking Controller-Observer for a Boost Inverter using
Feedback Linearization 342
*Ion Leandro Dos Santos, Tailan Orlando, Yohannes Amilcar Tekle Scherer, Telles Brunelli
Lazzarin, Hector Bessa Silveira*

Modeling and Control of a Cyclo-Active-Bridge Inverter for Single-Stage Three-Phase Grid
Interface 349
Tanuj Sen, Mian Liao, Yang Wu, Minjie Chen

Turn-On Transient Modeling of 10 kV SiC MOSFET Half-Bridge Power Module in LTspice 357
*Nianzun Qi, Jannick Kjær Jørgensen, Gao Liu, Zhixing Yan, Morten Rahr Nielsen, Asger
Bjørn Jørgensen, Hongbo Zhao, Stig Munk-Nielsen*

A Compact, Automated Sawyer-Tower System for Characterization of the High-Frequency, Soft-
Switching C_{oss} Loss of Wide Bandgap Devices 363
Katherine Liang, Malachi Hornbuckle, Juan Rivas-Davila

Enhancing Behind-the-Meter Visibility of Grid Edge PV Systems and Electric Vehicle Charging
Loads Through Integration of Compact Low-Cost Sensors 370
Mehrnaz Madadi, Paul Ohodnicki, Subhashish Bhattacharya

Supercapacitor based TMS Pulse Generator Design-Experimental Results Versus MATLAB
MOSFET Simulation Model 378
Soniya Raju, Nihal Kularatna, Marcus Wilson, Alistair Steyn-Ross

Application of Artificial Intelligence for Modeling SiC Power MOSFETs 385
Fredo Chavez, Danial Bavi, Sourabh Khandelwal

Multi-Objective Design Automation in Power Electronics using Bayesian Optimization Techniques 389
Tung-Tan Nguyen, Man-Hay Pong, Huang-Jen Chiu

Reduced Order Thermal Modelling of Multi-Chip Silicon Carbide Power Modules 395
Aamir Rafiq, Blake Nelson, Marshal Olimmah

Design and Evaluation of Dual-Resolver Emulation for Control System Verification in Aerospace
Actuation Applications 401
*Tomas Sadilek, Julian Opificius, Jason Wright, Alec Leslie, Jeremie Tuzizila, Cesar Alzate,
Hunter Burnett, Joshua Atkinson, Justin Stricula*

Un-Terminated Blackbox Modeling for Electric Machines...409
  Xinliang Yang, Vladimir Mitrovic, Qing Lin, Rolando Burgos

7.2 kW GaN-Based DAB Converter with 37 kW/L Power Density and High Efficiency...............................416
  Esmaeil Jalalabadi, Xiaoyu Wang, Jaksa Rubinic, Yang Jiao, Lucas Lu

A Novel Interleaving Method for High Power Integrated Electric Vehicle Charger with Three-Phase
Permanent Magnet Synchronous Motor 423
  Ryota Tanaka, Toshihiro Kai, Kenta Takishima, Yoshiyuki Nagai, Tetsuya Hayashi, Kantaro
  Yoshimoto

A Three-Phase CLLC Resonant Converter with Integrated Planar Magnetics for 22-kW On-Board
Chargers...429
  Tianlong Yuan, Zhangwei Xiang, Abdelrahman Ali, Feng Jin, Qiang Li, Wendell Da-Cunha-
  Alves, Xiaoshan Liu

Reconfigurable LLC Resonant Converter for Wide Voltage Range and Reduced Voltage Stress in
DC-Connected EV Charging Stations 436
  Yu Zuo, Xiaobing Shen, Bangli Du, Qingcheng Sui, Tim Geboers, Wilmar Martinez

Design and Control of GaN based Three-Phase / Single-Phase Combo Three-Level Flying
Capacitor PFC for OBC Applications...442
  Nidhi Haryani, Laszlo Huber, Anup Anurag, Juan Ruiz, Peter Barbosa

Optimization Strategy for Battery Electric Vehicle (BEV) DC Fast Charging (FC) in Cold
Environments...449
  Seif Sarofim, Cheng Feng Wang, Satyam Sa, Avram Kachura, Isaac Muscat, Olivier Trescases

DC-Link Voltage Reduction with Synergetic Common-Mode Voltage Control of Single-Phase Two-
Stage Non-Isolated EV Chargers...457
  Dongsu Lee, Juwon Lee, Jung-Ik Ha

DC-DC Converter Architecture for Fast Electric Vehicle (EV) Battery Charging Applications 464
  Shibaji Basu, Arjun Ivimey, Praveen Jain

Fast Simulator for the Estimation of Inverter DC-Link Temperature in e-Drives Subjected to Highly
Variable Working Cycles...472
  Simone Giuffrida, Fabio Mandrile, Radu Bojoi

A Monolithic Regulated 160 MHz Resonant DC-DC Converter 479
  Giacomo Ripamonti, Stefano Michelis, Georgios Bantemits, Pablo Daniel Antoszczuk, Khalil
  Khalife, Nils Hans Van Der Blij, Sokratis Koseoglou, Mattia Balutto, Francesco Driussi,
  Stefano Saggini

Reconfigurable Trans-Inductor Voltage Regulator with Improved Light Load Efficiency in Data
Center Applications 485
  Ziyao Wang, Zehui Li, Haoyu Wang

Fully Integrated Voltage Regulators (FIVRs) with Package In-Situ Coupled CoaxMIL Inductor for
High Power Density Microprocessor Applications 491
  Jaeil Baek, Beomseok Choi, Siddharth Kulasekaran, Huong Do, Brandon Marin, Jose
  Chavarria, Leigh Wojewoda, Kaladhar Radhakrishnan

Multiphase Lateral Flux Indirect Coupled Inductor for Vertical Power Delivery Voltage Regulator
Module 498
  Adhistira M. Naradhipa, Qiong Wang, Qiang Li

A High Density Three-Level Quadratic Buck Hybrid Converter for 48V-to-PoL Conversion......................... 505
    Kejia Wang, Si Yuan Sim, Yin Quen Choong, Xin Zhang, Sriharsh Pakala, Cheng Huang

Air-LEGO: A Magnetic-Free Ultra-Thin 24V-to-1V 120A VRM with Air-Coupled Inductors........................ 510
    Haoran Li, Wenliang Zeng, Youssef Elasser, Minjie Chen

A 15A 48V-Input Dual-Path Hybrid Dickson Converter with 6 mm³ Low Saturation Current
Inductors for Point-of-Load Conversion 518
    Hua Chen, Young-Seok Noh, Minxiang Gong, Vivek De, Arijit Raychowdhury

An Ultra-Fast Control Strategy and Pre-Current-Balancing Measures Prepared for Rapid Transients
in Constant On-Time Controllers 524
    Yijie Qian, Yuan Gao, Wenze Shu, Lingyun Li, Shen Xu, Weifeng Sun

Loosely Coupled Trans-Inductor Voltage Regulator (LC-TLVR) Inductor as Compensation Inductor
(Lc)... 530
    Pavan Kumar, Arturo Sanchez Hernandez

Novel Complex Permeability Model of Powder Magnetic Materials... 538
    Lukas Mueller, James Cox, Jun Wang, Enrique Garcia

Design Study Evaluating Impact of Gap Loss on Nanocrystalline Inductor Cores with Experimental
Validation 544
    Maurice Sturdivant, Brandon Grainger, Christopher Bracken, Paul R. Ohodnicki

A Permanent Magnet Variable Inductor for DC Fault Current Limiting Applications 552
    Mark Nations, Subhashish Bhattacharya

Design-Oriented Modeling and Multi-Objective Optimization of Two-Phase Coupled Inductors in
Multiphase PWM Converters 558
    Yicheng Zhu, Jiarui Zou, Robert C. N. Pilawa-Podgurski

MagNetX: Extending the Magnet Database for Modeling Power Magnetics in Transient 566
    Hyukjae Kwon, Shukai Wang, Haoran Li, Youssef Elasser, Gyeong-Gu Kang, Daniel Zhou,
    Davit Grigoryan, Minjie Chen

Non-Monotonic Influence of DC Bias on Ferrite Core Loss Up to 10 MHz with Sine Wave
Excitation 573
    Bohua Zhang, Martin Pfost

Comprehensive SPICE Model for Inductors Considering Magnetic Losses Under DC Bias Current 579
    Yuki Sato, Hirokazu Matsumoto, Junichi Kotani, Shohei Tomioka, Kenichiro Tanaka

Indented Core to Reduce and Desensitize Inductor's Fringing Losses without Increasing Volume................ 586
    Rajaie Nassar, Promit Datta, Guo-Quan Lu, Christina DiMarino, Khai Ngo

Coupled Inductor Analysis and Finite Element Modeling Assisted Design for Boost Extender
Topology... 594
    Vikas Kumar Rathore, Michael Evzelman, Mor Mordechai Peretz

Stability Analysis of Current-Limited Grid-Forming Inverters with Frequency Stabilization: An
Equivalent Impedance Approach... 602
    Bowen Yang, Gab-Su Seo

Revisit Active Power Oscillation in Multi-Virtual Synchronous Generators Gride 609
    Junjie Xiao, Pavol Bauer, Zian Qin

A Novel Current Control Technique for Off-Grid Single-Phase Inverters 616
Arpan Laha, Abirami Kalathy, Praveen Jain, Majid Pahlevani

Intelligent Low-Bandwidth Frequency Controller for VSGs at Economic Dispatch in Islanded
Microgrid 622
Shraf Eldin Sati, Ahmed Al-Durra, Hatem H. Zeineldin, Tarek H.M. El-Fouly, Ehab F. El-Saadany

Hardware-in-the-Loop of a Grid Forming Control Strategy Applied to a DC Off-Grid Green
Hydrogen Production System 629
*Diego Montoya-Acevedo, René Contreras-Barrios, Ángel Maureira-Riquelme, Esteban
Ibáñez-Muñoz, Catalina Gonzalez-Castaño, Carlos Restrepo*

Experimental Validation of a 40kW, 480V Point-to-Point DC Interlinks for Controller-Agnostic,
Interoperable Networked Microgrids 637
Maximiliano Ferrari, Michael Starke, John Smith, Joao Pereira, Misael Montejano

Andronov-Hopf Oscillator-Based Grid-Forming Converters with Embedded Disturbance Rejection
for Non-Ideal Loading Condition 645
Vikram Roy Chowdhury, Gab-Su Seo, Barry Mather

Estimation of Rectifier Output Current of the LLC Converter 651
Xin Wu, Yi Zhou, Haihong Long, Dehong Xu

A 100kHz Digitally Controlled 10kW, 2-Channel Solar MPPT Converter using 3-Level Topology
with >75W/in³ Power Density and >98.5% Peak Efficiency 658
Ranajay Mallik, Akshat Jain

A Bootstrapless KY-S-Hybrid Buck-Boost Converter with Full Range iLs Reduction and 400%
Line Transient Response Acceleration for AI-Mobile Application 664
Chuan-En Chang, Cheng-Ta Chuang, Hao-Ran Huang, Chieh-Ju Tsai, Ching-Jan Chen

Digital Control of a 600-V to 28-V 20-kW Two-Stage DC-DC Converter 670
*Shreyas B. Shah, Rachit Pradhan, Jiaqi Yuan, Mohamed Ibrahim, Ahmed Elezab, Samuel
Hemming, Giorgio Pietrini, Piranavan Suntharalingam, Mario F. Cruz, Ali Emadi*

Self-Calibrated Digital Current Emulation for High-Frequency Hysteretic Current-Mode Control in
GaN PFC Converters 676
Mohammad Shawkat Zaman, Olivier Trescases

High-Frequency Flying Capacitor Four-Level Drain Supply Modulator 682
Audrey Cheshire, Paul Flaten, Zoya Popoviœ, Dragan Maksimoviœ

Discontinuous Modulation Strategy for Voltage and Temperature Balancing of MMCs 689
Davide D'Amato, Stayner Nóbrega Barros, Jun-Hyung Jung, Marco Liserre

Damping Control and Improvement of Grid-Forming Inverter from a Wideband Stability
Perspective 696
Rui Kong, Subham Sahoo, Yubo Song, Frede Blaabjerg

A Grid-Forming Split-Phase Three-Leg Inverter with Unbalanced Loading and Active Power
Decoupling 703
Namwon Kim, Renata Kimpara, Michael Starke

Completely Decentralized Active and Reactive Power Control of Grid-Connected Cascaded H-
Bridge Inverters with Integrated Battery Storage 711
Soham Dutta, Brian Johnson

Small-Signal Modeling and Damping Design of Unfolding-Based Single Stage AC-DC Converter using the Extra Element Theorem 719
    Dakota Goodrich, Aditya Zade, Shubhangi Gurudiwan, Mahmoud Mansour, Regan Zane, Hongjie Wang

Methods to Enhance Cybersecurity of Multiple Inverters in Large Grid Connected PV / Battery Energy Storage Systems 727
    Hasan Ibrahim, Jaewon Kim, Peng-Hao Huang, Vishwam Raval, Prasad Enjeti

Optimal DC-DC Converter Topology and Control Algorithm for Fuel Cell Electric Vehicle with Series-Connected Supercapacitor 733
    Hyeon Soo Kim, Yun Seong Hwang, Seung Hyun Kang, Man Jae Kwon, Byoung Kuk Lee

Reliability-Constrained Design of a High-Gain Power Optimizer based on a Real Mission Profile 738
    Stefano Cerutti, Francesco Iannuzzo, Ariya Sangwongwanich, Tamás Kerekes, Mario Giuseppe Pavone, Francesco Gennaro, Natale Aiello, Francesco Musolino, Paolo Stefano Crovetti

Submodule Voltage Balancing Technique of Solar MMC for Firing the Switches using Integrated PWM Modules 746
    Ahmed Elsanabary, Saad Mekhilef, Mokhtar Aly, José Rodriguez

Single-Stage High-Frequency-Link Split-Phase Microinverter with High Voltage Gain based on Buck-Boost AC Chopper 751
    Xuewen Li, Jia Liu, Jinjun Liu

Fault Diagnosis and Tolerant Strategy for Triple-Port Hydrogen Converter using SSA-Optimized Random Forest Algorithm 757
    Shiqi Zhang, Yiyina Teng, Naizhe Diao, Xiaoqiang Guo, Vladimir Terzija, Lichong Wang

Resilient Operation for Grid-Connected Cascaded H-Bridge Multilevel Inverter with Improving PV Source Stress 761
    Jinli Zhu, Yuan Li, Hector Akuta, Jeonghun Kim, Uthandi Selvarasu, Shumeng Wang, Vikram Roy Chowdhury, Brad Lehman, Fang Z. Peng

A Medium Voltage Grid-Connected PV Inverter with a New Modular High Voltage Gain Converter Featuring Internal Modified Voltage Doubling Balancers 768
    Kajanan Kanathipan, Muhammad Ali Masood Cheema, John Lam

Split-Source Common-Ground Inverter for Photovoltaic Applications 775
    Mahmoud A. Gaafar, Mohamed Orabi, Samir Kouro, Ahmed Ibrahim, Eltaib Abdeen D. Ibrahim

Comprehensive Investigation and Proposal of a New Wireless Charging Road Structure using Low-Environmental-Impact Magnetic Concrete 782
    Shuntaro Inoue, Yuko Kano, Shin Tajima

Design of a Bidirectional High Power Inductive Power Transfer System with Auxiliary Winding for Automotive Applications 788
    Luis Ruiz Chamorro, Nikola Mirkoviœ, Alberto Delgado Expósito, Pedro Alou Cervera, Miroslav Vasiœ

Mutual Inductance and Load Identification Method based on the Voltage Transients of WPT Systems 795
    Xiaosheng Wang, C.Q. Jiang, Yibo Wang, Liping Mo

Digitally Controlled Misalignment-Tolerant Inductive Power Transfer System with Adaptive Hybrid Compensation for CC/CV Charging of E-Scooter 801
Niranjan Shrestha, V.S.R.Varaprasad Oruganti, Sheldon Williamson

On/Off Control of Modular Inductive Power Transfer System 809
Kunxiao Zhou, Guangdong Ning, Heyuan Li, Xinlin Wang, Minfan Fu

Receiver Side Regulation of LCC Wireless Power Transfer System with Variable Notch Filter 815
Hsin-Che Hsieh, Jih-Sheng Lai

84.7 Percent Peak Efficiency Stress Tolerant DC DC Buck Converter for Li Ion Battery Driven Standby Circuits in 18nm FDSOI 821
Gautam Dey Kanungo, Pijush Kanti Panja, Vikas Bugade, Kallol Chatterjee

Leveraging Ultrasound and Neural Networks for Non-Invasive Power Converter Efficiency Estimation 828
Youssof Fassi, Vincent Heiries, Jérôme Boutet, Julien Marianne, Sébastien Martin, Mathilde Chareyron, Clément Chambon, Sébastien Boisseau

A Load-Independent Multi-Relays Wireless Power Transfer with Self-Regulation and Single Compensation Network 834
Jong-Hun Kim, Najam Ul Hassan, Seogyong Jeong, Myeong-Ho Kim, Min-Sik Kim, Jee-Hoon Jung, Byunghun Lee, Se-Un Shin

A GaN-Based Single-Stage Solid-State Transformer Replacement for 40 VA Class 2 Line-Frequency Transformers 840
Allen T. Nguyen, Charles R. Sullivan

Survey of Components and Topologies for High-Efficiency and High-Power Density 48V DC-DC Converters 848
Joseph Winkler, Niklas Deneke, Bernhard Wicht

A Novel Solid-State Circuit Breaker using B-TRAN™ 854
Mudit Khanna, Ruiyang Yu, Milad Tayebi, Jiankang Bu, Jeffrey Knapp

Development of a Supercritical Fluid-Insulated Fast Mechanical Switch for MVDC Hybrid Circuit Breakers 860
Zhiyang Jin, Qichen Yang, Alfonso Cruz, Lukas Graber

Dynamic Impedance Matching for a Variable Reluctance Energy Harvesting Application with Constrained Space 868
Fernando Pérez, Alejandro Redondo, Airán Francés, Gabriel Mujica

Renewable Energy-Powered DC-Converted Refrigerator based on a Supercapacitor-Assisted Technique 874
Nirashi Polwaththa Gallage, Nihal Kularatna, Alistair Steyn-Ross, Dulsha Kularatna-Abeywardana

Design and Evaluation of Flexible Inductors for Wearable Power Electronics 880
Sean Logi, F. Selin Bagci, Katherine A. Kim

Design of Boost Power Factor Corrector and Asymmetrical Half-Bridge Flyback Converter for USB-PD Applications 887
Yun-Keng Cheng, Tsorng-Juu Liang, Kai-Hui Chen, Ming-Chang Tsou

Computationally Efficient Current Sensorless Predictive Control for PMSM Drive Fed by a Matrix Converter with CMV-Free Operation 895
Ali Sarajian, Ibrahim Harbi, Quanxue Guan, Davood Arab Khaburi, Ralph Kennel, José Rodriguez, Patrick Wheeler, Mokhtar Aly

PMSM Motor Drive with Current Direct Digital Control and Near 1st-Order Speed Control 900
Po-Chang Lee, Tsai-Fu Wu, Han Ku, Chien-Chih Hung, Jui-Yang Chiu

Fault-Tolerant Multilevel Converter for Multiphase Switched Reluctance Motor Drives based on q+2 Converter 906
Mahmoud A. Gaafar, Mohamed Orabi, Hao Chen, Mostafa Dardeer

Uncertainty-Aware Artificial Intelligence for Gear Fault Diagnosis in Motor Drives 912
Subham Sahoo, Huai Wang, Frede Blaabjerg

Neural Network based Digital Twin Health Monitoring of BLDC Motor Drives for Robots 919
Mohamed Y. Metwly, Benjamin Luckett, Landon Clark, JiangBiao He, Biyun Xie

MTPA Control using Predictive P&O Method for Dual Parallel Surface-Mounted Permanent Magnet Synchronous Motor Drives Fed by a Single Inverter 925
Jae-Seong Kim, Kyo-Beum Lee

A Novel I-f Startup Strategy with Smooth Transition to Sensorless Control for CSI-Fed PMSM Drives used in Submersible Pumps 930
Milad Bahrami-Fard, Majid Ghasemi Korrani, Babak Fahimi

Simulation-Assisted Design and Implementation of an Electrically Excited Synchronous Motor Drive System 938
Shih-Gang Chen, Jun-Ming Hsu, Chun-Yen Chen, Ming-Shi Huang

Implementation and Analysis of Direct Torque Control on High-Speed PMSMs: A Comparative Study of Commercial and Laboratory-Developed Motors 943
Md Moniruzzaman, Kishor Joshi, Md Rashedur Rahman, Md Khurshedul Islam, Seungdeog Choi, Masoud Karimi Ghartemani

A Ferrite based Carbon Reinforced Composite Wrapped IPM Rotor Design for High-Speed Traction Applications 951
Md Rashedur Rahman, Md Khurshedul Islam, Md Moniruzzaman, Seungdeog Choi, Han-Gyu Kim, Andrew Walters

A Novel Phase-Mode Controller for Resonant Converters 958
Claudio Adragna, Daniele Cazzaniga, Stefano Manzoni

A Regulated 36V-60V-Input VIN-Insensitive Resonant Switched-Capacitor Converter with Large Voltage Conversion Ratio 966
Yichao Ji, Jingyi Yuan, Lin Cheng

A Hybrid Switched Capacitor Converter Enabling Capacitive-Based Wireless Power Transfer for Battery Charging Applications 971
Jade Sund, Samantha Coday

A 48V to 50-110V Resonant Power-Bus Charger with Reduced Conduction Loss for MHz-Frequency Long-Range LiDAR Driver 978
Hangxiao Ma, Xuchu Mu, Yang Jiang, Weihang Zhang, Jincheng Zhang, Rui P. Martins, Pui-In Mak

A Trajectory Controlled 48-to-24 V Resonant Switched Capacitor Converter with 98.7% Efficiency and Ultrafast Dynamic Response 983
Hélène T.W. Ma Yang, Liang Wang, Haoyu Wang, Wai Tung Ng

Low Power, Non-Isolated, Extremely-High Step-Up, Quasi-Resonant Hybrid DC–DC Converter 990
Kumar Joy Nag, Aleksandar Prodić

Isolated Soft-Switching Flying-Capacitor based Quasi-Resonant Step-Up Converter 997
Kumar Joy Nag, Aleksandar Prodić

Accurate Small-Signal Phasor Transformation-Based Modeling of Secondary-Side Diode-Bridge Rectifiers for Battery Charging Applications 1004
Aditya Zade, Regan Zane

High-Efficiency Isolated Piezoelectric Transformers for Magnetic-Less DC-DC Power Conversion 1012
Sourav Naval, Wentao Xu, Mustapha Touhami, Jessica D. Boles

First Characterization of GaN Power Device and IC at Deep Cryogenic Temperatures Down to 100 mK 1020
Xin Yang, Matthew Porter, Zineng Yang, Zichen Xi, Liyang Jin, Liyan Zhu, Linbo Shao, Yuhao Zhang

Dynamic Environment-Aware Lifetime Prediction of SiC MOSFET Modules Through LSTM 1026
Md Zakir Hasan, Seungdeog Choi, Youssef Aider, Prashant Singh, Chun-Hung Liu

Guarding-Based C-V Characterization of 10 kV SiC MOSFET in Half-Bridge Module Configuration 1034
Nianzun Qi, Gao Liu, Zhixing Yan, Shaokang Luan, Pawel Piotr Kubulus, Yuan Gao, Stefan Meyer, Hongbo Zhao, Asger Bjørn Jørgensen, Stig Munk-Nielsen

Automated Characterization Platform for Comprehensive Dynamic R_{dson} Assessment of GaN HEMTs from 50 K to 400 K 1040
Tian Qiu, Zheyu Zhang, Purushottam Khadka, Ahmed Siraj, Dilip Rana

A Gate Driving Scheme for GaN Git with Enhanced Short Circuit Capability for Motor Drive Application 1047
Zongjie Zhou, Yan Cheng, Kevin J. Chen

Online Detection and Reduction of the Influence of Parameter Tolerance of Paralleled SiC MOSFETs in an EV Inverter Environment 1051
Hadiuzzaman Syed, Jochen Streit, Robert Kragl, Muhammad Muneeb Alam, Alberto Martinez-Limia, Karl Oberdieck, Ertuðrul Sönmez

Dynamic Current Sharing Issues with Paralleling SiC Power MOSFETs 1058
Ching-Yao Liu, Chen-Chan Lee, Jih-Sheng Lai

Integrated Short-Circuit Protection Design based on Dual-Channel Gate Driver for Series Connected Medium-Voltage SiC MOSFETs 1063
Rui Wang, Drazen Dujic

Long-Term High-Temperature Dynamic Gate Stress Reliability of a Last-Generation, Automotive-Grade, Planar 1200 V SiC MOSFET 1070
Giuseppe Mauromicale, Alessandro Sitta, Michele Fiore, Michele Calabretta, Francesco Iannuzzo

Innovative Gate Driver Structure Achieving Low Time Skew Across Isolation Barrier for Parallel Connected SiC Modules 1076
Louison Gouy, Anne-Sophie Descamps, Nicolas Ginot, Christophe Batard

Fully Integrated Closed-Loop Active Gate Driver IC with Real-Time Control of Gate Current Change Timing by Gate Current Sensing 1084
Yaogan Liang, Katsuhiro Hata, Makoto Takamiya

Analyze and Design of Digitally Load Current Modulated Active Gate Driver for GaN HEMTs based Buck DC-DC 1090
Wentao Liu, Zhina Lian, Taotao Wu, Xiaochuan Peng, Hao Min

Impact of Real-Time Variable Gate-Drive Strength on Drive Cycle Efficiency in SiC Inverter-Fed PMSM Traction Drives.. 1096
Matteo Pizzuto, Aiswarya Balamurali, Aniket Anand, Narayan C. Kar

Demonstration of Efficiency Increase of 350 V-to-13.3 V Isolated DC-DC Converters for Electric Vehicles by Active Gate Driving 1102
Yohei Sukita, Katsuhiro Hata, Hiroki Kondo, Kenichi Watanabe, Kenichi Nagayoshi, Makoto Takamiya

A Multi-Level Active Gate Driver for Achieving Thermal Balance in Parallel Connected Power MOSFETs..1108
Jingyuan Liang, Lingwei Sun, Wen Tao Cui, Wai Tung Ng, Motomitsu Iwamoto, Haruhiko Nishio

A Fast Short-Circuit Protection Method for Ohmic Gate P-GaN HEMT based on Gate Charge 1114
Yue Wu, Xi Jiang, Song Yuan, Xiaowu Gong, Zhaoheng Yan, Jiahong Chen, Yun Xu, Jinjie Liu

Comparison of Ultrafast-Rise-Time Gate Drivers for Wide-Bandgap Devices in Sub-Microsecond Pulsed Power Applications 1121
Soham Roy, Duy T. Nguyen, Neeraj Anantha, Alex J. Hanson

A Discrete Multilevel Active Gate Driver for GaN HEMTs to Optimize the Switching Behavior 1129
Celine Lawniczak, Martin Pfost

Attenuation of Fundamental Component of Differential Mode Noise using Active EMI Filter.....................1135
Guru Abhilash Mulumudi, Naveed Ishraq, Ayan Mallik

Graph Neural Network based Performance Modeling for the Dual Active Bridge Converter with Operational Generalization..1143
Weihao Lei, Fanfan Lin, Xinze Li, Xiaokun Bao, Xin Zhang

An Augmented State Space Modelling Approach for DC-DC Converter Start-Up in Closed Loop 1148
Waseah Anjum, Arkadeb Sengupta, Marco Liserre

The Utilization of a Parallel Computing Algorithm for Accelerating Switching-Level Modeling of Power Electronics Simulations in a T-Type PV Inverter 1153
Buck F. Brown III, Liwei Wang, Zheyu Zhang, Johan Enslin, Yi Li

A New Reduced Order Analytical Switching Model for eGaN HEMTs 1159
Ruqi Li, Douglas Arduini, Phen Lumod, Shobhana Punjabi, River Lin, Harold Gutierrez

Proposal of an Alternative Reverse Recovery Calculation Method...1167
Brian Deboi, Blake Nelson, Austin Curbow

Improvement of CM EMI Attenuation Ability of Transformer with Negative Capacitor 1173
    Qinghui Huang, Yiming Li, Yirui Yang, Shuo Wang, Yanwen Lai, Zhedong Ma

Damping Factor based PCB Parasitic Inductance Value Optimization to Minimize Voltage
Overshoot and Settling Time of Semiconductors 1179
    Reza Shahbazi, Yunting Liu

Hardware Implementation of Virtual Resistance based FRT Logic in Programmable 3-Level ANPC
Inverters...1184
    Mohammad Safayet Hossain, Shuvangkar Chandra Das, Paychuda Kritprajun, Amin Banaie,
    Tapas Barik, Deepak Ramasubramanian, Aboutaleb Haddadi, Evangelos Farantatos, Ulrich
    Muenz

Rad-Hard PSFB Controller for High-Voltage Space Applications 1190
    Reynaldo S. Gonzalez, Robert E. Bolaños

Modeling, Control and Digital Implementation of a Buck Converter Operating in Triangular
Current Mode for a Wide Output Voltage Range Space Application..........................1197
    Regina Ramos, Sara Pérez, Guillermo Núñez, Pedro Alou, Javier Torres

Thermal Model and Optimization of a Multi-Winding Transformer for Lunar Surface Power
Transmission...1203
    Zhining Zhang, Yuzhou Yao, Junchong Fan, Juchen Yang, Robert Guenther, Pengyu Fu, Jin
    Wang

Active Gate Driver Power Supply for High-Reliability Applications 1211
    Joseph P. Kozak, Juan Ramirez, Jesse Lin, Allison Orr, Alexander Martin, Hala Tomey

A Hybrid Energy Storage System for eVTOL Unmanned Aerial Vehicles using Supercapacitors.................1217
    Ali Alenezi, PengHao Huang, Prasad Enjeti

Evaluation of Retired Lithium-Ion Batteries for Second-Life Applications Through Electrochemical
Impedance Spectroscopy 1224
    Latha Anekal, Sheldon Williamson

Uninterruptable Non-Isolated Integrated Power Electronics Converter (UNIPEC) for Commercial
Truck Auxiliary Power Unit 1230
    Pouya Zolfi, Ahmad Alzahrani, Ayman El-Refaie

Investigation of Electrical Safety for Non-Isolated Single-Phase On-Board Chargers used in
BEV/PHEV 1237
    Soya Kataoka, Shohei Funatsu, Hiroaki Matsumori, Takashi Kosaka, Keisuke Nakamura,
    Subrata Saha

An 8-Level Flying Capacitor Multilevel Converter for Electric Aircraft Pulse Deicing 1242
    Nicole Stokowski, Andrew Freeman, Aidan Rodgers, Aria Delmar, Jonathan Sengstock, Alex
    Solecki, Andrew Stillwell

Impact of Position Measurement Delay Angle on Performance of PMSM Drives for Electric Power
Steering in a Wide Speed Range..1248
    Yingzhe Wu, Hengbin Zhang, Yuxiang Xue, Lisheng Wang, Hui Li, Shan Yin

Physical Parameter Estimation for a Two-Level VSI Three-Phase PMSM Electric Drivetrain 1255
    Bernard Steyaert, Ananda Tjakra Adisurja, Matthias Preindl

A Novel Two-Dimensional Random Switching Frequency PWM Method for Variable Frequency Drives 1261
    Mostafa Abarzadeh, Kevin Lee

Optimized Maximum Torque and Minimum Loss Fault-Tolerant Control Schemes for Dual Three-Phase PMSM 1267
    Syed Mohammad Maaz, Dong-Choon Lee

Wireless Actuation of Magnetic Robots with a Modular 60 mT 3-D Helmholtz Coil System 1274
    Konstantinos Manos, Yifan Rao, Tuo Zhao, Kevin Liu, Daniel Zhou, Calvin Nguyen, Eric Chen, Glaucio H. Paulino, Minjie Chen

A Versatile PHIL based Motor Emulator Testbench using a High-Performance Power Amplifier Testbench 1279
    Seyedeh Nazanin Afrasiabi, Rajendra Thike, Mathews Boby, K. S. Amitkumar

A 450V Three Phase GaN IPM Achieving 99.1% Efficiency in Smallest 12mm x12mm Package for 250W Power Delivery without Heatsink 1286
    Maik Kaufmann, Manu Balakrishnan, Stefan Herzer, Anand Chellamuthu, Hely Zhang

FEA based High-Frequency Synthesis for the Design and Optimization of GaN-Based Dual Three Phase Motor Drive System 1294
    Syed Imam Hasan, Alper Uzum, Ashraf Siddiquee, Yilmaz Sozer, Krishna Namburi

Evaluation of Passive Common-Noise Canceller Considering Both of Thermal Equilibrium and Common-Mode Noise Cancellation 1299
    Koji Mitsui, Kenshiro Katsura, Koki Notake, Koji Yamaguchi

Performance Evaluation of Isolated DC/DC Converters in Modularized Bridge Rectifier Solid-State Transformer 1305
    Zhenchao Li, Andrea Cervone, Drazen Dujic

Active and Reactive Power Flow Control of the Dual Active Bridge Converter 1311
    Lauryn Morris, Thomas W. Francois, Jonathan Saelens, Oroghene Oboreh-Snapps, Arnold Fernandes, Praneeth Uddarraju, Sophia A. Strathman, Jonathan W. Kimball

Comparative Analysis of Carbon Footprints and Material Usage of Solid-State Transformers and Low-Frequency-Transformer-Based MVac-LVdc Interfaces for High-Power EV Charging 1318
    Luc Imperiali, Rudy Wang, Anup Anurag, Peter Barbosa, Johann W. Kolar, Jonas Huber

Trade Study of Isolation Requirements and Magnetic Core Selection for Medium Frequency-Medium Voltage Transformers 1326
    Mohendro Kumar Ghosh, Mark A. Juds, Brandon Grainger, Ahmad El Shafei, Bogdan S. Borowy, Paul Ohodnicki

Comparative Evaluation of a Multilevel LLC Resonant Converter for a Modular DC/DC Stage in a Electrolyzer Power Supply 1334
    Samuel S. Queiroz, Levy F. Costa

Cost-Effectiveness Assessment of SiC MOSFET and Si IGBT Semiconductors in a Three-Level Resonant Converter for Solid-State Transformer 1341
    Samuel S. Queiroz, Levy F. Costa

Comparative Performance Analysis of Medium Voltage 3L-ANPC and 3L-DNPC Pole Enabled by Series-Connection of 10kV SiC MOSFETs and 10kV SiC JBS Diodes for Sine Triangle PWM Operation 1347
    Sanket Parashar, Shubham Rawat, Nithin Kolli, Raj Kumar Kokkonda, Subhashish Bhattacharya

A Zero Harmonic Distortion Master Converter for Medium Voltage Microgrids 1355
    Gabriel V. Ramos, Dener A. de L. Brandão, Thiago M. Parreiras, Danilo I. Brandão, Braz de J.C. Filho

An MILP Approach for Modeling and Analyzing the BESS for Smoothing Renewable Fluctuations Considering BESS Capacity Attenuation in the Bulk Power System with High Inverter-Based Resource Penetration 1363
    Hualong Liu, Wenyuan Tang

Thermal and Efficiency Characterization of Immersion Cooled SiC Traction Inverter 1368
    Yiju Wang, Reza Ilka, JiangBiao He

FPGA-Based Hybrid Simulator for Real-Time 3-D Temperature Monitoring of Power Converters 1375
    Xianghao Mo, Daniel Ríos Linares, Regina Ramos, Miroslav Vasiœ

A New Subassembly Concept for Enhanced Heat Dissipation and Reliability of Power Module 1383
    Yosuke Nakata, Yuji Sato, Shin Uegaki, Jun Fujita, Akihiko Furukawa, Masayoshi Tarutani

Stand-Alone R_{DS-ON} Sensor for In-Situ Prognostic, Protection and Reliability Enhancement of Power Converters 1388
    Zaheen Mustakin, Qiang Mu, Lucas Pereira, Jiale Zhou, Tiefu Zhao, Babak Parkhideh

Electrical Evaluation of a Modular High Voltage 3D Power Module using Direct Dielectric Liquid Cooling 1396
    Omar Sanjakdar, Yvan Avenas, Rachelle Hanna, Guillaume Piquet Boisson, Emmanuel Marcault, Antoine Philippe

Board Level Reliability of Gull-Wing, Micro-Leaded and Lead-Less Packaged MOSFETs in Automotive Environments 1403
    Christopher Liu, Vijayakrishna Satyamsetti, Xuanjing Wei, Christian Radici, Peter Vines, Wayne Lawson

Cost Effective and High Noise Immunity Methodology for Aging Evaluation of DC-Link Capacitors in Traction Inverters 1408
    Seyed Hossein Aleyasin, Fausto Stella, Radu Bojoi, Enrico Vico

A 3D Structure of Single-Sided Cooling Power Module with Low Thermal Resistance and Low Inductance 1414
    Hirofumi Hisamochi, Koki Notake, Yoshiaki Takahashi, Koji Yamaguchi

Aging of Y-Capacitor in an EMI Filter and Its Impact on Common-Mode Noises 1420
    Tahmid Ibne Mannan, Seungdeog Choi, Subarto Kumar Ghosh, Md Moniruzzaman

2200A/48V-to-1V Low-Profile Direct Power Converter with Standard PCB Transformer 1427
    Alejandro Figueroa, Pablo Mazariegos, Álvaro Cobos, Javier Goicoechea, Alejandro Castro, José A. Cobos

Single-Stage 48V-to-1V Regulator with a Half-Turn Transformer and Current-Doubler Rectifier 1433
    Xinmiao Xu, Qiang Li

Ultra-Low-Profile Single-Stage Voltage Regulator Module (VRM) for Next-Generation AI
Accelerators 1439
    Xufu Ren, Jinfeng Zhang, Zhenshuai Rong, Borong Hu, Teng Long

Novel TLVR Operation in Multi-Stage Voltage Regulator Module with Current Multipliers 1444
    *Kevin Zufferli, Roberto Rizzolatti, Mario Ursino, Simone Mazzer, Gerald Deboy, Stefano
    Saggini*

Interphase LC-Oscillation Suppression with Fast Line-Transient Response in 48-V Series-
Capacitor Buck Converters for Automotive Applications 1451
    W.L. Jiang, Y. Liu, N. Khan, J. Pigott, H.J. Bergveld, V. Chaturvedi, O. Trescases

An Approach to Compensate for Low Frequency DC-Link Voltage Ripple in High Power ANPC
Inverter 1459
    *Shaozhe Wang, Ankit Vivek Deshpande, Rolando Sandoval, Erick Pool-Mazun, Enrique
    Garza-Arias, Prasad Enjeti*

A Cascaded Multilevel Inverter System with Hot-Swapping and Fault Isolation Capability for
Improved Resiliency 1465
    *Uthandi Selvarasu, Vikram Roy Chowdhury, Shumeng Wang, Jinli Zhu, Mahshid Amirabadi,
    Yuan Li, Brad Lehman*

Layout Optimization for Parasitic Inductance Reduction of GaN-Based NPL.X Multilevel Inverter 1473
    Ali Halawa, Jinyeong Moon, Woongkul Lee

Topology Selection and Design Methodology for SiC based Solar Photovoltaic Medium Voltage
Direct Grid Connect Inverters 1481
    Jenson Joseph C. Attukadavil, Baylon G. Fernandes

EMI Modeling of PCB-Based Three-Level Active Neutral-Point-Clamped GaN Converter........................ 1489
    Mohammad Hassan Adeli, Necmi Altin, Erkan Deniz, Adel Nasiri

A Novel Layout for Improving Current Sharing of Paralleled SiC MOSFETs with TO-247 Package........... 1495
    *Che-Wei Chang, Matthias Spieler, Rolando Burgos, Ayman El-Refaie, Renato Amorim Torres,
    Dong Dong*

A Sensor-Less IGBT On-State Voltage Estimation Method using Inverter Control Variables 1501
    Shuyu Ou, Subham Sahoo, Ariya Sangwongwanich, Yongjie Liu, Frede Blaabjerg

A Novel Non-Intrusive Online Monitoring Method for Diagnosing the Lift-Off of Bonding Wires in
SiC MOSFETs 1507
    Keqi Song, Henry Shu-Hung Chung, Ho-Tin Tang

Optimizing MOSFET Selection for EMC-Critical Automotive Applications 1512
    Sacha J. Cazzitti, Christian Radici, Andrew J. Forsyth, Cheng Zhang, Peter Vines

Improving Dynamic Current Sharing Between Parallel MOSFETs by Optimizing Device
Parameters 1519
    Kunal Jha, Kapil Kelkar, Marina Hedenik, David Penof

A 21.6 kW/L Two-Phase Immersion-Cooled Isolated DC-DC Converter 1529
    Aleksandar Ristic-Smith, Kawsar Ali, Daniel Rogers

Extraction of Common Mode Parasitic Capacitance in Balance Filter for the Prediction of EMI
Noise Suppression 1537
    Qiuzhe Yang, Xingyu Chen, Zijian Wang, Qiang Li

A 660W, 96% Efficiency 3D Heterogeneously Integrated Digital DC/DC Power Module for Vertical Power Delivery 1544
Haoyu Wang, Xuliang Wang, Yan Wang, Xiaosen Liu

Planar Rogowski Coil-Based Switch Current Measurement for a 1.2 kV SiC MOSFET Embedded Die PCB... 1551
Matthias Spieler, Che-Wei Chang, Ayman M. El-Refaie, Dong Dong, Rolando Burgos

Effect of Magnetic Couplings on Conducted EMI of GaN-Based PFC Converter 1557
Tyler McGrew, Qiang Li

Optically-Controlled 3.3 kV SiC MOSFET with Fast Switching Speed and Low Optical Power 1564
Xin Yang, Guannan Shi, Liyang Jin, Yuan Qin, Matthew Porter, Che-Wei Chang, Xiaoting Jia, Dong Dong, Linbo Shao, Yuhao Zhang

Optimization Techniques for Parallel-Connected Devices in IPMs for Consumer Use 1569
Keisuke Kawamoto, Haruhiko Murakami, Teruaki Nagahara, Michael Rogers, Akiko Goto, Shoji Saito, Koichiro Noguchi

Investigating the Temperature Dependency and Operating Parameters of a Self-Driving Active Gate Driver 1576
Vin Loong Choo, Martin Pfost

Use of Switched-Capacitor Circuit to Generate Negative Gate-Source Voltage Pulses 1582
Ho-Tin Tang, Henry Shu-Hung Chung

An Optically Isolated Gate Driver with Simultaneous Data and Power Transmission Through a Miniaturized, Efficient Photonic Platform... 1590
Jiajun Li, Mariia Klymenko, Yanqiao Li, William Scheideler, Jason T. Stauth

Optimal Shared Energy Storage Capacity Configuration in Multi-Energy Microgrids Considering Battery Lifetime Loss based on Relaxation Techniques.. 1597
Hualong Liu, Wenyuan Tang

Virtual Resistance Control for an Active Battery Management System 1602
Alastair P. Thurlbeck, Ashraf Siddiquee, Mithat John Kisacikoglu, Yilmaz Sozer

Internal Voltage Source Saturation Impact on Stability Limits of Grid Forming Converter 1610
Divyanshu Bansal, Aravind G., L. Umanand

A Zero Harmonic Distortion Grid-Connected Grid-Forming Converter for Battery Energy Storage System 1615
Gabriel V. Ramos, Thiago M. Parreiras, Fangzhou Zhao, Xiongfei Wang, Braz de J.C. Filho

Single Cell Energy Router Justification for Three Phase Near Zero Energy Buildings 1622
Hossein Nourollahi Hokmabad, Tala Hemmati Shahsavar, Oleksandr Matiushkin, Tanel Jalakas, Oleksandr Husev, Juri Belikov

A Multi-UAV Charging Station Enabling Free Landing by Grid Pattern Transmitter................................... 1629
Jungho Kim, Hyunkyeong Jo, Seoktae Seo, Bonyoung Lee, Hyungki Min, Franklin Bien

Capacitor Design for Self-Resonant Coils for Long-Distance Wireless Power Transfer System................... 1635
Mostak Mohammad, Vandana Rallabandi, Omer C. Onar, Gui-Jia Su

A 10.4-kW High-Power-Transfer-Density Multi-MHz Capacitive Wireless Power Transfer System for EV Charging Utilizing Stacked-Inverter Stacked-Rectifier Architecture 1640
    Dheeraj Etta, Miguel Alvarez Dominguez, Sounak Maji, Syed Saeed Rashid, Khurram K. Afridi

Reduced-Fringing-Field Multi-MHz Capacitive Wireless Power Transfer System using Metasurface-Based Couplers with Active Field Cancellation 1646
    Syed Saeed Rashid, Dheeraj Etta, Matteo Ciabattoni, Francesco Monticone, Khurram K. Afridi

Living Object Detection in Wireless Power Transfer Systems using Remote Capacitive Bio-Signals Monitoring.. 1653
    Bruno M.G. Rosa, Paul D. Mitcheson

Modified N:1 Switched Capacitor Converter with Reduced Capacitor DC Bias Voltage for High Power Density 1659
    Taewoo Lee, Dam Yun, Sunghyuk Choi, Jung-Ik Ha

Wide Range Digital Control for Three-Level Buck Converters with Sensorless Flying-Cap Voltage Balancing.. 1666
    Hossein Hajisadeghian, Giovanni Bonanno

A Comparative Investigation of a New Continuous Voltage Conversion Ratio Approach in a Zero-Inductor Voltage Converter... 1673
    Sina Salehi Dobakhshari, Aamna Nasir Hameed, Binghui He, Mojtaba Forouzesh, Yan-Fei Liu

A 96.1% Peak Efficiency, 6.8 kW/in³, 48V-to-6V On-Package Intermediate Bus Converter with LV-GaN Power Transistors... 1681
    Mausamjeet Khatua, Nachiket Desai, Harish Krishnamurthy, Sheldon Weng, Jingshu Yu, Huong Do, Samuel Bader, Han Wui Then, Krishnan Ravichandran, James Tschanz, Kaladhar Radhakrishnan, Vivek De

A 48V to 2.4V-5V 95.8%-Peak-Efficiency 869W/in³-Power-Density Fibonacci Dual-Path Hybrid DC-DC Converter with Inductor Current Reduction and Low Output Resistance...................................... 1687
    Yichao Ji, Zeguo Liu, Lin Cheng

An Ultra-Fast Very Large Scale Interleaved Li-Fi Transmitter 1693
    Daniel H. Zhou, Konstantinos Manos, Minjie Chen

Isolated PWM DC-DC Converter with Single Magnetic Component, ZVS and Self-Balanced Switched-Capacitor Voltage 1701
    Pablo M. Gil, Juan Rodríguez, Diego G. Lamar

Analysis and Design of a Low-Complexity ZVS Buck-Boost Converter 1707
    Burkhard Ulrich

A High Conversion-Ratio Hybrid Series-Parallel DC-DC Converter with Pseudo-Soft-Charging and Inductor Current Frequency Multiplication.. 1715
    Avinash Maddela, Kishalay Datta, Jason T. Stauth

A Real-Time Variation Control of Deadtime in GaN-Based Bidirectional Buck-Boost Converter for Lithium-Ion Battery Formation System.. 1723
    Jong-Hun Lim, Go Woon Heo, Je-Yeong Lim, Dong Hwan Kim, Byoung Kuk Lee

A Space Vector PWM Strategy for Charging of Bootstrap Capacitor in Three-Level Neutral-Point-Clamped Inverter 1728

Anantha Hegde, Asamira Suzuki, Hirokazu Nakamura, Takamune Kabashima, Koji Higashiyama, Keiji Akamatsu

A Complementary Carrier based PWM Strategy for Average Current Sampling of Three-Phase Inverter using Single Current Sensor 1734

Byeong-Il Kim, Joon-Seok Kim, Yeongsu Bak, June-Seok Lee

Short-Circuit Ride-Through for a CRM-Based Soft-Switching Three-Phase Inverter 1741

Xingyu Chen, Gibong Son, Qiang Li

Modified Space Vector Modulation with Low Bandwidth Sensor to Reduce Losses in Soft Switching Three-Phase Inverters 1746

Md Didarul Alam, Nazmul Hassan, Iqbal Husain, Liming Liu, Hongrae Kim

A Feedforward Ripple Reduction Control Strategy based on a Hybrid GaN/Si Interleaved Inverter 1754

Mowei Lu, Jurgis Reinotas, Xiaoyang Tian, Stefan M. Goetz

IGBT Comparison for Optimized Switching Behavior in the SiC/Si-Hybrid Switch 1759

Adrian Amler, Thomas Heckel, Daniel Ruppert, Cornelius Rettner, Martin März

Forward Recovery and its Mitigation in Hybrid Si/SiC-Based DC–AC Converters 1767

Yan Zhou, Thomas Lehmeier, Adrian Amler, Martin März

Real-Time IGBT Module Ageing Characterization Through Temperature Monitoring 1774

Quirc Perez-Farre, Luis F. Gomez-Rivera, Carlos Lopez-Torres, Kai Dannehl, Antoni García-Espinosa, Alejandro Paredes-Camacho

Experimental Validation of Triangular SOA via Infrared Thermography of a MOSFET Die Operating in the Thermally Unstable Linear-Mode for Automotive Applications 1781

Yacine Ayachi Amor, Christian Radici, Kerry J. Abrams, Philip Ellis, Peter Vines, Wayne Lawson

Feasibility Study of the SuperIGBT: A Series-Connected High Voltage IGBT with a Single Gate 1786

Junhong Tong, Alex Q. Huang, Huanghaohe Zou, Zhiyuan Ma

Low Profile, Laminated Nife Transformers for Flyback Converters 1791

Xuan Wang, Reza Mounesi, Matthew Catanoso, Matthew Fox, Adel Nasiri, Mark G. Allen

Comprehensive Demonstration of New Magnetic Designs Utilizing Magnetic Anisotropy of the Cores for Integrated Magnetics 1797

Yota Takamura, Honami Nitta, Tatsuya Miyazaki, Kimito Yamanaka, Ryosuke Ishido, Akira Namba, Keisuke Fujisaki, Shigeki Nakagawa

A Two – Stage Artificial Neural Network (ANN) – based Design and Optimization of High Frequency Transformers for Dual Active Bridge Converter 1803

Lufan Zhou, Alberto Delgado Expósito, Adam Ruszczyk, Simon Round, Miroslav Vasiæ

Modeling and Optimizing Winding Arrangement for Gapped Planar Magnetics based on Artificial Neural Network 1810

Hanqing Cao, Bima Nugraha Sanusi, Ziwei Ouyang

Free-Shape Optimization of VHF Air-Core Inductors using a Constraint-Aware Genetic Algorithm 1816

Thomas Guillod, Charles R. Sullivan

Organic Direct Bonded Copper-Based Rapid Prototyping for Silicon Carbide Power Module Packaging 1824
Shuofeng Zhao, Joshua Major, Douglas DeVoto, Sarwar Islam, Xiaoling Li, Mike Tant, Faisal Khan, Sreekant Narumanchi

Discrete Power Device Packaging with Integrated Direct Two-Phase Cooling 1832
Jinpeng Cheng, Jinxiao Wei, Hao Feng, Li Ran

Investigation of Die Top-Side Re-Metallization for SiC-Based Double-Side Cooled Power Modules 1836
Narayanan Rajagopal, Christina DiMarino

Design of Low Parasitic Inductance GaN HEMT Flip-Chip Power Module 1844
Mohammad Dehan Rahman, Tanzila Akter, Abu Shahir Md Khalid Hasan, H. Alan Mantooth, Xiaoqing Song

A Scalable Dual-Orthogonal-Cooling Packaging Concept for Parallel-Series SiC Chips 1850
Ekaterina Muravleva, Youssef Abotaleb, Blake Anderson, Zichen Zhang, Boyi Zhang, Jerry Hudgins, Jun Wang

Parasitic Impact Analysis and Design of Hybrid EMI Filter for Active Clamp Flyback SMPS 1858
Tahmid Ibne Mannan, Seungdeog Choi, Masoud Karimi-Ghartemani

Overview of Dynamic Characterization of Switches for Three Phase Voltage Source, Current Source, and Matrix Converter Applications 1866
Sneha Narasimhan, Sathya Rupan Thirumoorthi, Subhashish Bhattacharya

Advanced Modeling Technique of Class-E Inverter Considering Low R_{on} of eGaN FETs and Different Design Procedures 1874
Manas Palmal, Jungwon Choi

PiezoNet and Data-Driven Models for Time-Domain Characterization of Piezoelectric Resonators 1882
Davit Grigoryan, Mian Liao, Haoran Li, Shukai Wang, Tanuj Sen, Matthew Tan, Minjie Chen

A New Gate Charge De-Embedding Method for Accurate On-Wafer Characterization of HV MOSFET Devices 1889
João R.R.O. Martins, Rachid Hamani, Vincent Quenette, Joerg Gessner

4 kW Auxiliary Power Module for Electric Vehicles Utilizing a Dual-Phase LLC DC-DC Converter 1892
Mojtaba Forouzesh, Xiang Yu, Yan-Fei Liu, Paresh C. Sen

New Reverse Mode Control Method of Phase-Shift Full-Bridge Converter for Bidirectional Auxiliary Power Module 1899
Jongyoon Chae, Dongmin Kim, Dongmin Choi, Gun-Woo Moon

In-Situ EV EIS with a High-Density Flying Capacitor Multi-Level Converter Supercapacitor System 1905
Avram Kachura, Gaël Vergès, Samantha K. Murray, Olivier Trescases

A Novel 500-kHz LLC-T Resonant Converter with Wide Output Range 1913
Zhengming Hou, Dong Jiao, Jih-Sheng Lai

High Efficiency Traction Drive Operation with a Partial Load Three-Phase Triangular Current Mode Modulation Concept 1919
Bhaskar Chatterjee, Jan Allgeier, Thomas Plum, Marc Hiller

Analysis of Maximum Power Transfer Limit for Linear Operation of Dual-Active-Bridge Converters 1927
    Radhika Sarda, Ezequiel Ramos Rodriguez, Gaowen Liang, Glen G. Farivar, Josep Pou, Vaisambhayana B. Sriram, Anshuman Tripathi

Enhanced Control for Integrated Active Power Decoupling in Single-Phase Three-Level Flying Capacitor PFC Converter 1935
    Gleisson Balen, Cristian Blanco, Ángel Navarro-Rodríguez, Pablo García, Rafael Peña-Alzola

Improving Transient Stability of PLL-Synchronized Grid-Following Inverters 1940
    Surya Prakash, Kalpana Beura, Mohamed Alkhatib, Omar Al Zaabi, Khalifa Al Hosani, Utkal Ranjan Muduli

Online Impedance-Based Analysis for Power System Stability Assessment using Transformer-Less and Filter-Less Switch-Mode Perturbation Generator 1946
    Tomoya Ide, Yuko Hirase, Cheng Huang, Takanori Isobe

PIR-R Control for Three-Phase Grid-Connected Inverter with Unbalanced Grid Current Correction 1953
    Haneen Ghanayem, Xingyu Yang, Mohammad Alathamneh, R.M. Nelms

Design and Placement of a Passive Clamp Snubber for Isolated SEPIC and Cuk Converters Working as Automatic Power Factor Correctors 1959
    Abraham López, Juan Rodríguez, Duberney Murillo-Yarce, Javier Sebastián, Diego G. Lamar

Current Sensorless Control Strategy for Single-Phase T-Type PFC Converter 1967
    Che-Yu Lu, Jia-En Zeng

Three-Phase Single-Stage Multiport AC-DC Converter with Integrated DC-DC Conversion Stages 1972
    Asad Hameed, Gerry Moschopoulos

High Efficiency AC-Adapter Realized by Voltage-Clamper with Mid-Voltage AHB Converter using Synchronous Rectification 1977
    Shuichiro Motoori, Akihiro Kawano, Toshiyuki Zaitsu, Riku Tatetsu, Kohei Sebata, Kazuki Miyanjou, Kimihiro Nishijima

Active Soft Switching Technique for Single Phase Series Capacitive Link Universal Rectifier 1983
    Anran Wei, Brad Lehman, Mahshid Amirabadi

A Multi Mode Control Algorithm for Totem-Pole Bridgeless PFC 1990
    Bosheng Sun, Sheng-Yang Yu, Amir Hussain

Protection Strategy for Flying Capacitor Totem-Pole PFC Under the AC Drop Transient 1995
    Yanqing Wu, Wending Zhao, Zhenhai Zhu, Xinke Wu

Three-Phase with Three Single-Phase Single-Stage Isolated AC-DC Converters for EV Charging Station Applications 2002
    Misha Kumar, Peter M. Barbosa, Juan M. Ruiz

400V SiC in Next-Generation 3-Level Flying Capacitor Bridgeless Totem-Pole PFC 2009
    Rytis Beinarys, Seamus O'Driscoll

Extended Smart-Link Quasi-Single-Stage 3-Phase AC-DC Power Supply Module for AI-Driving Data Centers 2014
    D. Biadene, J. Huber, J.W. Kolar, P. Mattavelli

A New Three-Phase Multi-Mode AC/DC LLC Converter with Output-Controlled Active Rectifier (with V2G and G2V Functions) for Fast DC Charging Application... 2022
Xiaoyi Xia, John Lam

Capacitorless Notch Resonant Converters for Miniaturized LLCLC Resonant Converters in Electric Vehicle Charging Applications 2029
Haitham Kanakri, Euzeli Cipriano Dos Santos Jr., Maher Rizkalla

Multiple-Core Transformer Design based on Half-Turn Structure in Two-Stage DC-DC Converter for Battery Storage System... 2035
Yilei Li, Bima Nugraha Sanusi, Pinhe Wang, Tianming Luo

Bidirectional DC-DC Converter Utilizing Coupled Inductors for Energy Storage System........................... 2043
Wen-Hsuan Lee, Jiann-Fuh Chen, Hsuan Liao, Kuo Fu Liao

Comparison of 2-Level and Quasi-2-Level Topologies in a Bidirectional Isolated DC-DC Converter for MVDC Networks... 2051
José Andrés Aguilar Croston, Jean-Yves Gauthier, Cyril Buttay, Maryam Saeedifard, Besar Asllani, Piotr Dworakowski

Sling Forward Converter for Offline Operation: Achieving High Efficiency and Wide Voltage Range Performance 2059
Nasherul Islam, Guozhu Chen, Honglei Miao, Fuxing Zhang

A Pulse Width Alternating Modulation Strategy for Three-Level Buck-Boost Converter 2066
Xinlong He, Caifeng Liu, Xudong Zou, Jiaao Zou, Tianyi Zhang, Yong Kang

ZVT Circuit Applied for Wide Input Range Isolated Converters 2070
Linguo Wang, Zhongyin Guo, Junjie Zhu, Bing Zhang, Zhiling Zuo, Xiaoguang Gao, Guangji Ma

Impact of Asymmetrical Leakage Inductance on a 380 V-12 V LLC Converter with Synchronous Rectifier for DCX Application 2075
Jinshu Lin, Shan Yin, Chen Song, Honglang Zhang, Minhai Dong, Limei Xu, Hui Li

Start-Up Techniques and Universal Closed Loop Control of Immittance Network based Resonant Converter 2082
Ripun Phukan, Misha Kumar, Randy Beckemeyer, Juan Ruiz, Peter Barbosa

Multi-Objective Efficiency-Oriented Optimization for DAB Converters Minimizing Current Stress and Backflow Power with Soft-Switching Assurance 2088
Kun Wang, Ian Laird, Jun Wang

An ISOP-PSFB PWM Converter based on Coupled Output Inductors and Phase-Shifted Modulation with Full ZVZCS Range 2096
Kang Hong, Guo Xu, Guangfu Ning, Mei Su

Design and Implementation of a GaN-Based Soft-Switched Series-Capacitor Buck Converter Operating at the CCM-DCM Boundary for High-Performance Computing Systems 2101
Ramin Rahimzadeh Khorasani, Kolman Puterman Ghitelman, Madhavan Swaminathan

Intrinsic Feedback Model for Coupled-Damped Self-Balancing of General Multiphase Hybrid Converters 2109
Haoran Xu, Weijia Hao, Desheng Zhang, Run Min, Qiaoling Tong, Xuecheng Zou

A High-Efficiency Switching Oscillation Suppression Strategy based on Damped Oscillation for Synchronous DC-DC Converter 2117

Hao Yuan, Chuan Ni, Zhengyu Ye, Wei Lu, Hui Xue, Ting Qian

Efficient and Streamlined Demodulation Strategy for High-Frequency Talkative DC-DC Converters 2125

Abdelmoumin Allioua, Hendrik Gockel, Gerd Griepentrog

A 90.9% Peak Efficiency KY Single-Inductor Bipolar-Output Converter with Conductance Modulation Controller for Active-Matrix Organic Light Emitting Diode Power Supply.............................. 2131

Sheng-Han Yu, Chieh-Ju Tsai, Hao-Ran Huang, Ching-Jan Chen

Constant-on-Time Control for Zero-Bias Trans-Inductor Voltage Regulators... 2138

Hank Zeng, Justin Lee, Rixin Lai, Hang Shao

An Improved PFM Control Scheme for Three-Level Buck Converter based on Ton Extension Achieving an 810% Frequency Reduction 2143

Yi-Chun Chang, Chieh-Ju Tsai, Ting-Lun Lee, Ching-Jan Chen

A Concept for Current Ripple or Transient Improvements in Multiphase Converters 2149

Alexandr Ikriannikov, Alex Gao

System Solutions and Design Trade-Offs to the Input Filter Interactions with Battery Chargers 2157

Xigen Zhou, Dan Mavencamp, Kuang-Yao Cheng

Modeling and Implementation of a Zero Bias TLVR 2162

Lei Wang, Travis Guthrie, Peyman Asadi, Mark Alexander, Kunrong Wang, Brandon Howell

cGANET-Enhanced Voltage Gain Modeling: Elevating CLLC Converter Accuracy..................................... 2167

Yu Zuo, Xiaobing Shen, Fanghao Tian, Jiaze Kong, Hans Wouters, Wilmar Martinez

Capacitive vs Inductive Coupling based DC-DC Converter Operating in MHz Switching Frequency Range... 2173

Saeid Pourjafar, Parham Mohseni, Oleksandr Husev, Ryszard Strezelecki, Oleksandr Matiushkin

LLC Converter Main Transformer Losses: Eliminating Air Gaps and Integrating Parallel External Inductors 2179

Yu-Chen Liu, Shang-Syun Wu

Small-Signal Phasor Modeling of T-Type Bridge-Based Single-Sided and Double-Sided LCC Resonant Converters for WPT Applications 2194

Aditya Zade, Shubhangi Gurudiwan, Regan Zane

A Hybrid Three-Port Topology for Urban Charging Stations... 2202

Mohammadreza Khodaparast Klidbari, Naser Souri, Zahra Sadat Habibolahi, Hamid Montazeri Hedeshi

Reconfigurable H5-Bridge based LLC-DAB Sigma Converter for EV Fast Charging Stations 2207

Huangsheng Xu, Mingde Zhou, Qishan Pan, Haoyu Wang

A Resonant Reset Forward Converter with Ultra-High Conversion Gain using Differential Transformation Technique (DTT)... 2213

Shubham Srivastava, Mandeep S. Rana, Santanu K. Mishra

Full-Range ZVS Modulation of Switched Capacitor Converter for Sensorless Voltage Balancing 2220

Md Tanvir Ahammed, Wensong Yu

Dimensional Parasitics Absorption in Capacitively-Isolated Ćuk Converter for Medium-Voltage High Step-Down Converters 2228
    Aakash Kamalapur, Jung-Soo Bae, Mark Cairnie, Rajaie Nassar, Jack Knoll, Dushan Boroyevich, Guo-Quan Lu, Christina DiMarino, Qiang Li, Khai D. T. Ngo

A 36-to-60V Input Dual-Phase 2MHz 93%-Efficiency ZVT Series-Parallel Hybrid Buck Converter using Single Auxiliary Inductor and Adaptive Time Multiplexing Control 2236
    Qi Cheng, Hoi Lee

Improved Efficiency in a 10 W Class-Φ_2 Converter Utilizing a Resonant Gate Drive 2241
    Malachi Hornbuckle, Katherine Liang, Juan Rivas-Davila

The Analysis and Design of a Resonant Capacitively-Isolated Cockcroft-Walton Converter 2249
    Elizabeth Rabenold, Raiphy Jerez, Samantha Coday

SHSC: Non-Isolated High-Density 4:1 IBC for 48 V Applications 2254
    Mario Ursino, Roberto Rizzolatti, Simone Mazzer

High-Performance Current Multiplier: A Hybrid Switched Capacitor Solution for High-Current Applications 2260
    Kevin Zufferli, Roberto Rizzolatti, Mario Ursino, Simone Mazzer, Gerald Deboy, Stefano Saggini

Representation and Design Methodology for Generalized Switched-Capacitor Converter Topologies 2268
    Seokwon Choi, Dam Yun, Jung-Ik Ha

A 48-V-to-1-V Gallium Nitride Switching Bus Converter for Processor Vertical Power Delivery with 2.7 mm Thickness and 3048 W/in³ Power Density 2276
    Jiarui Zou, Yicheng Zhu, Nathan M. Ellis, Logan Horowitz, Robert C. N. Pilawa-Podgurski

Ripple Reduction and Efficiency Improvement of Always-Dual-Path Hybrid DC-DC Converter based on Phase Shift Operation 2284
    Katsuhiro Hata, Shinsaku Tanaka, Toru Ashikaga, Yasuhiro Rikiishi

Ultralocal PQ Theory: A New Approach for Model-Free Predictive Direct Power Control of Shunt Active Power Filters 2290
    Mahdi S. Mousavi, Abolfazl Nassaji, Ibrahim Harbi, Behnam Nikmaram, S. Alireza Davari, Mokhtar Aly, José Rodriguez

Symmetrical Balanced Circuit for Common-Mode Noise Mitigation in LCL-T Resonant Converter 2296
    Ripun Phukan, Boyi Zhang, Juan Ruiz, Peter Barbosa

A Single-Phase Soft-Switching Buck-Boost Inverter 2303
    Lukas Wipprecht, Burkhard Ulrich

Low-Complexity Model Predictive Control Method for Driving Dual Induction Motors Fed by Five-Leg Inverter 2311
    Jun Young Lee, Eun Woo Lee, Dongho Choi, June-Seok Lee

Overvoltage Mitigation Filter using High-Frequency Cable Modeling in Long Transmission Lines for Silicon Carbide Inverter Systems 2317
    Yun-Jin Lee, Kyo-Beum Lee

Power Delivery Network (PDN) Design and Analysis to Achieve Low Impedance in Fast Edge Rate DC-DC Converters for EMI Compliance 2322
    Manraj Singh Ladhar, Sheldon Williamson

Enhancing the Performance of Dual Input Split Source Inverters using an Advanced Modulation Strategy 2327

Mustafa Abu-Zaher, Fang Zhuo, Mokhtar Aly, Mahmoud A. Gaafar, Mohamed Orabi, José Rodriguez, Alaaeldien Hassan, Jiachen Tian, Samir Kouro

A Novel GaN-HEMT Single-Phase Single-Stage Buck-Boost Micro-Inverter Topology for PV Applications 2332

Pengwei Li, Uiliam Kutrolli, Ali Bazzi

A Dynamic Current Sharing Method using Novel Clip Considering Mutual Inductance Coupling 2343

Zexiang Zheng, Jianwei Lv, Yiyang Yan, Baihan Liu, Yifan Zhang, Linhao Ren, Jiaxin Liu, Cai Chen, Yong Kang, Xiong Zhang, Hao Yu, Wei Jiang

Application-Oriented Test Setup for Measuring Dynamic Output and Transfer Characteristics of GaN-HEMTs 2348

Philipp Swoboda, Martin Fein, Simon Frank, Andreas Liske, Marc Hiller

Mitigating Gate Voltage Oscillation in Parallel SiC Power Modules for xEV 2356

Hideo Komo, Michael Rogers, Mark Steiner, Eric Motto, Koichi Taguchi, Chihiro Kawahara, Junichi Nakashima, Yasushige Mukunoki, Seiichiro Inokuchi, Rei Yoneyama

Switching Performance Comparison of Low-Voltage GaN and Si Devices 2361

Tianxiao Chen, Haoyang Liu, Pedro A.M. Bezerra, Eckart Hoene, Sibylle Dieckerhoff

Modeling of Switching Transients for Frequency-Domain CM EMI Analysis in Double Sided Cooling Power Modules 2369

Sijia Liu, Liu Yang, Heng Zhang, Yifan Zhang, Zexiang Zheng, Jianwei Lv, Jiaxin Liu, Cai Chen, Yong Kang, Yuebin Zhou, Daming Wang, Shuang Zhao

Leakage Current Detection Scheme for Aging Test of 10kV SiC MOSFET Power Module 2375

Peiyang Ding, Hong Zhang, Tianshu Yuan, Qiling Chen, Jiacheng Guo, Dingkun Ma, Peiyuan Sun, Ting Hou, Laili Wang

Physics-Informed Neural Network Approach for Early Degradation Trajectory Prediction of Power Semiconductor Modules 2380

Jie Kong, Yi Zhang, Yichi Zhang, Lukas Wick, Frederik Lillebæk Hansen, Dao Zhou, Huai Wang

Nonlinear Output Capacitance of Bidirectional Gallium Nitride Power Switches 2387

Michael Bosch, Jeremy Nuzzo, Dominik Koch, Mathias C.J. Weiser, Ingmar Kallfass

Novel Approach of Determining and Predicting SiC MOSFET's on Resistance from Device Case Temperature using Machine Learning 2393

Paul Bradford, Conner Deppe, Hongjie Wang

Comparison of Static Characteristics in GaN HEMTs Across 50K to 400K Considering Diverse Techniques and Statistical Variation 2400

Purushottam Khadka, Saumil Shivdikar, Zheyu Zhang, Tian Qiu, Ahmed Siraj

Compact Model of β-Ga$_2$O$_3$ Schottky Barrier Diode 2407

Abu Shahir Md Khalid Hasan, Mohammad Dehan Rahman, Tanzila Akter, Md Majharul Islam, Md Maksudul Hossain, Xiaoqing Song, H. Alan Mantooth

DC-Link Capacitor Board Design for Low Parasitic Inductance 2413

Mikayla Benson, Lifang Yi, Kangbeen Lee, Jinyeong Moon, Woongkul Lee

First Demonstration of a Gallium Oxide Power Converter 2419
  Joshua J. Piel, Elizabeth A. Sowers, Daniel M. Dryden, Thaddeus J. Asel, Adam T. Neal,
  Brenton A. Noesges, Shin Mou, Andrew J. Green

Optimized Integrated EMI Filter Design in SiC Power Modules with Terminal Inductor for Better
High-Frequency EMI Suppression 2426
  Yifan Zhang, Wenzhe Xu, Jianwei Lv, Yiyang Yan, Baihan Liu, Sijia Liu, Jiaxin Liu, Cai Chen,
  Yong Kang, Xiong Zhang, Hao Yu, Wei Jiang

Balanced Technique using Integrated Winding Coupled Inductor for High-Power Density Two-
Phase Interleaved Boost Converter 2431
  Yuta Imaeda, Jun Imaoka, Masayoshi Yamamoto, Hiroyuki Onishi

MagNetX: Foundation Neural Network Models for Simulating Power Magnetics in Transient 2438
  Shukai Wang, Hyukjae Kwon, Haoran Li, Youssef Elasser, Gyeong-Gu Kang, Daniel Zhou,
  Davit Grigoryan, Minjie Chen

Revisiting Models of Common Mode Inductors to Include the Magnetized Capacitance Effect 2446
  Rafael Bogo Portal Chagas, Marcelo Lobo Heldwein

A High Frequency Coupled Inductor with Distributed Air Gap for High Power DC-DC Converters 2453
  Muhammad Fasih Uddin, Ahmed H. Ismail, Peyman Darvish, Baher Abu Sba, Yue Zhao

High-Power Planar Transformer Design for Four-Port Converters 2461
  Arya Sadasivan, Behrooz Mirafzal

Optimal Design of Inductors with Aluminum Litz Wire for Inductive Power Transfer Systems 2468
  Jesús Acero, Claudio Carretero, Ignacio Lope, Óscar Lahuerta, José-Miguel Burdío

Analytic Design of Flat-Wire Inductors for High-Current and Compact DC-DC Converters 2474
  Sajjad Mohammadi, James L. Kirtley, Alireza Namadmalan

Insulation Dielectric Loss of High-Frequency Transformer Under Square Voltage Excitation with
Edge Oscillation 2482
  Zhanlei Liu, Lingyu Zhu, Yuntian Gao, Yongliang Dang, Cao Zhan, Shengchang Ji

Improved High-Speed Thermal Analysis based on Two-Step Simulation for High-Frequency
Transformers 2488
  Zheyuan Yi, Kai Sun, Qiang Li, Zengyang Liu

Core Material Characterization Under DC Bias Conditions 2495
  Jonas Mühlethaler, Fabrice Locher, Frédéric Mathieu, Edward Herbert

A Low-Cost Setup and Procedure for Measuring Losses in Inductors 2502
  Burkhard Ulrich

Effect of Temperature of Additively Manufactured Cores 2510
  Ken Johnson, Ali Bazzi

Extreme Temperature Permeability Engineered Soft Magnetics 2516
  Tyler W. Paplham, Alex M. Leary, Paul R. Ohodnicki Jr.

An Isolated RF Power Combining Approach with Multiple Decoupled Input Coils 2521
  Ziyang Xu, Yifan Zhao, Zhan Liu, Alex J. Hanson, Ming Liu

Simulation of a Custom Core, 15kV Isolated Gap Transformer Optimized for High Power Density 2527
  Andrew Galamb, Fei Teng, Srdjan Lukic

Low Interwinding Capacitance Design for PCB-Winding based Transformer in Self-Powered Gate Drive Power Supply for High-Voltage SiC MOSFET 2535
Yuan Zhou, Li Zhang, Yilun Chen, Tianxiang Yin, Lei Lin

Integrated 4-Level Dual-Phase Superimposed Quadratic Power Converter for High-Density Direct 48V/1V Conversion 2541
Prosenjit Ghosh, Jin Woong Kwak, Fei Zhou, D. Brian Ma

Compensation Method for Unbalance of the Multi-Channel Class E Power Amplifier using the Closed Loop Frequency Control 2547
Kyungmin Lee, Sungku Yeo

High Temperature Operation of Digital Gate Driver Integrated Into a Power Module 2551
Kazuma Saiga, Shohei Zaizen, Satoshi Nakano, Shigeru Kusunoki, Kiyoto Watabe, Katsuhiro Hata, Makoto Takamiya, Shin-Ichi Nishizawa, Wataru Saito

Evaluation Index-Based Multiphysics Coupling Model and Analysis Methodology for High-Reliable Power Supply Module 2556
Haoyu Wang, Xuliang Wang, Yan Wang, Xiaosen Liu

Electrical Characterization of Modular 3D Packaging Assembled with Compressed Metal Foams 2562
Paul Bruyere, Alexis Derbey, Betina Zynger-Capaverde, Yvan Avenas, Eric Vagnon, Jean-Luc Schanen, Jean-Michel Guichon, Omar Sanjakdar

Improvement in Short-Circuit Robustness of SiC-MOSFETs based Power Modules using Two-Level Turn-On (2LTO) 2569
Muhammad Muneeb Alam, Saad Khalid, Nisar Ahmed Khan, Ngoc Ho Tran, Sebastian Strache

GaN-Based Two Stage Point-of-Load (PoL) Converter with 2.5D Embedded Substrate Implementation 2576
Samuel Defaz, Yang Li, Fang Luo

Near-Field Coupling Mitigation of the Noise from High Voltage DC-Link Decoupling Capacitors in Voltage Source Converters 2582
Yuxuan Wu, Kushan Choksi, Samuel Defaz, Fang Luo

Advantages of Paralleling SiC MOSFETs in High-Performance Power Modules 2589
Steffen Beushausen, Dominik Alexander Ruoff, Wenqi Zhou, Karl Oberdieck

A SiC Half-Bridge Power Module based on Liquid Metal Packaging for High Performance and Low Thermal Stress 2597
Wei Mu, Ameer Janabi, Luke Shillaber, Borong Hu, Teng Long

Analysis and Modeling of Radiated EMI Considering Coupling Between Power Converter and Power Cable with LC-Type EMI Filter 2603
Qinghui Huang, Yingjie Zhang, Shuo Wang, Yirui Yang, Zhedong Ma, Yanwen Lai

Simple Prediction Method for Impacts of Switching Characteristics on EMI Noise of a Three-Phase PWM Inverter 2610
Shinobu Nagasawa, Toshiya Tadakuma, Keita Takahashi

Coaxially Nested 3.3 kV SiC MOSFET Packages with Uniform Interpackage Electric Field Distribution 2616
Jack Knoll, Mark Cairnie, Christina DiMarino

Thermal Modeling and Performance of a Bare-Die Embedded PCB for High Power Density Converters Design 2624
  Shahid Aziz Khan, Feng Zhou, Mengqi Wang, DucDung Le, Shivam Chaturvedi

Research on the Voltage Fluctuation Suppression Strategy in Weak Grid Under Pulsed Power Load Integration 2628
  Xi Chen, Jiazheng Zhang, Mingjun Bao

An Optimized Firmware-Based Cycle-by-Cycle Current Limiting Method for Power Electronic Converters in UPS 2634
  Teng Wu, Hong Liu

Frequency Stop-Band Management System for DC-DC Converters 2640
  Alessandro Bertolini, Alberto Cattani, Claudio Luise, Alessandro Gasparini

Multi-Stage Model Predictive Control with Enhanced Discrete-Time Models for Multilevel Inverters 2647
  Hoang Le, Apparao Dekka, Deepak Ronanki, Abdul R. Beig

Direct Effective Power Control (D-EPC) for LLC Resonant Converters Operating in Boost Mode using Event-Driven-Timer based Digital Controller 2654
  Yuto Yoshimura, Kenji Funatani, Kazuhiro Umetani, Toshiyuki Zaitsu, Akito Nakagaki, Masataka Ishihara, Eiji Hiraki

Mitigation Method of Resonance Between Paralleled On-Line UPS 2660
  Teng Wu, Zhenguo Huo, Shangxian Ning

An Extra-Element Small-Signal Model for a Current-Fed Resonant Dual-Active-Bridge Converter............ 2667
  Paolo Sbabo, Paolo Mattavelli, Giorgio Spiazzi, Andrea Petucco

Concurrent Charge Distribution and Time-Optimal Control for Unordered Single-Inductor Dual-Output Converter.. 2675
  Xuliang Wang, Haoyu Wang, Yang Liu, Yunxin Wang, Boran Zhang, Hongru Liu, Yan Wang, Xiaosen Liu

Circulating Current Control with Loss Reduction for Parallel Connected Inverters 2681
  Shun Endo, Takae Shimada, Masato Ando, Yuuichi Mabuchi, Masaki Miyamae, Naoki Takayama, Yohei Matsumoto, Naoto Onuma

Analysis of Power and Power Spectral Density for Quaternary Random Pulse Position Modulation 2687
  Hung-Chi Chen, Hsiang-Kai Wu, Chih-Chiang Wu

Bidirectional CLLC Converter using a Hybrid Control Method for Wide Voltage Range Applications.. 2692
  Jhih-Cheng Hu, Hong-Xuan Liao, Chien-Lung Liu, Wei Wang, Ming-Shi Huang

Design and Control of a High-Bandwidth Dual Active Bridge DC-DC Converter 2698
  Alper Uzum, Syed Imam Hasan, Yilmaz Sozer, Kenneth A. Loparo

Unified Model Predictive Control for DC-DC Buck Converters: From Start-Up to Steady-State Operation... 2703
  Zhengchen Guo, R.M. Nelms

A Novel IPPC Method for Precise Overload Protection and Burst Mode Operation in LLC Resonant Converters.. 2708
  Manikanta Pallantla, Ramkumar S

An Improved Current-Sensorless Model Predictive Voltage Control for Four-Leg Voltage Source Inverters.. 2713
Heng Guo, Yuxin Wei, Mengmeng Jing, Wenlong Ding, Bin Duan, Chenghui Zhang

A Highly Integrable, Modular and Multi-Functional Fault Monitoring Active Gate Driver with Parallel Buffers for a Global Enhanced Reliability of Gen. 3 SiC Power MOSFETs 2718
Mathis Picot-Digoix, Léo Seugnet, Frédéric Richardeau, Jean-Marc Blaquière, Sébastien Vinnac, Thanh-Long Le, Stéphane Azzopardi

A 24 – 16 V to 0.8 – 1.2 V Merged 4-Stage Hybrid-SC-SL Converter with 96.5% Peak Efficiency and Larger Than 50% iL Reduction.. 2725
Chien-Hao Tseng, Cheng-Ta Chuang, Chieh-Ju Tsai, Ching-Jan Chen

Innovation Active Gate Drive Method (Named TriC3™) for MOSFET Heat Reduction and EMI 2730
Hisashi Sugie

A KY Buck-Boost Converter with Extended Ramp Control Achieving 1500% Output Variation Reduction for Smooth Mode Transition 2735
Yu-Ting Hung, Chieh-Ju Tsai, Ching-Jan Chen, Chun-Yu Hsieh

An USB Cable based Extended Conversion Range L-First Hybrid-Converter using Valley-Virtual-Inductor-Current-Mode Control with Auto-Tracking Slope Compensation Against ±50% Inductance Variation 2741
Chun-I Li, Chieh-Ju Tsai, Ching-Jan Chen

Impact of Gate Resistor Configurations on Current Balancing in Paralleled SiC MOSFETs 2746
Yifu Zhang, Shashank Karanth, Emanuel Eni

Exploring the Potential of FPGA in High-Frequency Switching DC-DC Boost Converters using Model Predictive Control 2752
Qingcheng Sui, Bangli Du, Yu Zuo, Wilmar Martinez

A 7 Bit 5A 6.7 GHz Gate-Shaping Digital Gate Driver with Burst-Sampling ADC for Iterative Switching Optimization of SiC Power MOSFETs 2757
Tobias Zekorn, Kenny Vohl, Erik Wehr, Leon Weihs, Michael Hanhart, Ralf Wunderlich, Stefan Heinen

Decentralized Interleaving of Series-Stacked DC-DC Converters via Extremum-Seeking Control 2764
Ivan Petriœ, Vignesh Iyer, Shoudong Hu, Chirayu Rajpurohit, Bailey Sauter, Milan Iliœ, Luca Corradini, Dragan Maksimoviœ

Online Dead-Time Control for Half Bridges without Preliminary Training based on Switching Transient Steepness 2772
Lukas Knappstein, Niklas Falkenberg, Martin Pfost

Impedance-Based State-of-Health Estimation for Lithium-Ion Battery Management Systems 2779
Mohammad K. Al-Smadi, Jaber A. Abu Qahouq

Stability Analysis and Resonance Damping of LC Filter-Based Voltage Source Converter with Single-Loop Voltage Control.. 2785
Aravind G., Divyanshu Bansal, L. Umanand

Finite Control Set Model Predictive Control Combined with Online Junction Temperature Estimation for Reliability Enhancement of Voltage Source Inverters 2790
Qiang Mu, Jiale Zhou, Zaheen Mustakin, Lucas Pereira, Babak Parkhideh, Tiefu Zhao

Framework for Dynamic Control and Operation of Power Electronics Interfaces.. 2797
  Radha Sree Krishna Moorthy, Steven Campbell

Achieving Soft-Charging and Over 20% Input Current Ripple Reduction in a 48-to-6 V Dickson
Converter using 3-Phase Split-Phase Control.. 2805
  *Nagesh Patle, Rose A. Abramson, Sahana Krishnan, Jiarui Zou, Robert C. N. Pilawa-
  Podgurski*

Experimental Verification of Circuit-Losses Analysis-Model of DC-Output Converter Developed
using Approximated Equations from Measurement Data and Datasheet Data.. 2813
  Ryota Kondo, Tsuyoshi Funaki

Scattering Parameter Measurement System using Probes for Surface Mount Devices Operating in
the Frequency Range from 50 kHz to 1 GHz 2821
  *Ryoko Kishikawa, Masahiro Horibe, Tomokazu Shoji, Shigenori Yabuta, Toshi Ohi, Ryo
  Takeda, Takamasa Arai*

Optical Transformer Design with Additional Common-Mode Noise Reduction Winding for Flyback
DC-DC Converters 2828
  *Yusuke Irie, Shinichiro Eguchi, Yoichi Ishizuka, Toshiro Takeuchi, Akio Iwabuchi, Takahiro
  Koga, Toshiyuki Tanaka*

Enhanced Bus Voltage Stability Through Digital Twin-Enabled Adaptive Controller Tuning..................... 2833
  Matthew Belanger, Andy Wong, Kerry Sado, Enrico Santi

Modeling and Performance Characterization of Lithium-Ion Capacitor at Different Temperature
and Voltage Values.. 2840
  Mohammad K. Al-Smadi, Jaber A. Abu Qahouq, Sajad Saberi

Conveniently Identify Coils in Inductive Power Transfer System using Machine Learning.......................... 2846
  *Yifan Zhao, Mowei Lu, Ting Chen, Heyuan Li, Xiang Gao, Zhenbin Zhang, Minfan Fu, Stefan
  M. Goetz*

Accurate Modeling of LLC Resonant Converters with Enhanced Analytical Approach Considering
of Parasitic Capacitance 2851
  Dong Jiao, Zhengming Hou, Jih-Sheng Lai

High-Frequency Conditioning Circuits for Power-Related Information Extraction in Non-
Sinusoidal Power Electronic Systems 2857
  Haoyu Wang, Yuanxin Zhang, Di Mou, Alex Hanson, Shiqi Ji

Transconductance Model of the Dual Active Bridge Converter Under Single and Dual Phase Shift
Control... 2865
  Jared Cronin, Andrew Wunderlich, Enrico Santi

Lumped Parameter Modeling for Real-Time Thermal Regulation of Li-Ion Battery Packs........................... 2871
  Utkal Ranjan Muduli, Mohamed Shawky El Moursi, Khalifa Al Hosani, Ahmed Al-Durra

A Physics-Based Temperature Dependent Analytical Model for 2DEG Density in AlGaN/GaN
HEMT Devices.. 2877
  Kashfia Tajmim Nabila, Jerry L. Hudgins

Comparative Analysis of Stator-PM Machines: Design Optimization and Electromagnetic
Performance Evaluation 2883
  Maryam Salehi, Madhav Manjrekar

Elimination of Deadtime Effect on Resolver Offset Estimation using the Pulsating Current Command for Electric Vehicle Application 2889
Yingfeng Ji, Nurani Chandrasekhar

A Generic Load Emulator for Testing Motor Drives of E-Mobility 2894
Qingzheng Zhang, Kaiyuan Feng, Changsheng Hu, Dehong Xu

Design and Implementation of Power Assisted Control System for E-Bikes 2900
Che-Yu Lu, Tzu-Ping Cheng

A Hybrid PWM Strategy with Reduced Common-Mode Voltage and Extended Output Voltage Linearity for Adjustable Speed Drives 2907
Zhe Zhang, Kevin Lee

Single-Phase Open-Circuit Fault-Tolerant Control of Three-Phase PMSM Drives 2913
Yuichiro Minato, Yuki Nakata, Jun-Ichi Itoh

Multi-Vendor Encoder Position Sensing Interface using Programmable IP based Solution......... 2920
Rajul Bhambay, Dhaval Khandla, Pratheesh Gangadhar, Thomas Leyrer, Achala Ram, Manoj Koppolu, Archit Dev

Sensorless Control Method at Low-Speed Range using High-Frequency Voltage Injection for Synchronous Reluctance Motors Considering to Nonlinear Characteristic Due to Magnetic Saturation 2924
Sota Takizawa, Sari Maekawa

Hybrid Control Scheme for Permanent Magnet Gear Motor.. 2932
Bing Li, Takayoshi Matsuo, Ahmed Sayed-Ahmed, Yujia Cui, Jiangang Hu

Cost-Effective Fault Diagnosis for Motor and Inverter using Bootstrap Charging and Single DC Link Current Sensor 2937
Gyu Cheol Lim, Won Hyo Jeong, Kahyun Lee, Jung-Ik Ha

Improved PWM to Suppress Motor Overvoltage Caused by Voltage Reflection...................... 2943
Sung-Oh Kim, Kyo-Beum Lee

Analysis of Double Pulsing Effect in Motor Drives based on Vector Diagram........................... 2948
Byeong-Woo Kang, Kyo-Beum Lee

A Novel Speed Sensor-Less Control of a Solar-Powered PMSM Drive 2953
Abirami Kalathy, Arpan Laha, Praveen Jain, Majid Pahlevani

Design of a Compact Low-Loss MMC Double Submodule for MVDC and HVDC Applications 2960
Ali Sharaf Addin, Rainer Marquardt, Thomas Brückner

A Series-Type Dynamic Voltage Restorer Control Strategy to Cope with Voltage Swell............. 2968
Jiazheng Zhang, Hongyu Chen, Xi Chen, Mingjun Bao

Machine Learning Approach for Accurate Lithium-Ion Battery Temperature Prediction using Electrochemical Features Independent of Battery SOC and SOH... 2973
Vincent Masabiar Tingbari, Oluwaseun Isaiah Ekuewa, Anshul Nagar, Asad Abbas, Jamil Umar, Yuxin Zhang, Woonki Na, Jonghoon Kim

A Battery Strings Circulating Current Blocking Method for Battery Energy Storage Systems 2981
Haihong Long, Ziang Sun, Yucheng Fan, Xin Wu, Dehong Xu

A Hybrid Multilevel Converter-Based High-Gain Isolated DC/DC Converter for Grid-Tied Energy Storage Applications...................2986
  Pengyu Fu, Yizhou Cong, Jin Wang, Anant Agarwal

LCL Filter Parameter Selection using Graphical Method for a 13.8 kV ac 1.1 MVA 7-Level Flying Capacitor Grid-Connected Converter Utilizing Variable Switching Frequency...................2992
  Arthur Mendes, David Nam, Mingze Gao, Thimothy Thacker, Dong Dong, Rolando Burgos

Online Extraction of Electrochemical Impedance Spectroscopy Pattern based on EV Load Profile and Short Time Fourier Transform for Diagnosis of Lithium-Ion Battery Safety...................3000
  Miyoung Lee, Dongcheol Lee, Youngmin Bae, Jongchan An, Garam Yang, Woonki Na, Jonghoon Kim

Enhanced Incremental Capacity Analysis for Evaluating Battery Degradation Mechanisms of Optimized Fast Charging Methods...................3006
  Taehyeon Gong, Jaehyeong Lee, Sungjun Lee, Yura Kim, Bomyeong Ko, Woonki Na, Sungjin Choi, Jonghoon Kim

Co-Estimation of SOC and SOT in Lithium-Ion Batteries using an RLS-Based Heat Generation Model...................3012
  Seongkyu Lee, Eunjin Kang, Minhyeok Kim, Seunghyun Lee, Minwoo Song, Jaea Lee, Woonki Na, Jonghoon Kim

Three-Stage Adaptive Control Strategy for Stability Improvement of Grid-Connected Inverter in Weak Grid...................3018
  Longxiang You, Sicong Jin, Xin Zhang, Zuoshuai Wang, Sunqing Wang

Degradation Analysis of Offshore Bifacial PV Modules Under Multiple Climatic Stressors...................3024
  Aidha Muhammad Ajmal, Yongheng Yang

A Flexible Energy Management System for Solar Powered Electric-Bus Charging Stations...................3030
  Supun Amarathunga, Pasan Gunawardena, Xiaoting Wang, Yunwei Li

A Vienna Rectifier based Grid-Connected Powertrain for Hydrokinetic Turbine Systems...................3036
  Peidong Li, Md Tariquzzaman, Yue Cao

Condition Monitoring for DC-Link Capacitors and PV Arrays based on the Start-Up Process of the PV System...................3042
  Yongjie Liu, Ariya Sangwongwanich, Chen Liu, Xing Wei, Shuyu Ou, Tamás Kerekes, Jiahong Liu, Huai Wang

Electrically and Thermally Efficient Reliable Power Converter Design for Micro–Hydrokinetic Turbine...................3048
  Md Tariquzzaman, Peidong Li, Yue Cao

Comprehensive Evaluation of Cyber Attacks on Grid-Connected Smart Inverters...................3054
  Rishabh Singla, Vishwam Raval, Hasan Ibrahim, Jaewon Kim, Prasad Enjeti, Narsimha Reddy

Parallel Operation of Grid-Forming Converters based on Kuramoto Oscillators with Virtual Cable Emulation for Improved Power Sharing...................3059
  Vikram Roy Chowdhury, Gab-Su Seo, Barry Mather

Enhancing Hydrogen Production in Hybrid Standalone Microgrids...................3064
  Utkal Ranjan Muduli, Mohamed Shawky El Moursi, Khalifa Al Hosani, Ahmed Al-Durra

LSTM-Based Sub-Synchronous Oscillation Detection Scheme for Type 4 Wind Farm Interfaced with Weak AC Grid 3071
Omar Abu-Rub, Muhammad F. Umar, Jana A. Sheikh Ali, Yazan Qiblawey, Abdulrahman Alassi, Maryam Saeedifard, Mohammad B. Shadmand

A Study of Module Design Method to Suppress the Oscillation Occurs Between Parallel-Connected Power Devices 3077
Shinji Yato, Hiroto Sakai, Hideo Araki, Shumei Shimosako

A High-Efficient Hybrid Traction Inverter in Electric Vehicle Applications 3083
Yousefreza Jafarian, Omid Salari, Praveen Jain, Alireza Bakhshai, Mohamed Z. Youssef

Dual-Use of Onboard Chargers to Achieve Controllable DC Bus Voltage for Electric Vehicles 3089
Anuj Maheshwari, Elie Libbos, Arijit Banerjee

Isolated Single-Phase Onboard Chargers for BEV/PHEV using Active Power Decoupling Technology 3096
Yoshiki Amano, Keigo Nishimura, Hiroaki Matsumori, Takashi Kosaka, Kenichi Nagayoshi, Kenichi Watanabe

A Practical Use of xEVCap: The Modular and Standard DC-Link Capacitor Solution for the Main EV Powertrain Inverter 3100
David Olalla, Tomas Wagner, Fernando Rodriguez, Alberto Espinar

Optimized Bidirectional On-Board Charger using a Novel Unfolder-DAB Topology 3109
Héctor Sarnago, Ignacio Álvarez, Pablo Briz, Óscar Lucía

Critical Thermal Characterization of Next-Generation Solid-State Batteries for Automotive Battery Management Systems 3114
Chandan Chetri, Sheldon Williamson

Nanocrystalline CMC Inductors for EV Charging: Trade Studies and Testing Standardization 3119
Christopher Bracken, Mark A. Juds, Paul R. Ohodnicki, Bharadwaj Reddy Andapally, Jose Gato

Predicting Efficiency of On-Board and Off-Board EV Charging Systems using Machine Learning 3124
Mohamed Yasko, Fanghao Tian, Wilmar Martinez, Johan Driesen

High-Power and High-Speed Multi-Channel VCSEL Arrays with GaN Driver for Automotive LiDAR 3129
Yifu Liu, Sichao Li, Junlei He, Changyu Hu, Bill He, Karthik Krishnamurthy, Andy Shen

Double Pulse Test Platform for Hybrid SiC-IGBT Switch Characterization and Optimal Gate Control Strategy for EV Traction Inverters 3133
Rosario Attanasio, Harsha Ademane, Ryan Satterlee, Gianni Vitale

Critical Role of Individual Cell Temperature Monitoring in Mitigating Thermal Runaway and Reducing Accelerated Degradation in Lithium-Ion Batteries 3141
Mohit Sharma, Akash Samanta, William Locke, Sheldon Williamson

Loss-Optimized Design of a Triple Active Bridge DC-DC Converter for an Electric Vehicle Application 3147
Sreejith Chakkalakkal, Kyle Kozielski, Wesam Taha, Yicheng Wang, Aniket Anand, Ali Emadi

A Magnetic-Less DC/DC Converter with Pulse Charging for 800 V Powertrains from 400 V DC Fast Chargers 3155
Duc Dung Le, Shivam Chaturvedi, Shahid Aziz Khan, Mengqi Wang, Mohamed Elshaer

Boosting Charger Efficiency: A GaN-Based Flyback Converter with Energy Recycling 3160
Ahmad Nabizadah, Majid Ghasemi Korrani, Babak Fahimi

A Hybrid Three-Level Buck Converter with Flying Supercapacitor for High Load Current Surge
Capability using Peak Current Mode Control 3167
*Finlay Lodge, Rafael Peña-Alzola, Martin MacFadyen, Patrick Norman, Mark Sweet,
Graeme Burt*

Supercritical Carbon Dioxide (sCO.)-Cooled Current Source Inverter-based Integrated Motor Drive
for MW-Scale Electric Aviation Applications 3174
*Hang Dai, John Yagielski, Thomas Jahns, Kum-Kang Huh, Vandana Rallabandi, Libing
Wang, Tarak Saha, Wenda Feng, Bulent Sarlioglu*

The Challenge of Thermal Runaway in Soft Magnetic Materials for Inductive Power Transfer 3181
Yibo Wang, Ben Zhang, Weisheng Guo, Tianlu Ma, Sheng Ren, C.Q. Jiang

A Capacitively Coupled Alternative Electric Field Control for Freeze-Free based High Quality
Food Preservation.. 3187
Jaeyong Cho, Junhyeong Park, Sung-Bum Park, Daehyun Kim, Jinsoo Choi

The Characteristics of the Long Length Primary Loop and the Power Supply for the SCMaglev's
DWPT System 3194
Keisuke Yamamoto, Jun Enomoto, Shunsaku Koga, Junichi Kitano

A Wireless EV Charging System with a Double-Sided LCC Network using Variable Switching
Frequency and DC-Link Voltage Control.. 3200
Chae-Lyn Kim, Hyeonu Jo, Ju-A Lee, Dong Hyeon Sim, Byoung Kuk Lee

Class E/EF Inductive Power Transfer to Achieve Stable Output Under Variable Low Coupling................ 3206
Yifan Zhao, Mowei Lu, Heyuan Li, Zhenbin Zhang, Minfan Fu, Stefan M. Goetz

A Motorized Air-Core Variable Inductance Winding Structure 3212
Xindong Li, Sampath Jayalath, Cheng Zhang

Wireless Power Transfer System with Automatic Tuning Capability in Metallic Environment.................... 3220
*Renjie Zhang, Yue Wu, Delin Zhao, Yaohua Li, Yongbin Jiang, Yi Tang, Huan Yuan, Xiaohua
Wang, Mingzhe Rong*

Design of Wireless Power Transmitters for Enhanced Transmission Distance and Output Power 3227
Kaiyuan Wang, Shuang Zhao, Shuye Shang, Eric Ka-Wai Cheng, Siew-Chong Tan, Yun Yang

Optimization of Wireless Power Transfer Waveforms and In-Vivo Receivers for Implantable
Medical Devices 3232
Hanbing Liu, Xin Zan

Comparison of Compact Power Amplifier Designs for High Frequency Resonant Wireless Power
Transfer Systems at 6.78 MHz using High-Q Resonators... 3241
Manuel Rueß, Kilian Müller, Mathias C.J. Weiser, Ingmar Kallfass

Analysis and Design of Capacitive Coupling Wireless Power Transfer System using Load-
Independent Class-EF Inverter.. 3248
*Takumi Kobayashi, Yutaro Komiyama, Akihiro Konishi, Hiroaki Ota, Yuki Ito, Taichi Mishima,
Takeshi Uematsu, Kien Nguyen, Hiroo Sekiya*

Design and Optimization of a 600 W Wireless Drone Charger for High Gravimetric Power Density 3253
Arka Basu, Daniel Costinett

Stabilization Method for DC-Bus Oscillation in Dynamic Wireless Power Transfer Systems 3261
  Yuki Ochiai, Keisuke Kusaka

Unveiling Aliasing Effect on Resonant Pole Locations in Wireless Battery Chargers 3267
  Anwesha Mukhopadhyay, Daniel Costinett

Integrated Hybrid Inductive and Capacitive Power Transfer System with Asymmetrical PCB Self-
Resonator 3275
  Yao Wang, Zhen Sun, Xiangrong Zhang, Yun Yang, Shu Yuen Ron Hui

High Frequency Noise Reduction Method of the Class E Power Amplifier 3281
  Kyungmin Lee, Sungku Yeo

Single-Stage Three-Phase Buck-Matrix Rectifier with Series-Parallel Connected Transformers for
High-Power 48 V Data Center Power Supplies 3285
  Yuki Ishikura, Chinmay Bhagat

Sector Transition PWM Modulation Scheme for a Three-Phase Isolated Buck-Matrix Rectifier 3291
  Chinmay Bhagat, Yuki Ishikura

Adaptive Capacitance Circuit for Optimal Dynamic Impedance Matching in Variable Reluctance
Energy Harvesting Applications 3298
  Alejandro Redondo, Fernando Pérez, Sofía García, Gabriel Mujica, Airán Francés

Gallium Nitride (GaN) based Topology Comparison for Low Power Battery Charging Applications 3304
  Jai Aditya Chaudhary, Rosario Attanasio, Gianni Vitale

Server Motherboard Power Performance Study Under Immersion Cooling Environment 3312
  *Meng Wang, Haiyan Wang, Pavan Kumar, Haijin Zhang, Xiang Li, Fengwei Bian, Jianting
  Deng, Jiaqi Zhu, Yiming Lei*

Practical PCB Design Considerations for GaN HEMTs based Isolated DC-DC Converter 3316
  Gaureej Gauttam, Harish S. Krishnamoorthy, Sai Sushma Pasupuleti

Data-Driven Characterization and Forecasting of Metal-Oxide Varistor Degradation in DC Circuit
Breakers 3321
  Zhi Jin Zhang, Yang Liu, Lukas Graber, Maryam Saeedifard

A Thyristor-Based Fault Current Bypass Solid-State Circuit Breaker for DC Microgrid Applications 3328
  Jiale Zhou, Xiuhu Sun, Qiang Mu, Tiefu Zhao

Single-Stage Three-Phase AC-AC Isolated Inertialess Converter (IIC) for Industrial Drives 3334
  *Brad Houska, Decheng Yan, Aniruddh Marellapudi, Satish Belkhode, Joseph Benzaquen
  Sune, Deepak Divan*

Author Index

Design and Placement of a Passive Clamp Snubber for Isolated SEPIC and Cuk Converters Working as Automatic Power Factor Correctors

Abraham López, Juan Rodríguez, Duberney Murillo-Yarce, Javier Sebastián and Diego G. Lamar
University of Oviedo, Electronic Power Supply Systems Group. Campus de Viesques s/n, 33204 Gijón. Spain.
Email: lopezantunaa.fuo@uniovi.es

Abstract — DC/DC power converters with galvanic isolation and using only one power transistor need elements that limit the voltage peaks at the beginning of the transistor turn off. These elements are called clamp snubbers. Its placement and design are well known for power converters with two reactive elements (e.g. flyback). However, different placements can be considered for those clamp snubbers in converters with higher number of reactive elements. Moreover, if the converter works as a Resistor Emulator (RE) in a Power Factor Corrector (PFC), the snubber must take into account the continuous variation of some of the electrical variables. This paper presents the study of four different placements for a passive clamp snubber network in a SEPIC converter working as an automatic RE, i.e., working in the Discontinuos Conduction Mode (DCM) and with a constant duty cycle during a line period. The value of the clamp snubber resistor needed to achieve a specific clamp voltage for these four options is determined in this paper. Moreover, the four options are compared in terms of the dissipated power in the snubber resistor. Consequently, it is possible to determine which one is going to be the best snubber option, in terms of efficiency. This study has been carried out for a SEPIC topology, and it is also valid for the Cuk one. Finally, all the study developed in this paper has been validated considering PSIM simulations and experimental results using a SEPIC prototype working as an automatic PFC.

Keywords — *SEPIC converter, Clamp snubber, Power factor corrector, Resistor emulator.*

I. INTRODUCTION

Galvanic isolation between input and output is very common in AC/DC and in DC/DC converters, which implies the use of several-winding magnetic components. Introducing galvanic isolation in switching-mode DC/DC converters always implies some problems due to the abrupt interruption of the current passing through the leakage inductance that the magnetic component always has. If the voltage across the transistor is clamped to a certain value by the converter topology (e.g., half-bridge and full-bridge cases), then the leakage inductance only causes parasitic oscillations. However, it can cause serious damage for the transistors in topologies such us flyback, forward, push-pull, SEPIC, Cuk, Zeta and current-feed converters, if the voltage across the transistor is not properly clamped. In these cases, it is necessary to use a snubber network [1] to limit the voltage across the transistor and to reduce the parasitic oscillations. Snubbers can be active or passive networks. Active snubbers include transistors, diodes and capacitors [1],[2]. All the energy stored in the leakage inductance is recovered when active snubbers are used, thus

avoiding losses in the snubber. However, its use increases the converter complexity and cost due to the additional transistor used, including its control and its driving circuitry. Therefore, simple passive and dissipative snubber networks are preferred for low-power and low-cost converters. A very well-known passive clamp snubber is shown in Fig.1 for a flyback converter with leakage inductance L_k. This clamp snubber is in charge of limiting the drain-source peak voltage across the transistor. In this snubber, the electric charge injected into capacitor C_c, has to be compensated with the electric charge that flows from C_c to another point in the circuit (whose voltage must be lower). The balance of these electric charges is reached for a specific voltage across C_c, which in fact is the clamp voltage. The value of the clamp snubber resistor, R_c, and its placement in the circuit determines the clamp voltage. It should be noted that the position of resistor R_c is not irrelevant, because it determines whether a part of the energy stored in C_c is returned to the input port, or, if it is completely wasted. For example, one of the R_c terminals is connected to the positive terminal of V_g in Fig.1. With this placement of R_c, current i_{Rc} flows through both R_c and V_g. Therefore, there is a partial return of energy to the input voltage source V_g. If R_c were connected in parallel with C_c

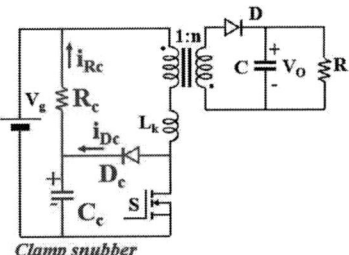

Fig. 1. Right connection of a passive clamp snubber for a flyback converter.

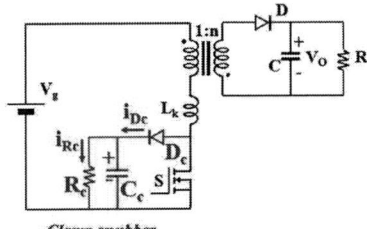

Fig. 2. Wrong connection of a passive clamp snubber for a flyback converter.

(Fig.2) this energy would not return to V_g. Therefore, the position of R_c in the circuit is not irrelevant.

A DC/DC converter can be placed between a line rectifier and a low frequency filter capacitor C_B, (as shown in Fig.3). In this case, an AC/DC converter with high power factor and low input harmonic distortion can be implemented. This is achieved by forcing the converter input current, averaged in a switching cycle, to be proportional to the converter input voltage. Once this behavior is achieved, the converter will be working as a Resistor Emulator (RE), performing a Power Factor Corrector (PFC). In the case of the converters belonging to the flyback family of converters (such as flyback, SEPIC, Cuk and Zeta converters), the easiest way to achieve RE behavior is by designing the converter as follows:

 a) The converter must always operate in Discontinuous Conduction Mode (DCM).

b) The converter switching frequency must be constant.

c) The converter duty cycle must remain almost constant during each line period [3].

The RE thus obtained is called "automatic RE" in this paper. The overall converter (line rectifier + RE) is called "automatic PFC".

Fig.3 shows a flyback converter working as an automatic PFC, with the same snubber placement as the one shown in Fig.1. However, the way to compute the value of the R_c will be different for several reasons:

a) The converter input voltage, called $v_g(\phi)$ in this case, is not constant now. It changes according to line angle ϕ, because it is a rectified version of the line voltage.

b) The energy stored in L_k also changes according to ϕ.

c) The connection of R_c could introduce some distortion in $i_g(\phi)$, especially in the line voltage zero crossing. This effect is, in practice, negligible.

The main objective of this paper is to evaluate the clamp snubber placement for the SEPIC (Fig.4) and Cuk converters operating as automatic PFC [4]-[7]. Therefore, they must always work in DCM, operating at constant switching frequency and maintaining their duty cycle almost constant each line period. Although the study is focused on the SEPIC topology, the study carried out is also valid for the Cuk one.

This paper is organized as follows: Section II describes the SEPIC converter working as an automatic PFC. Section III

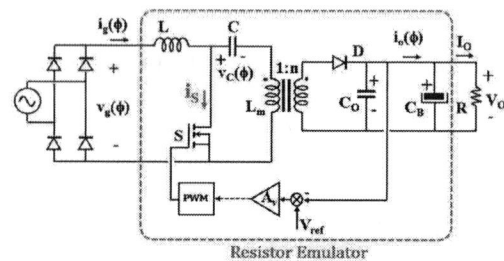
Fig.4. SEPIC converter working as an automatic PFC.

describes the four different options for a clamp snubber placement in the automatic SEPIC PFC. Simulation and experimental results are presented in Section IV. Finally, the conclusion is presented in Section V.

II. SEPIC CONVERTER WORKING AS AN AUTOMATIC PFC

Fig.4 shows a SEPIC converter working as an automatic PFC. In this converter the intermediate capacitor C is designed to have a negligible switching frequency ripple, but allowing variations of twice the line frequency [8]. Analyzing the average voltage (averaged in a switching period) in the loop made up of the rectifier output, inductance L, capacitor C and magnetizing inductance L_m, we easily obtain:

$$v_c(\phi) = v_g(\phi). \qquad (1)$$

The waveform corresponding to the current passing through transistor S, i_S, is represented in Fig.5. For the sake of clarity, the switching period T_s, and the line period T_L have not been represented in the proper scale. The line-rectifier output voltage can be expressed as follows:

$$v_g(\phi) = V_g|sin(\phi)|, \qquad (2)$$

where V_g is the peak line voltage. As the converter works as an automatic PFC, the low frequency component of $i_g(\phi)$ will be:

$$i_g(\phi) = I_g|sin(\phi)|, \qquad (3)$$

where I_g is the peak value of $i_g(\phi)$. It should be noted that $i_g(\phi)$ is the average value, averaged in a switching period, of the actual current passing through inductor L. Finally, the normalized conversion ratio at the peak line voltage is:

$$M = \frac{V_o}{nV_g} \qquad (4)$$

In order the SEPIC converter to achieve sinusoidal line current, the following conditions must be satisfied [8]-[10]:

a) In order to avoid the Continuous Conduction Mode (CCM), the converter duty cycle must verify:

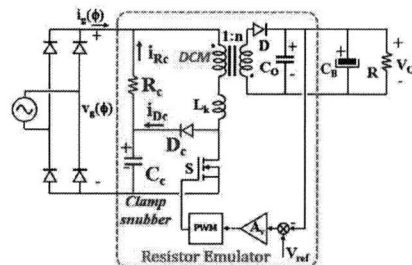
Fig.3. Flyback converter used as an automatic PFC.

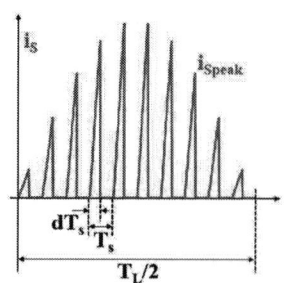
Fig.5. Transistor S current waveform.

979-8-3315-1612-3/25 $31.00 © 2025 IEEE

$$d < \frac{M}{M+1}. \tag{5}$$

b) In order to avoid operation in other discontinuous conduction modes, different from the classic one, L_m:

$$L_m < ML \tag{6}$$

According to [10], the value of I_g is:

$$I_g = \frac{d^2(L + L_m)T_s}{2LL_m}V_g \tag{7}$$

A power balance in a line half-period, allows us to relate I_g, V_o and R, as follows

$$I_g = \frac{2V_o^2}{V_g R} \tag{8}$$

As capacitor C has been designed to have constant voltage during a switching period, the average value of the current passing through it during a switching period will be zero in steady state. Therefore, the average value of $i_g(\phi)$, averaged in a switching period, will be the same as the average value of $i_S(\phi)$, also averaged in the same switching period. Therefore, the peak value of the current through transistor S, $i_{Speak}(\phi)$, can be easily related to its average value, due to the triangular shape of $i_S(\phi)$:

$$i_{Speak}(\phi) = \frac{2i_g(\phi)}{d} = \frac{2I_g}{d}|sin(\phi)| \tag{9}$$

The magnetic component that provides galvanic isolation in the converter shown in Fig.4 is a two-winding inductor, L_m being the inductance corresponding to the primary winding. The coupling between windings has been considered ideal so far. However, the actual coupling will not be ideal, resulting in the leakage inductance L_k. The presence of L_k justifies the use of a clamp snubber network in this converter.

III. PLACEMENT POSSIBILITIES FOR THE CLAMP SNUBBER IN A SEPIC-BASED AUTOMATIC PFC

Fig.6 shows the four options for the clamp snubber placement in an automatic SEPIC PFC. The main objective of the paper is to compute the power dissipated in resistor R_c for each snubber option, forcing the condition that the maximum voltage withstood by transistor S is the same in the four cases. Consequently, the option with lower value of power dissipated in R_c will the best.

The transistor off-state voltage, without L_k, is given by:

$$v_{SOFF}(\phi) = v_C(\phi) + \frac{V_o}{n} \tag{10}$$

The maximum value of $v_{SOFF}(\phi)$ takes place at $\phi = \frac{\pi}{2}$. Considering (1), (2) and (10), we obtain:

$$v_{SOFF}\left(\frac{\pi}{2}\right) = V_g + \frac{V_o}{n} \tag{11}$$

Due to L_k, the actual transistor off-state voltage will be higher than the value given by (11). In fact, the clamp snubber is in charge of limiting this voltage to a safe value. At this point parameter λ is introduced in order to determine the increase of the voltage across transistor S when both L_k and the clamp snubber are considered. The actual transistor off-state voltage, V_{Speak}, will be now:

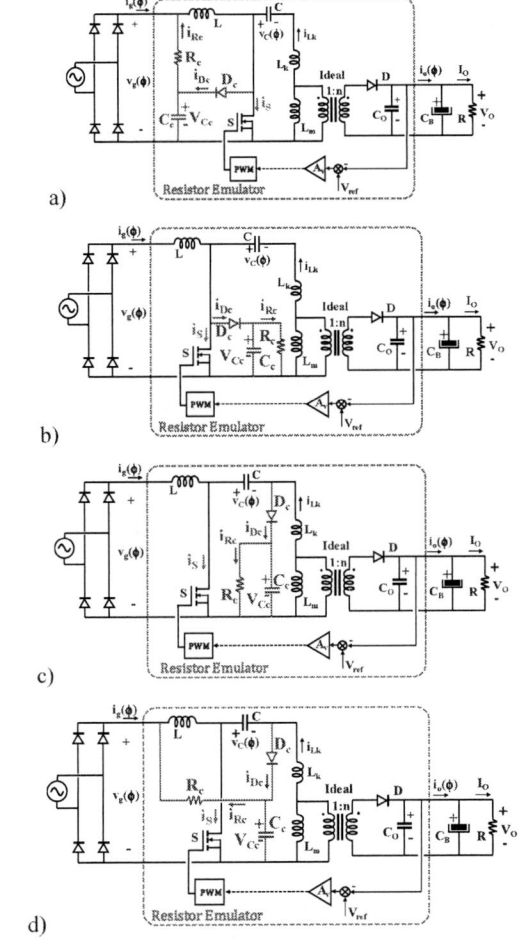

Fig.6. Different places for a clamp snubber in an automatic SEPIC PFC. a) Option A. b) Option B. c) Option C. d) Option D

$$V_{Speak} = \lambda v_{SOFF}\left(\frac{\pi}{2}\right) = \lambda\left(V_g + \frac{V_o}{n}\right) \tag{12}$$

According to (4), (12) can be rewritten as follows:

$$V_{Speak} = \lambda V_g(1 + M) \tag{13}$$

Fig.7 (a) shows an equivalent circuit to analyze the four clamp snubber options during the transistor off state. The value of voltage sources, v_1, v_2 and v_3 are given in Table 1 for each snubber option. As this Table 1 shows, the values of these voltage sources do not change in a switching cycle.

From Fig.7(a), the waveform corresponding to current $i_{Dc}(t, \phi)$ (current passing through diode D_c after turning the transistor off) can be easily computed. As Fig.7(b) shows, the value of $i_{Dc}(t, \phi)$ coincides with $i_{Speak}(\phi)$ just in the transistor turn off (at $t = dT_s$). As D_c starts conducting in this moment, the values of V_{Speak} and V_{Cc} verifies:

$$V_{Speak} = V_{Cc} + v_2\left(\frac{\pi}{2}\right). \tag{14}$$

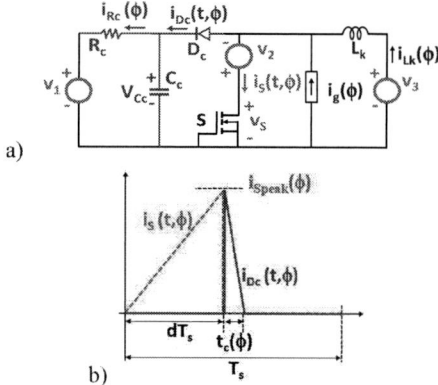

a)

b)

Fig.7. a) Equivalent circuit for the four snubber options during the transistor off state. b) $i_S(t,\phi)$ and $i_{Dc}(t,\phi)$ waveforms.

Table 1. Different values for v_1, v_2 and v_3 voltage sources depending on the snubber option

	v_1	v_2	v_3
Op. A	$v_g(\phi)$	0	$v_g(\phi)+V_o/n$
Op. B	0	0	$v_g(\phi)+V_o/n$
Op. C	0	$v_g(\phi)$	V_o/n
Op. D	$v_g(\phi)$	$v_g(\phi)$	V_o/n

Taking into account the actual values of L_k, current $i_{Dc}(t,\phi)$ reaches zero in a relatively short fraction of time, called $t_c(\phi)$ in Fig.7(b). Consequently, it is assumed that current $i_g(\phi)$ does not change significantly during $t_c(\phi)$ and, therefore, can be represented by a current source in Fig.7(a). Moreover, as the values of voltage sources, v_1, v_2 and v_3 remain constant during a switching cycle, the current waveform corresponding to $i_{Dc}(t,\phi)$ during the transistor off-state is a ramp, with a downwards slope. Obviously, diode D_c prevents a negative value for $i_{Dc}(t,\phi)$ (see Fig.7 (b)). Consequently, the value of $t_c(\phi)$ can be easily obtained from Faraday's law:

$$t_c(\phi) = \frac{L_k i_{Speak}(\phi)}{V_{Cc}-v_3}. \tag{15}$$

The average value of $i_{Dc}(t,\phi)$ in a switching period will be:

$$\langle i_{Dc}(t,\phi)\rangle_{T_s} = i_{Dc\_Ts}(\phi) = \frac{L_k (i_{Speak}(\phi))^2}{2T_s(V_{Cc}-v_3)}. \tag{16}$$

Considering (9), expression (16) can be rewritten as:

$$i_{Dc\_Ts}(\phi) = \frac{2L_k I_g^2 [sin(\phi)]^2}{d^2 T_s(V_{Cc}-v_3)}. \tag{17}$$

The average value of $i_{Dc\_Ts}(\phi)$ in a line half-period will be:

$$\langle i_{Dc\_Ts}(\phi)\rangle_\pi = I_{Dc} = \frac{2L_k I_g^2}{\pi d^2 T_s}\int_0^\pi \frac{[sin(\phi)]^2}{V_{Cc}-v_3} d\varphi. \tag{18}$$

From Fig.7(a), the value of $i_{Rc}(\phi)$ is:

$$i_{Rc}(\phi) = \frac{V_{Cc}-v_1}{R_c}. \tag{19}$$

Its average value in a line half-period will be:

$$\langle i_{Rc}(\phi)\rangle_\pi = \frac{1}{\pi}\int_0^\pi \frac{V_{Cc}-v_1}{R_c} d\phi \tag{20}$$

In steady state, the average currents given by (18) and (20) must be equal, thus obtaining the general equation that allow us to compute the value of R_c:

$$\frac{2L_k I_g^2}{d^2 T_s}\int_0^\pi \frac{[sin(\phi)]^2}{V_{Cc}-v_3} d\phi = \frac{1}{R_c}\int_0^\pi (V_{Cc}-v_1)\,d\phi. \tag{21}$$

Once value of R_c is known, the power dissipated in this resistor can be computed from the rms value of the voltage across it:

$$P_{Rc} = \frac{1}{\pi R_c}\int_0^\pi [(V_{Cc}-v_1)^2]\,d\phi. \tag{22}$$

Equations (21) and (22) are the key equations to compare the four clamp snubber options shown in Fig.6. In the following sub-section, a study of the four different clamp snubber options will be presented. It should be noted that the equivalent circuit and the waveforms of Fig.7 and the values given in Table 1 are also valid for the Cuk converter. Therefore, all the study carried out for the automatic SEPIC PFC is also valid for the Cuk case.

A. Option A of Clamp Snubber

As Table 1 shows, v_2 is zero in this case. According to (14), the value of V_{Cc} is equal to the value of V_{Speak}, given by (13). Taking into account the values of v_1 and v_3 (Table 1), (21) becomes:

$$\frac{2L_k I_g^2}{d^2 T_s V_g} H(\lambda, M) = \frac{V_g}{R_{c\_A}}[\pi\lambda(1+M)-2], \tag{23}$$

where R_{c_A} is the value of R_c for this snubber option and $H(\lambda, M)$ is defined as follows:

$$H(\lambda, M) = \int_0^\pi \frac{[sin(\phi)]^2}{\lambda(1+M)-M-|sin(\phi)|} d\phi. \tag{24}$$

The value of R_{c_A} obtained from (23) can be rewritten as:

$$R_{c\_A} = R_{base}\frac{[\lambda(1+M)-\frac{2}{\pi}]\pi}{H(\lambda,M)}, \tag{25}$$

where R_{base} is defined as:

$$R_{base} = \frac{d^2 T_s V_g^2}{2L_k I_g^2}. \tag{26}$$

According to (13) and (22), the power dissipated in R_{c_A} is:

$$P_{Rc\_A} = \frac{V_g^2}{R_{c\_A}}\left[\lambda^2(1+M)^2 + \frac{1}{2} - \frac{4}{\pi}\lambda(1+M)\right]. \tag{27}$$

B. Option B of Clamp Snubber

In this option, v_1 and v_2 are zero. Taking into account (13) and (14), (21) becomes in this case:

$$\frac{2L_k I_g^2}{d^2 T_s V_g} H(\lambda, M) = \frac{V_g}{R_{c\_B}}[\pi\lambda(1+M)] \tag{28}$$

where R_{c_B} is the value of R_c for this option. The value of R_{c_B} obtained from (28) can be rewritten as:

$$R_{c\_B} = R_{base}\frac{[\lambda(1+M)]\pi}{H(\lambda,M)}. \tag{29}$$

From (13), (14) and (22), the power dissipated in R_{c_B} is:

$$P_{Rc\_B} = \frac{V_g^2}{R_{c\_B}}[\lambda^2(1+M)^2]. \tag{30}$$

C. Option C of Clamp Snubber

In this case, the values of V_{Cc} and V_{Speak} are different. Taking into account (13) and (14), the value of V_{Cc} is:

$$V_{Cc} = v_{Speak} - v_g\left(\frac{\pi}{2}\right) = \lambda V_g(1+M) - V_g. \quad (31)$$

Taking into account (31), (21) becomes:

$$\frac{2L_kI_g^2}{d^2T_s}\int_0^\pi \frac{[sin(\phi)]^2}{V_g[(\lambda-1)(1+M)]}d\phi = \frac{V_g}{R_{c\_C}}\int_0^\pi[\lambda(1+M)-1]\,d\phi, \quad (32)$$

where R_{c_C} is the value of R_c for this option. Its value, obtained from (32), can be rewritten as:

$$R_{c\_C} = R_{base}2[\lambda(1+M)-1](1+M)(\lambda-1). \quad (33)$$

Once R_{c_C} is known, the power dissipated in this resistor can be easily computed from (22) and (31), the result being:

$$P_{Rc\_C} = \frac{V_g^2}{R_{c\_C}}(\lambda(1+M)-1)^2. \quad (34)$$

D. Option D of Clamp Snubber

In this case, (31) also gives the value of V_{Cc}. Taking into account this value and the values of v_1, v_2 and v_3, (21) becomes:

$$\frac{2L_kI_g^2}{d^2T_s}\int_0^\pi \frac{[sin(\phi)]^2}{V_g[(\lambda-1)(1+M)]}d\phi = \frac{V_g}{R_{c\_D}}\int_0^\pi[\lambda(1+M)-1-|sin(\phi)|]\,d\phi, \quad (35)$$

where R_{c_D} is the value of R_c for this option. Again, the value of R_{c_D} obtained from (35) can be rewritten as:

$$R_{c\_D} = R_{base}2\left[\lambda(1+M)-1-\frac{2}{\pi}\right](1+M)(\lambda-1). \quad (36)$$

As in the previous cases, the power dissipated in R_{c_D} can be calculated from (22) and (31), as follows:

$$P_{Rc\_D} = \frac{V_g^2}{R_{c\_D}}\left[(\lambda(1+M)-1)^2 + \frac{1}{2} - \frac{4}{\pi}(\lambda(1+M)-1)\right] \quad (37)$$

E. Comparison Between the Four Snubber Options

In order to compare the four snubber options, a base power value is selected as follows:

$$P_{base} = \frac{V_g^2}{R_{base}}. \quad (38)$$

This base value is used to normalize the dissipated power value corresponding to each option of clamp snubber. Therefore, (27), (30), (34) and (37) become:

$$PN_A = \frac{\lambda^2(1+M)^2+\frac{1}{2}-\frac{4}{\pi}\lambda(1+M)}{\left[\lambda(1+M)-\frac{2}{\pi}\right]\pi}H(\lambda,M), \quad (39)$$

$$PN_B = \frac{\lambda^2(1+M)^2+d\left(\frac{1}{2}+\frac{2M}{\pi}\right)}{\lambda(1+M)\pi}H(\lambda,M), \quad (40)$$

$$PN_C = \frac{\lambda(1+M)-1}{2(1+M)(\lambda-1)}, \quad (41)$$

$$PN_D = \frac{(\lambda(1+M)-1)^2+\frac{1}{2}-\frac{4}{\pi}(\lambda(1+M)-1)}{2\left[\lambda(1+M)-1-\frac{2}{\pi}\right](1+M)(\lambda-1)}, \quad (42)$$

where PN_x represents the normalized power dissipated in R_{c_x} when Option x has been selected as clamp snubber.

Fig.8 shows these normalized losses, for each option of clamp snubber, at different values of λ and M. As this Fig.8 shows, Option D is the one that exhibits the lowest power loss. However, for unusually low values of λ and M, Option C is the one that presents the minimum loss. By equaling (41) and (42), we can deduce the point where the power loss of these options coincide. The result is:

$$M_{C-D} = \frac{\pi+4}{4\lambda} - 1 \quad (43)$$

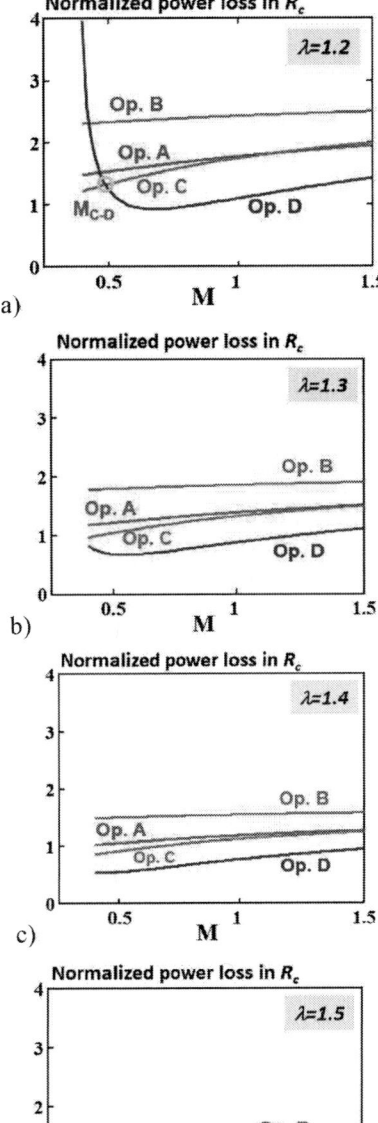

a)

b)

c)

d)

Fig.8. Normalized power loss in R_c for different values of λ and M: (a) $\lambda=1.2$. (b) $\lambda=1.3$. (c) $\lambda=1.4$. (d) $\lambda=1.5$.

Therefore, snubber power loss in Option D is lower than the corresponding loss in Option C if $M > M_{C-D}$, which is the common case. Otherwise, lower power loss is achieved in Option C. Moreover, the option that generates the highest power loss is Option B. Finally, power loss for Option A and for Option C are quite similar for all cases.

IV. EXPERIMENTAL AND SIMULATION RESULTS

In order to verify the analysis carried out to determine the value of resistor R_c and the power dissipated in this resistor for the four options of clamp snubber, an automatic SEPIC PFC has been used. First, the four clamp snubber options shown in Fig.6 have been simulated using PSIM. The values for the main SEPIC parameters are:

V_g = 169.71 V V_o = 48 V n = 0.318 M = 0.89

I_g = 0.742 A d = 0.413 T_s = 10 μs L_k = 1.46 μH

L = 3.19 mH L_m = 207 μH λ = 1.4

Bearing in mind these values, the maximum voltage withstood by transistor S, without considering the voltage spike due to L_k, would be 320.7 V (11). However, a voltage spike always appears due to L_k. The clamp snubber must limit this spike to a desired value, determined by λ. Taking into consideration the value selected for λ, the actual value of the maximum voltage withstood by transistor S is 450 V (12). The converter input power, P_g, calculated from the values of V_g and I_g, is 63 W.

A prototype, similar to the simulated one, has been built. In this case, the input power is 71.3 W and the output power is 64.7 W. In order to have an output voltage of 48 V, duty cycle d was adjusted to be 0.435. The values of snubber resistors R_{c_x}, computed from (25), (29), (33) and (36), are:

R_{c_A} = 107.8 kΩ R_{c_B} = 142.1 kΩ

R_{c_C} = 75.8 kΩ R_{c_D} = 46.5 kΩ

Table 2 shows a comparison between theoretical, simulation (using PSIM) and experimental results for each snubber option. These results have been obtained at the peak line voltage. The experimental values of V_{Speak} have been measured in the waveforms of Fig.9. As Table 2 shows, theoretical and simulated results fit very well, with errors lower than 7.2%. Regarding the experimental results, the values of the power loss and V_{Speak} are lower than those expected from the theoretical study. This is because the method proposed here does not consider the electric charge transferred to the transistor output capacitance, C_{oss}. A fraction of the energy variation in L_k is not transferred to capacitor C_c (and finally dissipated in R_c), but it is transferred to C_{oss} (and finally dissipated in the transistor when it turns on).

The discrepancy between theoretical and the simulated results can be predicted according to the considerations explained in [11]. Thus, total energy E_c stored in a nonlinear capacitor C_v when its voltage v changes from 0 V to a given voltage V_c can be computed as follows:

$$E_c = \int_0^{V_c} v \cdot C_v dv. \tag{44}$$

The transistor used in the converter prototype was a STP18N60DM2. Fig.10 shows the C_{oss}-V_{ds} curve for this transistor, obtained from its datasheet. From the information given in this Fig.10, the value of the energy stored in C_{oss} can be estimated according to (44). Thus, a numerical approach to (44) from 0 V to 400 V gives that the energy stored in C_{oss} is 2.974 μJ. This energy is wasted in the transistor during the transistor turn on, instead of being transferred to C_c. As the switching period is 10 μs, this energy means a power dissipation of 0.2974 W. This power dissipation must be discounted from the power transferred to C_c and finally dissipated in R_c. Therefore, the actual value of V_{Cc} (and, consequently, of V_{Speak})

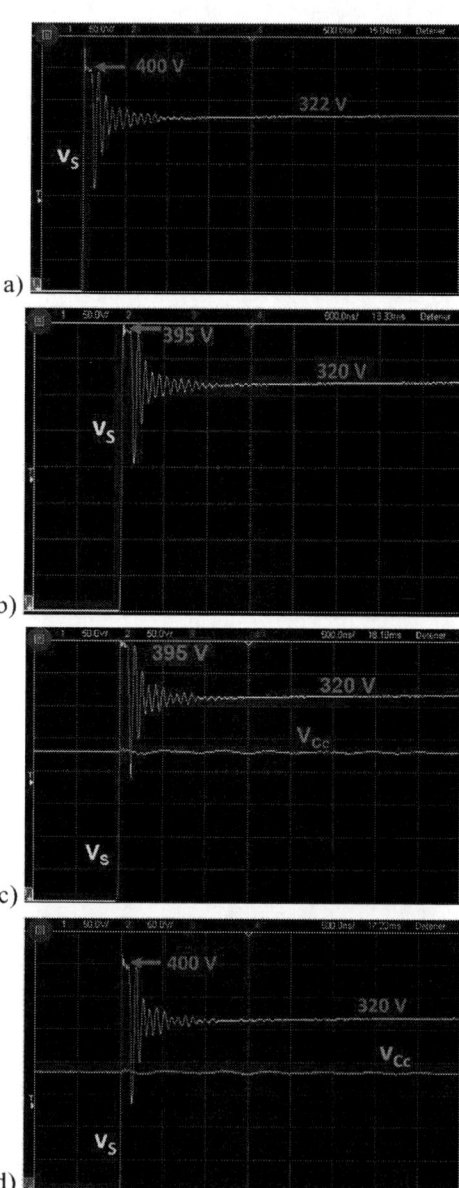

Fig.9. Experimental waveforms corresponding to the transistor drain-source voltage at the peak of the line voltage: a) Option A, b) Option B, c) Option C, and d) Option D.

Table 2. Comparison between theoretical, simulated and measured values for the four snubber options.

	A	B	C	D
Theoretical loss (W)	1.104	1.421	1.03	0.69
Simulated loss (W)	0.956	1.347	0.93	0.63
Measured loss (W)	0.634	1.10	0.749	0.411
V_{Speak} **simulated (V)**	425.20	423.30	426.80	429.10
V_{Speak} **measured (V)**	400	395	395	400

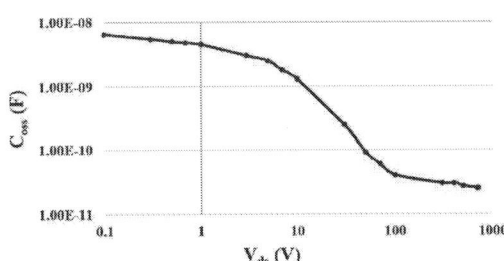

Fig.10. C_{oss}-V_{ds} curve from STP18N60DM2 datasheet.

and of P_{Rc} will be clearly lower than the one computed in the theoretical analysis.

To quantify this effect, parameter α is defined as follows:

$$\alpha = \frac{P_{Rc} - P_{Coss}}{P_{Rc}} \qquad (45)$$

where P_{Coss} is the power transferred to C_{oss}. The actual power dissipated in R_c taking into consideration P_{Coss} is called P_{Rc}^*. Its value is:

$$P_{Rc}^* = P_{Rc} - P_{Coss} = \alpha P_{Rc}. \qquad (46)$$

It should be noted that P_{Rc}^* is always lower than P_{Rc}. As R_c has been calculated without taking into account P_{Coss}, the actual value of V_{Cc}, V_{Cc}^*, will also be lower than V_{Cc}. According to (14), the actual value of V_{Speak}, V_{Speak}^*, will be:

$$V_{Speak}^* = V_{Cc}^* + v_2. \qquad (47)$$

From V_{Speak}^*, a new parameter λ^* can be defined as follows:

$$\lambda^* = \frac{V_{Speak}^*}{V_g(1+M)}. \qquad (48)$$

The procedure to compute V_{Speak}^* (and, therefore, λ^*) is based on adapting (22) to the fact that the actual power dissipated in R_c is P_{Rc}^*, instead of P_{Rc}. Thus, (22) becomes:

$$P_{Rc}^* = \alpha P_{Rc} = \frac{1}{\pi R_c} \int_0^\pi \left[(V_{Cc}^* - v_1)^2 \right] d\phi. \qquad (49)$$

This equation allows us to obtain the value of V_{Cc}^* for a given value of α (and, therefore, of P_{Coss}). Once V_{Cc}^* is known, (47) and (48) give us the values of V_{Speak}^* and λ^*.

The results obtained for the four options of clamp snubber applying this procedure are shown below. In all cases, λ_x^* and α_x are the values of λ^* and α for snubber option x. The results obtained are the following:

A. *Option A:*

Taking into account (27), (49) becomes:

$$\alpha_A \left[\lambda^2(1+M)^2 + \frac{1}{2} - \frac{4}{\pi}\lambda(1+M) \right]$$
$$= \lambda_A^{*2}(1+M)^2 + \frac{1}{2} - \frac{4}{\pi}\lambda_A^*(1+M). \qquad (50)$$

Solving this equation, we obtain:

$$\lambda_A^* = \frac{-b_A + \sqrt{b_A^2 - 4a_A c_A}}{2a_A} \qquad (51)$$

where:

$$a_A = (1+M)^2 \qquad (52)$$

$$b_A = -\frac{4}{\pi}(1+M) \qquad (53)$$

$$c_A = \frac{1}{2} - \alpha_A \left[\lambda^2(1+M)^2 + \frac{1}{2} - \frac{4}{\pi}\lambda(1+M) \right] \qquad (54)$$

In the prototype, $P_{Rc\_A} = 1.104$ W and $P_{Coss} \approx 0.3$ W. According to (45), $\alpha_A = 0.728$. From (51), $\lambda_A^* = 1.24$.

B. *Option B:*

Taking into account (30), (49) becomes:

$$\alpha_B \lambda^2(1+M)^2 = \lambda_B^{*2}(1+M)^2. \qquad (55)$$

This equation can be easily solved. The result is:

$$\lambda_B^* = \lambda \cdot \sqrt{\alpha_B}. \qquad (56)$$

In the prototype, $P_{Rc\_B} = 1.421$ W. Therefore, $\alpha_B = 0.789$. From (56), we obtain $\lambda_B^* = 1.243$.

C. *Option C:*

From (34), (49) becomes:

$$\alpha_C \cdot [\lambda(1+M) - 1]^2 = [\lambda_C^*(1+M) - 1]^2. \qquad (57)$$

The solution of (57) is:

$$\lambda_C^* = \frac{\sqrt{\alpha_C} \cdot [\lambda(1+M)-1]+1}{M+1}. \qquad (58)$$

In the prototype, $P_{Rc\_C} = 1.03$ W. Hence, $\alpha_C = 0.709$. Using (58), we obtain $\lambda_C^* = 1.262$.

D. *Option D:*

Using (37), (49) leads to:

$$\alpha_D \left[(\lambda(1+M) - 1)^2 + \frac{1}{2} - \frac{4}{\pi}(\lambda(1+M) - 1) \right]$$
$$= (\lambda_D^*(1+M) - 1)^2 + \frac{1}{2} \qquad (59)$$
$$- \frac{4}{\pi}(\lambda_D^*(1+M) - 1).$$

Solving (59), we obtain:

$$\lambda_D^* = \frac{-b_D + \sqrt{b_D^2 - 4a_D c_D}}{2a_D}, \qquad (60)$$

where:

$$a_D = (1+M)^2, \qquad (61)$$

$$b_D = -\left(\frac{4}{\pi} + 2\right)(1+M), \qquad (62)$$

$$c_D = \frac{3}{2} + \frac{4}{\pi} - \alpha_D \left[[\lambda(1+M) - 1]^2 + \frac{1}{2} \right. $$
$$\left. - \frac{4}{\pi}[\lambda(1+M) - 1] \right]. \qquad (63)$$

In the prototype, $P_{Rc\_D} = 0.69$ W. Therefore, $\alpha_D = 0.565$. From (60), we obtain $\lambda_D^* = 1.253$.

Table 3 shows a comparison between the value of V_{Speak} measured and V_{Speak}^* for the four snubber options. The values of V_{Speak}^* have been obtained from the values of λ_x^* already obtained and (48). As this Table 3 shows, excellent agreement between the results measured in the waveforms of Fig.9 and those predicted after taking into account P_{Coss} (the power transferred to C_{oss}) has been finally achieved.

Table 3. Comparison between P_{Rc}^* losses, V_{Speak} measured and V_{Speak}^* values for the four snubber options.

	A	B	C	D
V_{Speak} **measured (V)**	400	395	395	400
V_{Speak}^* **(V)**	397.78	398.82	404.86	401.86
P_{Rc}^* **(W)**	0.804	1.121	0.730	0.389

In some of the waveforms of Fig.9, some additional voltage spikes can be observed over the value marked for V_{Speak}. Thus, the spike shown in Fig.9(a) achieves about 420 V. These spikes are due to the parasitic inductance and resistance of diode D_c, capacitor C_c and the connection between these elements and the transistor. In the case of options C and D, the parasitic elements of capacitor C must be taken into consideration too. Obviously, a proper selection of the snubber devices and a proper design of the converter PCB is of primary concern. Fortunately, these additional spikes are quite narrow and do not exceed transistor voltage rating.

Finally, the fact of existing more options for the placement of the clamp snubber in the case of the SEPIC PFC represents a practical advantage of this topology over the use of the flyback PFC for the same application. In the case of the flyback PFC, an input filter made up of an inductor and a couple of capacitors must be placed at the flyback input to alleviate the differential-mode noise without deteriorating the power factor (see Fig.11). Therefore, the actual number of power elements (semiconductors and reactive elements) is similar in the case of both the flyback and the SEPIC PFC (see Fig.12). However, the SEPIC PFC has more options for the clamp snubber than the flyback PFC due to capacitor C. As demonstrated in this paper, lower power will be dissipated in snubber resistor R_c if Option D is selected. This option is not possible in the flyback topology.

V. CONCLUSIONS

The placement of a passive clamp snubber in isolated SEPIC and Cuk converters working as automatic PFC is not irrelevant. This work presents four different options for the placement of a passive clamp snubber, analyzing the value of snubber resistor R_c and of the power dissipated in this resistor for the four cases. The theoretical analysis has been normalized by selecting proper base values for resistance and power. According to the normalized values of the power loss in R_c, the option that exhibits the lowest power loss is option D, at least for common design values of λ and M. On the contrary, option B is the one that exhibits the highest power loss. Options A and C present similar values of power loss, higher than Option D, but lower than Option B. The conclusions of this study have been validated by simulation and with experimental results. The results obtained by simulation agree very well the theoretical study. Regarding the experimental results obtained for the maximum voltage across the transistor and the power dissipated in the snubber resistor, they are lower than the predicted by the study. However, this discrepancy can be overcome by taking into account the power transferred to the transistor output capacitance (C_{oss}). Once this power is discounted from the power dissipated in R_c, excellent agreement is observed between the new predicted results and the experimental results.

ACKNOWLEDGMENTS

This work has been carried out by funding from the Asturias government through the SV-PA-21-AYUD/2021/51931 project, and from the Spanish government through the PID2022-136969OB-I00, PID2021-127707OB-C21, MCINN-22-TED2021-130939B-I00 and MCINN-23-PID2022-136969OB-I00 projects.

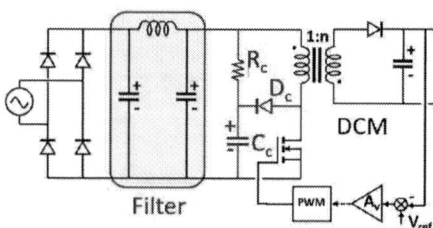

Fig.11. Flyback PFC with the best clamp snubber option.

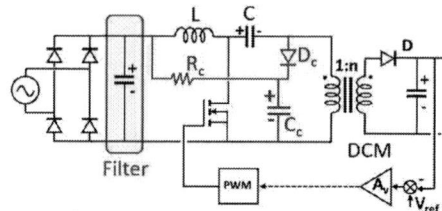

Fig.12. SEPIC PFC with the best clamp snubber option.

REFERENCES

[1] P.C. Todd, "Snubber circuits: Theory, design and application". Unitrode Power Supply Design Seminar, 1993, pp. 2.1-2.17.

[2] B. Carsten. "Design techniques for transformer active reset circuits at high frequencies and power levels". High Frequency Power Conversion (HFPC) Conference, 1990, pp. 235-246.

[3] R. Erickson, M. Madigan, and S. Singer, "Design of a simple high power-factor rectifier based on the flyback converter", in Fifth Annual Proceedings on Applied Power Electronics Conference and Exposition, 1990, pp. 792–801.

[4] M. Mahdavi and H. Farzanehfard, "Bridgeless SEPIC PFC rectifier with reduced components and conduction losses", IEEE Trans. Ind. Electron.,vol. 58, Sep. 2011.

[5] E. H. Ismail, "Bridgeless SEPIC rectifier with unity power factor and reduced conduction losses", IEEE Trans. Ind. Electron., vol. 56, pp. 1147–1157, Apr. 2009.

[6] A. J. Sabzali, E. H. Ismail, M. A. Al-Saffar, and A. A. Fardoun, "New bridgeless DCM Sepic and Cuk PFC rectifiers with low conduction and switching losses", IEEE Trans. Ind. Appl., vol. 47, no. 2, pp. 873–881, Mar./Apr. 2011.

[7] P. J. S. Costa, C. H. I. Font, and T. B. Lazzarin, "A family of single-phase voltage-doubler high-power-factor SEPIC rectifiers operating in DCM", IEEE Trans. Power Electron., vol. 32, no. 6, pp. 4279–4290, Jun. 2017.

[8] D. S. L. Simonetti, J. Sebastián and J. Uceda. "The discontinuous conduction mode SEPIC and Ćuk power factor pre-regulators: Analysis and design". IEEE Trans. on Industrial Electronics, vol. 44, n° 5, 1997, pp. 630-637

[9] J. Sebastián, J. A. Cobos, J. M. Lopera and J. Uceda. "The determination of the boundaries between continuous and discontinuous conduction modes in PWM dc-to-dc converters used as power factor preregulators". IEEE Trans. on Power Electronics, vol.10, n° 5, 1995, pp. 574-582.

[10] D. Murillo-Yarce, J. Rodriguez, F. Loose, M. Hernando and J. Sebastian, "Study of SEPIC and Ćuk converters working as automatic Power Factor Corrector when operating in unusual discontinuous conduction modes," 2024 IEEE Applied Power Electronics Conference and Exposition (APEC), Long Beach, CA, USA, 2024, pp. 42-49.

[11] D. Costinett, D. Maksimovic and R. Zane, "Circuit-oriented treatment of nonlinear capacitances in switched-mode power supplies". IEEE Trans. on Power Electronics, vol.30, n° 2, 2015, pp. 985-995.

Current Sensorless Control Strategy for Single-Phase T-Type PFC Converter

Che-Yu Lu
Department of Electrical Engineering
National United University, Miaoli, Taiwan
cylu@nuu.edu.tw

Jia-En Zeng
Department of Electrical Engineering,
National United University, Miaoli, Taiwan
chengjiang1126@gmail.com

Abstract — **This paper presents a current sensorless control strategy for a single-phase T-type power factor correction (PFC) converter. Unlike conventional multi-loop control schemes that require the measurement of input AC voltage, output DC voltage, and inductor current, the proposed control algorithm only uses feedback from the input and output voltages. This approach effectively regulates the output voltage and achieves power factor correction for the input current. Furthermore, the sensorless control strategy eliminates concerns regarding sampling inductor current, as it reduces the number of required sensors and simplifies the controller design. Simulation results demonstrate that the proposed current sensorless control strategy achieves a unity power factor operation and minimizes total harmonic distortion in the current.**

I. INTRODUCTION

Single-phase power factor correction converters are proverbially applied in various electronic products because of their distinguished merits, such as regulated output voltage, sinusoidal input current, lower harmonic distortions, and unity power factor. The performance of conventional two-level PFC converters [1], [2] could not be satisfying at high-voltage applications, and therefore, multilevel PFC converters have gained growing importance for industrial applications [3]-[10]. Multilevel converters have higher levels than conventional topologies, improved power ratings, decreased device numbers, high voltage gain, less switching stress, and low cost [3]-[5]. Their flexible use in medium voltage has raised their importance in industrial applications [6], [7].

The neutral point clamped (NPC) converter is introduced in [11], [12], and other multilevel topologies are presented, such as the flying capacitor (FC) [13], cascaded H-bridge (CHB) [14], and T-type converter [15]-[17]. The most used multilevel converter is the T-type converter, which performs better in low-voltage applications than NPC topologies because of its reduced switch losses and higher efficiency [15]. The T-type converter has several advantages over conventional multilevel converters, including a simple control strategy and enhanced reliability [16]. One control method with predictive observer for a single-phase T-type converter is presented in [17]. However, those control methods require numerous voltage or current transducers. Several current sensorless control methods for boost-type PFC converters are presented in [18]-[20] that withdraw extra current transducers and eliminate the sampling faults. Therefore, a single-phase T-type PFC converter with the current sensorless control strategy is presented in Fig. 1. The proposed control block diagram only requires grid voltage and two output capacitor voltages.

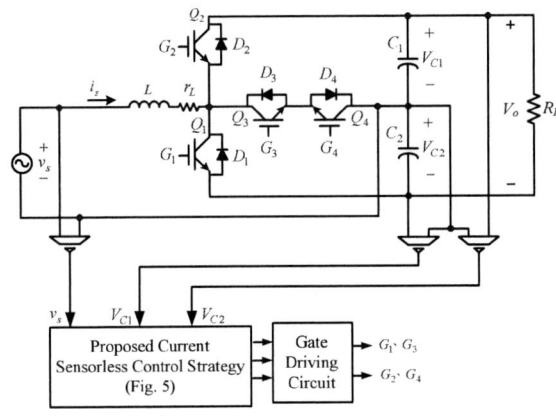

Fig. 1. The proposed current sensorless control strategy of single-phase T-type PFC converter.

II. ANALYSIS OF SINGLE-PHASE T-TYPE PFC CONVERTER

A. Operation States

The four switches $Q_1 \sim Q_4$ and body diodes $D_1 \sim D_4$ operate due to the respective ac-cycle based on the input voltage v_s. For T-type PFC converter topology, the switch pairs Q_1, Q_3 and Q_2, Q_4 are turned on within different half-cycle. In addition, the body diodes D_1, D_3 and D_2, D_4 are forward-biased within the opposite half-cycle. As drawn in Fig. 2, the single-phase T-type PFC converter operates at four switching states.

During the positive input voltage half-cycle $v_s > 0$, there are two switching states. The switches Q_1, Q_3 turn on and the diodes D_1, D_3 are cut-off during the state A, and thus the energy is stored in the capacitor C_2, as shown in Fig. 2(a). The switches Q_1, Q_3 turn off but the diodes D_1, D_3 are forward-biased during the state B. Hence, the boost inductor L releases the energy to the capacitor C_1, as shown in Fig. 2(b).

On the contrary, two other switching states exist during the negative input voltage half-cycle $v_s < 0$. As shown in Fig. 2(c), the switches Q_2, Q_4 turn on and diodes D_2, D_4 are cut-off within the state C. Hence, the capacitor C_1 release energy to the boost inductor L. The switches Q_2, Q_4 turn off but the diodes D_2, D_4 are forward-biased during the state D, as plotted in Fig. 2(d); thus, the capacitor C_2 absorbs energy from the boost inductor L.

979-8-3315-1612-3/25 $31.00 © 2025 IEEE

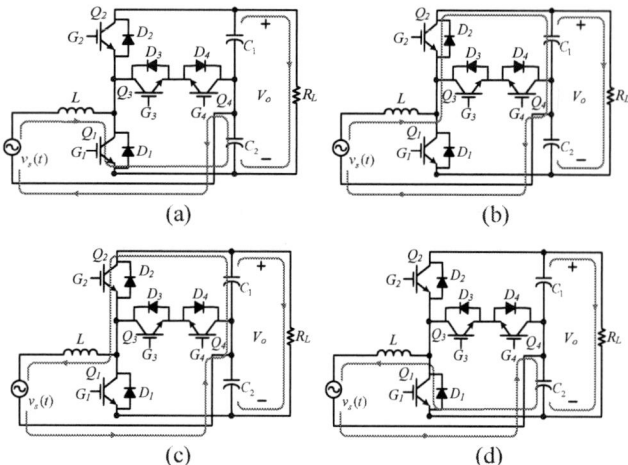

Fig. 2. Four switching states of single-phase T-type PFC converter: (a) state A; (b) state B; (c) state C; (d) state D.

B. Modeling of Single-Phase T-Type PFC Converter

The initial parameters are defined to model the single-phase T-type PFC converter. The boost inductance L, inductor resistance r_L, and dc-link voltage V_o are all regarded as constant values. Moreover, the switch's conduction voltage V_{on} and the diode's forward voltage V_F are assumed to be equal. The line frequency f is obviously lower than the switching frequency f_s. Therefore, the input voltage remains constant throughout each switching period T_s. The four equivalent circuits have been derived individually and are illustrated in Fig. 3(a) to 3(d), as the inductor current operates in continuous current mode (CCM).

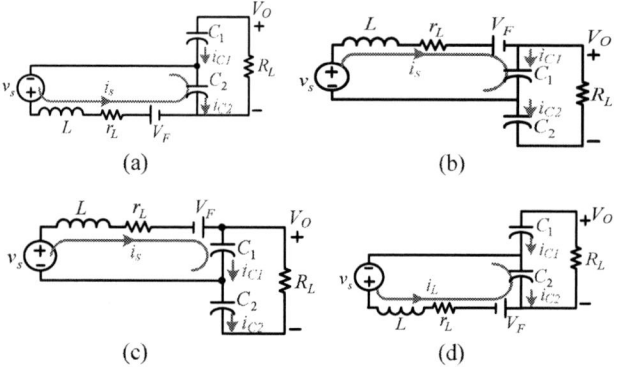

Fig. 3: Equivalent circuits of four switching states: (a) state A; (b) state B; (c) state C; (d) state D.

The inductor voltage v_L at four switching states uses Kirchhoff's voltage law to be written as in (1) to (4)

$$\text{State A: } v_{L,A} = v_s - i_s r_L - V_F + V_{C2} \tag{1}$$

$$\text{State B: } v_{L,B} = v_s - i_s r_L - V_F - V_{C1} \tag{2}$$

$$\text{State C: } v_{L,C} = v_s - i_s r_L + V_F - V_{C1} \tag{3}$$

$$\text{State D: } v_{L,D} = v_s - i_s r_L + V_F + V_{C2} \tag{4}$$

The input voltage v_s is replaced with an operator $sign(v_s)$ by (1) to (4), and that is expressed as

$$sign(v_s) = \begin{cases} 1, & v_s \geq 0 \\ -1, & v_s < 0 \end{cases} \tag{5}$$

The inductor voltages $v_{L,ON}$ and $v_{L,OFF}$ can be derived from (1) to (4) because the switches' turning-on and turning-off periods at different switching states are summarized.

$$v_{L,ON} = v_{L,A} + v_{L,C} = v_s - i_s r_L - sign(v_s)V_F - \frac{1}{2}(V_{C1} - V_{C2}) + \frac{sign(v_s)}{2}V_o \tag{6}$$

$$v_{L,OFF} = v_{L,B} + v_{L,D} = v_s - i_s r_L - sign(v_s)V_F - \frac{1}{2}(V_{C1} - V_{C2}) - \frac{sign(v_s)}{2}V_o \tag{7}$$

Eventually, a single-phase T-type PFC converter can be derived from (6) and (7), and its equivalent single-switch model is drawn in Fig. 4.

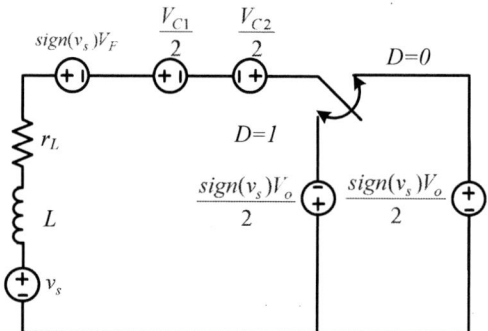

Fig. 4. Single-switch model of T-type PFC converter.

III. PROPOSED CURRENT SENSORLESS CONTROL STRATEGY

The proposed current sensorless control strategy is implemented because it meets the specifications without relying on a current transducer. By eliminating the need for such components, this approach streamlines the controller design and minimizes the dependence on feedback signals. Moreover, the sensorless control scheme not only adjusts the output voltage V_o to align perfectly with the voltage command V_o^* but also optimizes the input current i_s to be in phase with the input voltage v_s.

The average inductor voltage $\langle v_L \rangle$ can be derived as in (8) by using time averaging method, where the turning-on and turning-off time during one switching period are DT_s and $(1-D)T_s$, respectively.

$$\langle v_L \rangle = \langle v_s \rangle - \langle i_s \rangle r_L - sign(v_s)V_F - \frac{1}{2}(V_{C1} - V_{C2}) - \frac{sign(v_s)V_o}{2} \times (1-2D) \tag{8}$$

The average inductor current $\langle i_s \rangle$ is expressed as in (9) because the input current i_s is sinusoidal in phase with the input voltage v_s, where $\hat{V}_L$ is the amplitude of inductor voltage and ω is its angular frequency.

$$\langle i_s \rangle = \frac{\hat{V}_L}{\omega L} \sin(\omega t) \qquad (9)$$

Moreover, the average boost inductor voltage $\langle v_L \rangle$ is rewritten as in (10).

$$\langle v_L \rangle = L \frac{d \langle i_s \rangle}{dt} = L \frac{d}{dt} \left(\frac{\sin(\omega t)}{\omega L} \right) = \hat{V}_L \cos(\omega t) \qquad (10)$$

The control signal D can be expressed as in (11), which is derived from (8), (9), and (10).

$$D = \frac{1}{2} - \frac{1}{V_o^*} \left\{ \begin{array}{l} |v_s| - V_F - \frac{1}{2} sign(v_s)[V_{C1}(t) - V_{C2}(t)] \\ - \hat{V}_L \left[\cos(\omega t) sign(v_s) + \frac{r_L}{\omega L} \sin(\omega t) sign(v_s) \right] \end{array} \right\} \qquad (11)$$

Eventually, the $sign(v_s) \sin(\omega t)$ can be further predigested to $|\sin(\omega t)|$; therefore, the control signal D can be summarized as in (12).

$$D = \frac{1}{2} - \frac{1}{V_o^*} \left\{ \begin{array}{l} |v_s| - V_F - \frac{1}{2} sign(v_s)[V_{C1}(t) - V_{C2}(t)] \\ - \hat{V}_L \left[\cos(\omega t) sign(v_s) + \frac{r_L}{\omega L} |\sin(\omega t)| \right] \end{array} \right\} \qquad (12)$$

The power conversion efficiency may decline because of the high voltage boost ratio when traditional unipolar modulation is used, especially when the input voltage is 110V and the output voltage is 400V. To address this issue, the proposed current sensorless control strategy utilizes bipolar modulation, which is more straightforward to implement. The control scheme is illustrated in Fig. 5, where it operates the four switches during different half-cycles of the input voltage.

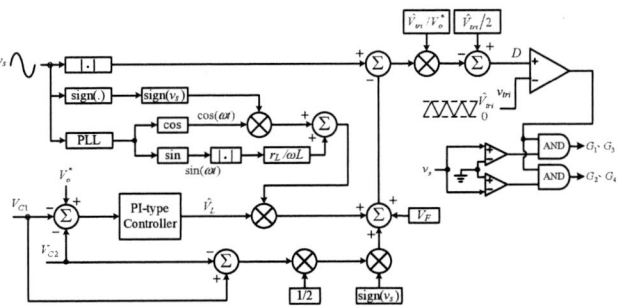

Fig. 5. Proposed current sensorless control strategy.

IV. VERIFICATION OF SIMULATION RESULTS

The parameters of the single-phase T-type PFC converter are provided in Table I. Additionally, simulation results are presented to validate the proposed current sensorless control strategy. The components used in the T-type PFC converter are standard commercially available products, with the IGBT model being IXXH50N60C3D1.

The input AC voltage v_s is 110 V at 60 Hz, while the output DC voltage V_o is set to 400 V, as depicted in Fig. 6. The simulated steady-state waveforms show that the input current i_s is sinusoidal and in phase with the input voltage v_s. The power factor (PF) remains close to unity for different load resistors: 160 Ω (1 kW) and 80 Ω (2 kW). Furthermore, the total harmonic distortion of the current (THD) is less than 5%. Additionally, the converter efficiency exceeds 96.5%, and further simulated results are detailed in Table II.

To evaluate the dynamic performance with the proposed current sensorless control strategy, the load resistor is varied between 160 Ω (1 kW) and 80 Ω (2 kW), as shown in Fig. 7. The settling time is approximately 100 ms, regardless of whether the load transitions from light to heavy or from heavy to light. Thus, the proposed current sensorless control strategy demonstrates satisfactory dynamic characteristics and system stability.

TABLE I. SIMULATED CIRCUIT PARAMETERS.

Input voltage (rms)	$v_s = 110\,\text{V}$
Output voltage	$V_o^* = 400\,\text{V}$
Load resistor	$R_L = 160\,\Omega(1\,\text{kW})$, $R_L = 80\,\Omega(2\,\text{kW})$
Switching frequency	$f_s = 50\,\text{kHz}$
Line frequency	$f = 60\,\text{Hz}$
Inductances	$L = 2\,\text{mH}$
Inductor resistances	$r_L = 0.1\,\Omega$
Capacitance	$C_1 = C_2 = 330 * 3\,\mu\text{F}$
Forward voltage	$V_F = 2\,\text{V}$

TABLE II. SIMULATED MEASUREMENT RESULTS

Output power	1 kW (R$_o$=160 Ω)	2 kW (R$_o$=80 Ω)
Power factor (PF)	0.987	0.986
Total harmonic distortion of current (THDi)	4.78 %	4.72 %
Copper loss	3.88 W	26.07 W
Conduction and Switching losses	18.7 W	37.7 W
Efficiency	97.79 %	96.86 %

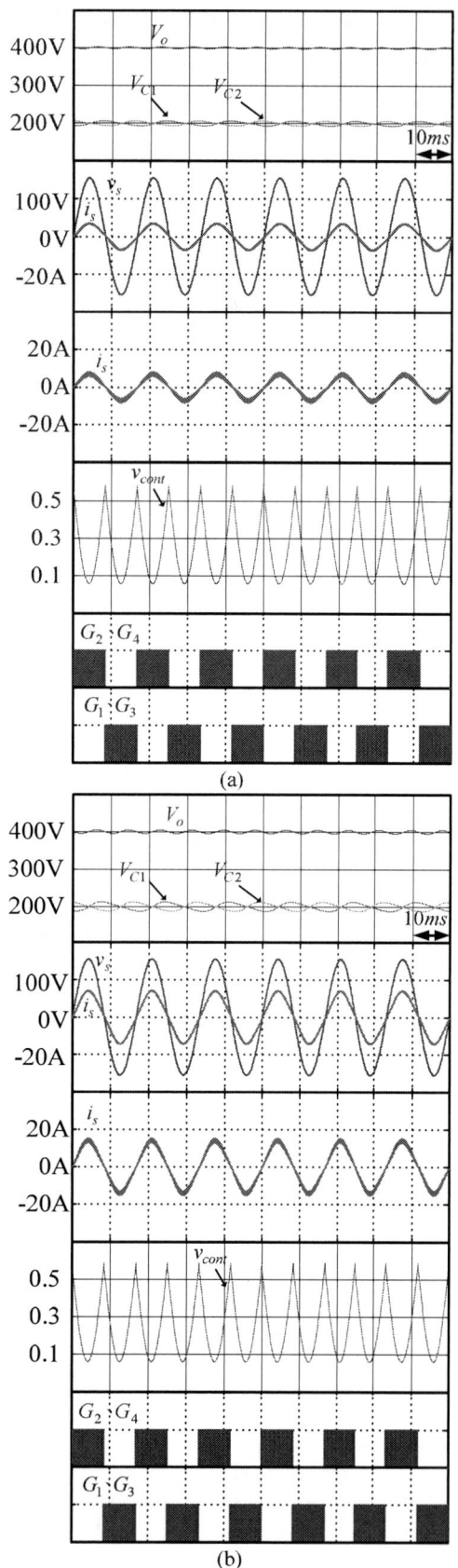

Fig. 6. Simulated steady-state waveforms under different output power (a) 1 kW (160 Ω); (b) 2 kW (80 Ω).

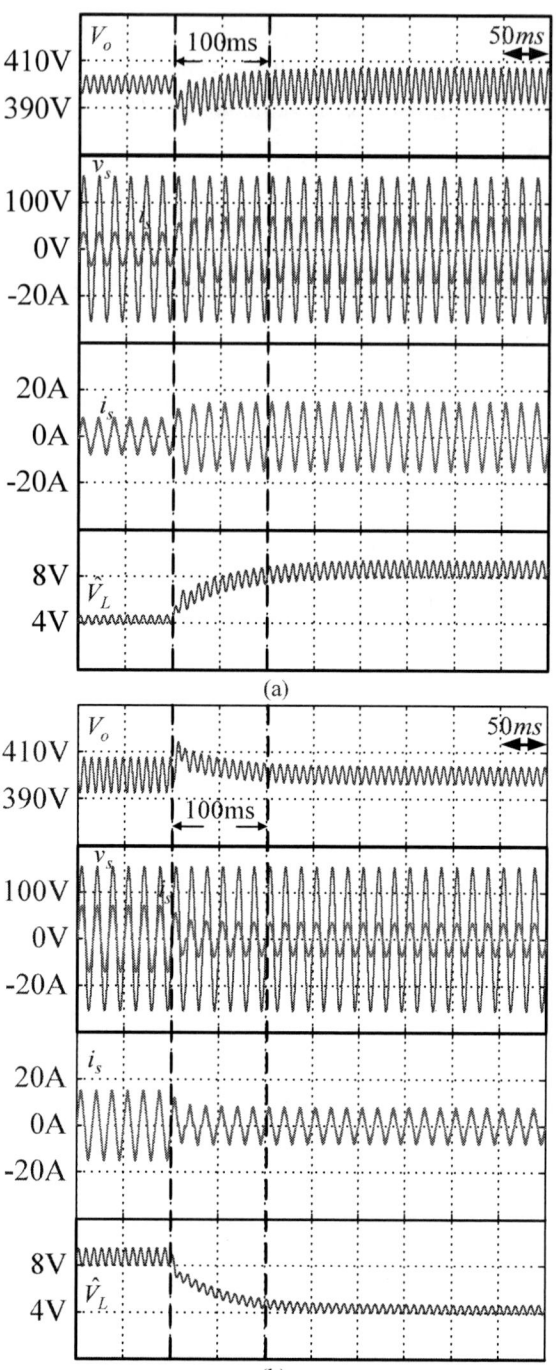

Fig. 7. Simulated transient waveforms with the varied load resistors (a) from 160 Ω (1 kW) to 80 Ω (2 kW); (b) from 80 Ω (2 kW) to 160 Ω (1 kW).

The simulation results indicate that the proposed current sensorless control strategy provides reliable steady-state and dynamic responses, including effective voltage regulation and power factor correction. The single-phase T-type PFC topology demonstrates high efficiency for low-voltage applications. Additionally, the total harmonic distortion (THD) of the current is fairly low, and the power factor (PF) is close to unity.

V. CONCLUSIONS

This paper introduces a current sensorless control strategy for a single-phase T-type power factor correction (PFC) converter. The simulation results demonstrate that the proposed control method achieves effective steady-state and dynamic performance. Moreover, this current sensorless control strategy successfully maintains a controllable output voltage, produces a sinusoidal inductor current, minimizes current harmonics, and achieves a high power factor. Consequently, this control scheme can be easily adapted with a multilevel T-type PFC converter.

REFERENCES

[1] Komurcugil and O. Kukrer, "A novel current-control method for three-phase PWM AC/DC voltage-source converters," *IEEE Trans. Ind. Electron.*, vol. 46, no. 3, pp. 544 -553, June 1999.

[2] J. Liang, H. Wang, and Z. Yan, "Grid voltage sensorless model-based predictive power control of PWM rectifiers basedon sliding mode virtual flux observer," *IEEE Access*, vol. 7, pp. 24007-24016, 2019.

[3] J. Rodriguez, J. S. Lai, and F. Z. Peng, "Multilevel Inverters: A Survey of Topologies, Controls, and Application," *IEEE Trans. Ind. Electron.*, vol. 49, no. 4, pp. 724–738, Aug. 2002.

[4] L. G. Franquelo, J. Rodriguez, J. I. Leon, S. Kouro, R. Portillo, and M. A. M. Prats, "The age of multilevel converters arrives," *IEEE Ind. Electron. Mag.*, vol. 2, no. 2, pp. 28–39, Jun. 2008.

[5] S. Kouro, M. Malinowski, K. Gopakumar, J. Pou, L. G. Franquelo, B. Wu, J. Rodriguez, M. A. Perez, and J. I. Leon, "Recent advances and industrial applications of multilevel converters," *IEEE Trans. Ind. Electron.*, vol. 57, no. 8, pp. 2553–2580, Aug. 2010.

[6] J. Rodriguez, S. Bernet, B. Wu, J. Pontt, S. Kouro, "Multilevel voltage-source-converter topologies for industrial medium-voltage drives," *IEEE Trans. Ind. Electron.*, vol. 54, no. 6, pp. 2930–2945, Dec. 2007.

[7] H. Abu-Rub, J. Holtz, J. Rodriguez, and B. Baoming, "Medium-voltage multilevel converters—State of the art, challenges, and requirements in industrial applications," *IEEE Trans. Ind. Electron.*, vol.57, no. 8, pp. 2581–2596, Aug. 2010.

[8] A. Nabae, I. Takahashi, and H. Akagi, "A new neutral-point-clamped PWM inverter," *IEEE Trans. Ind. Appl.*, vol. 17, no. 5, pp. 518–523, Sept./Oct., 1981.

[9] T. A. Meynard and H. Foch, "Multilevel conversion: high voltage choppers and voltage-source inverter," in *Proc. 23rd IEEE Power Electron. Spec. Conf.*, vol. 1, 1992, pp. 397–403.

[10] F. Z. Peng, J. S. Lai, J. W. McKeever, and J. VanCoevering, "A multilevel voltage-source inverter with separate dc sources for static var generation," *IEEE Trans. Ind. Appl.*, vol. 32, pp. 1130–1138, Sep./Oct. 1996.

[11] J. Dionisio, J. F. A. Silva, and E. G. A Jesus, "Fast-predictive optimal control of NPC multilevel converters," *IEEE Trans. Ind. Electron.*, vol. 60, no. 2, pp. 619-627, Feb. 2013.

[12] Y. Yin, J. Liu, J. A. Sanchez, L. Wu, S. Vazquez, J. I. Leon, and L. G. Franquelo, "Observer-based adaptive sliding mode control of NPC converters: an RBF neutral network approach," *IEEE Trans. Power Electron.*, vol. 34, no. 4, pp. 3831-3842, April 2019.

[13] C. A. Teixeira, D. G. Holmes, and B. P. McGrath, "Single-phase semibridge five-level flying-capacitor rectifier," *IEEE Trans. Ind. Appl.*, vol. 49, no. 5, pp. 2158-2166, Sept./Oct. 2013.

[14] C. Liu, F. Zhao, G. Cai, N. Huang, J. Wang, and M. Wang, "Novel individual voltage balancing control scheme for multilevel cascade active-front-end rectifier," *IET Power Electronics*, vol. 7, no. 1, pp. 50-59, 2014.

[15] M. Schweizer and J. W. Kolar, "Design and implementation of a highly efficient three-level T-type converter for low-voltage applications," *IEEE Trans. Power Electron.*, vol. 28, no. 2, pp. 899–907, Feb. 2013.

[16] M. Schweizer, T. Friedli, and J. W. Kolar, "Comparative evaluation of advanced three-phase three-level inverter/converter topologies against two-level systems," *IEEE Trans. Power Electron.*, vol. 60, no. 12, pp. 5515–5527, Dec. 2013.

[17] S. A. Khan, Y. Guo, and J. Zhu, "Model predictive observer based control for single-phase asymmetrical T-type AC/DC power converter," *IEEE Trans. Ind. Appl.*, vol. 55, no. 2, pp. 2033-2044, March/April 2019.

[18] C. Y. Li, N. C. Chao and H. C. Chen, "Design and implementation of four-switch current sensorless control for three-phase PFC converter," *IEEE Transactions on Ind. Electron.*, vol. 67, no. 4, pp. 3307-3312, Apr. 2020.

[19] H. C. Chen, C. Y. Lu, G. T. Li, "Design and implementation of three-phase current sensorless control for PFC bridge converter with considering voltage drops of power semiconductors," *IEEE Trans. Ind. Electron.*, vol. 65, no. 12, pp. 9234-9242, Dec. 2018.

[20] H. C. Chen, C. Y. Lu, G. T. Li, and W. C. Chen, "Digital current sensorless control for dual-boost half-bridge PFC converter with natural capacitor voltage balancing," *IEEE Trans. Power Electron.*, vol. 32, no. 5, pp. 4074–4083, May 2017.

Three-Phase Single-Stage Multiport AC-DC Converter with Integrated DC-DC Conversion Stages

Asad Hameed
Department of Electrical & Computer Engineering
Western University
London, ON, Canada
ahamee3@uwo.ca

Gerry Moschopoulos
Department of Electrical & Computer Engineering
Western University
London, ON, Canada
gmoschop@uwo.ca

Abstract— **A new three-phase multiport AC-DC converter with integrated DC-DC conversion stages is introduced in this paper. The converter features one three-phase AC port and three DC ports: one AC port and one DC port are bidirectional, while the remaining two DC ports are unidirectional. The proposed converter is formed by combining three sub-converters: a buck DC-DC converter, a three-phase full-bridge DC-DC converter, and a three-phase six-switch AC-DC voltage source converter (VSC). All sub-converters share the same switches to operate as a single-stage converter. The converter ensures a unity power factor at the AC port and offers galvanic isolation at one of the DC ports. All ports can be independently controlled, except for one DC port. The proposed converter topology is straightforward, using only six active switches and requiring a standard control system for operation. Additionally, when used in a microgrid, this converter can eliminate the need for two other converters, significantly reducing both the size and complexity of the system. This paper defines the workings, characteristics, and control of the proposed converter. Additionally, scaled-down prototype experimental results are provided to validate the converter.**

Keywords— *DC-DC conversion, full-bridge converter, microgrid converter, multiport converter, single-stage AC-DC conversion, transformers*

I. INTRODUCTION

Hybrid microgrids (HMGs) are gaining popularity due to the shift from centralized electric power generation toward smaller-scale systems. This transition is driven by environmental concerns, reduced power losses, improved energy efficiency, enhanced reliability, and growing energy demand [1]-[2]. Multiport converters allow multiple power sources and loads operating at different voltage levels to connect to an electrical network [3]. This integration significantly reduces the number of power electronic converters needed in HMG architectures. With recent technological advancements, multiport converters have become even more advantageous and cost-effective, offering fewer components, higher power density, compact design, and easier integration with centralized control [4].

In conventional multiport AC-DC converters, as shown in Fig. 1, a two-stage structure is commonly employed. This structure consists of a three-phase AC-DC converter followed by multiple DC-DC converters. Such a two-stage design reduces efficiency, increases costs, and adds to the converter's size, as power must pass through two separate conversion stages. Additionally, multiport converters often require complex designs with a high number of switches. To address these issues, researchers have developed single-stage multiport AC-DC converters that perform both AC-DC and DC-DC conversion simultaneously, eliminating redundant conversion stages and components. This approach reduces the number of switches, thereby enhancing overall conversion efficiency, lowering costs, and decreasing converter size [5]-[6]. However, previously

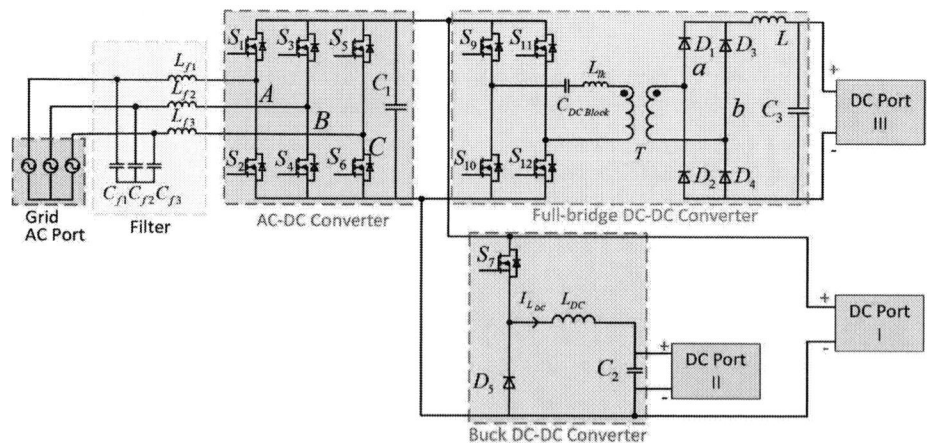

Fig. 1. Typical two-stage multiport converter.

979-8-3315-1612-3/25 $31.00 © 2025 IEEE

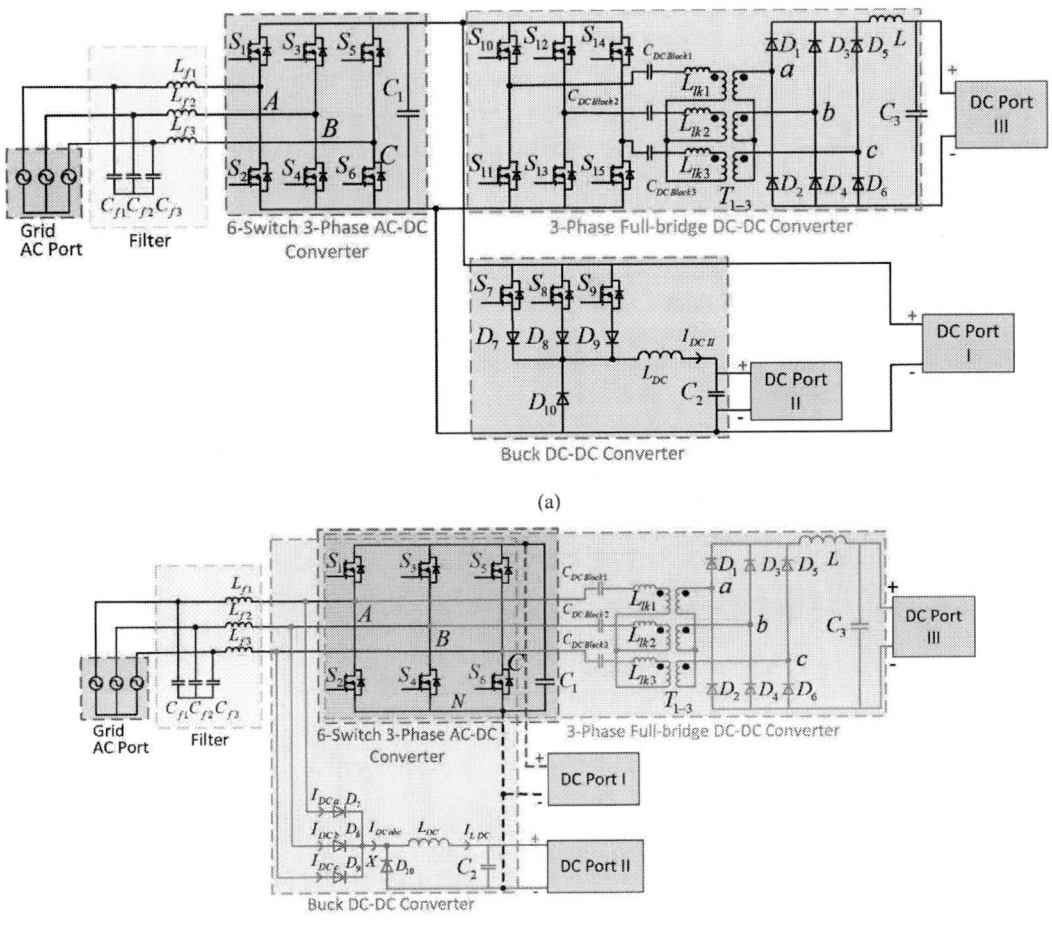

Fig. 2. (a) Topology derivation, (b) Proposed single-stage multiport AC-DC converter.

proposed multiport converters in the literature [7]-[15] still exhibit at least one of the following challenges:

- They have a complex framework that requires even more intricate gating signals, typically generated by high-speed FPGAs, to function properly.

- They lack any ports with galvanic isolation, limiting their utility in applications requiring electrical isolation between networks.

- They require a substantial number of conversion stages and switches to operate, which increases the likelihood of failure as the number of switches or stages rises.

- They manage power across multiple ports using complex or non-standard control methods. Such control systems often require high-speed and expensive microcontrollers or FPGAs.

The converter described in this paper combines several power converters into a single multiport device that operates as a single, cohesive unit. By sharing components, the proposed converter eliminates the need for redundant conversion stages and a high number of switches, thereby reducing overall cost. Its flexible architecture simplifies control implementation and centralizes operations, minimizing component count and improving power flow efficiency. The paper discusses the converter's features, control strategies, and operational principles. A scaled-down prototype is used for experimental validation, demonstrating the converter's functionality and viability.

II. CONVERTER OPERATION

Fig 2(b) illustrates the proposed multiport converter, which consists of three input inductors (L_{f1}-L_{f3}), three filter capacitors (C_{f1}-C_{f3}), six switches (S_1-S_6), four diodes (D_7-D_{10}), and a DC inductor (L_{DC}) connected to a bulk capacitor (C_2). A bulk capacitor (C_1) is connected across DC Port I, while DC blocking capacitors ($C_{DC\_Block\_1}$-$C_{DC\_Block\_3}$) are connected in series with the primary side of the transformers, linking to the three-phase legs. Additionally, the design includes three high-frequency isolation transformers (T_1-T_3) with an n:1 turns ratio, along with leakage inductances (L_{lk1}-L_{lk3}). A three-phase rectifier, composed of six diodes (D_1-D_6), a filtering inductor (L), and a smoothing capacitor (C_3), is located on the secondary side of the transformers to form DC Port III.

979-8-3315-1612-3/25 $31.00 © 2025 IEEE

The topology derivation of the proposed converter is shown in Fig. 2(a). The proposed converter design consists of three types of sub-converters: a three-phase bidirectional AC-DC converter, a three-phase unidirectional full-bridge DC-DC converter, and a buck DC-DC converter. This configuration enables the three converters to operate together, sharing the same switches. The AC-DC converter functions like a traditional three-phase, six-switch converter, transferring power between the AC Port and DC Port I.

The operation of the full-bridge DC-DC converter differs slightly from a standard three-phase full-bridge DC-DC converter. In the proposed design, the duty cycle of the square wave at the transformer primary varies based on a modified modulation signal, whereas in a standard full-bridge DC-DC converter, the duty cycle is adjusted according to the output voltage. The DC blocking capacitors filter out low-frequency and DC components, allowing only high-frequency components to pass through and be imposed across the transformer primaries. The three-phase rectifier on the transformers' secondary side rectifies the high-frequency square wave to DC, and the output low-pass filter, consisting of output inductor L and capacitor C_3, removes high-frequency ripples caused by switching frequency. In this three-phase full-bridge converter, low-frequency ripple from varying modulation signals is naturally suppressed due to the converter's three-phase nature. The full-bridge DC-DC converter transfers power from DC Port I to DC Port III through isolation transformers.

The operation of the buck DC-DC converter also differs slightly from a traditional buck converter. In a traditional converter, the PWM duty cycle is varied to adjust the output voltage, whereas in the proposed design, the maximum duty cycle of each phase leg is applied through three diodes (D_7-D_9), which function as a three-input OR gate. The variation in the bias signal adjusts the duty cycle applied to the DC inductor, causing the voltage at DC Port II to vary accordingly. The buck DC-DC converter transfers power from DC Port I to DC Port II through the DC inductor.

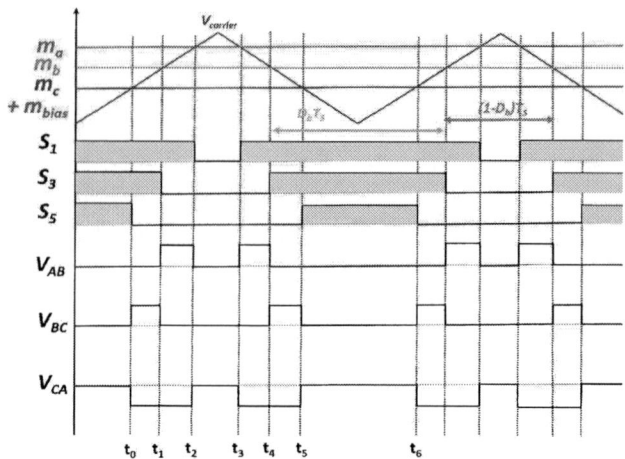

Fig. 4. Operation waveforms.

The AC port operates with a power factor close to one. Power flow at each port can be independently controlled, except for the galvanically isolated DC Port III. The six-switch VSC uses PWM techniques to generate three-phase sinusoidal voltages, with SPWM (Sinusoidal Pulse Width Modulation) employed in the proposed converter for simplicity. The control system adds a bias signal to the modulation signals to regulate the voltage level at DC Port II, as shown in Fig. 3. When the bias signal is zero, the modified modulation signal equals the modulation signal, oscillating around 0.5. By adding or subtracting the bias signal, the modified signal is offset from 0.5. Applying three-phase voltages at switching nodes A, B, and C allows control over active and reactive power exchange between the AC port and DC Port I. Fig. 4 illustrates the operational waveforms of the converter, including the high-side switch waveforms and the voltages between switching nodes V_{AB}, V_{BC}, and V_{CA}.

III. CONVERTER CONTROL

Fig. 5 illustrates the converter's control system. It uses a standard decoupled power control approach in the synchronous reference frame (dq) to manage power flow between the AC Port and DC Port I, ensuring sinusoidal AC current that is in phase with the voltage. The converter operates in three modes: grid-tied or grid-connected mode, rectifier mode, and stand-alone mode. In grid-tie mode, the d-axis reference current is set to control power flow between the AC Port and DC Port I. In rectifier mode, the voltage controller generates the d-axis reference current based on the DC Port I voltage. In stand-alone mode, the voltage controller generates I_d based on the feedback from V_d. Next, the dq-axis current errors are calculated for the current controllers to minimize these errors, and the current controllers' outputs are added to the feedforward components. Finally, the U_d and U_q values are transformed back to the abc reference frame to generate a modulation signal, M_{abc}. A voltage controller for DC Port II generates a bias signal by comparing the reference voltage of DC Port II with its actual voltage. This bias signal is then added to M_{abc}, creating a modified modulation signal that regulates the voltage at DC Port II. SPWM for the

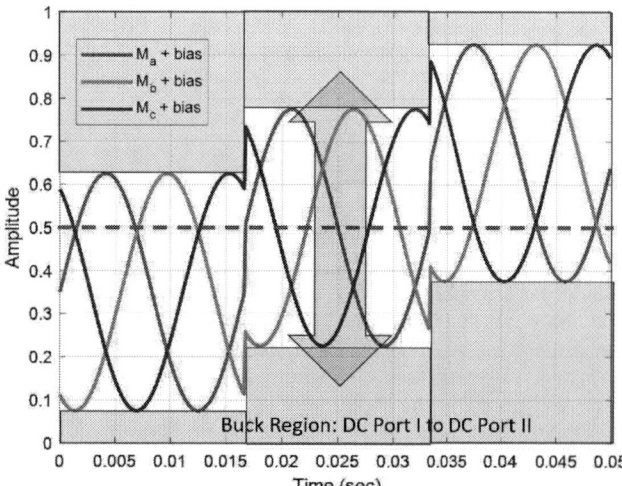

Fig. 3. Modified modulation signal.

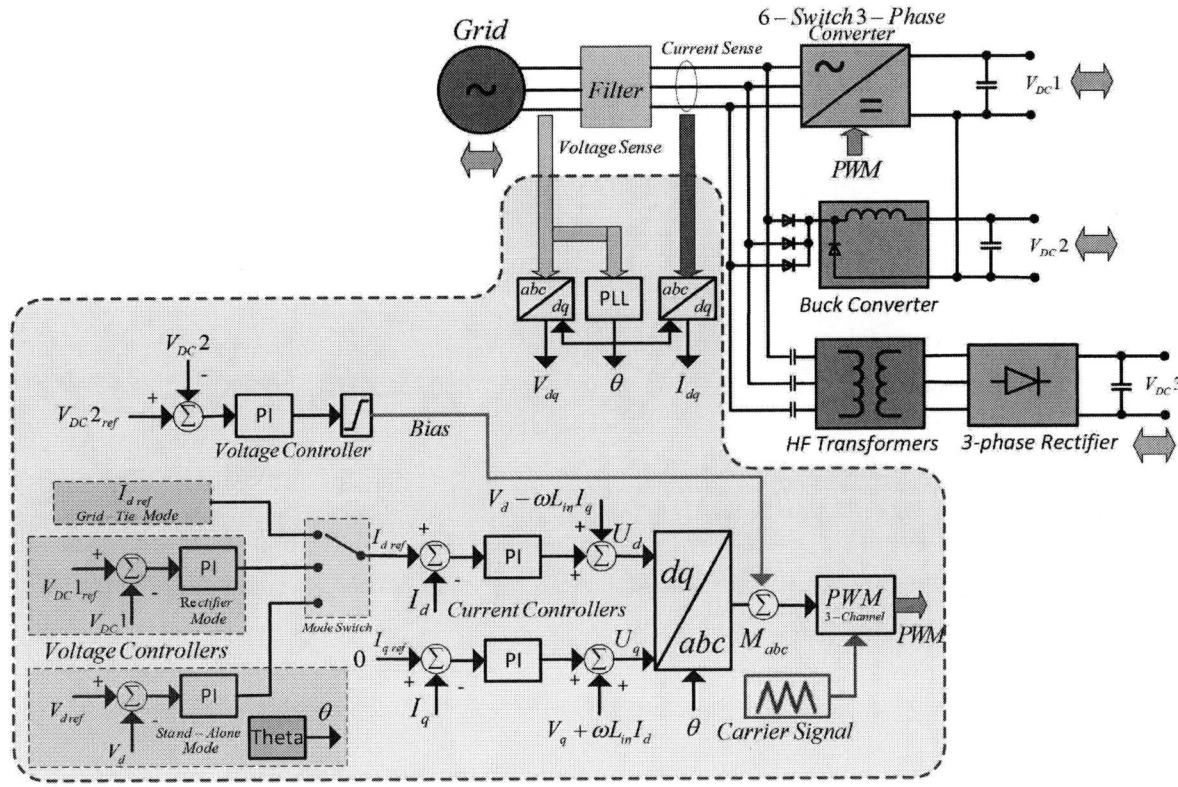

Fig. 5. Block diagram of the proposed converter control system.

switches is generated by comparing a triangle carrier with this modified modulation signal. Additionally, the controller removes low-frequency components caused by varying modulation signals. Importantly, the inclusion of the bias signal to create the modified modulation signal does not affect the voltages of the AC Port, DC Port I, and DC Port III.

IV. HARDWARE IMPLEMENTATION AND EXPERIMENTAL RESULTS

A simple scaled-down prototype was constructed in the laboratory to demonstrate the proposed converter's functionality. The prototype specifications are as follows: the AC Port voltage level is 208 V_{rms}, rated power is 2 kW, switching frequency is 40 kHz, and DC Ports I, II, and III have voltage levels of 380 V, 300 V, and 48 V, respectively. Active switches S_1-S_6 are SIHG47N65E MOSFETs, and passive switches D_1-D_{10} are APT60S20BG diodes. Capacitors C_1 and C_2 have a capacitance of 2000 μF, while C_3 has 3000 μF. Additionally, 1.5 mH inductors were used on the AC side, with a 7:1 turns ratio for the transformers. A 2 mH DC inductor and a 100 μH inductor (L) were also included, along with 4.7 μF DC blocking capacitors. The converter control system was implemented using a TI DSP TMS320F280049C, with a configuration similar to the system shown in Fig. 5.

A total load of 850 W is applied across all DC ports while the system operates in grid-connected mode. As shown in Fig. 6(a), the AC input phase 'a' current is in phase with the phase 'a'

voltage and is sinusoidal, indicating that the converter is effectively performing power factor correction (PFC). In Fig. 6(b), the voltages shown are across the three-phase VSC switching nodes, as well as the voltage before the DC inductor relative to the negative terminals of DC Ports I and II. The voltage before the DC inductor exhibits a longer pulse width compared to the other VSC switching node phases, due to the three diodes connected to the switching nodes. Fig. 6(c) shows the voltage on the primary side of the transformer, with a zoomed-in view demonstrating that the impressed voltage is a square wave with a frequency equal to the switching frequency. Finally, Fig. 6(d) illustrates the DC voltages at Ports I, II, and III, which exhibit slight low-frequency ripple. The input AC phase 'a' current is also shown in Fig. 6(d).

V. CONCLUSION

This paper presents a three-phase multiport AC-DC converter that integrates multiple DC conversion stages. The proposed converter uses a reduced-switch, single-stage topology, which can potentially eliminate the need for two additional converters in a microgrid. This reduction simplifies the system and significantly lowers its size, cost, and complexity. Galvanic isolation, high power density, standard controller compatibility, independent power management, and power factor correction make the converter suitable. The experimental results presented here confirm the practicality of the proposed converter by demonstrating its effective control using simple techniques.

REFERENCES

[1] M. Naghizadeh, H. S. Gohari, H. Hojabri and E. Muljadi, "New Single–Phase Three-Wire Interlinking Converter and Hybrid AC/LVDC Microgrid," in *IEEE Transactions on Power Electronics*, vol. 38, no. 4, pp. 4451-4463, April 2023, doi: 10.1109/TPEL.2022.3230787.

[2] D. S. D'antonio, O. López-Santos, A. Navas-Fonseca, F. Flores-Bahamonde and M. A. Pérez, "Multi-Mode Master-Slave Control Approach for More Modular and Reconfigurable Hybrid Microgrids," in *IEEE Access*, vol. 11, pp. 55334-55348, 2023, doi: 10.1109/ACCESS.2023.3280449.

[3] I. Ninma Jiya, H. Van Khang, N. Kishor and R. M. Ciric, "Novel Family of High-Gain Nonisolated Multiport Converters With Bipolar Symmetric Outputs for DC Microgrids," in *IEEE Transactions on Power Electronics*, vol. 37, no. 10, pp. 12151-12166, Oct. 2022, doi: 10.1109/TPEL.2022.3176688.

[4] R. Aravind, B. Chokkalingam and L. Mihet-Popa, "A Transformerless Non-Isolated Multi-Port DC–DC Converter for Hybrid Energy Applications," in *IEEE Access*, vol. 11, pp. 52050-52065, 2023, doi: 10.1109/ACCESS.2023.3280195.

[5] L. Zheng, R. P. Kandula and D. Divan, "Multiport Control With Partial Power Processing in Solid-State Transformer for PV, Storage, and Fast-Charging Electric Vehicle Integration," *in IEEE Transactions on Power Electronics*, vol. 38, no. 2, pp. 2606-2616, Feb. 2023, doi: 10.1109/TPEL.2022.3211000.

[6] F. Wu, K. Wang, G. Hu, Y. Shen and S. Luo, "Overview of Single-Stage High-Frequency Isolated AC–DC Converters and Modulation Strategies," in *IEEE Transactions on Power Electronics*, vol. 38, no. 2, pp. 1583-1598, Feb. 2023, doi: 10.1109/TPEL.2022.3210745.

[7] H. Wu, L. Zhu and F. Yang, "Three-Port-Converter-Based Single-Phase Bidirectional AC–DC Converter With Reduced Power Processing Stages and Improved Overall Efficiency," in *IEEE Transactions on Power Electronics*, vol. 33, no. 12, pp. 10021-10026, Dec. 2018, doi: 10.1109/TPEL.2018.2824242..

[8] J. Khodabakhsh and G. Moschopoulos, "Simplified Hybrid AC–DC Microgrid With a Novel Interlinking Converter," in *IEEE Transactions on Industry Applications*, vol. 56, no. 5, pp. 5023-5034, Sept.-Oct. 2020, doi: 10.1109/TIA.2020.2996537.

[9] S. Neira, J. Pereda and F. Rojas, "Three-Port Full-Bridge Bidirectional Converter for Hybrid DC/DC/AC Systems," in *IEEE Transactions on Power Electronics*, vol. 35, no. 12, pp. 13077-13084, Dec. 2020, doi: 10.1109/TPEL.2020.2990667.

[10] M. Rouhani and G. J. Kish, "Multiport DC–DC–AC Modular Multilevel Converters For Hybrid AC/DC Power Systems," *in IEEE Transactions on Power Delivery*, vol. 35, no. 1, pp. 408-419, Feb. 2020, doi: 10.1109/TPWRD.2019.2927324.

[11] C. Perera, J. Salmon and G. J. Kish, "Multiport Converter With Independent Control of AC and DC Power Flows for Bipolar DC Distribution," in *IEEE Transactions on Power Electronics*, vol. 36, no. 3, pp. 3473-3485, March 2021, doi: 10.1109/TPEL.2020.3016212.

[12] H. Wu, M. Han and K. Sun, "Dual-Voltage-Rectifier-Based Single-Phase AC–DC Converters With Dual DC Bus and Voltage-Sigma Architecture for Variable DC Output Applications," in *IEEE Transactions on Power Electronics*, vol. 34, no. 5, pp. 4208-4222, May 2019, doi: 10.1109/TPEL.2018.2864584.

[13] B. R. de Almeida, J. W. M. de Araújo, P. P. Praça and D. de S. Oliveira, "A Single-Stage Three-Phase Bidirectional AC/DC Converter With High-Frequency Isolation and PFC," in *IEEE Transactions on Power Electronics*, vol. 33, no. 10, pp. 8298-8307, Oct. 2018, doi: 10.1109/TPEL.2017.2775522.

[14] C. Perera, J. Salmon and G. J. Kish, "Multiport Converter With Enhanced Port Utilization Using Multitasking Dual Inverters," in *IEEE Open Journal of Power Electronics*, vol. 2, pp. 511-522, 2021, doi: 10.1109/OJPEL.2021.3109098.

[15] D. Ma, W. Chen, L. Shu, X. Qu, X. Zhan and Z. Liu, "A Multiport Power Electronic Transformer Based on Modular Multilevel Converter and Mixed-Frequency Modulation," in *IEEE Transactions on Circuits and Systems II: Express Briefs*, vol. 67, no. 7, pp. 1284-1288, July 2020.

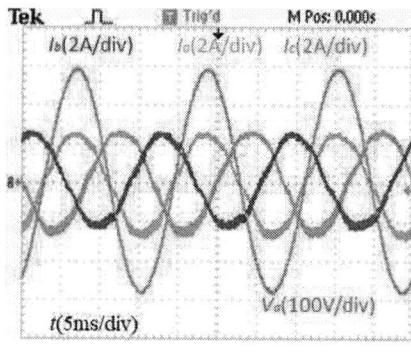

(a)

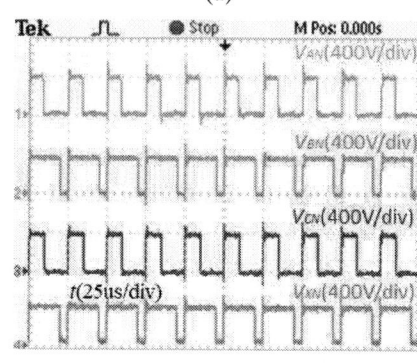

(b)

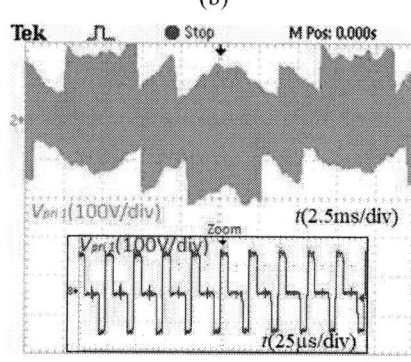

(c)

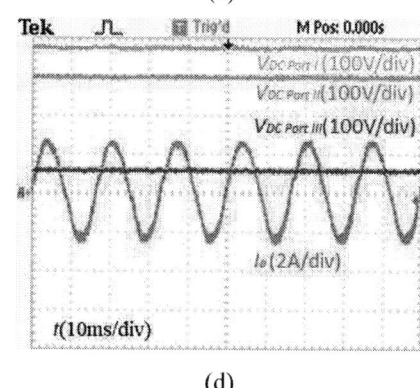

(d)

Fig. 6. (a) Three-phase currents and phase 'a' voltage, (b) voltage across the switching nodes and before the DC inductor, (c) Primary side voltages, (d) DC port voltages along with AC Port phase 'a' current.

High Efficiency AC-adapter realized by Voltage-clamper with Mid-voltage AHB converter using Synchronous Rectification

Shuichiro Motoori
ROHM Co., Ltd.
Yokohama, Japan
shuichiro.motori@rohm.co.jp

Akihiro Kawano
ROHM Co., Ltd.
Yokohama, Japan
akihiro.kawano@dsn.rohm.co.jp

Toshiyuki Zaitsu
ROHM Co., Ltd.
Yokohama, Japan
toshiyuki.zaitsu@rohm.co.jp

Riku Tatetsu
ROHM Co., Ltd.
Yokohama, Japan
riku.tatetsu@rohm.co.jp

Kohei Sebata
ROHM Co., Ltd.
Yokohama, Japan
kohei.sebata@rohm.co.jp

Kazuki Miyanjou
Sojo University
Kumamoto, Japan
g2411m06@m.sojo-u.ac.jp

Kimihiro Nishijima
Sojo University
Kumamoto, Japan
nisijima@cis.sojo-u.ac.jp

Abstract— **The advent of GaN devices has intensified the competition to miniaturize AC- adapter size. Utilizing the high performance of GaN-FET to reduce size, power supply topologies are transitioning from the traditional flyback topology to the resonant topologies such AHB (Asymmetric Half-bridge Flyback) configurations. A previously proposed new concept of two converters combining a voltage-clamper circuit with Mid-voltage (MV) AHB converter with GaN-FET successfully reduced the converter size by half. However, it still faced the issue that efficiency is not improved and is still at the same level as conventional AC-adapters. This paper introduces the incorporation of a Synchronous Rectification (SR) circuit in a MV-AHB converter with optimization of the soft-switching condition which achieves higher efficiency. The MV-AHB converter has achieved 96.4% efficiency at input voltage of 160Vdc, and output voltage of 20Vdc /3.25A(65W) with 500kHz, (up from the previous 91%). Finally, a complete AC-DC converter board was built by combining the AC input section (including voltage clamper) with the MV-AHB converter, resulting in an overall efficiency of 93% .**

Keywords— AC-adapter, miniaturization, voltage clamper, AHB, GaN, input capacitor, transformer

I. INTRODUCTION

AC-adapters for notebook PCs are an area in which there is strong demand for smaller size and higher efficiency, and the widespread use of high-voltage GaN-FETs is accelerating the competition to miniaturize AC-adapters [1]-[7]. However as shown in Fig. 1, both the input capacitor (which accounts for 40% of the total volume) and the transformer (which accounts for 45% of the total volume of components) are the major contributors that is preventing further miniaturization of AC adapters. Many challenges in circuit topologies and devices had to be addressed.

Fig.2 shows the traditional flyback topology. The use of 650V GaN-FET can decrease the Rds(on), aiding in the miniaturization of the circuit. However, low switching frequency in the range of 100kHz-200kHz results in the inability to reduce the transformer size in real market products. And the size of input electrolytic capacitor is still large due to requirement for high withstand voltage.

Fig. 3 shows the conventional High-Voltage (HV) AHB topology, which has recently attracted attention [2]-[7]. Compared to the Flyback converter, the Half-Bridge configuration of the AHB allows high frequency ZVS (Zero Voltage Switching) operation that allows easier downsizing of the transformer. But due to the operation at high voltages (400V), both the input capacitor Cin and the transformer cannot be made smaller.

To solve these problems simultaneously, a new concept of two stage converters combining a voltage-clamper circuit with MV-AHB converter, shown in Fig.4 was studied in 2016, and 1MHz operation with GaN-FET was proposed in 2024 [8] [9]. AC input voltage was clamped at 160Vdc (less than half of 400V), so the input capacitor size can be reduced to half and the MV-AHB converter is able to operate at high frequency which leads to a smaller size of transformer. However, the efficiency of MV-AHB converter was only 91% due to the diode rectification, which is same level as conventional AC-adapter, not enough to make it half the size as an AC adapter. In this paper, Synchronous Rectification (SR) is introduced for MV-AHB, resulting in a high efficiency of 96%. Finally, a complete AC-DC converter board was built by combining the AC input section with MV-AHB converter, resulting in an overall efficiency of 93%.

In section II, the MV-AHB converter with SR operation is described. In section III, the complete AC-DC converter operation and loss analysis are described. In section IV, the small size AC-DC demo board is presented. Conclusion and future work is discussed in section V.

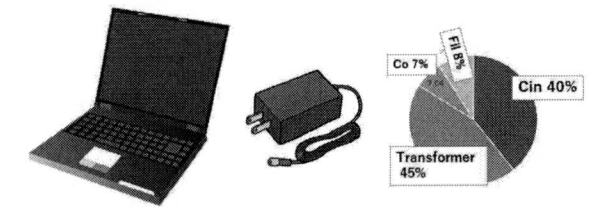

Fig.1. Volumue ratio of passive components in AC-adapter

979-8-3315-1612-3/25 $31.00 © 2025 IEEE

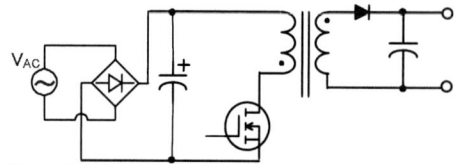

Fig.2. Traditional Flyback converter

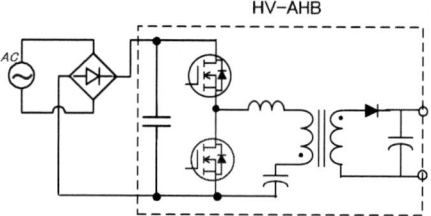

Fig. 3. Conventional HV-AHB converter

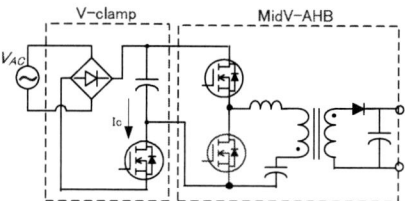

Fig. 4. Proposed Voltage-clamper+MV-AHB

II. MV-AHB CONVERTER WITH SR-FET

A. Designing MV-AHB converter

Fig. 5 shows the MV-AHB converter with SR-FET block diagram(a), and timing chart(b). Table I shows the specification of the circuit and components used.

In the previous case of "Diode Rectification" in Fig. 4 [9], the loss of the diode (Q_3) was dominant, and the loss of the snubber circuit was also large. As a result, the previous AHB converter loss in Fig.4 was 5.5W and the efficiency was only 91%. A synchronous Rectification (SR) operation can reduce the rectification loss dramatically, also, the snubber circuit can be removed by implementing a soft-switched operation as explained below.

In (1) and (2), the inductance L_m is set so that the negative peak current I_{np} of I_{Lm} achieves ZVS (Zero-Voltage-Switching) condition of Q_1 and Q_2. "n" is turns ratio, D is duty ratio, C_{oss1}, C_{oss2}, and $C_{oss\ SR}$ are the drain-source capacitance of Q_1, Q_2 and Q_3(SR-FET), respectively.

$$L_m = \frac{(V_{in}-n\cdot V_o)\cdot D\cdot T_{sw}}{2\cdot(I_{pri_{avg}}+I_{np})} = 19uH \quad (1)$$

Where, I_{np} is given as follows.

$$I_{np} = \frac{(C_{oss1}+C_{oss2}+C_{oss_{SR}}/n^2)\cdot V_{in}}{\Delta t_d} \quad (2)$$

In LLC resonant converter design, the inductance ratio (Lm/Lr) is often set to 3 or 5, leading to large secondary-side leakage inductance. The AHB converter dislikes large leakage inductance, as it causes ringing and surges due to half-wave resonance. Hence the inductance ratio should be set to a large value such as 50 which results in a smaller leakage inductance. L_r is given as follows.

$$L_r = \frac{L_m}{inductance\ ratio} = \frac{19uH}{50} \cong 0.4uH \quad (3)$$

The AHB converter, like the LLC converter, has a resonance capacitance C_r designed to achieve ZCS (Zero Current Switching) on the secondary side.

Therefore, C_r is given by the following equation.

$$C_r = \frac{1}{f_{sw}^2\cdot 4\pi^2\cdot L_r} = 0.25uF \rightarrow 0.3uF \quad (4)$$

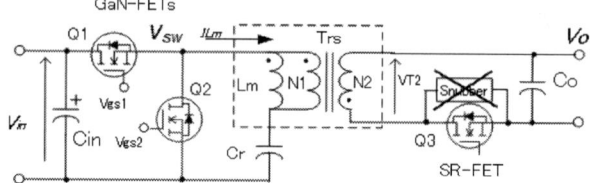

Fig. 5(a). MV-AHB with SR block diagram

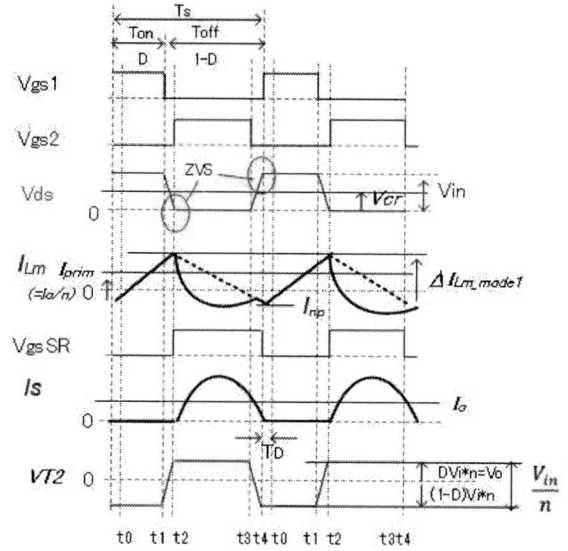

Fig. 5(b). Timing chart of key waveforms of AHB

TABLE I. MV-AHB CONVERTER SPECIFICATION AND COMPONENTS

Symbols	Value
Vin	160Vdc
Vo	20Vdc
Io	3.25A
Fsw	500kHz
Q1/Q2 (GaN-FET)	Rohm: GNP2130TCB (650V/130mΩ)
Q3 (SR-FET)	Infineon (80V/7.8mΩ)
Transformer	Coilfraft: ZF3187-AL (12T:4T, Lm=20uH, Lr=0.4uH)
Cin	Electrolytic cap (200V/47uF/0.5Ω) x2
Cr	Ceramic cap (0.4uF)
Co	Electrolytic cap (25V/820uF)
Controller	Rohm: ML62Q2045 [10]

979-8-3315-1612-3/25 $31.00 © 2025 IEEE

B. Experimental results of MV-AHB converter

Fig. 6 shows that the measured efficiency of the MV-AHB converter was improved dramatically from 91% to 96.4%. Fig. 7 shows the key waveforms. Even though there is no snubber circuit, no surge voltage is observed in the Vds waveform of the Q_3(SR-FET) due to the optimization of the soft-switching (ZVS) condition. Fig.7 shows the observed waveforms of the switch node (Q2_Vds) and inductor current I_{Lm} and SR-FET Vds. ZVS of Q2_Vds is achieved at primary side, and "no surge volage" is achieved at Q3(SR)_Vds.

III. COMPLETE AC-DC (VOLTGE-CLAMP+MV-AHB)

A. Over all AC-DC converter structure

The AHB converter, which achieved high efficiency through the use of the SR described in section II, was combined with the AC input section (including the voltage clamper stage) to function as an AC-DC converter. A test board of this AC-DC was implemented as shown in Fig. 8. Fig. 9 shows a block diagram of the test board which is depicted in a more practical manner than Fig. 4. A clamp FET Q_{clamp} is provided between the diode bridge and the input capacitor C_{in}. The AC input portion (includes the diode-bridge, voltage-clamp circuit and the input bulk capacitor) will operate in the capacitor input mode when the input voltage is low (eg. 100Vac), and in the voltage clamp mode with higher input voltage (eg. 200Vac). TABLE II shows the AC portion specification and components values. First, an input loss analysis is performed for each mode.

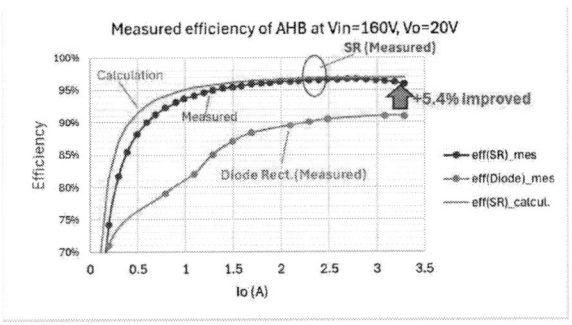

Fig. 6. Measured efficiency of (Mid-V) AHB converter

(Vin=160V, Vo=20V, 500kHz)

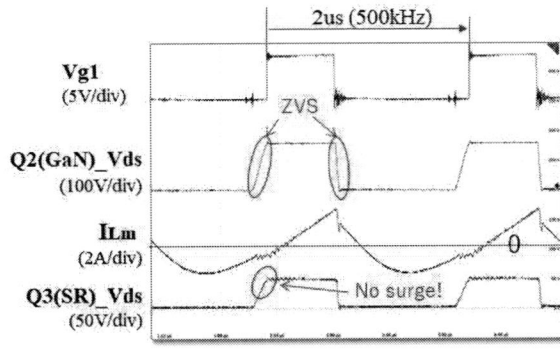

Fig. 7. Observed key waveforms of AHB converter

(Vin=160V, Vo=20V, Io=3.25A, 500kHz)

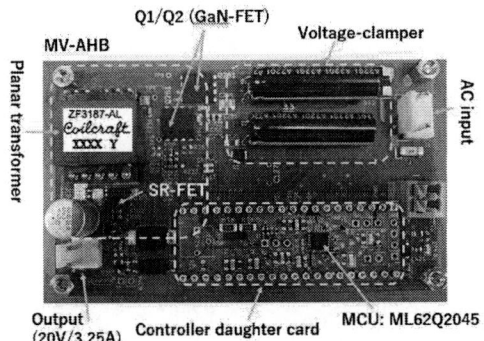

Fig. 8. Test board of AC-DC (Voltage-clamper+MV-AHB)

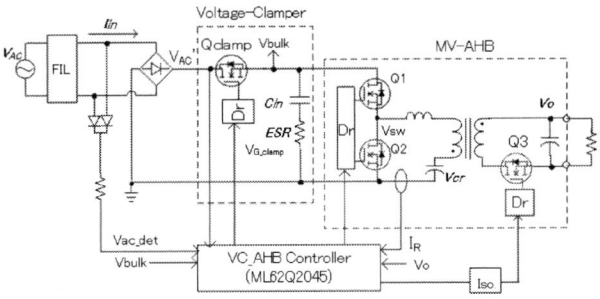

Fig. 9. Block diagram of test board AC-DC

TABLE II. AC PORTION SPECIFICATION AND COMPONENTS

Symbols	Value
V_{AC}	85V-240V
Q_{clamp}	600V/30A, 50mΩ
Cin (repost)	Electrolytic cap (200V/47uF/0.5Ω) x2

B. Loss Analysis of AC input portion at 100Vac input

When input voltage is $100V_{AC}$, the Q_{clamp} is fully-on for the entire duration in order to operate in the conventional capacitor input mode as shown in Fig. 10.

When it operates in the capacitor input mode, the loss analysis is the same as conventional AC-DC input portion. So, eqations from (5) to (15) are very basic formula for AC-DC design. Assuming the input conditions are $V_{p100Vac}=\sqrt{2}Vac$, $V_{th}=V_{p100Vac}*0.6$, the conduction angle $\alpha_{capmode}$ is given as follows.

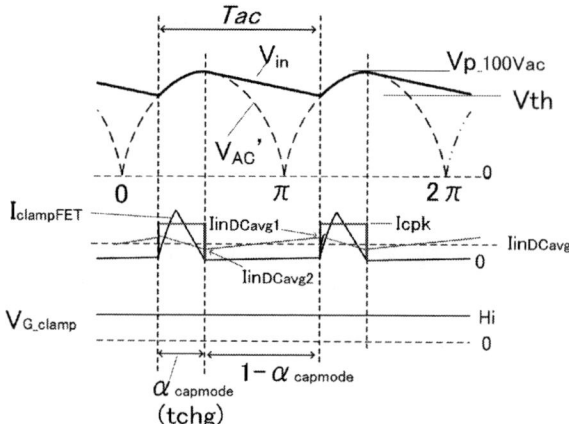

Fig. 10. Key waveforms of input at 100V_{AC} (Capacitor input mode)

$$\alpha_{capmode} = Acos\left(\frac{V_{th}}{V_{p100Vac}}\right)/\pi = 0.295 \quad (5)$$

The charging time tchg is given as follows, where Tac=10ms.

$$t_{chg} = \alpha_{capmode} \cdot T_{ac} = 3ms \quad (6)$$

So, the input capcitance $C_{inputcap\_mode}$ as "capacitor input mode" is calculated using the hold-up time formula as follows. This is the minmum value of Cin.

$$C_{inputcap\_mode} = \frac{2 \cdot P_{in} \cdot (T_{ac} - t_{chg})}{V_{p100Vac}^2 - V_{th}^2} = 84uF \quad (7)$$

Input average DC current Iin_{DCavg} is calculated as follows.

$$Iin_{DCavg1} = \frac{P_{in}}{\sqrt{2} \cdot V_{ac}}, \quad Iin_{DCavg2} = \frac{P_{in}}{\sqrt{2} \cdot V_{ac} * 0.6},$$

$$Iin_{DCavg} = \frac{Iin_{DCavg1} + Iin_{DCavg2}}{2} = 0.68A \quad (8)$$

- Bridge-diode loss at 100Vac P_{bridge_100} is calculated as follows.

$$P_{bridge\_100Vac} = 2 \cdot V_f \cdot Iin_{DCavg} = 1.36W \quad (9)$$

In order to calculate the clamp FET and the input capacitor ESR loss, Iin_{pk} and RMS curent are given as follows.

$$Iin_{pk} = \frac{Iin_{DCavg}}{\alpha_{capmode}} = 2.3A \quad (10)$$

$$Iin_{FET\_rms} = Iin_{pk}\sqrt{\alpha_{capmode}} = 1.25Arms \quad (11)$$

$$Iin_{rms} = Iin_{pk}\sqrt{\alpha_{capmode}(1 - \alpha_{capmode})} = 1.05Arms \quad (12)$$

- The calmp FET loss is calcuted as follows with R_{ds}=50mΩ.

$$P_{Clamp\_FET\_100} = Iin_{FET\_rms}^2 \cdot Rds = 0.08W \quad (13)$$

- The capacitor ESR loss is calculated as follows with ESR_{Cin}=250mΩ.

$$P_{Cin\_100} = Iin_{rms}^2 \cdot ESR_{Cin} = 0.28W \quad (14)$$

- Total loss in the AC input portion $P_{ACinput100}$ at $100Vac$ is given as follows.

$$P_{ACinput_{100}} = P_{bridge_{100}} + P_{Cin_{100}} + P_{ClampFET_{100}}$$
$$= 1.72W \quad (15)$$

C. Loss Analysis of AC input portion at 200Vac input
(Voltage-clamper is activated)

Detail explanation of the operation and deriving the equation of the voltage-clamper circuit was reported in [9]. So, here is a quick review and calculation of losses.

Fig. 11 shows the key input waveforms at *200Vac* where the voltage-clamper is activated. V_{AC}' is a rectified AC voltage, Ic is a charging current through Cin. When V_{AC}' is reaches V_1, Ic starts to flow to charge Cin. When V_{AC}' is reaches V_2, Q_{clamp} is turned off.

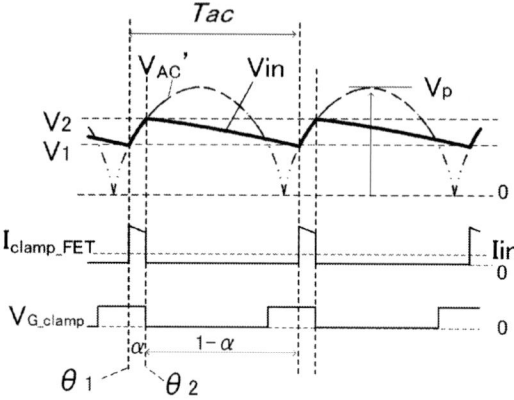

Fig. 11. Key waveforms of input at 200V_{AC} (Voltage-clamper activated)

θ_1 and θ_2 are set as the angle at V_1 and V_2, respectively. α is conduction angle between θ_1 and θ_2. Ripple ratio of input capacitor voltage V_{in} is expressed as k.

From (16)-(19), α is derived in (20). Input capacitor Cin is charged and discharged during period α and period (1- α), respectively. So, using the energy balance equation of (21), Cin is given in (22), where T_{ac} is half cycle period of AC input.

$$\theta_1 = sin^{-1}\left(\frac{V_1}{V_p}\right) \quad (16)$$

$$\theta_2 = sin^{-1}\left(\frac{V_2}{V_p}\right) \quad (17)$$

$$\alpha = \frac{1}{\pi}(\theta_1 - \theta_2) \quad (18)$$

$$k = \frac{V_2 - V_1}{V_2} \quad (19)$$

$$\alpha = \frac{1}{\pi}\left\{sin^{-1}\left(\frac{V_2}{V_p}\right) - sin^{-1}\left(\frac{V_2(1-k)}{V_p}\right)\right\} \quad (20)$$

$$\frac{1}{2}C_{in}(V_2^2 - V_1^2) = P_o(1 - \alpha)T_{ac} \quad (21)$$

$$C_{in} = \frac{2P_o(1-\alpha)T_{ac}}{k(2-k)V_2^2} \quad (22)$$

I_{cpk} is expressed in (23). In order to simplify the calculation, an approximation for dv/dt near t=0 of $sin(\omega t)$ is used, so I_{cpk} is derived in (25).

$$I_{cpk} = C_{in}\frac{dv}{dt} \quad (23)$$

$$\frac{dv}{dt} = V_p(sin(\omega t))' \cong V_p(\omega t)' = V_p\omega \quad (24)$$

$$I_{cpk} = V_pC_{in} \cdot \omega \quad (25)$$

979-8-3315-1612-3/25 $31.00 © 2025 IEEE

Under conditions of *Vac=200Vrms, Po=65W, V_2=160Vdc, k=0.25*, conduction angle of α, I_{cpk}, I_{in_avg} are calculated as follows.

$$\alpha = 0.052 \quad (26)$$

$$I_{cpk}=8.9A \quad (27)$$

$$I_{in\_avg} = \alpha \cdot I_{cpk} = 0.46A \quad (28)$$

- Bridge diode loss is calculated as follows.

$$P_{bridge} = 2 \cdot V_f \cdot I_{in} = 1.14W \quad (29)$$

- Clamp-FET Q_{clamp} loss is calculated as follows.

$$I_{cpk\_rms} = I_{cpk} \cdot \sqrt{\alpha} = 2.04Arms \quad (30)$$

$$P_{clampFET} = \left(I_{cpk_{rms}}\right)^2 \cdot Rds_{clampFET} = 0.31W \quad (31)$$

- Input capacitor Cin loss is calculated as follows.

$$I_{cpk\_Cin\_rms} = I_{cpk} \cdot \sqrt{\alpha(1-\alpha)} = 1.98Arms \quad (32)$$

$$P_{Cin} = I_{cpk\_Cin\_rms}^{\;2} \cdot ESR = 0.99W \quad (33)$$

- So, the loss of AC input portion $P_{ACinput}$ is given as follows.

$$P_{ACinput} = P_{bridge} + P_{clampFET} + P_{Cin} = 2.4W \quad (34)$$

D. Total loss of AC input portion and MV-AHB converter

TABLE III shows the results of calculated loss of total AC-DC converter, which consists of AC input portion and AHB converter. Fig. 12 shows the graph based on TABLE III. It is found that the AHB converter achieved high efficiency, but the AC input portion still has large loss. Especially the ESR loss in the electrolytic capacitor C_{in} at low frequency (eg. 50Hz) is still large.

TABLE III. Calculated loss(W) of all components in AC-DC total

Portion	Input voltage	100Vac	200Vac
AC input portion	Bridge diode	1.36	1.11
	ClampFET	0.08	0.31
	Cin (AL-cap)	0.28	0.98
AHB	Trans_core	0.5	0.5
	Trans_winding	1.1	1.1
	AHB_Q1(Hi)	0.334	0.334
	AHB_Q2(Lo)	0.398	0.398
	SR(80V/7.8m)	0.56	0.56
	other	0.4	0.4
	Total loss	5.012	5.692
	Po (rated)	65	65
	Efficiency	92.8%	91.9%

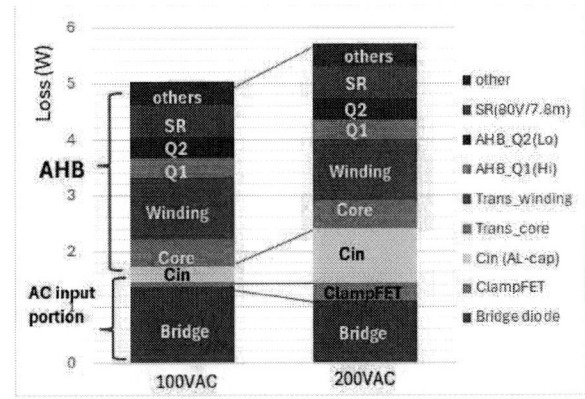

Fig. 12. Calculated loss analysis results graph at 500kHz AHB

IV. EXPERIMTANL RESULT OF TOTAL AC-DC CONVERTER

Fig. 13 shows the measured AC input portion waveforms includes the voltage clamp waveforms. It is found that the clamp circuit works well. Fig. 14 shows the measured overall efficiency as AC-DC converter when input voltage is 100Vac, and 200Vac. Total efficiency at 100Vac input voltage is 93% and at 200Vac input is 91%.

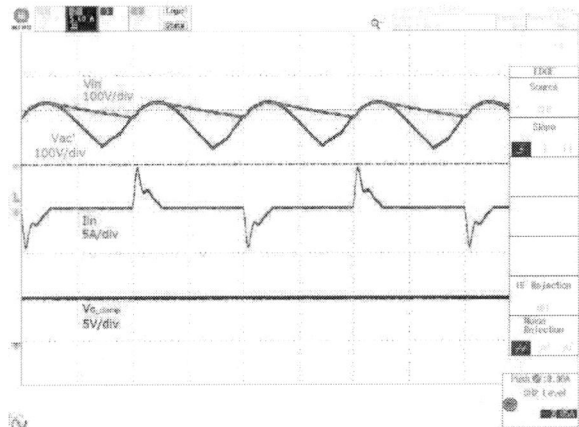

(a) Vin= AC100V, Load= 60W

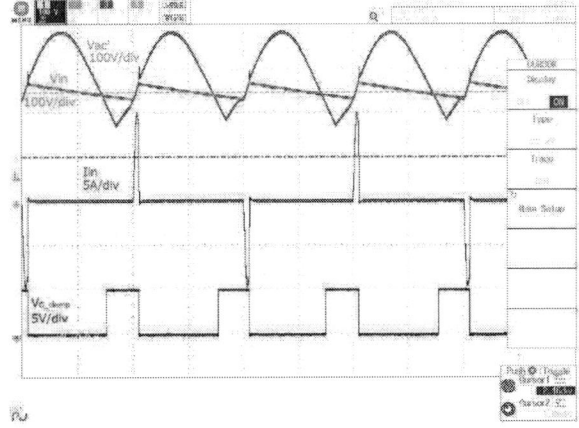

(b) Vin= AC200V, Load= 60W

Fig. 13. Observed switching waveforms of Voltage-clamper operations

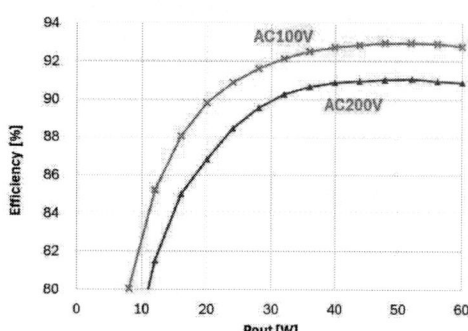

Fig. 14. Measured total efficiency at input Vac=100V and 200V, output Vout=20V with 500kHz AHB converter

V. CONCLUSIONS AND FUTURE WORK

A two-stage converter concept of the V-clamper with MV-AHB with GaN-FET converter was proposed for AC-adapter miniaturization. Regarding AHB converter portion, efficiency improvement was studied by incorporating the Synchronous Rectification (SR) with the optimization of soft-switching condition. The losses were reduced and the efficiency of the AHB converter was improved dramatically from 91% to 96.4% at Vin=160Vdc, Vout=20V, Io=3.25A(65W) at 500kHz.

In addition, AC-DC converter board was implemented (combined AC input portion with AHB converter) and overall efficiency was measured. As a result, 93% efficiency at 100Vac input, and 91% at 200Vac were achieved.

As next steps, the loss in AC input portion needs to be improved and AHB converter needs to be optimized in order to increase the total efficiency further.

REFERENCES

[1] X. Huang, J. Feng, W. Du, F. Lee, and Q. Li, "Design consideration of MHz active clamp flyback converter with GaN devices for low power adapter application," APEC2016 proc., pp.2334-2341.

[2] L. Huber, and M. Jovanovic, "Analysis, design, and performance evaluation of asymmetrical half-bridge flyback converter for universal-line-voltage-range applications", APEC2017 proc., pp.2481-2487.

[3] A. M. Garcia, M. Krueger, M. Schmid, J. Daimer, and M. Schlenk, "Hybrid-flyback and GaN enable ultra-high power density 240W USB-PD EPR adapter," APEC2023 proc., pp.1259-1264.

[4] G. Spiazzi, S. Buso, "Extended analysis of the asymmetrical half-bridge flyback converter," IEEE Transactions on Power Electronics, Vol. 36, No.7, pp.7956-7964, July 2021.

[5] M. Lie, Z. Ouyang, M. Andeson, "Analysis and optimal design of high-frequency and high-efficiency asymmetrical half-bridge flyback converter," IEEE Transaction on Industrial Electronics, Vol. 67, No. 10, pp.8312-8321, Oct., 2020.

[6] K. Cheng, T. Liang, H. Nghiep, and K. Chen, "Design and implementation of asymmetrical half-bridge flyback converter for USB power delivery applications", ECCE Asia 2023 proc., pp.593-600.

[7] Infineon data sheet, "XDP hybrid-flyback controller XDPS2201," https://www.infineon.com/cms/en/product/power/ac-dc-power-conversion/ac-dc-pwm-pfc-controller/llc-resonant-mode-controller/xdps2201/#!?fileId=5546d46277921c320177b3d4ea4a2055

[8] K. Nishijima, S. Takashita, T. Sato, S. Sohma, Y. Takeyama, "An isolated AC/DC Converter with two active-clamp circuits for low power apprications," IEICE Technical Report, Vol. 116, No.329, pp.51-55, 2016, Japan.

[9] S. Motoori, T.Zaitsu, A. Kawano, K. Tokumaru, K. Nishijima, "Miniaturization of AC-adapter realized by primary voltage-clamper with Mid-Voltage (150V) AHB converter", APEC2024 proc., pp.50-55.

[10] https://www.rohm.com/products/micon/logicoa/ml62q20xx-series/ml62q2045-nnngd_taping_-product

Active soft switching technique for single phase series capacitive link universal rectifier

Anran Wei
Department of Electrical and Computer
Engineering
Northeastern University
Boston, MA
wei.a@northeastern.edu

Brad Lehman
Department of Electrical and Computer
Engineering
Northeastern University
Boston, MA
lehman@ece.neu.edu

Mahshid Amirabadi
Department of Electrical and Computer
Engineering
Northeastern University
Boston, MA
m.amirabadi@northeastern.edu

Abstract— **Double line frequency pulsating power is a major challenge for single phase inverters and rectifiers. While traditional solution of bulky DC bus capacitor can filter out the inherent double frequency harmonic of single-phase inverters, its physical size and high failure rates are often undesirable, especially for applications that demand small volume and weight, high reliability and efficiency. In this paper a novel soft-switching single-phase AC to DC topology, which absorbs the pulsating power with a small film capacitor, is proposed. This eliminates the need for a bulky DC bus capacitor. The proposed topology uses an active auxiliary structure to achieve soft switching for all switches across the full load range. This lowers the power loss, volume and weight. In this paper simulation results and mathematical analysis are presented. Additionally, the behavior of the proposed rectifier is verified through experimental results.**

Keywords— soft-switching, single-phase rectifier, Universal converter

I. INTRODUCTION

Single phase inverters and rectifiers are widely used in renewable energy systems and electric vehicle (EV) battery chargers[1]. However, handling the inherent double line frequency power harmonic has always been challenging for these converters. Traditional single-phase AC to DC converters use bulky electrolytic bus capacitors to stabilize the bus voltage and filter the pulsating power, yet in many applications the massive volume and weight of the bus capacitor is undesirable[2]. Although different approaches have been developed to help reduce the double line frequency harmonics and thus improve output quality and power density, the constrain from the bus capacitor still exist and limits the performance of these converters[3], [4], [5]. .

Among different alternative solutions that were proposed to address this problem, active power decoupling / active filtering is one of the most trended and well developed solutions. By introducing extra switches and energy storage modules, the pulsating power causing mismatch between the instantaneous input and output powers can be absorbed and compensated[6]. In [7], [8], a ripple port structure is introduced to the system to serve as an active filter handling the double line frequency harmonics. By controlling the voltage across the buffer capacitor to be sinusoidal, the power mismatch is being transferred to the ripple port buffer. Although a larger voltage ripple across the

buffer capacitor helps reducing the required capacitance, a large capacitor is still necessary to maintain a stable bus voltage. Other active power decoupling strategies proposed in [9], [10], [11], [12] can also eliminate the double frequency pulsating power, yet they all require a number of extra switches and passive elements that are still considerably large.

Another effective solution is to use an inductive link topology. In[13], [14], a small inductor is connected in series with the input and output switch bridges and serves as an intermediate for transferring power. By allowing the link inductor to have a non-zero minimum current, this solution can absorb the pulsating power without any other components. The inductor, which has a positive current flow, serves as an energy storage that absorbs or releases energy as compensation for the mismatched input and output power. By introducing additional components soft switching can be achieved. Galvanic isolation can also be provided by adding a transformer to the link. Due to the dominant large link inductor the leakage inductance of the transformer does not cause any problems. However, there is a tradeoff between the link inductance and DC component of the link current, and as the power rating increases it may be difficult to optimize the link inductor. In practice, the link inductor has a limited energy capacity and usually additional active filter branches are needed for full elimination of the pulsating power[15].

Compared to inductive link converters, series capacitive link universal converters have higher power density [16]. In single phase AC to DC operation, as long as the link capacitor voltage has a sufficient DC bias, no additional energy storage is required

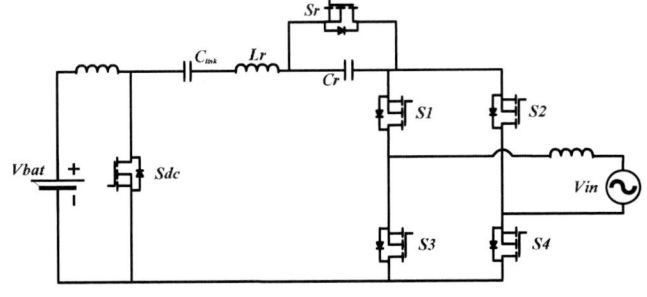

Figure 1: Proposed topology

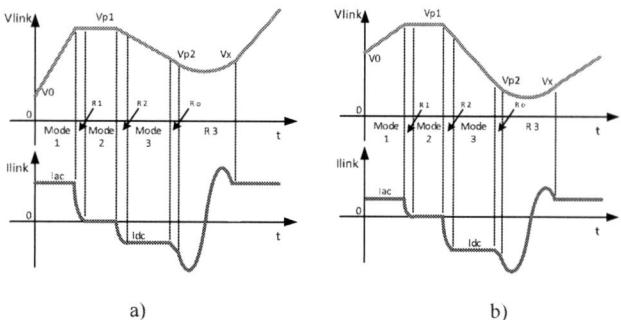

a)   b)

Figure 2: Link waveform when a) Pac>Pdc b) Pac<Pdc

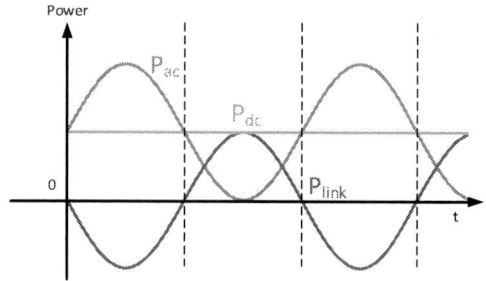

Figure 3: dynamic power of AC source, DC load and link capacitor

phase inverter and rectifier will suffer from extremely high peak current due to the relatively large value of link capacitance and high DC biased link voltage, which heavily offsets its merits and limits its application[17].

In this paper, a novel series capacitive link universal converter, which achieves full soft switching even when configured as a single-phase inverter or rectifier, is proposed. This converter adds an auxiliary switch and small passive components to the original series-capacitive link converter to achieve soft-switching, but the overall power density is not affected. The additional auxiliary circuit facilitates resonance and current swing in the link at the end of each switching cycle, preventing the unacceptable high peak current in the original LC circuit; thus, allowing the converter to benefit from the resonance inductor for full soft switching. Through this improvement, the applicability of series capacitive link universal converter is widely extended and can be a very attractive option for both household and industrial applications.

The principle of operation for the proposed converter is introduced in section II. Design and analysis are presented in section III. Simulation and experimental results are shown in section IV.

II. OPERATION PRINCIPLE

The proposed active soft switching series capacitive link converter, shown in Figure 1, uses the link capacitor C_{link} for transferring the power, while the series link inductance L_r, the resonance capacitor C_r and the auxiliary switch S_r facilitate soft-switching

for complete absorption of the double line frequency harmonics,. In general, the series capacitive link universal converter can benefit from full soft switching with adding only one small inductor to the link. However, soft-switching single-

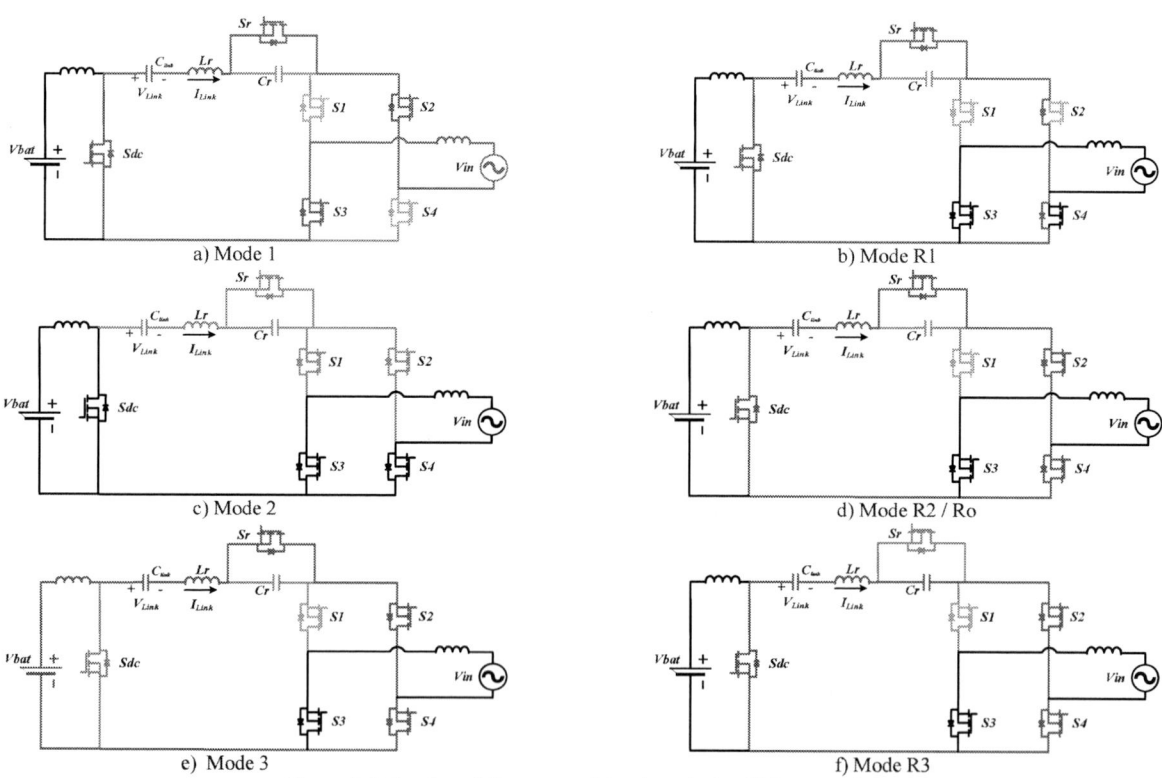

a) Mode 1

b) Mode R1

c) Mode 2

d) Mode R2 / Ro

e) Mode 3

f) Mode R3

Figure 4: Behavior of the proposed topology during different modes

Although the proposed converter is a universal topology the input and output of which can be single-phase / three-phase AC, or DC, the single phase AC to DC configuration is considered in this paper for explaining the principles of the operation. There are three main modes and three resonance modes in between that allow soft switching for the switches. An optional mode can be added before the beginning of the last resonance mode to boost the energy for guaranteed full soft switching. The detailed operation is explained in this section assuming the AC side source has positive current and voltage during the time span considered here, and the link voltage is controlled to have a positive DC bias. The absorption of pulsating power through link capacitor is explained through Figures 2 and 3, where link voltage and current waveforms under different power conditions are shown in Figure 2, and the instantaneous power from AC, DC and link are depicted in Figure 3. As shown in the figures the link capacitor actively absorbs or releases energy depending on the dynamic relationship between AC and DC side power. The mismatch between instantaneous input and output powers results in having different link capacitor voltages at the beginning and end of each switching cycle. If the instantaneous input power is higher than the output power, excess energy is absorbed by the link and only the required energy is delivered to the output, resulting in an overall energy surplus in the link capacitor. The net energy of the link is thus positive in this cycle with its end state voltage higher than beginning. Similarly, when the instantaneous input power is not enough for output, the insufficient amount of energy in each switching cycle will be provided by the link capacitor, resulting in an overall discharging of the link capacitor. Through this active balancing strategy the pulsating power can be completely absorbed and constant DC output power can be achieved. Detailed operation diagrams are shown in Figure 4.

Mode 1: In the first mode the AC source is connected across the link by turning on switches S_2, S_3, S_r and S_{dc}. The AC source charges the link capacitor, and the link voltage increases in mode 1. When C_{link} is sufficiently charged, S_2 is turned off and S_4 is turned on to enter the first resonance mode R1.

Mode R1: During this mode the link inductor (L$_r$) and capacitor (C$_{link}$) are shorted and resonate. Switch S_4 is turned on and its current has gradual increase due to the existence of resonance inductor L_r. the gradual increase of current in switch S_4 results in gradual decrease of current in S_2 and when the current of S_2 becomes zero, the switch can be turned off at zero current. Thus, these switches are all soft switched. Mode R1 ends at the end of switching transition.

Mode 2: The second main mode is a freewheeling mode for adjusting the switching frequency. During this mode both the DC and AC sides are disconnected from the link and no energy is being delivered or received. The link current that gradually decreased to zero in mode R1, remains zero in this mode and thus the link voltage will remain unchanged in Mode 2.

Mode R2: Mode 2 ends once S_2 is turned on. Similar to the condition in mode R1, it is turned on under soft switching. Due to the presence of resonance inductor L_r in the current path, the current of S_2 has a gradual increase in magnitude, and achieves soft turn on. When the switching transition complete the link inductor reaches DC side current, and the last main mode begins

with turning off switch S_{dc}, which is not carrying any current at the end of mode R2.

Mode 3: During mode 3 the link capacitor is connected across the output load and discharges its stored energy to the load. The AC side continues freewheeling during this mode. When the link capacitor is sufficiently discharged, the converter enters mode Ro or R3, depending on the relationship between input and output currents.

Mode Ro (Optional): If the absolute value of output current is equal or less than that of the input current, there will not be enough energy stored in the link in mode R3 to swing up and soft switching will not be achieved. To avoid this, in mode Ro switch S_{dc} is turned on after mode 3 ends and forms a closed loop for the link capacitor and inductor; thus, the link inductor is further discharged and the link current keeps decreasing. Once the link current reaches a value that guarantees a full soft switching mode, Ro ends and mode R3 begins.

Mode R3: If mode Ro is added prior to mode R3, only switch S_r is turned off, otherwise switch S_{dc} is also turned on simultaneously. Turning off S_r allows C_r to get involved in the resonance process. During this mode both the AC and DC sides are disconnected from the link. When mode R3 begins, C_r will be charged in negative direction and its voltage will start to decrease. The capacitance of C_r is much smaller than C_{link} and its voltage has a faster change rate. Once the absolute value of V_{cr} is larger than the link voltage, voltage across the link inductor will become positive and the link current will start to increase. Mode R3 ends when the link current swings up and reaches its peak value, which is slightly larger than the input current, and then decreases to input current. This satisfies the requirement for soft switching at the beginning of mode 1.

III. DESIGN AND ANALYSIS

A. General operation analysis

The proposed converter can be designed based on specified power, input voltage and output voltage. Due to the mismatch of instantaneous input and output powers, the instantaneous power of the link is also varying with a double line frequency harmonic:

$$P_{link}(t) = P_{ac}(t) - P_{dc}(t) = \frac{1}{2} V_{ac} I_{ac} \cos(2\omega t + \theta) \quad (1)$$

$$V_c(t) = \sqrt{V_{c0}^2 + \frac{V_{ac} I_{ac} \sin(2\omega t + \theta)}{2\omega C}} \quad (2)$$

where V_{ac} and I_{ac} are the amplitude of the AC source voltage and current, θ represents the phase angle between input voltage and current, which is zero if unity power factor is assumed. C is the link capacitance and V_{c0} is the DC offset voltage for the link capacitor. According to (1) and (2), the voltage of the link capacitor reaches its peak or valley point when $P_{link}(t)$ is zero, and $V_c(t) = V_{c0}$ when $P_{link}(t)$ is at its maximum or minimum value.

Since within each switching cycle the energy is transferred through the link capacitor, equations can be established assuming the instantaneous input and output powers are constant. The following equation is valid for the link capacitor:

$$P_{ac} - P_{dc} = \frac{1}{2} C_{link}(V_{start}^2 - V_{end}^2)f \quad (3)$$

where P_{ac} and P_{dc} are the instantaneous power at any given switching cycle, f represent the switching frequency and V_{start} and V_{end} refer to the voltage of link capacitor at the beginning and end of the switching cycle.

B. Control and design

Unlike the conventional single phase AC to DC rectifiers, where a large bus capacitor is used to suppress the bus voltage ripple, the proposed converter allows a much smaller link capacitor by having a higher link voltage ripple. By adding a DC bias to the link voltage, the link capacitor can store the necessary amount of energy for suppressing the pulsating power. The minimum DC bias voltage can be calculated using (4):

$$\frac{1}{2} C_{link} V_{c0,min}^2 = \int_0^{\frac{\pi}{2}} P_{link}(t) \quad (4)$$

Any DC bias voltage equal or higher than this minimum value should be enough to guarantee full suppression of the double frequency harmonic. In practical designs a slightly higher V_{c0} may be chosen.

The duration of the charging, discharging and freewheeling modes can be calculated based on instantaneous power and initial voltage of the link at the beginning of each switching cycle, according to (2). The proposed converter has decoupled charging and discharging modes; thus, the input or output port is only involved during their own charging or discharging mode. For a given condition, the charging mode duration of the n_{th} switching cycle can be calculated as:

$$t_{charge,n} = C_{link} \frac{\left(-V_{start,n} + \sqrt{V_{start,n}^2 + \frac{2P_{ac,n}}{C_{link}f}}\right)}{i_{ac,n}} \quad (5)$$

where $P_{ac,n}$ and $i_{ac,n}$ are the discretized instantaneous AC power and AC input current, which are considered constant values within the n_{th} switching cycle. The peak value of the link voltage during this cycle, $V_{peak,n}$, can be obtain:

$$V_{peak,n} = \sqrt{V_{start,n}^2 + \frac{2P_{ac,n}}{C_{link}f}} \quad (6)$$

Similarly, the duration of the discharging mode, as well as the V_{end} in the n_{th} switching cycle can be calculated:

$$t_{discharge,n} = C_{link} \frac{\left(V_{peak,n} + \sqrt{V_{peak,n}^2 - \frac{2P_{dc,n}}{C_{link}f}}\right)}{i_{dc,n}} \quad (7)$$

$$V_{end,n} = \sqrt{V_{peak,n}^2 - \frac{2P_{dc,n}}{C_{link}f}} \quad (8)$$

where $P_{dc,n}$ and $i_{dc,n}$ are the discretized instantaneous DC power and DC output current, which are considered constant values within the n_{th} switching cycle. With the calculated charging and discharging mode durations, the freewheeling mode time $t_{free,n}$ can be obtain:

$$t_{free,n} = \frac{1}{f} - t_{charge,n} - t_{discharge,n} - t_{swing,n} - t_{ro,n} \quad (9)$$

where $t_{swing,n}$ represents the duration of the final resonance mode R3, and $t_{ro,n}$ is the duration of the optional mode Ro. The detailed analysis for calculating $t_{swing,n}$ will be provided in next subsection. In practical designs the final resonance mode R3 and optional discharge mode Ro will be negligible compared to the main operation modes; thus, design constrain equations can be generated through (4) to (9), where switching frequency, link capacitance and link voltage DC bias can be selected based on system requirements subject to the target operation condition.

C. Active soft switching analysis

The proposed converter features full soft switching with the help of a small link inductor L_r, the value of which is decided by the link capacitor and duration of the final resonance mode R3. It is generally preferred to limit $t_{swing,n}$ to be less than 10% of the total cycle time for minimum impact on the overall system operation. In Mode R3, both DC and AC sides are shorted to create a resonance path for the link components, and the auxiliary switch S_r is turned off; thus, the resonance capacitor C_r is no longer shorted. The resonance happens between the three link components: link capacitor C_{link}, resonance inductor L_r and resonance capacitor C_r. The resonance mode allows the link current to swing up to a value equal or higher than the output current., A differential equation can be used to determine link current and resonance capacitor voltage during mode R3. Since C_r is much smaller than C_{link}, and the energy required in resonance mode for L_r is also negligible compared to the energy stored in the DC biased link capacitor, the link voltage is considered constant during the entire mode R3.

$$V_{link} + v_{Cr}(t) + L_r \frac{di_{link}(t)}{dt} = 0 \quad (10)$$

Since there is only one current path during mode R3, current flowing through the resonance capacitor C_r is exactly the same as that of L_r; thus:

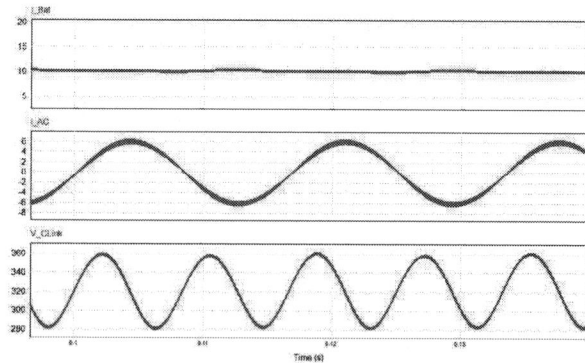

Figure 5: Input current, output current, and link voltage in the simulated converter

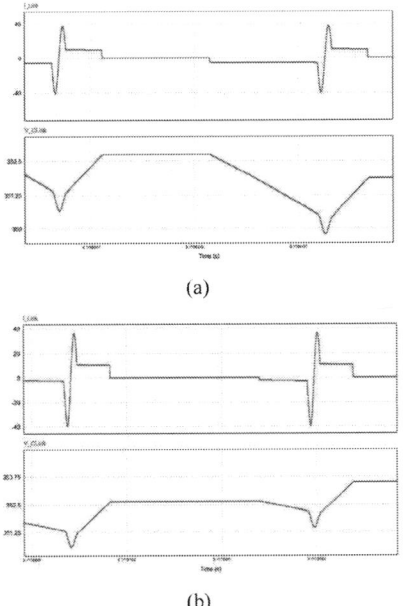

(a)

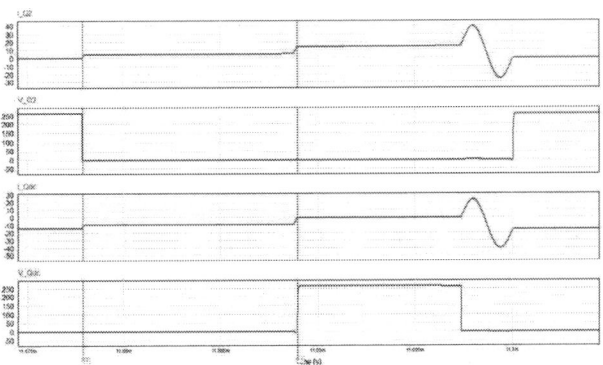

(b)

Figure 6: Link current and voltage in the simulated converter: (a) $P_{ac} < P_{dc}$ (b) $P_{ac} > P_{dc}$

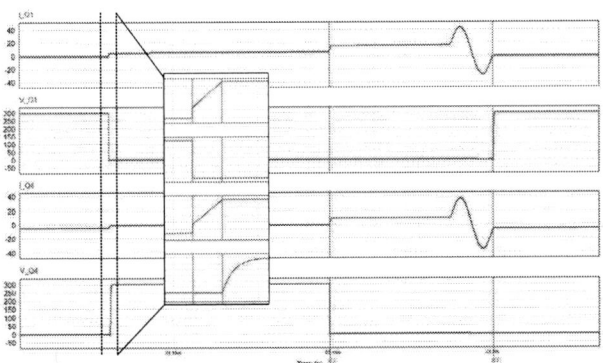

Figure 7: Switching transients for switches S2 and Sdc (ON & OFF)

Figure 8: Switching transients for switches S1 and S4 (ON & OFF)

$$i_{Cr}(t) = C_r \frac{dv_{Cr}(t)}{dt} = i_{link}(t) \tag{11}$$

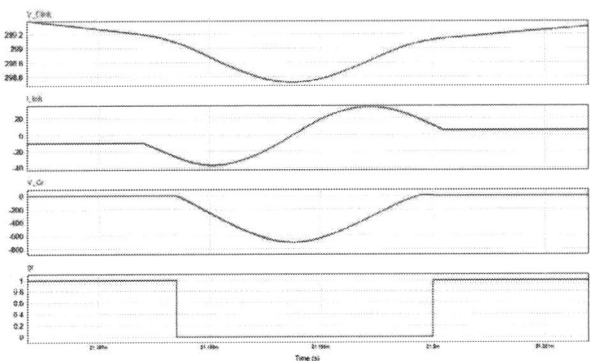

Figure 9: Zoomed in view of the key waveforms (link voltage, link current, Cr voltage and gate signal of Sr) during final resonant mode

$$L_r \frac{di_{link}(t)}{dt} = L_r C_r \frac{d^2 v_{Cr}(t)}{dt^2} \tag{12}$$

$$V_{link} + v_{Cr}(t) + L_r C_r \frac{d^2 v_{Cr}(t)}{dt^2} = 0 \tag{13}$$

The initial condition can be determined by analyzing the operation of the proposed converter. For soft switching of the auxiliary switch S_r, the voltage across resonance capacitor C_r should always start and go back to zero during mode R3; Also, due to the existence of L_r in the resonance path, link current is continuous and its value from the last discharging mode (or the optional mode Ro if it exists) can be included as $I_{Lr}(0)$ at the start of mode R3. If mode Ro is not present, $I_{Lr}(0)$ has the same value as $i_{dc,n}$.

Solving the second order linear differential equation gives the solution for resonance capacitor voltage $v_{Cr}(t)$:

$$v_{Cr}(t) = -V_{link} + V_{link} cos\left(\frac{1}{\sqrt{L_r C_r}}t\right) + \frac{I_{Lr}(0)}{C_r}\sqrt{L_r C_r}\sin(\frac{1}{\sqrt{L_r C_r}}t) \tag{14}$$

The link current can then be obtained through (11) and (14):

$$i_{Cr}(t) = i_{link}(t) = \frac{-V_{link}}{\sqrt{L_r C_r}} C_r sin\left(\frac{1}{\sqrt{L_r C_r}}t\right) + \frac{I_{Lr}(0)}{C_r} C_r cos(\frac{1}{\sqrt{L_r C_r}}t) \tag{15}$$

IV. SIMULATION AND EXPERIMENTAL RESULTS

A 480W converter with the parameters listed in Table I was designed and simulated. As can be seen in Figure 5, the proposed system can provide a smooth output current/voltage despite the input and output power mismatch, and the link voltage waveform carries the double line frequency harmonic. The zoomed in view for link voltage and current are shown in Figure 6. As can be seen the link current moves from positive to zero and then negative during the three main modes, and the link resonates at the end of each cycle to reverse the polarity of the link current that swings to a larger value than the next charging current.

TABLE I - SIMULATION AND EXPERIMENT PARAMETERS

Power	C_Link	L_r	C_r	V_ac_rms	V_dc	V_c0
480 W	50 µF	µH	40 nF	110 V	48 V	300 V

Switching transients in the proposed converter are shown in Figures 7 and 8. As shown in Figure 7, switch S_2 turns on with a slow increase of current due to the existence of L_r; thus, achieves zero current turn on. In the next mode the voltage across switch S_{dc} is clamped to zero until the resonance mode ends. At the end of the switching cycle , switch S_2 is turned off when its current swings back to zero. As can be seen in these figures, full soft switching is obtained for these switches across the shown switching cycle. In Figure 8, as shown in the zoomed-in window, switch S_1 benefits from zero current turn-on because of L_r, and voltage of switch S_4 is clamped to zero during the current transition; thus, S4 benefits from soft turn-on and off. The capability of all switches to achieve soft switching across the full switching cycle was verified through simulations.

The detailed operation waveform of the auxiliary circuit is shown in Figure 9. As can be seen, a short mode for link inductor discharging (mode Ro) is added before turning on S_r. This mode allows the link current to go further negative and provide enough energy for the final link resonance. When switch S_r is turned off, the resonance capacitor C_r starts to be charged. The link current stops decreasing when voltage of C_r reaches the link voltage.. Once the link current reverses its polarity, C_r, which was being charged before this time, will no longer be charged and its voltage starts to drop, resulting in a slowdown of current slew rate. Finally, when the voltage of C_r swings back to zero, it will be clamped to zero by the body diode of switch S_r. The link current is equal or higher than the charging current for the

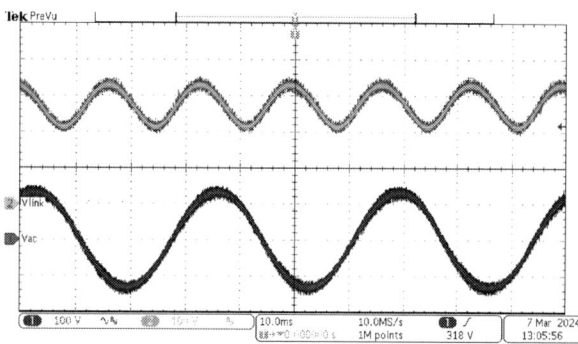

Figure 10: Experiment results on 400W DC to single phase AC inverter hard switching: Link voltage and output AC voltage

next switching cycle; thus, the switches can benefit from zero current turn off in next mode 1.

The estimated efficiency of the proposed soft-switching converter reaches 96% for the designed 480W system, while the efficiency of the hard-switching converter designed for the same system is 90%.

A prototype with the same parameter was fabricated. Experimental results of the hard switching configuration is shown in Figure 10. As can be seen the series capacitive link converter can absorb the double frequency pulsating power with the DC biased link voltage. To achieve soft switching, the proposed auxiliary circuit needs to be added to the circuit.

V. CONCLUSION

In this paper a novel active soft switching series capacitive link universal converter is presented. This converter is universal and can be configured as a rectifier, inverter or AC-AC converter. In this paper a single-phase AC to DC configuration is considered. By introducing an active soft switching structure, the proposed converter can handle the double frequency harmonic with only a small film capacitor while having soft switching for all switches in full load range. The proposed topology features high power density and efficiency; thus, can be an excellent candidate for renewable energy systems, on-board charging of electric vehicles, and many other applications.

REFERENCES

[1] T. Na, X. Yuan, J. Tang, and Q. Z. Hang, "A Review of on-Board Integrated Electric Vehicles Charger and a New Single-Phase Integrated Charger," *CPSS TPEA*, vol. 4, no. 4, Dec. 2019, doi: 10.24295/CPSSTPEA.2019.00027.
[2] S. S. Sayed and A. M. Massoud, "Review on State-of-the-Art Unidirectional Non-Isolated Power Factor Correction Converters for Short-/Long-Distance Electric Vehicles," *IEEE Access*, vol. 10, pp. 11308–11340, 2022, doi: 10.1109/ACCESS.2022.3146410.
[3] Y. Zhang *et al.*, "Low-frequency ripple-shaping controller for operation of non-inverting buck-boost converters near step-up step-down boundary," in *2018 IEEE Applied Power Electronics Conference and Exposition (APEC)*, San Antonio, TX, USA: IEEE, Mar. 2018, pp. 292–297. doi: 10.1109/APEC.2018.8341025.
[4] H. Wu, M. Han, and K. Sun, "Dual-Voltage-Rectifier-Based Single-Phase AC–DC Converters With Dual DC Bus and Voltage-Sigma Architecture for Variable DC Output Applications," *IEEE Transactions on Power Electronics*, vol. 34, no. 5, pp. 4208–4222, May 2019, doi: 10.1109/TPEL.2018.2864584.
[5] L. Zhang and X. Ruan, "Control Schemes for Reducing Second Harmonic Current in Two-Stage Single-Phase Converter: An Overview From DC-Bus Port-Impedance Characteristics," *IEEE Transactions on Power Electronics*, vol. 34, no. 10, pp. 10341–10358, Oct. 2019, doi: 10.1109/TPEL.2019.2894647.
[6] Y. Sun, Y. Liu, M. Su, W. Xiong, and J. Yang, "Review of Active Power Decoupling Topologies in Single-Phase Systems," *IEEE Transactions on Power Electronics*, vol. 31, no. 7, pp. 4778–4794, Jul. 2016, doi: 10.1109/TPEL.2015.2477882.
[7] S. Harb and R. S. Balog, "Single-phase PWM rectifier with power decoupling ripple-port for double-line-frequency ripple cancellation," in *2013 Twenty-Eighth Annual IEEE Applied Power Electronics Conference and Exposition (APEC)*, Mar. 2013, pp. 1025–1029. doi: 10.1109/APEC.2013.6520425.
[8] P. T. Krein, R. S. Balog, and M. Mirjafari, "Minimum Energy and Capacitance Requirements for Single-Phase Inverters and Rectifiers Using a Ripple Port," *IEEE Transactions on Power Electronics*, vol. 27, no. 11, pp. 4690–4698, Nov. 2012, doi: 10.1109/TPEL.2012.2186640.
[9] H. Li, K. Zhang, H. Zhao, S. Fan, and J. Xiong, "Active Power Decoupling for High-Power Single-Phase PWM Rectifiers," *IEEE Transactions on Power Electronics*, vol. 28, no. 3, pp. 1308–1319, Mar. 2013, doi: 10.1109/TPEL.2012.2208764.
[10] T. Shimizu and S. Suzuki, "A single-phase grid-connected inverter with power decoupling function," in *The 2010 International Power Electronics Conference - ECCE ASIA -*, Jun. 2010, pp. 2918–2923. doi: 10.1109/IPEC.2010.5543511.
[11] H. Han, Y. Liu, Y. Sun, M. Su, and W. Xiong, "Single-phase current source converter with power decoupling capability using a series-connected active buffer," *IET Power Electronics*, vol. 8, no. 5, pp. 700–707, 2015, doi: 10.1049/iet-pel.2014.0068.
[12] H. Zhou, L. He, and Z. Lin, "Low Frequency Current Ripple Suppression for Two-Stage Single-Phase Inverter Based on Impedance Editing," *IEEE Transactions on Industrial Electronics*, vol. 69, no. 12, pp. 13417–13427, Dec. 2022, doi: 10.1109/TIE.2021.3128913.
[13] H. Chen and D. Divan, "Soft-switching solid state transformer (S4T)," in *2016 IEEE Energy Conversion Congress and Exposition (ECCE)*, Sep. 2016, pp. 1–10. doi: 10.1109/ECCE.2016.7855455.

[14] H. Chen, A. Prasai, and D. Divan, "Dyna-C: A Minimal Topology for Bidirectional Solid-State Transformers," *IEEE Transactions on Power Electronics*, vol. 32, no. 2, pp. 995–1005, Feb. 2017, doi: 10.1109/TPEL.2016.2547983.

[15] L. Zheng, R. P. Kandula, and D. Divan, "Predictive Direct DC-Link Control for 7.2 kV Three-Port Low-Inertia Solid-State Transformer With Active Power Decoupling," *IEEE Transactions on Power Electronics*, vol. 37, no. 10, pp. 11673–11685, Oct. 2022, doi: 10.1109/TPEL.2022.3172957.

[16] M. Khodabandeh and M. Amirabadi, "A Single-Phase ac to Three-Phase ac Converter with a Small Link Capacitor," in *2018 IEEE Energy Conversion Congress and Exposition (ECCE)*, Sep. 2018, pp. 3942–3948. doi: 10.1109/ECCE.2018.8557639.

[17] X. Zhang, M. Amirabadi, and B. Lehman, "A Four-Mode Three-State (FMTS) Swinging Bus Controller for PV Micro-Inverters to Achieve Reactive Power Compensation and Remove Electrolytic Capacitor," in *2020 IEEE Applied Power Electronics Conference and Exposition (APEC)*, Mar. 2020, pp. 405–411. doi: 10.1109/APEC39645.2020.9124600.

A Multi Mode Control Algorithm for Totem-pole Bridgeless PFC

Bosheng Sun
Texas Instruments
Dallas, USA
b-sun@ti.com

Sheng-Yang Yu
Texas Instruments
Dallas, USA
seanyu@ti.com

Amir Hussain
Texas Instruments
Houston, USA
a-hussain@ti.com

Abstract—This paper presented a multi-mode control algorithm for totem pole bridgeless power factor correction (PFC): the PFC operates at continuous conduction mode (CCM) at heavy load or at AC peak, and triangular current mode (TCM) with zero voltage switching (ZVS) at light load or around AC zero-crossing. A processed zero current detection (ZCD) signal is used to determine the transition between CCM and TCM, a traditional average current mode controller generates pulse width modulation (PWM) duty cycle for both CCM and TCM, resulting in a smooth transition between each mode. The amount of inductor negative current, the deadtime between synchronous switch turns off and boost switch turns on, and the loop compensator parameters are dynamically adjusted to optimize efficiency and performance across the entire operation range. This multi-mode operation combines the advantages of both CCM and TCM operations, resulting in a high efficiency, high power density and low-cost totem pole bridgeless PFC solution.

Keywords—PFC, totem-pole, CCM, TCM, multi-mode

I. INTRODUCTION

With the increasing power consumption trend on server central processing unit (CPU) and graphics processing unit (GPU) load and limited server rack space [1], server power supply unit (PSU) is demanded to provide both high efficiency and high power density. The most recent Modular Hardware System – Common Redundant Power Supply (M-CRPS) [2] targets 3.6kW power with Titanium efficiency requirement in a 185mm x 39mm x 73.5mm form factor, which translates to 111W/in^3 power density while the leading-edge PSU products are still in 80~90W/in^3 level.

The power supply used in server and datacenters consists of two stages: a power factor correction (PFC) followed by an isolated DC/DC converter. Among all PFC topologies, totem-pole bridgeless PFC gives best efficiency, therefore it is dominant in server PSU. A totem-pole bridgeless PFC can operate at either continuous conduction mode (CCM) [3][4][5], or triangular ccurrent mode (TCM) [6][7][8], each has its advantages and disadvantages. Table I provides a high-level comparison between these two modes.

Ideally, the totem-pole bridgeless PFC could operate with multi-mode, as shown in Fig. 1: At heavy load or at the peak of an AC half cycle, the desired PFC input current is high and let the PFC operates at CCM mode. When load reduces or around AC zero-crossing area where the desired PFC input current is low, let the PFC switch to TCM mode and operates with zero voltage switching (ZVS). Compared to pure CCM mode, it has better efficiency at light load due to ZVS; compared to pure TCM mode, because the inductor current ripple is much lower, there is no need to use multi-phase interleaved operation,

therefore it significantly reduces the size and system cost. By combining the advantages of both CCM and TCM, it will be the ideal operation mode to meet both high efficiency and high power density requirements.

TABLE I. CCM/TCM COMPARISON FOR TOTEM-POLE PFC

	CCM Operation	**TCM Operation**
Pros	• Low peak-to-peak inductor current ripple • Simple control	• Zero voltage switching (ZVS)
Cons	• Hard switching – high switching loss	• High peak-to-peak inductor current ripple • Require multi-phase interleaved to reduce current ripple for high power applications – resulting in low power density and high cost • Complex control

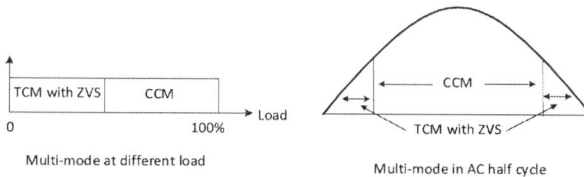

Fig. 1. CCM/TCM mulit-mode operation

However, to achieve CCM/TCM multi-mode control at different load, especially in each AC half cycle, there are three challenges:

1. The transition between CCM and TCM operation in each AC half cycle is a disturbance for the system, may cause input current distortion and affect total harmonic distortion (THD). Need a solution for a smooth transition.

2. Traditional CCM operation has a current loop, usually employing average current mode control [9]. Traditional TCM operation doesn't require current loop. Instead, it uses complex mathematical equations to calculate the required ON time [6]. These two different control methods will cause the duty cycle abruptly change at the boundary between CCM and TCM, this will also cause input current distortion and affect THD. Need a controller that works for both of CCM and TCM.

3. Once in TCM, how to achieve ZVS, especially at AC high line input.

979-8-3315-1612-3/25 $31.00 © 2025 IEEE

These challenges are addressed in this paper, a CCM/TCM multi-mode control algorithm is proposed with different methods to detect the boundary between CCM and TCM, and how a smooth transition between CCM and TCM is achieved through control logics.

II. PROPOSED CCM/TCM MULIT-MODE CONTROL ALGORITHM

Fig. 2 is the system diagram for the proposed CCM/TCM multi-mode control algorithm in a totem-pole bridgeless PFC. Q1 and Q2 are high frequency switches, Q3 and Q4 are low frequency switches switching at line frequency. At AC positive cycle where line voltage is higher than neutral, Q4 keeps on and Q3 keeps off, Q2 operates as PFC boost switch and Q1 operates as the synchronous switch. At AC negative cycle where line voltage is lower than neutral, Q3 keeps on and Q4 keeps off, Q1 operates as PFC boost switch and Q2 operates as the synchronous switch. The AC line voltage, AC neutral voltage, PFC output voltage and inductor current are sensed and sent to a PFC controller, a traditional average current mode controller is employed to generate pulse width modulation (PWM) signals.

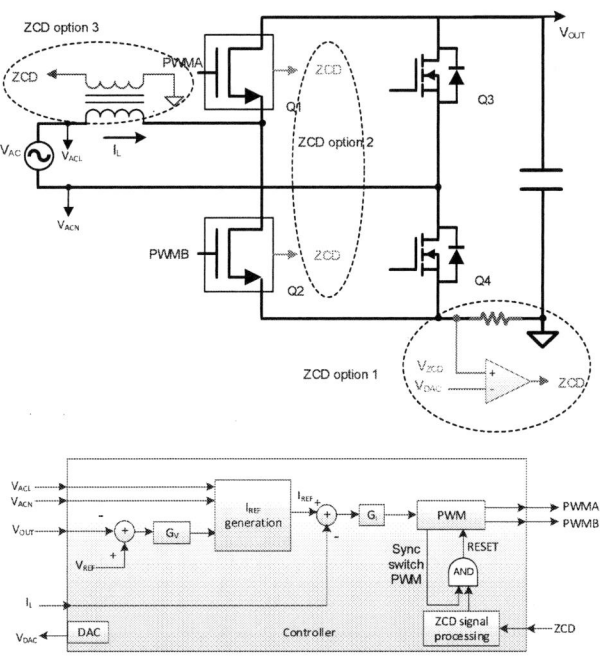

Fig. 2. System digagram for proposed CCM/TCM mulit-mode control algorithm

To achieve ZVS with TCM operation, the inductor current needs to drop below zero at the end of switching period, therefore the boost inductance needs to be less than the inductance used in a pure CCM PFC. This inductor zero current information can be detected with 3 different methods:

1. Place a resistor at PFC ground return path, the voltage drop across the resistor V_{ZCD} is compared to reference V_{DAC}, which is programmable by controller through its internal digital-to-analog converter (DAC). A zero current detection (ZCD) signal is generated when $V_{ZCD} > V_{DAC}$.

2. Place the current sensor in series with switch to sense switching current goes to 0.

3. Add a 2nd winding on the boost inductor to detect switch node oscillation transient as a ZCD signal.

The ZCD signal is then sent to the controller, and processed by the controller. Depending on how the ZCD signal is generated, the ZCD signal is processed differently.

A. ZCD method 1

Refer to Fig. 3. At light load or around AC zero-crossing area where the desired inductor current is low, when the synchronous switch turns on, inductor current starts to reduce and eventually reaches zero, then becomes negative. This negative current is detected by the comparator, a ZCD signal is generated when $V_{ZCD} > V_{DAC}$. By adjusting the V_{DAC} value, the amount of negative current can be dynamically adjusted. This negative current is used to discharge switch node voltage. Boost switch turns on when switch node voltage is discharged to 0, thus ZVS is achieved. For simplicity, a fixed V_{DAC} value can be used for the whole half AC cycle.

The ZCD signal is ANDed with synchronous switch PWM (1-D), to generate a RESET signal. The AND operation is to eliminate any suspicious ZCD signal which may be caused by noise.

RESET turns off PWM signal, resets PWM counter, a new switching period starts from here. Since PWM resets earlier before the end of nominal switching period, the real switching period $T_{TCM} < T_{nominal}$

A longer deadtime between synchronous switch turns off and boost switch turns on "t_turnon_delay" is used, this is to ensure the switch node is fully discharged before boost switch turns on, to achieve ZVS.

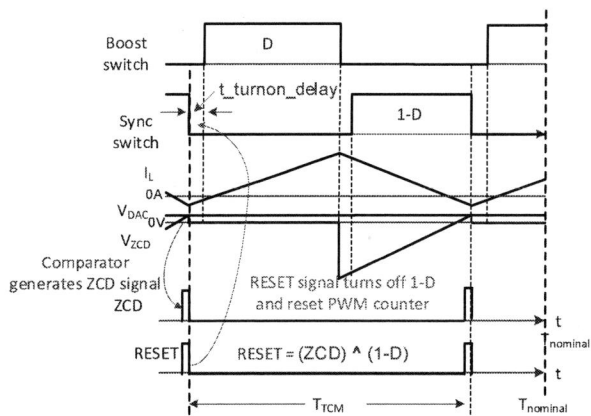

Fig. 3. Key waveforms in TCM for ZCD method 1

B. ZCD method 2 & 3

Refer to Fig. 4. At light load or around AC zero-crossing area where the desired inductor current is low, when the synchronous switch turns on, inductor current starts to reduce and eventually reaches zero. This zero current event is detected and a ZCD signal is generated.

979-8-3315-1612-3/25 $31.00 © 2025 IEEE   1991

Different from option 1, this ZCD signal stands for zero current, while the ZVS control needs negative current. To get negative current, the controller adds a delay "t_zcd_delay" to this ZCD signal. This delay makes the inductor current go to negative before turning off the synchronous switch. By adjusting the delay time, the amount of negative current can be dynamically adjusted. For simplicity, use fixed negative current $I_{NEGATIVE}$ for the whole AC half cycle, then the required "t_zcd_delay" can be calculated as (Fig. 5):

$$t\_zcd\_delay = \frac{L \times |I_{NEGATIVE}|}{V_{out} - V_{in}}, \qquad (1)$$

where L is the boost inductance, Vout is PFC instantaneous output voltage, Vin is PFC instantaneous input voltage.

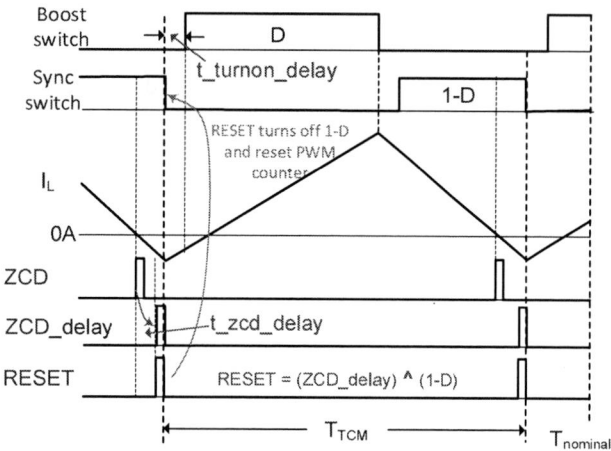

Fig. 4. Key waveforms in TCM for ZCD method 2&3

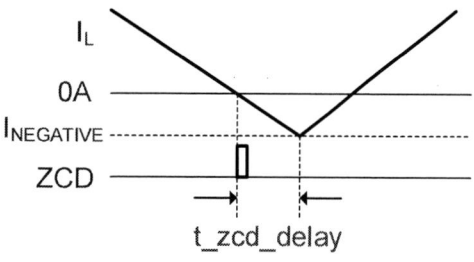

Fig. 5. Calculate required ZCD delay time

The delayed ZCD signal "ZCD_delay" is then ANDed with the synchronous switch PWM (1-D), to generate a RESET signal. The AND operation has two purpose:

1. Same as ZCD option 1, to eliminate any suspicious ZCD signal which may be caused by noise.

2. If the added ZCD delay time "t_zcd_delay" is too long, then the delayed ZCD signal "ZCD_delay" may appear after the end of switching period (the beginning of the next switching period), resetting PWM at this location is wrong. By ANDed with synchronous PWM (1-D), PWM won't reset if the aforementioned situation occurs.

RESET turns off PWM signal, resets PWM counter, a new switching period starts from here. Since PWM resets earlier before the end of nominal switching period, the real switching period $T_{TCM} < T_{nominal}$

A longer deadtime between synchronous switch turns off and boost switch turns on "t_turnon_delay" is used, this is to ensure the switch node is fully discharged before boost switch turns on, to achieve ZVS.

C. Haevy load operation

For all three ZCD methods, at heavy load or at AC peak where the desired inductor current is high, inductor current never goes to 0, therefore no ZCD signal is generated, then no RESET signal is generated, PWM resets naturally at the end of nominal switching period, $T_{TCM} = T_{nominal}$, it works as a tradition CCM totem pole PFC. This is shown in Fig. 6.

To improve efficiency, a shorter deadtime between synchronous switch turns off and boost switch turns on "t_turnon_delay" is used, this is to make the boost switch 3rd quadrant conduction time as short as possible (Especially the 3rd quadrant conduction causes extra power loss for Gallium Nitride device).

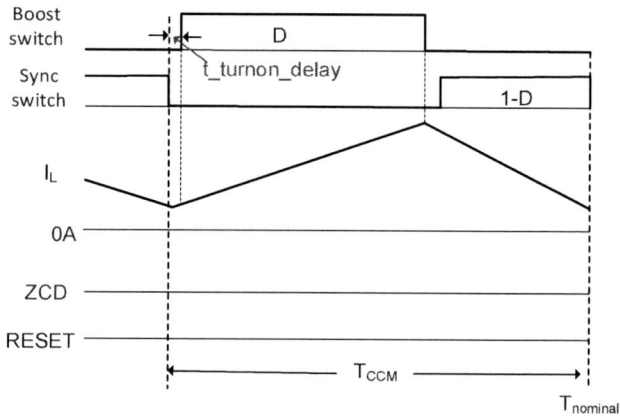

Fig. 6. Key waveforms in CCM

D. Transition between TCM and CCM

The operation mode TCM or CCM is based on whether the inductor current goes to zero or not, and inductor current naturally goes to zero when load reduces, therefore the transition between TCM and CCM is natural, resulting in a smooth transition. Moreover, in both TCM and CCM, the PWM duty cycle D is generated by the same average current mode controller, as shown in Fig. 1, resulting a smooth PWM duty cycle transition at the boundary between CCM and TCM, therefore, the proposed control method provides a smooth transition between TCM and CCM, not causing current distortion in each AC half cycle. This control method also makes TCM implementation much easier, as it does not need to use complex mathematical equations to calculate the required Ton time, a lower cost MCU with lower speed CPU is enough for the control implementation. To further optimize the performance for the entire operation range, different loop compensation may be used for TCM and CCM, to accommodate the fact that switching frequency is different at TCM and CCM.

III. TEST RESULTS

The above proposed CCM/TCM multi-mode operation algorithm is implemented in Texas Instruments 3.6kW totem pole bridgeless PFC reference design – PMP23537, as shown in Fig. 7. The high frequency switches are Texas Instruments LMG3427R030 GaN FET devices, it has an integrated zero current detection circuit. When the input PWM signal (IN) goes high, the ZCD circuit includes a blanking time t_{ZCD_Blank}, to prevent nuisance ZCD triggering during the turn-on transient. Following the blanking period, the ZCD circuit monitors the drain-to-source current. If the current is negative, a output pulse with a width of t_{WD_ZVD} is set on the ZCD pin after detecting the zero-crossing point, with a delay time of t_{zc_Det}, as indicated in the timing diagrams in Fig. 8. The t_{zc_Det} will be included in the ZCD delay time "t_zcd_delay" in Fig. 4. With this GaN device, no external zero current detection circuit is needed anymore, this significantly simplify the design. The controller uses Texas Instruments TMS320F280039C real time microcontroller, both top and bottom switches send ZCD signals to the controller. Depending on the AC polarity, the controller chooses the ZCD signal from the synchronous switch, adds delay and then ANDed with synchronous PWM (1-D). This delay and AND operations are implemented through TMS320F280039C internal Configurable Logic Block (CLB) module.

Fig. 7. A 3.6KW totem pole PFC reference design

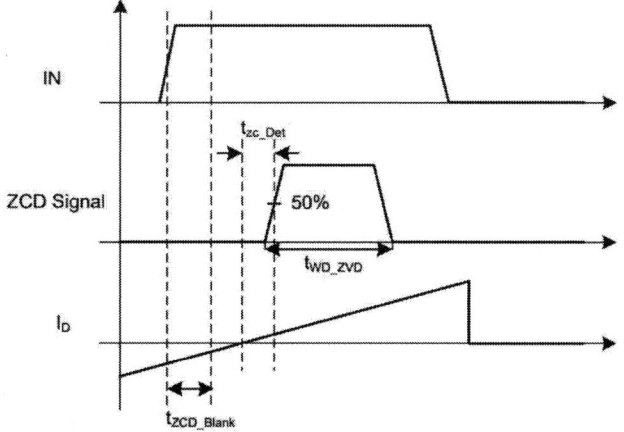

Fig. 8. LMG3427R030 ZCD timing diagram

Fig. 9 shows when the inductor current goes to negative, a ZCD signal is generated by LMG3427R030. The synchronous switch keeps on until 400ns later. This 400ns is the delay time added to the ZCD signal to generate negative current for ZVS. A new switching period starts when the synchronous switch turns off.

Fig. 10 is the inductor current waveform at light load (190W) and 230V input before turning on the TCM operation, PFC stays in CCM mode with a fixed switching frequency. There is excessive negative current which makes the root mean square (RMS) current high, causing extra conduction loss. The efficiency is 96.77%.

Fig. 11 shows the inductor current waveform under the same condition after turning on the TCM operation, it shows clearly that the inductor negative current is chopped to minimum to maintain ZVS operation. the efficiency is improved to 97.09%.

Fig. 12 is the inductor current waveform at a little heavier load (380W), PFC enters multi-mode: CCM (inductor current greater than 0) at AC peak, TCM (inductor current goes to negative) around AC zero-crossing area.

Fig. 13 shows the ZVS operation under TCM, the negative inductor current discharges the switch node, boost switch turns on after switch node voltage drops to zero, achieving ZVS.

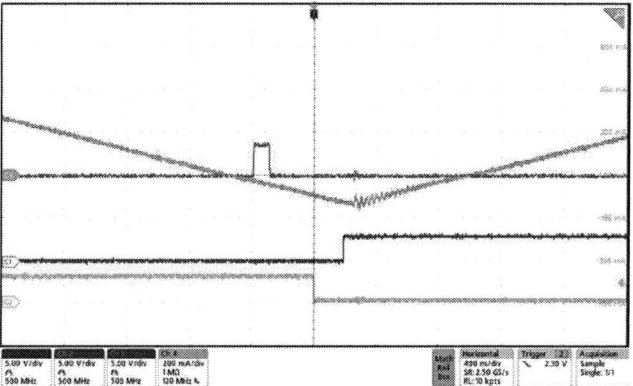

Fig. 9. Ch1: boost switch PWM; Ch2: synchronous switch PWM; Ch3: ZCD signal; Ch4: inductor current

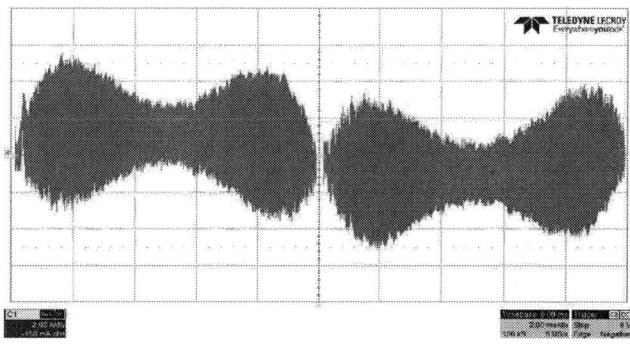

Fig. 10. PFC inductor current waveform before turn on TCM operation

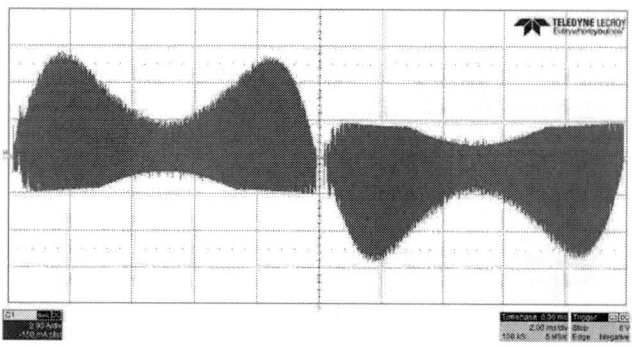

Fig. 11. PFC inductor current waveform after turn on TCM operation

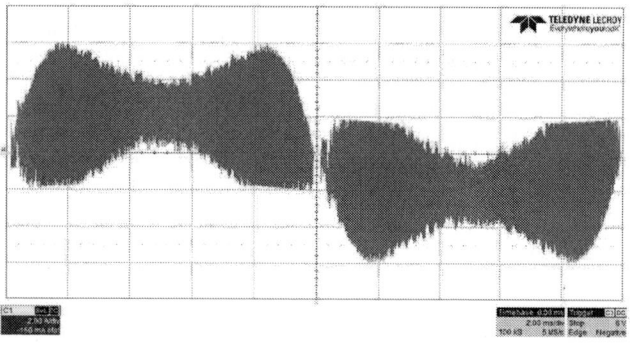

Fig. 12. PFC inductor current waveform at 380W load with CCM/TCM multi-mode operation

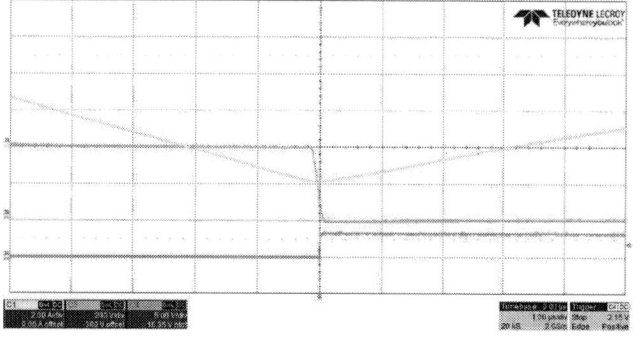

Fig. 13. ZVS under TCM. Ch1: Inductor current. Ch3: Switch node voltage. Ch4: Boost switch PWM

IV. CONCLUSION

A multi-mode control algorithm for totem pole bridgeless PFC is proposed in this paper: the PFC operates at CCM at heavy load or at AC peak, and TCM with ZVS at light load or around AC zero-crossing. It combines the advantages of both CCM and TCM operations, provides a high efficiency, high power density and low-cost totem pole bridgeless PFC solution.

REFERENCES

[1] R. Yin, "Five major trends in power-supply designs for servers," Texas Instruments Technical Articles. [Online]:

Avaliable: https://www.ti.com/lit/ta/ssztcw2/ssztcw2.pdf

[2] *Modular Hardware System – Common Redundant Power Supply (M-CRPS) Base Specification.* Open Compute Project, Version 1.02 RC2, Aug. 23, 2023. [Online].

Available:https://www.opencompute.org/wiki/Server/DC-MHS

[3] L. Zhou, Y. Wu, J. Honea and Z. Wang, "High-efficiency True Bridgeless Totem Pole PFC based on GaN HEMT: Design Challenges and Cost-effective Solution," Proceedings of PCIM Europe 2015; International Exhibition and Conference for Power Electronics, Intelligent Motion, Renewable Energy and Energy Management, Nuremberg, Germany, 2015, pp. 1-8.

[4] Z. Ye, A. Aguilar, Y. Bolurian, and B. Daugherty, "GaN FET Based CCM Totem-pole Bridgeless PFC," Texas Instruments Power Supply Design Seminar SEM600, Topic 6, 2014.

[5] B. Sun, S. -Y. Yu and T. Huh, "A Totem-pole PFC with Re-rush Current Control, Accurate E-metering, Low iTHD and High Power Density," 2024 IEEE Applied Power Electronics Conference and Exposition (APEC), Long Beach, CA, USA, 2024, pp. 14-21.

[6] B. Majmunović, B. A. McDonald, S. -Y. Yu and J. Strydom, "90∘-Valley Unified Controller for Zero-Voltage-Switching Quasi-Square-Wave (ZVS-QSW) Boost Converter," in *IEEE Transactions on Power Electronics*, vol. 39, no. 6, pp. 6930-6940, June 2024

[7] H. Zhou, W. Liu and E. Persson, "Evaluation of TCM and CrCM modulation for Totem Pole PFC," PCIM Europe 2016; International Exhibition and Conference for Power Electronics, Intelligent Motion, Renewable Energy and Energy Management, Nuremberg, Germany, 2016, pp. 1-7.

[8] M. Torrisi and S. Messina, "Three-Channel Interleaved Totem Pole PFC in Triangular Current Mode (TCM) with STM32G474 Microcontroller," PCIM Europe digital days 2021; International Exhibition and Conference for Power Electronics, Intelligent Motion, Renewable Energy and Energy Management, 2021, pp. 1-6.

[9] L. H. Dixon, "High Power Factor Preregulator for Off-Line Power Supplies." Unitrode Power Supply Design Seminar Manual SEM600, Topic 6, 1988.

Protection Strategy for Flying Capacitor Totem-Pole PFC under the AC Drop Transient

1st Yanqing Wu
College of Electrical Engineering
Zhejiang University
Hangzhou, China
yanqingwu@zju.edu.cn

2nd Wending Zhao
College of Electrical Engineering
Zhejiang University
Hangzhou, China
zhaowending@zju.edu.cn

3rd Zhenhai Zhu
College of Electrical Engineering and Automation
Fuzhou University
Fuzhou, China
HEARN.ZHU@DELTAWW.COM

4th Xinke Wu
College of Electrical Engineering
Zhejiang University
Hangzhou, China
wuxinke@zju.edu.cn

Abstract—The AC line transients are inevitable during the operation of the power factor correction (PFC). For 2-level PFCs, such transients can often pass smoothly without protection, as the large inductance of input filters can store enough energy to withstand short-term abnormalities. However, for the flying capacitor multilevel (FCML) PFC with a significantly smaller inductance, AC line transients could lead to rapid over-current issues and ruin the voltage balance of flying capacitors (FCs), resulting in over-voltage across switching devices and endangering the integrity of the converter. Among all different kinds of AC line transients, the AC drop transient is the most challenging one, in which the grid-side voltage suddenly drops to 0 and last for a certain period of time. In light of this issue, this paper presents a strategy for fast protection and quick recovery of the FCML PFC under the AC drop transient. It also analyzes the oscillation dynamics of the flying capacitor voltages (FCVs) during the recovery process and subsequently calculates the safe operating threshold of the switching devices. Finally, through a five-level FCML totem-pole prototype, this paper experimentally validates the feasibility of proposed PFC protection and recovery strategy.Furthermore, the oscillation dynamics of FCV is also tested under dc/dc boost mode.

Index Terms—Flying Capacitor Multilevel Converter(FCML), Bridgeless Totem-pole, Power Factor Correction(PFC), AC Line Transient, Over Current Protection(OCP)

I. INTRODUCTION

As the power demand of a single rack reaches $20kW$ or even more, there is an urgent need for a power supply unit (PSU) with higher power-density and efficiency [1], [2]. Multi-level converters, by significantly reducing inductor volt-second and utilizing low-voltage devices with lower NFoM value, can achieve higher power-density and higher efficiency. In recent studies, the FCML converter has demonstrated the highest power-density and efficiency among all kinds of multilevel converters for PFC applications [3], [4].

However, the FCML PFC may suffer from voltage balancing issues, especially during the AC line transients. Although

This work is supported by "Pioneer" and "Leading Goose" R&D Program of Zhejiang under Grant 2023C0101 and Power Management Innovation Consortium (PMIC).

FCML converters have some active balancing strategies [5]–[7] and the natural balancing mechanism [8]–[10] to ensure voltage balancing during steady-state operation, the control bandwidths of most mechanisms are too low to suppress the FCV oscillation dynamics caused by large AC line transients. Actual AC line transient test requirements are listed in TABLE I and the typical transient waveforms are depicted in Fig. 1. Among them, the worst scenario is that the grid voltage drops completely for 20ms when the PSU operates at full load condition, since the power gap between input and output here is the biggest, and the bus voltage drop is the fastest and the greatest. Under such working conditions, it is highly necessary to deal with the operation of the flying capacitor PFC. Previously, some scholars have proposed solutions for the short-circuit fault ride-through and lightning strike protection of FCML inverters [11]–[13]. However, almost no particularly suitable solution has been presented for the AC Drop transient of FCML PFC as of yet. Hence, this paper proposes an effective protection and recovery strategy for the AC Drop transient. Furthermore, we provide a model for evaluating the dynamics of FCVs in order to assess circuit reliability. This paper performs theoretical analysis and experimental verification on a 5-level FCML Totem-pole PFC as shown in Fig. 2.

TABLE I: AC Line Transient Test Standard

Transient Duration	Sag or Surge	Operating AC Voltage	Line Frequency	Output Load (up to)
20ms	-100%	Nominal Range	50/60Hz	100%
Continuous	-10%	Nominal Range	50/60Hz	100%
10s	-20%	Nominal Range	50/60Hz	100%
Continuous	+10%	Nominal Range	50/60Hz	100%
2s	+20%	Nominal Range	50/60Hz	100%

*Transients that within the standard time shall not allow for a shutdown

The remainder of this paper is structured as follows: the first section outlines the proposed protection and recovery

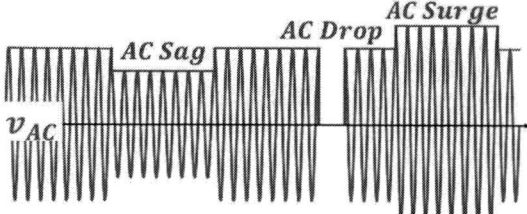

Fig. 1: Typical AC line transients.

strategies for the AC drop condition. Subsequently, an analysis of the FCV oscillation dynamics during the recovery process following the AC drop transient is presented to establish safe operating thresholds for switching devices. Following this, experimental validation of both the strategy and FCV oscillation under dc/dc boost conditions is conducted. Finally, we conclude with a discussion on future work.

II. PROPOSED AC DROP PROTECTION & RECOVERY STRATEGY FOR THE FCML TOTEM-POLE PFC

This paper takes the $20ms$ 100% AC Drop commencing at the phase angle of $90°$ scenario as an analytical example. Fig. 2 below shows the control block diagram of the protection and recovery strategy, while Fig. 3 presents the entire protection and recovery process with detailed waveforms. Next, the protection behaviors and effects in each time interval are explained:

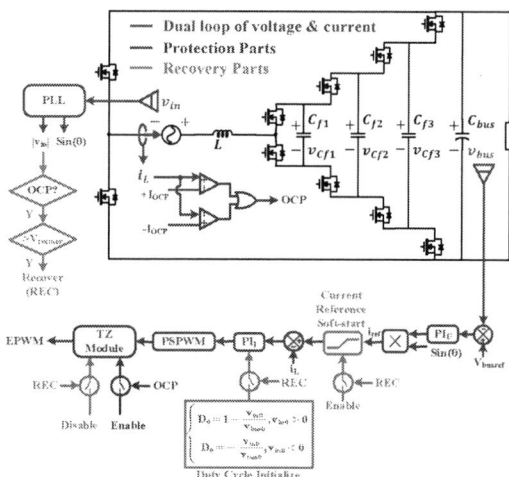

Fig. 2: The control block diagram of AC Drop protection and recovery strategy for FCML PFC.

- $[t_1, t_2]$ AC Drop transient occurs at t_1, causing the inductor current to drop rapidly.
- $[t_2, t_3]$ At t_2, the inductor current reaches the reverse over-current protection (OCP) threshold which is set according to the maximum value of the allowable voltage ripple of the FCs. Then, the comparator is set immediately, triggering the TZ submodule of the DSP to

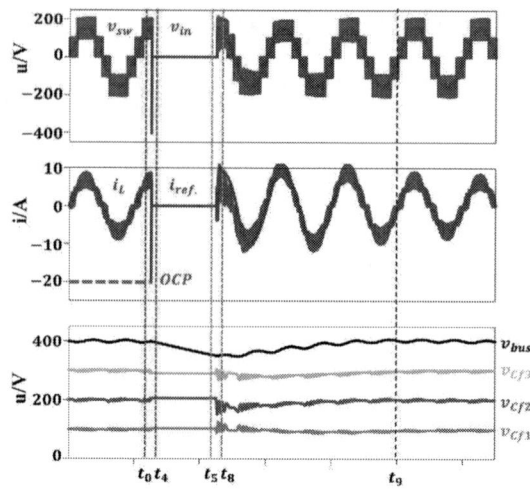

(a) The whole AC Drop protection & recovery process.

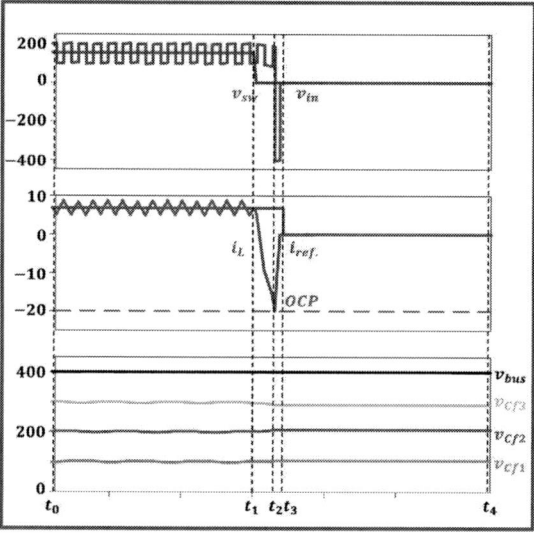

(b) The protection process $[t_0, t_4]$.

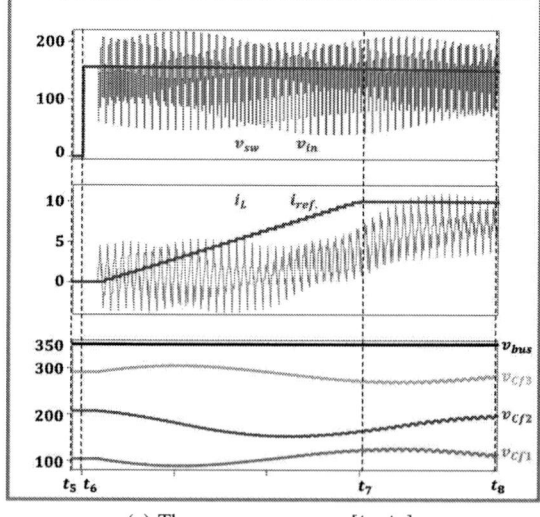

(c) The recovery process $[t_5, t_8]$.

Fig. 3: Critical simulation waveforms of proposed AC Drop protection & recovery strategy.

promptly shut off all switching devices. Subsequently, the inductor current rapidly drops to zero.

- $[t_3, t_6]$ During this period, the input voltage is zero, while the load is heavy, so the bus voltage continuously drops. Since all switching devices are turned off, there is no path for FCs to discharge, so the FCVs remain nearly invariant.

- $[t_6, t_7]$ v_{in} recovers at t_6. At this time, allow the current reference to soft-start from 0 and set the initial duty cycle of the lower bridge arm as:

$$\begin{cases} D_0 = 1 - \frac{v_{in0}}{v_{bus0}}, v_{in0} > 0 \\ D_0 = -\frac{v_{in0}}{v_{bus0}}, v_{in0} < 0 \end{cases} \quad (1)$$

Here, v_{in0} and v_{bus0} represent the sampled input voltage and bus voltage at this moment. Under this certain duty cycle of D_0, the average voltage of v_{sw} is the closest to v_{in}, minimizing the volt-second over the filter inductor and preventing over-current issue during the recovery stage. Notably, v_{bus0} is relatively low at this time due to v_{bus} drop previously and the FCVs are overall high. To reach a new steady state of FC voltage balancing, the FCVs will oscillate, which could cause over-voltage across the switching devices. The over-voltage situation will be analyzed in the second section.

- $[t_7, t_9]$ Thereafter, the oscillation of the FCVs gradually attenuates and the bus voltage undergoes a gradual restoration. Ultimately, the FCML totem-pole PFC completely resumes the normal operation state after t_9.

Preventing over-current to avoid severe FCV imbalance is the principle of the proposed protection and recovery strategy. The strategy utilizes comparators and TZ submodule for rapid OCP. It further avoids over-current in the recovery stage via proper initial duty cycle positioning and soft-start of the current reference. These measures are applicable to all transient protections involving protection and recovery stage. Therefore, the proposed method is applicable to other types of AC line transient. We will utilize this method to both AC drop and AC surge transient in the experiment section and compare it to cases without proposed method.

III. SAFE OPERATION THRESHOLD DURING RECOVERY STAGE

In the analysis of the previous section, during the $[t_6, t_7]$ interval, owing to the initially low v_{bus}, there will be a period of oscillation of FCVs during the recovery stage. The paper take the duty cycle in [0, 0.25] interval as an example to analyze the dynamic of the FCV and FCV dynamics in other duty cycle intervals can be analyzed through the same method. For the simplicity of the mathematical derivation, we hypothesize that parasitic parameters such as resistances in the circuit are disregarded, and that v_{in}, v_{bus}, and FCVs remain invariant within one switching cycle. The switch devices are controlled by a common duty cycle with the phase shift sequence of $S_{1a} \rightarrow S_{2a} \rightarrow S_{3a} \rightarrow S_{4a}$. Based on mode shown in Fig. 4, charge and discharge of FCs in 1 switching cycle can be calculated:

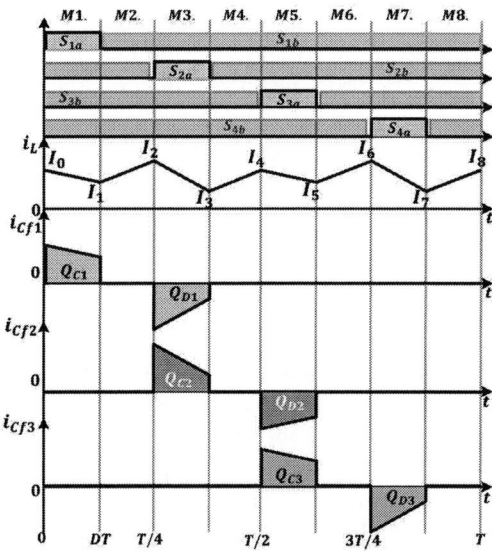

(a) Critical waveforms during 1 switching cycle.

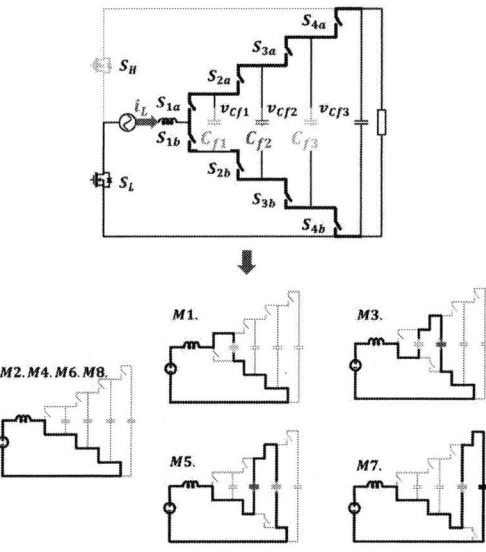

(b) The corresponding FCML mode circuits.

Fig. 4: The 1 switching cycle mode diagram in the [0, 0.25] duty cycle interval for 5L FCML totem-pole converter when S_L is on.

- In **M1.**, C_{f1} is charged and i_L decreases, so the charge of C_{f1} and change in current are:

$$\begin{cases} Q_{C1} = \frac{(I_0 + I_1)DT_s}{2} \\ I_1 - I_0 = \frac{(v_{in} - v_{Cf1})DT_s}{L} \end{cases} \quad (2)$$

where D is the duty cycle, T_s is the switching period and L is the inductance.

- In **M2.**, no FC is charged and i_L increases, so the change in current is:

$$I_2 - I_1 = \frac{v_{in} \left(\frac{1}{4} - D \right) T_s}{L} \quad (3)$$

- In **M3.**, C_{f1} is discharged, meanwhile C_{f2} is charged and i_L decreases, so the discharge of C_{f1} equals to the charge of C_{f2}, and the change in current are:

$$\begin{cases} Q_{D1} = Q_{C2} = \frac{(I_2+I_3)DT_s}{2} \\ I_3 - I_2 = \frac{(v_{in}+v_{Cf1}-v_{Cf2})DT_s}{L} \end{cases} \quad (4)$$

- In **M4.**, same mode to **M2.**, so the change in current is:

$$I_4 - I_3 = \frac{v_{in}\left(\frac{1}{4}-D\right)T_s}{L} \quad (5)$$

- In **M5.**, C_{f2} is discharged, meanwhile C_{f3} is charged and i_L decreases, so the discharge of C_{f2}, the charge of C_{f3} and the change in current are:

$$\begin{cases} Q_{D2} = Q_{C3} = \frac{(I_4+I_5)DT_s}{2} \\ I_5 - I_4 = \frac{(v_{in}+v_{Cf2}-v_{Cf3})DT_s}{L} \end{cases} \quad (6)$$

- In **M6.**, same mode to **M2.**, so the change in current is:

$$I_6 - I_5 = \frac{v_{in}\left(\frac{1}{4}-D\right)T_s}{L} \quad (7)$$

- In **M7.**, C_{f3} is discharged and C_{bus} is charged and i_L decreases, so the discharge of C_{f3} and the change in current are:

$$\begin{cases} Q_{D3} = \frac{(I_6+I_7)DT_s}{2} \\ I_7 - I_6 = \frac{(v_{in}+v_{Cf3}-v_{bus})DT_s}{L} \end{cases} \quad (8)$$

- In **M8.**, same mode to **M2.**, so the change in current is:

$$I_8 - I_7 = \frac{v_{in}\left(\frac{1}{4}-D\right)T_s}{L} \quad (9)$$

Since charge and discharge of each FC is get here, the voltage change of FCV during 1 switching cycle can be calculated by subtracting the corresponding charges and discharges:

$$\begin{cases} \Delta v_{Cf1} = \frac{Q_{C1}-Q_{D1}}{C_{f1}} \\ \Delta v_{Cf2} = \frac{Q_{C2}-Q_{D2}}{C_{f2}} \\ \Delta v_{Cf3} = \frac{Q_{C3}-Q_{D3}}{C_{f3}} \end{cases} \quad (10)$$

Where Δv_{Cfx} represents the voltage variation of C_{fx} in 1 switching cycle.

Additionally, for the purpose of further simplifying the mathematical derivation, we assume that the capacitance of the three FC stages are equal to C_f. After taking the above assumption, the ideal state-space model can be employed to depict the time-domain expression of the FCV oscillation:

$$\begin{bmatrix} \Delta v_{Cf1} \\ \Delta v_{Cf2} \\ \Delta v_{Cf3} \end{bmatrix} = k_1 \left(\begin{bmatrix} 0 & -2 & 0 \\ 2 & 0 & -2 \\ 0 & 2 & 0 \end{bmatrix} \begin{bmatrix} v_{Cf1} \\ v_{Cf2} \\ v_{Cf3} \end{bmatrix} + \begin{bmatrix} 1 \\ 1 \\ -1 \end{bmatrix} V_{bus} \right)$$
$$k_1 = \frac{D^2 T_s^2}{4LC_f} \quad (11)$$

The matrix structure in other duty cycle intervals is the same which are listed below:

- $D \in [0.25, 0.75]$:

$$\begin{bmatrix} \Delta v_{Cf1} \\ \Delta v_{Cf2} \\ \Delta v_{Cf3} \end{bmatrix} = k_2 \left(\begin{bmatrix} 0 & -2 & 0 \\ 2 & 0 & -2 \\ 0 & 2 & 0 \end{bmatrix} \begin{bmatrix} v_{Cf1} \\ v_{Cf2} \\ v_{Cf3} \end{bmatrix} + \begin{bmatrix} 1 \\ 1 \\ -1 \end{bmatrix} V_{bus} \right)$$
$$k_2 = \frac{(8(1-D)D-1)T_s^2}{32LC_f} \quad (12)$$

- $D \in [0.75, 1]$:

$$\begin{bmatrix} \Delta v_{Cf1} \\ \Delta v_{Cf2} \\ \Delta v_{Cf3} \end{bmatrix} = k_3 \left(\begin{bmatrix} 0 & -2 & 0 \\ 2 & 0 & -2 \\ 0 & 2 & 0 \end{bmatrix} \begin{bmatrix} v_{Cf1} \\ v_{Cf2} \\ v_{Cf3} \end{bmatrix} + \begin{bmatrix} 1 \\ 1 \\ -1 \end{bmatrix} V_{bus} \right)$$
$$k_3 = \frac{(1-D)^2 T_s^2}{4LC_f} \quad (13)$$

Therefore, the amplitude and phase characteristics of FCV oscillations are similar in different duty intervals for 5L FCML converter, merely the oscillation frequencies are different. So the worst oscillatory over-voltage value is the same in different duty cycle intervals, thus we can study the worst over-voltage of switching devices in just one duty cycle while applying the conclusion to other duty cycles. Thereafter, in order to get the worst over-voltage on switching devices, the time-domain expression of the FCV oscillations can be calculated by cumulative computation of over time, take $D \in [0, 0.25]$ as an example:

$$\begin{cases} v_{Cf1}(t) = \frac{1}{4} \begin{pmatrix} -v_{bus} + 2v_{Cf10} + 2v_{Cf30} \\ + (v_{bus}+2v_{Cf10}-2v_{Cf30})\cos\left[2\sqrt{2}\frac{D^2 T_s}{4LC_f}t\right] \\ +\sqrt{2}\,(v_{bus}-2v_{Cf20})\sin\left[2\sqrt{2}\frac{D^2 T_s}{4LC_f}t\right] \end{pmatrix} \\ v_{Cf2}(t) = \frac{1}{4} \begin{pmatrix} 2v_{bus} \\ -2(v_{bus}-2v_{Cf20})\cos\left[2\sqrt{2}\frac{D^2 T_s}{4LC_f}t\right] \\ +\sqrt{2}\,(v_{bus}+2v_{Cf10}-2v_{Cf30})\sin\left[2\sqrt{2}\frac{D^2 T_s}{4LC_f}t\right] \end{pmatrix} \\ v_{Cf3}(t) = \frac{1}{4} \begin{pmatrix} v_{bus} + 2v_{Cf10} + 2v_{Cf30} \\ -(v_{bus}+2v_{Cf10}-2v_{Cf30})\cos\left[2\sqrt{2}\frac{D^2 T_s}{4LC_f}t\right] \\ -\sqrt{2}\,(v_{bus}-2v_{Cf20})\sin\left[2\sqrt{2}\frac{D^2 T_s}{4LC_f}t\right] \end{pmatrix} \end{cases} \quad (14)$$

Where v_{Cfx0} represents the initial voltage of C_{fx}.

When we consider $v_{Cf10} = 100V$, $v_{Cf20} = 200V$, $v_{Cf30} = 300V$ and $v_{bus} \in [300, 400]V$, since v_{Cfx} remains nearly invariant while the bus voltage drops during the PFC protection stage. The maximum voltage stress of the switching devices v_{dsmax} can be calculated based on the max difference of the adjacent stage FCV as:

$$\begin{cases} v_{ds1\max} = \left(200 + 100\sqrt{3}\right) - (1+\sqrt{3})\frac{v_{bus}}{4} \\ v_{ds2\max} = 100 \\ v_{ds3\max} = 500 - v_{bus} \\ v_{ds4\max} = \left(-200 + 100\sqrt{3}\right) + (3-\sqrt{3})\frac{v_{bus}}{4} \end{cases} \quad (15)$$

The relationship between v_{dsmax} and v_{bus} is also shown in Fig. 5. In Fig. 5, v_{ds3max} appears to have the maximum over-voltage when v_{bus} drops. Besides, the more v_{bus} drops the greater v_{ds3max} and v_{ds1max} are. Hence, in the design of the 5L FCML PFC, the main emphasis should be placed on providing more sufficient voltage stress margins for S_{3a}, S_{3b} and S_{1a}, S_{1b}.

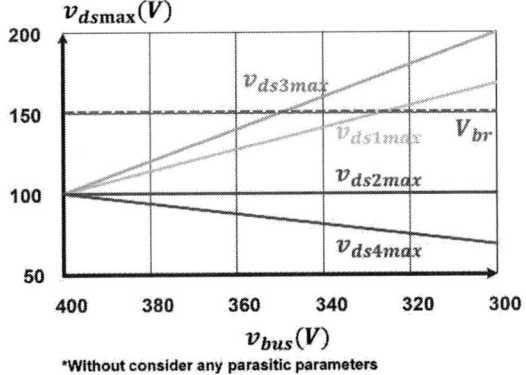

Fig. 5: $v_{ds\,max}vs.v_{bus}$(safe operating threshold is under $150V$).

IV. EXPERIMENT VERIFICATION OF PROPOSED PROTECTION STRATEGY&SAFETY OPERATION THRESHOLD

We employed a 5L FC totem-pole PFC prototype as depicted in Fig. 6 by short-circuiting one switching cell of a 6L prototype. The component details of the prototype are given in TABLE II.

The protection strategy of AC Drop was tested by using this 5L prototype. Under the working condition of a 260W load, an input of $110V_{ac}$, and an output of $400V_{dc}$, the AC Drop test results are shown in Fig. 7 which can achieve quick OCP and soft-start recovery stage. Moreover, some FCV oscillations can already be observed from Fig. 7b.

Fig. 6: 5L FC totem-pole bridgeless PFC prototype.

TABLE II: Component Details

Items	Details
MOSFET (150V,9.3mΩ)	Infineon BSC093N15NS5SC
Controller	DSP TMS320F280049
Inductor	29μH(Coilcraft XAL1513-153MED)
Flying Capacitor	13.2μF per level(CGK57NX7T2W225M)
Bus Capacitor	480μF

The paper also evaluated the protection of FCML totem-pole bridgeless PFC during AC surge transients with or without proposed soft-start recovery strategy. The AC surge transient here is a surge from $110V_{ac}$ to $160V_{ac}$ for $20ms$ with a $260W$ load and the output bus voltage is $400V_{dc}$. The AC Surge test results are shown in Fig. 8. In Fig. 8a, the PFC operation

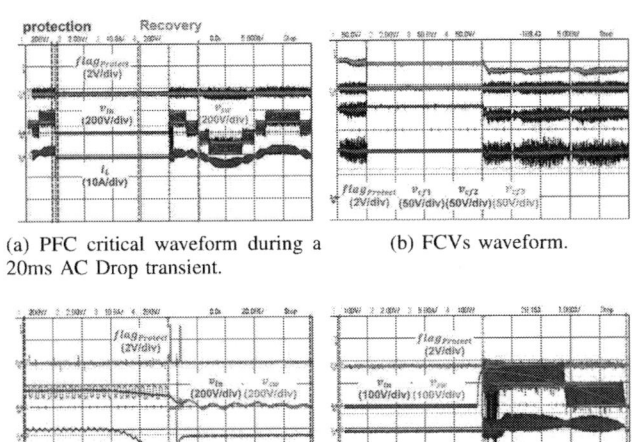

(a) PFC critical waveform during a 20ms AC Drop transient.

(b) FCVs waveform.

(c) The protection details.

(d) The recovery details.

Fig. 7: AC Drop protection & recovery experiment with proposed soft-start strategy.

waveform during AC surge transient without the proposed soft start recovery strategy is shown, where the inductor current experiences repeated over-current problem after the AC surge transient occurs. In Fig. 8b, it can be seen that the FCVs also experiences serious unbalance due to the repeated over-current, posing a serious threat to the safe operation of the converter. However, if the proposed recovery method is adopted, the repeated over-current problem will no longer exist after the AC surge transient as is shown Fig. 8c. Furthermore, as shown in Fig. 8d, the severe unbalancing of the FCV during the AC surge transient will not occur with assistance of proposed soft-start recovery strategy, thus ensuring the safety of the converter operation.

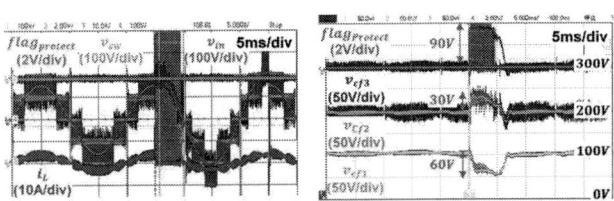

(a) PFC critical waveform without proposed method.

(b) FCVs waveform without proposed method.

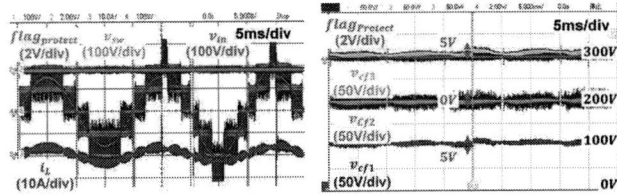

(c) PFC critical waveform with proposed method.

(d) FCVs waveform with proposed method.

Fig. 8: AC Surge protection & recovery experiment with/without proposed soft-start strategy.

979-8-3315-1612-3/25 $31.00 © 2025 IEEE

The paper also conducted verification of the FCV oscillation and over-voltage on devices in the recovery stage by operating the prototype in the synchronous rectification boost mode for the sake of simplicity. Under the operating condition of a $1200W$ load, with an input of $200V_{dc}$ and an output of $400V_{dc}$, the switching devices were turned off, and then they were turned on again when the bus voltage dropped to $300V_{dc}$. The v_{Cf} and v_{ds} test results are compared with the calculation results and presented in Fig. 9. Due to the presence of parasitic parameters, the attenuation of the FCV oscillation in the experiment is much faster. However, the frequency and phase of the oscillation are in good agreement with the calculation results which indicates that the proposed safe operating threshold can cover the worst operating conditions and thus is still of reference significance.

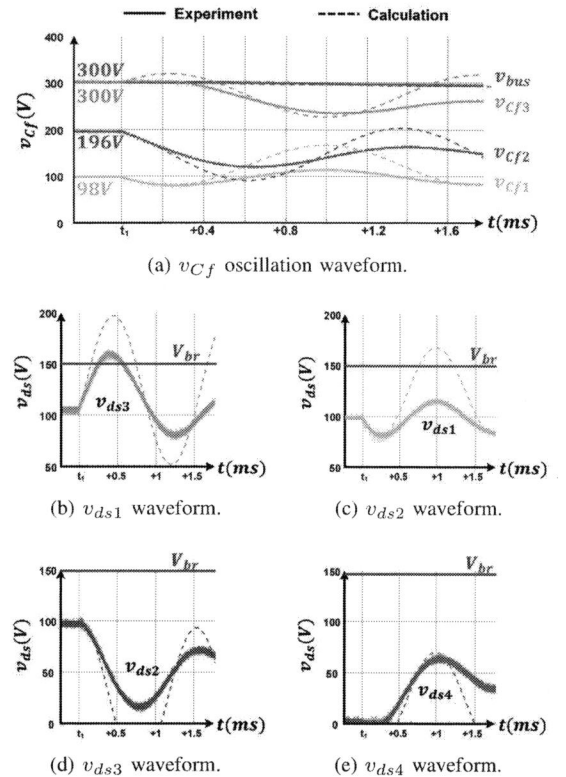

(a) v_{Cf} oscillation waveform.

(b) v_{ds1} waveform.

(c) v_{ds2} waveform.

(d) v_{ds3} waveform.

(e) v_{ds4} waveform.

Fig. 9: FCV oscillation waveform when v_{bus} drops from 400V to 300V under dc/dc boost mode, Experiment (solid line) vs. Calculation(dashed line).

V. CONCLUSION AND FUTURE WORK

This paper proposed a strategy for fast protection and quick recovery of the FCML PFC under the AC Drop transient and preventing over-current to avoid severe FCV imbalance is the principle. The strategy utilizes comparators for rapid OCP, averting unbalanced FCVs due to AC Drop transients. It further avoids over-current in the recovery stage via proper initial duty cycle positioning and soft-start of the current reference.

However, the FCV still oscillates during the recovery process due to the v_{bus} drop caused by the AC drop transient. The article analyzes the ideal state-space model of a 5L FCML converter. Based on the oscillation amplitude of the FCV, the safe operating thresholds of the switching devices are provided. The conclusion here is that the greater the v_{bus} drop is, the greater the voltage stress of the switching devices. Notably, the S_1 and S_3 switching pairs will undergo more severe over-voltage issue and require a larger margin of voltage stress for design purpose.

The experiment verified the feasibility of the protection strategy, not only in the AC drop transient, but also in the AC Surge transient. However, it was also observed that the oscillation of the FCV was smaller than the theoretical estimate. In future work, the parasitic parameters such as the parasitic resistance and capacitance will be substituted into the safe operating threshold model and the FCV oscillation under PFC operation mode needs to be further tested and analyzed. In addition to that, the authors are still working on finding some other ways such as use different control method and add clamping circuits to reduce the oscillation voltage.

ACKNOWLEDGMENT

This work is supported by "Pioneer" and "Leading Goose" R&D Program of Zhejiang under Grant 2023C0101 and the Power Management Innovation Consortium (PMIC) of Hangzhou Global Scientific and Technological Innovation Center, Zhejiang University. Thanks for the free materials supported by Coilcraft and Infineon.

REFERENCES

[1] Y. Lei, W.-C. Liu, and R. C. N. Pilawa-Podgurski, "An analytical method to evaluate flying capacitor multilevel converters and hybrid switched-capacitor converters for large voltage conversion ratios," in 2015 IEEE 16th Workshop on Control and Modeling for Power Electronics (COMPEL), Vancouver, BC, Canada: IEEE, Jul. 2015, pp. 1–7

[2] J. Wu, Y. Qi, F. Muhammad, and X. Wu, "A Unified Switch Loss Model and Design Consideration for Multilevel Boost PFC With GaN Devices," CPSS Trans. Power Electron. Appl., vol. 6, no. 4, pp. 349–358, Dec. 2021

[3] Q. Huang, Q. Ma, P. Liu, A. Q. Huang, and M. A. De Rooij, "99% Efficient 2.5-kW Four-Level Flying Capacitor Multilevel GaN Totem-Pole PFC," IEEE J. Emerg. Sel. Top. Power Electron., vol. 9, no. 5, pp. 5795–5806, Oct. 2021

[4] O. Lorenz and J. Sanchez, "Ultra low-profile flying capacitor 7-level 3kW PFC with optimized high frequency layout and active balancing using 100V GaN," in 2024 IEEE Applied Power Electronics Conference and Exposition (APEC), Long Beach, CA, USA: IEEE, Feb. 2024, pp. 22–28.

[5] X. Zhao, Y. Zhang, and Q. Guan, "A Balancing Control Method for Flying Capacitors in Five-Level Buck/Boost Converter with Synchronous Phase Shifting Decoupling," in 2021 IEEE 1st International Power Electronics and Application Symposium (PEAS), Shanghai, China: IEEE, Nov. 2021, pp. 1–6.

[6] A. Stillwell, E. Candan, and R. C. N. Pilawa-Podgurski, "Active Voltage Balancing in Flying Capacitor Multi-Level Converters With Valley Current Detection and Constant Effective Duty Cycle Control," IEEE Trans. Power Electron., vol. 34, no. 11, pp. 11429–11441, Nov. 2019

[7] S. D. Silva Carvalho, N. Vukadinovic, and A. Prodic, "Phase-Shift Control of Flying Capacitor Voltages in Multilevel Converters," in 2020 IEEE Applied Power Electronics Conference and Exposition (APEC), New Orleans, LA, USA: IEEE, Mar. 2020, pp. 299–304.

[8] Z. Xia, K. Datta, and J. T. Stauth, "State-Space Modeling and Control of Flying-Capacitor Multilevel DC–DC Converters," IEEE Trans. Power Electron., vol. 38, no. 10, pp. 12288–12303, Oct. 2023.

[9] R. Polichshuk, B. Reznikov, and A. Ruderman, "Six-Level DC-DC Flying Capacitor Converter Voltage Balancing Dynamics Analysis for Different Carrier-Based Modulation Strategies," in 2018 International Symposium on Power Electronics, Electrical Drives, Automation and Motion (SPEEDAM), Amalfi: IEEE, Jun. 2018, pp. 685–690

[10] Z. Ye, Y. Lei, Z. Liao, and R. C. N. Pilawa-Podgurski, "Investigation of Capacitor Voltage Balancing in Practical Implementations of Flying Capacitor Multilevel Converters," IEEE Trans. Power Electron., vol. 37, no. 3, pp. 2921–2935, Mar. 2022

[11] P. Papamanolis, D. Neumayr, and J. W. Kolar, "Behavior of the flying capacitor converter under critical operating conditions," in 2017 IEEE 26th International Symposium on Industrial Electronics (ISIE), Edinburgh, United Kingdom: IEEE, Jun. 2017, pp. 628–635

[12] N. Pallo, M. G. Taul, A. Stillwell, and R. C. N. Pilawa-Podgurski, "Short-Circuit Fault Ride-Through of Flying-Capacitor Multilevel Converters through Rapid Fault Detection and Idle-mode Operation", COMPEL 2022

[13] R. Beinarys, S. O'Driscoll, and T. T. Vu, "Flying Capacitor Voltage Imbalance Protection in Multilevel Bidirectional Inverters during Surge," in 2023 IEEE Applied Power Electronics Conference and Exposition (APEC), Orlando, FL, USA: IEEE, Mar. 2023, pp. 2068–2073

Three-Phase with Three Single-Phase Single-Stage Isolated AC-DC Converters for EV Charging Station Applications

Misha Kumar, Peter M. Barbosa and Juan M. Ruiz

Delta Electronics (Americas) Ltd., 5101 Davis Drive, Research Triangle Park, NC 27709, US

Abstract – **To achieve high power density, and high efficiency in EV charging station applications, in this paper, a three-phase with three single-phase single-stage ac-dc converters based on LLC resonant converter with variable switching frequency and variable delay-time control is proposed. With the proposed control, only one control variable, either the switching frequency or the delay-time, is controlled at a particular instant during ac line cycle in order to shape the input current to achieve PFC. The proposed topology and control ensure ZVS turn-on of switches while also limiting the switching frequency variation range in wide input/output voltage range. The detailed block-diagram of control is provided. Finally, the experimental waveforms for 10kW single-phase single-stage ac-dc converter are provided. At nominal operating condition, V_{AC} = 480Vrms, V_{DC} = 400V, measured efficiencies of 97.17% and 97.3% and iTHD of 3.04% and 4.47% are achieved at P_{DC} = 9.85 kW and P_{DC} = 6.59kW, respectively.**

I. INTRODUCTION

Today, to achieve high power density and high efficiency in EV charging station applications, single-stage ac-dc power conversion is becoming very attractive as it employs reduced number of components compared to that in conventional two-stage ac-dc power conversion. The reduction in the number of components helps to reduce the size and cost of the converter. In fact, the biggest reduction in size and cost comes from not employing bulky electrolytic capacitors in single-stage ac-dc converters [1],[2]. Additionally, the single-stage ac-dc converters also promise high efficiency compared to two-stage ac-dc converters as they process power only once from ac input to dc output. Moreover, the isolated single-stage ac-dc converters can leverage the energy stored in either the primary-side leakage inductor or the magnetizing inductor of the transformer to achieve zero voltage switching (ZVS) turn-on of the switching devices on the primary-side of the transformer. As a result, the switching losses are reduced, and the converter can be operated at higher switching frequencies, which further helps achieve high power density and high efficiency.

In literature, several approaches to achieve single-stage ac-dc power conversion are reported. One such approach is unfolding based single-stage ac-dc converters [1],[2]. These converters employ a three-phase unfolder which converts three-phase ac input voltage to three rectified soft dc-link voltages. These soft dc-link voltages are input to either an isolated T-type LCC network based resonant converter [1] or T-type LLC network based resonant converter [2] or T-type dual-active bridge (DAB) converter [3]. The converter modulates the duty-cycle of the switches so that input currents are sinusoidal to achieve power factor correction

(PFC) and output power is also regulated. The switching frequency of operation is either fixed as in [1], [3] or variable as in [2]. However, in [1] and [3], the ZVS turn on of all the switches of T-type converter cannot be achieved, especially in applications with wide input/output voltage ranges, which increases the switching loss and limits the efficiency and power density. Moreover, due to limited duty-cycle variation, a very wide output voltage range operation is not demonstrated in [1] and [3], as required by EV charging applications. In [2], even though, ZVS turn-on of switches is achieved, the range of switching frequency of operation is wide even at one particular input and output voltage condition shown in [2]. However, in EV charging station application, both input and output voltage ranges are very wide, which would further require the switching frequency to vary very widely. Very wide switching frequency variation results in challenging magnetics design, increased magnetic losses which limits the power density and efficiency.

Other approaches to achieve single-stage ac-dc power conversion, reported in the literature, are either based on dual DAB converter [4] or resonant converters [5]-[7]. In [4], ac-line voltage is rectified and boosted to twice its value by operating two interleaving legs on the primary-side of the transformer with 180° phase-shifted 50% duty-cycle switching pulses. The third leg connected to neutral operates as a low-frequency rectifier. The phase-shift of secondary-side switches vs. primary-side switches is controlled to shape the input ac current and regulate the power flow to the output. However, a DAB based topology suffers from high rms currents if the input and output voltage variations are very wide, which results in large conduction losses. In addition, a DAB based topology cannot achieve ZVS turn on of all switches over a wide input/output voltage range. Both of these factors limit the efficiency and power density. Resonant converter based single-stage topologies [5], [6] have lower rms currents and can achieve ZVS turn on of switches in wide input/output voltage range compared to DAB based topology. However, resonant converter topologies have very wide switching frequency variation, which limits the efficiency and power density, as explained before. To alleviate this issue, in [7], a dual-phase-shift and variable frequency modulation was proposed for series resonant dual active bridge (SR-DAB) converter. The converter control aims to maintain ZVS turn on of switches, limit switching frequency in a certain range, and achieve PFC while minimizing overall system losses. However, even with the proposed control, the switching frequency variation is almost 6 times, which is very wide. Moreover, the control is very complex since three control variables are controlled at a given instant during an ac line cycle. In [8] and [9], delay-

Table I : Specifications of 30kW EV charging station power supply

Description	Value
Input AC Voltage (phase-phase)	320-530 Vrms
Output Battery Voltage	150-500 V 150-300 V → CC region (100Amax)
Output Power	30kW

time control of secondary-side switches was proposed for dc-dc converters. The delay-time control helps in reducing the range of switching frequency of operation in wide input/output voltage range.

In this paper, a three-phase ac-dc converter with three single-phase single-stage ac-dc converters based on LLC resonant converter with variable switching frequency and variable delay-time control is proposed for EV charging station application. With the proposed control, only one control variable, either the switching frequency or the delay-time, is controlled at a particular instant during an ac line cycle in order to shape the input current to achieve PFC. The proposed topology and control ensure ZVS turn-on of switches while limiting the switching frequency variation range in wide input/output voltage range. The detailed block-diagram of control is provided. Finally, the experimental waveforms of a 10kW single-phase single-stage ac-dc converter are provided. At V_{AC} = 480Vrms, V_{DC} = 400V, measured efficiencies of 97.17% and 97.3%, measured iTHD of 3.04% and 4.47% are achieved at P_{DC} = 9.85 kW and P_{DC} = 6.59kW, respectively.

II. PROPOSED TOPOLOGY AND CONTROL FOR EV CHARGING STATION APPLICATION

Table I shows the specifications for a three-phase 30kW EV charging station power supply. The wide variation of ac input rms voltage and dc output voltage is evident from the specification. Figure 1 shows the block diagram of the proposed 30kW three-phase single-stage ac-dc converter implemented with three single-phase single-stage ac-dc converters. As shown in Fig.1, each single-phase single-stage ac-dc converter is connected to the phase-phase ac input voltage. The outputs of the three single-phase single-stage ac-dc converters are connected in parallel to provide the maximum load, therefore, each single-phase single-stage ac-dc converter carries 1/3rd of the maximum load, i.e.,

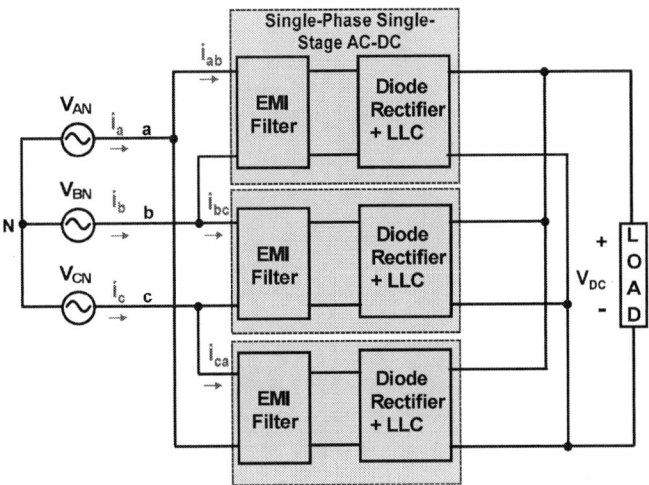

Fig. 1. Block diagram of proposed 30kW three-phase single-stage ac-dc converter implemented with three single-phase single-stage ac-dc converters.

P_{DCmax}/phase = 30kW/3 =10kW. Since in a balanced three-phase system, there is 120° phase shift between the three single phases, the output voltage/current ripple at the output of each single phase are also 120° phase shifted from each other which results in a very small net output voltage/current ripple at the load. Fig. 2 shows the circuit diagram of each 10kW single-phase single-stage ac-dc converter of Fig. 1. As shown in Fig. 2, the phase-phase ac input voltage is rectified with the help of a full-bridge diode rectifier so that the voltage across the soft DC link capacitor C_B is a rectified voltage with double the line frequency. This rectified voltage across capacitor C_B is the input to the LLC resonant converter, which transfers the power to a battery load. It should be noted that inductor L_F is a filter inductor which along with capacitor C_B, filters the high frequency current ripple at the input of LLC to achieve smooth line frequency varying ac source current i_{AC}. Fig. 2 shows the designed values of components and the selected components for experimental implementation. The series resonant frequency of LLC converter is $f_R = 1 / 2 \cdot \pi \cdot \sqrt{L_R \cdot C_R} \approx 100$kHz. The blocking capacitor C_{BI} on the secondary-side of the transformer has a larger value than the resonant capacitor and does not participate in the resonance. It is required to prevent saturation of transformer due to any mismatch in delays of

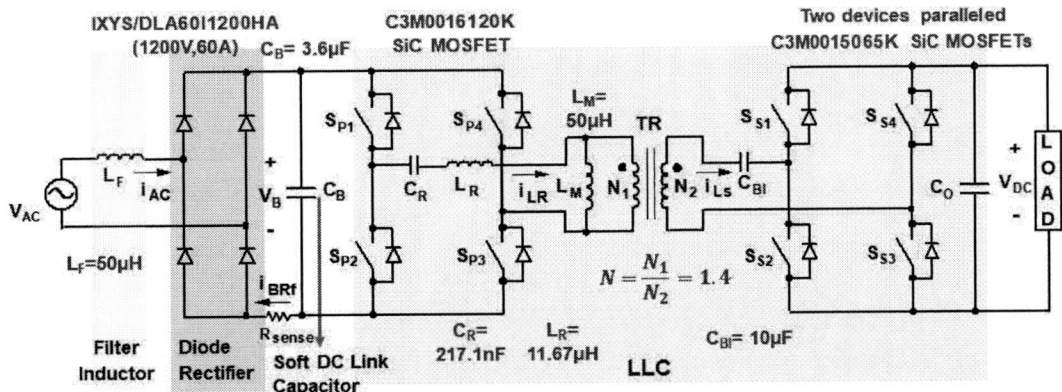

Fig. 2. Circuit diagram of each 10kW single-phase single-stage ac-dc converter of Fig. 1.

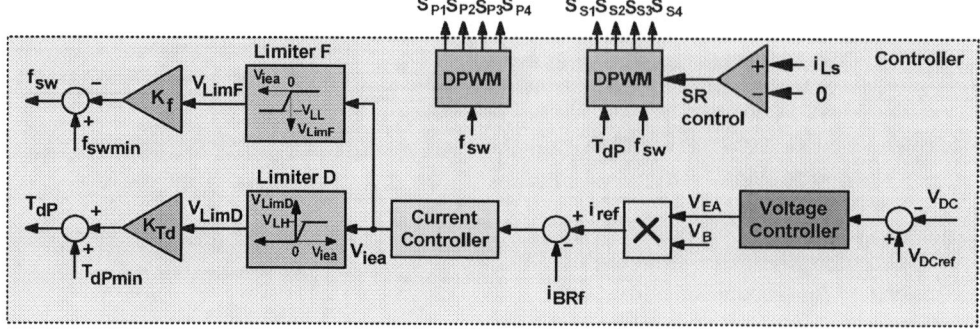

Fig. 3. Detailed block diagram of control of each single-phase single-stage ac-dc converter.

switching pulses. It should be noted that since the operation of all the three single-phase single-stage circuits are identical, for experimental evaluation, only one single-phase single-stage circuit is implemented. In order to handle the double line frequency of single-phase single-stage experimental circuit, C_O=10mF was selected. However, in three-phase experimental implementation, a very small value of C_O (<1mF) will be required, as explained before.

Figure 3 shows the detailed block diagram of control of each single-phase single-stage ac-dc converter. As seen in Fig. 3, the DC output voltage is regulated by subtracting sensed DC output voltage V_{DC} from DC output voltage reference V_{DCref}. This difference is passed through a voltage controller whose output is multiplied by the rectified voltage V_B across the capacitor C_B to obtain the reference current i_{ref}. The average value of input current of LLC i_{BRf} is measured directly with the help of sensing resistor R_{sense} as shown in Fig. 2. In Fig. 3, average current control is implemented by subtracting i_{BRf} from i_{ref} and passing their difference through a current controller. The output of the current controller V_{iea} controls either the percentage delay-time T_{dP} while the switching frequency f_{sw} is fixed at the minimum switching frequency f_{swmin} or controls the switching frequency f_{sw} while

the percentage delay time T_{dP} is fixed at its minimum value T_{dPmin}. It should be noted that T_{dP} is percentage of delay-time T_d with respect to switching period T_{sw}, i.e., $T_{dP} = T_d/T_{sw}$. The delay-time control helps to operate LLC converter in boost mode, i.e, when required gain of LLC, M= $(N_1/N_2){\cdot}V_{DC}/V_B$>1. Whereas the variable switching frequency control helps to operate LLC converter in buck mode, i.e, when M<1. It should also be noted that the control transitions automatically from delay-time control to variable switching frequency control and vice versa based on the sign of the current controller output V_{iea}. If V_{iea} is positive, the output of limiter F is zero which makes switching frequency equal to f_{swmin} and the output of limiter D is equal to V_{iea} which changes the percentage delay-time T_{dP}. Whereas, if V_{iea} is negative, the output of limiter, the output of limiter D is zero which makes $T_{dP}= T_{dPmin}$ and the output of limiter F is equal to V_{iea} which changes the switching frequency f_{sw}.

The proposed control is illustrated with the help of key Simplis simulation waveforms for single-phase single-stage ac-dc converter at V_{AC}=480Vrms, V_{DC}=400V, P_{DC}=10kW in Fig. 4 and its zoom-in waveforms in Figs. 5(a) and 5(b). It can be seen in Fig. 4 that PFC operation is achieved as the input current i_{AC} is approximately sinusoidal and in phase

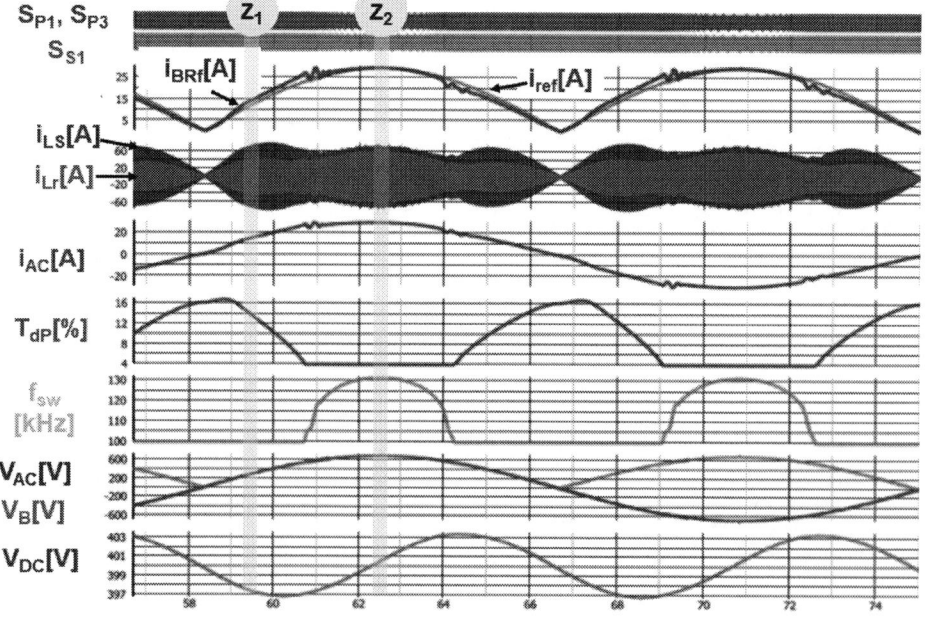

Fig. 4. Key simulation waveforms for single-phase single-stage ac-dc converter at V_{AC}=480Vrms, V_O=400V, P_O=10kW.

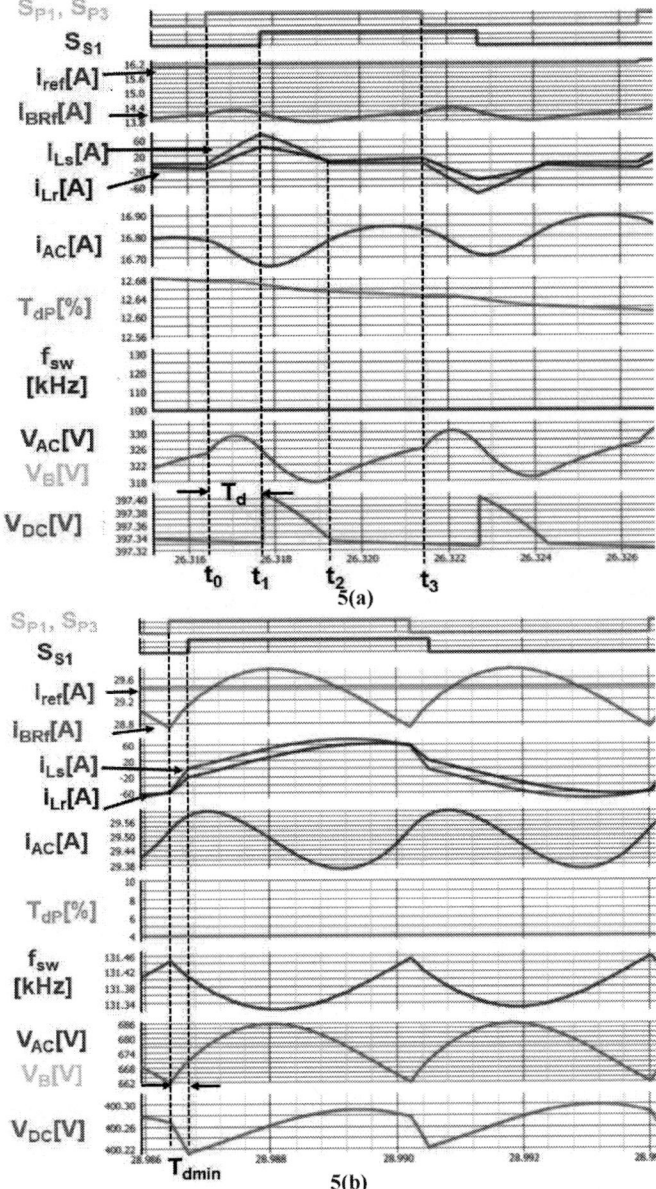

Fig. 5. Key zoom-in waveforms of simulation results in Fig. 4 during a) Z1 highlighted interval when M≥1 and b) Z2 highlighted interval when M<1.

when V_B starts decreasing from its peak value, f_{sw} decreases until it reaches its minimum value f_{swmin}. As, voltage V_B further decreases, the switching frequency is kept constant at f_{swmin} and the percentage delay-time T_{dP} is increased to perform the boost operation of the converter. T_{dP} reaches its maximum value when $V_B \approx 0V$ and the operation repeats in the next half-line cycle. It can be noted in Fig. 4 that at this operating condition, switching frequency varies from f_{swmin}=100kHz to ~131kHz at the peak of the ac line cycle.

To explain operation in detail, key zoom-in waveforms during Z_1 and Z_2 highlighted intervals of Fig. 4 are presented in Figures 5 (a) and 5 (b) when the converter operates in boost mode (M≥1) and buck mode (M<1), respectively. As shown in Fig. 5(a), primary switches S_{P1}, S_{P3} are operated with 50% duty-cycle pulses which operate at a constant switching frequency $f_{sw} = f_{swmin}$. Switches S_{P2}, S_{P4} are driven with complementary pulses as compared to S_{P1}, S_{P3}, respectively. Secondary-side switch S_{S1} is delayed with respect to S_{P1} by time delay T_d. Switch S_{S2} is operated with complementary pulse as compared to switch S_{S1}. The other secondary-side switches S_{S3} and S_{S4} are operated as synchronous rectifiers (SRs), i.e., when secondary-side current of transformer i_{LS} is positive, only switch S_{S3} is turned on and when i_{LS} is negative, only switch S_{S4} is turned on and when i_{LS} is zero, both S_{S3} and S_{S4} are turned off. As seen in Fig. 5(a), during time interval [t_0-t_1], switch S_{S1} is delayed with respect to S_{P1} by time delay T_d and the resonant inductor current i_{LR} and secondary-side current i_{LS} increase. This is because during this time, both the bottom secondary switches S_{S2} and S_{S3} are on, which results in zero voltage at secondary terminals of transformer. Therefore, the voltage across the resonant tank is the input voltage V_B, which makes the resonant inductor current i_{LR} increase and stores the energy at the resonant tank. At t = t_1, when switch S_{S1} turns on, the energy stored at the resonant tank is delivered to the output. During time interval [t_1-t_2], the voltage across the resonant tank is equal to V_B-$(N_1/N_2){\cdot}V_{DC}$, which is negative, therefore, i_{LR} decreases. At t = t_2, i_{LR} becomes equal to the magnetizing current i_{LM} of the transformer, and therefore, $i_{LS} = i_{LR}$ - i_{LM} = 0. At this instant, the synchronous rectifier S_{S3} turns off. Since, i_{LS}=0, the body diode of S_{S3} can no longer conduct and i_{LS} stays at zero from time interval [t_2-t_3]. At t=t_3, the primary-side switches S_{P2} and S_{P4} turn on with ZVS because resonant inductor current i_{LR} = i_{LM} is large enough to achieve ZVS. In Fig. 5(b), all switches of LLC are operated with conventional variable frequency control. Primary-side switches S_{P1}, S_{P3} are operated with 50% duty-cycle pulses with variable switching frequency. Switches S_{P2}, S_{P4} are driven with complementary pulses as compared to S_{P1}, S_{P3}, respectively. Secondary-side switches S_{S1} and S_{S3} are also operated with 50% duty-cycle, variable switching frequency pulses and are delayed with respect to primary switch S_{P1} by minimum delay-time T_{dmin} in order to turn on these secondary switches with ZVS after current i_{LS} changes direction. It should be noted in Fig. 5(b) that the switching frequency increases to ~131kHz to operate the

with the input voltage V_{AC}. Also, the output voltage V_O is well regulated. Voltage V_B is equal to the rectified ac input voltage V_{AC}. When $V_B \approx$ zero, i.e., it is much smaller than $(N_1/N_2){\cdot}V_{DC}$, M>>1, the converter operates in boost mode, therefore, the percentage delay-time T_{dP} is maximum and f_{sw} is fixed at its minimum value f_{swmin}. It should be noted that f_{swmin}=f_R=100kHz. As V_B starts increasing from zero, percentage delay-time T_{dP} starts decreasing until it reaches its minimum value T_{dPmin}. As voltage V_B increases, the requirement for boost operation slowly reduces, and therefore, the percentage delay time decreases. When T_{dP} = T_{dPmin}, it becomes constant and as voltage V_B increases further towards its peak value, f_{sw} starts increasing. During this time, voltage V_B > $(N_1/N_2){\cdot}V_{DC}$, i.e., M<1 and therefore, f_{sw} increases to operate the converter in buck mode. Further,

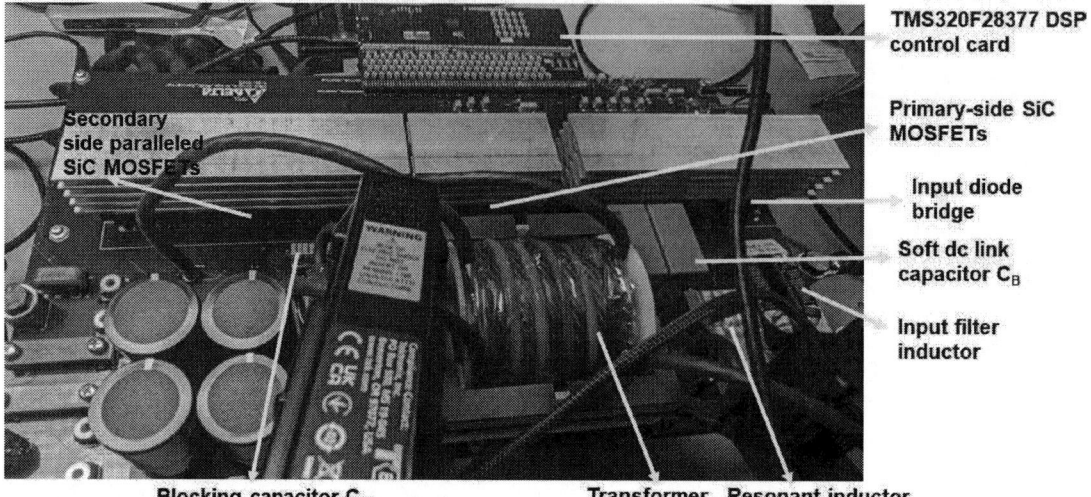

Fig. 6. Picture of experimental prototype for 10kW single-phase single-stage ac-dc converter.

converter in buck mode at this operating point in the ac line cycle.

III. EXPERIMENTAL CIRCUIT AND EXPERIMENTAL RESULTS

Figure 6 shows the picture of the experimental prototype for 10kW single-phase single-stage ac-dc converter. The designed values of components and selected devices are shown in Fig. 2. It should be noted that the transformer is built using EC 90X90X30 core in PE 22 core material from TDK. In order to optimize the core and winding loss, the number of turns on the primary side of transformer is 17 and secondary side of the transformer is 12. The transformer is designed with AWG 38 litz wire with 840 strands on the primary side and 1050 strands on the secondary side for a current density J = 7A/mm². The resonant inductor is designed using PQ50/50 core in 3C95 core material with 9 turns. The digital signal processor (DSP) based control is implemented using TI TMS320F28377 DSP.

Fig. 7 shows the experimental waveforms at V_{AC}=480 Vrms, V_{DC}=400 V, P_{DC}=9.85 kW. At this condition the measured efficiency of 97.17% and measured iTHD of 3.04% is achieved. It should be noted that in this operating condition, $V_{ACpk} > (N_1/N_2) \cdot V_{DC}$. Therefore, control operates with delay-time control and variable switching frequency

control as explained before. Figs. 8(a), (b), and (c) show zoom-in of waveforms of Fig. 7 when the converter operates

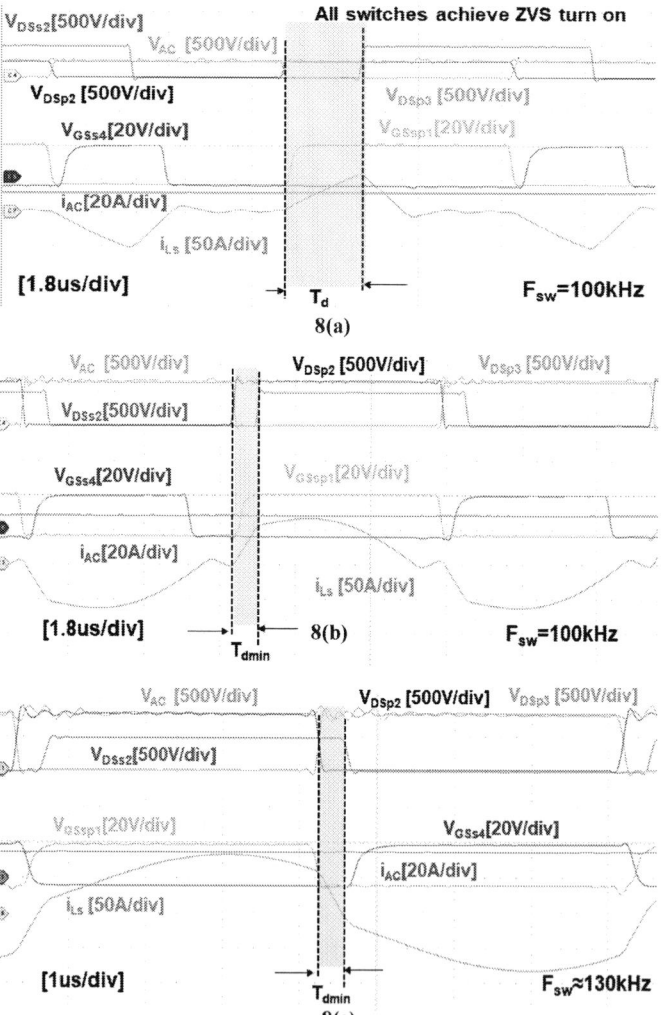

Fig. 8. Key zoom-in experimental waveforms of Fig. 7, a) near zero crossing of ac input voltage, b) at transition of control from delay-time to variable switching frequency control, and b) near peak of ac input voltage.

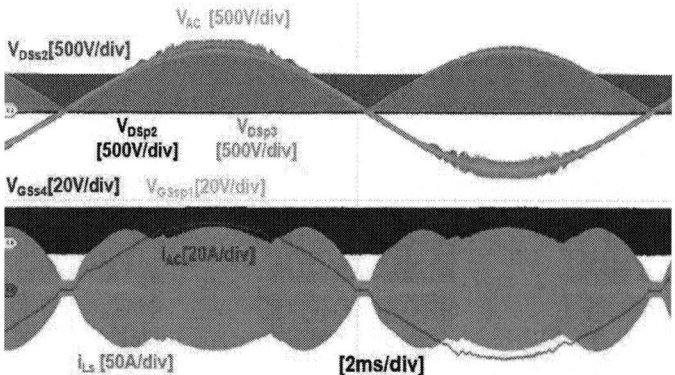

Fig. 7. Key experimental waveforms for single-phase single-stage ac-dc converter at V_{AC}=480Vrms, V_O=400V, P_O=9.85kW.

979-8-3315-1612-3/25 $31.00 © 2025 IEEE

in boost mode with only delay-time control near the zero crossing of the ac input voltage, when the converter operates at $V_B = (N_1/N_2) \cdot V_{DC}$ (M=1) at the transition of control from delay-time to variable switching frequency control, and when the converter operates in buck mode with only variable switching frequency control near the peak of the ac input voltage, respectively. As seen in Figs. 8(a), (b), and (c), all the switches on primary and secondary-side of transformer turn on with ZVS. The delay-time T_d between V_{DSs2} and V_{DSp2} is the largest in Fig. 8(a) near the zero crossing of the ac input voltage. In this condition, the switching frequency $f_{sw} = f_R = 100$kHz. As the ac input voltage increases and $V_B = (N_1/N_2) \cdot V_{DC}$, in Fig. 8(b), the delay-time T_d becomes minimum at T_{dmin}. The switching frequency at this condition is still at its minimum value 100kHz. As the input voltage increases further, the delay-time is kept constant at its minimum value and the switching frequency is increased as shown in Fig. 8(c). It should be noted that the minimum percentage delay-time $T_{dPmin} = T_{dmin}/T_{sw}$ is obtained at the peak of the line cycle to ensure that the secondary-side switches have enough delay with respect to primary-side switches so that the secondary-side switches can turn on with ZVS. This T_{dPmin} is kept constant for entire variable switching frequency mode operation and it ensures ZVS in entire variable switching frequency operation. In experiments minimum percentage delay-time $T_{dPmin} = 2.2\%$. The experimental waveforms have good correspondence with the simulation waveforms of Figs. 4 and 5.

Figure 9 shows experimental waveforms at V_{AC}=380 Vrms, V_{DC}=400 V, P_{DC}=9.95 kW. At this condition the measured efficiency of 96.36% and measured iTHD of 2.9% is achieved. It should be noted that in this operating condition, $V_{ACpk} < (N_1/N_2) \cdot V_{DC}$. Therefore, the control operates with delay-time control in entire half-line cycle. Figs. 10(a) and (b) show zoom-in of the waveforms of Fig. 9 when the converter operates at smaller ac input voltage levels and when the converter operates at the peak of the ac input voltage. As shown in Fig. 10(a), at smaller ac input voltage levels, a larger delay-time T_d is applied between V_{DSs2} and V_{DSp2} to meet a large boosting gain of the LLC converter. In Fig. 10(b), at the peak of the input voltage, a smaller delay-

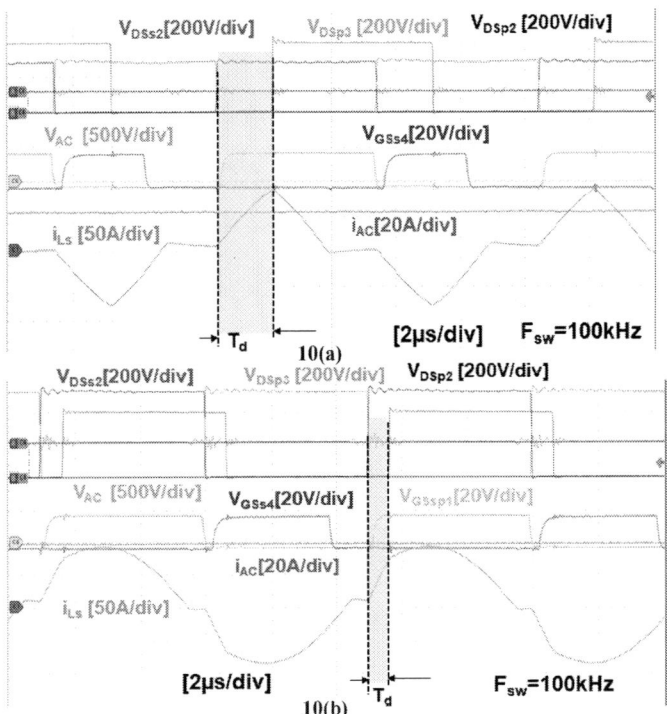

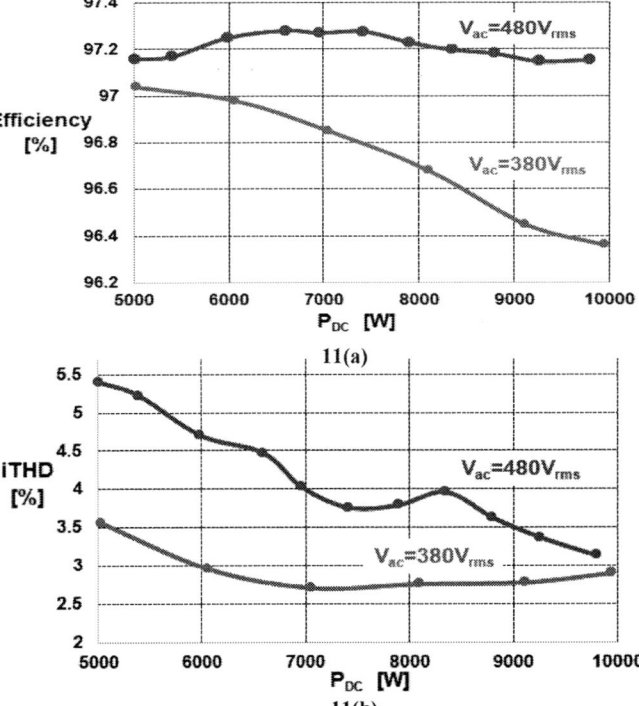

Fig. 10. Key zoom-in experimental waveforms of Fig. 9, a) near zero crossing of ac input voltage and b) near peak of ac input voltage.

time T_d is applied between V_{DSs2} and V_{DSp2} to meet a smaller boosting gain of the LLC converter. It should be noted that the switching frequency is fixed at the minimum switching frequency $f_{swmin} = f_R = 100$kHz in both Figs 10(a) and (b).

Figs. 11 (a) and (b) show measured efficiencies of the single-stage ac-dc converter and measured input ac current

Fig. 11 a) Measured efficiency and b) measured input ac current total harmonic distortion (THD) vs. output power P_{DC} at V_{DC}=400V for 10kW single-phase single-stage ac-dc converter.

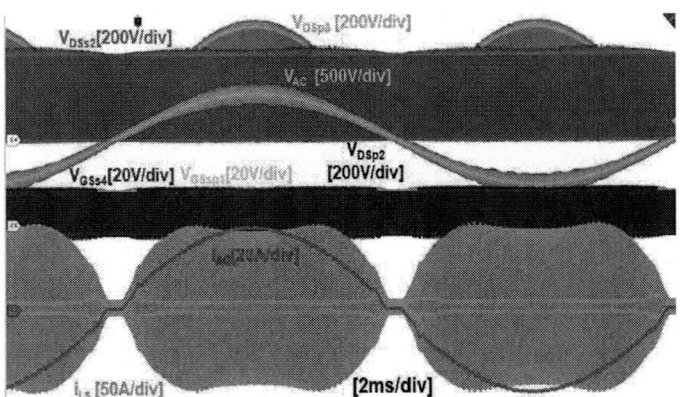

Fig. 9. Key experimental waveforms for single-phase single-stage ac-dc converter at V_{AC}=380Vrms, V_O=400V, P_O=9.95kW.

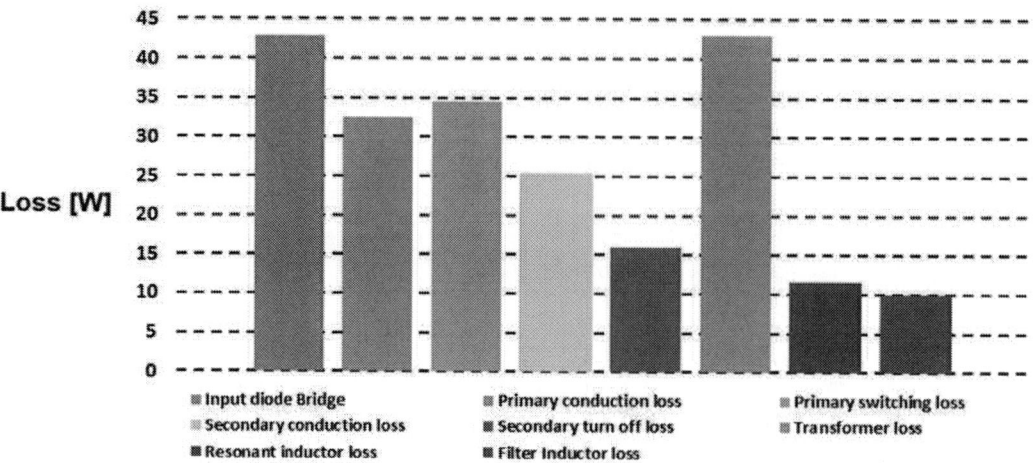

Fig. 12. Loss estimation based on Simplis simulation results at V_{AC}=480Vrms, V_{DC}=400V, P_{DC}=10kW for 10kW single-phase single-stage ac-dc converter.

total harmonic distortion (iTHD) vs. output power P_{DC}, respectively at output voltage V_O=400V at two different ac input voltages V_{AC}=380 Vrms and V_{AC}=480 Vrms. A peak efficiency of ~97.3% is achieved at V_{AC}=480 Vrms, V_{DC}=400 V, P_{DC}=6.59 kW. Measured iTHD < 5% is achieved at all output power levels P_{DC}> 5.4kW at both ac input voltages V_{AC}=480Vrms and 380Vrms at V_{DC}=400V.

Fig. 12 shows loss estimation based on the Simplis simulation results at V_{AC}=480Vrms, V_{DC}=400V, P_{DC}=10kW for 10kW single-phase single-stage ac-dc converter. As it can be seen from Fig. 12, the most dominant losses are in the input diode bridge, primary-side switch conduction and switch turn-off losses and the transformer loss. It should be noted that other losses like auxiliary power loss, fan loss, losses in the ESR of input and output capacitors are not included in the loss estimations. The efficiency could be further improved by ~0.3% if input diode bridge is replaced with low on-resistance R_{dson}=16mΩ SiC MOSFETs and by further design optimization of the transformer.

IV. CONCLUSION

In this paper, a three-phase with three single-phase single-stage ac-dc converters based on LLC resonant converter with variable switching frequency and variable delay-time control is proposed for EV charging station application. The proposed topology and control ensure ZVS of switches while also limiting switching frequency range. At V_{AC} = 480Vrms, V_O = 400V, measured efficiencies of 97.17% and 97.3%, iTHD of 3.04% and 4.47% are achieved at P_O = 9.85 kW and P_O = 6.59kW, respectively.

REFERENCES

[1] A. Zade, C. R. Teeneti, M. Mansour, S. Gurudiwan, B. Hesterman, H. Wang, and R. Zane, "A 21-kW unfolding-based single-stage ac-dc converter for wireless charging applications," *IEEE Journal of Emerging and Selected Topics in Power Electronics*, vol. 12, No.1, pp. 8-27, Feb. 2024.

[2] X. Li, J. Sun, L. Guo, M. Gao, H. Hu and M. Xu, "A Three-Phase Single-Stage ac/dc Converter Based on Swiss Rectifier and Three-Level LLC Topology," *in*

IEEE Transactions on Power Electronics, vol. 38, no. 2, pp. 1958-1972, Feb. 2023.

[3] C. R. Teeneti, R. Hatch, D. B. Yelaverthi, A. Kamineni, H. Wang and R. Zane, "Unfolder-Based Single-Stage AC-AC Conversion System for Wireless Charging Applications," *2020 IEEE Energy Conversion Congress and Exposition (ECCE)*, Detroit, MI, USA, 2020, pp. 5193-5198

[4] H. Kim, J. Park, S. Kim, R. M. Hakim, H. Belkamel and S. Choi, "A Single-Stage Electrolytic Capacitor-Less EV Charger With Single- and Three-Phase Compatibility," in *IEEE Transactions on Power Electronics*, vol. 37, no. 6, pp. 6780-6791, June 2022.

[5] M. Forouzesh, X. Zhou, Y.F. Liu, "Three-phase single-stage soft-switching ac-dc converter with power factor correction," U.S. Patent No. US 11,451,139 B2, Sept. 20, 2022.

[6] W. Liu, A. Yurek, B. Sheng, Y. Chen, Y. -F. Liu and P. C. Sen, "A Single Stage 1.65kW AC-DC LLC Converter with Power Factor Correction (PFC) for On-Board Charger (OBC) Application," 2020 *IEEE Energy Conversion Congress and Exposition (ECCE)*, Detroit, MI, USA, 2020, pp. 4594-4601.

[7] T. Chen, R. Yu and A. Q. Huang, "A Bidirectional Isolated Dual-Phase-Shift Variable-Frequency Series Resonant Dual-Active-Bridge GaN AC–DC Converter," in *IEEE Transactions on Industrial Electronics*, vol. 70, no. 4, pp. 3315-3325, April 2023.

[8] Y. Jang, M. M. Jovanović, *"System and Methods for Controlling Secondary Side Switches in Resonant Power Converters,"* U.S. Patent No. 9,490,704 B2, Nov. 8, 2016.

[9] G. Liu, Y. Jang, M. M. Jovanović and J. Q. Zhang, "Implementation of a 3.3-kW DC–DC Converter for EV On-Board Charger Employing the Series-Resonant Converter With Reduced-Frequency-Range Control," in *IEEE Transactions on Power Electronics*, vol. 32, no. 6, pp. 4168-4184, June 2017.

400V SiC in Next-Generation 3-Level Flying Capacitor Bridgeless Totem-pole PFC

Rytis Beinarys
ICERGi Limited
Dublin, Ireland
rytisbeinarys@icergi.com

Seamus O'Driscoll
Tyndall National Institute
Cork, Ireland
seamus.odriscoll@tyndall.ie

Abstract — **This paper proposes the design and implementation of a 3-level flying capacitor bridgeless totem-pole power factor correction (3L-FC-BTP-PFC) converter utilizing recently launched 400V Silicon Carbide (SiC) devices. It outlines key considerations and design challenges associated with adopting 400V SiC devices in place of a low-voltage series-connected silicon (Si) MOSFET arrangement. The proposed topology combines the benefits of a bridgeless totem-pole arrangement with the advantages of SiC technology. This combination enables superior performance in terms of higher efficiency, reduced gate driver complexity, increased system robustness, and enhanced power density. A comprehensive overview of the converter's operation, gate driver design, proposed power stage architecture, along with corresponding real-world experimental test results is presented.**

Keywords — silicon carbide (SiC), flying capacitor multilevel (FCML), low-cost isolated gate driver, hybrid flying capacitor voltage control, bridgeless totem-pole (BTP), power factor correction (PFC).

I. INTRODUCTION

As the demand for computational power in artificial intelligence (AI) applications continues to surge, server power supplies need to adapt to meet increasingly stringent industry requirements. Advanced GPUs are essential for AI training and are projected to require up to 3 kW per processor by the end of the decade. With AI server racks often housing multiple GPUs, the power supply unit (PSU) ratings are trending towards 8 kW and beyond [1]. This, in turn, drives the need for more efficient and robust power conversion solutions that can meet these requirements. With the recent announcement of the 400V CoolSiC™ MOSFET devices in the market [2], a significant advancement in multilevel power conversion systems can be achieved. Positioned between the existing 650V – 750V SiC device technology and the 150V – 200V Si MOSFET technology, 400V SiC switching devices in multilevel PFC design can offer a compelling balance of performance, cost, and design complexity. The evolution of power semiconductor technology has led to a paradigm shift in converter design, with increasing interest in adopting higher-voltage SiC MOSFETs to replace series-stacked low-voltage Si devices.

It is not uncommon to find series-connected Si MOSFETs utilized to achieve the necessary voltage handling capabilities in multilevel high-power applications [3], [4], but this approach introduces challenges such as increased circuit complexity, intricate voltage sharing mechanisms and higher switching losses due to higher component count in the power stage. On the other hand, the emergence of 400V SiC MOSFETs can address many of these limitations by offering superior material properties which allow for higher

breakdown voltage, ultra-low switching losses, faster switching speeds and fast commutation robust diode properties. The outlined advantages can enable simplified designs, reduced component count and improved efficiency in multilevel power converter systems such as single phase 3L-FC-BTP-PFC (Fig. 1). However, transitioning to SiC technology introduces its own set of design considerations, such as managing higher dv/dt, ensuring robust gate drive arrangements, and addressing electromagnetic interference (EMI) challenges. This paper explores the transition from stacked low-voltage Si MOSFETs to 400V SiC devices, highlighting the benefits, considerations, and challenges in adopting this advanced technology.

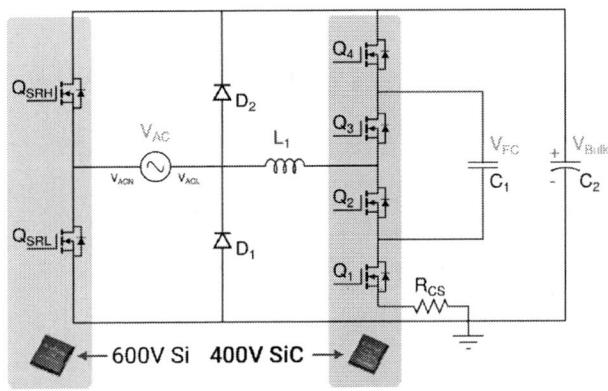

Fig. 1 Simplified circuit diagram of a 3L-FC-BTP-PFC utilizing 400V SiC MOSFET devices

II. TRANSITIONING FROM 150V SI TO 400V SIC IN BTP-PFC: DESIGN CHALLENGES AND CONSIDERATIONS

Well-known semiconductor manufacturers like *Infineon Technologies, Toshiba, Vishay,* etc., offer a wide range of moderate-frequency-capable Si MOSFETs primarily in the medium-voltage range (85V - 250V). However, 600V+ Si MOSFETs are generally limited to lower switching frequency applications such as synchronous rectifiers in PFC, solar inverters, etc. This is due to the large gate charge (Q_G) and poor reverse recovery performance (Q_{rr}) due to its slow intrinsic body diodes. While 600V – 750V SiC devices offer superior performance [5], they are not a viable solution in such cases due to their significantly larger V_{DS} rating which leads to unnecessarily increased system cost. This is where the recently announced 400V SiC devices from *Infineon Technologies* perfectly position themselves. It offers the ideal balance of high efficiency, fast switching capabilities and optimized cost for such applications.

979-8-3315-1612-3/25 $31.00 © 2025 IEEE

Table-I highlights some of the key parameters of 150V Si and 400V SiC MOSFET technology. The comparison is carried out between 2x 150V Si FETs (BSC093N15NS5) and a single 400V SiC MOSFET (IMT40R025M2H).

TABLE I.   150V Si MOSFETs vs 400V SiC MOSFET Comparison

Parameter	2x 150V Si	1x 400V SiC
Voltage blocking (V_{DS})	300V	400V
Reverse recovery charge (Q_{rr})	High	Extremely low
Switching speed (f_{sw})	Moderate	Faster
Thermal resistance (R_{th})	Higher	Lower
Gate driver design[a]	Complex	Simpler
Conduction losses ($T_j = 100°$)	Moderate	Moderate
Component Cost[b]	Lower	Higher

[a.] Number of gate driver circuits and the PCB layout

[b.] Cost based on 1,000 units from global component distributors

Considering the numerous advantages that SiC technology brings, the benefits far outweigh the higher initial cost per device. Additionally, since multilevel converters are typically used in medium- to high-power designs, where cost sensitivity is not as stringent, 400V SiC technology seems to be a perfect fit for the adoption.

III. Overview of Advanced 3L-FC-BTP-PFC Converter

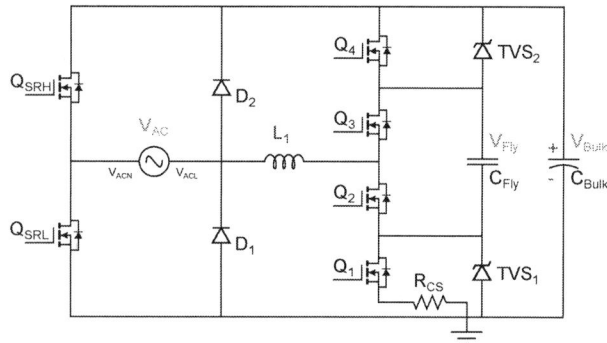

Fig. 2 Circuit diagram of 3L-FC-BTP-PFC with TVS diodes

Consider a simplified circuit of 3L-FC-BTP-PFC depicted in Fig. 2 above. The converter arrangement consists of the following components:

- Synchronous rectifier MOSFETs Q_{SRL} and Q_{SRH}.

- A pair of inrush current diodes D_1 and D_2, utilized exclusively during startup and latch up modes.

- The main PFC inductor L_1.

- HF boost leg with SiC MOSFETs Q_1, Q_2, Q_3, Q_4.

- Transient voltage suppressors TVS_1 and TVS_2 for flying capacitor initialization and protection.

- Additional flying capacitor C_{Fly} for voltage division and the output capacitor C_{Bulk}.

- Simple and cost-effective shunt-type pulsed inductor current sense resistor R_{CS}.

A. Control Technique: Phase Shift Modulation (PSM)

The HF string of SiC MOSFETs Q_{1-4} is controlled using a phase-shift modulation (PSM) in which devices Q_1 and Q_4 make up an outer phase and are driven in complementary fashion. The same applies to switches Q_2 and Q_3 forming an inner phase with an offset of 180°. The proposed modulation scheme allows for the flying capacitor voltage to track the reference voltage of $V_{bulk}/2$. During steady-state operation with a nominal PFC output voltage of 400V, each SiC MOSFET experiences a blocking voltage of approximately 200V. This occurs provided that the natural charge balance condition of the flying capacitor is satisfied.

B. Switching Frequency Control

The block diagram in Fig. 3 illustrates a variable switching frequency control architecture of a digital PFC controller. Central to this system is an ARM Cortex M0+ based microcontroller which encapsulates software defined control algorithms. Other key functional blocks include auxiliary voltage sense amplifiers for voltage measurement, isolated self-powered gate drivers and a switched-gain current sense amplifier to dynamically adapt current sensing based on the input current magnitude. This configuration allows for a reduced converter switching frequency, thereby minimizing switching losses and reducing the required drive power. The result is an optimized PFC system that achieves a balance between energy efficiency and improved current sense measurement accuracy at light to medium load operating conditions.

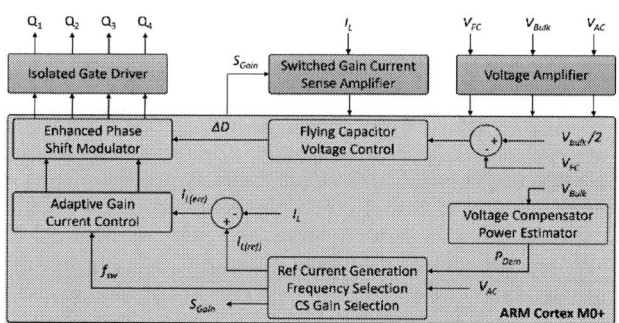

Fig. 3 Digital frequency control block diagram for 3L-BTP-PFC

C. Hybrid Flying Capacitor Voltage Balance Control and Pre-charge Mechanism

In the context of FCML power conversion systems, additional care must be taken to ensure that flying capacitor(s) charge balance is not compromised during high energy transient events such as start-up or input voltage surge. In addition, unbalanced voltage can occur during steady-state operation due to device tolerances, digital control timing delay, and hardware layout parasitic elements. The proposed hybrid flying capacitor voltage control is a simplified variant of the method which is employed in stacked 150V switching device PFC design [6]. The passive voltage balance enforcement circuitry requires only 2 additional components, namely TVS_1 and TVS_2 while the active part of it remained unchanged.

In this arrangement, TVS devices serve a secondary purpose, functioning as a flying capacitor voltage initialization mechanism during the power-up phase of the PFC converter. However, it is crucial to select the voltage clamping components carefully to ensure that the PFC converter operates without any adverse effects during both startup and steady-state conditions. In an ideal scenario, the TVS diodes should facilitate pre-charging of the flying capacitor to approximately half of the output voltage during the start-up sequence, while also assisting the outer SiC MOSFETs Q1 and Q4 in clamping excess energy during input voltage surges. However, with the nominal blocking voltage of each SiC MOSFET during steady-state operation of around 200V (just 50% of their V_{DS} rating) it becomes challenging to optimize the TVS diodes for both functions simultaneously. If TVS diodes with a breakdown voltage V_{BR} are selected close to $V_{Bulk}/2$, careful consideration must be given to their maximum peak pulse power (P_{PPM}) rating. These diodes will handle the bulk of the energy dissipation during critical events, such as the soft-start sequence and input voltage surges. Ensuring that P_{PPM} is sufficiently high to withstand these transient conditions is essential to prevent excess thermal stress and potential failure. However, with ICERGi's proprietary, fast (<20 μs) input voltage surge protection mechanism [7], the excellent pulsed avalanche energy handling capabilities of SiC MOSFETs, and a slightly overdesigned flying capacitor bank, alternative strategies can be employed to achieve the flying capacitor voltage pre-charge during startup.

IV. ISOLATED GATE DRIVER DESIGN

The shift from Si to SiC MOSFETs in the magnetically coupled floating gate driver circuit requires a thorough redesign to accommodate the distinct characteristics of SiC technology. While the switching profile for a SiC MOSFET is relatively similar to a Si MOSFET [8], several gate driver design challenges arise. First, SiC MOSFETs require a higher V_{GS}. The recommended unipolar gate drive voltage for 400V SiC is 18V, which is higher than the typical 12V used for Si devices. Second, achieving faster charge and discharge rates of the gate capacitances is essential to support high-speed switching. To satisfy these requirements, redesigning planar transformers is necessary. This process requires adjusting the turns ratio of the primary and secondary windings as well as resizing the PCB tracks. Moreover, the width of the primary side drive pulse must be carefully tuned to account for the external turn-on ($R_{ON,ext}$) and turn-off ($R_{OFF,ext}$) gate resistor values to achieve a slightly underdamped V_{GS} with minimal overshoots/undershoots during both turn-on and turn-off transitions. Precise dead time insertion (ideally variable) is also essential in this hard-switched topology to prevent shoot-through currents and enable reliable switching transitions.

A. Planar Transformer Implementation

The 400V CoolSiC™ MOSFETs can be reliably driven with a unipolar drive voltage $(0 - 18V)$ and does not require a negative turn-off drive voltage. The isolated gate driver transformer design consists of a single primary winding (2 turns) and two secondary windings for each SiC device: the turn-on winding (4 turns) and the turn-off winding (1 turn). Note, a logic-level pulse of 2V is sufficient to reliably pull the gate down to almost 0V during turn-off. The transformer turns ratio can be described by the following expression:

$$N_p : N_{s(on)} : N_{s(off)} = 2 : 4 : 1 \qquad (1)$$

With a primary side voltage ranging from 9V to 10V and the specified transformer turns ratios, both transitions can be satisfied based on the following expressions.

Turn-on:

$$V_p * \frac{N_{s(on)}}{N_p} \geq V_{S(on)} \rightarrow 9V * \frac{4}{2} \geq 18V \qquad (2)$$

Turn-off:

$$V_p * \frac{N_{s(off)}}{N_p} \geq V_{S(off)} \rightarrow 9V * \frac{1}{2} \geq 2V \qquad (3)$$

By reducing the number of FETs in boost leg from 8x 150V to 4x 400V SiC, a 4-layer PCB stack-up is sufficient to implement planar transformer windings. However, a 6-layer stack-up was used instead to test several winding constructions for both the primary and secondary sides. For all windings, the selected track width and gap are 0.10 mm and 0.15 mm, respectively. The primary-side windings are placed on external layers to handle the larger primary current. The secondary-side turn-on and turn-off windings are placed on four inner layers. To ensure strong magnetic coupling and minimize interwinding capacitance, the windings are routed with minimal interlayer overlap. Additionally, the top three layers and bottom three layers are designed symmetrically to provide greater flexibility in configuring the gate driver circuit, regardless of which side of the PCB it is placed on.

B. Gate Driver Optimisation

To accurately capture no-load switching transients of the SiC MOSFETs, waveforms were captured using a 200 MHz bandwidth, with measurements taken near the device terminals to accurately reflect gate drive loop behaviour and minimize parasitic influences. Test setup and some of the key parameters are outlined in Table-II. Based on Fig. 5, the tight gate drive loop with low parasitic inductance ensures that no substantial drive-on or drive-off oscillations are present at the drive terminals of each SiC device Fig. 4. However, more insightful test results can be obtained by measuring V_{GS} during PFC operating mode. Specifically, examining the floating gate drive waveforms which are crucial to verify that there are no parasitic turn-on events during hard-switched transitions.

TABLE II. GATE DRIVER OPTIMISATION PARAMETERS AND SETUP

Item	Value	Description
SiC MOSFET	IMT40R015M2H[a]	$400V, 111A\ (T_c), 15\ m\Omega$
Isolated Gate Driver	IC70-001	$20V, 2A\ (I_{sink}), 4\ A\ (I_{src})$
Transformer Core	EE7.3B	$R10KC$
Parameter	**Value**	**Description**
f_{sw}	65.2 kHz	Switching frequency
V_p	9V – 10V	Primary side voltage
$V_{s\,(on)}$	16V – 18V	Secondary side voltage
D	3.8%	Duty ratio
T_{pulse}	85ns – 165ns	Turn-on/-off pulse

[a]. Benchmark was carried out on a 15 mΩ component

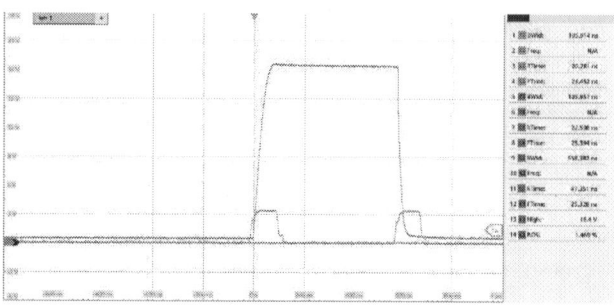

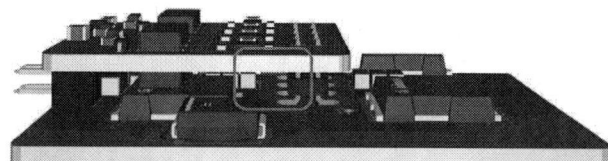

Fig. 4 Test conditions: $R_{ON} = 3.3\,\Omega, R_{OFF} = 2.2\,\Omega, T_{pulse} = 105\,ns, V_{pri} = 9V$ ($V_{GS}, T_{pulse(on)}, T_{pulse(off)}$)

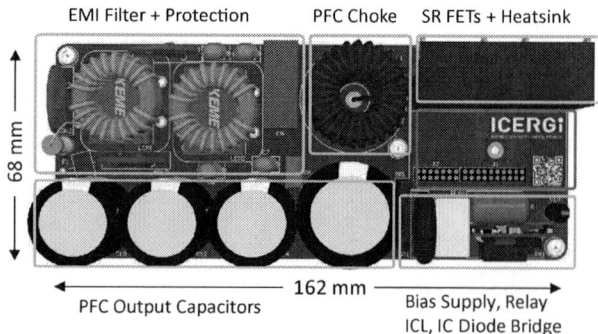

Fig. 5 Side profile of PFC control and boost leg power board assembly (area associated with the gate drive loop for Q_4)

The estimated gate driver loop inductances for the modular design shown in Fig. 5 ranges from 10 nH to 12 nH.

C. Potential Improvements

Several design changes can be adopted to enhance the gate driver design. To minimize parasitic gate drive loop inductance further and to optimize the circuit layout:

- Remove SMD connector from the gate drive loop which in turn can reduce the parasitic loop inductance in the order of several nH.
- Place gate drivers closer to the transistor devices to allow for even faster switching transitions.
- Widen and/or overlap the return path of the gate driver traces with the corresponding signal traces to significantly reduce parasitic inductance and minimize loop area.

V. HARDWARE DESIGN FOR TECHNOLOGY DEMONSTRATION

To evaluate and validate the adoption of 400V CoolSiC™ devices in multilevel PFC, 3.3 kW 3L-FC-BTP PFC prototype was designed and assembled (Fig. 6). The corresponding BOM of the design is presented in Table-III. Note that only some of the key components are listed.

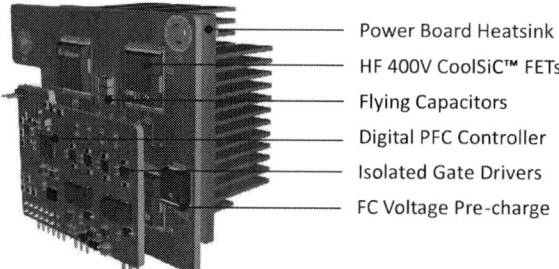

Fig. 6 3.3kW SiC-based 3L-BTP-PFC main board (area for the digital PFC module connection)

TABLE III. 3K3W SiC-BASED PFC BILL OF MATERIALS (BOM)

Component	Name	Description	Qty
Power Stage	C_{Bulk}	470µF 450V output capacitors	3
	$C_{Fly(1)}$	4.7µF 450V (film) FCs	2
	$C_{Fly(2)}$	68 nF 500V X7R (ceramic) FCs	6
	L_1	250µH inductor (MS109125-2)	2
	Q_{1-4}	400V 25mΩ HF SiC MOSFETs	4
	$Q_{SRL/H}$	600V 16mΩ SR MOSFETs	2
Input Protection	MOV_1	430V 8kA varistor (20 mm)	3
	F_1	20A 600VAC/500VDC fuse	1
	GDT	600V 15kA gas discharge tube[c]	1
EMI Filter	C_{X1-2}	1µF 310VAC X – capacitor	2
	C_{X3}	1.5µF 310VAC X – capacitor	1
	C_{Y1-4}	470pF 310VAC Y – capacitor	4
	L_{com1-2}	2mH common mode choke (16A)	2

The Digital PFC module hardware prototype is comprised of a control daughterboard and a power card that houses 4x IMT40R025M2H SiC devices (Fig. 7). A 45x45x20 mm heatsink is attached to the module to provide efficient bottom-side PCB cooling for all SiC devices.

Fig. 7 Digital PFC module prototype (*Note: Film-type flying capacitors are not shown*)

In the design phase of the 3L-FC-type boost leg, particular attention was paid to the layout and minimizing both the inner and the outer power loop areas. Poor design can lead to significant drain-source (V_{DS}) overshoots and failure to pass relevant conducted EMI standards.

VI. EXPERIMENTAL RESULTS

To validate the 3L-FC-BTP-PFC performance with 400V SiC devices, experimental results were collected with primary focus on efficiency metrics and operational behaviour. Waveforms during both startup (Fig. 8) and steady-state (Fig. 9) operations were captured and analysed.

A. Experimental Waveforms

During the startup of the PFC, close to the worst-case scenario is tested with $V_{IN} = 265V_{AC}$ and a phase angle near 90°. The flying capacitors are pre-charged to 140V, which is close to $V_{Bulk}/2$, thereby minimizing capacitor inrush current before the converter transitions to boost mode operation. Note that all waveforms have a 20 MHz bandwidth limit applied to each channel for clarity. Input and PFC choke currents are shown as 10V/div and 5V/div but should be interpreted as A/div instead.

979-8-3315-1612-3/25 $31.00 © 2025 IEEE

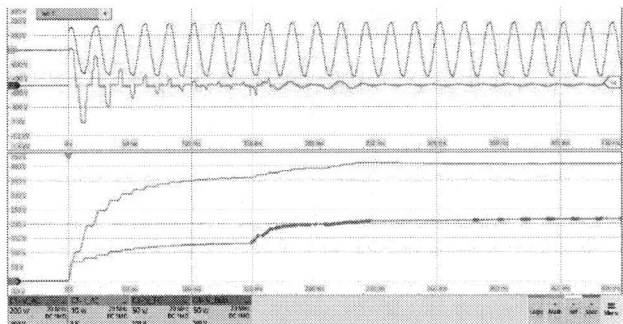

Fig. 8 Start-up sequence with flying capacitor voltage initialization ($V_{IN} = 265V_{AC}$, $P_{OUT} = 0W$)

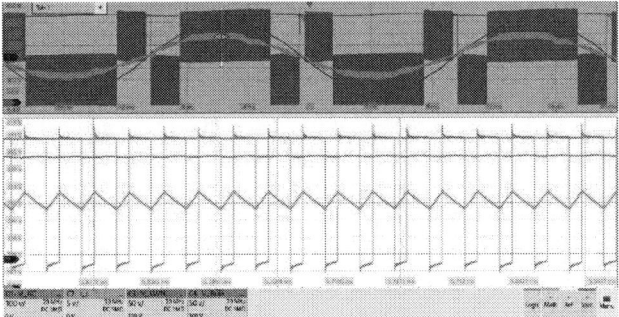

Fig. 9 Steady-state operation ($V_{IN} = 230V_{AC}$, $V_{OUT} = 385V_{DC}$, $f_{sw} = 45.8\ kHz$, $P_{OUT} = 50\%$)

B. Converter Efficiency

The PFC efficiency data outlined in Fig. 10 is measured up to 1650W using IMT40R025M2H SiC devices. It demonstrates peak efficiency exceeding 99.2%, with potential for further optimization. The efficiency measurements include both the two-stage line filter and the 12V bias supply but exclude the external DC fan. The converter was tested in an open-frame configuration. Natural convection was used for $V_{IN} = 230V_{AC}$, while $V_{IN} = 115V_{AC}$ had an approximate airflow of $0.34m^3/min$ (12 CFM).

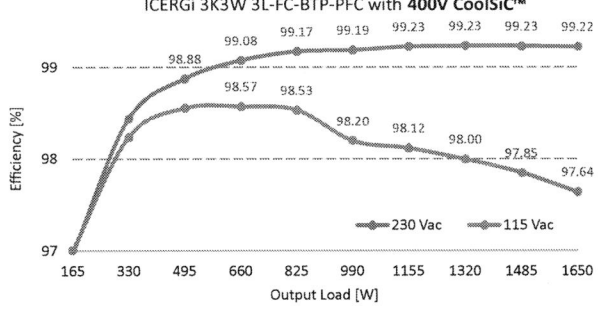

Fig. 10 PFC efficiency up to 1650W with IMT40R025M2H SiC

VII. CONCLUSION

In conclusion, the successful development of ICERGi's SiC gate drive solution has demonstrated robust gate driver capabilities. This achievement allows for a significant advancement in the multilevel power converter sector through the integration of 400V SiC technology, proprietary digital control, and proven gate drive techniques. The presented experimental results validate the successful integration and adoption of 400V SiC MOSFETs in a 3L-FC-BTP PFC applications. As a result, greater power density, simplified gate driver design, and enhanced reliability can be achieved, particularly in medium- to high-power single-phase PFC applications.

Some of the technological aspects and implementation details covered in this paper may be the subject of patent applications.

ACKNOWLEDGMENT

This material is based upon works supported by the Department of Enterprise, Trade and Employment and Enterprise Ireland through the Irish Government's Disruptive Technology Innovation Fund (Grant No. DT2020-0222).

REFERENCES

[1] S. Daryanani, "Improvements in Performance and Voltage Range of SiC MOSFETs," *Power Electronics News*, pp. 4-6, 2024.

[2] "400 V CoolSiC™ AI server deepdive," Infineon, 2024.

[3] P. B. G. Mostafa Khazraei, "Five-level Active Neutral Point Clamped Flying Capacitor Inverter Design Based on OptiMOS™ 5 150 V," Infineon Technologies AG, Munich, Germany, 2020.

[4] R. Beinarys, S. O'Driscoll and T. T. Vu, "Flying Capacitor Voltage Imbalance Protection in Multilevel Bidirectional Inverters during Surge," in *2023 IEEE Applied Power Electronics Conference and Exposition (APEC)*, Orlando, FL, USA, 2023.

[5] X. She, A. Q. Huang, Ó. Lucía and B. Ozpineci, "Review of Silicon Carbide Power Devices and Their Applications," in *IEEE Transactions on Industrial Electronics*, 2017.

[6] T. T. V. Rytis Beinarys, "Hybrid Voltage Balancing Control in 3-level Bridgeless Totem-pole PFC," in *Applied Power Electronics Conference and Exposition (APEC)*, Phoenix, AZ, USA, 2021.

[7] T. T. V. George Young, "Method of Controlling a Current Shaping Circuit in a Converter with Power Factor Correction". United States of America Patent US10944318B2, 26 March 2020.

[8] O. S. Corporation, "SiC MOSFETs: Gate Drive Optimization," 2023.

[9] D. Miller, "Surge Immunity Test Analysis for Modern Switching Mode Power Supplies," in *International Telecommunications Energy Conference (INTELEC)*, Austin, TX, USA, 2016.

Extended Smart-Link Quasi-Single-Stage 3-Phase AC-DC Power Supply Module for AI-Driving Data Centers

D. Biadene*, J. Huber[†], J.W. Kolar[†], and P. Mattavelli*

*Dept. of Management and Engineering, University of Padova, Italy
[†]Power Electronic Systems Laboratory, ETH Zürich, Switzerland

Abstract—In the rapidly evolving landscape of artificial intelligence (AI), the demand for computational power has surged, placing data centers at the forefront of technological progress. This increasing need for computational resources has led to a corresponding rise in energy consumption, currently estimated between 450 and 650 TWh/year and expected to reach 1000 TWh/year, driven by the emergence of next-generation 2 kW GPUs. Wide-bandgap (WBG) devices, including gallium nitride (GaN) and silicon carbide (SiC), offer the potential for up to a 1% efficiency improvement in applications such as state-of-the-art two-stage ac-dc converters, which currently achieve efficiencies of up to 98%. This paper presents a novel three-phase quasi-single-stage isolated ac-dc topology leveraging partial power processing to precisely control the input ac current and output dc voltage, ensuring unity power factor under constant power load conditions, while maintaining LLC converter operation at its resonant frequency despite input voltage variations. The eXtended Smart-Link (XS-Link) integrates both voltage and current regulation across the two stages of ac-dc converters, effectively improving efficiency.

Index Terms—ac-dc conversion, data centers, galvanic isolation, partial power processing, three-phase grid.

I. INTRODUCTION

The widespread and rapidly expanding use of artificial intelligence (AI) drives an unprecedented demand for computing power. Thus, large-scale data centers are at the forefront of technological advancements. The growth in computational power has been accompanied by a parallel increase in energy consumption. Current estimates place the annual global energy usage of data centers, including data computing and transmission, between 450 TWh and 650 TWh, accounting for up to 2.2% of the global electricity generation [1]. However, future hyperscale data centers are expected to contribute to more than a doubling of the energy consumption, reaching 1000 TWh/year, e.g., resulting from the next-generation of 2 kW GPUs [2]. New cutting-edge power supplies tailored for data center applications are thus needed, delivering exceptional efficiency, scalability, and reliability, while lowering operational costs and reducing environmental impact. For example, wide-bandgap (WBG) devices, such as gallium nitride (GaN) and silicon carbide (SiC), promise up to a 1% improvement in efficiency [2]. Due to their higher operating temperatures and greater thermal conductivity, they positively affect the liquid-based cooling system in terms of performance and volume [3].

The state-of-the-art power distribution architecture in modern data centers incorporates includes a shared three-phase low-frequency transformer (3Φ-LFT) that steps down the medium-voltage ac (MVAC) grid voltage (e.g., 13.4 kV) to low-voltage ac (LVAC, e.g., 400 V...690 V line-to-line rms), which is then distributed throughout the data center to the individual server racks [4]. At the rack level, a two-stage conversion system, consisting of an ac-dc converter followed by a dc-dc converter, is usually adopted. While single-phase ac-dc converters were predominant until recently, three-phase ac-dc converters are now the preferred choice due to higher power levels. This configuration supplies a 50 V backplane power bus of each CPU/GPU rack, providing up to 25 kW for the latest generation of processors. The front-end (FE) stage is designed to rectify the three-phase grid to an HVDC voltage (e.g., 800 V or ± 400 V), and provide power factor correction functionality through phase current control [5]. Then, the back-end (BE) dc-dc step-down stage must only provide voltage scaling and galvanic separation and regulate the output voltage, facing rapid fluctuations in the load power due to the high dynamics of the CPU/GPU current consumption depending on the computational load. The LLC converter is commonly used as the BE stage due to its outstanding performance. When operating at its optimum operating point (i.e., at the resonant frequency, known as DCX operation), the LLC converter shows constant voltage gain and guarantees zero voltage/current switching (ZVS/ZCS) commutations, drastically reducing the losses [6]–[9]. To overcome the limitation of fixed voltage gain, an additional partial power (PP) processing converter can be used to adjust the LLC input voltage [10], [11]. Alternatively, a dual active bridge (DAB) converter can be employed as a BE stage, providing performance comparable to the LLC converter but with increased control complexity [12]–[14]. Other approaches discussed in the literature feature single-phase single-stage ac-dc converters utilizing three-level converters [15], modified Cuk topologies [16], [17], and flying capacitors [18], as well as three-phase isolated single-stage ac-dc matrix converters with monolithic bidirectional switches [19] or anti-series connections of SiC transistors [20]; the latter achieves an efficiency of 99% but requires relatively complex modulation and commutation schemes. Finally, today, industrial state-of-the-art power supply units with power levels of up to 12 kW show rated efficiencies in the order of 97.5% [21].

979-8-3315-1612-3/25 $31.00 © 2025 IEEE

This paper proposes a novel isolated quasi-single-stage three-phase ac-dc power converter that has the potential to achieve efficiencies of 99%. The output voltage and input current regulation are integrated through a PP auxiliary converter (i.e., the "eXtended Smart dc-Link" or "XS-Link" concept), while ensuring that the LLC converter operates as DCX even under input voltage variations.

In the following, **Section II** introduces the XS-link concept and explains the operating principle. Then, **Section III** discusses design guidelines and the control method, before **Section IV** verifies the proposed XS-Link with detailed closed-loop circuit simulations. Finally, **Section V** concludes the paper.

II. Operating Principle

The proposed XS-Link concept is depicted in **Fig. 1**, alongside its predecessor, the "Smart dc-Link" (S-Link) introduced in [22].[1] In both cases, two-stage ac-dc power converters comprising an ac-dc front-end stage performing power factor correction (PFC) and an isolated dc-dc back-end stage that ensures the required galvanic isolation for the targeted application. For a first conceptual explanation of the operating principle, it is convenient to model the S/XS-Link circuit as a single current source in the S-Link configuration or as a pair of power-linked current sources in the XS-Link configuration, as shown in **Figs. 1a** and **Figs. 1b**, respectively. The interaction of the S/XS-Link with both FE and BE converters enables to select the best topology for each of these stages and operating them at their maximum-efficiency operating point. The considered hardware implementation of the proposed ac-dc converter based on the XS-Link circuit is reported in **Fig. 2**. **Fig. 1** reports the key voltage, current, and power waveforms to describe and highlight the main differences and capabilities of both concepts (S-Link and XS-Link), neglecting the switching-frequency components. In the following analysis, unity power factor and lossless power conversion is assumed.

A. AC-DC FE Stage

Under the assumption of unity power factor and symmetric grid voltages, the input three-phase voltages and currents of the ac-dc PFC stage are defined as

$$v_i(t) = \hat{V}_{ac} \cos(2\pi f_{ac} t + \varphi_i),$$
$$i_i(t) = \hat{I}_{ac} \cos(2\pi f_{ac} t + \varphi_i), \tag{1}$$

where $i = \{a, b, c\}$, $\varphi_i = \{0, -\frac{2}{3}\pi, \frac{2}{3}\pi\}$ are the phase angles, $\hat{V}_{ac}$ and $\hat{I}_{ac}$ are the peak line voltage and current, respectively, and f_{ac} is the grid frequency. To minimize losses, a three-phase rectifier employing an active third-harmonic current injection circuit is considered [23], i.e., an integrated active filter (IAF) rectifier, as shown in **Fig. 2**. The operating principle of the IAF is recalled next for completeness.

The dc-link $1\bar{1}$-port is connected through diodes (or switches) to the maximum line-to-line voltage, following the typical six-pulse-shaped voltage (see $v_{1\bar{1}}$ voltage in **Fig. 1**). Due to its periodicity, the analysis is reduced to its fundamental period $T_{ac}/6$. Within the interval $[0, T_{ac}/6]$, voltage $v_{1\bar{1}}$ is expressed as

$$v_{1\bar{1}}(t) = \max(v_a(t), v_b(t), v_c(t)) - \min(v_a(t), v_b(t), v_c(t))$$
$$= \sqrt{3}\hat{V}_{ac} \sin\left(2\pi f_{ac} t + \frac{\pi}{3}\right), \tag{2}$$

where $v_{1\bar{1}}(t) = v_a(t) - v_c(t)$ in this sector.

Using the hypothesis of a unity power factor, the absorbed power from the grid under balanced and sinusoidal conditions (i.e., $P_{ac} = p_{ac}(t) = p_a(t) + p_b(t) + p_c(t) = \frac{3}{2}\hat{V}_{ac}\hat{I}_{ac}$) must be equal to the power at the $1\bar{1}$-port, imposing the dc-link current i_1. Since lossless power conversion is assumed, the input and output power are equal, such that $p_{ac}(t) = p_{1\bar{1}}(t) = P_o$, allowing to define the dc-link current i_1 as

$$i_1(t) = \frac{P_o}{v_{1\bar{1}}(t)} = \frac{P_o}{\sqrt{3}\hat{V}_{ac} \sin\left(2\pi f_{ac} t + \frac{\pi}{3}\right)}. \tag{3}$$

The downstream converter stages must thus shape the dc-link current i_1 accordingly. Then, the third-harmonic current injection circuit must ensure symmetric three-phase currents at the input, such that $i_i(t) = G^* \cdot v_i(t)$ where $i \in \{a, b, c\}$ and G^* represents the equivalent input conductance. For this purpose, the smallest phase current (i.e., i_b in the considered time-interval $t \in [0, T_{ac}/6]$) is injected by modulating the L_j-inductor current with the auxiliary half-bridge $S_{j,x}$ and closing the related path of the bidirectional phase selector switch (PSS) $S_{abc,x}$.[2] Thus, within interval $[0, T_{ac}/6]$, $i_b = G^* \cdot v_b$.

To validate the aforementioned control strategy, the necessary i_1 current is derived. Consider negligible voltage drop across the L_j-inductor at low frequency (i.e., $\bar{v}_{kj} \simeq 0$), the half-bridge $S_{j,x}$ is modulated so that

$$v_{kn} \simeq v_b = d_j v_{1n} + (1 - d_j)v_{\bar{1}n} = d_j v_a + (1 - d_j)v_c \tag{4}$$

where d_j is the duty-cycle of the upper switch $S_{j,1}$. From (4), the definition of d_j is unique and given by $d_j = v_{bc}/v_{ac}$. Under the unity power factor assumption, the i_1 current is constrained to be equal to

$$i_1 = i_a - d_j i_j = i_a + \frac{v_{bc}}{v_{ac}} i_b = G^* \left(v_a + \frac{v_{bc}}{v_{ac}} v_b\right) = \frac{G^* \sum_i v_i^2}{v_{ac}}. \tag{5}$$

Recalling $P_{ac} = G^* \sum_i v_i^2$ and $v_{ac} = v_{1\bar{1}}$ in $t \in [0, T_{ac}/6]$, the equivalence between (3) and (5) is verified, highlighting that unity power factor is achieved only with a downstream converter stage that draws constant power [24].

In particular, the IAF FE stage combines low modulation complexity with high efficiency: the PSSs must commutate every $T_{ac}/6$, i.e., in each sector transition, reducing their switching loss contribution; further, since the inductor L_j always carries the smallest phase current, triangular current mode (TCM) modulation can be exploited to achieve ZVS of the $S_{j,x}$ switches for almost the entire sector.

---

[1] Note that the S-Link in [22] is used in combination with a Vienna rectifier ac-dc FE stage, and advantageously facilitates so-called 1/3-PWM operation of that Vienna rectifier, reducing switching losses. Here, the S-Link is combined with an IAF rectifier front-end, as discussed below.

[2] Note that the PSSs are switched at double-line frequency only.

979-8-3315-1612-3/25 $31.00 © 2025 IEEE

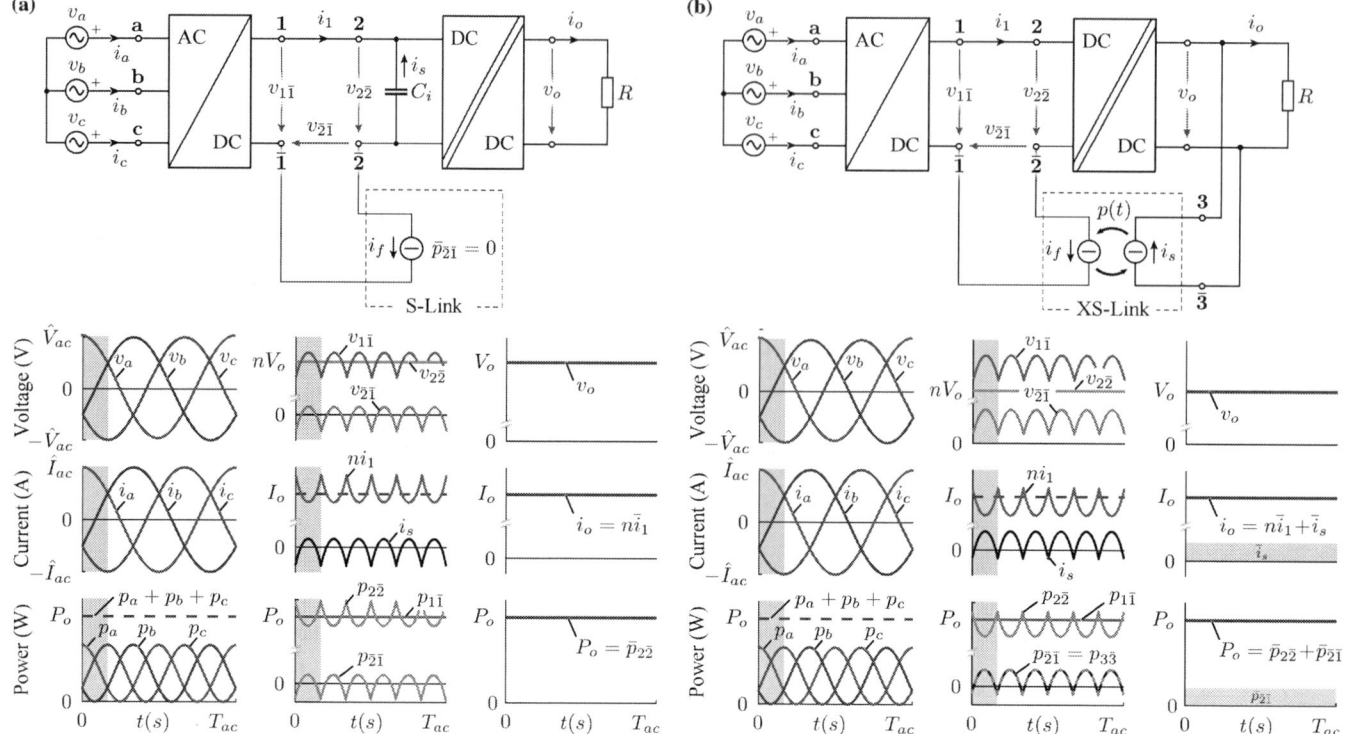

Fig. 1. Quasi-single-stage ac-dc power converters comprising an IAF PFC front-end and an isolated DCX-LLC dc-dc converter as BE stage. **(a)** The "S-Link" concept introduced in [22] can be utilized to achieve sinusoidal mains currents and constant power flow to the output; however, the output dc voltage is tied (fixed ratio) to the mains voltage amplitude. **(b)** Proposed eXtended Smart dc-Link (XS-Link) concept, where the partial-power stage can exchange active power with the dc output. This allows tight regulation of that output dc voltage to a fixed value independent of the mains voltage. The shown parameter n is the voltage gain of the isolated dc-dc converter, i.e., $n = v_{2\bar{2}}/v_o$.

B. DC-DC BE Stage

As in the FE stage, the simplicity of control and minimization of loss are the most significant decision elements for the chosen topology. For this reason, the LLC converter is selected for its well-known high-efficiency performance operating at its resonant frequency f_r (i.e., DCX mode) achieving ZVS for the primary-side switches and ZCS for the secondary-side ones. In this operating mode, the LLC acts like a dc-transformer where the voltage gain depends only on the selected transformer turn ratio $n_{12} = N_1/N_2$. The related circuit is reported in **Fig. 2** as the DCX-LLC block.

The two full bridges $S_{1,x}$ and $S_{2,x}$ are synchronously controlled and produce two identical square voltage waveforms across the L_rC_r resonant tank of amplitude $\pm v_{2\bar{2}} = \pm n_{12}v_o$. To increase DCX-LLC performance, the transformer can be designed as a matrix transformer offering excellent power density and better integration of the XS-Link stage.

It is important to note that the DCX-LLC stage operates in open-loop configuration, without any closed-loop control. As a result, the output voltage remains unregulated, and the voltage regulation must be implemented in the upstream converter stages.

C. S/XS-Link Circuit

The presented "S/XS-Link" concepts addresses the shortcomings of the two cascade converters: on one hand, the IAF FE

stage requires constant power load, i.e., an impressed current as defined in (3), and on the other, the DCX-LLC stage lacks output voltage regulation.

Given the strict constraint on the dc-link current i_1, the S/XS-Link stage interfaced with the $\bar{2}\bar{1}$-port must be designed as a current generator. For its realization, the inductor L_f is placed in series to the i_1 current path, and its current is controlled by the means of a full-bridge (FB) circuit $S_{4,x}$ connected to an energy storage capacitor C_h, as shown in **Fig. 2**.

As shown in [22], the S-Link circuit must process zero net power at its input port (i.e., $\bar{p}_{\bar{2}\bar{1}} = 0$)[3]. Within interval $[0, T_{ac}/6]$, the instantaneous input power expression at the $\bar{2}\bar{1}$-port from (2) and (3) is

$$p_{\bar{2}\bar{1}}(t) = p_{1\bar{1}} - p_{2\bar{2}} = P_o \left[1 - \frac{\bar{v}_{2\bar{2}}}{\sqrt{3}\hat{V}_{ac} \sin(2\pi f_{ac}t + \frac{\pi}{3})} \right] \quad (6)$$

and averaging over the considered sector, the net input power

---

[3]Note that the S-Link is designed solely to implement the circuit described here, without incorporating any additional ports. In contrast, the XS-Link circuit includes an extra port, enabling it to handle a net input power.

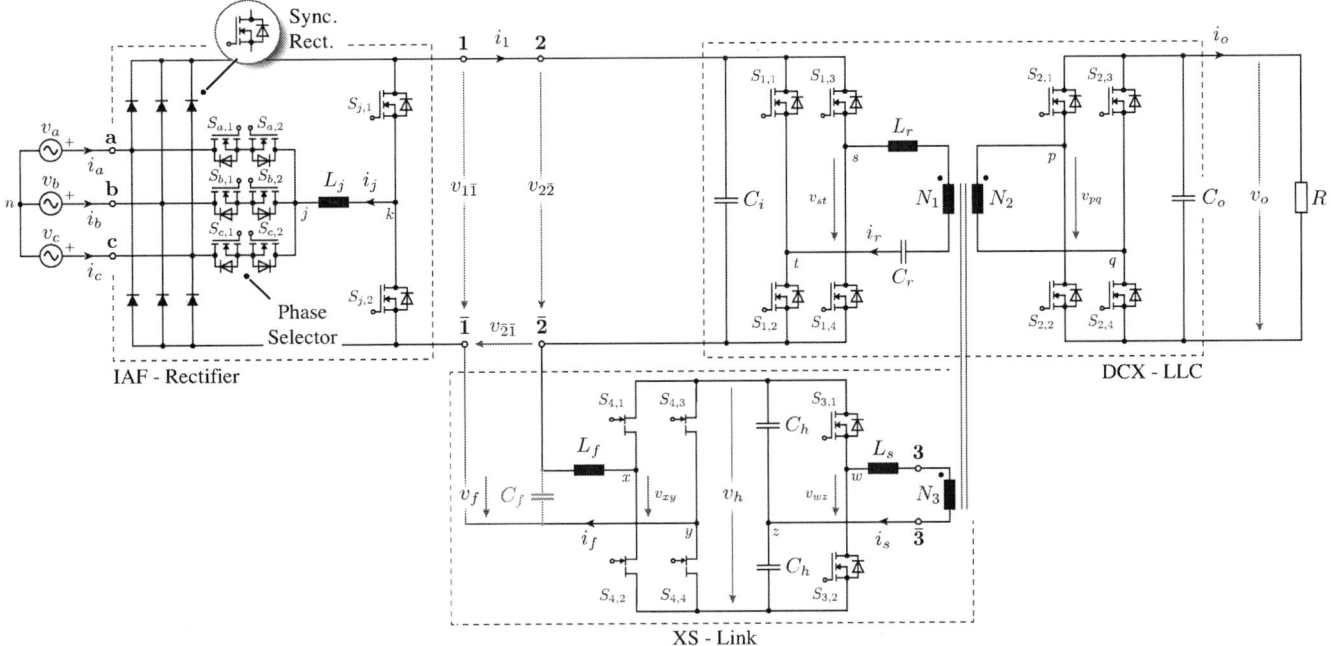

Fig. 2. Implementation example of the proposed quasi-single-stage isolated ac-dc power converter comprising an IAF PFC front-end, a DCX-LLC isolated dc-dc converter as back-end stage, and the proposed eXtended Smart dc-Link (XS-Link) partial-power converter (see also **Fig. 1b**. The exchange of power between the PPC stage and the dc output is implemented via a third winding on the DCX-LLC transformer as done in [10].

is found as

$$\bar{p}_{\bar{2}\bar{1}} = P_o \frac{6}{T_{ac}} \left[t - \frac{\bar{v}_{2\bar{2}}/\sqrt{3}\hat{V}_{ac}}{2\pi f_{ac}} \ln \left| \tan \left(\pi f_{ac} t + \frac{\pi}{6} \right) \right| \right]_0^{T_{ac}/6}$$

$$= P_o \left[1 - \frac{\bar{v}_{2\bar{2}}}{\hat{V}_{ac}} \frac{\sqrt{3}\ln 3}{\pi} \right]. \tag{7}$$

It is worth noting that the zero net power constraint can be satisfied only for a specific value of the $v_{2\bar{2}}$ voltage, fixing inevitably the the output voltage at

$$V_o = \frac{\bar{v}_{2\bar{2}}}{n_{12}} = \frac{1}{n_{12}} \frac{\pi}{\sqrt{3}\ln 3} \hat{V}_{ac}. \tag{8}$$

As emphasized in (8), the output voltage cannot be independently controlled relative to the grid voltage, meaning any fluctuation in the grid voltage directly impacts the output voltage V_o. However, as shown in **Fig. 1**, a buffer capacitor C_i must be added at the $2\bar{2}$-port to filter the ac component of i_1, such that $i_s = i_1 - I_o$. Otherwise, the output power should not be constant, which would violate the constant power load requirement required by the IAF stage.

To overcome the imposed voltage limitation at the $2\bar{2}$-port of the S-Link circuit, which enables an output voltage regulation scheme, the net-power setpoint of $p_{\bar{2}\bar{1}}$ must be moved from zero, as highlighted in (7). Therefore, the S-Link must be extended to enable partial processing of the input power $p_{1\bar{1}}$. Partial power (PP) architectures are well-suited for this purpose. As proposed in [10], a third port was added to the DCX transformer and connected through a phase shift modulated

(PSM) half-bridge. This created a power path for a voltage pre-regulator, which was connected in series with the DCX input port. Similarly, the S-Link is now equipped with a third port (i.e., $3\bar{3}$ in **Fig. 2**), interfacing an additional DCX transformer winding with a PSM half-bridge connected to the energy storage capacitor C_h, currently split into two. The obtained circuit is the implementation of the "XS-Link" concept.

As shown in **Fig. 1**, the additional circuit allows to share the input power $p_{1\bar{1}}$ among the DCX and the XS-Link without violating the constant power requirement at the IAF FE stage for output voltages differing from (8). The power processed by the XS-Link can be easily set with the additional half-bridge $S_{3,x}$ as in a DAB converter. Imposing a square voltage waveform of amplitude $\pm\frac{1}{2}v_h$ on $v_{wz}(t)$, leading in phase relative to voltage $v_{pq}(t)$ of φ_{32} angle, the power exchanged at $3\bar{3}$-port is regulated according to the following relation:

$$p_{3\bar{3}} = \frac{\frac{1}{2}v_h \cdot n_{32}v_o}{2\pi^2 f_r L_s} \varphi_{32}(\pi - |\varphi_{32}|), \tag{9}$$

where $n_{32} = N_3/N_2$ is the winding turn ratio between the transformer ports connected to the XS-Link and output, respectively, L_s is the leakage inductance of the third winding or a coupling impedance intentionally connected, and f_r is the resonance frequency of the DCX converter corresponding to the switching frequency of all bridge circuits connected to the DCX transformer (i.e., $S_{1,x}$, $S_{2,x}$, and $S_{3,x}$).

III. Design Guidelines

This section outlines general design considerations for determining the XS-Link parameters and the critical components of the front-end and back-end stages. Detailed design methodologies fall outside the scope of this paper. The reported guidelines are referred to the circuit implementation done in **Fig. 2**.

A. IAF Rectifier

The power semiconductor devices must be selected to block at least the maximum line-to-line grid voltage, i.e., $\sqrt{3}\hat{V}_{ac}$. Taking into account voltage fluctuations of $\pm 10\%$ relative to the nominal grid voltage and a 50% safety margin above the maximum operating voltage, the rated voltage for the switches and diodes of the IAF stage should be $V_R > 3/2(\sqrt{3}\hat{V}_{ac} + 10\%)$.

The injection inductor L_j is designed to guarantee ZVS of the $S_{j,x}$ switches for almost the entire sector. At the sector boundaries, i.e., at the crossings of the grid voltages, the duty cycle d_j saturates to 0 or 1, reducing the magnitude of the current ripple and loosing ZVS conditions.

B. DCX-LLC Converter

The voltage rating of the semiconductor devices connected to the $2\bar{2}$-port is selected following the same approach as the one used in the previous section. Likewise, the same safety margin can be considered for devices connected to the output port, that is, $V_R > 3/2\hat{V}_o$ with $\hat{V}_o$ the maximum allowed output voltage.

The resonant tank can be designed to limit the maximum voltage $\hat{V}_{Cr}$ reached by the resonant capacitor at the nominal power P_o, defining a minimum capacitance value equal to $C_r = P_o/(4n_{12}V_o f_r \hat{V}_{Cr})$. The related resonant inductor is derived from the switching frequency chosen for the DCX-LLC stage, which corresponds to the resonance one. The selection of the n_{12} turn-ratio value will be addressed in the following section as a key aspect of the XS-Link circuit.

The input and output capacitances, i.e., C_i and C_o in **Fig. 2**, respectively, are designed to filter the high-frequency components of the resonant current, limiting the voltage ripple at the DCX ports[4].

C. XS-Link

In the next step, all parameters for the XS-Link implementation are defined, starting with the n_{12} turn ratio value.

This parameter is responsible for the net power processed by the XS-Link, tuning the $v_{2\bar{2}}$ voltage, (i.e., $n_{12}v_o$) in (7). As demonstrated in [10], the PP approach is advantageous when the auxiliary circuit processes positive power, as the power handled by the PP circuit is directly transferred to the output port rather than recirculating between the input and output ports. For this reason, the turn ratio should be chosen as $n_{12} < \frac{3}{2}\hat{V}_{ac}/V_o$, accordingly to (6). However, lowering the $v_{2\bar{2}}$ voltage increases the average voltage at $v_{\bar{2}\bar{1}}$, at the expense of a higher voltage stress on the input full-bridge of the XS-Link circuit. Since the final target is the use of high-performance GaN MOSFETs to reduce the L_f inductor volume allowed

by very-high-switching frequency, the most significant design-driving parameter will be the maximum absolute voltage at $\bar{2}\bar{1}$-port. It follows that the turn ratio n_{12} is chosen close to the average voltage value of $v_{2\bar{2}}$ as $n_{12} \simeq \frac{3\sqrt{3}}{\pi}\hat{V}_{ac}/V_o$.

Once the turn ratio n_{12} is defined, the buffer capacitor voltage V_h is determined to keep the duty-cycle $d_{xy} = v_{\bar{2}\bar{1}}/V_h$ in its linear range (i.e. $d_{xy} \in [-1, 1]$) with a sufficient margin to ensure enough current control capability during transient (i.e. $|d_{xy}| < \hat{d}_{xy}$). Then, the buffer voltage is selected as $V_h = \hat{v}_{\bar{2}\bar{1}}/\hat{d}_{xy}$, and, consequently, the maximum voltage rating V_R of the XS-Link power semiconductor devices, keeping a reasonable safety margin like $V_R > 3/2V_h$.

The buffer capacitors C_h are designed to compensate for the power mismatch between the $\bar{2}\bar{1}$ and $3\bar{3}$-ports during grid voltage transient or output load step. Unlike the S-Link concept, where buffer capacitors serve as energy storage elements to absorb power fluctuations at the $\bar{2}\bar{1}$-port, the XS-Link circuit is power-transparent, meaning $p_{\bar{2}\bar{1}}(t) = p_{3\bar{3}}(t)$ as shown in **Fig. 1**. This enables a significant reduction in the minimum required stored energy and, consequently, in the C_h capacitance value.

The input inductor L_f is designed in a consistent manner with the selected modulation scheme of the full-bridge $S_{4,x}$. A three-level modulation can be employed, doubling the effective switching frequency of the $v_{xy}(t)$ voltage. Triangular current modulation can be also implemented to guarantee ZVS commutations at the cost of an increased rms value. The input capacitor C_f is dimensioned as a return path for the high-frequency components of the current i_f.

Since the XS-Link output stage operates equivalently to a DAB, the same design rules are applied: the turn ratio n_{23} is chosen closest to the voltage ratio V_h/V_o, while the leakage inductance L_s is selected to transfer the maximum rated power of the XS-Link circuit at maximum allowed phase-shift $\hat{\varphi}_{32}$ according to (9).

D. Control

Fig. 3 describes the closed-loop control structure of the quasi-single-stage three-phase ac-dc power module implementing the XS-Link concept. The output voltage Rv_o regulator sets an input power reference $p_{1\bar{1}}^*$, which is converted into the dc-link current i_1^* reference. This current is then tracked by the XS-Link current Ri_{Lf} regulator, incorporating appropriate feedforward terms (see **Fig. 3 a**). The additional voltage Rv_h regulator is used to maintain the buffer voltage v_h around its reference V_h^{*}[5] (**Fig. 3 b**), generating a power mismatch between the input and output processed by the XS-Link (i.e., $p_{\bar{2}\bar{1}}$ and $p_{3\bar{3}}$, respectively). Lastly, the input power reference $p_{1\bar{1}}^*$ is converted into a grid conductivity reference $G^* = \hat{I}_{ac}^*/\hat{V}_{ac}$, which is used to define the injection current reference i_j^*. This current is then tracked by the IAF current Ri_{Lj} regulator, incorporating appropriate feedforward terms (see **Fig. 3 c**). For the sake of completeness, the closed-loop control structure of the power module implementing the S-Link is reported in **Fig. 3 d**. In

---

[4]The input capacitance C_i of the S-Link must be over-designed to compensate also the low-frequency ac components of the input current i_1.

[5]The control bandwidth of Rv_h regulator is limited to not interfere with the wide-bandwidth current Ri_{Lf} regulator.

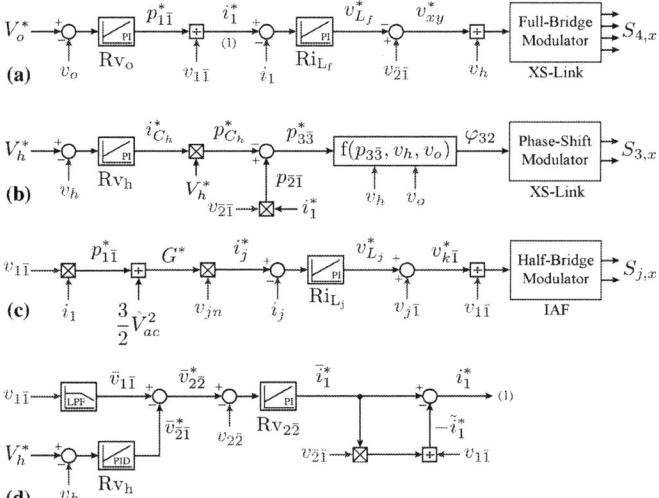

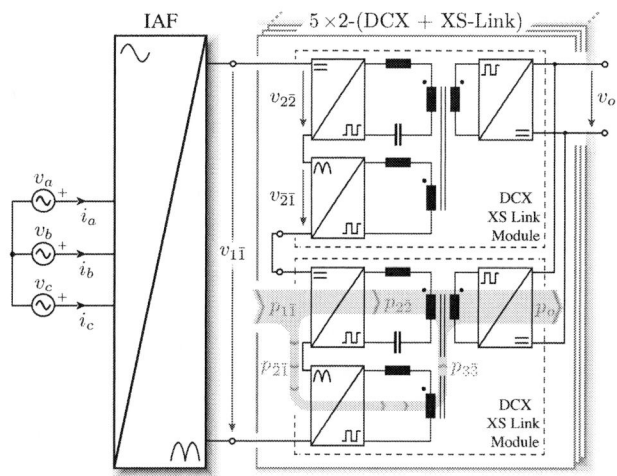

Fig. 3. Control structure of the XS-Link concept: **(a)** cascaded control structure with internal current i_1 control, and external output voltage v_o control; **(b)** buffer voltage v_h control; **(c)** injection current i_j control. The Ri_{Lj} regulator always controls the current only in one phase (j), which is selected by the phase selector switches (PSS); **(d)** i_1-current reference for the S-Link concept replacing i_1^* in **(a)**.

Fig. 4. Simulated structure of the quasi-single-stage three-phase ac-dc power module implementing the XS-Link concept. The simulation parameters are reported in **Tab. I**.

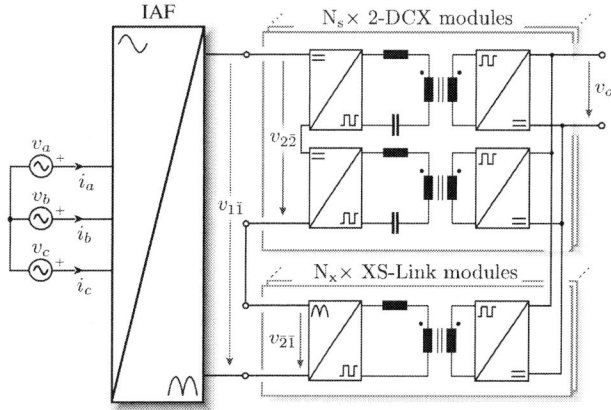

Fig. 5. Different arrangement of the quasi-single-stage ac-dc power module with N_s dc-dc stage in parallel composed of 2 DCX-LLC and N_x XS-Link module in parallel.

this case, the dc-link current i_1^* reference is obtained starting from its average value $\bar{i}_1^*$, which is provided by the $v_{2\bar{2}}$ voltage regulator $\text{Rv}_{2\bar{2}}$. The $v_{2\bar{2}}^*$ voltage reference is then built by the average value of the input voltage $\bar{v}_{1\bar{1}}$ and XS-Link $v_{\bar{2}\bar{1}}$ voltage reference. The latter is provided by the voltage Rv_h regulator.

IV. Verification

To validate the effectiveness of the proposed XS-Link concept, a performance evaluation is carried out using numerical simulations. These simulations confirm the operation of the proposed converter and the effectiveness of the control architecture, even in the critical transient conditions due to input voltage variations, and perform an evaluation of the conduction losses. To increase modularity, and limit the device voltage stress, the converter BE stage is split in 2.5 kW modules each. Two DCX + XS-Link modules are connected in series to share the six-pulse-shaped voltage, then paralleled to reach the required 25 kW. Conversely, the IAF rectifier is shared among all the BE modules. The considered module arrangement is depicted in **Fig. 4**.

A. Verification of Converter Operation

Tab. I lists the system specifications and key converter parameters considered in this section.

The main XS-Link converter waveforms simulated in PLECS for S-Link [22] (i.e., $p_{3\bar{3}} = 0$) and the proposed XS-Link operation (i.e., $p_{3\bar{3}} \approx p_{\bar{2}\bar{1}}$) are reported in **Fig. 6a** and **Fig. 6b**, respectively. In **Fig. 6-1** and **Fig. 6-2**, the line voltages and currents are reported, respectively, under a grid voltage steps of +10% and −10% with respect to the nominal value. In **Fig. 6-3**, the HV-side voltages are reported for a single DCX+XS-Link module: as mentioned, during the grid voltage steps, the XS-Link is able to keep the output voltage v_o constant, as well as

the input DCX voltage (i.e., $v_{2\bar{2}}$ - see **6b-3**). Unlike the S-Link, which, being unable to sustain a dc voltage offset at the $\bar{2}\bar{1}$-port, constraints the voltage $v_{2\bar{2}}$ to follow the grid voltage variation (see **6a-3**). In **Fig. 5-5**, the dc-link current i_1 and the DCX input current (i.e., i_o/n_{12}) are shown, demonstrating how the XS-Link ensures a seamless output current during grid transients. This is attributed to the XS-Link's ability to handle active power, as demonstrated in **Fig. 6-5**. By constraining $p_{3\bar{3}} = 0$ (refer to **6a-5**), the S-Link is limited to compensating only the ac component of the $v_{1\bar{1}}$ voltage. Notably, since constant power load is considered, the input power $p_{1\bar{1}}$ fluctuates accordingly with $p_{\bar{2}\bar{1}}$, causing a small variation in the output and grid currents (see **Fig. 6a-2.**). On the other side, when the XS-Link operates, the module is power-transparent keeping both input and output power constants (i.e., $p_{1\bar{1}} \simeq p_o$). Finally, the output voltage v_o and the buffer voltage v_h are shown in the **Fig. 6-6**.

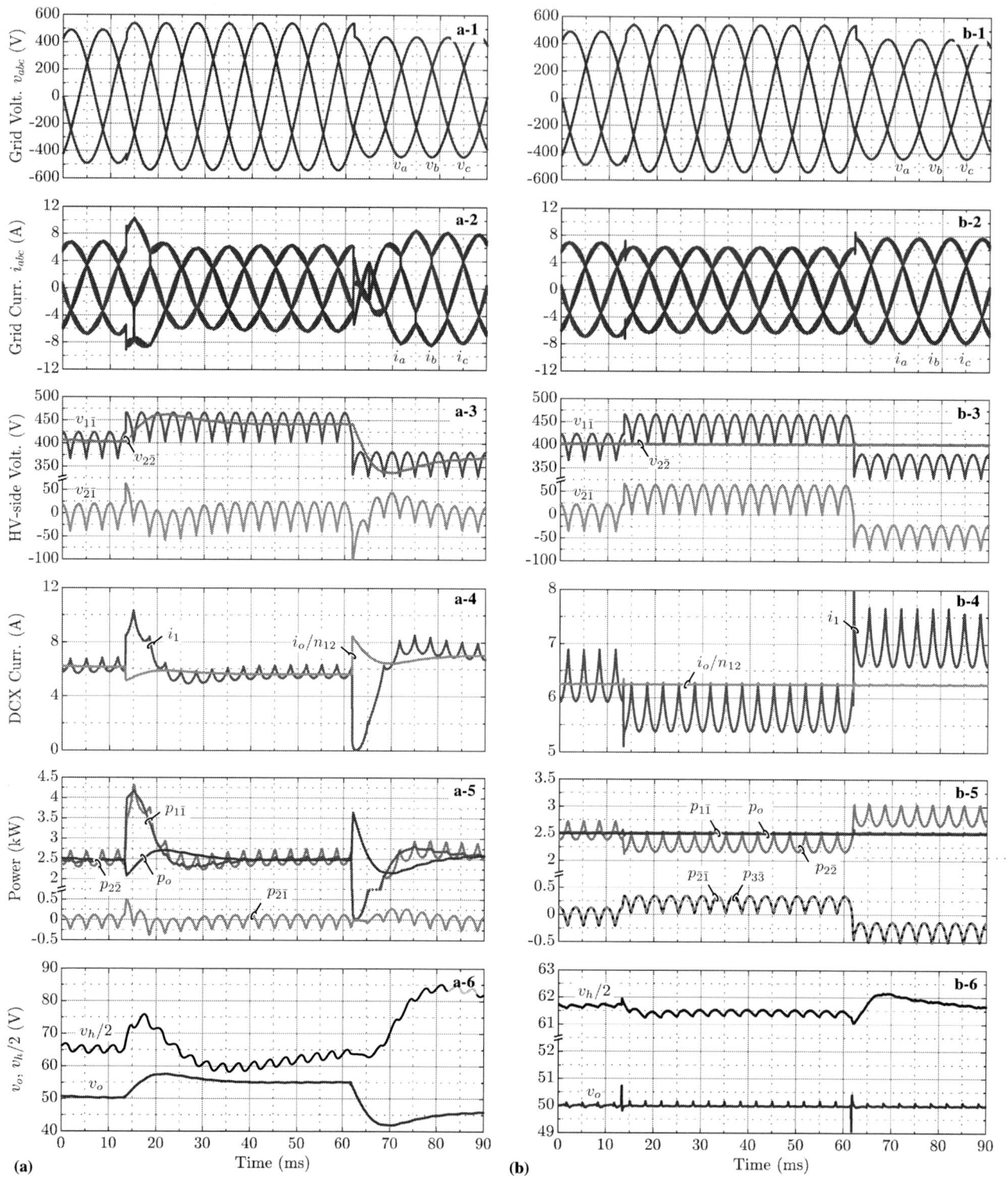

Fig. 6. Main XS-Link converter waveforms simulated in PLECS for **(a)** $p_{3\bar{3}} = 0$ (i.e., S-Link operation [22]) and **(b)** $p_{3\bar{3}} = p_{\bar{2}\bar{1}}$ (i.e., XS-Link operation), under grid voltage step transitions ($\pm 10\%$). The simulation parameters are reported in **Tab. I**. The vertical scale of **(4)**, **(5)** and **(6)** are different between S-Link**(a)** and XS-Link**(b)** operations to improve the visibility.

B. Performance Evaluation

To evaluate whether the proposed XS-Link approach can break the 99%-efficiency limit, the total conduction losses are estimated considering the following power transistors: Infineon CoolSiC 1200 V (10 mΩ) for $S_{abc,x}$ and $S_{j,x}$; Infineon CoolSiC

TABLE: I: Main system specifications for XS-Link

Parameter	Value	Parameter	Value
Grid volt.[1]	$V_{ac} = 600\,\text{V}\pm10\%$	DC volt.	$V_o = 50\,\text{V}\pm1\%$
Grid curr.	$\hat{I}_{ac} = 6.8\,\text{A}$	DC curr.	$I_o = 500\,\text{A}$
Grid freq.	$f_{ac} = 50\,\text{Hz}$	DC power	$P_o = 25\,\text{kW}$
$S_{j,x}$ freq.	$f_{sw,j} = 72\,\text{kHz}$	IAF ind.	$L_j = 86\,\mu\text{H}$
$S_{12,x}$ freq.	$f_{sw,12} = 144\,\text{kHz}$	DCX ind.	$L_r = 20\,\mu\text{H}$
DCX cap.	$C_o = 57\,\mu\text{F}$	DCX cap.	$C_r = 610\,\text{nF}$
DCX cap.	$C_i = 710\,\text{nF}$ (105 µF for S-Link)		
$S_{3,x}$ freq.	$f_{sw,3} = 144\,\text{kHz}$	XS ind.	$L_s = 1.5\,\mu\text{H}$
$S_{4,x}$ freq.	$f_{sw,4} = 288\,\text{kHz}$	XS volt.	$V_h = 125\,\text{V}$
XS ind.	$L_f = 8.9\,\mu\text{H}$	XS cap.	$C_f = 330\,\text{nF}$
XS cap.	$C_i = 67\,\mu\text{F}$ (100 µF for S-Link)		
Transf.	$N_1 : N_2 : N_3 = 8{:}1{:}1$		

[1] line-to-line rms voltage

650 V (7.6 mΩ) for $S_{1,x}$; Infineon OptiMOS 5 80 V (1.5 mΩ) for $S_{2,x}$; Infineon OptiMOS 5 200 V (7.8 mΩ) for $S_{3,x}$; and EPC2304 200 V (4.65 mΩ) for $S_{4,x}$. A junction temperature of 100° is considered for the evaluation of the on-state resistances. Then, the total conduction losses amount to 130 W (0.52%), allowing to allocate 120 W for the other loss contributions, e.g., switching losses, magnetics, EMI filter.

C. Alternative Realization Options

Thanks to the modularity allowed by the XS-Link concept, different arrangements of the quasi-single-stage three-phase ac-dc converter can be implemented, as shown in **Fig. 5**. In the reported solution, the ac-dc FE stage is kept unvaried, while the BE stage is substituted with N_s-paralleled modules of 2 series-connected DCX-LLC converter to split the dc-link voltage $v_{1\bar{1}}$ across the 2 dc-dc converters. Since single transformer integration is not possible with the used BE stage, N_x-paralleled modules of the XS-Link circuit are connected to the output with a dedicated full-bridge inverter which replaces the DCX-LLC output-side one of the arrangement of **Fig. 4**.

V. Conclusion

To address the rising energy demand driven by the increasing computational needs of next-generation datacenters, advancements in power distribution architectures utilizing wide-bandgap devices like GaN and SiC are crucial. This paper presents a three-phase quasi-single-stage isolated ac-dc converter, designed to step down from 600 V ac (line-to-line rms) to 50 V dc, which leverages the significant advantages of the partial power processing concept.

References

[1] Y. Chen *et al.*, "Data center power supply systems: From grid edge to point-of-load," *IEEE J. Emerg. Sel. Topics Power Electron.*, vol. 11, no. 3, pp. 2441–2456, Jun. 2023.

[2] "ONSEMI unveils complete power solution to improve energy efficiency for data centers," Jun. 2024.

[3] Q. Zhang *et al.*, "A survey on data center cooling systems: Technology, power consumption modeling and control strategy optimization," *J. Syst. Architect.*, vol. 119, p. 102253, Oct. 2021.

[4] J. Huber *et al.*, "Comparative evaluation of MVAC-LVDC SST and hybrid transformer concepts for future datacenters," in *Proc. Int. Power Electron. Conf. (IPEC/ECCE Asia)*, Himeji, Japan, May 2022, pp. 2027–2034.

[5] J. W. Kolar and T. Friedli, "The essence of three-phase PFC rectifier systems—Part I," *IEEE Trans. Power Electron.*, vol. 28, no. 1, pp. 176–198, Jan. 2013.

[6] A. Greifelt *et al.*, "Modular 11kW bidirectional onboard charger with SiC-MOSFET technology for mobile applications," in *Proc. Brazilian Power Electron. Conf. (COBEP)*, Juiz de Fora, Nov. 2017.

[7] K. Pande *et al.*, "Two-sstage on-board charger using bridgeless PFC and half-bridge LLC resonant converter with synchronous rectification for 48V e-mobility," in *Proc. IEEE Ind. Appl. Soc. Annu. Meet.*, Detroit, MI, USA, Oct. 2020.

[8] R. Gadelrab *et al.*, "LLC Resonant Converter with 99% Efficiency for Data Center Server," in *Proc. IEEE Appl. Power Electron Conf. Expo. (APEC)*, Phoenix, AZ, USA, Jun. 2021, pp. 310–319.

[9] P. R. Prakash *et al.*, "GaN-based 400V/48V DC-DC converter with 97% efficiency and PCB magnetics for automotive applications," in *Proc. IEEE Appl. Power Electron. Conf. Expo. (APEC)*, Orlando, FL, USA, Mar. 2023, pp. 3201–3208.

[10] D. Neumayr *et al.*, "P³DCT—partial-power pre-regulated DC transformer," *IEEE Trans. Power Electron.*, vol. 34, no. 7, pp. 6036–6047, Jul. 2019.

[11] X. Sang *et al.*, "Partial power LLC resonant converter with integrated transformer for wide input range," *CSEE J. Power Energy Syst.*, 2023.

[12] T. Chen *et al.*, "A single-stage bidirectional dual-active-bridge AC-DC converter based on enhancement mode GaN power transistor," in *Proc. IEEE Appl. Power Electron. Conf. Expo. (APEC)*, San Antonio, TX, USA, Mar. 2018, pp. 723–728.

[13] G. K. N. Kumar and A. K. Verma, "A two-stage interleaved bridgeless PFC based on-board Charger for 48V EV applications," in *Proc. IEEE Int. Conf. Smart Technol. Power Energy Control (STPEC)*, Bilaspur, Chhattisgarh, India, Dec. 2021.

[14] S. Chaithanya and M. Sindhu, "A PFC based onboard battery charger using isolated full-bridge DC-DC converter for electric vehicle application," in *Proc. IEEE IAS Global Conf. Emerg. Technol. (GlobConET)*, Arad, Romania, May 2022, pp. 581–586.

[15] S. Dusmez *et al.*, "A single-stage three-level isolated PFC converter," in *Proc. IEEE Energy Convers. Congr. Expo. (ECCE USA)*, Pittsburgh, PA, USA, Sep. 2014, pp. 586–592.

[16] N. Hasan and M. A. Saim, "A single stage off-board EV charger based on Cuk topology," in *Proc. 5th Int. Conf. Electr. Engin. Inf. Comm. Technol. (ICEEICT)*, Dhaka, Bangladesh, Nov. 2021.

[17] A. D. Kumar *et al.*, "A single-stage high step-down gain modified cuk-based HPF AC-DC converter for LVEVs chargers," in *Proc. IEEE Int. Conf. Power Electron. Drives Energy Syst. (PEDES)*, Jaipur, India, Dec. 2022.

[18] R. S. Bayliss and R. C. Pilawa-Podgurski, "An input inductor flying capacitor multilevel converter utilizing a combined power factor correcting and active voltage balancing control technique for buck-type AC/DC grid-tied applications," in *Proc. IEEE Workshop Control. Model. Power Electon. (COMPEL)*, Lahore, Pakistan, Jun. 2024.

[19] M. Vazzoler *et al.*, "Isolated active front-end with integrated bidirectional GaN switches for battery chargers," in *Proc. IEEE Appl. Power Electron. Conf. Expo. (APEC)*, Long Beach, CA, USA, Feb. 2024, pp. 1719–1726.

[20] L. Schrittwieser *et al.*, "99% efficient isolated three-phase matrix-type DAB buck–boost PFC rectifier," *IEEE Trans. Power Electron.*, vol. 35, no. 1, pp. 138–157, Jan. 2020.

[21] Infineon Technologies AG, "Infineon announces roadmap for state-of-the-art and energy-efficient power supply units in AI data centers," May 2024.

[22] D. Menzi *et al.*, "Novel S-Link enabling ultra-compact and ultra-efficient three-phase and single-phase operable on-board EV chargers," in *Proc. 24th IEEE Workshop Control. Model. Power Electron. (COMPEL)*, Ann Arbor, MI, USA, Jun. 2023.

[23] H. Yoo and S.-K. Sul, "A new circuit design and control to reduce input harmonic current for a three-phase AC machine drive system having a very small DC-link capacitor," in *Proc. Annu. IEEE Appl. Power Electron. Conf. Expo. (APEC)*, Palm Springs, CA, USA, Feb. 2010, pp. 611–618.

[24] L. Schrittwieser *et al.*, "99% efficient three-phase buck-type SiC MOSFET PFC rectifier minimizing life cycle cost in DC data centers," in *Proc. IEEE Int. Telecom. Energy Conf. (INTELEC)*, Austin, TX, USA, Oct. 2016.

A New Three-phase Multi-mode AC/DC LLC Converter with Output-controlled Active Rectifier (with V2G and G2V functions) For Fast DC Charging Application

Xiaoyi Xia
Dept. of Electrical Engineering and Computer Science
York University
Toronto, Canada
xxia@yorku.ca

John Lam, IEEE Senior Member
Dept. of Electrical Engineering and Computer Science
York University
Toronto, Canada
john.lam@yorku.ca

Abstract— **This paper proposes a new three-phase multi-mode AC/DC LLC resonant converter with an output-controlled active rectifier for electric vehicle (EV) fast DC charging applications. In the proposed approach, two low-frequency bidirectional switches at the primary side in each phase of the converter allow vehicle-to-grid (V2G, i.e. DC/AC mode) and grid-to-vehicle (G2V, i.e. AC/DC mode) modes. When the proposed converter operates in AC/DC mode, each phase consists of an integrated bridgeless boost power factor corrector and an LLC resonant converter at the primary side. Output voltage regulation is managed by switches on the high frequency transformer's secondary side output rectifier in each phase, with soft-switching operation achieved in all the switches. When the proposed three-phase converter operates in DC/AC mode, each phase consists of a half-bridge DC/DC resonant converter cascaded with a half-bridge grid-side inverter. The operation of the proposed converter is explained in this paper. Results from a 20kW, 480V$_{LLrms}$/350Vdc, 130kHz design, and a hardware test on a 130kHz, 250V-output proof-of-concept prototype are presented to validate the functionality of the proposed converter.**

Keywords—DC fast charging, resonant converter, output voltage regulation, zero voltage switching

I. INTRODUCTION

With the growing concern over greenhouse gas emissions and global warming, the attention to transferring the transportation method from internal combustion engine-based vehicles to electric vehicles (EV) has increased in the last ten years. Level 3 DC fast charging is mostly off-board system and converts AC from the grid into DC before it reaches the vehicles, bypassing the onboard converter to achieve much faster speeds [1]. Compared to the 5-10 hours needed for charging a medium-sized EV using level 1 and level 2 charging schemes, the level 3 DC fast charging station significantly reduces the charging time to around 15 to 55 minutes [2]. The converter must have bidirectional power flow capability coupled with high efficiency and power density to support both Vehicle-to-Grid (V2G) and Grid-to-Vehicle (G2V) functionalities in EV charging systems. Various DC/DC converter topologies have been introduced for bidirectional EV charging applications. [3]-[8]. However, most of the existing topologies are for level 1 and level 2 charging,

which are onboard systems with lower power ratings and longer charging times.

A new three-phase multi-mode AC/DC LLC converter is proposed for level 3 DC fast charging system that supports both V2G and G2V functionalities. Fig. 1 shows the proposed converter. The V2G and G2V modes can be selected using two low-frequency switches, S_A and S_B in each phase (see blue lines in Fig. 1), while soft switching operation of switches on the primary side of the transformer is achieved through the LLC resonant circuit. Power factor correction (PFC) and output voltage regulation are the two significant stages for the AC/DC converter utilized in the EV charging station [9]-[12]. In the proposed converter, in AC/DC mode, PFC in each phase is achieved by the input bridgeless boost inductor L_b operating in discontinuous conduction mode (DCM) and output voltage regulation is managed by adjusting the duty cycle of the rectifier switches S_7 on the secondary side. In DC/AC mode, each phase of the proposed converter becomes a half-bridge resonant converter cascaded with a half-bridge grid-side inverter. The simulation results on a 20kW, 480V$_{LLrms}$, 350V, 130kHz output system and some hardware waveforms on a proof-of-concept 130kHz, 250V output prototype are provided to validate the functionalities of the proposed converter.

II. OPERATION PRINCIPLES OF THE PROPOSED CONVERTER

In the proposed AC/DC biconverter topology, the two low-frequency switches S_A and S_B in each phase select the mode of the converter operation, either in V2G (DC/AC mode) or G2V (AC/DC mode). The operation modes of the proposed per-phase converter are shown in Fig. 2, and the operation principles of the V2G and G2V modes are introduced in this section. Since the converter mainly operates to transfer the power from the grid to vehicle [3] [13], the per-phase converter in AC/DC mode operation is analyzed in details in this paper.

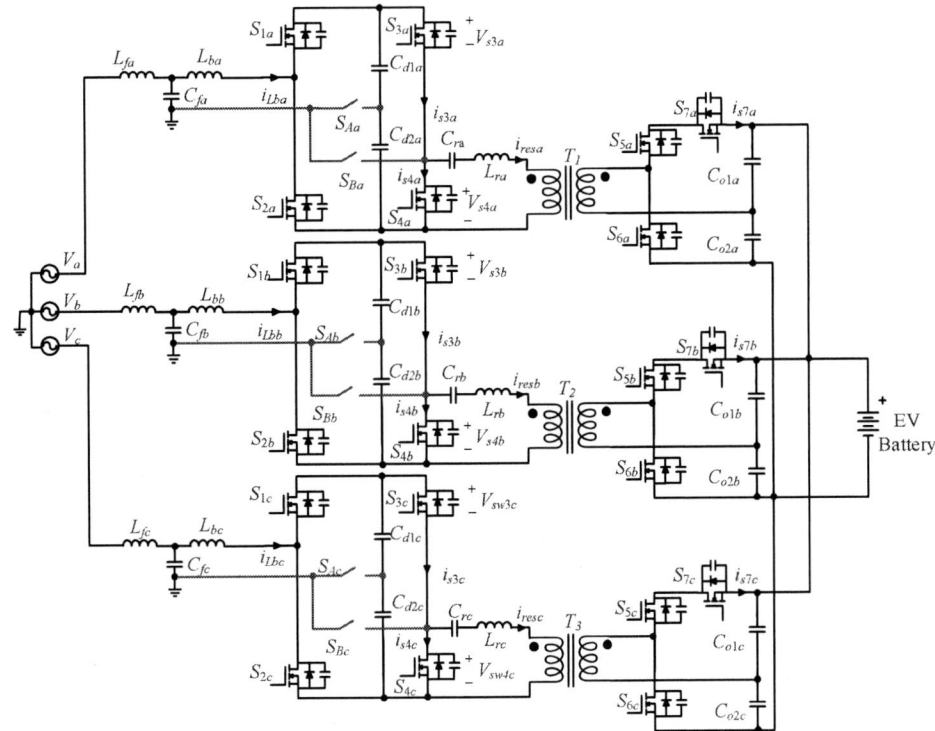

Figure 1:Proposed multi-mode three-phase AC/DC LLC resonant converter with bi-directional active output rectifier.

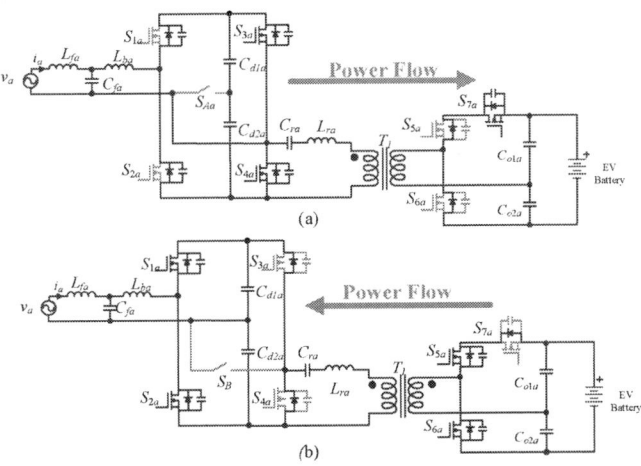

Figure 2: Operation modes of the proposed converter (shown with one phase only): (a) G2V mode with S_{Ba} on. (b) V2G mode with S_{Aa} on.

A. AC/DC Mode (G2V)

Fig. 2 (a) shows the converter (per-phase) operates in G2V mode with S_{Ba} on. In this mode, switches S_{1a} and S_{2a} are turned off, enabling the half-bridge diode rectifier to function and charge the DC link capacitor. The DC link capacitor C_{da}, is the equivalent capacitance formed by the series connection of C_{d1a} and C_{d2a}.

$$C_{da} = \frac{C_{d1a} \times C_{d2a}}{C_{d1a} + C_{d2a}} \quad (1)$$

PFC is achieved by the DCM operation mode of the boost inductor L_{ba} at the grid side. By cooperating with the input LC filter L_{fa} and C_{fa}, the input current properly follows the input voltage. To operate in DCM operation mode, the average current flows through the PFC inductor i_{Lba} needs to be lower than it's current ripple Δi_{Lba}, as shown in (1) - (3).

$$i_{Lba} = \frac{|v_{a(t)}|}{R_{ga}} \quad (2)$$

$$\Box i_{Lba} = \frac{v_{a(t)}}{L_{ba}} D_p T_{sw} \quad (3)$$

$$L_{ba} < \frac{\left(1 - \frac{2V_p}{V_{dc}} R_g\right)}{2f_{sw}} \quad (4)$$

where $v_{a(t)}$ is phase-a input voltage, R_g is the equivalent grid side resistance, V_{dc} is DC link voltage, and T_{sw} is the switching period. In (4), f_{sw} represents the switching frequency. These parameters are defined as follows, where ω_g is the grid angular frequency, P_{av} is the average input power and V_p is the peak of the input grid voltage.

$$v_{a(t)} = V_p \sin(\omega_g t) \quad (5)$$

$$R_g = \frac{v^2_{a(t)}}{P_{av}} \quad (6)$$

$$T_{sw} = \frac{1}{f_{sw}} \quad (7)$$

The average input current (for phase A) for the bridgeless PFC boost converter can be derived as (8) as a function of the duty cycle D_p of the half-bridge switches[14][15].

$$i_a = \frac{V_{dc} T_{sw} D_p^{\ 2} \sin(\omega_g t)}{2 L_{ba} \left(\dfrac{V_{dc}}{v_a} - \sin(\omega_g t) \right)} \qquad (8)$$

The DC/DC LLC resonant converter operates with switches S_3 and S_4 functioning at a 50% duty cycle, in conjunction with resonant components L_{ra} and C_{ra} and a high-frequency transformer. Switches S_{5a} and S_{6a} are in off condition with the anti-parallel diodes conducted. The duty cycle of switch S_{7a} is modulated to maintain a constant output voltage. The output voltage regulation is achieved by using a simple PI controller loop shown Fig. 3.

Fig. 4 and Fig. 5 illustrate the per-phase operation stages and waveforms of the proposed converter operating in G2V mode (S_{Ba} is on). There are six operation stages:

$t_0 < t < t_1$: During this stage interval, the snubber capacitor C_{s3a} is fully discharged to DC link capacitor C_{da} and resonant circuit. At beginning of t_0, the current resonant i_{resa} is negative, hence the diode D_{s3a} is on. The switch voltage v_{sw4a} is clamped to the DC link voltage V_{dca}.

$t_1 < t < t_2$: At this stage, the switch S_{3a} is on. The capacitor C_{s4a} starts charging. Since the negative switch current i_{s3a} flew through the antiparallel diode D_{s3a} in the previous stage, the zero voltage turn on of switch S_{3a} is achieved, providing zero losses during S_{3a} turn on. Now, i_{resa} is flowing the positive direction, transferring the power to resonant circuit and the load.

$t_2 < t < t_3$: At this stage, S_{3a} is turned off, C_{s4a} is discharging to the resonant circuit. C_{s3a} slowly charges until its voltage v_{sw3a} reaches the DC link voltage V_{dca}. Hence, the slow charging action of C_{s3a} allows near zero current turn off for S_{3a}.

$t_3 < t < t_4$: At the beginning of this interval, C_{s4a} is fully discharged and switch voltage v_{sw3a} is clamped to V_{dca}. The DC link capacitor current flows through a resonant circuit and the antiparallel diode D_{s4a} of switch S_{4a}.

$t_4 < t < t_5$: At this stage, S_{4a} is on. Since the negative switch current i_{sw4} flowed through D_{s4a} in the previous stage, the zero voltage turn on of S_{4a} was achieved.

$t_5 < t < t_6$: At this stage, S_{4a} is turned off, C_{s3a} is discharging to DC link capacitor C_{da} and resonant circuit. C_{s4a} is slowly charging until its voltage v_{sw4a} reaches the DC link voltage V_{dca}. Hence, near ZCS turn off for S_{4a} is achieved.

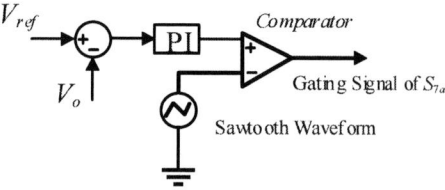

Figure 3:Block diagram of the output voltage regulation control in AC/DC mode.

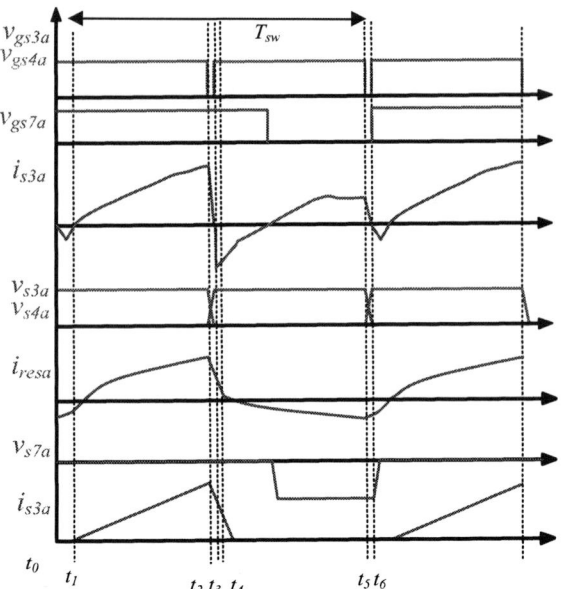

Figure 4: Operation switching waveforms of the proposed converter in AC/DC mode (one phase only).

The voltage gain of the resonant circuit and output rectifier in each phase is given by (9), where Q, ω_r, ω_n are given below. R_{Ds} is given in (10), with D_s representing the duty cycle of S_{7a} and R_o is the output load resistance. The voltage gain plot as a function of relative angular operation frequency for several Q values is shown in Fig. 6 and the two ZVS regions are illustrated in the diagram.

$$G = \frac{V_o}{v_{sw4a}} = \frac{1}{\sqrt{(1 + \frac{1}{k} - \frac{1}{\omega_n^2 k})^2 + (\frac{Q}{\omega_n} - Q\omega_n)^2}} \qquad (9)$$

$$Q = \frac{\omega_r L_{ra}}{R_{Ds}} = \frac{Z_R}{R_{Ds}}, \quad k = \frac{L_{ma}}{L_r}, \quad \omega_r = \frac{1}{\sqrt{C_{ra} L_{ra}}}, \quad \omega_n = \frac{\omega_s}{\omega_r}$$

$$R_{Ds} = \frac{R_o \sqrt{2}}{2\pi D_s} \left[1 - \cos(2\pi D_s) \right] \qquad (10)$$

B. DC/AC Mode (V2G)

Fig. 2 (b) shows the V2G mode with switch S_{4a} is on. In this mode, switch S_{7a} is off and its anti-parallel diode is conducted to enforce the power flow direction from EV battery to the grid side. Switches S_{5a} and S_{6a}, along with the resonant components L_{ra}, C_{ra} and the anti-parallel diodes of switches S_{3a} and S_{4a}, form a DC/DC isolated resonant converter. Soft switching operation of S_{5a} and S_{6a} can be achieved by operating the resonant current (i.e. i_{resa} in phase A) lagging behind the secondary voltage of transformer (T_1).

The grid-tied control is achieved by modulating the gating signals of the half-bridge inverter switches S_{1a} and S_{2a}. A standard grid-tied control scheme, incorporating with current and voltage control loops, can be employed to regulate the gating signals of S_{1a} and S_{2a} for the proposed topology [16]. The block diagram of the control scheme is shown in Fig. 7.

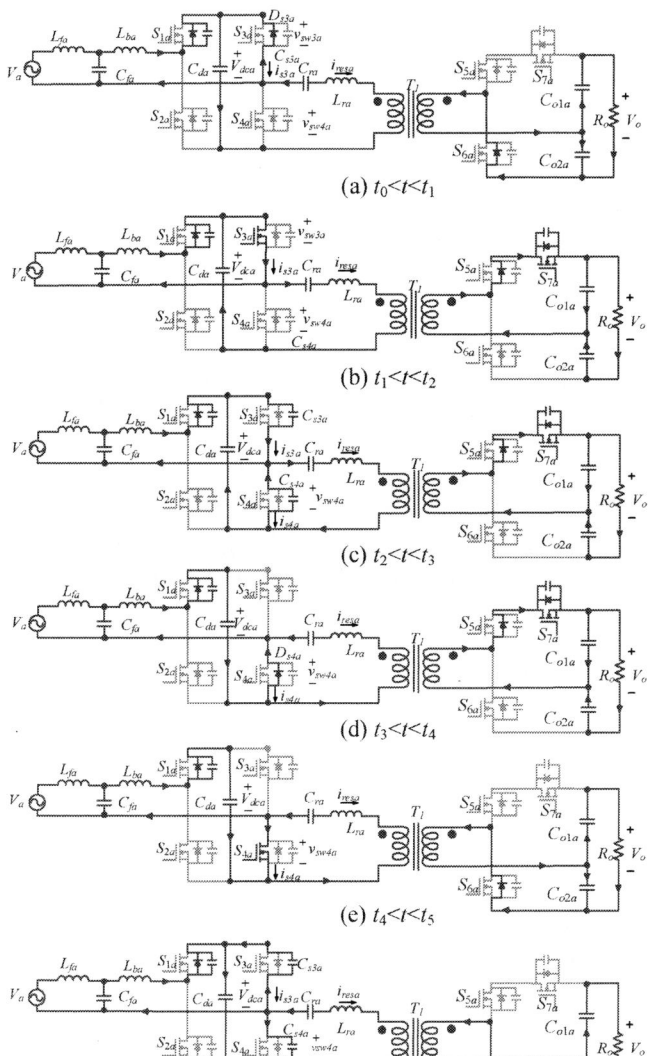

(a) $t_0 < t < t_1$

(b) $t_1 < t < t_2$

(c) $t_2 < t < t_3$

(d) $t_3 < t < t_4$

(e) $t_4 < t < t_5$

(f) $t_5 < t < t_6$

Figure 5: Operating stages of the proposed converter per phase (AC/DC mode)

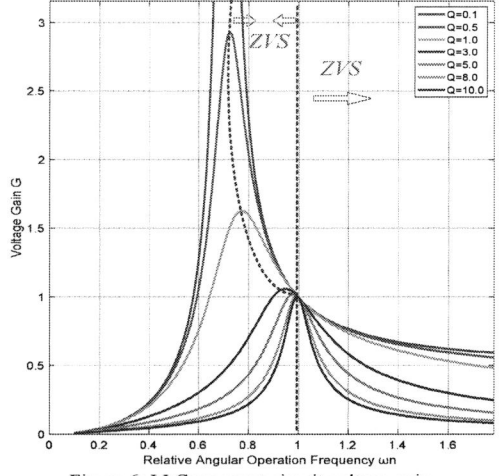

Figure 6: LLC resonant circuit voltage gain.

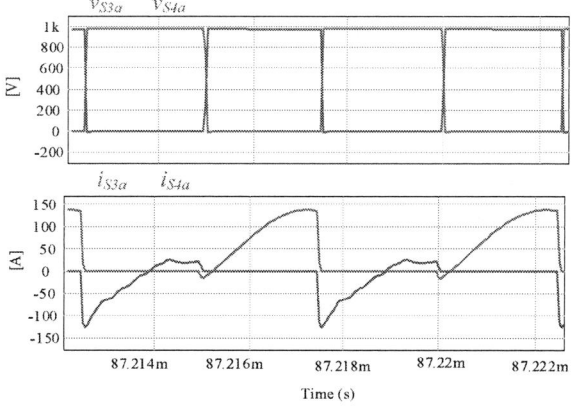

Figure 7 : Schematic block diagram of the grid-tied control in DC/AC mode (for one phase).

III. RESULTS AND PERFORMANCE

To verify the performance of the proposed converter, the proposed circuit is first simulated in PSIM with a switching frequency of 130kHz, a rated power of 20kW, the input AC voltage of $480V_{LLrms}$ and the output voltage of 350V. The system and converter parameters are shown in TABLE I. Simulation results are analyzed in this section.

TABLE I: System specification and per-phase converter parameters (simulation)

Power Rating	20kW
Switching Frequency	130kHz
Input Voltage	$480V_{rms}$
Output Voltage	350V
Input PFC Inductor	33µH
Resonant Inductance	27µH
Resonant Capacitance	1.5µF
Output Capacitance	150µF
Transformer Turns Ratio	2:1
SiC Switches $S_{1a} \sim S_{6a}$	NTH4L013N120M3S
SiC Switches S_{7a}	RFP4137PbF

Fig. 8 shows the voltage and current waveforms of switches S_{3a} and S_{4a}. ZVS turn-on and near ZCS turn-off is achieved by the resonant circuit. Fig. 9 shows the switch current and voltage waveforms of switch S_{7a}, confirming the achievement of soft-switching. Therefore, all the switches utilized in the converter in AC/DC mode achieve the soft-switching operation, significantly enhancing the efficiency of the converter. The DC output voltage at 350V is shown in Fig. 10.

Figure 8: Simulated waveforms in switch S_{3a} and S_{4a} (switching waveforms in each phase).

979-8-3315-1612-3/25 $31.00 © 2025 IEEE

PFC is achieved with the three-phase input current i_a, i_b and i_c in phase with input voltages v_a, v_b and v_c as shown in Fig. 11. The peak efficiency of 94.5% is achieved in the simulation. Fig. 12 shows the per-phase grid voltage and grid current in DC/AC mode. The grid current follows the grid voltage by controlling the gating signal of switch S_{1a} and S_{2a}. The input power factor is > 0.98.

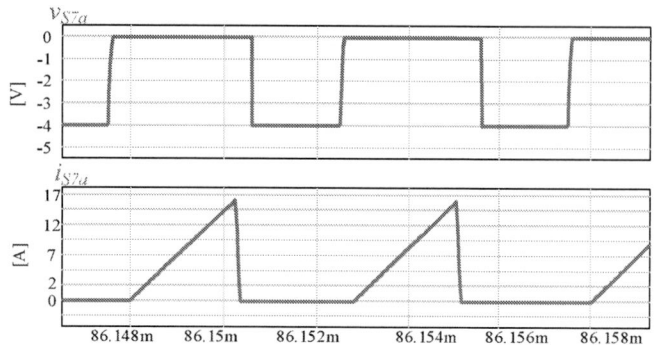

Figure 9: Simulated waveforms of voltage and current in switch S_{7a} (switching waveforms in each phase).

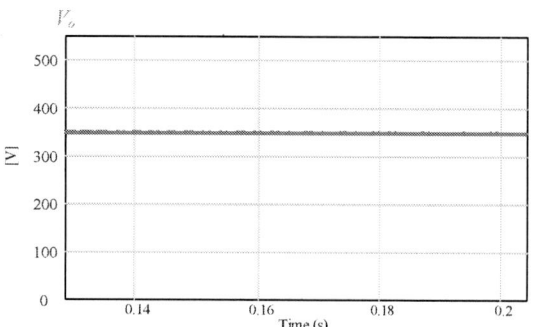

Figure 10: Simulated waveform of output voltage V_o.

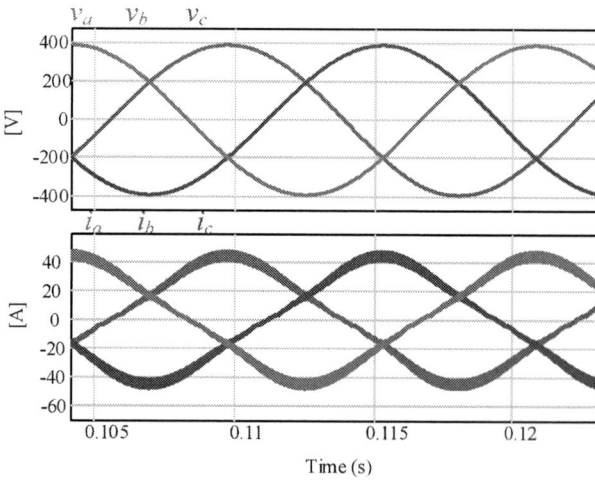

Figure 11: Simulated waveforms of three-phase input voltage and input current in AC/DC mode.

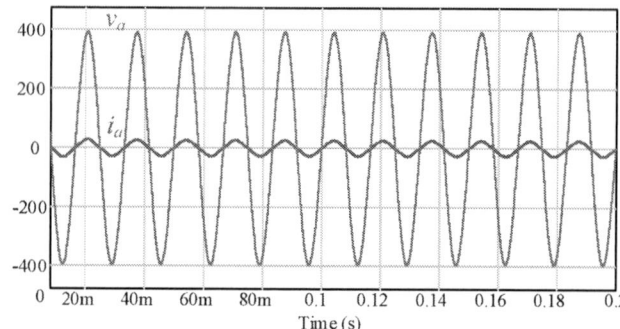

Figure 12: Simulated waveforms of grid current and grid voltage in DC/AC mode. (one phase only)

To further validate and evaluate the performance of the proposed AC/DC converter, a proof-of-concept laboratory-scale hardware prototype of the proposed circuit with a switching frequency of 130kHz and an output voltage of 250V was developed and tested in the laboratory. Fig. 13 – Fig. 20 present the waveforms of one-phase of the developed prototype.

Fig. 13 illustrates the output voltage V_o, the input voltage v_a, the PFC inductor current i_{Lba} and the input current i_a. The PFC inductor current is in DCM operation mode and the input current follows the AC input voltage.

Fig. 14 shows the DCM operation of the PFC inductor current in one phase. Fig. 15 - Fig. 17 show the waveforms of switching voltage and current waveforms in switches S_{3a} and S_{4a} with 50% duty cycle. Fig. 18 provides a zoomed-in view of the waveforms during a switch turn-on. ZVS turn-on is achieved for both switches, as the negative resonant current flows through their anti-parallel diodes prior to turn-on. Additionally, the snubber capacitor connected in parallel to the switches limits the voltage rise after turn-off, enabling near ZCS during turn-off for each switch.

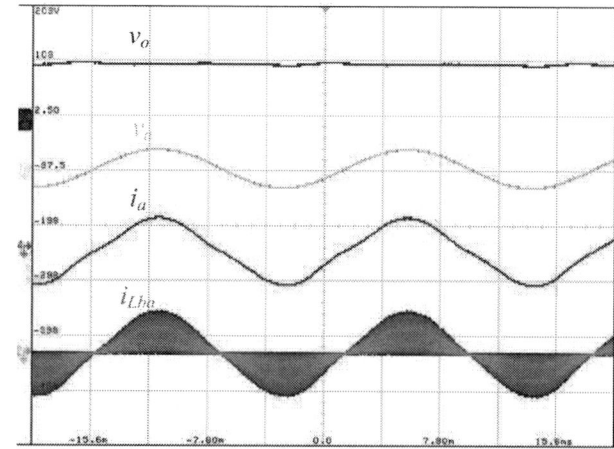

Figure 13: Experimental waveforms of per-phase circuit: output voltage (V_o), per phase input voltage (v_a), input current (i_a) and PFC inductor current (i_{Lba}).

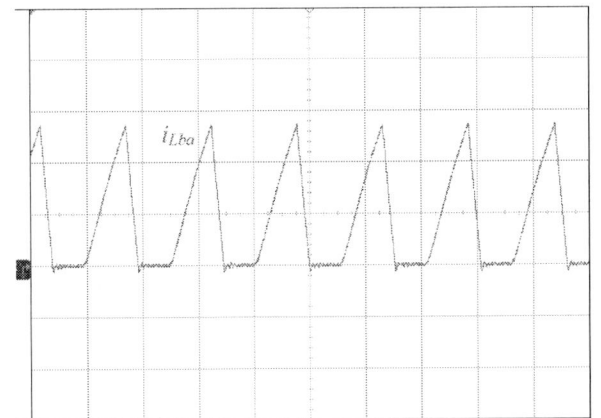

Figure 14: Experimental waveform of per-phase PFC inductor current for a few switching cycles

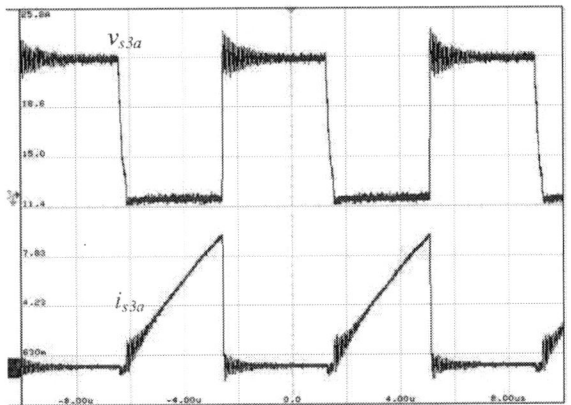

Figure 15: Experimental switching waveforms of switch S_{3a} during the peak of the positive input voltage in one phase of the prototype.

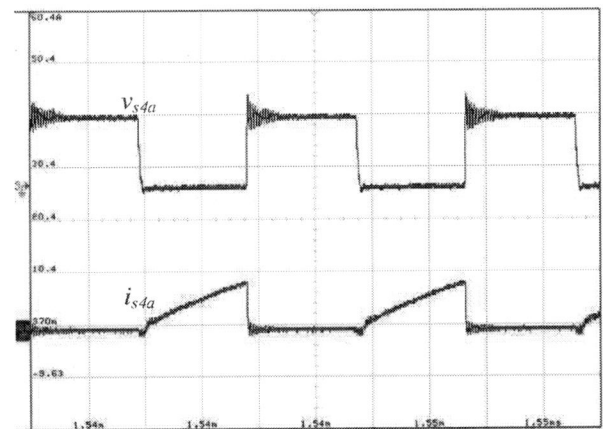

Figure 16: Experimental switching waveforms of switch S_{4a} (duty cycle of 50%) in one phase of the prototype.

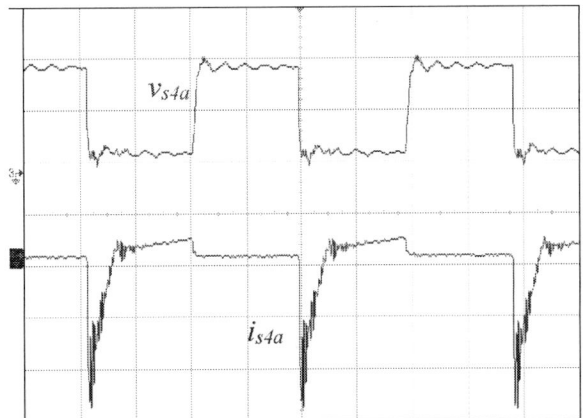

Figure 17: Experimental switching waveforms of switch S_{4a} during the lowest point of the negative input voltage in one phase of the prototype.

Fig. 19 and Fig. 20 show the voltage and current waveforms of switch S_{7a}.with 50% duty ratio and 70% duty ratio respectively. The operation of S_{7a} achieves ZVS during turn-on and ZCS during turn-off. It is observed that the current through S_{7a} reduced to zero prior to switch turn-off, occurring around $0.5T_s$. This behavior arises because power flows from the grid to the load only when both S_{3a} and S_{7a} are conducting, with S_{3a} also operating at a 50% duty cycle.

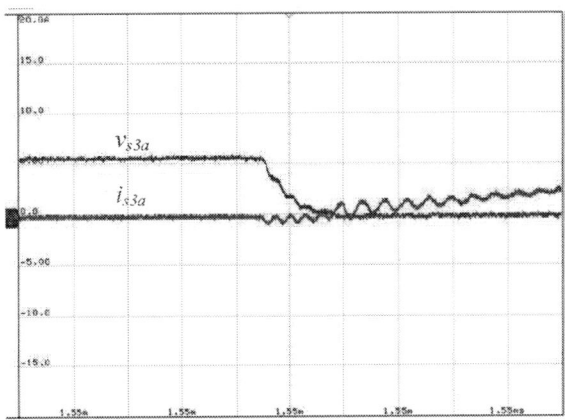

Figure 18: ZVS turn-on transition in one switch.

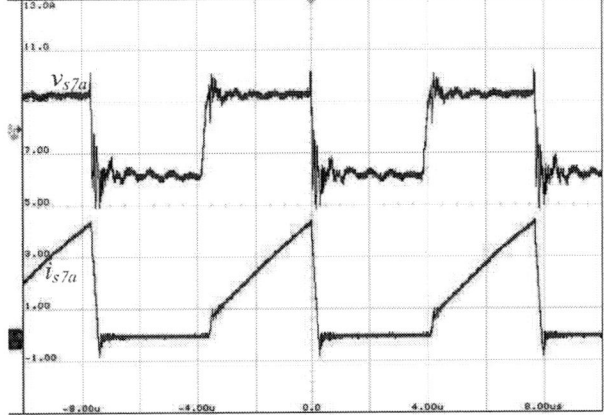

Figure 19: Experimental switching waveforms of switch S_{7a} (duty cycle of 50%).

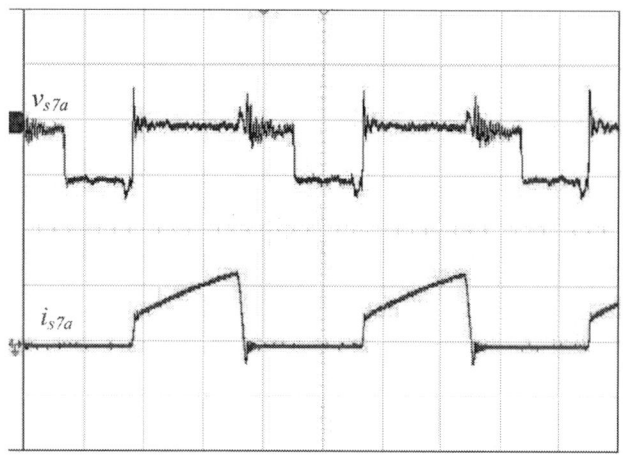

Figure 20: Experimental switching waveforms of switch S_{7a} (duty cycle of 70%).

CONCLUSIONS

A novel three-phase multi-mode AC/DC LLC bidirectional converter with an output-controlled rectifier, designed for DC fast charging applications, has been presented. The converter supports both V2G and G2V operations, controlled by two low-frequency switches. The operation principles of the converter are explained in detail, with particular emphasis on the AC/DC mode. All the switches utilized in the converter are able to achieve soft-switching operation. The simulation of the proposed converter operating at 130kHz, $480V_{LLrms}$, 350V was conducted and the results were provided. Finally, a proof-of-concept prototype of the proposed AC/DC converter with 130kHz and 250V output was developed and tested, the results and some hardware waveforms have been presented and discussed in this paper.

REFERENCES

[1] S. Srdic and S. Lukic, "Toward Extreme Fast Charging: Challenges and Opportunities in Directly Connecting to Medium-Voltage Line," in IEEE Electrification Magazine, vol. 7, no. 1, pp. 22-31, March 2019.

[2] H. Tu, H. Feng, S. Srdic and S. Lukic, "Extreme Fast Charging of Electric Vehicles: A Technology Overview," in IEEE Transactions on Transportation Electrification, vol. 5, no. 4, pp. 861-878, Dec. 2019.

[3] D. Das, N. Weise, K. Basu, R. Baranwal and N. Mohan, "A Bidirectional Soft-Switched DAB-Based Single-Stage Three-Phase AC–DC Converter for V2G Application," in IEEE Transactions on Transportation Electrification, vol. 5, no. 1, pp. 186-199, March 2019.

[4] H. -J. Chiu and L. -W. Lin, "A bidirectional DC–DC converter for fuel cell electric vehicle driving system," in IEEE Transactions on Power Electronics, vol. 21, no. 4, pp. 950-958, July 2006.

[5] U. R. Prasanna, A. K. Singh and K. Rajashekara, "Novel Bidirectional Single-phase Single-Stage Isolated AC–DC Converter With PFC for Charging of Electric Vehicles," in IEEE Transactions on Transportation Electrification, vol. 3, no. 3, pp. 536-544, Sept. 2017.

[6] Y. Du, S. Lukic, B. Jacobson and A. Huang, "Review of high power isolated bi-directional DC-DC converters for PHEV/EV DC charging infrastructure," 2011 IEEE Energy Conversion Congress and Exposition, Phoenix, AZ, USA, 2011, pp. 553-560.

[7] X. Zhou, S. Lukic, S. Bhattacharya and A. Huang, "Design and control of grid-connected converter in bi-directional battery charger for Plug-in hybrid electric vehicle application," 2009 IEEE Vehicle Power and Propulsion Conference, Dearborn, MI, USA, 2009, pp. 1716-1721.

[8] H. -J. Chiu and L. -W. Lin, "A bidirectional DC–DC converter for fuel cell electric vehicle driving system," in IEEE Transactions on Power Electronics, vol. 21, no. 4, pp. 950-958, July 2006.

[9] S. Srdic, X. Liang, C. Zhang, W. Yu and S. Lukic, "A SiC-based high-performance medium-voltage fast charger for plug-in electric vehicles," 2016 IEEE Energy Conversion Congress and Exposition (ECCE), Milwaukee, WI, USA, 2016.

[10] C. A. Suarez B, G. A. Melo and C. A. Canesin, "A 7.5 kW Off-board Three-Phase Fast Charger Prototype for Electric Vehicles," 2018 20th European Conference on Power Electronics and Applications (EPE'18 ECCE Europe), Riga, Latvia, 2018.

[11] J. Shabana and G. Renjini, "Analysis and design of bridgeless buck PFC rectifier with single inductor," 2014 International Conference on Circuits, Power and Computing Technologies [ICCPCT-2014], Nagercoil, India, 2014.

[12] M. Mahdavi and H. Farzanehfard, "Bridgeless SEPIC PFC Rectifier With Reduced Components and Conduction Losses," in IEEE Transactions on Industrial Electronics, vol. 58, no. 9, pp. 4153-4160, Sept. 2011.

[13] E. Azimi and Q. Wei, "Direct Connection of EV Fast Charging Station to Medium Voltage AC Grid," 2024 IEEE 10th International Power Electronics and Motion Control Conference (IPEMC2024-ECCE Asia), Chengdu, China, 2024.

[14] A. R. Prasad, P. D. Ziogas and S. Manias, "An active power factor correction technique for three-phase diode rectifiers," in IEEE Transactions on Power Electronics, vol. 6, no. 1, pp. 83-92, Jan. 1991.

[15] S. Derakhshan and J. Lam, "A Bridgeless, Stacked Switch-Pair-Based AC/DC On-Board High-Voltage EV Charger With Minimal Storage Capacitance Featuring Continuous Secondary-Side High-Frequency Rectifier Control," in IEEE Journal of Emerging and Selected Topics in Power Electronics, vol. 12, no. 4, pp. 3439-3455, Aug. 2024

[16] B. Wu, Power Conversion and Control of Wind Energy Systems. Hoboken, NJ, USA:Wiley/IEEE Press, 2011.

Capacitorless Notch Resonant Converters for Miniaturized LLCLC Resonant Converters in Electric Vehicle Charging Applications

Haitham Kanakri
Electrical Engineering Department
Purdue University-Indianapolis
Indianapolis, USA
hkanakri@purdue.edu

Euzeli Cipriano Dos Santos Jr.
Electrical Engineering Department
Purdue University-Indianapolis
Indianapolis, USA
edossant@purdue.edu

Maher Rizkalla
Electrical Engineering Department
Purdue University-Indianapolis
Indianapolis, USA
mrizkall@purdue.edu

Abstract—**This paper aims to establish the foundation for miniaturizing EV chargers by integrating stray capacitances into the LLC-LC converter operation. Capacitors are an indispensable part of power electronics converters that are widely employed in electric vehicle (EV) chargers. They are, however, particularly prone to wear-out mechanisms and failure modes, contributing to increased monitoring and replacement costs for these systems. Traditionally, stray capacitance in power electronics has been perceived as unavoidable and undesirable. However, this paper challenges this conventional view by advocating for the integration of stray capacitance as intrinsic components within the system's topology, particularly in LLC resonant converters utilizing the notch resonant converter on their secondary side. By employing alternative materials with high dielectric constants, such as calcium copper titanate (CCTO), specific parasitic components, notably intra-capacitance, are intentionally enhanced. This innovative approach reduces the dependence on external discrete capacitors and facilitates the development of highly reliable converters for electric vehicles (EVs). A comparison between the conventional LLC resonant converter and LLC-LC resonant converter is established. Using the proposed concept, the LLC-LC converter requires fewer passive components and demonstrates advantages in protection against short circuit and heavy load conditions. To validate the proposed concepts, we have conducted Ansys Maxwell and Ansys Simplorer co-simulations, supported by both simulation results and theoretical analysis. The proposed LC notch resonant converter, with $L_p = 539.97$ uH and $C_p = 9.09$ uF, showed good agreement between simulation data and the mathematical model detailed in the paper, thus validating the proposed design.**

Index Terms—**notch resonant filter, LLC-LC resonant converter, calcium copper titanate (CCTO), intra capacitance, miniaturizing EV chargers.**

I. INTRODUCTION

Electric vehicles (EVs) face multiple technical challenges in their widespread acceptance. These include prolonged charging times, range anxiety, and inefficiencies in current EV chargers [1]. Resonant converters have recently gained attention for their potential to address these issues. By enabling fast charging and reducing losses through zero voltage and current switching, resonant converters offer a solution to minimize

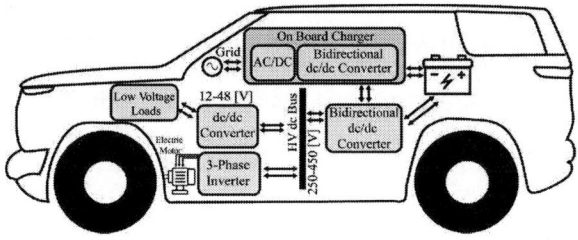

Fig. 1. The electric vehicle power train block diagram outlines the simplified system block diagram of a universal on-board two-stage battery charger.

range anxiety and improve EV charger efficiency [2]–[4]. EV chargers can be classified based on the output power they deliver, two-stage EV chargers are classified into Level-1, Level-2, and Level-3 chargers [5], [6]. Level-1 chargers operate on a common household circuit, supporting up to 120 V AC and up to 16 A [6]. Level-2 chargers specifically designed for rated use with up to 240 V AC, up to 60 A, and can handle power loads of up to 14.4 kW [6], whereas Level-3 chargers are rated greater than 14.4 kW. Level-3 chargers are typically fast chargers if they charge an average electric vehicle battery pack in 30 minutes or less [6].

The design of EV on-board charger illustrated in Fig. 1 faces crucial challenges such as high cost, and limited power density [1]. Moreover, to maintain the ripple magnitude within specified limits in the output of the ac-dc converter shown in Fig. 1, a suitable-value of dc-link capacitor is connected across its output. This capacitor stabilizes the ripple, ensuring consistent power delivery. The output of the dc-link capacitor is then directed to the isolated dc-dc converter, responsible for charging the EV battery. Galvanic isolation is provided by the isolated dc-dc converter, crucial for safety between the high-voltage side of the battery pack and the grid-sided dc-link. However, capacitors are susceptible to wear-out mechanisms and failure modes, which can result in increased monitoring and replacement costs for systems utilizing them [7]. However, their characteristics are greatly influenced by factors such as frequency, temperature, and aging time. As a result, their lifes-

979-8-3315-1612-3/25 $31.00 © 2025 IEEE

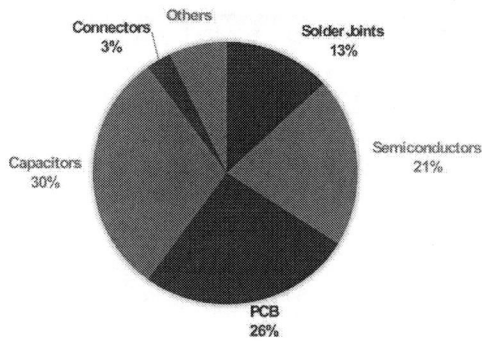

Fig. 2. The distribution of failures among major components in power electronics converters typically includes electrolytic capacitors with highest failure contribution of 30% [9].

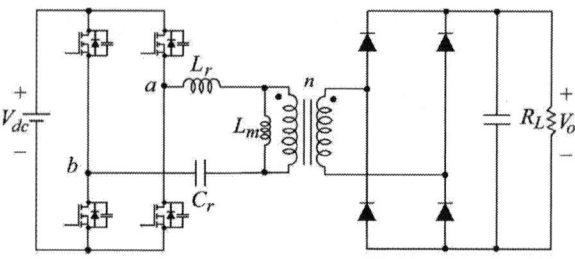

Fig. 3. The conventional LLC resonant converter consists of a full H-bridge, LLC resonant tank formed by resonant capacitor C_r, resonant inductor L_r and the magnetizing inductance of the transformer, and full bridge rectifier circuit.

pan significantly impacts the reliability of electrical systems [8]. As depicted in Fig. 2, capacitors contribute to the highest failure percentage of 30% among major components in power electronics converters, highlighting their significant impact on reliability degradation [9]. In the past several decades significant academic and industrial research efforts have focused on improving capacitor reliability [10]. This motivation stems from the rising need for highly dependable power electronic products in industrial and automotive products [11]. The LLC resonant converter shown in Fig. 3 is preferred in EV chargers due to its wide output voltage gain. It can operate in buck and boost modes, offering versatility. High efficiency is achieved through zero voltage switching (ZVS) and zero current switching (ZCS) capabilities [12]. However, the LLC resonant converter still faces challenges that need addressing. For instance, in the conventional LLC resonant converters, the voltage gain curve tends to flatten when the switching frequency f_s exceeds the resonant frequency f_r of the resonant tank [13]. Fig. 4 shows the LLC resonant inverter voltage gain that is given in equation (1) as a function of the normalized switching frequency f_n given in equation (3). Fig. 4(a) presents the LLC resonant inverter voltage gain at various load quality factor Q and fixed magnetic ratio $k = 6$, while Fig. 4(b) presents the LLC resonant inverter voltage gain at various magnetic ratio k and fixed quality factor $Q = 0.2$. The magnetics ratio k and the load quality factor Q are given in equations (2) and (3). This can lead to high voltage and current stresses during startup if the startup switching frequency isn't sufficiently high [13]. Another issue to address is the resonant converter loss of the ZVS when the switching frequency f_s goes down below the resonant frequency f_r. Consequently, if f_s is much greater than f_r, voltage regulation weakens due to rectifying diodes junction capacitance. Moreover the efficiency in LLC-based converters drops with increasing the difference between f_s and f_r [14].

$$Gain = \frac{1}{\sqrt{(1 + \frac{1}{k}(1 - \frac{1}{f_n^2}))^2 + Q^2(f_n - \frac{1}{f_n})^2}} \quad (1)$$

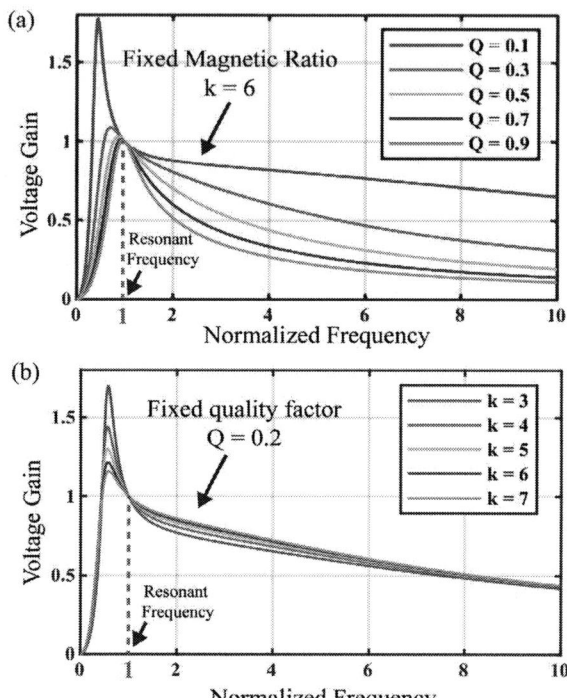

Fig. 4. LLC resonant inverter voltage gain as a function of the normalized switching frequency f_n. (a) At various load quality factor Q and fixed magnetic ratio $k = 6$. (b) At various magnetic ratio k and fixed quality factor $Q = 0.2$.

$$f_{rs} = \frac{1}{2\pi\sqrt{C_r L_r}}; Q = \frac{\sqrt{L_r/C_r}}{R_{ac}}; \quad (2)$$

$$R_{ac} = \frac{8}{\pi^2} n^2 R_L; k = \frac{L_m}{L_r}; f_n = \frac{f_s}{f_r} \quad (3)$$

Solutions proposed in technical literature for the LLC resonant converter challenges vary. These include employing reconfigurable circuits [12], modifying control and modulation strategies [15] and adjusting the resonant tank circuit and incorporating a notch resonant filter into the LLC converter structure [13]. The voltage gain of a converter shown in Fig. 5 can potentially drop to zero when it operates at the parallel resonant frequency of the notch filter and achieves higher voltage gain [13] as observed in Fig. 6.

In EV applications there is a consistent effort in shrinking

979-8-3315-1612-3/25 $31.00 © 2025 IEEE

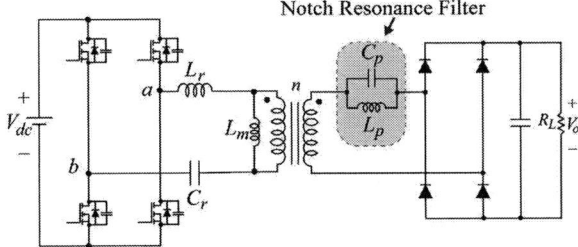

Fig. 5. The conventional LLC resonant converter with LC notch resonant filter attached on it's secondary side (LLC-LC) proposed in [13].

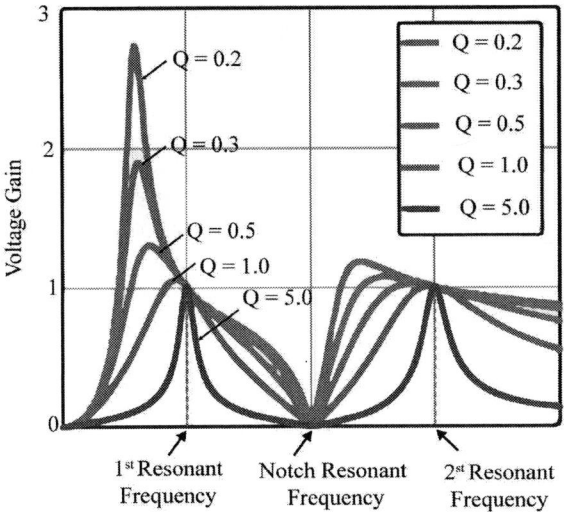

Fig. 6. The voltage gain of the conventional LLC resonant inverter with LC notch resonant filter attached on it's secondary side proposed in [13].

the power supply sizes while maintaining high power density [16]. For on board chargers, the demand persists to achieve greater output power with less physical volume [17]. Advancements in planar magnetic technology have contributed to the development of miniature magnetic components for LLC converters [18].

II. THE PROPOSED CAPACITORLESS PARALLEL LC NOTCH RESONANT FILTER

Notch resonance is established by incorporating an LC element in parallel, positioned in series between the input and output. Fig. 7 illustrates the fundamental notch resonant elements C_p, L_p and their corresponding voltage gain curves. The parallel LC in notch resonant elements transforms into an open circuit at the LC resonant frequency, causing the voltage gain to drop to zero at this frequency. This distinctive open circuit behavior of the NR elements generates a dip in the gain curve, as depicted in Fig. 7. Such a dip in the middle of the gain curve is advantageous for the operation of the LLC-LC converter, facilitating the reshaping of the voltage gain above the resonant frequency and the establishment of a zero voltage gain as illustrated in Fig. 6. The basic notch resonant voltage

gain G_{nr} is formulated in (4) and the corresponding notch resonant frequency f_{nr} is given in (5).

$$G_{nr} = \frac{R_L + s^2 R_L L_p C_p}{R_L + s L_p + s^2 R_L C_p L_p} \tag{4}$$

$$f_{nr} = \frac{1}{2\pi\sqrt{L_p C_p}} \tag{5}$$

The notch resonance structure resembles an inductor with stray capacitance in parallel, facilitating the integration of stray capacitance within the notch resonant structure, thus eliminating the need for external capacitors. Embracing passive integration and "domesticating" parasitic elements will create a new era in power converter technology. This paradigm shift promises higher power-density ratings, driving advancements in electric vehicles, compact power processing units, and renewable energy systems. This paper intentionally centers on augmenting parasitic components in the notch resonant structure to fulfill capacitance requirements in power converters. The hypothesis is that integrating high dielectric material-based thin-films, fabricated through nanotechnology, into planar magnetics, will enable the development of a capacitorless electronics converter family, achieving a two-fold increase in power density. This approach is a departure from the conventional industry practices in power converter design. This approach diverges from the conventional industry practices, where stray capacitances are typically treated as unwanted and undesired elements in power converter structures. The proposed electromagnetic device is investigated via simulation using Ansys PEmag and Ansys Twin-Builder (Simplorer), both based on finite element analysis solutions. This comprehensive analysis aims to demonstrate the advantages and potential performance enhancements offered by the newly proposed planar electromagnetic device in power converter applications. The winding resistance $R(\omega)$ and winding reactance $X(\omega)$, illustrated in Fig. 8(a), are calculated using the following equations [19]:

$$R(\omega) = c_0 + \sum_{i=1}^{h} \frac{\omega^2 c_{2i}}{\omega^2 + c_{2i+1}^2} \tag{6}$$

$$X(\omega) = \omega(c_1 + \sum_{i=1}^{h} \frac{c_{2i} c_{2i+1}}{\omega^2 + c_{2i+1}^2}) \tag{7}$$

Equations (6) and (7) reveal that the inductor winding can effectively be represented as a 2D Foster network, with the network's complexity dependent on the number of partial fractions in the generalized impedance model, as discussed in [19]. The coefficients c_0, c_{2i}, and c_{2i+1} represent partial fraction coefficients, which are determined through numerical estimation utilizing an impedance matching method aimed at minimizing estimation error criteria, as highlighted in [19]. Here, ω signifies the angular frequency, playing a crucial role in the characterization of the inductor winding and its corresponding electrical behavior. Fig. 8(a) shows that the proposed electromagnatic device can be effectively modeled as an inductor in parallel with the intra capacitance. The

979-8-3315-1612-3/25 $31.00 © 2025 IEEE

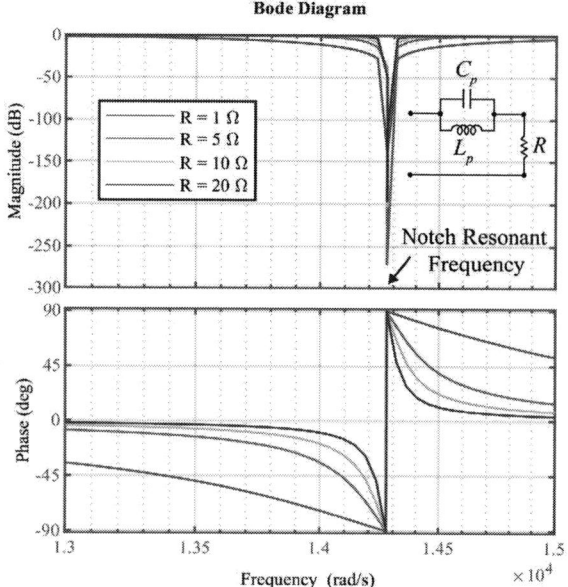

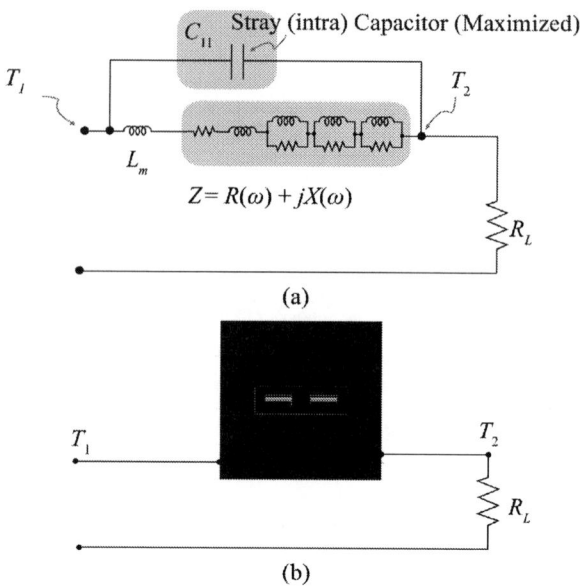

Fig. 7. Simple parallel notch resonance circuit with corresponding voltage gain curves at various load conditions. The notch resonant parameters are $L_p = 539.97$ uH and $C_p = 9.09$ uF.

Fig. 8. (a) The equivalent circuit of the proposed electromagnetic device functions as a notch resonant filter without needing an external capacitor. C_{11} represents the stray intra capacitance, which is maximized using CCTO material to serve as the notch resonant element. (b) The 2D cross-sectional view (FEA 2D block model) of the proposed device. It displays a clear depiction of the proposed device terminals T_1 and T_2 utilized in constructing the proposed notch filter circuit.

intra capacitance C_{11} can be determined based on the electric energy stored within a given volume E_s, as outlined in (8) and discussed in [20], where D and E are the electric flux density and intensity respectively.

$$Es = \frac{1}{2} \iiint_V \bar{D}.\bar{E}\, dv \qquad (8)$$

C_{11} is used to model the global energy stored in the inductor winding [20]. The solution for C_{11} is obtained by solving (9), via finite element analysis solver in Ansys Maxwell.

$$C_{11} = \frac{1}{V_1^2} \iiint_v E_1 D_1\, dv \qquad (9)$$

Fig. 8(b) shows the 2D cross-sectional view (FEA 2D block model) of the proposed device displays a clear depiction of the proposed device terminals T_1 and T_2 utilized in constructing the proposed notch filter circuit. Fig. 9(a) provides detailed illustration of the proposed electromagnetic device layer arrangement, with particular emphasis on the CCTO layer. Fig. 9(b) shows the detailed construction of the proposed electromagnetic device demonstrates the implementation of non-interleaving layer structure with series connected winding, and integrating a 0.01 mm thick CCTO layer to increase the intra-capacitance. Also, Fig. 9(b) shows the 3D representation of the proposed device, utilizing the Ferroxcube EE32/6/20 core composed of 3C90 ferromagnetic material with a total of 23 layers. It also incorporates copper, FR4, and CCTO layers, all employed in the construction of the device.

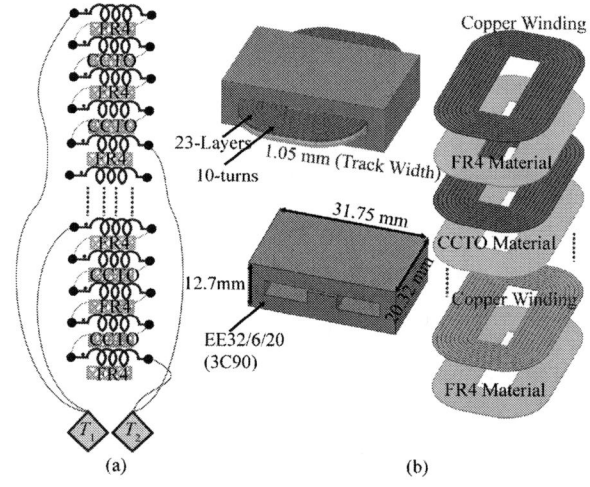

Fig. 9. (a) Detailed illustration of the proposed electromagnetic device layer arrangement, with particular emphasis on the CCTO layer. (b) The detailed construction of the proposed electromagnetic device. It demonstrates the implementation of non-interleaving layer structure with series connected winding, and integrating a 0.01 mm thick CCTO layer to increase the intra-capacitance. The 3D representation of the proposed device, including dimensions, utilizes a Ferroxcube EE32/6/20 core composed of 3C90 ferromagnetic material. A total of 23 layers, incorporating copper, FR4, and CCTO layers, are employed in the construction of the device.

III. DESIGN EXAMPLE, SIMULATION, EXPERIMENTAL RESULTS AND COMPARISON WITH CONVENTIONAL LLC RESONANT CONVERTER

The proposed concept of the capacitorless converter is validated through a comparison between the conventional LLC resonant converter voltage gain and the LLC-LC resonant converter incorporating the proposed notch resonance. Implementing the proposed concept will essentially boost the voltage gain of the LLC resonant converter within a minimal switching frequency range. Moreover, it will reduce the number of passive components used in the LLC-LC converter, and potentially will increase the converter efficiency. Moreover, the proposed concept will provide a protection against the short circuit output issue found in the conventional LLC converter. The conventional LLC converter, as depicted in Fig. 3, has design parameters listed in TABLE I. According to the design equation (2), the required resonant frequency to operate the LLC resonant converter at unity gain is 4.84 kHz. Selecting the resonant frequency as the switching frequency will ensure the maximum efficiency of the LLC converter, and unity gain will be achieved as the transformer turns ratio is equal to one. This ensures that the buck/boost gain is contributed from the LLC tank, ensuring fairness in comparison with the LLC-LC converter. Fig. 10 shows the experimental setup of the conventional LLC resonant converter highlighting the main components, while Fig. 11 and Fig. 12 show the simulation and experimental results of the conventional LLC converter respectively. Fig. 11 shows from top to bottom: v_{ab} is the output of the H-bridge inverter, i_r is the resonant current of the LLC tank, v_o is the output voltage and i_o is the output current given the load resistor = 10 Ω. Fig. 12 presents the same set of results verified experimentally. Note that the output voltage is below the expected voltage of 20 V due to the voltage drop across the full bridge diode rectifier and the winding resistance of the LLC transformer. The switching frequency can be decreased below the resonant frequency to boost the output voltage and achieve the desired output gain. Both converters, the LLC and LLC-LC, will exhibit similar behavior when operating at the frequency of the first resonant frequency indicated in Fig. 6. However, the advantage of the LLC-LC converter becomes apparent when the LLC resonant converter operates at a very heavy load or under short circuit conditions. In such situations, the LLC-LC converter provides short circuit output protection, thereby reducing the circulating energy of the resonant tank as well. The proposed LC notch resonant converter circuit components in this paper are in this paper is $L_p = 539.97$ uH and $C_p = 9.09$ uF. In this case the notch resonant frequency is equal to 3.1 kHz. Fig. 13 shows the resonant tank current when the proposed conventional LLC and LLC-LC operates at heavy load (near short circuit condition) with load resistor = 0.1 Ω. The advantage of utilizing the proposed concept is evident in its ability to maintain the resonant tank current at a lower value, thereby offering protection against short circuit conditions.

IV. CONCLUSION

This paper aims to establish the foundation for miniaturizing electric vehicle chargers by integrating stray capacitance into the operation of the LLC-LC converter. This innovative approach involves reducing or eliminating the dependency on the external capacitors typically utilized in the notch resonant

TABLE I
CONVENTIONAL LLC RESONANT CONVERTER DESIGN PARAMETERS

Input Voltage: V_{dc}	20 V
Output Voltage: V_o	20 V
Transformer's Turns Ratio: n	1:1
Resonant Capacitor: C_r	10.8 uF
Resonant inductor: L_r	100 uH

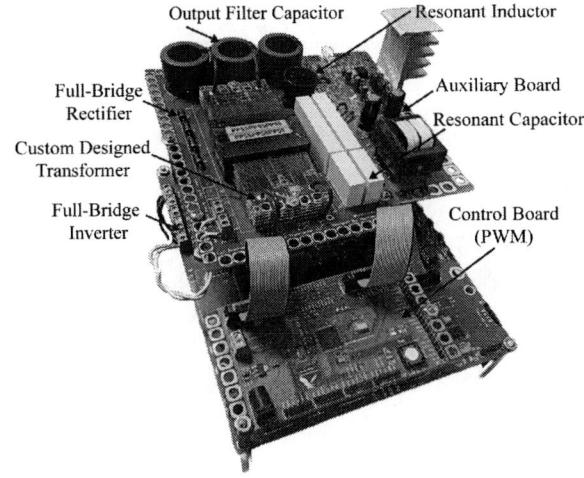

Fig. 10. Experimental setup highlighting the Conventional LLC resonant converter main components.

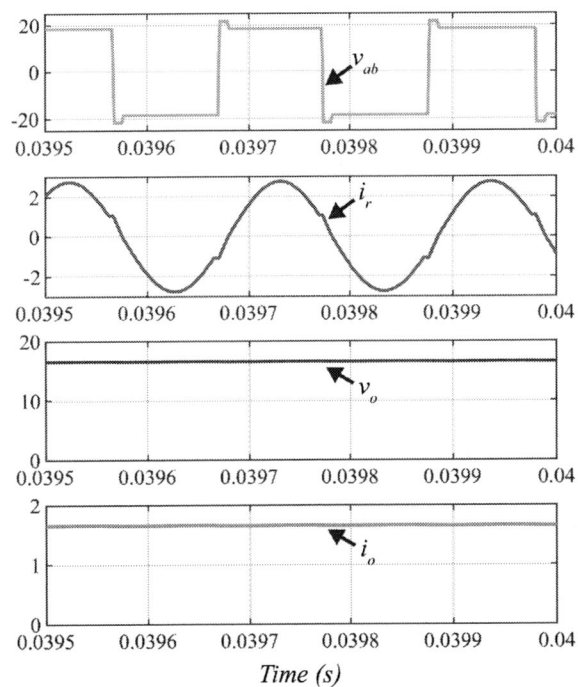

Fig. 11. Simulation results of the conventional LLC converter. From top to bottom: v_{ab} the output of the H-bridge inverter. i_r the resonant current of the LLC tank. v_o the output voltage. i_o is the output current given the load resistor = 10 Ω.

filter. Furthermore, it provides a comprehensive finite element analysis solution and presents a design example to exemplify

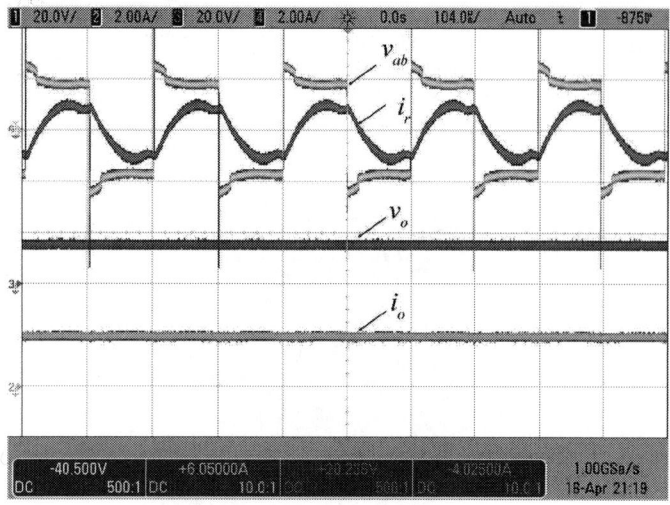

Fig. 12. Experimental results of the conventional LLC converter. From top to bottom: v_{ab} the output of the H-bridge inverter. i_r the resonant current of the LLC tank. v_o the output voltage. i_o is the output current given the load resistor = 10 Ω.

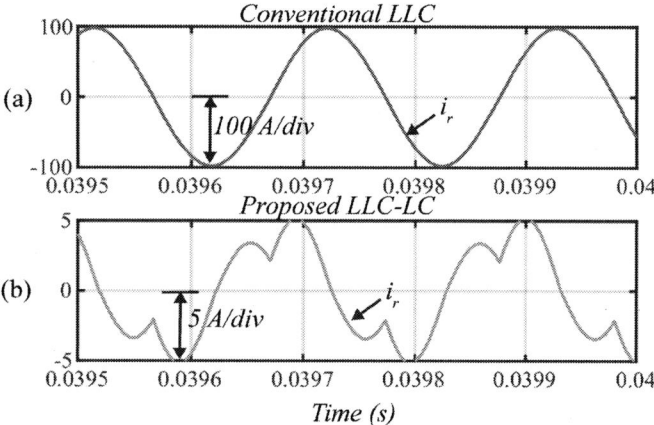

Fig. 13. A comparison between the conventional LLC and the proposed LLC-LC resonant tank current under heavy load condition (near short circuit). (a) Conventional LLC resonant converter. (b) Proposed LLC-LC converter.

the practical application of the proposed concept, with a focus on notch resonance design. A comparison between the conventional LLC resonant converter and the LLC-LC resonant converter is established. Utilizing the proposed concept, the LLC-LC converter necessitates fewer passive components and showcases advantages in protection against short circuit and heavy load conditions. Simulation and experimental results, supported by theoretical analysis, are provided to validate the proposed concepts. We conducted the system simulations using Ansys Maxwell and Ansys Simplorer, supported by experimental results and theoretical analysis. The proposed LC notch resonant converter, with $L_p = 539.97$ uH and $C_p = 9.09$ uF, demonstrated good agreement between the simulation data and the mathematical model detailed in the paper, thereby validating the proposed design. The cost comparison between

the conventional and proposed configurations will be the subject of future research.

REFERENCES

[1] S. Deshmukh, A. Iqbal, S. Islam, I. Khan, M. Marzband, S. Rahman, and A. M. Al-Wahedi, "Review on classification of resonant converters for electric vehicle application," *Energy reports*, vol. 8, pp. 1091–1113, 2022.

[2] Y. Xuan, X. Yang, W. Chen, T. Liu, and X. Hao, "A novel three-level cllc resonant dc–dc converter for bidirectional ev charger in dc microgrids," *IEEE Transactions on Industrial Electronics*, vol. 68, no. 3, pp. 2334–2344, 2021.

[3] M. Kwon and S. Choi, "An electrolytic capacitorless bidirectional ev charger for v2g and v2h applications," *IEEE Transactions on Power Electronics*, vol. 32, no. 9, pp. 6792–6799, 2017.

[4] C. Bai, C. Han, B.-H. Kwon, and M. Kim, "Highly efficient bidirectional series-resonant dc/dc converter over wide range of battery voltages," *IEEE Transactions on Power Electronics*, vol. 35, no. 4, pp. 3636–3650, 2020.

[5] J. Yuan, L. Dorn-Gomba, A. D. Callegaro, J. Reimers, and A. Emadi, "A review of bidirectional on-board chargers for electric vehicles," *IEEE Access*, vol. 9, pp. 51 501–51 518, 2021.

[6] S. S. Williamson, A. K. Rathore, and F. Musavi, "Industrial electronics for electric transportation: Current state-of-the-art and future challenges," *IEEE Transactions on Industrial Electronics*, vol. 62, no. 5, pp. 3021–3032, 2015.

[7] F. Blaabjerg, H. Wang, I. Vernica, B. Liu, and P. Davari, "Reliability of power electronic systems for ev/hev applications," *Proceedings of the IEEE*, vol. 109, no. 6, pp. 1060–1076, 2021.

[8] J. Torki, C. Joubert, and A. Sari, "Electrolytic capacitor: Properties and operation," *Journal of Energy Storage*, vol. 58, p. 106330, 2023.

[9] E. Wolfgang, "Examples for failures in power electronics systems," *ECPE tutorial on reliability of power electronic systems, Nuremberg, Germany*, pp. 19–20, 2007.

[10] S. Yang, A. Bryant, P. Mawby, D. Xiang, L. Ran, and P. Tavner, "An industry-based survey of reliability in power electronic converters," *IEEE Transactions on Industry Applications*, vol. 47, no. 3, pp. 1441–1451, 2011.

[11] H. Wang, M. Liserre, and F. Blaabjerg, "Toward reliable power electronics: Challenges, design tools, and opportunities," *IEEE Industrial Electronics Magazine*, vol. 7, no. 2, pp. 17–26, 2013.

[12] J.-W. Kim, B. Kim, D. Lee, Y. Cho, S. K. Ji, and D. Ryu, "Llc resonant converter for fast electric vehicle charging module with a reconfigurable bi-directional switch," *IEEE Transactions on Transportation Electrification*, pp. 1–1, 2024.

[13] H. Wu, X. Jin, H. Hu, and Y. Xing, "Multielement resonant converters with a notch filter on secondary side," *IEEE Transactions on Power Electronics*, vol. 31, no. 6, pp. 3999–4004, 2016.

[14] M. Kim, H. Jeong, B. Han, and S. Choi, "New parallel loaded resonant converter with wide output voltage range," *IEEE Transactions on Power Electronics*, vol. 33, no. 4, pp. 3106–3114, 2018.

[15] J.-Y. Lin, H.-Y. Yueh, Y.-F. Lin, and P.-H. Liu, "Variable-frequency and phase-shift with synchronous rectification advance on-time hybrid control of llc resonant converter for electric vehicles charger," *IEEE Journal of Emerging and Selected Topics in Industrial Electronics*, vol. 4, no. 1, pp. 348–356, 2023.

[16] J.-Y. Lee, "An el capacitorless ev on-board charger using harmonic modulation technique," *IEEE Transactions on Industrial Electronics*, vol. 61, no. 4, pp. 1784–1787, 2014.

[17] D. Patil and V. Agarwal, "Compact onboard single-phase ev battery charger with novel low-frequency ripple compensator and optimum filter design," *IEEE Transactions on Vehicular Technology*, vol. 65, no. 4, pp. 1948–1956, 2016.

[18] C. Fei, F. C. Lee, and Q. Li, "High-efficiency high-power-density llc converter with an integrated planar matrix transformer for high-output current applications," *IEEE Transactions on Industrial Electronics*, vol. 64, no. 11, pp. 9072–9082, 2017.

[19] R. Asensi, J. Cobos, O. Garcia, R. Prieto, and J. Uceda, "A full procedure to model high frequency transformer windings," in *Proceedings of 1994 Power Electronics Specialist Conference - PESC'94*, 1994.

[20] R. Prieto, R. Asensi, J. Cobos, O. Garcia, and J. Uceda, "Model of the capacitive effects in magnetic components," in *Proceedings of PESC '95 - Power Electronics Specialist Conference*, 1995.

Multiple-Core Transformer Design Based on Half-Turn Structure in Two-Stage DC-DC Converter for Battery Storage System

Yilei Li, Bima Nugraha Sanusi, Pinhe Wang, Tianming Luo

Department of Electrical and Photonics Engineering
Technical University of Denmark
Kgs. Lyngby, 2800, Denmark
s222480@dtu.dk, bnusa@dtu.dk, piwa@dtu.dk, tialu@dtu.dk

Abstract—**This paper aims to design a two-stage Buck-LLC DC/DC converter for battery energy storage system (BESS). Buck converter in the first stage features fast response for regulating, and together with second stage LLC-DCX, high power density and high efficiency can be achieved. To minimize the switching loss and boost the efficiency of GaN devices, dead time and magnetizing inductance are optimized. Moreover, different multiple-core planar transformers are proposed and compared in this paper. To achieve an improved winding design practice, winding loss models are built and simulated based on Finite Element Analysis (FEA). Proposed multiple-core transformer with half-turn structure is verified to be outstanding in eliminating proximity effect which reduces the winding loss by 27% compared to single-core design. A 384 V/12 V 240 W Buck-LLC converter is demonstrated, achieving a peak efficiency of 96%.**

Index Terms—**GaN device; half-turn structure; ISOP; LLC converter; magnetomotive force (MMF); matrix transformer; planar transformer (PT); soft-switching; winding loss**

I. INTRODUCTION

Battery energy storage system (BESS) has been gradually adopted in modern industrial use especially with an increasing implementation in power networks. BESS features various advantages not only for its lifespan, high energy density, but also significance in balancing the supply and demand with sufficient stability. By the end of 2024, U.S. battery capacity is hopefully expanding to more than 30 GW [1]. With rapid growth of battery energy storage system, power conversion for high capacity applications has vital importance in maintaining high efficiency before users.

In BESS configurations, battery pack connected in series which allows adding up the multiple cell units but remaining low current is widely used to achieve a better efficiency and thermal management. The current practice architecture of BESS from battery pack to low-voltage DC bus is illustrated in Fig. 1. For HV stacked battery solution, such as product from ZWAYN New Energy and QH Technology, a standard input of 384 V is commonly used for storage system. For industry user such as embedded CPU, communication system and advanced driver-assistance systems (ADAS), high input voltage before user is often converted and fed into a 12 V bus bar. Therefore, a high efficiency and robust DC/DC converter is needed.

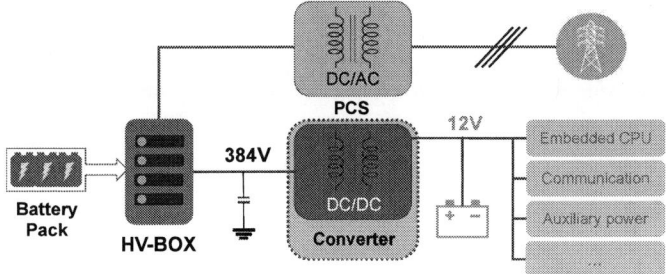

Fig. 1: BESS power conversion architecture

Traditional single-stage DC/DC converter is facing the challenge of large conversion ratio and high output current, which adds up the complexity of transformer design. In [2] and [3], it has been verified that ISOP converter is superior compared to conventional converter due to its high efficiency. Buck converter in first stage features fast dynamic response and provides a good pre-regulation for next stage. LLC has been proved to be an efficient topology with high power density which is capable to carry out zero-voltage-switching (ZVS) by fully utilizing its leakage inductance. Moreover, by designing LLC converter as a DC transformer (DCX) with a fixed converting ratio as well as optimal dead time and magnetizing inductance, the proposed Buck-LLC converter can further achieve its peak efficiency at resonant frequency.

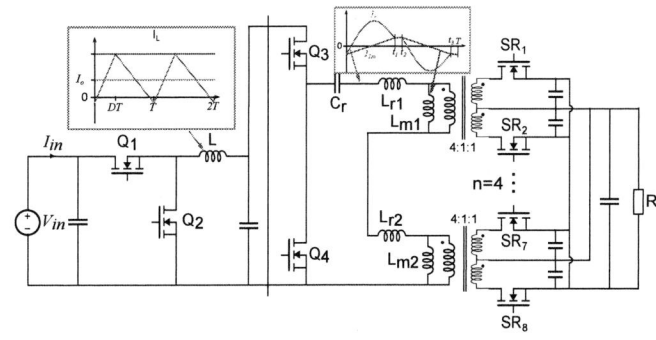

Fig. 2: Buck-LLC schematic

Other than traditional transformer with bulky volume, planar transformer has its natural merits in low profile, convenience for manufacturability as well as improved thermal characteristic [4]. Also, planar magnetic provides possibility of designing and controlling of parasitic parameters [5]. As in [6], [7] and [8], recent works have verified that matrix transformer helps to distribute the secondary side current more evenly, and it also avoids high power loss together with ISOP topology. Based on half-turn structure, which is an optimized winding layout which greatly limits the overall proximity effect, proposed Buck-LLC converter implements multiple-core transformer to magnetic design. Proposals are analyzed through FEA and winding loss comparison is made with single-core transformer. It is verified that proposed transformer design is superior than conventional single-core design with interleave structure in minimizing the winding loss and reducing the manufacturing cost with less PCB layers.

In this paper, Section II investigates the analysis of topology as well as optimization of resonant parameter including dead time and magnetizing inductance. Section III proposes and compares three different multiple-core transformer designs and corresponding MMF analysis. Section IV presents experiment results in a Buck-LLC converter prototype with a peak efficiency of 96%. The conclusion and summary is illustrated in Section V. Sufficient margin is considered through all design process.

II. PARAMETER OPTIMIZATION OF BUCK-LLC CONVERTER TOPOLOGY

A. Inductor of Buck

Fig. 2 shows the overview of Buck-LLC converter in this design. For first stage Buck converter which converts the voltage to 192 V at 200 kHz, to avoid large number of turns, half-bridge is applied to the topology and the ripple current is designed to achieve ZVS for both high side and low side MOSFETs.

To ensure soft-switching, the ripple current is designed to be zero-crossing for each period. By choosing the ripple factor K_r (ratio of peak current I_{pk} to ripple current Δi) to be 1. The ripple current Δi can be obtained by:

$$\begin{cases} \Delta i = 2I_o \\ I_{peak} = I_o + 0.5\Delta i \end{cases} \tag{1}$$

Under a duty cycle D within a range of 0.44 to 0.67, the inductance L needs to be design at worst case to obtain sufficient ripple current. Therefore, the inductance can be derived as expressed:

$$L_{min} = \frac{V_o(1 - D_{max})}{f_s \cdot \Delta i} \tag{2}$$

B. Dead Time Optimization of LLC

For second stage LLC converter, switching loss takes up a high proportion in total loss breakdown. Therefore, GaN devices which features ultrafast switching speed and lower output capacitance C_{oss} is applied to this design. Both primary

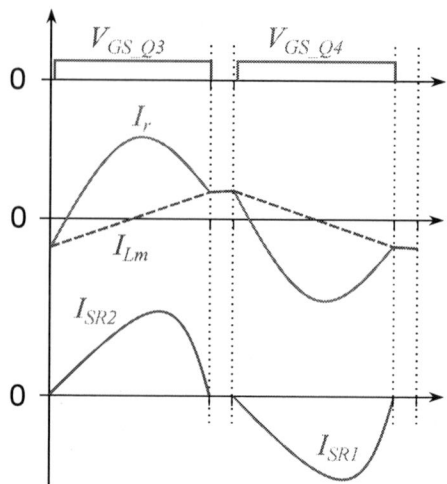

Fig. 3: Waveform of LLC resonant converter

side GaNFET and secondary side synchronous rectifier (SR) are supposed to completely achieve ZVS to minimize the switching loss. As in [9] and [10], dead time is introduced as a key factor of optimizing the FET loss in LLC converter.

As Fig. 3 shows, during dead time, both primary side GaNFET C_{oss_GaN} and secondary side C_{oss_SR} needs to be discharged by resonant current. Thus sufficient magnetizing current I_{L_m} and dead time t_d is needed to meet the balance of charge which can be expressed by:

$$I_{pk\_L_m} \cdot t_d = 2C_{oss\_GaN} \cdot V_{in} + \frac{1}{N} \cdot 2nC_{oss\_SR} \cdot 2V_o \tag{3}$$

By eliminating both V_{in} and V_o derived from turns ratio N as well as discharging magnetizing current $I_{pk\_L_m}$, the magnetizing inductance L_m can be obtained as follow:

$$L_m = \frac{t_d(\frac{T_s}{2} - t_d)}{4(2C_{oss\_GaN} + \frac{1}{N^2} \cdot 2nC_{oss\_SR})} \tag{4}$$

To limit the conducting loss caused by large output current on secondary side, the center tap structure is split to n phases and connected in parallel. Under a switching frequency of 500 kHz, optimal L_m is decided by dead time t_d.

In [11], an in-depth study foundation for the current calculation of half time period and corresponding simplification process of both for primary side and secondary side RMS current in half-bridge LLC converter is provided. The overall RMS value for resonant tank and SR can be derived as:

$$I_{RMS\_P\_total} = \sqrt{\frac{V_o^2 T_s^2 \pi^2}{8R_L N^2(T_s - 2t_d)^2} + \left(\frac{1}{2} + \frac{2t_d}{T_s}\right)\left(\frac{2C_{oss\_GaN}V_{in} + \frac{1}{N})2nC_{oss\_SR}*2V_o}{t_d}\right)^2} \tag{5}$$

$$I_{RMS\_S\_total} = N\sqrt{\frac{T_s - 2t_d}{2T_s}\left[\frac{V_o^2 T_s^2 \pi^2}{8R_L^2 N^2(T_s - 2t_d)^2} + \left(\frac{5}{6} - \frac{8}{\pi^2}\right)\left(\frac{2C_{oss\_GaN}V_{in} + \frac{1}{N})2nC_{oss\_SR}*2V_o}{t_d}\right)^2\right]} \tag{6}$$

At primary side, the GaNFET in half-bridge will experience a higher current than full-bridge topology thus higher conducting loss. However, more power switches require longer dead time and thus lower transferring efficiency for each period. Therefore, the overall conducting loss can be further optimized by calculating minimum t_d. The overall conducting loss can be expressed as:

$$P_{con} = 2(\sqrt{\tfrac{T_s - t_d}{T_s}} I_{RMS\_P\_total})^2 \cdot R_{ds\_on\_GaN} + 2n(\tfrac{I_{RMS\_S\_total}}{n})^2 \cdot R_{ds\_on\_SR} \tag{7}$$

For high output current, multi-phase center taps with single turn are implemented to lower down the conducting loss as [12] presents. The relationship between number of paralleling devices at secondary side n and conducting loss at $t_d = 50$ ns, 100 ns and 120 ns can be indicated as Fig. 4 shows.

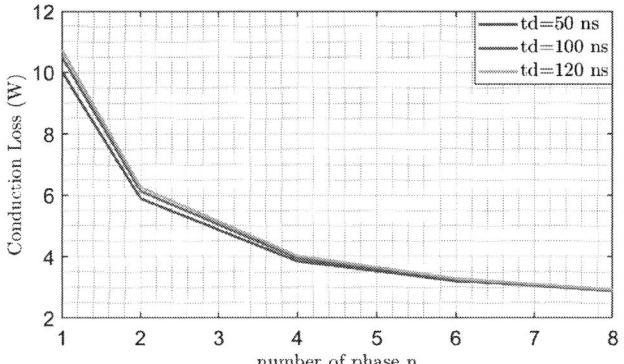

Fig. 4: Conducting loss under different number of n

By taking the derivative of t_d and equals to 0, the minimum conducting loss can be obtained at around $t_d = 80\ ns$. The optimized magnetizing inductance can also be calculated by inserting t_d into 4.

$$\frac{dP_{con}}{dt_d} = 0 \implies t_d = 81ns,\ L_m = 35\ uH \tag{8}$$

III. TRANSFORMER DESIGN AND WINDING LOSS COMPARISON

Due to large conversion ratio, LLC-DCX converter needs to handle high output current at secondary side. Except for dividing output current into 4 phases, a good structure of winding layout will also minimize the transformer loss and thus achieve high efficiency. Commonly, interleave structure is widely used in power converter transformer design, which greatly minimize the area that the magnetic field can penetrate through and hence reduce the magnetomotive force (MMF) between windings [13].

Meanwhile, with the proposal of planar transformer (PT), which integrates the winding into PCB manufacturing and increases power density, has also advanced the multiple-core application. Matrix transformer is one of the concept that divides one transformer core into array of element cores which is beneficial to split the large current and reduces the leakage

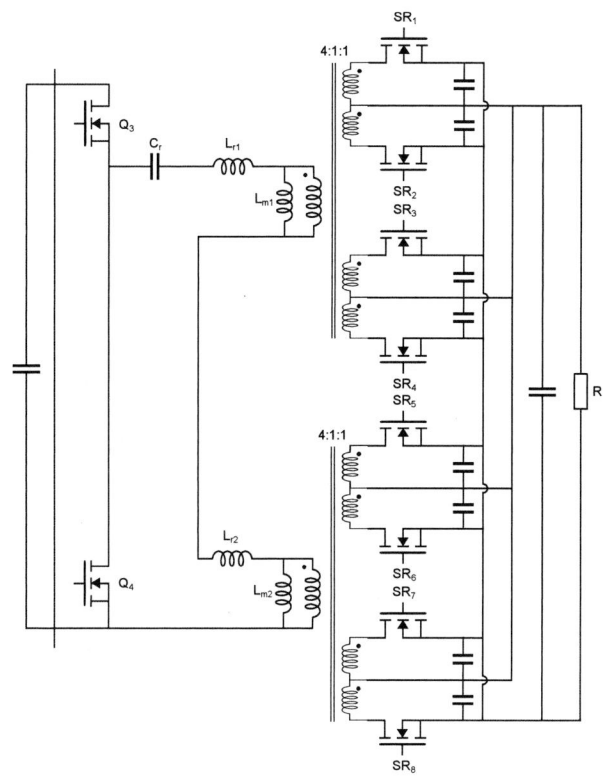

Fig. 5: Schematic of LLC converter

by splitting turns between each element core [14]. With the implementation of multiple-core design instead of simply paralleling and increasing the number of switches, unexpected termination loss and unbalanced current distribution can be avoided.

This section will give a detailed illustration on three multiple-core PT designs and comparison between traditional single-core design based on interleave and half-turn structure. ELP 32/6/20 core is used uniformly for all proposals.

A. Core Loss Model

For LLC-DCX, the conversion ratio is fixed at 16:1 and due to the larger output current, the secondary side winding is designed to be 1 turn for each phase to reduce R_{ac} caused by skin effect and proximity effect. Therefore, the peak flux density B_{peak} can be derived from (9).

$$dB = \frac{V_{in} \cdot D(T_s - t_d)}{N_p \cdot A_e} \implies B_{peak} = \frac{dB}{2} \tag{9}$$

Under a switching frequency of 500 kHz, N49 is chosen to be the core material with high permeability and good thermal stability. The core loss model can be derived based on Steinmetz equation as in (10) with coefficient $\alpha = 1.25$, $\beta = 2.85$ and $K_c = 31.5$ [15]:

$$P_{core} = V_e \cdot K_c \cdot f^{\alpha} \cdot B_{peak}^{\beta} \tag{10}$$

B. Winding Loss Model

As illustrated before, each secondary side winding is set to be 1 turn. Therefore, the primary side winding should be 8 turns with a conversion ratio of 8:1. For traditional single-core design with interleave structure, both 8 turns of primary side and secondary side winding are placed at center leg. Proposed multiple-core design divides the primary side winding into two part in series and each part of primary side winding is 4 turns, coupling two phases of secondary side winding. The schematic of LLC-DCX is shown as Fig. 5.

The winding loss model is based on Dowell's assumption for distribution of current density in one single layer of an infinitely foil conductor [16]. the ratio of AC resistance to DC resistance can be expressed as:

$$\frac{R_{ac\_m}}{R_{dc\_m}} = \frac{\Delta}{2}\left[\frac{sinh\Delta + sin\Delta}{cosh\Delta - cos\Delta} + (2m-1)^2 \cdot \frac{sinh\Delta - sin\Delta}{cosh\Delta + cos\Delta}\right] \quad (11)$$

Where Δ is the ratio of conductor thickness to skin depth. The first term on the right side of equation represents the skin effect, and the second term related to m represents the proximity effect, where magnetic field generated by adjacent conductor will interact with each other [17]. m is defined as:

$$m = \frac{F(h)}{F(h) - F(0)} \quad (12)$$

Where $F(h)$ and $F(0)$ is the MMF of conductor from one side of boundary to another.

1) Single-Core Transformer with Interleave: For interleave structure, m is restricted to 1 based on Dowell's equation. The MMF is canceled out between each two layers to constrain the effect on adjacent conductors as shown Fig. 6.

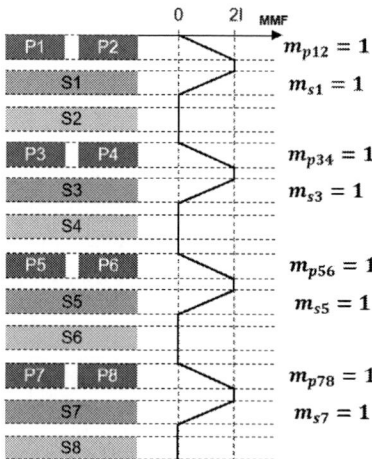

Fig. 6: MMF of single-core interleave structure

Simulation model was built based on Finite Element Analysis (FEA) in Ansys Maxwell to show the leakage inductance and winding loss. Fig. 7 shows the winding model of single-core design with interleave structure.

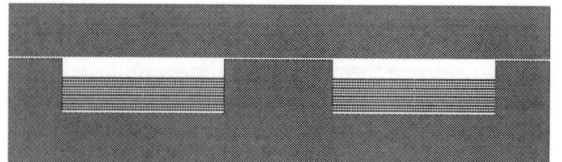

Fig. 7: Winding model of interleave single-core design

The leakage inductance L_r and magnetizing inductance L_m can be obtained from winding model simulation, and then the resonant capacitor can be calculated under the frequency of 500 kHz, the resonant parameter is shown in Table. I.

TABLE I: Resonant Parameter for Single-Core Design

Parameter	Symbol	Value
Magnetizing inductance	L_m	34.4 uH
Leakage inductance	L_r	0.38 uH
Resonant capacitor	C_r	267 nF

By using the resonant parameter to circuit simulation model, both primary and secondary side current waveform can be inserted to winding model in Ansys, which leads to an accurate result of overall winding loss shown in Fig. 8.

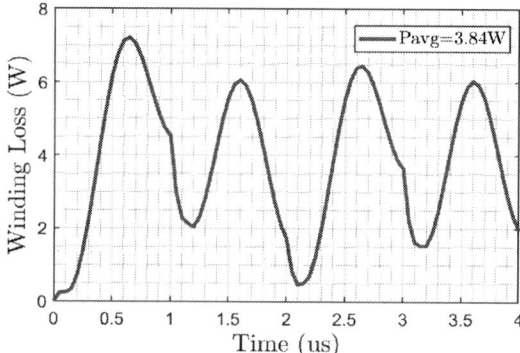

Fig. 8: Winding loss plot for single-core design

2) Multiple-Core Transformer Proposal 1: Half-turn is based on interleave structure and is an optimized winding layout which further limit the proximity effect. Paper [18] introduces a good way of splitting primary side current on top layer by adding another winding on bottom layer connected in parallel. In this way, except for top and bottom layer, the positive MMF can achieve zero-crossing with same amplitude to negative value and vice versa.

From (12), m can be calculated to be 0.5. Inserting $m = 0.5$ to (11), the second term of equation can be eliminated, which also means the proximity effect can be eliminated.

As shown in Fig. 9, based on half-turn structure, proposal 1 places the primary winding at center leg of both element cores with shortest conducting path as [19] presents on ELP core. For the secondary winding that conducting at the same time, each two windings are placed in different cores in adjacent layer. Instead of paralleling top layer of primary side winding, two turns of primary side winding are placed in the middle

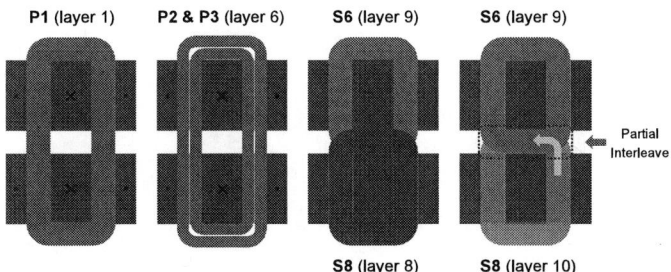

Fig. 9: Top view of part of winding layout of proposal 1

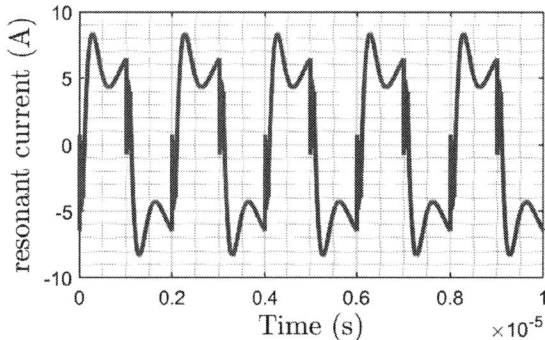

Fig. 11: Simulation current waveform of proposal 1

layer and the other two are placed at top layer and bottom layer. The MMF analysis can be drawn as shown in Fig. 10.

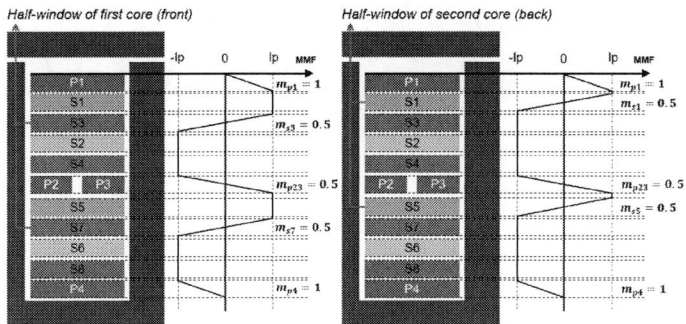

Fig. 10: MMF analysis of proposal 1

Proposal 1 implemented with multiple-core design not only shows strong advantage on eliminating the proximity effect to the greatest extent, but also achieves butterfly structure between secondary winding in the middle area of magnetic cores [20]. The current goes opposite direction between adjacent layers in this partial interleaving area, thus the magnetic flux in this area is canceled out. As a result, the loss can be further reduced. The leakage inductance is verified to be kept in a low value through simulation and corresponding resonant capacitance can be calculated as in Table. II.

TABLE II: Proposal 1 3D Model and Resonant Parameter

3D Model	Parameter	Symbol	Value
	Magnetizing inductance	L_m	34.5 uH
	Leakage inductance	L_r	16 nH
	Resonant capacitor	C_r	6.18 uF

Proposal 1 demonstrates an advantage in reducing the overall leakage inductance compared to single-core design. However, such a large resonant capacitor leads to an distorted resonant current at primary side, as shown in Fig. 11. During dead time, there is also a huge ringing caused by parasitic capacitance. Therefore, this proposal is not suitable for this specification.

3) Multiple-Core Transformer Proposal 2: The second proposal is also based on matrix transformer with two element cores. Proposal 2 moves secondary winding to the side leg of core. According to Farady's law, secondary side winding increases to 2 turns since the side leg cross-section A_e is only half of the area compared to center leg. The primary side winding remains the same as a shortest path coupling two cores. In this case, every two secondary windings that conduct at the same time are placed diagonally at the same layer. The overview winding placement is shown in Fig. 12.

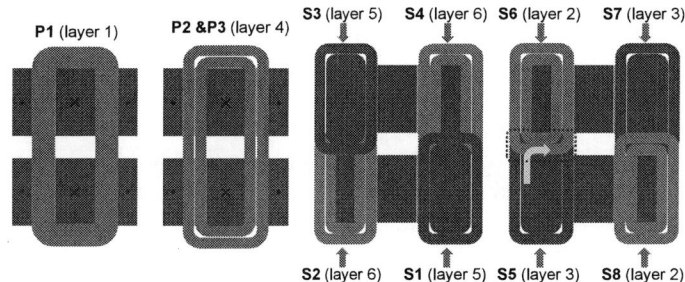

Fig. 12: Top view of part of winding layout of proposal 2

However, this design no longer follows the half-turn structure. Moreover, double turns of winding further increases the loss due to large current at secondary side. The MMF can be symmetrically analyzed for half window as shown in Fig. 13.

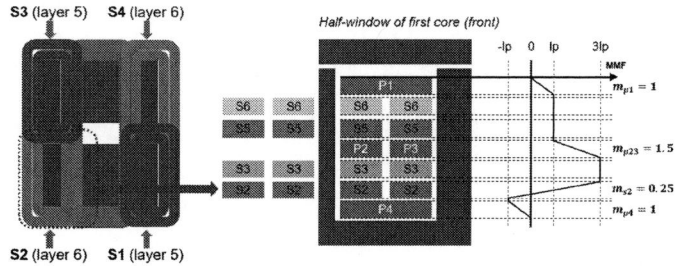

Fig. 13: MMF analysis of proposal 2

The 3D model and resonant parameter can be exported from simulation as shown in Table. III.

TABLE III: Proposal 2 3D Model and Resonant Parameter

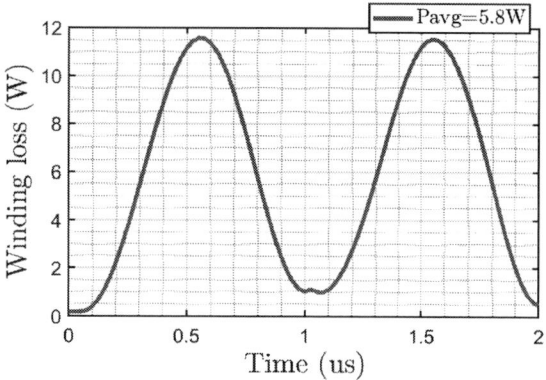

3D Model	Parameter	Symbol	Value
	Magnetizing inductance	L_m	34.5 uH
	Leakage inductance	L_r	1.8 uH
	Resonant capacitor	C_r	55 nF

The current waveform is exported from circuit simulation and inserted back to winding loss model. As shown in Fig. 14, the overall winding loss of proposal 2 is even higher than single-core design. Therefore, proposal 2 is also not suitable.

Fig. 14: Winding loss plot of proposal 2

4) Multiple-Core Transformer Proposal 3: Based on two previous designs, for LLC-DCX, it is not only important to have some leakage inductance to form the resonance but also noteworthy that high leakage inductance might bring high winding loss.

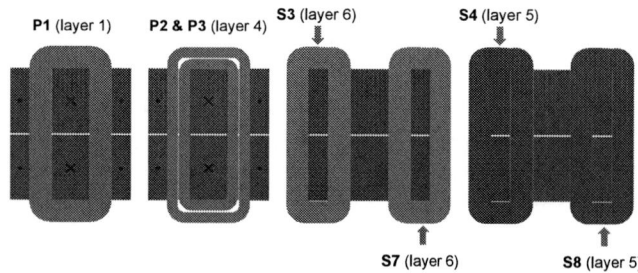

Fig. 15: Top view of part of winding layout of proposal 3

As Fig. 15 shows, proposal 3 removes the gap between two cores and attaches two cores together to make the secondary winding remain 1 turn since both primary and secondary side windings are coupling two cores with doubled effective cross-section A_e. This design shortened the conducting path both for primary side and secondary side winding and MMF analysis in Fig. 16 shows that it still follows the half-turn structure where the proximity effect is greatly eliminated.

As shown in Table. IV, proposal 3 keeps certain amount of leakage inductance but due to the reduced length of conducting

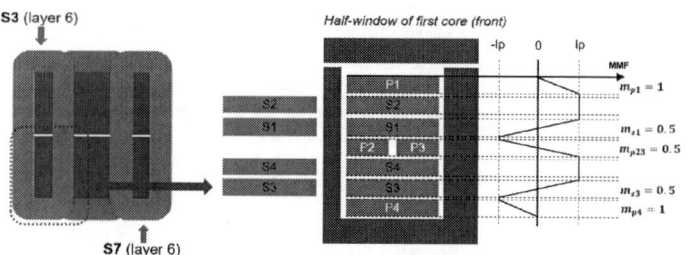

Fig. 16: MMF analysis of proposal 3

TABLE IV: Proposal 3 3D Model and Resonant Parameter

3D Model	Parameter	Symbol	Value
	Magnetizing inductance	L_m	34.5 uH
	Leakage inductance	L_r	1.64 uH
	Resonant capacitor	C_r	62 nF

path and achievement of half-turn structure, the winding loss is much improved, as shown in Fig. 17. Compared to conventional single-core design, the winding loss is reduced by 27% and the requiring winding layers reduces from 12 layers to 8 layers. Therefore, proposal 3 is implemented to the transformer design.

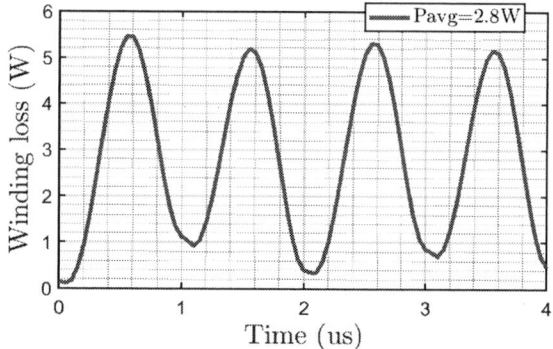

Fig. 17: Winding loss plot of proposal 3

IV. EXPERIMENT RESULT

A 200 kHz/500 kHz 384 V/12 V 240 W two-stage Buck-LLC converter prototype was built as shown in Fig. 18. MOSFET AIMBG120R080M1 is implemented to Buck converter for its high voltage endurance and fast switching for regulating. GaNFET GS66508B and SR BSC026N08NS5 are implemented to LLC primary and secondary side respectively, and both of them features low output capacitance and low on-state resistance. All primary and secondary side gate signals are provided by MCU TMS320F28379D. Secondary side SR signal is synchronized with primary side GaNFET.

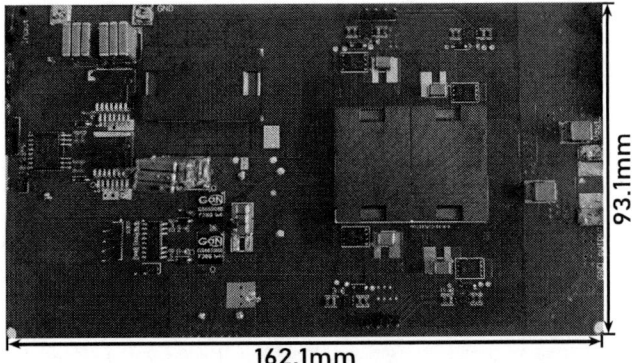

162.1mm / 93.1mm

Fig. 18: Experiment prototype

Fig. 21: Thermal image

The waveform of signal and drain-to-source voltages both for Buck MOSFET and LLC GaNFET are shown in Fig. 19 and 20. From where it can be clearly read that both first stage and second stage power switches have achieved ZVS within dead time as analyzed. The current waveform of inductor is also shown in figure below, which validates a correctness from topology analysis. The resonant current is sinusoidal and discharging current can be seen within dead time.

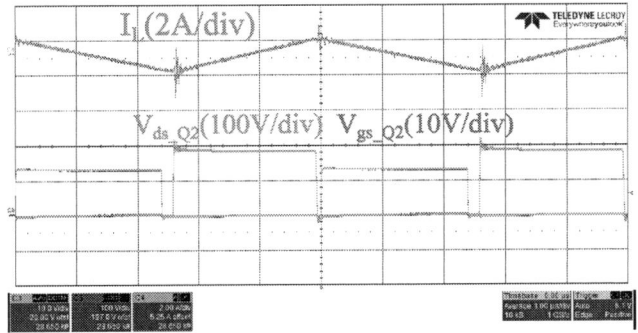

Fig. 19: Waveform of Buck I_L and Q_2

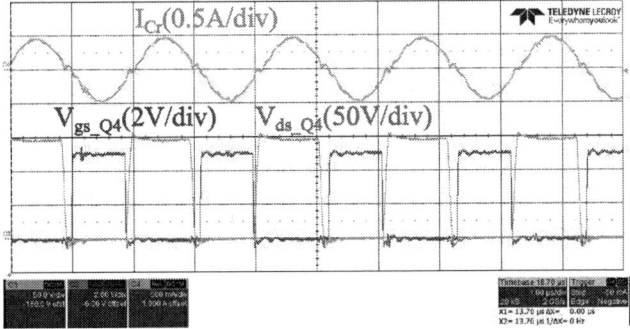

Fig. 20: Waveform of LLC I_{C_r} and Q_4

Thermal image is filmed under an input power of 240 W. As shown in Fig. 21, the maximum temperature from primary side GaNFET is 61.2 °C and both power switches and transformer are operating under a safe temperature range.

Due to some drawbacks of EMI design, some high frequency oscillation caused by parasitic capacitance in circuit may occur as the input power increases. To guarantee a safe operating, the power is limited under 240 W. By measuring the voltage and current both from input and output side, the efficiency curve can be drawn as shown in Fig. 22, indicating that the efficiency achieves a peak at 96%.

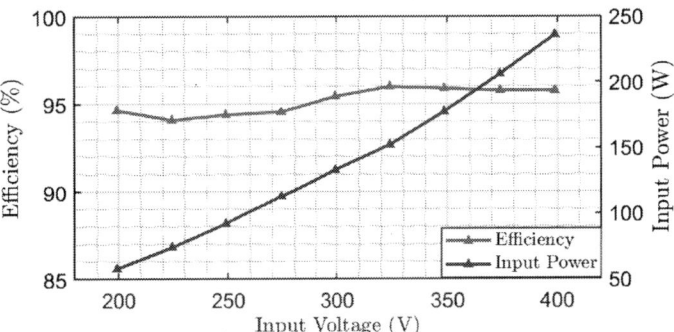

Fig. 22: Efficiency curve and input power plot

Compared to loss breakdown analyzed in Fig. 23, the experimental result basically fits the efficiency curve which reaches an efficiency of 96.47%. The main part of loss in power switches comes from turn-off loss under high voltage conversion on power switches but due to the achievement of ZVS, the switching loss is minimized. One main reason for the experimental result of peak efficiency being slightly lower is that due to the existence of some extra leakage inductance caused by uneven air gap, operating frequency is slightly declined to form a standard sinusoidal waveform. That increases part of the core loss which increases exponentially as frequency decreased. Besides, some ESR of capacitor and via termination loss also lead to some extra loss on PCB. But in general, this 384 V/12 V 240 W prototype is capable to validate the improvement and optimization of proposed Buck-LLC DCX converter with its multiple-core transformer with half-turn structure.

V. CONCLUSION

A 200 kHz/500 kHz 384 V/12 V 240 W two-stage Buck-LLC converter is designed for battery storage system. Both circuit topology and transformer are optimized for achieving

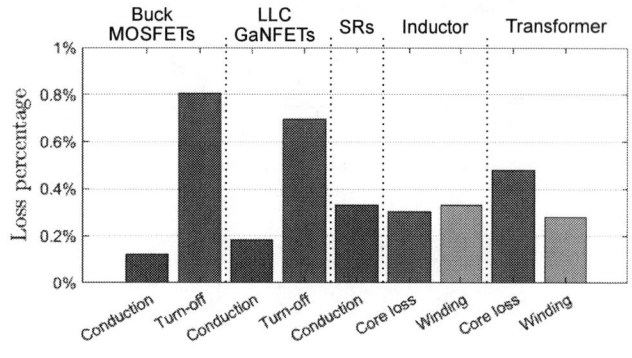

Fig. 23: Loss breakdown

higher efficiency. Compared to single-core transformer design with interleave structure, proposed design has greatly reduces winding loss by 27%. Experimental result shows that by implementing half-turn structure, the proximity effect has been eliminated to a fullest extent. It is also worth noting that for multiple-core transformer design, it requires less number of layers and the cost of manufacturing can be greatly reduced.

REFERENCES

[1] K. Antonio and A. Mey, "U.s. battery storage capacity expected to nearly double in 2024," 2024, accessed on: November 14, 2024. [Online]. Available: https://www.eia.gov/todayinenergy/detail.php?id=61202

[2] Y. Wang, C. Chen, B. Chen, Z. Wang, R. Ji, and M. Zhang, "Transformer Integration and Winding Design for ISOP-LLC Converter," in *2022 IEEE Energy Conversion Congress and Exposition (ECCE)*, 2022, pp. 1–5.

[3] Y. Liu, D. Hu, and H. Wu, "Magnetic Integration Design for Input-Series Output-Parallel LLC Resonant Converter," in *IECON 2023- 49th Annual Conference of the IEEE Industrial Electronics Society*, 2023, pp. 1–5.

[4] Z. Ouyang and M. A. E. Andersen, "Overview of Planar Magnetic Technology - Fundamental Properties," *IEEE Transactions on Power Electronics*, vol. 29, no. 9, pp. 4888–4900, 2014.

[5] J. Zhang, W. G. Hurley, and W. H. Wolfle, "Design of the planar transformer in llc resonant converters for micro-grid applications," in *2014 IEEE 5th International Symposium on Power Electronics for Distributed Generation Systems (PEDG)*, 2014, pp. 1–7.

[6] A. Nabih and Q. Li, "Low-Profile and High-Efficiency 3 kW 400 V-48 V LLC Converter with a Matrix of Four Transformers and Inductors for 48V Power Architecture for Data Centers," in *2021 IEEE Energy Conversion Congress and Exposition (ECCE)*, 2021, pp. 1813–1819.

[7] S. Zhu, Q. Qian, Q. Liu, and T. Wang, "Research on a High-Efficiency Half-Turn Matrix Planar Transformer Structure for LLC Resonant Converter," in *2020 21st International Conference on Electronic Packaging Technology (ICEPT)*, 2020, pp. 1–6.

[8] Z. Liu, J. Liang, and H. Wang, "Optimal Design of Hexagonal Matrix Transformer for 48V-1V Switched-Capacitor and Series-Resonant Converter," in *2024 IEEE 10th International Power Electronics and Motion Control Conference (IPEMC2024-ECCE Asia)*, 2024, pp. 1346–1351.

[9] P. Wang, B. N. Sanusi, T. Gabriel Zsurzsan, M. A. E. Andersen, and Z. Ouyang, "Vertical LLC Converter for High-Current Datacenter Application," in *2024 IEEE 10th International Power Electronics and Motion Control Conference (IPEMC2024-ECCE Asia)*, 2024, pp. 562–567.

[10] C. Wang, M. Li, Z. Ouyang, Z. Zhang, and M. A. E. Andersen, "High Step-Down Single-Stage DC-DC Converter with Improved Planar Matrix Transformer for High-Current Data Center Application," in *2022 IEEE Applied Power Electronics Conference and Exposition (APEC)*, 2022, pp. 709–715.

[11] W. Zhang, F. Wang, D. J. Costinett, L. M. Tolbert, and B. J. Blalock, "Investigation of Gallium Nitride Devices in High-Frequency LLC Resonant Converters," *IEEE Transactions on Power Electronics*, vol. 32, no. 1, pp. 571–583, 2017.

[12] M. Dai, X. Zhang, H. Li, D. Zhou, Y. Wang, and D. Xu, "LLC Converter With an Integrated Planar Matrix Transformer Based on Variable Width Winding," in *2019 22nd International Conference on Electrical Machines and Systems (ICEMS)*, 2019, pp. 1–4.

[13] B. Chen, "Analysis of Effect of Winding Interleaving on Leakage Inductance and Winding Loss of High Frequency Transformers," *Journal of Electrical Engineering Technology*, vol. 14, pp. 1211–1221, 2019.

[14] D. Huang, S. Ji, and F. C. Lee, "LLC Resonant Converter With Matrix Transformer," *IEEE Transactions on Power Electronics*, vol. 29, no. 8, pp. 4339–4347, 2014.

[15] W. G. Hurley, T. Merkin, and M. Duffy, "The Performance Factor for Magnetic Materials Revisited: The Effect of Core Losses on the Selection of Core Size in Transformers," *IEEE Power Electronics Magazine*, vol. 5, no. 3, pp. 26–34, 2018.

[16] Robert W. Erickson, Dragan Maksimovic, *Fundamentals of Power Electronics*. Springer Cham, 2020.

[17] A. Alabakhshizadeh and O.-M. Midtgård, "Optimum core dimension for minimizing proximity effect losses of an AC inductor for a galvanically isolated PV inverter," *2012 38th IEEE Photovoltaic Specialists Conference*, pp. 001 373–001 377, 2012.

[18] Z. Ouyang, O. C. Thomsen, and M. A. E. Andersen, "Optimal Design and Tradeoff Analysis of Planar Transformer in High-Power DC-DC Converters," *IEEE Transactions on Industrial Electronics*, vol. 59, no. 7, pp. 2800–2810, 2012.

[19] Q. Huang, Q. Ma, A. Q. Huang, and M. d. Rooij, "400V-to-48V GaN Modular LLC Resonant Converter with Planar Transformers," in *2021 IEEE Energy Conversion Congress and Exposition (ECCE)*, 2021, pp. 2129–2135.

[20] M. Li, Y. Liu, Z. Ouyang, and M. A. E. Andersen, "Butterfly Interleaving Winding Arrangements for Multiphase Coupled Inductors," *IEEE Transactions on Power Electronics*, vol. 38, no. 3, pp. 3315–3327, 2023.

Bidirectional DC-DC Converter Utilizing Coupled Inductors for Energy Storage System

Wen-Hsuan Lee
Department of Electrical Engineering,
National Cheng-Kung University
Tainan,Taiwan
ve6092119@gs.ncku.edu.tw

Jiann-Fuh Chen
Department of Electrical Engineering,
National Cheng-Kung University
Tainan,Taiwan
chenjf@mail.ncku.edu.tw

Hsuan Liao
Department of Electrical Engineering,
National Cheng-Kung University
Tainan,Taiwan
n28064046@gs.ncku.edu.tw

Kuo-Fu Liao
Department of Electrical Engineering,
National Cheng-Kung University
Tainan,Taiwan
n28074025@gs.ncku.edu.tw

Abstract—**In future power systems, the influence of diverse renewable energy sources will lead to power supply imbalance and energy intermittency. Energy storage systems with bidirectional operation mechanisms are needed to address these challenges. This paper proposes a non-isolated bidirectional DC-DC converter for energy storage systems. On the battery side, two sets of coupled inductors are used to achieve high voltage gain and reduce current ripple on the low-voltage side. The DC bus side employs a three-level neutral-point clamped structure to mitigate the problem of high switch voltage ratings. Finite element simulation is conducted on the circuit's magnetic components, and an improved magnetic core with two materials and varied gap distributions is introduced. A comparative analysis of the losses in the two magnetic components is also provided. Finally, following the design process outlined in this paper, a converter with a low-voltage side of 60 V, a high-voltage side of 400 V, a rated power of 300 W, and a switching frequency of 100 kHz was developed and tested. Experimental results demonstrate the feasibility of the proposed bidirectional topology, achieving a maximum efficiency of 93.72% in step-up stage and 92.33% in step-down stage under the condition of a 300 W output power.**

Keywords—**bidirectional DC-DC converter, coupled inductor, magnetic analysis, COMSOL Multiphysics®**

I. INTRODUCTION

In future power systems, the high penetration of renewable energy sources (RES) will lead to significant variability and unpredictability in electricity supply. Integrating RES with energy storage systems (ESS) can smooth frequency control, save fuel, reduce costs, and decrease emissions, thus improving energy conversion efficiency. Adjusting the power system's frequency or voltage to compensate for power fluctuations can alleviate supply imbalances [1]-[3]. ESS features bidirectional operation, allowing for rapid power supply and the ability to absorb and release energy. In distributed generation systems (DGS), a bidirectional DC-DC converter manages the voltage difference between the battery and the DC bus, ensuring stable system operation [3], [4], as shown in Fig. 1. Bidirectional DC-DC converters are categorized into isolated and non-isolated types. Isolated converters, such as Flyback, Forward, Half-Bridge, and Full-Bridge topologies, achieve high voltage conversion ratios through transformers, offering high safety and voltage gain. However, in high-power applications, their efficiency can decrease, size can increase, control can become complex, and costs can rise [5]-[10]. Non-isolated converters

include Cuk, Sepic/Zeta, coupled inductors, traditional Buck-Boost, and three-level topologies. Cuk and Sepic/Zeta converters reduce output voltage ripple and switch voltage stress but have lower efficiency [11], [12]. This paper proposes a non-isolated bidirectional DC-DC converter. Two coupled inductors are used on the battery side to achieve high voltage gain, while a three-level Neutral-Point-Clamped (NPC) structure is employed on the DC bus side to reduce switch-rated voltage issues. Three-level converters are suitable for high DC voltage applications due to their low switching voltage stress and smaller magnetic component size. Additionally, a detailed analysis of the magnetic flux density of the magnetic components used is conducted in this study, and a structure using two magnetic materials and different gap distributions on the magnetic core is proposed.

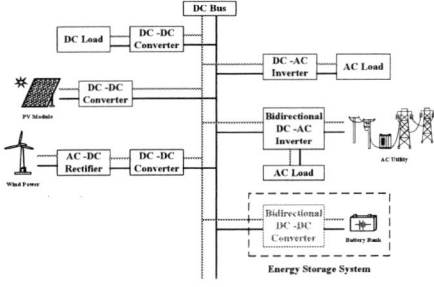

Fig. 1. Distributed Generation System [3], [4].

The paper is divided into five sections. Section II discusses the proposed converter's operational principles in both step-up and step-down stages. Section III covers the steady-state analysis and component design. Section IV presents the simulation and experimental results of the bidirectional converter obtained. Finally, Section V provides the conclusion of the paper.

II. ANALYSIS OF BIDIRECTIONAL DC-DC CONVERTERS

This paper discusses the operating principles and steady-state analysis of the proposed bidirectional DC-DC converter, whose topology is illustrated in Fig. 2. First, the circuit symbols and assumptions for the bidirectional converter are defined. Then, the operational principles of the converter are explained in two parts: step-up stage and step-down stage. This converter topology primarily consists of six switches (S_1-S_6), four capacitors (C_1-C_4), and two coupled inductors.

979-8-3315-1612-3/25 $31.00 © 2025 IEEE

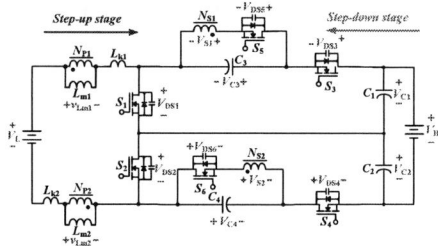

Fig. 2. The topology of the proposed bidirectional DC-DC converter circuit.

To simplify the subsequent analysis, the following assumptions are proposed before discussing the working principle:

1) The converter operates in steady state and at continuous conduction mode (CCM).
2) All switches, capacitors, and inductors are assumed to be ideal, with consideration given to the body diodes of all power switches.
3) The capacitors C_1 and C_2 on the high-voltage side are both sufficiently large to clamp the voltage at $\frac{V_{bus}}{2}$ with negligible ripple.
4) The inductance of the two magnetizing inductors is equal $L_{m1} = L_{m2}$, and the number of turns of the coupled inductor is defined as $n = \frac{N_s}{N_p}$.

A. Operating principle

1) Setp-up stage

The key waveforms for the proposed bidirectional converter in the step-up stage are shown in Fig. 3. In this stage, the primary roles are fulfilled by switches S_1 and S_2, with auxiliary roles played by S_3 and S_4, while S_5 and S_6 remain in the off state. The step-up stage is divided into four operational stages within each switching cycle.

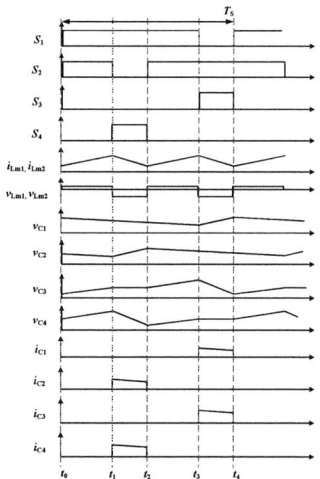

Fig. 3. Key waveform of the proposed bidirectional converter in step-up stage.

a) Mode I [t_0-t_1]: During the time interval between t_0 and t_1, S_1 and S_2 are turned on, and S_3 to S_6 are turned off. The

current path at this moment is illustrated in Fig. 4(a). During this period, L_{m1} and L_{m2} are charged by the input voltage V_L. Energy is released to C_3 and C_4 through secondary windings N_{S1} and N_{S2} by L_{m1} and L_{m2}, while energy is provided to V_H by C_1 and C_2 on the high side. This mode ends when S_2 turns off and S_4 turns on. In this mode, the voltages across the terminals of L_{m1} and L_{m2}, as well as C_3 and C_4, can be represented by equations (1) and (2), respectively.

$$v_{Lm1} + v_{Lm2} = V_L \tag{1}$$

$$v_{C3} = v_{C4} = \frac{n}{2} \cdot V_L \tag{2}$$

b) Mode II [t_1-t_2]: At $t=t_1$, S_1 is kept conductive, S_4 is turned on, S_3 is turned off, S_5 and S_6 are kept closed, as depicted in Fig. 4(b). C_2 is charged by L_{m1}, L_{m2}. Energy is discharged by C_4, and Lm1. This mode is concluded at $t=t_2$, with S_2 being opened and S_4 being cut off. The cross voltage of the two magnetizing inductances and the voltage across C_2 can be represented as equations (3) to (4).

$$v_{Lm1} + v_{Lm2} = V_L - v_{C2} + v_{C4} \tag{3}$$

$$v_{C2} = v_{C1} = \frac{1}{2} \cdot V_H \tag{4}$$

Additionally, the operation stage of Mode III is the same as that of Mode I, while Mode IV is similar to Mode II.

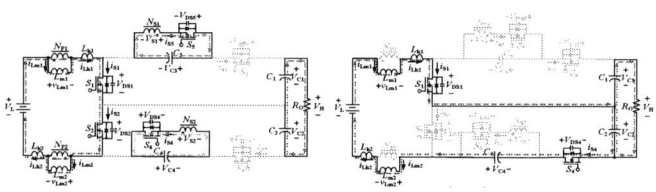

Fig. 4. Proposed bidirectional converter in step-up stage. (a) Mode I. (b) Mode II.

In step-up stage, the duty cycle is set at $0.5 < D_{step-up} < 1$, and there is a 180-degree phase difference in the control signals of S_1 and S_2, leading to their overlap. Modes I and III both S_1 and S_2 are turned on simultaneously, allowing voltage V_L to pass through and accumulate energy. Modes II and IV only one of S_1 or S_2 is turned on, resulting in an inductor voltage,. Using the formulas (1) and (3), the voltage conversion ratio in step-up stage can be derived as equation (6).

$$\frac{V_H}{V_L} = \frac{1}{1 - D_{step-up}} + n \tag{6}$$

2) Setp-down stage

The key waveforms for the proposed bidirectional converter in the step-down stage are shown in Fig. 5. In this stage, the primary switches are operated by S_3 and S_4, while S_5 and S_6 serve as auxiliary switches, and S_1 and S_2 remain closed throughout to facilitate conduction as parasitic diodes. Under the step-down stage, each switching cycle is divided into four operating stages for analysis.

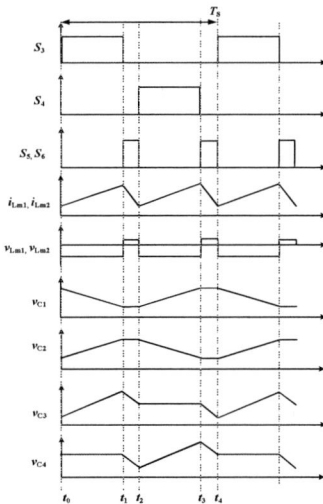

Fig. 5. Key waveform of the proposed bidirectional converter in step-down stage.

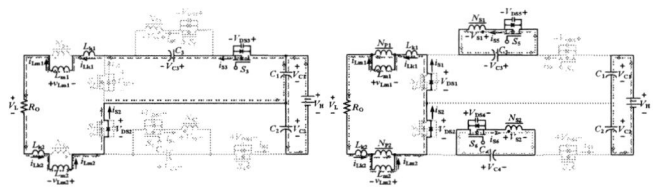

Fig. 6. Proposed bidirectional converter in step- down stage. (a) Mode I. (b) Mode II.

a) Mode I [t_0-t_1]: S_3 is turned on at t=t_0, and the parasitic diode of S_2 conducts. At this moment, S_1, S_4- S_6 are in the off state. C_3 is charged by high-voltage C_1 and input voltage V_H through L_{m1}, L_{m2}. Energy is provided to the low-voltage side load by L_{m1}, L_{m2}. The equivalent circuit diagram of this mode is depicted in Fig. 6(a). When S_5 and S_6 are turned on, and S_3 is turned off at t=t_1, this mode ends. The cross voltage on the magnetizing inductors is expressed by equation (7).

$$v_{Lm1} + v_{Lm2} = 2v_{Lm} = v_{C3} - V_L - v_{C1} \qquad (7)$$

b) Mode II [t_1-t_2]: Upon the activation of S_5 and S_6 at t=t_1, with main S_3 and S_4 in the off state, the high-voltage C_1 and C_2 are charged from the input voltage V_H. Energy transfer to C_3 and C_4 occurs through the secondary sides N_{S1} and N_{S2} of the coupled inductors. The load receives energy through the parasitic diodes of semiconductor devices S_1 and S_2, facilitated by L_{m1}, L_{m2}. The equivalent circuit diagram of this mode is depicted in Fig. 6(b). This mode at $t = t_2$ concludes with the activation of S_4 and the deactivation of S_5 and S_6. Equation (8) and (10) expresses the cross-voltage on the magnetizing inductors and C_3 and C_4.

$$v_{Lm1} + v_{Lm2} = V_L \qquad (8)$$

$$v_{C3} = \frac{v_{Lm1}}{n} \qquad (9)$$

$$v_{C4} = \frac{v_{Lm2}}{n} \qquad (10)$$

Furthermore, the operating method in Mode III resembles that of Mode I, while Mode IV aligns with Mode II.

In step-down stage, the duty cycle is set at $0< D_{step-down}<0.5$, and there is a 180-degree phase difference in the control signals of S_3 and S_4, leading to their overlap. Mode I and Mode III, only one of S_3 or S_4 is turned on, accumulating energy across the inductor terminals. Mode II and Mode IV, both S_5 and S_6 are turned on simultaneously, releasing energy from the inductor. Using the formulas (7) and (8), the equation for the magnetizing inductance in step-down stage can be derived as equation (12).

$$\frac{V_L}{V_H} = \frac{D_{step-down}}{nD_{step-down}+1} \qquad (12)$$

B. Design consideration of the coupled inductor

Assume the relationship between the average magnetizing inductor currents I_{Lm1} and I_{Lm2} and the average high-side current I_H under the BCM condition as equation (13) and (14).

$$I_{Lm\_BCM} = I_{Lm1} = I_{Lm2} = \frac{I_H(1-D_{step-up})}{2(1+n-n \cdot D_{step-up})} \qquad (13)$$

$$I_{Lm\_BCM} = \frac{1}{2} \cdot \frac{V_L}{L_m} \cdot D_{step-up}T_s = \frac{1}{2} \cdot \frac{V_H \cdot \frac{(1-D_{step-up})}{1+n-n \cdot D_{step-up}}}{L_m} \qquad (14)$$

Using the formulas (13) and (14), the equation for the magnetizing inductance in step-up stage can be derived as equation (15).

$$L_m = \frac{1}{2} \cdot \frac{V_H \cdot D_{step-up} \cdot (1-D_{step-up})^2}{I_{Lm\_BCM} \cdot f_s \cdot (1+n-n \cdot D_{step-up})} \qquad (15)$$

1) Core

The A_p method is related to the maximum energy stored in the inductor, defined as the product of the core window area W_a and the core cross-sectional area A_e, as shown in equation (16). Equation (17) is rewritten as equation (18).

$$A_P = A_e W_a = \frac{2 \text{Energy}(10^4)}{B_m J_m K_u} \qquad (16)$$

$$LI = NB_m A_e \qquad (17)$$

$$A_e = \frac{LI_{peak}^2}{NB_m I_{peak}} \qquad (18)$$

TABLE I. SOFT MAGNETIC MATERIALS.

Material	Ferrite	Nanocrystalline metal	Amorphous metal	MPPS	Ferrosilicon
Type	Epcos N87	Viroperm 500F	Metglas 2605	Micrometals 75μ	Unisil 23M3
Magnetic permeability, μ_e	2200	15000	10000~150000	75	5000~10000
B_{peak} (T)	0.49	1.2	1.56	0.6~1.3	2.0
ρ $(\mu\Omega \cdot m)$	10×10^6	1.15	1.3	10^6	0.48
Curie temperature (T_e/C)	210	600	399	665	745
P_{fe} $(\frac{mW}{cm^3})$	288 (0.2T, 50kHz)	312 (0.2T, 100kHz)	366 (0.2T, 25kHz)	1032 (0.2T,10kHz)	5.66 (1.5T,50kHz)
K_e	16.9	2.3	1.377	1798	3.388
α	1.25	1.32	1.51	1.02	1.7
β	2.35	2.1	1.74	1.89	1.9

Where the magnetic flux density B_m is measured in teslas, and the effective cross-sectional area of the core A_e is measured in cm^2. The window utilization factor K_u is defined as the ratio of the used window area W_a to the total winding area of the core W_c. W_c and W_a are in cm; I_{peak} is the inductor peak current; J_m is the current density, $\frac{A}{cm^2}$.

In addition to selecting the size and dimensions of the iron core, the properties of the magnetic materials can be optimized to enhance the converter's performance in various applications. Fiver common types of soft magnetic materials are introduced in Table I, with details provided on their permeability, maximum saturation magnetic flux density, resistivity, and core loss. Ferrites generally perform well in high-frequency applications due to their high resistivity, which reduces eddy current losses. Additionally, ferrite is characterized by a lower Curie temperature, while nanocrystalline materials exhibit excellent stability over a wide temperature range.

2) *Windings*

Before the wire area is calculated, the current density J_m should be defined, and the current density should be estimated using the Ap method according to equation (16). Due to the high operating frequency of the converter, the skin effect needs to be considered using the skin depth formula, as shown in equation (19) [13].

$$\delta = \frac{6.62}{\sqrt{f_s}} \quad (19)$$

where δ are in cm.

Additionally, the saturation flux density B_{peak} should be considered to design the maximum flux density B_m, and the number of turns for the primary and secondary windings should be calculated using equations (20) and (21).

$$N_p = \frac{L \cdot I_p}{A_e \Delta B_m} \quad (20)$$

$$N_s = \frac{L \cdot I_s}{A_e \Delta B_m} \quad (21)$$

The calculated number of turns for the primary and secondary windings, as given by equations (20) and (21), should be substituted into equation (19) to verify whether the coil's current density meets the skin depth requirements.

3) *Gap*

Adding an air gap to the magnetic core results in flattening the slope of the B-H curve, preventing core saturation, reducing the effective permeability of the core, and increasing the saturation current of the inductor [14]. The energy in the inductor is influenced by the core shape, size, winding method of the coil, number of turns, and the type of magnetic material used in the core. The inductance value is derived through the method of magnetic resistance, and based on the distribution of air gaps in the core and the use of different magnetic core materials, the equivalent magnetic resistance in the magnetic circuit is calculated [15].

a) Use single-gap to calculate the coupled inductance

Figure 7(a) depicts a cross-sectional view of a single air gap coupled inductor composed of two ETD-shaped magnetic cores with an air gap between the central pillars. The equivalent magnetic circuit diagram for the single air gap coupled inductor is shown in Fig. 7(b). $\Re_1$ is designated to represent the magnetic resistance on either side of the central pillars with the air gap. $\Re_2$ and $\Re_3$ denote the magnetic resistance of the outer pillars of the magnetic cores. $\Re_g$ is used to denote the magnetic resistance of the air gap. The generation of magnetic flux dependents on the magnetic potential mmf, given by (15). The total magnetic core resistance Re is calculated as the sum of $\Re_1$, $\Re_2$, and $\Re_3$, as expressed by equation (22). The formula for $\Re_g$ is provided by equation (23). The equivalent magnetic resistance $\Re_{eq(1gap)}$ in the magnetic circuit is derived as per equation (24). The formula for the inductance value of the single gap coupled inductor is represented by equation (35).

$$\Re_e = \frac{l_e - l_g}{\mu_r \mu_0 A_e} \quad (22)$$

$$\Re_g = \frac{l_g}{\mu_0 A_g} \quad (23)$$

$$\Re_{eq(1gap)} = \frac{l_e - l_g}{\mu_r \mu_0 A_e} + \frac{l_g}{\mu_0 A_g} \quad (24)$$

$$L_{(1gap)} = \frac{\mu_r A_e N^2}{l_g \left(\frac{\mu_r A_e - A_g}{\mu_r A_g} \right)} \quad (25)$$

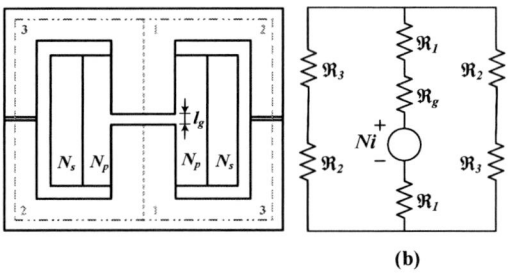

(b)

Fig. 7. Single-gap coupled inductor. (a) Cross-sectional view. (b) Equivalent magnetic circuit diagram.

b) Use dual-gap to calculate the coupled inductance

A coupled inductor is composed of two different magnetic materials and incorporates dual gaps. The outer frame of the core is made from one type of magnetic material, while the central pillars of the core are constructed from another type of magnetic material and bonded to the core frame with epoxy. Fig. 8(a) illustrates a cross-sectional view of a coupled inductor with dual gaps. It is comprised of two ETD-shaped magnetic cores, each with an air gap between the upper and lower sections of the central pillar, where the combined length of the dual gaps equals that of a single air gap. $\Re_1$ represents the magnetic resistance on both sides of the central pillars with the air gaps. $\Re_2$ and $\Re_3$ indicate the magnetic resistance of the outer pillars of the magnetic cores. $\Re_4$ signifies the magnetic resistance of the central pillar. $\Re_g$ denotes the magnetic resistance of the air gap. Fig. 8(b) displays the equivalent magnetic circuit diagram for the coupled inductor with dual gaps. The derivation concept for a coupled inductor with dual gaps is identical to that for a coupled inductor with only one air gap. The two different magnetic materials are designated as (1) representing the first magnetic material and (2) representing the second magnetic material. The total magnetic core resistance $\Re_e$ is the sum of $\Re_1$, $\Re_2$, $\Re_3$, and $\Re_4$, as expressed in equation (26). The formula for $\Re_g$ is given by equation (27). The derived equivalent magnetic resistance $\Re_{eq(dual-gaps)}$ in the magnetic circuit is provided by equation (28). The formula for the inductance value of the coupled inductor with dual gaps is represented by equation (29).

$$\Re_{(dual-gaps)} = \frac{l_{e(1)} - l_{2g}}{\mu_{r(1)} \mu_0 A_{e(1)}} + \frac{l_{e(2)} - l_{2g}}{\mu_{r(2)} \mu_0 A_{e(2)}} \quad (26)$$

$$\Re_g = \frac{l_{2g}}{\mu_0 A_{g(1)}} \quad (27)$$

$$\Re_{eq(dual-gaps)} = \frac{l_{e(1)} - l_{2g}}{\mu_{r(1)} \mu_0 A_{e(1)}} + \frac{l_{e(2)} - l_{2g}}{\mu_{r(2)} \mu_0 A_{e(2)}} + \frac{l_{2g}}{\mu_0 A_{g(1)}} \quad (28)$$

$$L_{(dual-gaps)} = \frac{\mu_0 (A_{e(1)} + A_{e(2)}) N^2}{l_{2g} \left(\frac{\mu_{r(1)} - 1}{\mu_{r(1)}} \right) + \left(\frac{l_{e(1)}}{\mu_{r(1)}} \right) + \left(\frac{l_{e(2)}}{\mu_{r(2)}} \right)} \quad (29)$$

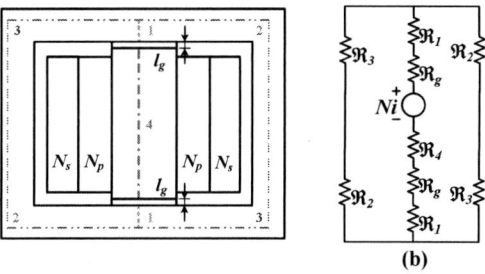

(b)

Fig. 8. Two different magnetic materials and dual-gaps coupled inductor. (a) Cross-sectional view. (b) Equivalent magnetic circuit diagram.

III. PARAMETERS DESIGN AND EXPERIMENTAL RESULTS

The parameter design and calculations of the proposed converter will be introduced. The specifications of the proposed bidirectional DC-DC converter are shown in Table II.

Table II The specifications of the proposed bidirectional DC-DC converter.

Circuit Parameters	Value
High voltage side voltage (V_H)	400 V
Low voltage side voltage (V_L)	60 V
Rated high voltage side current (I_H)	0.75 A
Rated low voltage side current (I_L)	5 A
Switching frequency (f_s)	100 kHz
Rated output power (P_o)	300 W

In the proposed converter, the ETD44/22/15 core model is chosen for the two coupled inductors, with parameters listed in Table III. The core material, 3C90 from FERROXCUBE, is widely used in magnetic applications, with a saturation flux density of 0.43 T at 25°C, decreasing to 0.35 T at 100°C. To prevent flux saturation and optimize efficiency, the operating point is set to 85% of the maximum flux density or 0.32 T. The flux density variation, ΔB_m, is designed as 0.22 T, calculated by subtracting the remanence from the maximum flux density. This ΔB_m value balances efficiency, thermal stability, and magnetic performance, keeping the core within its optimal operating range and minimizing energy loss.

Table III The parameters of core model ETD44/22/15

Symbol	Description	Value
B_{sat}	Saturation Flux Density	0.43 T @ 25°C/ 0.35 T @ 100°C
μ_r	Relative permeability	1800
A_e	Cross-section area	1.73 cm²
l_e	Mean path length (MPL)	10.3 cm
W_a	Window Area	2.14 cm²

IV. SIMULATION RESULTS & EXPERIMENTAL RESULTS

A prototype experiment was conducted to validate the theoretical analysis of the proposed bidirectional DC-DC converter. The feasibility of this converter will be verified using the SIMPLIS® simulation tool, while the magnetic flux will be simulated with the FEM software COMSOL Multiphysics®. Tables II, III, and IV outline the specifications and critical parameters of the experimental prototype.

Table IV Parameters of the key components

Items	Description
Turns ratios (n)	2
Magnetizing inductance (L_{m1}, L_{m2})	120 μH (core material: 3C90, bobbin: ETD44/22/15)
Capacitors (C_1, C_2)	220 μF/450 V (Electrolytic capacitor)
Capacitors (C_3, C_4)	120 μF/420 V (Electrolytic capacitor)
Power MOSFET ($S_1, S_2, S_3, S_4, S_5, S_6$)	NTP055N65S3H (V_{DSS}=650 V, I_D=47 A, $R_{ds,on}$=55 mΩ)

A. Finite element method

This study uses the multiphysics-coupled finite element software COMSOL Multiphysics for comprehensive simulation analysis. First, circuit's magnetic components model is constructed to examine the magnetic flux density distribution at different stages. An analysis of improved magnetic cores with two materials and varied gap distributions is also proposed. Finally, a comparative loss analysis of thermal simulations for the two magnetic components is discussed. The proposed bidirectional converter experiences a large current in step-down stage; therefore, Mode I in step-down stage is taken as an example to discuss the design of two coupled inductors in the actual circuit. These coupled inductors feature a single material and single air gap design, with the core made of ferrite material and a 0.4 mm air gap between the center posts. The cross-section and equivalent magnetic circuit diagrams of the simulated single-gap coupled inductor are shown in Fig. 9.

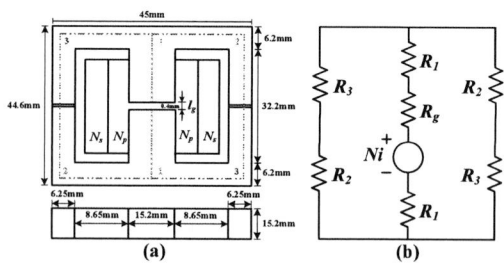

Fig. 9 Single-gap coupled inductor. (a) Coupled inductor cross-sectional view. (b) Equivalent magnetic circuit diagram.

The core losses as a function of frequency for ferrite and nanocrystalline materials are referenced in the literature [16]. In contrast, nanocrystalline materials exhibit an average

reduction in core losses at frequencies up to 300 kHz. Therefore, a design with two coupled inductors, each with magnetic cores made of different materials, is proposed. The core frame is made of ferrite material, while the center post core is composed of nanocrystalline material. The center post of the core is bonded to the core frame with epoxy resin, and the total air gap above and below the center post is 0.4 mm. The cross-sectional view and equivalent magnetic circuit diagram of the simulated double-gap coupling inductor are shown in Fig. 10.

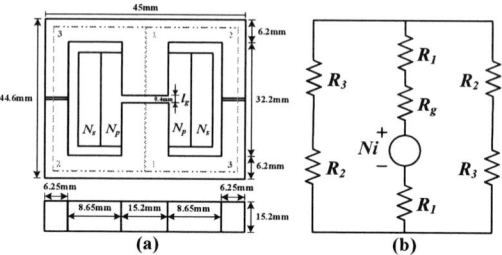

Fig. 10 Dual-gap coupled inductor. (a) Coupled inductor cross-sectional view. (b) Equivalent magnetic circuit diagram.

The core consists of ferrite 3C90 and nanocrystalline (NC) material, and its B-H curve is established, along with circuit simulation in a nodal manner, as shown in Figs. 11 and 12.

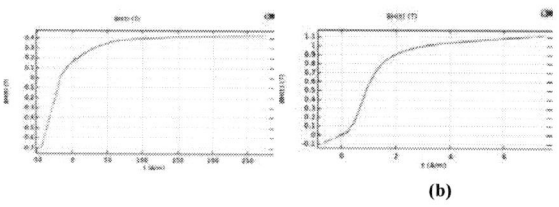

Fig. 11 The B-H curve. (a) Ferrite 3C90 material. (b) Nanocrystalline (NC) material.

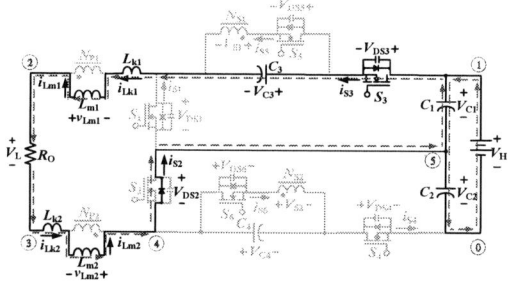

Fig. 12 Construct the circuit's equivalent circuit diagram using the node method (represented by Mode I in step-down stage).

1) Magnetic flux density simulation analysis

Finite element simulation software was used to model the two coupled inductors in the converter and to analyze the magnetic flux density distribution of improved magnetic cores made from two materials with varying gap distributions. For the step-down stage in Mode I, the maximum flux density and edge flux density of the two coupled inductors in the actual converter are 0.32 T and 0.2 T, respectively, as shown in Fig. 13, which meet the design specifications. In contrast, the maximum flux density and edge flux density of the proposed improved magnetic cores, also made from two different materials with varying gap distributions, are 0.2 T and 0.19 T, as shown in Fig. 14. The simulation results indicate that the magnetic components used in actual circuits have higher energy conversion efficiency; however, if lower edge effects and better high-frequency performance are required, the improved magnetic cores would be more suitable.

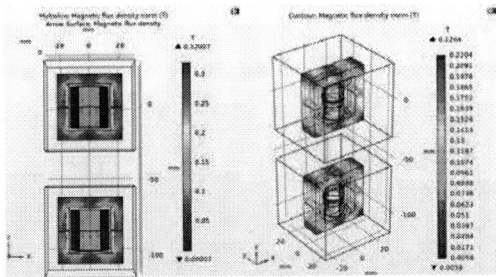

Fig. 13 Single-gap coupled inductor flux density analysis in step-down stage Mode I (a) Magnetic flux multilayer slice plot. (b) Magnetic flux contour plot.

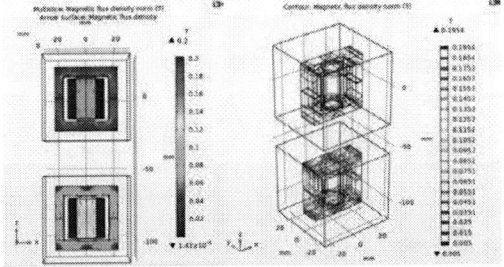

Fig. 14 Dual-gap coupled inductor flux density analysis in step-down stage Mode I. (a) Magnetic flux multilayer slice plot. (b) Magnetic flux contour plot.

2) Analysis of two different types of magnetic cores

The simulation conditions were set in the simulation software: input voltage of 400 V, output voltage of 60 V, rated power of 300 W, $Q_o = 5$, and ambient temperature of 25°C. Analyses were conducted at operating frequencies of 100 kHz and 300 kHz. Fig. 25 indicates that the single-material, single-gap core reaches a maximum temperature of 108°C at 100 kHz and 111°C at 300 kHz. Fig. 16 shows that the dual-gap core made of two different materials reaches a maximum temperature of 107.47°C at 100 kHz and 107.53°C at 300 kHz.

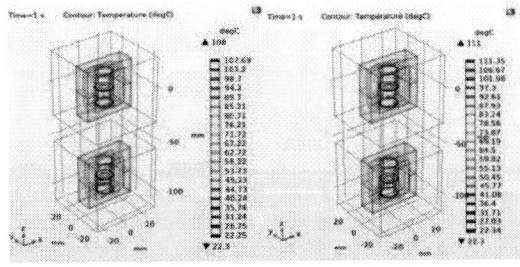

Fig. 15 Thermal simulation results of single gap core at 25°C at different frequencies in step-down stage Mode I. (a) 100 kHz. (b) 300 kHz.

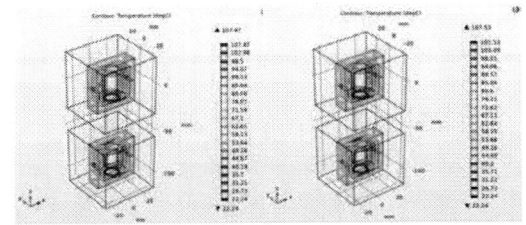

Fig. 16 Thermal simulation results of dual gap core at 25°C at different frequencies in step-down stage Mode I. (a) 100 kHz. (b) 300 kHz.

According to Fig. 15 and Fig. 16, the maximum temperature of the two types of cores at 100 kHz is quite similar. At 300 kHz, the maximum temperature of the dual-gap cores made of two distinct materials remains comparable to that at 100 kHz. In contrast, the maximum temperature of the single-material, single-gap cores increases significantly. The simulation results indicate that the dual-gap cores made of two different materials exhibit lower heat loss and better performance at high frequencies than the single-material, single-gap cores.

B. Experimental results and discussions

1) Step-up stage experimental results

Figure 17 and Fig. 18 show the experimental waveforms of the converter under 50 % load and full load in step-up stage and step-down stage, respectively. In step-up stage the main control switch signals, v_{gs1} and v_{gs2}, are phase-shifted by 180 degrees. S_1 and S_3 are complementary, as are switches S_2 and S_4. The i_{ds2} and i_{ds4} waveforms show that the charging and discharging of C_1 through C_4 occur in separate time intervals. In step-down stage the main control switch signals are v_{gs3} and v_{gs4}, with S_5 and S_6 acting as auxiliary switches, both turned on simultaneously. The i_{ds2} and i_{ds6} waveforms indicate that in step-down stage, the charging and discharging of C_1 through C_4 occur at different time intervals.

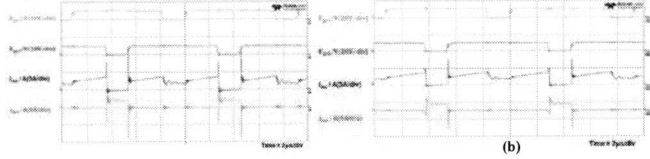

Fig. 17 The waveform of S_1 and S_2 condition in step-up stage. (a) 50 % load. (b) full load.

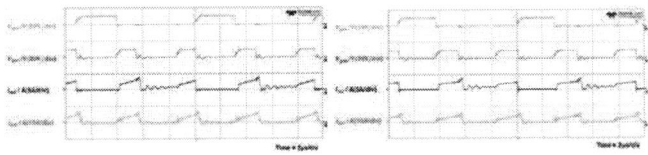

Fig. 18 The waveform of S_1 and S_2 condition in step-down stage. (a) 50 % load. (b) full load.

The efficiency curves for the step-up and step-down stages at different output power levels are shown in Fig. 18. The maximum efficiency of the step-up stage at 40% load is 93.72%, while the efficiency at full load is 92.68%. The maximum efficiency of the step-down stage at 50% load is 92.33%, while the efficiency at full load is also provided.

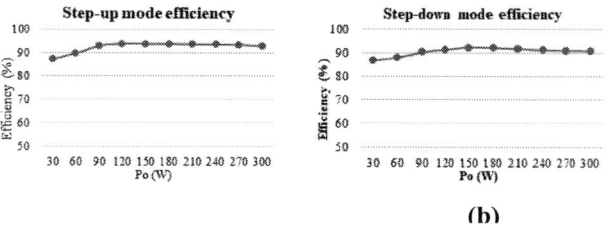

(b)

Fig. 18 Efficiency curve (a) step-up stage. (b) step-down stage.

In step-up stage, the full-load efficiency is 92.68 %, with a power consumption of 23.69 W. The switching loss is 5.12 W, the copper loss in the coupled inductors is 7.46 W, the core loss is 5.46 W, the capacitor loss accounts for 3.67 W, and the remaining losses account for 1.98 W. As shown in Fig. 19 (a), the copper loss and core loss of the inductors are the main reasons for the efficiency drop because the proposed converter uses two coupled inductors. In step-down stage, the full-load efficiency is 90.8 %, with a power consumption of 30.39 W. The switching loss is 4.66 W, the copper loss in the coupled inductors is 7.22 W, the core loss is 13.3 W, the capacitor loss accounts for 4.14 W, and the remaining losses are 1.07 W. As shown in Fig. 19 (b), because the proposed converter uses two coupled inductors, the core loss and copper loss of the inductors are the main reasons for the efficiency drop.

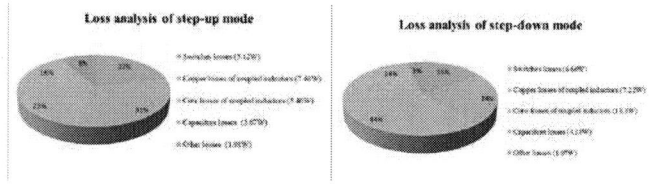

Fig. 19 Losses and proportion of various components. (a) step-up stage. (b) step-down stage.

V. CONCLUSIONS

This paper proposes a non-isolated bidirectional DC-DC converter to address the supply variability and unpredictability caused by renewable energy sources in future power systems. The converter utilizes a coupled inductor on the battery side to increase the bus voltage. It employs a three-level NPC structure on the DC bus side, effectively reducing the rated voltage stress on the switches. This approach improves system reliability and minimizes energy losses due to switching. A detailed analysis of the magnetic components is conducted, exploring various combinations of different magnetic materials. Magnetic flux and temperature distribution are confirmed through FEM simulation, which shows that at high frequencies, dual-gap cores made of two distinct materials exhibit lower heat loss and better performance compared to single-material, single-gap cores. This facilitates the efficient and rapid

selection of suitable air-gap inductors. Experimental results demonstrate the effectiveness of this converter, achieving a maximum efficiency of 93.72 % in the step-up stage at 40 % load and 92.33 % in the step-down stage at 50 % load.

REFERENCES

[1] A. Tayyebi, D. Groß, A. Anta, F. Kupzog and F. Dörfler, "Frequency Stability of Synchronous Machines and Grid-Forming Power Converters," in IEEE Journal of Emerging and Selected Topics in Power Electronics, vol. 8, no. 2, pp. 1004-1018, 2020.

[2] Y. P. Hsieh, J. F. Chen, L. S. Yang, C. Y. Wu, and W. S. Liu, "High-conversion-ratio bidirectional DC–DC converter with coupled inductor," IEEE Transactions on Industrial Electronics, vol. 61, no. 1, pp. 210-222, 2013.

[3] T. J. Liang and J. H. Lee, "Novel high-conversion-ratio high-efficiency isolated bidirectional DC–DC converter," IEEE transactions on industrial electronics, vol. 62, no. 7, pp. 4492-4503, 2014.

[4] Y. E. Wu and B. H. Pan, "High efficiency and voltage conversion ratio bidirectional isolated DC-DC converter for energy storage system," IEEE Access, vol. 10, pp. 55187-55199, 2022.

[5] S. B. Santra, D. Chatterjee, and T. J. Liang, "High gain and high-efficiency bidirectional DC–DC converter with current sharing characteristics using coupled inductor," IEEE Transactions on Power Electronics, vol. 36, no. 11, pp. 12819-12833, 2021.

[6] W. C. Liao, T. J. Liang, H. H. Liang, H. K. Liao, L. S. Yang, K. C. Juang, and J. F. Chen, "Study and implementation of a novel bidirectional DC-DC converter with high conversion ratio," in 2011 IEEE Energy Conversion Congress and Exposition, pp. 134-140, 2011.

[7] H. J. Choi, J. Y. Lee, J. Y. Sim, S. G. Cheon, C. U. Lee, and J. H. Jung, "Switching modulation method for current-fed dual-active-bridge converter to improve power conversion efficiency," in 2019 10th International Conference on Power Electronics and ECCE Asia (ICPE 2019-ECCE Asia), pp. 1-6, 2019.

[8] N. A. Kumar, R. Dhanalakshmi, and C. Gayithri, "Investigations on bidirectional resonant converters for renewable energy sources and energy storage systems," in 2016 International Conference on Advanced Communication Control and Computing Technologies (ICACCCT), pp. 396-401, 2016.

[9] Z. Hou, D. Jiao, B. C. Gutierrez, J. -S. Lai and P. -L. Chen, "Design of a 15kW High-Efficiency and High Power Density Bidirectional TCM Buck/Boost Converter," 2024 IEEE Applied Power Electronics Conference and Exposition (APEC), pp. 341-347, 2024.

[10] K. Bi, Y. Liu, Y. Zhu, and Q. Fan, "An impedance source modular DC/DC converter for energy storage system: analysis and design," International Journal of Electrical Power & Energy Systems, vol. 133, p. 107261, 2021.

[11] C. A. Ramos-Paja, D. G. Montoya, and J. D. Bastidas-Rodríguez, "Sliding-mode control of a CuK converter for voltage regulation of a dc-bus," Sustainable Energy Technologies and Assessments, vol. 42, p. 100807, 2020.

[12] M. Shahin, "Analysis of bidirectional SEPIC/Zeta converter with coupled inductor," in 2015 International Conference on Technological Advancements in Power and Energy (TAP Energy), pp. 103-108, 2015.

[13] C. T. McLyman: Transformer and Inductor Design Handbook, 4th ed. CRC Press, Boca Raton, 2011.

[14] A. Ayachit and M. K. Kazimierczuk, "Steinmetz equation for gapped magnetic cores," IEEE Magnetics Letters, vol. 7, pp. 1-4, 2016.

[15] H. Liao and J. F. Chen, "Design process of high-frequency inductor with multiple air-gaps in the dimensional limitation," The Journal of Engineering, vol. 2022, no. 1, pp. 16-33, 2022.

[16] R. A. Gomez, D. A. P. Fernandez, G. G. Oggier, J. C. Balda, and Y. Zhao, "Comparison of high-frequency ferrite and nanocrystalline core losses using identical geometries," in 2022 IEEE 13th International Symposium on Power Electronics for Distributed Generation Systems (PEDG), pp. 1-5, 2022.

Comparison of 2-Level and Quasi-2-Level Topologies in a Bidirectional Isolated DC-DC Converter for MVDC Networks

José Andrés Aguilar Croston
SuperGrid Institute and
INSA Lyon, Laboratoire Ampère
Villeurbanne, France
jose-andres.aguilar@supergrid-
institute.com

Jean-Yves Gauthier
INSA Lyon, CNRS, Universite
Claude Bernard Lyon 1, Ecole
Centrale de Lyon, Ampère,
UMR5005, 69621,
Villeurbanne, France
jean-yves.gauthier@insa-lyon.fr

Cyril Buttay
CNRS, INSA Lyon, Universite
Claude Bernard Lyon 1, Ecole
Centrale de Lyon, Ampère,
UMR5005, 69621,
Villeurbanne, France
cyril.buttay@insa-lyon.fr

Maryam Saeedifard
School of Electrical and
Computer Eng.
Georgia Institute of Tehcnology
Atlanta, USA
maryam@ece.gatech.edu

Besar Asllani
SuperGrid Institute
Villeurbanne, France
besar.asllani@supergrid-
institute.com

Piotr Dworakowski
SuperGrid Institute
Villeurbanne, France
piotr.dworakowski@supergrid-
institute.com

Abstract— **High-power DC-DC converters are essential for efficient power conversion in emerging DC networks. This paper presents a comparative analysis of bridge topologies suitable for medium–voltage (10-40 kV), isolated DC-DC converters, focusing on a 2-Level topology using series-connected SiC MOSFETs, and quasi-2-Level topologies, specifically the Flying Capacitor Converter and the Modular Multilevel Converter. Key performance indicators (KPIs) including semiconductor devices counts, voltage slew-rate during switching (dV/dt), energy stored in passive elements, and efficiency are analyzed. KPIs are evaluated across various voltage levels and two specific case studies. The Modular Multilevel Converter with quasi-2-level operation is demonstrated to be particularly appealing at higher voltage levels due to its modular design.**

Keywords—DC-DC converter, medium voltage, dual active bridge converter, multilevel, 2-level, quasi-2-level

I. INTRODUCTION

The rapid expansion of Distributed Energy Resources (DERs), Energy Storage Systems (ESSs), and DC loads—driven by global decarbonization efforts—requires an adaptation of the traditional electrical grid. DC grids are anticipated to play a pivotal role in this transformation. High-voltage DC (HVDC) systems are now well established and essential for deploying large offshore wind farms. Moreover, the increasing interest in electric vehicles (EVs), ESSs, and data centers is opening the door to Medium Voltage DC (MVDC) systems, which are currently the subject of extensive research [1], [2]. Central to MVDC systems are high-power DC-DC converters. Among the available options [3], DC-DC converters can be classified according to three criteria: galvanic isolation, power flow directionality and modularity.

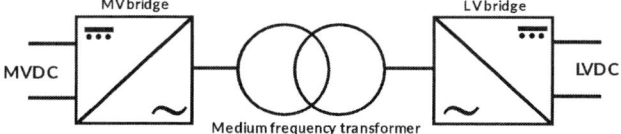

Fig. 1. Simplified single-line diagram of a monolithic isolated DC-DC converter.

For the interconnection between MVDC and Low Voltage DC (LVDC) networks, isolated converters, shown in Fig. 1, allow to decouple the MVDC constraints and the LVDC constraints in a similar way as it is done in AC with transformers. Step-up, unidirectional converters show promise for the MVDC collection networks which can be used for wind and photovoltaic [4], [5] generation. Due to the lack of semiconductors capable of sustaining the desired medium voltage levels (10-40 kV), bidirectional or step-down isolated converters can be developed using either a modular approach of input-series output-parallel (ISOP) isolated converters [6] or a monolithic approach with one transformer and a MV active bridge [7], [8], [9], [10], [11].

The modular ISOP approach enables the use of low-voltage, high-speed switching devices, though it requires multiple medium-frequency transformers (MFTs), each isolated to the medium voltage. Conversely, the monolithic approach requires an active bridge design capable of operating at medium voltage levels. The limited voltage rating of available semiconductor devices (up to 6.5 kV) is a major challenge in these applications. When the bus voltage level exceeds the maximum rating of a single semiconductor device there are two solutions: either a series connection of multiple devices [10], [12], [13] or a multilevel voltage-source topology. Fig. 2 shows several active

979-8-3315-1612-3/25 $31.00 © 2025 IEEE

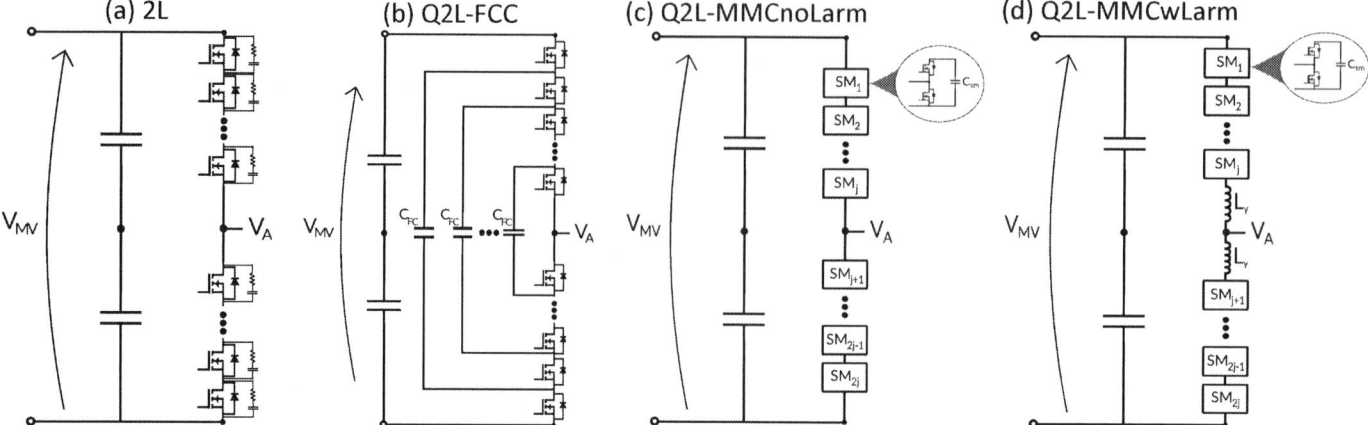

Fig. 2. Active bridge topologies suitable for MV applications: (a) MOSFET series connection (2L), (b) Flying Capacitor Converter (Q2L-FCC), (c) Modular Multilevel Converter without arm inductors (Q2L-MMCnoLarm) and (d) Modular Multilevel Converter with arm inductors (Q2L-MMCwLarm).

bridge topologies suitable for the MV side of a monolithic converter.

Voltage balancing and high dV/dt transients pose significant challenges for series-connected switching devices, especially at medium-voltage levels. In contrast, multilevel topologies—originally developed to address device voltage limitations—also improve the harmonic spectrum of the output, making them particularly suitable for DC-to-AC interfaces. However, these topologies require large passive components to generate multiple voltage levels, which can be a drawback. For isolated DC-DC converters using MFTs, where a sinusoidal AC waveform is not necessary, the harmonic spectrum is less critical.

Recent studies have explored a quasi-2-level (Q2L) or quasi-square wave (QSW) approach. Q2L operation consists in operating the transistors in a multilevel bridge in rapid succession, approximating the square wave output of a 2-level inverter. This approach has been demonstrated with Modular Multilevel Converters (MMCs) [7], [8], [14], [15], [16], Flying Capacitor Converters (FCCs) [11], [17], and Neutral Point Clamped (NPC) topologies [18]. Unlike traditional multilevel topologies, Q2L configurations do not require large passive components and enable the use of fast-switching power semiconductor devices, such as SiC MOSFETs, without exposing the AC output of a Q2L bridge to unacceptably high dV/dt transients.

Although some comparative studies of bridge topologies exist, few include Q2L operation [18], [19], [20], [21]. A comparison of bridge topologies with Q2L operation was conducted in [22], though it excluded 2-level (2L) topology. To

the best of the authors' knowledge, no existing studies compare a 2L bridge topology with prominent multilevel topologies in Q2L mode.

This paper identifies 2L and Q2L bridge topologies suitable for medium voltage, bidirectional, isolated DC-DC converters. Key performance indicators (KPIs) are established to facilitate the design and assess MV active bridges. The novelty of this work lies in its comparative analysis of 2L topology with multilevel topologies operating in Q2L, to determine the optimal choice based on KPIs calculated at different voltage levels.

The rest of the paper is structured as follows: Section II details the selected DC-DC converter topology, the corresponding modulation and operation of the MV bridge. Section III presents the MV bridge topologies, their properties and design rules. Section IV describes the models, defines the KPIs and compares bridge topologies for 2 voltage levels. Section V discusses the main findings and provides recommendations for choosing the MV bridge topology. The last section presents conclusions and perspectives.

II. CASE STUDY AND MV BRIDGE OPERATION

The selected DC-DC converter for this study is a single phase Dual Active Bridge (DAB) converter [23], shown in Fig. 3. Key parameters of the two specific cases considered in this paper are provided in Table I. The converter operates at a switching frequency (F_{SW}) of 10 kHz, with Silicon Carbide (SiC) MOSFETs chosen as the primary power semiconductors due to their high efficiency and fast switching capabilities. The voltage and current waveforms on the MV side of the DAB are shown in Fig. 4. For Q2L operation, the phase shift δ, is measured from the midpoint of the staircase transition to emulate the behavior of a conventional 2L bridge.

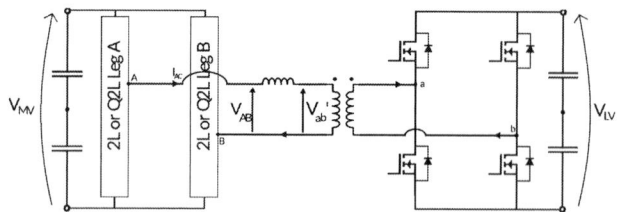

Fig. 3. Dual Active Bridge (DAB) converter based on 2L or Q2L operation. Q2L leg can be realized based on a multilevel topology such as FCC or MMC.

TABLE I. CASE STUDY PARAMETERS.

Case studies	Parameters		
	V_{MV}	V_{LV}	Nominal power
1st	40 kV (± 20 kV) ±10%	1.5 kV ±10%	4 MW
2nd	10 kV (± 5 kV) ±10%	1.5 kV ±10%	1 MW

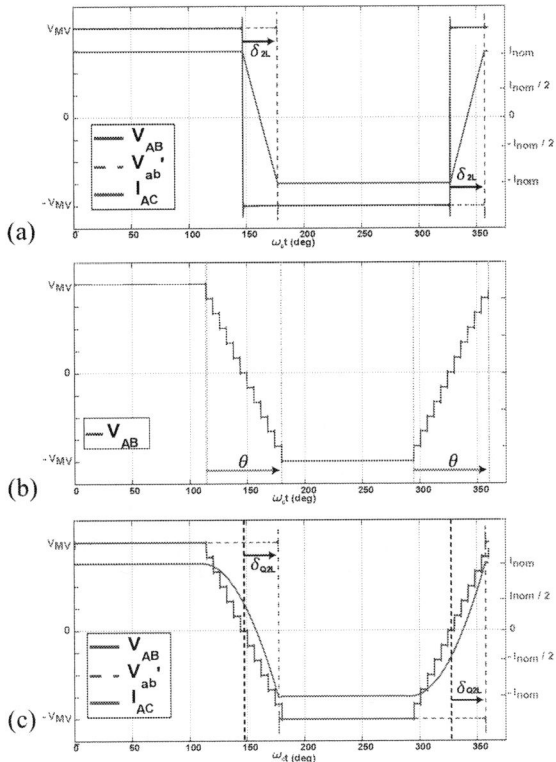

Fig. 4. DAB converter waveforms within one switching period with (a) 2L and (c) Q2L operation along with (b) the definition of transition time angle θ.

The chosen modulation strategy is the so-called single-phase shift (SPS). For the 2L bridge, V_{AB} (Fig. 3) can be considered as a square waveform, while for a Q2L bridge, V_{AB} appears as a staircase waveform (shown in Fig. 4). In this staircase, the transition time angle θ depends on the number of voltage levels (n) and the duration of each step (t_{delay}). To maintain similar operational characteristics between 2L and Q2L, θ must remain small relative to the fundamental period $T = 1/F_{SW}$. To achieve this, the duration of each step is limited to 200 ns duration. θ can be then calculated as:

$$\theta = (n - 2) \times t_{delay} \times 360 \times F_{sw} \qquad (1)$$

By setting t_{delay}, Q2L topologies allow to control the transition time angle (θ), and therefore the average dV/dt of the bridge over the entire switching sequence. The maximum dV/dt value, however, is larger and occurs at each intermediate voltage level in the staircase waveform. Considering a rise/fall time of 50 ns for each transistor, Fig. 5 illustrates the dV/dt variation across different V_{MV} values. Notably, dV/dt values in Q2L topologies remain independent of V_{MV}, highlighting a key advantage of Q2L operation. In 2L operation, dV/dt increases linearly with the voltage level. To manage extreme dV/dt values in 2L operation, switching times must be extended, which consequently raises switching losses. Due to its impact on transformer design and switching losses, the maximum dV/dt is selected as one of the KPIs for the final comparison of topologies.

III. MV BRIDGE TOPOLOGIES

The following arrangements (bridge topologies) are considered to form a MV bridge: series connection of SiC MOSFETs (the 2L topology), and FCC and MMC (operated with Q2L modulation). While the NPC topology can also be designed for Q2L operation, it is excluded from this study due to its complexity beyond three levels.

A. Series connection of SiC MOSFETs (2L)

As all MOSFETs are switched simultaneously, this is classified as a 2L topology. The bridge shown in Fig. 2 (a), employs RC snubbers for voltage balancing. The capacitance (here C=5 nF) is selected to be at least ten times larger than the C_{oss} of the SiC MOSFETs. Note that the addition of RC snubbers increases switching losses as demonstrated in [24].

B. Flying Capacitor Converter (Q2L-FCC)

Fig. 2 (b) illustrates the FCC topology [25], which is adapted for Q2L operation to allow the use of smaller capacitors [11]. According to the voltage balancing technique presented in [26], the flying capacitors value can be chosen following :

$$C_{fc} = \frac{\eta \times t_{delay} \times I_{peak}}{\Delta V} \qquad (2)$$

Here η is equal to the number of switching cycles required for voltage balancing, I_{peak} is the peak current and ΔV denotes the maximum permissible voltage ripple. A constant absolute ΔV across all switches implies that all capacitors have the same capacitance value, although their voltage ratings differ. The number of switching cycles required for voltage balancing η is equal to the number of switching devices in each arm, and it depends on the voltage level V_{MV}. As a result, the flying capacitance size increases significantly with higher voltage levels, driven by the series/parallel configurations needed to achieve the required C_{fc}. Although alternative voltage balancing strategies may reduce the required C_{fc}, achieving high V_{MV} levels remains challenging.

C. Modular Multilevel Converter (Q2L-MMC)

Shown in Fig. 2 (c) and (d), this topology consists of several identical submodules. Compared with its traditional implementation [27] for line-frequency AC to DC conversion, submodules designed for Q2L operation have much smaller capacitors. Two primary design variations are distinguished:

1) Quasi-2-level adaptation without arm inductors (Q2L-MMCnoLarm)

In this topology, shown in Fig. 2 (c), the only arm inductance present is the stray inductance [8]. Since the RMS currents differ

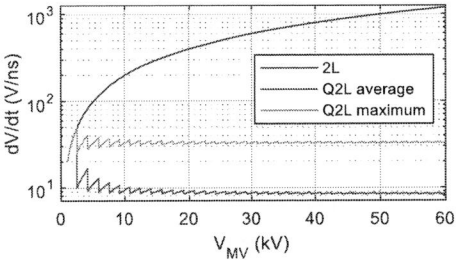

Fig. 5. dV/dt as a function of V_{MV} values for 2L and Q2L topologies.

significantly between the top and bottom switches of each submodule, asymmetric cells, such as those described in [12], can be used, where the top switch has a much lower current rating than the bottom switch. In this case, the submodule capacitance value can be calculated by:

$$C_{sm} = \frac{\eta(\eta+1) \times t_{delay} \times I_{peak} \times \xi}{2 \times \Delta V} \qquad (3)$$

Here, η is equal to the number of switching cycles required for voltage balancing, while ξ is an adjustable margin for precise ripple sizing, with a maximum value of 1. Notably, (3) provides a conservative sizing estimate assuming continuous charging or discharging over η cycles. In most applications, lower capacitance values may suffice. For this case, ξ is set to 1.

2) Classic approach with arm inductors (Q2L-MMCwLarm)

This approach, shown in Fig. 2 (d), includes arm inductors, which represent a significant part of the DAB inductor. Current waveforms differ from those in the Q2L adaptation without inductors [7]. In this setup, submodules are symmetric, meaning the top and bottom switches share identical current ratings. The submodule capacitance C_{sm} is calculated as follows:

$$C_{sm} = \frac{\eta \times I_{peak} \times \xi''}{\Delta V \times 2 \times F_{sw}} \qquad (4)$$

Where ξ'' is an adjustable margin maximized at a value of 1. This configuration uses continuous conduction in each arm, which means the submodule capacitor sizing depends on the switching frequency (F_{sw}) rather than t_{delay}. Since the current at the AC link is shared between the two arms of a leg, ξ'' reflects the current through the arm with inserted submodules. In this study, ξ'' is set to 0.3.

IV. COMPARISON BETWEEN 2L AND Q2L TOPOLOGIES

A variety of objectives must be considered when designing a DC-DC converter: cost, compact size, low complexity, DC-fault handling and high efficiency. Analytical models for the corresponding KPIs are presented below for each bridge.

A. Number of semiconductor devices

This first indicator is a primary determinant of converter cost. Fig. 6 presents the minimum number of 3.3 kV rated SiC MOSFETs required for one leg of each topology as a function of DC bus voltage, based on an operating voltage (V_{op}) of 1.7 kV, for reliability purposes, and a rated current of 100 A per device. The number of semiconductors ($\#_{SC}$) in a 2L and Q2L-FCC leg can be calculated following:

$$\#_{SC} = 2 \times ceil\left(\frac{V_{MV}}{V_{op}}\right) \qquad (5)$$

The lowest number of semiconductor devices is found in the 2L and Q2L-FCC topologies. Assuming identical chips with similar current ratings, this indicator enables the comparison and distinction of submodule designs between the two MMC topologies. The Q2L-MMCwLarm features a symmetrical submodule structure, resulting in twice the number of switches

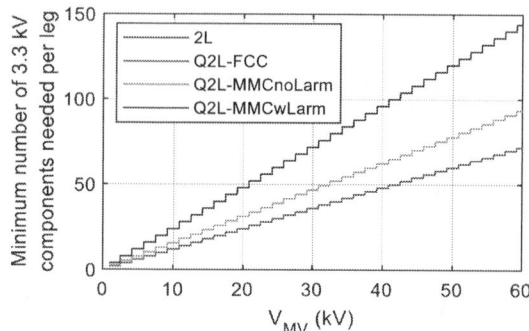

Fig. 6. Number of semiconductor switching components per leg for different V_{MV} values.

compared to the 2L topology. In contrast, the Q2L-MMCnoLarm employs an asymmetrical submodule design, requiring fewer chips than the Q2L-MMCwLarm. The Q2L-MMCnoLarm is considered to have 30% more chips than the 2L topology. This percentage is obtained after comparing the current rating of the bottom switching device and the RMS current seen by the top switching device in series with the submodule capacitor.

B. Energy stored in passive elements

This indicator represents the capacitors size within the converter. The energy stored in inductors is not considered because all cases are equivalent (they only differ in the way the required transformer inductance is distributed in the circuit). This indicator provides an estimate of the converter size.

This indicator is not only related to the size and cost of the capacitors: the energy stored in a single leg would supply a short circuit during a DC-fault. The converter must withstand the current in case of a fault. Therefore, it is important to keep stored energy at a minimum.

Fig. 7 illustrates the total capacitive energy in one 2L or Q2L leg across different V_{MV} values, with a voltage ripple of 200 V assumed for the Q2L legs.

For 2L and Q2L-MMC topologies the energy of passive elements (E_{pas}) in one leg for a fixed V_{MV} can be calculated with:

$$E_{pas} = N \times C_{top} \times (V_{sm})^2 \qquad (6)$$

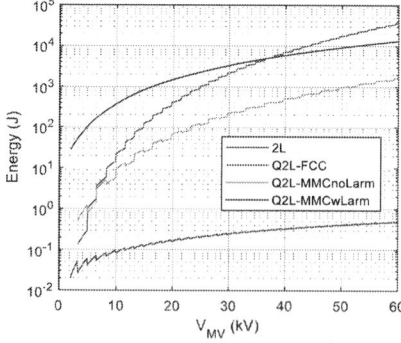

Fig. 7. Energy of passive elements installed within one leg for different topologies as a function of V_{MV}.

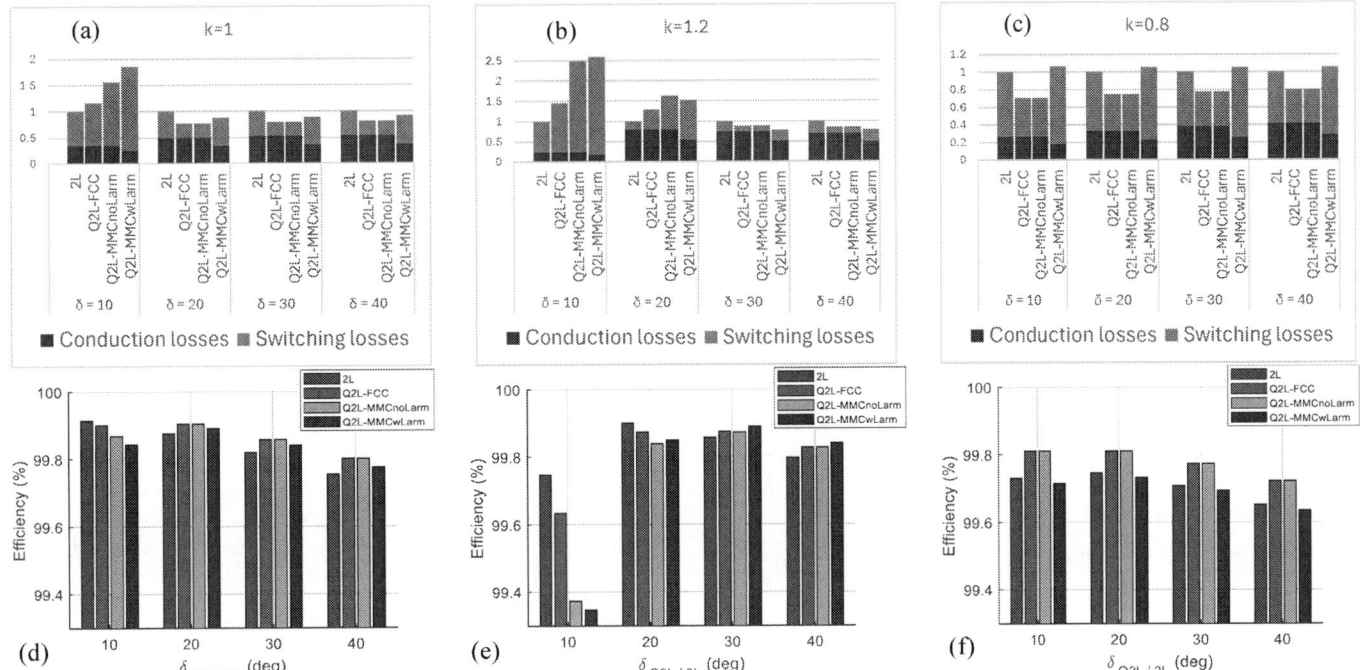

Fig. 8. MV bridge semiconductor losses repartition compared to Q2L-FCC (a), (b), (c), and total efficiency (d), (e), (f), for V_{MV} = 40 kV, different voltage ratios (k) and different active bridge topologies. Nominal operation with k = 1 is represented on (a) and (d). The maximum voltage ratio k = 1.2 is represented on (b) and (e). The minimum voltage ratio k = 0.8 is represented on (c) and (f).

Here V_{sm} = V_{MV} / N. Where N is the number of switching devices in one 2L or Q2L-FCC arm and the number of submodules in one arm for Q2L-MMC topologies. The capacitance C_{top} is the value of the snubber capacitor for a 2L topology and the value of C_{sm} for the Q2L-MMC topologies.

For the Q2L-FCC the energy of flying capacitors (E_{fcc}) in one leg for a fixed V_{MV} can be calculated following:

$$E_{fcc} = \frac{1}{2} \times C_{fc} \times \sum_{i=1}^{N-1} \left(\frac{V_{MV}}{N} \times i \right)^2 \quad (7)$$

The lowest energy is found in the series connection of MOSFETs with RC snubbers because of the small size of snubber capacitors compared to C_{fc} and C_{sm}. The Q2L-MMCnoLarm is the second lowest, however it is between 2 to 3 orders of magnitude higher, regardless of the voltage level. For voltage levels lower than 35 kV, the Q2L-MMCwLarm stores the most energy, while for higher voltages it is surpassed by the Q2L-FCC.

C. Losses distribution and efficiency

These indicators evaluate the performance of each topology in the DC-DC converter and indicate trends in the operation expenditure (OPEX) of the converter. In this work, only MV bridge semiconductor losses are considered since the transformer and LV bridge are assumed to be the same for all cases. This section aims to analyze the distribution of losses for each topology and to evaluate any efficiency differences.

Losses are calculated following the methods in [28] and with data directly from the datasheet of the SiC MOSFET considered

(GeneSiC G2R50MT33K [29]). The models are implemented using MATLAB Simulink. The DAB AC-link inductor (L) is designed for 2L operation at a nominal power flow with δ_{2L}=30°. The same inductor value is used for all subsequent cases.

Conduction losses are similar for the 2L, Q2L-FCC and Q2L-MMCnoLarm topologies. Conduction losses in Q2L-MMCwLarm differ due to the sharing of the AC-link current between the two arms in one leg. Given that conduction losses are proportional to the square of the current this results in lower conduction losses in Q2L-MMCwLarm.

Switching losses differ from 2L and Q2L operation due to the multiple switching instants at different current values in Q2L operation as shown in Fig. 4 (c). Fig. 8 (a), (b) and (c) illustrate the MV bridge semiconductor losses distribution for V_{MV} = 40 kV at different voltage ratios (k) to consider the extreme voltage variations of $\pm$ 10% presented in Table I. Where k = V_{LV} / ($V_{MV} \times n_t$) with a transformer turns ratio of 1: n_t. Losses are shown in per unit with total 2L losses as the reference.

Switching losses will also depend on the number of semiconductor devices, penalizing Q2L-MMC topologies with more semiconductor devices. However, depending on the current waveforms, the extra switches can operate under soft switching conditions and avoid decreasing efficiency.

In Q2L-MMC topologies, switches within each submodule operate as complements. In Q2L-MMCnoLarm, depending on the sign of the AC-link current, the semiconductor device in series with the submodule capacitor can achieve zero-voltage switching (ZVS) during turn on, while turn off occurs with

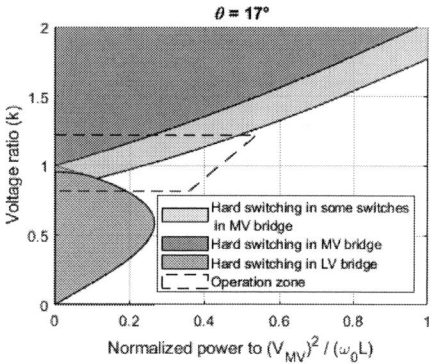

Fig. 9. Zero-voltage switching operating zones for a DAB converter with a Q2L bridge and a transition time angle (θ) of 17°.

minimal current, near zero. For the Q2L-MMCwLarm topology, the same principle applies for the turn-on but not for the turn-off. These characteristics contribute to lower switching losses in the Q2L-MMCnoLarm compared to Q2L-MMCwLarm.

To evaluate the differences in efficiency, Fig. 8 (d), (e) and (f) displays the MV bridge efficiency considering semiconductor losses of the different topologies. The ideal lossless power through the DAB converter is calculated following:

$$P = \frac{k \times (V_{MV})^2}{\omega_0 \times L} \delta_{Q2L/2L} \left(1 - \frac{|\delta_{Q2L/2L}|}{\pi}\right) \quad (8)$$

Efficiency is strongly influenced by the ZVS operation boundaries. The ZVS operation zones are shown in Fig. 9 where $\omega_0 = 2\pi F_{sw}$. Under nominal conditions with $k = 1$, SPS modulation enables ZVS operation in the 2L topology across all phase shifts (δ_{2L}). In contrast, during Q2L operation some switches operate outside the ZVS region at lower δ_{Q2L} values, represented by a gray zone in Fig. 9. This explains the higher efficiency of the 2L topology at $\delta = 10°$, as shown in Fig. 8 (d). For higher phase shifts, the Q2L topologies achieve greater efficiency due to the ZVS operation. Among the Q2L topologies, the Q2L-MMCwLarm has the lowest efficiency because of the parameters chosen in this case study, such as the switching frequency which results in higher switching losses.

With SPS modulation, the maximum voltage ratio $k = 1.2$ results in a loss of ZVS in the MV bridge at low δ values across all topologies. Q2L topologies, due to their multiple switching instants, exhibit a smaller ZVS region and thus higher switching

losses for low δ_{Q2L} values. Beyond a certain phase shift, however, Q2L topologies enter the ZVS operation zone, improving efficiency. At this voltage ratio, conduction losses dominate in the MV bridge, under which conditions the Q2L-MMCwLarm topology achieves the highest efficiency.

When the DAB converter operates with SPS and the minimum voltage ratio $k = 0.8$, soft switching of the MV bridge remains unaffected. Here, switching losses are predominant, favoring Q2L-FCC and Q2L-MMCnoLarm topologies.

Overall, the best efficiency will depend on the voltage ratio and phase shift. Nevertheless, the difference in efficiency is lower than 0.3 percentage points, for most of the presented cases in Fig. 8 (d), (e) and (f), which means that all topologies perform similarly but losses are distributed differently.

D. Transient behavior

Considering the DC-DC converter's goals of low complexity and DC-fault handling, it is essential to address transient behaviors, particularly during start-up and DC-fault conditions.

Start-up of the MV bridge considers the pre-charging of capacitors. In 2L the resistor in the RC snubber inherently limits the current inrush. Q2L topologies must pre-charge the capacitors to achieve the intermediate voltage levels with a limited current. In Q2L-FCC and both Q2L-MMC topologies, multiple capacitors must be charged during start-up. The start-up complexity can be assessed by the number of distinct voltage levels needed for capacitor charging. In Q2L-MMC topologies, each leg contains $2 \times N$ capacitors, each capacitor charged to the same voltage level V_{MV} / N, making start-up complexity independent of voltage level. Conversely, the Q2L-FCC topology has $N - 1$ capacitors, each at a unique voltage level. As a result, the complexity of start-up scales with N, therefore it depends on V_{MV}.

Regarding DC-fault handling, performance depends on the energy stored in the MV bridge and the capability to block fault currents in one arm. The 2L topology, with two orders of magnitude less energy stored than Q2L topologies, offers the best DC-fault performance. Among Q2L options, Q2L-MMC topologies perform better than Q2L-FCC, as this topology does not allow to prevent the flying capacitors to discharge in a DC-fault through the antiparallel diodes of the SiC MOSFETs.

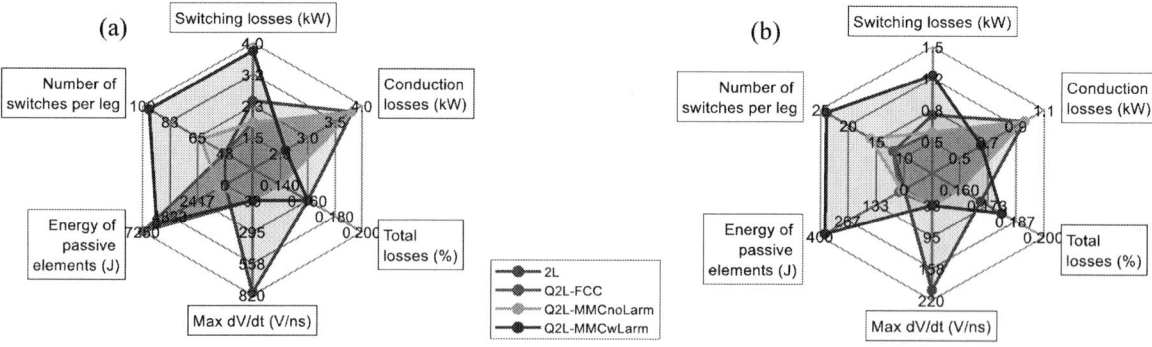

Fig. 10. Key performance indicators of different bridge topologies for 40 kV (a) and 10 kV (b) case studies.

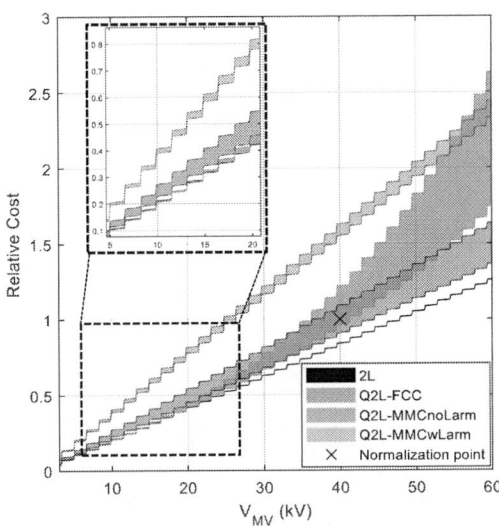

Fig. 11. Cost trend of different medium voltage active bridge topologies for a nominal current of 100 A. The voltage slew-rate during switching (dV/dt) is not monetized.

V. RESULTS

In Fig. 10 the KPIs for the two case studies are displayed. The losses and efficiency were calculated for nominal operation with a voltage ratio of 1. For all the listed KPIs, it is considered that the lowest value is the best one.

These results show that, depending on the voltage level V_{MV}, some topologies are more interesting than others. For both case studies it is shown that the biggest disadvantage of the 2L topology is the dV/dt. Q2L-FCC can be interesting at lower voltage levels because at higher voltage levels the energy of passive elements becomes an issue. The trend for the Q2L-MMC topologies is the same for both case studies, the Q2L-MMCnoLarm is the best candidate at 40 kV and is worth considering at 10 kV. In both case studies the Q2L-MMCwLarm has the most switching losses, a high stored energy and requires the largest amount of semiconductor devices.

Based on the list of components required for each topology, it is possible to estimate a final indicator to compare the cost of the different MV bridge topologies. Fig. 11 shows comparative results of relative cost, based on the cost of semiconductor devices, passive elements and losses for different voltage levels.

It must be noted that the dV/dt is not monetized. Therefore, 2L appears as the best option in the figure. The normalization was done with the cost of the Q2L-MMCnoLarm at 40 kV, following the design mentioned in previous sections. It is observed that the main drivers of the cost are the semiconductor devices and the capacitors in the case of the Q2L-FCC. The 2L topology estimate will depend on the semiconductor devices as shown in Fig. 11. The Q2L-FCC topology estimate depends on the semiconductor devices and on the capacitors used for the flying capacitors. Series/parallel arrangements of capacitors were considered in this case. The width of the zone is due to the price sensitivity range of one single capacitor. The Q2L-MMCnoLarm estimate depends on the sizing of the submodule capacitor and the dissymmetry of the switching devices in one submodule e.g. double of switching devices compared to 2L or only 30% more. The width of the zone considers different capacitor values and different number of switching devices in one submodule. Finally, the Q2L-MMCwLarm will depend mostly on semiconductor devices and sizing of submodule capacitors. The width of the zone is due to the sizing of the submodule capacitor. A final assessment showing the advantages and disadvantages of each topology is shown in Table II.

VI. CONCLUSION

As DC networks continue to evolve, high power DC-DC converters will play a crucial role in efficient power conversion. This paper focuses on the most promising bridge topologies as candidates for medium-voltage DC-DC converters, providing a detailed examination of their characteristics. Then, a list of key performance indicators is established to evaluate each topology across different voltage levels. These indicators include the number of semiconductor devices, energy of passive elements, losses distribution, efficiency, and transient behavior. Additionally, KPIs are calculated for two specific case studies (10 kV and 40 kV), and results were presented to highlight the advantages and disadvantages for each topology. Most KPIs were monetized, indicating an interest in the 2L topology, although its rapid dV/dt remains a significant limitation. Among the Q2L topologies, Q2L-MMCnoLarm stands out for its scalability, with only Q2L-FCC outperforming it at the lower end of the medium voltage range. If availability is a key factor, MMC topologies are the best option, as their modular design allows for bypassing, improving availability even in cases of reduced reliability.

TABLE II. ASSESSMENT OF DIFFERENT MV ACTIVE BRIDGE TOPOLOGIES FOR A BIDIRECTIONAL ISOLATED DC-DC CONVERTER FOR MVDC NETWORKS.

MV bridge topology	Advantages			Disadvantages
2L	-Most compact solution. -Least complex topology.			-High dV/dt impossible to regulate and dependent on V_{MV}.
Q2L-FCC	Q2L topology with the least number of semiconductor devices.		Q2L topologies allow to regulate and limit the average and maximum dV/dt respectively.	-High complexity for design and start-up, both depending on V_{MV}. -Difficult to block DC-fault leg current.
Q2L-MMCnoLarm	Q2L topology with the least stored energy in passive elements.	Q2L-MMC topologies feature a modular design that is independent of V_{MV}, enhancing system availability.		-High number of semiconductor devices.
Q2L-MMCwLarm	Topology with the lowest conduction losses.			-Highest number of semiconductor devices. -High switching losses.

979-8-3315-1612-3/25 $31.00 © 2025 IEEE

ACKNOWLEDGMENT

This work was supported by the French Government under the program Investissements d'Avenir (ANE-ITE-002-01)

REFERENCES

[1] International Council on Large Electric Systems, Ed., *Medium voltage DC distribution systems.* in Technical brochure / CIGRE, no. 875. Paris, France: CIGRE, 2022.

[2] V. Ciniello, R. Loggia, L. Martirano, and S. Lauria, "A Review of MVDC Smart Grids Technologies and Components," in *IEEE/IAS 60th Industrial and Commercial Power Systems Technical Conference (I&CPS)*, Las Vegas, NV, USA: IEEE, May 2024, pp. 1–10. doi: 10.1109/ICPS60943.2024.10563506.

[3] I. Alhurayyis, A. Elkhateb, and J. Morrow, "Isolated and Nonisolated DC-to-DC Converters for Medium-Voltage DC Networks: A Review," *IEEE J. Emerg. Sel. Top. Power Electron.*, vol. 9, no. 6, pp. 7486–7500, Dec. 2021, doi: 10.1109/JESTPE.2020.3028057.

[4] P. Dworakowski, P. Le Métayer, D. Dujic, and C. Buttay, "Unidirectional step-up isolated DC-DC converter for MVDC electrical networks," in *CIGRE Session 2022*, Paris, France, Aug. 2022. Accessed: Feb. 02, 2024. [Online]. Available: https://hal.science/hal-03771023

[5] P. Le Metayer *et al.*, "Break-even distance for MVDC electricity networks according to power loss criteria," in *23rd European Conference on Power Electronics and Applications (EPE'21 ECCE Europe)*, Ghent, Belgium: IEEE, Sep. 2021, pp. 1–9. doi: 10.23919/EPE21ECCEEurope50061.2021.9570416.

[6] S. P. Engel, M. Stieneker, N. Soltau, S. Rabiee, H. Stagge, and R. W. De Doncker, "Comparison of the Modular Multilevel DC Converter and the Dual-Active Bridge Converter for Power Conversion in HVDC and MVDC Grids," *IEEE Trans. Power Electron.*, vol. 30, no. 1, pp. 124–137, Jan. 2015, doi: 10.1109/TPEL.2014.2310656.

[7] S. Yin, Z. Zeng, S. Debnath, and M. Saeedifard, "Modeling and ZVS Operation of the Isolated Modular Multilevel DC–DC Converter With a Unified Trapezoidal Wave Modulation," *IEEE Trans. Power Electron.*, vol. 39, no. 7, pp. 8306–8322, Jul. 2024, doi: 10.1109/TPEL.2024.3384575.

[8] S. Milovanovic and D. Dujic, "Comprehensive Analysis and Design of a Quasi Two-Level Converter Leg," *CPSS Trans. Power Electron. Appl.*, vol. 4, no. 3, pp. 181–196, Sep. 2019, doi: 10.24295/CPSSTPEA.2019.00018.

[9] S. Mersche, D. Bernet, and M. Hiller, "Quasi-Two-Level Flying-Capacitor-Converter for Medium Voltage Grid Applications," in *IEEE Energy Conversion Congress and Exposition (ECCE)*, Baltimore, MD, USA: IEEE, Sep. 2019, pp. 3666–3673. doi: 10.1109/ECCE.2019.8913201.

[10] C. M. de VIENNE, P. Lefranc, B. Asllani, P.-O. Jeannin, and B. Lefebvre, "Experimental investigation of a 10 kV-70A switch with six SiC-MOSFETs in a series-connection configuration".

[11] A. Tcai, T. Wijekoon, and M. Liserre, "Evaluation of Flying Capacitor Quasi 2-level Modulation for MV Applications," 2021.

[12] J. Yu and R. Burgos, "Operation and Control of Converters Having Integrated Capacitor Blocked Transistor Cells," in *IEEE Energy Conversion Congress and Exposition (ECCE)*, Detroit, MI, USA: IEEE, Oct. 2020, pp. 2625–2632. doi: 10.1109/ECCE44975.2020.9235414.

[13] R. Elpelt, P. Friedrichs, R. Schorner, K. O. Dohnke, H. Mitlehner, and D. Stephani, "Serial connection of SiC VJFETs-features of a fast high voltage switch; Montage en serie de VJFET au SiC - caracteristiques d'un commutateur rapide pour hautes tensions," *Rev. Electr. Electron.*, Feb. 2004, Accessed: Jul. 12, 2024. [Online]. Available: https://www.osti.gov/etdeweb/biblio/20457550

[14] B. Zhao, Q. Song, J. Li, Y. Wang, and W. Liu, "Modular Multilevel High-Frequency-Link DC Transformer Based on Dual Active Phase-Shift Principle for Medium-Voltage DC Power Distribution Application," *IEEE Trans. Power Electron.*, vol. 32, no. 3, pp. 1779–1791, Mar. 2017, doi: 10.1109/TPEL.2016.2558660.

[15] I. A. Gowaid, G. P. Adam, S. Ahmed, D. Holliday, and B. W. Williams, "Analysis and Design of a Modular Multilevel Converter With Trapezoidal Modulation for Medium and High Voltage DC-DC Transformers," *IEEE Trans. Power Electron.*, vol. 30, no. 10, pp. 5439–5457, Oct. 2015, doi: 10.1109/TPEL.2014.2377719.

[16] I. A. Gowaid, G. P. Adam, A. M. Massoud, S. Ahmed, D. Holliday, and B. W. Williams, "Quasi Two-Level Operation of Modular Multilevel Converter for Use in a High-Power DC Transformer With DC Fault Isolation Capability," *IEEE Trans. Power Electron.*, vol. 30, no. 1, pp. 108–123, Jan. 2015, doi: 10.1109/TPEL.2014.2306453.

[17] P. Czyz, P. Papamanolis, T. Guillod, F. Krismer, and J. W. Kolar, "New 40kV / 300kVA Quasi-2-Level Operated 5-Level Flying Capacitor SiC 'Super-Switch' IPM," in *10th International Conference on Power Electronics and ECCE Asia (ICPE 2019 - ECCE Asia)*, Busan, Korea (South): IEEE, May 2019, pp. 813–820. doi: 10.23919/ICPE2019-ECCEAsia42246.2019.8796998.

[18] X. Lin and D. Dong, "Analysis of Quasi-Two-Level Modulation for Neutral-Point-Clamped Three-level Converter with 10 kV SiC MOSFETs," in *23rd European Conference on Power Electronics and Applications (EPE'21 ECCE Europe)*, Ghent, Belgium: IEEE, Sep. 2021, p. P.1-P.9. doi: 10.23919/EPE21ECCEEurope50061.2021.9570662.

[19] G. P. Adam, O. Anaya-Lara, G. Burt, and J. R. McDonald, "Comparison between flying capacitor and modular multilevel inverter".

[20] G. L. Goff, I. Chorfi, and C. Alonso, "Comparison of modular multilevel and flying capacitor converters made of wide bandgap switches for new MVDC grids," 2024.

[21] A. Campos, J. Paez, and P. Dworakowski, "Comparison of Modular Multilevel Converter and Neutral Point Clamped Converter Topologies for MVDC applications," in *25th European Conference on Power Electronics and Applications (EPE'23 ECCE Europe)*, Sep. 2023, pp. 1–9. doi: 10.23919/EPE23ECCEEurope58414.2023.10264438.

[22] S. C. Mersche, N. Katzenburg, P. Kiehnle, B. Schmitz-Rode, D. Schulz, and M. Hiller, "Comparison of Quasi-Two-Level Operation of a Flying Capacitor Converter with Quasi-Two-Level Operation of a Modular Multilevel Converter," in *2022 IEEE 7th Southern Power Electronics Conference (SPEC)*, Nadi, Fiji: IEEE, Dec. 2022, pp. 1–6. doi: 10.1109/SPEC55080.2022.10058325.

[23] R. W. A. A. De Doncker, D. M. Divan, and M. H. Kheraluwala, "A three-phase soft-switched high-power-density DC/DC converter for high-power applications," *IEEE Trans. Ind. Appl.*, vol. 27, no. 1, Art. no. 1, Feb. 1991, doi: 10.1109/28.67533.

[24] S. Liu, H. Lin, and T. Wang, "Comparative Study of Three Different Passive Snubber Circuits for SiC Power MOSFETs," in *IEEE Applied Power Electronics Conference and Exposition (APEC)*, Mar. 2019, pp. 354–358. doi: 10.1109/APEC.2019.8722302.

[25] T. A. Meynard and H. Foch, "Multi-level conversion: high voltage choppers and voltage-source inverters," in *PESC `92 Record. 23rd Annual IEEE Power Electronics Specialists Conference*, Toledo, Spain: IEEE, 1992, pp. 397–403. doi: 10.1109/PESC.1992.254717.

[26] J. A. Aguilar Croston, P. Dworakowski, and B. Asllani, "Generalized Switching Sequence for Voltage Balancing in a Flying Capacitor DC-DC Converter with Quasi-2-Level Modulation," in *PCIM Europe 2024; International Exhibition and Conference for Power Electronics, Intelligent Motion, Renewable Energy and Energy Management*, Jun. 2024, pp. 3145–3154. doi: 10.30420/566262446.

[27] A. Lesnicar and R. Marquardt, "An innovative modular multilevel converter topology suitable for a wide power range," in *IEEE Bologna Power Tech Conference Proceedings*, Bologna, Italy: IEEE, 2003, pp. 272–277. doi: 10.1109/PTC.2003.1304403.

[28] Y. H. Abraham, H. Wen, W. Xiao, and V. Khadkikar, "Estimating power losses in Dual Active Bridge DC-DC converter," in *2nd International Conference on Electric Power and Energy Conversion Systems (EPECS)*, Nov. 2011, pp. 1–5. doi: 10.1109/EPECS.2011.6126790.

[29] "Carbure de silicium (SiC) MOSFET - Semi-conducteur GeneSiC," Semi-conducteur GeneSiC, Inc. Accessed: Jul. 15, 2024. [Online]. Available: https://genesicsemi.com/sic-mosfet/

Sling Forward Converter for Offline Operation: Achieving High Efficiency and Wide Voltage Range Performance

Nasherul Islam
College of Electrical Engineering
Zhejiang University
Hangzhou, China
nasherul.islam@zju.edu.cn

Guozhu Chen
College of Electrical Engineering
Zhejiang University
Hangzhou, China
gzchen@zju.edu.cn

Honglei Miao
HNAC TECHNOLOGY CO.,
LTD.
RDC@cshnac.com

Fuxing Zhang
HNAC TECHNOLOGY CO.,
LTD.
zfx@nudt.edu.cn

Abstract—The Sling Forward Converter (SFC) is designed for offline DC-DC conversion across a wide voltage range, optimizing both input and output. The SFC reduces transformer size and switching losses by leveraging the transformer's magnetizing current for resetting and establishing Zero Voltage Switching (ZVS) conditions for the main power switch. This converter ensures the output inductor current remains independent of voltage variations, maintaining high, flat efficiency. MATLAB design and experimental results from a 200W prototype validate the SFC's stable performance, demonstrating consistent output and efficient ZVS operation across input voltages from 200V to 400V. The SFC offers a promising solution for applications demanding robust, wide-range voltage handling.

Keywords—Sling Forward Converter, Zero Voltage Switching, Forward Converter.

I. INTRODUCTION

As power electronics become established, the demand for converters that can efficiently and effectively handle wide voltage ranges is becoming crucial.

Flyback converters are recognized for their simplexes and cost-effectiveness, especially in low-power applications. They maintain energy in a transformer throughout the switch-on duration and release it throughout the switch-off duration. Nevertheless, at greater power degrees, flyback converters handle efficiency troubles due to boosted optimal existing anxiety and transmission losses. Furthermore, the power storage room in the transformer limits its efficiency over a large voltage variety [1]. Forward converters supply consistent energy transfer, reducing optimal present anxiousness compared to flyback converters. They use an output inductor to preserve consistent existing circulation, which boosts performance. Despite these advantages, forward converters handle performance drops due to altering losses and the details of handling transformer reset devices [2]. ZVS converters aim to reduce energy losses by making sure that switching over changes at zero voltage, hence lowering energy dissipation. These converters typically make use of robust circuits to acquire ZVS troubles. While ZVS converters supply higher performance, they frequently require complex control mechanisms and added components, which can improve system cost and intricacy [3]. The Sling Forward Converter (SFC) is a cutting-edge DC-DC converter created for offline applications.

It improves input and output voltage differences while keeping high efficiency. The SFC leverages a unique topology that utilizes the transformer's magnetizing current for resetting and developing Zero Voltage Switching (ZVS) problems, hence lessening transformer dimension and transforming losses. This method not only enhances the performance but also makes sure of steady efficiency throughout a wide voltage range.

The SFC incorporates the consistent energy transfer of forward converters with the ZVS capabilities of effective converters. This unique combination addresses the restraints of traditional topologies, using a much more reliable and reliable option for DC-DC conversion across a wide voltage array [4].

Over the years, researchers have improved forward converters in several ways. Wu and Xing (2014) presented a novel design that improves energy recycling and lessens system stress by better balancing the voltage [2]. These cells maximize voltage harmonizing of the demagnetizing resources, boosting voltage anxiety administration and power recycling. Lee (2017) introduced a forward-flyback converter that operates efficiently over various power varieties, using a voltage double framework for improved performance [5]. Lin (2015) defines an active-clamping forward converter with a non-linear step-down conversion ratio, which reduces current stress and anxiety on inductors and diodes. The SFC's high step-down conversion ability makes it ideal for applications needing substantial voltage reductions while keeping high efficiency [6]. The SFC's efficiency makes it suitable for various applications, including telecommunication systems, industrial power products, and renewable resource systems.

The SFC significantly boosts DC-DC conversion, giving high effectiveness, versatility, and resilient performance throughout a wide voltage range. By integrating continual power transfer with ZVS capacities, the SFC addresses the restrictions of regular converter locations, making it an enticing treatment for modern power digital device applications. Future research and development must improve the SFC's layout and increase its application range to identify its capacity in numerous commercial and organizational setups.

Project supported by The Key Research and Development Program of Hunan (2023GK2046)

II. OPERATION PRINCIPLES

A. Sling Forward Converter

The topology of an SFC is similar to that of a standard forward converter. It is shown in Fig. 1. The forward transformer has at least three windings: a primary winding Np, a secondary winding Ns, and a reset winding N_t. The difference from the regular forward converter is that the homonymous end of the reset winding N_t is connected through the drain and source of Q_2 to the primary ground. The non-homonymous end of reset winding is connected to the primary ground through a resonant capacitor C_r in parallel with a diode D_r. The reset and ZVS circuit of the sling forward converter is composed of the reset winding N_t, power switch Q_2, the resonant capacitor C_r and parallel diode D_r as shown in the block with a dashed line. Power switches Q_3 and Q_4 are used for synchronous rectification.

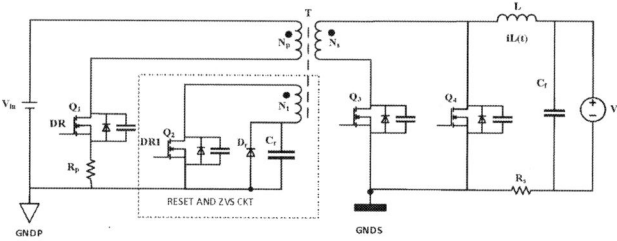

Fig. 1 Sling Forward Converter

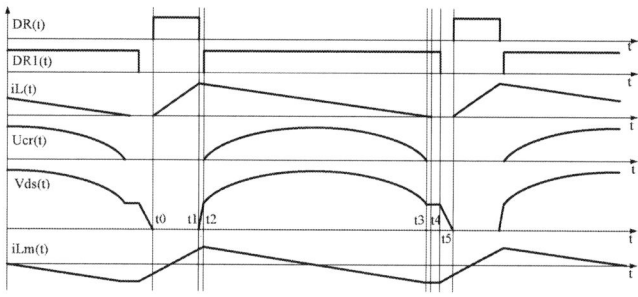

Fig. 2 Sling Forward Converter Operating Waveform

The illustrative waveforms of the sling forward converter are shown in Fig. 2. The current through the output inductor L is in discontinuous mode (DCM). The operation of the Sling Forward converter during different intervals is as follows:

Stage 1 [t0~t1]: As shown in Fig.2 and Fig.3 (a), the gate pulse of Q_1, $DR_{(t)}$, is at a high level, so Q_1 is on. The input DC voltage V_{in} appears across the primary winding N_p. The voltage induced on the secondary winding Ns is applied to the output inductor L and load through synchronous MOS Q_3 through the transformer's magnetic coupling. The current in the output inductor increases from zero to the peak value.

Stage 2 [t1~t2]: As shown in Fig.2 and Fig.3 (b), the gate pulse of Q_1, $DR_{(t)}$, goes low, turning Q_1 off. To keep the transformer's flux continuous, a current flows through the body diode of Q_2 into the homonymous end of the reset winding N_t. Thus, the capacitor C_r resonates with the magnetizing inductor L_m of the transformer. As the resonant current goes to zero, the body diode

of Q_2 turns on first, and then a gate pulse $DR_{1(t)}$ is applied to Q_2. Thus, Q_2 will turn on under ZVS condition.

Stage 3 [t2~t3]: As shown in Fig.2 and Fig.3 (c), due to the body diode of Q_2 and then Q_2 providing a path, the resonant capacitor C_r resonates with the magnetizing inductor L_m of the transformer. The voltage on the resonant capacitor C_r resonantly increases from zero to its peak value and then decreases to zero. The transformer completes the reset operation. The current in the magnetizing inductor L_m, $iL_m(t)$, resonantly varies from the positive peak to zero and then decreases to a negative peak. Thus, the current enters the homonymous end of the reset winding N_t and emerges.

Stage 4 [t3~t4]: As shown in Fig.2 and 3 (d), the current coming out from the homonymous end of the reset winding N_t flows through Q_2 and the diode D_r, shorting the reset winding N_t. Thus, the voltages across all the windings of the transformer are made zero. The voltage between the drain and source of Q_1 is equal to the input voltage V_{in}. During the time interval t3~t4, the current in the output inductor L decays. As the current decays to zero before t4, turning Q1 on under ZVS is possible.

Stage 5 [t4~t5]: As shown in Fig.2 and Fig.3 (e), at t4, the gate pulse $DR_{1(t)}$ of Q_2 becomes zero, Q_2 turns off, and the current through the reset winding N_t becomes zero. Due to the current through the output inductor L becoming zero, the current through the secondary winding N_s must go to zero. A current comes out of the homonymous end of the primary winding Np to keep the transformer's flux continuous. This current causes the voltage on the capacitor across the drain and the source of Q1 to decrease to zero. Thus, Q_1 can be turned on under ZVS.

The transformer is reset during a fixed-time TRESET in the Sling Forward Converter. The interval T_{RESET} is determined by the resonant capacitor C_r and the magnetizing inductor L_m of the transformer as:

$$T_{RESET} = \pi \cdot \sqrt{L_m \cdot C_r} \qquad (1)$$

The fixed-time T_{RESET} is independent of the input and output voltages.

The duty cycle of the Sling Forward Converter can be over 0.5. The peak voltage on the resonant capacitor C_r is proportional to the volt-second product of the input voltage and turn-on time τ of primary MOS Q_1 and inversely proportional to the resonant reset time:

$$U_{Cr\_Peak} = \frac{\pi \cdot \tau \cdot V_{IN}}{2 \cdot T_{RESET}} \qquad (2)$$

The maximum voltage on the main power switch Q_1 equals the sum of the peak voltage on the resonant capacitor C_r and the input DC voltage.

During transformer magnetization, the magnetizing current changes from a negative value to a positive value, and the operation of the magnetic core moves from the third quadrant to the first quadrant. This means that the transformer can be designed for twice the magnetic flux density Bs and thus half the number of turns, shrinking its size.

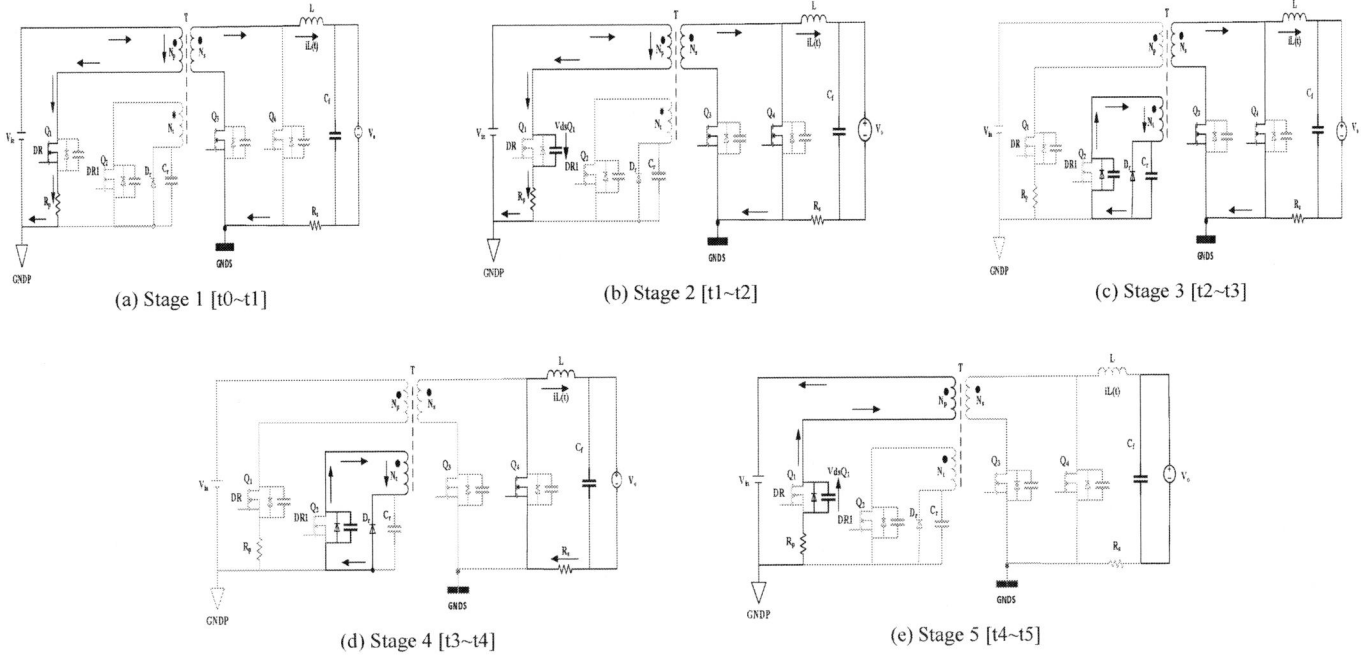

(a) Stage 1 [t0~t1] (b) Stage 2 [t1~t2] (c) Stage 3 [t2~t3]

(d) Stage 4 [t3~t4] (e) Stage 5 [t4~t5]

Fig. 3 Equivalent circuit of each stage

B. Closed loop PID Controller

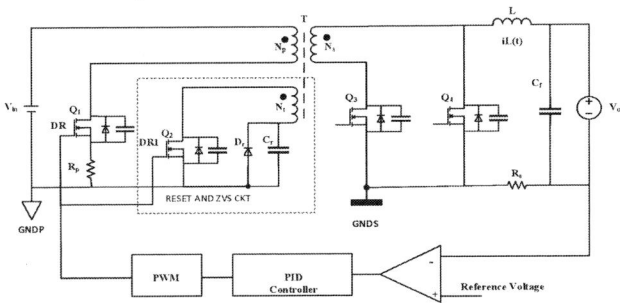

Fig. 4 Proposed SF-Converter with Closed Loop

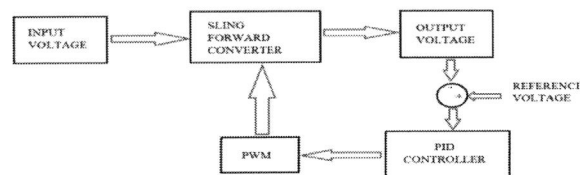

Fig. 5 Closed Loop Control of SF-Converter

TABLE I. PID PARAMETERS

K_p	K_i	K_d
0.02	50	0

The proposed Sling Forward (SF) Converter integrates a closed-loop control system to maintain a secure output voltage. The output of this system is kept constant by comparing the actual output voltage to a preset target voltage. PID (Proportional-Integral-Derivative) controllers adjust the PWM signal as needed to control the error signal, which is the difference between the two.

In the closed-loop control of the SF converter, the PID controller plays a vital role in maintaining a steady voltage. The controller continually tracks the output voltage and adjusts the PWM to lessen the difference between the actual outcome and the desired voltage. This vibrant modification helps accomplish specific and consistent manufacturing, which is essential for applications that need a secure power supply.

The table lays out the PID requirements used in the control system. These specifications are mainly chosen to guarantee the optimum efficiency of the SF converter in keeping a protected output voltage.

III. MATLAB DESIGN AND SIMULATION OF SFC WITH PID CONTROLLER

A. MATLAB design

The MATLAB design for the Sling Forward Converter (SFC) with a PID controller integrates several vital elements to achieve an efficient voltage policy. The system takes a DC input voltage ranging from 200V to 400V and converts it to a steady output voltage with the help of a robust circuit, transformer, and PID control mechanism. The PID controller is essential in preserving the wanted output by changing the Pulse Width Modulation (PWM) signals in reaction to voltage deviations.

979-8-3315-1612-3/25 $31.00 © 2025 IEEE

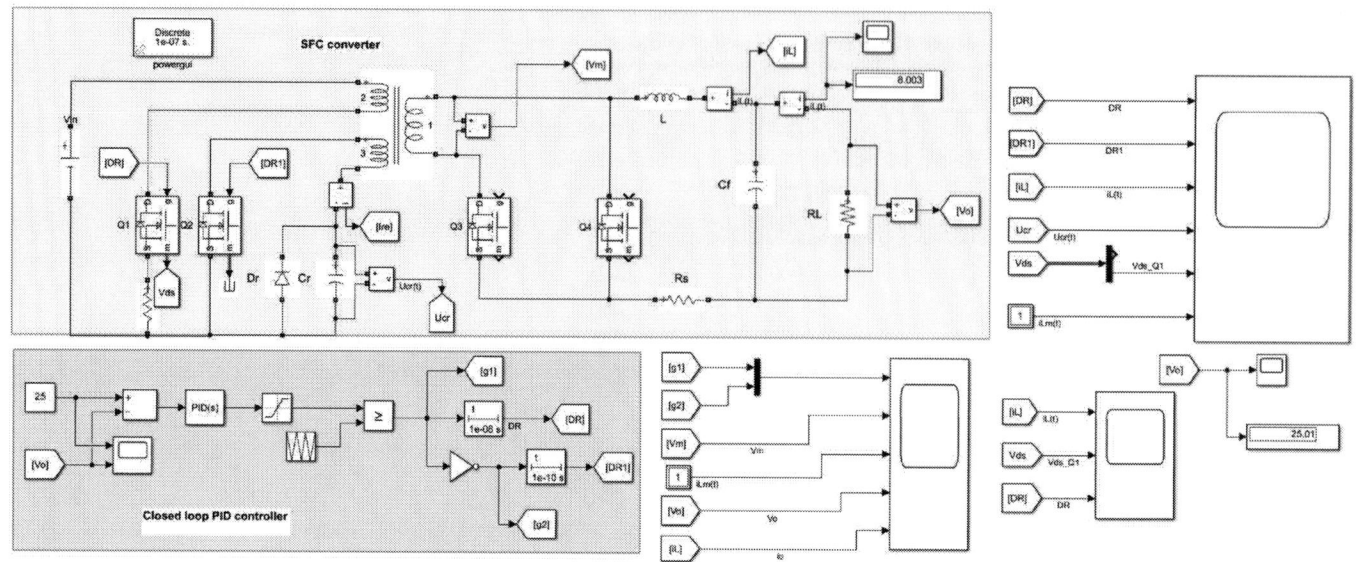

Fig. 6 SFC MATLAB design with PID controller

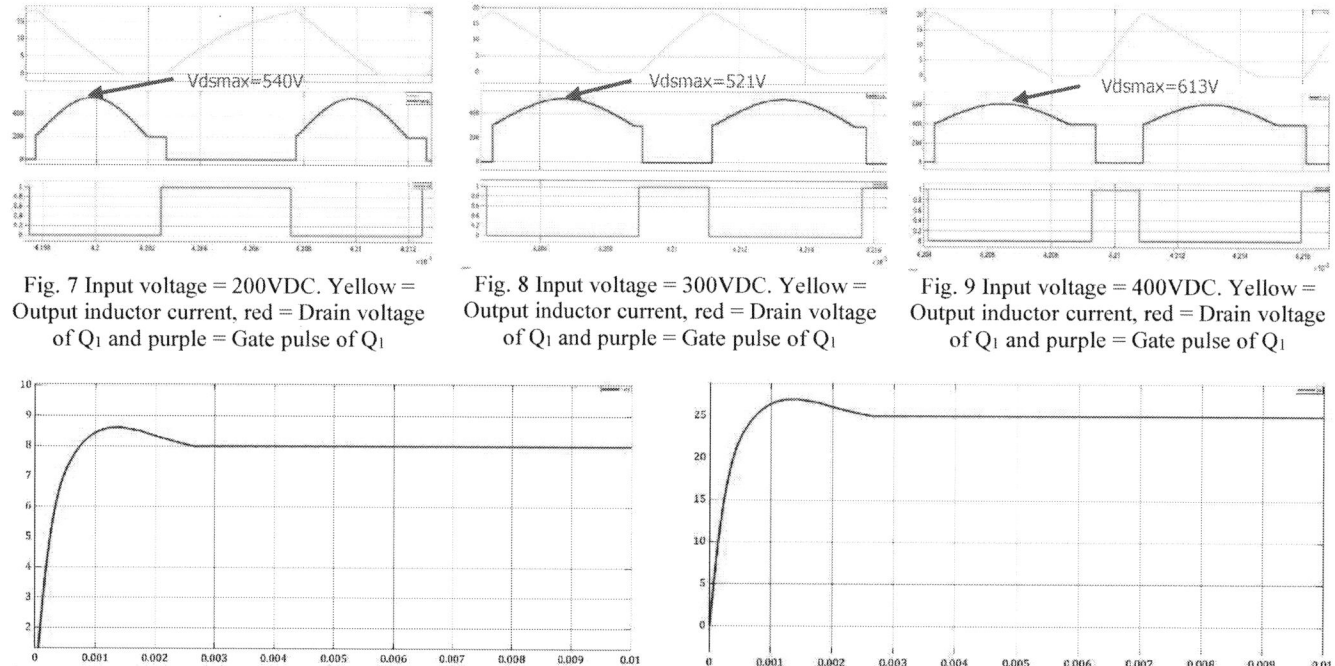

Fig. 7 Input voltage = 200VDC. Yellow = Output inductor current, red = Drain voltage of Q_1 and purple = Gate pulse of Q_1

Fig. 8 Input voltage = 300VDC. Yellow = Output inductor current, red = Drain voltage of Q_1 and purple = Gate pulse of Q_1

Fig. 9 Input voltage = 400VDC. Yellow = Output inductor current, red = Drain voltage of Q_1 and purple = Gate pulse of Q_1

Fig. 10 Output current of SFC when input voltage 400V

Fig. 11 Output voltage of SFC when input voltage 400V

B. Simulation Results

The simulation results demonstrate the SFC's performance under various input voltages. The figures below illustrate the waveforms of the output inductor current, the drain voltage of transistor Q1, and the gate pulse of Q1 at different input voltages.

Fig. 7: Input voltage = 200VDC. The yellow waveform represents the output inductor current, the red waveform shows the drain voltage of Q_1, and the purple waveform depicts the gate pulse of Q_1.

Fig. 8: Input voltage = 300VDC. Similar color coding applies: yellow for the output inductor current, red for the drain voltage of Q1, and purple for the gate pulse of Q_1.

Fig. 9: Input voltage = 400VDC. Again, the yellow, red, and purple waveforms represent the output inductor current, drain voltage of Q_1, and gate pulse of Q_1, respectively.

TABLE II. SPECIFICATIONS OF SLING FORWARD CONVERTER

Parameters	Values
Input voltage (DC)	200V-400 V
Resonant capacitor (C_r)	20e-9 F
R_p	4.5 Ω
R_s	0.4 Ω
Load resistance (R_L)	3.125 Ω
Output inductor (L)	5e-6 H
Output capacitor (C_f)	1000e-6 F
Magnetizing inductor (L_m)	6.25e-6 H
Transformer ratio (T_r)	n=250:250:65
Switching Frequency (f_s)	100-150 KHz
Output current (I_o)	7.288A-8.005A
Output voltage (V_o)	25.01 V-25.62 V

C. Wide Soft Switch Operating Range

TABLE III. TEST DATA OF SLING FORWARD CONVERTER

VDC(V)	Vo(V)	Io(A)	Po(W)
200	25.62	8.197	210.00
250	25.01	8.004	200.15
300	25.01	8.005	200.20
350	25.01	8.004	200.18
400	25.01	8.002	200.18

The SFC shows a wide soft switching operating range, as indicated by the MATLAB simulation data in Table III. The converter efficiently keeps a steady result voltage and current throughout varying input voltages. The following Fig. 10 and Fig. 11 reveal the output current and voltage of the SFC at an input voltage of 400V.

The MATLAB simulation of the SFC with the PID controller shows its capacity to maintain steady result voltage and current across input voltages. The integration of the PID controller makes certain specific laws, and the system's parts are optimized for efficient performance. The simulation data confirms the converter's integrity and performance in numerous operating conditions.

IV. EXPERIMENT RESULTS

A 200W prototype has been built to verify the performance of the sling forward converter. The bench waveforms for different input voltages are shown in Fig. 12, Fig. 13, and Fig. 14.

The output inductor's peak current remains 20A for the input DC voltage from 200V to 400V. The di/dt of the output inductor current changes with the input DC voltage.

The higher the peak voltage on the resonant capacitor C_r, the lower the input DC voltage. The input DC voltage changes from 200V~400V, and the main power MOS's V_{ds} is from 460V~600V.

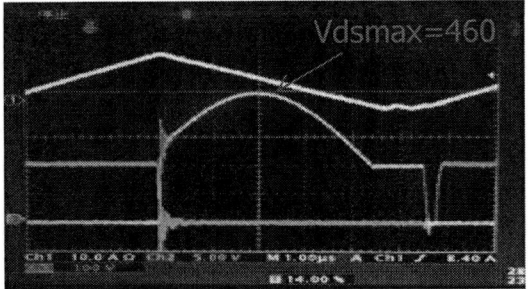

Fig. 12 Input voltage = 200VDC. Yellow = Output inductor current, Pink = Drain voltage of Q_1 and Blue = Gate pulse of Q_1

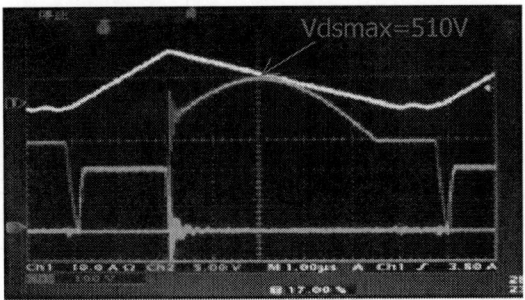

Fig. 13 Input voltage = 300VDC. Yellow = Output inductor current, Pink = Drain voltage of Q_1 and Blue = Gate pulse of Q_1

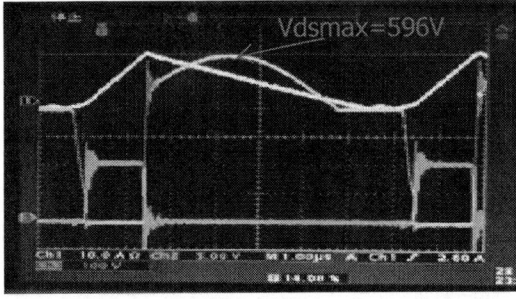

Fig. 14 Input voltage = 400VDC. Yellow = Output inductor current, Pink = Drain voltage of Q_1 and Blue = Gate pulse of Q_1

The main power MOS's V_{Ds} drop to zero, and the gate pulse comes. The main power MOS turns on in ZVS.

The bench test data and efficiency are shown in Table IV and Fig.

TABLE IV. THE BENCH TEST TATA OF EFFICIENCY

VDC(V)	P_{in} (W)	V_o (V)	I_o (A)	P_o (W)	E_{ff} (%)
200	221	25.1	8.07	202.56	91.65
250	220	25.1	8.07	202.56	92.07
300	219.5	25.1	8.08	202.81	92.40
350	219	25.1	8.07	202.56	92.49
400	219	25.1	8.06	202.31	92.38

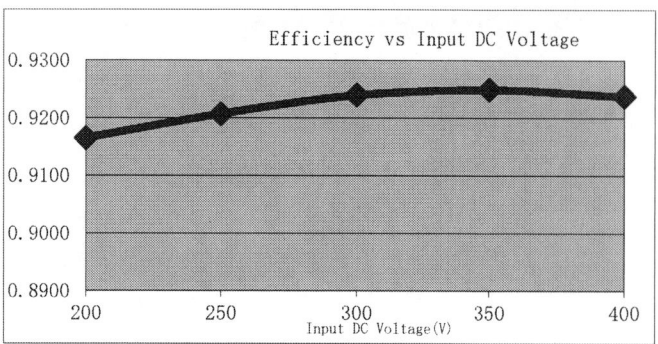

Fig. 15 Efficiency vs Input voltage

A. Analysis of Results

The examination results demonstrate that the Sling Forward Converter maintains a stable output voltage of around 25V and 8A across the wide input voltage range of 200V to 400V. This stability indicates the efficiency of the SFC design in controlling output specifications despite changes in input voltage. The performance dimensions, varying from 91.65% to 92.49%, verify the converter's high performance, which stays relatively level throughout the examined voltage array.

B. Influence of Resonant Capacitor (C_r) and Magnetizing Inductor (L_m) on Efficiency

- The resonant capacitor (Cr) and magnetizing inductor (Lm) play vital duties in attaining Zero-Voltage Switching (ZVS) and reducing energy losses. The vibration between Cr and Lm enables the voltage throughout the key power button (Q1) to get to know before it activates or goes off, considerably decreasing the power dissipated during altering transitions. This ZVS problem is essential for maintaining high performance.

- The size and values of Cr and Lm straight influence the resonance frequency and the duration of the ZVS condition. The optimum choice of these parts makes certain marginal changing losses and stable operation. In the SFC model, the chosen worth for Cr (20nF) and Lm (6.25 μH) gave efficient resonance, contributing to the observed high effectiveness and stable performance.

C. Comparison with Conventional Converters

The SFC displays better efficiency and stability throughout a large voltage variety than conventional converters like flyback and forward converters. Due to their energy storage capacity and release device, flyback converters often deal with high-existing stress and lower efficiency at different input voltages. The SFC appropriately deals with these restrictions with its ZVS capability and independent output inductor. In addition, the design guarantees that the output current remains steady irrespective of input voltage variants, a significant improvement over conventional topology. Table V compares the conventional converters and the proposed converter.

TABLE V. COMPARISON OF SCHEME PERFORMANCE WITH CONVENTIONAL CONVERTERS

Feature	LLC	Resonant Output (QR, ACF, AHB)	Forward Converter (FC)	Sling Forward Converter
Total load efficiency (optimized input/output voltage)	High (works under optimized input/output voltage)	High (in low input voltage scenario) Low (in high input voltage scenario)	Moderate (hard switching relationship)	High (ZVS)
Light load efficiency	Low (due to sizeable magnetizing current)	High	High	High
Current stress	Low (works close to optimal point) High (works far from optimal point)	High (peak current stress is affected by input/output voltage changes)	Low (peak current stress is independent of input voltage changes)	Low (peak current stress is independent of input voltage changes)
Synchronous rectification control	Difficult to control	Easy to control	Easy to control	Easy to control
Feedback control	Complex (supplementary network required)	QR simple (fixed frequency mode) ACF & AHB complex (requires additional circuitry)	Simple (traditional voltage mode control)	Simple (traditional fixed timing current mode control)
Transformer size	Automatic (magnetizing current allows for size reduction and works close to optimal second quadrant)	OR/ACF/AHB all have smaller transformer sizes (work close to optimal first quadrant)	Half-automatic (with traditional dual-feedback function, reduced size but requires added high-voltage MOSFETs and transformer design changes)	Automatic (optimized magnetizing current, reduced size in both first and third quadrants)
Reliability	High (MOSFETs have active body diodes, no direct conduction loss)	High (shared drive mechanism, no direct conduction loss)	High (shared drive mechanism, no direct conduction loss)	High (shared drive mechanism, no direct conduction loss)

979-8-3315-1612-3/25 $31.00 © 2025 IEEE

D. Discussion on Oscillations and Their Management

During the simulation, oscillations were observed in the outcome waveforms, particularly at higher input voltages over 350V. These oscillations were noted in the iL(t), Ucr(t), and Vds(t) waveforms. Such oscillations can occur between the circuit components and may affect the stability and efficiency of the converter.

The resonant capacitor (Cr), sense resistor Rs, and parasitic resistance Rp can be changed to manage these oscillations. Fine-tuning these values helps dampen the oscillatory behavior, ensuring stable operation across all input voltages. In the model examinations, adjusting Rp and Rs reduced oscillations without considerably influencing performance, demonstrating an efficient strategy for managing potential stability issues.

E. The Intended Application of SFC

The results confirm the Sling Forward Converter as a high-efficiency, stable solution for offline DC-DC conversion over a wide voltage range, which is made for systems that require high-efficiency, wide-voltage-range DC-DC conversion, like LED, battery charging circuits, industrial power products, etc. The ingenious use of resonant elements and ZVS conditions ensures marginal switching losses and constant efficiency, positioning the SFC as a premium alternative to standard converters for modern power electronics applications.

ACKNOWLEDGMENT

The authors thank Dr. Yuvarajan Subbaraya for his interest and constant encouragement of this work.

V. CONCLUSION

The Sling Forward Converter (SFC) demonstrates high efficiency, stable performance, and a reduced transformer size, addressing critical challenges in offline DC-DC conversion. By leveraging the transformer's magnetizing current for ZVS conditions, the SFC minimizes switching losses, ensuring consistent efficiency across a wide input voltage range. This makes the SFC particularly suitable for robust and efficient power conversion applications.

The experimental results confirm the SFC's potential, showing stable output and high efficiency from 200V to 400V input. Future work will optimize component parameters and explore real-world application scenarios, enhancing the SFC's reliability and performance in dynamic environments.

REFERENCES

[1] F. D. Tan, "The forward converter: from the classic to the contemporary," in APEC. Seventeenth Annual IEEE Applied Power Electronics Conference and Exposition (Cat. No.02CH37335), Dallas, TX, USA: IEEE, 2002, pp. 857–863. doi: 10.1109/APEC.2002.989344.

[2] H. Wu and Y. Xing, "Families of Forward Converters Suitable for Wide Input Voltage Range Applications," IEEE Trans. Power Electron., vol. 29, no. 11, pp. 6006–6017, Nov. 2014, doi: 10.1109/TPEL.2014.2298617.

[3] M. M. Jovanovic, D. C. Hopkins, and F. C. Y. Lee, "Evaluation and design of megahertz-frequency off-line zero-current-switched quasi-resonant converters," IEEE Trans. Power Electron., vol. 4, no. 1, pp. 136–146, Jan. 1989, doi: 10.1109/63.21882.

[4] H. Bai and C. Mi, "Comparison and evaluation of different DC/DC topologies for plug-in hybrid electric vehicle chargers," Int. J. Power Electron., vol. 4, no. 2, p. 119, 2012, doi: 10.1504/IJPELEC.2012.045627.

[5] H.-S. Lee, H.-J. Choe, S.-H. Ham, and B. Kang, "High-Efficiency Asymmetric Forward-Flyback Converter for Wide Output Power Range," IEEE Trans. Power Electron., vol. 32, no. 1, pp. 433–440, Jan. 2017, doi: 10.1109/TPEL.2016.2537930.

[6] J. Lin, W. Tzeng, C. Lin, C. Wang, and P. Liu, "Active-clamping forward converter with non-linear step-down conversion," IET Power Electron., vol. 8, no. 1, pp. 112–119, Jan. 2015, doi: 10.1049/iet-pel.2014.0170.

A Pulse Width Alternating Modulation Strategy for Three-Level Buck-Boost Converter

Xinlong He
Power Electronics and Energy Management Key Laboratory, Ministry of Education of China Huazhong University of Science and Technology
Wuhan, China
xinlonghe@hust.edu.cn

Caifeng Liu
Power Electronics and Energy Management Key Laboratory, Ministry of Education of China Huazhong University of Science and Technology
Wuhan, China
liucaifeng@hust.edu.cn

Xudong Zou
Power Electronics and Energy Management Key Laboratory, Ministry of Education of China Huazhong University of Science and Technology
Wuhan, China
xdzou@mail.hust.edu.cn

Jiaao Zou
Power Electronics and Energy Management Key Laboratory, Ministry of Education of China Huazhong University of Science and Technology
Wuhan, China
zoujiaao@hust.edu.cn

Tianyi Zhang
Power Electronics and Energy Management Key Laboratory, Ministry of Education of China Huazhong University of Science and Technology
Wuhan, China
tianyi_hust@hust.edu.cn

Yong Kang
Power Electronics and Energy Management Key Laboratory, Ministry of Education of China Huazhong University of Science and Technology
Wuhan, China
ykang@hust.edu.cn

Abstract—The three-level buck-boost(TLBB) converter is suitable for scenarios with a wide input voltage range and high voltage levels. However, conventional soft-switching modulation strategies face the problem of losing soft-switching when the duty cycle is large. This paper proposes a pulse-width alternating modulation strategy. By alternating the modulation of two switching sequences over two cycles, this strategy ensures the existence of the reverse current freewheeling stage across the entire input voltage range, achieving soft switching throughout the entire range. Additionally, it leverages the advantage of having more degrees of freedom to optimize the RMS value of the inductor current, resulting in a simplified calculation method for modulation parameters. Finally, a 300W prototype was built to verify the theoretical analysis.

Keywords—TLBB converter, soft-switching, inductor current modulation

I. INTRODUCTION

In fields such as wind power generation, PV systems, and EV/HEV/FEVs, DC-DC converters operating over a wide input voltage range and high voltage levels are required [1][2]. For example, a 1500V photovoltaic system is more cost-effective than a 1000V system [3]. Non-isolated DC-DC converters are increasingly used in these scenarios due to their compact structure and higher efficiency [4][5]. [6] introduces the three-level buck-boost (TLBB) converter, shown in **Fig. 1**. The converter's input side is a flying capacitor-type three-level unit, with the switching devices experiencing voltage stress at half the input voltage, making it ideal for high input voltage levels and wide voltage ranges. However, existing modulation strategies for this converter face limitations in soft-switching range and efficiency improvement [6][7].

To address these issues, this paper proposes a pulse-width alternating modulation strategy for the three-level Buck-Boost converter. This paper introduces the proposed modulation strategy and its principles in Section II. Section III optimizes the RMS value of inductor current, leveraging the advantages of the modulation strategy's degrees of freedom and deriving a simplified method for calculating modulation parameters. Section IV builds a 300W prototype to validate the effectiveness of the modulation strategy, followed by a summary in Section V.

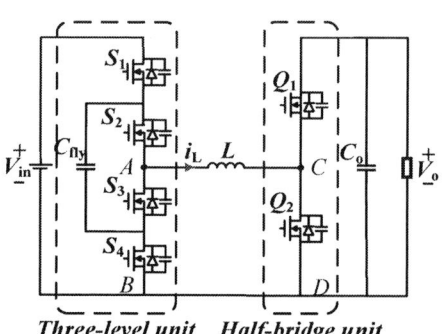

Fig. 1: Topology of the three-level buck-boost converter

II. PROPOSAL AND PRINCIPLES OF PULSE WIDTH ALTERNATING MODULATION STRATEGY

[7] identifies a flaw in the TLBB's conventional soft-switching method, as shown in **Fig. 2**. The switching transistors S_1 and S_4, S_2 and S_3, Q_1 and Q_2 conduct complementarily. For S_1, S_2, and Q_2 to achieve soft switching, reverse current is required. However, when the duty cycle of transistor S_1 increases to more than 1/2, the zero voltage stage

979-8-3315-1612-3/25 $31.00 © 2025 IEEE

of the voltage V_{ab} disappears. Consequently, the freewheeling stage of i_L vanishes, leading to the loss of soft switching for transistors S_1 and S_2 and an increase in the RMS value of the inductor current.

To address defects in the conventional soft-switching modulation strategy, an excessively large pulse width is divided into alternating modulation over two cycles. This approach prevents the loss of the freewheeling stage due to overlapping modulation pulses, which leads to the loss of soft switching. **Fig. 3** shows the key waveforms of the proposed pulse-width alternating modulation strategy. The duty cycle and switching times of S_1 in the first cycle are the same as those of S_2 in the second cycle. This strategy not only achieves soft switching across the entire range but also realizes self-balancing of the flying capacitor voltage within two cycles, ensuring stable converter operation.

III. OPTIMIZATION OF THE RMS VALUE OF INDUCTOR CURRENT

The proposed modulation strategy's six degrees of freedom enable optimizing the inductor current RMS value, resulting in a lower RMS value and simplified modulation parameter calculations.

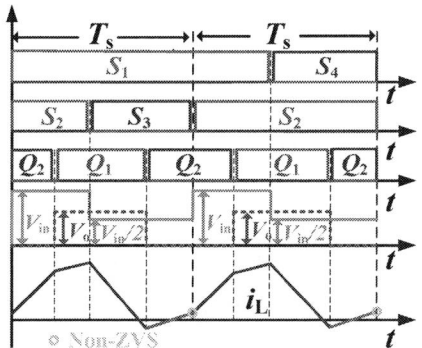

Fig. 2: Waveforms of conventional soft-switching modulation when soft switching is lost

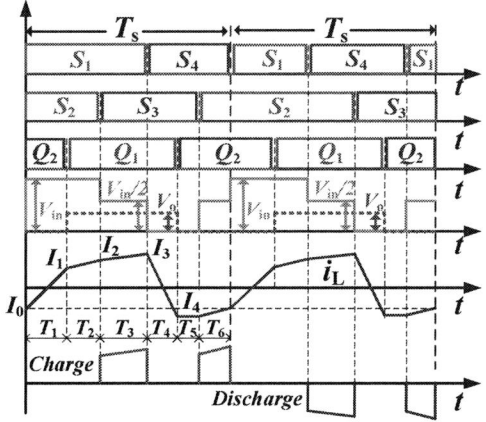

Fig. 3: Key waveforms of pulse-width alternating modulation

A. Simplification under Minimum RMS of Inductor Current Constraint

Using first cycle as an example (**Fig. 3**), the duration of each inductor current segment is T_1-T_6, with initial current values I_0-I_4. The T_6 phase increases negative current during the freewheeling stage, raising the inductor current RMS value and losses. Removing the T_6 phase and turning on S_1 and S_2 simultaneously does not affect soft switching. Therefore, the T_6 phase can be eliminated, and the I_0 value can be set to the critical soft-switching value I_{zvs}. Hence, it can be calculated that:

$$T_s = T_1 + T_2 + T_3 + T_4 + T_5 \tag{1}$$

$$\frac{V_o}{V_{in}} = \frac{T_1 + T_2 + 1/2T_3}{T_2 + T_3 + T_4} \tag{2}$$

$$P_{out} = \frac{V_o}{T_s}\left[\frac{I_1 + I_2}{2}T_2 + \frac{I_2 + I_3}{2}T_3 + \frac{I_3 - I_{zvs}}{2}T_4\right] \tag{3}$$

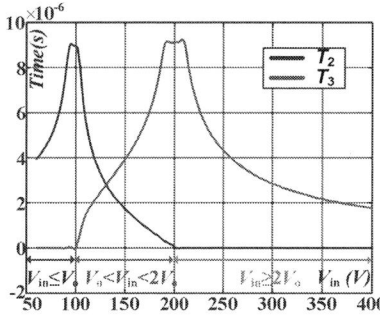

Fig. 4: Trend of T_2 and T_3 duration with input voltage variation at minimum inductor current RMS value

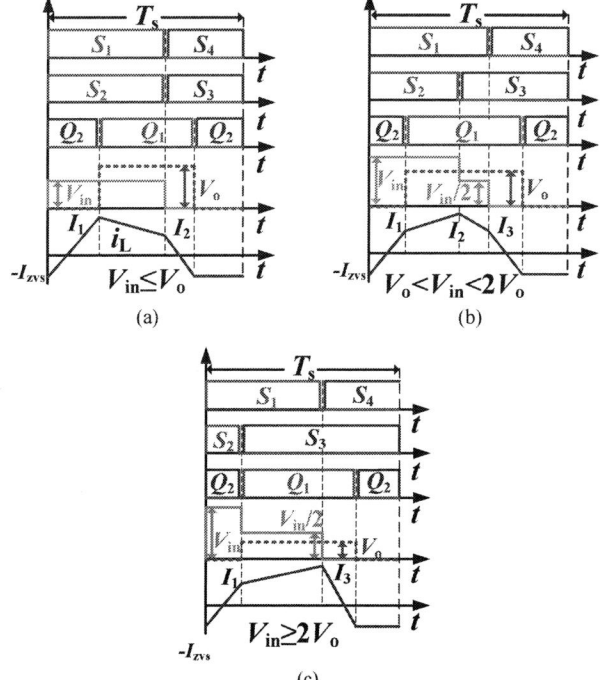

Fig. 5: Key waveforms of the simplified pulse-width alternating modulation

From (1)-(3), expressions for T_1, T_4, and T_5 in terms of T_2 and T_3 can be derived. Using the algorithm, the trend of the durations T_2 and T_3 with respect to the input voltage variation, when the RMS of inductor current is minimized under soft-switching conditions, can be obtained, as shown in **Fig. 4**. So, Minimum RMS of Inductor Current is always obtained at $T_3=0$ for $V_{in} \leq V_o$ and at $T_2=0$ for $V_{in} \geq 2V_o$. Thus, the modulation strategy can be further simplified. When $V_{in} \leq V_o$ and $V_{in} \geq 2V_o$, it simplifies to a four-segment inductor current modulation; when $V_o < V_{in} < 2V_o$, it simplifies to a five-segment inductor current modulation, with key waveforms illustrated in **Fig. 5**.

B. Further Simplification of Modulation Parameter Calculation

For $V_{in} \leq V_o$ or $V_{in} \geq 2V_o$, [8] shows that minimum RMS of inductor current is achieved when I_2 or I_1 equals I_{zvs}. When $V_o < V_{in} < 2V_o$, the algorithm shows that the RMS of inductor current is minimized when $I_1 = I_3 = I_{zvs}$, which is demonstrated below. Generally, the RMS value of current is proportional to its peak value. If $I_1 > I_{zvs}$ and the peak $I_2' = I_2$ as shown by the red line in Figure 6(a), the red shaded area is always smaller than the blue, indicating insufficient power transmission. To meet power requirements, I_2' must be greater than I_2, increasing the RMS value of the inductor current. The analysis for I_3 is similar, as shown in **Fig. 6(b)**.

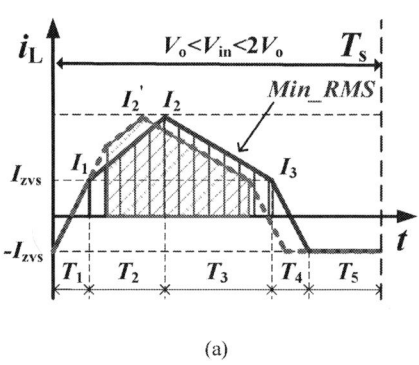

(a)

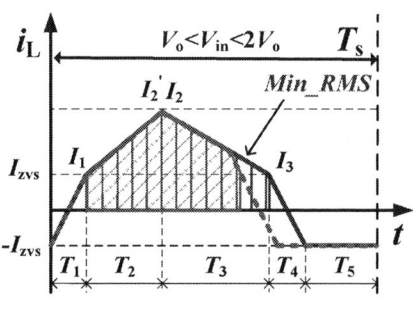

(b)

Fig. 6: Inductor current waveform when $V_o < V_{in} < 2V_o$

Therefore, for the simplified modulation method, (4) is the necessary condition for achieving the minimum effective value of the inductor current. By combining (1) to (4), the modulation parameters for different operating points (V_{in}, P) can be solved by controller.

$$\begin{cases} I_2 = I_{zvs} & V_{in} \leq V_o \\ I_1 = I_3 = I_{zvs} & V_o < V_{in} < 2V_o \\ I_1 = I_{zvs} & V_{in} \geq 2V_o \end{cases} \quad (4)$$

IV. EXPERIMENTAL VERIFICATION

To validate the proposed modulation strategy, a 300W TLBB converter prototype was built, as shown in **Fig. 7**. **Table I** lists key parameters. As illustrated in Fig. 8, the proposed modulation strategy effectively generates reverse current to achieve soft switching for the switches S_1, S_2, and Q_2, thereby reducing switching losses in high-frequency applications and improving the efficiency. In **Fig. 8(a)**, the brown waveform represents the flying capacitor voltage, which remains balanced under the proposed modulation strategy, ensuring normal operation of the converter in three-level mode. **Fig. 8 and Fig. 9** present the inductor current waveforms under different input voltage ranges, consistent with the previously mentioned theoretical analysis, thereby confirming the feasibility of the proposed modulation strategy. **Fig. 10** shows the inductor current waveform in critical mode at $V_{in} = V_o$. It can be observed that as the input power increases, the continuous conduction phase gradually disappears, resulting in a transition of the inductor current from four segments to three segments. If the input power is further increased, I_1 and I_3 will exceed the I_{zvs} limitation, thereby ensuring power output.

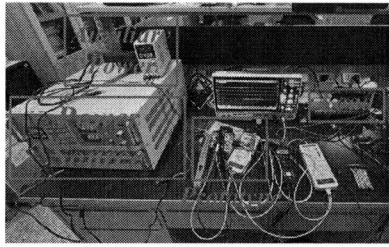

(a)

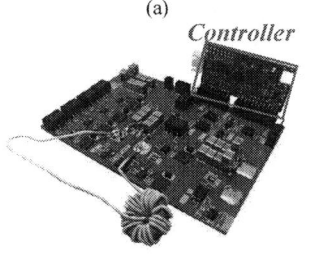

(b)

Fig. 7: (a) Experimental setup (b) Prototype and controller

TABLE I   CONVERTER PARAMETERS

Item	Parameter
Input voltage(V_{in})	60-400V
Output voltage(V_o)	100V
P_{rated}	300W
f_s	100kHz
L	15μH
S_1-S_4\Q_1-Q_2	GS-065-018-2-L

979-8-3315-1612-3/25 $31.00 © 2025 IEEE

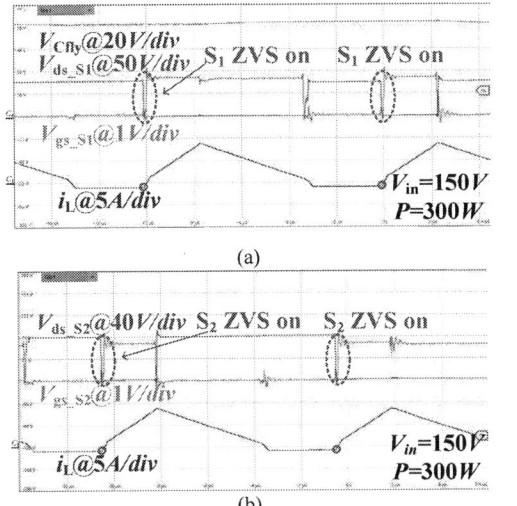

(a)

(b)

Fig. 8: Soft-switching waveforms of switches (a) S_1 and (b) S_2 ($V_o < V_{in} < 2V_o$)

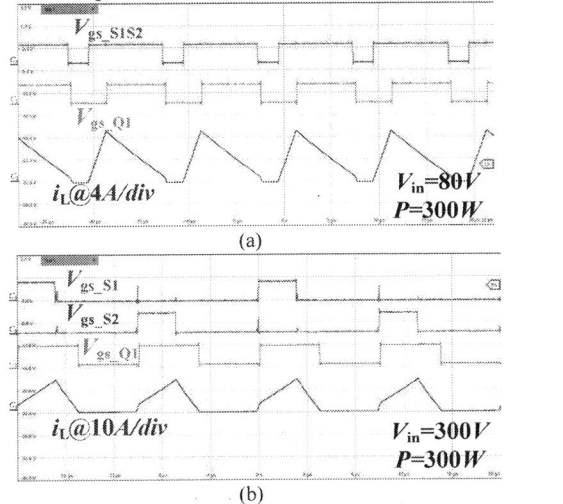

(a)

(b)

Fig. 9: Steady-state operation waveforms of the converter (a) $V_{in} < V_o$ (b) $V_{in} > 2V_o$

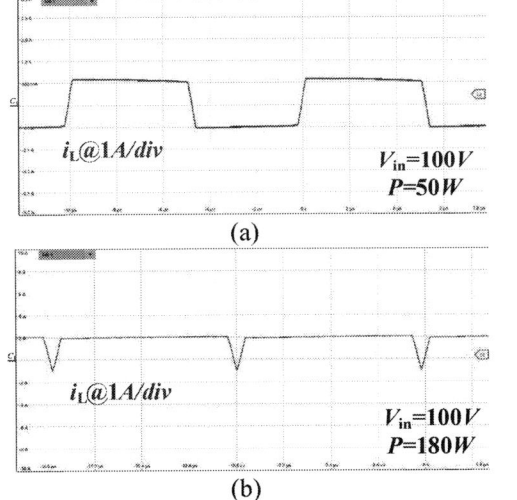

(a)

(b)

Fig. 10: Inductor current waveforms in critical mode when $V_{in} = V_o$ (a) $P = 50W$ (b) $P = 180W$

V. CONCLUSIONS

This paper proposes a pulse-width alternating modulation strategy based on the TLBB converter and optimizes the RMS value of inductor current, resulting in a simplified method for calculating modulation parameters. The correctness of the theoretical analysis is validated through a 300W prototype. Experimental results demonstrate that this strategy enables soft switching of the switching devices across the full range, confirming its feasibility and advantages for wide input voltage ranges and high voltage applications.

Future work will focus on developing closed-loop control strategies, optimizing the performance of experimental prototypes, and conducting further experimental validation. Additionally, the integration of variable frequency modulation with this strategy is also a worthwhile area of research.

REFERENCES

[1] S.-H. Lee, B.-S. Lee, D.-H. Kwon, J.-H. Ahn, and J.-K. Kim, "Two-Mode Low-Voltage DC/DC Converter With High and Wide Input Voltage Range," *IEEE Transactions on Industrial Electronics*, vol. 68, no. 12, pp. 12088–12099, Dec. 2021, doi: 10.1109/TIE.2020.3039214.

[2] R. Xie, Y. Chen, Z. Wang, S. Mei, and F. Li, "Online Periodic Coordination of Multiple Pulsed Loads on All-Electric Ships," *IEEE Transactions on Power Systems*, vol. 35, no. 4, pp. 2658–2669, Jul. 2020, doi: 10.1109/TPWRS.2019.2961147.

[3] E. Serban, M. Ordonez, and C. Pondiche, "DC-Bus Voltage Range Extension in 1500 V Photovoltaic Inverters," *IEEE Journal of Emerging and Selected Topics in Power Electronics*, vol. 3, no. 4, pp. 901–917, Dec. 2015, doi: 10.1109/JESTPE.2015.2445735.

[4] Y. Zhang, Y. Gao, J. Li, and M. Sumner, "Interleaved Switched-Capacitor Bidirectional DC-DC Converter With Wide Voltage-Gain Range for Energy Storage Systems," *IEEE Transactions on Power Electronics*, vol. 33, no. 5, pp. 3852–3869, May 2018, doi: 10.1109/TPEL.2017.2719402.

[5] Y. Koç, Y. Birbir, and H. Bodur, "Non-isolated high step-up DC/DC converters – An overview," *Alexandria Engineering Journal*, vol. 61, no. 2, pp. 1091–1132, Feb. 2022, doi: 10.1016/j.aej.2021.06.071.

[6] F. Li, R. Hao, H. Lei, X. You, C. Ke, and J. Wang, "Non-Inverting Three-Level Buck-Boost Converter for Wide Voltage Range Application," in *2018 IEEE Energy Conversion Congress and Exposition (ECCE)*, Sep. 2018, pp. 4870–4875. doi: 10.1109/ECCE.2018.8557724.

[7] C. Liu, S. Zhou, J. Zou, X. Zou, H. Peng, and Y. Kang, "Hybrid-Type Multilevel DAB Converter for High and Ultrawide Input-Voltage Range," *IEEE Transactions on Industrial Electronics*, pp. 1–7, 2024, doi: 10.1109/TIE.2024.3390749.

[8] L. Tian, X. Wu, C. Jiang, and J. Yang, "A Simplified Real-Time Digital Control Scheme for ZVS Four-Switch Buck–Boost With Low Inductor Current," *IEEE Transactions on Industrial Electronics*, vol. 69, no. 8, pp. 7920–7929, Aug. 2022, doi: 10.1109/TIE.2021.3104616.

ZVT Circuit applied for Wide Input Range Isolated Converters

Linguo Wang, Zhongyin Guo, Junjie Zhu, Bing Zhang, Zhiling Zuo, Xiaoguang Gao, Guangji Ma

ZTE Corporation, Shenzhen 518057, China

Email: wang.linguo@zte.com.cn

Abstract—This paper proposed an isolated zero voltage transition (ZVT) circuit for wide input range applications, which can achieve zero voltage switching (ZVS) for primary switches and zero current switching (ZCS) for secondary synchronous rectifiers in full input and load range. Compared with traditional ZVT circuits, the auxiliary inductor can be automatically fast reset, thus the auxiliary switch can also be turned off with ZCS and the rms value of auxiliary current is reduced accordingly. Besides, the auxiliary current can be further reduced by coupling to main inductor and the extra losses of auxiliary circuit are minimized. Finally, based on an existing full bridge topology, a standard 1/4 brick -48V bus converter prototype was built for verification. A peak efficiency up to 97.13% and maximum output power up to 1000W is achieved.

Keywords—soft switching, zero voltage transition (ZVT)

I. INTRODUCTION

Isolated power supplies with wide input range have important applications in telecommunication and other industrial area, due to battery voltage drifting, distributions voltage drops, etc. With the rapid growth in communication traffic and AI computing, the efficiency and power density are facing new challenges. However, the mainly adopted topology currently for such wide input application is still the traditional hard switched full bridge. Common isolated soft switching techniques like phase shift full bridge [1-3] and resonant converters [4-6] both have their limitations especially for wide input range application. For example, phase shift topology has circulating power and loss of duty cycle due to introduced inductor at primary side. Differing from pulse width modulation (PWM) converter, resonant converter like LLC has complicated gain characteristic, the optimized working zone is limited to around the resonant frequency and unity gain.

For non-isolation applications, typical soft switching technologies include the discontinuous conduction mode (DCM) or critical conduction mode (CRM) control [7-8] of inductor current and zero voltage transition (ZVT) circuit [9-10]. The DCM and CRM control is limited to small power due to large inductor current ripple and high turn-off loss. The ZVT circuit can keep the inductor current in continuous mode while only change the current direction at switching transient by injecting auxiliary current. However, the traditional ZVT topology as shown in Fig.1 introduces high extra loss of auxiliary circuit, especially the high turn-off loss of auxiliary switch.

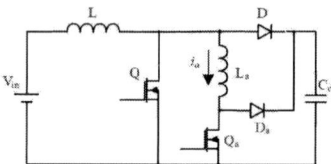

Fig. 1. Traditional ZVT circuit.

Newer researches on wide input range soft switching solution may include multi-resonant tank [13-15] or two-stage structure [16-17], which are based on more complex circuits and control, thus are not suitable for high performance product application. In industry, the improvement of efficiency and power density relies more on evolution of materials and processes such as third-generation semiconductor devices like GaN and SiC, low loss magnetic materials and high density PCB processes. Besides, for bus converters application, a tradeoff is often made by so-called semi-regulation converter, where the output is regulated at high input voltage and unregulated at low input voltage.

To address above issues, this paper proposed a novel ZVT circuit suitable for wide input range isolated converters and for product application. Compared with traditional isolated soft-switching technologies, this circuit can achieve soft switching under full input and load range with no efficiency compromise. Compared with traditional ZVT circuits, this circuit can solve the key issues like high turn-off loss of auxiliary switch by implementing a fast automatically reset scheme for the auxiliary inductor. Besides, by integrating the auxiliary inductor to the coupling winding of main inductor, the auxiliary current can be further reduced due to the coupling effect by turn ratio. The extra auxiliary components and losses are minimized. Applying to a traditional full bridge topology, zero voltage switching (ZVS) of all primary switches and zero current switching (ZCS) of secondary synchronous rectifiers (SR) switches can be realized.

II. THE PROPOSED ISOLATED ZVT CIRCUIT

In traditional ZVT circuit shown in Fig.1, the voltage over auxiliary inductor L_a is zero when main switch Q is turned on, thus auxiliary current i_a can only be reset by turning off the auxiliary switch Q_a. Q_a is then suffering high turn-off current and losses. Fig. 2 shows the proposed isolated ZVT circuit applied to a full bridge topology. In Fig. 2, the ZVT auxiliary switch Q_a and inductor L_a is connected to the output of secondary rectifiers in series with a coupling winding of main output inductor L. By coupling the voltage of main inductor to the auxiliary circuit, the voltage over auxiliary inductor L_a can be reversed when primary switches are turned on, thus the auxiliary current i_a can be automatically reset. Furthermore, by utilizing the voltage transition of the rectifier output together with the turn ratio of the auxiliary winding, a symmetric charging and discharging voltage over auxiliary inductor La can be designed. This is important to ZVT circuit, with fast charging and discharging feature, the auxiliary current will only be present a very short period, thus the extra conduction loss of the auxiliary circuit can also be minimized. Besides, with this coupling structure, the auxiliary inductor L_a can be integrated as leakage inductor, thus no extra magnetic component is needed.

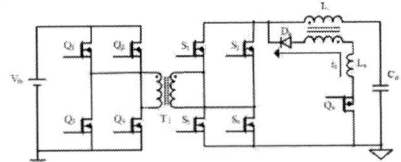

Fig. 2. Proposed isolated ZVT circuit.

ZVT of the proposed circuit is realized by turning on Q_a before primary switches when output inductor current is freewheeled in SR switches. Then the auxiliary current i_a is charged and injecting to the SR node. The current freewheeled in SR is decreased and SR switches can be turned off with ZCS. When the injected current is larger than output current, the current in SR is reversed to the primary. Then, the primary switches can be turned on with ZVS.

III. DETAILED OPERATING MODES

In this section, the detailed operating modes for soft switching realization is described. The key waveforms of each period are shown in Fig.3, where G_1~G_4, G_{1s}~G_{4s}, G_a is the respective drive signal for primary switches G_1~G_4, SR switches S_1~S_4 and auxiliary switch Q_a in Fig.2. The detailed equivalent circuits of each working modes are shown in Fig.4, where for better description, the Mosfet is using detailed model with inherent body diode and output capacitor C_{oss}, the integrated coupling inductor is using detailed model with primary winding, secondary winding, main inductor L and leakage inductor L_a.

A. Mode a, b (t_0~t_1)

The main inductor current i_L is freewheeled via SR switches S_1~S_4 when they are turned on (Mode a) or via the body diode when S_2, S_3 are turned off (Mode b). The switching node voltage V_{sw} of SR output is zero, thus when auxiliary switch Q_a is turned on prior to the primary switches, the charging voltage over auxiliary inductor L_a is

$$V_{La} = V_o \cdot N_a \tag{1}$$

Where Na is turn ratio of the auxiliary coupling winding to the main inductor. The auxiliary inductor current i_a starts

injecting to the SR switching node. Meanwhile, i_{ap} which is the coupling current of i_a with N_a times higher value is also injected to the SR. The current conducted in SR i_{sr} is decreased rapidly.

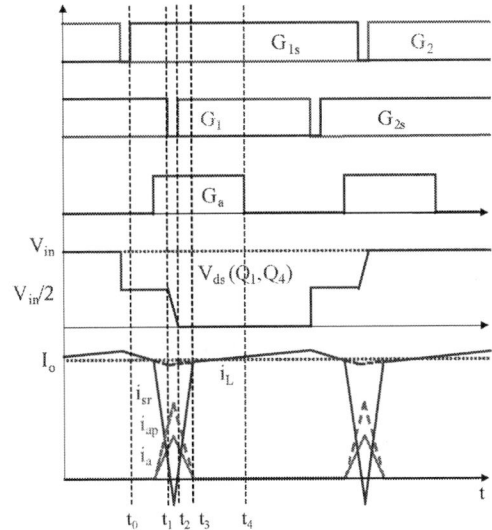

Fig. 3. Control scheme and key waveforms for ZVT realization.

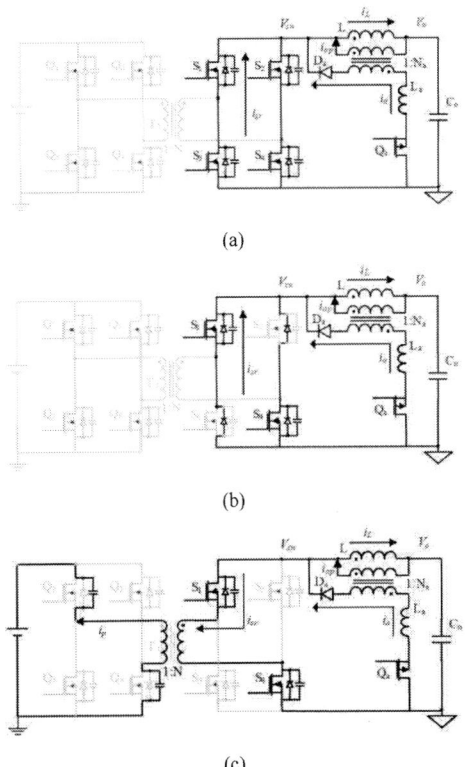

(a)

(b)

(c)

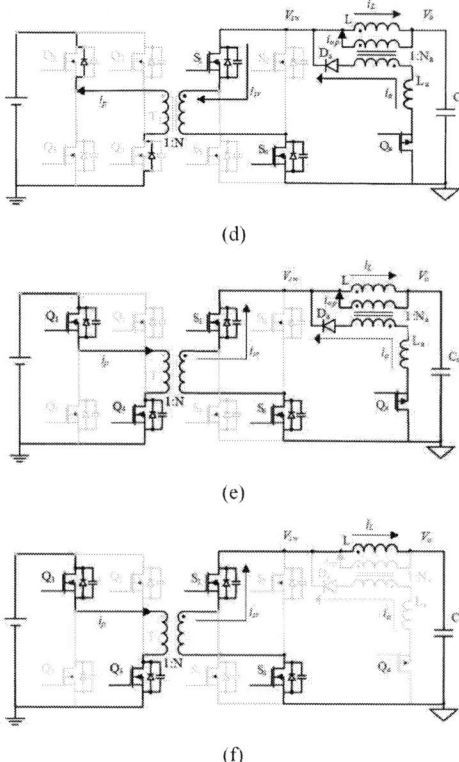

(d)

(e)

(f)

Fig. 4. Detailed operating modes.

B. Mode c, d (t₁~t₂)

When the injected current i_a plus i_{ap} equals to i_L, i_{sr} is decreased to zero, the SR switches S_2, S_3 can be turned off with ZCS, and the body diode has no reverse recovery issues. When i_a plus i_{ap} keeps increasing larger than i_L, i_{sr} turns negative. During the deadtime when both primary switches and S2, S3 are turned off, the reversed i_{sr} begins to discharge the Coss of primary switches (Mode c), the V_{ds} of Q_1, Q_4 is decreased accordingly. With the SR switching node voltage V_{sw} increased simultaneously, the voltage over auxiliary inductor L_a starts drop.

$$V_{La} = -\left(V_{sw} - V_o\right) \cdot N_a - V_{sw} \qquad (2)$$

When V_{La} falls to zero, i_a reaches to peak and keeps discharging C_{oss}. Then i_a starts discharging with V_{sw} keeps rising. When C_{oss} discharged to zero, the reversed primary current ip is conducting through body diode of Q_1, Q_4 (Mode d), keeps V_{ds} of Q_1, Q_4 zero. Q_1, Q_4 can be turned on with ZVS afterwards at t₂.

C. Mode e (t₂~t₃)

After Q_1, Q_4 turned on, switching node voltage V_{sw} equals to V_{in}/N, where N is the turn ratio of main transformer. The voltage over L_a is

$$V_{La} = -\left(\frac{V_{in}}{N} - V_o\right) \cdot N_a - \frac{V_{in}}{N} \qquad (3)$$

With this negative V_{La}, the auxiliary current i_a keeps discharging. Once i_a is decreased to zero, it's cut off by the auxiliary diode D_a. Thus, in this ZVT circuit, the auxiliary inductor can be automatically reset with no action of auxiliary switch Q_a.

D. Mode f (t₃~t₄)

Due to the auxiliary diode D_a, the auxiliary current i_a is kept zero in this mode. Then the auxiliary switch Q_a can be turned off with ZCS at any time during the turned-on period of primary switches. The main circuit is working as normal full bridge.

According to the symmetry of full bridge topology, primary switches Q_2, Q_3 can also be turned on with ZVS in the next half switching period. Thus, all primary switches Q_1~Q_4 can realize ZVS, and all secondary SR switches S_1~S_4 can realize ZCS in this circuit.

IV. HIGHLIGHT FEATURES

In addition to soft switching realization, the proposed ZVT circuit also shows below features, that can minimize the side effect brought by the auxiliary circuit, including extra power losses, components and control complexity. This is important to make a ZVT circuit for product ready use.

A. Fast automatically reset of auxiliary inductor

Though the automatically reset of auxiliary inductor can be realized by coupling with main inductor [11-12], the reset speed is slow in some application with asymmetric duty cycle. This increases the conduction period and the rms value of auxiliary current and hence the extra conduction loss. Especially in isolated application, for better power transfer efficiency purpose, the main conversion ratio should be handled by transformer, resulting a large duty cycle. In these cases, the charging and discharging voltage over main inductor is asymmetric. Taking a -48V input and +12V output isolated bus converter as an example, the turn ratio of full bridge transformer is designed to 3:1, then the voltage over main inductor is +4V and -12V. By reversed coupling, the voltage over auxiliary inductor is +12V and -4V, resulting a slow discharging speed. In this proposed ZVT circuit, the SR switching node voltage V_{sw} together with the turn ratio Na can be used to further compensate the unbalanced voltage over auxiliary inductor. By choosing $N=3$ and $N_a=2$, the voltage over auxiliary inductor V_{La} in (1), (3) can be designed to +24V and -24V. This makes a faster reset speed of auxiliary inductor and small rms value of auxiliary current.

B. Amplified injecting current

As shown in detailed coupling inductor model in Fig.4, the coupling current i_{ap} which is Na times larger of i_a is also injected to the switching node simultaneously. The equivalent injection current is

$$i_{inject} = \left(1 + N_a\right) \cdot i_a \qquad (4)$$

By $N_a=2$, that means only 1/3 of output current is required for the auxiliary inductor current i_a to achieve ZVT. This feature can further reduce the auxiliary conduction loss and make it easier for soft switching realization even at heavy load. The saturated current requirement of the auxiliary inductor is also reduced, a physical inductor can be eliminated by using leakage inductor instead.

C. Adaptive control for realizing soft switching

According to (2), the auxiliary current i_a is kept charging until the SR switching node voltage V_{sw} is charged to a certain value when $V_{La}=0$ by the reversed i_{sr} in mode c. This means, with various load the injection current can always be charged larger than the load current during deadtime period. The injection current value and ZVS realization is adaptive to load range. This feature makes the turn-on timing control of auxiliary switch flexible, together with the flexibility of turn-off timing control due to the fast automatically reset feature, the controller can keep simple.

All the above features make this ZVT circuit suitable for product application from form factor, whole efficiency aspects. In the next section, a prototype with original 1/4 brick standard package will be built for verification.

V. PROTOTYPE AND EXPERIMENTAL RESULTS

Based on a traditional full bridge bus converter, a ZVT prototype was built by adding auxiliary circuit in secondary side. With minimized additional ZVT components, the original 1/4 brick package is kept as shown in Fig.5. The additional auxiliary ZVT circuit only includes a P-type Mosfet as auxiliary switch and a diode for unidirectional conduction. The auxiliary inductor is integrated to main inductor by coupling with a 4-turn PCB winding. The ZVT components are located on the heatsink side in Fig.5 and can be seen in Fig.8.

Fig. 5. 1/4 brick ZVT full bridge prototype.

TABLE I.   DESIGN PARAMETERS

Parameters	Values
Input voltage	−36 ~ −40V
Output voltage	+12V
Output power	1000W
Form factor	1/4 brick (58.4*36.8*12.7mm)
Topology	Full Bridge

Fig.6 shows the key experimental waveforms with -48V input voltage and 40A load current. It can be observed that the auxiliary current i_a is rising sharply with auxiliary switch Q_a

turned on before primary switch Q_4. When the effective injection current is larger than load current, V_{ds} of Q_4 falls down before Q_4 is turned on by G_4 and ZVS of Q_4 is realized. After Q_4 turned on, i_a is discharged and then clamped to zero automatically. The auxiliary switch hereby using P-type Mosfet in this prototype is then turned off with ZCS.

In Fig.6, The fast automatically reset feature of auxiliary inductor can also be verified by the quite symmetrical rising and falling slope of i_a. The amplified injecting current feature can be verified by that the peak value of i_a is only about 15A while the load current is 40A.

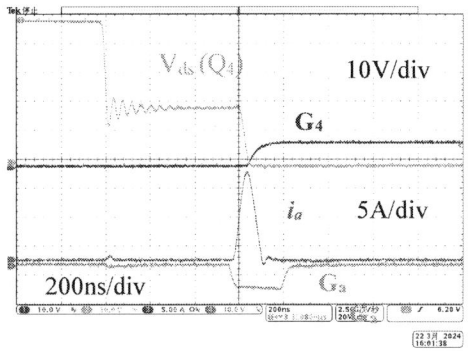

Fig. 6.   Waveforms of primary ZVS and injection current.

With the ZCS of SR switches and no reverse recovery of corresponding body diode, voltage spikes over SR switch S_3 is totally suppressed in Fig.7, while in hard switching full bridge it's a severe issue effecting efficiency and reliability.

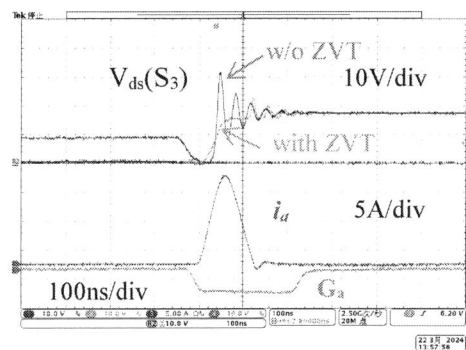

Fig. 7.   Waveforms of secondary voltage stress.

In addition to ZVS realization, the reverse recovery losses are also saved on primary switches. In Fig.8, the temperature on primary switches is 5 degrees lower than the hard switching version.

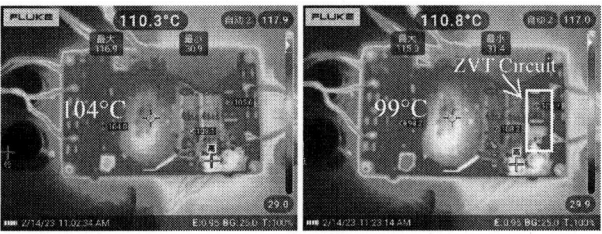

Fig. 8.   Thermal images of hard switching (left) and ZVT version (right).

Finally, the full load and input range efficiency curve is tested in Fig. 9. All extra losses introduced by auxiliary ZVT circuit is included. Under the -48V nominal input, a peak efficiency of 97.13% is achieved, with 0.5% improvement to the hard switching version.

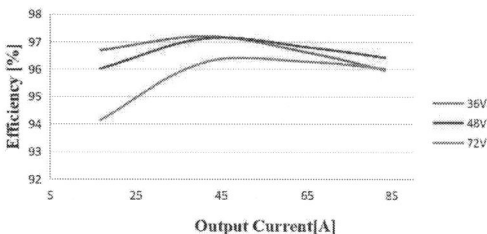

Fig. 9. Efficiency curve of full input and load range.

VI. CONCLUSIONS AND FUTURE WORK

This paper presented a ZVT circuit for isolated converters with wide input range. Compared to traditional isolated soft switching methods, ZVS and ZCS can be realized in full input and load range. Compared to traditional ZVT circuit, ZCS of the auxiliary switch can be realized, together with other features like fast reset and amplified injection current, integration of auxiliary inductor, the extra loss and the volume of auxiliary circuit are minimized for product ready implementation. Finally, a standard 1/4 brick bus converter prototype was built for verification. A peak efficiency up to 97.13% and maximum output power up to 1000W is achieved.

REFERENCES

[1] J. A. Sabate, V. Vlatkovic, R. B. Ridley, F. C. Lee and B. H. Cho, "Design considerations for high-voltage high-power full-bridge zero-voltage-switched PWM converter," in 1990 IEEE Applied Power Electronics Conference, 1990, pp. 275-284.

[2] P. F. Kocybik and K. N. Bateson, "Digital control of a ZVS full-bridge DC-DC converter," in 1995 IEEE Applied Power Electronics Conference, 1995, pp. 687-693

[3] Gwan-Bon Koo, Gun-Woo Moon and Myung-Joong Youn, "New zero-voltage-switching phase-shift full-bridge converter with low conduction losses," in IEEE Transactions on Industrial Electronics, vol. 52, no. 1, pp. 228-235, Feb. 2005

[4] R. L. Steigerwald, "A comparison of half-bridge resonant converter topologies," in 1987 IEEE Applied Power Electronics Conference, 1987, pp. 135-144

[5] Bo Yang, F. C. Lee, A. J. Zhang and Guisong Huang, "LLC resonant converter for front end DC/DC conversion," in 2002 IEEE Applied Power Electronics Conference, 2002, pp. 1108-1112

[6] Yuancheng Ren, Ming Xu, Julu Sun and F. C. Lee, "A family of high power density unregulated bus converters," in IEEE Transactions on Power Electronics, vol. 20, no. 5, pp. 1045-1054, Sept. 2005

[7] Jong-Hu Park and B. H. Cho, "The zero voltage switching (ZVS) critical conduction mode (CRM) buck converter with tapped-inductor," in 2003 IEEE Applied Power Electronics Conference, 2003, pp. 1077-1081

[8] Z. Zhou, H. Li and X. Wu, "A Constant Frequency ZVS Control System for the Four-Switch Buck-Boost DC-DC Converter With Reduced Inductor Current," in IEEE Transactions on Power Electronics, vol. 34, no. 7, pp. 5996-6003, July 2019

[9] G. Hua, C.-S. Leu, Y. Jiang and F. C. Y. Lee, "Novel zero-voltage-transition PWM converters", in IEEE Transactions on Power Electronics, vol. 9, no. 2, pp. 213-219, Mar. 1994

[10] E. Adib and H. Farzanehfard, "Zero-voltage-transition PWM converters with synchronous rectifier", in IEEE Transactions on Power Electronics, vol. 25, no. 1, pp. 105-110, Jan. 2010

[11] M. R. Mohammadi and H. Farzanehfard, "A New Family of Zero-Voltage-Transition Nonisolated Bidirectional Converters With Simple Auxiliary Circuit," in IEEE Transactions on Industrial Electronics, vol. 63, no. 3, pp. 1519-1527, March 2016

[12] X. -F. Cheng, C. Liu, D. Wang and Y. Zhang, "State-of-the-Art Review on Soft-Switching Technologies for Non-Isolated DC-DC Converters," in IEEE Access, vol. 9, pp. 119235-119249, 2021

[13] X. Zhang, J. Jing, Y. Guan, M. Dai, Y. Wang and D. Xu, "High-Efficiency High-Order CL-LLC DC/DC Converter With Wide Input Voltage Range", in IEEE Transactions on Power Electronics, vol. 36, no. 9, pp. 10383-10394, Sept. 2021.

[14] M. Chen, B. Chen, Y. Wang, P. Wang, M. Zhang and C. Che, "A High-efficiency Resonant DC-DC Converter with Wide Voltage Gain Range", in 2022 IEEE Energy Conversion Congress and Exposition (ECCE), pp. 1-6, 2022.

[15] R. M. Reddy, A. K. Jana and M. Das, "Novel Wide Voltage Range Multi-Resonant Bidirectional DC−DC Converter", in 2020 IEEE International Conference on Power Electronics Drives and Energy Systems (PEDES), pp. 1-6, 2020

[16] X. Sun, J. Qiu, X. Li, B. Wang, L. Wang and X. Li, "An improved wide input voltage buck-boost + LLC cascaded converter," 2015 IEEE Energy Conversion Congress and Exposition (ECCE), Montreal, QC, Canada, 2015, pp. 1473-1478

[17] J. Zhu, Q. Qian, C. Zhan, S. Lu, W. Sun, ZVS Buck-Boost LLC cascade converter with all soft switched switches, Journal | [J] International Journal of Electronics. Vol. 106, issue 6. 2019. pp 895-911.

Impact of Asymmetrical Leakage Inductance on a 380 V-12 V *LLC* Converter with Synchronous Rectifier for DCX Application

Jinshu Lin[1], Shan Yin[2], Chen Song[1], Honglang Zhang[1], Minghai Dong[1], Limei Xu[1], Hui Li[1]

[1] School of Aeronautics and Astronautics, University of Electronic Science and Technology of China, Chengdu, China
[2] Huawei Technologies Co., Ltd., Shenzhen, China
linjinshu@std.uestc.edu.cn

Abstract—The 380 V-12 V DC-DC converter utilizing an *LLC* resonant topology is considered a promising option for future data center architectures. To achieve high power density, the planar transformer is commonly employed, using leakage inductance for resonance, and operating in DCX mode. Additionally, magnetizing inductance is adjusted by introducing an air gap. However, this approach may disrupt the balance of leakage inductance distribution in the center-tapped (CT) transformer, leading to asymmetrical operation, especially under light load conditions. This study analyzes the impacts of asymmetrical leakage inductance in a CT transformer for a 380 V-12 V *LLC* resonant converter by examining the magnetic field energy distribution with different winding and core assembly methods. Moreover, a 600-kHz, 250-W 380 V-to-12 V *LLC* converter prototype was developed for this purpose. Results indicate that, for applications where leakage inductance serves as the resonant inductance, only symmetrical cores (such as the EE core used in this study) can ensure equal leakage inductance in a CT transformer. Furthermore, integrated magnetic solutions are also suggested, offering valuable insights for future high-density converter designs.

Index Terms—DC-DC, fractional-turn transformer, *LLC*

I. INTRODUCTION

The data center is an essential infrastructure to support the rapid advancements in emerging technologies such as AI, 5G, and cloud computing [1]. The accelerated evolution of these technologies has significantly increased electricity demand, emphasizing the need for power architectures with higher power density, greater efficiency, and increased current capacity. Currently, power supply efficiency from the grid to data centers is approximately 85%, revealing substantial potential for optimization [2]–[5]. Therefore optimizing data center power architecture is essential not only to address the increasing energy demands of advancing technologies but also to enhance overall operational efficiency.

The 380 V bus architecture directly steps down the input voltage of 380 V to a 12 V load, significantly reducing the number of conversion stages [6]–[8]. A fundamental 380 V-12 V isolated DC-DC converter, utilizing an *LLC* resonant converter, including three main modules: an inverter, an *LLC* resonant tank, and a rectifier, as shown in Fig. 1. The inverter converts the 380 V DC input into a square wave AC output for the LLC resonant tank. The half-bridge(HB) and full-bridge(FB) configurations are the most commonly

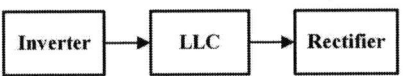

Fig. 1. Block diagram of fundamental 380 V-12 V isolated DC-DC converter.

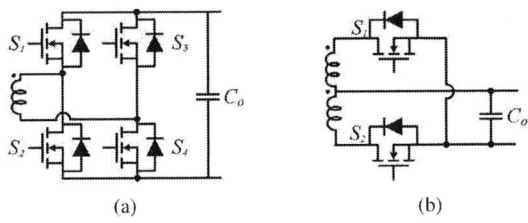

Fig. 2. Configurations of the rectifier: (a) SR, (b) FB.

employed topologies for this stage. Gallium nitride high-electron-mobility transistors (GaN HEMTs) are widely used in these inverters to achieve high-frequency operation, pushing the switching frequency to MHz-levels and enhancing power density [9]–[11]. The *LLC* resonant converter features soft-switching characteristics, enabling ZVS for the primary switches and zero-current switching (ZCS) for the secondary rectifier switches, which make it particularly suitable for high-frequency DC-DC applications to pursuit higher power density [12]–[16]. The rectifier module converts the sinusoidal-like current from the secondary side of the transformer into a DC output. In step-down converters, the full-bridge (FB) and CT topologies are the two most common rectifier configurations, as shown in Fig. 2.

From a topological perspective, the FB configuration allows each rectifier switch to endure only half the voltage stress, although it requires twice the number of switching rectifiers compared to the CT configuration [17]. From a practical design standpoint, for planar transformers, the FB structure requires at least three layers, as the secondary winding for the positive and negative half-cycles shares the same winding path. In contrast, the CT structure requires placing the secondary windings on different layers, resulting in a minimum of four layers. Most importantly, this configuration significantly impacts the distribution of stray parameters, particularly leakage inductance. As noted in [18], the leakage inductance between

979-8-3315-1612-3/25 $31.00 © 2025 IEEE

the primary and secondary sides of a transformer is not inherently equal. This imbalance occurs because the stored leakage energy in the primary winding area differs from that in the secondary winding area, which is primarily caused by the asymmetry of the transformer core (e.g., EI type).

In the context of an FB rectifier structure, this asymmetry leads only to a difference between the primary and secondary leakage inductances, while the total equivalent leakage inductance for the LLC resonant tank remains constant [19]. However, for the CT structure, the asymmetry impacts not only the primary and secondary sides but also the secondary windings during the positive and negative half-cycles. Consequently, the equivalent leakage inductance varies during the switching of LLC operating modes, significantly affecting the performance of the LLC resonant tank. Since the leakage inductance is intended to serve as the resonant inductance in DCX mode, where the switching frequency equals the resonant frequency, the asymmetrical leakage inductance results in unequal resonant frequencies, thereby degrading the efficient operation of the LLC resonant tank. The unequal leakage inductance between the primary and secondary sides in FB applications has been thoroughly researched in [18], [19]. However, for CT transformers, the additional effects on secondary leakage inductance during the positive and negative half-cycles have not yet been addressed in detail, which is critical given that the leakage inductance directly serves as the resonant inductance.

The rest of this paper is organized as follows: The design analysis of the 380 V-12 V high step-down DC-DC converter is presented in Section II. The variations in the H-field distribution within the transformer, caused by the assembly method between the PCB and the winding and resulting in asymmetrical leakage inductance, are investigated in Section III. Optimization solutions to eliminate the asymmetrical characteristics are discussed in Section IV. Experimental validations are illustrated in Section V. Finally, Section VI summarizes the main conclusions of this paper.

II. CIRCUIT ANALYSIS

The circuit diagram of a half-bridge 380 V-12 V DC-DC converter utilizing an LLC resonant converter with a CT transformer is shown in Fig. 3. In this configuration, the primary-side voltage v_{AB} is a square wave with a peak-to-peak amplitude of 380 V, having positive and negative peak values of +190 V and -190 V, respectively. To achieve a 12 V output, a turns ratio of 16:1 is required for the LLC resonant tank and transformer. Given that the gain of the LLC resonant tank is 1, a turn ratio of 16:1 is therefore required for the transformer. According to Equ. (1), the magnetizing inductance L_m significantly depends on the number of primary winding turns. In the 380 V bus converter, N_p^2 equals 256, whereas in a 48 V bus converter (with 4 primary winding turns), it is only 16. Therefore, at the same power level, the 380 V bus architecture results in a significantly higher magnetizing inductance L_m compared to the 48 V bus.

In order to achieve ZVS for the primary switches, sufficient magnetic field energy must be provided by L_m to discharge the

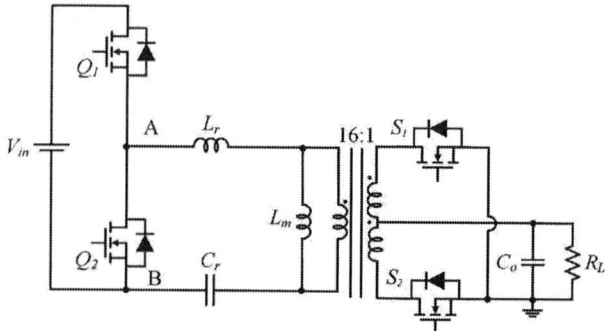

Fig. 3. Half-bridge LLC converter with CT transformer.

TABLE I
CORE PARAMETERS.

Parameters	Types or Value	Parameters	Types or Value
Material	PC 200	μ_r	800
l_g	65 mm	A_e	79 mm^2
B_s	530 mT	Window height	7 mm

C_{ds} of the primary switches. The relationship between L_m and C_{ds} is derived in Equ. (2), where C_{ds} is the output capacitor of the primary switch and f_{sw} is the switching frequency. To properly set the value of t_{dead}, since C_{ds} and f_{sw} are constant, L_m should be smaller. To adjust the value of L_m, an air gap is introduced to increase the magnetic reluctance of the core, as shown in Equ. (1), where N_p is the number of primary winding turns and R_g is the magnetic reluctance of the core. A smaller L_m is beneficial for achieving ZVS.

However, as shown in Equ. (3), this adjustment can lead to another issue: a larger I_m results in a smaller I_p, significantly affecting the energy transfer of the LLC resonant converter. Therefore, for the high-efficiency operation of the LLC resonant converter, a trade-off in selecting L_m must be carefully considered.

$$L_m = \frac{N_p{}^2}{R_g} \tag{1}$$

$$t_{dead} \geq 16 \times C_{ds} \times f_{sw} \times L_m \tag{2}$$

$$I_m = \frac{nV_o}{2fL_m} \tag{3}$$

Supposing core PC 200 EE 22/12/16 from TDK is selected, with key parameters shown in Table IV, the relationship between the air gap length and the magnitude of L_m is depicted in Fig. (4). When there is no air gap, L_m exceeds 400 μH, while the experiment value of leakage inductance should be less than 3μH, the difference between the L_r and L_m emphasizing the need to introduce an air gap to achieve ZVS. However, due to the presence of an air gap, the MMF between the primary and secondary windings cannot be fully canceled, as expressed by $N_pI_p - N_sI_s = \phi_mR$, where results in residual magnetic flux, and influences the magnetic field

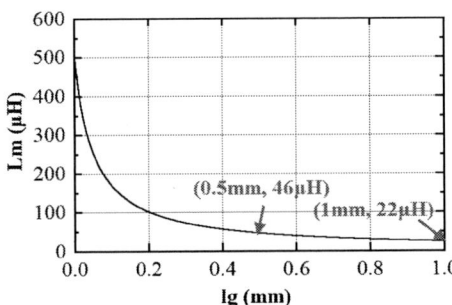

Fig. 4. Relationship between the air gap length and the magnitude of L_m.

Fig. 5. Winding arrangement.

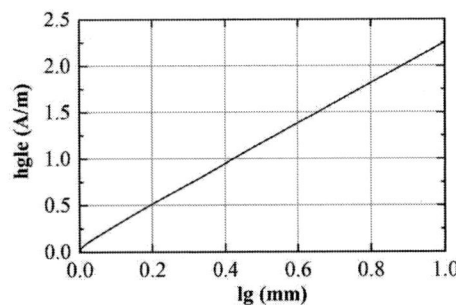

Fig. 6. The relationship between the $H_g l_g$ and the l_g.

TABLE II
PARAMETERS FOR EQU. (4)

Parameters	Value	Parameters	Value
n	16	μ_r	800
V_o	12 V	A_e	78 mm^2
N_p	16	f	1 MHz
l_c	65 mm		

distribution. The magnetomotive force of the air gap can be derived as:

$$H_g l_g = \frac{n V_0}{2 f N_p \mu_0 \mu_r A_c} \cdot \sqrt{l_g^2 \mu_r + l_c l_g} \tag{4}$$

Consider a four-layer CT winding arrangement, as illustrated in Fig. 5. Layers 1 and 4 are designated for the secondary windings, handling the positive and negative half-cycles, respectively, and are connected in parallel. Layers 2 and 3, each with 8 turns, are allocated to the primary winding and are connected in series, yielding a turn ratio of 16:1. Combining the winding arrangement, the relationship between $H_g l_g$ and the air gap length is shown in Fig. 6, with other values presented in Table II. As illustrated in the figure, $H_g l_g$ increases nearly linearly with the increase of l_g. Assuming the air gap length is 1 mm and the output current is 20 A, the MMF inside the air gap is approximately 2.2 A/m, while the MMF for the secondary winding, $N_s I_s$, is 20 A/m, roughly 10% of the secondary side. This affects both energy transfer and leakage inductance. When the output current is 10 A, the MMF for the secondary winding, $N_s I_s$, is 10 A/m, about 20% of the secondary side, which has a more significant impact. Therefore, it can be concluded that the presence of the air gap notably affects the leakage inductance, especially under light load conditions. More details will be presented in the following section.

III. ANALYSIS OF THE SYMMETRICAL LEAKAGE INDUCTANCE

EE and EI cores are the two most commonly used core types, as shown in Fig.7. The general manufacturing height of the basic four-layer PCB is 1.6 mm, which is considerably smaller than the window height of the specified EE core (EE 22/12/16). Therefore, the core window cannot be completely filled with the 4 layer PCB. In combination with the above winding structure, there are three typical assembling methods

for the EE core and two methods for the EI core, as shown in Fig. 8 and Fig. 9, respectively.

When there is no air gap, the magnetic flux originating from the primary and secondary windings ideally cancels out completely within the core. Consequently, the magnetic field intensity H can be considered as zero throughout the entire magnetic circuit. For the PCB, the boundaries of the top and bottom layers, which are adjacent to the core, exactly exhibit zero H.

For a 4-layer PCB with the default manufacturing process, the height of the signal layer is significantly smaller than that of the dielectric layer ($insu$). Additionally, the height of the central dielectric layer ($insu2$) is nearly twice that of the other two dielectric layers ($insu1$ and $insu3$). Considering the physical structure of the PCB, the MMF distributions for the transformer without an air gap during the positive and negative half-cycles can be derived, as shown in Fig. 10. Since the height of the dielectric layer is much greater than that of the signal layer, it is sufficient to consider only the dielectric layers when calculating the stored magnetic field energy inside the winding. Finally, according to Equ. (5), due to the complete symmetry with the same magnetic field intensity H in the same volume space, it can be concluded that this assembling method ensures equal leakage energy between the positive and negative half-cycles, thus achieving equal leakage inductance.

$$W = \frac{1}{2} \int_V H^2 \, dV \tag{5}$$

When the length of the air gap is non-zero, there will be residual magnetic flux within the core, as mentioned before. If assembled as shown in Fig. 8(a), the MMF distribution can be derived as shown in Fig. 11. Due to the presence of the air gap, the magnetic field density H on the layer bordering the air gap is no longer zero, and the zero point is located at the border of the air gap, close to the core. Thus, based on Equ. 5,

Fig. 7. Common Core Types: (a) EE, (b) EI.

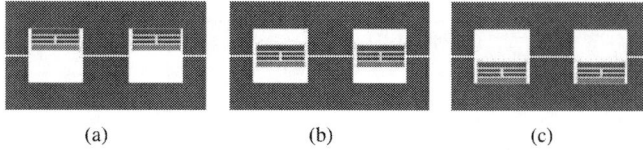

Fig. 8. Assembling methods with EE core, the order is as follows: (a)E core-winding-gap-E core, (b)E core-winding-air gap-winding-E core, (c)E core-air gap-winding-E core

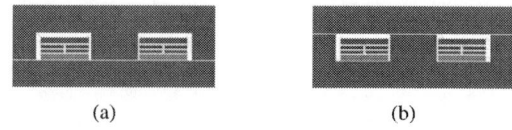

Fig. 9. Assembling methods with EI core, the order is as follows: (a)E core-winding-I core, (b)I core-winding-E core.

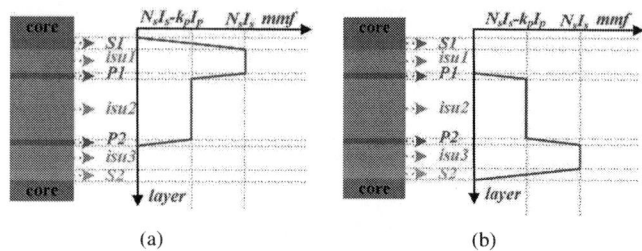

Fig. 10. The MMF distribution of EE core with l_g=0 mm at (a) positive half-cycle, (b) negative half-cycle.

it can be concluded that the magnetic field energy within the positive half-cycle will be smaller than that of the negative half-cycle, ultimately resulting in higher leakage inductance during the negative half-cycle.

For Fig. 8(b), the air gap is positioned to be entirely covered by the dielectric layer ($insu2$) along the longitudinal direction. When no air gap is present, H in $insu2$ remains constant, resulting in no significant variation. When an air gap is introduced, although it alters the H distribution, the variation occurs only within the central dielectric layer, and both half-cycles exhibit the same trend of change, thus ensuring equal magnetic field energy. Consequently, under this installation configuration, the magnetic field intensity achieves symmetry across both the positive and negative half-cycles, ensuring equal leakage inductance values for both cycles.

When the winding is assembled as shown in Fig. 8(c), its MMF distribution is symmetrical to that in Fig. 11. This also indicates an unequal leakage inductance distribution between the positive and negative half-cycles. For an EI core, which is also called an asymmetric core, the presence of the air gap causes H at the top and bottom border layers to differ. As a result, regardless of how the PCB and core are assembled, if a CT transformer is employed, the leakage inductance distribution for the positive and negative half-cycles will be unequal. Consequently, only an FB rectifier structure is suitable for achieving an integrated L_r.

From the above analysis of the three typical core and PCB assembly methods of EE core, a conclusion regarding the unequal leakage inductance distribution can be drawn. If the LLC resonant converter is to operate with the leakage inductance directly serves as the resonant inductance, the leakage inductance should be equal between the positive and negative half-cycles to ensure efficient converter operation. Thus, correct installation of the planar transformer is critical.

IV. SOLUTION AND INTEGRATED STACKED TRANSFORMER

As discussed previously, an improper assembly between the core and winding in a CT transformer can lead to unequal leakage inductance. Variations in L_r alter the input impedance of the LLC resonant tank, resulting in unequal energy circulation between the two half-cycles, which consequently impacts the efficient operation of the LLC resonant tank. The extent of the impact can be evaluated using Equ. (6), where D represents the difference in leakage inductance between the two half-cycles, and A denotes the average value of the two leakage inductances. This equation helps quantify the effect of uneven leakage inductance on the operation of the LLC converter. It illustrates that either a smaller Δ or a larger A leads to a lower ε, thereby mitigating the adverse impact of unequal leakage inductances, thus, three methods are proposed to solving the issue as follows:

$$\varepsilon = \left| \frac{\Delta}{A} \right| \qquad (6)$$

A. Assembling method

To minimize Δ, the most effective approach is to optimize the assembling method between the core, air gap, and PCB winding. For instance, the core should first be selected as a symmetrical type, such as an EE core. Additionally, the air gap and PCB should be arranged in a fully symmetrical and centralized configuration to ensure equal stored leakage magnetic field energy. This method is consistent with the configuration shown in Fig. 8(b), which, from a theoretical perspective, can ensure $\Delta = 0$.

B. Adding an inductor

To increase A, two methods can be considered. The first is to embed an external inductor into the LLC resonant circuit, which helps mitigate the overall impact and ultimately forms a conventional resonant tank. While adding an external inductor may not align with the primary goal of achieving high power density, it offers a design consideration when the assembly method is fixed, with a set leakage inductance difference. Additionally, this approach provides insight into how much inductance is required to alleviate the adverse effects to some extent.

979-8-3315-1612-3/25 $31.00 © 2025 IEEE

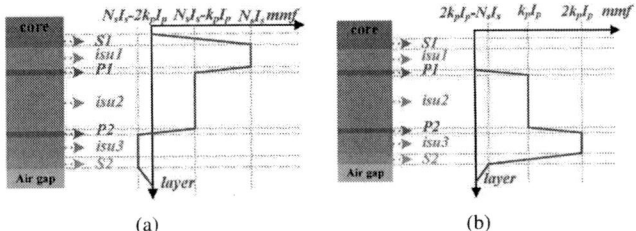

(a)   (b)

Fig. 11. The MMF distribution of EE core with air gap at (a) positive half-cycle, (b) negative half-cycle.

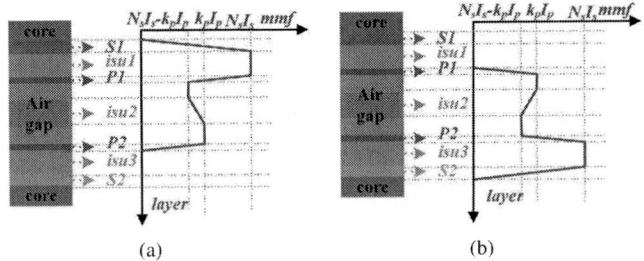

(a)   (b)

Fig. 12. The MMF distribution of EE core with air gap at (a) positive half-cycle, (b) negative half-cycle.

Using the same PCB arrangement and assembling method as shown in Fig. 5 and Fig. 8(a), the measured leakage inductance using an impedance analyzer (IM3570) is 1.6 μH for the positive half-cycle and 1.4 μH for the negative half-cycle, resulting in a difference of 0.2 μH. A simulation circuit is constructed in Plecs, where the leakage inductance difference is represented as an equivalent secondary leakage inductance, and the output load is set as fixed 1 ohm. For the primary side, the difference is 0.2 μH, and when reflected to the secondary side, it is divided by the square of the turns ratio n, i.e., n^2. Thus, the difference reflected to the secondary side is 7.8 nH.

The simulation results are shown in Table III. Since the impact of different L_r values is due to the varying impedance they introduce, the output secondary current for the positive half-cycle (I_p) and negative half-cycle (I_n) are used to investigate the extent of the asymmetrical leakage inductance impact. As the results show, when the resonant inductance increases to 4 μH (meaning the external inductance should be 4 - 1.4 = 2.8 μH), the difference in the secondary current between the positive and negative half-cycles is reduced to 1.32 A. Using the analysis method in Equ. 6, the ratio ε can be decreased to below 10%, indicating a significant improvement. This suggests that, despite a leakage inductance difference of approximately 0.2 μH, the total impact can be reduced to a negligible level.

C. Magnetic integration

Considering that asymmetrical leakage inductance arises from the inability to fully occupy the core window, magnetic integration can be achieved by increasing the distance between the primary and secondary winding layers. This adjustment

TABLE III
THE EFFECT OF THE VALUE OF THE i_{Lr} ON THE LLC

	1.6 μH	2 μH	3 μH	4 μH	5 μH
I_p (A)	15.53	15.48	14.06	12.66	11.48
I_n A	23.5	20.38	16.28	13.98	12.39
Δ A	7.93	4.9	2.22	1.32	0.91
ε	0.4053	0.2733	0.1463	0.0991	0.0762

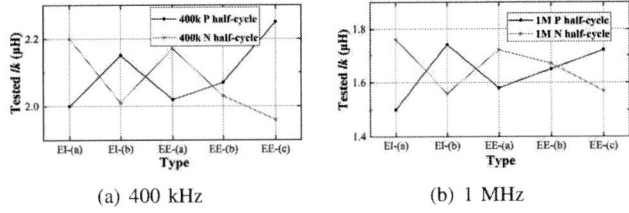

(a) 400 kHz   (b) 1 MHz

Fig. 13. The tested results.

allows for greater leakage energy storage while ensuring symmetric assembly and providing a larger leakage inductance, and the design options can refer to the [19].

V. EXPERIMENTAL VERIFICATION

A. impedance analyzer results

In the previous analysis, a four-layer transformer with a 16:1 high step-down ratio was examined. It was found that, under light load conditions, the positional relationship between the windings, core, and air gap affects the leakage inductance distribution of the CT transformer during the two half cycles. When the transformer leakage inductance is used directly as the resonant inductance to enhance power density, unequal leakage inductance distribution will result in unequal energy transfer between the positive and negative half cycles, thereby impacting the converter's efficient operation.

To further examine the above analysis, five different assembly structures, as shown in Fig. 8(a), Fig. 8(c), Fig. 8(b), Fig. 9(a), and Fig. 9(b), were considered. The leakage inductance distribution between the positive and negative half-cycles was precisely measured using an Impedance Analyzer (IM3570), and the tested results are shown in Fig. 13. As the test results demonstrate, in Fig. 8(a), the leakage inductance during the positive half-cycle is smaller than that during the negative half-cycle. In contrast, Fig. 8(c) presents a reversed but equal distribution compared to Fig. 8(a). Fig. 8(b) shows identical leakage inductance values for both half-cycles. The results in Fig. 9(a) align with those in Fig. 8(a), and Fig. 9(b) is consistent with the behavior observed in Fig. 8(a). In summary, the results support the theoretical analysis.

B. Experimental setup and results

To investigate the impact of asymmetrical leakage inductance distribution in practical implementation, a hardware prototype of a 380 V-12 V DC-DC converter was constructed, as shown in Fig. 14. The key parameters are summarized in Table IV. For the SR driver circuit, an adaptive driver method utilizing the UCC24612 was employed, which generates the

Fig. 14. Converter prototype.

TABLE IV
KEY PARAMETERS OF THE BUILT PROTOTYPE.

Parameters	Value	Parameters	Value
V_{in}	380 V	V_{out}	12 V
f_s	600 kHz	Core	PC200EE22/12/16
C_r	33 nF	gap length	1 mm
Primary switch	INN650D080BS	SR	IQE008N03LM5CG
Primary driver	UCC27511	SR driver	UCC24612

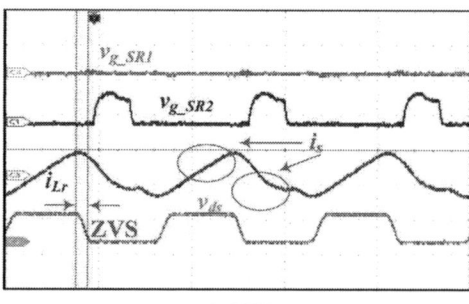

(a) 50W

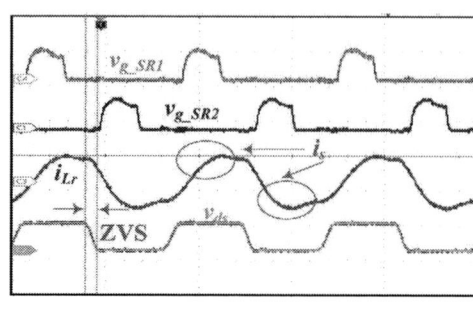

(b) 250W

Fig. 15. Measured waveforms with asymmetrical experimental performance of v_{g_SR1} (10 V/div) ,v_{g_SR2} (10 V/div), i_{Lr} (1 A/div), and v_s (500 V/div) at different switching frequencies. The time dimension is 1 μs/div.

TABLE V
PARAMETERS OF SR AND SR DRIVER.

	SR		SR Driver	
Parameters	Types/Value		Parameters	Types/Value
Device	IQE008N03LM5CG		Device	UCC24612
I_d	27 A		Frequency	1 MHz
V_{ds}	30 V		V_{THVGON}	-240 mV
$R_{DS(on),max}$	0.85 mΩ		$V_{THVGOFF}$	-9 mV

drive signal by monitoring the voltage across the SR. This method simplifies the SR driver circuit and eliminates the need for the SR drive signal to be referenced to a different ground than the primary switches.

In this implementation, the transformer's leakage inductance directly serves as the resonant inductance, with the assembling method following the installation shown in Fig. 8(c). Based on previous analysis and the results from the tests above, the leakage inductance during the positive half-cycle is expected to be higher than that during the negative half-cycle. The experimental waveforms are presented in Fig. 15. At light load (50 W), as depicted in Fig. 15(a), three distinct characteristics are observed: 1) ZVS is achieved for the primary switches; 2) The resonant current is asymmetrical; and 3) The synchronous rectifiers (SR1 and SR2) cannot be driven simultaneously, with SR1 failing to drive successfully during the positive half-cycle. When the power increases to 250 W, as shown in Fig. 15(b), the following characteristics are observed: 1) ZVS is achieved; 2) The resonant current remains asymmetrical; and 3) Both SR1 and SR2 can be driven successfully, in contrast to the light load condition.

Regarding 1), ZVS is primarily determined by sufficient magnetizing energy to discharge the C_{oss} of the switches, and by adjusting L_m, this result was achieved well. For 2), as proposed in the aforementioned analysis, this assembly method can cause asymmetrical leakage inductance distribution, resulting in a varying input impedance of the LLC resonant tank and an asymmetrical resonant current, as confirmed by experimental results.

As for 3), the selected SR is the IQE008N03LM5CG from Infineon, which has a low $R_{DS(on)}$ of 0.85 mΩ. The SR driver, UCC24612 from Texas Instruments (TI), is an adaptive driver that generates the drive signal by detecting the voltage across the SR, with power supplied through the output terminal. Therefore, to generate the normal drive signal V_{gs}, two necessary conditions must be fulfilled: 1) sufficient power supply, and 2) the attainment of the threshold voltage range.

In the experimental setup described above, the output voltage is 12 V, which is sufficient to fulfill condition 1. For condition 2, the voltage across the SR is given by $V_{ds,on} = R_{ds,on} \cdot I_d$, which is strongly correlated with the load current I_d. When the load current is large enough, both SRs can be driven to switch with the correct operation. However, at light load (e.g., 50W), the currents on both sides need to be equal in order to achieve a simultaneous switching state. As mentioned earlier, the asymmetric characteristics have a greater impact at light load; therefore, physical symmetry in the installation is crucial. With a symmetrical assembly method, as shown in Fig. 8(b), the waveforms in Fig. 16 show that both ZVS and simultaneous SR drive are achieved successfully.

VI. CONCLUSION

This study focused on the design of a 380 V-12 V DC-DC converter. To achieve ZVS, an air gap was introduced, resulting in asymmetrical leakage inductance in the CT transformer between the positive and the negative half-cycles. This asymmetry negatively affected the performance when the leakage inductance was intended to serve as the resonant inductance.

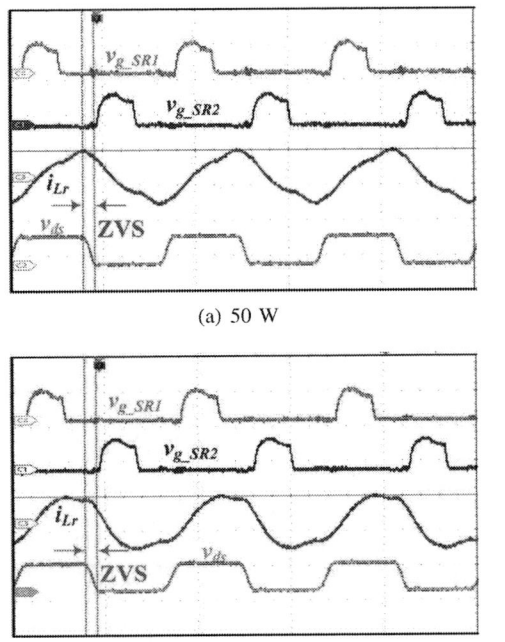

(a) 50 W

(b) 250 W

Fig. 16. Measured waveforms with symmetrical experimental performance of v_{g_SR1} (10 V/div) ,v_{g_SR2} (10 V/div), i_{Lr} (1 A/div), and v_s (500 V/div) at different switching frequencies. The time dimension is 1 μs/div.

According to the theoretical analysis, the impact of the asymmetry increased with the length of the air gap or a decrease in the load. It was concluded that only a fully symmetrical assembly between the winding and core could ensure equal leakage inductance with EE core. Furthermore, adding an external inductor or integrating magnetic components were alternative solutions to eliminate the asymmetrical effects between the two leakage inductances. A hardware prototype of a 600-kHz, 50-W/250-W, 380 V-12 V DC-DC converter was constructed to validate the analysis, and the experimental results agreed well with the theoretical predictions.

REFERENCES

[1] K. Tan, X. Song, C. Peng, P. Liu, and A. Q. Huang, "Hierarchical protection architecture for 380V DC data center application," in *2016 IEEE Energy Conversion Congress and Exposition (ECCE)*, Sep. 2016, pp. 1–8.

[2] W. Liu, Y. Guo, Q. Yang, C. Meng, and Y. Zhao, "Overview of energy-saving operation of data center under "double carbon" target," *Distribution & Utilization*, vol. 38, no. 09, pp. 49–55, 2021.

[3] F. C. Lee, "Power architecture for the next generation of data center," in *The China Power Supply Society Conference 2019*, Shenzhen, China, 2019.

[4] K. Wang, Q. Gao, G. Wei, and X. Yang, "Integrated fractional-turn planar transformer for MHz and high-current applications," *IEEE Trans. Power Electron.*, vol. 38, no. 6, pp. 7374–7384, Jun. 2023.

[5] C. Fei, F. C. Lee, and Q. Li, "High-efficiency high-power-density LLC converter with an integrated planar matrix transformer for high-output current applications," *IEEE Trans. Ind. Electron.*, vol. 64, no. 11, pp. 9072–9082, Nov. 2017.

[6] S. Li, E. Rong, Q. Min, and S. Lu, "A half-turn transformer with symmetry magnetic flux for high-frequency-isolated DC/DC converters," *IEEE Trans. Power Electron.*, vol. 33, no. 8, pp. 6467–6470, 2018.

[7] M. K. Ranjram and D. J. Perreault, "A 380-12 V, 1-kW, 1-MHz converter using a miniaturized split-phase, fractional-turn planar transformer," *IEEE Trans. Power Electron.*, vol. 37, no. 2, pp. 1666–1681, 2022.

[8] Y.-C. L. et al., "Quarter-turn transformer design and optimization for high power density 1-MHz LLC resonant converter," *IEEE Trans. Ind. Electron.*, vol. 67, no. 2, pp. 1580–1591, 2020.

[9] O. C. S. et al., "Optimized design of multi-MHz frequency isolated auxiliary power supply for gate drivers in medium-voltage converters," *IEEE Trans. Power Electron.*, vol. 35, no. 9, pp. 9494–9509, Sep. 2020.

[10] Y. Xiao, Z. Zhang, M. A. E. Andersen, and K. Sun, "Impact on zvs operation by splitting inductance to both sides of transformer for 1-MHz gan based dab converter," *IEEE Trans. Power Electron.*, vol. 35, no. 11, pp. 11 988–12 002, Nov. 2020.

[11] N. A. Dung, H.-J. Chiu, Y.-C. Liu, and P. J. Huang, "Analysis and implementation of a high voltage gain 1 MHz bidirectional DC-DC converter," *IEEE Trans. Ind. Electron.*, vol. 67, no. 2, pp. 1415–1424, Feb. 2020.

[12] B. Yang, F. C. Lee, A. J. Zhang, and G. Huang, "LLC resonant converter for front end DC/DC conversion," in *APEC. Seventeenth Annual IEEE Applied Power Electronics Conference and Exposition (Cat. No.02CH37335)*, Mar. 2002, pp. 1108–1112.

[13] A. Nabih and Q. Li, "Design of 98.8% efficient 400-to-48-V LLC converter with optimized matrix transformer and matrix inductor," *IEEE Trans. Power Electron.*, vol. 38, no. 6, pp. 7207–7225, Jun. 2023.

[14] P. R. Prakash, A. Nabih, and Q. Li, "Termination design optimization of high-current PCB-winding matrix transformers," *IEEE Trans. Power Electron.*, vol. 38, no. 4, pp. 4957–4971, Apr. 2023.

[15] S. Y. et al., "A 1-MHz GaN-based $LCLC$ resonant step-up converter with air-core transformer for satellite electric propulsion application," *IEEE Trans. Ind. Electron.*, vol. 69, no. 11, pp. 11 035–11 045, Nov. 2022.

[16] R. Wang, S. Mao, S. Yin, H. Liu, Q. Tan, and J. Fan, "Analysis and modeling of zero-voltage-switching condition for LCC resonant converter in above-resonance operation mode," *IEEE Trans. Power Electron.*, vol. 39, no. 9, pp. 10 950–10 961, Sep. 2024.

[17] M. K. Ranjram, I. Moon, and D. J. Perreault, "Variable-inverter-rectifier-transformer: A hybrid electronic and magnetic structure enabling adjustable high step-down conversion ratios," *IEEE Trans. Power Electron.*, vol. 33, no. 8, pp. 6509–6525, Aug. 2018.

[18] M. Noah, T. Shirakawa, K. Umetani, J. Imaoka, M. Yamamoto, and E. Hiraki, "Effects of secondary leakage inductance on the LLC resonant converter," *IEEE Trans. Power Electron.*, vol. 35, no. 1, pp. 835–852, Jan. 2020.

[19] S. A. Ansari, J. N. Davidson, and M. P. Foster, "Fully-integrated transformer with asymmetric primary and secondary leakage inductances for a bidirectional resonant converter," *IEEE Trans. Ind. Appl.*, vol. 59, no. 3, pp. 3674–3685, May 2023.

Start-Up Techniques and Universal Closed Loop Control of Immittance Network Based Resonant Converter

Ripun Phukan
MPEL, Delta Electronics
(Americas), Durham, USA
ripun.phukan@deltaww.com

Misha Kumar
MPEL, Delta Electronics
(Americas), Durham, USA
misha.kumar@deltaww.com

Randy Beckemeyer
MPEL, Delta Electronics
(Americas), Durham, USA
randy.beckemeyer@deltaww.com

Juan Ruiz
MPEL, Delta Electronics
(Americas), Durham, USA
Juan.Ruiz@deltaww.com

Peter Barbosa
MPEL, Delta Electronics (Americas), Durham, USA
peter.barbosa@deltaww.com

Abstract— **Immittance based resonant topologies have recently gained attention for on/off board charging due to their bidirectional nature. Thanks to the symmetry of the resonant tank, use of a simple phase shift control for power flow, simplified transformer design with high magnetizing impedance and no requirement for going into burst mode control at light loads. However, due to fixed frequency operation at or close to the resonant point, traditional start-up control techniques with variable frequency based resonant converters are not applicable. This paper therefore proposes techniques to handle start-up for immittance converters. Additionally, to cater to wide DC bus input voltage range, a multi-level stacked half bridge converter is used as primary side, for which a universal closed loop controller with active DC bus voltage balancing is proposed. The start-up and closed loop control scheme is verified on a 6.6 kW on board charger prototype.**

Keywords—Immittance converter, LCL-T, Period Doubling, Soft-Start

I. INTRODUCTION

LCL-T resonant converters are well known for its use in LED driver applications. Lately it has been extended to on-board charging applications as well [1]. Their benefits over conventional variable frequency based resonant converters include fixed switching frequency operation and simplified control. This comes at the expense of high circulating currents particularly during boost mode i.e. voltage step up operation. Furthermore, LCL-T converter offers better charging efficiencies (low circulation) at constant current (CC) and low battery voltage mode, which for the most part is critical in terms of efficiency curves.

The dynamics of the LCL-T resonant converter have been reported in prior art. [2, 3]. The LCL-T exhibits current source properties wherein the output current is controlled by changing the input voltage (using simple phase shift control of primary converter). Alternatively, the output voltage can be controlled by varying the phase shift between the primary and secondary side bridges. This leads to higher circulation currents. To allow full range Zero Voltage Switching (ZVS) of all switches in question, a phase shift between primary and secondary converters is employed; as demonstrated in [1]. However, A

simple phase shift control method requires further investigation. For instance, a better understanding of the possible start-up schemes is crucial. Particularly because traditional variable frequency and/or phase shift control start-up methods [4, 5, 6] do not work with immittance network-based converters. Additionally, with the incorporation of gallium nitride (GaN) power transistors and high DC link voltages, multi-level switching [7, 8, 9] circuits are gaining attention. A stacked half bridge circuit offers a simple modular implementation using two half bridge cells and is considered for demonstration in this paper. Therefore, dc-bus voltage balancing techniques of the primary side converter when using three-level stacked-half bridge architecture or other forms of multi-level circuits becomes important.

The paper is organized as follows. First the universal control architecture for the LCL-T converter is described. Some of the key points such as controller design, voltage balancing, and start-up control are discussed. A flowchart is presented that encompasses the critical features needed for successful operation of the converter.

II. PROPOSED UNIVERSAL CONTROL SCHEME

A. Overall System Architecture

The universal control scheme for the LCL-T resonant converter is presented in Fig 1. It comprises of three sub-blocks. The 1st sub-block is meant for period doubling modulation and start-up control, which is responsible for controlling the phase sequence of the primary side switches to ensure voltage balancing under steady state and control the phase sequence during start-up mode as well. Essentially, this part is handled by a dedicated interrupt service routine (ISR).

The 2nd block relies on voltage loop control for dynamic DC bus voltage balancing during events such as load transients, converter mismatch such as delays between gate signals or component tolerance. The 3rd sub-block controls the output battery current through a PI compensator loop. The closed loop current control is based on the primary side phase angle control (2). A fixed / variable load dependent phase shift between the primary and secondary side pole voltages is applied to maintain

ZVS on either side. The 2nd and 3rd blocks are combined into one ISR module. From the sensitivity of output battery current with respect to phase shift command, derived in (3), it is found that a simple PI controller can be used for control. The gain changes with the steady state phase shift angle. Here Ø denotes the primary side phase shift angle.

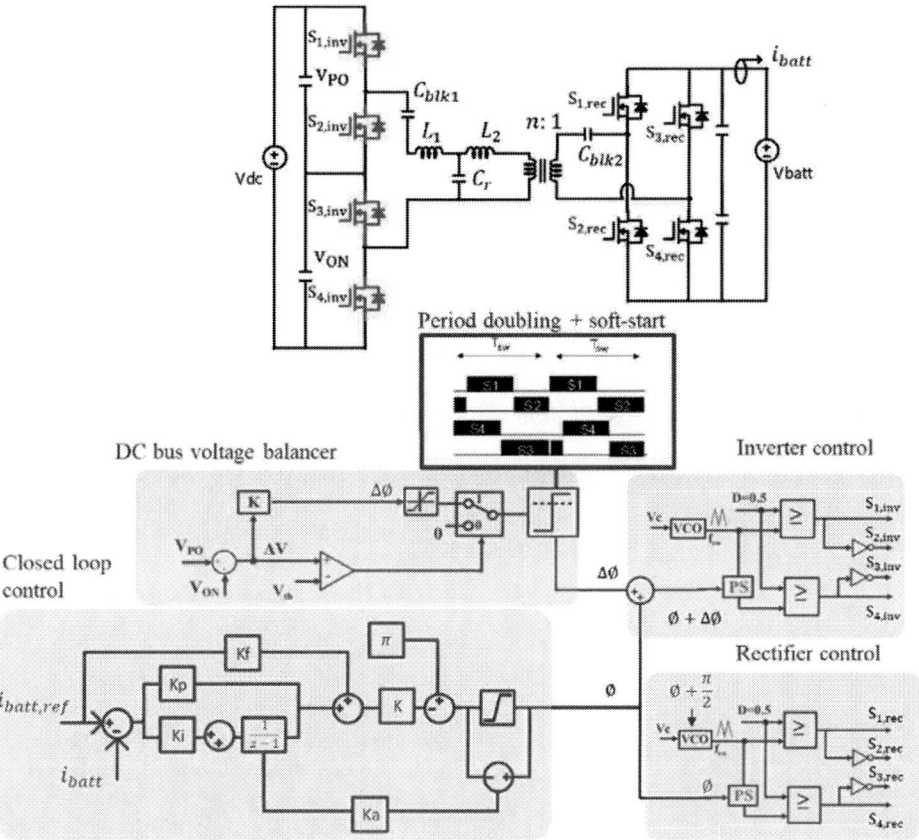

Fig. 1 Universal control scheme of LCL-T resonant converter, Kp and Ki – are the PI controllers, Kf is the feedforward term, Ka is the anti-wind up gain

$$I_{BATT} = \frac{4n \cos^3 \frac{\emptyset_{3level}}{2} V_{IN}}{\pi^2 X} \quad (1)$$

$$\frac{di_{BATT}}{d\emptyset} = \frac{-6nV_{in}}{\pi^2 X} \sin\left(\frac{\emptyset}{2}\right) \cos^2\left(\frac{\emptyset}{2}\right) \quad (2)$$

B. Current Mode Controller Design

Current mode control is implemented using a simple proportional and integral (PI) compensator. The plant transfer function, feedback gains and output current control loops are synthesized. The control block diagram is shown in Fig. 2. Here X – is the reactance of the tank, T_d is the computation delay, T_p is the period doubling rate, T_{RC} is the sensing time constant, R_c and R_L are the ESR and load resistor respectively and n is the transformer turns ratio.

A low bandwidth controller will suffice in the battery charging application due to slower system response. From the loop gain and step response, the PI parameters are optimized as indicated in Fig. 2.

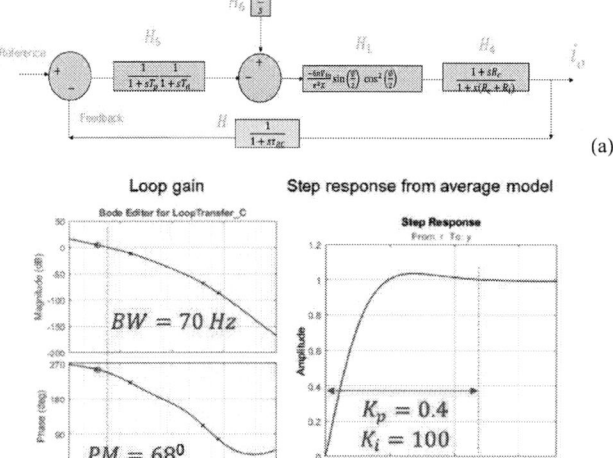

Fig. 2 Closed loop transfer function with tuned PI parameters (at 400 V_{in} and I_{batt} = 20 A)

C. DC Bus voltage balancing

The DC bus voltage balancing is done using cycle by cycle injection of small signal phase angle as illustrated in Fig 2(a). The DC bus voltages are naturally balanced without any active balancing through the implementation of a rotating sequence in the inverter switching patterns. S1 and S4 alternate the lead lag sequence every switching cycle. So the neutral point currents are naturally balanced in $2.T_{sw}$ hence the name period doubling modulation. While it may also be possible to implement the period doubling sequence every 2nd, 3rd or 4th switching cycle hence the name extended period doubling (EPD). The EPD is particularly useful when low computation power is required. Besides the period doubling scheme, an active phase injection depending on sensed voltage difference between upper and lower DC bus voltages (V_{PO} and V_{NO}) is used for active balancing. The effectiveness of active balancing is demonstrated through simulation results in Fig. 2(b), wherein an intentional delay in gating sequence is introduced at time t_2 leading to divergence of upper and lower voltages. At time t_3 the active balancing algorithm is enabled through a proportional controller which leads to convergence of the upper and lower capacitor voltages.

D. Soft start mechanism

Start-up control is essential to avoid inrush currents. The basic idea is to charge the primary end blocking capacitor C_{blk1} to ½ of the input DC bus voltage (V_{dc}) prior to charging the output bulk capacitor C_{dc}. Note that the blocking capacitor at the input side sustains a DC voltage bias across it and helps prevent transformer saturation. This is especially useful with the SHB architecture. The soft start procedure ensures minimum voltage drop and dv/dt across the LCL-T resonant tank. As a result, the inrush currents can be reduced. Once the capacitor voltage is equal to $V_{dc}/2$, the transition to closed loop control i.e., nominal operation, can be initiated. This approach ensures a smooth transition from start-up period to nominal operation.

There are two possible methods to achieve the soft start phenomenon. According to Fig. 4(a), this is achieved using 2-level modulation on inverter and rectifier voltages. Or using approach in Fig. 4(b), where the secondary side rectifier switches are shorting the transformer secondary windings and only the primary side has a 2-level modulation. Another difference between the two approaches is in the duty cycle ramp mode. While Fig. 4(a) relies on pulse skipping, Fig. 4(b) has continuous duty ramp. As a result, the DC output capacitor is charged along with the blocking capacitor in Fig. 4(a). Whereas the DC bulk capacitor is charged separately after the blocking capacitor is charged to ½ V_{dc} in Fig. 4(b). Hence in the latter case, start-up process is decoupled from output bulk cap charging, and the control is relatively simple.

In either case the transition points (the time instant when the system transitions from two level mode to three level mode) is determined either through simulations or through voltage feedback. An important point to note is that the starting phase angle immediately after transition time should be close to 180⁰. Doing this will ensure the lowest applied RMS voltage across the tank and a better soft start sequence.

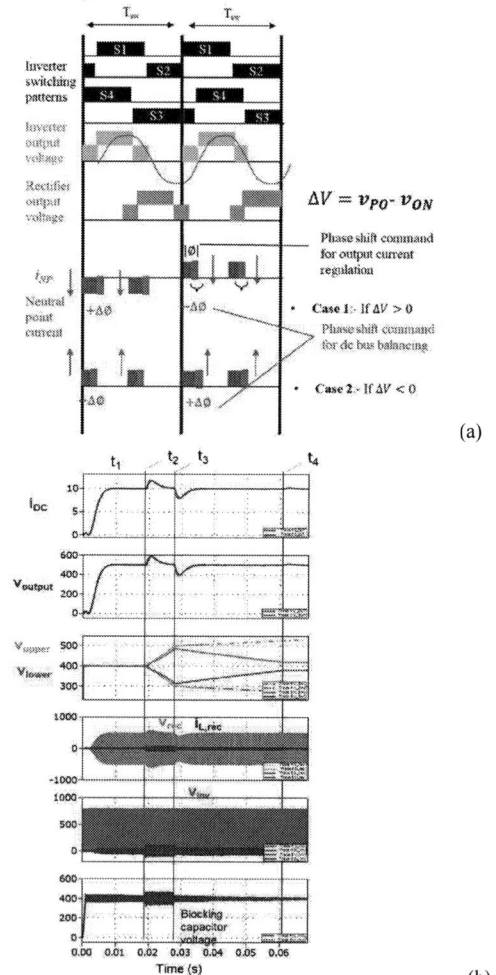

(a)

(b)

Fig. 3 Active balancing control and period doubling modulation for balancing of stacked half bridge capacitors

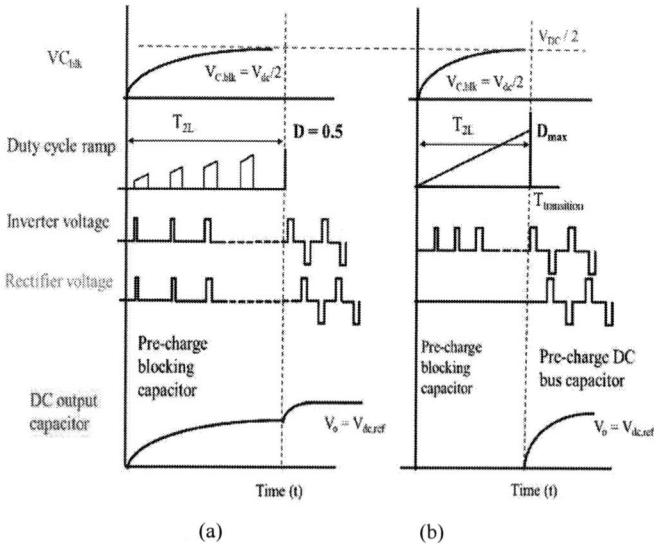

(a)   (b)

Fig. 4 Universal Control of LCL-T resonant converter, (a) Coupled pre-charge scheme, (b) decoupled pre-charge scheme

Under ultra-wide battery voltage range (150-900 V) a second blocking capacitor, C_{blk2}, is necessary as indicated in Fig. 5(a) to handle the DC bias on the secondary side of the transformer. In this case, a soft start slew rate control to minimize the peak transformer magnetization current is critical as indicated in Fig. 5. A slower ramp up in duty cycle will lead to a lower blocking capacitor V_{Cblk2} voltage under steady state Fig. 5(b). Thanks to a sufficiently longer free-wheeling period under low duty cycle mode, the blocking capacitor C_{blk2} has sufficient time to decay to a lower voltage level as indicated in Fig. 5(c). The average voltage is also lower compared to the case with larger duty cycle (red vs. black waveform in Fig. 5(c)). This voltage enforces a lower DC magnetizing current level in the transformer.

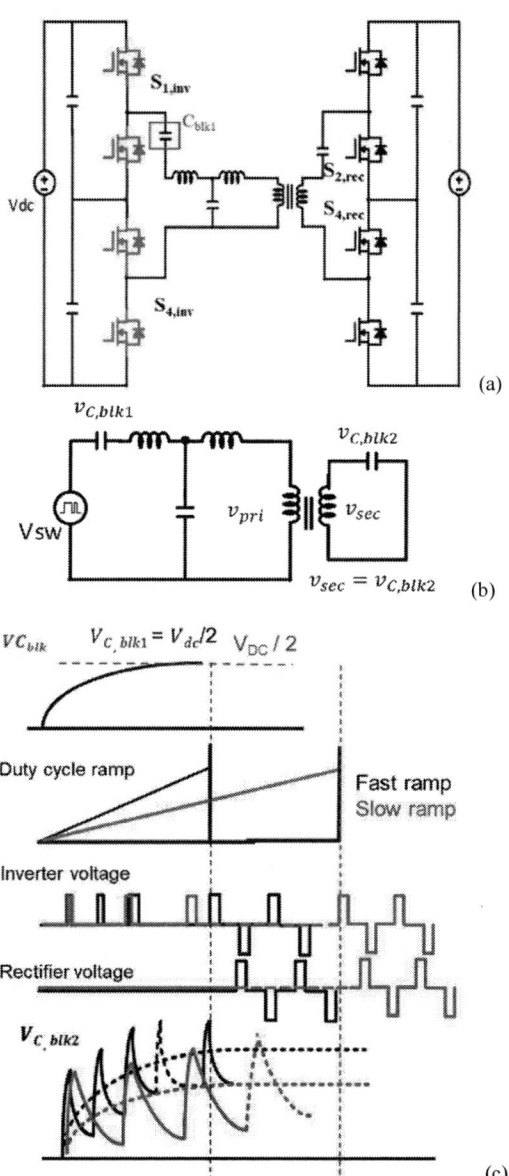

Fig. 5 (a) LCL-T architecture using SHB rectifier (b) Equivalent circuit during 2L mode (c) illustration of soft-start slew rate on transformer DC bias

E. Control flowchart

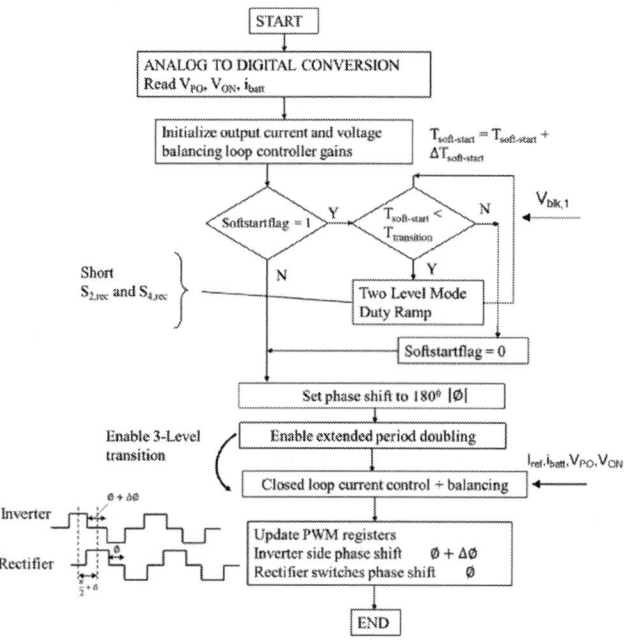

Fig. 6 Control flowchart of LCL-T resonant converter

The overall start up and control procedure is illustrated in Fig. 6.

III. EXPERIMENTAL EVALUATION

The GaN based hardware for demonstration is shown in Fig. 7 and the test parameters are indicated in Table I.

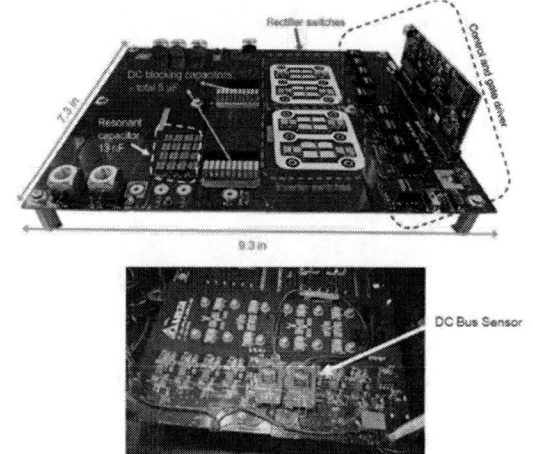

Fig. 7 6.6 kW, OBC hardware using LCL-T resonant converter structure

TABLE I. LCL-T TANK PARAMETERS

Parameters	Value
FET(s)	650 V/30 A *GaNSystems*
F_{sw}	500 kHz
L_r	7.3 uH
C_r	35 nF
C_{blk}	2 uF
n	2:1

The waveforms for period doubling, full scale start up and closed loop control are shown in Fig 8. From Fig. 8(a), the turn ON phase sequence of S1 and S4 are flipped every three switching cycles, i.e. S1 turn ON leads S4 turn ON for the first three cycles (T_{sw} = 2 μs) and then S4 turn ON leads S1 turn ON for the next three cycles, thereby repeating the sequence. This results in natural DC bus voltage balancing during steady state per the illustration in Fig. 3(a) due to identical neutral point current injections. The extended period doubling modulation with a period doubling frequency of 167 kHz is implemented. Note that the period doubling scheme is implemented at both the inverter and rectifier switches for the case with SHB converter on either sides. A phase shift of 90^0 is maintained between the inverter and rectifier pole voltages as per Fig. 8(a).

Fig. 8(b) shows the startup and closed loop control that builds the capacitor voltage during the first 600 us and then enables the closed loop control. This is shown in further detail in Fig. 9. Note that the start-up strategy from Fig. 4(b) is shown here. From the zoomed in views in Fig. 9, it is noticeable that the start-up is successfully enabled when the inverter pole voltage is almost entirely bipolar. This limits the inrush current to peak value of 20 A, well within the device SoA (Safe Operating Area), at V_{dc} = 400 V.

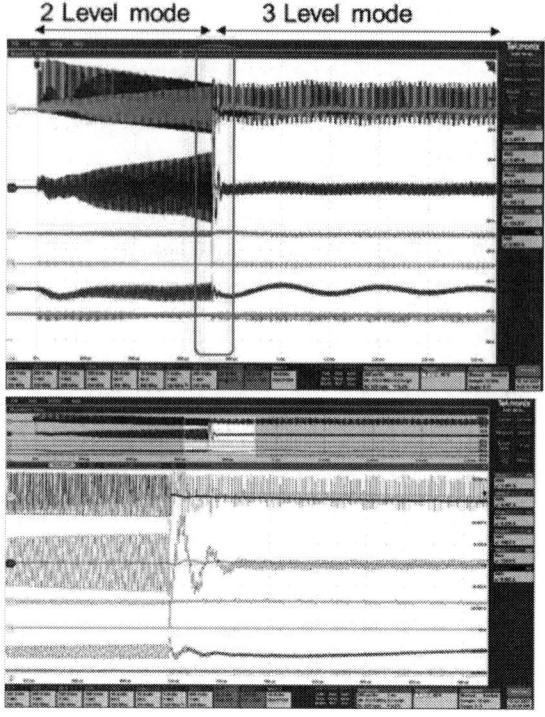

Fig. 9 Experimental results:- Zoom in waveforms during start-up

It is however interesting to note the impact of soft start slew rate on the inrush current as well. If the duty cycle ramp in Fig. 5 is too fast, this may result in potential saturation of the transformer due to fast build of DC magnetic bias, which stems from the output bulk capacitor. This is illustrated in the experimental results from Fig. 10 where the impact of slew-rate on the magnetizing current can be seen. The reason for this is primarily due to the presence of a blocking capacitor on the secondary winding of the transformer which stores DC charge during the soft-start phase as explained earlier section. The impact of slew rate control is validated through the experimental measurements in Fig. 10. A slow duty ramp control is analogous to pulse skipping control or burst mode control as well.

The low frequency oscillations in the magnetizing current are resulting from the transition from two level (2L) to three level (3L) modulation. This forces the DC magnetizing current to eventually drop to zero however accompanied by a low frequency resonance between the secondary slide blocking capacitor and magnetizing inductance. The natural damping component comes from transformer winding ac resistance, core material (resistive losses), Equivalent series resistances (ESR) of the resonant and blocking capacitors, and device $R_{ds(ON)}$. Nevertheless, the peak value of the resonant current is always lower than the DC magnetic bias in the circuit. There are state trajectory control techniques that allow transitioning between different modes with lower tank energy content [10], however, in the case of start-up, the trajectory definition is complicated. Therefore, such techniques are only suitable for steady state transitions.

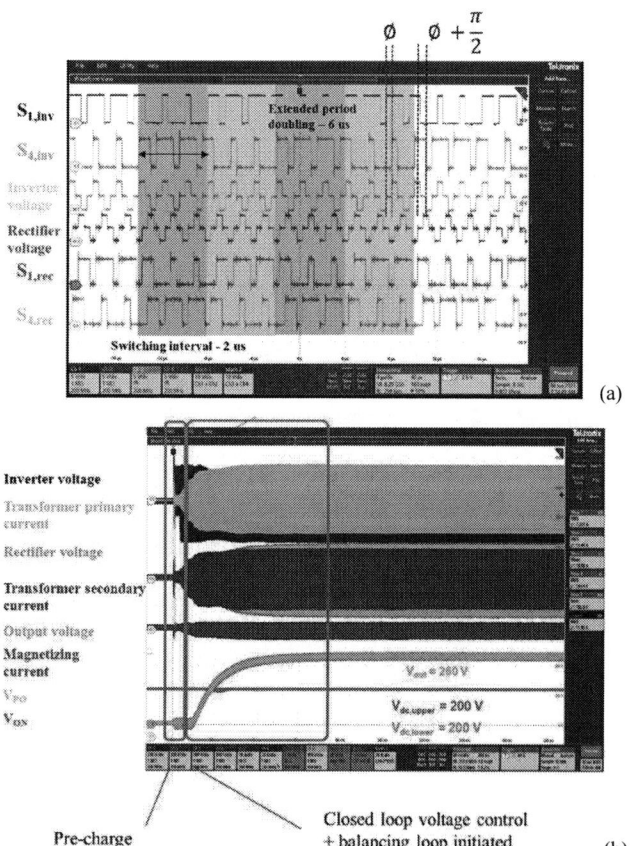

Fig. 8 Experimental results, (a) Period doubling modulation, (b) Start up and closed loop control

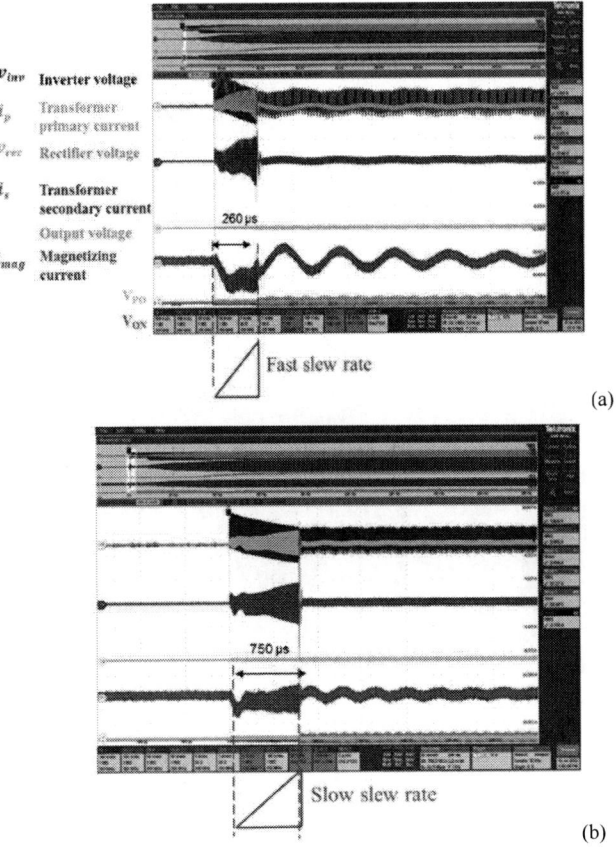

Fig. 10 Soft start slew rate and its impact on transformer magnetization current (a) waveforms under fast slew rate, (b) waveforms under slow slew rate

IV. CONCLUSIONS

In conclusion, a universal control strategy for immittance based resonant converters is proposed, which unifies start-up scheme, steady state and active DC bus voltage balancing, and current control strategies required for successful deployment of the immittance based resonant converters for on/off board charging applications. The overall control scheme is demonstrated on a 6.6 kW *GaN* based On-Board Charger. From the experimental results, it can be seen that the start-up sequence and closed loop current regulation is achieved successfully. The soft start slew rate is found to be critical component to control

the DC magnetizing flux across the transformer. Compared to the variable frequency type resonant converters, the immittance based converter is rather simpler to control under fixed frequency mode and presents lower control complexity as such. It can therefore be a potential candidate for wide range charging applications.

REFERENCES

[1] S. Mukherjee, J. M. Ruiz and P. Barbosa, "A High-Power Density Wide Range DC–DC Converter for Universal Electric Vehicle Charging," in IEEE Transactions on Power Electronics, vol. 38, no. 2, pp. 1998-2012, Feb. 2023, doi: 10.1109/TPEL.2022.3217092.

[2] M. Borage, S. Tiwari, and S. Kotaiah, "Analysis and design of an LCL-T resonant converter as a constant-current power supply," IEEE Trans. Ind. Electron., vol. 52, no. 6, pp. 1547–1554, Dec. 2005

[3] M. Borage, K. V. Nagesh, M. S. Bhatia, and S. Tiwari, "Resonant immittance converter topologies," IEEE Trans. Ind. Electron., vol. 58, no. 3, pp. 971–978, Mar. 2011.

[4] Qichao Chen, Jianze Wang, Yanchao Ji, and Shuang Liang, "Soft Starting Strategy of Bidirectional LLC Resonant DC-DC Transformer Based on Phase-Shift Control," 9th IEEE Conference on Industrial Electronics and Applications, 2016, pp. 318-322

[5] Ruichang Zheng, Bangyin Liu, and Shanxu Duan, "Analysis and Parameter Optimization of Start-Up Process for LLC Resonant Converter," IEEE Transactions on Power Electronics, vol. 30, no. 12, December 2015. pp. 7113-7122.

[6] Qiqi Li, Bangyin Liu, Ruichang Zheng, and Shanxu Duan, "Constant Resonant Current Limiting Strategy for LLC Converter Without Current Sensing," IEEE Transactions on Power Electronics, vol. 31, no. 9, Sept. 2016. pp. 6756-6764.

[7] S. -Y. Chen et al., "A Systematic Parasitic Capacitance Extraction Procedure for Three Level Neutral Point Clamped Inverter Modules," 2023 IEEE Applied Power Electronics Conference and Exposition (APEC), Orlando, FL, USA, 2023, pp. 2709-2714, doi: 10.1109/APEC43580.2023.10131260.

[8] R. Phukan, S. Ohn, S. Nielebock, G. Mondal, D. Dong and R. Burgos, "Comparison Between Interconnected Filter Blocks for Three-Phase AC–DC Interleaved Converters," in IEEE Journal of Emerging and Selected Topics in Power Electronics, vol. 10, no. 5, pp. 5956-5968, Oct. 2022, doi: 10.1109/JESTPE.2022.3179335.

[9] H. Akagi, "Multilevel Converters: Fundamental Circuits and Systems," in Proceedings of the IEEE, vol. 105, no. 11, pp. 2048-2065, Nov. 2017, doi: 10.1109/JPROC.2017.2682105.

[10] C. Zhang and P. Barbosa, "Modulation Transition Methods Based on Trajectory Control for LLC Resonant Converters Operating in Wide Input- and/or Output-Voltage Range," in IEEE Transactions on Power Electronics, vol. 37, no. 12, pp. 14103-14114, Dec. 2022, doi: 10.1109/TPEL.2022.3184631.

Multi-Objective Efficiency-Oriented Optimization for DAB Converters Minimizing Current Stress and Backflow Power with Soft-Switching Assurance

Kun Wang
Electrical Energy Management Group,
Department of Electrical and Electronic
Engineering, univerisity of Bristol,
Bristol, UK
Email: kun.wang@bristol.ac.uk,

Ian Laird
Electrical Energy Management Group,
Department of Electrical and Electronic
Engineering, univerisity of Bristol,
Bristol, UK
Email: ian.laird@bristol.ac.uk,

Jun Wang
Electrical Energy Management Group,
Department of Electrical and Electronic
Engineering, univerisity of Bristol,
Bristol, UK
Email: jun.wang@bristol.ac.uk

Abstract—Vehicle-to-Grid (V2G) electric vehicle chargers are becoming increasingly important due to their potential to enhance grid stability, support renewable energy integration, and provide economic benefits. This has led to a growing demand for dual active bridge (DAB) bidirectional DC-DC converters. While current stress and backflow power collectively impact DAB converter performance, existing optimization methods are mostly single-objective, focusing on just one of these two factors. In other words, there is rarely any co-optimization to minimize both current stress and backflow power simultaneously. To address this limitation, this work proposes a multi-objective, efficiency-oriented algorithm that minimizes current stress and backflow power simultaneously by optimizing triple phase shift (TPS) modulation parameters. This is achieved by combining time-domain analysis of the DAB under TPS modulation with an analytical power loss model into an automated, holistic digital optimization tool. Compared to conventional single-objective TPS methods that focus solely on current stress optimization, the proposed tool further improves power efficiency across certain power ranges. The validity of this multi-objective optimization method is verified through simulations and experiments.

Keywords—*Backflow power, current stress, dual active bridge (DAB) converter, multi-objective optimization, Pareto front, power losses model, triple phase shift (TPS).*

I. INTRODUCTION

Dual Active Bridge (DAB) galvanically isolated bidirectional DC-DC converters enable bidirectional energy flow and zero voltage switching (ZVS), offering advantages such as a simple structure, high efficiency, low cost, and high power density. In recent years, with the rise of technologies like electric vehicles, uninterruptible power supplies, and renewable energy generation, DAB converters have become increasingly popular for a wide range of applications.

During the charging process of electric vehicles, the battery voltage varies significantly with the state of charge. For example, a charging voltage can change from 200V to 420V [1]. To address this, the charger must maintain a constant current output across a wide range of voltage variations. Consequently, the DAB converter needs to achieve efficient energy transfer over a broad operating range. To enhance power efficiency, several modulation and control strategies have been introduced for DAB converters.

Among these strategies, phase shift modulation (PSM) methods are the most common, including single-phase shift (SPS), dual-phase shift (DPS), extended phase shift (EPS), and triple-phase shift (TPS). Typically, the duty cycles of these phase shift modulation methods are fixed at 50%. SPS is the simplest PSM method, but it suffers from a limited ZVS operating range and high current stress when the input and output voltages are not well-matched [2], [3]. Both EPS and DPS have two control variables. For the EPS scheme, it controls the duty ratio of the higher voltage side, which can be used to reduce backflow power [4]. DPS regulates the duty ratios of both sides but requires them to have identical values [5]. On the contrary, TPS modifies the duty ratios of both sides independently and regulates the phase shift between these two bridge voltages [6]-[14]. Ultimately, SPS, DPS, and EPS are special cases of TPS, which is the most general phase shift method and covers all possible scenarios.

Previous efficiency-oriented optimization methods based on TPS have primarily focused on minimizing either peak current or backflow power [6]-[10]. For example, [8], [9], and [10] focus exclusively on minimizing backflow power, while [11], [12], and [13] present various algorithms aiming at minimizing the inductor peak current. All these methods ([6]-[10]) address only a single objective. In contrast, [14] proposes an analytical multi-objective algorithm that seeks to minimize both backflow power and current stress, however since backflow power may not be a convex function, this approach can result in an inaccurate determination of the optimal three phase shift angles. Consequently, there remains a need for further research on multi-objective optimization for DAB converters. To address this gap, this article proposes an advanced Pareto-front multi-objective algorithm based on a non-dominated sorting genetic algorithm (NSGA-II), designed to optimize the TPS operation of a DAB converter. The goal is to minimize both current stress and backflow power while simultaneously guaranteeing ZVS. The remainder of this article is organized as follows. Section II

979-8-3315-1612-3/25 $31.00 © 2025 IEEE

defines the characteristics of twelve different operating modes under TPS modulation. Section III gives the power loss model of the DAB converter. The proposed Pareto-front multi-objective optimization algorithm is introduced in Section IV. Section V presents experimental verifications. Finally, Section VI concludes the article.

II. CHARACTERISTICS OF DAB UNDER TPS

A. DAB Topology and Operating modes under TPS

Fig. 1 illustrates the basic topology of a non-resonant DAB. It contains two bridges, a phase-shift inductor, and a high-frequency transformer. The two bridges contain eight switches: the upper switches Q_1, Q_2, Q_3, and Q_4, and their complementary switches Q_{1c}, Q_{2c}, Q_{3c} and Q_{4c}.

According to the TPS modulation strategy, the inner phase shift angles ϕ_p and ϕ_s are defined as the phase shift between Q_1 and Q_2, Q_3 and Q_4, respectively, as shown in Fig. 1. These two phase shift angles can modify the waveforms of the voltage V_p and V_s, respectively. However, the phase priority of the inner phase shift angles (i.e., Q_1 leads or lags Q_2 by the same angle) has no influence on the bridge voltage waveforms. Thus, both the two inner phase shift angles are confined to the range $[0, \pi]$. On the other hand, the polarity of the outer phase shift angle ϕ_o (phase shift between Q_1 and Q_3, as shown in Fig. 1) can alter the value of the inductor voltage. Therefore, the outer phase shift angle belongs to the range $[-\pi, \pi]$. In this article, phase shift ratios D_p, D_s and D are normalized by π to represent primary, secondary and outer phase shift angles, respectively. Therefore, D_p and D_s belong to the range $[0,1]$, while D belongs to the range $[-1, 1]$. Additionally, D_p and D_s are defined as the duty ratios of V_p and nV_s. Based on these phase shift ratios, twelve operating modes and their constraints can be mapped within two adjacent normalized cubes, as shown in Fig. 2 [1] [15] . The operating modes constraints are presented in TABLE I.

B. Inductor instantenaous and RMS current

According to the inductor voltage-second principle in steady state, the waveform of the inductor voltage is symmetric over one switching period. Based on the different voltage excitation combinations applied to the inductor, twelve distinct inductor voltage waveforms can be generated. From the equivalent circuit, the inductor current can be derived using the following equation:

$$V_L = V_p - nV_s = L\frac{di_L}{dt} \tag{1}$$

$$i_L(t) = \frac{1}{L}\int_{t_i}^{t} V_L(t)dt + i_L(t_i) \tag{2}$$

where t ∈ [t_i, t_{i+1}], t_n represents i^{th} switching instant.

The inductor current within a switching period is a piecewise function. Based on the continuity principle of inductor current, the inductor current expressions can be derived. Due to the symmetric waveform ($i_L(t_0) = -i_L(t_4)$), it is sufficient to solve for the values of the first half-cycle.

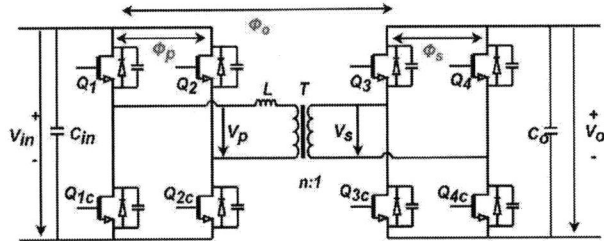

Fig. 1. Topology of non-resonant DAB

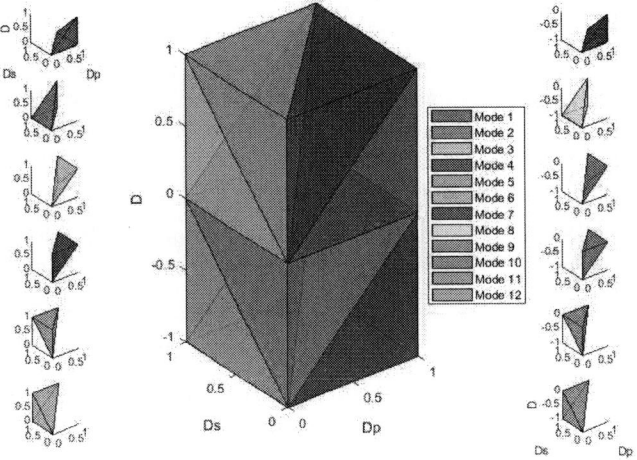

Fig. 2. Twelve operating modes and mode constraints.

TABLE I. OPERATING MODES CONSTRAINTS

Mode	Operating Mode Constraints
1	$0<D<1, D<D_p<1, 0<D+D_s<D_p$
2	$0<D<1, D<D_p<1, D_p<D+D_s<1$
3	$0<D<1, D<D_p<1, 1<D+D_s<2$
4	$0<D<1, 0<D_p<D, 0<D+D_s<1$
5	$0<D<1, 0<D_p<D, 1<D+D_s<1+D_s$
6	$0<D<1, 0<D_p<D, 1+D_s<D+D_s<2$
7	$-1<D<0, D+1<D_p<1, -1<D+D_s<-1+D_p$
8	$-1<D<0, D+1<D_p<1, -1+D_p<D+D_s<0$
9	$-1<D<0, D+1<D_p<1, 0<D+D_s<D_p$
10	$-1<D<0, 0<D_p<D+1, -1<D+D_s<0$
11	$-1<D<0, 0<D_p<D+1, 0<D+D_s<D_p$
12	$-1<D<0, 0<D_p<D+1, D_p<D+D_s<1$

RMS current determines the ohmic power loss. Based on the definition of RMS current, the switching instant time t_i, and the instantaneous inductor current value, the RMS inductor current for each operating stage [t_i, t_{i+1}] can be derived.

$$i_{rmsi} = \sqrt{\frac{1}{(t_{i+1}-t_i)}\int_{t_i}^{t_{i+1}} i_L^2(t)dt} \tag{3}$$

In a half period T_h, the RMS current can be expressed as:

$$i_{rms} = \sqrt{\frac{1}{T_h}\sum_{i=0}^{3} i_{rmsi}^2 \cdot (t_{i+1} - t_i)} \tag{4}$$

C. Transmission power

Assuming the components in the circuit are lossless and considering the half waveform symmetry of the inductor current and voltage, the average transmission power can be calculated by the following equation (5) for the first half period. The normalizing factor is the maximum transmission power as shown in equation (7).

$$P_{trs} = \frac{1}{T_h} \int_0^{T_h} V_p(t)\, i_L(t)\, dt \tag{5}$$

where T_h is the half period, $V_p(t)$ is the primary bridge output voltage, and $i_L(t)$ is the inductor current.

To decouple the input voltage V_{in}, output voltage V_o, turn ratio n, phase shift inductance L and switching frequency f_s, i_n is defined as the current normalizing factor and P_n is defined as the power normalizing factor. The voltage gain ratio k is defined as $k = \frac{V_i}{nV_o}$. In this study, the buck mode ($k > 1$) is chosen, as the analysis for boost mode is similar. The normalized current values and the expression for transmission power in each mode are illustrated in TABLE II.

$$i_n = \frac{n \cdot V_o}{8 \cdot f_s \cdot L} \tag{6}$$

$$P_n = \frac{n \cdot V_{in} \cdot V_o}{8 \cdot f_s \cdot L} \tag{7}$$

D. Backflow power

Generally, to ensure ZVS operation, converters primarily function in mode 3 of table II for heavy load situations. Therefore, this paper will use mode 3 as an example to examine the backflow power. As shown in Fig. 3, during the period $[t_0, t_0']$, the inductor current flows in the opposite direction to the primary full-bridge output voltage, indicating that the inductor's energy is flowing back to the primary-side capacitor. On the secondary side, during the interval $(t_0' - t_1)$, the inductor current is also in the opposite direction to the secondary full-bridge output voltage, resulting in energy flowing from the secondary side back into the inductor. In this paper, this portion of energy is referred to as backflow power.

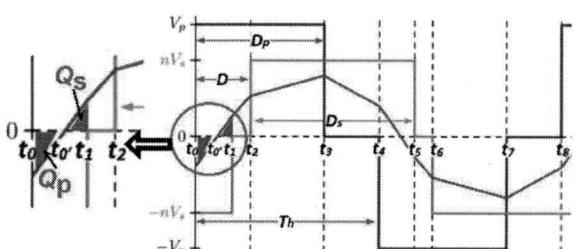

Fig. 3. The backflow power in operating mode 3. Q_p and Q_s are defined as the primary and secondary backflow power respectively.

The backflow energy generated on the primary and secondary sides of the converter leads to reactive power losses. For the same level of active power transmission, the smaller the backflow power, the higher the converter's power factor, which in turn improves conversion efficiency.

$$Q_p = \frac{U_{in}}{T_h} \int_0^{t_o'} |i_L(t)|\, dt \tag{8}$$

$$Q_s = \frac{nU_o}{T_h} \int_{t_o'}^{t_1} |i_L(t)|\, dt \tag{9}$$

$$P_{back} = Q_p + Q_s \tag{10}$$

where Q_p is the primary backflow power, Q_s is the secondary backflow power. T_h is half period, t_0' is the time inductor current is 0, $t_1 = (D + D_s - 1)\, T_h$.

In mode 5, 6, 7 and 8, the value of $V_p(t) \cdot V_s(t)$ remains non-positive, thereby preventing direct power transfer between the primary and secondary sides [6]. Based on the backflow power definition, the backflow power in these modes are larger than other operating modes. Therefore, these modes are ignored.

E. ZVS constraints

ZVS is achieved by charging and discharging the output capacitors of the semiconductors to ensure they turn on at zero voltage. There are two ZVS conditions in a DAB converter: the upper ZVS condition and the complementary ZVS condition, as shown in Fig. 4. During the transition before the switches turn on (Q_1 and Q_{1c} in Fig. 4 (a) and (b), respectively), the switches' output capacitance provide the current path. The device voltage will be zero when the current has fully discharged the output capacitances.

Fig. 4 (a) shows that when the inductor current is negative, it can discharge the output capacitance of Q_1 and charge the output capacitor of Q_{1c} simultaneously, the upper switches can achieve ZVS. Similarly, when the positive inductor current discharges and charges the output capacitance of Q_{1c} and Q_1, respectively, the lower switches can achieve ZVS, as shown in Fig. 4 (b). Therefore, the instantaneous inductor current value determines whether each switch achieves ZVS or not.

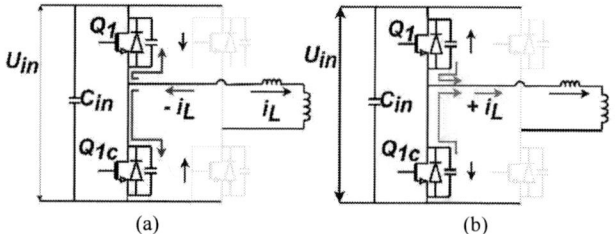

Fig. 4. ZVS processes. (a) Upper switches (b) Complementray switches.

If the inductor current is sufficient to charge and discharge the output capacitances of the switches, the turn-on voltage will be zero and the switches will achieve ZVS if they are then turned on. Since the output capacitances are small, only a minimal current is needed to charge and discharge them. Therefore, the ZVS condition can be simplified to depend only on the direction of the current flow. For the primary upper switches Q_1, Q_2, if i_L is negative at the switching instant, they can achieve ZVS. For the complementary switches Q_{1c}, Q_{2c}, when i_L is positive they can achieve ZVS. Similarly, the ZVS condition for secondary side switches can be derived. In summary, the following equations (11) can be used to judge the implementation of ZVS:

$$\text{For } Q_1, Q_2, Q_{3c}, Q_{4c}, \quad i_L(t) < 0;$$
$$\text{For } Q_{1c}, Q_{2c}, Q_3, Q_4, \quad i_L(t) > 0; \tag{11}$$

TABLE II. CHARACTERISTIC OF DIFFERENT OPERATING MODE

Mode 1	Mode 2	Mode 3
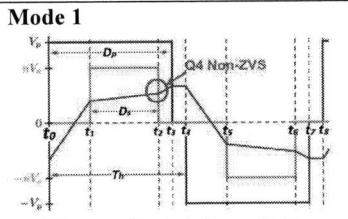	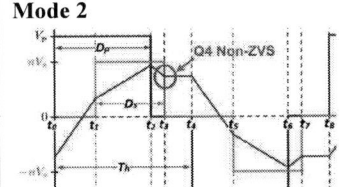	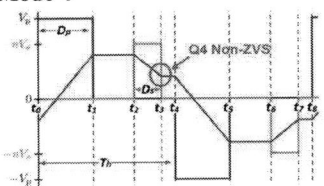
$t_1 = DT_h,\ t_2 = (D+D_s)T_h,\ t_3 = D_pT_h$ $P = 2(D_s{}^2 - D_pD_s + 2D_sD)$ $i_0 = -2(D_p{\cdot}k - D_s) < 0,$ $i_1 = 2(2{\cdot}D{\cdot}k - D_p{\cdot}k + D_s) > 0,$ $i_2 = 2(2{\cdot}D{\cdot}k - D_p{\cdot}k + 2D_s{\cdot}k - D_s) < 0,$ $i_3 = 2(D_p{\cdot}k - D_s) > 0.$	$t_1 = DT_h,\ t_2 = D_pT_h,\ t_3 = (D+D_s)T_h$ $P = 2(-D_p{}^2 - D^2 + D_pD_s + 2D_pD)$ $i_0 = -2(D_p{\cdot}k - D_s) < 0,$ $i_1 = 2(2{\cdot}D{\cdot}k - D_p{\cdot}k + D_s) > 0,$ $i_2 = 2(D_p{\cdot}k + 2{\cdot}D - 2{\cdot}D_p + D_s) > 0,$ $i_3 = 2(D_p{\cdot}k - D_s) < 0.$	$t_1 = (D+D_s-1)T_h,\ t_2 = DT_h,\ t_3 = D_pT_h$ $P = 2(-D_p{}^2 - D_s{}^2 - 2D^2 - 2DD_s + D_pD_s + 2D_pD + 2D + 2D_s - 1)$ $i_0 = -2(D_p{\cdot}k - 2 + 2{\cdot}D + D_s) < 0,$ $i_1 = 2(2{\cdot}D{\cdot}k - 2{\cdot}k - D_p{\cdot}k + 2{\cdot}D_s{\cdot}k + D_s) > 0,$ $i_2 = 2(2{\cdot}D{\cdot}k - D_p{\cdot}k + D_s) > 0,$ $i_3 = 2(D_p{\cdot}k + 2{\cdot}D - 2{\cdot}D_p + D_s) > 0\ (i_{peak})$

Mode 4	Mode 5	Mode 6
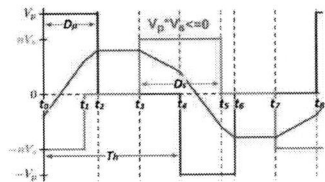	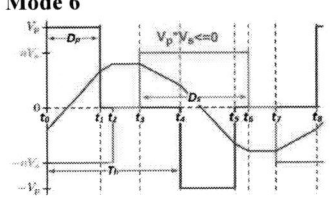	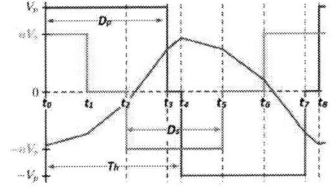
$t_1 = D_pT_h,\ t_2 = DT_h,\ t_3 = (D+D_s)T_h$ $P = 2(D_pD_s)$ $i_0 = -2(D_p{\cdot}k - D_s) < 0,$ $i_1 = 2(D_p{\cdot}k + D_s) > 0,$ $i_2 = 2(D_p{\cdot}k + D_s) > 0,$ $i_3 = 2(D_p{\cdot}k - D_s) < 0.$	$t_1 = (D+D_s-1)T_h,\ t_2 = D_pT_h,\ t_3 = DT_h$ $P = 2(-D_s{}^2 - D^2 + 2D + 2D_s - 2DD_s + D_pD_s - 1)$ $i_0 = -2(D_p{\cdot}k - 2 + 2{\cdot}D + D_s) < 0,$ $i_1 = 2(2{\cdot}D{\cdot}k - 2{\cdot}k - D_p{\cdot}k + 2{\cdot}D_s{\cdot}k + D_s) > 0,$ $i_2 = 2(D_p{\cdot}k + D_s) > 0,$ $i_3 = 2(D_p{\cdot}k + D_s) > 0.$	$t_1 = D_pT_h,\ t_2 = (D+D_s-1)T_h,\ t_3 = DT_h$ $P = 2(D_p{}^2 - D_pD_s + 2D_p - 2D_pD)$ $i_0 = -2(D_p{\cdot}k - 2 + 2{\cdot}D + D_s) < 0,$ $i_1 = 2(2 + D_p{\cdot}k - 2{\cdot}D + 2{\cdot}D_p - D_s) > 0,$ $i_2 = 2(D_p{\cdot}k + D_s) > 0,$ $i_3 = 2(D_p{\cdot}k + D_s) > 0.$

Mode 7	Mode 8	Mode 9
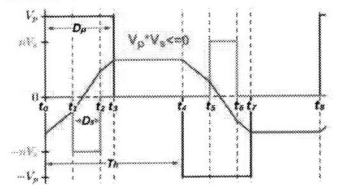	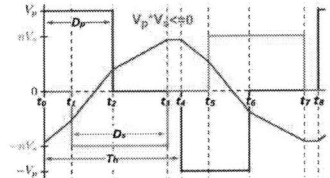	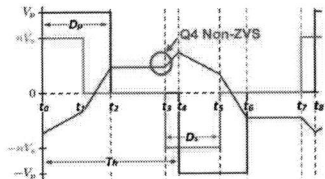
$t_1 = (D+1)T_h,\ t_2 = (D+D_s+1)T_h,\ t_3 = D_pT_h$ $P = -2(D_s{}^2 - D_pD_s + 2D_sD + 2D_s)$ $i_0 = -2(D_p{\cdot}k + D_s) < 0,$ $i_1 = 2(2{\cdot}k + 2{\cdot}D{\cdot}k - D_p{\cdot}k - D_s) < 0,$ $i_2 = 2(2{\cdot}k + 2{\cdot}D{\cdot}k - D_p{\cdot}k + 2{\cdot}D_s{\cdot}k + D_s) > 0,$ $i_3 = 2(D_p{\cdot}k + D_s) > 0.$	$t_1 = (D+1)T_h,\ t_2 = D_pT_h,\ t_3 = (D+D_s+1)T_h$ $P = -2(-D_p{}^2 - D^2 - 2D - 1 + D_pD_s + 2D_pD + 2D_p)$ $i_0 = -2(D_p{\cdot}k + D_s) < 0,$ $i_1 = 2(2{\cdot}k + 2{\cdot}D{\cdot}k - D_p{\cdot}k - D_s) < 0,$ $i_2 = -2(2 - D_p{\cdot}k + 2{\cdot}D - 2{\cdot}D_p + D_s) > 0,$ $i_3 = 2(D_p{\cdot}k + D_s) > 0.$	$t_1 = (D+D_s)T_h,\ t_2 = (D+1)T_h,\ t_3 = D_pT_h$ $P = -2(-D_p{}^2 - D_s{}^2 - 2D^2 - 2D - 2DD_s + D_pD_s + 2D_pD + 2D_s - 1)$ $i_0 = 2(2{\cdot}D - D_p{\cdot}k + D_s) < 0,\ (i_{peak})$ $i_1 = 2(2{\cdot}D{\cdot}k - D_p{\cdot}k + 2{\cdot}D_s{\cdot}k - D_s) < 0,$ $i_2 = 2(2{\cdot}k + 2{\cdot}D{\cdot}k - D_p{\cdot}k - D_s) < 0,$ $i_3 = -2(2 - D_p{\cdot}k + 2{\cdot}D - 2{\cdot}D_p + D_s) > 0.$

Mode 10	Mode 11	Mode 12
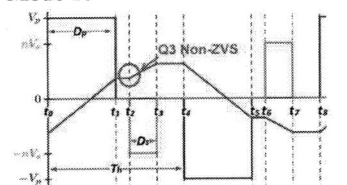	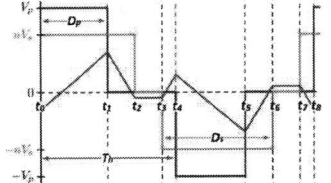	
$t_1 = D_pT_h,\ t_2 = (D+1)T_h,\ t_3 = (D+D_s+1)T_h$ $P = -2(D_pD_s)$ $i_0 = -2(D_p{\cdot}k + D_s) < 0,$ $i_1 = 2(D_p{\cdot}k - D_s) > 0,$ $i_2 = 2(D_p{\cdot}k - D_s) < 0,$ $i_3 = 2(D_p{\cdot}k + D_s) > 0.$	$t_1 = (D+D_s)T_h,\ t_2 = D_pT_h,\ t_3 = (D+1)T_h$ $P = -2(-D_s{}^2 - D^2 - 2DD_s + D_pD_s)$ $i_0 = 2(2{\cdot}D - D_p{\cdot}k + D_s) < 0,$ $i_1 = 2(2{\cdot}D{\cdot}k - D_p{\cdot}k + 2{\cdot}D_s{\cdot}k - D_s) < 0,$ $i_2 = 2(D_p{\cdot}k - D_s) > 0,$ $i_3 = 2(D_p{\cdot}k - D_s) < 0.$	$t_1 = D_pT_h,\ t_2 = (D+D_s)T_h,\ t_3 = (D+1)T_h$ $P = -2(D_p{}^2 - D_pD_s - 2D_pD))$ $(2D_0 + D_s < D_p\ i_{peak})$ $i_0 = 2(2{\cdot}D - D_p{\cdot}k + D_s) < 0,$ $(2D_0 + D_s > D_p\ i_{peak})$ $i_1 = 2(D_p{\cdot}k + 2{\cdot}D - 2{\cdot}D_p + D_s) > 0,$ $i_2 = 2(D_p{\cdot}k - D_s) < 0,\ i_3 = 2(D_p{\cdot}k - D_s) < 0.$

where $t_0 = 0$, $t_4 = T_h$

F. Feasible operating modes

Due to the ZVS constraints and the inductor current expressions presented in TABLE II. , mode 1, 2, 4, 10 and 11 cannot satisfy the ZVS condition, regardless of how the values of three phase shift ratios D_p, D_s & D, or the voltage gain k, are changed. Therefore, those modes can be ignored. Moreover, not all combinations in the other modes will meet the ZVS conditions. The ZVS constants can be considered as linear inequality constraints in the proposed optimization algorithm.

Based on the analysis above, only modes 3, 9 ,12 need to be further considered. The transmission power ranges for these three operating modes are derived and shown in F

TABLE III.   TRANSMISSION POWER RANGE IN DIFFERENT MODES

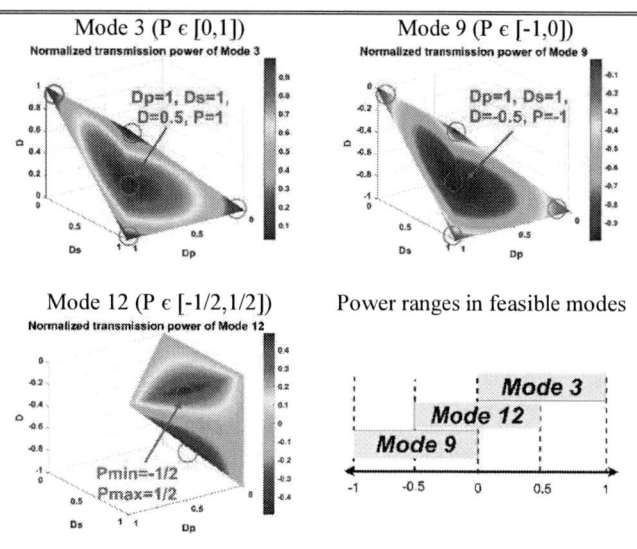

III. POWER LOSS ANALYSIS

This section derives a mathematical expression for the total power loss. Fig. 5 illustrates the power loss distribution in a DAB converter. The total power loss consists of switch loss, transformer loss, and inductor loss. Switch loss includes both conduction loss and switching loss, transformer loss comprises copper loss and iron loss, and inductor loss also includes copper loss (winding loss) and iron loss (core loss).

The RMS value of the inductor can be used to measure ohmic power loss, which includes the switch's conduction loss as well as the copper losses in the transformer and inductor.

Switching loss comprises turn-on loss and turn-off loss. According to the switching loss plots for each switch in their respective datasheets, both turn-on and turn-off switching losses can be determined using the following quadratic polynomial equation:

$$P_{sw} = f_s \cdot \frac{V_{wk}}{V_{test}} \cdot (A_0 I^2 + B_0 I + C_0) \tag{12}$$

where f_s is the switching frequency, V_{wk} is the switch working voltage in the application, V_{test} is the testing voltage in the datasheet. A_0, B_0 and C_0 are the expected coefficients.

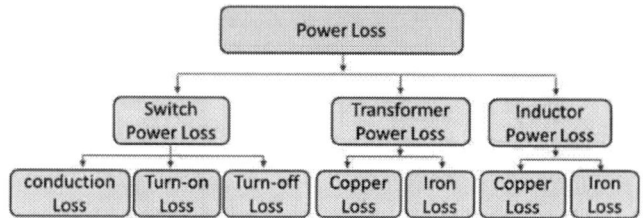

Fig. 5.   Power losses of DAB

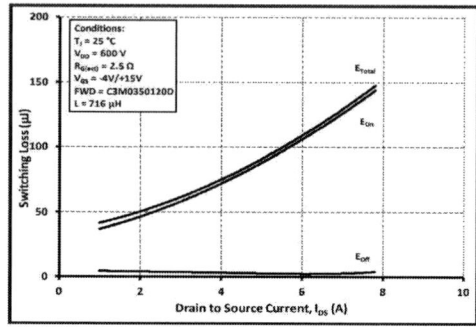

Fig. 6.   Clamped Inductive Switching Energy vs. Drain Current [18].

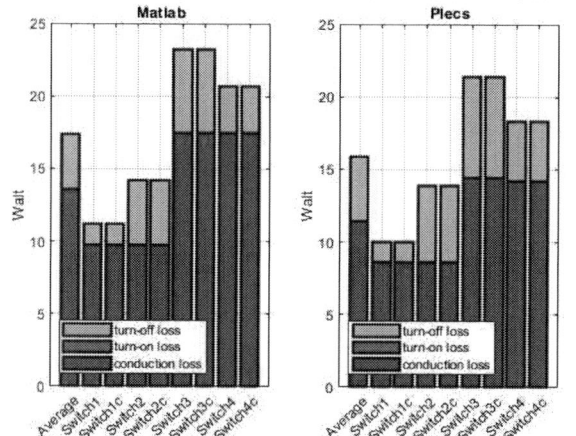

Fig. 7.   Comparison between power loss model estimation tool and PLECS.

For example, the plot from C3M0350120D MOSFET datasheet illustrates the relationship between switching loss and inductor current (with a test voltage V_{test} = 600V) in Fig. 6 [18]. By applying curving fitting tools in MATLAB, the coefficients of A_0, B_0 and C_0 can be derived.

Fig. 7 shows the comparison between the calculated results of the power loss model and the simulation results from PLECS. The conduction loss and switching loss of each switch during one switching period are depicted and compared. The power loss model can be utilized to estimate power loss and efficiency.

The iron loss in the inductor and transformer is primarily caused by stray eddy currents. The improved generalized Steinmetz equation can be used to calculate these iron losses [17].

$$k_i = \frac{K}{(2\pi)^{\alpha-1} \int_0^{2\pi} |cos\theta|^\alpha \, 2^{\beta-\alpha} d\theta} \quad (13)$$

$$P_{core} = \frac{2V_e}{T_r} \int_{t_0}^{t_0+\frac{t_r}{2}} k_i \, |\frac{dB_t(t)}{dt}|^\alpha \, (\Delta B_t)^{\beta-\alpha} \, dt \quad (14)$$

where V_e is the volume of the magnetic core, K, α, β are coefficients determined by the characteristics of the magnetic material, ΔB_t is the peak-to-peak flux density, $\frac{dB_t(t)}{dt}$ is the changing rate of magnetic flux density.

IV. PROPOSED MULTIOBJECTIVE ALGORITHM

The multi-objective optimization problem can be presented in the following standard form:

Minimize $f_1(x) = i_{peakn}(x)$

$\qquad f_2(x) = P_{backn}(x)$

Subject to $g_j(x) \leq 0, j = 1, 2, \ldots, J$ (15)

$\qquad h_m(x) = 0, m = 1, 2, 3, \ldots, M$

$\qquad x_{lb} \leq x_i \leq x_{ub}, i = 1, 2, \ldots, n$

where $f_1(x)$ is the normalized peak current and $f_2(x)$ is the normalized backflow power. x is the vector of the three phase shift ratios. The inequality constraints $g_j(x)$ contains operating mode constraints and ZVS region constraints, and the equality constraints $h_m(x)$ is the required power transmission. x_{lb} and x_{ub} are the lower and upper bounds on three phase shift ratios, respectively.

A. NSGA-II and Pareto front

NSGA-II is an advanced evolutionary algorithm designed to solve multi-objective optimization problems. It is widely used for finding the Pareto front for scenarios with multiple conflicting objectives. The Pareto front represents the set of non-dominated solutions, meaning that no other solutions in the search space perform better in all objectives [15]. Fig. 8 (a) and (b) illustrate the Pareto front when $k = 1.11$, $P_r = 0.9$ and $k = 2$, $P_r = 0.2$, respectively. The red circle in both Fig. 8 (a) and (b) represents the multi-objective optimization solution on the Pareto front, in contrast to the black circle, which highlights the solutions focused solely on minimizing peak current or backflow power.

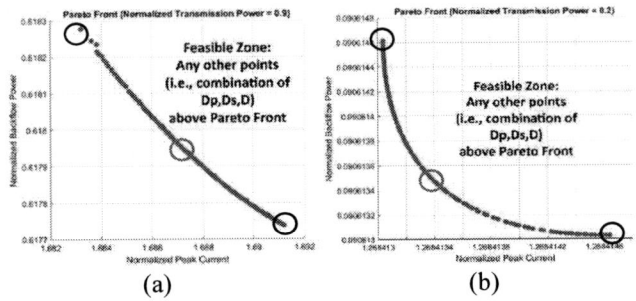

(a)   (b)

Fig. 8. Pareto-front (a) $P_r = 0.9$, k =1.11. (b) $P_r = 0.2$, k =2. The black circles are the optimal point of the minimization of peak current and backflow power. The red circles are the optimal points of the proposed multi-objective algorithm.

B. Flowchart of the proposed scheme

The flowchart of the proposed multi-objective optimization method is presented in Fig. 9. It consists of three main steps: First, based on the required normalized transmission power, NSGA-II is applied to construct the Pareto front in the suitable operating modes. Second, the power loss model is used to calculate the power efficiency of all possible solutions on the Pareto front to identify the local optimal solutions. Third, the local optimal solutions are sorted to find the global optimal solution.

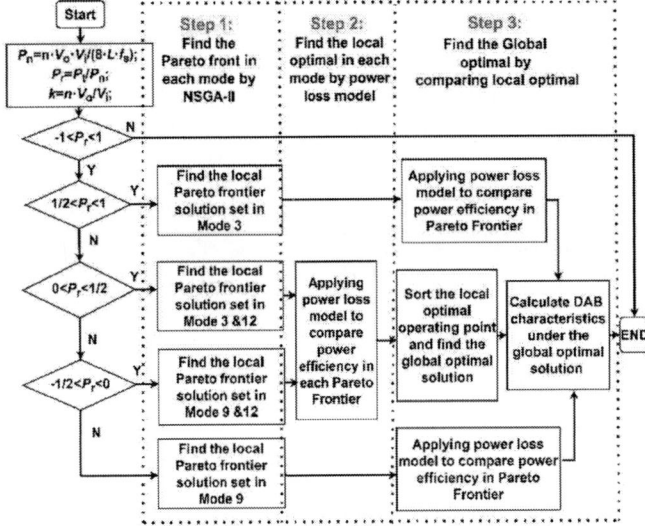

Fig. 9. Flow chart of the proposed Pareto front multi-objective optimization.

C. Close-loop control lookup table

Based on the strategy, the trends of the three phase shift ratios for different voltage gains are derived, as shown in Fig. 10. In the high-power range, the DAB converter primarily operates in mode 3. In the low-power range, modes 12 is the preferred choices, with the voltage gain influencing mode selection. The red line in Fig. 10 indicates that the outer phase shift ratio D is negative, signifying that the DAB converter operates in mode 12. As the voltage gain k increases, the range of mode 12 expands accordingly.

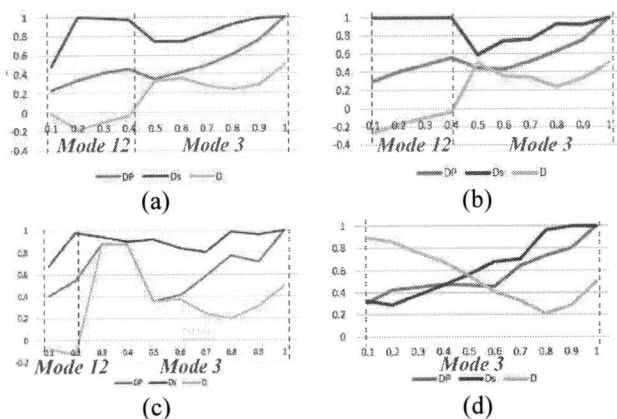

Fig. 10. The trend of three phase shift ratios in the whole power range (a) k = 2, (b) k = 1.67. (c) k = 1.33. (d) k = 1.11. The red line separates the two different work areas. When the voltage gain is equal to unity, the DAB converter will only work in mode 3. The mode 12 area will dispear.

V. EXPERIMENTAL VERIFICATIONS

A prototype system has been established to verify the proposed multi-objective efficiency-oriented algorithm, as illustrated in Fig. 11. The main DC source supplies input energy to the DAB converter, while the electrical load consumes and measures the converter's output energy. The system specifications are outlined in TABLE IV. The output voltage ranges from 100 V to 200 V, with a switching frequency of 100 kHz and a dead time of 100 ns.

TABLE IV.   PARAMETERS OF PROTOTYPE DAB CONVERTERS

Items	Symbol	Values
Input voltage	V_{in}	200 V
Output voltage	V_o	100 V - 200 V
Voltage gain	k	1 - 2
Transformer turns ratio	n	1:1
Phase shift inductance	L	200 µH
Switching frequency	f_s	100 kHz
Rated Power	P_{rate}	225 W
Dead time	t_{de}	100 ns
MOSFET	Q_1-Q_{4c}	C3M00350120D

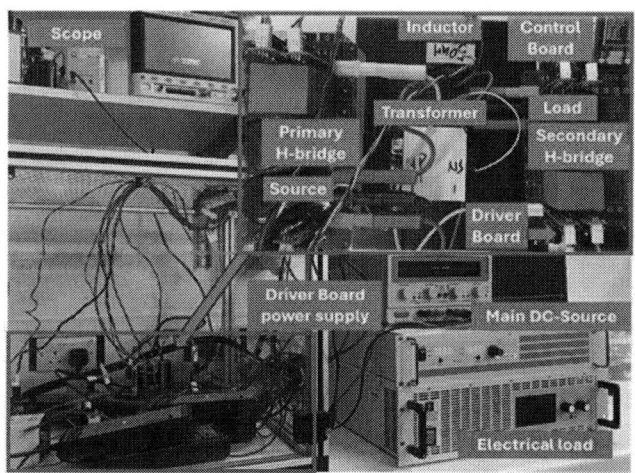

Fig. 11. Photo of the experimental prototype.

Fig. 12 presents a comparison between peak current optimization and the proposed multi-objective optimization results. Evidently, the current stress in the multi-objective scheme is lower than that in the single-objective approach. Additionally, the duty ratio of the secondary bridge voltage is approximately 50%, indicating operation in EPS modulation. The experimental results align well with the theoretical analysis, supporting the proposed approach.

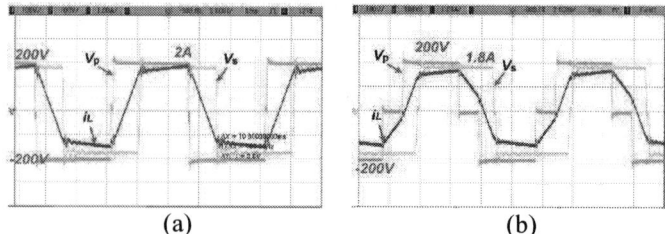

(a)   (b)

Fig. 12. Experimental waveforms under different normalized transmission power, when Vin = 200V, Vo =180V, k=1.11. (a) peak current single objective optimization when P_r=0.9. (b) proposed multi-objective optimization when P_r=0.9. Green and orange waveform is the primary and secondary bridge voltage, respectively. Purple waveform is the inductor current.

Fig. 13 illustrates the waveforms of bridge voltage and inductor current under varying conditions of normalized transmission power P_r and voltage gain k. The experimental results demonstrate that at lower transmission power levels, the Dual Active Bridge (DAB) converter operates in mode 12. As the transmission power increases, the converter transitions to mode 3. These findings align with the expected operational behavior of the DAB under different power levels.

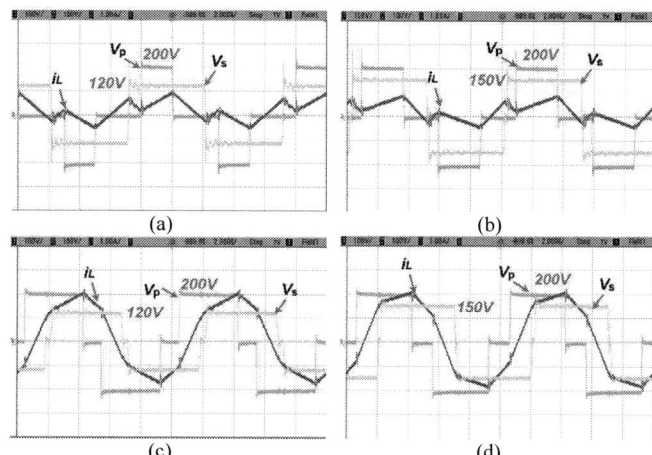

(a)   (b)

(c)   (d)

Fig. 13. Experimental waveforms under different voltage gain k. (a) Vin = 200V, Vo=120V, k =1.67, P_r=0.2. (b) Vin = 200V, Vo=150V, k =1.33, P_r=0.2. (c) Vin = 200V, Vo=150V, k =1.33, P_r=0.9. (d) Vin = 200V, Vo=150V, k =1.33, P_r=0.9. Green and orange waveform is the primary and secondary bridge voltage, respectively. Purple waveform is the inductor current.

The experimental results in Fig. 14. compare the power efficiency across five optimization methods when the voltage gain k = 1.11. Power efficiency is determined by the ratio of the input power of the electrical load to the output power of the main DC source. Notably, in the normalized transmission power range of 0.5 to 0.9, the proposed multi-objective optimization method achieves the highest power efficiency, with an average improvement of over 5% compared to prior single-objective optimization approaches. This indicates that the proposed multi-objective scheme effectively enhances power efficiency within a specific transmission power range.

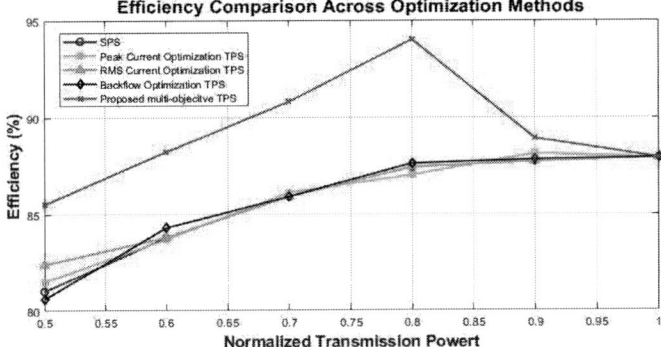

Fig. 14. Power efficiency comparison between different optimization method, when Vin = 200V, Vo = 180V, k = 1.11.

Fig. 15 compares the power efficiencies at various voltage gains k under the proposed multi-objective optimization modulation scheme. The results indicate that the efficiency remains consistent across different voltage gains.

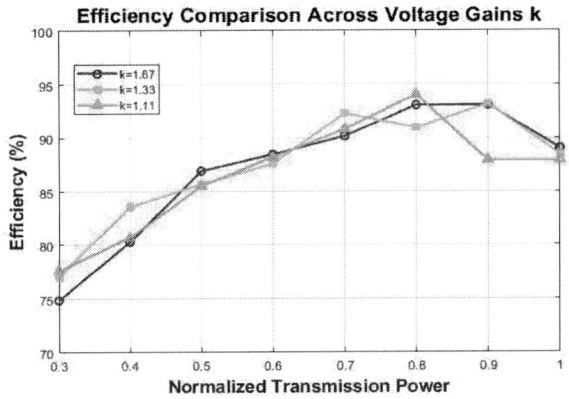

Fig. 15. Proposed multi-objective optimization method power efficiency comparison between different voltage gain, (Vin = 200V, Vo = 180V, k =1.11), (Vin = 200V, Vo = 150V, k=1.33), and (Vin = 200V, Vo = 120V, k = 1.67).

VI. CONCLUSION

This article proposes a multi-objective efficiency-oriented optimization algorithm for DAB converters, aimed at minimizing both backflow power and current stress while ensuring that all switches achieve ZVS. The main contributions of this work are as follows: a detailed time-domain analysis of operating modes is conducted, leading to the derivation of an accurate mathematical model; an implementable multi-objective optimization algorithm is developed and validated. Based on the experimental results, the proposed method achieves a 5% improvement in efficiency within specific power ranges from 0.5 to 0.9, compared to traditional single-objective peak current optimization using TPS modulation.

REFERENCES

[1] P. He and A. Khaligh, "Comprehensive analyses and comparison of 1kW isolated DCDC converters for bidirectional EV charging systems," IEEE Trans. Transport. Electrific., vol. 3, no. 1, pp. 147–156, Mar. 2017.

[2] F. Krismer and J. W. Kolar, "Accurate small-signal model for the digital control of an automotive bidirectional dual active bridge," IEEE Trans. Power Electron., vol. 24, no. 12, pp. 2756–2768, Dec. 2009.

[3] H. Bai, C. Mi, C. Wang, and S. Gargies, "The dynamic model and hybrid phase-shift control of a dual-active-bridge converter," in Proc. 34th Annu. Conf. IEEE Ind. Electron., 2008, pp. 2840–2845.

[4] B. Zhao, Q. Yu, and W. Sun, "Extended-phase-shift control of isolated bidirectional dc-dc converter for power distribution in microgrid," IEEE Trans. Power Electron., vol. 27, no. 11, pp. 4667–4680, Nov. 2012

[5] H. Bai and C. Mi, "Eliminate reactive power and increase system efficiency of isolated bidirectional dual-active-bridge dc–dc converters using novel dual-phase-shift control," IEEE Trans. Power Electron., vol. 23, no. 6, pp. 2905–2914, Nov. 2008.

[6] A. Tong, L. Hang, G. Li, X. Jiang and S. Gao, "Modeling and Analysis of a Dual-Active-Bridge-Isolated Bidirectional DC/DC Converter to Minimize RMS Current With Whole Operating Range," in IEEE Transactions on Power Electronics, vol. 33, no. 6, pp. 5302-5316, June 2018, doi: 10.1109/TPEL.2017.2692276.

[7] J. Tian, F. Wang, F. Zhuo and H. Deng, "Research on Multiple Duty Modulation Scheme in Dual-Active-Bridge-Based Energy Storage System," in IEEE Journal of Emerging and Selected Topics in Power Electronics, vol. 11, no. 3, pp. 3562-3573, June 2023, doi: 10.1109/JESTPE.2023.3255247.

[8] F. Xu, J. Liu and Z. Dong, "Minimum Backflow Power and ZVS Design for Dual-Active-Bridge DC-DC Converters," in IEEE Transactions on Industrial Electronics, vol. 70, no. 1, pp. 474-484, Jan. 2023, doi: 10.1109/TIE.2022.3156159.

[9] L. Deng, G. Zhou, Q. Bi and N. Xu, "Online Reactive Power Minimization and Soft Switching Algorithm for Triple-Phase-Shift Modulated Dual Active Bridge Converter," in IEEE Transactions on Industrial Electronics, vol. 70, no. 3, pp. 2543-2555, March 2023, doi: 10.1109/TIE.2022.3169713

[10] S. Shao, M. Jiang, W. Ye, Y. Li, J. Zhang and K. Sheng, "Optimal Phase-Shift Control to Minimize Reactive Power for a Dual Active Bridge DC-DC Converter," in IEEE Transactions on Power Electronics, vol. 34, no. 10, pp. 10193-10205, Oct. 2019, doi: 10.1109/TPEL.2018.2890292.

[11] Li, X. Zhang, F. Lin, C. Sun and K. Mao, "Artificial-Intelligence-Based Triple Phase Shift Modulation for Dual Active Bridge Converter With Minimized Current Stress," in IEEE Journal of Emerging and Selected Topics in Power Electronics, vol. 11, no. 4, pp. 4430-4441, Aug. 2023, doi: 10.1109/JESTPE.2021.3105522.

[12] M. Hebala, A. A. Aboushady, K. H. Ahmed and I. Abdelsalam, "Generic Closed-Loop Controller for Power Regulation in Dual Active Bridge DC-DC Converter With Current Stress Minimization," in IEEE Transactions on Industrial Electronics, vol. 66, no. 6, pp. 4468-4478, June 2019, doi: 10.1109/TIE.2018.2860535.

[13] J. Huang, Y. Wang, Z. Li and W. Lei, "Unified Triple-Phase-Shift Control to Minimize Current Stress and Achieve Full Soft-Switching of Isolated Bidirectional DC-DC Converter," in IEEE Transactions on Industrial Electronics, vol. 63, no. 7, pp. 4169-4179, July 2016, doi: 10.1109/TIE.2016.2543182

[14] B. Liu, P. Davari and F. Blaabjerg, "An Optimized Control Scheme to Reduce the Backflow Power and Peak Current in Dual Active Bridge Converters," 2019 IEEE Applied Power Electronics Conference and Exposition (APEC), Anaheim, CA, USA, 2019, pp. 1622-1628, doi: 10.1109/APEC.2019.8722273.

[15] Z. Xiao, Z. He, Z. Li, L. Zhu, and L. Wang, "Unified Description and Optimisation Method of Dual Active Bridge DC–DC Converters," in IEEE Transactions on Power Electronics, vol. 37, no. 10, pp. 11839-11854, Oct. 2022, doi: 10.1109/TPEL.2022.3177245

[16] K. Deb, A. Pratap, S. Agarwal and T. Meyarivan, "A fast and elitist multiobjective genetic algorithm: NSGA-II," in IEEE Transactions on Evolutionary Computation, vol. 6, no. 2, pp. 182-197, April 2002, doi: 10.1109/4235.996017.

[17] J. Muhlethaler, J. Biela, J. W. Kolar, and A. Ecklebe, "Core losses under the DC bias condition based on Steinmetz parameters," IEEE Trans. Power Electron., vol. 27, no. 2, pp. 953–963, Feb. 2012.

[18] CREE, "C3M0350120D Silicon Carbide Power MOSFET," C3M™ MOSFET Technology datasheet, 2024

An ISOP-PSFB PWM Converter Based on Coupled Output Inductors and Phase-Shifted Modulation with Full ZVZCS Range

Kang Hong
School of Automation
Central South University
Changsha, China
hkcsu102@csu.edu.cn

Guo Xu
School of Automation
Central South University
Changsha, China
xuguocsu@csu.edu.cn

Guangfu Ning
School of Automation
Central South University
Changsha, China
ningguangfu@csu.edu.cn

Mei Su
School of Automation
Central South University
Changsha, China
sumeicsu@mail.csu.edu.cn

Abstract—This paper proposes an Input-Series-Output-Parallel Phase-Shifted Full-Bridge (ISOP-PSFB) converter based on coupled output inductors. With the phase-shifted modulation, the coupled inductors enable mutual current interaction, reducing circulating current and eliminating duty cycle loss. Unlike conventional ISOP-PSFB converters that improve performance by adding auxiliary networks, the proposed converter adopts coupled inductors instead of the independent output inductors, combined with phase-shift modulation to achieve Zero Voltage and Zero Current Switching (ZVZCS) in the full operating range without the need for additional components. This paper provides the analysis of the proposed topology, operating principles, and characteristics. Finally, the simulation and experimental results have verified the effectiveness.

Keywords—Input-Series-Output-Parallel Phase-Shifted Full-Bridge, coupled output inductors, phase-shifted modulation, full ZVZCS Range

I. Introduction

With the rapid development and widespread application of renewable energy, the electrical energy grid has evolved rapidly as a key energy transmission medium [1]-[2]. In particular, for high-input-voltage and high-output-current DC converters in DC microgrid, the Phase-Shifted Full-Bridge (PSFB) converters are becoming increasingly popular due to their efficiency, reliability, and versatility in handling large power transfers [3]. These converters are particularly well-suited for applications in renewable energy systems, such as solar and wind power, where efficient energy conversion and transmission are critical [4]-[5].

However, traditional PSFB converters face significant challenges due to the large and fixed output inductance. These challenges include the generation of large primary circulating currents and the loss of Zero Current Switching (ZCS) in the lagging half-bridge as the load current increases [6]. As the load

current rises, the increased circulating current exacerbates conduction losses, while the failure to achieve ZCS reduces switching efficiency, both of which compromise the overall performance and efficiency of the converter [7]. To address these issues, some studies have attempted to improve the design of converter leakage [8] or output inductance by incorporating saturable coupled inductors [9], and adopt Input-Series-Output-Parallel (ISOP) structure [10]. However, these approaches often introduce complex auxiliary networks or control strategies [11], making it difficult to achieve Zero Voltage and Zero Current Switching (ZVZCS) in the full operating range of the converter [12]. In particular, for the ISOP structure, the addition of auxiliary networks significantly increases the circuit complexity, making it more challenging to implement in industrial applications [13]-[14].

In response to these challenges, this paper proposes an ISOP-PSFB converter based on coupled output inductors. Unlike the methods of adding auxiliary circuits to improve performance in [13] and [14], the proposed converter adopts coupled inductors with phase-shift modulation strategy instead of independent output inductors, which can achieve ZVZCS in the full load range, eliminate duty cycle losses, significantly reduce primary circulating current, and avoid the need for additional auxiliary components.

II. Topology and Modulation

A. The topology

The topology of the proposed ISOP-PSFB converter is depicted in Fig. 1, which comprises two PSFB modules connected in series on the input side. The primary side features eight MOSFETs, denoted as S_n (where n = 1-8), with each MOSFET's junction capacitance represented as C_{sn} (where n = 1-8). The transformers have equal turns ratios, denoted as N, and the leakage inductances are represented as L_k. On the secondary side, two diode rectifier circuits are connected in parallel, with D_n (where n = 1-4) representing the diodes. The output inductors of the two modules are magnetically coupled, with self-inductance values denoted as L_{o1} and L_{o2}, and mutual inductance denoted as M. Additionally, the output capacitor is represented by C_o, and the input and output voltages are designated as V_{in} and V_o, respectively.

This work was supported in part by the National Nature Science Foundation of China under Grant 52277210, in part by the Hunan Provincial Natural Science Foundation for Excellent Young Scholars of China under Grant 2023JJ20072, in part by the Innovation-Driven Project of Central South University under Grant 2023CXQD043, in part by the Changsha City Science and Technology Plan Project for Distinguished Young Scholars under Grant Kq2209004, and in part by the 2025 Graduate Innovation Project of Central South University (Independent Exploration). (Corresponding author: Guo Xu).

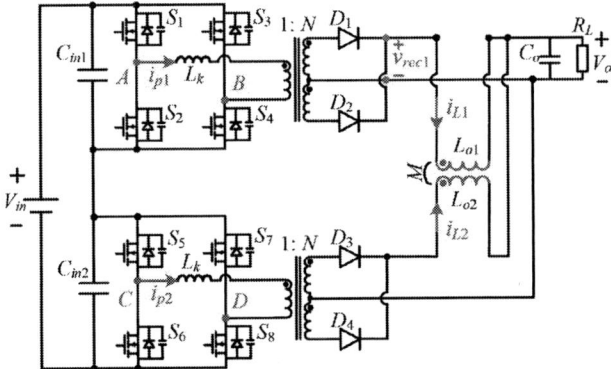

Fig. 1. The proposed topology.

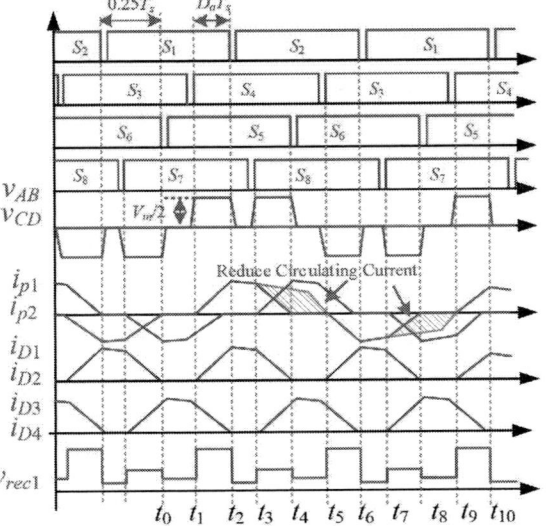

Fig. 2. The modulation strategy of the proposed topology.

B. The modulation strategy

The modulation strategy of the proposed ISOP-PSFB converter is illustrated in Fig. 2. The driving signals of the two PSFB modules are phase-shifted by 90°, with each module employing a phase-shift modulation strategy.

As depicted in Fig. 3, the time interval during which S_1 and S_4 are simultaneously turned on, as well as the corresponding period, are defined by the duty cycle D_a. For the coupled output inductors, as depicted in Fig. 3, the voltages across the two windings are denoted as v_{Lo1} and v_{Lo2}, while the currents flowing through these windings are represented as i_{L1} and i_{L2}, respectively.

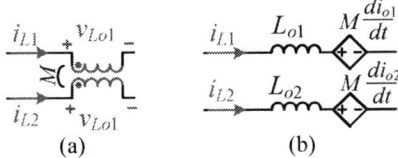

(a)   (b)

Fig. 3. The models of coupled output inductors. (a) The coupled output inductors. (b) The equivalent model.

The relationship between v_{Lo} and i_o can be expressed as:

$$\begin{cases} v_{Lo1} = L_{o1}\dfrac{di_{L1}}{dt} + M\dfrac{di_{L2}}{dt} \\ v_{Lo2} = L_{o2}\dfrac{di_{L2}}{dt} + M\dfrac{di_{L1}}{dt} \end{cases} \quad (1)$$

Owing to the coupling between the output inductors, variations in i_{L1} and i_{L2} will influence each other. The 90° phase shift between the two modules allows an increase in i_{L1} to rapidly decrease i_{L2}. By carefully selecting design parameters, the primary side circulating current can be minimized, facilitating the achievement of the Zero Current Switching (ZCS) as shown in the interval $[t_1, t_2]$ and $[t_3, t_4]$ in Figure 1(b).

III. CHARACTERISTICS ANALYSIS AND DESIGN

A. The gain

By analyzing the secondary rectifier-side voltage v_{rec} in each mode and applying the volt-second balance principle across the output inductor,

$$(0.25 - D_a)(v_{Lo(t0-t1)} + v_{Lo(t2-t3)}) + D_a(v_{Lo(t1-t2)} + v_{Lo(t3-t4)}) = 0 \quad (2)$$

After simplification, the converter gain Mv can be derived as follows:

$$M_v = \frac{2D_a L_k N^3 + 2D_a L_o N}{L_o + 4D_a L_o + M + L_k N^2 + 4D_a L_k N^2 - 4D_a M} \quad (3)$$

where the self-inductance of the two windings of the coupled inductor is designed to have the same value, denoted as L_o.

B. The ZVZCS range

For the leading half-bridge of the proposed PSFB converter, similar to traditional PFFB converter, the implementation of Zero Voltage Switching (ZVS) is facilitated by the secondary output inductor, which makes achieving ZVS relatively straightforward. Therefore, this paper primarily focuses on addressing the implementation of ZCS for the lagging half-bridge. Achieving ZCS in the lagging half-bridge is more challenging, particularly as the load current increases.

In the context where V_{in} ranges from 200V to 400V and V_o is 48V, the relationship between the duty cycle and the parameters of coupled inductors can be derived according to (3), which is illustrated in Fig. 4(a). To ensure the realization of ZCS in the lagging half-bridge, the primary current i_{p1} or i_{p2} must reach zero during the period when the V_{AB} or V_{CD} is zero, such as during the interval $[t_0, t_1]$ and $[t_2, t_3]$. The average value of this current waveform should be greater than the maximum load current I_{Load}, which can be expressed as:

$$\frac{1}{T_s}\left(\int_0^{T_s} i_{L1}dt + \int_0^{T_s} i_{L2}dt\right) > I_{Load} \quad (4)$$

The simulation results demonstrate that the proposed method can achieve ZVZCS across the entire load range, improving the efficiency and performance of the converter by minimizing switching and conduction losses.

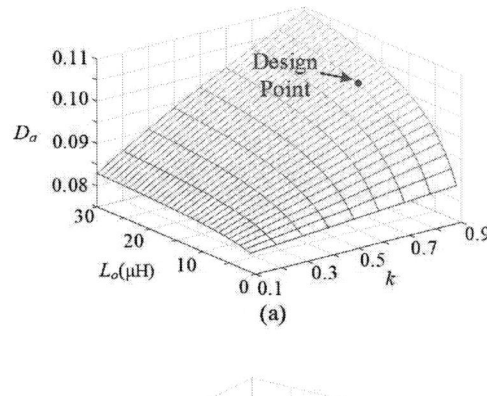

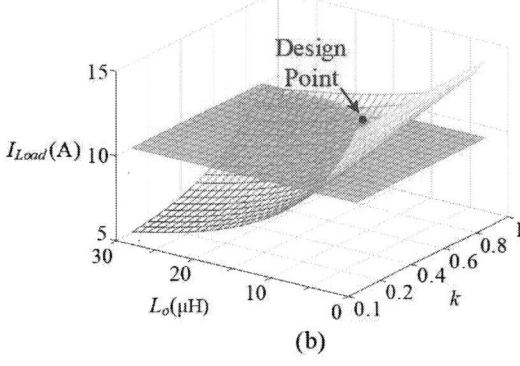

Fig. 4. The influence of coupled inductor parameters. (a) The D_a. (b) The maximum load current I_{Load}.

According to (4), the relationship between the maximum load current and the coupled inductors parameters under ZCS conditions is depicted in Fig. 4(b). The plane in the Fig. 4(b) indicates that the designed maximum load current is 10A. Considering the gain relationship, the output inductor self-inductance is determined to be 15μH, with a coupling coefficient of 0.8.

IV. SIMULATION AND EXPERIMENTAL RESULTS

The parameters V_{in} = 200-400V, V_o=48V, the maximum load current I_{Load} = 10A were selected. And $L_{o1} = L_{o2} = 15$μH, the coupling coefficient $k = 0.8$ were designed. Fig. 5 presents the simulation waveforms under 1A load current. It is observed that the primary current (i_{p1}/i_{p2}) decreases to zero before switching, thereby satisfying the ZCS condition for the lagging half-bridge. Additionally, the currents through the two secondary rectifier diodes (i_{D1}/i_{D2}) in the same module also reduce to zero before commutation, ensuring ZCS operation for the diodes. Moreover, no overlap in conduction between the two diodes occurs, demonstrating the absence of duty cycle loss in the converter under these conditions. Fig 6 and Fig 7 present the simulation waveforms under I_{Load} = 5A and I_{Load} = 10A, respectively. These results demonstrate that the primary current of the converter can reach zero in the half and full load, thereby achieving ZCS for both the lagging half-bridge and the secondary diodes, which can effectively reduce turn-off losses.

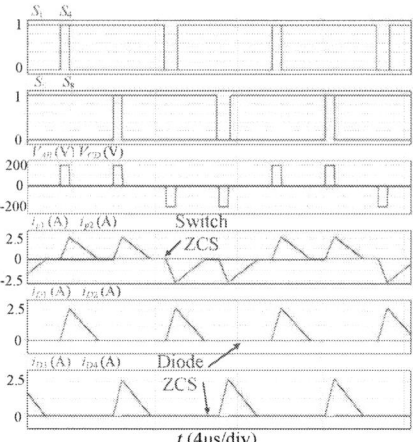

Fig. 5. Simulation waveforms under I_{Load} = 1A.

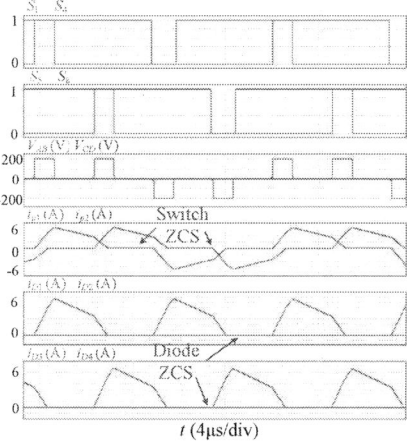

Fig. 6. Simulation waveforms under I_{Load} = 5A.

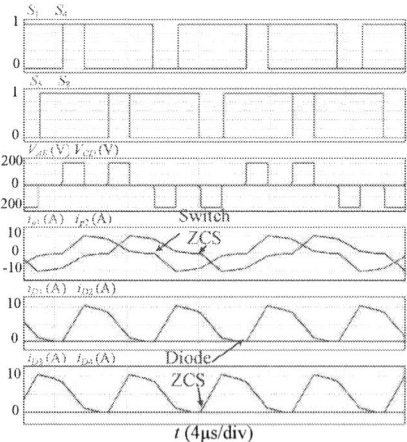

Fig. 7. Simulation waveforms under I_{Load} = 10A.

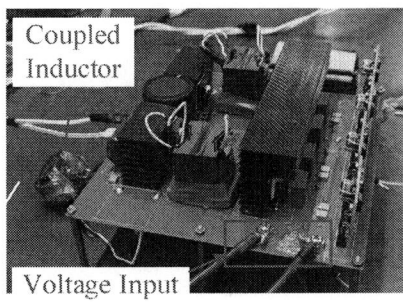

Fig. 8. A 480W experimental prototype.

As presented in Fig. 8, a 480W experimental prototype was developed to validate the effectiveness of the proposed ISOP-PSFB converter.

Fig. 9 - Fig 10 present the experimental waveforms under V_{in}=200V and V_{in}= 400V, respectively. From top to bottom, the waveforms depict the midpoint voltages V_{AB} and V_{CD} of the two modules, along with the primary currents i_{p1} and i_{p2}. It can be seen that there is a significant phase shift in the primary voltage and current waveforms of the two modules, which is consistent with the simulation results in Fig. 5-Fig. 7 and verifies the operation of the prototype.

The experimental results show that during the intervals when the midpoint voltage is zero, the primary currents in both modules reduce to zero, thereby satisfying the ZCS condition for the lagging half-bridge. This demonstrates the effectiveness of the proposed method in achieving ZCS operation and minimizing switching losses.

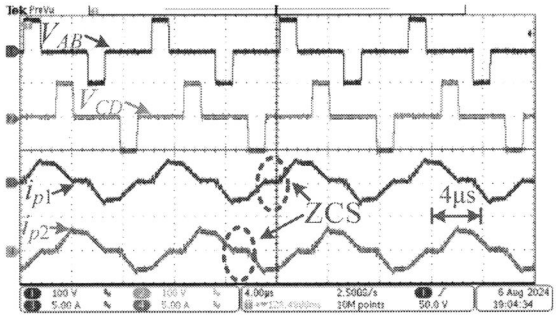

Fig. 9. Experimental waveforms under V_{in} = 200V.

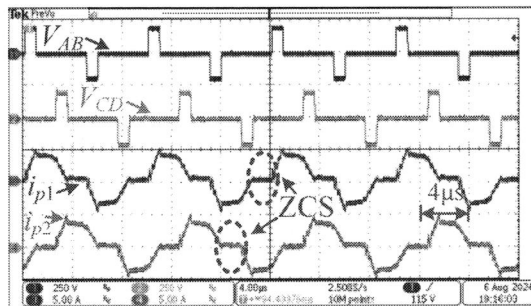

Fig. 10. Experimental waveforms under V_{in} = 400V.

V. CONCLUSIONS AND FUTURE WORK

In this paper, an ISOP-PSFB converter based on coupled output inductors was obtained and its modulation strategies were studied. Compared with the conventional ISOP-PSFB converters that improve performance by adding auxiliary networks, the proposed converter can achieve full-range ZVZCS by designing the coupled output inductors with phase-shifted modulation instead of using additional components. The currents flowing through the coupled output inductors can influence each other and rapidly decrease to zero. Therefore, the converter can eliminate duty cycle loss and significantly reduce the primary circulating current. This paper provided an analysis of the proposed topology, its operating principles, and characteristics. The simulation and experimental results were presented to verify the achievement of full-range ZVZCS.

REFERENCES

[1] M. Lu, M. Qin, W. Mu, J. Fang and S. M. Goetz, "A Hybrid Gallium-Nitride–Silicon Direct-Injection Universal Power Flow and Quality Control Circuit With Reduced Magnetics," in *IEEE Transactions on Industrial Electronics*, vol. 71, no. 11, pp. 14161-14174, Nov. 2024.

[2] M. Lu, M. Qin, J. Kacetl, E. Suresh, T. Long and S. M. Goetz, "A Novel Direct-Injection Universal Power Flow and Quality Control Circuit," in *IEEE Journal of Emerging and Selected Topics in Power Electronics*, vol. 11, no. 6, pp. 6028-6041, Dec. 2023.

[3] H. -P. Kieu, D. B. -H. Nguyen, D. Lee and S. Choi, "All-in-one Magnetic Structure for PSFB converter with Current Doubler Rectifier," *2024 IEEE Applied Power Electronics Conference and Exposition (APEC)*, Long Beach, CA, USA, 2024, pp. 875-880.

[4] C. -Y. Lim, J. -K. Han, M. -H. Park, K. -W. Kim and G. -W. Moon, "Phase-Shifted Full-Bridge DC-DC Converter with High Efficiency and Reduced Output Filter Using Center-Tapped Clamp Circuit," *2019 IEEE Applied Power Electronics Conference and Exposition (APEC)*, Anaheim, CA, USA, 2019, pp. 1710-1715.

[5] P. Rehlaender, R. Unruh, F. Schafmeister and J. Böcker, "Alternating Asymmetrical Phase-Shift Modulation for Full-Bridge Converters with Balanced Switching Losses to Reduce Thermal Imbalances," *2021 IEEE Applied Power Electronics Conference and Exposition (APEC)*, Phoenix, AZ, USA, 2021, pp. 1787-1795.

[6] S. Ren, N. Song, J. Song and N. Diao, "A Phase-Shifted Full-Bridge DC-DC Converter with Dual Clamping Circuits to Eliminate Voltage Oscillation for Electric Vehicles Charger," in *IEEE Transactions on Transportation Electrification*, doi: 10.1109/TTE.2024.3431954.

[7] Y. Shi, L. Feng, Q. Li and J. Kang, "High Power ZVZCS Phase Shift Full Bridge DC–DC Converter with High Current Reset Ability and No Extra Electrical Stress," in *IEEE Transactions on Industrial Electronics*, vol. 69, no. 12, pp. 12688-12697, Dec. 2022.

[8] Y. Gao, Y. Tang, H. Sun, Y. Guo, G. Liu and H. Yang, "Variable Saturation Inductor-Based Full Bridge Converter with Wide ZVS Range and Reduced Duty Cycle Loss," in *IEEE Transactions on Industrial Electronics*, vol. 69, no. 11, pp. 11055-11066, Nov. 2022.

[9] G. Liu et al., "An Improved Zero-Voltage and Zero-Current- Switching Phase-Shift Full-Bridge PWM Converter with Low Output Current Ripple," in *IEEE Transactions on Power Electronics*, vol. 38, no. 3, pp. 3419-3432, March 2023.

[10] Z. Pei et al., "Phase-Shift Full-Bridge (PSFB) Converter Integrated Double-Inductor Rectifier with Separated Resonant Circuits (SRCs) for 800-V High-Power Electric Vehicles," in *IEEE Journal of Emerging and Selected Topics in Power Electronics*, vol. 12, no. 1, pp. 269-282, Feb. 2024.

[11] G. Xu, K. Hong, G. Ning, W. Xiong and M. Su, "Dual-Transformer-Based Hybrid Phase-Shift PWM DC-DC Converter with Wide ZVZCS Range," in *IEEE Transactions on Circuits and Systems II: Express Briefs*, vol. 69, no. 12, pp. 5024-5028, Dec. 2022.

[12] L. Huang, Y. Zhou, J. Huang, J. Zeng and G. Chen, "Analysis and Design of ZVZCS Full-Bridge Converter with Reduced Components for Input-Series-Output-Parallel Application," in *IEEE Transactions on Industrial Electronics*, vol. 68, no. 8, pp. 6806-6817, Aug. 2021.

[13] Z. Guo, D. Sha, X. Liao and J. Luo, "Input-Series-Output-Parallel Phase-Shift Full-Bridge Derived DC–DC Converters with Auxiliary LC Networks to Achieve Wide Zero-Voltage Switching Range," in *IEEE Transactions on Power Electronics*, vol. 29, no. 10, pp. 5081-5086, Oct. 2014.

[14] H. Yu, Y. Shi, C. Wu, F. Wang and C. Xu, "Input Series Output Parallel Phase-Shifted Full-Bridge Converter Design for Full Range Soft-Switching Effect and Minimum Output Current Ripple," *2023 IEEE PELS Students and Young Professionals Symposium (SYPS)*, Shanghai, China, 2023, pp. 1-6.

Design and Implementation of a GaN-Based Soft-Switched Series-Capacitor Buck Converter Operating at the CCM-DCM Boundary for High-Performance Computing Systems

Ramin Rahimzadeh Khorasani
School of Electrical Engineering and Computer Science, Pennsylvania State University, University Park, USA
Ramin.Rahimzadeh@psu.edu

Kolman Puterman Ghitelman
School of Electrical Engineering and Computer Science, Pennsylvania State University, University Park, USA
kjp5754@psu.edu

Madhavan Swaminathan
School of Electrical Engineering and Computer Science, Pennsylvania State University, University Park, USA
mvs7249@psu.edu

Abstract—This paper presents a novel voltage regulator module (VRM) with zero-voltage switching (ZVS) operation, employing a multiphase-interleaved synchronous series-capacitor buck (SCB) converter operating at the boundary between discontinuous (DCM) and continuous (CCM) conduction modes. Building on our prior work introducing the SCB concept with low-valued filter inductances at this boundary mode and its integration into substrates and processor packages, this study provides a comprehensive circuit-level analysis, including mathematical modeling, operating modes, and ZVS soft-switching mechanism. A GaN-based 12V-to-1V prototype, operating at 0.5 MHz and 200W full load with scalable 3-phase SCB converters, is tested to validate the analysis. ZVS operation, coupled with low-inductance air-core inductors, eliminates core and switching losses. As a result, conduction losses dominate, necessitating low-voltage, low-resistance GaN switches for high efficiency in future designs. The converter's ZVS operation enables megahertz switching frequencies while maintaining low losses and improving power density in future implementations.

Keywords—*Voltage regulator module (VRM), series-capacitor buck converters, CCM-DCM boundary condition, zero voltage switching (ZVS), high performance computing systems*

I. INTRODUCTION

High-efficiency Voltage Regulator Modules (VRMs) with substantial voltage conversion ratios (>5:1) and high current capabilities (>100 A) are crucial for modern processors and data centers due to increasing demands from high-performance computing and AI [1]-[3]. The series capacitor buck (SCB) converter, introduced in 2004 [4]-[5], is commonly used in VRM architectures for its benefits, such as duty ratio extension, reduced switch voltage stress, automatic inductor current balancing, and soft charging of series capacitors [6]. However, high power density and minimal routing losses in high-current applications require MHz-range operation, which leads to excessive switching losses and reduced efficiency [1]-[2]. Recent VRMs utilizing conventional and hybrid SCBs in Continuous Conduction Mode (CCM) [7]-[14] show potential for direct 48V-to-1V and 12V-to-1V conversions but face efficiency limits at MHz frequencies due to hard switching. Similarly, multi-stage 12-48V/1V resonant architectures [15]-

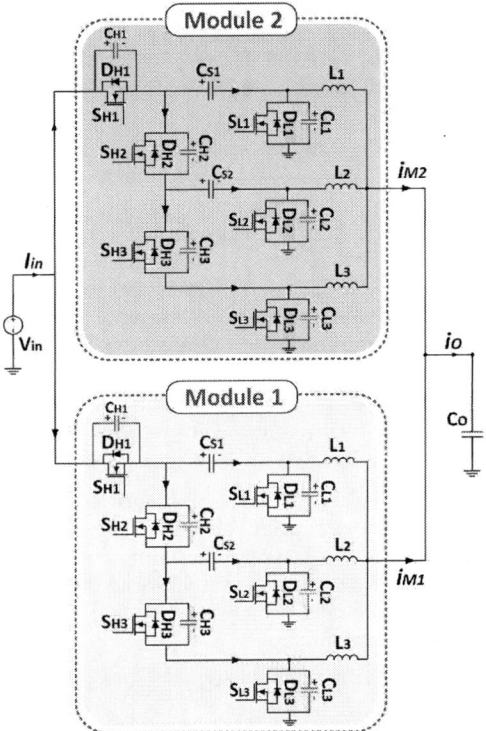

Fig.1. Soft-switched SCB converter from our prior work in [1] with added C_H and C_L snubber capacitors and a 3-phase, 2-module design at the CCM-DCM boundary.

[19] face efficiency challenges at these frequencies because of the additional switching stages or transformer losses. The SCB converter developed in [1] operates at the boundary between CCM and DCM, achieving zero switching losses at MHz frequencies without requiring additional components or transformers, while also reducing the inductance value needed for the output LC filter. However, our previous work did not address the detailed circuit-level analysis, design considerations, and implementation challenges. This paper builds on that foundation by providing a comprehensive circuit-level analysis, exploring the soft-switching mechanism, and addressing design

challenges related to high ripple currents—areas not covered in the earlier study. Additionally, we present a 12V-to-1V prototype with GaN switches and Coilcraft air-core inductors, designed for 200W and 0.5 MHz operation, to validate the effectiveness and practicality of the approach. Existing VRM prototypes in [7]-[9], [13], [15] use MOSFETs and have not demonstrated the operation of their converters with GaN switches.

II. MATHEMATICAL ANALYSIS AND CIRCUIT MODEL

The basic principles of the multi-phase series-capacitor buck (SCB) converter operating under continuous-discontinuous conduction mode (CCM-DCM) have been established in the literature [1]. This section provides a circuit-level analysis, including detailed operating modes, equivalent circuit models, mathematical expressions, and theoretical waveforms to exend that work. To streamline the analysis, a configuration with three phases and two modules is selected to capture all unique time intervals. **Fig. 1.** Illustrates the soft-switched SCB converter from [1], focusing on a single module. The second module operates identically with a 180-degree phase shift (**Fig. 2**). In configurations with more modules (n_M) and phases (n_p), each module is shifted by ($360/n_M$) degrees in phase, maintaining the same time intervals. The inherent drain-source capacitance of the switches (C_{oss}) can facilitate zero voltage switching (ZVS) at turn-off, if it is sufficiently large. In this work, additional snubber capacitors (C_{H1},...,C_{L3}) are incorporated into all switches to further reduce voltage and current overlap, enhancing soft-switching performance during turn-off. The schematic uses standard MOSFET symbols for simplicity, though GaN FETs are used in the prototype. GaN transistors conduct in reverse like a diode but without reverse recovery charge (QRR) [20], [21]. For simplicity, the ripple voltage across series capacitors C_{s1} and C_{s2} is assumed to be zero in this part, though it is considered in the design considerations.

Key waveforms and equivalent circuits for each operating interval have been extracted, as shown in Figures 2 and 3. The converter operates in 15 modes across three phases, and this paper explains six of these modes to demonstrate ZVS in Phase 2. In Figure 3, the changes between modes for each operating phase are highlighted in blue.

Interval 1 [t_0-t_1] (see Fig. 3, Mode 1): During this interval, S_{H1}, the top-side switch of phase 1, conducts current, while the bottom-side switches of phases 2 and 3, S_{L2} and S_{L3} respectively, freewheel the currents of inductors L_2 and L_3. This is the only interval during which power is transferred from the input source to the output. During this time, the current i_{L1} rises from zero to its peak value at t_1, while i_{L2} and i_{L3} decrease. letting $V_{DS(on)}$ denote the on-time Drian-Source voltage drop of the bottom side switch's during reverse conduction, the following equations apply:

$$I_{L1}(t) = \frac{V_{in}-V_{Cs1}-V_o}{L_1}\,(t-t_o) \tag{1}$$

$$I_{L1}(t_1) = \frac{V_{in}-V_{Cs1}-V_o}{L_1}\,(DT) \tag{2}$$

$$I_{L2}(t) = I_{L2}(t_0) - \frac{V_o+V_{DS(on)}}{L_2}\,(t-t_o) \tag{3}$$

$$I_{L3}(t) = I_{L3}(t_0) - \frac{V_o+V_{DS(on)}}{L_3}\,(t-t_o) \tag{4}$$

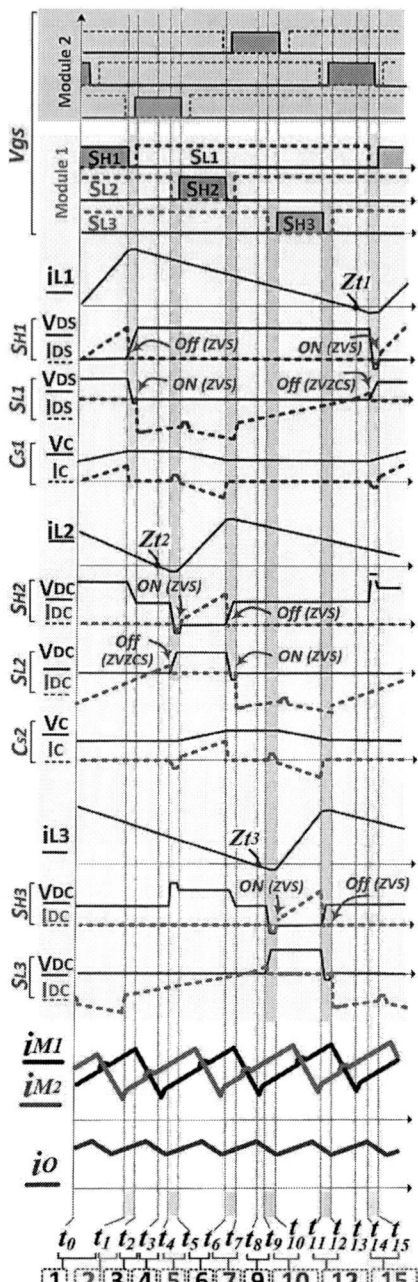

Fig. 2. Original theoretical waveform of the soft-switched SCB converter showing key operating modes and achieved Zero Voltage Switching. Colored intervals and numbers indicate ZVS for turn-on and turn-off across all three phases.

Interval 2 [t_1-t_2] (see Fig. 3, Mode 2 & 2'): At t_1, S_{H1} turns off under near-ZVS conditions due to the reduced overlap between the switch's voltage and current, aided by the C_{H1} capacitor and the C_{oss} of S_{H1}. During this interval, the peak L_1 current charges the C_{H1} capacitor and discharge C_{L1} capacitor until $V_{C_{L1}}$ reaches zero (**Fig. 3, Mode 2**), with i_{L1} remaining nearly constant. Depending on the dead-time between the top and bottom switches, $V_{C_{L1}}$ may turn negative, causing the body diode D_{L1} to conduct (**Fig. 3, Mode 2'**). This interval ends when

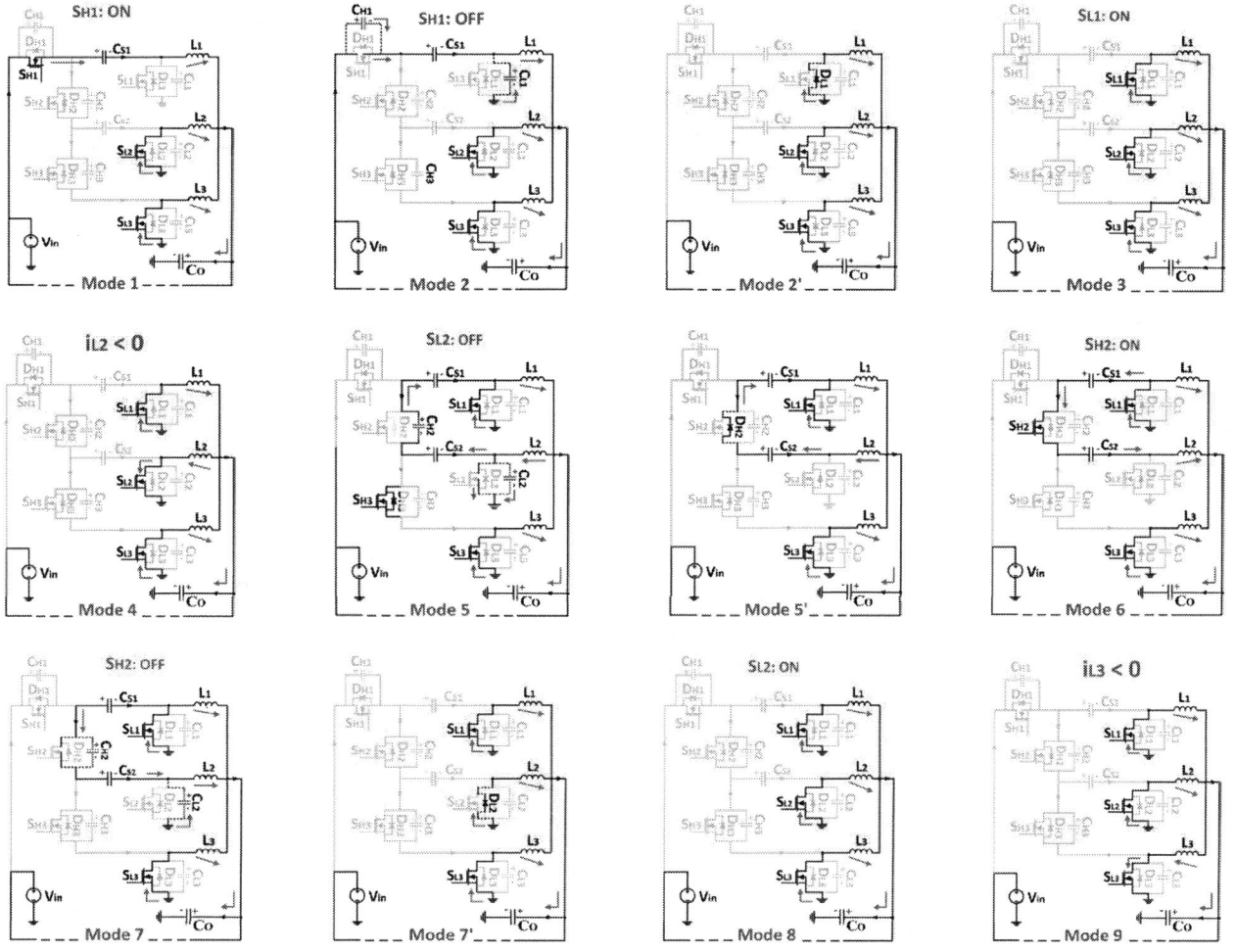

Fig.3. Equivalent circuit for each operating interval of the proposed converter. Only 9 of the 15 modes are shown here.

the S_{L1} switch turns on at t_2, by which time $V_{C_{L1}}$ must return to zero. For GaN power transistors, minimizing Mode 2' is essential to prevent dead-time losses (P_{dt}) [4], due to reverse current, while ensuring soft turn-off for S_{H1}.

$$V_{C_{L1}}(t) = V_{in} - V_{Cs1} - \frac{(V_{in}-V_{Cs1}-V_o)D \times T}{2L_1 \times C_{L1}} (t-t_1) \quad (5)$$

$$t_{21} = t_2 - t_1 = \frac{2L_1 \times C_{L1}(V_{in}-V_{Cs1})}{D \times T(V_{in}-V_{Cs1}-V_o)} \quad (6)$$

Interval 3 [t_2-t_3] (see Fig. 3, Mode 3): At the start of this interval, S_{L1} turns on under ZVS condition, allowing the inductor current I_{L1} to freewheel and begin decreasing. During this mode, the bottom-side switches S_{L2} and S_{L3} continue to freewheel the currents through the inductors in phases 2 and 3, respectively. At the end of this mode, the L_2 current reaches zero. The duration of this mode is crucial for the control circuit, as the zero-crossing point of I_{L2}(denoted Z_{t2} in brown in **Fig. 2**) occurs at t_3.

$$I_{L1}(t) = I_{L1}(peak) - \frac{V_o+V_{DS(on)}}{L_1} (t-t_2) \quad (7)$$

$$I_{L2}(t_1) = I_{L2}(t_2) - \frac{V_o+V_{DS(on)}}{L_2} (t-t_2) \quad (8)$$

$$I_{L3}(t) = I_{L3}(t_2) - \frac{V_o+V_{DS(on)}}{L_3} (t-t_2) \quad (9)$$

$$t_{32} = t_3 - t_2 = \frac{L_2 \times I_{L2}(t_2)}{V_o+V_F} \quad (10)$$

Interval 4 [t_3-t_4] (see Fig. 3, Mode 4): This interval begins when the I_{L2} current crosses Z_{t2} and starts flowing in reverse. This negative current is necessary for achieving ZVS condition for S_{H2} in the next interval. This mode ends when the control circuit turns off the S_{L2}, once I_{L2} reaches the required negative value (5-10% of the peak current) to quickly discharge the C_{H2} capacitor which is discussed in the design considerations. Note that I_{L2} and I_{L3} remain positive and continue to decrease, similar to previous intervals.

$$I_{L2}(t) = -\frac{V_o+V_{DS(on)}}{L_2} (t-t_3) \quad (11)$$

$$t_{43} = \frac{I_{L2}(4)}{V_o+V_o+V_{DS(on)}} \quad (12)$$

TABLE I
CURRENT AND VOLTAGE STRESS OF COMPONENTS IN SCB ARCHITECTURE WITH n_p PHASES AND n_m MODULES AT BOUNDARY CONDITIONS.

COMPONENTS	CURRENT RATING	VOLTAGE RATING
S_{H1}	$2I_o/(n_P \times n_M)^*$	V_{in}/n_p
$S_{H2}, ..., S_{Hn_p}.$	$2I_o/(n_P \times n_M)^*$	$2V_{in}/n_p$
$S_{L1}, S_{L2}, ..., S_{Ln_p}$	$\dfrac{2I_o}{n_p \times n_M} - \dfrac{1 - n_p(1 - D)}{L_1 \times n_p^2} DTV_{in}$	V_{in}/n_p
$C_1, C_2, ..., C_{(n_p-1)}$	$2I_o/(n_P \times n_M)$	$\dfrac{(n_p-1)}{n_p}V_{in}, \dfrac{(n_p-2)}{n_p}V_{in}, ..., \dfrac{1}{n_p}V_{in}$
$L_1, L_2, ..., L_{(n_p-1)}$	$2I_o/(n_P \times n_M)$	V_{in}/n_p

Interval 5 [t_5-t_6] (see Fig. 3, Mode 5 and 5'): This interval starts when the S_{L2} turns off at t_5 under zero-voltage and nearly zero current (ZVZCS) condition, since S_{L2} current is very small at t_5 and C_{L2} reduces overlap between the switch's voltage and current. This interval ends when the S_{H2} switch turns on at t_6, by which time $V_{C_{H2}}$ must return to zero. Minimizing Mode 5' is essential to prevent dead-time losses (P_{dt}), similar to mode 2'.

$$V_{C_{H2}}(t) = V_{Cs1} - V_{Cs2} - \frac{\left(V_o + V_{DS(on)}\right)t_{43}}{2L_2 \times C_{H2}}(t - t_5) \quad (13)$$

$$t_{65} = \frac{2L_2 \times C_{H2}(V_{Cs1} - V_{Cs2})}{D \times T\left(V_o + V_{DS(on)}\right)} \quad (14)$$

Interval 6 [t_5-t_6] (see Fig. 3, Mode 6): At the beginning of this interval, S_{H2} turns on under ZVS condition, which initiates an increase in the inductor current I_{L2} in a positive direction. The capacitor C_{s1}, which maintains a nearly constant voltage, serves as the input source for the second phase. During this mode, the bottom-side switches S_{L1} and S_{L3} continue to freewheel the currents through the inductors in phases 1 and 3, respectively, following the same equations as in previous intervals. The duration of this mode is $D \times T$, similar to mode 1, and I_{L2} reaches its peak value at the end of the interval.

$$I_{L2}(t) = \frac{V_{Cs1} - V_{Cs2}}{L_2}(t - t_5) \quad (15)$$

$$I_{L2}(t_6) = \frac{V_{Cs1} - V_{Cs2}}{L_2}(DT) \quad (16)$$

The remaining intervals associated with Phase 3 operate in the same modes as the preceding phases.

III. CONVERTER DESIGN AND CONSIDERATIONS

This section addresses the design challenges associated with the soft-switched SCB converter operating at the boundary condition. For each module with n_p inductors, current sensing through just one inductor is sufficient to detect the zero-crossing point, thereby enabling zero switching losses at turn-on. This is because the inductors within the same module naturally achieve auto volt-second balance. A high-frequency zero-cross detection circuit, such as the one introduced in [22], is well-suitable for MHz-range frequencies in this converter.

Unlike the CCM version, boundary condition operation can result in higher ripple currents in the capacitors and inductors. This design addresses these challenges while maintaining performance. The voltage conversion ratio (G) at the boundary condition matches that of the standard SCB operating in CCM, expressed as

$$G = \frac{V_o}{V_{in}} = \frac{D}{n_P} \quad (17)$$

In the SCB converters, the high-side switches operate in an interleaved mode introducing a phase shift between switches. To prevent overlap of high-side switches and ensure balance in the series capacitors and output inductors, a constraint is imposed on the maximum duty cycle (D), as described in [23]:

$$D \leq \frac{1}{n_P} \quad (18)$$

This constraint limits the maximum number of phases (n_P), which becomes a critical initial design parameter for SCB converters, irrespective of whether they operate in CCM or at the boundary condition. The maximum allowable number of phases is determined by:

$$n_p \leq floor\left(\sqrt{\frac{V_{in}}{V_o}}\right) \quad (19)$$

Here, the floor function returns the largest integer less than or equal to the computed value. This limitation, not addressed in [1], is critical for designing the proposed SCB converter at the boundary condition. It defines the maximum number of phases for various configurations, such as a 48-to-1V converter. For integrated voltage regulators (IVRs) operating at higher frequencies, this parameter must also comply with maximum on-time and frequency constraints to ensure high power density.

Additionally, when designing SCB inductors for operation at the boundary condition, allowing up to 5% of the peak current to flow in the negative direction can effectively discharge C_H snubber capacitors. This ensures zero switching losses while maintaining negligible circulating losses. The inductor value for boundary condition operation, guaranteeing ZVS-soft switching, is determined using the following equation:

$$L = \frac{0.95 \times n_p \times n_M \times V_o^2}{2P_{o_{(nom)}} \times f_{sw}}\left(1 - \frac{n_p \times V_o}{V_{in}}\right) \quad (20)$$

While the SCB operating at the boundary condition eliminates switching losses, it can introduce higher ripple currents to the output capacitor if not properly designed. This ripple current can be mitigated through careful design of interleaved modules and phases. For n_M interleaved modules, each comprising n_P phases, the ripple current for the module, i_{M1} and i_{M2} are given by:

$$\Delta I_M = \frac{2I_o}{n_P \times n_M}\left(1 - \frac{floor\left(n_p \times D\right)}{n_p \times D}\right)\left(1 + floor\left(n_p \times D\right) - n_p \times D\right) \quad (21)$$

TABLE II
DESIGN SPECIFICATIONS OF THE GAN-BASED PROTOTYPE

Parameter	Value	Component	Value
Input Voltage (V_{in})	12 V	Total Power	200 W (67 w for each module)
Output Voltage (V_o)	1 V	Number of interleaved Phases	$n_P = 3$
Switching Frequency (f_{sw}), nominal Duty Cycle (D)	0.5 MHz, D=0.26%	Number of interleaved Modules	$n_M = 3$

TABLE III
COMPONENTS USED IN THE DESIGNED PROTOTYPE

Component	Part number	Component	Part number
High-side switches (S_{H1}, S_{H2}, S_{H3})	GS61008T ($R_{DS(on)}$ = 7 mΩ)	Inductor L_1, L_2, L_3	22 nH Air core Coilcraft (1212VS-22N) (R_{DC} - 0.5 mΩ)
Low-side switches (S_{L1}, S_{L2}, S_{L3})	2-parallel switches (GS61008T)	Series capacitors C_1, C_2	GRM32EC72
Gate Driver	EiceDRIVER 1EDBx275F	Output capacitors	GRM219R60

The floor function returns the smallest integer greater than or equal to the number. Each module can be modeled as a single phase converter with an increased duty cycle of $D_{new} = n_P \times D$, effectively reducing output ripple current through n_m interleaved modules. Therefore, the output ripple current through the load capacitor is:

$$\Delta I_{CO} = \frac{2I_o}{n_P \times n_M} \left[1 - \frac{floor\,(D \times n_p \times n_m)}{D \times n_p \times n_m} \right] \left[1 - (n_p \times n_m)D + \right.$$

$$\left. + floor\,(D \times n_p \times n_m) \right] \tag{22}$$

This formula highlights the architecture's capability to reduce ripple current by incorporating an adequate number of phases and modules. A higher count of phases and modules results in lower ripple currents and enables the use of smaller output capacitors. However, while increasing the number of phases and modules improves ripple reduction, it can diminish efficiency under light loads due to the additional gate drivers required for the switches. Therefore, the maximum number of modules should be carefully selected to balance ripple reduction and efficiency at light loads. Finally, the series capacitors are designed accordingly. Selecting the optimal number of phases and modules can also minimize the capacitance required for these series capacitors.

$$C_s \geq \frac{1.05 \times I_o}{4 n_P \times n_M \times f_{sw} \times \Delta V_C} \tag{23}$$

In addition to selecting appropriate capacitor and inductor values, the current and voltage ratings of these passive components are critical for implementation. Similarly, the voltage and current ratings of the semiconductors must be calculated to ensure reliable operation. Table 1 summarizes the current and voltage stresses for all components in the SCB architecture under boundary conditions, as a function of the number of phases (n_P) and modules (n_M) in the converter design.

IV. EXPERIMENTAL RESULTS

The converter prototype is fabricated for a 200 W output power and a 0.5 MHz switching frequency, consisting of three modules, each containing three phases, with each module delivering 67 W of power, as specified in Table II and Table III. GS61008T GaN switches and the 1EDBx275F Infineon EiceDRIVER-1EDBx275F are used for the gate driver. Due to the small inductance requirement for this converter, air-core

TABLE IV
COMPARISON BETWEEN SCB AT CCM AND BOUNDARY OF CCM-DCM

PARAMETER	CONVENTIONAL SCB (CCM CONDITION)	PROPOSED SCB (BOUNDARY CONDITION)
Switching losses E_{on}, E_{off}, E_{oss}	Hi losses (Hard Switching) limited switching frequency	Nearly Zero (Soft Switching) Ability to operate at very high switching frequencies
Components Voltage Stress	Lower than a buck converter operating in CCM	The same as conventional SCB
Components Current Stress	Higher than a buck converter operating at CCM	Higher than CCM SCB (table I)
Filter Inductor	Large inductance	Low Inductance due to boundary conditions and high switching frequency
Control Method	adjustable on-time (PWM)	adjustable on-time and off-time / constant off-time

inductors are employed to eliminate core losses, contributing to further efficiency improvements.

The switches' waveforms, shown in **Fig. 4**, validate the theoretical analysis of the converter. **Fig. 5(a)** shows the current waveform for one of the inductors and the summed inductor currents of the phases in Module 1, representing the output current of the first module in **Fig. 5(b)**. This demonstrates continuous conduction mode (CCM) operation for each module, with each phase operating at the boundary between CCM and DCM to achieve ZVS soft switching for the switches. The peak current in each inductor reaches 50 A due to the three-phase configuration, supporting an average output current of 67 A per module. The ripple current for each module is reduced to approximately 15 A relative to the 67 A average. The interleaved operation of all three modules further minimizes the overall ripple current at the 200 A converter output, ensuring stable CCM operation and reducing ripple current for the output capacitors.

Table IV presents a comparison between the interleaved SCB converter at the boundary condition and the converter operating in CCM. The key advantage of the SCB at the boundary condition is its ability to provide soft switching and support Mega Hertz switching frequencies. This operating mode requires significantly smaller inductance values, which can be

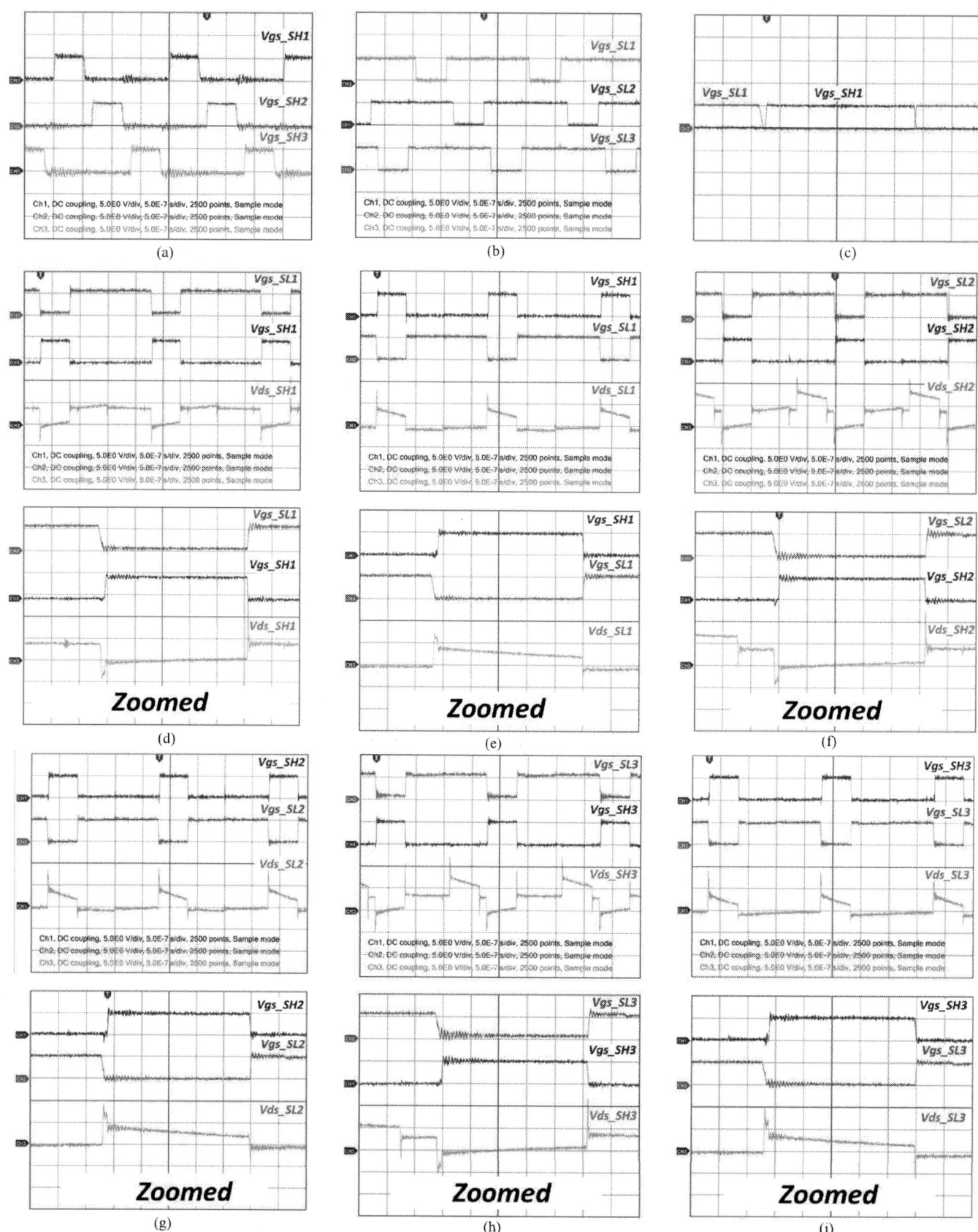

Fig. 4. Experimental waveforms for Module-1 at a 0.5 MHz switching frequency with an output power of 67 W (P_{Total}: 200 W): (a) Gate-Source voltage of the high-side switches. b) Gate-Source voltage of low-side switches c) Gate-Source voltage of Phase-1 showing the required delay between switching events. ZVS soft switching operation for all switches with zoomed figures: d) S_{H1} e) S_{L1} f) S_{H2} g) S_{L2} h) S_{H3} i) S_{L3}.

979-8-3315-1612-3/25 $31.00 © 2025 IEEE

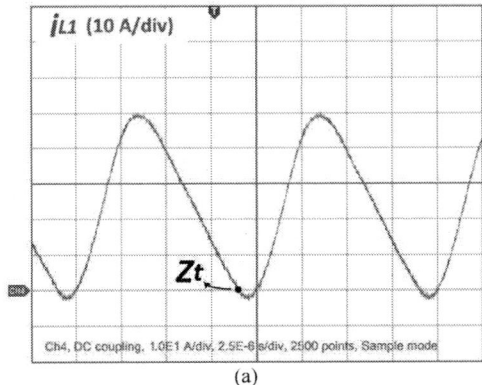

(a)

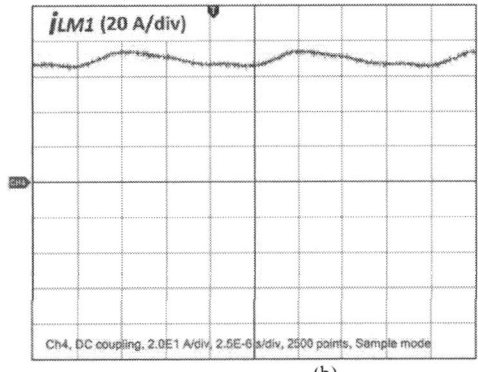

(b)

Fig. 5. Current waveforms: (a) Inductor current (i_{L1}) (b) Output current of the first module, i_{M1}. The current waveforms are captured at 100 kHz switching frequency instead of the nominal 500 kHz switching frequency due to the frequency limitations of the current probe.

further reduced at higher switching frequencies, making this converter an ideal candidate for integrated voltage regulator applications, as discussed in [1].

V. CONCLUSION

This paper presents a circuit-level analysis of an interleaved SCB converter operating at the CCM/DCM boundary for power delivery in CPU and GPU applications. The analysis is supported by a comprehensive mathematical framework addressing the design challenges under these conditions. Experiments on a GaN-based prototype with a 200W load, 12V-to-1V conversion ratio, and 0.5 MHz switching frequency, using three phases and three interleaved modules, demonstrate scalability for kilowatt loads. The prototype requires significantly smaller inductance values than existing VRMs, enabling the use of small air-core inductors and eliminating core losses. Simulations and experimental results confirm zero switching losses during switch turn-on and nearly zero switching losses during turn-off. However, due to high (RDS(on) = 7 mΩ) conduction losses from the 100V Infineon GaN switches, future designs can achieve higher efficiency with low-voltage GaN switches. With zero switching losses at the boundary condition and no need for additional auxiliary power components, this converter is an excellent candidate for both VRM and integrated voltage regulator (IVR) applications.

ACKNOWLEDGMENT

This work was supported in part by CHIMES: Center for Heterogeneous Integration of Micro Electronic Systems, one of the seven centers in JUMP 2.0, a Semiconductor Research Corporation (SRC) program sponsored by DARPA.

References

[1] R. R. Khorasani, X. Li, J. W. Kim, P. Murali, R. Sharma and M. Swaminathan, "Package Power Delivery Architecture for High Performance Computing Systems With a 1 kW IVR Operated in CCM-DCM Boundary Mode Condition," 2024 IEEE 74th Electronic Components and Technology Conference (ECTC), Denver, CO, USA, 2024, pp. 285-292, doi: 10.1109/ECTC51529.2024.00054.

[2] K. Radhakrishnan, M. Swaminathan and B. K. Bhattacharyya, "Power Delivery for High-Performance Microprocessors—Challenges, Solutions, and Future Trends," in IEEE Transactions on Components, Packaging and

Manufacturing Technology, vol. 11, no. 4, pp. 655-671, April 2021, doi: 10.1109/TCPMT.2021.3065690.

[3] M. Chen, S. Jiang, J. A. Cobos and B. Lehman, "Design Considerations for 48-V VRM: Architecture, Magnetics, and Performance Tradeoffs," 2023 Fourth International Symposium on 3D Power Electronics Integration and Manufacturing (3D-PEIM), Miami, FL, USA, 2023, pp. 1-9, doi: 10.1109/3D-PEIM55914.2023.10052608.

[4] Y. Jang and M. Jovanovi′c, "Non-isolated power conversion system having multiple switching power converters," U.S. Patent 10 972 632, Oct. 26, 2004.

[5] K. Nishijima, K. Harada, T. Nakano, T. Nabeshima and T. Sato, "A double step-down two-phase buck converter for VRM," 2005 European Conference on Power Electronics and Applications, Dresden, Germany, 2005, pp. 8 pp.-P.8, doi: 10.1109/EPE.2005.219347.

[6] P. S. Shenoy," Introduction to the Series Capacitor Buck Converter" Texas Instruments DC Solutions Application Report, SLVA750A–April 2016–Revised May 2016.

[7] J. Baek et al., "Vertical Stacked LEGO-PoL CPU Voltage Regulator," in IEEE Transactions on Power Electronics, vol. 37, no. 6, pp. 6305-6322, June 2022.

[8] Y. Elasser et al., "Mini-LEGO: A 1.5-MHz 240-A 48-V-to-1-V CPU VRM with 8.4-mm Height for Vertical Power Delivery," 2023 IEEE Applied Power Electronics Conference and Exposition (APEC), Orlando, FL, USA, 2023, pp. 1959-1966.

[9] Y. Zhu, J. Zou and R. C. N. Pilawa-Podgurski, "A 1500-A/48-V-to-1-V Switching Bus Converter for Next-Generation Ultra-High-Power Processors," in IEEE Transactions on Power Electronics, vol. 39, no. 9, pp. 11340-11355, Sept. 2024

[10] M. Gong, H. Chen, X. Zhang, R. Jain and A. Raychowdhury, "A 90.4% Peak Efficiency 48-to-1-V GaN/Si Hybrid Converter With Three-Level Hybrid Dickson Topology and Gradient Descent Run-Time Optimizer," in IEEE Journal of Solid-State Circuits, vol. 58, no. 4, pp. 1002-1014, April 2023,

[11] M. Gong et al., "A GaN-Based Reconfigurable Series-Parallel Hybrid Converter Supporting 48/24/12V Input and 0.8-1.2V Output with 83.7/87.8/90.7% Peak Efficiency," 2023 IEEE Applied Power Electronics Conference and Exposition (APEC), Orlando, FL, USA, 2023, pp. 912-918

[12] A. K. Delmar and A. Stillwell, "Current-Sourced Hybrid Switched-Capacitor Converter for Data Center Power Delivery," 2024 IEEE Applied Power Electronics Conference and Exposition (APEC), Long Beach, CA, USA, 2024, pp. 1357-1362, doi: 10.1109/APEC48139.2024.10509139.

[13] G. Roberts and A. Prodić, "Multilevel Series-Capacitor Buck Converter," 2024 IEEE Applied Power Electronics Conference and Exposition (APEC), Long Beach, CA, USA, 2024, pp. 1363-1370, doi: 10.1109/APEC48139.2024.10509279.

979-8-3315-1612-3/25 $31.00 © 2025 IEEE

[14] N. M. Ellis, Y. Zhu and R. C. N. Pilawa-Podgurski, "Gallium Nitride-based 48V-to-1V Point-of-Load (PoL) Converter for Aerospace Telecommunications and Computing Applications," 2024 IEEE Applied Power Electronics Conference and Exposition (APEC), Long Beach, CA, USA, 2024, pp. 1384-1388, doi: 10.1109/APEC48139.2024.10509202.

[15] M. H. Ahmed, C. Fei, F. C. Lee, and Q. Li, "Single-Stage HighEfficiency 48/1 V Sigma Converter With Integrated Magnetics," IEEE Transactions on Industrial Electronics, vol. 67, no. 1, pp. 192–202, 2020.

[16] P. R. Prakash et al., "A 2400 W/in3 1.8 V Bus Converter Enabling Vertical Power Delivery for Next-Generation Processors," 2024 IEEE Applied Power Electronics Conference and Exposition (APEC), Long Beach, CA, USA, 2024, pp. 910-917

[17] S. Zaffin et al., "A 60 A Switched Tank Converter with Buck-Boost Sigma Regulation for 48 V Bus Down-Conversion," 2024 IEEE Applied Power Electronics Conference and Exposition (APEC), Long Beach, CA, USA, 2024, pp. 63-69, doi: 10.1109/APEC48139.2024.10509131.

[18] A. Dago, M. Balutto, S. Saggini, M. Leoncini, S. Levantino and M. Ghioni, "A 260-A/48-V Bus Hybrid Resonant Converter with Large Conversion Ratio for Future Data Centers," 2024 IEEE Applied Power Electronics Conference and Exposition (APEC), Long Beach, CA, USA, 2024, pp. 83-87, doi: 10.1109/APEC48139.2024.10509193.

[19] B. A. McDonald, S. -Y. Yu and C. A. Rodriguez, "Practical Design Methodology for a High Efficiency LLC Converter," 2024 IEEE Applied Power Electronics Conference and Exposition (APEC), Long Beach, CA, USA, 2024, pp. 1734-1739, doi: 10.1109/APEC48139.2024.10509348.

[20] Bingyao Sun, "Does GaN Have a Body Diode? - Understanding the Third Quadrant Operation of GaN", Texas Instruments Application Report, SNOAA36, February 2019.

[21] Yichi Zhang, Fei Yang, "Maximizing the Performance of GaN with Ideal Diode Mode", Texas Instruments Application Report, SNOA932, OCTOBER 2020.

[22] C. Schaef et al., "8.5 A Fully Integrated Voltage Regulator in 14nm CMOS with Package-Embedded Air-Core Inductor Featuring Self-Trimmed, Digitally Controlled Variable On-Time Discontinuous Conduction Mode Operation," 2019 IEEE International Solid-State Circuits Conference - (ISSCC), San Francisco, CA, USA, 2019, pp. 154-156, doi: 10.1109/ISSCC.2019.8662294.

[23] G. Roberts and A. Prodić, "Modulation Improvements for High-Phase-Count Series-Capacitor Buck Converters," in IEEE Open Journal of Power Electronics, vol. 5, pp. 1071-1092, 2024, doi: 10.1109/OJPEL.2024.3417017.

979-8-3315-1612-3/25 $31.00 © 2025 IEEE

Intrinsic Feedback Model for Coupled-Damped Self-Balancing of General Multiphase Hybrid Converters

Haoran Xu
School of Integrated Circuits
Huazhong University of Science
and Technology
Wuhan, China
d202480677@hust.edu.cn

Weijia Hao
School of Integrated Circuits
Huazhong University of Science
and Technology
Wuhan, China
d202480674@hust.edu.cn

Desheng Zhang
School of Automation
Wuhan University of Technology
Wuhan, China
dszhang@whut.edu.cn

Run Min
School of Integrated Circuits
Huazhong University of Science
and Technology
Wuhan, China
minrun@hust.edu.cn

Qiaoling Tong
School of Integrated Circuits
Huazhong University of Science
and Technology
Wuhan, China
tongqiaoling@hust.edu.cn

Xuecheng Zou
Institute of Integrated Circuits
Henan Academy of Sciences
Wuhan, China
estxczou@hust.edu.cn

Abstract—This work develops a generalized analytical framework to demystify the current and voltage self-balancing in multiphase hybrid converters. Firstly, generalized mathematical criteria are explored to identify the self-balancing characteristics in steady-state. Moreover, the normalized oscillation equation of the balancing errors is derived, which reveals the coupled damping effect among phases. Finally, by performing the Laplace transform on the differential equations of the balancing errors, an intrinsic feedback model is established, which reveals the convergence speed and stability of the self-balancing process. The model indicates that small inductors, large equivalent parasitic resistors, and large flying capacitors are beneficial for fast and stable self-balancing. With the proposed framework, a new topology with self-balancing characteristics is predicted. The validity and generality of the developed theory are verified with experiments.

Keywords—DC-DC, hybrid converter, flying capacitor, modeling, feedback model, coupled damping, self-balancing

I. INTRODUCTION

Emerging AI computing needs of ASICs generate escalating peak currents under scaled-CMOS-compatible voltages (<1V) [1]. To ease the I^2R loss in power delivery, the DC bus voltage has increased from 12V to 48V [2]. To meet the high-conversion-ratio and high-current scenarios, multiphase hybrid converters were developed recently, such as the multiphase series capacitor buck (MPSCB) converter [3] and multiphase multilevel (MPML) converter [4] shown in Fig. 1 (a) and (b). By merging switched capacitors with switched inductors, multiphase hybrid converters can achieve higher efficiency and power density over conventional multiphase buck converters [5].

Furthermore, for conventional multiphase buck converters, inductor currents are not spontaneously balanced, which can lead to efficiency degradation, accelerated aging, and even thermal damage [6]. Besides, for some single-phase hybrid converters, flying capacitor voltages are not spontaneously balanced, which can lead to output ripple degradation and extra voltage stress on the devices [7]. However, for multiphase

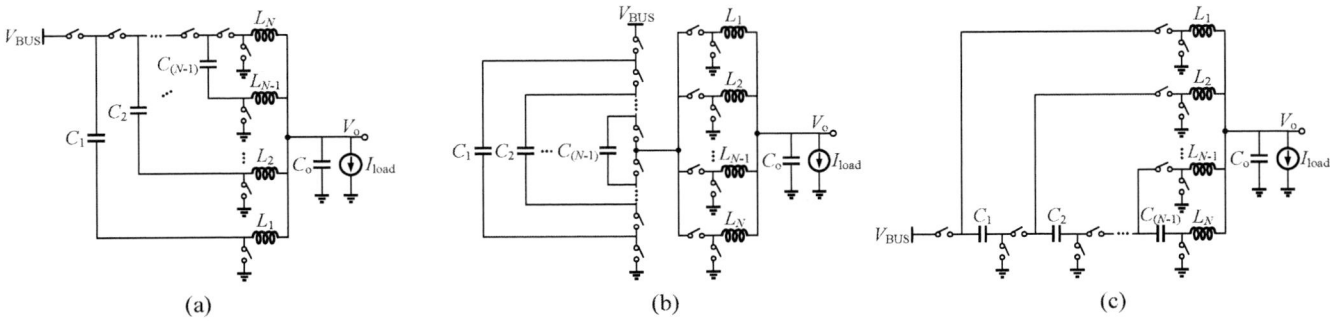

Fig. 1. Topologies of some multiphase hybrid converters with $(N-1)C N L$: (a) MPSCB, (b) MPML, and (c) MPSPCB.

979-8-3315-1612-3/25 $31.00 © 2025 IEEE

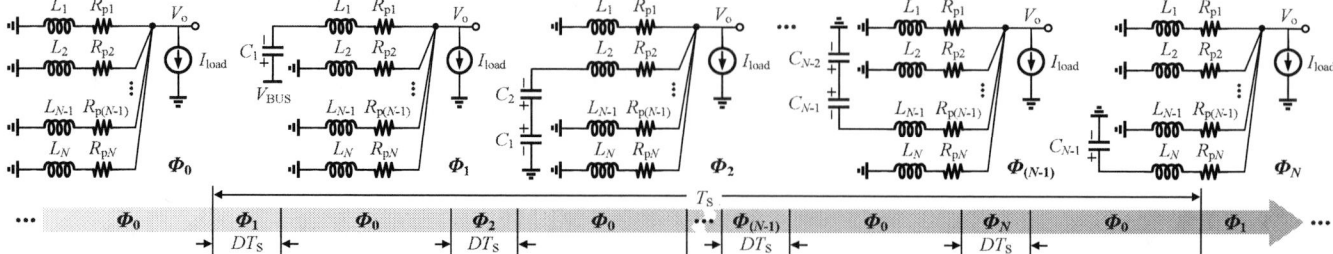

Fig. 2. Equivalent circuits of the multiphase hybrid converter with $(N\text{-}1)$CNL under different switching modes.

hybrid converters with N-1 capacitors and N inductors ($(N\text{-}1)$CNL), all inductor currents and flying capacitor voltages are spontaneously balanced [4, 8]. This delightful feature is known as self-balancing, which ensures high stability as well as simple control in practice. Although this feature has been revealed and validated for MPML [9, 10], the mathematical criteria and dynamic mechanism for general multiphase hybrid converters have not been fully demystified. Especially, an effective model to direct the topological parameter optimization for the self-balancing process is lacking.

Therefore, this work performs a generalized analysis of the self-balancing for the multiphase hybrid converters with $(N\text{-}1)$CNL. Firstly, taking MPSCB and MPML as examples, the mathematical criteria and normalized oscillation equation of the self-balancing are derived from the first-order and second-order differential equations for the $(N\text{-}1)$CNL, respectively. Moreover, an intrinsic feedback model is established in the s-domain, which can guide the optimization of the self-balancing process. Finally, with the proposed analytical framework, the self-balancing characteristics of the multiphase series-parallel capacitor buck (MPSPCB) converter shown in Fig. 1 (c) are predicted. The proposed analysis and prediction are verified on the experimental prototypes.

II. GENERALIZED ANALYSIS FOR SELF-BALANCING OF THE MULTIPHASE HYBRID CONVERTERS WITH (N-1)CNL

The equivalent circuits of the multiphase hybrid converter with $(N\text{-}1)$CNL under different switching modes are displayed in Fig. 2, which are valid for both MPSCB and MPML. There are N+1 switching modes ($\Phi_{0\sim N}$) within the switching period (T_S), where the $\Phi_{1\sim N}$ take the same duty-ratio (D) and shift at the same interval. The N-1 capacitors and N inductors are denoted by $C_{1\sim N\text{-}1}$ and $L_{1\sim N}$, respectively. The $R_{p1\sim N}$ represent the equivalent parasitic resistor of each current phase.

A. Steady-State Analysis for Mathematical Criteria

According to Fig. 2, the first-order differential equations of the flying capacitor voltages ($V_{C1\sim(N\text{-}1)}$) and inductor currents ($I_{L1\sim N}$) can be given by state-space averaging:

$$C_k \frac{dV_{Ck}(t)}{dt} = D\left(I_{Lk}(t) - I_{L(k+1)}(t)\right) \quad (1)$$

$$\begin{cases} L_1 \dfrac{dI_{L1}(t)}{dt} = D\left(V_{BUS} - V_{C1}(t)\right) - R_{p1}I_{L1}(t) - V_o \\[2mm] L_m \dfrac{dI_{Lm}(t)}{dt} = D\left(V_{C(m-1)}(t) - V_{Cm}(t)\right) - R_{pm}I_{Lm}(t) - V_o \quad (2) \\[2mm] L_N \dfrac{dI_{LN}(t)}{dt} = DV_{C(N-1)}(t) - R_{pN}I_{LN}(t) - V_o \end{cases}$$

where $1 \le k \le N$-1 and $2 \le m \le N$-1. According to the charge balance and volt-second balance principles, the differentiations of $V_{C1\sim(N\text{-}1)}$ and $I_{L1\sim N}$ are equal to 0 in steady-state. In addition, the sum of $I_{L1\sim N}$ is equal to the load current (I_{load}). With these considerations, the (1) and (2) in steady-state can be rewritten in matrix form:

$$D\underbrace{\begin{bmatrix} 1 & -1 & 0 & \cdots & 0 \\ 0 & 1 & -1 & \ddots & \vdots \\ \vdots & \ddots & \ddots & \ddots & 0 \\ 0 & \cdots & 0 & 1 & -1 \\ 1 & 1 & 1 & 1 & 1 \end{bmatrix}}_{A_1:\,(N-1)\times N} \times \begin{bmatrix} I_{L1} \\ I_{L2} \\ \vdots \\ I_{L(N-1)} \\ I_{LN} \end{bmatrix} = \underbrace{\begin{bmatrix} 0 \\ 0 \\ \vdots \\ 0 \\ I_{load} \end{bmatrix}^T}_{b_1:\,(N-1)\times 1} \quad (3)$$

$$D\underbrace{\begin{bmatrix} -1 & 0 & \cdots & 0 \\ 1 & -1 & \ddots & \vdots \\ 0 & 1 & \ddots & 0 \\ \vdots & \ddots & \ddots & -1 \\ 0 & \cdots & 0 & 1 \end{bmatrix}}_{A_2:\,N\times(N-1)} \times \begin{bmatrix} V_{C1} \\ V_{C2} \\ \vdots \\ V_{C(N-1)} \end{bmatrix} = \underbrace{\begin{bmatrix} V_o - DV_{BUS} \\ V_o \\ \vdots \\ V_o \\ V_o \end{bmatrix}}_{b_2:\,N\times 1}$$

$$+ \underbrace{\begin{bmatrix} R_{p1} \\ R_{p2} \\ \vdots \\ R_{p(N-1)} \\ R_{pN} \end{bmatrix}}_{b_3:\,N\times 1} \times \begin{bmatrix} I_{L1} \\ I_{L2} \\ \vdots \\ I_{L(N-1)} \\ I_{LN} \end{bmatrix}^T . \quad (4)$$

If the ranks of A_1 and its augmented matrix $[A_1, b_1]$ are equal to N, the (3) has only solution, which means that the $I_{L1\sim N}$ are self-balanced in steady-state:

$$r\{A_1\} = r\{[A_1, b_1]\} = N \quad (5)$$

$$\Rightarrow I_{L1} = I_{L2} = \ldots = I_{LN} = \frac{I_{load}}{N}. \quad (6)$$

As the $I_{L1\sim N}$ are solved, the (4) has only solution if the ranks of A_2 and its augmented matrix $[A_2, b_2]$ are equal to N-1.

$$r\{A_2\} = r\{[A_2, b_2 + b_3]\} = N - 1 \qquad (7)$$

Assuming all parasitic resistors are matched, it can be derived that the $V_{C1\sim(N-1)}$ of MPSCB and MPML are in an arithmetic progression.

$$\Rightarrow \frac{V_{C1}}{N-1} = \frac{V_{C2}}{N-2} = \ldots = V_{C(N-1)} = \frac{V_{BUS}}{N} \qquad (8)$$

Actually, the mismatch of parasitic resistors will lead to balancing errors in flying capacitor voltages. If the above criteria are met, it can be derived as:

$$
\begin{bmatrix}
V_{BUS} - V_{C1} \\
V_{C1} - V_{C2} \\
\ldots \\
V_{C(N-3)} - V_{C(N-2)} \\
V_{C(N-2)} - V_{C(N-1)}
\end{bmatrix}
-
\begin{bmatrix}
V_{C1} - V_{C2} \\
V_{C2} - V_{C3} \\
\ldots \\
V_{C(N-2)} - V_{C(N-1)} \\
V_{C(N-1)}
\end{bmatrix}
=
$$

$$
\frac{I_{load}}{N} \times
\begin{bmatrix}
R_{p1} - R_{p2} \\
R_{p2} - R_{p3} \\
\ldots \\
R_{p(N-2)} - R_{p(N-1)} \\
R_{p(N-1)} - R_{pN}
\end{bmatrix}
\qquad (9)
$$

which are related to the mismatch resistance and load current.

B. Time-Domain Analysis for Normalized Coupled-Damped Oscillation Equation

To demystify the general dynamic mechanism of the self-balancing, a time-domain analysis is conducted on the current and voltage balancing errors. Firstly, the balancing errors of $I_{L1\sim N}$ and $V_{C1\sim(N-1)}$ are defined as:

$$
\begin{cases}
\Delta I_{1i}(t) = I_{L1}(t) - I_{Li}(t) \\
\Delta I_{1N}(t) = I_{L1}(t) - I_{LN}(t) \\
\Delta I_{ij}(t) = I_{Li}(t) - I_{Lj}(t) \\
\Delta I_{iN}(t) = I_{Li}(t) - I_{LN}(t)
\end{cases}
\qquad (10)
$$

$$
\begin{cases}
\Delta V_{1i}(t) = (V_{BUS} - V_{C1}(t)) - (V_{C(i-1)}(t) - V_{Ci}(t)) \\
\Delta V_{1N}(t) = (V_{BUS} - V_{C1}(t)) - V_{C(N-1)}(t) \\
\Delta V_{ij}(t) = (V_{C(i-1)}(t) - V_{Ci}(t)) - (V_{C(j-1)}(t) - V_{Cj}(t)) \\
\Delta V_{iN}(t) = (V_{C(i-1)}(t) - V_{Ci}(t)) - V_{C(N-1)}(t)
\end{cases}
\qquad (11)
$$

where $2 \leq i \leq N-1$, $2 \leq j \leq N-1$ and $i \neq j$. For simplicity, it is assumed that all inductors, flying capacitors, and parasitic resistors in Fig. 2 are matched, whose inductance, capacitance, and resistance are denoted L, C_f, and R_p, respectively. With the assumption, the (1) and (2) can be rewritten as:

$$
\begin{cases}
C_f \dfrac{d\Delta V_{1i}(t)}{dt} = D(\Delta I_{2(i+1)}(t) + \Delta I_{i(i-1)}(t) - \Delta I_{1i}(t)) \\
C_f \dfrac{d\Delta V_{1N}(t)}{dt} = D(\Delta I_{2(N-1)}(t) - \Delta I_{1N}(t)) \\
C_f \dfrac{d\Delta V_{ij}(t)}{dt} = D(\Delta I_{(i-1)(j-1)}(t) + \Delta I_{(i+1)(j+1)}(t) - 2\Delta I_{ij}(t)) \\
C_f \dfrac{d\Delta V_{iN}(t)}{dt} = D(\Delta I_{i(i+1)}(t) + \Delta I_{(i-1)(N-1)}(t) - \Delta I_{iN}(t))
\end{cases}
\qquad (12)
$$

$$
\begin{cases}
L \dfrac{d\Delta I_{1i}(t)}{dt} = D\Delta V_{1i}(t) - R_p \Delta I_{1i}(t) \\
L \dfrac{d\Delta I_{1N}(t)}{dt} = D\Delta V_{1N}(t) - R_p \Delta I_{1N}(t) \\
L \dfrac{d\Delta I_{ij}(t)}{dt} = D\Delta V_{ij}(t) - R_p \Delta I_{ij}(t) \\
L \dfrac{d\Delta I_{iN}(t)}{dt} = D\Delta V_{iN}(t) - R_p \Delta I_{iN}(t)
\end{cases}
\qquad (13)
$$

By differentiating (13) and then substituting the voltage differentiations with (12), the second-order differential equations containing only the current balancing errors can be derived as:

$$
\begin{cases}
\dfrac{d^2 \Delta I_{1i}(t)}{dt^2} + \dfrac{R_p}{L} \cdot \dfrac{d\Delta I_{1j}(t)}{dt} \\
\quad + \dfrac{D^2}{LC_f}(\Delta I_{1i}(t) - \Delta I_{2(i+1)}(t) - \Delta I_{i(i-1)}(t)) = 0 \\
\dfrac{d^2 \Delta I_{1N}(t)}{dt^2} + \dfrac{R_p}{L} \cdot \dfrac{d\Delta I_{1N}(t)}{dt} \\
\quad + \dfrac{D^2}{LC_f}(\Delta I_{1N}(t) - \Delta I_{2(N-1)}(t)) = 0 \\
\dfrac{d^2 \Delta I_{ij}(t)}{dt^2} + \dfrac{R_p}{L} \cdot \dfrac{d\Delta I_{jk}(t)}{dt} \\
\quad + \dfrac{D^2}{LC_f}(2\Delta I_{ij}(t) - \Delta I_{(i-1)(j-1)}(t) - \Delta I_{(i+1)(j+1)}(t)) = 0 \\
\dfrac{d^2 \Delta I_{iN}(t)}{dt^2} + \dfrac{R_p}{L} \cdot \dfrac{d\Delta I_{jN}(t)}{dt} \\
\quad + \dfrac{D^2}{LC_f}(\Delta I_{iN}(t) - \Delta I_{i(i+1)}(t) - \Delta I_{(i-1)(N-1)}(t)) = 0
\end{cases}
\qquad (14)
$$

It is found that the current balancing errors will undergo coupled-damped harmonic oscillation during the self-balancing process. According to (14), the normalized dynamic equation of the coupled-damped harmonic oscillation can be given as:

$$\frac{d^2 x_1(t)}{dt^2} + 2\zeta\omega_o \frac{dx_1(t)}{dt} + \omega_o^2 x_1(t) + \gamma_1 x_2(t) + \gamma_2 x_3(t) = 0 \quad (15)$$

where $x_1(t)$, $x_2(t)$ and $x_3(t)$ can denote arbitrary current balancing errors. The ω_o and ζ denote the angular resonant frequency and damping factor of $x_1(t)$. The γ_1 and γ_2 denote the coupling coefficients of $x_2(t)$ and $x_3(t)$ to $x_1(t)$, respectively. For MPSCB and MPML, the specific parameters of the coupled-damped harmonic oscillation are given in TABLE I.

TABLE I. COUPLED-DAMPED HARMONIC OSCILLATION PARAMETERS FOR MPSCB AND MPML

x_1	x_2	x_3	ζ	ω_o	γ_1	γ_2
ΔI_{1i}	$\Delta I_{2(i+1)}$	$\Delta I_{i(i+1)}$	$\dfrac{R_p}{2D}\sqrt{\dfrac{C_f}{L}}$	$\dfrac{D}{\sqrt{LC_f}}$	$-\dfrac{D^2}{LC_f}$	$-\dfrac{D^2}{LC_f}$
ΔI_{LV}	$\Delta I_{2(N-1)}$	None	$\dfrac{R_p}{2D}\sqrt{\dfrac{C_f}{L}}$	$\dfrac{D}{\sqrt{LC_f}}$	$-\dfrac{D^2}{LC_f}$	0
ΔI_{ij}	$\Delta I_{(i-1)(j-1)}$	$\Delta I_{(i+1)(j+1)}$	$\dfrac{R_p}{2D}\sqrt{\dfrac{C_f}{2L}}$	$\dfrac{\sqrt{2}D}{\sqrt{LC_f}}$	$-\dfrac{D^2}{LC_f}$	$-\dfrac{D^2}{LC_f}$
ΔI_{iN}	$\Delta I_{i(i+1)}$	$\Delta I_{(i-1)(N-1)}$	$\dfrac{R_p}{2D}\sqrt{\dfrac{C_f}{L}}$	$\dfrac{D}{\sqrt{LC_f}}$	$-\dfrac{D^2}{LC_f}$	$-\dfrac{D^2}{LC_f}$

Since the $\zeta>0$, the current balancing errors will converge to zero during the oscillation and the convergence speed depends on the ζ and ω_o. According to (13), the voltage balancing errors can be given by the current balancing errors:

$$
\begin{cases}
\Delta V_{1i}(t) = \dfrac{L}{D}\cdot\dfrac{d\Delta I_{1i}(t)}{dt} + \dfrac{R_p}{D}\cdot\Delta I_{1i}(t) \\[2mm]
\Delta V_{1N}(t) = \dfrac{L}{D}\cdot\dfrac{d\Delta I_{1N}(t)}{dt} + \dfrac{R_p}{D}\cdot\Delta I_{1N}(t) \\[2mm]
\Delta V_{ij}(t) = \dfrac{L}{D}\cdot\dfrac{d\Delta I_{ij}(t)}{dt} + \dfrac{R_p}{D}\cdot\Delta I_{ij}(t) \\[2mm]
\Delta V_{iN}(t) = \dfrac{L}{D}\cdot\dfrac{d\Delta I_{iN}(t)}{dt} + \dfrac{R_p}{D}\cdot\Delta I_{iN}(t)
\end{cases} \tag{16}
$$

Thus, they will also converge to zero as the current balancing errors converge during the oscillation.

C. Frequency-Domain Analysis for Intrinsic Feedback Model

To quantitatively reveal the convergence speed and stability of the self-balancing, an s-domain modeling is also carried out for the multiphase hybrid converters. By performing the Laplace transform on (12) and (13), the s-domain equations of the current and voltage balancing errors are derived as:

$$
\begin{cases}
\Delta V_{1i}(s) = \dfrac{D}{sC_f}\big(\Delta I_{2(i+1)}(s) + \Delta I_{i(i-1)}(s) - \Delta I_{1i}(s)\big) \\[2mm]
\Delta V_{1N}(s) = \dfrac{D}{sC_f}\big(\Delta I_{2(N-1)}(s) - \Delta I_{1N}(s)\big) \\[2mm]
\Delta V_{ij}(s) = \dfrac{D}{sC_f}\big(\Delta I_{(i-1)(j-1)}(s) + \Delta I_{(i+1)(j+1)}(s) - 2\Delta I_{ij}(s)\big) \\[2mm]
\Delta V_{iN}(s) = \dfrac{D}{sC_f}\big(\Delta I_{i(i+1)}(s) + \Delta I_{(i-1)(N-1)}(s) - \Delta I_{iN}(s)\big)
\end{cases} \tag{17}
$$

$$
\begin{cases}
\dfrac{sL}{D}\cdot\Delta I_{1i}(s) + \dfrac{R_p}{D}\cdot\Delta I_{1i}(s) = \Delta V_{1i}(s) \\[2mm]
\dfrac{sL}{D}\cdot\Delta I_{1N}(s) + \dfrac{R_p}{D}\cdot\Delta I_{1N}(s) = \Delta V_{1N}(s) \\[2mm]
\dfrac{sL}{D}\cdot\Delta I_{ij}(s) + \dfrac{R_p}{D}\cdot\Delta I_{ij}(s) = \Delta V_{ij}(s) \\[2mm]
\dfrac{sL}{D}\cdot\Delta I_{iN}(s) + \dfrac{R_p}{D}\cdot\Delta I_{iN}(s) = \Delta V_{iN}(s)
\end{cases} \tag{18}
$$

Combining (17) and (18), the system diagram for the current and voltage balancing errors is displayed in Fig. 3. It reveals that the self-balancing is achieved by intrinsic feedback loops in the multiphase hybrid converter with $(N\text{-}1)C\!N\!L$. These loops are coupled to each other and the open-loop gains of each loop can be given as:

$$
G_{1i}(s) = G_{1N}(s) = G_{iN}(s) = -\frac{D^2}{sC_f(sL+R_p)}
$$

$$
G_{ij}(s) = -\frac{2D^2}{sC_f(sL+R_p)}. \tag{19}
$$

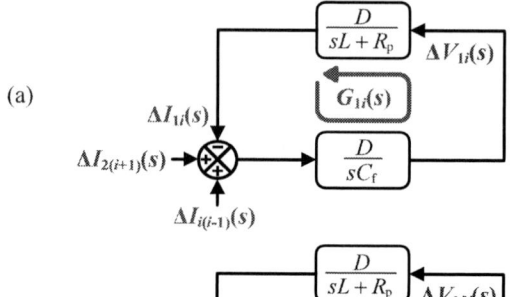

(a)

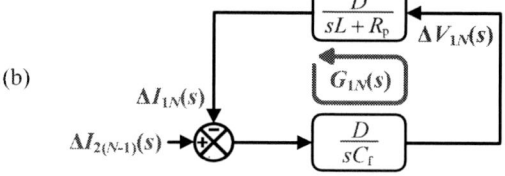

(b)

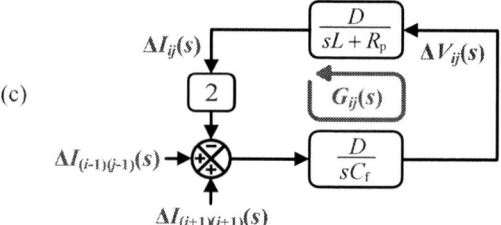

(c)

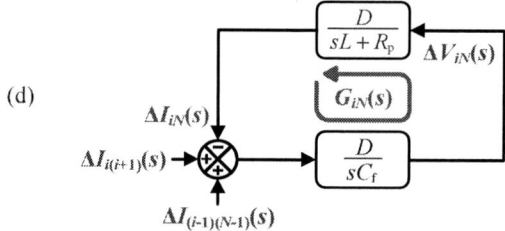

(d)

Fig. 3. System diagram for intrinsic feedback loops in the multiphase hybrid converter with $(N\text{-}1)C\!N\!L$.

Fig. 4 shows the bode plots of $G_{ij}(s)$ under different topological parameters. It suggests that large L will reduce the cross-over frequency (f_c) and phase margin (φ_m) of the loop. In contrast, large R_p and C_f will increase the φ_m. However, large R_p will also induce high I^2R loss. Thus, small L, small R_p, and large C_f are recommended for a fast and stable self-balancing process.

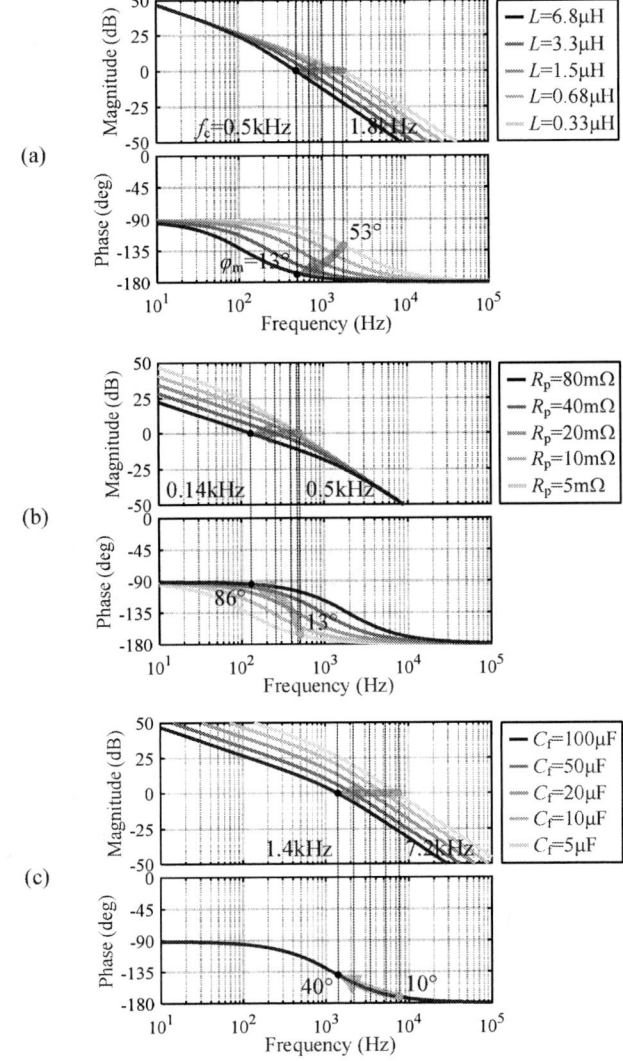

(a)

(b)

(c)

Fig. 4. Bode plots of the intrinsic feedback loops under different topological parameters.

III. PREDICTION FOR SELF-BALANCING OF MPSPCB

To demonstrate the generality of the proposed analytical framework, it is applied in a new multiphase hybrid converter with $(N-1)C/NL$, i.e. MPSPCB. The equivalent circuits of MPSPCB under different switching modes are displayed in Fig.

5. The self-balancing characteristics of MPSPCB are predicted as follows.

Firstly, to judge whether MPSPCB satisfies the mathematical criteria for self-balancing, the first-order differential equations of flying capacitor voltages and inductor currents are derived according to Fig. 5:

$$C_k \frac{dV_{Ck}(t)}{dt} = D\big(I_{L1}(t) - I_{L(k+1)}(t)\big) \tag{20}$$

$$\begin{cases} L_1 \dfrac{dI_{L1}(t)}{dt} = D\left(V_{BUS} - \displaystyle\sum_{k=1}^{N-1} V_{Ck}(t)\right) - R_{p1}I_{L1}(t) - V_o \\[2mm] L_m \dfrac{dI_{Lm}(t)}{dt} = DV_{C(m-1)}(t) - R_{pm}I_{Lm}(t) - V_o \end{cases} \tag{21}$$

where $1 \le k \le N-1$ and $2 \le m \le N$. According to the charge balance and volt-second balance principles, the (20) and (21) in steady-state can be rewritten in matrix form:

$$D \underbrace{\begin{bmatrix} 1 & -1 & 0 & \cdots & 0 \\ 1 & 0 & -1 & \ddots & \vdots \\ \vdots & \ddots & \ddots & \ddots & 0 \\ 1 & \cdots & 0 & 0 & -1 \\ 1 & 1 & 1 & 1 & 1 \end{bmatrix}}_{A_1:\ (N-1)\times N} \times \begin{bmatrix} I_{L1} \\ I_{L2} \\ \vdots \\ I_{L(N-1)} \\ I_{LN} \end{bmatrix} = \underbrace{\begin{bmatrix} 0 \\ 0 \\ \vdots \\ 0 \\ I_{load} \end{bmatrix}^T}_{b_1:\ (N-1)\times 1} \tag{22}$$

$$D \underbrace{\begin{bmatrix} -1 & -1 & \cdots & -1 \\ 1 & 0 & \ddots & \vdots \\ 0 & 1 & \ddots & 0 \\ \vdots & \ddots & \ddots & 0 \\ 0 & \cdots & 0 & 1 \end{bmatrix}}_{A_2:\ N\times(N-1)} \times \begin{bmatrix} V_{C1} \\ V_{C2} \\ \vdots \\ V_{C(N-1)} \end{bmatrix} = \underbrace{\begin{bmatrix} V_o - DV_{BUS} \\ V_o \\ \vdots \\ V_o \\ V_o \end{bmatrix}}_{b_2:\ N\times 1}$$

$$+ \underbrace{\begin{bmatrix} R_{p1} \\ R_{p2} \\ \vdots \\ R_{p(N-1)} \\ R_{pN} \end{bmatrix}}_{b_3:\ N\times 1} \times \begin{bmatrix} I_{L1} \\ I_{L2} \\ \vdots \\ I_{L(N-1)} \\ I_{LN} \end{bmatrix}^T . \tag{23}$$

Since the mathematical criteria in (5) and (7) are satisfied, the $I_{L1\sim N}$ and $V_{C1\sim(N-1)}$ have only solution in steady-state. This

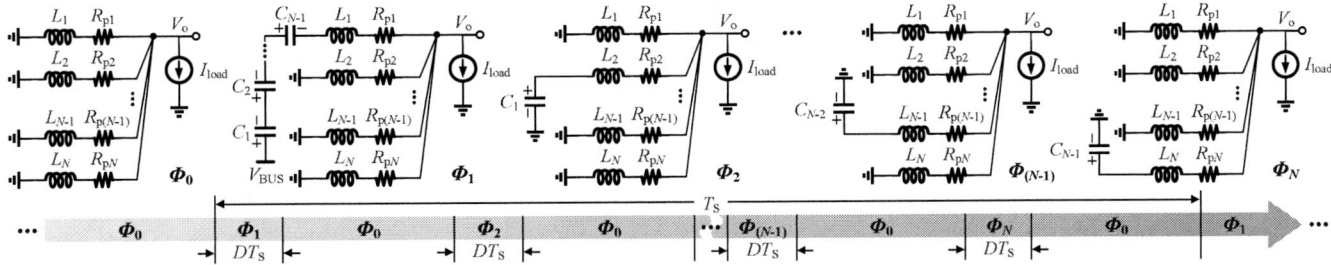

Fig. 5. Equivalent circuits of MPSPCB under different switching modes.

predicts that $I_{L1\sim N}$ and $V_{C1\sim(N-1)}$ are self-balanced in MPSPCB. In addition, assuming all parasitic resistors are matched, it can be derived that the $I_{L1\sim N}$ and $V_{C1\sim(N-1)}$ are all evenly shared:

$$I_{L1} = I_{L2} = ... = I_{LN} = \frac{I_{load}}{N}$$
$$V_{C1} = V_{C2} = ... = V_{C(N-1)} = \frac{V_{BUS}}{N}. \tag{24}$$

Secondly, to describe the self-balancing process of MPSPCB, the balance errors of $I_{L1\sim N}$ and $V_{C1\sim(N-1)}$ are defined as:

$$\begin{cases} \Delta I_{1i}(t) = I_{L1}(t) - I_{Li}(t) \\ \Delta I_{ij}(t) = I_{Li}(t) - I_{Lj}(t) \end{cases} \tag{25}$$

$$\begin{cases} \Delta V_{1i}(t) = \left(V_{BUS} - \sum_{k=1}^{N-1} V_{Ck}(t)\right) - V_{C(i-1)}(t) \\ \Delta V_{ij}(t) = V_{C(i-1)}(t) - V_{C(j-1)}(t) \end{cases} \tag{26}$$

where $2 \leq i \leq N$, $2 \leq j \leq N$ and $i \neq j$. From (20) and (21), the second-order differential equations of the current balancing errors can be derived as:

$$\begin{cases} \dfrac{d^2 \Delta I_{1i}(t)}{dt^2} + \dfrac{R_p}{L} \cdot \dfrac{d\Delta I_{1i}(t)}{dt} + \dfrac{D^2}{LC_f}\left(\Delta I_{1i}(t) + \Delta I_{1N}(t)\right) = 0 \\ \dfrac{d^2 \Delta I_{ij}(t)}{dt^2} + \dfrac{R_p}{L} \cdot \dfrac{d\Delta I_{ij}(t)}{dt} + \dfrac{D^2}{LC_f}\left(\Delta I_{ij}(t) - \Delta I_{(i-1)(j-1)}(t)\right) = 0 \end{cases} \tag{27}$$

It can be seen that the current balancing errors of MPSPCB also comply with the normalized coupled-damped oscillation equation in (15). The specific oscillation parameters for MPSPCB are given in TABLE II. Since the $\zeta > 0$, the current balancing errors will converge to zero during the oscillation.

TABLE II. COUPLED-DAMPED HARMONIC OSCILLATION PARAMETERS FOR MPSPCB

x_1	x_2	x_3	ζ	ω_o	γ_1	γ_2
ΔI_{1i}	ΔI_{1N}	None	$\dfrac{R_p}{2D}\sqrt{\dfrac{C_f}{L}}$	$\dfrac{D}{\sqrt{LC_f}}$	$\dfrac{D^2}{LC_f}$	0
ΔI_{ij}	$\Delta I_{(i-1)(j-1)}$	None	$\dfrac{R_p}{2D}\sqrt{\dfrac{C_f}{L}}$	$\dfrac{D}{\sqrt{LC_f}}$	$-\dfrac{D^2}{LC_f}$	0

By combining (21) and (26), the voltage balancing errors can be given by the current balancing errors. Thus, they will also converge to zero as the current balancing errors converge.

$$\begin{cases} \Delta V_{1i}(t) = \dfrac{L}{D} \cdot \dfrac{d\Delta I_{1i}(t)}{dt} + \dfrac{R_p}{D} \cdot \Delta I_{1i}(t) \\ \Delta V_{ij}(t) = \dfrac{L}{D} \cdot \dfrac{d\Delta I_{ij}(t)}{dt} + \dfrac{R_p}{D} \cdot \Delta I_{ij}(t) \end{cases} \tag{28}$$

Finally, the small-signal model of the intrinsic feedback loops in MPSPCB can be derived by performing the Laplace transform:

$$\begin{cases} \Delta V_{1i}(s) = \dfrac{D}{sC_f}\left(-\Delta I_{1N}(s) - \Delta I_{1i}(s)\right) \\ \Delta V_{ij}(s) = \dfrac{D}{sC_f}\left(\Delta I_{(i-1)(j-1)}(s) - \Delta I_{ij}(s)\right) \end{cases} \tag{29}$$

$$\begin{cases} \dfrac{sL}{D} \cdot \Delta I_{1i}(s) + \dfrac{R_p}{D} \cdot \Delta I_{1i}(s) = \Delta V_{1i}(s) \\ \dfrac{sL}{D} \cdot \Delta I_{ij}(s) + \dfrac{R_p}{D} \cdot \Delta I_{ij}(s) = \Delta V_{ij}(s) \end{cases} \tag{30}$$

The system diagram of (29) and (30) is displayed in Fig. 6, which is a coupled feedback system similar to Fig. 3. The open-loop gains of each loop can be given as:

$$G_{1i}(s) = G_{ij}(s) = -\frac{D^2}{sC_f(sL + R_p)}. \tag{31}$$

Thus, the topological parameter optimization in Section II-C is also suitable for MPSPCB.

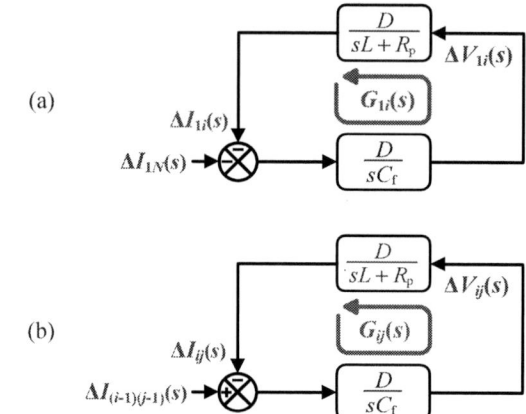

Fig. 6. System diagram for intrinsic feedback loops in MPSPCB.

IV. EXPERIMENT VERIFICATION

To verify the proposed theory and prediction, two different multiphase hybrid converter prototypes with 3C4L are built. Fig. 7 (a) and (b) show the prototypes of MPSCB and MPSPCB, respectively. To specify the equivalent parasitic resistances, the parasitic resistors are dominated by discrete resistors in series with the inductors. Moreover, the inductor currents can be detected from the voltages across the discrete resistors.

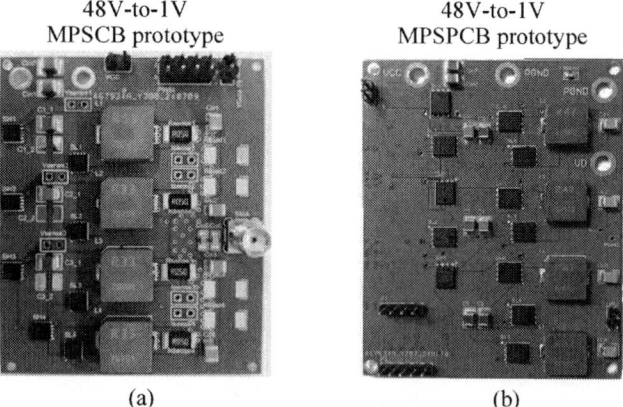

Fig. 7. Experimental prototypes of MPSCB and MPSPCB with 3C4L.

To investigate the effect of different topological parameters on the self-balancing, the test waveforms under L=330nH,

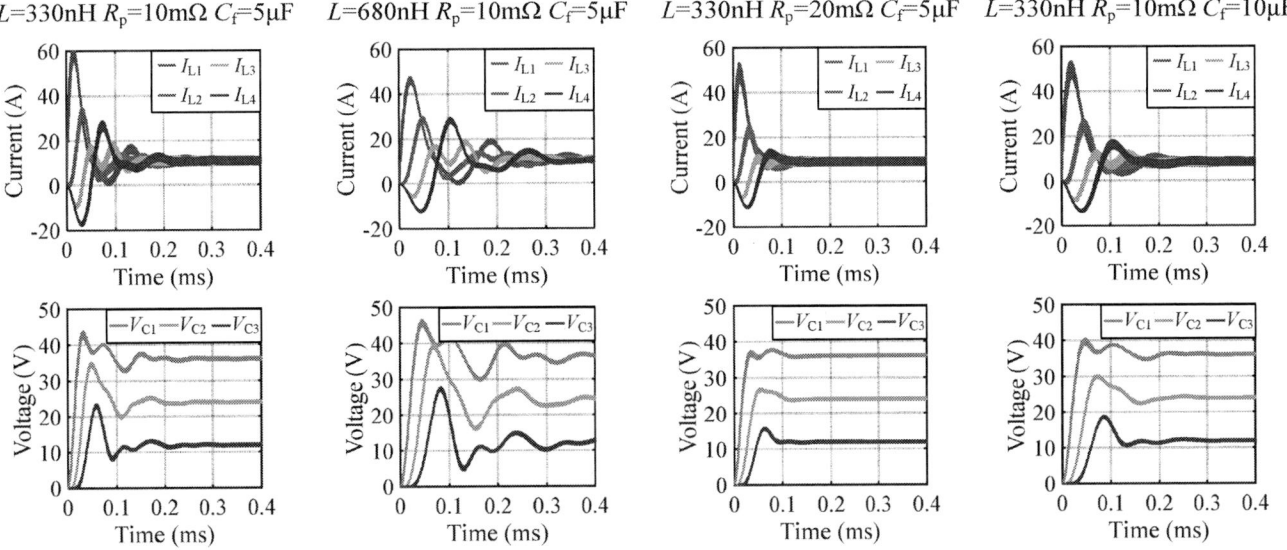

Fig. 8. Inductor current and flying capacitor voltage waveforms of MPSCB under different L, R_p, and C_f.

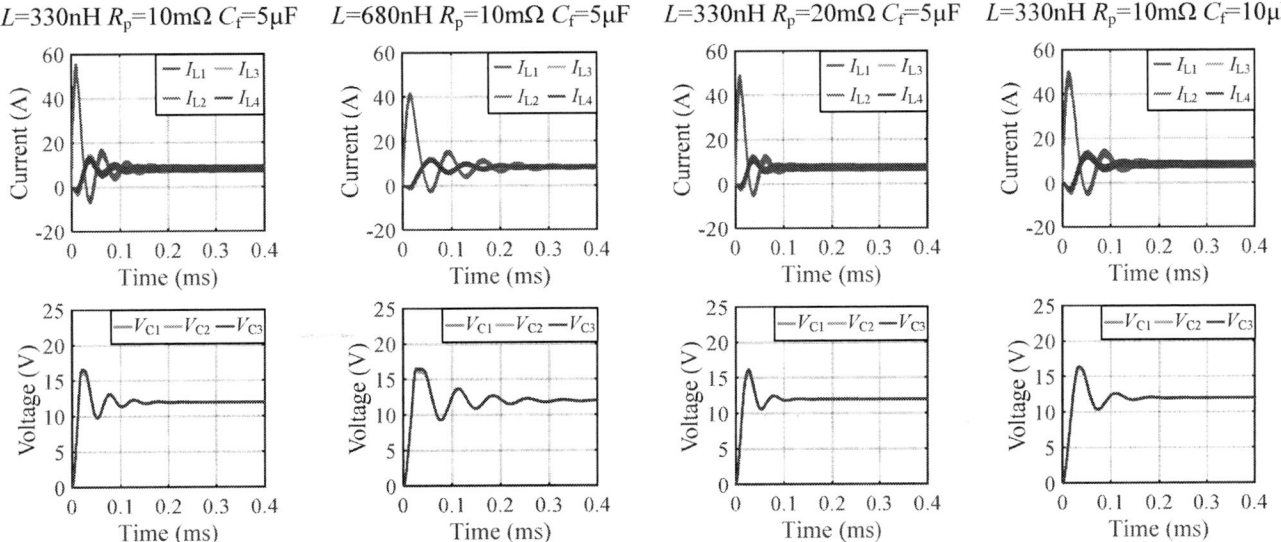

Fig. 9. Inductor current and flying capacitor voltage waveforms of MPSPCB under different L, R_p, and C_f.

R_p=10mΩ, and C_f=5μF are compared with the waveforms under increased L, R_p, and C_f, respectively. Fig. 8 shows the inductor current and flying capacitor voltage waveforms of the MPSCB during hard-starting. It can be seen that the inductor currents spontaneously reach balance in coupled-damped harmonic oscillation. Meanwhile, the flying capacitor voltages converge to their balance values as the inductor currents converge. When the L, R_p, and C_f increase, the oscillation cycle is prolonged. The increased L leads to increased oscillation cycles for self-balancing, whereas the increased R_p and C_f lead to decreased oscillation cycles. This is because the increased L is detrimental to the stability of the intrinsic feedback loops, whereas the increased R_p and C_f are favorable.

Fig. 9 shows the inductor current and flying capacitor voltage waveforms of MPSPCB during hard-starting. It can be seen that the inductor currents and flying capacitor voltages are also self-balanced and their steady-state values follow the prediction. During the self-balancing process, the inductor currents converge in coupled-damped harmonic oscillation and the flying capacitor voltages converge as well. Besides, the topological parameters have the same effect on the self-balancing of MPSPCB as that of MPSCB, which proves the generality of the proposed analytical framework.

V. CONCLUSIONS

This paper develops a generalized analytical framework for the current and voltage self-balancing of the multiphase hybrid converters with (N-1)CNL. The framework not only provides the mathematical criteria and normalized oscillation equation for the self-balancing, but also establishes an intrinsic feedback model for the balancing errors. The intrinsic feedback model indicates that small L, large R_p, and large C_f are favorable for a fast and stable self-balancing process. With the proposed framework, the self-balancing characteristics of MPSPCB are successfully predicted. Finally, the proposed theory and prediction are validated with the MPSCB and MPSPCB prototypes.

ACKNOWLEDGMENT

This work was supported by the Wuhan Youth Science and Technology Plan under Grant 2024040801020198 and by the National Natural Science Foundation of China under Grants 62074067 and 62374067.

REFERENCES

[1] H. Gan *et al.*, "Vertical Power Delivery for 1000 Amps Machine Learning ASICs," presented at the 2024 IEEE Applied Power Electronics Conference and Exposition (APEC), 2024.

[2] P. R. Prakash *et al.*, "A 2400 W/in3 1.8 V Bus Converter Enabling Vertical Power Delivery for Next-Generation Processors," presented at the 2024 IEEE Applied Power Electronics Conference and Exposition (APEC), 2024.

[3] X. Xu and Q. Li, "Analysis of Parasitic Stored Energy Loss and PCB Layout Optimization for 48V-to-1V Series-Capacitor Buck," presented at the 2024 IEEE Applied Power Electronics Conference and Exposition (APEC), 2024.

[4] Y. Zhou *et al.*, "A Direct 24- to 1-V Multilevel Multiphase Hybrid Buck Converter With Automatic VCF Balancing and Current Sharing Mechanism," *IEEE Journal of Emerging and Selected Topics in Power Electronics,* vol. 11, no. 6, pp. 5954-5968, 2023.

[5] P. S. Shenoy, M. Amaro, D. Freeman, and J. Morroni, "Comparison of a 12V, 10A, 3MHz buck converter and a series capacitor buck converter," presented at the 2015 IEEE Applied Power Electronics Conference and Exposition (APEC), Charlotte, NC, USA, 2015.

[6] J.-H. Cho *et al.*, "A Fully Integrated Multi-Phase Buck Converter With On-Chip Capacitor Dynamic Re-Allocation and Fine-Grained Phase-Shedding Techniques," *IEEE Journal of Solid-State Circuits,* vol. 57, no. 12, pp. 3840-3852, 2022.

[7] M. R. Khan, K. Wei, X. Zhang, and C. Huang, "A Single-Inductor 4-Phase Hybrid Switched-Capacitor Topology for Integrated 48V-to-1V DC-DC Converters," presented at the 2023 IEEE International Symposium on Circuits and Systems (ISCAS), 2023.

[8] K. I. Hwu, W. Z. Jiang, and P. Y. Wu, "An Expandable Four-Phase Interleaved High Step-Down Converter With Low Switch Voltage Stress and Automatic Uniform Current Sharing," *IEEE Transactions on Industrial Electronics,* vol. 63, no. 10, pp. 6064-6072, 2016.

[9] D. H. Zhou, A. Bendory, C. Li, and M. Chen, "Multiphase FCML Converter with Coupled Inductors for Ripple Reduction and Intrinsic Flying Capacitor Voltage Balancing," presented at the 2022 IEEE Applied Power Electronics Conference and Exposition (APEC), 2022.

[10] D. H. Zhou, J. Čeliković, D. Maksimović, and M. Chen, "Balancing Multiphase FCML Converters With Coupled Inductors: Modeling, Analysis, Limitations," *IEEE Transactions on Power Electronics,* vol. 39, no. 8, pp. 9268-9291, 2024.

A High-Efficiency Switching Oscillation Suppression Strategy Based on Damped Oscillation for Synchronous DC-DC Converter

Hao Yuan
Dept. of Electrical Engineering
Tongji University
Shanghai, China
2332151@tongji.edu.cn

Chuan Ni
LEN Technology Ltd.
Wuxi, China
jack.ni@len-technology.com

Zhengyu Ye
LEN Technology Ltd.
Shanghai, China
zhengyu.ye@len-technology.com

Wei Lu
LEN Technology Ltd.
Shanghai, China
wei.lu@len-technology.com

Hui Xue
LEN Technology Ltd.
Shanghai, China
hui.xue@len-technology.com

Ting Qian
Dept. of Electrical Engineering
Tongji University
Shanghai, China
tqian@tongji.edu.cn

Abstract—This paper proposes a high-efficiency strategy to suppress the switching-point oscillation by changing the state of the damping response. The detailed analysis is based on a synchronous rectification buck topology. In the new strategy, an additional MOSFET in a small package parallels the upper MOSFET. At the moment of turn-on of the additional MOSFET, an external resistor is introduced to increase the branch impedance and suppress oscillation. Thus, the equivalent dead time can be reduced to nearly zero while the switching safety is still ensured, and the reverse recovery of the body diode can also be reduced. As a result, this strategy allows the selection of MOSFETs with lower breakdown voltage and smaller parasitic parameters and further improves the performance of the converter. After detailed theoretical analysis, experimental results are presented to verify the effectiveness of the proposed strategy.

Keywords—*DC-DC converter, switching oscillation suppression, high efficiency*

I. INTRODUCTION

Buck converters are widely used in portable electronic devices and electric vehicles due to their high efficiency, simplicity, and stability. In applications, an increasing number of single-stage Buck solutions can directly step down 48V to the core voltages required by microprocessors [1]-[3]. However, because of its hard switching mechanism, Buck converters suffer from EMI problems and switching point oscillations. Currently, there are various methods to overcome these problems including optimizing circuit parameters, adding RC snubber, reducing switching speed, and etc [4]-[8]. Meanwhile, the balance between switching oscillation suppression and efficiency is also important. One of the traditional approaches to achieving such balance is by monitoring the drain current and changing the gate current as segmented driving [9].

This paper proposes a new strategy by changing the damping state of oscillations to suppress the Buck switching point voltage.

Based on the synchronous rectification Buck topology, a MOSFET with a small package is connected in parallel with the upper MOSFET. An external resistor is introduced to increase the branch impedance. The equivalent dead time can be reduced to nearly zero while the safety of the circuit is still ensured. The switching losses can be further reduced by minimizing the reverse recovery of the body diode. Therefore, the proposed strategy can achieve a minimum oscillation with high efficiency. This strategy allows the selection of MOSFETs with lower breakdown voltage and larger parasitic parameters and further improves the performance of the converter.

This paper begins with an analysis of the operation principle of the proposed strategy. Then it conducts a detailed theoretical analysis of the damping optimization, device selection, and loss analysis. Furthermore, it compares the effectiveness and efficiency advantages of this strategy with the conventional synchronous rectification Buck converter. Finally, experimental results are presented to verify the proposed strategy.

II. OPERATION PRINCIPLE

The switching-point oscillation phenomenon of the Buck converter usually occurs at the turn-on/off instant of the MOSFET, which is mainly caused by the parasitic inductance and the junction capacitance. Fig. 1 shows the equivalent circuit model of a conventional synchronous rectifier buck when the upper MOSFET is turned on. $R_{dsontra1}$ is the on-resistance of the upper MOSFET. $L_{ESLtra1}$ and $L_{ESLtra2}$ are the equivalent parasitic inductances. L_{LOOP1} is the equivalent PCB alignment inductance and C_{dstra2} is the junction capacitance of the lower MOSFET. Without any damping, these parasitics will keep charging and discharging continuously which externally manifests as an oscillation. Thanks to the resistive component, the amplitude of the oscillation will get smaller and smaller. However, the equivalent damping R is very small and hardly changes the oscillatory state. Therefore, we propose a strategy to introduce another external resistor to adjust the RLC damping oscillation state and suppress oscillations.

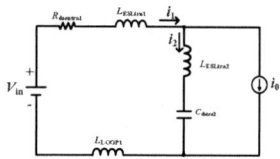

Fig. 1. Conventional synchronous rectification BUCK upper MOSFET turn-on moment equivalent diagram

Introducing an external structure is an important way of suppressing the switching point oscillations of the Buck converter. Fig. 2 shows two common strategies to suppress oscillations: Fig. 2(a) shows a strategy to suppress oscillations by increasing the external structure of the RC snubber, and Fig. 2(b) shows a strategy to suppress oscillations by reducing the driving resistor and thus reducing the turn-on speed of the MOSFET. Both approaches suffer from poor oscillation suppression and low efficiency, so we propose a high-efficiency strategy to suppress oscillations.

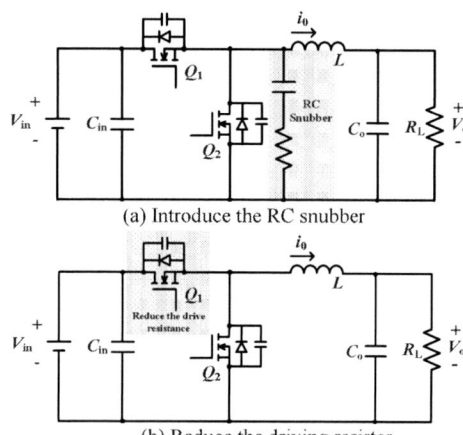

(a) Introduce the RC snubber

(b) Reduce the driving resistor
Fig. 2. Strategies for suppressing oscillations.

In Fig. 3, the Buck converter operates in CCM (Continuous Conduction Mode). Q_1 and Q_2 constitute the upper MOSFET of the synchronous rectification Buck, and Q_3 is the lower MOSFET of the synchronous rectification Buck. Q_1 is a small MOSFET, Q_2, and Q_3 are the large MOSFETs that carry the main power, and R_1 is the introduced resistance. L is the power inductor; C_{in} and C_o, respectively, are the input and output filter capacitance. By adjusting the on-time interval of Q_1 and Q_2, the damping oscillation time can be controlled. Due to the damping suppression, the equivalent dead time of Q_1 and Q_3 can also be greatly reduced to nearly zero, and efficiency can be optimized.

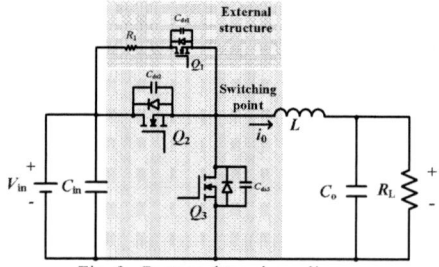

Fig. 3. Proposed topology diagram.

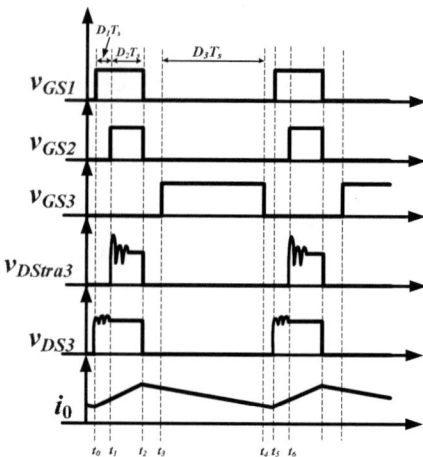

Fig. 4. Typical waveforms of the proposed topology.

Fig. 4 shows the typical waveforms, v_{GS1}, v_{GS2}, and v_{GS3} are the driving waveforms of $Q_1 \sim Q_3$; v_{DStra3} and v_{DS3} are the drain-source pole voltages of the lower MOSFETs of the conventional synchronous rectification Buck converter and the converter proposed in this paper, respectively. D_2, and D_3 are the duty cycles of Q_2 and Q_3. For the time variable, a complete switching cycle is defined as T_s. The period $t_0 \sim t_1$ is the conduction interval between Q_1 and Q_2, defined as $D_1 T_s$. The conduction times of Q_2 and Q_3 are defined as $D_2 T_s$ and $D_3 T_s$. The period $t_4 \sim t_5$ is the equivalent dead time of Q_1 and Q_3. Thus the converter proposed in this paper has a total of five modes, corresponding to the five phases of $t_0 \sim t_5$ in Fig. 4.

Fig. 5 gives a more detailed circuit modal analysis diagram. C_{ds1}, C_{ds2}, and C_{ds3} are the junction capacitance (drain-source pole capacitance) of $Q_1 \sim Q_3$, respectively. The following assumptions are made to simplify the working principle analysis. The discussion in this paper has a great relationship with the dead time, so the dead time can't be ignored. The conventional synchronous rectification Buck switching point oscillation occurs in the upper MOSFET turn-on moment, so the dead time of the upper MOSFET turn-off moment is not within the scope of this article which keeps dead time ($t_2 \sim t_3$) consistent. When $Q_1 \sim Q_3$ are turned on and off, the inductance of the power inductor L is much larger than the parasitic inductance between the branches of $Q_1 \sim Q_3$, and the effect of the parasitic inductance in the line on the power inductor current i_o is ignored, which can be regarded as a constant current source i_o at this moment. R_{dson1} and R_{dson2} are the Drain-Source On-Resistances of Q_1 and Q_2, which are treated as invariants. R_{dson} is so small that at the instant of the conduction interval between Q_1 and Q_2, C_{ds1} and C_{ds2} can be regarded as short circuits. Assuming that the capacitance values of C_o and C_{in} are large enough, they can be regarded as constant voltage sources V_o and V_{in}, respectively. To simplify the theoretical analysis, the PCB alignment inductance, the equivalent inductance of the input filter capacitor, and the bridge decoupling capacitor are ignored when analyzing the model equivalent diagram. In addition, the on-state voltage drops of the body diodes of the MOSFETs are all zero. Thus, the operations of the converter are described as follows.

979-8-3315-1612-3/25 $31.00 © 2025 IEEE   2118

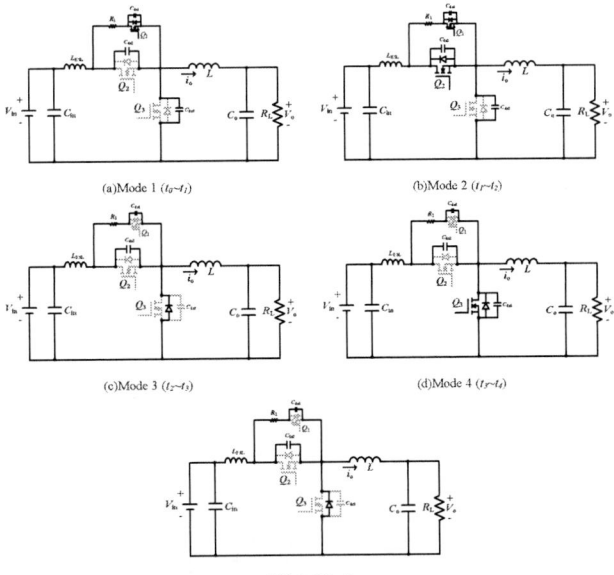

(a)Mode 1 $(t_0 \sim t_1)$ (b)Mode 2 $(t_1 \sim t_2)$

(c)Mode 3 $(t_2 \sim t_3)$ (d)Mode 4 $(t_3 \sim t_4)$

(e)Mode 5 $(t_4 \sim t_5)$

Fig. 5. Equivalent circuit diagram for each mode.

Mode 1 $(t_0 \sim t_1)$: As shown in Fig. 5(a), this mode starts from t_0. When Q_1 is turned on, the on-resistance of Q_1 and R_1 are introduced to suppress the oscillations. The equivalent circuit in this mode is shown in Fig. 6(a). The value of it at this point is:

$$i_o \approx i_1 \qquad \cdot (1)$$

i_1 is the inductor current at the moment of Q3 turn-off in the previous cycle.

After derivation, the equation of state for this mode is:

$$\frac{di_{L_{ESL}}}{dt} = \frac{1}{L_{ESL}}\left(v_{in} - v_{C_{ds2}} - v_{C_{ds3}}\right) \tag{2}$$

$$\frac{dv_{C_{ds2}}}{dt} = \frac{1}{C_{ds2}}\left(i_{L_{ESL}} - \frac{v_{C_{ds2}}}{R_1 + R_{dson1}}\right) \tag{3}$$

$$\frac{dv_{C_{ds3}}}{dt} = \frac{1}{C_{ds3}}\left(i_{L_{ESL}} - i_0\right) \tag{4}$$

Mode 2 $(t_1 \sim t_2)$: As shown in Fig. 5(b), this mode starts from t_1, Q_2 is turned on and the on-resistance of Q_2 is introduced. At this time, the damping oscillation has been completed. During this time the Buck converter carries out the transmission of the main power through Q_2. At this time, i_o rises linearly. The equivalent circuit in this mode is shown in Fig. 6(b). After derivation, the equation of state for this mode is:

$$\frac{di_L}{dt} = \frac{1}{L}\left(v_{C_{ds3}} - v_0\right) \tag{5}$$

$$\frac{di_{L_{ESL}}}{dt} = \frac{1}{L_{ESL}}\left(v_{in} - v_{C_{ds3}} - i_{L1}R'\right) \tag{6}$$

$$R' = \frac{R_1\left(R_{dson1} \cdot R_{dson2}\right)}{R_1 + R_{dson1} + R_{dson2}} \tag{7}$$

$$\frac{dv_{C_{ds3}}}{dt} = \frac{1}{C_{ds3}}\left(i_{L_{ESL}} - i_0\right) \tag{8}$$

Mode 3 $(t_2 \sim t_3)$: As shown in Fig. 5(c), this mode starts from t_2, Q_1, and Q_2 is turned off, into the dead time of the upper and lower MOSFETs. The Buck converter does not carry out the transmission of the main power during this period, i_o is renewed by the diode of the Q_3 body. The voltage at the ends of the power inductor L is:

$$v_L = -v_o \tag{9}$$

At this time, the equivalent circuit in this mode is shown in Fig. 6(c). i_o decreases slightly but remains almost unchanged, with a value of

$$i_o \approx i(t_2) \tag{10}$$

Where is the current value of inductor L at the moment t_2.

Mode 4 $(t_3 \sim t_4)$: As shown in Fig. 5(d), this mode starts from t_3, and Q_3 is turned on. The on-resistance of Q_3 is introduced, and during this time, it is renewed by Q_3, which realizes ZVS. At this time, the voltage at the ends of the power inductor L is:

$$v_L = -v_o - i_o R_{dson3} \tag{11}$$

At this time, i_o decreases. The equivalent circuit in this mode is shown in Fig. 6(d), and the i_o can be derived by combining the Laplace transform with Kirchhoff's law:

$$i_o = \frac{-V_{in}}{R_{dson3}} + \left(i(t_3) + \frac{V_{in}}{R_{dson3}}\right) \cdot e^{-\frac{R_{dson3}}{L}t} \tag{12}$$

We assume that R_{dson} is small and i_o is approximately equal to $i(t_3)$.

Mode 5 $(t_4 \sim t_5)$: As shown in Fig. 5(e), the mode starts from t_4. Q_3 is turned off and enters the equivalent dead time. During this time the converter does not transfer the main power, and it is renewed by the body diode of Q_3. The voltage across the power inductor L is:

$$v_L = -v_o \tag{13}$$

At this point, i_o decreases slightly but remains almost constant. The equivalent circuit in this mode is shown in Fig. 6(e), which is consistent with mode 3. Its value is:

$$i_o \approx i(t_4) \tag{14}$$

Where is the current value of inductor L at the moment t_4.

This mode continues until the end of the dead time and the next switching cycle is turned on. At this point, the operation of a complete switching cycle has been described.

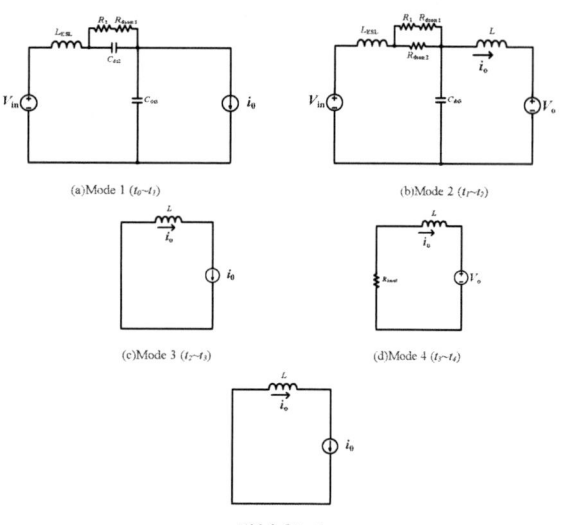

(a)Mode 1 (t_0~t_1)   (b)Mode 2 (t_1~t_2)

(c)Mode 3 (t_2~t_3)   (d)Mode 4 (t_3~t_4)

(e)Mode 5 (t_4~t_5)

Fig. 6. Simplified equivalent circuit diagram.

III. THEORETICAL ANALYSIS

A. Voltage Gain

The input voltage is defined as V_{in}, and the output voltage is defined as V_o. The voltage gain can be calculated as:

$$\frac{V_o}{V_{in}} = D_1 + D_2 \tag{15}$$

B. Optimisation of Damped Oscillatory States

In electrical circuits, *RLC-damped* oscillation is an oscillatory phenomenon involving a resistor (R), an inductor (L), and a capacitor (C). This oscillation usually occurs in an LC circuit in which a resistor is connected in series to provide damping. Due to the physical properties of the circuit components, there will always be some resistance, which will result in a gradual decrease in the energy of the oscillations, a phenomenon known as damping which prevents the oscillations from continuing indefinitely.

The type of oscillation in an *RLC* circuit depends largely on the damping factor (ζ) of the circuit, which is determined by the values of the resistor (R), inductor (L), and capacitor (C) in the circuit. The damping factor is defined as:

$$\zeta = \frac{R}{2}\sqrt{\frac{L}{C}} \tag{16}$$

RLC oscillations can be classified into three types: underdamped: when the damping factor is less than 1 ($\zeta<1$), the circuit will experience underdamped oscillations; critical Damping: when the damping coefficient is equal to 1 ($\zeta=1$), the circuit exhibits critical damping; overdamping: when the damping coefficient is greater than 1 ($\zeta>1$), the circuit exhibits overdamping.

The strategy proposed in this paper is to change the state of LC oscillation through the turn-on of Q_1 and the introduction of an external resistor R_1. As shown in Fig. 7(a), during the conduction interval, Q_1 conducts for damping the oscillations. After stabilizing the switching point, Q_2 is turned on, as shown in Fig. 7(b). At this point, the oscillation amplitude at the switching point is much smaller than that at the conduction of Q_1 and gradually stabilizes.

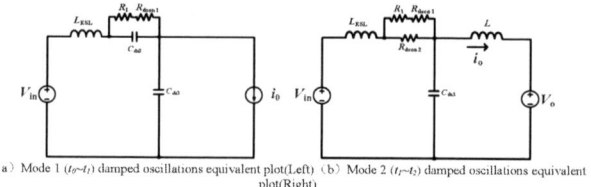

(a) Mode 1 (t_0~t_1) damped oscillations equivalent plot(Left) (b) Mode 2 (t_1~t_2) damped oscillations equivalent plot(Right)

Fig. 7. Equivalent circuit diagram of RLC damped oscillation.

According to the explanation above, the switching point oscillation of the conventional synchronous rectification Buck is generally more serious, and this optimization strategy performs damping oscillation by adding Q_1 and R_1. Ideally, we would like to perform a critical damping or overdamping state during the t_0~t_1, to eliminate the oscillation of the switching point.

C. Influence of Damped Oscillatory States

(1) Effect of conduction interval

The conduction interval (t_5~t_6) and the equivalent dead time (t_4~t_5) have a great influence on the oscillation spikes and the efficiency of the Buck converter. After introducing the external resistor R_1, it takes some time from oscillation to convergence. If the conduction interval is too small, the damping oscillation does not converge, the switching point has not yet regained stability, and the spike suppression effect is very poor; if the conduction interval is too large, after the convergence of the damping oscillation, Q_1 will be subjected to too much main power. The loss of R_1 can have a serious impact on efficiency. Therefore, the conduction interval should be chosen moderately, which will be verified later in the experiments.

(2) Effect of equivalent dead time

The dead time is to prevent the bridge arm from going straight through and causing the converter to short-circuit and thus cause danger. During the dead time, the converter does not transfer energy. Conventional synchronous rectification Buck is very sensitive to high-frequency occasions for dead time. If the dead time is too small, the oscillations may cause the bridge arm to pass through; if the dead time is too large, the amount of energy transferred will be limited, and the efficiency will be greatly reduced. The strategy suppresses the oscillation, thus the equivalent dead time can be reduced to nearly zero while ensuring the switching safety, and the switching losses can be reduced during the reverse recovery of the body diode. This improves the efficiency of the converter and compensates for the conversion of spikes into other forms of losses.

Since the damped oscillations take place during the conduction interval, the equivalent dead time minimally affects the amplitude of the oscillating spikes but can have a significant effect on the efficiency. Therefore, the conduction interval and the equivalent dead time are decoupled from each other and do not affect each other. It is also verified later in the experiments.

(3) The value of R_1

The value range of R_1 affects the damped oscillation state of the circuit. If R_1 is too small, the circuit is nearly undamped, and can not inhibit the switching point oscillation; if R_1 is too large, the branch current is too small, and the gate-source voltage of Q_1 does not reach its threshold voltage (V_{th}). Even if there is a current flow, the MOSFET will not be fully opened. The purpose of introducing an external resistor cannot be achieved. Therefore, the value of R_1 should be transformed according to the specific ringing noise requirements of the system and load changes, and the value should be moderate.

D. Component Selection

(1)Value of inductance L

According to IV, the range of values of L can be obtained by calculating from.

$$L = \frac{V_o}{f_s \cdot (0.2 \sim 0.4) \cdot i_0} \cdot \left(1 - \frac{V_0}{V_{in}}\right) = (7.5 - 15)uH \qquad (17)$$

To minimize the impact of current ripple on efficiency and oscillation, the value of L in this paper is 15uH.

(2) Selection of switching MOSFET

Voltage selection: The converter works in CCM, and the oscillation spike reaches $(1\sim1.5)\ V_{in}$, so to prevent the breakdown of the MOSFET, the voltage level of the selected MOSFET for the conventional circuit is 100V. When taking the strategy proposed in this paper, the voltage level of the selected MOSFET for the proposed circuit is 60V. The 60V withstand voltage of the MOSFETs can reduce the cost to compensate for the increased cost of adding Q_1.

On-resistance selection: According to the latter loss analysis calculations, the on-resistance is a large part of the loss of the entire circuit. Therefore, the on-resistance of Q_2 and Q_3 should be as small as possible, but when the on-resistance becomes small, the corresponding junction capacitance becomes large, which affects the oscillation state and the junction capacitance loss. Therefore, the MOSFET with an on-resistance of 26mΩ is selected for Q_2 and Q_3. In order to suppress the oscillation, the on-resistance of Q_1 should be larger, so the on-resistance of Q_1 is selected 330mΩ, which is in line with the parametric characteristics of MOSFETs in small packages.

E. Comparison of Driving Strategies

TABLE I.   COMPARISON OF DRIVING STRATEGIES

Driving Strategies	Flexible	Simple	Effective
Segmented Driving Strategy	Yes	No	Yes
The Proposed Driving Strategy	Yes	Yes	Yes

Segmented drive is a control strategy that responds to temperature and voltage variations by dividing the drive signal into multiple stages to optimize the performance of the switching device. While segmented drive can improve system efficiency and resilience, it also presents several challenges. Segmented drives require complex control logic to monitor and respond to changes in factors such as current and temperature, which increases system complexity and cost. Segmented drives require a high degree of real-time and processing power in the control system to ensure a fast and stable response. The design and implementation of segmented drives require consideration of various operating conditions, increasing the difficulty of debugging and maintaining the system. The proposed driving strategy is more flexible in that the dead time can be switched arbitrarily, which can be realized by simply adjusting the staggered conduction timing of the MOSFETs without the need for more complicated real-time monitoring.

Both the segmented driving strategy and the driving strategy proposed in this paper are effective and flexible, but the strategy proposed in this paper is simpler and low-cost.

IV. EXPERIMENTAL RESULTS

To verify the effect of the proposed strategy, a prototype is built by using the following specifications. The picture is shown in Fig. 8.

- Input Voltage: 48V
- Output Voltage: 12V
- Maximum Output Current: 5A
- Switching Frequency: 600kHz

TMS320F28377 (DSP) is utilized as the controller, it has a 200MHz mains frequency. Q_2 and Q_3 in large packages select the model number of SI4058DY (100V 10.3A 26mΩ SOP-8), Q_1 in small package selects the model number of SQ2398ES (100V 1.67A 300mΩ SOT-23). The driver chip is TI's single-channel isolation chip UCC5304, with a peak current output of up to 6A. The specific parameters are shown in Table II. For this experiment, we add decoupling capacitors(C_{in2}).

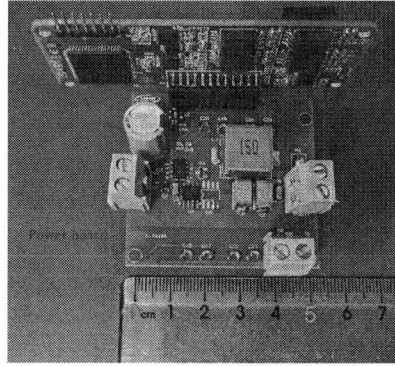

Fig.8. Experimental prototype diagram.

TABLE II.   PARAMETERS OF THE PROTOTYPE

Component	Value
L (μH)	15
C_{in1} (μF)	150
C_{in2} (μF)	1
R_{dson1} (mΩ)	300

R_{dson2} (mΩ)	26
C_{ds1} (pF)	28
$C_{ds2(3)}$ (pF)	280
C_o (μF)	220

When the conduction interval ($t_5 \sim t_6$) is fixed at 12ns, the equivalent dead time ($t_4 \sim t_5$) is varied, and different equivalent dead times have different effects on the efficiency, and the waveforms are shown in Fig. 9.

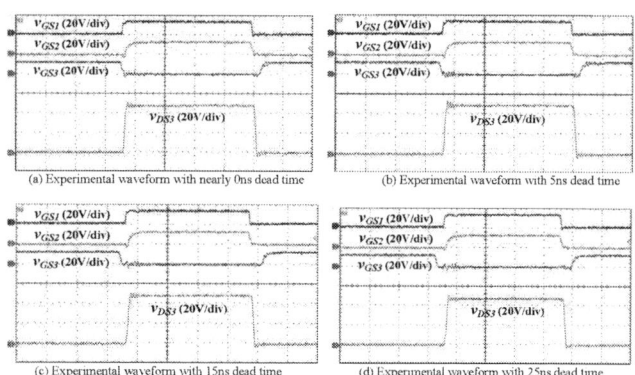

Fig. 9. Experimental waveforms with different dead times, full load (12V/5A)(**100ns/div**).

Fig. 9 shows the experimental waveforms at different equivalent dead times($t_4 \sim t_5$), where Figs. 9(a), 9(b), 9(c), and 9(d) show the equivalent dead time of nearly 0ns, 5ns, 15ns, and 25ns working waveforms respectively, and each waveform shows the gate-source voltages of Q_1, Q_2, and Q_3 (V_{GS1}, V_{GS2}, and V_{GS3}), and the drain-source pole voltage of Q_3 (V_{DS3}). The output voltage can be determined by adjusting the duty cycles D_1, D_2, and D_3.

Fig. 10 shows the experimental waveforms at different conduction intervals ($t_5 \sim t_6$) with the equivalent dead time of nearly 0ns. Fig. 10(a), 10(b), 10(c) and 10(d) show the conduction intervals of 2ns, 7ns, 12ns and 17ns, respectively, and the gate-source voltages of Q_1, Q_2 and Q_3(V_{GS1}, V_{GS2}, and V_{GS3}), and the drain-source pole voltage of Q_3(V_{DS3}), are shown in each waveform, respectively. It is worth noting that when the conduction interval is too small (2ns), the damping oscillation does not converge, so the oscillation is relatively serious, Q_1 does not play a role, and the oscillation also makes the gate-source voltage V_{GS1} affected, leading to the false shutdown, so in line with the theoretical analysis in the previous section, the conduction interval should be adjusted moderately to carry out the compromise between the efficiency and the oscillation suppression effect.

Fig. 11 shows the efficiency and oscillation spike curves with different equivalent dead times (experiment shown in Fig. 9). We find that the equivalent dead time has a greater impact on the efficiency, and has little effect on the suppression of oscillations. Therefore, we reduce the equivalent dead time to nearly 0ns, then carry out the later experiments(experiment shown in Fig. 10) to verify the impact of different conduction intervals on the state of damped oscillations.

Fig. 12 shows the efficiency and oscillation spike curves for circuits with different conduction intervals when the equivalent dead time is nearly 0ns(experiment shown in Fig. 10). We can

find that the time of the conduction interval affects the state of the damped oscillations and thus influences the suppression effect. When the conduction interval is set to 12 ns, we find that this parameter not only performs excellently in terms of the suppression effect but also achieves the ideal balance in terms of efficiency.

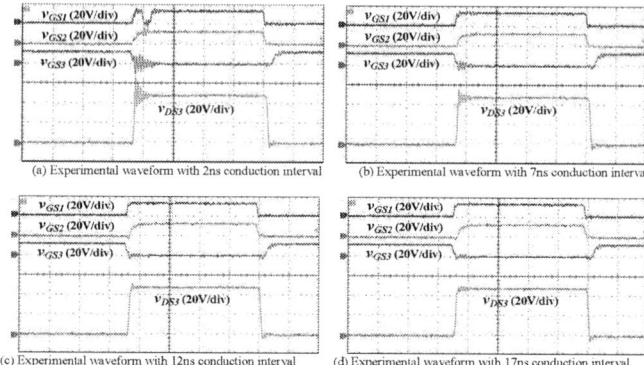

Fig. 10. Experimental waveforms with different conduction intervals at full load (12V/5A) with fixed dead time(**100ns/div**).

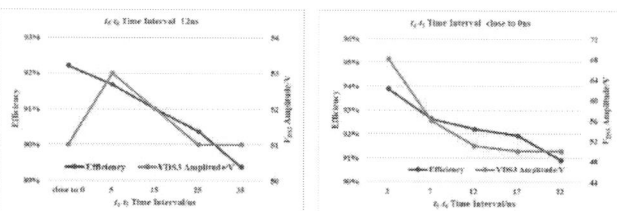

Fig. 11. Efficiency, oscillation spike graph for different dead time, full load (12V/5A)(Left). Fig. 12. Efficiency, oscillation spike graphs at full load (12V/5A) with different conduction intervals at fixed dead time (Right).

Fig. 13 shows the full load (12V/5A) efficiency and oscillation spike curves for different R_1 values with an equivalent dead time of 0ns and a fixed conduction interval of 12ns. Under this experimental condition, when the value of R_1 is 1.2Ω, we find that this parameter is not only excellent in the suppression effect but also in the efficiency to achieve the ideal balance. However, the value of R_1 should be determined according to the specific system oscillation and efficiency requirements.

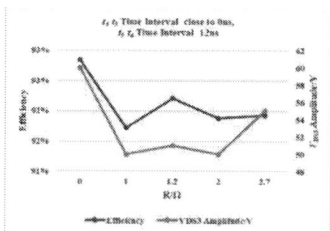

Fig. 13. Efficiency and oscillation spikes curves at full load (12V/5A) with fixed dead time, fixed on-time interval, and different magnitudes of R_1 values.

Fig. 14 shows the experimental waveforms of the conventional synchronous rectification Buck without the introduction of Q_1, where Figs. 14(a), 14(b), 14(c) and 14(d) show the operating waveforms at dead times of 12ns, 17ns, 27ns and 37ns, respectively. The Q_2 and Q_3 are the upper and lower

MOSFETs of the conventional synchronous rectification Buck, respectively. The gate-source pole voltages V_{GS2} and V_{GS3} of Q_2 and Q_3 and the drain-source pole voltage of Q_2, and V_{DS3}, are shown in each waveform respectively.

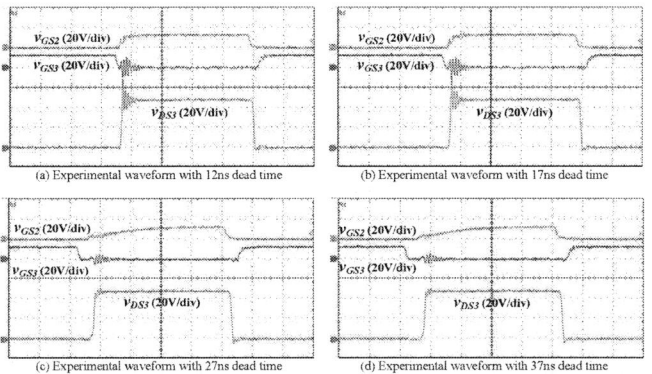

(a) Experimental waveform with 12ns dead time   (b) Experimental waveform with 17ns dead time

(c) Experimental waveform with 27ns dead time   (d) Experimental waveform with 37ns dead time

Fig. 14. Experimental waveforms of conventional synchronous rectification Buck at full load (12V/5A) with different dead times (**100ns/div**).

A common method of suppressing oscillations in a conventional synchronous rectification buck is to suppress spikes by increasing the drive resistance and reducing the speed of switching on. To illustrate the advantages of the strategy proposed in this paper, the efficiency of the conventional suppression strategy and the suppression strategy proposed in this paper are compared by controlling the variables in such a way that the effect of suppressing the oscillatory spikes is the same. Fig. 15(a), Fig. 15(b), Fig. 15(c), and Fig. 15(d) show the experimental waveforms that use the conventional suppression method with the same driving resistance, with the dead time of 12ns, 17ns, 27ns, and 37ns, respectively. The gate-source pole voltages V_{GS2}, V_{GS3} of Q_2 and Q_3, drain-source pole voltage V_{DS3} of Q_3, respectively, are shown in each waveform. (Note: For the accuracy of the experiments, for upper MOSFET, the speed of switching off is controlled by inverting the diode, and the switching speed is decoupled, the speeds of switching off are all the same)

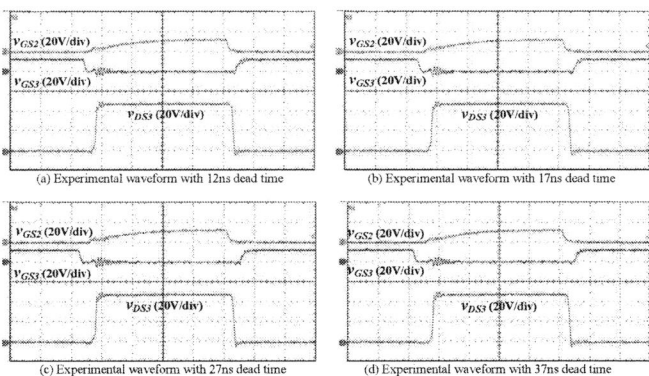

(a) Experimental waveform with 12ns dead time   (b) Experimental waveform with 17ns dead time

(c) Experimental waveform with 27ns dead time   (d) Experimental waveform with 37ns dead time

Fig. 15. Experimental waveforms of conventional suppression strategy at full load (12V/5A) with different dead time(**100ns/div**).

Fig. 16 shows the efficiency curves of the conventional synchronous rectification Buck, the strategy proposed in this paper (nearly 0ns for $t_4 \sim t_5$; 12ns for $t_5 \sim t_6$) and the conventional suppression strategy, the horizontal axis of the figure represents

the dead time of a conventional synchronous rectifier buck converter and the corresponding on-time of the upper MOSFET with conventional suppression strategy. Fig. 17 shows the comparison of the oscillation spikes in different dead times between the strategy proposed in this paper (nearly 0ns for $t_4 \sim t_5$; 12ns for $t_5 \sim t_6$) and the conventional synchronous rectification Buck. It can be found that this strategy does not sacrifice efficiency, in some cases the efficiency is higher than the conventional synchronous rectification Buck. The efficiency of this strategy is much higher than the efficiency of the conventional suppression strategy.

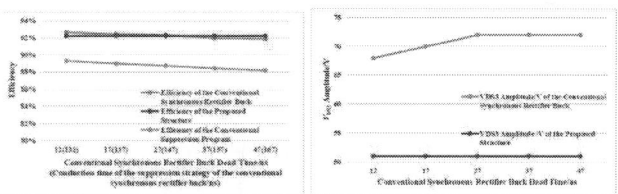

Fig. 16. Comparison of the efficiency of the proposed structure with the conventional synchronous rectification Buck and the conventional suppression strategy(Left). Fig. 17. Comparison of oscillatory spikes of the proposed structure with the conventional synchronous rectification Buck(Right).

V. LOSS ANALYSIS

In the experimental prototype, Q_1, Q_2, and Q_3, which are defined as Q_x (x=1, 2, and 3). For the switching losses, the drain-source pole voltages (v_{on-Qx} and v_{off-Qx}) and currents (i_{on-Qx} and i_{off-Qx}) at the turn-on and turn-off moments can be estimated according to (15). The switching losses [10] can be calculated as:

$$P_{sw-Qx} = P_{on-Qx} + P_{off-Qx} = 0.5v_{on-Qx}i_{on-Qx}t_{on-Qx} + 0.5v_{off-Qx}i_{off-Qx}t_{off-Qx} \tag{18}$$

Where P_{on-Qx} and P_{off-Qx} are the turn-on and turn-off losses, t_{on-Qx} and t_{off-Qx} and are the turn-on and turn-off times, respectively.

In addition, the switching MOSFET drain-source pole capacitance loss is:

$$P_{coss-Qx} = 0.5C_{oss-Qx}v_{on-Qx}^2 fs \tag{19}$$

where C_{oss-Qx} is the drain-source pole capacitance of the MOSFET(x=1,2 and 3).

The equivalent dead time ($t_4 \sim t_5$) is nearly 0ns, the voltage at the ends of the Q_1 branch can be viewed as V_{in}, and the losses of R_{dson1} and R_1 are:

$$P_{R-loss} = \frac{V_{in}^2}{R_1 + R_{dson1}} \tag{20}$$

The conduction loss of Q_2 and Q_3 is:

$$P_{con-Qx} = I_{RMS-Qx}^2 R_{dson-Qx} \tag{21}$$

979-8-3315-1612-3/25 $31.00 © 2025 IEEE   2123

Where $R_{dson-Qx}$ is the on-resistance and I_{RMS-Qx} is the RMS value of the current flowing through the switch(x=2 and 3), which can also be estimated by (5),(12).

The reverse recovery loss of the body diode of Q_3 is[11]:

$$P_{rec\_Q3} = f_s v_{off\_Q3} Q_{b\_Q3} \tag{22}$$

Where v_{off_Q3} is the reverse voltage at the moment of shutdown and Q_{b_Q3} is the value of charge in the reverse recovery current reduction time.

In this experiment, we improve the efficiency by reducing the equivalent dead time (t_4~t_5) and decreasing the loss in the reverse recovery of the body diode, but in the conduction interval (t_5~t_6), R_{dson1} and R_1 increase the losses, so we must compare the above losses. It can be found that the sum of the former losses is less than the latter. The convergence time of damped oscillation can be shortened by increasing the value of R_1 appropriately, and the efficiency loss in the convergence part of damped oscillation can be reduced. Through the optimization in this paper, the efficiency is improved in most of the conditions and it is higher than the conventional synchronous rectifier Buck.

VI. FUTURE WORK

- We can extend this strategy to chip integration to optimize the design of Buck converter analog chips by transforming the upper MOSFET into a dual-MOSFET integration.

- After the verification above, we further envision that the efficiency will be higher than the conventional synchronous rectifier buck over the full range of the new approach and manage to have an equivalent dead time of (-2~-5)ns while increasing the resistance of the introduced external resistor, this will stop the body diode of the lower MOSFET from renewing the current. The diode's reverse recovery loss will be reduced to close to zero on high-frequency occasions. The diode's reverse recovery loss is much smaller than that of the external resistor R_1. Therefore, the efficiency of the proposed strategy will be higher than that of the traditional synchronous rectifier buck in the whole range.

- When the upper MOSFET is turned off, the lower MOSFET utilizes a body diode to achieve zero-voltage switching (ZVS), which short-circuits the parasitic capacitance and reduces switching node oscillations. This portion of the dead time is not discussed in this paper, but future research could explore the suppression of oscillations by adjusting the MOSFET timing and the effect of this adjustment on efficiency.

- This strategy is theoretically more efficient and effective than the generic snubber device, and future work could include this part of the experiment in the comparison.

VII. CONCLUSION

This paper proposes a strategy to change the state of damped oscillations. The following conclusions are drawn from the theoretical analysis and experimental verification:

- The strategy proposed in this paper has advantages in terms of both switching oscillation suppression and efficiency.

- The strategy is effective in high-frequency occasions due to the reduced switching losses.

- The strategy allows the selection of MOSFETs with lower breakdown voltage and parasitic parameters to further improve the performance of the converter.

REFERENCES

[1] X. Xu and Q. Li, "Analysis of Parasitic Stored Energy Loss and PCB Layout Optimization for 48V-to-1V Series-Capacitor Buck," 2024 IEEE Applied Power Electronics Conference and Exposition (APEC), Long Beach, CA, USA, 2024, pp. 898-905.

[2] Y. Zhu, J. Zou and R. C. N. Pilawa-Podgurski, "A 1500-A/48-V-to-1-V Switching Bus Converter for Next-Generation Ultra-High-Power Microprocessors," 2024 IEEE Applied Power Electronics Conference and Exposition (APEC), Long Beach, CA, USA, 2024, pp. 890-897.

[3] G. Roberts and A. Prodić, "Multilevel Series-Capacitor Buck Converter," 2024 IEEE Applied Power Electronics Conference and Exposition (APEC), Long Beach, CA, USA, 2024, pp. 1363-1370.

[4] W. Han, Q. Cheng, C. Chen and H. Lee, "Conductive EMI Reduction Techniques for Soft-switched Half-bridge Buck Converters in Automotive Applications," 2022 IEEE Applied Power Electronics Conference and Exposition (APEC), Houston, TX, USA, 2022, pp. 646-649.

[5] N. Jia, X. Tian, L. Xue, H. Bai, L. M. Tolbert and H. Cui, "In-Package Common-Mode Filter for GaN Power Module with Improved Radiated EMI Performance," 2022 IEEE Applied Power Electronics Conference and Exposition (APEC), Houston, TX, USA, 2022, pp. 974-979.

[6] Y. Ishii et al., "Accurate Conducted EMI Simulation of a Buck Converter With a Compact Model for an SiC-MOSFET," 2020 IEEE Applied Power Electronics Conference and Exposition (APEC), New Orleans, LA, USA, 2020, pp. 2800-2805.

[7] K. Wang, H. Ma, X. Yang, X. Zeng and G. Li, "An optimized layout with split bus capacitors in eGaN-based integrated DC-DC converter module," 2014 International Power Electronics and Application Conference and Exposition, Shanghai, 2014, pp. 446-450.

[8] R. Shirai, S. Hayashi and K. Wada, "Accurate MOSFET Modeling Approach with Equivalent Series Resistance of Output Capacitance for Simulating Turn-OFF Oscillation," 2022 IEEE Applied Power Electronics Conference and Exposition (APEC), Houston, TX, USA, 2022, pp. 1762-1766.

[9] D. Zhang, K. Horii, K. Hata and M. Takamiya, "Digital Gate Driver IC with Real-Time Gate Current Change by Sensing Drain Current to Cope with Operating Condition Variations of SiC MOSFET," 2023 11th International Conference on Power Electronics and ECCE Asia (ICPE 2023 - ECCE Asia), Jeju Island, Korea, Republic of, 2023, pp. 374-380.

[10] D. Jauregui, B. Wang and R. Chen. Power Loss Calculation With Common Source Inductance, Consideration for Synchronous Buck Converters[R]. Texas Instrument, Appl. Report. SLPA009A, June. 2011.

[11] STMicroelectronics. Calculation of turn-off power losses generated by an ultrafast diode[R]. STMicroelectronics, Appl. Note. AN5028, 3 Oct. 2017.

Efficient and Streamlined Demodulation Strategy for High-Frequency Talkative DC-DC Converters

Abdelmoumin Allioua, Hendrik Gockel and Gerd Griepentrog
Institute for Power Electronics and Control of Drives
Technical University of Darmstadt
Darmstadt, Germany
abdelmoumin.allioua@lea.tu-darmstadt.de

Abstract—The integration of data transfer and power conversion in modern power electronics requires efficient encoding and decoding methods to achieve high transmission throughput. This paper introduces an analog Frequency-Shift Keying (FSK) receiver tailored for High-Frequency (HF) GaN-based talkative converters, capable of demodulating data from switching noise pulses at MHz frequencies. The proposed approach achieves high bit-rates while simplifying the circuit and enhancing reliability through the use of analog devices such as operational amplifiers, comparators, and monostables. By translating the switching noise—where the pulse repetition rates represent the transmitted information—into logic levels, the design offers a new and effective method for FSK demodulation in talkative converters. It successfully addresses challenges posed by rapid pulse rates while minimizing the need for costly digital processing components. Unlike prior methods limited to kHz frequencies with lower data rates, this development provides a robust solution for handling FSK demodulation in the MHz range, meeting the demand for high-speed and reliable data transfer in Power and Signal Dual Modulation (PSDM) systems.

Index Terms—PLC, PSDM, FSK Demodulation, DC-DC Convertes, GaN-HEMT, Hard switching, High switching frequency, High bit-rate.

I. INTRODUCTION

ADVANCEMENTS in power electronics increasingly focus on achieving integration, compactness, and miniaturization. One promising solution is "Power and Signal Dual Modulation (PSDM)"—also known as "Talkative Converters"—which enables simultaneous power delivery and data communication within a single converter framework. By leveraging power converters for dual purposes, PSDM enhances system efficiency, reduces costs, and simplifies architecture, making it an ideal choice for applications with stringent constraints on space, weight, and complexity.

While PSDM technology offers numerous advantages, achieving the high bit-rates needed for modern communication protocols remains a challenge. Previous studies [1]–[3] relied on low switching frequencies, limiting data rates to the kbit/s range and restricting PSDM's applicability in high-speed applications. Employing high switching frequencies addresses this limitation by increasing throughput, improving data handling, and enabling the miniaturization of converter components, aligning with the goal of compactness.

However, high switching frequencies pose challenges such as increased switching losses and Electromagnetic Interference (EMI). These can be reduced with wide-bandgap semiconductors such as Silicon Carbide (SiC) and Gallium Nitride (GaN), which decrease switching losses and improve efficiency, while techniques such as soft-switching [4] and hybrid EMI filters [5] effectively manage EMI. Within this High-Frequency (HF) context, an additional and critical challenge arises: effectively demodulating the HF switching noise into binary data without compromising power delivery or signal integrity.

Conventional Frequency-Shift Keying (FSK) receivers, including correlation matched-filters and asynchronous demodulators [2], [4], can adapt to PSDM decoding, requiring precise design for stable communication. Standard carrier waveforms enable reliable transmission, whereas using switching noise as signal carriers in talkative converters introduces complexity due to its HF content and amplitude variability.

Demodulating data signals from talkative converters has relied on analog and digital methods, both with limitations in HF environments. Analog techniques, such as Band-Pass Filters (BPFs) and envelope detection [6], are highly sensitive to noise and struggle to maintain signal integrity at elevated pulse rates. The consistent ringing frequency from the commutation cell further complicates binary state differentiation. Digital methods, though precise, require complex Analog Front-End (AFE) setups and high-speed Analog-to-Digital Converters (ADCs), adding cost and design complexity. A common digital approach, spectral detection, extracts binary information by analyzing the signal's frequency-domain components using the Discrete Fourier Transform (DFT) [7], where peaks at predefined mark and space frequencies are detected within each bit period to decode the binary sequence.

To overcome these challenges, this study presents a HF analog receiver for talkative converters, capable of interpreting pulses at switching frequencies up to 6 MHz—an advancement over prior low-frequency methods. Leveraging monostable multivibrators, recently applied in bearing current monitoring [8], the design enhances PSDM efficiency and functionality. Supporting Mbit/s data rates, it expands compatibility with communication protocols while seamlessly integrating with power conversion. Through rigorous testing and implementation, the proposed receiver offers a reliable, efficient, and cost-effective solution for high-speed data transfer in dual power and data transmission applications.

979-8-3315-1612-3/25 $31.00 © 2025 IEEE

II. BIT ENCODING TOPOLOGY

Binary data is encoded through HF switching signals generated by the talkative converter, leveraging the PSDM framework for seamless data integration.

A. Switching Pulse-Based Data Encoding

PSDM technologies are classified into two main types based on data integration: Single Carrier (SC) and Control Loop (CL). The PSDM-CL approach, shown in Fig. 1(a), incorporates data into the voltage control loop, providing greater modulation flexibility. In contrast, the PSDM-SC approach, illustrated in Fig. 1(b), embeds data within the voltage switching noise during power modulation [4]. Building on prior studies [3], PSDM-SC employs Binary Frequency-Shift Keying (BFSK) on the PWM carrier for data integration, enabling higher bit-rates while facing challenges in modulation flexibility.

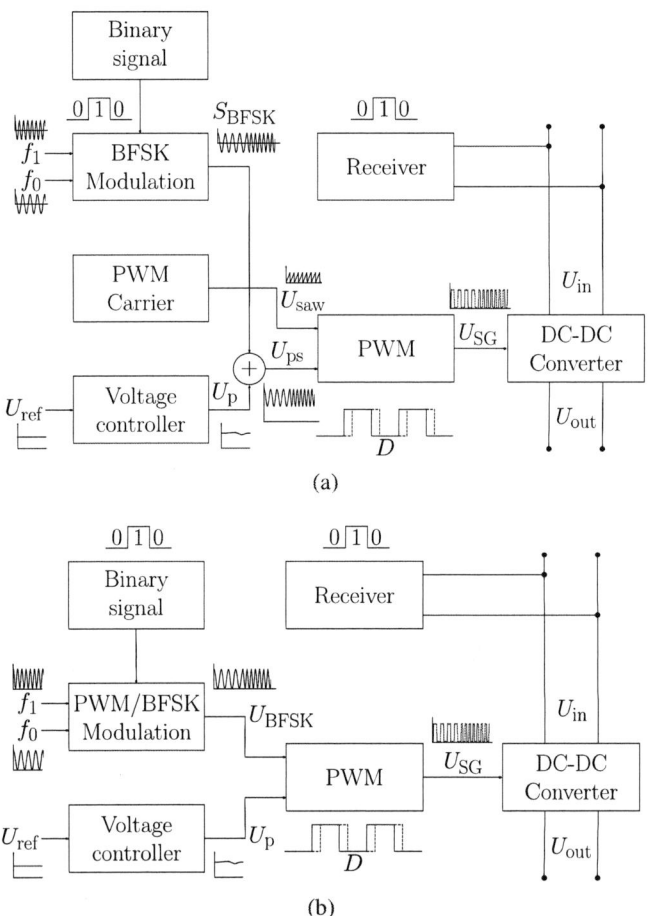

(a)

(b)

Fig. 1: PSDM methodology [4]: (a) PSDM-CL topology, (b) PSDM-SC approach.

PSDM systems can also be categorized by the voltage used for communication: input, output, or both. Powered Devices (PDs) use input DC bus U_{in} to communicate, while Power Sourcing Equipment (PSE) transmits data through the output voltage U_{out} [4].

B. Talkative Converter Overview

This study examines the PSDM-SC in PD-topology, where data is transmitted via inherent noise on the input DC bus, as shown in Fig. 2(a). A 60 W, 48/24 V hard-switching buck converter, depicted in Fig. 2(b), is utilized for its simplicity and cost-effectiveness, despite potential EMI challenges. Equipped with GaN-HEMTs, the converter supports HF operation and enables rapid switching events.

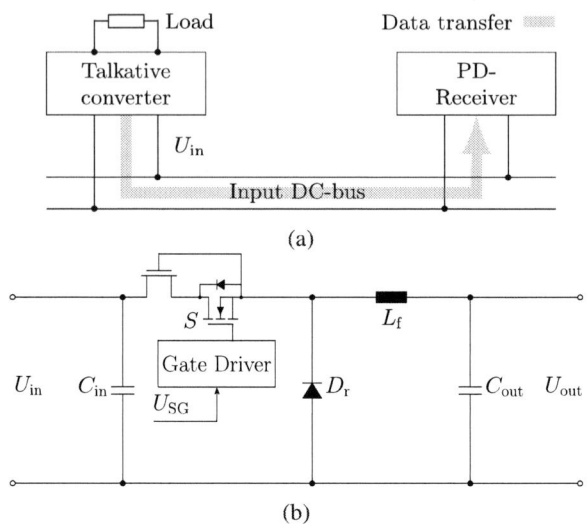

(a)

(b)

Fig. 2: PSDM system architecture: (a) PD transmission topology [4], (b) talkative converter circuit.

In this framework, BFSK-PWM encodes data within distinct switching pulses at predefined frequencies $f_0 = 4$ MHz and $f_1 = 6$ MHz, representing binary '0' and '1'. As illustrated in Fig. 3, the switching noise in U_{in} alternates between these frequencies, enabling simultaneous power transfer and data encoding for efficient demodulation at the receiver.

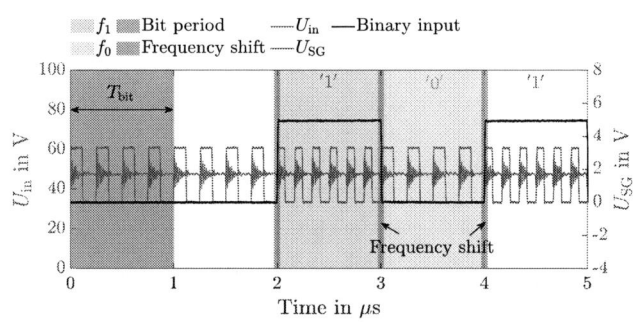

Fig. 3: Encoded data in PD-mode with PSDM converter.

III. RECEIVER METHODOLOGY

This methodology leverages switching noise peaks as binary signal carriers—representing 1 and 0—eliminating the need for additional signal conditioning. By interpreting pulse repetition rates as binary logic levels, it minimizes latency, enhances reliability in HF environments, and simplifies integration with microcontrollers for improved performance.

A. Receiver Operation Overview

The receiver operation, illustrated in Fig. 4, begins with the reception of encoded data as noise peaks on U_{in}. A capacitive decoupler isolates the AC content, preserving data integrity in U_{decin}, as shown in Fig. 5. If the signal level is insufficient, it is amplified and fed into a comparator, which generates trigger pulses U_{trigg}.

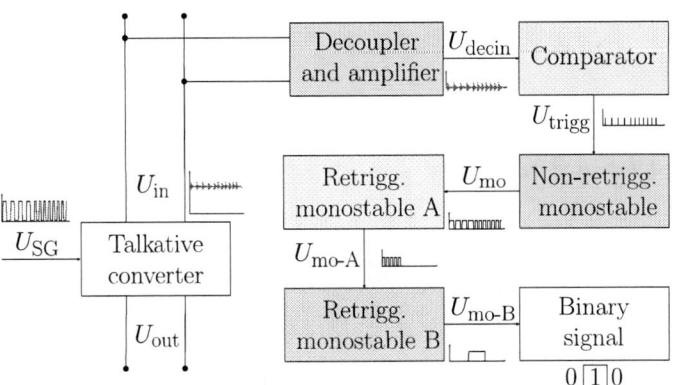

Fig. 4: Block diagram of the full decoding stage.

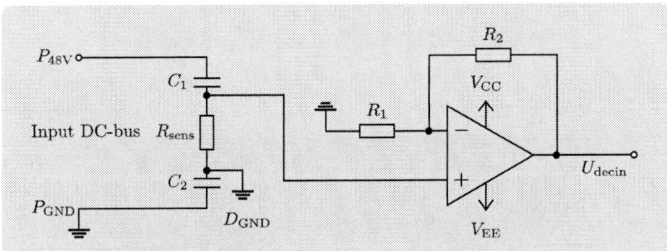

Fig. 5: Illustration of the decoupling and amplification stage.

These triggering pulses U_{trigg}, drive a non-retriggerable monostable circuit, producing U_{mo}, a square waveform representing BFSK-based PWM. The waveform U_{mo} is further processed by an inverting retriggerable monostable, labeled as 'monostable-A', which discriminates between the mark and space frequencies, generating U_{mo-A}. Lastly, a second retriggerable monostable, labeled as 'monostable-B', reconstructs the binary output signal U_{mo-B}, completing the demodulation process with reliable differentiation between binary '0' and '1'. As depicted in Fig. 6, the connection and functionality of the comparator and monostable circuits are clearly illustrated, showcasing the detailed integration of these key components in the receiver's design.

B. Managing Timing and Signal Integrity

Accurate timing within each stage of the decoding circuit is essential for the reliable operation of the proposed FSK demodulation method. These timings, illustrated in Fig. 7, must align with the carrier frequencies f_0 and f_1 and their respective switching periods, where $T_1 < T_0$, as commonly chosen in practice, the mark frequency f_1 is higher than the space frequency f_0.

The decoding process begins with the comparator detecting switching noise peaks and feeding its output to the first non-retriggerable monostable, referred to as 'mo.' This circuit maintains its output, y_{mo}, high for a duration $T_{h,mo}$, which is carefully chosen to be long enough to suppress multiple consecutive comparator pulses caused by switching noise ringing, setting its upper limit, while remaining shorter than T_1, ensuring reliable detection of all switching-on events.

Alternatively, a retriggerable monostable could also be employed, as long as its hold time adheres to the same constraints. However, a non-retriggerable monostable is generally safer, as it inherently ignores multiple comparator pulses, reducing the risk of timing errors.

Next, the retriggerable and inverting monostable 'mo-A' generates y_{mo-A}, which remains low during its hold time $T_{h,mo-A}$, triggered by a rising-edge at its input. This hold time ensures y_{mo-A} stays low during consecutive f_1 pulses, requiring $T_{h,mo-A} \geq T_1$, and recovers to high during f_0 pulses, necessitating $T_{h,mo-A} < T_0$. For f_0, the output alternates between low for $T_{h,mo-A}$ and high for T_{mo-H} as shown in Eq. (1).

$$T_{mo-H} = T_0 - T_{h,mo-A} \qquad (1)$$

The final stage, 'mo-B,' reconstructs the binary bit-stream. Its retriggerable and inverting function ensures y_{mo-B} remains high when the input is zero, corresponding to f_1. During f_0, the input exhibits the pulse pattern described previously, where the first detected pulse transitions y_{mo-B} to low for $T_{h,mo-B}$. For proper functionality, $T_{h,mo-B}$ must be long enough to be retriggered by subsequent pulses, requiring $T_{h,mo-B} \geq T_0$. Simultaneously, $T_{h,mo-B}$ must satisfy $T_{h,mo-B} < T_{bit}$, allowing the output to recover to the high-state within the bit period.

To analyze the latency, both transitions—from logical one to logical zero and vice versa—must be considered. For the transition from logical one to zero, the output of 'mo-B' changes when the rising-edge of the output of 'mo-A' is detected. This occurs during the hold time $T_{h,mo-A}$ after the first pulse of f_0. Additionally, the propagation delays, T_{pd}, of all participating monostables must be accounted for, resulting in the falling-edge latency described by Eq. (2).

$$T_{1 \to 0} = T_{h,mo-A} + T_{pd,mo} \\ + T_{pd,mo-A} + T_{pd,mo-B} \qquad (2)$$

Similarly, the output of 'mo-B' transitions from logical zero to one after $T_{h,mo-B} + T_{pd,mo-B}$ has elapsed since the last rising-edge of the output of 'mo-A'. Assuming no phase discontinuity occur during frequency switching, as discussed in [4], this last rising-edge of 'mo-A' should occur T_{mo-H} before the falling-edge of 'mo-A', and this takes place $T_{pd,mo} + T_{pd,mo-A}$ after the rising-edge of the input signal, resulting in the rising-edge latency expressed in Eq. (3).

$$T_{0 \to 1} = T_{h,mo-A} + T_{h,mo-B} - T_0 \\ + T_{pd,mo} + T_{pd,mo-A} + T_{pd,mo-B} \qquad (3)$$

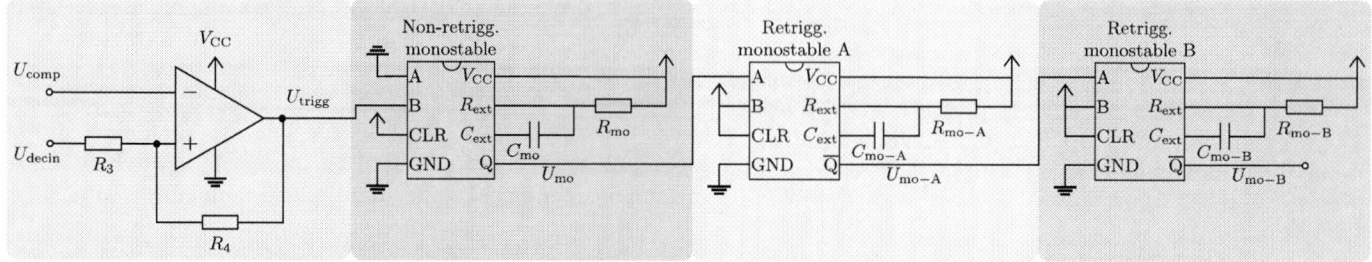

Fig. 6: Illustration of the comparator and the monostables stage.

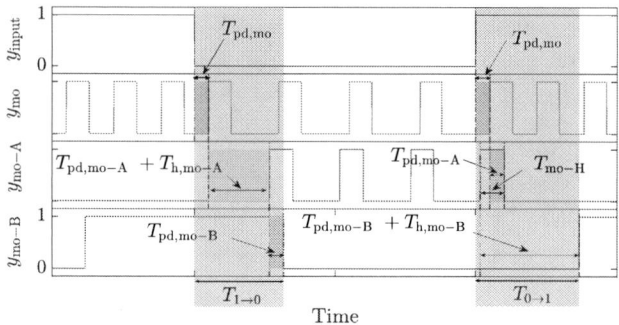

Fig. 7: Timing analysis of monostable stages in the decoding process, illustrating the case when $T_{1\to 0} \neq T_{0\to 1}$.

analyzed in Section IV-A for an in-depth analysis of the demodulation process.

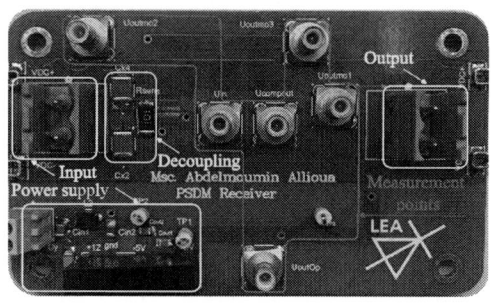

(a)

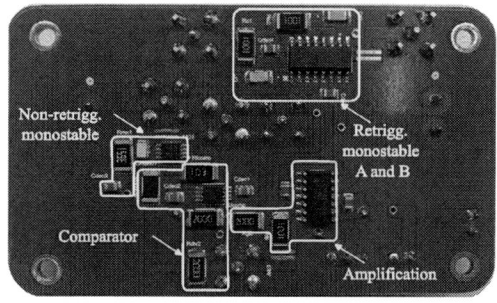

(b)

Fig. 8: Prototype setup of the PSDM receiver for performance validation: (a) PCB top and (b) bottom layer with receiver components.

For data integrity, both latencies must be equal ($T_{1\to 0} = T_{0\to 1}$). Any mismatch would result in differing durations for binary '0' and '1' bits, which becomes increasingly critical at higher bit-rates. For instance, a mismatch of around 10 ns in a 1000 ns bit period (1 Mbit/s) may be acceptable and manageable during the sampling process. However, the same 10 ns mismatch in a 100 ns bit period (10 Mbit/s) could significantly compromise data integrity. This requirement leads to the condition shown in Eq. (4).

$$T_{\text{h,mo}-\text{B}} = T_0 \qquad (4)$$

According to the previously mentioned lower limit of the hold time $T_{\text{h,mo}-\text{B}} \geq T_0$, this choice is feasible and satisfies the condition in Eq. (4). However, it poses a challenge, as the hold time of 'mo-B' must be finely tuned using $R_{\text{mo}-\text{B}}$ and $C_{\text{mo}-\text{B}}$ to precisely meet this requirement. Even a small deviation can result in a slightly longer latency for the transition from zero to one.

IV. PERFORMANCE ANALYSIS

A prototype of the proposed receiver for PSDM converters was constructed, as shown in Fig. 8. The prototype uses a dual monostable Integrated-Circuit (IC) for implementing both 'monostable-A and -B', ensuring compact and efficient circuitry. Performance evaluation involved taking measurements at each stage of the receiver, as depicted in Fig. 9, to assess its capability to accurately demodulate switching pulses into binary data. The waveforms captured in Fig. 9 are further

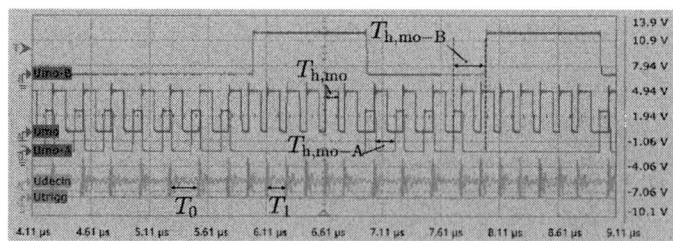

Fig. 9: Oscilloscope captures of receiver signal stages.

A. Evaluation of Demodulation

The receiver's performance was evaluated using a PSDM converter operating between 4 and 6 MHz. The AC content of U_{in} was measured and fed into the receiver through a signal generator. This approach enabled precise control over the signal amplitude to replicate various attenuation scenarios and determine the minimum detectable switching noise peaks, ensuring repeatability of the test conditions. The results, recorded at each stage of the receiver, provide a step-by-step view of the demodulation process.

Initially, as shown in Fig. 10, the receiver input U_{decin} after decoupling and U_{trigg} following comparison are presented. This illustrates how binary inputs, transformed into noise peaks at 4 and 6 MHz, are effectively captured and processed, demonstrating the receiver's capability to isolate and interpret frequency-based binary information.

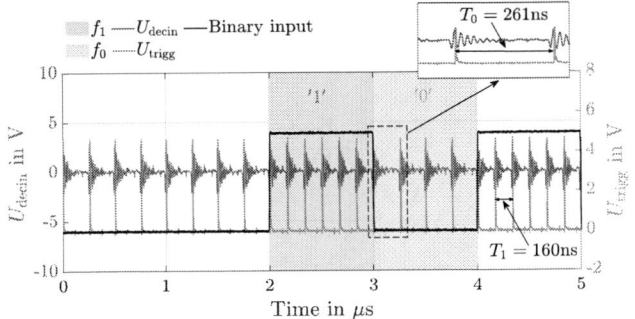

Fig. 10: Initial signal conditioning and trigger pulse generation.

In the subsequent stage, Fig. 11(a) highlights U_{mo}, the output from the first non-retriggerable monostable driven by U_{trigg}. Here, the initial trigger pulses are converted into a stable square waveform $U_{\text{FSK-PWM}}$ with $T_{\text{h,mo}} = 110$ ns, effectively preserving the binary input frequencies and marking a successful intermediate step in the demodulation process. Fig. 11(b) shows $U_{\text{mo-A}}$, the output from the retriggerable monostable-A with its hold time of $T_{\text{h,mo-A}} = 171$ ns. This stage is responsible for distinguishing between mark and space frequencies, selectively allowing only the space frequency to pass through, which helps isolate the intended binary states more accurately.

Finally, Fig. 12 presents $U_{\text{mo-B}}$, the final output from the retriggerable monostable-B. At this stage, the original binary data is fully reconstructed from the switching noise peaks, providing a clear distinction between the binary '0' and '1' states. The precise response of $U_{\text{mo-B}}$ validates the effectiveness of the decoding approach, as it successfully differentiates between the two operating frequencies, corresponding to the intended binary values. This final output confirms the receiver's capability to accurately demodulate HF switching pulses and reconstruct binary data, even in challenging HF environments, showcasing the robustness and precision of the proposed demodulation method.

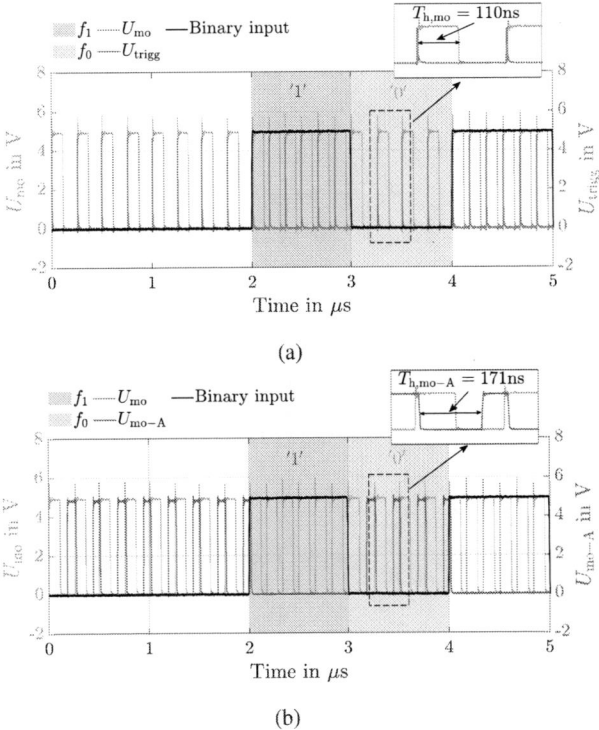

Fig. 11: Step-by-step demodulation stages: (a) Generation of stable square waveform U_{mo}, and (b) frequency discrimination using monostable-A to selectively pass space frequency through $U_{\text{mo-A}}$.

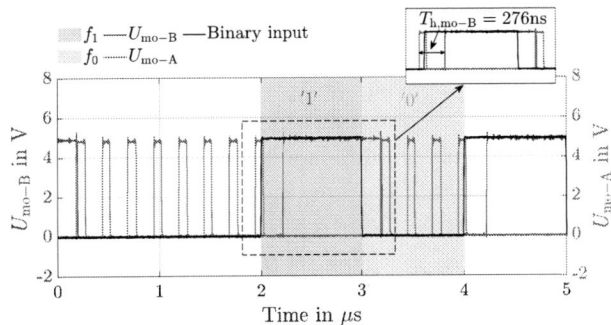

Fig. 12: Illustration of the recovered binary output.

B. Transmission Latency and Bit-Rate Testing

The receiver demonstrated precise binary data recovery and strong signal fidelity. A bit-stream of 92 bits was successfully received, confirming the reliability of the system. As shown in Fig. 13, the monostables introduce a transmission delay, or latency ($T_{1\rightarrow0} \approx T_{0\rightarrow1}$), with a maximum value of 220 ns. Despite this delay, the system consistently supports 1 Mbit/s bit rate with a bit period T_{bit} of 1 μs. While the latency has minimal impact on bit rate and reliability, it may vary with the distance between the transmitter and receiver. This highlights the importance of optimizing the communication channel to ensure robust performance in high-speed PSDM applications.

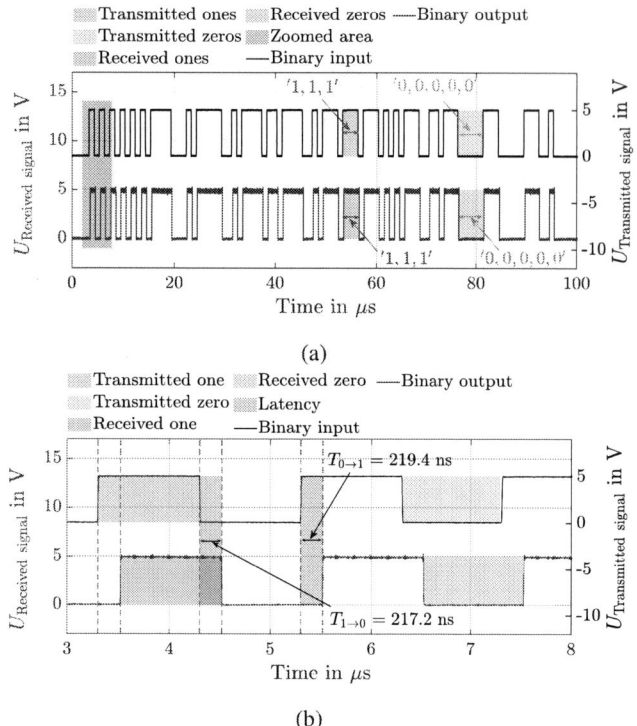

(a)

(b)

Fig. 13: Transmission latency analysis: (a) Overview of transmitted and receiver signals, (b) zoomed-in view.

C. Bit-Error Analysis

Accurate demodulation in the receiver depends on minimizing errors during signal processing. Errors can primarily arise from missed pulses at the comparator stage or additional pulses due to timing mismatches in monostable-B, which affect the correct reconstruction of binary data.

a) Missed Pulses: A missing pulse in U_{trigg} can result in errors depending on the frequency interval. During the mark frequency (f_1), a missed pulse may prevent $U_{\text{mo-A}}$ from staying low, causing an erroneous pulse to trigger monostable-B and resulting in a binary '1' error at $U_{\text{mo-B}}$. Conversely, during the space frequency (f_0), a missed pulse can create an additional pulse in $U_{\text{mo-A}}$, leading to a binary '0' error at $U_{\text{mo-B}}$.

b) Timing Mismatch: Errors may also occur if the hold time of monostable-B $T_{\text{h,mo-B}}$ is not precisely set:

- If shorter than T_0, short pulses may appear during the space frequency, causing errors in the binary '0' intervals.
- If longer than T_0, the bit period of binary '0' becomes inconsistent with binary '1,' leading to decoding inaccuracies.

By addressing these potential error sources through optimized comparator performance and precise timing configuration for monostable-B, the receiver can reliably decode binary data with minimal Bit-Error Rate (BER).

V. CONCLUSION AND FUTURE WORK

This study presented and validated a high-frequency receiver for PSDM converters, demonstrating its ability to reliably demodulate switching pulses into binary data while maintaining signal integrity. The design represents a significant advancement in integrating communication and power within compact power electronic systems, addressing the increasing demand for high-speed, dual-purpose applications.

The receiver achieved a transmission bit rate of 1 Mbit/s a measured maximum latency of 220 ns between the transmitted and received bit-streams. These results confirm the receiver's practicality and highlight the broader potential of PSDM technology in modern power electronics.

Future work will focus on enhancing system performance, investigating the effects of transmission distance on latency and BER, and exploring strategies to mitigate missing pulses and timing mismatches. Additionally, efforts will aim to broaden the system's functionality and scalability to support a wider range of applications.

REFERENCES

[1] I. Mandourarakis, E. Koutroulis and G. N. Karystinos, "Power Line Communication Method for the Simultaneous Transmission of Power and Digital Data by Cascaded H-Bridge Converters," In: *IEEE Transactions on Power Electronics*, vol. 37, no. 10, pp. 12793-12804, Oct. 2022.

[2] N. Bertoni, S. Bocchi, M. Mangia, F. Pareschi, R. Rovatti and G. Setti, "Ripple-based power-line communication in switching DC-DC converters exploiting switching frequency modulation," In: *2015 IEEE International Symposium on Circuits and Systems (ISCAS)*, Lisbon, Portugal, 2015, pp. 209-212.

[3] J. Wu, J. Du, Z. Lin, Y. Hu, C. Zhao and X. He, "Power Conversion and Signal Transmission Integration Method Based on Dual Modulation of DC–DC Converters," In: *IEEE Transactions on Industrial Electronics*, vol. 62, no. 2, pp. 1291-1300, Feb. 2015.

[4] A. Allioua and G. Griepentrog, "Power and Signal Dual Modulation with QR-ZVS DC/DC Converters using GaN-HEMTs," In: *2024 IEEE Applied Power Electronics Conference and Exposition (APEC)*, Long Beach, CA, USA, 2024, pp. 2164-2171.

[5] A. Allioua, D. Großmann, and G. Griepentrog, "Enhanced EMI Mitigation in High-Frequency DC-DC Converters through Hybrid Filtering Approaches," In: *2024 IEEE Southern Power Electronics Conference (SPEC)*, Brisbane, Australia, 2024, pp. 1-6.

[6] T. Kohama, S. Hasebe and S. Tsuji, "Simple Bidirectional Power Line Communication with Switching Converters in DC Power Distribution Network," In: *2019 IEEE International Conference on Industrial Technology (ICIT)*, Melbourne, VIC, Australia, 2019, pp. 539-543.

[7] R. Han and D. J. Rogers, "Zero-Additional-Hardware Power Line Communication for DC–DC Converters," In: *IEEE Transactions on Power Electronics*, vol. 37, no. 11, pp. 13107-13118, Nov. 2022.

[8] F. Schulte, M. Seidel and M. Pfost, "A High Bandwidth and Multilevel Counter Circuit for Bearing Current Evaluation," In: *PCIM Europe 2024; International Exhibition and Conference for Power Electronics, Intelligent Motion, Renewable Energy and Energy Management*, Nürnberg, Germany, 2024, pp. 3359-3365.

A 90.9% Peak Efficiency KY Single-Inductor Bipolar-Output Converter with conductance modulation controller For Active-Matrix Organic Light Emitting Diode Power Supply

Sheng-Han Yu
Graduate Institute of Electronics Engineering
National Taiwan University
Taipei, Taiwan
r10943043@ntu.edu.tw

Chieh-Ju Tsai
Graduate Institute of Electrical Engineering
National Taiwan University
Taipei, Taiwan
f04943123@ntu.edu.tw

Hao-Ran Huang
Graduate Institute of Electrical Engineering
National Taiwan University
Taipei, Taiwan
r13921025@ntu.edu.tw

Ching-Jan Chen
Graduate Institute of Electrical Engineering
National Taiwan University
Taipei, Taiwan
chenjim@ntu.edu.tw

Abstract—A single-inductor bipolar-output (SIBO) converter combined with KY boost converter and interleaved switched capacitor (SC) converter with conductance modulation controller is proposed for Active-Matrix Organic Light Emitting Diode (AMOLED) Power Supply. The KY converter demonstrates better efficiency compared to traditional boost converters through charge flow analysis. The SC converter used in this paper effectively addresses the interaction regulation issues present in conventional SIBO converters. By employing an interleaving architecture, the SC converter ensures continuous input current, preventing the positive output voltage from being affected by current fluctuations. To control the SC converter operating in fast switching limit (FSL) mode, conductance modulation is proposed. This approach adjusts the $R_{ds,on}$ value of the power MOSFET to maintain the negative output voltage at the desired level. The chip prototype is implemented in a 180 nm BCD process, The converter achieves a peak efficiency of 90.9%, operating with 1MHz switching frequency.

Keywords—Single Inductor Bipolar Output Converter, KY Converter, Switching Capacitor Converter, Conductance Modulation, Hybrid converter

I. INTRODUCTION

AMOLED displays offer several advantages over traditional passive OLEDs and LCDs, including self-emissive properties, wide viewing angles, high contrast ratios, and fast response times [1], [2]. Fig. 1 shows the traditional PMIC structure for AMOLED power supply. The conventional power converter architecture used for powering AMOLED displays typically includes two boost converters and one inverting buck-boost converter. These converters are responsible for generating the necessary positive voltage (ELV$_{DD}$), negative voltage (ELV$_{SS}$), and AV$_{DD}$ voltage required for the source driver integrated circuit (Source Driver IC) [1]-[3]. Despite their functionality, traditional AMOLED driving systems encounter challenges such as low efficiency, large volume, high chip area costs, and suboptimal power quality. To address these challenges, various chip manufacturers and power management IC developers in the industry are actively innovating their architectures. Their efforts aim to enhance efficiency, reduce

volume and costs, and improve overall power quality, optimizing AMOLED displays' performance in mobile devices.

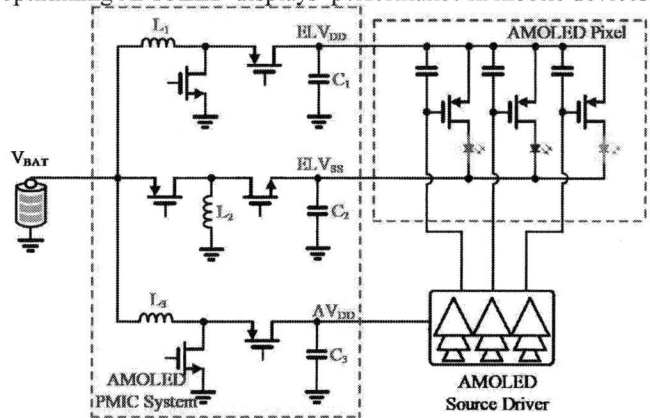

Fig. 1 Conventional PMIC architecture for AMOLED power supply

Recently, much research has centered on employing conventional single-inductor bipolar-output (SIBO) converter designs for AMOLED power supply. The primary objective has been to simplify the architecture by decreasing the number of large inductors from three sets to two. This effort places significant emphasis on optimizing transient response dynamics to achieve reductions in the sizes of both inductors and capacitors. The SIBO converter proposed in [1] employs a control method using time-division multiplexing to guarantee that the inductor current resets to zero at the end of each cycle. This control technique operates exclusively in Discontinuous Conduction Mode (DCM), resulting in higher inductor current ripple, output capacitor voltage ripple, slower transient response, and limitations on maximum output power. References [2-10] introduced a multi-phase continuous modulation mode SIBO converter. Because the inductor current energy is distributed within a single cycle, variations in load response for one output voltage can impact the regulation of the other output voltage, leading to suboptimal cross-regulation. In reference [11], the initial hybrid SIBO converter architecture was proposed, incorporating a flying capacitor. Despite this innovation, the converter continues to deliver power

979-8-3315-1612-3/25 $31.00 © 2025 IEEE

discontinuously to both outputs, thereby failing to resolve all the challenges encountered by earlier converter designs. Reference [12] expands upon the hybrid SIBO output converter architecture discussed earlier, employing a similar approach to float both positive and negative outputs. Nevertheless, this adaptation does not significantly enhance the architecture's limitations.

Core issue observed in current literature structures lies in the inability to fundamentally address the discontinuous behavior of the inductor current to either output terminal. This leads to drawbacks such as inefficiency, the need for large external inductor-capacitor components, and inadequate transient response. Otherwise, cross-regulation is also a big issue to affect the display quality [13][14]. Therefore, the proposed architecture in this thesis achieves continuous output at the positive terminal, while the negative output is isolated from the positive terminal by a flying capacitor, thereby eliminating the above drawbacks.

II. THE CONCEPT OF PROPOSED KY SIBO CONVERTER

Fig. 2 shows the system architecture of the proposed KY SIBO converter for AMOLED power supply. This converter integrates a KY boost converter with an interleaved switching capacitor (SC) converter. The design generates two output voltages: positive (ELVDD) and negative (ELVSS) voltage. It's important to understand that these two converters function independently. As a result, the duty cycles of the KY converter and the interleaved SC converter are not identical. Similarly, each converter has a separate control loop, with no overlap. It's worth noting that ELVDD serves a dual purpose: it's the positive output voltage of the SIBO converter and the input voltage for the subsequent interleaved SC converter, respectively. Consequently, maintaining a continuous input current for the interleaved SC converter is crucial.

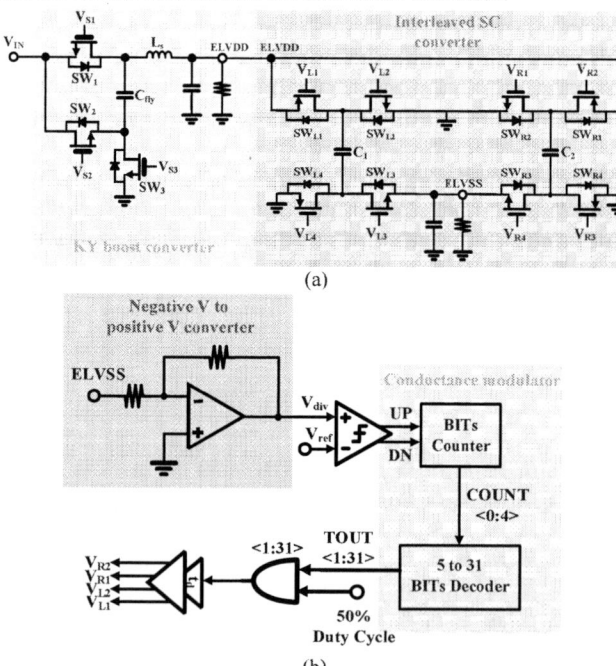

(a)

(b)

Fig. 2   Proposed KY SIBO converter(a) Power Stage (b) Control Stage

Fig. 3(a) shows the traditional SC converter structure, while Fig. 3(b) illustrates its operation. There is a discontinuous input current during state switching, which leads to an undesired ripple in the positive output voltage, ELVDD. Fig. 3(c) presents an interleaved SC converter, and Fig. 3(d) shows its operation. With interleaving, especially in Fast Switching Limit (FSL) mode, the input current ripple of the SC converter can be significantly reduced. This results in a substantial suppression of the input voltage ripple of the SC converter (KY converter's output voltage).

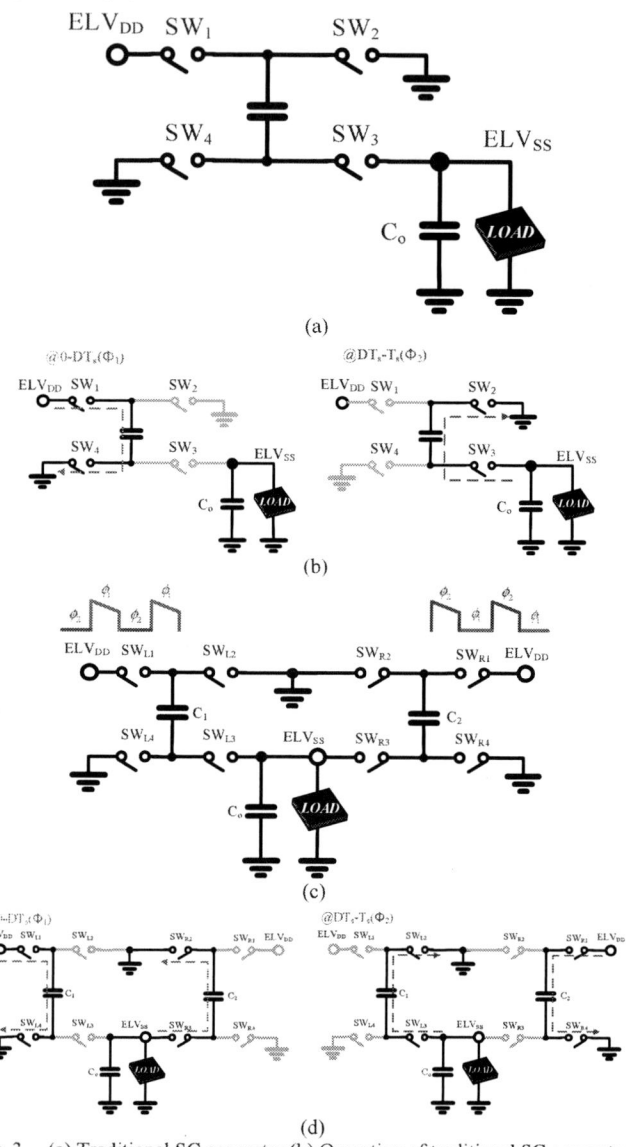

Fig. 3   (a) Traditional SC converter (b) Operation of traditional SC converter (c) Interleaving SC converter (d) Operation of interleaving SC converter

III. THE ANALYSIS OF THE KY CONVERTER FOR POSITIVE OUTPUT

The steady-state waveform of the proposed KY boost converter is shown in Fig. 4 (c) originally proposed in [15]. Because the inductor is connected to V_{out}, the small signal analysis [16][17] behaves more like that of a traditional buck converter rather than a boost converter[18]. This means that the control architecture of the KY converter can be simpler.

A. Charge flow analysis of KY converter and traditional boost converter

Fig. 4 illustrates the two states of the circuit during normal operation. In state $\varphi 1$, shown in Fig. 4 (a), V_{IN} flows through capacitor C_{fly}, which has been charged to V_{IN}, to transfer energy to the inductor. During this time, the voltage at node V_{LX} rises to $2V_{IN}$. In state $\varphi 2$, depicted in Fig. 4 (b), V_{IN} charges C_{fly} while simultaneously transferring energy to the inductor, resulting in V_{LX} being equal to V_{IN} due to its connection with V_{IN}.

The voltage conversion ratio of the converter can be calculated with charge flow analysis [19][20]. The result will be consistent with conducting with voltage-second balance. It is worth noting that the output capacitor is large enough. Then, the voltage conversion ratio is derived by the total input charge divided by the total output charge.

It follows that:

$$q_{in}^{\varphi 1} = q_L^{\varphi 1} = q_o^{\varphi 1} = I_L DT \tag{1}$$

$$q_O^{\varphi 2} = q_L^{\varphi 2} = I_L(1-D)T, q_{in}^{\varphi 2} = q_{in}^{\varphi 1} + q_L^{\varphi 2} \tag{2}$$

$$\text{Conversion ratio (M)} = \frac{q_{in}^{\varphi 1} + q_{in}^{\varphi 2}}{q_o^{\varphi 1} + q_o^{\varphi 2}} = 1 + D \tag{3}$$

Then, the conversion ratio of the traditional boost converter is calculated in (4), Which is consistent with calculation by voltage-second balance method.

$$M = \frac{q_{in}^{\varphi 1} + q_{in}^{\varphi 2}}{q_o^{\varphi 1} + q_o^{\varphi 2}} = \frac{I_L T}{I_L(1-D)T} = \frac{1}{1-D} \tag{4}$$

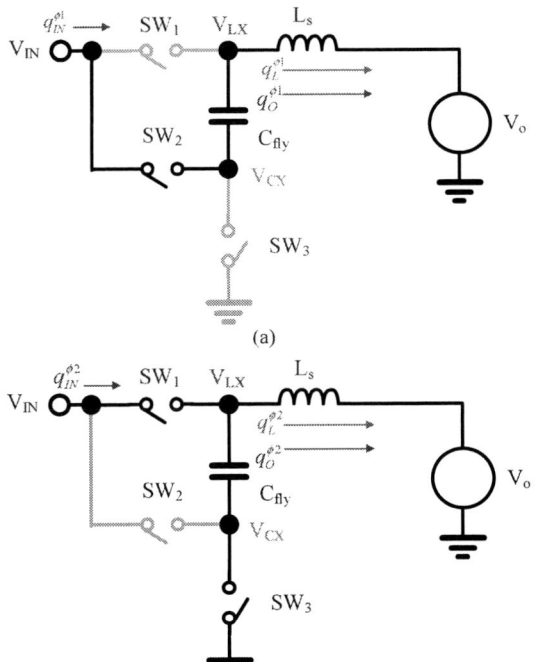

(a)

(b)

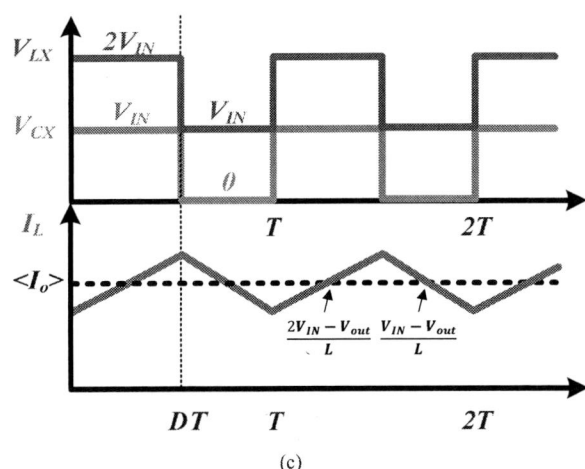

(c)

Fig. 4 The proposed KY converter: (a)State $\varphi 1$ (b)State $\varphi 2$ (c)Steady-State Waveform

B. Loss Analysis

A charge-flow analysis framework is outlined as follows: By calculating the charge that flows through the switches and their conduction time, we can determine the average current for each switch during its conduction period. Using the energy formula $E = I^2 R t_{on}$, we can then compute the losses for each switch over a switching cycle in FSL mode. Finally, we can establish the equivalent output resistance R_O by examining the relationship between I_L and I_O.

$$R_{o,FSL} = \frac{1}{1-D} R_{SW1} + D R_{SW2} + \frac{D^2}{1-D} R_{SW3} \tag{5}$$

The SSL loss analysis:

$$R_{o,SSL} = \frac{C(\Delta V)^2}{2CT} = \frac{(I_L DT)^2}{2CT} = \frac{D^2 T}{2C} \tag{6}$$

Combining (5) and (6):

$$R_{O,KY} = \sqrt{R_{o,FSL}^2 + R_{o,SSL}^2} \tag{7}$$

Fig. 5 presents the graph of Ro corresponding to switching frequency. In order to achieve better efficiency, FSL mode operation is determined to be utilized. Therefore, $R_{O,FSL}$ is much larger than $R_{O,SSL}$, $R_{O,total}$ is equal to $R_{O,FSL}$.

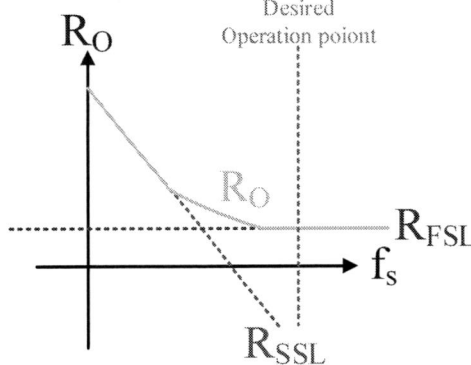

Fig. 5 Conduction loss consisting of switches loss and C_{fly} charging loss.

Then, R_O of traditional boost converter is calculated as follows:

$$R_{O,boost} = \left(R_{sw1}D + \frac{1}{1-D}R_{sw2}\right)M^2, M = \frac{1}{1-D} \quad (8)$$

To achieve better efficiency, $R_{O,KY}$ is supposed to be smaller than $R_{O,boost}$. It is noted that the same operation point is conducted. That is, the conversion ratio M and $R_{ds,on}$ of each switch is the same.

$$R_{O,KY} < R_{O,boost} \quad (9)$$

$$\left(\frac{1}{2-M} + (M-1) + \frac{(M-1)^2}{2-M}\right)R < \left(\frac{M-1}{M} + \frac{1}{1-\frac{M-1}{M}}\right)M^2R \quad (10)$$

$$1 < M < \sqrt{3} \quad (11)$$

From (11), better efficiency is achieved if the KY converter is operated in $1 < M < \sqrt{3}$.

IV. Operation of Conductance Modulation for negative output

In order to maintain continuous input current to minimize the amplitude ripple of positive output voltage, the SC converter must operate in FSL mode. If $R_{ds,on}$ is very small, this requires a large switching capacitor, which necessitates increasing $R_{ds,on}$. Consequently, it is necessary to control the output voltage of the SC converter to ensure that the load current does not affect the output voltage. Nevertheless, the proposed modulation method may result in relatively lower efficiency due to the increase in the value of $R_{ds,on}$, particularly under light load conditions.

Fig. 6(a) shows that the function blocks of proposed conductance modulation. The interleaved SC converter's control stage incorporates three key components: a negative-to-positive voltage converter, a comparator, and a conductance modulator. With conductance modulation, some power MOSFETs of interleaved SC converter (SW_{L1}, SW_{L2}, SW_{R1}, and SW_{R2}) are comprised of 32 segments. Fig. 6(b) shows the waveform of conductance modulation operation. The converter operates using conductance modulation through a series of steps:

1. negative ELVSS is converted to a positive value and scaled down (V_{div}).
2. V_{div} is compared against a predetermined reference voltage, generating either UP or DN signals.
3. In BITs counter, a high UP signal increments COUNT by one in binary, while a high DN signal decrements it. To prevent overflow, COUNT is capped at <11111> when it reaches this value and UP is high. Similarly, to avoid underflow, COUNT is limited to <00000> when it reaches this value and DN is high.
4. The COUNT signal is then fed into a BITs decoder, which determines the number of power MOSFET segments to activate, enabling precise conductance magnitude calculation. For instance, if the input signal B is 01101 (13 in decimal), TOUT becomes <0000000000000000001111111111111>.

Through above actions, the $R_{ds,on}$ of controlled power MOSFET is modulated by negative output negative feedback.

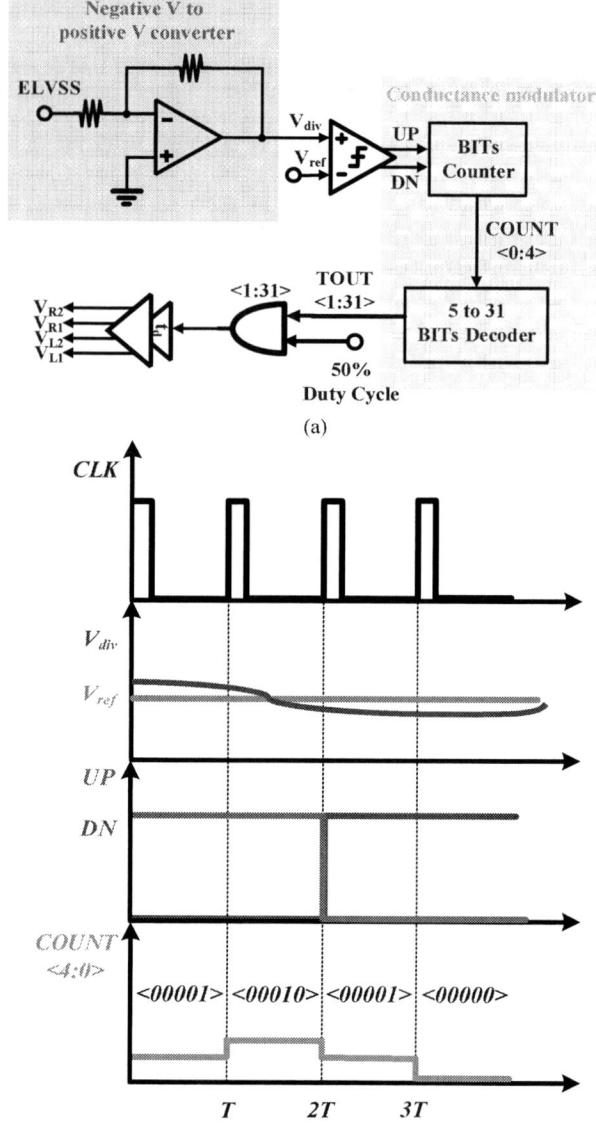

Fig. 6 (a)Block diagram of conductance modulation (b)Waveform of conductance modulation

V. Simulation Results

Fig. 7 (a), (b) shows the steady-state waveform of the proposed KY SIBO converter, V_{IN} =3.4 V, positive output voltage ELV_{DD} = 5 V, negative output voltage ELV_{SS} = -4.9 V. Loading current = 100 mA.

Fig. 7 (c) shows the load transient response of conductance modulation with a 250mA/50ns step-up current in ELVSS. Undershoot voltage of ELVDD is only 46mV, and recovery time within 1% of the output voltage is 38μs. COUNT is properly controlled from 00101 to 01011 according to ELVSS in Fig. 1Fig. 7 (b). Cross-Regulation in positive output is 46/250 mV/mA.

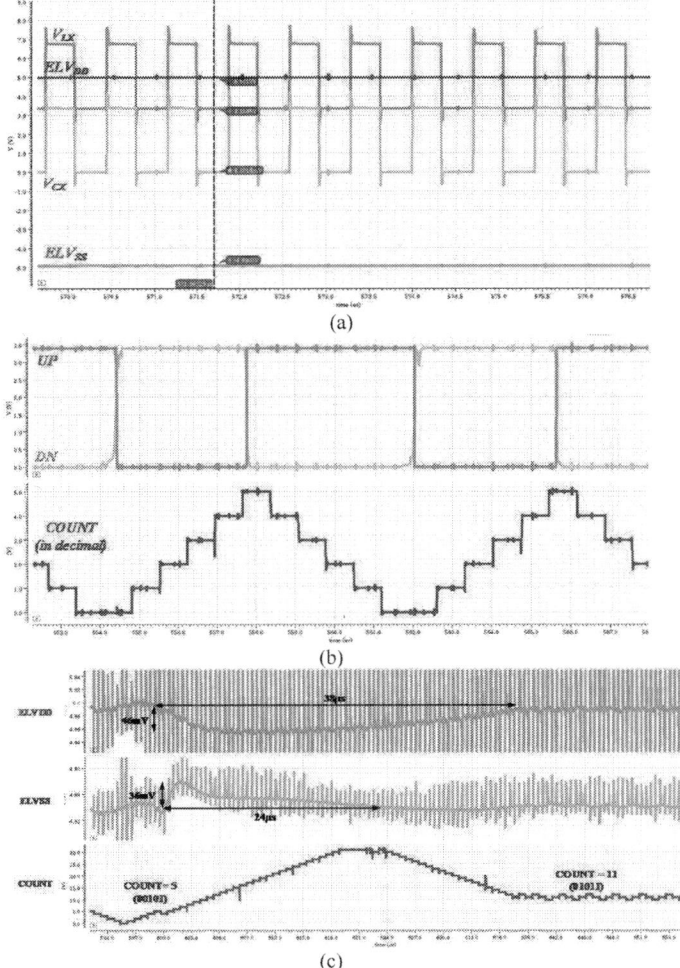

(a)

(b)

(c)

Fig. 7. Simulation waveform (a) Steady-State Operation in power stage signal (b) Steady-State (c) Load Transient at ELVSS from 30 mA to 280 mA in 50 ns

Fig. 8 (a) illustrates efficiency over the whole loading range. The efficiency of loading current from 30 mA to 350 mA is larger than 85 %, peak efficiency is 90.9 %. Fig. 9 shows the proposed KY-SIBO converter is fabricated into an IC using TSMC 180 nm CMOS high voltage mixed signal based generation II BCD technology. The chip micrograph is presented in Fig. 17, encompassing a total area of 3207 μm × 2506 μm, including all bonding pads.

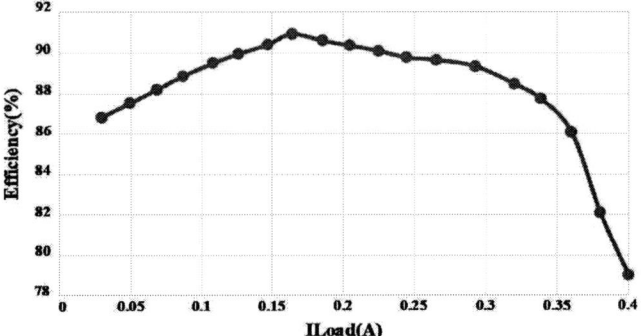

Fig. 8. The simulation efficiency over the loading range

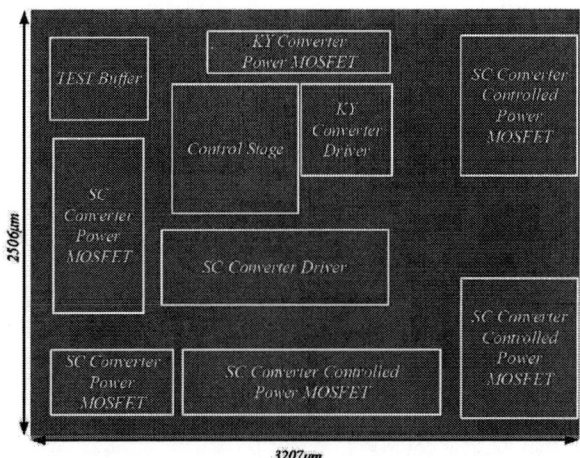

Fig. 9 Chip micrograph

VI. CONCLUSION

This paper presents an SIBO converter combining a KY converter and an interleaved SC converter with conductance modulation. It boosts efficiency within the KY converter and improves positive output voltage's ripple with the interleaving structure of the SC converter.

Table 1 shows comparison of the performance of the proposed converter with state-of-the-art SIBO and single-inductor multiple-output (SIMO) converters. With the inductor connected to the positive output, a smaller inductor can be adopted in the proposed converter. The proposed KY SIBO hybrid converter can achieve better efficiency compared with previous SIBO converters [7][12]. The proposed KY SIBO converter illustrates better cross-regulation suppression than [13][14].

ACKNOWLEDGMENT

The authors would like to thank the sponsorship from the National Science and Technology Council (NSTC), Taiwan and the educational subsidy for chip fabrication from TSRI, Taiwan. The authors would like to thank SIMPLIS Technologies for providing SIMPLIS simulation tool. Thanks also go to all the authors who previously worked on the dual-path architecture. This paper could not have been completed without everyone's contributions.

Table 1 Comparison table

	TCSII2018[7]		JSSC2021[12]		TCASI2022[13]	ISSCC2023[14]	This Work	
Technology	0.09μm CMOS		0.35μm 5V device		0.18μm CMOS	16nm FinFET	0.18μm BCD	
Structure	SIBO		SIBO		SIMO	SIMO	SIBO	
Input Voltage(V)	3.0~4.2		2.7~4.5		1.8	1.8	2.6~3.4	
Chip Area(mm²)	N/A		3.68		1.95	1.85	8.04	
Inductor (μH)	10		10		4.7	0.005/0.01	2.2	
Output Capacitor (μF)	10		4.7		1.5/3	20	4.7	
Flying Capacitor (μF)	N/A		4.7		2.3nF(on chip)	N/A	4.7	
Max Efficiency(%)	90.1		89.3		87.5	87.3	90.9	
Frequency (MHz)	0.5		1		500	10	1	
Output Voltage (V)	V_{OP}	V_{ON}	V_{OP}	V_{ON}	0.4-1.6	0.8-1.4	V_{OP}	V_{ON}
	4.8	-2.5	5.3	-4.7			5	-4.9
Line Regulation(mV/V)	1.25	2.5	18	2	N/A	N/A	12	1.5
Load Regulation(mV/mA)	0.03	0.06	0.1	0.06	N/A	N/A	0.03	0.02
Cross Regulation(mV/mA)	N/A		N/A		0.314	0.7	0.184	0.24

REFERENCES

[1] Paul C. P. Chao, C. -K. Chang, C. -K. Cheng, T. -H. Tran, T.Sauter, "A new single inductor bipolar multi-output (SIBMO) DC-DC converter driven by constant on-time with improved performance", in ACM Microsystem Technologies, vol. 22, No. 6, pp. 1405-1418, Jun. 2016.

[2] C. Chae, H. Le, K. Lee, G. Cho and G. Cho, "A Single-Inductor Step-Up DC-DC Switching Converter With Bipolar Outputs for Active Matrix OLED Mobile Display Panels," in IEEE Journal of Solid-State Circuits, vol. 44, no. 2, pp. 509-524, Feb. 2009.

[3] S. Wang, Y. Woo, Y. Yuk, G. Cho and G. Cho, "High efficiency Single-Inductor Boost/Buck Inverting Flyback converter with hybrid energy transfer media and multi level gate driving for AM OLED panel," Symposium on VLSI Circuits, 2010.

[4] W. Xu, Y. Li, Z. Hong, D. Killat and H. Schleifer, "A single-inductor multiple-bipolar-output (SIMBO) converter with fully-adaptive feedback matrix and improved light-load ripple," ESSCIRC, 2011, pp. 435-438.

[5] Y. -H. Lee et al., "A near-zero cross-regulation single-inductor bipolar-output (SIBO) converter with an active-energy-correlation control for driving cholesteric-LCD," IEEE Custom Integrated Circuits Conference (CICC), 2011.

[6] S. Hong, S. Park, T. Kong and G. Cho, "Inverting Buck-Boost DC-DC Converter for Mobile AMOLED Display Using Real-Time Self-Tuned Minimum Power-Loss Tracking (MPLT) Scheme With Lossless Soft-Switching for Discontinuous Conduction Mode," in IEEE Journal of Solid-State Circuits, vol. 50, no. 10, pp. 2380-2393, Oct. 2015

[7] B. Kwak, S. Hong and O. Kwon, "A Highly Power-Efficient Single-Inductor Bipolar-Output DC–DC Converter Using Hysteretic Skipping Control for OLED-on-Silicon Microdisplays," in IEEE Transactions on Circuits and Systems II: Express Briefs, vol. 65, no. 12, pp. 2017-2021, Dec. 2018.

[8] C. -K. Chang, C. -H. Su and P. C. . -P. Chao, "A new single inductor bipolar multiple output (SIBMO) on-chip boost converter using ripple-based constant on-time control for LCD drivers," IECON 2014 - 40th Annual Conference of the IEEE Industrial Electronics Society, Dallas, TX, USA, 2014, pp. 4122-4125

[9] C. -K. Chang, C. -H. Su, Y. -H. Kao, M. -H. Yu, T. Sauter and P. C. . -P. Chao, "A new single inductor bipolar multiple output (SIBMO) boost converter using pulse frequency modulation (PFM) control for OLED drivers and optical transducers," Sixteenth International Symposium on Quality Electronic Design, Santa Clara, CA, USA, 2015, pp. 552-555

[10] W. -C. Chen et al., "A single-inductor bipolar-output converter with 5 mV positive voltage ripple for active matrix OLED," 2011 IEEE Energy Conversion Congress and Exposition, Phoenix, AZ, USA, 2011, pp. 3229-3233

[11] S. Hong, "11.6 A 1.46mm2 Simultaneous Energy-Transferring Single-Inductor Bipolar-Output Converter with a Flying Capacitor for Highly Efficient AMOLED Display in 0.5μm CMOS," IEEE International Solid- State Circuits Conference - (ISSCC), 2020.

[12] F. Mao et al., "A Hybrid Single-Inductor Bipolar-Output DC-DC Converter With Floating Negative Output for AMOLED Displays," in IEEE Journal of Solid-State Circuits, Early access.

[13] Zhou Z, Tang N, Nguyen B, et al. An Inductor-First Single-Inductor Multiple-Output Hybrid DC-DC Converter With Integrated Flying Capacitor for SoC Applications[J]. IEEE Transactions on Circuits and Systems I-Regular Papers, 2022, 69(12): 4823-4836.

[14] Kim S, Krlshnarnurthy H K, Sofer S, et al. A 1.8W High-Frequency SIMO Converter Featuring Digital Sensor-Less Computational Zero Current Operation and Non-Linear Duty-Boost[M]. 2023: 10-2.

[15] K. I. Hwu and Y. T. Yau, "KY Converter and Its Derivatives," in IEEE Transactions on Power Electronics, vol. 24, no. 1, pp. 128-137, Jan. 2009.

[16] Y.-T. Hung, C.-J. Tsai, C.-J. Chen, C.-H. Hsu, C.-Y. Hsieh, "An All NMOS KY-Boost Converter with Double Injection Control for Fast Line and Load Transient Response," *IEEE Transactions on Circuits and Systems I: Regular Papers*, vol. 71, no. 11, pp. 5005-5016, Nov. 2024

[17] W. Zeng, C. Lam, S. Sin, F. Maloberti, M. Wong and R. P. Martins, "A 220-MHz Bondwire-Based Fully-Integrated KY Converter With Fast Transient Response Under DCM Operation," in IEEE Transactions on Circuits and Systems I: Regular Papers, vol. 65, no. 11, Nov. 2018.

[18] Y. -S. Hwang, Y. -T. Ku, A. Liu, C. -H. Chen and J. -J. Chen, "A New Efficiency-Improvement Low-Ripple Charge-Pump Boost Converter Using Adaptive Slope Generator With Hysteresis Voltage Comparison Techniques," in IEEE Transactions on Very Large Scale Integration (VLSI) Systems, vol. 23, no. 5, pp. 935-943, May 2015.

[19] Y. Yorozu, M. Hirano, K. Oka, and Y. Tagawa, "Electron spectroscopy studies on magneto-optical media and plastic substrate interface," IEEE Transl. J. Magn. Japan, vol. 2, pp. 740–741, August 1987 [Digests 9th Annual Conf. Magnetics Japan, p. 301, 1982].

[20] C.-J. Tsai, C.-H. Hsu, C.-J. Chen, Y.-T. Hung, C.-Y. Hsieh, "A Monolithic All-1.8V-Thin-Gate-NMOS KY-Boost Converter With Reused Flying-Capacitor Bootstrap Gate Driver Achieving 94.42% Peak Efficiency," IEEE Transactions on Power Electronics, vol. 39, no. 6, pp. 7238-7251, Jun. 2024

Constant-On-Time Control for Zero-Bias Trans-Inductor Voltage Regulators

Hank Zeng, Justin Lee, Rixin Lai, Hang Shao,
Monolithic Power Systems, Inc.
New Taipei City, Taiwan; San Jose, USA
{hank.zeng, justin.lee, rixin.lai, hang.shao}@monolithicpower.com

Abstract—Recently, the zero-bias trans-inductor voltage regulator (ZB-TLVR) was introduced to power high-performance processors for AI and cloud computing applications with superior transient performance and the potential for better power density and efficiency. This paper proposes a constant-on-time (COT) control algorithm for ZB-TLVR. A current-balance loop and ramp compensation were developed to address the challenges related to interleaving and phase current balancing. An 8-phase ZB-TLVR test platform was built, and the test results verified the performance of the proposed control approach.

Keywords—trans-inductor voltage regulator, constant-on-time control, zero-bias TLVR.

I. INTRODUCTION

Powering high-performance processors (e.g. CPUs and GPUs) has tremendous challenges due to the fast load transient [1-2] (current slew rate more than 1000A/µs) and stringent output voltage accuracy requirements. In addition, as the power consumption of the processors continues to grow, efficiency has become a key metric for the power solution. To meet performance and efficiency targets [3], both the power architecture and the control approach must be carefully designed. Recently, zero-bias trans-inductor voltage regulator (ZB-TLVR) topology has been presented to the industry (see Fig. 1a).

ZB-TLVR topology has a higher magnetic core utilization due to its significantly reduced bias current. It enables increased coupling inductance [4], which results in a lower ripple current and higher power efficiency. However, ZB-TLVR control is not well discussed for current multi-phase of voltage regulators application. The purpose of this paper is to propose a constant-on-time (COT) control [5-6] scheme customized for ZB-TLVR to achieve optimized control performance.

Section II presents the COT control principle and current-balance loop mechanism. Section III introduces the simulation result and test set-up, presents the challenges for current sharing and phase interleaving, and discusses the corresponding solutions. Section IV shows the transient performance and efficiency measurements.

II. CONSTANT-ON-TIME CONTROL AND CURRENT-BALANCE LOOP APPLICATON FOR ZB-TLVR

A. Constant-On-Time (COT) Control

Fig. 2 shows the block diagram of a proposed DC/DC buck converter with COT. With a COT architecture, the control structure is very simple compared to traditional voltage/current mode control. When using COT control, the sensed output voltage reference singnal is compared to the ramp voltage (V_{RAMP}), which is used to control and generate a fixed on time (t_{ON}) (see Fig. 3). V_{RAMP} is generated by an internal clock without inductor current information.

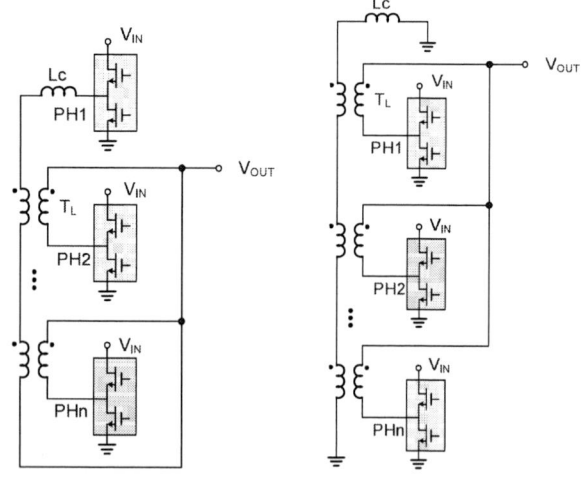

a) Zero-Bias TLVR   b) Traditional TLVR

Fig. 1. Zero-Bias TLVR and Traditional TLVR Topology

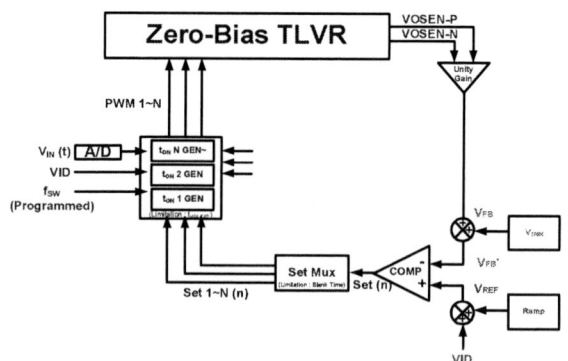

Fig. 2. COT Control Block Diagram

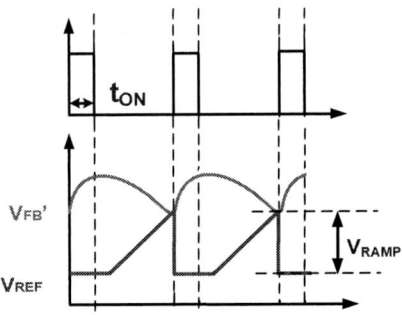

Fig. 3. COT Control

t_{ON} can be calculated with Equation (1):

$$t_{ON} = \frac{VID}{V_{IN}} \times \frac{1}{f_{sw}} \qquad (1)$$

Adjusting the slope slew rate affects the set signal's triggered timing (see Fig. 4). A slower slope makes it easier to trigger the set signal, which results in a larger pulse-width modulation (PWM) jitter and poor noise immunity. In contrast, a faster slope makes it difficult to trigger the PWM signal, which results in a smaller PWM jitter and good stability.

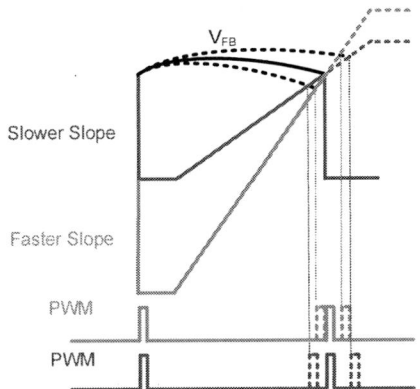

Fig. 4. Different Slew Rate Slope Behavior

Although a faster slope improves stability, it may lead to poor transient response (see Fig. 5 and Fig. 6). Because a faster slope contacted V_{FB} timing is slower than slower slope, which causes PWM pulling in frequency is too slower and worse undershoot occurs. When overshoot occurs, a faster slope makes it easier to trigger a PWM signal, so overshoot behavior is also poor with a fast slope.

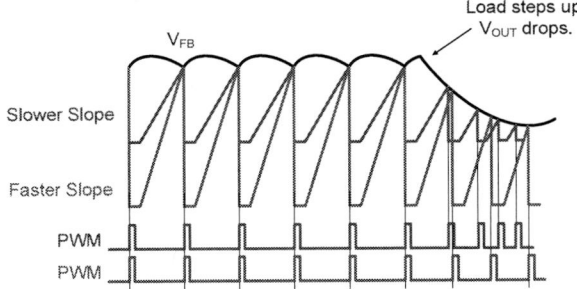

Fig. 5. Different Slew Rate Slope Behavior during Load Step-Up

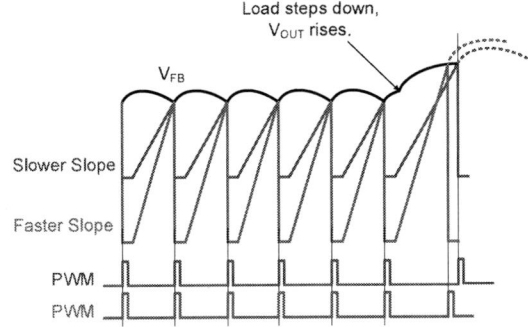

Fig. 6. Different Slew Rate Slope Behavior during Load Step-Down

B. Current-Balance Loop Mechanism

The sensed phase currents are compared to phase1 current (CS1) to obtain the error in the proportional integral (PI) loop to modulate the PWM on time. To achieve current balancing, the other PWM phases are based on PWM1 to increase or decrease the on time. Fig. 7 shows the current-balance loop control block diagram.

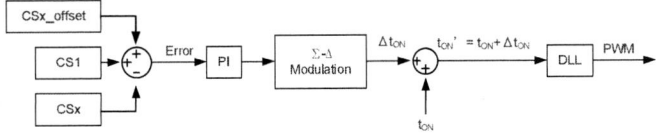

Fig. 7. Current-Balance Loop Block Diagram

III. SIMULATION AND TEST RESULT

A. SIMPLIS Simulation Result

The SIMPLIS simulation and the experimental prototype were both constructed in order to validate the proposed paper design. Fig. 8 and Fig. 9 show the SIMPLIS simulation bench and present the simulated waveform, including the phase 1 through phase 3 current, total current, V_{OUT}, and load current. Table 1 shows the key parameters of the ZB-TLVR.

TABLE I. KEY PARAMETERS OF THE ZERO-BIAS TLVR

V_{IN_NOM}	13.5V
V_{OUT_NOM}	0.9V
I_{OUT_MAX}	385A
f_{SW}	400kHz
Phase Number	8 phases
Load Line Resistance	0.4mΩ
Maximum Load Step	335A
Maximum Slew Rate of Load	1200A/μs
T_L	250nH
L_C inductance	100nH
Output Capacitance	4097μF
Controller	MP2985B
DrMOS	MP87290

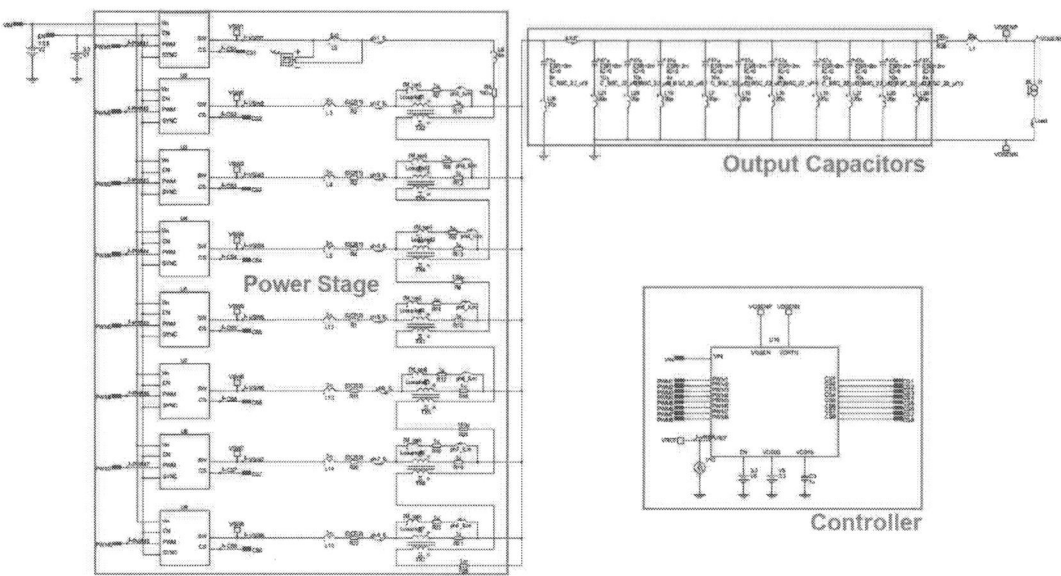

Fig. 8. SIMPLIS Simulation Block

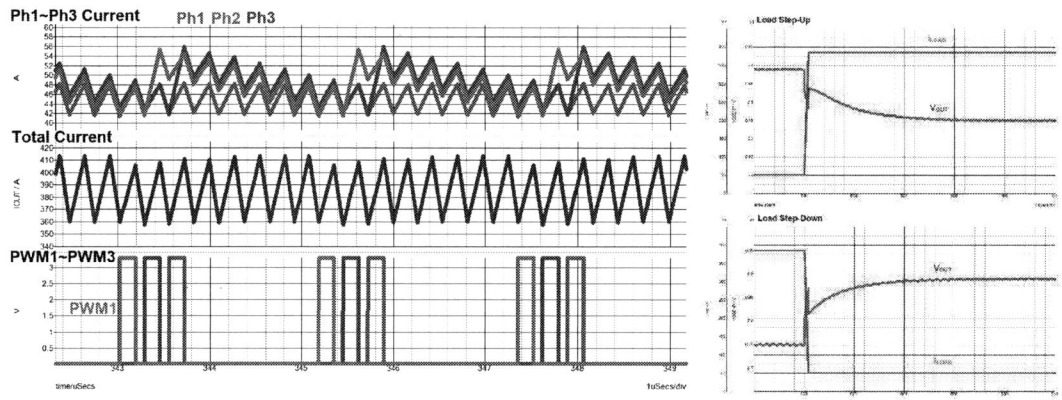

Fig. 9. SIMPLIS Simulation Results

According to the simulation results in Fig. 9, phase 1 experiences an unbalanced current because it is connected to the output with a longer path by the secondary side. This causes phase 1's impedance to be higher than the other phases. The next section will describe how a current-balance loop function can solve this problem while providing practical results.

Another phenomenon is that the total current does not have a symmetrical current ripple because the phase 1 current ripple of frequency is n times switching frequency that cause phase 1 current ripple is smaller than other phases; this behavior also affects the output waveform. In particular, it leads to non-ideal interleaving because the COT topology pulls in the next PWM signal (PWM2) to raise the output. However, suboptimal interleaving does not impact steady state or transient behavior.

B. Test Result

An 8-phase test set-up was built to verify the proposed control approach (see Fig. 10).

Fig. 10. The Experimental Platform for ZB-TLVR

Fig. 11 and Fig. 12 show the steady state test waveforms. Fig. 11 shows that the average current of phase 1 is lower than the other phases. This behavior is the same as the simulation result due to the ZB-TLVR structure. The unbalanced impedance causes an uneven current distribution. This issue can be addressed by adding a current-balance loop to adjust the on time of the different phases. Fig. 12 shows the test result with the current-balance loop.

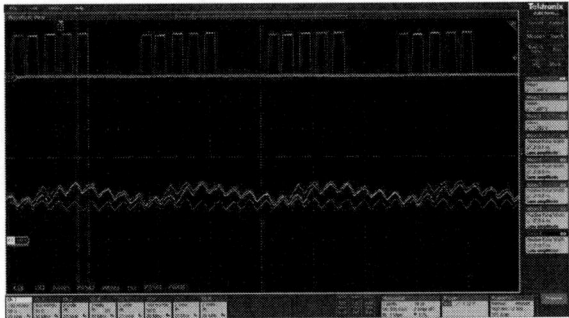

Probe 1: CS1; Probe 2: CS2; Probe 3: CS3

Fig. 11. CS Waveform (Disabled Current Balance Function)

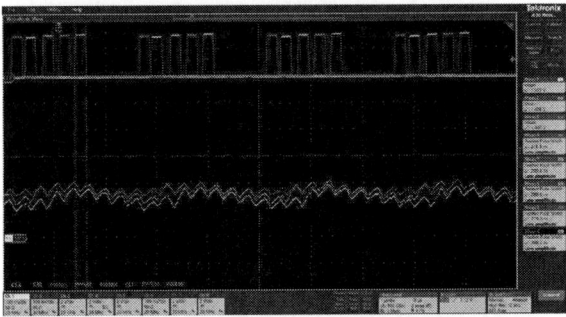

Probe 1: CS1; Probe 2: CS2; Probe 3: CS3

Fig. 12. CS Waveform (Enabled Current Balance Function)

Another problem found during the test is the non-ideal interleaving in steady state (see Fig. 13). The output voltage drops when phase 1 turns on because phase 1's current ripple is smaller than the other phases, resulting in an asymmetrical total current. COT control pulls in the next phase (phase 2) to compensate the output voltage. This process impacts the interleaving angle between phase 1 and phase 2. Interleaving can be improved by adjusting the ramp for the reference voltage (V_{REF}). Fig. 14 shows the test results with ramp compensation.

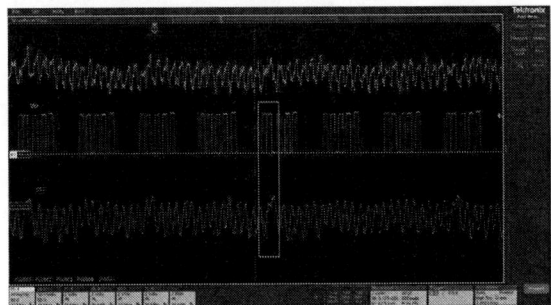

Probe 1: PWM1; Probe 2: PWM2; Probe 3: CS1

Fig. 13. Suboptimal Interleaving Behavior in Steady State

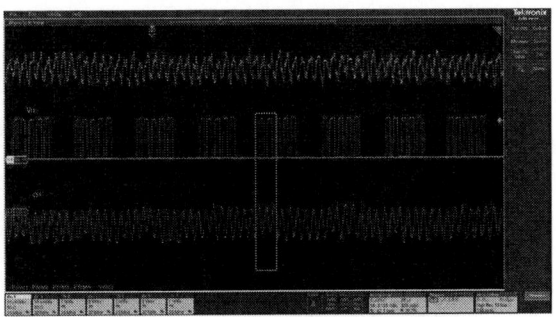

Probe 1: PWM1; Probe 2: PWM2; Probe 3: CS1

Fig. 14. Improved Interleaving in Steady State

IV. EFFICIENCY AND TRANSIENT RESULT

Due to the increased inductance (T_L) for ZB-TLVR applications, the switching frequency (f_{SW}) can be lower while keeping the ripple current within a reasonable range. Fig. 15 shows the efficiency curves of different TLVRs. A ZB-TLVR at 400kHz improves peak efficiency by about 1.26% compared to a traditional TLVR at 800kHz.

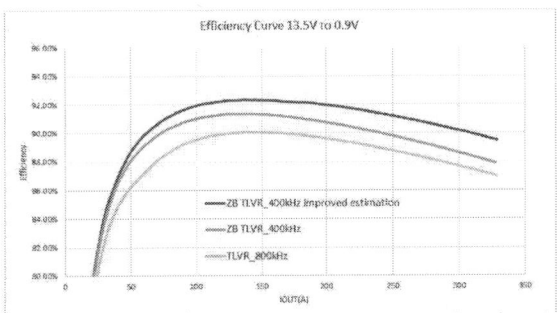

ZB-TLVR: 400kHz; Traditional TLVR: 800kHz

Fig. 15. Efficiency Curve

However, the efficiency curves are closer under heavy loads. This is because SW to output impedance difference. Fig. 16 shows the PCB differences between a ZB-TLVR and traditional TLVR. The ZB TLVR has extra PCB loss caused by a larger impedance between SW and the output.

From phase 2 through phase 8, the ZB-TLVR's PCB impedance is 0.87mΩ; from phase 1 through phase 8, the traditional TLVR's impedance is 0.35mΩ. The ZB-TLVR's PCB DCR loss rises as the output current (I_{OUT}) increases, which is the reason why the 'ZB TLVR_400kHz' curve has less efficiency improvement over the ' TLVR_800kHz' curve as the load goes higher, presented in Fig. 15. The authors anticipate more optimized TL designs in the near future, so that the impedance between the SW and the output of the ZB-TLVR can be similar or equal to that of a traditional TLVR. The 'ZB_TLVR_400kHz improved estimation' curve in Fig.15 is plotted based on such anticipation.

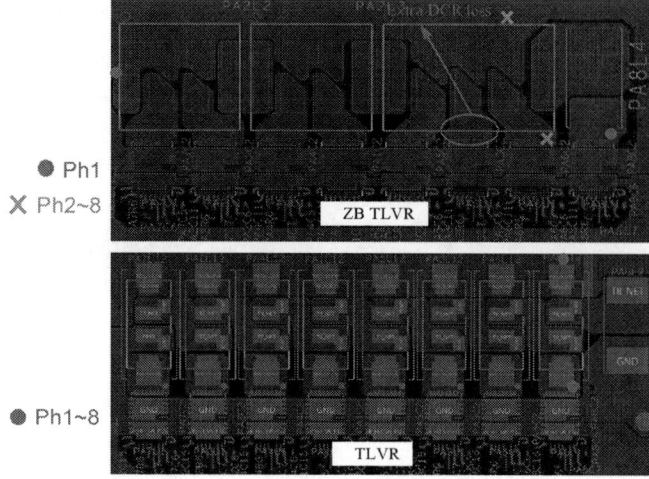

Fig. 16. PCB Comparison with ZB TLVR and Traditional TLVR

Fig. 17 and Fig. 18 show the transient behavior. When COT is implemented in a ZB-TLVR topology, the transient response mainly depends on the Lc inductance. Even when operating at a lower f_{SW}, the transient behavior at 400kHz is very similar to that of 800kHz.

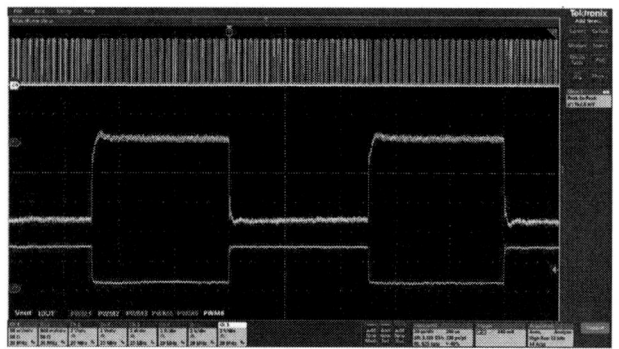

Probe 1: V$_{OUT}$; Probe 2: I$_{OUT}$ **(Pk-Pk: 163.8mV)**

Fig. 17. ZB TLVR 400kHz Transient Behavior at 10kHz, 50% Duty Cycle

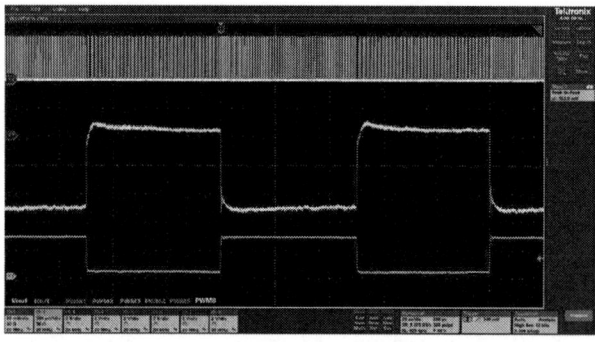

Probe 1: V$_{OUT}$; Probe2: I$_{OUT}$ **(Pk-Pk: 163mV)**

Fig. 18. TLVR 800kHz Transient Behavior at 10kHz, 50% Duty Cycle

V. CONCLUSION

This paper proposes and verifies a COT control algorithm for ZB-TLVR. Two challenges were introduced, as well as the corresponding methods to conquer these challenges. The first challenge is longer path and larger impedance from phase 1 SW to the output in a ZB-TLVR topology, which results in current balance issues. This can be significantly improved by enabling the proposed current balancing loop. The second challenge is the non-ideal PWM interleaving, which can be improved by adjusting the slope slew rate. Excellent stability, current balancing and transient performance can be achieved with the proposed approaches. Lastly, the test results proved that the ZB-TLVR topology with COT control achieves better efficiency when compared to the traditional TLVR topology.

REFERENCES

[1] Joseph G. Renauer, "Challenges in Powering High Performance Low Voltage Processors," in Proc.Appl. Power Electron. Conf., 1996, pp. 977-983.

[2] Nian Zhang, Chenchang Zhan, Guanghua Ye, Chuqi Chen, Xuenung Li, Jun Yi "Analysis of Multi-Phase Trans-inductor Voltage Regulator with Fast Transient Response for Large Load Current Applications, " IEEE Internatioal Symposiun on Circuits and Systems., 22-28 May 2021.

[3] S. Jiang, Xin Li, Woon-Seong Kwon, Cheng Chung Yang, Q. Wang, Nam Hoon Kim, Mikhail Popovich, H. Gan, Chenhao Nan, "Trans-Inductor Voltage Regulator for High Bandwidth Power Delivery, " United States Patent 20240120847A1., Apr. 2024.

[4] Hang Shao, Tao Zhao, Dianbo Fu, Daocheng Huang, Jinghai Zhou "Analysis Model and Desing Procedure of the Single-Secondary Trans-Inductor Voltage Regulator, " IEEE Energy Conversion Congress and Exposition., Oct.2021.

[5] L. Jiang, X. Wu and Q. Ouyang, "Constant on-time Switching Converter and Control Method Thereof " United States Patent 9356510, May 2016.

[6] K. Hariharan, Santanu Kapat, Siddhartha Mukhopadhyay, "Constant On-Time Multi-Mode Digital Control with Superior Performance and Programmable Frequency, " IEEE Applied Power Electronics Conference and Exposition., March 2019.

An improved PFM control scheme for Three-Level Buck Converter based on Ton Extension Achieving an 810% Frequency Reduction

Yi-Chun Chang
Graduate School of Advanced Technology
National Taiwan Univeristy
Taipei, Taiwan
mavissschang@gmail.com

Chieh-Ju Tsai
Graduate Institute of Electronics Engineering
National Taiwan University
Taipei, Taiwan
f04943123@ntu.edu.tw

Ting-Lun Lee
Department of Electrical Engineering
National Taiwan Univeristy
Taipei, Taiwan
r09921138@ntu.edu.tw

Ching-Jan Chen
Graduate Institute of Electronics Engineering
National Taiwan University
Taipei, Taiwan
chenjim@ntu.edu.tw

Abstract—**This work introduced a novel controlling technique of Pulse Width Modulation (PFM) on a three-level buck converter, to keep remarkable efficiency within wide range of duty cycle and loading. Since this interleaved signal converter with conventional constant on-time control cannot scale their switching frequency with load under discontinuous conduction mode when the conversion ratio(M) is larger than 0.5, an on-time extension method is applied. Additionally, a V_{LX} sensing scheme is proposed to achieve flying capacitor balancing in both CCM and DCM throughout the entire duty cycle. The chip prototype is designed using a TSMC 0.18um CMOS process. The simulation results show a 2MHz switching frequency with and a loading range from 20mA to 1A with > 90% efficiency.**

Keywords—Three-Level Buck Converter, Pulse Frequency Modulation, DCM Operation, On-Time Extension Control, Flying-Capacitor Balancing

I. INTRODUCTION

Advancements in mobile devices demand high-efficiency, compact SoC power solutions. Traditional Two-Level (2-L) buck converters require large inductors and increasing switching frequency reduces size but impacts efficiency. The Three-Level(3-L) buck converter, using four switches and a flying capacitor, minimizes area consumption and reduces losses, proving superior to 2-L buck converters [1]-[3].

In mobile device applications, which enter standby mode during periods of inactivity, maintaining high efficiency is crucial for extending battery life. Therefore, discontinuous conduction mode (DCM) operation under constant on-time (COT) control is a very effective method. However, the 3-L buck converter with conventional Constant On-Time (COT) control cannot scale its switching frequency with load under DCM operation when the conversion ratio ($M = V_O/V_{in}$) is larger than 0.5. This impedes the linear reduction of the switching frequency with the loading current, resulting in lower efficiency. Currently, only a few papers discuss and implement DCM operation in 3-L buck converters, which only focus on M < 0.5 [7]-[12]. In other words, DCM operation under M > 0.5 has not been implemented yet. The proposed on-time extension control ensures that the switching frequency adjusts appropriately to maintain high efficiency.

II. ANALYSIS OF THREE-LEVEL BUCK CONVERTER

A. Operation Principle of Three-Level Buck Converter

Fig. 1 illustrate the four operating states of the three-level buck converter. S_1, S_4 and S_2, S_3 are two pair of complementary switches having the same duty cycle signal but with a 180° phase shift. When the conversion ratio M<0.5, the switching voltage node (V_{LX}) varies from 0 to $V_{in}/2$, following a sequential operation in State B→A→B→D. Conversely, when M>0.5, V_{LX} varies from V_{in} to $V_{in}/2$, and the operation sequence follows in State D→A→D→C. It is important to note that the flying capacitor C_F is charges at State A and discharges at State C. While at conversion ratio(M) less than 0.5, these two states are the rising time of the inductor current, for

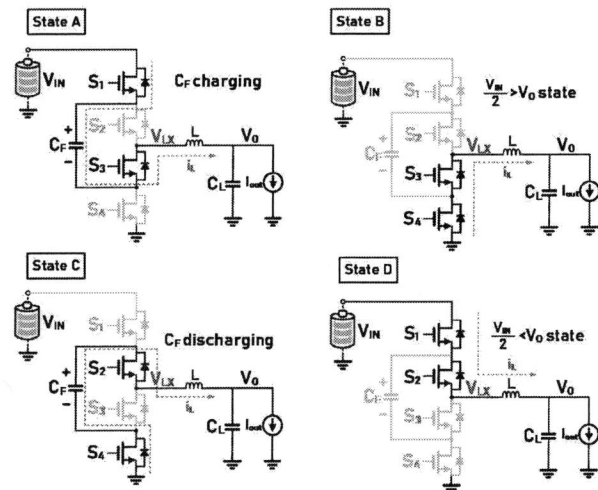

Fig. 1. Operating states of three-level buck converter

B. DCM Operation of Three-Level Buck Converter

In this section, the primary challenge of 3-L buck converter will be discussed. To dive in, the DCM conversion ratio for 3-L buck converter is derived in [7] with Fig. 2.

Take M<0.5 as example, applying the volt-second balance across the inductor and averaging its current:

$$\langle V_L \rangle = 0 \qquad (1)$$

$$\langle i_L \rangle = I_{out} \qquad (2)$$

(1) and (2) can be expressed as

$$\begin{cases} (V_{in}/2 - V_{out})T_{on} = V_{out} \cdot d_2 \\ (T_{on} + d_2)\dfrac{\dfrac{V_{in}/2 - V_{out}}{2L}T_{on}}{T_s/2} = I_{out} \end{cases} \qquad (3)$$

By pairing these two equations, M is calculated as:

979-8-3315-1612-3/25 $31.00 © 2025 IEEE

$$M = \frac{1}{2} \frac{T_{on}^2}{T_{on}^2 + I_o \cdot \dfrac{LT_s}{V_{in}/2}}, \qquad for\ M < 0.5 \quad (4)$$

Same procedure can be applied when M > 0.5:

$$M = \frac{1}{2} + \frac{(T_{on} - T_s/2)^2}{(T_{on} - T_s/2)^2 + I_{out} \cdot \dfrac{LT_s}{V_{in}/2}}, \qquad for\ M > 0.5 \quad (5)$$

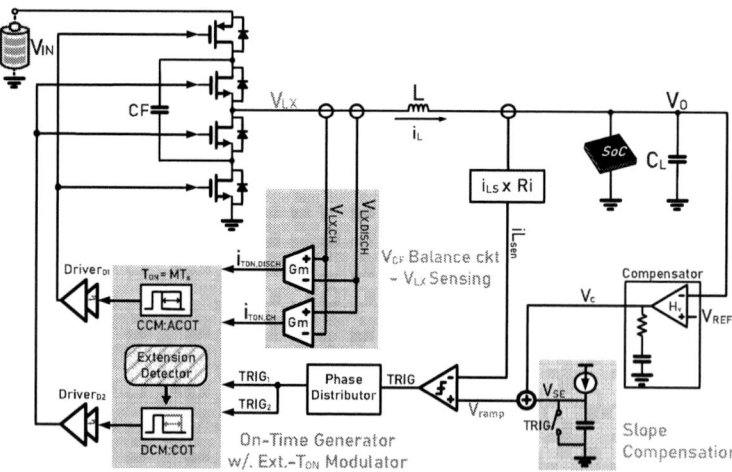

Fig. 2. Switching waveforms for (a) M<0.5 and (b) M>0.5

If V_{in}, V_{out}, L, and T_{on} are fixed, which is a conventional COT condition, according to (4), $I_{out} \times T_s$ is a constant in M < 0.5, indicating that the switching frequency decreases linearly with the load current. However, when M > 0.5 according to (5), the charging time of the inductor is determined by the overlapping time of the two interleaved duty cycles, causing $I_{out} \times T_s$ is not a constant, suppressing the ability of frequency scaling down with load under DCM. The theoretical concept is validated in Fig. 3, comparing the mathematical model of (4) and (5) with simulation results in DCM.

Furthermore, for a converter with N switches, the efficiency can be simply expressed as:

$$P_{eff} = P_{cond} + P_{switch} = N \cdot I_o^2 R_{on} + 2N \cdot \frac{1}{6} I_{max} V_{max} t_{on/off} \cdot f_{sw} \quad (6)$$

Switching loss (P_{switch}) is influenced by the maximum voltage

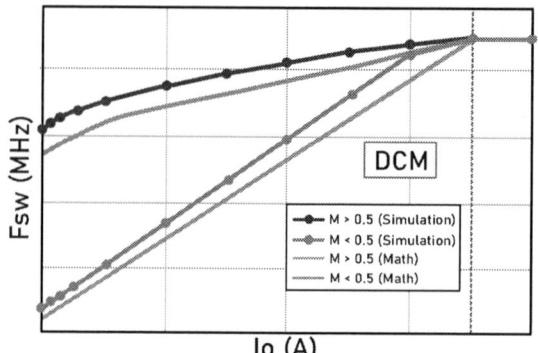

Fig. 3. Switching frequency vs. loading current for M < 0.5 and M > 0.5 in DCM

across the switch (V_{max}), switching frequency (f_{sw}), and the charge required for switching on and off. Conduction loss (P_{cond}) arises from resistive components in the circuit and is proportional to the square of the output current. At light load condition, I_O is relatively low, making conduction loss becomes less significant [5]. To achieve high efficiency across varying load conditions, it is essential to dynamically adjust the switching frequency during DCM. Current research on DCM operation in 3-L buck converters is limited and primarily focuses on conditions where M<0.5. In contrast, PFM schemes for M>0.5 remain largely unexplored.

III. PROPOSED THREE-LEVEL BUCK CONVERTER

A. Overall System Architecture and Operating Modes

Fig. 4 shows the architecture of the 3-L buck with proposed controller. The system scheme is similar to conventional COT [4]-[6], with an additional slope compensation, V_{CF} balance circuit and Extending-Ton modulator. In 3-L buck converter, aside from the ripple cancelling effect at M = 0.5, the non-fixed charging time of the inductor current when M > 0.5 also introduces stability issues [13], To address this issue, as illustrated in Fig. 5.(a), an external ramp (v_e) is added at compensator output (v_c) with an absolute slope value (S_e)

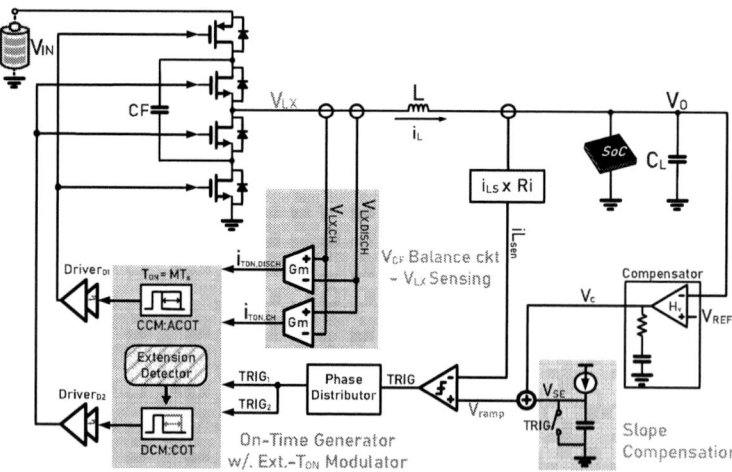

Fig. 4. The Proposed 3L Buck Converter

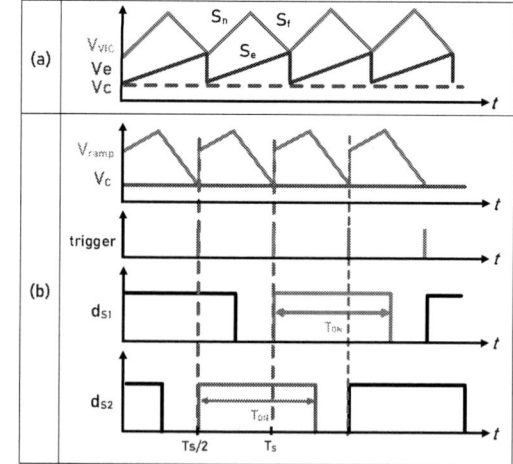

Fig. 5. (a)modulation waveform with slope compensation (b)Steady-state waveform at M>0.5

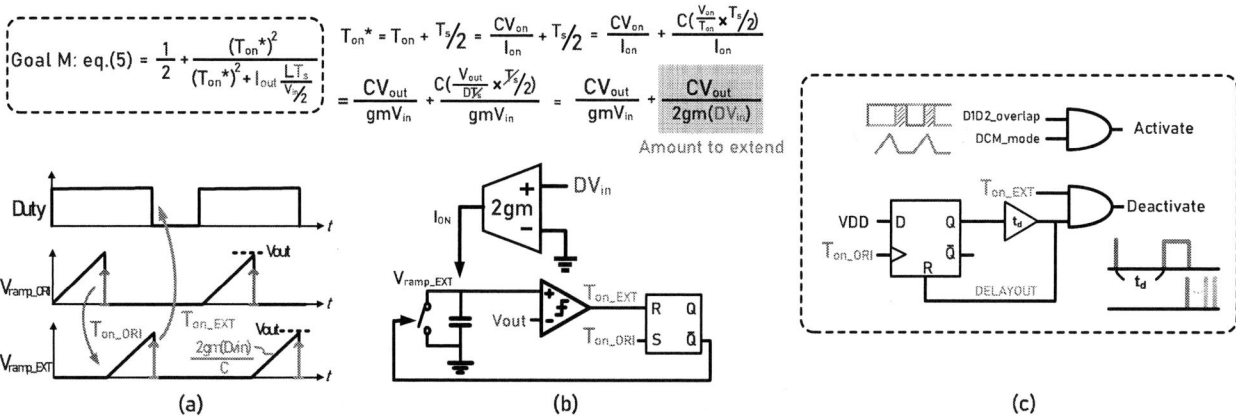

Fig. 6. (a)Modulation waveform in Extending mode (b)Implementation of Ext. ramp (c)Logic to De/Activate Ext.Ton Mod.

and compared with virtual inductor current (v_{VIC}). The steady-state modulation waveform at M>0.5 is depicted in Fig. 5.(b), where $V_{ramp} = v_{VIC} - v_e$. As V_{ramp} touches v_c, the trigger will be distributed to d_{S1} or d_{S2} to generate a constant T_{on}.

B. Implementation of Extending Ton modulator

As mentioned in previous section, the primary challenge in reducing frequency with decreasing load in a 3L buck converter while applying COT control lies in the inability to maintain a linear relationship between switching frequency and load current in M > 0.5. Therefore, the Extending Ton Modulator (Ext.-Ton Mod.) is proposed. According to (5), requires $(T_{on} - \frac{T_s}{2})$ to remain constant. While the load current decreases, the switching period T_s increases, indicating that the on-time should also be extended to keep $(T_{on} - \frac{T_s}{2})$ constant. Fig. 6 shows the derivation and the implementation of the extended ton (T_{on}^*), which can be substitute to the original T_{on} to cancel the term related to T_s. When the system detects DCM with overlapping time between two duty cycles, the Ext.-Ton Mod. activates. Reaching the target voltage (V_{on}), the original on-time generator triggers the extended on-time generator to start charging, with the extended on-time generator, which then controls the final duty cycle turn-off signal.

For a smooth transition, the additional on-time gradually

increases until it reaches the target value. Conversely, to deactivate Ext.-Ton Mod., the additional on-time gradually decreases until the turn-off signal of the two on-time generators are closed enough, shown in Fig. 6. (c), allowing the system to return to regular mode.

Fig. 7 compared the waveforms between with and without the Extending Ton Modulator (Ext.-Ton Mod.). As illustrated in Fig. 7.(b), the constant Ton duty cycle restrains the ability to decrease switching frequency among the loads. The adjustment in Fig. 7.(c) ensures that the charging time of the inductor current, the period during which the converter supplies energy to the output, remains constant. Consequently, this method maintains high efficiency across a wide range of operating conditions and has potential applications in any converter with interleaved duty signals.

C. Flying Capacitor Balance - V_{LX} sensing

As discussed in [21]-[22], non-idealities deviate V_{CF} from Vin/2. For example, the charge distribution between the flying capacitor and its parasitic capacitor, the time mismatch between driving signals. The method employed in [15] can result in some voltage deviation from the desired steady-state value, since each charging and discharging time interval is the same. [8] is proposed by a senior member in the laboratory, while the technique effectively address the V_{CF} balance issue in CCM, they do not delve into the problem in DCM.

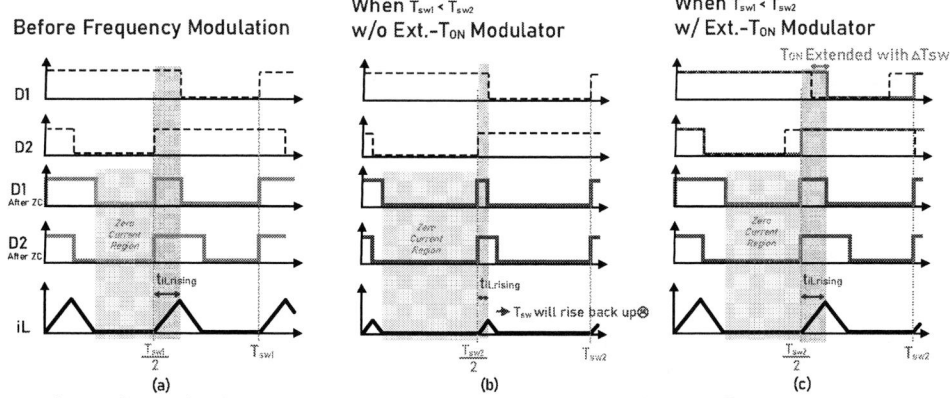

Fig. 7. Switching waveforms for 3L buck converter at DCM operation during (a)initial waveform, (b)when Io decreases w/o Ext.-Ton Mod. (c)when Io decreases w/ Ext.-Ton Mod.

979-8-3315-1612-3/25 $31.00 © 2025 IEEE

To eliminate the error, a V_{CF} calibration loop is required. Fig. 4 analyzes how single-ended V_{LX} sensing achieves V_{CF} balance from the architecture in the diagram. If $V_{CF} > V_{in}/2$, then $V_{LX,DISCH} > V_{LX,CH}$. The voltage difference between these two is processed through a VCCS to reduce the on-time for the charging path in the T_{ON} generator, which gradually balance V_{CF} back to $V_{in}/2$. The same procedure can be easily adapted for the opposite condition when $V_{CF} < V_{in}/2$.

Fig. 8 demonstrates the operation of full range V_{LX} sensing capacitor voltage balance. Be aware that in DCM, the rising time determines the falling time. In case when M>0.5, the falling time of the inductor current becomes the charge/discharge time of V_{CF}. Therefore, we must reverse the mechanism that extends (shortens) the charging and discharging time to achieve voltage balance of the flying capacitor in this scenario.

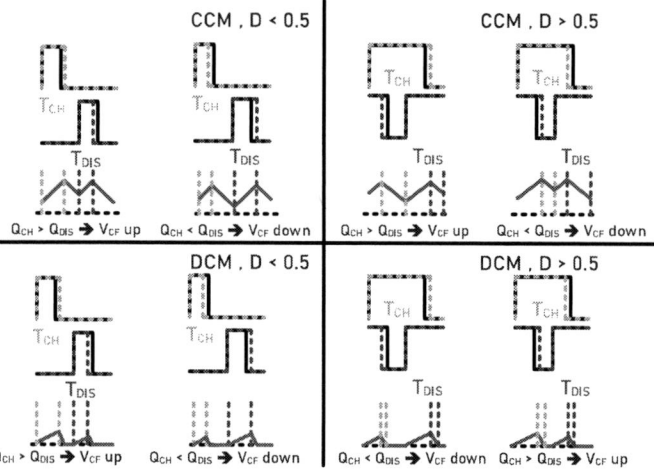

Fig. 8. The operation of V_{LX} sensing for V_{CF} balance

IV. SIMULATION RESULT AND CONCLUSION

The proposed current mode COT with Extended-Ton Control 3-L Buck Converter has been implemented in a 0.18um CMOS process with 2MHz switching frequency during CCM and variable frequency during DCM. Fig. 9 shows the switching frequency versus loading current when M=0.74, meanwhile achieving 810% frequency reduction at Io=20mA.

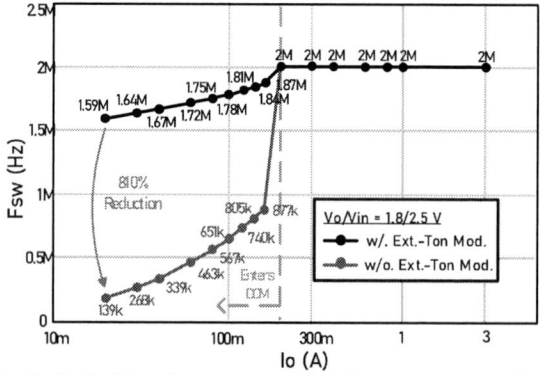

Fig. 9. Switching frequency vs. load current comparing with and without Ext.-Ton Modulator for M = 0.72

This function operates effectively without additional mode switching, even when M≅0.5, which is the ripple cancellation point often associated with stability issues, as shown in Fig. 10. This results in high efficiency for 3L-buck converter in a wide conversion ratio, as demonstrated in Fig. 11, achieving peak efficiency 98.2% @Vin = 2.5V, Vo = 1.8V, Io = 200mA, with overall efficiency > 90% from Io 20mA to 1A.

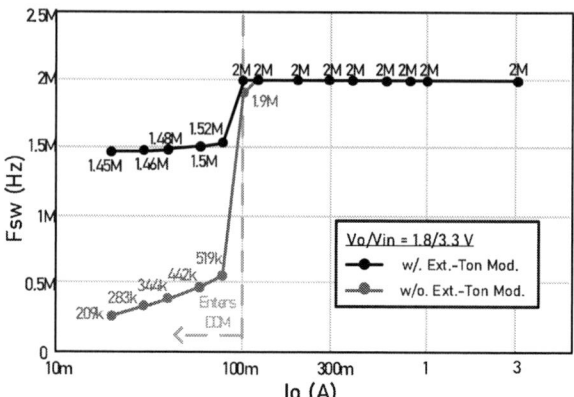

Fig. 10. Switching frequency vs. load current comparing with and without Ext.-Ton Modulator for M = 0.54

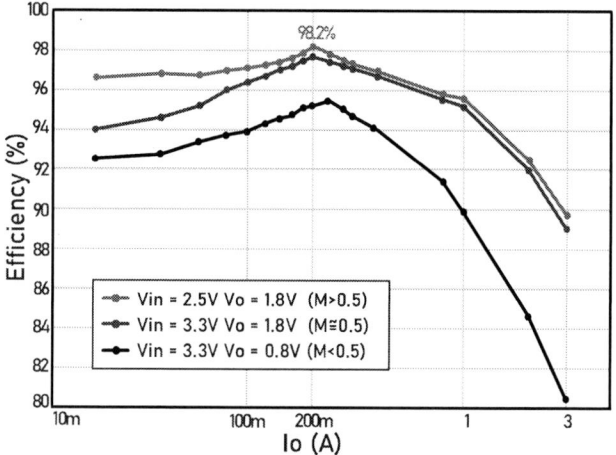

Fig. 11. Efficiency vs. the load current under various conversion ratio(M)

TABLE 1 provides comparison with other state-of-the-art three-level buck converters. Most of the papers published in these prominent conferences focus on improving transient response, [10], [19] and present notable techniques for V_{CF} calibration [10], [19]-[21], while [18] optimizes light load efficiency by establishing ZVS mechanism. Additionally, [10], [18], [20] utilizes adaptive constant-on-time control, which inherently supports pulse-frequency-modulation (PFM) only at M<0.5. In contrast, the proposed 3-L buck converter not only incorporates a robust scheme for V_{CF} calibration, but also significantly reduces switching frequency across a wide range of conversion ratios ($M < 0.5, M \cong 0.5, M > 0.5$) as load decreases. The chip fabrication is currently in progress.

ACKNOWLEDGMENT

The author gratefully acknowledges the sponsorship of MediaTek Inc., Taiwan, and the support for chip fabrication provided by TSRI, Taiwan. Appreciation is also extended to SIMPLIS Technologies for supplying simulation tools. Sincere thanks are given to the researchers whose previous work on three-level buck converter architectures laid the foundation for this paper. This work would not have been possible without the invaluable contributions of all involved.

TABLE 1

	[21]JSSC' 18	[18]ISSCC' 16	[19]ISSCC' 17	[20]VLSIC' 18	[10]TCSI' 22	This work
Process	65nm	500nm	28nm	250nm	65nm	180nm
L(H) / Cout(F)	100n / 10n	1.5μ / 4.7μ	2.2μ / 2μ	0.22μ / 2.4μ	47n / 47n	470n / 28μ
Cfly(F)	5n	1μ	Not shown	1μ	5n	14μ
Cfly type	on chip	Not shown	On chip	Off chip	on chip	off chip
Cfly balance	V_{CF} calibration	Auto calibration	AFC calibration	V_{CF} calibration	V_{CF} calibration	VLX sensing calibration
Vin(V)	5	12 - 100	3 - 4.5	2.6 - 5	2.4	2.5 - 3.3
Vo(V)	0.6 - 4.2	10	1.45, 1.2, 0.8	0.34 - 4.5	0.4 - 2	0.8 - 1.8
Io(A)	50m - 0.7	0.1 - 0.5	0.2 - 1	0.2 - 0.8	5m - 500m	20m - 3
Fsw,max(Hz)	50M	2M	3M	11.8M	25M	2M
PFM @DCM	x	x	x	M < 0.5	M < 0.5	Wide Duty (0.24-0.72)

REFERENCES

[1] D. Reusch, "High frequency, high power density integrated point of load and bus converters," Ph.D. dissertation, Virginia Tech, USA, 2012

[2] George Lakkas, "MOSFET power losses and how they affect power-supply efficiency," Analog Appl. J., Enterprise Systems, 2016.

[3] M. Halamicek, T. Moiannou, N. Vukadinović and A. Prodić, "Capacitive Divider Based Passive Start-up Methods for Flying Capacitor Step-down DC-DC Converter Topologies," 2018 International Power Electronics Conference (IPEC-Niigata 2018 - ECCE Asia), Niigata, Japan, 2018, pp. 831-837

[4] C. -Y. Hong, C. -J. Tsai and C. -J. Chen, "A gm-ramped interleaving technique with adaptive-extended TON control (AETC) scheme for multi-phase buck converter achieving fast load response," in Proc. IEEE Appl. Power Electron. Conf. Expo. (APEC), 2020

[5] W. Chen et al., "Pseudo-constant switching frequency in on-time controlled buck converter with predicting correction techniques," IEEE Trans. Power Electron., vol. 31, no. 5, pp. 3650-3662

[6] C.-H. Tsai, S.-M. Lin, and C.-S. Huang, "A fast-transient quasi-V2 switching buck regulator using AOT control with a load current correction (LCC) technique," IEEE Trans. Power Electron., vol. 28, no. 8, pp. 3949–3957

[7] Nenad Vukadinovic, "Low-Power Multi-Level Flying Capacitor Converters - Modeling and Control", PhD. Dissertation, University of Toronto,2018

[8] Y. -Y. Lin, " A Bidirectional Three-Level Converter with Single Point Sensing Technique for Flying Capacitor Balance," Master. dissertation, NTUEE, 2022

[9] Y. Yamauchi, T. Sai, T. Sakurai, and M. Takamiya, "Modeling of 3-level buck converters in discontinuous conduction mode for stand-by mode power supply," in Proc. IEEE Int. Symp. Circuits Syst. (ISCAS), May 2017, pp. 1–4.

[10] Sijie Pan and Philip K. T. Mok, "A 25 MHz Fast Transient Adaptive-On/Off-Time Controlled Three-Level Buck Converter," IEEE Trans. Circuits Syst. I, Reg papers

[11] G. Villar and E. Alarcon, "Inductor-current zero-crossing detection mixed-signal CMOS circuit for a DCM-operated 3-level switching power converter," IEEE International Symposium on Circuits and Systems, May. 2008, pp. 2606-2609.

[12] G. Villar, E. Alarcon, F. Guinjoan and A. Poveda, "Automatic dead-time adjustment CMOS mixed-signal circuit for a DCM-operated 3-level switching power converter," IEEE International Symposium on Circuits and Systems, May, 2008, pp. 3053-3056.

[13] E. Abdelhamid, G. Bonanno, L. Corradini, P. Mattavelli, and M. Agostinelli, "Stability properties of the 3-level flying capacitor buck converter under peak or valley current programmed control," IEEE Trans. Power Electron., vol. 34, no. 8, pp. 8031-8044, Aug. 2019.

[14] M. Du and H. Lee, "A 5-MHz 91% peak-power-efficiency buck regulator with auto-selectable peak- and valley-current control," IEEE Custom Integrated Circuits Conference 2010, San Jose, CA, 2010.

[15] S. -J. Lee et al., "30.5 A 95.3% 5V-to-32V Wide Range 3-Level Current Mode Boost Converter with Fully State-based Phase Selection Achieving Simultaneous High-Speed VCF Balancing and Smooth Transition," 2023 IEEE International Solid-State Circuits Conference (ISSCC), San Francisco, CA, USA, 2023, pp. 446-448

[16] S. Bari, Q. Li and F. C. Lee, "A new fast adaptive on-time control for transient response improvement in constant on-time control," in IEEE Transactions on Power Electronics, vol. 33, no. 3, pp. 2680-2689, March 2018

[17] K. D. T. Ngo, S. K. Mishra and M. Walters, "Synthetic-ripple modulator for synchronous buck converter," in IEEE Power Electronics Letters, vol. 3, no. 4, pp. 148-151, Dec. 2005.

[18] J. Xue and H. Lee, "12.5 A 2MHz 12-to-100V 90%-efficiency self-balancing ZVS three-level DC-DC regulator with constant-frequency AOT V2 control and 5ns ZVS turn-on delay," 2016 IEEE International Solid-State Circuits Conference (ISSCC), San FFrancisco, CA, USA, 2016, pp. 226-227

[19] L. -C. Chu et al., "10.5 A three-level single-inductor triple-output converter with an adjustable flying-capacitor technique for low output ripple and fast transient response," 2017 IEEE International Solid-State Circuits Conference (ISSCC), San Francisco, CA, USA, 2017, pp. 186-187

[20] Y. Karasawa, T. Fukuoka and K. Miyaji, "A 92.8% Efficiency Adaptive-On/Off-Time Control 3-Level Buck Converter for Wide Conversion Ratio with Shared Charge Pump Intermediate Voltage Regulator," 2018 IEEE Symposium on VLSI Circuits, Honolulu, HI, USA, 2018, pp. 227-228

[21] X. Liu, C. Huang and P. K. T. Mok, "A high-frequency three-level Buck converter with real-time calibration and wide output range for fast-DVS," IEEE J. Solid-State Circuits, vol. 53, no. 2, pp. 582-595, Feb. 2018

[22] X. Liu, P. K. T. Mok, J. Jiang, and W.-H. Ki, "Analysis and design considerations of integrated 3-level buck converters," IEEE Trans. Circuits Syst. I, Reg. Papers, vol. 63, no. 5, pp. 671–682, May 2016

A Concept for Current Ripple or Transient Improvements in Multiphase Converters

Alexandr Ikriannikov
Cloud and Communications Power
Analog Devices Inc
San Jose, USA
Alexandr.Ikriannikov@analog.com

Alex Gao
Cloud and Communications Power
Analog Devices Inc
San Jose, USA
Alex.Gao@analog.com

Abstract—Separate coupled inductors can be linked together with secondary windings to improve performance, similar to a previous concept of the trans-inductor voltage regulator. The secondary windings have an AC current, in sync with the ripple in the main power phases. The paper proposes to simplify the estimate of the current ripple in the arbitrary connection of several coupled inductors. It is also proposed to appropriately route the already existing secondary current for the ripple cancellation at the converter output. Alternatively, the secondary AC can be used to multiply the transient current step and therefore improve the transient performance of a voltage regulator. The experimental results confirm the operation of the proposed concepts.

Keywords—*linked coupled inductors, multiphase converters, voltage regulators, current ripple, transient.*

I. INTRODUCTION

Magnetically coupled inductors (CL) are known to significantly outperform conventional discrete inductors (DL) in the sense that a faster current slew rate is obtained for a certain current ripple in a steady state. This allows to either improve efficiency for a given transient performance, or significantly improve the transient and reduce output capacitance without lowering the efficiency [2, 4, 6, 10-12, 18-19, 22, 24]. An alternative solution where the phases are linked with individual secondary transformer windings was initially proposed in [7-8] and was later termed the "trans-inductor voltage regulator" (TLVR) where the tuning inductor L_c sets the transient and affects the current ripple (Fig. 1a) [13-16, 20, 23, 25, 27, 30]. The monolithic CL with the same number of coupled phases was mathematically and experimentally shown to significantly outperform the TLVR [22, 24]. One of the biggest TLVR drawbacks is the full current rating of the effective mutual inductance. In other words, each TLVR is based on a DL that has to be rated for the full load current per phase, plus the current ripple and transient margin on top of it. The transformer is fundamentally an AC device, and therefore the DC offset is not shared and not cancelled out between the phases. This leads to much smaller effective mutual inductance between the TLVR phases as compared to the CL.

However, the TLVR theoretically allows a very large number of phases to be linked, while the monolithic CL has manufacturing limitations for the aspect ratio of the ferrite core. Linking many TLVR phases causes significant high voltage concerns and potential difficulty with safety ratings [22] but

generally allows to contain the guaranteed efficiency hit in TLVR solution as compared to the DL [20, 24]. While the CL is the best performing solution for the same number of coupled phases, [22, 24], the low height requirement will most likely limit how long the CL can be manufactured or in other words, how many phases can be magnetically coupled in a single core.

To address this drawback, the boosted CL (BCL) was introduced, [17, 22]. The idea is to use the secondary transformer windings to link together mechanically separate CLs (Fig. 1b). When the secondaries of all the phases in the same CL are connected in series, the net mutual flux from each main winding is zero. In other words, the linking windings that connect BCLs are across only the total leakage of the CL, from all phases together $N_{ph} \times L_k$. L_k is the leakage per phase, and N_{ph} is the number of phases in that particular CL.

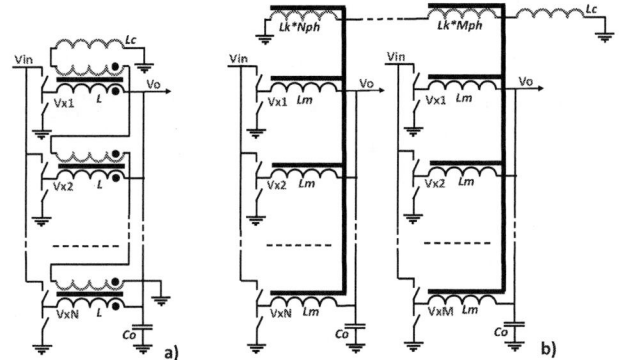

Fig. 1. Multi-phase buck converter with (a) a TLVR, and (b) a BCL.

The BCL allows to link several CLs together, so the advantage of current ripple cancellation (for a given transient set by tuning inductor L_c) can be utilized, while it may be impossible to manufacture a single monolithic CL with that total number of phases. The analytical result for the current ripple estimates in a BCL was presented in [25], but it was limited to a particular configuration of two-phase building blocks, which is generally the simplest system with a lowest ripple cancellation. An exact analytical derivation of the current ripple in the BCL solution with an arbitrary number of building blocks that could have a different number of phases would be much more complicated. The two types of interactions between phases, purely magnetic coupling in each CL and linking the building

979-8-3315-1612-3/25 $31.00 © 2025 IEEE   2149

blocks with connections of secondary windings are the main source of this analytical challenge.

This paper will propose a simplified ballpark estimate for the current ripple in an arbitrary BCL system and then introduce the concept of current ripple cancellation at the output with the related analytical analysis. Since in high current applications with a large number of phases the total current and therefore output voltage ripple are typically less critical than transient considerations, an alternative arrangement will be introduced to multiply the output transient current during the load step.

II. BCL MODELS AND CURRENT RIPPLE

As was described above, the net mutual L_m flux from each main phase that is seen by the BCL secondary is zero (because that secondary is coupled to all the phases at the same time). The original BCL model was then derived from the standard CL model by coupling the secondary winding to each phase leakage (Fig. 2a). The parasitic inductances L_a and L_b are effectively stray inductances that illustrate non-ideal coupling between main windings and the secondary.

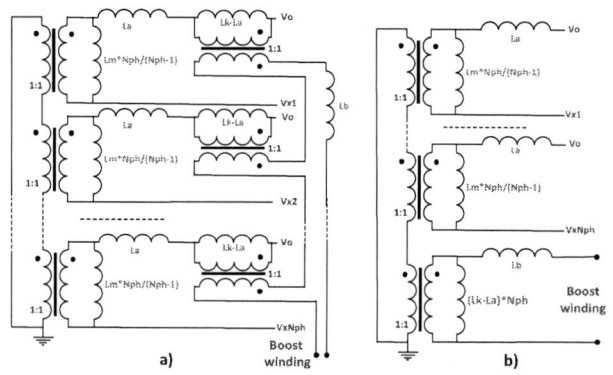

Fig. 2. BCL models (a) original, and (b) proposed with lumped total leakage.

The model can be rearranged into a more convenient form, as shown in Fig.2b, where all the leakage is lumped into a single element. This is similar to the model shown in [29], albeit for a different system and operation.

Starting with the conventional buck topology with the DL, the current ripple in each phase can be found as (1) [3], where the duty cycle is $D=V_o/V_{in}$, V_o is the output voltage, V_{in} is the input voltage, L is the inductance value, and F_s is the switching frequency.

Coupling the inductors magnetically introduces current ripple cancellation, with the resulting CL current ripple for any number of phases and any operation condition shown in [10] and then reworked in [22, 24]. Connecting several BCLs with the tuning inductor L_c (Fig. 1b) is generally similar to a system with a single monolithic CL, but the problem is that two different coupling mechanisms are introduced.

The proposal is to make a simplification of the BCL system so as to use already derived equations for the CL current ripple. For this, all the phases are assumed to be coupled in the same (single) way with L_m mutual inductance, not only for phases in the same BCL building block but also for phases that belong to

different BCLs. This effectively represents the system as a single monolithic CL where all the phases are coupled in the same way. Linking different BCLs with secondaries across (ideally) $N_{ph} \times L_k$ (Fig. 1a) is clearly different from magnetic coupling between phases with a mutual inductance L_m. However, a typical goal for a high-performance CL is to have $L_m \gg L_k$, [22, 24]. Electrically linking BCLs with secondaries across the total leakage $N_{ph} * L_k$ would potentially approach this objective if a number of magnetically coupled phases in the BCL building block is reasonably high and then $N_{ph} \times L_k \gg L_k$.

Thus, the standard leakage inductance L_k in the original CL model needs to be recalculated into an effective leakage L_{k_bcl} of the BCL system. Looking at a general connection of BCLs in Fig. 1b and inserting the BCL model from Fig.2, L_{k_bcl} can be derived. In each phase, parasitic L_a (Fig.2) always remains an independent leakage, while the (L_k-L_a) part ends up in parallel with a phase fraction of L_c and a sum of all parasitic inductances L_b on the secondary (Fig. 3). Assuming the same BCLs are linked, (6) can be derived, where all the L_b values are assumed the same and are just summed as $N_{ph} \times L_b$. If BCLs with different numbers of phases are linked, then the L_b values will be different for different BCLs and (6) can be easily adjusted to account for this.

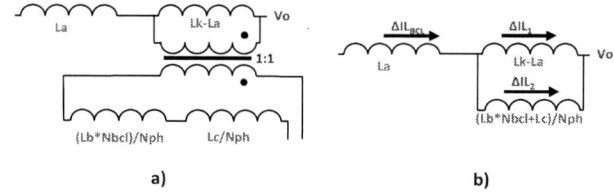

Fig. 3. Recalculating the effective leakage L_{k_bcl} in each phase: (a) splitting the L_b and L_c contributions for each phase, and (b) an equivalent current divider in each phase.

The leakage inductance L_k and the mutual inductance L_m are "original" values measured in the BCL. The figure-of-merit (FOM) is expressed as (3), where N_{ph} is the number of coupled phases, ρ is a coupling coefficient (4), and j is a running index (5) which simply defines an applicable interval of the duty cycle.

$$\Delta IL_{DL} = \frac{Vin-Vo}{L} \cdot \frac{D}{Fs} \quad (1)$$

$$\Delta IL_{BCL} = \frac{V_{in}-V_o}{L_{k\_BCL}} \cdot \frac{D}{Fs} \cdot \frac{1}{FOM(D,N_{ph},\rho,k)} \quad (2)$$

$$FOM = \frac{\left(1+\frac{\rho}{\rho+1}\frac{1}{N_{ph}-1}\right)}{1-\frac{(N_{ph}-2\cdot j-2)+\frac{j\cdot(j+1)}{N_{ph}\cdot D}\frac{N_{ph}\cdot D\cdot(N_{ph}-2\cdot j-1)+j\cdot(j+1)}{N_{ph}\cdot(1-D)}}{\frac{N_{ph}-1}{\frac{\rho}{\rho+1}}}} \quad (3)$$

$$\rho = \frac{L_m}{L_{k\_BCL}} \quad (4)$$

$$j = trunc(N_{ph} \cdot D) \quad (5)$$

$$L_{k\_BCL} = L_a + (L_k - L_a) // \left(\frac{L_b N_{bcl}+L_c}{N_{ph}}\right) \quad (6)$$

It is necessary to remember that the proposed current ripple estimate enables a very quick result for the arbitrary configuration of potentially different BCLs, but it is based on simplification and will be prone to errors. The estimate should be useful to attain a ballpark number for the current ripple and to investigate overall trends, but then the actual current ripple can be obtained - for example, from a simulation with chosen parameters and in exact conditions of interest.

For comparison, the TLVR current ripple for any duty cycle range is shown as (9), [16, 23], where the maximum number of phases occurring simultaneously in the on-state is $N_{SimOnMax}$ (7). D_{HF} is the on-time calculation for the total ripple current (8). Assuming that the connected TLVR value splits into $L = L_k + L_m$, the TLVR coupling coefficient k is defined by (10). L_c is an external tuning inductor, as shown in Fig.1a.

$$N_{SimOnMax} = Roundup(N_{ph} \cdot D, 0) \qquad (7)$$

$$D_{HF} = \frac{t_{overlap}}{(1/N_{ph} \cdot F_s)} = N_{ph} \cdot D - INT(N_{ph} \cdot D) \qquad (8)$$

$$\Delta I_{TLVR} = \frac{V_{in}}{F_s} \left[\frac{1}{L_m} (1 - D)D + \frac{k^2 \cdot D_{HF}}{L_c} \left(\frac{N_{SimOnMax}}{N_{ph}} - D \right) \right] \qquad (9)$$

$$k = \frac{L_m}{L_m + L_k} \qquad (10)$$

To the first degree, the transient in both the BCL and the TLVR is set by L_c and is therefore equal. The ideally small parasitics L_a and L_b in the BCL and L_k in the TLVR will also affect the transient and may result in a simple adjustment of L_c if the transient is too affected by parasitics.

A developed $h=12$mm four-phase BCL is shown in Fig. 4. It is designed to be compatible with the TLVR footprint, so the same multiphase layout can be loaded either with BCL or TLVR alternatives. The windings are lifted on leads by 0.5mm so they are not shorting to the pads in the middle for the TLVR secondary. The system with eight phases in total was targeted.

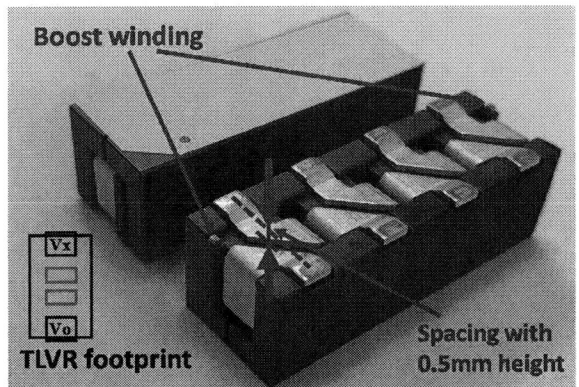

Fig. 4. Developed four-phase BCL, compatible with the TLVR footprint.

The four main parameters for the models in Fig.2 were extracted from the four BCL measurements in different connection configurations as $L_k=89$nH, $L_m=211$nH, $L_a=7.5$nH, and $L_b=35$nH. This is measured for the BCL with $Nph=4$. Notice that parasitics L_a and L_b are most likely different for the BCL

with a different number of phases. The BCL was designed to have a comfortable $I_{sat}=110$A/phase for the aggressive transient performance.

The BCL solution is compared to the TLVR in the same size and with the same transient current slew rate. To be as optimistic as possible with TLVR efficiency, TLVR=150nH was used to fit in the same solution footprint and dimensions. Notice that this leads to only $I_{sat}=65$A/ph, which still allows for the system to run properly, but is insufficient for the aggressive transient steps, especially with a fast repetition rate. The current ripple vs V_o for the $V_{in}=12$V, $F_s=400$KHz conditions is plotted in Fig. 5. A typical value $L_c=120$nH was used for both BCL and TLVR systems.

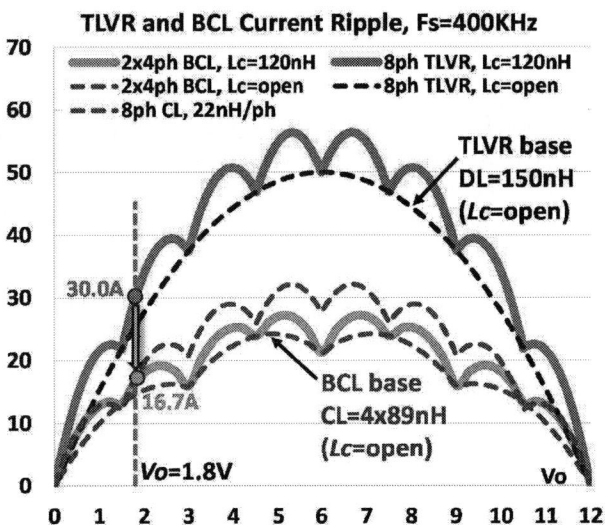

Fig. 5. Current ripple for eight-phase systems: eight TLVR=150nH and two BCL=4×89nH. $V_{in}=12$V, $F_s=400$KHz. Proposed BCL estimate: eight phase CL.

TLVR=150nH is fundamentally based on DL=150nH, so when L_c is open circuit, the phases operate as independent DLs (dashed black curve in Fig. 5). When L_c is sufficiently small (120nH), the TLVR phases are linked and operate similar to the eight-phase CL: the current ripple cancellation forms (N_{ph}-1) notches (red curve, calculated from (9)). However, the current ripple can only increase from the DL base. This is guaranteed to decrease efficiency, but a sufficiently high N_{ph} would generally mitigate the current ripple increase and therefore efficiency impact [22, 24].

Interestingly enough, the BCL current ripple behaves in a similar fashion. The baseline ripple corresponds to the ripple in the CL building block, in this case a single four-phase CL=4×89nH (purple curve in Fig. 5; L_c is open circuit). When L_c is reduced to 120nH, the two BCLs are linked and the system starts to behave similarly to eight-phase CL (green curve, simulated with the BCL model from Fig. 2 and the measured BCL parameters described above). The BCL ripple can also only increase from the baseline curve, similar to TLVR behavior.

The proposed analytical estimate for the current ripple in the BCL system is shown in Fig. 5 as a dashed blue curve. It is calculated from (2), where the effective leakage value L_{k_bcl} is recalculated as (9) and ends up as 22nH/ph (for $L_c=120$nH). The estimate provides a good ballpark value for the current ripple in an easy way, although the purposeful simplification leads to the difference between the estimated and simulated curves.

Notice in Fig. 5 that since the current ripple in four-phase CL as a baseline for the BCL is significantly lower than the current ripple in the DL as a baseline for the TLVR, the BCL ripple remains significantly lower than the TLVR ripple when L_c is connected. For example, in the condition of interest V_o=1.8V, the TLVR has a ripple of 30A, while the BCL ripple is only 16.7A (1.8× advantage, with a large I_{sat} advantage on top of that).

This also illustrates an interesting change in design objective. A typical CL is designed with considerations of both the current ripple and the transient for particular application. This often leads to excessive I_{sat} for L_k as leakage is taken down for the faster transient while still enabling a reasonably low current ripple. In the BCL, the transient is mostly set by the external L_c. Then the objective for the BCL is to design the largest possible L_k for the application defined I_{sat} per phase, so the lowest current ripple base (Fig. 5) is established for the highest efficiency. Notice also that the larger L_k is favorable for linking different BCLs (Fig. 1b) as better coupling between different BCLs enables better current ripple cancellation.

III. PERFORMANCE COMPARISON

A good way to compare different magnetics solutions in terms of current ripple (that affects efficiency) and transient tradeoff was shown in [22]. The FOM is defined as a ratio between the relative change of the current slew rate in the transient, normalized by a benchmark DL, to the relative change in the current ripple, also normalized by the same DL. Given that the DL current slew rate is the same in the steady state or transient (so the FOM for DL is always 1), equation can be simplified as (11), where SR_{tr} is the current slew rate in the transient, ΔIL is the current ripple, D is the duty cycle, and T_s is a switching period. The transient slew rate is simply a voltage (V_{in}-V_o) applied during the loading transient divided by the transient inductance in each phase: for the BCL, it is L_{k_bcl}, calculated as 22nH from (6) when L_c=120nH, and for the TLVR it is L_c/N_{ph} plus measured L_k=7nH.

$$FOM = \frac{SR_{tr}}{SR_{tr\_DL}} / \frac{\Delta IL}{\Delta IL_{DL}} = \frac{SR_{tr}}{\Delta IL} \cdot D \cdot T_s \quad (11)$$

To put things into perspective, the current ripple plot in Fig. 6 compares some additional eight-phase systems. The same F_s=400KHz was used for all curves for comparison purposes.

Matching the I_{sat}=110A/ph spec of BCL would lead to TLVR=100nH and a corresponding current ripple of 42.6A for the 1.8V output. The efficiency impact caused by the high current ripple or increased F_s would be very significant. Choosing TLVR=150nH would improve the current ripple to 30.0A but lead to I_{sat}=65A/ph, which allows the system to still run in the full load range but creates issues for the aggressive transient with a fast repetition rate. Linking two BCLs=4×89nH (L_m=211nH) leads to much better current ripple of 16.7A and therefore significant efficiency improvement.

If it is possible to manufacture a single eight-phase NCL=8×22nH structure [22, 24], then the leakage plate would

be removed, and in these given dimensions, L_m would be reoptimized to ~400nH, resulting in a further current ripple reduction to 10.9A. The latter potentially enables further F_s reduction for even more efficiency gain, while the current ripples of TLVR options above generally imply an increase of F_s from 400KHz.

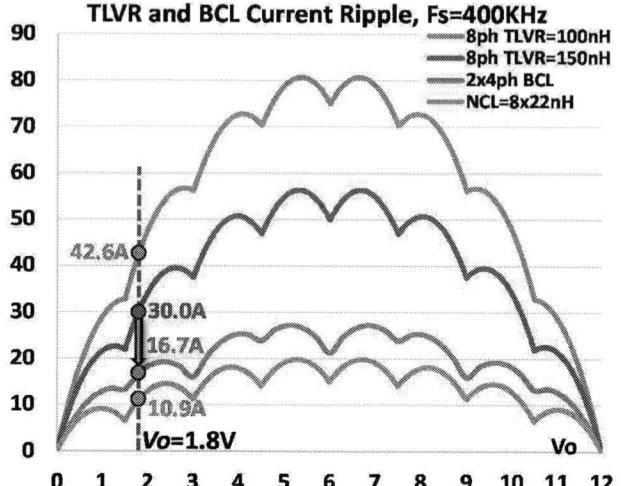

Fig. 6. Current ripple for the eight-phase systems: V_{in}=12V, F_s=400KHz, and L_c=120nH for the TLVR and BCL.

Fig. 7 illustrates the FOM of these different magnetics options. The DL baseline for the TLVR is always equal to 1. Linking TLVR phases improves the transient at a faster rate than it increases the current ripple (especially with a sufficiently large N_{ph}), so the FOM is increased. TLVR=100nH has a small FOM increase, which could be improved with TLVR=150nH at the expense of the saturation rating. The BCL starts with a much higher base of CL=4×89nH (L_c open), and when linked into eight-phase system: the FOM is increased significantly further. If it is possible to make a monolithic 8-phase NCL=8×22nH, optimized for coupling, then the FOM would increase even more.

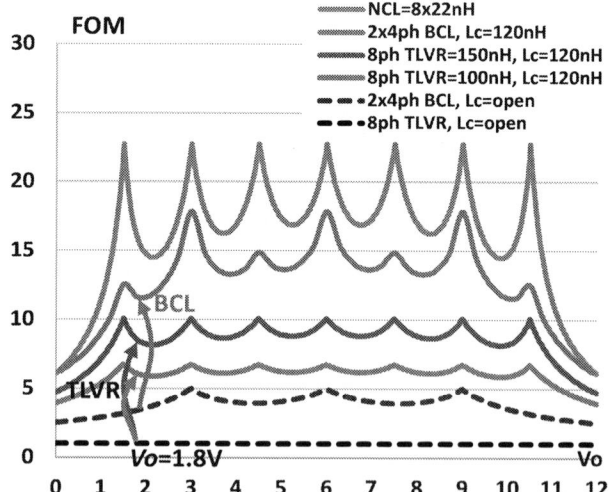

Fig. 7. FOM for the 8-phase systems: V_{in}=12V, F_s=400KHz.

Notice that all the options in Fig. 6 are purposefully set for the same transient and set at the same F_s to emphasize the current ripple difference. However, the FOM can be generally compared between systems that have different transients and F_s, etc.

To complete the performance comparison, the eight-phase board in Fig. 8 was developed with a common footprint (CF) for magnetics that can accept either the TLVR or the BCL. The board demonstrated better efficiency when loaded with two BCLs=4×89nH as compared to the TLVR=150nH option (Fig. 9). If the TLVR=100nH parts are loaded to match I_{sat}=110A and meet the repetitive transient spec with the same output capacitance C_o, the efficiency would drop much further. This is expected from the current ripple estimates in Fig. 6.

Fig. 8. Developed eight-phase board with a common footprint that accepts both the TLVR and the BCL.

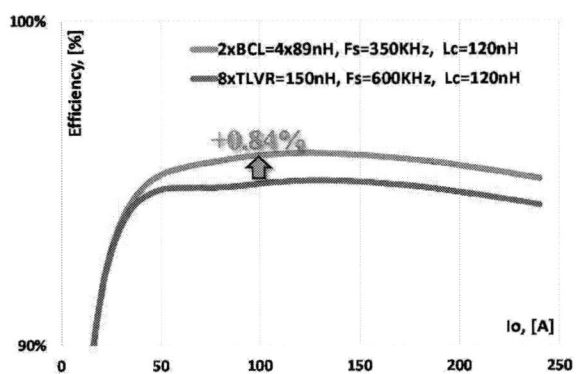

Fig. 9. Measured efficiency: eight TLVRs=150nH and two BCLs=4×89nH.

IV. OUTPUT CURRENT RIPPLE CANCELLATION

Both the BCL and the TLVR are linked on the secondary side with a tuning inductor L_c (Fig. 10a) if L_c inductance is sufficiently low. As the secondaries are coupled to all the phases, L_c has an AC synchronized with peaks from all the phases, which results in a similar shape to the total output current ripple that goes into the output capacitance C_o.

The idea is then illustrated in Fig. 10b: the L_c current is sent to the output out of phase to cancel the current ripple. The blocking cap C_b ensures that the V_o DC level remains unaffected. The transformer is proposed to replace the inductor L_c. If the inductance on the transformer primary (with the secondary shorted) is arranged to match the original L_c value, then the internal BCL waveforms will be preserved. This implies the same phase current ripple, same efficiency, etc. The transformer is necessary to step up the AC, as the original L_c current I_{BCL} in Fig. 10a ideally approaches the ripple in a single phase (depending on the coupling), while at the output, all the phases add their ripples to the total I_o. So ideally (if coupling to the BCL secondary is very high), the transformer needs to have a $N_{tr}=N_{ph}$ turns ratio to match the I_{BCL} ripple to I_o for optimal ripple

cancellation. Current ripple cancellation circuit was also proposed in [5], but since it is a single-phase circuit and a passive secondary winding is effectively shorting out the mutual inductance – the current ripple and therefore efficiency can be significantly affected.

The N_{ph}:1 ratio could be considered for the BCL itself, but it could be difficult to arrange. The additional benefit of arranging the N_{ph}:1 ratio separately in a small external transformer that replaces the inductor L_c is that in systems with different BCL building blocks and a different number of BCLs – only that small transformer would be modified, while the BCL design would not change from system to system.

From Fig. 10b, it is easy to see that the actual output current I_{O_TOT}, (12), is now a sum of original I_o and the added I_{BCL}, depending on the polarity of the transformer connection. Now the AC components of these two currents are needed in analytical form to find the resulting ripple cancellation.

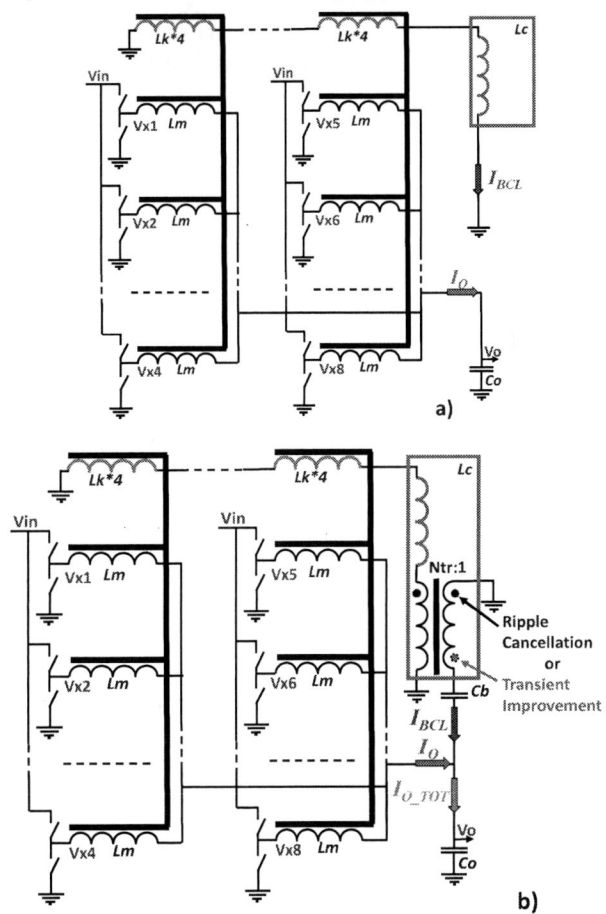

Fig. 10. BCL system: a) original, b) proposed circuit with the output current ripple cancellation (or transient improvement if polarity is switched).

For the conventional multiphase buck converter with the DL, the total output current ripple ΔI_o can be found as (13) [1, 9].

$$I_{O\_TOT} = I_O \pm I_{BCL} \tag{12}$$

$$\Delta I_O = \frac{V_o}{L \cdot F_s} \cdot \left(1 - \frac{j}{N_{ph} \cdot D}\right) \cdot \left(1 + j - N_{ph} \cdot D\right) \tag{13}$$

It has been established that the CL with the same leakage value as the DL inductance would have the same total output current ripple. In other words, the amount of coupling between phases does not affect the total output current ripple to the first degree. The BCL system is effectively a mix of two different coupling mechanisms: one is direct magnetic coupling between the phases in each monolithic BCL, and another is between BCL building blocks via the electrical connection of the secondaries and L_c. One hypothesis could be that similar to the direct magnetic coupling not affecting the total output current ripple in a single CL, the same could be assumed for a system with two different coupling arrangements. Then L_{k_BCL} (6) can be substituted into (13).

The BCL system was simplified above as effectively a system with the same magnetic coupling between all the phases to estimate the phase current ripple (2) with effective leakage inductance in each phase as L_{k_bcl} (6). With this simplification, the current in each phase would look like a typical current in the CL phase (Fig. 11), where in one switching cycle N_{ph} ripple peaks would appear that relate to the leakages in different phases. The simulation in Fig. 11 has the conditions N_{ph}=8, V_{in}=12V, V_o=0.75V (non-overlapping duty cycles for clarity), and F_s=400KHz. The current ripple in one phase is shown, together with the related PWM signal. Fig. 11 clearly shows a (highlighted) triangular current ripple component associated with L_m, which raises the "native" ripple peak associated with the duty cycle in the local phase to be higher than the reflected ripple peaks.

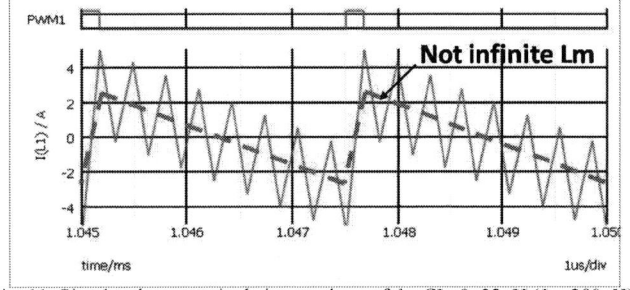

Fig. 11. Simulated current ripple in one phase of the CL=8×22nH (L_m=280nH): V_{in}=12V, V_o=0.75V F_s=400KHz.

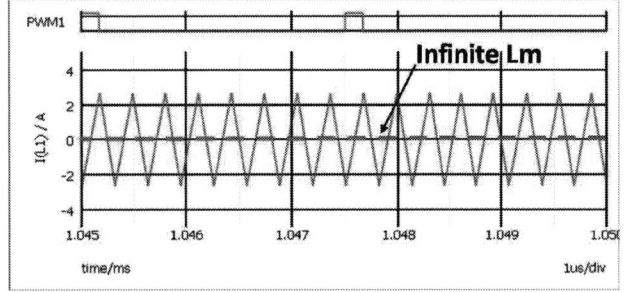

Fig. 12. Simulated leakage only current ripple in one phase of the CL=8×22nH (L_m=infinite): V_{in}=12V, V_o=0.75V F_s=400KHz.

Assuming that the coupling and mutual inductance value do not affect the total output current ripple, a similar hypothesis can be applied to the current ripple in the BCL secondary I_{BCL}, coupled to all the phases at the same time, where all phase AC effectively mix in a similar fashion. The component of the phase current of interest that relates to the leakage can therefore be

obtained from the phase current ripple ΔIL_{BCL} (2) by eliminating the ripple component from L_m. This can be easily done by modifying (2) for the infinite coupling coefficient or just inserting some very large L_m into the equations. The phase current waveform without the L_m component is shown in Fig.12 for comparison - the same scale as the original waveform in Fig. 11. The "leakage only ripple component" ΔIL_{BCL_k} is then defined as (14).

$$\Delta IL_{BCL\_k} = \Delta IL_{BCL}(L_m = infinite) \tag{14}$$

The next objective is to calculate I_{BCL} (the AC in the secondary winding of the L_c). Consider the current divider in Fig. 3b. The AC associated with the transformer (BCL secondary) can be found as (15), while (16) can also be established. The current of interest I_{BCL} is on the transformer secondary (17) and can be solved from (15-17) as (18).

$$\Delta I_2 = \Delta IL_{BCL\_k} - \Delta I_1 \tag{15}$$

$$\frac{\Delta I_1}{\Delta I_2} = \frac{(L_b \cdot N_{bcl} + L_c)/N_{ph}}{L_k - L_a} \tag{16}$$

$$I_{BCL} = \Delta I_2 \cdot N_{tr} \tag{17}$$

$$I_{BCL} = \frac{\Delta IL_{BCL\_k} \cdot N_{tr}}{1 + \frac{L_b \cdot N_{bcl} + L_c}{N_{ph}(L_k - L_a)}} \tag{18}$$

$$I_{O\_TOT} = I_O - I_{BCL} \tag{19}$$

The plot is Fig. 13 compares simulated (dots) and calculated (solid lines) output currents: the output current ripple from the main phases I_O, I_{BCL}, and the resulting I_{O_TOT}=I_O-I_{BCL} (19), when the two are subtracted for the ripple cancellation.

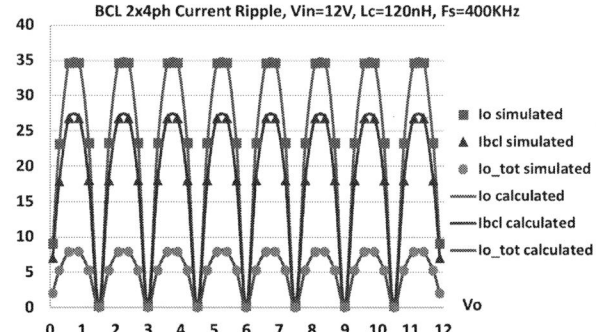

Fig. 13. Simulated and calculated output currents in the eight-phase (two BCLs=4×89nH) system: V_{in}=12V, F_s=400KHz, L_c=120nH.

In an interesting outcome, the calculated and simulated ripple currents at the output significantly match in Fig. 11, while the proposed analytical estimate (2) for the phase current ripple in the BCL shows a certain deviation (Fig. 5). The accurate correlation for the calculated output currents based on a "less accurate" estimate for the phase current ripple illustrates that the assumed hypothesis that the output currents in the BCL system are not dependent on the internal coupling between phases is correct, similar to a case of a single monolithic CL. This is different from the BCL phase current ripple (2), which is actually quite dependent on different coupling mechanisms, hence why the proposed simplification of coupling affected the accuracy in this case (Fig. 5). The fully derived analytical output ripple cancellation (18-19) can now be easily analyzed. The first observation from Fig. 13 is that the current ripple cancellation is not ideal. While the total output ripple is decreased (4.4× at V_o=1.8V), it is not as perfect as expected. The reason is that the coupling between the primary and secondary of the BCL is not ideal; in other words, mutual inductance $(L_k-L_a)\times N_{ph}$ is not infinite, and the other inductances (L_a, L_b, and L_c) are not insignificant. Fig. 14 shows the total output current ripple as a function of the transformer turns ratio for different L_c values. The ideal transformer turns ratio $N_{tr}=N_{ph}$ can be a starting point, but the larger turns ratio is generally needed to compensate for the non-ideal coupling and achieve better ripple cancellation at output. The smaller the L_c, the better the coupling, and the closer the transformer ratio for the maximum ripple cancellation to the ideal $N_{tr}=N_{ph}$.

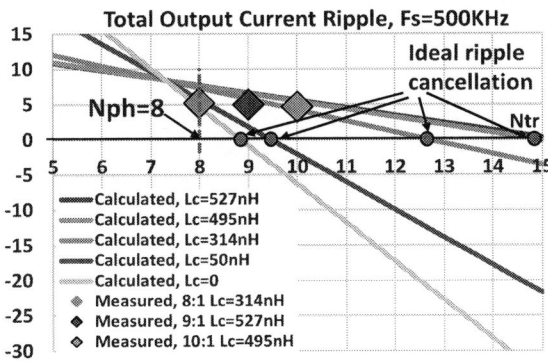

Fig. 14. Total output current ripple in two BCLs=4×89nH vs N_{tr} for different L_c values. V_{in}=12V, V_o=0.75V F_s=500KHz. Calculated and measured.

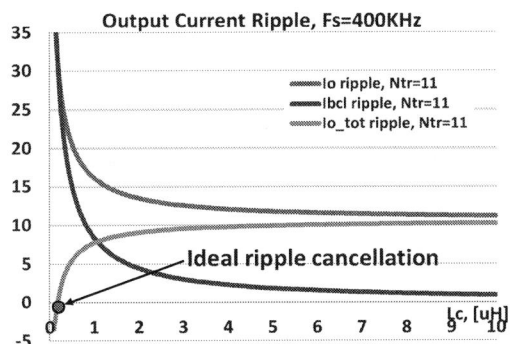

Fig. 15. Output current ripple in two BCLs=4×89nH vs L_c for N_{tr}=11. V_{in}=12V, V_o=0.75V F_s=400KHz.

Fig. 15 shows the output ripple currents as a function of L_c. Making L_c smaller generally improves the coupling between separate BCLs and the L_c, until the zero current ripple is achieved at output and then the ripple flips polarity. Increasing L_c effectively disconnects BCL blocks from each other, while I_{BCL} decreases and stops cancelling the output ripple.

V. EXPERIMENTAL RESULTS.

The eight-phase board from Fig. 8 was hacked with a daughter card that has current sense resistors for I_O, I_{BCL}, and I_{O_TOT}, C_b and a footprint for the L_c transformer (Fig. 16). The measured output currents are shown in Fig. 17 for a L_c=8:1 transformer on an RM5 core with 314nH value on the primary (when the secondary is shorted). The I_{BCL} AC ripple is subtracted from the main windings current I_O and results in I_{O_TOT} that goes to the output. Significant 2.8× reduction in the output current

ripple is achieved (from 13.8A to 5A). Lowering stray inductance in the 8:1 L_c transformer would improve the ripple cancellation further (the L_c prototype has excessive 314nH on the primary). Another improvement can result from optimizing the transformer turns ratio for a given L_c value.

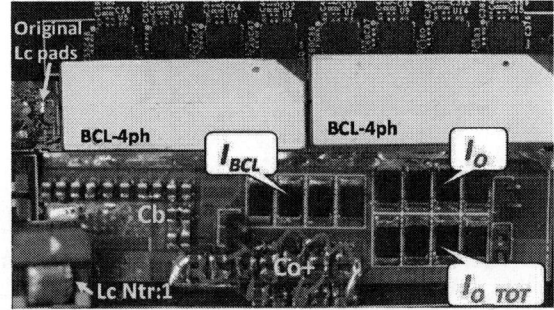

Fig. 16. Eight-phase board with current sensors, C_b and L_c transformer.

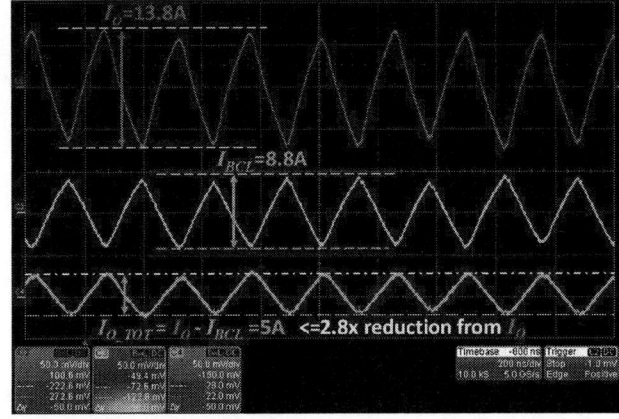

Fig. 17. Measured output ripple cancellation (V_{in}=12V, V_o=0.75V, F_s=500KHz, L_c=314nH 8:1).

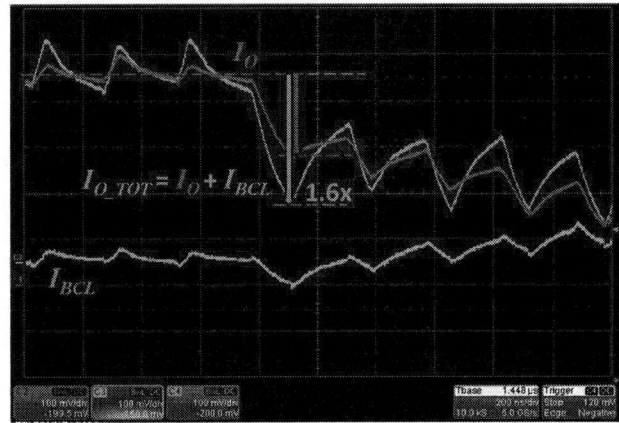

Fig. 18. Transient currents for 50A load step (V_{in}=12V, V_o=0.75V, F_s=500KHz).

Alternatively, the connection of the BCL secondary current I_{BCL} in Fig. 10 can be reversed and added to the I_O during the transient step. Fig. 18 shows the measured currents during the 50A unloading step: the change of I_{O_TOT} is 1.6× larger than the initial I_O step, which generally enables size and cost reduction in the output capacitance tank. Extra gain shows up in the transfer functions only above the corner frequency (14).

979-8-3315-1612-3/25 $31.00 © 2025 IEEE

$$F_{Lc} = \frac{N_{tr}}{2\pi\sqrt{L_c C_b}} \tag{14}$$

More of a transient related analysis cannot be inserted here and must be investigated separately.

VI. Conclusions

A model for the BCL was developed to be used in simulations of the multiphase converters. The proposed simplified analytical estimate for the phase current ripple in the BCL system showed good ballpark accuracy.

The concept of adding the already existing secondary BCL current to the output allows to either significantly cancel the total output current ripple or multiply the transient output current of the converter. An analysis was done for the fundamentals of the current ripple cancellation at output, demonstrating a match with the simulated results. The proposal was verified in the prototype where $2.8\times$ reduction in the output current ripple was demonstrated. Alternatively, $1.6\times$ improvement in the transient current step was shown. The latter proposal can be beneficial in high current systems where the output capacitance depends more on the transient specs rather than the output current ripple.

The proposed concepts could be implemented in the TLVR solution or a single BCL, similar to the shown multiple BCL system. A patent has been filed. Assitional references can be found in [29].

The future work should include the L_c transformer with PCB windings to have a better control over the L_c leakage value and repeatability in manufacturing, and the transient improvement concept needs to be investigated and detailed.

References

[1] B. Miwa, "Interleaved conversion techniques for high density power supplies," Ph.D. thesis, 1992, Massachusetts Institute of Technology, USA.

[2] A. M. Schultz and C. R. Sullivan, "Voltage converter with coupled inductive windings, and associated methods," U.S. Patent 6,362,986, filed March 22, 2001.

[3] R. W. Erickson and D. Maksimović, Fundamentals of Power Electronics, 2nd ed., Kluwer Academic Publishers, 2001.

[4] J. Li, C. R. Sullivan, and A. Schultz, "Coupled inductor design optimization for fast-response low-voltage DC-DC converters," in Proc. IEEE Applied Power Electronics Conference, 2002, vol.2, pp. 817–823.

[5] M. Schutten; R. Steigerwald; and J. Sabate, "Ripple current cancellation circuit," in Proc. IEEE 2003 Applied Power Electronics Conference, 2023.

[6] J. Li, A. Stratakos, C. R. Sullivan, and A. Schultz, "Using coupled inductors to enhance transient performance of multi-phase buck converters," in Proc. IEEE Applied Power Electronics Conference, 2004, vol. 2, pp. 1289–1293.

[7] M. Xu, Y. Ying, Q. Li, and F. C. Lee, "Novel coupled-inductor multi-phase VRs," in Proc. IEEE Applied Power Electronics Conference, pp. 113-119, 2007.

[8] Z. Lu and W. Chen, "Multi-phase inductor coupling scheme with balancing winding in VRM applications," in Proc. IEEE Applied Power Electronics Conference, 2007.

[9] T. Hegarty, "Benefits of multi-phasing buck converters," EE Times, November 2007.

[10] Y. Dong, "Investigation of multiphase coupled-inductor buck converters in point-of-load applications," Ph.D. thesis, 2009, Virginia Polytechnic Institute and State University, USA.

[11] G. Zhu, B. McDonald, and K. Wang, "Modeling and analysis of coupled inductors in power converters," in Proc. IEEE Applied Power Electronics Conference, 2009.

[12] J. Imaoka, Y. Ishikura, T. Kawashima, and M. Yamamoto, "Optimal design method for interleaved single-phase PFC converter with coupled inductor," in Proc. IEEE Energy Conversion Congress and Expo, 2011.

[13] N/A, "Fast multi-phase trans-inductor voltage regulator", Technical Disclosure Commons, May 09 2019, available online: https://www.tdcommons.org/dpubs_series/2194.

[14] S. Jiang, X. Li, M. Yazdani, and C. Chung, "Driving 48V technology innovations forward - hybrid converters and trans-inductor voltage regulator (TLVR)," industry session in IEEE Applied Power Electronics Conference, 2020.

[15] J. Zhou, "Scalable 2-stage 48V to PoL power delivery for data centers," industry session in IEEE Applied Power Electronics Conference, 2020.

[16] Infineon Technologies, "Multiphase buck converter with TLVR output filter," eeNews Europe, February 2021. Available at: https://www.eenewseurope.com/Learning-center/multiphase-buck-converter-tlvr-output-filter

[17] D. Yao, A. Ikriannikov, A. Pizzutelli, and T. Paing, "Boosted coupled inductors and associated systems and methods," U.S. Patent 11,756,725, filed May 2021.

[18] M. Chen and C. R. Sullivan, "Unified models for coupled inductors applied to multiphase PWM converters," IEEE Transactions on Power Electronics, vol. 36, no. 12, pp. 14155–14174, Dec. 2021.

[19] Y. Elasser, J. Baek, C. R. Sullivan, and M. Chen, "Modeling and design of vertical multi-phase coupled inductors with inductance dual model," in Proc. IEEE Applied Power Electronics Conference, 2021, pp. 1717–1724.

[20] D. Wiest and Y. Zhou "Trans-inductor voltage regulator (TLVR): circuit operation, power magnetic construction, efficiency and cost trade-offs," industry session in IEEE Applied Power Electronics Conference, 2022.

[21] J. Baek, Y. Elasser, K. Radhakrishnan, H. Gan; J. Douglas, et al., "Vertical stacked LEGO-PoL CPU voltage regulator," IEEE Transactions on Power Electronics, vol. 37, no. 6, pp. 6305 - 6322, June 2022.

[22] A. Ikriannikov, "Evolution and comparison of magnetics for the multiphase DC-DC applications," industry session in IEEE Applied Power Electronics Conference, March 2023.

[23] A. Fard, S. Naidu, H. Tamdem, and B. Vafakhah, "Trans-inductors versus discrete inductors in multiphase voltage regulators: An analytical and experimental comparative study," in Proc. IEEE Applied Power Electronics Conference, March 2023.

[24] A. Ikriannikov and D. Yao, "Converters with multiphase magnetics: TLVR vs CL and the novel optimized structure," in Proc. PCIM Europe, May 2023.

[25] F. Zhu, X. Lou, and Q. Li, "Design comparison of different coupled inductor concepts for voltage regulators," in Proc. IEEE Applied Power Electronics Conference, March 2023.

[26] A. Ismail, Z. Ma, A. Al-Hmoud, and Y. Zhao, "A high frequency coupled inductor design for high power density DC-DC converters," in Proc. IEEE 2023 Applied Power Electronics Conference, March 2023.

[27] P. Kumar, J. Tippetts, S. Naidu, and P. Brusco, "Efficiency impact of phase firing order in dual-sided power entry with trans-inductor voltage regulators (TLVR)," in Proc. IEEE Applied Power Electronics Conference, Feb. 2024.

[28] N. Ellis, Y. Zhu, and R. Pilawa-Podgurski, " A Gallium Nitride-Based 48V-to-1V Point-of-Load (PoL) Converter for Aerospace Telecommunications and Computing Applications," in Proc. IEEE Applied Power Electronics Conference, Feb. 2024.

[29] A. Ikriannikov and L. Lipcsei, "A Novel Concept of Injected Coupled Inductors," in Proc. IEEE Applied Power Electronics Conference, Feb. 2024.

[30] S. Jiang, "Next TLVR innovations: topologies, magnetics and control," industry session in IEEE Applied Power Electronics Conference, Feb. 2024.

System Solutions and Design Trade-offs to the Input Filter Interactions with Battery Chargers

Xigen Zhou
Battery Charging Products
Texas Instruments
Dallas, Texas, USA

Dan Mavencamp
Battery Charging Products
Texas Instruments
Dallas, Texas, USA

Kuang-Yao (Brian) Cheng
Battery Charging Products
Texas Instruments
Dallas, Texas, USA

Abstract— **An input filter is often required for the on-board battery chargers to meet the stringent electromagnetic interference (EMI) requirements. This creates a high-Q resonant tank and may cause input current oscillation if not properly designed. The adapter with high-Q resonant tank generates a peak output impedance at resonant frequency. The charger is a current-limiting DCDC converter with a large duty cycle, it creates a very low input impedance. When the peak output impedance adapter meets the low input impedance battery charger, it forms an unstable region, and induces input current oscillation.**

This paper studies a practical 4-cell 100W notebook charger, uses impedance analysis to illustrate the input filter interactions with the battery charger, discusses the design trade-offs of the input filters, and proposes three system solutions to avoid the input current oscillation. Experimental results show the input current oscillation is avoided. Furthermore, new dual random spread spectrum (DRSS) technique mitigates the EMI noise without the input filters. This provides an alternative and cost-effective path to address the EMI requirements without the input filter interactions.

Keywords—input filter, input current oscillation, impedance analysis, system interaction, EMI, DRSS.

I. INTRODUCTION

In a typical notebook computer (Fig. 1), AC power adapter connects to the utility line and provides power to the whole computer. A battery charger sits on the board, manages the battery charging and system power delivery.

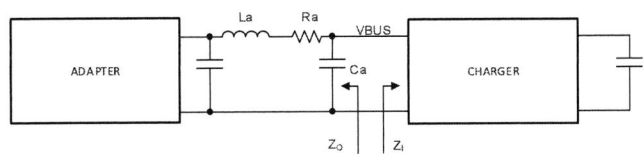

Fig. 1. A typical notebook architecture with an adapter and a charger

To comply with EMI requirements, a low pass filter is needed to filter out the high frequency noise, for example, a π filter consists of two capacitors and a ferrite bead/inductor. The filter inductor and the capacitors create a high Q resonant tank, which is prone to system oscillation under the noise excitation or perturbance.

On the other hand, high power battery chargers in buck mode with large duty cycles are capable of creating very low input impedances. High speed converters with extended bandwidth extends the low input impedance to a high frequency range, which is unfavorable for system stability.

Previous studies show the output impedance of the adapter Z_O (including the input filter and input capacitors) needs to be lower than the input impedance of the charger Z_i to guarantee a stable system design ($Z_O < Z_i$) [1-3]. This criterion serves as the theoretical basis for the system stability analysis.

This article uses a practical 4-cell 100W notebook charger as a case study, analyzes the adapter and the charger impedances for the first time, and reveals the input filter system interaction as the root cause of the input current oscillation. Three system solutions are proposed to avoid the input current oscillation, with design trade-offs discussed. With novel DRSS technology [4], modern battery chargers achieve low EMI without the need of the input filters, thus avoid the input filter system interactions.

The output impedance of the adapter is investigated in section 2. The input impedance of the battery charger is modeled in section 3. The input filter system impact is studied in section 4. Three different solutions are proposed and verified experimentally in section 5. EMI noise spectrums are compared to show the DRSS effectiveness without input filters in section 6.

II. ADAPTOR OUTPUT IMPEDANCE

The adapter usually incorporates a low pass filter between the charger and the adapter power supply. This filter inductor has an inductance and resistance. The impedance that the charger sees is a function of the filter, the output impedance of the adapter power supply, and the capacitance placed in front of the charger. If we assume the adapter power supply is well decoupled, and has sufficient bandwidth, we can write a simple form of the impedance Z_O in (1).

$$Z_o = \frac{\left[\frac{1}{S \cdot C_a}\right] \cdot [R_a + L_a \cdot S]}{\left[\frac{1}{S \cdot C_a} + R_a + L_a \cdot S\right]} \qquad (1)$$

Now insert $\omega = \sqrt{\frac{1}{L_a \cdot C_a}}$, and assume that Ra is small compared to $\sqrt{(L_a/C_a)}$, the peak impedance for the adapter $Z_{o(pk)}$ is derived in (3). The resonant frequency F_w is calculated in (4).

$$Z_o = \frac{R_a + J\sqrt{(L_a/C_a)}}{J \cdot R_a \cdot \sqrt{(C_a/L_a)}} \qquad (2)$$

$$Z_{o(pk)} = \frac{1}{R_a} \cdot \frac{L_a}{C_a} \qquad (3)$$

$$F_w = \frac{1}{2\pi \cdot \sqrt{L_a \cdot C_a}} \qquad (4)$$

where Ra is the filter resistance and output resistance of the adapter power supply, La is the filter inductance and cable inductance of the adapter power supply, and Ca represents the input capacitance in front of the charger.

III. CHARGER INPUT IMPEDANCE

The charger is a family of switching converters with wide input range, wide output range and multiple control loops.

The input source of the charger includes USB type A port, travel adapter and USB type-C adapter. The input range could be from 5V to 36V.

The battery is a multi-chemistry battery depending on the applications, commonly 3-cell or 4-cell Lithium-Ion (Lion) batteries for notebook applications.

One feature of the battery charger is safety. It usually regulates the charge current and charge voltage with input current limit and input voltage limit.

The multi-cell buck-boost battery charger is gaining increasing popularity because of its wide input range. For example, Texas Instruments' BQ25720 is able to charge 1-to 4-cell batteries from 3.5V to 26V input range.

The buck-boost battery charger operates in three operation modes. It is in buck mode when the input voltage is higher than the battery voltage; in boost mode when the input voltage is lower than the battery voltage; and in buck-boost mode when the input voltage is close to the battery voltage.

Furthermore, the buck-boost charger regulates the charge current in constant current(CC) mode or charge voltage in constant voltage (CV) mode. When the input power exceeds the input current limit, the charger starts to regulate the input current. The operation mode and regulation loop affect the charger input impedance and make the impedance analysis a complicated task.

In previous studies[5-7], the switching regulators were modeled with voltage loop in regulation. In this study, multiple loops were included in the simulation.

The open loop charger input impedance is modeled as a LC tank with a resistive load.

$$Z_{i(ol)} = \left(s \cdot L_O + \frac{R_L}{1 + s \cdot C_O \cdot R_L}\right) \cdot [V_{BUS}/V_{SYS}]^2 \qquad (5)$$

where R_L is the load resistance, L_O is the output inductance of the charger, and C_O represents the output capacitance of the charger.

The open loop input impedance is the lowest at the resonant frequency.

As shown in Fig. 2, when the output voltage is regulated, the regulation loop increases the converter bandwidth beyond the double pole frequency, thus helps to shift the $Z_{i(ol)}$ curve to a higher impedance.

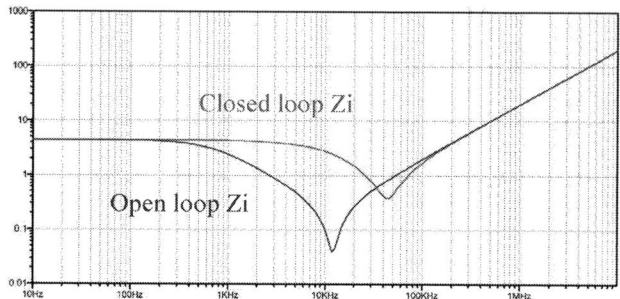

Fig. 2. Loop modulated charger input impedance

When the charger, a high-power switching converter, is operating in buck mode with large duty cycles, it creates very low input impedance. The input impedance for a buck converter inside the bandwidth of the converter loop, is defined as:

$$Z_i = R_L \cdot [V_{BUS}/V_{SYS}]^2 \qquad (6)$$

Advanced power path chargers have high speed converters with extended bandwidth. Beyond the bandwidth of the converter the input impedance will climb.

IV. SYSTEM INTERACTION

As illustrated in Fig. 3, when we overlap the adapter output impedance curve and charger input impedance curve, there could be an unstable region if the adapter output impedance Z_O is higher than the charger input impedance Z_i. This usually happens when the resonant frequency Fw is close to the converter bandwidth Fs.

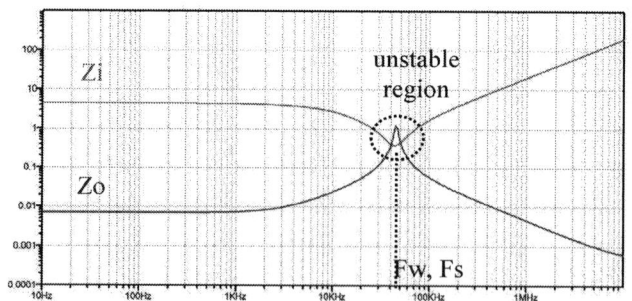

Fig. 3. Unstable region when adapter output impedance exceeds the charger input impedance

Another explanation of the system interaction is from the equivalent circuit. As shown in Fig. 4, the adapter is equivalent to a voltage source with internal impedance Z_O, while the charger is equivalent to a load Zi. When the charger is regulated as a constant power source, the input impedance is a negative value. In other words, when the charger input voltage increases, the charger input current decreases accordingly.

When the Z_O increases to the $|Zi|$ value, the adapter current will reach an infinite value, resulting in input current oscillation.

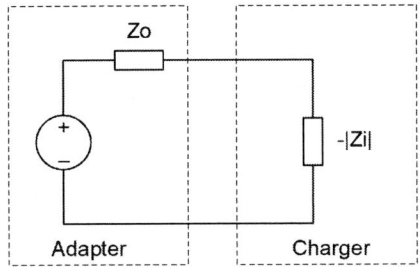

Fig. 4. Adapter and charger equivalent circuit highlighting negative Zi.

To meet the stability criteria, Zi should be higher than $Z_{O(PK)}$[1], otherwise, we will see input impedance instability issues. The calculations to find the intercept point where Z_O just intersects Zi are derived in (7).

$$R_L \cdot \left[\frac{V_{BUS}}{V_{SYS}}\right]^2 \geq \frac{L_a}{C_a} \cdot \frac{1}{R_a} \tag{7}$$

Now solve for La,

$$L_a \leq R_L \cdot R_a \cdot \left[\frac{V_{BUS}}{V_{SYS}}\right]^2 \cdot C_a \tag{8}$$

This means the adapter inductance La should be small enough to avoid the system interaction between the adapter and the charger. The upper limit of the La is shown in (8).

V. CASE STUDY AND SOLUTIONS

In a practical 100W 4-cell notebook battery charger application, the design parameters are listed in Tab. 1. The system runs a heavy load, and pushes the input current to reach its current limit. The original design deploys a 0.33uH inductor for EMI purpose. The input is a low-impedance 130W power adapter with around 10mΩ internal impedance. Zi-Zo impedance analysis reveals an unstable region in the original design in Fig. 5(a). It has been confirmed by the experimental results. The original system shows input current oscillation when the charger is operating full load with 5A input current limit. The input current oscillates from zero to 10A as shown in Fig. 5(b).

TABLE I.   100W BATTERY CHARGER DESIGN PARAMETERS

Input voltage (V)	Input current limit (A)	System voltage (V)	System load (A)
20	5	16.5	5.5

Input filter (µH)	Input capacitance (µF)	Output capacitance (µF)	Inductor (µH)
0.33	7x10	8x10	2.2

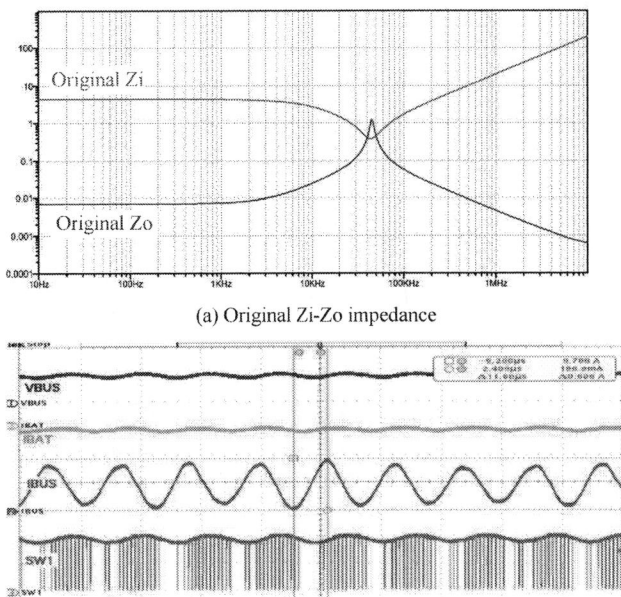

(a) Original Zi-Zo impedance

(b) Original system input current oscillation

Fig. 5. Zi-Zo impedance analysis shows an unstable region in the original design, which causes input current oscillation under full load conditions.

There are three ways to avoid the unstable region.

A. Increase adapter capacitance Ca

To reduce the output impedance Z_O, one method is to increase the adapter capacitance Ca by adding a 22uF bulk capacitor. As illustrated in Fig. 6, this method effectively lowers the resonant peak $Z_{O(PK)}$ and the resonant frequency Fw.

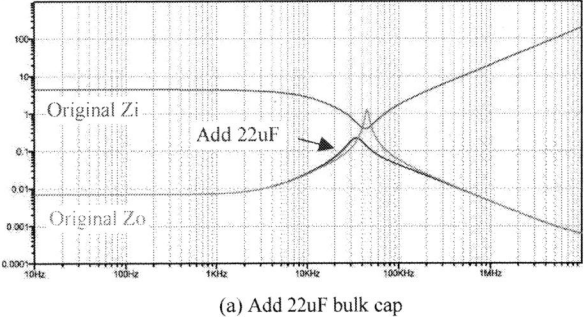

(a) Add 22uF bulk cap

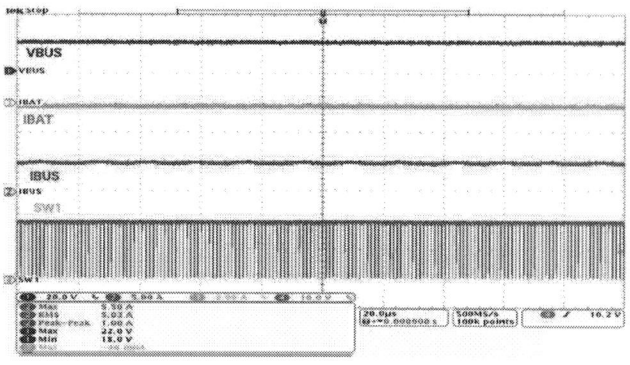

(b) No input current oscillation

Fig. 6. By increasing adapter capacitance Ca, the input current oscillation is eliminated.

B. Lower the charger bandwidth

By lowering the charger bandwidth Fs, the input impedance curve shifts to a lower frequency, thus avoid intersecting with the Z_O curve. Fig.7 shows the charger input impedance change and resulting waveform.

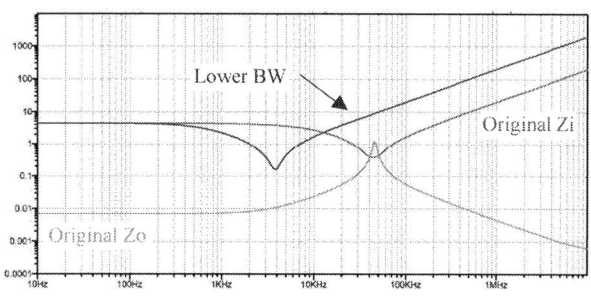

(a) Lower charger bandwidth

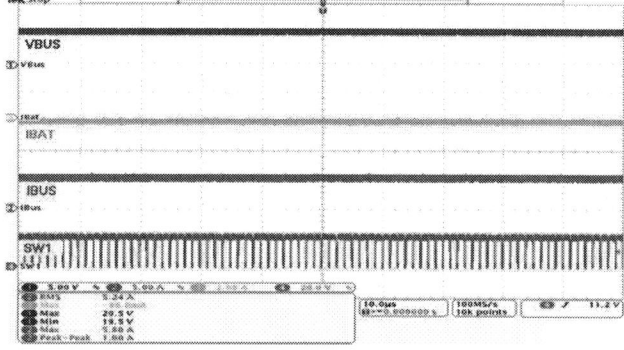

(b) No input current oscillation

Fig. 7. By lowering the charger bandwidth, the input current oscillation is avoided.

C. Reduce the adapter inductance La

The ferrite bead in the π-filter is the culprit of the system interaction. It significantly lowers the resonant tank frequency to near the frequency of the charger bandwidth. By removing the ferrite bead, the output impedance resonant peak is further away

from the charger bandwidth, thus avoid the unstable region. Fig. 8 illustrates the impedance changes from above method.

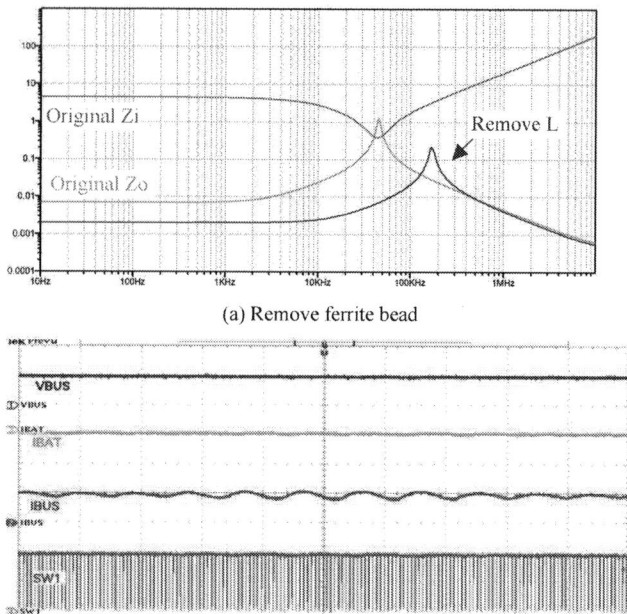

(a) Remove ferrite bead

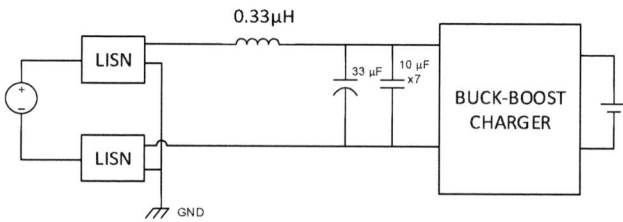

(b) No input current oscillation

Fig. 8. By reducing the adapter inductance La, the input current oscillation is suppressed.

In summary, the input current oscillation is gone by either (a) adding 22uF bulk cap, (b) lowering the charger bandwidth or (c) removing the ferrite bead.

VI. SHAPING EMI WITH DRSS

There is a small problem when removing the ferrite bead. It defeats the EMI filtering purpose. However, dual random spread spectrum (DRSS) is a new feature in modern battery chargers, which is able to change the switching frequency pattern thus re-shaping the EMI noise spectrum.

The conductive EMI noise was evaluated with 0.33uH filter inductor in the original battery charger. Fig. 9 shows the test setup under 20V input, 12.2V, 65W output conditions. Fig. 10 shows the original system noise spectrum without input filter.

Fig. 9. Test setup to evaluate the input filter-attenuated EMI noise spectrum

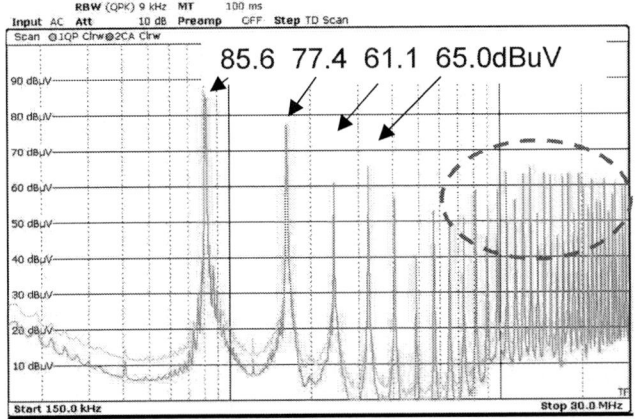

Fig. 10. Original system EMI noise spectrum without input filter

When the input filter is deployed, the noise spectrum in Fig. 11 shows 5-10dBuV noise reduction in the low frequency range (<4MHz), however, the high frequency (>5MHz) noises are about the same level. That means the filter inductor does not attenuate the high frequency noises which are dominated by common-mode noises.

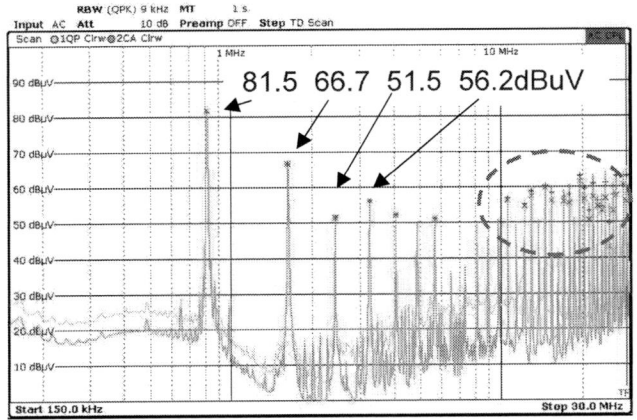

Fig. 11. Reduced EMI noise spectrum by 0.33uH filter inductor

For comparison, the filter inductor is removed, and the charger turns on the DRSS to reduce the EMI noise. The noise spectrum in Fig. 12 shows 5-10 dBµV reduction in the low frequency (<4MHz) noise peaks, moreover, the high frequency (>5MHz) noises are significantly reduced by 15-20 dBµV.

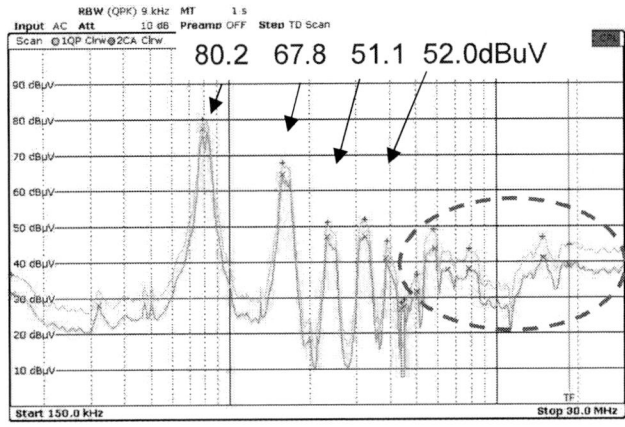

Fig. 12. Reduced EMI noise spectrum by 6% DRSS frequency dithering.

This comparison shows DRSS technique is an effective way to attenuate the EMI noises without affecting the system stability. By combining both DRSS and filter inductor, the EMI noise could be further reduced.

VII. SUMMARY

Input filter interactions with battery chargers are studied in a practical 4-cell 100W charger application. Impedance analysis is first applied to the high-bandwidth battery charger. Input filter design trade-offs are discussed. Additional capacitance or low charger bandwidth are proposed solutions to avoid the input current oscillation when an input filter is used. Alternatively, the DRSS technique can be used to achieve the EMI performance without the input filters.

REFERENCES

[1] Mohammed Alfayyoumi, Ali H. Nayfeh, Dusan Borojevic, Input Filter Interactions in DC-DC Switching Regulators, IEEE-PESC, 1999

[2] Yungtaek Jang, Robert W. Erickson, Physical Origins of Input Filter Oscillations in Current Programmed Converters, IEEE Trans. On Power Electronics, Vol 7, No 4, Oct 1992.

[3] Mosaddique Nawaz Hussain, Vivek Agarwal, A Novel Feedforward Stabilizing Technique to Damp Power Oscillations Caused by DC–DC Converters Fed From a DC Bus, IEEE Journal of Emerging and Selected Topics in Power Electronics, Vol 8, No. 2, June 2020.

[4] Paul Curtis, Eric Lee, EMI Reduction Technique Dual Random Spread Spectrum, Application Notes, Texas Instruments, Nov, 2020.

[5] Robert W Erikson, Dragan Maksimovic, Fundamentals of Power Electronics, Springer, Third Edition, 2020

[6] R. D. Middlebrook, Input Filter Considerations in Design and Application of Switching Regulators, IEEE, 1976.

[7] Charles Zhang, Anlysis and Design of Input Filter for DC-DC Circuit, Application Note, Texas Instruments, 2017.

Modeling and Implementation of a Zero Bias TLVR

Lei Wang
Renesas electronics America
(Austin)
Performance Core Power
(America)
Austin, USA
email address or ORCID

Travis Guthrie
Renesas electronics America
(Austin)
Performance Core Power
(America)
Austin, USA
email address or ORCID

Peyman Asadi
Renesas electronics America
(Austin)
Performance Core Power
(America)
Austin, USA
email address or ORCID

Mark Alexander
Renesas electronics America
(Austin)
Performance Core Power
(America)
Austin, USA
email address or ORCID

Kunrong Wang
Renesas electronics America
(Austin)
Performance Core Power
(America)
Austin, USA
email address or ORCID

Brandon Howell
Renesas electronics America
(Austin)
Performance Core Power
(America)
Austin, USA
email address or ORCID

Abstract—An alternative version of TLVR, named Zero Bias TLVR, has been proposed in recent years. This new topology employs magnetic flux cancellation in all windings, thus enabling a much lower saturation requirement and higher density. At the same time, this topology also introduces an asymmetry, since the external coupled phase has a different inductance value compared to that of the other phases. A modeling approach is developed here to analyze the circuit, and the transient inductance of each phase (or the current slew rate of each phase), is derived to facilitate the control loop design. Simulation and experiments are presented to verify this modeling approach.

Keywords—Tran-Inductance Voltage regulator (TLVR), Zero bias TLVR, modeling

I. INTRODUCTION

TLVR has been an attractive topology choice in recent years [2-5] for high power and high-speed CPU/GPU applications. It provides faster transient response, since its equivalent inductance is much smaller than a traditional VR due to the magnetic coupling effect. However, one of its shortcomings is that there is no magnetic flux cancellation in all trans-inductors, and thus each one of them has to be designed with a high saturation point based on real peak current. This requirement increases the magnetic core size. To address that issue, an alternative version, known as zero-bias TLVR, was introduced recently [1]. This topology connects the previous secondary Lc windings to the output point, making it operate much like a regular phase. This change causes the special secondary Lc phase to have the same DC current as other phases, thus enabling continuous flux cancellation. As a result, the saturation current requirements and physical size of all the inductors are greatly reduced.

Although TLVR has been thoroughly investigated in previous studies [2-5], this new version, zero bias TLVR, has

not been fully analyzed. Its asymmetry needs to be studied to help develop an appropriate control strategy, thus avoiding any closed loop instability. In this paper, a modeling approach is presented, and the multi-winding transformer is modeled using the inductance-dual method [6-8]. The circuit behavior during transient is investigated. The transient inductance (or current slew rate) of each phase is derived, and this will enable a correct control system design to help mitigate any trial-and-error debugging. Simulation and experiments have been conducted to validate this modeling approach and control loop programming.

II. MODELING

Zero-bias TLVR consists of multi-winding grouped inductors, instead of discrete transformers to reduce the total resistance of the special secondary Lc phase as shown in Fig.1. In this example configuration, group one has fours windings while group two has three windings. An inductance-dual method [6-8] is adopted to model the grouped inductors. If the leakage inductance is much smaller than the magnetizing inductance, then the model for the grouped inductors in Fig. 1 could be simplified as shown in Fig.2. In which Lk_total_GP1 and Lk_total_GP2 represent the total leakage inductance for either group, Lm_pri represent the magnetizing inductance for each primary winding, Lm_sec_GP1 and Lm_sec_GP2 represent the magnetizing inductance for each group's secondary winding.

The advantage of this model is that it translates magnetic coupling into electric circuits, thus much easier for simulation and analysis.

One of the important parameters in the circuit is the transient inductance for each phase, and this will mandate certain aspects of the control loop design. Unlike traditional TLVR, which has a symmetrical topology for each phase, zero bias TLVR has an asymmetrical topology. This is because the

Lc phase, which is coupled with all the other phases, also connect to Vout. It is perceivable that this Lc phase has different equivalent inductance versus others, therefore it is necessary to derive the transient inductance parameter for all of the phases and observe the difference.

The equivalent circuit is shown in Fig. 2, during the occurrence of a transient event, two or more phases could be overlapped. Below example assumes all the phases are turned on and overlapping but the analysis applies to other cases too. the voltage at each phase's SW point is a positive pulse voltage, then the phase current slew rate, can be derived using the superimposing method by analyzing the effect of the individual pulse sources one by one as shown in (1).

In following analysis SW1 phase is used as the example while SWn phase is the special Lc phase. V1 is the pulse voltage applied on SW1 phase, I1-In are the currents generated by each phase's voltage respectively.

$$Z_{T1} = V_1/(I_1 + I_2 + \cdots + I_n) \qquad (1)$$

In Fig.3, only the pulse source from SW1 phase is considered, while all other pulses are treated as short circuit, in that way, all the other phases' winding also being considered as short circuit. The generated current will take three paths. The first will take its own Lm_pri path, the second will take this group's secondary magnetizing inductance Lm_sec_GP1, the third will travel through all those shorted windings but go through all the leakage inductance and that Lc. Its current is shown in (2).

$$I_1 = \frac{V_1}{s(L_c+L_{ktot})//L_{m\_sec\_GP1}//L_{m\_pri1}} \qquad (2)$$

In which Lktot represents the combination of all the groups' leakage inductance.

In Fig.4, only the pulse source from SW2 is considered, which is still in the same group as SW1, while all other pulses are treated as a short circuit. Since only the current running through phase SW1 is evaluated here, this current will only take two paths. The first will travel through all those shorted windings including SW1 and all the leakage inductance and that Lc, the second will take this group's secondary magnetizing inductance Lm_sec_GP1. Its current is shown in (3). This scenario also applies to SW3, which is in the same group of SW1.

$$I_2 = \frac{V_1}{s(L_c+L_{ktot})//L_{m\_sec\_GP1}} \qquad (3)$$

In Fig.5, only the pulse source from SW4 is considered, which is in a different group other than SW1, while all other pulses are treated as a short circuit. Again, only the current running through phase SW1 is evaluated here, this current will only take one path. it will travel through all those shorted windings including SW1 and all the leakage inductance and that Lc. And its current is shown (4). This scenario also applies to SW5, or any other phase/pulse which is not in the same group of SW1.

$$I_4 = \frac{V_1}{s(L_c+L_{ktot})} \qquad (4)$$

In Fig.6, only the pulse source from special Lc phase SWn is considered, while all other pulses are treated as a short circuit. Again, only the current running through phase SW1 is evaluated here, this current will also only take one path. it will travel through all those shorted windings including SW1 and all the leakage inductance and that Lc. Its current is shown in (5).

$$I_n = \frac{V_1}{s(L_c+L_{ktot})} \qquad (5)$$

By combining all the currents together, the total equivalent impedance and inductance seen by SW1 phase, a regular phase, with all the phases overlapped is shown in (6) and (7).

$$Z_{T1} = s[\frac{1}{N}(L_c + L_{ktot})//\frac{1}{M}L_{m\_sec\_GP1}//L_{m\_pri1}] \qquad (6)$$

$$L_{T1} = \frac{1}{N}(L_c + L_{ktot})//\frac{1}{M}L_{m\_sec\_GP1}//L_{m\_pri1} \qquad (7)$$

Where M is the number of phases in that first group, N is the number of the total phases of this VR. From (7), it can be seen this inductance will be different for each phase if they are in a different group with different M, for example, phase 1 is in group one and M is 3, but phase 4 is in group two and M is 2; or if the Lm_sec_GP is different for each group. But the difference could be minor.

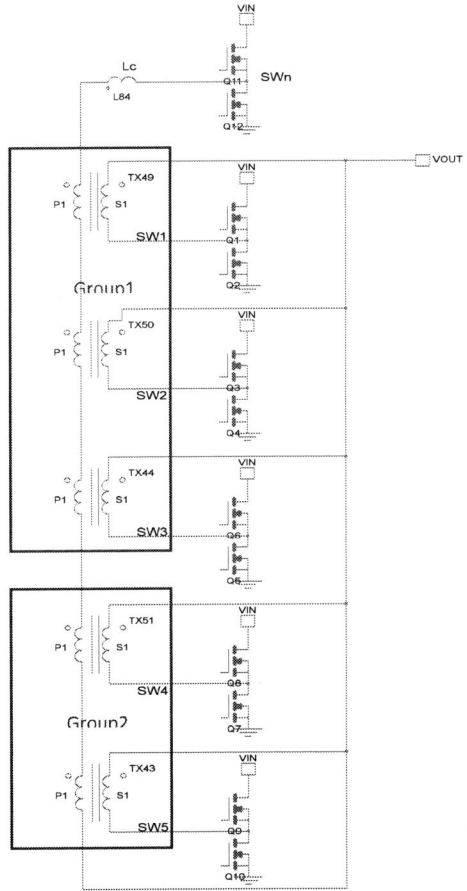

Fig. 1. Block diagram of Zero bias TLVR

979-8-3315-1612-3/25 $31.00 © 2025 IEEE

By following same process, the equivalent impedance or inductance seen by SWn phase, the special Lc phase, can be also derived. No matter which pulse source is considered, the current running through phase SWn will take the path of all those shorted windings and all the leakage inductances. Thus, the transient inductance can be calculated from (8). It can be seen that this inductance is bigger than all the other regular phases.

$$L_{Tn} = \frac{1}{N}(L_c + L_{ktot}) \qquad (8)$$

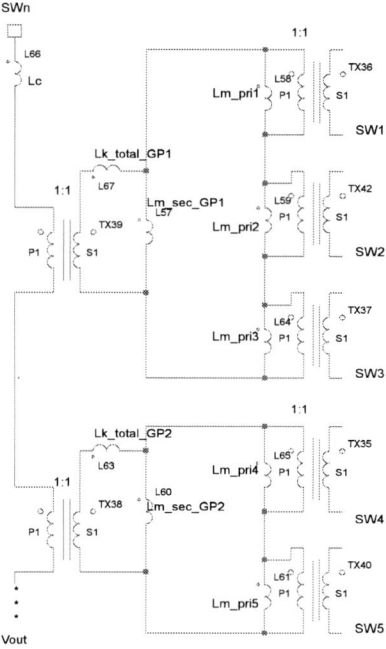

Fig. 2. Model of grouped inductor

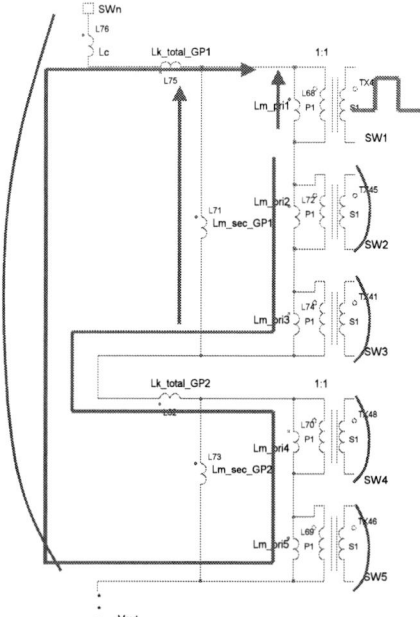

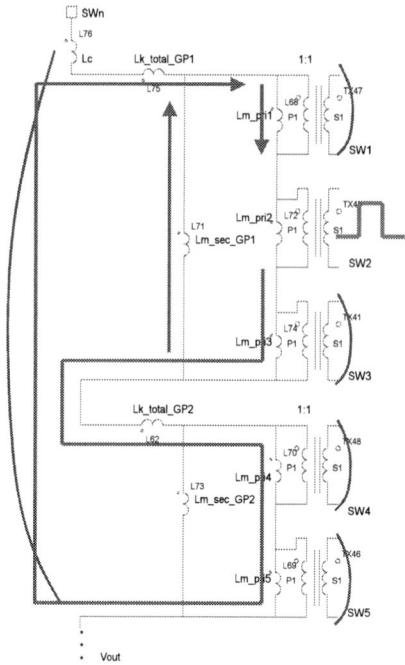

Fig. 3. The current flowing chart with only SW1 source

Fig. 4. The current flowing chart with only SW2 source

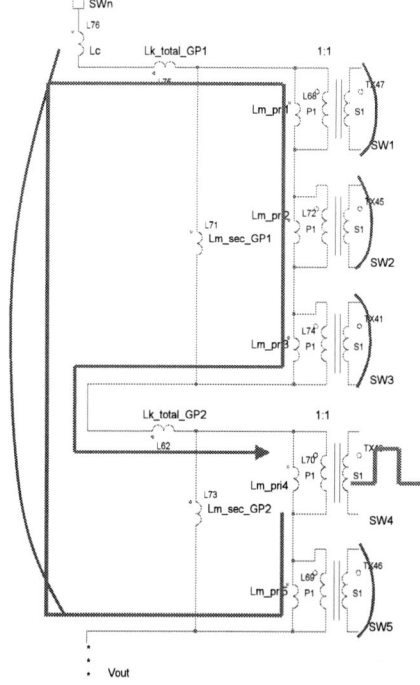

Fig. 5. The current flowing chart with only SW4 source

The asymmetry will be addressed in the control scheme by adopting different control parameters for each phase.

III. SIMULATION

A simulation model of this zero bias TLVR has been developed in Simplis. It consists of **8** phases: one special Lc phase, a 3-phase grouped inductor, and two 2-phase grouped inductors. Input voltage is 12V, output voltage is 1V. the primary winding inductance of phase 1-7 is 300nH, Lc is 150nH. Output current is from 100A to 300A. Switching frequency is 600kHz.

The steady state waveforms are shown in Fig. 7. It can be seen that special Lc phase current is flat and different from others, it only contains N times the switching frequency elements and a DC component, it does not has a sawtooth element like other phases, from the analysis in section two, it has a different equivalent inductance than other phases.

In Fig.7, I_PH2 and I_PH5 represent regular phase current from phase 2 and 3, while I_Lc_PHn represents the current of the Lc phase. I_Lm_Pri and I_Lm_Sec represent the current in Lm_pri and Lm_sec in the equivalent model respectively.

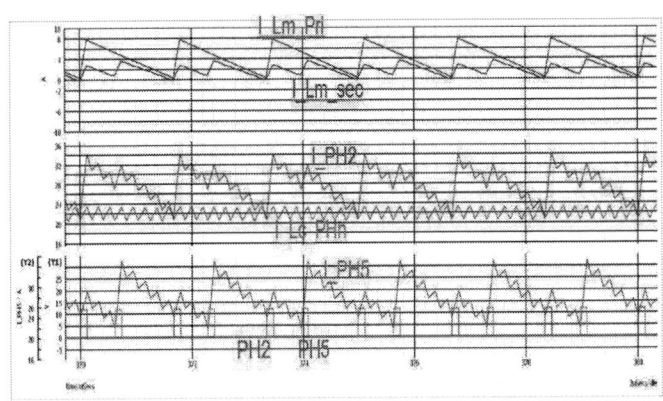

Fig. 7. Steady state phase currents of Zero bias TLVR

IV. EXPERIMENTS

A test board for the zero-bias TLVR, has been built and tested in the lab. Operating conditions are as below: input is 12V, output is 1V, switching frequency is 600khz, total 8 phases, peak current 400A. The steady state ripple currents are shown in Fig. 8, by measuring its peak to peak ripple, which matches the inductance predicted by our equations and simulation.

The transient response waveforms are shown in Fig. 9-10, load step is from 0A to 240A. It can be seen that the phase currents are well balanced. The Lc phase current is measured through current probe, but the regular phase current is measured using power stage's current sense output, which shows a huge discontinuity during some moments, this is caused by its hold-circuitry. Overall, it proves that a control scheme considering different phase inductance provides good current balance.

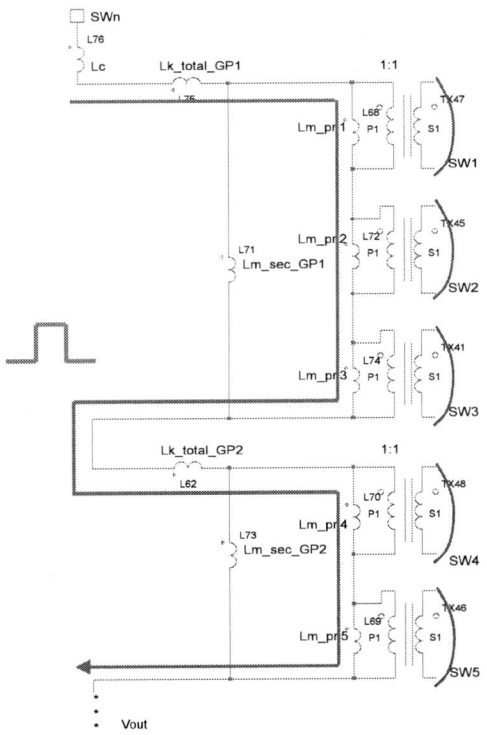

Fig. 6. The current flowing chart with only SWn source

By measuring the phase current slew rate, it is confirmed that it matches with theoretical calculation from section II.

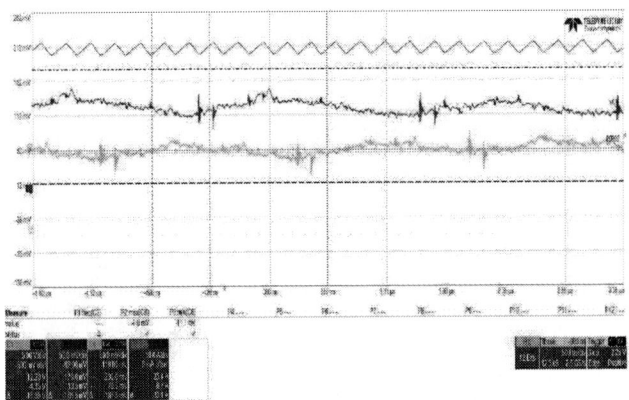

Fig. 8. Steady state phase currents of Zero bias TLVR(experiment)

Red: I_Lc Blue: I_PH2 Green: I_PH5 (10A/div)

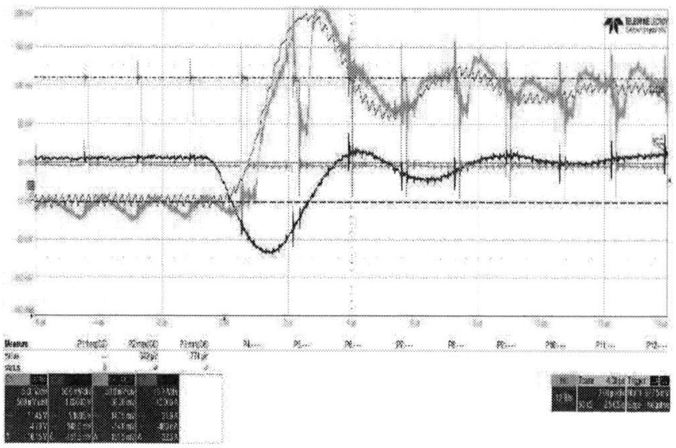

Fig. 11. Experimental setup of a Zero bias TLVR

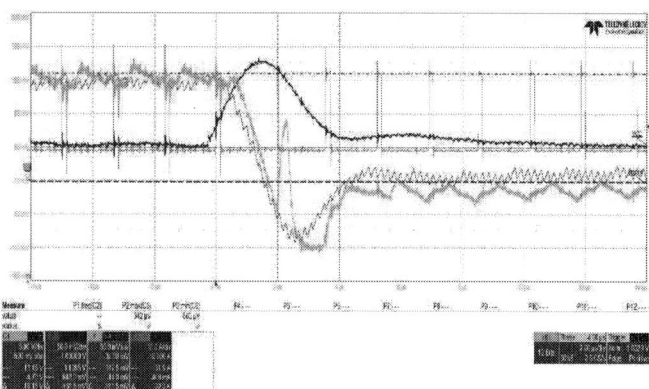

Fig. 9. Load step up case (0A-240A) of Zero bias TLVR

Red: I_Lc Green: I_PH2 (10A/div) Blue: Vout (50mV/div)

Fig. 10. Load release case (240A-0A) of Zero bias TLVR

Red: I_Lc Green: I_PH2 (10A/div) Blue: Vout (50mV/div)

V. Conclusions

The modeling of a new version of TLVR, zero bias TLVR, has been developed in this paper. The asymmetry of this topology has been fully investigated, and it is clear that the Lc phase has a different transient inductance versus other phases, Both inductances are derived from the model. Since the Lc phase has a different current slew rate comparing to others, the control loop programming needs to be adjusted to accommodate that phenomenon. Simulation and experiments have been done to validate this modeling, the effectiveness of the proposed control loop method is demonstrated.

Acknowledgment

The authors would like to thank Jorge Reyes and Francois Oudomphong for their kind help in experiments, and also Cyntec inc. for their support in providing inductor samples.

References

[1] S. Jiang, H. Gan, G. Sizikov, C. Chung, and S. Ko, "Next TLVR innovations: topologies, magnetics and control," *IEEE APEC*, 2024.

[2] P. Kumar, J. Tippetts, S. Naidu, and P. Brusco, "Efficiency impact of phase firing order in dual-sided power entry with trans-inductor voltage regulators(TLVR)," in *IEEE Applied Power Electronics Conference*, 2024.

[3] S. Krishnamurthy, D. Wiest, and Y. Zhou, "Trans-inductor voltage regulator (TLVR): circuit operation, power magnetic construction, efficiency and cost trade-offs," in *IEEE*, PCIM Europe 2022.

[4] G. Zhu, B. McDonald, and K. Wang, "Modeling and analysis of coupled inductors in power converters," in *IEEE Trans. On power electronics*, 2011.

[5] A. Ikriannikov, and D. Yao, "Converters with Multiphase magnetics: TLVR vs CL and the novel optimized structure," in *IEEE*, PCIM Europe 2023.

[6] M. Chen, and C. R. Sullivan, "Unified models for coupled inductors applied to multiphase PWM converters," in *IEEE Trans. On power electronics*, vol. 36, No. 12. 2021.

[7] S. Hamamsy, and E. Chang, "Magnetics modeling for computer-aided design of power electronics circuits," in *IEEE*, PESC 1989.

[8] G. Ludwig, and S. Hamamsy, "Coupled inductance and reluctance models of magnetic components," in *IEEE Trans. On power electronics*, 1991.

[9] I. Park, and S. Kim, "Modeling and analysis of multi-interphase transformers for connecting power converters in parallel," in *IEEE*, PESC 1997.

[10] F. Zhu, Q. Li, and F. Lee, "Modeling and analysis of multi-phase coupled inductor structures for voltage regulators," in *IEEE*, ECCE 2021.

cGANET-Enhanced Voltage Gain Modeling: Elevating CLLC Converter Accuracy

Yu Zuo
Dept. of Electrical Engineering
KU Leuven - EnergyVille
Leuven, Belgium
yu.zuo@kuleuven.be

Xiaobing Shen
Dept. of Electrical Engineering
KU Leuven - EnergyVille
Leuven, Belgium
xiaobing.shen@kuleuven.be

Fanghao Tian
Dept. of Electrical Engineering
KU Leuven - EnergyVille
Leuven, Belgium
fanghao.tian@kuleuven.be

Jiaze Kong
Dept. of Electrical Engineering
KU Leuven - EnergyVille
Leuven, Belgium
jiaze.kong@ kuleuven.be

Hans Wouters
Dept. of Electrical Engineering
KU Leuven - EnergyVille
Leuven, Belgium
hans.wouters@kuleuven.be

Wilmar Martinez
Dept. of Electrical Engineering
KU Leuven - EnergyVille
Leuven, Belgium
wilmar.martinez@ kuleuven.be

Abstract— The voltage gain model of resonant converters is crucial for parameter design and optimization. Common methods, such as Fundamental Harmonic Approximation (FHA), lack sufficient accuracy, and time-domain analysis (TDA) is computationally intensive and varies across different operating modes. Therefore, this paper proposes a voltage gain prediction model based on conditional Generative Adversarial Networks (cGANET). The new voltage gain model is developed by training the cGANET with a dataset of voltage gain data from simulations under various conditions, enabling accurate predictions of the complex, non-linear interactions within the resonant converter without relying on simplifying assumptions or mode-specific formulas like FHA and TDA. Leveraging its powerful extrapolation and interpolation abilities, the cGANET model maintains high accuracy for voltage gain within and beyond the training range, even with limited training data, and adapts to new conditions without extensive reconfiguration. To validate the proposed model, a prototype CLLC converter was constructed and tested across various switching frequencies. The results demonstrates that the proposed model improves accuracy by 60% under the worst conditions compared to FHA, beneficial for enhancing design and optimization of resonant converters.

Keywords— CLLC resonant converter, Wide voltage range, Analysis methodologies, Conditional generative adversarial networks, Artificial intelligence (AI)

I. INTRODUCTION

The bidirectional symmetric CLLC DC-DC converter is widely applied in various fields, including electric vehicles, renewable energy systems, and industrial power supplies [1-3]. To operate over a wide voltage range, Pulse Frequency Modulation (PFM) is often employed. Achieving wide voltage range operation requires balancing the trade-off between inductance ratio, quality factor, and switching frequency [4-5]. A larger inductance ratio and quality factor can enhance efficiency but necessitate a broader operating frequency range to achieve specific voltage gain, which can complicate

TABLE I. ANALYSIS OF VOLTAGE-GAIN ERRORS AMONG TYPCIAL MODELS

	Error Analysis
FHA	1. Only the fundamental harmonic is considered. 2. Resonant waveforms are continuous sinusoids to get equivalent output resistance Req
TDA [20]	1. Magnetizing inductance is constant in O mode. 2. Numerical solutions are sensitive to initial conditions.
STDA [21]	1. It inherits the characteristics of TDA. 2. Analytical Solutions are independent of output load.
SA	1. Some parasitic parameters of components are neglected.

magnetic design and optimization. Therefore, selecting an appropriate inductance ratio, magnetizing inductance, and quality factor is crucial to ensure soft switching across the entire voltage and load range. Consequently, an accurate voltage gain model is essential for designing symmetric CLLC resonant converters [6-7].

Common analysis methodologies include Fundamental Harmonic Approximation (FHA), Extended Fundamental Harmonic Approximation (eFHA) [8], Time-Domain Analysis (TDA) [9-11], Simplified Time-Domain Analysis (STDA) [12-13], and Simulation Analysis (SA). FHA simplifies the analysis of resonant converters by considering only the fundamental frequency harmonics of the switching waveform, assuming higher-order harmonics have a negligible impact. This method provides quick insights with simplified mathematical expressions, yielding analytical solutions for voltage gain and key parameters, facilitating initial design and analysis. However, FHA fails to accurately predict converter behavior when the switching frequency deviates significantly from the resonant frequency, or when higher-order harmonics are non-negligible [14]. eFHA incorporates the effects of multiple harmonics, offering a more accurate predictive model. It provides precise predictions of efficiency, voltage gain, and power loss across a wide frequency and load range [15]. However, this increased accuracy comes with greater computational complexity, making it less practical for real-time applications.

TDA offers an intricate voltage-gain model of the converter's performance by resolving various differential equations across different operational modes within a cycle [16]. Due to numerous variables and their interdependent relationships, simplifications are necessary to facilitate numerical solutions, such as assuming a constant resonant current in O mode and disregarding dead time. Although TDA provides highly accurate results and detailed insights into both transient and steady-state performance, it is computationally demanding, which poses challenges for real-time application implementation [17]. STDA aims to balance accuracy and complexity by introducing more simplifications to derive a gain model without solving complex differential equations. It offers more accuracy than FHA by considering non-linearities and dynamic interactions while maintaining moderate computational complexity [18]. However, some assumptions and simplification of STDA is only suitable for specific applications, lacking universality.

Simulation analysis (SA) employs software tools to model and simulate converter performance under various conditions, using numerical techniques to solve circuit equations and visualize behavior. This method easily accommodates different scenarios, component variations, and control strategies, providing graphical outputs for intuitive understanding. However, its application in parameter design relies heavily on the engineer's experience and requires iterative parameter tuning to meet expectations, which can be time-consuming and impractical [19]. Other methods, such as phase plane analysis or state space analysis, focus on the dynamic processes of voltage and current but do not directly impact parameter design and are therefore not the focus of this paper.

In this paper, a highly accurate voltage gain model for the CLLC resonant converter based on conditional Generative Adversarial Networks (cGANET) is proposed. The contributions and advantages of the proposed model include: 1) By effectively modeling complex, non-linear interactions within the converter, the cGANET-based model surpasses traditional methods that rely on simplifying assumptions, providing a more comprehensive and precise understanding of the converter's voltage gain model. 2) The remarkable extrapolation and interpolation abilities of cGANET enable precise voltage gain predictions across the entire voltage and load range even beyond training range, ensuring robust performance under varying operating conditions. 3) The model's capacity to learn from diverse operational data provides flexibility and adaptability to new conditions without requiring extensive reconfiguration, making it applicable to other resonant converters such as LLC and LCC. Therefore, the accurate cGANET-based voltage gain model could facilitate improved design accuracy, and optimization. The structure of this paper is as follows: Section II explains the working principles and outcomes of the proposed cGANET-based voltage gain model. Section III validates the feasibility and advantages through experimental results. Finally, the conclusion summarizes the paper.

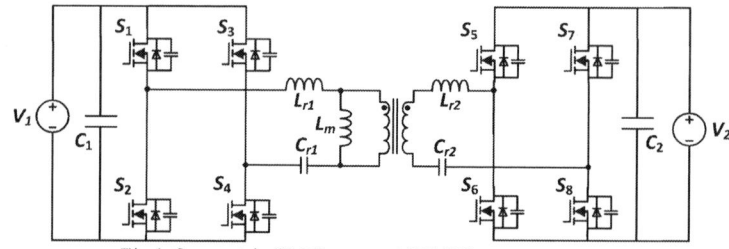

Fig.1. Symmetric CLLC resonant DC-DC converter

II. PRELIMINARY WORK BEFORE TRAINING

A. Setup for Training Data Acquisition

The circuit topology used in this study is a bidirectional symmetric CLLC resonant converter, as shown in Fig. 1. It adopts a dual-active bridge (DAB) configuration with CLLC resonant circuits, featuring a full-bridge structure on both the primary and secondary sides. To ensure unified bidirectional operation, the transformer is designed with a primary-to-secondary turns ratio of 1, and the magnetizing inductance is denoted as L_m. The resonant parameters are symmetrically designed on both sides, with $L_{r1}=L_{r2}$ and $C_{r1}=C_{r2}$. Therefore, the series resonant frequencies on both the primary and secondary sides are equal, denoted as (f_r). In the converter, V_1 and V_2 represent the DC voltages on the primary and secondary sides, while C_1 and C_2 are the filter capacitors at the power source terminals.

$$f_{r1} = \frac{1}{2\pi\sqrt{L_{r1}C_{r1}}} = \frac{1}{2\pi\sqrt{L_{r2}C_{r2}}} \tag{1}$$

$$f_{r2} = \frac{1}{2\pi\sqrt{(L_{r1}+L_m)C_{r1}}} \tag{2}$$

Training data is critical. For the voltage gain model of the symmetric CLLC resonant converter, accuracy is paramount. Table I presents an analysis of gain errors across commonly used models, including the FHA, eFHA, TDA, STDA, and SA. Firstly, FHA is based on simplifications that only consider the fundamental harmonic, neglecting higher harmonics. It assumes that resonant waveforms are continuous sinusoids, leading to an equivalent output resistance for the resonant tank. This assumption holds true primarily when the switching frequency is equal to the series resonant frequency. As a result, FHA maintains comparative accuracy at the series resonant point but deteriorates when the switching frequency deviates from this working point. TDA assumes that magnetizing inductance remains constant when magnetizing components participate in resonance (i.e., O mode), which is not always the case. It also notes that numerical solutions are sensitive to initial conditions, which can affect the consistency of results. STDA inherits characteristics from TDA and specifies that the duration of P mode is equal to half of the series resonant period. Consequently, its analytical solutions are independent of the output load, a condition that only applies in specific scenarios. Comparatively, SA achieves the highest accuracy as it does not rely on simplifications and assumptions. However, like the other models, SA also suffers from a common limitation that affects accuracy: only some parasitic parameters of actual components are considered. Overall, the SA simulation data proves to be

TABLE II SIMULATION SETTING

Name	Parameters
Input voltage range (V_I)	400V
Series resonant frequency (f_{r1})	100kHz
Transformer turns ratio (n)	1:1
Resonant capacitor (C_{r1}, C_{r2})	47.3nF
Inductance ratio ($L_m^{'}/L_{r1}$)	2-8
Quality factor (Q)	0.01-0.81
Switching frequency range (f_s)	50-100kHz
Capacitors (C_1, C_2)	47uF
Switches' junction capacitance(C_{oss})	160pF
Switches' deadtime	110ns
Simulation time step	100ns

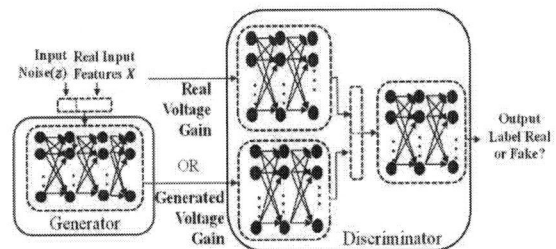

Fig.2. cGANET's principle and process

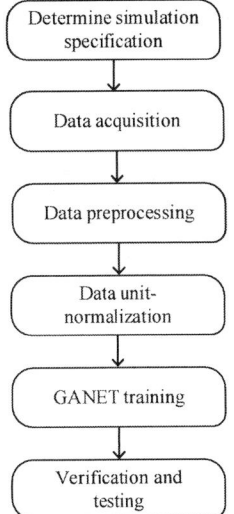

Fig.3. Flowchart of cGANET-based voltage gain model prediction

more accurate and reliable than that of other models. The proposed model utilizes simulation data which closely reflects real-world conditions and is readily available in substantial volumes.

B. Data acquisition for training

The simulation setup involves a PSIM and Hyper study co-simulation environment, modeling the symmetric CLLC converter circuit, as shown in Fig. 1. The resonant frequency is set to 100 kHz and switching frequency range from 50kHz to 150kHz, suitable for ferrite magnetic components. Various parameters such as inductance ratio, quality factor, and frequency, as detailed in TABLE II, are used to generate 1890 sets of data under different operating conditions. This dataset is divided into training and test sets. The training data is used to build and train the model, while the test data is used to evaluate and test the model's performance once it has been completed. In this paper, the training data includes inductance ratios from 3 to 8, which are commonly used for wide voltage range applications. The test data includes inductance ratios of 2, 9, and 10, which are used to evaluate the extrapolation capability of the trained model. While including more parameters like conduction losses, reverse diode voltage drops, and transformer parasitic capacitances could yield more accurate training data, it would reduce the generalizability of the voltage gain model, so these factors were not considered. Note that the training data does not differentiate between whether the converter achieves soft switching or the various operating modes it uses, such as PO, NP, PON, etc. After data acquisition, training involve filtering out non-monotonic interval data in data preprocessing, particularly for operating frequencies below fr2, normalizing the data, and structuring it for training purposes.

III. PRINCCIPLES AND IMPLEMENTATION OF CGANET-BASED VOLTAGE GAIN MODEL PREDICTION

A. Principle of cGANET-based voltage gain model

cGANET enhance standard GANET architectures by integrating conditional inputs, allowing for targeted data generation that aligns with specific CLLC converter operational modes or settings. In cGANET, the generator uses both noise and conditional labels to create realistic voltage gain predictions tailored to specific conditions, while the discriminator evaluates the authenticity of these predictions against real data, also considering the same

conditions, as shown in Fig.2 [22]. This conditional input enables the model to focus on particular scenarios, thereby improving the accuracy and applicability of predictions across different states of the converter. The following function used for training the cGANET to predict voltage gain of CLLC is formulated as:

$$\min_G \max_D V(D,G) = E_{x \sim Pdata}[logD(x)]$$
$$+E_{z \sim Pg(z)}[1 - logD(G(z))] \quad (3)$$

Where $E_{x \sim Pdata}$ represents the mathematical expectation of real voltage gain data distribution while $E_{z \sim Pg(z)}$ is the mathematical expectation of input noise distribution. x denotes input variables, L_n, Q and f_n. D(x) represents the discriminator's estimate of the probability a given voltage gain G, corresponding to input x. G(z) denotes the generator's output voltage gain given noise z and conditioning variables while $D(G(z))$ means the probability that the probability that generated data G(z) is regarded as real by the discriminator.

B. Training process

This section details the implementation of the CGANET-based voltage gain model prediction, illustrated in Fig.3. The process begins with setting simulation specifications for data collection and model training. Data preprocessing follows, which refines the dataset by eliminating inconsistencies and

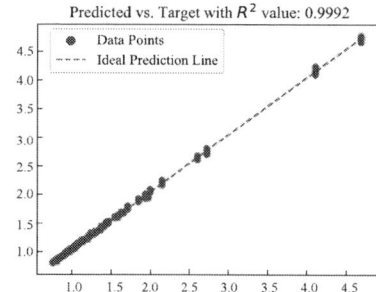

Fig.4. Training voltage gain model interpolation evaluation

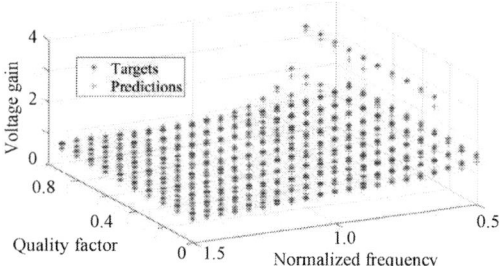

Fig.5. Training voltage gain model extrapolation evaluation

TABLE III. SETUP OF cGANET HYPERPERAMETERS

Parameters	Value
Batch size	16
Activation	ReLu
Learning rate	0.0002
Epochs	2000
Optimizer	Adam

TABLE IV. PROTOTYPE PARAMETERS

Components	Parameters
Input voltage range (V_{in})	200V
Transformer turns ratio (n)	1:1
Magnetizing inductance (L_m)	98uH
Resonant inductance (L_{r1}, L_{r2})	54uH
Resonant capacitance (C_{r1}, C_{r2})	47.3nF
Switching frequency range (f_s)	80-160kHz
Capacitors (C_1, C_2)	10uF
Rated output current	2A
Active's deadtime	200ns

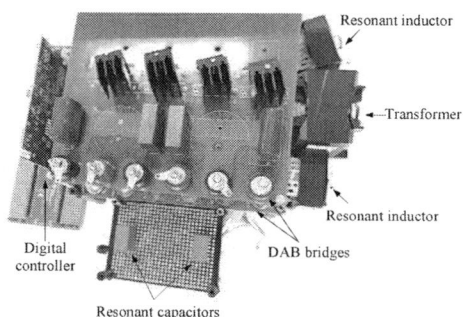

Fig. 6. Converter prototype

applying unit-normalization to standardize inputs. The cGANET is trained on this dataset to generate accurate voltage gain predictions, enhancing model reliability. Initial tests helps establish hyperparameter ranges with short, manual training sessions as indicated in Table III. The training involves an iterative process where the discriminator works to accurately identify real versus generated voltage gains, and the generator strives to produce indistinguishable voltage gain predictions. This process is repeated until the generator can reliably predict voltage gains across varied voltages and loads, utilizing the cGAN's capabilities to navigate complex, non-linear interactions within the converter.

C. Result Analysis

After training is completed, it is needed to assess both its accuracy and extrapolation capabilities, which could be used in CLLC design and optimization. Fig.4 compares the predicted values from the cGANET-trained model against the target values from the test data. The data points closely align with the ideal prediction line, as indicated by the average relative error value of 0.9992. This suggests that the cGANET model has achieved a high level of accuracy in predicting the target values, demonstrating its strong performance and reliability. Fig.5 demonstrates the extrapolation capability of the trained model with plotting voltage gain predictions against actual target values across a range of quality factors and normalized frequencies. The close alignment of predicted values with target values illustrates the model's ability to accurately extrapolate voltage gain beyond the training data, effectively capturing the underlying trends and non-linear relationships within the

converter's performance. This high level of agreement between predictions and targets underscores the model's robustness and reliability in predicting voltage gain across various operating conditions.

IV. EXPERIMENAL RESULTS VERIFICATION

To validate the cGANET-based CLLC voltage gain model, a prototype was constructed, as depicted in Fig. 6. The series resonant frequency was established at 100 kHz, featuring a symmetrical structure with a 1:1 transformer turn ratio. The prototype underwent testing across a range of switching frequencies from 80 kHz to 160 kHz to observe different voltage gains. Detailed results of these tests are presented in Table IV.

Fig.7 displays the experimental waveforms at switching frequencies below [Fig. 7(a)(b)], equal to [Fig.7(c)(d)], and above [Fig.7(e)(f)] the series resonant frequency. These include voltage and current of resonant tank, output voltage, and soft-switching waveforms. In order to verify the accuracy, voltage gains from test under different switching frequency are compared with outputs from cGANET-based voltage gain

979-8-3315-1612-3/25 $31.00 © 2025 IEEE

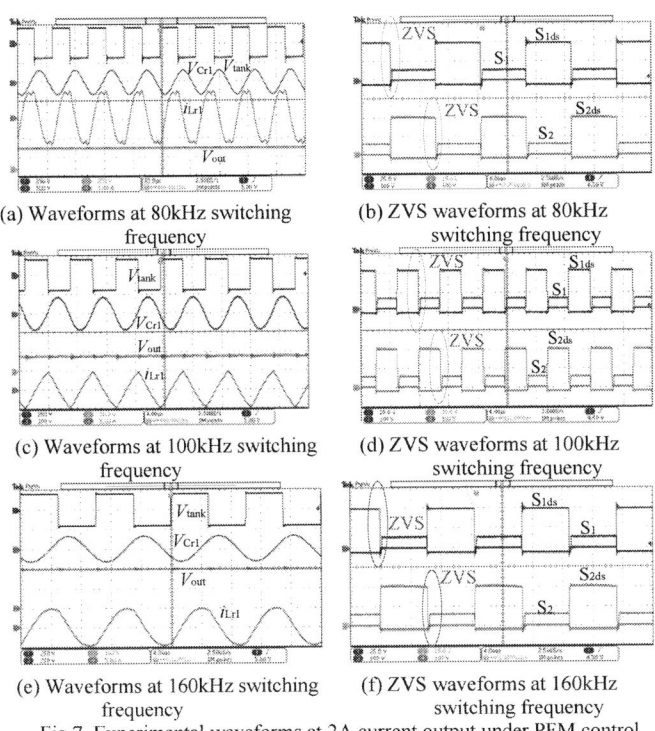

(a) Waveforms at 80kHz switching frequency

(b) ZVS waveforms at 80kHz switching frequency

(c) Waveforms at 100kHz switching frequency

(d) ZVS waveforms at 100kHz switching frequency

(e) Waveforms at 160kHz switching frequency

(f) ZVS waveforms at 160kHz switching frequency

Fig.7. Experimental waveforms at 2A current output under PFM control

(a) Voltage gain curves in different models

(b) Relative errors curves in different models

Fig.8 Voltage gain Comparisons among different models

TABLE V COMPARISONS BETWEEN DIFFERENT VOLTAGE GAIN ANALYSIS METHODOLOGIES

	FHA	eFHA[8]	TDA[11, 20]	STDA [13,21]	SA	Proposed
Domain	Frequency	Frequency	Time	Time	Time	Frequency
Computation	★	★★★	★★★★	★★★	—	★★★★
Analytical expression	Yes	Yes	Rarely	Yes	No	No
Accuracy	★	★★★	★★★★	★★★	★★★★	★★★★
Design and optimization practicality	★★★	★★★	★★★	★★★	★★	★★★★
Validity range	LC,LLC,LCC, CLLC,etc.	LC,LLC,LCC, CLLC,etc.	LC, LLC,LCC, CLLC,etc.	—	LC, LLC,LCC, CLLC,etc.	LC, LLC,LCC, CLLC,etc.

model, plus FHA and SA methodologies, as shown in Fig.8. Besides, the relative errors of voltage gain predictions from different models were compared against the test data to evaluate their accuracy. Positive relative error values indicate that the model overestimates test results, while negative values indicate underestimation. Since the cGANET model's training data is derived from simulation results, the outcomes of the SA and the proposed model are very similar, and they exhibit higher accuracy compared to FHA, especially at frequencies far from 1. The simulation results tend to overestimate the voltage gain due to the ideal components used. When the CLLC resonant converter operates over a wide frequency range, the proposed model's accuracy improves by around 60%. Future work could achieve even closer alignment with experimental results by incorporating component parameters into the training data.

Finally, a comparison of different methodologies for voltage gain model analysis, including FHA, eFHA, TDA, STDA, SA, and a proposed methodologies are presented in TABLE V. These methods are evaluated based on domain, computational load, presence of analytical expressions, accuracy, practicality

in design and optimization, and their validity range. FHA and eFHA operate in the frequency domain and provide analytical expressions with lower computational demands compared to TDA and STDA, but their accuracy is a main drawback. TDA, STDA, and SA, while highly accurate, produce results specific to particular operating modes and require extensive derivation or simulation across various conditions. n contrast, the proposed solution, despite necessitating a complex training process, offers high accuracy for voltage gain across a wide range due to its exceptional interpolation and extrapolation abilities. Notably, this method can also be applied to other resonant converters, such as LLC and LCC, to accurately model the voltage gain, making it highly suitable for applications requiring precise parameter design and optimization.

V. CONCLUSION

This paper introduces a novel voltage gain model for CLLC converters based on cGANET. Utilizing the robust extrapolation and interpolation capabilities of cGANET, the model delivers highly accurate voltage gain predictions by effectively modeling the complex, non-linear dynamics within the CLLC resonant converter, even with limited training data. This approach surpasses traditional methods like FHA, which suffers from accuracy issues, and time-domain analysis, which is computationally demanding and varies across operating modes. The cGANET model's interpolation and extrapolation capabilities enable it to adapt to new conditions with high accuracy, eliminating the need for extensive reconfiguration and significantly benefiting parameter design and optimization. Experimental validation using a prototype CLLC converter confirmed the model's effectiveness across a wide range of switching frequencies, demonstrating a 60% improvement in accuracy under the worst conditions compared to FHA. This method lays a beneficial groundwork for future advancements in resonant converter design and optimization. By incorporating actual component parameters into the training data, there is potential for even greater precision, ensuring closer alignment with experimental outcomes.

REFERENCES

[1] H. Li et al., "Bidirectional Control With Fitting Model-Based Synchronous Rectification and Input Ripple Current Feedforward for SiC Bidirectional CLLC EV Charger," in IEEE Transactions on Industrial Electronics, vol. 70, no. 9, pp. 9136-9146, Sept. 2023, doi: 10.1109/TIE.2022.3212382.

[2] R. Emamalipour and J. Lam, "A Multi-Mode Full-Bridge/Modified-Stacked- Switches Structured CLLC Resonant Converter for Energy Storage Applications," in IEEE Transactions on Power Electronics, vol. 39, no. 5, pp. 5967-5981, May 2024, doi: 10.1109/TPEL.2024.3364233.

[3] H. Wouters, H. Pervaiz, T. Geboers, Y. Zuo, W. -R. Lin and W. Martinez, "Interleaved PCB Winding Planar Transformer for Electric Vehicle Charging CLLC Converters," 2024 IEEE Applied Power Electronics Conference and Exposition (APEC), Long Beach, CA, USA, 2024, pp. 3216-3223, doi: 10.1109/APEC48139.2024.10509410.

[4] C. Zhao, F. Jin, Z. Li, Y. -H. Hsieh, F. C. Lee and Q. Li, "Design Consideration for CLLC Converter with High Power and Wide Gain Range," 2023 IEEE Applied Power Electronics Conference and Exposition (APEC), Orlando, FL, USA, 2023, pp. 1-6, doi: 10.1109/APEC43580.2023.10131583.

[5] M. Li, Z. Ouyang and M. A. E. Andersen, "High frequency LLC resonant converter with magnetic shunt integrated planar transformer," 2018 IEEE Applied Power Electronics Conference and Exposition (APEC), San Antonio, TX, USA, 2018, pp. 2678-2685, doi: 10.1109/APEC.2018.8341395.

[6] F. Jin, A. Nabih, T. Yuan and Q. Li, "A High-Efficiency High-Density Three-Phase CLLC Resonant Converter With a Universally Derived Three-Phase Integrated Transformer for On-Board-Charger Application," in IEEE Transactions on Power Electronics, vol. 39, no. 4, pp. 4350-4366, April 2024, doi: 10.1109/TPEL.2024.3354679.

[7] R. Wei, L. Ding and Y. Li, "Efficiency-Oriented Optimized Design and Control of Hybrid FSBB–CLLC Converters With Partial Power Processing Capability," in IEEE Transactions on Power Electronics, vol. 39, no. 5, pp. 6364-6375, May 2024, doi: 10.1109/TPEL.2024.3363629.

[8] A. Sankar, A. Mallik and A. Khaligh, "Extended Harmonics Based Phase Tracking for Synchronous Rectification in CLLC Converters," in IEEE Transactions on Industrial Electronics, vol. 66, no. 8, pp. 6592-6603, Aug. 2019, doi: 10.1109/TIE.2018.2874348.

[9] B. Li, M. Chen, X. Wang, N. Chen, X. Sun and D. Zhang, "An Optimized Digital Synchronous Rectification Scheme Based on Time-Domain Model of Resonant CLLC Circuit," in IEEE Transactions on Power Electronics, vol. 36, no. 9, pp. 10933-10948, Sept. 2021, doi: 10.1109/TPEL.2020.3044297.

[10] Z. Xiao, Z. He, F. Deng and Y. Tang, "Power Loss Minimization of CLLC Resonant Converters via Time Domain Analysis," 2023 IEEE Energy Conversion Congress and Exposition (ECCE), Nashville, TN, USA, 2023, pp.

3392-3399, doi: 10.1109/ECCE53617.2023.10362154.

[11] R. Wei, L. Ding, R. Liu and Y. Li, "An Intuitive and Noniterative Design Methodology for CLLC Chargers Employing Simplified Operation Modes Model," in IEEE Transactions on Power Electronics, vol. 38, no. 6, pp. 7771-7784, June 2023, doi: 10.1109/TPEL.2023.3251284.

[12] Y. Zuo, H. Niu, R. Zhang and X. Pan, "The Modified FHA and Simplified Time-doamin Analysis methodologies for LLC Resonant Converter," 2021 IEEE 12th Energy Conversion Congress & Exposition - Asia (ECCE-Asia), Singapore, Singapore, 2021, pp. 56-61, doi: 10.1109/ECCE-Asia49820.2021.9479161.

[13] Y. Cao, M. Ngo, D. Dong and R. Burgos, "A Simplified Time-Domain Gain Model for CLLC Resonant Converter," 2021 IEEE Energy Conversion Congress and Exposition (ECCE), Vancouver, BC, Canada, 2021, pp. 3079-3086, doi: 10.1109/ECCE47101.2021.9596002.

[14] Z. Li et al., "An Accurate, Universal and Fast Time Domain Model for Different Types of Resonant Converters by Considering Parasitic Capacitors and Deadtime," in IEEE Transactions on Power Electronics, doi: 10.1109/TPEL.2024.3437251.

[15] Y. Fang, B. M. H. Pong and R. S. Y. Hui, "An Enhanced Multiple Harmonics Analysis Method for Wireless Power Transfer Systems," in IEEE Transactions on Power Electronics, vol. 35, no. 2, pp. 1205-1216, Feb. 2020, doi: 10.1109/TPEL.2019.2925050.

[16] R. Gu, D. Zhang, J. Duan, R. Qu and A. Li, "Fixed-Frequency Circulant Phase-Shift Controlled Current-Fed Resonant Converter With Wide Voltage Gain Range," in IEEE Transactions on Industrial Electronics, vol. 71, no. 8, pp. 8782-8792, Aug. 2024, doi: 10.1109/TIE.2023.3322011.

[17] X. Li, L. Guo, S. Chen, T. Lang, D. Lu and H. Hu, "Time Domain Analysis of Three-Phase Single-Stage AC/DC Resonant Converter Using Numerical Calculation," in IEEE Transactions on Power Electronics, vol. 37, no. 6, pp. 6857-6872, June 2022, doi: 10.1109/TPEL.2022.3140547.

[18] J. Niu, Y. Tong, Q. Ding, X. Wu, X. Xin and X. Wang, "Time Domain Simplified Equations and its Iterative Calculation Model for LLC Resonant Converter," in IEEE Access, vol. 8, pp. 151195-151207, 2020, doi: 10.1109/ACCESS.2020.3016975.

[19] Z. Xiao, X. Li and Y. Tang, "A Lightweight Artificial Neural Network Start-up Controller for CLLC Resonant Converters," in IEEE Transactions on Power Electronics, doi: 10.1109/TPEL.2024.3436847.

[20] L. Wang, H. Chen and K. Sun, "A Synchronous Rectification Method with Switching Delay for CLLC Converters to Achieve Secondary-side ZVS," 2022 IEEE Energy Conversion Congress and Exposition (ECCE), Detroit, MI, USA, 2022, pp. 1-6, doi: 10.1109/ECCE50734.2022.9947373.

[21] JL. Jiao, L. Li, C. Wang, S. Zhang, B. Liu and X. Fang, "High-Precision Time-Domain Analysis Method Based on the Superposition Principle for CLLC Converter in Above-Resonant-Frequency Mode," in IEEE Transactions on Power Electronics, vol. 39, no. 11, pp. 14550-14564, Nov. 2024, doi: 10.1109/TPEL.2024.3433504.

[22] X. Shen, Y. Zuo and W. Martinez, "Conditional Generative Adversarial Network Aided Iron Loss Prediction for High-Frequency Magnetic Components," in IEEE Transactions on Power Electronics, vol. 39, no. 8, pp. 9953-9964, Aug. 2024, doi: 10.1109/TPEL.2024.3397041.

Capacitive vs Inductive Coupling Based DC-DC Converter Operating in MHz Switching Frequency Range

Saeed Pourjafar, Parham Mohseni
Department of Electrical Power Engineering and Mechatronics
Tallinn University of Technology
Tallinn, Estonia
saeid.pourjafar@taltech.ee, ,
parham.mohseni@taltech.ee

Oleksandr Husev, Ryszard Strzelecki
Department of Power Electronics and Electrical Machines
Gdańsk University of Technology
Gdansk, Poland
oleksandr.husev@ pg.edu.pl,
ryszard.strzelecki@pg.edu.pl

Oleksandr Matiushkin
Department of Electrical, Electronic and Control Engineering
University of Extremadura
Badajoz, Spain
oleksandr.matiushkin@taltech.ee

Abstract— In this paper galvanically isolated dc-dc interfaces are studied. Capacitive coupling is studied as an alternative solution to magnetic isolation. The recommended converter configuration is combination of a full bridge structure in input side and a full bridge rectifier as output side. Instead of using magnetic isolation, capacitive isolation has been utilized in the proposed topology to provide the galvanic isolation and also the required energy for the load. The proposed converter provide benefits including decreased size, weight, and losses, consequently improving both high power density and high efficiency in compared to the inductive coupling converters. The comprehensive performance comparison between capacitive coupling and inductive coupling converters, have been carried out in this work. An experimental prototype of the suggested converter operated in 1.5 MHz switching frequency and rated power equal to 6 kW is presented to prove the correctness of the theoretical surveys.

Keywords — Isolated dc-dc converter, capacitive coupling, high power density, magnetic isolation.

I. INTRODUCTION

Recently there's a growing demand for isolated dc-dc converters, which are essential in various applications like data centers and electric vehicles [1]. These converters ensure safe and efficient power transfer between different parts of the system [2]. Traditionally, Multi-Frequency Transformers (MFTs) have been the solution for achieving this isolation. MFTs have long been a basis in the demand for efficient power conversion and the design and construction of MFTs inherently impose limitations on their performance characteristics. Traditional MFTs often exhibit bulkiness and weight, which can be impractical in applications where space and weight constraints are critical considerations [3] – [4]. Additionally, the reliance on the conventional materials and manufacturing techniques may hinder the optimization of MFTs for specific operating conditions, which leads to limit their efficiency and performance [5].

Given the rising need for more compact and efficient dc-dc converters, researchers are actively exploring alternative isolation techniques to overcome the limitations associated with MFTs [6] - [7]. One promising approach is the adoption of capacitive coupling, which offers a fresh perspective on power isolation that diverges from the traditional reliance on transformers. Capacitive-coupled converters are emerging as a feasible solution for developing compact and efficient power conversion systems. Capacitive coupling provides

distinct advantages over inductive methods, particularly in reducing size and weight, while simultaneously enhancing efficiency [8] - [10]. This transition towards capacitive coupling represents a potential breakthrough in power conversion technology, offering a pathway to more compact and effective solutions suitable for a variety of applications [10]. Figs. 1a and 1b provide a comparative illustration of capacitor power transfer applications, highlighting the potential space and efficiency improvements [11]. Additionally, a broader overview of capacitive power transfer applications is depicted in Fig. 1c [12].

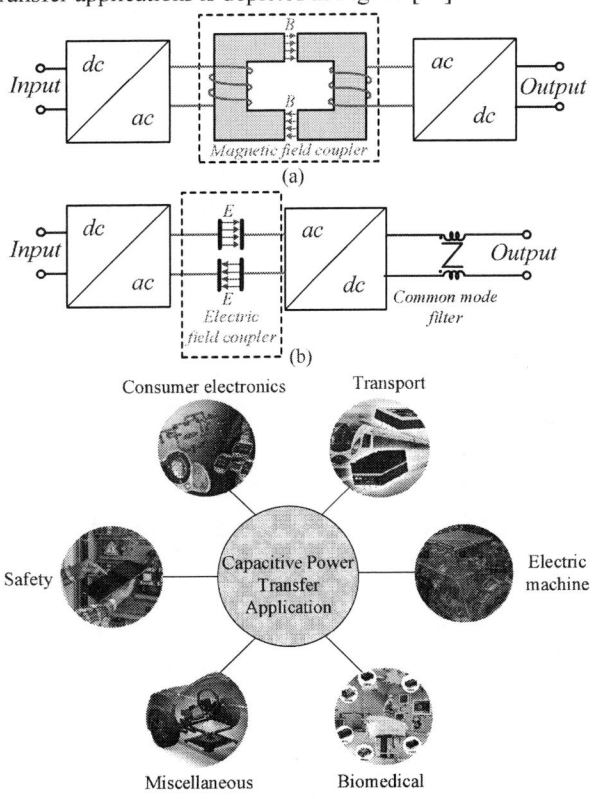

Fig. 1. The overview of power transfer, (a) magnetic power transfer (MPT), (b) Capacitive power transfer (CPT), (c) The brief overview capacitive power transfer applications [13].

However, while significant progress has been made, challenges remain in fully understanding and optimizing the design of the capacitive-coupled converters. Ensuring galvanic isolation without compromising the functionality of Capacitive Power Transfer (CPT) converters is essential to maintain safety standards and prevent potential electrical hazards [13].

This manuscript presents an advanced isolated capacitive-coupled dc-dc converter design that achieves high power density with minimized size and volume. Chapter II covers the proposed converter topology, outlining the various operating, and guidelines for selecting appropriate components. Chapter III provides a comprehensive performance comparison between capacitive coupling and inductive coupling converters, highlighting the relative advantages and trade-offs of each solution. Chapter IV details the experimental validation of the design, including results obtained at a switching frequency of 1.5 MHz and a power output of 6 kW. Lastly, Chapter V summarizes the key findings and conclusions of this study.

II. Proposed Solutions

The considered solutions are presented in Fig. 2. The first solution that is illustrated Fig. 2a represents a classical LLC converter with inductive isolation, while the second, which has been showed in Fig. 2b employs capacitive isolation. Both configurations use air-core inductors or transformers to reduce core losses and improve high-frequency performance. In the capacitive isolation case, instead of utilizing two series inductances across the isolation barrier, a single coupled inductor is used to reduce the system's overall size and volume, enhancing both power density and efficiency. We emphasize, that in case of air core mutual inductors, the size of inductor does not depend on saturation current, which in enhances benefits from coupling. We can almost double equivalent inductance without size increasing.

Key parameters influencing the operation of these setups include the capacitance value C of the isolation capacitors and the maximum allowable dc and ac voltage ratings, V_{DCmax} and V_{ACmax}, respectively. The capacitance C determines the converter's resonant frequency, which is critical to ensuring effective energy transfer in the capacitive-coupled converter. For a capacitive-coupled LLC converter, the resonant frequency f_r can be approximated as:

$$ f_r = \frac{1}{2\pi\sqrt{LC}}, \qquad (1) $$

where L is the inductance value of the coupled inductor and C represents the total effective capacitance across the isolation barrier. This resonant frequency should ideally match the switching frequency of the converter for maximum efficiency. The voltage ratings, V_{DCmax} and V_{ACmax} define the isolation capability of the converter. The DC voltage rating ensures safe operation under steady-state conditions, while the ac voltage rating protects against peak transients, which could arise due to switching or load transients.

For isolation design, the maximum peak voltage V_{peak} across the isolation barrier should satisfy:

$$ V_{peak} \prec V_{AC_{max}}. \qquad (2) $$

These alternative configurations can include combinations of dc-ac and ac-dc bridges on both the primary and secondary sides, which are connected through a network of reactive components, as shown in Fig. 2.

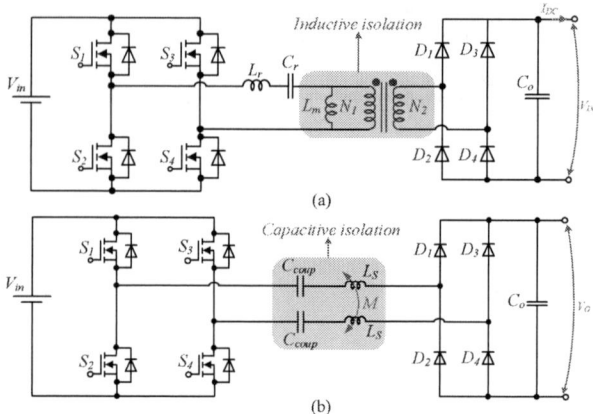

(a)

(b)

Fig. 2. The converter configuration of each structures, (a) inductive isolation, (b) capacitive isolation.

This network of reactive components plays a key role in determining the shape of the switch current waveforms, thereby affecting how the power converter performs. By adjusting these waveforms, the converter can be tailored to operate in either a resonant mode, producing a sinusoidal waveform, or a non-resonant mode, creating a trapezoidal waveform, depending on the requirements of the specific application, which has been indicated in Fig. 3.

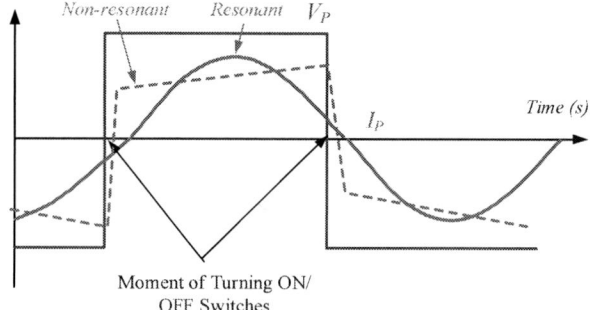

Fig. 3. The primary diagram of isolated dc-dc converters utilizing capacitive coupling, (a) resonant network current waveform, (b) The current and voltage waveform of a proposed solution [17].

Considering very high switching frequency, only GaN semiconductors are considered for application. In case of GaN transistors utilizing in ZCS condition it is well-known that power losses are defined by $Eoss$ and $Eqoss$ losses where the $Eoss$ loss is introduced by the capacitance self-discharging current of the switch device itself and $Eqoss$ loss is introduced by the capacitance charging current from the opposite switch device [14], [15].

Fig. 4 shows simplified theoretical evaluation of conduction losses, turn OFF losses and losses related to $Eoss/Eqoss$ energy versus switching frequency for Zero Voltage Switching (ZVS) and Zero Current Switching (ZCS) conditions for 5 kW rated power. In both cases the value of magnetizing inductance is assumed to be large enough to provide ideal resonant current waveforms. Transistors GS66516T are used for losses estimation.

P_{OSS} are losses related to the capacitance self-discharging current of the switch device $Eoss$ and the capacitance charging current $Eqoss$. This type of losses is associated only with ZCS case and are not considered in ZVS case. At the same time, losses, related to the turn OFF losses (P_{SW}) are considered only for ZVS. It is assumed, that turn off current in ZVS case has vale not larger to the 10% of the peak current due to the sufficient magnetizing inductance.

Also, the conduction losses in case of ZVS cases have slightly higher value (P_{CON}) compared to the ZCS.

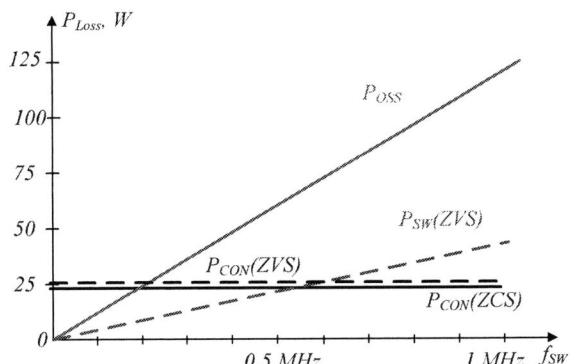

Fig. 4. Theoretical losses breakdown analysis for considered solutions as a frequency function in case of GaN transistors utilization in ZVS vs ZCS condition.

It can be seen, that after 200 kHz switching frequency, the losses related to the switching frequencies (P_{SW} and P_{OSS}) dominate over conduction losses. The conclusion is that even in case of GaN semiconductors, the ZCS is not enough and at high frequency (200+ kHz) the switching losses related to $Eoss$ and $Eqoss$ dominate over other losses. It shows clearly that ZVS condition becomes more important compared to ZCS in case of very high switching frequency. In [16] it was clearly shown that for a design and optimization of a converter system with ZVS, the inductance present is not enough as ZVS condition requires the presence of an impressed current of an inductive component which charges/discharges the output capacitances of the MOSFETs within a bridge leg during the interlocking time of the associated gate signals.

As a conclusion, our solutions operate only in ZVS conditions that is achieved by means of non-zero current in the resonant inductor L_r. It generates a circulating current that discharges the parasitic output capacitance C_{oss} of the transistor before it turns ON. When the switch is ready to turn ON, the voltage across its drain-source terminals has already dropped to zero due to this discharge. Consequently, the switch can turn ON without experiencing high voltage across it, effectively eliminating the C_{oss} losses, which typically arise from charging and discharging the output capacitance during each switching cycle [17].

A. Design consideration of capacitive coupling based converters

The primary considerations for capacitive coupling applications are the capacitance value and the maximum dc and ac voltage rating. The capacitance value determines the operational mode of the capacitive-coupled dc-dc converter, distinguishing between resonant and non-resonant modes. Meanwhile, the voltage rating indicates the isolation capability of the capacitor, enabling its use in specific applications. Off-the-shelf capacitors typically have voltage ratings ranging from 200 V to 1.0 kV across various technologies such as ceramic, film, and aluminum electrolytic capacitors. The isolation within a capacitor is achieved through dielectric layers, which vary in dielectric

strengths and thicknesses as outlined in Table I [18] – [21].

Additionally, other important features to consider include the temperature coefficient, leakage current, and tolerance levels, which can significantly impact the performance and reliability of the capacitor in diverse applications. According to the data provided in Table I, film capacitors can be emerged as the preferred choice for the particular capacitive coupling isolated application. Hence, they are often selected for integration into the design, which can align perfectly with the converter's requirements [22].

According to Table V, the dielectric strength of a capacitor can surpass that of transformer insulation material by multiple times. However, the thickness of dielectric layer in capacitors is significantly lower, up to a factor of 10^4, compared to the distance between the primary and secondary sides in a transformer, which spans centimeters. Capacitors can withstand voltages ranging from hundreds to thousands of volts, depending on the technology used. Hence, capacitive isolation may be adequate to replace the transformer (MFT) in certain applications. The common insulation material parameters of the MFT have been depicted in Table II [23] – [24].

TABLE I
CAPACITORS DIELECTRIC MATERIALS [18] – [21]

Capacitor model	Material of dielectric	Dielectric strength (KV/mm)	Thickness of material (μm)
Electrolytic	Al_2O_3	710	120
Film	Polypropylene (PP)	640	4-14
Film	Polyester (PET)	570	4-14
Film	Polycarbonate	528	4-14
Ceramic	$BaTiO_3$	90	1-10

TABLE II
THE COMMON INSULATION MATERIAL PARAMETERS OF THE MFT [23] – [24]

Material of Isolation	Dielectric strength (KV/mm)
DMD Paper	22 – 49
Silicone	4 – 28
NOMEX Paper	27
Micares	8 – 24
PVC	9.8 – 19
HDPE	19
Epoxy	16
Transformer oil	10 – 15
Air	3

III. PERFORMANCE COMPARISON OF THE CAPACITIVE COUPLING AND INDUCTIVE COUPLING CONVERTER

The converter configuration of both capacitive coupling and inductive coupling-based structures for comparison has been indicated in Fig. 2. Besides, the performance comparison of the capacitive coupling and inductive coupling-based converters has been depicted in Fig. 5. It's important to highlight that this comparative analysis has been implemented by considering the specifications illustrated in Table III. In addition, it has been considered that both converters are operating in soft switching conditions. To ensure a comprehensive and fair comparison, other dynamic switch configurations are also kept consistent across all converters. It's worth noting that such converters typically incorporate a voltage multiplier rectifier on their secondary side, which is a common feature among most of isolated converters.

TABLE III
DESIGN PARAMETER OF COMPARISON FOR ALL CONVERTERS

Parameters	Value
Input voltage range (V_{in})	350 V
Output voltage (V_o)	350 V
Maximum rated power (P_o)	5 kW
Switching frequency	500 kHz
Maximum input current	15 A
Maximum voltage ripple of capacitors	3 %

The following features have been considered for this comparison [25]:

- Capacitor size ($\sum_{i=1}^{n} E_{C_i} = \frac{1}{2} C_i V_{C_i}^2$)

- Magnetic size ($\sum_{j=1}^{m} E_{L_j} = \frac{1}{2} L_j I_{L_j}^2$)

- Switch voltage stress ($VSR = \dfrac{V_{stress}}{V_{out}}$)

- Semiconductor conduction losses of the proposed converters ($P_{conduction} = \sum_{i=1}^{N} R_{on} I_{RMSi}^2$)

- Switching losses ($SL \cong \sum_{i=1}^{N_S} \left\langle \hat{i}_{Si} \cdot \hat{v}_{Si} \right\rangle_T$)

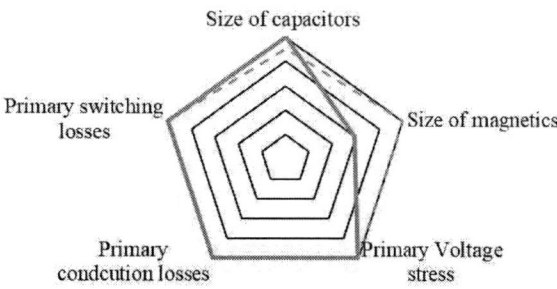

——— Capacitive isolation – – – Inductive isolation

Fig. 5. Overall performance comparison of the capacitive isolation and inductive isolation-based converters.

In the equations provided above, E_C represents the accumulated energy in capacitors, E_L stands for the accumulated energy in inductors, V_C denotes the voltage across the capacitors, I_L represents the current through the inductors, and n and m respectively refer to the number of capacitors and the number of inductors. Variable VSR is the voltage stress ratio. I_{RMS} refers to the Root Mean Square (RMS) current flowing through the semiconductor device while it's conducting, and R_{on} represents the ON-state resistance of the semiconductor device. In switching losses

ratio (S_L) equation v_{Si} and i_{Si} are the average value of voltage and current throughout a fundamental period T.

Based on the findings presented in Fig. 5, it is evident that both converters have almost the same performance characteristics such as equal switching voltage stress, conduction losses and equal switching losses. However, what makes structures based on capacitive isolation superior to inductive isolation is the small size and volume. This is mainly because they don't need bulky transformers and their inductive elements are smaller. As a result, these systems can pack more power into a smaller space and cost less to build. Therefore, according to this comparison it can be conclude that the capacitive isolated based converters can provide higher power density with lower overall system cost.

IV. EXPERIMENTAL RESULTS

In this chapter the experimental tests are performed to verify and estimate the capability of the proposed solution. Fig. 6 displays the experimental prototype of the designed dc-dc converter with capacitive coupling. This prototype has been developed to meet the specifications outlined in Table IV, including the power rating, switching frequency, and input/output voltage range.

Table IV
EXPERIMENTAL PROTOTYPE SPECIFICATIONS

Parameter	Value
Input voltage range v_{IN}	300-400 V
Output voltage v_o	350 V
Max input power P_{INMAX}	6 kW
Max input current i_{INMAX}	15 A
Max output current i_{OMAX}	16 A
Switching frequency f_{SW}	1.5 MHz
Switches $S_1..S_4$	GS66516T-TR
Output diodes	SCS230AE2HRC11
Capacitor C_{coup}	42 nF
Air transformer (L_p, L_s, M)	2.9 µH, 3 µH, 1.3

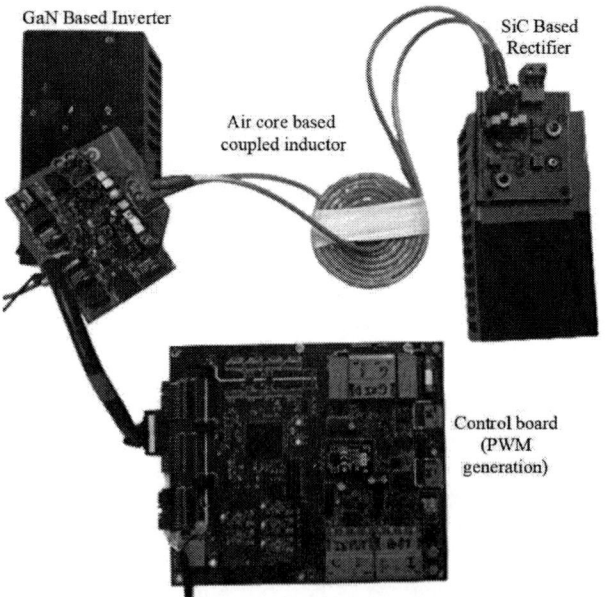

Fig. 6. Experimental prototype of the proposed converter based on the capacitive coupling and using air core based coupled inductor.

The setup includes key components such as the primary and secondary windings, coupled inductor, rectification stage, and high-frequency switches. The layout is designed for easy access to critical points for waveform measurements and thermal analysis. GaN semiconductors are chosen as high-frequency switching transistors for $S_1 - S_4$, taking advantage of their fast switching speed and low on-resistance to achieve high efficiency at elevated frequencies. For the rectification stage, SiC diodes (SCS230AE2HRC11) are utilized due to their high thermal stability and fast recovery characteristics, which reduce losses during high-speed operation. Together, these components ensure the converter can handle rapid switching transitions, supporting ZVS and minimizing conduction losses. Furthermore, this design is optimized for thermal performance, enabling stable operation even at high power levels. Moreover, the control board, built on the TMS32028379DPTPT microcontroller, is used to provide open-loop switching signalsare performed to verify and estimate the capability of the proposed solution.

Fig. 7a presents the main waveforms of the converter, including the drain-source voltage of switches S_1 and S_2, as well as the primary (I_p) and secondary (I_s) currents of the coupled inductor. This figure clearly shows that the converter operates in ZVS conditions for both turn-on and turn-off transitions, with minimal voltage stress on the transistors at switching instances. The ZVS mode ensures that the switches transition with zero voltage, eliminating C_{oss} losses and thereby reducing heat generation. The primary and secondary currents also display the expected resonant waveforms, validating the effectiveness of the resonant circuit in shaping current and voltage profiles to achieve efficient power transfer. Figures 7b and 7c show the operating temperature profiles of the main switch S_1 and the air-core coupled inductor, respectively, when delivering approximately 6 kW of output power. The temperature of switch S_1 remains well within safe operational limits, demonstrating effective thermal management and heat dissipation. Similarly, the air-core coupled inductor's

temperature profile indicates stable operation without significant thermal stress, suggesting the design's robustness in high-power conditions. These results confirm that the thermal behavior of both components does not negatively impact the converter's overall performance, even under high load conditions.

In Figs. 8a and 8b, numerical results from the precision power analyzer (YOKOGAWA WT1800) are presented, illustrating the power and efficiency measurements of the converter. At an output power of 5 kW, the converter achieves an efficiency of 95%, while at a slightly higher output power of 5.46 kW, the efficiency is 94.55%. These relatively high efficiency values demonstrate the converter's capability to minimize losses across the full range of operation, largely due to the ZVS operation mode and the use of low-loss components such as GaN transistors and SiC diodes. The stability of efficiency across different load levels highlights the effectiveness of the converter's design. Fig. 8c illustrates the size of the coupled inductor in magnetic coupling and capacitive coupling-based converter. According to this figure, the inductive element within the capacitive coupling design has been significantly reduced under equivalent test conditions, resulting in increased power density and reduced overall cost for the converter.

Fig. 9 shows losses breakdown analysis in the 5 kW power point for inductive and capacitive coupling. It can be seen that at nominal power point the GaN transistors contribute by 34 W which is less than 1 % of total power. The major losses still come from air inductors/transformer. Fig. 7c shows that despite the neglecting skin effect, the proximity effect is significant and current distribution in the wires are not equal, which in turns cause the extra heating and losses. Due to this, the capacitive coupling has slightly higher efficiency. Another point that should be emphasized is that other losses mostly related to PCB and wires are also significant and can be optimized.

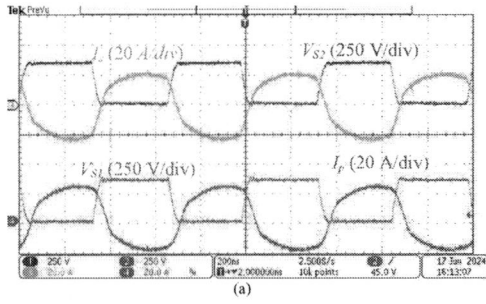 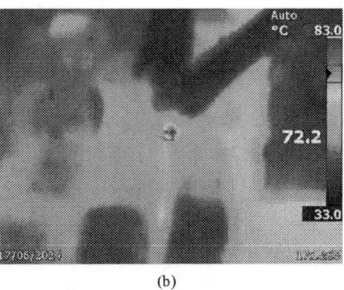 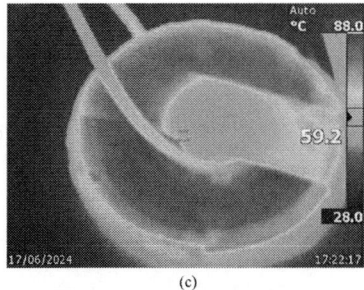

| (a) | (b) | (c) |

Fig. 7. Experimental results, (a) Experimental waveforms of the main components, (b) Thermal measurement of switch S_1, (c) Thermal measurement of air core coupled inductor.

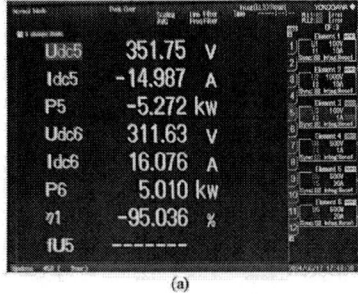 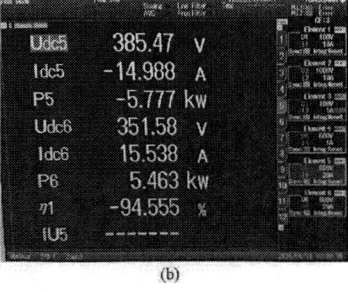 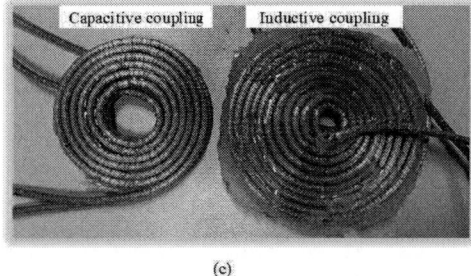

| (a) | (b) | (c) |

Fig. 8. Experimental results, (a) Numerical result in 5 kW, (b) Numerical result in 5.5 kW, (c) Air core difference in inductive coupling and capacitive coupling.

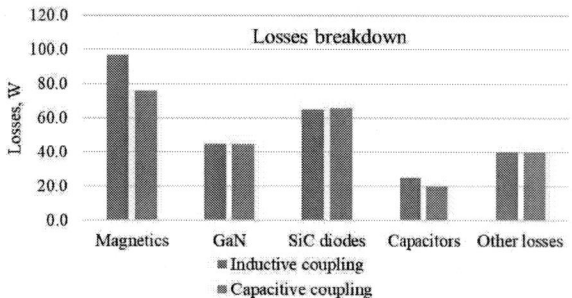

Fig. 9. Losses breakdown analysis for inductive and capacitive coupling.

V. CONCLUSIONS

This paper introduced a high-power density dc-dc converter utilizing capacitive coupling, demonstrating significant improvements over traditional inductive methods by being smaller, lighter, and more efficient. Design considerations for capacitive-based converters, and also the performance comparison of the capacitive coupling and inductive coupling converter in same test condition also discussed in this literature. The experimental prototype, operating at a frequency of 1.5 MHz and delivering 6 kW of power, validates the effectiveness of this approach. Experimental results demonstrate that the proposed capacitive coupling-based approach with coupled resonance inductors achieves suitable efficiency, stable thermal performance, and effective ZVS operation for high-power, high-frequency applications. The reduced size of the inductive element at higher frequencies offers a clear advantage in applications where space and weight are critical.

ACKNOWLEDGMENT

This research was supported by the Polish National Center of Science in frame of the project Sonata BIS: 2023/50/E/ST7/00097 and Estonian Research Council grant PRG675, postdoctoral grant PUTJD1209.

REFERENCES

[1] K. -C. Tseng, H. -L. Hsu, Y. -H. Su and C. -A. Cheng, "A Novel Isolated High-Step-Up Interleaved Converter for Renewable Energy Systems," *IEEE Journal of Emerging and Selected Topics in Industrial Electronics*, vol. 4, no. 4, pp. 1235-1243, Oct. 2023

[2] S. Pourjafar, H. Afshari, P. Mohseni, O. Husev, O. Matiushkin and N. Shabbir, "Comprehensive Comparison of Isolated High Step-up DC-DC Converters for Low Power Application," *IEEE Open Journal of Power Electronics*, vol. 5, pp. 1149-1161, 2024

[3] Y. Cao, K. Ngo and D. Dong, "A Scalable Electronic-Embedded Transformer, a New Concept Toward Ultra-High-Frequency High-Power Transformer in DC–DC Converters," *IEEE Transactions on Power Electronics*, vol. 38, no. 8, pp. 9278-9293, Aug. 2023

[4] S. Pourjafar, P. Mohseni, O. Matiushkin, O. Husev and D. Vinnikov, "Novel Isolated High Step-up DC-DC Converter with Wide Input Voltage Regulation Range," *2023 IEEE 64th International Scientific Conference on Power and Electrical Engineering of Riga Technical University (RTUCON)*, Riga, Latvia, 2023, pp. 1-6

[5] M. Mogorovic and D. Dujic, "100 kW, 10 kHz medium-frequency transformer design optimization and experimental verification," *IEEE Transactions on Power Electronics*, vol. 34, no. 2, pp. 1696-1708, Feb. 2019.

[6] C. Liu, A. P. Hu, G. A. Covic and N. -K. C. Nair, "Comparative Study of CCPT Systems With Two Different Inductor Tuning Positions," *IEEE Transactions on Power Electronics*, vol. 27, no. 1, pp. 294-306, Jan. 2012

[7] E. Culurciello and A. G. Andreou, "Capacitive Inter-Chip Data and Power Transfer for 3-D VLSI," *IEEE Transactions on Circuits and Systems II: Express Briefs*, vol. 53, no. 12, pp. 1348-1352, Dec. 2006

[8] J. Dai and D. C. Ludois, "A Survey of Wireless Power Transfer and a Critical Comparison of Inductive and Capacitive Coupling for Small Gap Applications," *IEEE Transactions on Power Electronics*, vol. 30, no. 11, pp. 6017-6029, Nov. 2015

[9] W. Chen, P. Han and A. M. Bazzi, "Evaluation of H-bridge and half-bridge resonant converters in capacitive-coupled wireless charging," *2017 IEEE Applied Power Electronics Conference and Exposition (APEC)*, Tampa, FL, USA, 2017, pp. 3738-3742

[10] A. Toebe, P. H. B. Löbler, L. Schuch and C. Rech, "A New Capacitive Coupled Step-up DC-DC Converter," *2023 IEEE Energy Conversion Congress and Exposition (ECCE)*, Nashville, TN, USA, 2023, pp. 105-111

[11] Z. Cao, Y. Gu, D. Zhang, "A Novel Magnetic Integrated Boost-forward Converter," *in proc. of IEEE/IAS Industrial and Commercial Power System Asia (I&CPS Asia)*, Chengdu, China, Jul. 18-21, 2021.

[12] S. Pourjafar, O. Husev and C. Roncero-Clemente, "Review and Outlook of Isolated Capacitive Coupling Based Converters," *2024 IEEE 18th International Conference on Compatibility, Power Electronics and Power Engineering (CPE-POWERENG)*, Gdynia, Poland, 2024, pp. 1-6

[13] M. Z. Erel, K. C. Bayindir, M. T. Aydemir, S. K. Chaudhary and J. M. Guerrero,"A Comprehensive Review on Wireless Capacitive Power Transfer Technology: Fundamentals and Applications," *IEEE Access*, vol. 10, pp. 3116-3143, 2022

[14] R. Hou, J. Lu and D. Chen, "Parasitic capacitance Eqoss loss mechanism, calculation, and measurement in hard-switching for GaN HEMTs," *2018 IEEE Applied Power Electronics Conference and Exposition (APEC)*, San Antonio, TX, USA, 2018, pp. 919-924.

[15] R. Miftakhutdinov, "New aspects on analyzing ZVS conditions for converters using super-junction Si and wide bandgap SiC and GaN power FETs," *2014 16th European Conference on Power Electronics and Applications*, Lappeenranta, Finland, 2014, pp. 1-9.

[16] M. Kasper, R. M. Burkart, G. Deboy and J. W. Kolar, "ZVS of Power MOSFETs Revisited," in *IEEE Transactions on Power Electronics*, vol. 31, no. 12, pp. 8063-8067, Dec. 2016.

[17] O. Husev, O. Matiushkin, P. Mohseni, F. Canales and D. Vinnikov, "Feasibility Study of High-Power Density of Modified Isolated CLLC dc-dc Interface with Wide Range of Voltage/Current Regulation," *PCIM Europe 2024; International Exhibition and Conference for Power Electronics, Intelligent Motion, Renewable Energy and Energy Management*, Nürnberg, Germany, 2024, pp. 893-902

[18] I. Rytoeluoto and K. Lahti, "Effect of film thickness and electrode area on the dielectric breakdown characteristics of metallized capacitor flms," *Proceedings of the Nordic Insulation Symposium*, 02 2018

[19] A. Yoshida, C. Kuji, T. Hasebe and M. Ozawa, "A Novel Aluminum Electrolytic Capacitor Suitable for High-frequency Power Converters," *2018 20th European Conference on Power Electronics and Applications (EPE'18 ECCE Europe)*, Riga, Latvia, 2018, pp. P.1-P.10.

[20] L. Qi, L. Petersson, and T. Liu, "Review of recent activities on dielectric films for capacitor applications," *Journal of International Council on Electrical Engineering*, vol. 4, no. 1, pp. 1-6, Jan 2014.

[21] Y. Tang, D. Fu, J. Kan, T. Wang, "Dual Switches DC/DC Converter With Three-Winding-Coupled Inductor and Charge Pump," *IEEE Transaction on Power Electronics*, vol. 31, no. 1, Jan. 2016.

[22] E. A. TDK, "Film capacitors - general info," Application Note, 2018

[23] G. Ortiz, J. Biela and J. W. Kolar, "Optimized design of medium frequency transformers with high isolation requirements," *IECON 2010 - 36th Annual Conference on IEEE Industrial Electronics Society*, Glendale, AZ, USA, 2010, pp. 631-638

[24] M. A. Bahmani, T. Thiringer and M. Kharezy, "Design Methodology and Optimization of a Medium-Frequency Transformer for High-Power DC–DC Applications," *IEEE Transactions on Industry Applications*, vol. 52, no. 5, pp. 4225-4233, Sept.-Oct. 2016

[25] O. Husev, O. Matiushkin, T. Jalakas, D. Vinnikov and N. V. Kurdkandi, "Comparative Evaluation of Dual-Purpose Converters Suitable for Application in DC and AC Grids," *IEEE Journal of Emerging and Selected Topics in Power Electronics*, vol. 12, no. 2, pp. 1337-1347, April 2024

LLC Converter Main Transformer Losses: Eliminating Air Gaps and Integrating Parallel External Inductors

1st Yu-Chen Liu
Department of Electrical Engineering
National Taipei University of Technology
Taipei, Taiwan
ycliu@mail.ntut.edu.tw

2nd Shang-Syun Wu
Department of Electrical Engineering
National Taipei University of Technology
Taipei, Taiwan
t112318020@ntut.org.tw

Abstract—Magnetic components are critical in power electronics design, particularly in high-efficiency power conversion applications. Although the introduction of an air gap can effectively control inductance and achieve zero-voltage switching (ZVS), it also causes fringing effects, resulting in additional losses, increased AC resistance, inconsistent coupling between the primary and secondary windings, uneven current distribution, and hotspot formation. While distributed air gaps have been developed to alleviate these issues, different circuits require varying numbers of distributed gaps. Additionally, this design faces manufacturing challenges, such as increased complexity in winding layout to avoid air gaps and the need to increase the magnetic column height to meet design requirements. To address these challenges, this paper proposes using an external inductor to replace traditional air gap magnetizing inductance, enabling air-gap-free energy transfer in the main transformer. A 380 V/12 V input, 500 W output LLC resonant converter serves as the validation platform. Simulation and experimental results show that this approach reduces air gap losses by 7.2 W and effectively mitigates the issue of uneven winding current distribution. Furthermore, copper losses related to the air gap are reduced by 49.6 %, demonstrating a significant improvement in energy efficiency. This paper provides a solution for designing high-power density converters, significantly enhancing system efficiency and reliability, and offering valuable insights for future power electronics applications.

Keywords—magnetic component, zero-voltage switching (ZVS), fringing effect, current distribution, distributed air gap, power electronic

I. INTRODUCTION

With the rapid advancement of power electronics converter technology, effectively managing the size and losses of passive components has become one of the core challenges for enhancing overall system performance and achieving high power density [1-3]. During the optimization process of magnetic element design, engineers must balance various factors to meet performance objectives. This is particularly critical in high-efficiency and high-power-density applications, where design considerations encompass magnetizing inductance requirements, operating voltage and current, switching frequency, and multiple parameters of magnetic materials, such as winding current density, power loss, saturation flux density, permeability, and core size. Furthermore, specific application scenarios require addressing effects like skin effect, proximity effect, and fringing flux to minimize unnecessary losses and electromagnetic interference, while ensuring the reliability and optimal performance of magnetic components.

Magnetic elements in power electronics applications are often designed with air gaps in their magnetic circuits to achieve desired electrical characteristics [4-6]. However, as power conversion technology advances to increasingly higher frequencies, the effects of AC losses caused by air gaps become significantly more pronounced, impacting efficiency and system performance. The presence of an air gap not only alters the linearity of the B-H curve but also significantly reduces the effective inductance, as shown in Fig. 1. As the winding current increases, the magnetic field intensity within the core rises, pushing the magnetic material's operating point closer to saturation [7]. Additionally, the fringing flux effect exacerbates leakage flux, increasing eddy current losses in the core and the windings near the air gap [8]. These issues decrease overall energy efficiency and generate higher electromagnetic noise and losses, leading to non-uniform flux density distribution and imposing constraints on system performance.

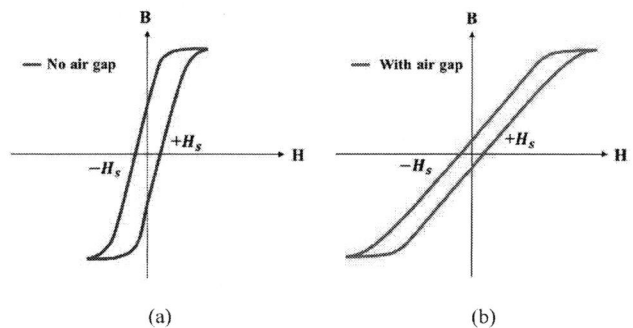

Fig. 1. Hysteresis loops (a) With air gaps (b) Without air gaps

Parallel winding current sharing issues in planar transformers are commonly encountered [9–10], which are related to the magnetic field distribution and influenced by factors such as the distance between winding PCBs and core structure. Additionally, the air gap alters the magnetic field distribution, thereby affecting current distribution. In modular stacked transformers with multi-layer PCBs, the presence of an air gap exacerbates current sharing imbalances. This paper uses finite element analysis (FEA) simulations to compare the losses and current sharing issues in parallel windings with and without an air gap. The results show that the air gap worsens the current sharing imbalance, especially in modular stacked transformers with multi-layer PCB structures. This highlights the adverse impact of the air gap on current sharing and emphasizes the need to consider the air gap's influence when designing transformers.

Air-gapped transformers still face challenges like reduced energy transfer efficiency, additional leakage losses [11], and discontinuities in the winding structure [12]. Although distributed air gap designs can alleviate these issues in some cases [13-14], their practicality in high-frequency applications is still constrained by manufacturing technology and economic considerations. This paper proposes an innovative method to isolate magnetizing flux and employs a parallel design strategy to mitigate the adverse effects of air gaps on transformer performance. We derive voltage gain equations and simplify the system using an equivalent circuit model analogous to traditional transformer structures, validating the feasibility and stability of the proposed design approach. Additionally, this research investigates the impact of fringing flux effects on transformer performance, particularly concerning current sharing among synchronous rectification switches. We will conduct comprehensive electromagnetic and thermal performance simulations using finite element analysis tools (e.g., ANSYS Maxwell) and thermal simulation software (e.g., Icepak) to identify optimal design parameters. The effectiveness of the proposed method will be verified through experimental validation. Ultimately, this work aims to develop a highly efficient DC/DC isolated converter based on an air-gap-free design concept, offering an innovative solution to advance power electronics technology.

This paper provides an in-depth analysis of the transformer air gap issue and its improvement process, offering a detailed examination of the associated technologies. The structure of the paper is as follows: Section II discusses the impact of transformer air gaps on converter performance, including edge effects leading to leakage flux, eddy current losses in windings, and current-sharing issues. Section III elaborates on the gain curve of the proposed topology and analyzes its resulting effects and advantages. Section IV introduces the concept of parallel inductors, presenting optimization design methods supported by simulation results. Section V compares the electrical characteristics and losses of transformers with and without air gaps using simulation software, accompanied by thermal characterization results. Section VI showcases hardware results of a 500 W LLC converter, which successfully addresses the current-sharing problem caused by the air gap. The converter achieves a peak efficiency of 96.2 %, and with the proposed improvements, the efficiency at full load is increased by 1.89 %. Finally, Section VII concludes the paper.

II. DEFINING THE LLC CIRCUIT STRUCTURE

This paper adopts the LLC resonant converter. Given that the specification has a high step-down ratio, the current in the low-voltage side windings of the transformer becomes relatively large, thereby increasing conduction losses. To address the issue of conduction losses in low-voltage, high-current applications,

this paper proposes using a multi-center-tapped parallel configuration, as shown in Fig. 2, to effectively share the current stress and significantly reduce losses.

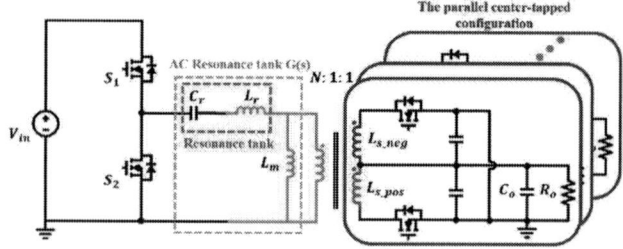

Fig. 2. The traditional LLC resonant converter topology

A. Analysis of the Winding Structure

Since the secondary side adopts a parallel center-tapped configuration to reduce current stress, this section provides a preliminary analysis to determine the relationship between conduction losses of the primary and secondary windings and the number of parallel center-tapped groups. The aim is to find a balance between cost and loss. Fig. 3 illustrates the model of the transformer and windings. Fig. 3 (a) shows a top view of the primary winding arrangement, while Fig. 3 (b) presents a top view of the secondary winding. To minimize the number of stacked winding layers, the primary side is wound using a single-layer, four-turn Litz wire structure. The secondary side employs a two-layer PCB winding method, with each layer consisting of a single turn. Focusing on the primary structure for analysis, the DC resistance of a single-layer, single-turn winding is denoted as R_{DC}. Since the secondary side's line width is four times that of the primary side, the DC resistance is one-fourth of the primary's. The actual I_{rms} currents I_p and I_s for the primary and secondary windings are derived using (1) to (2). As the number of parallel connections P_p on the primary side or P_s on the secondary side increases, the RMS current decreases, showing an inverse relationship with P_p and P_s.

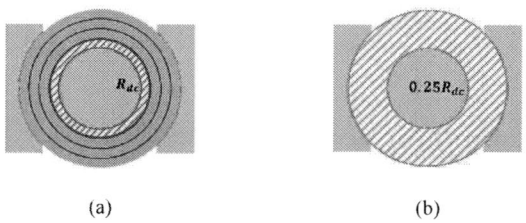

(a)   (b)

Fig. 3. Transformer winding model (a) Primary side (b) Secondary side

$$I_p = I_{rms} \cdot \frac{1}{P_p} \tag{1}$$

$$I_s = I_{rms} \cdot N_n \cdot \frac{1}{P_s} \tag{2}$$

Next, we will calculate the total DC resistance R_p and R_s of the primary and secondary windings after equalization. Since the number of turns and impedance are directly proportional, this relationship is defined through the parameter l_s. Ultimately, the DC resistance and corresponding conduction losses are calculated based on the total number of turns and the number of parallel groups, The final results of (3) and (4) are derived as follows:

$$R_p = P \cdot (R_{dc} \cdot l_s \cdot 32) \tag{3}$$

$$R_s = P \cdot (0.25 R_{dc} \cdot l_s \cdot 2) \tag{4}$$

The optimization of winding arrangement methods is presented in Table I, which lists the conduction losses of the equalized windings. These results are then combined with the conduction losses of the synchronous rectifier to further calculate the total conduction losses under different parallel group numbers, as shown in Fig. 4. Fig. 4 demonstrates that, as the number of parallel groups increases, the total conduction losses decrease significantly. However, the benefits of increasing the parallel groups begin to diminish after a certain point. Therefore, the optimal design region is chosen near the inflection point where the losses begin to decrease. For parallel group numbers of 3 and 4, both configurations demonstrate relatively excellent performance.

TABLE I.   COPPER CONDUCTION LOSS IN TRANSFORMER WINDINGS

	1P1S	1P2S	1P3S	1P4S	1P5S	1P6S
N_n			32:2			
P_p			1			
P_s	1	2	3	4	5	6
I_p			$1/16 \cdot I_{rms}$			
I_s			$1 \cdot I_{rms}$			
R_p			$32 R_{dc}$			
R_s	$2R_{dc}$	$1R_{dc}$	$0.67R_{dc}$	$0.5R_{dc}$	$0.4R_{dc}$	$0.33R_{dc}$
P_{dc}	2.152	1.152	0.795	0.625	0.525	0.455

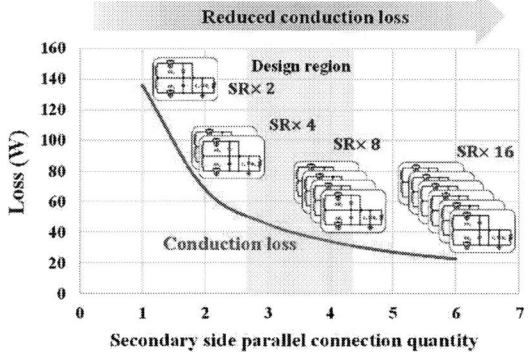

Fig. 4. Total conduction loss of individual parallel connections at the center-tapped.

Based on the preliminary plan, we selected winding structures with 3 and 4 parallel groups, as shown in Fig. 5 . It is worth noting that the structure with 3 parallel groups has an asymmetrical winding issue, which may lead to various problems, such as uneven magnetic field distribution, poor

current sharing between windings, and higher losses. To avoid these potential issues, we ultimately selected the structure with 4 center-tapped groups as the circuit architecture for this work, as shown in Fig. 5.

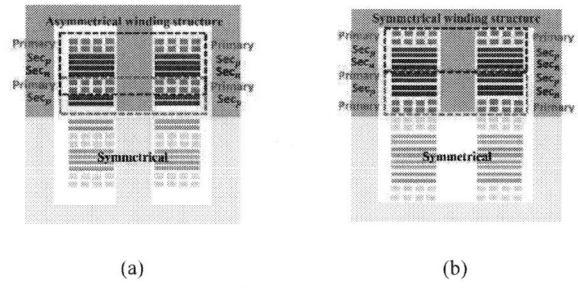

(a)   (b)

Fig. 5. Winding arrangement structure (a) Three sets of center-tapped (b) Four sets of center-tapped

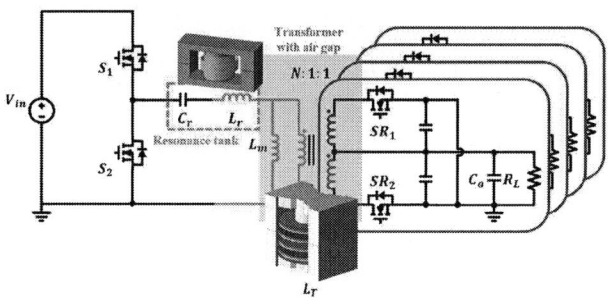

Fig. 6. The LLC resonant converter topology adopted in this paper

B. Effect of Leakage Flux on Winding Losses

The air gap is crucial in LLC resonant converters [15-17], used to adjust the excitation inductance and achieve zero-voltage switching (ZVS). However, to enhance power density, it is necessary to increase the operating frequency of magnetic components. This will cause windings to be increasingly affected by air gap edge effects in high-frequency environments, which has become one of the key challenges in designing high-power-density transformers. Using finite element analysis (FEA) with Ansys, the edge effects caused by the air gap lead to an uneven distribution of leakage flux around the gap, as shown in Fig. 7, which generates significant eddy currents within the windings. These eddy currents exacerbate the AC losses in the windings, especially in high-current, low-voltage applications, resulting in additional energy loss, incomplete magnetic field transmission, and reduced transformer efficiency. Furthermore, these losses are often concentrated in areas where the windings intersect with the core, as illustrated in Fig. 8, making thermal management more complex and challenging. The formation of localized hotspots not only decreases the reliability of the windings but also accelerates material aging, shortening the device's lifespan, which is particularly detrimental to long-term stable operation. Therefore, mitigating the thermal issues caused by air gap edge effects is one of the key factors in improving transformer performance. In subsequent sections, we will explore methods to reduce core-induced losses by improving winding layouts and optimizing conductor structures. Additionally, we will apply precise thermal characterization techniques to analyze the most affected thermal regions, enabling targeted design improvements to reduce thermal losses and enhance the thermal stability of the transformer, effectively addressing the issues caused by eddy currents.

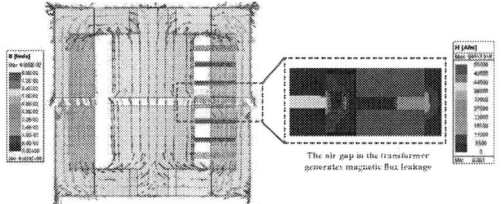

Fig. 7.   Fringing effects caused by the air gap

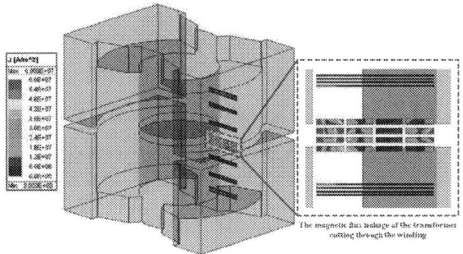

Fig. 8.   Leaked magnetic flux induces eddy currents in the winding

C. Effect of Leakage Flux on Current Sharing

The presence of air gaps significantly impacts the performance of transformers and circuits, especially in parallel windings or parallel circuit structures, where this effect is more pronounced. The magnetic field's uneven distribution caused by the air gap leads to poor current sharing in parallel windings or circuits, resulting in an uneven current density distribution, as shown in Fig. 9. This non-uniform current distribution causes certain windings or circuit components to carry higher currents, leading to additional copper losses and localized hotspots, which further affect the overall system stability.

As the current density increases, the copper losses also increase, particularly in high-power-density applications, where these additional losses have a more significant impact on system efficiency. Furthermore, the air gap exacerbates the current stress on synchronous rectifier switches, as the uneven current distribution affects the switch conduction time, increasing the switching losses of the synchronous rectifier. This not only reduces the efficiency of the switching components but also can trigger thermal effects, further affecting the system's reliability and stability. These issues become more prominent under high load or extreme operating conditions. Moreover, the air gap introduces the issue of asymmetric leakage inductance between parallel structures. During the positive and negative half-cycle operation, the asymmetric magnetic field caused by the air gap leads to an uneven distribution of leakage inductance between different windings or parallel structures, causing impedance mismatch between parallel structures. This further exacerbates the uneven current distribution. The asymmetry in leakage inductance complicates the system's impedance characteristics and adversely affects the system's stability during load changes. This could result in unnecessary current fluctuations or even unstable oscillations or resonant phenomena, further damaging

the dynamic performance of the system. To effectively reduce the negative impact of air gaps on parallel structure performance, comprehensive optimization during the design phase is essential. This includes improving the winding structure to enhance current sharing and precisely adjusting the air gap design to minimize the impact of leakage inductance asymmetry. Distributed air gap technology is an effective solution, as it reduces edge effects by altering the air gap distribution. However, despite the fact that this technology can alleviate the effects of air gaps to some extent, the improvements are still limited by process costs and capabilities. Therefore, subsequent chapters will delve into the specific applications and limitations of these technologies.

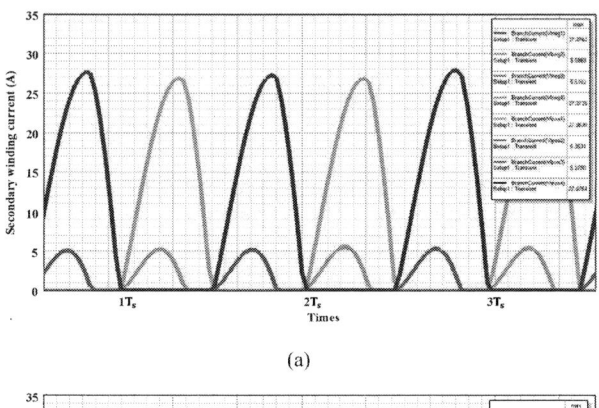

(a)

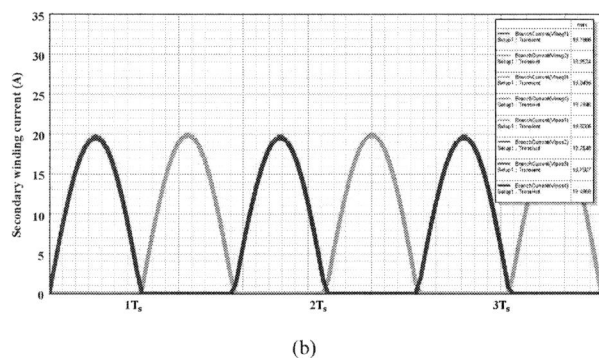

(b)

Fig. 9.   Rectified current of each winding group (a) With air gap (b) Without air gap

III. Circuit Topology And Gain Curve

In summary, the presence of air gaps introduces numerous challenges to the design and performance of magnetic components, such as uneven current distribution, increased copper losses, localized hotspots, and reduced system stability. To address these difficulties, this paper proposes an innovative topology, as shown in Fig. 10. This design strategically separates the magnetizing inductance from the transformer, so the transformer no longer relies on air gaps to regulate inductance. Consequently, the transformer's primary role is limited to isolation and energy transfer, while the magnetizing inductance can be managed through other precise control methods, effectively mitigating the various adverse effects caused by air gaps. This decoupled design offers significant advantages. Firstly, it notably reduces copper losses in the transformer, enhancing system efficiency, particularly in high-

979-8-3315-1612-3/25 $31.00 © 2025 IEEE

frequency applications. Secondly, the topology helps optimize thermal management, minimizing the formation of hotspots and thereby improving the reliability and lifespan of the components. Moreover, this design effectively alleviates the issue of asymmetric leakage inductance in parallel structures, enhancing current sharing and balance, and ultimately improving the overall stability and performance of the system."

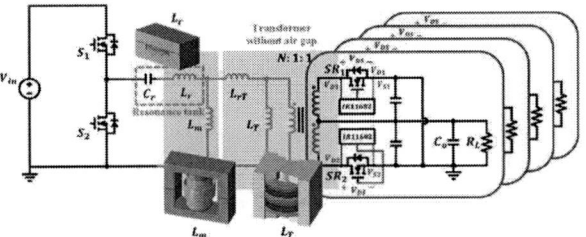

Fig. 10. The topology proposed in this paper

The transformer model proposed in this paper is applied to LLC resonant converters, facilitating the conversion into the linear circuit shown in Fig. 11 for subsequent circuit characteristic analysis. In this model, C_r and L_r represent the resonant tank, L_m denotes the additional excitation inductance introduced in this paper, L_T stands for the excitation inductance of the gapless transformer, and L_{rt} represents the leakage inductance of the gapless transformer.

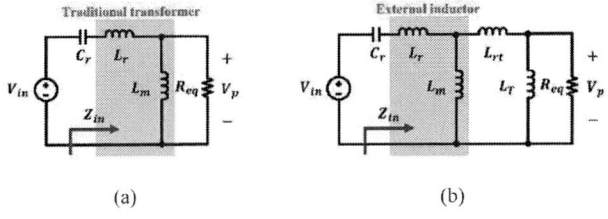

Fig. 11. LLC equalization model (a) Traditional (a) Proposed

Using (5), the relationship between the coupling coefficient and gain is shown in Fig. 12. Leakage inductance primarily affects gain; as transformer coupling decreases and leakage inductance increases, circuit gain rises, aiding hold time design. In region II, gain can be adjusted with less frequency variation, unlike in region I, allowing parameter adjustments to meet design requirements. Using the First Harmonic Approximation (FHA) method, we analyze the gain curve of the proposed topology and compare it with that of a traditional resonant converter. By establishing an equivalent model and deriving the gain transfer function based on (6) and (7), we explore the characteristics and advantages of this topology. The gain curve of the proposed topology exhibits significant differences compared to conventional resonant converters. Through FHA analysis, we gain a clearer understanding of the new topology's gain behavior across various frequency ranges. This analysis enables us to further assess its efficiency and stability improvements in different application scenarios.

$$G_t(s) = \frac{V_o}{V_{in}} \quad (5)$$

$$R_{DCX}(s) = \left(j\omega L_m \parallel \left(j\omega L_{rt} + \left(j\omega L_T \parallel R_{eq}\right)\right)\right) \quad (6)$$

$$G_t(s) = \frac{R_{DCX}(s)}{j\omega L_r + \frac{1}{j\omega C_r} + R_{DCX}(s)} \quad (7)$$

Fig. 14 is the gain characteristics of the LLC converter as a function of frequency, highlighting the impact of different coupling coefficients k on gain behavior. The blue curve represents the conventional LLC transformer design with k = 1, corresponding to a scenario without leakage inductance, and serves as the reference gain curve of a traditional transformer. The green, purple, and red curves correspond to the proposed topology with coupling coefficients of = 0.999, k = 0.995, and k = 0.97, respectively. The paper reveals that as the k value decreases, the gain in the low-frequency region (Region II) significantly increases, extending the stable operating range and enhancing system performance. The graph also marks the nominal operating point, clearly showing the differences between the traditional design and the improved topology. This validates the effectiveness of the proposed approach in enhancing both gain and efficiency. Such improvements are particularly beneficial for applications requiring a wider gain range, providing greater design flexibility and optimization potential for LLC converters.

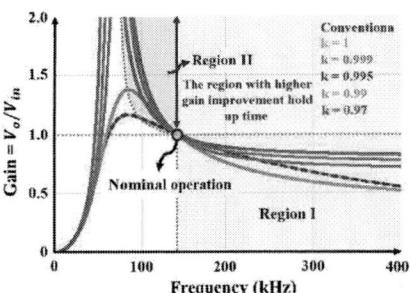

Fig. 12. Gain curves with different coupling coefficients.

IV. PARALLEL INDUCTORS DESIGN

In inductor design, the absence of interleaving in the windings and the presence of higher magnetomotive force (MMF) stress between layers make the number of layers significantly impact the winding structure, especially in terms of AC effects and conduction losses. As the number of layers increases, designers gain more flexibility in adjusting the winding structure, and the increased winding area helps improve current distribution, thereby reducing copper losses and enhancing overall efficiency. It also improves current sharing, further boosting efficiency in high-current applications. However, increasing the number of layers can also introduce several disadvantages, such as increased parasitic capacitance, more pronounced AC effects, higher conduction losses, and increased design and manufacturing complexity. Specifically, under high-frequency operation, as shown in Fig. 13, the AC resistance of a five-layer inductor demonstrates strong AC effects as the layer thickness changes, which leads to higher high-frequency losses and reduces system efficiency. Overall, while multi-layer PCB designs contribute to better heat dissipation and efficiency, careful adjustment of parameters is necessary to balance various influencing factors and minimize losses. Therefore, when opting for multi-layer designs, it is important to consider the impact of the number of layers on the overall performance.

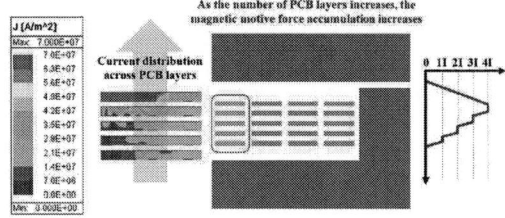

Fig. 13. Magnetomotive force and current density distribution

The losses of magnetic components primarily include core losses and winding losses. Therefore, adjusting the winding width and the core cross-sectional area is particularly important when designing PCB transformers. Within a given layout area constraint, an optimal combination of winding width P and core cross-sectional area r can be identified to minimize losses. By using a parameterized model to adjust P and the core parameter r and then calculating the transformer losses, Fig. 14 presents contour plots. Fig. 14 provides a visual representation of the overall loss distribution as the layout area changes, effectively helping to pinpoint the ideal values of P and r for a specific layout area. For an operating frequency of 140 kHz, the resonant inductor was optimized using a single-layer, four-turn structure stacked into four layers. The geometric optimization function was applied across various ranges of transformer layout areas, repeatedly assessing their impact on transformer losses, as shown in Figure 16, which presents the parameterized analysis results for 16 turns. The results indicate an inverse relationship between losses and layout area; however, once the layout area surpasses a certain critical point, the reduction in losses becomes less pronounced. In Fig. 15, the "design region" highlights the optimal design range, demonstrating the ideal balance achieved at this point. This optimization technique enables efficient design, enhancing performance and effectively reducing power loss.

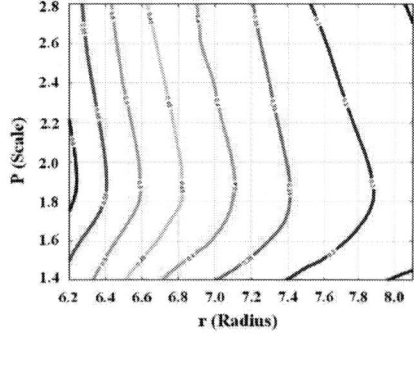

(c)

Fig. 14. Total loss and footprint

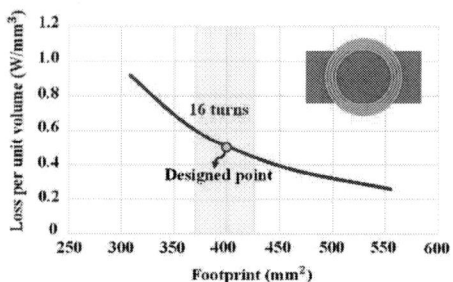

Fig. 15. Loss curves for different number of turns

V. COMPARISON WITH TRADITIONAL TRANSFORMER

Air gaps in magnetic components can lead to uneven current distribution, increase eddy current losses, and cause significant thermal stress within PCB windings. Our proposed solution effectively mitigates the issue of uneven current distribution and simplifies thermal management. Specifically, an air-gap-free transformer eliminates the challenges associated with uneven current distribution, ensuring a more uniform temperature distribution across the windings. As shown in Fig. 16, thermal simulation results indicate that under forced air cooling, our air-gap-free design achieves approximately a 34% temperature reduction compared to traditional transformer structures. This significant temperature reduction is attributed to the complete elimination of hot spots commonly caused by air gap designs. The edge effects caused by a single air gap are the most severe, resulting in higher winding losses compared to designs with four distributed air gaps or no air gap at all. Our analysis further demonstrates that the air-gap-free transformer achieves the most uniform winding temperature distribution, reducing the risk of localized overheating and significantly enhancing the overall reliability of the component. In terms of current-carrying capability, the air-gap-free transformer continues to outperform other designs. In summary, the air-gap-free design not only simplifies the manufacturing process but also provides more efficient heat dissipation, making it highly suitable for high-power-density applications with stringent thermal management requirements. In conclusion, this paper highlights the importance of strategic air gap management in transformer design. By optimizing air gap design and winding layout, it is possible to minimize thermal stress and losses while achieving superior performance.

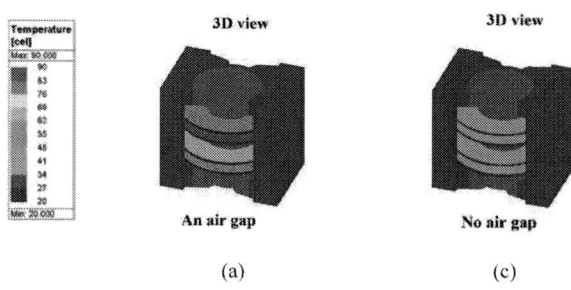

Fig. 16. Thermal simulation results (a) With air gap (b) Without air gap

The final loss distribution results are presented in Table II. Our proposed solution offers a practical and effective approach, significantly improving transformer efficiency and reliability, which is crucial for advanced power electronic systems.

TABLE II.

SIMULATION RESULTS OF TRANSFORMER TOTAL LOSSES DUE TO AIR GAP

Item	Spec		Loss (W)			
	Inductance	Turns	Copper	Core	SR	Total
Traditional transformer	120 μH	32:2:2	12.75	0.27	1.51	19.07
This work transformer	10.1 mH	32:2:2	4.94	0.53	1.21	12.59
This work inductor	120 μH	16	1.21	0.65		

VI. EXPERIMENTAL RESULTS

This paper presents a converter with additional magnetizing inductance, designed with specifications of 500W power, 380V input, and 12V output, aiming to enhance power density, efficiency, and reliability. The paper delves into the optimization of magnetic components within a multi-layer PCB structure. Fig. 17 illustrates the LLC resonant converter circuit employing the air-gap-free transformer used in this research. The main circuit board utilizes a four-layer PCB to form the power loop for all essential components, including transformer windings, resistors, capacitors, and other devices. The transformer's primary winding is designed as a single-layer, four-turn coil distributed across two layers, constructed using four PCBs. To reduce costs and achieve an interleaved winding arrangement, the secondary winding is designed as a single-layer, one-turn coil spread across four layers. These layers correspond to the positive and negative half-cycles of the winding and are connected in parallel using four PCBs. All PCBs share the same stack-up structure and are electrically connected through soldered PCB connectors. Additionally, an inductor is connected in parallel at the transformer's input to implement the proposed external magnetizing inductance design. Table III provides the detailed design parameters and components used in this paper.

TABLE III.

PARAMETERS OF THE 380 V/12 V, 500 W LLC CONVERTER

Design Parameters	Values
Input Voltage, V_{in}	380 V
Output Voltage, V_o	12 V
Output Power, P_o	500 W
Primary Devices	IPP60R180CM8
Synchronous Rectifiers	BSC010N04LS
Gate Driver of Synchronous Rectifiers	IR11682S
Resonant Frequency	143 kHz
Core Material	3C96
Magnetizing Inductance, L_{mt}	2.7 mH
Resonant Inductance, L_r	25 μH
Resonant Capacitoers, C_r	50 nC
Parallel External Inductors Inductance, L_m	134 μH

(a)

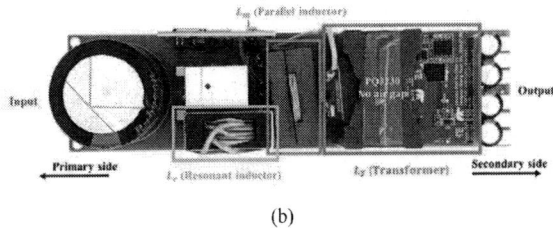

(b)

Fig. 17. Hardware 500W 380 V/12 V prototype (a) 3D View of prototype (b) Top view of the without air gap transformer

Experimental results demonstrate significant advantages of the air-gap-free transformer design in terms of thermal management efficiency and current-sharing capability. Fig. 18 highlights the current-sharing performance of the secondary circuits. In this design, four parallel secondary circuits are controlled by the synchronous rectifier IC IR11682, which adjusts the V_{DS} voltage difference among the secondary circuits by monitoring the current through the SR switches, enabling efficient current sharing. This configuration ensures load balance among the circuits, preventing overheating or losses caused by uneven current distribution, thereby improving the system's performance and reliability. Fig. 18(a) shows a traditional transformer with an air gap, while Fig. 18(b) depicts the air-gap-free transformer. As the circuit operates in the second resonance region, the waveform observed in the figure indicates that the circuit is functioning in the decoupling region during oscillation. Additionally, the results indicate that traditional transformers with air gaps exhibit significantly uneven current distribution among the parallel windings near the air gap. In contrast, the air-gap-free transformer achieves a more uniform current distribution among its parallel windings. This confirms that the air-gap-free transformer design successfully enhances current sharing among the four central junctions, improving the operational consistency of the secondary circuits.

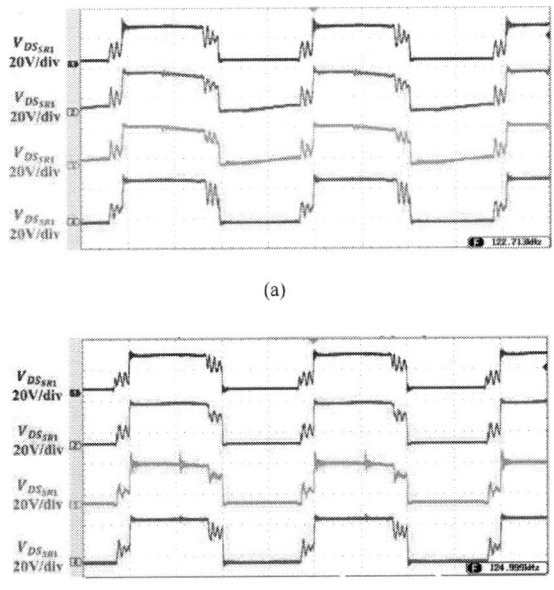

Fig. 18. The V_{DS} of the positive half-cycle in SR (a) With air gap (b) Without air gap

Fig. 19 presents the thermal imaging results of the circuit under full-load conditions, showing highly uniform heat distribution during operation, particularly in the central junction area. Compared to the air gap transformer in Fig. 19(a), the air-gap-free transformer in Fig. 19(b) demonstrates a lower temperature profile, highlighting the superior thermal management performance of the proposed design. These results further validate the ability of the proposed design to effectively control thermal stress under high-power operation, maintaining stable operating temperatures under both steady-state and dynamic conditions.

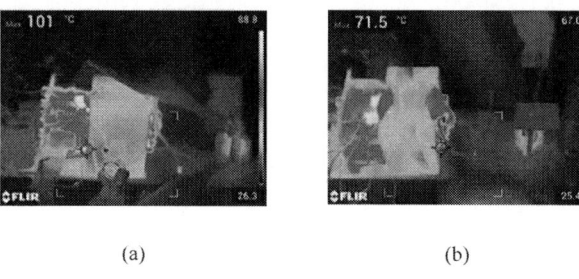

(a)   (b)

Fig. 19. Thermal performance at full load (500 W) (a) With air gap (b) Without air gap

Finally, the efficiency comparison of the two designs is shown in Fig. 20. The paper indicates that the proposed air-gap-free transformer design significantly reduces conduction losses in individual components and windings, resulting in an overall efficiency improvement of 1.89 % under full-load conditions.

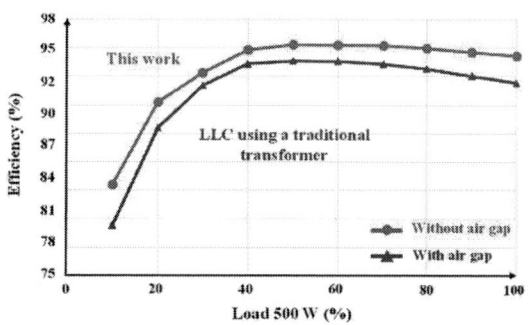

Fig. 20. Efficiency Comparison

VII. CONCLUSION

This paper suggests a way to increase circuit conversion efficiency and output winding current sharing, and this is through the use of external magnetizing inductance. The effectiveness of an air-gap-free design is analyzed through simulation and the optimization of the external magnetizing inductance is done with the assumption of an air-gap-free primary transformer. The paper achieves a 500 W resonant converter with an input voltage of 380 V and an output voltage of 12 V. The simulation shows that this method can reduce the loss by 6.5 W and the experimental result is in accordance with

the simulation, efficiency of the system is measured to be 96.2 % and overall efficiency is improved by 1.89 % at full load. From the above results, it can be concluded that when incorporated with an air-gap-free transformer and a precise current-sharing control, not only does the thermal management of the converter is enhanced but also the operation of the multiple parallel connected secondary circuits. These improvements not only contribute to the enhancement of efficiency of the converter but also greatly enhance the lifetime of the system, especially in the high-power density applications. The work presented in this paper offers useful information for the design of high efficiency and high reliability power.

REFERENCES

[1] S. Wang, F. C. Lee and Q. Li, "Improved Balance Technique for Common-Mode Noise Suppression of PCB-Based PFC," IEEE Trans., vol. 37, no. 4, pp. 4174-4182

[2] R. Yu, Q. Huang, T. Chen, A. Q. Huang, and T. Ribarich, "High-frequency and high-density design of all GAN power supply unit," in Proc. Int. Exhib. Conf. Power Electron., Intell. Motion, Renew. Energy Energy Manage., 2018, pp. 1–5

[3] R. Yu, T. Chen, P. Liu, and A. Q. Huang, "A 3-D winding structure for planar transformers and its applications to LLC resonant converters," IEEE Trans. Emerg. Sel. Topics Power Electron., vol. 9, no. 5, pp. 6232–6247, Oct. 2021.

[4] W. A. Roshen, "Fringing field formulas and winding loss due to an air gap," IEEE Trans. Magn., vol. 43, no. 8, pp. 3387–3394, Aug. 2007.

[5] M. K. Kazimierczuk, High-Frequency Magnetic Components. New York, NY, USA: Wiley, 2009.

[6] J. D. Pollock, W. Lundquist, and C. R. Sullivan, "Predicting inductance roll-off with DC excitations," in Proc. IEEE Energy Convers. Congr. Expo., Sep. 2011, pp. 2139–2145.

[7] A. Ayachit and M. K. Kazimierczuk, "Steinmetz equation for gapped magnetic cores," IEEE Magn. Lett. vol. 7, May 2016.

[8] N. H. Kutkut and D. M. Divan, "Optimal air-gap design in highfrequency foil windings," in Proc. IEEE 12th Annu. Appl. Power Electron. Conf. Expo., Atlanta, GA, USA, Feb. 1997, pp. 381–387.

[9] W. Chen, Y. Yan, Y. Hu, and Q. Lu, "Model and design of PCB parallel winding for planar transformer," IEEE Trans. on Magn., vol.39, no. 5, pp. 3202–3204, Sep. 2003.

[10] Y. Cai, M. H. Ahmed, Q. Li and F. C. Lee, "Optimized design of integrated PCB-winding transformer for MHz LLC converter," in Proc. IEEE Applied Power Electron. Conf. and Expo., pp. 1452–1458, March. 2019.

[11] C. W. T. McLyman, Transformer and Inductor Design Handbook. New York, NY, USA: Marcel Dekker, 2004, pp. 1.21–1.26

[12] J. -R. Sibue, J. -P. Ferrieux, G. Meunier and R. Periot, "Modeling of Losses and Current Density Distribution in Conductors of a Large Air-Gap Transformer Using Homogenization and 3-D FEM," in IEEE Transactions on Magnetics, vol. 48, no. 2, pp. 763-766, Feb. 2012

[13] S. Hao, Z. Zhang, J. Li and J. Han, "Analysis of Distributed Air Gap Parameters of Differential Mode Inductor Considering Core Loss and Saturation," In IEEE 2019 International Conference on Electrical Machines and Systems (ICEMS), Harbin, China, 2019, pp. 1-5

[14] R. Jez, "Influence of the Distributed Air Gap on the Parameters of an Industrial Inductor," in IEEE Transactions on Magnetics, vol. 53, no. 11, pp. 1-5, Nov. 2017

[15] M. Ilic and D. Maksimovic, "Interleaved Zero-Current-Transition Buck Converter," in IEEE Transactions on Industry Applications, vol. 43, no. 6, pp. 1619-1627, Nov.-dec. 2007.

[16] S. Jiang, C. Nan, X. Li, C. Chung and M. Yazdani, "Switched tank converters," 2018 IEEE Applied Power Electronics Conference and Exposition (APEC), San Antonio, TX, 2018, pp. 81-90,

[17] B. Lu, W. Liu, Y. Liang, F. C. Lee, and J. D. van Wyk, "Optimal design methodology for LLC resonant converter," in Proc. 21st Annu. IEEE Appl. Power Electron. Conf. Expo., Mar. 2006, pp. 533–

Gap in pagination due to withheld paper.

Pages 2187-2193

Small-Signal Phasor Modeling of T-Type Bridge-Based Single-Sided and Double-Sided *LCC* Resonant Converters for WPT Applications

Aditya Zade, Shubhangi Gurudiwan, and Regan Zane
Department of Electrical and Computer Engineering, Utah State University
Logan, Utah – USA 84341
Email: aditya.p.zade@usu.edu

Abstract—Wireless power transfer systems often utilize dc-dc resonant converters with single-sided or double-sided *LCC* resonant networks for effective power conversion. With the increasing prevalence of *LCC* networks, this paper investigates a dc-dc resonant converter comprising a T-type bridge, single-sided and double-sided *LCC* resonant tanks, and an isolated diode-bridge rectifier connected to a battery load, and conducts a thorough small-signal analysis. The study employs phasor transformation-based small-signal modeling, which provides clear and intuitive insights into the model derivation. The precise model presented here enhances the design and stability of closed-loop feedback control, resulting in more reliable and efficient power delivery. The proposed analysis is validated through simulations conducted on a 20 kW T-type bridge-based dc-dc converter that incorporates an *LCC* resonant network. Additionally, hardware testing of a 4 kW, 85 kHz dc-dc resonant converter prototype further substantiates the accuracy of the small-signal modeling process for perturbation frequencies up to 55 kHz.

Index Terms—Battery chargers, dc-dc converters, electric vehicles (EVs), *LCC*, phasor transformation, resonant converters, small-signal modeling, T-type, wireless power transfer (WPT).

I. INTRODUCTION

Resonant power converters are widely used for dc-dc conversion at high switching frequencies due to their soft-switching capabilities, high efficiency, high power density, and low electromagnetic interference [1]. These advantages make them suitable for applications such as telecom, energy storage, wireless power transfer (WPT), battery charging, and LED drivers [2]. In particular, WPT applications have gained significant attention in recent years for high-power charging systems. WPT provides a safe and convenient solution for charging electric vehicles (EVs) by eliminating the need for high-voltage, high-current, and moving mechanical connectors, which are prone to wear and fatigue [3]–[6].

WPT systems often use dc-dc resonant converters with single-sided or double-sided *LCC* resonant networks to generate a current source for charging EV batteries [6]–[8]. Given the widespread use of *LCC* networks in dc-dc converters, this study employs a T-type bridge-based dc-dc converter with single-sided and double-sided *LCC* resonant networks, as

This material is based in part upon work supported by the United States Department of Energy (DOE) under Award Number DE-EE0008803 Falcon and by the National Science Foundation (NSF) through the ASPIRE Engineering Research Center under Grant 1941524.

shown in Fig. 1, and conducts a thorough small-signal analysis of the system. In this study, the dc-dc resonant converter is connected to a grid-tied Unfolder. The Unfolder switches operate at a maximum of twice the grid frequency, converting each negative segment of the ac input voltages into a positive polarity [9]–[11]. With the Unfolder operating in open-loop, the small-signal modeling centers on the T-type bridge-based dc-dc converter, which is closed-loop controlled to regulate the battery load current and ensure grid-side power factor correction (PFC).

The higher-order single-sided or double-sided *LCC* resonant tank, designed with an appropriate quality factor, leads to sinusoidal current or voltage waveforms on both the primary and secondary sides of the dc-dc converter. In this context, the use of a phasor transformation-based small-signal modeling approach, as discussed in [12], [13], is deemed appropriate. Previously, generalized state-space averaging (GSSA) has been used for modeling *LCC* tank-based resonant converters [8], [14]. However, GSSA involves complex mathematical analysis and lacks intuitive clarity in the modeling steps, and prior works have not verified *LCC* tank-based modeling in hardware. In contrast, the phasor transformation-based modeling proposed in this paper offers an intuitive understanding of each step, serving as a valuable tool for both analysis and closed-loop control design.

The proposed small-signal modeling analysis is validated through simulations on a 20 kW T-type bridge-based dc-dc converter incorporating an *LCC* resonant network. Additionally, hardware testing on a 4 kW, 85 kHz battery charger prototype confirms the accuracy of the modeling process for perturbation frequencies up to 55 kHz.

II. BRIEF OVERVIEW OF CIRCUIT CONFIGURATION, MODULATION STRATEGY, AND CLOSED-LOOP CONTROL

In Fig. 1, a circuit diagram of the T-type bridge-based dc-dc converter is illustrated, consisting of either a single-sided or double-sided *LCC* resonant tank with isolation. As the topology is designed for WPT applications, where communicating control signals to the isolated side is challenging, a diode-bridge rectifier is used on the secondary side to connect to the battery load. The dc-dc converter in this study is connected to a grid-tied Unfolder operating in open-loop mode, switching

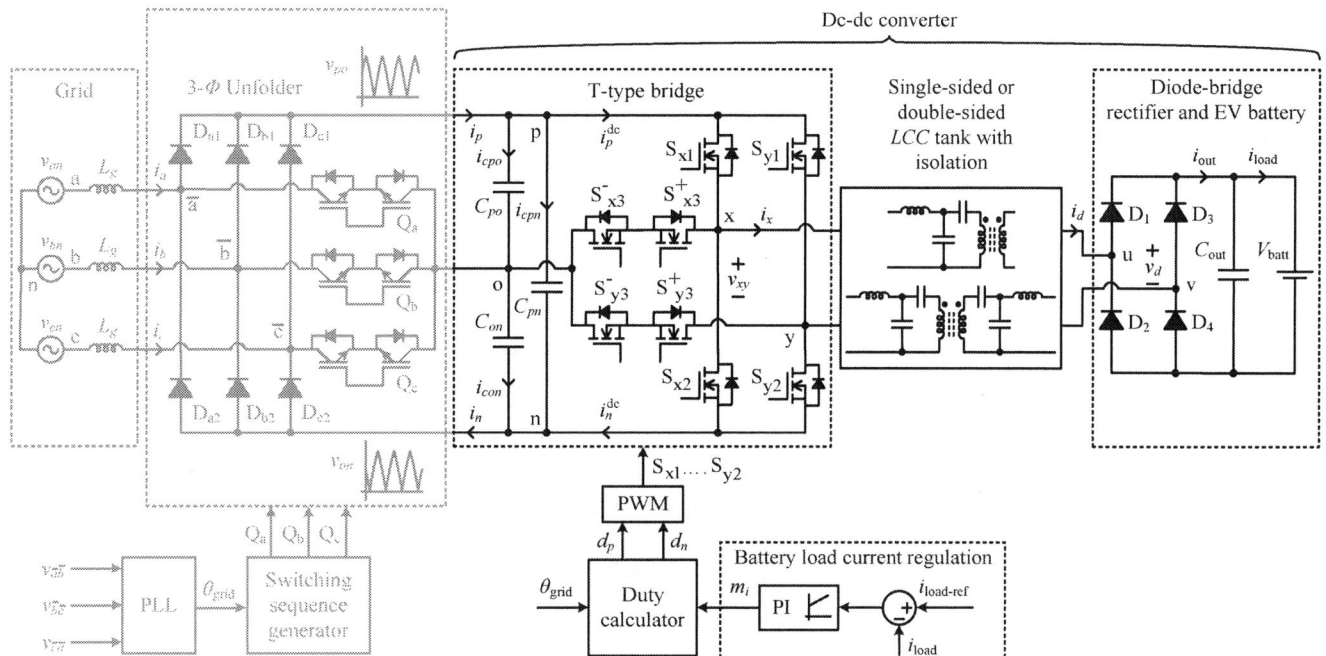

Fig. 1: Circuit diagram of a T-type bridge-based dc-dc converter featuring a single-sided or double-sided LCC tank with isolation and a diode bridge in an unfolding-based battery charger. The closed-loop control regulates the battery load current.

at a frequency up to twice the grid frequency. The Unfolder rectifies each negative segment of the ac input voltages into a positive polarity, producing two time-varying dc link voltages, v_{po} and v_{on}, which pulsate at three times the grid frequency at the input of the subsequent T-type bridge, as shown in Fig. 1.

The closed-loop control architecture is specifically designed for the high-frequency (HF) T-type bridge-based dc-dc converter to regulate the battery load current. The duty calculator, as depicted in Fig. 1, accurately calculates the duty ratios (d_p and d_n) based on the modulation index (m_i) obtained from the closed-loop control and the grid angle (θ_{grid}), which is calculated using the phase-locked loop (PLL). The duty calculator, along with the closed-loop control, ensures the desired load current while maintaining PFC on the grid side. A detailed description of the circuit configuration, modulation strategy, and closed-loop control is provided in [9]. Since the Unfolder operates in open-loop mode, the small-signal analysis in this paper focuses exclusively on the closed-loop controlled T-type bridge-based dc-dc conversion system, incorporating either a single-sided or double-sided LCC resonant tank.

III. Notation

In this paper, small-signal quantities are denoted by lowercase letters with a hat notation (such as $\hat{i}_{load}$), while steady-state quantities are indicated using capital letters (for instance, I_{load}). Quantities specified with only lowercase letters, without a hat (e.g., $i_{load} = I_{load} + \hat{i}_{load}$), represent full signal quantities, which include both steady-state and small-signal components. Phasors are indicated by an arrow (as in $\vec{v}_{xy}$), while magnitudes (peak values) are denoted by double bars (such as $\|\vec{i}_x\|$). Additionally, magnitudes are also identified by 'mag' in the

subscript. Finally, angled brackets (e.g., $\langle i_{out} \rangle$) are used for average values.

IV. Derivation of the Phasor Transformer

The application of the phasor transformer to model switching converters within the resonant dc-dc system has been previously investigated for representing an H-bridge [13]. The derivation established in the prior work has been extended here to represent the T-type bridge and the diode bridge utilized in this analysis. The implementation of a single-sided or double-sided higher-order LCC resonant tank with an appropriate quality factor facilitates the achievement of sinusoidal current or voltage waveforms on both the primary and secondary sides of the dc–dc conversion system. Consequently, the subsequent analysis considers only the fundamental components of the tank voltage and current quantities. The fundamental component of the voltage at the output of the T-type bridge, denoted as v_{xy} and shown in Fig. 2(a), is derived from the two dc input voltages, v_{po} and v_{on}, along with their corresponding duty ratios, d_p and d_n:

$$v_{xy}(t) = v_{xy\text{-}po}(t) + v_{xy\text{-}on}(t)$$
$$= \frac{4}{\pi} \left[v_{po} \sin\left(\frac{\pi d_p}{2}\right) + v_{on} \sin\left(\frac{\pi d_n}{2}\right) \right] \cos\left(\omega_s t\right) , \tag{1}$$

where

$$v_{xy\text{-}po}(t) = \frac{4}{\pi} v_{po} \sin\left(\frac{\pi d_p}{2}\right) \cos\left(\omega_s t\right) , \tag{2}$$

$$v_{xy\text{-}on}(t) = \frac{4}{\pi} v_{on} \sin\left(\frac{\pi d_n}{2}\right) \cos\left(\omega_s t\right) , \tag{3}$$

979-8-3315-1612-3/25 $31.00 © 2025 IEEE   2195

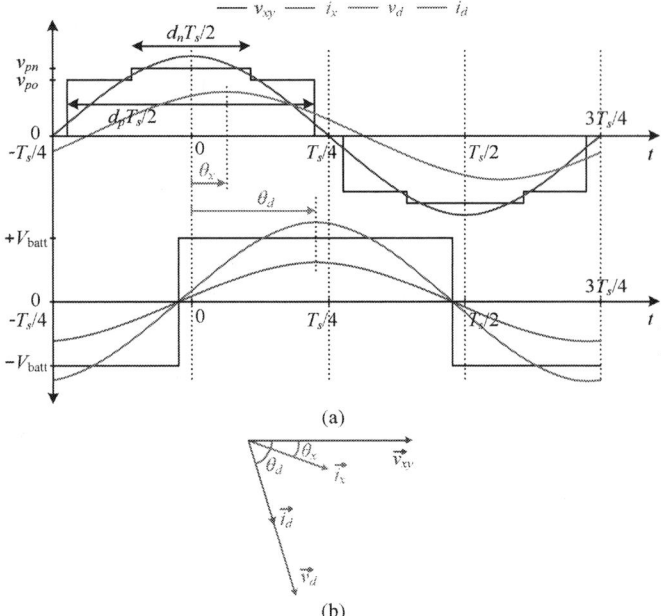

Fig. 2: (a) Fundamental components of the T-type bridge output voltage, v_{xy}, and current, i_x, as well as the diode bridge input voltage, v_d, and current, i_d. The operating region where $d_p > d_n$ is shown, and the two quasi-square voltages generated at the output of the T-type bridge using these duty ratios are center-aligned. (b) Phasor diagram of the voltage and current quantities in the T-type bridge-based dc-dc converter.

and ω_s represents the angular switching frequency of the T-type bridge, which remains constant (f_s = 85 kHz) for this application in accordance with the WPT standard J2954, established by the Society of Automotive Engineers (SAE) [15]. By considering the voltage ratio between the ac and dc sides, two effective turns ratios (or switching functions) for the T-type bridge can be defined as

$$s_{po}(t) = \frac{v_{xy\text{-}po}}{v_{po}} = \frac{4}{\pi}\sin\left(\frac{\pi d_p}{2}\right)\cos\left(\omega_s t\right), \quad (4)$$

$$s_{on}(t) = \frac{v_{xy\text{-}on}}{v_{on}} = \frac{4}{\pi}\sin\left(\frac{\pi d_n}{2}\right)\cos\left(\omega_s t\right), \quad (5)$$

or, equivalently, in phasor form at a constant angular frequency of ω_s, as

$$\overrightarrow{s_{po}} = \frac{4}{\pi}\sin\left(\frac{\pi d_p}{2}\right), \quad (6)$$

$$\overrightarrow{s_{on}} = \frac{4}{\pi}\sin\left(\frac{\pi d_n}{2}\right). \quad (7)$$

By combining the above equations, the voltage at the output of the T-type bridge can be expressed in phasor form using the switching functions as

$$\overrightarrow{v_{xy}} = \overrightarrow{v_{xy\text{-}po}} + \overrightarrow{v_{xy\text{-}on}} = \overrightarrow{s_{po}}\,v_{po} + \overrightarrow{s_{on}}\,v_{on}. \quad (8)$$

Similarly, the voltage at the input of the diode bridge can be expressed in phasor form as

$$\overrightarrow{v_d} = \overrightarrow{s_d}\,V_{\text{batt}}, \quad (9)$$

where

$$\overrightarrow{s_d} = \frac{4}{\pi}e^{-j\theta_d}. \quad (10)$$

Here, θ_d represents the phase lag of the diode bridge input voltage phasor, $\overrightarrow{v_d}$, relative to the T-type bridge output voltage phasor, $\overrightarrow{v_{xy}}$, as shown in Fig. 2(b). It is important to note that the phases of the fundamental components of the diode bridge input voltage, v_d, and the diode bridge input current, i_d, always align exactly in both steady-state and small-signal conditions. The overlap of these two quantities is illustrated in the phasor diagram provided in Fig. 2(b). A comprehensive discussion of this phenomenon is given in [16]. The diode bridge input current phasor is therefore expressed as

$$\overrightarrow{i_d} = \|\overrightarrow{i_d}\|e^{-j\theta_d}. \quad (11)$$

By employing the definitions of tank ac voltages derived in (8) and (9) in relation to the switching functions, the T-type bridge-based dc-dc conversion system can be represented using phasor transformers for both single-sided and double-sided LCC resonant tanks, as depicted in Fig. 3. Using the phasor transformers-based circuit diagram, the output current of the diode bridge can be calculated as

$$i_{\text{out}} = \langle i_{\text{out}}\rangle + i_{\text{out}}^{\text{HF}} = \Re\left[\overrightarrow{s_d}\,e^{j\omega_s t}\right]\Re\left[\overrightarrow{i_d}\,e^{j\omega_s t}\right]$$
$$= \Re\left[\frac{\overrightarrow{s_d}^{*}}{\sqrt{2}}\frac{\overrightarrow{i_d}}{\sqrt{2}}\right] + \Re\left[\frac{\overrightarrow{s_d}}{\sqrt{2}}\frac{\overrightarrow{i_d}}{\sqrt{2}}e^{2j\omega_s t}\right]. \quad (12)$$

The battery load current, which is the averaged component of the output current of the diode bridge, can now be written as

$$i_{\text{load}} = \langle i_{\text{out}}\rangle = \Re\left[\frac{\overrightarrow{s_d}^{*}}{\sqrt{2}}\frac{\overrightarrow{i_d}}{\sqrt{2}}\right] = \Re\left[\left(\frac{4}{\sqrt{2}\pi}e^{-j\theta_d}\right)^{*}\frac{\|\overrightarrow{i_d}\|}{\sqrt{2}}e^{-j\theta_d}\right]$$
$$= \frac{2}{\pi}\|\overrightarrow{i_d}\|. \quad (13)$$

Similarly, the input dc currents of the T-type bridge can be calculated as

$$i_p = \langle i_p^{\text{dc}}\rangle = \Re\left[\frac{\overrightarrow{s_{po}}^{*}}{\sqrt{2}}\frac{\overrightarrow{i_x}}{\sqrt{2}}\right] = \frac{2}{\pi}\|\overrightarrow{i_x}\|\cos(\theta_x)\sin\left(\frac{\pi d_p}{2}\right), \quad (14)$$

$$i_n = \langle i_n^{\text{dc}}\rangle = \Re\left[\frac{\overrightarrow{s_{on}}^{*}}{\sqrt{2}}\frac{\overrightarrow{i_x}}{\sqrt{2}}\right] = \frac{2}{\pi}\|\overrightarrow{i_x}\|\cos(\theta_x)\sin\left(\frac{\pi d_n}{2}\right), \quad (15)$$

where $\|\overrightarrow{i_x}\|$ and θ_x are the magnitude and phase angle, respectively, of the T-type bridge output current phasor, $\overrightarrow{i_x}$, as depicted in Fig. 2(b).

V. SMALL SIGNAL ANALYSIS

The output voltage of the T-type bridge, as defined in (1), can be expressed in phasor form as

$$\overrightarrow{v_{xy}} = \frac{4}{\pi}\left[v_{po}\sin\left(\frac{\pi d_p}{2}\right) + v_{on}\sin\left(\frac{\pi d_n}{2}\right)\right]\angle 0, \quad (16)$$

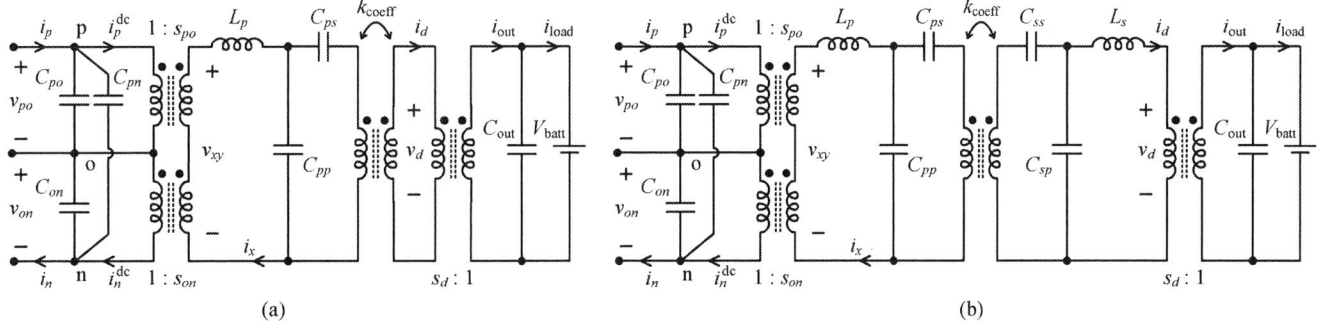

Fig. 3: T-type bridge-based dc-dc converter with a primary T-type bridge and secondary diode bridge modeled using phasor transformers, featuring (a) a single-sided LCC resonant tank with isolation, and (b) a double-sided LCC resonant tank with isolation.

which can be further solved by substituting the following expressions for v_{po}, v_{on}, d_p, and d_n during $0 < \theta_{grid} \leq \pi/3$:

$$v_{po} = V_{g\text{-mag}} \sin\left(\theta_{grid} + 2\pi/3\right), \quad (17)$$

$$v_{on} = V_{g\text{-mag}} \sin\left(\theta_{grid}\right), \quad (18)$$

$$d_p = (2/\pi) \sin^{-1}\left(m_i \sin\left(\theta_{grid} + \pi/2\right)\right), \quad (19)$$

$$d_n = (2/\pi) \sin^{-1}\left(m_i \sin\left(\theta_{grid} + \pi/6\right)\right), \quad (20)$$

where $V_{g\text{-mag}}$ represents the peak value of the line-to-line ac input voltage of the Unfolder, and m_i is the modulation index that controls the magnitude of the duty ratios. A detailed discussion of the above equations is provided in [9]. By combining (16)–(20), the T-type bridge output voltage phasor, $\vec{v_{xy}}$, can be expressed as

$$\vec{v_{xy}} = \frac{2\sqrt{3}}{\pi} m_i V_{g\text{-mag}} \angle 0 = \frac{2\sqrt{3}}{\pi} m_i V_{g\text{-mag}}. \quad (21)$$

As the modulation index m_i controls the magnitude of the T-type bridge output voltage, which defines the load current i_{load} charging the battery, a small-signal analysis is conducted on the plant transfer function derived between the small-signal load current $\hat{i}_{load}$ and the small-signal modulation index $\hat{m}_i$:

$$G_{plant} = \frac{\hat{i}_{load}}{\hat{m}_i}. \quad (22)$$

The small-signal analysis is first performed for the single-sided LCC tank and then extended to the double-sided LCC tank.

A. Single-Sided LCC Tank

1) Small-signal phasor-transformed circuit: To derive the small-signal phasor-transformed circuit, the tank inductors and capacitors are phasor-transformed as described in [12]. The inductor is represented as a series combination of the dynamic impedance (sL) and the steady-state impedance ($j\omega_s L$) while the capacitor is represented as a parallel combination of the dynamic impedance $\left(\frac{1}{sC}\right)$ and the steady-state impedance $\left(\frac{1}{j\omega_s C}\right)$. The output voltage of the T-type bridge is linearized around the steady-state value of the control input m_i, allowing the steady-state and small-signal components of $\vec{v_{xy}}$ to be expressed as

$$\vec{v_{xy}} = \vec{V_{xy}} + \hat{\vec{v_{xy}}} = \vec{V_{xy}} + \left(\frac{\partial \vec{v_{xy}}}{\partial m_i}\right)\hat{m}_i$$

$$= \frac{2\sqrt{3}}{\pi} M_i V_{g\text{-mag}} + \frac{2\sqrt{3}}{\pi} \hat{m}_i V_{g\text{-mag}}. \quad (23)$$

Therefore,

$$\frac{\hat{\vec{v_{xy}}}}{\hat{m}_i} = \frac{2\sqrt{3}}{\pi} V_{g\text{-mag}}. \quad (24)$$

Now, the diode bridge input voltage, written as

$$\vec{v_d} = \frac{4}{\pi} V_{batt} e^{-j\theta_d}, \quad (25)$$

can be expressed in terms of the small-signal component by taking the partial derivative with respect to the control input, m_i. The phase lag, θ_d, undergoes small-signal deviations due to perturbations in m_i. The magnitude, which is a function of the battery voltage (V_{batt}), is treated as a constant, as the battery voltage shows a negligible deviation in the small-signal domain due to the significantly higher time constant of the State of Charge [17]. A detailed discussion of this phenomenon is provided in [16]. Therefore, the steady-state and small-signal components of the diode bridge input voltage can be expressed as

$$\vec{v_d} = \vec{V_d} + \hat{\vec{v_d}} = \vec{V_d} + \left(\frac{\partial \vec{v_d}}{\partial m_i}\right)\hat{m}_i$$

$$= \frac{4}{\pi} V_{batt} e^{-j\Theta_d} - j\frac{4}{\pi} V_{batt} e^{-j\Theta_d} \hat{\theta}_d. \quad (26)$$

Therefore,

$$\frac{\hat{\vec{v_d}}}{\hat{m}_i} = -j\frac{4}{\pi} V_{batt} e^{-j\Theta_d} \frac{\hat{\theta}_d}{\hat{m}_i}. \quad (27)$$

The overall linearized steady-state and small-signal model of the T-type bridge-based dc-dc converter with a single-sided LCC tank is illustrated in Fig. 4. The isolation is represented using the cantilever model, incorporating leakage and magnetizing inductances. To enhance the accuracy of the model, the parasitic resistances of inductors and capacitors are introduced as series and parallel resistances, respectively, as done while presenting the simulation and experimental results.

2) Small-signal diode bridge input current: As derived in (13), the load current i_{load} is a function of the magnitude of the diode bridge input current i_d. Therefore, to obtain the plant transfer function, the small-signal deviations in the diode bridge

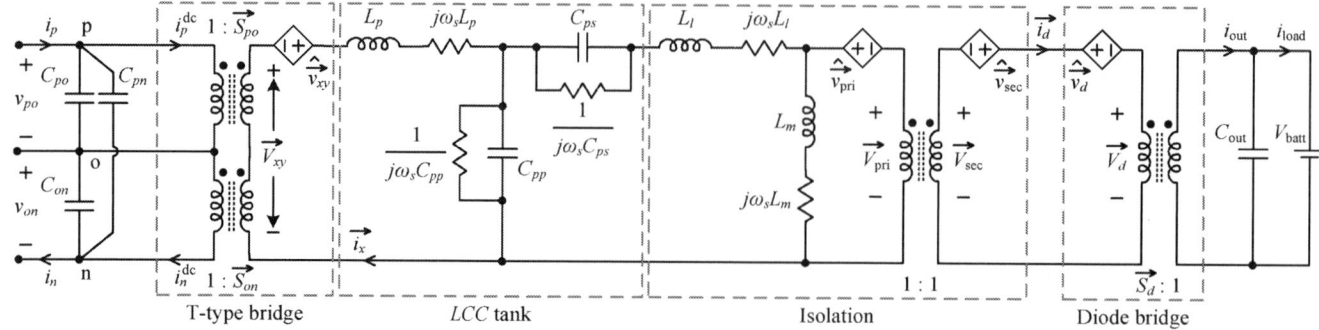

Fig. 4: Combined steady-state and small-signal phasor-transformed model of the T-type bridge-based dc-dc system, with a single-sided LCC tank, a diode bridge connected to a battery, and isolation represented by the cantilever model. Parasitic resistances are not shown but are included in the derived expressions.

input current phasor, $\vec{\hat{i}}_d$, are calculated in relation to perturbations introduced in the control input, $\hat{m}_i$. The small-signal diode bridge input current phasor is calculated using Fig. 4 as

$$\vec{\hat{i}}_d = \frac{\left(\vec{\hat{v}}_{xy} - \vec{\hat{v}}_d(f_1 f_3 - f_4)\right)}{f_2 f_3}$$

$$= \frac{\left(\frac{2\sqrt{3}}{\pi}\hat{m}_i V_{g\text{-mag}} + j\frac{4}{\pi}V_{\text{batt}} \, e^{-j\Theta_d}\hat{\theta}_d(f_1 f_3 - f_4)\right)}{f_2 f_3},$$

(28)

where f_1, f_2, f_3, and f_4 are functions of the tank parameters:

$$f_1 = \frac{Z_{ps} + Z_l + Z_m}{Z_m},$$

(29)

$$f_2 = Z_{ps} + Z_l,$$

(30)

$$f_3 = 1 + \frac{Z_p}{Z_{pp}} + \frac{Z_p}{Z_{ps} + Z_l},$$

(31)

$$f_4 = \frac{Z_p}{Z_{ps} + Z_l},$$

(32)

$$Z_p = sL_p + j\omega_s L_p + r_{L_p},$$

(33)

$$Z_{pp} = \frac{r_{C_{pp}}}{1 + r_{C_{pp}}(sC_{pp} + j\omega_s C_{pp})},$$

(34)

$$Z_{ps} = \frac{r_{C_{ps}}}{1 + r_{C_{ps}}(sC_{ps} + j\omega_s C_{ps})},$$

(35)

$$Z_l = sL_l + j\omega_s L_l + r_{L_l},$$

(36)

$$Z_m = sL_m + j\omega_s L_m,$$

(37)

where r_{L_p}, $r_{C_{pp}}$, $r_{C_{ps}}$, and r_{L_l} represent the parasitic resistances of the tank inductors and capacitors. Equation (28) can be modified into a transfer function using (24) and (27) as

$$\frac{\vec{\hat{i}}_d}{\hat{m}_i} = \frac{\left(\frac{\vec{\hat{v}}_{xy}}{\hat{m}_i} - \frac{\vec{\hat{v}}_d}{\hat{m}_i}(f_1 f_3 - f_4)\right)}{f_2 f_3}$$

$$= \frac{\left(\frac{2\sqrt{3}}{\pi}V_{g\text{-mag}} + j\frac{4}{\pi}V_{\text{batt}} \, e^{-j\Theta_d}\frac{\hat{\theta}_d}{\hat{m}_i}(f_1 f_3 - f_4)\right)}{f_2 f_3}.$$

(38)

An alternative, more useful form of the result (38) is derived by splitting this equation into its linearized magnitude and phase

components using a first-order Taylor series approximation. This method allows the complex phasor transfer function to be approximated by two transfer functions with real coefficients, representing the magnitude and phase deviations of the diode bridge input current. These transfer functions are calculated as follows:

$$\frac{\hat{i}_{d\text{-mag}}}{\hat{m}_i} = \frac{\Re\left[\frac{\vec{\hat{i}}_d}{\hat{m}_i}\right]\Re[\vec{I}_d] + \Im\left[\frac{\vec{\hat{i}}_d}{\hat{m}_i}\right]\Im[\vec{I}_d]}{\|\vec{I}_d\|},$$

(39)

$$\frac{\hat{\theta}_d}{\hat{m}_i} = \frac{\Re\left[\frac{\vec{\hat{i}}_d}{\hat{m}_i}\right]\Im[\vec{I}_d] - \Im\left[\frac{\vec{\hat{i}}_d}{\hat{m}_i}\right]\Re[\vec{I}_d]}{\|\vec{I}_d\|^2}.$$

(40)

It is important to note that (38) and (40) are interdependent and must be solved as simultaneous equations.

3) Small-signal load current: The output current of the diode bridge is a dc quantity and, therefore, cannot be represented in phasor form. A time-domain analysis is conducted to determine the complete output current, which includes both the dc and HF components. The output current in the time domain can be calculated using the circuit model shown in Fig. 4 as

$$i_{\text{out}}(t) = s_d(t)i_d(t)$$

$$= \frac{4}{\pi}\cos(\omega_s t - \theta_d)\, i_{d\text{-mag}}\cos(\omega_s t - \theta_d)$$

$$= \frac{2}{\pi}(I_{d\text{-mag}} + \hat{i}_{d\text{-mag}})(1 + \cos(2\omega_s t - 2\Theta_d - 2\hat{\theta}_d)).$$

(41)

The battery load current, which is the averaged component of the output current of the diode bridge, can now be written as

$$i_{\text{load}} = \langle i_{\text{out}}\rangle = I_{\text{load}} + \hat{i}_{\text{load}} = \frac{2}{\pi}I_{d\text{-mag}} + \frac{2}{\pi}\hat{i}_{d\text{-mag}}.$$

(42)

Therefore, the small-signal deviations in i_{load} that directly depend on $\hat{i}_{d\text{-mag}}$ and $\hat{\theta}_d$ are

$$\frac{\hat{i}_{\text{load}}}{\hat{i}_{d\text{-mag}}} = \frac{2}{\pi},$$

(43)

$$\frac{\hat{i}_{\text{load}}}{\hat{\theta}_d} = 0,$$

(44)

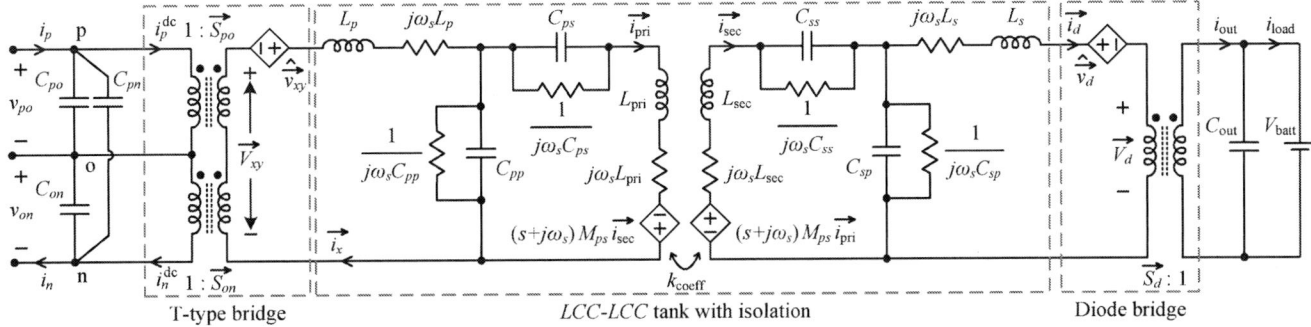

Fig. 5: Combined steady-state and small-signal phasor-transformed model of the T-type bridge-based dc-dc converter, featuring a double-sided LCC tank and a diode bridge connected to a battery load. Isolation is represented by a two-coupled inductor model, with k_{coeff} as the coupling coefficient.

respectively.

Now, by combining (39), (40), (43), and (44), the plant transfer function can be calculated as

$$G_{\text{plant}} = \frac{\hat{i}_{\text{load}}}{\hat{m}_i} = \frac{\hat{i}_{\text{load}}}{\hat{i}_{d\text{-mag}}} \frac{\hat{i}_{d\text{-mag}}}{\hat{m}_i} + \frac{\hat{i}_{\text{load}}}{\hat{\theta}_d} \frac{\hat{\theta}_d}{\hat{m}_i}. \quad (45)$$

B. Double-Sided LCC Tank

The three steps outlined for the small-signal modeling of the single-sided LCC tank are similarly applied to the double-sided LCC tank.

1) Small-signal phasor-transformed circuit: Similar to the single-sided LCC tank, the linearized steady-state and small-signal model of the T-type bridge-based dc-dc converter with a double-sided LCC tank is derived by representing the tank inductors and capacitors in terms of dynamic and steady-state impedances. The T-type bridge output voltage is linearized as given in (23), and the diode bridge input voltage is linearized as derived in (26). The complete steady-state and small-signal model is depicted in Fig. 5, with isolation modeled using the two-coupled inductor approach commonly employed in WPT systems.

2) Small-signal diode bridge input current: As explained earlier, since the load current i_{load} is a function of the diode bridge input current i_d, the small-signal deviations in the diode bridge input current phasor, $\vec{\hat{i}}_d$, are calculated in relation to perturbations introduced in the control input, $\hat{m}_i$, as provided in (46) at the bottom of the page. In (46),

$$f_{p1} = Z_{ps} + Z_{\text{pri}}, \quad (47)$$

$$f_{p2} = 1 + \frac{Z_p}{Z_{pp}}, \quad (48)$$

$$f_{s1} = \frac{Z_{sp} + Z_{\text{sec}} + Z_{ss}}{Z_{sp}}, \quad (49)$$

$$f_{s2} = Z_{\text{sec}} + Z_{ss} + Z_s \left(\frac{Z_{sp} + Z_{\text{sec}} + Z_{ss}}{Z_{sp}} \right), \quad (50)$$

$$f_{s3} = 1 + \frac{Z_s}{Z_{sp}}, \quad (51)$$

$$f_{ps} = sM_{ps} + j\omega_s M_{ps}, \quad (52)$$

$$Z_p = sL_p + j\omega_s L_p + r_{L_p}, \quad (53)$$

$$Z_{pp} = \frac{r_{C_{pp}}}{1 + r_{C_{pp}}(sC_{pp} + j\omega_s C_{pp})}, \quad (54)$$

$$Z_{ps} = \frac{r_{C_{ps}}}{1 + r_{C_{ps}}(sC_{ps} + j\omega_s C_{ps})}, \quad (55)$$

$$Z_{\text{pri}} = sL_{\text{pri}} + j\omega_s L_{\text{pri}} + r_{L_{\text{pri}}}, \quad (56)$$

$$Z_{\text{sec}} = sL_{\text{sec}} + j\omega_s L_{\text{sec}} + r_{L_{\text{sec}}}, \quad (57)$$

$$Z_{ss} = \frac{r_{C_{ss}}}{1 + r_{C_{ss}}(sC_{ss} + j\omega_s C_{ss})}, \quad (58)$$

$$Z_{sp} = \frac{r_{C_{sp}}}{1 + r_{C_{sp}}(sC_{sp} + j\omega_s C_{sp})}, \quad (59)$$

$$Z_s = sL_s + j\omega_s L_s + r_{L_s}, \quad (60)$$

$$M_{ps} = k_{\text{coeff}} \sqrt{L_{\text{pri}} L_{\text{sec}}}, \quad (61)$$

where r_{L_p}, $r_{C_{pp}}$, $r_{C_{ps}}$, $r_{L_{\text{pri}}}$, $r_{L_{\text{sec}}}$, $r_{C_{ss}}$, $r_{C_{sp}}$, and r_{L_s} are the parasitic resistances of the tank inductors and capacitors. As described earlier for the single-sided LCC tank, (46) can be split into its linearized magnitude and phase components using a first-order Taylor series approximation, as provided in (39) and (40). This method allows the complex phasor transfer function to be approximated by two transfer functions with real coefficients, representing the magnitude $\left(\frac{\hat{i}_{d\text{-mag}}}{\hat{m}_i} \right)$ and phase $\left(\frac{\hat{\theta}_d}{\hat{m}_i} \right)$ deviations of the diode bridge input current relative to the perturbations in the control input, $\hat{m}_i$.

3) Small-signal load current: Similar to the single-sided LCC tank, by combining the magnitude and phase deviations of the small-signal diode bridge input current phasor, $\vec{\hat{i}}_d$, relative to the perturbations in the control input, $\hat{m}_i$ (as provided by (39) and (40)), with the small-signal deviations in the load current, $\hat{i}_{\text{load}}$, which directly depend on the magnitude

$$\frac{\vec{\hat{i}}_d}{\hat{m}_i} = \frac{\frac{\vec{\hat{v}}_{xy}}{\hat{m}_i} - \frac{\vec{\hat{v}}_d}{\hat{m}_i}\left(\left(\frac{f_{p1}f_{s1}}{f_{ps}} - \frac{f_{ps}}{Z_{sp}} \right) f_{p2} + \frac{Z_p f_{s1}}{f_{ps}} \right)}{\left(\frac{f_{p1}f_{s2}}{f_{ps}} - f_{ps}f_{s3} \right) f_{p2} + \frac{Z_p f_{s2}}{f_{ps}}} = \frac{\frac{2\sqrt{3}}{\pi} V_{g\text{-mag}} + j\frac{4}{\pi} V_{\text{batt}} e^{-j\Theta_d} \frac{\hat{\theta}_d}{\hat{m}_i}\left(\left(\frac{f_{p1}f_{s1}}{f_{ps}} - \frac{f_{ps}}{Z_{sp}} \right) f_{p2} + \frac{Z_p f_{s1}}{f_{ps}} \right)}{\left(\frac{f_{p1}f_{s2}}{f_{ps}} - f_{ps}f_{s3} \right) f_{p2} + \frac{Z_p f_{s2}}{f_{ps}}}. \quad (46)$$

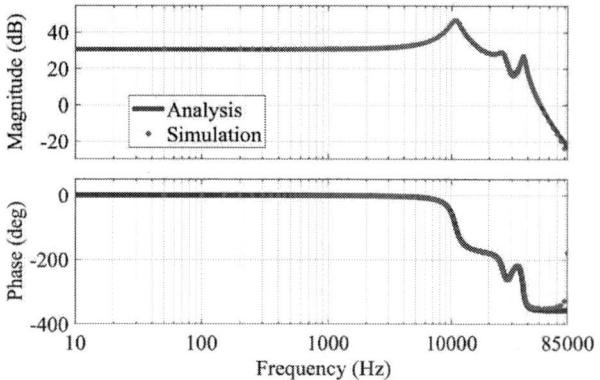

Fig. 6: Comparison of Bode plots of the plant $\left(G_{\text{plant}} = \frac{\hat{i}_{\text{load}}}{\hat{m}_i}\right)$ of the T-type bridge-based dc-dc converter with a double-sided LCC resonant tank, obtained from the phasor transformation-based small-signal modeling and PLECS simulation multitone analysis.

TABLE I: Simulation parameters for the T-type bridge-based dc-dc conversion system with a double-sided LCC tank.

Parameter	Value	Parameter	Value
P_{out}	20 kW	$C_{ps}, r_{C_{ps}}$	9.9 nF, 10 kΩ
f_s	85 kHz	$C_{ss}, r_{C_{ss}}$	9.9 nF, 10 kΩ
$V_{g\text{-mag}}$	$480\sqrt{2}$ V	$C_{sp}, r_{C_{sp}}$	46.7 nF, 10 kΩ
v_{po}, v_{on}	379.4 V, 297.4 V	L_s, r_{L_s}	75.1 μH, 70 mΩ
V_{batt}	650 V	k_{coeff}	0.44
L_p, r_{L_p}	64.5 μH, 70 mΩ	$L_{\text{pri}}, r_{L_{\text{pri}}}$	419.2 μH, 150 mΩ
$C_{pp}, r_{C_{pp}}$	54.4 nF, 10 kΩ	$L_{\text{sec}}, r_{L_{\text{sec}}}$	427.9 μH, 150 mΩ

deviation, $\hat{i}_{d\text{-mag}}$, and phase deviation, $\hat{\theta}_d$, in the diode bridge input current (as given by (43) and (44)), the plant transfer function can be calculated as

$$G_{\text{plant}} = \frac{\hat{i}_{\text{load}}}{\hat{m}_i} = \frac{\hat{i}_{\text{load}}}{\hat{i}_{d\text{-mag}}} \frac{\hat{i}_{d\text{-mag}}}{\hat{m}_i} + \frac{\hat{i}_{\text{load}}}{\hat{\theta}_d} \frac{\hat{\theta}_d}{\hat{m}_i}. \quad (62)$$

VI. SIMULATION AND EXPERIMENTAL RESULTS

A 20 kW T-type bridge-based dc-dc converter featuring a double-sided LCC resonant tank with isolation and a diode bridge connected to a battery load is simulated using

PLECS multitone analysis. This analysis applies a multitone signal, composed of multiple sinusoidal signals, to encompass all investigated frequencies simultaneously. The simulation parameters are listed in TABLE I. Parasitic resistances of inductors and capacitors are expressed in terms of series and parallel resistances, respectively. The Bode plots of the plant $G_{\text{plant}} = \frac{\hat{i}_{\text{load}}}{\hat{m}_i}$ obtained from small-signal modeling closely align with the simulation results in Fig. 6, thereby validating the accuracy of the modeling approach for the dc-dc converter with a double-sided LCC tank.

To validate the phasor transformation-based small-signal modeling in hardware, tests are conducted on the T-type bridge-based dc-dc conversion system illustrated in Fig. 7. This system comprises a single-sided LCC tank with isolation and a secondary-side diode-bridge rectifier connected to a battery load. Two dc-voltage sources, manufactured by REGATRON, are connected to the inputs of the T-type bridge, while a dc-voltage sink (NHR 9300) emulates a battery at the output. Experimental parameters are provided in TABLE II. Parasitic resistances of inductors and capacitors are expressed in terms of series and parallel resistances, respectively. The modeling verification is carried out for the plant transfer function, $G_{\text{plant}} = \frac{\hat{i}_{\text{load}}}{\hat{m}_i}$, at an output power of 4 kW. In this verification, the modulation index, m_i, is sinusoidally perturbed by $\pm$ 0.05 around a steady-state value of 0.95. The perturbation is performed across a range of frequencies: 50 Hz, 100 Hz, 500 Hz, 1 kHz, 3.7 kHz, 5 kHz, 10 kHz, 20 kHz, 30 kHz, 32.5 kHz, 36.6 kHz, 40 kHz, 42.5 kHz, and 55 kHz. The control input, m_i, is updated at twice the switching frequency (170 kHz) using an up-down carrier, implemented with the TMS320F28379D microcontroller. The resulting magnitude and phase of the perturbation-frequency component present in the load current, i_{load}, are measured to derive the Bode plots, as depicted in Fig. 8. The figure illustrates a comparison between the experimental Bode plots of G_{plant} and those derived from the small-signal modeling. The close similarity between the two sets of plots confirms the accuracy and reliability of the modeling process.

Moreover, a step response of the diode bridge output current, i_{out}, has been measured during the hardware testing by

TABLE II: Experimental parameters.

Parameter	Value
P_{out}, f_s	4 kW, 85 kHz
$V_{g\text{-mag}}, V_{\text{batt}}$	$208\sqrt{2}$ V, 316 V
v_{po}, v_{on}	208 V, 76 V
L_p, r_{L_p}	29.3 μH, 116.4 mΩ
$C_{pp}, r_{C_{pp}}$	119.9 nF, 14.9 kΩ
$C_{ps}, r_{C_{ps}}$	112.4 nF, 15.6 kΩ
L_l, r_{L_l}	37.1 μH, 31.5 mΩ
L_m	802.2 μH

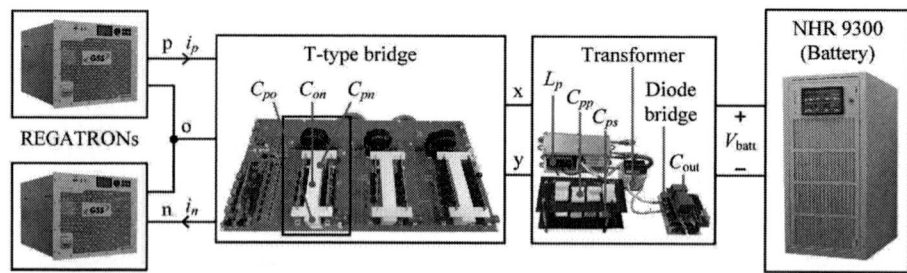

Fig. 7: A 4 kW hardware setup of a dc-dc conversion system featuring a T-type bridge, a single-sided LCC tank, and a diode bridge on the secondary side of the isolation. Two dc power supplies from REGATRON serve as dc input sources, while the NHR 9300 functions as a battery load. Small-signal perturbations are applied to the modulation index m_i using the TMS320F28379D microcontroller.

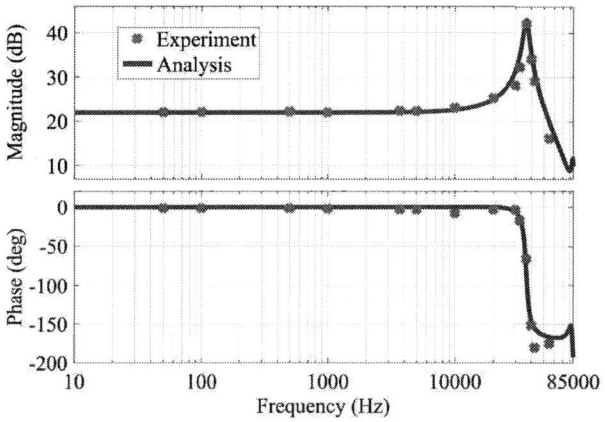

Fig. 8: Comparison of Bode plots of the plant $\left(G_{\text{plant}} = \dfrac{\hat{i}_{\text{load}}}{\hat{m}_i}\right)$ of the T-type bridge-based dc-dc converter with a single-sided LCC tank, obtained from the phasor transformation-based small-signal modeling and hardware testing.

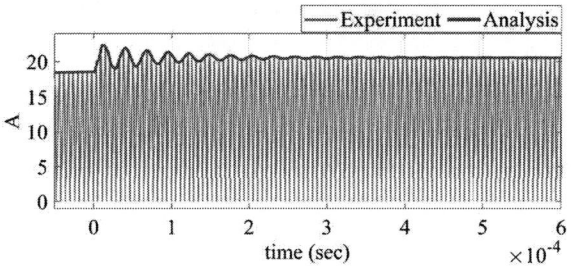

Fig. 9: Experimental waveform of the diode-bridge rectifier output current, i_{out}, for the T-type bridge-based dc-dc converter with a single-sided LCC resonant tank, superimposed on the envelope predicted by the small-signal model during a step change in the modulation index, m_i, from 0.9 to 1.

introducing a step change in the modulation index from 0.9 to 1. The envelope observed during this step change in the hardware has been compared with the step response obtained from the small-signal modeling, as presented in Fig. 9. The results demonstrate that the modeling precisely predicts the envelope of the diode bridge output current during this step change, thereby confirming the accuracy of the modeling.

VII. CONCLUSION

This paper presents a comprehensive small-signal modeling that utilizes phasor transformation for a T-type bridge-based dc-dc converter. This system features both single-sided and double-sided LCC resonant tanks, with a secondary side isolated diode bridge connected to a battery load, representing a typical configuration used in WPT applications. The modeling procedure is divided into distinct steps, each of which is clarified to facilitate a clear and intuitive understanding of the methodology employed. The precise model presented here enhances the design and stability of closed-loop control, resulting in more reliable and efficient power delivery. The proposed analytical model is validated through simulations of a 20 kW T-type bridge-based dc-dc converter with a double-sided LCC resonant tank. Furthermore, the small-signal modeling is also verified through hardware testing of a 4 kW, 85 kHz prototype

consisting of a single-sided LCC tank, confirming its accuracy for perturbation frequencies up to 55 kHz.

REFERENCES

[1] R. L. Steigerwald, "A comparison of half-bridge resonant converter topologies," in IEEE Transactions on Power Electronics, vol. 3, no. 2, pp. 174-182, April 1988, doi: 10.1109/63.4347.

[2] G. A. Mudiyanselage, N. Keshmiri and A. Emadi, "A Review of DC-DC Resonant Converter Topologies and Control Techniques for Electric Vehicle Applications," in IEEE Open Journal of Power Electronics, vol. 4, pp. 945-964, 2023, doi: 10.1109/OJPEL.2023.3331180.

[3] Wave IPT—Wirelessly Charging Electric Vehicles, accessed on Sep. 16, 2015. [Online]. Available: www.waveipt.com.

[4] G. A. Covic and J. T. Boys, "Inductive Power Transfer," in Proceedings of the IEEE, vol. 101, no. 6, pp. 1276-1289, June 2013, doi: 10.1109/JPROC.2013.2244536.

[5] S. -Y. R. Hui, Y. Yang and C. Zhang, "Wireless Power Transfer: A Paradigm Shift for the Next Generation," in IEEE Journal of Emerging and Selected Topics in Power Electronics, vol. 11, no. 3, pp. 2412-2427, June 2023, doi: 10.1109/JESTPE.2023.3237792.

[6] S. Li and C. C. Mi, "Wireless Power Transfer for Electric Vehicle Applications," in IEEE Journal of Emerging and Selected Topics in Power Electronics, vol. 3, no. 1, pp. 4-17, March 2015, doi: 10.1109/JESTPE.2014.2319453.

[7] A. Mahesh, B. Chokkalingam and L. Mihet-Popa, "Inductive Wireless Power Transfer Charging for Electric Vehicles–A Review," in IEEE Access, vol. 9, pp. 137667-137713, 2021, doi: 10.1109/ACCESS.2021.3116678.

[8] R. Tavakoli and Z. Pantic, "Analysis, Design, and Demonstration of a 25-kW Dynamic Wireless Charging System for Roadway Electric Vehicles," in IEEE Journal of Emerging and Selected Topics in Power Electronics, vol. 6, no. 3, pp. 1378-1393, Sept. 2018, doi: 10.1109/JESTPE.2017.2761763.

[9] A. Zade et al., "A 21-kW Unfolding-Based Single-Stage AC–DC Converter for Wireless Charging Applications," in IEEE Journal of Emerging and Selected Topics in Power Electronics, vol. 12, no. 1, pp. 8-27, Feb. 2024, doi: 10.1109/JESTPE.2023.3309588.

[10] S. GURUDIWAN, A. Zade, H. Wang and R. Zane, "Accurate ZVS Analysis of a Full-Bridge T-Type Resonant Converter for a 20-kW Unfolding-Based AC-DC Topology," in IEEE Open Journal of Power Electronics, vol. 5, pp. 692-708, 2024, doi: 10.1109/OJPEL.2024.3400256.

[11] B. Zhang, S. Xie, Z. Li, P. Zhao and J. Xu, "An Optimized Single-Stage Isolated Swiss-Type AC/DC Converter Based on Single Full-Bridge With Midpoint-Clamper," in IEEE Transactions on Power Electronics, vol. 36, no. 10, pp. 11288-11297, Oct. 2021, doi: 10.1109/TPEL.2021.3073742.

[12] C. T. Rim and G. H. Cho, "Phasor transformation and its application to the DC/AC analyses of frequency phase-controlled series resonant converters (SRC)," in IEEE Transactions on Power Electronics, vol. 5, no. 2, pp. 201-211, April 1990, doi: 10.1109/63.53157.

[13] D. Seltzer, L. Corradini, D. Bloomquist, R. Zane and D. Maksimović, "Small signal phasor modeling of dual active bridge series resonant DC/DC converters with multi-angle phase shift modulation," 2011 IEEE Energy Conversion Congress and Exposition, Phoenix, AZ, USA, 2011, pp. 2757-2764, doi: 10.1109/ECCE.2011.6064139.

[14] A. P. Hu, "Modeling a contactless power supply using GSSA method," 2009 IEEE International Conference on Industrial Technology, Churchill, VIC, Australia, 2009, pp. 1-6, doi: 10.1109/ICIT.2009.4939571.

[15] Wireless Power Transfer for Light-Duty Plug-In Electric Vehicles and Alignment Methodology, International Standard SAE J2954, 2019. [Online]. Available: https://www.sae.org/standards/content/j2954-201904/

[16] A. Zade and R. Zane, "Accurate Small-Signal Phasor Transformation-Based Modeling of Secondary-Side Diode-Bridge Rectifiers for Battery Charging Applications," 2025 IEEE Applied Power Electronics Conference and Exposition (APEC), Atlanta, GA, USA, 2025.

[17] M. Cacciato, G. Nobile, G. Scarcella and G. Scelba, "Real-Time Model-Based Estimation of SOC and SOH for Energy Storage Systems," in IEEE Transactions on Power Electronics, vol. 32, no. 1, pp. 794-803, Jan. 2017, doi: 10.1109/TPEL.2016.2535321.

A Hybrid Three-Port Topology for Urban Charging Stations

Mohammadreza Khodaparast Klidbari[1], Naser Souri[2], Zahra Sadat Habibolahi[3], Hamid Montazeri Hedeshi[4]

[1]Faculty of Electrical Engineering, K. N. Toosi University of Technology, Tehran, Iran.
[2]The Bradley Department of Electrical and Computer Engineering, Virginia Tech, United States
[3]Department of Electrical Engineering, Iran University of Science and Technology, Tehran, Iran
[4]Faculty of Electrical and Computer Engineering, University of Tabriz, Tabriz, Iran

Emails: [1]khpmohammadreza@email.kntu.ac.ir, [2]nsouri@vt.edu, [3]zahra_habibolahi78@elec.iust.ac.ir, [4]h.montazeri.h@gmail.com

Abstract—**Electric vehicles are rapidly gaining popularity as a sustainable alternative to conventional gasoline. In urban areas, chargers with different ratings can accommodate the diverse needs of electric vehicles. However, the available multiport topologies have variable switching frequencies. This paper introduces a hybrid multiport isolated DC-DC converter for urban charging stations, incorporating fast and slow charging ports with a fixed switching frequency. It provides isolation and enables soft switching on the primary side of the converter without circulating current on its secondary side. The primary side does not need feedback, which reduces complexity. The second stage generates a wide output voltage range to charge the electric vehicle battery by employing a switch. In addition, the proposed topology offers reduced component count and simple control with fixed-frequency operation. This paper provides the concept and the operation modes. Experimental results are provided to validate its features. The prototype converter achieves 96% peak efficiency.**

Index Terms—**DC-DC converters, EV charging station, topology, multi-port converter, ZVS.**

I. INTRODUCTION

Increasing greenhouse gas emissions from the automotive industry is a concerning issue. Electric and hybrid electric vehicles have received significant attention in recent years with advancements in battery technology [1], [2]. Their potential to lower greenhouse gas emissions and lessen reliance on fossil fuels promotes sustainable transportation. Therefore, charging stations are increasing in number in urban areas. Charging stations can benefit from an AC or DC grid. Fig. 1 shows a DC-based charging station structure supplying from both AC and DC sources. DC grids have fewer power conversion stages than AC grids, reducing power conversion losses and potentially increasing efficiency as electric vehicle batteries are also powered with DC [3]–[5].

Slow charging stations, commonly known as level 2, are ideal for overnight charging or suitable for parking periods, while fast charging is ideal for long journeys and rapid energy replenishment in suburban areas, ensuring that drivers can conveniently and efficiently charge their vehicles. Therefore, hybrid charging stations enhance flexibility, accommodating various charging speeds for a diverse range of EVs [6], [7]. Hybrid charging stations can benefit from multiport topologies, as they can integrate fast and slow chargers in one package,

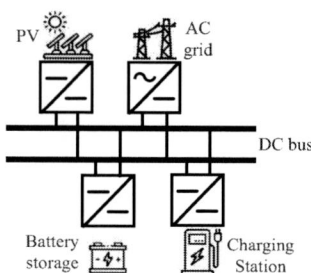

Fig. 1. Charging station topology.

as shown in Fig. 2. In addition, multiport topologies usually offer a lower number of components, which reduces the cost.

Among multiport topologies, LLC provides a simple structure with zero-voltage switching (ZVS) capability on its primary side and a wide range of output voltage [8]. ZVS ensures that the voltage across the switch is zero when it turns on or off, reducing the switching losses. The main challenge for LLC topology is that the switching frequency is not fixed and varies according to the output voltage. On the other hand, to achieve a high gain, the switching frequency should be less than the resonant frequency. However, reducing the switching frequency affects the transformer and inductor size. In contrast, considering a higher switching frequency increases the switching losses. In addition, adjusting the voltage gain using the switching frequency complicates the transformer design and can lead to a reduction in efficiency [9], [10].

For EV applications, it is recommended that the input and output are galvanically isolated so that battery protection is not affected by the charging system. However, [11], [12] propose topologies that isolation is not respected. References [7], [13], [14] present an isolated three-port topology with two LLC resonant tanks that can operate at a high switching frequency with lower switching losses due to soft-switching operation. However, the design of these topologies is complex as they use a three-winding transformer. In addition, this type of transformer introduces a circulating current, which increases the losses and stress over the components. Therefore, an additional control algorithm is required to minimize this current. Reference [15] proposes a topology that the frequency is fixed during some modes. However, it changes in other

979-8-3315-1612-3/25 $31.00 © 2025 IEEE

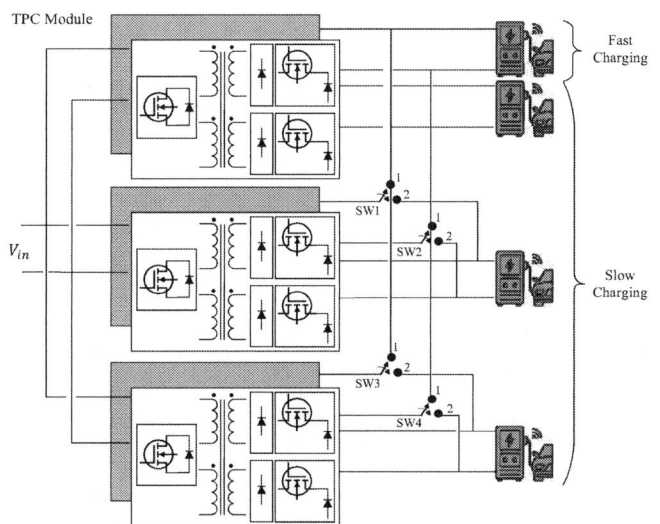

Fig. 2. A hybrid charging station block diagram.

modes. Therefore, [16] proposes a new topology that uses a fixed frequency. However, this topology has a single port structure. Moreover, it has a large number of components, which increases the cost. To overcome the existing drawbacks, a new multiport SiC-based isolated topology is proposed in this paper to address the aforementioned issues. Table I shows a comparison of different topologies. This comparison is based on the control method, the number of ports, the circulating current, the number of components, and the bidirectional or unidirectional structure.

The contribution of this paper is as follows.

- Proposing a new single-stage multiport topology for hybrid charging stations with fixed frequency.

Section II provides the proposed topology, its features, and its operation modes. Section III provides simulation studies to evaluate the switching waveforms. Section IV provides experimental studies to validate the effectiveness of the proposed topology. In the last section, a conclusion is provided.

II. PROPOSED MULTIPORT TOPOLOGY

This section studies the concept of the proposed charger topology and the operating modes. Fig. 3 shows the proposed LLC topology for a charging station. This topology consists of two stages and two outputs, one of which is fast charging, and the other is slow charging. The primary side of this converter is a full bridge converter, and the switching frequency is close to the resonance frequency. The transformer isolates the primary and secondary sides. The power is shared between the secondary sides through this transformer. ZVS occurs for the primary switches, reducing the switching losses. However, the charging algorithm is implemented on the secondary side. The proposed topology utilizes a two-winding transformer instead of a three-winding transformer, resulting in eliminating the circulating current and reducing complexity. To achieve ZVS during dead time, the magnetizing current of the transformer I_m must be capable of charging and discharging the capacitors

C_{oss}. The current that passes through each of the output capacitors C_{oss} is calculated as follows.

$$I_{\text{coss}} = C_{\text{oss}} \frac{\Delta V}{\Delta t} \qquad (1)$$

where $I_{\text{coss}} = (I_{\text{lm1}} + I_{\text{lm2}})/2$ and $I_{\text{lm1,2}} = V_{\text{out1,2}}/4L_m f_s$, and f_s is the switching frequency. Therefore, the duration for charging the capacitor C_{oss} should be less than the dead time to achieve zero-voltage switching. Fig.4 shows the circuit in different operation modes for the proposed topology. These modes are discussed below.

A. Operation Mode I: $t_1 \leq t < t_2$

Fig. 4(a) shows the equivalent circuit in this operation mode during t_1 to t_2. In this mode, S_2 and S_3 are ON at the primary side. At the beginning of this interval, the voltage across the MOSFET is zero as the body diode conducts before t_1, which indicates a ZVS. In this mode, I_{lr1} and I_{lr2} are positive and greater than I_{lm1} and I_{lm2}. As a result, diodes D_1, D_4, D_5, D_8 are ON and V_{lm1} and V_{lm2} are equal to V_{o1} and V_{o2}, respectively.

B. Operation Mode II: $t_2 \leq t < t_3$

In this mode, S_2 and S_3 are ON and S_1 and S_4 are OFF as shown in Fig. 4(b). In this mode, I_{lr2} reaches I_{lm2} at t_2, and the secondary side of T_2 is zero. In this case, the magnetizing inductor at the primary side of this transformer resonates with the resonant tank, which is calculated from $f_{\text{sr2}} = \frac{1}{2\pi\sqrt{(l_r + l_m)c_r}}$.

C. Operation Mode III: $t_3 \leq t < t_4$

In this mode, all switches are OFF, as shown in Fig. 4(c). I_{lr1} is greater than I_{lm1} in this mode, helping I_{lm1} to charge and discharge the capacitors C_{oss} to achieve ZVS. The resonant tank current passes through the output capacitor of the switches, and based on the current direction, it charges the capacitor of the switches S_2 and S_3, and discharges S_1 and S_4. Finally, the capacitors C_{oss} are fully charged and discharged, and the current passes through the body diodes $S_{1,4}$.

D. Operation Mode IV: $t_4 \leq t < t_5$

In this mode, the body diodes of S_1 and S_4 are ON, and D_1 and D_2 conduct as shown in Fig. 4(d). The direction of the inductors current remains unchanged during this interval. In addition, I_{lr1} and I_{lr2} pass through the body diodes of S_1 and S_4. The V_{DS} is, therefore, zero for S_1 and S_4.

E. Operation Mode V: $t_5 \leq t < t_6$

In this mode, S_1 and S_4 are ON and D_2, D_3, D_6, and D_7 conduct as shown in Fig. 4(e). In the previous mode (IV), the body diodes of S_1 and S_4 are ON. Therefore, the corresponding switches turn on with ZVS.

Fig. 5. shows the current and voltage waveforms of the diodes, switches, and the magnetizing current during the operation modes. G_1, G_4, V_{s3}, and V_{s4} are the gate drive commands and the switch voltages for S_1 and S_4 from the top, respectively. In this figure, I_{Lm1}, I_{Lm2}, I_{lr1}, and I_{Lr2} are the magnetizing current and inductor current for the transformers,

TABLE I
COMPARISON OF DIFFERENT TOPOLOGIES.

Reference	Control method	Circulating current	Port	Switching	Direction	Transformers	Inductors	Diodes	Switches
Proposed	Fixed F_{sw}, PWM	No	Multiport	ZVS	Unidirectional	2	4	8	6
[16]	Variable F_{sw}, PWM	No	Single port	ZVS	Unidirectional	3	3	8	6
[17]	Variable F_{sw}, PSH	Yes	Multiport	ZVS	Unidirectional	2	2	4	6
[18]	Variable F_{sw}, PWM	Yes	Multiport	ZVS	Unidirectional	1	2	8	8
[19]	Fixed F_{sw}, PS-PWM	Yes	Multiport	Hard	Unidirectional	1	1	8	8
[20]	Variable F_{sw}, PSH	No	Multiport	ZVS	Unidirectional	1	2	0	12

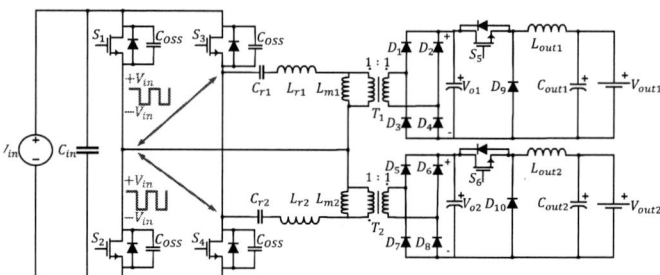

Fig. 3. Proposed multiport LLC topology for a hybrid charging station.

TABLE II
SYSTEM PARAMETERS.

Parameter		Value	Parameter		Value
Input voltage	V_{in}	150 V	Output voltage	V_O	30-137 V
Magnetizing inductance	$L_{m1,2}$	240 uH	Resonant Inductor	$L_{r1,2}$	31 uH
Resonant capacitor	$C_{r1,2}$	60 nF	Resonant frequency	F_r	116 kHz

respectively. During the charging interval, I_{D1}, I_{D3}, I_{D5}, and I_{D7} show the output rectifier currents, respectively.

F. Resonant Tank Design

This section discusses the designs of L_r and C_r. The transformer leakage inductor L_r affects the resonant tank gain and the resonance frequency. To reduce the effect of leakage inductance on the resonance frequency, L_r should be large enough. The capacitor value is designed based on the resonance frequency and the maximum voltage stress across the capacitor in the resonant tank, which is calculated from:

$$V_{cr}^{max} = I_{pri}^{peak} \sqrt{\frac{L_r}{C_r}} \tag{2}$$

where I_{pri}^{peak} is the primary peak current. The resonance frequency is calculated using $F_{r1,2} = 1/(2\pi\sqrt{L_r L_c})$. In practice, the resonant component values may not be equal for both transformers.

III. SIMULATION RESULTS

Simulation studies are performed in Plexim software to see the switches waveform and the zero-voltage switching operation under different rating power. In the simulation study, both slow and fast chargers are working simultaneously. The simulation tests are done according to Fig. 2, which shows one input and multiple outputs. For simplicity, only one of the parallel converters is included in the simulation studies.

In this simulation study, one of the ports is examined under different rating powers to see the currents and voltages under ZVS. This study assumes that the first port provides power to the load and the second port is deactivated. In these cases, the converter delivers 9 kW, 18 kW, and 26 kW to validate under different ratings. The voltages and currents under these three

rating powers are 303 V, 30 A, 425 V, 42.5 A, 510 V, and 51 A, respectively. The simulation results for these case studies are shown in Fig. 6. In this figure, I_{s3} and V_{s3} are the current and voltage for one switch of the primary side, and V_{o1} is the output voltage of the rectified voltage on the secondary side. As shown in this figure, in all cases the proposed topology delivers power under zero-voltage switching.

IV. EXPERIMENTAL RESULTS

A prototype is built in the lab to evaluate the proposed converter practically. The specification of the converter is shown in Table II. Fig. 7 shows the experimental setup. Two case studies are provided to show the converter efficiency and waveforms, as below.

A. Case I: $P_{out1} = 460$ W and $P_{out2} = 80$ W

In this case, the converter is examined to see the waveform and validate ZVS. The output power of the first port is assumed to deliver 460 W power to the load, and the second port provides 80 W to the load. In this test, the gate-source and gate driver voltages are captured. The result is shown in Fig. 8(a). As can be seen, the switches achieve zero-voltage switching. In this case study, the converter achieves 96% efficiency.

B. Case II: $P_{out1} = 80$ W and $P_{out2} = 450$ W

In this case, the converter is examined at another rating power. The output power of the first port is assumed to provide 80 W power to the load, and the second port delivers 450 W to the load in the case. In this test, the gate-source and gate driver voltages are captured. The result is shown in Fig. 8(b). As can be seen, the switches achieve zero-voltage switching. In this case study, the converter achieves 96% efficiency using SiC-based switches.

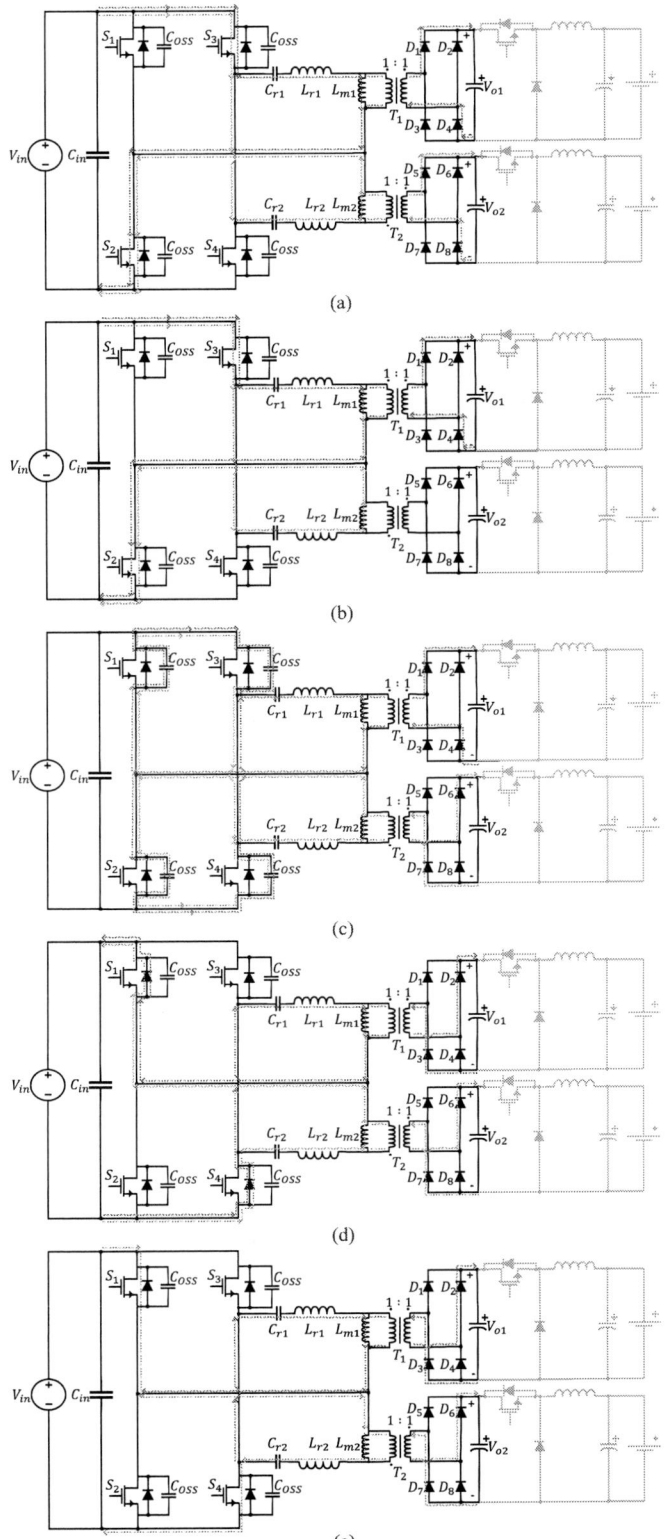

Fig. 4. Operation modes of the topology: (a) mode I, (b) mode II, (c) mode III, (d) mode IV, (e) mode V.

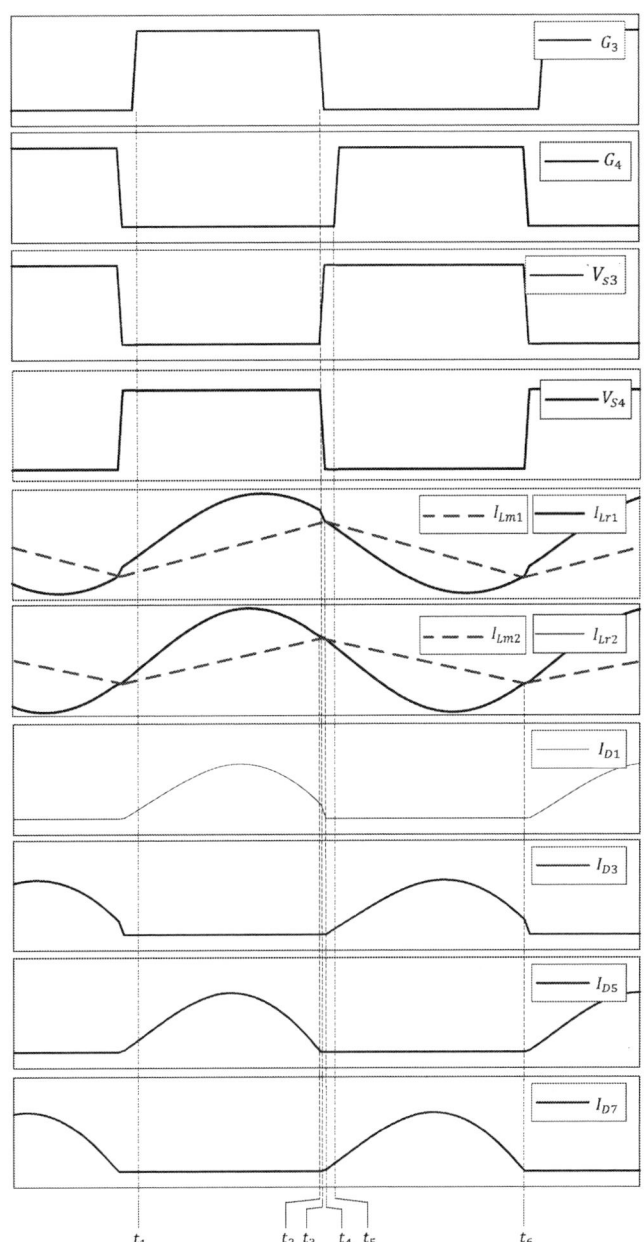

Fig. 5. Voltages and currents of the switches and diodes in steady-state operation.

V. CONCLUSION

This paper proposes a hybrid three-port isolated DC-DC converter, which is based on the LLC resonant topology.

This topology is designed specifically for hybrid EV charging stations. The proposed converter features both fast and slow charging ports. The topology uses an LLC resonant tank, providing isolation and enabling soft switching. By eliminating the need for the primary-side feedback signals, the complexity of the control circuit is significantly reduced. The proposed converter offers a lower number of components, which reduces the cost. In addition, it offers a simple control under fixed-frequency operation and ZVS. The efficacy of the converter is demonstrated through a 600 W prototype, validating the operation and its potential benefits for urban EV charging infrastructure. The converter achieves 96% efficiency.

979-8-3315-1612-3/25 $31.00 © 2025 IEEE

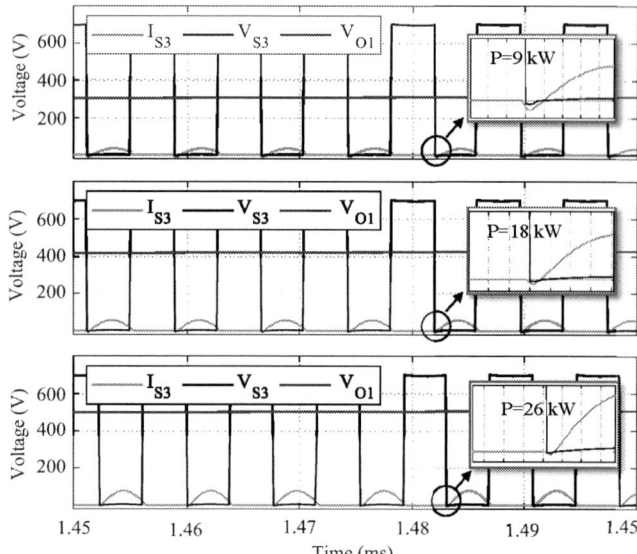

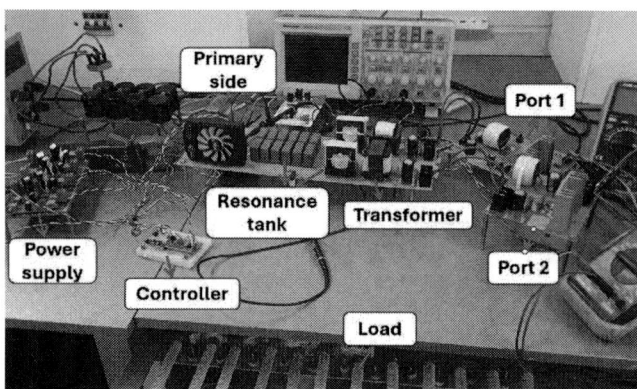

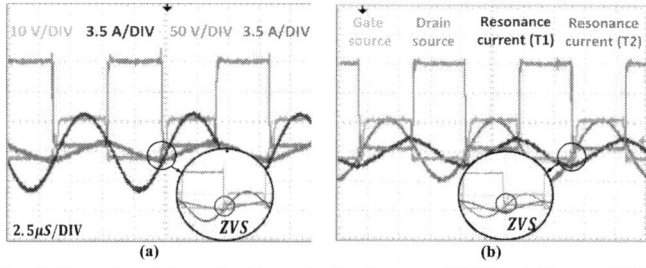

Fig. 6. Voltages and current of a switch at different rating power.

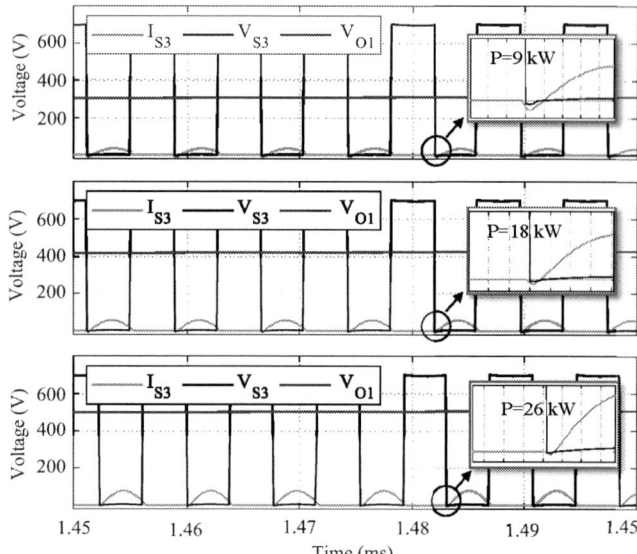

Fig. 7. Experimental setup.

Fig. 8. Experimental results: (a) results for $P_{out1} = 460$ W and $P_{out2} = 80$ W, (b) results for $P_{out1} = 80$ W and $P_{out2} = 450$ W.

ACKNOWLEDGMENT

The authors would like to thank Kv Material, s.r.o company, for the financial support at the conference.

REFERENCES

[1] J. Chevinly, S. S. Rad, E. Nadi, B. Proca, J. Wolgemuth, A. Calabro, H. Zhang, and F. Lu, "Gallium nitride (GaN) based high-power multi-level H-bridge inverter for wireless power transfer of electric vehicles," in *IEEE Transportation Electrification Conference and Expo (ITEC)*, Chicago, IL, June 2024.

[2] S. H. Nazaralizadeh, A. Banerjee, P. Srivastava, and Famouri, "Battery energy storage systems: A review of energy management systems and health metrics," *Energies*, vol. 17, no. 5, p. 1250, March 2024.

[3] N. Souri and A. Mehrizi-Sani, "Accurate current sharing in a DC micro-grid using modified droop control algorithm," in *Industrial Electronics Society (IES)*, Chicago, IL, June 2024.

[4] M. Ghavaminejad, E. Afjei, and M. Meghdadi, "Double-input/double-output buck-zeta converter," in *29th Iranian Conference on Electrical Engineering (ICEE)*, Tehran, Iran, May 2021.

[5] M. Ghasemi, A. Honarbakhsh, M. Saradarzadeh, and M. Hamzeh, "Ultra-wide voltage range control of DC-DC full-bridge converter with hysteresis controller," in *13th Power Electronics, Drive Systems, and Technologies Conference (PEDSTC)*, February 2022.

[6] S. A. Assadi, Z. Gong, N. Coelho, M. S. Zaman, and O. Trescases, "Modular multiport electric-vehicle DC fast-charge station assisted by a dynamically reconfigurable stationary battery," *IEEE Transactions on Power Electronics*, vol. 38, no. 5, pp. 6212–6223, January 2023.

[7] H. Krishnaswami and N. Mohan, "Three-port series-resonant DC–DC converter to interface renewable energy sources with bidirectional load and energy storage ports," *IEEE Transactions on Power Electronics*, vol. 24, no. 10, pp. 2289–2297, August 2009.

[8] I. Kougioulis, A. Pal, P. Wheeler, and M. R. Ahmed, "An isolated multiport DC–DC converter for integrated electric vehicle on-board charger," *IEEE Journal of Emerging and Selected Topics in Power Electronics*, vol. 11, no. 4, pp. 4178–4198, May 2023.

[9] N. D. Dao, D.-C. Lee, and Q. D. Phan, "High-efficiency SiC-based isolated three-port DC-DC converters for hybrid charging stations," *IEEE Transactions on Power Electronics*, vol. 35, no. 10, pp. 10455–10465, February 2020.

[10] M. Mohebifar, N. Rostami, E. Babaei, and M. Sabahi, "Dual-output step-down soft switching current-fed full-bridge DC-DC converter," in *14th International Conference on Electrical Engineering/Electronics, Computer, Telecommunications and Information Technology (ECTI-CON)*, Phuket, Thailand, June 2017.

[11] J. Y. Yong, V. K. Ramachandaramurthy, K. M. Tan, and J. Selvaraj, "Experimental validation of a three-phase off-board electric vehicle charger with new power grid voltage control," *IEEE Transactions on Smart Grid*, vol. 9, no. 4, pp. 2703–2713, October 2018.

[12] N. Kumar, S. K. Mazumder, and A. Gupta, "SiC DC fast charger control for electric vehicles," in *IEEE Energy Conversion Congress and Exposition (ECCE)*, Portland, OR, December 2018.

[13] Y. Wang, F. Han, L. Yang, R. Xu, and R. Liu, "A three-port bidirectional multi-element resonant converter with decoupled power flow management for hybrid energy storage systems," *IEEE Access*, vol. 6, pp. 61331–61341, September 2018.

[14] M. Phattanasak, R. Gavagsaz-Ghoachani, J.-P. Martin, B. Nahid-Mobarakeh, S. Pierfederici, and B. Davat, "Control of a hybrid energy source comprising a fuel cell and two storage devices using isolated three-port bidirectional DC-DC converters," in *Eighth International Conference and Exhibition on Ecological Vehicles and Renewable Energies (EVER)*, May 2013.

[15] Z. Shi, Y. Tang, Y. Zhang, Y. Guo, H. Sun, and L. Jiang, "A secondary-side semiactive 3-phase interleaved resonant converter employing multimode modulation scheme for fast EV charger applications," *IEEE Transactions on Power Electronics*, vol. 37, no. 11, pp. 13385–13397, June 2022.

[16] F. Liu, Y. Chen, and X. Chen, "Comprehensive analysis of three-phase three-level LC-type resonant DC-DC converter with variable frequency control—series resonant converter," *IEEE Transactions on Power Electronics*, vol. 32, no. 7, pp. 5122–5131, September 2017.

[17] X. Gao, H. Wu, and Y. Xing, "A multioutput llc resonant converter with semi-active rectifiers," *IEEE Journal of Emerging and Selected Topics in Power Electronics*, vol. 5, no. 4, pp. 1819–1827, June 2017.

[18] Y. Shi and X. Yang, "Wide load range ZVS three-level DC–DC converter: Four primary switches, capacitor clamped, two secondary switches, and smaller output filter volume," *IEEE Transactions on Power Electronics*, vol. 31, no. 5, pp. 3431–3443, August 2016.

[19] S. Rivera, F. Flores-Bahamonde, H. Renaudineau, T. Dragicevic, and S. Kouro, "A buck-boost series partial power converter using a three-port structure for electric vehicle charging stations," in *IEEE 12th Energy Conversion Congress and Exposition - Asia (ECCE-Asia)*, July 2021.

[20] S. S. Chakraborty, S. Dey, and K. Hatua, "Design of a three-winding transformer for power decoupling of a three-port series resonant converter for an integrated on-board EV charger," *IEEE Transactions on Power Electronics*, vol. 38, no. 11, pp. 14262–14273, August 2023.

Reconfigurable H5-bridge Based *LLC*-DAB Sigma Converter for EV Fast Charging Stations

Huangsheng Xu, Mingde Zhou, Qishan Pan and Haoyu Wang
School of Information Science and Technology
ShanghaiTech University, Shanghai, China
Shanghai Engineering Research Center of Energy Efficient and Custom AI IC
wanghy@shanghaitech.edu.cn

Abstract—In fast charging stations, the electric vehicle (EV) charger needs to adapt to the traction batteries of different models. The ultra-wide voltage range brings significant challenges to the optimal design of the dc/dc converter in the EV charger. To cope with this issue, we propose a Sigma converter based on a reconfigurable H5 bridge. The converter consists of two submodules, the *LLC* submodule and the DAB submodule, with a shared H5 bridge. By reconfiguring the H5 bridge, the *LLC* submodule exhibits a trapezoidal gain characteristic. On this basis, the DAB submodule realizes the residual gain through a hybrid modulation. The *LLC* submodule delivers the majority of power and always operates at the optimal operating point, while the voltage regulation is majorly realized by the DAB submodule. The circulating current of the *LLC* submodule contributes to the zero-voltage switching (ZVS) of the DAB, and the partial power processing structure makes the power path simpler and more efficient. In addition, the modulation method is capable of battery current control, making it suitable for EV charging applications. A 3.3kW prototype that converts 800V input to a 150-900V ultra-wide output is designed, simulated, and experimentally tested to validate the concept.

Keywords—DAB, dc transformer (DCX), H5-bridge, LLC, wide voltage range

I. INTRODUCTION

For the booming EV industries, off-board chargers play an important role in linking the traction batteries to the ac power grid. Due to the coexistence of 400V and 800V electric vehicle (EV) architectures in the current market, the dc/dc stage in the off-board EV charger needs to have an ultra-wide voltage gain range to be compatible with both voltage frameworks as shown in the system diagram in Fig. 1.

The practical application of EV charger usually adopts a two-stage architecture [1] - [5]. The frontend ac/dc stage achieves power factor correction, while the backend dc/dc stage achieves galvanic isolation and charging regulation. The dc/dc converter plays a crucial role in adapting to a wide voltage range on the battery side. DAB and resonant isolated converter are the main candidate topologies for the dc/dc stage due to their electrical isolation capability, soft switching characteristics, bi-directional power transfer capability, simple topology and mature control methods. However, they all have some limitations in terms of performance degradation in wide-gain range applications.

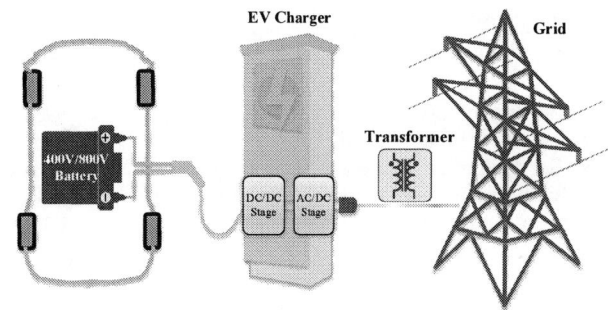

Fig. 1. The system block diagram corresponding to the application in EV charging.

The soft-switching performance of DAB is load and gain range dependent, and ZVS can only be reliably achieved at a heavy load and limited gain range. Under light load and ultra-wide gain range applications, to ensure ZVS, the inductance of DAB must be reduced, which introduces a large amount of reactive circulating current and reactive power, resulting in large circulating current losses in the circuit [6] - [10].

Resonant-isolated converters, such as *LLC* and *CLLC*, have high efficiency and a load-independent gain when the switching frequency is equal to the resonant frequency. However, when the gain of the converter is varied over a wide gain range, the efficiency of the converter will decrease rapidly. In applications with ultra-wide voltage gain ranges, a wider frequency range is required to realize wide-range voltage regulation, and then the problems of efficiency degradation and loss of soft-switching become more and more severe as the switching frequency deviates further from the resonant frequency [11] - [12].

To develop the high-performance dc/dc stage, two-stage solutions are investigated in the literature [13] - [14]. The two-stage converter implements voltage regulation and galvanic isolation separately, which can be optimized respectively. However, the overall system efficiency degrades due to the multiplication effect. To further improve the efficiency, quasi-single-stage topologies combining a DCX and a voltage regulator have been studied [15] - [16]. However, there is still some power flowing through a two-stage channel, which harms the efficiency the system's efficiency.

979-8-3315-1612-3/25 $31.00 © 2025 IEEE
2207

To achieve a true single-stage structure, this paper replaces the commonly used DCX plus PWM converter in quasi-single-stage structures with a DAB converter. Meanwhile, to further extend the gain range, an H5-bridge structure is introduced. Therefore, an *LLC*-DAB Sigma converter based on the H5 bridge is constructed. Six operating modes adapting to an ultra-wide output voltage range can be realized by changing the configurations of the H5-bridge. The *LLC* submodule always operates at the resonant frequency, which indicates good efficiency performance; the DAB submodule is regulated in a narrow voltage gain range with hybrid modulation to suppress the circulating current and extend the ZVS range. Moreover, the circulating current of the *LLC* submodule aids the DAB's primary switch in achieving ZVS under all load conditions, ameliorating the issue of ZVS loss at light load.

The rest of this paper is organized as follows. Section II introduces the proposed *LLC*-DAB Sigma converter. Section IIIdiscusses parameter design, including gain design and ZVS parameter design. Section IV demonstrated the simulation and experimental simulation verification. Finally, Section V summarizes this paper.

II. TOPOLOGY DESCRIPTION AND OPERATION PRINCIPLES

A. Topology Description

The schematic of the proposed converter is illustrated in Fig. 2. The H5 bridge consists of five switches S_1-S_5. The H5 bridge converts the dc voltage V_A to two ac voltages, v_{ab} and v_{bc} [17]. They are fed to two impedance networks RT_1 and RT_2 with different parameters.

The H5-bridge with v_{ab}, RT_1, and the active full-bridge FB_1 forms an *LLC* submodule. The *LLC* converter always operates at the optimal operating point, with a fixed voltage gain as DCX. The H5 bridge with v_{bc}, RT_2, and the active full-bridge FB_2 forms a DAB submodule. Both the primary and secondary bridges of the DAB can operate in half-bridge mode due to the presence of the DC blocking capacitors C_P and C_S.

By configuring the H5-bridge switches among ON state, OFF state, and frequency modulation (FM) state, six configurations can be achieved, namely half bridge *LLC*, half-bridge DAB, half-bridge *LLC* plus half bridge DAB, full bridge *LLC* plus half-bridge DAB, half-bridge *LLC* plus full bridge DAB, and full-bridge *LLC* plus full-bridge DAB, as plotted in Fig. 3. By changing the configuration, the RMS values of v_{ab} and v_{bc} can be adjusted, thus the six-step voltage gain curve can be derived, expanding the entire voltage gain range.

B. DAB Modulation Method

The DAB submodule needs to handle the remaining power and regulate the voltage within a certain range, so the optimization of DAB modulation method plays a decisive role in the performance of the converter. This paper adopts a hybrid modulation method of pulse width modulation (PWM) and phase shift modulation (PFM). The v_{bc} of the DAB

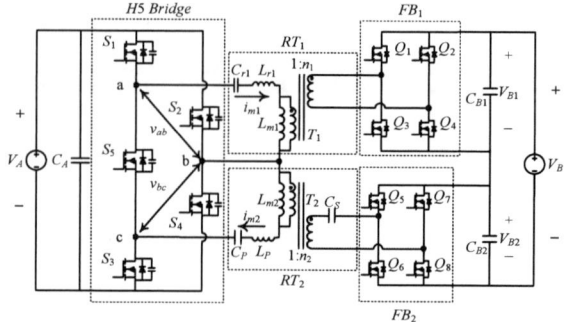

Fig. 2. Schematic of the proposed converter.

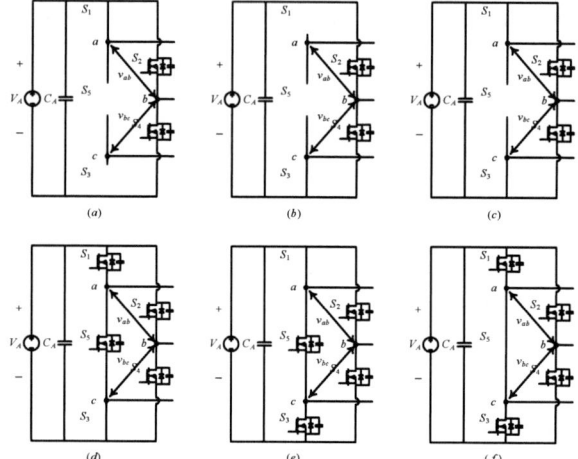

Fig. 3. Six configurations of the H5-bridge: (a) Configuration 1; (b) Configuration 2; (c) Configuration 3; (d) Configuration 4; (e) Configuration 5; (f) Configuration 6.

submodule is controlled by a complementary driving signal with a 50% duty cycle. The FB_2 operates in asymmetric pulse width modulation (APWM) to continuously adjust the dc bias of the dc blocking capacitor, so V_{B2} can be regulated. The output power is adjusted by the phase shift of the H5-bridge side and FB_2 side. Taking the operation mode of a single half-bridge DAB submodule in configuration 2 as an example, the equivalent topology is shown in Fig. 4, and the key waveforms are shown in Fig. 5.

In Fig. 5, S_2 and S_4 are driven in a complementary fashion with a 50% duty cycle, incorporating a defined dead time. The duty cycle of Q_5 and Q_7 varies by the battery side voltage. The duty cycle of Q_8 exceeds 50%. D_1 is the turning-off delay between Q_8 and Q_5. It is between 0 and 0.5.

D_1 is controlled to satisfy the following expression:

$$\frac{V_A}{2} = n_2(1-D_1)V_B \tag{1}$$

The normalized voltage gain is defined as

$$M = \frac{1}{1-D_1} = \frac{2n_2V_B}{V_A} \tag{2}$$

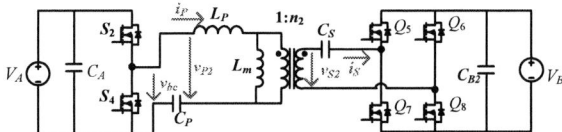

Fig. 4. Equivalent topology of configuration 2.

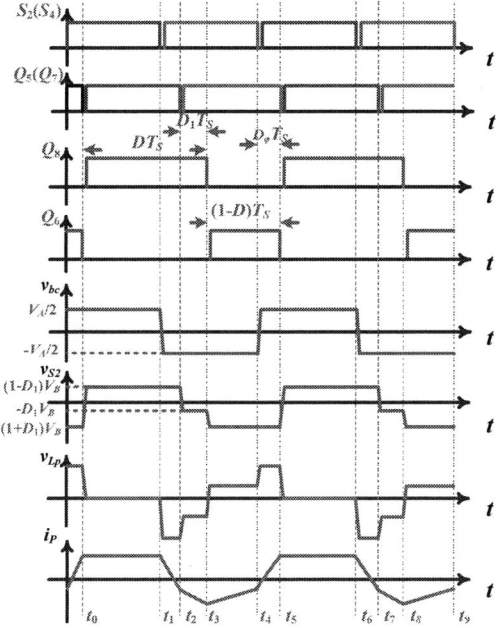

Fig. 5. Key waveform of DAB under hybrid modulation.

For 0.5 < M <1, the hybrid modulation realizes a general voltage match with 0 <D_1 < 0.5. In addition, for M= 0.5 or 1, the highest and lowest gain points of the DAB submodule are completely voltage matched points, with the smallest circulating current, widest ZVS range, and optimal efficiency.

The external phase shift D_φ modulates the power transmitted. The range of D_φ is between 0 and 0.5. The output power of the half-bridge DAB submodule over a switching period is expressed as

$$P_O = \frac{V_A V_B T_S (-2D_1^2 - 4D_1 D_\varphi - 8D_\varphi^2 + 4D_\varphi + D_1)}{8n_2 L_k} \quad (3)$$

III. DESIGN CONSIDERATIONS

The two submodules of the proposed converter proposed in this paper can be decoupled into two separate parts for design, suitable for traditional DAB and *LLC* design methods. However, partial power processing techniques may pose challenges in gain design, and the ZVS enhancement effect between the two submodules and the H5-bridge configuration may result in differences in ZVS analysis. This section will focus on introducing gain analysis and ZVS analysis.

A. Gain Analysis

Partial power processing technology divides the overall gain of the converter into primary gain and residual gain. Meanwhile, the transmission power is also divided into main power and residual power. In the proposed topology, the main transmitted power is processed through the *LLC*

TABLE I
VOLTAGE GAIN RANGE OF THE CONVERTER

Configuration	$v_{ab, RMS}$	$v_{bc, RMS}$	Minimum gain	Maximum gain
Configuration 1	$V_A/2$	0	0.5n	0.5n
Configuration 2	0	$V_A/2$	0.5n	n
Configuration 3	$V_A/2$	$V_A/2$	n	1.5n
Configuration 4	V_A	$V_A/2$	1.5n	2n
Configuration 5	$V_A/2$	V_A	1.5n	2.5n
Configuration 6	V_A	V_A	2n	3n

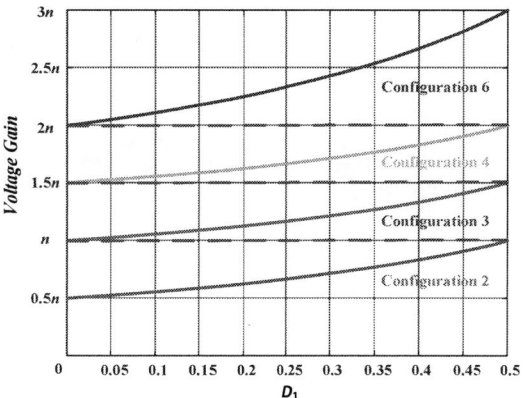

Fig. 6. Voltage gain curves of the proposed topology.

submodule, and the DAB submodule processes the remaining power. By changing the configuration of the H5 bridge, the *LLC* resonant submodule for the main power transmission can be configured with a laddered voltage gain characteristic. The DAB submodule processes the remaining voltage gain.

According to the DAB modulation method mentioned earlier, the range of *M* is between 0.5 and 1. That is $0.5 = M_{min} \le M \le M_{max} = 1$.

To ensure continuous gain and minimize the voltage regulation range of DAB, the voltage regulation range of DAB is set to match the voltage gain of *LLC* during gain design. That is

$$n_2 \times (M_{max} - M_{min}) = n_2 \times (1 - 0.5) = \frac{1}{2} \times n_1 \quad (4)$$

Therefore, it can be concluded that the transformer turns ratio of DAB and *LLC* is the same, defined as *n*. That is $n_1 = n_2 = n$. The voltage gain range of the converter under different configurations is shown in Table I.

The converter only operates in four configurations of 2, 3, 4, and 6 to achieve a gain range of 0.5n to 3n, as plotted in Fig. 6, and the turns ratio of two transformers can be derived from this gain range.

Taking the application of EV charging proposed in this article as an example, the output voltage range required for two voltage platforms should be able to cover 250V to 500V and 500V to 900V. So when the input voltage is 800V, n is 0.375, which means the gain range of the converter is 0.375 to 1.125, and the output voltage range is 150V to 900V.

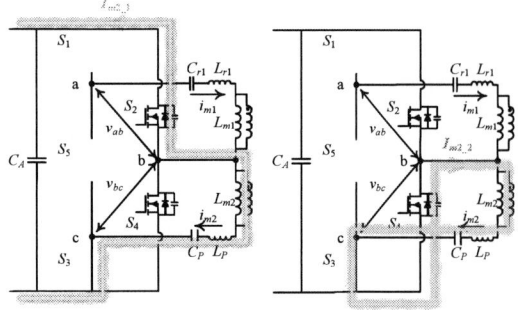

Fig. 7. The flow direction of circulating current in configuration 2 during the dead time.

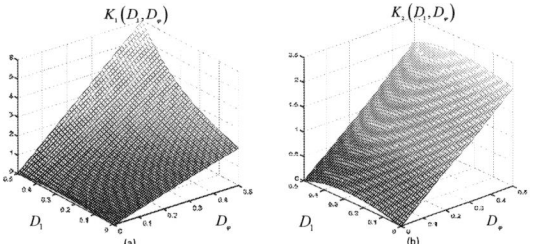

Fig. 8. The impact of D_1 and D_φ on the magnitude of circulating current.

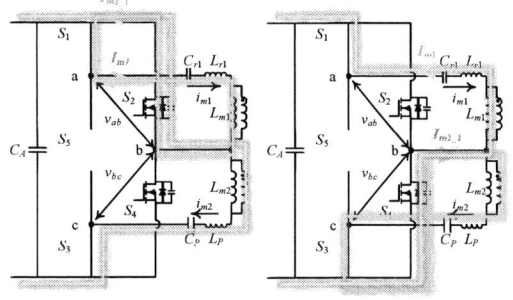

Fig. 9. The flow direction of circulating current in configuration 2 during the dead time.

B. ZVS Analysis of DAB

Based on capacitor charging balance and the waveform diagram as plotted in Fig. 5, it can be deduced that when the DAB submodule works alone, that is, configuration 2, the inductor current expression at different time points is

$$\begin{cases} i_P(t_0) = I_B\left(-2D_1^2 - 4D_1D_\varphi + 4D_\varphi + D_1\right) \\ i_P(t_1) = I_B\left(-2D_1^2 - 4D_1D_\varphi + 4D_\varphi + D_1\right) \\ i_P(t_2) = I_B\left(-2D_1^2 + 4D_1D_\varphi - 4D_\varphi + D_1\right) \\ i_P(t_3) = I_B\left(6D_1^2 + 4D_1D_\varphi - 4D_\varphi - 3D_1\right) \\ i_P(t_4) = I_B\left(-2D_1^2 - 4D_1D_\varphi - 4D_\varphi + D_1\right) \\ I_B = nV_OT_S / 4L_P \end{cases} \tag{5}$$

To simplify the analysis, C_{oss} is neglected. Therefore, when the MOSFET is turned on and the current is negative, it is considered that ZVS has been achieved. If MOSFET

wants to achieve ZVS, the inductor current needs to meet the following conditions:

$$\begin{cases} Q_5,\ Q_8:\ -2D_1^2 - 4D_1D_\varphi + 4D_\varphi + D_1 > 0 \\ Q_7:\ -2D_1^2 + 4D_1D_\varphi - 4D_\varphi + D_1 < 0 \\ Q_6:\ 6D_1^2 + 4D_1D_\varphi - 4D_\varphi - 3D_1 < 0 \\ S_2:\ -2D_1^2 - 4D_1D_\varphi - 4D_\varphi + D_1 < 0 \\ S_4:\ -2D_1^2 - 4D_1D_\varphi + 4D_\varphi + D_1 > 0 \\ 0 < D_1,\ D_\varphi < 0.5,\ D_1 + D_\varphi < 0.5 \end{cases} \tag{6}$$

Considering the range of control variables, the ZVS inequalities of Q_5, Q_6, Q_8, and S_4 are always satisfied, that is Q_5, Q_6, Q_8, and S_4 can always achieve ZVS. In addition, ZVS conditions for Q_7 and S_2 can be simplified as

$$\begin{cases} S_2:\ \dfrac{1}{4}\dfrac{D_1(1-2D_1)}{1+D_1} \le D_\varphi \\ Q_7:\ \dfrac{1}{4}\dfrac{D_1(1-2D_1)}{1-D_1} \le D_\varphi \end{cases} \tag{7}$$

According to (7), when D_1 approaches 0.3, ZVS for Q_7 is the most difficult to achieve, and when D_1 approaches 0.225, ZVS for S_2 is the most difficult to achieve. When the transmission power increases, the ZVS range of Q_7 and S_2 expands. Under boundary conditions, DAB operates in single-phase shift modulation (SPS) mode, where ZVS conditions for all MOSFETs become easy to realize.

C. Design of Dead Time

In order to achieve ZVS of the primary MOSFET, the circulating current should be large enough to fully charge and discharge the output capacitor of the MOSFET during the dead zone period [18]. T_d is the duration of the dead zone. C_{oss} is an energy-normalized output capacitor.

$$I_m t_d \ge 2C_{oss}V_A \tag{8}$$

In configuration 2, DAB submodule works alone, where S_2 and S_4 form a half bridge. The circulating current when S_2 S_4 are turned on is I_{m2_1} and I_{m2_2} respectively, and their flow directions are as plotted in Fig. 7. According to (1) and (5), the value of the circulating current can be derived as follows:

$$\begin{cases} \left|I_{m2\_1}\right| = \dfrac{V_A T_S \left|2D_1^2 + 4D_1D_\varphi + 4D_\varphi - D_1\right|}{8L_P(1-D_1)} = \dfrac{V_A T_S}{8L_P} \cdot K_1(D_1, D_\varphi) \\ \left|I_{m2\_2}\right| = \dfrac{V_A T_S \left|-2D_1^2 - 4D_1D_\varphi + 4D_\varphi + D_1\right|}{8L_P(1-D_1)} = \dfrac{V_A T_S}{8L_P} \cdot K_2(D_1, D_\varphi) \end{cases} \tag{9}$$

where K_1 and K_2 are circulating current factors, whose values are related to D_1 and D_φ. The impact of D_1 and D_φ on the magnitude of circulating current is shown in Fig. 8.

According to (9), it can be concluded that

$$\begin{cases} t_{d\_on} \ge \dfrac{16C_{oss}f_sL_P}{K_1(D_1, D_\varphi)} \\ t_{d\_off} \ge \dfrac{16C_{oss}f_sL_P}{K_2(D_1, D_\varphi)} \end{cases} \tag{10}$$

When the phase shift D_φ is small, K_1 and K_2 approach 0, requiring a significant dead time to achieve ZVS. In other words, ZVS is difficult to achieve when DAB is at light load. When the phase shift D_φ is large, K_1 and K_2 are greater than 1, and the circulating current of DAB is greater than that of *LLC*. In other words, when DAB is under heavy load, it is easier to achieve ZVS compared to *LLC*.

In configuration 3, I_m of both submodules charge and discharge C_{oss} of MOSFETs, as plotted in Fig. 9.

$$(I_{m1} + I_{m2})t_d \geq 2C_{oss}V_A \quad (11)$$

The circulating current I_{m1} provided by the *LLC* submodule is expressed as follows

$$I_{m1} = \frac{V_{B1}}{4nL_{m1}f_s} = \frac{V_A}{8L_{m1}f_s} \quad (12)$$

Inserting (10), and (12) into (11), we can obtain

$$\begin{cases} \left(\dfrac{V_A}{8L_{m1}f_s} + \dfrac{V_A K_1}{8L_P f_s}\right) t_{d\_on} \geq 2C_{oss}V_A \\ \left(\dfrac{V_A}{8L_{m1}f_s} + \dfrac{V_A K_2}{8L_P f_s}\right) t_{d\_off} \geq 2C_{oss}V_A \end{cases} \quad (13)$$

Therefore, the ZVS conditions of the primary MOSFETs in configuration 3 can be obtained, and the following are given:

$$\begin{cases} t_{d\_on} \geq 16C_{oss}f_s \dfrac{L_{m1}L_P}{K_1 L_{m1} + L_P}, \ Configuration \ 3 \\ t_{d\_off} \geq 16C_{oss}f_s \dfrac{L_{m1}L_P}{K_2 L_{m1} + L_P}, \ Configuration \ 3 \end{cases} \quad (14)$$

The same analysis method is applied to several other configurations, so that the range of dead time values for configurations 4, 5, and 6 can be obtained as follows:

$$\begin{cases} t_{d\_on} \geq 16C_{oss}f_s \dfrac{L_{m1}L_P}{K_1 L_{m1} + 2L_P}, \ Configuration \ 4 \\ t_{d\_off} \geq 16C_{oss}f_s \dfrac{L_{m1}L_P}{K_1 L_{m1} + 2L_P}, \ Configuration \ 4 \\ t_{d\_on} \geq 16C_{oss}f_s \dfrac{L_{m1}L_P}{2K_1 L_{m1} + L_P}, \ Configuration \ 5 \\ t_{d\_off} \geq 16C_{oss}f_s \dfrac{L_{m1}L_P}{2K_1 L_{m1} + 1L_P}, \ Configuration \ 5 \\ t_{d\_on} \geq 8C_{oss}f_s \dfrac{L_{m1}L_P}{K_1 L_{m1} + L_P}, \ Configuration \ 6 \\ t_{d\_off} \geq 8C_{oss}f_s \dfrac{L_{m1}L_P}{K_1 L_{m1} + L_P}, \ Configuration \ 6 \end{cases} \quad (15)$$

D. Design of Inductor L_P

The design of L_P is crucial in the DAB converter as it determines the power transfer capability and the power conversion efficiency. Large L_P leads to a poor power transfer capability. Therefore, L_P should be designed to meet the output power requirements with a 90% design margin as

$$L_P \leq \frac{0.9nT_s V_A V_{B2}}{2P_{rate}} \quad (20)$$

where P_{rate} is the rated power of the DAB submodule [6].

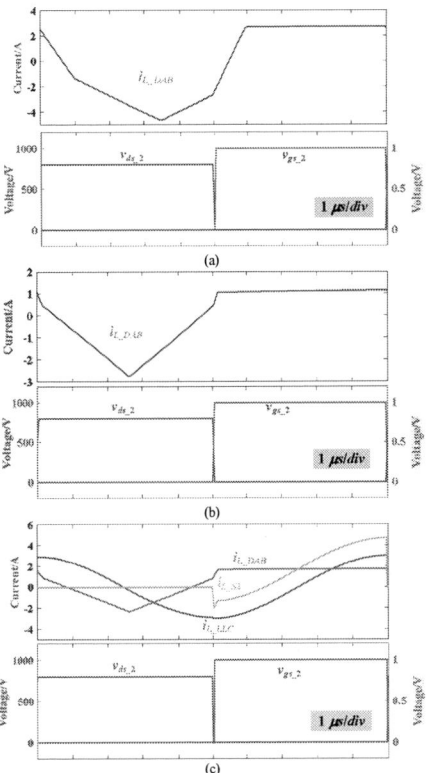

Fig. 10. Simulation result: (a) Primary current and ZVS performance of S_2 in configuration 2 with and $D_\varphi = 0.1$ and $D_1 = 0.25$; (b) Primary current and ZVS performance of S_2 in configuration 2 with $D_\varphi = 0.02$ and $D_1 = 0.25$; (c) Primary current and ZVS performance of S_2 in configuration 3 with $D_\varphi = 0.02$ and $D_1 = 0.25$.

E. Design of the Blocking Capacitor C_P and C_S

Blocking capacitors are utilized to maintain dc voltage bias. Firstly, because the H5 bridge on the primary side can switch configurations, the DAB submodule has two operating modes for the primary side: half bridge and full bridge. Only when the primary side operates in a half bridge, C_P acts as the dc blocking capacitor to maintain dc voltage bias. So the rated voltage of C_P only needs to be greater than $0.25V_A$, equivalent to 200V. Meanwhile, due to the hybrid modulation method of DAB, the secondary side of DAB also has two operating modes: half bridge and full bridge. When the secondary side of DAB operates in half-bridge mode and the primary side operates in full-bridge mode, C_S experiences the maximum voltage stress of $0.5V_{B2}$, equivalent to 300V.

Secondly, blocking capacitors should be large enough to maintain a relatively constant voltage during charging or discharging. C_P and C_S are selected at 5μF to ensure that the voltage change rate is less than 0.01.

IV. SIMULATION AND EXPERIMENTAL RESULTS

To validate the proposed concept, a prototype of a 3.3kW converter was designed, simulated and experimentally tested. The input voltage is 800V and the output voltage is 150 - 900V. The turns n is 3/8. The leakage inductance of *LLC* is 26.6μH, the magnetic inductance is 263.17μH, and the resonant capacitance is 93nF (resonant frequency $f_r = 100$ kHz). The DAB inductor L_P is 200μH, and the dc blocking capacitors C_P and C_S are both 5μF.

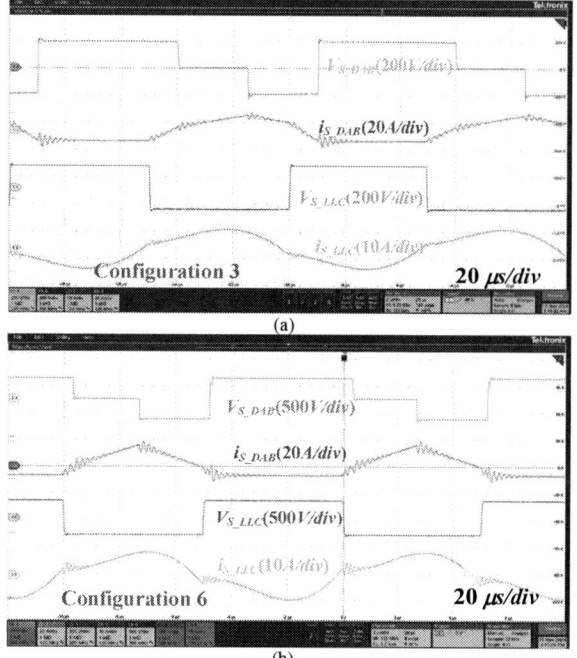

Fig. 11. Experimental steady state waveforms: (a) Configuration 3 with $V_B = 350$V; (b) Configuration 6 with $V_B = 700$V.

Fig. 10 shows the switching process of S_2 under different working states. Fig. 10 (a) shows that in configuration 2, $D_\varphi = 0.1$ and $D_1 = 0.25$. At this point, meeting the ZVS condition as in (7), S_2 achieves ZVS. And Fig. 10 (b) is also the steady-state waveform under configuration 2, $D_1 = 0.25$. But at this point, D_φ is too small to meet the ZVS condition proposed in (7), and the circulating current is opposite to discharge direction of the C_{oss} of S_2, making it impossible for S_2 to achieve ZVS. The converter in Fig. 10 (c) operates in configuration 3. Although $D_\varphi = 0.02$ and $D_1 = 0.25$ are the same as (b), the circulating current is equal to the inductor current of the DAB submodule plus the magnetizing current of the *LLC* submodule, which is independent of the load. Therefore, S_2 can still achieve ZVS at this time, which is the ZVS enhancement performance between *LLC* and DAB.

Fig. 11 (a) shows the steady-state waveform of the design prototype in configuration 3 when the output voltage is 350V. Fig. 11 (b) shows the steady-state waveform of the design prototype in configuration 6 when the output voltage is 700V. Both submodules can operate normally regardless of whether the primary side is a full bridge or a half bridge. According to the waveform, it can be observed that the output voltage gains of DAB and *LLC* submodules do not interfere with each other, which is consistent with the gain analysis in the paper. And thanks to the hybrid modulation method, the peak current of the DAB submodule is greatly reduced.

V. CONCLUSION

This paper introduces an *LLC*-DAB Sigma converter. The topology is thoroughly described, alongside a detailed analysis of its steady-state operation. A hybrid modulation method is employed to enhance the control of the DAB submodule, which is crucial for voltage regulation. The paper derives the converter's gain range and conducts a ZVS analysis. The proposed topology and modulation technique effectively address the challenge of achieving both high

efficiency and a wide voltage gain range. To validate the converter's performance and feasibility, a 3.3 kW prototype was developed with an input voltage of 800V and an adjustable output voltage ranging from 150V to 900V.

REFERENCES

[1] A. Verma and B. Singh, "Multi-Objective Reconfigurable Three-Phase Off-Board Charger for EV," *IEEE Tran. Ind. Appl.*, vol. 55, no. 4, pp. 4192-4203, July-Aug. 2019.

[2] R. Gadelrab, Y. Yang, B. Li, F. Lee, and Q. Li, "High-frequency high-density bidirectional EV charger," in *Proc. IEEE Transp. Electrific. Conf. Expo.*, 2018, pp. 687–694.

[3] A. Khaligh and M. D. Antonio, "Global trends in high-power on-board chargers for electric vehicles," *IEEE Trans. Veh. Technol.*, vol. 68, no. 4, pp. 3306–3324, Apr. 2019.

[4] V. Monteiro, J. C. Ferreira, A. A. Nogueiras Meléndez, C. Couto and J. L. Afonso, "Experimental Validation of a Novel Architecture Based on a Dual-Stage Converter for Off-Board Fast Battery Chargers of Electric Vehicles," *IEEE Trans. Veh. Technol.*, vol. 67, no. 2, pp. 1000-1011, Feb. 2018.

[5] R. Gadelrab, Y. Yang, B. Li, F. Lee, and Q. Li, "High-frequency high-density bidirectional EV charger," in *Proc. IEEE Transp. Electrific. Conf. Expo.*, 2018, pp. 687–694.

[6] J. Deng and H. Wang, "A Hybrid-Bridge and Hybrid Modulation-Based Dual-Active-Bridge Converter Adapted to Wide Voltage Range," *IEEE J. Emerg. Sel. Topics Power Electron.*, vol. 9, no. 1, pp. 910-920, Feb. 2021.

[7] D. Shu, H. Wang, and M. Zhou, "Universal control scheme to achieve seamless dynamic transition of dual-active-bridge converters using zero-current prediction," *IEEE Trans. Ind. Electron.*, vol. 69, no. 6, pp. 5826-5834, Jun. 2022.

[8] D. Mou et al, "Hybrid duty modulation for dual active bridge converter to minimize RMS current and extend soft-switching range using the frequency domain analysis," *IEEE Trans. Power Electron.*, vol. 36, no. 4, pp. 4738–4751, Apr. 2021.

[9] D. Mou et al, "Optimal asymmetric duty modulation to minimize inductor peak-to-peak current for dual active bridge DC–DC converter," *IEEE Trans. Power Electron.*, vol. 36, no. 4, pp. 4572–4584, Apr. 2021.

[10] S. Shao et al, "Modeling and advanced control of dual active bridge DCDC converters: A review," *IEEE Trans. Power Electron.*, vol. no. 37, no. 2, pp. 1524–1547, Feb. 2021.

[11] M. Zhou, D. Shu, and H. Wang, "An H5-bridge based laddered *CLLC* DCX with variable DC-link for PEV charging applications," *IEEE Trans. Power Electron.*, vol. 37, no. 4, pp. 4249-4260, Apr. 2022.

[12] D. Shu and H. Wang, "An ultra-wide output range *LLC* resonant converter based on adjustable turns ratio transformer and reconfigurable bridge," *IEEE Trans. Ind. Electron.*, vol. 68, no. 8, pp. 7115-7124, Aug. 2021.

[13] M. Fu, C. Fei, Y. Yang, Q. Li, and F. C. Lee, "A gan-Based DC–DC module for railway applications: Design consideration and high-frequency digital control," *IEEE Trans. Ind. Electron.*, vol. 67, no. 2, pp. 1638–1647, Feb. 2020.

[14] Y. C. Liu, C. Chen, K. D. Chen, Y. L. Syu, and N. A. Dung, "High-frequency and high-efficiency isolated two-stage bidirectional DC–DC converter for residential energy storage systems," *IEEE J. Emerg. Sel. Topics Power Electron.*, vol. 8, no. 3, pp. 1994–2006, Mar. 2020.

[15] G. Fan, X. Wu, T. Liu and Y. Xu, "High-Efficiency High-Density MHz Cellular DC/DC Converter for On-Board Charger," *IEEE Trans. Power Electron.*, vol. 37, no. 12, pp. 15666-15677, Dec. 2022.

[16] Y. Cao, M. Ngo, N. Yan, D. Dong, R. Burgos and A. Ismail, "Design and Implementation of an 18-kW 500-kHz 98.8% Efficiency High-Density Battery Charger With Partial Power Processing," *IEEE J. Emerg. Sel. Topics Power Electron.*, vol. 10, no. 6, pp. 7963-7975, Dec. 2022.

[17] C. Li, M. Zhou and H. Wang, "An H5-Bridge-Based Asymmetric LLC Resonant Converter With an Ultrawide Output Voltage Range," *IEEE Trans. Ind. Electron.*, vol. 67, no. 11, pp. 9503-9514, Nov. 2020.

[18] U. Kundu, K. Yenduri, and P. Sensarma, "Accurate ZVS analysis for magnetic design and efficiency improvement of full-bridge," *IEEE Trans. Power Electron.*, vol. 32, no. 3, pp. 1703–1706, Mar. 2017

A Resonant Reset Forward Converter with Ultra-High Conversion Gain using Differential Transformation Technique (DTT)

Shubham Srivastava
Centre for Automotive Research &
Tribology (CART)
Indian Institute of Technology, Delhi
New Delhi, India
ctz238443@iitd.ac.in

Mandeep S. Rana
Centre for Automotive Research &
Tribology (CART)
Indian Institute of Technology, Delhi
New Delhi, India
mandp@iitd.ac.in

Santanu K. Mishra
Centre for Automotive Research &
Tribology (CART)
Indian Institute of Technology, Delhi
New Delhi, India
skmishra@iitd.ac.in

Abstract—**Achieving ultra-high step-down using high frequency transformers is difficult as the minimum duty cycle limit and leakage inductance of the transformer limits the operation. In order to circumvent these issues a DTT based Resonant Reset Forward Converter is proposed in this paper. The converter is characterized under steady state and a step by step design method to implement the DTT is proposed. Experimental verifications carried out with a 600 V to 12 V/ 200 W converter demonstrates the operation. The peak efficiency recorded is 85 %.**

Keywords—Resonant Reset Forward Converter, Differential Transformation Technique (DTT)

I. INTRODUCTION

Due to increasing demand for power, many critical power distribution systems use medium-/high-voltage distribution bus for efficient power delivery. Auxiliary power supply (APS) takes power from these high voltage distribution bus to bias electronic systems. This conversion from high voltage to bias supply grade voltage level happens in multiple stages as shown in Fig. 1. The step-down stage connected to the high voltage bus normally has galvanic isolation for protection of the electronic circuit against faults. There is a constant endeavor to reduce the number of conversion stages in similar systems. In this paper, a resonant reset forward converter with differential transformation

technique (DTT) is proposed to achieve ultra-high conversion gain in just one conversion stage.

In order to appreciate the need of the proposed circuit, it can be seen that if a very low duty cycle is applied to conventional forward converter in Fig. 2 (a), it leads to duty cycle issues in Fig. 2 (b). On the contrary, if a very high step-down ratio transformer is used, it leads to higher leakage inductance or larger transformer size (to keep $\frac{L_{lk}}{L_m}$ minimum) as shown in Fig. 3. Therefore, a multi-stage approach as shown in Fig. 1 is employed. A review of various high stepdown ratio converter used in literature [1-4] is tabulated in Table I. In this paper, it will be shown that by using DTT based transformers, all the drawbacks can be eliminated, and an ultra-high stepdown forward topology can be implemented in a single stage.

II. PROPOSED TOPOLOGY & OPERATION

The schematic of the proposed resonant reset forward converter with DTT based transformer is shown in Fig. 4. In DTT a single input single output transformer uses three windings; one primary and two secondaries. The secondaries are wound differentially on the core as shown. Due to this connection, the secondary side sees a voltage given by

$$v_{s1} - v_{s2} = \left(\frac{a}{n_1}\right)v_p \tag{1}$$

The resonant forward converter [5] uses a reset capacitor (C_p) across the MOSFET (S_m) and doesn't have a dedicated reset leg as in conventional forward topology. This small

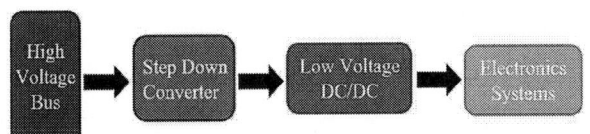

Fig. 2. Block Diagram of the isolated DC/DC converter system.

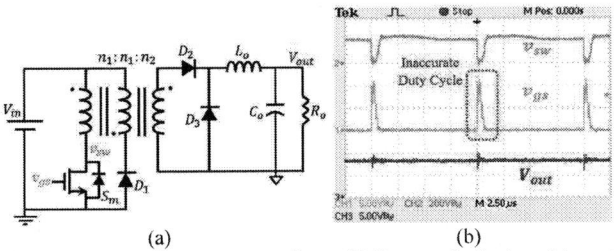

(a)   (b)

Fig. 1. (a) Forward converter topology, (b) Duty cycle limit problem at high conversion ratio ($V_{in} = 200\,V, V_{out} = 8\,V, D = 0.05, n_1{:}n_2 = 4{:}3$).

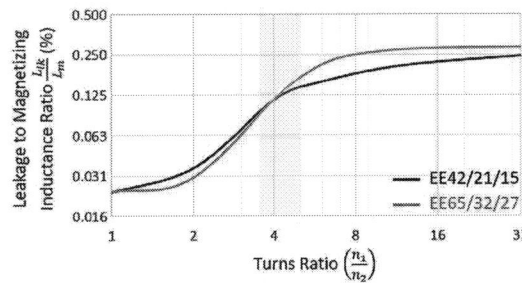

Fig. 3. Variation of leakage to magnetizing inductance ratio $\left(\frac{L_{lk}}{L_m}\right)$ with turns ratio of the transformer.

TABLE I - COMPARISON OF PROPOSED TOPOLOGY WITH THE STATE-OF-THE-ART TOPOLOGIES

Topology	Application	Voltage Ratio (V_{in}/V_{out})	Output Power	Component Count					Peak Efficiency	Switching Frequency
				Switches	Transformers/ Coupled Inductors	Capacitors	Diodes	Inductors		
Modular Flyback [1]	APS in SST	1000/15 V	125 W	9	1	8	1	3	77 %	17 kHz
Flyback [2]	Charge balance in battery management systems	300/15 V	35 W	1	1	1	1	0	~84 %	60 kHz
Flyback [3]	APS in automotive	600/15 V	60 W	1	1	2	1	1	~80 %	140 kHz
Flyback [4]	APS	1000/14 V	56 W	2	1	2	1	0	84.6 %	35 kHz
Proposed	APS	600/12 V	200 W	1	1	2	2	1	85 %	50 kHz
		600/9 V	150 W	1	1	2	2	1	83 %	
		600/5 V	75 W	1	1	2	2	1	75 %	

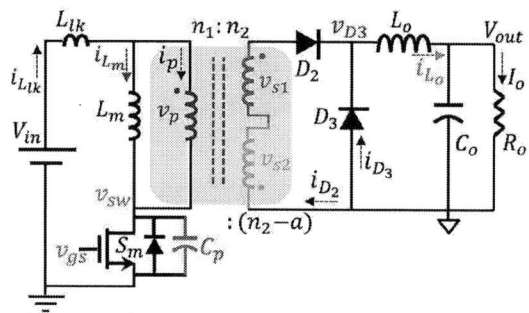

Fig. 4. Schematic of DTT based forward topology.

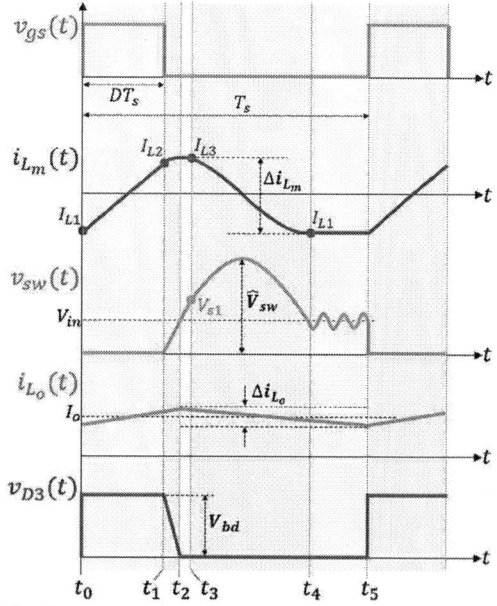

Fig. 5. Steady state waveforms

capacitor (C_p) across the MOSFET (S_m), resonates with the leakage (L_{lk}) and magnetizing inductance (L_m) of the

transformer to reset the flux in the core. The secondary side of the forward converter is similar to a conventional forward topology with diodes $(D_2$ and $D_3)$, output inductor (L_o), capacitor (C_o), and load resistance (R_o).

III. OPERATING INTERVALS

The steady state waveform of the converter in a single switching cycle is shown in Fig. 5. There are five distinct operating intervals in this converter. The equivalent circuit of the converter in all these intervals are given in Table II. The following assumptions are made for the steady state analysis:

1) Small ripple approximation in inductor current i_{L_o}, i.e., $\Delta i_{L_o} = 0$.

2) Input voltage V_{in}, output voltage V_{out}, and output current I_o are all constants in a switching cycle.

3) The operation of switch (S_m), and diodes $(D_2$ and $D_3)$ is ideal.

4) Initial conditions are: $i_{L_m}(t_0) = I_{L1}$, $v_{sw}(t_0) = 0$, and $i_{L_o}(t_0) = I_o$.

a) *Interval I* $(t_0 < t < t_1)$:

Interval I is initiated by turning ON the switch (S_m). The DC input voltage V_{in} is applied across the primary winding as shown in the equivalent circuit during interval I, ref to Table II. The current i_{L_m} will rise linearly from I_{L1} to I_{L2} given as,

$$i_{L_m}(t) = \frac{V_{in}}{(L_m + L_{lk})}(t - t_0) + I_{L1} \qquad (2)$$

The value of current i_{L_m} at the end of this interval will be,

$$I_{L2} = \frac{V_{in}}{(L_m + L_{lk})}DT_s + I_{L1} \qquad (3)$$

where, D is the duty cycle, and T_s is the switching period for applied gate pulse v_{gs}.

TABLE II - OPERATING INTERVALS IN STEADY STATE

	Operating circuit	Primary Side	Secondary Side
Interval I	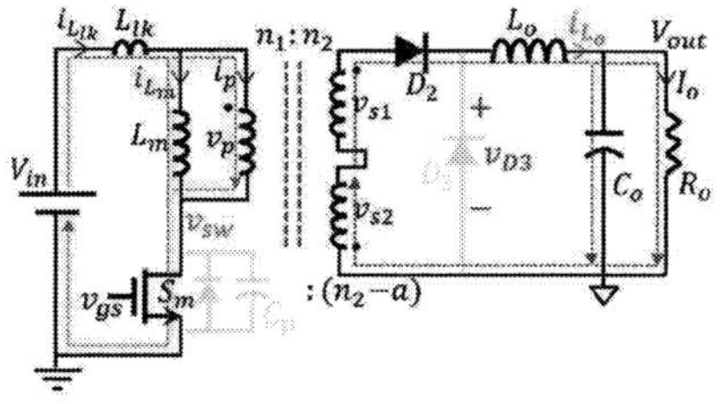	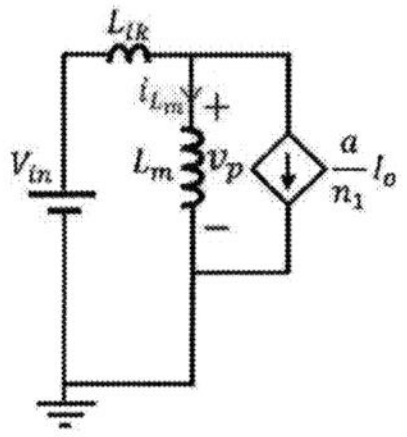	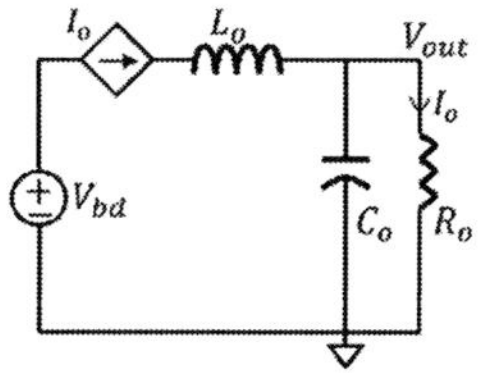
Interval II	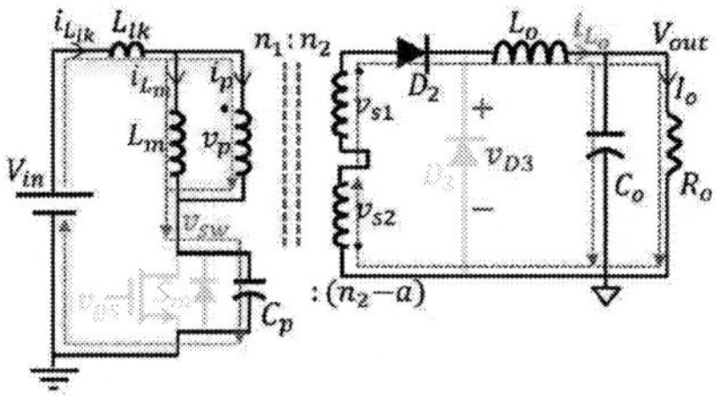	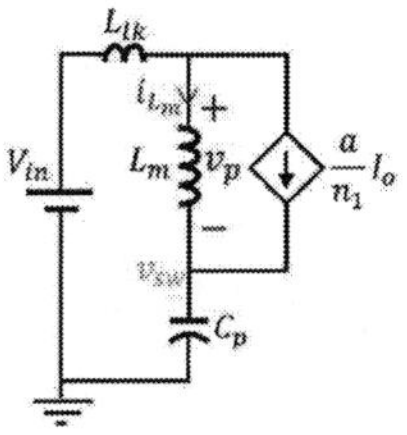	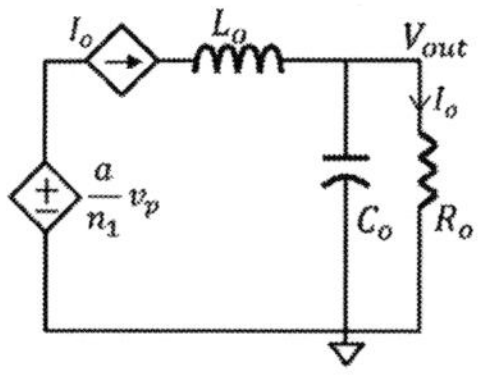
Interval III	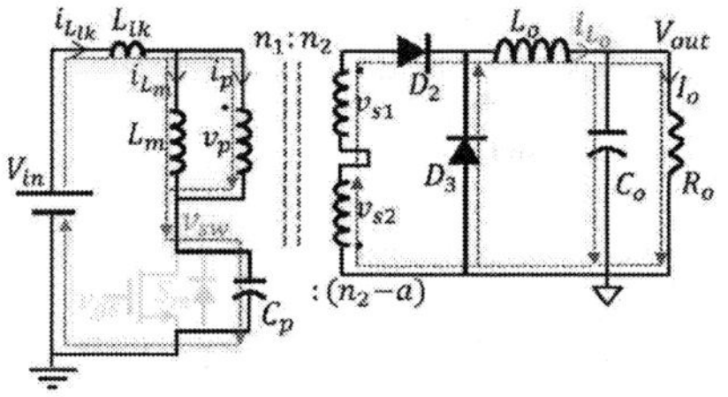	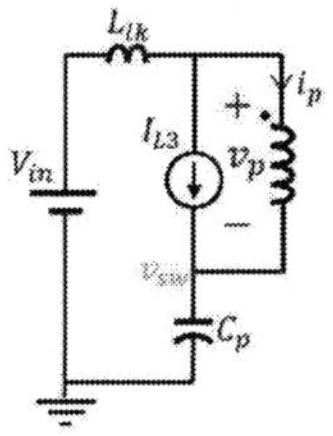	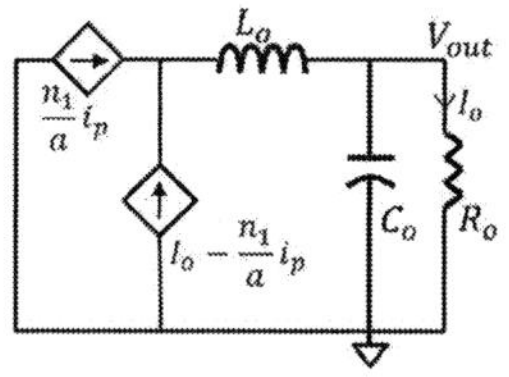
Interval IV	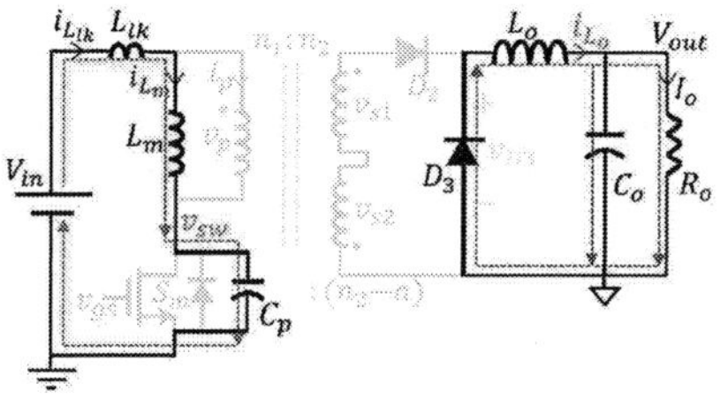	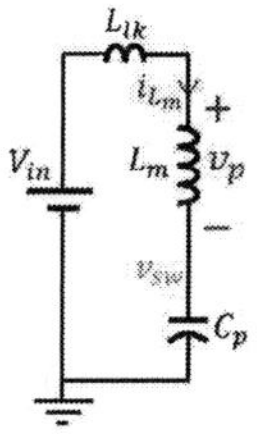	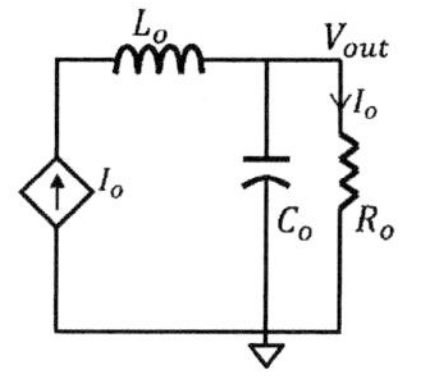
Interval V	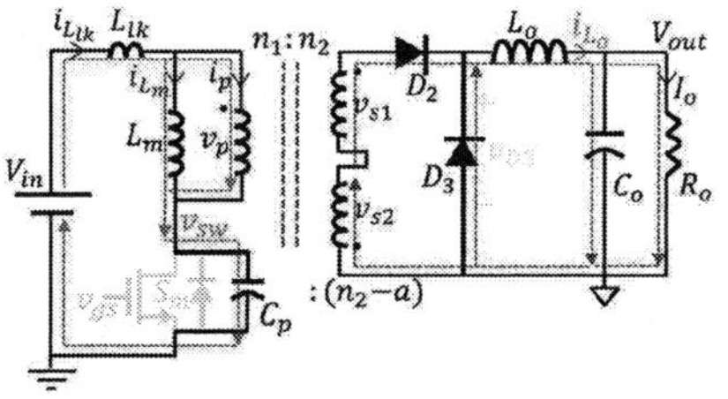	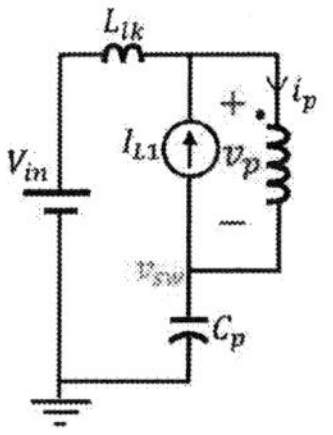	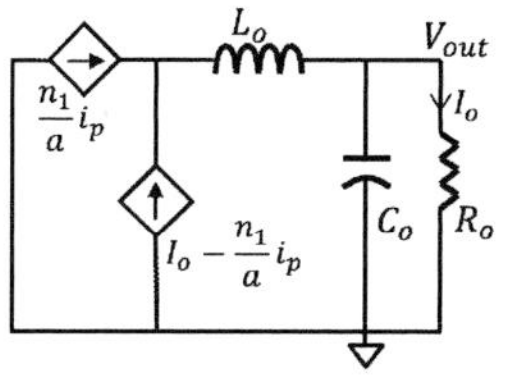

In the secondary side, current i_{D_3} becomes zero as diode (D_3) is reverse biased. The diode (D_2) carries the load current I_o throughout this interval. The blocking voltage for diode (D_3) reflected from the primary side is given as,

$$v_{D3}(t) = V_{bd} = \left(\frac{a}{n_1}\right)\frac{V_{in}}{\left(1 + \frac{L_{lk}}{L_m}\right)} \tag{4}$$

b) *Interval II ($t_1 < t < t_2$):*

Interval II is initiated as gate-to-source voltage v_{gs} falls to zero to turn OFF the switch (S_m). As per the equivalent circuit during interval II, ref to Table II, the current established in the primary winding will start flowing through capacitor (C_p). The expressions for current i_{L_m} and voltage v_{sw} are given as,

979-8-3315-1612-3/25 $31.00 © 2025 IEEE

$$i_{L_m}(t) = \frac{V_{in}}{Z_r} \sin \omega_r (t - t_1) + \left(\frac{a}{n_1} I_o + I_{L2}\right) \cos \omega_r (t - t_1)$$
$$- \frac{a}{n_1} I_o \tag{5}$$

$$v_{sw}(t) = V_{in}\{1 - \cos \omega_r (t - t_1)\}$$
$$+ \left(\frac{a}{n_1} I_o + I_{L2}\right) Z_r \sin \omega_r (t - t_1) \tag{6}$$

where, $\quad Z_r = \sqrt{\frac{(L_m + L_{lk})}{C_p}}$, and $\omega_r = \frac{1}{\sqrt{(L_m + L_{lk})C_p}} \tag{7}$

The voltage v_{sw} starts to increase as per (6), and the interval ends as soon as it becomes equal to input voltage V_{in}. The duration of this interval is given as,

$$(t_2 - t_1) = \frac{1}{\omega_r} tan^{-1} \left\{ \frac{V_{in}}{\left(\frac{a}{n_1} I_o + I_{L2}\right) Z_r} \right\} \tag{8}$$

The value of current i_{L_m} at the end of this interval will be,

$$I_{L3} = \frac{\sqrt{V_{in}^2 + \left(\frac{a}{n_1} I_o + I_{L2}\right)^2 (Z_r)^2}}{Z_r} - \frac{a}{n_1} I_o \tag{9}$$

In the secondary side, voltage v_{D3} gradually falls to zero as shown in Fig. 5. The diode (D_2) continues carrying the load current I_o in this interval.

c) *Interval III* $(t_2 < t < t_3)$:

In interval III, voltage v_{sw} increases further beyond input voltage V_{in} causing the current in leakage inductance (L_{lk}) to decrease gradually. The leakage inductance (L_{lk}) carries the summation of currents i_{L_m}, and i_p as shown in Fig. 4. The change in current i_{L_m} is very negligible compared to current i_p therefore, i_{L_m} is considered almost constant ($= I_{L3}$) throughout this interval as shown in the equivalent circuit, ref to Table II. The expression for current i_p can be obtained as,

$$i_p(t) = \left(\frac{a}{n_1} I_o + I_{L3}\right) \cos \omega_{lk}(t - t_2) - I_{L3} \tag{10}$$

where, $\omega_{lk} = \frac{1}{\sqrt{L_{lk}C_p}}$. As soon as the current i_p falls to zero, this interval ends. The duration of this interval is given as,

$$(t_3 - t_2) = \frac{1}{\omega_{lk}} cos^{-1} \left\{ \frac{1}{\left(1 + \frac{a}{n_1} \frac{I_o}{I_{L3}}\right)} \right\} \tag{11}$$

The switch node voltage v_{sw} reaches to value V_{s1} given by,

$$V_{s1} = V_{in} + Z_{lk} I_{L3} \sqrt{\left(1 + \frac{a}{n_1} \frac{I_o}{I_{L3}}\right)^2 - 1} \tag{12}$$

where, $Z_{lk} = \sqrt{\frac{L_{lk}}{C_p}}$.

In the secondary side, both the diodes (D_2 and D_3) share the load current I_o such that their respective currents i_{D_2}, and i_{D_3} are given as,

$$i_{D_2}(t) = \frac{n_1}{a} i_p(t) \tag{13}$$

$$i_{D_3}(t) = I_o - \frac{n_1}{a} i_p(t) \tag{14}$$

d) *Interval IV* $(t_3 < t < t_4)$:

The equivalent circuit during interval IV is shown in Table II. In this interval, resonance between capacitor (C_p) and the combination of inductances (L_m and L_{lk}), resets the flux in the core. The expressions for current i_{L_m} and voltage v_{sw} are given as,

$$i_{L_m}(t) = \left\{\frac{V_{in} - V_{s1}}{Z_r}\right\} \sin \omega_r (t - t_3) + I_{L3} \cos \omega_r (t - t_3) \tag{15}$$

$$v_{sw}(t) = V_{in} + I_{L3} Z_r \sin \omega_r (t - t_3)$$
$$+ (V_{s1} - V_{in}) \cos \omega_r (t - t_3) \tag{16}$$

The peak voltage across the switch is a critical parameter to design and it is calculated as,

$$\hat{V}_{sw} = V_{in} + \sqrt{(I_{L3} Z_r)^2 + (V_{s1} - V_{in})^2} \tag{17}$$

This interval lasts till the voltage v_{sw} becomes equal to input voltage V_{in} following (16). The duration of this interval is given as,

$$(t_4 - t_3) = \frac{1}{\omega_r} \left[\pi - tan^{-1} \left\{ \frac{(V_{s1} - V_{in})}{I_{L3} Z_r} \right\} \right] \tag{18}$$

The value of current i_{L_m} at the end of this interval resets to value I_{L1} given as,

$$I_{L1} = -\frac{\sqrt{(V_{s1} - V_{in})^2 + (I_{L3} Z_r)^2}}{Z_r} \tag{19}$$

In the secondary side, current i_{D_2} becomes zero as diode (D_2) is reverse biased. The diode (D_3) carries the load current I_o throughout this interval. The peak blocking voltage $v_{D2,peak}$ across the diode (D_2) will be given as,

$$v_{D2,peak} = \left(\frac{a}{n_1}\right) \frac{(\hat{V}_{sw} - V_{in})}{\left(1 + \frac{L_{lk}}{L_m}\right)} \tag{20}$$

e) *Interval V* $(t_4 < t < t_5)$:

In interval V, the average of switch node voltage v_{sw} is clamped to V_{in}. However, leakage inductance (L_{lk}) and capacitor (C_p) resonates generating LC oscillations in voltage

979-8-3315-1612-3/25 $31.00 © 2025 IEEE

v_{sw} as shown in Fig. 5. The equivalent circuit during interval V is shown in Table II. The governing equations are expressed as,

$$i_{L_{lk}}(t) = I_{L1} \cos \omega_{lk}(t - t_4) \tag{21}$$

$$v_{sw}(t) = V_{in} + I_{L1} Z_{lk} \sin \omega_{lk}(t - t_4) \tag{22}$$

The equivalent schematic on the secondary side of the converter is same as in interval III, ref to Table II. The expressions for currents i_{D_2}, and i_{D_3} are given as (13), and (14), respectively as both the diodes (D_2 and D_3) conduct during this interval. The end of this interval marks the completion of one switching cycle.

The DC output voltage V_{out} can be obtained by applying the principle of volt-sec balance across the output inductor (L_o). The expression for V_{out} is given as follows,

$$V_{out} = \left\{ \frac{DV_{in}}{(L_m + L_{lk})f_{sw}} + \sqrt{\left(\frac{V_{in}}{Z_r}\right)^2 + \left(\frac{a}{n_1}I_o + I_{L2}\right)^2} \right.$$
$$\left. - \left(\frac{a}{n_1}I_o + I_{L2}\right)\right\} \left(\frac{a}{n_1}\right)(L_m f_{sw}) \tag{23}$$

The term $\frac{V_{in}}{Z_r}$ is less dominant as compared to $\left(\frac{a}{n_1}I_o + I_{L2}\right)$, hence the expression can be simplified as,

$$V_{out} = \left(\frac{a}{n_1}\right)\frac{DV_{in}}{\left(1 + \frac{L_{lk}}{L_m}\right)} \tag{24}$$

As can be noted, the leakage and magnetizing inductance have a considerable effect on the conversion ratio. Higher the ratio between leakage and magnetizing inductance, lower is the conversion ratio.

IV. Differential Transformer Design

The design of the differential transformer is extremely crucial in the proposed topology, and it is different from a conventional transformer design. Based on the curve shown in Fig. 3, it is observed that choosing $n_1{:}n_2$ in between 3.5~5 suffices both the requirements of optimum size of the transformer and nominal leakage inductance.

Since, the operation of the converter is dependent on the value of magnetizing inductance (L_m); therefore, a coupled inductor based approach is followed for the transformer design. Also, the peak-to-peak ripple $\left(\Delta i_{L_m} = I_{L3} - I_{L1}\right)$ in current i_{L_m} is significant as shown in Fig. 5. Hence, AC flux density cannot be neglected in this coupled inductor based transformer design. The three major objectives of the design methodology are:

1) Obtain desired value of the magnetizing inductance (L_m).

2) Limit the maximum operating flux density (B_{max}) in the core below its saturation.

3) Limit the power density of the converter by choosing the minimum possible size of the core.

TABLE III - SUMMARIZATION OF TRANSFORMER DESIGN

Step-1: Calculate the value of Kg

$$K_g \geq \frac{\rho(L_m)^2(i_{tot})^2(i_{L_{m,max}})^2}{(B_{max})^2 P_{cu} K_u}$$

Choose a core to satisfy this inequality.

Step-2: Core Parameters

Note down the values of following core parameters:

$$MLT, A_c, W_A, \text{ and } l_m$$

Step-3: Determine air-gap length l_g

$$l_g = \frac{\mu_o L_m (i_{L_{m,max}})^2}{(B_{max})^2 A_c} - \frac{l_m}{\mu_r}$$

If $l_g < 10$ µm, then consider a minimum value of $l_g = 10$ µm.

Step-4: Determine the turns n_1 in primary winding

$$n_1 = \sqrt{\frac{L_m \left(\frac{l_m}{\mu_r} + l_g\right)}{\mu_o A_c}}$$

Round-off the obtained value to the nearest whole number.

Step-5: B_{max} check

$$B_{max} = \frac{L_m}{n_1 A_c}\left(I_{L_m} + \frac{\Delta i_{L_m}}{2}\right)$$

Check if $B_{max} < B_{sat}$, then proceed to next step. Otherwise, choose a larger core and repeat from step-2.

Step-6: Determine the turns in secondaries

$$a = \left(\frac{a}{n_1}\right) n_1$$

$$n_2 = \left(\frac{n_2}{n_1}\right) n_1$$

Step-7: Selection of the wire size

Choose appropriate wire for the three windings based on their current rating [6].

Step-8: Convergence test

$$K_u = \frac{n_1 A_{wp} + n_2 A_{ws1} + (n_2 - a)A_{ws2}}{W_A}$$

Typically, if $K_u < 0.5$, then the design converges. Otherwise, choose a larger core and repeat from step-2.

Nomenclature:

ρ = resistivity of copper, i_{tot} = total rms current, ref to primary, $i_{L_{m,max}}$ = peak value of magnetizing current ($= \max[|I_{L_1}|, |I_{L_3}|]$), Δi_{L_m} = peak-to-peak ripple in magnetizing current, I_{L_m} = average value of magnetizing current, B_{max} = maximum flux density, B_{sat} = saturation flux density, P_{cu} = total copper loss, K_u = winding fill factor, MLT = mean length per turn of the core, A_c = cross-sectional area of the core, W_A = window area of the core, l_m = magnetic path length of the core, μ_o = permeability of air, μ_r = relative permeability of the core, l_g = air-gap length, A_{wp} = conductor area of primary winding, A_{ws1} = conductor area of secondary1 winding, and A_{ws2} = conductor area of secondary2 winding.

The methodology follows an iterative process with initial step of selection of the core using core geometry constant (K_g) method as per [7]. Table III shows the step-by-step transformer design method. The value of K_g was found to be 3.33E-02 cm⁵ in step-1, ref to Table III. Based on this value, after several iterations to achieve, the required value of magnetizing inductance (L_m), and winding fill factor (K_u) less than 0.5, EE42/21/20 core was selected for the transformer design. The dimensions and properties of the core can be found in [8]. The values of all the obtained parameters for transformer are tabulated in Table IV.

V. EXPERIMENT RESULTS

The proposed converter is validated using a laboratory prototype as per the electrical specifications in Table IV. The steady state waveforms of the converter are obtained for 200 W load. Fig. 6 shows the gate pulse v_{gs} having a duty cycle $D = 0.3$ for a step down from 600 V input to 12 V output. The average inductor current is around 16 A with a peak to peak ripple (Δi_{L_o}) of 14 A.

Fig. 7 shows one switching cycle of the switch node voltage (v_{sw}). The five intervals shown in Fig. 5 are distinctly identified in the experiment. The duration of interval II and III comes out

TABLE IV - SPECIFICATIONS OF THE CONVERTER

Parameter	Specification
Input Voltage (V_{in})	600 V
Output Voltage (V_{out})	12 V
Output Power (P_o)	200 W
Nominal Duty Cycle (D)	0.3
Switching Frequency (f_{sw})	50 kHz
Capacitor (C_p)	0.8 nF
Leakage Inductance (L_{lk})	0.15 mH
Magnetizing Inductance (L_m)	11.6 mH
Turns in primary winding (n_1)	50
Turns in secondary1 winding (n_2)	14
Difference in secondaries' turns (a)	4

to be 240 ns and 500 ns which closely matches as obtained from (8) and (11), respectively. In interval IV, the switch node voltage v_{sw} attains its peak value ($= \hat{V}_{sw}$), observed to be 1.44 kV, which is in close agreement with (17). The duration of interval IV is 6 μs, which closely matches as obtained from (18). In interval V, the angular frequency of oscillation $\omega_{lk} = 2.96 * 10^6$ rad/s, closely matches for the values of leakage inductance (L_{lk}) and capacitor (C_p) as defined in Table IV.

The partial-ZVS switching is proven in Fig. 8. It can be seen that the switch node voltage v_{sw} falls to zero before the current $i_{L_{lk}}$ starts to rise. This leads to low loss and higher efficiency of this topology compared to the existing options given in Table I.

A comparative analysis for the efficiency achieved in high step-down operation for existing topologies is presented in Table I. Generally, a resonant reset forward design at the given power level has an efficiency of nearly 80% [5]. Fig. 9 shows the efficiency of the proposed converter verses load current I_o for a constant 12 V output voltage. A peak efficiency of 85% is

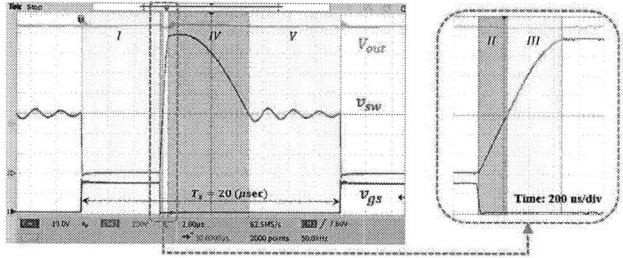

Fig. 7. Time duration for five distinct operating intervals in steady state.

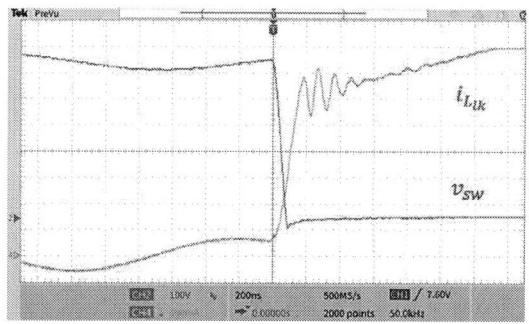

Fig. 8. ZVS during turn-on of switch (S_m).

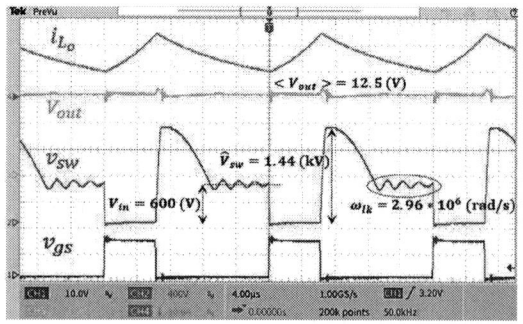

Fig. 6. Experiment results showing waveforms of i_{L_o}, V_{out}, v_{sw}, and v_{gs} for the proposed converter at $V_{in} = 600$ V.

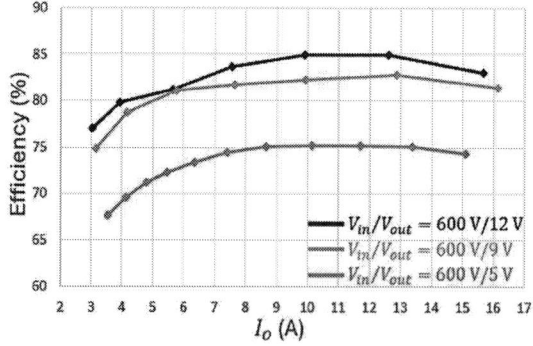

Fig. 9. Efficiency curve for the proposed converter.

observed for the output current $I_o = 12\ A$, which is comparatively higher than the state-of-the-art topologies.

The performance of the designed hardware is also evaluated by lowering the duty cycle. As the step down ratio increases the peak efficiency of the converter decreases. The efficiency of the proposed converter for two other high step-down ratios in voltage, (V_{in}/V_{out}): (i) 600 V/ 9 V, and (ii) 600 V/5 V is also shown in Fig. 9.

VI. CONCLUSION

An ultra-high step-down resonant reset forward converter is proposed and implemented in this paper. The converter utilized a novel differential transformation technique (DTT), to achieve this high conversion ratio. The modified transformer design achieved this high step-down ratio with minimal leakage. Also, the converter avoided the duty cycle limit in obtaining 12 V output from 600 V input. The five distinct operating intervals of the topology during steady state are discussed in detail along with the characterization of the performance. The paper also included a detailed methodology for the design of the differential transformer. The proposed topology is validated using an experimental prototype for up to 200 W output power. The maximum efficiency of the proposed converter is recorded to be 85 %.

REFERENCES

[1] J. Liu, S. Zhong, J. Zhang, Y. Ai, N. Zhao and J. Yang, "Auxiliary Power Supply for Medium-/High-Voltage and High-Power Solid-State Transformers," in *IEEE Transactions on Power Electronics*, vol. 35, no. 5, pp. 4791-4803, May 2020.

[2] Do-Hyun Kim, Sol Moon, Chan-In Kim and Joung-Hu Park, "High efficiency step-down flyback converter using coaxial cable transformer," Proceedings of the 7th International Power Electronics and Motion Control Conference, Harbin, 2012, pp. 1855-1858.

[3] Texas Instruments, "Automotive 40-V to 1-kV Input Flyback Reference Design Supporting Regenerative Braking Test," Design Guide: TIDA-01505, August 2019.

[4] Texas Instruments, "350-V to 1000-V DC Input, 56-W Flyback Isolated Power Supply Reference Design," Test Report: PMP41009, April 2022.

[5] Intersil, "Resonant Reset Forward Converters for Low Power Applications," ISL6721EVAL3Z Application Note, Sept. 2009.

[6] Solaris, "American Wire Gauge Conductor Size Table," [Online]. Available: https://www.solaris-shop.com/content/American%20Wire%20Gauge%20Conductor%20Size%20Table.pdf

[7] R. W. Erickson and D. Maksimovic, Fundamentals of Power Electronics, 2nd ed. Dordrecht, Netherlands: Kluwer Academic Publishers Group, 2001.

[8] TDK Electronics AG, Ferrites and accessories, [Online]. Available: https://www.tdk-electronics.tdk.com/inf/80/db/fer/e_42_21_20.pdf

Full-Range ZVS Modulation of Switched Capacitor Converter for Sensorless Voltage Balancing

Md Tanvir Ahammed
FREEDM Systems Center
North Carolina State University
Raleigh, USA
mahamme@ncsu.edu

Wensong Yu
FREEDM Systems Center
North Carolina State University
Raleigh, USA
wyu2@ncsu.edu

Abstract—This paper presents a new sensorless full-range zero-voltage switching (ZVS) modulation technique for a resonant switched-capacitor converter (RSCC) using a constant non-symmetrical duty cycle control and demonstrated as a DC-link voltage balancing circuit in split-phase inverter applications. A DC-link voltage balancer is required in a split-phase inverter to compensate for load imbalances between the two output phases, and it typically operates under slightly unbalanced or light-load conditions. While the existing phase-shift modulation of the RSCC can achieve high efficiency with ZVS operation, it has inherent non-ZVS regions and a notable efficiency decline at light loads. A decline in light load efficiency is also evident with the typical zero-current switching (ZCS) modulation of the RSCC. As a solution, in the proposed sensorless ZVS modulation, the minimum required ZVS current is maintained across the entire load range using a constant non-symmetrical duty cycle control, ensuring ZVS operation for all switching devices even during load transients. The proposed ZVS modulation is verified in simulation and experimental results from a 5-kW hardware prototype confirmed the full-range ZVS operation of the RSCC and efficiency improvement of the split-phase inverter for both balanced and unbalanced load conditions across the entire load range.

Index Terms—Resonant switched-capacitor converter, non-symmetrical duty, sensorless modulation, split-phase inverter, light-load efficiency, zero-voltage switching (ZVS).

I. INTRODUCTION

Split-phase inverters are used to generate two different single-phase voltages (e.g., 120V and 240V), where the phase voltages are equal in amplitude and frequency but 180° out of phase, and the line-to-line voltage is twice the phase voltage [1]. The DC-link of the split-phase inverter uses two series-connected DC capacitors to split the DC-link into two. While the total DC-link voltage across the two capacitors equals the overall DC-link voltage, the voltage on each capacitor may deviate from half of the DC-link voltage due to load imbalances on the inverter side. This imbalance can complicate the regulation of the split AC voltages and potentially cause overvoltage across the switching devices. To address this issue, a DC-link voltage balancer circuit is typically incorporated to compensate for the load imbalance. The conventional buck-boost-based DC-link voltage balancer, using a two-level leg [1], [2], is the most popular one, primarily because it requires fewer semiconductor devices. However, its application is often often constrained by the need for bulky inductors [2]. Another

commonly used topology is the neutral-point-clamped (NPC) leg-based DC-link voltage balancer, which has been proposed and demonstrated for applications in split-phase inverters [1] and multiphase NPC converters [3]. Despite its advantages, the NPC leg-based voltage balancer typically requires additional hardware or complex control strategies [3].

In recent years, resonant switched-capacitor converters (RSCCs) have gained significant attention for their capability to enable bidirectional power flow while simultaneously achieving exceptionally high efficiency and power density [4]. The inductor size in RSCCs is significantly smaller than that of a conventional buck converter, as the converter primarily stores energy in the switched capacitor, similar to the switched-capacitor converter (SCC) [5]. Another notable advantage of RSCCs is the control flexibility, allowing operation in either zero-current switching (ZCS) mode or zero-voltage switching (ZVS) mode. The RSCCs naturally achieve zero-current switching (ZCS) for all switches when the switching frequency is near the resonant tank frequency [6], while ZVS for all switches is achieved with phase-shift control when the switching frequency is much higher than the resonant tank frequency [5]. As a result, RSCCs are widely employed in various applications [4] - [10], including battery cell voltage balancing systems [6], DC-link voltage balancing [7], DC microgrids [8], point-of-load power supplies [9], step-down [4] and step-up [10] applications.

An RSCC with phase-shift control has been proposed and demonstrated in a multilevel inverter for split DC voltage balancing [7], which improves conversion efficiency through ZVS operation. However, in phase-shift control, the switching frequency is significantly higher than the resonant frequency and the RSCC can't achieve ZVS at light loads. As a consequence, the frequency-dependent losses, such as switching losses caused by the output capacitance of MOSFETs become notable, significantly diminishing the efficiency of the RSCC when operating under no-load to light-load conditions. The efficiency drop at light loads is also noticeable in the typical ZCS mode of operation [6]. However, the ZCS mode maintains higher efficiency than the ZVS mode with phase-shift control under light-load conditions, as the switching frequency in ZCS mode is significantly lower. A full-range ZVS modulation has been demonstrated for the RSCC with the inductor placement

979-8-3315-1612-3/25 $31.00 © 2025 IEEE

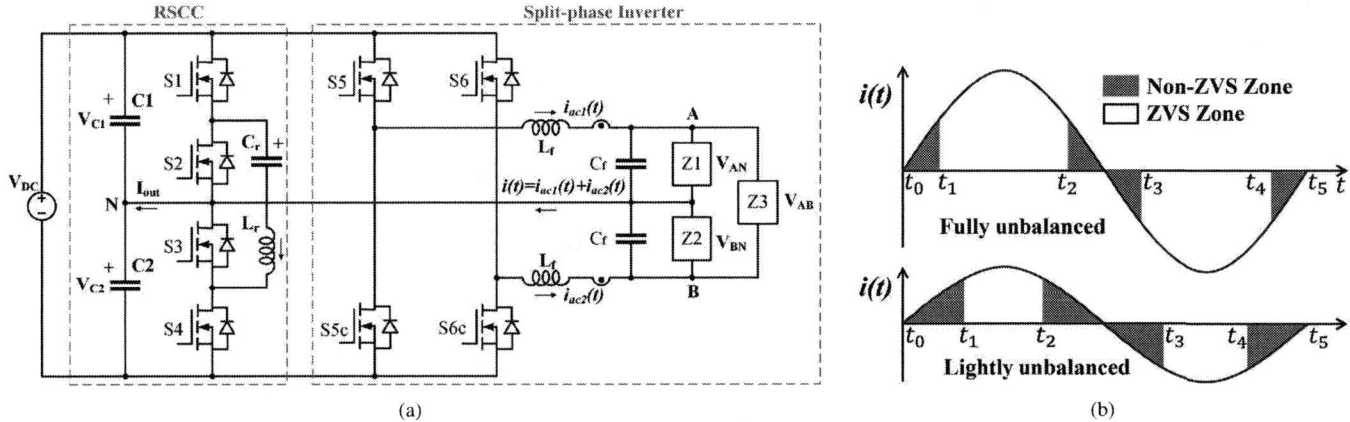

Fig. 1. (a) Circuit configuration of the split-phase inverter. (b) Load current seen by the RSCC.

at the output node [4]. However, a drawback of such inductor placement is that the resonant capacitor ripple voltage, which is load-dependent and will appear across the switching devices. Therefore, special consideration is required while selecting switching devices. Conversely, placing the inductor in series with the resonant capacitor can shield the switch stress from the resonant capacitor ripple voltage.

This paper presents a sensorless full-range ZVS modulation for the RSCC where the resonant inductor is connected in series with the resonant capacitor. The approach employs a constant switching frequency and a constant non-symmetrical duty cycle. The resonant tank circuit determines the switching frequency, while the duty cycle control ensures ZVS operation for all switches. The constant non-symmetrical duty cycle introduces additional resonant current boost subintervals, enabling the minimum required ZVS current to be maintained across the entire load range, from no load to full load, thereby achieving ZVS for all MOSFETs within the RSCC. A mathematical model is developed to derive the conditions necessary to satisfy ZVS requirements. The effectiveness of the approach is validated through both simulation and experimental results. Experimental tests conducted on a 5-kW hardware prototype confirm the efficiency improvements of the split-phase inverter. The rest of the paper is organized as follows. Section II describes the RSCC-based DC-link voltage balancing circuit. The proposed full-range ZVS modulation is described in Section III. Section IV presents the experimental results and finally, conclusions are made in Section V.

II. RSCC-BASED VOLTAGE BALANCING CIRCUIT

The circuit configuration of the proposed split-phase inverter with the RSCC as a DC-link voltage balancer is shown in Fig. 1(a). The split-phase inverter (within the blue dashed lines) comprises two half-bridges: one formed by switches $S5$ and $S5c$, and the other by $S6$ and $S6c$. Both are modulated using sinusoidal pulse width modulation (SPWM), with the second inverter's carrier and modulating signals phase-shifted $180°$ from the first. This results in phase voltages V_{AN} and V_{BN} that are $180°$ out of phase, yielding a line-line voltage V_{AB}

equal to their sum. The DC-link consists of capacitors C_1 and C_2. While their combined voltage is V_{DC}, unbalanced loads can cause voltage deviations from the ideal $V_{DC}/2$ per capacitor. To address this, the RSCC is incorporated to maintain a balanced voltage of $V_{DC}/2$ across each capacitor, ensuring well-regulated split-phase voltage generation even under fully unbalanced load conditions.

The configuration of the RSCC-based voltage balancing circuit is shown in Fig. 1(a) (inside the red dashed line). It consists of two half-bridges: switches $S1$ and $S2$ form the upper half-bridge, while switches $S3$ and $S4$ form the lower half-bridge. The midpoint of each bridge is connected through a series resonant circuit comprising L_r and C_r. The four switching devices facilitate bidirectional power flow between capacitors C_1 and C_2. The instantaneous load current seen by the RSCC is the summation of the two inverter currents as given in (1).

$$i(t) = i_{ac1}(t) + i_{ac2}(t) \qquad (1)$$

This implies that the RSCC's load current is $60Hz$, with amplitude determined by the split-phase inverter's load. If one inverter is at no load and the other is operating at its rated load, the load current seen by the RSCC will be at its maximum value. In contrast, if both inverters are loaded equally, the load current seen by the RSCC will be zero. The load current seen by the RSCC under unbalanced load conditions is shown in Fig. 1(b), revealing four non-ZVS zones within a $60Hz$ fundamental cycle. Under balanced load conditions, ideally, $i(t) = 0$, meaning there will be no ZVS regions for the RSCC.

Therefore, the full-range ZVS modulation for the RSCC can significantly improve the efficiency of the RSCC-based DC-link voltage balancer by ensuring ZVS operation throughout the entire fundamental cycle, thereby further enhancing the overall efficiency of the split-phase inverter.

III. PROPOSED FULL-RANGE ZVS MODULATION

The PWM scheme of the existing phase-shift modulation (PSM), ZCS modulation, and the proposed ZVS modulation for the RSCC are shown in Fig. 2(a), 2(b), and 2(c), respectively. Fig. 2 is drawn with the assumption that $V_{C1} > V_{C2}$.

979-8-3315-1612-3/25 $31.00 © 2025 IEEE

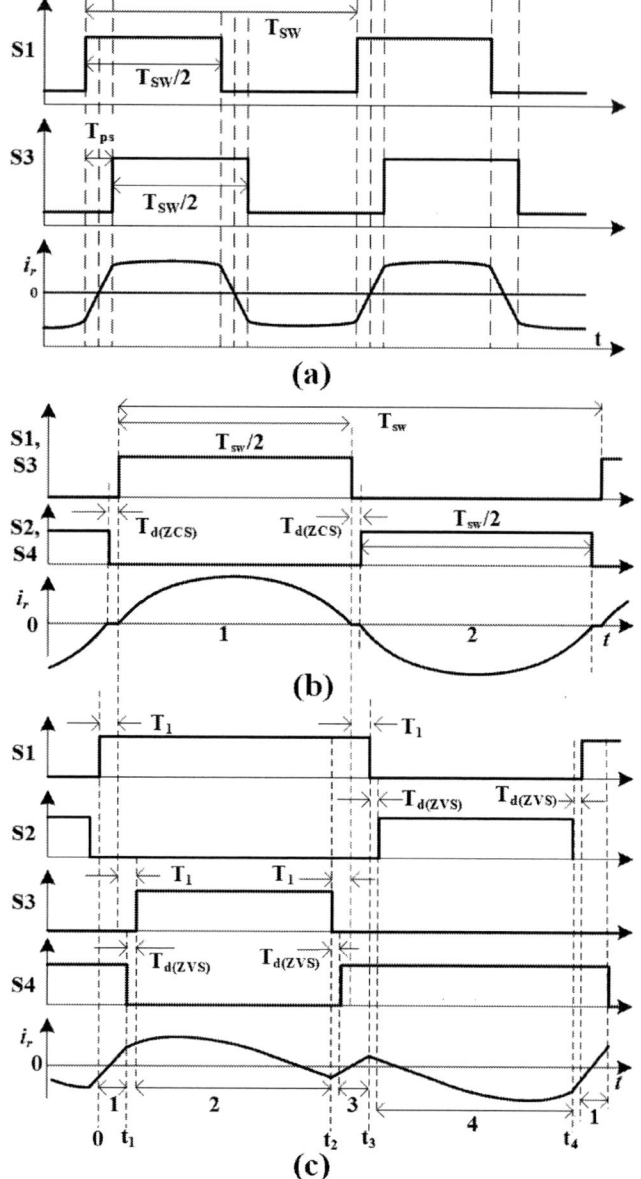

Fig. 2. (a) PSM. (b) ZCS modulation. (c) Proposed ZVS modulation.

In PSM, all switches ($S1$, $S2$, $S3$, $S4$) are operated at constant 50% duty with a constant switching frequency which is much higher than the resonant frequency [5]. The power transfer is achieved by regulating the phase-shift time (T_{PS}). To accomplish this, the DC-link capacitor voltages V_{C1} and V_{C2} are sensed and the voltage difference is fed to a PI (proportional-integral) controller, which generates the reference phase-shift time (T_{PS}). Unlike PSM, the proposed ZVS modulation of the RSCC is very similar to the typical ZCS modulation [6] where natural voltage balancing is achieved without any closed loop control. In the ZCS modulation, the switching frequency (f_{SW}) is constant and kept slightly higher than the resonant tank frequency to account for tolerances in the tank circuit, and all switches are operated at constant 50%

duty as illustrated in Fig. 2(b). Similar to the ZCS modulation, the switching frequency in the proposed modulation is also constant and slightly higher than the resonant tank frequency (f_r). The gating sequence and the resonant current (i_r) waveforms are shown in Fig. 2(c). Compared to the ZCS modulation (Fig. 2(b)), the proposed ZVS modulation introduces additional boost subintervals to ensure a minimum ZVS current. To achieve this, the duty cycle of the upper and lower half-bridges is no longer fixed at 50%; instead, it is made non-symmetrical. The duty cycle of $S1$ is extended from $T_{SW}/2$ to $T_{SW}/2 + 2T_1$ while the duty cycle of $S3$ is reduced by the same amount, setting it to $T_{SW}/2 - 2T_1$. Switches $S2$ and $S4$ are operated in complementary logic to $S1$ and $S3$, respectively with a dead band time of $T_{d(ZVS)}$. With this modification, the proposed modulation introduces two additional resonant current boost stages during Mode-1 and Mode-3, which are load-independent and can ensure the minimum required ZVS current across all loading conditions. During Mode-2, C_1 is discharged by the resonant current i_r, while in Mode-4, C_2 is charged by i_r. As a result, the proposed ZVS modulation inherently achieves voltage balancing without relying on any sensing or feedback control.

A. Steady-state Circuit Analysis

The transition of switching modes including the direction of the resonant current during the dead time, the charge and discharge process of MOSFETs' output capacitance (C_{oss}), and the ZVS turn-on operation of the MOSFETs are illustrated in Fig. 3.

1) Mode-1 $[0 \leq t \leq t_1]$: Switches $S1$ and $S4$ are on, V_{DC} is applied across the resonant tank circuit. The resonant current $i_r(t)$ is expressed in (2).

$$i_r(t) = I_a \sin\omega_r t + I_m \sin\omega_r t \qquad (2)$$

$$I_a = \frac{V_{DC} - \langle V_{Cr} \rangle}{Z} \qquad (3)$$

$$I_m = \frac{V_{DC}/2 - V_{Cr0}}{Z} \qquad (4)$$

$$Z = \sqrt{R_{cir}^2 + (\omega_r L_r)^2} \qquad (5)$$

Where I_a is the amplitude of the additional current introduced by the proposed ZVS modulation which is load-independent. I_m is the amplitude of the resonant current seen in typical ZCS modulation ($f_{SW} = f_r$ with 50% duty) which is load dependent. $\langle V_{Cr} \rangle = V_{DC}/2$ is the average voltage of the resonant capacitor C_r; $\omega_r = 1/\sqrt{L_r C_r}$; $R_{cir} = 2R_{ds} + R_{LC}$; R_{ds} is the on-state resistance of MOSFET, and R_{LC} is the parasitic resistances of the resonant tank circuit. V_{Cr0} is the initial voltage of the resonant capacitor C_r at $t = 0$. If I_{out} is the average rectified load current then V_{Cr0} can be expressed as given in (6).

$$V_{Cr0} = \frac{V_{DC}}{2} - \frac{I_{out}}{4 f_{SW} C_r} \qquad (6)$$

979-8-3315-1612-3/25 $31.00 © 2025 IEEE

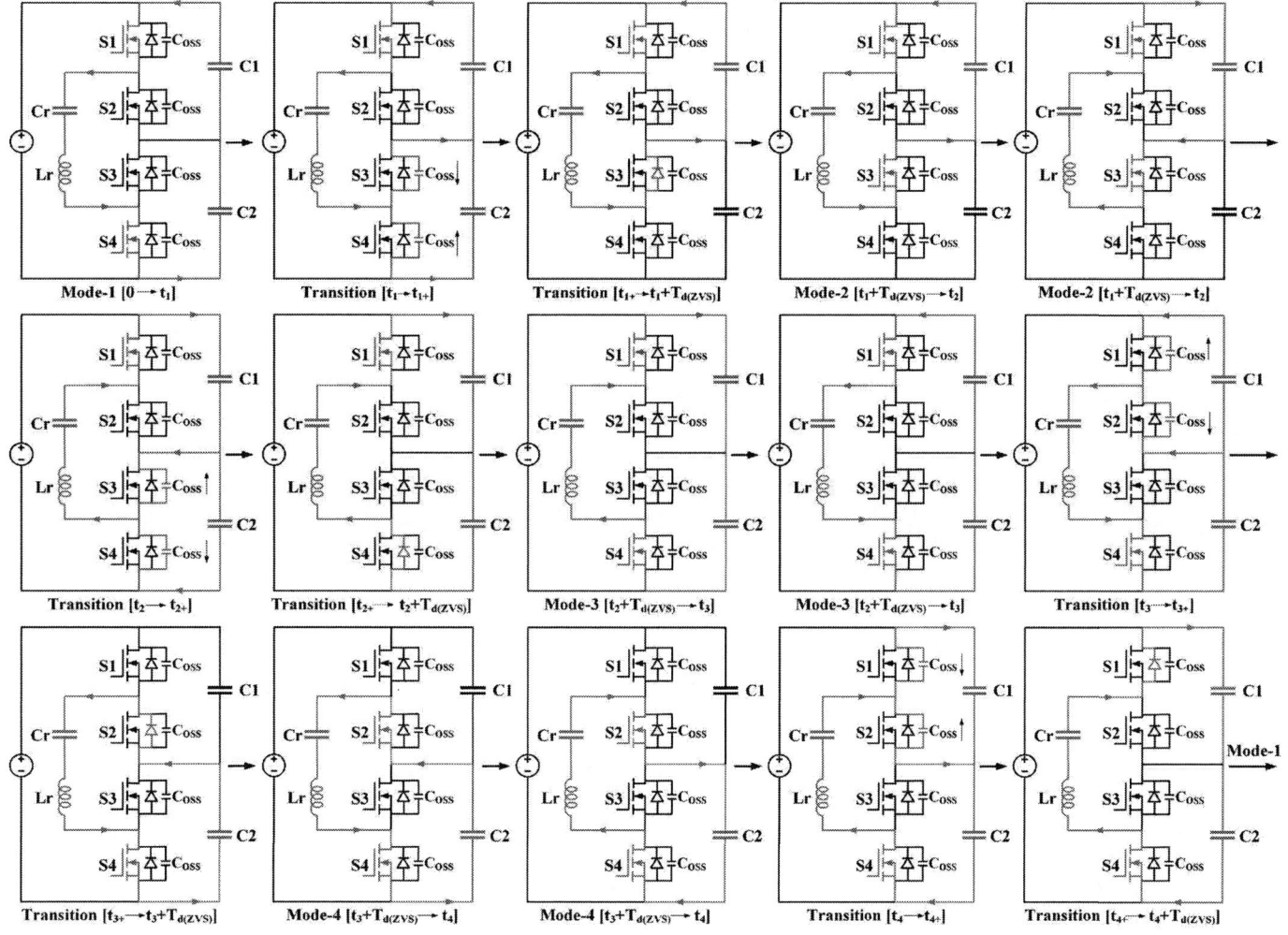

Fig. 3. Switching mode transition with the proposed full-range ZVS modulation.

$ZVS - transition[t_1 \leq t \leq t_1 + T_{d(ZVS)}]$: At $t = t_1$, $S4$ is turned off and the resonant current $i_r(t)$ starts discharging the output capacitance (C_{oss}) of switch $S3$ while charging the C_{oss} of switch $S4$ at $t = t_{1+}$. This causes the drain-source voltage of $S4$ to increase towards the DC blocking voltage of $V_{DC}/2$, while the drain-source voltage of $S3$ decreases toward zero. Once the drain-source voltage of $S3$ drops below its body diode's forward voltage, $i_r(t)$ starts conducting through the body diode of $S3$. Finally, at $t = t_1 + T_{d(ZVS)}$, $S3$ is turned on with ZVS.

2) Mode-2 $[t_1 + T_{d(ZVS)} \leq t \leq t_2]$: Switches $S1$ and $S3$ are on. The resonant current $i_r(t)$ starts discharging C_1 while charging C_r. The resonant current during this mode is expressed in (7). Where, $I_{a1} = I_a sin\omega_r T_1$.

$$i_r(t) = I_{a1} cos\omega_r t + I_m sin\omega_r t \tag{7}$$

$ZVS - transition[t_2 \leq t \leq t_2 + T_{d(ZVS)}]$: At $t = t_2$, $S3$ is turned off and the resonant current $i_r(t)$ starts charging the C_{oss} of switch $S3$ while discharging the C_{oss} of switch $S4$ at $t = t_{2+}$. As a result, the drain-source voltage of $S3$

rises toward $V_{DC}/2$, while the drain-source voltage of $S4$ falls toward zero. When the drain-source voltage of $S4$ drops below its body diode's forward voltage, $i_r(t)$ starts conducting through the body diode of $S4$. Finally, at $t = t_2 + T_{d(ZVS)}$, $S4$ achieves ZVS turn on.

3) Mode-3 $[t_2 + T_{d(ZVS)} \leq t \leq t_3]$: Switches $S1$ and $S4$ are on. The resonant current $i_r(t)$ during this mode is expressed in (8).

$$i_r(t) = I_m sin\omega_r t - I_a sin\omega_r t \tag{8}$$

$ZVS - transition[t_3 \leq t \leq t_3 + T_{d(ZVS)}]$: At $t = t_3$, $S1$ is turned off, $i_r(t)$ starts charging the C_{oss} of switch $S1$ while discharging the C_{oss} of switch $S2$ at $t = t_{3+}$. As a result, the drain-source voltage of $S1$ rises toward the DC blocking voltage, while the drain-source voltage of $S4$ decreases toward zero. As the drain-source voltage of $S2$ drops below its body diode's forward voltage, $i_r(t)$ starts conducting through the body diode of $S2$. Finally, at $t = t_3 + T_{d(ZVS)}$, $S2$ achieves ZVS turn on.

4) *Mode-4* $[t_3 + T_{d(ZVS)} \leq t \leq t_4]$: Switches $S2$ and $S4$ are on. The resonant current $i_r(t)$ starts discharging C_r to C_2. The resonant current during this mode is expressed in (9).

$$i_r(t) = I_m sin\omega_r t - I_{a1}cos\omega_r t \quad (9)$$

$ZVS - transition[t_4 \leq t \leq t_4 + T_{d(ZVS)}]$: At $t = t_4$, when $S2$ turns off, just after the turn off of $S2$ at $t = t_{2+}$, the resonant current $i_r(t)$ begins charging the C_{oss} of switch $S2$ while while simultaneously discharging the C_{oss} of switch $S1$. This causes the drain-source voltage of $S2$ to increase towards the DC blocking voltage of $V_{DC}/2$, while the drain-source voltage of $S1$ decreases toward zero. Once the drain-source voltage of $S1$ falls below its body diode's forward voltage, $i_r(t)$ starts flowing through this diode. This allows $S1$ to be turned on with ZVS at $t = t_4 + T_{d(ZVS)}$.

The proposed modulation has been verified in PSIM simulation and illustrated in Fig. 4 for the worst-case load current of 30 A for the RSCC. As seen in (2), (7)-(9), the resonant current ($i_r(t)$) has two current components: one is load-independent current I_a introduced by the proposed ZVS modulation and another one is the load-dependent current I_m. In the existing ZCS modulation, $T_1 = 0$, so, I_a current is zero, only I_m contributes to the resonant current. In contrast, with the proposed ZVS modulation, the resonant current is the summation of both I_a and I_m. Since I_a depends solely on T_1, the proposed ZVS modulation can guarantee the required ZVS current even under no load conditions. The rectified average load current (I_{out}) of the RSCC with the proposed ZVS modulation is given in (10) and the *rms* current is given in (11).

$$I_{out} = \frac{2}{\pi(1 - 2T_1/T_{sw})}I_{m(max)} \quad (10)$$

$$I_{rms} = \sqrt{\frac{I_{m(max)}^2 + I_{a1}^2}{2}} \quad (11)$$

where $I_{m(max)} = I_m sin\frac{\omega_r T_{SW}}{4}$. In typical ZCS modulation, $T_1 = 0$ therefore, the proposed ZVS modulation results in

slightly higher conduction losses compared to the ZCS modulation. However, by fully eliminating C_{oss} losses, the proposed ZVS modulation significantly enhances the efficiency of the RSCC-based DC-link voltage balancer for split-phase inverter applications under slightly unbalanced load to balanced load conditions. Additionally, the complete elimination of turn-on switching losses in MOSFETs with the proposed modulation enables the use of low ON-state resistance (R_{ds}) MOSFETs with relatively large output capacitance (C_{oss}).

B. ZVS Requirements

The minimum current required to achieve the ZVS turn-on for MOSFET is given in (12). Where C_{oss} is the output capacitance of the MOSFET.

$$i_{ZVS(min)} = V_{DC}\sqrt{\frac{C_{oss}}{L_r}} \quad (12)$$

With the proposed modulation, to achieve ZVS turn-on for all switches within the RSCC under all load conditions, T_1 should be selected such that at the end of Mode-2 (13) is satisfied.

$$|i_r(t = t_2)| \geq i_{ZVS(min)} \quad (13)$$

Using (7), the resonant current $i_r(t)$ at $t = t_2 = T_{SW}/2 - T_1$ is obtained in (14).

$$i_r(t = t2) = (I_a sin\omega_r T_1)cos\omega_r t_2 + I_m sin\omega_r t_2 \quad (14)$$

Using the parameters listed in Table I, the mathematical model derived earlier has been verified with PSIM simulation in Fig. 5 for the worst-case load current $I_{out} = 30A$. As seen in Fig. 5, there is a close match between the mathematical modeling and simulation. Furthermore, Fig. 5 confirms that with the specified T_1 and the C_{oss} values provided by the manufacturer, the proposed ZVS modulation meets the ZVS requirements under the worst-case scenario.

C. Simulation Verification of Voltage Balancing and ZVS Operation

The dynamics of the RSCC-based DC-link voltage balancer with the proposed sensorless ZVS modulation are verified

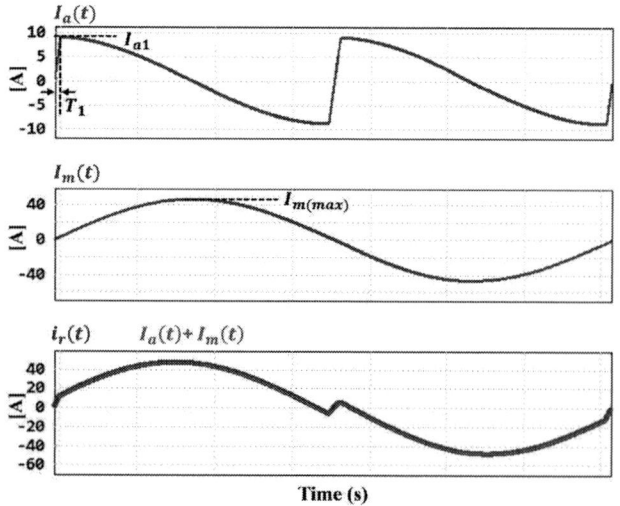

Fig. 4. Simulation verification of the proposed ZVS modulation.

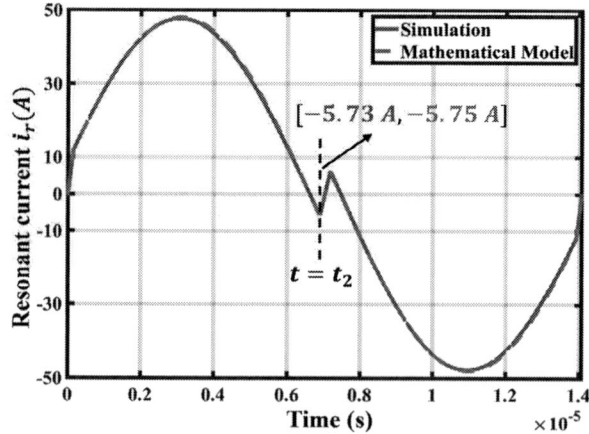

Fig. 5. Verification of the mathematical modeling with simulation.

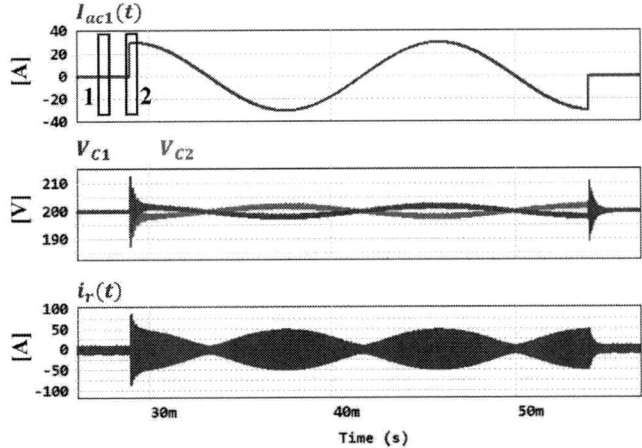

Fig. 6. Voltage balancing dynamics of the RSCC.

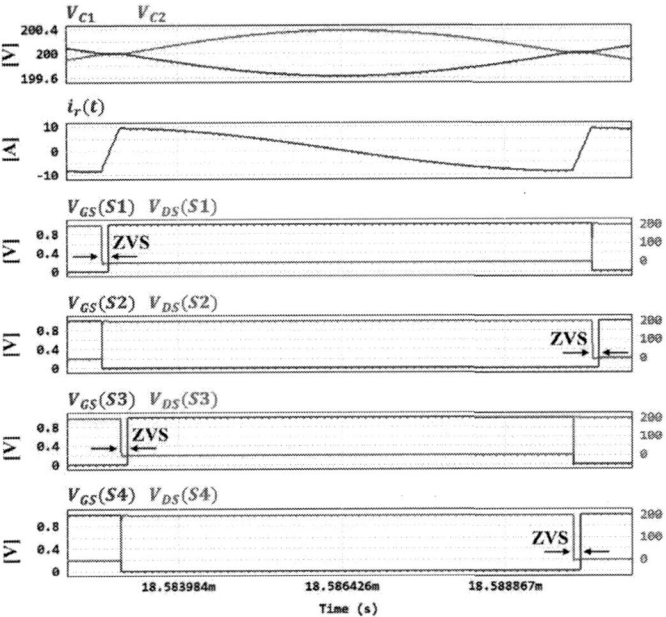

Fig. 7. ZVS operation of RSCC under no-load (or balanced load).

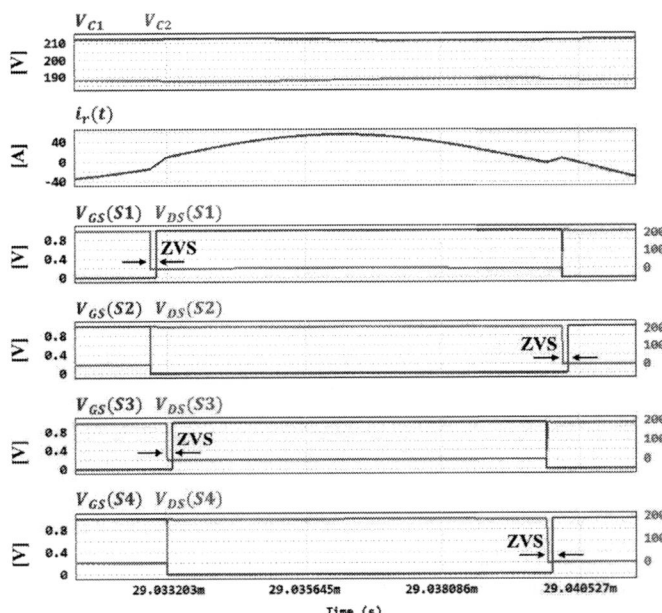

Fig. 8. ZVS operation of RSCC under transient periods.

achieve ZVS turn-on under no-load (also true for balanced load) conditions. Furthermore, Fig. 8 confirms that the ZVS turn-on is also maintained even during the transient period of load imbalance, highlighting the robustness of the proposed modulation strategy.

IV. EXPERIMENTAL RESULTS

The specifications of the split-phase inverter with the RSCC as the DC-link voltage balancer are listed in Table I and a 5-kW hardware prototype design is shown in Fig. 9. The power stage has been designed onto an insulated metal substrate (IMS) with NTBG025N065SC1, D2PAK-7L SiC MOSFETs. The bottom part of the IMS power board is attached to the heat sink using thermal paste and screws. The power stage is then interfaced with the control board via male-female header connectors.

The steady-state experimental waveforms of the split-phase inverter under the worst-case condition (fully unbalanced load)

in $PSIM$ simulation under the worst-case condition: a step change in AC load from no-load to a full unbalanced load of 2.5 kW. As illustrated in Fig. 6, the DC-link capacitor voltages temporarily deviated from their nominal values of $V_{DC}/2$. The resonant current $i_r(t)$ increased proportionally to the voltage difference, aiding in balancing the DC-link capacitor voltages. As seen in Fig. 6, the DC-link capacitor voltage recovered quickly within a few milliseconds without requiring any sensor or feedback control. It is worth mentioning that the inverter is closed-loop controlled and such voltage deviation for a short time will not create any problems for the AC voltage regulation. The zoomed switching waveforms for case-1 and case-2 (Fig. 6) are shown in Fig. 7 and Fig. 8, respectively. As seen in Fig. 7, the proposed ZVS modulation maintains the required ZVS current at no load, enabling all switches to

TABLE I
DESIGN PARAMETERS

Parameters	Value
DC input voltage (V_{DC})	400 V
Rated output AC voltage	120/240 V rms, 60 Hz
Rated power	2.5 kW + 2.5 kW = 5.0 kW
DC-link Capacitor(C_1, C_2)	25 μF
Resonant inductor(L_r)	3.15 μH
Resonant capacitor(C_r)	1.6 μF
Resonant frequency(f_r)	71 kHz
Switching frequency of RSCC (f_{SW})	71.1 kHz, $T_{d(ZVS)}$ = 100 ns
T_1	0.01T_{SW}
Inverter switching frequency	60 kHz
Inverter filter inductor(L_f)	250 μH
Inverter filter capacitor(C_f)	5.6 μF

Fig. 9. 5-kW hardware prototype design.

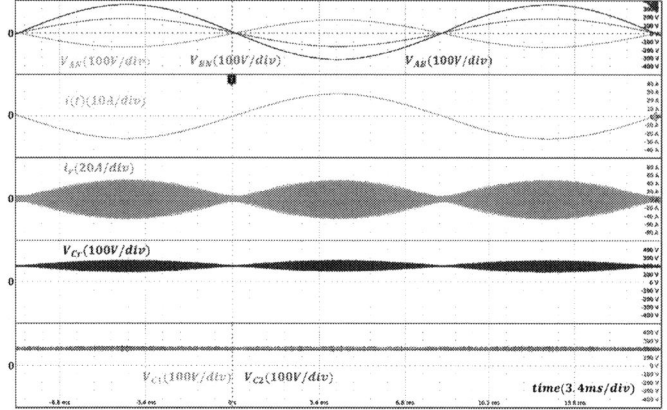

Fig. 10. Steady-state waveforms under a 2.5 kW unbalanced load.

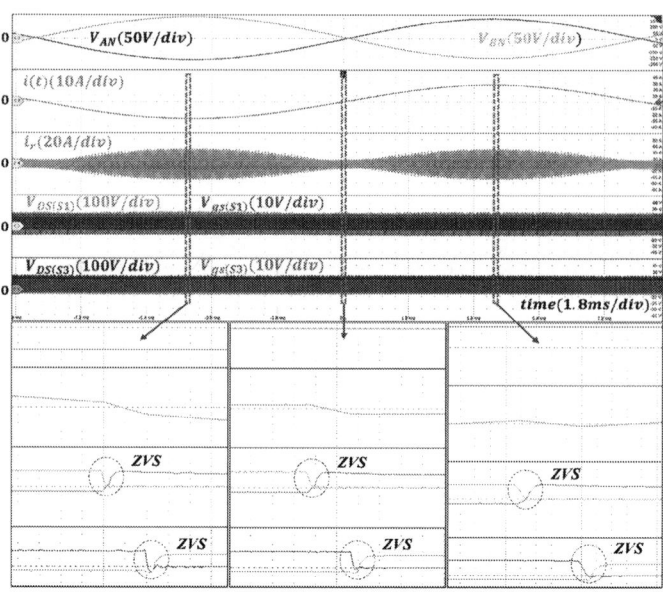

Fig. 11. Verification of full-range ZVS.

for the RSCC are presented in Fig. 10. A rated load of 2.5 kW is connected to the inverter's output terminal V_{AN} while no-load is connected to the V_{BN} terminal. As shown in Fig. 10, the split-phase voltages V_{AN} and V_{BN} are well-regulated, with a phase difference of $180°$, and the line-to-line voltage satisfies $V_{AB} = V_{AN} + V_{BN}$. Additionally, the DC-link capacitor voltages V_{C1} and V_{C2} are well-balanced at 200 V, confirming the effectiveness of the proposed ZVS modulation for the RSCC-based DC-link voltage balancing circuit.

The ZVS switching waveforms are presented in Fig. 11. Observing the drain-source voltage and gate-source voltage of the top switches in each half-bridge of the RSCC ($S1$ and $S3$), it can be seen that $V_{DS(S1)}$ and $V_{DS(S3)}$ drops to zero before $S1$ and $S3$ turned on. These ZVS waveforms confirm that the proposed modulation meets all ZVS requirements while maintaining a constant non-symmetrical duty cycle and switching frequency under all conditions, including the three critical scenarios: reverse maximum load current, zero load current, and forward maximum load current. The ZVS waveforms for the complementary switches ($S2$ and $S4$) are not shown; however, these switches are also turned on with ZVS. Additionally, Fig. 11 illustrates that the switches $S1 - S4$ experience a voltage stress of $V_{DC}/2$, effectively shielded from the resonant capacitor ripple voltage (ΔV_{Cr}).

The efficiency plot of the split-phase inverter under different unbalanced load conditions (a resistor load bank is connected to the inverter's V_{AN} terminal and no loads are connected to

the V_{BN} terminal) are measured using YOKOGAWA WT3000 Power Analyzer and shown in Fig. 12. To highlight the efficiency improvements achieved with the proposed ZVS modulation, the plot also includes efficiency measurements using the existing PSM and ZCS modulation for the RSCC-based DC-link voltage balancer. In the PSM, the switching frequency is 120 kHz and for the ZCS modulation, the switching frequency is 71.1 kHz (same as the switching frequency used for the proposed ZVS modulation). Since the switching frequency in PSM is much higher than the resonant tank frequency and can't achieve ZVS at light loads, the efficiency drops as the load decreases. However, in the moderate to full load range, the ZVS region expands, and efficiency with

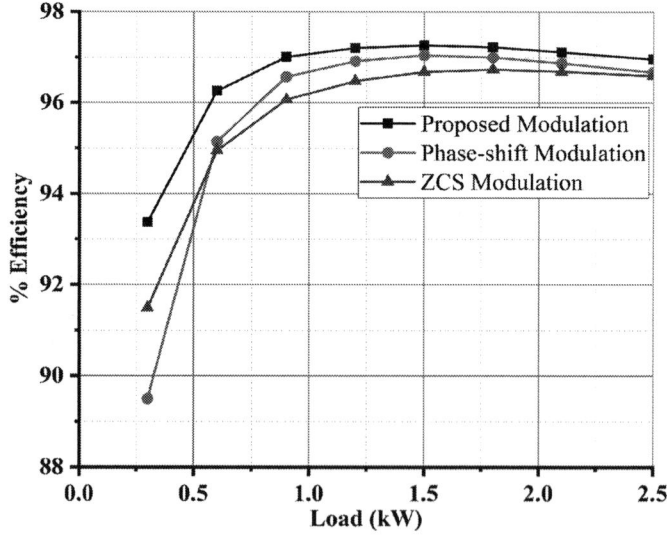

Fig. 12. Efficiency plot at unbalanced load.

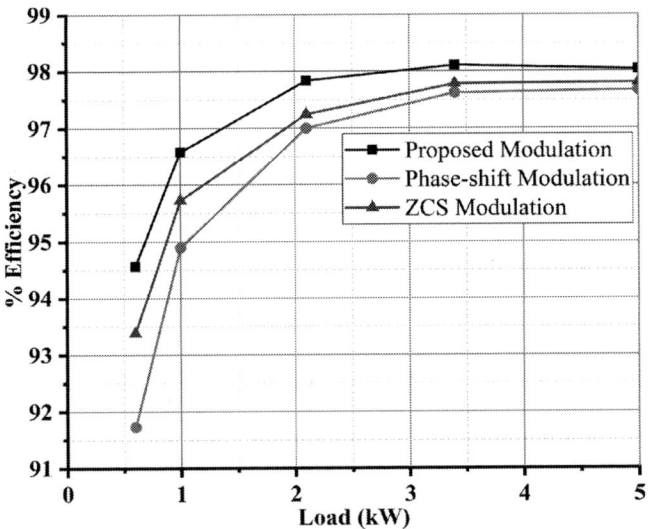

Fig. 13. Efficiency plot at balanced load.

PSM improves. In contrast, the ZCS modulation operates at a lower switching frequency (71.1 kHz), resulting in higher efficiency under light loads compared to PSM. However, ZCS modulation does not eliminate losses associated with the MOSFETs' output capacitance (C_{oss}), leading to slightly lower efficiency than PSM in the moderate to full load range. As seen in Fig. 12, the proposed ZVS modulation maintains consistently higher efficiency across the entire load range by ensuring ZVS operation under all conditions.

Fig. 13 shows the efficiency plot of the split-phase inverter under balanced load conditions. A resistor load bank is connected to the inverter's V_{AB} terminal. Under balanced load conditions, the load current for the RSCC is zero, meaning that the losses associated with the RSCC are mainly the no-load losses caused by the MOSFETs' output capacitance (C_{oss}). The plot shown in Fig. 13 compares the measured efficiency of three modulation techniques: PSM, ZCS modulation, and the proposed ZVS modulation for the RSCC. At balance load, RSCC with PSM doesn't have any ZVS zones, and its high switching frequency (120 kHz) results in the lowest efficiency. In contrast, ZCS modulation operates at a reduced switching frequency (71.1 kHz), leading to improved efficiency compared to PSM under balanced load conditions. As seen in Fig. 13, the proposed ZVS modulation has the highest efficiency across the entire load range. This improvement is attributed to the modulation's ability to maintain the required ZVS current by incorporating additional resonant current boost subintervals, even under no-load conditions.

V. CONCLUSION

This paper proposed a new sensorless ZVS modulation for the RSCC-based DC-link voltage balancing circuit for split-phase inverter applications to improve efficiency across the entire load range. In a split-phase inverter, the DC-link voltage balancer typically operates under slightly unbalanced (light-load) to balanced load conditions (no load). The existing PSM

and ZCS modulation can't achieve high efficiency at these load ranges. The proposed sensorless ZVS modulation addresses this issue by maintaining the minimum required ZVS current across the full load range using a constant non-symmetrical duty cycle control. This approach ensures ZVS operation for all switching devices, even during load transients. The non-symmetrical duty cycle introduces an additional degree of freedom in controlling the resonant current, which is critical for achieving ZVS. A mathematical model is developed to derive the conditions necessary to satisfy ZVS requirements and verified with simulations. The ZVS operation under no load and during load transient periods was also verified with simulations. The experimental results from 5-kW hardware prototype verify that the proposed modulation for the RSCC meets all ZVS requirements while maintaining a constant non-symmetrical duty cycle and frequency across all scenarios, including the three corner cases: reverse maximum load current, zero load current, and forward maximum load current. Finally, the efficiency plot confirms the efficiency improvement of the split-phase inverter for both balanced and unbalanced load conditions across the entire load range with the proposed sensorless ZVS modulation for the RSCC-based DC-link voltage balancer.

REFERENCES

[1] B. S. Gehrke, C. B. Jacobina, R. P. R. de Sousa, I. R. F. M. P. da Silva, N. B. de Freitas and M. B. d. R. Corrêa, "Single-Phase Three-Wire Power Converters Based on Two-Level and Three-Level Legs Using a Space-Vector PWM-Based Voltage Balancing," in IEEE Transactions on Industry Applications, vol. 57, no. 3, pp. 2654-2665, May-June 2021.

[2] Y. Zhu, H. Wen and Y. Yang, "A Novel Single-Phase Three-Wire Three-Level Power Inverter with a Balance Bridge," 2023 IEEE 2nd International Power Electronics and Application Symposium (PEAS), Guangzhou, China, 2023, pp. 1326-1330.

[3] A. Cervone, G. Brando and O. Dordevic, "Hybrid Modulation Technique With DC-Bus Voltage Control for Multiphase NPC Converters," in IEEE Transactions on Power Electronics, vol. 35, no. 12, pp. 13528-13539, Dec. 2020.

[4] T. Ge, Z. Ye and R. C. N. Pilawa-Podgurski, "Geometrical State-Plane Analysis of Resonant Switched-Capacitor Converters: Demonstration on the Cascaded Multiresonant Converter," in IEEE Transactions on Power Electronics, vol. 38, no. 9, pp. 11125-11140, Sept. 2023.

[5] K. Sano and H. Fujita, "Performance of a High-Efficiency Switched-Capacitor-Based Resonant Converter With Phase-Shift Control," in IEEE Transactions on Power Electronics, vol. 26, no. 2, pp. 344-354, Feb. 2011.

[6] Y. Yuanmao, K. W. E. Cheng and Y. P. B. Yeung, "Zero-Current Switching Switched-Capacitor Zero-Voltage-Gap Automatic Equalization System for Series Battery String," in IEEE Transactions on Power Electronics, vol. 27, no. 7, pp. 3234-3242, July 2012.

[7] K. Sano and H. Fujita, "Voltage-Balancing Circuit Based on a Resonant Switched-Capacitor Converter for Multilevel Inverters," in IEEE Transactions on Industry Applications, vol. 44, no. 6, pp. 1768-1776, Nov.-dec. 2008.

[8] Satvik, W. Yu, D. Wang and S. Chen, "Switched Capacitor Converter with Flexible Voltage Gain and 99.2% Efficiency Utilizing Auto-transformer," 2020 IEEE Energy Conversion Congress and Exposition (ECCE), Detroit, MI, USA, 2020, pp. 165-172.

[9] A. Amin, M. Shousha, A. Prodić and B. Lynch, "A transformerless dual active half-bridge DC-DC converter for point-of-load power supplies," 2015 IEEE Energy Conversion Congress and Exposition (ECCE), Montreal, QC, Canada, 2015, pp. 133-140.

[10] K. K. Law, K. W. E. Cheng and Y. P. B. Yeung, "Design and analysis of switched-capacitor-based step-up resonant converters," in IEEE Transactions on Circuits and Systems I: Regular Papers, vol. 52, no. 5, pp. 943-948, May 2005.

Dimensional Parasitics Absorption in Capacitively-Isolated Ćuk Converter for Medium-Voltage High Step-Down Converters

Aakash Kamalapur, Jung-Soo Bae, Mark Cairnie, Rajaie Nassar, Jack Knoll
Dushan Boroyevich, Guo-Quan Lu, Christina DiMarino, Qiang Li, and Khai D. T. Ngo
Center for Power Electronics Systems
Virginia Polytechnic Institute and State University
Blacksburg, VA 24061, United States
Email: {aakashkamalapur, jamesbae, mcairnie, rajaienassar, knolljs, dusan, gqlu, dimaricm, lqvt, kdtn}@vt.edu

Abstract—The demand for medium-voltage direct current (MVdc) power distribution in land-constrained residential areas has highlighted the need for compact power processors and architectures. High-density power distribution requires co-packaging of power passive components and switches to enhance power densities while addressing insulation and thermal management for medium-voltage (MV) systems. However, such designs inherently introduce parasitics within the converter. This work focuses on absorbing these parasitics as manifestations of non-linear insulation and coaxial structural constraints. The capacitively-isolated Ćuk converter (CICC) is proposed as a high step-down solution for modular converters. The CICC mitigates parasitics, enabling integration into input-series output-parallel (ISOP) architectures, such as solid-state transformers (SSTs). By using the dielectric for both insulation and energy transfer, the CICC eliminates transformer-based cells in SSTs. Additionally, this work achieves a significant reduction in the blocking capacitor size required for zero-voltage switching (ZVS). The proposed topology is validated using a custom-packaged 2 kV-to-400 V, 50 kW coaxial modular converter operating at 100 kHz, achieving a power density of 8 MW/m³. These results demonstrate the CICC's potential to address the challenges of high-density, modular MV power conversion.

Index Terms—dc-to-dc converters, modular converters, multi-level converters, Ćuk converter, CIĆuk, medium-voltage, state-plane, soft-switching, MVdc-to-LVdc, high step-down conversion.

I. INTRODUCTION

The growing demand for medium-voltage electrical distribution, driven by factors such as the expansion of electric vehicle (EV) charging infrastructure, urbanization-induced real estate constraints, increased reliance on distributed renewable energy, the need for efficient energy storage, and the rising frequency of severe weather events, has led to a significant shift in distribution systems. This shift has progressed from low-frequency transformer (LFT) based systems to the initially proposed power-electronics-based medium-voltage processors [1], ultimately culminating in the formalized system of solid-state transformers (SST) [2].

SSTs offer notable advantages, including bidirectional power flow control, advanced protection, and diagnostic capabilities. However, as input or output port voltages increase, the

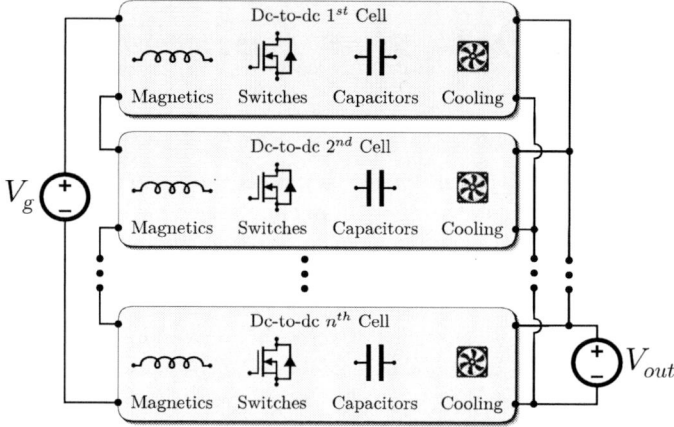

Fig. 1. Ideal layout of an input-series output-parallel (ISOP) system with n cells assumes that the dc-to-dc stage is isolated either galvanically or dielectrically. Any number of stages can be cascaded to achieve the desired conversion ratio, V_{out}/V_g for each cell.

non-linear insulation requirements introduce significant challenges in the design of high-frequency transformers (HFTs) within individual SST cells, reducing the achievable power density to as low as 3.8 MW/m³ [3]. This limitation exemplifies one aspect of the broader complexity in power electronics design. Another critical concern is the rise of *dimensional parasitics*, which are inherent in input-series output-parallel (ISOP) converters—a common architecture in SST designs, as shown in Fig. 1. These parasitics often manifest as inductive elements in the circuit, as illustrated in Fig. 2, and play a pivotal role in the design of medium-voltage converters [4]. Furthermore, these parasitics are accompanied by coupling capacitances [5], which, while contributing to increased common-mode currents, have minimal impact on steady-state converter operation. This raises a fundamental question: Can these *dimensional parasitics* be effectively absorbed into the converter's operation, transforming them from constraints to avoid into functional energy storage or transfer elements?

This research aims to develop solutions for absorbing dimen-

979-8-3315-1612-3/25 $31.00 © 2025 IEEE

Dimensional parasitics in an input-series output-parallel (ISOP) architecture can arise due to the length and spacing between cells, which result from thermal and insulation considerations when positioning them apart.

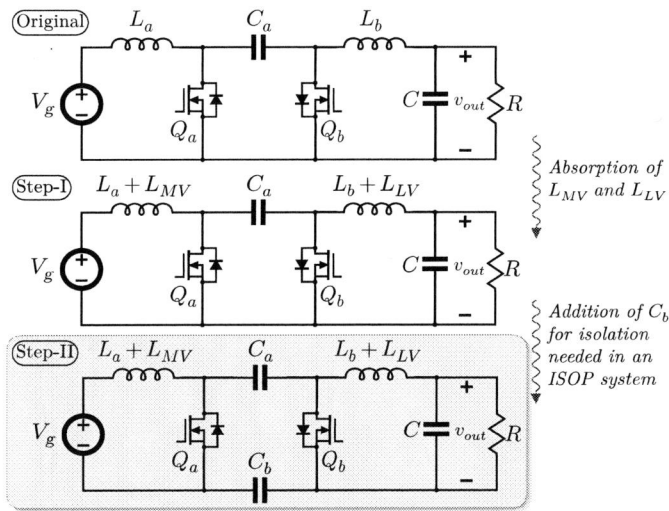

Fig. 2. Generalized representation of the ISOP system from Fig. 1, where the connections between cells in the medium-voltage (MV) and low-voltage (LV) buses are left as ports to signify the manifestation of *dimensional parasitics*. Four types of dimensional parasitics arise as functions of spacing within and between cells, governed by insulation and thermal constraints. These are defined as the input-port bus impedance (Z_{MV}) or parasitic (L_{MV}); the output-port bus impedance (Z_{LV}) or parasitic (L_{LV}); the input-port return impedance ($Z_{MV,\text{return}}$) or parasitic ($L_{MV,\text{return}}$); and the commutation-loop impedance (Z_{CL}) or parasitic (L_{CL}).

sional parasitics in individual cells of an MV ISOP system while avoiding cascaded dc-to-dc stages to minimize the component count. Based on isolation requirements, the dielectric material—typically used as an encapsulant for local insulation within and between cells—is also utilized for isolation, enabling capacitive isolation similar to the approach in systems proposed in [6], [7]. To achieve single-stage operation per dc-to-dc cell with capacitive isolation, a classic Ćuk converter is employed, with the coupling capacitor split into the return path, as illustrated in Fig. 3.

This simple modification facilitates the study of absorbing *dimensional parasitics*, combined with an examination of the steady-state time evolution of the converter, while enabling zero-voltage switching (ZVS). These measures address the inherent limitations of the Ćuk converter. This work provides a formal and qualitative analysis of the converter by identifying and absorbing *dimensional parasitics* in the ISOP architecture, as detailed in Section II. This approach allows the converter to be evaluated independently of *dimensional parasitics*, leading naturally to Section III, which applies the state-plane method to analyze the converter's multi-resonant behavior.

Section IV validates the effectiveness of absorbing dimensional parasitics and corroborates the theoretical framework established in Section III, while demonstrating the efforts detailed in Section II. Finally, Section V summarizes the findings and outlines ongoing research directions to further advance this work.

II. ABSORBING DIMENSIONAL PARASITICS

The generalized schematic for an ISOP system is shown in Fig. 2, which extends the representation in Fig. 1. It incorporates *dimensional parasitics* as port impedances at the MV or the input port Z_{MV} and LV or the output port Z_{LV}, representing the impedance in the input and output

Fig. 3. First modification, (Step-I), facilitates the absorption of L_{MV} and L_{LV}. The second modification, (Step-II), involves splitting C_a into another capacitor, C_b, to achieve dielectric isolation instead of galvanic isolation. This approach leverages the dielectric material commonly used for insulation in transformers to provide complete solid-state insulation. Together, these modifications enable the converter to operate effectively as a cell in an ISOP system.

port connections, respectively. These impedances depend on the cell length and the spacing between cells, influenced by insulation and thermal considerations. Additionally, the schematic illustrates another parasitic in the input-port return path, $Z_{MV,\text{return}}$, which is a function of the cell length and is repeated n times, corresponding to the number of cells. The $Z_{MV,\text{return}}$ parasitic has no distinct impact from Z_{MV}, as both are effectively in series. For clarity and to avoid redundancy, the port impedance is henceforth considered as part of Z_{MV}.

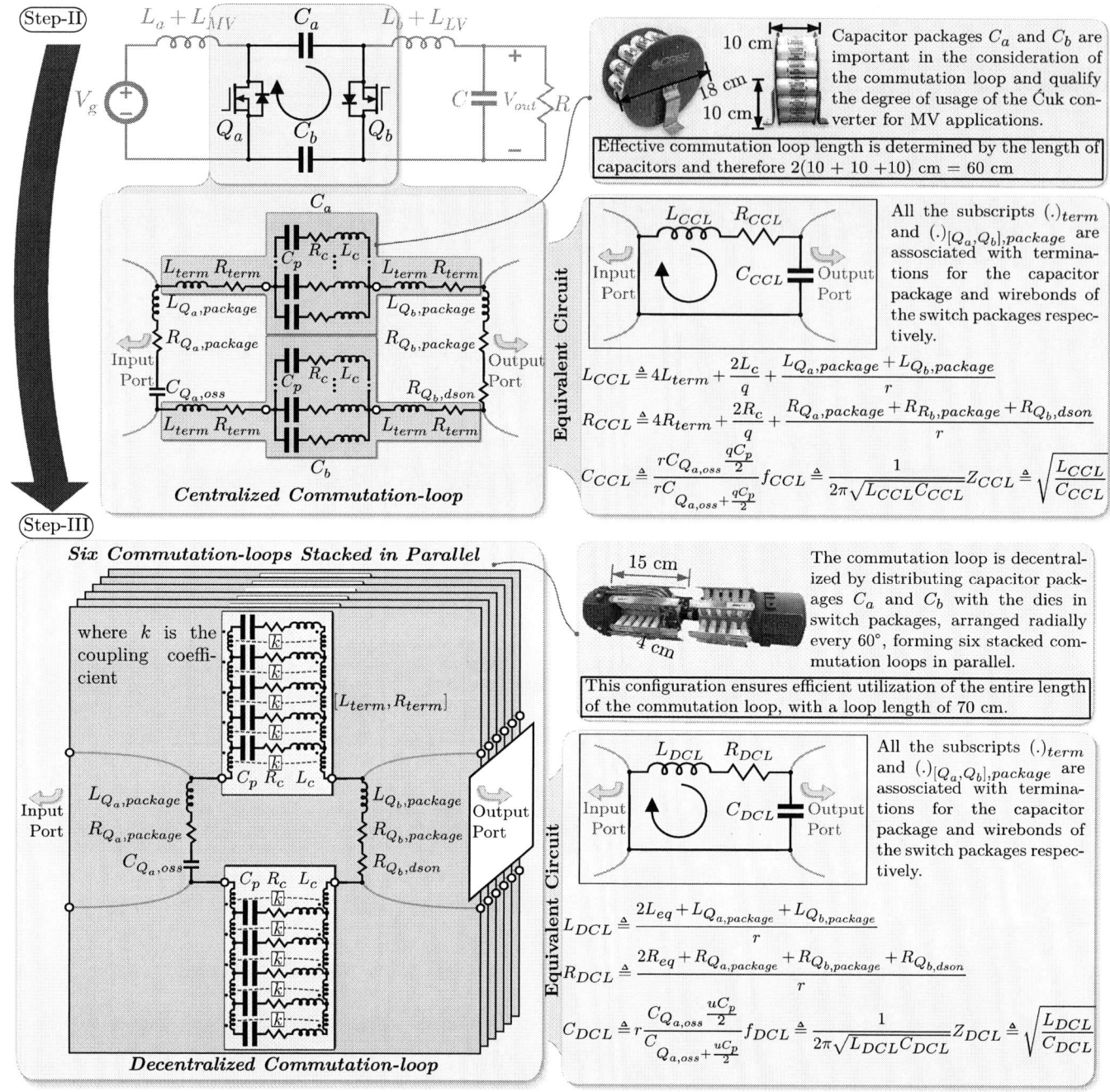

The resistors R_{term}, $R_{Q_{a,package}}$, and $R_{Q_{b,package}}$ exhibit both dc and ac resistances. The dc resistances primarily affects the efficiency of the converter, while the ac resistances plays a critical role in damping high-frequency differential-mode (DM) currents associated with f_{CCL}, thereby mitigating peak voltage and current stresses on the switches.

Fig. 4. Transition from (Step-II) in Fig. 3 to (Step-III) is achieved by distributing the capacitor C_p throughout the commutation loop. By designing six individual cells of commutation-loops and connecting their input and output ports in parallel, allows to decentralize the commutation loop from (Step-II) and reduce the commutation-loop inductance from $L_{CCL} = 82$ nH to $L_{DCL} = 3.6$ nH.

The final *dimensional parasitic* is Z_{CL}, which represents the impedance of the commutation loop within individual dc-to-dc cells of the ISOP system. For simplicity in referring to these *dimensional parasitics* throughout the text, they are represented as $L_{[LV,MV,CL]}$, where the Z identifier is replaced with L to denote their inductive nature, aligning with the circuit-oriented analysis developed in subsequent sections of this work.

To absorb *dimensional parasitics*, L_{MV} and L_{LV}, the Ćuk converter [8] shown in Fig. 3 was selected due to its input inductor, which effectively reduces both differential and common-mode currents by maintaining continuous input current. However, the Ćuk converter is not inherently suitable for ISOP integration as a single cell due to its lack of isolation and requires modifications.

One approach involves absorbing the parasitics L_{MV} and L_{LV} into the energy storage elements and splitting the capacitor C_a into C_a and C_b for solid-state insulation, as shown in Fig. 2 as an evolution from the *original* circuit to *step-I* and *step-II*, respectively. Interestingly, the dielectric in these split capacitors facilitates energy transfer during steady-state operation while also providing high-voltage insulation. This modification results in a fully solid-state dc-to-dc converter optimized for the first two *dimensional parasitics*, L_{LV} and L_{MV}, reducing potential energy losses and mitigating increased switch stresses seen in basic converter topologies such as buck, boost, or even buck-boost, while improving efficiency.

In a similar effort to absorb *dimensional parasitics*, a critical question arises: How can the commutation loop be effectively managed? If left unresolved, significant ringing and voltage spikes across switches Q_a and Q_b may occur, leading to reduced converter efficiency and potential damage.

To address this, the commutation loop in the Ćuk converter, particularly for MV applications, is defined in Fig. 4. The commutation loop corresponds to the condition when the synchronous rectifier Q_b turns on at high current, causing ringing and voltage spikes. In Ćuk converters, capacitors C_a and C_b typically exhibit a large equivalent series inductance (ESL) exceeding $100\,\text{nH}$, which significantly contributes to this ringing.

The commutation loop, as outlined in (Step-II), comprises two capacitor packages, C_a and C_b, each containing $q = 30$ capacitors C_p in parallel, with individual equivalent series resistance (ESR) R_c and ESL L_c. Additionally, termination inductances L_{term} and resistances R_{term} exist at either end of the loop. The active switch Q_a and synchronous rectifier Q_b packages include drain and source inductances $L_{Q_a,\text{package}}$, $L_{Q_b,\text{package}}$, and corresponding resistances $R_{Q_a,\text{package}}$ and $R_{Q_b,\text{package}}$.

Each switch package, as depicted in Fig. 4, is assumed to contain $r = 6$ dies in parallel. While the dc resistances of terminations and packages are typically small (less than $1\,\text{m}\Omega$) and negligible relative to the on-resistance of the switches, their ac resistances can be considerably higher due to skin and proximity effects at high frequencies. The ac resistance and inductance values for the termination dimensions in Fig. 4, and for wire bonds with a diameter of $380\,\mu\text{m}$, can be calculated using closed-form expressions provided in [9]–[11].

Using the definitions for the commutation loop in (Step-II) from Fig. 4, the equivalent commutation-loop impedances are

derived as follows:

$$L_{CCL} \triangleq 4L_{term} + \frac{2L_c}{q} + \frac{L_{Q_a,\text{package}} + L_{Q_b,\text{package}}}{r} \quad (1)$$

$$R_{CCL} \triangleq 4R_{term} + \frac{2R_c}{q}$$
$$+ \frac{R_{Q_a,\text{package}} + R_{Q_b,\text{package}} + R_{Q_b,\text{dson}}}{r} \quad (2)$$

$$C_{CCL} \triangleq \frac{rC_{Q_a,\text{oss}}\frac{qC_p}{2}}{rC_{Q_a,\text{oss}} + \frac{qC_p}{2}} \quad (3)$$

$$f_{CCL} \triangleq \frac{1}{2\pi\sqrt{L_{CCL}C_{CCL}}} \quad (4)$$

The values of these parameters are $L_{CCL} = 82\,\text{nH}$, $C_{CCL} = 1.19\,\text{nF}$, and $R_{CCL} = 25\,\text{m}\Omega$. The resonant frequency $f_{CCL} = 17\,\text{MHz}$ is calculated and verified using the frequency response of the equivalent circuit shown in Fig. 5a. For R_{CCL}, the ac resistances of the package and terminations are considered at the resonant frequency f_{CCL}.

To reduce the loop inductance L_{CCL}, the commutation loop is decentralized into six parallel commutation loops, as depicted in Fig. 4. The split capacitors and switches are distributed throughout the loop, resulting in the equivalent loop impedances:

$$L_{DCL} \triangleq \frac{2L_{eq} + L_{Q_a,\text{package}} + L_{Q_b,\text{package}}}{r} \quad (5)$$

$$R_{DCL} \triangleq \frac{2R_{eq} + R_{Q_a,\text{package}} + R_{Q_b,\text{package}} + R_{Q_b,\text{dson}}}{r} \quad (6)$$

$$C_{DCL} \triangleq r\frac{C_{Q_a,\text{oss}}\frac{uC_p}{2}}{C_{Q_a,\text{oss}} + \frac{uC_p}{2}} \quad (7)$$

$$f_{DCL} \triangleq \frac{1}{2\pi\sqrt{L_{DCL}C_{DCL}}} \quad (8)$$

where $u = 5$ is the number of capacitors C_p in parallel in one commutation-loop stack. The effective inductance L_{eq} and resistance R_{eq} at high frequencies are determined by shorting the capacitor C_p and calculating the resulting impedances. The values of these parameters are $L_{DCL} = 3.5\,\text{nH}$, $C_{DCL} = 1.19\,\text{nF}$, and $R_{DCL} = 25\,\text{m}\Omega$, with a resonant frequency of the decentralized commutation loop $f_{DCL} = 78\,\text{MHz}$, as verified in Fig. 5a.

Simulation instances of the Ćuk converter with both commutation loops are created for a soft-switching case with $V_g = 2000\,\text{V}$, $V_{out} = 400\,\text{V}$, $L_a = 30\,\mu\text{H}$, $L_b = 13\,\mu\text{H}$, $C_a = C_b = 660\,\text{nH}$, $C = 120\,\mu\text{F}$, and $R = 3.2\,\Omega$. The voltage stress across the switch Q_a, depicted in Fig. 5b, clearly shows that the resonant frequency during Q_b turn-on matches the calculated values in (4) and (8). The results demonstrate that decentralizing the commutation loop significantly reduces voltage stress.

Finally, the modifications to absorb *dimensional parasitics* lead to a variant of the Ćuk converter, henceforth referred to as the capacitively-isolated Ćuk converter (CIĆC), illustrated in Fig. 3.

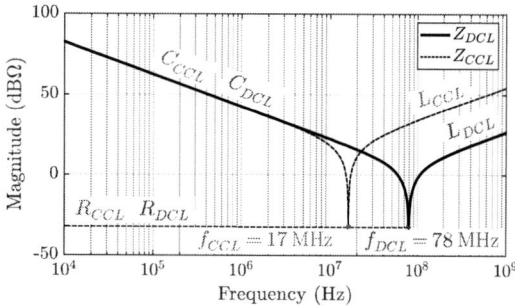

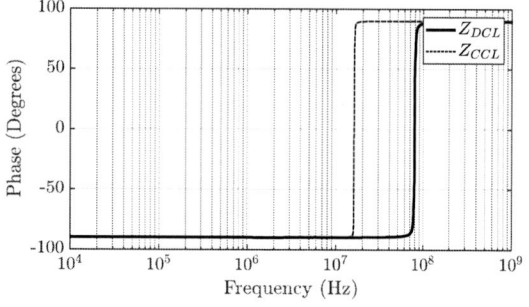

(a)

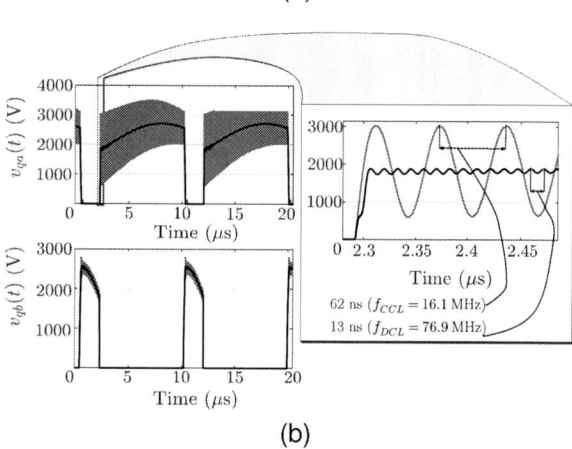

(b)

Fig. 5. (a) Frequency-response comparison of the commutation-loop impedances Z_{CCL} and Z_{DCL}, aligning with the calculated resonant frequencies. (b) Time-response comparison of the voltage across switches Q_a and Q_b during Q_b turn-on. The red waveform represents the centralized commutation loop, while the black waveform corresponds to the decentralized loop. The commutation-loop modeled in simulation closely matches the analytical model developed in (1) and (8), validating its use as a blueprint for minimizing commutation loops in converters designed for medium-voltage systems.

III. CAPACITIVELY-ISOLATED ĆUK CONVERTER

The steady-state operation and time evolution of the CIĆC with ZVS is illustrated in Fig. 6a. The CIĆC operates similarly to quasi-square-wave converters [12]–[14], but with capacitors C_a and C_b resonating with inductors due to their small size. This design improves the power density of the converter by reducing the size of the capacitors from the typical μH range to the nH range, causing ripple in the blocking voltages of the active switch and synchronous rectifier Q_a and Q_b.

As shown in Fig. 6a, the CIĆC operates in four subintervals. Power inversion and rectification occur during subintervals **(I)** and **(III)**, while zero-voltage switching (ZVS) is achieved during subintervals **(II)** and **(IV)**. The capacitors C_a and C_b have identical values, C, and the output capacitances of switches Q_a and Q_b are also assumed to be identical, denoted as $C_{oss,Q_a} = C_{oss,Q_b} = C_{oss}$.

The dynamics of the converter are described using three families of normalization equations: power inversion (10), soft-switching (11), and power rectification (12), all normalized with (9). These dynamics are expressed as state-plane trajectories [15], illustrated in Fig. 6b and Fig. 6c. The state-plane diagrams represent the normalized voltages of Q_a and Q_b relative to the input and output inductor currents of L_a and L_b. The scalar coordinates, J_x and M_x, are also normalized using (9).

$$V_{base} \triangleq V_g$$
$$\mathfrak{M} \triangleq \frac{V_{out}}{V_g} \tag{9}$$

$$\omega_{o,in} \triangleq \frac{1}{\sqrt{L_a\left(0.5C + C_{oss}\right)}}$$
$$R_{o,in} \triangleq \sqrt{\frac{L_a}{0.5C + C_{oss}}} \tag{10}$$
$$I_{base,in} \triangleq \frac{V_{base}}{R_{o,in}}$$

$$\omega_{o,ss} \triangleq \frac{1}{\sqrt{2\left(L_a\|L_b\right)C_{oss}}}$$
$$R_{o,ss} \triangleq \sqrt{\frac{\left(L_a\|L_b\right)}{2C_{oss}}} \tag{11}$$
$$I_{base,ss} \triangleq \frac{V_{base}}{R_{o,ss}}$$

$$\omega_{o,out} \triangleq \frac{1}{\sqrt{L_b\left(0.5C + C_{oss}\right)}}$$
$$R_{o,out} \triangleq \sqrt{\frac{L_b}{0.5C + C_{oss}}} \tag{12}$$
$$I_{base,out} \triangleq \frac{V_{base}}{R_{o,out}}$$

A. Subinterval-I (t_α)

This subinterval undergoes power inversion, detailed in Fig. 6a. During this period, the active switch Q_a is turned on, while the synchronous rectifier Q_b remains off until the duty or control signal is deactivated. The input port exhibits a linear increase in current $i_{L_a}(t)$, while the output port resonates between L_b and $0.5C + C_{oss}$, leading to resonances in $v_{Q_b}(t)$ and $i_{L_b}(t)$ with a dc bias of V_{out}.

The state-plane diagram for subinterval-I (Fig. 6c) provides the timing t_α and voltage stress $V_{Q_b,pk}$, derived using (13) and (14), respectively.

979-8-3315-1612-3/25 $31.00 © 2025 IEEE

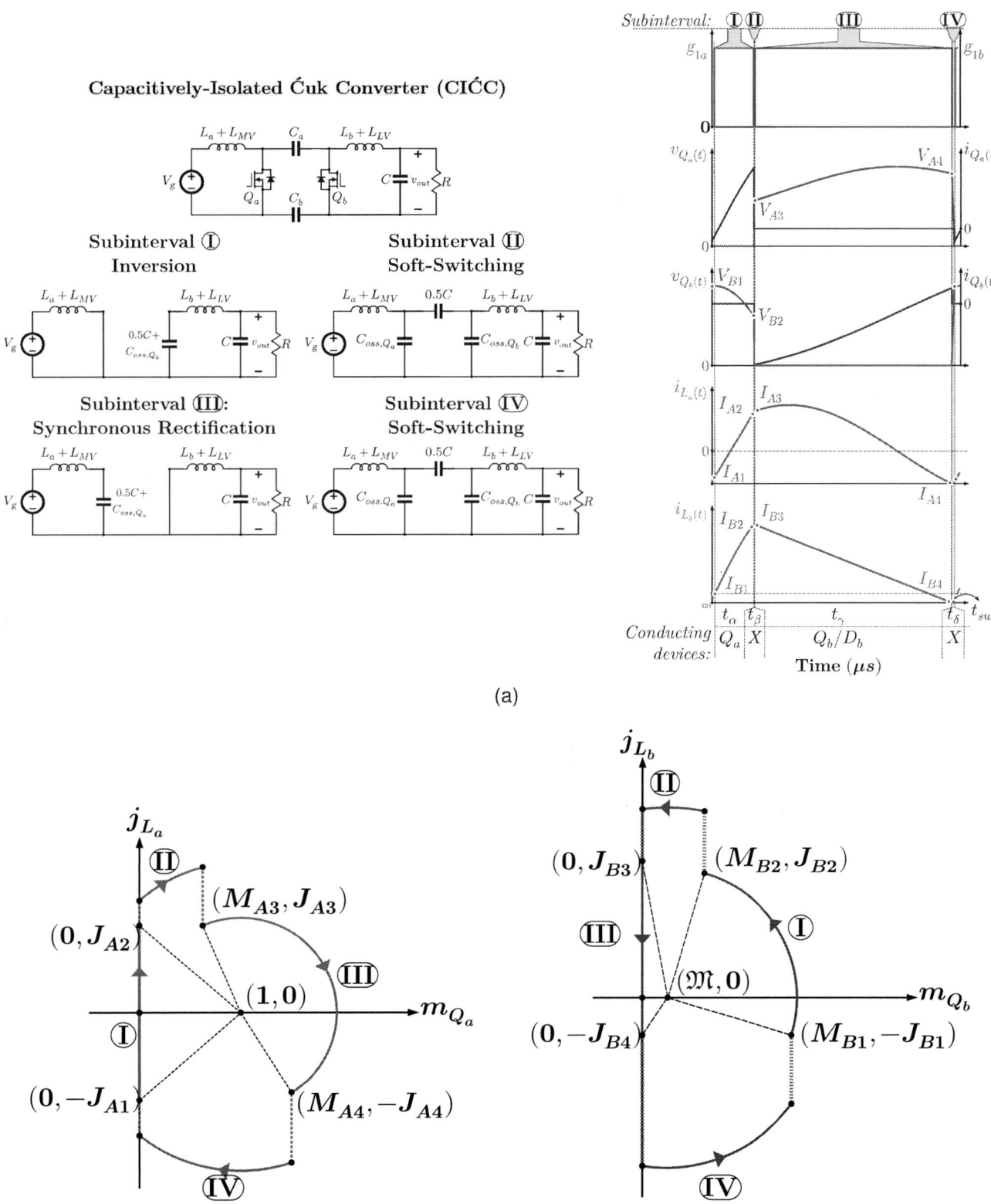

Fig. 6. Four subintervals and qualitative waveforms of the CIĆC in (a) are as follows: the first row illustrates the gating signals g_{1a} and g_{1b} for switches Q_a and Q_b; the second and third rows show the blocking voltages and conduction currents of Q_a and Q_b, respectively; and the fourth and fifth rows depict the inductor currents $i_{L_a}(t)$ and $i_{L_b}(t)$, respectively. The input and output inductors resonate with $0.5C + C_{oss,Q_b}$ and $0.5C + C_{oss,Q_a}$ during the inversion and rectification subintervals I and III, enabling soft-switching in II and IV. The converter's (b) input and (c) output ports are subsequently represented as state-plane diagrams normalized using (9) and (12).

979-8-3315-1612-3/25 $31.00 © 2025 IEEE   2233

Converter Hardware	2 kV Test Bench

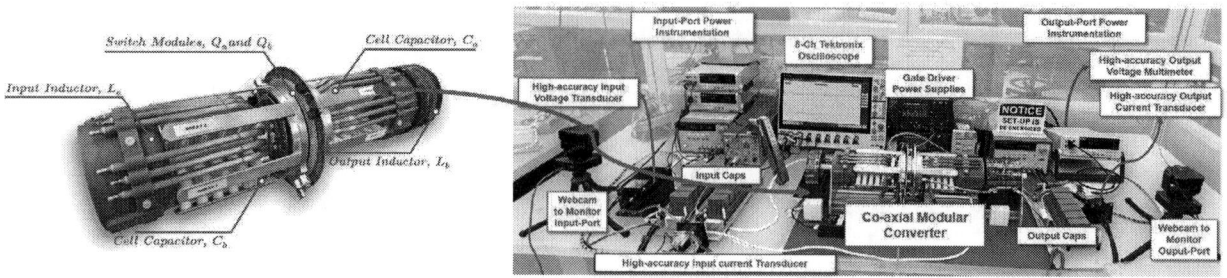

Fig. 7. 2 kV-to-400 V, 50 kW co-axial cell [16] with its full-load operating condition test-bench.

$$t_\alpha = D t_{sw} = \left(\pi - \arctan\left(\frac{M_{B1} - \mathfrak{M}}{J_{b1}} \right) \right.$$
$$\left. - \arctan\left(\frac{M_{B2} - \mathfrak{M}}{J_{B2}} \right) \right) \omega_{o,out} \quad (13)$$

$$V_{Q_{1b,pk}} = \left(\sqrt{(M_{B1} - \mathfrak{M})^2 + J_{B2}^2} + \mathfrak{M} \right) V_{base} \quad (14)$$

B. Subinterval-II (t_β)

In this subinterval (Fig. 6a), the converter enters a dead-band where the stored energy in both inductors discharges the equivalent C_{oss}, enabling ZVS for Q_{1b}. The state-plane diagram for subinterval-II (Fig. 6b) provides the timing t_β, determined using (15).

$$t_\beta = \left(\arctan\left(\frac{1}{J_{A2}\frac{R_{o,ss}}{R_{o,in}} + J_{B2}\frac{R_{o,ss}}{R_{o,out}}} \right) \right.$$
$$\left. + \arctan\left(\frac{M_{A3} - 1}{J_{A3}\frac{R_{o,ss}}{R_{o,in}} + J_{B3}\frac{R_{o,ss}}{R_{o,out}}} \right) \right) \omega_{o,ss} \quad (15)$$

C. Subinterval-III (t_γ)

This subinterval involves synchronous rectification (Fig. 6a), where the active switch Q_b remains off for a period t_γ while the synchronous rectifier Q_b is on, continuing until nearly the end of the switching period t_{sw}. The output port experiences a linear increase in current $i_{L_b}(t)$, while the input port resonates between L_a and $0.5C + C_{oss}$, resulting in $v_{Q_a}(t)$ and $i_{L_a}(t)$ exhibiting resonances with a dc bias of V_g. The state-plane diagram for subinterval-III (Fig. 6b) provides the timing t_γ (16) and the voltage stress $V_{Q_{1a,pk}}$ (17).

$$t_\gamma = \left(\pi - \arctan\left(\frac{M_{A4} - 1}{J_{a4}} \right) \right.$$
$$\left. - \arctan\left(\frac{M_{A3} - 1}{J_{A3}} \right) \right) \omega_{o,in} \quad (16)$$

$$V_{Q_{1a,pk}} = \left(\sqrt{(M_{A4} - 1)^2 + J_{A4}^2} + 1 \right) V_{base} \quad (17)$$

D. Subinterval-IV (t_δ)

In this subinterval (Fig. 6a), the converter enters a dead-band where the stored energy in the inductors discharges the equivalent C_{oss}, enabling ZVS for Q_a.

The state-plane diagram for subinterval-IV (Fig. 6c) provides the timing t_δ as given in (18).

$$t_\delta = \left(\arctan\left(\frac{\mathfrak{M}}{J_{B4}\frac{R_{o,ss}}{R_{o,out}} + J_{A4}\frac{R_{o,ss}}{R_{o,in}}} \right) \right.$$
$$\left. + \arctan\left(\frac{M_{B1} - \mathfrak{M}}{J_{A1}\frac{R_{o,ss}}{R_{o,in}} + J_{B1}\frac{R_{o,ss}}{R_{o,out}}} \right) \right) \omega_{o,ss} \quad (18)$$

The equations (9)-(18), combined with simulations, are utilized to guide the component design of a coaxial modular converter cell [16] rated for 2 kV–to–400 V, 50 kW. Notably, the CIĆC shares key features with stacked active bridge converters [6], [7], but its unique bridgeless configuration, characterized by fewer switches, enhances both efficiency and reliability.

IV. Experimental Results

A 2 kV-to-400 V, 50 kW modular co-axial converter [16], employing custom co-axial components with $L_a = 30\,\mu H$, $L_b = 13\,\mu H$, and $C_a = C_b = 660\,nF$, as shown in Fig. 7, was tested at 2 kV-to-400 V, 50 kW, achieving 97.2% efficiency at 100 kHz. Experimental results, depicted in Fig. 8, indicate that the voltage stresses across the devices are negligible and closely align with the simulation results in Fig. 5b and the qualitative waveforms in Fig. 6a, demonstrating the effectiveness of decentralizing the commutation loop.

Additionally, it is noteworthy that the co-axial modular converter is designed for passive cooling [16], with the maximum case temperature ($T_{case,max}$) of the custom co-axial switch modules reaching 150 °C.

V. Conclusion

The impact of *dimensional parasitics* is critical in the design of high step-down medium-voltage power processors and has been carefully incorporated into the development of

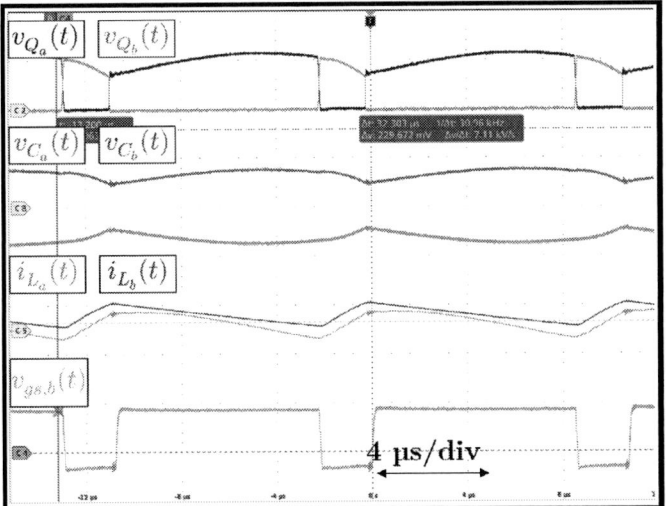

Fig. 8. Experimental waveforms of the CIĆC during steady-state operation. The first row displays the blocking voltages of the active switch $v_{Q_a}(t)$ (black) and the synchronous rectifier $v_{Q_b}(t)$ (cyan). The second row shows the capacitor voltages $v_{C_a}(t)$ (magenta) and $v_{C_b}(t)$ (green), corresponding to the split capacitors. The third row depicts the inductor currents $i_{L_a}(t)$ (orange) and $i_{L_b}(t)$ (blue), demonstrating the resonance in the converter's operation. The fourth row represents the gate-source voltage $v_{gs,b}(t)$ (green) of the synchronous rectifier Q_b.

the CIĆC. Experimental waveforms for a 2 kV-to-400 V, 50 kW operating point with 97.2% efficiency have been validated against simulations. Building on the design methodology to absorb *dimensional parasitics* and the analysis of the CIĆC presented in this work, future research will focus on further improving efficiency through the integration of advanced co-axial custom switches and passive components. While the CIĆC effectively absorbs these parasitics, its small-signal control-to-output voltage response is anticipated to exhibit fourth-order dynamics, similar to the SEPIC converter [17], posing potential stability challenges that are currently under investigation. Although not the primary focus of this work, the CIĆC's applicability to ac-to-dc, ac-to-ac, dc-to-ac, and bidirectional power conversion will be explored in future research. The results presented here validate the CIĆC cell's viability for use in ISOP systems designed for high step-down medium-voltage conversion.

VI. Acknowledgement

The information, data, or work presented herein was funded in part by the Advanced Research Projects Agency-Energy (ARPA-E), U.S. Department of Energy, under Award Number DE-AR0001568 in the OPEN-2021 program monitored by Dr. Peter de Bock. The views and opinions of authors expressed herein do not necessarily state or reflect those of the United States Government or any agency thereof.

References

[1] E. Ronan, S. Sudhoff, S. Glover, and D. Galloway, "A power electronic-based distribution transformer," *IEEE Transactions on Power Delivery*, vol. 17, no. 2, pp. 537–543, 2002.

[2] L. Yang, T. Zhao, J. Wang, and A. Q. Huang, "Design and analysis of a 270kw five-level dc/dc converter for solid state transformer using 10kv sic power devices," in *2007 IEEE Power Electronics Specialists Conference*, 2007, pp. 245–251.

[3] Z. Li, E. Hsieh, Q. Li, and F. C. Lee, "High-frequency transformer design with medium-voltage insulation for resonant converter in solid-state transformer," *IEEE Transactions on Power Electronics*, vol. 38, no. 8, pp. 9917–9932, 2023.

[4] X. Lin, D. Nam, N. Yan, J. Stewart, A. Mendes, D. Dong, R. Burgos, and D. Boroyevich, "Design and testing of a 13.8 kv/1.1mva 3-phase 5-level flying capacitor converter with 10 kv sic mosfets," in *2023 IEEE Applied Power Electronics Conference and Exposition (APEC)*, 2023, pp. 158–164.

[5] A. Barzkar, B. Fan, H. Song, J. Stewart, R. Burgos, D. Dong, and D. Boroyevich, "Modeling of conducted emi emissions in 10 kv sic mosfet based power electronics building blocks," in *2023 IEEE Energy Conversion Congress and Exposition (ECCE)*, 2023, pp. 2967–2974.

[6] B. Majmunovic and D. Maksimovic, "400–48-V Stacked Active Bridge Converter," *IEEE Trans. Power Electron.*, vol. 37, no. 10, pp. 12 017–12 029, Oct. 2022.

[7] D. Neuner, M. Hartmann, J. W. Kolar, and J. Huber, "Analysis of a solid-state transformer employing capacitively isolated series-stacked converter cells and a single medium-frequency transformer," in *2023 IEEE 24th Workshop on Control and Modeling for Power Electronics (COMPEL)*, 2023, pp. 1–8.

[8] S. Cuk and R. Middlebrook, "A new optimum topology switching DC-to-DC converter," in *1977 IEEE Power Electronics Specialists Conference*. Palo Alto, CA, USA: IEEE, Jun. 1977, pp. 160–179.

[9] C. Hoer and C. Love, "Exact inductance equations for rectangular conductors with applications to more complicated geometries," *Journal of Research of the National Bureau of Standards Section C: Engineering and Instrumentation*, vol. 69C, no. 2, p. 127, April 1965.

[10] "Round wire ac resistance calculator," https://chemandy.com/calculators/round-wire-ac-resistance-calculator.htm, accessed: Nov. 30, 2024.

[11] "Learnemc resistance calculator - wire with rectangular cross-section," https://learnemc.com/ext/calculators/resistance/rect.html, accessed: Nov. 30, 2024.

[12] K. D. T. Ngo, "Generalization of resonant switches and quasi-resonant dc-dc converters," in *1987 IEEE Power Electronics Specialists Conference*, 1987, pp. 395–403.

[13] V. Vorperian, "Quasi-square-wave converters: topologies and analysis," *IEEE Trans. Power Electron.*, vol. 3, no. 2, pp. 183–191, Apr. 1988.

[14] D. Maksimovic and S. Cuk, "A general approach to synthesis and analysis of quasi-resonant converters," *IEEE Transactions on Power Electronics*, vol. 6, no. 1, pp. 127–140, 1991.

[15] R. Oruganti and F. C. Lee, "Resonant Power Processors, Part I—State Plane Analysis," *IEEE Trans. on Ind. Applicat.*, vol. IA-21, no. 6, pp. 1453–1460, Nov. 1985.

[16] M. Cairnie, A. Kamalapur, R. Nassar, J. Knoll, Q. Yuchi, J.-S. Bae, C. DiMarino, K. Ngo, G.-Q. Lu, Q. Li, D. Boroyevich, J. Zhou, Y. Cao, D. DeVoto, and B. Kekelia, "Modular coaxial power converter for high-density integration into medium-voltage cables," in *PCIM Europe 2024; International Exhibition and Conference for Power Electronics, Intelligent Motion, Renewable Energy and Energy Management*, 2024, pp. 599–607.

[17] R. W. Erickson and D. Maksimović, *Techniques of Design-Oriented Analysis: Extra Element Theorems*. Cham: Springer International Publishing, 2020, pp. 625–673. [Online]. Available: https://doi.org/10.1007/978-3-030-43881-4_16

A 36-to-60V Input Dual-Phase 2MHz 93%-Efficiency ZVT Series-Parallel Hybrid Buck Converter Using Single Auxiliary Inductor and Adaptive Time Multiplexing Control

Qi Cheng[1,2] and Hoi Lee[1]

[1]Department of Electrical and Computer Engineering, University of Texas at Dallas Richardson, TX 75080, USA
[2]Texas Instruments Inc., Santa Clara, CA 95095, USA

Abstract— This paper presents a dual-phase series-parallel hybrid buck (SPHB) converter featuring a passive saving (PS) ZVT scheme and an adaptive time-multiplexing (ATM) controller. The aim is to intensify both power density and efficiency at light loads without reducing the switching frequency. With the PS-ZVT scheme, the converter can share the auxiliary inductor across two phases. Meanwhile, the ATM controller adjusts the peak auxiliary inductor current injected into switch nodes via the adaptive control of the interleaved auxiliary switches. Designed in a 0.5μm HV CMOS process, the proposed converter operates at a fixed 2MHz frequency, achieves over 85% efficiency across 0.2A-to-5A load, and has its power density improved by 1.6 times compared with the prior art.

Keywords— Adaptive-time-multiplexing, hybrid DC-DC converter, zero-voltage switching, zero-voltage transition

I. INTRODUCTION

Demands for 10s-of-Watt wide-input-range switched-mode power converters have surged due to increased load requirements in 48V-bus systems for automotive and industrial applications [1]. Conventional hard-switching (HS) buck converters [2, 3] encounter notable switching power losses under high input voltages, limiting the switching frequency (FS) to the range of 100s-of-kHz. Soft-switching (SS) technique in [4, 5] can mitigate the major switching power loss and enhance the power efficiency by enabling zero-voltage-switching (ZVS) turn-on of power FETs. However, the presence of fixed large continuous current ripple in the power inductor with considerable conduction loss impairs the converter light-load power efficiency. To address this challenge, methods such as zero-voltage-transition (ZVT) [6, 7] and adaptive ZVT [8] were reported to reduce the conduction loss by employing a discontinuous auxiliary inductor current and decreasing its current-ripple magnitude with the load current. However, when the load current is beyond 2A, the converter in [8] needs to switch to the HS mode to avoid excessive RMS current in the ZVT-conventional buck (CB) power stage. This however necessitates the use of bulky power inductors with low DC resistance to limit the inductor conduction loss under high load currents. To improve both power density and light-load efficiency, this paper introduces a dual-phase series-parallel hybrid buck converter with a passive-saving (PS) ZVT scheme

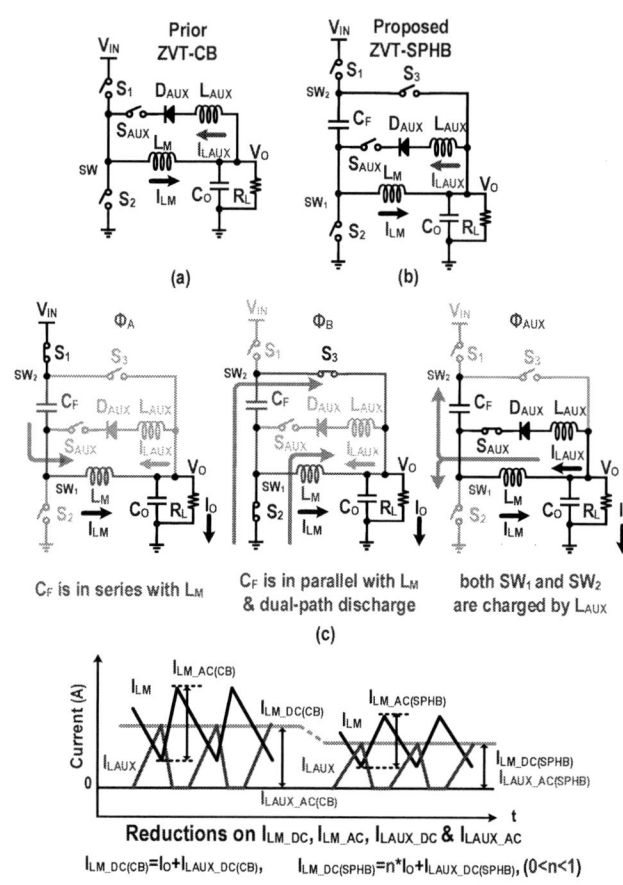

Fig. 1: (a) Prior ZVT-CB converter in [8]. (b) Structure, (c) operations, and (d) RMS current reduction advantage of the proposed ZVT-SPHB converter.

and an adaptive time-multiplexing (ATM) controller. This paper is organized as follows. Section II reviewed the advantages of proposed ZVT-series parallel hybrid buck (SPHB) converter over ZVT-CB buck converter. Section III details the design of proposed dual-phase PS-ZVT converter including the adaptive time multiplexing controller and the discontinuous mode auxiliary inductor current sensor. Simulation verifications and

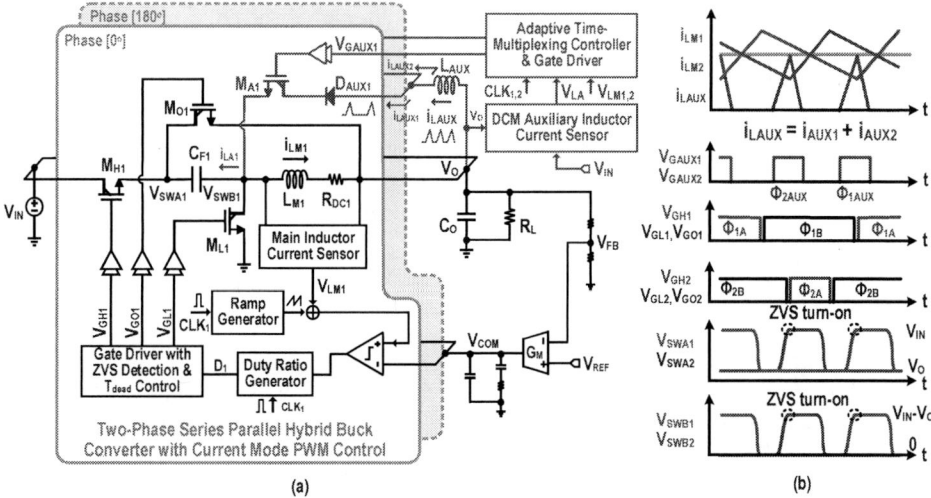

Fig. 2: (a) Structure and (b) operation of the proposed dual-phase ZVT-SPHB converter.

performance comparisons among the proposed and state-of-the-art wide-input-range buck converters are provided in Section IV. Finally, conclusions are given in Section V.

II. ZERO VOLTAGE TRANSITION SERIES-PARALLEL HYBRID BUCK CONVERTER

Figs. 1(a) and 1(b) show the structures of the conventional ZVT-CB converter [8] and the proposed single-phase ZVT-SPHB converter, respectively. Compared with the ZVT-CB converter, the proposed converter requires an additional flying capacitor C_F and one more power switch S_3. The proposed converter operates in two main states, Φ_A and Φ_B, with an auxiliary operation state, Φ_{AUX} that is active only during the transition from Φ_B to Φ_A. Detailed operations are depicted in Fig. 1(c). Energy transfer components C_F and L_M are first in series to increase I_{LM} in Φ_A, then in parallel to decrease I_{LM} in Φ_B. In the Φ_{AUX} state, the auxiliary inductor current charges switch nodes SW_1 to $(V_{IN}-V_O)$ and SW_2 to V_{IN}, realizing ZVS turn-on of S_1 during the Φ_B-to-Φ_A transition. On the other hand, during the Φ_A-to-Φ_B transition, the peak I_{LM} discharges SW_1 to ground and SW_2 to V_O, providing ZVS turn-on of S_2 and S_3. Since C_F is tied between converter output and ground during Φ_B, V_{CF} is refreshed to V_O in each switching cycle and ensures self-balancing of C_F in the steady state.

The proposed converter offers several benefits. (1) Both C_F and S_3 provide part of the load current in Φ_B to lower the average inductor current $I_{LM\_DC(SPHB)}$ of L_M. Voltage V_{CF} reduces the volt-seconds across L_M and L_{AUX}, decreasing both current ripples $I_{LM\_AC(SPHB)}$ and $I_{LAUX\_AC(SPHB)}$ of L_M and L_{AUX}, and the conduction losses of switches and inductors. (2) The maximum voltage stress V_{DS} of all power switches $S_1 - S_3$ is reduced to $V_{IN}-V_{CF}$ ($=V_{IN}-V_O$) instead of V_{IN} in the CB converter, enabling the use of lower-voltage-rating MOSFETs with smaller device figure-of-merit ($=R_{ON}*Q_G$). This reduces switch losses and saves silicon area. (3) The V_O/V_{IN} of the proposed converter equals $D_{SPHB}/(1+D_{SPHB})$, where D_{SPHB} is the duty ratio. For the same V_{IN}-to-V_O conversion ratio, D_{SPHB} is always higher than that of the CB converter. Larger switch duty ratio in the

proposed converter enables higher frequency operation, which consequently reduces the required passive values and lowers the converter volume.

III. PROPOSED PASSIVE-SAVING CONVERTER

To enhance the maximum output power/load current, a dual-phase ZVT-SPHB could provide a much lower total conduction loss of the main inductors than the single-phase design. Nevertheless, a conventional dual-phase ZVT scheme would require a designated L_{AUX} for each phase. In a bid to save converter volume and intensify the power density, the PS-ZVT scheme is proposed to allow two ZVT auxiliary branches to share one auxiliary inductor between two phases.

A. Passive-Saving Dual-Phase ZVT-SPHB Converter

Fig. 2(a) illustrates the architecture of the dual-phase PS-ZVT-SPHB converter, where each phase comprises one set of SPHB power stage and a ZVT auxiliary branch. By connecting the shared L_{AUX} to both anodes of D_{AUX1} and D_{AUX2}, the auxiliary current (i_{AUX1} or i_{AUX2}) flows into one switching node (V_{SWB1} or V_{SWB2}) in a phased manner by controlling the auxiliary switches M_{A1} and M_{A2}. Fig. 2(b) illustrates the basic operations of the proposed converter. To realize ZVS turn-on of M_{H1} and M_{H2}, M_{A1} and M_{A2} are turned on during Φ_{1AUX} and Φ_{2AUX}, respectively. The auxiliary inductor current i_{AUX} alternates through D_{AUX1} and D_{AUX2}, allowing i_{AUX1} and i_{AUX2} to charge V_{SWA1} and V_{SWA2} to V_{IN} via flying capacitor C_{F1} and C_{F2}, respectively. In the proposed converter, on-chip HV gate drivers featuring ZVS detection and dynamic dead-time were designed to properly drive power transistors for minimizing the switching power loss. Output voltage regulation is realized through a two-phase peak current mode PWM controller with Type-II compensation.

B. Adaptive Time-Multiplexing ZVT Controller

To complement the PS-ZVT scheme, we have developed an adaptive time-multiplexing (ATM) controller alongside a discontinuous-mode (DCM) auxiliary current sensor. This sensor retrieves peak auxiliary inductor current information in

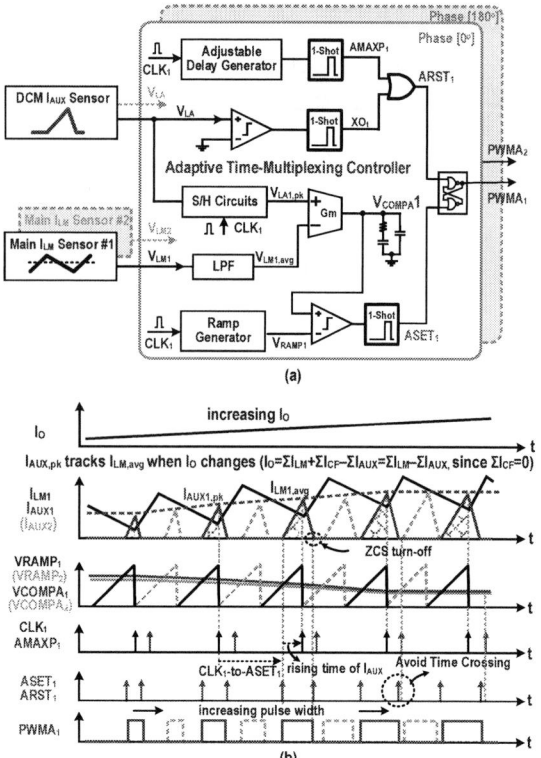

Fig. 3: (a) Structure and (b) illustration of the proposed adaptive time-multiplexing controller.

each phase, allowing for the regulation of the turn-on/off instants of auxiliary switches M_{A1} and M_{A2}. This, in turn, minimizes power loss of the ZVT branch under different loads. Fig. 3(a) shows the structure of the proposed ATM controller, while Fig. 3(b) illustrates timing waveforms of auxiliary current modulation with increasing load current.

The ATM controller takes input voltages of V_{LM1}, V_{LM2}, and V_{LA}, representing inductor current information from two main inductor current sensors and the shared DCM auxiliary inductor sensor (as depicted in Fig. 2(a)). Taking phase [0°] as an example, the ATM controller operates as follows: a sample-and-hold block senses the peak value of V_{LA} ($V_{LA1,pk}$) via the system clock CLK_1 of phase [0°]. The type-II compensator compares the average signal of V_{LM1} ($V_{LM1,avg}$) with $V_{LA1,pk}$ to generate a stable voltage V_{COMPA1}. Meanwhile, a ramp signal V_{RAMP1} that is synchronized with the CLK_1 is used to compare with V_{COMPA1} and generate $ASET_1$ to initiate the turn-on instant of the auxiliary switch M_{A1} via $PWMA_1$ when V_{RAMP1} crosses V_{COMPA1}. Essentially, V_{COMPA1} is transformed to CLK_1-to-$ASET_1$ delay based on the difference between $V_{LA1,pk}$, and $V_{LM1,avg}$. Since the CLK_1 period is fixed, the delay from $ASET_1$ to the next CLK_1 is predictable, enabling determination of the rising time and the magnitude of auxiliary current $i_{AUX1,pk}$. The compensator's high DC gain gradually adjusts $V_{LA1,pk}$ towards $V_{LM1,avg}$ due to virtual short and modulates $V_{LA1,pk}$ according to $i_{LM1,avg}$. Given that $V_{LM1,avg}$ represents the average current of the main inductor L_{M1} and the load current I_O is linearly proportional to I_{LM1} in the SPHB buck converter (according to the equation in Fig. 3(b)), the peak auxiliary current i_{AUX} can be regulated

with the load current I_O. Furthermore, the proposed ATM controller ensures proper control of the turn-off instant for M_{A1} with zero-current switching, thereby minimizing the switching loss of M_{A1}. When i_{LAUX1} decreases to 0, an XO_1 pulse triggers $ARST_1$ to turn off the auxiliary switch M_{A1}. To prevent output $PWMA_1$ signal logic errors, an extra pulse $AMAXP_1$ is introduced in the ATM controller as a preventive measure against cross-turn-on of M_{A1} and M_{L1}.

C. DCM Auxiliary Inductor Current Sensor

Different from the auxiliary current sensor in [8], which requires a second order RC sensing network [9], a high-voltage error amplifiers and a complex signal flipping and ripple cancellation scheme, the proposed auxiliary sensor emulates DCM auxiliary inductor current using simple circuitry confined to the low-voltage (5V) domain. As shown in Fig. 4(a), with the

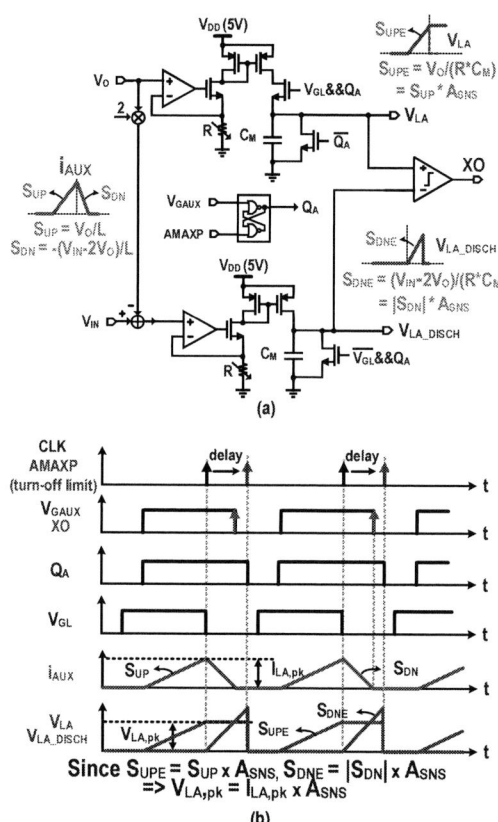

Fig. 4: (a) Schematic and (b) waveforms of the proposed DCM auxiliary current sensor.

Fig. 5: Chip micrograph.

979-8-3315-1612-3/25 $31.00 © 2025 IEEE 2238

knowledge of input/output voltage and the value of the auxiliary inductor, i_{LAUX} can be emulated via two ramp signals, V_{LA} and V_{LA_DISCH}, whose slopes (S_{UPE} and S_{DNE}) are proportional to up- and down-slope (S_{UP} and S_{DN}) of i_{AUX}, respectively.

For instance, the rising slope of i_{AUX} is V_O/L, and the up-slope of V_{LA} is $V_O/(R*C_M)$ which can be realized by a mirrored current source from a voltage-to-current converter charging a capacitor C_M. Therefore, the current sensing gain (A_{SNS}) is expressed as $A_{SNS}=S_{UPE}/S_{UP}=L/(R*C_M)$. Similar circuitry is utilized to generate V_{LA_DISCH} with the same $A_{SNS}=S_{DNF}/|S_{DN}|$ when R and C_M are perfectly matched. Fig. 4(b) shows the key timing waveforms of the auxilairy current sensor, V_{LA} is sampled at system CLK and held as $V_{LA,\ pk}$ until AMAXP (turn-off limit of V_{GAUX}) asserts. At CLK, V_{LA_DISCH} starts to charge with the slope of S_{DNE}. When V_{LA_DISCH} crosses V_{LA}, the XO signal asserts to indicate that the auxiliary current is decreased to zero, prompting M_A to turn off and minimize power loss.

IV. PERFORMANCE VERIFICATION

The proposed dual-phase ZVT-SPHB buck converted was implemented in a 0.5-μm HV CMOS technology. Fig. 5 shows the chip micrograph with a total die area of 9 mm² including 8 on-chip 55-V MOSFETs. The converter supports an input range of 36-to-60V, produces a nominal 12V output, and can deliver a load current of up to 5 A. With two main inductors of 4.7μH, the proposed converter operates at a fixed F_S of 2MHz. To enable ZVT operation, one 0.68μH auxiliary inductor and two Schottky diodes are employed to form the auxiliary branches.

Figs. 6 (a) and 6(b) depict the results of 48-V input cases under different loads. As the load current increases from 0.4 A to 4 A, the turn-on duration of the M_{A1} (M_{A2}) also increases from 50 ns to 220 ns, resulting in a rise in the peak auxiliary current $i_{AUX,\ pk}$ from 0.37 A to 2.16 A. This demonstrates the adaptive adjustment of $i_{AUX,pk}$ based on the load current level by the proposed ATM controller. Fig. 7 shows the simulated gate-drive voltages of M_{HI} (M_{H2}) and major switch-node waveforms of the proposed converter. In this figure, as the $V_{DS,MH}$ ($=V_{IN}-V_{SWA}$) of high-side power FET M_H drops to 0 before V_{GH} reaches the threshold voltage to turn on the device, full ZVS of M_H is achieved under the input voltage of 60 V. Similarly, full ZVS of M_L is also achieved as $V_{DS,ML}$ ($=V_{SWB}$) drops to 0 before V_{GL} is driven to turn-on M_L. Fig. 8 illustrates the power efficiency of the converter that operating at the 2MHz across different input voltage of 36V, 48V and 60V, spanning the load range of 0.2 A to 5 A. The peak power efficiencies are 94.7%, 93.1%, and 91%, respectively. Particularly, for the 48V input condition, the power efficiency remains above 85% across the entire load range.

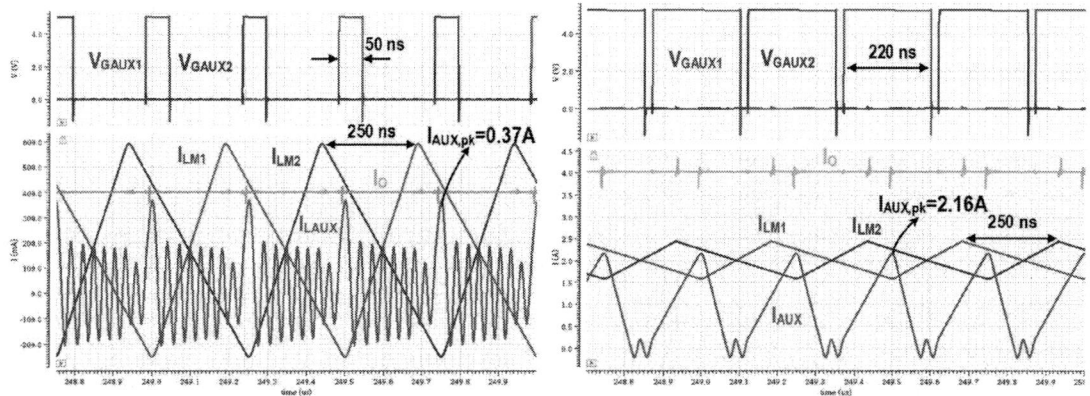

Fig. 6: Main waveforms of the proposed converter with V_{IN}=48V (a) I_O=0.4A, and (b) I_O=4A.

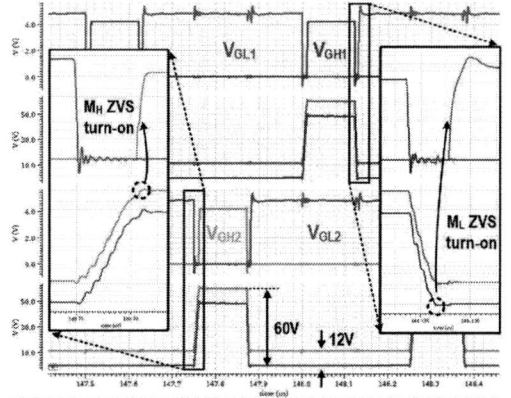

Fig. 7: Main waveforms of ZVS operation when V_{IN} = 60V.

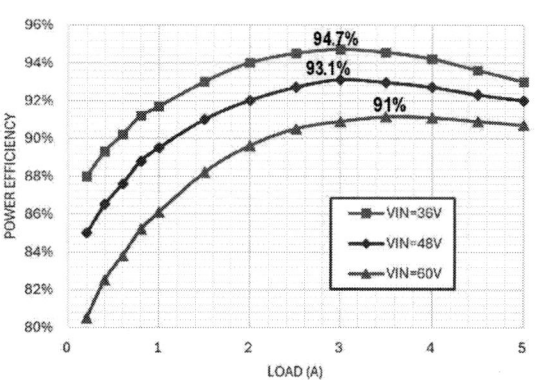

Fig. 8: Power efficiency of the proposed converter.

TABE I: Performance Comparisons Of Different Wide-Input-Range Buck Converters

	LMR38020 [3]	ISSCC [5]	ISSCC [8]	This work
Process	N.A.	0.5μm 120V CMOS	0.5μm 120V CMOS	**0.5μm 70V CMOS**
Topology	Buck	2/3-level Buck	ZVT-Buck	**2-Phase ZVT-SPHB**
Power Transistors	On-chip MOSFET	On-chip MOSFET	Off-chip eGaN	**On-chip MOSFET**
Input Voltage (Output Voltage)	24 – 64V (12V)	12 – 100V (10V)	48 – 80V (12V)	**36 – 60V (12V)**
Max. Output Power (Switching Method)	24W (HS)	5W (SS)	24W (SS) 60W (HS)	**60W (SS)**
Fs (MHz)	0.4	1	2	**2 (per phase)**
Inductor (DCR)	15μH (21.8mΩ)	1.5μH (N.A.)	1.5μH (4.3mΩ) 0.82μH (3.3mΩ)	**2 x 4.7μH (17.5mΩ) 0.82μH (7.7mΩ)**
Converter (Magnetic) Vol. (in³)	0.05 (0.0406)	N.A.	0.05 (0.0255)	**0.031 (0.02)**
Energy-Transfer Capacitors	N.A.	1 μF	N.A.	**2 x 1 μF**
Output Capacitors	3 x 22 μF	4.7μF	4 x 4.7 μF	**2 x 4.7 μF**
Worst-Case ΔVo (mV)	N.A.	100	100	**60**
Power Density(W/in³)*	480.12	N.A.	1191.75	**1923**
Peak Efficiency @ 48V-V_{IN}	95% (1A) @0.4MHz	90% (0.5A) @1MHz	92.5% (5A) @2MHz	**93.1% (3A) @2MHz (per phase)**
Full-Load Efficiency @ 48V-V_{IN}*	93% (2A) @0.4MHz	90% (0.5A) @1MHz	92.5% (5A) @2MHz	**92% (5A) @2MHz (per phase)**
Light-Load Efficiency @ 48V-V_{IN}	67% (0.1A) @0.4MHz	75% (0.1A) @1MHz	92.4% (0.5A) @2MHz	**85% (0.2A) @2MHz (per phase)**

*Power density = Max. output power / converter box volume; where converter box volume is the product of the max. height and area of all power-stage components.

Table I provides the performance comparisons of the proposed converter with state-of-the-art wide-input-range DC-DC converters. With the proposed passive saving scheme, this work achieves the smallest magnetic volume and highest power density. With the proposed ATM control, the proposed dual-phase ZVT-SPHB operation converter can achieve soft-switching while delivering output power of up to 60 W.

V. Conclusions

A dual-phase ZVT series-parallel hybrid buck converter with a single auxiliary inductor and ATM control is proposed and verified in this paper. The proposed converter operates at 2 MHz and achieves a peak efficiency of 93.1% with a 48-V input voltage. Thanks to the proposed PS-ZVT scheme, it minimizes magnetic volume while maximizing output power. The ATM controller modulates and injects the peak auxiliary inductor current for ZVS turn-on into the switched nodes of both phases in a time-multiplexed manner, enhancing power efficiency. Throughout a wide load range of 0.2 A to 5 A, the converter maintains efficiency above 85% and enhances the power density by 1.6 times compared with the prior art.

References

[1] L. Kou and J. Lu, "A GaN and Si hybrid solution for 48V-12V automotive DC-DC application," in *Proc. IEEE Energy Convers. Congr. Expo.*, pp. 2858 – 2864, Oct. 2020.

[2] Texas Instruments, LM46002, "3.5-V to 60-V, 2-A synchronous step-down voltage converter," datasheet, Apr. 2019.

[3] Texas Instruments, LMR38020, "4.2-V to 80-V, 2-A synchronous buck converter with 40-μA I_Q," Datasheet, Jan. 2022.

[4] I. H. Oh, "A soft-switching synchronous buck converter for zero voltage switching (ZVS) in light and full load conditions," in *Proc. IEEE Appl. Power Electron. Conf. Expo.*, pp. 1460 – 1464, Feb. 2008.

[5] J. Xue and H. Lee, "A 2MHz 12-to-100V 90%-efficiency self-balancing ZVS three-level DC-DC regulator with constant-frequency AOT V^2 control and 5ns ZVS turn-on delay," in *IEEE ISSCC Dig. Tech. Papers*, pp. 226 – 227, Feb. 2016.

[6] C. Nan and R. Ayyanar, "A high frequency zero-voltage-transition (ZVT) synchronous buck converter for automotive applications," in *Proc. IEEE Energy Convers. Congr. Expo.*, pp. 1 – 6, Sep. 2016.

[7] E. Adib and H. Farzanehfard, "Zero-voltage-transition PWM converters with synchronous rectifier," IEEE Trans. Power Electron., vol. 25, no. 1, pp. 105 – 110, Jan. 2010.

[8] Q. Cheng, L. Cong and H. Lee, "A 48-to-80V input 2MHz adaptive ZVT-assisted GaN-based bus converter achieving 14% light-load efficiency improvement," in *IEEE ISSCC Dig. Tech. Papers*, pp. 196 – 197, Feb. 2020.

[9] Y. Yan and E. Gu, "A precise sub-milliohm DCR current sensing scheme for area-efficient multiphase peak current mode controller", in *Proc. IEEE Appl. Power Electron. Conf. Expo.*, pp. 927 – 933, Jun. 2021.

Improved Efficiency in a 10 W Class-Φ_2 Converter Utilizing a Resonant Gate Drive

Malachi Hornbuckle
Electrical Engineering
Stanford University
Stanford, CA 94305
malachih@stanford.edu

Katherine Liang
Electrical Engineering
Stanford University
Stanford, CA 94305
katliang@stanford.edu

Juan Rivas-Davila
Electrical Engineering
Stanford University
Stanford, CA 94305
jmrivas@stanford.edu

Abstract—GaN power devices are popular for VHF power conversion due to their favorable features such as small capacitances, low threshold voltages, and compact size. Here, we present the design of a class-Φ_2 power converter that uses a GaN switching device, analyzing its performance delivering 10 W at 50 MHz. We conducted a comparative study between two different gate drive topologies for this converter: a conventional gate driver IC and a resonant gate drive topology. The conventional implementation achieves a total efficiency of 78.7%, while the resonant topology reaches 84.1%, demonstrating the importance of minimizing gating losses to design high efficiency VHF converters.

Index Terms—Resonant Power Converter, Zero-Voltage Switching, Phi2 Converter, RF Power, GaN Power Device, Gate Drive

I. INTRODUCTION

Very high frequency (VHF) power supplies are used in a variety of applications including semiconductor fabrication, MRI, and radar [1]–[3]. Switch-mode converter topologies, which utilize a power semiconductor device as a switch, are an appealing option for such power supplies since their operation is theoretically lossless. Increased development of power devices built in Gallium Nitride (GaN), a wide-bandgap material, has provided opportunities for pushing the frequencies, power levels, and efficiencies achievable with these topologies due to favorable device characteristics such as lower on-resistances, smaller parasitic capacitances, and higher voltage ratings [4], [5]. Despite challenges associated with VHF switch-mode power supplies, such as gating losses and C_{OSS} losses, the benefits of GaN's material properties can enable the design of high-efficiency VHF power supplies.

This work presents the design and implementation of a class-Φ_2 dc-ac converter using a GS61004B GaN transistor to deliver 10 W of power to a 50 Ω load at 50 MHz. Gating losses, which occur in every switching cycle and therefore scale with frequency, can significantly impact overall efficiency at VHF. To investigate this impact, two versions of the same converter are constructed with different strategies for driving the gate of the GaN device. Through this comparison, a 5.4% improvement in overall efficiency is observed in the version with a less power-intensive gate drive design. The converter operates with a single 12 V dc supply and receives a 10 dBm 50 Ω referenced input signal at 50 MHz. It is desired that this input signal drive an impedance close to

50 Ω, resulting in a low input standing wave ratio (SWR). Low harmonic distortion in the output signal is desired, so filtering techniques are used to attenuate harmonic content by 50 dB relative to the 50 MHz switching frequency.

In this paper we first discuss the two gate drive strategies tested in the class-Φ_2 design, as well as test results from other gate driving methods implemented on stand-alone GS61004B devices. Second, the design of our class-Φ_2 topology is discussed, including our chosen switching device, output filtering, and component values. We then analyze the performance of the two gate drives in converters, noting in particular the efficiency improvement in the converter with a resonant gate drive (RGD) compared to the one with the conventional gate drive (CGD) from a commercial IC. Finally, we estimate sources of loss in the converters and consider how these losses could be reduced further.

II. GATE DRIVE DESIGN

A. Power Source

A buck converter with APC63200 is used to step down the 12 V source supplying the drain to a 5 V source for the gate driver circuit. This buck converter is labeled in Fig. 6 as "5 V Buck". The 10 dBm input signal drives a TLV3502 comparator to generate a 5 V pulse width modulated (PWM) square wave which serves as the input to the rest of the gate drive. A 50 Ω resistor in parallel with the high input impedance of the comparator provides a reasonable match to the input signal and low SWR.

B. Drivers

A variety of drivers are tested on stand-alone GS61004B devices, including the LMG1025 driver, 551MILFT clock buffer, SN74LVC2G04 digital inverter, and TLV3502 comparator. Each driver receives its input signal from the single output of a TLV3502 comparator operating at 50% duty cycle. In order to achieve fast rising and falling edges, all four outputs of the clock buffer are connected to the gate of the switching device. Similarly, the outputs of four inverters are connected, and six comparator outputs are connected, forming drivers capable of delivering more current. The resulting waveforms and power consumptions of these drivers are shown in Fig. 1.

979-8-3315-1612-3/25 $31.00 © 2025 IEEE

The theoretical loss incurred from charging and discharging the gate capacitance with a driver is $f_s V_{gate} Q_{gate}$ [6]. Based on the GS61004B datasheet, we estimate the gate requires a total of 2.75 nC to reach 5 V, leading to an expected power dissipation just under 700 mW. Experimentally, we measure power levels that are above and below this level. The output capacitance of the driver itself presents an additional capacitance that must be driven in addition to the gate, resulting in more power consumption. In particular, the LMG1025 driver (900 mW) seems to suffer from this effect as its output capacitance is likely large to support its high current rating. Parasitic inductance from packages and PCB traces can also affect performance, introducing a small amount of resonance to the system (which reduces power consumption) and even increasing the amplitude of the voltage waveform seen at the gate. This may help explain our observations regarding the 551MILFT driver (545 mW) and the SN74LVC2G04 driver (660 mW).

C. Resonant Topologies

Along with testing different drivers, we test resonant topologies that can reduce the power consumed by a gate drive circuit. Many resonant gate drive topologies have been proposed, and in this work we measure the power consumptions and gate waveforms of the topologies shown in Fig. 2. For the topologies in Fig. 2a-c, the SN74LVC2G04 digital inverter is used as a driver due to its low input power and the large voltage swing it exhibits in Fig. 1d.

1) Long Topology (Fig. 2a): This topology seen in [7], which we will refer to as the "Long topology", utilizes a series inductor L_{Res} to resonantly charge the gate capacitance, with clamping diodes D_{Clamp} to prevent the gate voltage from exceeding 5 V or dropping below 0 V. In our implementation of this topology, Coilcraft 0805HP inductors and BAT60AE6327HTSA1 diodes were used. A larger inductance limits the power consumption of this topology, but slows down the waveform.

2) Gu Topology (Fig. 2b): We design this multiresonant topology, which we will refer to as the "Gu topology", using the methodology described in [8]. In this methodology, the driver is treated as a square voltage waveform source, and the resonant elements L_F, L_{MR}, and C_{MR} are selected based on the desired voltage amplitude at C_G in the fundamental, second, and third harmonics. The amplitude of the gate waveform and the power required can be adjusted by changing passive element values. Coilcraft 0805HP inductors are used as L_F and L_{MR} in our implementation.

3) Fujita Topology (Fig. 2c): This parallel type gate drive from [9] includes a parallel inductance L_{Res} tuned close to resonance with C_G. As a result, L_{Res} supplies some of the current that charges and discharges C_G to reduce the current, and therefore power, supplied by the driver. C_B is a dc blocking capacitor providing a level shift to the average voltage of the gate node. We realize this design using a hand-wound inductor to allow greater tuning flexibility. As the resonant frequency of the parallel LC tank approaches

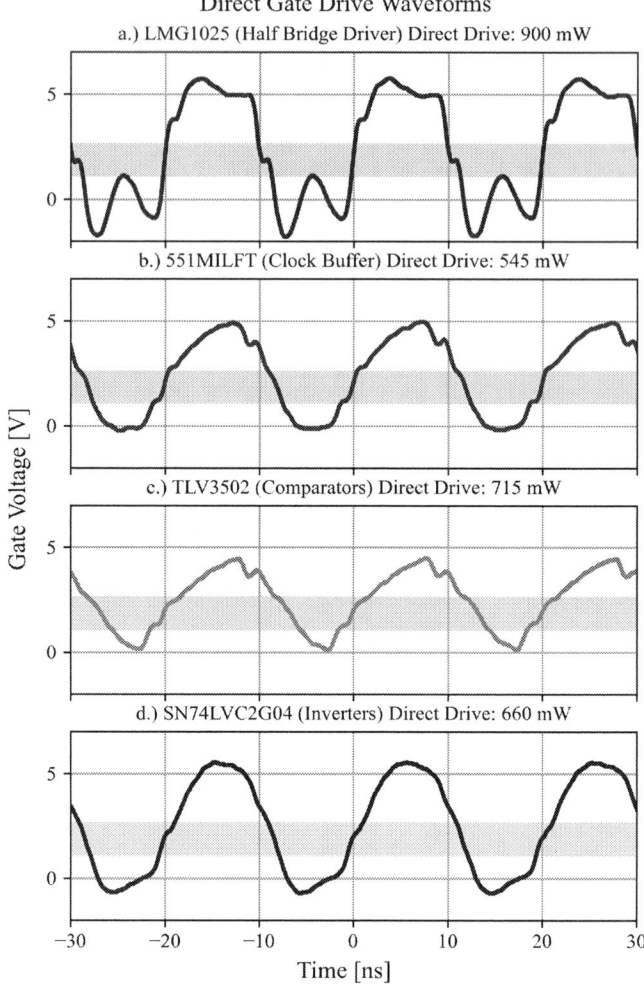

Fig. 1. Waveforms and measured powers from drivers tested on standalone GS61004B gates, with yellow shaded regions indicating the published expected range for the threshold voltage V_{TH}. The waveforms shown are generated by the a.) LMG1025 driver, b.) 551MILFT clock buffer, c.) TLV3502 comparators, and d.) SN74LVC2G04 digital inverters.

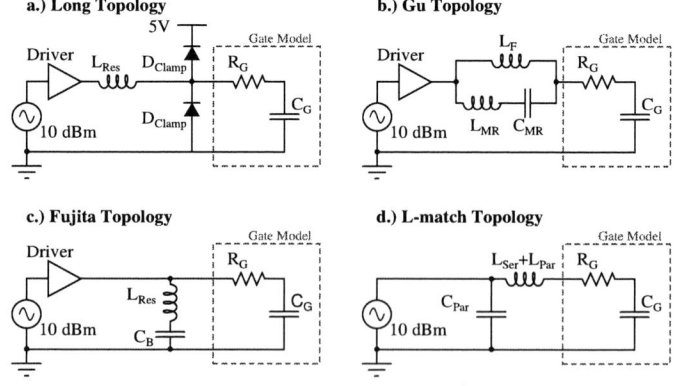

Fig. 2. Tested resonant gate drive topologies.

the switching frequency, more current is recycled by L_{Res}, reducing the power consumption.

4) L-match Topology (Fig. 2d): The last resonant gate drive we test is inspired by the series type drive in [9]. Assuming a relatively small R_G, a series inductor L_{Ser} can be added to the circuit to resonate with C_G at the switching frequency and produce a low impedance. We add extra inductance L_{Par} to L_{Ser}, and then C_{Par} in parallel to the existing circuit to resonate with L_{Par} at the switching frequency, effectively forming the L-match network described in [10] and discussed further below. The resulting impedance presented to the input 10 dBm signal is close to enough to 50 Ω to produce a satisfactory input SWR. Due to the circuit's resonance, a large current should flow between the C_{Par} and C_G branches at the switching frequency, yielding a sizable voltage swing across the gate capacitance.

5) Test Results: Waveforms from each of the resonant topologies discussed above are shown in Fig. 3. For the Long, Gu, and Fujita topologies, two waveforms are displayed to highlight the dependence of waveform shape and power consumption on component values. In the Long topology waveforms of Fig. 3a, we observe that a larger inductance tends to reduce power consumption, but reduces waveform amplitude. The Gu topology waveforms in Fig. 3b indicate that the circuit can be tuned for a higher gate waveform amplitude, but it requires increased power. From the Fujita topology waveforms in Fig. 3c, we see that decreasing L_{Res} to push the circuit's resonant frequency closer to 50 MHz yields lower power consumption. After first-pass tuning, the L-match topology resonates and produces the largest waveform amplitude near 35 MHz, as shown in Fig. 3d. We observe that this amplitude is likely insufficient to effectively drive the gate, so effort is not spent tuning the circuit to 50 MHz and the L-match topology is excluded from consideration in this application.

D. Selected Gate Drives

Fig. 4 compares the input power levels of the tested gate drive strategies. The LMG1025 is selected for the CGD due to its design for gate drive applications and sharp rising and falling edges. This benefit comes at the expense of input power, as it is measured to consume 900 mW driving a stand-alone switching device. The inverter-driven Fujita topology is chosen for use in the RGD converter. With an estimated L_{Res} value of 45 nH it is only measured to require only 340 mW while maintaining large voltage swings across the switching device gate capacitance. The duty cycle of the gate voltage waveform provided by the Fujita topology also responds well to the duty cycle of the PWM signal driving the inverters, allowing duty cycle to be easily tuned to help the class-Φ_2 topology reach zero voltage switching (ZVS). These CGD and RGD are integrated with our converter design to provide a direct performance comparison driving a switching device in an active circuit.

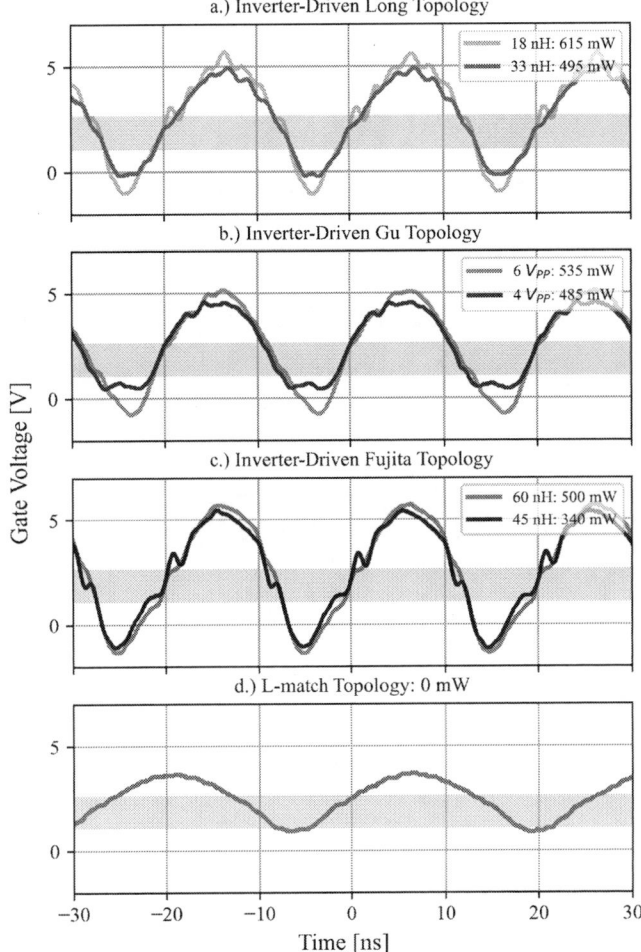

Fig. 3. Resonant gate drive waveforms and powers on stand-alone GS61004B gates, with yellow shaded regions indicating the published expected range for the threshold voltage V_{TH}. The waveforms shown are generated by the a.) Long topology, b.) Gu topology, c.) Fujita topology, and d.) L-match topology.

III. CONVERTER DESIGN

A. Switching Device and Topology

An important step in designing a converter for operation in the VHF range is choosing a suitable switching device. Most switch-mode VHF converter topologies operate under ZVS to minimize the switching losses that are repeated with every cycle, and therefore scale directly with frequency [11], [12]. Achieving ZVS requires circuit designers to account for the parasitics associated with the switching device(s), particularly the output capacitance C_{OSS}, to ensure that the drain-source voltage V_{DS}, otherwise known as the drain waveform, reaches 0 V before the switch is turned on. Though its voltage and current ratings are larger than needed for a 10 W application, the Enhancement-mode GaN power transistor GS61004B from Infineon is chosen as the switching device due to the thermal characteristics of its package. Relevant device parameters are shown in Table I.

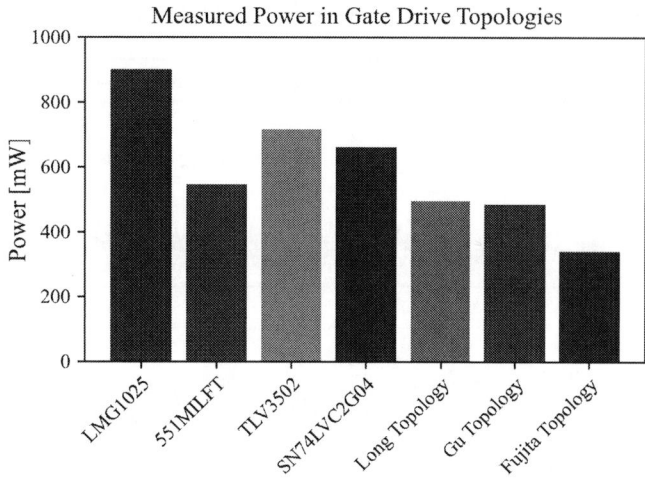

Fig. 4. Comparison of the measured power for different gate drives tested, including four different drivers and three resonant topologies driven by SN74LVC2G04 digital inverters. The L-match topology is not included as it requires no power beyond the 10 dBm input signal.

TABLE I
GS66104B SWITCHING DEVICE PARAMETERS

Device	C_{OSS}	C_{ISS}	$V_{DS,max}$	$R_{DS,on}$
GS66104B	200 pF[a]	260 pF	100 V	16 mΩ

[a] Approximate value over the 0-30 V range of V_{DS} in this work.

In this work we implement the class-Φ_2 topology, shown in Fig. 5 inside the "Converter Topology" box. This topology is an offshoot of the class-Φ topology, which utilizes a quarter-wavelength transmission line to eliminate even harmonic components from the drain waveform [13]. Instead, the class-Φ_2 uses a multiresonant network of passive elements (L_F, L_{MR}, C_{MR}, and $C_P + C_{OSS}$) to mimic the impedance of the class-Φ's quarter-wavelength transmission line up to the third harmonic of the switching frequency [14]. This topology also includes a blocking capacitor C_B to provide dc isolation to the output, as well as a series inductor L_S which assists with ZVS by making the load seen by the converter more inductive [14]. At 50 MHz switching frequency, the class-Φ_2 topology is chosen over the multi-switch class-D and class-DE topologies shown in [11] due to the simplicity of driving a single ground-referenced switching device at high frequencies. Other resonant topologies like the class-E [11] and class-F [15] similarly utilize a single switching device and have great potential for this application. However, the class-Φ_2 is chosen due to its strong second harmonic attenuation, which improves the purity of the spectrum at the output, as well as its fast transient speeds and relatively small number of lumped elements.

B. Output Impedance Transformation and Filtering

Since the class-Φ_2 drain waveform approximates a square wave alternating between 0 V and $2V_{in}$, its rms fundamental

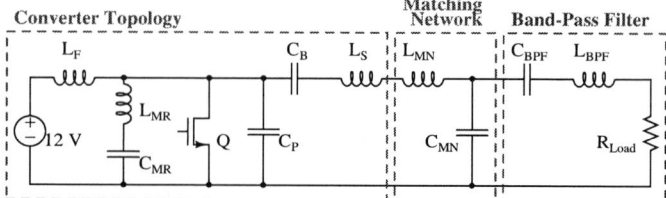

Fig. 5. Schematic for Φ_2 converter topology, plus matching network and band-pass filter.

component is roughly that of a square wave [14]:

$$V_{D,rms} \approx \frac{2\sqrt{2}V_{in}}{\pi} \quad (1)$$

Given that V_{in} = 12 V, if this rms voltage were applied directly to a 50 Ω load, it would only provide a power of 2.33 W. As a result, the matching network shown in Fig. 5 is included in the design to reduce the load impedance seen by the converter. A low-pass L-match network [10] was used for the impedance transformation due to its 40 dB per decade roll-off above the switching frequency, providing attenuation of harmonics in the output spectrum. Equation (1) implies that the converter must see a load of less than 11.7 Ω in order to deliver 10 W at the fundamental frequency. However, this analysis ignores L_S, which is in series with the converter output and therefore reduces the rms voltage seen by the load. As a result, the L-match network is designed to transform the 50 Ω load into 6 Ω, allowing the converter to comfortably deliver 10 W.

The LC band-pass filter (BPF) shown in Fig. 5, which rolls off at 20 dB per decade both above and below the switching frequency, is included to further attenuate non-fundamental spectral components in the output waveform. Its quality factor can be varied to fine-tune the attenuation of lower frequency noise (such as from the buck converter discussed above) and higher frequency harmonics. A higher BPF quality factor increases the attenuation of undesired spectral components, but also increases the series resistance in L_{BPF}, thereby increasing conduction losses.

C. Component Value Selection

The input wave-shaping network consisting of L_F, L_{MR}, C_{MR}, and $C_F = C_P + C_{OSS}$ is largely determined using (2), which comes from the transmission line-based method from [14].

$$L_F = \frac{1}{9\pi^2 f_s^2 C_F}, \; L_{MR} = \frac{1}{15\pi^2 f_s^2 C_F}, \; C_{MR} = \frac{15}{16}C_F \quad (2)$$

Here, f_s is the switching frequency and we assume that C_{OSS} behaves as roughly 200 pF over the relevant voltage range. First, a small value of 25 pF is chosen for C_P to minimize the circulating currents needed to charge and discharge the drain node while accommodating possible variance in the value of C_{OSS} and any PCB parasitics. Next, C_B and L_S were selected. C_B is simply a dc blocking capacitor, and was selected as a large 100 nF to provide dc isolation between the converter and the load. L_S is important for two main reasons. First, it causes the load driven by the converter to be inductive, meaning its current waveform will lag in time behind its voltage waveform. This causes L_S to continue pulling current out of the drain node capacitance ($C_P + C_{OSS}$) as the drain voltage approaches 0 V, helping to enable ZVS. Second, L_S creates a voltage divider, splitting the voltage seen at the drain between itself and the converter load and therefore determining the power delivered to the load. Analysis from [14] yields an expression that can be used to determine the value of L_S required to deliver 10 W into a 6 Ω converter load:

$$L_S = \frac{\sqrt{R_L}}{\omega}\sqrt{\frac{||V_{D,rms}||^2}{P_o} - R_L} = 18.5 \text{ nH} \quad (3)$$

In (3), $\omega = 2\pi f_s$ is the angular switching frequency and P_o is the output power. The values for L_{MN} and C_{MN} are chosen based on the impedance transformation equations in [10] to produce the desired 6 Ω load impedance. The BPF is tuned to a quality factor near one, which is sufficient to attenuate all output spectral components by at least 50 dB relative to the carrier frequency of 50 MHz (see Fig. 10). In the LTSpice simultion of the converter, L_F is reduced to slightly below the value calculated by (2) to more easily reach ZVS. The relatively large C_{OSS} of the switching device requires a large circulating current to charge and discharge at high frequencies, and decreasing L_F is a simple way to increase this circulating current. Final component values for the topology in Fig. 5 are displayed in Table II.

IV. Performance Comparison

A. Measured Performance and Analysis

The constructed CGD and RGD converters are displayed in Fig. 6 and Fig. 7 respectively. Both converters are built with the same components for all elements in Fig. 5 and similar tuning in an attempt to isolate the differences in performance caused by their gate drives. Using the current probe loops seen at the top of the PCBs in Fig. 6 and Fig. 7, we are able to measure the current entering the gate and drain sides of the device independently during steady state 10 W output operation, allowing us to calculate the power and efficiency data shown in Table III. The RGD converter has a clear advantage in overall efficiency, requiring nearly 1 W less input power to produce the same 10 W output as the CGD converter.

A comparison of the final gate waveforms in operational converters is provided in Fig. 8. The CGD consumes roughly

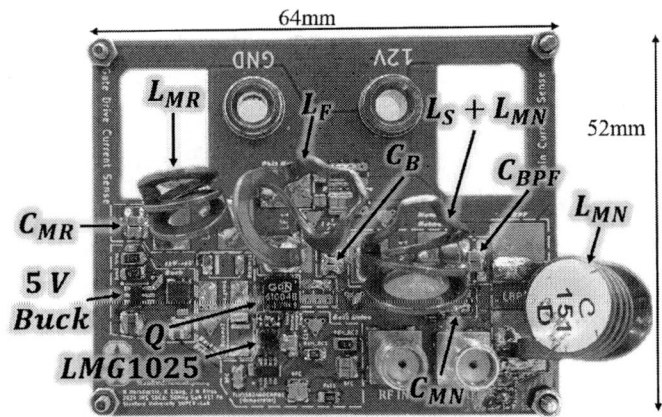

Fig. 6. CGD converter, with important components annotated.

Fig. 7. RGD converter, with important elements of the RGD annotated.

1.1 W, about twice that of the 0.51 W required by the RGD. These powers are slightly larger than those measured on stand-alone switching devices due to inefficiencies in the buck converter IC supplying the gate drive and dynamics in the operating converter that are not captured by driving a stand-alone device.

The characteristic second harmonic attenuation of the class-Φ_2 topology is evident in the symmetry of the drain waveforms of Fig. 9, as well as in the spectra of Fig. 10. Here, the strength of the second harmonic is roughly equal to that of the third,

TABLE II
COMPONENT VALUES

Component	Value	Component	Value	Component	Value
L_F	29 nH	C_P	25 pF	C_{MN}	170 pF
L_{MR}	36 nH	C_B	100 nF	C_{BPF}	64 pF
C_{MR}	70 pF	L_S+L_{MN}	70 nH	L_{BPF}	160 nH

TABLE III
CGD AND RGD CONVERTER PERFORMANCE COMPARISON AT 10 W OUTPUT

Spec	CGD	RGD
Total Input Power [W]	12.70	11.89
Total Efficiency	78.7%	84.1%
Gate Power [W]	1.10	0.51
Drain Power [W]	11.60	11.38
Drain Efficiency	86.2%	87.9%

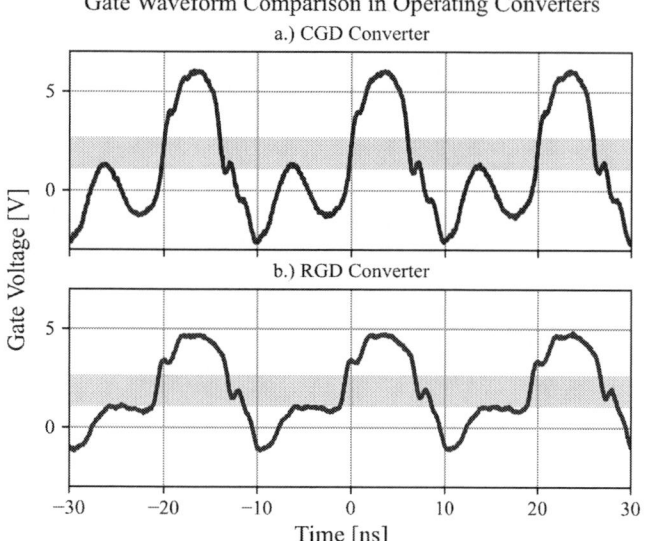

Fig. 8. Comparison of the gate waveforms of both converters, in steady state operation outputting 10 W. The yellow shaded region indicates the published expected range for the threshold voltage V_{TH}.

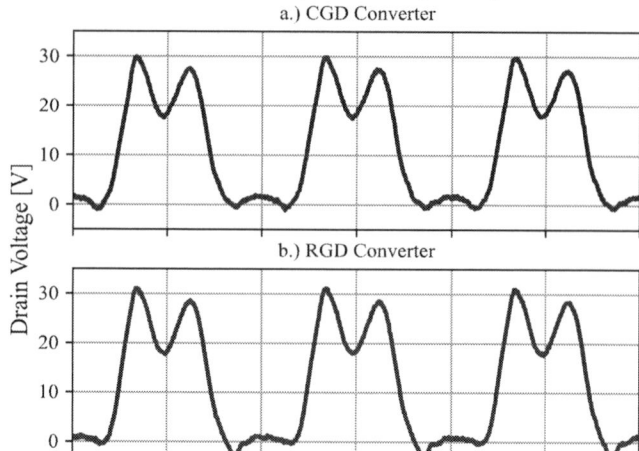

Fig. 9. Comparison of the drain waveforms of both converters, in steady state operation outputting 10 W.

even though it is attenuated less by the low-pass filtering in the output network, implying that the drain waveform has a much smaller second harmonic component.

The similarity between the drain efficiencies in Table III, as well as the drain waveforms in Fig. 9 and the spectra in Fig. 10, gives us confidence that the drain sides of the CGD and RGD converters are operating similarly. The difference in drain efficiency between the two converters is likely due to small differences in tuning, but can also be partly attributed to greater conduction losses in the CGD switching device due to increased temperature. The steady state temperature of the CGD switching device is around 70°C compared to 45°C for the RGD switching device. A possible explanation for this temperature discrepancy is the increased amplitude and oscillations of the CGD waveform in Fig. 8a, implying higher current flow in the gate resistance and more power dissipation in the CGD device. Overall, we find that the RGD significantly improves overall efficiency without adversely affecting performance on the drain side of the converter.

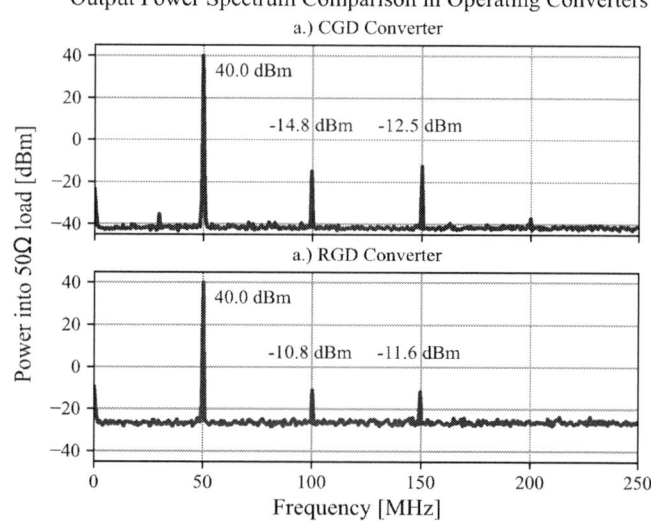

Fig. 10. Experimental output spectrum measurements for both versions of the converter.

B. Parasitics and Simulation

In the drain waveform measurements of Fig. 9 it is difficult to precisely determine if ZVS is achieved in the two converters due to the small "bump" in drain voltage when the switching device is turned on. However, the duty cycle of each converter is tuned to maximize efficiency, and it is clear that the switches turn on under very small drain-source voltages. An LTspice simulation of the converter closely matches the experimental drain waveform when parasitic inductance in the power current loop is accounted for, as seen in Fig. 11. Roughly 0.5 nH is added in series with the switching device (directly below Q in Fig. 5) in this simulation. Probing directly from the

source to the drain of the switching device package does not appreciably change the drain waveform, so we hypothesize that this parasitic inductance can be largely attributed to the device itself.

C. Estimation of Losses

With our simulation we can also estimate sources of loss in the converter. Based on the high quality factors of SMD capacitors compared to inductors, we predict that the greatest power losses are in the inductors and the switching device. Using the plots of quality factor provided by Coilcraft for its 1212VS and 2014VS inductors, we estimate the equivalent

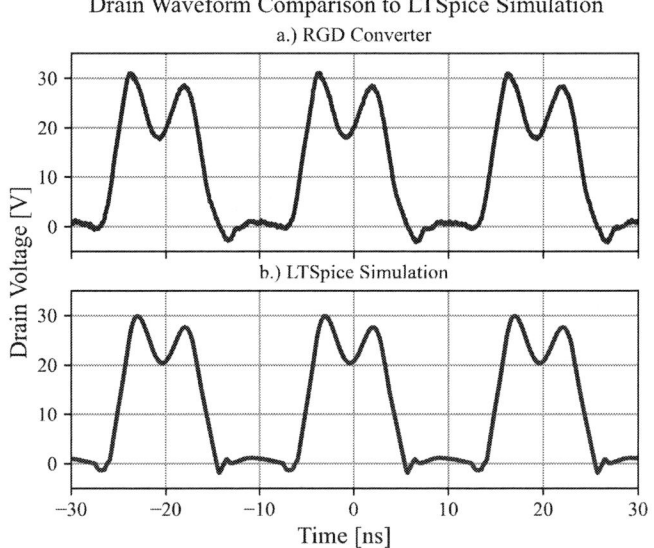

Fig. 11. Comparison of the drain waveforms of our LTspice simulation and the RGD converter.

series resistance (ESR) of our inductors at each harmonic of the switching frequency as $R_{ESR} = \omega L / Q$. With the rms current and resistance of each inductor at each harmonic frequency, we estimate the losses in each inductor as $P_{loss} = I_{rms}^2 R_{ESR}$. Conduction losses in the switching device are calculated in simulation using the manufacturer's SPICE model. C_{OSS} losses in our switching device are important at this frequency, and their mechanisms and characteristics are an active research area [16]. As was done in [17], we estimate our C_{OSS} loss per cycle as $0.2E_{OSS}$, where E_{OSS} is calculated from the device datasheet as the peak energy stored in C_{OSS} each cycle. We estimate that $C_{OSS} \approx 0.8$ nJ, leading to a loss estimation of 800 mW:

$$P_{COSS\ Loss} \approx 0.2 f_s E_{OSS} = 800 \text{ mW} \qquad (4)$$

Table IV reports our drain side loss calculations. These losses total to approximately 1.52 W and closely match the measured drain-side losses of 1.60 W and 1.38 W from Table III. C_{OSS} losses are the most dominant, and have been found to scale with the size of a switching device's output capacitance [18], implying that they could be reduced in a device with a smaller die size. Although a smaller die would increase on-resistance, device conduction losses are

TABLE IV
ESTIMATING SOURCES OF LOSS

Source	Loss	Source	Loss	Source	Loss
L_F	153 mW	L_{MR}	250 mW	$L_S + L_{MN}$	148 mW
L_{BPF}	45 mW	Q Cond.[a]	127 mW	C_{OSS}	800 mW

[a]Conduction losses in Q, the switching device.

currently much smaller than C_{OSS} losses in this converter, so this trade-off could be worthwhile. Additionally, the large resonant currents required to discharge C_{OSS} contribute to considerably to conduction losses losses in L_F, L_{MR}, and the switching device. A switching device with less output capacitance is therefore expected to reduce these losses as well. These findings highlight the importance of selecting appropriate switching devices when designing high-efficiency, high-frequency power converters.

V. CONCLUSION

This work demonstrates the impact of gating losses on the efficiency of a 50 MHz class-Φ_2 dc-ac converter utilizing a GaN switching device. Two implementations of the same converter are constructed to directly compare the performances of a conventional IC-based gate drive and a resonant gate drive technique. With the drain sides of the converters operating similarly, the resonant gate drive provides a 5.4% overall efficiency advantage compared to the conventional gate drive. Key aspects of the converter's design, performance, and sources of loss are discussed, along with a possible strategy for further improving the efficiency in future work.

REFERENCES

[1] K. Suzuki, S. Okudaira, N. Sakudo, and I. Kanomata, "Microwave plasma etching," *Japanese Journal of Applied Physics*, vol. 16, no. 11, p. 1979, nov 1977. [Online]. Available: https://dx.doi.org/10.1143/JJAP.16.1979

[2] B. Gruber, M. Froeling, T. Leiner, and D. W. Klomp, "Rf coils: A practical guide for nonphysicists," *Journal of Magnetic Resonance Imaging*, vol. 48, no. 3, pp. 590–604, 2018. [Online]. Available: https://onlinelibrary.wiley.com/doi/abs/10.1002/jmri.26187

[3] F. H. Raab, "GaN-FET class-E amplifier for 60-MHz radar," in *2020 50th European Microwave Conference (EuMC)*, 2021, pp. 1099–1102.

[4] D. J. Perreault, J. Hu, J. M. Rivas, Y. Han, O. Leitermann, R. C. Pilawa-Podgurski, A. Sagneri, and C. R. Sullivan, "Opportunities and challenges in very high frequency power conversion," in *2009 Twenty-Fourth Annual IEEE Applied Power Electronics Conference and Exposition*, 2009, pp. 1–14.

[5] R. Sun, J. Lai, W. Chen, and B. Zhang, "GaN power integration for high frequency and high efficiency power applications: A review," *IEEE Access*, vol. 8, pp. 15 529–15 542, 2020.

[6] Y. Ren, M. Xu, J. Zhou, and F. Lee, "Analytical loss model of power MOSFET," *IEEE Transactions on Power Electronics*, vol. 21, no. 2, pp. 310–319, 2006.

[7] Y. Long, W. Zhang, D. Costinett, B. B. Blalock, and L. L. Jenkins, "A high-frequency resonant gate driver for enhancement-mode GaN power devices," in *2015 IEEE Applied Power Electronics Conference and Exposition (APEC)*, 2015, pp. 1961–1965.

[8] L. Gu, Z. Tong, W. Liang, and J. Rivas-Davila, "A multiresonant gate driver for high-frequency resonant converters," *IEEE Transactions on Industrial Electronics*, vol. 67, no. 2, pp. 1405–1414, 2020.

[9] H. Fujita, "A resonant gate-drive circuit capable of high-frequency and high-efficiency operation," *IEEE Transactions on Power Electronics*, vol. 25, no. 4, pp. 962–969, 2010.

[10] T. H. Lee, *Planar Microwave Engineering: A Practical Guide to Theory, Measurement, and Circuits*. Cambridge University Press, 2004.

[11] D. Hamill, "Class DE inverters and rectifiers for dc-dc conversion," in *PESC Record. 27th Annual IEEE Power Electronics Specialists Conference*, vol. 1, 1996, pp. 854–860 vol.1.

[12] S. Kee, I. Aoki, A. Hajimiri, and D. Rutledge, "The class-E/F family of ZVS switching amplifiers," *IEEE Transactions on Microwave Theory and Techniques*, vol. 51, no. 6, pp. 1677–1690, 2003.

[13] J. W. Phinney, D. J. Perreault, and J. H. Land, "Radio-frequency inverters with transmission-line input networks," *IEEE Transactions on Power Electronics*, vol. 22, no. 4, pp. 1154–1161, 2007.

979-8-3315-1612-3/25 $31.00 © 2025 IEEE

[14] J. M. Rivas, Y. Han, O. Leitermann, A. D. Sagneri, and D. J. Perreault, "A high-frequency resonant inverter topology with low-voltage stress," *IEEE Transactions on Power Electronics*, vol. 23, no. 4, pp. 1759–1771, 2008.

[15] F. Raab, "Maximum efficiency and output of class-F power amplifiers," *IEEE Transactions on Microwave Theory and Techniques*, vol. 49, no. 6, pp. 1162–1166, 2001.

[16] M. Guacci, M. Heller, D. Neumayr, D. Bortis, J. W. Kolar, G. Deboy, C. Ostermaier, and O. Häberlen, "On the origin of the Coss-losses in soft-switching GaN-on-Si power HEMTs," *IEEE Journal of Emerging and Selected Topics in Power Electronics*, vol. 7, no. 2, pp. 679–694, 2019.

[17] J. Rademacher, X. Zan, and A. Avestruz, "High power, high efficiency wireless power transfer at 27.12 MHz using CMCD converters," in *2021 IEEE Energy Conversion Congress and Exposition (ECCE)*, 2021, pp. 5698–5703.

[18] G. Zulauf, S. Park, W. Liang, K. N. Surakitbovorn, and J. Rivas-Davila, "Coss losses in 600 V GaN power semiconductors in soft-switched, high- and very-high-frequency power converters," *IEEE Transactions on Power Electronics*, vol. 33, no. 12, pp. 10 748–10 763, 2018.

The Analysis and Design of a Resonant Capacitively-Isolated Cockcroft-Walton Converter

Elizabeth Rabenold, Raiphy Jerez, and Samantha Coday
Department of Electrical Engineering and Computer Science, Massachusetts Institute of Technology
Email: {rabenold, coday}@mit.edu

Abstract—**Capacitive isolation is a potential alternative to magnetics-based isolation in emerging applications, such as in partial power processing converter topologies, and in applications where weight and component volume are limiting factors. This work presents a capacitively-isolated Cockcroft-Walton converter capable of isolation through the flying capacitors. Generalized equations for mid-range flying capacitor voltages and switch voltages for converters of even level count are detailed. Experimental results validate the topology and analysis with a hardware prototype demonstrating 120 V input, 93.87% efficiency, and up to 60 V of isolation.**

I. INTRODUCTION

Hybrid switched-capacitor converters have been shown to be competitive with traditional magnetics-based converters in both efficiency and passive component volume due to their utilization of energy-dense capacitors for energy processing [1]. However, one important feature that is yet underdeveloped in hybrid switched-capacitor converters is isolation between the input and output voltages. Traditional implementations of galvanic isolation employ transformers to create the isolation barrier. It is important to note, though, that transformers are costly in both component volume and weight. In applications where passive volume is a key design constraint, such as in data centers and in electric aircraft, the dielectric material of capacitors may instead be used to achieve isolation [2] [3]. While typical capacitors do not offer the same safety or reliability as magnetic isolation, capacitor-based isolation can still be utilized in alternative applications, such as in enabling partial power processing or in composite converters. This work is motivated by converter applications that require extreme conversion ratios and output regulation. In such applications, series-input, parallel-output stacked converters can achieve high power density [4] [5] [6], though such topologies require isolation to allow for a series-input configuration. An example system architecture for a series-input parallel-output composite converter is shown in Fig. 1. Previous work introduced a capacitively-isolated Dickson converter [7] [8], which utilizes capacitors for energy processing and as the isolation barrier for a composite converter, thereby enabling the high conversion ratios required for tethered robots in space [9]. Existing literature [10] demonstrates the viability of the Cockcroft-Walton converter for extreme conversion ratios. This work proposes the introduction of a capacitor-based isolation barrier into the Cockcroft-Walton converter to support isolated power transfer in applications requiring extreme conversion ratios. The proposed topology, shown in Fig. 2, relies on distributed

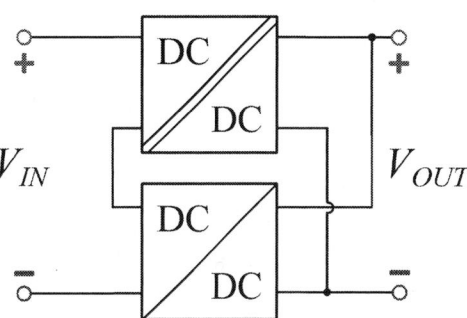

Fig. 1: Example system diagram for a partial power processing converter.

inductors to achieve full soft charging, and uses capacitors to create an isolation barrier between the converter's input and output return path. For this topology, the level count is equal to the conversion ratio. This work presents the design, analysis, and characterization of a 6:1 capacitively-isolated Cockcroft-Walton prototype, with hardware results validated at a high-side voltage of 120 V and a demonstrated 60 V of isolation.

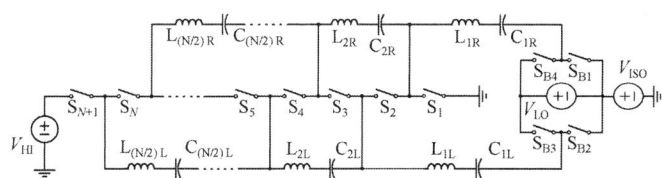

Fig. 2: Schematic of an N-level capacitively-isolated Cockcroft-Walton converter.

II. ANALYSIS

Figure 3 shows the two-phase operation of a 6:1 variant of the proposed topology. An odd level count introduces significant changes to analysis and design of the converter and will not be presented in this work. However, analysis presented holds for any even level conversion ratio. Charge flow analysis [11] of the 6:1 converter yields a normalized charge vector of (1) for capacitor vector (2).

$$q = [3, 3, 2, 2, 1, 1] \tag{1}$$

$$C = [C_{1L}, C_{1R}, C_{2L}, C_{2R}, C_{3L}, C_{3R}] \tag{2}$$

979-8-3315-1612-3/25 $31.00 © 2025 IEEE

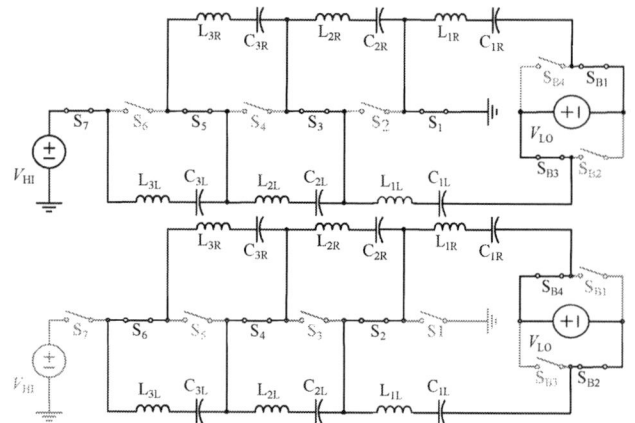

Fig. 3: (a) Phase one operation of a 6:1 isolated Cockcroft-Walton converter under two-phase control. (b) Phase two operation.

The charge at each stage can be generalized for an N-level version of the converter as (3) where k is the index of the left and right capacitors as defined in Fig. 2. The peak-to-peak capacitor ripple for both capacitors in stage k is defined as (4).

$$q_k = \frac{N}{2} - k + 1 \qquad (3)$$

$$\Delta v_{pp,k} = \frac{q_k}{C_k} \qquad (4)$$

Equal capacitor ripple is chosen to simplify analysis of the converter. Therefore, flying capacitor values are calculated as (5), where each value is scaled by the capacitance at the $\frac{N}{2}$ stage, C_0. The resultant capacitor vector normalized by C_0 for this 6:1 converter is then (6).

$$C_k = C_0(\frac{N}{2} - k + 1) \qquad (5)$$

$$C = [1, 1, 2, 2, 3, 3] \qquad (6)$$

Accordingly, inductors in each stage are selected to maintain resonant operation. The resonant frequency is presented in (7), where L_0 and C_0 are the values of the inductors and capacitors, respectively, at stage $\frac{N}{2}$ of the converter. As a result, the inductor value at each stage is equal to (8).

$$f_{res} = \frac{1}{2\pi\sqrt{L_0 C_0}} \qquad (7)$$

$$L_k = \frac{L_0}{\frac{N}{2} - k + 1} \qquad (8)$$

The inductor vector for this converter, normalized by L_0, is as follows.

$$L = [1, 1, \frac{1}{2}, \frac{1}{2}, \frac{1}{3}, \frac{1}{3}] \qquad (9)$$

The mid-range capacitor voltages can be generalized to the following, where the isolation voltage, V_{ISO}, is defined as the dc voltage offset between the low side of the rectifier and the input ground.

$$v_{k,L} = \begin{cases} -V_{ISO} & \text{for } k = 1 \\ \frac{2V_{HI}}{N} & \text{for } k = [2, \frac{N}{2} - 1] \end{cases} \qquad (10)$$

$$v_{k,R} = \begin{cases} \frac{V_{HI}}{N} - V_{ISO} & \text{for } k = 1 \\ \frac{2V_{HI}}{N} & \text{for } k = [2, \frac{N}{2} - 1] \end{cases} \qquad (11)$$

The switch voltages are presented in (12), where m is the index of the switch as described in Fig. 2. Note, the switch stress decreases as a function of level count, enabling the usage of lower voltage switches, which have improved figures of merit [12] [13] [14]. Moreover, as is typical of resonant tank converters [15], the switch voltage does not experience increased stress with capacitor ripple and can be sized independent of load and isolation voltage.

$$v_{ds,m} = \begin{cases} \frac{V_{HI}}{N} & \text{for } m = [1, N+1] \\ \frac{2V_{HI}}{N} & \text{for } m = [2...N] \\ \frac{V_{HI}}{N} & \text{for } m = B[1-4] \end{cases} \qquad (12)$$

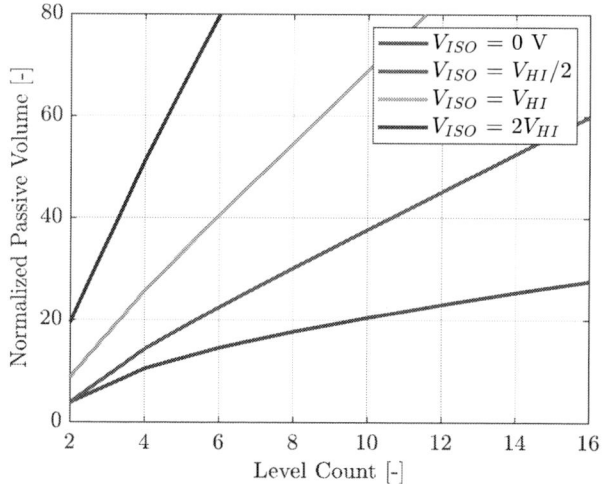

Fig. 4: Passive volume as a function of level count and isolation voltage.

Previous work [16] introduces a method for optimizing capacitor and inductor sizing for hybrid switched-capacitor converters operating at and above resonance. This approach computes the peak energy stored in each passive component and then uses the approximate energy density of capacitors and inductors, as found in [17], to determine the total volume of the passive components. Figure 4 shows this method applied to the capacitively-isolated Cockcroft-Walton converter demonstrated in this work. As the level count, conversion ratio, or isolation voltage is increased, the energy storage required of the capacitors also increases, thereby increasing the total

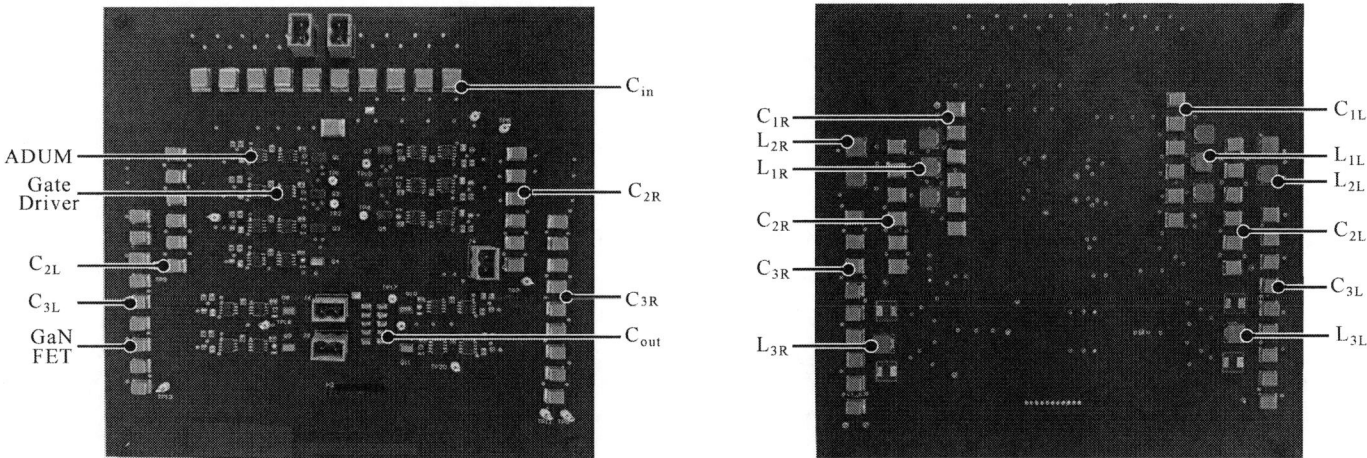

Fig. 5: (a) Front side of the 6:1 Cockcroft-Walton prototype. (b) Back side of the prototype.

passive component volume.

III. HARDWARE DESIGN

A 6:1 capacitively-isolated resonant Cockcroft-Walton converter was built to validate the proposed topology and analysis. The operating conditions are shown in Table I. The value of C_0 was informed by the optimization method [16] with consideration for level count and conversion ratio, and then selected from among commercially available capacitors of discrete values. The value of L_0 was chosen for resonant operation at 200 kHz, per (7). The passive components are listed in Table II. The prototype board uses a two-sided design for improved space utilization: switching cells and bulk input and output capacitor banks are on the top side of the board as seen in Fig. 5a, while the flying capacitors are on both sides of the board, and the inductors are solely on the bottom side, as in Fig. 5b. To minimize the commutation loop, the string switches were placed in a U-shape, such that S_1 and S_7 (Fig. 3) are close to the input. For the initial experimental prototype, signal and logic power isolation were achieved using ADUM5241s. Future implementations of this topology will investigate using cascaded bootstrap circuits to improve the overall system efficiency and to decrease board area [18]. The distributed inductors can be replaced with a lumped inductor on both input sides of the bridge. This would allow for a reduced passive component volume relative to the version of the converter with distributed inductors, owing to inductor scaling laws as presented in [19]. However, removing the resonant tanks at each stage of the converter will remove the converter's soft charging capabilities, so a converter with lumped inductance under two-phase control would incur hard charging losses. To avoid this, split-phase control can be implemented on the converter.

IV. EXPERIMENTAL RESULTS

The measured efficiency of the 6:1 converter across various loads at 120 V input voltage and 0 V of isolation voltage is demonstrated in Fig. 6. The converter's peak efficiency was

Description	Parameter	Value
Output Voltage	V_{OUT}	20 V
Input Voltage	V_{IN}	120 V
Operating Frequency	f_{res}	193 kHz
Isolation Voltage	V_{ISO}	60 V

TABLE I: Converter specifications.

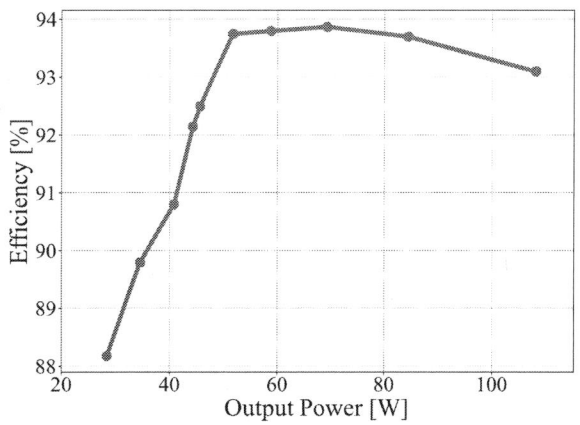

Fig. 6: Measured hardware prototype efficiency as a function of output power, with fixed 120 V input.

found to be 93.87% at a load of 5 Ω and an output power of 69.4 W. The efficiency was measured with a Yokogawa WT5000 power analyzer and does not include gate drive losses.

Figure 7 demonstrates the isolation capability of the converter by presenting the measured flying capacitor voltages at a V_{ISO} of 0 V, 30 V, and 60 V. This figure validates (10) and (11). Furthermore, the sinusoidal nature of the flying capacitor voltages indicates resonant operation, affirming the passive component vectors in (6) and (9) Additionally, the sinusoidal waveforms in Fig. 7 indicate soft charging of the

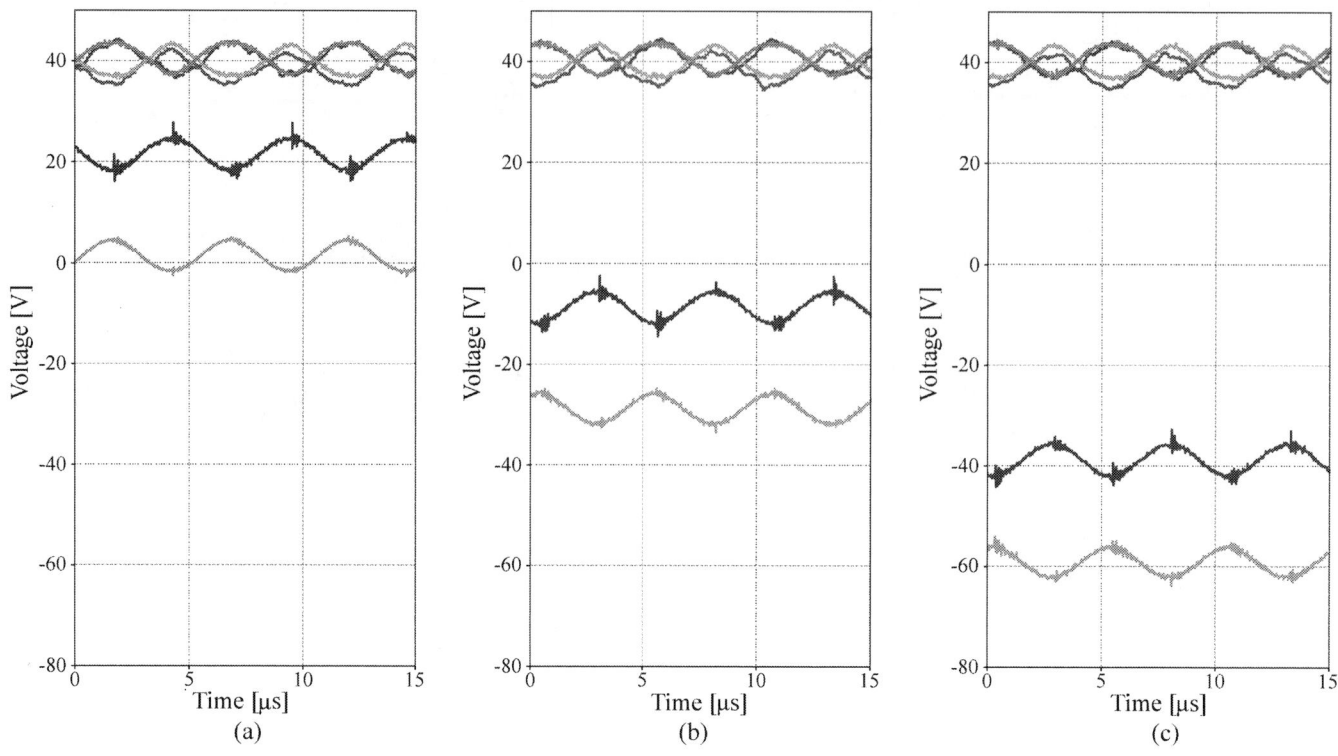

Fig. 7: (a) Measured flying capacitor voltages with an applied isolation voltage of 0 V, (b) 30 V, and (c) 60 V.

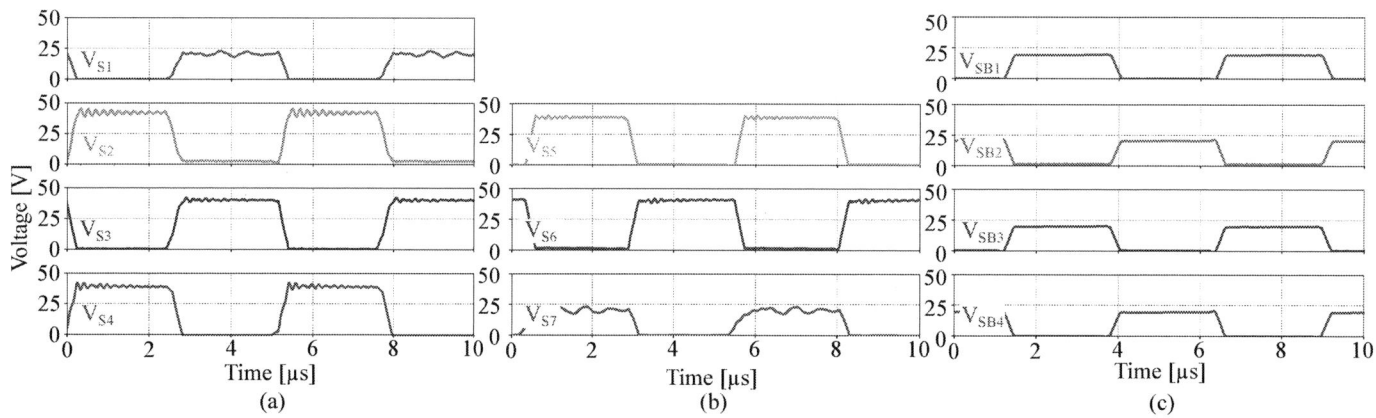

Fig. 8: (a) Switch voltages measured at an input voltage of 120 V.

Component	Part Number	Description
C_0	C5750NP02W104J280KA	100 nF, 450 V
L_0	IHLP2525CZERR33M01	330 nF, 20 A
C_{in}	KTS501B105M76N0T00	1μF, 500 V
C_{out}	CL31B105KCHNNNE	1μF, 100 V
$S_{[1-7]/[B1-B4]}$	EPC2304ENGRT	200 V, 5 mΩ
Gate Driver	NCP81074ADR2G	5 V, 10 A
Isolator	ADUM5241ARZ	Power and Signal

TABLE II: Components for the hardware prototype.

capacitors owing to the inductors distributed at each stage of the converter. The ripple on the flying capacitors is equal in magnitude, further supporting the component scaling vectors. The generalized switch voltage equations presented in (12) are experimentally validated in Fig. 8, where all 11 switching devices are measured at an input voltage of 120 V.

V. CONCLUSION

This work has presented the analysis, design, and implementation of a 6:1 capacitively-isolated Cockcroft-Walton converter capable of achieving isolation between the input and output return path. Contributions include generalized equations for mid-range flying capacitor voltages for even level counts, as well as generalized switch voltage equations and verification of the topology under various levels of isolation. These equations are validated in hardware, and experimental results are shared. The prototype converter reaches a peak efficiency of 93.87% at an input voltage of 120 V and output power of 69.4 W, thereby proving the suitability of this converter for future implementation in partial power processing converters.

REFERENCES

[1] Z. Ye, S. R. Sanders, and R. C. N. Pilawa-Podgurski, "Modeling and comparison of passive component volume of hybrid resonant switched-capacitor converters," *IEEE Transactions on Power Electronics*, vol. 37, no. 9, pp. 10903–10919, 2022.

[2] P. Granello, F. Pellitteri, R. Miceli, and L. Schirone, "Highly efficient capacitive galvanic isolation for ev charging stations," in *2022 International Symposium on Power Electronics, Electrical Drives, Automation and Motion (SPEEDAM)*, pp. 389–394, 2022.

[3] S. Pourjafar, O. Husev, and C. Roncero-Clemente, "Review and outlook of isolated capacitive coupling based converters," in *2024 IEEE 18th International Conference on Compatibility, Power Electronics and Power Engineering (CPE-POWERENG)*, pp. 1–6, 2024.

[4] J. Baek, Y. Elasser, K. Radhakrishnan, H. Gan, J. P. Douglas, H. K. Krishnamurthy, X. Li, S. Jiang, C. R. Sullivan, and M. Chen, "Vertical stacked lego-pol cpu voltage regulator," *IEEE Transactions on Power Electronics*, vol. 37, no. 6, pp. 6305–6322, 2022.

[5] M. H. Ahmed, C. Fei, F. C. Lee, and Q. Li, "Single-stage high-efficiency 48/1 v sigma converter with integrated magnetics," *IEEE Transactions on Industrial Electronics*, vol. 67, no. 1, pp. 192–202, 2020.

[6] C. Li, Y. E. Bouvier, A. Berrios, P. Alou, J. A. Oliver, and J. A. Cobos, "Revisiting "partial power architectures" from the "differential power" perspective," in *2019 20th Workshop on Control and Modeling for Power Electronics (COMPEL)*, pp. 1–8, 2019.

[7] J. Dickson, "On-chip high-voltage generation in mnos integrated circuits using an improved voltage multiplier technique," *IEEE Journal of Solid-State Circuits*, vol. 11, no. 3, pp. 374–378, 1976.

[8] A. Jackson, N. M. Ellis, and R. C. Pilawa-Podgurski, "A capacitively-isolated dual extended lc-tank hybrid switched-capacitor converter," in *2022 IEEE Applied Power Electronics Conference and Exposition (APEC)*, pp. 1279–1283, 2022.

[9] S. Coday, E. Krause, M. E. Blackwell, N. M. Ellis, A. Barchowsky, and R. C. Pilawa-Podgurski, "Design and implementation of a gan-based capacitively-isolated hybrid dickson switched-capacitor dc-dc converter for space applications," in *2023 IEEE Applied Power Electronics Conference and Exposition (APEC)*, pp. 3154–3159, 2023.

[10] L. Müller and J. W. Kimball, "High gain dc–dc converter based on the cockcroft–walton multiplier," *IEEE Transactions on Power Electronics*, vol. 31, no. 9, pp. 6405–6415, 2016.

[11] M. D. Seeman and S. R. Sanders, "Analysis and optimization of switched-capacitor dc–dc converters," *IEEE Transactions on Power Electronics*, vol. 23, no. 2, pp. 841–851, 2008.

[12] J. T. Stauth, "Pathways to mm-scale dc-dc converters: Trends, opportunities, and limitations," in *2018 IEEE Custom Integrated Circuits Conference (CICC)*, pp. 1–8, 2018.

[13] R. A. Abramson, S. J. Gunter, D. M. Otten, K. K. Afridi, and D. J. Perreault, "Design and evaluation of a reconfigurable stacked active bridge dc/dc converter for efficient wide load-range operation," in *2017 IEEE Applied Power Electronics Conference and Exposition (APEC)*, pp. 3391–3401, 2017.

[14] M. Guacci, D. Bortis, and J. W. Kolar, "High-efficiency weight-optimized fault-tolerant modular multi-cell three-phase gan inverter for next generation aerospace applications," in *2018 IEEE Energy Conversion Congress and Exposition (ECCE)*, pp. 1334–1341, 2018.

[15] C. Schaef, J. Rentmeister, and J. T. Stauth, "Multimode operation of resonant and hybrid switched-capacitor topologies," *IEEE Transactions on Power Electronics*, vol. 33, no. 12, pp. 10512–10523, 2018.

[16] N. M. Ellis, N. C. Brooks, M. E. Blackwell, R. A. Abramson, S. Coday, and R. C. N. Pilawa-Podgurski, "A general analysis of resonant switched-capacitor converters using peak energy storage and switch stress including ripple," *IEEE Transactions on Power Electronics*, vol. 39, no. 7, pp. 8363–8383, 2024.

[17] N. C. Brooks, J. Zou, S. Coday, T. Ge, N. M. Ellis, and R. C. N. Pilawa-Podgurski, "On the size and weight of passive components: Scaling trends for high-density power converter designs," *IEEE Transactions on Power Electronics*, vol. 39, no. 7, pp. 8459–8477, 2024.

[18] Z. Ye, Y. Lei, W. C. Liu, P. S. Shenoy, and R. C. Pilawa-Podgurski, "Improved bootstrap methods for powering floating gate drivers of flying capacitor multilevel converters and hybrid switched-capacitor converters," *IEEE Transactions on Power Electronics*, vol. 35, pp. 5965–5977, June 2020.

[19] C. R. Sullivan, B. A. Reese, A. L. F. Stein, and P. A. Kyaw, "On size and magnetics: Why small efficient power inductors are rare," in *2016 International Symposium on 3D Power Electronics Integration and Manufacturing (3D-PEIM)*, pp. 1–23, 2016.

SHSC: Non-Isolated High-Density 4:1 IBC For 48 V Applications

Dr. Mario Ursino, Dr. Roberto Rizzolatti, Simone Mazzer

Infineon Technologies, Villach, Austria

Email: mario.ursino, roberto.rizzolatti,
simone.mazzer@infineon.com

Abstract—**The rack-level 54/48 V bus architecture is an established standard in modern Data Centers, both in classical server applications and accelerator boards. The digital load supply (xPU/ASIC) is almost always provided through a Voltage Regulator Module (VRM), which is a multi-phase buck converter implementing the last stage of the power conversion. With modern technology, Voltage supply for the VRM is at its sweet spot between 6 V and 12 V, enabling the use of wide-spread, commercial power stages. To provide this Voltage rail to the VRM, a first stage connected to the 48 V bus is needed (Intermediate Bus Converter – IBC): as regulation is provided with high performance by the VRM, the IBC can be unregulated to enable very high power density. In this work the Authors present an unregulated, non-isolated 4:1 fixed-ratio Hybrid Switched Capacitor converter called SHSC. Detailed operation is shown, together with experimental results for a 40x18x8 mm discrete implementation providing up to 1.3 kW with a power density of 3.7 kW/in^3.**

Index Terms—**Power conversion, Data centers, 48 V power distribution, Resonant converters, Switched-tank converters, Hybrid resonant converters.**

I. INTRODUCTION

In the power conversion chain of a typical data center, approximately half of energy is wasted in power conversion and cooling [1]. The need to reduce CO2 emissions demands a continuous improvement at system and converter levels. 48 V voltage power distribution at rack level represents an accepted and widespread solution, that was recently proposed by vendors and research centers [2], which are already moving to the next step with a 400 V bus bar [3]. Typically, from 48 V bus voltage a two-stage approach is adopted where the first stage is commonly a resonant converter such as an LLC or a STC, while the second stage is usually a multiphase buck topology with a 12 V input [4]. LLC resonant converters are widely used as unregulated Intermediate Bus Converters (IBCs) [5], [6] due to Zero Voltage Switching (ZVS) operation at primary side and zero current switching (ZCS) operation at secondary side. GaN implementations are also becoming widespread [7], [8]. Combining the advantages of STCs and LLCs, hybrid resonant converters are now becoming widespread [9] for high stepdown (above 4:1) applications. As the 12 V rail is still the most used in server applications and many accelerator cards, the 4:1 conversion is still the most desirable to maximize overall system efficiency. Accordingly, most solutions provide a fixed 4:1 conversion ratio [10], [11], [4]. Furthermore, the use of advanced materials and topologies, such as the cascaded res-

onant converter, has led to significant improvements in power density and efficiency [11]. Additionally, the development of new control strategies and modulation techniques has enabled the optimization of converter performance and efficiency [12]. Recent studies have also demonstrated the feasibility of high-efficiency 48 V to 12 V conversion using hybrid resonant converters [13], switched-capacitor converters [14], and GaN-based resonant converters [15].

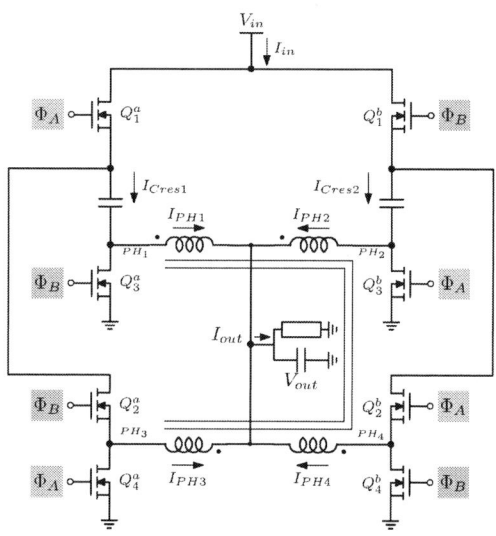

Fig. 1. Symmetric Hybrid Switched capacitor Converter (SHSC) topology.

In this work, a novel 4:1 fixed-ratio IBC is proposed, with the name of SHSC, from Symmetrical Hybrid Switched Capacitor converter, featuring resonant operation, ZVS and near-ZCS on all switches, a minimum bill of materials (BoM), and very high power density and efficiency, similarly to [9]. Resonant operation is achieved with the combination of resonant capacitors and a 4-windings *autotransformer* (TX in short), where each winding has the same number of turns N, enabling for straightforward interleaving and high-frequency resistance (R_{ac}) minimization thanks to very high coupling. With SHSC, each resonant capacitor bank (C_{res1} or C_{res2}) is individually connected to a single TX winding.

This is beneficial to decrease charge redistribution among the two banks, when compared to other dual-phase topologies. Moreover, in the same comparison, the combination of stand-

979-8-3315-1612-3/25 $31.00 © 2025 IEEE

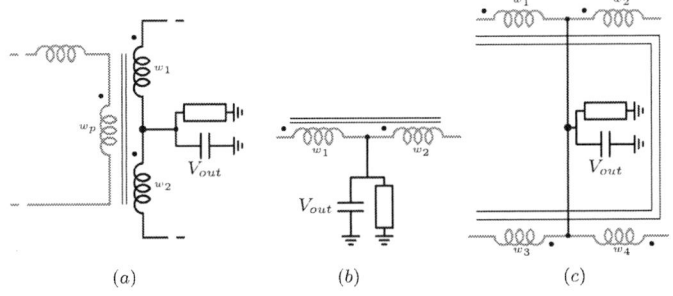

Fig. 2. (Auto)transformers comparison of different topologies. Center-tapped LLC (a), [7] (b) and SHSC, this work (c). Orange windings are always delivering energy to the load, to highlight the difference of non-isolated topologies with center-tapped LLC (a).

alone tanks and separated windings contributes to minimize AC-resistance R_{ac} in the TX [12]. TX volume and technology remains the same of a same-power classic hybrid, non-isolated converter: the 2-windings solution of [7] can be compared with a same-volume version of SHSC TX, as shown in Figure 3.

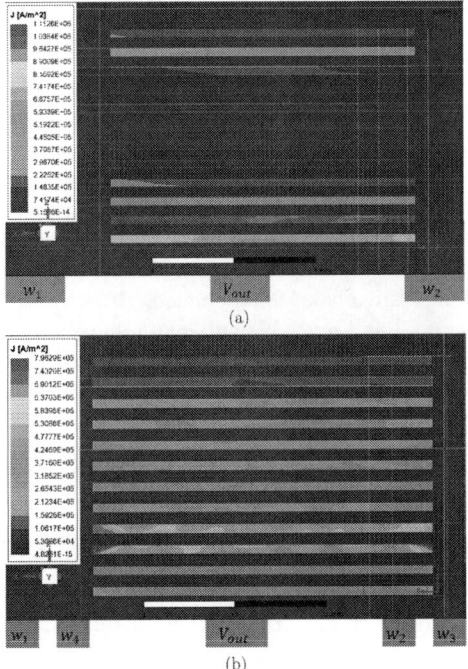

Fig. 3. Current density distribution in the winding area of a E22/6/16/R transformer used as in [7] (a) and in SHSC (b, this work). First harmonic at $2 \times f_{sw}$ (most relevant) is shown. $P_{out} = 1.2$ kW, $V_{in} = 54$ V.

In SHSC, each winding has a halved number of paralleled windings, the number of connectors is doubled, the winding turns and the ferrite core are unchanged. SHSC has the same TX volume and BoM of [7] for any practical 48 V application, i.e., where windings have at least two paralleled windings, which is true for any practical application. The same holds for the number of Synchronous Rectifiers (SRs): a practical implementation with 3 paralleled SRs is BoM-equivalent to a $2 \times Q_3 + 1 \times Q_4$ SHSC. To summarize, the addition

of 4 TX connectors mitigates capacitor charge redistribution problems and minimizes the equivalent AC-resistance of the windings. Figure 3 shows this effect on a 12 layers, 3 oz autotransformer arranged in the two configurations of [7] (a) and SHSC (b), with an E 22/15/6 core. In Figure 3 (a), the TX is composed of two winding types, w_1 and w_2, replicated 6 times and connected in parallel. In (b), the same set of layers is assigned to w_1, w_2, w_3 and w_4 windings, replicated 3 times. These transformer 3D models are designed targeting a discrete solution with a *turn-key*, Surface-Mounted Device (SMD) planar transformer in the range of 1.2 kW.

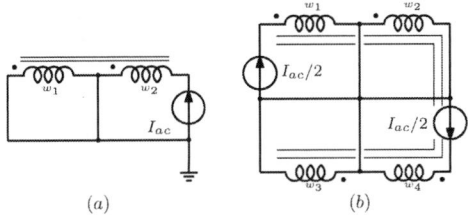

Fig. 4.

The two solutions are simulated with AC excitations as shown in Figure 4. In this test model, the converter operates at $f_{sw} = 500$ kHz with an output power of 1.2 kW. The first harmonic is therefore at $f_{sw} = 1$ MHz. As Figure 3 shows, the current density is better distributed in SHSC. In (a), the current crowds near the connectors, while in (b) the effect is mitigated by separating the two capacitor banks at topology-level. The model of Figure 4 can be further improved by implementing a time-domain simulation, also including resonant capacitor models and PCB stray inductance. In general, this analysis shows a qualitative benefit associated with AC losses. With this model, the AC resistance is decreased by 42% when using SHSC.

II. CONVERTER ARCHITECTURE

SHSC operates, similarly to [9], in two symmetrical configurations separated by a short dead-time (DT). The resonant tanks are alternatively connected between input (V_{in}, which is the 48 V rail) and the V_{out} output, delivering to the output rectified sinusoidal currents. Each resonant tank is composed by the leakage inductance of the TX and two resonant capacitor banks C_{res1} and C_{res2}. The leakage energy is dominated by the TX windings, but is also associated with PCB connections and MOSFETs stray inductance. ZVS is obtained on all MOSFETs in all load conditions with the magnetizing energy of the TX, which can be precisely selected with a proper *gap*. In general, the converter is operated slightly above resonance. Combined with a very low leakage inductance, this yields a very high robustness against component mismatch and resonant frequency variation.

Figure 5 can be used to understand converter timings and main waveforms. SHSC operates in two symmetric sub-intervals ϕ_α and ϕ_β, separated by a fixed-length dead-time used for ZVS. During ϕ_α (Figure 6), Q_1^a, Q_2^b, Q_3^b, and Q_4^a are ON. The first resonant tank, composed by C_{res1}

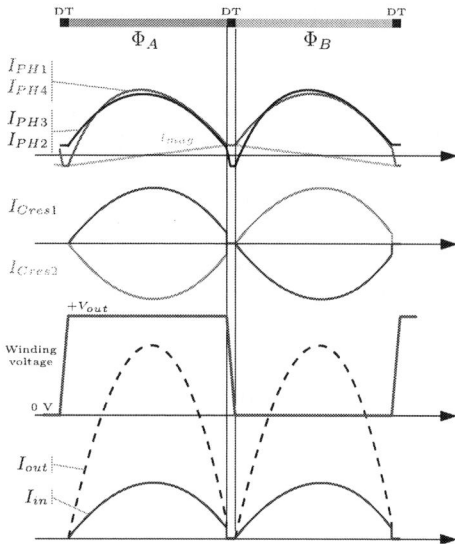

Fig. 5. SHSC timings and main waveforms.

and distributed leakage energy, has a positive current I_{Cres1} flowing through winding w_1 and delivered to the output. Synchronous-rectifier (SR) Q_3^b connects C_{res2} to the TX through Q_2^b, composing the second resonant tank that delivers another resonant current to the output ($I_{Cres2} = -I_{Cres1}$) by mean of winding w_4. The other two windings, w_2 and w_3, are connected to GND through SRs Q_3^b and Q_4^a, allowing for a total of $2 \times I_{Cres1}$ to be induced to the output. From this sub-interval, it is also clear that Q_3 need to carry twice the current of the counterparts Q_4. In this sub-interval, as in the next one, $I_{out} \simeq 4 \cdot I_{in}$.

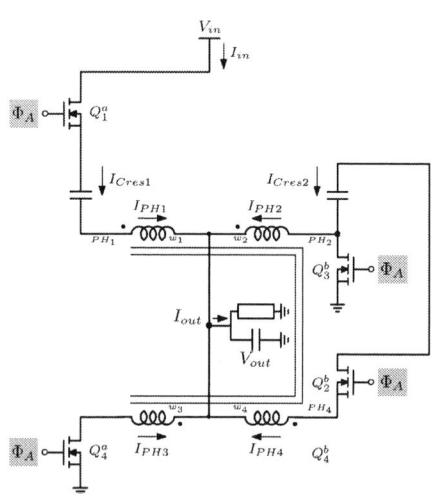

Fig. 6. SHSC equivalent circuit during ϕ_α sub-interval.

During ϕ_α and ϕ_β, a triangular current is superimposed to the resonant waveforms flowing in the TX windings. This is associated with the magnetizing flux linearly increasing and decreasing inside the core. By observing the winding voltage shown in Figure 5, its magnitude can be simply derived by

the considering one winding voltage, which is almost constant during these phases. When DT occurs, the magnetizing current drives the C_{oss} of each MOSFETs to achieve ZVS operation.

During ϕ_β (Figure 7), Q_1^b, Q_2^a, Q_3^a, and Q_4^b are ON. Operation is fully symmetrical, and the resonant capacitors banks are balanced with an equal and opposite current. After ϕ_β, another DT occurs and the cycle is complete.

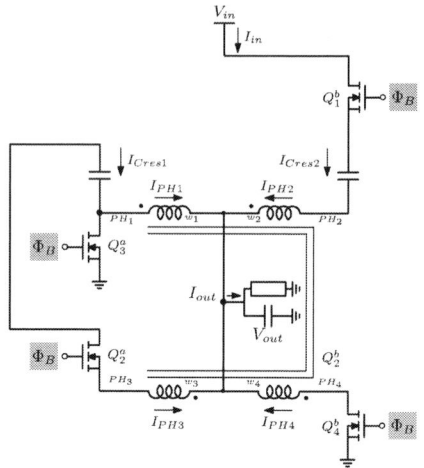

Fig. 7. SHSC equivalent circuit during ϕ_β sub-interval.

In addition to the previous analysis, Figure 8 can be used to derive the total conversion ratio. As the turn-ratio of each transformer winding is the same, one half-cycle (ϕ_β in this case) can be used to derive it. Considering two active SRs, two windings are electrically connected to V_{out}, while the two remaining windings have an equal induced voltage due to the high coupling. In steady-state, voltages on C_{res} is forced to $V_{in}/2$. This yields a total conversion ratio of $1/4$.

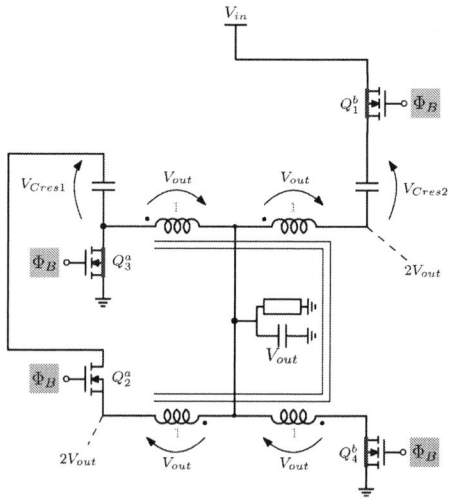

Fig. 8. SHSC equivalent circuit during ϕ_α sub-interval with windings and resonant capacitors voltage.

III. EXPERIMENTAL RESULTS

SHSC was built for two different applications: a 1.3 kW discrete solution, shown in Figure 9 and a 1 kW module, in Figure 10. Table I summarizes the two versions' specifications and power components.

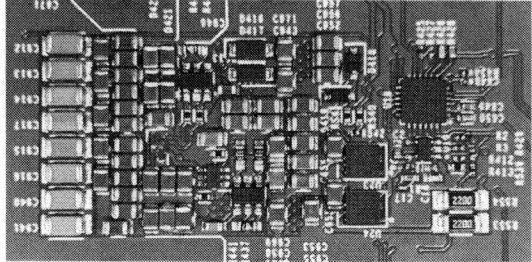

Fig. 9. 1.3 kW SHSC discrete implementation. Top-side (top figure) and bottom-side (bottom figure).

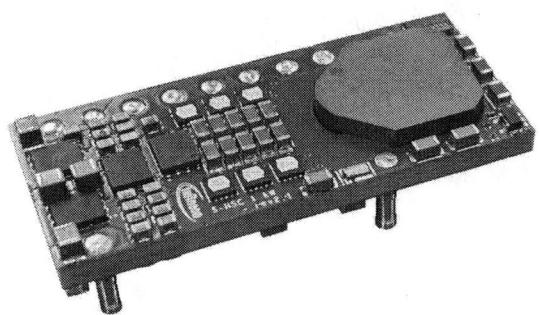

Fig. 10. 1 kW SHSC module implementation.

Both implementations are based on a custom-shaped ferrite core TX. In the discrete implementation the TX is an SMD component, while in the module it is embedded in the PCB cross-section. For both, the connectors for windings w_1 to w_4 are on the vertices and V_{out} is in the middle part, as shown in Figure 11 for the SMD version. Power MOSFETs are placed on the top side, to enable optimal cooling solutions. In the discrete version, all the MLCC capacitors are placed on the bottom side, while in the module some of them occupy the top side as well. The module version includes 12 V to 7.2 V and a 7.2 V to 3.3 V on-board auxiliary converters, while in the discrete implementation they are outside of the rectangular layout. This supplies are considered in the efficiency data of Figure 12.

The discrete implementation has a TDP efficiency of 97.8% (1.3 kW, V_{in} = 54 V) and a peak efficiency of 98.3% (600 W, V_{in} = 54 V). The module counterparts reaches

TABLE I
SHSC SPECIFICATIONS

	Discrete	Module
V_{in}	40÷60 V	40÷60 V
V_{out}	10÷15 V	10÷15 V
P_{out}	1.3 kW TDP, 3.6 kW peak	1 kW TDP, 3.2 kW peak
f_{sw}	300 kHz	500 kHz
L_k (w_1, others shorted)	13 nH	11 nH
L_{mag} (w_1, others open)	989 nH	1.05 uH
Size	40×18×8 mm	60×25.4×10 mm
Power density	3.7 kW/in³ (TDP)	1.07 kW/in³ (TDP)
PCB technology	12 layers, 1 oz all	12 layers, 2oz ext. 3 oz int.
Core	Custom, DMR51W ferrite	Custom, DMR51W ferrite
Driving voltage	10 V	7.2 V
Q_1, Q_2	4×BSC023N08NS5SC	4×BSC025N08LS5
Q_3	1×BSC007N04LS6SC	4×IQE013N04LM6CGSC
Q_4	1×IQE013N04LM6SC	2×IQE013N04LM6CGSC
C_{in}	8×GRM31CZ72A475KE11	6×GRM31CZ72A475KE11
C_{res}	28×GRM21BZ71H475KE15	28×C2012X7R1H475K125AC
C_{out}	19×GRM21BZ71E106KE15	23×GRM31CC71E226ME15
Drivers	4×1EDN7550B	4×1EDN7550U
	2×2EDN7534G	2×2EDN7534G
Controller	XDPP1100-Q024	XDPP1100-Q024

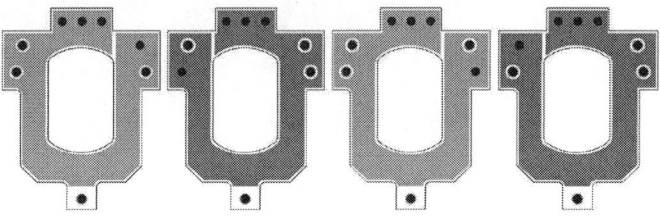

Fig. 11. SHSC autotransformer windings in the SMD implementation. From left to right: w_2, w_1, w_3, w_4 are connected at the vertices. Each winding is terminated in the north-center part, connected to V_{out}. The south-center pin is for mechanical stability.

97.55% (1 kW, V_{in} = 54 V) and 98.5% (450 W, V_{in} = 54 V). The two versions are designed to operate in two different load scenarios: the discrete SHSC is designed to be highly efficient at full-load, while the module has a higher peak efficiency, with decreased full-load performance.

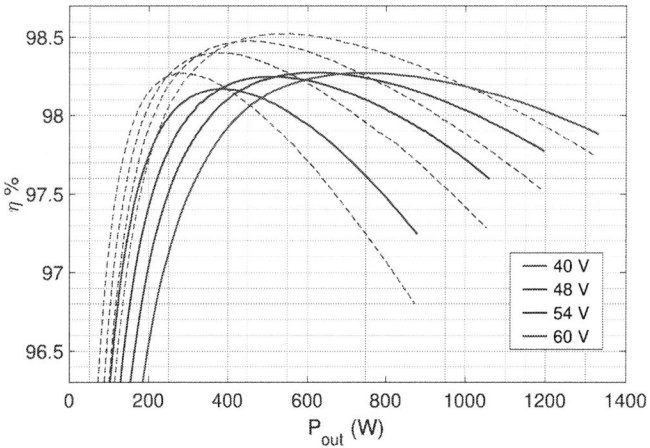

Fig. 12. Total converter efficiency (including auxiliary supply) for the two implementations: discrete (solid lines) and module (dashed lines) at different input voltages.

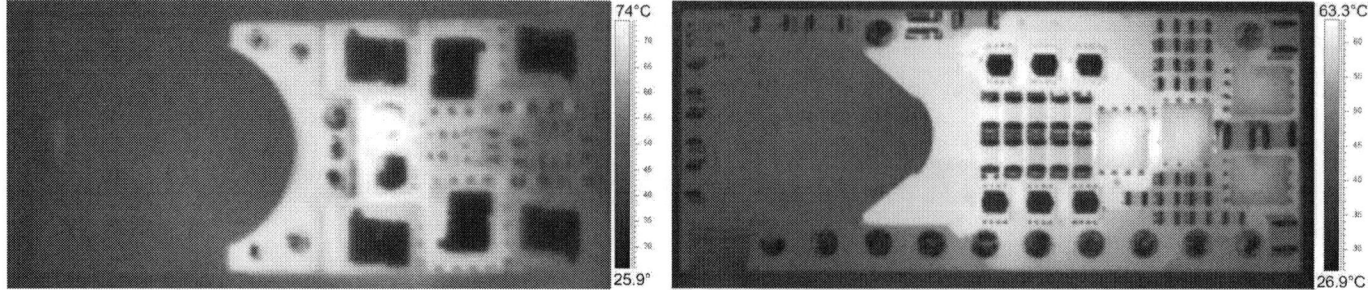

Fig. 13. Thermal images of SHSC at full load. Discrete implementation (left) operating at 1.3 kW, $V_{in} = 54$ V, and module (right) at 1 kW, $V_{in} = 54$ V. Both images are at thermal steady-state, with 3 m/s air cooling. In both versions Q_3 and Q_4 groups are the closest to the TX to minimize the termination loss. From the middle to the right Q_2 and Q_1 groups are placed.

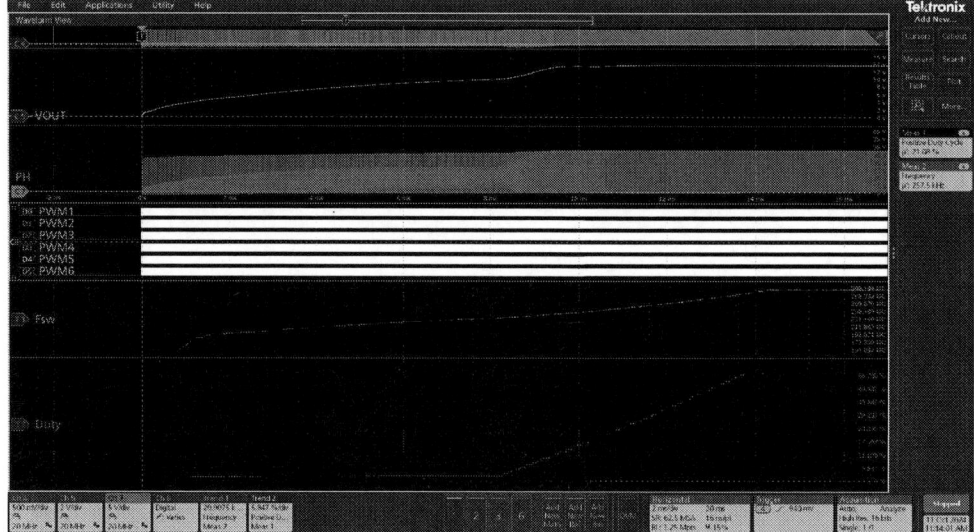

Fig. 14. SHSC startup at $V_{in} = 54$ V. Top-to-bottom: PWM1, V_{out}, PH1, PWM signals, f_{sw}, PWM1 duty cycle (equal for all).

Fig. 15. Load transient from 1.3 kW (TDP) to 3.6 kW (peak power) in the discrete version. Top-to-bottom: V_{out}, V_{in}, PH1 and PH2, P_{out}, I_{out}.

979-8-3315-1612-3/25 $31.00 © 2025 IEEE

Figure 13 shows thermal images for both versions of SHSC operating at TDP, in both cases. In the discrete SHSC, a clear hot-spot is visible for the small SRs (Q_3). Their power dissipation density is similar to Q_4, as they deliver half the current but have double the $R_{ds,ON}$, the hot-spot is caused by the increased junction-to-PCB thermal resistance compared to Q_3, which are on the sides of the layout. In a heat-plated application, this behavior does not manifest as the junction-to-pad (exposed) is much lower than the forced-air case. In the module, the SRs (both Q_3 and Q_4) are on the sides and well-connected to the GND planes, showing an almost uniform temperature. In this case, a different hot-spot is present in the middle part of Q_2: here the thermal *crowding* is more severe than in the discrete solution, as the module layout increases Q_2 junction-to-PCB (main board) thermal resistance. Also in this case, a cold plate can be used to increase temperature uniformity and decrease it in value. For the discrete solution, a maximum temperature of 74 °C is observed on Q_3, while a maximum of 63.3% is observed on the module near Q_2. Both values are well below safe operating margins, which are usually above 100°C, even without heat-sink.

Converter start-up is shown in Figure 13. Although SHSC is a non-regulated converter, it is able to start in the full V_{in} range without an external eFuse, relying on duty-cycle and frequency (f_{sw} modulation. Contrary to LLC-derived topologies, SHSC commences with a very small duty-cycle and f_{sw}. The controller ramps-up these values based on a preset V_{out}/V_{in} ratio ramp. This is combined with a linear pre-charge phase (not shown) that brings the C_{res} banks to $V_{in}/2$ before the modulation phase. The pre-charge, set to a fixed 5 ms duration, is necessary to protect the 40 V MOSFETs during the first modulation pulses. In [9] this phase is not present as the small duty-cycle, together with the presence of the primary windings, is sufficient to keep the SRs well-within their Safe Operating Area (SOA).

Figure 15 shows SHSC main waveforms during a load transient from 1.3 kW (TDP) to 3.6 kW (peak power) in the discrete version. Observing the waveforms of channel C5 (PH1 and PH2 of Figure 1) the load-independent ZVS can be observed. At peak power, which is 2.7× TDP, the turn-off voltage on the SRs is still well-withing the MOSFET SOA. The voltage drop can be used to calculate the output resistive impedance of this converter, which is approximately 2.9 mΩ.

IV. CONCLUSIONS

This work presents a novel fixed-ratio, resonant, hybrid soft-switched Intermediate Bus Converter called Symmetrical Hybrid Switched Capacitor converter (SHSC) for 48 V data centers. SHSC relies on a fully-interleaved, extremely efficient autotransformer to yield an unregulated 12 V intermediate rail to supply modern VRMs. When compared to other hybrid solutions, the splitting of the resonant tank allows to further decrease the R_{ac} related to winding losses.

The converter is implemented in two versions, discrete and module, for the classic V_{in} = 40 V to 60 V range. The discrete implementation has a TDP efficiency of 97.8%

(1.3 kW, $V_{in} = 54$ V) and a peak efficiency of 98.3% (600 W, $V_{in} = 54$ V). The module reaches 97.55% (1 kW, $V_{in} = 54$ V) and 98.5% (450 W, $V_{in} = 54$ V) with forced air cooling. SHSC can reach very high power densities, 3.7 kW/in^3 (TDP) for the discrete and 1.07 kW/in^3 (TDP) for the module. The proposed topology can support very high over-power transients due to its low leakage inductance, which ensures low stress on power MOSFETs. The use of magnetizing inductance, integrated in the autotransformer, enables load-independent ZVS on all MOSFETs.

REFERENCES

[1] S. Oliver. (July 2012) From 48 v direct to intel vr12.0: Saving 'big data' $500,000 per data center, per year. [Online]. Available: https://www.mouser.com/pdfdocs/wp_VR12.pdf

[2] Z. Wu, Y. He, J. Zheng, B. Cheng, J. Zhang, M. Wu, W. Lv, Y. He, P. Yue, N. Ahuja, and Q. Qiao, "High power density rack design optimal for hyper scale data center carbon emission reduction," in *2023 22nd IEEE Intersociety Conference on Thermal and Thermomechanical Phenomena in Electronic Systems (ITherm)*, 2023, pp. 1–6.

[3] G. Cloud. (2024) 2024 ocp global summit keynote. Blog post. [Online]. Available: https://cloud.google.com/blog/topics/systems/

[4] S. Jiang, S. Saggini, C. Nan, X. Li, and C. Chung, "Switched tank converters," *IEEE Transactions on Power Electronics*, vol. 34, no. 6, pp. 5048–5062, 2019.

[5] M. H. Ahmed, M. A. de Rooij, and J. Wang, "High-power density, 900-w llc converters for servers using gan fets: Toward greater efficiency and power density in 48 v to 6/12 v converters," *IEEE Power Electronics Magazine*, vol. 6, no. 1, 2019.

[6] S. Qin, H. D. Tamdem, B. Tsai, Z. Huang, S. Tian, F. Huang, A. Guo, W. Yuan, and M. Cao, "All-gan hemts and planar transformer based 48v voltage regulator common module (vrcm) design," in *2023 18th International Microsystems, Packaging, Assembly and Circuits Technology Conference (IMPACT)*, 2023, pp. 259–262.

[7] Renesas, "Comparison of silicon mosfet vs gan for 48v input intermediate bus conversion (ibc) based on multiphase buck controllers," in *2024 IEEE Applied Power Electronics Conference and Exposition (APEC)*, 2024.

[8] L. A. D. Ta, N. D. Dao, and D.-C. Lee, "High-efficiency hybrid llc resonant converter for on-board chargers of plug-in electric vehicles," *IEEE Transactions on Power Electronics*, vol. 35, no. 8, pp. 8324–8334, 2020.

[9] R. Rizzolatti, C. Rainer, S. Saggini, and M. Ursino, "High density hybrid switched capacitor converter for data-center application," in *2021 IEEE Applied Power Electronics Conference and Exposition (APEC)*, 2021.

[10] K. Hata, S. Tanaka, Y. Rikiishi, and T. Matsumoto, "48 v-to-12 v always-dual-path hybrid dc-dc converter for inductor current reduction," in *2022 IEEE Energy Conversion Congress and Exposition (ECCE)*, 2022, pp. 1–6.

[11] Z. Ye, Y. Lei, and R. C. N. Pilawa-Podgurski, "A 48-to-12 v cascaded resonant switched-capacitor converter achieving 4068 w/in3 power density and 99.0% peak efficiency," in *2021 IEEE Applied Power Electronics Conference and Exposition (APEC)*, 2021, pp. 3421–3428.

[12] L. W. J. Y. Saggini, "Wireless charging with split resonant capacitors," Patent US20 240 223 012A1, Aug 10, 2021.

[13] R. Rizzolatti, S. Saggini, M. Ursino, and L. Corradini, "A high-density hybrid resonant converter for 48v-to-12v conversion," *IEEE Transactions on Power Electronics*, vol. 37, no. 5, pp. 4121–4133, 2022.

[14] Y. Li, X. Lyu, D. Cao, S. Jiang, and C. Nan, "A 98.5% efficient 48v-to-12v switched-tank converter for data center applications," *IEEE Transactions on Power Electronics*, vol. 37, no. 4, pp. 3421–3433, 2022.

[15] M. Ahmed, M. A. de Rooij, and J. Wang, "Gan-based 48v-to-12v resonant converter for high-efficiency data center power supplies," *IEEE Transactions on Power Electronics*, vol. 37, no. 7, pp. 5221–5233, 2022.

High-Performance Current Multiplier: a Hybrid Switched Capacitor Solution For High-Current Applications

Kevin Zufferli, Roberto Rizzolatti, Mario Ursino,
Simone Mazzer, Gerald Deboy
Infineon Technologies, Villach, Austria
Email: kevin.zufferli, roberto.rizzolatti, mario.ursino,
simone.mazzer, gerald.deboy@infineon.com

Stefano Saggini
DPIA, University of Udine, Italy
Email: stefano.saggini@uniud.it

Abstract—The rapid growth of AI applications in data centers has led to a surge in the development of new hardware accelerators, resulting in increased power demands and energy consumption. To ensure this trend remains feasible and sustainable, it is crucial to focus on enhancing efficiency, especially within the energy conversion chain. The key enablers for achieving this objective include the precise integration of topology and magnetic structure, as well as the implementation of advanced 3D designs to enhance two critical parameters: efficiency and current density. This paper introduces a compact ultra-low profile Current Multiplier module, exploiting the advantages offered by the Hybrid Switched Capacitor (HSC) topology. Experimental results show the effectiveness of the topology achieving Electrical Current Design (EDC) density of 2.2 A/mm^2.

Index Terms—Data Center, Current Multiplier, HSC, AI, 48V.

I. INTRODUCTION

The escalating need for advanced data storage solutions has triggered a substantial rise in energy consumption within data centers [1]. Consequently, there has been a notable migration from conventional 12V DC power distribution systems to 48V DC systems in server racks over the past few years [2]. The rapid advancement of artificial intelligence within data centers has led to a substantial increase in current demand from graphics processing units (GPUs) and dedicated ASICs, placing power delivery design under high pressure. Conventional 48V architectures typically involve the cascaded connection of two stages: a regulated stage and an unregulated stage. The Intermediate Bus Architecture (IBA) [3] comprises a 48V Intermediate Bus Converter (IBC) [4]–[6] that performs unregulated conversion to power the Voltage Regulator Module (VRM) converter , which regulates the output voltage with high bandwidth. While this architecture offers high-speed control to load transients, it falls short in terms of current density to PoL. In contrast, the Fractional Power Architecture (FPA) [2] can be seen as an inverted IBA, where the regulated stage (Modular Current Driver – MCD) is connected to the input,

while the fixed-ratio stages (Modular Current Multipliers – MCMs) are connected to the output. This approach enables the exploitation of the high current Thermal and Electrical Design Current (TDC-EDC) density of MCMs, but regulating performance becomes a critical concern. As Current Multipliers act as a low-pass filter, they can cut the bandwidth of the system, potentially limiting its overall performance. However, as the current demand continues to surge, the confined space dedicated to power conversion is becoming increasingly inadequate for state-of-the-art VRM. In this context, Current Multipliers offer a promising solution, enabling designers to create more efficient and compact power conversion systems that can meet the growing demand for high current.

State-of-the-art Current Multipliers have adopted the use of soft switching DCX converters, specifically LLC resonant converters, to achieve high efficiency and performance. In these converters, the conduction losses of the MOSFETs are primarily determined by their on-resistance ($R_{ds,on}$). However, the primary source of losses is actually the transformer used in the DCX converters [7]. The transformers play a critical role in the converter's performance, and their losses can significantly impact the overall efficiency of the system. To minimize these losses, it is essential to study the AC resistance of them and choose a suitable topology. Since AC resistance increases with the frequency of the harmonic components in the winding current, it is essential to choose a topology that minimizes high-frequency components in high-current applications.

This paper presents a novel design for a Hybrid Switching Capacitor (HSC)-based Current Multiplier, which offers a significant improvement over existing state-of-the-art LLC solution. The proposed design is based on a careful selection of the topology, which takes into account the optimal excitation of the transformer and the reduced stress on the Synchronous Rectifiers.

979-8-3315-1612-3/25 $31.00 © 2025 IEEE

II. COMPARATIVE ANALYSIS OF HSC AND LLC CENTER-TAPPED CONVERTERS FOR CURRENT MULTIPLIER APPLICATIONS

Current Multipliers are distinguished by the high currents they must handle to power PoL converters, making it crucial to select an architecture that minimizes stress on components within the circuit. The current state-of-the-art solution is the LLC center-tapped converter, which inherently can provide isolation (a feature not necessary in data center applications). However, as revealed by the analysis in [8], this approach falls short in terms of magnetic and synchronous rectifier performance compared to an HSC-based design, depicted in Fig. 1. A fair

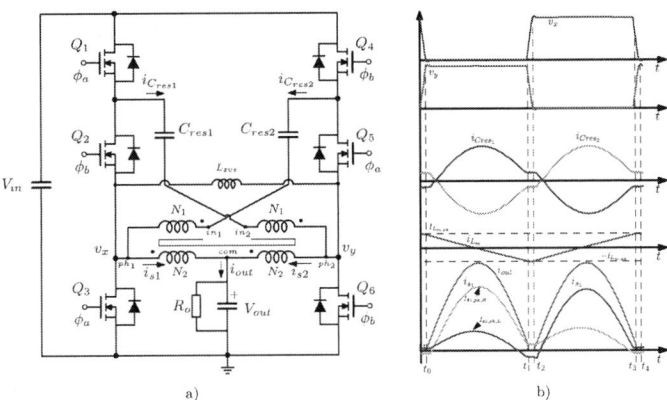

Fig. 1: a) HSC topology b) Main waveforms.

comparison, as presented in [8], demonstrates that the LLC secondary windings and Synchronous Rectifiers (SR) carry higher currents compared with HSC, resulting in overall higher losses. In this work, an additional analysis is conducted to provide deeper insights about transformer losses which involves the actual AC resistance of the windings. More specifically, for a rectified wave of current with two different peaks in T_{SW} as depicted in Figure 1b, the general Fourier expansion of the current flowing in a secondary is:

$$I_s(t) = \frac{I_{pH} + I_{pL}}{\pi} + \frac{I_{pH} - I_{pL}}{2} \sin(2\pi f_{sw}t)$$
$$- \sum_{2n=2}^{\infty} \frac{I_{pH} + I_{pL}}{\pi} \frac{2}{n^2 - 1} \cos(2n\pi f_{sw}t) \quad (1)$$

where I_{pH} is the higher peak and I_{pL} is the lower peak, while the total rms current is:

$$I_{s_{rms}} = \frac{\sqrt{I_{pH}^2 + I_{pL}^2}}{2} \quad (2)$$

Considering the same input current for both the topologies and the same conversion ratio $\left(\frac{N_1}{N_2}_{LLC} = 4 + 2\frac{N_1}{N_2}_{HSC} \right)$, the values of current peaks for LLC are $I_{pH} = \left(4 + 2\frac{N_1}{N_2}_{HSC} \right) I_{in}$ and $I_{pL} = 0$; for

HSC instead $I_{pH} = \left(2 + 2\frac{N_1}{N_2}_{HSC} \right) I_{in}$ and $I_{pL} = 2I_{in}$. Given the non-sinusoidal behavior of the secondary windings, different AC-resistance are excited in the frequency spectrum. Consequently, total copper losses depend on the contribution of each spectrum resistance, as described by the following equation:

$$P_{loss} = R_{dc}I_{dc}^2 + \sum_{n=1}^{N} R_{ac_n}I_{rms_n}^2 \quad (3)$$

where N is the number of total harmonics. Then to determine the total resistance of the secondary windings, copper losses needs to be divided by the total rms current flowing through them:

$$R_{sec} = \frac{P_{loss}}{I_{s_{rms}}^2} = k_{dc}R_{dc} + \sum_{n=1}^{N} k_{ac_n}R_{ac_n}$$
$$k_{dc} + \sum_{n=1}^{N} k_{ac_n} = 1 \quad (4)$$

Considering (1) and (2), the k_x values of (4) can be determined. As shown in Fig. 2, the comparison reveals that the HSC topology assigns greater importance to the resistance DC component, depending on the conversion ratio. This, in turn, enhances the total efficiency and current capability as the resistance increases with frequency. As a result, the HSC topology achieves a lower total R_s compared to the LLC topology.

III. ULTRA-COMPACT 3D PCB ARCHITECTURE

Conventionally, DC-DC discrete modules consist of one or more PCBs interconnected by pinout copper pillars. These PCBs run parallel to the base board. The primary purpose of the pinout is to facilitate signal and power transfer between the base board and the module. However, pinouts contribute to space inefficiency due to their relatively large size compared to the module. This large size ensures that the current density of a single pin is not too high for thermal purposes. Additionally, given the high currents handled by the CM, they introduce undesired copper losses.

To address these issues, an effective approach is to replace the copper pillars with vertical PCBs that are integral parts of the entire converter. These PCBs can also accommodate components mounted on them, further increasing the potential power density since they won't occupy space on the horizontal PCB. The connection between different PCBs is ensured through edge plating of the vertical PCBs allowing the copper layers to extend to the boundary of the PCB and be plated and molded to the base board or horizontal PCB. By doing so, copper losses can be minimized while maximizing efficiency and power/current density of the whole module.

In the specific case at hand, a 6:1 HSC is implemented using four distinct PCBs: three vertical and one horizontal.

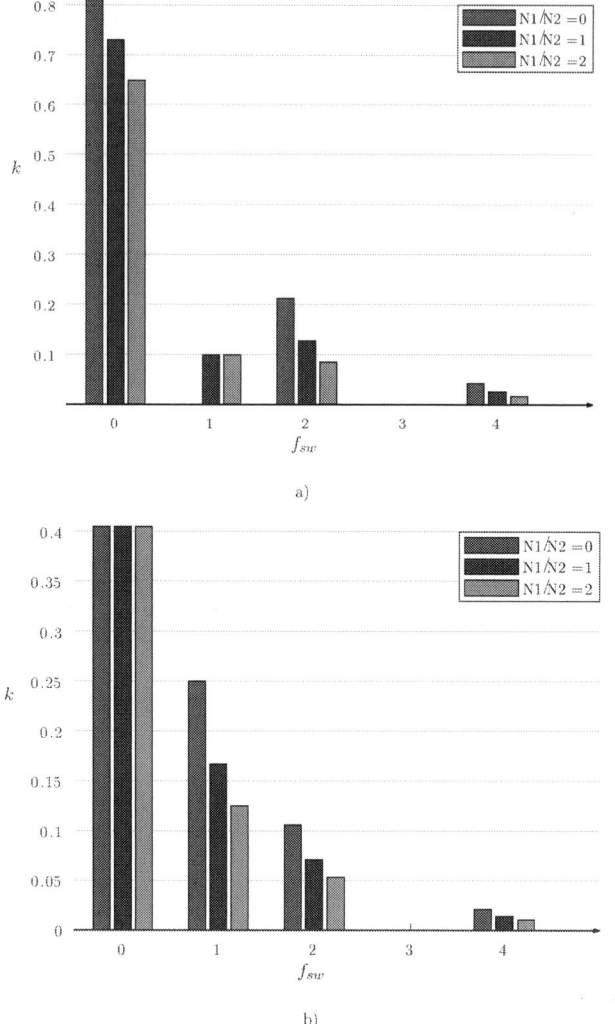

Fig. 2: k_x values comparison: a) HSC b) LLC-CT.

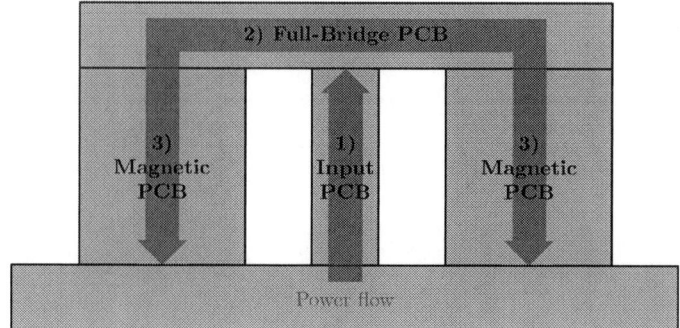

Fig. 3: Current Multiplier architecture and power flow.

Referring to Fig. 3, we can differentiate these PCBs based on their intended purposes:

1) **Input PCB**: a vertical PCB connected to input voltage V_{in}, drivers voltage V_{drive} and PWM for the drivers. The components mounted are Synchronous Rectifier Drivers and input capacitance C_{in}

2) **Full-Bridge PCB**: a horizontal PCB where the top side houses the molded input full-bridge MOSFETs (Q_1, Q_2, Q_4 and Q_5), their drivers and resonance capacitors C_{res}. The bottom side is dedicated solely to connecting vertical PCBs.

3) **MagneticPCB**: a vertical PCB where the core and windings of the MTA is located. Synchronous Rectifiers (Q_3 and Q_6,) and output capacitors C_{out} are also molded onto this PCB.

In Fig. 3 two Magnetic PCBs are depicted, indicating that this HSC is implemented using two distinct MTA and a single Full-Bridge. This design choice is based on the power dissipation characteristics of each power switch.

In the HSC, ZVS and ZCS for all MOSFETs is feasible, resulting in negligible switching losses. Consequently, we can approximate the total dissipated power by each switch as the sum of conduction and driving losses.

$$P_{MOS} = P_{drv} + P_{cond} + P_{sw} \simeq$$
$$\simeq P_{drv} + P_{cond} = V_{drv}Q_g f_{sw} + R_{ds_{(on)}} I_{rms}^2 \quad (5)$$

As the input voltage is significantly lower than 15V, we can leverage Infineon's latest top-performing source-down MOSFET. This MOSFET boasts an exceptionally low on-resistance, denoted as $R_{ds_{(on)}}$. When multiplied with the square of the root mean square current I_{rms}^2, it becomes the dominant factor in equation (5). In high-current applications, minimizing conduction losses is crucial for achieving higher current capability, as these losses contribute significantly to inefficiencies and heat generation.

To achieve maximum current capability, it is essential to balance losses across each switch, ensuring no bottlenecks are introduced. When using the same MOSFETs for both FB ($Q_{1,2,4,5}$) and SR ($Q_{3,6}$) stages, the primary difference lies in the number of parallel MOSFETs employed. Considering the topology, the rms currents of MOSFETs can expressed as follows:

$$I_{FB,rms} = \frac{I_{out,DC}\pi(3 + 2\frac{N_1}{N_2})}{8(\frac{N_1}{N_2} + 2)} \quad (6)$$

$$I_{SR,rms} = \frac{I_{out,DC}\pi}{8(\frac{N_1}{N_2} + 2)} \quad (7)$$

Here, N_{FB} and N_{SR} represent the number of parallel MOSFETs connected to each FB and SR MOSFETs. To match conduction losses between FB and SR MOSFETs, we find that:

$$N_{SR} = \left(3 + 2\frac{N_1}{N_2}\right) N_{FB} \sqrt{\frac{R_{ds_{on_{SR}}}}{R_{ds_{on_{FB}}}}} \quad (8)$$

In the specific context, where $N_1 = N_2$ and $R_{ds_{on_{SR}}} = R_{ds_{on_{FB}}}$, the objective is to achieve maximum power density. Consequently, N_{FB} is set to 1. Referring to equation (8), the optimal number of SR MOSFETs for each HSC leg is calculated as $N_{SR} = 5$. However, due to the negative impact on power and current density caused by an odd number of SRs (which would double the module's height), $N_{SR} = 4$ was chosen. Additionally, including a total of eight SRs on a single Magnetic PCB would lead to space inefficiencies, as well as increased winding loop length and AC resistance. To address this, using two Magnetic PCBs appears to be the best approach for improving efficiency, current capability, and distributing the current across the entire module footprint. In this scenario, a single Magnetic PCB can accommodate two SRs at the top and two at the bottom.

Once the number of MOSFETs is determined, drivers and capacitors are strategically positioned around them to achieve maximum power density. Additionally, in the low-voltage domain, the minimum manufacturable thicknesses of the core and prepreg are sufficient to ensure isolation between layers. Therefore, selecting dielectric materials capable of withstanding high temperatures becomes another critical consideration.

IV. MAGNETIC DESIGN

The design of the magnetic component inside the HSC plays a crucial role. Not only does it influence the overall dimensions of the converter, but it also directly impacts its final efficiency. In the current application, minimizing the size of this component is essential. Additionally, thoughtful analysis of the winding configuration is necessary.

More specifically, when aiming for maximum power density, the choice of ultra-low-profile custom magnetics becomes crucial to avoid wasting space. Consequently, operating at a high switching frequency within the range of 500 kHz to 1 MHz is meaningful for reducing the volume of the magnetic core. Additionally, to ensure good thermal performance at full load, it is essential to minimize the DC resistance R_{dc} and AC resistance R_{ac} of the autotransformer's windings. Given these considerations, a planar MTA is the most suitable component for achieving these goals as it reduces AC eddy currents, and therefore minimizes R_{ac}.

A. Magnetic Core

Typically, DCX magnetic designs involve the use of three-column-based cores, as depicted in Fig. 4a . In this configuration, the magnetic flux is initially generated in the central leg with area A_c, and then it splits into the two lateral legs with area A_l. This design approach aims to distribute the flux more effectively along the central leg, resulting in reduced core losses and minimizing local saturations.

For low-voltage, high-current applications, the impact of core loss is minimal compared to copper losses, primarily because the volume of the magnetic component is smaller. A favorable approach is to prioritize space for windings

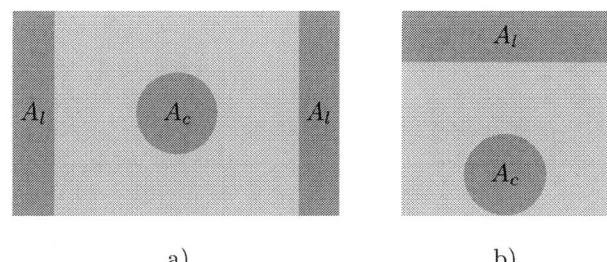

a)   b)

Fig. 4: Typical DCX magnetic cores: a) three-column-based b) two-column-based.

over the magnetic core. With this consideration, a two-column-based core can be chosen, as depicted in Fig. 4b.

As a starting point, A_c was calculated considering a switching frequency f_{sw} of 800 kHz. The material suitable for this frequency is the DMR51W soft ferrite from DMEGC. Based on the hysteresis cycle, the chosen maximum flux density B_{max} is 150 mT. The output voltage is 0.8 V. The result is expressed by the following equation:

$$A_c = \frac{V_{out}T_{on}}{2B_{max}} \simeq \frac{V_{out}}{4B_{max}f_{sw}} = 1.66 \; cm^2 \qquad (9)$$

Taking this into account, the magnetic core drawings are depicted in Fig. 5. The lateral leg was chosen larger than the central one due to available space, reducing flux flow and minimizing core loss.

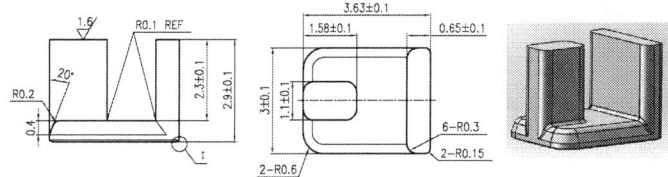

Fig. 5: Magnetic core drawings for CM.

The evaluation of core loss is essential to understand its impact on design. While the Steinmetz Equation (SE) is typically estimated for sinusoidal magnetic fields, we can analytically calculate core loss using the Modified Steinmetz Equation (MSE) [9]:

$$P_{MSE} = (kf_{eq}^{\alpha-1}B_M^{\beta})f \qquad (10)$$

The equivalent frequency f_{eq} is given by:

$$f_{eq} = \frac{2}{\Delta B^2 \pi^2} \int_0^T \left(\frac{dB}{dt}\right)^2 dt \qquad (11)$$

Here, B_M represents the maximum magnetic flux density, ΔB is is the peak-to-peak magnetic flux density, and k, α, and β are the Steinmetz parameters from SE. The excitation frequency is denoted as f.

For a triangular magnetic field (as in the HSC), the total core loss can be expressed as:

$$P_{core} = \left(\frac{8}{\pi^2}\right)^{\alpha-1} kf^{\alpha}B_M^{\beta}V_e \qquad (12)$$

Here, $V_e = A_e l_e$ represents the effective volume of the magnetic core, where A_e and l_e are respectively the effective area and length. Accurately calculating A_e is challenging because l_e corresponds to the actual path of the flux. Referring to Fig. 5, an approximated value of A_e can be extrapolated considering the area and length of different segments. This results in $A_e = 1.844\ mm^2$, and consequently $B_M = 135.6\ mT$. Taking into account $V_{tot} = 2.85\ mm^3$ and DMR51W Steinmetz coefficients $\alpha = 1.188$, $\beta = 6.4$ and $k = 276\ kW/m^3$, the total core loss is $P_{core} = 300\ mW$.

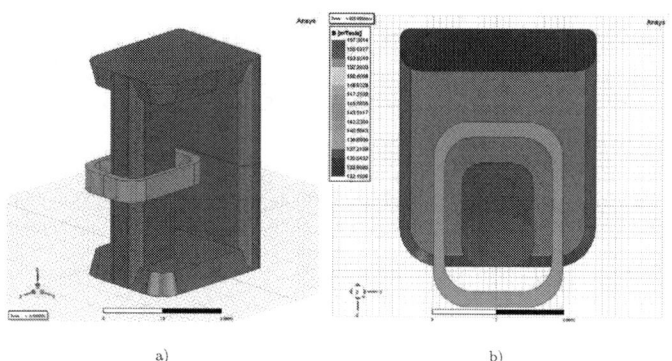

Fig. 6: Transient simulation: a) 3D model, b) peak flux in the two legs.

To effectively estimate core losses, ANSYS Maxwell can be employed, utilizing Steinmetz coefficients from the material. By conducting transient simulations, we can replicate the voltage behavior of windings with a single turn, as the windings arrangement does not affect core losses. The simulation model is depicted in Fig. 6a, while Fig. 6b illustrates the peak magnetic flux in the two legs. Fig. 7 displays the core loss curve over the period, with an average value of 325 mW and an 8% error compared to the analytical calculation.

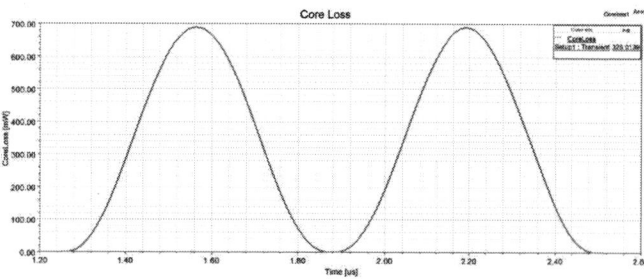

Fig. 7: Current Multiplier single core loss.

Another parameter dependent on the core shape is the magnetizing inductance, which needs to be below a certain value to achieve zero-voltage switching (ZVS) in all switches of the HSC. However, due to the complex structure of the full converter, it was not possible to adjust the core gap to achieve the desired L_{zvs}. Nevertheless, magnetizing inductance is not the sole contributor to ZVS;

energy stored in L_k also assists in moving the Q_{oss} charge, especially when the current flow is high. Additionally, because of the low switch voltages, ZVS does not significantly impact losses. Consequently, the core will be gapless, with a magnetizing inductance of $L_m = 300\ nH$. Simulation results demonstrate that achieving ZVS is feasible with the inclusion of this magnetizing inductance.

B. Windings Layout

Design of MTA is challenging: all the windings must surround only one leg to achieve maximum coupling coefficient and consequently minimum leakage. In order to do this, it is necessary to split each phase of the converter (ϕ_x) in two AC loops. The first loop is connected to V_{in} and charges the resonance capacitors, with a current I_{in} flowing also through a primary. The second discharges the resonance capacitors, with the same current flowing through the other primary. These two currents are then summed up in a secondary that contributes in generating flux flowing through the core. For the sake of explanation, the first loop will be named *charging loop* and the second *discharging loop*, and an example for the first phase (ϕ_a) is depicted in Fig. 8.

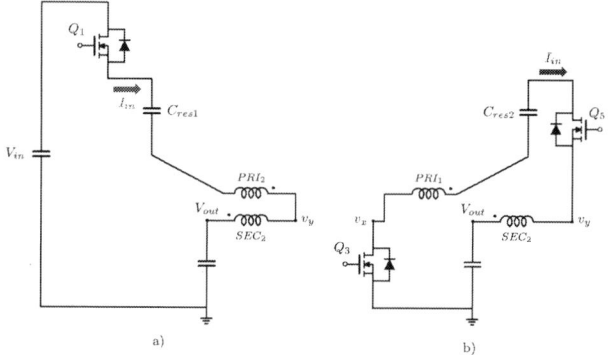

Fig. 8: HSC loops: a) charghing loop, b) discharging loop.

As the currents of the two loops must flow surrounding only the central leg, the PRI_1 to v_y connection through Q_5 must to be taken into account as it can led to a surrounding of the lateral leg.

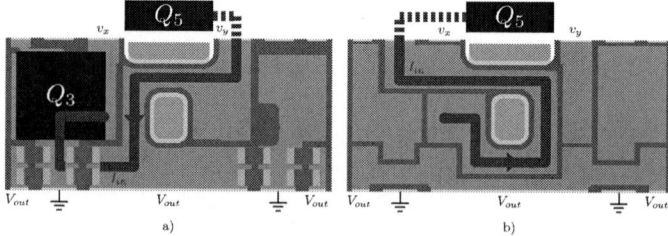

Fig. 9: Discharging Loop: a) SEC_2 winding surrounding lateral leg, b) PRI_1 winding surrounds central leg.

In fact, considering the first traditional design solution in Fig. 9, it can be denoted that with this type of connection, the PRI_1 winding is surrounding the central leg while the SEC_2 winding is surrounding the lateral leg, then increasing the leakage. Additionally, the current is flowing in the Full-Bridge PCB, presenting less copper than Magnetic PCB and increasing AC resistance.

With this consideration, a novel paradigm of windings structure was investigated. Instead of connecting Q_5 directly to the drain of Q_6 (v_y), the MOSFET source could be connected *between* the PRI_2 winding. In this manner, PRI_x will be connected to a total of three nodes. The new current loops are depicted in Fig. 10. In this way, the

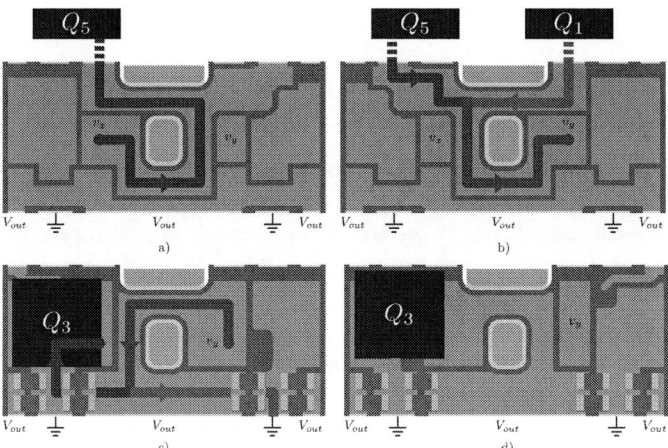

Fig. 10: ϕ_a Current Loops: a) PRI_1 b) PRI_2 c) SEC_2 d) SEC_1.

discharging loop will surround only the central leg, and Full-Bridge PCB will not contribute as windings.

Another crucial strategy for reducing leakage inductance and AC resistance involves properly interleaving different windings. It's crucial to simulate different configurations and select the one that not only reduces leakage inductance and AC resistance but also meets other design criteria such as efficiency, size, and cost-effectiveness. Fig. 11 illustrates the most effective configuration identified through such simulations.

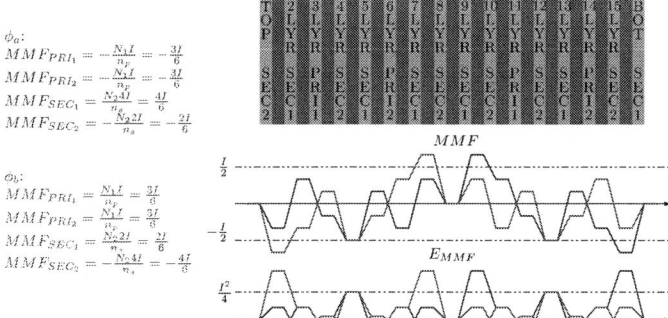

Fig. 11: Current Multiplier Interleaving.

Taking all these factors into consideration, MTA parasitics such as AC resistance R_{ac} and leakage inductance L_k can be extrapolated in order to build the transformer

model for SPICE simulations. To achieve this, FEM simulations in the frequency domain are performed using ANSYS Maxwell. Since the MTA is not purely sinusoidal, its parasitics cannot be accurately represented using the traditional RL (resistor-inductor) series model applied to the primary windings. Instead, an approximate model is shown in Fig. 12a. While the primary current exhibits sinusoidal behavior, its parasitic impedance follows an RL series model, as depicted in Fig 12b. However, the secondary current has different spectral components, leading to a winding model for the parasitics, as illustrated in Fig. 12c [10].

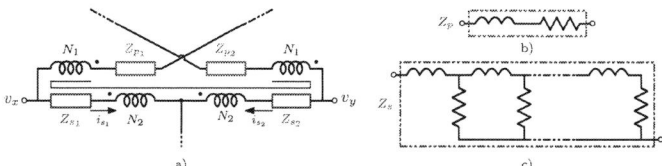

Fig. 12: Parasitics model of MTA: a) Main circuit, b) Z_p model, c) Z_s model.

Furthermore, the secondary currents flow in the opposite direction relative to the flux generated by windings, effectively compensating for any leakage contributions from the secondaries in the frequency domain. This compensation ensures that the value of L_k in Z_p remains unaffected.

However, this compensation mechanism does not apply to copper losses, which are proportional to the square of the current. Copper losses depend on the contribution of each spectrum resistance, as described by (4). Referring to Fig. 2 with a 6:1 HSC ration ($N_1/n_2 = 1$), the main contributor is the DC resistance and the other contributors are the f_{sw} and $2f_{sw}$ resistances, then frequency analysis with FEM simulation is possible considering $N = 2$.

Fig. 13 shows the different steps in order to extrapolate parasitics of the MTA. In Fig. 13a the primaries are excited with the same current and a secondary is shorted, in order to extrapolate the leakage L_k at f_{sw} evaluating the magnetic energy in the total volume (e.g. leakage energy as it is not exploited for generating current in the shorted secondary). In Fig. 13b the primaries are excited with the same current again but everything is forced in order to excite the secondaries with the same current as primaries. In this way, R_{pri} and R_{sec} (@f_{sw}) can be evaluated, considering also the eddy effects induced in both primaries and secondaries.

In Fig. 13c the secondaries are excited with the same current: in this way approximately no flux is generated and no current will be induced in the primaries. Then the whole resistance of secondaries R_{sec} is evaluated with $N = 2$ and referred then to the primaries as R'_{sec}. Hence the model of the MTA will assume the form of a traditional transformer model with the $R_{ac} = R_{pri} + R'_{sec}$ and L_k series. The outcome of the aforementioned par-

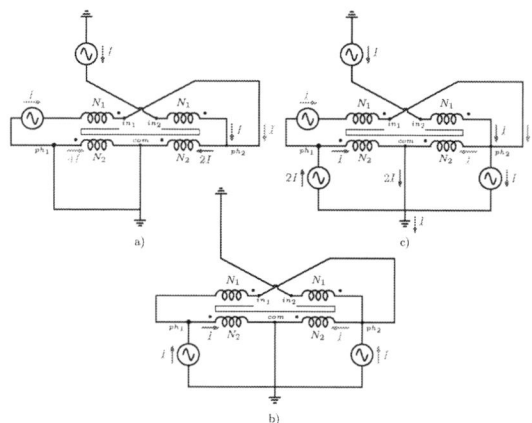

Fig. 13: Simulations methodology: a) L_k evaluation at f_{sw}, b) R_{ac} evaluation at f_{sw}, c) R_{sec} evaluation at DC and f_{sw}.

asitic analysis yields a total $R_{ac} = 25.3$ $m\Omega$ and total $L_k = 30$ nH for the individual autotransformer unit. By incorporating these values into the SPICE simulation, the model can more closely reflect the real-world behavior of the autotransformer, exploiting the available SPICE models of MOSFETs, thereby improving the reliability and validity of the simulation results for further analysis and design optimization.

V. EXPERIMENTAL RESULTS

To validate the proposed approach, the prototype described in the previous chapters was built achieving 15x12x8mm. Converter specifications are reported in Table I. This module is intended to provide PoL output. Consequently, it operates with an input voltage ranging from 6V. The operative switching frequency of the module is set at 700kHz.

TABLE I: HSC-based Current Multiplier prototype specifications.

V_{in}	3.9÷6 V
V_{out}	0.65÷1 V
P_{out}	200 W
f_{sw}	700 kHz
L_{mag} (magnetizing. single core)	300 nH
N1:N2 (turns ratio)	1:1
Size	15×12×8 mm
Power density	2.27 kW/in³
PCB technology	**Input PCB:** 4 layers, 2 oz ext., 4 oz int.
	Full-Bridge PCB: 6 layers, 3 oz ext., 4 oz int.
	Magnetic PCB: 16 layers, 3 oz ext., 4 oz int.
Core	Custom, DMR51W ferrite
Q_1, Q_2, Q_4, Q_5 FETs	IQE004NE1LM7CGSC, 15 V, 450 $\mu\Omega$
Q_7, Q_8 FETs (4 per phase)	IQE004NE1LM7CGSC, 15 V, 450 $\mu\Omega$
C_{res} (6 per phase)	GRM21BC81H475KE11
C_{out}	16×GRM155D70G106ME18
HSC drivers	6×1EDN7550U

The molds between PCBs and base boards was possible through the using of different solder pastes exhibiting distinct profile temperatures. In Fig. 14 it is possible to distinguish the different PCBs described in the previous

sections. The top side cooling of FETs enables the module to be attached to an heatsink in order to increase the current capability.

Fig. 14: Prototype of Current Multiplier.

The development of a demo board was a crucial step in verifying the capabilities of the module. This demo board was designed to test the module's performance and provide a platform for comparing the results with simulation data. The demo board's design is illustrated in Fig. 15.

Fig. 15: Current Multiplier demo board.

Efficiency curves at different frequencies are depicted in Fig. 16 which reveals a peak efficiency of 94.8% when powering an output voltage of 0.8V. Furthermore, the converter's output resistance per surface area has been characterized and found to be 0.216mΩ*mm2.

Thermal performance are shown in Fig. 17. It derives a Thermal Design Current (without heatsink) of 130A corresponding to a corresponding current density of 0.72A/mm2. Furthermore, the Electrical Design Current reaches 400A with a corresponding current density of 2.2A/mm2.

VI. CONCLUSIONS

This paper proposes a Hybrid Switched Capacitor-based Current Multiplier for data center applications. In-depth harmonic current content analysis flowing in the secondary windings reveals the topology's suitability for

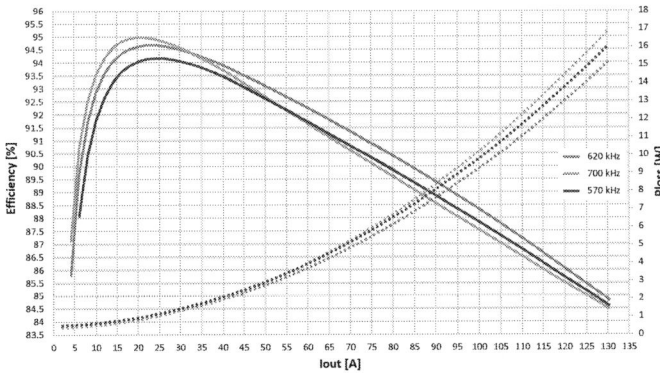

Fig. 16: Current Multiplier efficiency with 0.8V output voltage at different frequencies.

Fig. 17: Thermal performance of Current Multiplier at 130A TDC.

the low voltage high current domain, outperforming the state-of-the-art LLC topology. The ultra-compact module design achieves high Thermal Design Current (TDC) and Electrical Design Current (EDC) densities of 0.72A/mm2 and 2.2A/mm2, respectively, meeting the high current capability requirements of AI devices.

REFERENCES

[1] IEA, "Electricity 2024," https://www.iea.org/reports/electricity-2024, 2024.

[2] S. Oliver, "From 48 v direct to intel vr12.0: Saving 'big data' $500,000 per data center, per year," https://www.mouser.com/pdfdocs/wp_VR12.pdf, 2012.

[3] R. White, "Emerging on-board power architectures," in *Eighteenth Annual IEEE Applied Power Electronics Conference and Exposition, 2003. APEC '03.*, vol. 2, 2003, pp. 799–804 vol.2.

[4] B. Yang, F. Lee, A. Zhang, and G. Huang, "Llc resonant converter for front end dc/dc conversion," in *APEC. Seventeenth Annual IEEE Applied Power Electronics Conference and Exposition (Cat. No.02CH37335)*, vol. 2, 2002, pp. 1108–1112 vol.2.

[5] R. Rizzolatti, C. Rainer, S. Saggini, and M. Ursino, "Ultra-low profile hybrid switched capacitor converter with matrix multi-tapped autotransformer," in *2021 IEEE Applied Power Electronics Conference and Exposition (APEC)*, 2021, pp. 869–874.

[6] S. Jiang, S. Saggini, C. Nan, X. Li, C. Chung, and M. Yazdani, "Switched tank converters," *IEEE Transactions on Power Electronics*, vol. 34, no. 6, pp. 5048–5062, 2019.

[7] Y. Cai, M. H. Ahmed, Q. Li, and F. C. Lee, "Optimized design of integrated pcb-winding transformer for mhz llc converter," in *2019 IEEE Applied Power Electronics Conference and Exposition (APEC)*, 2019, pp. 1452–1458.

[8] R. Rizzolatti, C. Rainer, S. Saggini, and M. Ursino, "High density hybrid switched capacitor converter for data-center application," in *2021 IEEE Applied Power Electronics Conference and Exposition (APEC)*, 2021, pp. 1288–1293.

[9] H. Hein, S. Yue, and Y. Li, "Comparative core loss calculation methods for magnetic materials under harmonics effect," *IOP Conference Series: Materials Science and Engineering*, vol. 486, no. 1, p. 012019, jun 2019. [Online]. Available: https://dx.doi.org/10.1088/1757-899X/486/1/012019

[10] L. H. Dixon, "Eddy current losses in transformer windings and circuit wiring," in *SEM600 Unitrode Seminar*, 1988, pp. R2–1.

Representation and Design Methodology for Generalized Switched-Capacitor Converter Topologies

Seokwon Choi
Department of Electrical and Computer Engineering
Seoul National University
Seoul, Republic of Korea
skc411@snu.ac.kr

Dam Yun
Department of Electrical and Computer Engineering
Seoul National University
Seoul, Republic of Korea
dam0710@snu.ac.kr

Jung-Ik Ha
Department of Electrical and Computer Engineering
Seoul National University
Seoul, Republic of Korea
jungikha@snu.ac.kr

Abstract—This paper presents a design methodology for generalized switched-capacitor (SC) converter topologies, structured around an explicit equation-based representation for topological circuit models. The proposed topology equations incorporate key principles for SC converter designs, including Kirchhoff's voltage law (KVL), Kirchhoff's current law (KCL), and charge balance. Specific rules and design constraints for SC converter topology generation are detailed in this paper. The optimization process and the derivation of performance metrics for both passive and active components are illustrated through detailed examples. Following the outlined design procedure, the methodology enables the realization of a general two-phase step-down SC converter with an arbitrary voltage conversion ratio. As a practical example, this paper introduces a novel 6:1 SC converter topology that reduces the theoretical passive component volume by 75% compared to the conventional Dickson converter. The proposed topology is demonstrated through a 48V-to-8V hardware prototype, which achieves a peak efficiency of over 98%, along with a 66% reduction in passive component volume and a 44% increase in loss. The experimental results validate the representation model applied in the theoretical performance estimation.

Keywords—Switched-capacitor (SC) converters, topology synthesis, circuit optimization, generalized topology

I. INTRODUCTION

With the development of high power density switched-capacitor (SC) converters, the minimization of passive component volume and switch stress has become critical considerations, often serving as the key performance criteria for evaluating converter topologies [1]. These factors align with a traditional cost-scaling model of SC converters, which involves the calculation of output impedance in Fig. 1 [2]. Specifically, the slow-switching limit (SSL) impedance and its associated constraint in [2] represent the capacitor volume, while the fast-switching limit (FSL) impedance model corresponds with the optimization of active device power loss under a constant electric field scaling model in a cascode configuration [3].

Extensive research on SC converter topology has aimed to identify high-performance circuits across various specifications. In response, numerous efforts have been made to synthesize

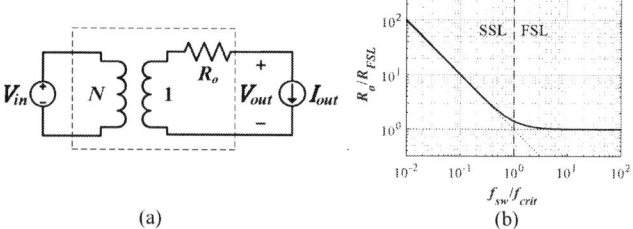

Fig. 1. (a) Output Impedance model of an ideal SC converter. (b) Output impedance as a function of frequency for an ideal SC converter.

generalized topologies. One common approach involves assembling predefined cells [4], [5], which facilitates the recognition of structures compatible with a wide range of conversion ratios. However, this method often requires an excessive number of capacitors, limiting the optimization of passive components. Another approach develops iterative wiring methods based on graphical models [6], [7]. This method can be effective in computing all possible topologies for a specified number of capacitors [8], though the voltage conversion ratio can only be determined after a topology is defined. These studies generally derive topologies with the minimum number of capacitors, which may exclude more efficient designs from consideration. Meanwhile, Zhu et al. have proposed a general network model to explore N:1 SC converter topologies through parallel simulation [9], yet this model tends to generate numerous redundant and equivalent topologies, which results in a large time complexity [8]. In addition, topologies with overlapping voltage nodes are often overlooked.

This paper addresses these limitations by introducing a novel representation method for SC converter topologies. An equation-based approach is proposed with a set of well-defined rules to facilitate efficient depiction and synthesis of converter circuit designs. Series and parallel connections are denoted using intuitive mathematical symbols, specifically "+" and "=" or "‖," respectively. The symbols correspond to voltage nodes and play a crucial role in connection vector derivation, circuit validity, and performance calculation. Each component is assigned a predefined DC bias value to satisfy fundamental circuit laws and support the generation of converters with

979-8-3315-1612-3/25 $31.00 © 2025 IEEE

arbitrary conversion ratios. The charging phase of each component can be identified and utilized to maintain charge balance.

The design methodology incorporates essential constraints to ensure a valid SC converter topology, with respect to the derivation of a unique charge flow. Consequently, the approach simplifies the design process of the converter. The connection vectors derived from the representation model are instrumental in calculating theoretical performance, specifically both the SSL and FSL impedances. In addition, passive and active components are optimized to minimize energy storage capability and reduce switch stress.

To further validate the methodology, a new 6:1 two-phase step-down SC converter topology is proposed. This paper provides a performance comparison between the novel topology and conventional designs, focusing on component volume and conversion efficiency. The proposed design concept is evaluated through simulations and experimental hardware prototypes.

II. AN EQUATION-BASED REPRESENTATION METHOD FOR TOPOLOGICAL MODELS

A. Rules of Equation Representations

The core concept of the method is to depict the connections between capacitors and voltage terminals, where series connections are represented by "+" and parallel connections by "=" or "∥." Numerical values indicate the DC bias voltage relative to the minimum input or output terminal voltage. Each equation represents one phase of the SC converter, where validity indicates that Kirchhoff's voltage law (KVL) is met.

The circuit topology is represented by the following rules:

1) Component Enumeration: Each number denotes the DC bias of a component. Input and output voltages are labeled with subscript 'in' and 'out,' respectively. Capacitors are labeled alphabetically in descending order of DC bias to prevent duplicate topologies.

2) Connection Representation: Series connections are indicated by "+," while parallel connections are denoted by "=" or "∥." The "∥" notation has higher precedence over "+," and brackets may be used to group expressions. The component with the highest DC bias appears on the left-hand side of the equation, which indicates that it is grounded.

3) Voltage Sum Validity: The voltage sum on each side of "=" and "∥" must be balanced to comply with KVLs. For example,

$$4_{in} = 2_a + 2_b \parallel (1_c + 1_{out}) \qquad (1)$$

illustrates KVL equations $V_{in} = V_a + V_b$ and $V_b = V_c + V_{out}$.

4) Additional Voltage Node: If there is no component with the highest DC bias grounded, an additional unscripted node is introduced to represent the highest voltage node. This node must connect to two or more components for charge balance. For example,

$$3 = 2_a \parallel 2_b + 1_c \parallel 1_{out} \qquad (2)$$

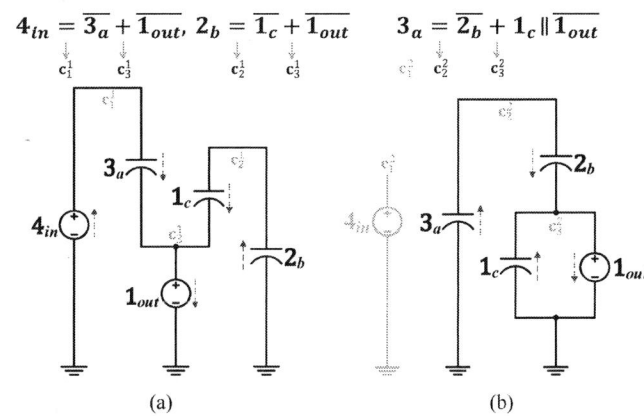

$$4_{in} = \overline{3_a} + \overline{1_{out}}, \; 2_b = \overline{1_c} + \overline{1_{out}} \qquad 3_a = \overline{2_b} + 1_c \parallel \overline{1_{out}}$$

(a) (b)

Fig. 2. 4:1 Dickson converter represented using the proposed method. (a) Phase 1: the input terminal directly charges the capacitors. (b) Phase 2: the input terminal is disconnected.

denotes that a voltage node of $3V_{out}$ is connected to the positive terminals of capacitors a and b, ensuring proper charge distribution.

5) Leaf Equations: An additional equation is introduced when connections cannot be fully represented in a single equation, called a *leaf equation*. Common elements may appear in both equations, and grounds in leaf equations are assumed to be shared. For example, a phase of a 4:1 Dickson converter, as shown in Fig. 2 (a), can be represented as

$$4_{in} = 3_a + 1_{out}, \; 2_b = 1_c + 1_{out}, \qquad (3)$$

with two leaf equations that share the output terminal.

6) Charging Phase: The charging phase of a component is indicated with an overbar, while the discharging phase is left unmarked. Components independently connected in series share the equivalent charging condition. One possible charging phase of (1) is represented as

$$4_{in} = \overline{2_a} + 2_b \parallel (\overline{1_c} + \overline{1_{out}}), \qquad (4)$$

where capacitors a, c and the output terminal are charged.

7) Phase Equation: Each equation corresponds to one phase of an SC converter, with two or more phases generating an SC converter topology. For example, the 4:1 Dickson converter topology illustrated in Fig. 2 can be represented as

$$\begin{cases} \varphi_1: \; 4_{in} = \overline{3_a} + \overline{1_{out}}, \; 2_b = \overline{1_c} + \overline{1_{out}} \\ \varphi_2: \; 3_a = \overline{2_b} + 1_c \parallel \overline{1_{out}}. \end{cases} \qquad (5)$$

where leaf equations are applied in phase 1 (φ_1).

B. SC Converter Topology Design Constraints

A valid SC converter topology should meet the following design constraints:

1) Charge Balance in Steady-State: Each SC converter topology must include both a charging and a discharging phase for each capacitor to ensure charge balance in steady-state. This indicates that in a two-phase SC converter topology, each phase equation must include all capacitors.

2) Kirchhoff's Current Law (KCL) Constraints: Each "=" and "+" symbol represents all voltage nodes except the ground node. To satisfy KCL, the sum of charge flow on each side of "=" or "+" (including all components in parallel) must be balanced. For example, (2) (in phase j) requires that

$$
\begin{aligned}
0 &= q_a^j + q_b^j, \\
q_a^j + q_b^j &= q_c^j + q_{out}^j.
\end{aligned}
\tag{6}
$$

3) Redundant Connections: Identical capacitor connections across all phases should be avoided to prevent redundancy.

The following lemmas are established to support the validity of SC converter topologies.

Lemma 1. *Each "=" (except for additional nodes) and "||" denotes a voltage loop. For a topology with a total of k capacitors, the total number of loops $\sum l_j$ must satisfy $\sum l_j \geq k + 1$ for the topology to be valid.*

Proof. Under steady-state conditions, there are $(k + 1)$ variables representing the DC bias of each capacitor and the output terminal relative to the input voltage.

In order to derive a unique solution for the DC bias from the KVLs, there should be $(k + 1)$ linearly independent KVLs, which correspond to voltage loops.

Lemma 2. *If $\sum l_j = k + 1$, a unique charge flow for each component can be determined, independent of other conditions (e.g., the capacitance values or the duty ratio).*

Proof. For a planar graph, the equation $v - e + f = 2$ holds, where v represents the number of voltage nodes, e is the number of components (edges), and f equals $l + 1$ with l being the number of (voltage) loops.

There are a total of $e = k + 2$ components, including capacitors and the input and output terminals. In an m-phase converter topology, the total number of charge flow variables is $m(k + 2)$.

The number of KCLs is $\sum_{j=1}^{m}(v_i - 1) = m(k + 2) - \sum_{j=1}^{m} l_j$, excluding the ground node, as its which the corresponding KCL is a linear combination of KCLs for other nodes.

Additionally, there are k charge balance equations (one for each capacitor) and one normalization equation.

Therefore, if $\sum l_j = k + 1$, a unique solution for the charge flow can be obtained.

C. Two-Phase Step-Down SC Converter Constraints

From this point onward, the paper focuses on *two-phase step-down SC converter topologies*. In this setup, the design constraint for charge balance requires that each phase equation includes all capacitors. The following constraints are defined to simplify the design process and prevent invalid topologies:

- The input terminal is designated as the highest voltage node, while the lowest voltage node is the ground. Both the input and output terminals are referenced to ground.

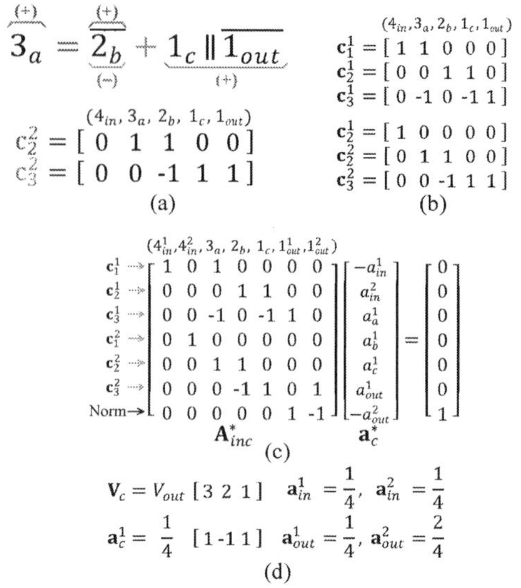

Fig. 3. Calculation of the capacitor charge multiplier vector from the topology in Fig. 2. (a) Example derivation of connection vectors. (b) Connection vectors of each voltage node. (c) Matrix equation. (d) Derived charge multipliers.

- Input and output capacitors are assumed to be very large, with negligible voltage drops across switches.

- In order to prevent duplicate topologies, parallel-connected components are ordered based on their final component values, with the largest placed first, followed by subsequent values in descending order. For instance, components that sum up to 3 are ordered as "3, (1+2), (2+1), (1+1+1)."

III. THEORETICAL EFFICIENCY CALCULATION AND COMPONENT OPTIMIZATION

As demonstrated in [2], the conversion loss of an idealized SC converter can be represented by an output impedance model that accounts for resistive losses. In this model, the charge flow in the capacitors is calculated under the slow-switching limit (SSL) condition, while switch on-resistance is considered dominant under the fast-switching limit (FSL) condition. This leads to the following expressions:

$$
\begin{aligned}
R_{SSL} &= \sum_{i \in caps} \frac{(a_{c,i})^2}{C_i f_{sw}}, \\
R_{FSL} &= \sum_{i \in sw} \frac{R_i (a_{r,i})^2}{D_i},
\end{aligned}
\tag{7}
$$

where $a_{c,i}^j = q_{c,i}^j f_{sw}/I_{out}$ and $a_{r,i}^j = q_{r,i}^j f_{sw}/I_{out}$ represent the charge multipliers for the capacitors and switches, respectively. The total output impedance can be approximated as $R_{OUT} \approx \sqrt{R_{SSL}^2 + R_{FSL}^2}$, resulting in the behavior shown in Fig. 1 (b).

The capacitor optimization follows a fixed constraint on total energy storage capability, $E_{tot} = \sum_i 1/2 \left(v_{c,i(rated)}\right)^2 C_i$. This corresponds to the normalized passive component volume described in [1]. For switch optimization, the sum of the switch V·A product (or G·V^2 metric), $A_{tot} = \sum_i G_i \left(v_{r,i(rated)}\right)^2$, is used

$c_1^1 = [1\ 1\ 0\ 0\ 0]: 4V_{out}$ $a_{r1}^1 = a_{in}^1$
$c_2^1 = [0\ 0\ 1\ 1\ 0]: 2V_{out}$ $a_{r2}^1 = -a_b^1$
$c_3^1 = [0\ 1\ 0\ 1\ 1]: 1V_{out}$ $a_{r3}^1 = a_a^1 + a_c^1$
$c_4^1 = [1\ 0\ -1\ 0\ 1]: 0V_{out}$ $a_{r4}^1 = a_b^1$

(a)

$c_1^2 = [1\ 0\ 0\ 0\ 0]: 4V_{out}$
$c_2^2 = [0\ 1\ 1\ 0\ 0]: 3V_{out}$ $a_{r1}^2 = -a_a^2$
$c_3^2 = [0\ 0\ -1\ 1\ 1]: 1V_{out}$ $a_{r2}^2 = a_b^2,\ a_{r3}^2 = -a_c^2$
$c_4^2 = [1\ 1\ 0\ 1\ 1]: 0V_{out}$ $a_{r4}^2 = a_a^2 + a_c^2$

(b)

$$a_r = \frac{1}{4}[1\ 1\ 2\ -1|1\ 1\ 1\ -2]$$
$$V_r = V_{out}[1\ 2\ -1\ 1|2\ -1\ 1\ 1]$$

(d)

Fig. 4. Switch optimization for the topology in Fig. 2. Each designed switch is shown with bidirectional arrows, and merged terminals are highlighted. Connection vectors and required switches for (a) Phase 1 and (b) Phase 2. (c) Example switch configuration in phase 2. (d) Derived switch charge multipliers.

as a fixed constraint. Minimizing the output impedance under these constraints results in the optimal values for each capacitance and switch on-resistance as:

$$C_i = \left|\frac{a_{c,i}}{v_{c,i}}\right| \frac{2E_{tot}}{\Sigma_k |a_{c,k} v_{c,k}|},$$
$$R_i = \left|\frac{v_{r,i}}{a_{r,i}}\right| \frac{\Sigma_k |a_{r,k} v_{r,k}|}{A_{tot}}. \tag{8}$$

Recent works have demonstrated that optimizing the $G \cdot V^2$ metric produces similar results to switch width optimization for minimizing the power loss associated with switching devices [3], [10]. This power loss is expressed as the sum of conduction and switching loss, assuming constant electric field scaling in a cascode configuration of switches.

Consequently, the performance metric for SSL and FSL limit can be expressed as

$$M_{SSL} = \Sigma_i \left|\frac{a_{c,i} v_{c,i}}{v_{out}}\right|,$$
$$M_{FSL} = \Sigma_i \left|\frac{a_{r,i} v_{r,i}}{v_{out}}\right|. \tag{9}$$

It is known that the series-parallel converters achieve the theoretical Wolaver limit for passive components, represented by $M_{SSL} = (N-1)/N$. This indicates that the design minimizes SSL output impedance for a given passive component volume. In contrast, ladder-type converters, including Dickson and Cockroft-Walton converters, reach the theoretical FSL Wolaver limit, $M_{FSL} = 4(N-1)/N$, which maximizes conversion efficiency under equivalent switch stress conditions.

A. Capacitor Charge Multiplier and SSL Metric Derivation

The efficiency of an N:1 SC converter, represented by the proposed method, can be derived by defining incidence matrices, A_{inc} [7], [8]. This paper simplifies the derivation of the incidence matrix and the capacitor charge multiplier vector.

Assuming steady-state conditions for each capacitor, where $a_{c,i}^1 + a_{c,i}^2 = 1$, the charge multiplier vector can be defined as

$$a_c = \left[a_{in}^j \mid a_{c,i}^1 \mid a_{out}^j\right]^T$$
$$= \left[a_{in}^1\ a_{in}^2\ a_{c,1}^1 \cdots a_{c,k}^1\ a_{out}^1\ a_{out}^2\right]^T, \tag{10}$$

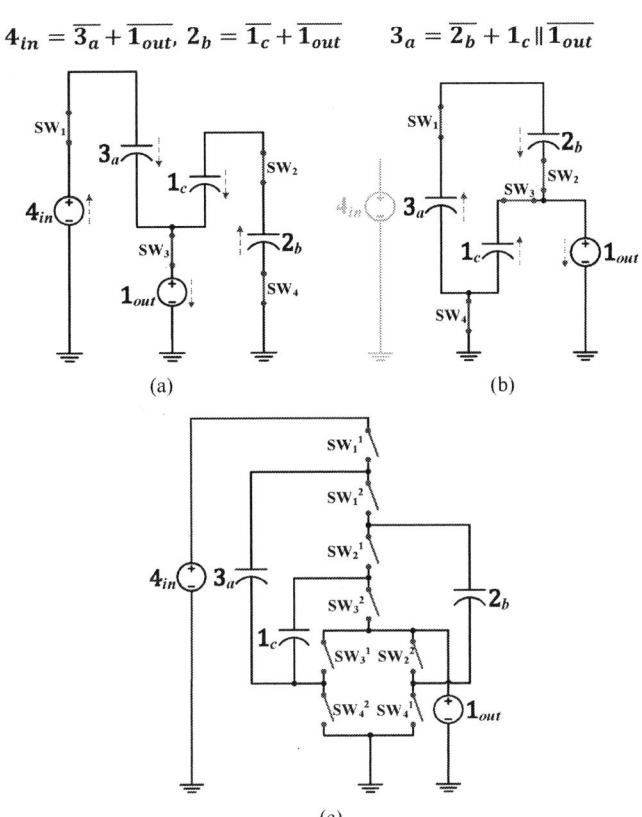

$$4_{in} = \overline{3_a} + \overline{1_{out}},\ 2_b = \overline{1_c} + \overline{1_{out}} \qquad 3_a = \overline{2_b} + 1_c \| \overline{1_{out}}$$

(a)   (b)

(c)

Fig. 5. Complete SC converter topology derived from Fig. 2. (a) Phase 1 circuit representation. (b) Phase 2 circuit representation. (c) Full SC converter topology.

with $(k+4)$ variables, where k is the number of capacitors.

The customized incidence matrix A_{inc}^* is a square matrix consisting of $(k+3)$ connection vectors indicating the connections of each capacitor terminal, along with one normalization row that denotes $a_{out}^1 + a_{out}^2 = 1$.

A connection vector represents the connection of each component and corresponds to a node (or a KCL constraint), where 1 indicates a positive terminal connection, and -1 indicates a negative terminal, as shown in Fig. 3. The symbols "+" and "=" relate to voltage nodes, as described in Section II. Therefore, connection vectors can be derived by searching for these symbols along with the parallel connections surrounding them. Specifically, the symbol "=" denotes that the positive terminals of components on both sides are connected, while "+" indicates that the negative terminal of the component on the left side is connected to the positive terminal of the component on the right side. An example of a connection vector derivation is illustrated in Fig. 3 (a). The input terminal is not connected to other components in phase 2, making its own vector.

Thus, the formation of this matrix is given by

$$A_{inc}^* = \begin{bmatrix} c_n^{1*} \\ \hline c_n^{2*} \\ \hline c_{Norm} \end{bmatrix}, \tag{11}$$

TABLE I. THE COMPARISON OF CONVENTIONAL 6:1 SC CONVERTER TOPOLOGIES

Topology	Phase Equations	# Capacitors	# Switches	M_{SSL}	M_{FSL}
Dickson	$6_{in} = \overline{5_a} + \overline{1_{out}}, 4_b = \overline{3_c} + \overline{1_{out}}, 2_d = \overline{1_e} + \overline{1_{out}}$ $5_a = \overline{4_b} + 1_e \parallel \overline{1_{out}}, 3_c = \overline{2_d} + 1_e \parallel \overline{1_{out}}$	5	10	2.500	3.333
Cockroft-Walton	$6_{in} = \overline{2_a} + 2_b \parallel \overline{2_c} + 2_d \parallel (\overline{1_e} + \overline{1_{out}})$ $5 = 2_a \parallel \overline{2_b} + 2_c \parallel \overline{2_d} + 1_e \parallel \overline{1_{out}}$	5	10	2.500	3.333
Ladder	$6_{in} = \overline{1_a} + 1_b\parallel\overline{1_c} + 1_d\parallel\overline{1_e} + 1_f\parallel\overline{1_g} + 1_h\parallel\overline{1_i} + \overline{1_{out}}$ $5 = 1_a\parallel\overline{1_b} + 1_c\parallel\overline{1_d} + 1_e\parallel\overline{1_f} + 1_g\parallel\overline{1_h} + 1_i\parallel\overline{1_{out}}$	9	12	4.167	3.333
Series-Parallel	$6_{in} = \overline{1_a} + \overline{1_b} + \overline{1_c} + \overline{1_d} + \overline{1_e} + \overline{1_{out}}$ $1 = 1_a \parallel 1_b \parallel 1_c \parallel 1_d \parallel 1_e \parallel \overline{1_{out}}$	5	16	0.833	6.667
Cascaded Monolithic [14]	$6_{in} = \overline{3_a} + \overline{2_b} + 1_e\parallel\overline{1_{out}}, 2_c = \overline{1_d} + 1_e\parallel\overline{1_{out}}$ $3_a = \overline{2_c} + 1_e\parallel\overline{1_{out}}, 2_b = \overline{1_e} + 1_d\parallel\overline{1_{out}}$	5	16	1.500	4.333
New Topology	$6_{in} = \overline{4_a} + 2_b \parallel (\overline{1_c} + 1_d \parallel \overline{1_{out}})$ $4_a = \overline{2_b} + \overline{1_d} + 1_c \parallel \overline{1_{out}}$	4	13	1.500	4.000

$$6_{in} = \overline{4_a} + 2_b \parallel (\overline{1_c} + 1_d \parallel \overline{1_{out}}) \qquad 4_a = \overline{2_b} + \overline{1_d} + 1_c \parallel \overline{1_{out}}$$

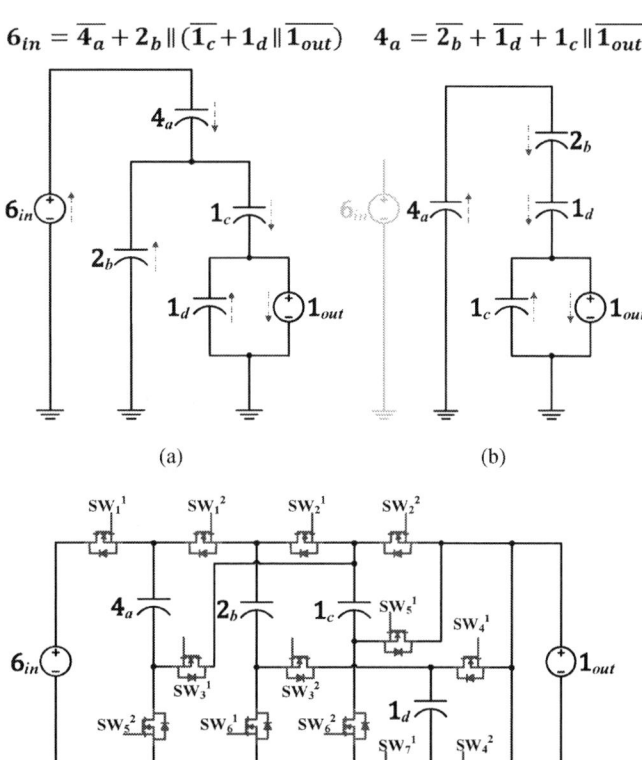

Fig. 6. Proposed 6:1 SC converter topology. (a) Phase 1. (b) Phase 2. (c) Entire converter circuit.

with connection vectors expanded to represent two phases. By applying the customized charge multiplier vector,

$$\mathbf{a}_c^* = \begin{bmatrix} -a_{in}^1 & a_{in}^2 & a_{c,1}^1 & \cdots & a_{c,k}^1 & a_{out}^1 & -a_{out}^2 \end{bmatrix}^T, \qquad (12)$$

the matrix equation

$$\mathbf{A}_{inc}^* \mathbf{a}_c^* = \begin{bmatrix} 0 & 0 & \cdots & 0 & 1 \end{bmatrix}^T \qquad (13)$$

is satisfied, which allows $\mathbf{a}_{c,i}^1$ to be derived using the inverse of $\mathbf{A}_{inc}^*$. This implies that the rank of $\mathbf{A}_{inc}^*$ must be $(k + 4)$ for a valid converter topology.

Using the derived $\mathbf{a}_{c,i}^1$ values, the optimal capacitance values SSL metric M_{SSL} can be calculated as given in (8) and (9).

B. Switch Charge Multiplier and FSL Metric Derivation

Following the derivation of capacitor charge multiplier vector $\mathbf{a}_{c,i}^1$, the optimal placement of switching devices can be determined. Each connection in a phase, including ground connections, or a group of non-zero elements in a connection vector, corresponds to a configuration of closed switch(es). Terminals connected in both phases are merged into a single terminal, as no switch is needed between them. For example, in Fig. 4, the negative terminals of capacitors a and c, as well as the input and output, are connected in both phases and colored to indicate that they are merged.

For each voltage node with s connected terminals, $(s - 1)$ switches are required across s^{s-2} possible spanning trees [8]. To minimize the FSL Metric, the total switch stress (V·A product) of the $(s - 1)$ switches is calculated across s^{s-2} configurations and compared. The blocking voltage $v_{r,i}$ can be identified by examining the other phase. The switch placement that minimizes switch stress is then selected, as shown in Fig. 4 (c).

By synthesizing the switches for each phase, the complete SC converter topology generated from the phase equations in (5) is obtained, as shown in Fig. 5. The configuration in Fig. 5 (c) is equivalent to the conventional topology.

IV. THE PROPOSED 6:1 SC CONVERTER TOPOLOGY

A. Conventional 6:1 Circuit Topologies

Data center and telecom power supplies traditionally deliver power to the blade level at high voltages, typically around 48 V, often using a two-stage power conversion architecture to manage power efficiently across components [11], [12]. In this architecture, a 6:1 converter can serve as an intermediate bus converter that steps down the voltage to 8 V [13]. While 4:1 converter is widely used and extensively studied, there is a

TABLE II. PARAMETERS OF THE CONVERTER HARDWARE PROTOTYPES

Topology	E_{tot}^{simul} [J]	E_{tot}^{exp} [J]	A_{tot}^{simul} [MV·A]	A_{tot}^{simul} [MV·A]	Capacitor Volume [mm³]	Switch Volume [mm³]
New Topology	0.021	0.0207	0.21	0.239	221	141
Dickson	0.084	0.0834	0.21	0.234	655	109

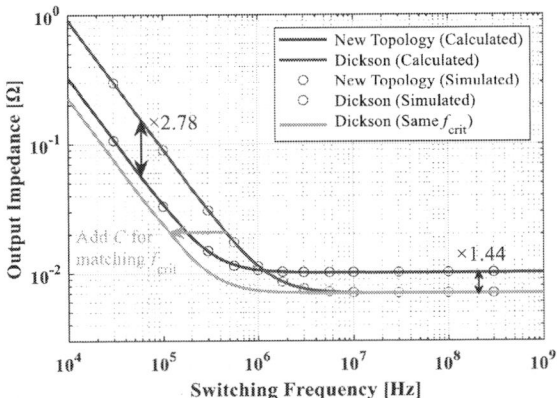

Fig. 7. Output impedance simulated through Simulink.

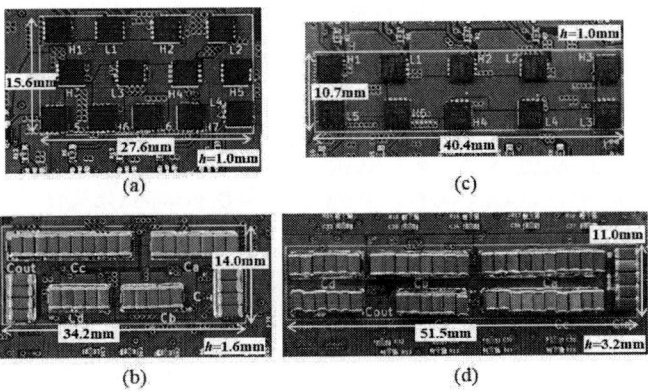

Fig. 8. Hardware prototypes of the converters. (a) The top side of the new topology converter board shows active components, and (b) the bottom side shows passive components. (c) The top side of the Dickson converter board, and (d) the bottom side.

notable lack of conventional topologies available for 6:1 switched-capacitor converters.

Conventional topologies are analyzed and summarized in Table 1 based on the representation criteria and performance metrics described above. The series-parallel topology achieves the theoretical minimum SSL metric, whereas ladder-type topologies reach the theoretical minimum FSL metric.

This paper proposes a novel 6:1 two-phase step-down SC converter topology founded on the outlined design concepts. The theoretical performance metric of this topology lies in the middle of the series-parallel and Dickson converters, achieving a balance between passive component volume and switch stress. Additionally, this topology demonstrates improved switch stress while maintaining the same SSL metric compared to a previous study on 6:1 SC converters [14].

B. The Proposed 6:1 Converter Topology

Fig. 6 illustrates the proposed 6:1 SC converter topology, represented by:

$$\begin{cases} \varphi_1 : \ 6_{in} = \overline{4_a} + 2_b \parallel (\overline{1_c} + 1_d \parallel \overline{1_{out}}) \\ \varphi_2 : \ 4_a = \overline{2_b} + \overline{1_d} + 1_c \parallel \overline{1_{out}}. \end{cases} \quad (14)$$

The design includes 4 capacitors and 13 switches. With consistent output charge between the two phases, the output current remains relatively stable, which allows the use of a smaller output capacitance. The maximum charge flow through any active component reaches two-thirds of the active output charge flow, minimizing resistive loss.

Compared to the conventional Dickson converter, which achieves minimal switch stress, the proposed converter topology reduces the SSL metric by 40%, with a 20% margin relative to the FSL Wolaver limit. This trade-off allows a reduction in passive component volume compared to the Dickson converter while maintaining a similar level of component loss. In order to achieve an equivalent critical frequency, where SSL loss equals FSL loss as shown in Fig. 1 (b), the new topology requires only one-quarter of the theoretical passive component volume of the Dickson converter. It is crucial to minimize the operating frequency, as higher frequencies in the FSL region lead to increased overall power loss driven by switching loss. Specifically, the overall power loss is proportional to the square root of the switching frequency when optimized [15].

V. EXPERIMENTAL RESULTS

A. Simulation Results

A simulation through Simulink and a hardware prototype are implemented to validate the design concept by comparing the performance of the new topology converter with the conventional Dickson converter. A 48V-to-8V SC converter is designed for the simulation and experiment.

For the design of a hardware prototype, the passive component values are chosen to achieve a critical frequency of 350 kHz. The total switch stress, $A_{tot} = \sum G_i v_{r,i}^2$, which serves as the FSL constraint, is set at 0.21 MV·A. In this configuration, the SSL constraint (defined as the total capacitor energy storage capability $E_{tot} = \sum 1/2 \, C_i v_{c,i}^2$) is set at 0.021 J. The Dickson converter requires 0.084 J for the SSL constraint to match the critical frequency. Hence, the capacitance values and switch on-resistance values are designed as given in (8).

Fig. 7 illustrates the theoretical and simulated output impedance of the two topologies. Under equivalent conditions, the Dickson converter has a critical frequency of 1.4 MHz, while the new topology achieves 350 kHz. Additional capacitance is added to the Dickson converter to match the critical frequency,

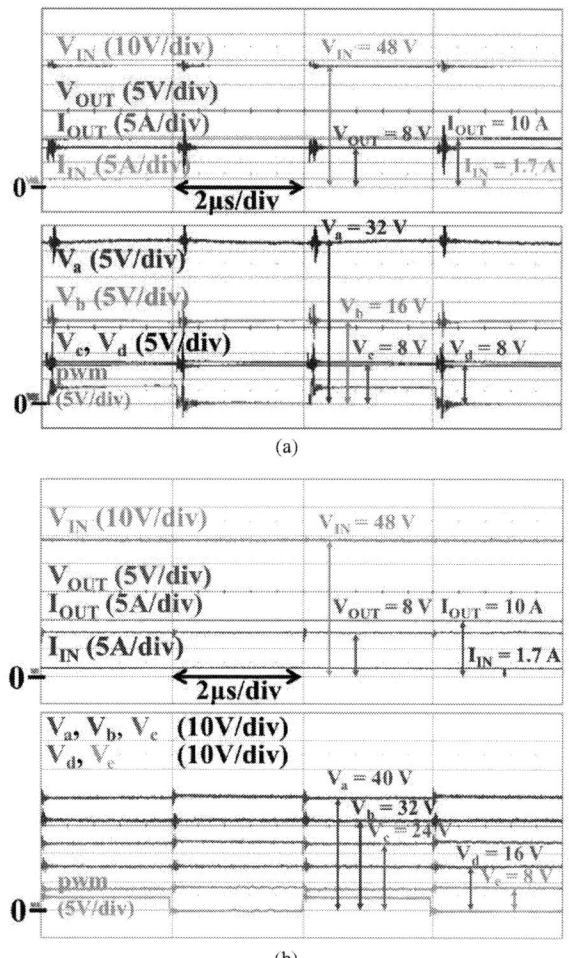

(a)

(b)

Fig. 9. Measured output waveforms of the capacitor, input, and output voltage, along with input and output current in 250 kHz switching frequency and an 80 W load (a) New Topology. (b) Dickson.

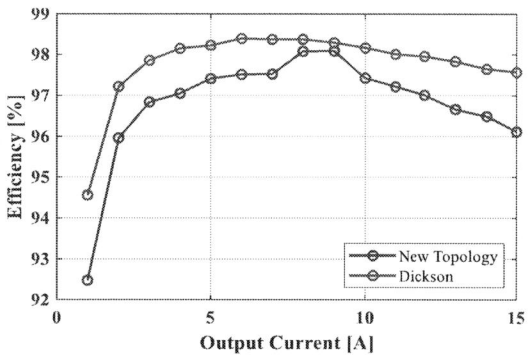

Fig. 10. Measured conversion efficiency with 250 kHz switching frequency.

resulting in an SSL constraint that is four times larger. At a 120W load condition, maximum conversion efficiency of 98.7% is achieved in the Dickson topology, slightly higher than the 98.1% efficiency of the new topology. The validity of the model in Fig. 1 (b) is confirmed through simulation results.

B. Experimental Results

The achieved converter parameters are presented in Table 2. TDK multilayer ceramic chip capacitors of the C3216 series are used for the passive components. Capacitors are selected according to the criteria that require a rated voltage greater than 1.5 times the DC bias, a critical frequency above 1 MHz, and the largest capacitance under the DC bias condition. The number of capacitors is determined based on the capacitance values specified in (8). The active components are chosen from Infineon OptiMOS™ power MOSFETs with PFQN packages that meet the requirements of the rated voltage exceeding 1.5 times the blocking voltage. The selected MOSFETs have drain-source on-resistance values closest to the values calculated from (8).

Due to the discrete nature of product parameters, differences exist between the design constraints used in the simulation and those in the hardware experiment. Specifically, while the nominal capacitor energy storage capability is 4.03 times greater in the Dickson converter prototype, the capacitor volume in the conversion module is 2.96 times larger. This discrepancy is a result of the rated voltage not perfectly matching the DC bias of each capacitor, which affects the effective energy storage. In addition, the active component volume has an unintentional gap due to the limited switch options.

The measured voltage and current waveforms at 66% load conditions are shown in Fig. 9. The operating frequency is set at 250 kHz, which optimizes conversion efficiency over a wide range of load conditions. It is noticeable that the voltage of each capacitor aligns closely with the DC bias ratio used in the converter design, as indicated by the numerical labels. The input and output current are filtered to ensure accurate measurements, demonstrating stable power conversion at an 80 W load.

Fig. 10 illustrates the measured conversion efficiency across the full load range. The hardware prototype for the proposed topology achieves a peak efficiency of 98.1% and a full-load efficiency of 96.1%. By comparison, the Dickson converter board achieves a slightly higher peak efficiency of 98.4%, with a full-load efficiency of 97.6%. On average, the power loss of the proposed converter prototype is 1.44 times greater than that of the Dickson converter, which matches the theoretical model depicted in Fig. 7. This confirms the validity of the theoretical predictions and highlights the trade-off between efficiency and component size inherent in the new topology design.

VI. CONCLUSION

This paper presents a novel design methodology for representing SC converter topologies using accessible equations and introduces a new approach to the synthesis of generalized N:1 SC converters. The new designs can be compared to existing topologies since the proposed method leverages performance metrics established in previous studies.

A 6:1 topology developed on the design concept is realized as a hardware prototype. The proposed topology optimizes the use of passive components and achieves better volume efficiency than conventional designs such as the Dickson converter. Specifically, it reduces the passive component volume by a factor of 2.96 compared to the Dickson converter, while achieving 98.1% peak conversion efficiency with an average power loss that is 1.44 times larger. A detailed comparison highlights the trade-offs between the passive component volume and the conversion efficiency.

Furthermore, the SC topology representation and design methodology introduced in this work establish a foundation for automated topology synthesis. This study can provide a basis for a new approach to synthesizing arbitrary N:M SC topologies.

REFERENCES

[1] Z. Ye, S. R. Sanders, and R. C. N. Pilawa-Podgurski, "Modeling and Comparison of Passive Component Volume of Hybrid Resonant Switched-Capacitor Converters," *IEEE Trans. Power Electron.*, vol. 37, no. 9, pp. 10903–10919, Sep. 2022, doi: 10.1109/TPEL.2022.3160675.

[2] M. D. Seeman and S. R. Sanders, "Analysis and Optimization of Switched-Capacitor DC–DC Converters," *IEEE Trans. Power Electron.*, vol. 23, no. 2, pp. 841–851, Mar. 2008, doi: 10.1109/TPEL.2007.915182.

[3] M. H. Kiani and J. T. Stauth, "Optimization and comparison of hybrid-resonant switched capacitor DC-DC converter topologies," in *2017 IEEE 18th Workshop on Control and Modeling for Power Electronics (COMPEL)*, Stanford, CA, USA: IEEE, Jul. 2017, pp. 1–8. doi: 10.1109/COMPEL.2017.8013321.

[4] M. S. Makowski, "A canonical switched capacitor DC-DC converter," in *2014 IEEE 15th Workshop on Control and Modeling for Power Electronics (COMPEL)*, Santander, Spain: IEEE, Jun. 2014, pp. 1–8. doi: 10.1109/COMPEL.2014.6877113.

[5] Chi-Wah Kok, Oi-Ying Wong, Wing-Shan Tam, and Hei Wong, "Design strategy for two-phase switched capacitor step-up charge pump," in *2009 IEEE International Conference of Electron Devices and Solid-State Circuits (EDSSC)*, Xi'an: IEEE, Dec. 2009, pp. 423–428. doi: 10.1109/EDSSC.2009.5394227.

[6] R. Karadi, "Synthesis of switched-capacitor power converters: An iterative algorithm," in *2015 IEEE 16th Workshop on Control and Modeling for Power Electronics (COMPEL)*, Vancouver, BC, Canada: IEEE, Jul. 2015, pp. 1–4. doi: 10.1109/COMPEL.2015.7236477.

[7] D. O. Larsen, M. Vinter, and I. Jorgensen, "Systematic Synthesis of Step-Down Switched-Capacitor Power Converter Topologies," *IEEE Trans. Circuits Syst. II Express Briefs*, vol. 66, no. 5, pp. 863–867, May 2019, doi: 10.1109/TCSII.2019.2909442.

[8] Z. Gu *et al.*, "Synthesizing Step-Down Switched Capacitor Power Converter Topologies," *IEEE Trans. Circuits Syst. Regul. Pap.*, vol. 71, no. 3, pp. 1465–1479, Mar. 2024, doi: 10.1109/TCSI.2023.3336718.

[9] J. Zhu, Y. Yang, D. Zheng, and Y. Deng, "A Generalized Topology Synthesis for the Switched-Capacitor Converter," *IEEE Trans. Ind. Electron.*, vol. 70, no. 10, pp. 10024–10033, Oct. 2023, doi: 10.1109/TIE.2022.3224191.

[10] S. R. Pasternak, M. H. Kiani, J. S. Rentmeister, and J. T. Stauth, "Modeling and Performance Limits of Switched-Capacitor DC–DC Converters Capable of Resonant Operation With a Single Inductor," *IEEE J. Emerg. Sel. Top. Power Electron.*, vol. 5, no. 4, pp. 1746–1760, Dec. 2017, doi: 10.1109/JESTPE.2017.2730823.

[11] P. Sandri, "Increasing Hyperscale Data Center Efficiency: A Better Way to Manage 54-V\/48-V-to-Point-of-Load Direct Conversion," *IEEE Power Electron. Mag.*, vol. 4, no. 4, pp. 58–64, Dec. 2017, doi: 10.1109/MPEL.2017.2760113.

[12] W. C. Liu, Z. Ye, and R. C. N. Pilawa-Podgurski, "A 97% Peak Efficiency and 308 A/in³ Current Density 48-to-4 V Two-Stage Resonant Switched-Capacitor Converter for Data Center Applications," in *2020 IEEE Applied Power Electronics Conference and Exposition (APEC)*, New Orleans, LA, USA: IEEE, Mar. 2020, pp. 468–474. doi: 10.1109/APEC39645.2020.9124488.

[13] Y. Zhu, Z. Ye, T. Ge, R. Abramson, and R. C. N. Pilawa-Podgurski, "A Multi-Phase Cascaded Series-Parallel (CaSP) Hybrid Converter for Direct 48 V to Point-of-Load Applications," in *2021 IEEE Energy Conversion Congress and Exposition (ECCE)*, Vancouver, BC, Canada: IEEE, Oct. 2021, pp. 1973–1980. doi: 10.1109/ECCE47101.2021.9596032.

[14] T. Van Daele, E. De Pelecijn, T. Thielemans, M. Steyaert, and F. Tavernier, "A Fully-Integrated 6:1 Cascaded Switched-Capacitor DC-DC Converter Achieving 74% Efficiency at 0.1W/mm²," in *2019 15th Conference on Ph.D Research in Microelectronics and Electronics (PRIME)*, Lausanne, Switzerland: IEEE, Jul. 2019, pp. 49–52. doi: 10.1109/PRIME.2019.8787774.

[15] P. H. McLaughlin, J. S. Rentmeister, M. H. Kiani, and J. T. Stauth, "Analysis and Comparison of Hybrid-Resonant Switched-Capacitor DC–DC Converters With Passive Component Size Constraints," *IEEE Trans. Power Electron.*, vol. 36, no. 3, pp. 3111–3125, Mar. 2021, doi: 10.1109/TPEL.2020.3017123.

A 48-V-to-1-V Gallium Nitride Switching Bus Converter for Processor Vertical Power Delivery with 2.7 mm Thickness and 3048 W/in³ Power Density

Jiarui Zou[†], Yicheng Zhu[†], Nathan M. Ellis[‡], Logan Horowitz[†], and Robert C. N. Pilawa-Podgurski[†]

[†] Department of Electrical Engineering and Computer Sciences, University of California, Berkeley
[‡] Department of Electrical and Computer Engineering, University of California, Santa Cruz
Email: [†]{jiarui.zou, yczhu, logan_h_horowitz, pilawa}@berkeley.edu, [‡]nathan.ellis@ucsc.edu

Abstract—**High-performance processors demand over 1000 A of current at a core voltage of 1 V or lower when running computationally intensive tasks, demanding improved 48 V to point-of-load (PoL) power conversion solutions with high efficiency and power density. In pursuit of improved performance for 48-V-to-PoL conversion, a switching bus converter (SBC) was recently proposed to achieve single-stage vertical power delivery and pushed the performance boundary of existing solutions. This work presents an ultra-low-profile, high-density implementation of the SBC, with high switching frequency enabled by Gallium Nitride field-effect transistors and an advanced packaging structure featuring recessed coupled inductors for vertical current flow. A 2.7 mm thick low-profile SBC hardware prototype was designed and built for 48-V-to-1-V conversion. The prototype was tested up to 405 A of output current at an output voltage of 1 V, achieving a high power density of 3048 W/in³. At 0.8 V and 420 A, it reached a power density of 2529 W/in³, demonstrating promising performance compared to state-of-the-art solutions.**

I. INTRODUCTION

Due to the rapid growth in computationally demanding tasks, such as training artificial intelligence models, the current demand for modern high-performance processors has surged to over 1000A [1]. At these increased current levels, the i^2R power distribution network (PDN) losses of existing lateral power delivery (LPD) architectures are severe. This challenge is further exacerbated by the growing number of peripheral devices and associated interconnects, such as the high-bandwidth memory (HBM), surrounding the processor core, which limits the space for power delivery circuits in the LPD architecture. With an increased dc bus voltage from 12 V to 48 V and an operating processor core voltage of 1 V or lower, voltage regulation modules (VRMs) for next-generation processors are required to achieve a much larger step-down ratio for 48 V to point-of-load (PoL) power conversion.

To address these challenges, multiple 48-V-to-PoL power converters have been proposed in academia for improved efficiency and power density [2]–[13]. Recently, a switching bus converter (SBC) [14], [15] and its dual-inductor variant [16] were proposed for single-stage vertical power delivery (VPD) as a higher-performance alternative to current LPD solutions. Compared to the state of the art, the SBC achieved excellent efficiency and power density, pushing the performance boundaries for 48-V-to-PoL conversion.

This work discusses the design and implementation of a low-profile, high-density SBC prototype for processor VPD. With a rack-level power consumption of up to 132 kW and

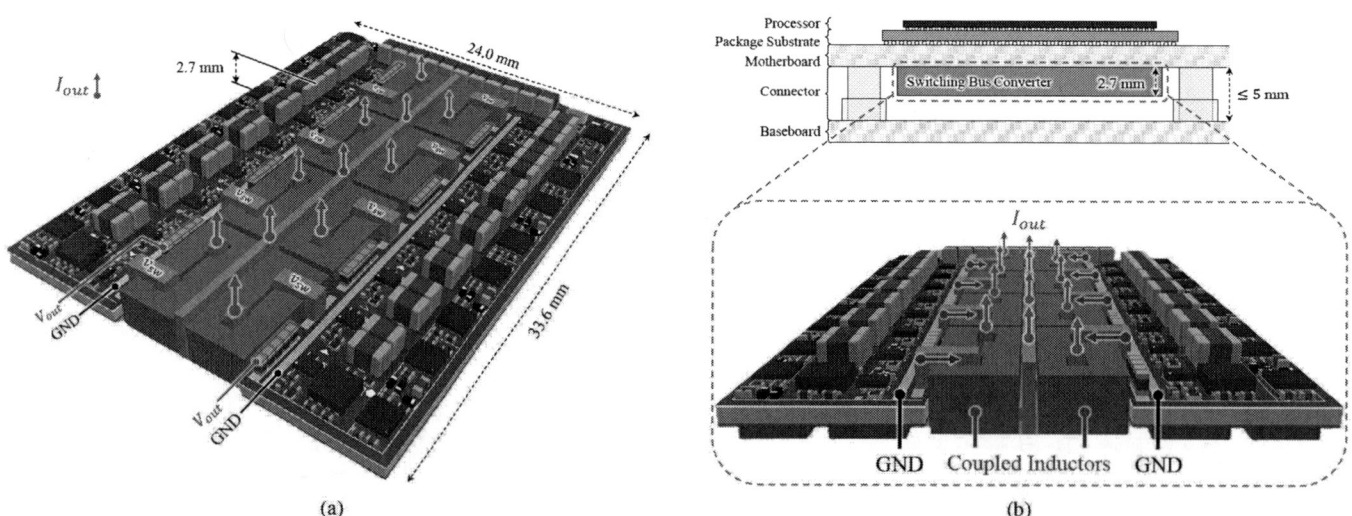

Fig. 1: 3D rendering of the converter illustrating the VPD concept. Dimension: 33.6 mm × 24.0 mm × 2.7 mm.

each 1U server (1.75 inches thick) consuming over 5 kW, space within the server—particularly height—is at a premium for the latest generation designs [17], [18]. Components such as computing and networking modules, liquid cooling cold plates with associated tubing, and various power supplies compete for space [19]. In some processor VPD applications, due to the limited vertical space ($\leq$ 5mm typically) between the motherboard and the baseboard, power converters are typically required to meet stringent height constraints to achieve VPD, making low-profile solutions highly desirable.

Another critical consideration for converters designed for processor VPD is thermal management within these constrained spaces, which further reduces the available space for the converter themselves. Converters intended for processor VPD must allocate volume for thermal management solutions, such as heat spreaders or heatsinks, to cool heat-generating components like transistors and inductors [20]. To achieve a compact form factor for the cooling solution, the power converter should ideally be designed for single-sided cooling, allowing a cooling solution to contact all heat-generating components on one side, thereby reducing both the volume and complexity of the thermal solution.

Using Gallium Nitride (GaN) field-effect transistors (FETs) and advanced packaging structures with recessed coupled inductors, a 2.7 mm thick ultra-thin hardware prototype (SBC-GaN) was designed and built for 48-V-to-PoL processor VPD. As illustrated in Fig. 1, this prototype can be placed on the backside of the motherboard directly under the processor die, thereby reducing the PDN size and losses.

The prototype was tested up to 405 A of output current at an output voltage of 1 V and achieved a power density of 3048 W/in³. The prototype's operation was further verified at an output of 0.8 V and 420 A with a high power density of 2529 W/in³, demonstrating its robust performance across different operating points.

II. TOPOLOGY AND IMPLEMENTATION

In the switching bus converter illustrated in Fig. 2, a 2-to-1 switched-capacitor (SC) front end is connected to two subsequent series-capacitor buck (SCB) modules via two switching buses [15]. Due to the superior energy density of capacitors compared to inductors [21], it is advantageous to design the SC network to achieve a larger conversion ratio so that the inductor volume of the following buck-type stage, which typically dominates the overall converter volume, can be reduced [22]. A larger SC stage conversion ratio can be achieved by adding more branches in the SCB modules. However, the maximum achievable output voltage decreases as the branch count of the SCB module increases. Therefore, the number of SCB branches is typically selected as the maximum allowable value to ensure the required output voltage over the entire load range. This work uses the 8-branch configuration to achieve the required 48-V-to-1-V conversion.

To significantly reduce the converter height compared to prior SBC implementations [14], [15], this work (SBC-GaN) 1) increases the switching frequency to 1 MHz by using GaN

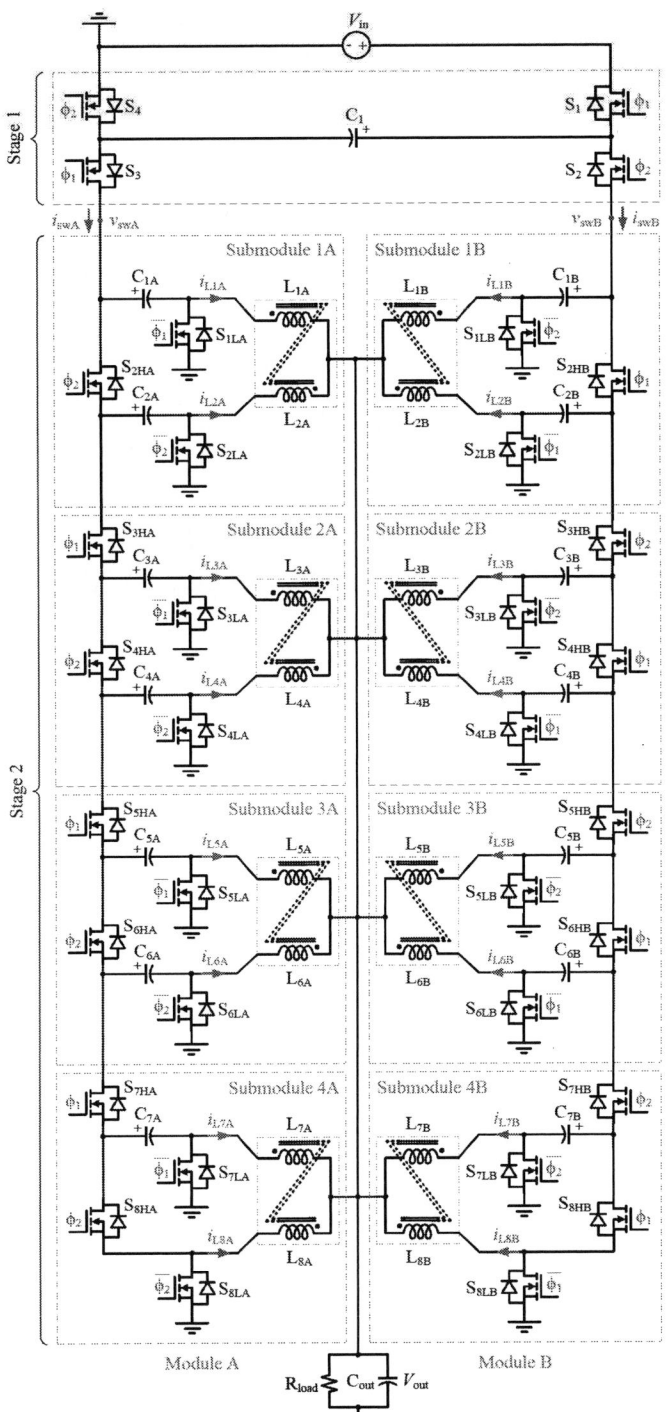

Fig. 2: Schematic drawing of the SBC.

FETs instead of Si MOSFETs, 2) uses a thin PCB with a thickness of 0.8 mm, and 3) recesses the coupled inductors into the PCB to minimize the total height.

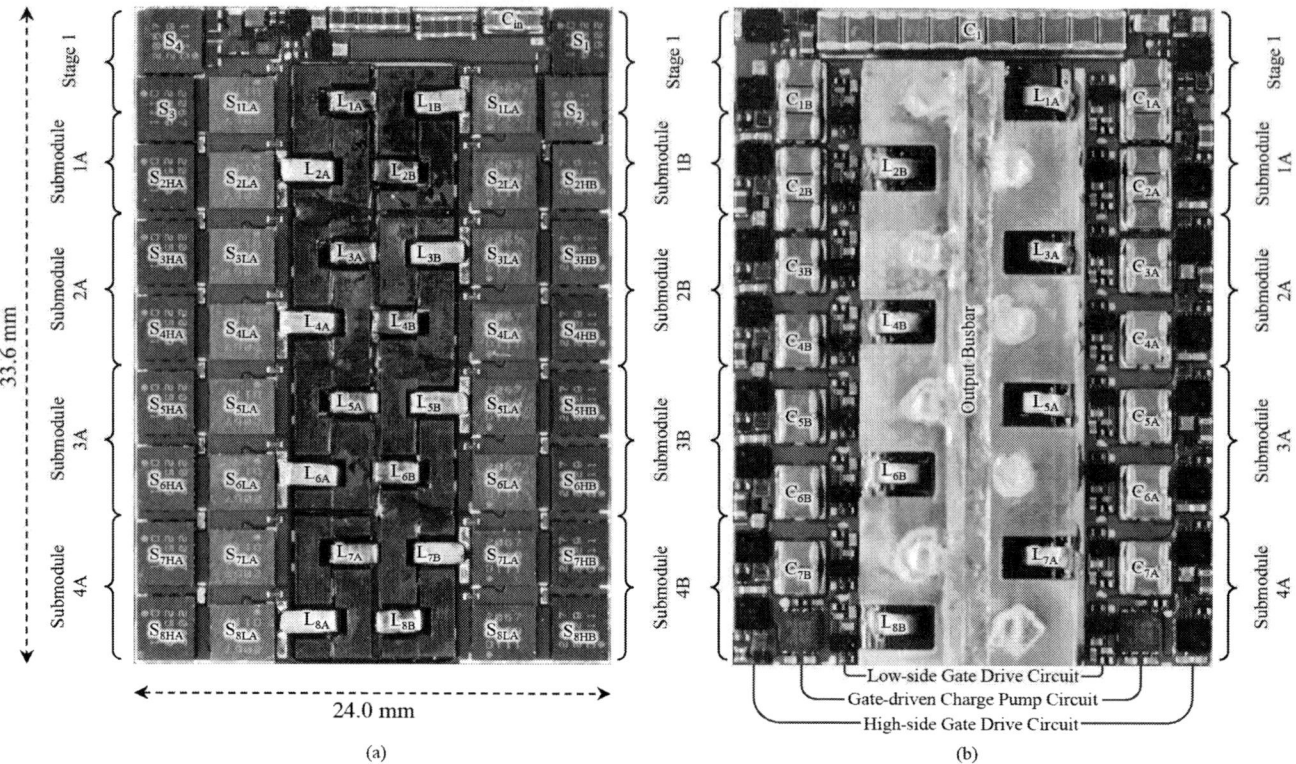

Fig. 3: Photographs of the hardware prototype. (a) Top view. (b) Bottom view. Output capacitors included in the box volume are covered by the output busbar and can be seen in Fig. 5(b). Dimension: 33.6 mm × 24.0 mm × 2.7 mm.

TABLE I: Main components used in the hardware prototype

Component	Part number	Parameters
Transistor	EPC2067	40V 1.55mΩ GaN E-HEMT
Flying capacitor C_1	Murata GRM21BR6YA106KE43L	X5R 35V 10μF ×11 in parallel
Flying capacitor $C_{1A/B-2A/B}$	Murata GRT21BR61E226ME13L	X5R 25V 22μF ×3 in parallel
Flying capacitor $C_{3A/B-7A/B}$	Murata GRT21BR61E226ME13L	X5R 25V 22μF ×2 in parallel
Input capacitor C_{in}	Murata GRM21BD72A225KE01	X7T 100V 2.2μF ×20 in parallel
Output capacitor C_{out}[†]	Murata GRM158R61A226ME15D	X5R 10V 22μF ×40 in parallel
	Murata GRM155R61A106ME11	X5R 10V 10μF ×315 in parallel
Gate driver in Stage 1	EPC uP1966E	0.4Ω pull-down resistance, 0.7Ω pull-up resistance
High-side gate driver in Stage 2	EPC uP1966E	0.4Ω pull-down resistance, 0.7Ω pull-up resistance
Low-side gate driver in Stage 2	Texas Instruments LMG1020	7-A peak source current, 5-A peak sink current

* The capacitance listed in this table is the nominal value before dc derating.
† All output capacitance implemented in the hardware prototype fits within the converter box volume without increasing assembly difficulty, as detailed in Section III-D.

III. SBC-GaN CONVERTER DESIGN

A. Prototype Implementation

By taking advantage of the fast switching capability of GaN FETs, the size of passive components, including flying capacitors and output inductors, can be significantly reduced compared to lower frequency Si designs. Compared to the previous SBC prototype [15], a high switching frequency of 1 MHz allowed the flying capacitor size to be reduced by 50%, contributing to a more compact and low-profile design.

However, compared to silicon MOSFETs, GaN FETs exhibit a much higher reverse conduction voltage drop when the device is off [23]. As shown in Fig. 2, an external diode was anti-paralleled with each transistor to reduce losses caused by the reverse conduction of GaN FETs, providing a low-voltage-drop current path during dead time [24]. Moreover, the anti-parallel diodes mitigate the overcharging effect in the gate drive bootstrap circuit caused by the transistors' high reverse conduction voltage drop, which can increase the local gate drive voltage seen by each GaN FET and may result in overvoltage gate breakdown [25].

B. Coupled Inductor Design

To realize decreased inductor volume and transient inductance [26], custom two-phase coupled inductors were designed for the SBC-GaN converter prototype. In addition, several packaging structures were employed to further decrease the

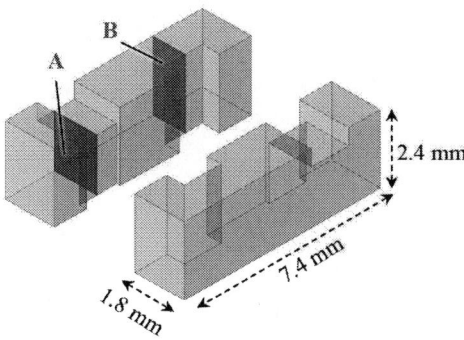

Fig. 4: Asymmetrical EE coupled inductor core (top view). Colored regions indicate cross-sectional areas.

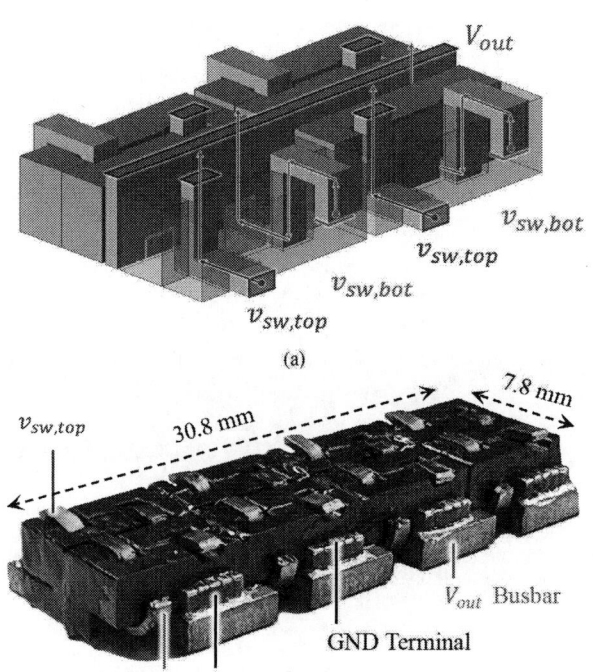

(a)

(b)

Fig. 5: (a) Winding current visualization (bottom view). (b) Photograph of the coupled inductors assembly (top view).

volume of the output inductors. Since inductors normally dictate the height of the converter, all coupled inductors in the SBC-GaN prototype were recessed into the PCB as illustrated in Fig. 1. As shown in Fig. 3(a) and Fig. 9, this recessed design additionally enables placement of the inductors at the same level as the GaN FETs, allowing both components to be cooled by a heat spreader (not shown) on the top side of the converter, thereby reducing the volume and complexity of the cooling solution.

To achieve inverse coupling of the two phases, a twisted winding configuration, shown in Fig. 5(a), was chosen to improve manufacturability at the small scale of the designed coupled inductor core, as depicted in Fig. 4. However, the twisted winding arrangement causes the switch node pairs in each submodule, shown in Fig. 2, to be positioned at opposite sides of the PCB.

To aid the visualization of the winding current in the inductor assembly, Fig. 5(a) labels the top-side switch node in purple ($v_{sw,top}$), the bottom-side switch node in red ($v_{sw,bot}$), and the output node in blue. Copper windings directly connected top-side switch nodes to the output node. All bottom switch nodes were connected to the output via twisted windings, allowing current to flow in the opposite direction through the core. To decrease conduction loss, a central output busbar was placed between the inductors to aggregate all inductor current from the bottom switch nodes.

As shown in Fig. 4, each coupled inductor was constructed using two custom-made E-cores to include winding cutouts that recess the copper winding into the core. This custom E-core ensures all volume occupied by coupled inductors is fully utilized by either the core material or the copper, maximizing volume utilization. The winding cutout, however, reduces the cross-sectional area of the core as illustrated in Fig. 4. If the cutout were made to a symmetrical E-core, the cross-section of the core with the cutout would be subjected to higher flux density during converter operation and thus would become a bottleneck of the inductor core. To achieve an optimal design, the flux density distribution should be uniform across the core [27]. Therefore, an asymmetrical E core shown in Fig. 4 was designed to ensure that the cross-sectional areas of the blue region A and the purple region B were equal.

TABLE II: Key performance metrics and operating conditions of the custom two-phase coupled inductor

Parameter	Value
Coupling coefficient	-0.83
Per-phase steady-state inductance	83.4 nH
Per-phase transient inductance	26.2 nH
Per-phase dc resistance	0.191 mΩ
Nominal output voltage	1 V
Switching frequency	1 MHz
Nominal duty ratio	0.34
Per-phase peak-to-peak current ripple	8.7 A
Per-phase average current at full load	25 A
Per-phase saturation current	35 A
Height	2.4 mm

C. Coupled Inductor Performance

The two-phase coupled inductors were manufactured using DMR53 Mn-Zn ferrite material from DMEGC [28]. Key performance metrics and operating conditions of the custom two-phase coupled inductor are listed in Table II. The inductor achieves an 83.4 nH per-phase steady-state inductance to maintain a peak-to-peak current ripple of 8.7 A and features a low per-phase transient inductance of 26.2 nH to improve transient response during load steps [26].

While the inductor is designed for a nominal output voltage of 1 V and a nominal duty ratio of 0.34, it must be robust against variations in load voltage (duty ratios), load steps, and manufacturing tolerances. The designed coupled inductor is configured to operate across an output voltage range of 0.6 V to 1.4 V, with a 10% manufacturing tolerance.

Fig. 6: Experimental setup for automated efficiency measurement. Equipment list: ① Equipment control interface. ② Keysight E36312A dc power supply for powering the control and gate drive circuitry. ③ GW Instek PSU 60-25 dc power supply (60 V 25 A). ④ Chroma 63206A-60-1000 dc electronic load (60 V 1000 A). ⑤ Keysight InfiniiVision MSOX4024A oscilloscope. ⑥ Vertically mounted hardware prototype. ⑦ FLIR T540 thermal camera. ⑧ CUI Devices CBM-979533B-154 44 CFM dc blower. ⑨ Yokogawa WT3000E precision power analyzer for input/output voltage and input current measurement. ⑩ Prototype bottom-side real-time thermal monitoring. ⑪ Prototype top-side real-time thermal monitoring.

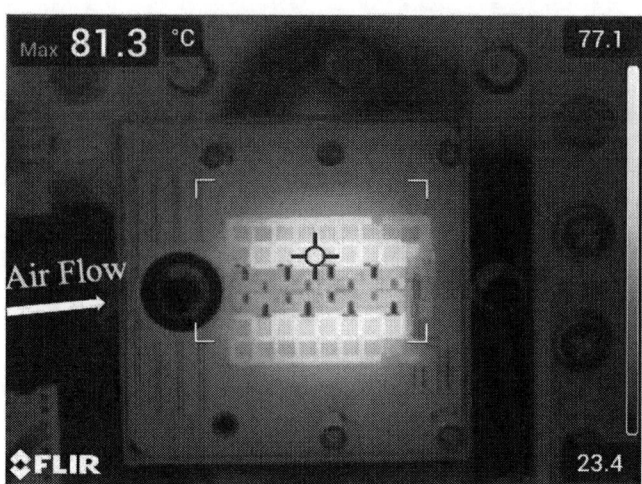

Fig. 7: Steady-state thermal image (top view) of the prototype operating at an output of 1 V and 250 A, cooled by a 44 CFM blower-style fan (CUI Devices CBM-979533B-154).

D. Output Capacitors

For processor power delivery, an adequate amount of output capacitance is needed to meet stringent transient performance requirements. As illustrated in Fig. 1 and Fig. 5(b), vertical output capacitor banks were inserted between the switch nodes located on the backside of the PCB, with the PCB-facing terminal connected to ground, fully utilizing the space provided by the alternating switch nodes of the coupled inductors.

The current SBC-GaN prototype includes $40 \times 22\,\mu F$ 0402 vertical output capacitors within the converter area shown in Fig. 1 and Fig. 5(b), along with an additional $315 \times 10\,\mu F$ 0402 MLCCs adjacent to the converter output, with 180 on the top side of the converter and 135 on the bottom, as shown in Fig. 9 and Table I. In total, the prototype includes $4030\,\mu F$ of output capacitance.

Similar to a previous SBC prototype designed for processor VPD [15], the SBC-GaN prototype can take advantage of the space between the gate drive circuits and the motherboard to place all output capacitance within the converter box volume.

Referring to Fig. 3(b), $65 \times 22\,\mu F$ 0402 MLCCs, such as the Murata GRM158R61A226ME15D, can be placed in parallel on top of each column of the high-side gate driver circuits, while an additional 60 such MLCCs can be placed on top of each column of the bottom-side gate driver circuits.

Along with the 40 vertical output capacitors adjacent to the coupled inductors, a total of $290 \times 22\,\mu F$ 0402 MLCCs can be placed within the converter box volume for a total output capacitance of $6380\,\mu F$— over 50% more output capacitance than is currently implemented in the hardware prototype.

TABLE III: Hardware specifications and test conditions

Parameter	Value
Nominal input voltage	48 V
Nominal output voltage	1 V
Nominal duty ratio	0.34
Switching frequency	1 MHz
Gate drive voltage of Stage 1	5 V
High-side gate drive voltage of Stage 2	5.4 V
Low-side gate drive voltage of Stage 2	5 V
Length	33.6 mm
Width	24.0 mm
Height	2.7 mm
Output Voltage @ 1 V	
Full-load output current	405 A
Per-phase current	25.31 A
Output Voltage @ 0.8 V	
Full-load output current	420 A
Per-phase current	26.25 A

IV. EXPERIMENTAL RESULTS AND PERFORMANCE COMPARISON

A hardware prototype of the 48-V-to-1-V SBC-GaN, shown in Fig. 9, was constructed to validate the functionality and performance of the proposed implementation. To evaluate the prototype's performance across various operating points, the converter was connected to an automated efficiency measurement test bench shown in Fig. 6.

A. Experimental Setup

To ensure precise efficiency measurements, the Yokogawa WT3000E power analyzer was used to provide high-precision measurements for input voltage, input current, and output voltage. Simultaneously, the Chroma 63206A-60-1000 dc electronic load maintained an accurate, specified current draw from the converter and provided output current readings to the equipment control interface.

As shown in Fig. 6, the prototype was cooled by a single 44 CFM blower fan directed at the top side, where the GaN

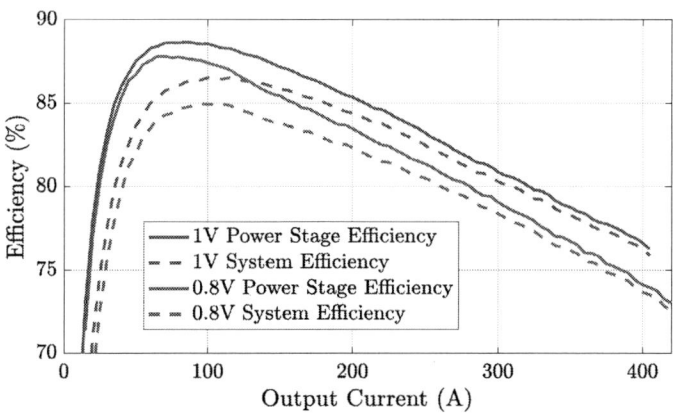

Fig. 8: Measured power stage efficiency (excluding the gate drive loss) and system efficiency (including the gate drive loss) of the 48-V-to-PoL SBC-GaN prototype. The output voltage was maintained within ±1% of the nominal output voltage (1 V & 0.8 V).

Fig. 9: Photograph of the top side of the hardware prototype, with a U.S. penny included for perspective. The board is mounted to copper output busbars for convenient output connections.

FETs are located. To assess the converter's thermal performance in real-time, the converter was mounted vertically, with two FLIR T540 thermal cameras positioned to capture images of the top and bottom sides of the prototype.

B. Experimental Results

The hardware specifications and test conditions for the converter are listed in Table III. During efficiency measurements, the output voltage was maintained within ±1% of the nominal output. The measured power stage efficiency and system efficiency across the load range are shown in Fig. 8, and the performance of the hardware prototype is summarized in Table IV.

At an output voltage of 1 V, the converter was tested up to a maximum output current of 405 A, achieving a power density of 3048 W/in^3 at full load. Power density was calculated based on the total box volume of the converter, with dimensions detailed in Table III and Fig. 1. At full load, the converter has an output current density of 0.502 A/mm^2, despite its slim 2.7 mm profile.

The prototype reached a peak power stage efficiency (excluding gate drive losses) of 88.7% and a full-load efficiency of 76.7%. System efficiency, accounting for gate drive and auxiliary circuit losses, was 86.5% at peak and 76.3% at full load.

Throughout testing, the switching frequency was held at 1 MHz, with the output voltage controlled within ±1% of the nominal 1 V. Fig. 7 provides a steady-state thermal image of the converter operating at 250 A under single-sided cooling from a 44 CFM blower.

The converter was further tested at an output voltage of 0.8 V to assess operational robustness, reaching an output current of 420 A with a power density of 2529 W/in^3 and current density of 0.521 A/mm^2. The power stage efficiency was 87.8% at peak and 73.0% at full load, while system efficiency reached 85.0% at peak and 72.6% at full load.

C. Performance Comparison

Table VI and Table V present a performance comparison between this work and other state-of-the-art 48-V-to-1-V and 48-V-to-0.8-V academic works. The prototype achieved a high power density and maintained a low profile, albeit with a trade-off in reduced full-load efficiency. According to Table V, for an output voltage of 1 V, this design achieves approximately 50% higher power density than the second-highest design. The prototype is also the thinnest design with a single-sided cooling solution, with a thickness around 40% less than the second-lowest design [15].

V. Conclusion

This work presents a 48-V-to-1-V switching bus converter (SBC) design that achieves high power density and a low profile of 2.7 mm, making it ideal for height-constrained vertical power delivery (VPD) applications. Custom two-phase coupled inductors were recessed into the PCB to meet stringent VPD form factor requirements. Leveraging the fast-switching capabilities of GaN FETs and various packaging structures, the hardware prototype achieved a high power density of 3048 W/in^3 and a current density of 0.502 A/mm^2 at a 405 A output and 1 V. Additionally, at an output of 0.8 V and 420 A, the converter reached a power density of 2529 W/in^3 and a current density of 0.521 A/mm^2. This work offers a compelling solution for applications requiring extremely high power density and single-sided cooling capability within height-constrained spaces.

VI. Acknowledgements

The authors acknowledge the financial support from the Berkeley Power and Energy Center (BPEC).

TABLE IV: Measured performance of the hardware prototype

Output Voltage	Full-Load Current	Per-Phase Current	Current Density	Power Density[†]	Power Stage Efficiency (Peak / Full-load)	System Efficiency[‡] (Peak / Full-load)
1.0 V	405 A	25.31 A	0.502 A/mm^2	3048 W/in^3	88.7% / 76.7%	86.5% / 76.3%
0.8 V	420 A	26.25 A	0.521 A/mm^2	2529 W/in^3	87.8% / 73.0%	85.0% / 72.6%

[†] The box volume is measured as the smallest rectangular box that can contain the converter, including the gate drive circuitry.
[‡] Gate drive loss is included in the calculation of system efficiency.

TABLE V: Performance comparison between this work and the state-of-the-art 48-V-to-1-V academic works

Year	Reference	Full-Load Current	Switching Frequency	Height	Current Density	Power Density[†]	Power Stage Efficiency (Peak / Full-load)	System Efficiency[‡] (Peak / Full-load)
2025	This work (SBC-GaN)	405 A	1 MHz	2.7 mm	0.502 A/mm^2	3048 W/in^3	88.7% / 76.7%	86.5% / 76.3%
2024	SBC-VPD [15] (8-branch)	400 A 300 A	600 kHz	4.7 mm	0.501 A/mm^2 0.375 A/mm^2	1745 W/in^3 1309 W/in^3	90.2% / 85.1% 90.2% / 88.0%	88.2% / 84.4% 88.2% / 87.0%
2024	Dual-inductor SBC* [16]	50 A	1 MHz	2.6 mm	0.321 A/mm^2	2020 W/in^3	91.9% / 88.0%	90.1% / 87.3%
2024	20-to-1 SBC [14]	1500 A	220 kHz	6.2 mm	0.287 A/mm^2	759 W/in^3	94.1% / 86.0%	92.7% / 85.7%
2024	Mini-LEGO [13]	240 A	1.5 MHz	8.4 mm	0.714 A/mm^2	1390 W/in^3	87.1% / 84.1%	84.1% / 82.3%
2023	MSC [10]	220 A	400 kHz	7.0 mm	0.265 A/mm^2	607 W/in^3	92.9% / 86.3%	91.1% / 85.8%
2022	Dickson^2 [9]	270 A	280 kHz	6.45 mm	0.142 A/mm^2	360 W/in^3	93.8% / 88.4%	91.6% / 87.7%
2022	VIB [8]	450 A	417 kHz	11.0 mm	0.156 A/mm^2	232 W/in^3	95.2% / 89.1%	93.3% / 88.1%
2022	LEGO [7]	450 A	1 MHz	16.65 mm	0.298 A/mm^2	294 W/in^3	91.1% / 85.7%	88.4% / 84.8%
2020	Crossed-coupled QSD buck** [6]	40 A	125 kHz	N.A.	N.A.	100 W/in^3	94.5% / 91.1%	N.A. / N.A.
2020	Sigma [5]	80 A	1 MHz	4.0 mm	0.127 A/mm^2	420 W/in^3	94.0% / 92.5%	N.A. / N.A.

[†] The box volume is measured as the smallest rectangular box that can contain the converter, including the gate drive circuitry.
[‡] Gate drive loss is included in the calculation of system efficiency.
* The Dual-inductor SBC requires dual-side cooling.
** The power density of the cross-coupled QSD buck converter is calculated with the power component volume.

TABLE VI: Performance comparison between this work and the state-of-the-art 48-V-to-0.8-V academic works

Year	Reference	Full-Load Current	Switching Frequency	Height	Current Density	Power Density[†]	Power Stage Efficiency (Peak / Full-load)	System Efficiency[‡] (Peak / Full-load)
2025	This work (SBC-GaN)	420 A	1 MHz	2.7 mm	0.521 A/mm^2	2529 W/in^3	87.8% / 73.0%	85.0% / 72.6%
2024	SBC-VPD [15] (10-branch)	600 A 400 A	600 kHz	4.7 mm	0.614 A/mm^2 0.410 A/mm^2	1713 W/in^3 1142 W/in^3	89.9% / 80.1% 89.9% / 86.0%	87.1% / 79.5% 87.1% / 84.8%
2024	Mini-LEGO [13]	210 A	1.5 MHz	8.4 mm	0.625 A/mm^2	975 W/in^3	86.0% / 84.1%	82.2% / 81.8%
2022	VIB [8]	425 A	417 kHz	11.0 mm	0.147 A/mm^2	175 W/in^3	94.5% / 88.6%	N.A. / N.A.
2022	LEGO [7]	450 A	1 MHz	16.65 mm	0.298 A/mm^2	235 W/in^3	89.8% / 84.7%	86.8% / 83.5%

[†] The box volume is measured as the smallest rectangular box that can contain the converter, including the gate drive circuitry.
[‡] Gate drive loss is included in the calculation of system efficiency.
[§] For a comprehensive list of references, please refer to [15].

REFERENCES

[1] NVIDIA Corporation, "NVIDIA H100 Tensor Core GPU Datasheet," 2023, Accessed: Nov. 11, 2024. [Online]. Available: https://resources.nvidia.com/en-us-tensor-core/nvidia-tensor-core-gpu-datasheet

[2] Vicor Inc., *Factorized Power Architecture and VI Chips: Flexible, High Performance Power System Solutions*, 2013. [Online]. Available: http://www.vicorpower.com/documents/whitepapers/fpa101.pdf.

[3] S. Saggini, O. Zambetti, R. Rizzolatti, M. Picca, and P. Mattavelli, "An Isolated Quasi-Resonant Multiphase Single-Stage Topology for 48-V VRM Applications," *IEEE Transactions on Power Electronics*, vol. 33, no. 7, pp. 6224–6237, 2018.

[4] G.-S. Seo, R. Das, and H.-P. Le, "A 95%-efficient 48v-to-1v/10a vrm hybrid converter using interleaved dual inductors," in *2018 IEEE Energy Conversion Congress and Exposition (ECCE)*, 2018, pp. 3825–3830.

[5] M. H. Ahmed, C. Fei, F. C. Lee, and Q. Li, "Single-Stage High-Efficiency 48/1 V Sigma Converter With Integrated Magnetics," *IEEE Transactions on Industrial Electronics*, vol. 67, no. 1, pp. 192–202, 2020.

[6] M. Halamicek, T. McRae, and A. Prodić, "Cross-Coupled Series-Capacitor Quadruple Step-Down Buck Converter," in *2020 IEEE Power Electronics Conference and Exposition (APEC)*, 2020, pp. 1–6.

[7] J. Baek, Y. Elasser, K. Radhakrishnan, H. Gan, J. P. Douglas, H. K. Krishnamurthy, X. Li, S. Jiang, C. R. Sullivan, and M. Chen, "Vertical Stacked LEGO-PoL CPU Voltage Regulator," *IEEE Transactions on Power Electronics*, vol. 37, no. 6, pp. 6305–6322, 2022.

[8] Y. Chen, P. Wang, H. Cheng, G. Szczeszynski, S. Allen, D. M. Giuliano, and M. Chen, "Virtual Intermediate Bus CPU Voltage Regulator," *IEEE Transactions on Power Electronics*, vol. 37, no. 6, pp. 6883–6898, 2022.

[9] Y. Zhu, T. Ge, Z. Ye, and R. C. Pilawa-Podgurski, "A Dickson-Squared Hybrid Switched-Capacitor Converter for Direct 48 V to Point-of-Load Conversion," in *2022 IEEE Applied Power Electronics Conference and Exposition (APEC)*, 2022, pp. 1272–1278.

[10] P. Wang, Y. Chen, G. Szczeszynski, S. Allen, D. M. Giuliano, and M. Chen, "MSC-PoL: Hybrid GaN-Si Multistacked Switched Capacitor 48V PwrSiP VRM for Chiplets," *IEEE Transactions on Power Electronics*, pp. 1–20, 2023.

[11] L. Han, H. Liu, S. Zhang, and Y. Chen, "Two-stage 48v-to-16v-to-1v hybrid-switched-capacitor converter for high performance computing," in *2023 IEEE 2nd International Power Electronics and Application Symposium (PEAS)*, 2023, pp. 367–373.

[12] X. Lou and Q. Li, "Single-Stage 48 V/1.8 V Converter With a Novel Integrated Magnetics and 1000 W/in3 Power Density," *IEEE Transactions on Industrial Electronics*, vol. 71, no. 7, pp. 6601–6611, 2024.

[13] Y. Elasser, J. Baek, K. Radhakrishnan, H. Gan, J. P. Douglas, H. K. Krishnamurthy, X. Li, S. Jiang, V. De, C. R. Sullivan, and M. Chen, "Mini-LEGO CPU Voltage Regulator," *IEEE Transactions on Power Electronics*, vol. 39, no. 3, pp. 3391–3410, 2024.

[14] Y. Zhu, J. Zou, and R. C. N. Pilawa-Podgurski, "A 1500-A/48-V-to-1-V Switching Bus Converter for Next-Generation Ultra-High-Power Processors," *IEEE Transactions on Power Electronics*, vol. 39, no. 9, pp. 11 340–11 355, 2024.

[15] Y. Zhu, J. Zou, N. M. Ellis, S. Kudva, M. Mosa, C. T. Gray, and R. C. Pilawa-Podgurski, "A compact 48-v-to-sub-1-v switching bus converter with 4.7-mm height for processor vertical power delivery," in *2024 IEEE Energy Conversion Congress and Exposition (ECCE)*, 2024.

[16] N. M. Ellis, Y. Zhu, and R. C. Pilawa-Podgurski, "Gallium Nitride-based 48V-to-1V Point-of-Load (PoL) Converter for Aerospace Telecommunications and Computing Applications," in *2024 IEEE Applied Power Electronics Conference and Exposition (APEC)*, 2024, pp. 1384–1388.

[17] Vertiv. (2024) Vertiv codevelops with nvidia complete power and cooling blueprint for nvidia gb200 nvl72 platform. Accessed: Nov. 11, 2024. [Online]. Available: https://www.vertiv.com/en-emea/about/news-and-insights/news-releases/vertiv-codevelops-with-nvidia-complete-power-and-cooling-blueprint-for--nvidia-gb200-nvl72-platform/

[18] NVIDIA Corporation, "NVIDIA GB200 NVL72," 2024, Accessed: Nov. 11, 2024. [Online]. Available: https://www.nvidia.com/en-us/data-center/gb200-nvl72/

[19] C. Johnstone. (2024) How power density is changing in data centers and what it means for liquid cooling. Accessed: Nov. 11, 2024. [Online]. Available: https://jetcool.com/post/how-power-density-is-changing-in-data-centers/

[20] H. Gan, S. Jiang, S. Teng, S. Yamamoto, V. Chivukula, B. Edwards, C. Chung, J. Chen, M. Mohideen, G. Sizikov, and X. Li, "Vertical power delivery for 1000 amps machine learning asics," in *2024 IEEE Applied Power Electronics Conference and Exposition (APEC)*, 2024, pp. 906–909.

[21] J. Zou, N. C. Brooks, S. Coday, N. M. Ellis, and R. C. Pilawa-Podgurski, "On the Size and Weight of Passive Components: Scaling Trends for High-Density Power Converter Designs," in *2022 IEEE 23rd Workshop on Control and Modeling for Power Electronics (COMPEL)*, 2022, pp. 1–7.

[22] Y. Zhu, N. M. Ellis, and R. C. N. Pilawa-Podgurski, "Comparative performance analysis of regulated hybrid switched-capacitor topologies for direct 48 v to point-of-load conversion," *IEEE Open Journal of Power Electronics*, vol. 5, pp. 1735–1755, 2024.

[23] E. A. Jones, F. Wang, and B. Ozpineci, "Application-based review of gan hfets," in *2014 IEEE Workshop on Wide Bandgap Power Devices and Applications*, 2014, pp. 24–29.

[24] Texas Instrument, *Optimizing Efficiency Through Dead Time Control With the LMG1210 GaN Driver*, 2018, Accessed: Nov. 11, 2024. [Online]. Available: https://www.ti.com/lit/an/snva815a/snva815a.pdf?ts=1731198446119&ref_url=https%253A%252F%252Fwww.google.com%252F

[25] A. Mazany, *Implementing Bootstrap Overcharge Prevention in GaN Half-bridge Circuits*, 2023, Accessed: Nov. 11, 2024. [Online]. Available: https://www.ti.com/lit/an/snvaa94/snvaa94.pdf?ts=1733709982284&ref_url=https%253A%252F%252Fwww.google.com%252F

[26] P.-L. Wong, P. Xu, P. Yang, and F. Lee, "Performance improvements of interleaving vrms with coupling inductors," *IEEE Transactions on Power Electronics*, vol. 16, no. 4, pp. 499–507, 2001.

[27] Y. Zhu, J. Zou, and R. C. N. Pilawa-Podgurski, "Design-oriented modeling and multi-objective optimization of two-phase coupled inductors in multiphase pwm converters," in *2025 IEEE Applied Power Electronics Conference and Exposition (APEC)*, 2025.

[28] DMEGC, *DMR53 Material Characteristics*, 2019, Accessed: Nov. 11, 2024. [Online]. Available: https://dongyangdongci.oss-cn-hangzhou.aliyuncs.com/uploads/20230401/DMR53%20Material%20Characteristics.pdf

Ripple Reduction and Efficiency Improvement of Always-Dual-Path Hybrid DC-DC Converter Based on Phase Shift Operation

Katsuhiro Hata[1], Shinsaku Tanaka[2], Toru Ashikaga[2], and Yasuhiro Rikiishi[2]

[1]Shibaura Institute of Technology, Tokyo, Japan
[2]Sanken Electric Co., Ltd., Saitama, Japan
E-mail: khata@shibaura-it.ac.jp

Abstract—**This study proposes a novel phase-shift operation for an always-dual-path hybrid (ADPH) DC-DC converter with soft charging, utilizing a sub-inductor to improve performance. The proposed operation enhances efficiency by achieving zero voltage switching (ZVS) and reducing body diode losses, while significantly lowering output voltage ripple by optimizing the sub-inductor current waveform. To maintain high efficiency and low ripple across a wide load range, a dynamic phase-shift adjustment method is introduced, linearly adjusting the phase shift delay with the load current. Experimental results show that the proposed approach outperforms conventional operation in both efficiency and ripple reduction. This method holds promise for applications such as data centers, robotics, and other systems requiring high power density, efficiency, and stable operation.**

Keywords—hybrid DC-DC converter, always-dual-path, soft charging, phase shift, ripple reduction

I. INTRODUCTION

Hybrid step-down DC-DC converters that combine a buck converter and a switched capacitor DC-DC converter can overcome the trade-off between efficiency and form factor commonly encountered with traditional buck converters [1–5]. Additionally, to accommodate increased output current in various applications, dual-path hybrid converters have been proposed to reduce inductor current by supplying the output current through both inductor and capacitor paths, thereby reducing conduction losses caused by inductor's equivalent series resistance [6–9]. To further reduce inductor current, always-dual-path hybrid (ADPH) converters that consistently operate in dual-path mode have also been proposed, which can dramatically reduce inductor conduction losses and achieve high efficiency in high-current output applications [10–13].

While these converters raise concerns about spike currents from flying capacitors, soft charging operation using a sub-inductor has been introduced to mitigate this issue [14, 15]. These ADPH converters with soft charging can achieve smaller size and higher efficiency than conventional buck converters by appropriately selecting passive components, including the sub-inductor [14]. They can also achieve zero voltage switching (ZVS) and enhance efficiency by optimizing the dead time based on the sub-inductor current behavior [15]. However, previous studies have primarily focused on the main circuit states, and the commutation behavior during the dead time, along with the detailed operation of each switch, have not been sufficiently explored.

This study examines the commutation process during the dead time period and proposes an operating method to achieve proper ZVS for each switch. This approach not only enhances efficiency but also reduces output voltage ripple by optimizing the sub-inductor current waveform. Additionally, the relationship between efficiency and output voltage ripple, influenced by changes in the on-timing of each switch, is

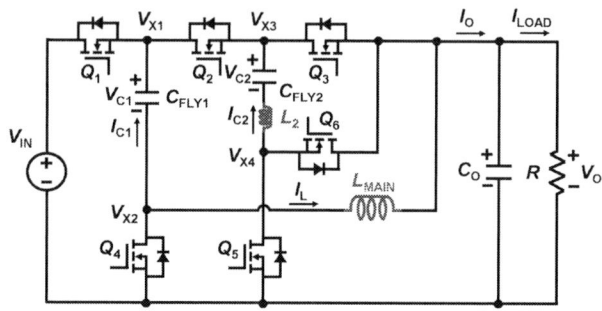

Fig. 1. ADPH converter with soft charging using a sub-inductor L_2.

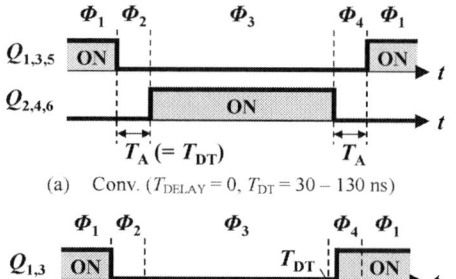

(a)   Conv. ($T_{DELAY} = 0$, $T_{DT} = 30 - 130$ ns)

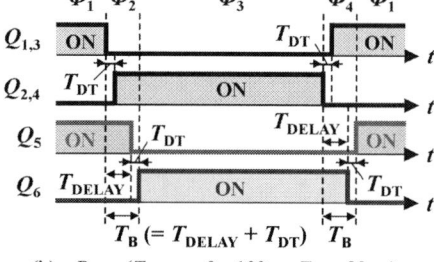

(b)   Prop. ($T_{DELAY} = 0 - 100$ ns, $T_{DT} = 30$ ns)

Fig. 2. Operating methods for the ADPH converter with soft charging.

experimentally demonstrated. Furthermore, a method to adapt to a wide load current range without performance degradation is proposed, and its feasibility is demonstrated through experimental verification.

II. PROPOSED OPERATION FOR ALWAYS-DUAL-PATH HYBRID CONVERTER WITH SOFT CHARGING

A. Circuit topology and sub-inductor design considerations

Fig. 1 illustrates the circuit diagram of the ADPH converter with soft charging, incorporating a sub-inductor L_2. While there are several options for placing the sub-inductor, a circuit topology that minimizes the sub-inductor RMS current while achieving ZVS is preferable [15]. Furthermore, the minimum value of the sub-inductor required to sufficiently suppress spike currents depends on the switching frequency and the flying capacitor's capacitance. Therefore, the resonant operation for each circuit state should be carefully designed.

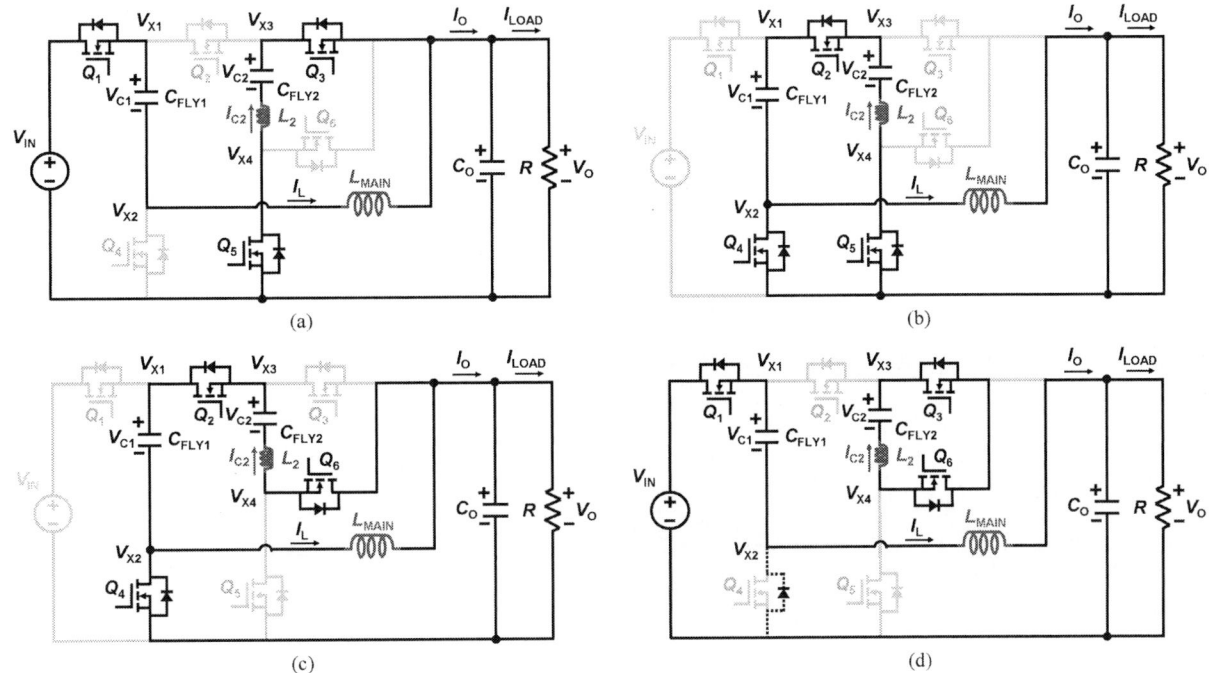

Fig. 3. Circuit states in the ADPH converter. (a) State 1: Φ_1 and Φ_4 ($I_{C2} > 0$). (b) State 2: Φ_2 ($I_{C2} > 0$). (c) State 3: Φ_2 ($I_{C2} < 0$) and Φ_3. (d) State 4: Φ_4 ($I_{C2} < 0$).

B. Proposed operation for zero voltage switching (ZVS)

Operating methods for the ADPH converter with soft charging are illustrated in Fig. 2. In the conventional method, the odd- and even-numbered switches are alternately operated with a common dead time T_{DT} (i.e., T_A). However, during this process, some switches fail to achieve ZVS due to the commutation behavior, and the conduction loss of the body diode increases because of the extended dead time.

On the other hand, the proposed method achieves ZVS for all switches except Q_1 by shifting the phase between the Q_1–Q_4 group and the Q_5–Q_6 group. Additionally, the conduction loss of the body diode is reduced by minimizing the dead time T_{DT}. The phase shift is controlled using the turn-off delay time T_{DELAY} between the Q_1–Q_4 group and the Q_5–Q_6 group. The sum of T_{DELAY} and T_{DT} is defined as T_B, which represents the time difference between the earliest switch turn-off and the latest switch turn-on during a main circuit state transition.

If T_{DELAY} exceeds T_{DT}, there will be an overlap period where the odd- and even-numbered switches do not operate alternately. However, this does not affect the commutation process. While phase shift operation has been proposed for other converters [16, 17], this paper demonstrates its applicability and effectiveness for the ADPH converter with soft charging.

C. Circuit states considering the commutation process

The detailed circuit states involved in the commutation process are illustrated in Fig. 3. The primary circuit states are shown in Figs. 3(a) and 3(c), while the intermediate circuit states occurring during the commutation process are depicted in Figs. 3(b) and 3(d). It is important to note that these circuit states are influenced not only by the switching events (on and off) but also by the directions of both the main inductor current and the sub-inductor current.

When the switches are turned off in the circuit state of Fig. 3(a), the body diodes of the switches shown in Fig. 3(b) conduct, maintaining the current paths of the inductors. During this commutation, Q_2 and Q_4 are turned on in Fig. 3(b) to achieve ZVS. As the current direction of the sub-inductor with small inductance reverses, the circuit transitions to the state shown in Fig. 3(c), where turning on Q_6 ensures a smooth transition between the main circuit states while maintaining ZVS.

Similarly, when the switches are turned off in the circuit state of Fig. 3(c), the body diodes of the switches shown in Fig. 3(d) conduct, maintaining the inductor current paths. During this commutation, turning on Q_3 ensures ZVS for Q_3, but ZVS for Q_1 cannot be achieved because the main inductor current flows through the body diode of Q_4 instead of Q_1. To minimize the conduction loss of Q_4's body diode, Q_1 should be turned on with minimum dead time. Once the sub-inductor current reverses, the circuit transitions to the state in Fig. 3(a), where turning on Q_5 ensures ZVS for Q_5. This sequence ensures proper transitions between the main states, achieving ZVS for all switches except Q_1.

Furthermore, to reduce the body diode conduction loss of Q_5 in Fig. 3(b) and Q_6 in Fig. 3(d), delaying turn-off until the sub-inductor current reverses enables the proposed operation shown in Fig. 2(b), leading to improved efficiency. During this delay, commutation from Q_5 in Fig. 3(b) to the body diode of Q_6 in Fig. 3(c), and from Q_6 in Fig. 3(d) to the body diode of Q_5 in Fig. 3(a), is prevented, even if the sub-inductor current reverses. This allows control over the sub-inductor current waveform, which can potentially reduce the output voltage ripple. However, this does not always improve overall efficiency, as the time during which the sub-inductor current is not delivered to the load increases. Therefore, the delay time should be appropriately designed according to the required specifications.

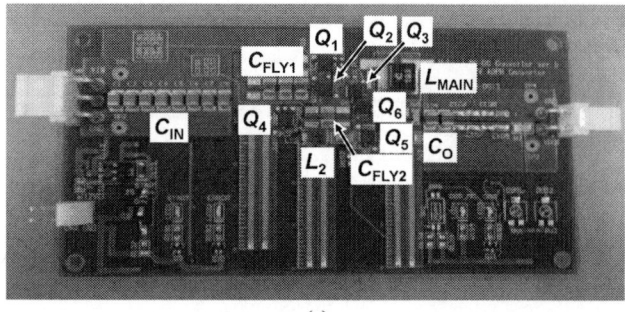

(a)

MCU (MD6603, Sanken Electric, Co., Ltd.)

(b)

Fig. 4. Prototype of the ADPH converter. (a) Top view. (b) Bottom view.

TABLE I. CIRCUIT COMPONENTS AND PARAMETERS.

Parts	Manufacture and Part number	Parameters
Q_{1-6}	Infineon, IQE050N08NM5CG	80 V, 5 mΩ × 6
C_{IN}	Murata, GRM32EC72A106KE05L	100 V,10 μF (2.5 μF @ 48 V_{DC}) × 8
C_{FLY1}	Murata, GCM32EC71H106KA03L	50 V, 10 μF (6.3 μF @ 24 V_{DC}) × 3
C_{FLY2}	Murata, GCM32EC71E226KE36K	25 V, 22 μF (17 μF @ 12 V_{DC}) × 1
C_O	Murata, GCM32EC71E226KE36K	25 V, 22 μF (17 μF @ 12 V_{DC}) × 2
L_{MAIN}	Coilcraft, XGL6030-103ME	10 μH, 38.0 mΩ
L_2	Wurth Elektronik, 744340300055	55 nH, 0.27 mΩ

TABLE II. EXPERIMENTAL CONDITIONS.

Parameters	Value
Input voltage V_{IN}	48 V
Output voltage V_O	12 V
Output current I_O	8 A_{MAX}
Switching frequency f_{SW}	300 kHz
Gate resistance $R_{G,ON}$	47 Ω for Q_1, 10 Ω for Q_{2-6}
Gate resistance $R_{G,OFF}$	0 Ω for Q_{1-6}
Time T_A, T_B	30 – 130 ns

III. EXPERIENTAL VERIFICATION

A. Experimental setup

The prototype of the ADPH converter with soft charging was implemented on a 114 mm × 60 mm 4-layer PCB, as shown in Fig. 4. The design targets a 48 V-to-12 V bus converter with a rated load current of 8 A. The circuit components and parameters are summarized in Table I, while the experimental conditions are listed in Table II.

To suppress spike currents, a switching frequency of 300 kHz was selected. For Q_1, which cannot achieve ZVS, the on-gate resistance was set to a relatively high value of 48 Ω, while it was set to 10 Ω for the other switches.

The transition times for T_A (in the conventional operation) and T_B (in the proposed operation) were set within a range of 30–130 ns. In the conventional operation, T_{DT} was set to 30–130 ns, while in the proposed operation, it was minimized to 30 ns. Additionally, T_{DELAY}, which is exclusive to the proposed operation, was configured within a range of 0–100 ns. These timing parameters were controlled using a microcontroller (MD6603, Sanken Electric Co., Ltd.).

B. Measured results

The efficiency measured under varying T_A and T_B for the conventional and proposed operations is shown in Fig. 5. In the conventional operation, a common dead time is applied to all switches, causing the ZVS condition of the switches to vary under different operating conditions. This results in multiple inflection points as T_A changes, leading to complex efficiency characteristics. In contrast, in the proposed operation, changes in T_B have a smaller impact on maintaining the ZVS for each switch, resulting in a smoother efficiency curve. The peak efficiency was 96.88 % for the conventional operation and 97.25 % for the proposed operation, representing a modest improvement of 0.37 percentage points.

Fig. 6 shows the measured output voltage ripple. While the conventional operation uses a common dead time without phase shifting, adjusting the on-timing of each switch via T_A provides only limited control over the sub-inductor current waveform, resulting in only a slight reduction in the voltage ripple. In contrast, the proposed operation employs phase shifting, offering greater flexibility in controlling the sub-inductor current waveform and achieving a significantly larger reduction in ripple voltage.

This result is further supported by the sub-inductor RMS current measurements shown in Fig. 7, evaluated under varying T_A and T_B. The proposed operation effectively reduces the sub-inductor RMS current compared to the conventional operation and demonstrates better control of the sub-inductor current waveform. In the conventional operation, efficiency peaks when the sub-inductor RMS current reaches its minimum, and at this same T_A, the output voltage ripple is also nearly minimized. In contrast, for the proposed operation, the minimization of the sub-inductor RMS current coincides with the lowest output voltage ripple. However, the condition for achieving maximum efficiency requires a T_B different from that which minimizes the ripple, indicating the necessity of careful design adjusted to specific requirements. These experimental results validate the discussions and analyses presented in the previous chapter.

To clearly illustrate the changes in efficiency and voltage ripple with variations in T_A and T_B, the results are summarized in Fig. 8. As previously discussed, in the conventional operation, maximum efficiency and minimum voltage ripple are achieved under nearly identical conditions. In contrast, the proposed operation provides greater design flexibility, allowing for prioritization of either efficiency (Prop. 1) or ripple reduction (Prop. 2).

Figs. 9 and 10 compare the efficiency and ripple voltage under these conditions, respectively. The results demonstrate that the proposed operation outperforms the conventional operation in both efficiency and ripple reduction, with a more pronounced improvement in ripple voltage. These findings validate the feasibility of the proposed phase-shift operation and emphasize its effectiveness, particularly in reducing output voltage ripple.

979-8-3315-1612-3/25 $31.00 © 2025 IEEE

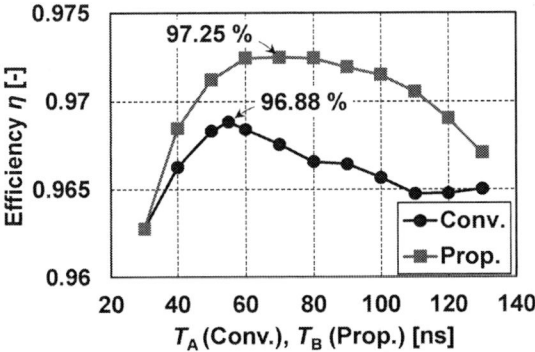

Fig. 5. Measured efficiency under different operating conditions with varying transition times at a rated load of 8 A_{OUT}.

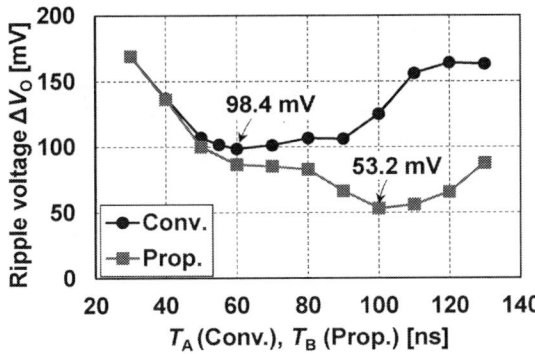

Fig. 6. Measured ripple voltage under different operating conditions with varying transition times at a rated load of 8 A_{OUT}.

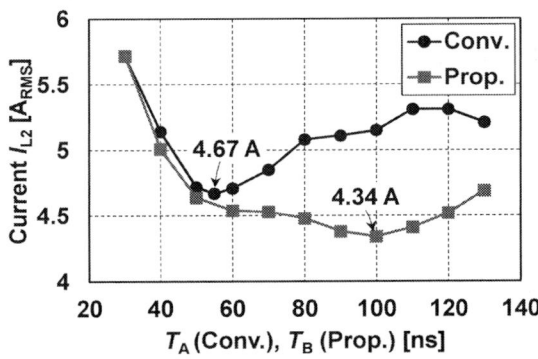

Fig. 7. Measured sub-inductor RMS current under different operating conditions with varying transition times at a rated load of 8 A_{OUT}.

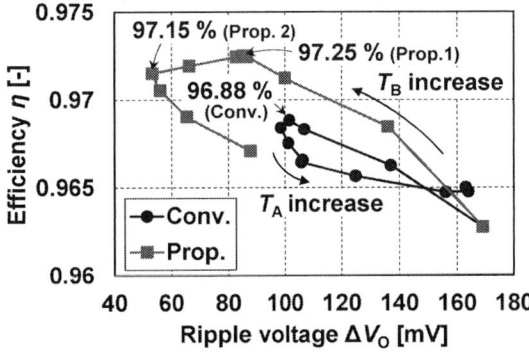

Fig. 8. Efficiency and ripple voltage measured under different operating conditions with varying transition times at a rated load of 8 A_{OUT}.

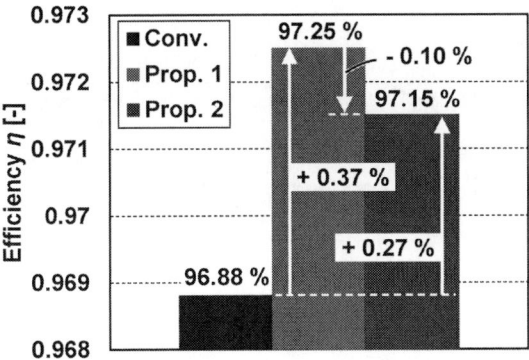

Fig. 9. Comparison of efficiency at optimum points between conventional and proposed operations at the rated load of 8 A_{OUT}.

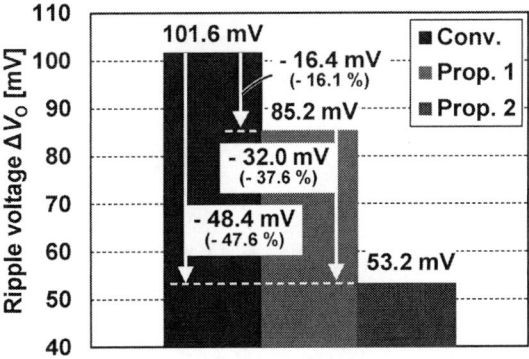

Fig. 10. Comparison of ripple voltage at optimum points between conventional and proposed operations at the rated load of 8 A_{OUT}.

C. Measured waveforms

Fig. 11 shows the waveforms of the sub-inductor current and output voltage at the optimum points for both conventional and proposed operations, measured at the rated load of 8 A_{OUT}. In conventional operation (Conv.), the on-times of Q_1 and Q_6 do not overlap, preventing the sub-inductor current waveform from forming a square wave. As a result, the ripple component in the output voltage cannot be sufficiently reduced. In contrast, in the proposed operation, the on-times of Q_1 and Q_6 overlap. Compared to efficiency-focused priority (Prop. 1), the ripple-focused priority (Prop. 2) increases T_{DELAY}, aligning the sub-inductor current waveform closer to a square wave. This adjustment effectively suppresses the output voltage ripple.

IV. LOAD CURRENT CHARACTERISTICS

In the proposed operation, the output voltage ripple is effectively suppressed by controlling the sub-inductor current waveform through phase shifting. However, the T_{DELAY} needed to shape the sub-inductor current closer to a square wave depends on the load current. Assuming that current sharing between the main and sub-inductors is determined by the charge balance of the flying capacitors and remains constant regardless of the load current, the RMS current of the sub-inductor is expected to vary linearly with the load current. Consequently, if the sub-inductor current is controlled to form a square wave, the transition time between the main circuit states should change linearly with the amplitude of the sub-inductor current waveform. Based on this principle, an

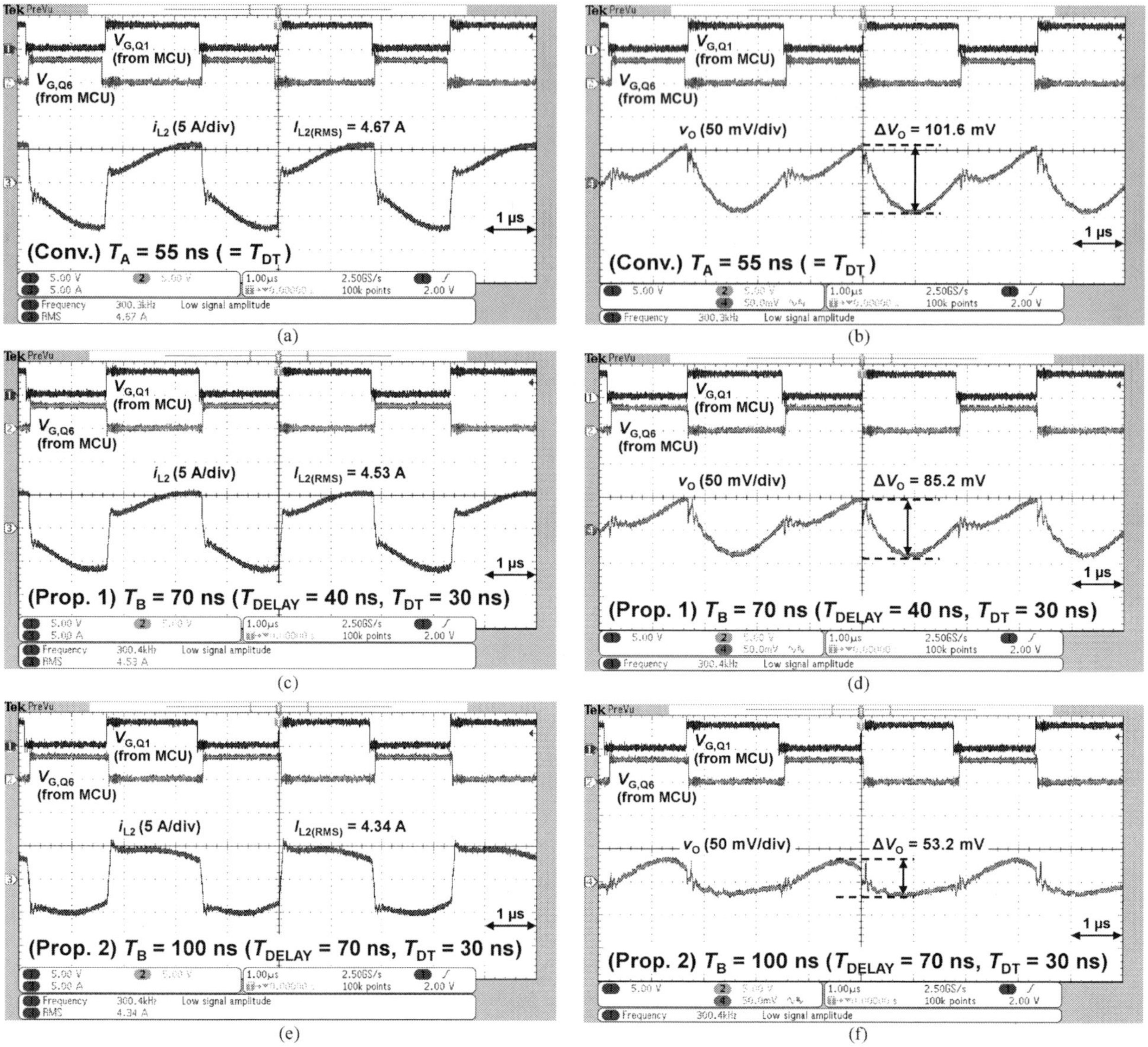

Fig. 11. Sub-inductor current and output voltage waveforms measured at the optimum points of conventional and proposed operations under a rated load of 8 A_{OUT}. (a) i_{L2} (Conv.). (b) v_O (Conv.). (c) i_{L2} (Prop. 1). (d) v_O (Prop. 1). (e) i_{L2} (Prop. 2). (f) v_O (Prop. 2).

operation that linearly adjusts T_{DELAY} (i.e., T_B) according to the load current is proposed (Prop. 3), as shown in Fig. 12.

The efficiency, output voltage ripple, and sub-inductor RMS current, measured across varying values of the load current under different operating conditions, are shown in Figs. 13, 14, and 15, respectively. Except at light loads, Prop. 1 and Prop. 3 achieved the maximum efficiency, while Prop. 2 showed a significant efficiency drop as the load current decreased. Similar trends were observed for the output voltage ripple and sub-inductor RMS current.

As shown in Fig. 12, as the load current decreases, the difference in operating conditions between Prop. 2 (where T_B is fixed) and Prop. 3 (where T_B is adjusted with the load current) increases. This highlights the importance of controlling the sub-inductor current waveform, which has a significant impact on both efficiency and ripple. Notably, when the load current is around 4 A, Prop. 1 and Prop. 3 use nearly the same T_B, resulting in the greatest reduction in output voltage ripple. The results show that Prop. 3 minimizes the

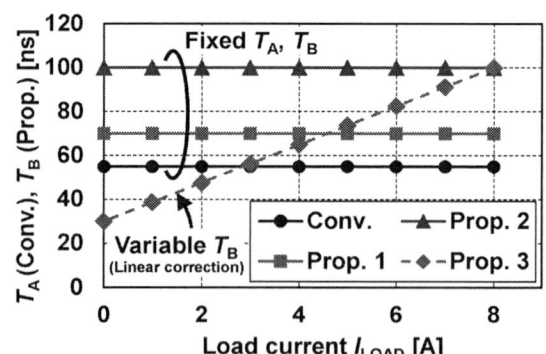

Fig. 12. Timing settings for load current in conventional and proposed operations. The transition time of conv., prop. 1, and prop. 2 is constant with respect to the load current, while prop. 3 varies linearly with the load current through linear correction.

sub-inductor RMS current across all load current conditions, offering superior performance compared to the other methods.

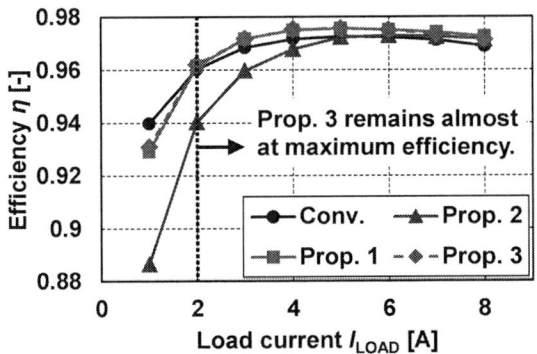

Fig. 13. Measured efficiency vs. load current for different operations.

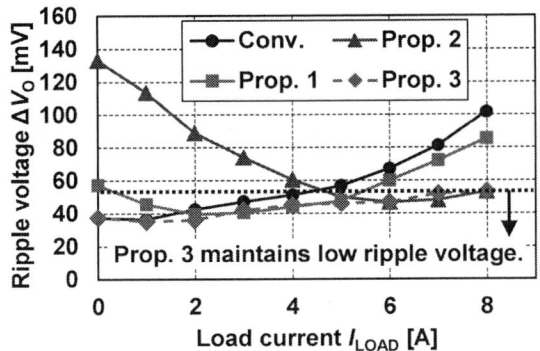

Fig. 14. Measured ripple voltage vs. load current for different operations.

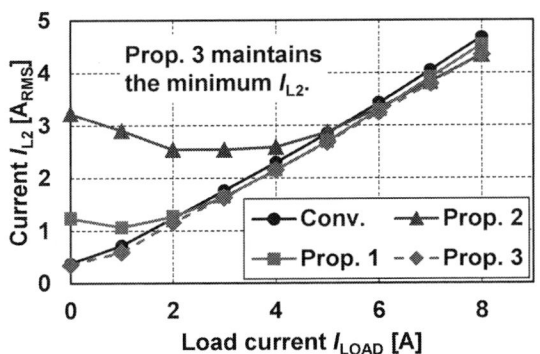

Fig. 15. Measured sub-inductor current vs. load current for different operations.

V. CONCLUSIONS

This study introduced a phase-shift operation for the ADPH converter with soft charging, incorporating a sub-inductor to enhance performance. The proposed method improves efficiency by achieving ZVS (Zero Voltage Switching) and reducing body diode losses, while significantly lowering output voltage ripple by optimizing the sub-inductor current waveform. To maintain high efficiency and low ripple across a wide load range, a dynamic phase-shift adjustment based on load current was proposed and experimentally validated. This approach minimized the sub-inductor RMS current and ensured optimum performance under varying load conditions. These results demonstrate the potential of phase-shift techniques to enhance both efficiency and ripple performance in power converters for demanding applications.

Future work will focus on deriving the optimal phase shift through theoretical analysis and implementing phase-shift control to evaluate its performance under load variations in real-world applications.

REFERENCES

[1] M. Halamicek, T. McRae, N. Vukadinović and A. Prodić, "Modulation Scheme for an Effective Increase in the Number of Levels of DC-DC Multi-Level Flying Capacitor Converters," in *Proc. IEEE Appl. Power Electron. Conf. Expo. (APEC)*, 2019, pp. 45–49.

[2] Y. Lei, W. Liu and R. C. N. Pilawa-Podgurski, "An Analytical Method to Evaluate and Design Hybrid Switched-Capacitor and Multilevel Converters," *IEEE Trans. Power Electron.*, vol. 33, no. 3, pp. 2227–2240, March 2018.

[3] F. Bez, G. Bonanno, L. Corradini and C. Garbossa, "Control technique for reliable operation of the synchronous series capacitor tapped inductor converter," in *Proc. IEEE Appl. Power Electron. Conf. Expo. (APEC)*, 2018, pp. 113–120.

[4] O. Kirshenboim and M. M. Peretz, "High-Efficiency Nonisolated Converter With Very High Step-Down Conversion Ratio," *IEEE Trans. Power Electron.*, vol. 32, no. 5, pp. 3683–3690, May 2017.

[5] P. S. Shenoy, M. Amaro, J. Morroni and D. Freeman, "Comparison of a Buck Converter and a Series Capacitor Buck Converter for High-Frequency, High-Conversion-Ratio Voltage Regulators," *IEEE Trans. Power Electron.*, vol. 31, no. 10, pp. 7006–7015, Oct. 2016.

[6] G. Seo and H-P. Le, "S-Hybrid Step-Down DC-DC Converter-Analysis of Operation and Design Considerations," *IEEE Tran. Ind. Electron.*, vol. 67, no. 1, pp. 265–275, Jan. 2020.

[7] Y. Huh, S. Hong and G. Cho, "A Hybrid Structure Dual-Path Step-Down Converter With 96.2% Peak Efficiency Using 250-mΩ Large-DCR Inductor," *IEEE J. Solid-State Circuits*, vol. 54, no. 4, pp. 959–967, April 2019.

[8] K. Hata, Y. Yamauchi, T. Sai, T. Sakurai, and M. Takamiya, "48V-to-12V Dual-Path Hybrid DC-DC Converter," in *Proc. IEEE Appl. Power Electron. Conf. Expo. (APEC)*, 2020, pp. 2279–2284.

[9] S. Zhen, R. Yang, D. Wu, Y. Cheng, P. Luo and B. Zhang, "Design of Hybrid Dual-Path DC-DC Converter with Wide Input Voltage Efficiency Improvement," in *Proc. IEEE International Symposium on Circuits and Systems (ISCAS)*, 2021, pp. 1–5.

[10] K. Hata, Y. Jiang, M. -K. Law, and M. Takamiya, "Always-Dual-Path Hybrid DC-DC Converter Achieving High Efficiency at Around 2:1 Step-Down Ratio," in *Proc. IEEE Appl. Power Electron. Conf. Expo. (APEC)*, 2021, pp. 1302–1307.

[11] K. Hata, S. Tanaka, Y. Rikiishi, and T. Matsumoto, "48 V-to-12 V Always-Dual-Path Hybrid DC-DC Converter for Inductor Current Reduction," in *Proc. IEEE Energy Conversion Congress and Exposition (ECCE)*, 2022, pp. 1–6.

[12] G. Cai, Y. Lu, and R. P. Martins, "An SC-Parallel-Inductor Hybrid Buck Converter With Reduced Inductor Voltage and Current," *IEEE J. Solid-State Circuits*, vol. 58, no. 6, pp. 1758–1768, June 2023.

[13] W. Jung, M. Kim, H. Park, S.-M. Yoo, J.-H. Yang, M. Choi, J. Shin, and H.-M. Lee, "A 95.4% Hybrid Always-Dual-Path Recursive Step-Down Converter Using Adaptive Switching Level Control With 288 mΩ Large-DCR Inductor," *IEEE Trans. Power Electron.*, vol. 39, no. 2, pp. 2258–2269, Feb. 2024.

[14] K. Hata, S. Tanaka, T. Ashikaga, and Y. Rikiishi, "Always-Dual-Path Hybrid DC-DC Converter with Soft Charging for High Efficiency with Reduced Passive Components," in *Proc. IEEE Appl. Power Electron. Conf. Expo. (APEC)*, 2024, pp. 1371–1376.

[15] K. Hata, S. Tanaka, T. Ashikaga, and Y. Rikiishi, "Topology Comparison and Dead-Time Optimization for Efficiency Enhancement of Always-Dual-Path Hybrid DC-DC Converter with Soft Charging," in *Proc. IEEE Energy Conversion Congress and Exposition (ECCE)*, 2024, pp. 2576–2582.

[16] K. Sano and H. Fujita, "Performance of a High-Efficiency Switched-Capacitor-Based Resonant Converter With Phase-Shift Control," *IEEE Trans. Power Electron.*, vol. 26, no. 2, pp. 344–354, Feb. 2011.

[17] J. Zhu and D. Maksimovic, "A family of transformerless stacked active bridge converters," in *Proc. IEEE Appl. Power Electron. Conf. Expo. (APEC)*, 2019, pp. 19–24.

979-8-3315-1612-3/25 $31.00 © 2025 IEEE

Ultralocal PQ Theory: A New Approach for Model-Free Predictive Direct Power Control of Shunt Active Power Filters

Mahdi S. Mousavi*, Abolfazl Nassaji[†], Ibrahim Harbi[‡],
Behnam Nikmaram*, S. Alireza Davari*, Mokhtar Aly[§], Jose Rodriguez[§]
*Department of Electrical Engineering, Shahid Rajaee Teacher Training University, Tehran, Iran
[†]Department of Electrical Engineering, University of Science and Culture, Tehran, Iran
[‡]Faculty of Engineering, Technical University of Munich, Munich, Germany
[§]Facultad de Ingeniería, Arquitectura y Diseño, Universidad San Sebastián, Bellavista 7, Santiago 8420524, Chile
Email: m.mousavi@sru.ac.ir

Abstract—**This paper presents a novel PQ (active and reactive power) theory based on an ultralocal model to enable model-free predictive direct power control (MF-PDPC) for shunt active power filters (SAPFs). The proposed PQ theory defines general dynamic equations for active and reactive powers using voltages from two corresponding nodes and incorporates unknown disturbance terms, which are estimated in real-time through the designed PQ observers. Observer convergence is validated using the Lyapunov theory. With this PQ theory, an MF-PDPC approach for SAPFs is developed that operates independently of system parameters. Unlike traditional model predictive direct power control, which relies on current predictions, the proposed MF-PDPC method directly predicts active and reactive powers at the next sampling instant. Simulation results confirm the effectiveness of the proposed approach.**

Index Terms—**Model-free, PQ, ultralocal model, SAPF, direct power control, predictive, active power filter.**

I. INTRODUCTION

Active power filters (APFs) have emerged as effective grid-connected inverters for mitigating harmonics and providing reactive power compensation. Compared to passive filters, APFs offer improved dynamic response and accuracy, making them a preferred solution for power quality enhancement in various applications [1], [2]. Recently, the focus on advanced control methods for APFs has increased, as these methods can significantly reduce harmonic distortions [3]. Notable control approaches include model predictive control (MPC) [4], fuzzy control (FC) [5], sliding mode control (SMC) [6], and neural network control (NNC) [7]. Of these, MPC stands out as a modern and robust technique, offering advantages such as fast response, straightforward implementation, and high steady-state accuracy.

An enhanced MPC technique with signal correction for LC-coupled hybrid APFs (LC-HAPFs) was presented in [4]. In [8], a finite set MPC (FS-MPC) approach was applied to a shunt APF (SAPF) using two parallel inverters. A comparative analysis in that study showed this configuration, combined with the control method, provides benefits such as improved

transient response during load changes, reduced total harmonic distortion (THD), and lower switching losses. However, this approach also involves a high computational load. A new deadbeat direct power control (DB-DPC) with simple delay compensation for APFs was proposed in [9], featuring fast response, simplicity, and improved robustness against parameter variations. In [10], a deadbeat control strategy with repetitive predictions was introduced.

Despite these advancements, the methods above may face challenges related to system modeling. Variations in APF model parameters can result in issues such as increased THD, higher steady-state error, and suboptimal control in the MPC approach [11].

To address the inherent limitations of MPC as a model-based method, the concept of model-free predictive control (MFPC) has been introduced in various applications, including motor drives [12] and grid-connected inverters [13]. Generally, MFPC establishes the relationship between the current and voltage of the inverter using ultralocal models or generalized time series [14]. As a result, MFPC is primarily a current control approach and is not typically applied directly to power control systems.

This paper introduces a new theory called ultralocal active and reactive powers (PQ) to leverage the advantages of model-free control in power management. This theory is used to develop a model-free predictive direct power control (MF-PDPC) strategy for grid-tied inverters. The proposed ultralocal PQ theory directly calculates the powers between two nodes in the grid using measured voltages from each node and unknown PQ disturbances, which need to be estimated in real time. To achieve this, two extended state observers (ESOs) are defined—one for active power and one for reactive power.

While the ultralocal PQ theory and ESOs are presented generally for grid-tied inverters, SAPFs are utilized as a case study. Consequently, the ultralocal PQ theory is implemented to realize the MF-PDPC approach for SAPFs. To validate the proposed method, it is applied to a SAPF in MATLAB simulations. The simulation results demonstrate that the pro-

979-8-3315-1612-3/25 $31.00 © 2025 IEEE

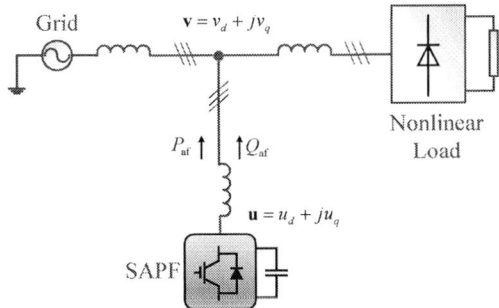

Fig. 1: Schematic of SAPF

posed theory performs well compared to the standard direct power control approach, particularly under uncertainties such as parameter variations.

II. SHUNT ACTIVE POWER FILTER

A. Basic Principles

SAPFs are essential power electronics devices designed to compensate for reactive power and current harmonics in the grid, which are often caused by nonlinear loads. The schematic of the SAPF connected to the grid with a nonlinear load is illustrated in Fig. 1. In this configuration, a two-level voltage source inverter (VSI) is employed, along with series inductances at the inverter output. These inductances possess relatively small internal resistances, which are included in the mathematical model. By applying Kirchhoff's Voltage Law (KVL) to the path from the grid to the VSI, the following vector equation can be derived.

$$\mathbf{v}_{abc}(t) = \mathbf{u}_{abc}(t) - R_f \mathbf{i}_{af,abc}(t) - L_f \frac{d\mathbf{i}_{af,abc}(t)}{dt}. \quad (1)$$

In the above equation, $\mathbf{v}_{abc} = [v_a\ v_b\ v_c]^\mathsf{T}$ is the grid phase voltages, $\mathbf{u}_{abc} = [u_a\ u_b\ u_c]^\mathsf{T}$ denotes the SAPF voltages, $\mathbf{i}_{af,abc} = [i_{af,a}\ i_{af,b}\ i_{af,c}]^\mathsf{T}$ is the SAPF currents, and (R_f, L_f) represent the series inductance and resistance at the inverter's output.

B. Conventional Model Predictive Direct Power Control

To perform the conventional MPC for an APF, the controlled variables should be predicted using the dynamic equation of the APF. The current prediction is the basic format of the MPC, and it can be expressed in the dq frame by the following equations [9].

$$
\begin{aligned}
i_{af,d}(k+1) &= \left(1 - \frac{t_s R_f}{L_f}\right) i_{af,d}(k) - t_s \omega i_{af,q}(k) \\
&\quad + \frac{t_s}{L_f}\left(u_d(k) - v_d(k)\right), \\
i_{af,q}(k+1) &= \left(1 - \frac{t_s R_f}{L_f}\right) i_{af,q}(k) + t_s \omega i_{af,d}(k) \\
&\quad + \frac{t_s}{L_f}\left(u_q(k) - v_q(k)\right).
\end{aligned}
\quad (2)
$$

In the above equation, $(i_{af,d}, i_{af,q})$ are the SAPF's current components; (u_d, u_q) depict the inverter's voltage components; (v_d, v_q) are the grid's voltage components; ω is the angular frequency of the grid; t_s is the sampling time; and (k) and $(k+1)$ show the sampling instants.

The instantaneous active and reactive powers of the SAPF in the dq frame are obtained as follows:

$$
\begin{aligned}
P_{af} &= v_d i_{af,d} + v_q i_{af,q}, \\
Q_{af} &= v_q i_{af,d} - v_d i_{af,q}.
\end{aligned}
\quad (3)
$$

When the dq frame is synchronized with the grid voltage, the d-component of the grid's voltage is constant, and the q-component is zero:

$$
\begin{aligned}
v_d(k+1) &= v_d(k) = v_d, \\
v_q &= 0.
\end{aligned}
\quad (4)
$$

Therefore, the active and reactive power can be predicted as follows:

$$
\begin{aligned}
P_{af}(k+1) &= v_d i_{af,d}(k+1), \\
Q_{af}(k+1) &= -v_d i_{af,q}(k+1).
\end{aligned}
\quad (5)
$$

Merging (2) and (5) leads to:

$$
\begin{aligned}
P_{af}(k+1) &= \left(1 - \frac{t_s R_f}{L_f}\right) P_{af}(k) + t_s \omega Q_{af}(k) \\
&\quad + \frac{t_s}{L_f}\left(u_d(k)v_d(k) - v_d^2(k)\right), \\
Q_{af}(k+1) &= \left(1 - \frac{t_s R_f}{L_f}\right) Q_{af}(k) - t_s \omega P_{af}(k) \\
&\quad - \frac{t_s}{L_f}\left(v_d u_q(k)\right).
\end{aligned}
\quad (6)
$$

The above equations predict the active and reactive power in the next sampling instant. To apply the deadbeat predictive control, the reference PQ values are set equal to the future instants, i.e., $P_{af}(k+1) = P^*$ and $Q_{af}(k+1) = Q^*$. Then, (6) is rearranged to obtain the voltage components:

$$
\begin{aligned}
u_d(k) &= v_d + \frac{L_f\left(P^* - P(k)\right)}{v_d t_s} + \frac{R_f}{v_d}P(k) - \frac{L_f\omega}{v_d}Q(k), \\
u_q(k) &= \frac{L_f\left(Q(k) - Q^*\right)}{v_d t_s} - \frac{R_f}{v_d}Q(k) - \frac{L_f\omega}{v_d}P(k).
\end{aligned}
\quad (7)
$$

Finally, the achieved voltage signals in the above equation are applied to the inverter using the modulation techniques.

III. PROPOSED ULTRALOCAL PQ THEORY

A. Ultralocal PQ Model Definition

This paper presents new dynamic equations for active and reactive powers using the model-free control theory. Therefore, the active and reactive powers are obtained directly from general model-free equations instead of using the classic current and voltage mathematical equations as in (1). According to [15], any system with input u and output y can be expressed by $y^{(n)} = F + \alpha u$ where n is the order of the system, F is an unknown term to be estimated in real-time, and α is an adjustable gain. Exploiting this theory, the active and reactive

979-8-3315-1612-3/25 $31.00 © 2025 IEEE

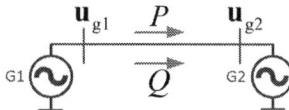

Fig. 2: Power flow

powers that flow between two voltage sources as in Fig. 2 can be expressed as follows:

$$\frac{dP}{dt} = F_P + \alpha \left(\mathbf{u}_{g1} \cdot \mathbf{u}_{g2} \right)$$
$$\frac{dQ}{dt} = F_Q + \alpha \left(\mathbf{u}_{g1} \times \mathbf{u}_{g2} \right) \tag{8}$$

where $\mathbf{u}_{g1}$ and $\mathbf{u}_{g2}$ are the voltage vectors of the nodes, $(\times)$ denotes the cross product and $(\cdot)$ is the dot product. The unknown variables F_P and F_Q in (8) represent all disturbances and uncertain parts in the proposed theory. Also, α is an input coefficient that can be designed heuristically.

B. Real-time Estimation of F_P and F_Q

To employ a control system for the ultralocal PQ model (8), it is necessary to obtain real-time values of F_P and F_Q. The following state observer is presented to estimate active power and the unknown term F_P:

$$\begin{cases} e_P = P - \hat{P} \\ \dfrac{d\hat{P}_{af}}{dt} = \hat{F}_P + \alpha \left(\mathbf{u}_{g1} \cdot \mathbf{u}_{g2} \right) + \gamma_1 G(e_P) \\ \dfrac{d\hat{F}_P}{dt} = \gamma_2 e_P \end{cases} \tag{9}$$

where superscript ($\hat{\ }$) denotes the estimated variables, (γ_1, γ_2) are gains of the observer, and $G(.)$ is the error function. In this study, the hyperbolic tangent of error, $\tanh(e_P)$, is utilized to limit the effect of the higher error values on the active power estimation. It is worth mentioning that the hyperbolic tangent is similar to the signum function utilized in the sliding mode control, but it does not suffer from the chattering problem of the signum function.

For the reactive power and the unknown term F_Q, the state observer is proposed as follows:

$$\begin{cases} e_Q = Q - \hat{Q} \\ \dfrac{d\hat{Q}_{af}}{dt} = \hat{F}_Q + \alpha \left(\mathbf{u}_{g1} \times \mathbf{u}_{g2} \right) + \eta_1 G(e_Q) \\ \dfrac{d\hat{F}_Q}{dt} = \eta_2 e_Q \end{cases} \tag{10}$$

where (η_1, η_2) are the gains of the observer.

The Lyapunov stability theory is employed here to ensure the stability of the proposed observers. To avoid repetition, the stability analysis is only given for the active power, and it can be expressed similarly for the reactive power observer.

Subtracting (9) from (8) leads to the following error system:

$$\begin{cases} \dot{e}_P = e_F - \gamma_1 \tanh(e_P) \\ \dot{e}_F = -\gamma_2 e_P \end{cases} \tag{11}$$

where $e_F = F_P - \hat{F}_P$. A Lyapunov function is defined for the above system as written below:

$$V = \frac{1}{2} \left(\gamma_2 e_P^2 + E^2(e_P, e_F) \right) \tag{12}$$

where $E^2(e_P, e_F) = e_F - \gamma_1 \tanh(e_P)$, and for simplicity will be shown by E in the following. The derivative of the Lyapunov function is expressed below:

$$\begin{aligned} \dot{V} &= \gamma_2 e_P \dot{e}_P + E\dot{E} \\ &= \gamma_2 e_P E + (\dot{e}_F - \gamma_1 \dot{e}_P \text{sech}^2(e_P))E \\ &= \gamma_2 e_P E + (-\gamma_2 e_P - \gamma_1 E \text{sech}^2(e_P))E \\ &= -\gamma_1 E^2 \text{sech}^2(e_P) \end{aligned} \tag{13}$$

It can be seen that $\dot{V} < 0$ for the positive gain γ_1. Thus, the proposed observers for the active and reactive powers are convergent. This means that the proposed ultralocal models in (8) can be utilized in the control system using the observers in (9) and (10).

IV. CASE STUDY: ULTRALOCAL PQ THEORY FOR SAPFs

A. Ultralocal PQ Model of the SAPF

Using the new PQ theory presented in (8), the active and reactive powers of the SAPF shown in Fig. 1 can be dynamically achieved as follows:

$$\frac{dP_{af}}{dt} = F_P + \alpha(u_d v_d + u_q v_q)$$
$$\frac{dQ_{af}}{dt} = F_Q + \alpha(u_d v_q - u_q v_d) \tag{14}$$

where u_d, and u_q show the dq components of the SAPF voltage in the synchronous frame. Similarly, v_d, and v_q are the dq components of the grid voltage.

In the synchronized system given in (4), the ultralocal PQ equations of the SAPF can be simplified to:

$$\frac{dP_{af}}{dt} = F_P + \alpha u_d v_d$$
$$\frac{dQ_{af}}{dt} = F_Q - \alpha u_q v_d \tag{15}$$

The same simplification applies to the observers. The ESO for active and reactive powers of the SAPF can be expressed as follows, respectively.

$$\begin{cases} e_P = P - \hat{P} \\ \dfrac{d\hat{P}_{af}}{dt} = \hat{F}_P + \alpha u_d v_d + \gamma_1 \tanh(e_P) \\ \dfrac{d\hat{F}_P}{dt} = \gamma_2 e_P \end{cases} \tag{16}$$

979-8-3315-1612-3/25 $31.00 © 2025 IEEE

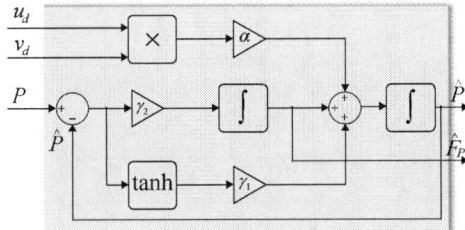

Fig. 3: State observer for the active power and unknown function F_P

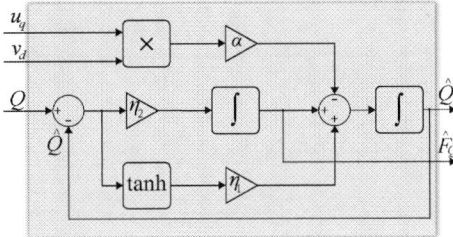

Fig. 4: State observer for the active power and unknown function F_Q

$$\begin{cases} e_Q = Q - \hat{Q} \\ \dfrac{d\hat{Q}_{\text{af}}}{dt} = \hat{F}_Q - \alpha u_{\text{q}} v_{\text{d}} + \eta_1 \tanh(e_Q) \\ \dfrac{d\hat{F}_Q}{dt} = \eta_2 e_Q \end{cases} \quad (17)$$

The block diagrams of the proposed observers are illustrated in Fig. 3 and Fig. 4. In the following, the proposed ultralocal PQ theory is utilized to present an MF-PDPC for the SAPF.

B. Model-free Predictive DPC for SAPFs Using the Proposed Ultralocal PQ Model

To perform the predictive control method using the proposed PQ ultralocal model, the forward Euler discretization technique is used on (15) as follows:

$$\begin{aligned} P_{\text{af}}(k+1) &= \hat{P}_{\text{af}}(k) + t_s\left(\hat{F}_P(k) + \alpha u_{\text{d}}(k) v_{\text{d}}(k)\right) \\ Q_{\text{af}}(k+1) &= \hat{Q}_{\text{af}}(k) + t_s\left(\hat{F}_Q(k) - \alpha u_{\text{q}}(k) v_{\text{d}}(k)\right) \end{aligned} \quad (18)$$

In the above equation, the estimated values are taken from the proposed observers in (16) and (17). The MF-PDPC is achieved by finding the voltage components from (18) as expressed below:

$$\begin{aligned} u_{\text{d}}(k) &= \frac{P^* - \hat{P}(k) - t_s \hat{F}_P}{\alpha t_s v_{\text{d}}} \\ u_{\text{q}}(k) &= \frac{-Q^* + \hat{Q}(k) + t_s \hat{F}_Q}{\alpha t_s v_{\text{d}}} \end{aligned} \quad (19)$$

where P^* and Q^* are the references for the active and reactive powers, and according to the deadbeat control, it is assumed that $P(k+1) = P^*$ and $Q(k+1) = Q^*$.

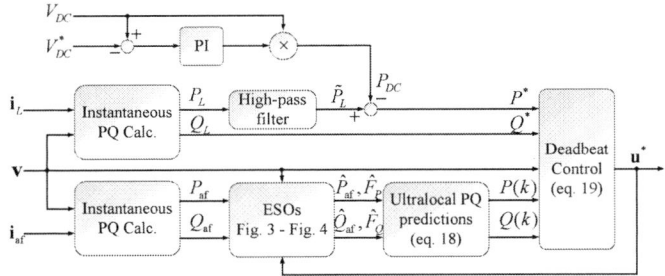

Fig. 5: Block diagram of the proposed MF-PDPC for SAPFs

Comparing (19) and (7) shows that the proposed MF-PDPC has a simpler structure and, unlike conventional predictive DPC, does not rely on system parameters.

The overall block diagram of the proposed MF-PDPC is illustrated in Fig. 5. In SAPF applications, references are generally derived from load measurements. As shown in Fig. 5, the instantaneous active power of the load P_L is calculated, and its harmonic components $\tilde{P}_L$ are extracted using a high-pass filter. The SAPF then generates all harmonic powers, ensuring the grid supplies only the fundamental active power.

To maintain the desired voltage level on the SAPF's DC link, the DC link voltage V_{DC} is regulated to match a reference value V_{DC}^* using a proportional-integral (PI) controller. The grid supplies the necessary active power to keep the DC link capacitor charged at the reference voltage. This power demand is subtracted from the harmonic power compensation requirement to define the reference for the active power that the SAPF must generate. To achieve a unity power factor on the grid side, the reactive power reference is set to match the total reactive power demand of the load.

V. RESULTS AND DISCUSSION

In this section, the performance of the proposed MF-PDPC strategy is evaluated using MATLAB/Simulink. The simulated system is a three-phase, 380 V, 50 Hz network supplying a nonlinear load composed of a diode bridge rectifier with an RC load on its DC side, as well as a 30 kVA load with a 0.8 power factor. The SAPF connects to the network at the point of common coupling (PCC) through inductors. Table I lists the power circuit parameters.

Fig. 6 illustrates the PQ profiles of the load, SAPF, and grid, where the nonlinear load experiences a sudden increase.

TABLE I: System specifications

Name	Symbol	Value
Rated line voltage	V_{L-L}	380 V
Frequency	f	50 Hz
DC link voltage	V_{DC}	700 V
Filter inductance	L_f	4 mH
Filter resistance	R_f	0.1 Ω
Nonlinear load capacitance	C_L	400 μF
Nonlinear load resistance	R_L	40 Ω
Linear load	S	30 kVA
Power factor of the linear load	$\cos\phi$	0.8

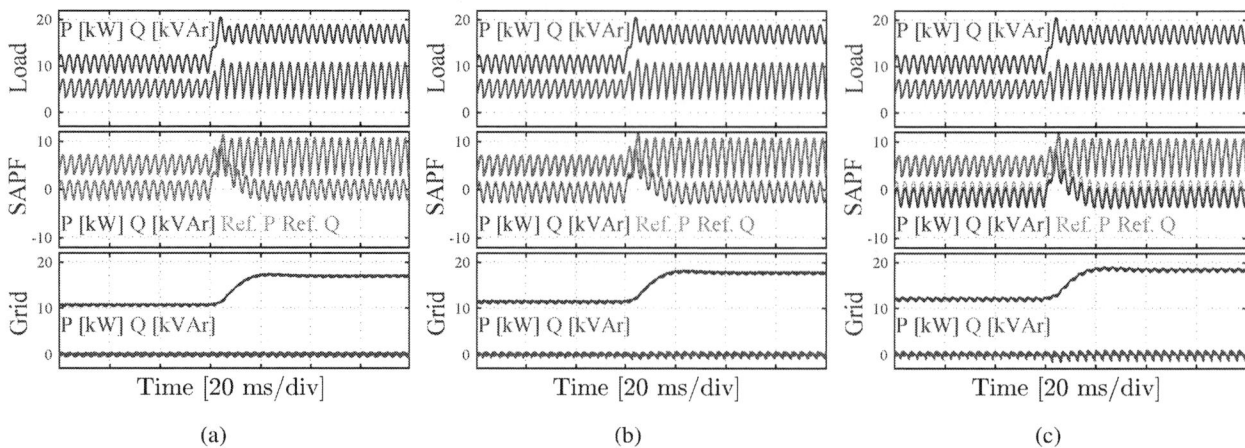

Fig. 6: Active and reactive power profiles of the load, SAPF, and grid when a nonlinear load is suddenly increased. (a) Proposed MF-PDPC method, (b) Conventional deadbeat DPC, (c) DPC with 50% inductance mismatch.

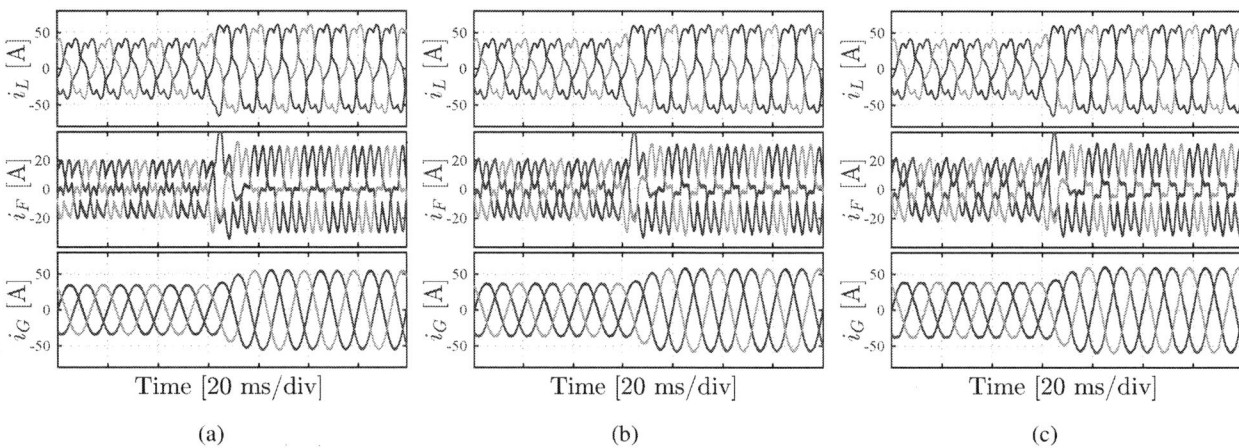

Fig. 7: Current waveforms of the load, SAPF, and grid under increased nonlinear load conditions. (a) Proposed MF-PDPC method, (b) Conventional deadbeat DPC, (c) DPC with 50% inductance mismatch.

The performance of the proposed MF-PDPC method (Fig. 6a) is compared with a conventional deadbeat DPC method [9] (Fig. 6b) and DPC under parameter mismatch (Fig. 6c), where the inductance is increased by 50%. Results indicate that the SAPF effectively compensates for harmonic and reactive power, ensuring that the grid only supplies active power to the loads. This compensation is more accurate in the proposed method, as evidenced by reduced PQ distortion on the grid side. Fig. 6a shows that the SAPF output closely follows the reference values while tracking performance degrades in the conventional deadbeat DPC [9] under inductance mismatch.

These results demonstrate the superior performance of the MF-PDPC approach, which performs well in both steady-state and transient conditions. Fig. 7 presents the current waveforms for the load, SAPF, and grid. The load current is heavily distorted, with a THD of 15.43% under high load. The SAPF successfully compensates for these harmonics, resulting in a nearly sinusoidal grid current.

Fig. 8 further illustrates the harmonic spectra, with Fig. 8a

displaying the load current spectrum and Fig. 8b comparing the harmonic spectra of the grid current for the three control methods. The proposed MF-DPDC achieves the lowest grid current THD at 1.68%, compared to 2.02% for conventional deadbeat DPC and 2.97% for DPC with parameter mismatch.

These findings highlight the effectiveness of the proposed method, which is robust to parameter variations thanks to the ultralocal PQ theory introduced in this work.

VI. CONCLUSION

This paper proposed a new PQ theory, named ultralocal PQ theory, for the grid-tied inverters to construct a model-free predictive PQ control method. While the existing MFPCs are utilized as a current control approach, the proposed method applies the MFPC as a DPC method. Hence, the benefits of model-free control and direct power control are obtained simultaneously. In this work, the proposed method has been applied to a SAPF as the case study. However, future works can investigate the proposed ultralocal PQ theory in various applications of grid-tied inverters.

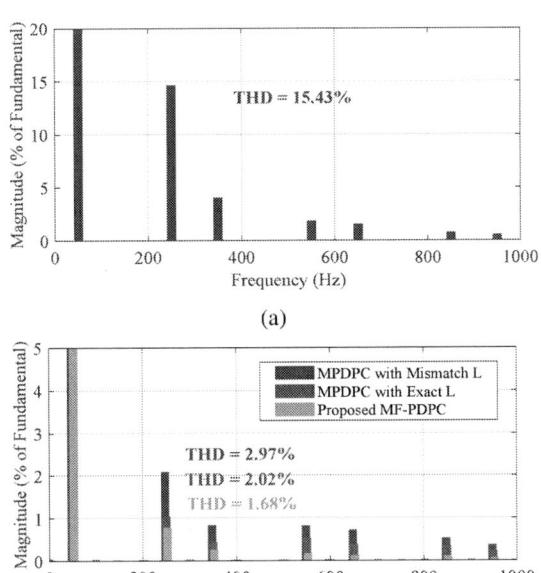

(a)

(b)

Fig. 8: Harmonic spectra (a) Load current (b) Grid current using three different methods

ACKNOWLEDGMENTS

J. Rodriguez acknowledges the support of ANID through projects FB0008, 1210208, and 1221293. This work was also supported by ANID, Chile FONDECYT 1230250, FONDECYT Iniciacion 11230430, and SERC-Chile ANID/FONDAP/1523A0006.

REFERENCES

[1] D. Li, T. Wang, W. Pan, X. Ding, and J. Gong, "A comprehensive review of improving power quality using active power filters," *Electric Power Systems Research*, vol. 199, p. 107389, 2021.

[2] W.-K. Sou, P.-I. Chan, C. Gong, and C.-S. Lam, "Finite-set model predictive control for hybrid active power filter," *IEEE Transactions on Industrial Electronics*, vol. 70, no. 1, pp. 52–64, 2023.

[3] J. Fei and L. Liu, "Real-time nonlinear model predictive control of active power filter using self-feedback recurrent fuzzy neural network estimator," *IEEE Transactions on Industrial Electronics*, vol. 69, no. 8, pp. 8366–8376, 2022.

[4] P.-I. Chan, W.-K. Sou, and C.-S. Lam, "Improved model predictive control with signal correction technique of lc-coupling hybrid active power filter," *IEEE Journal of Emerging and Selected Topics in Power Electronics*, vol. 10, no. 4, pp. 4650–4664, 2022.

[5] J. Fei, Z. Wang, and Y. Fang, "Self-evolving recurrent chebyshev fuzzy neural sliding mode control for active power filter," *IEEE Transactions on Industrial Informatics*, vol. 19, no. 3, pp. 2729–2739, 2023.

[6] C. Gong, W.-K. Sou, and C.-S. Lam, "Second-order sliding-mode current controller for lc-coupling hybrid active power filter," *IEEE Transactions on Industrial Electronics*, vol. 68, no. 3, pp. 1883–1894, 2021.

[7] Y. Chu, S. Fu, S. Hou, and J. Fei, "Intelligent terminal sliding mode control of active power filters by self-evolving emotional neural network," *IEEE Transactions on Industrial Informatics*, vol. 19, no. 4, pp. 6138–6149, 2023.

[8] L. L. d. Souza, N. Rocha, D. A. Fernandes, R. P. R. d. Sousa, and C. B. Jacobina, "Grid harmonic current correction based on parallel three-phase shunt active power filter," *IEEE Transactions on Power Electronics*, vol. 37, no. 2, pp. 1422–1434, 2022.

[9] M. Pichan, M. Seyyedhosseini, and H. Hafezi, "A new deadbeat-based direct power control of shunt active power filter with digital implementation delay compensation," *IEEE Access*, vol. 10, pp. 72 866–72 878, 2022.

[10] J. Chen, H. Shao, and C. Liu, "An improved deadbeat control strategy based on repetitive prediction against grid frequency fluctuation for active power filter," *IEEE Access*, vol. 9, pp. 24 646–24 657, 2021.

[11] N. Yang, S. Zhang, X. Li, and X. Li, "A new model-free deadbeat predictive current control for pmsm using parameter-free luenberger disturbance observer," *IEEE Journal of Emerging and Selected Topics in Power Electronics*, vol. 11, no. 1, pp. 407–417, 2023.

[12] M. S. Mousavi, S. A. Davari, V. Nekoukar, C. Garcia, and J. Rodriguez, "Computationally efficient model-free predictive control of zero-sequence current in dual inverter fed induction motor," *IEEE Journal of Emerging and Selected Topics in Power Electronics*, vol. 11, no. 2, pp. 1332–1344, 2023.

[13] R. Heydari, H. Young, F. Flores-Bahamonde, S. Vaez-Zadeh, C. González-Castaño, S. Sabzevari, and J. Rodríguez, "Model-free predictive control of grid-forming inverters with *lcl* filters," *IEEE Transactions on Power Electronics*, vol. 37, no. 8, pp. 9200–9211, 2022.

[14] F. Wang, Y. Wei, H. Young, D. Ke, D. Huang, and J. Rodríguez, "Continuous-control-set model-free predictive control using time-series subspace for pmsm drives," *IEEE Transactions on Industrial Electronics*, vol. 71, no. 7, pp. 6656–6666, 2024.

[15] M. Fliess and C. Join, "Model-free control," *International journal of control*, vol. 86, no. 12, pp. 2228–2252, 2013.

979-8-3315-1612-3/25 $31.00 © 2025 IEEE

Symmetrical Balanced Circuit for Common-Mode Noise Mitigation in LCL-T Resonant Converter

Ripun Phukan
MPEL, Delta Electronics (Americas)
Durham, USA
ripun.phukan@deltaww.com

Boyi Zhang
MPEL, Delta Electronics (Americas)
Durham, USA
boyi.zhang@deltaww.com

Juan Ruiz
MPEL, Delta Electronics (Americas)
Durham, USA
juan.ruiz@deltaww.com

Peter Barbosa
MPEL, Delta Electronics (Americas)
Durham, USA
peter.barbosa@deltaww.com

Abstract — **This paper presents a common mode (CM) noise mitigation technique of a three-element resonant converter at its DC input terminal. Using a novel balanced LCL-T tank structure, that ensures symmetry for CM noise propagation and decouples the effect of secondary side converter, the DC input CM emissions in the conducted EMI compliance range (150 kHz to 30 MHz) are minimized significantly. Unlike traditional methods, the proposed technique is not limited to upper limit of common mode Y capacitors and relies on simple wheatstone bridge balancing principle. As such, the CM filter size can be significantly smaller. The proposed circuit is demonstrated on a GaN based three level LCL-T converter used in On-Board charger applications switching at a fixed frequency of 500 kHz.**

Keywords— Conducted emissions, LCL-T resonant converter, Symmetry, Wheatstone bridge

I. INTRODUCTION

LCL-T resonant converters are well known for its use in LED driver applications and lately have been extended to on-board charging applications as well [1]. The dynamics of the LCL-T resonant converter have been reported in prior art. [2,3]. Dual phase shift control is typically employed. The LCL-T converter exhibits current source properties wherein the output current is controlled by changing the input voltage (using simple phase shift control of primary converter). To allow full range ZVS of both primary and secondary side switches a phase shift between primary and secondary converters is employed, as demonstrated in [1]. However, the phase shift results in CM emissions propagation from primary to secondary side of the transformer. CM displacement currents propagate from transformer primary to secondary side and the magnitude itself is proportional to the phase shift between the two sides. This in effect is found to be even severe with LCL-T structure when compared to traditional resonant DC-DC topologies such as LLC and its variants [4] and/or boost-SRC [5] topologies due to the additional phase shift requirement between primary and secondary converters. A common practice would be to introduce AC line filters to solve the EMC filtering requirements [6]. In another approach DC side filters [7] or shielding techniques [8] are proposed. AC filters would occupy a significant portion of the total converter volume. While shielding may solve some of the issues, but its use with LCL-T

tank may result in severe eddy current effects for which segmented shielding must be incorporated [9].

Balancing technique is well known and its efficacy with front end power factor correction (PFC) and inverter circuits was demonstrated in [10, 11] respectively. The key is to determine the switching states leading to CM current in any given circuit. The next step is to realize a balanced wheatstone bridge configuration to cancel CM noise flowing to the Line Impedance Stabilization Network (LISN) terminal. However this is to be done using circuit modifications that would not affect the differential mode (DM) currents significantly. In this paper, first the LCL-T resonant converter equivalent CM-EMI model is developed. Thereafter, a CM-EMI suppression circuit is proposed using a balanced LCL-T resonant tank circuit. In conclusion, experimental confirmation of the proposed circuit is shown.

II. CM-EMI MODEL OF LCL-T RESONANT CONVERTER

At first the conventional LCL-T resonant converter is shown in Fig. 1(a), adopted from [1]. The stacked half bridge structure on the primary side is introduced to meet the 800 V bus category while the full bridge architecture on the secondary side meets the 400 V battery specification. The LCL-T operates under fixed switching frequency which is set equal to the resonant frequency given by (1). Here L_1 and L_2 are the LCL-T tank inductors while C_1 is the LCL-T tank capacitor, f_r is the tank resonant frequency and f_{sw} is the switching frequency of the converter.

$$f_r = f_{sw} = \frac{1}{2\pi\sqrt{CL}}; \; L_1 = L_2 = L; C_1 = C \qquad (1)$$

The universal CM-EMI model is derived in Fig. 1. Using super-position principle, the final lumped circuit can be obtained. The nodal capacitances ($C_{px,inv}$ and $C_{px,rec}$) are obtained from measurements taken through a vector network analyzer (VNA) while the voltage sources are obtained through V_{ds} measurements of the active switches on primary and secondary converters. Note that $C_{p,battery}$ represents the stray parasitic capacitance from the battery chassis to ground.

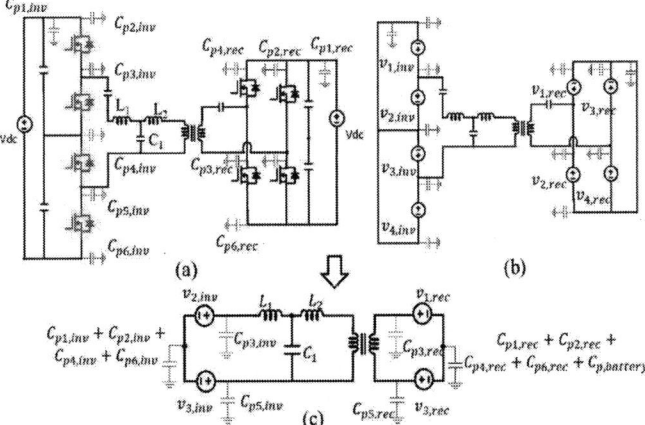

Fig. 1 CM-EMI model of the LCL-T resonant converter, (a) LCL-T resonant tank model with device parasitics to ground, (b) Equivalent representation by replacing switches with voltage sources and (c) CM equivalent lumped single line circuit – the tank parasitics are not shown for simplicity

From the model, it is evident that the voltage $v_{2,inv}$ and $v_{3,inv}$ are critical sources as shown in Fig. 1(c). The full system model in Fig. 1(a) and CM-EMI model in Fig. 1(b) was verified in Fig. 2. The equivalent model and full system model emissions were verified in Fig. 2(a) and then the lumped model simulated measurements were verified through noise measurements taken at the LISN terminal shown in Fig. 2(b). From the noise characterization, it is obvious that we can see the dominant resonant mode f_{sw} and its multiples in addition to the beat frequency (f_b), which comes from the update rate of period doubling modulation [5], mandatory for steady state DC bus voltage balancing. Therefore, the noise spectrum seems rich in lower and mid order harmonics (< 1MHz) for which a smart filter scheme is necessitated. The impact of beat frequency and phase shift is discussed in the following section.

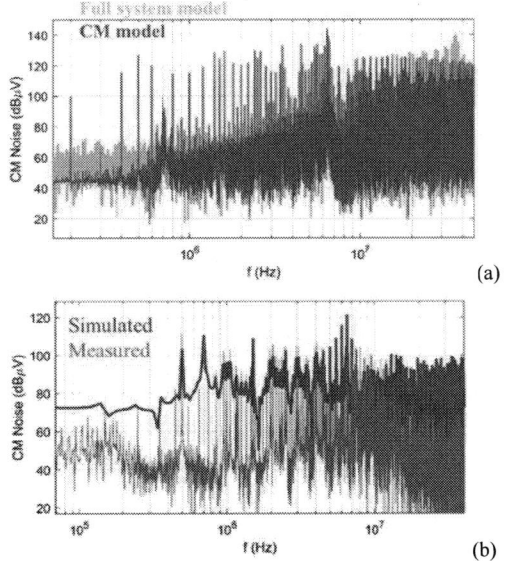

Fig. 2 (a) Comparison between full system model in Fig 1(a) and lumped CM equivalent circuit Fig. 1(c), (b) simulated and experimental noise measurements comparison for f_{sw} = 500 kHz and $f_{period\_doubling}$ = 166.6 kHz.

A. Impact of period doubling modulation

The period doubling modulation was proposed to balance the top and bottom DC bus capacitors, V_{PO} and V_{NO}, in the circuit as shown in Fig. 3. The idea is to rotate the phase sequence of the primary side switches S1-S4 such that the total neutral point current injection within each (hence the name period doubling) or multiple of switching cycles (extended period doubling) remains the same. As shown in Fig. 3(a), period doubling feature results in the expected DM resonant tank currents and voltages. However, it leads to additional CM component. This is illustrated in Fig. 4 in detail where the impact of beat frequency on the CM emission is studied. The explanation lies in the time domain voltage profile information for the noise sources $v_{rec,2}$ and $v_{rec,3}$ which will alternate lead and lag sequence every other switching cycle.

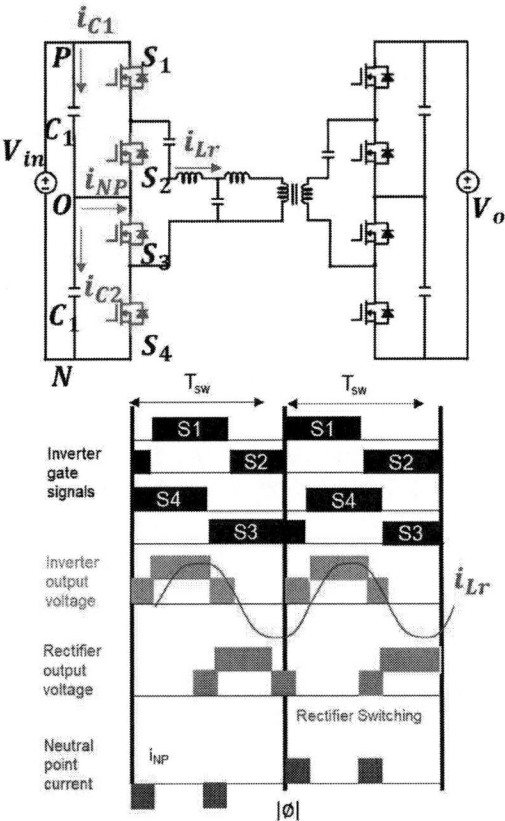

Fig. 3 Period doubling modulation to balance neutral point current i_{NP} in subsequent cycles and balance the capacitor voltages

If the period doubling is extended to occur in multiple switching cycles such as 2, 3 and 4 switching cycles, then it results in higher harmonic content c.f. Fig. 4. The case where no period doubling occurs is the best case. For example when using a full bridge based primary converter with the 400 V bus architecture. In this case the requirement for period doubling is alleviated.

979-8-3315-1612-3/25 $31.00 © 2025 IEEE

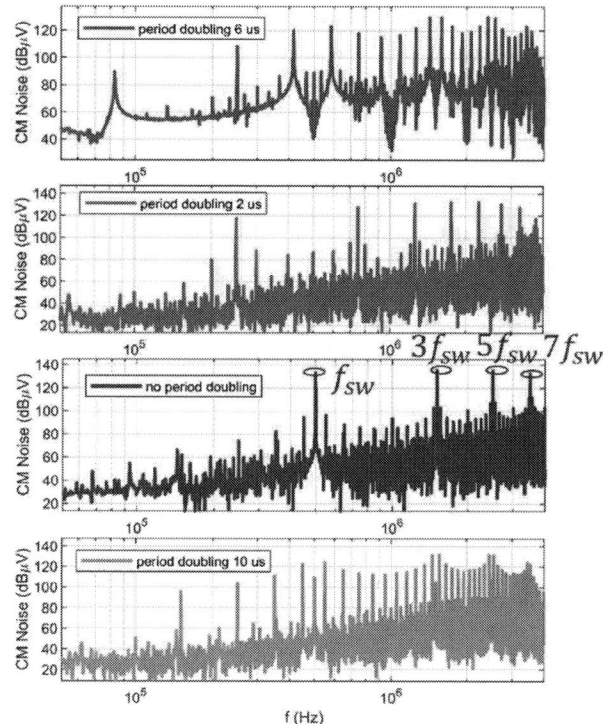

Fig. 4 Period doubling modulation impact on common mode emissions, (a) beat frequency component seen on the EMI spectrum.

This issue may however arise in duty controlled phase shifted bull bridge converters when using the period doubling feature to redistribute the thermal stress on the devices.

With *GaN* devices, the idea is to push the switching frequencies to be higher, resulting in faster period doubling rate requirements. Eventually, this necessitates the use of a powerful digital signal processor (DSP) to achieve the functionality. Extending the period doubling rate results in lower controller complexity and requirements but leads to rich lower and mid frequency harmonics.

B. Impact of primary and secondary side phase shift

Unlike variable frequency type converters, LCL-T converter requires an additional 90^0 phase shift between the primary and secondary side bridges to meet the ZVS conditions. In the CM sense, this results in significant CM currents propagating from the primary to secondary side of the transformer. A fair comparison under same power rating and input / output voltage conditions was conducted between the LCL-T converter and series resonant converter (SRC). The comparison is done under boost mode. From Fig. 5, the LCL-T exhibits almost 10~20 dB higher CM noise content compared to the conventional SRC due to different phase shift angles.

Note that SRC has a higher frequency under boost mode due to delay time effect for gain boost. The effective switching frequency must be increased to compensate (minimize the power transfer period) for the extra energy stored in the resonant inductor due to the phase shift.

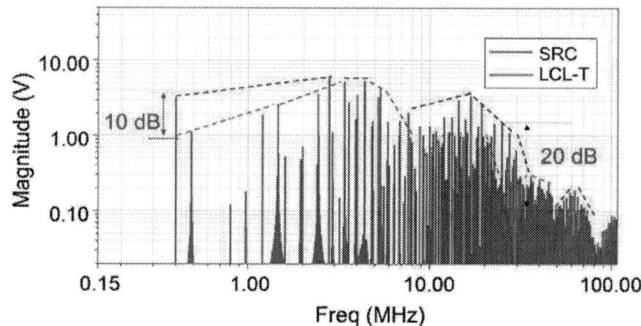

Fig. 5 Illustration of phase shift impact between SRC (15^0) and LCL-T (100^0) under boost mode 420 V_{in} – 460 V_o. F_{res} = 400 kHz

C. Impact of asymmetric line impedance

Another important aspect to consider, is the unbalanced tank impedances. This is more generic to any resonant topology, or simply any isolated DC-DC converter. The key is to ensure impedance on both the positive and negative lines of the converter for better CM immunity. For instance, the CM equivalent model in Fig. 6(a) offers higher impedance than Fig. 6(b). The following condition holds true: $V_{CM1} > V_{CM2}$.

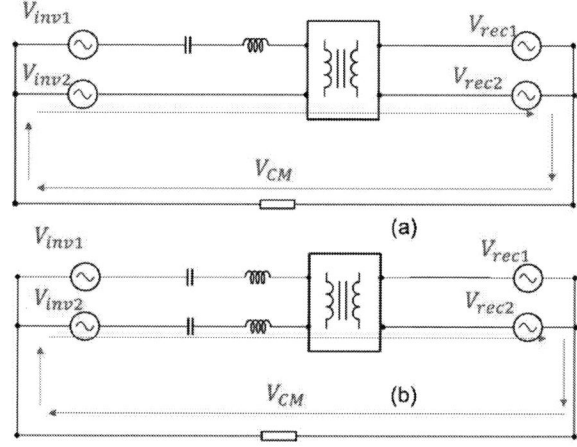

Fig. 6 Illustration of balanced impedance using series resonant converter as an example, (a) unbalanced, (b) balanced

In summary, the LCL-T structure presents major challenges in filter design owing to modulation sequence which includes voltage balancing and current control, along with impedance distribution.

III. PROPOSED SYMMETRICAL BALANCED LCL-T CONVERTER

This section presents details on the proposed EMI filter circuit. Note that the presented changes do not affect the DM waveforms in any form and is meant to improve the CM immunity of the resonant tank. From the previous analyses, the approach for filter design must include ways to decouple the effect of the secondary side converter on the primary side converter in order to mitigate the effects of phase shift modulation on CM noise. Secondly, the proposed circuit must include ways to mitigate the effect of device parasitics to ground on the CM noise. Finally, the circuit must be symmetric.

A. Circuit synthesis

Eventually, the proposed symmetrical resonant network is shown in Fig. 7(d). By splitting the first and second tank inductors L_1 and L_2 as shown in Fig. 7(a), then adding a baby LC filter (L_3 and C_{cm}) on the DC bus shown in Fig. 7(b) and connecting the mid-point of the split resonant tank capacitor C_1 to the LC filter via. a discrete inductor L_4 in Fig. 7(c), the final circuit can be deduced. Additionally, a high CM impedance transformer is realized using either Litz wire type transformers or using planar PCB winding based transformer + shield windings. The shield winding is connected to the secondary side to ensure the effect of secondary stage converter does not affect the noise on the primary side. Moreover, the proposed structure is symmetric thereby improving the CM noise immunity.

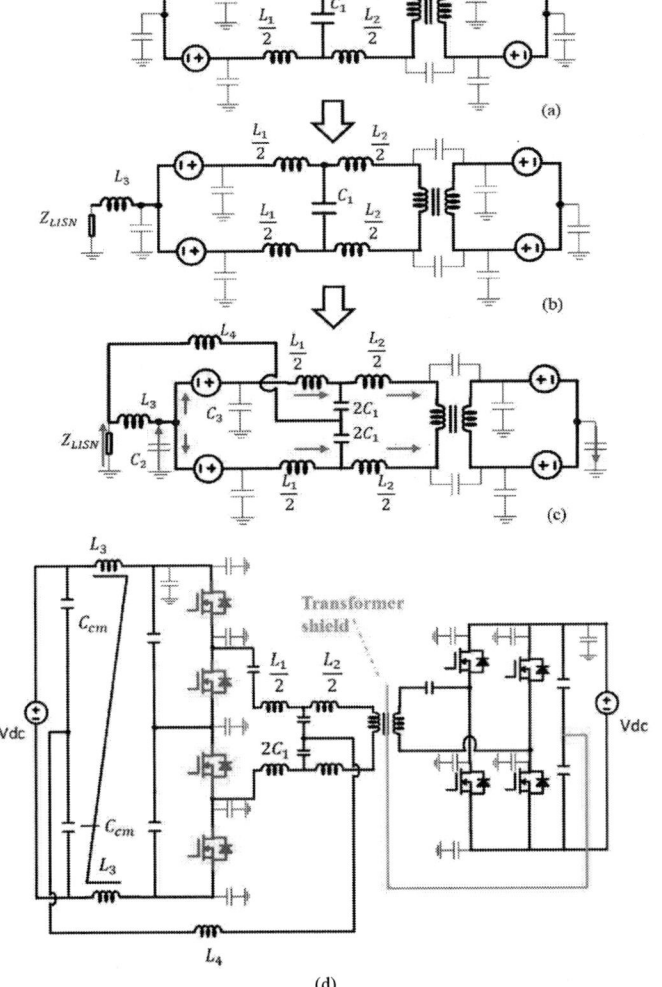

(a)

(b)

(c)

(d)

Fig. 7 Synthesis of a symmetrical LCL-T network, (a) Split inductances, (b) addition of DC LC filter, (c) split resonant cap and introduce return line L4, (d) full system layout of the circuit

B. Balancing conditions

An equivalent balanced bridge configuration can be deduced as shown in Fig. 8. In reality, there will be two balanced circuits working in tandem. Each circuit generated from either of the sources $v_{2,inv}$ and $v_{3,inv}$. Analyzing this circuit in detail, the balancing condition is given by (2).

$$\frac{jL_3\omega}{\frac{-j}{(2C_1||C_{cm})\omega}+j2L_4\omega+\frac{jL_1\omega}{2}} = \frac{C_3}{C_2+C_{add}} \quad (2)$$

By adjusting the series resonant frequency of the equivalent impedance in the denominator Z_{22} to be greater than f_{min} (3), the balancing criterion is found to be purely dependent on circuit parameters and is reduced to (4).

$$f_{series} = \frac{1}{2\pi\sqrt{(\frac{L_1}{2}+2L_4)(2C_1||C_{cm})}} > f_{min} \quad (3)$$

$$f_{min} > 150\ kHz$$

$$\frac{L_3}{\frac{L_1}{2}+2L_4} = \frac{C_3}{C_2+C_{add}} \quad (4)$$

Where C_2 and C_3 are defined as shown in (5).

$$C_3 = C_{p3,inv}$$
$$C_2 = C_{p1,inv} + C_{p2,inv} + C_{p4,inv} + C_{p6,inv} \quad (5)$$

The value of f_{min} would depend on the lower frequency limit of the EMI standard. Typically this limit is set to 150 kHz. The simplified circuit is shown in Fig. 8. Note that, with C_{add} in circuit, the value of L_3 can be smaller, implying lower turns and lower impact of EPC. The noise spectrum comparison under balanced and unbalanced conditions is shown in Fig. 9, where three cases are simulated. The two unbalanced case show that when the ratio in (4) is either higher or lower than the required balanced ratio, the noise spectrum is high. This validates the balancing principle.

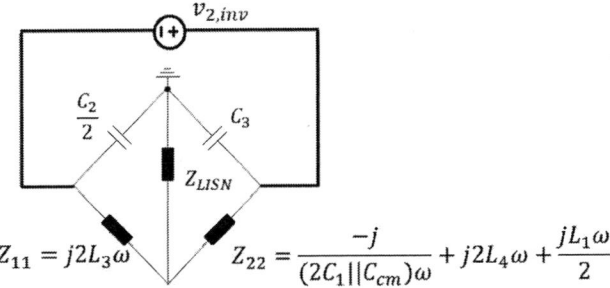

$$Z_{11} = j2L_3\omega \qquad Z_{22} = \frac{-j}{(2C_1||C_{cm})\omega} + j2L_4\omega + \frac{jL_1\omega}{2}$$

Fig. 8 Equivalent balancing circuit

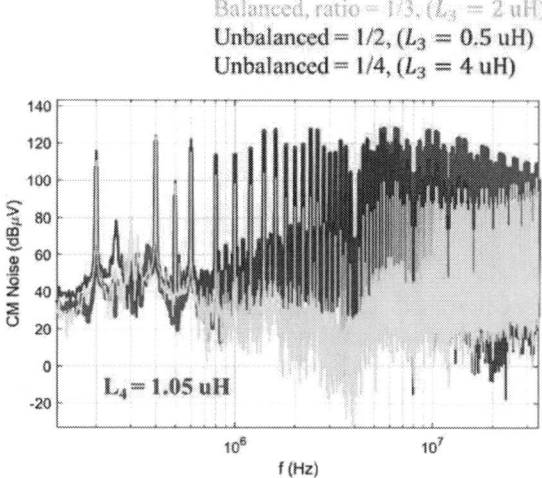

Fig. 9 Balanced circuit performance using simulations.

From the structure, the EPC of L_2 does not matter if the transformer CM impedance is large enough as long as the transformer inter-winding capacitance is small. Furthermore, L_2 can be a single core and does not need to be split as the high transformer impedance will mask its effect.

In cases where the transformer shield or Litz wire type transforemrs are not used, the balancing ratio is given by (5), derived from the circuits in Fig.10.

$$\frac{Z_1 + j2L_4\omega}{j2L_3\omega} = \frac{Z_2}{\frac{-j2}{C_2\omega}} \quad (6)$$

Where;

$$Z_1 = \frac{\frac{jL_1\omega}{2}Z_{LOAD}}{\frac{jL_1\omega}{2} + Z_{LOAD} - \frac{j}{C_3\omega}} \quad (7)$$

$$Z_2 = \frac{-Z_{LOAD}\frac{j}{C_3\omega}}{\frac{jL_1\omega}{2} + Z_{LOAD} - \frac{j}{C_3\omega}} \quad (8)$$

$$Z_3 = \frac{\frac{-L_1}{2C_3}}{\frac{jL_1\omega}{2} + Z_{LOAD} - \frac{j}{C_3\omega}} \quad (9)$$

The terms Z1 – Z3 are a complex function of transformer load impedance. The balancing criterion is rather complicated and would depend on the battery bank parasitics. Hence, it is critical to ensure the effect of secondary stage rectifier is bypassed using high CM impedance transformer. This way a load independent filter is possible.

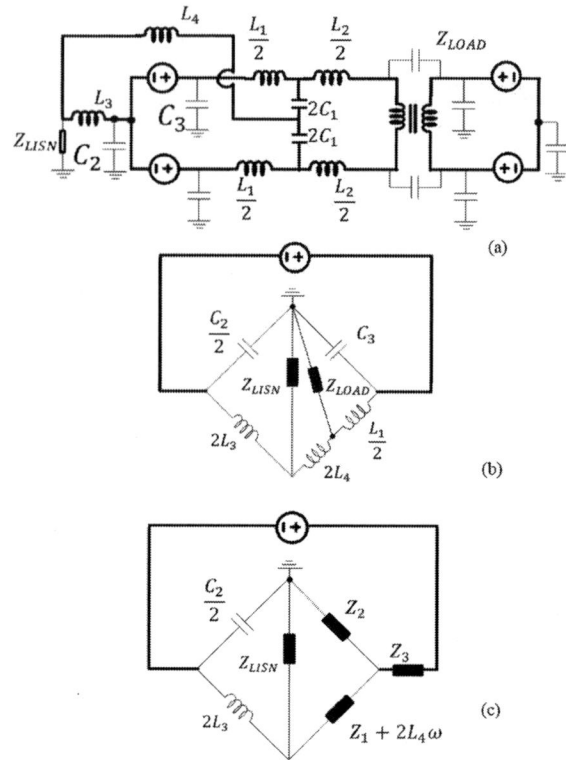

Fig. 10 Balanced circuit under low transformer CM impedance, (a) single line lumped diagram, (b) representative wheatstone bridge circuit, (c) simplified wheatstone bridge circuit

C. Comparison with conventional EMI filter solution

A fair comparison between a single stage DC side CM filter and the proposed balanced topology approach is conducted. The filter parameters are calculated under the same peak requirement of approximately 40 dB in Fig. 11.

Using $C_{cm} = 100\ nF$ and respecting the touch current and safety limit, the corner frequency f_c is calculated as shown in (5).

$$f_c = \frac{1}{2\pi\sqrt{L_{cm}C_{cm}}} \quad (10)$$

The results in the common mode inductance L_{cm} to be, $L_{cm} = 160\ \mu H$. It is noticeable from the corner frequency calculations, that the required CM inductance in a traditional filter solution is much larger than the proposed solution. Thanks to the absence of leakage current limit and therefore maximum limit on CM capacitor, the proposed solution presents a small overall filter size. In addition, since the balancing principle solely relies on parameter ratio to achieve CM current cancellation, this inherently results in smaller filter size.

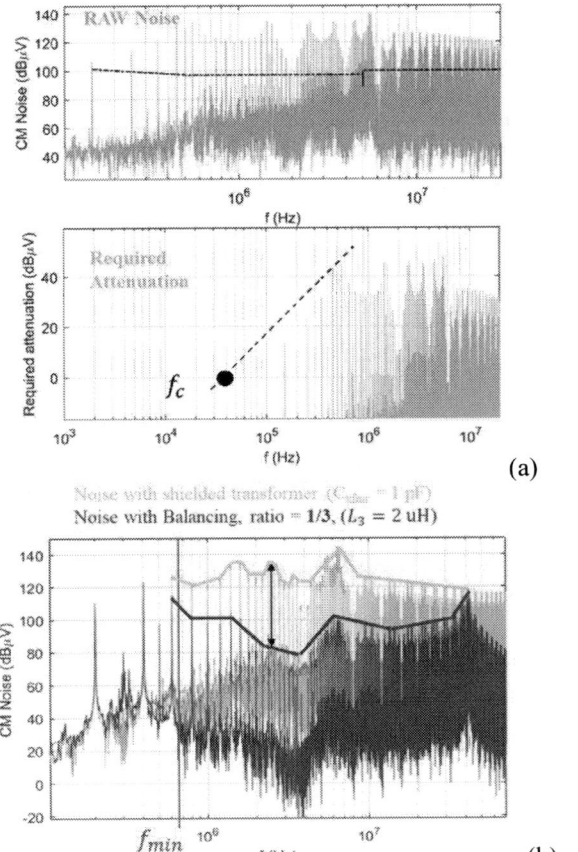

Fig. 11 (a) Conventional filter sizing, (b) Sizing using proposed filter under same attenuation.

TABLE I.   COMPARISON

Proposed		Conventional	
Parameters	**Value**	**Parameters**	**Value**
L_3	2 uH	L_{cm}	160 uH
L_4	1.05 uH		
C_{cm}	500 nF	C_{cm}	100 nF

IV. EXPERIMENTAL RESULTS

The 6.6 kW *GaN* based OBC charger under test is shown in Fig. 5, [1]. The system parameters are listed in Table II.

TABLE II.   LCL-T TANK PARAMETERS

Parameters	Value
FET(s)	650 V/30 A *GaNSystems*
F_{sw}	500 kHz
L_r	7.3 uH
C_r	35 nF
C_{blk}	2 uF
n	1:1

The test setup is shown in Fig. 12(a). The parameters and components were picked for the balancing network based on equations in (1, 2, and 3), RMS current rating and peak volt-second, shown in Table III.

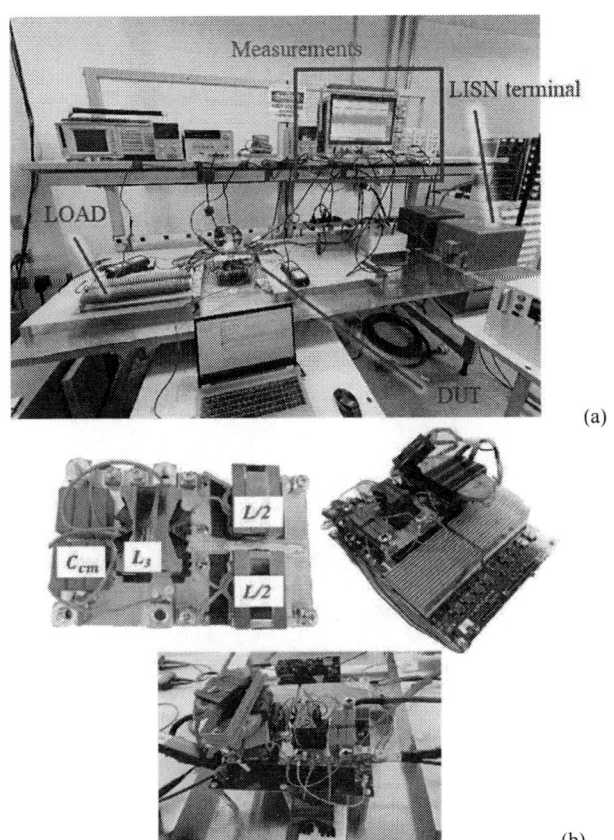

Fig. 12 (a) Experimental setup, (b) hardware of balancing circuit (Note: L_4 is mounted externally and not shown here)

TABLE III.   LCL-T TANK FILTER PARAMETERS

Parameters	Value
Balancing inductor L_4	13.8 uH, n = 7, lg = 0.35 mm, $B_{pk,L4}$ = 0.21 T, B65883A ferrite core
Balancing inductor L_3	15.7 uH, n = 8, lg = 0.35 mm, $B_{pk,L3}$ = 0.24 T, B65883A ferrite core
C_{cm}	500 nF
f_{min}	182 kHz

The effectiveness of the proposed approach is dependent on the balanced inductor impedance characteristics. In this case, ferrite-based PQ cores were adopted for demonstration purposes. The EMI filtering range was dictated by the permeability characteristics as seen from Fig. 13(a). Based on the impedance characteristics the balancing criterion beyond 6-8 MHz, is not met, resulting in poor attenuation as seen from measured spectrum. The spectrum was obtained through post processing of DC side CM currents through the oscilloscope measurements. Note that in this case, ferrite core was selected due to its low sensitivity to load current. This limitation can be eliminated by using stray air core inductors or selecting better core permeability characteristics.

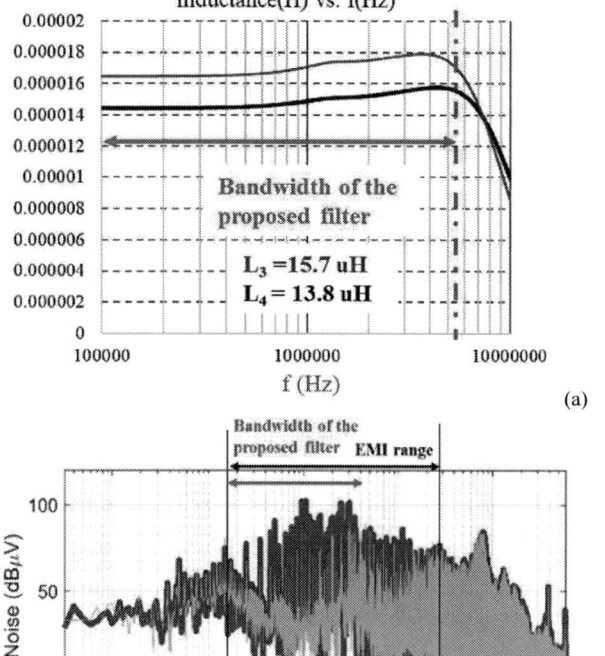

Fig. 13 (a) Tuned and measured impedance for L_3 and L_4 branch (b) Measured spectrum comparison between no-filter and proposed balanced filter structure

V. Conclusions

This paper presents a technique to suppress the low and mid frequency common mode harmonics in an immittance based resonant converter, specifically the LCL-T resonant converter. By introducing a symmetric architecture, adding a feedback path from the resonant cap mid-point to an LC filter branch, the LCL-T resonant converter can be reduced to a balanced circuit which allows one to compensate for common mode noise at the DC input terminal using a balanced wheatstone bridge network.

The proposed filter can be smaller compared to a conventional LC filter added at the DC input terminal.

References

[1] S. Mukherjee, J. M. Ruiz and P. Barbosa, "A High-Power Density Wide Range DC–DC Converter for Universal Electric Vehicle Charging," in IEEE Transactions on Power Electronics, vol. 38, no. 2, pp. 1998-2012, Feb. 2023, doi: 10.1109/TPEL.2022.3217092.

[2] M. Borage, S. Tiwari, and S. Kotaiah, "Analysis and design of an LCL-Tresonant converter as a constant-current power supply," IEEE Trans. Ind. Electron., vol. 52, no. 6, pp. 1547–1554, Dec. 2005

[3] M. Borage, K. V. Nagesh, M. S. Bhatia, and S. Tiwari, "Resonant immittance converter topologies," IEEE Trans. Ind. Electron., vol. 58, no. 3, pp. 971–978, Mar. 2011.

[4] Z. Fang, T. Cai, S. Duan, and C. Chen, "Optimal design methodology for LLC resonant converter in battery charging applications based on time-weighted average efficiency," IEEE Trans. Power Electron., vol. 30, no. 10, pp. 5469–5483, Oct. 2015

[5] Y. Jang, M. M. Jovanović, J. M. Ruiz, M. Kumar and G. Liu, "Implementation of 3.3-kW GaN-based DC-DC converter for EV on-board charger with series-resonant converter that employs combination of variable-frequency and delay-time control," 2016 IEEE Applied Power Electronics Conference and Exposition (APEC), Long Beach, CA, USA, 2016, pp. 1292-1299, doi: 10.1109/APEC.2016.7468035.

[6] R. Phukan et al., "Design of an Indirectly Coupled Filter Building Block for Modular Interleaved AC–DC Converters," in IEEE Transactions on Power Electronics, vol. 37, no. 11, pp. 13343-13357, Nov. 2022, doi: 10.1109/TPEL.2022.3179346.

[7] A. Vedde, M. Neuburger and H. Reuss, "An optimized high-frequency EMI filter design for an automotive DC/DC-converter," 2021 National Power Electronics Conference (NPEC), Bhubaneswar, India, 2021, pp. 01-06, doi: 10.1109/NPEC52100.2021.9672512.

[8] P. R. Prakash, A. Nabih and Q. Li, "Design Optimization of PCB-Winding Matrix Transformer for 400V/12V Unregulated LLC Converter," 2021 IEEE Energy Conversion Congress and Exposition (ECCE), Vancouver, BC, Canada, 2021, pp. 1777-1784, doi: 10.1109/ECCE47101.2021.9595190.

[9] Z. Ge, H. Wu and Y. Liu, "Low-Loss Segmented Shielding Technique for PCB-Winding Planar Transformers," in IEEE Transactions on Power Electronics, vol. 38, no. 1, pp. 12-16, Jan. 2023

[10] P. Kong, S. Wang and F. C. Lee, "Common Mode EMI Noise Suppression for Bridgeless PFC Converters," in IEEE Transactions on Power Electronics, vol. 23, no. 1, pp. 291-297, Jan. 2008, doi: 10.1109/TPEL.2007.911877.

[11] Z. Qing et al., "Common-Mode EMI Noise Reduction With Improved Balance Technique for DC–AC Converters," in IEEE Transactions on Power Electronics, vol. 39, no. 11, pp. 14318-14329, Nov. 2024, doi: 10.1109/TPEL.2024.3429165.

A Single-Phase Soft-Switching Buck-Boost Inverter

Lukas Wipprecht
Electronics & Drives
Reutlingen University
Reutlingen, Germany
lukas.wipprecht@web.de

Burkhard Ulrich
Electronics & Drives
Reutlingen University
Reutlingen, Germany
burkhard.ulrich@reutlingen-university.de

Abstract— **This article presents a single-phase, single-stage unidirectional inverter derived from a non-inverting buck-boost dc/dc converter combined with a full bridge unfolder circuit. Three different operating modes to achieve zero-voltage switching (ZVS) are discussed and evaluated against each other, demonstrating the feasibility of a soft-switching operation for this converter. The circuit structure and operating principles are discussed in detail. A modular 400 W prototype utilizing GaN devices switching at 500 kHz was built to verify the design, using an input dc voltage range of 100 V to 400 V and delivering an ac output voltage of 230 V at 50 Hz.**

Keywords— soft-switching, inverter, control modes

I. INTRODUCTION

The changing landscape of electrical energy production, shifting from centralized generation to decentralized systems characterized by multiple smaller energy production sites, has increased the demand for lower power single-phase grid-tied inverters. These inverters connect decentralized dc sources or energy storage devices to the low-voltage ac grid. Only a unidirectional power flow may be required in some applications, such as photovoltaic generators, fuel cells, or small-scale wind turbines. Also, in this type of application, the dc input voltage of the inverter can vary significantly due to load and time dependencies.

For many applications, buck-derived voltage source inverters are commonly used [1], but these require additional power conversion stages when the dc input voltage drops below the ac grid's peak voltage; otherwise, an additional dc/dc converter stage, such as a boost converter needs to be connected in series to step up the voltage to be able to use the buck-derived inverter. This results in the requirement of two or more high-frequency switching stages connecting in series to process the energy fed into the grid, possibly increasing complexity and losses.

Therefore, buck-boost type inverters have emerged to overcome these limitations and are considered a beneficial alternative [2]-[5], capable of reducing the number of required high-frequency switching stages. Most solutions described in the literature focus on the discussion of hard-switching converters. E.g., both the Aalborg inverter proposed [2], and the buck-boost converter with unfolder presented in [3] operate with hard-switching in continuous conduction mode (CCM), using a fixed-frequency pulse-width modulation (PWM) control. Although these solutions mitigate the problem of multiple high-frequency switching stages connected in series, the hard-switching operation might result in additional switching losses and increased electromagnetic emissions (EMI).

In contrast, a soft-switching operation of power converters, particular zero-voltage switching (ZVS), is beneficial for reducing switching losses and mitigating EMI. However, soft-switching operational modes often have restrictions regarding the converter control, such as variable frequency operation, adding control complexity. For dc/dc converters, simple low-complexity operation principles such as the clamp-switch approach [7]-[10] have been proposed, but these have yet to be applied to inverters.

Therefore, this article aims to describe a new single-stage buck-boost inverter design, which integrates a soft-switching operation based on a clamp-switch approach, demonstrating the feasibility of a ZVS operation of a single-phase inverter at a constant switching frequency. This approach could lead to reduced switching losses and mitigate EMI, which constitute critical challenges in practical designs.

II. PROPOSED INVERTER SYSTEM AND VARIANTS

A. Inverter System Overview

Fig. 1 illustrates the system of a single-phase inverter, comprised of a buck-boost dc/dc converter followed by a low-frequency full bridge unfolder circuit, as considered in this article, which is, on a systems level, similar to the system proposed in [3]. The input capacitor, C_1, is a low-frequency decoupling capacitor designed to handle power fluctuations at twice the line frequency. In contrast, C_2 is a small high-frequency link capacitor that filters out the high-frequency

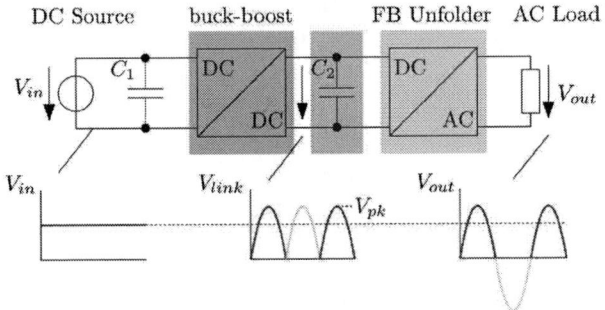

Fig. 1.   Proposed inverter system overview, consisting of a buck-boost dc/dc converter and a full-bridge (FB) unfolder circuit, similar to the circuit structure in [3].

components of the inductor current. The dc/dc converter structures employed here are derived from a non-inverting buck-boost converter and are designed to generate a (rectified) half-sine wave voltage, V_{link}, across C_2. Due to the buck-boost operation, the amplitude V_{pk} of the rectified sine wave V_{link} can be larger than the input voltage V_{in}. The full-bridge unfolder circuit, operating at the line frequency, converts the voltage V_{link} to an ac voltage, applied to an ac load, which is considered here to be a pure resistive load. Although the proposed system concept shown in **Fig. 1** is similar to that in [3], there are some key differences, which will be discussed in the following sections. A significant difference lies in the circuit structures used to implement the buck-boost converter and the control system implementation. The converter here is designed to operate with ZVS soft-switching at a constant frequency for at least part of the mains cycle, thereby reducing switching losses and mitigating EMI.

B. Circuit Variants And Clamp-Switch ZVS Operation

Two different implementations of the non-inverting buck-boost converter shown in **Fig. 2a** and **Fig. 2b** are considered here. **Fig. 2a** presents a three-switch variant, where a diode at the input side of the buck-boost converter is replaced by the active switch Q_2, and **Fig. 2b** shows a four-switch variant of the non-inverting buck-boost converter. In both cases, a clamp-switch approach similar to [7]-[10] is employed to achieve ZVS at a constant switching frequency $f_{sw} = 1/T_{sw}$. The clamp-switch ZVS approach will be reviewed in the context of the converter variants used here utilizing the following **Fig. 3**. In **Fig. 3a**, the typical inductor current waveform i_L from a converter operating with a clamp-switch operation is shown. The current i_L resembles the waveforms for a converter operating in discontinuous conduction mode (DCM), with one crucial difference: the current will freewheel at a negative value during the time interval when in a DCM-operated converter, both the main switch and the diode are off. This freewheeling process is achieved by placing a clamp-switch, which is usually a series connection of two semiconductor switches parallel across the inductor. In its simplest form, this switch consists of a MOSFET (or a GaN HEMT) and a diode. The operating principle is as

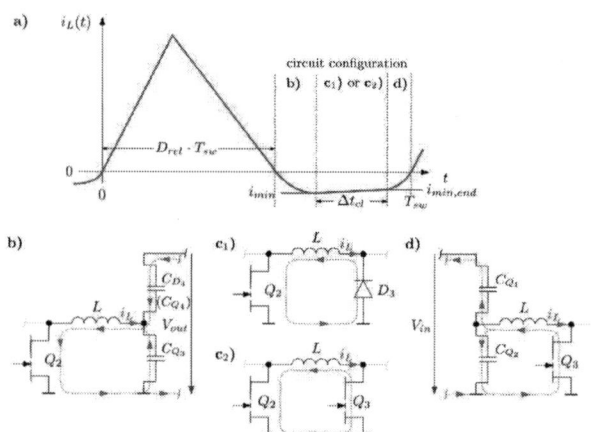

Fig. 3. Clamp-switch operation of the proposed converters: a) inductor current i_L, b) circuit configuration after diode D_4 turn-off to c_1) and c_2) circuit configurations during clamp interval and c) circuit configuration after clamp-switch (Q_2) turn-off.

follows: at the zero-crossing of the inductor current, a resonant circuit is formed from L, and the device capacitances connected to the right switch node (cf. to **Fig. 3b**). This causes the inductor current to undergo a sinusoidal transition, and i_L will become negative. The resonant transition ends at the negative peak value i_{min}. At this time, the voltage at the right switch node reaches approximately 0 V, and diode D_3 gets forward-biased. Therefore, the current can now freewheel through the series connection of Q_2 and D_3, as shown in **Fig. 3c$_1$**. It is possible (but not necessary) to also turn on Q_3 at the time instant when D_3 turns on or somewhat later; this results in the circuit configuration depicted in **Fig. 3c$_2$**. Therefore, both circuit configurations will lead to a freewheeling path for the inductor current i_L. During the freewheeling phase, the current i_L will decay from i_{min} to $i_{min,end}$. The difference in the circuit configurations regarding this decay is that the turn-on of Q_3 can lead to a lower applied voltage across L and, therefore, to a slower decay. Assuming that the inductor is still negative at the value $i_{min,end}$ at the end of the freewheeling phase, then the turn-off of Q_2 will initiate a resonant charging process of the capacitors C_{Q1} and C_{Q2} at the left switch node as shown in **Fig. 3c**. If the absolute value $|i_{min,end}|$ at the turn-off of Q_2 is sufficiently large, i.e.:

$$L \cdot i_{min,end}^2 \geq \left(C_{Q1} + C_{Q2} \right) \cdot V_{in}^2, \tag{1}$$

then a ZVS turn-on of Q_1 is possible. A more detailed discussion of the basic clamp-switch converter operation using the three-switch buck-boost converter (cf. to variant 1 in **Fig. 2**) can be found in our other work [10], where also the effect of increasing $|i_{min,end}|$, and therefore, extending the ZVS range by adding the capacitor C_{ZVS} is discussed. The focus here is on extending this approach to other circuit variants and discussing an inverter application.

C. Design Considerations

Regarding the operation of the buck-boost converter in the single-phase proposed inverter applications, using the clamp-switch approach poses unique demands on the selection of the passive components C_1, C_2 and L.

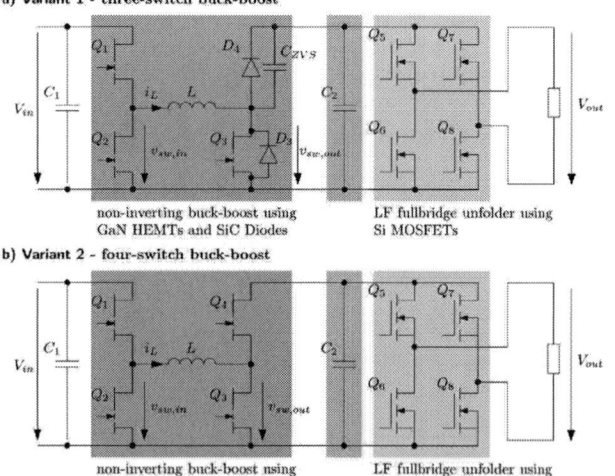

a) Variant 1 - three-switch buck-boost

non-inverting buck-boost using GaN HEMTs and SiC Diodes | LF fullbridge unfolder using Si MOSFETs

b) Variant 2 - four-switch buck-boost

non-inverting buck-boost using GaN HEMTs | LF fullbridge unfolder using Si MOSFETs

Fig. 2. Circuit variants under study: a) three-switch buck-boost converter and b) four-switch buck-boost converter.

Input capacitor C_1: Assuming a source supplying pure dc power, the input capacitor C_1 has to be designed to handle the power fluctuations at double the ac frequency of the output voltage and, therefore, will have to be sized according to the power requirements. A large-sized capacitor will, therefore, be required regardless of the converter's switching frequency.

Inductor value L: For operating the converter using a ZVS clamp-switch approach, such a low inductance value L, is required so that the converter would operate in DCM if Q_2 were not turned on. This requirement must be met depending on input voltage and load variations. The worst case will occur at minimum input voltage $V_{in,min}$ and maximum load $P_{out,max}$. This leads to the following requirement for the inductance value:

$$L \le \frac{D_{max}^2 \cdot V_{in,min}^2}{4 \cdot f_{sw} \cdot P_{out,max}}. \tag{2}$$

Here, D_{max} represents the maximum attainable duty cycle, which can be expressed as:

$$D_{max} = \frac{D_{rel}}{1 + \frac{V_{in,min}}{V_{pk}}}. \tag{3}$$

$D_{rel} < 1$ in (3) is the maximum attainable time duration where a current flows through the inductor in DCM operation (cf. to **Fig. 3**), which ensures that enough time in the period is left for the resonant transitions. V_{pk} is the amplitude of the output voltage.

Output capacitor C_2: The unidirectional nature of the proposed inverter, due to the always positive average current through D_4 or Q_4, respectively, means that capacitor C_2 must be carefully selected, as the control loop can only increase its

voltage. Therefore, a trade-off regarding the voltage ripple of V_{link} on C_2 and the total harmonic distortion (THD) of the output current must be considered. Assuming a resistive load R_L connected to the output of the inverter, the fall time of V_{link} and V_{out} is limited by the exponential discharge of the RC circuit formed from C_2 and R_L. This means that when the sine wave approaches the zero-crossing, at some point, the waveform changes to the exponential discharge waveform of an RC circuit. This has significant implications for the voltage ripple and THD, as C_2 cannot be increased indefinitely until a desired low ripple is achieved because this would result in distorting the sinusoidal waveform with an earlier transition to the RC discharge, resulting in a higher THD value. Therefore, to design the converter, an equation was derived using the derivatives of both the sinusoidal and exponential part at the transition point to allow the calculation of C_2 for a defined transition voltage:

$$C_2 = \frac{(V_{c,off}/V_{pk})}{R_L} \cdot \frac{1}{2 \cdot \pi \cdot f \cdot \cos(\pi - arcsin(V_{c,off}/V_{pk}))}, \tag{4}$$

where $V_{c,off}$ is the desired voltage threshold at which no control is possible, and f is the line frequency. After choosing $V_{c,off}$, depending on the THD, the ripple voltage can only be decreased using a higher switching frequency.

III. CONVERTER OPERATION MODES

Three operational modes are derived from the two circuit variants and the two approaches to clamp the inductor current in the freewheeling phase; their major waveforms are visualized in **Fig. 4**.

Mode I is the simplest and uses the three-switch configuration in combination with clamping using diode D_3. The

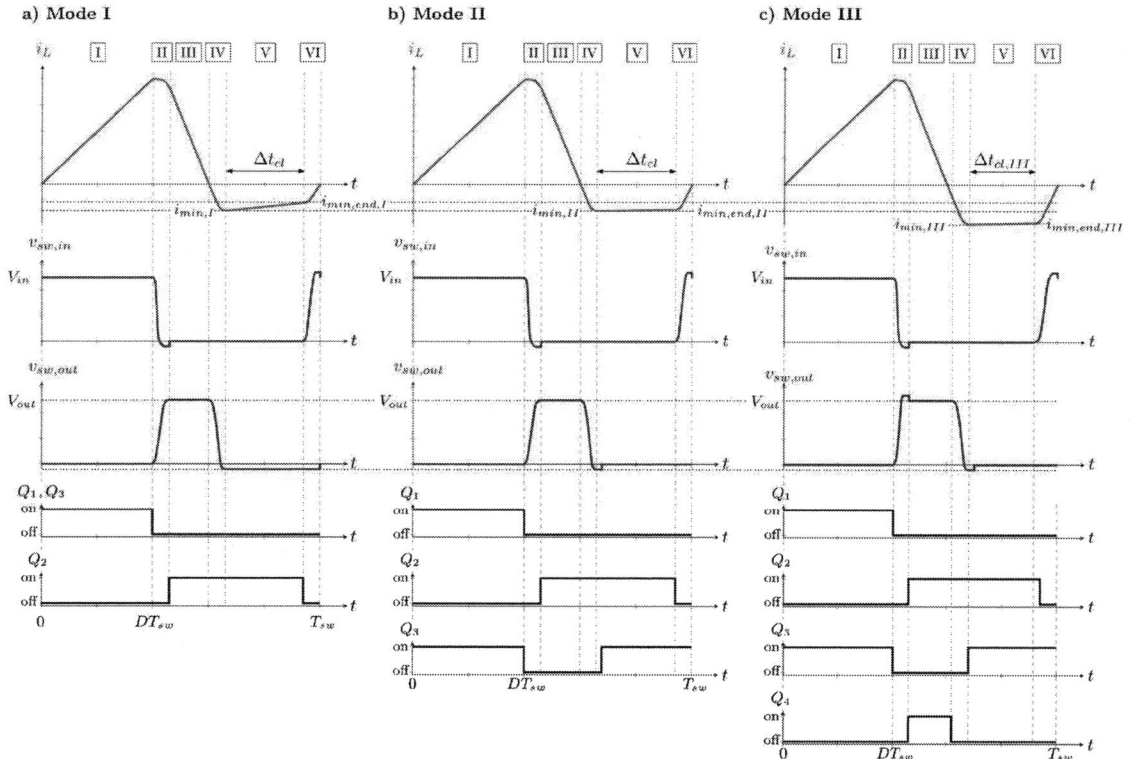

Fig. 4. Idealized waveforms for i_L, $v_{sw,in}$, $v_{sw,out}$ and switch control signals for all three modes.

corresponding waveforms are shown in **Fig. 4a**. The primary advantage of mode I is its simplicity. The same control signals are used for Q_1 and Q_3, while Q_2 operates with the inverted drive signal and added dead time. Therefore, no changes to the control scheme are needed compared to a standard PWM operation without ZVS.

Mode II also uses the three-switch configuration; however, instead of utilizing diode D_3 for freewheeling, transistor Q_3 is turned on for a potentially lower voltage drop. This effect can be observed in **Fig. 4b**, as the inductor current decays slower during the clamp interval. The disadvantage is that an additional signal needs to be calculated, which means that a more complex control scheme is required. However, an advantage could be that a separate diode is no longer required as the body-diode (reverse-conduction) carries the freewheeling current until the transistor is turned on.

Mode III offers the most advanced features by employing a four-switch configuration. Compared to mode II, this brings the advantage that a second possibility to achieve a negative current in the inductor exists by leaving Q_4 turned on after the inductor current reaches zero, which can be seen in **Fig. 4c** as the resonant transition into the freewheeling phase already starts at a negative current. Therefore, the freewheeling current can be increased to enable ZVS over a broader operational range, while being limited to the minimum required to maintain ZVS. Although in modes I and II, the freewheeling current can be increased by adding a capacitor C_{ZVS} in parallel to D_4 (cf. to **Fig. 2a**), this addition of C_{ZVS} also leads to an increase in current, and therefore conduction losses, in operating points, where a ZVS is possible without an additional capacitance. Consequently, in mode III, a ZVS operation with potential lower losses is possible compared to the other modes because the amount of negative current can be controlled.

The following Table I summarizes the approaches:

TABLE I.   MODE COMPARISON

Mode	Circuit	Clamping Method	Advantage	Disadvantage
I	3-switch	Diode D3	Simple control	Limited ZVS range, highest losses
II	3-switch	Active Q3	Reduced voltage drop during Δt_{cl}	More complex control
III	4-switch	Active Q3 and Q2	Greater ZVS range and possibly lower losses	Increased control and circuit complexity

IV. COMPARATIVE EVALUATION

Comparing the three modes regarding efficiency across different parameters, such as varying input voltage and load conditions, requires accounting for the constantly changing operating point. This variation arises from the proposed inverter application, where the output voltage continuously fluctuates.

A circuit simulation using SPICE is deemed unsuitable for this analysis for two main reasons. First, to cover the entire range of operating points during the mains period, the required simulation timespan must be at least 5 ms (one-quarter of the mains period). This period and the proposed switching frequency of 500 kHz would result in prohibitively long simulation times, given the many anticipated operating points and the small timesteps needed to accurately simulate effects like switching losses. Second, accurately modeling factors like inductor losses and calculating the ZVS range within a circuit simulation is challenging.

A comprehensive Python program was developed to address these limitations for a detailed converter analysis and design. The basic structure of the program and the computations it performs are depicted in the flowchart in **Fig. 5**. The main process is shown in black at the top, with detailed steps illustrated below.

The analysis begins by configuring parameters such as the input voltage range and output power. Initial calculations include determining the inductance value L and coil parameters and the equivalent (charge-related) capacitances for all switches. These capacitances are based on datasheet diagrams and corrected using measurements on the prototype to account for the additional stray capacitances. Additionally, a virtual double-pulse test, using manufacturer-supplied models, was conducted for the GaN HEMTs to estimate their turn-off energy, a parameter that is missing from the respective device datasheets. This simulation is used to evaluate the turn-off energy as a function of voltage, current, and parallel capacitance, and the interpolated results are incorporated into the program.

Using the provided and calculated data, the ZVS behavior is calculated for the entire quarter period. This involves determining the required current to achieve ZVS and the current at the end of the clamping phase based on the value of the equivalent capacitances. These calculations establish if ZVS is feasible.

In the next step, the inductor current at the actual operating point is approximated using linear and sinusoidal segments and sampled at 100 points for detailed analysis of the high-frequency switching period. Using the so-calculated current waveform, the conduction losses in the GaN-HEMTs and/or SiC-Diodes are calculated, as well as the inductor losses, which are divided into core losses and copper losses. A fast-fourier transform (FFT) is

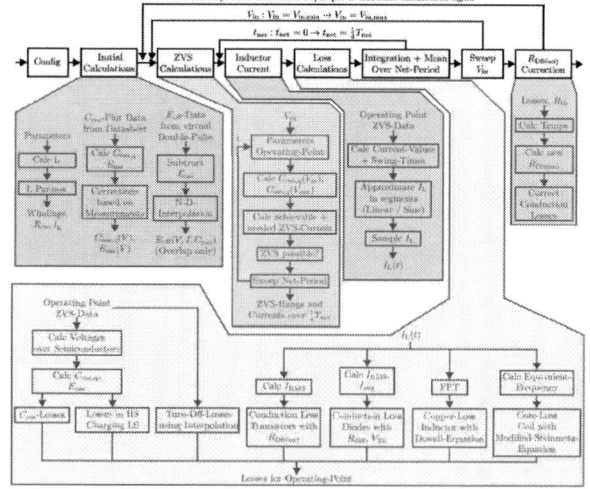

Fig. 5.   Program structure for analyzing the proposed buck-boost inverter circuits.

performed in the case of the copper losses to calculate the individual harmonics of the inductor current. Then, the Dowell equation [11] is used to calculate the ac resistance and the losses of the winding at each individual frequency, and the results are summed up to get the total winding losses. The modified Steinmetz equation approach [12] is used to calculate the core losses to take into account the non-sinusoidal flux density. Therefore, an equivalent frequency is calculated based on the inductor current waveform, and, together with data from the core manufacturer, the core losses during the high-frequency switching period are estimated.

The switching losses are also considered. Using the data for ZVS behavior at the current operating point and the equivalent capacitances, the drain-source voltage over the transistors in the respective turn-on instant is calculated so that the losses associated with the discharge of the output capacitance C_{oss} in the case of a partial ZVS can be appropriately calculated. In the case of transistor Q_1, the losses due to charging the output capacitance of the low-side transistor Q_2 are also considered. For all GaN-HEMTs, the turn-off losses are determined using the interpolated data of the virtual double-pulse test.

The loss calculation mentioned above is repeated for 50 distinguished high-frequency switching periods during one-quarter of the mains period, considering the constantly changing operating point. The calculated losses are averaged to estimate the losses for a full mains cycle.

After this, a second loop sweeps over the input voltage range by starting at the ZVS calculations with the next voltage value. The respective data of each cycle is saved for later analysis.

In the final step, the junction temperature of the HEMTs and diodes is approximated using the losses and the experimentally determined thermal resistance of the prototype converter. A new estimate for the $R_{DS(on)}$ value is calculated using the temperature increase, and the conduction losses are corrected accordingly.

The conduction losses in the unfolder bridge are also calculated and added to get the total converter losses.

As the losses will lead to an increase in the actual duty cycle values in a closed-loop operating converter, the program jumps back once to the start of the calculations to repeat everything with all losses added to the output power to consider this effect to gain a better estimation of the total losses.

The program always performs all the described calculations for all three described operation modes.

Because the program runs much faster than a SPICE simulation, it could also aid in designing the coil by automatically trying different form/material combinations. However, it must be clarified that the program's calculations are based on several simplifying approximations and, therefore, not all effects of the actual inverter prototype are considered.

Fig. 6 and **Fig. 7** depict the results of the calculations above. In **Fig. 6**, the upper three diagrams show the calculated losses for the main components, and the lower graphs show the derived efficiency. The results show that mode II achieves the highest efficiency. Mode III has a lower calculated efficiency. However, it should be noted that the operating points on the calculations for mode III are done so that a ZVS is always achieved. This necessitates a larger negative inductor current and, therefore, causes higher conduction losses in the components, which can offset the reduction in switching losses. Therefore, to fully leverage the additional option to control the freewheeling current to increase the efficiency, a more comprehensive control system implementation and calculation have to take place, which optimize the freewheeling current for the highest efficiency instead of achieving ZVS.

In contrast, for modes I and II, there is no possibility of adjusting the ZVS current; therefore, at least for parts of mains frequency, depending on V_{out}, there will be turn-on switching losses. However, it can be concluded that from an efficiency point of view, the best operating point is not necessarily the one

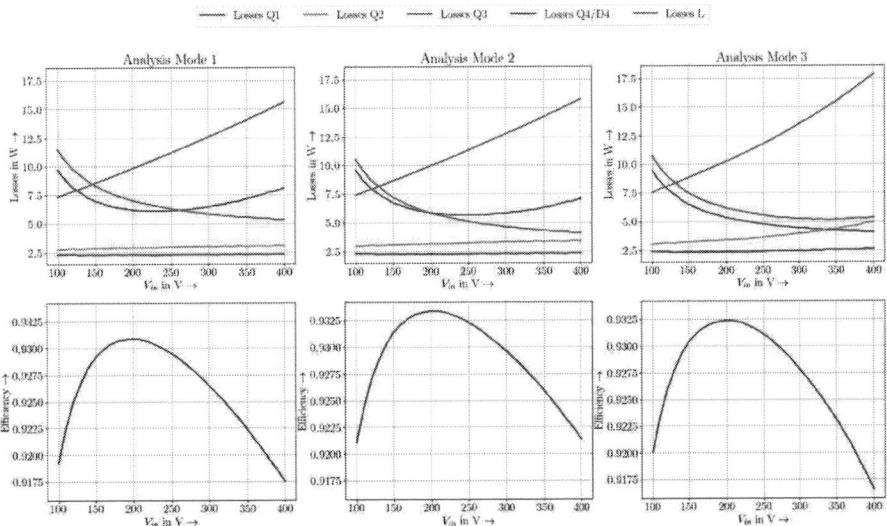

Fig. 6. Losses and efficiency comparison dependent on the input voltage V_{in} for the three mode at $V_{out} = 230$ V ($f = 50$ Hz) and with $P_{out} = 400$ W.

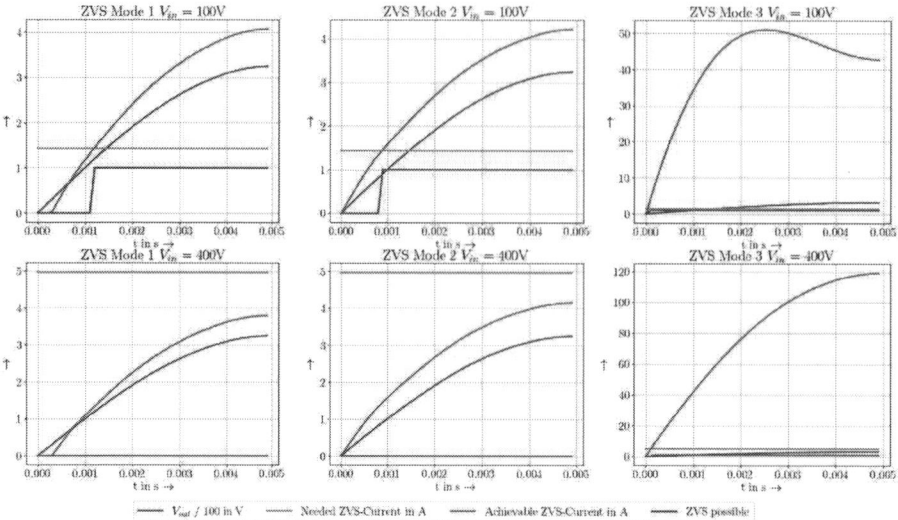

Fig. 7. ZVS behavior for all variants during the first-quarter of a mains-cycle at $V_{in} = 100$ V (upper graphs) and $V_{in} = 400$ V (lower graphs).

that achieves ZVS. Note that all these calculations depend on the characteristics of the actual semiconductor devices used. Therefore, the situation could change if other types of switches were used. In **Fig. 7**, the ZVS behavior of the converter modes is depicted over one-quarter of the mains period. The upper three graphs show the situation for $V_{in} = 100$ V and the lower graphs for $V_{in} = 400$ V. The following quantities are shown: the output voltage (blue), the required current to achieve ZVS (orange), the achievable current for ZVS (green), which is either determined by V_{out} and the capacitances in mode I and II, or adjustable in mode III, and the red trace indicates as a logic variable if ZVS is possible. These graphs show that a ZVS for modes I and II will only be achievable over parts of the mains period and for lower input voltages. In contrast, for mode III, under almost any conditions, ZVS can be achieved.

V. PROTOTYPE AND EXPERIMENTAL RESULTS

A. Inverter System Overview

A prototype inverter system, as shown in **Fig. 8**, has been built to verify the proposed operating method. The prototype consists of a main printed circuit board which contains the

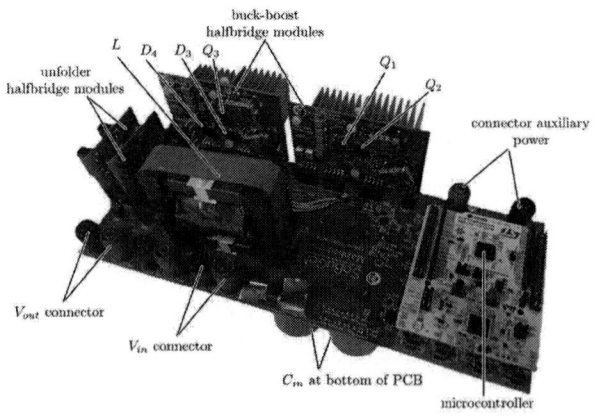

Fig. 8. Prototype inverter.

inductor L, the input buffer capacitor C_{in} and provides sockets for plugging in the output half-bridge modules for the unfolder, the high-frequency switching half-bridge modules for the buck-boost converter using GaN HEMTs and a connector for a microcontroller evaluation board, which is used to control the inverter system. As the prototype uses exchangeable half-bridge modules, the converter system can be reconfigured by exchanging these to be adapted to the different circuit variants. The specifications of the converter are summarized in **Table II**, and **Table III** provides information on the main components used in the converter.

TABLE II.   CONVERTER SPECIFICATIONS

parameter	value
switching frequency f_{sw}	500 kHz
input voltage range V_{in}	100 V … 400 V
nominal output voltage V_{out}	230 V (AC) / 50 Hz
maximum output power P_{out}	400 W

TABLE III.   MAIN CONVERTER COMPONENTS

component	value
transistors Q_1, Q_2, Q_3 and Q_4	GAN080-650EBE (NEXPERIA) / 650 V / 80 mΩ / GaN HEMT
transistors Q_5, Q_6, Q_7 and Q_8	TK190U65Z (TOSHIBA) / 650 V / 190 mΩ / Si MOSFET
diodes D_3 and D_4	TRS12V65H (TOSHIBA) / 650 V / 12 A / SiC Schottky diode
inductor L	3.6 µH / 6 windings: 4 x parallel litz wire 120x0.1 mm / core ETD59 / material 3F36 / airgap 6.5 mm
microcontroller evalboard	STM32 NUCLEO G474RE
input bulk capacitor C_1	4 x 560 uF / 450 V / Aluminum electrolytic capacitor / Wuerth Electronics / 861141486024
output capacitor C_2	7 x 220 nF / 630V / ceramic X7R / size 2220

B. Control System

The inverter is controlled by the STM32 microcontroller on the evaluation board shown on the right side of **Fig. 8**. A

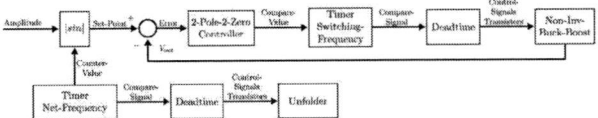

Fig. 9. Block diagram of control structure.

straightforward voltage mode control principle is implemented to control the voltage across C_2, such that it follows a rectified sinusoidal signal. For testing purposes, only a resistive load R_L is considered. A block diagram of the main, fully software-implemented control structure is shown in **Fig. 9**. A rectified sinusoidal signal, with programmable amplitude, is derived from a timer operating with a period equal to the frequency of the ac signal. This rectified sinusoid acts as the set point value, which is compared to the measured output V_{out} to generate the error signal at the input of the controller. A discrete-time two-pole two-zero compensator implements the control law, calculating the compare value (i.e., the duty cycle) of a timer at the switching frequency to generate the buck-boost transistors control signals, which, after inserting the required dead times, are provided to the gate drivers of the buck-boost switching transistors. The control signals of the unfolder bridge are also derived from the timer at the mains frequency. For the operation in mode I, the shown control structure is already sufficient. For an operation in mode II, the generation of the control signal for transistor Q_3 requires additional calculations; this can be either done by using the measured input and output voltage of the buck-boost converter or, if the output resistance is known, by a direct calculation. Also, a current measurement of the inductor current can be used to derive the control signal, but this is not implemented here. For operation in mode III, which is not considered further in the experiments, additional calculations or measurements of the inductor current would also be required to generate the control signals.

C. Experimental Results

The following **Fig. 10** and **Fig. 11** show major experimental converter waveforms. These waveforms were recorded in operation mode I with $V_{in} = 250$ V, $V_{out} = 230$ V, and for a resistive load $R_L = 200\ \Omega$. The measured THD of the current waveform in this operating point is $THD_I = 0.73\%$. **Fig. 10a** shows the inverter input and output voltage, the output current, and the voltage V_{link} at the output of the buck-boost converter

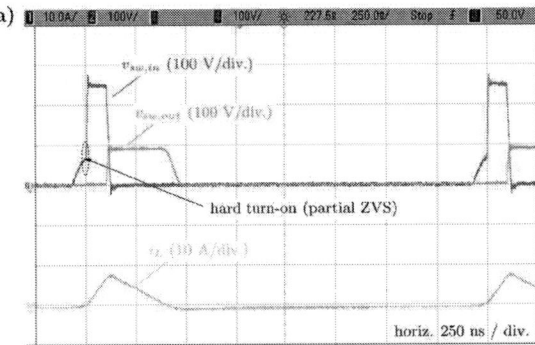

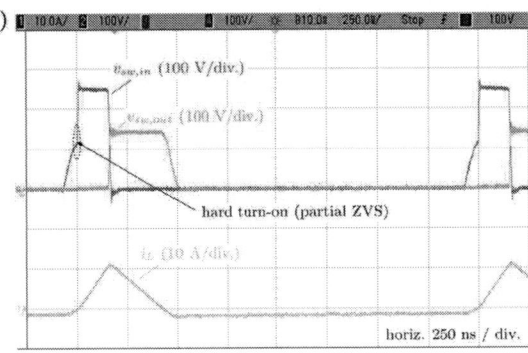

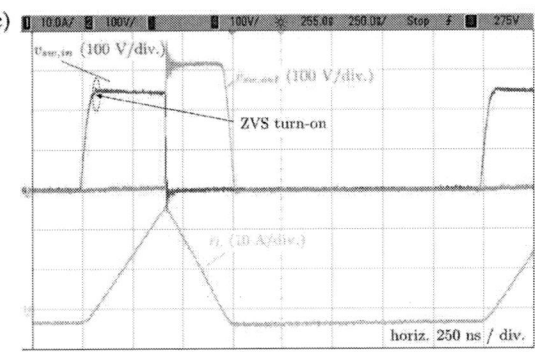

Fig. 12. Detail views of the inductor current i_L, the input switch node voltage $v_{SW,in}$ and output switch node voltage $v_{sw,out}$ for different time instants across the line cycle. a) $V_{out} = 90$ V, b) $V_{out} = 140$ V and c) $V_{out} = 320$ V.

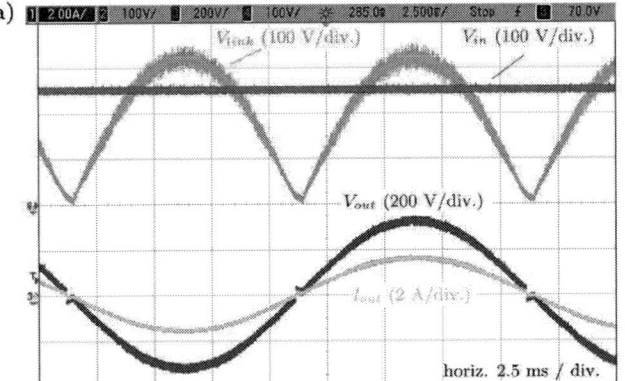

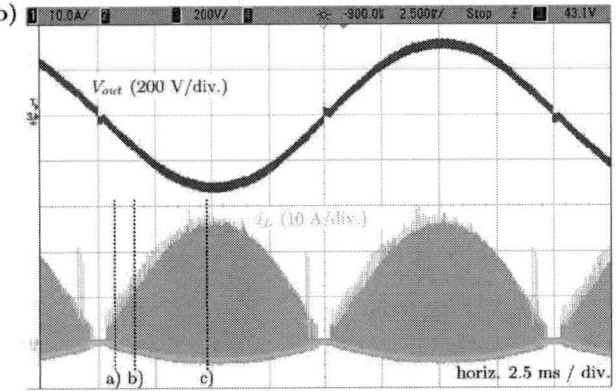

Fig. 10. Experimental waveforms: a) input voltage V_{in}, buck-boost output voltage V_{link}, converter output V_{out} and output current I_{out} over full line cycle and b) output voltage V_{out} and inductor current i_L over line cycle.

979-8-3315-1612-3/25 $31.00 © 2025 IEEE

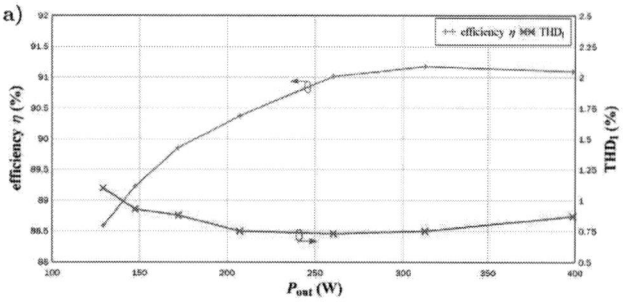 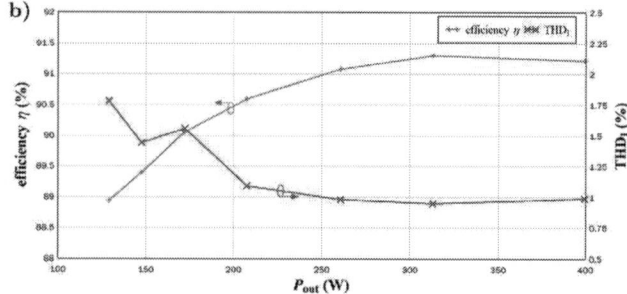

Fig. 13. Measured efficiency and THD of current in a) mode I and b) mode II, measured at $V_{in} = 250$ V and $V_{out} = 230$ V.

over an entire line cycle. As can be seen, the converter operates as proposed, generating a sinusoidal voltage at the load. **Fig.10b** depicts the output voltage and the inductor current, showing the variation in the current during the mains cycle. **Fig. 11** shows detailed views of both switch node voltages ($v_{sw,in}$ and $v_{sw,out}$) and the inductor current for three distinct time instants (cf. to points marked in **Fig. 10** by a), b) and c)). As can be seen by observing the voltage $v_{sw,in}$ this operating mode, the converter only achieves a ZVS at the input side half-bridge at high instantaneous output voltage values, as the current i_{min} is lower at lower values of V_{out}. Therefore, a hard turn-on of Q_1 in **Fig. 11a** and **Fig. 11b** is visible, but as can be seen, the dead time is chosen so that the switching losses are minimized, as the turn-on of Q_1 occurs at the peak of the resonant voltage. Therefore, a minimal voltage drop across Q_1 occurs. Nevertheless, the results indicate a proper inverter operation.

Finally, in **Fig. 12**, the measured efficiency and current THD_I for a varying load resistance at $V_{in} = 250$ V are depicted. As can be seen, the efficiency in mode II is slightly higher than in mode I, reaching a peak efficiency of about 91.3% at $P_{out} = 312$ W. The efficiency is lower than anticipated by the calculation of the design program discussed in the preceding chapter. There are various reasons for this, e.g., the calculations ignore some losses, such as PCB resistances and losses in the capacitors. Also, some loss mechanisms in the calculations are rough estimates, such as the switching losses derived from a simulation model or the simplified calculation of the inductor losses using Dowell's equation, which is only valid for a 1D winding geometry. Note that the prototype is neither optimized for size nor efficiency. Other semiconductor devices with lower resistances should be considered to optimize the efficiency.

VI. CONCLUSION

A novel soft-switching single-stage buck-boost inverter, based on the clamp-switch ZVS approach traditionally used in DC/DC converters, has been proposed and demonstrated. The inverter's operation is analyzed across three different modes, comparing their performance using metrics such as efficiency and ZVS behavior. Measurement results from a 400 W prototype, utilizing GaN HEMTs and operating at 500 kHz, confirm the feasibility of achieving a low THD using the proposed soft-switching operation.

However, the prototype's efficiency is primarily limited by conduction losses in the switches and inductor losses, both represent areas for improvement. Addressing these losses through better switch selection and magnetic component

optimization could further enhance the converter's performance. For this purpose, the Python-based program described in this work provides a valuable tool for optimizing the inverter topology.

REFERENCES

[1] Nimrod Vázquez, Joaquín Vaquero López, "Inverters", chapter 11 in Power Electronics Handbook (Fourth Edition), Editor: Muhammad H. Rashid, Butterworth-Heinemann, 2018, Pages 289-338, ISBN 9780128114070, https://doi.org/10.1016/B978-0-12-811407-0.00011-8.

[2] W. Wu and F. Blaabjerg, "Aalborg inverter — A new type of "Buck in Buck, Boost in Boost" grid-tied inverter," 2013 Twenty-Eighth Annual IEEE Applied Power Electronics Conference and Exposition (APEC), Long Beach, CA, USA, 2013, pp. 460-467, doi: 10.1109/APEC.2013.6520250.

[3] Yao Z, Cai L. Single-stage buck-boost off-grid inverter with feedforward control. Int J Circ Theor Appl. 2023; 51(6): 2705-2715. doi:10.1002/cta.3555

[4] C. Roncero-Clemente, O. Husev, O. Matiushkin, D. Vinnikov and F. Blaabjerg, "Reactive Power Injection Capability of Buck–Boost Inverter With Unfolding Circuit," in IEEE Transactions on Power Electronics, vol. 37, no. 10, pp. 11876-11886, Oct. 2022, doi: 10.1109/TPEL.2022.3179784.

[5] R. O. Caceres, W. M. Garcia and O. E. Camacho, "A buck-boost DC-AC converter: operation, analysis, and control," 6th IEEE Power Electronics Congress. Technical Proceedings. CIEP 98 (Cat. No.98TH8375), Morelia, Mexico, 1998, pp. 126-131, doi: 10.1109/CIEP.1998.750672.

[6] Fesenko, Artem, Oleksandr Matiushkin, Oleksandr Husev, Dmitri Vinnikov, Ryszard Strzelecki, and Piotr Kołodziejek. "Design and Experimental Validation of a Single-Stage PV String Inverter with Optimal Number of Interleaved Buck-Boost Cells", 2021, Energies 14, no. 9: 2448. https://doi.org/10.3390/en14092448.

[7] O. Knecht, D. Bortis and J. W. Kolar, "ZVS Modulation Scheme for Reduced Complexity Clamp-Switch TCM DC–DC Boost Converter," in IEEE Transactions on Power Electronics, vol. 33, no. 5, pp. 4204-4214, May 2018, doi: 10.1109/TPEL.2017.2720729.

[8] J. Prager and P. Vinciarelli, "Loss and noise reduction in power converters", US Patent US 6,522,108, 18-Feb-2003.

[9] B. Ulrich, "Improved Clamp-Switch Boost Converter with Extended ZVS range," 2021 IEEE Applied Power Electronics Conference and Exposition (APEC), Phoenix, AZ, USA, 2021, pp. 1747-1754, doi: 10.1109/APEC42165.2021.9487229.

[10] B. Ulrich, " Analysis and Design of a Low-Complexity ZVS Buck-Boost Converter",2025 IEEE Applied Power Electronics Conference and Exposition (APEC), Atlanta, GA, USA, 2025, in press.

[11] P. L. Dowell, "Effects of eddy currents in transformer windings", Proc. Inst. Electr. Eng., vol. 113, no. 8, pp. 1387-1394, Aug. 1966.

[12] M. Albach, T. Durbaum and A. Brockmeyer, "Calculating core losses in transformers for arbitrary magnetizing currents a comparison of different approaches", PESC Record. 27th Annual IEEE Power Electronics Specialists Conference, Baveno, Italy, 1996, pp. 1463-1468 vol.2, doi: 10.1109/PESC.1996.548774.

Low-Complexity Model Predictive Control Method for Driving Dual Induction Motors Fed by Five-leg Inverter

Jun Young Lee
School of Electroncis and Electrical Engineering
Dankook University
South Korea
eepdp@dankook.ac.kr

Eun Woo Lee
School of Electroncis and Electrical Engineering
Dankook University
South Korea
LEW@dankook.ac.kr

Dongho Choi
School of Electroncis and Electrical Engineering
Dankook University
South Korea
dhchoi@dankook.ac.kr

June-Seok Lee
School of Electroncis and Electrical Engineering
Dankook University
South Korea
ljs@dankook.ac.kr

Abstract—This paper proposes a low-complexity model predictive control (L-MPC) method for a five-leg inverter (FLI) that independently drives dual three-phase induction motors (IMs). In the FLI system, a conventional MPC method considers all 32 candidate voltage vectors and selects the one with the lowest cost. This method causes a high computational burden and may limit the control period depending on the performance of the controller. However, low computation can be achieved using the L-MPC method considering only 4 candidate voltage vectors. In the L-MPC method, two IMs are prioritized, and the priority of both IMs alternates every control period. Priority indicates the optimization order. High-priority motor selects the voltage vector with the lowest cost among 3 effective vectors and 1 zero vector based on the sector where the reference voltage vector is located. Low-priority motor is automatically determined based on the previously selected voltage vector from high-priority motor. L-MPC method considers fewer candidate voltage vectors than the conventional method in the control period. Therefore, L-MPC method results in the high switching frequency by reducing the computation time, which results in reduction of current THD. The validity of the proposed method is verified by simulations and experimental results.

Keywords—Five-leg inverter(FLI), Model predictive control(MPC), Induction motor(IM), computation, current THD.

I. INTRODUCTION

Over the past decade, AC motors have been utilized in a variety of industrial fields. Generally, one three-leg inverter is used to operate one AC motor in motor drive systems. As a result, driving two motors required two three-leg inverter. In dual motor drive systems, several studies have been developed to reduce the number of switches [1]-[3].

Five-leg inverter (FLI) is one of several topologies for dual-motor drive systems. The FLI is composed of only 10 switches to operate dual-motor, which has advantages over the system that uses two three-leg inverters in terms of size, capital cost, and switching losses. Fig. 1 shows the circuit configuration of the FLI with two identical induction motors (IMs). In the FLI which consists of five legs of two switches, four legs (*leg A, leg B leg E leg D*) are connected to phases *a* and *b* of each IM, and one leg (*leg C*) is connected in common to phase *c* of both IMs.

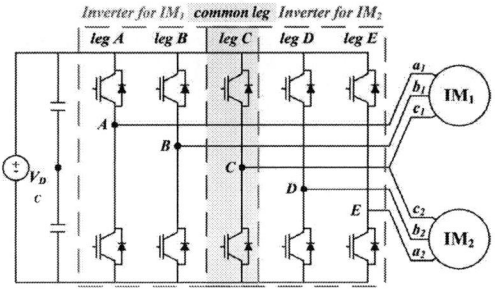

Fig. 1. Circuit of Five-Leg Inverter.

One of the FLI control methods [4]-[10], model predictive control (MPC) method has the advantage of fast dynamic characteristics, but the disadvantage is that it requires a lot of computation. The conventional method is the full-set MPC (FMPC) method, which considers all candidate voltage vectors in selecting a voltage vector of the lowest cost. Considering switches in each leg operate complementarily, the possible number of switching states is 32. In other words, 32 candidate voltage vectors are considered in the FMPC method, which has the significant drawback of a long computation time.

This paper proposes the low-complexity MPC (L-MPC) method to achieve low computational cost. L-MPC is performed by the same mechanism as the FMPC, but only considers a smaller number of candidate voltage vectors. In the L-MPC method, two IMs are prioritized to determine the optimization order. The priority of IMs changes alternately in every control period. First, a high-priority motor considers 4 candidate voltage vectors in selecting a voltage vector with the lowest cost during a control period. The 4 candidate voltage vectors consist of 3 effective vectors based on the sector where the reference voltage is located and the rotation direction of the reference voltage, and 1 zero vector. After that, the low-priority motor is automatically determined to zero vector depending on the previously selected voltage vector in the high-priority motor. L-MPC considers fewer candidate voltage vectors than the FMPC in the control period. Therefore, the proposed method achieves the high switching frequency by reducing the computation time, which results in reduction of current THD. The validity of the proposed method is demonstrated by simulations and experimental results.

979-8-3315-1612-3/25 $31.00 © 2025 IEEE

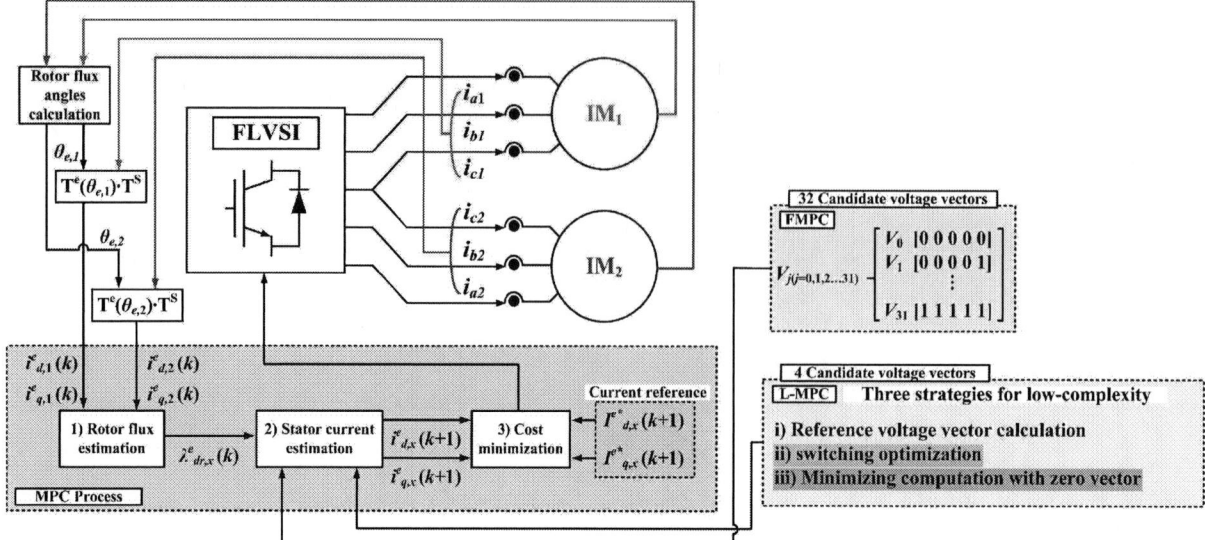

Fig. 2. Control block diagram of the MPC process

II. CONVENTIONAL MPC METHOD

Fig. 2 shows a control block diagram of the MPC process. Where θ_e is the synchronous angle, k is the k_{th} sampling value in the discrete domain, $i^e_{d,x}$ ($x = 1, 2$) and $i^e_{q,x}$ ($x = 1, 2$) are d-q axis stator currents, $\lambda^e_{dr,x}$ ($x = 1, 2$) is the d-axis rotor flux, superscript '*' and 'e' are the reference and the synchronous reference frame, respectively, and subscript 'x' is IM$_1$ and IM$_2$. The MPC process consists of the following three steps.

1) Rotor flux estimation

In step 1), rotor flux estimation is peformed. As rotor flux oriented control is applied, $\lambda^e_{dr,x}$ is needed in the stator current estimation formula and is calculated as

$$\lambda^e_{dr,x}(k) = \frac{L_{r,x}\lambda^e_{dr,x}(k-1) + L_{m,x}T_s R_{r,x}i^e_{d,x}(k)}{L_{r,x} + T_s R_{r,x}}, (x = 1,\ 2), \quad (1)$$

where $R_{s,x}$, $L_{s,x}$, $R_{r,x}$, $L_{r,x}$, $L_{m,x}$, and T_s represent the stator resistance and inductance, rotor resistance and inductance, mutual inductance, and control period, respectively.

2) Stator current estimation

It is the process of estimating the value of the $(k+1)_{th}$ stator current along a voltage vector for motor drive control. From candidate voltage vectors, the $i^e_{d,x}(k+1)$, $i^e_{q,x}(k+1)$ are calculated by following (2).

$$i^e_{d,x}(k+1) = T_s K_1 \Big[\big\{ T_s^{-1}K_1^{-1} - K_2 \big\} i^e_{d,x}(k)$$
$$+ \omega_{e,x} K_1^{-1} i^e_{q,x}(k) + L_{m,x} R_{r,x} \lambda^e_{dr,x}(k) + L^2_{r,x} v^e_{d,x}(k) \Big]$$
$$i^e_{q,x}(k+1) = T_s K_1 \Big[\big\{ T_s^{-1}K_1^{-1} - T_s K_1 K_2 \big\} i^e_{q,x}(k) \quad (2)$$
$$- \omega_{e,x} K_1^{-1} i^e_{d,x}(k) - L_{m,x} L_{r,x} \omega_{r,x} \lambda^e_{dr,x}(k) + L^2_{r,x} v^e_{q,x}(k) \Big],$$

$$\left(\begin{array}{l} x = (1,\ 2), \\[2mm] K_1 = \dfrac{1}{(L_{s,x}L^2_{r,x} - L_{r,x}L^2_{m,x})}, \\[3mm] K_2 = R_{s,x}L^2_{r,x} + R_{r,x}L^2_{m,x} \end{array} \right),$$

where K_1 and K_2 are the estimation factor, ω_e, ω_r, $v^e_{d,x}$, and $v^e_{q,x}$ are synchronous angular speed, rotor angular speed, and d-q axis voltage vector, respectively.

3) Cost minimization

This process is to select the optimal voltage vector using a cost function to minimize current ripple.

$$G_{c,2} = w_2 \left[\begin{array}{l} \big\{ i^{e*}_{d,2}(k+1) - i^e_{d,2}(k+1) \big\}^2 \\[2mm] + \big\{ i^{e*}_{q,2}(k+1) - i^e_{q,2}(k+1) \big\}^2 \end{array} \right]$$

$$G_{c,2} = w_2 \left[\begin{array}{l} \big\{ i^{e*}_{d,2}(k+1) - i^e_{d,2}(k+1) \big\}^2 \\[2mm] + \big\{ i^{e*}_{q,2}(k+1) - i^e_{q,2}(k+1) \big\}^2 \end{array} \right] \quad (3)$$

$$\Rightarrow \min \big\{ G_{c,1} + G_{c,2} \big\},$$

where $G_{c,x}$, w_x are the cost function and weighting factors. The $G_{c,x}$ means the error between the current estimate and the reference current. Through the step 3), the V_j with the minimum $G_{c,x}$ is selected.

As shown in Fig. 2, all possible voltage vectors are considered as candidate voltage vectors in the FMPC. The FLI with two switches on each leg that operate complementarily has 32 candidate voltage vectors. During each T_s, a candidate voltage vector is selected that minimizes the $G_{c,x}$ of IM1 and IM2. As a result, the FMPC, which considers all candidate voltage vectors, reduces the size in terms of current ripple, but is limited in the switching frequency due to high computation.

III. PROPOSED L-MPC METHOD

The L-MPC process in Fig. 2 is used equally in the proposed L-MPC method. However, in the L-MPC method, 2) step stator current estimation is performed by applying three strategies.

i) Reference voltage vector calculation

In the motor drive systems, $i^{e*}_{d,x}$, $i^{e*}_{q,x}$ are used to control the motor. The current references can be calculated as $v^{e*}_{d,x}$, $v^{e*}_{q,x}$ using the induction motor's voltage equation.

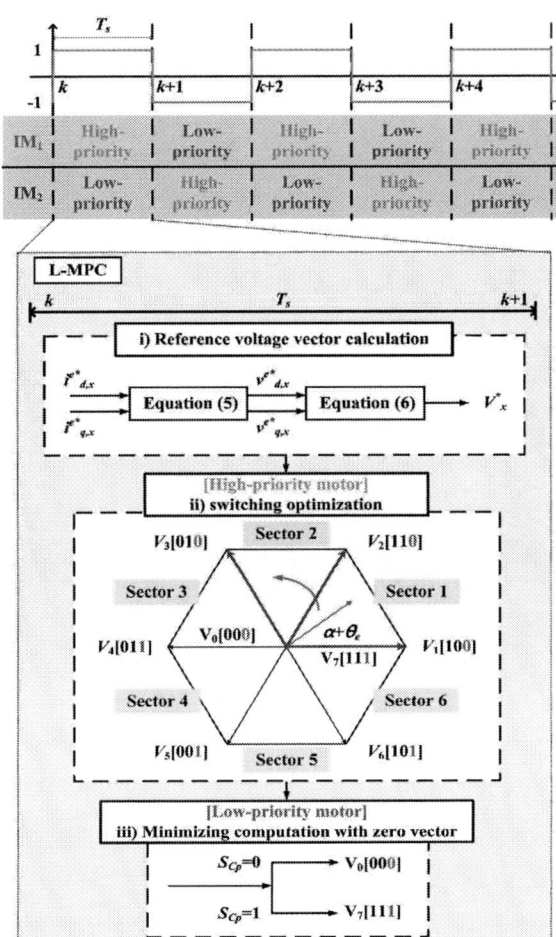

Fig. 3. Mechanism of the L-MPC Method

$$v_{d,x}^{e*} = R_{s,x}i_{d,x}^{e*} + p\left(\sigma L_{s,x}i_{d,x}^{e*} + \frac{L_{m,x}}{L_{r,x}}\lambda_{dr,x}^{e}\right) - \omega_{e,x}\sigma L_{s,x}i_{q,x}^{e*},$$

$$v_{q,x}^{e*} = R_{s,x}i_{q,x}^{e*} + p\sigma L_{s,x}i_{q,x}^{e*} + \omega_{e,x}\left(\sigma L_{s,x}i_{d,x}^{e*} + \frac{L_{m,x}}{L_{r,x}}\lambda_{dr,x}^{e}\right), \quad (4)$$

$$\left(x=1,\ 2,\ \sigma=1-\left\{L_{m,x}^2 / (L_{s,x}\cdot L_{r,x})\right\}, \lambda_{dr,x}^{e}=L_{m,x}\cdot i_{d,x}^{e*}\right),$$

where p and σ represent the differential operator and the leakage coefficient. If the rotor flux is expressed as (5) in the synchronous coordinate system and assuming no variation in current and flux in steady state, (4) can be expressed as

$$v_{d,x}^{e*} = R_{s,x}i_{d,x}^{e*} - \omega_{e,x}\sigma L_{s,x}i_{q,x}^{e*},$$

$$v_{q,x}^{e*} = R_{s,x}i_{q,x}^{e*} + \omega_{e,x}\left(\sigma L_{s,x} + \frac{L_{m,x}^2}{L_{r,x}}\right)i_{d,x}^{e*}. \quad (5)$$

From $v_{d,x}^{e*}$, $v_{q,x}^{e*}$, (x=1, 2), V_x^* can be expressed as

$$V_x^* = \sqrt{\left(v_{d,x}^{e*}\right)^2 + \left(v_{q,x}^{e*}\right)^2}\cos\left(\alpha+\theta_{e,x}\right),$$

$$\left(x=1,\ 2\ , \alpha=\cos^{-1}\left(v_{d,x}^{e*} / \sqrt{\left(v_{d,x}^{e*}\right)^2 + \left(v_{q,x}^{e*}\right)^2}\right)\right), \quad (6)$$

where α represents reference power factor angle.

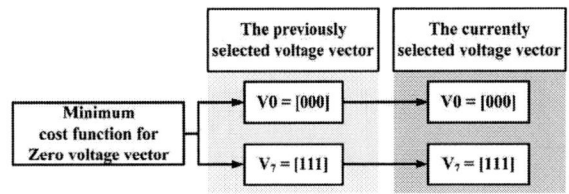

Fig. 4. Zero vector selection in the L-MPC Method

ii) Switching optimization

Before optimization with candidate voltage vectors takes place, the priority of the two IMs must be determined. The prioritization(f_{order}) of the IMs is conducted as shown in (7) below.

$$f_{order} = \begin{cases} 1, & \therefore \text{High-priority} = \text{IM}_1, \text{Low-priority} = \text{IM}_2 \\ -1, & \therefore \text{Low-priority} = \text{IM}_1, \text{HIgh-priority} = \text{IM}_2 \end{cases}. \quad (7)$$

As expressed in (7), f_{order} is the priority of the IMs whose value is switched sequentially every T_s. The high-priority motor is optimized first, and in the next T_s, the low-priority motor is promoted to high-priority and optimized accordingly. If IM1 is the high-priority motor with the f_{order} of 1 during the k_{th} T_s, then during the $(k+1)_{th}$ T_s, IM$_1$ becomes the low-priority motor with the f_{order} of -1 and IM$_2$ becomes the high-priority motor with the f_{order} of 1.

In a voltage vector diagram divided into 6 sectors as shown in Fig. 3, the candidate voltage vectors of the high-priority motor are determined by a sector according to the $(\alpha+\theta_e)$ during T_s. Depending on the sector where V_x^* is located and the rotation direction of V_x^*, the 3 effective voltage vectors (V_1, $V_2 \dots V_6$), and 1 zero voltage vector (V_0, V_7) are considered as the candidate voltage vector. Among the 4 candidate voltage vectors, the voltage vector with the minimum $G_{c,x}$ is selected through equation (3). If the $G_{c,x}$ of the zero vector is minimal as shown in Fig. 4, the one with the least switching among V_0, V_7 is selected according to the previous switching state.

$$V_x^* \begin{cases} if\ (0° \le \alpha+\theta_{e,x} < 60°) \to \text{Sector 1,} \\ if\ (60° \le \alpha+\theta_{e,x} < 120°) \to \text{Sector 2,} \\ if\ (120° \le \alpha+\theta_{e,x} < 180°) \to \text{Sector 3,} \\ if\ (180° \le \alpha+\theta_{e,x} < 240°) \to \text{Sector 4,} \\ if\ (240° \le \alpha+\theta_{e,x} < 300°) \to \text{Sector 5,} \\ if\ (300° \le \alpha+\theta_{e,x} < 360°) \to \text{Sector 6.} \end{cases} \quad (8)$$

(8) shows the corresponding sector based on the angle at which V_x^* is located. For example, if the angle of V_x^* is 30°, it will be located in Sector 1. In the case of counterclockwise rotation, the effective vector (V_1, V_2, V_3) adjacent to V_x^* and the zero vector (V_0, and V_7) are considered.

iii) Minimizing computation with zero vectors.

Since the *leg C* is connected to both IMs in the FLI, the switching state (S_{Cp}) of *leg C* is determined by the voltage vector selected during high-priority motor optimization. In other words, the voltage vector of the low-priority motor is automatically selected as either V_0 or V_7 depending on the S_{Cp}. For example, high-priority motor is IM$_1$ and the low-priority motor is IM$_2$. If S_{Cp} of voltage vector selected in high-priority

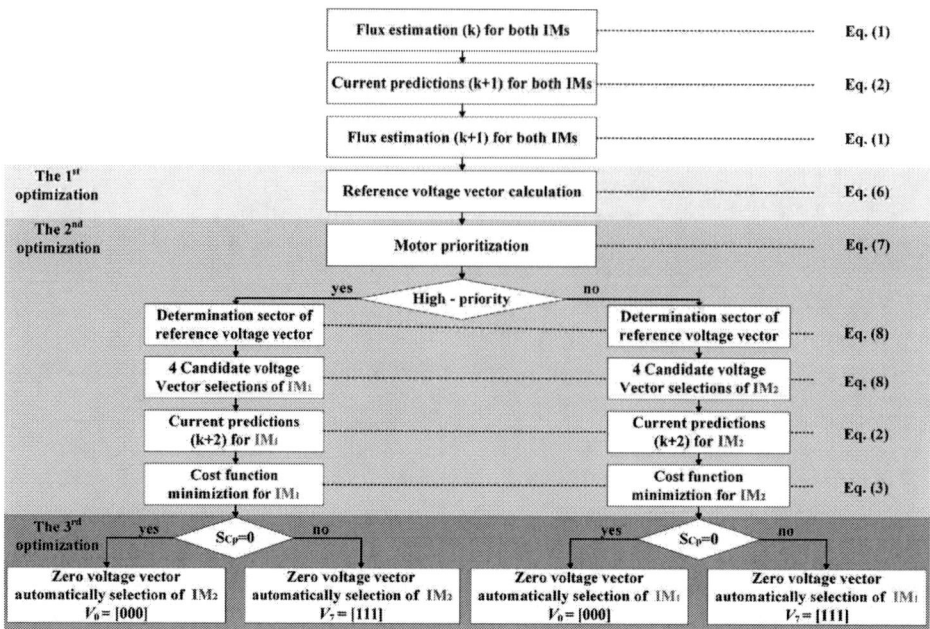

Fig. 5. Flow chart of L-MPC Method for implementation in real systems

TABLE I. COMPUTATION COS COMPARISON OF MPC METHODS

Parameters	MPC Methods	
	FMPC	L-MPC
Number of cost functions	32	4
Number of Stator current estimations	128	16

TABLE II. SIMULATION PARAMETERS

Parameters	Mark	Value	Unit
Stator resistance	$R_{s,x}$	1.96	Ω
Rotor resistance	$R_{r,x}$	2.74	Ω
Mutual inductance	$L_{m,x}$	210	mH
Stator inductance	$L_{s,x}$	221	mH
Rotor inductance	$L_{r,x}$	221	mH
Pole	P	4	-
Control period	T_s	250 / 100	μs
DC-Link voltage	V_{dc}	250	V
Rated current	i	4.6	A

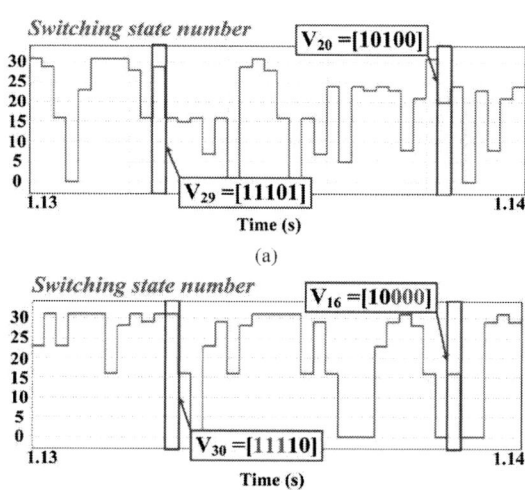

Fig. 6. Simulation results of switching state. (a) FMPC (b) L-MPC

motor is 0, then V_0 is selected for voltage vector of the low-priority motor. If S_{Cp} of voltage vector selected in high-priority motor is 1, then V_7 is selected for voltage vector of the low-priority motor.

Due to the aforementioned three strategies, the FMPC method has 32 cost functions and 128 stator current estimations, while the L-MPC method has 4 cost functions and 16 stator current estimations, resulting in 8 times less computation compared to the FMPC method as shown in Table I. The FMPC causes high computation. In contrast, the L-MPC has low computation. Therefore, compared to the FMPC, a higher switching can be achieved, which implies an improvement in current THD.

Fig. 5 shows the flowchart for implementing the L-MPC method. This flowchart aligns with the L-MPC explanations provided in the earlier subsections. However, in practical systems, due to the inability to perform measurement, computation, and execution simultaneously, stator currents

must be predicted two steps in advance, as described. Consequently, an additional step for rotor flux estimation and stator current prediction is incorporated at the beginning of the flowchart.

IV. SIMULATION RESULTS

The proposed L-MPC method was verified by simulation using PSIM and experimental results. The simulation parameters are listed in Table II, and both IM_s utilize the same motor. Additionally, IM_1 and IM_2 were operated at a speed of 150 rpm. In this section, first, the basic operating principle of the L-MPC method is explained using simulation results. Second, the current ripple characteristics of the L-MPC method are compared with the conventional FMPC method. Finally, it is demonstrated that increasing the switching frequency reduces the current THD.

Fig. 6 shows the switching states of FMPC and L-MPC. In the FMPC, it can be observed that one of the V_j states is

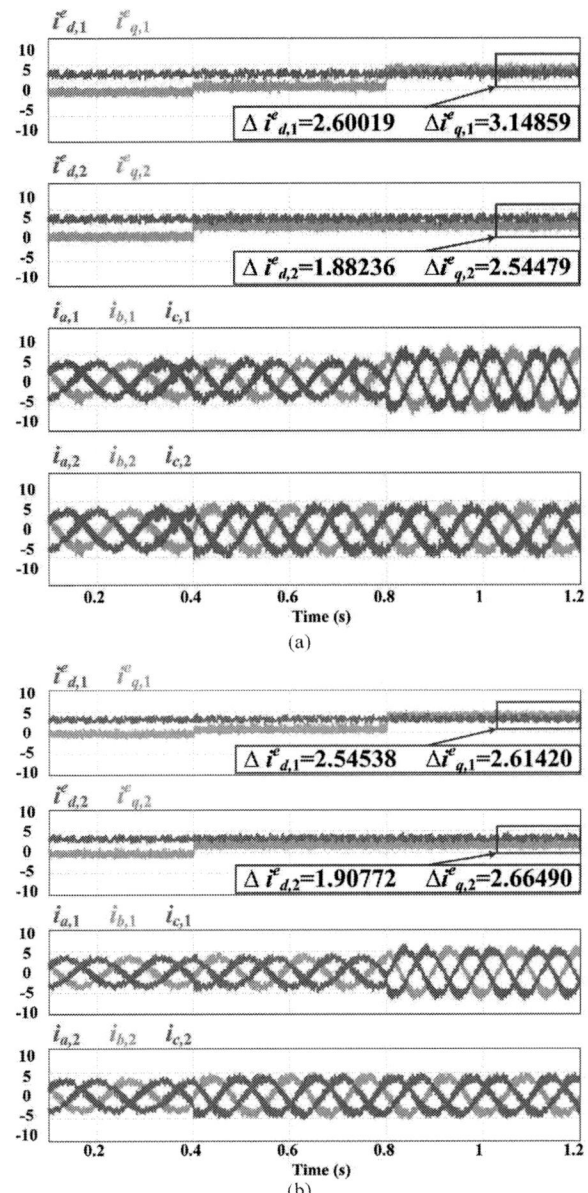

Fig. 7. Simulation results of *d-q* axis stator currents at $T_s = 250\mu s$
(a) FMPC (b) L-MPC

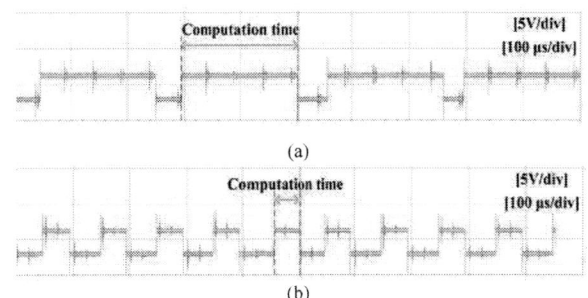

Fig. 8. Experimental results of computation time
(a) FMPC (b) L-MPC

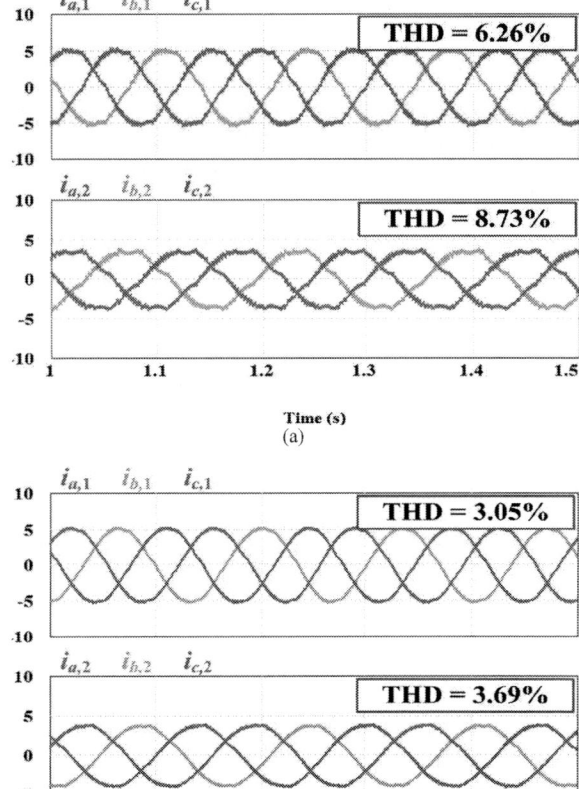

Fig. 9. Simulation results of *3-phase* currents
(a) $T_s = 250 \mu s$ (b) $T_s = 100 \mu s$

selected. In contrast, in L-MPC, a zero vector is selected for IM with low-priority, resulting in the switching states being determined based on S_{Cp}, as illustrated in Fig. 6.

Fig. 7 shows the simulation results of $i^e_{d,x}$ and $i^e_{q,x}$ for the FMPC and the L-MPC under identical $i^e_{d,x}$ and $i^e_{q,x}$ conditions. In this scenario, $i^{e*}_{d,1}$ and $i^{e*}_{d,2}$ are maintained at the rated value of 3.31 [A], while $i^{e*}_{q,1}$ changes from 0 [A] to 1 [A] and 4 [A], and $i^{e*}_{q,2}$ changes from 0 [A] to 2 [A]. The steady-state current ripple ($\Delta i^e_{d,x}$, $\Delta i^e_{q,x}$) for each condition is also presented. As shown in Fig. 7, under steady-state conditions where $i^e_{q,1}$ is 4 [A] and $i^e_{q,2}$ is 2 [A], the difference in $\Delta i^e_{d,x}$ and $\Delta i^e_{q,x}$ between the two MPC methods is a maximum of 0.05 for $i^e_{d,x}$ and a maximum of 0.5 for $i^e_{q,x}$. This demonstrates that the steady-state ripple performance of the L-MPC is comparable to that of the FMPC.

Fig. 8 shows the computation signal for each MPC method. TMS320F28355 was used as the processor to implement the proposed method. The computation time is the time to process the whole algorithm. The overall computation time for the FMPC method takes 203us on average, while L-MPC method takes 46us. As a result, the L-MPC method can use a higher switching frequency than the FMPC method.

Fig. 9 shows the three-phase stator currents of IM$_1$ and IM$_2$. The simulation conditions are identical to those in Fig. 7, with $i^{e*}_{d,1}$ and $i^{e*}_{d,2}$ set to 3.31 A, $i^{e*}_{q,1}$ to 4 A, and $i^{e*}_{q,2}$ to 2

A, representing steady-state currents. As shown in Fig. 8, the computation time of L-MPC is 46us, allowing the switching frequency to be increased to 10kHz. As the switching frequency increases, a reduction in THD can be observed. Specifically, the THD for IM_1 is reduced by approximately 51%, while the THD for IM_2 is reduced by about 58%.

V. CONCLUSION

This paper proposes the L-MPC method to drive dual induction motors using FLI. The conventional FMPC method for FLI has difficulty in using high switching frequency due to the large amount of computation. However, the L-MPC considers fewer candidate voltage vectors, resulting in lower computation. In the proposed method, 3 effective voltage vectors and 1 zero voltage vector are considered as the candidate voltage vector of high-priority motor based on the sector where the reference voltage vector is located. The switching state of the common leg determines the voltage vector of the low-priority motor. For this property, the L-MPC method has less computation, unlike the FMPC method which considers 32 candidate voltage vectors. Compared to the FMPC method, the L-MPC method has a similar current ripple in steady-state, 87.5% less amount of computation, and about 77% less computation time. As a result, the L-MPC method can use a high switching frequency due to a small amount of computation, which can reduce the current THD. The current THD for IM_1 is reduced by approximately 51%, while the current THD for IM_2 is reduced by about 58%.

ACKNOWLEDGMENT

This work is supported by the Korea Agency for Infrastructure Technology Advancement (KAIA) grant funded by the Ministry of Land, Infrastructure and Transport (Grant RS-2024-00417481).

REFERENCES

[1] E. Ledezma, B. McGrath, A. Munoz and T. A. Lipo, "Dual AC-drive system with a reduced switch count," *IEEE Trans. Ind. Appl.*, vol. 37, no. 5, pp. 1325-1333, Sept.-Oct. 2001.

[2] Y. Song, J. Sun, Y. Zhou, Y. Liu, H. Luo and J. Zhao, "Minimization of Capacitor Voltage Difference for Four-Leg Inverter Dual-Parallel IM System," *IEEE Trans. Power Electron.*, vol. 37, no. 4, pp. 3969-3979, April 2022.

[3] Y. Lee and J. -I. Ha, "Control method for mono inverter dual parallel interior permanent magnet synchronous machine drive system," *2015 IEEE Energy Conversion Congress and Exposition (ECCE)*, Montreal, QC, Canada, 2015, pp. 5256-5262.

[4] M. Jones, S. N. Vukosavic, D. Dujic, E. Levi, and P. Wright, "Five-leg inverter PWM technique for reduced switch count two-motor constant power applications," *IET Electric Power Appl.*, vol. 2, no. 5, pp. 275-287, Sep. 2008.

[5] Y. -S. Lim, J. -S. Lee and K. -B. Lee, "Advanced Speed Control for a Five-Leg Inverter Driving a Dual-Induction Motor System," *IEEE Trans. Ind. Electron.*, vol. 66, no. 1, pp. 707-716, Jan. 2019.

[6] E. Lee, J. -H. Lee, D. -H. Kim, D. Choi and J. -S. Lee, "An Extended DPWM Method for Switching Loss Reduction With Dual Induction Motors Fed by Five-Leg Inverter," *IEEE Trans. Ind. Appl.*

[7] Y. Mei, Z. Yi, and Z. Li, "A Model Predictive Control Strategy for Dual Induction Motor Drive System Fed by Five-leg Inverter," in *Proc. International Conference on Electrical Machines and Systems (ICEMS)*, Oct. 2014, pp. 2959–2963.

[8] C. S. Lim, N. A. Rahim, W. P. Hew, and E. Levi, "Model predictive control of a two-motor drive with five-leg inverter supply," *IEEE Trans. Ind. Electron.*, vol. 60, no. 1, pp. 54-65, Jan. 2013.

[9] C. Lim, E. Levi, M. Jones, N. A. Rahim, and W. Hew, "A Comparative Study of Synchronous Current Control Schemes Based on FCS-MPC and PI-PWM for a Two-Motor Three-Phase Drive," *IEEE Trans. Ind. Electron.*, vol. 61, no. 8, pp. 3867-3878, Aug. 2014.

[10] D. Choi, J. -S. Lee, Y. -S. Lim and K. -B. Lee, "Priority-Based Model Predictive Control Method for Driving Dual Induction Motors Fed by Five-Leg Inverter," *IEEE Trans. Power Electron.*, vol. 38, no. 1, pp. 887-900, Jan. 2023.

Overvoltage Mitigation Filter Using High-Frequency Cable Modeling in Long Transmission Lines for Silicon Carbide Inverter Systems

Yun-Jin Lee
Department of Electrical and Computer Engineering
Ajou University
Suwon, South Korea
lhgg0614@ajou.ac.kr

Kyo-Beum Lee
Department of Electrical and Computer Engineering
Ajou University
Suwon, South Korea
kyl@ajou.ac.kr

Abstract—This paper analyzes a design of overvoltage mitigation filter using high-frequency cable modeling in long transmission lines for silicon carbide inverter systems. The SiC inverter systems operating with long transmission lines may experience overvoltage caused by voltage reflection, leading to insulation breakdown. The proposed method to address the problem of overvoltage occurring on long transmission lines involves connecting a RLC filter at the end of the cable to reduce the overvoltage. The RLC filter is calculated by deriving the characteristic impedance using the parameters of the cable modeling. The validity of cable modeling and designing overvoltage mitigation filter is verified through simulation results.

Keywords—SiC MOSFET, filter, cable modeling, overvoltage, reflected voltage

I. INTRODUCTION

The SiC inverter systems are widely used to increase global energy consumption and the demand for power efficiency. The SiC inverter systems have been applied in various fields such as power converter systems, battery charging systems for electric vehicles, and renewable energy systems [1]–[2]. The SiC-based inverters have enabled high-frequency switching operation in inverter-cable-high impedance load systems. However, voltage reflection occurs if the impedances of the inverter, long transmission line, and load are mismatched [3]. Overvoltage greater than the inverter output voltage is generated when the voltage reflection occurs [4]–[5]. The overvoltage increases stress on the loads such as induction motor insulation and the overcurrent affects the power devices.

The causes of voltage reflection have been analyzed using transmission theory to mitigate the problems caused by overvoltage [6]–[7]. In addition to impedance mismatches, overvoltage also occurs due to the fast rise time of the SiC inverters. Various hardware methods are being studied to suppress overvoltage by connecting filter to the inverter output side or the load input side [8]–[10]. The proposed filter suppresses overvoltage by connecting a filter to the inverter output side to increase the rise time [11]. Characteristic impedance may be derived through high-frequency cable modeling to design an optimal filter. The parameters for high-frequency cable modeling are measured using the impedance analyzer.

This paper investigates the design of overvoltage mitigation filter using high-frequency cable modeling in long transmission lines for silicon carbide inverter systems. The method for reducing overvoltage using filter involves connecting the filter to the output side of the SiC inverter systems to increase the rise time. The method for designing filter parameters may be determined by deriving the characteristic impedance through cable modeling. The validity of the proposed method is demonstrated by the simulation.

II. HIGH-FREQUENCY CABLE MODELING CONSIDERING LONG TRANSMISSION LINES

The voltage reflection due to long transmission lines in the SiC inverter systems causes overvoltage, leading to insulation breakdown. The voltage reflection is analyzed through cable modeling that considers high-frequency in long transmission lines. High-frequency cable modeling allows for more accurate modeling by considering dielectric loss, proximity effect, and skin effect. This is the first reason why high-frequency cable modeling for occurring overvoltage due to the voltage reflection is used in this paper.

A. High-frequency cable modling

Accurate modeling of power cable is essential for designing filters to mitigate overvoltage caused by voltage reflections. The conventional second-order section model is commonly used to represent cable as shown in Fig. 1. The advantage of the conventional cable model is simplicity. However, the disadvantage of the conventional cable model is that it doesn't represent frequency-dependent phenomena such as dielectric, proximity effect, and skin effect. The parameters of the traditional model are challenging to determine precisely because the resonant frequency of the cable varies based on its properties and length.

As shown in Fig. 2, high-frequency cable modeling allows for more accurate modeling, unlike the conventional simple model. Futher improvements are proposed to include dielectric losses, proximity effects, and skin dffects, which are not considered in models using a first-oreder series branch. The parameters of the high-frequency cable model in Fig. 2 are identified from the cable SC and OC impedances Z_{SC} and Z_{OC}. In the parallel branch, R_{p2} and C_{p2} represent dielectric losses, while in the series branch, L_{s2} and R_{s2} represent proximity and skin effects. The parameters in Table I are values measured directly using an impedance analyzer and can be derived through the calculation method of Eq. (1) and Eq. (2). Based on SC results, the inductance Eq. (1) is determined from the measurements in low and high-frequency regions as follows [12]:

$$
\begin{cases}
L_{s1} = L_{SC1\_HF} \\
L_{s2} = \dfrac{L_{SC2\_LF} - L_{SC2\_HF}}{2} \\
R_{s1} = \left| Z_{SC1\_LF} \right| \cos(\theta_{SC1\_LF}) \\
R_{s2} = \dfrac{R_{SC2\_HF} - R_{SC2\_LF}}{2}
\end{cases}
\tag{1}
$$

Similarly, based on OC results, the capacitance Eq. (2) is determined from the measurements as follows:

$$
\begin{cases}
C_{p1} = C_{OC\_HF} \\
C_{p2} = C_{OC\_LF} - C_{p1} \\
R_{p1} = \left| Z_{OC\_LF} \right| \left[\cos(\theta_{SC\_LF}) \right]^{-1}, \\
R_{p2} = \left[\left(R_{p1//p2} \right)^{-1} - \left(R_{p1} \right)^{-1} \right]^{-1}
\end{cases}
\tag{2}
$$

where L_{s1} is series inductance, L_{s2} is parallel inductance in series, R_{s1} is series resistance, R_{s2} is parallel resistance in series, C_{p1} is parallel capacitance, C_{p2} is series capacitance in parallel, R_{p1} is parallel resistance, R_{p2} is series resistance in parallel.

B. Voltage reflection theory for overvoltage

Using high-frequency cable modeling, Fig. 3 shows the cable connected between the inverter and the load. The parameters of the proposed cable model in Fig. 3 are determined based on the short-circuit (SC) impedance Z_{SC} and open-circuit (OC) Z_{OC} of the cable. The impedance Z and admittance Y of the long transmission lines with high-frequency cable modeling are given by Eq. (3) and (4), respectively.

$$
Z = R_{s1} + j\omega L_{s1} + (R_{s2} // j\omega L_{s2}),
\tag{3}
$$

$$
Y = R_{p1} // \frac{1}{j\omega C_{p1}} // (R_{p2} + \frac{1}{j\omega C_{p2}}).
\tag{4}
$$

The propagation delay constant γ of the transmission line is calculated through the square root of the product of the cable series impedance Z and parallel admittance Y. This relationship determines the speed at which signals propagate through the line. The expression for γ is given in Eq. (5).

$$
\gamma = \sqrt{ \{ R_{s1} + j\omega L_{s1} + (R_{s2} // j\omega L_{s2}) \} \times \{ R_{p1} // \frac{1}{j\omega C_{p1}} // (R_{p2} + \frac{1}{j\omega C_{p2}}) \} }.
\tag{5}
$$

The cable characteristic impedance Z_0 is given by the ratio

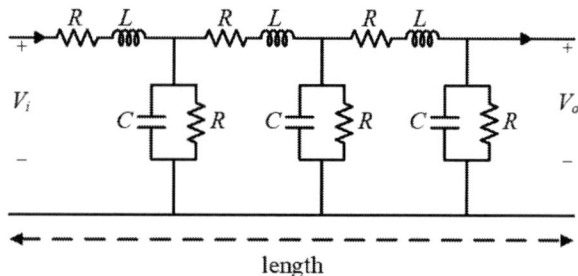

Fig. 1. Simplified long transmission line model.

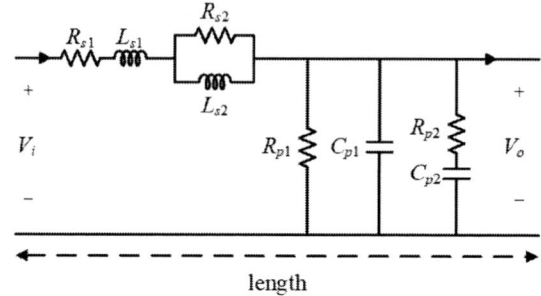

Fig. 2. High-frequency cable model.

TABLE I. CABLE PARAMETERS

Parameter	Value
L_{s1}	332 [nH]
L_{s2}	36.6 [nH]
R_{s1}	17.5 [mΩ]
R_{s2}	650 [mΩ]
C_{p1}	43.8 [pF]
C_{p2}	12.3 [pF]
R_{p1}	47.9 [MΩ]
R_{p2}	55.9 [kΩ]

of the series impedance to the propagation delay constant, as shown in Eq. (6).

$$
Z_0 = \sqrt{ \frac{ R_{s1} + j\omega L_{s1} + (R_{s2} // j\omega L_{s2}) }{ (R_{p1} // \frac{1}{j\omega C_{p1}}) // (R_{p2} + \frac{1}{j\omega C_{p2}}) } }.
\tag{6}
$$

Based on the method of measuring cable impedance considering high frequencies, the characteristic impedance is expressed as shown in (7). Assuming no resistance

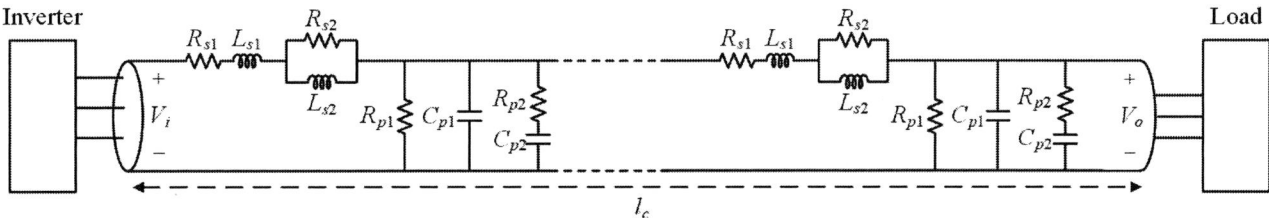

Fig. 3. Inverter-high-frequency cable-load circuit.

corresponding to losses, the final impedance Z_0 is expressed as shown in Eq. (8).

$$Z_0 = \sqrt{\frac{R_{s1} + j\omega L_{s1} + (R_{s2} // j\omega L_{s2})}{(R_{p1} // \frac{1}{j\omega C_{p1}}) // (R_{p2} + \frac{1}{j\omega C_{p2}})}}$$

$$\approx \sqrt{\frac{R_{s1} + R_{s2} + j\omega L_{s1}}{R_{p1} // R_{p2} // \frac{1}{j\omega C_{p2}}}}$$

$$\approx \sqrt{\frac{j\omega L_{s1}}{j\omega C_{p1}}}, \tag{7}$$

$$\therefore Z_0 = \sqrt{\frac{L_{s1}}{C_{p1}}}. \tag{8}$$

The voltage reflection occurring at the load end due to impedance mismatch between the cable impedance and the load impedance moves towards the inverter output side. However, the voltage reflection occurs again at the inverter side, producing a forward voltage towards the load end Eq. (9) and (10) represent the reflection coefficients at the inverter end and the load end.

$$\Gamma_L = \frac{Z_L - Z_0}{Z_L + Z_0}, \tag{9}$$

$$\Gamma_S = \frac{Z_S - Z_0}{Z_S + Z_0}, \tag{10}$$

where Z_L and Z_S are the load impedance and the inverter impedance, respectively. Eq. (11) and Eq. (12) represent the time for the inverter output voltage wave to travel from the inverter output to the load terminal.

$$t_p = \frac{l_c}{v}, \tag{11}$$

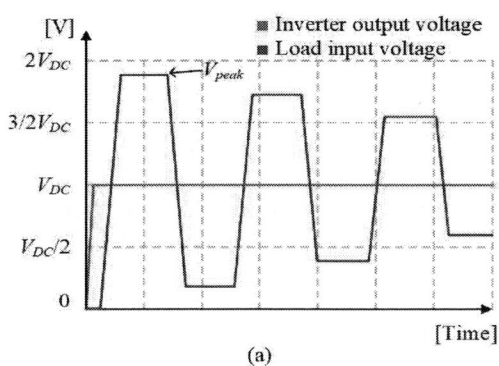

(a)

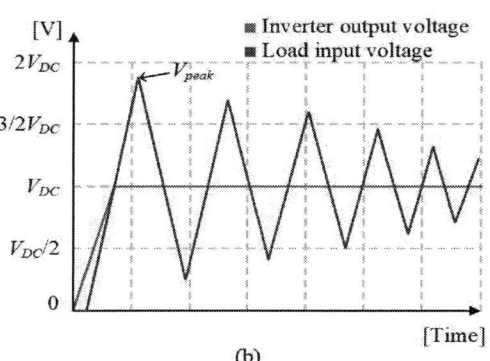

(b)

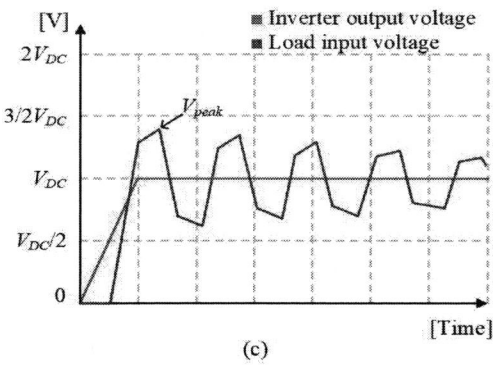

(c)

Fig. 4. Overvoltage of load input voltage depending on the rise time of inverter output voltage: (a) t_p=0.33μs and rise time=0.1μs, (b) rise time=$2t_p$, and (c) rise time=$3t_p$.

$$v = \frac{1}{\sqrt{L_{s1}C_{p1}}}, \qquad (12)$$

where v is pulse velocity, l_c is cable length, L_{s1}, C_{p1} and t_p are cable inductance per unit length, cable capacitance per unit length and cable propagation delay, respectively.

III. DESIGN OF OVERVOLTAGE MITIGATION FILTER

The voltage reflection in the inverter-cable-load system varies based on the rise time and propagation delay time. The rise time that is less than twice the propagation delay time results in the load voltage magnitude nearly twice that of the DC voltage. There is several method to suppress overvoltage on the load side due to the inverter output. One such method involves connecting a RLC filter to the inverter output to reduce the voltage rise time.

A. Effect of overvoltage rise time

The inverter and the load are connected through several meters of cable, which determines the propagation delay time. The output impedance of the inverter is negligible compared to the characteristic impedance of the cable. Using Eq. (9) and (10), the inverter reflection coefficient is given by $K_S=-1$, and the load reflection coefficient is given by $K_L=1$. Fig. 4 shows the overvoltage occurring at the load input due to the rise time of the inverter output voltage. Using a typical SiC inverter rise time of 0.1 μs, which is less than the propagation time of the cable, the peak voltage can be predicted as shown in Eq. (13).

$$V_{peak} = (1+K_L) \cdot V_{dc} = 2V_{dc}. \qquad (13)$$

B. High-frequency cable modeling parameters for filter design

A filter is used to suppress overvoltage, placing the inductor directly at the inverter output and connecting the resistor and capacitor in parallel with the cable. Fig. 5 illustrates the dv/dt suppressing second-order RLC filter. The RLC filter is designed to have a rise time corresponding to three times the propagation delay, and the quality factor is selected to be 0.5. The filter output voltage is given by Eq. (14).

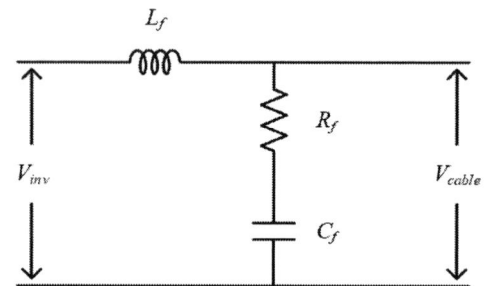

Fig. 5. Overvoltage suppressing second-order RLC filter topology.

TABLE II. SIMULATION PARAMETERS

Parameter	Value
DC-link voltage	300 [V]
RL load	10 [Ω], 2 [mH]
Resistor filter	87.06 [Ω]
Inductance filter	26.22 [μH]
Capacitance filter	13.84 [nF]
Resonant frequency	1.66 [MHz]

$$V_o(t) = V_{dc} \cdot (1 - e^{-\omega_o t} + \omega_o t e^{-\omega_o t}) \cdot u(t). \qquad (14)$$

The equations for determining the resonant frequency required for filter design are given by Eq. (15) through (17).

$$t_r = 3t_p, \qquad (15)$$

$$t_{pk} = 1.77058 \times t_r, \qquad (16)$$

$$\omega_o = \frac{2}{t_{pk}}, \qquad (17)$$

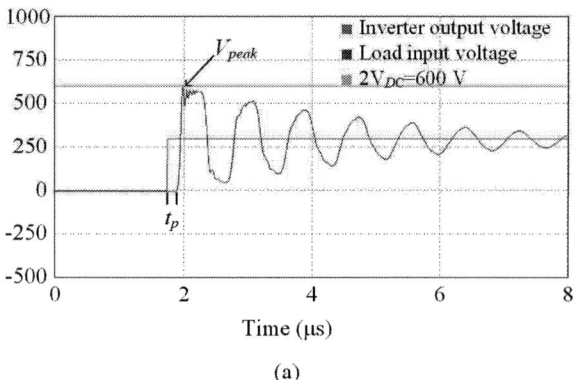

(a)

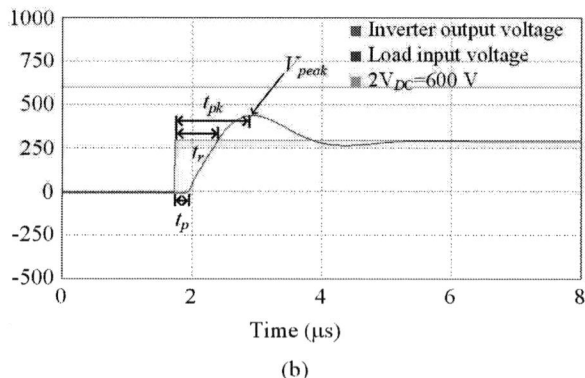

(b)

Fig. 6. Inverter-high-frequency cable-load circuit: (a) without a filter and (b) with the RLC filter.

where t_{pk} is the time to reach the peak voltage, ω_o is the resonant frequency.

The filter resistor is expressed as Eq. (18) through Eq. (8) to achieve impedance matching with the cable. The parameters of the filter designed to suppress overvoltage are determined as follows, with Eq. (19) and (20) derived from Eq. (14).

$$R_f = Z_0, \tag{18}$$

$$L_f = \frac{R_f}{2\omega_o}, \tag{19}$$

$$C_f = \frac{1}{\omega_o^2 L_f}, \tag{20}$$

where R_f is the resistance of the filter, L_f is the inductance of the filter, C_f is the capacitance of the filter.

IV. SIMULATION RESULTS

The performance of the filter designed to mitigate overvoltage in the SiC inverter systems was verified through simulation. The simulation was conducted under the conditions specified in Table II. To observe the overvoltage occurring in the long transmission, the simulation was performed over 30 [m]. The results were compared before and after applying the filter. Fig. 6(a) the peak voltage was shown to be 605.31V, and after applying the filter, the peak voltage Fig. 6(b) was reduced to 433.33V, representing a 28.4 [%] decrease. It was confirmed that the overvoltage magnitude decreased to below $3.1V_{dc}$, meeting the NEMA (National Electrical Manufacturers Association) regulations. Additionally, the rise time measured through simulation was found to be more than three times the propagation delay of the cable. The advantage of the overvoltage reduction filter design, which considers rise time, is that it can be used with inverter systems having various rise times. As a result, the RLC filter using high-frequency cable modeling demonstrated high performance in suppressing overvoltage.

V. CONCLUSIONS

This paper analyzed the design of overvoltage mitigation filter using high-frequency cable modeling in long transmission lines for silicon carbide inverter systems. The cable parameters were measured using an impedance analyzer, and the cable modeling was conducted through simulations based on these measurements. The characteristic impedance for designing the RLC filter was obtained from cable modeling. The RLC filter was used to suppress overvoltage at the output of the inverter, resulting in the rise time of $1\,\mu s$. The validity of cable modeling and designing overvoltage mitigation filter was demonstrated by the simulations.

REFERENCES

[1] M. Bragard, N. Soltau, S. Thomas, and R. W. De Doncker, "The balance of renewable sources and user demands in grids: power electronics for modular battery energy storage systems," *IEEE Trans. Power Electron.*, vol. 25, no. 12, pp. 2049–3056, Dec. 2010.

[2] J. Zhu, H. Kim, H. Chen, R. Erickson, and D. Maksimovic, "High-efficiency SiC traction inverter for electric vehicle applications," in *Proc. IEEE Appl. Power Electron. Conf. Expo. (APEC)*, San Antonio, Tx, USA, pp. 1428–1433, Mar. 2018.

[3] W. Zhou, M. Diab, X. Yuan, and C. Wei, "Mitigation of motor overvoltage in SiC-based drives using soft-switching voltage slew-rate (dv/dt) profiling," *IEEE Trans. Power Electron.*, vol. 37, no. 8, pp. 9612–9628, Aug. 2022.

[4] S. Amarir and K. Al-Haddad, "Mathematical analysis and experimental validation of transient over-voltage higher than 2 per unit along industrial ASDM long cables," in *Proc. IEEE Power Electron. Spec. Conf.*, pp. 1846–1851, 2008.

[5] H. Xiong, A. Louie, R. Liu, J. Zhang, and A. von Jouanne, "Finite element analysis modeling and experimental verification of reflected wave phenomena in variable speed machine drive cables," in *Proc. IEEE Int. Electric Machines Drives Conf.*, pp. 1–8, 2017.

[6] K. K. Yuen and H. S. Chung, "Use of synchronous modulation to recover energy gained from matching long cable in inverter-fed motor drives," *IEEE Trans. Power Electron.* vol. 29. no. 2, pp. 883–893, Feb. 2014.

[7] A. V. Jouanne and P. N. Enjeti, "Design considerations for an inverter output filter to mitigate the effects of long motor leads in ASD applications," *IEEE Trans. Ind. Appl.*, vol. 33, pp. 1138–1145, Sep./Oct. 1997.

[8] H. Akagi and I. Matsumura, "Overvoltage mitigation of inverter-driven motors with long cables of different lengths," *IEEE Trans. Ind. Appl.*, vol. 47, no. 4, pp. 1741–1748, Jul./Aug. 2011.

[9] E. Velander et al., "An ultralow loss inductorless dv/dt filter concept for medium-power voltage source motor drive converters with SiC devices," *IEEE Trans. Power Electron.*, vol. 33, no. 7. pp. 6072–6081, Jul. 2018.

[10] W. Zhou, Z. Zhang, X. Yuan, and M. Diab, "Design and implementation of a dv/dt filter for motor overvoltage mitigation in SiC-based adjustable speed drives," in *2024 IEEE Applied Power Electronics Conference and Exposition (APEC)*, Long beach, CA, USA, pp. 1046-1053, 2024.

[11] A. Moreira, T. Lipo, G. Venkataramanan, and S. Bernet, "High-frequency modeling for cable and induction motor overvoltage studies in long cable drives," *IEEE Trans. Ind. Appl.*, vol. 38, no. 5, pp. 1297–1306, Sep./Oct. 2002.

[12] L. Wang, C. N. M. Ho, F. Canales, and J. Jatskevich, "High-frequency cable and motor modeling of long-cable-fed induction motor drive systems," in *Proc, IEEE ECCE. Atlanta, GA, USA*, pp. 846–852, Sep. 2010.

Power Delivery Network (PDN) Design and Analysis to Achieve Low Impedance in Fast Edge Rate DC-DC Converters for EMI Compliance

Manraj Singh Ladhar, *Member, IEEE;* and Sheldon Williamson, *Fellow, IEEE*
Smart Transportation Electrification and Energy Research (STEER) Group
Faculty of Engineering and Applied Science, Ontario Tech University
Oshawa, Ontario, Canada
manraj.ladhar@ontariotechu.net; sheldon.williamson@ontariotechu.ca

Abstract— This paper examines optimal design principles and practices essential for the printed circuit board (PCB) layout of switching regulators with fast dv/dt and di/dt edge rates to achieve electromagnetic interference (EMI) compliance standards. Reflections, crosstalk, and transmission line management techniques along with power delivery network/system (PDN/PDS) design are implemented for enhanced signal and power integrity. 4-layer board designed with embedded interplane capacitance, controlled impedance traces, proper signal termination and crosstalk management exhibits lowest noise emissions. The experimental results show good consistency with electromagnetic field theory and simulations. The paper also outlines effective design rules and practices essential for optimizing EMI/EMC performance and power integrity during the layout process of switching regulators in PCBs. By employing robust transmission line design techniques and power delivery system (PDS) design principles, a reliable PCB design for the DC/DC converter is achieved, ensuring compliance with EMI standards, and maintaining power integrity.

Keywords—PCB Layout Design, Transmission Lines, Signal Integrity, Power Integrity, EMI/EMC.

I. INTRODUCTION

The industry requirements for smaller footprint, efficient, and cost-effective dc-dc converters drives the critical need for fast dv/dt edge rates for switching regulators. This necessitates the requirement of high-speed printed circuit board (PCB) design capabilities. However, ensuring optimal signal and power integrity while maintaining compliance standards for stringent electromagnetic interference (EMI) remains a significant challenge. Many products fail to pass compliance testing in time for scheduled product launch. The complexity of managing noise emissions and designing power delivery networks in such high-speed systems becomes a significant hurdle. This paper aims to address these critical challenges by applying signal integrity principles and PDN design for power integrity [1,2]

This paper aims to address the critical challenges of passing EMI compliance testing in high-speed PCBs by applying electromagnetic field theory, signal integrity principles and power delivery network design for enhanced power integrity. To address these potential concerns in design phase of a product and meet the required goal, the following objectives are highlighted:

- To design and develop a 500W bidirectional DC-DC converter with fast edge rates, focusing on signal and power integrity.

- To implement and validate transmission line principles on PCBs, controlled impedance routing, management of signal reflections, terminations, and crosstalk.

- To attain target impedance at high frequencies in power delivery network/system (PDN/PDS) on PCBs and measure using a vector network analyzer (VNA).

II. BIDIRECTIONAL DC-DC CONVERTER

A. Design Calculations and Schematic Capture

The LT8228 controller is designed to support bidirectional operation by using two active switches (High-Side and Low-Side) and a single inductor, forming a bidirectional buck-boost topology. The MOSFETs M2 and M3 serve as the primary synchronous switching components, with a fixed dead time applied to their gate signals. Capacitors C1 and C2 function as the output bulk filter capacitors.

The MOSFETs M1A, M1B, M2A, and M2B are n-channel MOSFETs arranged in a common source back-to-back configuration to enable bidirectional current control between V-high and V-low directions. The Cbst and Dbst components form the bootstrapping circuit used to drive the gates of the high side MOSFETs, aided by an internal charge pump that allows M1 and M2 to operate with a 100% duty cycle. The DRXN pin determines whether the LT8228 operates in buck or boost mode: pulling it low selects boost mode, while pulling it high, with a logic-level voltage, selects buck mode. Feedback resistors RFB1A, RFB1B, RFB2A, and RFB2B are employed for closed-loop regulation of the output voltage.

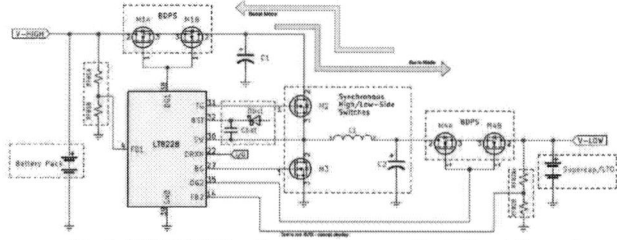

Fig. 1. LT8228 Bidirectional Synchronous Buck-Boost Converter

The calculation of output voltage is given by the equation (1), and the calculation of the inductor value is given by equations (2) and (3).

$$V_o = Vfb \times \left(1 + \frac{R_{FB1A}}{R_{FB1B}}\right) \qquad (1)$$

979-8-3315-1612-3/25 $31.00 © 2025 IEEE

$$L_{Buck} > V2 \times \left(\frac{V1_{MAX} - V2}{f \times \Delta I_L \times V1_{MAX}} \right) \qquad (2)$$

$$L_{Boost} > V2 \times \left(\frac{V1 - V2}{f \times \Delta I_L \times V1} \right) \qquad (3)$$

The critical value of inductor L for continuous conduction mode must meet the requirements of both the operating frequency (f) and the peak inductor ripple current (IL) in both boost mode and buck mode of operation. Therefore, the chosen value of inductor L must satisfy both these conditions to ensure stable and efficient operation of the circuit.

The schematic design of the LT8228 Bidirectional DC/DC Buck/Boost converter was developed using Altium Designer. The LT8228 functions as a current-mode controlled bidirectional synchronous DC/DC converter, capable of maintaining constant current and constant voltage mode regulation, which is essential for li-ion battery charging algorithms. It includes a bidirectional power switch (BDPS) featuring two back-to-back connected N-Channel MOSFETs that can be activated upon over-voltage and over-current detection. A current sense shunt resistor is utilized to monitor the voltage drop across it to measure the output current.

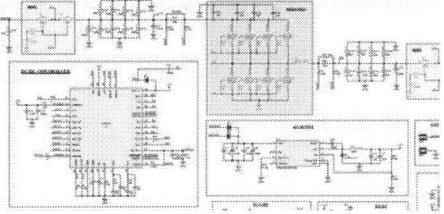

Fig. 2. LT8228 Schematic Capture

The high-side and low-side MOSFETs are driven by an internal totem-pole driver with bootstrap circuitry, which can handle a maximum gate charge pulse of 8A. The HS/LS FETs are connected in parallel to achieve an output current of 35A in buck mode and 10A in boost mode, with a total output power rating of 500W. The inductor L1 has a value of 10µH, a peak saturation current rating of 60A, and a self-resonant frequency of 14MHz. The LT8228 controller operates at a switching frequency of 150KHz, determined by the resistor on the RT pin.

B. PCB Design

The PCB design of the LT8228 controller board was created using Altium Designer. The board is designed as a 4-layer board with a SIG-GND-GND-SIG stack-up configuration. Figure 3 below illustrates the layer stack configuration, including the selected core, prepreg thickness, and copper weight.

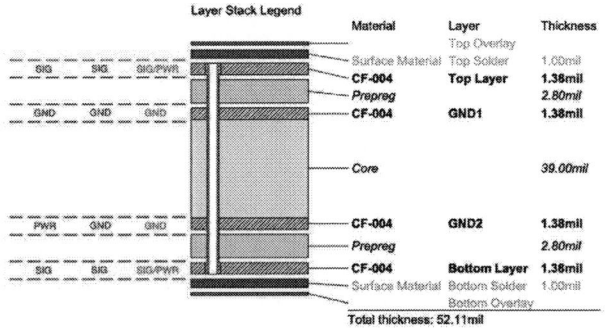

Fig. 3. PCB Stackup Configuration

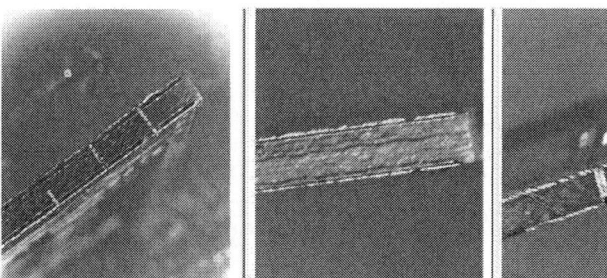

Fig. 4. 4-Layer PCB Layer Stackups – 8 mil and 3 mil thick prepreg, 2-Layer PCB 62 mil thick core

The above fig. 4 shows 3 different layer stack configurations, 4 Layer with PDN/PDS, 4 Layer Microstrip and 2 Layer Board. The chosen layer stack configuration is considered optimal for achieving the best EMC and Signal/Power Integrity performance in a 4-layer board. This is because every signal trace on the outer layers references the GND on the inner layers. Consequently, all signal traces can be routed as controlled impedance microstrip traces, which helps manage crosstalk and reflections effectively. If the power distribution system/network (PDS/PDN) is designed to provide good power bus impedance and decoupling for the ICs across a range of frequency bands, the EMI/EMC performance of a 4-layer board can be significantly improved [3-5]. Figure 5 below presents both a 2D and a 3D realistic view of the PCB design.

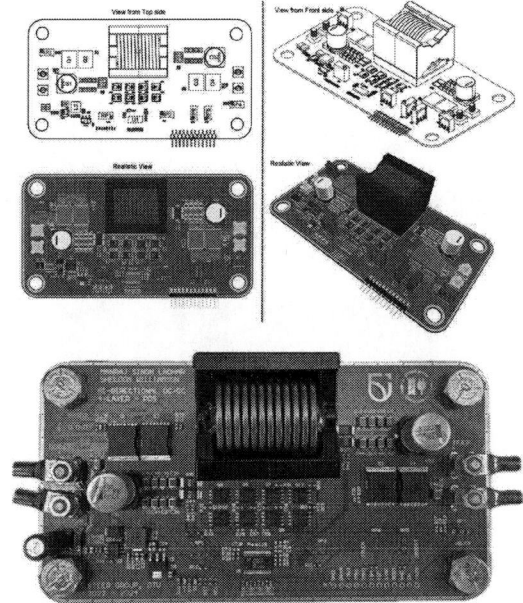

Fig. 5. LT8228 Bidirectional DC/DC Converter 2D/3D View

This PCB was designed following industry-standard design rules and best practices for Signal and Power Integrity. It includes careful routing of traces, strategic placement of vias, and a well-designed PDS/PDN to minimize EMI, reflections, and crosstalk.

III. DESIGN RULES FOR SIGNAL AND POWER INTEGRITY

A. Signal Integrity – Microstrip Traces

All traces on a printed circuit board (PCB) must be designed as transmission lines. A transmission line is composed of two conductors separated by a dielectric medium. In a PCB, this structure includes signal traces and planes. Specifically, a transmission line in a PCB context refers to signal traces routed over a continuous adjacent reference plane, with dielectric material between them [5-9]. Figure 6 illustrates a microstrip trace routed on a PCB.

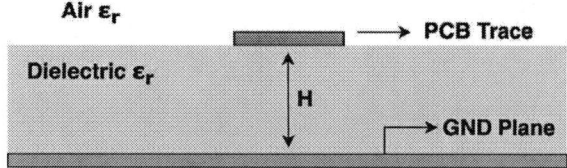

Fig. 6. A Microstrip Trace on PCB

The equation (4) for impedance is given below:

$$Zo = \sqrt{\frac{Ro+j\omega Lo}{Go+j\omega Co}} \qquad (4)$$

Fig. 6. depicts the geometry of a microstrip trace routed on a PCB, shown in a 2D front view. A microstrip trace is defined as a trace routed over a reference plane on the top or bottom outer layers of a PCB, interacting with air dielectric on one side and FR4 dielectric on the other side [10,11]. The characteristic impedance of this trace is determined by three variables: the width of the trace, the dielectric constant of the medium surrounding the trace, and the height above the reference plane. This impedance is crucial for maintaining signal and power integrity [8-10].

B. Field Solver Simulation – Electric and Magnetic Fields

2D/3D field solvers are capable of simulating the electric and magnetic fields associated with a microstrip trace routed on a PCB. These fields carry the electromagnetic energy of the signal and impact electromagnetic interference (EMI), power distribution, and crosstalk [11]. A thorough understanding and effective management of these fields are essential for designing a robust PCB that complies with industry standards. Fig. 7. shows a simulation of a 50-ohm microstrip transmission line.

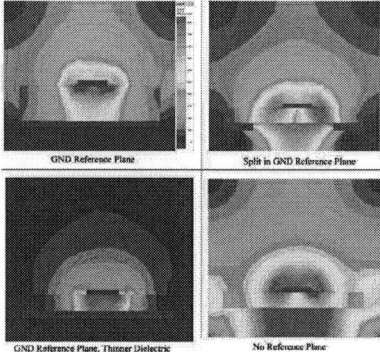

Fig. 7. Microstrip Transmission Line – EM Field Simulation

Fig. 7. illustrates the simulation results of the electric and magnetic fields of a microstrip trace. The magnetic field strength is represented by circular lines, while the electric field strength is shown using a color legend. The simulation involved a sinusoidal signal with a frequency of 1 MHz and an amplitude of 1 V, routed over a ground (GND) reference plane with a common FR4 dielectric material in between, possessing a dielectric constant (Dk) of 4.3.

The field solver simulation results illustrate how electromagnetic fields propagate along a PCB trace. As previously explained, all PCB traces should be routed as

transmission lines. This routing is necessary because the combination of the trace and the reference plane forms a waveguide for electromagnetic energy, providing both forward and return paths. This waveguide enables the electromagnetic fields to travel through the dielectric with a characteristic impedance. The simulation results indicate that the concentration of these fields is highest directly under the trace, highlighting the path of least impedance. Additionally, at high frequencies, most of the return current follows this path of least impedance, which is directly beneath the trace on the surface of the adjacent layer, known as the reference plane 11-13].

C. Crosstalk and Reflection on a PCB Transmission Line

Any discontinuity in the adjacent reference plane layer causes these electromagnetic fields to spread causing EMI and impedance discontinuities which result in signal reflections, increased crosstalk, losses, and distortion [14,15].

As the frequency of the signal increases, or if the edge rates (rise times) of signals decrease, the impedance equation becomes dominated by the inductive and capacitive properties of the interconnect. Consequently, the current follows the path with the highest capacitance and least inductance, forming a waveguide created by the combination of the trace and reference plane [16,17]. Fig. 8. is a comparison of measured crosstalk, reflections between 3 different PCB layouts.

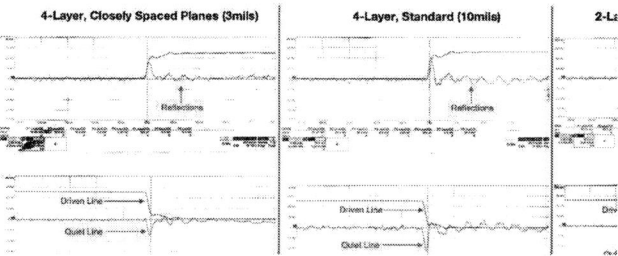

Fig. 8. Measured Crosstalk and Reflections on PCBs

Any discontinuities in the reference plane result in an impedance change causing reflections and force the return current to take a longer path, potentially sharing return paths with other signals. This can result in noise issues, which manifest as electromagnetic interference (EMI), crosstalk and undesired signal reflections on the transmission line. PCBs with poorly designed return paths, transmission lines with poorly managed impedances and terminations, inadequate power distribution systems (PDS/PDN) often fail to comply with EMI and SI standards and suffer from signal and power integrity problems [16-18].

D. PCB Design for EMC and Power Integrity

An effective power distribution system (PDS) provides decoupling to integrated circuits (ICs) experiencing fast current switching transients and edge rates across all necessary frequency ranges on a PCB. Signals with fast edge rates and rise times have harmonic frequencies that range from 200 MHz to 2 GHz [9,10].

Surface-mount device (SMD) ceramic capacitors with the lowest equivalent series inductance (ESL) have resonant frequencies in the range of 80-100 MHz. These capacitors are inadequate for decoupling at frequencies required by modern CMOS logic ICs with fast switching transients. Therefore, instead of using arbitrary decoupling capacitor values, a carefully calculated PDS/PDN design is necessary [10].

979-8-3315-1612-3/25 $31.00 © 2025 IEEE

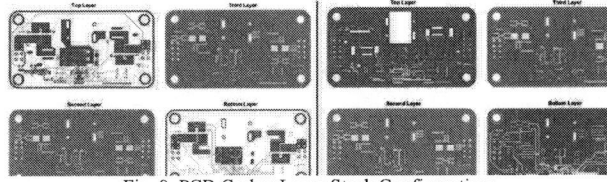

Fig. 9. PCB Gerber Layer Stack Configurations

Fig. 9. illustrates 3 different PCB layer stack configurations, 4-layer without PDS, 4-layer with PDS and 2-layer respectively. The 4-layer PCB with PDS is designed to optimize power integrity and electromagnetic compatibility (EMC) in the bidirectional DC/DC converter. The inner layers are solid ground (GND) planes that serve as reference planes for signals on the top and bottom layers.

Consequently, each signal trace routed on the outer layers is a controlled impedance microstrip trace, effectively managing impedance, crosstalk, and reflections. Additionally, due to layer count limitations, the outer layers are equipped with copper pours of Vdd. These pours form plane capacitance with the adjacent reference plane, separated by a very thin 2.8 mil FR4 dielectric, assisting the smd decoupling capacitors on PCB [1,9,10].

E. Power Distribution System (PDS) – Solution for EMI/EMC Radiated and Conducted Emissions

The plane capacitance created by closely spaced power and ground plane sandwiches exhibits very low inductance, typically in femtohenries, compared to ceramic capacitors with ESL in nanohenries. This configuration results in low power bus impedance and efficient decoupling at very high frequencies ranging from 100 MHz to 2 GHz [1,10].

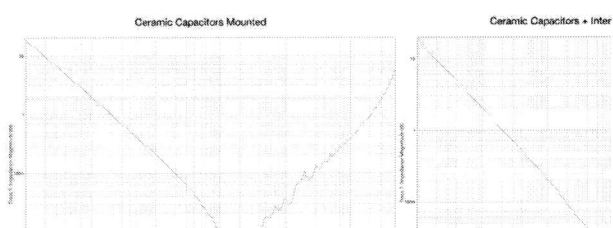

Fig. 10. PDS/PDN Impedance at High Frequencies Measured with a Vector Network Analyzer (VNA)

Table 1. Power Bus Impedance Measurement using VNA

PCB	Frequency	Impedance
4-Layer with PDN/PDS	30.91 MHz	245mΩ (0.245Ω)
4-Layer with 0805 SMD Capacitors	30.89 MHz	1821mΩ (1.821Ω)

Table 2. gives us the impedance measurement comparison between the PCBs with and without the interplane capacitance. Plane capacitance, supplemented by an adequate number of decoupling capacitors, is crucial for supplying the rapid current transients demanded by modern CMOS logic ICs and gate drivers. Without sufficient PCB plane capacitance, the Vdd rail of the power supply can experience ringing when logic gates switch with fast rise times, leading to conducted and radiated electromagnetic interference (EMI) noise. The

following fig. 11. Depicts the ringing on Vdd that is a major cause of EMI.

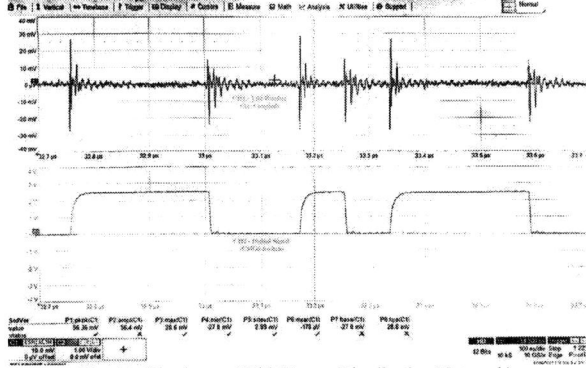

Fig. 11. Ringing on Vdd (Power Distribution Network)
CH1 – Vdd Ringing (AC Coupled) CH2 – Digital Signal (CMOS Switch)

The Vdd noise is a primary contributor to EMC failures in many PCB designs [1,10]. Therefore, optimizing plane capacitance and decoupling strategies is essential for maintaining proper power integrity and meeting EMC standards in high-speed digital circuits. The near-field radiated emissions are measured on the 3 different types of PCB layouts/stack-ups.

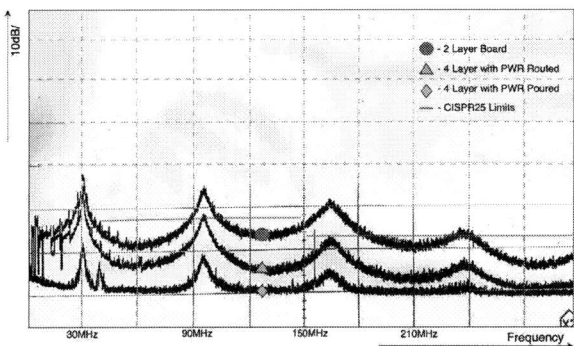

Fig. 12. PCB Near-Field Radiated Emissions Comparison

The above fig. 12. illustrates the comparison between the measurement results observed. The two-layer board depicts highest EMI noise, followed by 4-layer board without interplane capacitance. The 4-layer board with well-designed PDS/PDN delivers lowest EMI noise and passes automotive compliance standards.

IV. CONCLUSION

The design of DC/DC converters incorporating modern logic ICs and gate drivers, which exhibit fast edge rates, rise times, and current transients, necessitates adherence to design rules for (EMC), signal and power integrity.

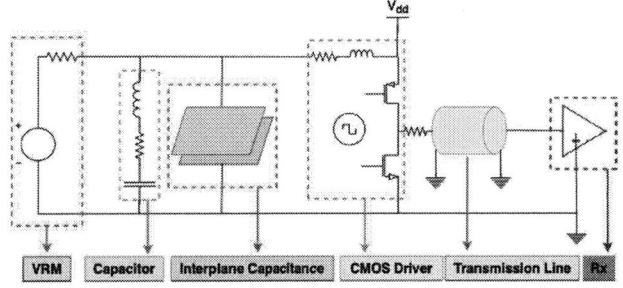

Fig. 13. PCB Near-Field Radiated Emissions Comparison

Field solver simulations of a microstrip trace reveal that most electromagnetic fields propagate along the transmission line, with return currents flowing beneath the signal trace on the adjacent reference plane.

Consequently, a carefully designed and calculated Power Distribution Network/Power Delivery System (PDN/PDS) design is implemented in 4-layer PCB bidirectional DC/DC converter to enhance electromagnetic interference, electromagnetic compatibility (EMI/EMC) performance, power integrity, and immunity to gate drive switching noise and crosstalk.

The impedance of the PDN is measured using a VNA at high frequencies. It is concluded that the 4-layer PCB which has interplane capacitance incorporated into the structure and stackup of the PCB provides lower impedance and better decoupling characteristics at high frequencies as compared to the boards which consists of only ceramic decoupling capacitors. Hence, calculating the required interplane capacitance based on signal switching transients and highest harmonic frequencies propagating on PCB is crucial for optimal EMI/EMC performance and reduced noise.

REFERENCES

[1] E. Bogatin, "Essential Principles of Signal Integrity," in *IEEE Microwave Magazine*, vol. 12, no. 5, pp. 34-41, Aug. 2011, doi: 10.1109/MMM.2011.941411.

[2] J. N. Tripathi, L. Niu and T. Hubing, "Signal Integrity and Power Integrity," in *IEEE Electromagnetic Compatibility Magazine*, vol. 12, no. 3, pp. 35-43, 3rd Quarter 2023, doi: 10.1109/MEMC.2023.10364702.

[3] T. H. Hubing, "Common PCB layout errors that cause products to fail to meet automotive EMC requirements," in *IEEE Electromagnetic Compatibility Magazine*, vol. 8, no. 3, pp. 86-91, 3rd Quarter 2019, doi: 10.1109/MEMC.2019.8878242.

[4] Z. Sun, J. Liu, X. Xiong, D. Kim, D. Beetner and V. Khilkevich, "Characterization of a Microstrip Line Referenced to a Meshed Return Plane Using 2-D Analysis," in IEEE Transactions on Signal and Power Integrity, vol. 3, pp. 13-20, 2024, doi: 10.1109/TSIPI.2023.3339445.

[5] T. H. Hubing, "Autonomous Vehicles Will Transform the Field of Automotive EMC," *2018 IEEE Symposium on Electromagnetic Compatibility, Signal Integrity and Power Integrity (EMC, SI & PI)*, Long Beach, CA, USA, 2018, pp. 1-29, doi: 10.1109/EMCSI.2018.8495279.

[6] T. H. Hubing, "Grounding," *2018 IEEE Symposium on Electromagnetic Compatibility, Signal Integrity and Power Integrity (EMC, SI & PI)*, Long Beach, CA, USA, 2018, pp. 1-39, doi: 10.1109/EMCSI.2018.8495282.

[7] T. H. Hubing, J. L. Drewniak, T. P. Van Doren and D. M. Hockanson, "Power bus decoupling on multilayer printed circuit boards," in *IEEE Electromagnetic Compatibility Magazine*, vol. 9, no. 2, pp. 66-77, 2nd Quarter 2020, doi: 10.1109/MEMC.2020.9133248.

[8] L. Smith and E. Bogatin, "Principles of Power Integrity for PDN Design:," 2018 IEEE Symposium on Electromagnetic Compatibility, Signal Integrity and Power Integrity (EMC, SI & PI), Long Beach, CA, USA, 2018, pp. 1-74, doi: 10.1109/EMCSI.2018.849541.

[9] J. Joo et al., "Modeling of a Voltage Regulator Module for Power Integrity: Power Supply Induced Jitter," in IEEE Transactions on Signal and Power Integrity, vol. 3, pp. 110-125, 2024, doi: 10.1109/TSIPI.2024.3416088.

[10] Lee W. Ritchey. 2008. Right the First Time. Speeding Edge, USA.

[11] J. B. Nitsch, R. Rambousky and S. Tkachenko, "Introduction of Reflection and Transmission Coefficients for Nonuniform Radiating Transmission Lines," in *IEEE*

[12] Eric Bogatin, *Bogatin's Practical Guide to Transmission Line Design and Characterization for Signal Integrity Applications*, Artech, 2020.

[13] Eric Bogatin, *Bogatin's Practical Guide to PCB Design for New Product Development*, Artech, 2021.

[14] Q. Wang, X. Huang and B. Yin, "Crosstalk analysis and suppression of microstrip lines on PCB board," *2022 IEEE 5th Advanced Information Management, Communicates, Electronic and Automation Control Conference (IMCEC)*, Chongqing, China, 2022, pp. 875-879, doi: 10.1109/IMCEC55388.2022.10019868.

[15] R. Mudavath, B. R. Naik and B. Gugulothu, "Analysis of Crosstalk Noise for Coupled Microstrip Interconnect Models in High-Speed PCB Design," *2019 International Conference on Electronics, Information, and Communication (ICEIC)*, Auckland, New Zealand, 2019, pp. 1-5, doi: 10.23919/ELINFOCOM.2019.8706385.

[16] E. Song, J. Cho, J. Kim, Y. Shim, G. Kim and J. Kim, "Modeling and Design Optimization of a Wideband Passive Equalizer on PCB Based on Near-End Crosstalk and Reflections for High-Speed Serial Data Transmission," in *IEEE Transactions on Electromagnetic Compatibility*, vol. 52, no. 2, pp. 410-420, May 2010, doi: 10.1109/TEMC.2010.2042452.

[17] E. Bogatin, "Essential Principles of Cross Talk and Mitigation Strategies," *2018 IEEE Symposium on Electromagnetic Compatibility, Signal Integrity and Power Integrity (EMC, SI & PI)*, Long Beach, CA, USA, 2018, pp. 1-29, doi: 10.1109/EMCSI.2018.8495242.

[18] A. Rao, S. Sawant, E. Bogatin and M. Piket-May, "Impact of Copper Pour on Crosstalk: Measurement and Simulation Correlation," *2021 IEEE 30th Conference on Electrical Performance of Electronic Packaging and Systems (EPEPS)*, Austin, TX, USA, 2021, pp. 1-3, doi: 10.1109/EPEPS51341.2021.960911.

Enhancing the Performance of Dual Input Split Source Inverters Using an Advanced Modulation Strategy

Mustafa Abu-Zaher[1,2], Fang Zhuo[1], Mokhtar Aly[3], Mahmoud A. Gaafar[4], Mohamed Orabi[4], José Rodriguez[3], Alaaeldien Hassan[5], Jiachen Tian[1], and Samir Kouro[6]

[1] State Key Laboratory of Electrical Insulation and Power Equipment, Xi'an Jiaotong University, Xi'an ,710049, China
[2] Electrical Department, Faculty of Technology and Education, Sohag University, Sohag 82524, Egypt
[3] Facultad de Ingeniería, Arquitectura y Diseño, Universidad San Sebastián, Bellavista 7, Santiago 8420524, Chile
[4] APEARC, Faculty of Engineering, Aswan University, Aswan, 81542, Egypt
[5] Department of Electrical Power and Machines Engineering, Faculty of Engineering, South Valley University, Qena, Egypt
[6] Department of Electronic Engineering, Universidad Tecnica Federico Santa Maria, Valparaiso, Chile
Mostafa abu-Zaher@techedu.sohag.edu.eg, zffz@mail.xjtu.edu.cn, mokhtar.aly@aswu.edu.eg, mgaafar@apearc.aswu.edu.eg,
morabi@apearc.aswu.edu.eg, jose.rodriguezp@uss.cl, alaaeldien@eng.svu.edu.eg,
tianjiachen1995@stu.xjtu.edu.cn, samir.kouro@ieee.org

Abstract— **The increasing demand for compact, efficient, and reliable power conversion structures, especially for photovoltaics, has promoted the development of sophisticated single-stage inverter topologies such as split source inverter (SSI). In this study, improving the performance of dual input split source inverters (DSSI) is targeted at utilizing a sophisticated modulation scheme. The DSSI integrates dual input sources with a split-source inverter structure, providing continuous DC input current, reduced voltage stress, and stable DC-link voltage, which are essential requirements in such applications. The proposed control strategy uses sinusoidal pulse width modulation (SPWM) and optimized simple boost SPWM (SB-SPWM) to control the output voltage and improve the system efficiency; by implementing these modulation techniques, the system offers better control of voltage gain and, in turn, exhibits better overall control of the inverter. The proposed SB-SPWM method is validated through a set of experiments with the conventional SPWM scheme. It demonstrates the improved efficiency and reliability of the proposed approach in meeting the stringent requirements of modern power applications.**

Keywords—split source inverter, Z-source, sinusoidal pulse width modulation, simple boost sinusoidal pulse width modulation.

I. INTRODUCTION

Requirements of reliability, high performance, and compact sizes of power electronics conversion systems have become essential for promoting the planned energy transition in different industrial sectors, especially from two-stage-based structures [1] to single-stage-based structures [2]. The main objective is to provide reductions in the overall system costs, footprints, and power losses. The development of single-stage solutions leads to having more efficient energy transfer/conversion with inherent functions to reduce the total number of required components, especially the passive ones. This, in turn, enhances the system reliability index and reduces the required shutdowns. This trend is especially significant in renewable energy, electric vehicles, and portable electronics sectors, where volume, weight, and

efficiency are paramount. Single-stage topologies and their corresponding modulation techniques have been conducted and introduced in [3] and [4]. Among these, the three-phase split-source inverter (SSI) has emerged as a promising substitute for the conventional Z-source inverter (ZSI). The SSI-based solutions overcome several challenges related to the ZSI systems, including the associated discontinuous input current problem and the increased voltage stresses at high voltage gains requirements. The SSI-based solutions provide input current in DC form without discontinuity, leading to better reliability for the source side. Furthermore, it minimizes voltage stresses on the switches, particularly at higher overall voltage gains, thereby contributing to improved durability and operational efficiency [5]. The SSI's design facilitates a continuous DC-link voltage, allowing for the use of high-frequency decoupling capacitors across the inverter legs to reduce voltage spikes caused by parasitic inductances [6].

The increasing utilization of renewable energy, photovoltaic (PV) systems in particular, has clearly increased the need for enhanced power conversion technologies that are suitable for managing variable power sources. Standard inverter layouts also fail to balance well with the inherent variability of power originating from the fluctuations of the environmental conditions; this is mainly proven in regions characterized by variations such as partial shading and variation from one season to the other. Negative impacts on the system's reliability, stability, and even efficiency can result in these cases with sophisticated control requirements. Accordingly, developing new and compact inverter solutions with the possibility of multi-input connection has become essential. The developed solutions have to adjust the system operation based on the continuously changing conditions while maximizing the extracted energy from the system. Several attempts in the literature have been presented in this area through developing multi-input-based power inverter systems to connect different sources using integrated inverter solutions [7]–[9].

Dual Input Split Source Inverter (DSSI), a new type of inverter as considered in the recent literature [10]–[12], is a combination of SSI with dual input flexibility. This design

979-8-3315-1612-3/25 $31.00 © 2025 IEEE

unites the robust operational features of the conventional SSI while incorporating the flexibility of the dual-input configuration. Due to its versatility in handling multiple input sources, the DSSI provides enhanced power control and efficiency in different applications. Moreover, this dual-input arrangement allows for the optimization of energy usage in addition to enhancing system flexibility to system load fluctuations, which are issues chiefly experienced in conventional SSI applications. The introduction of DSSI is a major development in the progress of inverter systems, opening new opportunities to improve capabilities in contemporary energy systems.

In this regard, the modulation index is used as the determinant feature influencing the operation of these sophisticated inverter systems. It contributes significantly to decisions concerning the output waveform, efficiency, and stability of inverters. Recent studies show that adjusting the modulation index [13], [14] can effectively improve the harmonic characteristic and decrease voltage demand in components, improving the inverter reliability and efficiency. Sinusoidal Pulse Width Modulation (SPWM) is a widely adopted modulation technique in inverter configuration that modulates the output voltage generated by comparing a sinusoidal reference signal with a triangular carrier signal. This method can control the pulse widths of the output waveform to reduce harmonic distortion and improve power quality and operational efficiency. However, Simple Boost Pulse Width Modulation (SB-SPWM) has recently been presented as a new development in this area since it incorporates sinusoidal references together with the duty cycle to offer the required outlay construction and boost capacity. This refined technique not only ensures more accurate control of the output waveform but also optimizes voltage regulation and power consumption under varying load conditions, an issue that cannot be resolved by basic SPWM. A comparative analysis through experimental comparison and evaluation of SPWM and SB-SPWM has been carried out in this paper in order to determine their effects on the inverter topologies.

II. COMPREHENSIVE ANALYSIS OF DUAL INPUTSPLIT SOURCE INVERTER DSSI

The DSSI, shown in Fig. 1, has a unique configuration characterized by a primary leg, which has three switches, while the two additional legs each contain two switches. This configuration enables the integration of dual input sources, each with its voltage and current: for the upper source V_{in1}, and i_{L1} for the lower source, V_{in2}, i_{L2}. The upper source is connected to the top switches of the converter leg through an inductor L_2 and diodes D_{21}, D_{22}, and D_{23}, respectively. On the other hand, the lower source is connected through diodes D_{11}, D_{12}, and D_{13}, respectively, with the connection of an inductor L_1. The arrangement of these components facilitates improved flexibility in using power supply from different sources to increase the efficiency of the inverter. Additionally, capacitor C serves the DC-link coupling, stabilizes the inverter output voltage V_{inv}, and performs satisfactorily across full load ranges. It also enriches the configuration of the system and demonstrates the flexibility of the inverter in various operational conditions.

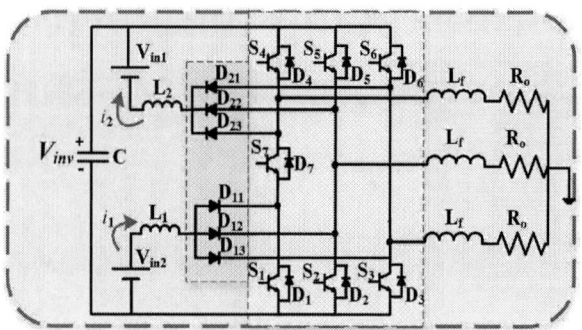

Fig. 1: The configuration of DSSI.

A. Optimized Control Using SPWM Modulation

SPWM occupies a leading control area of power electronics as one of the critical functions for controlling and improving the inverter's performance. The fundamental operation of SPWM involves the sinusoidal modulating signal, denoted as (V_a, V_b, V_c), as described by:

$$\begin{cases} V_a = M\sin(\omega t) \\ V_b = M\sin(\omega t - \frac{2\pi}{3}) \\ V_c = M\sin(\omega t + \frac{2\pi}{3}) \end{cases} \tag{1}$$

where the symbols M and ω represent the magnitude and frequency of such signals equivalently. Moreover, the incorporation of boosting techniques within this framework is governed by variable duty cycles denoted as D_{up}, D_{down}, defined as:

$$\begin{cases} D_{up} = \frac{1}{2} + \frac{3\sqrt{3}}{4\pi}M \\ D_{down} = -(\frac{1}{2} + \frac{3\sqrt{3}}{4\pi}M) \end{cases} \tag{2}$$

These duty cycles dynamically adjust to modulate the power delivered to the load, thereby offering fine-grained control over the inverter's output, which is then compared to a high-frequency triangular carrier signal, as shown in Fig. 2. The intersections between the sinusoidal modulation signals and the triangular carrier wave command the switching instances of the inverter's transistors, thereby determining the pulse widths in the output waveform.

The formation of a variable duty cycle involves some inconvenient features. This is a major disadvantage because the switching frequency must be controlled to stabilize at a suitable output when the load gets tough, and this is realized by changing the duty cycle intermittently. This can lead to additional switching losses, which negatively impact on the overall efficiency of the inverter. Moreover, variable duty cycles can cause fluctuating stress on power electronic components, potentially leading to reduced lifespan due to thermal and electrical stresses.

979-8-3315-1612-3/25 $31.00 © 2025 IEEE

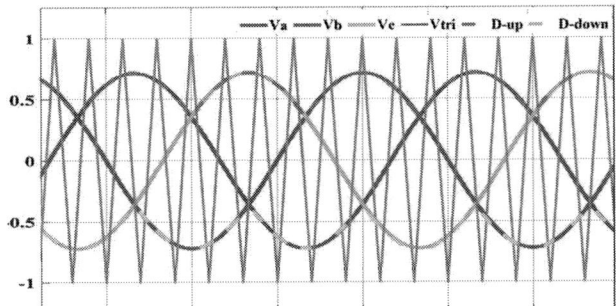

Fig. 2: Reference signals of SPWM.

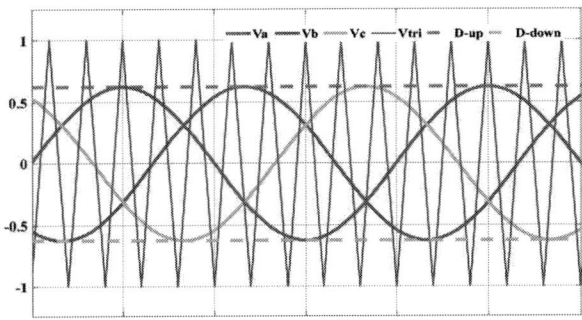

Fig. 3: Reference signals of SB-SPWM.

B. Simple Boost SPWM Modulation (SB-SPWM)

The Simple Boost SPWM Modulation (SB-SPWM) is an advanced control method proposed as an improvement over regular SPWM, it offers better efficiency, enhanced stability of the DC-link voltage, and higher voltage gains. This method is particularly effective through its design; three sinusoidal reference signals as described by

$$\begin{cases} V_a = M \sin(\omega t) \\ V_b = M \sin(\omega t - \dfrac{2\pi}{3}) \\ V_c = M \sin(\omega t + \dfrac{2\pi}{3}) \end{cases} \quad (3)$$

and two constant duty cycles; D_{up} and D_{down} as described in:

$$\begin{cases} D_{up} = M \\ D_{down} = -M \end{cases} \quad (4)$$

These duty cycles are critical signals, as they control the entire period of the switching intervals and thus regulate the voltage gain. The constancy of D_{up} and D_{down} is a defining feature of SB-SPWM from other methods as shown in Fig. 3. This constancy eliminates the need for a dynamic duty cycle, which makes the DC-link voltage smoother and has less voltage ripple.

The SB-SPWM operates by comparing the three sinusoidal reference signals against a high-frequency triangular carrier wave. The intersections of the reference signals with the carrier wave determine the precise switching points for the inverter's transistors, thus modulating the output waveform. This structure also gives the operator greater control over the pulse amplitude and the extent of the waveform variations from the ideal sine waveform. The continuous operation of the duty cycles makes it possible to have the optimum timing of operation ON/OFF for the transistors, thus minimizing grating switching, which leads to energy loss.

Structurally, the SB-SPWM configuration is designed to provide a stable and high-performance output by integrating the modulation references with the topology's boosting mechanism.

The modulation reference signals, in combination with the duty cycles, contribute to a balanced energy flow through the inverter's stages, which is particularly advantageous in applications requiring high power density and reliability. By maintaining consistent D_{up} and D_{down} values, the SB-SPWM reduces the stress on the inverter's components, allowing for more efficient thermal management and extended component longevity.

III. EXPERIMENTAL INVESTIGATION

The experimental results demonstrate notable differences between SPWM and SB-SPWM, particularly in terms of waveform stability, boosting capability, and overall system performance. Table. I outlines the specified parameters utilized in the test bench setup. As shown in Fig. 4, the input source voltage and current waveforms highlight that SPWM's variable duty cycle introduces a higher ripple. This ripple not only compromises the quality of the waveform but also undermines system stability and reduces efficiency by causing input power oscillations. The continuously changing nature of the duty cycle in SPWM requires frequent modulation, which negatively impacts power supply regulation and increases switching losses, which increases the stress across the components and, hence, reduces the lifetime operation on the available inverter. Conversely, SB-SPWM has a constant duty ratio. This stability in the duty cycle is vital in controlling the switching losses since a reduction in the voltage oscillations reduces the occurrence of energy transfer across the inverter.

TABLE. I
PARAMETERS OF THE TEST BENCH PROTOTYPE

Symbol	Quantity	Value
V_{in1}, V_{in2}	Input Voltage	40 V
f_o	Fundamental AC frequency	50 HZ
L_1, L_2	Inductance of boosting circuits	4 mH
L_f	output filter inductance	6 mH
f_{sw}	Switching frequency	10 KHZ
C	Capacitance of the DC-link	220 μF
IGBT	IKW20N65ET7	
Diode	DHG30I600HA	

979-8-3315-1612-3/25 $31.00 © 2025 IEEE

SPWM	SB-SPWM

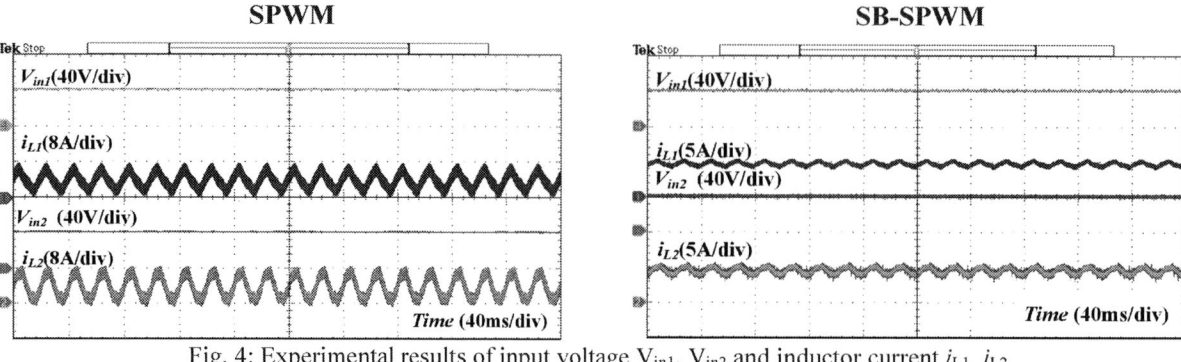

Fig. 4: Experimental results of input voltage V_{in1}, V_{in2} and inductor current i_{L1}, i_{L2}.

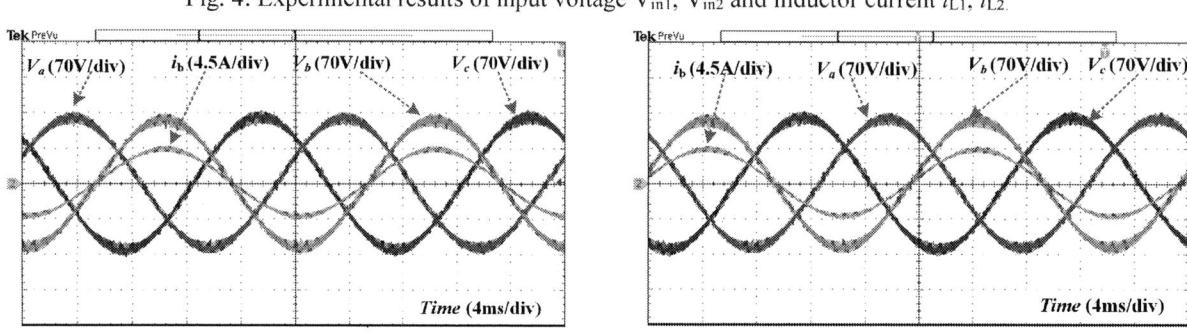

Fig. 5: Experimental results of three phase voltages with emphasis on phase b current.

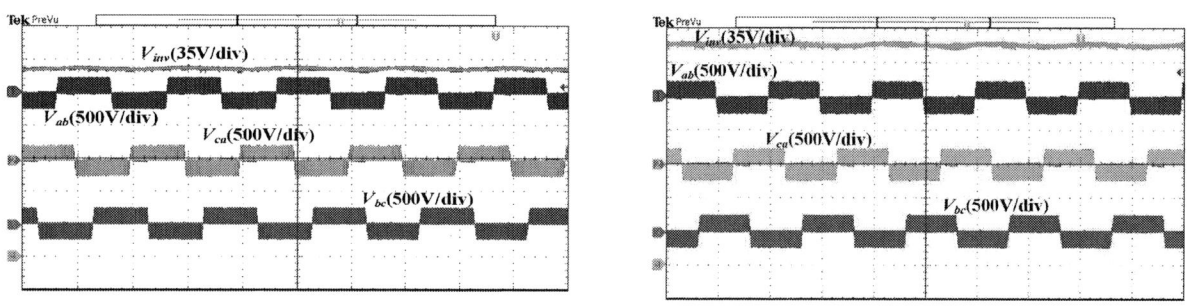

Fig. 6: Experimental results of the DC-link voltage (V_{inv}), line voltage V_{ab}, V_{ca} and V_{bc}.

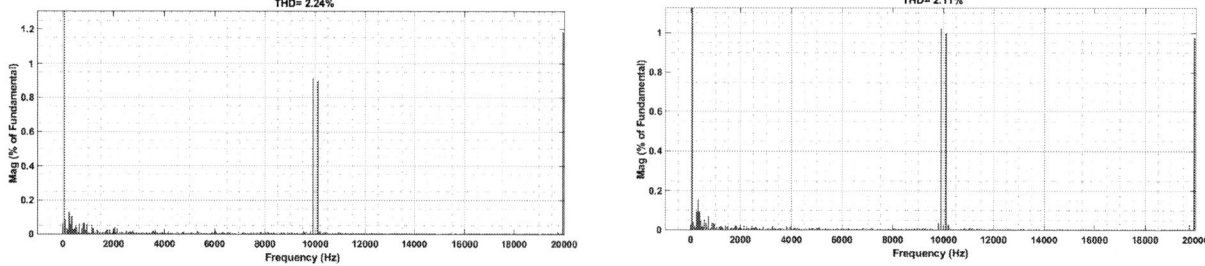

Fig. 7: FFT of the output current i_b.

The analysis of Fig. 5 demonstrates the output voltage and current characteristics for SPWM and SB-SPWM, the steady duty cycle in SB-SPWM facilitates a significantly higher boosting ratio. This increase in the boost ratio offers greater flexibility in regulating the output voltage. Fig. 6 shows the improved performance of the proposed SB-SPWM compared to the conventional SPWM, where the measurements obtained for the DC-link voltage for SPWM reaches 190V, whereas the DC-link voltage of the proposed technique, SB-

SPWM reaches 225V. This increase clearly shows that SB-SPWM has a greater boosting capability compared to the conventional SPWM method. The constant duty cycle in SB-SPWM not only allows for exact fine tuning according to conditions but also minimizes the losses resulting from the variable duty cycle across switching. Therefore, the application of SB-SPWM is highly effective for achieving improved levels of reliability and performance in systems that require a high level of consistency and stability.

979-8-3315-1612-3/25 $31.00 © 2025 IEEE

The Fast Fourier Transform (FFT) of the output current structure it is necessary for the analysis due to the direct impact of the modulation techniques on Total Harmonic Distortion (THD). Evaluating THD provides deeper insight into the effectiveness of any modulation technique in minimizing the levels of harmonics across the waveform. As can be deduced from the FFT analysis in Fig. 7, the THD of the i_b has a THD of 2.24 % under SPWM, while it has 2.11% under SB-SPWM in response to delivering the same output current. That is why the presented comparison emphasizes the specifics and enhanced performance of SB-SPWM in terms of harmonic distortion control and the general quality of the obtained waveform in the inverter.

As a result, the experimental data demonstrate that SB-SPWM distinctly offers superior benefits over conventional SPWM by providing a steady and high boosting ratio, reduced switching losses, and low voltage ripple. These attributes of SB-SPWM make it particularly suitable for high-performance applications where a high boosting ratio and high performance in voltage regulation in terms of efficiency, stability, and flexibility are desirable. By comparing SPWM and SB-SPWM, SB-SPWM is considered a more robust and resilient solution for the latest inverter, demonstrating significant enhancements in DC-link voltage stability and overall system reliability.

IV. CONCLUSIONS AND FUTURE WORK

This paper demonstrates that for contemporary applications, the SB-SPWM technique offers significant advantages over conventional SPWM. It effectively minimizes ripple in the input current, reduces waveform oscillations, and achieves a higher boost ratio, thereby enhancing the overall performance of the system. Unlike the conventional SPWM, the proposed SB-SPWM method delivers improved system efficiency and stability. The results clearly illustrate the superiority of SB-SPWM, as evidenced by its ability to achieve a DC-link voltage of 225V compared to 190V for conventional SPWM in applications requiring precise voltage control, efficiency, and reliability. Finally, the findings of this study highlight that SB-SPWM is a highly favorable modulation strategy for modern power electronic systems, particularly considering the growing emphasis on efficiency and stability in cutting-edge technologies and applications.

ACKNOWLEDGMENTS

This work was supported by ANID FONDECYT Iniciacion 11230430, SERC-Chile ANID/FONDAP/1523A0006, AC3E (ANID/BASAL/AFB240002), 1210208 and 1221293.

REFERENCES

[1] S. I. Nanou, A. G. Papakonstantinou, and S. A. Papathanassiou, "A generic model of two-stage grid-connected PV systems with primary frequency response and inertia emulation," *Electr. Power Syst. Res.*, vol. 127, pp. 186–196, 2015, doi: 10.1016/j.epsr.2015.06.011.

[2] L. Wu, Z. Zhao, and J. Liu, "A single-stage three-phase grid-connected photovoltaic system with modified MPPT method and reactive power compensation," *IEEE Trans. Energy Convers.*, vol. 22, no. 4, pp. 881–886, 2007, doi: 10.1109/TEC.2007.895461.

[3] M. Zhu, K. Yu, and F. L. Luo, "Switched inductor Z-source inverter," *IEEE Trans. Power Electron.*, vol. 25, no. 8, pp. 2150–2158, 2010, doi: 10.1109/TPEL.2010.2046676.

[4] X. Zhu, B. Zhang, and D. Qiu, "A High Boost Active Switched Quasi-Z-Source Inverter With Low Input Current Ripple," *IEEE Trans. Ind. Informatics*, vol. 15, no. 9, pp. 5341–5354, 2019, doi: 10.1109/tii.2019.2899937.

[5] A. Abdelhakim, P. Mattavelli, and G. Spiazzi, "Three-Phase Split-Source Inverter (SSI): Analysis and Modulation," *IEEE Trans. Power Electron.*, vol. 31, no. 11, pp. 7451–7461, 2016, doi: 10.1109/TPEL.2015.2513204.

[6] M. Zhu, D. Li, P. C. Loh, and F. Blaabjerg, "Tapped-inductor Z-source inverters with enhanced voltage boost inversion abilities," *2010 IEEE Int. Conf. Sustain. Energy Technol. ICSET 2010*, vol. 2, pp. 1–6, 2010, doi: 10.1109/ICSET.2010.5684446.

[7] A. R. Dehghanzadeh, V. Behjat, and M. R. Banaei, "Double input Z-source inverter applicable in dual-star PMSG based wind turbine," *Int. J. Electr. Power Energy Syst.*, vol. 82, pp. 49–57, Nov. 2016, doi: 10.1016/j.ijepes.2016.02.017.

[8] Y. M. Chen, Y. C. Liu, S. C. Hung, and C. S. Cheng, "Multi-input inverter for grid-connected hybrid PV/wind power system," *IEEE Trans. Power Electron.*, vol. 22, no. 3, pp. 1070–1077, May 2007, doi: 10.1109/TPEL.2007.897117.

[9] N. Beniwal, H. D. Tafti, G. G. Farivar, S. Ceballos, J. Pou, and F. Blaabjerg, "A Control Strategy for Dual-Input Neutral-Point-Clamped Inverter-Based Grid-Connected Photovoltaic System," *IEEE Trans. Power Electron.*, vol. 36, no. 9, pp. 9743–9757, Sep. 2021, doi: 10.1109/TPEL.2021.3063745.

[10] M. Abu-Zaher, F. Zhuo, M. Orabi, A. Hassan, and M. A. Gaafar, "Dual-input configuration of three-phase split-source inverter for photovoltaic systems with independent maximum power point tracking," *Electr. Power Syst. Res.*, vol. 232, no. November 2023, p. 110375, 2024, doi: 10.1016/j.epsr.2024.110375.

[11] M. Abu-Zaher, M. A. Gaafar, and M. Orabi, "Dual-Input Three Phase Split-Source Inverter," *IEEE Conf. Power Electron. Renew. Energy, CPERE 2023*, pp. 1–6, 2023, doi: 10.1109/CPERE56564.2023.10119585.

[12] M. Abu-Zaher, F. Zhuo, M. Aly, J. Tian, and M. Ahmed, "Simplified Predictive Control Strategy for Dual-Input Three-Phase Split Source Inverter with Minimized Computational Burdens," *IEEE J. Emerg. Sel. Top. Power Electron.*, vol. 12, no. 5, pp. 4703–4715, 2024, doi: 10.1109/JESTPE.2024.3450206.

[13] A. Abdelhakim, P. Mattavelli, P. Davari, and F. Blaabjerg, "Performance Evaluation of the Single-Phase Split-Source Inverter Using an Alternative DC-AC Configuration," *IEEE Trans. Ind. Electron.*, vol. 65, no. 1, pp. 363–373, 2018, doi: 10.1109/TIE.2017.2714122.

[14] A. Abdelhakim, F. Blaabjerg, and P. Mattavelli, "Modulation Schemes of the Three-Phase Impedance Source Inverters-Part I: Classification and Review," *IEEE Trans. Ind. Electron.*, vol. 65, no. 8, pp. 6309–6320, 2018, doi: 10.1109/TIE.2018.2793255.

A Novel GaN-HEMT Single-Phase Single-Stage Buck-Boost Micro-Inverter Topology for PV Applications

Pengwei Li, Uiliam Kutrolli, *Member, IEEE*, Ali Bazzi, *Senior Member, IEEE*

University of Connecticut, Storrs, CT, USA.

Abstract—To enhance the input voltage compatibility of single-stage inverters used in medium-power solar energy products, a novel single-phase, single-stage inverter topology is proposed in this paper. The switching states and modulation methods are analyzed. Based on steady-state analysis, its transfer gain is derived, and a linearized modulation method is proposed. The proposed topology is implemented with GaN-HEMT devices on a 300W prototype with 100kHz switching frequency. Selected experimental results verify the proposed topology and modulation method.

Index Terms—Single Stage Inverter, Buck-Boost, Solid State Transformer, Input-Output Feasibility.

I. INTRODUCTION

Due to the significant expansion of the renewable energy market and the maturity of solar energy technologies, solar power applications are no longer limited to large-scale solar farms or residential microgrids [1]. Today, solar energy is increasingly being integrated into distributed energy systems, as shown in Figure 1, creating an "energy internet" where each terminal can contribute power to the grid [2]. Additionally, advances in solar harvesting have enabled its application across various outdoor platforms, such as electric vehicles, emergency rescue equipment, and remote off-grid installations [3], [4]. Despite their diverse settings, these applications often share common requirements: medium power levels (200W–500W), high power density, high efficiency, a wide input voltage range, and grid-tied/islanded compatibility.

Achieving both high efficiency and high-power density requires careful consideration of the inverter's topology as well as the semiconductor devices used. Wide-bandgap (WBG) semiconductors, including silicon carbide (SiC) and gallium nitride (GaN), have seen extensive adoption in these applications due to their ability to operate at high switching frequencies. These frequencies enable designers to reduce the size of passive components, particularly inductors, thus minimizing the inverter's volume and weight while significantly improving power density and efficiency [5]. This compact design is particularly advantageous in solar applications that require portability or need to integrate seamlessly with constrained spaces.

In solar photovoltaic (PV) applications, where the input voltage can vary widely based on solar irradiance conditions and load demand, the inverter must accommodate these variations to maintain a consistent AC output voltage and frequency [6]. Traditional approaches, such as boost converters

combined with voltage source inverters (VSIs) [7], current source inverters (CSIs) [8], or impedance source inverters [9], typically require multiple semiconductor devices and complex pulse-width modulation (PWM) control schemes. Additionally, these configurations often impose greater electrical stress on DC-link capacitors, limiting the inverter's durability and operational efficiency.

Single-stage DC-AC inverters with wide gain have been proposed to address these challenges in wide input voltage applications [10]-[12]. For example, paralleling two boost converters at the input can generate an AC output with a differential voltage; however, this setup increases the number of passive components and complicates control, hindering high-performance operation [10]. Another approach involves placing a boost converter in the AC stage of a traditional H-bridge inverter to enable wide gain. Unfortunately, this requires additional semiconductor devices to achieve bidirectional operation, resulting in higher costs and switching losses [11]. Similarly, available grid-connected single-phase buck-boost inverters for PV arrays may utilize high-frequency modulation in a single stage but still rely on many passive components, which can increase cost and system complexity [12].

Considering these limitations, this paper proposes a novel single-phase, single-stage inverter topology derived from the buck-boost converter. This topology is designed to achieve a stable AC output across a broad range of DC input voltages, making it well-suited for dynamic solar applications. By incorporating GaN high-electron-mobility transistors (HEMTs), the proposed converter can operate at a switching frequency of 100kHz, effectively reducing the size of passive components and enhancing power density. The paper further analyzes the modulation strategies and switching states within the topology, as well as the continuous conduction mode (CCM) condition, which helps to minimize electrical stress on semiconductor devices. This analysis supports the hardware design process to ensure reliable performance at the target power level.

A 300W prototype has been constructed to experimentally validate the proposed inverter topology. Experimental results demonstrate that the topology meets the required specifications, providing efficient and stable power conversion with a high degree of modulation flexibility. These findings highlight the proposed inverter's potential as a robust and cost-effective solution for medium-power solar applications, offering both high power density and adaptability for a range of on/off-grid interfacing needs

979-8-3315-1612-3/25 $31.00 © 2025 IEEE

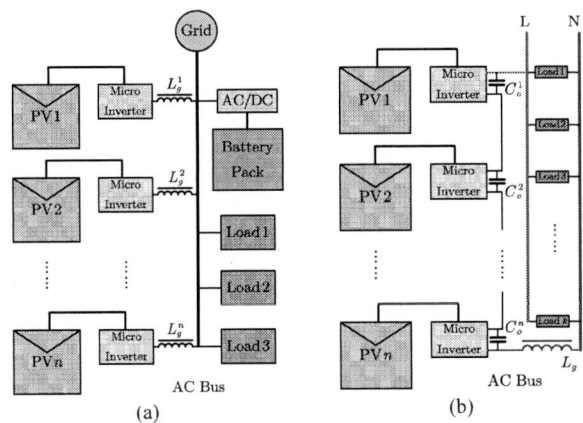

Fig. 1. Two different interconnected micro-inverters in a microgrid.

II. COUPLED-INDUCTOR BUCK-BOOST INVERTER TOPOLOGY

A. Topology Derivation

In [11], an inverter is composed of the two parallel boost converters with their differential voltage as the AC output. Since this circuit has a higher voltage on the output capacitor and there are two inductors, the volume and weight cannot be minimized. Therefore, a new solution is to use a buck-boost converter as a basic element to output differential AC voltage as shown in Fig. 2. The buck-boost converter consists of the input DC source, an inductor, a single pole double throw (SPDT) switch, and a capacitor. The SPDT switch is implemented with a MOSFET and a diode. Note that the output voltage on the capacitor in an ideal buck-boost converter has the inversed polarity referred to its input. i.e.:

$$V_o = -\frac{D}{1-D} V_{in}, \tag{1}$$

where D is the duty cycle of the SPDT switch on the throw tied to V_{in}.

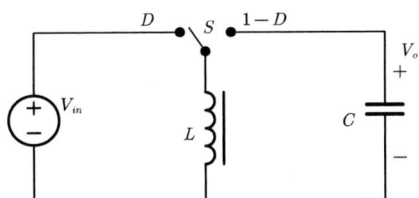

Fig. 2. Basic buck-boost converter.

Combining two basic buck-boost converters shown in Fig. 2 into an AC output inverter, shown in Fig. 3, the inverter output voltage is the differential value of the two outputs:

$$v_{ac} = v_1 - v_2. \tag{2}$$

Substitute (1) explicitly assigned with foot index into (2), the duty cycle defined AC voltage is:

$$v_{ac} = \left(\frac{d_2}{1-d_2} - \frac{d_1}{1-d_1}\right) V_{dc}, \tag{3}$$

where d_1 and d_2 represent the duty cycle of switches S_1 and S_2, respectively. S_{1N} and S_{2N} are complementary switches of S_1 and S_2, respectively. Note that several aspects in Figure 3 can be optimized into Figure 4(a). First, the two capacitors C_1 and C_2, and the AC load resistor R are composed into a loop. Thus, these two capacitors can be optimized into one capacitor C paralleling with the load resistor R. Second, the two separated inductors L_1 and L_2 can be replaced with a coupled inductor L_m with a turns ratio of 1:1 to improve the utilization of the magnetic core in the converter.

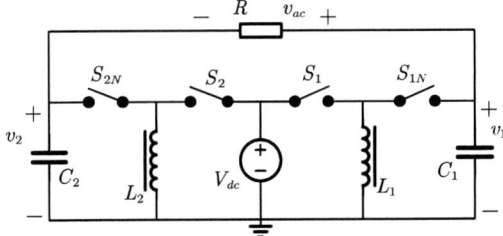

Fig. 3. Differential AC output of two parallel buck-boost converters.

However, in this topology, a diode must be added with the DC source to block the reversing current from the trans-coupled inductor. This diode may not be necessary when a PV source is used. This process is shown in Figure 4(b) and (c) in detail.

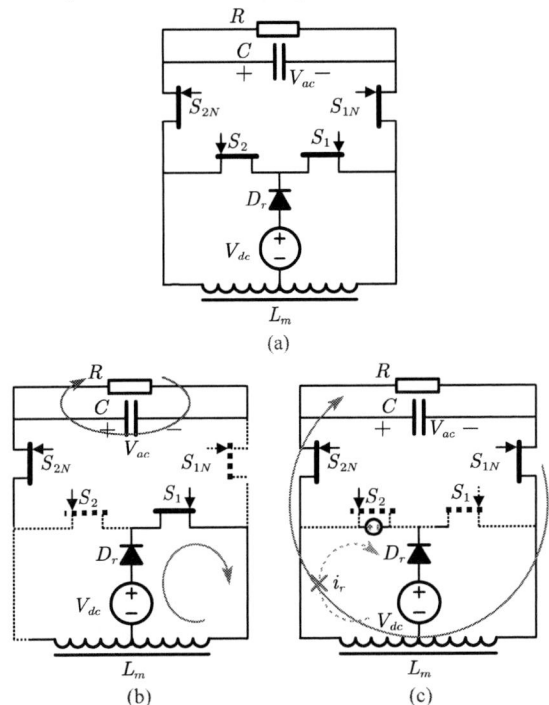

Fig. 4. Trans-coupled buck-boost inverter (a) derived from Fig. 3; (b) Inductor charging interval; (c) inductor discharging interval.

Remark: The output voltage is V_{ac} as shown in Fig. 4 for topology derivation, while its absolute value $v_o = |V_{ac}|$ is being used for simplicity in the next few sections to analyze the circuit since the positive and negative half-cycle are symmetrically related.

B. Circuit Analysis

Since the topology is symmetrical, only half-cycle analysis is sufficient to demonstrate the operation of the inverter. During this process, consider the voltage ripple in C and voltage drop through the semiconductor device are negligible. When devices S_1 and S_{1N} are modulated with high-frequency complementary PWM, switch S_2 is turned off and S_{2N} is turned on constantly. In Figure 4(b), the coupled inductor is magnetized through half

979-8-3315-1612-3/25 $31.00 © 2025 IEEE

of the winding tied to S_1. At the same time, the AC output capacitor feeds the resistive load as a voltage source. The inductor satisfies the voltage equation (4).

$$L'_m \frac{\Delta i_{L'_m}}{D_1 T_s} = V_{dc} \qquad (4)$$

Similarly, the demagnetizing process of the inductor L_m is shown in Figure 4(c). Note that the inductor value is $L'_m = 0.25 L_m$ because half of the winding is being utilized at a given instant. The other inductor equation (5) can be obtained as:

$$L_m \frac{\Delta i_{L_m}}{(1-D_1) T_s} = V_o . \qquad (5)$$

For the trans-coupled inductor, consider a PWM period, its incremental flux is zero to satisfy the volt-second balancing principle:

$$\Delta \lambda_{L_m} + \Delta \lambda_{L'_m} = 0 \qquad (6a)$$

That is:

$$N \Delta i_{L_m} + 2N \Delta i_{L'_m} = 0 \qquad (6b)$$

From equation (4) ~ (6a), the input-output gain of the circuit is:

$$G = \frac{V_{ac}}{V_{dc}} = \frac{2D_1}{1-D_1} \qquad (7)$$

Equation (7) represents the average gain from DC input to AC output when S_1 and S_{1N} are modulated. Ideally, the output voltage can be arbitrarily high as duty cycle D approaches 1. Considering the other half cycle modulated by complementary switches S_2 and S_{2N}, the average AC output over a PWM period can be expressed explicitly:

$$v_{ac}(t) = \left(\frac{2d_1}{1-d_1} - \frac{2d_2}{1-d_2} \right) V_{dc} , \qquad (8)$$

where the small letters d_1 and d_2 used in (8) indicate that those variables are time-variant according to the sinusoidal output requirements. Moreover, there is only one non-zero term in (8) to have the effective AC output.

In Figure 4(c), even when switch S_2 is turned off and according to the GaN HEMT reverse conduction characteristic, there will be a current loop i_r if the voltage on the other half winding L_m high than $V_{dc}+V_r$. V_r is the reverse conduction voltage in the GaN HEMT device [13]. In other words, the DC source and the reversing conduction path through S_2 limited the output voltage to no more than $2(V_{dc}+V_r)$. To increase the instant AC output voltage, a diode is added in Figure 4, which can block the dashed loop current marked in Figure 4(c).

C. Modulation and Linearization

Equation (10) presents the relationship between the output voltage of the proposed topology and the input voltage during either the positive or negative half cycle. Even though there is a dynamic process during the duty cycle variation, assume that the dynamic response of the circuit is fast enough compared to the variation of duty cycle d_x. The average output voltage can be specified by the slow time-variant gain $g(t)$:

$$g(t) = \frac{v_o(t)}{V_{dc}} = \frac{2d_x}{1-d_x} \qquad (9)$$

The duty cycle d_x can be solved in terms of input and output voltages:

$$d_x = \frac{v_o(t)}{2V_{dc} + v_o(t)} \qquad (10)$$

Therefore, in the case of open-loop operation, the duty cycle can be parameterized into (11a) and (11b) by replacing output voltage v_o with $V_m \sin(\omega t)$. If it is operated in closed-loop condition, v_o shall be a control parameter u_c, which is the output of the closed-loop controller, like a proportional-resonant controller for AC output.

$$d_1(t) = \begin{cases} \dfrac{V_m \sin(\omega t)}{2V_{dc} + V_m \sin(\omega t)} & 0 < \omega t < \pi, \\ 0 & \pi < \omega t < 2\pi \end{cases} \qquad (11a)$$

$$d_2(t) = \begin{cases} 0 & 0 < \omega t < \pi, \\ \dfrac{-V_m \sin(\omega t)}{2V_{dc} - V_m \sin(\omega t)} & \pi < \omega t < 2\pi \end{cases} \qquad (11b)$$

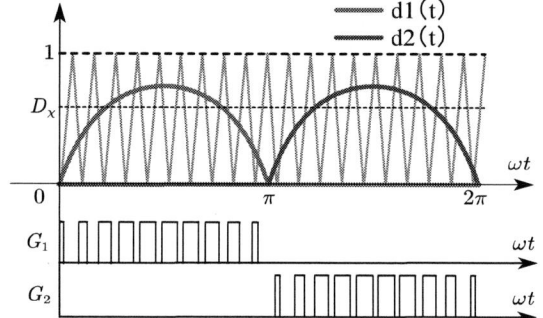

Fig. 5. Modulation scheme of the proposed buck-boost inverter.

III. CIRCUIT EVALUATION

A. Voltage Rating

Compared to the input and output voltage, the deadtime reversing voltage drop V_r and the on-state voltage drop $v_{ds\_on} = i_d R_{on}$ are negligible. Hence, in Fig. 7(a), assuming that the switches are ideal, the voltage on switch S_{1N} can be obtained through the load current loop. Because the voltage on the coupled inductor is 2x the DC source voltage, the drain to source voltages of S_{1N} and S_2 over this duration are:

$$v_{ds}^{1N} = 2V_{dc} + v_o \qquad (12a)$$

$$v_{ds}^2 = 2V_{dc} \qquad (12b)$$

In Fig. 4(c), since the diode is in reverse bias, the voltage stress on switches S_1 and S_2 are:

$$v_{ds}^1 = V_{dc} + \frac{v_o}{2} \qquad (12c)$$

$$v_{ds}^2 = V_{dc} - \frac{v_o}{2} \qquad (12d)$$

However, the diode is used to block the reverse current when output voltage exceeds double input voltage. From the blocked current loop in Fig. 4(c), its voltage stress is:

$$v_{ds}^r = \frac{v_o}{2} - V_{dc} \qquad (12e)$$

Consider the worst-case scenario for the design rating, the voltage stress on these can be plotted in Figure 6 according to the input and output voltage ratio g.

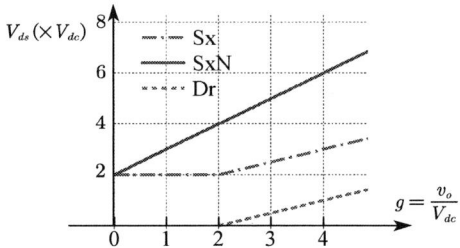

Fig. 6. Voltage stress of semiconductor devices.

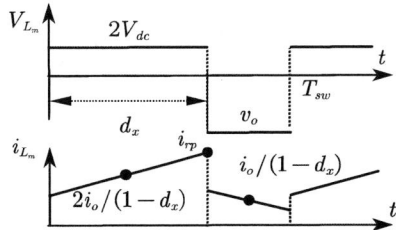

Fig. 7. Principle of selected switches.

B. Current Continues Mode Analysis

To verify if the proposed buck-boost inverter, tests are performed in continuous conduction mode (CCM) and discontinuous conduction mode (DCM). The instantaneous current ripple and average current should be compared over the energy release intervals in Figure 7. Assume that the circuit is operated in CCM, the middle point of the current ripple during switching state Figure 4(c) is:

$$i_{av} = \frac{i_o}{1-d_x} = \frac{v_o}{R_L(1-d_x)} \quad (13)$$

On the other side, the current ripple of this interval is:

$$\Delta i_{Lm} = T_s(1-d_x)\frac{v_o}{L_m} \quad (14)$$

Therefore, the circuit is operated in CCM if and only if:

$$\Delta i_{Lm}(\omega t) \leqslant 2i_{av}(\omega t), \quad 0 \leqslant \omega t \leqslant \pi. \quad (15)$$

Substitute (13) and (14) into (15), and simplify the equation into (16):

$$\frac{L_m}{R_L} \geqslant \frac{T_s}{2}(1-d_x(\omega t))^2 \quad (16)$$

From the duty cycle analysis in Figure 5. The largest right-side value in (16) is $T_s/2$ at the voltage zero crossing. Hence, to satisfy the CCM operation over the entire inversion cycle, the time constant $\tau = L_m/R_L$ should be larger than half of the PWM period.

$$\tau = \frac{L_m}{R_L} \geqslant \frac{T_s}{2} \quad (17)$$

Note that the definition of CCM in this context is that the coupled inductor current doesn't reach zero when the entire PWM cycle completes. That is, the energy in the inductor is not fully exhausted over the entire T_s. CCM operation can guarantee that the analysis in Section II can be applied.

IV. EXPERIMENTS

As shown in Figure 8, a 120VAC, 300W output prototype is built to verify the proposed buck-boost inverter. An external controller STM32F103 was interfaced to modulate the GaN HEMTs with 100kHz PWM signals. The prototype was tested in the open loop condition with different input voltages and fixed output voltage and load. Table 1 lists the detailed parameters and configuration of the test setup.

TABLE 1. Specification of the experimental setup.

Item	value	Item	value
GaN device	GS0650605TA	V_{dc}	50~150 V
V_{gs_on}	6.2V	V_{o_rms}	120V
V_{gs_off}	-3V	I_o	1~5 A
f_{sw}	100kHz	L_m	160 μH
f_o	50Hz	R_L	200 mΩ
C_{dc}	164 μF	C_o	1.5 μF

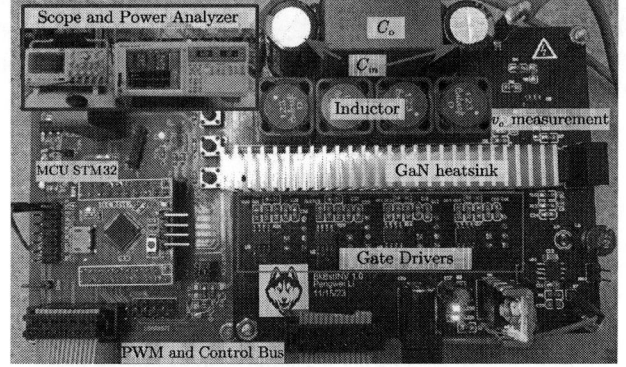

Fig. 8. Prototype built and test setup.

With different input voltages at 60V, 80V, and 120V by adjusting the modulation amplitude and maintaining the output voltage at 50Hz 120VAC, the output voltage and PWM signals are recorded as shown in Figure 9. The scope channels show the PWM of S_1, S_2, diode conduction state D_r, and output voltage v_o from CH1 to CH4, respectively. The 14kHz noise-filtered waveform shows the average duty cycle in the full range of 3.3V, which validates the consistency of analysis in Figure 5. In Figure 9 (a) and (b), since the input voltage is low and the output AC peak voltage exceeds double the input voltage, the reverse blocking diode D_r was reverse biased around the peak region to block the reversing current. However, in (c) and (d), the diode D_r was not reversed because the output voltage is lower than the reversing threshold. It also shows that there is output voltage distortion when input voltage is too low, because of the modulation value reached to the region where the parasitic parameter distorts the original transfer function. On the other side, note that the half-cycle modulation waveforms are not sinusoidal, which is a function of the sinusoidal (11a) and (11b) analyzed in this paper and defined in the modulation controller.

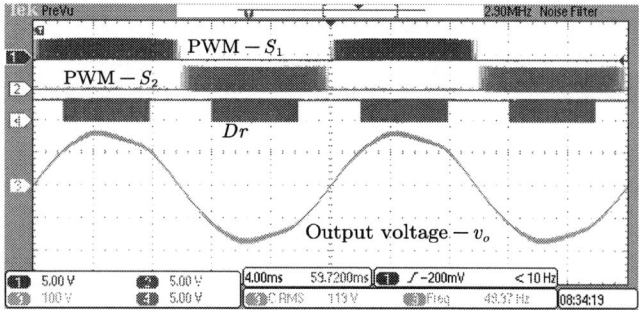

(a). 60V input without filtering.

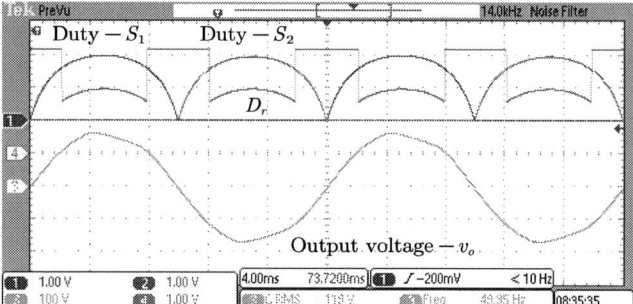

(b). 60V input with filtering.

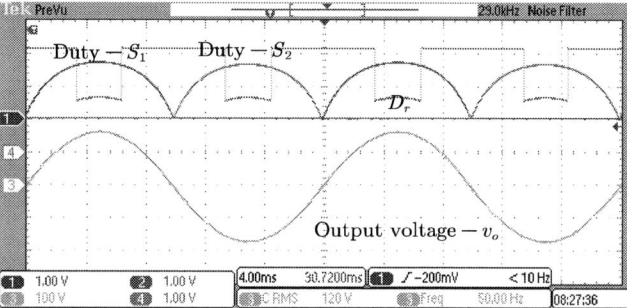

(c). 80V input with filtering.

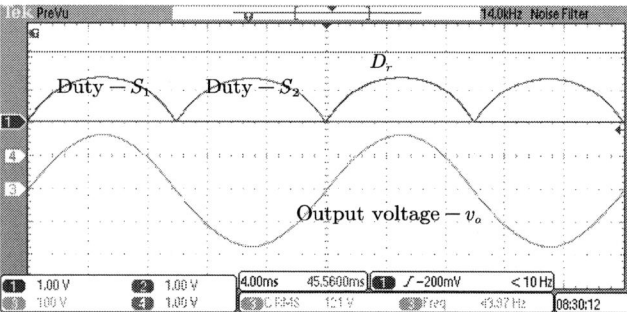

(d). 120V input with filtering.

Fig. 9. Experimental results.

V. CONCLUSION

This paper presents a novel single-phase, single-stage buck-boost inverter topology with a coupled inductor, specifically designed to improve input voltage compatibility and modulation flexibility for medium-power solar energy applications. To validate the proposed topology, A 300W prototype implemented with GaN HEMTs and operating at a 100kHz switching frequency was constructed with smaller passive elements including a couple inductor and an output capacitor. Experimental results confirm the feasibility and effectiveness of the design, demonstrating its potential for high-performance solar interfacing with broad voltage range adaptability. This topology is shown to be a promising solution for advancing power optimization in solar energy systems.

REFERENCES

[1] L. Kazmerski, S. Kurtz and J. Vasi, "Photovoltaics: Impact on People and Society," in IEEE Electron Devices Magazine, vol. 2, no. 3, pp. 6-15, Sept. 2024, doi: 10.1109/MED.2024.3396614.

[2] H. Ma, Z. Guo, C. Hao, G. Qin and J. Lei, "Specific Implementation of a Household Distributed Energy Resource System Based on a Bottom-Up Energy Internet Architecture," 2023 IEEE International Conference on Energy Internet (ICEI), Shenyang, China, 2023, pp. 74-79, doi: 10.1109/ICEI60179.2023.00023.

[3] T. Z. Qi, "A Review of Solar DC Microgrids Design for Smart Farming in a New Zealand Lifestyle Block," 2023 IEEE Fifth International Conference on DC Microgrids (ICDCM), Auckland, New Zealand, 2023, pp. 1-5, doi: 10.1109/ICDCM54452.2023.10433636.

[4] P. Li and A. M. Bazzi, "Efficiency Optimization Method for Parallel Converters in Fault-tolerant Microgrids," 2022 International Power Electronics Conference (IPEC-Himeji 2022- ECCE Asia), Himeji, Japan, 2022, pp. 1898-1902, doi: 10.23919/IPEC-Himeji2022-ECCE53331.2022.9807112.

[5] M. -Y. Fan, Y. Jiang, G. -Y. Yang, Y. -L. Jiang and H. -Y. Yu, "Very-Low Resistance Contact to 2D Electron Gas by Annealing Induced Penetration Without Spikes Using TaAl/Au on Non-Recessed i-AlGaN/GaN," in IEEE Electron Device Letters, vol. 41, no. 10, pp. 1484-1487, Oct. 2020, doi: 10.1109/LED.2020.3020232.

[6] W. Wu, J. Ji and F. Blaabjerg, "Aalborg Inverter - A New Type of "Buck in Buck, Boost in Boost" Grid-Tied Inverter," in IEEE Transactions on Power Electronics, vol. 30, no. 9, pp. 4784-4793, Sept. 2015, doi: 10.1109/TPEL.2014.2363566.

[7] K. Ogura, T. Nishida, E. Hiraki, M. Nakaoka and S. Nagai, "Time-sharing boost chopper cascaded dual mode single-phase sinewave inverter for solar photovoltaic power generation system," 2004 IEEE 35th Annual Power Electronics Specialists Conference (IEEE Cat. No.04CH37551), Aachen, Germany, 2004, pp. 4763-4767 Vol.6, doi: 10.1109/PESC.2004.1354841.

[8] Marignetti, F.; Di Stefano, R.L.; Rubino, G.; Giacomobono, R. Current Source Inverter (CSI) Power Converters in Photovoltaic Systems: A Comprehensive Review of Performance, Control, and Integration. Energies 2023, 16, 7319. https://doi.org/10.3390/en16217319

[9] O. Ellabban and H. Abu-Rub, "Z-Source Inverter: Topology Improvements Review," in IEEE Industrial Electronics Magazine, vol. 10, no. 1, pp. 6-24, March 2016, doi: 10.1109/MIE.2015.2475475.

[10] A. Ashraf Gandomi, K. Varesi and S. H. Hosseini, "DC-AC buck and buck-boost inverters for renewable energy applications," The 6th Power Electronics, Drive Systems & Technologies Conference (PEDSTC2015), Tehran, Iran, 2015, pp. 77-82, doi: 10.1109/PEDSTC.2015.7093253.

[11] S. Dutta and K. Chatterjee, "A Coupled Inductor-Based Buck–Boost Type Grid Connected Transformerless PV Inverter Having the Ability to Control Two Subarrays Simultaneously," in IEEE Transactions on Industrial Electronics, vol. 67, no. 7, pp. 5543-5553, July 2020, doi: 10.1109/TIE.2019.2931512.

[12] W. Wu, J. Ji and F. Blaabjerg, "Aalborg Inverter - A New Type of "Buck in Buck, Boost in Boost" Grid-Tied Inverter," in IEEE Transactions on Power Electronics, vol. 30, no. 9, pp. 4784-4793, Sept. 2015, doi: 10.1109/TPEL.2014.2363566.

979-8-3315-1612-3/25 $31.00 © 2025 IEEE

Gap in pagination due to withheld paper.

Pages 2337-2342

A Dynamic Current Sharing Method using Novel Clip Considering Mutual Inductance Coupling

Zexiang Zheng
School of Electrical and Electronic Engineering Huazhong University of Science and Technology
Wuhan, China
zzx_huster@hust.edu.cn

Jianwei Lv
School of Electrical and Electronic Engineering Huazhong University of Science and Technology
Wuhan, China
jianweil@hust.edu.cn

Yiyang Yan
School of Electrical and Electronic Engineering Huazhong University of Science and Technology
Wuhan, China
yanyiyang@hust.edu.cn

Baihan Liu
School of Electrical and Electronic Engineering Huazhong University of Science and Technology
Wuhan, China
loubeckham@hust.edu.cn

Yifan Zhang
School of Electrical and Electronic Engineering Huazhong University of Science and Technology
Wuhan, China
z_yifan@hust.edu.cn

Linhao Ren
School of Electrical and Electronic Engineering Huazhong University of Science and Technology
Wuhan, China
ringo@hust.edu.cn

Jiaxin Liu
School of Electrical and Electronic Engineering Huazhong University of Science and Technology
Wuhan, China
liujx711@hust.edu.cn

Cai Chen
School of Electrical and Electronic Engineering Huazhong University of Science and Technology
Wuhan, China
caichen@hust.edu.cn

Yong Kang
School of Electrical and Electronic Engineering Huazhong University of Science and Technology
Wuhan, China
ykang@hust.edu.cn

Xiong Zhang
GAC Aion New Energy Automobile Co., Ltd.
Guangzhou,
zhangxiong@aion.com.cn

Hao Yu
GAC Aion New Energy Automobile Co., Ltd.
Guangzhou, China
yuhao@aion.com.cn

Wei Jiang
GAC Aion New Energy Automobile Co., Ltd.
Guangzhou, China
jiangwei@aion.com.cn

Abstract—**Imbalanced parasitic parameters can cause uneven dynamic currents among parallel chips, leading to uneven losses and junction temperatures, which in turn affect system lifetime. In addition, clip bonding is a low thermal resistance and highly reliable packaging method. Therefore, this paper proposes a novel clip to optimize the dynamic current sharing performance. The positions of the proposed clip's pins can be changed to adjust imbalanced self-inductances and mutual inductances within the power module. A design method for the positions of the pins is presented. Based on ANSYS Q3D, the optimized clip can be effectively designed with the maximum allowable value. Compared to existing methods, the proposed method considers the unbalanced mutual inductances and does not require changing the chip layout. Simulation results validate the effectiveness of the proposed clip. Finally, two half-bridge power modules were fabricated, and double pulse tests were conducted to further verify the proposed method.**

Keywords—Clip, SiC MOSFET, Power module, Dynamic current

I. INTRODUCTION

Silicon carbide (SiC) MOSFETs are increasingly being adopted in transportation due to their superior electrical characteristics. Limited to the current carrying capability, SiC MOSFETs are commonly used in parallel within power modules [1]. However, the unbalanced parasitic parameters and chip parameters can cause imbalanced dynamic currents among parallel MOSFETs, which results in thermal imbalance, ultimately affecting system reliability [2].

Numerous current balancing methods have been proposed. In [3]-[6], symmetrical layouts were proposed to achieve more balanced currents. However, modifying the layout requires changes to the external busbar connection structure, which limits the applicable scenarios. In [7-9], the connection points of bonding wires were modified to improve dynamic current sharing performance. In [10], a clip with extra modification paths (MPs) was proposed. The length of the MPs is adjusted to achieve balanced power source inductances, leading to balanced dynamic currents. On the other hand, integrated components can also improve dynamic current sharing performance. In previous research, capacitances and coupling inductances were integrated to mitigate imbalanced currents [2], [11], [12]. However, the above research mainly focused on addressing imbalance self-inductances. The impact of mutual inductance coupling of package structure is also significant and cannot be ignored [13] [14]. Even though theoretical models have been researched in [13] and [14], passive dynamic current sharing methods considering mutual inductance coupling fully are still rare.

On the other hand, the bonding form of packaging is also a research hotspot. Conventional Al bonding wires are widely used due to mature technology and low cost. However, the high di/dt of SiC devices causes significant voltage overshoot due to the large parasitic inductance introduced by Al bonding wires. Additionally, the high operating temperature of SiC devices increases the risk of bonding wire failure [15]. To address these issues, Cu clip bonding packaging is considered a good solution for high reliability and low thermal resistance [16], [17].

In this paper, a clip structure for multichip power modules is introduced. Based on a dynamic current sharing model considering mutual inductance coupling, the proposed clip is optimized to improve the dynamic current sharing

979-8-3315-1612-3/25 $31.00 © 2025 IEEE

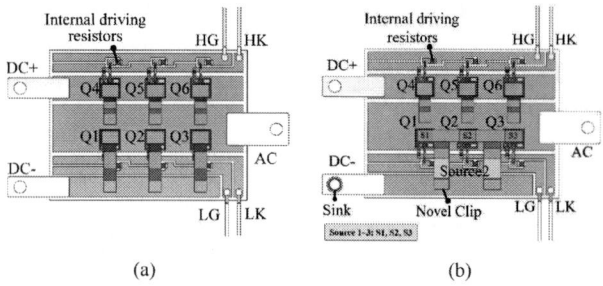

(a)   (b)

Fig. 1. (a) Conventional clip bonding power module. (b) Proposed clip bonding power module.

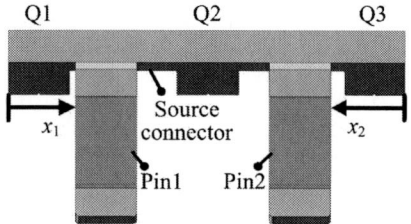

Fig. 2. Structural diagram of the proposed clip.

performance of parallel SiC MOSFETs effectively. Both simulations and experiments have validated the effectiveness of the proposed method.

II. PROPOSED CLIP STRUCTURE

A clip-bonding power module with a conventional layout is shown in Fig. 1(a), which is similar to EconoDUAL. In the power loop, one side of the conventional Clip is soldered to the power source of the SiC MOSFET, while the other side is soldered to the copper layer of the substrate to establish an electrical connection. Each MOSFET has a dedicated power clip. Similarly, the driving loops of the MOSFETs are also connected using a clip.

The proposed clip is implemented as shown in Fig. 1(b). The proposed Clip consists of a source connector, pin 1, and pin 2, as shown in Fig. 2. The source connector connects the power source of MOSFETs and pins. Pin1 and pin2 are the electrical connection points for the power sources. Compared to the conventional clip, the biggest advantage of the proposed clip is that the positions of pin1 and pin2 can be freely adjusted.

The specific positions of pin1 and pin2 can be represented by x_1 and x_2, respectively. The proposed clip can change the self-inductances and mutual inductances within the power module by adjusting the positions of pin1 and pin2. Therefore, it is necessary to consider all self-inductances and mutual inductances to optimize the clip. Next, the optimized process of finding satisfactory x_1 and x_2 is considered.

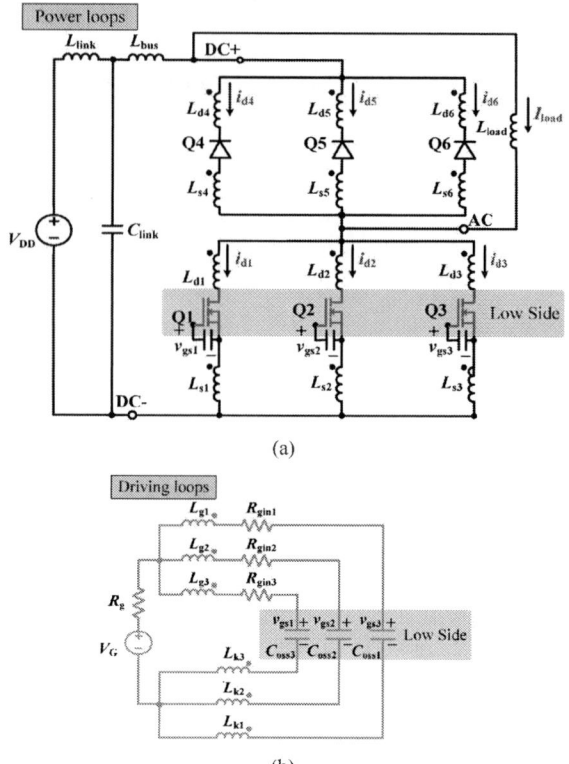

(a)

(b)

Fig. 3. Parasitic circuit model considering self-inductances and mutual inductances. (a) Power loop. (b) Driving loop.

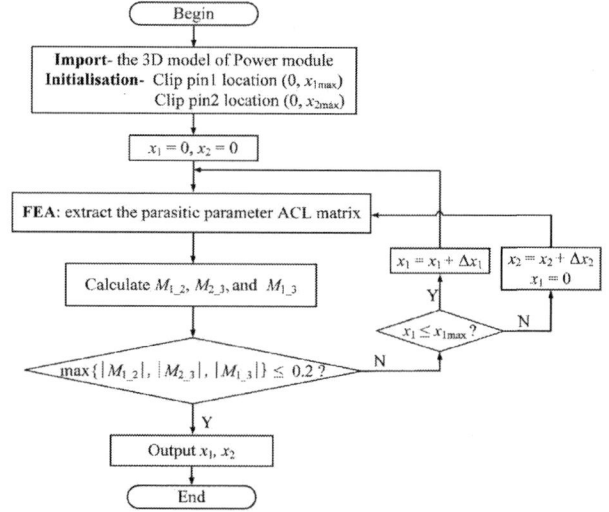

Fig. 4. Solution block diagram of clip structure optimization.

$$\begin{cases} M_{\text{Hd1\_2}} = -(M_{\text{Ls1Ld4}}+M_{\text{Ls1Ld5}}+M_{\text{Ls1Ld6}})+(M_{\text{Ls2Ld4}}+M_{\text{Ls2Ld5}}+M_{\text{Ls2Ld6}})-(M_{\text{Lg1Ld4}}+M_{\text{Lg1Ld5}}+M_{\text{Lg1Ld6}})+(M_{\text{Lg2Ld4}}+M_{\text{Lg2Ld5}}+M_{\text{Lg2Ld6}}) \\ M_{\text{Hs1\_2}} = -(M_{\text{Ls1Ls4}}+M_{\text{Ls1Ls5}}+M_{\text{Ls1Ls6}})+(M_{\text{Ls2Ls4}}+M_{\text{Ls2Ls5}}+M_{\text{Ls2Ls6}})-(M_{\text{Lg1Ls4}}+M_{\text{Lg1Ls5}}+M_{\text{Lg1Ls6}})+(M_{\text{Lg2Ls4}}+M_{\text{Lg2Ls5}}+M_{\text{Lg2Ls6}}) \\ M_{\text{Ld1\_2}} = -(M_{\text{Ls1Ld1}}+M_{\text{Ls1Ld2}}+M_{\text{Ls1Ld3}})+(M_{\text{Ls2Ld1}}+M_{\text{Ls2Ld2}}+M_{\text{Ls2Ld3}})-(M_{\text{Lg1Ld1}}+M_{\text{Lg1Ld2}}+M_{\text{Lg1Ld3}})+(M_{\text{Lg2Ld1}}+M_{\text{Lg2Ld2}}+M_{\text{Lg2Ld3}}) \\ M_{\text{Ls1\_2}} = -(L_{\text{s1}}+M_{\text{Ls1Ls2}}+M_{\text{Ls1Ls3}})+(M_{\text{Ls2Ls1}}+L_{\text{s2}}+M_{\text{Ls2Ls3}})-(M_{\text{Lg1Ls1}}+M_{\text{Lg1Ls2}}+M_{\text{Lg1Ls3}})+(M_{\text{Lg2Ls1}}+M_{\text{Lg2Ls2}}+M_{\text{Lg2Ls3}}) \end{cases}$$

(2)

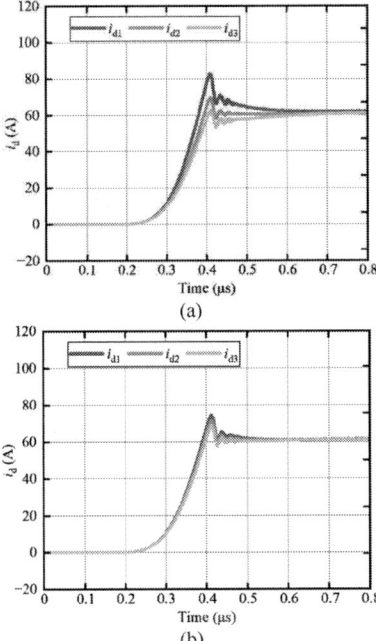

Fig. 5. Simulation turn-on waveforms. (a) Conventional module. (b) Optimized module.

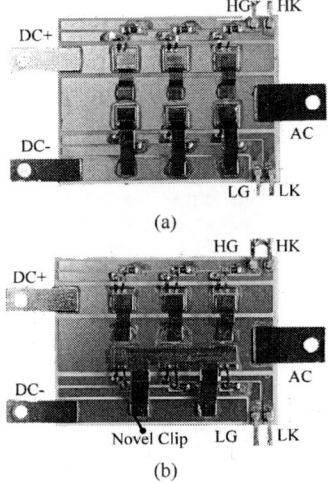

Fig. 6. Fabricated power modules. (a) Conventional module. (b) Optimized module.

III. PROPOSED CLIP OPTIMIZATION

In the section, the pin positions of the proposed clip are optimized based on the dynamic current sharing model considering mutual inductance coupling [14].

A half-bridge parasitic circuit with three chips in parallel is shown in Fig. 3. The proposed heterogeneous clip makes the parasitic circuit of the power module different from the parasitic circuit in Fig. 3. However, in ANSYS Q3D, DC- can be set as the sink and the power sources of Q1~Q3 can be set as source 1-3, as shown in Fig. 1(b). Then, the power module

with the heterogeneous clip can be directly converted into the parasitic circuit shown in Fig. 3.

According to the dynamic current sharing design guide in [14], the current sharing coefficient M_{12} of Q1 and Q2 of the two chips can be expressed as (1) and (2).

$$M_{1\_2} = M_{Hd1\_2} + M_{Hs1\_2} + M_{Ld1\_2} + M_{Ls1\_2} \tag{1}$$

where L_{s1} and L_{s2} represent the power source self-inductances of Q1 and Q2. M_{xy} represents the mutual inductance between the branches x and y. For example, M_{Ls1Ld3} represents the mutual inductance between L_{s1} and L_{d3}. Similarly, current sharing coefficients M_{23} and M_{13} can also be derived in the same way. Detailed derivation steps and analysis of the equations are covered in [14]. When M_{12}, M_{23}, and M_{13} are close to 0 at the same time, Q1-Q3 can achieve balanced dynamic currents. It can be used as a guide for the optimization design of pin1 and pin2.

The solution procedure for optimizing the positions of pin1 and pin2 is shown in Fig. 4. First, the 3D model of the power module is imported into ANSYS Q3D. The positions of pin 1 and pin 2 are parameterized as x_1 and x_2. When pin 1 is at the far left and pin 2 at the far right, x_1 and x_2 are both set to 0. Initialization determines the maximum values for x_1 and x_2. The parasitic inductance matrix of the power module is extracted at various x_1 and x_2. Subsequently, current sharing coefficients (M_{12}, M_{23}, and M_{13}) are calculated based on the parasitic inductances. If the maximum absolute value of the current sharing coefficient is less than the maximum allowable value (σ), the result is considered to have achieved the final goal.

Obviously, the selection of σ determines both the final dynamic current sharing performance and the computational time required for optimization. In this paper, σ is set to 0.2. When their maximum absolute value does not exceed 0.2, x_1 and x_2 are selected as the optimized position. Finally, $x_1 = 3.4$ mm, and $x_2 = 4.0$ mm are selected. The maximum current sharing coefficient (M_{13}) is reduced from the initial 3.01 to 0.16.

In order to verify the effectiveness, the conventional and optimized parasitic inductance networks are extracted by Q3D and imported into LTSIPCE. Double pulse simulation is performed to verify the dynamic current sharing performance. The simulated turn-on waveforms of Q1-Q3 in the conventional and optimized power modules are shown in Fig. 5. The maximum dynamic current difference of the conventional power module is 20.0 A. However, with the optimized clip, the maximum dynamic current is only 5.2 A. The effectiveness of the proposed clip is verified. In addition, the dynamic current sharing characteristic can be further optimized by reducing σ, but this may increase the optimization calculations.

IV. FABRICATION AND EXPERIMENTAL VERIFICATION

To further verify the effectiveness of the proposed clip, a double pulse test is performed. First, the conventional and optimized power modules are fabricated. The fabrication process consists of two steps. In the first step, SiC MOSFETs, gate resistors, and terminals are soldered. In the second step,

979-8-3315-1612-3/25 $31.00 ©2025 IEEE

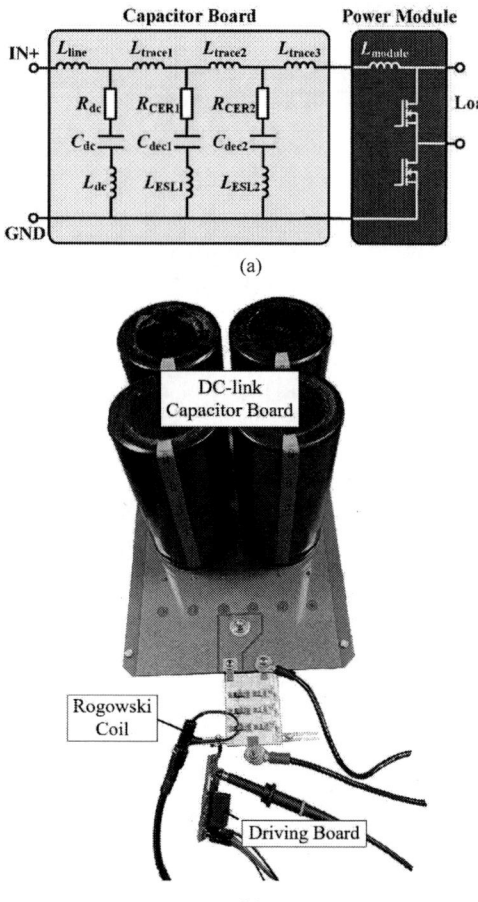

(a)

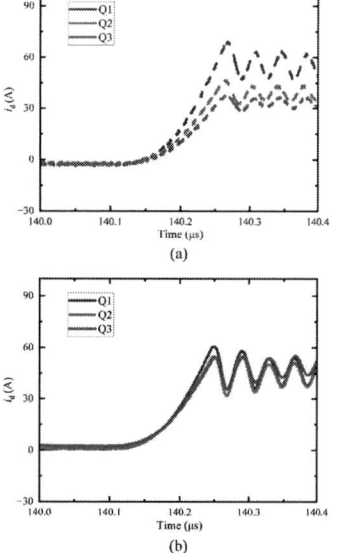

(b)

Fig. 7. Double pulse test. (a) Schematic diagram. (b) Test setups.

(a)

(b)

Fig. 8. Comparison of turn-on current waveforms. (a) Conventional module. (b) Optimized module.

the clips for the gate drive and power loops are soldered onto the MOSFETs. The conventional and optimized power modules are fabricated as shown in Fig. 6.

The schematic of the double pulse test for low-side MOSFETs is shown in Fig. 7(a). The power module is mounted on a test board equipped with multi-stage decoupling capacitors. The test tig is shown in Fig. 7(b). The driver board outputs a +18V/-2V signal to control the turn-on and turn-off of the low-side MOSFETs. The body diodes of the high-side MOSFETs serve as freewheeling diodes. The gate and source of the high-side MOSFETs are shorted. The current flowing through the single chip is measured by the Rogowski coil. The turn-on dynamic currents of the conventional and optimized power modules are compared in Fig. 8. Manufacturing tolerances in the clip and parameter inconsistencies between chips may introduce some errors, but the effectiveness of the proposed clip is still well-validated. The low-side MOSFETs with optimized Clip achieve better dynamic current sharing performance.

V. Conclusions

In this paper, a dynamic sharing method using a novel clip has been presented. The positions of the proposed clip's pins can be adjusted to modify various parasitic parameters within the power module. This method does not require changing the chip layout or adding additional components, which is easier to implement. Furthermore, compared with previous methods, mutual inductance coupling is considered in the optimization process. The optimal positions of the clip's pins are obtained based on a dynamic current sharing model. The specific optimization process is demonstrated and verified. An electromagnetic simulation was conducted to verify the effectiveness of the proposed method. Finally, two comparative power modules were fabricated and tested to further verify the proposed method.

References

[1] J. Millán, P. Godignon, X. Perpiñà, A. Pérez-Tomás, and J. Rebollo, "A Survey of Wide Bandgap Power Semiconductor Devices," IEEE Transactions on Power Electronics, vol. 29, no. 5, pp. 2155-2163, May 2014.

[2] J. Lv, C. Zhang, C. Chen and Y. Kang, "A Dynamic Current Sharing Method in Multi-chip SiC Power Module Using Stacked DBC Bridges and Decoupling Capacitors Based on the Original Simple Module Layout," 2021 IEEE Workshop on Wide Bandgap Power Devices and Applications in Asia (WiPDA Asia), Wuhan, China, 2021, pp. 184-188.

[3] H. Li, S. Munk-Nielsen, S. Bęczkowski, and X. Wang, "A Novel DBC Layout for Current Imbalance Mitigation in SiC MOSFET Multichip Power Modules," IEEE Transactions on Power Electronics, vol. 31, no. 12, pp. 8042-8045, Dec. 2016.

[4] B. Zhao, P. Sun, Q. Yu, Y. Cai and Z. Zhao, "Layout-Dominated Dynamic Imbalanced Current Analysis and Its Suppression Strategy of Parallel SiC MOSFETs," in IEEE Transactions on Device and Materials Reliability, vol. 21, no. 3, pp. 394-404, Sept. 2021.

[5] M. Wang, F. Luo, and L. Xu, "A Double-End Sourced Wire-Bonded Multichip SiC MOSFET Power Module With Improved Dynamic Current Sharing," in IEEE Journal of Emerging and Selected Topics in Power Electronics, vol. 5, no. 4, pp. 1828-1836, Dec. 2017.

[6] S. Weihua, L. Ran, Z. Zheng, L. Xiaoling, and P. Mawby, "Design and evaluation of SiC multichip power module with low and symmetrical inductance," J. Eng., vol. 2019, no. 17, pp. 3573–3577, 2019.

[7] S. Bęczkowski, A. B. Jørgensen, H. Li, C. Uhrenfeldt, X. Dai and S. Munk-Nielsen, "Switching current imbalance mitigation in power

modules with parallel connected SiC MOSFETs," in 2017 19th European Conference on Power Electronics and Applications (EPE'17 ECCE Europe), Warsaw, Poland, 2017, pp. P.1-P.8.

[8] C. Zhao, L. Wang, F. Zhang, and F. Yang, "A Method to Balance Dynamic Current of Paralleled SiC MOSFETs With Kelvin Connection Based on Response Surface Model and Nonlinear Optimization," in IEEE Transactions on Power Electronics, vol. 36, no. 2, pp. 2068-2079, Feb. 2021.

[9] B. Zhang, R. Wang, P. Barbosa, Y. -H. Tsai, W. -S. Wang and W. -S. Lai, "Common Source Inductance Compensation Technique for Dynamic Current Balancing in SiC MOSFETs Parallel Operations," 2023 IEEE Applied Power Electronics Conference and Exposition (APEC), Orlando, FL, USA, 2023, pp. 358-365.

[10] L. Wang, T. Zhang, F. Yang, D. Ma, C. Zhao, Y. Pei, "Cu Clip-Bonding Method With Optimized Source Inductance for Current Balancing in Multichip SiC MOSFET Power Module," in IEEE Transactions on Power Electronics, vol. 37, no. 7, pp. 7952-7964, July 2022.

[11] Z. Miao, Y. Mao, G. -Q. Lu and K. D. T. Ngo, "Magnetic Integration Into a Silicon Carbide Power Module for Current Balancing," in IEEE Transactions on Power Electronics, vol. 34, no. 11, pp. 11026-11035, Nov. 2019,.

[12] Y. Mao, Z. Miao, C. -M. Wang and K. D. T. Ngo, "Balancing of Peak Currents Between Paralleled SiC MOSFETs by Drive-Source Resistors and Coupled Power-Source Inductors," in IEEE Transactions on Industrial Electronics, vol. 64, no. 10, pp. 8334-8343, Oct. 2017.

[13] Y. Ge, Z. Wang, Y. Yang, C. Qian, G. Xin, and X. Shi, "Layout-Dominated Dynamic Current Balancing Analysis of Multichip SiC Power Modules Based on Coupled Parasitic Network Model," in IEEE Transactions on Power Electronics, vol. 38, no. 2, pp. 2240-2251, Feb. 2023.

[14] Z. Zheng, C. Chen, J. Lv, Y. Yan, J. Liu, and Y. Kang, "A Dynamic Current Sharing Model of Multichip Parallel SiC MOSFETs Considering Layout-Dominated Mutual Inductance Coupling," in IEEE Transactions on Power Electronics, vol. 39, no. 9, pp. 11060-11073, Sept. 2024.

[15] C. Durand, M. Klingler, D. Coutellier, and H. Naceur, "Power cycling reliability of power module: A survey," IEEE Trans. Device Mater. Rel., vol. 16, no. 1, pp. 80–97, Mar. 2016.

[16] H. Chang, J. Bu, G. Kong, and R. Labayen, "300A 650V 70 um thin IGBTs with double-sided cooling," in Proc. IEEE 23rd Int. Symp. Power Semicond. Devices ICs, May 2011, pp. 320–323.

[17] Y. Zhu, H. Chen, K. Xue, M. Li, and J. Wu, "Thermal and reliability analysis of clip bonding package using high thermal conductivity adhesive," in Proc. IEEE 15th Electron. Packag. Technol. Conf., Dec. 2013, pp. 259–263.

Application-Oriented Test Setup for Measuring Dynamic Output and Transfer Characteristics of GaN-HEMTs

1st Philipp Swoboda
Elektrotechnisches Institut (ETI)
Karlsruhe Institute of Technology (KIT)
Karlsruhe, Germany
philipp.swoboda@kit.edu

2nd Martin Fein
Elektrotechnisches Institut (ETI)
Karlsruhe Institute of Technology (KIT)
Karlsruhe, Germany
martin.fein@kit.edu

3rd Simon Frank
Elektrotechnisches Institut (ETI)
Karlsruhe Institute of Technology (KIT)
Karlsruhe, Germany
s.frank@kit.edu

4th Andreas Liske
Elektrotechnisches Institut (ETI)
Karlsruhe Institute of Technology (KIT)
Karlsruhe, Germany
andreas.liske@kit.edu

5th Marc Hiller
Elektrotechnisches Institut (ETI)
Karlsruhe Institute of Technology (KIT)
Karlsruhe, Germany
marc.hiller@kit.edu

Abstract—Apart from the excellent switching and conduction properties of power GaN-HEMTs, trapping effects are still an issue degrading the device's performance. As conventional methods have been proven inadequate for device characterization with respect to trapping, new application-oriented characterization methods have been developed in recent years. In this paper, an application-oriented test setup for measuring output and transfer characteristics is presented, which extends an existing setup for the measurement of the dynamic on-resistance. Initial results indicate that there is a significant change in characteristics dependent on the applied off-state drain voltage bias. Additionally, several typical trapping phenomena such as the dynamic on-resistance, threshold voltage shift and current collapse were observed.

Index Terms—Dynamic on-state resistance, gallium nitride high-electron-mobility transistors (GaN-HEMT), power semiconductor device characterization, conduction-losses, double pulse test (DPT), threshold voltage shift, transfer and output characteristics

I. INTRODUCTION

Trapping effects such as threshold voltage shift or $V_{\mathrm{gs,th}}$-shift, the dynamic on-resistance $dR_{\mathrm{ds,on}}$ and current collapse are still present in today's power GaN-HEMTs. Those effects are mainly stimulated by applying a gate or a drain voltage bias to the device, when it is blocking. Previous investigations have shown that the impact of those effects on the device's performance depends on the magnitude and duration of the applied voltage bias [1]–[4]. In many cases, those effects are reversible, if a relaxation interval is inserted in which the biases are removed.

This work was carried out within the GaNius-programme for "Energy Efficient Power Electronics" funded by DFG (Deutsche Forschungsgemeinschaft). Name of the partial project: *"Modelling and characterization of GaN-HEMT devices with respect to effects of charge carrier trapping"*

Conventional methods for device characterization, as used for Si-based devices, are often inadequate, as the device is operated under idealistic conditions and not in an application-oriented scenario. The influence of trapping is neglected, since no preconditioning with drain or gate bias is applied to the device under test (DUT) prior to the measurement. To tackle that issue, several application-oriented characterization methods have been developed in recent years. In the majority of publications, the dynamic on-resistance $dR_{\mathrm{ds,on}}$ was investigated, mostly in a modified double pulse test (DPT) setup, which is close to an actual power electronic converter [1], [5]–[7].

Further, the $V_{\mathrm{gs,th}}$-shift phenomenon gained academic attention in recent years [2], [3]. In [2], the standardized method according to IEC60747-8-4 [8] for evaluation of the threshold voltage $V_{\mathrm{gs,th}}$ was expanded by additional circuitry to allow investigations on the $V_{\mathrm{gs,th}}$-shift phenomenon under consideration of drain- and gate-bias. The applied drain-bias V_{ds} during the $V_{\mathrm{gs,th}}$-measurement was kept at a constant low voltage of 2.4 V, which is much lower than a typical V_{ds} in a real hard-switching application. This, however, allows a more isolated investigation of the $V_{\mathrm{gs,th}}$-shift, as V_{ds}-dependent effects, such as Drain-Induced Barrier Lowering (DIBL), do not occur. Strong $V_{\mathrm{gs,th}}$-shifts of up to 500 mV were observed both under applied drain and gate bias in [2], with time constants of up to the hour range.

The test circuit presented in [3] allows $V_{\mathrm{gs,th}}$-measurements, whilst a V_{ds}-bias of up to 400 V is applied. Thereby, the DUT was switched under zero load current, which is more close to a real application and includes V_{ds}-dependent effects like DIBL. The results show a significant time-dependent $V_{\mathrm{gs,th}}$-shift.

979-8-3315-1612-3/25 $31.00 © 2025 IEEE

Further, the output characteristics was measured using a device analyzer. Thereby, a V_{ds}-bias of $400\,\mathrm{V}$ was applied prior to the measurement. The result was a poorer performance of the device, indicated by an increased $R_{ds,on}$, a knee-point that was shifted downwards and a decreased saturation current. However, no previous work has been found that investigates those "dynamic" output and transfer characteristics in a converter-like setup, as it was done especially in the $dR_{ds,on}$-investigations [1], [4].

In this paper, a test setup is presented, which is capable of measuring both transfer and output characteristics with respect to trapping, which includes the evaluation of the threshold voltage $V_{gs,th}$. It extends an existing test setup to investigate the dynamic on-resistance $dR_{ds,on}$ [4]. The measurement is carried out in a converter-like setup, whereby the DUT is switched under an inductive load, as often the case in a real power electronics application. Similar to [3], the threshold voltage $V_{gs,th}$ can be measured as a part of the transfer characteristics under fully applied blocking voltage V_{ds} of up to $400\,\mathrm{V}$. This makes the characterization very close to the application.

The paper is structured as follows: First, the test circuit with its auxiliary circuitry around the DUT is described. This is followed by a description of the proposed measurement procedures for transfer and output characteristics. Subsequently, a short description of the practical test setup is provided. Finally, first measurements are shown and their results discussed.

II. DESCRIPTION OF THE TEST CIRCUITS

A. Overview

The main part of the test setup is an advanced DPT-circuit, consisting of a full bridge with one additional device LS1 connected in parallel to the DUT (Fig. 1). As with most power electronic circuits, an inductive load L_{DPT} is used, which is connected between the two switching nodes. The additional

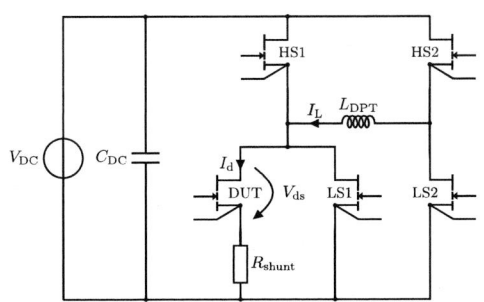

Fig. 1: Modified Double Pulse Circuit

device LS1 is required for the test procedures, to i.a. reduce the self-heating of the DUT. Its further purposes are described in the following sections.

In contrast to a conventional DPT setup, the given setup allows to control the applied drain-bias $V_{ds,IDLE}$ on the DUT,

when no measurement is performed. By switching HS1 and HS2 on, $V_{ds,IDLE}$ becomes the DC-link voltage V_{DC}, whereas when LS1 and LS2 are turned on, $V_{ds,IDLE}$ becomes zero. This ability is crucial to control the trapping effects of GaN-HEMTs. The test circuit was first introduced in [4], where it

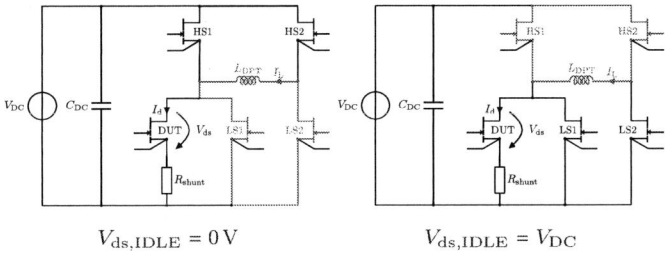

Fig. 2: Idle states for drain-bias preconditioning

was used for $dR_{ds,on}$-measurements. In this paper, the circuit is expanded by additional circuitry, which enables output and transfer characteristics measurements.

B. Adaptive Gate Driver

As the procedure for transfer characteristics requires a slowed-down hard-switching event, the rise time at the gate needs to be adjustable down to several hundreds of nanoseconds. Further, for measuring the output characteristics at different magnitudes of the gate voltage V_{gs}, the gate driver must be capable of adjusting its output voltage. At the same time, parasitic turn-on of the DUT must be prevented, which requires a low-impedance path from the gate to the source in the off-state.

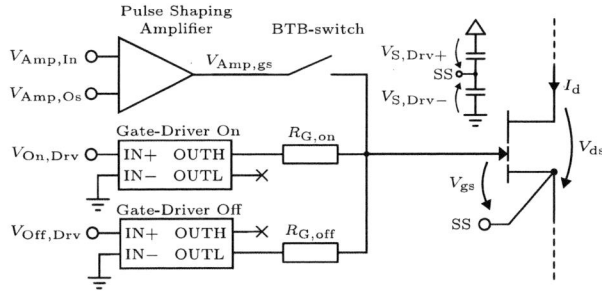

Fig. 3: Adaptive Gate Driver

To meet those requirements, an adaptive gate driver is required which is depicted in Fig. 3. It consists of a pulse shaping amplifier (PSA) and two commercially available gate-drivers [9], which are supplied by two adjustable bipolar power supply rails $V_{S,Drv+}$ and $V_{S,Drv-}$. The gate-drivers feature two outputs each, one for turn-on (OUTH) and another one for turn-off (OUTL). By combining their logic inputs $V_{On,Drv}$ and $V_{Off,Drv}$ and by tieing their outputs together via the gate resistors $R_{G,on}$ and $R_{G,off}$, a gate-driver with tristate-output is formed. The gate can either be left floating or be tied to $V_{S,Drv+}$ or $V_{S,Drv-}$ via the gate resistors (see TABLE I). Although the two gate-drivers can provide a low-impedance path from the gate to the supply rails, they cannot adjust

TABLE I: Truth Table of the formed Tristate Gate-Driver

$V_{On,Drv}$	$V_{Off,Drv}$	V_{gs}
H	H	$V_{S,Drv+}$
L	L	$V_{S,Drv-}$
L	H	Floating
H	L	Forbidden

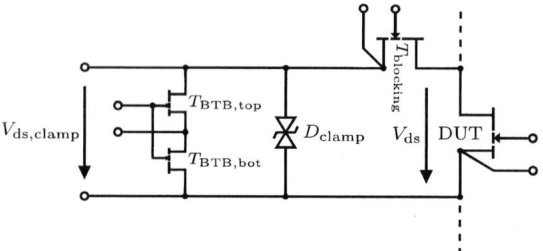

Fig. 5: Clamp Circuit with High Accuracy

V_{gs} in magnitude and speed. This feature is provided by a PSA (Fig. 4), which can be connected to the gate-node of the DUT using two back-to-back (BTB)-connected transistors. Its output voltage $V_{Amp,gs}$ is controlled by the two inputs $V_{Amp,In}$ and $V_{Amp,Os}$. Both of them are generated by a digital-to-analog converter (DAC). To filter out the stairs of

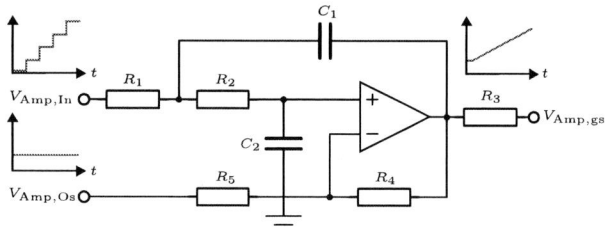

Fig. 4: Pulse Shaping Amplifier

the DAC-signal, $V_{Amp,In}$ is smoothed by a Sallen-Key low-pass filter [10], which is composed of R_1, R_2, C_1 and C_2. Due to its low magnitude, $V_{Amp,In}$ does not fit the required range of V_{gs}, which can be for instance from $-6\,\mathrm{V}$ to $6\,\mathrm{V}$. Thus, amplification and offset shift is required. This feature is provided by the non-inverting amplifier part of the PSA, formed by R_4 and R_5. The offset is thereby controlled by $V_{Amp,Os}$. As $V_{Amp,Os}$ is a DC-signal, there is no need for filtering. The DC transfer function of the PSA is

$$V_{Amp,gs} = V_{Amp,In} \cdot \left(1 + \frac{R_4}{R_5}\right) - V_{Amp,Os} \cdot \frac{R_4}{R_5} . \quad (1)$$

Note that it is not possible that both the PSA and the gate-drivers are active at the same time. If the gate of the DUT needs to be charged with a reduced and adjustable speed to a certain magnitude, the two gate-driver outputs OUTH and OUTL need to be left floating and the BTB-switch needs to be turned on, which connects the PSA to the gate. Outside the measurement phase or when the inductor is charging, the gate potential is pulled against the $V_{S,Drv-}$-potential, providing a low-impedance path, whilst the BTB-switch is turned off.

C. Clamp Circuit

As the DUT can see blocking voltages of up to $400\,\mathrm{V}$ while the on-state voltages are no higher than $15\,\mathrm{V}$, a clamp circuit is required. This enables measurements of the on-state drain voltage V_{ds} with high accuracy, which is required for the output characteristics. The actively controlled clamp circuit is depicted in Fig. 5 and was already described in [4]. The

only difference is the choice of the TVS-diode D_{clamp}, which has now a higher clamp voltage V_{clamp} of $19\,\mathrm{V}$, extending the measurement range to about $\pm 15\,\mathrm{V}$. With the higher measurement range, the output characteristics can be acquired for V_{ds} of up to $15\,\mathrm{V}$.

D. High-Voltage Comparator

A high-voltage comparator is crucial for the operation of the actively controlled clamp circuit and the measurement procedure for the output characteristics. The simplified circuitry is depicted in Fig. 6. The task of the comparator is to check whether V_{ds} has fallen below a certain threshold $V_{Comp,th}$, such as $12\,\mathrm{V}$. If that is not the case, the clamp circuit must

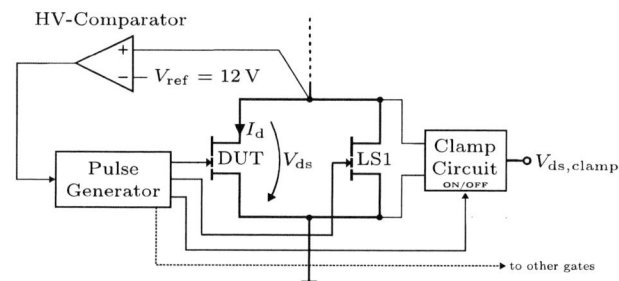

Fig. 6: Simplified Circuitry of the High-Voltage Comparator

not turn-on, to prevent damage to the clamp circuit or to the measurement equipment. This situation can be caused, for instance, when V_{gs} is below $V_{gs,th}$. In this case, V_{ds} will not drop. Further, during an output characteristics measurement, the comparator has to check whether V_{ds} exceeded the threshold, to turn-off the clamp circuit and to turn-on the bypass switch LS1. More details about that are given in the following section.

III. MEASUREMENT PROCEDURES

A. Transfer Characteristics

The procedure for measuring transfer characteristics is depicted in Fig. 7. The individual steps of the procedure are separated in time by the vertical grey bars and numbered as follows:

① Inductor charging phase
② Freewheeling phase
③ Measurement phase
④ Idle state

979-8-3315-1612-3/25 $31.00 © 2025 IEEE

First the inductor L_{DPT} is charged by switching on HS2 and LS1 ①. The duration of this phase defines the maximum drain current I_{d} of interest in the transfer characteristics curve. Second, a freewheeling phase ② is inserted by only turning HS2 on, while HS1 is left in the off-state. As the current I_{L} is positive, the current will flow through HS1 in reverse. Finally, the measurement phase ③ is initiated by applying a V_{gs}-ramp with adjustable ramp-time t_{ramp} and adjustable final value $V_{\mathrm{gs,max}}$ to the DUT. The current slowly commutates from HS1 to the DUT as depicted in Fig. 7, while V_{gs} is increasing to its final value $V_{\mathrm{gs,max}}$.

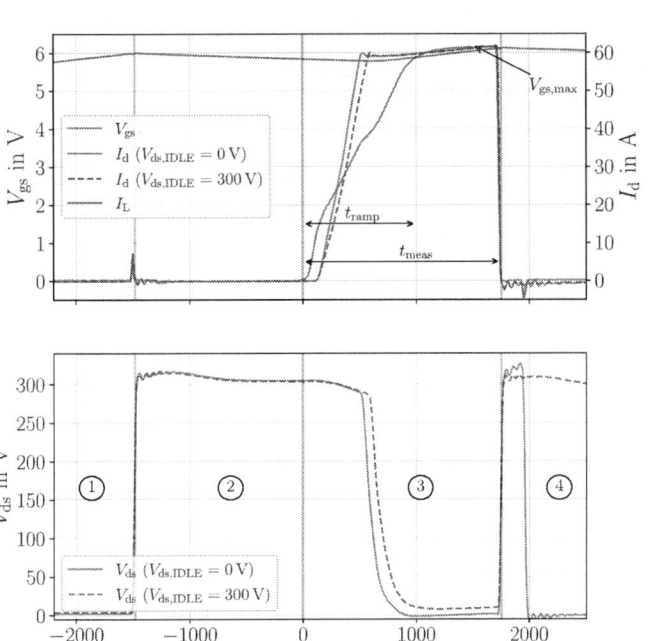

Fig. 7: Procedure for Transfer Characteristics Measurement

The choice of the ramp-time t_{ramp} is crucial for the accuracy of the measurement. For an ideal measurement, the channel current I_{ch} equals the measurable drain current I_{d} and the internal gate voltage $V_{\mathrm{gs,int}}$ equals the measurable gate voltage at the terminals V_{gs}. Due to the parasitic capacitances, inductances and resistances of the transistor (see Fig. 8), this can never be the exact case, as their currents and voltage drops affect the measurable quantities I_{d} and V_{gs}. Those parasitic effects, however, can be weakened significantly by slowing down the switching speed. Slower switching speeds also allow to filter the signals, which reduces the noise of the measurement. In contrast, when the switching speed is set too slow, the self-heating of the DUT can become excessive, as the switching energy heavily increases due to larger overlap in time of V_{ds} and I_{d}. Therefore, the optimal t_{ramp} has to be examined by performing circuit simulations including the thermal model of the DUT.

From Fig. 7 another effect can be seen: The drain voltage V_{ds} is smaller than V_{DC} while the current I_{d} is increasing. This voltage drop is mainly caused by the resistive elements

in the loop, such as the current shunt, the Equivalent Series Resistance (ESR) of the DC-link capacitor and the tracks on the PCB. As soon as the drain current I_{d} has reached the inductor current I_{L}, the main drop of V_{ds} is initiated, at which V_{DC} is gradually taken over by HS1.

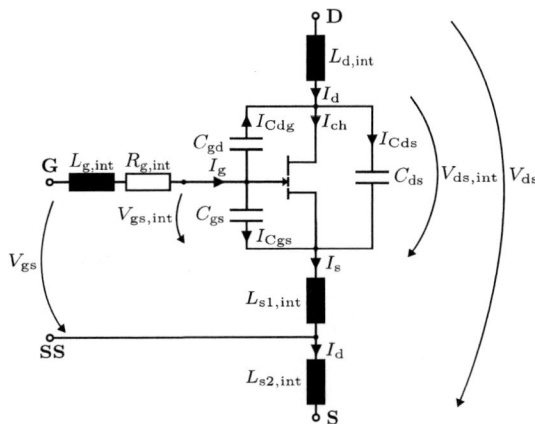

Fig. 8: Equivalent Circuit of a GaN-HEMT with parasitics

It is apparent that the procedure is very close to a real hard-switching turn-on event. The only difference is that the switching speed of an actual application is much higher. By plotting the drain current I_{d} against the gate-source voltage V_{gs}, the transfer characteristics at $V_{\mathrm{ds}} \approx V_{\mathrm{DC}}$ is obtained. After the measurement phase with the duration t_{meas} is finished, either HS1 and HS2 or LS1 and LS2 will be switched on ④, to define $V_{\mathrm{ds,IDLE}}$ to be V_{DC} or 0 V, until the next measurement is conducted.

B. Output Characteristics

The procedure for the output characteristics is depicted in Fig. 9. In short, it consists of the following steps:

① Charge the gate of the DUT to the desired V_{gs}
② Measurement interval
③ Optional: Turn on LS1, if V_{ds} exceeds 12 V
④ Idle state (Upper freewheeling state in the given case)

The idle state ④ is left by applying a gate-source voltage V_{gs} to the DUT ①. If V_{gs} is above $V_{\mathrm{gs,th}}$, the drain-source voltage V_{ds} of the DUT will drop. Subsequently, HS2 is turned on, which causes the inductor current I_{L} to rise. The maximum possible drain current I_{d} and on-time of the DUT is defined by the duration of the measurement interval t_{meas}, when the DUT and HS2 is turned on at the same time ②. While I_{d} is increasing, the drain-source voltage V_{ds} of the DUT increases as well due to the $R_{\mathrm{ds,on}}$. If the current I_{d} is high enough for a certain V_{gs}, the saturation region is reached, which leads to a faster increase of V_{ds}. In case V_{ds} exceeds the comparator threshold $V_{\mathrm{Comp,th}}$ of 12 V, the inductor current I_{L} will be bypassed through switch LS1, which prevents V_{ds} to increase further and terminates the measurement phase. This optional step ③ is necessary to avoid destruction of the DUT due to excessive self-heating. Fig. 9 shows both cases, with and without the optional step.

979-8-3315-1612-3/25 $31.00 © 2025 IEEE

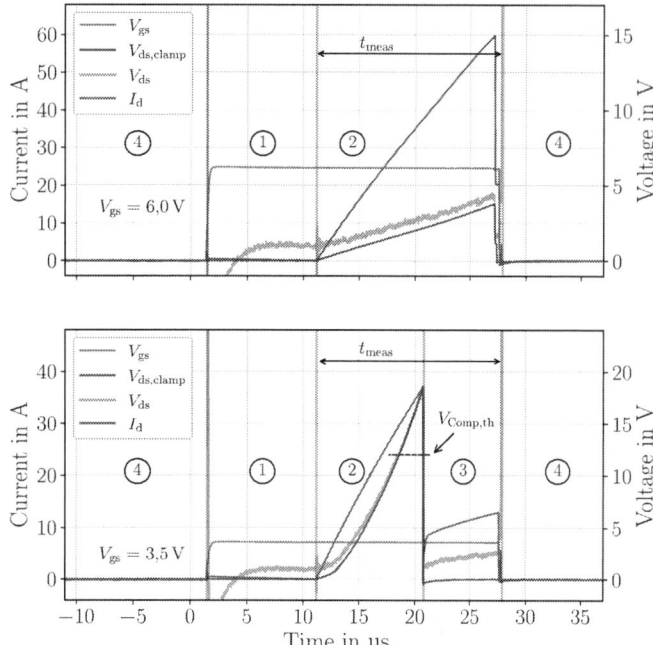

Fig. 9: Raw Data of Output Characteristics Measurement at $V_{\mathrm{ds,IDLE}} = 300\,\mathrm{V}$

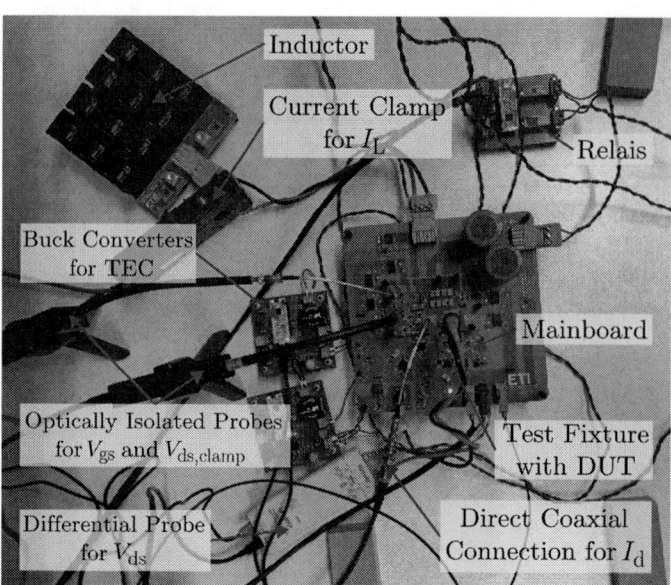

Fig. 10: Picture of the Test Setup

In the lower graph for $V_{\mathrm{gs}} = 3.5\,\mathrm{V}$, the drain current I_{d} becomes high enough to reach the saturation region and exceeds $V_{\mathrm{Comp,th}}$. The sudden drop of I_{d} close to $V_{\mathrm{Comp,th}}$ indicates that LS1 was switched on. Both switches are operating in parallel in this case. In the upper graph with $V_{\mathrm{gs}} = 6.0\,\mathrm{V}$, the DUT remains in the linear region and the measurement is terminated, when the interval t_{meas} has elapsed.

By plotting I_{d} against V_{ds} within the measurement interval ②, the output characteristics at the desired gate-source voltage V_{gs} is obtained. After the measurement has finished, $V_{\mathrm{ds,IDLE}}$ is applied at the DUT ④. In contrast to the procedure for the transfer characteristics, the parasitic influence on the measurement, except self-heating, is negligibly small. This arises from the fact, that the slopes of I_{d} and V_{ds}, that cause parasitic voltage drops and currents, are several magnitudes slower than it is the case with the transfer characteristics measurement. Nevertheless, it is reasonable to perform circuit simulations, to assess self-heating.

IV. PRACTICAL REALIZATION OF THE TEST SETUP

As already mentioned, the given test setup depicted in Fig. 10 is an expansion of the one introduced in [4]. The major part is the mainboard, on which the exchangeable test fixture is mounted. Further, it contains the control circuit, which includes the circuits introduced in section II. On the test fixture, the DUT is mounted together with the test circuit depicted in Fig. 1. It further employs coaxial connectors for

the connection of the measurement equipment to measure the signals of interest I_{d}, V_{gs}, $V_{\mathrm{ds,clamp}}$ and V_{ds}. Both V_{gs} and $V_{\mathrm{ds,clamp}}$ are measured using two optically isolated differential probes of the type *SigOFIT MOIP02P* from *Miscsig* [11]. The current is measured using shunt-resistors with a total resistance of $40\,\mathrm{m\Omega}$, which are mounted on the test fixture. As the voltage drop on the shunt resistors can be very small, the voltage is measured directly with the oscilloscope via a series termination and a coaxial cable to achieve a good measurement accuracy. For additional monitoring of the applied drain-bias, the drain source voltage V_{ds} is measured using a conventional differential probe. Not shown in Fig. 10 is the oscilloscope, which is a *WavePro 404HD* with $4\,\mathrm{GHz}$ bandwidth, 12-bit resolution and $0.5\,\%$ full-scale gain accuracy from *LeCroy* [12].

Due to the strong temperature coefficient of GaN-HEMTs, it is important to conduct the measurements at a defined temperature. This is achieved by using a thermoelectric cooler (TEC), which allows a temperature control with high dynamic [4]. The TEC is powered by two back-to-back connected buck converters, which are also shown in Fig. 10. The controlled variable is the case temperature T_{c} of the DUT, which is measured with a thermally well-connected NTC-resistor.

The voltage V_{DC} is provided by a high voltage power supply, whose poles are disconnected during a measurement using two relais, to reduce common-mode disturbances.

V. MEASUREMENTS AND DISCUSSION

A. Measurement Conditions

The following measurements are carried out using the GS66508B GaN-HEMT with a Schottky p-GaN gate from GaN-Systems. It has a current rating of $30\,\mathrm{A}$ and a voltage rating of 650 V. In the off-state of the DUT, the gate-source voltage V_{gs} is set to 0 V. The charging time of the inductor L_{DPT}, which is t_{charge} for the transfer characteristics and t_{meas} for the output characteristics, is always set such that the desired current I_d is reached. The inductance L_{DPT} has a nominal value of 90 µH. Further, the case temperature T_c is kept constant at 25 °C using the TEC-based temperature control. These conditions apply to all measurements.

Whenever the drain-bias $V_{ds,IDLE}$ is changed, a relaxation time of 20 min is introduced to ensure that the traps are in a steady state. However, since trapping effects are also caused by any gate stress V_{gs}, the influence of this on the treshold voltage $V_{gs,th}$ is the first part of the investigation. The threshold voltage $V_{gs,th}$ is defined at the threshold, when the drain current I_d exceeds 100 mA. For all transfer characteristics measurements or $V_{gs,th}$-measurements, the ramp time t_{ramp} is set to 1 µs.

The transient junction temperature rise ΔT_j due to self-heating was simulated with the thermal model from the datasheet [13]. It was shown, that ΔT_j completely decayed within one second after a measurement. Thus, if the interval between two consecutive measurements t_{IDLE} is chosen larger than one second, T_j is not expected to accumulate.
For the output characteristics, ΔT_j was always below 10 K for currents I_d of up to 60 A. For the transfer characteristics, ΔT_j was 0.5 K at 5 A, 10 K at 30 A and 30 K at 60 A. Consequently, except for the $V_{gs,th}$-measurements at 5 A, self-heating can not be neglected. However, as the pulse pattern applied to the DUT is identical with and without applied drain-bias $V_{ds,IDLE}$, this is not expected to affect the general outcome of the result.

B. Consideration of the Threshold Voltage Shift

Any voltage V_{gs} applied to the gate results in a $V_{gs,th}$-shift. To acquire the characteristics, charging of the gate is essential, which in turn means that trapping at the gate or $V_{gs,th}$-shift always occurs.
For reproducible measurements, these influences must be taken into account and by an appropriate timing of the measurements, the $V_{gs,th}$-shift can be kept at an almost constant level. The measurements need to be conducted periodically, with a constant measurement time t_{meas}, when a voltage is applied to gate and a constant resting time t_{IDLE}, when V_{gs} is zero.

Fig. 11 shows measurements of the time dependency of $V_{gs,th}$ at different $V_{gs,max}$, t_{IDLE} and t_{meas}. The measurements were conducted at a very low V_{DC} of 5 V with zero drain bias $V_{ds,IDLE}$, to minimize the impact of drain-induced trapping and DIBL. In addition, the current I_d was limited to 5 A, to minimize self-heating and to allow a reduced full-scale range of the oscilloscope, resulting in a more accurate determination of $V_{gs,th}$. Between every measurement curve, the DUT rests for 10 min, to restore the initial state of $V_{gs,th}$. This can be

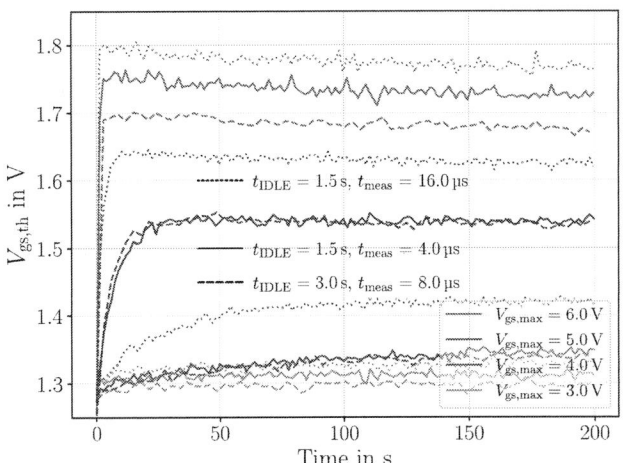

Fig. 11: Time dependency of $V_{gs,th}$-shift

recognized by the fact that all curves start at $V_{gs,th} \approx 1.27$ V. It can be seen that the greatest change of $V_{gs,th}$ for all curves occurs within the first hundred seconds. After that, $V_{gs,th}$ changes only in low tens of mV-steps. Further, the higher $V_{gs,max}$, the higher the $V_{gs,th}$-shift and the smaller the trapping time constant. For $V_{gs,max}$ smaller than 3.0 V, the $V_{gs,th}$-shift becomes negligibly small and is therefore not depicted in Fig. 11. When comparing different combinations of t_{IDLE} and t_{meas}, a different steady state $V_{gs,th}$ was observed in most of the cases. Two measurements are therefore only comparable, if their t_{IDLE} and t_{meas} are identical.

In order to obtain reproducible measurements, it is crucial that $V_{gs,th}$ is in a steady state. For the transfer characteristics measurements this state can be achieved, by always setting $V_{gs,max}$ to 6 V. In this case, $V_{gs,th}$ reaches its steady state within a couple of measurements or seconds. By only considering the last of ten consecutive measurements, the transfer characteristics measurement becomes reproducible.
The output characteristic measurement, on the other hand, is associated with greater effort, as the measurements have to be conducted at different on-state V_{gs}. Between each output characteristics curve, defined at a certain V_{gs}, a resting time of 10 min has to be inserted. Further, much more consecutive measurements have to be conducted, until $V_{gs,th}$ has reached its steady state. With $t_{IDLE} = 1.5$ s and 60 consecutive measurements, reproducible results can be produced, by only considering the last measurement.

C. Transfer Characteristics

Two major effects, that significantly affect the transfer characteristics, are part of the investigation: The DIBL-effect and trapping induced by applying a drain-bias $V_{ds,IDLE}$ in the off-state. All measurements are conducted periodically, consisting of ten consecutive measurements, as described earlier, with $t_{meas} = 4.0$ µs and $t_{IDLE} = 1.5$ s.

In Fig. 12 the influence of V_{ds} on the steady state $V_{gs,th}$ at different $V_{ds,IDLE}$ is shown. As expected, the DIBL-effect occurs and significantly affects the threshold voltage $V_{gs,th}$. Between 100 V and 400 V, this effect is noticeably stronger, if a drain-bias $V_{ds,IDLE}$ is applied in the pauses between the measurements, as the absolute sensitivity $\|dV_{gs,th}/dV_{ds}\|$ is much higher. However, this is not the case within the first 50 V. In this case, the curve for $V_{ds,IDLE} = 0\,V$ shows a stronger slope.

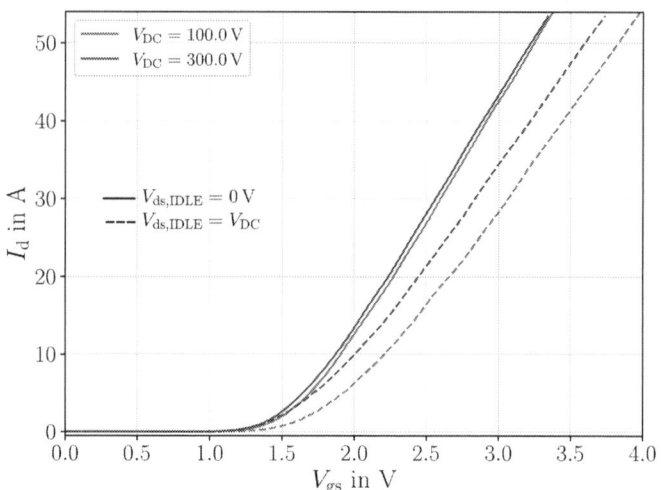

Fig. 13: Transfer Characteristics Comparison

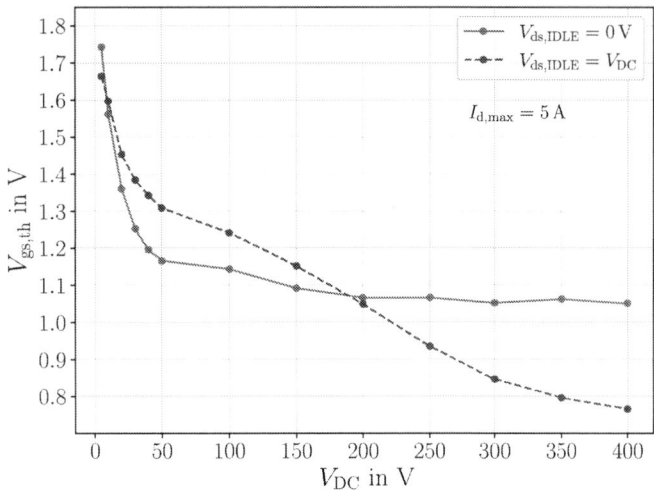

Fig. 12: Influence of V_{ds} on $V_{gs,th}$ (DIBL-effect)

Important to mention is that $V_{gs,th}$ at zero drain bias $V_{ds,IDLE}$ and $V_{DC} = 5\,V$ is 1.72 V, which is very close to the typical value of 1.7 V from the datasheet of the DUT [13]. Here, $V_{gs,th}$ was defined at $I_d = 7\,mA$, while the DUT was operated as diode-connected transistor.

The actual transfer characteristics at chosen V_{DC} and different $V_{ds,IDLE}$ is shown in Fig. 13. It can be clearly seen, that the curves for a V_{DC} of 300 V increase earlier than the curves for 100 V. This indicates the DIBL-effect and coincides with the results shown in Fig. 12. As shown in Fig. 12, this effect appears to be stronger, if a drain-bias $V_{ds,IDLE}$ is applied in the off-state of the DUT. Further, the slope or transconductance g_m of the curves differs dependent on $V_{ds,IDLE}$ and is lower, if a bias $V_{ds,IDLE}$ is applied. The transconductance g_m, however, does not appear to be significantly affected by V_{DC}.

If the average transconductance $\hat{g}_m$ is considered between 10 A and 50 A, the result for $V_{ds,IDLE}$ of V_{DC} is about 25 A V^{-1}. For a $V_{ds,IDLE}$ of 0 V, one obtains about 30 A V^{-1}.

Consequently, a reduced transconductance leads to a lower current rise during switching, as it can also be seen in Fig. 7. In [3], switching transients are shown with and without applied drain-bias $V_{ds,IDLE}$, that yield a slower current rise, after a $V_{ds,IDLE}$ was applied for 30 min. Further, a turn-on loss increase of 20 % was reported.

D. Output Characteristics

For the output characteristics measurements, a maximum drain-current I_d of 60 A was chosen, while the resting-time t_{IDLE} between two periodical measurements was set to 1.5 s. As described earlier, the result was extracted only from the last measurement, after 60 measurements were performed. In Fig. 14 the output characteristics at $V_{DC} = 100\,V$ with

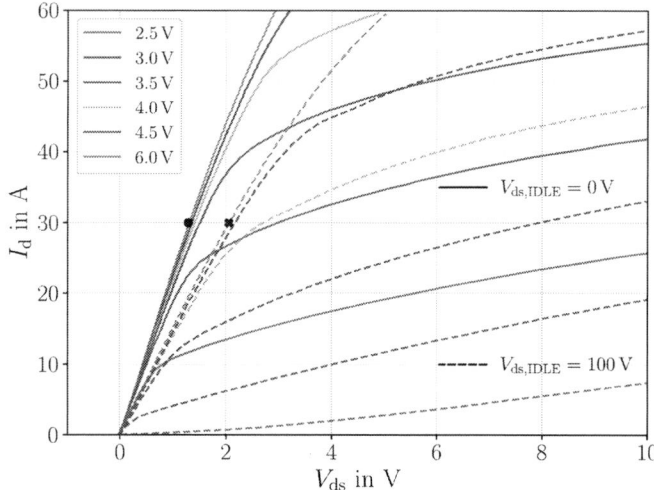

Fig. 14: Output Characteristics Comparison at $V_{DC} = 100\,V$

and without applied drain-bias $V_{ds,IDLE}$ are shown. It can be clearly seen, that the curves for an applied drain-bias $V_{ds,IDLE}$ indicate a poorer performance than the curves without bias. First, the slope of the linear region is reduced. From particular interest is the slope at $V_{gs} = 6\,V$, as this refers to the on-resistance $R_{ds,on}$. Without applied bias $V_{ds,IDLE}$, the on-resistance $R_{ds,on}$ at the rated current $I_d = 30\,A$ is 46.2 mΩ ($\bullet$), which comes close to the value of 50 mΩ from the datasheet [13]. With the bias $V_{ds,IDLE}$ applied, the on-resistance $R_{ds,on}$ is 76.2 mΩ ($\ast$), which results in an increase

979-8-3315-1612-3/25 $31.00 © 2025 IEEE

of 65 %.

Second, the knee point of the characteristics is shifted towards lower currents and the saturation current is decreased. For a V_{gs} of 2.5 V, the linear region seem to disappear completely and the curve looks similar to the characteristics of a diode. The decrease of the saturation current correlates with the transfer characteristics measurements, as g_m is reduced with applied drain-bias $V_{ds,IDLE}$. It can be assumed, that the shifted knee point, the disappearing linear region at low V_{gs} and the reduced g_m refers to the current collapse effect [14].

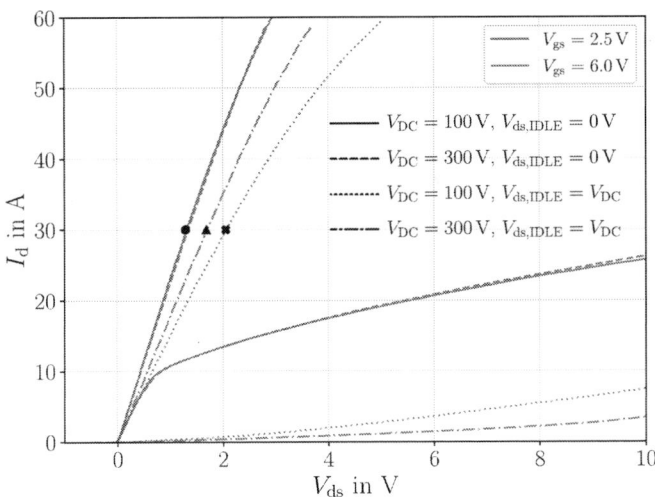

Fig. 15: Output Characteristics Comparison at different V_{DC}

In Fig. 15 the output characteristics at two different V_{DC}, 100 V and 300 V, is shown. As expected, the curves without drain-bias $V_{ds,IDLE}$ coincide. In this case, the DIBL-effect does not occur, as the DUT is switched under zero voltage. At an applied bias $V_{ds,IDLE}$ of 300 V, the $R_{ds,on}$ is 59.3 mΩ (▲), which is 28 % more than the original value at $V_{ds,IDLE} = 0$ V. This voltage-dependency indicates the effect of the dynamic on-resistance $dR_{ds,on}$, which is reported in earlier investigations [1], [4]. Both studies reported that the $dR_{ds,on}$ is reduced towards higher V_{DC}. It should be also noted, that all three $R_{ds,on}$-values are very close to the results from our previous $dR_{ds,on}$-investigation [4]. In this case, the same DUT was used, but from a different batch.

VI. CONCLUSIONS AND FUTURE WORK

Within this contribution, new methods are provided to measure the transfer and output characteristics of GaN-HEMTs in an environment that is close to an actual power converter. Those methods were already realized in a practical test setup and results were provided. The results show, that trapping stimulated by gate and drain bias significantly affects the device's characteristics. It could be shown that the effect of the dynamic on-resistance $dR_{ds,on}$ and the threshold voltage shift are present in the investigated DUT. Further, we assume that the current collapse effect still occurs, because of the i.a. reduced transconductance g_m and the shifted knee

point, after a drain bias $V_{ds,IDLE}$ was applied for 20 min.

Future studies could focus on the variation between different batches, different DUTs with different gate structures and the influence of the temperature on the characteristics. Thanks to the modular design of the test setup and the already existing temperature control using a TEC, the prerequisites for those investigations are given.

In addition, the switching losses and conduction-losses could also be investigated. An initial study already reported increased switching losses under the influence of trapping [3]. If the variation between the different batches is moderate, compact models can be developed using the obtained data, that include the influence of trapping.

REFERENCES

[1] M. Goller, C. Schwabe, N. Thoenelt, G. Li, J. Lutz, and T. Basler, "Precise determination of dynamic RDSon in AlGaN/GaN power HEMTs under soft switching condition," in *PCIM Europe 2022; International Exhibition and Conference for Power Electronics, Intelligent Motion, Renewable Energy and Energy Management*, 2022, pp. 1–8.

[2] B. Kohlhepp, D. Breidenstein, and T. Dürbaum, "Test setup to study threshold voltage shift of GaN-HEMTs under short- and long-term gate and drain bias," *IEEE Journal of Emerging and Selected Topics in Power Electronics*, pp. 1–1, 2024.

[3] F. Yang, C. Xu, and B. Akin, "Characterization of threshold voltage instability under off-state drain stress and its impact on p-gan hemt performance," *IEEE Journal of Emerging and Selected Topics in Power Electronics*, vol. 9, no. 4, pp. 4026–4035, 2021.

[4] P. Swoboda, S. R. Frank, A. Liske, and M. Hiller, "Improved test circuit for characterization of the dynamic on-resistance of GaN-HEMTs over a wide operating range," in *2024 Energy Conversion Congress & Expo Europe (ECCE Europe)*, 2024, pp. 1–8.

[5] E. A. Jones and A. Pozo, "Hard-switching dynamic Rds,on characterization of a GaN FET with an active GaN-based clamping circuit," in *2019 IEEE Applied Power Electronics Conference and Exposition (APEC)*, 2019, pp. 2757–2763.

[6] R. Li, X. Wu, S. Yang, and K. Sheng, "Dynamic on-state resistance test and evaluation of GaN power devices under hard- and soft-switching conditions by double and multiple pulses," *IEEE Transactions on Power Electronics*, vol. 34, no. 2, pp. 1044–1053, 2019.

[7] R. Hou, Y. Shen, H. Zhao, H. Hu, J. Lu, and T. Long, "Power loss characterization and modeling for GaN-based hard-switching half-bridges considering dynamic on-state resistance," *IEEE Transactions on Transportation Electrification*, vol. 6, no. 2, pp. 540–553, 2020.

[8] "Discrete semiconductor devices - metal-oxide semiconductor field-effect transistors (MOSFETs) for power switching applications," British Standards Institution, Standard, 2004, IEC 60747-8-4:2004.

[9] *UCC27511A Single-Channel High-Speed Low-Side Gate Driver With 4A Peak Source and 8A Peak Sink*, Texas Instruments, 2024, SLUSD95A, available at: https://www.ti.com/lit/ds/symlink/ucc27511a.pdf (Oct 2024).

[10] R. P. Sallen and E. L. Key, "A practical method of designing RC active filters," *IRE Transactions on Circuit Theory*, vol. 2, no. 1, pp. 74–85, 1955.

[11] *SigOFIT Optical-fiber Isolated Probe Datasheet*, Shenzhen Micsig Technology Co., Ltd., Shenzhen, Guangdong, March 2023, available at: https://www.micsig.com/SigOFIT (June 2024).

[12] *WavePro HD High Definition Oscilloscope*, Teledyne LeCroy, Chestnut Ridge, NY, December 2018, available at: https://de.teledynelecroy.com/waveprohd (June 2024).

[13] *GS66508B Bottom-side cooled 650 V E-mode GaN transistor Datasheet*, GaN Systems, 2020, REV 200402, available at: https://gansystems.com/gan-transistors/gs66508b (Oct 2024).

[14] G. Meneghesso, F. Rampazzo, P. Kordos, G. Verzellesi, and E. Zanoni, "Current collapse and high-electric-field reliability of unpassivated GaN/AlGaN/GaN HEMTs," *IEEE Transactions on Electron Devices*, vol. 53, no. 12, pp. 2932–2941, 2006.

Mitigating Gate Voltage Oscillation in Parallel SiC Power Modules for xEV

Hideo Komo
Semiconductor & Device Division
Mitsubishi Electric US, INC.
Export, PA, USA
hideo.komo@meus.com

Michael Rogers
Semiconductor & Device Division
Mitsubishi Electric US, INC.
Export, PA, USA
michael.rogers@meus.com

Mark Steiner
Semiconductor & Device Division
Mitsubishi Electric US, INC.
Export, PA, USA
mark.steiner@meus.com

Eric Motto
Semiconductor & Device Division
Mitsubishi Electric US, INC.
Export, PA, USA
eric.motto@meus.com

Koichi Taguchi
Power Device Works
Melco Semiconductor Engineering Corp.
Fukuoka, Japan
taguchi.koichi@zd.mitsubishielectric.co.jp

Chihiro Kawahara
Advanced Technology R&D Center
Mitsubishi Electric Corpolation
Amagasaki, Japan
kawahara.chihiro@ak.mitsubishielectric.co.jp

Junichi Nakashima
Advanced Technology R&D Center
Mitsubishi Electric Corpolation
Amagasaki, Japan
nakashima.junichi@dr.mitsubishielectric.co.jp

Yasushige Mukunoki
Advanced Technology R&D Center
Mitsubishi Electric Corpolation
Amagasaki, Japan
mukunoki.yasushige@cj.mitsubishielectric.co.jp

Seiichiro Inokuchi
Power Device Works
Mistubishi Electric Corporation
Fukuoka, Japan
inokuchi.seiichiro@bk.mitsubishielectric.co.jp

Rei Yoneyama
Power Device Works
Mitsubishi Electric Corpolation
Fukuoka, Japan
yoneyama.rei@db.mitsubishielectric.co.jp

Global vehicle electrification is accelerating to combat global warming. Recently, Silicon Carbide (SiC) Metal-Oxide-Semiconductor Field-Effect Transistors (MOSFETs) are increasingly used in electric vehicle inverters for their low power loss at fast switching speeds compared to conventional Silicon Insulated Gate Bipolar Transistors (IGBTs). Although SiC MOSFETs are usually used in parallel to gain power, these fast-switching speeds can lead to gate voltage oscillations across paralleled SiC chips, potentially leading to failures such as gate drive malfunction and gate oxide breakdown. This paper addresses the oscillation caused by source voltage fluctuations and unbalanced gate drive circuit characteristics. We propose a method to mitigate these issues using a symmetrical module structure which equalizes stray inductance between modules.

Keywords—SiC MOSFET, Gate voltage oscillation, Parallel

I. Introduction

In the pursuit of higher efficiency and the fight against global warming, the adoption of electric vehicles is rapidly accelerating worldwide. Reducing system cost and improving the cruising distance of xEVs (such as Hybrid Vehicles, Plug-in Hybrid Vehicles, and Battery Electric Vehicles) are key factors in a successful vehicle design. To optimize these factors, improving inverter efficiency is one of the most important challenges because it directly affects battery capacity. Within the scope of the power module, which is a key component in the inverter, key areas for improvement include reductions in chip power loss and total module size. So far, Silicon Insulated-Gate-Bipolar-Transistors (Si IGBTs) have been used in power modules, and their performance improvements have been aimed at enhancing inverter efficiency. However, in recent years, Silicon Carbide (SiC) Metal-Oxide-Semiconductor Field-Effect Transistors

(MOSFETs) have begun to gain popularity for further improving inverter efficiency due to their low power losses compared to conventional Si IGBTs. SiC MOSFETs exhibit not only low conduction loss (thanks to approximately ten times higher dielectric breakdown electric field strength than Si [1]) but also low switching loss because they are unipolar devices. However, since SiC is still in the developmental stage, it has various challenges, such as including crystallographic defects during the wafer manufacturing process [2]. To minimize the number of defects in SiC MOSFETs, the chip size must be small, and they must be used in parallel, which leads to an increase in the power module size. On the other hand, paralleling small SiC MOSFET power modules introduces challenges for high-speed switching operation. In this paper, we will present the cause of the technical issues and technical solutions for paralleling small SiC MOSFET power modules.

II. Gate Voltage Oscillation Due To High-Speed Switching Operation

Although SiC power modules can operate with faster switching speeds than IGBT power modules, it can result in a voltage oscillation between the gate and source terminals [3]. The gate voltage oscillation occurs during the state transition at high switching speeds. The oscillation might result in failures, such as gate drive malfunction due to abnormal gate voltage, or gate oxide breakdown due to excessive voltage. These failure modes represent limitations for the system efficiency improvements that can be realized by higher switching speeds. Therefore, it is important to reduce the oscillation to make full use of SiC modules [4]. The oscillation primarily results from imbalanced source currents across each SiC chip. Figure 1 illustrates the resonance half-bridge circuits in parallel SiC chips. In the case of high-side switching, when the high-side (MOSFET1 and MOSFET2) is turned on and the low-side (MOSFET3 and MOSFET4) is turned off, the switching ringing occurs in loop (a) and loop (b) due to the imbalanced source currents on the low-side. This imbalance is caused by differences in circuit constants, as well as variations in chip properties. Simultaneously, the switching ringing influences the resonance circuit between the high-side SiC chips via C_{iss} (loop (c) and loop(d)). Consequently, these events lead to oscillation in the high-side gate voltage. Therefore, reducing the source voltage fluctuation and balancing circuit characteristics within the resonance circuit are necessary to prevent the gate oscillation.

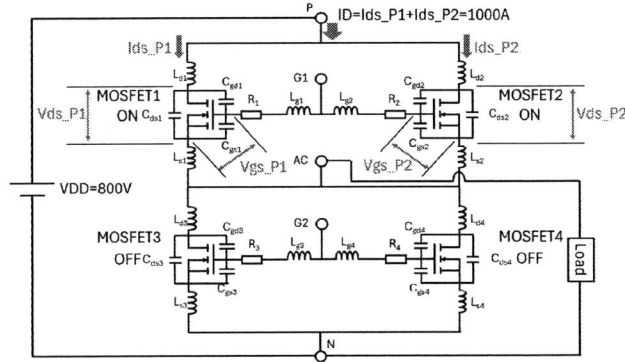

Fig. 2. The simulation circuit.

TABLE I. THE IDEAL CIRCUIT CONSTANTS

Symbol	L_{d1}	L_{d3}	L_{s1}	L_{s3}	L_{g1}	L_{g3}	R_1	R_3
	L_{d2}	L_{d4}	L_{s2}	L_{s4}	L_{g2}	L_{g4}	R_2	R_4
Value	15nH	1nH	1nH	15nH	10nH	10nH	4Ω	4Ω
	15nH	1nH	1nH	15nH	10nH	10nH	4Ω	4Ω

TABLE II. THE SIMULATION ITEMS

No.	Simulation Item	Conditions (Constants not specified are ideal values)
1	Ideal circuit	The circuit constants are the same as Table 1
2	Varying High-side L_d and L_s	L_{d1}=18nH, L_{d2}=15nH, L_{s1}=1.2nH, L_{s2}=1.0nH
3	Varying Low-side L_d and L_s	L_{d3}=18nH, L_{d4}=15nH, L_{s3}=1.2nH, L_{s4}=1.0nH
4	Varying High-side L_g	L_{g1}=12nH, L_{g2}=10nH
5	Varying High-side R	R_1=4.8Ω, R_2=4.0Ω

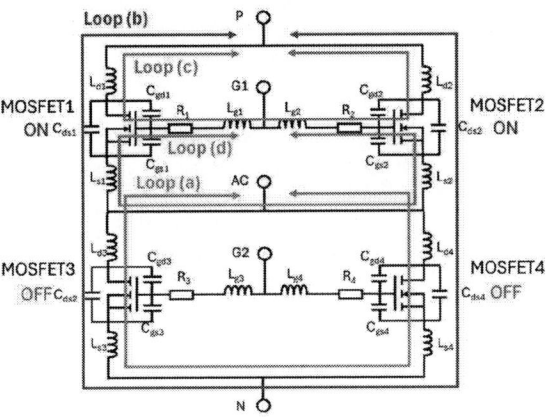

Fig. 1. A resonance hald-bridge circuit between SiC chips.

In this paper, to compare the impacts of each circuit constant derived from a power module configuration on gate voltage oscillation in a half-bridge configuration with two paralleled SiC MOSFETs, high-side switching simulations were conducted under two conditions using the simulation circuit shown in Figure 2. The first condition was ideal, with all factors having the same values between the two paralleled SiC MOSFETs (Table I). The second condition was realistic, assuming constant chip properties while varying each factor related to the module configuration, such as L_d, L_s, L_g, and R, by 20% between the two SiC MOSFETs, as shown in Table II. The ideal values of these factors have been determined from the general power module, and the operation conditions of the simulation are VDD=800V, ID=1000A, Vgs=-5V, +20V, Tj=25°C, Rgon=4Ω (apart from R in Figure 1). Figures 3, 4, 5, 6, and 7 show the results of V_{gs}, V_{ds}, and I_{ds} waveforms for the high-side SiC MOSFETs in the simulations. The waveforms in the ideal circuit show no oscillation in the gate voltages, whereas the other conditions exhibit gate voltage oscillations. Of these factors, the most sensitive circuit constants for gate voltage oscillation are the low-side L_d and L_s. Compared to the high-side, the stray inductance difference between the SiC-MOSFETs on the low-side causes greater fluctuations in the source voltage of the high-side SiC-MOSFETs. It is most important to reduce the difference in Low-side L_d and L_s between paralleled SiC MOSFETs to mitigate gate voltage oscillation.

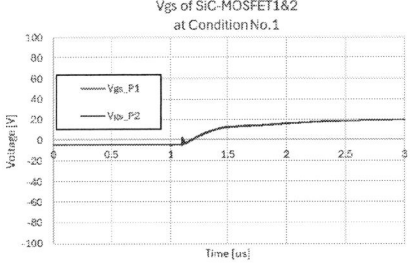

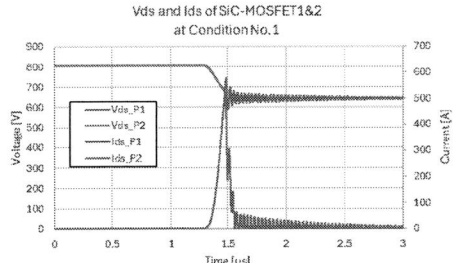

Fig. 3. Simulation waveforms in the ideal circuit.

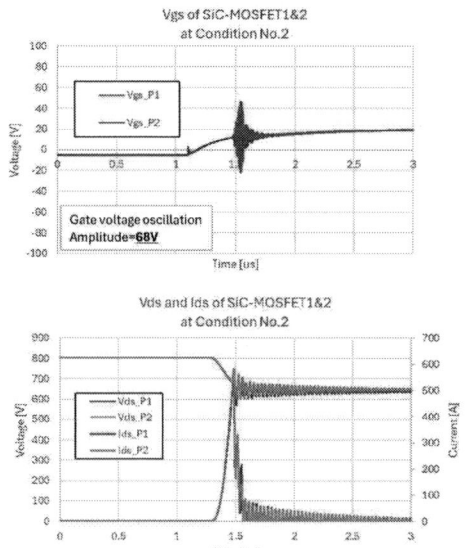

Fig. 4. Simulation waveforms in the circuit varying High-side L_d and L_s.

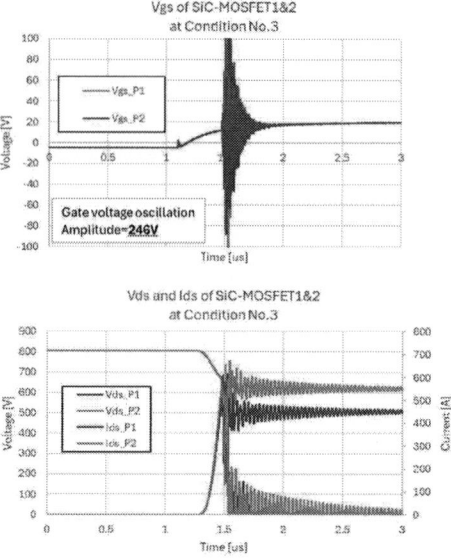

Fig. 5. Simulation waveforms in the circuit varying Low-side L_d and L_s.

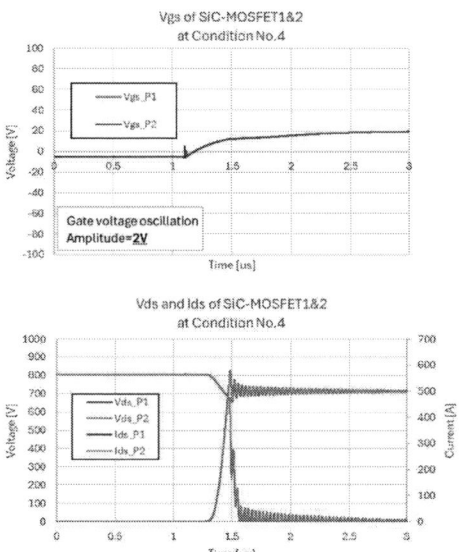

Fig. 6. Simulation waveforms in the circuit varying High-side L_g.

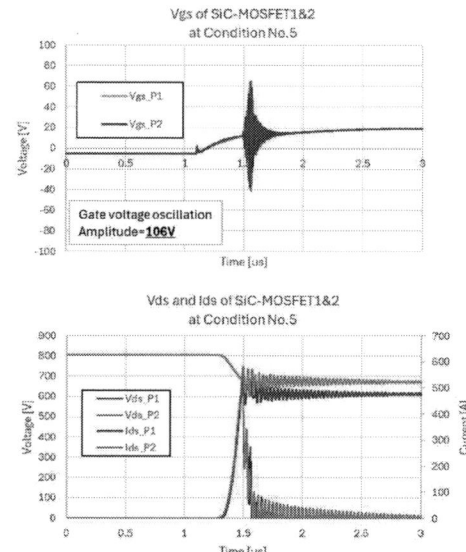

Fig. 7. Simulation waveforms in the circuit varying High-side R

III. GATE VOLTAGE OSCILLATION MITIGATION FOR PARALLELED SiC POWER MODULES

SiC power modules typically include many SiC chips in parallel within a package. This design choice is due to the small size of SiC chips, which results from their crystal defects as mentioned section I. However, this approach makes SiC modules bigger, creating difficulties in optimizing the power module size for different motor powers. On the other hand, in general, to parallelize small SiC modules causes gate voltage oscillation because stray inductance differences between the modules occur easily. To address this challenge, this paper presents a newly developed compact SiC power module with a 2-in-1 (half bridge) configuration (Figure 8).

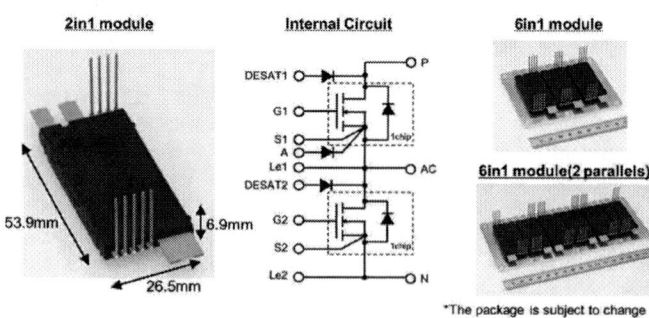

Fig. 8. Proposed module outline.

979-8-3315-1612-3/25 $31.00 © 2025 IEEE   2358

This compact module is designed to be easily paralleled for applications requiring increased power with a small footprint [5]. To minimize the module size, we have developed the multifunctional Si chip (Figure 9). The module size is determined by the internal layout of the module such as signal patterns, gate resistors, and a thermistor. By incorporating these components, and a DESAT diode into the multifunctional Si chip, we achieved a significant 15 % size reduction for the SiC module compared to conventional structures.

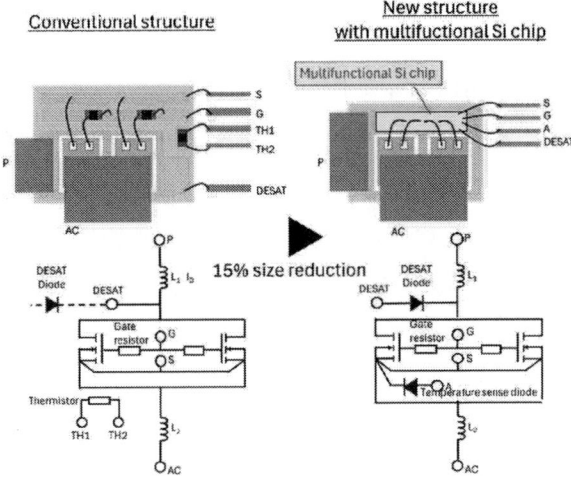

Fig. 9. Package minimization by multifunctional Si chip.

Despite the gate voltage oscillation issue inherent in parallel use, the new module incorporates equalization of stray inductance between modules through a symmetrical structure to mitigate the problem.

While the new module can be paralleled as needed for higher power output, measures must be taken to prevent gate oscillation due to the difference in stray inductance in the resonance circuit among the modules. In an actual parallel system, the stray inductance difference between the modules generally occurs at the P-bus bar and N-bus bar between the power modules and the smoothing capacitor. Figure 10 shows the bus bar configuration. To minimize the stray inductance on the bus bars, they are usually laminated with an insulator to cancel the magnetic field generated by the currents. However, when the two identical modules are parallelized, a difference in stray inductance occurs between them because the cancellation of the magnetic field generated by the currents on the bus bar differs partially between the path from the capacitor to the P-terminal and the path from the N-terminal of the power module to the capacitor, as shown in Figure 11. This also means that the difference in the high-side L_d and the low-side L_s between two SiC-MOSFETs occurs in the circuit of Figure 1. Therefore, to prevent differences in stray inductance on the bus bar, it is important to consider the layout so that the cancellation of the magnetic field is even. By developing a symmetrical structural module for parallel use, the cancellation of the magnetic field has been equalized between the current paths on the bus bar, as shown in Figure 12. To confirm the minimization of gate voltage oscillation with the configuration, switching evaluations on the high-side for the asymmetrical and symmetrical structures have been conducted. The test conditions were VDD=870V, ID=1000A, and

Tj=25°C. As shown in Figure 13, the symmetrical structure establishes a resonance circuit equilibrium and significantly reduces gate oscillation between the modules, decreasing the amplitude of Vgs from 70.9V to 10.6V compared to the asymmetrical structure.

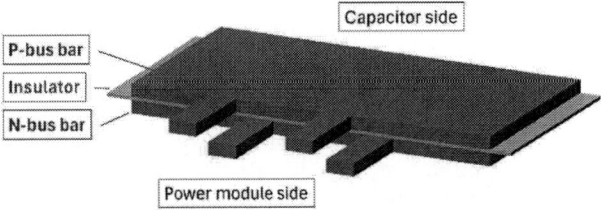

Fig. 10. Bus bar configuration.

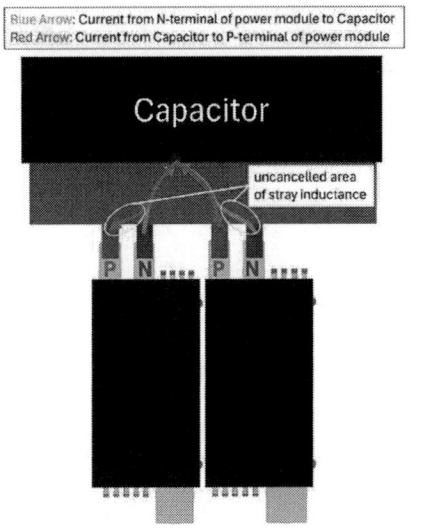

Fig. 11. Uncancelled area of stray inductance in asymmetrical structure.

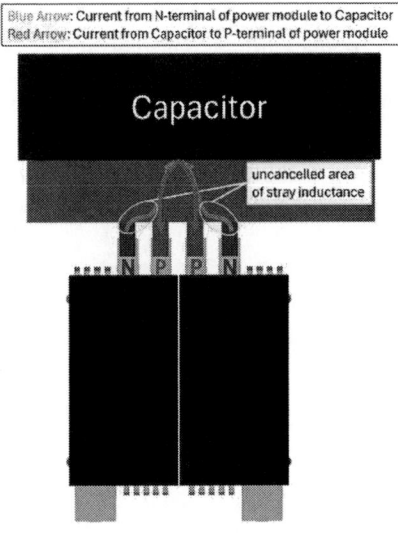

Fig. 12. Uncancelled area of stray inductance in symmetrical structure.

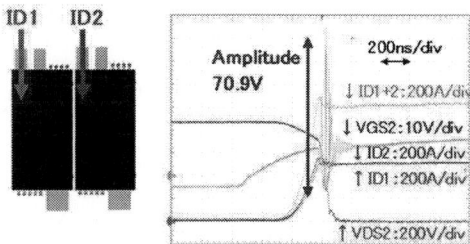

Asymmetrical structure

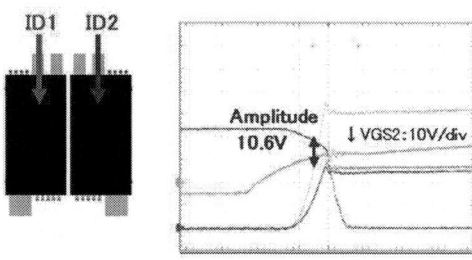

Symmetrical structure

Fig. 13. Waveforms comparison between the modules with and without symmetrical structure.

IV. CONCLUSION

In this paper, we investigated the circuit parameters that most contribute to gate voltage oscillation in a general module structure with two parallel SiC-MOSFETs through simulations. The simulations suggested that the low-side stray inductances, L_d and L_s, have the most significant impact on gate voltage oscillation.

To minimize the gate voltage oscillation in parallel use, the author presented a newly developed compact SiC power module with a 2-in-1 (half bridge) configuration. This compact module is designed to be symmetrically paralleled for applications requiring increased power with a small footprint. Despite the gate voltage oscillation issue inherent in parallel use, the new module incorporates equalization of stray inductance between modules through a symmetrical structure to mitigate the problem.

The experimental results demonstrate that a symmetrical structure minimizes gate voltage oscillation compared to the asymmetrical structure.

REFERENCES

[1] T. Kimoto, "High-voltage SiC power devices for improved energy efficiency," in Proc. Jpn. Acad., Ser. B Vol.98, pp 161-189, 2022.

[2] P. Chen, W Miao, T. Ahmed, Y. Pan, C. Lin, S. Chen, H. Kuo, B. Tsui, and D. Lien, "Defect Inspection Techniques in SiC" in Nanoscale Research Letters, 2022.

[3] F. Xiu, and L. Chen, "Suppressing Gate Voltage Oscillation in Paralleled SiC MOSFETs for HEV/EV Traction Inverter Application," in 2019 IEEE Energy Conversion Congress and Exposition (ECCE), pp. 3548-3553.

[4] R. K. Thirukoluri, and R. Paul, "High-Frequency Oscillations in SiC MOSFET Power Modules During Turn-on Switching Transient – Analysis Based on Simulations and Mitigation Methods," in PCIM Europe 2024, 11-13 June 2024, Nuremberg, pp. 1860-1866.

[5] T. Tokorozuki, H. Komo, K. Nishimura, R. Yoneyama, and G. Majumdar, "Technological Approaches to High-Power Density SiC Power Module for Automotive," in PCIM Europe 2024, 11-13 June 2024, Nuremberg, pp. 849-854.

Switching Performance Comparison of Low-Voltage GaN and Si Devices

Tianxiao Chen[1,3], Haoyang Liu[3], Pedro A. M. Bezerra[3], Eckart Hoene[2], and Sibylle Dieckerhoff[1]

[1]Technische Universität Berlin, Berlin, Germany
[2]Fraunhofer IZM, Berlin, Germany
[3]Huawei Technologies Duesseldorf GmbH, Nuremberg, Germany

Abstract—This paper presents a comprehensive comparison of the switching performance of low-voltage (100 V) Gallium Nitride (GaN) transistors and Silicon (Si) Metal-Oxide-Semiconductor Field-Effect Transistors (MOSFETs), featuring some of the lowest $R_{DS(on)}$ devices on the market. Key performance parameters, including turn-on and turn-off switching losses, gate driver losses, and lumped $R_{DS(on)}$, are evaluated using an indirect non-invasive method. Results show that GaN devices achieve at least 42% lower turn-off losses, 45% lower turn-on losses, and 71% lower gate driver losses compared to Si counterparts. However, GaN devices are highly sensitive to dead-time (DT) control, requiring precise tuning to fully exploit their low turn-off losses. Additionally, GaN devices exhibit a lumped dynamic $R_{DS(on)}$ 3–4 times higher than static values at 1 MHz, posing concerns for high-conduction-loss applications. These findings underscore the trade-offs between GaN and Si devices in optimizing low-voltage converter designs.

Index Terms—Switching losses, GaN, WBG, soft switching, dead-time

I. INTRODUCTION

In the rapidly evolving fields of telecommunications, data centers, and AI-related applications, the demand for high-efficiency power converters is critical. These systems typically operate on a 48 V bus voltage, requiring 100 V devices to handle input voltages up to 75 V [1]. To accommodate the large current consumption in these applications, two primary categories of power converter topologies are commonly employed: (1) transformer-based converters, which are extensively used in applications requiring galvanic isolation and rely on Zero Voltage Switching (ZVS) to minimize turn-on losses [2][3][4], making hard turn-off losses and optimal DT critical design considerations (cf. Fig 1). DT refers to the intentional delay between the turn-off of one switch and the turn-on of another in a half-bridge or full-bridge

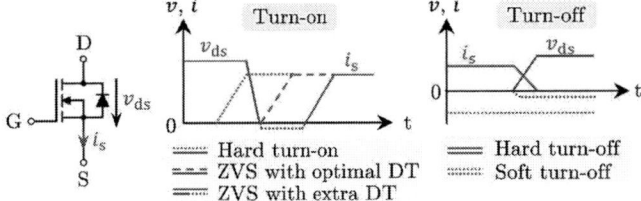

Fig. 1. The types of switching losses are defined by the simplified waveforms of the drain-to-source voltage, v_{ds}, and the current, i_s.

circuit, preventing shoot-through and enabling ZVS by allowing load current to discharge the drain-to-source voltage. (2) Hybrid switched-capacitor converters, where some switches experience hard turn-on [5][6], making hard turn-on losses a significant concern. Accurate evaluation of the switching performance of candidate devices is therefore essential for achieving optimal system design. While GaN devices have demonstrated superiority in terms of power density and efficiency over Si counterparts in 650 V applications [7], their advantages in low-voltage applications, particularly under ZVS conditions, remain less evident and necessitate further investigation.

Current methods for estimating switching losses of low-voltage devices face several limitations. Analytical calculations, while theoretically comprehensive, are complex and often fail to account for all circuit parasitics [8][9]. The double pulse test, a widely used approach, lacks accuracy for fast-switching devices and is challenging to implement non-intrusively [10]. Indirect methods, such as calorimetric measurements, provide high accuracy but are time-consuming and require detailed thermal network modeling [11][12]. Electrical indirect methods, like the Opposition Method (OM), estimate switching losses by measuring total input DC power and accounting for other losses, enabling faster and accurate measurements. However, previous studies have mainly focused on high-voltage devices and/or insufficiently address conduction losses in PCB layouts by ignoring the harmonics [10][13][14].

Applying OM to GaN devices is further complicated by the need to estimate Dynamic on-resistance (D-R_{DS}), for which limited data exist for low-voltage devices [15][16]. Moreover, some measurements based on pulsed measurement methods may yield inaccurate results according to some researchers [17][18].

This work enhances the accuracy of the OM by integrating the Finite Element Method (FEM) to extract frequency-dependent parasitic resistances within the PCB layout and applying Fourier Transform analysis to current waveforms for precise loss calculations. It also introduces a straightforward approach to estimate lumped D-R_{DS} of GaN devices across various operating frequencies within the OM setup, largely improving the accuracy of switching loss estimations. Furthermore, this study provides a comprehensive comparison of the switching performance of $100\,V$ GaN and Si devices, evaluating key parameters such as turn-on and turn-off switching losses, gate driver losses, and lumped D-R_{DS}. Results highlight that GaN devices offer substantial advantages, including up to 45% lower turn-on losses, 42% lower turn-off losses, and 71% lower gate driver losses compared to Si counterparts. However, GaN devices are found to be highly sensitive to DT control, with precise tuning required to fully exploit their low turn-off losses. Additionally, GaN devices exhibit a lumped dynamic D-R_{DS} 3–4 times higher than static values at $1\,MHz$, which raises concerns in high-conduction-loss applications.

The underlying principles of the proposed method are outlined in Section II. Section III details the setup design and impedance extraction process, including PCB layout, litz-wire inductor, and D-R_{DS} of GaN devices. Section IV presents the experimental results and discussions. The paper concludes in Section V.

II. PRINCIPLES OF PROPOSED METHOD

A. Working Principles of OM

The OM employs an indirect approach to estimate switching losses (P_{sw}) by first measuring the total power consumption (P_{in}) of the system and then subtracting other losses (P_{other}), as expressed in:

$$P_{sw} = P_{in} - P_{other}. \tag{1}$$

As shown in Fig. 2, the system comprises a DC-link and a full-bridge circuit operating in two distinct modes to generate waveforms that allow extraction of turn-off energy (E_{off}) and turn-on energy (E_{on}), respectively. P_{in}

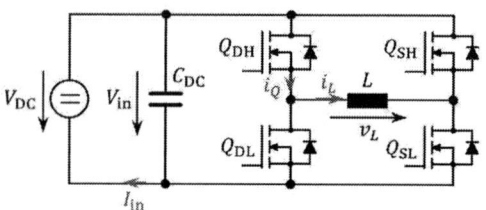

Fig. 2. The circuit of the OM.

is obtained by measuring input current (I_{in}) and voltage on the DC-link (V_{in}). To achieve accurate estimation of losses in the system, litz-wire air-core inductors are used to eliminate core losses. Consequently, P_{other} are primarily conduction losses. These include conduction losses in the litz-wire inductor, switching devices, PCB layout, DC-link capacitors, i.e.,

$$P_{other} = P_{cond} = P_{ind} + P_{Q,cond} + P_{PCB} + P_{cap}. \tag{2}$$

In Mode 1 (cf. Fig. 3(a)), the two half-bridges (D and S) operate with approximately a 50% duty cycle and a controlled phase shift (PS) to create a trapezoidal current in the inductor. A larger PS results in a higher switching current, also determined by the DC-link voltage and inductance. A smaller inductance is preferred to produce a current waveform closer to triangular, reducing the root mean square (RMS) current and thus minimizing conduction losses. In this mode, all four devices operate with ZVS during turn-on and hard switching during turn-off. The turn-off loss energy is then calculated as:

$$E_{off} = \frac{P_{in} - P_{other}}{4f_{sw}}. \tag{3}$$

Since P_{other} primarily consists of AC conduction losses, and the AC resistances of the litz-wire inductor, PCB, and DC-link capacitors are strongly frequency-dependent, their resistances are extracted across the frequency spectrum. The currents are then transformed into the frequency domain using Fourier Transform. The root mean square (RMS) values of the harmonics are multiplied by the corresponding resistances at each frequency to calculate the AC conduction losses. The $R_{DS(on)}$ of the Si devices are considered constant, while for GaN devices, which exhibit dynamic R_{DS} effects that are also frequency-dependent, the lumped $R_{DS(on)}$ are extracted separately and will be discussed later.

In Mode 2 (cf. Fig. 3(b)), the two half-bridges also operate with a duty cycle close to 50%, but one of the half-bridges introduces a slight extra duty cycle (ΔD) to generate a DC bias current. A large inductance is used, and no phase shift is applied to suppress current ripple and reduce AC conduction losses. In this mode,

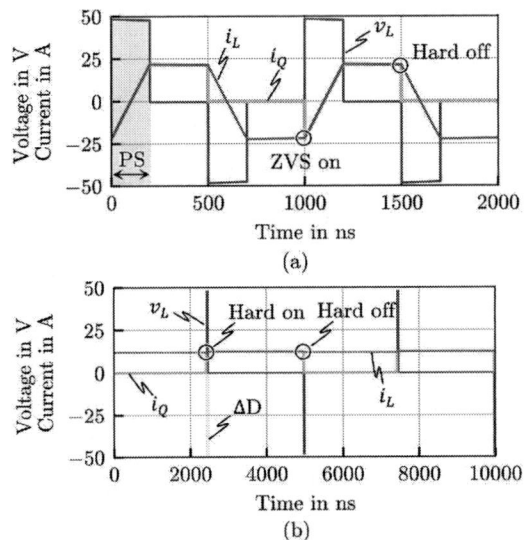

(a)

(b)

Fig. 3. Waveforms of the OM and switching events of Q_{DH}: (a) Mode 1, (b) Mode 2.

two devices experience hard turn-on and hard turn-off, while the other two operate with ZVS turn-on and soft turn-off (cf. Fig. 1). The switching loss energy in this mode includes contributions from two times of E_{on} and E_{off}, with E_{on} extracted as:

$$E_{on} = \frac{P_{in} - P_{other}}{2f_{sw}} - E_{off}. \quad (4)$$

Given the current ripple is kept to a minimum, the conduction losses in the inductor are primarily due to DC conduction losses. However, AC currents flow in parts of the PCB and the switching devices, their harmonics are considered when calculating AC conduction losses.

B. Estimation of Lumped Dynamic On Resistance

The extraction of D-R_{DS} is crucial for accurately estimating switching losses, as D-R_{DS} is typically much larger than the static on resistance $R_{DS,DC}$, particularly at high switching frequencies. For convenience, the ratio of the lumped D-R_{DS} to $R_{DS,DC}$ is defined as the dynamic resistance factor, K_D, given by:

$$K_D = \frac{D\text{-}R_{DS}}{R_{DS,DC}}. \quad (5)$$

To estimate K_D, a straightforward method is proposed that can be directly performed in the OM setup. The approach operates the system in Mode 1 and the following relationship holds:

$$P_{in} = P_{ex} + 4I_{Q,RMS}^2 K_D R_{DS,DC}, \quad (6)$$

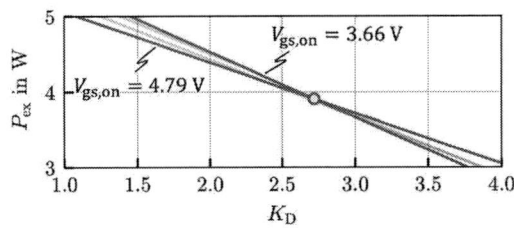

Fig. 4. Identification of K_D of GaN devices by changing $V_{gs,on}$.

where P_{ex} refers to the total losses, excluding conduction losses in the GaN devices.

Next, the turn-on gate voltage is adjusted from $V_{gs,on,1}$ to $V_{gs,on,2}$, which changes the lumped resistance from $K_D R_{DS,DC,1}$ to $K_D R_{DS,DC,2}$ (with $R_{DS,DC}$ as a function of $V_{gs,on}$ according to the datasheet). By keeping other control parameters constant, the switching losses and conduction losses in other components (P_{ex}) remain unchanged. Therefore, the difference in P_{in} is directly related to the change in conduction losses in the GaN devices, as given by:

$$P_{in,1} - P_{in,2} = 4I_{Q,RMS}^2 K_D \left(R_{DS,DC,1} - R_{DS,DC,2}\right). \quad (7)$$

Given that $I_{Q,RMS}$ is known, K_D can be determined by solving equation (7). Alternatively, to integrate multiple measurement points, Equation (6) can be rearranged to:

$$P_{ex} = P_{in} - 4I_{Q,RMS}^2 K_D R_{DS,DC}. \quad (8)$$

For a fixed $V_{gs,on}$, K_D is swept from low to high values, and the corresponding P_{ex} is calculated and plotted, as shown in Fig. 4. Multiple curves are plotted with different $V_{gs,on}$, and since P_{ex} remains constant across different $V_{gs,on}$ values, the point of intersection of the curves represents the K_D of the device.

This method provides a simple and practical approach for estimating K_D in GaN devices within the OM setup, offering insight into the lumped $R_{DS(on)}$, which is critical for estimating switching losses. While error is expected due to the underlying assumptions:

1) Constant K_D across $V_{gs,on}$ Variations: It is assumed that K_D does not change with variations in $V_{gs,on}$.
2) Consistency of Static Resistance: The static resistance $R_{DS,DC}$ as a function of $V_{gs,on}$ for the tested devices is assumed to be consistent with the values provided in the datasheet.
3) Consistency of Resistance vs. Temperature: The static resistance $R_{DS,DC}$ as a function of temperature for the tested devices is assumed to be consistent with the values provided in the datasheet. $R_{DS,DC}$ is then scaled based on the measured

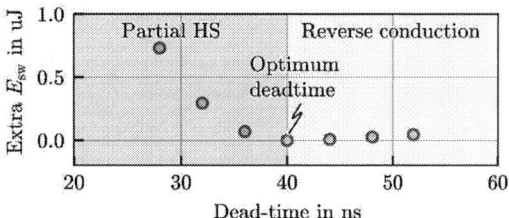

Fig. 5. The optimal DT is identified by sweeping DT values to minimize extra switching losses.

temperature on the top case, as observed from the high-frame-rate thermal camera.

4) Mean Value of V_{gs}: The $V_{gs,on}$ values are measured during device operation, and a slight voltage drop occurs during turn-on of the device. The mean value of $V_{gs,on}$ is selected for calculation and plotting.

C. Optimal Dead-Time

To ensure ZVS during turn-on, a specific dead-time is required to fully discharge the output capacitor. As shown in Fig. 5, too short a DT can lead to incomplete reloading and increased losses due to partial hard switching [19], while too long a DT results in additional losses from reverse conduction, either through the body diode in Si or the channel in GaN [20]. This issue is particularly critical for GaN devices, as their fast switching speeds can easily cause full hard switching without sufficient DT, while excessive DT increases losses due to the higher reverse conduction voltage of GaN compared to Si, especially with negative gate voltages used for secure turn-off.

Thus, optimizing the DT is essential for minimizing losses. Although theoretically, it can be calculated from parasitic capacitances in the datasheet, such calculations are often inaccurate due to small-signal extraction [21][22]. In this work, the optimal DT is determined by sweeping through DT values and selecting the one that minimizes total losses P_{in} (cf. Fig. 5).

III. IMPEDANCE EXTRACTION

To accurately estimate conduction losses in the system, it is essential to precisely extract the impedance of all components, including the lumped D-R_{DS} of GaN devices, litz-wire inductors, PCB layout, and DC-link capacitors.

This work investigates four devices with some of the lowest and comparable $R_{DS(on)}$ on the market, summarized in Table I. These include two Si devices from the same manufacturer but different generations, (Si G1 and Si G2) and two GaN devices from different manufacturers (GaN D1 and GaN D2, both Schottky

Fig. 6. PCB layouts of the test boards.

TABLE I
100 V DEVICES CONSIDERED IN THIS WORK.

Parameter	Si G1	Si G2	GaN D1	GaN D2
$V_{DS,max}$ (V)	100	100	100	100
$R_{DS(on),max}$ (mΩ)	2.7[1]	2.2[1]	2.7[2]	2.2[2]
Q_g (nC)	89[1]	73[1]	13[2]	18[2]
Q_{gd} (nC)	18[1]	11.9[1]	2.5[2]	1.8[2]
Q_{oss} (nC)	114	135	77	71
Package size (mm^2)	5×6	5×6	4.5×2.3	4.5×2.3

[1] $V_{gs} = 10$ V.
[2] $V_{gs} = 5$ V.

gate p-GaN HEMT), all with similar characteristics. The PCB layouts, shown in Fig. 6, use the same Infineon gate driver (2EDF7275K) for both Si and GaN devices, with zero external gate resistance to maximize slew rate, typical for low-voltage applications. For the GaN devices, a negative drive voltage is applied via the EZdrive circuit [23].

The proposed method for extracting lumped D-R_{DS} and K_D is applied to three devices, tested at 500 kHz, 1 MHz, and 1.5 MHz, all operating in Mode 1 at a load current of approximately 15 A. The results, shown in Fig. 7, indicate that at 1 MHz, GaN D1 exhibits a K_D of 3.3, while GaN D2 shows 4.3. These values provide a reference for estimating switching losses, though some error is expected, as discussed in Sec. II. For instance, Si G2, which is believed not to have dynamic R_{DS} effects, shows a K_D of 1.3 at 1 MHz. These measurements suggest that the lumped D-R_{DS} of GaN devices can be significantly higher at high frequencies, much more than previously reported values for low-voltage devices, which are typically below 2, even at 1 MHz [15][16]. Notably, a high K_D ratio of 4–6 has been reported for several 650 V devices in soft-switching operation at 3 MHz [18].

Two different litz-wire inductors are employed in this work, with No.1 having 210 nH inductance used in Mode 1, No.2 having 4 µH inductance used in Mode 2. The impedance of the litz-wire inductors is measured

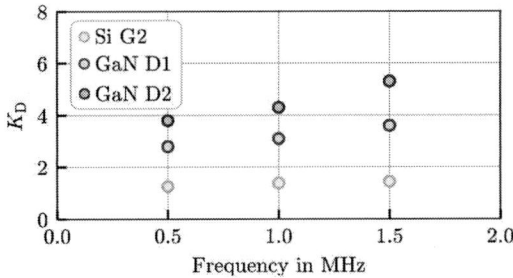

Fig. 7. Extracted K_D of Si G2, GaN D1, and GaN D2 at the frequencies of 0.5 MHz, 1 MHz, and 1.5 MHz under ZVS condition at 48 V.

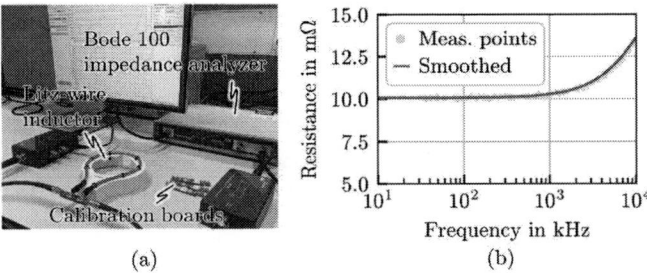

Fig. 8. Experimental setup for impedance measurement: (a) measurement on a litz-wire inductor, (b) measurement result of a $10\,\mathrm{m\Omega}$ 0603 resistor with 1% tolerance.

using a Bode 100 impedance analyzer with a shunt-through (4-wire) connection, as shown in Fig. 8(a). Specialized calibration boards minimize measurement errors from connection resistance, with the inductors directly soldered onto the board. To verify low-resistance measurement accuracy, a $10\,\mathrm{m\Omega}$ 0603 resistor with 1% tolerance is measured (cf. Fig. 8(b)), yielding a resistance of $10.08\,\mathrm{m\Omega}$ below 400 kHz, after which it rises due to the skin effect. The litz-wire inductors show a resistance of $8.9\,\mathrm{m\Omega}$ for No.1 at 1 MHz, and $4.6\,\mathrm{m\Omega}$ for No.2 at 10 Hz.

The frequency-dependent resistances of the PCB layout are also critical, especially due to the significantly low switching losses in low-voltage devices. These resistances are extracted using FEM simulations for different current loops. Fig. 9 shows the current distribution in the PCB layout when Q_{SL} and Q_{DH} are both on. The resistances of different nets are extracted accordingly. At a frequency of 1 MHz, the combined resistance of the VDC and PGND nets is $4.3\,\mathrm{m\Omega}$, while the combined resistance of the SW-D and SW-S nets is $1.2\,\mathrm{m\Omega}$. The PCB resistance is also measured with the impedance analyzer; however, the measured loop resistance is approximately $3\,\mathrm{m\Omega}$ higher than the simulated value. This discrepancy may be due to additional resistances from the device footprint connections and variations in the excitation injection points. As it is challenging to separate the AC resistances for the different nets from the

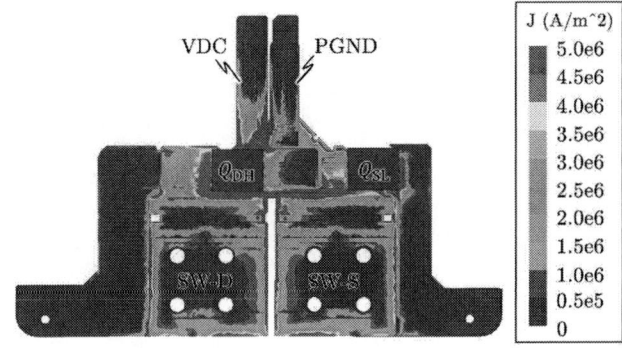

Fig. 9. Current distribution of the PCB layout in Ansys with AC excitation of 1 A.

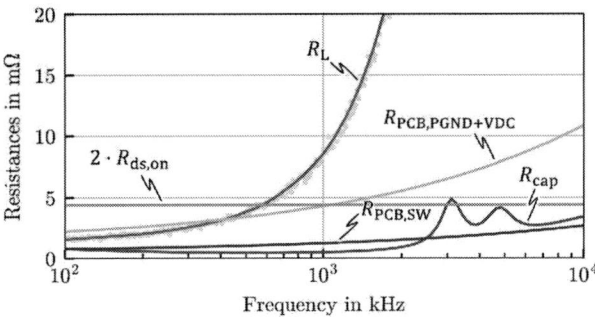

Fig. 10. Extracted frequency-dependent resistances of litz-wire inductor, PCB layout, $R_{\mathrm{DS(on)}}$, and DC-link capacitors.

measurement results, the simulation results are used in the final calculations.

The frequency-dependent resistance of the DC-link capacitor is provided by the manufacturer. Then all the resistances are obtained and depicted in Fig. 10. The results show that the PCB layout resistances are of the same order of magnitude as R_L and $R_{\mathrm{DS(on)}}$. This highlights the importance of accurately estimating and accounting for losses caused by current harmonics in the PCB layouts.

IV. EXPERIMENTAL SETUP AND MEASUREMENT RESULTS

Fig. 11 shows the system design, where the test board interfaces with both a DC power supply and an auxiliary supply via a motherboard. Input current I_{in} is measured using a high-precision current shunt (0.1% tolerance), while input voltage V_{in} is monitored at the DC-link capacitors with a digital multimeter. Depending on the operating mode, either a 210 nH or 4 µH inductor is soldered onto the test board. Control signals, generated by the controller board, vary the switching frequency from 200 kHz to 1.5 MHz with DT resolution of 1 ns. A high-frame-rate thermal camera tracks temperature rise in components. Cooling fans are used to maintain moderate temperature increases. In case of significant

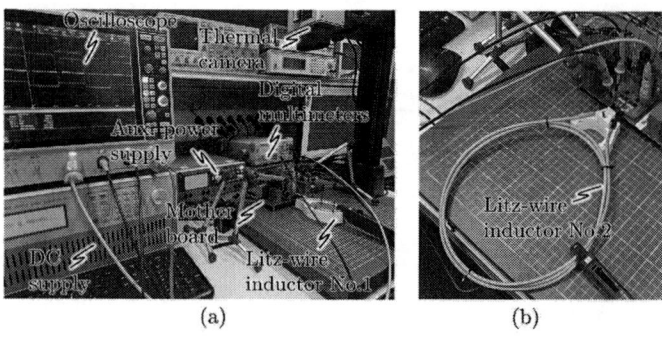

(a)   (b)

Fig. 11. Experimental implementation of the OM: (a) setup for Mode 1 with a 210 nH litz-wire inductor (No.1), (b) setup for Mode 2 with a 4 µH litz-wire inductor (No.2).

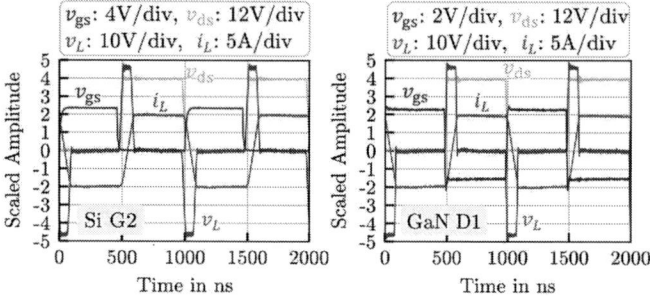

Fig. 12. Measured waveforms when operating in Mode 1 for both Si and GaN devices at 1 MHz, 48 V, and 10 A.

losses, temperature changes are recorded, and scaling factors are applied to minimize error. Since these devices are commonly used with a 48 V bus voltage, the optimal DT extraction and switching loss measurements are performed at 48 V for comparison.

A. Turn-off Losses

Turn-off losses are measured with operation Mode 1, the measured waveforms of Si and GaN devices are shown in Fig. 12. To identify the optimal DT for the desired load current, DTs are adjusted from a low to a high value. The DT that results in the lowest P_{in} is considered the optimal DT. The measurement results of the optimal DTs for the four devices at 48 V are illustrated in Fig. 13. Due to significantly lower values of Q_g, Q_{gd}, and Q_{oss}, the GaN devices require much shorter DTs to complete the recharging of the output capacitor. At a 10 A load current, the optimal DT for GaN devices is reduced from 40 ns to approximately 10 ns compared to Si counterparts. This reduction is especially beneficial for high-frequency operations, as the effective duty cycle is increased with less DT required. However, within the same semiconductor material, no device exhibits significant advantages.

A practical consideration when comparing Si and GaN devices is the sensitivity of DT. In real applications, DT is difficult to maintain optimal due to load current

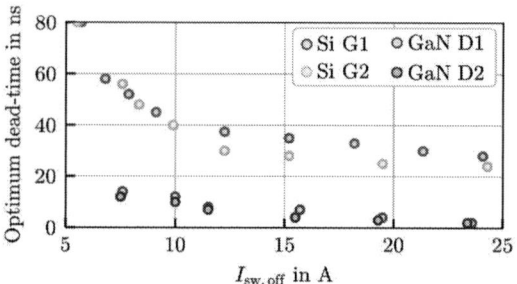

Fig. 13. Comparison of optimal DTs at 48 V for various load currents.

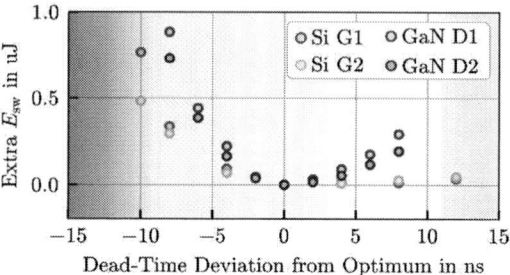

Fig. 14. DT sensitivity of considered devices at 48 V and 10 A load current.

variations or safety margins for mass-produced products. Either excessive or insufficient DT can significantly increase switching losses. A comparison of DT sensitivity for four devices at 48 V and 10 A, shown in Fig. 14, reveals that GaN devices are more sensitive to DT mismatches. When DT is too short, the high slew rate of GaN devices results in larger remaining v_{DS} during turn-on, leading to higher partial hard turn-on losses compared to Si devices. When DT is too long, GaN devices experience higher reverse conduction voltage, particularly with negative $V_{gs,off}$. For example, at 10 A, the reverse conduction voltage in GaN devices is -4.4 V, while for Si it is only -0.7 V. This highlights the need for precise DT control in systems using GaN devices, ideally adjustable based on load current. In contrast, Si devices can tolerate a safety margin in DT (e.g., 10 ns) without significant efficiency loss.

With optimal DTs set for all load currents, turn-off losses are estimated for all devices at 48 V and load currents ranging from 5 A to 25 A. The results, shown in Fig. 15(a), present turn-off losses E_{off} at switching frequencies of 1 MHz, 1.2 MHz, and 1.5 MHz, with each frequency represented as a separate point for each load current. The close agreement of E_{off} values across frequencies indicates measurement accuracy. GaN devices show significantly lower turn-off losses, due to their lower parasitic capacitances. Turn-off losses are 42% lower compared to Si G2 and 85% compared to Si G1. Additionally, Si G2 achieves much lower turn-off losses than Si G1 with even lower $R_{DS(on)}$, despite both

979-8-3315-1612-3/25 $31.00 © 2025 IEEE   2366

being Si-based. The reason is explained in the literature [24] and is out of scope of this paper.

Another interesting comparison involves examining the total system losses when Si and GaN devices operate under the same conditions, allowing for a direct comparison of their overall performance. While GaN devices exhibit lower turn-off losses, their lumped D-R_{DS} is significantly higher than that of Si devices. With almost identical layout designs, same litz-wire inductor, switching frequency, and load current, the power difference in P_{in} is primarily due to the switching devices. A comparison at 1 MHz with a trapezoidal load current of 10 A is shown in the last row of Table II. The results show that Si G2 yields the lowest P_{in}, indicating that when both conduction and switching losses are considered, Si G2 outperforms both GaN D1 and GaN D2. Given the less dead-time sensitivity of Si devices, this suggests that for soft-switching applications, where conduction losses dominate, Si devices like Si G2 could be more suitable than GaN devices. However, when driver losses are considered, the total losses from GaN devices remain lower. Therefore, determining which device is more suitable becomes very case-dependent. Other key device parameters are summarized in Table II.

B. Turn-on Losses

After measuring the turn-off losses, turn-on losses are assessed in Mode 2 at a switching frequency of 200 kHz to prevent excessive heating. By adjusting ΔD, the load current is varied from very low values up to approximately 15 A. The DT is set to a value that ensures minimum losses are caused by mismatches during the two ZVS-on events. The results, shown in Fig. 15(b), reveal that GaN devices have 45% lower turn-on losses than Si G2 and 69% lower than Si G1. Si G2 also shows a 45% reduction compared to Si G1. Furthermore, GaN devices exhibit much lower driver losses. For hard-switching and high-frequency applications, GaN devices clearly outperform Si devices, making them the preferred choice.

V. CONCLUSIONS

This work compares the switching performance of low-voltage GaN and Si MOSFETs, focusing on key parameters such as turn-on and turn-off losses, gate driver losses, and lumped $R_{DS(on)}$. GaN devices show significant advantages, with at least 42% lower turn-off losses, 45% lower turn-on losses, and 71% lower gate driver losses compared to Si counterparts, making them ideal for high-frequency hard-switching applications. However, GaN devices require precise DT control

TABLE II

COMPARISON OF KEY PARAMETERS OF THE DEVICES IN THIS WORK CONSIDERING A 10 A LOAD CURRENT.

Parameter	Si G1	Si G2	GaN D1	GaN D2
Lumped R_{DS} (mΩ)[1]	2.11	1.90	8.3	7.8
E_{off} (µJ)	1.44	0.34	0.21	0.14
Opt. DT (ns)	40	40	12	10
E_{on} (µJ)	14.4	7.94	4.48	3.09
E_{drv} (µJ)[1]	0.81	0.64	0.16	0.21
P_{in} (W)[2]	7.25	2.74	3.48	2.95

[1] Considering $V_{gs} = 9.5$ V for Si devices and $V_{gs} = -2.8$ V-4.5 V for GaN devices.

[2] The total input power is measured at 1 MHz with a 10 A trapezoidal load current and 48 V DC-link voltage, operating in Mode 1 (ZVS for all devices). The same litz-wire inductor and similar layouts are used for each device.

to fully exploit these benefits, and measured by the proposed method, their higher dynamic $R_{DS(on)}$ at high frequencies poses challenges for high-conduction-losses applications. Si devices offer lower $R_{DS(on)}$ and are less sensitive to DT mismatches, making them more suitable for soft-switching applications or systems where conduction losses dominate. Thus, the choice between GaN and Si devices should be guided by specific application requirements, with GaN preferred for high-frequency, hard-switching applications and Si for more conduction-loss-sensitive designs.

REFERENCES

[1] S. Li, "Intermediate bus converters for high efficiency power conversion: A review," in *2020 IEEE Texas Power and Energy Conference (TPEC)*, IEEE, 2020, pp. 1–6.

[2] M. H. Ahmed, C. Fei, F. C. Lee, and Q. Li, "48-v voltage regulator module with pcb winding matrix transformer for future data centers," *IEEE Transactions on Industrial Electronics*, vol. 64, no. 12, pp. 9302–9310, 2017.

[3] S. Saggini, O. Zambetti, R. Rizzolatti, M. Picca, and P. Mattavelli, "An isolated quasi-resonant multiphase single-stage topology for 48-v vrm applications," *IEEE Transactions on Power Electronics*, vol. 33, no. 7, pp. 6224–6237, 2017.

[4] X. Ren, J. Zhang, P. Xu, J. Song, and T. Long, "Comparative evaluation of transformer-based fixed-ratio dc-dc converters for 48v data centers," in *PCIM Europe 2024; International Exhibition and Conference for Power Electronics, Intelligent Motion, Renewable Energy and Energy Management*, VDE, 2024, pp. 803–811.

[5] Y. Zhu, J. Zou, and R. C. Pilawa-Podgurski, "A 1500-a/48-v-to-1-v switching bus converter for next-generation ultra-high-power microprocessors," in *2024 IEEE Applied Power Electronics Conference and Exposition (APEC)*, IEEE, 2024, pp. 890–897.

[6] P. Wang, Y. Chen, G. Szczeszynski, S. Allen, D. M. Giuliano, and M. Chen, "Msc-pol: Hybrid gan-si multistacked switched capacitor 48v pwrsip vrm for chiplets," *IEEE Transactions on Power Electronics*, 2023.

979-8-3315-1612-3/25 $31.00 © 2025 IEEE

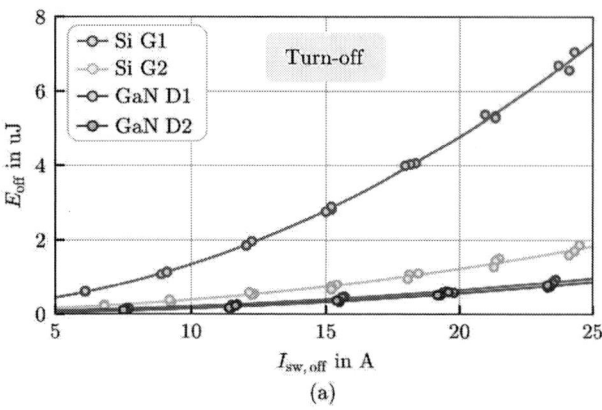 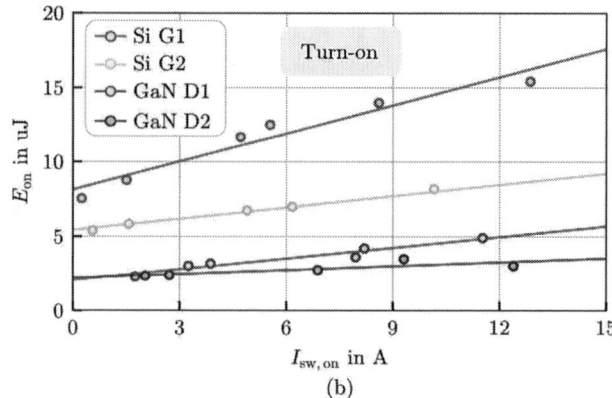
(a)   (b)

Fig. 15. Measurement of switching energy at 48 V across various load currents: (a) turn-off losses, measured at switching frequencies of 1 MHz, 1.2 MHz, and 1.5 MHz; (b) turn-on losses, measured at a frequency of 200 kHz.

[7] G.-J. Su, "Comparison of si, sic, and gan based isolation converters for onboard charger applications," in *2018 IEEE Energy Conversion Congress and Exposition (ECCE)*, 2018, pp. 1233–1239. DOI: 10.1109/ECCE.2018.8558063.

[8] A. Hu and J. Biela, "Verification and application of an analytical switching loss model for a sic mosfet and schottky diode half-bridge," in *2022 International Power Electronics Conference (IPEC-Himeji 2022-ECCE Asia)*, IEEE, 2022, pp. 2599–2606.

[9] D. Christen and J. Biela, "Analytical switching loss modeling based on datasheet parameters for mosfet s in a half-bridge," *IEEE Transactions on Power Electronics*, vol. 34, no. 4, pp. 3700–3710, 2018.

[10] A. Kuczmik, S. Hoffmann, and E. Hoene, "Double pulse vs. indirect measurement: Characterizing switching losses of integrated power modules with wide bandgap semiconductors," in *CIPS 2020; 11th International Conference on Integrated Power Electronics Systems*, VDE, 2020, pp. 1–8.

[11] M. Guacci, J. A. Anderson, K. L. Pally, *et al.*, "Experimental characterization of silicon and gallium nitride 200 v power semiconductors for modular/multi-level converters using advanced measurement techniques," *IEEE Journal of Emerging and Selected Topics in Power Electronics*, vol. 8, no. 3, pp. 2238–2254, 2019.

[12] D. Neumayr, M. Guacci, D. Bortis, and J. W. Kolar, "New calorimetric power transistor soft-switching loss measurement based on accurate temperature rise monitoring," in *2017 29th International Symposium on Power Semiconductor Devices and IC's (ISPSD)*, IEEE, 2017, pp. 447–450.

[13] B. Cougo, H. Schneider, and T. Meynard, "Accurate switching energy estimation of wide bandgap devices used in converters for aircraft applications," in *2013 15th European Conference on Power Electronics and Applications (EPE)*, IEEE, 2013, pp. 1–10.

[14] Q. Yang, A. Nabih, R. Zhang, Q. Li, and Y. Zhang, "A converter based switching loss measurement method for wbg device," in *2023 IEEE Applied Power Electronics Conference and Exposition (APEC)*, IEEE, 2023, pp. 8–13.

[15] G. Zulauf, M. Guacci, J. M. Rivas-Davila, and J. W. Kolar, "The impact of multi-mhz switching frequencies on dynamic on-resistance in gan-on-si hemts," *IEEE Open Journal of Power Electronics*, vol. 1, pp. 210–215, 2020. DOI: 10.1109/OJPEL.2020.3005879.

[16] T. Foulkes, T. Modeer, and R. C. Pilawa-Podgurski, "Quantifying dynamic on-state resistance of gan hemts for power converter design via a survey of low and high voltage devices," *IEEE Journal of Emerging and Selected Topics in Power Electronics*, vol. 9, no. 4, pp. 4036–4049, 2020.

[17] H. Zhu and E. Matioli, "Accurate measurement of dynamic on-resistance in gan transistors at steady-state," *IEEE Transactions on Power Electronics*, vol. 38, no. 7, pp. 8045–8050, 2023.

[18] B. J. Galapon, A. J. Hanson, and D. J. Perreault, "Measuring dynamic on resistance in gan transistors at mhz frequencies," in *2018 IEEE 19th Workshop on Control and Modeling for Power Electronics (COMPEL)*, 2018, pp. 1–8. DOI: 10.1109/COMPEL.2018.8460051.

[19] M. Kasper, R. M. Burkart, G. Deboy, and J. W. Kolar, "Zvs of power mosfets revisited," *IEEE Transactions on Power Electronics*, vol. 31, no. 12, pp. 8063–8067, 2016.

[20] E. A. Jones, F. Wang, and B. Ozpineci, "Application-based review of gan hfets," in *2014 IEEE Workshop on Wide Bandgap Power Devices and Applications*, IEEE, 2014, pp. 24–29.

[21] N. Perera, G. Kampitsis, R. Van Erp, *et al.*, "Analysis of large-signal output capacitance of transistors using sawyer–tower circuit," *IEEE Journal of Emerging and Selected Topics in Power Electronics*, vol. 9, no. 3, pp. 3647–3656, 2020.

[22] T. Chen, P. A. Bezerra, Z. He, G. Li, and E. Hoene, "Systematic derivation and experimental verification of a compact loss model for soft-switching half-bridges," in *2023 25th European Conference on Power Electronics and Applications (EPE'23 ECCE Europe)*, IEEE, 2023, pp. 1–9.

[23] L. Kou and J. Lu, "A gan and si hybrid solution for 48v-12v automotive dc-dc application," in *2020 IEEE Energy Conversion Congress and Exposition (ECCE)*, IEEE, 2020, pp. 2858–2864.

[24] R. Siemieniec, S. Mazzer, E. Kahrimanovic, *et al.*, "Next step in power mosfet evolution boosts application efficiency," in *2024 IEEE Applied Power Electronics Conference and Exposition (APEC)*, 2024, pp. 1767–1774. DOI: 10.1109/APEC48139.2024.10509054.

Modeling of Switching Transients for Frequency-Domain CM EMI Analysis in Double Sided Cooling Power Modules

1st Sijia Liu
School of Electrical and Electronic Engineering
Huazhong University of Science and Technology
Wuhan, China
sijia_l@hust.edu.cn

2nd Liu Yang*
State Key Laboratory of HVDC Electric Power Research Institute of China Southern Power Grid
Guangzhou, China
Yangliu3@csg.cn

3rd Heng Zhang
School of Electrical and Electronic Engineering
Huazhong University of Science and Technology
Wuhan, China
m202372269@hust.edu.cn

4th Yifan Zhang
School of Electrical and Electronic Engineering
Huazhong University of Science and Technology
Wuhan, China
z_yifan@hust.edu.cn

5th Zexiang Zheng
School of Electrical and Electronic Engineering
Huazhong University of Science and Technology
Wuhan, China
zzx_huster@hust.edu.cn

6th Jianwei Lv
School of Electrical and Electronic Engineering
Huazhong University of Science and Technology
Wuhan, China
jianweil@hust.edu.cn

7th Jiaxin Liu
School of Electrical and Electronic Engineering
Huazhong University of Science and Technology
Wuhan, China
liujx711@hust.edu.cn

8th Cai Chen*
School of Electrical and Electronic Engineering
Huazhong University of Science and Technology
Wuhan, China
caichen@hust.edu.cn

9th Yong Kang
School of Electrical and Electronic Engineering
Huazhong University of Science and Technology
Wuhan, China
ykang@hust.edu.cn

10th Yuebin Zhou
State Key Laboratory of HVDC Electric Power Research Institute of China Southern Power Grid
Guangzhou, China
Zhouyb@csg.cn

11th Daming Wang
State Key Laboratory of HVDC Electric Power Research Institute of China Southern Power Grid
Guangzhou, China
Wangdm3@csg.cn

12th Shuang Zhao
School of Electrical Engineering and Automation
Hefei University of Technology
Hefei, China
shuang.zhao@hfut.edu.cn

Abstract—**This paper provides detailed models and analysis of the impact of parasitic capacitances on CM EMI performance in the double sided cooling (DSC) power modules. The effect of parasitic capacitances on CM EMI in DSC modules are analyzed first. Then, small-signal models for different switching stages are established to better understand the mechanisms from the perspective of interference source. Finally, based on the above analysis, several design guidance of the DSC module is given for better CM EMI performance. Moreover, an optimized DSC module integrated with a capacitor is proposed. Simulation results show that the optimized module can achieve a maximum reduction of 13 dBµV in CM EMI.**

Keywords—***Common-mode(CM) EMI, frequency-domain model, parasitic capacitances, double sided cooling power modules.***

I. INTRODUCTION

With the advancement of power electronics technology, there is an increasing demand for greater performance in terms of switching frequency, power density, and other characteristics of power electronic devices. As the core component of power electronics, power semiconductor devices play a key role on the development of power electronics technology. The emergence of wide bandgap (WBG) power semiconductors has introduced both new opportunities and challenges in the field of power electronics. WBG devices exhibit faster switching frequencies and can operate at higher voltages and temperatures, significantly pushing the development of power electronic systems towards higher frequency and higher power density[1]. However, the high switching speed also bring new challenges in the packaging design of power modules. The parasitic parameters introduced by the packaging can cause overvoltage and oscillations, which provide conducted path for electromagnetic interference (EMI), further exacerbating the EMI issues within the system[2].

The primary approach for EMI optimization in power electronic systems involves optimizing both the EMI source and the conducted paths[3]. The analysis of the EMI interference source require modeling the switching transients of the WBG devices. Reference [4] proposes a time-domain model to analyze the oscillation behavior of SiC MOSFETs with RC snubbers, focusing on reducing voltage overshoot and improving current sharing. In [5], two equivalent RLC circuit models are developed to analyze the switching oscillations in SiC MOSFETs and guidance is given for RC snubber design to suppress the switching oscillation. For analysis of the conducted path, a EMI model which considers parasitic inductances and

979-8-3315-1612-3/25 $31.00 © 2025 IEEE

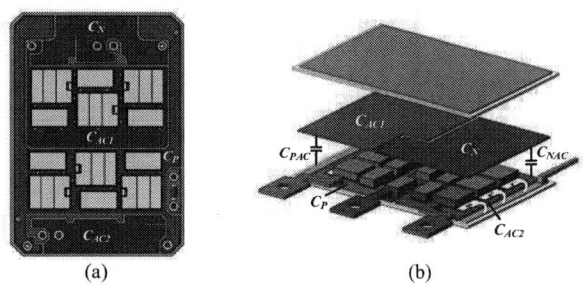

Fig. 1. Structures of two modules. (a) Conventional wire bonded power module. (b) Double sided cooling power module.

TABLE I.   TABLE TYPE STYLES

	Material	Thickness (mm)	Relative permittivity
Bare Die	Si	0.16	-
Copper of DBC	Copper	0.365	-
Ceramic of DBC	ALN	0.33	8.8

TABLE II.   PARASITIC EXTRACTION RESULTS OF THE TWO MODULES

Power Module Structures	Conventional Wire bonded Module	Double Sided Cooling Module
C_{AC}	238.14pF	471.78pF
C_P	240.54pF	182.05pF
C_N	71.36pF	248.84pF
C_{PAC}	0.03pF	17.6pF
C_{NAC}	0.03pF	18pF

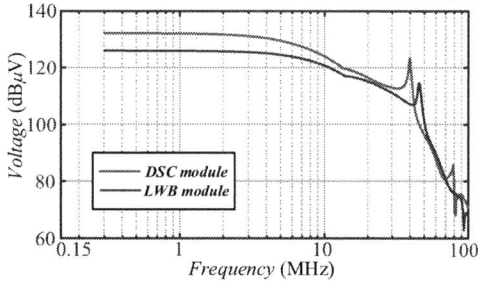

Fig. 2. Comparsion of CM EMI between two modules.

parasitic capacitances is needed[6]. The work in [7] proposes a EMI model from the perspective of conducted paths and gives a solution to reduce CM current by optimizing power module layout and parasitic inductances. In [8], a EMI noise model is proposed based on a boost converter and a general balance concept is drawn to cancel CM EMI. Reference [9] proposes a detailed frequency-domain model as an efficient approach for analyzing EMI behavior and validated its reliability through experiments.

Through the analysis of the interference source and conducted paths, several packaging optimization designs are proposed to achieve lower EMI performance. Reference [10] utilizes the multi-layer substrate to shield the output grounding capacitance, effectively reducing parasitic inductances and

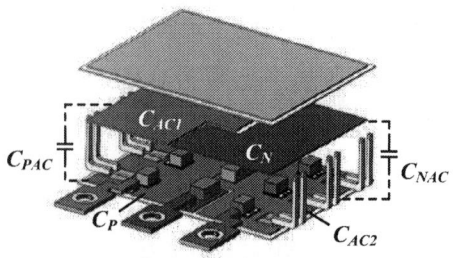

Fig. 3. Parasitic capacitances of double sided cooling power module.

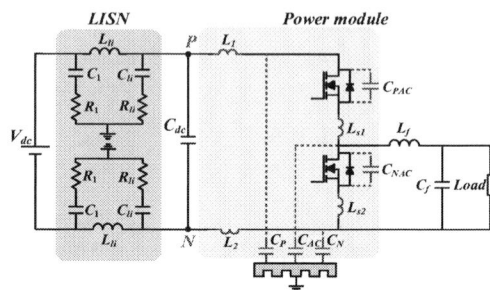

Fig. 4. Schematic diagram of a buck circuit with parasitic parameters.

achieve lower EMI in high-speed switching applications. In [11], a hybrid stacked substrate packaging structure with optimized copper patterns is proposed, which effectively parasitic capacitance by 70% and CM EMI by 6 dBμV without compromising switching performance.

Among the current studies, research on EMI in double sided cooling (DSC) power modules is still limited. The use of double sided cooling substrates lead to unique parasitic parameter distributions. Therefore, there is a need for further investigation into the EMI issues associated with such modules.

This paper provides detailed models and analysis of the impact of parasitic capacitances on CM EMI performance in the DSC power modules. In section II, a comparison is given first and then the main parasitic capacitances in DSC modules are analyzed. In section III, small-signal models for different switching stages are established and validated through simulations. Finally, several design guidelines are given and an optimized DSC module is proposed for lower EMI performance.

II.   CM EMI IN DOUBLE SIDED POWER MODULES

Due to the structural differences between DSC modules and conventional wire bonded modules, it is essential to investigate the impact of parasitic parameters introduced by the DSC module packaging on CM EMI. This section first compares the CM EMI characteristics of two typical power modules. Then, the influence of several key parasitic capacitances on CM EMI is analyzed.

A. Comparsion of CM EMI in Two Typical Modules

Two typical packaging structures are used as study cases in this part. The structures of two power modules are shown in Fig. 1 The DSC power module investigated in this study is based on a commercial DSC power module with model number of FS820R08A6P2B. For ease of comparison, a conventional wire bonded power (CWB) module from the same manufacturer with model number of FF400R07A01E3 is utilized in this study.

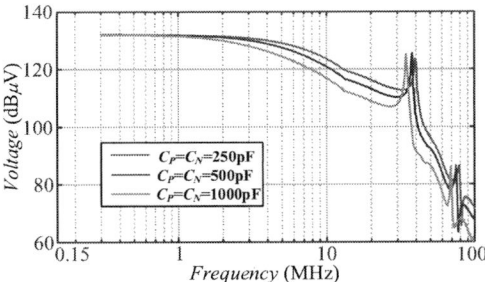

Fig. 5. The impact of C_P and C_N on CM EMI.

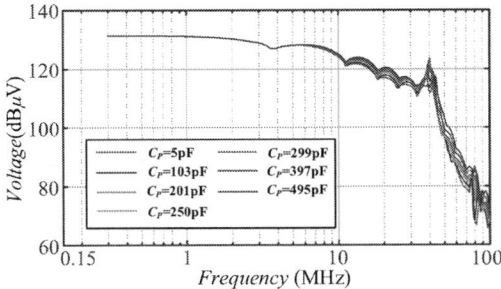

Fig. 6. The mismatch effect of C_P and C_N on CM EMI.

To control variables, both power modules use the same chip model, copper layer thickness, and substrate thickness. Detailed parameters are provided in Table I. Five main parasitic capacitances are extracted by ANSYS Q3D, which are shown in Table II. From Table II, it can be observed that the DSC power module has a C_{AC} of 471.78 pF, which is two times as much as that of the conventional wire bonded module, indicating a severe CM EMI problem in the DSC module. Moreover, DSC module has relatively large C_P and C_N, which plays a similar role to that of the Y capacitors in an EMI filter. Thus, due to the large parasitic capacitances, DSC module suffer more complex CM EMI situations. Fig. 2 shows the simulation results of the CM EMI for both modules, which also clearly demonstrates that the DSC module exhibits more severe CM EMI. Therefore, to further elucidate the impact of these parasitic capacitances on the CM EMI, several simulations have been conducted in LTSPICE. A SiC MOSFET bare die SPICE model is from manufacturer with part number of S4661.

B. Analysis of Parasitic Capacitances in DSC Modules

This section studies the CM EMI of a typical DSC power module. Fig. 3 shows the structure of the DSC power module. This module structure is based on the configuration in Fig. 1(b), with similar copper layer layout and dimensions. Five parasitic capacitances are considered, of which distribution in the power module are also shown in Fig. 3. To better understand the CM EMI characteristics of the DSC power module, a buck circuit with multiple parasitic parameters is used as a study case. The configuration of the buck circuit is shown in Fig. 4.

Fig. 5 shows the impact of C_P and C_N on CM EMI. In this figure, a 6 dBμV reduction on CM EMI is achieved by increasing the two capacitances from 500pF to 1000pF in the low-frequency band. However, a spectral peak appears in high-frequency band and shift to lower frequency as the C_P and C_N increase. Although the amplitude of CM EMI in low-frequency band is suppressed, the amplitude of spectral spikes in high-

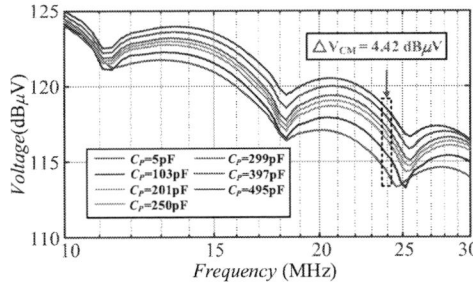

Fig. 7. CM EMI from 10MHz to 30MHz range.

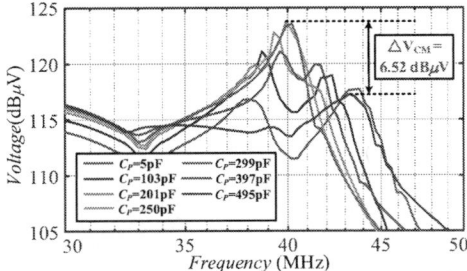

Fig. 8. CM EMI from 30MHz to 50MHz range.

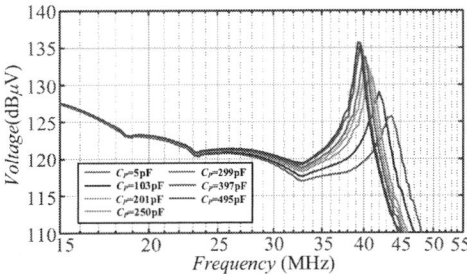

Fig. 9. Voltage spectrums of active switch during turn-off stage.

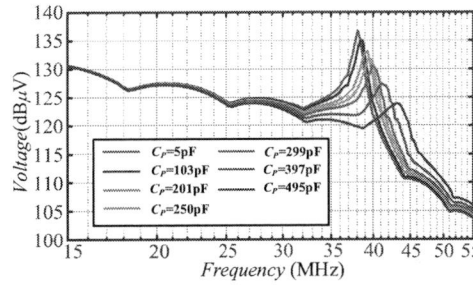

Fig. 10. Voltage spectrums of freewheeling switch during turn-on stage.

frequency band remains unchanged and instead shifts to lower frequency.

The mismatch effect of C_P and C_N on CM EMI is also conducted in the simulations. By changing the ration of C_P and C_N while keeping the sum of C_P and C_N and other parameters unchanged, the partial shifting of the spectrum depending on the frequency ranges can be observed as shown in Fig. 6. The entire spectrum can be divided into three parts. The first part begins at 150kHz and ends at 10MHz. The spectrum in this frequency band remains no change. The second part begins at 10MHz and ends at 30MHz as shown in Fig. 7. By changing C_P from 5pF to 495pF, an increase of the spectrum amplitude in this frequency range is caused with maximum difference of

979-8-3315-1612-3/25 $31.00 © 2025 IEEE

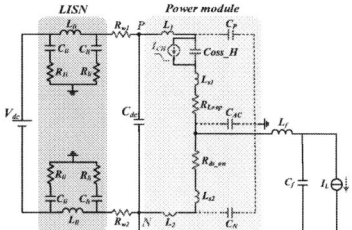

Fig. 11. Large-signal model for turn-off stage.

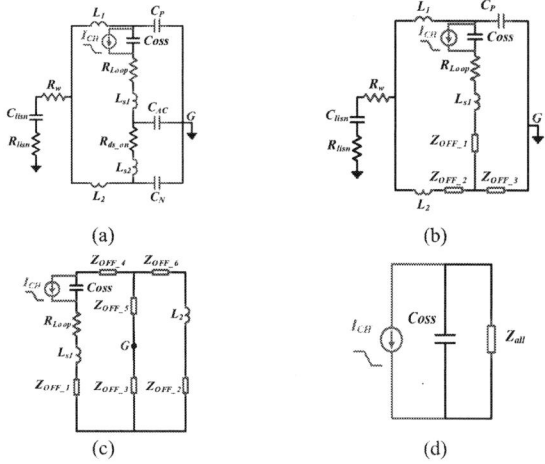

Fig. 12. Simplified process of the small-signal model for turn-off stage.(a) Small-signal model. (b) The first simplification process. (c) The second simplification process. (d) Equivalent circuit of the small-signal model after simplification.

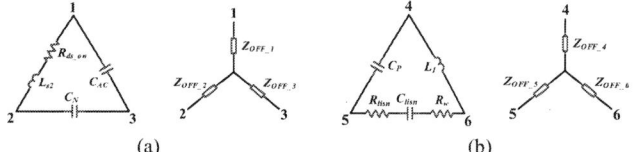

Fig. 13. △-Y impedance transformation for turn-off stage. (a) The first transformation process. (b) The second transformation process.

4.42dBμV. The third part begins at 30MHz and end at 50MHz, in Fig. 8. Spectral peaks appear in this frequency range. As C_P increases from 5pF to 495pF, the spectral spikes initially merge into a single peak and then split into two separate peaks, with a highest amplitude when C_P and C_N are balanced.

Since electronic devices should comply with Electro-Magnetic Compatibility (EMC) standards, the spectral spikes will determine whether the devices can pass these standards or not. Thus, it is necessary to analyze the mechanism behind it.

III. MODELING AND ANALYSIS OF SPECTRAL PEAKS ON CM EMI

Studies have shown that CM interference is caused by the interaction between rapidly changing voltages and parasitic capacitance during the dynamic process of the switching device. Thus, the devices are usually replaced by a voltage source in CM equivalent model. The voltage source is composed of two parts: a trapezoidal wave and a high-frequency voltage oscillation. The voltage oscillation will create a peak at the corresponding frequency in the spectrum, as shown in Fig. 9 and Fig. 10, which will propagate to LISN through CM conducted path. The

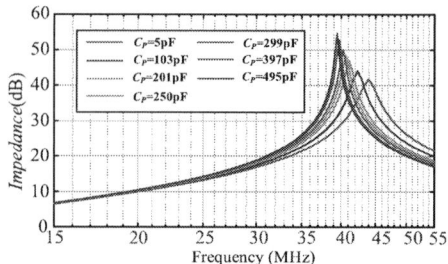

Fig. 14. Input impedances for turn-off stage.

TABLE III.   COMPARISON OF SIMULATIONS AND THEORETICAL MODEL RESULTS FOR THE TURN-OFF STAGE

Turn-on comparison	Frequency (MHz)		
C_P (pF)	V_{DS}	V_{Model}	Differences
5	43.8	43.9	0.1
103	42	42.3	0.3
201	40.8	41.3	0.5
250	40.5	40.7	0.2
299	40.2	40.3	0.1
397	39.6	39.7	0.1
495	39.3	39.4	0.1

frequency of the peaks and their shifting trends in each switch spectrum correspond to those observed in the CM EMI spectrum. Therefore, it is essential to study the mechanisms behind the generation and shifting trends of spectral peaks. Considering the symmetry of upper and lower switches, the analysis only focuses on one case where the upper switch acts as the active switch in the buck circuit.

A. Modeling of Turn-off Stage

To simplify the analysis, this study disregards the parasitic inductance between parallel branches and represents the three switches as an equivalent source for evaluation. The large-signal equivalent module based on buck circuit for turn-off stage is shown in Fig. 11 [12]. Since the voltage oscillation of the active switch occurs during the current falling stage of the turn-off process, the switch is replaced by a current source with a specific di/dt to serve as the excitation in the large-signal equivalent model. Based on the large-signal equivalent model, the small-signal oscillation equivalent model can be easily derived, as shown in Fig. 12. The small-signal oscillation equivalent module can be simplified by using △-Y transformation for two times, which is shown in Fig. 13. From Fig. 12(d), the switching oscillation of the active switch in the turn-off stage is caused by the resonance of RLC network under the current excitation. The spectrum of the input impedance is shown in Fig. 14. To further verify the accuracy of the model, a comparison is presented between the simulation and model results, as shown in the Table III. From Table III, the switching oscillation frequency from the LTSPICE transient simulation closely matches the oscillation frequency of the turn-off small-signal equivalent model, with a maximum error of 1.2%. The resonant frequency and the shifting trend of the peaks are consistent with the voltage spectrum shown in Fig. 9, indicating the accuracy of the small-signal model for the turn-off stage.

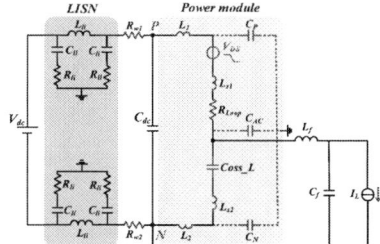

Fig. 15. Large-signal model for turn-on stage.

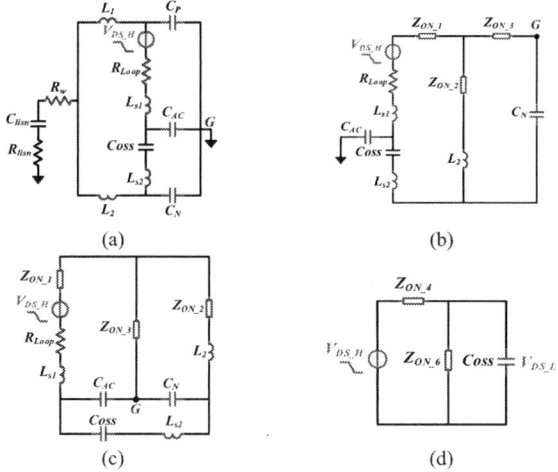

(a)   (b)

(c)   (d)

Fig. 16. Simplified process of the small-signal model for turn-on stage.(a) Small-signal model. (b) The first simplification process. (c) The second simplification process. (d) Equivalent circuit of the small-signal model after simplification.

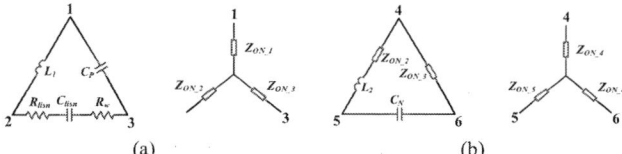

(a)   (b)

Fig. 17. △-Y impedance transformation for turn-on stage. (a) The first transformation process. (b) The second transformation process.

B. Modeling of Turn-on Stage

The large-signal equivalent module for turn-on stage is shown in Fig. 15. During the turn-on stage, the voltage oscillation mainly focuses on the freewheeling switch. The voltage oscillation of the freewheeling switch occurs during the voltage rising stage of the active switch. Thus, the active switch is modeled as a voltage source with an specific dv/dt value. Similarly, the small-signal oscillation equivalent model can be derived and simplified by using △-Y transformation, as shown in Fig. 16 and Fig. 17 respectively. From Fig. 16(d) the switching oscillation of the freewheeling switch in the turn-on stage is caused by the resonance of RLC network under the voltage excitation. Similarly, the amplitude-frequency characteristic of the input impedance during turn-on stage can be obtained through MATLAB. The spectrum of the input impedance is shown in Fig. 18. The comparison between the simulation and model results are given in Table IV. As shown in the Table IV, the switching oscillation frequency from the LTSPICE transient simulation closely matches the oscillation frequency of small-signal equivalent model for the turn-on

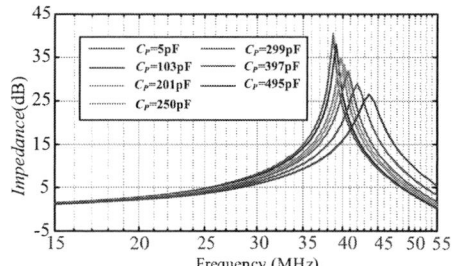

Fig. 18. Input impedances for turn-on stage.

TABLE IV.   COMPARISON OF SIMULATIONS AND THEORETICAL MODEL RESULTS FOR THE TURN-OFF STAGE

Turn-on comparison	Frequency (MHz)		
C_P (pF)	V_{DS}	V_{Model}	Differences
5	38.1	38.7	0.6
103	38.7	39.1	0.4
201	39.3	39.7	0.4
250	39.6	40.2	0.6
299	39.9	40.7	0.8
397	41.1	42	0.9
495	43.2	43.7	0.5

stage, with a maximum error of 2.2%. The resonant frequency and the shifting trend of the peaks are consistent with the voltage spectrum shown in Fig. 10, indicating the accuracy of the small-signal model for the turn-on stage.

From the above analysis, it can be observed that as C_P increases from 5pF to 495pF, the voltage spectral peaks of the active switch shift towards lower frequency while its amplitudes increase. Conversely, the freewheeling switch exhibits the opposite trend. This explains why two peaks initially appear in the CM EMI spectrum, then merge into a single peak, and finally separate into two peaks again. In summary, the case with $C_P = 5$pF demonstrates the best CM EMI characteristics, with the lowest CM EMI amplitude within the 10 MHz to 30 MHz frequency range and minimal spectral peaks compared to other cases. To better meet EMC standards, the amplitude of the spectral peaks should be minimized and its frequency should be shifted outside the frequency range of conducted EMI. Therefore, for module design, it is better to maximize the sum of C_P and C_N, with smaller C_P and larger C_N.

IV. MODULE OPTIMIZATION AND SIMULATION VERIFICATION

To improve the EMI performance of the DSC module, optimization is required. For achieving lower CM EMI, the negative copper layer should be maximized. However, this increases the module size, which hinders the realization of high power density. Therefore, in this study, a capacitor is chosen to be integrated directly into the DSC power module to achieve larger C_N. From Fig. 5, when the sum of C_P and C_N is four times greater than its original value, a significant suppression on CM EMI is achieved. Therefore, based on the extraction results in Table II, a Multilayer Ceramic Capacitor (MLCC) with a value of 1500 pF and rated at 1000V is selected for integration , which

979-8-3315-1612-3/25 $31.00 © 2025 IEEE

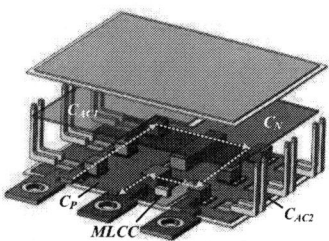

Fig. 19. Structure of the proposed module with a integrated capacitor.

TABLE V.   PARASITIC EXTRACTION RESULTS OF THE TWO MODULES

Power Module Structures	Original Module	Proposed Module
C_{AC}	435pF	417.16pF
C_P	170.85pF	171pF
C_{NY}	-	1500pF
C_N	218.07pF	221.01pF

part number is C1206C152KDRACTU. This capacitor is surface mounted, making it easy for integration into the DSC module. The proposed module with integrated capacitor is shown in Fig. 18. Parasitic parameters of the original module and proposed module are extracted through ANSYS Q3D and shown in Table V. From Table V, it can be observed that the C_{AC} of the proposed module is slightly reduced, while other parameters remain nearly the same as the original module.

To further validate the EMI characteristic of the proposed module, a simulation is carried out based on the buck circuit. Multiport network SPICE model is exported from ANSYS Q3D and imported into LTSPICE for simulations. The spice model of the bare die used in the simulation is from manufactory with part number of S4661. The relevant parameters are as follows: input voltage 400V, duty cycle 0.5, output power 8kW, switching frequency 300kHz. The simulation result is shown in Fig. 19. It can been seen that the CM EMI is effectively suppressed across the entire frequency range of the spectrum. The CM EMI amplitude achieves a maximum reduction of 13dBµV in the 10 MHz to 30 MHz frequency range, which demonstrates that the proposed module exhibits lower CM EMI.

V. CONCLUSION

This paper provides detailed models and analysis of the impact of parasitic capacitances on CM EMI performance in DSC power module. The parasitic parameters of two typical power modules are compared first. Then, the influence of five key parasitic capacitances on CM EMI is analyzed. Both the value and the asymmetry of the parasitic capacitance have a significant impact on the CM EMI. Small-signal models are used to explain the effects of dc-side grounding capacitance asymmetry on EMI spectral peaks. The spectral peaks in the CM EMI spectrum can be easily explained through the models. For better CM EMI performance, it is better to maximize the sum of C_P and C_N, with smaller C_P and larger C_N. Finally, based on the above analysis, an optimized module is proposed, in which a capacitor is integrated. Through simulation verification, the proposed module achieves a maximum reduction of 13 dBµV in CM EMI, demonstrating the effectiveness of the

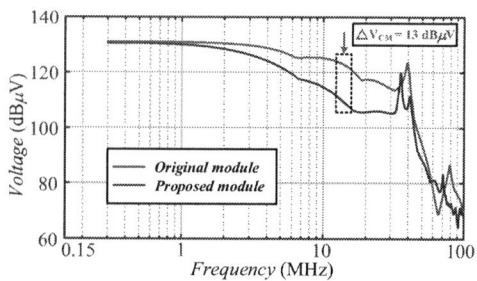

Fig. 20. Simulation results of CM EMI for two modules.

optimization method.

REFERENCES

[1] Y. Wu, N. He, L. Yu, D. Xu, S. Igarashi and T. Fujihira, "Effectiveness Analysis of SiC MOSFET Switching Oscillation Damping," in *2020 IEEE 9th International Power Electronics and Motion Control Conference (IPEMC2020-ECCE Asia)*,2020, pp. 20-27.

[2] X. Chen, W. Chen, X. Yang, Y. Ren and L. Qiao, "Common-Mode EMI Mathematical Modeling Based on Inductive Coupling Theory in a Power Module With Parallel-Connected SiC MOSFETs," in *IEEE Transactions on Power Electronics*, vol. 36, no. 6, pp. 6644-6661, June 2021.

[3] T. Huber, A. Kleimaier, S. Polster and O. Mathieu, "Low Inductive SiC Power Module Design Using Ceramic Multilayer Substrates,"in *PCIM Europe 2018; International Exhibition and Conference for Power Electronics, Intelligent Motion, Renewable Energy and Energy Management*, 2018, pp. 1-8.

[4] Lemmon A N, Cuzner R, Gafford J, et al. Methodology for characterization of common-mode conducted electromagnetic emissions in wide-bandgap converters for ungrounded shipboard applications[J]. IEEE Journal of Emerging and Selected Topics in Power Electronics, 2017, 6(1): 300-314.

[5] Y. Ding, S. Mao, H. Liu, Q. Tan, S. Yang and P. Liu, "Modeling of SiC MOSFETs Switching Oscillation for Dynamic Optimization with RC Snubber in the Half-bridge Circuit,"in *2023 IEEE Energy Conversion Congress and Exposition (ECCE)*, 2023, pp. 5970-5975

[6] T. Liu, R. Ning, T. T. Y. Wong and Z. J. Shen, "Modeling and Analysis of SiC MOSFET Switching Oscillations," in *IEEE Journal of Emerging and Selected Topics in Power Electronics*, vol. 4, no. 3, pp. 747-756, Sept. 2016.

[7] Y. Zhang et al., "Comprehensive Analysis and Optimization of Parasitic Capacitance on Conducted EMI and Switching Losses in Hybrid-Packaged SiC Power Modules," in *IEEE Transactions on Power Electronics*, vol. 38, no. 11, pp. 13988-14003, Nov. 2023.

[8] S. Wang, P. Kong and F. C. Lee, "Common Mode Noise Reduction for Boost Converters Using General Balance Technique," in *IEEE Transactions on Power Electronics*, vol. 22, no. 4, pp. 1410-1416, July 2007.

[9] Jih-Sheng Lai, Xudong Huang, E. Pepa, Shaotang Chen and T. W. Nehl, "Inverter EMI modeling and simulation methodologies," in IEEE Transactions on Industrial Electronics, vol. 53, no. 3, pp. 736-744, June 2006.

[10] S. Tanimoto and K. Matsui, "High Junction Temperature and Low Parasitic Inductance Power Module Technology for Compact Power Conversion Systems," in *IEEE Transactions on Electron Devices*, vol. 62, no. 2, pp. 258-269, Feb. 2015.

[11] Y. Xie, Z. Huang, C. Chen and Y. Kang, "An EMI performance improved stacked substrate packaging structure with ultra-low parasitics for SiC half-bridge power module," *PCIM Europe 2019; International Exhibition and Conference for Power Electronics, Intelligent Motion, Renewable Energy and Energy Management*, 2019, pp. 1-6.

[12] Chen Z. Electrical integration of SiC power devices for high-power-density applications[D]. Virginia Polytechnic Institute and State University, 2013.

Leakage Current Detection Scheme for Aging Test of 10kV SiC MOSFET Power Module

Peiyang Ding
School of Electrical Engineering
Xi'an Jiaotong University
Xi'an, Shaanxi, China
pyding@xjtu.edu.cn

Hong Zhang
School of Electrical Engineering
Xi'an Jiaotong University
Xi'an, Shaanxi, China
mhzhang@mail.xjtu.edu.cn

Tianshu Yuan
School of Electrical Engineering
Xi'an Jiaotong University
Xi'an, Shaanxi, China
yuantianshu@stu.xjtu.edu.cn

Qiling Chen
School of Electrical Engineering
Xi'an Jiaotong University
Xi'an, Shaanxi, China
qiling_chen@stu.xjtu.edu.cn

Jiacheng Guo
School of Electrical Engineering
Xi'an Jiaotong University
Xi'an, Shaanxi, China
1272419245@stu.xjtu.edu.cn

Dingkun Ma
School of Electrical Engineering
Xi'an Jiaotong University
Xi'an, Shaanxi, China
3121104252@stu.xjtu.edu.cn

Peiyuan Sun
School of Electrical Engineering
Xi'an Jiaotong University
Xi'an, Shaanxi, China
spy2018@stu.xjtu.edu.cn

Ting Hou
China Southern Power Grid Electric
Power Research Institute
Guangzhou, Guangdong, China
houting@csg.cn

Laili Wang
School of Electrical Engineering
Xi'an Jiaotong University
Xi'an, Shaanxi, China
llwang@mail.xjtu.edu.cn

Abstract—**Existing aging experiments primarily focus on low-voltage MOSFETs and medium-voltage (MV) Insulated Gate Bipolar Transistors (IGBTs), while lacking of experimental data for MV SiC MOSFETs. Due to their device characteristics, MV SiC MOSFETs exhibit significantly lower leakage currents compared to IGBTs, and the interference caused by MV environments makes traditional monitoring methods unsuitable. This paper designs a high-precision leakage current monitoring circuit capable of covering a range from nA to µA and sets up a MV High-Temperature Reverse Bias (HTRB) testing platform for the aging testing of SiC MOSFET power modules. Finally, HTRB test is conducted on 6.5kV and 10kV SiC MOSFET power modules, with real-time monitoring of leakage current variations, and successfully yields relevant aging characteristics.**

Keywords—*SiC MOSFET, medium-voltage, leakage current, HTRB, aging characteristics*

I. INTRODUCTION

Medium-voltage (MV) silicon carbide (SiC) metal-oxide-semiconductor field-effect transistor (MOSFET) devices have garnered significant attention within the field of power electronics due to their outstanding performance characteristics. Notably, these devices exhibit a high breakdown voltage, rapid switching speeds, and minimal leakage currents, which collectively enhance their efficiency and suitability for a wide range of demanding applications [1]. Despite these advantages, the comprehensive understanding of their reliability and long-term operational stability remains incomplete. Consequently, in-depth studies are imperative to fully evaluate their reliability testing parameters and address existing knowledge gaps [2].

High-Temperature Reverse Bias (HTRB) testing is a critical stress evaluation method that involves applying both elevated temperature and reverse voltage to assess the reliability of semiconductor devices and identify potential defects [3]. Accurate leakage current measurement is pivotal

in this process [4][5]. MV SiC MOSFET devices typically demonstrate leakage currents in the range of tens of nanoamps, markedly lower than those observed in Insulated Gate Bipolar Transistors (IGBTs) and other low-voltage semiconductor devices. Given the elevated risk of breakdown under MV conditions, effective electromagnetic interference (EMI) suppression and robust MV protection mechanisms are essential.

However, performing HTRB tests on high-voltage SiC power devices presents several technical challenges. These include achieving high-precision current sampling, ensuring voltage isolation exceeding 10 kV, and maintaining rapid response times. Specifically, the leakage current I_{DS} of SiC chips generally remains in the nanoamp range, rarely reaching the milliamp level, even after extended periods of accelerated aging. Additionally, the potential for breakdown under high-voltage conditions necessitates extremely fast sampling response times to ensure accurate data acquisition.

In conventional medium and low-voltage domains, mature commercial solutions for HTRB testing are available. However, these systems typically support test supply voltages up to only 2kV, which is inadequate for high-voltage MOSFET evaluations. High-precision current sampling methods are generally categorized into contact and non-contact approaches. Non-contact sampling, utilizing technologies such as Hall sensors or magnetic rings, provides electrical isolation between the measurement and control circuits and ensures high voltage isolation. Nevertheless, these methods suffer from drawbacks, including complex circuitry, substantial weight, and limited accuracy, often failing to achieve microamp-level precision.

Three principal techniques for current sampling include current sampling chips, current mirrors, and sampling resistor amplifiers[1][6]. Current sampling chips currently lack the necessary voltage isolation standards for high-voltage environments. The current mirror circuit is relatively straightforward but falls short in sampling precision, while the sampling resistor amplifier method meets accuracy requirements but faces challenges related to amplifier voltage isolation. Consequently, achieving microamp-level sampling accuracy, voltage isolation above 10 kV, and second-level

This work was supported by the Open Fund Projects of China Southern Power Grid Electric Power Research Institute under Grant DPGCSG-2024-KF-12..

response times remain significant hurdles for HTRB testing of high-voltage SiC devices.

Furthermore, electrical aging platforms must address the operational demands of high-voltage SiC power devices, which often function under conditions of high voltage and high switching frequencies. Despite this, there are no specific standards tailored to the reliability testing of such devices. Existing electrical aging test platforms, both domestically and internationally, are primarily designed for silicon-based IGBTs, whose leakage currents are considerably higher than those of SiC MOSFETs. While the International Electrotechnical Commission (IEC) standards for MOSFETs provide a foundational framework for reliability testing, current commercial HTRB test equipment for small and medium power levels is limited by inadequate voltage capabilities, insufficient measurement precision, and incomplete functionality.

This paper presents a leakage current monitoring circuit and the development of an aging test platform, which was utilized to conduct HTRB tests on MV SiC MOSFET power modules. Subsequently, the feasibility of the platform was validated through HTRB testing of 6.5kV and 10kV SiC MOSFET power modules, providing a reliable basis for understanding the aging characteristics of MV SiC MOSFETs.

II. CURRENT DETECTION SCHEME

Fig 1 shows the overall design of the circuit. The leakage current is sampled and amplified, and then displayed on the host computer after Analog-to-Digital Converter (ADC) [6].

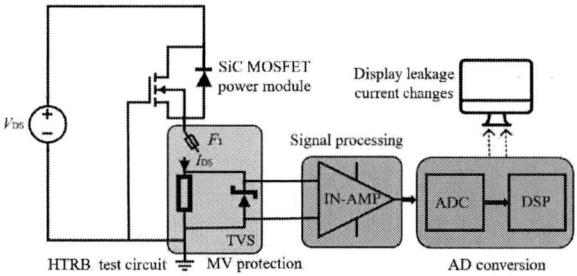

Fig. 1. Overall design diagram of current monitoring

A. Current sampling design

For high-precision current sampling, temperature fluctuations in resistance must be given special consideration. In this design, multiple high-precision resistors with a temperature coefficient of only 20ppm/°C are connected in parallel to form a resistor network, which is beneficial to balance heat distribution and reduce resistor temperature rise.

The instrumentation amplifier (IN-AMP) is configured between the MV protection section and the ADC. The IN-AMP serves several key functions: it uses its high input impedance to isolate the preceding and subsequent circuit stages, matches the voltage output from the sampling resistor with the input of the ADC, and reduces errors caused by common-mode signals at the input by leveraging its high Common-Mode Rejection Ratio (CMRR) [7]. Furthermore, because the selected IN-AMP chip has a maximum input

offset voltage of ±10µV, a sampling resistor of at least 1kΩ is required to minimize sampling errors.

B. MV protection design

To safeguard the measurement system, isolation between high- and low-voltage circuits is required, as the measurement circuit has a low voltage withstand rating. This isolation prevents irreversible damage to the measurement system from transient overvoltages. The device implements an overvoltage protection circuit using fuses and transient voltage suppressor (TVS) diodes. The TVS diodes absorb energy from transient overvoltages, thereby preventing energy surges from impacting the data acquisition system. In the event of a short circuit in the device under test, the TVS diodes are the first to respond by absorbing energy, providing the fuse with sufficient time to operate and subsequently disconnect the circuit[8].

C. DSP program configuration

In MV test circuits, the effects of EMI are particularly severe due to the MV environment, leading to significant fluctuations in AD conversion and affecting the accuracy of leakage current data collection. In this design, digital filtering is applied to the AD-sampled data to eliminate impulsive and periodic interferences. Since HTRB tests typically last for 1000 hours, the program is configured to precisely capture leakage current changes and their corresponding time intervals, ensuring accurate data recording and storage in the appropriate arrays for long-term unattended operation.

III. ACCURACY ANALYSIS

A. The error of the sampling resistor

The inherent noise of a resistor refers to the noise generated by itself, including thermal noise and excess noise[9]. Thermal noise, also known as white noise, is caused by the thermal vibration of electrons in a conductor and exists in all electronic devices and transmission media. It is the result of temperature changes and is not affected by frequency changes. The thermal noise formula of a resistor is shown in (1), where K is the Boltzmann constant, T is the ambient temperature (Kelvin temperature), B is the current bandwidth, and R is the resistance.

$$V = \sqrt{4kTBR} \tag{1}$$

The sampling resistor in the design is 100Ω. Assuming T= 400K and B = 50Hz in the experiment, the thermal noise equivalent voltage is approximately 10.51nV.

B. The error of the amplifier

The current sampling circuit uses the LTC2053 chip as an amplifier, and its basic characteristics have a variety of errors:

- Common-mode signal suppression: CMRR is a key indicator to measure the amplifier's ability to suppress common-mode signals. It is used to describe the device's ability to resist the influence of common-mode signals. The relevant calculation formula is shown in formula (2).

$$CMR(dB) = 20\log_{10}(CMRR) \tag{2}$$

The common-mode rejection ratio of a general op amp is at least 80dB, while the CMRR of the LTC2053 chip can reach 116dB. Then the common-mode amplification factor of the entire circuit will be less than 10^{-8}, and the common-mode signal error can be ignored.

- Offset Voltage and its temperature drift error: According to the chip data sheet, the maximum Offset Voltage is 10μV, which is equivalent to a measurement error of 10nA. When the leakage current is 10nA, the op amp input voltage is 10μV, which is consistent with the maximum Offset Voltage. 10nA is also the minimum resolution of the entire current sampling circuit, and this error cannot be eliminated. The input voltage temperature drift parameter is less than ±50nV/°C. Whenever the ambient temperature changes by 1°C, the voltage error can reach ±50nV, which has little effect on the input voltage. A heat sink is installed around the chip to maintain the circuit at a stable temperature and further reduce the error.

- Noise error: The typical noise of this chip is 2.5μ$V_{\text{P-P}}$ (0.01Hz to 10Hz). For the HTRB test, its leakage current is basically a DC value, and this error can be ignored.

- Gain error: The maximum gain error of LTC2053 is 0.01%, and the typical gain error is 0.001%. For full-scale measurement, its maximum equivalent current measurement error is 6nA.

- Gain nonlinearity error: According to the chip manual, the gain nonlinearity is 10ppm/FSR, and the equivalent measurement error is 6nA.

C. The error of A/D Converter module

The circuit's AD conversion uses Texas Instruments' DSP28335. During chip manufacturing and testing, Texas Instruments calibrated the ADC's gain, offset, and linearity together with the buffered DAC. These correction settings are embedded in TI's reserved OTP memory as functions that can be called in C language[10].

- The *Device_cal()* function is used to copy the ADC and DAC offset correction values from the OTP memory to their respective correction registers.

- The *CalAdcXINL()* function is used to copy the linear correction value from the OTP memory to their respective correction registers.

- Each possible combination of resolution and signal mode requires a different offset correction value. The *GetAdcOffsetTrimOTP(Uint16)* function carries an input value of the resolution and signal mode corresponding to the ADC. This function returns the corresponding offset correction value from the OTP memory, and then the user moves this value to the ADC offset correction register.

- The correction function *Device_cal()* can call the *ADC_setOFFSETTRIM()*, *ADC_setINLTRIM()*, and *DAC_setDACTRIM()* functions in the C2000ware component. These functions retrieve the correction values from the corresponding storage locations reserved by TI in the OTP memory, which are copied to the corresponding storage locations along with the analog module register addresses during testing.

To reduce errors, load the corresponding factory correction values ADC during the use of the DSP to ensure that they operate within the technical parameters of the data sheet. The boot ROM will call the calibration functions without user intervention.

D. Error Summary

For the entire current detection system, the error mainly comes from the amplifier itself, the temperature drift of the device and the environmental noise. For the AD conversion module in the DSP, its error can be effectively compensated by calling the function and configuring the program. Assuming that the system errors are uncorrelated, there are s single undetermined coefficient errors and q single random errors[11]. According to formula (3), the standard deviation can be expressed as:

$$e = \sqrt{\sum_{i=1}^{s} u_i^2 + \sum_{i=1}^{q} \sigma_i^2} \tag{3}$$

It can be calculated that the maximum error of the sampling circuit is less than 15nA. Taking into account the environmental noise, the maximum error of the whole system is less than 50nA.

IV. EXPERIMENTAL VERIFICATION

A. Current sampling accuracy verification

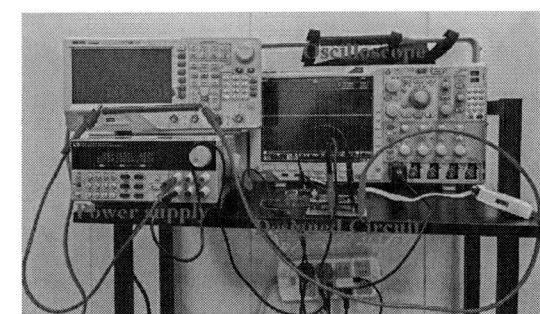

Fig. 2. Current sampling verification experimental equipment.

To preliminarily verify the accuracy of the current sampling circuit, an auxiliary power supply was connected in series with a resistor to generate a small current signal as the circuit input. By varying the input voltage, the leakage current data displayed on the host computer was recorded. The experimental equipment is shown by Fig. 2.

The sampling resistor and the series resistor are measured by a high-precision multimeter, and the theoretical output current values under different given voltages are calculated respectively. The experimental results are shown in Fig. 3, where U is the supply voltage, I_T is the theoretical current value, and I_M is the actual sampling current value. The test range is from 40nA to 5.71μA. According to Fig. 3(a) and Fig. 3(b),the output sampling value is accurate and can effectively record the current with this precision. There is a slight deviation of 0.02μA compared with the theoretical value as shown in Fig. 3(c), which is less than the maximum error result of 50nA analyzed. At the same time, the difference is stable and subsequent aging tests can be carried out.

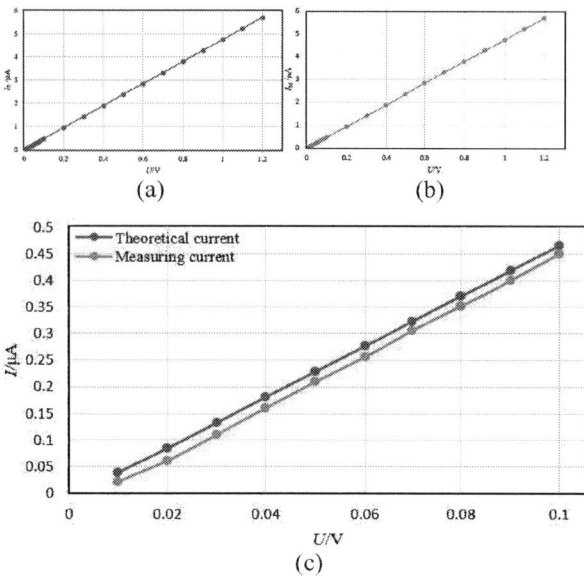

(a)   (b)

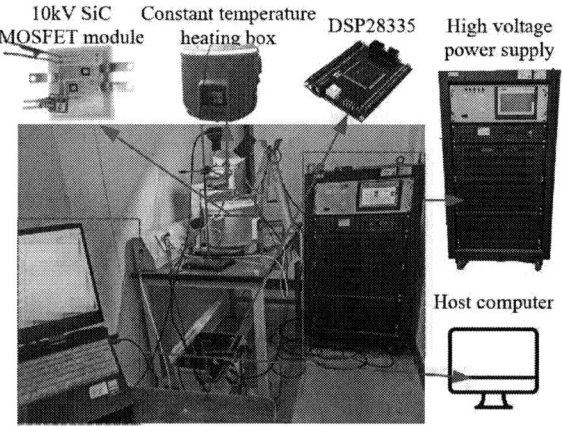

(c)

Fig. 3. Sample Circuit Test Results , (a) Theoretical current and given voltage relationship, (b) Relationship between detection current and given voltage, (c) Comparison of two current values when the given voltage is 0 to 0.1V.

B. MV SiC MOSFET power module aging platform verification

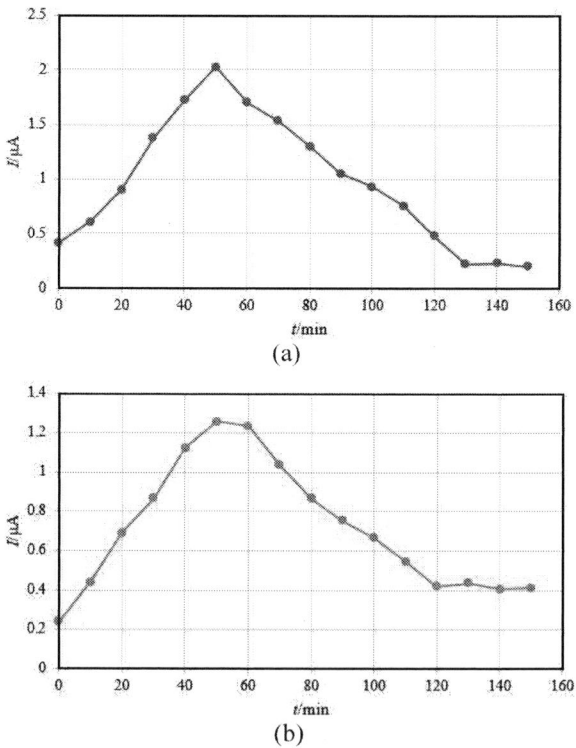

Fig. 4. MV HTRB testing platform.

Fig. 4 shows the test platform based on the current sensing scheme, which is used to test SiC MOSFET power module. Fig. 5 illustrates the two module tested in this experiment: (a) depicts a 10kV SiC power module, while (b) presents a 6.5kV SiC power module.

To test the performance of the platform, the 10kV SiC MOSFET power module was subjected to HTRB test under the test conditions of V_{DS} = 8kV (80% V_{DSmax}) at 175°C[13][14]. The leakage current of the power module after stabilization was less than 100nA and didn't change significantly for a period of time.

This also proves that the test platform is capable of achieving aging tests with high voltage levels and high-precision current sampling.

(a)   (b)

Fig. 5. Test Module, (a) 10kV SiC power module, (b) 6.5kV SiC power module.

C. Research on MV SiC MOSFET power module aging characteristics

In order to explore the aging characteristics of MV SiC power modules, this paper selected two 6.5kV modules for further experiments, which was subjected to HTRB test under the test conditions of V_{DS} = 5.2kV (80% V_{DSmax}) at 170°C. Fig. 6 is the current experimental result.

Fig. 6. Sa Transient test results of 6.5kV SiC power module , (a) power module A, (b) power module B

In the experiment, it was found that the leakage current of the power module increased first and then decreased within one to two hours before the start, and finally stabilized as the test time increased. The leakage current of A 6.5kV module in the figure rose to a maximum of 2μA in the first hour, then gradually decreased and finally stabilized, and

979-8-3315-1612-3/25 $31.00 © 2025 IEEE   2378

basically stabilized below 200nA after 3 hours. The change trend of module B is similar to that of the first module, which rose to a maximum of 1.2μA at the beginning of the experiment and then fluctuated around 400nA. As shown in Fig. 7, the power module has been continuously tested for more than 168h and is still stabilizing at this value.

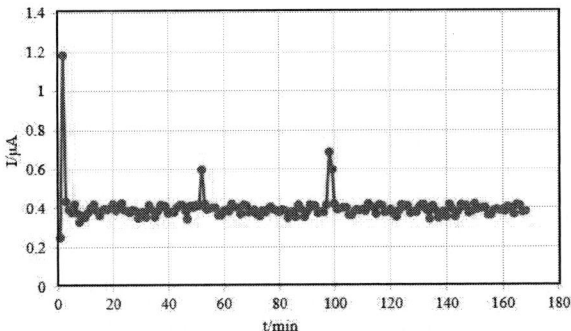

Fig. 7. Evolution of drain leakage current by HTRB stress of 6.5kV SiC MOSFET power module of VDS = 5.2kV at 170℃ for 168h.

This leakage current change trend is a common phenomenon in low-voltage SiC MOSFET aging experiment[12][15]. Relevant researchers pointed out that when the HTRB test begins, the device is subjected to high temperature and high voltage stress. Under the action of the electric field, the free charges in the terminal and the passivation layer are relocated, which will cause charge accumulation on the contact surface between the terminal and the passivation layer. The breakdown voltage of the device is sensitive to the charge concentration. Once the charge accumulation reaches a certain concentration, the breakdown voltage of the device will decrease significantly, resulting in an increasing trend in the leakage current, which also triggers a transient increase in the leakage current.

As the test time increases, the internal free charges move under the action of the external electric field, the accumulated charge concentration decreases, and the breakdown voltage of the device gradually recovers. After a period of time, the breakdown voltage returns to the initial value and remains unchanged, so the leakage current tends to be stable again and remains stable at this value for a long time.

V. CONCLUSION

This paper proposes a high-precision leakage current monitoring circuit covering the range of nA to μA, and analyzes and optimizes its error. Among them, a low-temperature drift resistor network and a high-isolation instrument amplifier are selected, and the filtering algorithm is improved to achieve high-precision current sampling with a sampling error of less than 50nA; TVS tubes and fuses are used to achieve circuit voltage isolation of more than 10kV; and the long-term unattended experimental mode is achieved through the configuration of the DSP program.

Then, based on this current detection circuit, an HTRB test platform was built, 10kV SiC MOSFET power modules was tested. The leakage current of the power module after stabilization was less than 100nA. In order to explore the aging characteristics of MV SiC power modules, further experiments were conducted on two 6.5kV modules. The

transient leakage current has a trend of first increasing, then decreasing and finally stabilizing, which verifies the initial aging characteristics of the MV SiC device.

In the future, more experimental tests will be conducted on 10kV SiC power modules to reach the complete HTRB test standards and further reveal their aging characteristics.

REFERENCES

[1] Zhang, Weiping, et al. "Analysis of SiC MOSFET switching performance and driving circuit," *2018 IEEE International Power Electronics and Application Conference and Exposition* (PEAC). IEEE, 2018.

[2] Chen Jie, et al. "Failure Mechanism Analysis of SiC MOSFETs under Different Aging Test Methods," *Transactions of China Electrotechnical Society* 35.24 , 2020, pp. 5105-5114.

[3] Dong, Shaohua, et al. "Research on HTRB Test Methods and Phenomena," Smart Grid 4.12 , 2016, pp. 1200-1203.

[4] Du Zechen, et al. "Impact of High-Temperature Reliability Testing on the Characteristic Parameters of SiC MOSFETs," *Semiconductor Technology* , 2021.

[5] Wang Yanhao, et al. "Review of Accelerated Aging Models and Test Conditions for High-Temperature Reliability Testing of Power Devices," *Journal of North China Electric Power University: Natural Science Edition* 50.5 , 2023, pp.68-77.

[6] Yin Shaoxuan, et al. "Design of a Compact High-Precision Microcurrent Measurement Module," *Marine Electric & Electronic Technology* 41.4 , 2021, pp. 4.

[7] Li, Zhuo-ran, et al. "Design and Implementation of Highly Accurate Current Measurement Acquisition Card," *2019 14th IEEE Conference on Industrial Electronics and Applications* (ICIEA). IEEE, 2019.

[8] Deng, Erping, et al. "Development of a High-Temperature Gate Bias Testing Device for High-Voltage High-Power Devices," *Electric Power* 52.9 ,2019, pp. 48-53.

[9] Li Jiaming, et al. "Teaching Exploration of Measuring Boltzmann Constant Using Thermal Noise of Resistors," *Physical Experiment* 43.08, 2023, pp. 29-35..

[10] Wu Limin. "A high-precision AD acquisition system based on dual DSP," *Microcomputers and Applications* 36.08, 2017, pp.16-18.doi:10.19358/j.issn.1674-7720.2017.08.006.

[11] Dou Jianhua. "Application of Error Theory to Circuit Tolerance Analysis," *Journal of Hefei University of Technology (Natural Science Edition)* 02, 1996, pp.83-87.

[12] L. Xie et al., "Remaining Useful Lifetime Prediction Method of Power Modules Based on the Aging Characteristic Parameters," *IEEE Transactions on Power Electronics*, doi: 10.1109/TPEL.2024.3459070.

[13] Hoffmann, Felix, et al. "Long term high temperature reverse bias (HTRB) test on high voltage SiC-JBS-diodes." *2018 IEEE 30th International Symposium on Power Semiconductor Devices and ICs* (ISPSD). IEEE, 2018.

[14] Yang, Li, and Alberto Castellazzi. "High temperature gate-bias and reverse-bias tests on SiC MOSFETs," *Microelectronics Reliability* 53.9-11, 2013, pp.1771-1773.

[15] Deng, Erping, et al. "Development of a 6kV/180℃ High-Temperature Reverse Bias Testing Device for High-Voltage High-Power Devices," *Electric Power* 54.2, 2021, pp.133-139.

Physics-informed Neural Network Approach for Early Degradation Trajectory Prediction of Power Semiconductor Modules

Jie Kong[†], Yi Zhang[†], Yichi Zhang[†], Lukas Wick[*], Frederik Lillebæk Hansen[*], Dao Zhou[†], and Huai Wang[†]

[†]*AAU Energy, Aalborg University, Aalborg, Denmark*
E-mail: {jiek, yiz, yzhang, zda, hwa}@energy.aau.dk

[*]*Neurospace, Hørning, Denmark*
E-mail: {lukas.wick, frederik.lillebaek}@neurospace.io

Abstract—DC power cycling tests in semiconductor modules induces repetitive thermal-mechanical stresses that accumulate as fatigue over time. This paper proposes a physics-informed neural network method to reduce the reliability testing time for power semiconductor modules. The main objective of this study is to reduce the testing time while maintain a satisfactory degradation trajectory prediction accuracy using physics-informed data-driven methods. The impact of testing noise and inconsistencies from device to device is attenuated. On-state saturation voltage temperature-dependence compensation and physics based loss term regularization technique are applied in Long Short-Term Memory (LSTM) architechture, which can enhance the accuracy of degradation curve prediction under early degradation. A total of 18 IGBT devices were tested in the power cycling experiments. The proposed degradation curve prediction model can achieve an average end-of-life (EOL) prediction accuracy of 90% using approximately 40% of the early degradation testing data, which can reduce the testing time by about 60%.

Index Terms—DC power cycling, Neural Network, Degradation trajectory prediction, Reliability, End-of-life

I. INTRODUCTION

Power semiconductor devices, such as IGBT, are one of the most critical components for power electronics converters. Their reliability impacts the performance of electric vehicles, renewable energy, and data center, etc [1], [2].

To ensure the reliability of power semiconductor devices, power cycling is an effective and standard method to evaluate the device's lifetime and degradation mechanism [3], [4]. Conventional power cycling methods cost intensive testing hours to reach the EOL point, while the valuable degradation process along the power cycling test does not attract sufficient attentions [5]. For instance, a new standard power module has been tested more than 1 million cycles (i.e., above three-month testing time) in [6], where only the thermal and electrical characteristics around the EOL has been investigated. Thus, a research hypothesis of this work is that if the early degradation process can be utilized to predict the EOL point, the intensive testing hours can be reduced.

Utilizing early degradation process to predict the EOL is a longstanding dream in literature. The statistical methods have been used to predict the EOL of capacitors and power semiconductor devices [7], [8]. However, the lifetime prediction based on statistical characteristics of early degradation data is strongly affected by test noise, temperature coupling and variances from device to device. Machine learning techniques [9], [10], such as neural networks [11], can train models to learn from sensor data and identify the meaningful feature related to degradation. Recurrent Neural Networks (RNN) [12], 1-Dimensional Convolutional Neural Networks (1D CNN) [13] and LSTM [14], can effectively process time-series data to capture complex patterns of device performance degradation [15]. However, purely data-driven approaches are limited by the size of dataset and often do not have good generalisation performance. Therefore, by combining machine learning with the physical models, hybrid physics-informed data-driven methods is possible to more accurately predict the remaining life of IGBTs [16], thereby enhancing the reliability and efficiency of power electronic systems [17].

To address the challenge of the effect by temperature coupling and reduce dependance on large testing dataset, this study proposes a physics-informed neural network (PINN) method which can enhance the accuracy of degradation trajectory curve prediction and EOL value estimation. The main contribution of this study is the on-state saturation voltage temperature-dependence compensation and the physics based loss term regularization technique are applied in LSTM neural network architechture, which can maintain a satisfactory degradation curve prediction accuracy using early degradation testing data during DC power cycling.

The rest of the paper is organized as follows: Section II presents the power cycling data base. Section III describes the proposed degradation curve prediction method and Section IV analyze and discuss the results obtained. Section V presents the conclusions and future research directions.

II. POWER CYCLING DATA BASE

A. DC Power Cycling Tests

The DC power cycling test is a typical method for evaluating the reliability of power semiconductor device. The purpose of power cycling test is to perform reliability assessment, which is aimed to determine the power cycles to failure and

979-8-3315-1612-3/25 $31.00 © 2025 IEEE

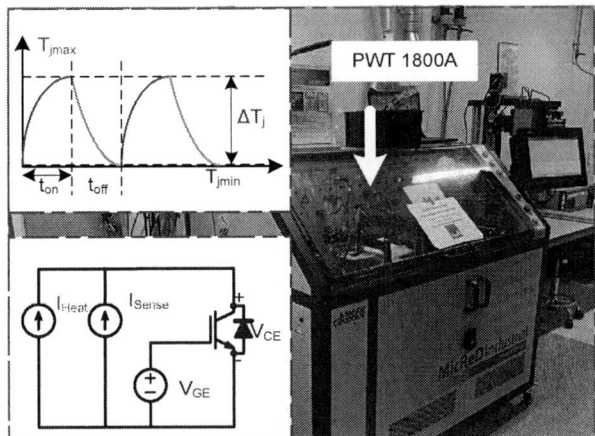

Fig. 1. The overview of DC power cycling experimental platform. (a) Principle of DC power cycling. (b) Power Tester.

TABLE I
DETAILS OF TESTING CONDITIONS AND ACTUAL LIFETIME CYCLE TO FAILURE (DEVICE TYPE: FS25R12KT3).

No.	Test Condition	DUTs and Actual cycle to failure	
i	ΔT_j=100 °C, T_{jmax}=150 °C, t_{on}=t_{off}=1.5s	N1 (54075)	N2 (64558)
		N3 (56091)	N4 (51341)
		N5 (49542)	N6 (44189)
ii	ΔT_j=100 °C, T_{jmax}=125 °C, t_{on}=t_{off}=1.5s	N7 (88218)	N8 (87154)
		N9 (54841)	N10 (86793)
		N11 (70866)	N12 (89298)
iii	ΔT_j=90 °C, T_{jmax}=150 °C, t_{on}=t_{off}=1.5s	N13 (63368)	N14 (66880)
		N15 (76667)	N16 (71748)
		N17 (48648)	N18 (71852)

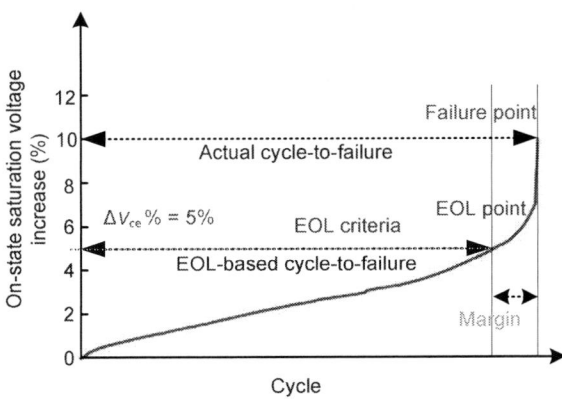

Fig. 2. (a) Typical device degradation pattern overtime (b) IGBT end of life criteria.

failure modes of the device over a long period of operation. The physics of failure mode examined in this paper addresses the degradation including the lift-off of bond wires and the reformation of the metallization layer, which are known as prevalent failure mechanisms.

The testing circuit and the photo of the testing platform are shownin Fig. 1. Experimental test is time-consuming, as it collects monitoting data on IGBT performance under long-term power cycles, and then these high quality data can be used to establish and later validate degradation estimation models.

B. Health Indicator and End of Life Criteria

In practical application, key parameters related to the device performance are selected as health indicators to identify potential developing degradation of the specific device in service.

Junction temperature, thermal resistance, on-state voltage drop and on-state resistance are the commonly used indicators for health monitoring and lifetime prediction of IGBT. For example, an increase in thermal resistance indicates a decline in heat dissipation performance, possibly due to internal material aging or degradation of thermal interface materials. Typically, the on-state voltage drop increases with the aging of IGBT dies, which can be measured online and has a clear correlation with the EOL. By monitoring these indicators, the health status of the IGBT can be evaluated, and its remaining useful lifetime (RUL) can be predicted. The on-state saturation voltage $V_{ce,sat}$ is measured, where 5% voltage increase is regarded as the EOL according to the standard [18], as depicted in Fig. 2. In addition, the heating current I_{heat}, the maximum junction temperature T_{jmax} of each cycle are recorded along the testing. These records will be utilized to enhance the predictability of the proposed method.

The fluctuation range and rate of change in the junction temperature of the IGBT significantly affects its lifespan. Since temperature has an impact on the health indicators during power cycling tests, decoupling the effect of temperature can improve the accuracy of degradation curve prediction.

C. Dataset Description

As listed in the Table I, 18 IGBT power modules (1200V, 25A) have been tested under three different conditions, which consist of different testing parameters includes T_{jmax}, ΔT_j, t_{on}, t_{off}. There are 6 samples under the same testing condition with different cycle ranges to failure. Actual cycle to failure for different samples under DC power cycling aging conditions are shown in brackets in the Table I. It can be seen that different samples have different voltage curves even under the same testing conditions, indicating the variation from sample to sample. Moreover, the different testing conditions also lead to various voltage profiles, which can be seen from Fig. 3 with different colors.

III. PROPOSED DEGRADATION CURVE PREDICTION METHOD

A. On-state Saturation Voltage Temperature-dependence Compensation

The on-state saturation voltage data shown in Fig. 3(a) are loading current, junction-temperature, and degradation dependent. For a specific power cycling condition, the current

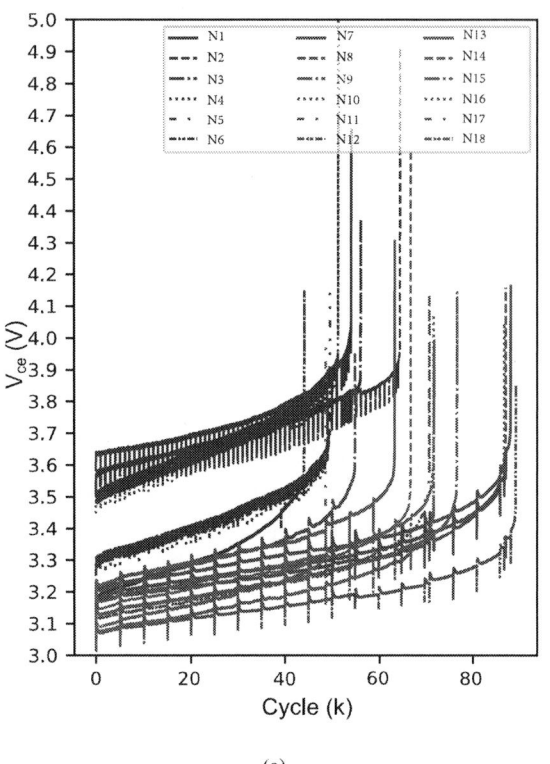

(a)

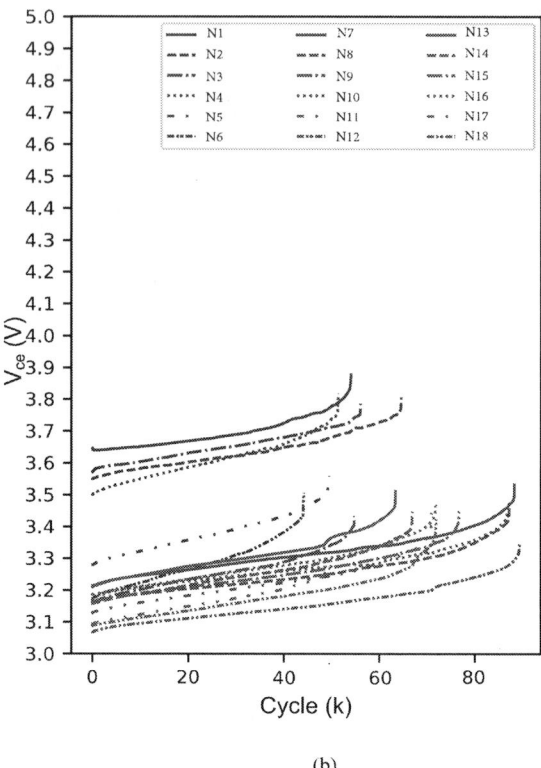

(b)

Fig. 3. The overview of DC power cycling dataset. (a) Raw dataset (b) Dataset after data cleaning and window smoothing.

is constant along the testing, while the IGBT junction temperature could increase from the initial value during the test due to component degradation, e.g., on-state voltage increase, thermal resistance increase. Therefore, for each of the data curve, it contains the impact from both IGBT degradation and the junction temperature change. It is necessary to compensate the temperature-dependence for degradation and lifetime analysis.

The on-state saturation voltage of IGBT semiconductor devices is linearly dependent on temperature at the same aging state,

$$V_{\text{ce}}(T) = (T - T_{\text{ref}})\, K + V_{\text{ce,ref}} \qquad (1)$$

where $V_{\text{ce,ref}}$ is the on-state saturation voltage measured at reference temperature T_{ref} and K indicates the temperature coefficient. The measured voltage $V_{\text{ce}}(t, T)$ at temperature T depends on the resistivity and the heating current, $V_{\text{ce}}(0)$ is the initial voltage drop when the current is zero:

$$V_{\text{ce}}(t, T) = V_{\text{ce}}(0) + R(t, T)I(t) \qquad (2)$$

To remove the dependence of temperature, the $V_{\text{ce}}(t, T)$ at temperature T can be decoupled to the voltage at reference temperature T_{ref} in a short time period,

$$V_{\text{ce,filter}}(t, T) = V_{\text{ce}}(t, T) - (T - T_{\text{ref}})K \qquad (3)$$

For the DC power cycling data, if $V_{\text{ce,0}} = V_{\text{ce}}(t_0, T_0)$, $V_{\text{ce,1}} = V_{\text{ce}}(t_1, T_1)$ are measured, T_0 and T_1 are different when $t_0 \approx t_1$, Therefore, the temperature coefficient can be caculated below:

$$K = \frac{\Delta V}{\Delta T},\ K_0 = \frac{(V_{ce,1} - V_{ce,0})}{(T_1 - T_0)} \qquad (4)$$

where T_1, $V_{ce,1}$ at the top of the first peak and T_0, $V_{ce,0}$ as minimum value around the first peak are chosen to caculate K_0. The temperature decoupling coefficients are then updated at the new aging state through a sliding window. The data decoupling results based on physical knowledge is illustrated in Fig. 4. After data cleaning and window smooting, the spikes in the voltage waveform fluctuating with the maximum junction temperature are attenuated and the waveform quality becomes smoother, compared to Fig. 3(b).

B. Physics Based Loss Term Regularization

Incorporating a physics-informed loss function into machine learning model can indeed yield strong regularization effects, particularly by embedding prior knowledge about the physical system into the training process. By leveraging regularization terms that enforce the physical laws in loss, which can help the model reduce the solution space and converge quickly. The physical knowledge learning from the previous study is that the first and second derivative of the V_{ce} overtime is positive. The regulariztion effect can be realized by penalizing the deviation from the expected physical behavior.

979-8-3315-1612-3/25 $31.00 © 2025 IEEE

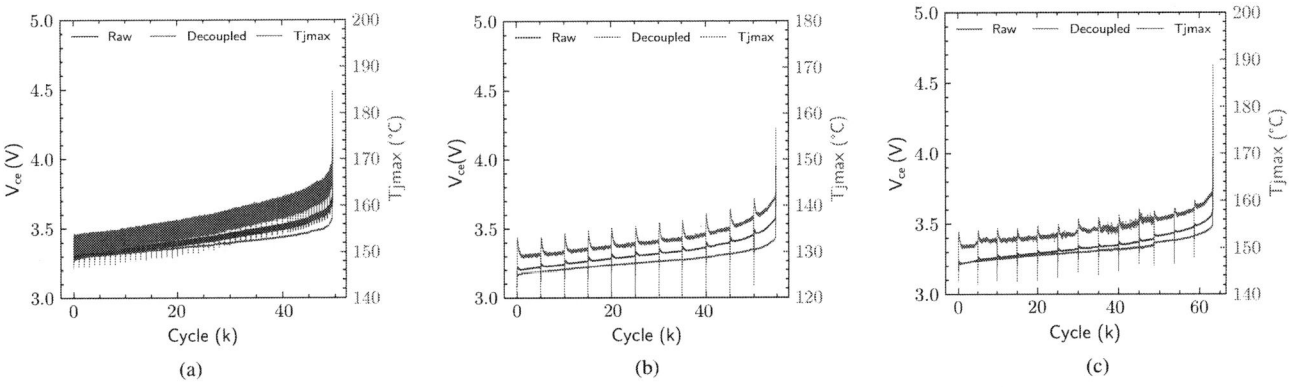

Fig. 4. Data cleaning by decoupling with T_{jmax}. (a) N5 (b) N9 (c) N13.

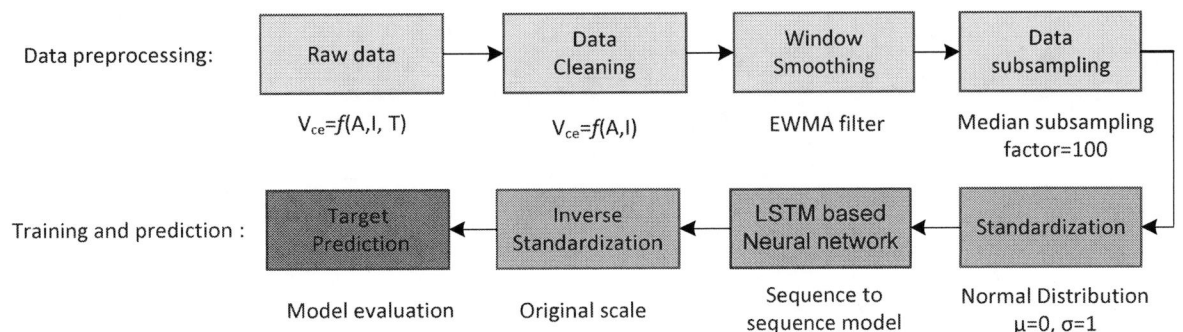

Fig. 5. Overall framework of PINN lifetime prediction model. In $f(A, I, T)$ and $f(A, I)$, A means aging related variable, I means the heating current, T means the temperature. EWMA means exponentially weighted moving average [19].

$$\text{Loss}(\hat{y}) = \frac{1}{n} \sum_i (y_i - \hat{y}_i)^2 + \frac{\alpha}{n-1} \sum_i \text{Re} lu \left(\frac{-d\hat{y}_i}{di} \right) + \frac{\beta}{n-2} \sum_i \text{Re} lu \left(\frac{-d^2\hat{y}_i}{di^2} \right)$$

(5)

where, the first item is the mean suqare error (MSE), which is a basic loss function for fitting the data, n is the number of the data points for a traning sample, y_i is the V_{ce} data label and $\hat{y}_i$ is the output of the PINN model at the i_{th} iteration. The second term is that the voltage is monotonically increasing during the aging process. The third item is that the rate of change of voltage during the power cycling test is also monotonically increasing. The regularization parameter α and β are used to balance the data fitting and satisfying the physical constraints and the regularization parameter needs to be carefully chosen. It is worthwhile to note that α is set as 1 and β is set as 0.25 in this paper.

C. LSTM Neural Network Architecture

LSTM neural network is a special kind of RNN, which aims to solve the problem of gradient vanishing and gradient explosion of the traditional RNNs when dealing with long time dependencies. LSTMs can effectively remember or forget the past information through the introduction of a gating mechanism, which ensures that the network will process the long sequence data with a stable performance. To evaluate the power cycling testing sequence data for degradation curve estimation, the feature engineering part employs an LSTM architechture, as depicted in Fig. 5. The LSTM based neural network comprises 2 layers. The first layer and the second layer is an LSTM layer with 128 neurons, followed by 3 fully connected layers.

IV. RESULTS AND DISCUSSION

To estimate the EOL point of an IGBT under DC power cycling, a commonly used method involves analyzing a sequence of on-state voltage observations from the early stage cycles. These observations are input into a data-driven PINN model, which then generates a sequence of probabilistic voltage estimation. This future voltage sequence forms a degradation trajectory. The EOL is determined as the cycle number at which this trajectory crosses a predefined voltage threshold, typically set at 5% increase of the initial on-state voltage. The RUL also can be calculated by subtracting the current cycle number from the EOL estimation. After data pre-processing, it can be noted that the amplitude range of the voltage has changed, which will result in the EOL calculated based on a 5% increase in voltage being different from the original data,

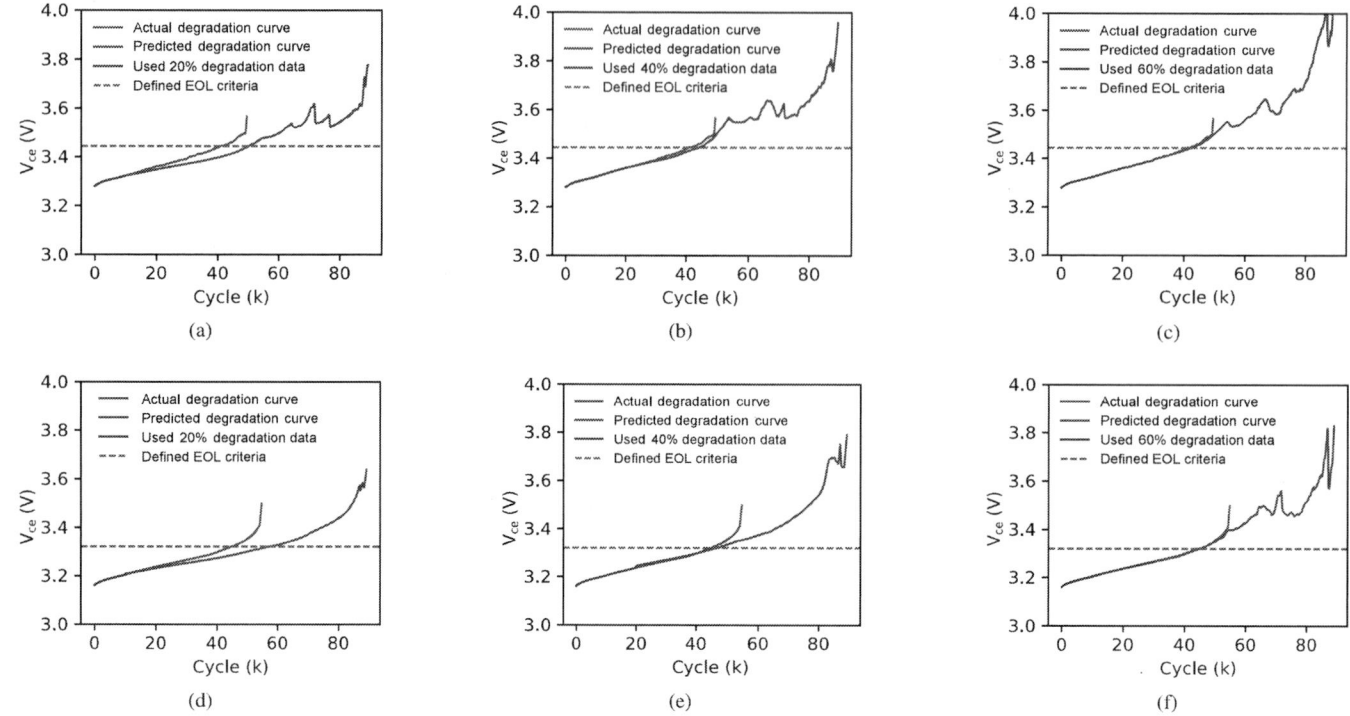

Fig. 6. The prediction of on-state voltage degradation curve by using PINN. (a)-(c) N5. (d)-(f) N9.

TABLE II
LIFETIME PREDICTED PERCENTAGE ERROR (%) WITH PHYSICS BASED LOSS REGULARIZATION.

DUTs		Monitored Number of Cycles					
		First 10000 Cycles (approximately 20% degradation)		First 20000 Cycles (approximately 40% degradation)		First 30000 Cycles (approximately 60% degradation)	
No.	Real EOL	Predicted EOL	Percentage error	Predicted EOL	Percentage error	Predicted EOL	Percentage error
N1	53400	89200	67.04%	83600	56.55%	73400	37.45%
N3	55400	41600	-24.91%	55600	0.36%	54000	-2.53%
N5	40600	50000	23.15%	44500	9.61%	42300	4.19%
N7	72700	63000	-13.34%	67600	-7.02%	74200	2.06%
N9	44800	57000	27.23%	46900	4.69%	44800	0.00%
N11	59700	73200	22.61%	69800	16.92%	68300	14.41%
N13	51700	57300	10.83%	47800	-7.54%	50500	-2.32%
N15	65600	58400	-10.98%	62200	-5.18%	69200	5.49%
N17	48300	59600	23.40%	56900	17.81%	50800	5.18%
Mean error (Absolute value)		24.83%		13.96%		8.18%	

and therefore the setting of the EOL voltage threshold needs to be evaluated in practical applications.

In this paper, model training stage is performed using cross-validation methods, where different combinations of training dataset can be trained to obtain different degradation curve estimation models. As shown in Fig. 6, the results of the degradation curve prediction for the device under test that did not participate in the model training are plotted, with the horizontal axis indicating the power cycling test time and

the vertical axis being the on-state saturation voltage drop. It can be summarized that the prediction accuracy of the voltage profile increases as the monitoring time increases in the early operation stages.

The prediction results are listed in the Table II. The EOL value corresponding to the extrapolated voltage profile exceeding the EOL threshold criterion for different monitoring times are calculated separately. Fig. 7 visualize the variance degree between the true EOL and the predicted EOL value, it can be

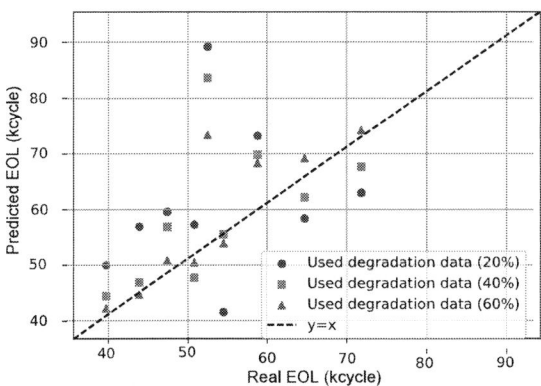

Fig. 7. Variance between the true EOL and the predicted EOL.

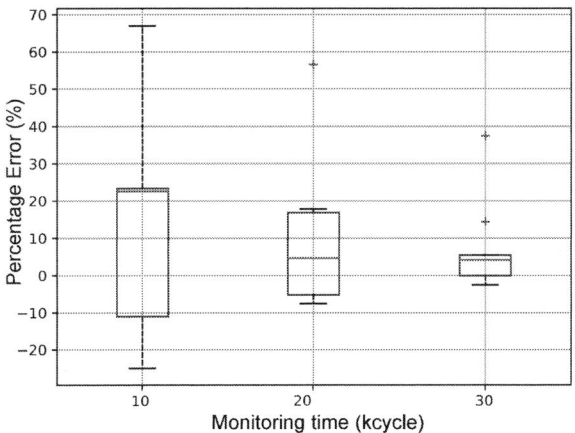

Fig. 8. Boxplot results with physics based loss regularization.

seen how well the predicted values match the real values for different testing samples with various monitoring time. The degradation curve prediction accuracy is compromised with the reduced testing time, as the error would increase if less test-to-failure data is applied for the training. The residual statistics are shown in Fig. 8. When the degradation data from the first 10,000 cycles are used, the lifetime prediction yields a 24.83% average error for the EOL criterion compared to the target value. When the degradation data from the first 20,000 cycles are used, the average error decreases to 13.96%. Finally when 60% of degradation data are used, the average error further reduces to 8.18%.

V. CONCLUSIONS

This paper highlights the physics informed neural network method for degradation curve prediction of IGBT semiconductor modules using early degradation data under DC power cycling test. The PINN method can enhance the accuracy of degradation trajectory curve prediction while reducing the impact of testing noise and minimizing the reliance on large datasets. The EOL value estimation has a 24.83% error when the degradation data of the first 10,000 cycles are

used (20% of degradation data). The error drops to 13.96% when the monitoring time increaces to 20,000 cycles (40% of degradation data). When the degradation data from the first 30,000 cycles (60% of degradation data) is applied, the error further reduces to 8.18%. It is evident that, this physics-informed early lifetime prediction approach can shorten the reliability testing time of power semiconductor devices while maintain a satisfactory degradation curve prediction accuracy. In the future, the generalization ability and interpretability of the data-driven model can be further improved by using uncertainty quantification based on probabilistic model.

REFERENCES

[1] S. Yang, D. Xiang, A. Bryant, et al. "Condition Monitoring for Device Reliability in Power Electronic Converters: A Review," *IEEE Trans. Power Electron.*, vol. 25, no. 11, pp. 2734-2752, Nov. 2010.

[2] H. Oh, B. Han, P. McCluskey, C. Han, and B. D. Youn, "Physics-of-Failure, Condition Monitoring, and Prognostics of Insulated Gate Bipolar Transistor Modules: A Review," *IEEE Trans. Power Electron.*, vol. 30, no. 5, pp. 2413-2426, May 2015.

[3] L. R. GopiReddy, L. M. Tolbert, and B. Ozpineci, "Power Cycle Testing of Power Switches: A Literature Survey," *IEEE Trans. Power Electron.*, vol. 30, no. 5, pp. 2465-2473, May 2015.

[4] A. Hanif, Y. Yu, D. DeVoto, and F. Khan, "A Comprehensive Review Toward the State-of-the-Art in Failure and Lifetime Predictions of Power Electronic Devices," *IEEE Trans. Power Electron.*, vol. 34, no. 5, pp. 4729-4746, May 2019.

[5] J. Lutz, "Power cycling – methods, measurement accuracy, comparability," *11th International Conference on Integrated Power Electronics Systems.*, Berlin, Germany, pp. 1-8, 2020.

[6] Y. Zhang, R. Wu, F. Iannuzzo, and H. Wang, "Aging investigation of the latest standard dual power modules using improved interconnect technologies by power cycling test," *Microelectronics Reliability*, vol. 138, 2022.

[7] Y. Zhang, Z. Wang, S. Zhao, F. Blaabjerg, and H. Wang, "Lifetime prediction of the film capacitor based on early degradation information," *2021 IEEE Applied Power Electronics Conference and Exposition (APEC)*, pp. 407-412, 2021.

[8] L. Meng, Y. Chen, and Z. Zhou, "Segmental Degradation RUL Prediction of IGBT Based on Combinatorial Prediction Algorithms," *IEEE Access*, vol. 10, pp. 127845-4746, 2022.

[9] Roman, D., Saxena, S., Robu, V. et al. Machine learning pipeline for battery state-of-health estimation. *Nat Mach Intell.* 3, 447–456, 2021.

[10] O. Fink, Q. Wang, M. Svensén, et al. "Potential, challenges and future directions for deep learning in prognostics and health management applications," *Eng. Appl. Artif. Intel.*, vol. 92, 103678, 2020.

[11] A. Vaccaro, P. Magnone, A. Zilio, and P. Mattavelli, "Predicting Lifetime of Semiconductor Power Devices Under Power Cycling Stress Using Artificial Neural Network," *IEEE J. Emerg. Sel. Top. Power Electron.*, vol. 11, no. 6, pp. 5626-5635, Dec. 2023.

[12] M. Catelani, L. Ciani, R. Fantacci, G. Patrizi, and B. Picano, "Remaining Useful Life Estimation for Prognostics of Lithium-Ion Batteries Based on Recurrent Neural Network," *IEEE Trans. Instrum. Meas.*, vol. 70, pp. 1-11, 2021.

[13] H. Cheng, Z. Xie, Y. Shi, and N. Xiong, "Multi-Step Data Prediction in Wireless Sensor Networks Based on One-Dimensional CNN and Bidirectional LSTM," *IEEE Access*, vol. 7, pp. 117883-117896, 2019.

[14] A. Vaccaro, D. Biadene, and P. Magnone, "Remaining Useful Lifetime Prediction of Discrete Power Devices by Means of Artificial Neural Networks," *IEEE Open J. Power Electron.*, vol. 4, pp. 978-986, 2023.

[15] B. Zraibi, C. Okar, H. Chaoui, and M. Mansouri, "Remaining Useful Life Assessment for Lithium-Ion Batteries Using CNN-LSTM-DNN Hybrid Method," *IEEE Trans. Veh. Technol.*, vol. 70, no. 5, pp. 4252-4261, May 2021.

[16] Lu, Z., Guo, C., Liu, M. et al., "Remaining useful lifetime estimation for discrete power electronic devices using physics-informed neural network," *Sci Rep.*, vol. 13, 10167, 2023.

979-8-3315-1612-3/25 $31.00 © 2025 IEEE

[17] Y. Fassi, V. Heiries, J. Boutet, and S. Boisseau. Toward Physics-Informed Machine-Learning-Based Predictive Maintenance for Power Converters—A Review. *IEEE Trans. Power Electron.*, vol. 39, no. 2, pp. 2692-2720, Feb. 2024.

[18] ECPE Guideline: AQG 324–Qualification of Power Modules for Use in Power Electronics Converter Units in Motor Vehicles, 2021.

[19] Sheldon M. Ross, Introductory Statistics (Fourth Edition). Academic Press, 2017.

Nonlinear Output Capacitance of Bidirectional Gallium Nitride Power Switches

Michael Bosch, Jeremy Nuzzo, Dominik Koch, Mathias C.J. Weiser and Ingmar Kallfass

Institute of Robust Power Semiconductor Systems
University of Stuttgart
Stuttgart, Germany
E-Mail: Michael.Bosch@ilh.Uni-Stuttgart.de

Abstract—In this work, the characterization of the non-linear output capacitance of bidirectional GaN transistors is presented for the first time. An experimental characterization in an Unclamped Inductive Switching Circuit of the nonlinear output characterization of a discrete bidirectional GaN transistor consisting of two 650 V anti-serial unidirectional GaN transistors in common-drain and common-source configuration is shown. Besides the evaluation of the dissipated energy due to the non-linear output capacitance the influence of the negative gate source voltage is investigated. While the dissipated energy decreases with a more negative gate source voltage for a unidirectional switch the shows the inverse behavior. At 400 V the common drain configuration has with 17 µJ almost twice as much dissipated energy as the common source counterpart. Furthermore, an analysis into the origin of the losses will be given, which can be used to improve the modeling of bidirectional GaN transistors. Thus, a better understanding for the switched operation of a bidirectional switch is gained.

Index Terms—Gallium Nitride, Nonlinear Capacitance, Bidirectional Switch

I. Introduction

Modern power electronic systems which are operating at switching frequencies in the MHz-range have to simultaneously maintain a high efficiency and power density. With their low parasitic capacitances gallium nitride high electron mobility transistors (GaN HEMTs) enable such high switching frequencies and fast switching operations with never-before-seen performances. This increase can be reached by reducing switching losses via establishing zero-voltage soft-switching operation. By bringing the transistor in the on-state when the voltage between drain and source of the transistor, which also corresponds to the voltage over the output capacitance COSS, is close to zero the overlap between voltage and current, and thus, the switching losses can be minimized. Therefore, a characterization of the non-linearity of the output capacitance is needed. In [1]–[6] different approaches like the Sawyer-Tower (ST), Non-Linear Resonance and the Unclamped Inductive Switching (UIS) to extract the dissipated energy related to the non-linear output capacitance (NLOC) of unidirectional power transistors are described. In [3] the resistive nature of the parasitic capacitance between the drain substrate are suspected to be the origin of these losses. Recent advancements in the field of bidirectional power switches (BDS) [7], [8] and their integration in novel topologies like the T-Type-Inverter, Matrix Converter and Vienna Rectifier [8]–[10] underline the importance to improve their efficiency and power density. Given their role in blocking and conducting in both directions, exact knowledge about the NLOC is necessary to establish and optimize soft switching operation in both directions.

II. Measurement Setup and Methodology

To measure the NLOC the method of the UIS as proposed in [4] is utilized and modified for the application with a BDS. This base setup has multiple advantages over other methods such as modularity and ease of adaptation of relevant circuit and test parameters like switch voltage, current, resonance frequency. Also compared to the small signal analysis of the output capacitance the characterization is more application-near by measuring the large signal behavior. In contrast to the calorimetric approach to measure NLOC also knowledge about the exact voltage waveform can be gained. Fig. 1 shows the

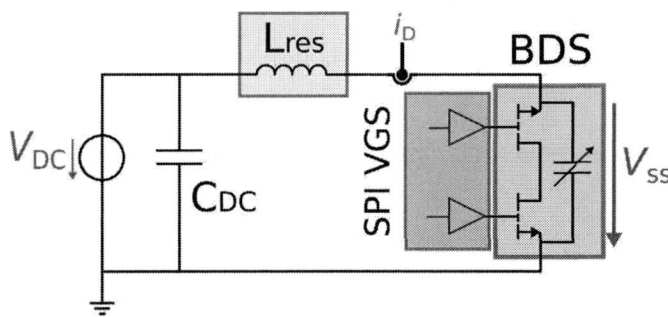

Fig. 1. Equivalent circuit of the UIS modified to a BDS

SPI-controlled UIS circuit, which allows to sweep the gate-source voltage for each gate of the BDS. By applying the principle described in [2], [4] the NLOC can be extracted. From the basic equation (eq. 1) which describes this circuit electrically a term to extract the capacitance in relation to the measured voltage over the switch can be derived.

$$LC(v_{SS}(t))\frac{d^2 v_{SS}(t)}{dt^2} + v_{SS} = L\frac{di(t)}{dt} \qquad (1)$$

When the switch in the circuit is turned on a current and thus a flux is building up in the inductance. At a specified current

979-8-3315-1612-3/25 $31.00 © 2025 IEEE

(I_O) the switch is turned off. Integrating eq.1 at the time point of turning off the current leads to equation 2.

$$LC(v_{SS}(t))\frac{dv_{SS}(t)}{dt} + \int_{t_0}^{t} v_{SS}(t)\,dt = Li(t = t_0) \quad (2)$$

From this point on the energy stored in form of the magnetic flux (Φ) in the inductance is then transferred into an induced voltage at the output capacitance (equation 3).

$$\Phi = Li(t = t_0) = \int_{t_0}^{t_{max}} v_{SS}\,dt \quad (3)$$

Replacing (3) in (2) and rearranging it gives a term for the output capacitance in dependence of the voltage which can be fit into the general equation for the energy stored in a capacitance (4).

$$E_C = \int_T v_{SS}(t)C(v_{SS})\,dv_{SS} \quad (4)$$

Calculating the integral between the starting point ($t = t_0$) and the peak voltage ($t = t_{max}$) results in the energy stored in the output capacitance during the charging process (5). Similar to this the energy of the discharging process can be calculated with the limits from the peak voltage to the next zero crossing (6).

$$E_{COSS}^{Charge} = \frac{1}{2L}\left(\int_{t_0}^{t_{max}} v_{SS}(t)\,dt\right)^2 \quad (5)$$

$$E_{COSS}^{Discharge} = \frac{1}{2L}\left(\int_{t_{max}}^{t_1} v_{SS}(t)\,dt\right)^2 \quad (6)$$

As can be seen in Fig.2 the area underneath the curve from the starting point to the maximum value is larger than the area from the peak value to the zero crossing. This is related to the lossy output capacitance. As can be seen in (5) and (6) these areas can be related to the energies for charging and discharging and thus the difference between these is the dissipated energy (E_{DISS})(7).

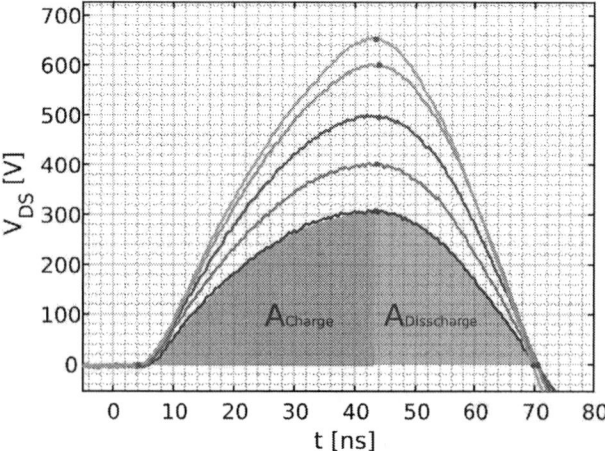

Fig. 2. Exemplary evaluation of the NLOC

$$E_{DISS} = E_{COSS}^{Charge} - E_{COSS}^{Discharge} \quad (7)$$

The only value that is assumed to be constant is the resonance inductance itself. In this work air coils with a self-resonance frequency of over 2 GHz are used. As can be seen in Fig. 3 the windings have quite a large space between each other to reduce the parasitic capacitance of the coil. As DUT a commercial 25 mOhm, 650 V enhancement mode GaN HEMT is employed. To create a BDS two of these transistors are connected with their drains in the anti-serial common drain configuration (CD). Similar, a common source configuration (CS) is created by connecting both source terminals. Last but not least, a DUT with two parallel paths in the common drain configuration (CDp) is tested. This is done to create a comparison to a later monolithic integrated switch with the same $R_{DS,ON}$. For the extraction of the switching voltage waveforms a passive HV-probe with a bandwidth of 800 MHz is used.

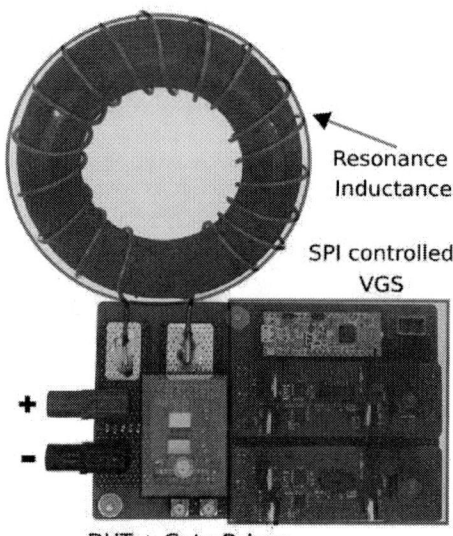

Fig. 3. Measurement setup with the DUT, SPI automatized Gate Supply and Resonance Inductance

III. MEASUREMENT RESULTS

First a reference measurement is taken to compare the measured energies to the datasheet and validate the test bench. For this measurement a 1 µH inductance was used and the negative gate source voltage was 0 V. In Fig. 4 the energy stored in the output capacitance according to the datasheet is compared to the one measured during the charging process. It can be seen that there is only a slight deviation in the lower voltage regions between both curves. The slight differences can be explained by the different techniques to measure the energy stored in the output capacitance. Whereas this work uses the UIS the standardized method is still the small signal measurement.

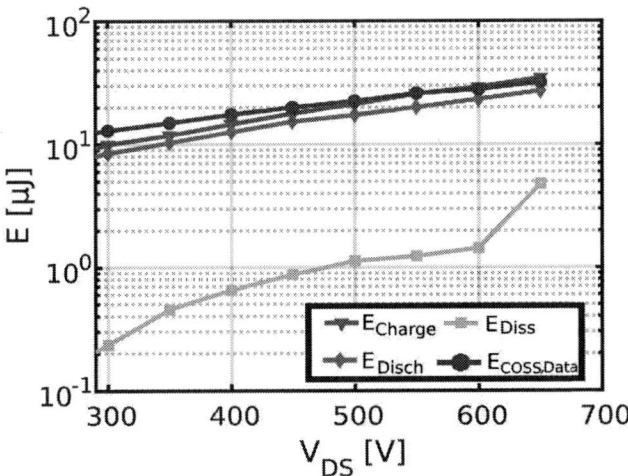

Fig. 4. Exemplary Waveforms of the UIS with a BDS and the

A. Dissipated Energy in dependency of the voltage

In Fig. 5 E_{DISS} of a UDS in dependency at V_{DS} the negative V_{GS} and measured with different resonance inductivities is shown. A clear correlation to a higher voltage and a higher E_{DISS} can be made. There is also a clear correlation between a lower inductance and higher E_{DISS}. At 650 V E_{DISS} with an inductance of 1 μH is 24 μJ lower than with 300 nH. Furthermore, a dependency on the negative V_{GS} can be observed. With a more negative V_{GS} for all inductivities E_{DISS} is decreased. The lower the inductance the smaller the relative difference. Whereas, the deviation between E_{DISS} of $V_{\mathrm{GS}} = -1\ V$ and $V_{\mathrm{GS}} = -3\ V$ the 1 μH inductance often is higher than the factor of 2, this factor decreases with lower inductance. When

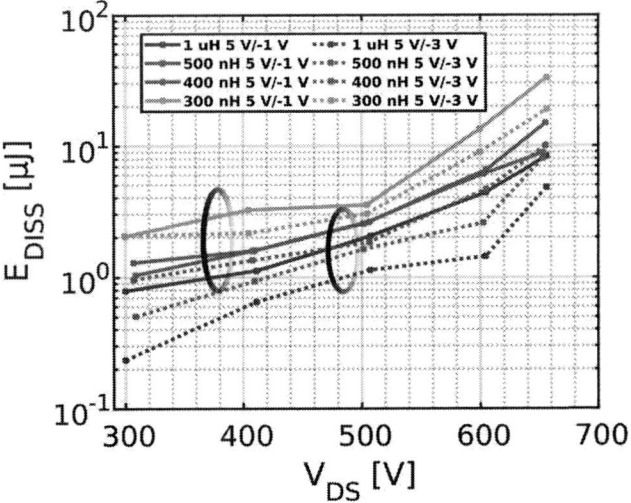

Fig. 5. Dissipated Energy of an UDS in dependence of V_{DS} and the negative V_{GS}

having a look on the CS BDS configuration in Fig. 6 some correlations are changing. Still E_{DISS} is rising with higher

voltages, and a lower inductance increases E_{DISS}. But here a more negative V_{GS} does increase E_{DISS}. Similar to the UDS this correlation is also connected to the inductance but vice versa. The different E_{DISS} for 1 μH do not increase as much as for the lower inductivities. E_{DISS} for the CD configuration

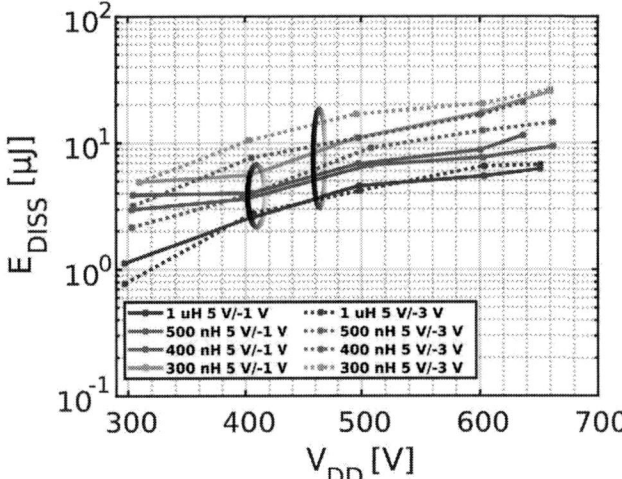

Fig. 6. Dissipated Energy of a BDS CS configuration in dependence of V_{DD} and the negative V_{GS}

which is shown in Fig. 7 behaves in the same way as for the CS configuration. Though, E_{DISS} for lower voltages is much higher independently of V_{GS} than those of the CS counterpart. For example at 300 V, a negative V_{GS} of -3 V and a resonance inductance of 300 nH the CD configuration is showing a E_{DISS} which is 17 μJ and therefore more than 3 times larger. Whereas at higher voltages of about 600 V the energies of the different negative V_{GS} for the smaller inductivities converge. The value of the resonance inductance as well as V_{GS} are having more influence on E_{DISS} compared to the CS configuration. As can be seen in Fig. 8 the parallelization of the CD configuration leads to a rise of E_{DISS} compared to the single path under every circumstance. Here also the measurements with the higher inductivities are showing much higher E_{DISS}. In case of 1 μH and a negative V_{GS} of -1 V at 650 V E_{DISS} is with 38 μJ more than 10 times larger compared to the 2.5 μJ of the single path. Also the relative influence of V_{GS} is not as significant as for the other configurations.

B. E_{DISS} and the influence of dv/dt_{OFF}

To explain the measured behavior and perform a root case analysis a closer look on the slope of the falling voltage (dv/dt_{OFF}) is presented in this section. Since the energy of a capacitance is determined by the voltage, its time derivative is a valid source to explain the behavior. dv/dt_{OFF} was calculated from 90% of the peak value to 10% of the peak value. For the UDS (Fig. 9) a higher dv/dt_{OFF} results in a higher E_{DISS}. Similar to the dependency on the peak voltage over the switch a smaller inductance also results in higher dv/dt_{OFF}. Therefore there is a clear relation between a higher

979-8-3315-1612-3/25 $31.00 © 2025 IEEE

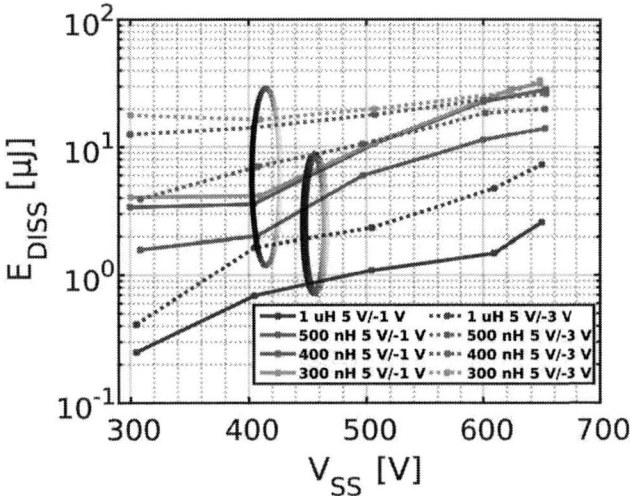

Fig. 7. Dissipated Energy of a BDS CD configuration in dependence of V_{SS} and the negative V_{GS}

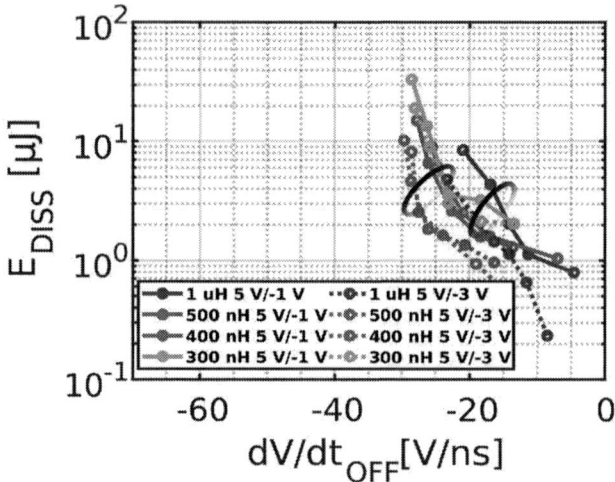

Fig. 9. Dissipated Energy of an UDS in dependence of dv/dt_{OFF} and the negative V_{GS}

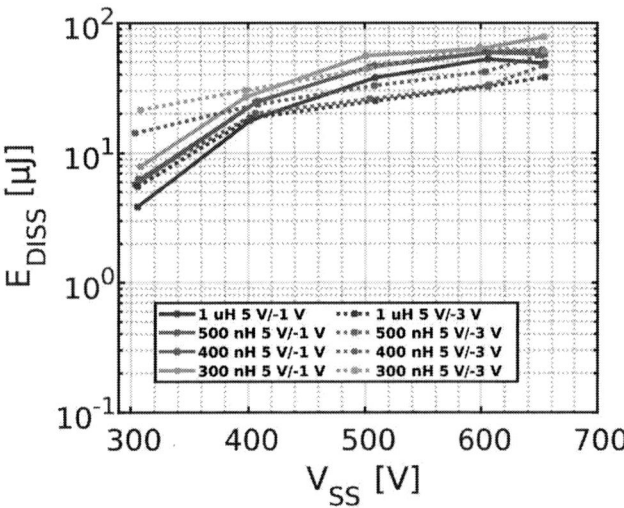

Fig. 8. Dissipated Energy of a BDS parallel CD configuration in dependence of V_{SS} and the negative V_{GS}

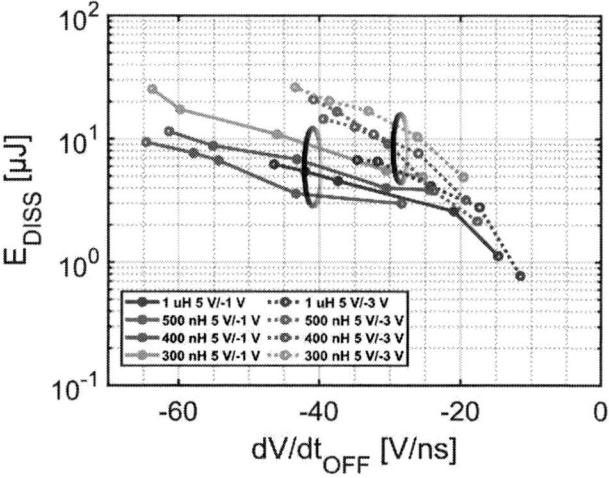

Fig. 10. Dissipated Energy of a BDS in CS configuration in dependence of dv/dt_{OFF} and the negative V_{GS}

dv/dt_{OFF}. For the more negative V_{GS} dv/dt_{OFF} increases but E_{DISS} decreases for each inductance. Relative to the BDS configurations the absolute values of dv/dt_{OFF} are the lowest. The BDS in CS configuration shows much higher dv/dt_{OFF} (Fig.10). Analogous to the UDS this correlates with a higher E_{DISS}. Also a lower inductance is not automatically related to a higher dv/dt_{OFF} for the negative V_{GS} of -1 V. In contrast to the UDS dv/dt_{OFF} of the BDS in CS configuration for the negative V_{GS} of -3 V does result in a slower voltage slope and a higher E_{DISS}. The same behavior can be observed for the BDS in CD configuration (Fig.11). Also here V_{GS} induces more losses due to a lower dv/dt_{OFF}. The deviation between the two V_{GS} in dependency of the voltage over the switch can also be observed for dv/dt_{OFF}. The observation that the

negative V_{GS} does not have a huge influence on E_{DISS} in the time domain for the BDS parallel CD configuration is also congruent to what can be seen in the voltage domain. Moreover it can bee seen that a smaller dv/dt_{OFF} induces more losses for the BDS configurations.

C. Comparison to the State of the Art

In [5], [12] the losses were measured with a calorimetric method. Even though the losses for a UDS in [12] were less than those in this work, the general observation that with a higher dv/dt_{OFF} E_{DISS} rises is confirmed. The smaller losses can be explained by the smaller and therefore less lossy output capacitance. In [5] also the influence of the charge and discharge current was investigated, but this inherently correlates with dv/dt_{OFF} since the current of a capacitance is determined by the time derivative of the capacitance voltage.

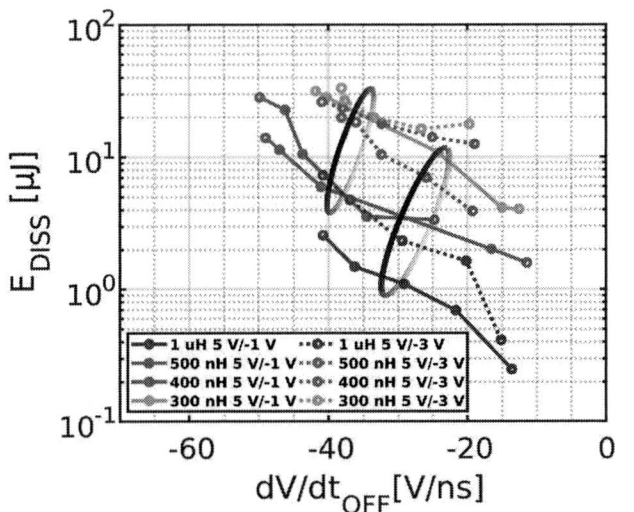

Fig. 11. Dissipated Energy of a BDS in CD configuration in dependence of dv/dt_{OFF} and the negative V_{GS}

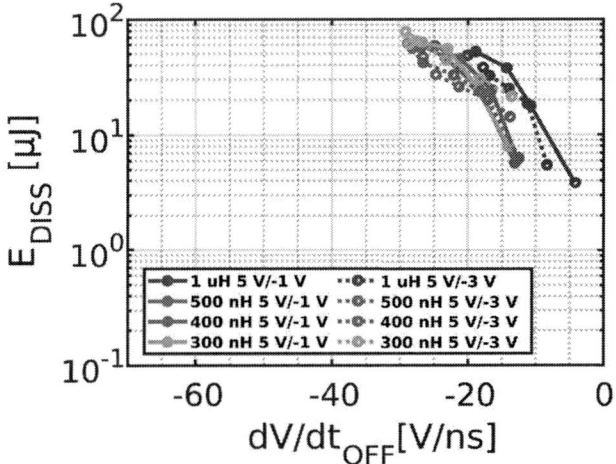

Fig. 12. Dissipated Energy of a parallel BDS in CD configuration in dependence of dv/dt_{OFF} and the negative V_{GS}

Whereas with the measured current there can be distinguished between the influence of dv/dt_{OFF} and the non-linearity of the capacitance.

D. Possible physical root-cause for the Losses

In conclusion the loss behavior for the UDS as well as for the different BDS can be correlated to the slope of the falling voltage. Interestingly the negative V_{GS} has a contrary effect on the loss behavior for the UDS and the BDS configurations. As an explanation to this the classical view of the output capacitance has to be modified. The combination of the parasitic capacitance between drain and gate (C_{GD}) and that between drain and source (C_{DS}) is only sufficient to explain the correlations in the UDS. In [3] the lossy nature of the dielectric between drain and substrate are assumed to be the

main reason for the NLOC losses. As can be seen in this work there is also a dependency of E_{DISS} to the negative V_{GS} which was not taken into account in [3]. By applying a more negative V_{GS} the charge on C_{GD} gets cleared more quickly thus the Miller-plateau is shorter. Since the length of the Miller-plateau defines the time of voltage commutation a shorter time means a higher voltage slope. Therefore, the assumption here is that besides the lossy dielectric between drain and substrate the influence of less gate traps can be seen here. The explanation for the observations on the BDS have to lead to another than those of the UDS. Unlike the UDS the discrete BDS configurations are having two gate electrodes and don't share the same substrate like a monolithic integrated BDS. So one potential reason for the higher general losses can be that there are two lossy dielectrics from the quasi drain to the substrate. However, this does not give a coherent explanation to the contrary influence of a more negative V_{GS}. What can be seen from the dv/dt_{OFF} of the different BDS is that the slope is reduced. Analogous to the UDS this shift results in higher losses. The assumption here is that the second gate electrode affects the coupling capacitance between input and output in a way that the voltage commutation is slowed down which results in higher losses in the lossy dielectrics to the substrate. For the parallelization this means that due to the added parasitic capacities the general losses resulting from slower voltage commutation and a lossy substrate are inherently much higher.

IV. CONCLUSION AND OUTLOOK

In this work, for the first time the characterization in an unclamped inductive switching test of the non-linear output capacitance of a bidirectional GaN switch is presented. Additionally the influence of the negative gate source voltage on the dissipated energy was made. Whereas the UDS shows lower losses with a more negative gate source voltage all the BDS configurations are showing an increase. For a single gate source voltage the general observation is that a higher voltage slope results in higher losses. The more negative gate source voltage of the BDS configurations the slower the slope, which results in higher losses. Therefore the assumption was made that besides the lossy dielectric the quasi-reverse-transfer-capacitance slows down the transition and therefore induces more losses in the lossy dielectrics. For a future work the investigation of the loss behavior of a monolithic integrated BDS with the same substrate would be interesting. Also the influence of the temperature and the biasing of the substrate can be part of further investigations.

V. ACKNOWLEDGMENT

This work was supported by the German BMWK - Federal Ministry of Economics and Climate Protection (BMWK) within the projects GaN4EmoBiL (FKZ: 01MV23003A).

REFERENCES

[1] G. Zulauf, S. Park, W. Liang, K. N. Surakitbovorn and J. Rivas-Davila, "COSS Losses in 600 V GaN Power Semiconductors in Soft-Switched, High- and Very-High-Frequency Power Converters," in IEEE

Transactions on Power Electronics, vol. 33, no. 12, pp. 10748-10763, Dec. 2018, doi: 10.1109/TPEL.2018.2800533

[2] M. Samizadeh Nikoo, A. Jafari, N. Perera and E. Matioli, "Measurement of Large-Signal COSS and COSS Losses of Transistors Based on Nonlinear Resonance," in IEEE Transactions on Power Electronics, vol. 35, no. 3, pp. 2242-2246, March 2020, doi: 10.1109/TPEL.2019.2938922

[3] M. Guacci et al., "On the Origin of the Coss -Losses in Soft-Switching GaN-on-Si Power HEMTs," in IEEE Journal of Emerging and Selected Topics in Power Electronics, vol. 7, no. 2, pp. 679-694, June 2019, doi: 10.1109/JESTPE.2018.2885442

[4] Q. Song, Q. Li and Y. Zhang, "Output Capacitance Loss in Widebandgap and Superjunction Power Transistors: Impact of Switching Voltage and Current," 2023 IEEE 10th Workshop on Wide Bandgap Power Devices & Applications (WiPDA), Charlotte, NC, USA, 2023, pp. 1-4, doi: 10.1109/WiPDA58524.2023.10382205.

[5] D. Bura, T. Plum, J. Baringhaus and R. W. De Doncker, "Hysteresis Losses in the Output Capacitance of Wide Bandgap and Superjunction Transistors," 2018 20th European Conference on Power Electronics and Applications (EPE'18 ECCE Europe), Riga, Latvia, 2018, pp. P.1-P.9.

[6] N. Perera, A. Jafari, R. Soleimanzadeh, N. Bollier, S. G. Abeyratne and E. Matioli, "Hard-Switching Losses in Power FETs: The Role of Output Capacitance," in IEEE Transactions on Power Electronics, vol. 37, no. 7, pp. 7604-7616, July 2022, doi: 10.1109/TPEL.2021.3130831.

[7] C. Kuring, O. Hilt, J. Böcker, M. Wolf, S. Dieckerhoff and J. Würfl, "Novel monolithically integrated bidirectional GaN HEMT," 2018 IEEE Energy Conversion Congress and Exposition (ECCE), Portland, OR, USA, 2018, pp. 876-883, doi: 10.1109/ECCE.2018.8557741.

[8] X. Wen, H. Kasai, K. Lee, M. Noshin, J. Chun and S. Chowdhury, "First Experimental Demonstration of Monolithic Bidirectional Switch Using GaN Current Aperture Vertical Electron Transistor (CAVET)," 2024 Device Research Conference (DRC), College Park, MD, USA, 2024, pp. 1-2, doi: 10.1109/DRC61706.2024.10605251

[9] Q. Yu, D. Klikic, V. Aulagnier, E. Persson and J. Cerce, "Evaluation of Monolithic AC GaN Switch in A Vienna Rectifier for UPS," 2024 IEEE Applied Power Electronics Conference and Exposition (APEC), Long Beach, CA, USA, 2024, pp. 767-773, doi: 10.1109/APEC48139.2024.10509374

[10] F. Vollmaier, N. Nain, J. Huber, J. W. Kolar, K. K. Leong and B. Pandya, "Performance Evaluation of Future T-Type PFC Rectifier and Inverter Systems with Monolithic Bidirectional 600 V GaN Switches," 2021 IEEE Energy Conversion Congress and Exposition (ECCE), Vancouver, BC, Canada, 2021, pp. 5297-5304, doi: 10.1109/ECCE47101.2021.9595422.

[11] H. Ueno et al., "A 3-Phase T-type 3-Level Inverter using GaN Bidirectional Switch with Very Low On-State Resistance," PCIM Europe 2019; International Exhibition and Conference for Power Electronics, Intelligent Motion, Renewable Energy and Energy Management, Nuremberg, Germany, 2019, pp. 1-4.

[12] J. Weimer, R. Schnitzler, D. Koch and I. Kallfass, "Thermal Impedance Calibration for Rapid and Noninvasive Calorimetric Soft-Switching Loss Characterization," in IEEE Transactions on Power Electronics, vol. 38, no. 7, pp. 8472-8485, July 2023

Novel Approach of Determining and Predicting SiC MOSFET's On Resistance From Device Case Temperature Using Machine Learning

Paul Bradford*, Conner Deppe, and Hongjie Wang
Department of Electrical and Computer Engineering
Utah State University, Logan, Utah 84321
*Email: paul.bradford@usu.edu

Abstract—Silicon-carbide (SiC) MOSFET devices are increasing in popularity in high-power converter applications. Device on-resistance (R_{dson}) is an important indicator for SiC MOSFET health status. Increments in R_{dson} over device lifetime indicate imminent device failure and result in decreased system efficiency. Direct and accurate measurement of SiC MOSFET device R_{dson} in high-power applications is difficult. Another approach is to estimate/predict the device R_{dson} from other, more easily measurable quantities, however, little work has been done on this approach in the literature. This leaves a significant technical gap in measuring/predicting device R_{dson} and slows down the device health status monitoring and power converter reliability research. To address the technical gaps, this work proposes a novel approach to predicting device R_{dson} from thermal cycle count and instantaneous temperature using machine learning regression models. The actual hardware data collected from accelerated lifetime tests of high-power SiC MOSFETs are used to train, test, and validate the proposed machine-learning regression models. The developed models, coupled with cycle counting algorithms, and device case thermal measurements, provide accurate live estimates of R_{dson} and can be used to predict changes in R_{dson} over expected mission profiles during power converter design.

Index Terms—Machine Learning, SiC MOSFET, EV Charging, Accelerated Lifetime Test, Reliability, Health Monitoring, Remaining Useful Life

I. INTRODUCTION

As the demand for high-power converters increases in the fields of renewable energy and electric transportation, new technologies are emerging to meet these demands. To ensure the safety, reliability, and sustainability of these equipment, methods of monitoring and predicting device wear-out and failure are important. For SiC MOSFET, which is widely used in high-power and high-performance power converters, one key indicator of device utility and wear-out is R_{dson}[1, 2] . During the operation of power converters, power cycling and current conduction lead to thermal cycling and the expansion and contraction of the semiconductor materials inside the SiC device. These forces break down the connections between the various layers within the device and degrade the electrical conductivity within those layers [3]. This degradation and the resultant reduced conductivity of the device has both immediate and long-term effects on the device lifetime and

This work is based in part upon work supported by the National Science Foundation (NSF) through the ASPIRE Engineering Research Center under Grant EEC-1941524 and CAREER Award under Grant 2239169.

R_{dson} [4, 5, 6]. Because of this, R_{dson} is a preferred predictor of the device's remaining useful lifetime (RUL) [7, 8]. Additionally, as SiC MOSFETs age, large changes in R_{dson} will affect the performance of the converters and applications in which these devices are used. While R_{dson} is a nice indicator of device lifetime, research gaps exist in both R_{dson} monitoring and modeling. Directly monitoring device R_{dson} is difficult as current and voltage measurements must be made for each device. A simpler method of R_{dson} measurement would facilitate RUL monitoring of SiC MOSFETs in the field, and enable more efficient operation of power converter equipment.

Work has been done in the field of power converter design for reliability [9, 10, 11], but much of this work relies on Coffin-Manson based lifetime models of SiC MOSFET devices. These models are based on research done on Si IGBT modules decades ago [12], and are difficult to obtain. The model coefficients must be provided by the SiC MOSFET manufacturer [13] or obtained by fitting the correct variation of the model [14] on data collected from long lifetime tests. Alternatively, the Machine Learning (ML) model proposed in this work, can be used in conjunction with thermal modeling tools and cycle counting algorithms to directly model the behavior of SiC MOSFET R_{dson} during expected mission profiles. This type of design-time model of R_{dson} behavior in SiC MOSFET devices can replace the traditional lifetime models in converter design for reliability. The ML R_{dson} model can easily be updated to new devices and a model of R_{dson} behavior during an entire device mission profile provides additional information about the direct impacts of design changes on converter lifetime.

This paper addresses these research gaps by presenting ML models that accurately predict R_{dson} in SiC MOSFETs from MOSFET case temperature measurements over the application lifetime. A preliminary data analysis is performed on data collected from SiC MOSFET accelerated lifetime test (ALT) hardware in Section II. For proof of concept, a preliminary Random Forest (RF) model is trained to indicate the potential of ML models and to demonstrate the importance of using both cycle count and temperature data in Section III. In Section IV, several different models are trained on the same data to evaluate overall performance. A discussion of these results and example implementations of the ML models are addressed in

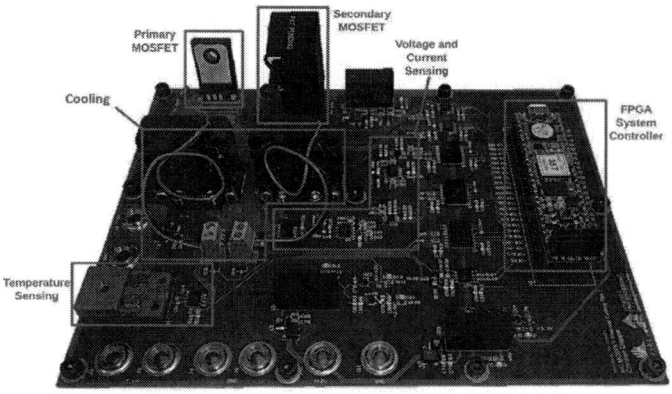

Fig. 1. ALT hardware used for thermal testing of SiC MOSFET devices in this study [15].

Section V. Finally, a conclusion is presented in Section VI.

II. DATA ANALYSIS AND VARIABLE DEPENDENCE

Data used in this work to inform and test the ML models were collected on accelerated lifetime test (ALT) hardware presented in [15]. On this testbed, devices are thermally cycled until failure while device drain-to-source voltage (v_{ds}), on-state current (i_{on}), and case temperature (T_c) are measured synchronously at 5 Hz rate and the average across the 5 measurements in one second for each signal is recorded, respectively. R_{dson} is calculated from the voltage and current using ohms law and device junction temperature (T_j) is calculated using

$$T_j = T_c + v * i * R_{th}, \tag{1}$$

from case temperature and the case thermal resistance (R_{th}) reported in the device data sheet.

The devices are heated by driving the gate-source voltage high to place the device in the on state and passing a high current through the devices. The R_{dson} losses resulted from the conducting current heat the MOSFETs outside the typical operating temperature range. The devices are then cooled by forced air to a set level. The cycle is repeated until the device fails in open or closed circuit. In total 11 NVH4L020N120SC1 SiC MOSFET devices were tested on the platform and data from 6 are used to inform the models in this paper. The ALT hardware is depicted in Fig. 1.

During the data collection, the devices were turned on with a 20 V gate-source voltage and were heated with 18 A of current. The devices were heated to either $225°C$ or $245°C$ depending on the test. Between the six devices tested, a total of 3,797,379 data points are collected at various temperatures and cycle counts when current is flowing through the device that allows a valid R_{dson} measurement. When the device temperature and R_{dson} are plotted, as illustrated in Fig. 2, a generally increasing relationship is revealed. This observation is supported by the R_{dson} relationship given in the device datasheet. When measured across the full device lifetime, range of device on resistances are observed at at a given temperature. When the

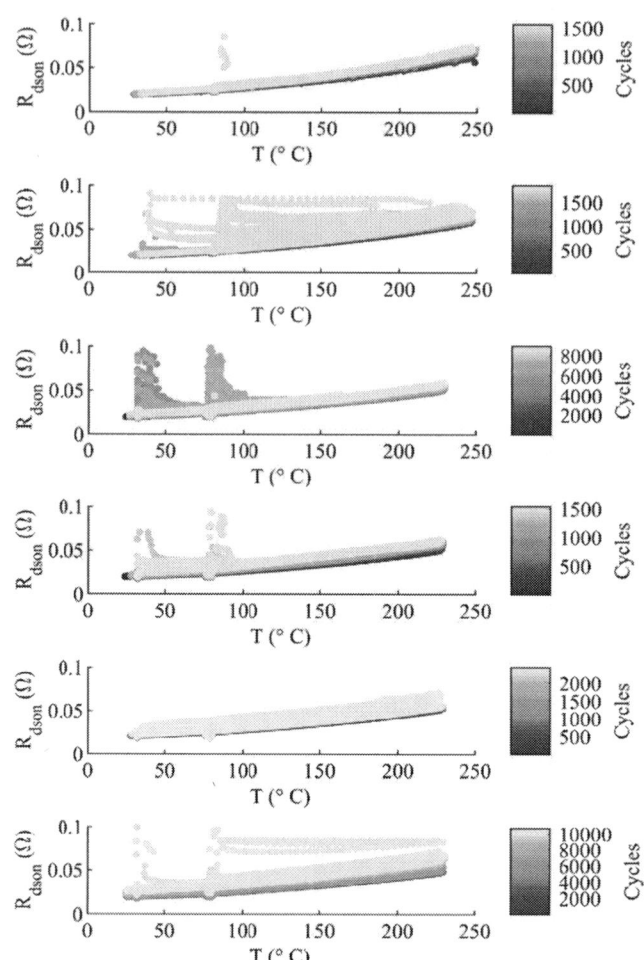

Fig. 2. R_{dson} vs temperature for 6 NVH4L020N120SC1 SiC MOSFET devices.

TABLE I
CAPTION

Mutual Information	Temperature	Cycle Number	R_{dson}
Temperature	x	0	1.163
Cycle Number	0	x	0.035
R_{dson}	1.163	0.035	x

data points are colored by their corresponding thermal cycles, as shown in Fig. 2, it can be seen that the additional spread is due to device aging.

This relationship is further exposed through mutual information analysis. Mutual information is calculated using k-nearest-neighbor algorithms to determine entropy estimates between two variables [16]. Mutual information scores reveal the dependence between two variables. A mutual information score of zero indicates complete independence between variables, and higher mutual information scores indicate higher dependence. The mutual information scores between each of the variables of interest measured during the ALT are shown in Table 1.

A substantial mutual information score between temperature and R_{dson} is apparent, as well as a less significant, yet still present, mutual information score between cycle count and R_{dson}. This is because temperature has a substantial affect on device R_{dson}. Device age, reflected in the cycle count, has a less obvious affect on R_{dson}. The mutual information between temperature and cycle count is 0 showing complete independence. This indicates that these two variables do not contain much of the same information and will each contribute independently to R_{dson} prediction. The non-zero mutual information scores between R_{dson} and temperature and between R_{dson} and cycle count combined with the zero mutual information score between the two predictive variables indicates that cycle count and temperature together will form a more accurate R_{dson} predictor than either variable alone.

Some outliers, particularly at lower temperatures, can be seen in the thermal data. These are likely due to measurement noise and difficulty obtaining accurate R_{dson} and thermal measurements as the device cools. The device junction cools slightly slower than the device case and heats slightly faster, while the measurements are filtered to compensate for this discrepancy this may cause the testbed to read higher on-resistances at certain points during cool-down. Ultimately, the noise and outliers present in the data shown in Fig. 2 were kept to evaluate the ML model's robustness against noise.

III. PRELIMINARY ML ASSESSMENT

The visual indications and the mutual information scores demonstrate the relationship between R_{dson}, thermal cycle count, and device junction temperature. ML regression tools are a flexible, noise resistant method of predicting complex nonlinear relationships to build an accurate R_{dson} model with the potential to adapt this model as more data becomes available.

To understand the impact of using both temperature, and cycle count in an R_{dson} model, a preliminary Random Forest (RF) based model is trained to predict R_{dson} from thermal cycle counts and temperature alone and with both variables together.

The RF algorithm is selected for this preliminary analysis due to its robustness against noise, and broad applicability to many types and sizes of datasets [17]. The RF algorithm works by growing large numbers of decision trees on subsets of the data. Each decision tree splits the data into multiple nodes and makes predictions about the data at each split. The ends of the trees, where final predictions appear, are called leaf nodes. The RF algorithm obtains its final prediction by averaging the predictions at the leaf nodes to make a final prediction. The decision boundaries are tuned during the training of the RF algorithm where the models predictions are compared with ground truth data collected from the testbed. A depiction of the RF algorithm is shown in Fig. 3.

To preserve data for testing and validation of the RF predictor, the 3,797,379 data points are split into 3 groups, a training dataset, a test dataset, and a validation dataset. For the validation dataset, 20% of the original data is held out. For

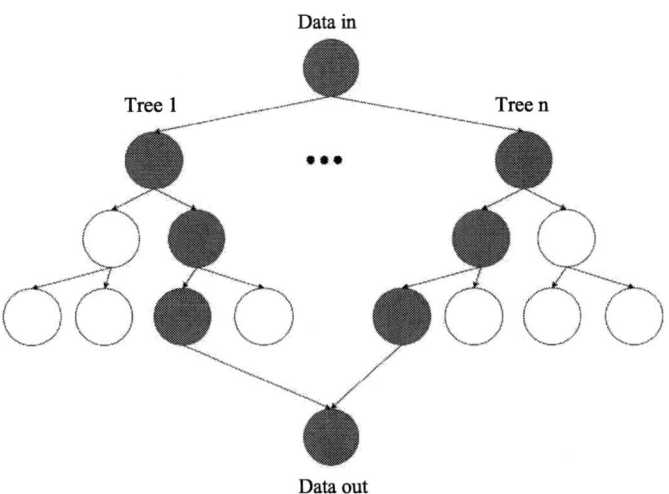

Fig. 3. Illustration of the random forest algorithm operation. Random predictions about the data are made at each decision tree node within the multi-tree forest. The decisions at one node in a tree inform decisions further down the tree and the end predictions in each tree are combined to form the final forest predictions. The decision boundaries are tuned in training.

the testing dataset, 25% of the remaining data is reserved and the other 75% of the non-validation data is used as a training dataset. This preliminary RF model uses 100 trees with no limit on the maximum tree depth.

When trained on the test data, the mean absolute percent error (MAPE) of the RF model is 3.5% on the test dataset. When trained to predict R_{dson} from only cycle count data, the model MAPE is 22.1%. When trained to predict R_{dson} from only temperature data, the model MAPE is 7.1%. The MAPE is calculated using

$$MAPE = \frac{1}{n_{samples}} \sum_{i=0}^{n_{samples}} \frac{|y_i - \hat{y}_i|}{|y_i|}. \qquad (2)$$

These results demonstrate the importance of both cycle count and temperature when predicting R_{dson}.

IV. MODEL COMPARISON

After proving its potential with the 100 tree RF model, variations of five other ML algorithms: histogram based gradient boosting, decision tree bagging, K-Nearest Neighbors (KNN) regression, Single Layer Neural Network, and Multi-Layer Neural Network, along with other sizes of RF models are trained and tested on the same dataset to compare model efficacy.

Each different algorithm has its associate hyper parameters and to get a full picture of model utility multiple hyper parameter combinations are tested for each algorithm type. The hyper parameters of interest are selected for each of these algorithms and a grid search is run across all hyper parameter combinations for each algorithm. The hypermeter and model with the best test scores within the grid search is kept. The best models of each algorithm are compared to understand which algorithm provides the best R_{dson} model. Different trained models were scored based on their prediction coefficient of determination R^2 calculated as

$$R^2 = 1 - \frac{\sum (y_i - \hat{y}_i)^2}{\sum (y_i - y_{avg})^2}. \quad (3)$$

Each of the different models and the hyper parameter spaces explored are explained below:

Histogram gradient boosting is a computationally efficient form of decision tree gradient boosting [18]. Where predictions from multiple decision trees are generated and the model is trained by leveraging the areas of most potential improvement from previous trees predictions. For histogram gradient boosting the tunable hyper parameters are: learning rate, maximum number of iterations, and L2 regularization level. The learning rates explored in the grid search are: 0.1, 0.5, and 1. The maximum number of iterations explored in the grid search are: 50, 100, and 150. The L2 regularization levels explored in the grid search are: 0 and 0.1.

A bagging regressor [19] works by fitting large numbers of less accurate, and less complex ML algorithms to small random subsets of data and averaging their predictions to produce a more accurate prediction. In this work the Decision tree based bagging was explored. Random forest [17] is similar type of bagging regressor similar to decision tree based bagging. The only hyper parameter tuned for decision tree bagging regressor is number of estimators. The numbers of estimators explored are: 5, 10, and 15.

K-nearest neighbors regression [20, 21] works by making a prediction based on the weighted average of the k closest samples in the training dataset. The weighting of these neighbors is determined during model training. For K-Nearest Neighbors number of neighbors is the only hyper parameter for which different values are tested. The values for number of neighbors explore are: 3, 5, 10, and 25.

Both single-layer and multi-layer neural networks [22] are explored in this work. Neural networks make predictions on data by passing input data into a series of layers of interconnected neurons which perform nonlinear transformations to the data yielding a prediction output at the last layer. The coefficients of the transformations are modified in gradient based training to produce the most accurate predictions based on training data. For the single-layer neural network the hyper parameters that are tuned are: hidden layer size and activation function. The hidden layer sizes explored are: 50, 100, and 150. The activation functions tested are: relu, tanh, and logistic.

For the multi-layer neural network the hyper parameters that are tested are also: number of layers and activation function. Because a single layer, 50 node neural network was already tested only 2 and 3 layer neural network are explore here. The activation functions tested are: relu, tanh, and logistic. Each layer of the multi-layer neural network is 50 nodes.

In addition to these models, RF algorithms with different numbers of estimators are also tested. The RF number of estimators tested are: 50, 100, 150, and 200.

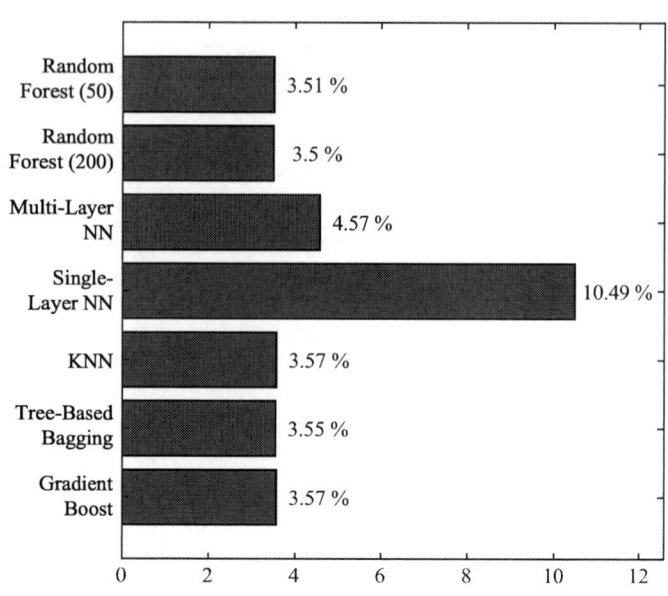

Fig. 4. Side-by-side comparison of ML model MAPE.

V. RESULTS

The most accurate models of each type with their hyper parameters are:

- Histogram gradient boosting regressor
  - Learning Rage: 1
  - Maximum iterations: 100
  - L2 regularization: 0.1
  - Best MAPE: 3.57%
- Tree based bagging regressor
  - Number of estimators: 15
  - Best MAPE: 3.55%
- K nearest neighbors regressor
  - Number of neighbors: 25
  - Best MAPE: 3.57%
- Single-layer neural network regressor
  - Single layer size: 150
  - Activation function: logistic
  - Best MAPE: 10.49%
- Multi-layer neural network regressor
  - Number of layers: 3
  - Activation function: logistic
  - Best MAPE: 4.57%
- Random forest
  - Number of layers: 200
  - Best MAPE: 3.50%

The results of training and testing multiple ML models as R_{dson} predictors are presented in Fig. 4.

A. Discussion

The large (200 tree) RF algorithm performed slightly better than the other algorithms in the study. It is worth noting, however, that the improvements gained by growing a larger random forest are not necessarily substantial and a smaller RF algorithm may be more computationally efficient and accurate enough. For example, the 50 tree RF had a MAPE of 3.51%.

Additionally, these models have been built and trained on a preliminary dataset. New data that becomes available from the second version of the ALT hardware and other test sources may reveal over-fitting in the RF algorithm which scored the best in this scenario making other models, with different regularization schemes more useful in future R_{dson} modeling. For example, a large neural network with dropout may perform better on a larger dataset, featuring tests at additional temperatures, and higher noise. Overall, more data, in a wider array of circumstances and with different devices needs to be collected and processed to fully validate these models and to make full use of ML as an R_{dson} predictor.

B. Implementation

The potential applications for this model include predicting the behavior of R_{dson} in devices prior to the implementation and live monitoring of R_{dson} once the devices are employed in power applications. Designers using thermal simulation tools such as complex fluid dynamics software, finite element analysis, or thermal equivalent circuit models [23] and cycle counting algorithms such as rain flow [24] can model device temperature over time and cycle counts. This information can be used in the ML model to predict the behavior of R_{dson} in new, untested power converter topologies prior to implementation. This enables design of converter topologies, and control schemes with reliability in mind. The model can also be used to monitor the R_{dson} of devices online by using online cycle counting algorithms [25] to collect temperature and cycle count information from the devices during use. Live monitoring of R_{dson} can enable preventative maintenance and reliability-oriented scheduling and deployment of high-power SiC MOSFET applications such as EV charging.

There are two main applications for the proposed ML based R_{dson} model: one is R_{dson} modeling, and the other is real-time R_{dson} estimation. R_{dson} modeling enables charger designers to predict R_{dson} behavior in SiC based power converters based on preliminary thermal models and expected mission profiles. A designer can combine thermal models of the designed power converter with expected mission profiles to predict long-term thermal behavior of the converter. This long-term thermal behavior can be passed into rain flow cycle counting algorithms. With the thermal models and rain flow algorithms, predicted thermal cycle count and component temperatures can be obtained and these can be used in the RF model to predict device R_{dson} change over time at each time of interest. A sample diagram of this process is shown in Fig. 5.

Using the ML model to perform online prediction of R_{dson}, SiC MOSFET based charger component R_{dson} and lifetime can be continuously monitored by logging the case

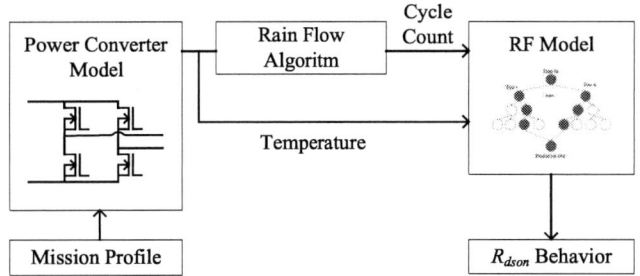

Fig. 5. Sample implementation of the RF modeling of R_{dson} behavior

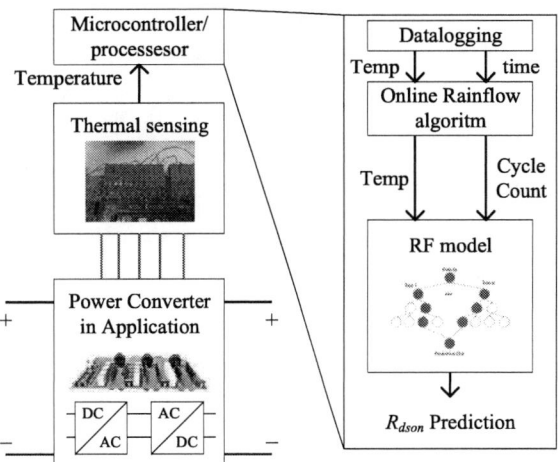

Fig. 6. Sample implementation of the R_{dson} estimation algorithm in power converter operation

temperature of the SiC devices. From the case temperature, junction temperature is calculated as outlined in (1). The logged junction temperatures are fed into an online rain flow counting algorithm [25] and cycle counts are extracted and the temperature and cycle count are passed into the RF algorithm at each time step. The ML model generates an accurate R_{dson} prediction which can be used to directly monitor device health or be fed into a lifetime model to predict RUL. A sample implementation of RF algorithm for R_{dson} prediction is shown in Fig. 6.

VI. CONCLUSION

In this work a novel ML method for determining SiC MOSFET R_{dson} based on device junction temperature and cycle count is presented. An analytical basis for the use of both temperature and cycle count as R_{dson} predictors is given and tested with a preliminary RF model. Additional ML models are also tested and compared. Prediction errors near 3.5% are achieved using multiple ML algorithms trained on temperature and cycle count data. The potential applications for these ML models in power electronics design and monitoring are also outlined.

Preliminary data is limited and work is already underway to collect additional data on SiC MOSFET degradation and R_{dson} behavior. A second revision of the ALT hardware is under development which addresses some of the issues with testbed breakdown and allow for the collection of additional SiC MOSFET lifetime data at various power levels in in configurations. Ultimately, the models will be verified in live power converter operation.

REFERENCES

[1] Enes Ugur and Bilal Akin. "Aging assessment of discrete SiC MOSFETs under high temperature cycling tests". In: *2017 IEEE Energy Conversion Congress and Exposition (ECCE)*. 2017, pp. 3496–3501. DOI: 10.1109/ECCE.2017.8096624.

[2] Masoud Farhadi, Bhanu Teja Vankayalapati, and Bilal Akin. "Reliability Evaluation of SiC MOSFETs Under Realistic Power Cycling Tests". In: *IEEE Power Electronics Magazine* 10.2 (2023), pp. 49–56. DOI: 10.1109/MPEL.2023.3271621.

[3] Ze Ni et al. "Overview of Real-Time Lifetime Prediction and Extension for SiC Power Converters". In: *IEEE Transactions on Power Electronics* 35.8 (Aug. 2020), pp. 7765–7794. ISSN: 1941-0107. DOI: 10.1109/TPEL.2019.2962503.

[4] Enes Ugur et al. "Degradation Assessment and Precursor Identification for SiC MOSFETs Under High Temp Cycling". In: *IEEE Transactions on Industry Applications* 55.3 (2019), pp. 2858–2867. DOI: 10.1109/TIA.2019.2891214.

[5] F. Blaabjerg, K. Ma, and D. Zhou. "Power electronics and reliability in renewable energy systems". In: *2012 IEEE International Symposium on Industrial Electronics*. 2012 IEEE International Symposium on Industrial Electronics. ISSN: 2163-5145. May 2012, pp. 19–30. DOI: 10.1109/ISIE.2012.6237053.

[6] Ujjwal Karki and Fang Zheng Peng. "Effect of Gate-Oxide Degradation on Electrical Parameters of Power MOSFETs". In: *IEEE Transactions on Power Electronics* 33.12 (2018), pp. 10764–10773. DOI: 10.1109/TPEL.2018.2801848.

[7] Serkan Dusmez, Hamit Duran, and Bilal Akin. "Remaining Useful Lifetime Estimation for Thermally Stressed Power MOSFETs Based on on-State Resistance Variation". In: *IEEE Transactions on Industry Applications* 52.3 (2016), pp. 2554–2563. DOI: 10.1109/TIA.2016.2518127.

[8] José R. Celaya et al. "Prognostics approach for power MOSFET under thermal-stress aging". In: *2012 Proceedings Annual Reliability and Maintainability Symposium*. 2012, pp. 1–6. DOI: 10.1109/RAMS.2012.6175487.

[9] Alessandro Soldati et al. "Electric-vehicle power converters model-based design-for-reliability". In: *CPSS Transactions on Power Electronics and Applications* 3.2 (2018), pp. 102–110. DOI: 10.24295/CPSSTPEA.2018.00010.

[10] Yongheng Yang, Ariya Sangwongwanich, and Frede Blaabjerg. "Design for reliability of power electronics for grid-connected photovoltaic systems". In: *CPSS Transactions on Power Electronics and Applications* 1.1 (2016), pp. 92–103. DOI: 10.24295/CPSSTPEA.2016.00009.

[11] Jinkui He et al. "Design for Reliability of SiC-MOSFET-Based 1500-V PV Inverters With Variable Gate Resistance". In: *IEEE Transactions on Industry Applications* 58.5 (2022), pp. 6485–6495. DOI: 10.1109/TIA.2022.3183029.

[12] M. Held et al. "Fast power cycling test of IGBT modules in traction application". In: *Proceedings of Second International Conference on Power Electronics and Drive Systems*. Vol. 1. 1997, 425–430 vol.1. DOI: 10.1109/PEDS.1997.618742.

[13] *Load-cycling capability of HiPak*. URL: https://search.abb.com/library/Download.aspx?DocumentID=5SYA2043&DocumentPartId=&DocumentRevisionId=A (visited on 06/22/2023).

[14] Yi Zhang et al. "Impact of lifetime model selections on the reliability prediction of IGBT modules in modular multilevel converters". In: *2017 IEEE Energy Conversion Congress and Exposition (ECCE)*. 2017, pp. 4202–4207. DOI: 10.1109/ECCE.2017.8096728.

[15] Conner Deppe and Hongjie Wang. "Silicon Carbide MOSFETs Aging Testing Platform for EV Chargers Using Power Cycling". In: *2024 IEEE 10th International Power Electronics and Motion Control Conference (IPEMC2024-ECCE Asia)*. 2024, pp. 655–660. DOI: 10.1109/IPEMC-ECCEAsia60879.2024.10567447.

[16] Alexander Kraskov, Harald Stögbauer, and Peter Grassberger. "Estimating mutual information". In: *Physical Review. E, Statistical, Nonlinear, and Soft Matter Physics* 69.6 (June 2004), p. 066138. ISSN: 1539-3755. DOI: 10.1103/PhysRevE.69.066138.

[17] Leo Breiman. "Random Forests". In: *Machine Learning* 45.1 (Oct. 1, 2001), pp. 5–32. ISSN: 1573-0565. DOI: 10.1023/A:1010933404324.

[18] Guolin Ke et al. "LightGBM: A Highly Efficient Gradient Boosting Decision Tree". In: *Advances in Neural Information Processing Systems*. Ed. by I. Guyon et al. Vol. 30. Curran Associates, Inc., 2017.

[19] Leo Breiman. "Bagging Predictors". In: *Machine Learning* 24.2 (Aug. 1, 1996), pp. 123–140. ISSN: 1573-0565. DOI: 10.1023/A:1018054314350.

[20] T. Cover and P. Hart. "Nearest neighbor pattern classification". In: *IEEE Transactions on Information Theory* 13.1 (1967), pp. 21–27. DOI: 10.1109/TIT.1967.1053964.

[21] Charles J. Stone. "Consistent Nonparametric Regression". In: *The Annals of Statistics* 5.4 (1977), pp. 595–620. DOI: 10.1214/aos/1176343886.

[22] Geoffrey E. Hinton. "Connectionist learning procedures". In: *Artificial Intelligence* 40.1 (1989), pp. 185–

234. ISSN: 0004-3702. DOI: https://doi.org/10.1016/0004-3702(89)90049-0.

[23] Paul Braford et al. "Power electronics and reliability in renewable energy systems". In: *2024 IEEE Energy Conversion Conference and Expo.* 2024 IEEE Energy Conversion Conference and Expo. 2024.

[24] *ASTM E1049-85(2017) - Standard Practices for Cycle Counting in Fatigue Analysis.* URL: https://webstore.ansi.org/standards/astm/astme1049852017?msclkid=ba2bccbe2eb91d9a0f92ab8b31b66058 & utm_source = bing&utm_medium=cpc&utm_campaign=Campaign%20%231&utm_term=ASTM%20E1049-85%202017& utm_content=ASTM-E (visited on 11/13/2024).

[25] Mahera Musallam and C. Mark Johnson. "An Efficient Implementation of the Rainflow Counting Algorithm for Life Consumption Estimation". In: *IEEE Transactions on Reliability* 61.4 (2012), pp. 978–986. DOI: 10.1109/TR.2012.2221040.

Comparison of Static Characteristics in GaN HEMTs Across 50K to 400K Considering Diverse Techniques and Statistical Variation

Purushottam Khadka
Electrical, Computer, and Systems Engineering
Rensselaer Polytechnic Institute
Troy, NY, USA
khadkp@rpi.edu

Saumil Shivdikar
ECSE
Rensselaer Polytechnic Institute
Troy, NY, USA
shivds@rpi.edu

Zheyu Zhang
ECSE
Rensselaer Polytechnic Institute
Troy, NY, USA
zhangz49@rpi.edu

Tian Qiu
Electrical, Computer, and Systems Engineering
Rensselaer Polytechnic Institute
Troy, NY, USA
qiut2@rpi.edu

Ahmed Siraj
Electrical, Computer, and Systems Engineering
Rensselaer Polytechnic Institute
Troy, NY, USA
siraja@rpi.edu

Abstract—GaN HEMTs are prominent candidates for space applications, such as lunar missions, which require reliable operation over a wide temperature range from 50K to 400K. Previous studies have examined the performance of GaN HEMTs at cryogenic temperatures (typically down to 77K), but they often focus on a limited selection of models and sample sizes. To address this gap, a comprehensive analysis of diverse GaN HEMTs is necessary, considering their performance across a broad temperature range and statistical variations due to manufacturing inconsistencies. This paper provides a comparison of GaN HEMTs based on different voltage ratings, technologies and packaging approaches. A key observation is that the on-state resistance of the studied Gate Injection Transistor (GIT) device exhibits a parabolic variation with temperature, with resistance at 50K being 1.25 times higher than at room temperature. Additionally, the statistical variation in the static parameters of a cascode GaN device across six samples reveals that the variation increases as the temperature decreases, indicating that the statistical variation over a wide temperature range cannot be overlooked.

Index Terms—static characterization, GaN HEMT, GIT (Gate Injection Transistor), D-mode GaN, EPC, GaN Systems, technology, on-state resistance, cryogenic temperature, leakage current

I. INTRODUCTION

Wide-bandgap devices, such as SiC and GaN, are among the most promising alternatives to traditional Si MOSFETs and IGBTs. These devices are increasingly being adopted to replace conventional Si-based converters due to their ability to enhance efficiency and support high power density. This superior performance stems from several key advantages: a high critical electric field, faster switching speeds, superior thermal conductivity, lower on-state resistance, and reduced power losses [1].

Among these technologies, GaN HEMTs exhibit enhanced mobility and low on-state resistance, particularly at low temperatures. Unlike Si and SiC, GaN HEMTs avoid carrier freeze-out at cryogenic temperatures because they rely on a two-dimensional electron gas (2DEG) formed at the AlGaN/GaN heterojunction. This 2DEG is induced by spontaneous and piezoelectric polarization fields rather than by inversion, enabling GaN HEMTs to achieve high conductivity in cryogenic environments. In contrast, silicon and SiC devices suffer from carrier freeze-out at low temperatures due to insufficient thermal energy to ionize dopant atoms, which results in reduced conductivity. Additionally, SiC devices are further impacted by interface traps, which increase on-state resistance and threshold voltage under cryogenic conditions [2]–[4]. Silicon MOSFETs, on the other hand, experience a reduction in breakdown voltage at low temperatures. This is because the mean free path of carriers increases, allowing them to gain more energy for a given electric field before collisions, thus resulting in a reduced avalanche breakdown voltage [5].

Furthermore, GaN HEMTs are being considered for space applications that require reliable operation over a broad temperature range, such as lunar missions from 50K to 400K [6]. While extensive characterization of GaN HEMTs under cryogenic conditions has been conducted in previous studies, these studies often involve small sample sizes and temperature measurements that typically extend only down to 77K [7]–[9]. Some research attempts to compare GaN devices based on gate technology; however, such comparisons are limited by small sample sizes, lack of characterization at 50K, and

979-8-3315-1612-3/25 $31.00 © 2025 IEEE

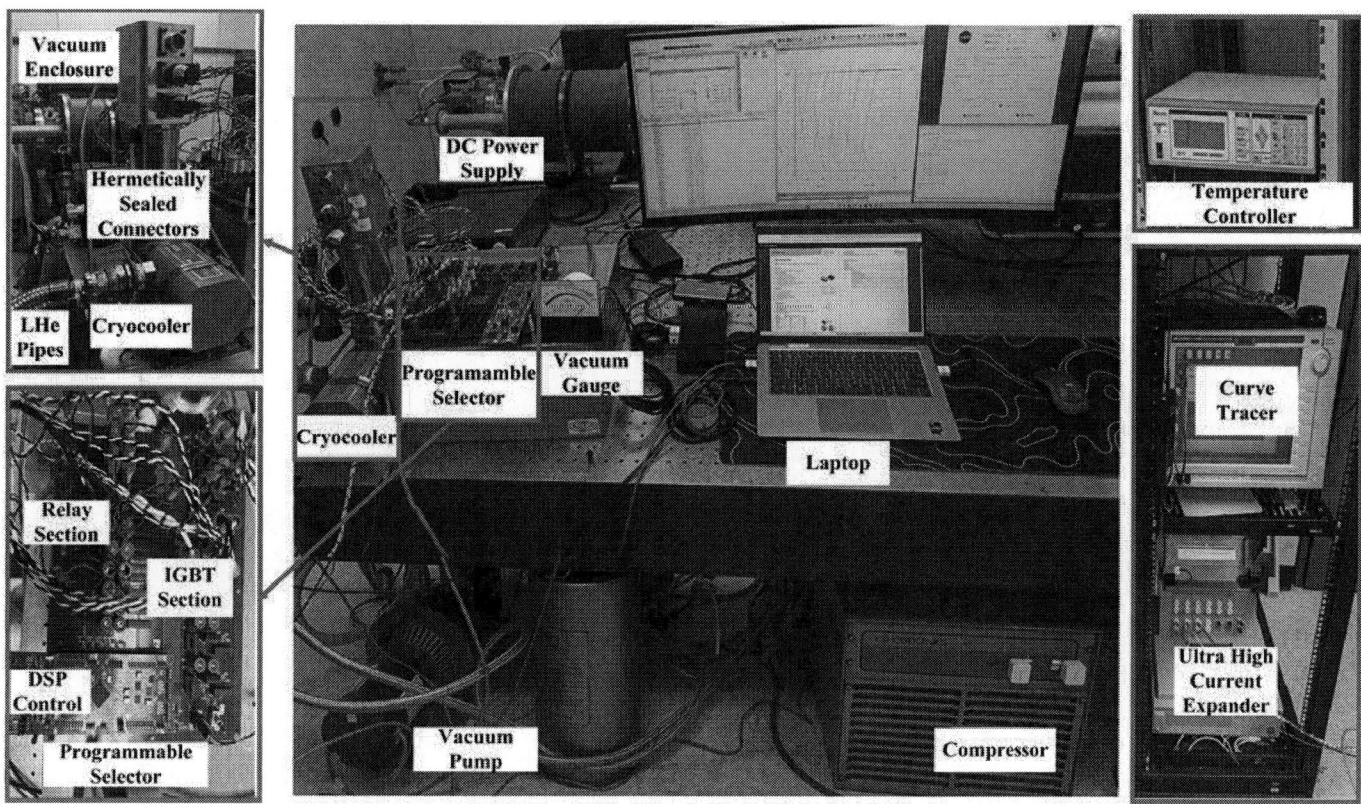

Fig. 1. Hardware overview of static characterization platform.

insufficient consideration of statistical variations [10]–[13]. For instance, Nela investigated the static characteristics of GaN devices down to 4.2K, but this study was restricted to only two samples and focused exclusively on devices in the 600V–650V range [14]. Many papers have attempted to model the behavior of GaN devices for a particular technology, focusing either on high-temperature variations or limited temperature ranges. However, these studies often focus on specific technologies, which do not comprehensively represent GaN device performance under cryogenic conditions.

The limitations of current state-of-the-art studies highlight the need for a comprehensive comparative analysis of GaN HEMTs across a wide temperature range, considering the often-overlooked impact of statistical parameter variations among samples. This paper addresses these gaps by systematically examining the temperature-dependent behaviors (from 50K to 400K) of static parameters in a diverse array of GaN HEMTs, with holistic consideration of GaN semiconductor chips, packaging technologies, voltage ratings, and statistical variations.

First, the hardware setup for the static characterization platform is overviewed. Then, the impacts of different chip, packaging technologies and voltage ratings dependence on temperature are discussed, respectively. Next, the statistical variation among different samples of the same device is analyzed. Finally, key takeaways and future work are summarized.

II. AUTOMATED STATIC CHARACTERIZATION PLATFORM

A fully automated platform, consisting of a cryocooler, curve tracer, programmable selector, temperature controller, and a computer (as shown in Fig. 1), has been developed to characterize the static behaviors of GaN HEMTs across a temperature range from 50K to 400K. The detailed setup description of the static characterization platform can be found in [15], and the lead optimization inside the cryocooler for this setup can be found in [16]. Multiple devices can be characterized simultaneously on this platform. The devices under test are placed inside the cryocooler, and the temperature is varied from 50K to 400K using the temperature controller. The vacuum inside the cryocooler is maintained by a vacuum pump, and the pressure is monitored using a vacuum gauge.

The devices are characterized using the curve tracer. To switch between devices for characterization, the relay board employs relays and IGBT switches. The relays are used for low-current signals, such as gate and sense wires, while the IGBTs handle high-current force wires. These components are controlled by a microcontroller. The entire automation process—device selection, temperature adjustment, static characterization initiation, data saving, and data processing—is managed using a MATLAB GUI.

Table I summarizes all the GaN devices from various vendors that were characterized across the 50K to 400K temperature range with complete data.

III. DEPENDENCE ON VOLTAGE RATINGS

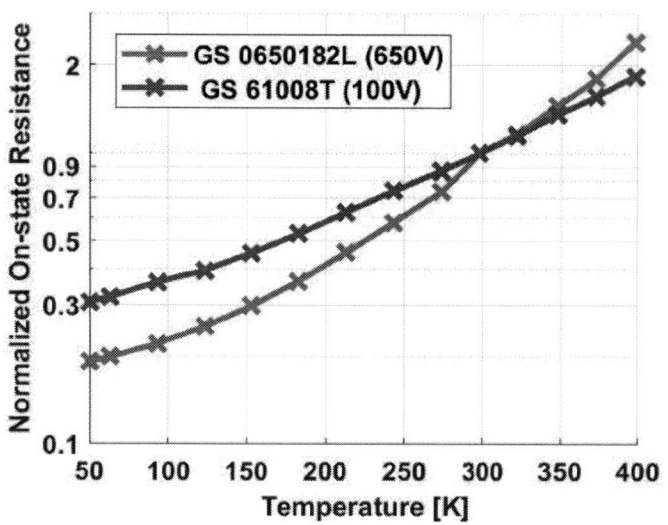

Fig. 2. Normalized on state resistance of GS0650182L [650V] and GS61008T [100V].

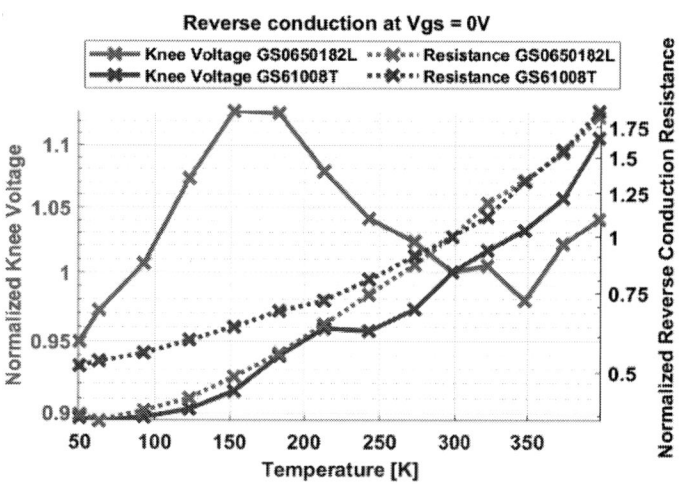

Fig. 3. Normalized knee voltage and reverse conduction resistance of GS0650182L [650V] and GS61008T [100V].

In this paper, we have used normalized values of static parameters such as on-state resistance, knee voltage, and leakage current. The normalized value of a static parameter is defined as the ratio of the parameter at a given temperature to its corresponding value at room temperature.

Fig. 2, Fig. 3, and Fig. 4 compare the variation in static parameters between the GS0650182L and GS61008T devices, both of which utilize the same enhancement-mode GaN HEMT chip technology from the same manufacturer. The GS0650182L is rated for 650V, while the GS61008T is rated for 100V.

Fig. 2 shows that both devices exhibit a decrease in normalized on-state resistance as the temperature decreases. This

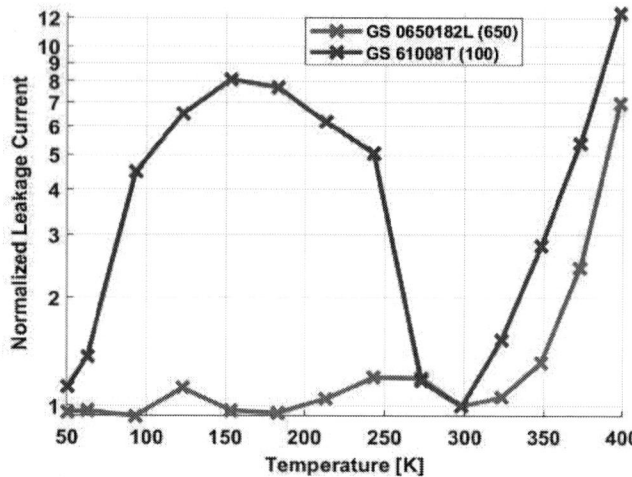

Fig. 4. Normalized leakage current of GS0650182L [650V] and GS61008T [100V].

trend is primarily attributed to two factors: the reduction in resistance of the 2DEG regions due to improved electron mobility at lower temperatures, and the decrease in metal contact and package resistance with lower temperatures.

The GS0650182L (650V) exhibits a steeper reduction in normalized on-state resistance compared to the GS61008T (100V). At 50K, the normalized on-state resistance of the 650V device is 0.19, significantly lower than that of the 100V device (0.31). This observation suggests that higher voltage-rated devices are more sensitive to temperature variations.

The normalized knee voltage of the GS61008T during reverse conduction consistently decreases with temperature. In contrast, the variation in the normalized knee voltage for the GS0650182L is inconsistent, as shown in Fig. 3.

Fig. 4 shows that the leakage current for the 650V device remains stable at lower temperatures, whereas the leakage current for the 100V device varies significantly. This behavior indicates that the leakage current for the 100V device is more sensitive to temperature variations and suggests greater susceptibility of its breakdown voltage (BV) to temperature changes.

IV. DEPENDENCE ON CHIP TECHNOLOGIES

Fig. 5, Fig. 6, and Fig. 7 illustrate the variation in static parameters for the IGOT60R070D1, GS0650182L, and TP65H070G4LSG devices, all of which belong to the 600–650 V family, originate from different vendors, and utilize distinct chip technologies.

The p-GaN HEMT (GS0650182L) was discussed in the previous section, where it was observed that the on-state resistance decreases with temperature. Additionally, the knee voltage during reverse conduction exhibits inconsistent behavior, while the leakage current shows minimal variation at lower temperatures.

For the GIT device (IGOT60R070D1), the variation in normalized on-state resistance with temperature follows a

TABLE I
COMMERCIALLY AVAILABLE GaN HEMTs: THEIR VOLTAGE AND RESISTANCE RATINGS AT ROOM TEMPERATURE (25°C), TECHNOLOGY TYPE, AND PACKAGING.

Sr. No	Vendor	Part Number	Ratings (@25 °C) Voltage / Resistance	Technology	Package
1	GaN Systems	GS61008T	100V / 7mΩ	E-mode GaN HEMT (p-GaN)	GaNPx packaging
2	EPC	EPC2215	200V / 6mΩ	E-mode GaN HEMT (p-GaN)	LGA
3	EPC	EPC2034C	200V / 8mΩ	E-mode GaN HEMT (p-GaN)	BGA
4	EPC	EPC2307	200V / 8.2mΩ	E-mode GaN HEMT (p-GaN)	QFN
5	Infineon	IGOT60R070D1	600V / 55mΩ	GIT, E-mode GaN HEMT (p-GaN gate with Ohmic Gate)	PG-DSO
6	Transphorm	TP65H070LSG	650V / 72mΩ	D-mode GaN HEMT w/ Si for normally-off cascode	PQFN
7	GaN Systems	GS-065-018-2-L	650V / 78mΩ	E-mode GaN HEMT (p-GaN)	PDFN packaging

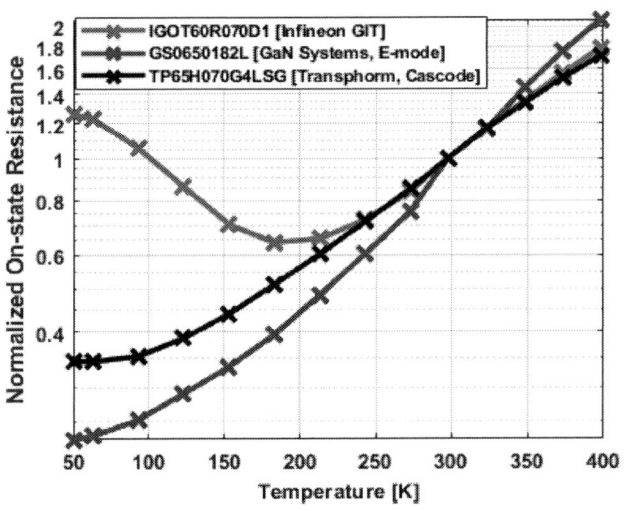

Fig. 5. Normalized on state resistance of IGOT60R070D1 [p-GaN with Ohmic gate], GS0650182L [p-GaN], TP65H070G4LSG [Cascode].

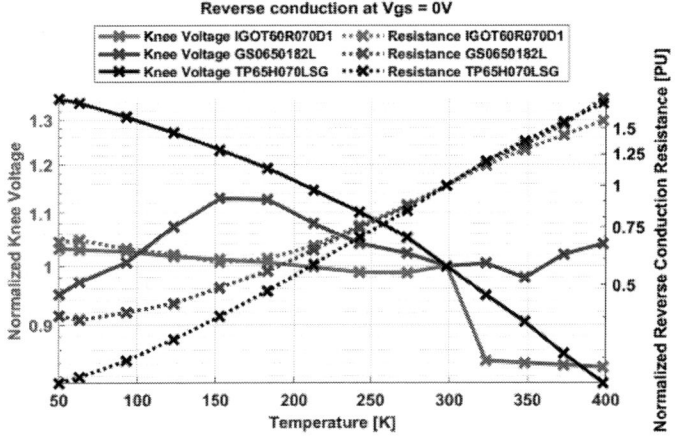

Fig. 6. Normalized knee voltage and reverse conduction resistance of IGOT60R070D1 [p-GaN with Ohmic gate], GS0650182L [p-GaN], TP65H070G4LSG [Cascode].

parabolic trend. The resistance decreases as the temperature drops below room temperature, reaching a minimum value at around 183K, and then begins to increase. This behavior can be attributed to the carrier freeze-out of dopants in the p-GaN layer. At 50K, the normalized on-state resistance is approximately 1.25 times higher than at room temperature, indicating reduced efficiency for power converter applications at low temperatures. Despite this, the drain leakage current remains stable under cryogenic conditions. Additionally, the knee voltage for this device increases as the temperature decreases.

For the cascode device (TP65H070G4LSG), the on-state

resistance decreases with temperature, consistent with the expected behavior of Si and GaN devices. However, compared to GIT devices, the decrease is less steep. Below 100K, the slope of the resistance-temperature curve becomes nearly constant, suggesting limited improvement in conductivity due to carrier freeze-out in the silicon MOSFET component of the device. Furthermore, the leakage current for the cascode device increases at cryogenic temperatures, aligning with the behavior observed in silicon devices, where leakage current typically rises as temperature decreases.

V. DEPENDENCE ON PACKAGING TECHNOLOGIES

Fig. 8, Fig. 9, and Fig. 10 illustrate the variation in static parameters for the EPC 2215, EPC 2034C, and EPC 2307

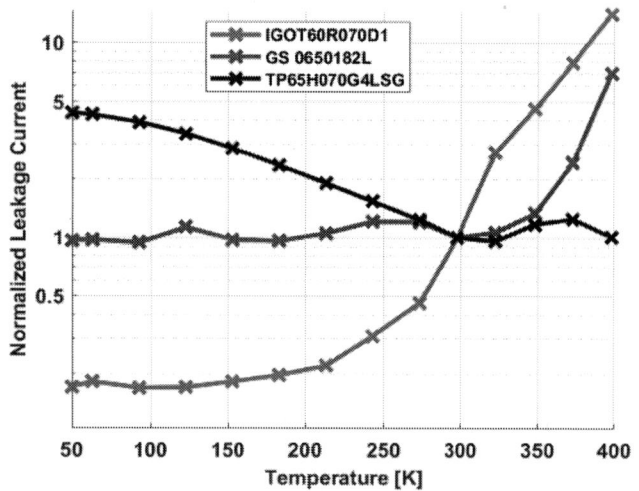

Fig. 7. Normalized leakage current of IGOT60R070D1 [p-GaN with Ohmic gate], GS0650182L [p-GaN], TP65H070G4LSG [Cascode].

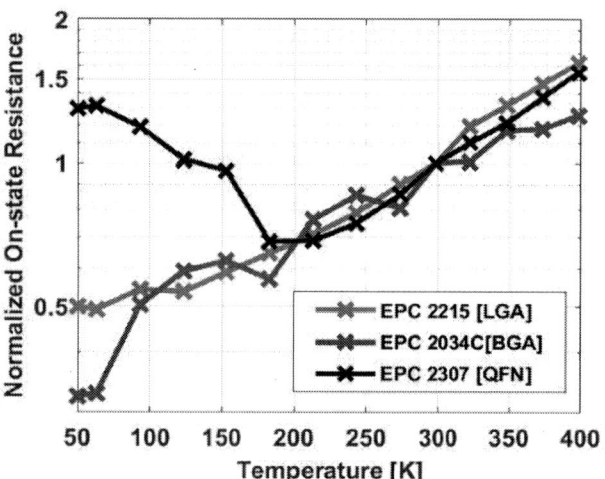

Fig. 8. Normalized on-state resistance of EPC 2215 [LGA], EPC 2034C [BGA], EPC 2307 [QFN].

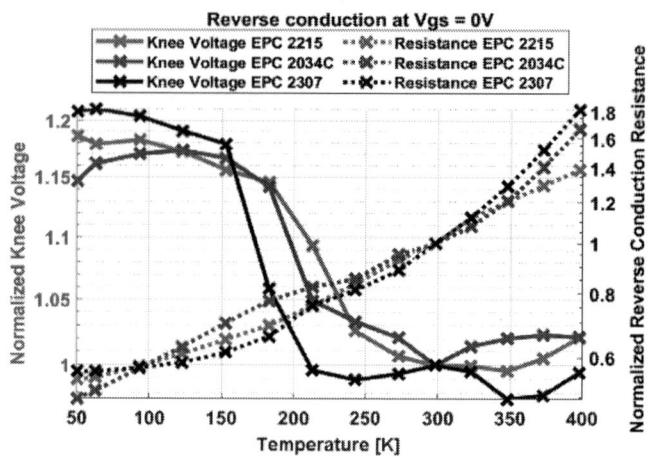

Fig. 9. Normalized knee voltage and reverse conduction resistance of EPC 2215 [LGA], EPC 2034C [BGA], EPC 2307 [QFN].

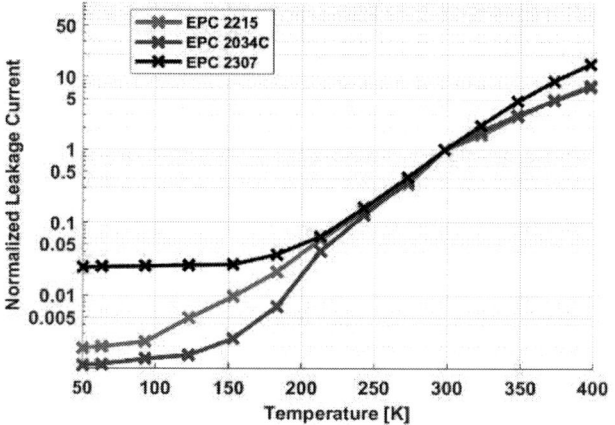

Fig. 10. Normalized leakage current of EPC 2215 [LGA], EPC 2034C [BGA], EPC 2307 [QFN].

devices. All three devices have the same voltage rating of 200V and are manufactured by the same vendor (EPC) but differ in packaging: LGA, BGA, and QFN, respectively.

The normalized on-state resistance for the EPC 2215 and EPC 2034C follows a consistent decreasing trend with temperature, as expected for p-GaN devices. However, the EPC 2307, which uses a QFN package, exhibits an anomalous trend. Its normalized on-state resistance decreases with temperature down to 183K, where it reaches a minimum, and then begins to increase. At 50K, the on-state resistance for the EPC 2307 is approximately 1.31 times higher than at room temperature, which is atypical for p-GaN-based devices.

The variation in reverse conduction resistance across the three devices consistently shows a decreasing trend with temperature. The reverse knee voltage demonstrates little sensitivity at temperatures near room temperature but becomes increasingly sensitive as the temperature drops further.

The parabolic nature of the $R_{DS(on)}$ trend for the EPC 2307 is unusual for EPC devices. Whether this behavior is due to the packaging or other factors requires further investigation.

VI. QUANTIFICATION OF STATISTICAL VARIATION

Fig. 11, Fig. 12, and Fig. 13 illustrate the statistical variation in the static parameters of the TP65H070G4LSG (cascode device) across six samples. The data indicate that the variation in static parameters, such as $R_{DS(on)}$, knee voltage, and reverse conduction resistance, is greater at 50K compared to room temperature. Additionally, the variation in leakage current becomes most pronounced at 400K. At 50K, the tolerances for the static parameters are as follows: on-state resistance ($R_{DS(on)}$) ± 1.11 mΩ, knee voltage ± 0.02 V, and leakage current ± 1.98 μA, relative to their respective mean values of 18.64 mΩ, 1.23 V, and 9.59 μA.

The results indicate that the statistical variation among samples from the same device cannot be overlooked, as it

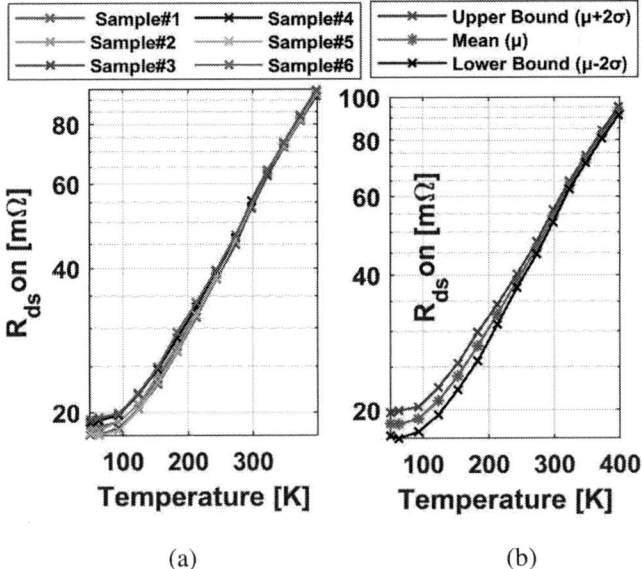

(a)   (b)

Fig. 11. On state resistance of TP65H070G4LSG. (a) Individual values of the 6 devices. (b) Calculated confidence interval.

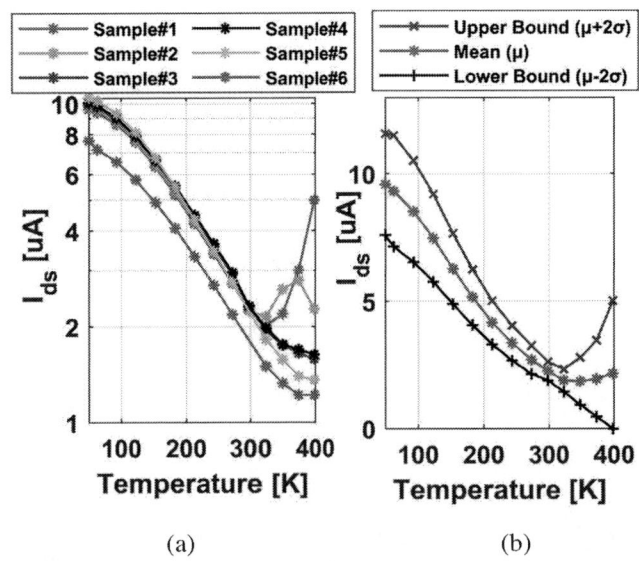

(a)   (b)

Fig. 13. Leakage current of TP65H070G4LSG. (a) Individual values of the 6 devices. (b) Calculated confidence interval.

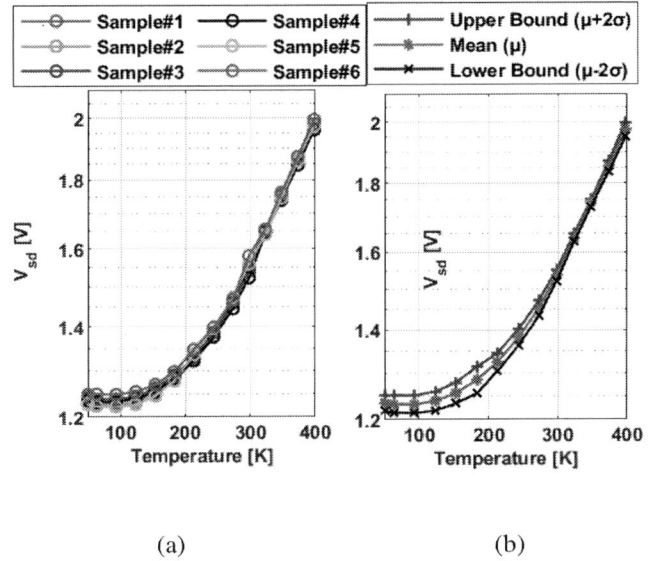

(a)   (b)

Fig. 12. Knee voltage of TP65H070G4LSG. (a) Individual values of the 6 devices. (b) Calculated confidence interval.

is particularly important for high-power modules integrating multiple chips. Deviations in static parameters can lead to imbalances in current sharing, potentially causing premature failure. Ensuring uniformity in these parameters is essential for robust, efficient, and reliable circuit performance.

VII. CONCLUSION

In this paper, we analyzed the static performance of various commercially available GaN HEMTs over a wide temperature range, from 50K to 400K. The comparison included devices with different voltage ratings, chip technologies, and pack-

aging approaches. At 50K, the 600V GaN device from GaN Systems exhibited greater temperature sensitivity in $R_{DS(on)}$ compared to the 100V device from the same manufacturer. Despite this, the 600V device demonstrated a lower normalized on-state resistance and more consistent leakage current behavior at cryogenic temperatures.

For 650V GaN devices employing different technologies (p-GaN HEMT, p-GaN with Ohmic gate (GIT), and cascode configurations), we observed notable distinctions in temperature-dependent performance. The on-state resistance of the p-GaN HEMT decreased consistently as the temperature dropped. In contrast, GIT device exhibited a parabolic trend, with its $R_{DS(on)}$ at 50K being approximately 1.25 times higher than at room temperature. This indicates higher conduction losses and reduced efficiency under cryogenic conditions, making it less suitable for power conversion at low temperatures.

For devices from the same 200V family manufactured by EPC, we investigated the impact of packaging (QFN, BGA, and LGA). Significant performance differences were observed, with the EPC2307 (QFN package) exhibiting an unusual $R_{DS(on)}$ trend compared to other EPC devices. Whether this behavior is due to packaging or other factors requires further investigation.

Finally, we found that the statistical variation in device performance at 50K was more pronounced than at room temperature. This increased variability is critical for applications involving parallel operation of GaN HEMTs under cryogenic conditions, as it directly affects reliability and uniformity of performance.

In the future, we plan to further investigate the unusual behavior of the EPC2307 and expand our analysis by characterizing additional samples to better understand statistical variations. Additionally, we aim to perform dynamic characterization and develop a comprehensive model for GaN devices that accounts for wide temperature ranges (50K to 400K) and statistical variations.

ACKNOWLEDGMENT

This material is based on work supported by NASA under Award No. 80NSSC24K0179. The authors would also like to thank Mr. Richard C. Oeftering, Dr. Nicholas R. Uguccini, Dr. Brian A. Holler, and Mr. Frederick W. Van Keuls from NASA Glenn Research Center for their contributions to the design of the cryocooler platform. Additionally, we extend our gratitude to Prof. Ishwara Bhat from Rensselaer Polytechnic Institute for his assistance with the operation and maintenance of the cryogenic apparatus.

REFERENCES

[1] M. Beheshti, Wide-bandgap Semiconductors: Performance and Benefits of GaN Versus SiC. Texas Instruments, [Online]. Available: https://www.ti.com/lit/an/slyt801.pdf

[2] H. Gui, X. Yan, J. Wu, and J. Shen, "Review of Power Electronics Components at Cryogenic Temperatures," IEEE Transactions on Power Electronics, vol. 35, no. 5, pp. 5144–5156, May 2020.

[3] Y. Liao, A. Elwakeel, Y. Xiao, R. Peña Alzola, M. Zhang, W. Yuan, A. J. Cruz Feliciano, and L. Graber, "Review of semiconductor devices and other power electronics components at cryogenic temperature," iEnergy, vol. 3, no. 2, pp. 95–107, 2024.

[4] "Cryogenic Cooling Technology using SiC and GaN Devices," Utmel, Jul. 7, 2023. [Online].

[5] H. Ye, C. Lee, J. Raynolds, P. Haldar, M. J. Hennessy, and E. K. Mueller, "Silicon power MOSFET at low temperatures: A two-dimensional computer simulation study," Cryogenics, vol. 47, no. 4, pp. 243-251, 2007.

[6] EPC Space, "Space Level Qualified 40 to 300 V Rad Hard eGaN Power Transistors," [Online].

[7] R. Ren, H. Xiao, Z. Liu, X. Huang, and Q. Li, "Characterization of 650 V Enhancement-mode GaN HEMT at Cryogenic Temperatures," Proc. IEEE Energy Convers. Congr. Expo. (ECCE), Portland, OR, USA, vol. 2018, pp. 891–897, Sep. 2018.

[8] Z. Zhang, J. Lin, X. Zhou, and Y. Pan, "Characterization of Wide Bandgap Device for Cryogenically-Cooled Power Electronics in Aircraft Applications," Proc. AIAA/IEEE Electr. Aircraft Technol. Symp., Cincinnati, OH, USA, Jul. 2018.

[9] R. Ren, H. Xiao, Z. Liu, X. Huang, and Q. Li, "Characterization and Failure Analysis of 650-V Enhancement-Mode GaN HEMT for Cryogenically Cooled Power Electronics," IEEE J. Emerg. Sel. Topics Power Electron., vol. 8, no. 1, pp. 66–76, Mar. 2020.

[10] Z. Liu, X. Huang, F. C. Lee, and Q. Li, "Package Parasitic Inductance Extraction and Simulation Model Development for the High-Voltage Cascode GaN HEMT," IEEE Trans. Power Electron., vol. 29, no. 4, pp. 1977–1985, Apr. 2014.

[11] M. Mehrabankhomartash, J. Wang, H. Zhang, S. Xie, and F. Blaabjerg, "Static and Dynamic Characterization of 650 V GaN E-HEMTs in Room and Cryogenic Environments," Proc. IEEE Energy Convers. Congr. Expo. (ECCE), Vancouver, BC, Canada, vol. 2021, pp. 5289–5296, Oct. 2021.

[12] Z. Liu, X. Huang, W. Zhang, F. C. Lee, and Q. Li, "Evaluation of High-Voltage Cascode GaN HEMT in Different Packages," Proc. IEEE Appl. Power Electron. Conf. Expo. (APEC), Fort Worth, TX, USA, vol. 2014, pp. 168–173, Mar. 2014.

[13] X. Geng, N. Wieczorek, C. Kuring, O. Hilt, M. Wolf, and S. Dieckerhoff, "Comparison of the Static Characteristics of GaN HEMTs With Different Gate Technologies and the Impact on Modeling," Proc. IEEE 10th Workshop Wide Bandgap Power Devices and Appl. (WiPDA), Charlotte, NC, USA, vol. 2023, pp. 1–6, Dec. 2023.

[14] L. Nela, N. Perera, C. Erine, and E. Matioli, "Performance of GaN Power Devices for Cryogenic Applications Down to 4.2 K," IEEE Trans. Power Electron., vol. 36, no. 7, pp. 7412–7416, Jul. 2021.

[15] P. Khadka, C. Wan, Z. Zhang, Q. Tian, and A. S. Siraj, "Automated static characterization platform for multi-GaN devices across cryogenic and high-temperatures," presented at ECCE 2024, Arizona, USA, unpublished.

[16] T. Qiu, Z. Zhang, and P. Khadka, "Conduction lead design and optimization for cryogenic characterization of GaN HEMTs," presented at ECCE 2024, Arizona, USA, unpublished.

Compact Model of β-Ga₂O₃ Schottky Barrier Diode

Abu Shahir Md Khalid Hasan
Department of Electrical Engineering
and Computer Science
University of Arkansas
Fayetteville, Arkansas, United States
ah162@uark.edu

Md Majharul Islam
Department of Electrical Engineering
and Computer Science
University of Arkansas
Fayetteville, Arkansas, United States
mi031@uark.edu

Mohammad Dehan Rahman
Department of Electrical Engineering
and Computer Science
University of Arkansas
Fayetteville, Arkansas, United States
mr117@uark.edu

Md Maksudul Hossain
Department of Electrical Engineering
and Computer Science
University of Arkansas
Fayetteville, Arkansas, United States
mh080@uark.edu

H. Alan Mantooth
Department of Electrical Engineering
and Computer Science
University of Arkansas
Fayetteville, Arkansas, United States
mantooth@uark.edu

Tanzila Akter
Department of Electrical Engineering
and Computer Science
University of Arkansas
Fayetteville, Arkansas, United States
tanzilaa@uark.edu

Xiaoqing Song
Department of Electrical Engineering
and Computer Science
University of Arkansas
Fayetteville, Arkansas, United States
songx@uark.edu

Abstract— **Gallium oxide is emerging as a leading ultra-wide bandgap material for future applications. Metals with large work functions have been used with β-Ga₂O₃ (the most stable polytype of gallium oxide) to form the Schottky barrier diode. In this work, a β-Ga₂O₃ Schottky barrier diode has been characterized using the Keysight B1505A curve tracer. The IV characteristics of the β-Ga₂O₃ die are measured with four probe measurements. A diode spice model for β-Ga₂O₃ Schottky barrier diode has been developed based on diffusion mechanism and charge control continuity equation. A series resistance is also included in the diode spice model. The model has shown good agreement on both linear and logarithmic scale. Transient simulation for half and full bridge rectifier circuit has been performed with the model.**

Keywords— *β-Ga₂O₃, Schottky barrier diode, continuity equation, spice model, full bridge rectifier*

I. INTRODUCTION

Wide bandgap (WBG) power devices are increasingly being used in applications such as transportation electrification, motor drives, energy storage, and renewable energy due to their faster switching speeds, high voltage operation and lower conduction resistance. To further enhance the performance of power electronics converters, including achieving higher efficiency and power density, ultra-wide bandgap (UWBG) power semiconductor devices have attracted growing interest from researchers. Among the most promising UWBG semiconductors is gallium oxide (Ga₂O₃), which offers exceptional intrinsic material properties [1]. These include a significantly larger bandgap energy—4.4 times greater than that of silicon, and approximately 1.5 times greater than that of silicon carbide (SiC) [2-3]. These properties enable gallium oxide to achieve higher breakdown voltages, lower conduction resistances, and, importantly, superior operational temperature capabilities in semiconductor devices. However, a significant drawback of this material is its difficulty in achieving p-type conductivity [4-5], which limits its use in bipolar devices

This work was supported by the U.S. National Science Foundation (NSF) Center on Grid Connected Advanced Power Electronic Systems (GRAPES) under Grant 1939144.

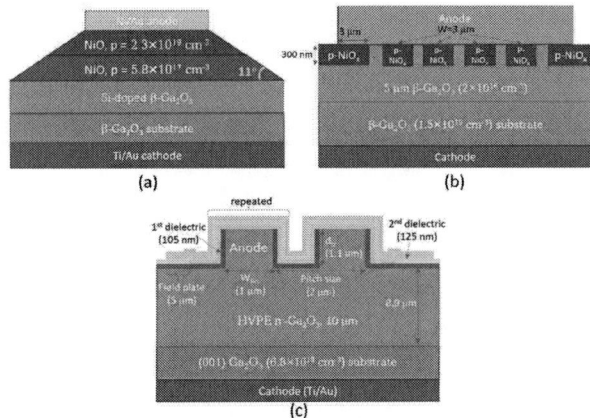

Fig. 1: Device structure of (a) Hetero-Junction Ga₂O₃ P-N diode [30], (b) Hetero-Junction Barrier Ga₂O₃ Schottky diode [32], and (c) Field-plated Ga₂O₃ Schottky diode [31]

such as p–n junctions and bipolar junction transistors (BJT) [6]. Developing unipolar devices using Ga₂O₃ is more straightforward, as it is not hindered by the aforementioned limitation.

Due to the difficulty of developing the p-type layer in Ga₂O₃ devices, nickel oxide based has been developed to perform as p-type layer. A group of researchers have reported to fabricate a P-N diode with NiO-Ga₂O₃ [as shown in Fig. 1(a)] with 1.9 mΩcm² on resistance, 20A current and 1.95kV breakdown voltage [7]. Ga₂O₃ has been used to develop Schottky barrier diodes (SBDs). For SBDs, high work-function metals like nickel and platinum are essential for forming a Schottky contact with β-Ga₂O₃. Metal deposition is typically carried out using electron-beam evaporation (EBE) [8]. Prior to the deposition of the metal for Schottky contact, the Ga₂O₃ surface is etched. However, this process can cause surface damage to the material due to the strong Ga₂O₃ bonds, potentially resulting in the degradation of the SBD. Cornell University has reported a trench structure [as depicted in Fig. 1(c)] for SBD with 10.5 mΩcm² on resistance and breakdown voltage of 2.89 kV [9]. Another group has reported a scattered p-type NiOₓ layer as shown in Fig. 1(b) to fabricate a Ga$_2$O$_3$ SBD with 1.94 mΩcm^2 on resistance and breakdown voltage of 1.34 kV [10].

Ga$_2$O$_3$ based SBDs are in an early phase of study with limited prototypes, still in the initial phase of development. Technology Computer-Aided Design (TCAD) is a highly cost-effective tool for exploring innovative ideas and structures in β-Ga$_2$O$_3$ devices, particularly given the current expense and developing phase of Ga$_2$O$_3$ technology. But TCAD based models are not suitable for power electronics circuits simulations. Although TCAD based model for β-Ga$_2$O$_3$ SBD [8] has been reported, no compact model has been presented for this device. In this work, a compact model of β-Ga$_2$O$_3$ SBD has been developed based on diffusion mechanism and charge control continuity equation. A β-

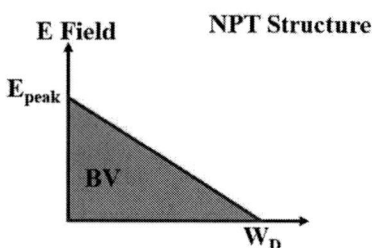

Fig. 2: Electric field strength with depletion layer width

Ga$_2$O$_3$ SBD has been characterized and fit with the diode model. Section II describes the advantages and limitations of gallium oxide for power device devices. Section III describes the model formulation for the β-Ga$_2$O$_3$ SBD while Section IV explains the characterization & model fit results. Section V gives the simulation time estimation for the β-Ga$_2$O$_3$ SBD model.

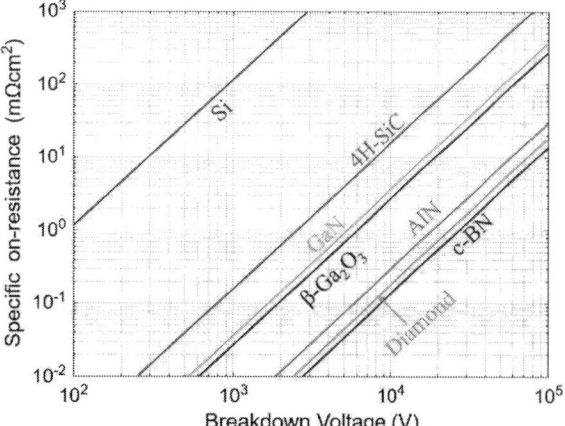

Fig. 3: Specific on resistance with breakdown voltage for different materials

II. ADVANTAGES & LIMITATIONS OF GALIIUM OXIDE

The primary advantage of Ga$_2$O$_3$ as a material for power devices lies in its high critical electric field. In vertical power devices, the drift layer sustains the blocking voltage and is typically lightly doped, with a thickness ranging from a few microns to several tens of microns. Under reverse bias in the commonly used non-punch-through (NPT) design, the relationship between the electric field and drift layer width is illustrated in Fig. 2.

From Fig. 3, it can be observed that Ga$_2$O$_3$ demonstrates a superior R$_{on,sp}$ ~ breakdown voltage relationship compared to SiC and GaN. Theoretically, for a given breakdown voltage, a Ga$_2$O$_3$ power device can achieve a lower R$_{on,sp}$ than those made from Si, SiC, or GaN. This advantage is primarily due to the high critical electric field of Ga$_2$O$_3$, positioning it as a competitive material choice for power devices.

Gallium oxide, with an ultra-wide bandgap of 4.5 to 4.8

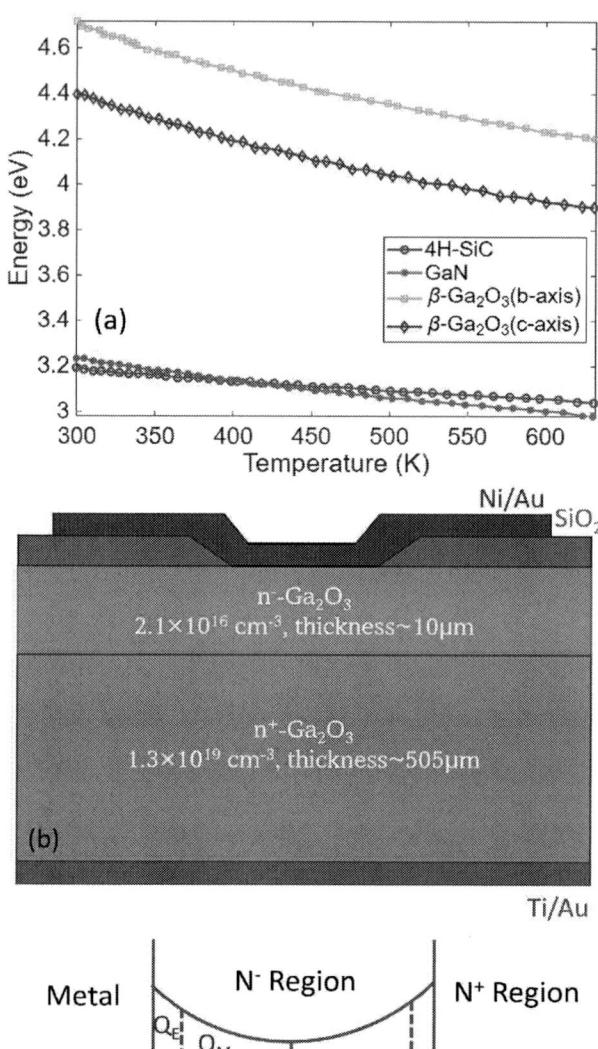

Fig. 4: (a) Bandgap energy with temperature for 4H-SiC, GaN, β-Ga$_2$O$_3$ (b-axis) and β-Ga$_2$O$_3$ (c-axis); (b) Cross-sectional device structure of the β-Ga$_2$O$_3$ Schottky Barrier Diode (c) charge storage location in the β-Ga$_2$O$_3$ SBD

eV, exhibits an extremely low intrinsic carrier concentration (n$_i$), approximately 10^{10} times lower than that of silicon, even at elevated temperatures. Additionally, Ga$_2$O$_3$ has a high melting point of 1800°C, and stable Schottky contacts for Ga$_2$O$_3$ devices, capable of operating at temperatures up to 500°C, have been successfully fabricated by numerous research groups [11, 12]. These properties make Ga$_2$O$_3$ highly suitable for high-temperature applications.

limitation of Ga_2O_3 for power devices is its low thermal conductivity, approximately 10–39 $Wm^{-1}K^{-1}$. This low thermal conductivity increases the junction-to-case thermal resistance in Ga_2O_3 power devices, particularly when conventional bottom-side cooling packages are used. As a result, efficient heat dissipation in Ga_2O_3 power devices may present a challenge.

III. MODEL DESCRIPTION

The bandgap comparison of 4H-SiC, GaN and β-Ga_2O_3 with temperature dependence has been depicted in Fig. 4(a). Although the bandgap energy drops down with temperature for both β-Ga_2O_3 (b-axis) and β-Ga_2O_3 (c-axis), it still has a better profile than 4H-SiC and GaN from 300K to 600K. The temperature dependent bandgap model [8] can be formulated as,

$$E_{gGaO} = \begin{cases} -1.26 \times 10^{-3}T_K + 5.10 & (b\ axis) \\ -1.40 \times 10^{-3}T_K + 4.90 & (c\ axis) \end{cases} \quad (1)$$

The device structure of large-area β-Ga_2O_3 SBD is illustrated in Fig. 4(b) where 2-inch n+ β-Ga_2O_3 substrate was used for fabrication. 10μm epitaxial layer of n- β-Ga_2O_3 (Si-doped~2.1×10^{16} cm^{-3}) were grown before forming the metal contacts. The diffusion mechanism, used by Lauritzen and Ma [9] in the simple diode model, has been used here to model the β-Ga_2O_3 diode current. The instantaneous diffusion current results from the redistribution of the charges between two storages Q_E and Q_M at the anode can be defined as

$$I_{GaO} = \frac{4ead_A(n_E - n_M)}{(w + \Delta)} \quad (2)$$

$$Q_E = ea\Delta(n_E - n_{io}) \quad (3)$$

$$Q_M = eaw(n_M - n_{io}) \quad (4)$$

where e, a and d_A are the electron charge, junction area and diffusion constant respectively. n_E and n_M are the average electron concentrations, while Δ and w are the widths corresponding to the two charge storages Q_E and Q_M, correspondingly and n_{io} is the equilibrium electron concentration. The charge storage locations have been depicted in Fig. 4(c). Symmetric charge has been assumed to simplify the analysis. Δ is designed to be much smaller than w to limit the current flow between the external lead and Q_E. Considering $\Delta \to 0$, the relation in (2) becomes,

$$I_{GaOt} = \frac{Q_0 - Q_M}{T_M} \quad (5)$$

where,

$$Q_0 = eaw(n_E - n_{io}) \quad (6)$$

$$T_M = \frac{w^2}{4d_A} \quad (7)$$

The continuity equation in case of Q_M can be represented by,

$$0 = \frac{\delta Q_M}{\delta t} + \frac{Q_M}{\tau} - \frac{Q_0 - Q_M}{2T_M} \quad (8)$$

where τ is the recombination lifetime. The mathematical formulation to relate the junction voltage V_d with Q_0 can be described as,

$$n_E - n_{io} = n_{io}\left[exp^{\frac{V_d - r_s I_{GaOt}}{2V_T}} - 1\right] \quad (9)$$

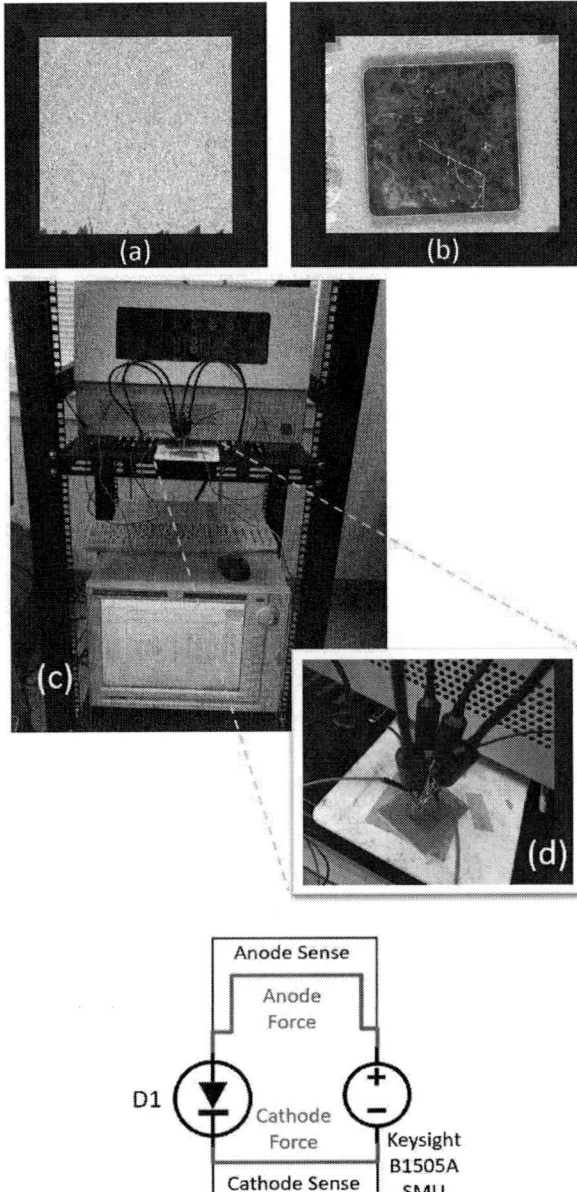

Figure 5: Bare die of β-Ga_2O_3 Schottky Barrier Diode (a) cathode side and (b) anode side. (c) Test setup for IV characterization with curve tracer, (d) zoomed-in version and (e) circuit diagram of the probing

The primary drawback for power device applications is the absence of p-type Ga_2O_3. Due to the relatively flat valence band in momentum space, Ga_2O_3 has a large effective hole mass, leading to the formation of localized polarons rather than free-moving holes [13,14]. Consequently, conducting p-type Ga_2O_3 is not feasible, limiting edge termination options and complicating the fabrication of certain devices such as junction barrier Schottky (JBS) diodes and metal-oxide-semiconductor field-effect transistors (MOSFETs). Another

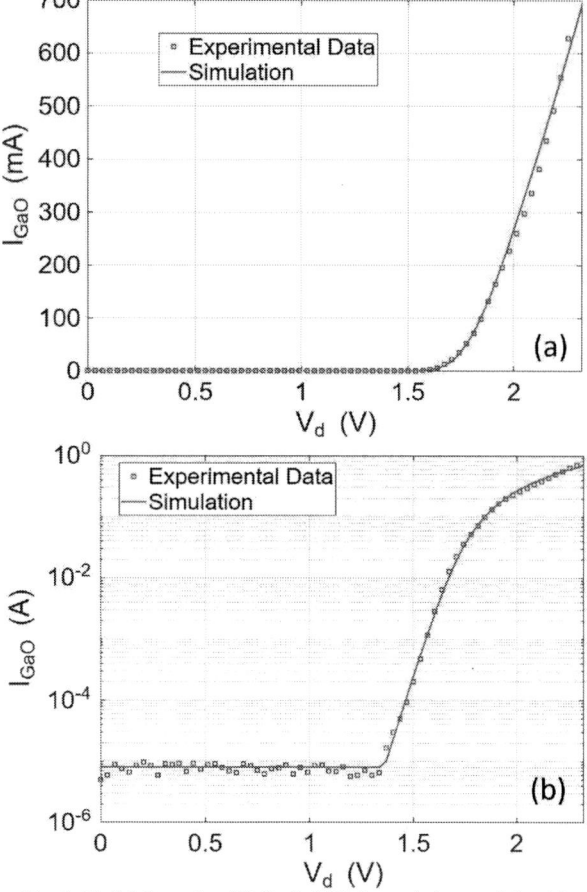

Fig. 6: Model fit results of β-Ga₂O₃ SBD forward characteristics (a) current in linear scale and (b) current in logarithmic scale at room temperature

where r_s is the temperature dependent series resistance. By multiplying either side of (9) by eaw to derive Q_0 with the following relationship

$$eaw(n_E - n_{io}) = Q_0 = \frac{i_S \tau}{2}\left[exp^{\frac{V_d - r_s I_{GaOt}}{2V_T}} - 1\right] \quad (10)$$

where,

$$i_S = \frac{2eawn_{io}}{\tau} \quad (11)$$

Using (7), (8) and (10), the forward current in steady-state condition can be formulated as,

$$I_{GaO} = \frac{i_S}{\left(1 + \frac{2T_M}{\tau}\right)}\left[exp^{\frac{V_d - r_s I_{GaOt}}{2V_T}} - 1\right] \quad (12)$$

IV. CHARACTERIZATION & MODEL FIT RESULTS

Fig. 5(a) and 5(b) shows the microscopic capture of the backside (cathode) and front side (anode) of the β-Ga₂O₃ SBD. The test setup for the IV measurements has been shown in Fig. 5(c) and 5(d). Two different probes (force and sense) have been used for both anode and cathode terminals to connect with the Keysight B1505A curve tracer. The circuit diagram of the test setup is depicted in Fig. 5(e). The diode was placed on a gold-plated PCB to probe simultaneously the anode and cathode. The current measurement was accurate

up to 10μA as shown in Fig. 6(b) and at $V_d \sim 1.33$V, the current seems to cross the 10μA mark.

The model is written in verilog-A and the parameters have been extracted using the IC-CAP software. The model fit

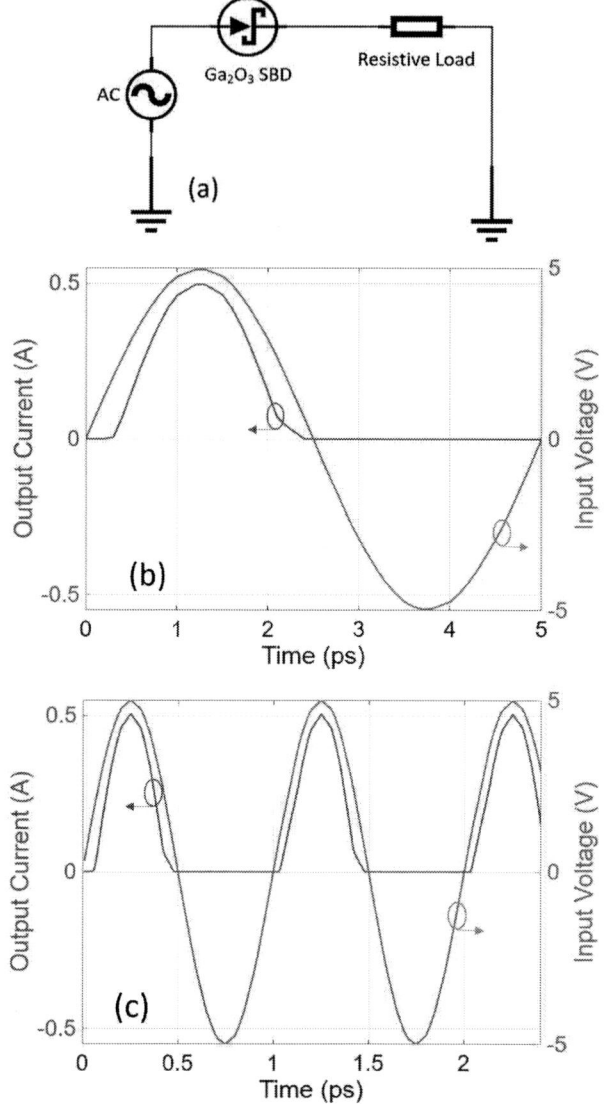

Fig. 7: (a) The circuit diagram of the rectifier circuit with one β-Ga₂O₃ SBD and resistive load, output current (peak~0.5 A) through the β-Ga₂O₃ SBD (b) at 200kHz and (c) 1 MHz

result for the β-Ga₂O₃ SBD model has been illustrated in Fig. 6(a) and 6(b). The measured data are plotted as blue squares while solid orange lines are used to plot the model simulations. The model has shown good accuracy not only on the linear current scale but also on the logarithmic current scale.

The model has been used to perform transient simulations using the HSPICE simulator. At first, a rectifier with a single β-Ga₂O₃ SBD circuit (shown in Fig. 7(a)) has been simulated with resistive load and the output current for the 200 kHz 5V sinusoidal input voltage has been plotted in Fig. 7(b). The maximum output current of circuit is 0.497A with 5Ω

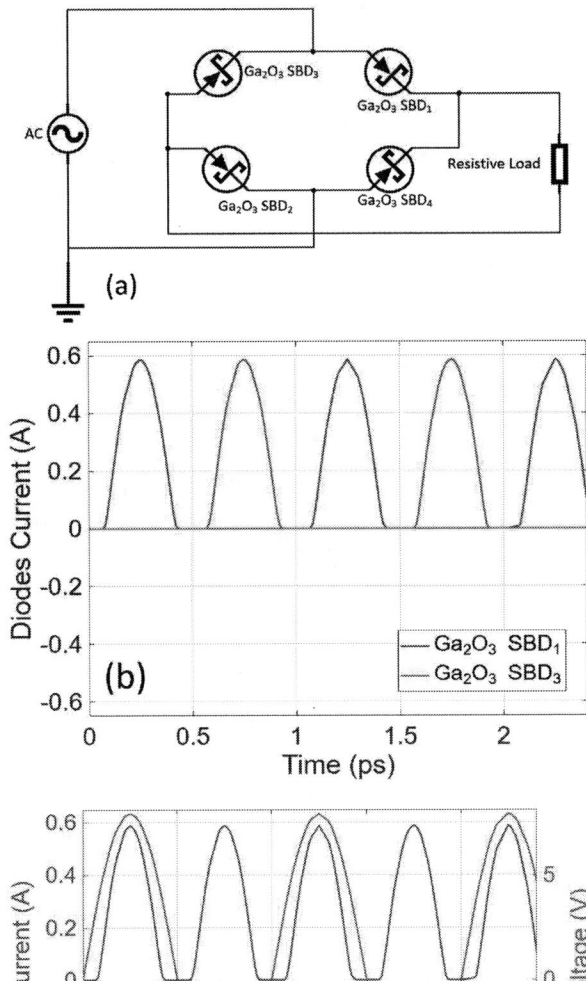

(a)

(b)

(c)

Fig. 8: (a) The circuit diagram of the full bridge rectifier circuit with four β-Ga$_2$O$_3$ SBD and resistive load, (b) the current through SBD$_1$ and SBD$_3$ (c) output current (peak~0.6 A) through the resistive load at 1 MHz

resistive load. The circuit has been also tested up to 1 MHz and the transient results have been demonstrated in Fig. 7(c).

Another full bridge rectifier circuit as depicted in Fig. 8(a) with four β-Ga$_2$O$_3$ SBDs has been simulated with a resistive load. The currents SBD$_1$ and SBD$_3$ have been plotted in Fig. 8(b). The two diodes are complementing each other in the alternate half cycle of the ac signal and both of them the peak value 0.587 A with 5Ω resistive load and 1 MHz 8V sinusoidal input voltage. The output current through the resister has been shown in Fig. 8(c).

V. SIMULATIONS TIME ANALYSIS

The simulation of the circuits shown in Fig. 7 and Fig. 8 has been done in Intel(R) Xeon(R) CPU E5-2650 at 2.30 GHz.

0.01 µs step time has been used for both cases with maximum simulation time of 2.4 µs. The simulation were in linux64 operating system with HSPICE Q-2020.03-3 and Newton-Raphson method has been utilized for the DC convergence. The relevant information to simulation time in the two circuits have been presented in Table I. In all cases the peak memory used is 410.4 megabytes. For single β-Ga$_2$O$_3$ SBD rectifier the total CPU time was 0.11 seconds for 200kHz and 0.12 seconds for 1MHz frequencies for the same number of data points. In case of the full bridge rectifier circuit, four β-Ga$_2$O$_3$ SBDs have been utilized and the total CPU time was 0.23 seconds. So although the number of β-Ga$_2$O$_3$ SBDs increases four times, the simulation time has creased only twice which proves the model is good for multi-diode simulations.

TABLE I. SIMULATION INFORMATION

Circuit	Component Description			Peak Memory used (mega-bytes)	Simulation Time (Total CPU Time)
	Number of β-Ga$_2$O$_3$ SBDs	Current Level	Freq.		
Basic Rectifier Ciruit	1	0.497A	200 kHz	410.14	0.11 seconds
			1 MHz	410.14	0.12 seconds
Full Bridge Rectifier	4	0.587A	1 MHz	410.14	0.23 seconds

VI. CONCLUSIONS AND FUTURE WORK

The forward IV characteristics of β-Ga$_2$O$_3$ SBD has been successfully characterized up to 700 mA at room temperature. The diode model has shown good agreement with the experimental data for both linear and logarithmic current scale up to 10 µA. This spice model can be used for different circuit simulations. Here, the model has been simulated in basic rectifier circuit in two different frequencies (200 kHz and 1 MHz). A full bridge rectifier consists of four SBDs has been also worked without any convergence issues. The total CPU time for the full bridge rectifier circuit ensures that the model can be comfortably used for circuits with multiple diodes. The temperature scaling and CV characteristics modeling would be presented in the future work.

ACKNOWLEDGMENT

The opinions, findings, conclusions, or recommendations exhibited in this work are solely those of the author(s) and do not essentially act for the views of the National Science Foundation. The author(s) would like to express sincere gratitude to GRAPES for providing the opportunity to collaborate with other researchers.

REFERENCES

[1] O. Slobodyanl, J. Flicker, J. Dickerson, A. Binder, T. Smith, R. Kaplat, and M. Hollis, "Analysis of the dependence of critical electric field on semiconductor bandgap", *Journal of Materials Research*, 2022, https://doi.org/10.1557/s43578-021-00465-2.

[2] M. D. Rahman and X. Song, "Investigation of the Impact of Low Thermal Conductivity on Gallium Oxide Power Module Packaging," *2023 IEEE 10th Workshop on Wide Bandgap Power Devices & Applications (WiPDA)*, Charlotte, NC, USA, 2023, pp. 1-5, doi: 10.1109/WiPDA58524.2023.10382165.

[3] A. S. M. K. Hasan, M. M. Hossain, P. C. Heris and H. A. Mantooth, "Single Event Upset in Depletion-Mode Gallium Oxide MOSFETs at

the Breakdown Region," *2024 IEEE Applied Power Electronics Conference and Exposition (APEC)*, Long Beach, CA, USA, 2024, pp. 2461-2467, doi: 10.1109/APEC48139.2024.10509030.

[4] E. Chikoidze, et al. "Enhancing the intrinsic p-type conductivity of the ultra-wide bandgap Ga_2O_3 semiconductor," *Journal of Materials Chemistry C*, vol. 7, no. 33, pp. 10231–10239, 2019.

[5] A. Kyrtsos, M. Matsubara and E. Bellotti "On the feasibility of p-type Ga_2O_3," *Applied Physics Letters*, vol. 112, no. 3, pp. 032108, Jan. 2018, https://doi.org/10.1063/1.5009423

[6] H. Xue, Q. He, G. Jian, S. Long, T. Pang and M. Liu "An overview of the ultrawide bandgap Ga_2O_3 semiconductor-based Schottky barrier diode for power electronics application" *Nanoscale research letters*, vol. 13, no. 290, Sep. 2018.

[7] H. Gong *et al.*, "1.37 kV/12 A NiO/β-Ga2O3 Heterojunction Diode With Nanosecond Reverse Recovery and Rugged Surge-Current Capability," in *IEEE Transactions on Power Electronics*, vol. 36, no. 11, pp. 12213-12217, Nov. 2021, doi: 10.1109/TPEL.2021.3082640.

[8] M. Labed, et al. "Modeling a Ni/β-Ga$_2$O$_3$ Schottky barrier diode deposited by confined magnetic-field-based sputtering." *Journal of Physics D: Applied Physics*, vol. 54, no. 11, pp. 115102, Jan. 2021. doi: 10.1088/1361-6463/abce2c

[9] W. Li, K. Nomoto, Z. Hu, D. Jena and H. G. Xing, "Field-Plated Ga2O3 Trench Schottky Barrier Diodes With a BV2/ Ron,sp of up to 0.95 GW/cm2," in *IEEE Electron Device Letters*, vol. 41, no. 1, pp. 107-110, Jan. 2020, doi: 10.1109/LED.2019.2953559.

[10] H. H. Gong, X. X. Yu, Y. Xu, X. H. Chen, Y. Kuang, Y. J. Lv, Y. Yang, F.-F. Ren, Z. H. Feng, S. L. Gu, Y. D. Zheng, R. Zhang, and J. D. Ye, "β-Ga$_2$O$_3$ vertical heterojunction barrier Schottky diodes terminated with p-NiO field limiting rings," *Applied Physics Letters*, vol. 118, no. 20, p. 202102, May 2021 https://doi.org/10.1063/5.0050919

[11] S. Oh, G. Yang, and J. Kim, "Electrical characteristics of vertical Ni/β-Ga2O3 Schottky barrier diodes at high temperatures," *ECS Journal of Solid State Science and Technology*, vol. 6, no. 2, p. Q3022, Sep. 2016. doi: 10.1149/2.0041702jss

[12] B. Wang, M. Xiao, X. Yan, H. Y. Wong, J. Ma, K. Sasaki, H. Wang, and Y. Zhang, "High-voltage vertical Ga2O3 power rectifiers operational at high temperatures up to 600 K," *Applied Physics Letters*, vol. 115, no. 26, p. 263503, Dec. 2019. doi: 10.1063/1.5132818.

[13] H. Von Wenckstern, "Group-III sesquioxides: growth, physical properties and devices," *Advanced Electronic Materials*, vol. 3, no. 9, p. 1600350, 2017.

[14] S. J. Pearton, J. Yang, P. H. Cary, F. Ren, J. Kim, Marko J. Tadjer, and M. A. Mastro, "A review of Ga2O3 materials, processing, and devices," *Applied Physics Reviews*, vol. 5, no. 1, p. 011301, Mar. 2018. https://doi.org/10.1063/1.5006941

[15] L. Cheng, J. Y. Yang, and W. Zheng, "Bandgap, Mobility, Dielectric Constant, and Baliga's Figure of Merit of 4H-SiC, GaN, and β-Ga$_2$O$_3$ from 300 to 620 K" in *ACS Applied Electronic Materials*, vol. 4, no. 8, pp. 4140-4145, Aug. 2022, doi:10.1021/acsaelm.2c00766.

[16] P. O. Lauritzen and C. L. Ma, "A simple diode model with reverse recovery," in *IEEE Transactions on Power Electronics*, vol. 6, no. 2, pp. 188-191, April 1991, doi: 10.1109/63.76804.

DC-Link Capacitor Board Design for Low Parasitic Inductance

Mikayla Benson
Department of Electrical and Computer Engineering
Michigan State University
East Lansing, MI, USA
benson80@msu.edu

Lifang Yi
Department of Electrical and Computer Engineering
Florida State University
Tallahassee, FL, USA
lyi@fsu.edu

Kangbeen Lee
Elmore Family School of Electrical and Computer Engineering
Purdue University
West Lafayette, IN, USA
lee4097@purdue.edu

Jinyeong Moon
Department of Electrical and Computer Engineering
Florida State University
Tallahassee, FL, USA
j.moon@fsu.edu

Woongkul Lee
Elmore Family School of Electrical and Computer Engineering
Purdue University
West Lafayette, IN, USA
wklee@purdue.edu

Abstract— DC-link capacitors absorb ripple current and provide stable DC voltage in power converter applications. As switching speed and frequency increase, the parasitic inductance of the DC-link capacitor causes transient overvoltage, which can lead to premature failures of the semiconductor and capacitor. Therefore, the parasitic inductance of the DC-link board should be minimized to increase the resonance frequency above the switching frequency. This paper explores the effects of parasitic inductance on the capacitor voltage and explores three different DC-link board layouts to determine the best configuration for decreased parasitic inductance. The board designs explored the presence of a bottom layer of copper and the alignment of adjacent capacitors. FEM simulation was used with varying numbers of parallel capacitors. It was found that increasing the number of capacitors will cause an initial decrease in parasitic inductance, then begin increasing. With an increased number of capacitors in parallel, the presence of a copper bottom layer with direct parallel capacitors allowed for a 50% decrease in parasitic inductance, and interleaving parallel capacitors offered an overall 93% decrease in parasitic inductance. Experimental results were conducted to validate that the interleaving of parallel capacitors allows for additional reduction in parasitic inductance, which showed a 59% decrease in parasitic inductance from direct parallel oriented capacitors and a 77% increase in resonant frequency.

Keywords—Capacitors, DC-link capacitor, Parasitic Inductance, PCB Design

I. INTRODUCTION

The primary function of DC-link capacitors is to absorb ripple current and stabilize the DC bus voltage, effectively smoothing out fluctuations and maintaining a more consistent voltage level for power electronics. The recent development and adoption of wide-bandgap (WBG) semiconductors, such as SiC and GaN, have enabled much higher switching speeds, allowing for greater efficiency and power density in converters [1]-[2]. These higher speeds, however, demand stricter control over circuit parasitics, as even minor inductances in the layout can result in substantial voltage spikes and ringing. With advances in semiconductor packaging, components are now smaller and

designed to reduce internal parasitic inductances. However, as these packaging constraints tighten, the external connections to commutation loops become increasingly critical in minimizing overall parasitics [2]-[4]. Increased inductance in these circuits can lead to transient overvoltage conditions, which may exceed the breakdown voltage of the semiconductors, causing potential reliability issues and even damage due to oscillatory switching behavior and electromagnetic interference (EMI). This oscillating behavior, often caused by interactions between the WBG devices and the surrounding parasitic elements, can degrade switching performance, reduce efficiency, and shorten the lifespan of the semiconductors if not adequately mitigated [1].

Many studies exist to reduce parasitic inductance through PCB layout optimization for power converter topologies. Extensive research has focused on reducing parasitics in power loops for WBG semiconductors by minimizing loop areas and optimizing current paths, as investigated in [2]-[4]. Techniques such as multilayer PCB structures, via placement strategies, and the use of ground planes have all shown promise in lowering parasitic effects, as seen in [7]. The connection of DC-link capacitors through meticulously designed busbars has also been a significant area of study. Research in [5]-[6] has demonstrated that busbars with minimized loop inductance, optimized conductor geometry, and careful layer stacking can substantially lower inductance and improve thermal management in high-frequency switching applications. Moreover, there has been a detailed exploration of the types, sizes, and arrangements of capacitors to reduce overall parasitic inductance in power converters. Studies like [7]-[8] have analyzed the trade-offs between different capacitor technologies, such as film and ceramic. Additionally, hybrid DC-link capacitor boards have been designed, utilizing multiple types of capacitors in [12]-[13]. However, the specific orientation and placement of the DC-link capacitors themselves, factors that can critically influence the impedance of the circuit, have not been discussed in depth. This is a gap in the literature, as the capacitor orientation could affect coupling effects, especially in compact, high-density power designs. Addressing these orientation

979-8-3315-1612-3/25 $31.00 © 2025 IEEE

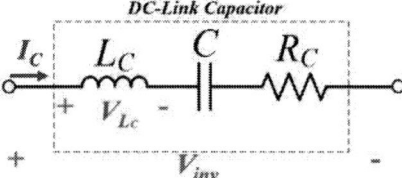

Fig. 2. Equivalent circuit of DC-link capacitors.

strategies could further advance the field by reducing parasitic inductance and enhancing the overall efficiency of WBG-based power converters. Therefore, this paper aims to reduce the parasitic inductance of the overall DC-link capacitor board by improving the layout and configuration of parallel capacitors on PCB boards.

This study will first discuss the fundamental effects of parasitic inductance on capacitor voltage and explore the interaction of the capacitor parasitics impedance of a DC-link capacitor when paralleled with a DC source. Utilizing PCB design, this paper aims to reduce the overall impedance of the DC-link capacitor by exploring the layout and configuration of paralleled capacitors on PCBs. This paper will also explore the effects of a copper ground layer and the direction of current flow through adjacent capacitors. FEM simulation will be used to approximate the total parasitic inductances for each board design with varying capacitor values. The DC-link capacitor boards will be constructed to experimentally validate the orientation of capacitors and their effect on parasitic inductances.

II. DISCUSSION OF PARASITICS IN DC-LINK CAPACITORS

A. Effects of Parastics on DC-Link Capacitors

The parasitics in a DC-link capacitor affect the operation of the capacitor in power electronic applications. At low frequencies, capacitors are purely capacitive and resistive. However, as switching frequencies are increased, the parasitic components of the capacitor become increasingly important. The parasitics of capacitors can be modeled as a series circuit containing a resistive, inductive, and capacitive component, as shown in Fig. 2. The series impedance of the DC-link capacitor can be described in (1).

$$Z_C(f) = R_C + \frac{1}{j2\pi f C} + j2\pi f L_C \qquad (1)$$

where Z_C is the impedance of the capacitor, R_C is the parasitic resistance, C is the capacitance, L_C is the parasitic inductance, and f is the frequency. This equation shows that as the frequency increases, the impedance from the capacitance decreases, and the impedance contribution of the parasitic inductance increases while the parasitic resistance remains constant. Therefore, it is crucial to consider the effects of the parasitic inductance on the performance of the capacitor.

A fast-switching current was applied to the equivalent circuit of the capacitor to determine the effects of the parasitic inductance magnitude. The capacitance and resistance were kept constant at 200 µF and 5 mΩ, respectively, while the inductance was changed, with 0.2 nH, 2 nH, and 20 nH to show the effects of increased inductance. The capacitor current and voltage of the

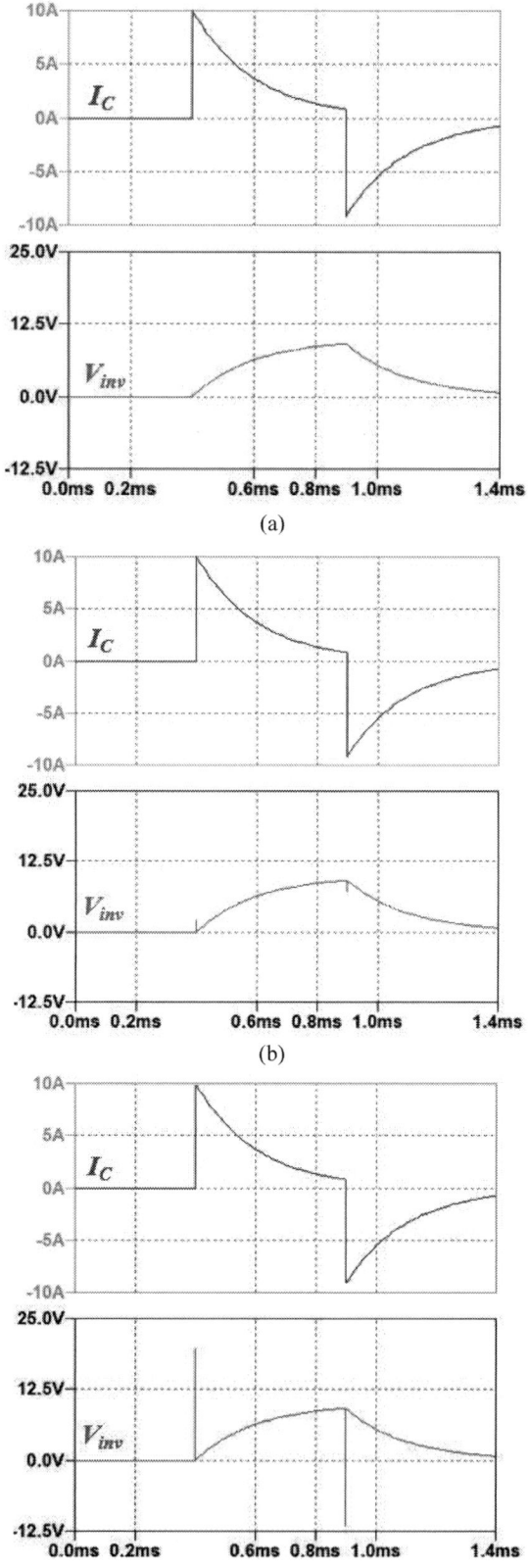

Fig. 1. Capacitor current and voltage for a single pulse with (a) 0.2 nH, (b) 2 nH, and (c) 20 nH of parasitic inductance.

979-8-3315-1612-3/25 $31.00 © 2025 IEEE   2414

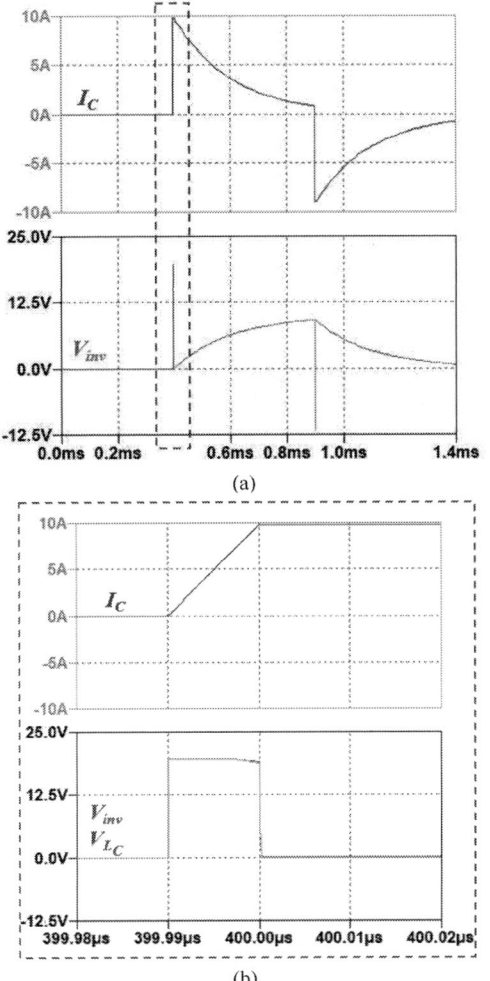

(a)

(b)

Fig. 3. Capacitor current and voltage response compared to inductor voltage during the voltage spike with high parasitic inductance (20 nH) for (a) a single pulse and (b) a rising edge.

equivalent circuit were measured for a single pulse in Fig. 1. In Fig. 1(a), with 0.2 nH of parasitic impedance, there is no voltage overshoot. In contrast, in Fig. 1(b) and Fig. 1(c), with 2 nH and 20 nH of parasitic impedance, respectively, a voltage overshoot appears as a result of the spike in current, with the magnitude of the voltage spike increasing with parasitic impedance. The voltage spike from the 20 nH parasitic impedance test was analyzed in Fig. 3, with the inductor voltage overlayed on the circuit voltage. The capacitor current shows a rapid rise time of the current at 10 ns, and the inductor voltage matches perfectly to the total capacitor voltage, confirming this spike is directly related to the parasitic inductance. Therefore, in fast-switching applications, the parasitic inductance of DC-link capacitors should be minimized to reduce voltage overshoot.

B. Impedance Characteristics of DC-Link Capacitors

DC-link capacitors are placed parallel to a DC source in power electronic applications. The DC source adds additional parasitics, with a series of resistive and inductive structures, which can be modeled as shown in (2).

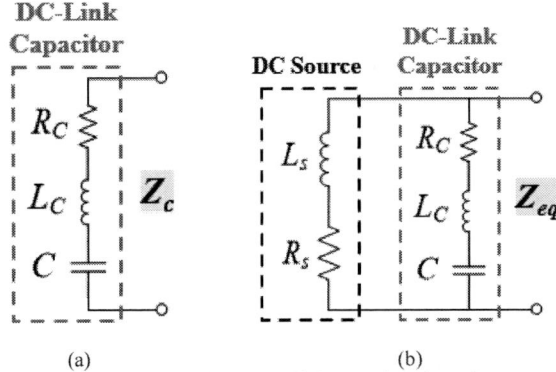

(a)   (b)

Fig. 4. Equivalent circuit of (a) DC-link capacitor impedance, and (b) combined dc-link and DC source equivalent impedance.

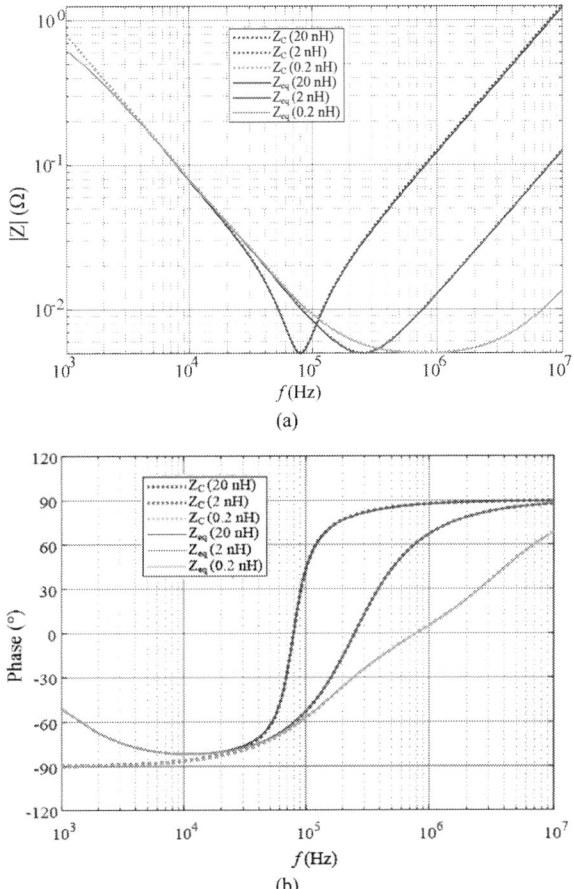

(a)

(b)

Fig. 5. Impedance measurement for (a) impedance magnitude and (b) phase angle of the DC-link capacitor and DC source equivalent circuit for parasitic inductances of 0.2 nH, 2 nH, and 20 nH.

$$Z_S(f) = R_S + j2\pi f L_S \qquad (2)$$

where Z_S is the impedance of the source, R_S is the parasitic resistance, L_S is the parasitic inductance, and f is the frequency. The impedance can be calculated through the combination of (1) and (2) to determine the equivalent impedance or the paralleled DC source and capacitor, as shown in (3).

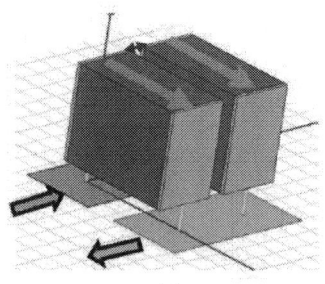 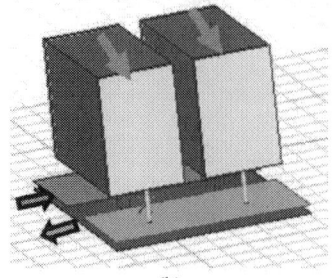 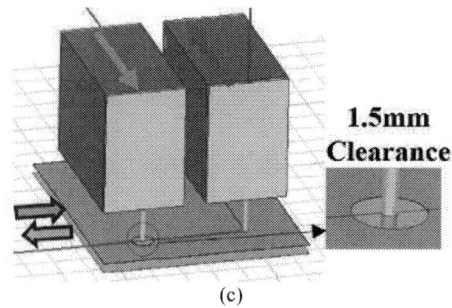

| | (a) | (b) | (c) |

Fig. 6. DC-Link board design with (a) direct parallel A, (b) direct parallel B, and (c) interleaving parallel.

TABLE I. TOTAL INDUCTANCE FOR EACH DC-LINK CAPACITOR BOARD LAYOUT WITH VARYING CAPACITANCE.

No. of Capacitors	1	2	4	8	10	20
Total Capacitance	5 µF	10 µF	20 µF	40 µF	50 µF	100 µF
Direct parallel (A)	23.30 nH	18.80 nH	21.81 nH	35.59 nH	43.54 nH	85.10 nH
Direct parallel (B)	23.26 nH	15.55 nH	13.32 nH	17.44 nH	20.62 nH	38.80 nH
Interleaving parallel	22.26 nH	9.78 nH	5.32 nH	4.09 nH	4.17 nH	5.92 nH

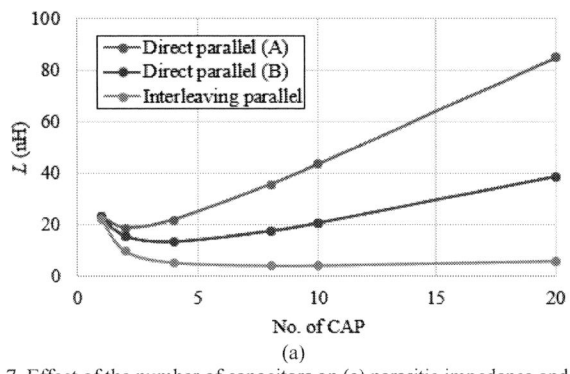 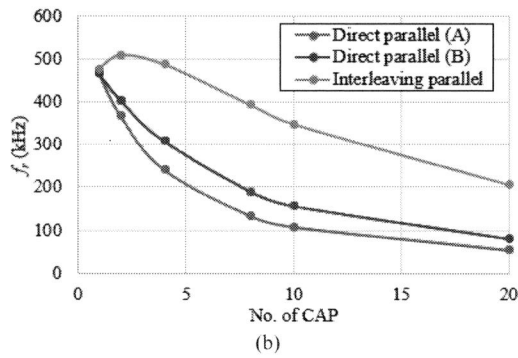

| | (a) | (b) |

Fig. 7. Effect of the number of capacitors on (a) parasitic impedance and (b) resonant frequency for different DC-link capacitor board layouts.

$$Z_{eq}(f) = (Z_S(f)^{-1} + Z_C(f)^{-1})^{-1} =$$
$$\frac{R_C R_S + j2\pi f(L_S R_C + L_C R_S) + \frac{R_S}{j2\pi f C} + \frac{L_S}{C} - (2\pi f)^2 L_C L_S}{R_C + R_S + \frac{1}{j2\pi f C} + j2\pi f(L_S + L_C)} \quad (3)$$

The impedance characteristics of the DC-link capacitor and the combined DC-link and DC source circuit can be analyzed, and their equivalent circuits are shown in Fig. 4. The source is modeled with a resistance of 1 Ω and an inductance of 1 µH. The impedance measurements are shown in Fig. 5 for the capacitor only impedance measurement and the capacitor in parallel with the DC source, with capacitor parasitic inductances varying. The results of the capacitor only and paralleled circuit align closely for each parasitic inductance test point, with the source impedance only slightly affecting the overall impedance measurement in the low frequency range. Therefore, the effect of the source impedance on the DC-link impedance is negligible at fast switching frequencies.

III. EXPLORATION OF BOARD LAYOUT FOR REDUCED PARASITICS IN DC-LINK CAPACITORS

A. Design of DC-Link Capacitor Boards

DC-link capacitor boards typically consist of numerous parallel capacitors to achieve the desired overall capacitance. The configuration of the individual capacitors and copper connections with the PCB can be designed to decrease the overall parasitic inductance of the DC-link capacitor board. To explore the effects of the dc-link capacitor board layout, three variations were created, as shown in Fig. 6. In Fig. 6(a)-(b), the capacitors are aligned in direct parallel, meaning the positive and negative terminals of each capacitor are aligned, causing the current flow through the capacitors to occur in the same direction. Fig. 6(a) has a single power pour for the positive and negative terminals with no bottom layer of copper, whereas Fig. 6(b) has a bottom copper layer. In Fig. 6(c), the capacitors are aligned in interleaving parallel, meaning the positive and negative terminal locations are alternated, causing opposing current flow direction in adjacent capacitors, with the positive and negative copper located on different layers of the PCB and cover the entire board.

B. Estimation of DC-Link Board Impedance Through FEM Simulation

FEM simulation was used to calculate the parasitic inductance of each board design with varying numbers of capacitors to determine the effects of board design on the overall parasitic inductance. The capacitor used in the simulation was KEMET C4AQIBU4500M1YJ, with a nominal capacitance of 5 µF. A single row of capacitors was tested at increasing numbers of capacitors. The overall inductance and resonant frequency results are shown in Fig. 7, with the total parasitic

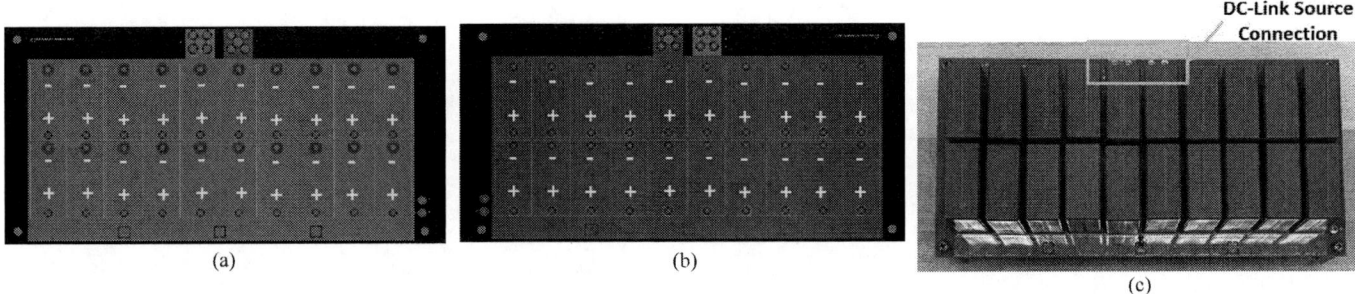

Fig. 8. DC-Link board design of direct parallel for (a) top layout, (b) bottom layout, and (c) assembled board.

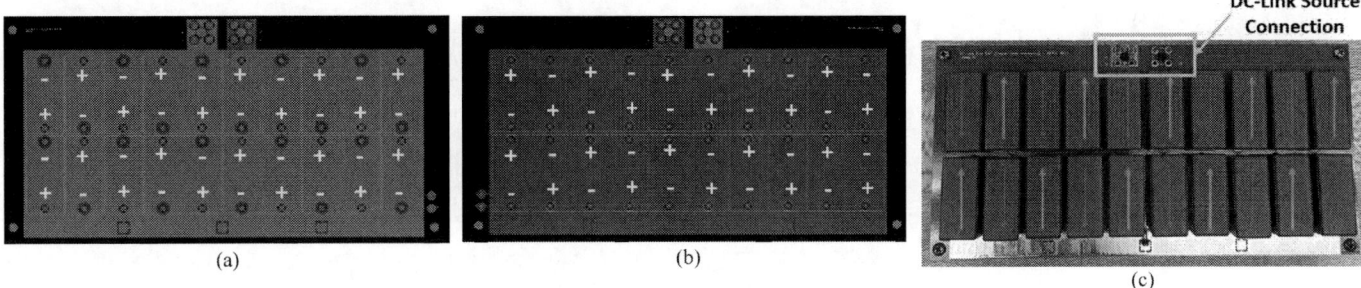

Fig. 9. DC-Link board design of interleaving parallel for (a) top layout, (b) bottom layout, and (c) assembled board.

inductance for each case summarized in Table I. Fig. 7(a) shows that as the number of capacitors increases, the total inductance initially decreases but then increases. For the direct parallel A layout, the parasitic inductance has a maximum of 19% reduction from the board with a single capacitor when the board consists of two parallel capacitors. However, the 100 μF board increases the parasitic capacitance up to 265% of the parasitic capacitance of a single capacitor. The direct parallel B layout has a peak reduction of 43% with four parallel capacitors from the parasitics of a single capacitor. Similar to the direct parallel A board, increasing the number of capacitors brings the 100 μF board to a 67% increase in overall parasitic inductance. In the interleaved parallel board design, the maximum reduction in parasitic inductances reached eight parallel capacitors with an 82% decrease. However, with this board design, the increase in parasitic inductance with additional capacitors is much slower, resulting in a 73% decrease in the 100 μF board from the single capacitor case. Overall, the direct parallel A layout has the highest inductance, whereas the direct parallel B layout shows that a copper bottom layer reduces the parasitic inductance. Furthermore, the interleaving parallel layout decreases the inductance further through the alternating current between adjacent capacitors.

The resonant frequency in Fig. 7(b) shows that the frequency decreases with the increase in the number of capacitors. In all boards, the resonant frequency of one capacitor is approximately 475 kHz. With the 100 μF board design, the direct parallel A board has the largest reduction of approximately 87%, followed by the direct parallel B board with a reduction of 81%. The interleaved parallel board has over two times higher resonant frequency than the direct parallel boards, with only a 58% reduction. Overall, each board design shows a decrease in resonant frequency, with the interleaving parallel achieving higher resonant frequencies than the direct parallel boards. Therefore, the interleaving parallel layout is able to successfully

TABLE II. MEASURED INDUCTANCE AND RESONANT FREQUENCY FOR EACH DC-LINK CAPACITOR BOARD.

Board Type	Inductance, Z (mΩ)	Resonant Frequency, f_r (kHz)
Direct Parallel B	5.283	111.11
Interleaving Parallel	2.152	197.42

reduce the inductance of the DC-link capacitor board and increase the resonant frequency compared to the direct parallel board designs.

IV. EXPERIMENTAL VALIDATION OF BOARD IMPEDANCE

DC-link boards were constructed and measured to confirm the simulation results. A direct parallel and an interleaving parallel boards, as shown in Fig. 8 and Fig. 9, respectively, were created with 20 capacitors. Both boards were designed with full layer copper pours, with the positive leg poured on the top layer (Fig. 8(a) and Fig. 9(a)) and the negative leg poured on the bottom layer (Fig. 8(b) and Fig. 9(b)). The boards have identical dimensions, capacitor spacing, and dc-link source connectors. The variation in the board design results from the configuration of capacitors, with the capacitors in the direct parallel B board all oriented with current flowing in the same direction, as shown in Fig. 9(c), and the capacitors in the interleaving parallel board oriented with current flowing in opposing directions, as seen in Fig. 9(c).

The boards were measured with an impedance analyzer, and the impedance versus frequency graphs are shown in Fig. 10. The minimum impedance value and the corresponding resonant frequency for each board type are shown in Table II. Compared to the direct parallel B board layout, the interleaving parallel layout achieves a 59% reduction in inductance and a 77% increase in resonant frequency. These results confirm that with identical board structure and connections, the interleaving parallel capacitor board layout can significantly reduce

979-8-3315-1612-3/25 $31.00 © 2025 IEEE

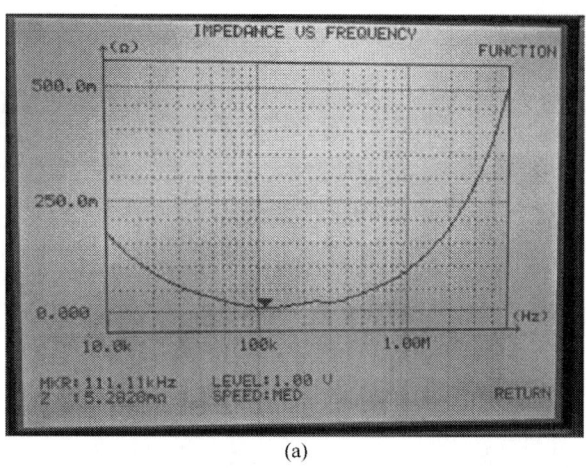 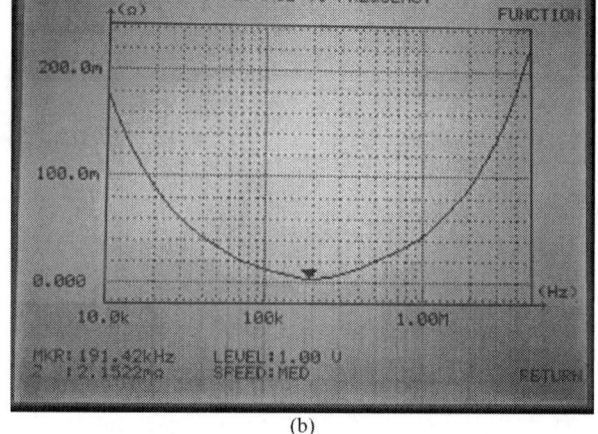

(a)   (b)

Fig. 10. Impedance measurement of DC-link board designs with (a) direct parallel B and (b) interleaving parallel.

inductance and increase resonant frequencies compared to a direct parallel layout.

V. CONCLUSIONS

DC-link capacitors are essential for absorbing ripple currents and maintaining stable DC voltage in power converters. However, as switching speeds increase, the parasitic inductance of these capacitors can lead to transient overvoltage, risking semiconductor damage. To address this, minimizing parasitic inductance in the DC-link board is crucial. This paper investigates how parasitic inductance impacts capacitor voltage. Three different DC-link board layouts were designed to identify the optimal configuration for reduced inductance. The designs tested include variations with a copper bottom layer and different alignments of adjacent capacitors. A simulation was conducted with multiple values of paralleled capacitors. Results showed that initially, adding more capacitors lowers parasitic inductance, but beyond a certain point, it begins to rise again. Using a copper bottom layer with directly parallel capacitors, direct parallel B, achieved a 50% reduction in parasitic inductance while interleaving parallel capacitors led to an overall 93% reduction from the baseline design, direct parallel A, with no bottom copper layer. PCB boards with direct parallel and interleaved parallel capacitors were designed with identical dimensions, copper pours, capacitor spacing, and DC source connections to determine the effect of capacitor orientation. Experimental results confirmed that interleaving parallel capacitors could further reduce parasitic inductance, achieving a 59% reduction compared to directly oriented capacitors and a 77% increase in resonant frequency.

REFERENCES

[1] H. Wang and F. Blaabjerg, "Reliability of capacitors for DC-link Appl. in power electronic converters—an overview," *IEEE Trans. on Ind. Appl.*, vol. 50, no. 5, pp. 3569–3578, Sep. 2014.

[2] H. Wang, H. Wang, G. Zhu, and F. Blaabjerg, "An Overview of Capacitive DC-Links-Topology Derivation and Scalability Analysis," *IEEE Trans. Power Electron.*, vol. 35, no. 2, pp. 1805–1829, Feb. 2020.

[3] D. Reusch and J. Strydom, "Understanding the effect of PCB layout on circuit performance in a high-frequency gallium-nitride-based point of load converter," *IEEE Trans. on Power Electron.*, vol. 29, no. 4, pp. 2008–2015, Apr. 2014.

[4] T. Ibuchi and T. Funaki, "A study on parasitic inductance reduction design in GaN-based power converter for high-frequency switching operation," in *2017 International Symposium on Electromagnetic Compatibility - EMC EUROPE*, Sep. 2017, pp. 1–5.

[5] W. Kangping, M. Huan, L. Hongchang, G. Yixuan, Y. Xu, Z. Xiangjun, and Y. Xiaoling, "An optimized layout with low parasitic inductances for GaN HEMTs based DC-DC converter," in *2015 IEEE Appl. Power Electron. Conf. and Expo. (APEC)*, Mar. 2015, pp. 948–951.

[6] K. Wang, L. Wang, X. Yang, X. Zeng, W. Chen, and H. Li, "A multiloop method for minimization of parasitic inductance in GaN-based high-frequency DC–DC converter," *IEEE Trans. on Power Electron.*, vol. 32, no. 6, pp. 4728–4740, Jun. 2017.

[7] M. Kumar, R. K. Yakala, S. Pramanick, and B. K. Panigrahi, "Improved Interconnect Layout of DC Link Capacitor Bank to Minimize Parasitic Inductance and its Effect on Performance of SiC MOSFET," in *2020 IEEE International Conference on Power Electronics, Smart Grid and Renewable Energy (PESGRE2020)*, Jan. 2020, pp. 1–6.

[8] R. Alizadeh, M. Schupbach, T. Adamson, J. C. Balda, Y, Zhao, and S. Long, "Busbar design for distributed DC-link capacitor banks for traction applications," in *2018 IEEE Energy Convers. Congress and Expo. (ECCE)*, Sep. 2018, pp. 4810–4815.

[9] M. Kumar, R. K. Yakala, S. Pramanick, and B. K. Panigrahi, "Improved interconnect layout of DC link capacitor bank to minimize parasitic inductance and its effect on performance of SiC MOSFET," in *2020 IEEE Int. Conf. on Power Electron., Smart Grid and Renewable Energy (PESGRE2020)*, Jan. 2020, pp. 1–6.

[10] H. Wen, W. Xiao, and X. Wen, "Comparative evaluation of DC-link capacitors for electric vehicle application," in *2012 IEEE Int. Symp. on Ind. Electron.*, May 2012, pp. 1472–1477.

[11] D. Cittanti, F. Stella, E. Vico, C. Liu, J. Shen, G. Xiu, "Analysis, design, and experimental assessment of a high power density ceramic DC-link capacitor for a 800 V 550 kVA electric vehicle drive inverter," *IEEE Trans. on Ind. Appl.*, vol. 59, no. 6, pp. 7078–7091, Nov. 2023.

[12] H. Wang, C. Li, G. Zhu, Y. Liu, and H. Wang, "Model-Based Design and Optimization of Hybrid DC-Link Capacitor Banks," *IEEE Transactions on Power Electronics*, vol. 35, no. 9, pp. 8910–8925, Sep. 2020.

[13] D. Wang, M. Preindl, F. Peng, J. Ye, and A. Emadi, "DC-bus design with hybrid capacitor bank in single-phase PV inverters," in *IECON 2017 - 43rd Annual Conference of the IEEE Industrial Electronics Society*, Beijing: IEEE, Oct. 2017, pp. 2425–2430

First Demonstration of a Gallium Oxide Power Converter

Joshua J. Piel[1], Elizabeth A. Sowers[1], Daniel M. Dryden[1], Thaddeus J. Asel[2], Adam T. Neal[2],
Brenton A. Noesges[2], Shin Mou[2], Andrew J. Green[1]

Sensors Directorate[1], Materials and Manufacturing Directorate[2]
Air Force Research Laboratory
Wright-Patterson AFB, OH, USA
{joshua.piel.1, elizabeth.sowers, andrew.green.25}@us.af.mil

Abstract—Ultra-wide bandgap (UWBG) semiconductor materials are predicted to greatly improve the performance of next-generation power electronics. β-Ga$_2$O$_3$ is a promising UWBG material due to its large critical electric field strength, readily accessible n-type doping, and the availability of native substrates. However, no power converters using only β-Ga$_2$O$_3$ devices have been demonstrated yet because the field has prioritized low periphery devices with large unnormalized on-resistances. In this paper, we present the first demonstration of a dc-dc converter using β-Ga$_2$O$_3$ lateral field effect transistors with large periphery. We first fabricate and package the transistors then place them into the high and low side switch of a buck converter circuit. The devices are tested at 45 V input voltage, 2.7 W output power, and the converter achieves an efficiency of over 90% under various operating conditions.

Index Terms—ga2o3, gallium oxide, buck converter, mosfet, power conversion, ultra wide bandgap, wide bandgap, switching performance

I. INTRODUCTION

There is an ever-increasing demand for smaller, lighter, and more efficient power electronics. Semiconductor switching devices are critical for the operation of power electronics and can be a bottleneck for their performance. Ultra-wide bandgap (UWBG) semiconductor materials show strong potential for reducing loss and increasing the breakdown voltage and switching frequency of power semiconductor devices, as predicted by several figures of merit [1], [2]. Beta gallium oxide (β-Ga$_2$O$_3$) is one such UWBG material, with a bandgap of 4.8 eV and an estimated critical electric field strength (E_c) of 8 MV/cm. β-Ga$_2$O$_3$ is predicted to have improved performance over the wide bandgap materials GaN and SiC. The material shows particularly strong potential to impact switching voltages above 10 kV due to its high demonstrated E_c enabling thin epitaxial stacks for vertical device topologies. Additionally, native β-Ga$_2$O$_3$ substrates can be grown from the melt, leading to significantly reduced costs compared to GaN and SiC substrates. Drawbacks of β-Ga$_2$O$_3$ include a lack of p-type doping, which limits the available device architectures, and low thermal conductivity, which increases the need for advanced packaging and cooling solutions. Understanding the full impact of low thermal conductivity requires further investigation of the switch's behavior during operation.

Over the last decade, significant progress and improvements have been made to β-Ga$_2$O$_3$ power devices. Lateral depletion mode [3], lateral enhancement mode [4], and vertical [5] field effect transistors (FETs) have been fabricated. Lateral FETs have reached breakdown voltages exceeding 8 kV [6], and electric field strengths greater than the material limit of SiC have been measured in lateral devices [3]. Work has also been done to characterize the switching performance of β-Ga$_2$O$_3$ diodes and FETs, though almost all of this work has focused on diodes. The clamped inductive switching test, or double pulse test, which replicates the switching transient present in many dc-dc converters, has been extensively performed for β-Ga$_2$O$_3$ Schottky barrier and heterojunction diodes [7]–[17]. The unclamped inductive switching test has been used to observe avalanche breakdown in β-Ga$_2$O$_3$/p-NiO heterojunction diodes [18], [19]. β-Ga$_2$O$_3$ diodes have also been tested in dc-dc converter topologies including the buck converter [20], the boost converter [21]–[25], and a Cockcroft-Walton voltage multiplier [26]. In these tests, all active switch components are implemented with other semiconductor materials such as Si or GaN solutions. Switching characterization of β-Ga$_2$O$_3$ FETs has been limited to double pulse testing [27], [28] and resistive load testing [29], and to date β-Ga$_2$O$_3$ FETs have not yet been demonstrated or evaluated operating in a dc-dc converter.

In this paper, we present the first dc-dc converter which implements β-Ga$_2$O$_3$ based field effect transistors, marking a significant step towards the realization of UWBG semiconductor technology. First, we will describe the fabrication and process flow we followed to fabricate the devices. Next, we will describe how the devices are packaged for converter level testing. Finally, we will discuss the integration scheme and module design resulting in analysis of β-Ga$_2$O$_3$ device performance in a step-down dc-dc converter.

II. DEVICE FABRICATION AND PACKAGING

The devices were fabricated on a Si-doped β-Ga$_2$O$_3$ epitaxial layer grown by molecular beam epitaxy on an Fe-doped semi-insulating β-Ga$_2$O$_3$ substrate purchased from Novel Crystal Technologies. The substrate is a single β-Ga$_2$O$_3$ crystal grown using the edge-fed growth method. The epitaxial deisgn consisted of a 65 nm channel thickness targeting a Si concentration of 3e17 cm^{-3}. The following procedure was followed to fabricate the devices:

979-8-3315-1612-3/25 $31.00 © 2025 IEEE

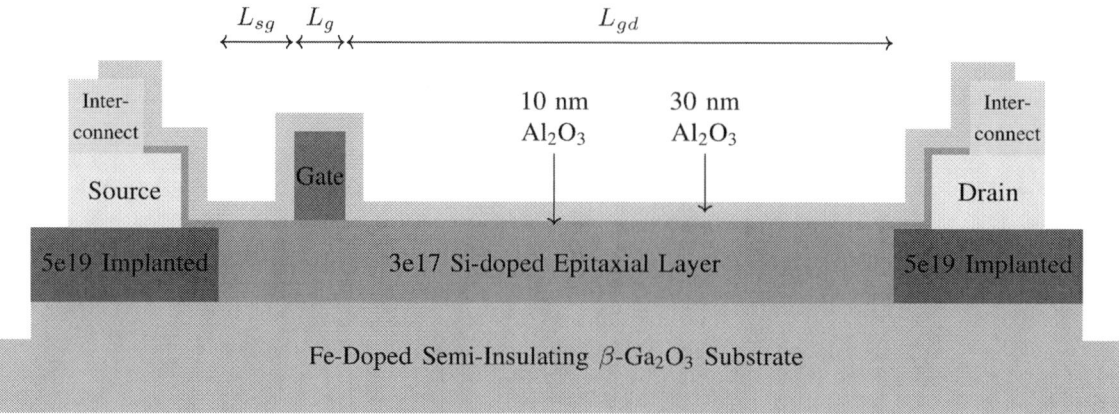

Fig. 1: Cross sectional view of the fabricated devices. L_{sg} = 1.15 μm, L_g = 800 nm, and L_{gd} = 6.15 μm.

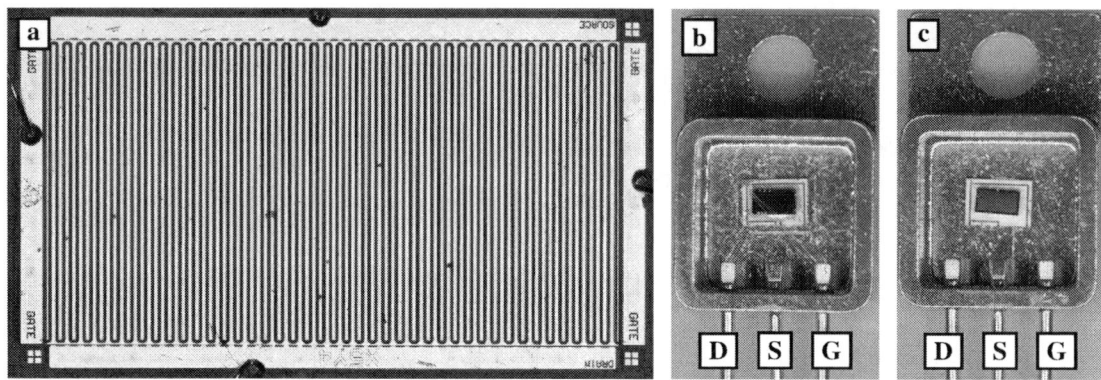

Fig. 2: a) Optical image of a 2.3mm x 1.3mm lateral β-Ga$_2$O$_3$ device with an 94 mm gate periphery. b) Image of high-side β-Ga$_2$O$_3$ packaged device and c) low-side β-Ga$_2$O$_3$ packaged device into a gold plated TO-220. The gate (G), source (S), and drain (D) pins are labeled.

1) Highly doped contact regions were formed using Si ion implantation. The implant profile was a 65 nm deep box with 5e19 cm^{-3} ion concentration.
2) Ohmic contacts were formed by depositing 20 nm Ti and 380 nm Au by electron beam evaporation.
3) Device mesa isolation was performed using a BCl$_3$ inductively coupled plasma etch. The etch depth was 150 nm to ensure isolation of the active region.
4) 10 nm of Al$_2$O$_3$ was deposited by atomic layer deposition (ALD) to form the gate dielectric. A 5 minute buffered oxide/piranha etch preclean was used before deposition, and afterwards Al$_2$O$_3$ was cleared from the field via a wet etch.
5) 20 nm of Ti and 480 nm of Au was deposited by electron beam evaporation to form a metal interconnect layer.
6) 20 nm of Ni and 380 nm of Au was deposited by electron beam evaporation to form gates. The gate layer was optically defined.
7) 30 nm Al$_2$O$_3$ was deposited by ALD for final passivation.

The devices have a gate periphery of 94 mm, which is the largest periphery reported in the literature to date. The devices have a source-gate spacing (L_{sg}) of 1.15 μm, a gate length (L_g) of 800 nm, and a gate-drain spacing (L_{gd}) of 6.15 μm. A device cross-section cartoon can be seen in Fig. 1, and an image of a completed device can be seen in Fig. 2a. Based on the L_{gd} and a conservative average breakdown field estimate of 2 MV/cm [30], the breakdown voltage of these devices is predicted to be about 1200 V.

After fabrication, the substrate was diced to separate discrete devices; these were then screened for low on-resistance (R_{on}) and low off-state current. Two devices were selected, and we will now refer to these as the high side (HS) device and the low side (LS) device. The selected devices were each bonded onto an Au-plated alumina mounting plate using silver epoxy. The mounting plates were then similarly bonded into a TO-220 package, and the device contacts were connected to the package leads with 1 mil Au wire bonds. The mounting plates were used to aid the wire bonding process by raising the devices and allowing connection of the gate pads on opposite sides of the device. Fig. 2b and 2c show images of the packaged devices.

Fig. 3 shows the measured drain-to-source voltage (v_{ds}) and current (i_{ds}) curves of the packaged devices. HS has

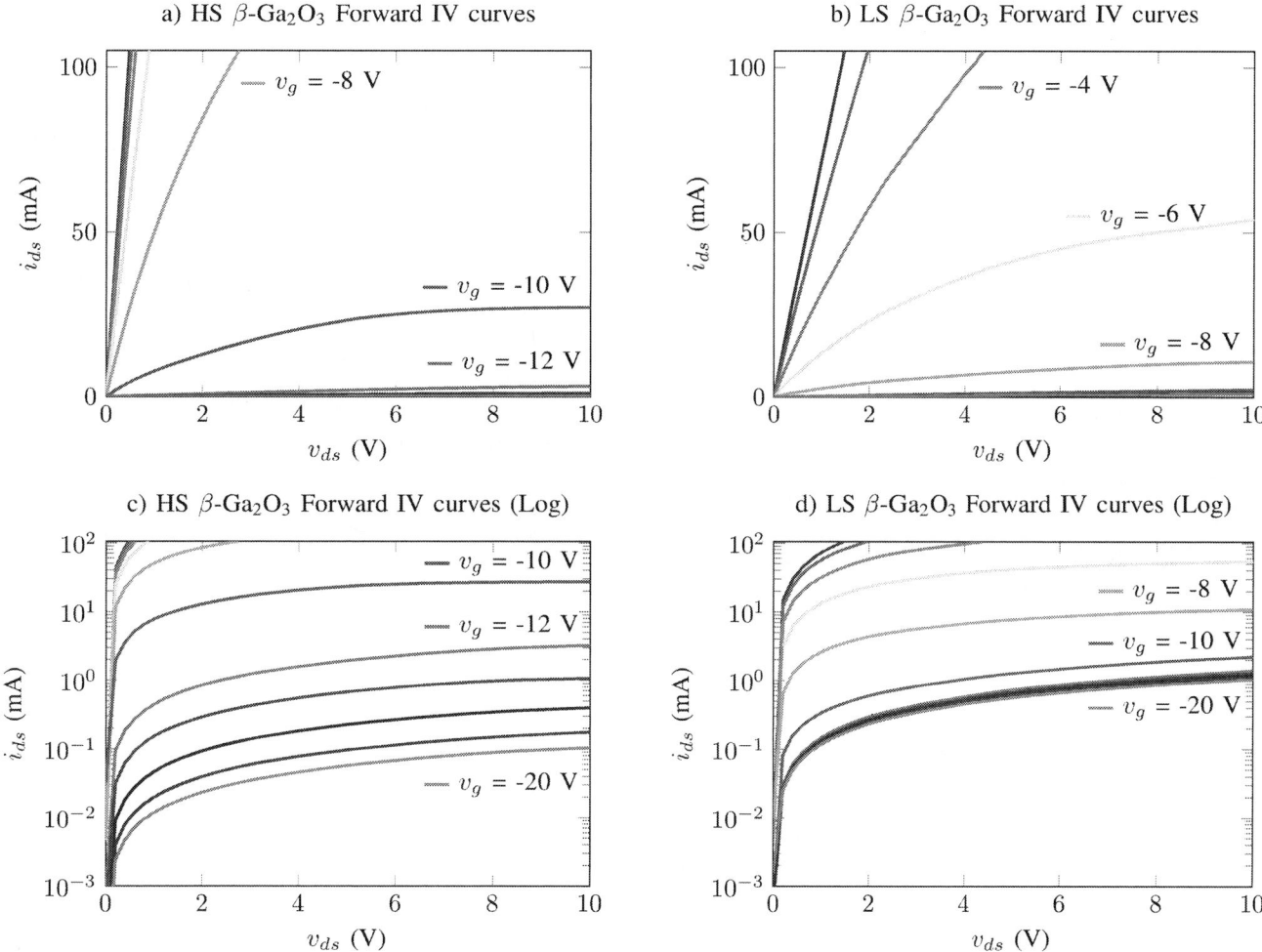

Fig. 3: v_{ds}-i_{ds} curves for Gox A (high-side) and Gox B (low-side) switches in linear and log scale. The high-side and low-side switches have an R_{on} of 4.6 Ω and 13.6 Ω, respectively. v_{gs} ranges from 0 V to -20 V, with a step of -2 V between curves.

an R_{on} of 4.6 Ω and has leakage current under 100 μA at -20 V gate-source voltage (v_{gs}), while LS has an R_{on} of 13.6 Ω and has leakage current of around 1 mA at v_{gs} = -20 V, indicating that HS has better gate control. We note that achieving high yield on devices with periphery this large is difficult in a research fabrication environment, and that only 8% of fabricated devices could successfully modulate the gate with a drain current on-off ratio greater than 10^3. The fabrication process implemented a minimum amount of processing steps to increase yield, and the layout was designed to eliminate vias and interconnect overpasses. While these steps enabled fabrication simplicity, there is risk that high frequency switching could result in non-uniform gate control throughout the large area device. At the testing conditions described below, we did not observe this behavior.

III. CONVERTER ARCHITECTURE

A standard step-down dc-dc converter was constructed to test the fabricated devices. The circuit schematic is shown in Fig. 4, hardware specifications and component values are

Component	Value/Part Number	Voltage Rating
Si Driver Switch	IRF530NPBF	100 V
Gate Driver IC	IR21844PBF	
Bootstrap Diode	VS-ETL0806-M3	600 V
Inductor (L)	10 mH	
Output Capacitor (C)	1 μF	250 V
Load Resistor (R)	400 Ω	

TABLE I: Hardware Specifications

shown in Table I, and an image of the assembled converter is shown in Fig. 5. The converter is rated to handle up to 400 V input voltage (V_{in}), 200 V output voltage (V_{out}), 100 W output power (P_{out}), and low 100s of kHz switching frequency (f_{sw}). The inductor and capacitor values were selected to adequately filter output ripple at a worst-case scenario of f_{sw} = 10 kHz and 10 mA of output current. The gate drive is implemented with a cascode configuration because the β-Ga$_2$O$_3$ transistors are depletion mode and have a negative threshold voltage. This is accomplished by placing the β-Ga$_2$O$_3$ devices in series with a low voltage enhancement mode device; in this case, we choose a commercial Si MOSFET. The gate of the β-Ga$_2$O$_3$

979-8-3315-1612-3/25 $31.00 © 2025 IEEE

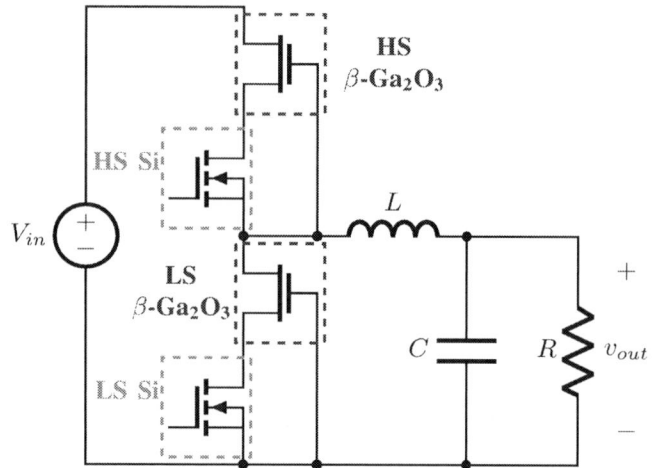

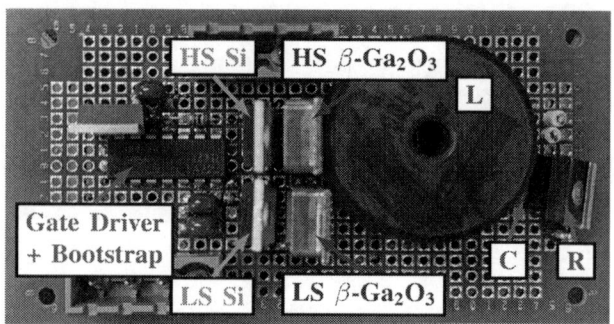

Fig. 5: Image of the assembled test dc-dc converter.

Fig. 4: Circuit diagram of the test dc-dc converter. For the cascode connection, each β-Ga$_2$O$_3$ device is in series with a Si driver switch, and the β-Ga$_2$O$_3$ gate is tied to the Si source.

Operating waveforms at V_{in} = 45 V, f_{sw} = 50 kHz, D = 0.8

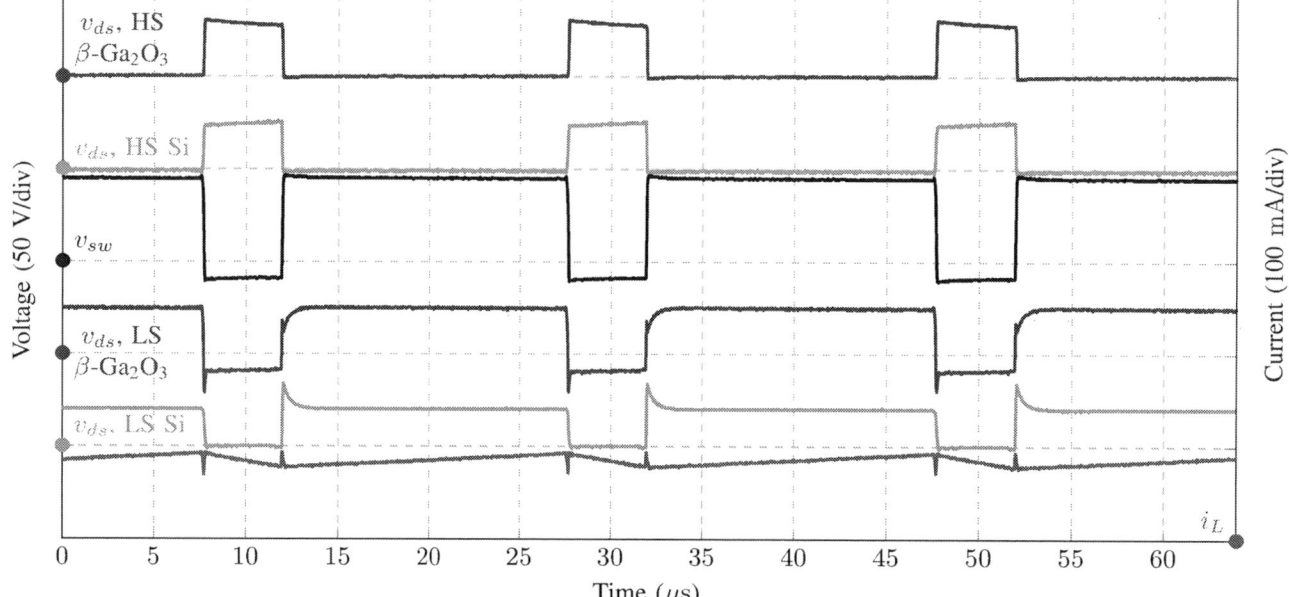

Fig. 6: Experimental operating waveforms at V_{in} = 45 V, f_{sw} = 50 kHz, D = 0.8, and R = 400 Ω. The output power is 2.7 W and the efficiency is 90.6%. v_{ds} of each switch, the switch node voltage v_{sw}, and inductor current i_L are plotted. The waveforms are offset similar to an oscilloscope, zero level of each waveform is marked with a colored circle on the voltage- or current- axis.

device is connected to the source of the Si MOSFET, and the β-Ga$_2$O$_3$ device is turned off when its source is brought positive by the Si MOSFET. The Si switches are driven by a bootstrapped gate driver IC and were conservatively chosen to have 100 V breakdown to protect against failure from potentially large β-Ga$_2$O$_3$ threshold voltages. A function generator is used to control the switching frequency and duty cycle in open loop, and digital multimeters are used to

measure the true input and output voltage and current, allowing calculation of the power stage efficiency.

IV. EXPERIMENTAL RESULTS

The converter with the devices was tested between input voltages (V_{in}) of 25 V to 45 V, up to a maximum output power of 2.7 W, at f_{sw} between 25 kHz to 50 kHz, and at duty cycles (D) between 0.3 and 0.8 with a constant load

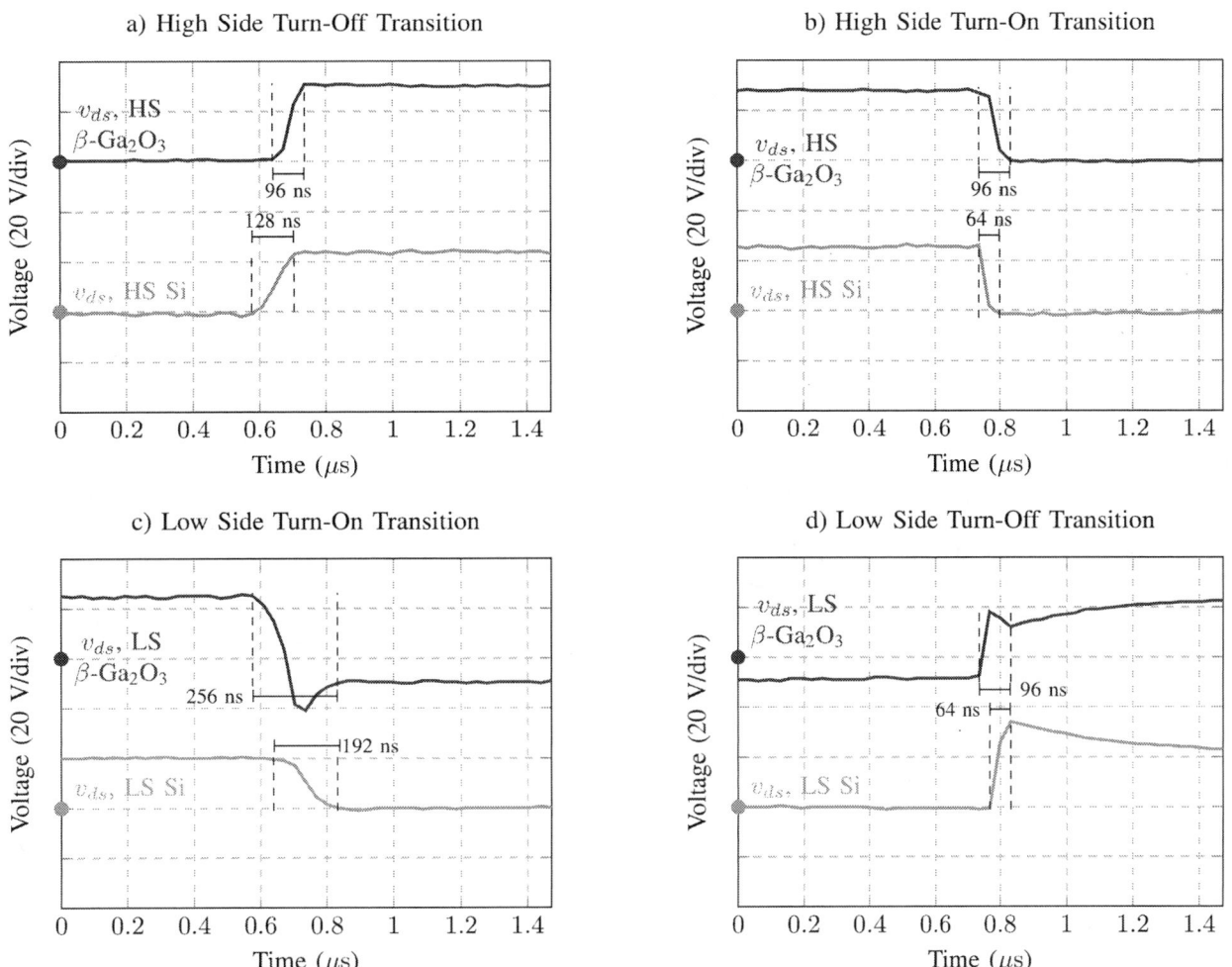

Fig. 7: Turn-on and turn-off transitions for the high side and low side devices. The β-Ga$_2$O$_3$ device v_{ds} waveforms are colored blue and the Si device v_{ds} waveforms are colored green. The time for each switching transition is annotated on the plot.

resistance (R) of 400 Ω. The maximum voltage and power levels were limited by a premature breakdown mechanism between the β-Ga$_2$O$_3$ device contact pads. Fig. 6 shows an example operating waveform measured at V_{in} = 45 V, f_{sw} = 50 kHz, and D = 0.8. The waveforms shown are the v_{ds} for both the high side and low side β-Ga$_2$O$_3$ and Si devices, the converter switch node voltage, and the inductor current. The waveforms are offset from 0, mimicking the display of an oscilloscope. The waveforms clearly show the devices are turning off properly under the cascode gate drive as the blocking voltage is being shared between the Si driver device and the β-Ga$_2$O$_3$ device. The Si device should block the threshold voltage of the β-Ga$_2$O$_3$ device, while the β-Ga$_2$O$_3$ device will block the remaining voltage.

Fig. 6 reveals asymmetries between the low and high-side switch. The LS β-Ga$_2$O$_3$ device is observed to have a significantly degraded R_{on} of approximately 100 Ω compared to the measurement from Fig. 3, while the HS β-Ga$_2$O$_3$ device is consistent with the dc measurements. The HS β-Ga$_2$O$_3$ device has a threshold voltage around -25 V, aligning with

the dc measurements from Fig. 3 that show the device is not fully pinched off at v_{gs} = -20 V. Additionally, v_{ds} of the HS β-Ga$_2$O$_3$ device slowly decreases during the off state, indicating leakage current may be charging the Si driver switch capacitance. The low-side switch has a lower and more stable threshold voltage around -20 V, also aligning with the dc measurements, but it is much slower to turn on, leading to a high voltage stress on its Si driver switch.

Fig. 7 shows a zoomed in view of the turn-off and turn-on transitions of the HS and LS devices at the operating point from Fig. 6. The switching times of each transition are annotated on each plot, and, besides the turn-on transition for the LS switch, all β-Ga$_2$O$_3$ switching times are less than 100 ns. The Si MOSFETs and gate driver ICs were not optimized for fast switching, so it is likely the switching times can be significantly improved. Again, further asymmetries between the HS and LS devices can be observed. The HS β-Ga$_2$O$_3$ device is well-behaved but the LS β-Ga$_2$O$_3$ device experiences significant overshoot on the turn-on transition and has a slow ramp to reach steady state on the turn-off transient.

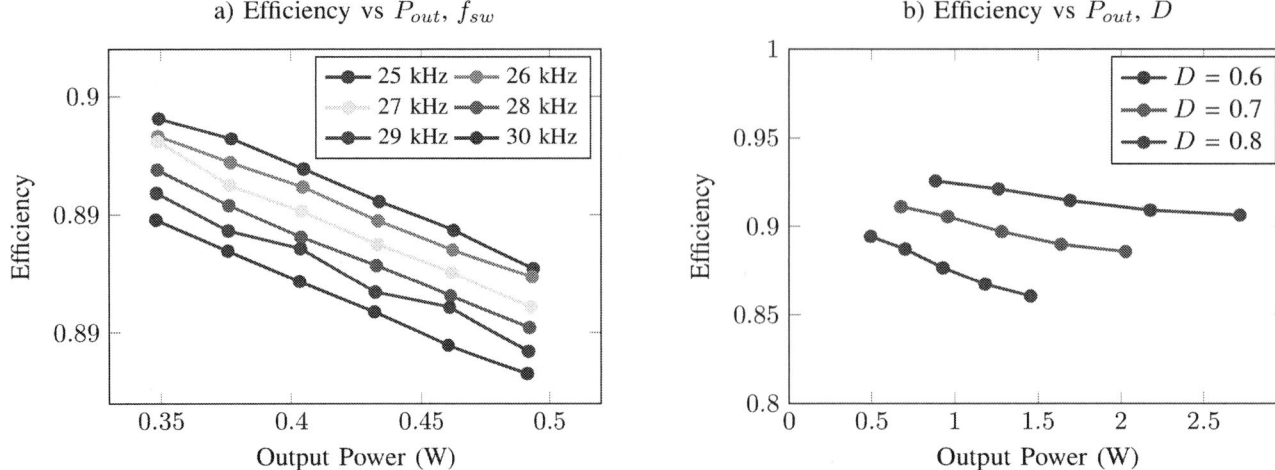

Fig. 8: Plots of converter efficiency vs P_{out}. a) compares the efficiency across different f_{sw}. D is held at 0.5 and V_{in} is varied between 25 V and 30 V. b) compares the efficiency across different D. f_{sw} is held at 50 kHz and V_{in} is varied between 25 V and 45 V. In both cases, the $R = 400\ \Omega$.

Fig. 8 plots the converter efficiency against the output power at different values of f_{sw} and D. In the first plot, f_{sw} and V_{in} are varied while D is held constant at 0.5. In the second plot, f_{sw} is held constant at 50 kHz and V_{in} and D are varied. The efficiency change with frequency, shown by the spacing between the lines in Fig. 8a, is nearly linear, suggesting switching loss is well behaved with frequency in this operating region. Fig. 8b illustrates that at smaller values of D, where the converter spends more time conducting current through the LS switch, there is a significant decrease in efficiency because of the LS β-Ga$_2$O$_3$ switch's unexpectedly high R_{on}.

Because of the large device area of the β-Ga$_2$O$_3$ devices, it is expected that there will be differences in device performance resulting from the challenging fabrication process. Here, the HS and LS devices show significant differences in both on state, off state, and switching characteristics. The large area fabrication process must be optimized for large area devices to achieve the breakdown voltage and R_{on} specifications that have been reached for small periphery devices.

V. Conclusion

For the first time, large-area lateral β-Ga$_2$O$_3$ field effect transistors have been fabricated, packaged and tested at the converter level. The devices were able to transfer 2.7 W at over 90% efficiency. Basic operating conditions, such as device on-resistance, switching performance, and efficiency over different output powers, frequencies, and duty cycles were analyzed. Looking forward, high-performing β-Ga$_2$O$_3$ devices will need to be optimized and implemented into similar designs to demonstrate the technology's value. Future work will emphasize large area device yield and and obtain more consistent large-area β-Ga$_2$O$_3$ devices for testing in realistic power electronics applications. While the device design was rated to operate at over 1000 V, extrinsic breakdown limited the operating power of the final converter. As high-power density converters are demonstrated in future iterations, close consideration will be required to examine thermal limitations for the technology. Solving these challenges will be key to improving the efficiency and power density of future power electronics.

References

[1] B. Baliga, "Power semiconductor device figure of merit for high-frequency applications," *IEEE Electron Device Letters*, vol. 10, no. 10, pp. 455–457, 1989.

[2] A. Huang, "New unipolar switching power device figures of merit," *IEEE Electron Device Letters*, vol. 25, no. 5, pp. 298–301, 2004.

[3] A. J. Green, K. D. Chabak, E. R. Heller, R. C. Fitch, M. Baldini, A. Fiedler, K. Irmscher, G. Wagner, Z. Galazka, S. E. Tetlak, A. Crespo, K. Leedy, and G. H. Jessen, "3.8-mv/cm breakdown strength of movpe-grown sn-doped β -ga2o3 mosfets," *IEEE Electron Device Letters*, vol. 37, no. 7, pp. 902–905, 2016.

[4] K. D. Chabak, N. Moser, A. J. Green, J. Walker, Dennis E., S. E. Tetlak, E. Heller, A. Crespo, R. Fitch, J. P. McCandless, K. Leedy, M. Baldini, G. Wagner, Z. Galazka, X. Li, and G. Jessen, "Enhancement-mode ga2o3 wrap-gate fin field-effect transistors on native (100) β-ga2o3 substrate with high breakdown voltage," *Applied Physics Letters*, vol. 109, no. 21, p. 213501, 11 2016. [Online]. Available: https://doi.org/10.1063/1.4967931

[5] Z. Hu, K. Nomoto, W. Li, Z. Zhang, N. Tanen, Q. T. Thieu, K. Sasaki, A. Kuramata, T. Nakamura, D. Jena, and H. G. Xing, "Breakdown mechanism in 1 ka/cm2 and 960 v e-mode β-ga2o3 vertical transistors," *Applied Physics Letters*, vol. 113, no. 12, p. 122103, 09 2018. [Online]. Available: https://doi.org/10.1063/1.5038105

[6] S. Sharma, L. Meng, A. F. M. A. U. Bhuiyan, Z. Feng, D. Eason, H. Zhao, and U. Singisetti, "Vacuum annealed β-ga2o3 recess channel mosfets with 8.56 kv breakdown voltage," *IEEE Electron Device Letters*, vol. 43, no. 12, pp. 2029–2032, 2022.

[7] H. Gong, F. Zhou, W. Xu, X. Yu, Y. Xu, Y. Yang, F.-f. Ren, S. Gu, Y. Zheng, R. Zhang, H. Lu, and J. Ye, "1.37 kv/12 a nio/ β-ga2o3 heterojunction diode with nanosecond reverse recovery and rugged surge-current capability," *IEEE Transactions on Power Electronics*, vol. 36, no. 11, pp. 12 213–12 217, 2021.

[8] A. Takatsuka, H. Miyamoto, K. Sasaki, and A. Kuramata, "Fabrication of ampere-class p-cu2o/n − β-ga2o3 trench heterojunction barrier schottky diodes and double-pulse evaluation," in *2023 35th International Symposium on Power Semiconductor Devices and ICs (ISPSD)*, 2023, pp. 342–345.

[9] J. Yang, F. Ren, Y.-T. Chen, Y.-T. Liao, C.-W. Chang, J. Lin, M. J. Tadjer, S. J. Pearton, and A. Kuramata, "Dynamic switching characteristics of 1 a forward current β-ga2o3 rectifiers," *IEEE Journal of the Electron Devices Society*, vol. 7, pp. 57–61, 2019.

[10] J.-S. Li, C.-C. Chiang, X. Xia, C.-T. Tsai, F. R. Y.-T. Liao, and S. J. Pearton, "Dynamic switching of 1.9 a/1.76 kv forward current nio/ β-ga2o3 rectifiers," *ECS Journal of Solid State Science and Technology*, vol. 11, no. 10, p. 105003, oct 2022. [Online]. Available: https://dx.doi.org/10.1149/2162-8777/ac942c

[11] X. Lu, X. Zhang, H. Jiang, X. Zou, K. M. Lau, and G. Wang, "Vertical β-ga2o3 schottky barrier diodes with enhanced breakdown voltage and high switching performance," *physica status solidi (a)*, vol. 217, no. 3, p. 1900497, 2020. [Online]. Available: https://onlinelibrary.wiley.com/doi/abs/10.1002/pssa.201900497

[12] J. Yang, C. Fares, F. Ren, Y.-T. Chen, Y.-T. Liao, C.-W. Chang, J. Lin, M. Tadjer, D. J. Smith, S. J. Pearton, and A. Kuramata, "Switching behavior and forward bias degradation of 700v, 0.2a, β-ga2o3 vertical geometry rectifiers," *ECS Journal of Solid State Science and Technology*, vol. 8, no. 7, p. Q3028, jan 2019. [Online]. Available: https://dx.doi.org/10.1149/2.0061907jss

[13] Y.-T. Chen, J. Yang, F. Ren, C.-W. Chang, J. Lin, S. J. Pearton, M. J. Tadjer, A. Kuramata, and Y.-T. Liao, "Implementation of a 900 v switching circuit for high breakdown voltage β-ga2o3 schottky diodes," *ECS Journal of Solid State Science and Technology*, vol. 8, no. 7, p. Q3229, may 2019. [Online]. Available: https://dx.doi.org/10.1149/2.0421907jss

[14] M. Xiao, B. Wang, J. Liu, R. Zhang, Z. Zhang, C. Ding, S. Lu, K. Sasaki, G.-Q. Lu, C. Buttay, and Y. Zhang, "Packaged ga2o3 schottky rectifiers with over 60-a surge current capability," *IEEE Transactions on Power Electronics*, vol. 36, no. 8, pp. 8565–8569, 2021.

[15] F. Zhang, X. Zheng, Y. He, X. Wang, Y. Hong, X. Zhang, Z. Yuan, Y. Wang, X. Lu, J. Yin, Y. Gao, X. Ma, and Y. Hao, "The impact of anode p+ islands layout on the performance of niox/ β-ga2o3 hetero-junction barrier schottky diodes," *IEEE Transactions on Electron Devices*, vol. 70, no. 11, pp. 5603–5608, 2023.

[16] Y. Wei, X. Luo, Y. Wang, J. Lu, Z. Jiang, J. Wei, Y. Lv, and Z. Feng, "Experimental study on static and dynamic characteristics of ga2o3 schottky barrier diodes with compound termination," *IEEE Transactions on Power Electronics*, vol. 36, no. 10, pp. 10 976–10 980, 2021.

[17] J. Yang, P. C. IV, F. Ren, Y. Liao, C.-W. Chang, J. Lin, M. Tadjer, S. J. Pearton, D. J. Smith, and A. Kuramata, "DC and dynamic switching characteristics of field-plated vertical geometry [beta]-Ga2O3 rectifiers," in *Oxide-based Materials and Devices X*, D. J. Rogers, D. C. Look, and F. H. Teherani, Eds., vol. 10919, International Society for Optics and Photonics. SPIE, 2019, p. 1091916. [Online]. Available: https://doi.org/10.1117/12.2515006

[18] F. Zhou, H. Gong, M. Xiao, Y. Ma, Z. Wang, X. Yu, L. Li, L. Fu, H. H. Tan, Y. Yang, F.-F. Ren, S. Gu, Y. Zheng, H. Lu, R. Zhang, Y. Zhang, and J. Ye, "An avalanche-and-surge robust ultrawide-bandgap heterojunction for power electronics," *Nature Communications*, vol. 14, no. 1, p. 4459, 2023. [Online]. Available: https://doi.org/10.1038/s41467-023-40194-0

[19] H. Gong, F. Zhou, M. Xiao, Z. Yang, F.-F. Ren, S. Gu, H. Lu, R. Zhang, Y. Zhang, and J. Ye, "Enhanced avalanche (2.1 kv, 83 a) in nio/ga2o3 heterojunction by edge termination optimization," *IEEE Electron Device Letters*, vol. 45, no. 8, pp. 1421–1424, 2024.

[20] F. Wilhelmi, Y. Komatsu, S. Yamaguchi, Y. Uchida, T. Kase, S. Kunori, and A. Lindemann, "Switching properties of 600 v ga2o3 diodes with different chip sizes and thicknesses," *IEEE Transactions on Power Electronics*, vol. 38, no. 7, pp. 8406–8418, 2023.

[21] F. Wu, J. Wen, J. Liu, Q. Li, Z. Han, W. Hao, X. Zhou, G. Xu, and S. Long, "Reliability of 1.5 × 1.5 mm2 β-ga2o3 power diodes and application in dc–dc converter," *physica status solidi (b)*, vol. n/a, no. n/a, p. 2400438. [Online]. Available: https://onlinelibrary.wiley.com/doi/abs/10.1002/pssb.202400438

[22] H. Gong, F. Zhou, X. Yu, W. Xu, F.-F. Ren, S. Gu, H. Lu, J. Ye, and R. Zhang, "70-μm-body ga2o3 schottky barrier diode with 1.48 k/w thermal resistance, 59 a surge current and 98.9% conversion efficiency," *IEEE Electron Device Letters*, vol. 43, no. 5, pp. 773–776, 2022.

[23] W. Guo, G. Jian, W. Hao, F. Wu, K. Zhou, J. Du, X. Zhou, Q. He, Z. Yu, X. Zhao, G. Xu, and S. Long, "-gao field plate schottky barrier diode with superb reverse recovery for high-efficiency dc–dc converter," *IEEE Journal of the Electron Devices Society*, vol. 10, pp. 933–941, 2022.

[24] W. Guo, Z. Han, X. Zhao, G. Xu, and S. Long, "Large-area β-ga2o3 schottky barrier diode and its application in dc–dc converters," *Journal of Semiconductors*, vol. 44, no. 7, p. 072805, jul 2023. [Online]. Available: https://dx.doi.org/10.1088/1674-4926/44/7/072805

[25] W. Guo, G. Jian, F. Wu, K. Zhou, G. Xu, X. Zhou, Q. He, X. Zhao, and S. Long, "A dc-dc converter utilizing $\beta - Ga_2O_3$ schottky barrier diode," in *2021 5th IEEE Electron Devices Technology & Manufacturing Conference (EDTM)*, 2021, pp. 1–3.

[26] F. Wu, Y. Wang, G. Jian, G. Xu, X. Zhou, W. Guo, J. Du, Q. Liu, S. Dun, Z. Yu, Y. Lv, Z. Feng, S. Cai, and S. Long, "Superior performance β-ga2o3 junction barrier schottky diodes implementing p-nio heterojunction and beveled field plate for hybrid cockcroft–walton voltage multiplier," *IEEE Transactions on Electron Devices*, vol. 70, no. 3, pp. 1199–1205, 2023.

[27] C. Kuring, K. Tetzner, A. Popp, S. Heucke, O. Hilt, S. B. Anooz, J. Würfl, and S. Dieckerhoff, "Switching behavior and dynamic on-resistance of lateral β-ga2 o3 mosfets up to 400 v," in *2021 IEEE 8th Workshop on Wide Bandgap Power Devices and Applications (WiPDA)*, 2021, pp. 52–57.

[28] J. Böcker, K. Tetzner, S. Heucke, O. Hilt, E. Bahat-Treidel, S. Dieckerhoff, and J. Würfl, "Dispersion effects in on-state resistance of lateral ga2o3 mosfets at 300 v switching," *Electronics Letters*, vol. 56, no. 16, pp. 838–840, 2020. [Online]. Available: https://ietresearch.onlinelibrary.wiley.com/doi/abs/10.1049/el.2020.1286

[29] H. Dong, S. Long, H. Sun, X. Zhao, Q. He, Y. Qin, G. Jian, X. Zhou, Y. Yu, W. Guo, W. Xiong, W. Hao, Y. Zhang, H. Xue, X. Xiang, Z. Yu, H. Lv, Q. Liu, and M. Liu, "Fast switching β-ga2o3 power mosfet with a trench-gate structure," *IEEE Electron Device Letters*, vol. 40, no. 9, pp. 1385–1388, 2019.

[30] D. M. Dryden, K. J. Liddy, A. E. Islam, J. C. Williams, D. E. Walker, N. S. Hendricks, N. A. Moser, A. Arias-Purdue, N. P. Sepelak, K. DeLello, K. D. Chabak, and A. J. Green, "Scaled t-gate β-ga2o3 mosfets with 2.45 kv breakdown and high switching figure of merit," *IEEE Electron Device Letters*, vol. 43, no. 8, pp. 1307–1310, 2022.

Optimized Integrated EMI Filter Design in SiC Power Modules with Terminal Inductor for Better High-Frequency EMI Suppression

Yifan Zhang
School of Electrical and Electronic Engineering Huazhong University of Science and Technology
Wuhan, China
z_yifan@hust.edu.cn

Wenzhe Xu
School of Electrical and Electronic Engineering Huazhong University of Science and Technology
Wuhan, China
wenzhe_xu@hust.edu.cn

Jianwei Lv
School of Electrical and Electronic Engineering Huazhong University of Science and Technology
Wuhan, China
jianweil@hust.edu.cn

Yiyang Yan
School of Electrical and Electronic Engineering Huazhong University of Science and Technology
Wuhan, China
yanyiyang@hust.edu.cn

Baihan Liu
School of Electrical and Electronic Engineering Huazhong University of Science and Technology
Wuhan, China
loubeckham@hust.edu.cn

Sijia Liu
School of Electrical and Electronic Engineering Huazhong University of Science and Technology
Wuhan, China
m202271950@hust.edu.cn

Jiaxin Liu
School of Electrical and Electronic Engineering Huazhong University of Science and Technology
Wuhan, China
liujx711@hust.edu.cn

Cai Chen
School of Electrical and Electronic Engineering Huazhong University of Science and Technology
Wuhan, China
caichen@hust.edu.cn

Yong Kang
School of Electrical and Electronic Engineering Huazhong University of Science and Technology
Wuhan, China
ykang@hust.edu.cn

Xiong Zhang
GAC Aion New Energy Automobile Co. Ltd.
Guangzhou, China
zhangxiong@aion.com.cn

Hao Yu
GAC Aion New Energy Automobile Co. Ltd.
Guangzhou, China
yuhao@aion.com.cn

Wei Jiang
GAC Aion New Energy Automobile Co. Ltd.
Guangzhou, China
jiangwei@aion.com.cn

Abstract—**Using fast-switching wide-bandgap devices such as SiC MOSFETs can reduce switching losses but also bring more serious common-mode (CM) electromagnetic interference (EMI) problems. Integrating EMI filters directly in SiC power modules is an effective way to solve this problem. However, the parasitic inductance of the Y capacitor branch will affect the filtering effect of the integrated filter. To solve those problems, this paper analyzed the influence of these parameters on the insertion loss of the integrated filters. Then, a terminal inductance structure is proposed according to the results of the analysis, using the mutual effect to cancel the parasitic inductance. The optimized SiC power module with an integrated EMI filter has been designed to achieve lower CM EMI in the high-frequency range.**

Keywords—integrated EMI filter, SiC power module, CM EMI

I. INTRODUCTION

Wide bandgap semiconductor devices have attracted more and more attention because of their fast-switching characteristics. SiC MOSFETs' opening speed can reach 10 times that of silicon-based devices [1]. These devices can work at a switching frequency of hundreds of kHz. In such a fast-switching action, the high-amplitude dv/dt and di/dt generated by the switching waveform will generate more serious EMI problems than silicon-based devices through the coupling of subtle parasitic parameters, which is especially serious in the high-frequency range[2].

To further suppress the EMI generated by the power module of the wide-bandgap device, reasearch [3] put Y capacitors on the DBC (Direct Bond Copper) of the Si IGBT power module. Research [4] integrated the EMI filter on the top of the GaN power module. Research [5] integrated the

CM EMI filter directly into the DBC of the SiC power module. With low connection impedance, the integrated filters can significantly reduce the EMI in the high-frequency range. However, due to space constraints, the filter parameters are limited, making it challenging to integrate multi-stage filters into the power module for further EMI suppression. Additionally, the pursuit of higher power density brings components closer together, exacerbating coupling issues, which have not been thoroughly analyzed in integrated EMI filters. Therefore, in the integrated power modules, it is essential to conduct a detailed analysis of the effects of parasitic coupling on EMI. Structural optimizations and innovations based on this analysis can help mitigate adverse effects and enhance EMI filter performance in the high-frequency range. To address this problem, this paper proposes an optimized integrated structure of EMI filters for the SiC power module, which utilizes the mutual inductance cancellation effect of the terminal inductor to reduce the parasitic inductance of the ground path and achieve EMI reduction.

The paper is structured as follows: Part II analyses the influence of parasitic and coupling parameters on the performance of integrated CM EMI filters. Then, the method of extracting the parasitic inductance parameters of the Y capacitor and the influence of the parasitic inductance on the filter performance are presented. Finally, the principle of mutual coupling cancellation of parasitic inductance is carried out. Part III describes the terminal CM inductor structure using coupling to reduce the parasitic inductance, and based on this structure, the power module layout with terminal CM inductor is designed, and the superiority of reducing CM EMI in high-frequency range is verified by simulation. The last part is the conclusion of this paper.

979-8-3315-1612-3/25 $31.00 © 2025 IEEE

II. ANALYSIS OF PARASITIC AND COUPLING EFFECTS' INFLUENCE ON IL OF INTEGRATED EMI FILTERS

A. Analysis of the Influence of Parasitic Parameter of Filter Elements on Insertion Loss

Firstly, this analysis is based on existing SiC power module architectures with integrated CM EMI filters [5]. The CM EMI filter structure in the power module is shown in Fig. 1. In the insertion loss analysis, both the noise source and the load side are connected by a 50Ω impedance. L_a, L_b, and L_c in the figure are the parasitic inductance values of the branches respectively. In this structure, L_c is 1.64nH.

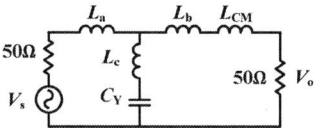

Fig. 1. CM equivalent circuit of integrated EMI filter for insertion loss analysis

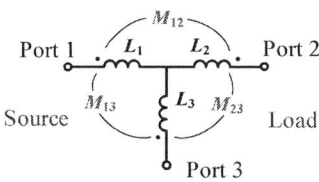

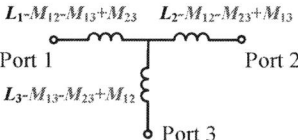

Fig. 2. Equivalent circuits for inductively coupled simplifications of T-circuits

Then the structure in Fig. 2 is obtained by considering the parasitic inductance and the coupling parameters of the integrated power module. The CM inductance is 6.1µH and the Y-capacitance is 10nF.

When $L_3=0$, L_c is completely canceled. At this time, the value of M_{13} is shown in (1).

$$M_{13}\vert_{cancellation} = L_3 + M_{12} - M_{23} = 1.01229nH \quad (1)$$

The equivalent circuit results are brought into the filter insertion loss calculation results. The insertion loss of the filter is obtained as shown in equation (1).

$$IL = \frac{As^3 + Bs^2 + Cs + D}{Es^2 + D} \quad (2)$$

Among them,

$$A = C_Y(-M_{12}{}^2 + 2M_{12}M_{13} + 2M_{12}M_{23} - 2L_3M_{12} - M_{13}{}^2 + 2M_{13}M_{23} - 2L_2M_{13} - M_{23}{}^2 - 2L_1M_{23} + L_1L_2 + L_1L_3 + L_2L_3) \quad (3)$$

$$B = C_Y L_1 R_{50} + C_Y L_2 R_{50} + 2C_Y L_3 R_{50} - 2C_Y M_{13} R_{50} - 2C_Y M_{23} R_{50} \quad (4)$$

$$C = C_Y R_{50}{}^2 + L_1 + L_2 - 2M_{12} \quad (5)$$

$$D = 2R_{50} \quad (6)$$

$$E = 2C_Y R_{50}(L_3 + M_{12} - M_{13} - M_{23}) \quad (7)$$

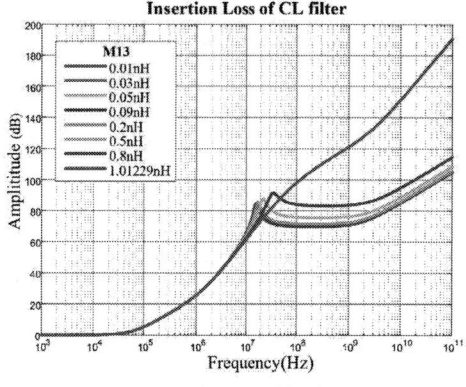

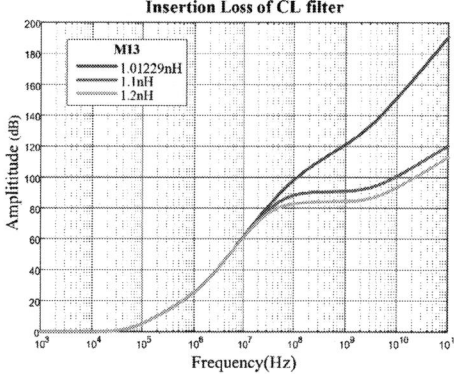

Fig. 3. Changes in insertion loss of CL-type EMI filters by varying coupling parameters.

The effect of the coupling parameters on the insertion loss of the filter is calculated using MATLAB. Since only M_{13} is easy to change using the structure in the actual position of the power module, the effect of M_{13} change on insertion loss is calculated here. The results are shown in Fig. 3. It can be found that the insertion loss of the EMI filter can be increased by appropriately increasing the forward coupling parameter, but the insertion loss decreases when the coupling coefficient exceeds $M_{13}\vert_{cancellation}$. The filter structure that utilizes coupling to achieve complete cancellation of parasitic inductance can increase the insertion loss by 30 dB in the 20-100 MHz range.

B. The Parasitic Parameter Extraction Method of Y Capacitors

The parasitic parameters of the MLCC capacitor include the equivalent series inductance and the equivalent series resistance, as shown in Fig. 4. As for the modeling and simulation method of MLCC capacitor, it is pointed out in the existing literature[6]-[7] that the equivalent structure model of MLCC capacitor can be established by ANSYS HFSS and the parasitic parameters of the capacitor can be simulated. Fig. 5 shows the established finite element simulation model of MLCC capacitor.

The model of the selected capacitor is C4532C0G2J103JT000N and the parameter is 10nF. The high-frequency equivalent circuit of the capacitor is shown in Fig. 6. The high-frequency impedance curve in its data sheet is provided and the high-frequency parameters are listed in the table. Through simulation, the parasitic parameter values extracted by us are also listed in Table I , which are consistent with the parameters in the datasheet.

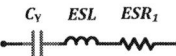

Fig. 4. High-frequency parameters of MLCC capacitors

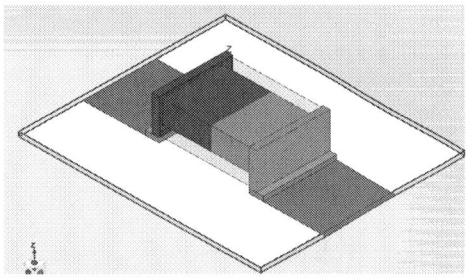

Fig. 5. Finite element simulation model of MLCC capacitor

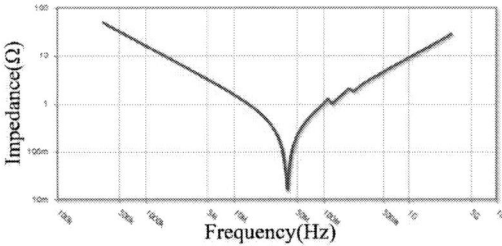

Fig. 6. Impedance curve in MLCC capacitor datasheets

TABLE I. PARASITIC PARAMETERS OF MLCC CAPACITANCE

Parameters	Parameter value	
	Actual value	*Simulation results*
C_Y	9.8212 nF	10.08 nF
ESL	1.6403 nH	1.6346nH
ESR	15.556 mΩ	17.5 mΩ

C. The Coupling Parameter Extraction Method

In order to make use of the coupling inside the power module, the coupling situation needs to be analyzed and modeled. For the inductance and coupling of conductors, the simulation method is as follows. The single-layer flat conductor model is first considered. As shown in Fig. 7 (a).

The size of the self-inductance of a flat conductor of length l, width w, and thickness t is given by equation (8)[8]. According to the results in the literature, the size of the thickness dimension is much smaller than the other two parameters, and its effect can be neglected. Reducing the length and increasing the width can reduce the self-inductance. However, due to the need to consider factors such as core layout, bonding and terminal connections, and module size, there is less freedom to reduce the self-inductance by changing the size of the conductor layer for a single-layer structure.

$$L = \frac{\mu_0 l}{2\pi}(\ln\frac{2l}{t+w}+\frac{1}{2}) \qquad (8)$$

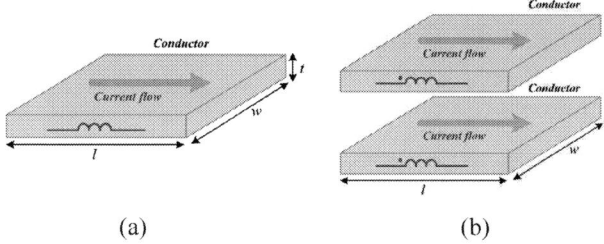

(a)   (b)

Fig. 7. Single and double conductor construction

The coupling can be effectively enhanced and utilized by using a multilayer conductor back. As shown in Fig. 7 (b) double-layer conductor, the length is l, the width is w, the distance between the conductor is d, and the thickness of the conductor is ignored, when the upper and lower two layers of the current flows in the reverse direction, the total inductance of the conductor can be calculated by equation (9). The first part of the equation is the self-inductance of the conductor, and the second part is the mutual inductance between the upper and lower layers of the conductor, and the mutual inductance takes a positive value when the current flows in the same direction, so the mutual inductance can be used to achieve the mutual cancellation of the branch inductance in Fig. 2, thus reducing the total inductance.

$$L = \frac{\mu_0 l}{2\pi}(\ln\frac{2l}{t+w}+\frac{1}{2})$$
$$-\frac{\mu_0 l}{2\pi}\left[\ln\frac{2l}{d}+\frac{1}{2}-2\frac{d}{w}\tan^{-1}\frac{w}{d}-\frac{1}{2}(1-\frac{d^2}{w^2})\ln(1+\frac{w^2}{d^2})\right] \qquad (9)$$

III. THE TERMINAL INDUCTANCE AND THE INTEGRATED POWER MODULE DESIGN

Based on the above results, the coupling parameter M_{13} can reduce the Y capacitor branch inductance brought about by the parasitic parameters of the MLCC capacitor itself. At the same time, the impedance balance needs to be maintained as much as possible. As needed, a terminal CM inductor was designed as shown in Fig. 8. As shown in the figure, the bottom of the power terminals and the Y capacitor form a coupling structure. Finite element simulations yield a coupling coefficient M_{13} of 0.63nH for this structure. The core is sculpted into a toroidal shape using ferrite, and the CM inductor of this structure is placed above the Y-capacitor to realize the coupling effect and reduce the parasitic inductance of the Y capacitor by closing the distance between the terminals and the Y capacitor. The overall structure of the optimized SiC power module with an integrated EMI filter is shown in Fig. 9.

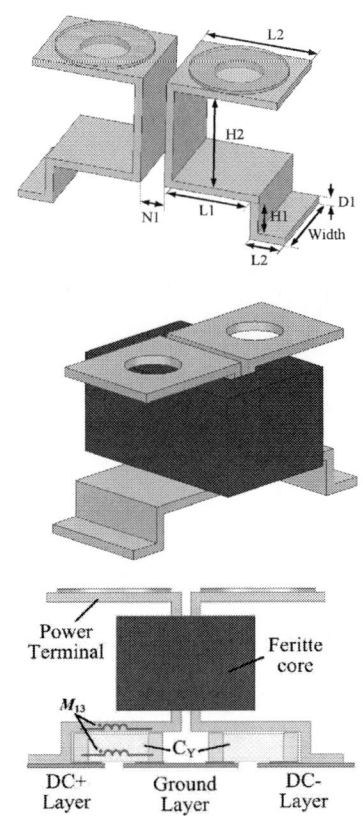

Fig. 8. : Construction and dimensional parameters of terminal inductors and cores

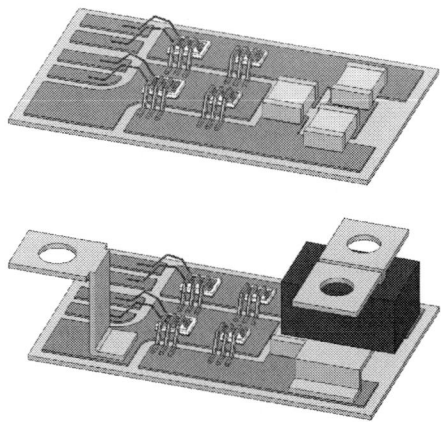

Fig. 9. : The optimized SiC power modules with integrated EMI filter

Based on this structure in Fig. 9, ANSYS finite element simulations are built and its parasitic coupling parameters are

calculated. Then the spice simulation is built based on these parameters. The circuit simulation results show that the module using the coupling structure to offset the parasitic inductance reduces the CM EMI by up to 20 dB over the original module in the 30-100MHz band.

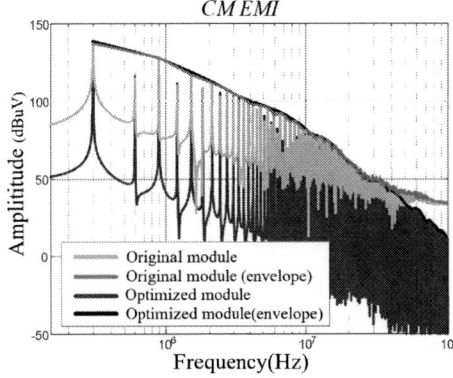

Fig. 10. : The CM EMI simulation result

IV. CONCLUSIONS

In this paper, the effect of parasitic inductance on the filtering effect is analyzed by circuit insertion loss analysis and finite element simulation. Then, the parasitic parameter and the coupling parameter's simulation method of components are analyzed, and the principle of parasitic inductance cancellation by mutual coupling effect is explained. Finally, the structure of the integrated EMI filter is optimized on the basis of the sic power module structure of the integrated EMI filter. The article finally proposes a terminal CM inductor structure, which effectively achieves the impedance cancellation of the Y capacitor branch by strengthening the mutual inductance effect and effectively improves the filtering effect of the integrated filter and is verified by simulation.

REFERENCES

[1] J. Millán, P. Godignon, X. Perpiñà, A. Pérez-Tomás and J. Rebollo, "A Survey of Wide Bandgap Power Semiconductor Devices," in *IEEE Transactions on Power Electronics*, vol. 29, no. 5, pp. 2155-2163, May 2014, doi: 10.1109/TPEL.2013.2268900.\

[2] X. Gong and J. A. Ferreira, "Comparison and Reduction of Conducted EMI in SiC JFET and Si IGBT-Based Motor Drives," *IEEE Transactions on Power Electronics*, vol. 29, pp. 1757-1767, 2014.

[3] R. Robutel *et al.*, "Design and Implementation of Integrated Common Mode Capacitors for SiC-JFET Inverters," in *IEEE Transactions on Power Electronics*, vol. 29, no. 7, pp. 3625-3636, July 2014, doi: 10.1109/TPEL.2013.2279772.

[4] N. Jia, X. Tian, L. Xue, H. Bai, L. M. Tolbert and H. Cui, "Integrated Common-Mode Filter for GaN Power Module With Improved High-Frequency EMI Performance," in *IEEE Transactions on Power Electronics*, vol. 38, no. 6, pp. 6897-6901, June 2023, doi: 10.1109/TPEL.2023.3248092.

[5] Y. Zhang, Y. Xie, W. Xu, Y. Yan, C. Chen and Y. Kang, "Design and Integration Methodology of CM Stripline Inductors and EMI Filters in SiC Power Modules for Low Conductive CM EMI," 2024 IEEE 10th International Power Electronics and Motion Control Conference (IPEMC2024-ECCE Asia), Chengdu, China, 2024, pp. 2253-2258, doi: 10.1109/IPEMC-ECCEAsia60879.2024.10567189.

[6] R. He *et al.*, "Modelling Strategy for Film Capacitors in EMI Filters," *2019 IEEE International Symposium on Electromagnetic*

Compatibility, Signal & Power Integrity (EMC+SIPI), New Orleans, LA, USA, 2019, pp. 259-265, doi: 10.1109/ISEMC.2019.8825302.

[7] P. González-Vizuete, F. Fico, A. Fernández-Prieto, M. J. Freire and J. B. Mendez, "Calculation of Parasitic Self- and Mutual-Inductances of Thin-Film Capacitors for Power Line Filters," in *IEEE Transactions on Power Electronics*, vol. 34, no. 1, pp. 236-246, Jan. 2019, doi: 10.1109/TPEL.2018.2824658.

[8] Z. Piatek, B. Baron, T. Szczegielniak, D. Kusiak, and A. Pasierbek, "Self inductance of long conductor of rectangular cross section," *Przeglad Elektrotechniczny Electr. Rev.*, vol. 88, no. 8, pp. 323–326, 2012

Balanced Technique Using Integrated Winding Coupled Inductor for High-Power Density Two-Phase Interleaved Boost Converter

Yuta Imaeda
Department of Electrical Engineering
Nagoya University
Nagoya, Japan
imaeda.yuta.y0@s.mail.nagoya-u.ac.jp

Jun Imaoka
Insitute of materials and Systems for
Sustainability (IMaSS)
Nagoya University
Nagoya, Japan
imaoka.jun.n9@f.mail.nagoya-u.ac.jp

Masayoshi Yamamoto
Insitute of materials and Systems for
Sustainability (IMaSS)
Nagoya University
Nagoya, Japan
m.yamamoto@imass.nagoya-u.ac.jp

Hiroyuki Onishi
Power Stage Product Design Division
ROHM Co., Ltd
Kyoto, Japan
hiroyuki,onishi@rohm.co.jp

Abstract— **This paper proposes a two-phase boost converter using an integrated winding coupled inductor (IWCI) that achieves high-power density and common-mode (CM) noise reduction. As an attractive feature of the proposed circuit using IWCI, two output diodes are eliminated compared to the conventional circuit using loosely coupled inductor (LCI) and symmetry-balanced technique. The CM noise of the proposed converter is effectively reduced by fulfilling the balancing conditions derived from the CM equivalent circuit. Additionally, components are inherent to the circuit, and it can be done without compromising other design parameters or requirements. To confirm the validity of the derived balancing conditions, experimental evaluations are carried out by using a 300W prototype of the proposed circuit using Gallium Nitride (GaN) devices. The experimental results demonstrated a significant common mode noise reduction effect, achieving up to 50 dBμV across a frequency range of 1 MHz to 30 MHz compared to conventional circuits.**

Keywords—Two-phase interleaved DC/DC converter, coupled inductor, balanced technique, common-mode noise, high-frequency, GaN

I. INTRODUCTION

The switched-mode power supplies (SMPS) requires high power density and high efficiency [1]-[2]. By using wide-bandgap devices such as gallium nitride (GaN), high-frequency operation becomes more feasible, allowing for the downsizing of magnetic components [3]. However, applying high switching frequency increases conducted noise propagating between wires in circuits, which affects other circuits or components as electromagnetic interference (EMI) as reported in [4]-[5]. Although EMI filters using passive components can attenuate the conducted noise, they occupy significant space in power converters and lead to low power density of converters. Therefore, conducted noise reduction methods without increasing power converter volume are required.

As noise reduction methods that avoid the use of EMI filters, Other approaches such as auxiliary windings and balanced technique have been studied. Fig. 1 shows the noise reduction methods by modifying boost converter configuration reported in [6]-[12]. In Fig. 1, the input voltage is defined as V_{in}. Similarly, the input capacitor is C_{in}, the output capacitor is C_{out}, the main switch is S_1, the output diode is D, and the boost inductor is L. The parasitic capacitance between the drain terminal of S_1 and frame ground (FG) is defined as C_d and the parasitic capacitance between the source terminal of S_1 and FG is denoted as C_s. In general, the common-mode (CM) noise flows through C_d due to pulsating voltage potential at the drain terminal of S_1. Fig. 1(a) shows the circuit of the boost converter applying an auxiliary

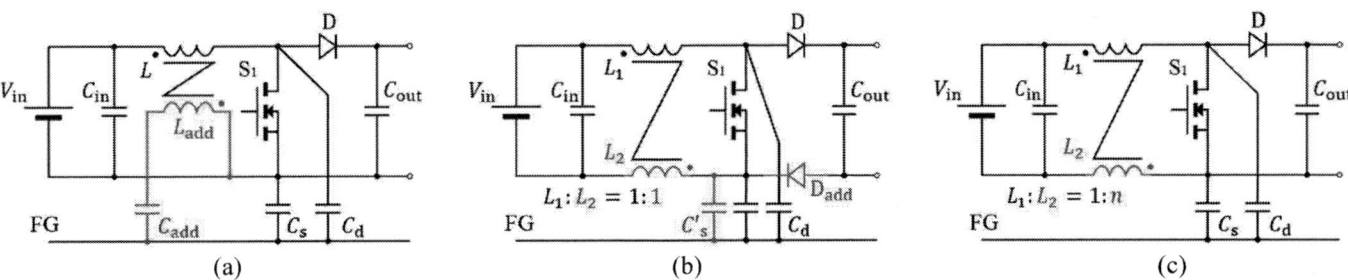

(a)　　　　　　　　　　　　(b)　　　　　　　　　　　　(c)

Fig. 1. Noise reduction methods within the internal circuit of single-phase boost converter (a) Auxiliary winding (b) Symmetry circuit (c) Balanced technique (adjust winding ratio).

windings L_{add} and additional capacitor C_{add} as proposed in [6]-[8]. Auxiliary windings L_{add} are magnetically coupled on the boost inductor L to generate the opposite direction current of the original CM current. As a result, the original CM current flowing toward through C_d is effectively canceled out, thereby reducing the CM noise. Fig. 1(b) shows the balanced boost converter considering the symmetry of the impedance [9]-[10]. The boost inductance ratio is split in a 1:1 ratio, with one side placed on the source side of S_1. Additionally, an external capacitor C'_s and an output diode D_{add} are added to the source side, and the impedance is designed symmetrically. In this paper, a circuit designed with symmetrical impedance using external components is defined as a "Symmetry circuit". As the operation of the symmetry circuit, the drain voltage and source voltage vary complementarily due to the divided inductor. Therefore, the CM noise generated from C_s cancels out the noise from C_d, reducing CM noise. Fig. 1(c) shows the balanced boost converter achieved by adjusting the winding ratio shown in [11]-[12]. In this method, the boost inductor is splitted according to the capacitances C_d and C_s. The source voltage fluctuates complementarily with the drain voltage, and the noise current flowing from C_d to FG is canceled out by the noise current from C_s, thereby reducing. Although this approach requires a complex design with high precision, it effectively reduces CM noise without the need of additional diode, i.e., D_{add}. However, although the noise reduction technology shown in Figures 1(a)-(c) is used in single-phase boost converters, the application of the single-phase boost converter is unsuitable from the perspective of increasing output power.

Therefore, this paper proposes a two-phase interleaved boost converter using an integrated magnetics and balanced technique that can achieve both high-power density and CM noise reduction. As a magnetic structure type used in the two-phase interleaved boost converter, Integrated Winding Coupled Inductor (IWCI) is adopted. The proposed circuit applies the balanced technique by placing part of the central leg winding on the source side of the switch. As an attractive feature of the proposed circuit, the conventional circuit using loosely coupled inductors (LCI) and symmetry balanced technique requires

additional output diodes, whereas the proposed circuit solves this issue by using IWCI. Additionally, adjusting the winding split ratio to fulfill the derived balanced condition significantly reduces CM noise, which enables the downsizing of the EMI filter.

The structure of this paper is as follows. Section II introduces previous studies on high power density circuits that combine coupled inductors and the balanced technique. Section III describes the topology of the proposed circuit and its advantages compared to conventional designs. Section IV derives the circuit balancing conditions from the CM equivalent circuit of the proposed design. Section V presents simulation-based analysis to verify the consistency of the derived balancing conditions. Section VI evaluates the noise reduction effect using test circuit equipped with GaN devices. Section VII is concludes the paper.

II. PREVIOUS RESEARCH COMBINING BALANCED TECHNIQUE WITH COUPLED INDUCTOR

This chapter introduces previous research aimed at achieving high power density. Multi-phase interleaved converters have been extensively studied as a topology to achieve high power density in SMPS. Applying each switch gate signals shifted by 360°/number of phases, the interleaved multi-phase circuit reduces the ripple current of the input and output currents, enabling downsizing of the input and output capacitors as shown in [13]-[14]. Additionally, by coupling multiple-phase inductors into a single core, further high-power density can be achieved. Table I summarizes the characteristics of each method of coupled inductors proposed in [15]-[20]. Various methods for coupled inductors have been proposed, and by selecting the appropriate one based on the circuit's requirements, high power density can be achieved. From Table I, the LCI places the windings on the outer leg of the EE core with negative coupling as shown in [18]. Due to the negative coupling, the DC flux of each phase cancels out, reducing the maximum flux. Therefore, the required cross-sectional area to prevent magnetic saturation is reduced, allowing for the downsizing of the inductor. Additionally, the central leg is used

TABLE I

STRUCTURE AND CHARACTERISTICS OF A COUPLED INDUCTOR DESIGNED FOR A TWO-PHASE INTERLEAVED CONVERTER

Loosely Coupled Inductor [18] (LCI)	Closely Coupled Inductor [19] (CCI)	Integrated Winding Coupled Inductor [20] (IWCI)
✓ Canceling the DC flux enables downsizing. ✓ The coupling coefficient is adjust using the air gap.	✓ Two magnetic cores compose the structure. ✓ High efficiency is achieved by using different material.	✓ Canceling the DC flux enables downsizing. ✓ Ripple current is reduced by the winding on the central leg.

as a path for leakage flux, and an arbitrary coupling coefficient can be achieved by adjusting the air gap at central leg. Then, the closely coupled inductor (CCI) uses two magnetic cores as reported in [19].This method reported that high efficiency operation can be achieved by selecting appropriate core materials for boost inductors and closely coupled inductors. Ideally, no DC flux is generated in the core in a closely coupled inductor. In case of IWCI shown in Table I, windings are placed on all legs of the EE core as analyzed in [20]. The advantages of IWCI in comparsion with other magnetic coupling methods are the reduction of the maximum flux by negatively coupling the outer leg winding, similar to LCI, and the increase in the number of windings common to each phase by adding windings to the central leg. This enables the reduction of the core volume and the ripple current value of each phase under conditions where the duty ratio is farther than 0.5. In ref. [20], it is shown that IWCI has better downsizing performance than LCI for duty cycles other than 0.5. By combining the above techniques with noise reduction, significant high power density can be achieved, including the reduction of EMI filters.

In relation to studies combine coupling inductors and noise reduction, a low-noise and high-power density circuit topology with the balanced technique applied to a two-phase boost converter with LCI is proposed in [21]. The circuit configuration and magnetic core stracture are respectively shown in Fig. 2 (a) and (b). In the circuit shown in Fig. 2 (a), symmetry balanced technique is applied by dividing the LCI into a 1:1 winding ratio and placing it on the source side of S_1 and S_2. By adding an external capacitor and output diode on the source side of S_1 and S_2, the balanced condition can be easily achieved even in a two-phase circuit. However, adding the diodes D_3 and D_4 on the return path is undesirable as it increases the number of components and losses.

III. PROPOSED CIRCUIT AND NOISE REDUCTION

A. Configuration of the Proposed Circuit

Fig. 3(a) shows the circuit diagram of the proposed circuit, and Fig. 3(b) presents the structure of the coupled inductor used in the proposed circuit. The proposed circuit uses IWCI in a two-

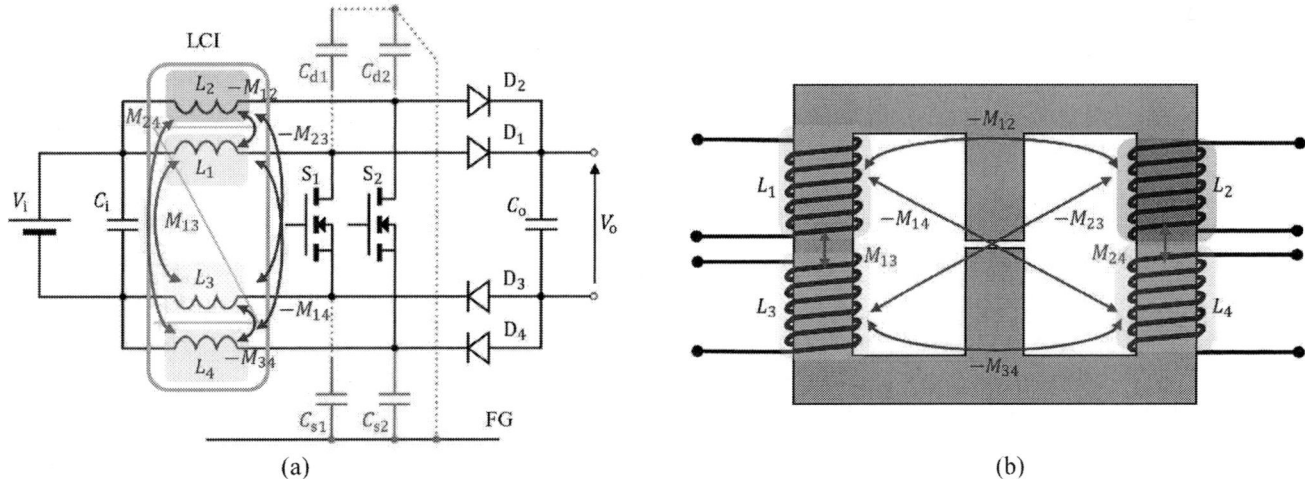

Fig. 2. (a) Circuit diagram of a symmetric interleaved two-phase boost converter using loosely coupled inductor. (b) Structural diagram of a loosely coupled inductor symmetrically split with a 1:1 winding split ratio.

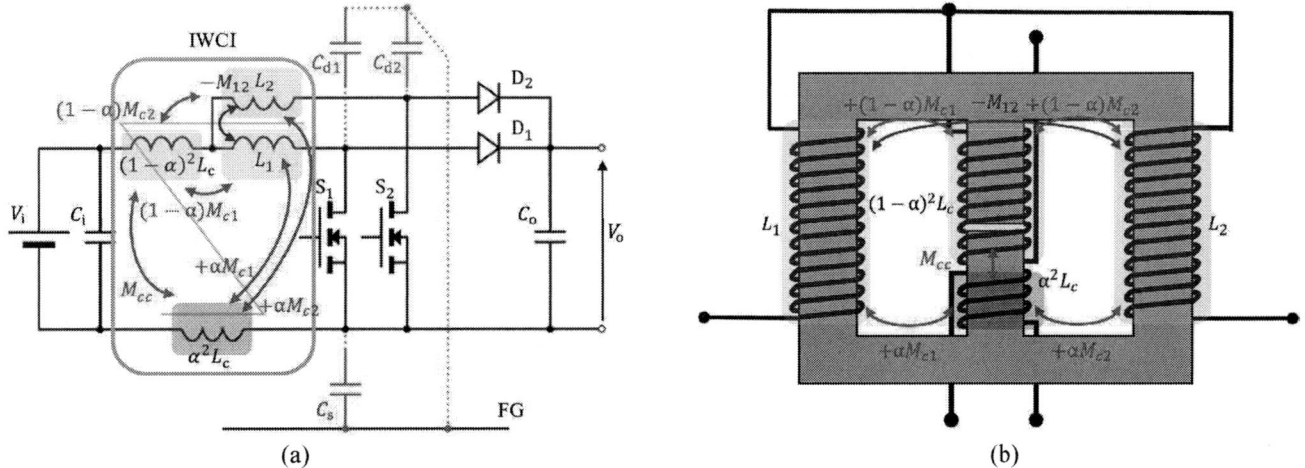

Fig. 3. (a) Circuit diagram proposed balanced interleaved two-phase boost converter using integrated winding coupled inductor (b) Structural diagram of integrated winding coupled inductor with a central leg split according to parasitic capacitance.

phase boost converter. The proposed circuit applies the balanced technique by placing part of the central leg winding on the source side of the switch. The winding split ratio α (ranging from 0 to 1) is defined as the ratio of windings placed on the source side of the switch relative to the total number of windings on the central leg. By adjusting α according to the values of C_d and C_s, the balanced technique condition can be fulfilled without affecting the inductor ripple current. In the LCI method, since there are paths on the source side of the switch for each phase, an additional diode is necessary. On the other hand, in the proposed IWCI method, the path on the source side is combined into a single path, so additional diodes are not necessarily required. This leads to a reduction in component count and enhanced efficiency compared to the conventional LCI method.

B. Noise Reduction Mechanism

In this subsection, CM noise reduction mechanism is explained. Fig. 4 shows the voltage variation in the proposed circuit during operation (Duty = 0.69). Where v_{gs1} and v_{gs2} correspond to the gate signals input to each switch. To enable interleaved driving, the phases are input with a 180° shift. Literals v_{d1}, v_{d2} are the voltages between the drain terminals of the switches in each phase and FG, and v_s represents the voltage between the source terminal of the switch and FG. The modes are also defined as follows.

$$\begin{cases} \text{Mode1 } S_1 = \text{ON} , \ S_2 = \text{OFF} \\ \text{Mode2 } S_1 = \text{OFF}, \ S_2 = \text{ON} \\ \text{Mode3 } S_1 = \text{ON} , \ S_2 = \text{ON} \end{cases} \quad (1)$$

A voltage fluctuation at where occurs due to switching at each mode switching timing. The CM noise current generated in the proposed circuit at this time can be expressed by (2).

$$i_{CM} = C_{d1}\frac{dv_{d1}}{dt} + C_{d2}\frac{dv_{d2}}{dt} + C_s\frac{dv_s}{dt} \quad (2)$$

The source voltage of the proposed circuit fluctuates complementarily with the drain voltage. The inductors on the outer legs mutually influence each other through magnetic coupling. Based on the above, the switching patterns for each

mode are explained. When switching from Mode 3 to Mode 1, SW_2 turns off, causing v_{d2} to increase and v_s to decrease. Additionally, since L_1 and L_2 are negatively coupled, an increase in v_{d2} causes v_{d1} to decrease. In the case of symmetric phases, switching between the other modes can be considered in the same way. Based on Fig. 4, in the switching of all modes, there are nodes where the voltage increases and nodes where it decreases. Therefore, by adjusting the winding split ratio, coupling coefficient, and parasitic capacitance, $i_{CM} = 0$ can be achieved.

IV. DERIVATION OF THE BALANCING CONDITION

This chapter presents the derivation of the balancing condition equation for achieving zero CM noise current in the proposed circuit. Fig. 5 shows the equivalent circuit for CM noise in the proposed circuit. Since CM noise is a high-frequency component, the input and output capacitors with large capacitance are regarded as short circuits [22]. Additionally, the switch voltages shown by v_{n1} and v_{n2} are considered as a noise voltage source, and the diode is considered as a noise current source. CM noise is generally measured using a line impedance stabilization network (LISN). The currents flowing through the outer leg inductors L_1 and L_2 are defined as i_1 and i_2, respectively, and the current flowing through the central leg inductor L_c is defined as i_c. The mutual inductance between the central leg and the outer legs is defined as M_c, and the mutual inductance between the outer legs is defined as M_o. Additionally, the winding split ratio of the central leg is denoted as α. If the coupling coefficient between the central leg windings is 1, the mutual inductance can be expressed by (3).

$$M_{cc} = k\sqrt{(1-\alpha)^2 L_c \cdot \alpha^2 L_c} = \alpha(1-\alpha)L_c \quad (3)$$

If each phase is symmetrical, when the balancing condition of one phase is satisfied, the other phase also satisfies the balancing condition.

The modified CM equivalent circuit, when considering a single switch as a noise voltage source, is shown in Fig. 6. According to the superposition theorem, other voltage sources are considered short circuits, and current sources are considered open circuits. The voltages at both ends of the LISN are defined

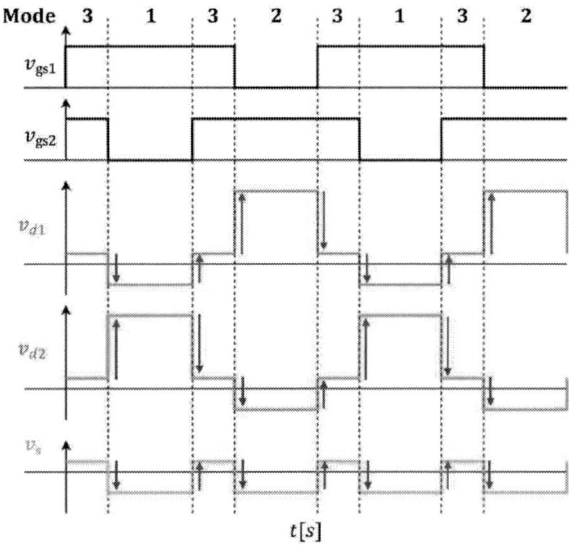

Fig. 4. Operating of the proposed circuit (duty ratio = 0.69).

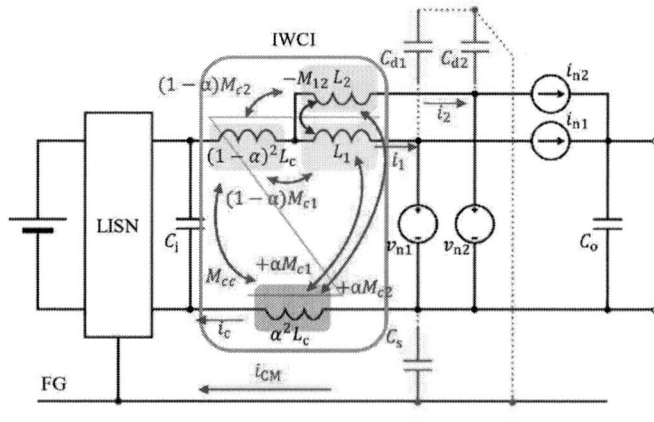

Fig. 5. CM equivalent circuit diagram of the proposed circuit.

as v_a and v_b, and the voltage at the connection point to each phase are defined as v_A. In this case, to achieve $i_{CM} = 0$, the condition $v_a = v_b$, must be fulfilled. From Fig. 6, by applying Kirchhoff's law to v_a, v_A, and v_n, the system of equations can be expressed as follows.

$$\begin{cases} v_a = \alpha L_c \dfrac{di_c}{dt} + \alpha M_c \dfrac{di_1}{dt} + \alpha M_c \dfrac{di_2}{dt} \\[2mm] v_A = M_c \dfrac{di_c}{dt} + M_o \dfrac{di_1}{dt} - L_2 \dfrac{di_2}{dt} \\[2mm] v_A = L_c \dfrac{di_c}{dt} + M_c \dfrac{di_1}{dt} + M_c \dfrac{di_2}{dt} \\[2mm] v_n = v_A + M_c \dfrac{di_c}{dt} + L_1 \dfrac{di_1}{dt} - M_o \dfrac{di_2}{dt} \end{cases} \quad (4)$$

Assuming the phases are symmetric, it follows that $L = L_1 = L_2$ and $C_d = C_{d1} = C_{d2}$. Additionally, when $i_{CM} = 0$ is satisfied, the current flowing through the outer leg inductors is entirely directed through the central leg inductor, so the relation $i_c = i_1 + i_2$ holds. Furthermore, the voltage v_b can be expressed by (5) the following equation using the voltage divider rule.

$$v_b = \frac{C_d}{2C_d + C_s} \times v_n \quad (5)$$

Based on the above, transforming (4) and (5) gives the following expression (6).

$$\begin{cases} v_n = (L + M_o)\left(\dfrac{di_1}{dt} + \dfrac{di_2}{dt}\right) \\[2mm] v_n = (L + L_c + 2M_c)\dfrac{di_1}{dt} + (L_c + 2M_c - M_o)\dfrac{di_2}{dt} \\[2mm] v_n = \alpha(L_c + M_c)\left(\dfrac{di_1}{dt} + \dfrac{di_2}{dt}\right) \times \dfrac{2C_d + C_s}{C_d} \end{cases} \quad (6)$$

By eliminating v_n, i_1, and i_2 from this system of equations, α can be expressed by the following (7).

$$\alpha = \frac{L + 2L_c + 4M_c - M_o}{L_c + M_c} \times \frac{C_d}{2C_d + C_s} \quad (7)$$

Equation (7) is the balancing condition for the proposed circuit, and by adjusting the winding split ratio to satisfy this equality, noise can be minimized to the greatest extent without affecting other design requirements, such as inductor ripple current.

V. SIMULATION VALIDATION OF CM NOISE REDUCTION

A. Purpose of the Verification

In this chapter, CM noise is measured through simulation to verify the noise reduction effect of the balanced technique. The simulation model is built using PLECSver4.7.7. A comparison is made between a conventional circuit, in which the central leg winding in a two-phase boost converter using 1WC1 is placed only on the drain side, and a proposed circuit, in which the central leg winding is split and placed on the source side to enable the balanced technique. Furthermore, by adjusting the winding split ratio α of the central leg winding to satisfy the balancing condition derived in (7), we verify whether significant noise reduction is achieved.

B. Simulation Conditions

In this simulation, the circuit is operated under conditions where the number of central leg windings is increased so that the winding split ratio can be finely adjusted. The simulation conditions are as follows : $V_{in} = 101\text{V}$, Output voltage $V_{out} = 325\text{V}$, Output power $P_{out} = 1\text{kW}$, Duty ratio = 0.69, Switching frequency $f_{sw} = 50\text{kHz}$, Outer leg winding inductance $L = 192\mu\text{H}$, Central $L_c = 709\mu\text{H}$, Outer-to-outer mutual inductance $M_o = 135\mu\text{H}$, Outer-to-central mutual inductance $M_o = 142\mu\text{H}$, $C_d = 470\text{pF}$ $C_s = 470\text{pF}$, The noise reduction effect of the balanced technique targets the noise generated by the parasitic capacitance between the switch and FG. Therefore, other parasitic components are not considered. By substituting the above parameters into (7), $\alpha = 0.8$ is obtained. For this verification, α is varied from 0.7 to 0.9 for the analysis. The central leg winding consists of 60 turns. When $\alpha = 0.7$, 42 turns are placed on the source side and 18 turns are placed on the drain side. Similarly, when $\alpha = 0.8$, 12 turns are placed on the source side and 48 turns are placed on the drain side. Likewise, when $\alpha = 0.9$, 54 turns are placed on the source side and 6 turns are placed on the drain side.

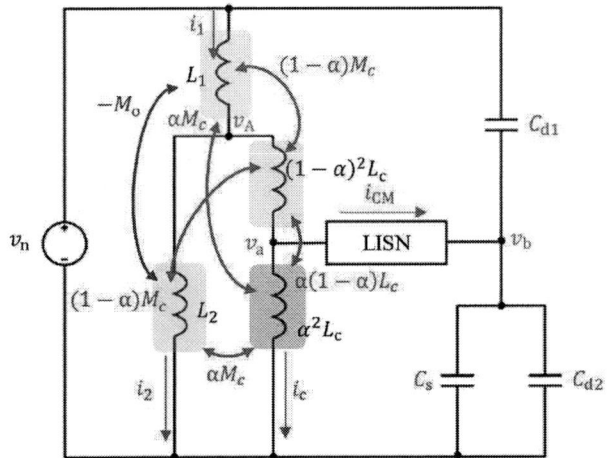

Fig. 6. Wheatstone bridge circuit centered on LISN with one switch as a noise voltage source.

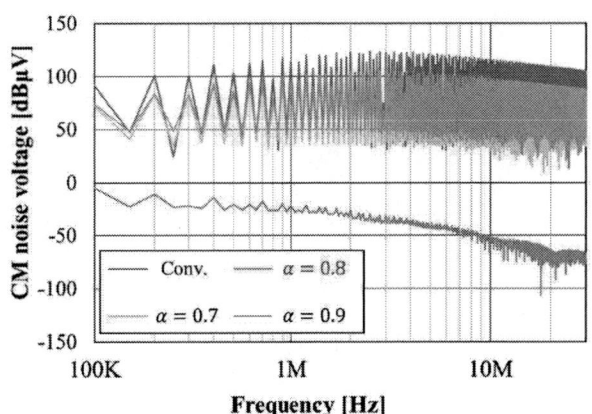

Fig. 7. Simulation results of CM noise with varying α (Balanced conditions are satisfied when $\alpha = 0.8$).

979-8-3315-1612-3/25 $31.00 © 2025 IEEE

C. Simulation Results

The measurement results of the CM noise in the steady state of the circuit are shown in Fig. 7. The results show that in the proposed circuit, CM noise is reduced by more than 12 dB for all winding split ratios α, confirming the reduction effect of the Balanced technique. At the balancing condition of $\alpha = 0.8$, the CM noise is attenuated by 120 dB, indicating that almost no noise current is flowing. Therefore, the validity of the balance condition derived in (7) was confirmed.

VI. Experimental Verification Results

This chapter presents the experimental results obtained using actual equipment and demonstrates the effectiveness of the proposed circuit in reducing CM noise. Fig. 8 shows the prototype board used for the experiment and the structure of the IWCI. The magnetic core of the IWCI uses "ML91s EE21.8-11.4-15.8 made from PROTERIAL," with an air gap of 0.8 mm at central legs The winding consists of six PCB layers, and the winding split ratio α can be adjusted by inserting separation layers between them. Table II presents the operating conditions and the parameters of the IWCI. In this verification, an external capacitor is added to ensure the consistency of the balanced technique conditions. Also, by substituting the parameters above into (7), $\alpha = 0.167$ is obtained. Therefore, the experiment measures CM noise for three patterns: non-balanced $\alpha = 0$ (conventional), balanced $\alpha = 0.167$, and $\alpha = 0.333$, which does not satisfy the balancing condition (7). Fig. 9 provides a comparison of the measured CM noise spectra. The bold lines represent the envelope of each CM noise spectrum. With $\alpha = 0.167$, the noise reduction reaches a maximum of 50 dB and an average of approximately 30 dB compared to the conventional circuit. In contrast, with $\alpha = 0.333$, there is no noticeable noise reduction in the 1 to 3 MHz range, and the reduction is approximately 15 dB above 3 MHz. This is because the external capacitor is large, causing a significant deviation from the balanced condition equation due to changes in the winding split ratio. Therefore, the experimental results confirmed the validity of the balanced condition equation and the noise reduction effect of the proposed circuit.

TABLE II
SPECIFICATIONS OF PROPOSED CIRCUIT

Parameters		Value
Input Voltage	V_{in}	40V
Output Voltage	V_o	129V
Operating Frequency	f_{sw}	1MHz
Output Power	P_{out}	300W
Main Switch (GNP1150TCA-Z)	SW	650V/11A
Output diode (SCS210AJHRTLL)	D	650V/10A
Central leg Inductance	L_c	5.04μH / 6turns
Outer leg Inductance 1	L_1	100.1μH / 12turns
Outer leg Inductance 2	L_2	100.4μH / 12turns
Central to Outer leg Mutual Inductance	M_c	4.61μH
Outer to Outer leg Mutual Inductance	M_o	85.59μH
External Drain side Parasitic Capacitance	C_d	100pF
External Source side Parasitic Capacitance	C_s	2420pF

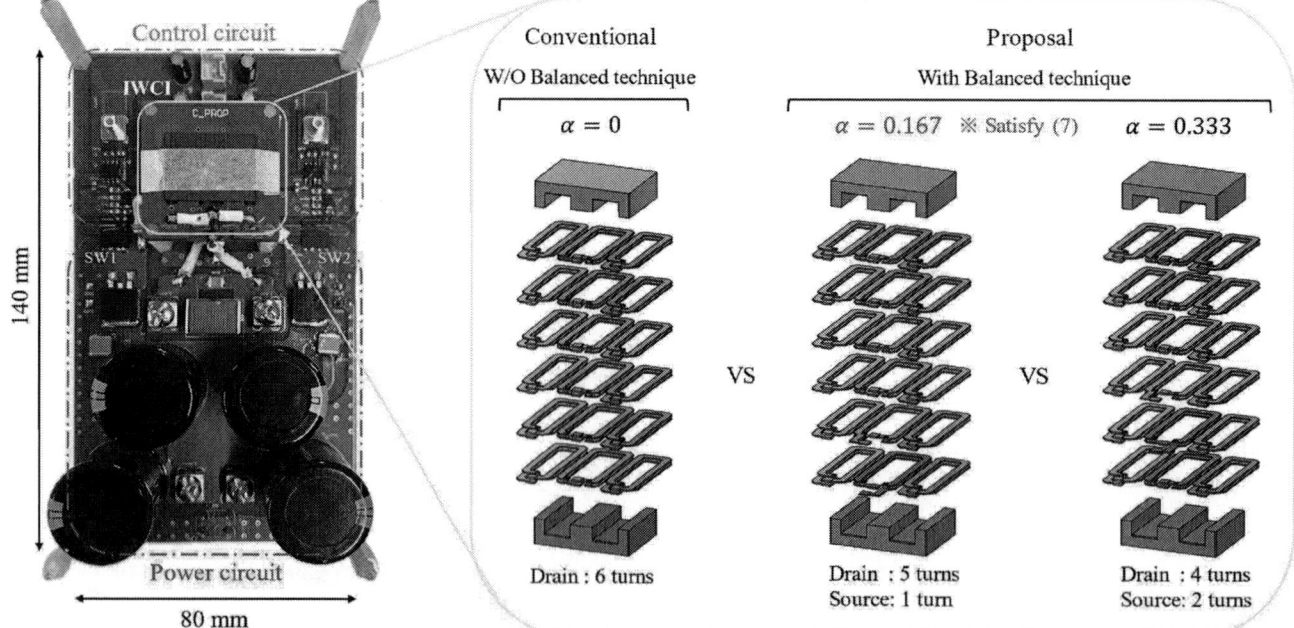

Fig. 8. Prototype circuit of the proposed two-phase interleaved boost converter and integrated winding coupled inductor with adjustable winding split ratio.

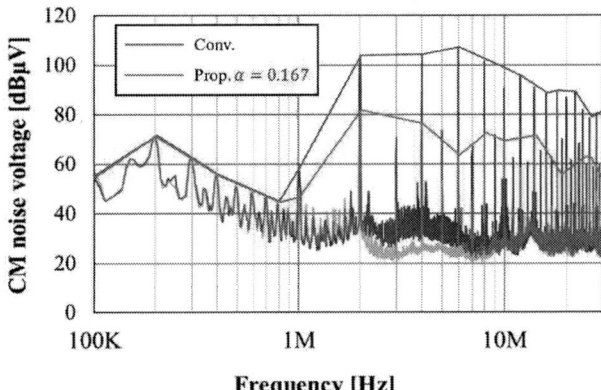

(a)

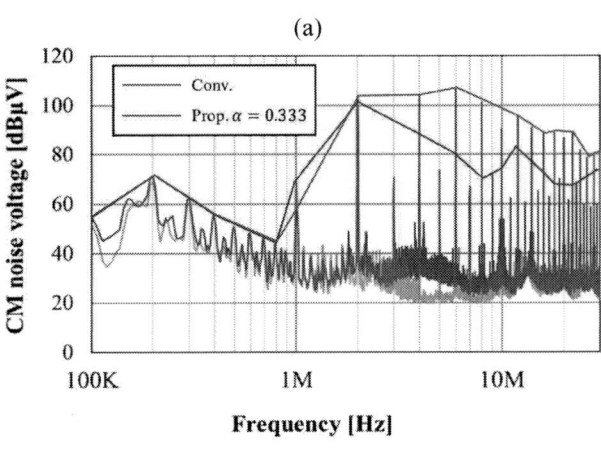

(b)

Fig. 9. CM noise measurement results from the prototype circuit (a) α =0.167 satisfied balanced condition (b) α =0.333 no satisfied balanced condition.

VII. CONCLUSION

In this paper, we propose an interleaved two-phase boost converter with a noise reduction mechanism based on the balanced technique using IWCI. The noise reduction principle of the proposed circuit was theoretically analyzed, and the balancing condition was derived. We validated the noise reduction effect through simulations and experiments. In addition to the downsizing effect of the IWCI, the elimination of unnecessary EMI filter volume achieves significant improvements in power density.

REFERENCES

[1] CC BY 4.0 IEA (2024), "Global EV Outlook 2024", https://www.iea.org/reports/global-ev-outlook,-pp110-118, 2024.

[2] IPCC, 2023: Summary for Policymakers. "Climate Change 2023: Synthesis Report". Contribution of Working Groups I, II and III to the Sixth Assessment Report of the Intergovernmental Panel on Climate Change [Core Writing Team, H. Lee and J. Romero (eds.)]. IPCC, 2023, pp. 1-34

[3] H. A. Mantooth, M. D. Glover, and P. Shepherd, "Wide bandgap technologies and their implications on miniaturizing power electronic systems," *IEEE J. Emerg. Sel. Topics Power Electron.*, vol. 2, no. 3, pp.374- 385, Sep. 2014.

[4] D. Han, S. Li, W. Lee, W. Choi and B. Sarlioglu, "Trade-off between switching loss and common mode EMI generation of GaN devices-

analysis and solution," *in Proc. of 2017 IEEE Applied Power Electron. Conf. and Expo. (APEC)*, 2017, pp. 843-847.

[5] N. Oswald, P. Anthony, N. McNeill and B. H. Stark, "An Experimental Investigation of the Tradeoff between Switching Losses and EMI Generation With Hard-Switched All-Si, Si-SiC, and All-SiC Device Combinations," *IEEE Trans. Power Electron.*, vol. 29, no. 5, pp. 2393-2407, May. 2014

[6] Wu Xin, M. H. Pong, et.al., "Novel boost PFC with low common mode EMI: modeling and design," *in Proc. IEEE Applied Power Electron. Conf. Expo. (APEC)*, 2000, pp. 178-181.

[7] M. Kchikach, Z.M. Qian, X. Wu, M.H. Pong, "The influences of parasitic capacitances on the effectiveness of anti-phase technique for common mode noise suppression", *in Proc. IEEE Int Conf. Power Electronics Drive Systems*, vol.1, pp.115-120, 2001.

[8] M. Sasaki, J. Imaoka, M. Yamamoto, A. Nakano, K. Fuse , "Common mode conductive noise cancellation for multiphase converter using auxiliary winding ," *Electr Eng Jpn.*, no. e23439, pp. 1-11. 2023.

[9] M. Shoyama, G. Li, and T. Ninomiya, "Balanced Switching Converter to Reduce Common-Mode Conducted Noise," *IEEE Trans. Ind. Electron.*, vol. 50, no. 6, 2003, pp. 1095-1099.

[10] Q. Lin, S. Zhang, B. Zhang and M. Shoyama, "Balanced Technique for Common Mode EMI Suppression in Transformer-less PV Systems," *in Proc. 2020 IEEE 9th International Power Electronics and Motion Control Conference (IPEMC2020-ECCE Asia)*, 2020, pp. 1900-1904.

[11] S. Wang, P. Kong and F.C. Lee, "Common mode noise reduction for boost converters using general balance technique", *in Proc. 2006 37th IEEE Power Electron. Specialists Conf.*, pp. 1-6, 2006.

[12] S. Wang, F. C. Lee and Q. Li, "Improved Balance Technique for Common-Mode Noise Suppression of PCB-Based PFC," *IEEE Trans. Power Electron.*, vol. 37, no. 4, pp. 4174-4182, April. 2022.

[13] J. Imaoka, M. Yamamoto, Y. Nakamura, and T. Kawashima, "Analysis of output capacitor voltage ripple in multi-phase transformer-linked boost chopper circuit," *IEEJ J. Ind. App.*, vol. 2, no. 5, pp. 252-260, 2013.

[14] Y. Wang, P. Cheng, X. Ma, et al., "Design of miniaturized and lightweight coupling inductors for interleaved parallel DC/DC converters," *J. Power Electron.* 21, pp. 1439-1450, 2021.

[15] Pit-Leong Wong, Peng Xu, P. Yang and F. C. Lee, "Performance improvements of interleaving VRMs with coupling inductors," in IEEE Transactions on Power Electronics, vol. 16, no. 4, pp. 499-507, July 2001.

[16] W. Huang and B. Lehman, "A compact coupled inductor for interleaved multiphase DC–DC converters", *IEEE Trans. Power Electron.*, vol. 31, no. 10, pp. 6770-6775, Oct. 2016.

[17] M. Pavlovsky, G. Guidi, and A. Kawamura, "Assessment of coupled and independent phase designs of interleaved multiphase buck/boost DC-DC converter for EV power train", *IEEE Trans. on Power Electron.*, vol. 29, no. 6, pp. 2693-2704, 2014

[18] J. Imaoka, K. Okamoto, M. Shoyama, et.al.,"A Magnetic Design Method Considering DC-Biased Magnetization for Integrated Magnetic Components Used in Multiphase Boost Converters", *IEEE Trans. on Power Electron.*, vol. 33, no. 4, pp. 3346-3362, 2018.

[19] M. Hirakawa *et al.*, "High power DC/DC converter using extreme close-coupled inductors aimed for electric vehicles," *in Proc. the 2010 International Power Electronics Conference (IPEMC2010-ECCE Asia)*, 2010, pp. 2941-2948.

[20] J. Imaoka, K. Umetani, S. Kimura,et.al.,"Magnetic Analysis, Design, and Experimental Evaluations of Integrated Winding Coupled Inductors in Interleaved Converters," *IEEJ Journal of Ind. App.*, vol.5, no3, pp276-288, 2016.

[21] T. Nagai, M. Sasaki, J. Imaoka, M. Yamamoto, A. Nakano, "Balanced Two-phase Interleaved Boost Converter with Integrated Magnetics for Common-Mode Noise Reduction," *IEEJ Journal of Ind. App.*, vol. 12, no. 4, pp. 664-675, 2023

[22] S. Choi, Y. Yin, J. Shin, J. Imaoka and M. Yamamoto, "Common-Mode Noise Reduction for Bridgeless Flyback PFC Rectifier with Balance Technique," in Proc. IEEE Appl. Power Electron. Conf. Expo., 2024, pp. 1478-1483

MagNetX: Foundation Neural Network Models for Simulating Power Magnetics in Transient

Shukai Wang°, Hyukjae Kwon°, Haoran Li°, Youssef Elasser‡,
Gyeong-Gu Kang°, Daniel Zhou°, Davit Grigoryan°, and Minjie Chen°
°Princeton University, Princeton, NJ, United States
‡Nvidia Research, Durham NC, United States
Email: {sw0123, hk1715, minjie}@princeton.edu

Abstract—This paper introduces a foundation neural network framework for modeling power magnetics in transient, based on MagNetX[1] – a new extension of the MagNet database which includes extensive measurement data in transient. Provided with flux density $B(t)$ and field intensity $H(t)$ waveforms, the model uses partial memories and the next-state flux density excitation to predict the response of the field intensity in the next time step. The model is in time domain and is frequency independent. An example sequence-to-scalar LSTM neural network was designed, trained, and tested. This modeling framework can greatly enhance the modeling and design of power magnetics operating in transient condition, such as in PFCs and power amplifiers.

Index Terms—power magnetics, hysteresis loop, machine learning, data-driven method, neural network, transformer.

I. INTRODUCTION

Power magnetics are key components for storing energy in power electronic circuit designs, but they tend to be lossy and occupy a large volume, leading to less efficient and dense converter designs. Equation 1 depicts the relationship between volumetric core loss P_v and B-H loop. Material core losses exhibit highly nonlinear characteristics due to the intricate hysteresis mechanisms inherent in magnetic materials.

$$P_V = \frac{1}{T} \int_{B(0)}^{B(T)} H(t)\, dB(t) \tag{1}$$

The impact of frequency, temperature, dc-bias and other factors challenge existing loss and hysteresis models [1]. Most existing hysteresis models can be divided into two groups: traditional, or data driven method. Traditional magnetic hysteresis models include the Stoner-Wohlfarth model [2], the Jiles-Atherton model [3], and the Preisach model [4]. All three models can approximately trace the hysteresis curve but struggle to accurately capture magnetic behavior under specific operating conditions, especially temperature and dc-bias. While extended models and improved Steinmetz equations have been developed to address these limitations [5], most remain confined to static hysteresis analysis, and lacking the capability to model transient condition where the excitation behavior rapidly changes [6]–[9].

Most converter topologies, however, introduces transient conditions when the load behavior changes. In certain topology

[1]MagNetX GitHub Repository: https://github.com/PaulShuk/MagNetX

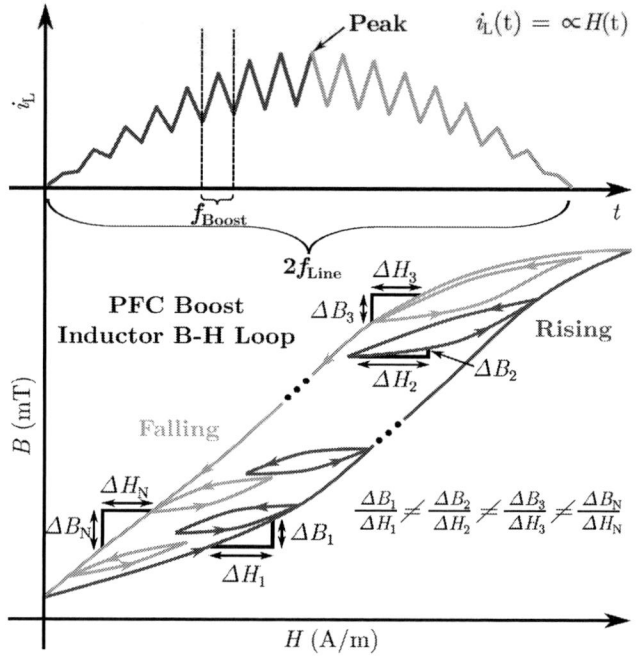

Fig. 1. Example of PFC-Boost multi-cycle transient B-H hysteresis: inductor current waveform and dynamic cycle transient condition with numerous minor loops.

such as power factor correction boost converter–PFC Boost, contains composite inductor voltage and current waveform that is highly dynamic [10], [11]. Figure 1 depicts an example PFC boost inductor current as well as B-H loop waveform. Since the front-end provides a low frequency period ($2f_{Line}$), and back-end switches at much higher frequency ($f_{Boost} = 100f_{Line}$), the inductor voltage and current are composite waveforms operating between two periods. Different from traditional static B-H loop, the slopes of the curve are dynamic in nature and the B-H position is heavily dependent on the past states. This highly complex hysteresis behavior renders the modeling of inductors in these topologies challenging.

The recent advanced data-driven method utilizing machine learning with neural network (NN) architectures have garnered much attention. Though there are techniques that predict hysteresis curves with simple neural network structures [12], or in combination with Preisach state memory [13], more advanced neural network structures such as LSTM [14] and Transformer

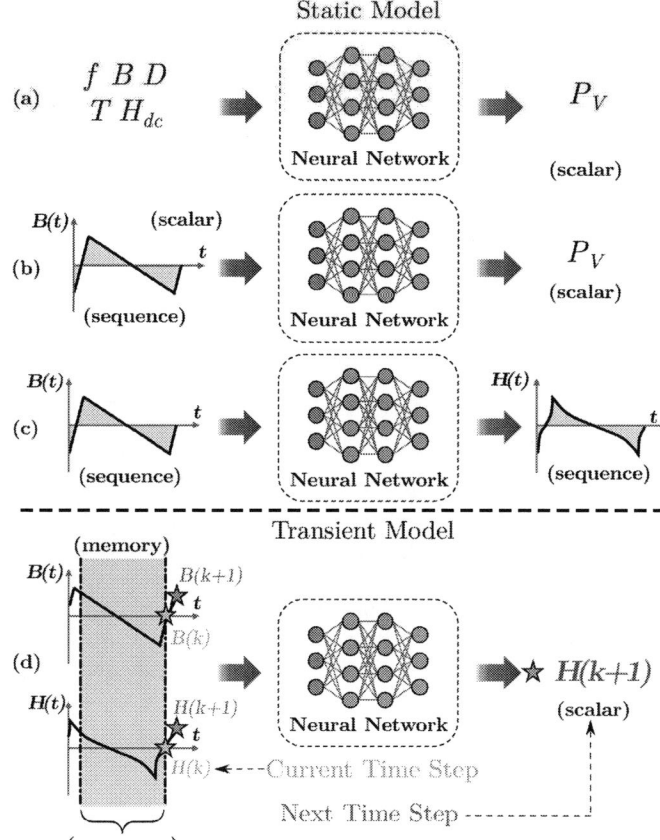

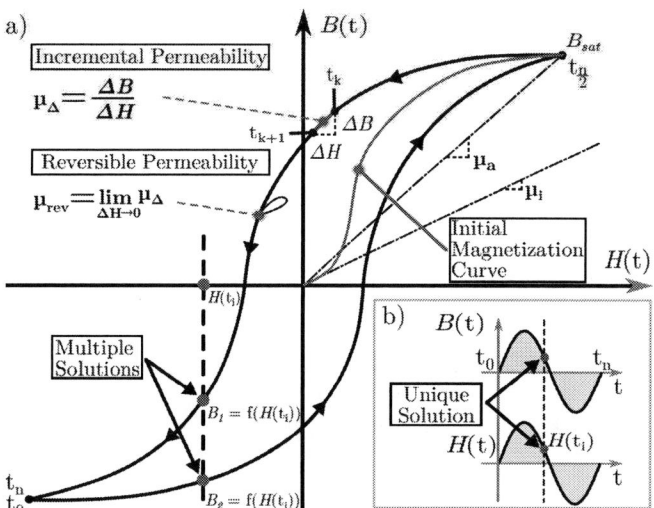

Fig. 3. Definitions of permeability on the B-H loop: a) static hysteresis loop on the B-H plane; b) B and H waveform on time-series scale.

Fig. 2. A variaty of machine learning models: a) scalar-to-scalar; b) sequence-to-scalar; c) sequence-to-sequence; d) dual sequence-to-scalar models.

[15] based architecture have proven to be solving multi-variable non-linear regression problems with high accuracy [16]. Aiming to characterize core loss and hysteresis behaviors, MagNet[2] – an open-sourced core loss and static hysteresis database, built and deployed its first AI model for multiple soft ferrite materials utilizing both LSTM [17] and Transformer [18]. The MagNet project has created models for multiple purposes.

Figure 2a depicts a simple scaler-to-scaler neural network structure that maps scalar inputs of various operating parameters to volumetric core loss. However, this method is limited by its waveform types, hence not great for model generalization. As a result, time sequence $B(t)$ is introduced to the NN structure in Fig. 2b, which is capable of capturing arbitrary waveform type in its core loss model. The MagNet Challenge [19] yielded significantly more efficient and accurate neural network models for core loss modeling, advancing the Pareto frontier of modeling techniques. Lastly, the sequence-to-sequence model depicted in Fig. 2c is developed to map time sequence $B(t)$ to $H(t)$ for better exploration of a comprehensive steady-state hysteresis behavior [17], [18]. Figure 1a shows a seq-to-seq testing result of mapping a steady state full cycle B to its corresponding H sequence. The transformer model performed well at predicting correlations between two steady-state time sequences. All three models are trained based

---

[2]Princeton MagNet Project Website: https://mag-net.princeton.edu/

on the steady-state full-cycle data provided in the MagNet database. In practice, however, power converters operate in both steady-state and transient conditions, rendering static models unable to correctly predict the transitional behavior between the steady-states. The existing models can not track a sequence with ever-changing duty cycles, or a full cycle with many minor loops as shown in Fig. 1b.

To model the transient behavior of power magnetics, a foundation neural network framework is needed which can generalize magnetic B-H behavior with minimal input constraints, eliminating the need to predefine specific parameters such as waveform type, frequency, or duty cycle. The key factor that governs the behavior of hysteresis curve is the precise relationship between $B(t)$ and $H(t)$ at any given time step. A pair of $B(t)$ and $H(t)$ sequences have a strong correlation and both exhibit memory effects. The excitation B in the subsequent state has a direct influence on the corresponding H, with their relationship additionally dependent on the system's previous state and the magnetic memory accumulated in recent states. Therefore, we extend the MagNet database to include transient data (the MagNetX database described in [20]), and propose a generalized MagNetX machine-learning framework for modeling the dynamic hysteresis behavior of materials. The new model is shown in Fig. 2d, predicting the future state based on the input of both B and H historical sequences.

The rest of the paper is organized as follows: Section II introduces the principles modeling B-H relations in dynamic state; Section III pictures the structure and the data flow of the LSTM-based Encoder-Projector architecture, modified from the existing MagNet sequence-to-sequence architecture; Section IV discusses how the training data is constructed; Section V shows the experimental results. Finally, Section VI concludes this paper.

II. TRANSIENT STATE MODELING

Ferromagnetic Material exhibits non-ideal hysteresis B-H loops shown in Fig. 3a. Permeability (μ) is the fundamental parameter governing magnetic hysteresis behavior, following

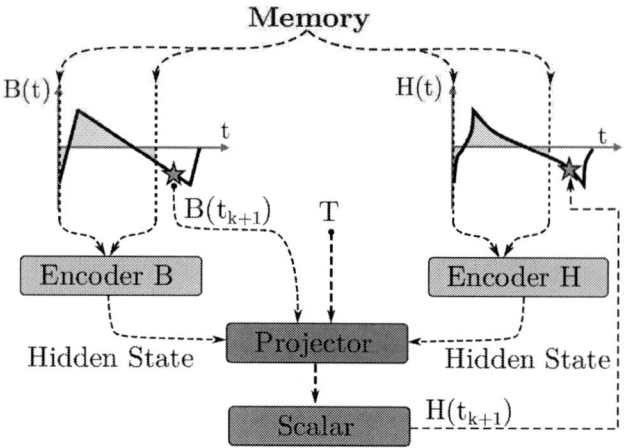

Fig. 4. Progression flow of the proposed neural network architecture. The system takes in both B and H memory sequences, and outputs the scalar H.

Algorithm 1 LSTM-based Sequence-to-Scalar Model

Input:
> Segmented Flux Density $B_{mem}(t)$, Segmented Field Strength $H_{mem}(t)$, Next Step Flux Density $B(t_{k+1})$ Temperature T, Next Step Field Strength $H(t_{k+1})$ (only available in training);

Output:
> Next Step Magnetic Field Strength $H(t_{k+1})$;

1: Initialize hideen states h_0 and cell states c_0;
2: $X_1 \leftarrow B(t_1)$;
3: $X_2 \leftarrow H(t_1)$;
4: **for** $i = 1$ to L **do**
> $h_i, c_i \leftarrow \text{LSTM}_1 (h_{i-1}, c_{i-1}, B(t_i))$ **[Encoder B]**

5: **for** $i = 1$ to N **do**
> $h_i, c_i \leftarrow \text{LSTM}_1 (h_{i-1}, c_{i-1}, H(t_i))$ **[Encoder H]**

6: $Y' \leftarrow \text{FNN}(X_1, \; X_2, \; B(t_{k+1}), \; T)$;   **[Projector]**
7: **return** $H(t_{k+1})$;

$B = \mu H$. However, numerous permeability types exist, with each defining distinct characteristics of the hysteresis curve, capturing different aspects of magnetic material behavior. Figure 3 presents the definition of various permeability types. Manufacturer often tends to provide either the initial permeability (μ_i), or amplitude permeability. Both are good metrics used for characterizing and classifying different materials, but lacks the ability to delineate the B-H loop on their own. On the hysteresis curve, the slope changes dynamically throughout the excitation cycle.

One could try generating a unique mapping between B and H to characterize dynamics for each hysteresis pair. Figure 3a has demonstrated, however, that if we apply an excitation of $H(t_i)$, a direct mapping could result in more than two non-identical solutions such as B_1 and B_2. To generate a unique mapping, we invoke the third dimension, time. As shown in Fig. 3b, time stamps can be used as tokens for each $B(t_k)$ - $H(t_k)$ pair which results in a unique solution. With each time sequence $B(t)$ and $H(t)$ bearing a total of n time steps, one may calculate a corresponding μ_Δ that yields a unique slope at each step. To calculate μ_Δ, or slope of the curve, information on the current time step $B(t_k)$, $B(t_k)$, $H(t_k)$, $B(t_{k+1})$, and $H(t_{k+1})$ needs to be known. The μ_Δ in this case is calculated as

$$\mu_\Delta = \frac{1}{\mu_0} \frac{\Delta B}{\Delta H} = \frac{1}{\mu_0} \frac{B(t_{k+1}) - B(t_k)}{H(t_{k+1}) - H(t_k)}, \quad (2)$$

where ΔB and ΔH are incremental changes of $B(t)$ and $H(t)$ from k^{th} time step to $(k+1)^{th}$ in a period. μ_Δ represents the slope of the B-H curve between the kth and $(k+1)^{th}$ point. If the current time step excitation $B(t_k)$, $H(t_k)$ and μ_Δ are known, then given a new excitation $B(t_{k+1})$ in the next time step, the $H(t_{k+1})$ would be obtained by

$$H(t_{k+1}) = H(t_k) + \frac{B(t_{k+1}) - B(t_k)}{\mu_0 \mu_\Delta} \quad (3)$$

at any given time in a dynamic system. There are works that have attempted to utilize simple series interlinked feedforward neural networks to predict the μ_Δ [21], [22]. However, the model is restricted to the transient sequence that has a strong resemblance to sinusoids. The framework should be able to utilize the input information and correctly map out slope μ_Δ

needed to obtain $H(t_{k+1})$. In the proposed setup, $H(t_{k+1})$ is directly used as the output of the system, since measurement data could introduce noises that result in a near-infinite μ_Δ slope between the current and next state. Thus, to better predict the corresponding response of the future state, the NN structure should at minimum include the information of the current time step B-H pair.

In addition, each state has strong dependencies on the waveform shape and trends of the immediate past. The presence of any distortion, ripple, large spikes, or noises may have different degrees of impact on the result of the next time step. Scalar inputs of the current state are not enough to represent the true behavior of the next time step. Therefore, the input of both B and H should both be sequences of most recent memories to properly describe the behavior of the next state. Thus the overall framework of the machine learning model revolves around utilizing a small portion of the immediate memory of both B-H pair for each time step to predict the immediate next state. The sequence-to-scalar framework is first used to predict all the $H(t_{k+1})$, then forming the whole sequence with all the scalar outputs as the fundamental building blocks.

III. MEMORY-BASED LSTM NEURAL NETWORK STRUCTURE

The LSTM is a commonly used NN framework that is best suited for sequential inputs solving complex regression problems [14]. Stock market predictions are good examples of applying LSTM networks to time series data with complicated dynamics [23]. The memory-based prediction of arbitrary waveforms is conceptually similar to the prediction of stock prices, both involving complex, nonlinear time-series prediction challenges that require capturing historical patterns and state-dependent dynamics. Therefore, as shown in Fig. 4, we select memory sequence $B(t)$ and $H(t)$ as input sequences. The output is the next state field intensity $H(t_{k+1})$. From Eq. 2, the future step $B(t_{k+1})$ is necessary for informing the $H(t_{k+1})$. In addition, variable such as temperature externally

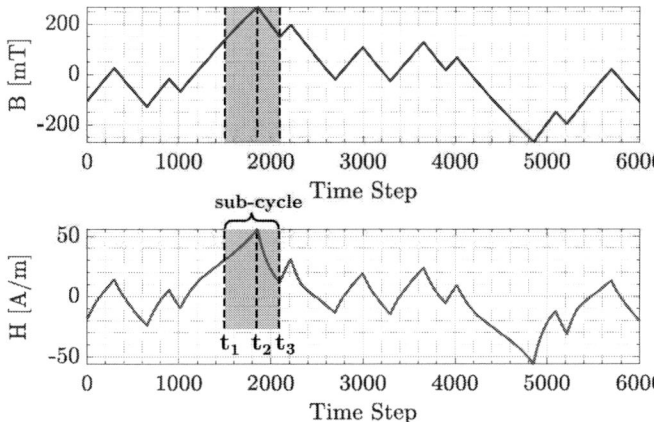

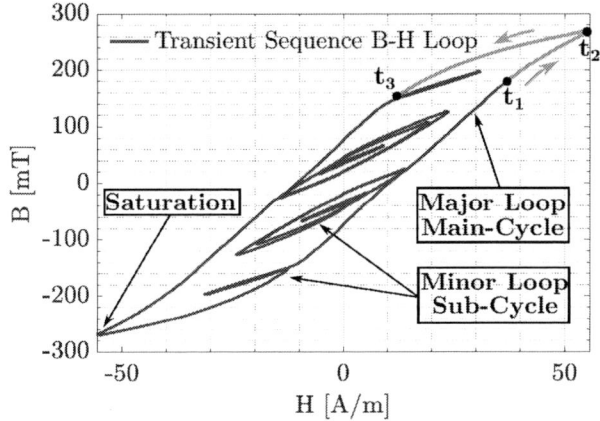

Fig. 5. Example measurement $B(t)$ and $H(t)$ data sequence. Waveform contains 10 cycles of data that is sampled with 6,000 points.

Fig. 6. A single B-H loop, with many transient minor loops, echoing the behavior found in Fig 5.

influences the hysteresis behavior as well. Therefore, a projector module is placed past the encoder stage, where the scalar input of the $B(t_{k+1})$, and temperature (T) are implemented. Other variables such as frequency, dc-bias, and duty cycle are not necessary as explicit inputs since the time series data $B(t)$ and $H(t)$ intrinsically encode their information.

In the proposed NN structure, memory segments $B(t)$ and $H(t)$ are individually fed through separate LSTM modules to address the temporal correlation of the most recent memories to the distant past. LSTM module will first characterize the individual sequence by memorizing only the important information such as the waveform's turning points, peaks, and troughs. Then, the output of the LSTM cells provides the updates of the hidden states and is loaded into the feedforward neural network (FNN). With the provided $B(t_{k+1})$ and T as scalar inputs to the FNN, the projector performs the learning process to determine the output $H(t_{k+1})$. The detailed architecture can be described in Fig. 7. The training flow of the model are detailed in the pseudo code in Algorithm 1.

IV. DATA COLLECTION AND PROCESSING

High quality machine learning model always rely on high quality database. MagNet database collects data using the four wire method [24], and controls frequency under 500 kHz to minimize probe delay error. A T-type inverter is used as the power stage to provide excitations to the windings of the core under test (CUT). The previously open-sourced data only contain steady-state voltage $v(t)$ and current $i(t)$ pairs. In order to model the transient hysteresis, MagNet database is updated with new triangular current dynamic data for 10 materials [20]. The corresponding data collection sequence is developed aiming at providing transient data measurement. The transient data is created by creating sequences of randomly changing duty cycles during each period of switching. The voltage current sequences are inferred to corresponding B-H pairs.

Figure 5 shows an example data collection of B and H waveforms under triangular (current) excitation. Step changes are applied by varying random sequence of duty cycle changes. The CUT is excited with 10 transient sub-cycles forming the main cycle. The sub-cycles are excited at 50 kHz, whereas the total excitation period is 5 kHz. Each sub-cycle is represented as the minor loops on the B-H curve in Fig. 6. The sub-cycles together form a major cycle, mimicking transient hysteresis conditions. As current increases, the CUT undergo saturation which can be observed in the H waveform as well as the B-H loop. The exampled waveform is post processed by down sampling to 600 points per sub-cycle.

There are various number of combinations when applying the voltage duty cycle step changes. Magnetics react differently when voltage excitation changes from 20% to 50% verses from 80% to 50%. Same set of duty cycles step changes at 50 kHz and 500 kHz also appear to be different. The model is likely to benefit from a highly diverse and extensive combinations of step changes. In the final database, each sequential data contains 100 normalized distributed random duty cycle step changes to enrich the diversity of magnetic response to various transient states. A total of 13,587 sequences of data is collected for 3C90, including frequencies ranging from 50 kHz to 800 kHz at three temperatures (25, 50, 70 °C). Given the horizontal resolution of the oscilloscope, 100 sub-cycle sequence achieves a good balance between waveform degradation and combination richness. Figure 8 showcases one of such sequence. Each long sequence is downsampled to 2,000 points for the whole period, or 20 points per sub-cycle. This is the maximum amount of downsampling before the loss of critical information. In addition, the data is prone to having noises and sharp spike due to circuit network's capacitive effect during testing. Therefore, a post-processed software filter was also applied to eliminate data with extremely large slopes to avoid confusion with true step changes in training.

Figure 8 showcases an example downsampled data sequence being prepared for training. In this example, The CUT employs material Ferroxcube 3C90. The original sequence length is sampled with 2000 points which is represented by $B(t_{k+1})$ and $H(t_{k+1})$ respectively. For training such a sequence, the model takes each $H(t_{k+1})$ point as an output. This implies that each $H(t_{k+1})$ is associated with a corresponding $B(t_{k+1})$ scalar input and both B_{mem} and H_{mem} segmented memory as sequence inputs. The model's input encompasses both the current excitation point $B(t_k)$ and $H(t_k)$ and several preceding states near the present time step, with the number of past states determined by the strength of the material's memory effect. Since a single 100 sub-cycle period is continuous from

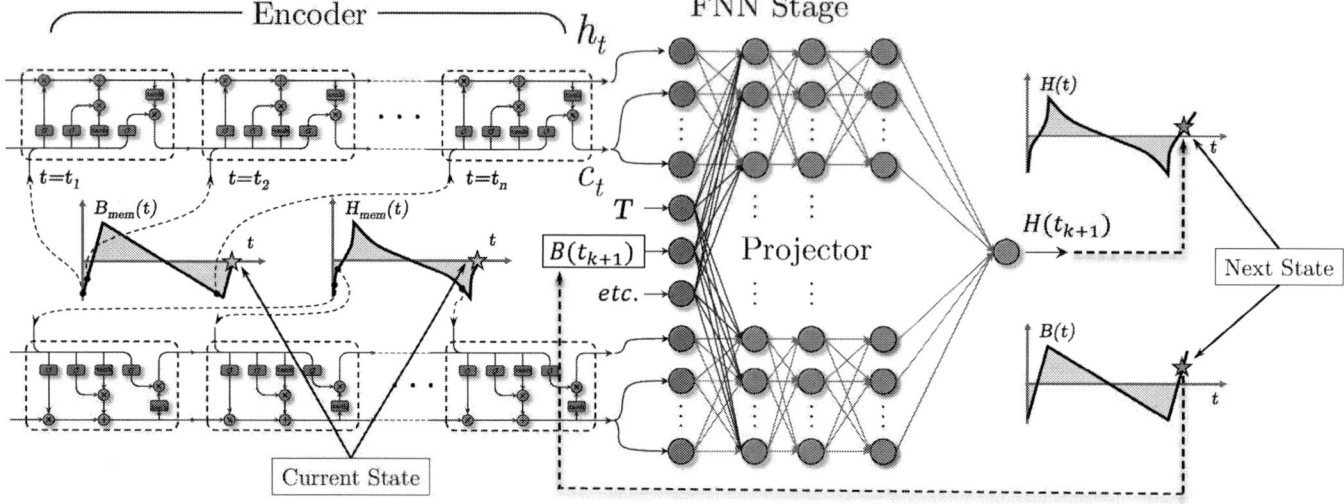

Fig. 7. Overview of the dual encoder, projector architecture. The model takes in two segmented memory inputs and each point on the sequence is fed sequentially into concatenated LSTM cell. The projector then takes in the hidden, and cell states into the projector. The next state $B(t_{k+1})$ and $H(t_{k+1})$ are siphoned from the original sequence and fed into the projector and output.

the end to the beginning, a segment length equivalent to the memory window size is copied and appended in front of the sequence for the training of the first few points. The specific segmented memory points chosen for material 3C90 is selected as 60. When selecting window range, there are a couple important considerations.

(1) **Memory Length:** Theoretically, the magnetic remanence lasts indefinitely, making every historical point relevant to future excitation behavior. However, given the data storage limit, it is acceptable to "remember" only the most recent few cycles depending on the gravity of the memory effect tailored to each specific material.

(2) **Peak Values:** The local peaks and troughs are of high significance in determining future excitation states, exerting greater influence than other historical points in the sequence. It may be crucial to selectively store the positive and negative peaks values from distant past sub-cycles, rather than the entire waveform data.

The memory is selected based on the memory effect of the highest frequency (500 kHz as tested) as it sees more dynamic changes in the past number of points than any other frequencies. Given that 60 points are chosen for each segmented memory sequences, the selected material only retains memory of the past 3 cycles. As seen in Fig 8, each sub-cycle in the memory window inherit a different duty cycle excitation.

V. NEURAL NETWORK TRAINING AND TESTING

For training the neural network, 600 sequences comprised of frequencies from 50 kHz, 200 kHz, and 500 kHz at 25°C are selected. Since each sequence contains 2000 output points for training, the total sets of output data is 1,200,000, with 2,400,000 input sequences. The final post processed data size is 2.0 GB. The LSTM-based encoder-projector sequence-to-scalar model is synthesized with the PyTorch framework. Hyper-parameters of the network structure are determined and optimized based on experimental training results. For the

model proposed in this work, both LSTM cells have 32 cell states and 32 hidden states. The output of the LSTM cells are fed into the projector which comprises 3 layers with 64 neurons in the first two layers, and 32 neurons in the third layer. The total number of parameters trained for the model is 19,587. The mean-squared error (MSE) between the predicted scalar output $H_{pred}(t_{k+1})$ and the measured $H_{meas}(t_{k+1})$ is selected as the loss function for the backpropagation during training. The proposed framework was trained with 3,000 epochs using the Princeton Research Computing Della Cluster on the Nvidia A100 80GB GPU devices. The total training took around 4 hours.

There are two steps to evaluating the model. Step 1: The model is to be tested with the measured dataset of H_{mem}, to validate the robustness of the model at predicting the next time step point. Step 2: The H_{mem} is updated iteratively, incorporating the newly predicted value from the previous iteration. If the first step demonstrates highly accurate prediction, only then will the second step be considered. The trained LSTM model in this paper is evaluated on the test set based on the implementation of the first step. The testing dataset employs the 600 sequences at 25 °C. All data are tested and recorded as scalar outputs, before all 2,000 points are synthesized to form the sequence again. Figure 9 depicts the model prediction of the dataset compared to the raw dataset. The model makes prediction closely matches with the measurement result, even during the saturation period. A zoomed in view provides a more detailed look at the prediction accuracy. The figure on the very right represents the final converged results. As the training proceeds, the model gradually converges and the discrepancy between the predicted and the measured hysteresis loops is minimized, eventually achieving a good match.

In addition, a reference $H(t)$ is calculated using Eq. 2 by replacing μ_Δ with initial permeability μ_i, which is depicted in the Fig. 9. The μ_i used for material 3C90 is 2300. As observed, at very low excitation levels, the incremental permeability,

979-8-3315-1612-3/25 $31.00 © 2025 IEEE   2442

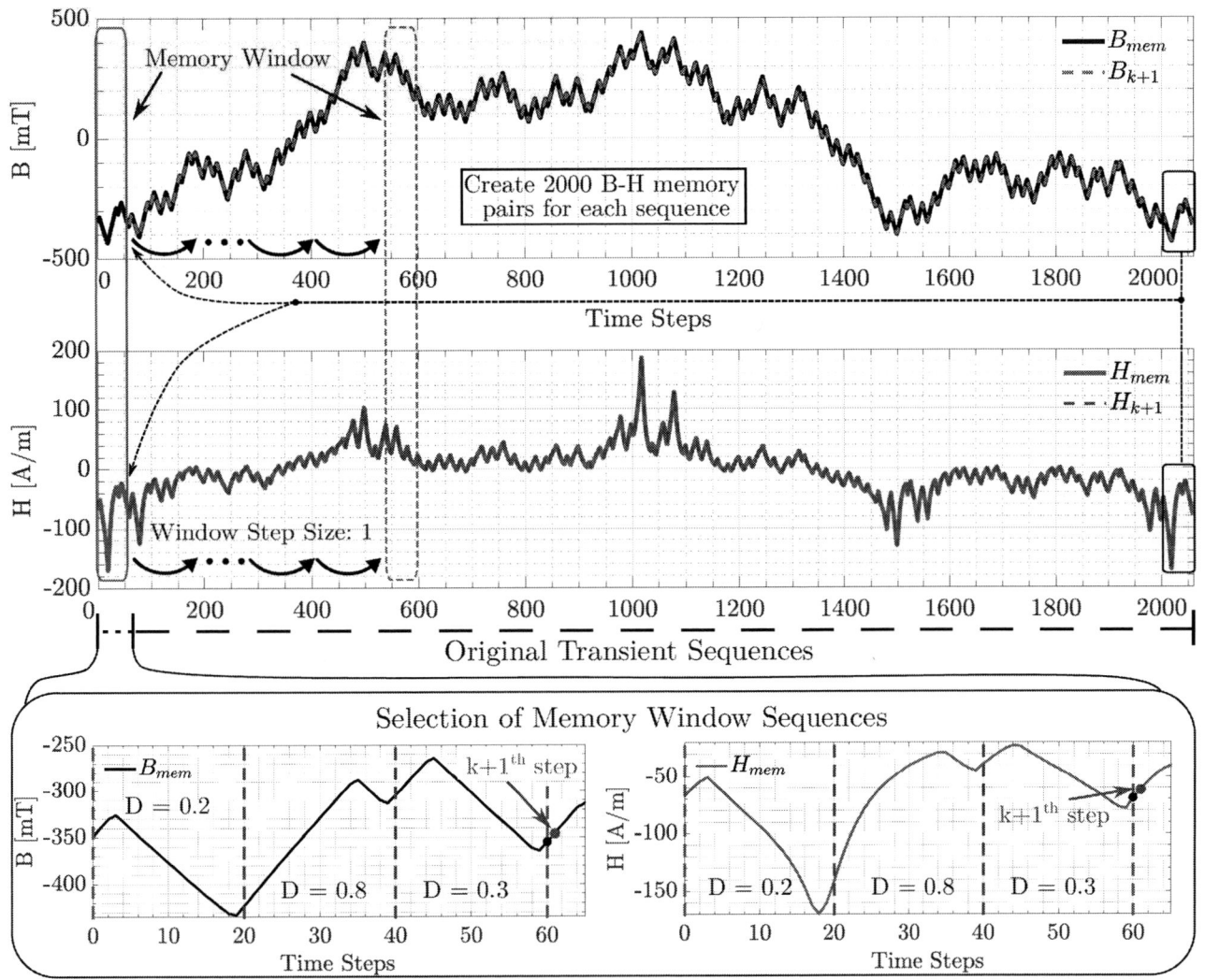

Fig. 8. Updated MagNet database (the MagNetX database) including transient sequences. Each large cycle contains 100 step change duty cycles. The last window size is copied and appended to the beginning. Each window is sized with 60 point memories.

μ_Δ is much larger than 2300, hence the calculation with μ_i induces a larger slope of in the H sequence, deviating from true measurements. On the contrary, when the core reaches saturation, μ_i becomes greater than μ_Δ, resulting in the calculated H sequence being unable to reach the magnitude it would otherwise in a true saturation condition.

Even though the model predicts scalar outputs, the accuracy of the model is still quantified by calculating the relative error between the predicted sequence $H_{pred}(t)$ and the measured sequence $H_{meas}(t)$ of the entire cycle at once. The relative error of the sequence is calculated as:

$$\text{Relative Err. of Sequence} = \frac{\text{rms}(H_{pred} - H_{meas})}{\text{rms}(H_{meas})}$$

$$= \frac{\sqrt{\frac{1}{n}\sum_{t=t_1}^{t_n}\left(H_{pred}(t) - H_{meas}(t)\right)^2}}{\sqrt{\frac{1}{n}\sum_{t=t_1}^{t_n}\left(H_{meas}(t)\right)^2}} \quad (4)$$

Figure 10 demonstrates the distribution of relative errors in the predicted $H(\text{t})$ sequences in the test set generated by the

transformer-based model. As shown in the figure, the model is capable of accurately predicting the $H(\text{t})$ sequence with an average, and 95th percentile of the sequence relative error of 3.81% and 6.552% respectively. These statistics on the prediction results provided good evidence that the proposed model is capable of making accurate predictions for the hysteresis loops under various operating conditions.

In order to validate the model's capabilities of handling various frequency, another test dataset is created. Compared to the training data which contains 600 sequences operating at frequencies 50 kHz, 200 kHz, and 500 kHz, the test data contains 600 data sequences of 80 kHz, 125 kHz, and 320 kHz. The test dataset introduces 100% of new data. Figure 12 presents one of such prediction, measurement sequence at 200 kHz. The predicted result still managed to achieve reasonable results, with predictions having slight deviations from the measurement data. Another characteristic of the model is that the prediction always has slight delay at sharp transitions where the slope of H changes from positive to negative. Such delay could be minimized by increasing the resolution of the

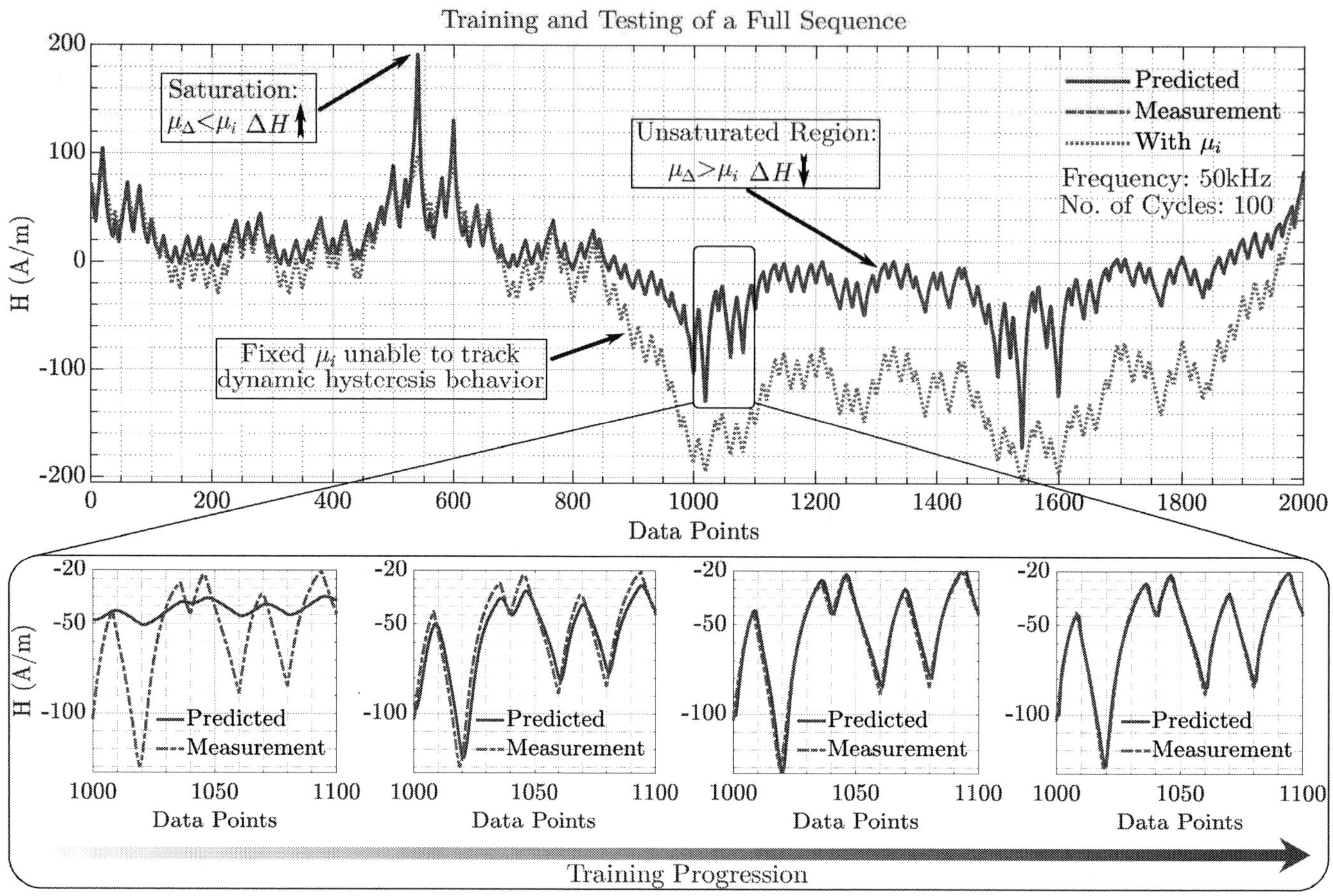

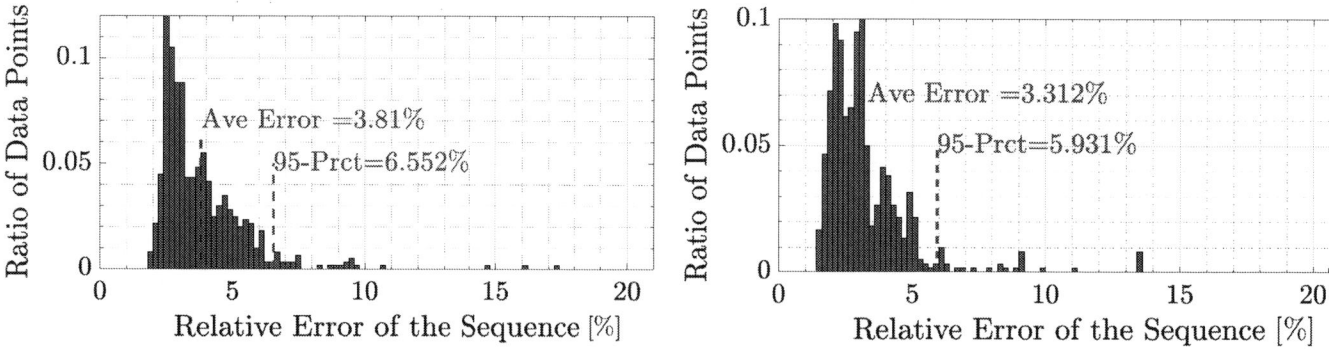

Fig. 9. The testing of the full sequence. Sequence includes low flux excitations as well as saturation point. A comparison of predicted and calculated waveform using initial permeability μ_i. The neural network model shows much better prediction results compared to a fixed permeability approximation.

Fig. 10. Relative error of the sequence between predicted and measured result. The average error is calculated based on the result of the testing data and training data of the same frequency.

Fig. 11. Relative error of the sequence comparing the training dataset and new test dataset on frequencies that the original training dataset does not cover.

sequence. The average error and $95\%^{th}$ error are 3.312% and 5.931% respectively. The comparison indicates that the performance of the model with data of different frequency is comparable to the data tested with the original training data. The model demonstrates reasonable generalization capability.

VI. CONCLUSIONS

This paper introduces a foundation neural network framework for modeling magnetic hysteresis under transient excitations, and presents an LSTM-based encoder-projector

sequence-to-scalar neural network architecture. The developed model takes its first step towards a foundational model that is capable of modeling complex hysteresis behaviors in transient. An example LSTM neural network model was designed, trained, and tested, validating the effectiveness of the proposed architecture. This breakthrough sparked a new approach to modeling complex magnetic behaviors, aiming to accurately predict B–H relationships by incorporating both material characteristics and memory effects, moving beyond the limitations of fixed frequency periodic steady state assumptions.

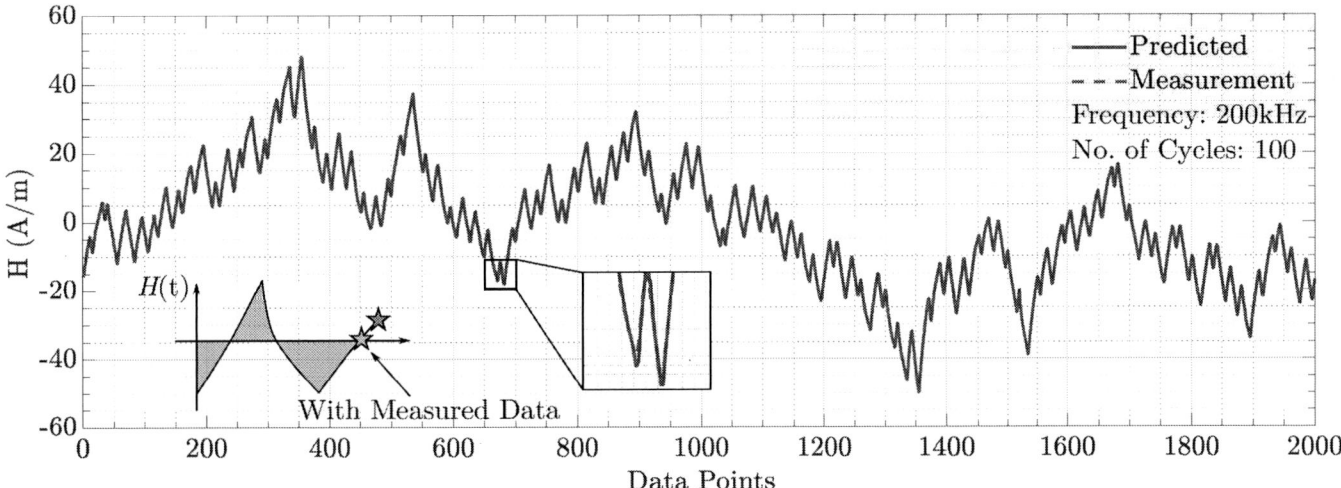

Fig. 12. Test dataset with different frequency values as the training dataset. The neural network model was able to accurately predict the material response even though the training data and testing data were collected at different frequencies.

ACKNOWLEDGEMENTS

This work was jointly supported by the National Science Foundation under Award #2344664, ITG Electronics, TSMC, pSemi, and the Princeton Andlinger Center for Energy and the Environment.

REFERENCES

[1] D. Serrano, H. Li, S. Wang, T. Guillod, M. Luo, V. Bansal, N. K. Jha, Y. Chen, C. R. Sullivan, and M. Chen, "Why magnet: Quantifying the complexity of modeling power magnetic material characteristics," *IEEE Transactions on Power Electronics*, vol. 38, no. 11, pp. 14 292–14 316, 2023.

[2] E. C. Stoner and E. P. Wohlfarth, "A mechanism of magnetic hysteresis in heterogeneous alloys," *Philosophical Transactions of the Royal Society of London. Series A, Mathematical and Physical Sciences*, vol. 240, no. 826, pp. 599–642, 1948. [Online]. Available: https://royalsocietypublishing.org/doi/abs/10.1098/rsta.1948.0007

[3] D. Jiles and D. Atherton, "Theory of ferromagnetic hysteresis," *Journal of Magnetism and Magnetic Materials*, vol. 61, no. 1, pp. 48–60, 1986. [Online]. Available: https://www.sciencedirect.com/science/article/pii/0304885386900661

[4] F. Preisach, "Über die magnetische Nachwirkung," *Zeitschrift für Physik*, vol. 94, no. 5-6, pp. 277–302, May 1935.

[5] H. H. Cui, S. Dulal, S. B. Sohid, G. Gu, and L. M. Tolbert, "Unveiling the microworld inside magnetic materials via circuit models," *IEEE Power Electronics Magazine*, vol. 10, no. 3, pp. 14–22, 2023.

[6] A. Raghunathan, Y. Melikhov, J. E. Snyder, and D. C. Jiles, "Modeling the temperature dependence of hysteresis based on jiles–atherton theory," *IEEE Transactions on Magnetics*, vol. 45, no. 10, pp. 3954–3957, 2009.

[7] R. Szewczyk, A. Bieńkowski, and J. Salach, "Extended jiles–atherton model for modelling the magnetic characteristics of isotropic materials," *Journal of Magnetism and Magnetic Materials*, vol. 320, no. 20, pp. e1049–e1052, 2008, proceedings of the 18th International Symposium on Soft Magnetic Materials. [Online]. Available: https://www.sciencedirect.com/science/article/pii/S0304885308005556

[8] H. Cui and K. D. T. Ngo, "Transient core-loss simulation for ferrites with nonuniform field in spice," *IEEE Transactions on Power Electronics*, vol. 34, no. 1, pp. 659–667, 2019.

[9] M. Chen, S. Chakraborty, and D. J. Perreault, "Multitrack power factor correction architecture," *IEEE Transactions on Power Electronics*, vol. 34, no. 3, pp. 2454–2466, 2019.

[10] J. P. M. Figueiredo, F. L. Tofoli, and B. L. A. Silva, "A review of single-phase pfc topologies based on the boost converter," in *IEEE/IAS International Conference on Industry Applications*, 2010, pp. 1–6.

[11] T. Jappe, M. Lohn, and S. Mussa, "Gan-based single-phase bridgeless pfc boost rectifier," *The Journal of Engineering*, vol. 2019, 04 2019.

[12] Z. Zhao, F. Liu, S. L. Ho, W. N. Fu, and W. Yan, "Modeling magnetic hysteresis under dc-biased magnetization using the neural network," *IEEE Transactions on Magnetics*, vol. 45, no. 10, pp. 3958–3961, 2009.

[13] C. Serpico and C. Visone, "Magnetic hysteresis modeling via feed-forward neural networks," *IEEE Transactions on Magnetics*, vol. 34, no. 3, pp. 623–628, 1998.

[14] S. Hochreiter and J. Schmidhuber, "Long short-term memory," *Neural Comput.*, vol. 9, no. 8, p. 1735–1780, Nov. 1997. [Online]. Available: https://doi.org/10.1162/neco.1997.9.8.1735

[15] A. Vaswani, N. Shazeer, N. Parmar, J. Uszkoreit, L. Jones, A. N. Gomez, Ł. Kaiser, and I. Polosukhin, "Attention is all you need," in *Advances in neural information processing systems*, 2017, pp. 5998–6008. [Online]. Available: http://arxiv.org/abs/1706.03762

[16] Y. LeCun, Y. Bengio, and G. Hinton, "Deep learning," *Nature*, vol. 521, pp. 436–44, 05 2015.

[17] H. Li, D. Serrano, S. Wang, and M. Chen, "Magnet-ai: Neural network as datasheet for magnetics modeling and material recommendation," *IEEE Transactions on Power Electronics*, vol. 38, no. 12, pp. 15 854–15 869, 2023.

[18] H. Li, D. Serrano, T. Guillod, S. Wang, E. Dogariu, A. Nadler, M. Luo, V. Bansal, N. K. Jha, Y. Chen, C. R. Sullivan, and M. Chen, "How magnet: Machine learning framework for modeling power magnetic material characteristics," *IEEE Transactions on Power Electronics*, vol. 38, no. 12, pp. 15 829–15 853, 2023.

[19] M. Chen *et al.*, "Magnet challenge for data-driven power magnetics modeling," *IEEE Open Journal of Power Electronics*, pp. 1–16, 2024.

[20] H. Kwon, S. Wang, H. Li, Y. Elasser, G.-G. Kang, D. Zhou, D. Grigoryan, and M. Chen, "Magnetx: Extending the magnet database for modeling power magnetics in transient," in *2025 IEEE Applied Power Electronics Conference and Exposition*, 2025.

[21] A. Laudani, G. M. Lozito, and F. Riganti-Fulginei, "Dynamic hysteresis modelling of magnetic materials by using a neural network approach," *2014 AEIT Annual Conference - From Research to Industry: The Need for a More Effective Technology Transfer, AEIT 2014*, 01 2015.

[22] A. Laudani, G. M. Lozito, F. R. Fulginei, and A. Salvini, "Modeling dynamic hysteresis through fully connected cascade neural networks," in *International Forum on Research and Technologies for Society and Industry Leveraging a better tomorrow (RTSI)*, 2016, pp. 1–5.

[23] A. Moghar and M. Hamiche, "Stock market prediction using lstm recurrent neural network," *Procedia Computer Science*, vol. 170, pp. 1168–1173, 2020, the 11th International Conference on Ambient Systems, Networks and Technologies (ANT) / The 3rd International Conference on Emerging Data and Industry 4.0 (EDI40) / Affiliated Workshops. [Online]. Available: https://www.sciencedirect.com/science/article/pii/S1877050920304865

[24] S. Wang, H. Li, D. Serrano, T. Guillod, J. Li, C. Sullivan, and M. Chen, "A simplified dc-bias injection method for characterizing power magnetics using a voltage mirror transformer," *IEEE Transactions on Power Electronics*, vol. 39, no. 6, pp. 6608–6612, 2024.

Revisiting Models of Common Mode Inductors to Include the Magnetized Capacitance Effect

Rafael Bogo Portal Chagas
Dept. of Research and Development
Nidec Global Appliance
Joinville, Brazil
rafaelbogochagas@gmail.com

Marcelo Lobo Heldwein
Chair of High-Power Converter Systems
Technical University of Munich
Munich, Germany
marcelo.heldwein@tum.de

Abstract—Common mode (CM) inductors are filter components that are key to achieving electromagnetic compatibility
(EMC) in power electronics systems. Their design depends upon
comprehensive system models, which require the best knowledge
of parasitic effects. Typically, a filter design is based on equivalent
circuits, and the quality of the component models defines the
outcome and applicable design safe margins. Thus, high-accuracy
models of every filter component are desirable. Past works
proposed models for CM inductors. However, recently, the notion
of magnetized capacitance has been proposed to model the effect
of the energy stored in a magnetic core of power inductors and the
consequent generation of electric fields that also store energy. The
impact of such an effect on CM inductors has yet to be reported.
Thus, this work addresses the question of how significant this
effect is in the modeling of CM inductors. The method used
here incorporates state-of-the-art models for the well-known
electromagnetic effects occurring in CM inductors, including
the magnetized capacitance model. Experimental results on
inductors built with different Mn–Zn and Ni–Zn ferrite cores are
compared with the presented theoretical models. It is found that
the magnetized capacitance improves the parallel capacitance
modeling for Mn–Zn, with reduced errors in experimental results.
However, it has less effect on Ni–Zn cores. A critical evaluation
of the results is presented.

Index Terms—Common Mode Inductors, Magnetized Capacitance, Parallel Capacitance, Mn–Zn, Ni–Zn

I. INTRODUCTION

In terms of cost and time, it is more efficient for equipment designers to consider electromagnetic interference (EMI)
issues from the beginning of a system design, minimizing
the testing, rework, and costs of late solutions. However,
considering EMI early on in a project requires extensive
knowledge of the system at hand, the emissions and propagation paths that might exist in its intended environment, and
the immunity and emission reduction techniques [1]. On static
converters coupled via cables, the dominant mechanism of
radiated emissions is caused by common mode (CM) currents,
which should be limited to achieve EMC standards compliance
[2]. EMI filters are essential in this context. A CM inductor is
typically required, and its frequency behavior in the frequency
range of interest determines a successful filter design [3].

At high frequencies (HFs), the behavior of a magnetic
component differs significantly from its behavior at low frequencies, characterized by parasitic non-idealities. The capacitances in an inductor cause it to become capacitive at high

frequencies [4], [5]. The characteristics of magnetic materials
strongly depend on their operating points, such as operating
frequency and temperature [6]. The losses in the windings
depend on nonlinear effects like skin effect and proximity
effect, which, together with losses in the core, raise the
component's temperature and alter its operating point [7],
[8]. Other effects, such as couplings and stray fields [9], can
also change the high-frequency characteristics of a magnetic
component and the system in which it is placed. These effects
are typically considered dominant and are used to model
CM inductors. Recently, an internally magnetized capacitance
effect within a power inductor core has been proposed [10],
which models the impact of the magnetic energy stored in a
magnetic core and consequently generates associated electric
fields that store energy. The impact of such an effect on a CM
inductor's behavior has yet to be reported.

This work addresses the research question of how significant
this effect is in modeling CM inductors. A study is presented
here that considers the HF characteristics of CM inductors,
their parasitic components, and the resulting equivalent circuit
models. In particular, the prediction of their high-frequency
impedance and self-resonance frequency, with the addition
of the magnetized capacitance effect. Experimental results on
inductors built with different Mn–Zn and Ni–Zn ferrite cores
are compared with the presented theoretical models. A critical
evaluation of the results is presented.

II. COMMON-MODE INDUCTOR EQUIVALENT MODELS

The CM inductor is commonly built with windings that
have the same number of turns (see Fig. 1(a)) but wound
in the direction that makes the magnetic fields induced by
differential-mode (DM) current components (i_{DM}) mutually
cancel, leaving mainly the magnetic fields caused by CM
currents (i_{CM}). This way, it can conduct elevated DM currents
without saturating, making it possible to build CM inductors
made of materials with high permeability that cannot be used
for DM inductors [6].

Fig. 1 shows the schematic of a CM inductor connected to
phases a and b of a circuit. Considering the self-inductances
$L_{Sa} = L_{Sb} = L_S$, it is possible to find equivalent inductances
through the description of the voltages across the CM inductor
windings. The CM equivalent circuits can be derived from the

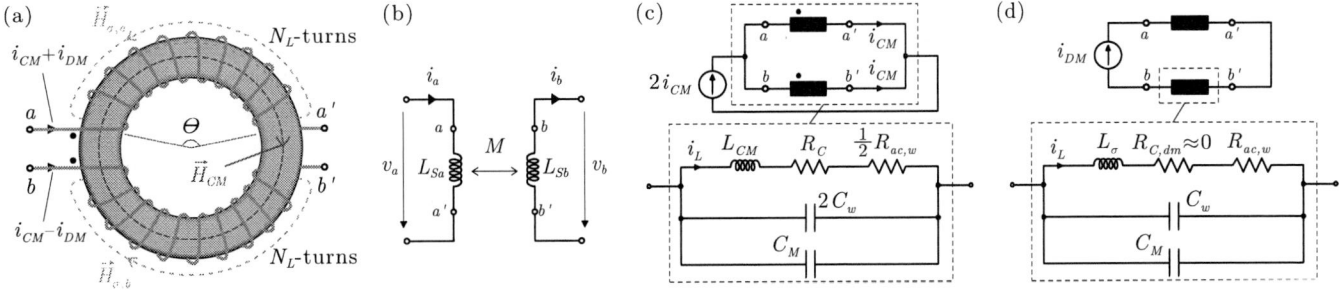

Fig. 1. Common mode inductor: (a) typical single-phase configuration and simplified distribution of currents and magnetic fields; (b) ideal equivalent circuit; (c) CM equivalent circuit including parasitic effects, and (d) DM equivalent circuit including main parasitic effects.

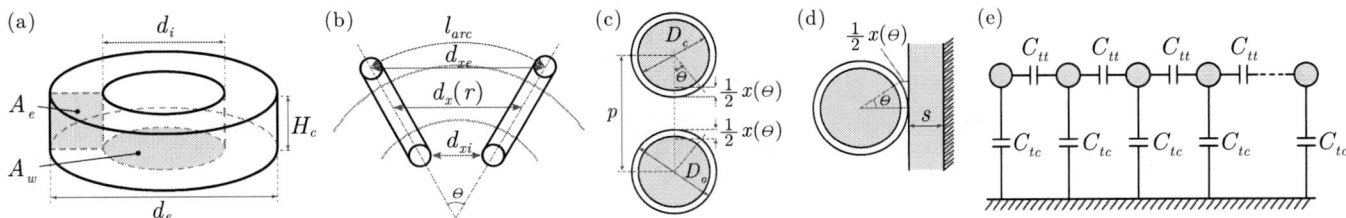

Fig. 2. Common mode inductor: (a) toroidal magnetic core geometry; (b) definition of distances between turns; (c) geometric definitions to compute turn-to-turn capacitances; (d) geometric definitions to compute turn-to-core capacitances, and; (e) parasitic capacitances equivalent circuit for single layer windings.

definition of CM currents, i.e., $i_a = i_b = i_{CM}$, which leads to $v_a = v_b = v_{CM}$. Thus,

$$\begin{bmatrix} v_a \\ v_b \end{bmatrix} = \begin{bmatrix} L_S & M \\ M & L_S \end{bmatrix} \cdot \frac{d}{dt} \begin{bmatrix} i_a \\ i_b \end{bmatrix}, \qquad (1)$$

where M is the mutual inductance between the windings, defined as $M = k_{CM} L_S$, with k_{CM} the magnetic coupling coefficient between the two windings, and L_S the self inductance of a winding. The common mode equivalent inductance can be derived from (1), leading to

$$L_{CM} = \frac{v_{CM}}{\frac{di_{CM}}{dt}} = L_P + M = L_S \frac{1 + k_{CM}}{2} \approx L_S. \qquad (2)$$

The inter-winding coupling is perfect in an ideal CM inductor, resulting in $k_{CM} = 1$. As such, it is possible to say that the CM inductance of the CM inductor is equal to the self-inductance of each winding. On the other hand, considering that only DM currents circulate through the CM inductor, i.e., $i_a + i_b = 0$, and defining $i_a = i_{DM}$ leads to the identification of the equivalent DM inductance. Using the previous procedure to derive the equivalent DM inductance (L_σ) of a CM inductor, one finds that

$$L_\sigma = L_P \cdot (1 - k_{CM}). \qquad (3)$$

resulting in a null DM inductance for ideal inductors.

On the other hand, when observing a non-ideal choke inductor, the windings are not perfectly coupled, leading to non-concatenated magnetic flux. As a result, the coupling factor k_{CM} becomes less than one and varies with the core

permeability. Consequently, leakage inductances L_σ exist in each of the windings, influencing differential mode signals and, depending on their magnitude, may saturate the core [6].

The intra-winding capacitance (C_w) is essential in the impedance's HF behavior [4], [5]. The core (R_C) and winding (R_w) resistances are also highly relevant, with the core resistance being capable of attenuating high-frequency noise emissions [2]. The magnetic core magnetized capacitance [10] effect has yet to be reported in the literature to model the coupling between magnetic and electric fields. Thus, this work proposes HF models for the CM inductor as shown in Fig. 1(c) for CM and in Fig. 1(d) for DM components.

A. Common Mode Inductance

The common mode inductance L_{CM} is a relevant parameter that represents the inductor's behavior at lower frequencies and is frequently used to calculate the attenuation of an EMI filter for conducted emissions. Since the CM signals have a low amplitude when compared to DM signals, many authors use the small-signal modeling to predict the self-inductance of CM inductors, a domain in which the complex permeability model is valid [2], [6].

Fig. 2(a) shows the geometry of a toroidal core with a square cross-section. For this geometry, the self-inductance of each winding can be calculated through (4), where N_L represents the number of turns, μ_0 is the vacuum permeability, and μ' is the core material's real part of the complex permeability.

$$L_S = A_L N_L^2 \qquad (4)$$

$$A_{L\_toroidal} = \frac{\mu_0 \mu' H_c}{2\pi} \cdot \ln\left(\frac{d_e}{d_i}\right) \qquad (5)$$

979-8-3315-1612-3/25 $31.00 © 2025 IEEE

In general, CM inductors should filter high-frequency noise. The absolute permeability, however, decreases with increasing frequency, making the magnetic component less inductive and reducing the filtering effects. Different materials have different characteristics, with some being more able to filter higher frequencies than others [2]. As such, it is necessary to consider this variation in the design of EMI filters. Other operating conditions affect the inductance, such as the temperature and the DC level of the signal. This data is often found on the datasheets of the magnetic materials.

B. Leakage Inductance

The leakage inductance L_σ can reach non-negligible levels due to the high permeability of the materials used on CM inductors and can be used as part of DM mode filters. Initially proposed in [9], who modeled the leakage inductance as a rod inductor due to the similarity between the magnetic field paths, and later improved in [6], the leakage inductance of the toroidal core can be approximated through

$$L_\sigma = \mu_0 \mu_{ef} N_L^2 \frac{A_e}{l_{ef}}, \qquad (6)$$

where μ_{ef} represents the equivalent permeability due to a magnetic path that partially flows through the air, defined by

$$\mu_{ef} = 2.5 \Gamma^{1.45}, \qquad (7)$$

$$\Gamma = \frac{l_e}{2} \sqrt{\frac{\pi}{A_e}}, \qquad (8)$$

l_{ef} is the equivalent magnetic path length,

$$l_{ef} = \sqrt{\left[\frac{d_e}{2} \left(\frac{\theta}{4} + 1 + \sin\left(\frac{\theta}{2}\right) \right) \right]^2 + \left[d_i \left(\frac{\theta}{4} - 1 + \sin\left(\frac{\theta}{2}\right) \right) \right]^2}, \quad (9)$$

θ represents the winding' coverage angle as defined in Fig. 1(a), and A_e is the area of the cross-section of the core.

Since the leakage magnetic fields of each winding do not cancel each other, they can saturate the core [6]. To avoid this saturation, the leakage magnetic flux density inside the core must be lower than the core's saturation. Thus,

$$B_\sigma = \mu_0 \mu_{ef} H_\sigma = \mu_0 \mu_{ef} \frac{N_L I_{DM}}{l_{ef}} < B_{sat}, \qquad (10)$$

where I_{DM} represents the differential mode currents and B_{sat} represents the saturation magnetic flux density, defined in the materials datasheet. This expression shows that materials with higher B_{sat} can contribute to the miniaturization of CM inductors.

Another method of finding the leakage inductance is presented in [11], where neural networks trained by finite element method (FEM) simulations are used to improve estimations.

C. Core Equivalent Resistance

The core's equivalent resistance presents a way of attenuating emissions that is an alternative to self-inductance. While the self-inductance reduces with frequency, the equivalent resistance of typical ferrites has low values at lower frequencies, significant values in the range between hundreds of kHz and MHz, and reduces again at the frequencies above [2].

The CM inductor's core equivalent resistance R_c can be calculated through a small signal complex permeability model, using the imaginary part of the permeability μ'', provided in the core material's datasheet [2], resulting in,

$$R_c(f) = \omega \frac{\mu_0 \mu''(f) N_L^2 H_c}{2\pi} \ln\left(\frac{d_e}{d_i}\right). \qquad (11)$$

The core losses can be used to dissipate the energy that would otherwise be emitted into the environment. To reduce emissions using losses, the core material has to fulfill two criteria [2], namely: the magnitude of the complex permeability needs to be constant or increasing and cannot be sharply decreasing, and the impedance must be dominated by the μ''. Comparing Mn–Zn to Ni–Zn ferrites, Ni–Zn tends to have higher core losses at higher frequencies, a suitable material for resistive noise attenuation [2].

D. Magnetized Capacitance

The magnetized capacitance C_M represents the electrical energy stored inside the core, induced by the inductor's magnetic fields, hence the name *magnetized*. In [10], the authors argue that the typical magnetic component models depend on electrostatic assumptions that do not hold on higher frequencies. When these assumptions are discarded, it is possible to calculate an equivalent capacitance. The magnetized capacitance can explain why inductors with a low number of turns have a seemingly constant resonance frequency, which does not depend on the number of turns. This effect might be significant for CM filter devices since CM inductors are typically wound with few turns. To model C_M, the authors use Ampère's circuital law to find the magnetic field induced inside the core and the volumetric stored energy. The electric field induced by this magnetic field is found using Faraday's Law and the volumetric energy stored in the electric fields. Representing this electric energy as a capacitance, the authors find that

$$C_M = \frac{\varepsilon l_e}{8\pi N_L^2} \qquad (12)$$

where ε is the core's electric permittivity and l_e is the magnetic path length.

Equation (12) shows that C_M reduces with increased turns. Thus, C_M is more relevant in components with fewer turns. From [12], Mn–Zn ferrites typically have $\varepsilon \approx 10^5$ at 1 MHz, while Ni–Zn ferrites have $\varepsilon \approx 25$. Consequently, C_M tends to be lower on Ni–Zn cores, partially explaining why this material typically has higher self-resonance frequencies than Mn–Zn. Every core has a natural resonance frequency caused by the interaction between its internal magnetic and electric

979-8-3315-1612-3/25 $31.00 © 2025 IEEE

fields [10], which depends on the core's geometry and electrical characteristics. This frequency can be computed through

$$f_{res,M} = \frac{1}{2\pi\sqrt{L \cdot C_M}} = \frac{\sqrt{2}}{\pi r_i \sqrt{\mu\varepsilon}}, \tag{13}$$

where r_i is the internal radius of the cross-section of the core, while these equations were developed for DM inductors, they also hold for common-mode inductors.

E. Winding Capacitance

The winding capacitance C_w represents the capacitances between the windings and between the windings and the core or shielding. In the case of the toroidal cores, typical expressions for calculating winding capacitances, such as found in [4], [5], cannot be used since these expressions consider that the distance between turns is constant and small. As shown in Fig. 2(b), the distance between turns of a toroidal core is variable, being equal to d_i for the inner part, equal to d_e for the external part, and radius-dependent on the intermediary part $d_x(r)$.

The turn-to-turn capacitance is calculated [13] through

$$C_{tt} = \frac{2\varepsilon_0 l_t A}{\sqrt{\left(\frac{1}{\varepsilon_r}\ln\left(\frac{D_o}{D_c}\right) + \frac{p}{D_o}\right)^2 - 1}}, \tag{14}$$

where,

$$A = \tan^{-1}\left(\sqrt{1 + \frac{2}{\frac{1}{\varepsilon_r}\ln\left(\frac{D_o}{D_c}\right) + \frac{p}{D_o} - 1}}\right), \tag{15}$$

l_t is the conductor length, ε_0 the permittivity of free space and ε_r the relative permittivity of the insulation material. For toroidal inductors, capacitances can be computed by dividing the capacitance of each turn into the capacitances of its inner, outer, and intermediate sections and associating them in parallel, resulting [14] in

$$C_{tt\_toroid} = C_{tt}(d_e) + C_{tt}(d_i) + 2C_{tt}\left(\frac{d_e + d_i}{2}\right). \tag{16}$$

The turn-to-core capacitance, as seen in Fig. 2), can be computed [14] with

$$C_{tc} = \frac{8\varepsilon_0 l_t B}{\sqrt{\left(\frac{2}{\varepsilon_r}\ln\left(\frac{D_o}{D_c}\right) + \frac{2s}{D_o} + 1\right)^2 - 1}}, \tag{17}$$

where,

$$B = \tan^{-1}\left(\sqrt{1 + \frac{2}{\frac{2}{\varepsilon_r}\ln\left(\frac{D_o}{D_c}\right) + \frac{2s}{D_o}}}\right) \tag{18}$$

and s is the coating insulation thickness of the core.

The capacitive network is to be solved to find the total winding capacitance, as seen in Fig. 2(e), where the core is considered a single node. For winding configurations different from a single layer of turns with a core, the capacitive network must be adjusted to reflect the winding configuration [15].

F. Winding Resistance

The winding resistance R_w is a significant source of losses in an inductor and is an essential parameter in the design of magnetic components. When working at higher frequencies, the winding is subject to non-linear effects, namely the skin effect and the proximity effect, resulting in AC resistance sometimes orders of magnitude higher than its DC resistance. It depends on the configuration of the electric conductors and has been modeled in different ways. For the sake of simplicity, but not of precision, the model used here applies Dowell's equation [16]. Improved models can be found and are compared in [17].

III. EXPERIMENTAL RESULTS

To validate the common-mode inductor modeling, the impedances of four inductors were measured and compared to the impedances computed with the described models.

A. Setup description

The tested cores were manufactured by Hengdian Group DMEGC Magnetics [18]–[20] and their construction details are described in Table I. Photographs of the inductor are presented in Figs. 3 through 6). The measurements were made using an Agilent 4294A impedance analyzer as shown in Fig. 7.

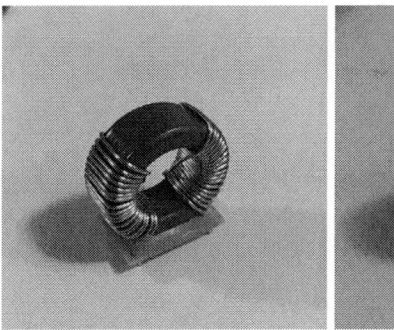 

Fig. 3. Inductor #1. Fig. 4. Inductor #2.

Fig. 5. Inductor #3. Fig. 6. Inductor #4.

TABLE I
DESCRIPTION OF THE MAIN FEATURES OF THE TESTED INDUCTORS.

Inductor number	Material	Number of Turns	Conductor Diameter [mm]	Core Dimensions ($d_e \times d_i \times H_c$) [mm]	L_S [mH]
1	Mn–Zn R10K	18	1.10	36 x 23 x 15	3.88
2	Mn–Zn R5K	19	1.00	38 x 22 x 14	2.76
3	Mn–Zn R5K	21	1.00	50 x 25 x 20	6.11
4	Ni–Zn DN85H	25	0.95	50 x 25 x 20	1.34

TABLE II
COMPARISON BETWEEN MODEL AND MEASUREMENT FOR PARALLEL CAPACITANCE C_P AND RESONANCE FREQUENCY f_{res}.

Inductor	C_P [pF] (with C_M)	C_P [pF] (without C_M)	C_P [pF] (measurement)	Model Error (with C_M)	Model Error (without C_M)	Model Log Error (with C_M)	Model Log Error (without C_M)
1	19.54	9.47	15.97	22.35%	-40.70%	-8.76%	22.70%
2	18.33	9.14	18.58	-1.35%	-50.81%	0.59%	30.81%
3	23.21	13.80	25.40	-8.62%	-45.67%	3.92%	26.50%
4	13.10	13.10	4.65	181.72%	181.72%	-44.98%	-44.98%

Inductor	f_{res} [kHz] (with C_M)	f_{res} [kHz] (without C_M)	f_{res} [kHz] (measurement)	Model Error (with C_M)	Model Error (without C_M)	Model Log Error (with C_M)	Model Log Error (without C_M)
1	537	719	573	-6.28%	25.48%	2.82%	-9.86%
2	612	825	696	-12.07%	18.53%	5.59%	-7.38%
3	386	490	392	-1.53%	25.00%	0.67%	-9.69%
4	1112	1113	1981	-43.87%	-43.82%	25.08%	25.04%

It was considered that the wire had an enamel with a thickness of 25.7 μm and a relative permittivity of 3.5 for the calculation of the mathematical model impedance. The coatings of the core were assumed to be 0.1 mm, and the cores' permittivity values were $\varepsilon_{Mn-Zn} = 10^5$ and $\varepsilon_{Ni-Zn} = 25$. The achieved results are presented in Figs. 8 through 11.

An important figure of merit is the parallel capacitance C_P, extracted from the measurements and computed with (20), where the resonance frequency ω_{res} is defined as the frequency in which the impedance has a phase equal to zero. The self-inductance is the measured low-frequency L_S value corrected to the frequency using the nominal μ' variation and the core resistance R_C was calculated through (11). Thus, the parallel capacitance value estimated from the measured impedances was computed with

$$C_P = \frac{L_S(\omega_{res})}{\omega_{res}^2 L_S(\omega_{res})^2 + (R_c(\omega_{res}) + R_{ac\_w}(\omega_{res}))^2}, \quad (19)$$

while the developed analytical model assumed that this parallel capacitance is given by the sum,

$$C_P = C_M + C_w. \quad (20)$$

Comparisons for the parallel capacitance and resonance frequencies obtained through the model and experimental results are presented in Table II, where the errors were calculated with,

$$\text{Error}(x) = \frac{x_{model} - x_{measurement}}{x_{measurement}} \quad (21)$$

$$\text{Log Error}(x) = \log_{10}\left(\frac{x_{model}}{x_{measurement}}\right). \quad (22)$$

To evaluate the relevance of the magnetized capacitance C_M, the impedance model was also computed by neglecting the effects of C_M, i.e., considering $C_M = 0$.

B. Discussion on the achieved results

The calculated magnetized capacitances C_M for inductors 1 to 4 are 10.7 pF, 9.2 pF, 9.4 pF and 0.0017 pF. On the other hand, the winding capacitances C_w calculated are 4.7 pF, 4.6 pF, 6.9 pF and 6.5 pF. The total winding capacitances are the capacitance of a single winding multiplied by the number of turns, resulting in capacitances around the same order of magnitude as the magnetized capacitances, except for the Ni–Zn core. It is possible to observe that the model including C_M presented lower error than the model that neglected it for all Mn–Zn samples, with C_P errors reducing from -50.81% to -1.37% in case of inductor #2 and reducing from -45.67% to -8.62% for inductor #3. The same holds for the self-resonance frequency errors.

The Ni–Zn inductor result, however, showed a minor improvement with the inclusion of the magnetized inductance. The core for inductor #4 has a lower permittivity, resulting in a magnetized impedance that is orders of magnitude below the winding impedance and does not impact the parallel

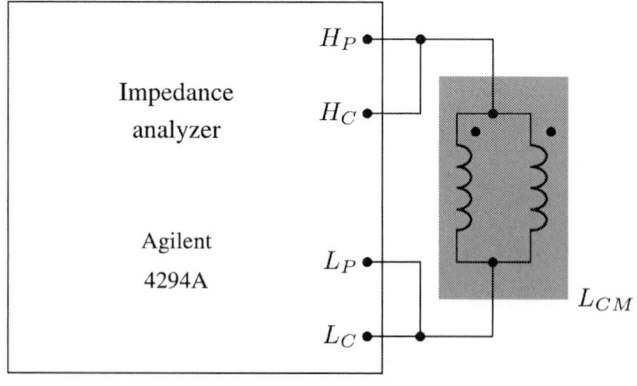

Fig. 7. Common-mode impedance measurement setup.

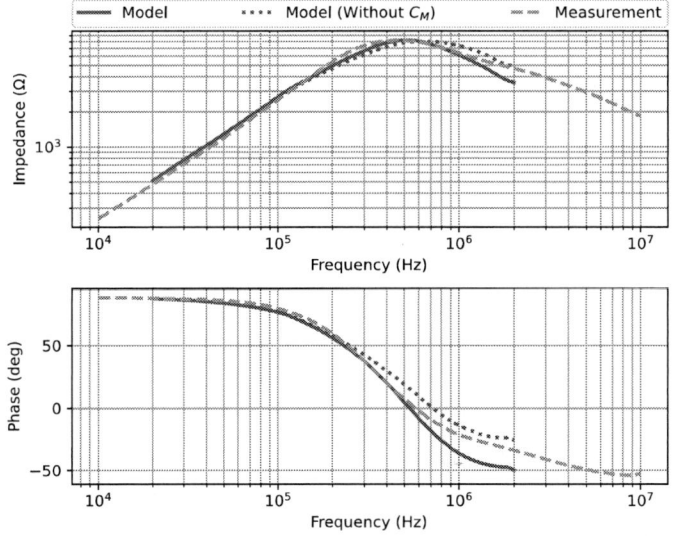

Fig. 8. Impedance measurement results for inductor #1.

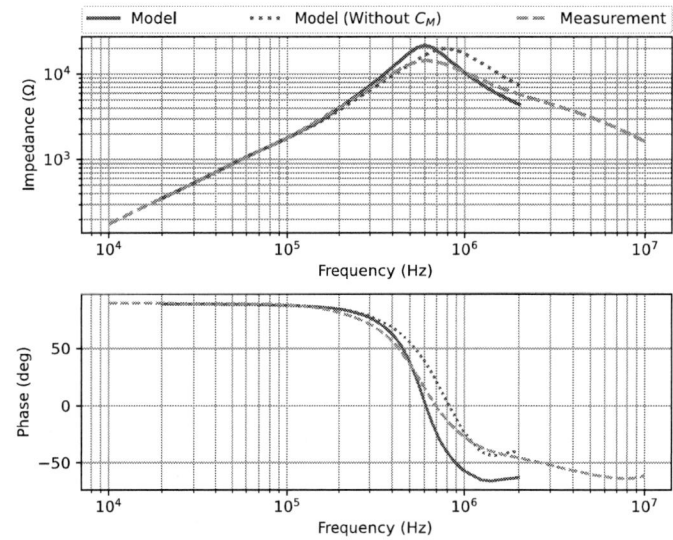

Fig. 9. Impedance measurement results for inductor #2.

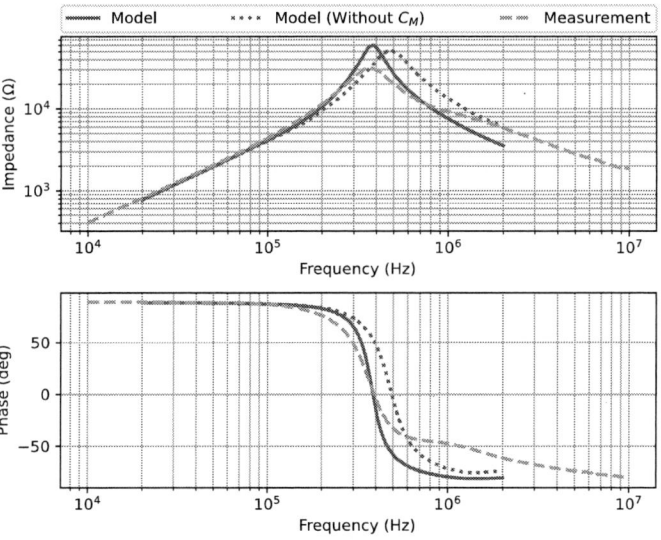

Fig. 10. Impedance measurement results for inductor #3.

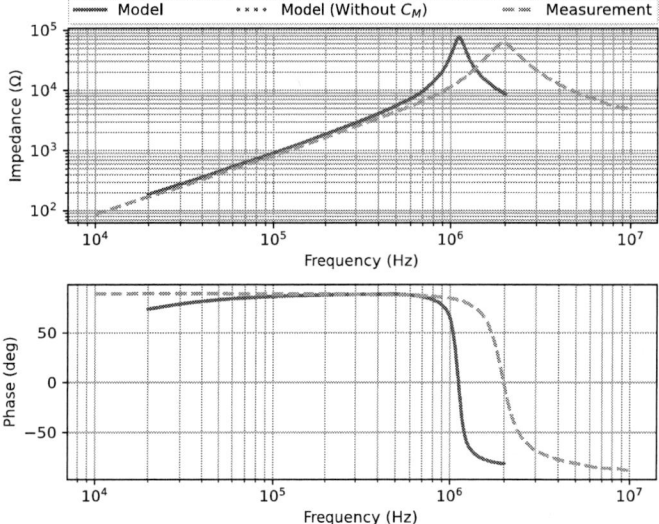

Fig. 11. Impedance measurement results for inductor #4.

capacitance C_P. Moreover, it is possible to observe that the Ni–Zn core's resistivity is high, above $10^5 \Omega \cdot m$ [20]. As such, the typical winding capacitance C_w model from Fig. 2(e) needs to be more adequate, as it considers that the core is a single node. To improve the modeling of the capacitance of Ni–Zn cores, it could be necessary to develop a model that includes the effects of core resistivity between the windings and core capacitances. As for the resonance frequency, the Ni–Zn inductor also showed higher errors than the Mn–Zn inductors.

From these results, the magnetized capacitance has a relevant influence on the total capacitance of the Mn–Zn inductors, while it could be considered much less critical for Ni–Zn inductors. This could help explain why Ni–Zn inductors

typically show self-resonance at frequencies higher than the Mn–Zn, together with the self-inductance that tends to be constant at higher frequencies.

IV. CONCLUSIONS

This work has extended the modeling of CM inductors, focusing on the impact of magnetized capacitances on Mn–Zn and Ni–Zn ferrites. Impedance models were validated through experimental results. Due to its high permittivity, it was found that C_M plays a significant role in the parallel capacitance of Mn–Zn inductors. Including the magnetized capacitance has improved the analytical models of these inductors, reducing the errors for the calculated total parallel capacitance (C_P) and the CM inductor self-resonance frequency (f_{res}).

979-8-3315-1612-3/25 $31.00 © 2025 IEEE

While it is possible to model Mn–Zn inductors with reasonable error, improvement is still necessary for Ni–Zn inductors, whose resonance frequency and parallel capacitance values showed little influence from C_M, as it has a lower permittivity. In addition, it is benefitial to develop a better model for the winding capacitance of the inductor, considering the core's resistivity.

ACKNOWLEDGMENT

The authors would like to thank the Power Electronics Institute (INEP) at the Federal University of Santa Catarina (UFSC) in Florianópolis, Brazil, for its support and infrastructure and Nidec Global Appliance for providing the inductor samples.

REFERENCES

[1] F. A. Kharanaq, A. Emadi, and B. Bilgin, "Modeling of conducted emissions for emi analysis of power converters: State-of-the-art review," *IEEE Access*, vol. 8, pp. 189 313–189 325, 2020.

[2] J. Yao, Y. Li, H. Zhao, and S. Wang, "Design of cm inductor based on core loss for radiated emi reduction in power converters," in *2019 IEEE Applied Power Electronics Conference and Exposition (APEC)*, 2019, pp. 2673–2680.

[3] F. Salomez, A. Videt, and N. Idir, "Modeling and minimization of the parasitic capacitances of single-layer toroidal inductors," *IEEE Transactions on Power Electronics*, vol. 37, no. 10, pp. 12 426–12 436, 2022.

[4] A. Massarini, M. Kazimierczuk, and G. Grandi, "Lumped parameter models for single- and multiple-layer inductors," in *PESC Record. 27th Annual IEEE Power Electronics Specialists Conference*, vol. 1, 1996, pp. 295–301 vol.1.

[5] A. Massarini and M. Kazimierczuk, "Self-capacitance of inductors," *IEEE Transactions on Power Electronics*, vol. 12, no. 4, pp. 671–676, 1997.

[6] M. L. Heldwein, "Emc filtering of three-phase pwm converters," Doctoral Thesis, ETH Zurich, Zürich, 2008.

[7] H. Li, D. Serrano, T. Guillod, E. Dogariu, A. Nadler, S. Wang, M. Luo, V. Bansal, Y. Chen, C. R. Sullivan, and M. Chen, "Magnet: An open-source database for data-driven magnetic core loss modeling," in *2022 IEEE Applied Power Electronics Conference and Exposition (APEC)*, 2022, pp. 588–595.

[8] N. Rasekh, J. Wang, and X. Yuan, "A novel in-situ measurement method of high-frequency winding loss in cored inductors with immunity against phase discrepancy error," *IEEE Open Journal of the Industrial Electronics Society*, vol. 2, pp. 545–555, 2021.

[9] M. Nave, "On modeling the common mode inductor," in *IEEE 1991 International Symposium on Electromagnetic Compatibility*, 1991, pp. 452–457.

[10] R. Zhang, S. Wang, T. Long, J. Qiu, K. Liu, and H. Zhao, "The magnetized capacitance, first resonant frequency, and electromagnetic analysis of inductors with ferrite cores," *IEEE Transactions on Industrial Electronics*, pp. 1–11, 2023.

[11] Z. Dong, R. Ren, B. Liu, and F. Wang, "Data-driven leakage inductance modeling of common mode chokes," in *2019 IEEE Energy Conversion Congress and Exposition (ECCE)*, 2019, pp. 6641–6646.

[12] Ferroxcube, *Soft Ferrites: Ferrite Materials Survey*. Ferroxcube, 2008.

[13] A. Ayachit and M. K. Kazimierczuk, "Self-capacitance of single-layer inductors with separation between conductor turns," *IEEE Transactions on Electromagnetic Compatibility*, vol. 59, no. 5, pp. 1642–1645, 2017.

[14] B. Bertoldi, "Systematic procedures for the design of passive components applied to a high performance three-phase rectifier," Master Thesis, Universidade Federal de Santa Catarina, Florianopolis, 2021.

[15] N. B. Chagas and T. B. Marchesan, "Analytical calculation of static capacitance for high-frequency inductors and transformers," *IEEE Transactions on Power Electronics*, vol. 34, no. 2, pp. 1672–1682, 2019.

[16] P. L. Dowell, "Effects of eddy currents in transformer windings," in *Proceedings of the IEE*, vol. 113, no. 8, 1966, pp. 1287–1394.

[17] Y. Zhao, Z. Ming, and B. Han, "Analytical modelling of high-frequency losses in toroidal inductors," *IET Power Electronics*, vol. 16, no. 9, pp. 1538–1547, 2023.

[18] H. G. D. MAGNETICS", *Mn-Zn ferrite series 34: R5K Material Characteristics*. DMEGC, 2023.

[19] ——, *Mn-Zn ferrite series 36: R10K Material Characteristics*. DMEGC, 2023.

[20] ——, *Ni-Zn ferrite series: Material Characteristics*. DMEGC, 2023.

A High Frequency Coupled Inductor with Distributed Air Gap for High Power DC-DC Converters

Muhammad Fasih Uddin, Ahmed H. Ismail, Peyman Darvish, Baher Abu Sba and Yue Zhao
Power Electronic Systems Laboratory at Arkansas (PESLA)
Department of Electrical Engineering, University of Arkansas
Fayetteville, AR, 72701, USA
Email: yuezhao@uark.edu

Abstract— **In coupled inductor designs, minimizing AC resistance and achieving an optimal coupling coefficient are crucial for the performance of high frequency, high power DC-DC converters. The location and length of air gaps not only influence the loss profile but also affect the coupling coefficient, directly impacting the system's performance. This article is focused on a coupled inductor design using custom planar ferrite blocks and distributed air gaps with 3D printed bobbins for mechanical stability. The proposed design achieves over 30% reduction in AC losses by reducing fringing effects and offers the flexibility to adjust the coupling coefficient, while ensuring a cost-effective and repeatable assembly process. System level advantages of the proposed method are verified using 2D FEM simulations. Finally, experimental results for a two-phase interleaved boost converter application have been presented in this paper to validate the effectiveness of the proposed inductor design.**

Keywords—Coupled Inductors, AC Losses, Coupling Coefficient, Interleaved Boost Converter

I. INTRODUCTION

The power electronics industry is continuously pushing towards higher efficiency and higher power density. With recent advancements in Wide Bandgap Devices (WBG), the switching frequency of power electronics converters has increased significantly. Power loss distribution indicates that active switches are no longer the main limitations on the performance of high frequency power converters. In comparison with active devices, passive components such as inductors still face many challenges. Inductors play a critical role in energy storage and conversion in power electronics systems, but are often bulky and inefficient at high power densities. One promising solution to overcome these limitations is the use of coupled inductors [1]. Coupled inductors can improve the performance and dynamic response of DC-DC converters by leveraging magnetic coupling between windings and achieving flux cancellation within the core [2]-[6]. These inherent characteristics enable more efficient utilization of magnetic materials at higher power densities, enabling a more compact and efficient design.

In design of high frequency coupled inductors, the AC resistance and coupling coefficient (k) are of paramount importance, as they directly impact efficiency and performance of DC-DC converters [7]. Recently, planar magnetics with printed circuit board (PCB) windings have emerged as a popular choice for such applications due to their compact geometry [8], [9]. However, their implementation comes with notable challenges. The high cost and limited flexibility in controlling winding parameters can hinder design optimization. Furthermore, the high length-to-thickness ratio of the PCB winding cross-section exacerbates the sensitivity of AC resistance to leakage fringing flux. These issues impose significant constraints, limiting the efficiency and scalability of planar magnetics.

Several solutions have been proposed to reduce the AC resistance, including use of parallel and orthogonal air gaps [10]. Reference [11] introduced the concept of redistributing the gaps from core legs to the core plates, which not only reduces AC resistance, but also improves core losses. However, the proposed solutions necessitate machining of ferrites to make custom cores, complicating the design and assembly process, and limiting scalability [13], [14]. Alternatively, [12] proposed use of vertical copper foils which offers several advantages such as reduced cost, control over inductor parameters and high reproducibility.

Furthermore, the selection of optimal magnetic structure and coupling coefficient is regarded as an important task in the design of coupled inductors, especially for interleaved buck/boost converters [7]. Previous research in design of coupled inductor has paid less attention to the design elements such as core material, core structure and air gap placement for achieving optimal coupling coefficient necessary for the performance of DC-DC converters. Based on these concepts, a coupled inductor core structure with distributed air gaps is proposed in this paper. The design effectively minimizes the AC losses while providing the provision to tune the coupling coefficient. A 3D printed bobbin is utilized to provide mechanical stability to the core structure and facilitate seamless assembly, thereby ensuring scalability of the design.

This paper is organized into three main sections. Section II highlights the advantages of the proposed coupled inductor structure, focusing on its improved AC resistance and flexibility in adjusting the coupling coefficient. These performance metrics are benchmarked against the conventional coupled inductor design with a discrete air gap. Section III evaluates the practical application of the proposed coupled inductor in a two-phase interleaved boost converter. Finally,

979-8-3315-1612-3/25 $31.00 © 2025 IEEE

Section IV concludes the paper, summarizing the findings and implications of the proposed design.

II. PROPOSED COUPLED INDUCTOR DESIGN

Coupled inductors take advantage of flux cancellation in the core to improve performance in DC-DC converters. The details of flux cancellation depend on the geometry of core and how windings are wound depending on the degree of coupling required. Fig. 1 shows an exemplary 2D diagram of a coupled inductor using EI cores. It is expected that the coupled inductor will have zero flux in the center leg of the E-core due to flux cancellation. This phenomenon is favorable for reduction of core losses and prevention of core saturation, but it has an adverse impact on winding losses due to the fringing field caused by the air gaps. In planar cores, due to reduced height, fringing flux becomes densely concentrated in the winding window, leading to substantial copper losses, particularly in high-current and high-frequency application. Moreover, in coupled inductors, the location and length of the air gaps not only has a profound effect on the loss profile of the inductor but can also influence the coupling coefficient (k). It has a direct impact on the performance of the DC-DC converter, thus the selection of right value of k for the inductor is really important. Unfortunately, not every value of k can be achieved since it depends on the physical geometry of core. In order to achieve desirable coupling coefficient, complex custom core geometries are made by machining the ferrites which is not favorable for mass production.

Fig. 2(a) shows the proposed coupled inductor structure with distributed air gap while Fig. 2(b) shows the conventional coupled inductor. The proposed structure is assembled using planar ferrite blocks enabling distribution of a single discrete air gap into multiple smaller gaps along the path of flux in the core. Moreover, the use of planar ferrite blocks allows for tuning the coupling coefficient by adjusting the reluctance and cross-sectional area of the core legs, while maintaining the same volume. In order to simplify the assembly process, a 3D printed bobbin design is presented which provides the required mechanical stability to the proposed core structure.

A. Magnetic Circuit Analysis

The equivalent magnetic structure for the proposed coupled conductor is shown in Fig. 3(a). It consists of ferrite blocks that are stacked together to form the core structure. Each ferrite block is separated by an air gap of length $l_g/3$. Two coils, each consisting of an equal number of turns N, are wound around the outer legs of the core. The width of the block in the central leg of the core is a, which is the same as in the conventionally gapped inductor. The width of the block in the outer leg is c, which for simplicity is considered to be equal to $a/2$. The depth of ferrite block is b and is the same as the depth of the conventionally gapped inductor core. The fringing effect caused by the air gap causes the reluctance of the path to decrease due to increase in the effective cross-sectional area of the air gap. One simple method to account for the fringing effect is to treat it as a parallel reluctance as shown in Fig. 3, as described in [12].

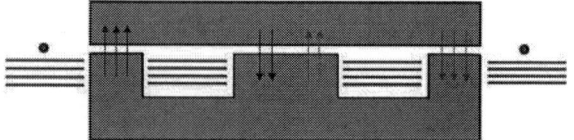

Fig. 1. Example of a planar coupled inductor using EI core

From the magnetic circuit in Fig 3(a), self inductance and mutual inductance can be derived as:

$$
\begin{cases}
L_s = \dfrac{N^2\left(R_{c\text{-}eff} + R_{o\text{-}eff}\right)}{R_{o\text{-}eff}\left(2R_{c\text{-}eff} + R_{o\text{-}eff}\right)} \\[3mm]
M = \dfrac{N^2 R_{c\text{-}eff}}{R_{o\text{-}eff}\left(2R_{c\text{-}eff} + R_{o\text{-}eff}\right)} \\[3mm]
k = \dfrac{M}{L}
\end{cases}
\tag{1}
$$

$$
R_{c\text{-}eff} = \frac{l_g}{3\mu_0 A_{c\text{-}eff}} + \frac{l_g}{3\mu_0 A_{c\text{-}eff}} + \frac{l_g}{3\mu_0 A_{c\text{-}eff}} = \frac{l_g}{\mu_0 A_{c\text{-}eff}}
\tag{2}
$$

$$
A_{c\text{-}eff} = (ab) + \frac{2l_g}{3}\left(a + b + \frac{2l_g}{3}\right)
\tag{3}
$$

Similarly, the expression for $R_{o\text{-}eff}$ is derived as:

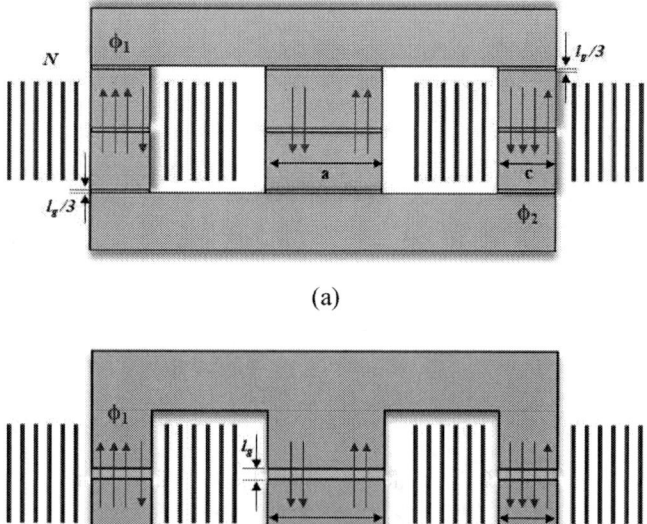

(a)

(b)

Fig. 2. Coupled inductor structure with (a) distributed air gap (proposed) and (b) discrete air gap (conventional)

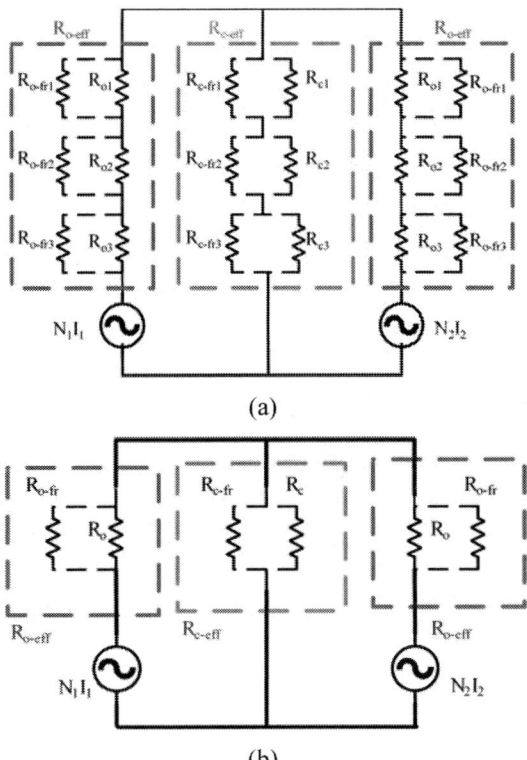

(a)

(b)

Fig. 3. Equivalent magnetic circuit models (a) proposed inductor and (b) conventional inductor

$$R_{o-eff} = \frac{l_g}{3\mu_0 A_{o-eff}} + \frac{l_g}{3\mu_0 A_{o-eff}} + \frac{l_g}{3\mu_0 A_{o-eff}} = \frac{l_g}{\mu_0 A_{o-eff}} \quad (4)$$

$$A_{o-eff} = \left(\frac{ab}{2}\right) + \frac{l_g}{3}\left(a + 2b + \frac{4l_g}{3}\right) \quad (5)$$

Similar derivation for the conventional coupled inductor from Fig. 3(b) leads to:

$$\begin{cases} L_s = \dfrac{N^2\left(R_{c-eff(conv)} + R_{o-eff(conv)}\right)}{R_{o-eff(conv)}\left(2R_{c-eff(conv)} + R_{o-eff(conv)}\right)} \\[3mm] M = \dfrac{N^2 R_{c-eff(conv)}}{R_{o-eff(conv)}\left(2R_{c-eff(conv)} + R_{o-eff(conv)}\right)} \end{cases} \quad (6)$$

$$\begin{cases} R_{c-eff(conv)} = R_c \parallel R_{c-fr} = \dfrac{l_g}{\mu_0 A_{c-eff(conv)}} \\[3mm] R_{o-eff(conv)} = R_o \parallel R_{o-fr} = \dfrac{l_g}{\mu_0 A_{o-eff(conv)}} \end{cases} \quad (7)$$

$$\begin{cases} A_{c-eff(conv)} = ab + 2l_g\left(a + b + 2l_g\right) \\[3mm] A_{o-eff(conv)} = \left(\dfrac{ab}{2}\right) + l_g\left(a + b + 4l_g\right) \end{cases} \quad (8)$$

Equation (2) and (4) define the effective reluctance of the central and outer leg for the proposed inductor structure, which maintains a total air gap length equal to that of a conventional gapped inductor as shown in (7). However, the effective cross-sectional areas due to the fringing effect differ in the two inductors designs. Equations (3) and (5) illustrate the effective cross-sectional area of the central and outer legs of the proposed core. Due to the distribution of the air gap in three positions along the leg of the core, the effective cross-sectional area only increases by a factor of $2l_g/3$ along the width and depth of the core. In contrast to this, (8) shows that the effective cross-sectional area of the conventional design increases by a factor of $2l_g$. This shows that for the same length of total air gap l_g, the inductance in the proposed structure is going to be lower as compared to the conventional design. Hence, the length of each individual air gap has to be decreased to reach the same value of self inductance [14], which is another advantage of the proposed design.

B. Reducing AC Losses

AC losses due to skin effect and proximity effect limit the performance of inductors in high frequency and high power applications. In high power applications, large air gaps are required to prevent core saturation, which exacerbates the AC losses. The use of parallel and orthogonal air gaps [10] improves AC resistance by aligning the fringing flux with the windings. However, this approach requires precise ferrite machining to position and optimize the air gap. This complicates the core fabrication process for large scale production. Alternatively, the vertical foil winding design presented in [12] achieves similar benefits by changing the winding orientation. While this approach reduces AC losses and eliminates the complexity of air gap machining, it still faces challenges with fringing flux due to discrete air gap. This leads to current crowding and significant AC losses in the turns near the air gap. Increasing the distance between the turns and the air gap can cause a slight improvement in the loss profile, but it stops improving after a certain distance. Additionally, the distance between the core and windings is limited by the available winding space in planar cores.

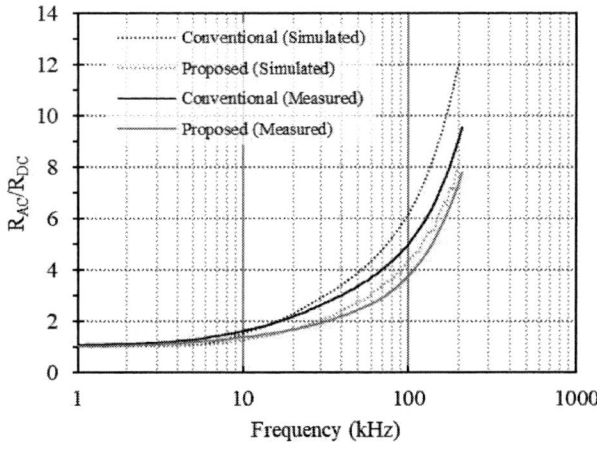

Fig. 4. Ratio of R_{AC} to R_{DC} for both proposed and conventional inductors

The proposed structure distributes the discrete gap into a number of smaller individual gaps, and utilizes foil windings, as shown in Fig. 2(a). By using ferrite blocks to assemble the structure, the three discrete air gaps shown in Fig. 2(b) have been distributed into nine smaller gaps. This distribution of air gaps into smaller sections reduces the intensity of fringing field near each individual gap, resulting in a significant decrease in AC resistance. This even distribution minimizes the penetration of fringing fields in both windings and core and helps maintain a more uniform magnetic field distribution both in core and the winding window. To quantify the advantages of distributed gapped coupled inductor over a conventionally gapped inductor, ANSYS 2D FEM simulation were conducted for both the configurations. The parameters were kept the same for both inductors except for the air gap configuration. The length of each individual air gap in the proposed structure was adjusted to achieve the same value of inductance as the conventionally gapped inductor, as discussed in the previous section. The AC resistance for both configurations was extracted from ANSYS across a wide frequency spectrum. Fig. 4 shows the ratio of AC resistance to DC resistance for both the designs. At 150 kHz, the simulated AC resistance of the proposed structure is 31% lower than that of the conventional design. This difference increases as the frequency rises. The measured AC resistance for the proposed structure shows a 27.7% reduction at 150 kHz, which aligns closely with the ANSYS simulation results.

Fig. 5 compares the flux distribution in the core for the proposed and conventional designs under the same operating conditions. In the proposed design, the flux distribution within the core is more uniform, in contrast to the conventional design, where flux tends to concentrate around the air gaps. This enhanced uniformity is attributed to better management of the flux path, which reduces the intensity of fringing flux and mitigates its impact, thereby preventing the formation of localized hot spots in the core. Fig. 6(a) and Fig. 6(b) highlights the fringing flux effects in the winding window and the current crowding near the air gaps for the proposed and the conventional designs, respectively. It can be seen that the current distribution in the windings has considerably improved in the proposed design.

C. Tuning the Coupling Coefficient

The coupling coefficient in coupled inductors is a key parameter that directly impacts the performance of DC-DC converters by defining the efficiency of magnetic linkage between the two windings [3]. Optimizing this coefficient is essential to improve the converter's overall efficiency, transient response, and steady-state behavior [2]. However, achieving a specific coupling coefficient is challenging because it relies heavily on factors such as the core geometry, the positioning and orientation of windings, and the placement and size of air gaps. These design aspects often necessitate careful machining of ferrite materials and a complex assembly process to achieve the desired value of coupling coefficient. Once the core is modified to achieve a particular value of k, it is generally fixed, making any further adjustments difficult or impractical. This design rigidity emphasizes the importance of precise initial tuning and manufacturing to ensure the desired magnetic coupling for optimal converter performance.

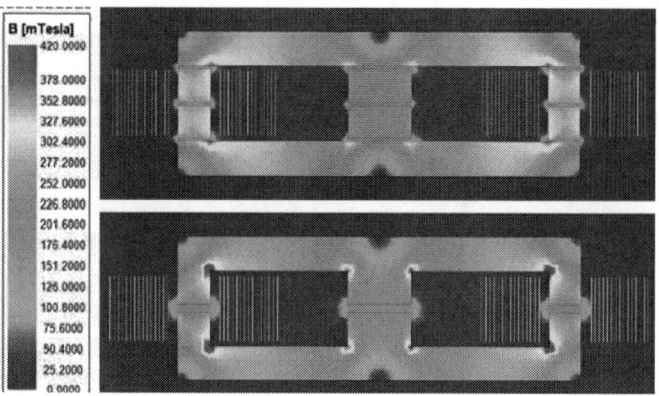

Fig. 5. Comparison of flux distribution in the cores of proposed (top) and conventional (bottom) designs

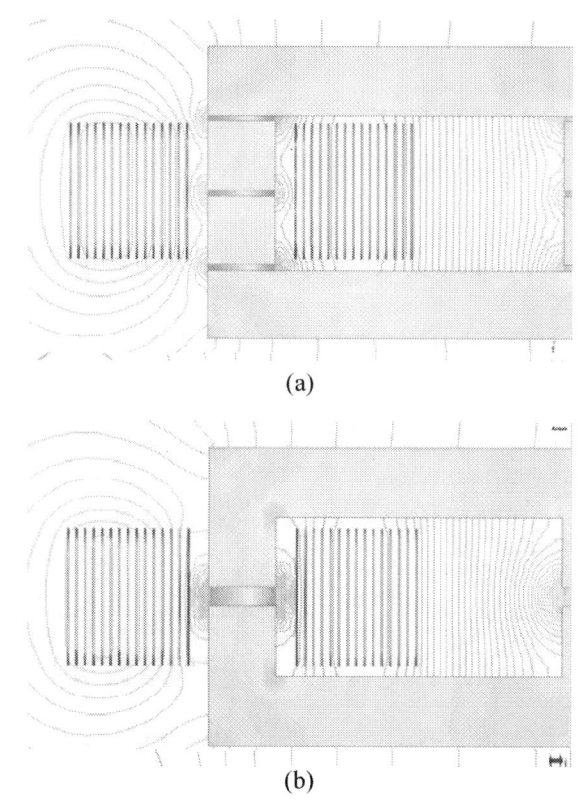

(a)

(b)

Fig. 6. Comparison of fringing flux in the winding window for (a) proposed design and (b) conventional design

The proposed distributed-gap structure offers a practical solution for fine tuning the coupling coefficient without requiring extensive machining of ferrites or a complex assembly process. From the magnetic circuit illustrated in Fig. 3, the coupling coefficient is calculated as follows:

$$k = \frac{M}{L} = \frac{R_{c-eff}}{R_{c-eff} + R_{o-eff}} \qquad (9)$$

This expression can also be written, in terms of the parameters of ferrite blocks in the central and outer legs of the core, as:

$$k = \frac{A_{o-eff}}{A_{o-eff} + A_{c-eff}} \tag{10}$$

$$k = \frac{c + l_g}{a + c + 2l_g} \tag{11}$$

Equation (9) demonstrates that the coupling coefficient k is determined by the reluctance ratio between the core legs, with variations in this ratio directly impacting magnetic coupling. (10) and (11) further establish a direct relationship between k and the widths of the core's central and outer legs. By adjusting the widths of ferrite blocks in the proposed structure, the coupling coefficient k can be effectively controlled, allowing fine-tuning without extensive alterations.

ANSYS simulations were conducted using various combinations of available ferrite block widths. The combinations are presented as a function of the ratio of width of core legs. The coupling coefficients and self inductances for each core configuration is summarized in Table I. To ensure consistency, the number of turns N was adjusted in each case to maintain the same self inductance. These simulations confirm that modifying the ratio of widths in the central and outer legs allows control over the coupling coefficient simply by selecting and positioning ferrite blocks as required.

TABLE I. CORE CONFIGURATIONS FOR DIFFERENT COUPLING COEFFICIENTS

Core Construction	L, (µH)	k
$c = a/2$ ($a = 10.4$ mm)	53	0.28
$c = a$ ($a = 10.4$ mm)	52	0.42
$c = a$ ($a = 5.2$ mm)	49	0.36
$c = 10.4$ mm ($a = 0$, i.e.,no center leg)	51	0.52

III. EXPERIMENTAL RESULTS AND DISCUSSION

To evaluate and compare the performance of the proposed coupled design with the conventional design, a test case involving two-phase interleaved boost converter is considered. The following section details the implementation and results of the proposed design within this context.

A. Coupled Inductors in Two-Phase Boost Converter

Fig. 7 shows the schematic layout for a typical multiphase boost converter, a four-phase example is shown. For two-phase converters, the inductors should be distributed among the phases so as to maintain a 180° phase shift between the currents in a coupled inductor pair. This is to maximize the flux cancelation in a single core. As shown in Fig. 8, the mutual flux in a coupled inductor pair affects the voltage across each inductor, causing changes in the current slopes during four distinct periods. This results in different equivalent inductances, i.e., L_{eq1}, L_{eq2} and L_{eq3}, for each period. L_{eq1} and L_{eq3}, which correspond to $D < 0.5$ and $D > 0.5$ respectively, directly influence the peak and ripple inductor current.

From [2], L_{eq1} or L_{eq3} can be written as:

$$L_{eq1} \text{ (or } L_{eq3}) = \alpha \times L \tag{12}$$

$$\begin{cases} for \quad D < 0.5 \quad \alpha = \dfrac{1 - k^2}{1 + k\dfrac{D}{D'}} \\[4mm] for \quad D > 0.5 \quad \alpha = \dfrac{1 - k^2}{1 + k\dfrac{D'}{D}} \end{cases} \tag{13}$$

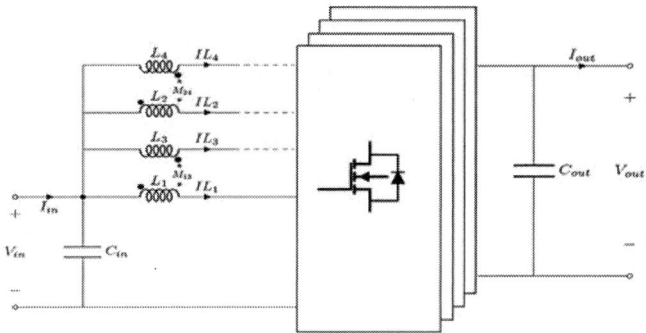

Fig. 7. Schematic of a multiphase DC-DC converter with coupled inductors

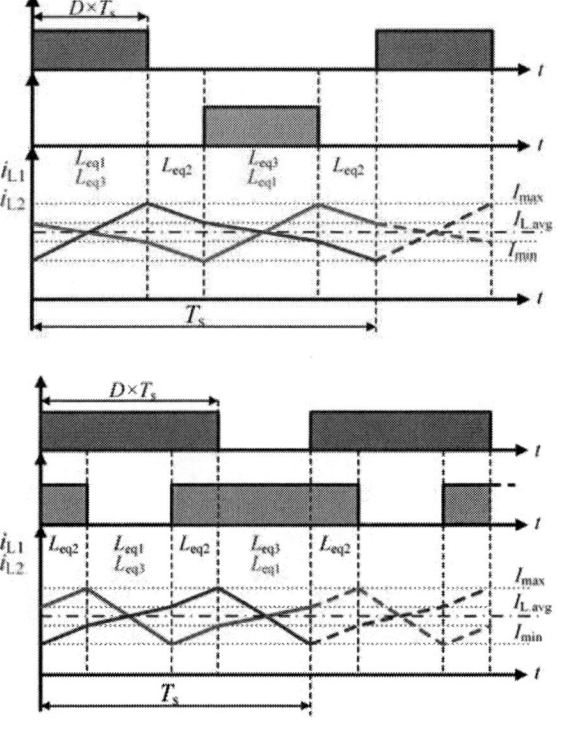

Fig. 8. Inductor current waveform with coupled inductor for $D < 0.5$ (top) and $D > 0.5$ (bottom)

979-8-3315-1612-3/25 $31.00 © 2025 IEEE 2457

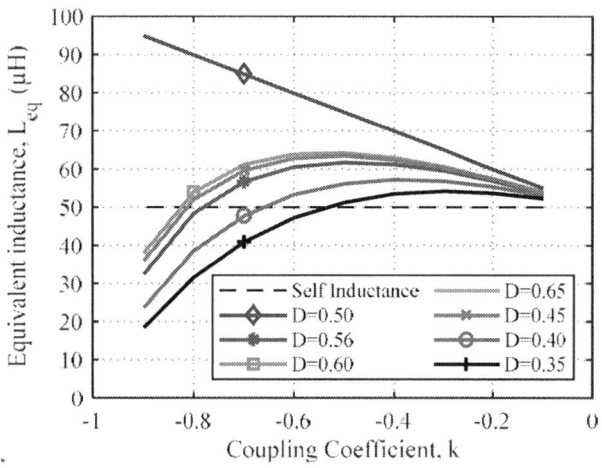

Fig. 9. Relationship between equivalent inductance (L_{eq1} or L_{eq3}) and coupling coefficient (k) at different duty cycles.

If the coupling factor k is negative, i.e., inverse coupling, the value of α can exceed 1, depending on the operating duty cycle D. This means that L_{eq1} or L_{eq3} will be greater than the self-inductance L. Since the coupling factor k is fixed once the inductor is designed, it should be chosen during the design stage to ensure the desired performance across the entire duty cycle range. Fig. 9 illustrates the relationship between the equivalent inductance of the coupled inductor and the coupling coefficient across various duty cycles typical in converter applications. The plot reveals that for both duty cycles range, above and below 0.5, the equivalent inductance initially increases as the coupling coefficient k rises from 0.1 to 0.5. However, beyond a coupling coefficient of 0.5, the inductance begins to decrease, eventually falling below the inductor's self-inductance for a higher value of k. This behavior highlights an optimal range for the coupling coefficient, between 0.3 and 0.5.

In contrast to the steady state operation, where a large inductance value is preferred, large inductance values slow down the dynamic response of the converter. However, for the inverse coupled inductor, L_{eq2} is responsible for the dynamic response of the converter [2], where $L_{eq2} = (1+k)\,L$, and since k is negative, L_{eq2} is always smaller than L. Therefore, the converter can have faster dynamic response than that with uncoupled inductors.

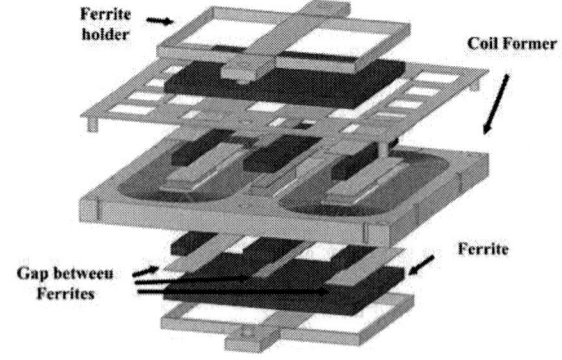

Fig. 10. Exploded 3D model of the proposed coupled inductor

Fig. 11. Photograph of the coupled inductor prototypes

TABLE II. BOOST CONVERTER KEY PARAMETERS

Parameter	Symbol	Value
Output Power	P_{out}	20 kW
Output Voltage	V_{out}	800 V
Input Voltage	V_{in}	$350-500$ V
Max. Input Current	$I_{in\,max}$	36 A
Number of Phases	n	2
Switching Frequency	f_{sw}	150 kHz
Inductor Current Ripple	Δi_L	50%
Self Inductance	L	50 μH
Inductor Coupling Coefficient	k	-0.33

Building on the earlier discussion, a 20 kW, two-phase interleaved boost converter was assembled to verify the proposed inductor design. Table II summarizes the system and inductor parameters. The next section discusses the prototyping of the required inductors.

B. Inductor Prototyping

The proposed coupled inductor configuration is shown in Fig. 10, featuring a 3D-printed bobbin that provides mechanical stability to the core. The bobbin has precisely defined slots that accommodate the ferrite blocks with a small tolerance, ensuring a secure fit. To create the required air gaps, small 3D-printed shims are placed between the ferrite blocks in both the central and outer legs of the core. The vertical windings, made from uninsulated copper foils, are shaped and inserted into the spiral slots of the bobbin. The bobbin is fabricated using ABS material, which has a dielectric strength of approximately 12 kV/mm [2], and a 0.5 mm separator is used between winding turns for insulation.

The prototypes for both the proposed and conventional inductor designs, each with the same number of turns, are presented in Fig 11. In the conventional design, the air gap length (l_g) is 2.00 mm, while in the proposed design, the air gap is divided into three sections, with one-third of the gap length ($l_g/3$) being 0.45 mm, as detailed in Section I. As a result, the total air gap length in one leg of the proposed design is 1.35 mm, compared to 2.00 mm in the conventional design. Both the inductors have the same inductance of 50 uH and a resonant frequency of 7 MHz as measured by the impedance analyzer, shown in Fig. 12. The coupling coefficient for this prototype is chosen to be 0.33 based on the results in Fig. 9.

C. Experimental Results

The tests were conducted at two duty cycles: one below 0.5, corresponding to V_{in} = 450 V, and the other above 0.5, corresponding to V_{in} = 350 V. For each duty cycle, the power was varied from 7 kW to 14 kW, with the upper limit constrained by the power supply's current capacity. Input power, output power, and inductor losses were measured using a Yokogawa WT5000 power analyzer to ensure precise and reliable measurements. Fig. 13 shows the test bench setup of the boost converter for prototype verification. The coupled inductor current waveform for the proposed design at $D > 0.5$ is shown in Fig. 14.

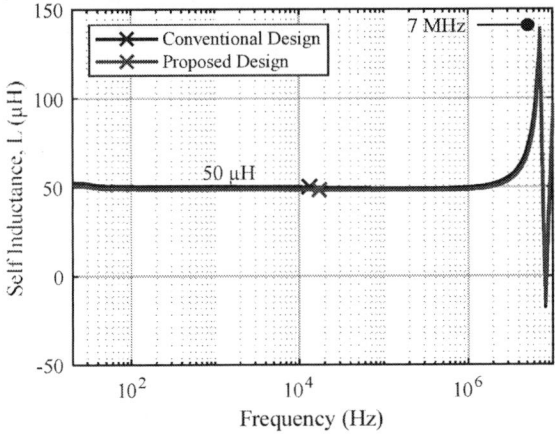

Fig. 12. Inductance measurement using impedance analyzer

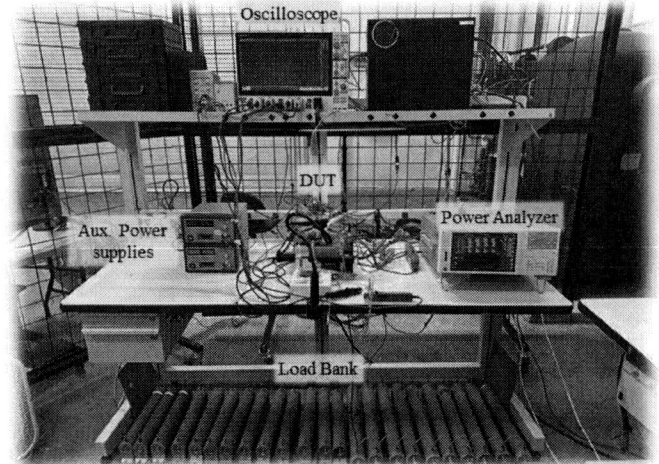

Fig. 13. Test bench setup for experimental verification

Fig. 15 compares the total inductor losses as a function of output power for the proposed coupled inductor design and the conventional design. In all test cases, the converter

components, apart from the inductors, were kept identical. Since both prototypes have the same inductance values, operating under identical input voltages and duty cycles produced equivalent current waveforms and RMS currents. As a result, the losses associated with the power stage and other components remained nearly constant across all cases. Consequently, the observed difference in the measured losses can be attributed primarily to the reduction in AC winding losses in the inductors. To enhance the clarity of the results, the inductor losses were isolated from the overall system power losses and presented separately.

The experimental results for duty cycles below 0.5 (V_{in} = 450 V) indicate that the proposed coupled inductor design achieves a 31% reduction in AC winding losses, aligning closely with predictions from ANSYS FEM simulations. At an output power of 13 kW, the conventional design exhibits AC losses of approximately 16 W, while the proposed design reduces these losses to around 11 W.

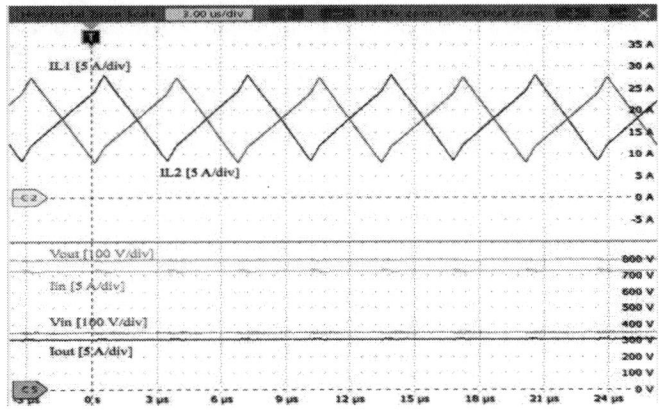

Fig. 14. Inductor current waveforms at 13 kW, 800 V output

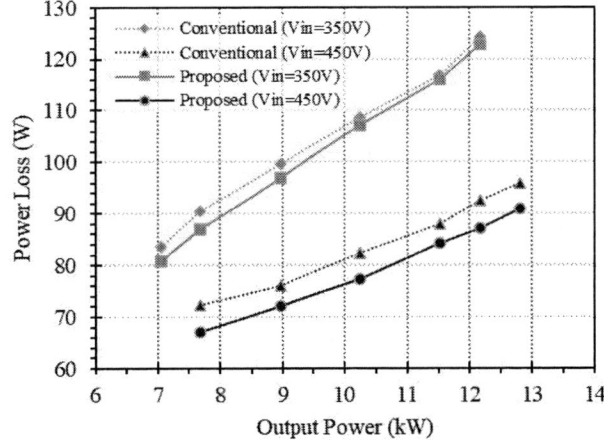

Fig. 15. Comparison of measured inductor losses in the proposed inductor design and conventional inductor design for different input voltages

In contrast, for the V_{in} = 350 V case, the reduction in AC winding losses is less pronounced, as core losses dominate the total losses in the coupled inductor at this operating point.

These results highlight that while the proposed design effectively minimizes AC winding losses, its system-level benefits are limited in applications where DC current is high, and AC current is relatively small. However, in applications where coupled inductors operate under higher AC current conditions, the proposed design would offer more significant system-level advantages.

IV. CONCLUSION

In this article, a coupled inductor structure is proposed and evaluated within a two-phase interleaved boost converter. In coupled inductors, the flux cancellation in the core can cause uneven fringing field in the winding window that affect the AC losses more adversely. Moreover, the selection of k in the design of coupled inductor can have a direct impact on the overall performance of the converter. The proposed design extends the use of vertical copper foil presented in [13] and offers a significant reduction in AC resistance by the use of segmented air gap. Through the distribution of air gaps, the effect of fringing flux in the winding window is mitigated, and offers a better current distribution in the windings, as well as better flux distribution in the core. Simulation and experimental results demonstrate a significant reduction in AC losses.

The proposed structure also simplifies the inductors fabrication process compared to designs previously presented in the literature. Due to the flexible nature of the core structure, the coupling coefficient can be easily tuned by using different combinations of the widths of the ferrite blocks used to assemble the core. Experimental results shows that the use of the proposed design can decrease the AC losses in inductors by more than 30%, which can contribute significantly in high power converters.

REFERENCES

[1] V. K. Rathore, M. Evzelman, and M. M. Peretz, "Coupled Inductor Design Methodology for Optimization of Boost Extending Topology," in *2024 IEEE Applied Power Electronics Conference and Exposition (APEC)*, Feb. 2024, pp. 1389–1395. doi: 10.1109/APEC48139.2024.10509217.

[2] A. H. Ismail, Z. Ma, A. Al–Hmoud, and Y. Zhao, "An Interleaved Multi-Phase Boost Converter with Coupled Inductors for High Power Density," in *2022 IEEE Energy Conversion Congress and Exposition (ECCE)*, Oct. 2022, pp. 01–07. doi: 10.1109/ECCE50734.2022.9947586.

[3] P.-L. Wong, P. Xu, P. Yang, and F. C. Lee, "Performance improvements of interleaving VRMs with coupling inductors," *IEEE Trans. Power Electron.*, vol. 16, no. 4, pp. 499–507, Jul. 2001, doi: 10.1109/63.931059.

[4] F. Zheng, Y. Pei, Y. Liu, L. Wang, X. Yang, and Z. Wang, "Design Coupled Inductors for Interleaved Converters Using a Three-Leg Core," *IEEE Trans. Magn.*, vol. 44, no. 12, pp. 4697–4705, Dec. 2008, doi: 10.1109/TMAG.2008.2002300.

[5] U. Asif, M. A. Aslam, S. Ahmed, and M. F. Uddin, "Design and Implementation of 125 and 243 Level Cascaded H-Bridge Multilevel Inverter using Binary Search Algorithm," in *2022 10th International Conference on Smart Grid (icSmartGrid)*, Jun. 2022, pp. 40–45. doi: 10.1109/icSmartGrid55722.2022.9848592.

[6] Z. Wang, C. Li, Z. Zheng, H. Jiang, and Z. Huang, "A DC Flux Cancellation Design Method for Inverse-Coupled Inductors Used in Interleaved Boost Converters," *IEEE Trans. Transp. Electrification*, vol. 10, no. 2, pp. 2956–2964, Jun. 2024, doi: 10.1109/TTE.2023.3302803.

[7] T. Kang, A. Gandomkar, J. Lee, and Y. Suh, "Design of Optimized Coupling Factor for Minimum Inductor Current Ripple in DC-DC Converter Using Multiwinding Coupled Inductor," *IEEE Trans. Ind. Appl.*, vol. 57, no. 4, pp. 3978–3989, Jul. 2021, doi: 10.1109/TIA.2021.3079289.

[8] G. Y. Sizov, Z. Vrankovic, and G. L. Skibinski, "Novel PCB Integrated Magnetic Component Design for Reduced AC Power Losses," in *2019 IEEE Energy Conversion Congress and Exposition (ECCE)*, Sep. 2019, pp. 4146–4153. doi: 10.1109/ECCE.2019.8912228.

[9] S. Wang, P. H. Pham, Q. Li, A. Nabih, and P. R. Prakash, "PCB Winding-Based Coupled Inductor for a High-Frequency DC/DC Converter with 99% Efficiency," in *2023 IEEE Applied Power Electronics Conference and Exposition (APEC)*, Mar. 2023, pp. 420–425. doi: 10.1109/APEC43580.2023.10131439.

[10] S. Mukherjee, Y. Gao, and D. Maksimović, "Reduction of AC Winding Losses Due to Fringing-Field Effects in High-Frequency Inductors With Orthogonal Air Gaps," *IEEE Trans. Power Electron.*, vol. 36, no. 1, pp. 815–828, Jan. 2021, doi: 10.1109/TPEL.2020.3002507.

[11] Y. Liu *et al.*, "Optimized Air-Gap Configuration for an Integrated Coupled Inductor With Lower Height and Reduced Core/Winding Losses," *IEEE Trans. Ind. Appl.*, vol. 60, no. 2, pp. 2980–2990, Mar. 2024, doi: 10.1109/TIA.2023.3321265.

[12] A. H. Ismail, Z. Ma, A. Al-Hmoud, and Y. Zhao, "A High Frequency Coupled Inductor Design for High Power Density DC-DC Converters," in *2023 IEEE Applied Power Electronics Conference and Exposition (APEC)*, Mar. 2023, pp. 3275–3280. doi: 10.1109/APEC43580.2023.10131629.

[13] J. Zhou, H. Liu, T. Zhao, and X. Guo, "A Novel One-Dimension Space Vector Strategy for Multilevel Cascaded Inverters," in *2023 IEEE Applied Power Electronics Conference and Exposition (APEC)*, Mar. 2023, pp. 251–256. doi: 10.1109/APEC43580.2023.10131653.

[14] J. Knowles, "The origin of the increase in magnetic loss induced by machining ferrites," *IEEE Trans. Magn.*, vol. 11, no. 1, pp. 44–50, Jan. 1975, doi: 10.1109/TMAG.1975.1058549.

High-Power Planar Transformer Design for Four-Port Converters

Arya Sadasivan
Department of Electrical and Computer Engineering
Kansas State University
Manhattan, KS, USA
arya09@ksu.edu

Behrooz Mirafzal
Department of Electrical and Computer Engineering
Kansas State University
Manhattan, KS, USA
mirafzal@ksu.edu

Abstract— **This paper analyzes a high-power planar transformer designed for GaN-based bidirectional DC-DC four-port converters. The multiport configuration introduces challenges in achieving uniform magnetic coupling among the windings as their number increases. This study addresses these magnetic design complexities, proposing solutions to enhance power density and magnetic coupling while minimizing core losses. Interleaving and flux path management designs are applied to develop a planar transformer featuring four windings with a 3(400)/800V and 3(5kW)/15kW rating. Finite Element (FE) simulations validate the design's performance, and experimental evaluation of a fabricated prototype confirms its magnetic characteristics.**

Index Terms— **Planar transformer, GaN-based H-bridge converters, multiport converters.**

I. INTRODUCTION

With rising global energy consumption and growing environmental concerns, the development of sustainable and energy-efficient systems is essential for meeting climate targets and low-carbon energy infrastructures. For these reasons, nearly net-zero energy buildings (NZEBs) play a critical role, helping to reduce energy demand, greenhouse gas emissions, and reliance on nonrenewable resources. These buildings are designed to balance annual energy production and consumption by utilizing renewable energy sources (RES), like solar and battery systems.

In NZEBs, power management involves integrating multiple energy generation, storage, and consumption systems to ensure a balanced supply and demand. When on-site renewable generation exceeds consumption, surplus energy is stored in batteries or exported to the grid [1]. Conversely, during periods of low on-site generation (e.g., cloudy days), energy is drawn from batteries or the grid to meet demand. Such systems can enhance grid stability by providing ancillary services to the grid, such as frequency regulation, voltage stabilization, and fast dynamic responses, by utilizing the rapid response capabilities of battery energy storage systems (BESS) [2]. For instance, in-demand responses, the loads, and batteries are strategically managed to either absorb energy or discharge it into the grid based on real-time grid conditions. Furthermore, with advanced autonomous controllers such as nano or microgrids, the inverters can seamlessly transition between grid-following and grid-forming modes [3]-[6]. This operational flexibility helps address scenarios of high demand or limited supply. These capabilities stabilize the grid and reduce peak loads and support the transition to a more sustainable and resilient energy system. In this context, NZEBs can be technically classified as nano grids.

Advances in power electronics, particularly the development of fast semiconductor switches capable of high-speed switching and substantial power handling, have significantly enabled the integration of multiple sources and loads in a compact form for NEZBs. However, a major challenge lies in the inherent mismatch between the voltage-current characteristics of RES and those of connected loads. Additionally, battery storage systems are typically designed with standardized voltage and current ratings to ensure safety, compatibility, and adherence to regulatory standards [1], [7]. Various power converter topologies have been proposed to address these mismatches in the literature [8]-[12]. They are generally categorized into multiple individual converters, DC-link battery circuit topologies, and AC-side battery circuits. All these approaches employ combinations of conventional converters—such as buck, boost, buck-boost, and inverter circuits—to regulate voltage and current and meet the requirements of RES, energy storage systems (ESS), and loads [13]. These topologies are often application-specific; for instance, some configurations are particularly well-suited for high-voltage battery systems. The use of multiple converters proportional to the number of power sources leads to redundant components, lower utilization of devices, reduced power density and efficiency, and higher costs [13]-[14]. As a result, there is a growing need for integrated and optimized power converter solutions that simplify system architecture and enhance overall performance.

Recent investigations have shown that multiport converters are promising solutions for achieving compact and energy-efficient systems. These converters typically share passive and active components, incorporating advanced control strategies such as time-sharing control and duty-cycle modulation [14]. Multiport converters are broadly classified based on the degree of isolation between their ports: non-isolated, partially-isolated, and fully-isolated topologies. The non-isolated topologies are known for their compactness and high efficiency, achieved by eliminating isolating components such as transformers. However, certain high-voltage applications demand complete isolation between the ports to ensure broader voltage gain capabilities, enhanced safety, and seamless integration of ports operating at different voltage levels. In such cases, partially and fully-isolated topologies become essential, balancing efficiency with the functional requirements of the system.

In multiport converters, to enhance system performance, wide-bandgap (WBG) devices such as silicon carbide (SiC) and gallium nitride (GaN) are increasingly used as switching elements [15]-[18]. These devices enable operation at significantly higher switching frequencies (100 kHz to 1 MHz), which improves efficiency, reduces system size, and increases

979-8-3315-1612-3/25 $31.00 © 2025 IEEE

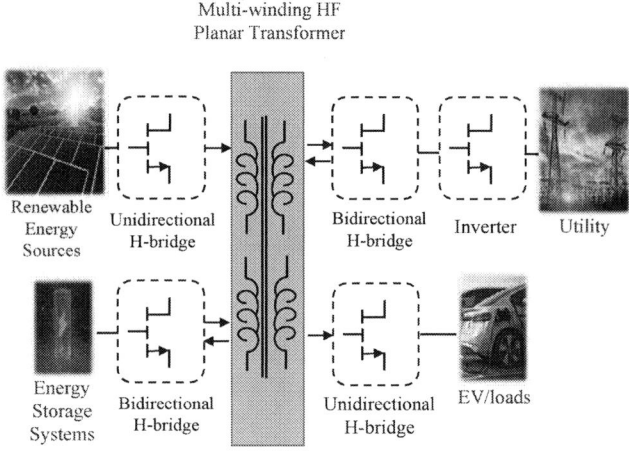

Fig. 1. Fully-isolated multiport bidirectional DC-DC converter.

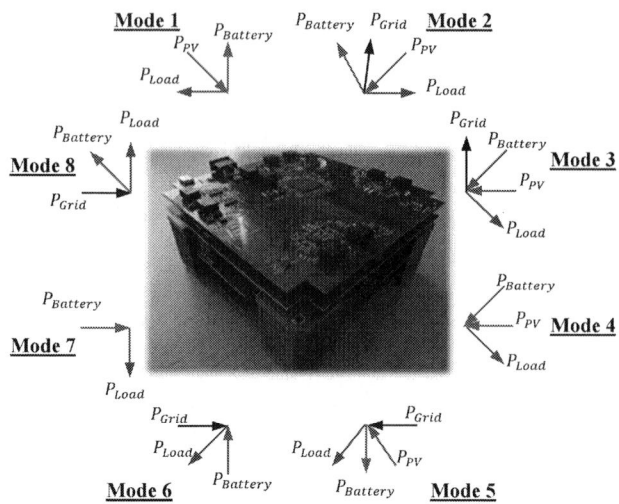

Fig. 2. Operating modes of a fully-isolated four-port converter for grid-integrated renewable energy systems.

power density. However, these advancements impose stringent requirements on transformer design, as traditional low-frequency transformers are inadequate for high-frequency applications in terms of compactness, efficiency, and performance. To address this challenge, innovations in PCB-based windings and high-frequency core materials have led to the development of planar transformers [19]. Achieving an optimal planar transformer design requires careful consideration of several parameters, including operating frequency, magnetic flux distribution, core losses, copper losses, and leakage inductance [19]-[22]. Various planar transformer design strategies, such as matrix transformers, fractional-flux or fractional-turn transformers, and planar transformers with engraved ferrite cores or adjustable leakage inductances, have been explored in the literature [23]-[27]. However, most of these designs are limited to low-power applications and have a reduced number of windings, making them unsuitable for fully isolated multiport converters in NZEB applications.

This article proposes a four-winding planar transformer optimized for use in a fully isolated four-port converter. The design targets an operating frequency of 100 kHz and incorporates several key considerations, such as the choice of core material, uniform flux distribution, and a winding configuration that minimizes interconnections and vias. The winding terminals are distributed to accommodate the four ports of the converter easily. Furthermore, the core structure is designed with reduced height and increased surface area to enhance the thermal performance of the transformer. The design incorporates flux cancellation techniques, commonly used in matrix transformers, and winding interleaving methods to minimize core losses and leakage inductance. An ER-core structure is used as a benchmark to study leakage inductance and assess the performance of the proposed design through ANSYS Maxwell finite element analysis.

In the following, Section II discusses the power flow analysis of the four-port converter. Section III covers the design aspects of high-frequency transformer cores and the strategies for flux management in a multi-winding transformer. Section IV presents the coupling effects between windings and provides the finite element analysis of the proposed transformer. Section V details the experimental analysis conducted on the prototype. Finally, Section VI concludes this paper by summarizing the findings and outcomes of the presented work.

II. POWER FLOW IN FOUR-PORT CONVERTER

In this section, the power flow dynamics in a RES-based NZEB are analyzed to develop a four-port converter configuration. In buildings and industrial sites aiming for net-zero or nearly-zero energy, the goal is to maximize energy efficiency which necessitates addressing losses caused by inefficient and bulky converters. The use of a multiport converter offers significant advantages over multiple individual converters, as it reduces redundancy in component usage, streamlines the control system, and enhances energy management by efficiently balancing power flow across different energy sources.

Fig. 1 illustrates a fully isolated four-port converter, where ports 1,2,3, and 4 are coupled via a four-winding transformer for power transfer. The multi-winding transformer provides electrical isolation between the ports and voltage level matching with a suitable turns ratio. Typically, fully isolated multiport converters are implemented using half-bridge or full-bridge topologies, both of which can support bidirectional power flow when equipped with bidirectional switches such as SiC moseft, GaN, etc. However, the half-bridge topology is limited by a duty cycle constraint of 50%. This limitation impacts the implementation of soft-switching, thereby affecting the overall efficiency and performance of a half-bridge-based multiport converter [13]. Hence, for the implementation of the four-port fully isolated converter for NZEBs systems, the full-bridge topology is considered, as shown in Table I, with specifications.

Fig. 2 illustrates the various operating modes required for the four-port converter in a NZEB application. The power generated by the RES, typically photovoltaic (PV), is primarily used to meet the local load demand. After satisfying the load demand, any surplus power is directed to the battery, the grid, or both, based on the battery's state of charge (SoC) and the grid operating conditions. To ensure efficient and reliable operation, the converter must support multiple operating modes. The net power equation for the system is given by:

$$P_{net} = P_{PV} + P_{Load} - P_{Battery} + P_{Grid} \qquad (1)$$

where, P_{PV} is the power generated by the PV source, which is always unidirectional, P_{Load} represents the power demand of the connected local loads, $P_{Battery}$ denotes the power exchanged with the battery (positive when charging and negative when discharging), and P_{Grid} is the power exchanged with the grid (positive when importing and negative when exporting).

The operating modes described can be effectively achieved using a fully isolated converter with appropriate phase angle control. However, this requires the design and implementation of a high-power, high-frequency multi-winding transformer to support the compact and high-power-density design of the converter.

III. CORE STRUCTURE AND FLUX MANAGEMENT IN MULTI-WINDING TRANSFORMER

This section discusses the design considerations for a multi-winding planar transformer in a fully isolated four-port converter. In multiport converters, WBG devices enable switching frequencies in the MHz range, significantly reducing the size and weight of passive components and enhancing the power density. However, achieving a compact and efficient multi-winding transformer capable of operating at such high switching frequencies is a critical challenge. In the proposed high-power, fully isolated four-port converter, the switching frequency is restricted to 100 kHz to minimize the frequency-dependent losses, such as switching losses, core losses, and losses in magnetic and winding components.

The primary consideration in the design of a high-frequency transformer is the selection of a suitable core material. Conventional low-frequency transformers commonly use cores made of iron alloys due to their high magnetic permeability and saturation flux density. However, their relatively high electrical conductivity makes them unsuitable for high-frequency applications because of increased eddy current losses. Therefore, ferrite materials with low electrical conductivity are preferred for power transformers operating at frequencies up to 1 MHz. Among ferrite materials, MnZn-based ferrites are favored over NiZn-based ferrites due to their superior permeability [22]. A core material equivalent to Ferroxcube 3C95 has been selected for this application. For the selected core material, at a switching frequency. $f_{sw} = 100 \ kHz$, and a peak flux density, $B_{max} = 310 \ mT$, the core loss per unit volume (P_v) is approximately $900 \ kW/m^3$. The number of turns for the transformer winding can be calculated using the following equation,

$$N = \frac{V}{k f_{sw} B_{max} A_e} \qquad (2)$$

where k is a waveform-dependent constant. Typically, for sinusoidal excitation, $k = 4.44$, as the magnetizing flux, ψ, increases from zero to its maximum value, ψ_m, in one-quarter of the cycle ($T/4$), and the form factor is 1.11. However, for square wave excitation, $k = 2$, as ψ reaches to ψ_m in half the

cycle ($T/2$), and the form factor is 1. Hence, with the above parameters, and for an effective core cross-sectional area, $A_e = 804.247 mm^2$, the number of turns for the proposed multiwinding transformer is 8 for 400 V ports and 16 for the 800 V port.

Once the effective cross-sectional area of core and the number of turns of windings are known, selecting an appropriate core structure is crucial for optimizing the design. This includes minimizing core and winding losses, effectively managing magnetic flux, ensuring efficient winding interconnections, and mitigating issues like EMI and crosstalk. The EI core is the most commonly used design among various core structures. However, previous studies indicated that for the same effective cross-sectional area, the current path length in a rectangular core pillar is longer than in a circular core pillar. This increased path length leads to higher winding losses in rectangular pillars. Fig. 3 illustrates the current distribution for rectangular and circular core pillars for a cross-sectional area of $804.247 mm^2$. The winding length required for two turns is $322.273 \ mm$ in the rectangular pillar and $256.096 \ mm$ in the circular pillar, resulting in slightly higher winding losses in the rectangular core pillar than in the circular core pillar. Fig. 4 illustrates the winding loss and core loss for both cases. The core and winding specifications were kept identical in both scenarios to evaluate the impact of core geometry on winding length and, consequently, on winding loss. Based on these findings, the ER core is used in this paper as the benchmark for evaluating the performance of the proposed core structure. The ER core not only features a circular pillar but also offers better magnetic coupling between windings by allowing the primary and secondary windings to be placed on the same pillar in a fully interleaved configuration.

Fig. 5 illustrates the flux distribution in the conventional ER-core, considering both primary and secondary windings are uniformly distributed along the center leg. The radius of the core is maintained at 16 mm to achieve the desired cross-sectional area for carrying the maximum magnetizing flux (ψ_m). From the figure, it is evident that the flux path utilizes the two side legs, each with a cross-sectional area of $A_e/2$, to have a complete the magnetic circuit path. However, in this configuration, the two side legs of the core are not used for windings, leading to an underutilization of the core material. The total footprint area and height of the core structure needed for the proposed four-winding planar transformer are $5400 \ mm^2$ and $90 \ mm$, respectively. A higher core height is required to accommodate the horizontal arrangement of PCB-based windings on a single pillar while maintaining sufficient clearance. This clearance accounts for the thickness of the PCBs, the minimum spacing between PCBs, and the required gaps between the core and the top and bottom PCBs. Since this transformer design includes four PCB-based windings that can function as either primary or secondary depending on the mode

TABLE I
FOUR-PORT CONVERTER SPECIFICATIONS

	Voltage (V)	Current (A)	Power (kW)
Port 1,3,4	400 V	18.75	7.5
Port 2	800V	18.75	15

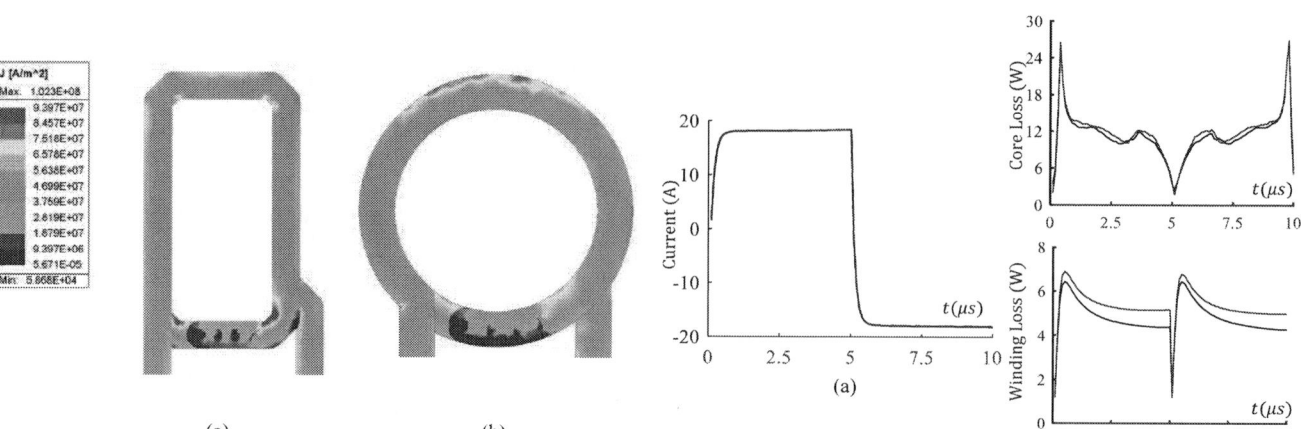

Fig. 3. Current distribution in windings for various core cross-sections with the same cross-sectional area, A_e (a) rectangular core pillar, (b) circular core pillar.

Fig. 4. Current through windings and losses over a single switching cycle for rectangular and circular core cross-sections with equal cross-sectional area (A_e), demonstrating reduced winding losses for the circular core pillar.

of operation, they are labeled as coil 1, coil 2, coil 3, and coil 4 for simplicity.

As mentioned earlier, core loss can be mitigated to some extent through careful material selection and by effectively utilizing the core area in the design of high-frequency transformers. In ER cores, the side legs are often underutilized, contributing to inefficiency. To address this, this paper explores an effective flux path management technique known as flux cancellation for designing the core structure of the proposed four-winding planar transformer. In a transformer, the core is primarily intended to carry magnetizing flux. The flux responsible for transferring load energy is inherently canceled per the basic transformer operating principle. In this context, flux cancellation refers to optimizing the handling of magnetizing flux in the core without disrupting the overall flux path.

In the literature, flux cancellation is primarily used in matrix transformer configurations, where the core is excited by identical voltage-second, resulting in the same flux generation in each core [24]. To further illustrate the concept of flux cancellation, Fig. 5(b) illustrates the flux distribution within the core structure of the proposed four-winding planar transformer. In this design, each core leg is designed to handle the maximum magnetizing flux, ψ_m. The turns of Coil 1 are evenly distributed, ensuring that current flows through them in the direction indicated in Fig. 5(b). A similar pattern is maintained for the remaining coils. This arrangement allows each core leg to carry the flux generated by the surrounding coil turns while simultaneously providing a return path for flux from adjacent legs. Consequently, this core structure achieves better core utilization compared to an ER core. Furthermore, this technique ensures that the top and bottom cores handle only half of the total magnetizing flux, further enhancing the core design's efficiency. The total footprint area and height of the proposed core structure are 11236 mm^2 and 42 mm, respectively.

The proposed core structure increases the footprint area by 2.08 times compared to the ER core. However, the core height is reduced to 0.533 times that of the ER core. The larger footprint provides more surface area, enhancing thermal dissipation. Moreover, this design enables the four coil terminals to be distributed more effectively, simplifying integration with the four ports of the converter. In contrast, an ER core requires additional PCBs to accommodate the turns of the four coils in a vertical arrangement within a two-layer PCB. This complexity leads to a design heavily reliant on vias and terminals, making the ER-core structure more intricate. Additionally, it results in significant conduction losses due to the alternating current flowing through vias and terminals across multiple PCB layers.

Figs. 6 and 7 illustrate the 3D structure and detailed winding arrangement for both the ER core and the proposed core structure. In the proposed core structure, the turns of the coils are distributed among the four core pillars, requiring only a 2-layer PCB to accommodate the 8 turns needed to achieve 400 V. Consequently, 4 coils are realized using five PCBs. A copper thickness of 2 oz is used to ensure compatibility with the required current rating.

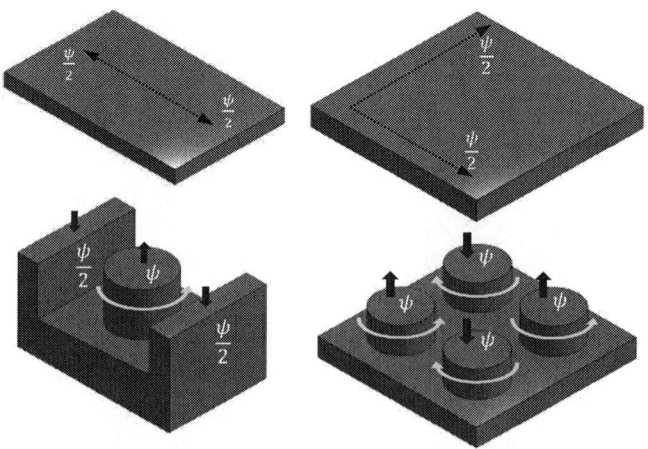

Fig. 5. Flux distribution in (a) ER core, (b) Proposed planar transformer structure

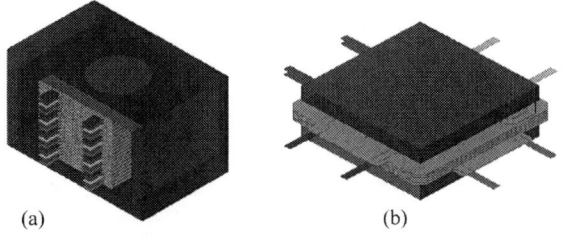

Fig. 6. 3-D demonstration of core and winding configuration (a) ER core, (b) Proposed planar transformer.

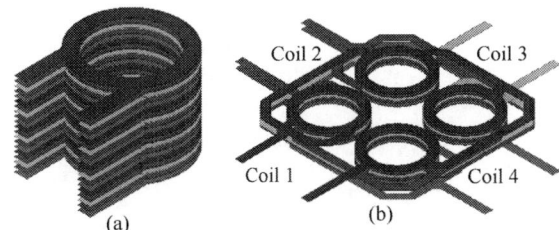

Fig. 7. Interleaved winding configuration for a four-winding transformer (a) ER core, (b) Proposed planar transformer.

IV. ANALYSIS OF WINDING COUPLING EFFECTS AND FINITE ELEMENT SIMULATION

A four-winding planar transformer is proposed for the fully-isolated four-port converter to achieve a high-power density. In these converters, the leakage inductance of transformer windings significantly impacts power transfer, along with the phase displacement of the control signals in individual full-bridge cells [28]. For example, the power flow between two phase-shift-controlled full-bridge cells is given by

$$P_{12} = \frac{\theta_2(\pi - \theta_2)V_1V_2}{2\pi^2 f_s(L_1 + L_2)} \quad (3)$$

where V_1 and V_2 are the magnitudes of voltage, θ_2 is the phase shift of the control signals of two full-bridge cells, f_s is the switching frequency, L_1 and L_2 are the leakage inductance of transformer windings connected to the full-bridge cells. "From (3), it is evident that power transfer between the ports can be controlled by appropriately adjusting the phase shift angle. For instance, when $\theta_2 > 0$, power flows from port 1 to port 2, whereas when $\theta_2 < 0$, power flows from port 2 to port 1. Therefore, selecting a suitable combination of phase shift angles for the various ports is important, following the operational modes described in Section II.

The leakage inductance of the windings plays a crucial role in determining the amount of power transferred between ports. According to Eq. (3), higher leakage inductance results in lower power transfer. Therefore, minimizing leakage inductance is essential for improving transformer efficiency. In a two-winding transformer, the primary and secondary windings are typically interleaved to achieve maximum coupling and lower leakage inductance. However, in a multi-winding transformer, the number of turns for each winding varies depending on the voltage and current ratings of the respective ports. As a result, multiple winding configurations exist, leading to more complex coupling relationships between the windings.

For the proposed four-winding transformer with a winding ratio of 8:8:8:16, various winding configurations are analyzed using finite element analysis (FEA). The study is performed on an ER-core, where all windings are arranged on a single pillar, providing optimal coupling between the windings and reducing leakage inductance. Fig. 7(a) illustrates the winding configuration that achieves the best coupling and the lowest leakage inductance for ER-core. Fig. 7(b) illustrates the winding configuration employed in the proposed core structure.

The self and mutual inductance matrix, and coupling coefficient values are obtained through ANSYS Maxwell FEA.

The total leakage inductance seen from each coil when the other coils are shorted (hereafter called *reflected leakage inductance*) can be calculated from the following inductance matrix.

$$\begin{bmatrix} v_1 \\ v_2 \\ v_3 \\ v_4 \end{bmatrix} = \begin{bmatrix} L_{11} & M_{12} & M_{13} & M_{14} \\ M_{21} & L_{22} & M_{23} & M_{24} \\ M_{31} & M_{32} & L_{33} & M_{34} \\ M_{41} & M_{42} & M_{43} & L_{44} \end{bmatrix} \frac{d}{dt} \begin{bmatrix} i_1 \\ i_2 \\ i_3 \\ i_4 \end{bmatrix} \quad (4)$$

If G is defined as follows [18]:

$$G = L^{-1} = \begin{bmatrix} g_{11} & g_{12} & g_{13} & g_{14} \\ g_{21} & g_{22} & g_{23} & g_{24} \\ g_{31} & g_{32} & g_{33} & g_{34} \\ g_{41} & g_{42} & g_{43} & g_{44} \end{bmatrix} \quad (5)$$

Then, *reflected leakage inductances* can be obtained from $L_{coil1} = 1/g_{11}$, $L_{coil2} = 1/g_{22}$, $L_{coil3} = 1/g_{33}$, $L_{coil4} = 1/g_{44}$. Table II and Table III present the inductance matrix obtained through FEA and the corresponding reflected leakage inductances, respectively. In addition to the reflected leakage inductance,

TABLE II
INDUCTANCE MATRIX OF MULTI WINDING PLANAR TRANSFORMER (UNIT: mH)

Core Structure		Coil 1	Coil 2	Coil 3	Coil 4
ER	Coil 1	2.69382	5.38661	2.69326	2.69307
	Coil 2	5.38661	10.77391	5.38686	5.38664
	Coil 3	2.69326	5.38686	2.69369	2.69328
	Coil 4	2.69307	5.38664	2.69328	2.69386
		Coil 1	Coil 2	Coil 3	Coil 4
Proposed	Coil 1	2.33841	4.67478	2.33768	2.33674
	Coil 2	4.67478	9.35079	4.67519	4.67519
	Coil 3	2.33768	4.67519	2.33835	2.33710
	Coil 4	2.33674	2.33710	2.33710	2.33845

TABLE III
REFLECTED LEAKAGE INDUCTANCE OF MULTIWINDING PLANAR TRANSFORMER (UNIT: μH)

Core Structure	L_{coil1}	L_{coil2}	L_{coil3}	L_{coil4}
ER	0.681	0.628	0.313	0.684
Proposed	1.104	1.34	0.731	1.254

Fig. 8. Magnetic flux path in the proposed core structure, illustrating how the magnetizing flux completes the path through the adjacent core legs by applying the flux cancellation technique.

Fig. 9. Magnetic flux density plot, validating that the entire core area is effectively utilized without significant areas of magnetic saturation.

coupling coefficient analysis also provides insight into the power-sharing dynamics between ports, especially when certain ports are left open. Table IV summarizes the coupling coefficient values for the proposed winding configuration. Although the core and winding structures exhibit slightly higher leakage inductance, the leakage inductance pattern remains consistent; Coil 3 demonstrates the lowest reflected leakage

TABLE IV
COUPLING COEFFICIENT OF MULTI WINDING PLANAR TRANSFORMER

Core Structure		Coil 1	Coil 2	Coil 3	Coil 4
ER	Coil 1	1	0.999872	0.999815	0.999715
	Coil 2	0.999872	1	0.999942	0.999871
	Coil 3	0.999815	0.999942	1	0.999814
	Coil 4	0.999715	0.999871	0.999814	1
		Coil 1	Coil 2	Coil 3	Coil 4
Proposed	Coil 1	1	0.999715	0.999699	0.999280
	Coil 2	0.999715	1	0.999818	0.999711
	Coil 3	0.999699	0.999818	1	0.999445
	Coil 4	0.999280	0.999711	0.999445	1

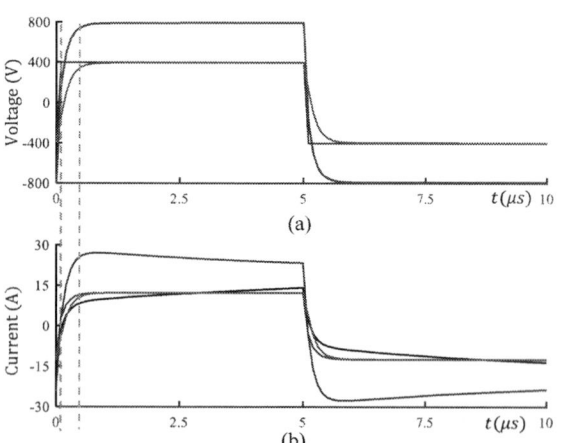

Fig. 10. Ansys Maxwell transient simulation (a) Voltage (b) Current, with loads 10kW and 5kW connected to coils 2 and 4, respectively, illustrating the phase shift between voltage and current in coil 3.

indutance in both scenarios. Moreover, Coil 3 achieves better coupling with other windings, making it suitable for efficient power transfer from the RES.

FEA is also used to validate the magnetic performance of the design at a switching frequency of $100kHz$, ensuring that the core operates within the permissible saturation limits and also the voltages induced in the windings as per the design specifications. Fig. 8 shows the magnetic flux path through the core structure. By using the proposed winding configuration, the flux gets uniformly distributed over the entire cross-section of the core. The flux cancellation method discussed in section III helps to reduce the top and bottom core plate thickness to half. From the magnetic flux density plot, depicted in Fig. 9, it is validated that the entire core area is effectively utilized without significant areas of magnetic saturation.

Figs. 10 (a) and (b) show the voltage and current outputs obtained from Ansys Maxwell transient simulations under loaded conditions. In this simulation setup, Coil 1 and 3 are connected to the input sources, while Coil 2 and 4 are connected to loads of 10 kW and 5 kW, respectively. The simulation is conducted without phase angle control; as a result, power transfer from the sources to the loads depends solely on the coupling or reflected leakage inductance of the coils. As previously noted, Coil 3 has lower reflected leakage inductance, indicating better coupling with the other windings. Hence, the source connected to the corresponding winding shares most of the load demand. The efficiency curve with different output powers is shown in Fig. 11. For the proposed design, the efficiency is around 94% at $15kW$.

V. EXPERIMENTAL RESULTS

Fig. 12 illustrates the the proposed $100kHz$, $15kW$ four-port planar transformer. The core material used is equivalent to 3C95, which is suitable for power transformers at frequencies up to $0.5\,MHz$. Each PCB is designed with 8 turns of circular copper windings, evenly distributed across the four core pillars, and implemented on a two-layer PCB with a copper thickness

TABLE IV
MEASURED REFLECTED LEAKAGE INDUCTANCE (UNIT: μH)

	L_{coil1}	L_{coil2}	L_{coil3}	L_{coil4}
Simulation	1.104	1.34	0.731	1.254
Experiment	1.6	1.872	0.997	1.644

979-8-3315-1612-3/25 $31.00 © 2025 IEEE

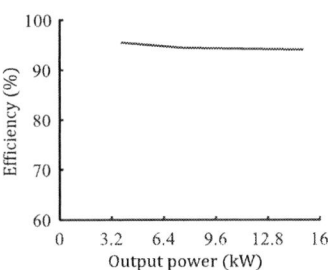

Fig. 11. Efficiency curve of proposed multiwinding transformer design.

Fig. 12. Prototype of proposed four-winding planar transformer.

of $70\mu m$. Table IV compares the simulated and measured reflected leakage inductances of each coil. The experimental leakage inductance values were obtained using an LCR meter. In the measurements, the coil under test was connected to the instrument while the other coils were shorted.

VI. CONCLUSION

A high-power planar transformer design tailored for four-port converters, which has been presented in this paper, has addressed the challenges of integrating multiple sources and loads with varying voltage-current characteristics. The design has effectively minimized core losses and leakage inductance by utilizing advanced techniques such as flux cancellation and winding interleaving, enhancing overall efficiency and performance. The prototype testing and experimental results for reflected leakage inductance have validated the feasibility of the design, achieving the desired voltage ratios with minimal losses. This work has advanced the design of efficient, compact multiport converters, with a focus on applications demanding galvanic isolation and high-power density.

ACKNOWLEDGMENT

We would like to acknowledge the support of ABC Taiwan Electronics Corp, Dr. Cedric Hsu, and Epower Technology LLC.

VII. REFERENCES

[1] C. A. Hill, M. C. Such, D. Chen, J. Gonzalez and W. M. Grady, "Battery energy storage for enabling integration of distributed solar power generation," *IEEE Trans. Smart Grid*, vol. 3, pp. 850-857, Jun. 2012.

[2] J. M. Carrasco et al., "Power-electronic systems for the grid integration of renewable energy sources: a survey," *IEEE Trans. Ind. Electron.*, vol. 53, no. 4, pp. 1002-1016, Jun. 2006.

[3] B. Mirafzal and A. Adib, "On grid-interactive smart inverters: features and advancements," *IEEE Access*, vol. 8, pp. 160526-160536, 2020.

[4] Q. C. Zhong, W.-L. Ming and Y. Zeng, "Self-synchronized universal droop controller," *IEEE Access*, vol. 4, pp. 7145-7153, 2016.

[5] F. Sadeque, M. Gursoy, D. Sharma and B. Mirafzal, "Autonomous control of inverters in microgrid," *IEEE Trans. Ind. Appl.*, vol. 60, no. 3, pp. 4313-4323, May-June 2024.

[6] B. Mirafzal, Power Electronics in Energy Conversion Systems, New York, USA, McGraw Hill, 2022.

[7] A. Adib et al., "E-Mobility — Advancements and Challenges," *IEEE Access*, vol. 7, pp. 165226-165240, 2019.

[8] C. Wang and M. H. Nehrir, "Power management of a stand-alone wind/photovoltaic/fuel cell energy system," *IEEE Trans. Energy Convers.*, vol. 23, no. 3, pp. 957-967, Sept. 2008.

[9] B. Wang, M. Sechilariu, and F. Locment, "Intelligent DC microgrid with smart grid communications: Control strategy consideration and design," *IEEE Trans. Smart Grid*, vol. 3, no. 4, pp. 2148–2156, Dec. 2012.

[10] Y. Riffonneau, S. Bacha, F. Barruel, and S. Ploix, "Optimal power flow management for grid connected PV systems with batteries," *IEEE Trans. Sustain. Energy*, vol. 2, no. 3, pp. 309–320, Jul. 2011.

[11] H. Mahmood, D. Michaelson and J. Jiang, "A Power Management Strategy for PV/Battery Hybrid Systems in Islanded Microgrids," *IEEE J. Emerg. Sel. Topics Power Electron.*, vol. 2, no. 4, pp. 870-882, Dec. 2014.

[12] S. J. Chiang, K. T. Chang, and C. Y. Yen, "Residential photovoltaic energy storage system," *IEEE Trans. Ind. Electron.*, vol. 45, no. 3, pp. 385–394, Jun. 1998.

[13] A. K. Bhattacharjee, N. Kutkut and I. Batarseh, "Review of multiport converters for solar and energy storage integration," *IEEE Trans. Power Electron.*, vol. 34, no. 2, pp. 1431-1445, Feb. 2019.

[14] Z. Wang, Q. Luo, Y. Wei, D. Mou, X. Lu and P. Sun, "Topology analysis and review of three-port DC-DC converters," *IEEE Trans. Power Electron.*, vol. 35, no. 11, pp. 11783-11800, Nov. 2020.

[15] N. D. Dao, D. -C. Lee and Q. D. Phan, "High-efficiency SiC-based isolated three-port DC/DC converters for hybrid charging stations," *IEEE Trans. Power Electron.*, vol. 35, no. 10, pp. 10455-10465, Oct. 2020.

[16] R. Faraji, H. Farzanehfard, G. Kampitsis, M. Mattavelli, E. Matioli and M. Esteki, "Fully soft-switched high step-up non-isolated three-port DC–DC converter using GaN HEMTs," *IEEE Trans. Ind. Electron.*, vol. 67, no. 10, pp. 8371-8380, Oct. 2020.

[17] M. Mu and F. C. Lee, "Design and optimization of a 380–12 V high-frequency, high-current LLC converter with GaN devices and planar matrix transformers," *IEEE J. Emerg. Sel. Topics Power Electron.*, vol. 4, no. 3, pp. 854-862, Sept. 2016.

[18] Z. Zhang, B. He, D. -D. Hu, X. Ren and Q. Chen, "Multi-winding configuration optimization of multi-output planar transformers in GaN active forward converters for satellite applications," *IEEE Trans. Power Electron.*, vol. 34, no. 5, pp. 4465-4479, May 2019.

[19] C. Quinn, K. Rinne, T. O'Donnell, M. Duffy and C. O. Mathuna, "A review of planar magnetic techniques and technologies," in *Proc. 16th Annu. IEEE Appl. Power Electron. Conf. Expo.*, 2001, pp. 1175–1183.

[20] R. Petkov, "Optimum design of a high-power, high-frequency transformer," *IEEE Trans. Power Electron.*, vol. 11, pp. 33-42, Jan. 1996.

[21] Z. Ouyang, O. C. Thomsen and M. A. E. Andersen, "Optimal design and tradeoff analysis of planar transformer in high-power DC–DC converters," *IEEE Trans. Ind. Electron.*, vol. 59, no. 7, pp. 2800-2810, July 2012.

[22] A. Sadasivan, F. Fateh and B. Mirafzal, "Design of planar transformers for multiport converters: A Study," *2024 IEEE Kansas Power and Energy Conference (KPEC)*, Manhattan, KS, USA, 2024.

[23] S. M. Djuric and G. M. Stojanovic, "A compact planar transformer with an improved winding configuration," *IEEE Trans. Magn.*, vol. 50, no. 11, pp. 1-4, Nov. 2014.

[24] D. Huang, S. Ji and F. C. Lee, "LLC resonant converter with matrix transformer," *IEEE Trans. Power Electron.*, vol. 29, no. 8, pp. 4339-4347, Aug. 2014.

[25] C. Fei, F. C. Lee and Q. Li, "High-efficiency high-power-density LLC converter with an integrated planar matrix transformer for high-output current applications," *IEEE Trans. Ind. Electron.*, vol. 64, no. 11, pp. 9072-9082, Nov. 2017.

[26] Y. -C. Liu et al., "Quarter-turn transformer design and optimization for high power density 1-MHz LLC resonant converter," *IEEE Trans. Ind. Electron.*, vol. 67, no. 2, pp. 1580-1591, Feb. 2020.

[27] K. Wang, Q. Gao, G. Wei and X. Yang, "Integrated fractional-turn planar transformer for MHz and high-current applications," *IEEE Trans. Power Electron.*, vol. 38, no. 6, pp. 7374-7384, June 2023.

[28] C. Zhao, S. D. Round and J. W. Kolar, "An isolated three-port bidirectional DC-DC converter with decoupled power flow management," *IEEE Trans. Power Electron.*, vol. 23, no. 5, pp. 2443-2453, Sept. 2008.

Optimal design of inductors with aluminum litz wire for inductive power transfer systems

Jesús Acero
Dept. of Electronic Eng. and Communications
I3A, University of Zaragoza
Zaragoza, Spain
jacero@unizar.es

Claudio Carretero
Dept. of Applied Physics
I3A, University of Zaragoza
Zaragoza, Spain
ccar@unizar.es

Ignacio Lope
Dept. of Applied Physics)
I3A, University of Zaragoza
Zaragoza, Spain
nlope@unizar.es

Óscar Lahuerta
Dept. of Electronic Eng. and Communications
University of Zaragoza
Zaragoza, Spain
olahuerta@unizar.es

José-Miguel Burdío
Dept. of Electronic Eng. and Communications
I3A, University of Zaragoza
Zaragoza, Spain
burdio@unizar.es

Abstract— This work presents a design method for magnetic couplers with aluminum litz-wire for wireless vehicle chargers. This alternative is considered due to its potential benefits, such as lightweight and cost savings. The work not only proposes the replacement of copper wire with aluminum but also proposes an analytical method to obtain the optimal aluminum volume that maximizes the efficiency of the inductive link. The optimal aluminum volume is expressed as the product of the number of turns of the magnetic coupler by the cable's number of strands. The optimization strategy pursues the maximization of the product of the magnetic coupling coefficient and the quality factor. This product is considered a figure of merit in the proposed method. The method is used to design an aluminum inductive link fitting the standard SAE J2954. The performance of a prototype of 12 kW is evaluated and compared with a conventional copper litz-wire implementation.

Keywords— Ac resistance, electromagnetic induction, electromagnetic simulation, litz wire, wireless power transfer.

I. INTRODUCTION

Shortly copper will be one of the most demanded materials due to the progressive electrification of modern societies (electrical mobility, smart grids, data centers, etc.). Worldwide copper consumption could increase by 50% by 2050. Prices have risen by 30% by 2024, exceeding 9,000 euros per ton. In the 21st century, copper shortages are a destabilizing threat to international stability. Shortages will place value chains under unprecedented stress. This is reminiscent of what we experienced in the 20th century with oil, but with copper this may be accentuated by an even higher geographic concentration when it comes to refining industrial products. Current development projects will not be sufficient to compensate for the lack of supply [1]. Copper is needed for electric vehicles,

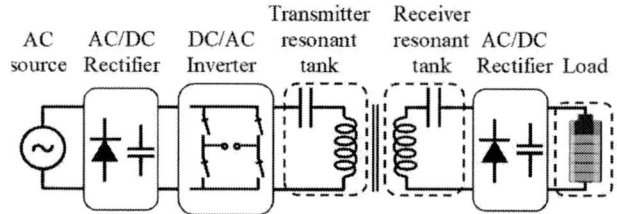

Fig. 1. Block diagram of an inductive power transfer system.

heat pumps, and batteries in the energy transition. The development of data centers for artificial intelligence must be added to this.

Cost and weight savings can be potentially achieved with aluminum wiring in power electronic applications [2], [3], [4], [5]. However, aluminum wiring did not take off until a few years ago in the power electronics ambit because of several problems. Among these problems were the need for strands of sufficiently small diameter, the crimping of the terminals, and oxidation of the aluminum. In recent years, improvements in manufacturing processes and the use of solderable enamels have made it possible to overcome these problems, and currently, aluminum litz wires are often part of the product portfolio of wire manufacturers [6].

Fig. 1 shows the typical block diagram of a wireless battery charger. The magnetic coupler occupies a central role in this scheme. In the electric vehicle field, some standards regulate the structure, size and other manufacturing parameters of magnetic couplers according to power classes in order to achieve interoperability between OEMs [7]. In this standard, the use of copper litz wires is specifically mentioned. Currently, an intense research activity is being carried out on magnetic couplers [8], [9], [10], [11], [12]. However, the current literature contains few works dealing with cabling losses or new cabling arrangements.

This work was partly supported by Projects PID2022-136621OB-I00 and CPP2021-008938, TED2021-129274B-I00, PDC2023-145837-I00, co-funded by MICIU/AEI/10.13039/501100011033, by "ERDF A way of making Europe", by the "European Union NextGeneration EU/PRTR", by the DGA-FSE and by BSH Home Appliances Group.

979-8-3315-1612-3/25 $31.00 © 2025 IEEE

This work explores the implementation of aluminum magnetic couplers for inductive power transfer (IPT) systems as an alternative to copper implementations. A design procedure that considers kQ as a figure of merit [13] is followed, being k the magnetic coupling and Q the quality factor of the inductive link. This factor is related to the efficiency of the electromagnetic power transfer. The optimal aluminum volume that maximizes the efficiency is obtained by maximizing the kQ factor [14] for the conductor volume. The kQ maximization process combines FEA simulations and analytical calculations of the skin and proximity resistances of ideal litz wires [15], and results are experimentally compared with a conventional copper litz-wire implementation.

II. FINITE-ELEMENT MODEL OF AN INDUCTIVE POWER TRANSFER SYSTEM

As mentioned above, the optimal design procedure is based on maximizing the kQ factor. Implementing this procedure requires the following electrical magnitudes: the magnetic coupling coefficient k, the coil self-inductances, and the coil frequency-dependent resistances. The first two magnitudes can be directly derived from finite-element simulations. In contrast, the coil frequency-dependent resistance can affordably be calculated using the combination of finite-element results and the analytical loss model of cylindrical conductors.

Fig. 2 shows the finite-element model of an IPT system. This model is implemented in COMSOL Multiphysics. The simulation model includes the two coils, ferrites, and aluminum shielding. The model definition is based on the following assumptions:

- The wire doesn't modify the form and magnitude of the magnetic fields. Consequently, the wire structure is replaced by an ideal free-loss constant current density.

- Coils are identical.

- The shielding is modeled by an Impedance Boundary Condition (IBC). Consequently, the number of mesh nodes is reduced.

- At the interest frequency range, below 100 kHz, parasitic capacitors are neglected.

According to the first two assumptions, each coil is modeled as an impressed constant current density of value:

$$J = \frac{I}{S_w} = n \cdot \frac{I}{S_c} \tag{1}$$

where I is the current amplitude and S_w is the cross-sectional area of the wire, n is the number of turns of the coils, and S_c is the cross-sectional area of the coil. This formulation is advantageous because magnitudes per turn (pt) are obtained when n is set to 1.

The voltage of the ith-coil is obtained as follows:

$$V_i = -\oint \mathbf{E}_i(\mathbf{r}) \cdot d\mathbf{r} \tag{2}$$

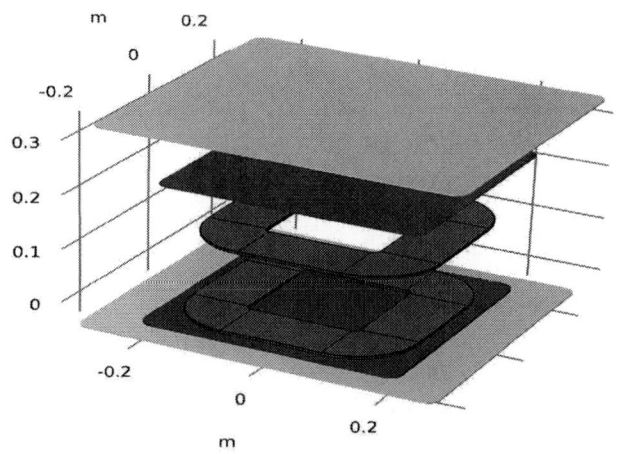

Fig. 2. Image of the 3D finite-element simulation model of the IPT system.

According to the ideal loss-free coil model, the real part of (2) represents only the power dissipated in the shielding, and the imaginary part is related to the self and mutual inductances including the effect of the ferrite and shielding. Consequently, both self and mutual inductances can be straightforwardly derived as follows:

$$L_i = \frac{\text{Im}\left(\left.\frac{V_i}{I_i}\right|_{I_j=0}\right)}{\omega}; M = \frac{\text{Im}\left(\left.\frac{V_i}{I_j}\right|_{I_i=0}\right)}{\omega}; \quad i,j=1,2 \tag{3}$$

As mentioned above, magnitudes per turn are obtained if n equals 1. The relationship between the magnetic field and inductances and their corresponding magnitudes per turn are:

$$\begin{aligned}
\mathbf{H}_i(\mathbf{r}) &= n_i \cdot \overline{\mathbf{H}}_i(\mathbf{r}) \\
L_i &= n_i^2 \cdot \overline{L}_i \\
M &= n_i \cdot n_j \cdot \overline{M} \\
l_{c,i} &= n_i \cdot \text{MLT}_i
\end{aligned} \tag{4}$$

where $\mathbf{H}_i(\mathbf{r})$ is the magnetic field generated by the ith coil at an arbitrary position. The magnetic field is required for calculating the *proximity losses* in coils. More specifically, the FEA tool is used to obtain the average of the squared magnetic field in the coil volume normalized per unit of current and per turn in the ith winding. This magnitude is:

$$\left\langle \overline{H}_i^2 \right\rangle = \frac{1}{V_c} \int_{V_c} \overline{H}_i^2 \cdot dv_c \tag{5}$$

where V_c is the coil volume.

Moreover, in (4) $l_{c,i}$ is the length of the ith-coil cable and MLT is the mean length of the coil's turn.

According to (4) and considering identical coils:

$$k = \frac{M}{\sqrt{L_i \cdot L_j}} = \frac{\bar{M}}{\bar{L}_i} \qquad (6)$$

III. AC-LOSS MODEL OF LITZ-WIRE COILS

The ac-loss model of litz wire coils considers the dc resistance and the skin and proximity ac effects. In this case, the ac-loss model is developed considering the following assumptions:

- The transposition pattern of the litz wire is ideal and, therefore, equalizes the resistance contributed by each strand. Equivalently, all strands can be considered as equivalent.

- Assuming that the litz wire doesn't modify the form and magnitude of the magnetic fields, the loss model is based on the analytical expressions of infinitely long isolated round strands.

The dc and skin resistance of a strand of radius r_s and electrical conductivity σ per unit of length is [15]:

$$R_{\text{dc·skin}} = \frac{1}{\pi r_s^2 \sigma} \Phi_{\text{skin}} \left(r_s / \delta \right) \qquad (7)$$

where δ is the skin depth. The function $\Phi_{\text{skin}}(r_s/\delta)$ represents the increase of the strand resistance due to the skin effect. The asymptotic approximation of $\Phi_{\text{skin}}(r_s/\delta)$ at low and high frequencies (equivalent to low and high values of is) are $\Phi_{\text{skin,LF}} \simeq 1$, and $\Phi_{\text{skin,HF}} \simeq r_s/2\delta$.

From the point of view of the dc by skin resistance, a coil can be considered as the parallel connection of n_s strands and the series connection of n tuns. Consequently, the dc by skin resistance of the ith coil is:

$$R_{\text{dc·skin},i} = \frac{n_i}{n_{s,i}} \cdot \text{MLT}_i \cdot \frac{1}{\pi r_s^2 \sigma} \cdot \Phi_{\text{skin}} \left(r_s / \delta \right) \qquad (8)$$

The previous expression can be normalized per turn and strand taking $n_i=1$, $n_{s,i}=1$. In this case:

$$\bar{R}_{\text{dc·skin},i} = \text{MLT}_i \cdot \frac{1}{\pi r_s^2 \sigma} \cdot \Phi_{\text{skin}} \left(r_s / \delta \right) \qquad (9)$$

and therefore:

$$R_{\text{dc·skin},i} = \frac{n_i}{n_{s,i}} \cdot \bar{R}_{\text{dc·skin},i} \qquad (10)$$

Similarly, the proximity resistance per unit of length of an isolated strand immersed in a uniform ac field whose amplitude normalized per unit of ampere is denoted as $\bar{H}$ is [15]:

$$R_{\text{prox}} = \frac{4\pi}{\sigma} \cdot \Phi_{\text{prox}} \left(r_s / \delta \right) \cdot \bar{H}^2 \qquad (11)$$

In this case $\Phi_{\text{prox}}(r_s/\delta)$ captures the inductive nature of the proximity resistance. Consequently, Φ_{prox} increases with the frequency, nulling at dc. The asymptotic approximation of $\Phi_{\text{prox}}(r_s/\delta)$ at low and high frequencies is and $\Phi_{\text{prox,HF}} \simeq r_s/\delta$.

The total proximity resistance of a coil can be calculated as the sum of the proximity resistance of its n_s strands. In this case, it should be considered that strands placed at different turns can be under various magnetic fields. For this reason, the average of the square magnetic field shown in (5) is used. Moreover, considering also the total length of the cable, the proximity resistance of the ith coil is:

$$R_{\text{prox},i} = n_{s,i} \cdot n_i^3 \cdot \text{MLT}_i \cdot \frac{4\pi}{\sigma} \cdot \Phi_{\text{prox}} \left(r_s / \delta \right) \cdot \left\langle \bar{H}_i^2 \right\rangle \qquad (12)$$

As in the previous case, this resistance can be normalized by $n_i=1$, $n_{s,i}=1$. Therefore:

$$\bar{R}_{\text{prox},i} = \text{MLT}_i \cdot \frac{4\pi}{\sigma} \cdot \Phi_{\text{prox}} \left(r_s / \delta \right) \cdot \left\langle \bar{H}_i^2 \right\rangle \qquad (13)$$

And consequently:

$$R_{\text{prox},i} = n_{s,i} \cdot n_i^3 \cdot \bar{R}_{\text{prox},i} \qquad (14)$$

IV. OPTIMUM CONDUCTOR VOLUME OF AN IPT SYSTEM

In an IPT system with series-series compensation, the maximum efficiency of the inductive energy transfer η_{max} is approximated by [13]:

$$\eta_{\text{max}} \simeq 1 - \frac{2}{kQ} \qquad (15)$$

being k the magnetic coupling coefficient and Q the quality factor of the system defined as:

$$Q = \sqrt{Q_1 \cdot Q_2} \qquad (16)$$

where Q_i is the quality factor of the ith coil. Regarding (16), the higher the kQ value is, the higher the maximum efficiency can be achieved. This relationship between the kQ factor and efficiency applies not only to the case of series-series compensation but also to IPT systems with other compensation networks [16], [17], [18], [19], [20].

Assuming that, at the interest frequency range the losses of both ferrite and shielding are enough smaller than the winding losses, Q_i can be written as follows [14]:

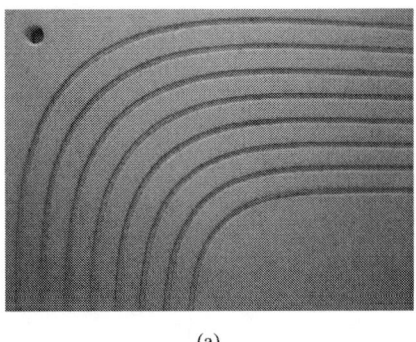 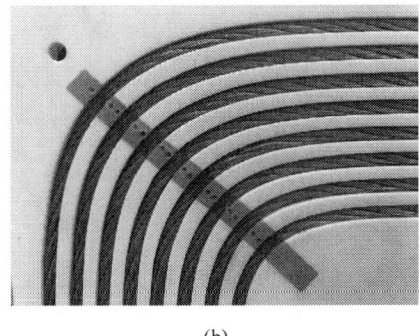 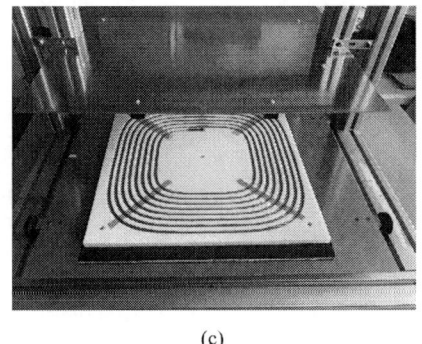

(a)   (b)   (c)

Fig. 3. Images of the corners of the prototypes with the different cables. (a) Aluminum litz wire. (b) Standard copper litz wire. (c) Aluminum inductor in the chassis used for the experimental verification.

$$Q_i = \frac{\omega L_i}{R_i} = \frac{\omega L_i}{R_{\text{dc·skin},i} + R_{\text{prox},i}} = \frac{\omega \cdot n_i^2 \cdot \overline{L}_i}{\dfrac{n_i}{n_{s,i}} \cdot \overline{R}_{\text{dc·skin},i} + n_{s,i} \cdot n_i^3 \cdot \overline{R}_{\text{prox},i}}$$

$$= \frac{\omega \cdot \overline{L}_i}{\dfrac{1}{n_i \cdot n_{s,i}} \cdot \overline{R}_{\text{dc·skin},i} + n_{s,i} \cdot n_i \cdot \overline{R}_{\text{prox},i}} \qquad (17)$$

Considering identical power pads ($n_1=n_2=n$, $n_{s,1}=n_{s,2}=n_s$, $\overline{R}_{\text{dc·skin},1} = \overline{R}_{\text{dc·skin},2} = \overline{R}_{\text{dc·skin}}$, $\overline{R}_{\text{prox},1} = \overline{R}_{\text{prox},2} = \overline{R}_{\text{prox}}$), and (6), kQ adopts the following form:

$$kQ = \frac{\omega \cdot \overline{M}}{\dfrac{1}{n_s \cdot n} \cdot \overline{R}_{\text{dc·skin}} + n_s \cdot n \cdot \overline{R}_{\text{prox}}} \qquad (18)$$

The maximum kQ is obtained for the value $n_s \cdot n$ that maximizes the (18). This value corresponds to the optimal product of the number of turns by the number of strands that maximizes the efficiency of the electromagnetic transfer. This value is:

$$\frac{\partial (kQ)}{\partial (n_s \cdot n)} = 0 \Rightarrow (n_s \cdot n)_{\text{opt}} = \sqrt{\frac{\overline{R}_{\text{dc·skin}}}{\overline{R}_{\text{prox}}}} \qquad (19)$$

The product $n_s \cdot n$ is proportional to the winding volume (the conductor volume of each coil is $n_s\, n\, \text{MLT} \cdot \pi \cdot r_s^2$). Therefore, $(n_s \cdot n)_{\text{opt}}$ represents the optimal conductor volume that maximizes the efficiency of the magnetic link.

This criterion is applied to design an IPT system for a battery charger according to the standard SAE J2954. Considering the power class WPT3 (up to 15 kW), the coil dimension is 380 × 380 mm. This dimension and the self-inductance set in the standard (comprised between 39 µH and 39.5 µH) results in $n = 8$ and MLT = 1.082 m.

The criterion adopted for selecting the diameter of aluminum strands is the skin depth at the interest frequency. The standard conductivity of the aluminum used for cables is $\sigma = 30$ MS/m.

For this material, the skin depth at 85 kHz is $\delta = 315$ µm. Moreover, also considering the current strand diameter portfolio, aluminum strands of diameter $\phi_s = 200$ µm, are selected. For this diameter, the ac current is uniformly spread in the section in the strand. According to this diameter and applying (19) at f=85 kHz, $\overline{R}_{\text{dc·skin}} = 1.15$ Ω, and $\overline{R}_{\text{prox}} = 0.0728$ µΩ. Accordingly, $(n_s \cdot n)_{\text{opt}}$=3944 and therefore n_s=493.

For simplicity of manufacturing, the optimum number of strands of the aluminum wire is set to n_s=500 instead of 493.

V. EXPERIMENTAL VALIDATION

The selected cable is compared with a standard copper litz wire with $n_s = 200$ strands of diameter $\phi_s = 200$ µm. Some images of the tested inductors with the selected cables are shown in Fig. 3. The diameter of the aluminum cable is considerably larger than that of the standard copper because it has a higher number of strands. The dc resistances of both cables per unit of length are $R_{\text{dc,Al}} = 2.122$ mΩ and $R_{\text{dc,Cu}} = 2.744$ mΩ. Consequently, the dc resistance of the optimal aluminum cable is 22% less than that of the copper standard litz wire.

Considering the density of copper and aluminum, and its current commodity price per unit of volume, the aluminum winding is 17% lighter than the copper winding, which is 4.73 times more expensive than the aluminum winding.

Fig. 4 shows the calculated and measured resistance and inductance versus frequency. The observed mismatching between the calculated and measured resistances points that the model of ideal litz wire with equivalent strands has some limitations at the medium frequency range. These results indicate that the assumption of equivalent strands is not strictly true. Moreover, the discrepancies observed between calculations and measurements at the lower frequency range point that some strands could not be connected in both prototypes. In any case, the resistance of the aluminum winding is lower in the considered frequency range. The inductance of the aluminum is slightly lower than that of the copper winding due to the larger cable diameter.

Fig. 5 compares the calculated and measured values of the magnetic coupling coefficient and kQ factors versus the horizontal misalignment d at 85 kHz. In general, the predictions

979-8-3315-1612-3/25 $31.00 © 2025 IEEE   2471

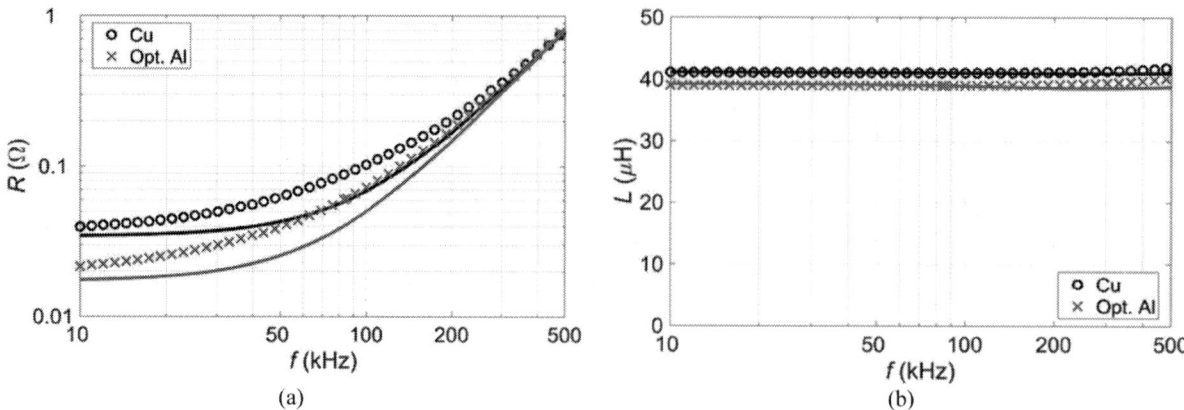

Fig. 4. Calculated and measured impedance of the tested prototypes with the frequency. Dots correspond to measurements and the continuous line corresponds to simulation. (a) Resistance. (b) Inductance.

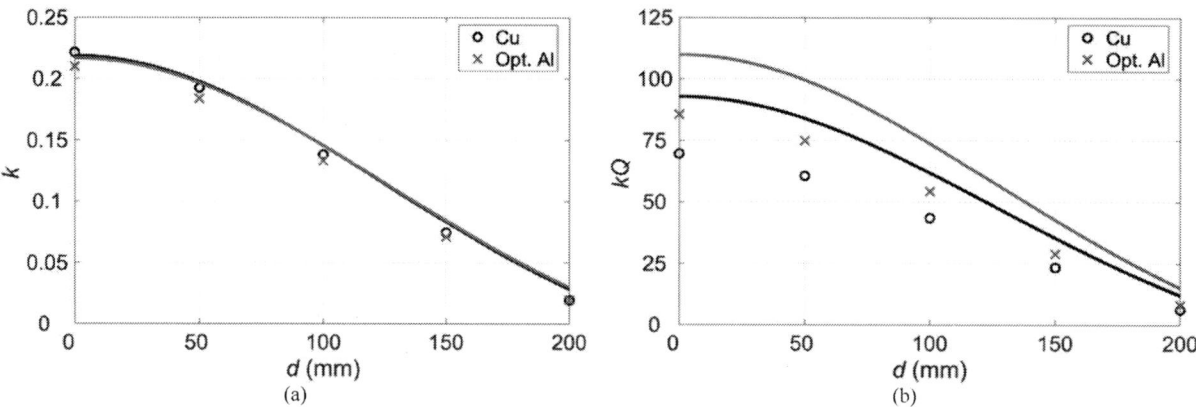

Fig. 5. Calculated and measured results with respect to misalignment at 85 kHz. Dots correspond to measurements and the continuous line corresponds to simulation. (a) Coupling factor. (b) kQ factor.

of the kQ factor are overestimated with both cables because the calculated resistances were lower than the measurements. Moreover, mismatches can be observed between the measurements and the predictions. The mismatching is amplified because the resistance divides in the formula of the Q. In any case, predictions and measurements follow similar tendencies for the considered misalignment span and both results point that an optimal aluminum design can outperform a standard copper implementation.

The calculated and measured electromagnetic efficiency is shown in Table I. This Table shows the electromagnetic efficiency, or equivalently, the relationship between the received power and the transmitted power in the coils. According to the results presented, the efficiency of the optimal aluminum cable is almost one point higher than that of the copper cable. These results are similar to those observed in the simulation.

Finally, some waveforms acquired transferring P_o=12.57 kW are shown in Fig. 6.

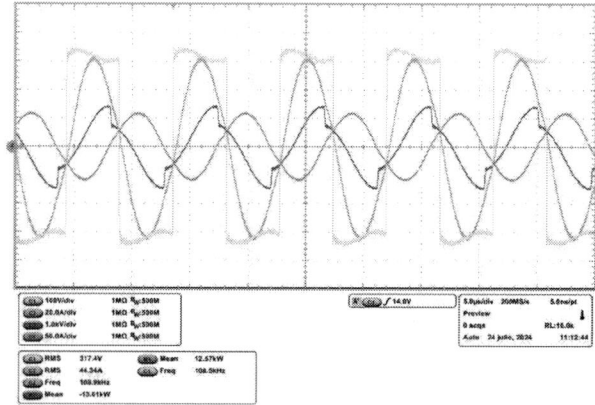

Fig. 6. Transmitter and receiver voltages and currents at P_o=12.57 kW with aluminum inductors.

TABLE I.	EFFICIENCY COMPARISON	
	Copper litz	*Aluminum litz*
η_{calc}	96.1%	96.8%
η_{meas}	95.2%	96%

VI. CONCLUSION

Cost and weight savings can be achieved when copper is substituted by aluminum in the litz wires used in IPT systems. Moreover, performance is not necessarily penalized if an optimal design based on the maximum kQ factor is adopted. Results show that the parasitic series resistance with aluminum can be lower than copper standard implementations at the interest frequency range, which is reflected by a higher kQ factor and higher efficiency of the electromagnetic power transfer.

REFERENCES

[1] "LME Copper | London Metal Exchange." Accessed: Jul. 22, 2024. [Online]. Available: https://www.lme.com/Metals/Non-ferrous/LME-Copper#Summary

[2] C. R. Sullivan, "Aluminum windings and other strategies for high-frequency magnetics design in an era of high copper and energy costs," *IEEE Trans Power Electron*, vol. 23, no. 4, pp. 2044–2051, Jul. 2008, doi: 10.1109/TPEL.2008.925434.

[3] R. Wojda, V. P. Galigekere, J. Pries, and O. Onar, "Copper-Clad Aluminum Windings as an Alternative Conductor for High-Power Electric Vehicle Wireless Charging," 2020. [Online]. Available: http://energy.gov/downloads/doe-

[4] J. Acero, I. Lope, C. Carretero, H. Sarnago, and J. M. Burdio, "Comparative Evaluation of Different Cables for Magnetic Couplers in Inductive Power Transfer Systems," *IEEE Trans Ind Appl*, 2023, doi: 10.1109/TIA.2023.3344686.

[5] T. Maezawa, H. Zhou, M. Sato, Y. Bu, and T. Mizuno, "Low Loss on a Litz Aluminum Wire Coil Using Magnetic Tape for Automotive Wireless Power Transmission," *IEEE Trans Magn*, vol. 57, no. 9, 2021, doi: 10.1109/TMAG.2021.3097880.

[6] "ELEKTRISOLA." Accessed: Oct. 12, 2024. [Online]. Available: https://www.elektrisola.com/en

[7] "J2954: Wireless Power Transfer for Light-Duty Plug-in/Electric Vehicles and Alignment Methodology - SAE International." Accessed: Dec. 28, 2021. [Online]. Available: https://www.sae.org/standards/content/j2954_202010/

[8] H. Wang, Y. Wu, Z. Shen, W. Pan, X. Chen, and Y. Zhang, "High-Misalignment-Tolerant Dual-Channel Inductive Power Transfer System Based on Cross-Shaped Reversed-Winding-Incorporated Solenoid Pad," *IEEE Transactions on Industrial Electronics*, vol. 71, no. 10, pp. 12499–12509, 2024, doi: 10.1109/TIE.2024.3360620.

[9] Y. Li, Y. Feng, and Y. Guo, "Free-Positioning and Omnidirectional Wireless Power Transfer Using Self-Decoupled Planar Transmitter Coils," *IEEE Transactions on Industrial Electronics*, vol. 71, no. 7, pp. 8091–8101, Jul. 2024, doi: 10.1109/TIE.2023.3308129.

[10] U. Gordhan and S. Jayalath, "Comparative Analysis of Wireless Power Transfer Couplers for Unmanned Aerial Vehicles and Drones," *IEEE Open Journal of Power Electronics*, vol. 5, pp. 618–633, 2024, doi: 10.1109/OJPEL.2024.3395174.

[11] L. Zhu, L. Wang, C. Zhao, J. Shen, M. Wu, and L. Pei, "Design and Optimization of Unequal-Pitch SelfResonant Helical Coils for High-Efficiency MidRange Wireless Power Transfer," *IEEE Trans Power Electron*, 2024, doi: 10.1109/TPEL.2024.3413346.

[12] Z. Chen, Z. Li, Z. Lin, J. Li, and Y. Zhang, "Mutual Inductance Calculation of Rectangular Coils at Arbitrary Position With Bilateral Finite Magnetic Shields in Wireless Power Transfer Systems," *IEEE Trans Power Electron*, 2024, doi: 10.1109/TPEL.2024.3425713.

[13] G. Vandevoorde and R. Puers, "Wireless energy transfer for stand-alone systems: a comparison between low and high power applicability," *Sens Actuators A Phys*, vol. 92, no. 1–3, pp. 305–311, Aug. 2001, doi: 10.1016/S0924-4247(01)00588-X.

[14] J. Acero, I. Lope, C. Carretero, and J. M. Burdio, "Analysis of Winding Loss and optimization of Inductive Power Transfer Coils," *IEEE International Symposium on Industrial Electronics*, vol. 2020-June, pp. 1435–1441, Jun. 2020, doi: 10.1109/ISIE45063.2020.9152283.

[15] C. Carretero, J. Acero, R. Alonso, C. Carretero,) 84 Carretero, and A. Acero, "TM-TE Decomposition of Power Losses in Multi-Stranded Litz-Wires Used in Electronic Devices," *Progress In Electromagnetics Research*, vol. 123, pp. 83–103, 2012, doi: 10.2528/PIER11091909.

[16] K. Lee and S. H. Chae, "Power Transfer Efficiency Analysis of Intermediate-Resonator for Wireless Power Transfer," *IEEE Trans Power Electron*, vol. 33, no. 3, pp. 2484–2493, Mar. 2018, doi: 10.1109/TPEL.2017.2698638.

[17] W. Zhang and C. C. Mi, "Compensation topologies of high-power wireless power transfer systems," *IEEE Trans Veh Technol*, vol. 65, no. 6, pp. 4768–4778, Jun. 2016, doi: 10.1109/TVT.2015.2454292.

[18] S.-Y. Ron Hui, Y. Yang, and C. Zhang, "Wireless Power Transfer: A Paradigm Shift for the Next Generation," *IEEE J Emerg Sel Top Power Electron*, vol. 11, no. 3, 2023, doi: 10.1109/JESTPE.2023.3237792.

[19] W. Zhang, S. C. Wong, C. K. Tse, and Q. Chen, "Analysis and comparison of secondary series-and parallel-compensated inductive power transfer systems operating for optimal efficiency and load-independent voltage-transfer ratio," *IEEE Trans Power Electron*, vol. 29, no. 6, pp. 2979–2990, Jun. 2014, doi: 10.1109/TPEL.2013.2273364.

[20] Y. Yin, H. Li, and M. Fu, "Inductive Coupler Analysis Based on Scattering Parameters With Nonstandard Terminal Impedance," *IEEE Journal of Emerging and Selected Topics in Industrial Electronics*, vol. 3, no. 4, pp. 1168–1176, Aug. 2022, doi: 10.1109/JESTIE.2022.3199673.

979-8-3315-1612-3/25 $31.00 © 2025 IEEE

Analytic Design of Flat-Wire Inductors for High-Current and Compact DC-DC Converters

Sajjad Mohammadi
Electrical Engineering and Computer Science
MIT
Cambridge, USA
sajjadm@mit.edu

James L. Kirtley
Electrical Engineering and Computer Science
MIT
Cambridge, USA
kirtley@mit.edu

Alireza Namadmalan
Power Electronics & Magnetics Lab
Bourns Electronics Ireland (BEI)
Cork, Ireland
alireza.namadmalan@bourns.com

Abstract—**This paper presents analytic study and design considerations of flat wire inductors with distributed gaps for high-power and compact DC-DC Converters. The focus is eddy-current loss components within the conductors due to fringing and leakage fluxes. A magnetic equivalent circuit (MEC) is proposed in which eddy currents are modeled by MMFs opposing the primary flux as well as frequency dependent reluctances, which finally leads to a frequency dependent inductance describing the behavior of the inductor at high frequencies. Three formulations for DC resistance depending on the required accuracy are developed. Calculations of the AC resistance based on vector potential obtained from FEM are provided. To provide an insight into the optimized design of such inductors, components of the magnetic flux and induced eddy currents along with sensitivity of the main inductor quantities such as DCR, ESR, loss components and inductance values to the design parameters are investigated. Finally, an inductor is prototyped and experimentally tested to verify the design.**

Keywords—*Finite Elements, Eddy-current, Power Inductors, Magnetic Equivalent Circuit (MEC).*

I. INTRODUCTION

Recent advances in high-frequency and wide-band-gap semiconductors, such as GaN and SiC switches, have prioritized the miniaturization of power converters [1], where magnetic components, especially high-frequency inductors are usually the limiting components in terms of size and losses [2]-[5]. Typically, there are applications where current of high-power inductors include both high-frequency and low-frequency (or DC) components. These applications include DC-DC inductors, power factor correction (PFC) inductors, output filter inductors, and chokes [4]-[7]. The main challenge in these designs lies in selecting and designing the winding to manage both low and high-frequency conduction losses effectively [8].

Litz wire is a common solution to this challenge, yet its poor filling factor of about 50%, or sometimes less, limits its effectiveness for high power density applications. Compared to solid copper wires, litz wires leads to bulkier inductors with higher direct current resistance (DCR), negatively impacting low-frequency or DC performance. However, poorly designed solid wires can lead to increased equivalent series resistance (ESR) at high frequencies [8]. Various solutions using solid copper wires aim to optimize the balance between DCR and ESR at high frequencies. For example, a design using single-layer winding of round wires can reduce ESR for high-frequency applications [9]. Although it works well for toroidal shapes, it presents high ESR for multi-layer designs [10].

Solid wires with spiral shape, although promising, are mainly suitable for PCB inductors [2]. Solid flat wires, although provide a better thermal conductivity and DCR, require careful design optimization [7]. Among solid wire options, flat spiral and helix shapes offer superior performance and offer more practical manufacturing on printed circuit boards (PCBs), helping to reduce the size of the magnetic components. Spiral shapes, though effective, have proximity effect issues and are more suitable for high-frequency planar transformers [2].

Beside superior benefits for solid flat wire inductors, still they require further research in terms of accurate modelling [7], and [10]. To do design optimization of such inductors, e.g. using heuristic methods presented in [12], accurate models are needed to derive analytical equations for AC and DC losses in the solid flat wire coils. The model should consider the frequency-dependent effects of eddy-currents, reluctance paths and fringing effect of the core's gaps, which has not been proposed for such inductors. As an example, [13]-[14] has presented such a model for eddy currents in the lamination and magnets of an actuator, where eddy current impacts are represented as frequency-dependent reluctance and inductance.

The main contribution of this paper is presenting an analytic study and design considerations of high-power inductors with helical flat wires and distributed gaps with a focus on eddy-current effects in the coil. A magnetic equivalent circuit (MEC) including magneto-motive forces or frequency-dependent reluctances is offered which explain the inductance reduction at higher frequencies. Equations for DC and AC resistances for three cases are developed. FEM is employed in the analyses. Flux and eddy current components as well as the impact of design parameters on the performance indices of the inductor are investigated. Finally, an inductor with PQ 40/40 ferrite core is prototyped and tested to experimentally verify the design.

II. INDUCTOR TOPOLOGY

Fig. 1 shows an inductor including a PQ 40/40 core with a distributed gap and a flat-wire coil. As the fringing fluxes almost take a circular path with an effective radius of the gap length, as

979-8-3315-1612-3/25 $31.00 © 2025 IEEE

illustrated in Fig. 2, distributing a large gap into multiple small gaps leads to less penetration of the fringing flux into the window, resulting in less eddy current loss.

III. COUPLED MAGNETIC-ELECTRIC CIRCUIT

Figs. 3 (a)-(b) show flux lines, flux density and field intensity within the inductor for a DC current of 5 A with no eddy-current effect. It is a two-dimensional FEM with a symmetry along ϕ axis. We have flux fringing at the gaps and a leakage through the window. Field intensity distribution shows that most of the stored energy is mainly within the air-gap, and then the fringing regions.

Accordingly, magnetic equivalent circuit shown in Fig. 4(a) can be developed in which R_g, R_{ci}, R_f, and R_{lw} are gap reluctance, i^{th} core reluctance, fringing reluctance and window leakage reluctance. As shown in Fig. 4(b), eddy currents induced in the conductors by fringing or window leakage fluxes can be modeled by magneto-motive forces F_{ef} and F_{ew}, respectively, which generate fluxes ϕ_{ef} and ϕ_{ew} in the opposite direction of the original flux generated by the coil in the fringing region and the window, respectively. As represented in Fig. 4(c), it can be shown that these MMFs can be modeled by frequency dependent reluctances $R_{ef}(j\omega)$ and $R_{elw}(j\omega)$ in series with the primary zero-frequency reluctance. These frequency-dependent reluctances go up at higher frequencies which explains why inductance goes down by a reduction in the total flux linked by the coil at higher frequencies [13]-[14].

$$\varphi(j\omega) = \frac{Ni_L}{R_t(j\omega)} = \frac{Ni_L}{R_0 + \underbrace{R_0 Q(j\omega)}_{R_e(j\omega)}} = \frac{Ni_L}{R_0\left(1 + Q(j\omega)\right)} \quad (1)$$

where R_t is the total reluctance, R_0 is the total reluctance at zero frequency and R_e is the total reluctance added by eddy currents which is zero at zero frequency with Q as an auxiliary term:

$$\omega \to 0 \Rightarrow Q(j\omega) \to 0; \; R_e(j\omega) \to 0; \; R_t(j\omega) \to R_0; \; \varphi(j\omega) \to \varphi_0 \quad (2)$$

where ϕ_0 is the primary flux at zero frequency:

$$\varphi_0 = \frac{Ni_L}{R_0} \quad (3)$$

The coupled electric-magnetic circuit for both zero-frequency case and the frequency-dependent case are shown in Fig. 5. The governing equations can be expressed as:

$$\begin{cases} V_L = R_L I_L + j\omega N\varphi \\ NI_L = R_t\varphi \end{cases} \Rightarrow \begin{bmatrix} R_L & j\omega N \\ -N & R_t \end{bmatrix} \begin{bmatrix} I_L \\ \varphi \end{bmatrix} = \begin{bmatrix} V_L \\ 0 \end{bmatrix} \quad (4)$$

There is a codependency between electrical and magnetic circuits. From the magnetic circuit, flux is returned to the electrical circuit, and from the electrical circuit, the current is returned to the magnetic circuit. By solving the above system of equation, the terminal impedance can be obtained as:

$$Z(j\omega) = \frac{V_L}{I_L} = R_L + j\omega L(j\omega); \; L(j\omega) = \frac{N^2}{R_t(j\omega)} = \frac{L_0}{1 + Q(j\omega)} \quad (5)$$

It can be seen that the inductance $L(j\omega)$ is L_0 at zero frequency, and at higher frequencies, as Q and subsequently the reluctance goes up, the inductance goes down.

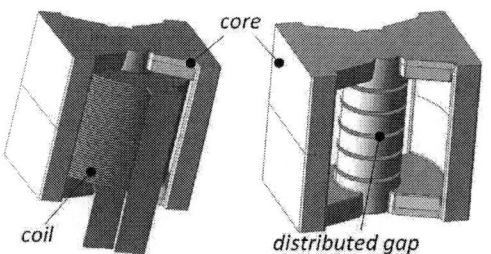

Fig. 1. Topology (top) and exploded view (bottom) of the proposed coupler.

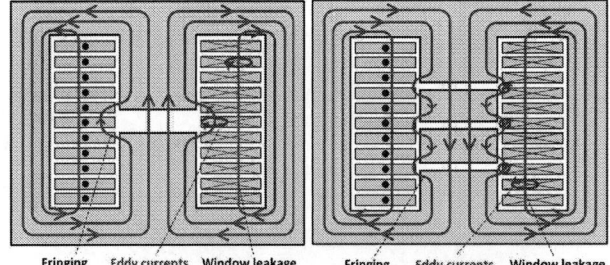

Fig. 2. Flux paths for inductors with single and distributed gaps.

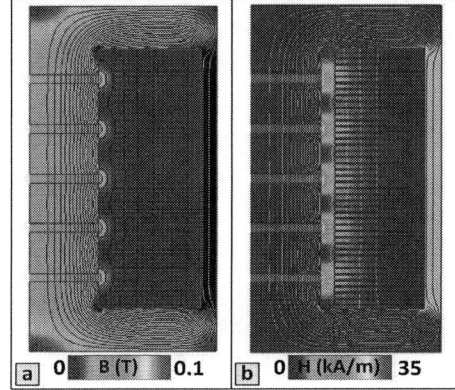

Fig. 3. Flux lines, flux density and field intensity with the core and window.

IV. DC RESISTANCE AND LOSS

Three models for calculating DCR are represented here that can be employed depending on the inductor topology and the required accuracy.

A. Coil with Helical Shape

Fig. 6 shows the geometry of the coil in which R_w, D_w, t_w, h_w, s, N are inner radius of the coil, radial length of the coil, thickness or z-height of the flat wire, total height of the coil, distance between turns, and number of turns, respectively. Moreover, D_{righ} and D_{left} are clearance between coil and core, shown in Fig. 6. A differential area of $dA = t_w \, dr$ for a differential radius of dr is employed in the integration. Using the length of helix $l(r)$ with radius r, N, and a total height of $h_w = Nt_w + (N-1)s$, the conductance is obtained as:

$$G_{DC} = \frac{1}{R_{DC}} = \int_{r=r_w}^{r=r_w+D_w} \frac{\sigma t_w \, dr}{l(r)} \quad (6)$$

where the total length of the coil with N turns is as follows:

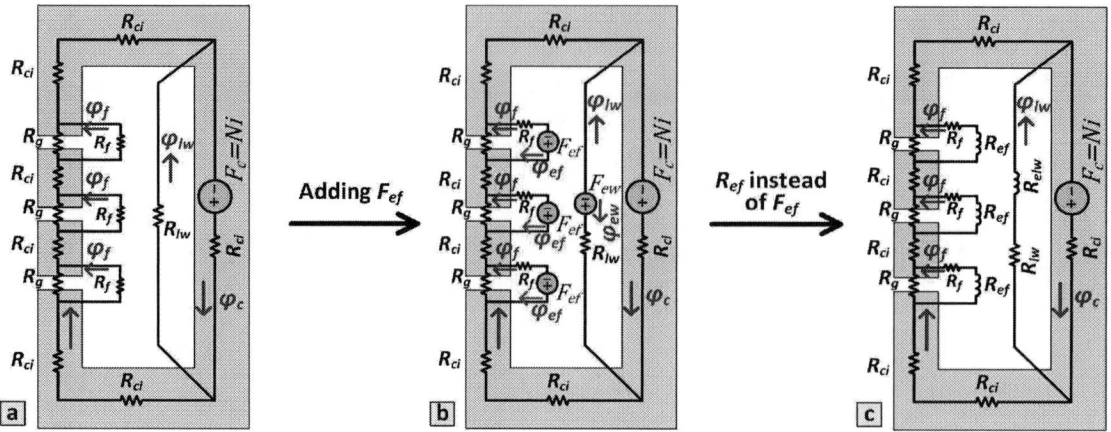

Fig. 4. MEC development: (a) eddy-currents ignored, (b) MMFs F_e representing eddy-currents an opposing flux, and (c) frequency-dependent reluctances R_e.

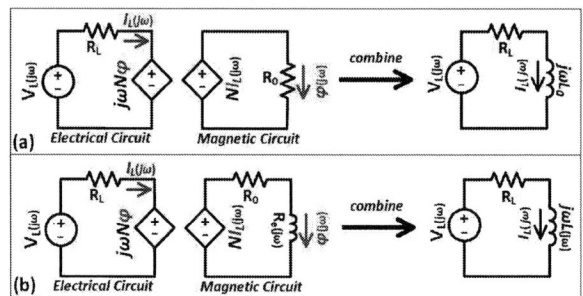

Fig. 5. Coupled electric-magnetic circuit: (a) low frequency case with no frequency dependency, and (b) frequency dependency due to eddy currents.

$$l(r) = 2\pi N \sqrt{r^2 + \left(\frac{h_w}{2\pi N}\right)^2} \qquad (7)$$

Finally, DCR is obtained as:

$$R_{DC} = 2\pi N / \left\{ \sigma t_w \ln \left(\frac{r_w + D_w + \sqrt{(r_w + D_w)^2 + \left(\frac{h_w}{2\pi N}\right)^2}}{r_w + \sqrt{r_w^2 + \left(\frac{h_w}{2\pi N}\right)^2}} \right) \right\} \qquad (8)$$

This is the general and the most accurate relationship for calculation of the DCR of this type of coil which is particularly useful for small number of turns where $h_w/2\pi N$ is not small compared to $r_w + D_w$.

B. Planar Circular Approximation

For large number of turns, coil length can be approximated by discrete circular turns as follows. It can also be obtained from approximation of the previous equation if $h_w/2\pi N$ is small enough and negligible. The conductance can be derived as:

$$G_{DC} = \frac{1}{R_{DC}} = \int_{r=r_w}^{r=r_w+D_w} \frac{\sigma t_w \, dr}{l(r)} \qquad (9)$$

where the total length of the coil with N turns is as follows:

$$l(r) = 2\pi Nr \qquad (10)$$

Finally, DCR is obtained as:

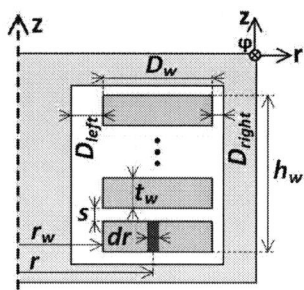

Fig. 6. Coil dimensions.

$$R_{DC} = 2\pi N / \left\{ \sigma t_w \ln \left(\frac{r_w + D_w}{r_w} \right) \right\} \qquad (11)$$

C. Average Radius Approximation

An initial estimation of DCR can be obtained by cconsidering an average radius $R_{av} = r_w + D_w/2$ as follows:

$$G_{DC} = \frac{1}{R_{DC}} = \int_{r=r_w}^{r=r_w+D_w} \frac{\sigma t_w \, dr}{l(r)} \qquad (12)$$

where the total length of the coil with N turns is as follows:

$$l(r) = 2\pi N R_{av} \qquad (13)$$

Finally, DCR is obtained as:

$$R_{DC} = \frac{2\pi N R_{av}}{\sigma t_w D_w} \qquad (14)$$

V. AC RESISTANCE AND CONDUCTION LOSSES

This section is devoted to the calculations of AC resistance of the inductor at high frequencies. According to Gausses' law $\nabla . B = 0$, a magnetic vector potential A can be defined as $B = \nabla \times A$. Employing the identity $\nabla \times \nabla \times A = \nabla(\nabla . A) - \nabla^2 A$ and Coulomb's gauge condition $\nabla . A = 0$ in Ampere's law yields:

$$\nabla \times H = J \xrightarrow{H = B/\mu_0} \nabla \times \left(\frac{\nabla \times A}{\mu_0} \right) = J \rightarrow \nabla^2 A = -\mu_0 J \qquad (15)$$

In the two-dimensional domain with symmetry in the ϕ direction, the vector potential A_ϕ and the currents J_ϕ are in the

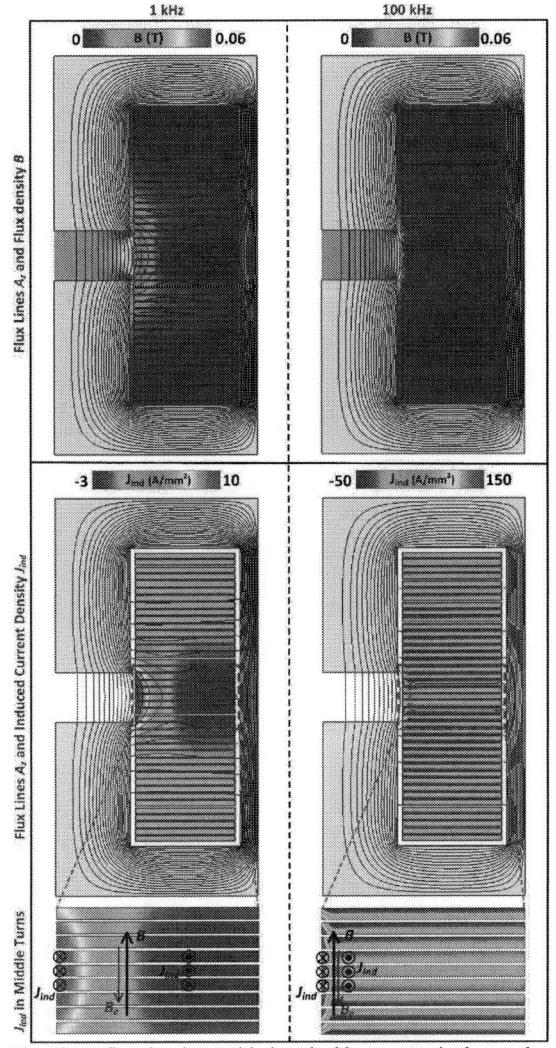

Fig. 7. Flux lines, flux density, and induced eddy currents in the conductors at frequencies of 1 kHz and 100 kHz for inductor with single large gap.

ϕ direction, while magnetic fields B and H are in the rz-plane. We have:

$$\nabla^2 A_\varphi(r,z,t) = -\mu_0 J_\varphi(r,z,t) \tag{16}$$

It is a 2-D problem with φ-directed current J_φ that can be decomposed into the source $J_s(t)$ due to inductor terminal current, and also induced eddy-current density J_{eddy}. As follows:

$$J_\varphi(r,z,t) = J_s(t) + J_{eddy}(r,z,t) \tag{17}$$

The inductor terminal current I_L can be obtained by integrating J_s which is uniformly distributed over a conductor area as follows:

$$I_L(t) = \iint_{A_w} J_s(t) \, dr \, dz = t_w D_w J_s(t) \tag{18}$$

Using Faraday's law and Ohm's law $J_\varphi = \sigma E_\varphi$, induced eddy current density can be obtained in terms of A_φ as follows:

$$\nabla \times E_\varphi(r,z,t) = -\frac{\partial B_{rz}(r,z,t)}{\partial t} \xrightarrow{B = \nabla \times A} E_\varphi(r,z,t) = -\frac{\partial A_\varphi(r,z,t)}{\partial t} \tag{19}$$

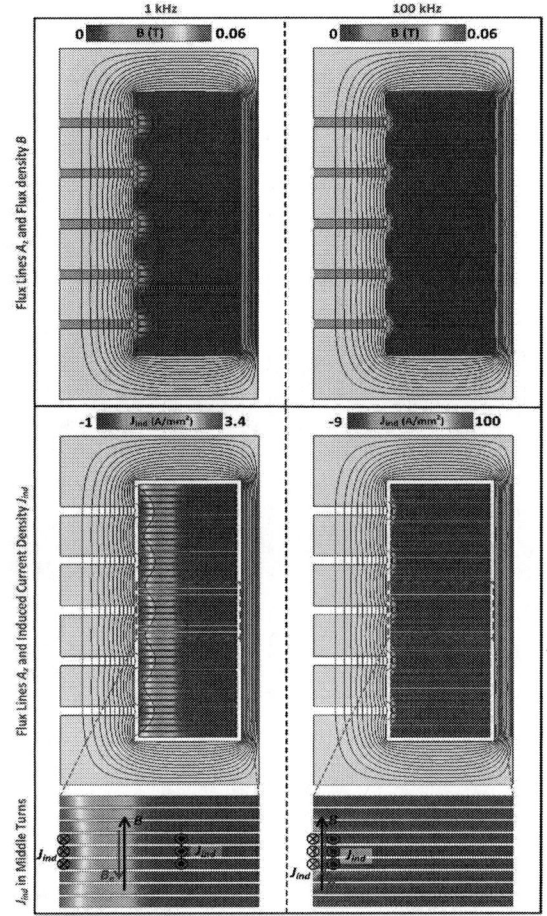

Fig. 8. Flux lines, flux density, and induced eddy currents in the conductors at frequencies of 1 kHz and 100 kHz for distributed gap inductor if conductors are close to the gap (small D_{left}).

Then,

$$J_{eddy}(r,z,t) = -\sigma \frac{\partial A_\varphi(r,z,t)}{\partial t} \tag{20}$$

In the phasor domain, we obtain:

$$J_{eddy}(r,z,j\omega) = -j\omega\sigma A_\varphi(r,z,j\omega) \tag{21}$$

The AC resistance R_{ac} is the effective resistance seen at terminals that can be obtained from dissipated power as:

$$P = R_{ac} I_L^2 = \iiint_{volume} \frac{\left|J_\varphi(r,\varphi,z)\right|^2}{\sigma} dr \, d\varphi \, dz = \frac{2\pi}{\sigma} \iint_{Area} \left|J_\varphi(r,z)\right|^2 dr \, dz \tag{22}$$

Substituting J_φ in terms of its components leads to:

$$R_{ac} = \frac{2\pi}{\sigma} \frac{\iint_{Area} \left|J_s + J_{eddy}(r,z)\right|^2 dr \, dz}{I_L^2} \tag{23}$$

VI. EDDY-CURRENT DESOMPOSITION AND DESIGN CONSIDERATIONS

In this section, eddy current components induced in the coil due to leakage or fringing fluxes are analyzed to provide a deeper insight into the design of the inductor with minimized

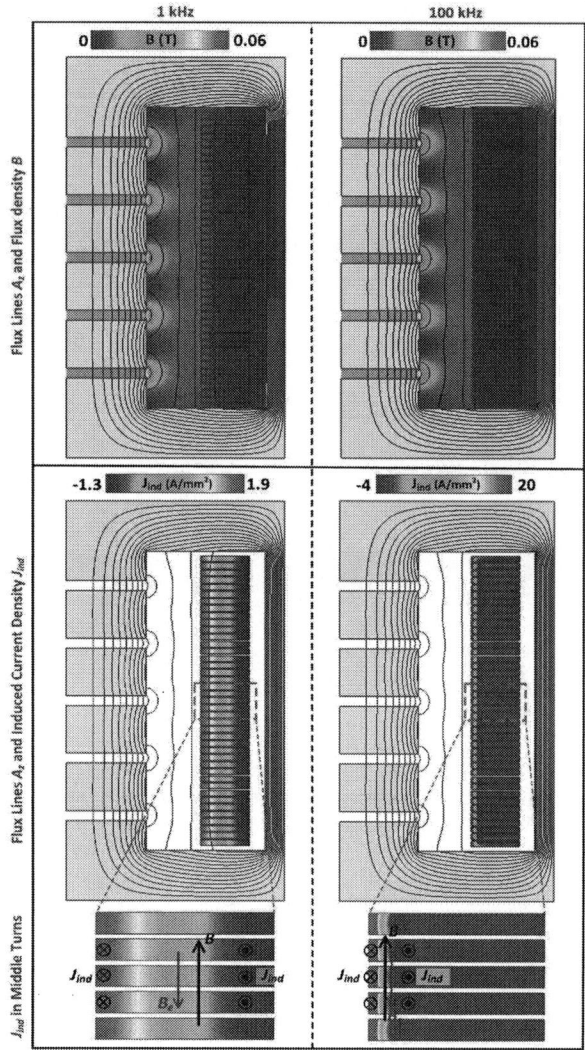

Fig. 9. Flux lines, flux density, and induced eddy currents in the conductors at frequencies of 1 kHz and 100 kHz for distributed gap inductor if conductors are away from the gap (large D_{left}).

conductive loss. Figs. 7-10 provide flux lines, flux density and induced eddy currents at a low frequency of 1 kHz and a high frequency of 100 kHz under various scenarios. Also, Fig. 11 shows the impact of design parameters on inductors quantities including DC and AC resistances, loss components, and inductance, while there is 5 A peak as AC current for different frequencies and constant DC current of 15 A.

As shown in Figs. 7-8, the fringing fluxes at the gap of an inductor with a single large gap flows into a large volume of conductors, while the fringing fluxes in the inductor with distributed gap are limited to smaller areas, leading to smaller eddy current losses in the conductor. It should be noted that, in order to show the closed loops of the induced eddy currents in $+\phi$ and $-\phi$ directions, the current density plots illustrate only the induced eddy currents J_{eddy}, not the total current J_ϕ. It can be seen that the vertical field B due to gap fringing fluxes or the window leakage flux which are in the z direction induce circulating eddy currents in the r-φ plane.

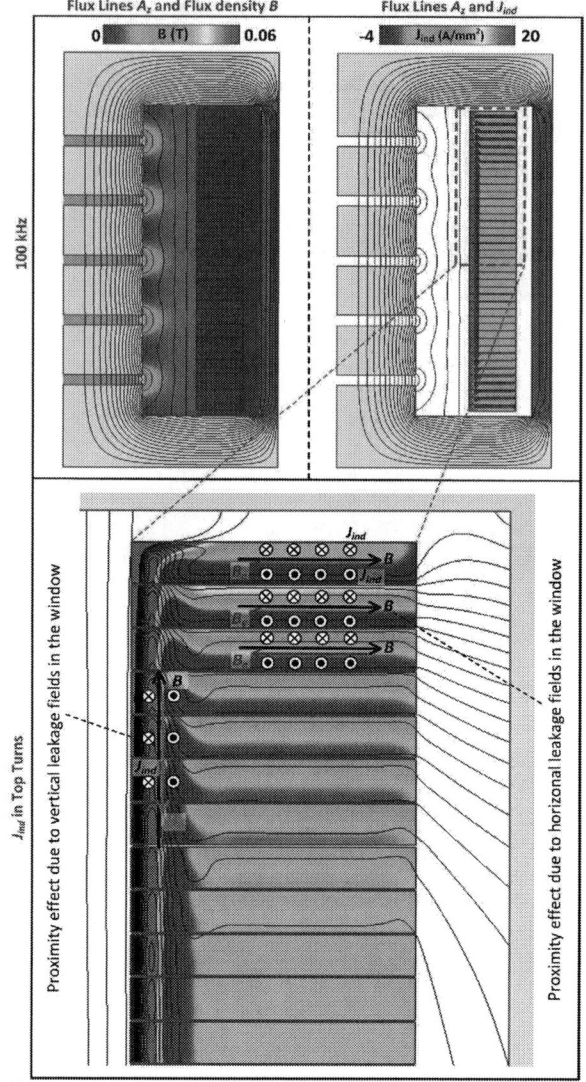

Fig. 10. Flux lines, flux density, and induced eddy currents in the conductors at frequencies of 1 kHz and 100 kHz for distributed gap inductor if conductors are very close to each other (small s).

As expected, the reaction fields B_e generated by induced eddy currents are in the opposite direction of the initial field B which explains how eddy currents reduce the flux linked by the coil λ and thus the inductance $L = \lambda/i$ at higher frequencies. It can also be observed that, at 100 kHz, induced currents penetrate less into the depth of the conductor and mostly concentrate at the inner radius of the coil with a higher magnitude.

As shown in Fig. 9, there is a leakage flux within the window. The larger the total air-gap or the higher the core saturation, the larger the window leakage. In other words, if the reluctance of the main flux path within the core goes up, a larger portion of flux tends to flow into the window. These vertical fluxes cause eddy currents circulating in the r-φ plane which will be limited to the inner radius of conductors at higher frequencies, and also the window leakage flux will be confined to the inner section of the window while they also penetrate into the coil area at the lower frequencies.

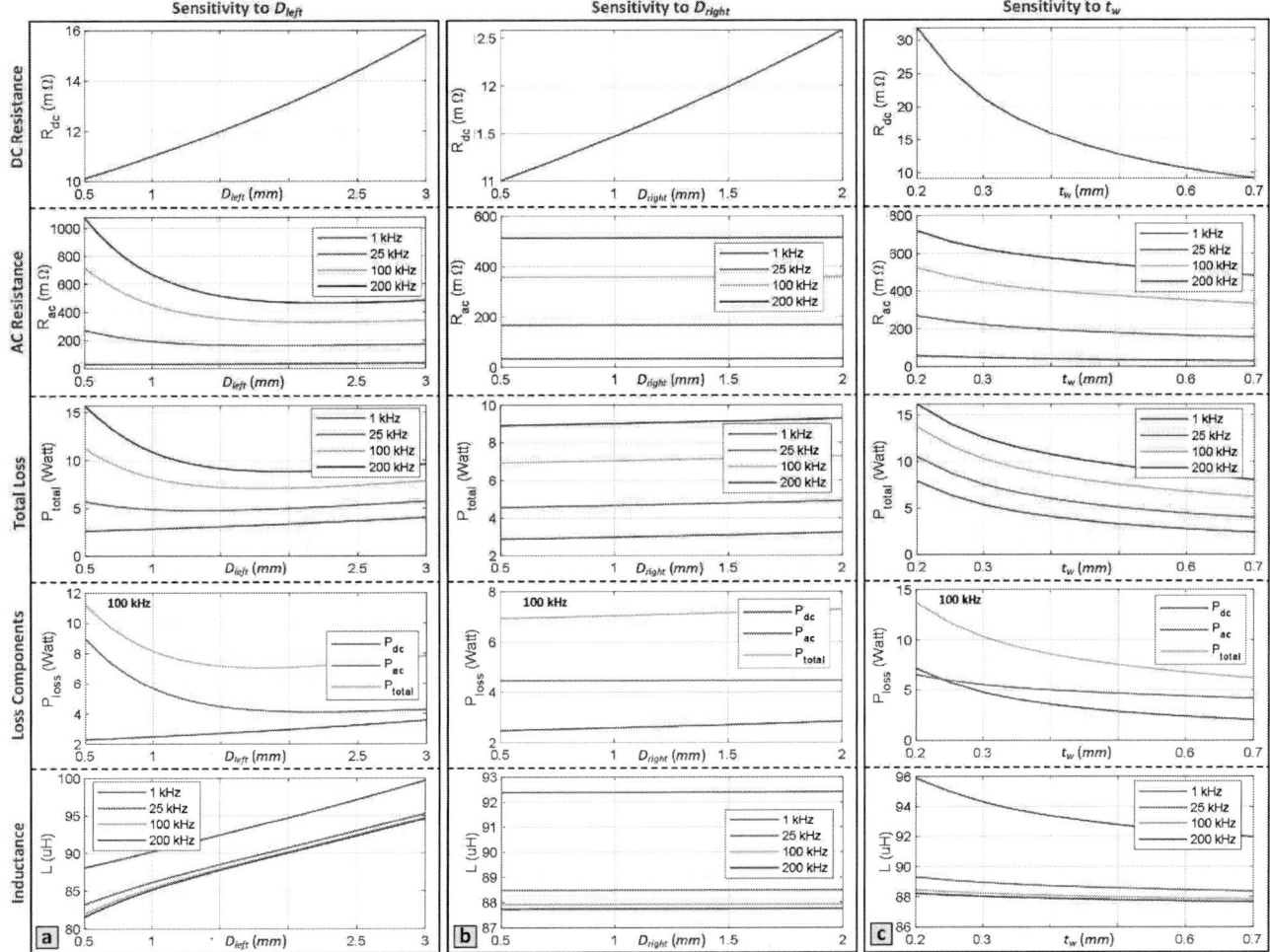

Fig. 11. Sensitivity of DC resistance, AC resistance, loss components, and inductance to design variables.

Fig. 10 represents finite element analysis for a case with small vertical distance s between the turns to study proximity between coil turns. Due to the horizontal leakage fluxes in the window, eddy-currents are induced in the z-φ plane.

It is seen in Fig. 11(a) that, as the coil distance from the gaps D_{left} is increased, the induced eddy currents are decreased, and therefore, AC resistance and AC losses go down, while DC resistance and DC loss go up due to smaller coil cross section. It shows an optimal point to achieve the minimum total conduction loss. As leakage flux within the window is rejected by eddy currents in the conductors, this leakage flux exists only within the area between the inner radius of the coil and the center leg of the core. Therefore, the inductance linearly goes up with D_{left} as this area goes up linearly. This fact shows D_{left} that is a key parameter to change L when N >> 1.

As shown in Fig. 11(b), the DC resistance and DC loss go up with increasing the distance between the outer radius of the coil and the core D_{right}. However, as the eddy currents are mostly induced on the left side of the coil near the gaps, the AC resistance and AC loss are not affected by increasing the coil distance from the right side. Additionally, the inductance is not affected by D_{right} as fringing and window leakage fluxes are rejected by the inner radius of the coil. Fig. 11(c) suggests that, by increasing the coil thickness t_w, the DC resistance and loss goes down, and also AC resistance and AC loss go down as the skin effect goes down. However, inductance goes down due to flux rejection by the induced eddy currents.

VII. PROTOTYPE AND EXPETRIMENTAL RESULTS

As shown in Fig. 12(a), a flat wire inductor with five distributed gaps of 1 mm each using a PQ 40/40 ferrite core is prototyped and tested, as shown in Fig. 12 (b), to verify the analytical modeling. The prototype is for an application customer spec of 87 μH ±10 % with 30 % drop at 34 A under 100 °C. The main specifications are $N = 41$, $t_w \approx 0.58$ mm, $D_w \approx 8.0$ mm, $s \approx 0.13$ mm and $r_w \approx 9.0$ mm, based on PQ 40/40 core with N95 material.

Inductance and total ESR values versus frequency are shown in Fig. 13 (a) and (b) using WYNE KERR 6500B impedance analyzer. As seen, at operating frequency of 100 kHz, L is measured about 82.8 μH with 5.8 % discrepancy from value of 87.9 μH obtained by FEM. Furthermore, total ESR is measured about 500 mΩ, which here is the combination of winding's AC resistance, R_{ac}, discussed in previous sections, and the ESR

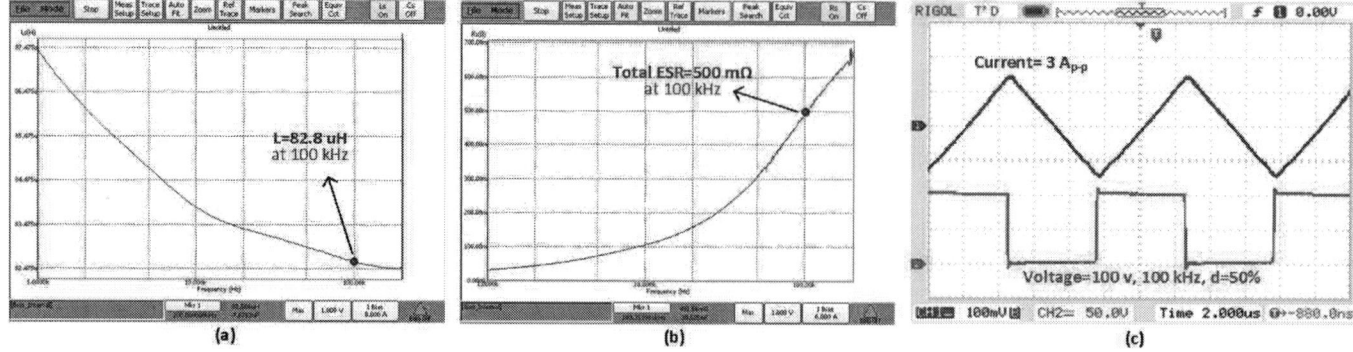

Fig. 13. Measurements of (a) inductance versus frequency, (b) AC resistance versus frequency and (c) voltage on S1 and inductor's current waveform at 100 kHz with duty cycle of =50 %.

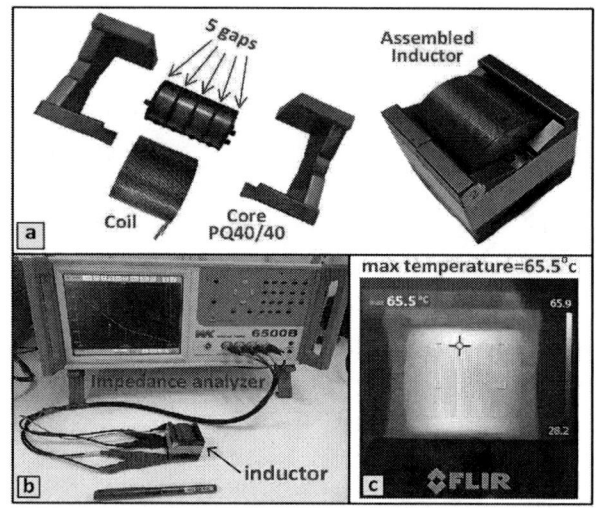

Fig. 12. Experiment: (a) prototype, (b) measurement setup, and (c) thermal camera.

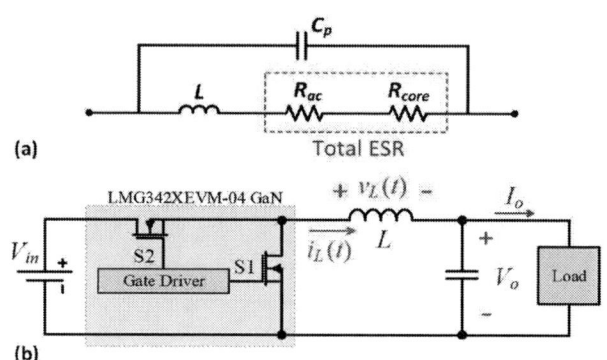

Fig. 14. Test circuit: buck converter.

reflected from the PQ40/40 core, R_{core}, due to hysteresis and eddy losses. As shown in Fig. 14(a), the lumped model of the inductor can be simplified for operating frequencies below the first resonant frequency, where C_p represent the total parasitic capacitor of the inductor which models the first resonant frequency of the component, f_r:

$$f_r = \frac{1}{2\pi\sqrt{LC_p}} \qquad (24)$$

To exclude R_{core}, first ESR of a reference inductor made by 170x0.05 mm litz wire, with $N = 41$ and the same core is measured. Regarding FEM simulations, AC and DC resistances of the litz wire are very close to each other for 100 kHz, less than 10 % deviation. Hence the measured ESR of the reference inductor is approximately equal to its DCR plus the extra ESR reflected from the core, i.e., R_{core} [10].

Using WYNE KERR 6500B impedance analyser and for 100 kHz, the ESR reflected from core is estimated about 75 mΩ, by excluding DCR of the litz wire from the total ESR measured by the impedance analyser. Hence, for the flat wire prototype, ESR of the winding, R_{ac}, is estimated about 425 mΩ, with 19 % deviation from 357mΩ of 2D FEM results (discrepancy is due to the 3D impacts, especially large leakage fluxes in the two

conductor regions without a surrounding core which causes more eddy currents losses). The DCR of the flat wire inductor is measured 12.4 mΩ, with 3.3% deviation from 12 mΩ obtained by equations (1) and (2). Due to the low DCR value, to have an accurate measurement, DCR is measured by applying DC current of 10 A and measuring the voltage drops on the inductor's terminals.

Furthermore, the inductor has been utilized in a synchronous buck converter using LMG342XEVM-04 GaN daughter card from Texas Instruments. The converter's topology is selected as a buck converter, shown in Fig. 14(b). The buck converter has been tested from 1 kHz to 200 kHz with input voltage V_{in} from 10 V up to 200 V. The load is selected a variable resistive load capable of handling 0 up to 20 A.

Fig. 13 (c) shows the voltage over switch S1 and the inductor's current ripple at $f_s = 100$ kHz, duty cycle of 50 % and input voltage of 100 V. In this condition, the current probe is connected through AC-coupling condition, showing only the ac ripple current.

Values reported for the ac inductance, included in Fig. 11, have been verified by measuring the peak-to-peak ripple current, I_{pp}, of the inductor and knowing the output voltage, V_O, and f_s under duty cycle of 50 %:

$$L = \frac{V_O}{2I_{pp}f_s} \qquad (25)$$

Comparing the values derived by measurements and the values presented in Fig. 11, the maximum deviation was reported about 7.5 % for 1, 25, 100 and 200 kHz operating conditions. To illustrate the thermal performance, steady-state temperature of inductor with f_s = 100 kHz, 50 % duty cycle and DC current of about 15.0 A is shown Fig. 12(c). With a temperature rise of 40 °C from the ambient temperature, 25 °C, the inductor temperature of 65 °C is still well below 100 °C which is satisfactory.

Finally, it is important to calculate the AC conduction losses of the winding for DC-DC converters. Generally, the current waveforms are triangle, as seen in Fig. 13 (c) and not a sinusoid, while the maximum ripple happens at 50 % duty cycle. Using Fourier series, voltage on the inductor $v_L(t)$ and the hth harmonic of the current I_h are derived by:

$$v_L(t) = \sum_{h=1}^{\infty} \frac{4V_O}{\pi h} \sin(2\pi h f_s t) \qquad (26)$$

$$I_h = \frac{2V_O}{(\pi h)^2 L f_s} \qquad (27)$$

where V_o is the load voltage, shown in Fig. 14 (b). Equations (26) and (27) are accurate when $hf_s < f_r$. Hence, the total ac conduction losses, P_{ac}, is derived by:

$$P_{ac} = \frac{1}{2} \sum_{h=1}^{\infty} R_{ac}(hf_s) I_h^2 \qquad (28)$$

where $R_{ac}(hf_s)$ is the ESR at hth harmonic. As discussed in [10], for the flat wire inductors, ESR is proportional to $\sqrt{f}$; hence (28) can be simplified by:

$$P_{ac} = \frac{1}{2} R_{ac} \sum_{h=1}^{\infty} \sqrt{h} I_h^2 \qquad (29)$$

By approximating (29) up to $h = 25^{\text{th}}$ harmonic, P_{ac} can be simplified more in (30), where R_{ac} is the winding's ESR measured at the switching frequency of the converter, f_s:

$$P_{ac} \approx 1.027 \left(\frac{2R_{ac}V_O^2}{\pi^4 L^2 f_s^2} \right) = 1.027 \left(\frac{8R_{ac}I_{pp}^2}{\pi^4} \right) \qquad (30)$$

Hence, P_{ac} can finally be simplified based on peak-to-peak current of inductor, I_{pp} and measured ESR at f_s.

VIII. CONCLUSION

Analytic study and design considerations for flat wire inductors are presented. Magnetic equivalent circuits including frequency-dependent reluctances and inductance are introduced. Three formulations for DCR are derived which can be used depending on the coil topology and the required accuracy. Calculations for ESR and AC losses are presented. Effectiveness of distributed gaps in terms of reduced fringing flux and thus eddy current losses is shown. Eddy current components are scrutinized. Sensitivity of inductor quantities such as DCR, ESR, loss components and inductance to design parameters are studied. The optimum design and the balance between DCR and ESR are discussed. Finally, an inductor is prototyped, where the experimental results are on par with FEM and show the satisfactory performance of the design.

There is a growing need for interpretable, generalizable and low-cost AI models, while physics-based models face their own challenges. In a future work, such physics-based analytical models can be integrated into the architecture, loss function, or optimization constraints of AI/ML models as physics-informed artificial intelligence (PIAI) to leverage both data-driven and physics-based techniques.

REFERENCES

[1] F. C. Lee, S. Wang and Q. Li, "Next Generation of Power Supplies—Design for Manufacturability," *IEEE Journal of Emerging and Selected Topics in Power Electronics*, vol. 9, no. 6, pp. 6462-6475, Dec. 2021.

[2] M. Sato, Y. Hattori, M. Ueda, Y. Bu and T. Mizuno, "Improved Performance of a Flat-Wire Coil with Magnetic Composite Material for Wireless Power Transfer," *IEEE Magnetics Letters*, vol. 12, pp. 1-5, 2021.

[3] Jiankun Hu and C. R. Sullivan, "AC resistance of planar power inductors and the quasi-distributed gap technique," *IEEE Transactions on Power Electronics*, vol. 16, no. 4, pp. 558-567, July 2001.

[4] R. Jez, "Influence of the Distributed Air Gap on the Parameters of an Industrial Inductor," *IEEE Transactions on Magnetics*, vol. 53, no. 11, pp. 1-5, Nov. 2017.

[5] C. Kjeldsen, C. Østergaard, M. Nymand and R. Ramachandran, "Procedure to compare different design methods for implementation-ready high power inductors," *2019 IEEE 13th International Conference on Compatibility, Power Electronics and Power Engineering (CPE-POWERENG)*, Sonderborg, Denmark, 2019, pp. 1-6.

[6] Y. Zhang, X. Guo, G. Wang and Q. Xiao, "A Novel Power Inductor and Its Application for Compact, Large Current DC-DC Converters," *2021 IEEE Applied Power Electronics Conference and Exposition (APEC)*, Phoenix, AZ, USA, 2021, pp. 2865-2869.

[7] L. Wang, W. Liu, D. Malcom and Y. -F. Liu, "An Integrated Power Module Based on the Power-System-In-Inductor Structure," *IEEE Transactions on Power Electronics*, vol. 33, no. 9, pp. 7904-7915, Sept. 2018.

[8] R. P. Wojda and M. K. Kazimierczuk, "Winding Resistance and Power Loss of Inductors with Litz and Solid-Round Wires," *IEEE Transactions on Industry Applications*, vol. 54, no. 4, pp. 3548-3557, July-Aug. 2018.

[9] D. K. Saini, A. Ayachit, A. Reatti and M. K. Kazimierczuk, "Analysis and Design of Choke Inductors for Switched-Mode Power Inverters," *IEEE Transactions on Industrial Electronics*, vol. 65, no. 3, pp. 2234-2244, March 2018.

[10] A. Namadmalan, "Modeling of High Power Inductors Based on Solid Flat Wires for Compact DC-DC Converters," *2024 IEEE Kansas Power and Energy Conference (KPEC)*, Manhattan, KS, USA, 2024, pp. 1-5.

[11] R. Rajamony, S. Wang, R. Navaratne and W. Ming, "Multi-Objective Design of Single-Phase Differential Buck Inverters with Active Power Decoupling," *IEEE Open Journal of Power Electronics*, vol. 3, pp. 105-114, 2022.

[12] A. Namadmalan, B. Jaafari, A. Iqbal and M. Al-Hitmi, "Design Optimization of Inductive Power Transfer Systems Considering Bifurcation and Equivalent AC Resistance for Spiral Coils," *IEEE Access*, vol. 8, pp. 141584-141593, 2020.

[13] S. Mohammadi, W. R. Benner, J. L. Kirtley, and J. H. Lang, "An Actuator with Magnetic Restoration, Part I: Electromechanical Model and Identification," *IEEE Transactions on Energy Conversion*, 2024, doi:10.1109/TEC.2024.3387390.

[14] S. Mohammadi, "Modeling, Design, Identification, Drive, and Control of a Rotary Actuator with Magnetic Restoration," *Ph.D dissertation, Department of Electrical Engineering and Computer Science.*, *Massachusetts Institute of Technology*, 2022.

Insulation Dielectric Loss of High-Frequency Transformer Under Square Voltage Excitation With Edge Oscillation

Zhanlei Liu
State Key Lab of Electrical Insulation
and Power Equipment
Xi'an Jiaotong University
Xi'an, China
lzl0283@stu.xjtu.edu.cn

Lingyu Zhu
State Key Lab of Electrical Insulation
and Power Equipment
Xi'an Jiaotong University
Xi'an, China
zhuly1026@xjtu.edu.cn

Yuntian Gao
State Key Lab of Electrical Insulation
and Power Equipment
Xi'an Jiaotong University
Xi'an, China
2203211439@stu.xjtu.edu.cn

Yongliang Dang
State Key Lab of Electrical Insulation
and Power Equipment
Xi'an Jiaotong University
Xi'an, China
dyl877759724@stu.xjtu.edu.cn

Cao Zhan
Center for Power Electronics Systems
Virginia Tech
Blacksburg, VA, USA
caozhan@vt.edu

Shengchang Ji
State Key Lab of Electrical Insulation
and Power Equipment
Xi'an Jiaotong University
Xi'an, China
jsc@xjtu.edu.cn

Abstract—**High-frequency transformer (HFT) withstands high-frequency non-sinusoidal voltage and high temperature stresses. The edge oscillation (EdgOsc) in square voltage (SquVol) can significantly increase the insulation dielectric loss of potted epoxy resin. This paper investigates the epoxy resin insulation dielectric loss under SquVol with EdgOsc. The waveform of SquVol with EdgOsc can be described by an analytical function. The permittivities of epoxy resin sample are measured by broadband dielectric spectroscopy tester. The insulation dielectric loss under SquVol with EdgOsc can be calculated by Fourier superposition method with frequency-dependent ε''. Calculation results show that EdgOsc can increase the insulation dielectric loss by more than 10 times. In addition, insulation dielectric loss can increase by 0.5 to 1 times when the temperature increases from 25°C to 125°C. The contribution of EdgOsc in SquVol should be taken into considerations in insulation dielectric loss calculation.**

Keywords—*High-frequency transformer, epoxy resin, insulation dielectric loss, edge oscillation.*

I. INTRODUCTION

High-frequency transformer has attracted wide attentions in distribution grid [1], electric locomotive [2], [3], energy storage station [4] and renewable energy grid connection [5]. With the emergence and development of high-voltage IGBTs and SiC MOSFETs [6]-[9], high-voltage and high power density HFTs are becoming increasingly important [10]. To ensure the insulation safety of HFT and meanwhile reduce the HFT size, epoxy resin potting schemes [11], [12] are usually adopted for high-voltage HFTs.

Despite enhanced insulation performance, the low thermal conductivity of epoxy resin reduces the heat dissipation efficiency and may lead to local hotspots inside the HFT, bringing insulation risks. In addition, the insulation dielectric loss of HFT under high-frequency non-sinusoidal excitation is significant. It can account for 17% of total power loss in HFT as mentioned in [13], which is usually overlooked in the design of HFT. The insulation dielectric loss can increase the total power loss of HFT and may also lead to local hotspots inside the HFT. Therefore, accurate estimation of insulation dielectric loss in epoxy resin potting HFT is significant.

The insulation dielectric loss under non-sinusoidal excitation can be calculated by summation of dielectric losses under all harmonic excitations for linear dielectric materials [14]. For typical electric field strength in HFT insulation, the linear dielectric material assumption can be satisfied [15]. The effect of dielectric relaxation of epoxy resin on insulation dielectric loss is investigated in [16]. The dielectric loss closely corresponds to the α-relaxation and β-relaxation. The effect of spike voltage in square voltage on the dielectric loss is investigated in [17]. It was found that spike voltage will significantly increase the dielectric loss. The calculation of dielectric loss under PWM voltage is investigated in [18]. The impact of frequency-dependent permittivity is also discussed. The dielectric loss of a HFT employed in a 25kW and 48kHz MV DC-DC converter is measured. It is discovered that the ratio of insulation dielectric loss to total HFT power loss reaches 17%. However, the practical excitation voltage in HFT may be square voltage (SquVol) with edge oscillation (EdgOsc) [19]. The EdgOsc in SquVol contains rich high-frequency harmonics and can increase the insulation dielectric loss. The effect of EdgOsc on insulation dielectric loss is not clear yet and needs further comprehensive research.

This paper focuses on the insulation dielectric loss of high-frequency transformer under SquVol with EdgOsc. In Section II, the practical excitation voltage waveform in HFT is analyzed. An analytical function is proposed to describe the SquVol with EdgOsc. In Section III, the permittivities of epoxy resin sample is measured. The calculation method of insulation dielectric loss under SquVol with EdgOsc is introduced. In Section IV, the characteristics of insulation dielectric loss under SquVol with EdgOsc are investigated. Section V concludes this paper.

II. ANALYSIS OF PRACTICAL VOLTAGE WAVEFORM IN HFT

The typical application topology of high-voltage HFT is dual-active-bridge (DAB) DC-DC converter as shown in Fig. 1. HFT acts as an isolation component in the DAB converter. The parasitic inductance and stray capacitance in switching devices and HFT can compose a resonant circuit. The fast switching transient of switching devices can induce high-frequency oscillations in the resonant circuit. Thus, the

979-8-3315-1612-3/25 $31.00 © 2025 IEEE

practical excitation voltage waveform in HFT is not ideal square wave. Instead, the practical excitation voltage waveform may be SquVol with EdgOsc as shown in Fig. 2. The EdgOsc in SquVol contains a large amount of high-frequency harmonics. These high-frequency harmonics will contribute considerable high-frequency dielectric loss and cannot be neglected in insulation dielectric loss calculation.

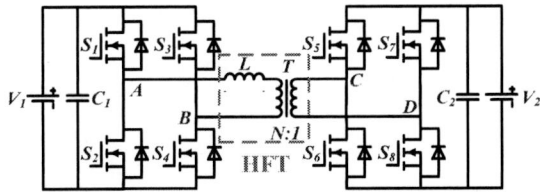

Fig. 1. Dual-active-bridge DC-DC converter.

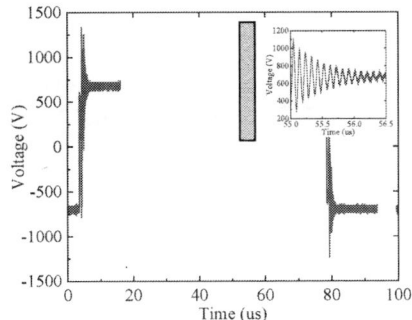

Fig. 2. Practical square voltage with edge oscillation.

Since the high-frequency resonant circuit is composed of resistance, inductance and capacitance, sine function with exponentially decaying amplitude can be used to describe the edge oscillation. The practical square voltage waveform can be expressed as

$$
y = \begin{cases} F \cdot \text{square}(t) + Ae^{-bt} \sin(\dfrac{2\pi}{T} \cdot t) * \text{square}(t) & 0 < t \leq \dfrac{T}{2} \\ F \cdot \text{square}(t) + Ae^{-b(t-\frac{T}{2})} * \\ \quad \sin(\dfrac{2\pi}{T} \cdot (t - \dfrac{T}{2})) * \text{square}(t) & \dfrac{T}{2} < t \leq T \end{cases}
$$

(1)

where F is square voltage amplitude, square(t) is unit square voltage function, A is edge oscillation amplitude, b is edge oscillation decaying factor, T is edge oscillation period.

Taking F=700, A=700, b=2.5×10⁶, T=1×10⁷ and square voltage rising time t_r=0.1μs as an example, the simulated square voltage waveform by equation (1) is shown in Fig. 3. Equation (1) can accurately describe the practical square voltage waveform with edge oscillation. The high-frequency harmonics of the SquVol with EdgOsc calculated by Fourier decomposition method are shown in Fig. 4. The harmonics of square voltage decreases with frequencies. The center frequency of harmonics of edge oscillation voltage are about 10MHz. The maximum harmonic amplitude of edge oscillation voltage is about 11.2V. Since the insulation dielectric loss resistance at high frequency is small, the contribution of edge oscillation voltage to the total insulation dielectric loss cannot be neglected. In this paper, equation (1) is used to investigate the calculation method

and characteristics of insulation dielectric loss under SquVol with EdgOsc.

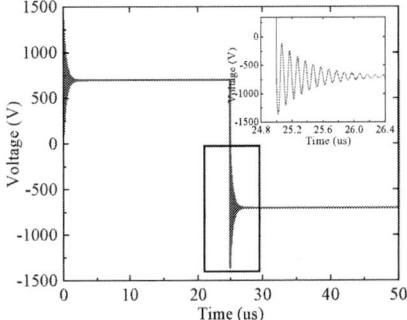

Fig. 3. Simulated square voltage with edge oscillation.

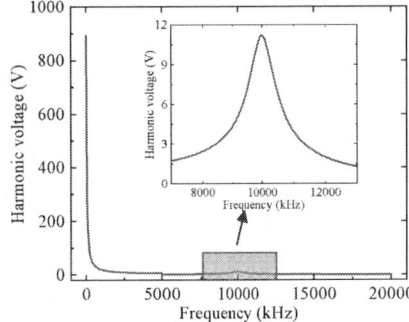

Fig. 4. Simulated square voltage with edge oscillation.

III. CALCULATION METHOD OF DIELECTRIC LOSS UNDER SQUVOL WITH EDGOSC

The insulation dielectric loss of epoxy resin mainly includes conductivity loss and relaxation loss. The conductivity loss comes from conduction current under external electric field. The relaxation loss comes from the power loss related to dipole turning and interface polarization during the medium polarization process. The conductivity loss can be characterized by conductivity σ and the relaxation loss can be characterized by imaginary permittivity ε''. Insulation dielectric loss under high-frequency excitation mainly comes from relaxation loss and can be characterized by imaginary permittivity ε''. To calculate the insulation dielectric loss under high-frequency non-sinusoidal excitations, imaginary permittivity ε'' should be measured and analyzed first. In this section, the epoxy resin sample under test is introduced. The real and imaginary permittivities are measured and analyzed. The insulation dielectric loss calculation method under high-frequency non-sinusoidal excitations is presented.

A. Preparation of epoxy resin sample

To test the dielectric loss of epoxy resin, a toroidal epoxy resin sample with a diameter of 40mm and thickness of 1mm is made, as shown in Fig. 5. The selected epoxy resin is 910-1803K&B-002 series A/B epoxy resin potting material produced by Jiaxing Libeide New Material Technology Co.,Ltd. from China. This epoxy resin potting material can be cured at medium to low temperatures. The insulation strength of the cured epoxy resin is ≥18kV/mm and the thermal conductivity is 0.7-0.9 W/(m·K). This epoxy resin potting material is suitable for the insulation packaging of HFT.

To reduce the contact resistance between test electrodes and epoxy resin sample, silver electrodes were sprayed on both sides of the sample. The diameter of the sprayed silver electrodes is 30mm.

Fig. 5. Epoxy resin sample.

B. Permittivity test of epoxy resin sample

The broadband dielectric spectroscopy tester (Novocontrol Concept 80), as shown in Fig. 6, is utilized to measure the permittivity of epoxy resin at different temperatures and frequencies. The variations of real permittivity with temperature and frequency are shown in Fig. 7. Real permittivity decreases with frequency and increases with temperature. The variations of imaginary permittivity with temperature and frequency are shown in Fig. 8. From 10kHz to 2MHz, ε'' is almost constant with frequency. From 2MHz to 20MHz, ε'' increases rapidly with frequency, indicating that insulation dielectric loss increases rapidly at high frequency.

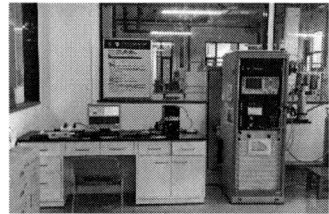

Fig. 6. Broadband dielectric spectroscopy tester.

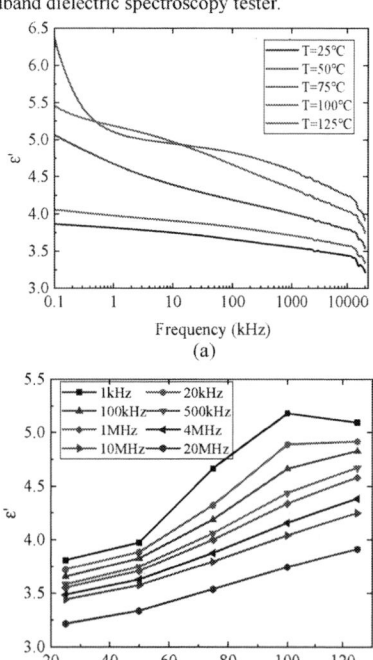

Fig. 7. Variations of real permittivity ε' with (a) frequency and (b) temperature.

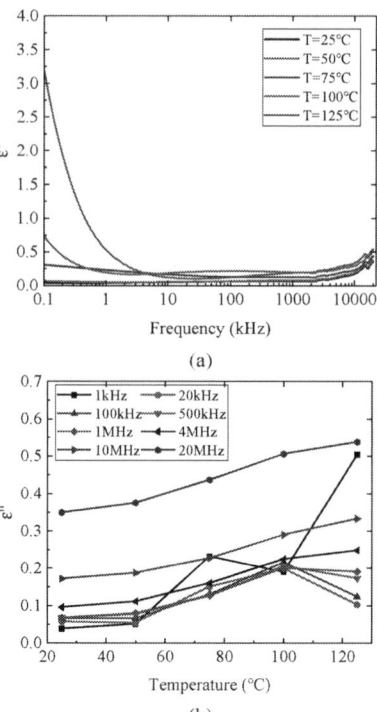

Fig. 8. Variations of imaginary permittivity ε'' with (a) frequency and (b) temperature.

The insulation dielectric loss resistance can be calculated by

$$R = \frac{1}{2\pi f C_0 \varepsilon''} \tag{2}$$

where f is frequency, C_0 is vacuum capacitance, ε'' is imaginary permittivity.

The variations of insulation dielectric loss resistance with frequency and temperature are shown in Fig. 9. As illustrated in Fig. 9(a), insulation dielectric resistance decreases with frequency. When frequency is over 2MHz, insulation dielectric resistance decreases faster with frequency, indicating that calculating insulation dielectric loss using imaginary permittivity data below 2MHz will underestimate the dielectric loss. As illustrated in Fig. 9(b), when frequency is below 1MHz, dielectric loss resistance first decreases and then increases with temperature. The dielectric loss resistance is minimum at about 100°C. When frequency is above 1MHz, dielectric loss resistance decreases with temperature. As the typical operating temperature of HFT is around 100°C, insulation dielectric loss under rated load condition will be higher than no-load condition.

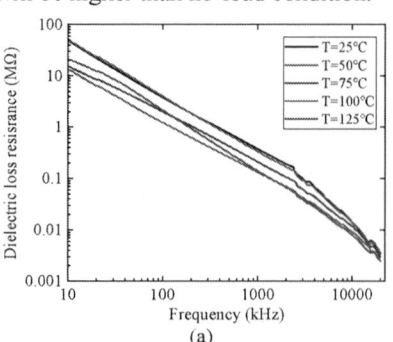

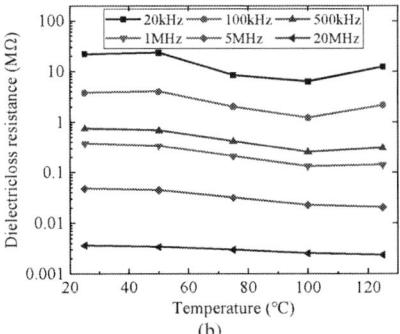

(b)

Fig. 9. Variations of dielectric loss resistance with (a) frequency and (b) temperature.

C. Dielectric loss calculation method

The insulation dielectric loss of epoxy resin under non-sinusoidal excitation can be calculated by summation of dielectric loss under all harmonics for linear dielectric materials. Firstly, Fourier decomposition is implemented on the waveform of SquVol with EdgOsc to obtain the harmonics. Secondly, equation (3) can be used to calculate the insulation dielectric loss based on Fourier superposition principle. Since ε'' increases rapidly with frequency over 2MHz, frequency-dependent ε'' should be adopted in equation (3).

$$P = \sum_{n=1}^{\infty} P_n = C_0 \sum_{n=1}^{\infty} \varepsilon^{*}\left(nf,T\right)\left(2\pi fn\right)V_{n,RMS}^{2} \quad (3)$$

where C_0 is vacuum capacitance, ε'' is imaginary permittivity, f is fundamental frequency, n is harmonic order, T is temperature. $V_{n,rms}$ is RMS of n-th harmonic voltage.

IV. CHARACTERISTICS OF DIELECTRIC LOSS UNDER SQUVOL WITH EDGOSC

A. Impact of harmonic superposition order on insulation dielectric loss

Since square voltage waveform contains a large number of harmonics, the harmonic superposition order can have significant impact on the calculated insulation dielectric loss. Taking electric strength E=1kV/mm and square voltage with different rising times as study cases, the impacts of harmonic superposition order on calculated insulation dielectric loss are shown in Fig. 10. It can be seen that the contributions of high-frequency harmonics to the total dielectric loss cannot be neglected. Superposition of harmonics dielectric loss until cutoff frequency is needed for accurate dielectric loss calculation under square voltage excitation.

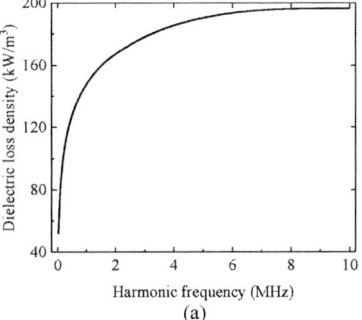

(a)

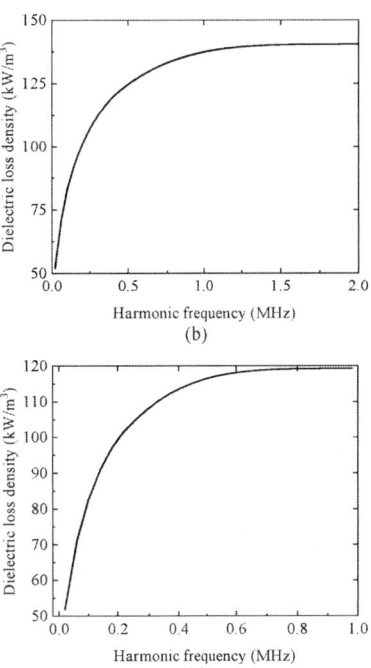

Fig. 10. Calculated insulation dielectric loss with harmonic superposition order for square voltage with rising time of (a) 0.1μs, (b) 0.5μs and (c) 1μs.

B. Impact of square voltage rising time on insulation dielectric loss

The rising time of square voltage determines the cutoff frequency and high-frequency harmonics, therefore having impact on insulation dielectric loss. Taking square voltage with electric strength E=1kV/mm as a study case, the impacts of square voltage rising time on calculated insulation dielectric loss are shown in Fig. 11. The insulation dielectric loss decreases with square voltage rising time. When square voltage rising time is reduced from 1μs to 0.1μs, insulation dielectric loss is almost doubled. With the development of fast-switching MOSFETs, the rising time of square voltage in converters decreases and the total insulation dielectric loss will increase.

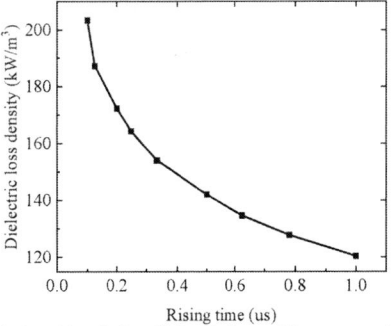

Fig. 11. Calculated insulation dielectric loss with square voltage rising time.

C. Impact of EdgOsc in SquVol on insulation dielectric loss

The EdgOsc can significantly increase the insulation dielectric loss under SquVol. In this paper, a loss ratio k is introduced to describe the characteristic of dielectric loss under SquVol with EdgOsc. The loss ratio k is defined as the ratio of insulation dielectric loss under SquVol with

EdgOsc to insulation dielectric loss under SquVol without EdgOsc and can be expressed as

$$k = P_{squ\_edg} / P_{squ} \qquad (4)$$

where P_{squ_edg} is the dielectric loss under SquVol with EdgOsc, P_{squ} is the dielectric loss under square voltage without EdgOsc.

Taking a square voltage of 20kHz with rising time of 0.5μs as an example, the variations of loss ratio k with A, b and T are shown in Fig. 12. As illustrated in Fig. 12(a), loss ratio k increases with oscillation amplitude A. As illustrated in Fig. 12(b), loss ratio k decreases with oscillation decaying factor b. When b is over 3×10^6, k decreases slowly with oscillation decaying factor b. As illustrated in Fig. 12(c), loss ratio k decreases with oscillation period T. When T is over 2.5×10^{-6}, k decreases slowly with oscillation period T. EdgOsc can increase the insulation dielectric loss under SquVol excitation by up to 10 times. With the emergence of fast-switching power devices, EdgOsc amplitude in SquVol and EdgOsc equivalent frequency improves, increasing the insulation dielectric loss under practical voltage excitations.

The variations of loss ratio k with temperature are shown in Fig. 13. The insulation dielectric loss ratio k increases with temperature. Insulation dielectric loss ratio k at 100°C is about 1.5-1.6 times of that at 25°C, and the insulation dielectric loss ratio k at 125°C is about 1.6-1.8 times of that at 25°C. Temperature can significantly increase the dielectric loss. The insulation dielectric loss of epoxy resin in HFT under load condition will be higher than no-load condition.

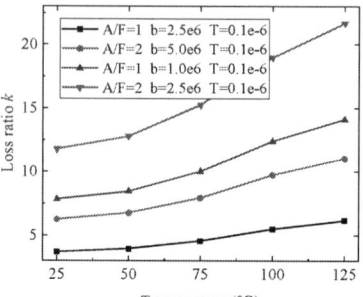

Fig. 13. Variations of insulation dielectric loss ratio k with temperature.

Taking a square voltage of 20kHz with rising time of 0.1μs as an example, the variations of loss ratio k with A, b and T are shown in Fig. 14. Since square voltage rising time is increased, insulation dielectric loss under square voltage excitation increases. Therefore, insulation dielectric loss ratio k for rising time of 0.1μs is lower than rising time of 0.5μs. Nevertheless, EdgOsc can still increase the insulation dielectric loss under square voltage excitation by up to 7 times. The contribution of EdgOsc to insulation dielectric loss still demands considerations.

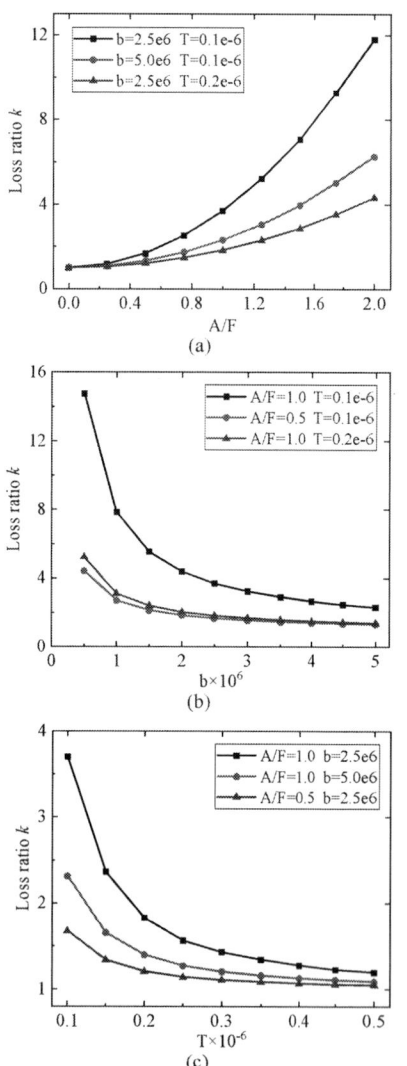

Fig. 12. Variations of insulation dielectric loss ratio k with (a) A, (b) b and (c) T when rising time of square voltage is 0.5μs.

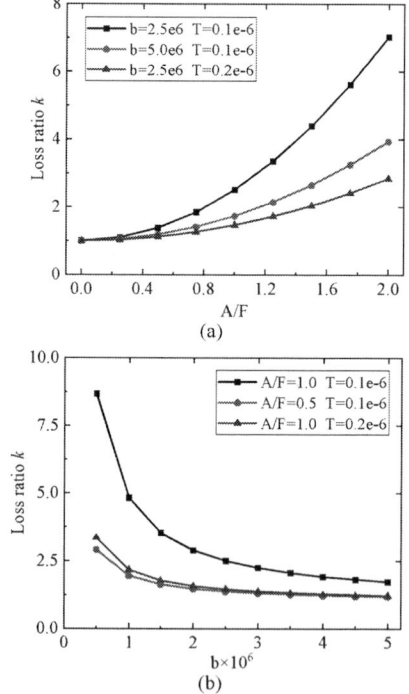

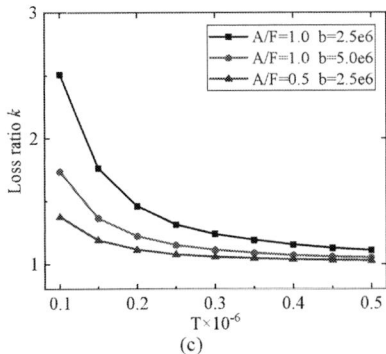

(c)

Fig. 14. Variations of insulation dielectric loss ratio k with (a) A, (b) b and (c) T when rising time of square voltage is 0.1 μs.

V. CONCLUSION

This paper proposes calculation method and analyzes characteristics of insulation dielectric loss of HFT under SquVol with EdgOsc. The main conclusions are listed as follows.

1) From 10kHz to 2MHz, ε'' is almost constant with frequency. From 2MHz to 20MHz, ε'' increases rapidly with frequency, indicating that insulation dielectric loss increases rapidly at high frequency.

2) The practical voltage in the HFT is SquVol with EdgOsc. An analytical formula is proposed to describe the square waveform with edge oscillation. The Fourier superposition method with frequency-dependent ε'' is proposed to calculate the insulation dielectric loss under SquVol with EdgOsc.

3) The EdgOsc can significantly increase the insulation dielectric loss by more than 10 times. Temperature can increase the dielectric loss by 0.5 to 1 time when it increases from 25°C to 125°C. The contribution of edge oscillation to insulation dielectric loss cannot be neglected.

REFERENCES

[1] B. Zhao, Q. Song, J. Li, W. Liu, G. Liu, and Y. Zhao, "High-frequency link DC transformer based on switched capacitor for medium-voltage DC power distribution application," *IEEE Trans. Power Electron.*, vol. 31, no. 7, pp. 4766–4777, Jul. 2016.

[2] C. Zhao et al., "Power Electronic Traction Transformer—Medium Voltage Prototype," *IEEE Trans. Ind. Electron.*, vol. 61, no. 7, pp. 3257-3268, July 2014.

[3] C. Gu, Z. Zheng, L. Xu, K. Wang and Y. Li, "Modeling and Control of a Multiport Power Electronic Transformer (PET) for Electric Traction Applications," *IEEE Trans. Power Electron.*, vol. 31, no. 2, pp. 915-927, Feb. 2016.

[4] J. Lu, Y. Wang, X. Li and C. Du, "High-Conversion-Ratio Isolated Bidirectional DC–DC Converter for Distributed Energy Storage Systems," *IEEE Trans. Power Electron.*, vol. 34, no. 8, pp. 7256-7277, Aug. 2019.

[5] M. R. Islam, Y. Guo and J. Zhu, "A High-Frequency Link Multilevel Cascaded Medium-Voltage Converter for Direct Grid Integration of Renewable Energy Systems," *IEEE Trans. Power Electron.*, vol. 29, no. 8, pp. 4167-4182, Aug. 2014.

[6] S. Madhusoodhanan et al., "Harmonic Analysis and Controller Design of 15 kV SiC IGBT-Based Medium-Voltage Grid-Connected Three-Phase Three-Level NPC Converter," *IEEE Trans. Power Electron.*, vol. 32, no. 5, pp. 3355-3369, May 2017.

[7] L. Zhang et al., "Design Considerations for High-Voltage Insulated Gate Drive Power Supply for 10-kV SiC MOSFET Applied in Medium-Voltage Converter," *IEEE Trans. Ind. Electron.*, vol. 68, no. 7, pp. 5712-5724, July 2021.

[8] C. Zhan et al., "A Novel Sensor-Reduction Condition Monitoring Approach for MMC Submodule IGBTs Based on Statistics of Inferred On-State Voltage," *IEEE J. Emerg. Sel. Topics Power Electron.*, vol. 12, no. 1, pp. 1068-1077, Feb. 2024.

[9] C. Zhan et al., "Intelligent Condition Monitoring of Multiple Thermal Degradation of IGBT Modules Based on Case Temperature Matrix," *IEEE Trans. Power Electron.*, vol. 39, no. 10, pp. 12490-12501, Oct. 2024.

[10] T. Yuan, F. Jin, Z. Li, C. Zhao and Q. Li, "Design of an Integrated Transformer With Parallel Windings for a 30-kW LLC Resonant Converter," *IEEE Trans. Power Electron.*, vol. 38, no. 11, pp. 14317-14333, Nov. 2023.

[11] W. Wang et al., "An Improved Design Procedure for a 10 kHz, 10 kW Medium-Frequency Transformer Considering Insulation Breakdown Strength and Structure Optimization," *IEEE J. Emerg. Sel. Topics Power Electron.*, vol. 10, no. 4, pp. 3525-3540, Aug. 2022.

[12] Z. Li, E. Hsieh, Q. Li and F. C. Lee, "High-Frequency Transformer Design With Medium-Voltage Insulation for Resonant Converter in Solid-State Transformer," *IEEE Trans. Power Electron.*, vol. 38, no. 8, pp. 9917-9932, Aug. 2023.

[13] T. Guillod, R. Faerber, D. Rothmund, F. Krismer, C. M. Franck and J. W. Kolar, "Dielectric Losses in Dry-Type Insulation of Medium-Voltage Power Electronic Converters," *IEEE Journal of Emerging and Selected Topics in Power Electronics*, vol. 8, no. 3, pp. 2716-2732, Sept. 2020.

[14] B. Sonerud, T. Bengtsson, J. Blennow, and S. M. Gubanski, "Dielectric heating in insulating materials subjected to voltage waveforms with high harmonic content," *IEEE Trans. Dielectr. Electr. Insul.*, vol. 16, no. 4, pp. 926–933, Aug. 2009.

[15] R. Eichhorn, *Engineering Dielectrics: Electrical Properties of Solid Insulating Materials, Molecular Structure and Electrical Behavior*. Philadelphia, PA, USA: ASTM, 1983.

[16] Z. Wu, B. Lin, J. Fan, J. Zhao, Q. Zhang, and L. Li, "Effect of dielectric relaxation of epoxy resin on dielectric loss of medium-frequency transformer," *IEEE Trans. Dielectr. Electr. Insul.*, vol. 29, no. 5, pp. 1651–1658, Oct. 2022.

[17] W. Wang, J. He, Y. Liu, X. Wang and S. Li, "Effects of Spike Voltages Coupling With High dV/dt Square Wave on Dielectric Loss and Electric-Thermal Field of High-Frequency Transformer," *IEEE Access*, vol. 9, pp. 137733-137743, 2021.

[18] T. Guillod, R. Faerber, D. Rothmund, F. Krismer, C. M. Franck and J. W. Kolar, "Dielectric Losses in Dry-Type Insulation of Medium-Voltage Power Electronic Converters," *IEEE J. Emerg. Sel. Topics Power Electron.*, vol. 8, no. 3, pp. 2716-2732, Sept. 2020.

[19] T. -S. Li, M. Ngo, R. Burgos and D. Dong, "Modeling and Analysis of Voltage Overshoot in Bidirectional Phase-Shift Full Bridge Converters," *2024 IEEE Sixth International Conference on DC Microgrids (ICDCM)*, Columbia, SC, USA, 2024, pp. 1-7.

Improved High-Speed Thermal Analysis Based on Two-Step Simulation for High-Frequency Transformers

Zheyuan Yi
[1]State Key Laboratory of Power System
Department of Electrical Engineering
Tsinghua University
Beijing, China
[2]Delta Electronics Shanghai Co. Ltd.
Shanghai, China
zheyuan.yi@deltaww.com

Kai Sun
State Key Laboratory of Power System
Department of Electrical Engineering
Tsinghua University
Beijing, China
sun-kai@mail.tsinghua.edu.cn

Qiang Li
Center for Power Electronics Systems
The Bradley Department of Electrical and Computer Engineering
Virginia Polytechnic Institute and State University
Blacksburg, Virginia, USA
lqvt@vt.edu

Zengyang Liu
State Key Laboratory of Power System
Department of Electrical Engineering
Tsinghua University
Beijing, China
liuzy23@mails.tsinghua.edu.cn

Abstract—**Thermal analysis is a crucial and time-consuming step in the optimization design process of high-frequency transformers (HFTs). This paper proposes an improved thermal analysis method based on heat dissipation-conduction two-step simulation, significantly reducing the computation time while ensuring the accuracy of thermal analysis. Simulation results validate the accuracy and of the proposed method and demonstrate that the proposed method saves 82% of computation time for a single thermal analysis case compared to direct CFD. Experimental results further verify the accuracy of the proposed method. An exemplary application of the proposed method in the optimization design of the HFT illustrates a 98% reduction in computation time for thermal analysis during the optimization process.**

Keywords—high-frequency transformer, thermal analysis, surface heat transfer coefficient, CFD simulation, optimization

I. INTRODUCTION

High-frequency transformers (HFTs) are essential components in isolated power converters [1][2], widely used in various applications [3]–[5]. Given the significant effect of temperature on the losses, insulation performance, and reliability of HFTs, thermal analysis is a critical step in the optimization design process. However, the complex structure of HFTs makes thermal analysis particularly challenging and time-consuming, especially when large-scale optimization is involved. As a result, accurate, universal, and high-speed methods are needed for efficient thermal analysis of HFTs.

Currently, there are three primary types of thermal analysis methods used for HFTs, each with advantages and disadvantages, as summarized in Table I.

Thermal Resistance Network is the simplest and most commonly used method for thermal analysis of HFTs, which remains integral to the thermal modeling and design of HFTs nowadays [6]–[8]. However, compared to line-frequency transformers, HFTs have more complex structures and operating conditions, making thermal resistance network modeling more challenging for them. The selection of network nodes and the calculation of thermal resistances critically impact the accuracy of thermal resistance network modeling and are also the issues that researchers are most concerned about. Reference [9] specifically models the

This work was supported by the National Key Research and Development Program (2021YFB1507000，2021YFB1507002).

TABLE I
ADVANTAGES AND DISADVANTAGES OF EXISTING THERMAL ANALYSIS METHOD FOR HFTS

Existing Method Type	Advantages	Disadvantages
Thermal resistance network[6]—[11]	High calculation speed	Low universality, leading to insufficient accuracy in complex structures
CFD simulation[12]	High accuracy	Low calculation speed
Thermal steady-state simulation[13][14]	Balancing speed and accuracy	Requiring external settings of surface heat transfer conditions

common rectangular corner structure of the magnetic core in HFTs, accounting for high-frequency effect of surface thermal resistance to improve the accuracy of thermal resistance calculation. Reference [10] attempts to simplify and circumvent direct thermal resistance modeling by using artificial neural networks (ANN) to derive the relationship between design parameters (input variables) and temperature (output variables), and proposes corresponding methods to gather datasets and conserve computational resources. Reference [11] developed thermal network model without empirical or numerical adjustment coefficients in order to increase model accuracy, and use numerical methods to solve the temperatures. Generally, the thermal resistance network offers high calculation speed but lacks universality, leading to insufficient accuracy for complex structures such as HFTs.

With the rapid development of computer technology, numerical simulation has also been gradually widely used in thermal analysis of HFTs. The most commonly used type of numerical simulation is computational fluid dynamics (CFD) simulation, which accounts for all three types of heat transfer—conduction, convection, and radiation—making it a recognized high-precision thermal analysis method [12]. However, CFD simulation is hindered by low calculation speed, especially with complex structures, as single-point simulations typically take several tens of minutes to several hours. Another type of simulation is Finite Element Analysis (FEA)-based thermal steady-state simulation, which models only heat conduction, requiring the user to set parameters for heat exchange with the external environment. Although thermal steady-state simulation is faster than CFD, its accuracy is affected if heat dissipation parameters are not accurately set. Nonetheless, thermal steady-state simulation

979-8-3315-1612-3/25 $31.00 © 2025 IEEE

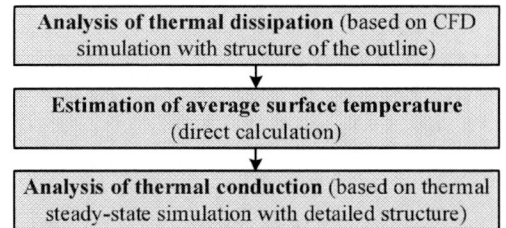

Fig. 1. Overall process of proposed thermal analysis method based on two-step simulation.

provides a balance between speed and accuracy, outperforming thermal resistance networks in simulating internal temperature distribution. Reference [13] and [14] provide ways to simplify specific HFT structure in FEA thermal simulation, intending to save computation resources consumed by solving internal temperature distribution, but it is still difficult to solve the fluid flow.

In summary, existing methods cannot fully meet all requirements for thermal analysis methods. Effective thermal design involves both heat dissipation and conduction, offering critical insights into thermal analysis. This paper proposes an improved thermal analysis method based on a heat dissipation-conduction two-step simulation process, significantly reducing computation time while maintaining accuracy of thermal analysis. Section II explains the principles of the proposed method. Simulation and experimental validation results are discussed in Sections III and IV, respectively. Section V presents an example application of the proposed method in the HFT optimization process, and Section VI concludes with final remarks.

II. HIGH-SPEED THERMAL ANALYSIS BASED ON TWO-STEP SIMULATION

Considering that the thermal design of HFTs primarily consists of two aspects—heat exchange with the external environment and internal heat conduction, this concept can be applied to the thermal analysis of HFTs, dividing the analysis into two parts: surface thermal dissipation analysis and internal heat conduction analysis. According to this idea and combining the benefits of the aforementioned CFD simulation and FEA-based thermal steady-state simulation, this paper proposes a thermal analysis method based on two-step simulation, delivering both speed and accuracy. The overall process is shown in Fig. 1, with specific implementation steps detailed below.

Step 1: Analysis of Thermal Dissipation

The first step, thermal dissipation analysis, utilizes CFD simulation to obtain the surface heat transfer coefficient, h, for the transformer. To enhance calculation speed, the CFD simulation model focuses only on the outline of the transformer, capturing its shape, size, and surface materials, and is known as the outline-structure model. This simulation model is used to analyze heat exchange between the transformer and the fluid environment.

For simplicity, the transformer surface is divided into regions based on heat exchange characteristics, calculating an average h for each region based on CFD simulation results as follows:

$$h_i = \frac{1}{A_i} \iint_{S_i} \frac{q}{T_s - T_0} \, dA, \tag{1}$$

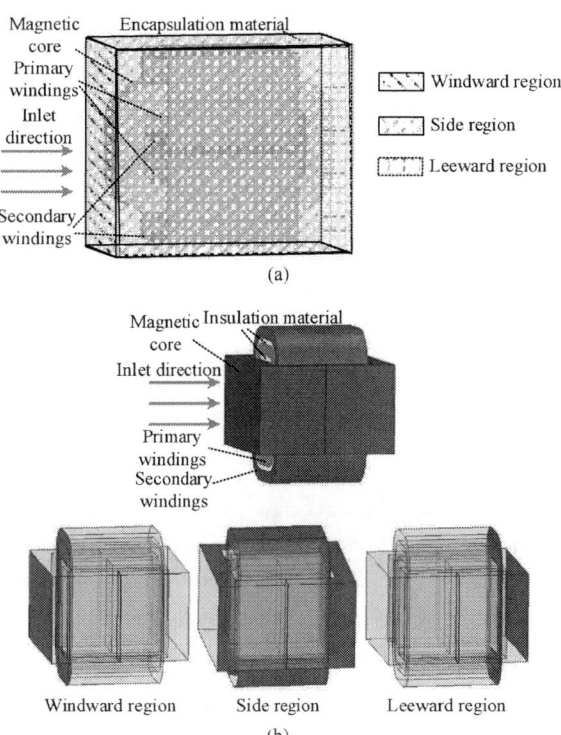

Fig. 2. Examples of dividing the surfaces of a transformer in thermal dissipation analysis. (a) Surface division of a cubic encapsulated transformer. (b) Surface division of an un-encapsulated transformer with EE-shaped magnetic cores.

where h_i is the average surface heat transfer coefficient of the i-th region (S_i), q is the heat flux rate per unit area, T_s is surface temperature, T_0 is ambient temperature, and A_i is the total area of region S_i.

In the most common forced-air cooling conditions, for instance, a simple method of dividing the transformer surfaces is to divide them based their relationship with air inlet direction. Fig. 2 shows how the surfaces of transformers with two typical structures are divided. Fig. 2 (a) shows a cubic transformer, which is the typical shape of encapsulated transformers, and Fig. 2 (b) shows an un-encapsulated transformer with EE-shaped magnetic cores. The surfaces of the two example transformers in Fig. 2 are both divided into a windward region perpendicular to the air inlet direction and facing the air inlet, a leeward region perpendicular to the air inlet direction and facing away from the air inlet, and a side region parallel to the air inlet direction. Beside the surface heat transfer coefficient of the windward, leeward, and side regions calculated by CFD simulation, the bottom surface region is in contact with the placement and dissipates heat to the environment through heat conduction, and the heat dissipation coefficient of this region is directly estimated by thermal conduction resistance analysis. This part of heat transfer usually only accounts for a small proportion, so the error caused by the estimation is very small. Another example is liquid cooling condition, where the transformer surfaces can be divided according to the flowing path and temperature of the fluid, which is not considered in detail in this paper.

To address random effects from meshing and residuals in CFD simulation, multiple repeated simulations can be conducted to reduce random error. For varying thermal analysis conditions, computing resources can be saved by setting concerned variables to some specific values within a range, using preliminary simulations to establish data that can be fitted and interpolated. For one thing, when analyzing an individual HFT with complex external structure features, environmental parameters like air inlet speed can be used as variables. For another, in large-scale optimization of HFTs with different size values, especially for transformers with a simple external structure like cubic encapsulated transformers shown in Fig. 2 (a), primary outline dimensions like length, width, and height can be variables. Whatever the selected variable is, it is only necessary to perform one-time preliminary simulations to fit and obtain the relationship between the surface heat transfer coefficients and the selected variables before the large-scale computation, and interpolate the preliminary simulation results across a large number of target conditions.

Step 2: Estimation of Average Surface Temperature

Before thermal conduction analysis, the average surface temperature, $T_{s\text{-avg}}$, is estimated as follows:

$$T_{s-\text{avg}} = T_0 + \frac{P_t}{\sum_i h_i A_i}, \qquad (2)$$

where P_t is the total loss of the transformer.

If the average surface temperature calculated according to (2) exceeds the maximum allowable temperature limit of the transformer, it is determined that the current design scheme does not meet the temperature rise limit and is eliminated early, reducing the number of required thermal steady-state simulations.

Step 3: Analysis of Thermal Conduction

This approach achieves faster computation by separating external and internal heat simulations.

Thermal conduction analysis, based on FEA-based thermal steady-state simulation, accurately calculates the internal temperature distribution to assess whether the temperature limit is met. In this step, it is necessary to analyze the detailed structure of the transformer, while it is no longer necessary to model the environment. Instead, the surface heat transfer coefficient is set for each region of the transformer surface according to the results from the first step. The maximum temperature of each material in the HFT are determined by steady-state simulation and compared against each material's allowable limits.

In the proposed thermal analysis process, compared to direct full CFD simulation, potential errors primarily stem from simplifications made in the thermal analysis steps. These error sources include: (1) discrepancies between the simplified temperature distribution in the outline-structure model used for thermal dissipation analysis and the actual distribution, which can impact the accuracy of surface heat transfer coefficients; and (2) inaccuracies due to the use of average surface heat transfer coefficients for each region instead of detailed surface heat transfer coefficient distributions. According to reference [15], when thermal radiation contributes minimally to overall heat transfer, that is, with cooling methods other than natural air cooling, the relationship between temperature rise and loss is approximately linear, meaning that the surface heat transfer coefficient is not significantly affected by surface temperature. This allows error source (1) to be largely ignored. The impact of simplification (2), however, depends on a reasonable

TABLE II
INFORMATION OF SIMULATION SETUP

Ambient temperature		300 K
Thermal conductivity	Core	4 W/(m·K)
	Winding	387.6 W/(m·K)
	Insulating material	0.2 W/(m·K)
Simulation model		Viscous: laminar Radiation: discrete ordinates
Number of iterations		500

division of surface areas; for it to be negligible, surface heat dissipation distribution should remain relatively uniform across each region, so its effect on the internal temperature rise of the transformer remains minimal. This will be further validated by simulation results presented in Section III.

When applying the above process to HFT thermal analysis and design, the results of three analysis steps provide insights for possible thermal design improvements. For instance, if the average surface temperature estimated in Step 2 is high and approaches or exceeds the maximum temperature limit, it is necessary to reduce losses or switch to a cooling method with a higher surface heat transfer coefficient. On the other hand, if the estimated surface temperature is well below the maximum temperature limit but thermal conduction analysis shows that the internal temperature exceeds the limit, enhancing internal heat conduction becomes important, which could involve using materials with better thermal conductivity or implementing specific structure design to channel internal heat outward.

III. SIMULATION VALIDATION

To validate the principles and effectiveness of the proposed thermal analysis method, an exemplary HFT model is used for simulation. The simulation model and dimensions of the HFT are provided in Fig. 3, and the simulation setup details are shown in Table II.

When using the proposed improved thermal analysis method to analyze the HFT model shown in Fig. 3, its surfaces are divided according to their relationship with the air inlet direction, as described in Section II. Given that the transformer structure in Fig. 3 includes a hollow channel to enhance heat dissipation, the surface of this channel exhibits unique heat transfer characteristics. Consequently, the hollow channel surface is classified separately as a "hollow region," resulting in a total of four surface regions for the HFT model, highlighted in red in Fig. 3. For the simulation, the surface heat transfer coefficient of the bottom region (in contact with the HFT's placement) is set at 11 W/(m²·K).

The CFD simulations are conducted using ANSYS Fluent software in this section. The average surface heat transfer coefficient for each HFT surface region is calculated based on (1), after obtaining the heat flow and temperature distribution via CFD simulation. Here, air inlet speed is chosen as the variable, allowing the relationship between surface heat transfer coefficient and air inlet speed to be simulated and fitted. This approach helps reduce random errors from meshing and residuals without increasing computational demand, as discussed in Section II.

To confirm the validity of using the outline-structure model instead of the detailed-structure model for the simulation in the first step of the proposed method (heat dissipation analysis), it is first assessed whether surface

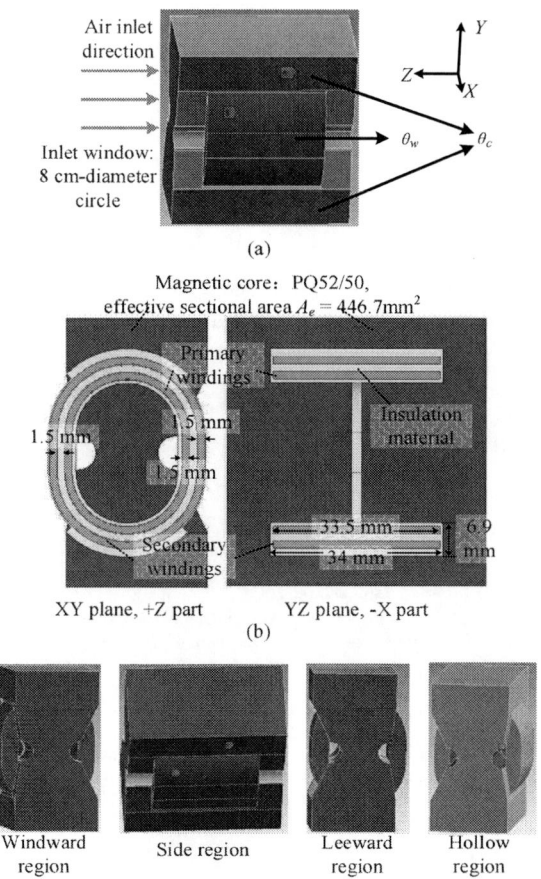

Fig. 3. Model of the HFT used in simulation validation. (a) Simulation model. (b) Sectional plane and dimensions. (c) Region division of transformer surfaces.

temperature affects the surface heat transfer coefficient. This check establishes that, despite differences in temperature distribution between the outline-structure and detailed-structure models, they can approximately share the same surface heat transfer coefficient. Fig. 4 (a) illustrates the simulated surface heat transfer coefficient across different total transformer losses (28W and 10W) applied to the detailed transformer model, which reflect different transformer surface temperatures. The CFD simulation results indicate that, across a wide range of air inlet speeds, the surface heat transfer coefficient remains almost unchanged, validating the above assumption.

Furthermore, a direct comparison of surface heat transfer coefficients between the outline-structure and detailed-structure models, derived from CFD simulation, shows close alignment, as seen in Fig. 4 (b). The maximum relative error between the two models is within ±10%, supporting the simplification of the detailed-structure model into the outline-structure model for thermal dissipation analysis.

The surface heat transfer coefficient derived from the outline-structure model simulation serves as the target output of the proposed method's first step. To streamline subsequent thermal analysis steps, the relationship between the surface heat transfer coefficient and air inlet speed is fitted, with the fitting curves and equations presented in Fig. 4 (c). A linear function is used to fit the surface heat transfer coefficients of the side, leeward, and hollow regions against air inlet speed,

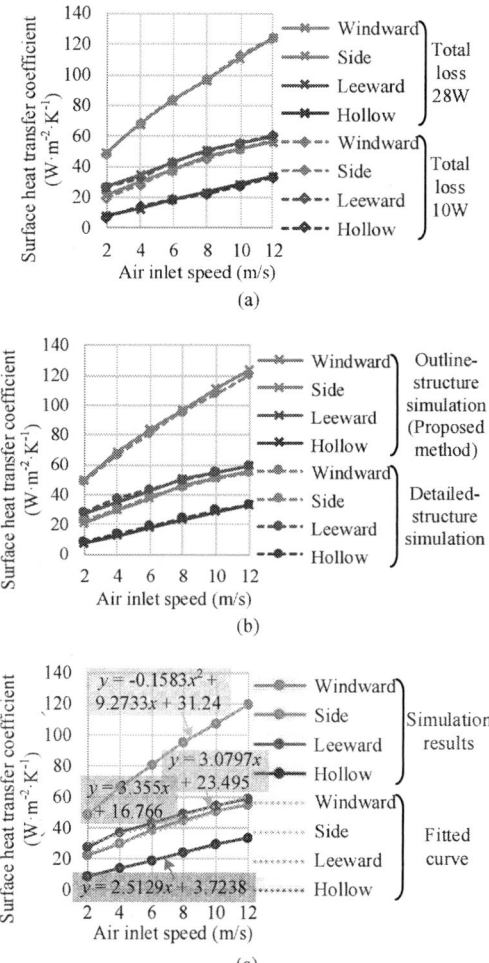

Fig. 4. CFD simulation results of surface heat transfer coefficient. (a) Comparison of results with different transformer total losses, representing different surface temperatures. (b) Comparison of results with outline-structure model and detailed-structure model. (c) Fitted curves and fitted equations calculated based on results with outline-structure model.

yielding R^2 values of 0.9867, 0.9833, and 0.9985, respectively. A quadratic function fits the windward region, achieving an R^2 value of 0.9997. These high R^2 values confirm an excellent fit. In later thermal analysis steps, these fitted equations can replace the interpolation of discrete surface heat dissipation coefficient values, enhancing the efficiency of temperature estimation or simulation.

The second step in the proposed thermal analysis method involves estimating the average surface temperature. The surface areas for each transformer region—windward, side, leeward, and hollow regions—are calculated as 15.19 cm², 72.72 cm², 15.19 cm², and 16.53 cm², respectively. Using these surface areas and the heat transfer coefficients obtained in the first step, the average surface temperature rise of the transformer, denoted as $\theta_{\text{surf-avg}}$, is estimated as the difference between the results of (2) and the ambient temperature. The results, depicted as a green dashed line in Fig. 5 (a), show the relationship between $\theta_{\text{surf-avg}}$ and varying air inlet speeds with a 10 W total loss.

The third step, thermal conduction analysis, is performed using thermal steady-state simulation through ANSYS Mechanical software in this section. For simplicity while

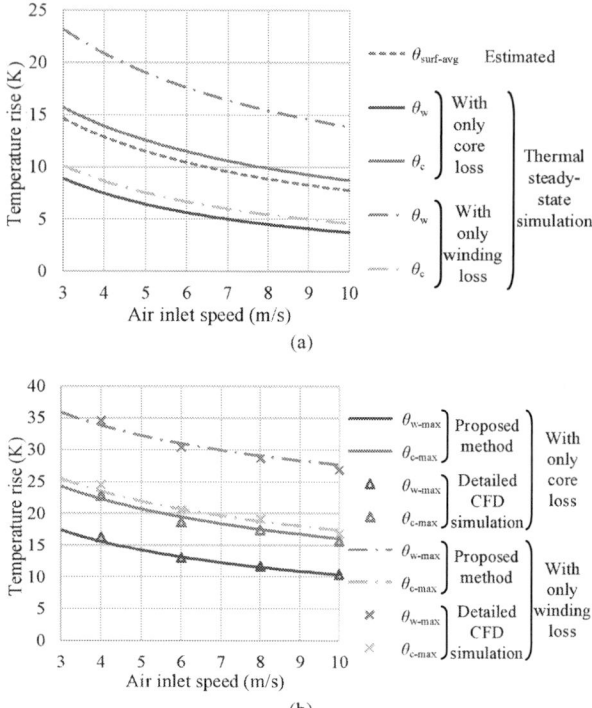

(a)

(b)

Fig. 5. Temperature rises of the modeled HFT obtained in different ways. (a) Surface temperature rises estimated by (2) or simulated by the proposed method based on thermal steady-state simulation. (b) Maximum temperature rises simulated with CFD simulation based on detailed-structure model or with the proposed method.

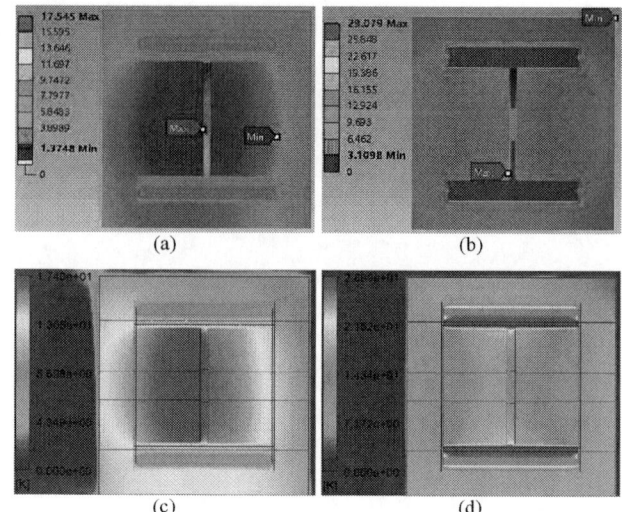

(a)   (b)

(c)   (d)

Fig. 6. Simulated cross-sectional temperature rise distributions. (a) With the proposed method (steady-state simulation), under open-circuit condition. (b) With the proposed method (steady-state simulation), under short-circuit condition. (c) With CFD simulation on the detailed-structure model, under open-circuit condition. (d) With CFD simulation on the detailed-structure model, under short-circuit condition.

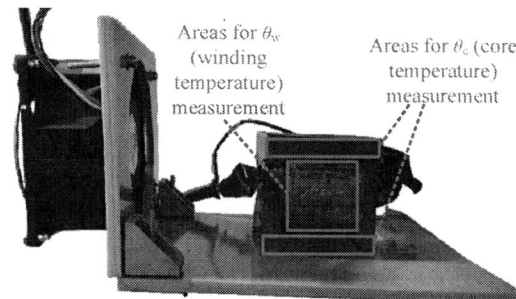

Fig. 7. Photos of the tested HFT and testing platform.

covering extreme cases, simulations are conducted with only core loss or only winding loss, assigning a 10 W loss under each condition. This analysis also considers various air inlet speeds. The average temperature rises of the representative surface regions of the core and winding, labeled θ_c and θ_w (shown in Fig. 3 (a)), are obtained from thermal steady-state simulation results, and are compared to $\theta_{surf\text{-}avg}$ in Fig. 5 (a). The results indicate that $\theta_{surf\text{-}avg}$ falls between θ_c and θ_w, validating $\theta_{surf\text{-}avg}$ estimated in Step 2 as a useful reference temperature.

In addition to the proposed method, the temperature rises of the HFT are also simulated using CFD simulations on the detailed-structure model. The maximum temperature rises in the winding ($\theta_{w\text{-}max}$) and the core ($\theta_{c\text{-}max}$) are calculated, as shown in Fig. 5 (b), with the cross-sectional temperature rise distribution illustrated in Fig. 6. The results demonstrate that both the proposed method and the detailed CFD simulation produce similar temperature rise distributions and maximum temperatures, confirming the high accuracy of the proposed method.

Using the same computer and comparable meshing settings, the average time for a single-point CFD simulation on the detailed-structure model is around 1 hour, whereas it takes only about 10 minutes for the CFD simulation on the outline-structure model and just 1 minute for thermal steady-state simulation on the detailed-structure model. Therefore, the proposed improved thermal analysis method reduces computation time by approximately 82%, from around 60 minutes to 11 minutes, compared to direct CFD simulation.

IV. EXPERIMENTAL VALIDATION

The accuracy of the proposed thermal analysis method is further verified through experimental testing. Photographs of the tested HFT and the testing platform are shown in Fig. 7, with specifications provided in Table III.

Similar to the simulation setup, experiments are conducted under short-circuit and open-circuit conditions, corresponding approximately only winding losses and only core losses, respectively. For simplicity, only the surface temperature rises are measured in the experiment using a thermal camera, which provides the average temperature rises for the representative surface regions of the core (θ_c) and the winding (θ_w) (see Fig. 7). The difference between θ_c and θ_w reflects the thermal conduction characteristics within the HFT.

Two sample thermal images, illustrating the tests under short-circuit and open-circuit conditions, are shown in Fig. 8. During the experiment, it is challenging to adjust the loss to an exact value, so the loss is controlled at approximately 10W. As indicated by the CFD simulation results in Section III, transformer loss has minimal impact on surface heat dissipation coefficients, so for varying transformer loss levels, temperature rises should be proportional to the loss. Based on this hypothesis, the calculated and experimental results are processed as unified temperature rises—specifically, the ratios of temperature rise to total loss. The results, presented

TABLE III
INFORMATION OF EXPERIMENTAL SETUP

Transformer specification	
Core material	HE7G Mn-Zn ferrite from HEC company
Winding specification	Primary: 0.07 mm*1450 litz wire, 7 turns
	Secondary: 0.07 mm*1100 litz wire, 9 turns
Fan specification	
Model	Delta FFB0812SHE-R00
Size	80 mm length * 80 mm height * 38 mm width
Rated air flow	293.25 CFM

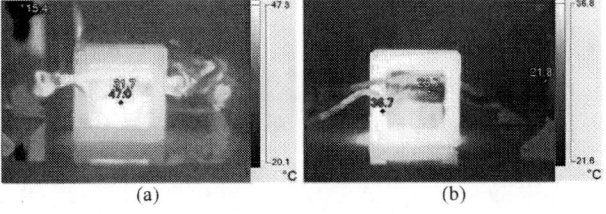

Fig. 8. Sample thermal images in experimental validation. (a) Short-circuit condition with 7.8m/s air inlet speed and 10.86W total loss. (b) Open-circuit condition with 7.9m/s air inlet speed and 10.90W total loss.

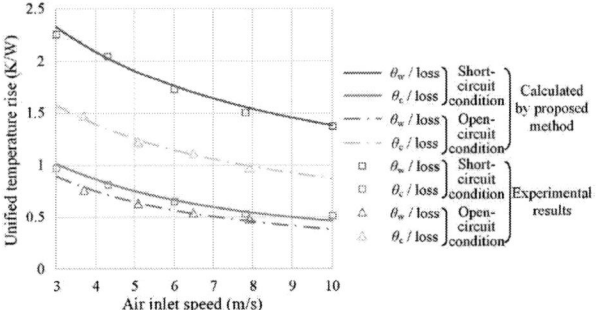

Fig. 9. Comparison of unified average surface temperature rises of the HFT calculated from the proposed method and the measured values.

in Fig. 9, show a maximum relative error of 9.4% between the unified temperature rises calculated from the proposed method and the measured values under both open- and short-circuit conditions, validating the accuracy of the proposed method.

V. APPLICATION IN OPTIMIZATION DESIGN

During the optimization process for HFTs, evaluating the performance of numerous design schemes is typically necessary, regardless of the optimization algorithm used. To accurately assess the temperatures for each design scheme, a CFD simulation on a detailed-structure model would be ideal; however, this approach is extremely time-consuming. Consequently, the proposed thermal analysis method offers a more efficient alternative, saving computational resources during HFT optimization.

This section provides an example demonstrating the time-saving effect of the proposed thermal analysis method in a HFT optimization scenario. The optimized HFT is encapsulated in the structure shown in Fig. 2 (a), with design specifications and parameters outlined in Table IV and Fig. 10, respectively.

In Fig. 10, the parameters highlighted in red are variable. However, instead of using these parameters directly, five independent variables with clear physical significance are selected for optimization: operating frequency f, rated current

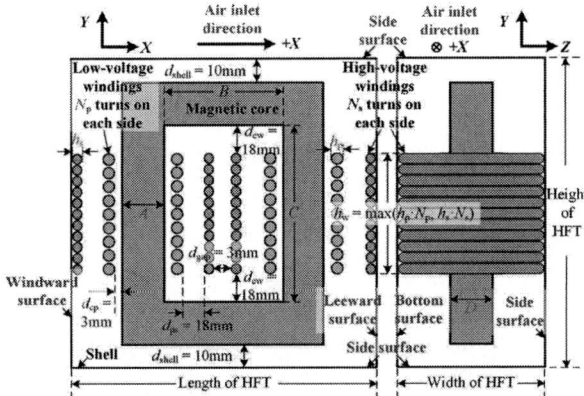

Fig. 10. Parameters and dimensions of exemplary design case of HFT.

Pre-simulation (obtain data for interpolation)

Analysis of thermal dissipation (based on CFD simulation on outline-structure model, with limited dimension values)

↓

Optimization and iteration (evaluate different schemes)

Estimation of average surface temperature; analysis of thermal conduction (based on thermal steady-state simulation on detailed-structure model)

Fig. 11. Procedure of applying proposed thermal analysis method in optimization design of the HFT.

TABLE V
LIMITS AND OPTIMIZED VALUES OF INDEPENDENT VARIABLES IN THE OPTIMIZATION DESIGN CASE OF HFT

Variable	Lower limit	Uppeer limit	Optimized value
f	30 kHz	80 kHz	37.7 kHz
J_w	1 A/mm²	6 A/mm²	3.14 A/mm²
P_{cv}	20 kW/m³	90 kW/m³	61.8 kW/m³
N_s	5	12	8
r_{rect}	0.2	5	0.2

density J_w in the litz wire, core loss density P_{cv}, number of turns in the secondary winding on each side N_s, and cross-sectional width-to-thickness ratio of the magnetic core $r_{rect} = A / D$. The upper and lower limits for these variables are provided in Table V. The HFT is optimized in tandem with the converter, with total converter efficiency considered. The objective of the optimization is to minimize HFT volume while meeting the requirements of at least 97% converter efficiency and the maximum HFT temperature specified in Table IV.

The genetic algorithm [16] is employed to optimize the HFT, evaluating a total of 8,020 design schemes to achieve a converged optimal design scheme, as detailed in Table V. Using the conventional approach of CFD simulation on a detailed-structure model for thermal analysis, each of these 8,020 schemes would need to be simulated, with each simulation taking approximately 1 hour. Consequently, the total computation time for thermal analysis would be around 8,020 hours with conventional approach.

With the proposed thermal analysis method, however, thermal dissipation analysis can be conducted only once in a pre-simulation phase using a limited set of dimension values before the optimization process begins, as shown in Fig. 11. Since the HFT in this case is encapsulated and has a relatively simple structure, pre-simulation can be limited to the primary dimension variables: length, width, and height of the HFT, with five values chosen for each to maintain enough accuracy. Therefore, the pre-simulation requires only 125 CFD simulations on the outline-structure model, each taking approximately 10 minutes, totaling around 1,250 minutes for this step. During the optimization process, the surface heat transfer coefficients are then interpolated from these pre-simulation results, and only average surface temperature estimation and thermal conduction analysis via thermal steady-state simulation are performed for each design scheme, taking about 1 minute per scheme and totaling 8020 minutes. This results in a total computation time of approximately 155 hours with the proposed thermal analysis method—98% less than the 8,020 hours required by direct CFD simulation on a detailed-structure model.

Additionally, the proposed method estimates average surface temperature before conducting thermal conduction analysis on each scheme, allowing for the early elimination of schemes that obviously do not meet the temperature rise limit, reducing the number of thermal steady-state simulations. In this case, 277 schemes are eliminated at this preliminary stage, saving their thermal steady-state simulations, which accounted for 3.5% of the total schemes.

VI. CONCLUSION

This paper presents an improved thermal analysis method for HFT based on a heat dissipation-conduction two-step simulation. In the first step, surface heat transfer coefficients are derived using CFD simulations on an outline-structure model to provide input for subsequent steps. The second step involves estimating the average surface temperature through direct calculation, allowing for the preliminary elimination of solutions that obviously exceed temperature rise limits. The final step uses thermal steady-state simulations on a detailed-structure model to accurately determine the temperature rise distribution inside the HFT. Compared to CFD simulations on a detailed-structure model, the proposed thermal analysis method maintains high accuracy while significantly enhancing computational speed. Simulation results validate the accuracy of this method, demonstrating a reduction of 82% in computation time for thermal analysis of a single case compared to direct CFD simulation. Experimental results further confirm the method's accuracy. Additionally, an example of applying the proposed method to HFT optimization design shows a 98% reduction in computation time for thermal analysis within the optimization process.

ACKNOWLEDGMENT

The authors would like to acknowledge the supports by the National Key Research and Development Program (2021YFB1507000，2021YFB1507002).

REFERENCES

[1] C. Zhao et al., "Power Electronic Traction Transformer — Medium Voltage Prototype," IEEE Transactions on Industrial Electronics, vol. 61, no. 7, pp. 3257-3268, July 2014.

[2] M. Leibl, G. Ortiz and J. W. Kolar, "Design and Experimental Analysis of a Medium-Frequency Transformer for Solid-State Transformer Applications," in IEEE Journal of Emerging and Selected Topics in Power Electronics, vol. 5, no. 1, pp. 110-123, March 2017.

[3] S. Inoue, H. Akagi, "A Bidirectional DC-DC Converter for an Energy Storage System with Galvanic Isolation", IEEE Transactions on Power Electronics, vol. 22, no. 6, pp. 2299-2306, Nov. 2007.

[4] B. Whitaker et al., "A high-density high-efficiency isolated on-board vehicle battery charger utilizing silicon carbide power devices", IEEE Trans. Power Electron., vol. 29, no. 5, pp. 2606-2617, May 2014.

[5] G. AlLee and W. Tschudi, "Edison redux: 380 VDC brings reliability and efficiency to sustainable data centers," IEEE Power Energy Mag., vol. 10, no. 6, pp. 50–59, Nov. 2012.

[6] M. Rascon, J. Ara, R. Madsen, J. Navas, M. Perez, and F. San Miguel, "Thermal analysis and modelling of planar magnetic components," in APEC 2001. Sixteenth Annual IEEE Applied Power Electronics Conference and Exposition (Cat. No.01CH37181), in 3*, vol. 1. Mar. 2001, pp. 97–101 vol.1.

[7] Z. Shen, Y. Shen, and H. Wang, "Thermal Modelling of Planar Transformers Considering Internal Power Loss Distribution," in 2019 IEEE 4th International Future Energy Electronics Conference (IFEEC), in 3*. Nov. 2019, pp. 1–5.

[8] Y. Gao et al., "Modeling and Design of High-Power, High-Current-Ripple Planar Inductors," IEEE Transactions on Power Electronics, vol. 37, no. 5, pp. 5816–5832, May 2022.

[9] R. Shafaei, M. Ordonez, and M. A. Saket, "Three-Dimensional Frequency-Dependent Thermal Model for Planar Transformers in LLC Resonant Converters," IEEE Transactions on Power Electronics, vol. 34, no. 5, pp. 4641–4655, May 2019.

[10] D. Santamargarita, G. Salinas, M. Vasic, and E. Bueno, "Trade-Off Between Accuracy and Computational Time for Magnetics Thermal Model Based on Artificial Neural Networks (ANN)," in 2021 IEEE Design Methodologies Conference (DMC), in 4*. Jul. 2021, pp. 1–6.

[11] L. A. Militão, B. Bertoldi, A. G. L. Furlan, M. L. Heldwein, and J. Riso Barbosa, "Thermal Network Models for Toroidal and E-Core Inductors in Still Air," IEEE Transactions on Power Electronics, vol. 38, no. 12, pp. 15879–15892, Dec. 2023.

[12] M. Ngo, Y. Cao, K. Nguyen, D. Dong, and R. Burgos, "Computational Fluid Dynamic Analysis and Design of an Air Duct Cooling System for 18 kW, 500 kHz Planar Transformers," in 2021 IEEE Applied Power Electronics Conference and Exposition (APEC), in 4*. Jun. 2021, pp. 1496–1504.

[13] G. Salinas, A. Delgado, J. A. Oliver, and R. Prieto, "Fast FEA Thermal Simulation of Magnetic Components by Winding Equivalent Layers," in 2018 IEEE Energy Conversion Congress and Exposition (ECCE), Sep. 2018, pp. 7380–7385.

[14] G. Salinas López, A. D. Expósito, J. Muñoz-Antón, J. Á. O. Ramírez, and R. P. López, "Fast and Accurate Thermal Modeling of Magnetic Components by FEA-Based Homogenization," IEEE Transactions on Power Electronics, vol. 35, no. 2, pp. 1830–1844, Feb. 2020.

[15] Z. Yi, Z. Liu, K. Sun, and B. Su, "A Simple Power Loss Evaluation Method for High-Frequency Transformers Based on Surface Temperature Measurement Within Wide Operation Range," in 2024 IEEE Applied Power Electronics Conference and Exposition (APEC), Feb. 2024, pp. 402–409.

[16] K. Deb and S. Tiwari, "Omni-optimizer: A generic evolutionary algorithm for single and multi-objective optimization," European Journal of Operational Research, vol. 185, no. 3, pp. 1062–1087, Mar. 2008.

Core Material Characterization under DC Bias Conditions

Jonas Mühlethaler[*Ω], Fabrice Locher[*], Frédéric Mathieu[*], and Edward Herbert[†]

[*]*Lucerne University of Applied Sciences and Arts (HSLU), Lucerne, Switzerland*
[†]*Power Sources Manufacturers Association (PSMA), Mendham NJ, United States*
[Ω]*Frenetic Switzerland GmbH, Zürich, Switzerland*

Abstract— **This paper investigates the impact of a DC premagnetization on permittivity, conductivity, permeability, as well as core losses across various frequencies in the ferrite N87 (TDK). The aim is to better understand the reasons behind the increase in loss due to a DC premagnetization and to explore how dimensional effects, such as bulk eddy currents or dimensional resonance, are influenced by a DC magnetic field. Experimental results show that the application of a DC bias primarily influences magnetic core losses through modifying the material's complex permeability, while exhibiting no significant effect on its permittivity and conductivity. To account for the observed increase in core losses under DC bias conditions, we introduce a small-signal DC Bias Loss Factor, the "*DC Permeability Loss Estimator*" (DPLE). This factor can be extracted from simple measurements and can help predict core losses under DC Bias condition.**

Keywords — Core Losses, DC Bias, Magnetic Components, Permeability. Permittivity, and Conductivity.

I. INTRODUCTION

The analysis of core losses is crucial for designing optimal power electronics converters. However, this analysis is complex and will become even more challenging as switching frequencies increase. At higher frequencies, new effects such as bulk eddy current losses and dimensional resonance will play significant roles [1]. To better understand core losses and prepare for the future of power electronics, a deeper physical understanding of these losses is necessary. Currently, there are still challenges about core losses, such as a description of DC Bias losses [2], or the modeling of effects like dimensional resonance and bulk eddy currents [3], which are often not included in common core loss models. To address these gaps, recent research has focused on physical modeling to better understand core losses using basic physical parameters like permeability, permittivity, and conductivity [4][5].

This paper aims to measure and discuss how material properties such as permittivity, permeability, and conductivity are affected by a DC magnetic field (the "DC Bias"). This is done on the material N87 (TDK) to investigate the following two research questions: **RQ1:** Is there an alternative approach to describe the loss increase in the case of DC premagnetization that is not requiring extensive measurements (as e.g. was the case in [2]). **RQ2:** How are dimensional effects such as bulk eddy currents or dimensional resonance affected by a DC magnetic field? The research project's objective is to fill the gap in understanding the impact of a DC field on core material properties, ultimately contributing to a better understanding on core losses, hence, enabling a more efficient power electronics design.

A test setup for characterizing core materials has been built, as will be described in Section II. Section III presents the initial results and interesting observations. Section IV provides the derivation of a new Core Loss Estimation Model: we propose an approach to estimate the increase in core losses resulting from DC premagnetization, based on a simple measurement principle.

II. THE TEST SETUP

Permittivity, conductivity, and permeability under the influence of a DC magnetic flux (i.e. DC premagnetization) at different frequencies is an area that has not yet been thoroughly explored. The conduction of measurements of these parameters would provide valuable insights into the origin of core losses. For this work, specimens are either prepared to measure permittivity and conductivity, or permeability. All measurements were conducted at room temperature (at around 25°C). The measurement principles are given in the following Subsections.

A. Permittivity and Conductivity

A setup to measure permittivity and conductivity under DC Bias has been developed, as shown in Fig. 1 and Fig. 2. The test setup consists of a magnetic fixture with a coil around one leg, allowing the injection of a DC current to create a DC magnetic field in the air gap. The specimen, or material under test, is placed in this air gap. The fixture has a thermal sensor between bobbin and winding to prevent overheating. A gaussmeter is placed in the air gap of the fixture in order to measure the magnetic flux, i.e. the DC bias premagnetization. The equipment used is given in TABLE I.

TABLE I   MEASUREMENT EQUIPMENT

Equipment	Model
Impedance Analyzer	Bode 100 from Omicron (Vector Network Analyzer configured as an Impedance Analyzer)
Voltage Source	DC Power Supply NGSM 32/10
Gauss meter	Oumefar Gauss Meter, cross-checked with Model GM-2 by AlphaLab Inc.
Thermal Sensor	PT100 / Multimeter Metra Hit 18S

The specimens are built out of toroidal cores with a central hole, designed to limit the dimensional scale at which resonance can occur. Fig. 3 shows the equivalent circuit of the plate capacitor-shaped specimen with a central hole that is used to measure permittivity and conductivity. The greyish material between the two plates in Fig. 3(a) is the ferrite under test. The relative permittivity ε_r of the material under test is represented as a complex number as follows:

$$\varepsilon_r = \varepsilon'_r - j\varepsilon''_r = \varepsilon'_r - j\frac{\sigma}{\omega\varepsilon_0} \tag{1}$$

In the above formula, ε_0 is the vacuum permittivity, σ the conductivity of the material, ω the angular frequency,

Fig. 1 - Photo of the Test Setup

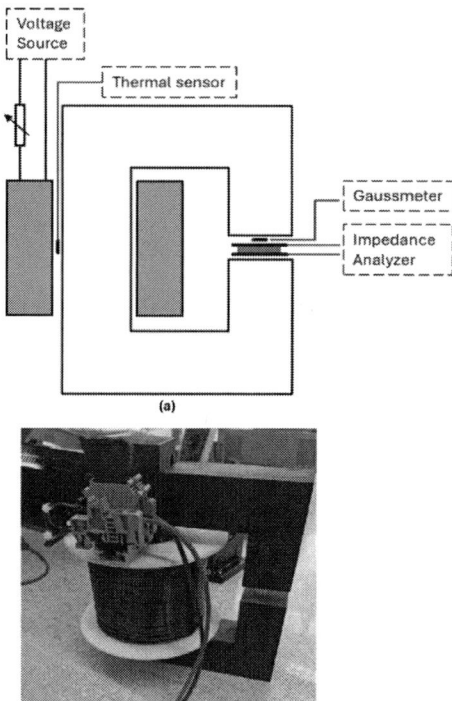

Fig. 2 – (a) Schematic of Test Setup to measure permittivity and conductivity under DC Bias condition, (b) Foto of fixture

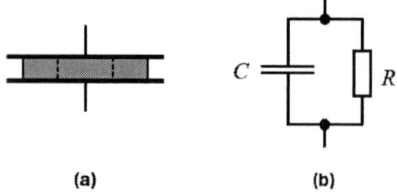

Fig. 3 - (a) Illustration of the built plate capacitor with the ferrite under test in between, (b) the conisdered equivalent circuit for calculating conductivity and permittivity.

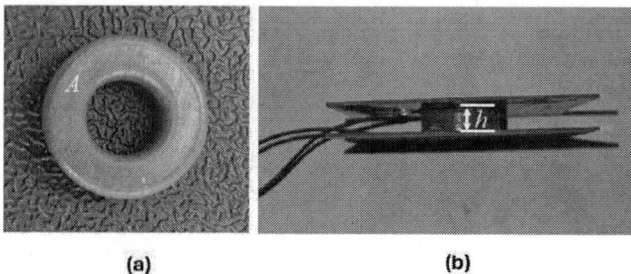

Fig. 4 – (a) Photo of specimen, with cross-sectional area A illustrated. (b) Built specimen, i.e. a plate capacitor with ferrite as dielectric with dimensions height h illustrated.

The impedance of a plate capacitor is given as

$$Z(\omega) = \frac{1}{j\omega\varepsilon_0\varepsilon_r\frac{A}{h}} \tag{2}$$

where A is the cross-sectional area of the specimen, and h is the height of the specimen as illustrated in Fig. 4. Hence, by measuring the impedance, the complex permeability becomes

$$\varepsilon_r = \varepsilon'_r - j\varepsilon''_r = \frac{1}{j\omega\varepsilon_0 Z\frac{A}{h}} \tag{3}$$

The formula (1) and (3) lead to the parameters to be measured as

$$\varepsilon'_r = \mathrm{Re}\left(\frac{1}{j\omega\varepsilon_0 Z\frac{A}{h}}\right) \tag{4}$$

$$\sigma = -1 \cdot \mathrm{Im}\left(\frac{1}{j\omega\varepsilon_0 Z\frac{A}{h}}\right)\omega\varepsilon_0 \tag{5}$$

The specimen's inductivity has been neglected as it is not dominant in the frequency range considered. The same measurements principles have been used e.g. in [6].

B. Permeability

A toroidal specimen with magnetic length l_e and magnetic cross-sectional area A_e is used for permeability measurements. The complex permeability can be calculated from impedance measurements as [3]

$$\mu_r = \mu'_r - j\mu''_r = \frac{L_s l_e}{\mu_0 N^2 A_e} - j\frac{R_s l_e}{\mu_0 \omega N^2 A_e} \tag{6}$$

The series parameters of the impedance R_s and L_s are directly given by the Bode 100. For these measurements, the DC bias is introduced through an additional wire carrying a DC current. To prevent voltage induction in this DC auxiliary winding, an approach as illustrated in Fig. 5(a) was implemented. By reversing the winding direction in the second toroid, a net zero flux through the DC auxiliary winding is ensured. Further details on this approach can be found in [8][9]. It is important to note that when calculating the permeability using Equation (6), the parameters (i.e., effective cross-sectional area (A_e), effective magnetic path length (l_e), inductance (L_s), and number of turns (N)) should be based on only one core as depicted in Figure 5. This means one should use half of the total measured inductance L_s, R_s and HF turns (N), along with the A_e and l_e values corresponding to a single core

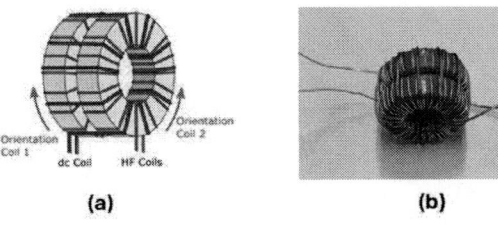

(a)   **(b)**

Fig. 5 – (a) Illustration of measurement setup for permeability measurement (figure taken from [8]) (b) Photo of two toroids to measure permeability.

Before the actual measurements, numerous tests have been conducted to validate the setup and ensure accurate results. In the following the most important aspects to be considered are discussed.

C. Specimen preparation

The specimens for the permittivity and conductivity measurements are built out of toroids with a central hole. Fig. 4 shows how the upper surface has been sanded flat to create a smooth surface. However, the ferrite core and PCB copper surfaces are not perfectly smooth. If a small air gap forms between the PCB and the core, it can create a small capacitor, which may dominate the overall capacitance of the setup. One solution is to silver or gold plate the ferrite material [6], however, this approach is costly as it requires special equipment. Tightly compressing the two capacitor plates to minimize air gaps reduces air gaps as well. However, applying mechanical pressure can alter the permittivity [7], an effect that must be kept at a minimum. A cost effective solution is to use conductive epoxy resin to fill the air gaps between PCB and core. We have used CW2460, a silver-based conductive epoxy resin. This method produced stable results across core samples of varying sizes, leading to high confidence in the measurement outcomes. Please note, after assembling a sample, it is necessary to allow enough time for the epoxy to reach optimal conductivity and adhesion. In our case, we waited approximately 96 hours.

D. Dimensional Resonance:

The electric and magnetic effects are coupled in ferrites due to their very high permittivity and high permeability [10][11][12]. To ensure accurate measurement of the material parameters, it is crucial for the field distribution within the samples to be homogeneous. This requires the wavelength and the skin depth to be significantly larger than the smallest specimen dimension perpendicular to the magnetic and electric field [6][10]. At half the wavelength $\lambda/2$, a standing wave is created in the sample, which distorts the results.

According to [10], the skin depth δ and wavelength λ can be calculated as follows:

$$\delta = \frac{1}{\alpha} \quad (7)$$

and

$$\lambda = \frac{2\pi}{\beta} \quad (8)$$

where α and β are defined as

$$\beta - j\alpha = 2\pi f \sqrt{(\mu' - j\mu'')(\epsilon' - j\epsilon'')} \quad (9)$$

Fig. 8 shows the half wavelength $\lambda/2$ as well as the skin depth as a function of frequency, calculated based on Equation (7), (8), and (9). The underlying values for permittivity were taken from [6], and for permeability from the data sheet N87 [13]. The specimens built have smaller dimensions than critical length according to Fig. 6, hence, the measurements should not be affected by dimensional resonance and skin effect up to a frequency of ~1MHz.

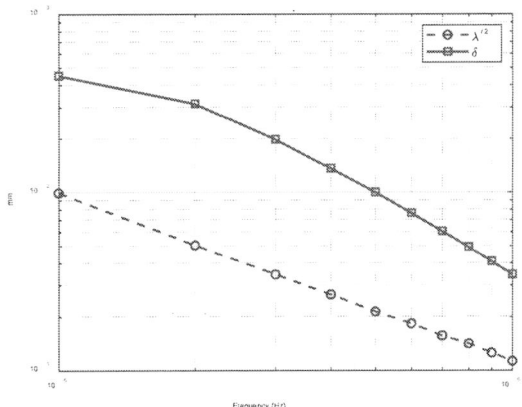

Fig. 6 – Half of wavelength $\lambda/2$ and skin depth δ as a function of frequency. Underlying values for permittivity taken from [6], and for permeability from the data sheet N87 [13].

TABLE II   SPECIMEN FOR PERMITTIVITY AND CONDUCTIVITY

ID	Material	Height [mm]	Outer diam. [mm]	Inner diam. [mm]
PIT1	N87	3.85	10.8	5.25
PIT2	N87	6.7	21.2	8.95

III. RESULTS

Measurements of permittivity, conductivity, and permeability have been conducted and results will be discussed in this Section.

A. Permittivity and Conductivity

Measurements for permittivity and conductivity were conducted on toroid-shaped specimens, as previously described and illustrated. TABLE II presents the parameters of the specimens. Fig. 7 shows the measured permittivity, and

Fig. 8 the conductivity. When we examine the range below 1 MHz, we observe that a DC bias has hardly any influence on the permittivity and conductivity. This is an important conclusion of this work, as it shows that dimensional resonance and skin depth are influenced by a pre-magnetization only through changes in permeability (and not permittivity and conductivity). The region above 1 MHz is shaded gray because it is a range in which MnZn ferrites are seldom used; for example, N87 has a maximum frequency of approximately 1 MHz before it loses its magnetic properties [13]. Furthermore, depending on the specimen's size, dimensional resonance can occur. In the region above that, the results are correspondingly not meaningful to extract permittivity and conductivity (for example, the local conductivity in the material will hardly reach such a high peak as shown Fig. 8, but instead a standing wave / dimensional resonance builds up, which effectively short-circuits the core; see previous discussion). Nevertheless, this region is depicted here because the dimensional resonance is nicely visible. The small core PIT1 exhibits resonance above 10 MHz, while the core PIT2 exhibits resonances at lower frequencies. A comparison of the specimen dimensions with Fig. 6 shows that this is plausible.

One can see how the resonance in PIT2 shifts to higher frequencies with increasing DC bias. This is expected, as the permeability decreases with increasing DC bias. An interesting observation is that in PT1, even below the resonance frequency, there is a variation in permittivity with changes in DC pre-magnetization. This is evident in Fig. 9. The underlying cause of this change in permittivity is currently unclear to the authors and requires further study. To conclude this subsection, a few things are to be noted: the results differ between PIT1 and PIT2, and they also differ from the results in [6]. However, it should be noted that all results are of the same order of magnitude. Differences may arise from varying pressure (see [7]), the use of epoxy resin instead of gold plating, or differing success in uniformly applying the epoxy, which is more challenging with larger cores like PIT2. Nevertheless, the results showed that a DC pre-magnetization does hardly affect permittivity and conductivity, something that, to the authors' knowledge, has not been investigated before.

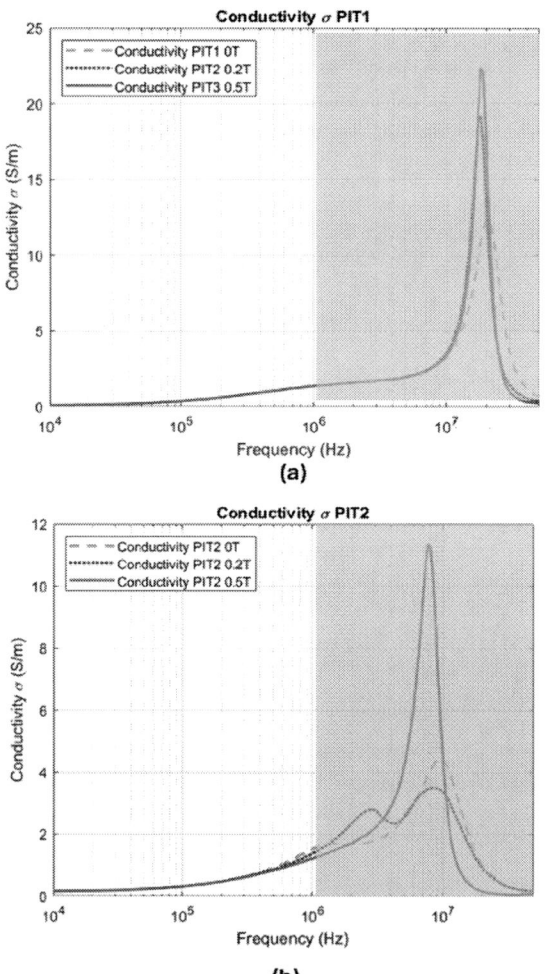

Fig. 8 - Measured Conductivity of N87 with and without DC Bias on all specimens. (a) results of PIT1, (b) results of PIT2.

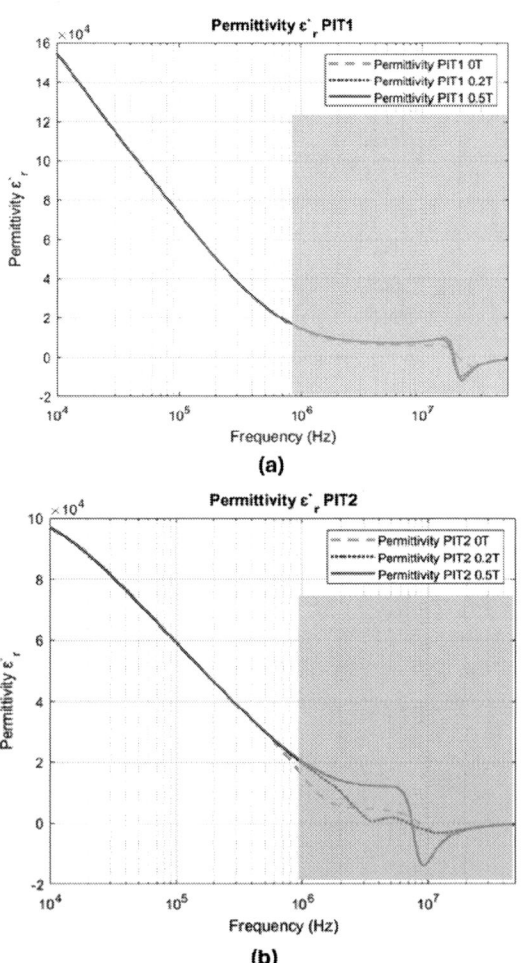

Fig. 7 - Measured Permittivity of N87 with and without DC Bias on all specimens. (a) results of PIT1, (b) results of PIT2.

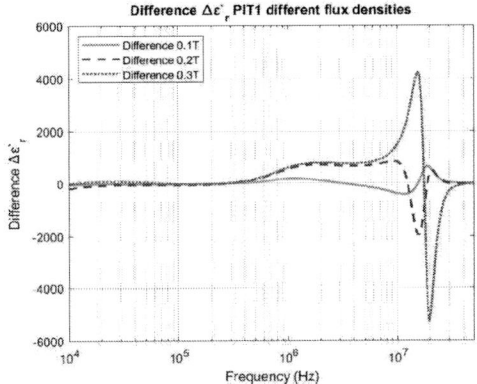

Fig. 9 – Difference in Permittivity of N87 under DC Bias Condition.

TABLE III   SPECIMEN FOR PERMEABILITY MEASUREMENTS

ID	No. of turns HF	No. of turns LF	Material	A_e [mm²]	l_e [mm]
PAD1	40 per core	16	N87	33.63	43.55

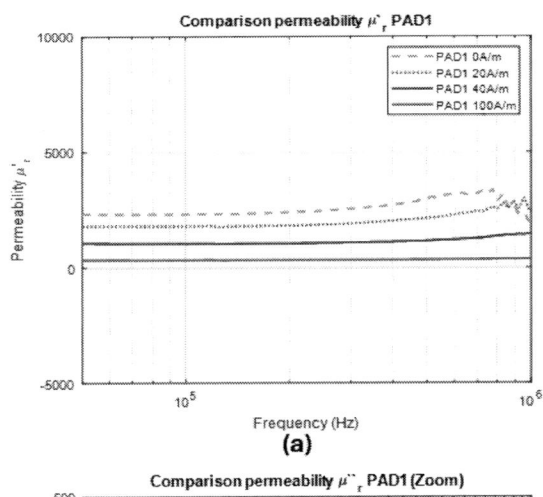

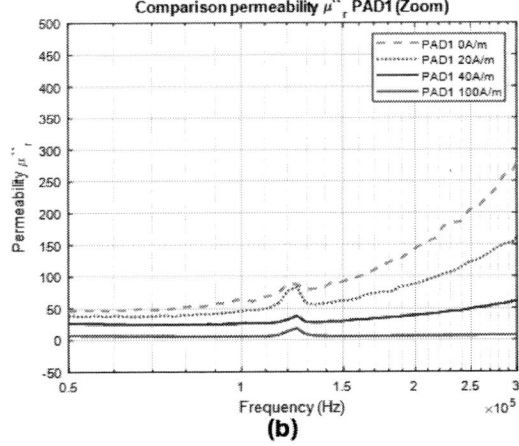

Fig. 10 - Measured Permeability of N87 with and without DC Bias. (a) real permeability, (b) imaginary permeability.

B. Permeability

Measurements for permeability were conducted on one toroid-shaped specimen, as previously described and illustrated. TABLE III presents the parameters of the specimen. Fig. 10 shows the measured permeability as function of frequency, and Fig. 11 the permeability as function of magnetic field strength H_{DC}. Generally, the permeability results are as expected. It will be discussed in the following chapter how to estimate core losses under DC magnetization based on these results.

IV. DERIVATION OF NEW CORE LOSS ESTIMATION MODEL

At the beginning of this paper, we formulated the first research question as whether there is an approach to describe the loss increase in the case of DC premagnetization that is not requiring extensive measurements. Conventional methods typically necessitate excitation circuits with controlled DC current [2][15]. In the following, we aim to develop a method to account for core losses under DC bias using simple measurement of the complex permeability, as previously described. Since permittivity and conductivity remain unaffected by DC bias, and permeability measurements are straightforward to perform, a loss model that relies solely on complex relative permeabilities would be highly advantageous.

According to [4], core losses can be calculated as

$$P_c = \frac{1}{2}\text{Re}\{j\omega\hat{\underline{\theta}}\hat{\underline{\emptyset}}\} \tag{9}$$

with $\underline{\theta}$ is the magneto-motive force and $\underline{\emptyset}$ the magnetic flux. The magneto-motive force can be calculated as

$$\hat{\underline{\theta}} = \hat{H}l_e = \frac{\hat{B}l_e}{u_0(\mu' - j\mu'')} \tag{10}$$

where l_e and A_e are the effective magnetic length and area, respectively. $\hat{B}$ is the magnetic flux density, and $\hat{H}$ the magnetic field strength. With $\hat{\underline{\emptyset}} = \hat{B}A_e$, core losses can be calculated as

$$P_c = -\frac{1}{2}\frac{\hat{B}^2 l_e A_e}{u_0}\text{Im}\left\{\frac{1}{(\mu' - j\mu'')}\right\} \tag{11}$$

Hence, the losses are proportional to

$$P_c \propto -\text{Im}\left\{\frac{1}{(\mu' - j\mu'')}\right\} = \frac{\mu''}{(\mu'^2 + \mu''^2)} \tag{12}$$

This proportionality of core losses to the right side of equation (12) has been derived in [16] as well, although in a different way. If we now normalize the losses to the losses at $H_{DC} = 0$, we will get a factor that accounts for the loss increase under DC Bias condition, at least for small signals:

$$P_c(H_{DC}) = P_c(H_{DC} = 0) \cdot \frac{\frac{\mu''(H_{DC})}{(\mu'(H_{DC}))^2 + (\mu''(H_{DC}))^2}}{\frac{\mu''(H_{DC}=0)}{(\mu'(H_{DC}=0))^2 + (\mu''(H_{DC}=0))^2}} \tag{13}$$

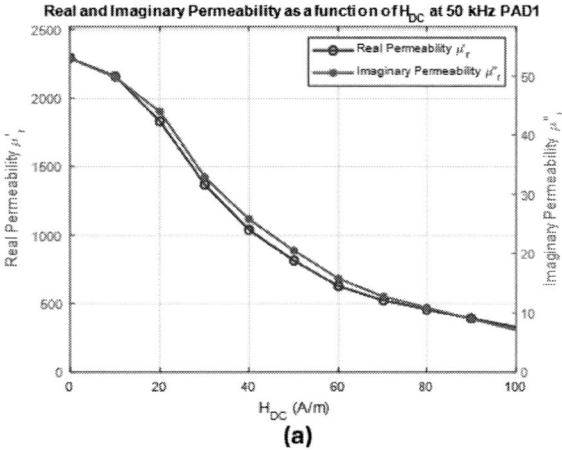

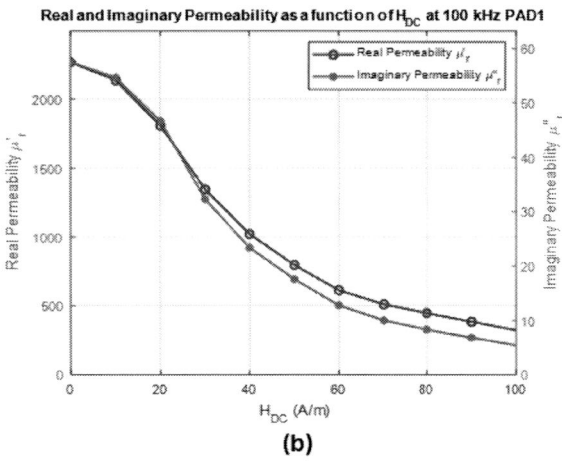

Fig. 11 - Measured Permeability of N87 as a function of DC Bias. (a) at 50 kHz, (b) at 100 kHz.

According to Fig. 11, the relative change of the imaginary permeability seems to follow the relative change of the real permeability. Hence, when assuming that $\mu' \gg \mu''$ and $\mu' \propto \mu''$ (with the proportionality factor K), the right factor in Equation (13) becomes

$$\frac{\frac{K \cdot \mu'(H_{DC})}{\mu'(H_{DC})^2}}{\frac{K \cdot \mu'(H_{DC}=0)}{\mu'(H_{DC}=0)^2}} = \frac{\mu'(H_{DC}=0)}{\mu'(H_{DC})} \qquad (14)$$

We call the newly derived factor the **DC Permeability Loss Estimator (DPLE)**, i.e.

$$\textbf{DPLE}(H_{DC}) = \frac{\mu'(H_{DC}=0)}{\mu'(H_{DC})} \qquad (15)$$

It can be used as follows:

$$P_c(H_{DC}) = P_c(H_{DC} = 0) \cdot \textbf{DPLE}(H_{DC}) \qquad (16)$$

The measured DEPLE for N87 is presented in Fig. 13 and is also superimposed over an image from publication [2] in Fig. 14. The results show good agreement, particularly at lower

DC Bias. DC premagnetization is often completely ignored in core loss calculations due to the challenge of measuring it. Using the DEPLE allows for an estimation without the need for extensive/complicated measurements.

This approach has been demonstrated on a single material. As will be shown in the following sections, further validation on different materials and operating points, as well as a development to account for the influence of ΔB (magnetic flux density variation) is planned.

Before concluding this Section, we would like to briefly explain why we believe that estimating losses using the real permeability is advantageous compared to using the imaginary permeability. Firstly, measuring the inductance L_s is simpler than measuring the resistance R_s (see Equation (6)). This is because even a small phase error in the current/voltage measurement can lead to significant errors in the determination of R_s, given that the reactance of R_s is much smaller than the reactance of L_s. Therefore, measuring the real permeability is easier and more reliable. Additionally, the real permeability is more commonly provided in datasheets.

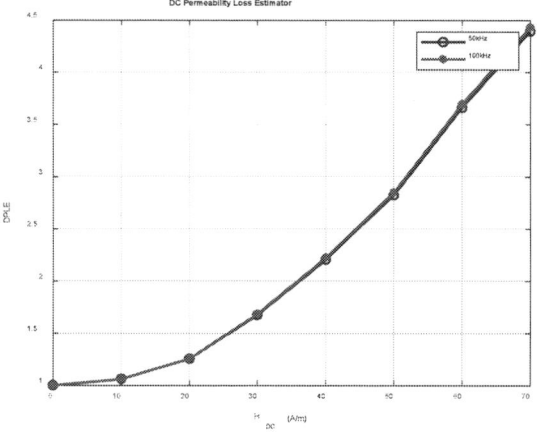

Fig. 12 – The DC Permeability Loss Estimator (DPLE) of the material N87

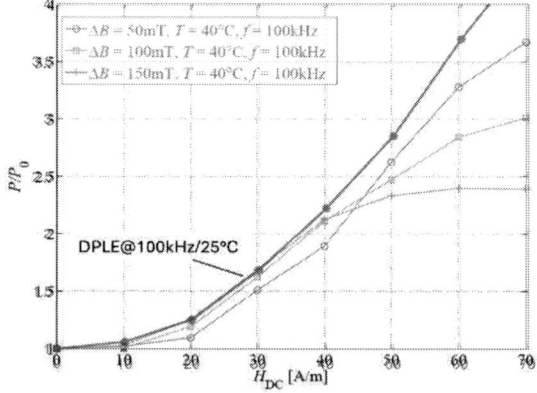

Fig. 13 - The DC Permeability Loss Estimator (DPLE) of the material N87 superimpose to Fig. 6 in [2].

V. Conclusions and Outlook

This paper investigated the impact of a DC premagnetization on permittivity, conductivity, permeability, as well as core losses across various frequencies in the ferrite N87 (TDK).

This has been done to investigate two research questions: **RQ1:** Is there an alternative approach to describe the loss increase in the case of DC premagnetization that is not requiring extensive measurements. **RQ2:** How are dimensional effects such as bulk eddy currents or dimensional resonance affected by a DC magnetic field?

Regarding RQ2: An important conclusion of this work was that permittivity and conductivity are mostly not affected by a DC pre-magnetization; hence, dimensional resonance and skin depth are influenced only through changes in permeability due to a DC bias. However, at higher frequencies (> 1Mhz, see Fig. 9) a change in permittivity due to a DC premagnetization was observed. The underlying cause should be further studied; for this, the setup must be prepared for higher frequencies first. This includes preparation and analysis of the specimen for higher frequencies (e.g. PCBs, etc.), as well as a gold plating of the ferrites.

Regarding RQ1: The introduced DC Permeability Loss Estimator shows promise for initial estimation of core losses under DC bias. Its derivation was straight forward; however, the authors are not aware of a similar analysis. The considerations made so far are only valid for MnZn ferrites (specifically for N87), and further studies should extend this approach to additional materials, for example its applicability to materials with higher eddy current losses, such as iron powder cores. It is also important to think of an extension of the method to account for operating points with higher flux variations (e.g., $\Delta B = 50/100/150 \, mT$).

ACKNOWLEDGMENT

We would like to express our sincere gratitude to PSMA for their financial support, insightful ideas, and invaluable partnership throughout our research.

REFERENCES

[1] A. Nabih, R. Gadelrab, Q. Li and F. C. Lee, "Dimensional Effects of Core Loss and Design Considerations for High Frequency Magnetics," 2021 IEEE Energy Conversion Congress and Exposition (ECCE), Vancouver, BC, Canada, 2021, pp. 5488-5495.

[2] J. Mühlethaler, J. Biela, J. W. Kolar and A. Ecklebe, "Core Losses Under the DC Bias Condition Based on Steinmetz Parameters," in IEEE Transactions on Power Electronics, vol. 27, no. 2, pp. 953-963, Feb. 2012.

[3] Skutt, Glenn Richard. High-frequency dimensional effects in ferrite-core magnetic devices. Diss. Virginia Polytechnic Institute and State University, 1996.

[4] T. Dimier and J. Biela, "Non-Linear Material Model of Ferrite to Calculate Core Losses With Full Frequency and Excitation Scaling," in IEEE Transactions on Magnetics, vol. 59, no. 7, pp. 1-10, July 2023, Art no. 6300610.

[5] Dobák, S., Beatrice, C., Tsakaloudi, V., & Fiorillo, F. (2022). Magnetic losses in soft ferrites. Magnetochemistry, 8(6), 60

[6] T. Guillod, W. V. R. Roberts and C. R. Sullivan, "Characterization and Impact of Large-Signal Dielectric Properties in MnZn Ferrites," 2024 IEEE Applied Power Electronics Conference and Exposition (APEC), Long Beach, CA, USA, 2024, pp. 384-390

[7] M. Kącki, M. S. Ryłko, J. G. Hayes and C. R. Sullivan, "Measurement Methods for High-Frequency Characterizations of Permeability, Permittivity, and Core Loss of Mn-Zn Ferrite Cores," in IEEE Transactions on Power Electronics, vol. 37, no. 12, pp. 15152-15162, Dec. 2022

[8] A. Schröder, A. Savca and D. Bormann, "High Frequency Permeability Measurements and Modeling of Magnetic Powder Cores Under DC Bias" in IEEE Transactions on Power Electronics, vol. 39, no. 12, pp. 16361-16370, Dec. 2024.

[9] M. Szewczyk, K. Kutorasiński, J. Pawłowski, W. Piasecki and M. Florkowski, "Advanced Modeling of Magnetic Cores for Damping of High-Frequency Power System Transients," in IEEE Transactions on Power Delivery, vol. 31, no. 5, pp. 2431-2439, Oct. 2016

[10] M. Kącki, M. S. Ryłko, J. G. Hayes and C. R. Sullivan, "Analysis and Experimental Investigation of High-Frequency Magnetic Flux Distribution in Mn-Zn Ferrite Cores," in IEEE Transactions on Power Electronics, vol. 38, no. 1, pp. 703-716, Jan. 2023

[11] T. Dimier and J. Biela, "Semi-Analytical Non-Linear Physical Model of Core Losses in Ferrite Ring Cores," in IEEE Transactions on Magnetics, vol. 59, no. 11, pp. 1-5, Nov. 2023

[12] G. R. Skutt and F. C. Lee, "Characterization of dimensional effects in ferrite-core magnetic devices," PESC Record. 27th Annual IEEE Power Electronics Specialists Conference, Baveno, Italy, 1996, pp. 1435-1440 vol.2

[13] TDK Datasheet, Ferrites and accessories - SIFERRIT material N87, February 2023

[14] B. N. Sanusi, M. Zambach, C. Frandsen, M. Beleggia, A. Michael Jørgensen and Z. Ouyang, "Investigation and Modeling of DC Bias Impact on Core Losses at High Frequency," in IEEE Transactions on Power Electronics, vol. 38, no. 6, pp. 7444-7458, June 2023

[15] C. A. Baguley, U. K. Madawala and B. Carsten, "A New Technique for Measuring Ferrite Core Loss Under DC Bias Conditions," in IEEE Transactions on Magnetics, vol. 44, no. 11, pp. 4127-4130, Nov. 2008

[16] J. W. M. De Araújo, R. N. M. Oliveira, D. Sgró and D. Oliveira, "On the Estimation of the Magnetic Behavior of Sendust Cores Under Different DC-Bias Conditions," 2023 IEEE 8th Southern Power Electronics Conference and 17th Brazilian Power Electronics Conference (SPEC/COBEP), Florianopolis, Brazil, 2023, pp. 1-8

A Low-Cost Setup and Procedure for Measuring Losses in Inductors

Burkhard Ulrich
Electronics & Drives
Reutlingen University
Reutlingen, Germany
burkhard.ulrich@reutlingen-university.de

Abstract—**Inductors are critical components in power electronic converters, determining their efficiency and size. A converter's performance is limited by the losses associated with inductors. In lower power applications, commercially available off-the-shelf inductors are usually used, but their losses are often difficult to predict solely based on datasheet values. Therefore, a direct measurement is typically required; here, a separate measurement of the dc and an ac loss is proposed to simplify the measurement process. This paper presents a low-cost setup based on affordable equipment, such as a self-built current probe, and uses simple compensation methods that allow developers to efficiently and cost-effectively measure losses in inductors. The setup is designed to emulate conditions comparable to real-world applications. The proposed procedure and setup are validated through experimental results, and potential sources of error and methods for compensation are discussed.**

Keywords—magnetics, components, losses, measurement, current probe

I. INTRODUCTION

Inductors are vital components of many power electronic systems, determining their size and efficiency. Commercial off-the-shelf components are often used for convenience for lower power applications due to their availability and price. However, these components are sometimes poorly specified by the respective manufacturers; in some cases, only the dc resistance is provided, and there is no or insufficient information about core or high-frequency winding losses. This lack of information makes it difficult for designers to accurately predict the overall losses when using such components. Also, for off-the-shelf inductors, parameters like winding geometry, core material, and the associated core losses cannot easily be assessed to calculate (or simulate) the overall losses by a designer who wants to use such a standard part. To properly use such a component, a designer will, therefore, need to estimate the losses either through measurement, trial and error testing in the actual hardware or eventually have to choose another component from a vendor that provides more detailed information (e.g., [1]-[3]). However, even if the manufacturer supplies more information, especially on the ac losses, a designer cannot be sure if the data is extrapolated from other measurements, and sometimes there will be no data available at desired operating points; for example, when using an inductor in an application with large ripple such as discontinuous conduction mode (DCM) or some sort of resonant converters. In these cases, measurements with

an inductor current excitation similar to the waveforms occurring in real-world applications are beneficial for measuring the losses.

Various measurement procedures and setups to measure inductor losses have been reported or are commercially available [4]-[17]. These methods can be broadly separated into calorimetric/thermal and electrical measurements. Calorimetric methods [14],[15] do not directly measure the losses but the generated heat generated by the inductive component, either through a direct temperature/heat measurement [14] or using the principle of thermal equilibrium [15]. These methods are considered precise and reliable but usually require long measuring times, although fast calorimetric methods have also been proposed [16]. Electrical measurement methods, in contrast, try to assess the power loss by directly measuring the electrical quantities, such as device voltage and current. Many methods have been proposed, including measurements employing impedance analyzers using sinusoidal ac-signals [17], resonant circuits [7], power analyzers [9], special test circuits [6],[8], and in-circuit/in-situ measurements [5],[13]. These measurements usually allow a faster measurement than thermal methods but are prone to errors, such as phase-shift discrepancies (an exemption being resonant measuring methods), when measuring high-frequency signals; therefore, they are more prone to error. Unfortunately, most measurement methods are complex, require special equipment [12], or are costly, making them inaccessible to many designers.

Therefore, in this article, a measurement setup is proposed to assess the losses of an inductor using two separate measurements, with the help of commonly available and low-cost laboratory equipment and a procedure to enable an accurate assessment of the losses. The proposed method is designed to offer a rough estimation of overall inductor losses within a reasonable error margin. However, it is essential to note that while the method aims to provide an estimation of the overall losses, the trade-off is the inability to distinguish between core and winding losses. Nevertheless, the method aims to provide practical insights for designers prioritizing simplicity and cost-effectiveness over ultra-high precision when accessing losses in inductive components.

II. MEASUREMENT METHOD

The measurement method used here can be considered a type of direct electrical power measurement, which uses the

979-8-3315-1612-3/25 $31.00 © 2025 IEEE

measured inductor voltage and current to calculate the power losses. But instead of trying to measure directly the power losses, a separate measurement of the ac and dc losses is proposed. Due to their physical origin, inductor losses can be separated into winding and core losses [10],[11]. While core losses are pure ac losses, winding losses are due to both dc and ac effects. Therefore, the inductor losses P_L can be represented as dc losses (P_{DC}) plus ac losses (P_{AC}), where the ac losses combine ac winding and core loss as:

$$P_L = P_{DC,Winding} + \underbrace{P_{AC,Winding} + P_{Core}}_{P_{AC}} = P_{DC} + P_{AC}. \quad (1)$$

In contrast to the physical origin of the losses, from an electrical point of view, the losses of an inductor (as any two-terminal device) can be expressed as the average of the instantaneous power at the device terminals. For example, considering an inductor in a power converter such as depicted in **Fig. 1**, if the inductor current and voltage are denoted as i_L and v_L and assumed as periodic quantities (with a fundamental period of T_S), then the inductor losses can be expressed as an n-cycle ($n \in \mathbb{N}$) average:

$$P_L = \frac{1}{n \cdot T_S} \int_0^{n \cdot T_S} v_L(t) \cdot i_L(t) dt$$
$$= \frac{1}{n \cdot T_S} \int_0^{n \cdot T_S} \underbrace{(V_{L,DC} + v_{L,AC}(t))}_{v_L(t)} \cdot \underbrace{(I_{L,DC} + i_{L,AC}(t))}_{i_L(t)} dt. \quad (2)$$

The rearrangement in the second row of (2) is because the terminal voltage v_L and current i_L can be expressed as a constant dc part ($V_{L,DC}$ and $I_{L,DC}$) and an ac part ($v_{L,AC}$ and $i_{L,AC}$) when using a Fourier series representation:

$$v_L(t) = \underbrace{V_{L,0}}_{V_{L,DC}} + \underbrace{\sum_{n=1}^{\infty} V_{L,n} \cdot \cos\left(\frac{2\pi}{T_s} \cdot t - \phi_{V,n}\right)}_{v_{L,AC}(t)} \text{ and } \quad (3a)$$

$$i_L(t) = \underbrace{I_{L,0}}_{I_{L,DC}} + \underbrace{\sum_{n=1}^{\infty} I_{L,n} \cdot \cos\left(\frac{2\pi}{T_s} \cdot t - \phi_{I,n}\right)}_{i_{L,AC}(t)}. \quad (3b)$$

Using (3a) and (3b) in (2), the losses P_L can be simplified to:

$$P_L = \underbrace{V_{L,DC} \cdot I_{L,DC}}_{=P_{DC}} + \underbrace{\frac{1}{n \cdot T_S} \int_0^{n \cdot T_S} v_{L,AC}(t) \cdot i_{L,AC}(t) dt}_{=P_{AC}}. \quad (4)$$

By comparing (1) and (4), the basic idea of the measurement principle proposed here can be derived: the losses P_L can be measured by determining the dc and ac losses separately. This leads to the following process:

   a.   The inductor under test will be exciting with a current i_L similar to practical conditions.

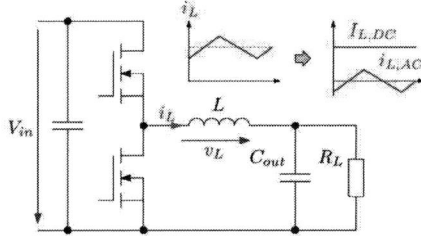

Fig. 1. Buck converter with inductor under test and separation of inductor current into dc and ac parts

   b.   The dc part P_{DC} of the losses can be assessed simply by knowing the dc resistance R_{DC} and measuring the dc component of the inductor current using a multimeter, as then P_{DC} can be calculated as

$$P_{DC} = I_{L,DC}^2 \cdot R_{DC}. \quad (5)$$

   c.   The ac part P_{AC} will be assessed as a one-port power measurement of the inductor, using the ac components $v_{L,AC}(t)$ and $i_{L,AC}(t)$ of the inductor terminal quantities, which can be measured using ac-coupled probes on an oscilloscope and then evaluating the integral

$$P_{AC} = \frac{1}{n \cdot T_s} \int_0^{n \cdot T_s} v_{ac}(t) \cdot i_{ac}(t) dt. \quad (6)$$

This method is, therefore, very similar to a classical power measurement, but there is a significant difference; the use of an ac coupling eliminates problems due to dc offsets and also allows the use of simple and more low-cost ac current probes, which can be self-built by the designer. Therefore, only a less expensive current probe will be sufficient to take the measurements, and the requirements for the measurement equipment are relaxed. The measurement method intends to measure total inductor losses but does not differentiate between core and winding loss. These cannot be directly inferred from the measurement results.

Note that the separation of the power into a dc and an ac component is no contradiction to the fact that the inductor losses will be nonlinear. As the inductor is excited by real-world signals, the nonlinearity will be reflected in the terminal voltage and currents.

The measurement method shares some ideas of the direct one-port power measurement (cf. the two-wire method in [11]) and resembles the partially described method in [4]. However, a more detailed description of a low-cost measurement setup, signal evaluation, and a discussion of potential sources of errors are provided here.

III. MEASUREMENT SETUP AND PROCEDURE

A. Measurement Setup

Fig. 2 shows a measurement setup, and the major equipment used in the practical setup (cf. to section V) is listed on the right side. The setup consists of a half-bridge power stage printed circuit board (PCB) supplied by an adjustable dc lab power supply (V_{in}), forming an open loop operating buck converter with floating output and the inductor (device under test DUT) connected to ground. The half-bridge PCB contains the transistors, the high-frequency decoupling capacitance and the necessary gate drivers. The half-bridge transistors' control signals are generated using a function generator. A power potentiometer is used as variable load R_{LOAD}, and two multimeters measure I_{DC} (of inductor current) and V_{out}. A differential voltage probe and an ac current probe sense the inductor voltage and current, which are finally measured by a USB oscilloscope in an ac input coupled mode. In this way, only the ac components of v_L and i_L are captured. The setup is chosen as it is to enhance the measurement in the following ways:

979-8-3315-1612-3/25 $31.00 © 2025 IEEE

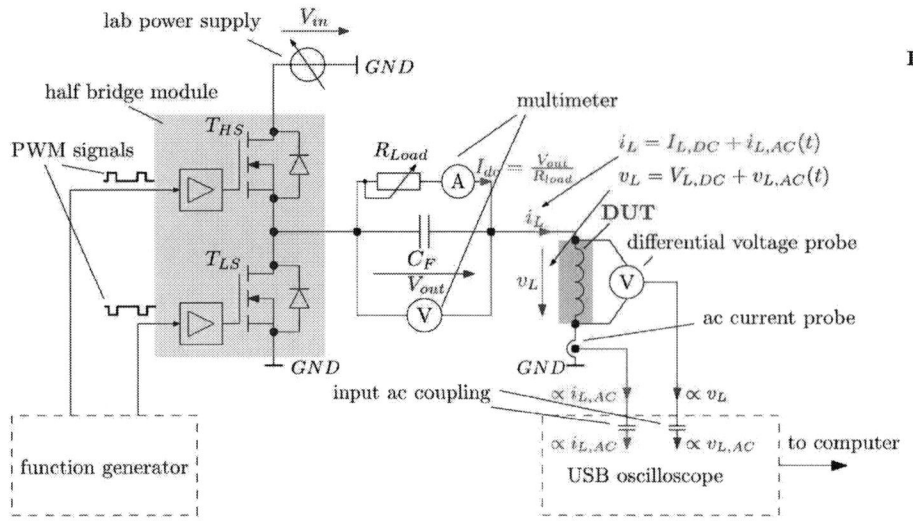

Fig. 2. Proposed measurement setup and list of used equipement in practical measurement environment.

- The inductor is arranged so that an isolated current probe and a differential voltage probe are connected on one side to GND. This is done to reduce noise coupling in the measurement signal.
- The converter is operated in an open loop manner by the function generator using a pulse-width modulated (PWM) signal generated from a stable source with a clock signal independent of the acquisition signal of the USB oscilloscope. This step is crucial, as these types of data acquisition devices often also have signal generator capabilities. However, the use of two different, not synchronized, clocks in the PWM signal generation and the data acquisition allows an averaging and especially oversampling in the data acquisition process, reducing the requirements for the sampling rate.
- An isolated USB oscilloscope and isolated or differential signal measurements are used to break ground loops.
- A USB oscilloscope also has the advantage of typically having a higher vertical resolution (here, a 14-Bit ADC is used).

B. Current Transformer

The used current probe is a self-built current transformer, built in a similar fashion as described in [18]. It uses a ferrite toroid from Fair Rite (Part No. 5977002721 [19]), and $n_s = 20$ secondary turns. The winding is terminated with a burden resistor directly a BNC connector. The burden resistance $R_B = 2.2\,\Omega \| 39\,\Omega$ is chosen, such that the effective burden resistance $R_{B,eq}$ is about $2\,\Omega$, when connected to a $50\,\Omega$ load via BNC cable. Additionally, a copper foil with a gap is used as an electrical field shield, which is connected to the GND connection of the BNC connector. This screening is critical regarding noise reduction in the current measurement. **Fig. 3** shows the actual current transformer construction and a schematic representation. In **Fig. 3a** the transformer construction can be seen, before the copper shield is applied, **Fig. 3b** shows the assembled transformer and **Fig. 3c** shows a schematic representation when connected to an oscilloscope.

The primary design characteristics of the current probe can be calculated as follows:

Ideal sensor transfer function

$$\frac{v_{\mathrm{isense}}}{i_p} = \frac{1}{n_S} \cdot \underbrace{R_B \| R_{\mathrm{term}}}_{R_{B,\mathrm{eq}}} = 0.1\,\frac{\mathrm{V}}{\mathrm{A}} \qquad (7)$$

Lower cut-off frequency

$$f_{\mathrm{cut,off}} = \frac{1}{2\cdot\pi} \cdot \frac{R_{B,\mathrm{eq}}}{n_S^2 \cdot A_L} \approx 354\,\mathrm{Hz} \qquad (8)$$

Maximum dc current

$$I_{\mathrm{DC,max}} = \frac{B_{\mathrm{max}}}{\mu_0 \cdot \underbrace{\mu_R}_{\approx\mu_i}} \cdot l_e \approx 10.6\,\mathrm{A} \qquad (9)$$

The core parameters for the A_L value, $A_L = 2250$ nH and the mean magnetic path length $l_e = 8.9$ cm, have been taken from the datasheet of the toroid [19]. Instead of the relative permeability, the initial permeability $\mu_i = 2000$ has been taken from the materials datasheet [20], and $B_{\mathrm{max}} = 0.3$ T has been

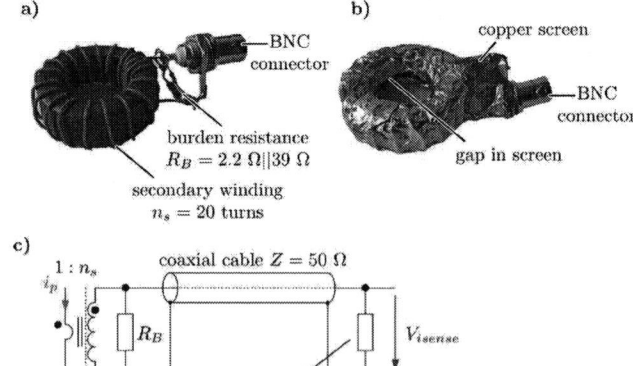

Fig. 3. Self-built current transformer as ac probe. Views during construction a) before and b) after adding a copper screen. c) schematic representation.

assumed in (9). The higher cutoff frequency depends on the parasitic capacitances and is analytically not easily accessible, but using a function generator, a probe bandwidth of more than 10 MHz was estimated (cf. to **Fig. 4**).

As with any other probe, this will also introduce a measurement error. **Fig. 4** and **Fig. 5** compare the self-built to a dedicated commercial pulse current monitor probe from Pearson Electronics (model 6298, which is similar to model 2877 [21]). In **Fig. 4**, the relative gain, i.e., the output signal of the self-built current probe divided by the commercial probe's output signal (with 1% accuracy [21]), is depicted in the range of 10 kHz to 10 MHz. The maximum error compared to the Pearson current monitor is about 4.7% at 10 MHz. Although the error will further increase to 9.6% at 20 MHz, the curve in **Fig. 5** indicates that for measuring inductors with switching frequencies up to 1 MHz, the self-built current probe should be sufficiently accurate if the first ten harmonics in the current waveform are considered. This is also indicated by the signals shown in **Fig. 5**. Here, the current from 1 MHz sawtooth signal applied to 50 Ω is measured with both probes, using oversampling (100 times) and averaging (50 times) to reduce noise. As can be seen, the signal of the self-built probe is almost identical to the signal from the commercial current sensor when using a signal similar to practical ac current waveforms observed in dc-dc converters. Therefore, it can be concluded that although the self-built probe will introduce an error in the measurement results, its use is feasible and can drastically reduce the overall system cost. Nevertheless, if a higher accuracy is needed, then more expensive commercial current monitors or dedicated ac current probes are required.

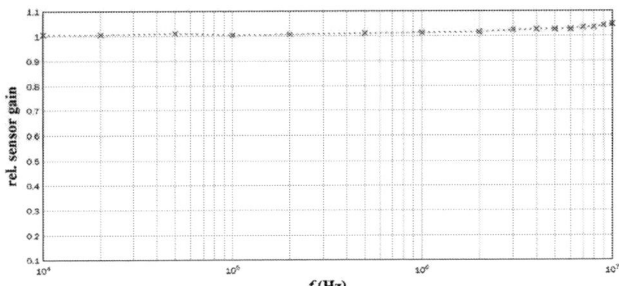

Fig. 4. Measured relative sensor gain for sinusoidal current (amplitude 150 – 190 mA depending on frequency) between 10 kHz and 10 MHz.

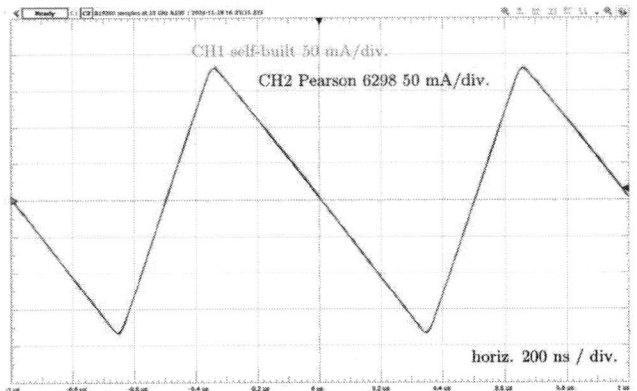

Fig. 5. Triangular current measurement with 30% duty cycle at 1 MHz; comparison of self-built probe to Pearson model 6298.

C. Measurement Procedure

The following measurement procedure is used:

1. Prior to taking any measurement, the ac voltage and current probes should be properly deskewed. E.g., using a function generator loaded with $R_L = 50$ Ω, the voltage and current are simultaneously measured to make a proper time-aligned measurement.

2. The desired operating point to measure the losses is set by adjusting the input voltage (V_{in}), as well as the duty cycle (D) and the switching frequency (f_s) from the signal generator. Adjusting the load resistance R_L ensures that the required V_{out} and $I_{L,DC}$ are reached.

3. DC losses are calculated using (5), where $I_{L,DC}$ is measured by a multimeter in series to the load resistance. Due to a large output filter capacitance C_F, the multimeter will capture the dc component of the inductor accurately when used in this position.

4. Inductor ac voltage $v_{L,AC}$ and current $i_{L,AC}$ are captured using the USB oscilloscope.

5. Using the oscilloscope's math functionality and cursor tools to measure the average values of $v_{L,AC}$ and $i_{L,AC}$ over n cycles, then use these averages to zero the offset of the voltage and current measurement channels.

6. A math channel is set up to continuously calculate $p = v_{L,AC} \cdot i_{L,AC}$ and integrate over n cycles to measure the average power. The final ac loss is determined by averaging the results of multiple acquisitions to enhance the accuracy.

IV. SOURCES OF ERROR AND COMPENSATION

A. Overview of Error Influences

Several factors affect the accuracy of the power measurement results in this setup:

- Accuracies/tolerances of the voltage and current probes and multimeters used
- Resolution and accuracy of the oscilloscope
- Propagation delays of the probes
- Temperature dependence of the inductor dc resistance
- Bandwidth limitations of the equipment
- Proper synchronizing when integrating to calculate the average power
- Noise due to the switching actions and parasitic magnetic and electrical coupling into the measured signals

Besides the error of the current probe (see preceding section) and the voltage probe's error of approximately ±2%, there are two additional factors which are particularly critical and will therefore be discussed in the following sections in more detail:

1. Compensation for the different probe propagation delays through a proper deskew procedure is needed, as this leads to an erroneous loss term being measured.

2. Determining the dc losses using (5) will systematically underestimate this part of the losses due to the temperature dependence of R.

Besides these effects, care should be taken to calculate the average power during signal acquisition over an interval, which is precisely an integer number of switching cycles.

Also, a thermal measurement can be done to cross-check the measurement results. Oversampling and averaging during signal acquisition will enhance the measurements and reduce noise.

B. Measurement Error Due To Probe Propagation Delay

The most significant source of error in the power measurement is the delay between the voltage and current probes. If not accounted for, this delay can significantly alter the measurement results. **Fig. 6** and **Fig. 7** illustrate this problem using a simplified case, when the inductor under test can be modeled as a series connection of a resistor R and an ideal inductor L (cf. to **Fig. 6**). It is assumed that the ac inductor current $i_{L,AC}$ is a triangular waveform defined by parameters Δi_L, $T_s = 1/f_s$) and D as depicted in **Fig. 7**. Using the proposed measurement principle the ac voltage $v_{L,AC}(t)=v_L(t) + v_R(t)$ and the current $i_{L,AC}$ are measured. However, the acquired waveforms will differ by a time delay Δt introduced by the used probes, resulting in an erroneous instantaneous ac loss given by:

$$p_{AC}(t) = i_{L,AC}(t) \cdot \left(\underbrace{v_L(t - \Delta t) + v_R(t - \Delta t)}_{v_{L,AC}(t-\Delta t)} \right)$$
$$= \underbrace{i_{L,AC}(t) \cdot v_L(t - \Delta t)}_{\neq 0 \text{ if } \Delta t \neq 0} + i_{L,AC}(t) \cdot v_R(t - \Delta t). \quad (10)$$

model inductor under test

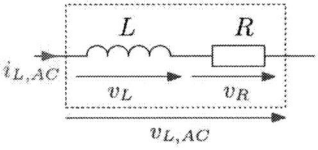

Fig. 7. Inductor model considered for ideal error calculation

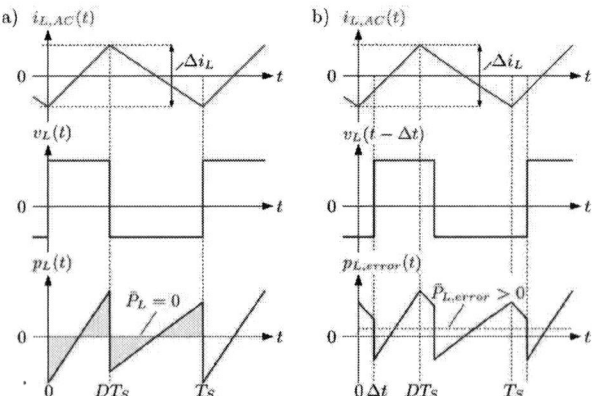

Fig. 6. Inductor current and voltage waveform with instantaneous power across ideal inductor part of inductor model a) without delay in voltage measurement and b) with delay Δt in measurement of v_L

The first term $i_{L,AC}(t) \cdot v_L(t - \Delta t)$ in (10) represents the instantaneous power of the inductive component. For an ideal inductor, the average value of this term will be equal to zero. However, with an introduced delay of Δt, this term will have a non-zero average value. **Fig. 7a** and **Fig. 7b** illustrate this phenomenon by comparing the ideal instantaneous power waveform of the inductor (with the average value $\bar{P}_L$ in **Fig. 7a**) to one affected by the probe delay Δt, resulting in an erroneous average $\bar{P}_{L,error}$. This additional term will add to actual ac power loss, leading to an inflated measurement result. For small time delays Δt, the error can be approximated by:

$$\bar{P}_{L,error}(\Delta t) \approx \frac{1}{2} \frac{L \cdot \Delta i_L^2 \cdot \Delta t}{(1-D)^2 \cdot D^2 \cdot T_s^3} \quad (11)$$

Assuming the triangular current waveform $i_{L,AC}$ (cf. to **Fig. 7**) and a constant resistance value R, the relative error E introduced by this mechanism, due to the delay Δt, is:

$$E \approx 12 \cdot \frac{L}{R} \frac{f_s^2}{(1-D) \cdot D} \cdot \Delta t. \quad (12)$$

Evaluating (12) shows that the error sensitivity increases with switching frequency f_s, for inductors with a high L/R ratio (i.e., low ac losses) and for very low or very high duty cycle values. Solving (12) for Δt leads to an expression for the allowed delay time, depending on the maximum acceptable error E_{max}:

$$\Delta t \leq \frac{1}{12} \cdot \frac{R}{L} \frac{(1-D) \cdot D}{f_s^2} \cdot E_{max}. \quad (13)$$

Fig. 8 depicts evaluations of (13) vs. the switching frequency f_s for a maximum error of $E_{max} = 10\%$, a duty cycle of $D = 0.3$, and different values for the L/R ratio. The results show that for measurements to be accurate within the required 10% margin, the maximum allowable timing delay can be in the nanosecond range for high switching frequencies. Therefore, a compensation (deskew) of the relative probe delay Δt is crucial to achieve accurate measurements.

A simple calibration method can be applied to correct the probe delay. For the ac probes in this setup, voltage and current are simultaneously measured across a purely resistive load connected to a signal generator. A periodic signal, such as a square wave, is applied to the resistor, and the following process is used:

1. The dc offset in the measurement signals is removed. For this, the average signal values over n cycles are measured and then subtracted.

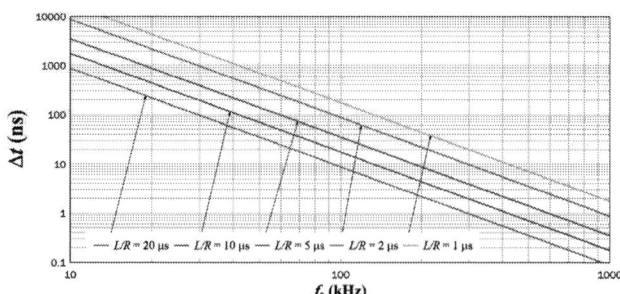

Fig. 8. Maximum allowable delay time Δt vs. frequency f_s for achieving a relative error $E < 10\%$ for different values of L/R at $D = 0.3$

2. The delay is adjusted. Without dc offset, this is easily done on the waveforms' positive or negative slope. Therefore, the time delay Δt between the waveforms is determined at the signal zero crossings. The measured waveforms are then time-shifted to align the voltage and current waveforms accurately.

For improved accuracy of this calibration and reduced noise sensitivity, an oversampling using an equivalent-time sampling and a signal averaging can be applied during the process.

Note that also the second term in (10) will introduce a measurement error for a non-zero delay Δt, but this effect is a much lower error, then the introduced additional erroneous power term described by (11).

C. Measurement error in P_{DC} due to temperature rise

A second systematic error occurs due using R_{DC} in calculating dc losses according to (5). Two main factors influence this error: first, the R_{DC} is subject to manufacturing tolerances, and second, it is highly temperature-dependent. Assuming a copper winding, an increase in temperature of 1 °C raises R_{DC} (and consequently the anticipated dc losses) by approximately 0.393%. Therefore, simply using the datasheet values for R_{DC} in (5) may lead to inaccuracies at elevated temperatures, either from significant self-heating due to losses or from higher ambient temperatures T_A. For higher accuracy, an iterative approach is recommended. First, the total inductor losses P_L will be estimated using measurement results calculated using (5) and (6). Second, using a simple thermal model, the anticipated stationary temperature rise ΔT is calculated as:

$$\Delta T = P_L \cdot R_{th} = (P_{DC} + P_{AC}) \cdot R_{th}. \quad (14)$$

Where R_{th} represents the thermal resistance of the inductor. It is crucial to use the specific R_{th} of the actual application setup, especially if the cooling conditions differ significantly from those used to measure the temperature rise in the inductor's datasheet. With this value of ΔT an increased winding temperature $T_{wind} = T_A + \Delta T$ and therefore, a new value for winding resistance $R_{DC,new}$ can be calculated:

$$R_{DC,new} = R_{DC}(T_{wind}) = R_{DC}(T_A) \cdot (1 + \alpha \cdot \Delta T), \quad (15)$$

where α is the temperature coefficient of copper. Using this new resistance value in (5) a better estimate of of the dc losses can be carried out.

V. EXPERIMENTAL RESULTS

Fig. 9 shows a practical setup on a workbench (not shown lab power supply and function generator) to measure off-the-shelf inductors in dc-dc converters with input voltages lower than 60 V and currents up to 10 A (DC). The used equipment has been listed in **Fig. 2**. Note that current and voltage limitations are not inherent to the measurement method but to the actual equipment used. E.g., the lab power supply and the half-bridge transistors limit the maximum voltage to about 60 V and the design of the self-built current transformer limits the dc current value. Therefore, by exchanging these components, the measurement method is also applicable to higher power inductors at higher voltage and current levels if precautions are

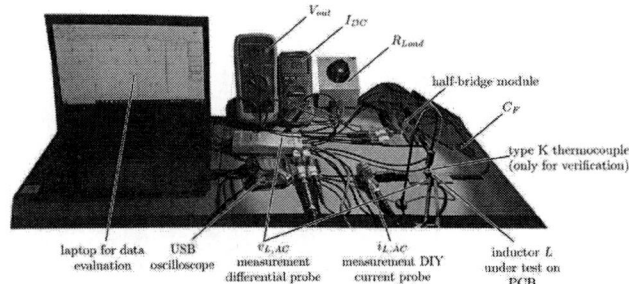

Fig. 9. Practical setup on workbench

taken. The additional type K thermocouple shown in the measurement setup is used here to verify the experimental results and is not needed to determine the losses.

Fig. 10 shows exemplary waveforms taken from the measurements of an inductor for two different operating points. **Fig. 10a** depicts waveforms for an off-the-shelf 9.2 µH inductor (Wuerth Electronics Part No. 7443551920) operating in DCM, and in **Fig. 10b** waveforms for the same inductor operating in CCM are shown. **Fig. 11** shows exemplary results for this inductor in a synchronous buck converter operating with $V_{out} = 3.3$ V and $I_{DC} = 1.4$ A, first at $f = 100$ kHz (cf. **Fig. 11a**) and second for $f = 50$ kHz (cf. **Fig. 11b**) with $V_{out} = 3.3$ V. In these cases, an operating point indicating the usefulness of the measurement setup is shown. Here, the ac losses dominate (cf. **Fig. 11**) over the dc losses. Although the losses can be obtained using the tool [1], here, these data underestimate the losses by

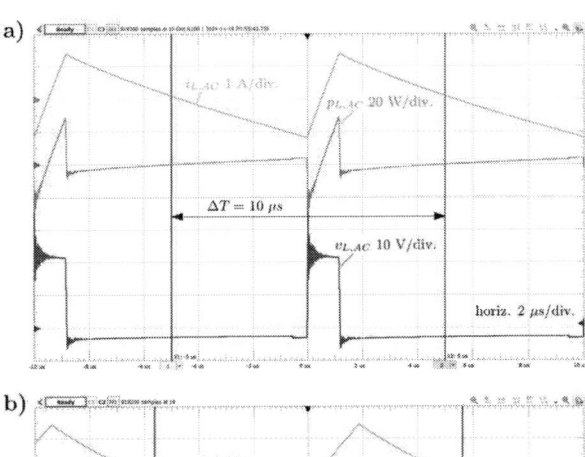

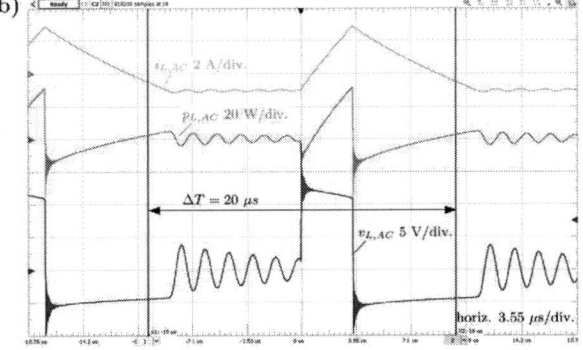

Fig. 10. Example inductor ac waveforms measured using the setup: a) CCM operation $V_{in} = 30$ V, $V_{out} = 3.3$ V, $I_{DC} = 1.4$ A, $f = 100$ kHz and $L = 9.2$ µH and b) DCM $V_{in} = 20$ V, $V_{out} = 5$ V, $I_{DC} = 1$ A, $f = 50$ kHz and $L = 9.2$ µH

a)

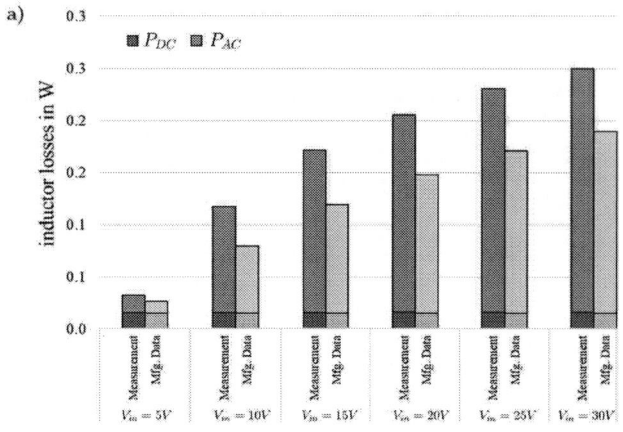

b)

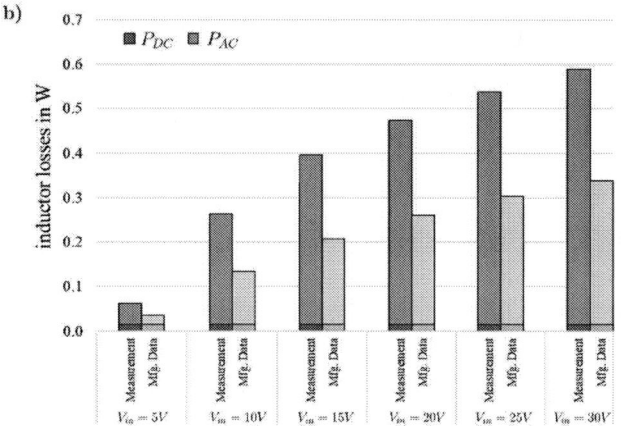

Fig. 12. Comparison of measurements to vendor supplied loss data [1] for a inductor in a synchronous buck converter with V_{out} = 3.3 V, I_{DC} = 1.4 A a) for f = 100 kHz and b) for f = 50 kHz

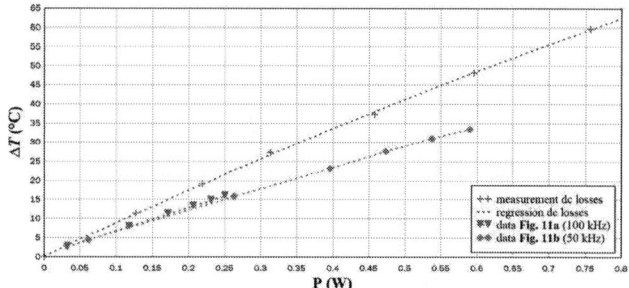

Fig. 11. Measured temperature rise vs. losses for the test inductor for different operating conditions.

These results underscore the necessity of accurately measuring the actual losses in the inductor through a direct power measurement and emphasize the importance of a dedicated measurement setup, as proposed in this article.

VI. CONCLUSION

A low-cost setup for measuring inductor losses under real-world excitations is proposed, enabling designers of inductors to practically test the occurring losses in the lab using standard equipment. The measurement method and setup are described. Using a low-cost USB oscilloscope, together with a self-built current probe, whose design is described in detail, reduces the overall cost of the equipment. Potential sources of error, such as a proper deskewing of the probes and measurement errors due to a temperature rise of the inductor's resistor, are identified and discussed. Practical measurements on an off-the-shelf inductor indicate the approach's usefulness.

REFERENCES

[1] REDEXPERT, Wuerth Electronics, [online], Available: https://redexpert.we-online.com/ (Accessed Aug. 12, 2024)

[2] "Power Inductor Finder and Analyzer", Coilcraft, [online], Available: https://www.coilcraft.com/en-us/tools/power-inductor-finder/, (Accessed Aug. 12, 2024)

[3] "Inductor Loss Calculation Tool", TDK, [online], Available: https://product.tdk.com/en/search/inductor/inductor/smd/core_loss_simu lation, (Accessed Aug. 10, 2024)

[4] R. Bramanpalli, "Accurate Calculation of AC Losses of Inductors in Power Electronic Applications," PCIM Europe 2016; International Exhibition and Conference for Power Electronics, Intelligent Motion, Renewable Energy and Energy Management, Nuremberg, Germany, 2016, pp. 1-8.

[5] L. Yi and J. Moon, "Accurately Disentangling Core and Winding Losses in Experimental, In-situ Magnetic Loss Measurement for Power Electronic Circuits and Applications," in IEEE Journal of Emerging and Selected Topics in Power Electronics, doi: 10.1109/JESTPE.2024.3426283.

[6] B. Liu, W. Chen, J. Wang and Q. Chen, "A Practical Inductor Loss Testing Scheme and Device With High Frequency Pulsewidth Modulation Excitations," in IEEE Transactions on Industrial Electronics, vol. 68, no. 5, pp. 4457-4467, May 2021, doi: 10.1109/TIE.2020.2984985.

[7] M. Mu and F. C. Lee, "A new high frequency inductor loss measurement method," 2011 IEEE Energy Conversion Congress and Exposition, Phoenix, AZ, USA, 2011, pp. 1801-1806, doi: 10.1109/ECCE.2011.6064003.

[8] H. Xu et al., "A Loss Measurement Approach of Power Magnetic Components under Practical Power Electronics Conditions," 2022 IEEE Transportation Electrification Conference and Expo, Asia-Pacific (ITEC

a factor of 30% - 90% depending on the respective operating point. An attempt was made to cross-check the losses using a thermal measurement; however, this approach is more complex for the reasons explained below. A thermal verification measurement was conducted to evaluate the thermal resistance of the inductor on the test PCB to perform the cross-check. This was done by applying a DC current to heat the winding, with the resulting temperature rise shown in **Fig. 12**. For comparison, the corresponding temperature rises from the measurements in **Fig. 11** are also depicted. It is important to note that the temperature rise under DC conditions differs from that in **Fig. 11**. This discrepancy arises due to differences in heat flow paths between core and winding losses. In the DC measurement, all generated heat originates from the winding, leading to a higher measured thermal resistance compared to cases where core losses also contribute. Even if the loss data from the vendor tool [1] would be used to estimate the temperature rise based on the measurements from **Fig. 12**, the heating of the inductor would still be underestimated, albeit to a lesser extent. However, such a comparison provides limited insight, as the thermal behavior differs significantly when losses are distributed between the core and winding. Consequently, it was not possible to verify the losses through a thermal measurement in this case.

979-8-3315-1612-3/25 $31.00 © 2025 IEEE

Asia-Pacific), Haining, China, 2022, pp. 1-6, doi: 10.1109/ITECAsia-Pacific56316.2022.9941831.

[9] K. Hayashi," Real Operating Loss Measurement of Low-Loss Inductors Using High-Precision Wideband Power Analyzer and Current Sensor", HIOKI, online: https://www.hioki.com/download/38355, accessed Aug. 12, 2024

[10] W.G. Hurley and W.H. Woelfle, Transformers and Inductors for Power Electronics: Theory, Design and Applications, Wiley, Hoboken, NJ, 2013

[11] A. Van den Bossche and V. Valchev, Inductor and Transformers for Power Electronic; CRC Press: Boca Raton, FL, USA, 2005.

[12] "MADMIX-Power Inductor Measurement", Mindcet, [online], Available: https://www.mindcet.com/measurement-systems/madmix, (Accessed Aug. 12, 2024)

[13] "In-circuit Measurements of Inductors and Transformers in Switch Mode Power Supplies", Application Note, 07/18, [online] Available: https://download.tek.com/document/In-circuit%20Measurements%20of%20Inductors_App-Note_55W-61401-2.pdf, (accessed: Oct. 30, 2024)

[14] F. Grecki and U. Drofenik, "Calorimetric medium frequency loss measurement of the foil inductor winding," 2021 IEEE 19th International Power Electronics and Motion Control Conference (PEMC), Gliwice, Poland, 2021, pp. 611-614, doi: 10.1109/PEMC48073.2021.9432598.

[15] J. Wang, P. Fan, S. Lin, F. Huang and W. Chen, "Comparative Calorimetric Method for Magnetic Component Losses Measurement With High Frequency," in IEEE Journal of Emerging and Selected Topics in Power Electronics, vol. 12, no. 5, pp. 5112-5122, Oct. 2024, doi: 10.1109/JESTPE.2024.3417354.

[16] P. Papamanolis, T. Guillod, F. Krismer and J. W. Kolar, "Transient Calorimetric Measurement of Ferrite Core Losses," 2020 IEEE Applied Power Electronics Conference and Exposition (APEC), New Orleans, LA, USA, 2020, pp. 195-201, doi: 10.1109/APEC39645.2020.9124132.

[17] S. Prabhakaran and C. R. Sullivan, "Impedance-analyzer measurements of high-frequency power passives: techniques for high power and low impedance," Conference Record of the 2002 IEEE Industry Applications Conference. 37th IAS Annual Meeting (Cat. No.02CH37344), Pittsburgh, PA, USA, 2002, pp. 1360-1367 vol.2, doi: 10.1109/IAS.2002.1042734.

[18] K.Wyatt, "Troubleshooting Radiated Emissions Using Low-Cost Bench Top Methods", [online] Available: https://interferencetechnology.com/troubleshooting-radiated-emissions-using-low-cost-bench-top-methods/, (Accessed Aug. 11, 2024)

[19] "Toroids (5977002701)", Fair-Rite [online] Available: https://fair-rite.com/product/toroids-5977002701/, (Acessed Nov. 19, 2024)

[20] "77 material Datasheet", Fair-Rite [online] Available: https://fair-rite.com/77-material-data-sheet/, (Acessed Nov. 19, 2024)

[21] "Pearson Current Monitor Model 2877", Datasheet, Pearson Electronics, [online] Available: https://pearsonelectronics.com/pdf/2877.pdf, (Acessed Nov. 19, 2024)

Effect of Temperature of Additively Manufactured Cores

Ken Johnson
Department of *Electrical and Computer Engineering*
University of Connecticut
Storrs, United States
kenneth.r.johnson@uconn.edu

Ali Bazzi
Department of *Electrical and Computer Engineering*
University of Connecticut
Storrs, United States
bazzi@uconn.edu

Abstract— Staying updated on emerging technologies with the potential to revolutionize industries is essential. Additive manufacturing offers a promising alternative to traditional methods. This technique allows for the three-dimensional printing of various metals, replicating complex structures with maintained integrity. Despite its widespread adoption, additive manufacturing's application in electrical machines remains relatively unexplored. To harness its potential in constructing electrical machines, specialized alloys like electrical steel are used to mitigate performance-limiting factors such as hysteresis losses. Controlling the silicon percentage in electrical steel is critical for defining its magnetic properties. (*Abstract*)

Keywords—Additive Manufacturing (AM), Selective Laser Melting (SLM), Laser Powder Bed Fusion (LBPF), Magnetic Flux Density (B), Relative Permeability (μR) (*key words*)

I. INTRODUCTION

A comprehensive nonlinear model is crucial for assessing new magnetic materials for electric machinery, with ideal core materials possessing high relative permeability and low coercive force to minimize energy losses. Thus, materials with favorable B-H curves are vital for optimizing electric motor performance, illustrating the relationship between magnetic flux density (B) and magnetic field strength (H). Several researchers have explored core materials produced via Selective Laser Melting (SLM), claiming comparable hysteresis losses to commercially machined cores [1]. We aimed to replicate these findings using a similar SLM process but with Fe-Si powder containing 3.3% silicon, contrasting with Fe-Si-Cr powder with 4% silicon [2]. We used a IPG Photonics one of a kind SLM machine under the alias AM-RAD Laser Powder Bed Fusion System. The purpose of developing SLM and a manufacturing technique for electrical motors is the fact that the advent of AM brings with it many advantages that are sought after in electric motor design that are not available through current typical industry standards of motor core production. A few of these advantages can be described as, flexibility in 3-D geometries due to the flexibility of 3-D printing, modular designs can be presented that allow for easier repairs and assembly and reduced requirements on laminated structures. The gain that is most sought after is the reduced machine space and manufacturing cost that AM provides, since less machines are required to produce the same end product. The cost of cutting or stamping the laminations is often a significant portion of the production cost of motors [9].

TABLE I -POWDER INFORMATION

Vendor: Hoeganaes Specialty powders	Material: Fe-3.66wt%Si
Lot Number: 202159001	Product Code: 560108
Apparent Density: 4.04g/cm^3	Tap Density: 4.82g/cm^3
Manufacture process: Gas atomized	

II. PERFORMANCE OF COMERCIAL CORES

A. Discussion of Material

First, let me explain the goal of this paper. Efficiency of electric motors rely heavily on minimizing losses of the machine [11]. The losses of these motors typically include Copper losses, losses in air gap, mechanical losses due to friction and Core losses, there are others but this paper will focus on the Core losses. Core Losses can be split into Hysteresis loss, Eddy Current Loss and stray Loss. Hysteresis Loss can be measured by using Maxwell's equations to form a B-H curve, The area of this curve is directly related to Hysteresis loss of the core material. When measuring B-H characteristics of the material B which is magnetic flux density is plotted on the y-axis of the graph, this tells you how much of the magnetic field is being absorbed within the material. H has units of amp turns per meter and can be interpreted as magnetic field strength of the exciting magnetic field. At some point the material cannot produce more B field regardless of H field, this is when the material is in saturation and cannot absorb more magnetic field within the material. When it comes to material used in alternating current motor designs, B-H curves that show saturation occurring early under low H field, while dissipating that magnetic field quickly with little residual flux density are sought after because of the low Hysteresis loss and ability to change polarity quickly and efficiently. A typical material that is used silicon steel commonly referred to as electrical steel. A specific grade to this electrical steel we would like to achieve is M250 steel [7], which is one of the industry standards of Silicon Steel used in magnetic cores. Other more developed low loss electrical steels exist and are topic of many researchers to improve efficiency of the motors, an example of these high efficiency materials is 50PN400 [12]. For materials used in motors, the material used is non-grain oriented. For materials used in machines like transformers grain-oriented steels are used in transformers due to the superior magnetic properties in the direction of the oriented grains [6]. The downside of metallic crystal grain

979-8-3315-1612-3/25 $31.00 © 2025 IEEE

orientation is that is hinders the magnetic properties in the direction opposite the orientation of the aligned grains. This is the reason for using the non-grain-oriented material in electric machines, since operation direction would likely need to change direction, which if the material had grain orientation one direction would be vastly less efficient than the other. The focus of this writing is to describe the typical process of characterizing core material that has been additively manufactured while including the procedure of improving core losses within this AM steel. Our goal is to meet the low hysteresis loss characteristics of grain-oriented silicon steels, with non-oriented crystalline structure since the SLM technology does not allow for orienting these grains in any specific direction, since the orientation of these grains is highly dependent on the thermal cycling of the material during and after the melt process. Although this paper characterizes inductor cores and not characterization of electric motors, the findings on AM core material being developed would carry over into designing the cores of electric motors. Since AM has already developed the ability to 3-D print complex geometries with relative ease, using slicing software to print the exact shapes required for electric machines as shown by other research from [7] and [8], more research is needed in controlling the core losses of this material [2]. Once a printing condition that prints non-grain oriented with core losses comparable to commercially available electrical steels gets established in the industry AM manufacturing will likely surpass already developed industry standards due to the lower price, freedom in designing geometries and reduced production equipment requirements. This paper will strive to further mature the development of controlling the core losses of the additively manufactured core materials for use in electric motors.

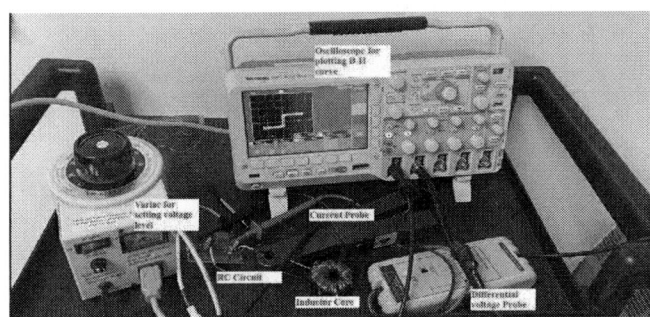

Figure 2: Setup

B. Measuring Characteristics of Silicon Steel

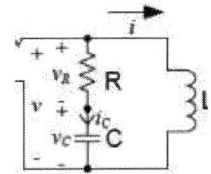

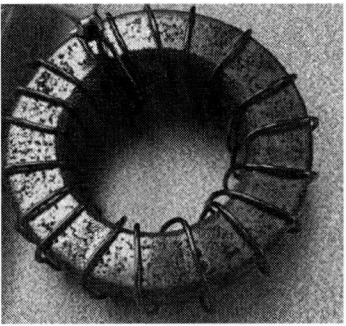

Figure 1. Test schematic Figure 4. Printed

Toroid

Using toroidal cores with matching dimensions as a reference for comparing material performance. We measured hysteresis losses of the Silicon Steel cores using an RLC circuit [4] shown in Figure 1. The core material, wrapped with 65 turns of wire, was connected in parallel with the R-C branch to assess hysteresis loss according to the provided theory, where N represents the number of turns, e(t) is induced voltage from faraday's law, values I denote peak current, V_c represents peak voltage across the capacitor and l signifies the mean path length of the core [5]. 65 turns were chosen to allow the AM cores to reach saturation shown later. The relative permeability is calculated using the slope of the B-H plot, where the values are points on the line of the linear region of the graph. The resistance was chosen to be 100k ohms and the capacitance value of the capacitor was determined as 100µ Farad. This RLC circuit will work with any core material as long as the expected operating frequency is around 60 Hertz, in which all the testing in this paper is evaluated with a pure sinusoidal 60 Hertz input voltage. This sinusoidal input voltage was adjusted by a Variac to vary the input to the RLC circuit until the core being tested reached into saturation. The current into the core was measured on the x axis of an oscilloscope with the voltage across the capacitor, representing the integral of the back emf of the core significant to faraday's law, on the y axis to provide a curve proportional to the B-h curve, this data is then extracted from the oscilloscope and processed in MATLAB to provide the actual B-H curve of the core material being tested. The Testing setup is shown in Figure 2.

$$H = \frac{N*i}{l}, \quad B = \frac{1}{N*A} \; e(t)dt, \; B = \frac{(RC)}{(NA)} * vc, \; \mu R = \frac{B}{\mu 0 * H}, \; Reluctance = \frac{length}{\mu 0 * \mu R * A}, \; L = \frac{N^2}{Reluctance}$$

979-8-3315-1612-3/25 $31.00 © 2025 IEEE

The results of a commercial silicon steel toroid core include the B-H curve and the parameter values calculated from the equations above [4]. The silicon steel core has characteristics of an efficient core material for electric motors since it is showing soft magnetic characteristics as is shown in Figure 3. Soft magnets have very steep permeability and very little residual magnetism. We would like to be able to reproduce similar material properties using AM techniques where the SLM machine uses LPBF to 3-D print the core material avoiding the use of stamping machines or laser cutters hopefully making the process of manufacturing motor cores possible with only one machine. If AM methods can provide core material with desirable hysteresis losses, then the creation of 3-D printed electric machines is possible.

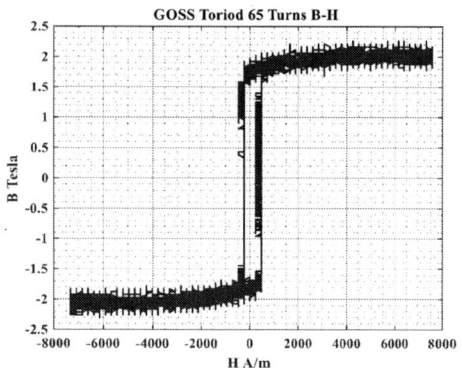

Figure 3. Silicon Steel B-H Curve

TABLE II - SILICON STEEL RESULTS

Goss Toroid Values						
B(max)	H(max)	μ_R	Reluctance	Flux	L (inductance)	Hysteresis Loss
2.06 Tesla	7448.16 Amp Turns/m	3978.19	171.813k Amp Turns/weber	4.77mWeber	24.6mHenry	.24Watts

III. PRINTING CONDITIONS

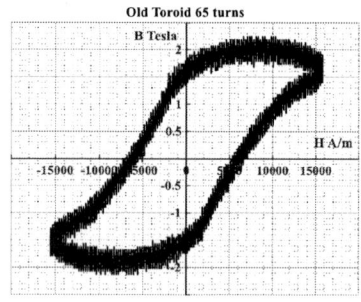

Figure 5. BH Toroid₁

Using the SLM machine we first needed to determine the printing conditions that would provide the highest relative permeability. To do this we printed out several test conditions and picked what provided a good balance of porosity and surface finish. When printing using LBPF it is required to look at how the material forms while being melted, the powdered metal is supposed to bead up and make straight lines with even distribution of material. This does not always happen since the material is melting and can form as irregular shapes.

TABLE III -OLDER TOROID RESULTS

Toroid₁ Values						
B(max)	H(max)	μ_R	Reluctance	Flux	L(inductance)	Hysteresis Loss
1.96 Tesla	15646.2 Amp Turns/m	198.94	3.8M Amp Turns/weber	430μWeber	1.11 μHenry	2.89 Watts

These are two printing conditions we decided try out and see how they would compare to the silicon steel core. Sine the AM technology has not been explored by many other researchers it was important to see if any other groups have been successful. As mentioned before a group of researchers had published information on their success of SLM methods and proved to have similar results to commercially available silicon steel cores [2]. Since other researchers were able to produce desirable characteristics using SLM we printed our first core. A picture of the core

TABLE V -NEWER TOROID RESULTS

Toroid₂ Values						
B(max)	H(max)	μ_R	Reluctance	Flux	L(inductance)	Hysteresis Loss
2.15 Tesla	16888 Amp Turns/m	206.9	3.9 M Amp Turns/weber	209 μWeber	1.08 μ Henry	3.57 Watts

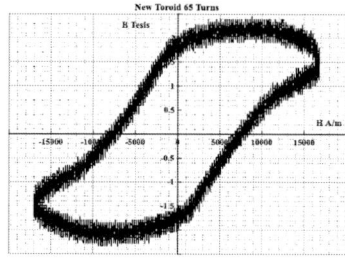

Figure 6. BH Toroid₂

is available in Figure 4, this picture is to show what the core looks like not what the testing conditions were. In our testing we used insulation to avoid shorting the core and had more turns of wire. Overall, our first printing condition was undesirable since it had a relative permeability of less than ten times the value of [2] and commercial cores [3]. A new toroid was printed at a reduced laser temperature to prevent potential silicon burn-off during SLM. The melting point of Fe-3Si is at least 1700-

1800°C during printing smoke emission was observed indicating alloy vaporization. To avoid burning silicon content, the temperature was lowered.

TABLE VI - BODE 100 IMPEDANCE MEASURMENT

Impedance	
Toroid1	Toroid2
308mΩ∠100.9	285mΩ∠100.5

Literature highlights relative permeability as a key metric for evaluating magnetic performance and enabling easy comparison among cores to assess hysteresis losses [10]. In our printed cores there was not much difference between the permeability of the cores. Another way of defining performance of the cores is by looking at impedance. Using a bode 100 the impedance of the two printed cores were measured. This resulted in showing how similar the impedance of the two cores were regardless of the change in printing conditions.

IV. PURE POWDER RESULTS

TABLE VII - NON ADDITIVE MANUFACTURED RESULTS

Pure Powder Toroid Values					
B(T)	H (A/m)	μ_R	Reluctance	Flux	L (inductance)
1.73 Tesla	15894.5 A/m	66.31	11.4 M Amp Turns /Weber	146 (μWeber)	370 (μHenry)

Preliminary findings show that toroid shapes printed in this study have significantly lower relative permeability than commercial counterparts. Our focus is on optimizing silicon retention and metal crystallization in our SLM machine to meet commercial electrical steel standards. Despite some improvement in core performance with minor adjustments in printing temperature, challenges persist at lower temperatures due to powder balling issues, leading to uneven metal structure. FeSi3 powder is characterized without heat manipulation using a nonmagnetic container enclosing the toroid shape filled with compressed powder. The powder in a 2.5mm PLA container does not reach saturation which is likely due to large pores in the powder thus providing poor magnetic properties. This shows that the SLM process is beneficial since it reduces the porosity of the powder and thus provides better magnetic properties than the pure powder. Although unlikely to exceed 1.6T in these findings, a consistent correlation is evident despite voltage probe noise affecting accuracy. The inaccuracy in the results is because the method of measuring an RLC circuit introduces such noise that affects results. Our additive manufacturing results indicate subpar magnetic properties, yet reducing temperatures during the SLM process does not show any benefit. While literature indicates that 3% silicon steel achieves a relative permeability close to 4000 under similar conditions, our SLM cores exhibit a relative permeability less than one-tenth that of commercial cores and those achieved by others. Addressing issues like porosity and crystalline structure control during SLM by adjusting temperature is crucial for further enhancement. Mitigating problems such as delamination and powder balling at lower temperatures significantly extends printing time. In future studies, we aim to optimize printing parameters to match those in the literature and

conduct temperature testing for various atmospheres. It's worth noting that our SLM machine's oxygen content may differ from that achieved by Tiismus, Hans, et al., as their machine reaches 0.35% oxygen content while ours is set to less than 0.0025%. Working on controlling the surface finish by using a printing condition closer to the conditions discussed in [2]. It is likely that adjusting the conditions to allow the layer thickness to be smaller will allow better surface finish control for each layer reducing porosities and therefore the reluctance of the core which should provide better performance even without having the atmospheric content similar to Tiismus, Hans, et al. [2].

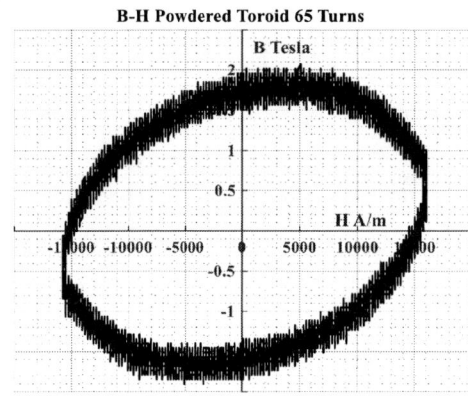

Figure 7. BH Pure Powder Toroid₃

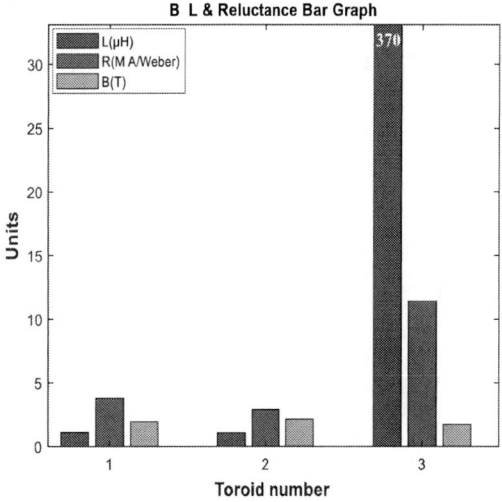

Figure 8. B L & Reluctance Bar Graph

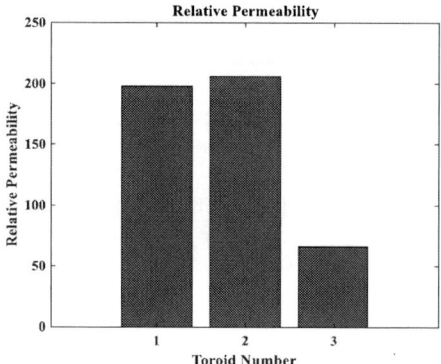

Figure 9 μ_R Bar Graph

provide modularity to design of electric motors. Since having isolation between layers is known to reduce magnetic losses, having modular pieces that allow for electric isolation while allowing each piece to attach rigidly to the neighboring pieces will provide a different method to maturing the AM field for electric machines, while still using pure FeSi AM manufacturing techniques that will work for every SLM machine. Once this method is proven the same method will be used to develop a fully AM FeSi stator for a motor to determine how performance I hindered using AM over proven laminated cores.

V. CONCLUSION

The hysteresis loss AM cores printed from the IPG SLM machine were quite excessive and shows that much more research is needed in additive manufacturing of electrical machines to learn how to control the losses of core material produced by AM techniques. Other researchers including [1], [2] and [7] show that hysteresis loss of AM silicon steel can get close to commercially laminated silicon steel core material, but controlling eddy current losses of such a manufacturing process has not been possible for pure silicon steel cores due to the inability to electrically isolate each layer, and eddy current increases proportionally to the height of non-isolated material. Other research in this topic ignores measurement of eddy current losses have or estimate by the use of FEM simulations, which does not provide accurate representation of whether or not the material from AM is suitable for use in electric machines. The research done by us provide physical measurements and actual metrics to describe the material performance using multiple devices and mathematical theory. Some companies have developed ways in isolating each layer with a layer of insulation between each additively manufactured layer but show that the eddy current losses are still more than commercial techniques. This will not work with most SLM machines due to the requirement of having two hoppers one for the FeSi powder and another for the insulation layers. This also complicates design since different conditions are required for different materials. After all, unless a SLM machine is built to accommodate the insulation layers and is able to print the silicon steel to retain its soft magnetic performance after the printing process, while keeping the dimensions highly accurate, it will not be able to provide the performance necessary to compete with commercially available techniques. Further more the SLM machine used in this research was not able to print the performance described by other researchers using the same printing conditions showing that the performance is highly dependent on powder choice and SLM capabilities of the machine. With the help of this paper researchers who are interested in further developing this technology can determine exactly how to measure the performance of such a material to determine what is needed to provide a material suitable for electric motors. Future work included in our research will include using modular layers to try and reduce eddy currents and

REFERENCES

[1] Selema, Ahmed, et al. "Metal Additive Manufacturing for Electrical Machines: Technology Review and Latest Advancements." MDPI, Multidisciplinary Digital Publishing Institute, 31 Jan. 2022, www.mdpi.com/1996-1073/15/3/1076.

[2] Tiismus, Hans, et al. "Hysteresis Measurements and Numerical Losses Segregation of Additively Manufactured Silicon Steel for 3D Printing Electrical Machines." MDPI, Multidisciplinary Digital Publishing Institute, 18 Sept. 2020, www.mdpi.com/2076-3417/10/18/6515.

[3] "3% Silicon Steel Core Material." 3% Silicon Steel Core Datasheet, U.S. Department of Energy - National Energy Technology Laboratory, Sept. 2008, www.netl.doe.gov/sites/default/files/netl-file/Core-Loss-Datasheet---3-percent%5B1%5D.pdf.

[4] G. Paltanea, V. Manescu, P. C. Andrei, C. Grumeza and S. Marinescu, "Laboratory set-up to evaluate the B-H relationship in soft magnetic materials," 2017 10th International Symposium on Advanced Topics in Electrical Engineering (ATEE), Bucharest, Romania, 2017, pp. 420-423, doi: 10.1109/ATEE.2017.7905145.

[5] C. Urabinahatti, S. S. Ahmad and G. Narayanan, "Magnetic Characterization of Ferromagnetic Alloys for High-Speed Electric Machines," in IEEE Transactions on Industry Applications, vol. 56, no. 6, pp. 6436-6447, Nov.-Dec. 2020, doi: 10.1109/TIA.2020.3023868.

[6] "Selection of Electrical Steels for Magnetic Cores." *AK Steel Corporation*, AK Steel Corporation, 7 July 2007, www.brown.edu/Departments/Engineering/Courses/ENGN1931F/mag_cores_dataAKSteel-very%20good.pdf.

[7] T. Vaimann, H. Tiismus and A. Kallaste, "Concepts of Additively Manufactured Electrical Machines and Components," 2023 23rd International Scientific Conference on Electric Power Engineering (EPE), Brno, Czech Republic, 2023, pp. 1-6, doi: 10.1109/EPE58302.2023.10149252.

[8] T. Mix et al., "Additive Manufacturing of Low Loss Electrical Steel Sheets for High Efficiency Electrical Devices," in IEEE Transactions on Transportation Electrification, vol. 9, no. 4, pp. 5226-5231, Dec. 2023, doi: 10.1109/TTE.2023.3298488.

[9] V. Hundak, T. Cox, G. Vakil and C. Gerada, "Mass Production Costing of Induction Machines for Automotive Applications," 2018 IEEE Transportation Electrification Conference and Expo (ITEC), Long Beach, CA, USA, 2018, pp. 500-505, doi: 10.1109/ITEC.2018.8450116. keywords: {Manufacturing;Torque;Induction machines;Optimization;Costing;Mathematical model;Windings},

[10] M. M. Rahaman and K. S. Sandhu, "Energy Efficient magnetic materials for Electrical Machines," 2019 5th International Conference on Advanced Computing & Communication Systems (ICACCS), Coimbatore, India,

2019, pp. 642-646, doi: 10.1109/ICACCS.2019.8728342. keywords: {Soft magnetic materials;Magnetic hysteresis;Metals;Magnetic cores;Eddy currents;Magnetic flux;Induction Machine;Transformer;MATLAB;Silicon-Iron Alloy;Soft Magnetic Composites;Amorphous Magnetic Material.},

[11] M. Najgebauer, J. Szczyglowski and A. Kaplon, "Soft magnetic materials for energy-efficient electric motors," 2015 Selected Problems of Electrical Engineering and Electronics (WZEE), Kielce, Poland, 2015, pp. 1-4, doi: 10.1109/WZEE.2015.7394038. keywords: {Steel;Magnetic cores;Induction motors;Loss measurement;Magnetic properties;Amorphous magnetic materials;efficiency of electric motors;motor losses;non-oriented steels;6.5% Fe-Si steels;Fe-based amorphous alloys},

[12] S. Lee and J. Yun, "Influence of electrical steel characteristics on efficiency of industrial induction motors," 2017 20th International Conference on Electrical Machines and Systems (ICEMS), Sydney, NSW, Australia, 2017, pp. 1-4, doi: 10.1109/ICEMS.2017.8056353. keywords: {Core loss;Steel;Induction motors;Magnetic flux density;Harmonic analysis;Load modeling;Stator cores;Core loss;Electrical steel;Induction motor;Stray load loss},

Extreme Temperature Permeability Engineered Soft Magnetics

Tyler W. Paplham
Mechanical Engineering & Materials Science
University of Pittsburgh
Pittsburgh, USA
tyler.paplham@pitt.edu

Alex M. Leary
Materials and Structures Division
NASA Glenn Research Center
Cleveland, USA
alex.m.leary@nasa.gov

Paul R. Ohodnicki, Jr.
Mechanical Engineering & Materials Science
University of Pittsburgh
Pittsburgh, USA
pro8@pitt.edu

Abstract— **Applications may require high temperature power components, either due to temperature rise or high ambient temperatures. For temperatures greater than 300 C and frequencies greater than 1 kHz, novel nanocrystalline soft magnetic materials based upon Co- and FeCo-alloy systems become uniquely relevant due to the limitations of permalloys, ferrites, and amorphous alloys. Controlling material permeability directly (i.e. "permeability engineering") through advanced manufacturing to produce an induced magnetic anisotropy within the core rather than gapping is also desired to avoid instabilities of binders and resins used in the gapping process. However, the temperature stability of permeability engineered nanocrystalline alloys has not been well explored in the literature. Here we present measurement comparisons of current materials for high temperature and frequency, show impact of high temperature and large excitation on stabilities of anisotropies in relevant materials, and discuss limitations and opportunities for future materials.**

Keywords—power magnetics, soft magnetic materials, high-temperature

I. INTRODUCTION

Components for electronic systems may require high temperature capability if heat generated in the volume exceeds the heat flux through the surface of the component, or if ambient temperature is high and available cooling is limited. In the first case, common in systems seeking high power density, relative temperature rise above ambient is determined by component geometry/materials, electrical input (magnitude, frequency, etc.), and thermal boundary conditions. High ambient temperatures for the second case are found in drilling applications [1], use near engines [2], and various aerospace applications [3]. A component may be cooled during use, but this adds costs and high temperatures after shutdown can occur from thermal soak back when cooling is no longer available. Practical application of inductive components requires suitable performance from all materials used in the core including the soft magnetic material, binders or impregnation used to consolidate powder and laminated cores, and bobbins/cases that provide mounting support and decouple winding stress from the core. Binders or impregnation are important in discrete or distributed (powder) air gapped cores that must maintain dimensional stability for constant permeability. Many standard electronic components are rated to 125 °C, with a wide selection available from both active and passive components that meet

reliability demands [4]. High power density applications can benefit from inductive components with higher temperature ratings. Factors which affect the high temperature performance of soft magnetic materials include: (1) direct effects on the magnetic exchange interactions which control Curie temperatures, (2) indirect effects on secondary magnetic properties (e.g., magnetostriction, magnetocrystalline anisotropy), (3) effects on stresses internal to the core material or applied to cores through mounting fixtures, windings, etc., (4) oxidation/corrosion on core material surfaces, (5) stability of relevant magnetic phases and interfaces, and (6) stability of binders and core packaging materials. The temperature dependent effects can be categorized into those that are reversible or irreversible after temperature cycling. Reversible changes to core properties are possible when both the magnetic phases and microstructure do not change and the mechanical stresses are reversible. Irreversible impacts on grain size and structure as a result of temperature cycling must be considered. Kinetics of property changes must also be considered to determine the impact on the larger system.

This work describes the current state of the art for soft magnetic materials with intrinsic permeability for use in high temperature (> 300 °C) and high frequency (> 1 kHz) applications through a combination of alloy chemistry and details of thermal processing conditions. Given our target of extreme applications temperatures approaching 500 °C, such as in aerospace and planetary exploration, the performance is measured at ambient temperatures through 500 °C and mechanisms limiting performance are classified and discussed. Emphasis is placed on FeCo-based and Co-based nanocrystalline alloys, which are differentiated from Fe-based and FeSi-based alloys by their exceptionally high Curie temperatures and improved mechanical integrity [5]. Additionally, these materials do not require an air gap to control permeability and provide designers with new options for achieving desired inductance values without the limitations and weaknesses associated with the integration of a gap through binders or impregnating resins. Instead, their permeability along the magnetizing direction can be tuned directly through an induced anisotropy [6]. This concept, often referred to as "permeability engineering", is performed during initial thermal processing of the nanocrystalline alloy, by annealing under either a saturating magnetic field (which may be oriented parallel or transverse to the ribbon axis) or a large tensile stress

---

Funding support by the NASA HOTTech program award number #80NSSC22K0415 and the AMPED Consortium.

along the ribbon axis. As shown schematically in Fig. 1, field annealing generally produces an easy axis of magnetization along the applied field direction. Microstructurally, this results from preferential orientation of particular chemical bonds in the alloys [7]. Stress annealing can produce more complex magnetic anisotropies, such as easy planes normal to the stress axis depending on composition and applied stress tensor; the exact nature of the relationship between applied stress magnitude and permeability is also material dependent. In Co-based alloys, the stress-induced anisotropy is believed to be associated with oriented structural defects in the nanocrystalline phase (these may also play a role in field-annealed Co-based alloys) [8]. FeCo-based alloys are typically not stress annealed for inductive applications as a longitudinal easy axis with high permeability results. Generally, stress-induced anisotropies are much stronger than field-induced anisotropies in Co-based systems, with a consequently larger range of permeabilities accessible from ~10-10,000 depending on chemistry [8]. The stability of this anisotropy over extended periods at high temperature, ultimately responsible for measured value of permeability and hence inductance, is thereby critical for a component operating in a high-temperature environment but has not yet been well-explored in the literature. Here, the basic impact of temperature, induced anisotropy type, and applied excitation amplitude (including those of significant magnitude with respect to the saturation field) are considered for relevant materials systems.

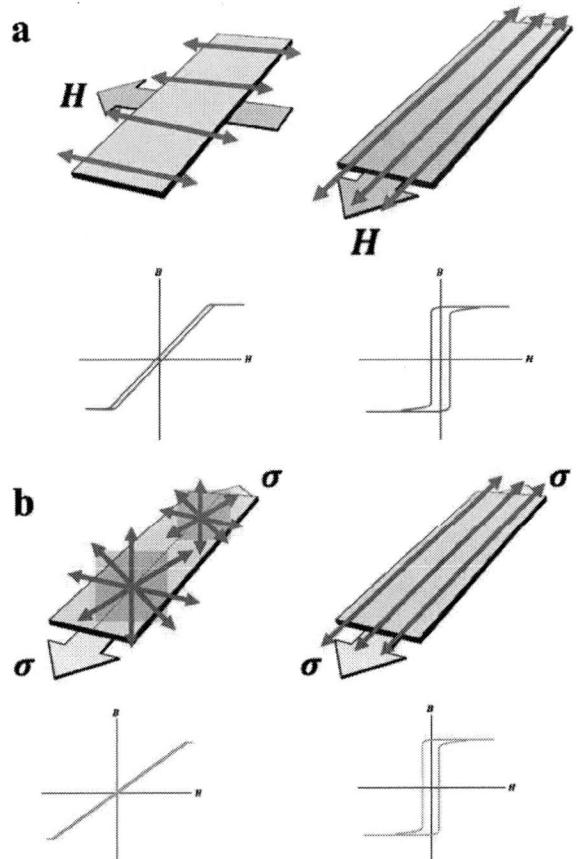

Fig. 1.   Schematic of the formation of easy axes (red arrows) and representative hysteresis loops under the influence of an applied magnetic field H or tensile stress σ in nanocrystalline alloys (loops not to scale).

II. METHODS

A system to measure core properties at high ambient temperatures in air was constructed. Sample cores include a variety of commercially sourced cores and custom nanocomposite cores produced by NASA Glenn Research Center (GRC). Vented ceramic bobbins housed sample cores to limit external stresses and were wound with Ni wire (Ceramawire). Temperature dependent core properties were measured using the wattmeter method [9] using a Keysight DSOX3034 oscilloscope, 1147B probe measured current in the primary winding, and N2790A differential probe measured voltage in the open secondary winding. An AE Techron 7224 wideband amplifier supplied a sinusoidal voltage with zero dc bias to the primary winding and applied voltage was adjusted to maintain a constant flux density, as measured by the secondary winding. The core was placed in a convection furnace to maintain uniform temperatures and a heating rate of 2 °C/min to limit temperature differences between core and ambient. Temperature was measured with a Type K thermocouple at the center of the wound toroid.

III. RESULTS AND DISCUSSION

Fig. 2 presents measured relative permeabilities μ_r and core losses under a 50 mT, 50 kHz excitation as a function of temperature for several representative core materials. The supermalloy (Magnetics FeNiMo) and the MnZn ferrite (Ferroxcube 3C97) both have relatively high phase stability across the measured temperature range, but other factors limit their high temperature magnetic performance. The supermalloy has a Curie temperature T_c = 410 °C. Observed fluctuations in the permeability up to this temperature are attributed to factor (2) listed above, specifically temperature dependence of the magnetostriction or magnetocrystalline anisotropy. By design, these quantities are nearly zero in supermalloy leading to extremely high permeabilities, but temperatures at which the magnetostriction or anisotropy have an exact zero crossing lead to further peaks in permeability. This is most commonly observed at the Curie temperature, where the anisotropy constants go to zero at some power $n > 1$ of the magnetization, called a Hopkinson peak [10]. Direct studies of the temperature dependent anisotropy constants in supermalloy could not identified, but the broad peak just above 200 °C corresponds to a zero crossing in nickel, the primary constituent of the alloy [11]. Supermalloy is also limited to frequencies < 10 kHz due to eddy current losses. The MnZn ferrite is limited to T < 220 °C due to its low Curie temperature, and shows no lower-temperature peaks. Mn-based ferrites, including similar Ferroxcube compositions, do tend to exhibit a peak in permeability at lower temperature, however this effect is greatly reduced with Zn addition and is absent in NiZn ferrites, again due to influences on zero crossings of the anisotropy [12]. The chosen nanocrystalline alloys exhibit very stable permeability through 500 °C. This is in part due to the high Curie temperatures of these alloys, satisfying the first factor listed above, but also due to the relatively flat temperature dependence of the induced anisotropy, which dominates the magnetization process in these alloys. Unlike in the other materials where the factors which dictate permeability can have significant variation with temperature, the mechanisms controlling permeability in these alloys (oriented chemical bonds/structural defects) are

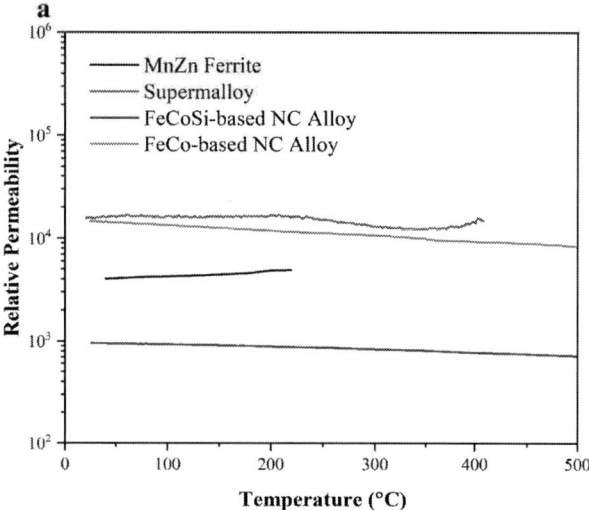

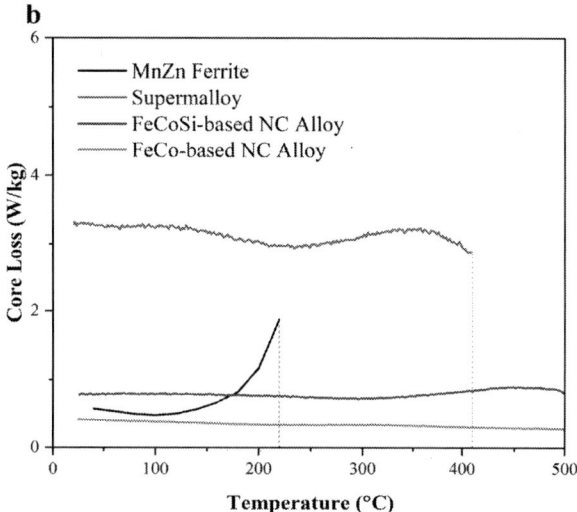

Fig. 2. Permeability (a) and core loss (b) measured under 50 mT, 50 kHz sinusoidal excitation (zero dc bias) as a function of temperature.

induction B_m that reaches the saturating field B_s when $\theta = \pi/2$ and the magnetization aligns with the applied field direction. Fig. 3(b) shows the permeability evolution over time at 500 °C for the FeCo-based nanocrystalline cores subjected to applied signals of magnitudes ranging from $B_m < 0.05B_s$ up to $B_m = 0.5B_s$. For all cases, the permeability tends to approach a new limiting value with time; this value can be smaller than the original permeability for small signals, while larger amplitudes tend to result in increasingly large, positive changes in permeability. Larger field amplitudes also require longer times for permeability to settle to the new limiting value. Uniform magnetic anisotropies parallel to H create high permeabilities, so the permeability increase with large excitation suggests weakening of the existing transverse anisotropy field (easy axis perpendicular to H) or the creation of a separate longitudinal anisotropy field parallel to H. Mixed global magnetic anisotropies increase losses and will be an expected signature of this high temperature effect. For small amplitudes, θ is small and the magnetization therefore lies close to the original transverse easy axis; at high temperatures, the original anisotropy may then be strengthened through a self-field anneal effect rather than weakened, leading to a drop in hard axis permeability as seen for the 50 mT excitation in Fig. 3(b). Cores are also exposed to air during test, so stress related to surface oxidation may create magnetic anisotropies that change the permeability. The dependence of the kinetics on field strength, i.e., the increased settling time at higher field strength is not well understood at present and will be a focus of future work. In addition, the role of the waveform should be considered, as many power applications involve non-sinusoidal signals, potentially with a dc bias. There is evidence to suggest that a dc bias may decrease stability [13], with the simple interpretation that a steady-state component of H along the ribbon axis will more readily stabilize a longitudinal anisotropy associated with large permeability. However, the magnitude of this impact on the long-term kinetics of the anisotropy are as yet unknown, and the results of Fig. 3 suggest that there will be a strong sensitivity to the magnitude of the dc component. Spatial inhomogeneities in the magnitude and orientation of flux with respect to the magnetizing direction, which may be imposed by core geometry, may create differences in the local stability of the anisotropy, however as only a single effective permeability value is measured for the core via the current technique, this effect has not been observed in experiment.

Lastly, the sensitivity of the kinetic stability of the core to ambient temperature and induced anisotropy type (i.e. field-induced vs. stress-induced) is presented in Fig. 4 for Co-based nanocrystalline cores excited by a 50 mT, 5 kHz longitudinal signal. The cores had initial relative permeabilities of $\mu_r = 343$ (transverse field anneal) and $\mu_r = 38$ (stress anneal). At 400 °C, the two types of induced anisotropy exhibit permeability changes of opposite sign, with the field annealed alloy showing a decrease of ~0.6% in permeability (representing an enhancement of the induced anisotropy) while the stress anneal shows an increase of ~3.6% (representing a degradation of induced anisotropy). Dips in the permeability curves spaced ~24 hours apart are expected to be an artifact of daily changes in ambient laboratory conditions.

very stable. In the case of oriented chemical bonds, destabilization requires sufficiently high temperature to allow for atomic diffusion, generally > 400 °C for practical time scales. The oriented structural defects are likely even more stable [8], although this is the subject of ongoing research. Provided the other factors are not significant, the low core losses will also be stable in these nanocrystalline alloys.

The relative strength of the *in-situ* magnetic field applied during high-temperature operation can influence induced anisotropy stability, as shown in Fig. 3 for a set of FeCo-based nanocrystalline cores ($B_s \sim 1.4$ T). These cores were annealed with a transverse magnetic field, to produce a low frequency, low temperature permeability of $\mu_r \sim 900$. Fig. 3(a) shows the relative geometry of the initial easy axis in the core with magnetization (M_s) oriented along the transverse direction. During excitation, the field produced by the primary winding (H), rotates the magnetization through angle θ, creating a dB/dt component parallel to H that induces an emf in the secondary winding. This voltage is integrated to yield the measured

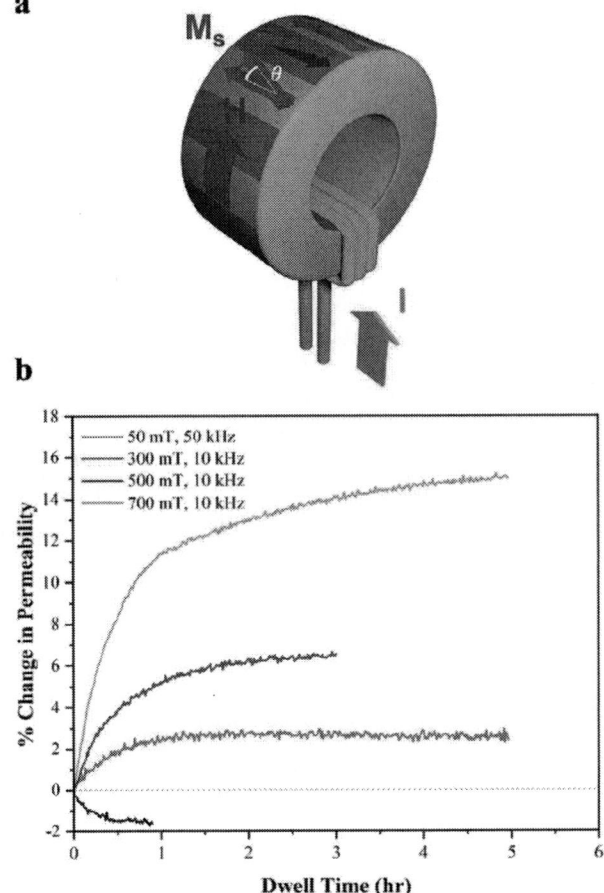

Fig. 3. (a) Schematic of domain magnetization rotation when the applied field is along the hard axis in a toroidal inductor. (b) Evolution of permeability with time in a FeCo-based nanocrystalline core under sinusoidal excitations of different magnitude.

At 500 °C, the evolution of permeability occurs more rapidly due to the enhanced kinetics (see different time scales between Fig. 4(a) and Fig 4(b)). The stress annealed alloy exhibits a significantly larger increase in permeability (18.5%) which has not fully stabilized by the end of the measurement period. The field-annealed alloy, on the other hand, exhibits an initial decrease of approximately the same magnitude as at the lower temperature, but then begins to climb after approximately three hours as visible in the inset to Fig. 4(b). This non-monotonic behavior suggests that there may be competing influences from microstructural changes developing at different kinetic time scales, such as atomic diffusion acting to increase or decrease number of preferentially oriented bonds, effects associated with coarsening of the crystalline phase, changes in the density or orientation distribution of defects in the crystalline phase, etc. It is also possible that the differences in stability of field-induced vs. stress-induced anisotropies is related to differences in domain structure between the two classes of processing conditions [14]. More specifically, closure domains are typically formed at the surface of the stress-annealed alloy due to the in-plane easy axis anisotropy, while in-plane stripe domain structures largely absent of surface closure domains are characteristic of transverse field-annealed samples. Surface closure domains, in which the magnetization is not directed along the in-plane transverse direction of the ribbon, may serve as convenient centers for growth of domains oriented along the ribbon axis, resulting in the destruction of the transverse anisotropy. These surface domains are absent in the field-annealed alloy, which may result in enhanced stability of the originally established domain structure, anisotropy and permeability.

IV. CONCLUSIONS AND FUTURE WORK

High-temperature soft magnetic properties were measured at frequency range relevant to power electronic application (kHz) and traditional materials were compared to nanocrystalline cores with induced magnetic anisotropy. Preliminary results show that nanocrystalline cores have potential for applications through 500 °C, but the operating temperature, type of induced anisotropy, and excitation field were all shown to influence the kinetic stability of the cores. The present results show a significant improvement compared to MnZn ferrites, limited to

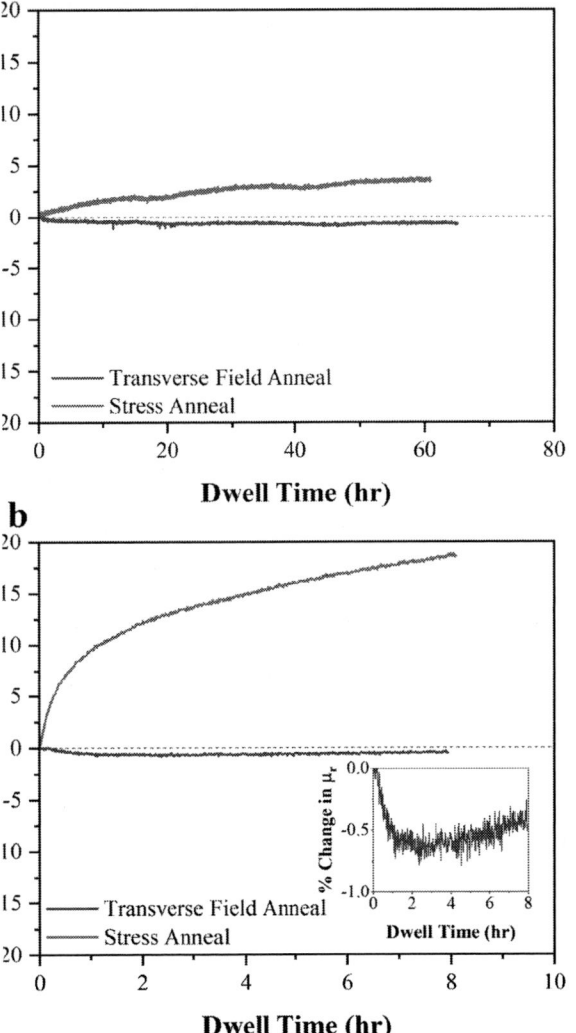

Fig. 4. Evolution of permeability in field- or stress-annealed Co-based nanocrystalline cores at ambient temperatures of (a) 400 °C and (b) 500 °C isothermal annealing while measuring small signal permeability.

less than 220 °C, and supermalloy, limited to less than 410 °C, with higher losses compared to the nanocrystalline alloys. Future work will attempt to establish functional relationships between the anisotropy (permeability) stability and ambient temperature, applied excitation, and initial value of induced anisotropy (permeability) to enable materials design for further stability improvement, and will set limits for the kinetic effects of property changes at high temperature to establish potential for use by circuit designers.

ACKNOWLEDGMENT

The authors acknowledge the support of the NASA High Operating Temperature Technology (HOTTech) and NASA Research Opportunities in Space and Earth Science (ROSES) programs. Partial support is also provided by the Advanced Magnetics for Power and Energy Development (AMPED) industry consortium.

REFERENCES

[1] S. P. Rountree, S. Berjaoui, A. Tamporello, B. Vincent, and T. Wiley "High temperature measurement while drilling systems and applications," In 5th International HiTEC Conference, 2000.

[2] W. C. Nieberding and J. A. Powell, "High-temperature electronic requirements in aeropropulsion systems," IEEE Trans. Industrial Elec., vol. IE-29, 1982.

[3] P. F. McCluskey, T. Podlesak, and R. Grzybowski, High Temperature Electronics. CRC Press, 2018.

[4] W. R. Johnson, J. L. Evans, P. Jacobsen, J. R. Thompson, and M. Christopher, "The changing automotive environment: High-temperature electronics," IEEE Trans. Electronics Packaging Manuf., vol. 27, pp. 164-176, 2004.

[5] P. R. Ohodnicki, S. Y. Park, H. K. McWilliams, K. Ramos, D. E. Laughlin, and M. E. McHenry, "Phaasee evolution during crystallization of nanocomposite alloys with Co:Fe ratios in the two-phasee region of the binary Fe-Co phase diagram," J Appl. Phys., vol. 101, 2007.

[6] K. Byerly, P. R. Ohodnicki, S. R. Moon, A. M. Leary, V. Keylin, M. E. McHenry et al. "Metal amorphous nanocomposite (MANC) alloy cores with spatially tuned permeability for advanced power magnetics applications," JOM, vol. 70, pp. 879-891, 2018.

[7] L. Néel, "Anisotropie magnétique superficielle et surstructures d'orientation," J Phys. Le Radium, vol. 15, pp. 225-239, 1954.

[8] A. M. Leary, V. Keylin, A. Devaraj, V. DeGeorge, P. R. Ohodnicki, and M. E. McHenry, "Stress induced anisotropy in Co-rich magnetic nanocomposites for inductive applications," J Mat. Res., vol. 31, pp. 3089-3107, 2016.

[9] International Electrotechnical Commission, "Magnetic materials—Part 6: Methods of measurement of the magnetic properties of magnetically soft metallic and powder materials at frequencies in the range 20 Hz to 100 kHz by the use of ring specimens," IEC 60404-6, 2018.

[10] J. Hopkinson, "Magnetic and other physical properties of iron at a high temperature," Philos. Trans. R. Soc. London A, pp. 443-465, 1889.

[11] M. Kersten, "Zur Wirking der Versetzungen auf die Anfangspermeabilität von Nickel im rekristallisierten und im plastich verformten Zustand," Ann. Phys., vol. 455, pp. 337-344, 1957.

[12] B. Kubota and T. Nishikawa, "On the temperature dependency of initial permeability for Ni-Zn ferrites," Zairyoshiken, vol. 10, pp. 472-479, 1962.

[13] P. R. Ohodnicki, D. E. Laughlin, M. E. McHenry, V. Keylin, and J. Huth, "Temperature stability of field induced anisotropy in soft ferromagnetic Fe, Co-based amorphous and nanocomposite ribbons," J Appl. Phys., vol. 105, 2009.

[14] P. R. Ohodnicki, A. M. Leary, R. R. Bowman, R. D. Noebe, G. E. Feichter, K. Byerly et al., "Methods of modifying a domain structure of a magnetic ribbon, manufacturing an apparatus, and magnetic ribbon having a domain structure," US Patent 20220298615A1, 2022.

An Isolated RF Power Combining Approach with Multiple Decoupled Input Coils

Ziyang Xu
Dept. of Electrical and Computer Engineering
The University of Texas at Austin
Austin, United States
ziyangxu@utexas.edu

Yifan Zhao
School of Electrical Engineering
Shandong University
Jinan, China
yifan.zhao@mail.sdu.edu.cn

Zhan Liu
UMich-SJTU Joint Institute
Shanghai Jiao Tong University
Shanghai, China
worldhuman@sjtu.edu.cn

Alex J. Hanson
Dept. of Electrical and Computer Engineering
The University of Texas at Austin
Austin, United States
ajhanson@utexas.edu

Ming Liu
Dept. of Electrical Engineering
Shanghai Jiao Tong University
Shanghai, China
mingliu@sjtu.edu.cn

Abstract—To meet high RF power levels, it is common to combine the output power from multiple power amplifiers. Conventional isolated power combining networks are bulky and inefficient, while efficient lossless combiners lack necessary isolation for industrial applications. In this paper, an isolated RF power combining approach is proposed. A polygonal coupling coil structure is further proposed, which can allow an arbitrary number of input ports while eliminating cross-coupling between them by flux cancellation. A capacitor decoupling network is also introduced to further improve decoupling performance at high frequencies. The proposed system maintains ZVS operation of all amplifiers with mismatched gate signals. An experimental prototype with compact isolation coupler of 10cm×10cm×1.3cm is demonstrated, which combines four 13.56MHz Class-E power amplifiers and outputs 50W-300W power with a stable efficiency of 65%.

Index Terms—radio frequency, power combining, power amplifier, decoupling, coil design

I. INTRODUCTION

Radio frequency (RF) power amplifiers are widely used across various fields, including communications, wireless power transfer, and industrial plasma generation. Traditionally, linear amplifiers such as Class A, B, and AB have been preferred due to their design simplicity and linearity. However, as demand for higher efficiency grows, switched-mode power amplifiers—such as Class D, E, EF, and Φ—are gaining popularity [1]–[4], which can achieve higher efficiency with a more compact size. However, as operating frequencies increase beyond 10 MHz, the output power of a single amplifier becomes restricted by various practical limitations, including significant parasitic parameters, thermal design, the voltage/current ratings of high-frequency switching devices [5]. These factors collectively limit the achievable power output of individual amplifiers at higher frequencies. Consequently, high power levels are typically achieved by combining the output power from multiple amplifiers.

Conventional RF power combining network can be divided into two categories: isolated and non-isolated, as shown in

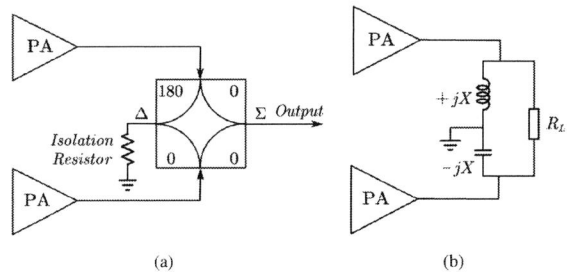

Fig. 1. (a) Conventional isolated combiner (b) non-isolated combiner

Fig. 1. Conventional isolated RF power combination networks, such as commonly-used Wilkinson combiner, are bulky and inefficient, in which all differential portion of energy is lost as heat on the isolation resistor [6], [7]. Conventional non-isolated combining methods use power combiner networks to combine output power of power amplifiers, which can solve the mismatch problem and achieve improved efficiency, but they still have disadvantages of high loss and weight [8], [9].

A new RF power combining method for high power applications was proposed in [10], which directly combines output power of power amplifiers without combiner networks and achieves relatively high efficiency. However, this setup doesn't show potentials for galvanic isolation, which is crucial in many industrial applications. Also, electrical isolation also offers a promising solution to address signal mismatch issues at high frequencies. At these frequencies, mismatched signals are often unavoidable due to parasitic elements in circuits and variations in digital chip characteristics. By eliminating direct electrical connections between amplifiers, isolation can significantly reduce interference among amplifiers, even when signals are mismatched. As a result, an isolated and efficient power combining approach is desirable, which also provides the potential for modular RF power transmission and easier control implementations.

To fill the technique gap, this paper proposes an isolated RF power combining approach with multiple decoupled input coils, with inspiration taken from the field of wireless power transfer [11], [12]. We introduce a polygonal coupling coil structure capable of combining power from multiple amplifiers while simultaneously eliminating cross-coupling between input ports. We also present a capacitor-based decoupling network to further enhance input port decoupling, which effectively mitigates amplifier mismatches and robustly maintains their zero-voltage-switching (ZVS) conditions. An experimental prototype with a compact isolation coupler (10cm×10cm×1.3cm), which combines four 13.56 MHz Class-E power amplifiers, validates the proposed methods.

II. POLYGONAL COUPLING COIL DESIGN FOR RF POWER COMBINATION

To realize isolated power combination of different amplifiers, a well-designed magnetic coupler is of great importance. It should be compact and efficient to align with the advantages of switched-mode power amplifiers. Additionally, it should allow arbitrary number of amplifiers to be combined while minimizing the interference between them.

Conventional transformer designs employ either multiple magnetic cores to avoid input ports coupling, increasing the system size and relying on limited high-frequency core selections, or a multi-winding design that tightly couples all ports [13], which would introduce the circulating currents caused by unavoidable mismatched signals of RF power amplifiers and cause the amplifiers to deviate from their desired operating condition. To further illustrate the influence of coupling input ports, impedance analysis is employed as follows. Fig. 2 shows the commonly-used mutual inductance model, in which the input impedance Z_i of ith input port can be derived as shown in (1),

$$
\begin{aligned}
Z_i =& \frac{V_i}{I_i} = j\omega L_i + \sum_{k \neq i} j\omega M_{ik}\frac{I_k}{I_i} + j\omega M_{io}\frac{I_o}{I_i} \\
=& j\omega L_i + \sum_{k \neq i} \alpha_{ki}\omega M_{ik}\angle(90° + \theta_{ki}) + j\omega M_{io}\frac{I_o}{I_i} \\
=& \sum_{k \neq i} \alpha_{ki}\omega M_{ik}\cos(90° + \theta_{ki}) + \\
& j\omega(L_i + \alpha_{ki}\sin(90° + \theta_{ki})) + j\omega M_{io}\frac{I_o}{I_i}
\end{aligned}
\tag{1}
$$

In (1), V_i and I_i represent the voltage and current at the i-th input port, while V_o and I_o correspond to the voltage and current at the output port. L_i denotes self-inductance of the ith input coil, and M_{ik} and M_{io} denote mutual inductance between the ith coil and the kth coil, and between the ith coil and the output coil, respectively. Additionally, α_{ki} and θ_{ki} are defined to represent the amplitude and phase influence of the kth port on the ith port respectively. It can be seen that when input ports are coupled, both amplitude and phase mismatches in the currents cause variations in the port impedance. This results in circulating currents between input ports, which do not contribute to power delivery to the load but instead flow between the ports. Consequently, the equivalent load seen by power amplifiers is significantly affected, leading power amplifiers to lose their ZVS operations, which can ultimately damage the system.

Drawing inspiration from wireless power transfer coils—specifically, planar coils without cores—this paper proposes a polygonal planar coupling coil structure, offering high design flexibility at high frequencies. Unlike typical coupling coils used for wireless power delivery, the coupling conditions of the coils in this design are fixed, enabling precise designs with high accuracy requirements.

The structure and design procedure of the proposed polygonal coupling coils are shown in Fig. 3, where the blue coils represent input coils connected to the power amplifiers, and the yellow coil represents the output coil connected to the final load. By canceling the positive and negative flux generated by adjacent coils, the coupling between adjacent input coils is eliminated. Additionally, the coupling between non-adjacent input coils can be significantly reduced by designing the rectangular coils to be thinner and keeping them spaced apart. Theoretically, the cross-coupling between input coils can be minimized to nearly zero.

The coil parameters design basically follows the general design procedure within certain constraints. The shape of output coil is determined with the number of input power ports, while the size is limited by the requirement of system. The line width and number of turns in the coils must be balanced according to the voltage and current level of the system. Increasing the number of turns enhances the mutual inductance (necessary for higher voltages) while limiting the width of lines and increasing coil resistance (detrimental at high currents). The size and positioning of the input coils can then be optimized using finite element analysis (FEA) tools

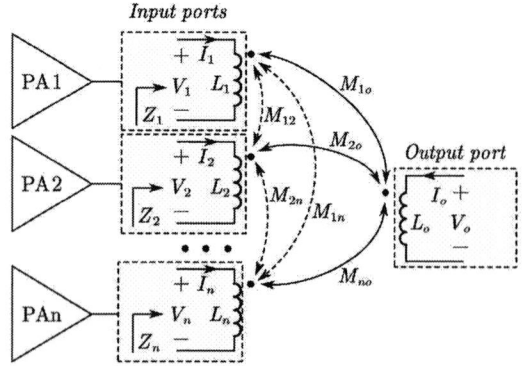

Fig. 2. Mutual inductance model of multiple input coils connected with power amplifiers and single output coil connected with final load.

979-8-3315-1612-3/25 $31.00 © 2025 IEEE

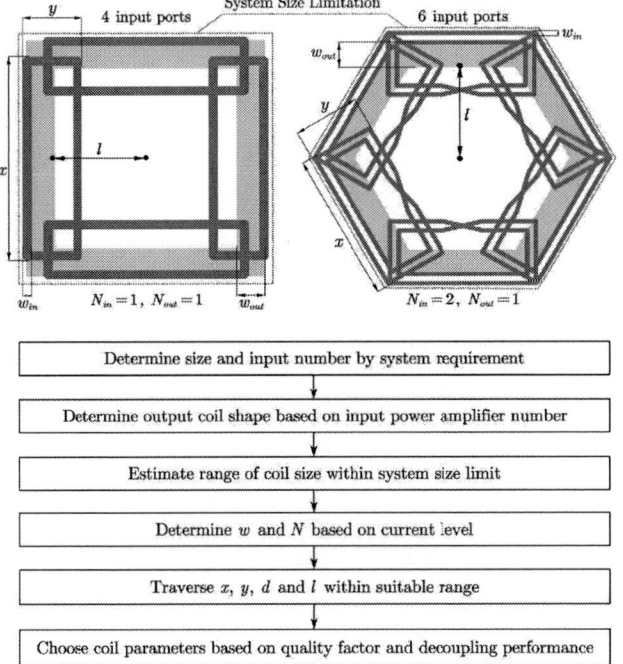

Fig. 3. The structure and design procedure of the proposed polygonal coupling coils.

to achieve a higher quality factor and improved decoupling performance as mentioned earlier.

III. CAPACITOR DECOUPLING NETWORK

Due to the significant limitations and challenges in the electromagnetic simulation of high-frequency coils, which cannot fully achieve decoupling of input coils, this paper also proposes a capacitor decoupling network to further enhance the decoupling performance. This network can also be applied to a wide range of multi-port systems that use ordinary magnetic couplers to decouple any ports [14], [15].

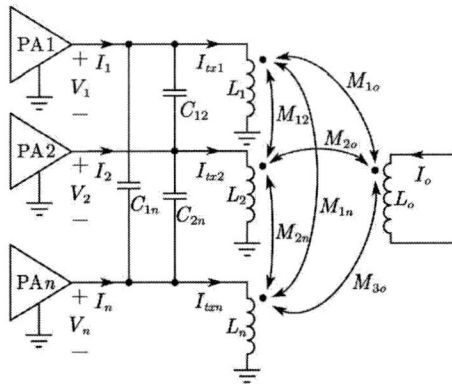

Fig. 4. Capacitor decoupling network for n input coils and a single output coil.

Fig. 4 shows the proposed capacitor decoupling network with with n input coils and a single output coil. By introducing

decoupling capacitors among input coils, the circuit now can be viewed as two cascaded stages. The port characteristics of each network can be determined separately as follows,

$$
\begin{bmatrix} I_1 \\ I_2 \\ \vdots \\ I_n \\ I_o \end{bmatrix} - \begin{bmatrix} I_{tx1} \\ I_{tx2} \\ \vdots \\ I_{txn} \\ I_o \end{bmatrix} = j\omega \begin{bmatrix} \sum_{i\neq 1} C_{1i} & -C_{12} & \cdots & -C_{1n} & 0 \\ -C_{21} & \sum_{i\neq 2} C_{2i} & \cdots & -C_{2n} & 0 \\ \vdots & \vdots & \ddots & \vdots & \vdots \\ -C_{n1} & -C_{n2} & \cdots & \sum_{i\neq n} C_{ni} & 0 \\ 0 & 0 & \cdots & 0 & 0 \end{bmatrix} \begin{bmatrix} V_1 \\ V_2 \\ \vdots \\ V_n \\ V_o \end{bmatrix}
$$

(2)

$$
\begin{bmatrix} V_1 \\ V_2 \\ \vdots \\ V_n \\ V_o \end{bmatrix} = j\omega \begin{bmatrix} L_1 & M_{12} & \cdots & M_{1n} & M_{1o} \\ M_{21} & L_2 & \cdots & M_{2n} & M_{2o} \\ \vdots & \vdots & \ddots & \vdots & \vdots \\ M_{n1} & M_{n2} & \cdots & L_n & M_{no} \\ M_{o1} & M_{o2} & \cdots & M_{on} & L_o \end{bmatrix} \begin{bmatrix} I_{tx1} \\ I_{tx2} \\ \vdots \\ I_{txn} \\ I_o \end{bmatrix}
$$

(3)

In (2) and (3), I_i represents the output current of ith power amplifier, which is also the input current of the network. I_{txi} represents the current flowing into the ith input coil. While there is no capacitor added to the output side, I_o is used to denoted the output port current. C_{ij} represents the capacitor value added between the ith and jth port while M_{ij} is the original mutual inductance of coils. Notably, if symmetry is perfectly achieved between different input ports, the capacitor decoupling network would be inherently open.

The matrices (2) and (3) can be abbreviated into symbolic form as follows,

$$
\begin{aligned} \boldsymbol{I} - \boldsymbol{I_{tx}} &= j\omega \boldsymbol{C} \boldsymbol{V} \\ \boldsymbol{V} &= j\omega \boldsymbol{M} \boldsymbol{I_{tx}} \end{aligned}
$$

(4)

Then, the port characteristics of the entire network to be derived as follows,

$$
\boldsymbol{V} = j\omega(\boldsymbol{M}^{-1} - \omega^2 \boldsymbol{C})^{-1}\boldsymbol{I}
$$

(5)

It can be seen from (5) that by adjusting the value of the capacitor network, the port characteristics of the entire network can be altered, allowing specific couplings to be eliminated. More intuition can be gained by using an equivalent circuit model of the coupled magnetic component such as the extended cantilever model (Fig. 5). For instance, ports can be decoupled by choosing capacitors which are parallel resonant with equivalent inductance L_{ij} in the extended cantilever model.

When evaluating the difference between two decoupling approaches of the mutual inductance model and cantilever model mathematically, a duality becomes apparent. Specifically, the first model is based on the **Z** impedance matrix, while the second is based on the **Y** admittance matrix, resulting in different capacitor values. In this paper, the impedance matrix-based approach is employed to achieve decoupling of the input coils.

979-8-3315-1612-3/25 $31.00 © 2025 IEEE

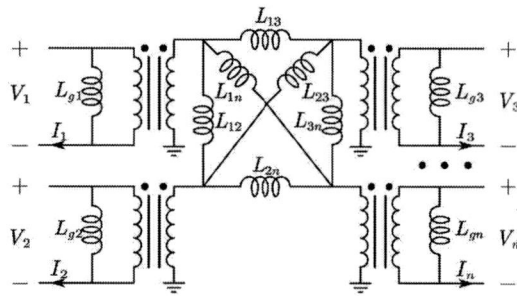

Fig. 5. An extended cantilever model with n ports.

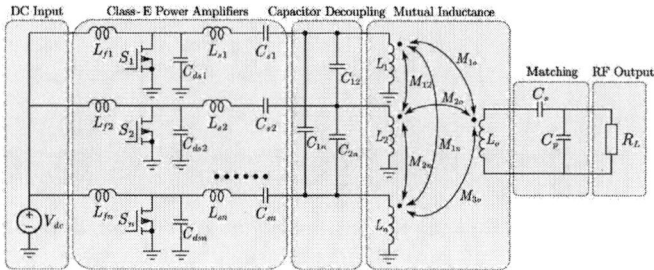

Fig. 6. System diagram of RF power combining with proposed capacitor decoupling network.

Fig. 6 shows the diagram of the proposed RF power combining system, which combines the output power of n power amplifiers to a single output through an $(n+1)$-port magnetic coupler along with the capacitor decoupling network. To validate the effectiveness of the proposed capacitor decoupling network countering mismatched signals, simulations of four Class-E power amplifiers under two mismatched conditions are conducted. Variations in gate signal timing of up to 2 ns are introduced, which is considered as a reasonable approximation for typical mismatches of gate driving signal chain elements that would appear at 13.56MHz. The switch voltage V_{ds} and output current I_{PA} of four power amplifiers adding gate signal delays of 0ns, 0.5ns, 1ns, 2ns and gate signal delays of 0ns, 0.1ns, 0.2ns, 2ns are shown without the capacitor decoupling network (Fig. 7) and with the capacitive network (Fig. 8).

It is evident that with a properly-designed capacitor decoupling network, the ZVS operation of all four power amplifiers can be maintained under both mismatched conditions. In contrast, without the decoupling network, ZVS operation is lost. However, one drawback of the decoupling network is that it slightly degrades the current sharing performance.

IV. EXPERIMENTAL RESULTS

The experimental prototype is shown in Fig. 9, featuring coupling coils at the center with four surrounding subcircuits. For compact isolation, 10 x 10 cm PCB coils are used, with two 0.5 cm-thick Fair-Rite 67 magnetic layers added for EMI shielding. The system parameters are listed in Table I. To validate the proposed system and methods, four identically designed Class-E power amplifiers are used. In this system,

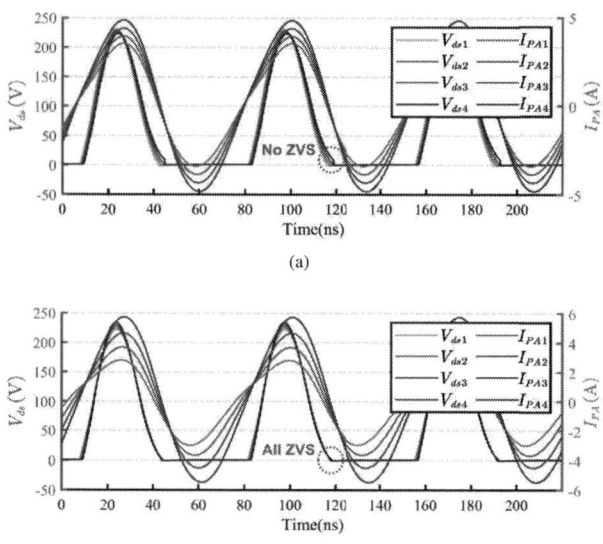

Fig. 7. V_{ds} and I_{PA} waveforms of four power amplifiers adding gate signal delays of 0ns, 0.5ns, 1ns, 2ns (a) with (b) without capacitor decoupling network

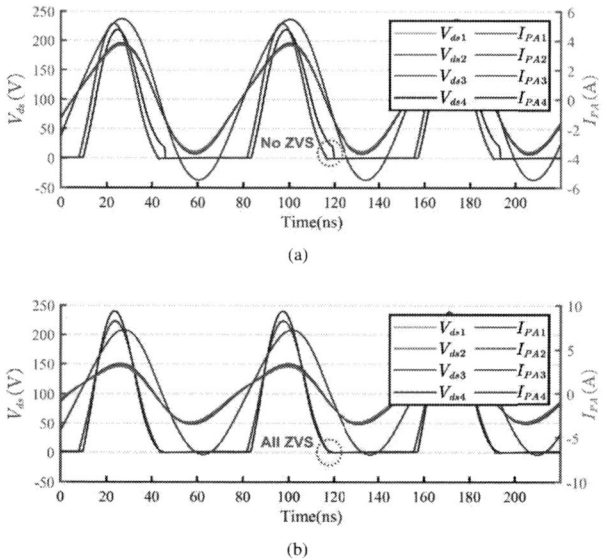

Fig. 8. V_{ds} and I_{PA} waveforms of four power amplifiers adding gate signal delays of 0ns, 0.1ns, 0.2ns, 2ns (a) with (b) without capacitor decoupling network

TABLE I
SYSTEM PARAMETERS

Parameters	Value
System operating frequency	13.56MHz
Duty cycle of power amplifiers	0.5
Input inductor L_{fi}	144nH
Parallel capacitor C_{dsi}	347pF
Series resonant network L_{si}, C_{si}	880nH, 108pF
Self-inductance of input, output coil	300nH, 738nH
Coupling coefficient between input, output coils	0.308
Cross-coupling coefficient between input coils	0.1
Decoupling capacitor C_{ij}	39.2pF
Matching network capacitor C_s, C_p	310pF, 235pF
RF load R_L	50Ω

Fig. 9. Experimental prototype

finite-inductor Class-E power amplifiers are used to counter the influence of potential load variations caused by misalignment or parameter variations.

Fig. 10 shows the measured drain-to-source waveforms of the switches of all four power amplifiers. While the V_{ds} waveforms differ slightly as expected due to the mismatched signals and components' parameter deviations, all of them exhibit ZVS operation, verifying the effectiveness of decoupling coils and capacitor network. Experiments without the proposed capacitor decoupling network are also conducted, where mismatches led to hard switching of power amplifiers and cannot operate properly.

The system's overall efficiency maintains stable around 65% with output power ranging from 50W to 300W. The losses can be attributed to two main sources: the power amplifiers and the coupler. Testing reveals that the efficiency of a single power amplifier in the experiment is around 85%, indicating that the efficiency of the coupling coils is approximately 80%. The majority of the coil losses are caused by the added magnetic shielding material, which introduces significant magnetic losses at high frequencies. The loss analysis also provides valuable insights for enhancing the system's efficiency. Class-E power amplifiers can be optimized by incorporating considerations for high-frequency switching devices, while the coupler's performance can be improved through refined simulations that accurately account for the effects of magnetic shielding.

V. CONCLUSION

This paper proposes an isolated RF power combining approach with multiple decoupled input coils. Compared to conventional isolated RF power combiners, this approach improves efficiency and compactness while providing the necessary isolation for industrial applications. A polygonal coupling coil structure is proposed, allowing power combination of any number of input ports while simultaneously eliminating cross-coupling between input ports. A capacitor decoupling network is also proposed for decoupling input ports, which can also maintain ZVS operation of all amplifiers with mismatched

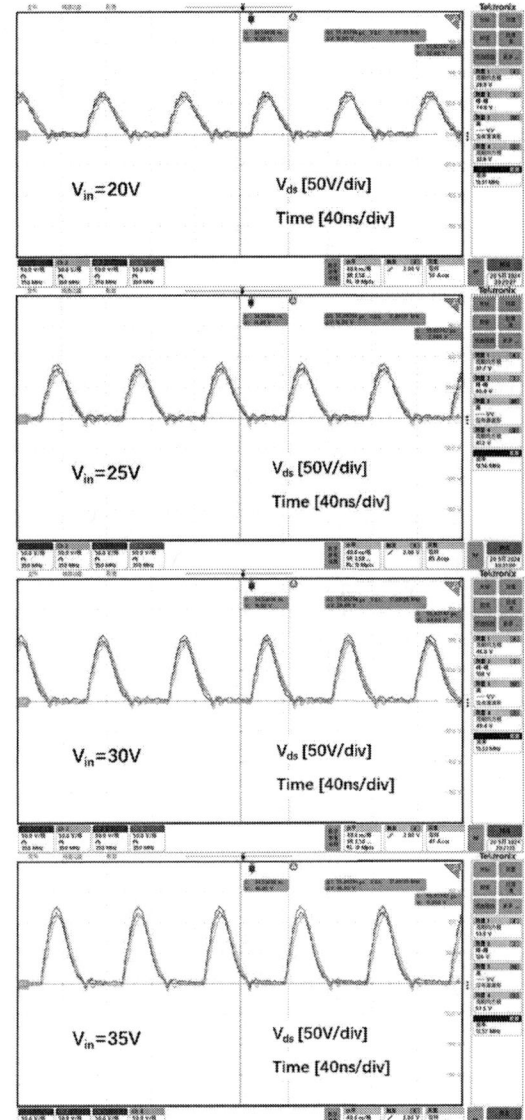

Fig. 10. Experimental V_{ds} waveforms of all four power amplifiers

gate signals. The capacitor network also shows the potential to be employed in a variety of applications involving multi-port magnetic couplers. An experimental prototype consisting of four 13.56 MHz Class-E power amplifiers was built and validated the effectiveness of the proposed decoupling methods.

REFERENCES

[1] S. . -A. El-Hamamsy, "Design of high-efficiency RF Class-D power amplifier," in IEEE Transactions on Power Electronics, vol. 9, no. 3, pp. 297-308, May 1994, doi: 10.1109/63.311263.

[2] M. Hayati, S. Roshani, S. Roshani, M. K. Kazimierczuk and H. Sekiya, "Design of Class E Power Amplifier with New Structure and Flat Top Switch Voltage Waveform," in IEEE Transactions on Power Electronics, vol. 33, no. 3, pp. 2571-2579, March 2018, doi: 10.1109/TPEL.2017.2698834.

[3] S. Aldhaher, D. C. Yates and P. D. Mitcheson, "Modeling and Analysis of Class EF and Class E/F Inverters With Series-Tuned Resonant Networks," in IEEE Transactions on Power Electronics, vol. 31, no. 5, pp. 3415-3430, May 2016, doi: 10.1109/TPEL.2015.2460997.

[4] L. Gu, G. Zulauf, Z. Zhang, S. Chakraborty and J. Rivas-Davila, "Push–Pull Class Φ_2 RF Power Amplifier," in IEEE Transactions on Power Electronics, vol. 35, no. 10, pp. 10515-10531, Oct. 2020, doi: 10.1109/TPEL.2020.2981312.

[5] D. J. Perreault et al., "Opportunities and Challenges in Very High Frequency Power Conversion," 2009 Twenty-Fourth Annual IEEE Applied Power Electronics Conference and Exposition, Washington, DC, USA, 2009, pp. 1-14, doi: 10.1109/APEC.2009.4802625.

[6] F. H. Raab et al., "RF and microwave power amplifier and transmitter technologies—Part 1," High Freq. Electron., vol. 2, no. 3, p. 22–36, 2003.

[7] L. Guo et al., "A High-Isolation Eight-Way Power Combiner," in IEEE Transactions on Microwave Theory and Techniques, vol. 68, no. 3, pp. 854-866, March 2020, doi: 10.1109/TMTT.2019.2951113.

[8] H. Zhang et al., "Modeling and Design of High-Power Non-Isolating RF Power Combiners based on Transmission Lines," 2022 IEEE Applied Power Electronics Conference and Exposition (APEC), Houston, TX, USA, 2022, pp. 259-266, doi: 10.1109/APEC43599.2022.9773770.

[9] D. J. Perreault, "A New Power Combining and Outphasing Modulation System for High-Efficiency Power Amplification," in IEEE Transactions on Circuits and Systems I: Regular Papers, vol. 58, no. 8, pp. 1713-1726, Aug. 2011, doi: 10.1109/TCSI.2011.2106230.

[10] K. Surakitbovorn and J. M. Rivas-Davila, "A Simple Method to Combine the Output Power From Multiple Class-E Power Amplifiers," in IEEE Journal of Emerging and Selected Topics in Power Electronics, vol. 10, no. 2, pp. 2245-2253, April 2022, doi: 10.1109/JESTPE.2020.3011658.

[11] Y. Wu, C. Deng, M. Zhang, S. Roy and A. J. Hanson, "Dual-PWM Control of Inductive Power Transfer Systems for High Efficiency Over Wide Load Ranges," 2024 IEEE Applied Power Electronics Conference and Exposition (APEC), Long Beach, CA, USA, 2024, pp. 1920-1926, doi: 10.1109/APEC48139.2024.10509365.

[12] Y. Wu, L. Zhou, S. Liu, R. Mai, J. Tian and S. Goetz, "A Highly-Efficient and Cost-Effective Reconfigurable IPT Topology for Constant-Current and Constant-Voltage Battery Charging," 2021 IEEE Applied Power Electronics Conference and Exposition (APEC), Phoenix, AZ, USA, 2021, pp. 451-455, doi: 10.1109/APEC42165.2021.9487242.

[13] H. Qin, H. Zhang, M. Liu and C. Ma, "Comparison of Different Multi-winding Transformer Models in Multi-port AC-coupled Converter Application," IECON 2021 – 47th Annual Conference of the IEEE Industrial Electronics Society, Toronto, ON, Canada, 2021, pp. 1-6, doi: 10.1109/IECON48115.2021.9589283.

[14] M. Chen and C. R. Sullivan, "Unified Models for Coupled Inductors Applied to Multiphase PWM Converters," in IEEE Transactions on Power Electronics, vol. 36, no. 12, pp. 14155-14174, Dec. 2021, doi: 10.1109/TPEL.2021.3088083.

[15] Y. Chen, P. Wang, Y. Elasser and M. Chen, "Multicell Reconfigurable Multi-Input Multi-Output Energy Router Architecture," in IEEE Transactions on Power Electronics, vol. 35, no. 12, pp. 13210-13224, Dec. 2020, doi: 10.1109/TPEL.2020.2996199.

Simulation of a Custom Core, 15kV Isolated Gap Transformer Optimized for High Power Density

Andrew Galamb
FREEDM Systems Center
North Carolina State University
Raleigh, USA
aegalamb@ncsu.edu

Fei Teng
FREEDM Systems Center
North Carolina State University
Raleigh, USA
fteng@ncsu.edu

Srdjan Lukic
FREEDM Systems Center
North Carolina State University
Raleigh, USA
smlukic@ncsu.edu

Abstract—Designing medium voltage transformers with a high power density requires a multidisciplinary effort, combining electric field stress control, magnetics design, and thermal mitigation. In this paper, an interleaved, isolated core-gap transformer is optimized for power density and losses using a genetic algorithm. The transformer construction is based on a 15 kV partial discharge (PD) free isolation structure that separates the high voltage and low voltage cores. Included is an overview of the Pareto front designs, showing the trade-offs fundamental to the isolated gap topology. The result is three custom core transformer design points that show the trade-offs of a planar isolation system. An experimental prototype is constructed using off-the-shelf components to validate the simulated results.

Index Terms—transformer, dual active bridge, high voltage, partial discharge free, genetic algorithm, optimization, high power density

I. MOTIVATION AND BACKGROUND

DC chargers that can charge with power levels above 350 kW are categorized as extreme fast chargers (XFC), and are the latest generation of vehicle chargers [1]. As the power capability of electric vehicle batteries increases, more of the vehicle chargers installed will have increased power capacity to reduce charging times. By deploying a Solid State Transformer (SST), the typically required service upgrades and low frequency transformer installation can be replaced with a single modular unit that connects directly to the medium voltage (MV) grid. SSTs also allow for bidirectional power flow, power factor correction, and a direct DC bus output that allows for easier conversion to the desired charging voltage and current.

An AC/DC active front end (AFE) is used to connect the SST to the low frequency grid and generated the high voltage direct current (HV DC) bus. This HV DC bus connects to a dual active bridge (DAB) converter to allow for control of power flow in both directions. The DAB includes a transformer for galvanic isolation and steps the voltage down from the HV DC bus to the low voltage direct current (LV DC) bus.

As the grid voltage that the SST is connected to increases, the voltage withstand ability of the semiconductors must also increase. Once the grid voltage has surpassed the useful range of the semiconductor switches that are readily available, the modules can be cascaded in an input-series output-parallel (ISOP) connection, and the voltage and current stresses can be shared equally between the modules. This allows the differential mode voltage of each module to be reduced, while the sum of the voltages allows for connection to a MV grid. The differential mode voltage of the transformer is much lower, and reduces the insulation requirements of the windings to manageable levels, however, the common mode voltage is on the order of magnitude of the MV grid voltage. The high frequency transformer (HFT) of the DAB is used to provide galvanic isolation between the MV grid and the LV DC bus, which is important for safety and operational concerns.

II. ISOLATION APPROACHES AND POWER DENSITY

Transformer are designed with various isolation methods, which each have their own advantages and disadvantages. This section will describe the categories that the different transformer isolation systems can be sorted into and highlight a few examples from the literature in each category with exceptional power density or ingenuity.

An example of an air isolated transformer design can be seen in [2]. The 200 kW, 15 kHz design has a unity turns ratio with a voltage rating of 1300 V on both the primary and secondary of the DAB converter. The design philosophy with the unique parallel-concentric winding topology was to minimize the leakage inductance, which can be a source of loss in nanocrystalline cores. The use of air as an insulator significantly reduces the window area utilization for the design, but allows for optimal air flow to reach the windings. The specially designed bobbins allow for a high air flow over the windings to achieve and impressive 7.5 A/mm^2 current density without overheating. A low loss FINMET core was used to reduce core losses, and a power density of 19.23 kW/L was achieved with a partial discharge (PD) inception of 5.3 kV rms.

An example of a much high isolation transformer design can be seen in [3]. The 240 kVA, 10 kHz design uses a combination of epoxy and air isolation to achieve a 35 kV PDIV. This hybrid isolation strategy allows for the epoxy to take the electric field when it is concentrated close to the high voltage winding, and as the field spreads out, air is used to support the lower electric field densities. This hybrid approach allows for air cooling channels between the primary and secondary windings. Although the paper does

979-8-3315-1612-3/25 $31.00 © 2025 IEEE

not specify the Litz wire used, the current densities would likely exceed 3 A/mm². A nanocrystalline core is used to reduce the losses of the transformer at the 10 kHz switching frequency. This transformer achieved the highest PDIV of the transformers reviewed, but the power density of this design is quite low at 3.6 kW/L. The excess space between the windings increased the volume of the transformer, showing the challenge of designing a power dense, high voltage transformer using air as isolation.

The transformer proposed in [4] uses unshielded epoxy for the isolation. The 80 kW, 43 kHz transformer, and boasts a 42 kV hipot success. The planar design allowed for impressive thermal dissipation, maintaining a hot spot temperature of 102 °C at 594 W of loss, and the current density of the winding was pushed to 4.7 A/mm². The transformer used a wireless power transfer style construction to provide a gap for isolation that, when submerged in epoxy, provided robust isolation. The hipot results showed significant leakage current suggesting that the transformer was close to failure. PD results were not shared, but with the distance between the primary and secondary, a PDIV of less than 15 kV is expected.

An example of a shielded epoxy design is in [5]. The 100 kW, 20 kHz transformer incorporates dual shields surrounding both the primary and secondary winding, achieving a PDIV of 13 kV rms. With 10 mm of epoxy surrounding the windings, the current density was kept at 3.5 A/mm² to avoid overheating. Terminating the wires such that there are no discharges in air required that the transformer have the large, extended terminals which occupied a significant portion of the volume. Even with these challenges, a power density of 11.2 kW/L was achieved.

The shielded epoxy transformer with the greatest power density of the transformers surveyed that included PD results was in [6]. The 15 kW, 200 kHz transformer had a PDIV of 5.4 kV rms due to the shielded epoxy construction. At the switching frequency of 200 kHz, ferrite has much lower losses than nanocrystalline core, making it the choice for power density at higher frequency. A resonant converter was used to drive the transformer at the high frequency with the large leakage inductance inherent with high voltage designs. The current density of windings was an impressive 10.7 A/mm². The superior current density along with a high switching frequency resulted in the power density of 35 kW/L. This transformer design shows the ability of frequency to reduce the size of a transformer.

This paper will optimize a transformer that can utilize air isolation current densities while achieving shielded epoxy levels of isolation.

III. EXISTING POWER DENSITY VS ISOLATION

A thorough literature review was conducted to determine the state of the art power density for a given isolation voltage and plotted in Fig. 1. The transformers covered in this graph follow a general trend that as the isolation voltage increases, the power density decreases.

Table I shows the references for each point on the graph, and their respective isolation method and how the isolation was tested.

TABLE I
TRANSFORMERS IN THE LITERATURE THAT PROVIDE INFORMATION REGARDING POWER DENSITY.

Chart #	Reference	Power Density (kW/L)	Isolation (kV)	Isolation Test	Isolation Method
1	[7]	8.2	2	None	Air
2	[8]	11.5	6	None	Air
3	[8]	15	6	None	Air
4	[3]	3.6	35	PD	Unshielded Epoxy
5	[9]	5.1	42	None	Shielded Epoxy
6	[10]	8	30	Hipot	Air
7	[11]	8.2	4	PD	Air
8	[12]	8.4	12	PD	Shielded Epoxy
9	[13]	9.6	20	Hipot	Unshielded Epoxy
10	[14]	10.5	35	Hipot	Unshielded Epoxy
11	[15]	19.2	5.3	PD	Air
12	[5]	11.2	15	PD	Shielded Epoxy
13	[6]	35	5.4	PD	Shielded Epoxy
14	[16]	4.2	45	Hipot	Unshielded Epoxy
15	[17]	50	1	None	Air
16	[4]	21.1	42	Hipot	Unshielded Epoxy
17	[18]	29.6	4	simulated	Air
18	[19]	13.5	7.8	PD (100pC)	Shielded Epoxy
19	[2]	8.5	9.5	PD	Shielded Epoxy
20	[20]	10.6	14	PD (300pC)	Unshielded Epoxy
21	[21]	17.7	1	None	Air
22	[22]	18	7	PD	Air

Figure 1 shows how few designs in the research space can achieve a power density of above 20 kW/L while maintaining isolation above 20 kV. The low isolation designs focus on cooling, allowing for high core flux densities and high current densities without overheating. For the higher isolation designs, especially the ones that conduct rigorous PD tests, focus on the isolation design and the thermal impedance added by the isolation reduces the possible current densities. Only [4] showed above 20 kV isolation and above 20 kW/L power density, although the isolation was only tested with a hipot test, not PD.

IV. TOPOLOGY INVESTIGATED

The focus of this section will be the interleaved, core-type topology due to its high coupling coefficient for a given footprint area. The form factor of the interleaved topology should allow for it to be substituted in for a traditional transformer design in a module while maintaining a good coupling coefficient. 4mm between cores and 10mm between widings

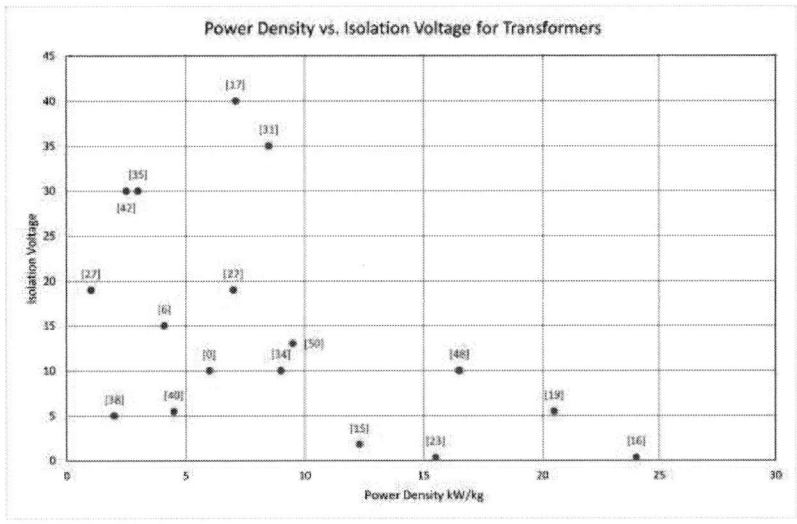

Fig. 1. Power density plotted against isolation voltage for transformers published in IEEE.

are neceseary for the insulation system. Further description of the isolation system is out of the scope of this paper.

A. Ansys Setup

An overview of the interleaved, isolated gap transformer geometry is shown in Fig. 2.

The design targets for the optimization are shown in Table II. The voltage is chosen based on a DAB converter with 1.2 kV silicon carbide (SiC) MOSFETs and the power level is based on the ampacity of a known Litz wire multiplied by the assumed voltage. The switching frequency range was chosen based on the most power dense designs in the literature, with 80 kHz being used to see if the benefits of power density continue to improve beyond 60 kHz.

The geometric variables that can be optimized by the genetic algorithm can be seen in Table III. These variables are chosen such that a variety of designs are generated within the desired design space.

The winding geometry for the transformer was simplified to include the volume that the winding would occupy. Then the windings excitations were set to stranded so that the current density was evenly distributed, and the the material was copper

TABLE II
TARGETS FOR ISOLATED GAP TRANSFORMER TO BE OPTIMIZED.

Parameter	Target/Nominal
Power	70 kW
Voltage	800 V
Power Density	>10 kW/L
Efficiency	>99.7%
Frequency	40-80 kHz
Magnetizing Current	$<15\% I_{load}$
Current Density	<6 A/mm^2

TABLE III
VARIABLES USED FOR OPTIMIZATION.

Variable	Range
Leg Length	30-480 mm
Leg Width	40-160 mm
Leg Height	14-22 mm
Winding Width	20-80 mm
Top Core Thickness	5-10 mm
Winding Layers	1-2
Number of Turns	10-20
Core Gap	4 mm
Winding Gap	10 mm

with Ansys' built-in Litz wire material type. This allows the simulation to calculate the AC losses of the winding, including skin effect and proximity effect. The Steinmetz parameters for the core loss of the N87 ferrite material were extracted from the datasheet and used to simulate the losses.

The transformer excitation was set to current mode and the load resistance of the transformer was set so that the desired voltage of 800 V was achieved on the primary.

V. GENETIC OPTIMIZATION

In this section, a genetic algorithm is used to optimize the geometry of the core-type, interleaved, isolated gap transformer.

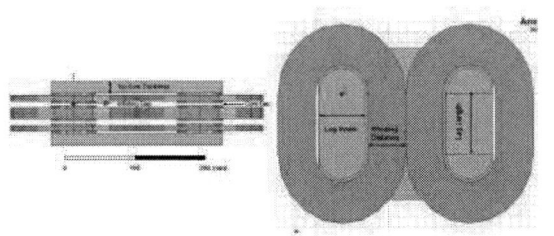

Fig. 2. View of side and top of transformer labeled for pertinent geometric variables.

A. Genetic Algorithm Setup

Figure 3 shows the optimization process for the isolated gap transformer. First, the targets need to be set, such as the output power, voltage, turns ratio, target power density, target efficiency, and magnetizing current as seen in Table II. The variables that can modified need to be identified and constrained so the desired goals can be achieved. In this design, the frequency, number of turns, leg length, leg width, and top core thickness are given a range of acceptable values as seen in Table III.

A screening algorithm within MATLAB is used to generate the initial population of 550 designs which are evenly spread through the design space. Ansys then simulates the output power, efficiency, expected power density, self inductance, leakage inductance, and loss distribution of each of the initial designs. MATLAB calculates the cost of each design and ranks the design in order of ascending cost.

The lowest cost designs are kept for the next generation, and the higher cost designs are culled and replaced with crossovers or mutations of the low cost individuals. Crossovers take a random weighted average of the traits of two randomly selected parents to produce an offspring. Mutations take a low cost design and change a random trait to a random value. An additional 50% population reduction occurred after the initial screening to improve convergence. The genetic algorithm stops after the cost of the top individual does not increase for a generation, or the maximum number of generations is achieved at 8.

B. Cost Function

The genetic algorithm requires a cost function to determine the designs that are better than the others. If the inductance constraints are met, then the cost is a function of normalized transformer volume, and normalized loss, as seen in Eq. 1.

$$C = \frac{Volume_{tx}}{15L} + \frac{Loss}{900W} \quad (1)$$

C. Output Constraints

The output constraints of the magnetizing inductance and leakage inductance are enforced by increasing the cost of the design by 100 and adding additional cost based on how far the inductance is from the target value. The value of 100 was chosen to ensure that any design that is outside of the output constraints is below any designs that achieve the output constraints.

The output constraints for leakage and magnetizing inductance are determined by the requirements of the DAB at the switching frequency. To achieve stable power transfer, the leakage inductance of the transformer must be set at a value that provides adequate power transfer while being large enough to allow for accurate and stable control. The equation for the leakage inductance and power flow relationship is given by:

$$L_{lk} = \frac{V_1 V_2 (\pi - \phi)\phi}{2n f_s P_{out}\pi^2} \quad (2)$$

The upper end of the leakage inductance range determined for the DAB assumed a phase shift of 60° to ensure that there is sufficient control margin. The lower bound was set at 45° to allow for power flow control stability.

The magnetizing current is given by the following equation:

$$I_{mag} = \frac{V_i}{2\pi f_s (L_m + L_{lk})} \quad (3)$$

This was set at 15% of the load current to ensure the current stress on the switching devices due to the magnetizing current is low.

If the inductance output constraints are not met, then the leakage and magnetizing inductance determine the cost of the designs as seen in Eq. 4.

$$C = 100 + (240mH * \frac{40kHz}{F_s} - L_s) + |19uH * \frac{40kHz}{F_s} - L_l| \quad (4)$$

100 is added to ensure that any design that meets the output constraints will be favored over a design that does not meet the constraints. Additional cost is added based on the self inductance being below the required self inductance that to achieve the desired magnetizing current of less than 15% of the full load current. Additional cost is also added based on the deviation of the leakage inductance from the required leakage inductance for a DAB converter.

D. Genetic Algorithm Results

The design space explored by the optimization algorithm can be seen in Figs. 4-6. The plots show that there is little loss in efficiency when optimizing for just power density. The lowest cost design of each frequency is highlighted with a red circle. In all simulated frequencies, the power density tends to increase as the number of turns increases, since the the inductance is proportional to the number of turns squared. At a certain point the losses start increasing due to the increased copper losses from the increased number of turns.

The lowest cost design at 40-80 kHz can be seen in Figs. 7-9.

Table IV summarizes the characteristics of the three pictured designs.

E. Genetic Algorithm Discussion

As the switching frequency of the designs increased, the method by which the design generated the leakage inductance changed. At 40 kHz, the potential for proximity losses are low, so the optimal design uses two winding layers to reduce the mean turn length of the design, and the flux passing through the inner most layers do not significantly increase the winding loss. This design had the thickest top ferrite plate to manage the higher flux conditions at the lower frequency.

With the 60 kHz design, the requirements for leakage and self inductance are lower, so the optimal design switches to using one winding layer. This increases the mean turn length, but the flux passing through the single winding layer results

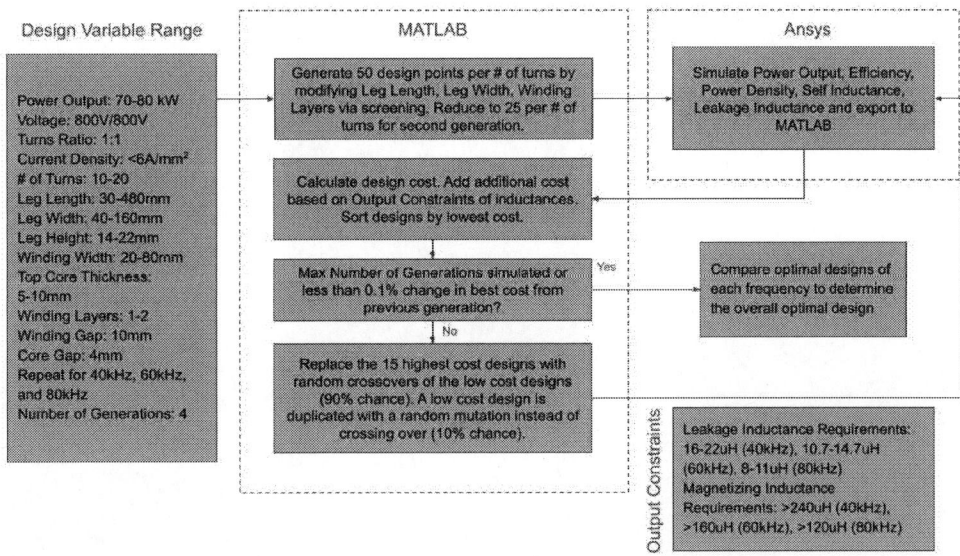

Fig. 3. Flow chart depicting the process for Pareto front generation.

in lower proximity losses. The 60 kHz design has the most copper area of the designs, so it has the highest losses.

For the 80 kHz design, proximity losses can be the highest so a single winding layer design dominates. At this frequency, the inductances are small enough that they can be achieved with a lower number of turns, which reduced the copper loss and increased the core loss. This design had the thinnest top ferrite plate to manage the lower flux conditions at the higher frequency.

The improvement up to 80 kHz suggests that designs for switching frequencies of 100 kHz and beyond may also provide increased power density and loss reduction benefits.

VI. EXPERIMENTAL RESULTS

To ensure the accuracy of the simulations, a transformer using off-the-shelf ferrite plates was simulated and built to measure the inductances and gather temperature data. Comparisons are made between the simulated performance and the experimental performance.

A. Simulated Performance

The transformer model is shown in Fig. 10. The winding area is shown in the copper color, and the ferrite material is in gray. The windings were simplified to solid shapes so that the simulations would run more efficiently and converge faster.

The methodology for simulating this design is the same as used for the genetic optimization. The Ansys Eddy Current Solver was used to simulate the transformer at 25 kW with the desired input and output voltages of 750 V and 2150 V. Table V shows the results of the simulation.

The following figures were from the Eddy Current simulation of the transformer at 25 kW and 40 kHz. Figure 11 shows the B field magnitude of the bisecting plane. The highest flux density in the transformer is in the top and bottom plates. The flux spreads out as it goes through the air gaps, reducing the reluctance of the gaps.

Figure 12 show the core loss distribution of the transformer in W/m³. The majority of the core loss is concentrated on the top plate, with the legs of the transformer experiencing an order of magnitude less loss.

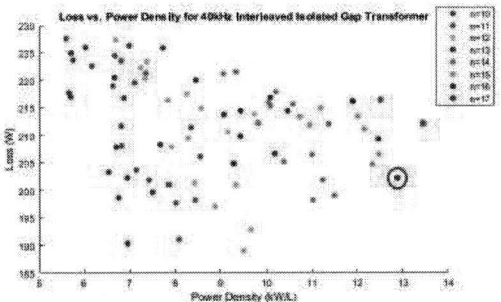

Fig. 4. Plot of Loss vs. Power Density for 40 kHz Designs.

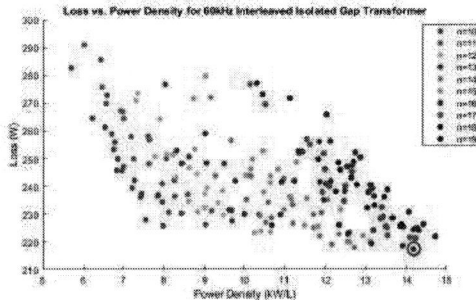

Fig. 5. Plot of Loss vs. Power Density for 60 kHz Designs.

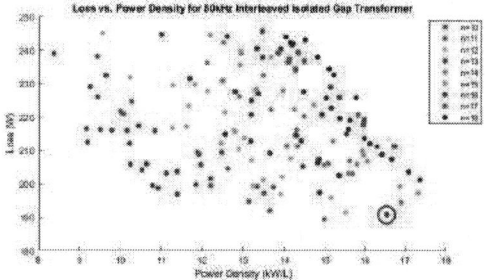

Fig. 6. Plot of Loss vs. Power Density for 80 kHz Designs.

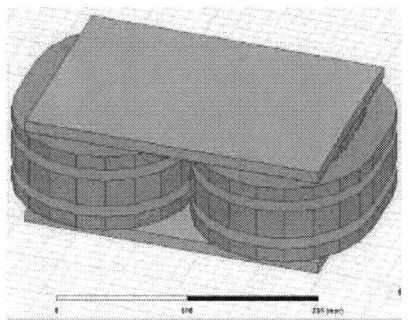

Fig. 7. Lowest Cost Design at 40 kHz.

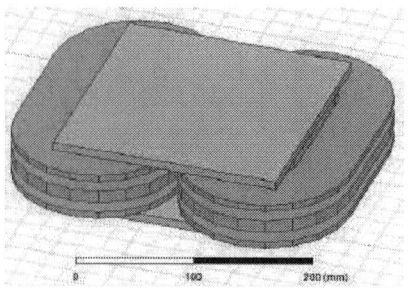

Fig. 8. Lowest Cost Design at 60 kHz.

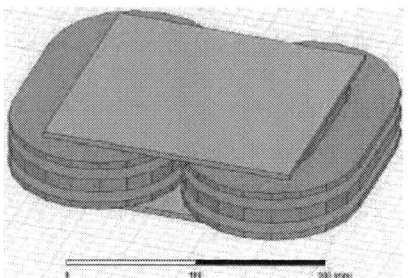

Fig. 9. Lowest Cost Design at 80 kHz.

B. Experimental Prototype

The built transformer can be seen in Fig. 13 and Fig.14. Special attention was paid to ensure that sufficient airflow would be able to cool the top and bottom cores, as well as the windings. The side view in Fig. 13 shows the isolation barriers and the additional space to ensure that windings don't overheat.

VII. LCR METER TESTING

The transformer was testing using the LCR meter to measure the leakage inductance, self inductance, and resistances. The self inductance was measured with the opposite side terminals open, and the leakage inductance and AC resistance were measured with the opposite side shorted. All tests were performed using 40 kHz as the test frequency.

The simulated and measured self inductance match very well with less than three percent difference between simulated and actual results. The results of the leakage inductance was less than ten percent off, which can be accounted for with added inductance in the experimental prototype due to long wire leads and imperfections in the winding location. Differences in the winding resistances can also be attributed to the long leads required for testing and the resistance of the termination.

VIII. TRANSFORMER POWER TESTING

The transformer was testing at 25 kW using a DAB converter. The setup can be seen in Fig. 14. A power supply fed the low voltage side bridge on the right of the figure with 750 V. The transformer in the center was excited from the

TABLE IV
RESULTS FROM SIMULATIONS OF SELECTED DESIGNS.

Parameter	40 kHz Design	60 kHz Design	80 kHz Design
Efficiency (%)	99.72	99.70	99.71
Power Density (kW/L)	12.87	14.18	16.55
Self Inductance (μH)	248	184	126
Magnetizing Current (% of Full Load)	12.83	11.53	12.63
Leakage Inductance (μH)	20.1	10.9	8.1
Core Loss (W)	16	13	33
Copper Loss (W)	189	204	175
Number of Turns	16	16	13
Core Width (mm)	93	50	60
Core Length (mm)	45	108	90
Number of Layers	2	1	1
Top Thickness	10	8	5

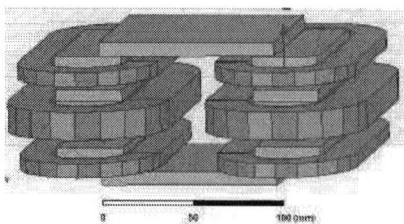

Fig. 10. Transformer as modeled in Ansys.

TABLE V
SIMULATED PERFORMANCE OF THE PROTOTYPE.

Parameter	Simulated Value
Output Power	25.5 kW
Frequency	40 kHz
Turns Ratio	48:17
Leakage Inductance	160μH
Coupling Coefficient	0.89
Core Loss	20.0 W
Copper Loss	40.0 W

TABLE VI
TRANSFORMER PARAMETERS MEASURED BY AN LCR METER AT 40 KHZ
COMPARED TO THE SIMULATED VALUES.

Parameter	LCR Meter Results	Simulated Values	Percent Difference
Primary Self Inductance	785 μH	775 μH	1.27%
Secondary Self Inductance	90.7 μH	92.9 μH	2.43%
Primary Leakage Inductance	175 μH	160.4 μH	8.34%
Secondary Leakage Inductance	20.7 μH	19.1 μH	7.7%
Primary Resistance	200 mΩ	193 mΩ	3.5%
Secondary Resistance	23.5 mΩ	20.5 mΩ	12.77%

low voltage side and transferred the power to the high voltage side at 2150 V and capacitor bank, seen on the left. Finally, the resistive load was used to sink the power. The modules and transformer are cooled by a fan blowing at 3 m/s wind speed.

Figure 15 shows the operational waveforms of the transformer at 40 kHz and 25 kW. Channel 1 is the low voltage side current, channel 2 is high voltage side current, channel 4 is low voltage side voltage, and channel 5 is high voltage side voltage.

The top plate reached a temperature of 61.7°C when operated at 25 kW. The temperature of the top and bottom ferrites was relatively uniform as seen in Figs.16 and 17.

IX. CONCLUSION

A genetic optimization was performed, with an emphasis on constraining the output designs to the required leakage and magnetizing inductances required by a DAB converter. The design space for each frequency was explored, and the trends

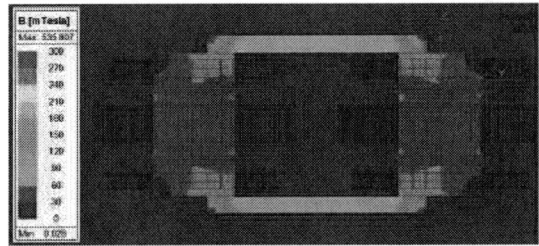
Fig. 11. Plot of B field magnitude on the bisecting plane.

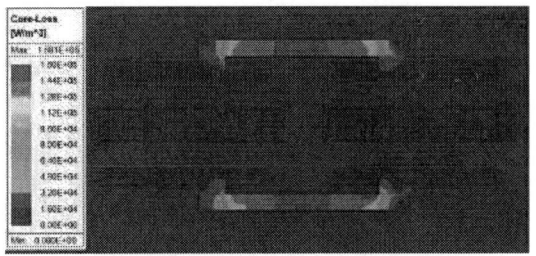
Fig. 12. Plot of core loss density in the bisecting plane.

Fig. 13. Side view of built transformer.

Fig. 14. DAB converter test setup to test transformer.

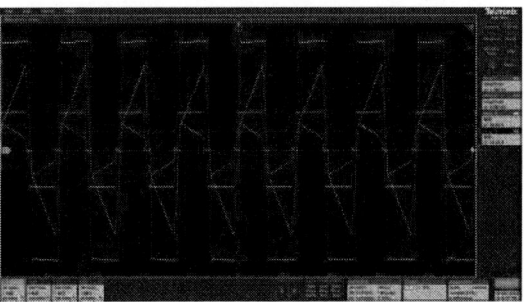
Fig. 15. Waveform of transformer operating at 25 kW.

Fig. 16. Thermal image from the top of transformer operating at 25 kW.

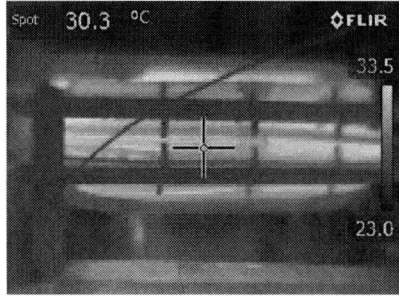

Fig. 17. Thermal image of the side of transformer operating at 25 kW.

of each were discussed. Three designs of different frequencies were outlined, with the 80 kHz, 16.55 kW/L transformer achieving the best power density. Future work may determine if the converter losses dominate over the transformer losses at frequencies above 80 kHz. A prototype was built and matched the simulated inductances within 10% variance. A power test was conducted to ensure that the transformer was thermally stable.

ACKNOWLEDGMENT

I would like to thank Erik Bishop for help with mechanical design and construction of the transformer.

REFERENCES

[1] H. Tu, H. Feng, S. Srdic, and S. Lukic, "Extreme fast charging of electric vehicles: A technology overview," *IEEE Transactions on Transportation Electrification*, vol. 5, no. 4, pp. 861–878, 2019.

[2] Z. Gao, H. Li, and F. Wang, "A medium-voltage transformer with integrated leakage inductance for 10 kv sic-based dual-active-bridge converter," in *2022 IEEE 9th Workshop on Wide Bandgap Power Devices Applications (WiPDA)*, 2022, pp. 221–226.

[3] T. B. Gradinger, U. Drofenik, and S. Alvarez, "Novel insulation concept for an mv dry-cast medium-frequency transformer," in *2017 19th European Conference on Power Electronics and Applications (EPE'17 ECCE Europe)*, 2017, pp. P.1–P.10.

[4] S. Lu, D. Kong, S. Xu, L. Luo, and S. Li, "A high-efficiency 80-kw split planar transformer for medium-voltage modular power conversion," *IEEE Transactions on Power Electronics*, vol. 37, no. 8, pp. 8762–8766, 2022.

[5] R. Lu, C. Li, J. Yu, C. Li, W. Li, and X. He, "Modeling and design of insulation structure for high power density medium voltage high-frequency transformers," *IEEE Journal of Emerging and Selected Topics in Power Electronics*, vol. 11, no. 6, pp. 6005–6015, 2023.

[6] Z. Li, Y.-H. Hsieh, Q. Li, F. C. Lee, and M. H. Ahmed, "High-frequency transformer design with high-voltage insulation for modular power conversion from medium-voltage ac to 400-v dc," in *2020 IEEE Energy Conversion Congress and Exposition (ECCE)*, 2020, pp. 5053–5060.

[7] G. Ortiz, M. Leibl, J. W. Kolar, and O. Apeldoorn, "Medium frequency transformers for solid-state-transformer applications — design and experimental verification," in *2013 IEEE 10th International Conference on Power Electronics and Drive Systems (PEDS)*, 2013, pp. 1285–1290.

[8] M. A. Bahmani, T. Thiringer, and M. Kharezy, "Optimization and experimental validation of medium-frequency high power transformers in solid-state transformer applications," in *2016 IEEE Applied Power Electronics Conference and Exposition (APEC)*, 2016, pp. 3043–3050.

[9] T. Tang, J. Ferreira, S. Mao, W. Wang, and M. G. Niasar, "Design of a medium frequency transformer with high insulation level for dual active bridge dc-dc converter," in *2019 10th International Conference on Power Electronics and ECCE Asia (ICPE 2019 - ECCE Asia)*, 2019, pp. 1–8.

[10] B. Chen, X. Liang, and N. Wan, "Design methodology for inductor-integrated litz-wired high-power medium-frequency transformer with the nanocrystalline core material for isolated dc-link stage of solid-state transformer," *IEEE Transactions on Power Electronics*, vol. 35, no. 11, pp. 11 557–11 573, 2020.

[11] M. Mogorovic and D. Dujic, "100 kw, 10 khz medium-frequency transformer design optimization and experimental verification," *IEEE Transactions on Power Electronics*, vol. 34, no. 2, pp. 1696–1708, 2019.

[12] C. Zhu, "High-efficiency, medium-voltage-input, solid-state-transformer-based 400-kw/1000v/400a extreme fast charger for electric vehicles," *U.S. Department of Energy Office of Scientific and Technical Information*, 2023. [Online]. Available: https://www.osti.gov/biblio/1987553

[13] D. Rothmund, T. Guillod, D. Bortis, and J. W. Kolar, "99

[14] R. Lu, J. Yu, D. Feng, K. Bu, Z. Yang, C. Li, and W. Li, "Modeling and design of a medium frequency transformer with high isolation and high power-density," in *2021 IEEE Applied Power Electronics Conference and Exposition (APEC)*, 2021, pp. 1694–1700.

[15] Z. Guo, R. Yu, W. Xu, X. Feng, and A. Q. Huang, "Design and optimization of a 200-kw medium-frequency transformer for medium-voltage sic pv inverters," *IEEE Transactions on Power Electronics*, vol. 36, no. 9, pp. 10 548–10 560, 2021.

[16] E. S. Lee, J. H. Park, M. Y. Kim, and J. S. Lee, "High efficiency integrated transformer design in dab converters for solid-state transformers," *IEEE Transactions on Vehicular Technology*, vol. 71, no. 7, pp. 7147–7160, 2022.

[17] P. Yao, X. Jiang, P. Xue, S. Li, S. Lu, and F. Wang, "Design optimization of medium-frequency transformer for dab converters with dc bias capacity," *IEEE Journal of Emerging and Selected Topics in Power Electronics*, vol. 9, no. 4, pp. 5043–5054, 2021.

[18] E. A. Younis, O. Zayed, A. Elezab, M. Ibrahim, and M. Narimani, "Transformer design for solid-state transformer (sst)-based ev charging station applications," in *2023 IEEE Transportation Electrification Conference Expo (ITEC)*, 2023, pp. 1–6.

[19] Z. Gao, D. Li, R. Chen, H. Bai, L. Tolbert, and F. Wang, "Design of a 50-kw medium frequency medium voltage transformer for 10-kv sic-based dual active bridge converter," in *2023 IEEE Energy Conversion Congress and Exposition (ECCE)*, 2023, pp. 5839–5845.

[20] Z. Guo, S. Rajendran, J. Tangudu, Y. Khakpour, S. Taylor, L. Xing, Y. Xu, X. Feng, and A. Q. Huang, "A novel high insulation 100 kw medium frequency transformer," *IEEE Transactions on Power Electronics*, vol. 38, no. 1, pp. 112–117, 2023.

[21] Z. Zhao, Y. Wu, F. Diao, N. Lin, X. Du, and Y. Zhao, "A high-power, high-frequency matrix core transformer design for medium voltage dual active bridge," in *2022 IEEE Applied Power Electronics Conference and Exposition (APEC)*, 2022, pp. 582–587.

[22] A. El Shafei, S. Ozdemir, N. Altin, and A. Nasiri, "Development of a high power, medium frequency transformer for medium voltage applications," in *2023 IEEE Energy Conversion Congress and Exposition (ECCE)*, 2023, pp. 891–898.

Low Interwinding Capacitance Design for PCB-Winding Based Transformer in Self-Powered Gate Drive Power Supply for High-Voltage SiC MOSFET

Yuan Zhou
School of Electrical and Electronic Engineering
Huazhong University of Science and Technology
Wuhan, China
zhouyuan5677@hust.edu.cn

Li Zhang
School of Electrical and Electronic Engineering
Huazhong University of Science and Technology
Wuhan, China
zhangli_frank@hust.edu.cn

Yilun Chen
School of Electrical and Electronic Engineering
Huazhong University of Science and Technology
Wuhan, China
chenyilun123@ hust.edu.cn

Tianxiang Yin
School of Electrical and Electronic Engineering
Huazhong University of Science and Technology
Wuhan, China
dpblk@hust.edu.cn

Lei Lin
School of Electrical and Electronic Engineering
Huazhong University of Science and Technology
Wuhan, China
linlei@hust.edu.cn

Abstract—**High-performance gate drive power supply (GDPS) is the critical component to guarantee the reliable and safe operation of the gate driver for SiC semiconductor power devices. To meet the demand for suppressing the common-mode interference (CMI) resulting from the high dv/dt, it is of significance to design GDPSs with low interwinding capacitance. In this paper, design considerations, including winding layout design and transformer geometric design, are proposed as a guidance for designing the printed-circuit-board-winding (PCB-winding) based transformer in the GDPS. The doing so results in an optimized transformer design with ultra-low interwinding capacitance of 0.86pF. Finally, a prototype is fabricated in the lab to verify the feasibility of the proposed work.**

Keywords—*Gate drive power supply (GDPS), interwinding capacitance, PCB-winding based transformer, medium voltage (MV), silicon carbide (SiC) MOSFET*

I. INTRODUCTION

The 10kV/15kV silicon-carbide (SiC) MOSFETs, with the advantages of high blocking voltage, fast switching speed, and low switching loss, are promising to change the landscape of medium-voltage (MV) power conversion [1], [2], [3], [4]. As the energy source for the gate driver, the gate drive power supply (GDPS) is of great significance in maintaining the reliable operation of the MV converter [5], [6].

Nonetheless, the high working voltage and high dv/dt of 10kV/15kV SiC MOSFET impose two major challenges for the high-performance GDPS design. One is the high voltage insulation design [7], and the other is the low interwinding capacitance (C_{ps}) design. In recent years, the self-power-supplied GDPS (S-GDPS) has been proposed in [8], effectively lowering down the insulation voltage from the system voltage to half of the submodule capacitor voltage. Therefore, the remaining challenge for building a reliable GDPS for high-

This work is supported by National Natural Science Foundation of China (52307203), Natural Science Foundation of Hubei Province of China (2023AFB361), and Grants from Delta Power Electronics Science and Education Development Program of Delta Group (DREG2023003).

voltage SiC MOSFET is the methodology of lowering down the interwinding capacitance of the GDPS (C_{ps} <5pF). Considering the fact that quite a number of GDPSs will be used in a MV converter, the ease of manufacturing is of significant importance as well from the application perspective.

Blocking the electrical coupling path can be a promising approach. For the transformer in [9], the magnetic core is split into two parts to cut off the coupling path, lowering down C_{ps} to 4.85pF. On the basis of the air core transformer, a GDPS, with C_{ps} of 2.5pF, is proposed in [10]. Nonetheless, the excessively increased leakage inductance would be a problem. Reducing the winding turn number can be another solution. In [11], a current transformer, with a single primary turn, is proposed to achieve C_{ps} of 1.67pF. Though effective, the potential single-point fault issue might limit the practical application. By using PCB-winding based transformer, an external powered GDPS, with C_{ps} of 1.85pF, is proposed in [12]. However, the power density is low because of the enlarged core volume for meeting the insulation clearance requirement. By contrast, a compact PCB-winding based transformer is demonstrated in [13]. Nonetheless, such a transformer design, with C_{ps} of 14pF, can be hardly satisfactory for developing the GDPS with low C_{ps}.

Considering that the transformer is the critical element to determine the interwinding capacitance of the GDPS, it is of archival value to propose a complete design procedure for having low-interwinding-capacitance transformer used in S-GDPS for high-voltage SiC MOSFETs. The rest of the paper is organized as follows. Section II describes the interwinding capacitance model for the PCB-winding based transformer. Thereafter, different winding layouts are compared to determine the optimized layout in Section III. After that, the relationship, between C_{ps} and the transformer geometric parameters, is revealed, based on which, a detailed transformer geometric parameter design approach is proposed in Section IV. The experimental results are presented in Section V to verify the effectiveness of the proposed design approach. Finally, Section VI concludes this paper.

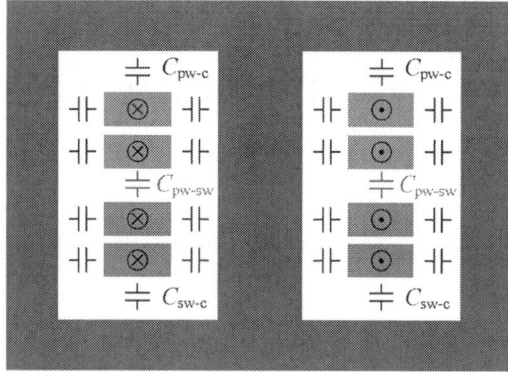

■Primary Windings ■Secondary Windings ■Core

Fig. 1. Transformer interwinding capacitance model.

	(a)		(b)		(c)
P_1		P_1		P_1	
P_2		S_1		S_1	
P_3		S_2		P_2	
P_4		P_2		S_2	
S_1		P_3		P_3	
S_2		S_3		S_3	
S_3		S_4		P_4	
S_4		P_4		S_4	

Fig. 2. Three winding layouts. (a) Non-interleaved. (b) Alternatively-interleaved. (c) Fully-interleaved.

II. INTERWINDING CAPACITANCE MODEL FOR PCB-WINDING BASED TRANSFORMER

In this paper, the transformer turn ratio is set to 1. This symmetrical structure simplifies the layout design and transformer optimization. In addition, this symmetry is beneficial to the high voltage insulation as the E-field can be distributed uniformly, reducing the risk of partial discharge (PD).

The transformer interwinding capacitance model is given in Fig. 1. As shown, C_{ps} can be written as

$$C_{ps} = C_{pw-sw} + \frac{C_{pw-c} C_{sw-c}}{C_{pw-c} + C_{sw-c}} \tag{1}$$

where C_{pw-sw} is the winding-to-winding capacitance, and C_{pw-c} and C_{sw-c} are respectively the lumped primary and secondary winding-to-core capacitances. The coupling capacitance can be roughly estimated as

$$C = \varepsilon_r \varepsilon_0 \cdot A / d \tag{2}$$

where A and d are the overlapping area and distance between the conductors, respectively, and ε_r is determined by the insulation material.

Eq. (2) indicates that, either reducing the overlapping area or increasing the distance is beneficial to minimizing the coupling capacitances. An insulation material with low permittivity is preferred, but the dielectric strength should be taken into account. Based on these analyses, design considerations for designing transformers with low C_{ps} are illustrated as follows.

TABLE I.   SIMULATION RESULTS WITH DIFFERENT WINDING LAYOUTS

	C_{ps}(pF)	C_{pw-sw}(pF)	C_{pw-c}(pF)	C_{sw-c}(pF)
Non-interleaved	6.21	5.53	1.35	1.34
Alternatively-interleaved	18.58	17.91	1.48	1.12
Fully-interleaved	28.06	27.39	1.34	1.34

III. ANALYSIS OF WINDING LAYOUT

In conventional transformers, interleaved winding layouts are widely used to improve the coupling factor, however, this design may lead to increased C_{ps}. Figs. 2(a), (b), and (c) present three kinds of winding layouts, namely, non-interleaved layout, alternatively-interleaved layout and fully-interleaved layout.

As shown in Fig. 2, with the interleaving degree between the primary and secondary windings increasing, the overlapping area gets larger, resulting in increased C_{pw-sw}. The simulation results are given in Table I, with an 8-layer PCB and parameters given in Table III (referring to Section IV-D). As shown, C_{ps} increases to 3 or 5 times with interleaved layouts. Therefore, a non-interleaved winding layout is preferred to realize low C_{ps}.

However, as shown in Table I, with the non-interleaved layout, the C_{ps} doesn't meet the requirement yet (<5pF). To further reduce C_{ps}, the primary and secondary windings should be physically separated into 2 PCBs, such that the distance between the two windings can be enlarged, while the insulating material is substituted from FR4 ($\varepsilon_r \approx 4.7$) to air.

The main challenge of employing this layout is how to select appropriate geometric parameters to attain low C_{ps}, sufficient insulation strength, and a low profile. This problem will be addressed by a dedicated transformer geometric parameter design approach proposed in Section IV.

IV. ANALYSIS OF TRANSFORMER GEOMETRIC PARAMETERS

The influence of the transformer geometric parameters on C_{ps} will be analyzed in this section. It is found that C_{ps} has strong nonlinearity to the geometric parameters. To avoid the derivation of complex analytical expression, a finite-element-method (FEM) -based approach is proposed for designing the geometric parameters. Firstly, the key factors that influence C_{ps} will be pointed out and their effects on C_{ps} will be analyzed theoretically. Then a parameter-sweeping FEM simulation is utilized to verify the analysis. Meanwhile, the optimized value can be determined to facilitate the parameter design. Finally, the PD-free parameter design will be examined by simulation.

Considering the complexity of the transformer structure, the magnetic core is divided into three parts to clarify the analysis, namely, the column, the wings, and the bases, as shown in Fig. 3. Hence, we will have

$$C_{w-c} = C_{w-cl} + C_{w-wg} + C_{w-bs} \tag{3}$$

where C_{w-cl}, C_{w-wg} and C_{w-bs} are the winding-to-column, winding-to-wing, and winding-to-base capacitance, respectively. The main transformer geometric parameters that influence C_{ps} include the distance between the windings and the core base d_{w-bs}, the distance between the primary and secondary windings d_{pw-sw}, and the radius of the trace R_t.

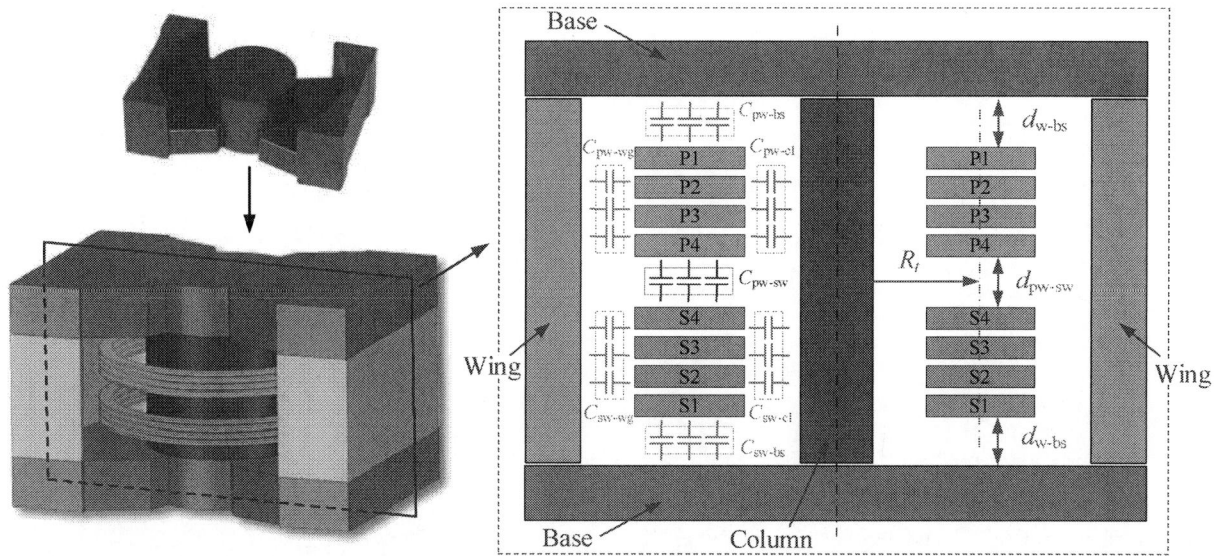

Fig. 3.   Detailed PCB transformer structure.

A. Distance Between Windings and Core Base

Fig. 4 shows the simulation results of coupling capacitances against $d_{w\text{-}bs}$ with $d_{pw\text{-}sw}$=2.0mm and R_t=6.5mm. The relationship between the coupling capacitances and $d_{w\text{-}bs}$ is demonstrated as follows.

As shown in Fig. 4, $C_{pw\text{-}sw}$ keeps constant with the variation of $d_{w\text{-}bs}$. The reason lies in that, $C_{pw\text{-}sw}$ is dominated by the capacitance of the two adjacent turns of the primary and secondary windings, which are specifically the P4 and S4 in Fig. 3. Particularly, $d_{w\text{-}bs}$ exerts no influence on the P4-to-S4 capacitance.

It is obvious that $C_{w\text{-}bs}$ will decrease as $d_{w\text{-}bs}$ increases. Please keep in mind that large $d_{w\text{-}bs}$ results in a bulky magnetic core, for the sake of leaving sufficient window height to accommodate the windings. This finally causes reduced power density of the GDPS. Hence, $d_{w\text{-}bs}$ cannot increase unlimitedly.

In addition, increased $d_{w\text{-}bs}$ doesn't necessarily bring small C_{ps}. Regarding $C_{w\text{-}cl}$ and $C_{w\text{-}wg}$, $d_{w\text{-}bs}$ is irrelevant to either the overlapping area or the distance between the windings and the core. However, the core base has a shielding effect of suppressing the electrical coupling between the windings and the core legs (the column or the wings). As $d_{w\text{-}bs}$ increases, this shielding effect gradually weakens. Therefore, $C_{w\text{-}cl}$ and $C_{w\text{-}wg}$ will increase as seen in Fig. 4.

Finally, when $d_{w\text{-}bs}$ increases, C_{ps} decreases dramatically at the very beginning but stays almost unchanged afterward, as shown in Fig. 4. This implies that, $d_{w\text{-}bs}$ should be optimized at the knee point of the curve to balance C_{ps} and the power density. In this case, for example, $d_{w\text{-}bs}$ can be set to 2.0mm.

B. Distance Between Primary and Secondary Windings

Fig. 5 shows the simulation results of coupling capacitances against $d_{pw\text{-}sw}$ with $d_{w\text{-}bs}$=2.0mm and R_t=6.5mm. It can be found that the relationship between every partial capacitance and $d_{pw\text{-}sw}$ shows high symmetry with that of $d_{w\text{-}bs}$.

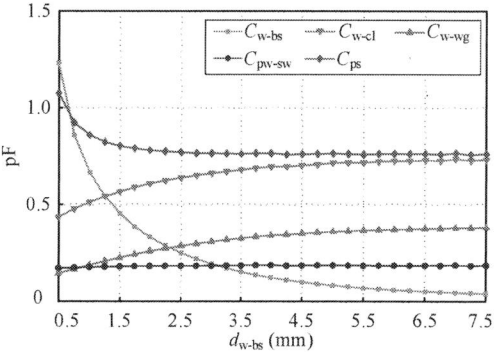

Fig. 4.   Simulation results of coupling capacitances against $d_{w\text{-}bs}$.

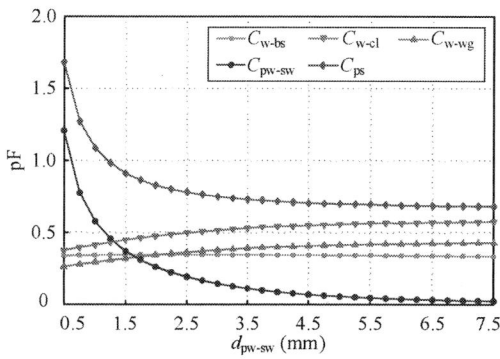

Fig. 5.   Simulation results of coupling capacitances against $d_{pw\text{-}sw}$.

Likewise in Section IV-A, $C_{w\text{-}bs}$ remains constant with varied $d_{pw\text{-}sw}$, as shown in Fig.5. This is because $d_{pw\text{-}sw}$ has no influence on the capacitance of the core base to P1 or S1. As $d_{pw\text{-}sw}$ increases, $C_{pw\text{-}sw}$ decreases in an inversely proportional manner, leading to an initial decline for C_{ps}. Meanwhile, $C_{w\text{-}cl}$ and $C_{w\text{-}wg}$ gradually increase because of the weakening shielding effect. Therefore, C_{ps} remains almost constant same when $d_{pw\text{-}sw}$ is fairly large.

TABLE II.	GDPS SPECIFICATIONS
Parameter	Value
Input Voltage	48V
Output Voltage	24V
Output Power	2W
Switching Frequency	300kHz
Insulation Voltage	3kV

TABLE III.	PCB TRANSFORMER PARAMETERS
Parameter	Value
Magnetic Material	DMR95
Core Size	PQ20/xx
Primary Turns N_p	4
Secondary Turns N_s	4
Thickness of Trace t_t	2oz/ft^2
Width of Trace	15mil

In order to compromise C_{ps} and the power density, $d_{pw\text{-}sw}$ should be optimized at the knee point of the curve. Take this design as an example, $d_{pw\text{-}sw}$ is selected to be 2.5mm.

C. Radius of the Trace

Figs. 6 and 7 show the simulation results of coupling capacitances against R_t with $d_{pw\text{-}sw}$=2.0mm and $d_{w\text{-}bs}$=2.0mm respectively.

For the selected core with fixed window width w_w, as shown in Figs. 6 and 7, $C_{w\text{-}cl}$ decreases inversely proportionally to R_t while $C_{w\text{-}wg}$ increases inversely proportionally to w_w-R_t. As a result, defining $C_{w\text{-}legs}$ as $C_{w\text{-}cl}$ + $C_{w\text{-}wg}$, $C_{w\text{-}legs}$ will initially decrease and then increase with increased R_t.

Considering the toroidal PCB trace with width of trace w_t, the surface area of the trace A_t is $2\pi w_t R_t$. When R_t increases, A_t, the overlapping area of the windings and the base as well as the overlapping area of the primary and secondary windings, will increase. Hence, $C_{w\text{-}bs}$ and $C_{pw\text{-}sw}$ increase with growing R_t as shown in Figs. 6 and 7.

As for the variation trend of C_{ps}, theoretically, there are two scenarios. The first case is that, either $C_{w\text{-}bs}$ or $C_{pw\text{-}sw}$ takes the dominant part in C_{ps} when $d_{w\text{-}bs}$ or $d_{pw\text{-}sw}$ is small such that C_{ps} will monotonically increase as shown in Fig. 6(a) or in Fig. 7(a). The other case is that, $C_{w\text{-}legs}$ is the dominant part in C_{ps} when $d_{w\text{-}bs}$ and $d_{pw\text{-}sw}$ are large as shown in Fig. 6(b) or in Fig. 7(b). Identically to $C_{w\text{-}legs}$, C_{ps} will initially decrease and then increase.

With the aid of the simulation results, an optimum R_t can be found to achieve a minimum value of C_{ps}.

D. Design Example

The specification of a flyback-type GDPS is given in Table II and Table III, respectively. Fig. 8 shows the simulation results with different parameters. To obtain a possibly small C_{ps} without causing a bulky core, the optimized parameters are determined as: $d_{w\text{-}bs}$=2.5mm, $d_{pw\text{-}sw}$=4.0mm, R_t=6.4mm, with C_{ps}=0.86pF. Fig. 9 shows the FEM simulation result of E-field intensity distribution. As shown, the maximum E-field intensity E_{max} is 1.46kV/mm. Considering the dielectric strength of air is 3kV/mm, PD risk is eliminated in this design.

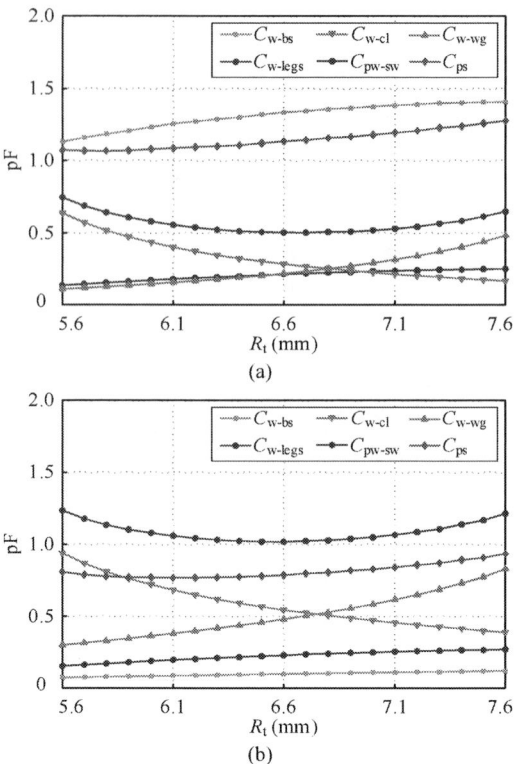

Fig. 6. Simulation results of coupling capacitances against R_t with $d_{pw\text{-}sw}$=2.0mm and different $d_{w\text{-}bs}$. (a) $d_{w\text{-}bs}$=0.5mm. (b) $d_{w\text{-}bs}$=5.0mm.

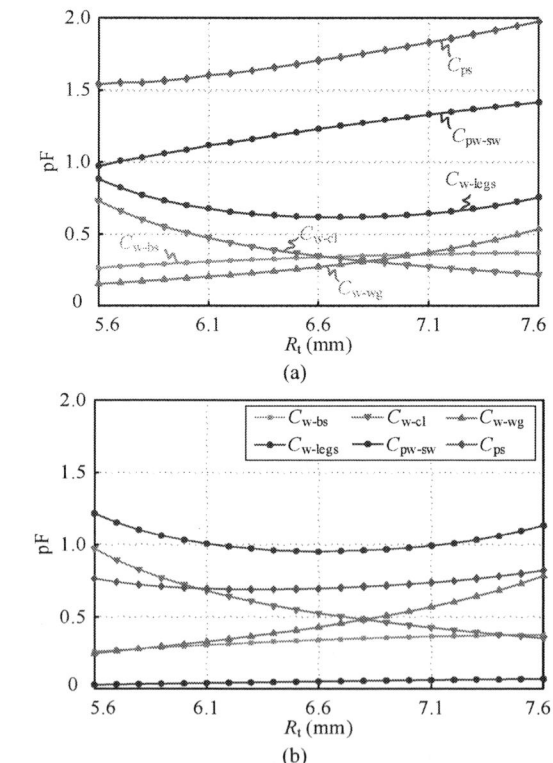

Fig. 7. Simulation results of coupling capacitances against R_t with $d_{w\text{-}bs}$=2.0mm and different $d_{pw\text{-}sw}$. (a) $d_{pw\text{-}sw}$=0.5mm. (b) $d_{pw\text{-}sw}$=5.0mm.

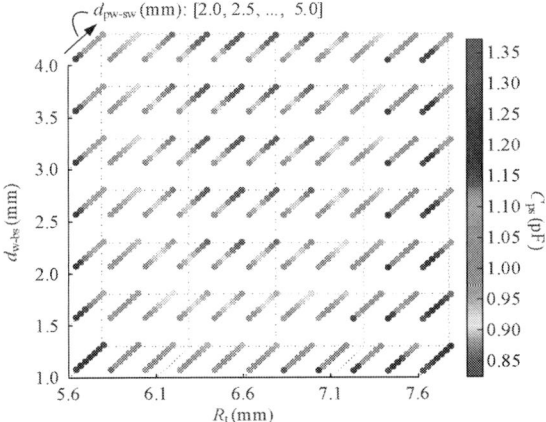

Fig. 8. Simulation results of C_{ps} with different transformer parameters.

Fig. 9. Simulation result of E-field intensity distribution.

V. EXPERIMENTAL RESULTS

To validate the feasibility of the proposed work, a PCB transformer with optimized geometric parameters given in Section IV-D is fabricated and tested in the lab. As a reference group, a conventional 8-layer PCB transformer with a non-interleaved layout in Fig. 2(a) is also fabricated.

Fig. 10 shows the prototype photo of the PCB transformers. The C_{ps} measurement setup by OMICRON Lab Bode100 is presented in Fig. 11. The test results are given in Fig. 12 and Table IV. Considering the rising/falling time of high-voltage SiC MOSFET is about 200ns [7], the measurement point is selected as 1.75MHz [14].

As can be seen, the measurement results match well with the simulation results. Please note that the measurement results are slightly higher than the simulation results. Such errors come from the parasitic capacitances introduced by manual manufacturing error and test leads.

As seen from Fig. 12, compared with the conventional 8-layer PCB transformer, C_{ps} of the optimized PCB transformer is dramatically reduced, demonstrating the effectiveness of the proposed winding layout and FEM-based transformer geometric parameters design approach.

VI. CONCLUSIONS

To mitigate the potential risk by the common mode current resulting from the high dv/dt of SiC MOSFETs, it is necessary to reduce the interwinding capacitance of the GDPS for high-voltage SiC MOSFET. In this paper, design considerations are proposed for designing low-interwinding-capacitance PCB-winding-based transformer used in the GDPS. Different winding layouts are compared to determine the optimized winding layout. In addition, the relationship between the interwinding capacitance and the PCB transformer geometric parameters is analyzed, based on which, a FEM-based geometric parameters design approach is proposed. Taking all the design considerations into account, a complete and practical PCB-winding based transformer design procedure is proposed to achieve low interwinding capacitance. Finally, the effectiveness of the proposed work is validated by experimental results.

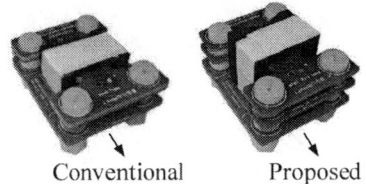

Fig. 10. Prototype photo of the PCB transformers.

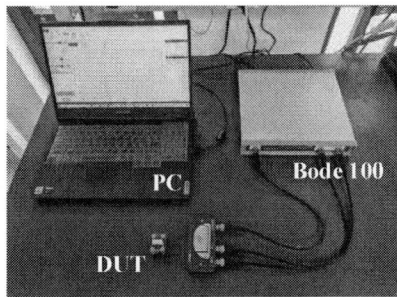

Fig. 11. C_{ps} measurement setup.

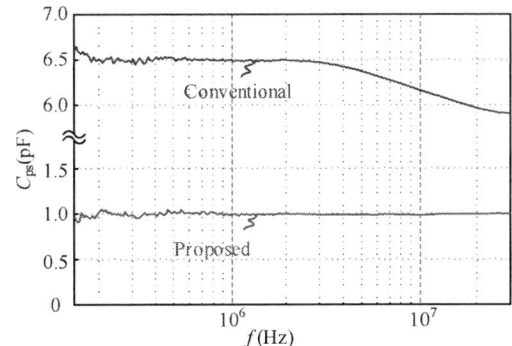

Fig. 12. C_{ps} measurement results.

TABLE IV.   TEST RESULTS OF INTERWINDING CAPACITANCE

	Simulation (pF)	Measurement (pF) @1.75MHz
Conventional	6.21	6.48
Proposed	0.86	0.96

References

[1] A. N. Lemmon, R. C. Graves, R. L. Kini, M. R. Hontz, and R. Khanna, "Characterization and modeling of 10-kV silicon carbide modules for naval applications," *IEEE J. Emerg. Sel. Top. Power Electron.*, vol. 5, no. 1, pp. 309–322, Mar. 2017.

[2] D. Rothmund, D. Bortis, and J. W. Kolar, "Highly compact isolated gate driver with ultrafast overcurrent protection for 10 kV SiC MOSFETs," *CPSS Trans. Power Electron. Appl.*, vol. 3, no. 4, pp. 278–291, Dec. 2018.

[3] H. Li, Z. Gao, and F. Wang, "Medium-voltage isolated auxiliary power supply design for high insulation capability, ultra-low coupling capacitance, and small size," *IEEE Trans. Power Electron.*, vol. 38, no. 6, pp. 7226–7240, Jun. 2023.

[4] L. Li, Y. Gan, T. Yuan, D. Ma, Y. Nie, P. Sun, X. Dong, K. Gao, and L. Wang, "Design and optimization of high performance gate driver for medium-voltage SiC power modules," in *Proc. IEEE Appl. Power Electron. Conf. Expo.*, 2024, pp. 2800–2805.

[5] R. Wang, A. B. Jørgensen, W. Liu, H. Zhao, Z. Yan, D. N. Dalal, and S. M.- Nielsen "Auxiliary power supply startup evaluation and improvement of the input-series system with small submodule capacitances," in *Proc. IEEE Appl. Power Electron. Conf. Expo.*, 2023, pp. 1323–1330.

[6] Y. Li, A. J. Watson, and M. Kaya, "A Class-E-based multichannel auxiliary power supply with individual hysteresis control," *IEEE J. Emerg. Sel. Top. Power Electron.*, vol. 11, no. 3, pp. 3335–3347, Jun. 2023.

[7] S. Ji, Z. Zhang, and F. Wang, "Overview of high voltage SiC power semiconductor devices: development and application," *CES Trans. Elect. Mach. Syst.*, vol. 1, no. 3, pp. 254–264, Sep. 2017.

[8] Y. Zhou, L. Zhang, Y. Chen, T. Yin, and L. Lin, "Architecture of self-power-supplied auxiliary power supply for high-voltage SiC MOSFET," in *Proc. IEEE Energy Conv. Cong. Expo.*, 2024, pp. 2933–2937.

[9] A. Anurag and P. Barbosa, "High-voltage isolated power supply structure for gate drivers of medium-voltage SiC devices," *IEEE Trans. Power Electron.*, vol. 38, no. 6, pp. 6907–6911, Jun. 2023.

[10] J. Sabate, E. Delgado, and M. Harfman-Todorovic, "Gate driver power supply for medium voltage SiC MOSFETs with air core transformer," in *Proc. IEEE Energy Conv. Cong. Expo.*, 2022, pp. 1–6.

[11] J. Hu, J. Wang, R. Burgos, B. Wen, and D. Boroyevich, "High-density current-transformer-based gate-drive power supply with reinforced isolation for 10-kV SiC MOSFET modules," *IEEE J. Emerg. Sel. Top. Power Electron.*, vol. 8, no. 3, pp. 2217–2226, Sep. 2020.

[12] L. Zhang, S. Ji, S. Gu, X. Huang, J. E. Palmer, W. Giewont, F. Wang, and L. M. Tolbert, "Design considerations for high-voltage insulated gate drive power supply for 10-kV SiC MOSFET applied in medium-voltage converter," *IEEE Trans. Ind. Electron.*, vol. 68, no. 7, pp. 5712–5724, Jul. 2021.

[13] E. Serban, M. A. Saket, and M. Ordonez, "High-performance isolated gate-driver power supply with integrated planar transformer," *IEEE Trans. Power Electron.*, vol. 36, no. 10, pp. 11409–11420, Oct. 2021.

[14] V. Nguyen, L. Kerachev, P. Lefranc, and J. Crebier, "Characterization and analysis of an innovative gate driver and power supplies architecture for HF power devices with high dv/dt," *IEEE Trans. Power Electron.*, vol. 32, no. 8, pp. 6079–6090, Aug. 2017.

Integrated 4-Level Dual-Phase Superimposed Quadratic Power Converter for High-Density Direct 48V/1V Conversion

Prosenjit Ghosh, Jin Woong Kwak, Fei Zhou, D. Brian Ma
Department of Electrical and Computer Engineering
The University of Texas at Dallas
Richardson, TX 75080, USA
Email: {prosenjitkumar.ghosh, jinwoong.kwak,fei.zhou;brian.ma}@utdallas.edu

Abstract—Recent hybrid power converter topologies have encountered significant design and performance challenges for high step-down DC-DC power conversion. To minimize power passives and achieve high power density, both high switching frequency (f_{SW}) operation and single-stage conversion are preferred; however, the resulting high step-down ratio imposes extremely short ON-time (T_{ON}), which limits the maximum achievable f_{SW}. To address these, this paper presents a novel 4-level dual-phase superimposed quadratic (4-LDPSQ) converter topology, which significantly extends T_{ON} and enhances efficiency by reducing voltage stresses on switches and optimizing RMS currents with parallel multipath power delivery. A prototype of this work, designed in a 0.18µm HV-BCD process, demonstrates direct 48V-to-1V conversion with a maximum I_O of 5A at 5MHz, a peak efficiency of 88.28%, and power density of 69mW/mm³, representing ×9.35 power density improvement and ×13.9 T_{ON} extension over a conventional half-bridge (HB) topology.

Keywords—Quadratic converter, inductor-first buck converter, high step-down conversion, high-density converter, non-isolated converter, point-of-load converter.

I. INTRODUCTION

Fast development of high-density and high-efficiency data centers and automotive electronics have imposed fast-growing demands for direct DC-DC power conversion between 48V supply rails and sub-1.8V at point-of-load (PoL) [1]. However, conventional buck converters experience high voltage swings at the switching node (V_{SW}) and escalated switching loss (P_{SW}) leads to very low efficiency [2-3]. Several advanced hybrid converter topologies have recently been reported in the search

for more competent candidates [4-7]. As power demands grow exponentially, minimizing power losses and enhancing power density have emerged as top priorities in the design process.

The widely used HB converters in Fig. 1(a), cannot suffice due to inherent topological limitations. For example, to perform 48V-to-1V conversion, its duty ratio is limited to ~2.1%, which results in extremely narrow T_{ON} and limits high f_{SW}. Meanwhile, the low-side switch S_L remains ON during most switching cycles, inducing substantial conduction loss. A series of hybrid converter topologies have been proposed to overcome such challenges. A double step-down (DSD) converter in Fig. 1(b) subdivides V_{IN} by half with C_{F1}. It extends T_{ON} by ×2 and reduces the voltage swings by half [8], thus significantly reducing the switching losses. Thanks to its dual-phase operation where each inductor carries $I_O/2$, the overall conduction loss is also reduced and the inductor current ripples (ΔI_L) cancel each other, minimizing output ripple (ΔV_O). realize the desirably high f_{SW}. A superimposed quadratic buck (SQ) converter in Fig. 1 (c) further extends T_{ON} by ×6.9 compared to HB topology and allows lower voltage-rated devices to handle most of the I_O [9]. However, the frontend half-bridge switches S_{H1} and S_{L1} still deal with excessive switching voltage swings. Meanwhile, its I_O delivery ability is limited at heavy loads due to using only two inductive paths with unbalanced currents.

In order to solve the aforementioned problems, this paper presents a novel 4-level dual-phase superimposed quadratic (4-LDPSQ) converter topology. By stepping down the input voltage by $V_{IN}/4$, the converter effectively reduces voltage stress

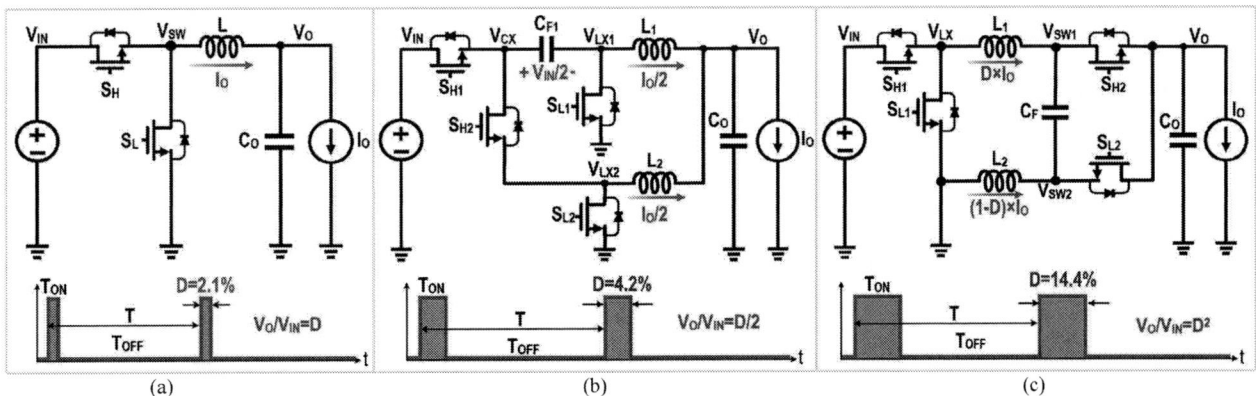

Fig. 1. Topologies and duty ratios of (a) half-bridge (b) double step-down, and (c) superimposed quadratic converter for direct 48V-to-1V power conversion.

979-8-3315-1612-3/25 $31.00 © 2025 IEEE

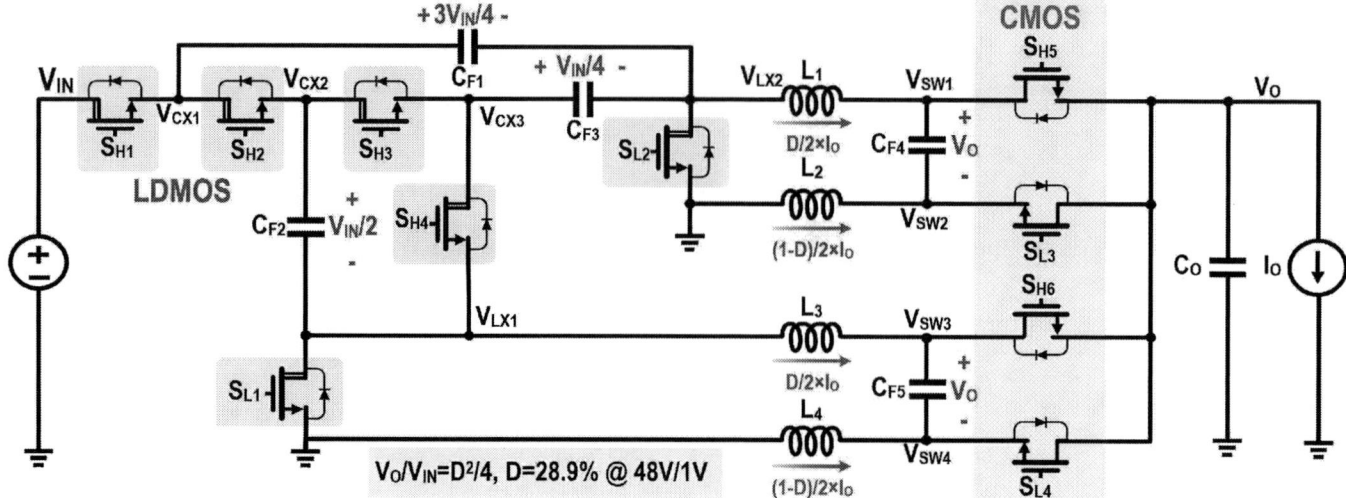

Fig. 2. Proposed topology of 4-level dual-phase superimposed quadratic (4-LDPSQ) buck converter.

on the switching node. This enables longer T_{ON}, supporting operation at higher f_{SW}, which reduces the size of passive components and enhances power density. Additionally, the converter's parallel path power delivery technique improves overall efficiency.

The remainder of this paper is organized as follows: Section II introduces the converter topology and outlines its operational strategies. Section III presents a duty-ratio-based inductor optimization approach to effectively showcase its performance. Section IV describes the performance verification of a converter and validates the design through simulation results, including a comparison with HB, DSD, and SQ converter topologies. Finally, this research study is concluded in section V.

II. 4-LEVEL DUAL-PHASE SUPERIMPOSED QUADRATIC TOPOLOGY

The proposed 4-level dual-phase superimposed quadratic (4-LDPSQ) buck converters as shown in Fig. 2, introduce a series-capacitor (SC) structure as the frontend, which merges with a power-delivery improved multi-path SQ stage. The SC frontend subdivides high V_{IN} by $\times 1/4$ using flying capacitors C_{F1-3}, providing a $\times 1/4$ step-down ratio. The steady-state operation of the 4-LDPSQ converter presented in Fig. 3, consists of four switching states that control the magnetization and demagnetization of its inductors to achieve efficient power conversion.

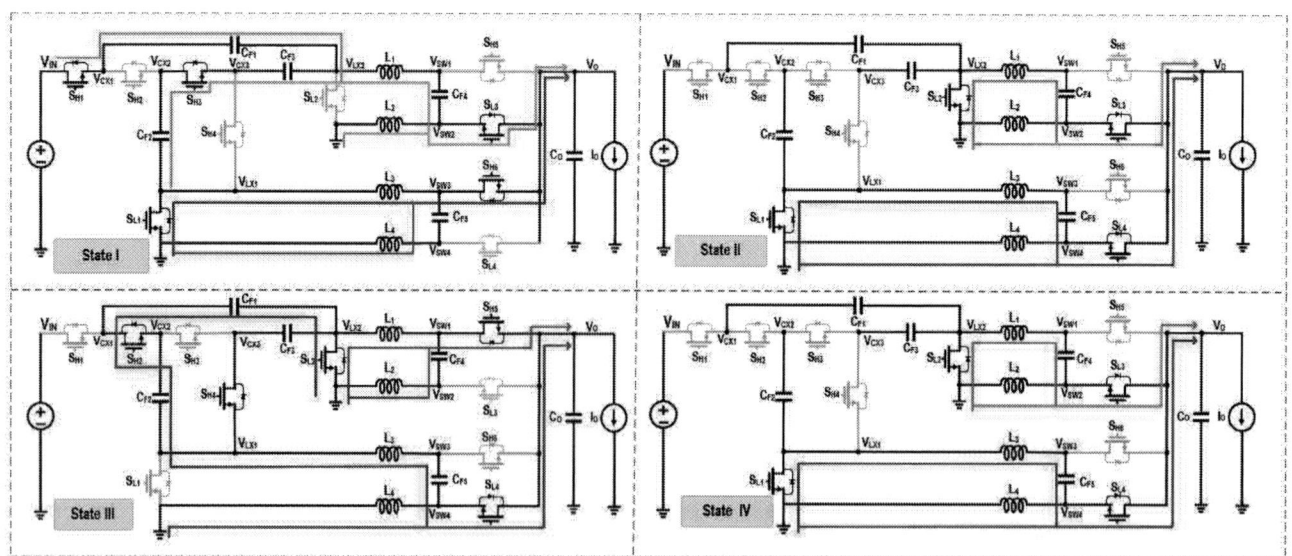

Fig. 3. Key steady state operation of 4-LDPSQ buck converter.

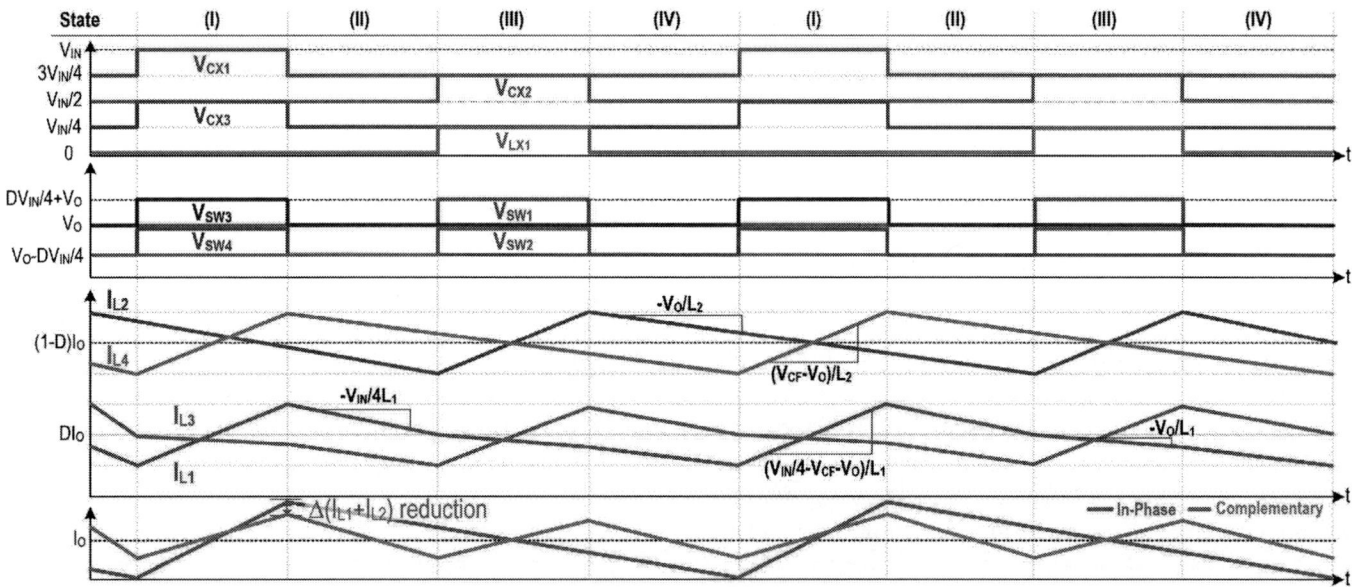

Fig. 4. Key waveforms of 4-LDPSQ buck converter.

In State I: switches S_{H1}, S_{H3}, S_{H6}, and S_{L3} are ON, resulting inductors L_1 and L_3 are magnetized while L_2 and L_4 are demagnetized.

In state II and state IV: all high-side switches S_{H1-4} are OFF, and all low-side switches S_{L1-4} are ON, leading to the demagnetization of all inductors (L_{1-4}).

In state III: the switches S_{H2} and S_{H4} are turned ON, causing L_2 and L_4 to are magnetized while L_1 and L_3 are demagnetized.

In steady state, the volt-second balance for I_{L1} gives,

$$\left(\frac{V_{IN}}{4} - V_O\right)DT + (-V_O - V_{CF})(1 - D)T = 0 \quad (1)$$

and the volt-second balance I_{L2} leads,

$$(V_{CF} - V_O)DT - V_O(1 - D)T = 0 \quad (2)$$

Now solving for V_{CF},

$$V_{CF} = DV_{IN} \quad (3)$$

By substituting V_{CF} in Eqn. (1), (2) and solving for V_O/V_{IN},

$$M = \frac{D^2}{4} \quad (4)$$

With the proposed conversion ratio, the proposed topology delivers 48V-to-1V step-down with duty ratio of 28.9%, extending T_{ON} by $\times13.9$ compared to the HB and enabling much higher f_{SW} operation.

III. DUTY-RATIO-BASED INDUCTOR OPTIMIZATION

Two strategies are employed to minimize current ripples. First, a complementary switching scheme operates the frontend switches (S_{H5}-S_{L3}) and the backend switches (S_{H6}-S_{L4}) 180° out of phase to achieve current ripple cancellations as shown in Fig. 4. It is crucial to ensure identical current ripple in both inductors to achieve maximum ripple cancellation, as shown in Fig. 5. Therefore, inductor optimization is performed based on the duty ratio. The sizes of L_1 and L_2 are optimized based on duty ratio further minimizing overall inductance and associated DCR loss. The inductor current ripple ΔI_{L1} can be expressed as

$$\Delta L_1 = \frac{\left(4 - \dfrac{DV_{IN}}{4} - \dfrac{D^2 V_{IN}}{4}\right)}{L_1} \times DT \quad (5)$$

and the inductor current ripple ΔI_{L2} can be expressed as:

$$\Delta L_2 = \frac{(V_{CF} - V_O)}{L_2} \times DT \quad (6)$$

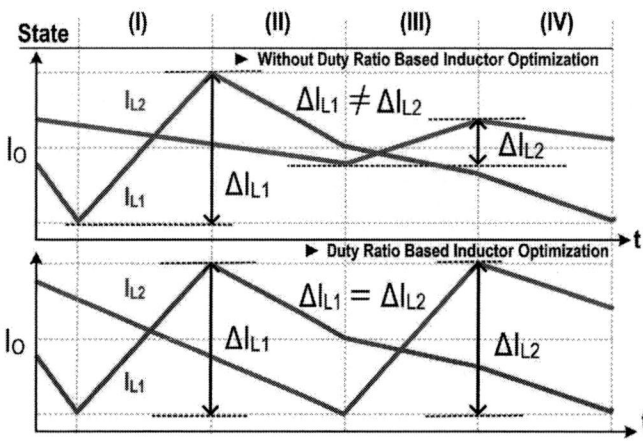

Fig. 5. Duty ratio-based inductor optimization.

To find the optimal inductance for L_2 to achieve maximum ripple cancellation,

$$L_2 = \frac{4D - D^2}{4 - 4D - D^2} L_1 \qquad (7)$$

which implies that L_2 must be chosen according to the desired duty ratio for current ripple minimization. For example, in 48V/1V conversion, $M = \frac{D^2}{4}$, where $D \approx 0.28$, and L_2 is given by,

$$L_2 = 0.38 \times L_1 \qquad (8)$$

Consequently, for 48V/1V conversion, the size of L_2 can be reduced by ~62% compared to L_1 and yet achieve a reduced current ripple size, which further improves efficiency and power density. In Fig. 6, the normalized current ripple is presented and compared between conventional HB, DSD, SQ, and the proposed 4-LDPSQ buck topologies. The results demonstrate that the proposed 4-LDPSQ converter achieves the smallest inductor current ripple compared to all topologies. This lower ripple improves system stability and minimizes the output filter size, further optimizing overall converter performance.

To maintain high efficiency at elevated switching frequencies, it is essential to keep switching losses to a minimum. The proposed 4-LDPSQ topology achieves this by leveraging a switched-capacitor (SC) frontend stage, which effectively reduces the voltage ratings required for the power switches. By lowering these voltage ratings, the design enables the use of lower-voltage-rated devices with significantly lower $R_{ON} \times Q_G$. As shown in Fig. 7, these lower-rated devices contribute to decreased conduction and switching losses, further enhancing efficiency. To deliver high-output current I_O, the 4-LDPSQ topology strategically distributes I_O across four inductors (L_1–L_4). In this arrangement, L_1 and L_3 each deliver $D \times I_O/2$, while L_2 and L_4 provide $(1-D) \times I_O/2$. This division of load current across four inductors in parallel helps balance the power delivery, reducing stress on individual components and improving overall converter reliability and performance. Furthermore, the proposed 4-LDPSQ is extended by the quadratic conversion ratio. For direct 48V/1V conversion, the T_{ON} is extended by a factor of $\times 14.8$, $\times 6.9$ and $\times 2$ compared to HB, DSD, and SQ buck converter. This extended T_{ON} enables operation at a higher f_{SW} which minimizes the size of passive components and enhances power density.

IV. PERFORMANCE VERIFICATION

The proposed 4-LDPSQ converter is designed in a 0.18μm HV-BCD process and verified by fully transistor-based post-layout Cadence simulations. Meanwhile, chip tape-out is planned for further validation. Fig. 8 examines the steady-state operation of the proposed converter. The switching node

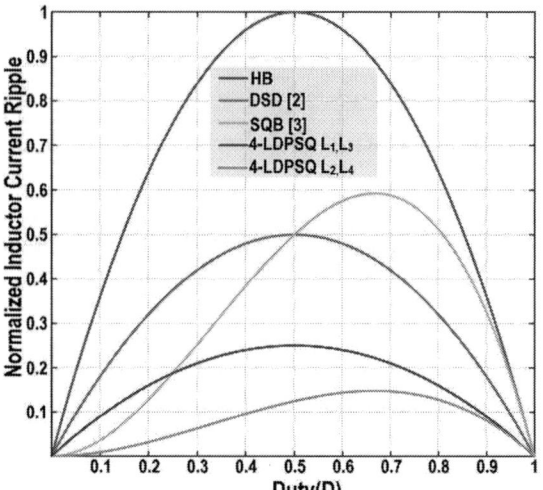

Fig. 6. Inductor current ripple ΔI_L comparison.

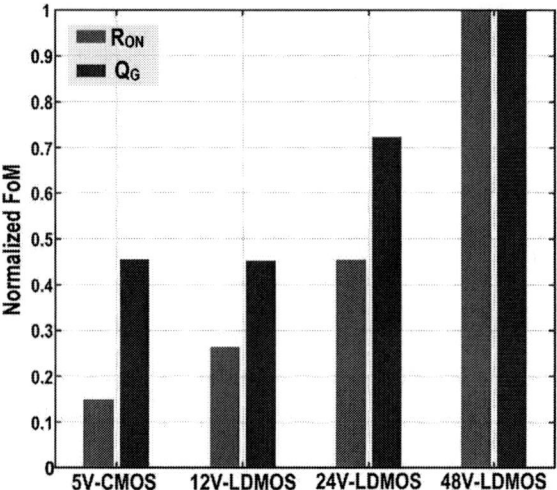

Fig. 7. R_{ON} and Q_G comparison between power transistors [10].

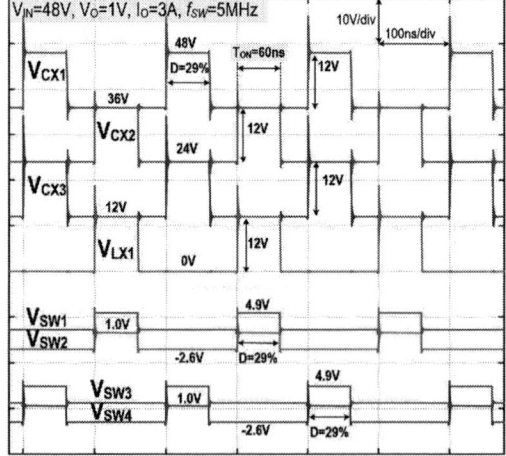

Fig. 8. Waveform of switching node voltages (V_{CX1}-V_{SW4}).

voltage drop of $V_{IN}/4$ at the frontend stage helps reduce the voltage stress on the backend switches (V_{SW1-4}). Additionally, Fig. 9 shows the inductor current waveform where L_2-L_4 are the smallest inductors and deliver ~70% of the output current with smaller DCR loss. Fig. 10 compares the total power passive volumes and power densities of the HB, DSD, and SQ buck topologies. With ×13.9 longer T_{ON}, the 4-LDPSQ converter operates at the highest f_{SW} of 5 MHz, which allows the smallest inductance as listed in Table I. With the same T_{ON} of 60ns, the HB converter operates at f_{SW} of 333 kHz, DSD at 666 kHz, and the SQ at 2.5 MHz. The HB topology shows the largest total passive volume, resulting in a low power density of 7.37 mW/mm³, while the DSD and SQ topologies offer moderate improvements in power density and reduced passive volume. However, the proposed 4-LDPSQ topology stands out with the smallest passive volume and the highest power density at 69 mW/mm³, approximately 9.35 times greater than the HB topology. Fig. 11. compares the power efficiency curves of the proposed converter. Delivering up to 5A, the prototype achieves a peak efficiency of 88.28%, offering the highest efficiency and power density among the compared designs.

V. CONCLUSION

To overcome the performance challenges of the hybrid topologies for high step-down (i.e.,48V/1V), a novel 4-level dual-phase superimposed quadratic (4-LDPSQ) buck converter is developed. The converter reduces voltage stress on the switching node by $V_{IN}/4$, enabling the use of lower voltage-rated switches at the backend, which enhances efficient power delivery. To further optimize the performance of the proposed topology, a complementary switching scheme and duty-ratio-based inductor size optimization technique are introduced. The proposed 4-LDPSQ converter significantly improves T_{ON} compared to traditional topologies, achieving 14.8 times higher than HB topology for 48V/1V conversion. Furthermore, by operating at an elevated switching frequency the converter achieves a power density 9.35 times higher than the conventional HB buck converter. The simulation results demonstrate a peak efficiency of 88.28% with power delivery up to 5W, showcasing the best performance among all compared converters.

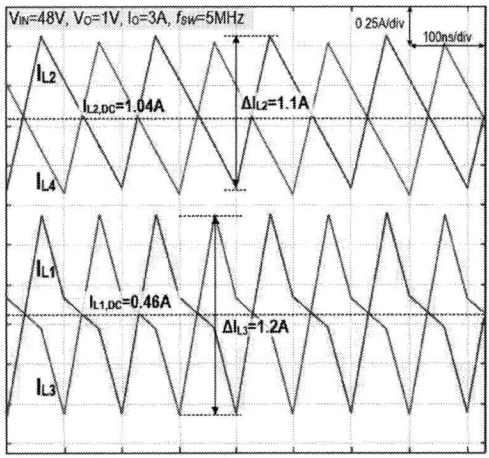

Fig. 9. Inductor current waveforms of proposed 4-LDPSQ converter.

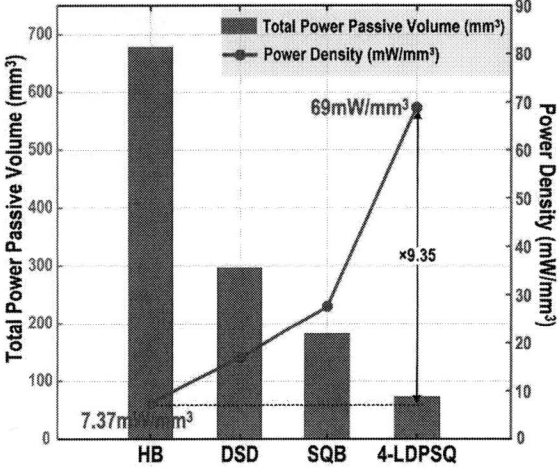

Fig. 10. Power densities of HB, DSD, SQ, and 4-LDPSQ converters.

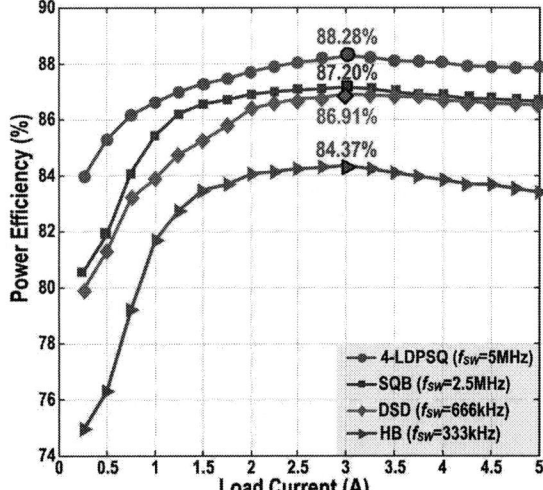

Fig. 11. Efficiencies of HB, DSD, SQ and 4-LDPSQ converters.

Table I. Comparison between HB, DSD, SQ, and 4-LDPSQ.

Topology	HB	DSD [8]	SQB [9]	This work
Specification	V_{IN}=48V, V_O=1V, T_{ON}=60ns, ΔI_L=1A, ΔV_O=10mV			
f_{SW} (T_{ON}=60ns)	333kHz	666kHz	2.5MHz	5MHz
V_O/V_{IN} (M)	D	D/2	D^2	$D^2/4$
Inductance	L=3.3uH	L_1, L_2=1.6uH	L_1=2.2uH L_2=330nH	L_1, L_3=450nH L_2, L_4=180nH
Capacitance	C_O=40uF	C_{F1}=10uF C_O=30uF	C_{F1}=20uF C_O=10uF	C_{F1}=600nF, C_{F2}=800nF C_O=7uF
Device Type	2×48V LDMOS	1×48V LDMOS 3×24V LDMOS	2×48V LDMOS 2×8V LDMOS	4×24V LDMOS, 2×12V LDMOS 4×5V CMOS
Power Density	7.37mW/mm³	16.88mW/mm³	27.44mW/mm³	69mW/mm³
Peak Efficiency @ 48V/1V	84.37% @ 3A	86.91% @ 3A	87.20% @ 3A	88.28% @ 3A
Power Switch Area	3.53mm²	1.60mm²	0.72mm²	1.41mm²

REFERENCES

1. X. Li et al., "Google 48V Power Architecture," 2017 IEEE Applied Power Electronics Conference & Exposition (APEC), March 2017.

2. J. W. Kwak and D. Brian Ma, "Comparative Topology and Power Loss Analysis on 48V-to-1V Direct Step-Down Non-Isolated DC-DC Switched-Mode Power Converters," 2020 IEEE Energy Conversion Congress and Exposition (ECCE), Detroit, MI, USA, 2020, pp. 943-949.

3. E. Aklimi and et. al, "Hybrid CMOS/GaN 40-MHz Maximum 20 V Input DC–DC Multiphase Buck Converter," IEEE J. of Solid-State Circuits, vol. 52, no. 6, pp. 1618–1627, Jun. 2017.

4. N. M. Ellis and R. C. N. Pilawa-Podgurski, "A Symmetric Dual-Inductor Hybrid Dickson Converter for Direct 48V-to-PoL Conversion," 2022 IEEE Applied Power Electronics Conference and Exposition (APEC), Houston, TX, USA, 2022, pp. 1267-1271.

5. M. Gong et al., "A GaN-Based Reconfigurable Series-Parallel Hybrid Converter Supporting 48/24/12V Input and 0.8-1.2V Output with 83.7/87.8/90.7% Peak Efficiency," 2023 IEEE Applied Power Electronics Conference and Exposition (APEC), Orlando, FL, USA, 2023, pp. 912-918.

6. Y. Chen, H. Cheng, D. M. Giuliano and M. Chen, "A 93.7% Efficient 400A 48V-1V Merged-Two-Stage Hybrid Switched-Capacitor Converter with 24V Virtual Intermediate Bus and Coupled Inductors," 2021 IEEE Applied Power Electronics Conference and Exposition (APEC), Phoenix, AZ, USA, 2021, pp. 1308-1315.

7. W. C. Liu, Z. Ye and R. C. N. Pilawa-Podgurski, "A 97% Peak Efficiency and 308 A/in3 Current Density 48-to-4 V Two-Stage Resonant Switched-Capacitor Converter for Data Center Applications," 2020 IEEE Applied Power Electronics Conference and Exposition (APEC), New Orleans, LA, USA, 2020, pp. 468-474.

8. K. Nishijima, et al., "Analysis of Double Step-Down Two-Phase Buck Converter for VRM," INTELEC 05 - Twenty-Seventh International Telecommunications Conference, Berlin, Germany, 2005, pp. 497-502.

9. J. W. Kwak, et al. "Superimposed Quadratic Buck Converter for High-Efficiency Direct 48V/1V Applications," 2023 IEEE Applied Power Electronics Conference and Exposition (APEC), Orlando, FL, USA, 2023, pp. 2253-2259.

10. TSMC, "0.18 Micron Technology," TSMC, [Online]. Available: https://www.tsmc.com/english/dedicated Foundry/technology/logic/l_018micron.

Compensation method for unbalance of the multi-channel class E power amplifier using the closed loop frequency control

Kyungmin Lee

Samsung Research

Samsung Electronics

Seoul, South Korea

km1019.lee@samsung.com

Sungku Yeo

Samsung Research

Samsung Electronics

Seoul, South Korea

sungku.yeo@samsung.com

Abstract— **Class E power amplifiers are widely used in communication fields, particularly for repeaters, due to their high efficiency. In the field of communications, they transmit tens of watts of power at frequencies greater than several tens of MHz. To use a Class E power amplifier for applications requiring high frequency and high output, multiple PAs must be operated in parallel. When operating in parallel, individual channels may experience imbalance caused by errors or parasitic components in passive elements such as inductors and capacitors, which can lead to reduced lifespan. This paper presents a method for detecting channel imbalances and resolving them through frequency compensation, thereby eliminating the imbalance in paralleled-driven power amplifiers and extending their lifespan. The proposed circuit was verified by implementing an eight-channel multi-power amplifier with a capacity of 2kW.**

Keywords— class E power amplifier, multi-channel

I. INTRODUCTION

Power amplifiers serve the purpose of amplifying AC voltages for specific applications. There are two types of power amplifiers: linear power amplifier (LPA), such as classes A, B, and AB, that operate in non-switching, linear regions; and switching power amplifier (SPA), which drive switches in saturated regions and perform switching operations. LPAs exhibit low efficiency but generate less noise compared to switching PAs, which offer higher efficiency but also carry the potential risk of generating switching noises during operation [1]-[3].

In the domain of power amplifiers requiring kilowatts of power and capable of amplifying signals beyond ten MHz, LPAs are typically employed. High frequency and high power SPA have not been developed much because they do not have fast switching devices and filtering techniques for switching noise are difficult when there is a lot of switching noise. Nevertheless, recent advancements in wide bandgap (WBG) materials and devices have made it possible to explore the feasibility of developing technologies based on switching modes [4]-[7].

This paper proposes a method that enables the operation of power amplifiers exceeding kilowatts and ten megahertz simultaneously. For this purpose, we selected the class E power amplifier topology, known for its high-efficiency power circuits. In order to develop a high power PA, it is necessary to drive multiple channels in parallel [8]-[10]. In low frequency power circuits at the kHz level, it is easy to drive circuits in parallel, but as the frequency increases,

driving in parallel becomes more difficult. Various factors such as channel mismatch, imbalance and resonant components (inductor and capacitor) errors must be addressed. In particular, when there is an imbalance between channels, electrical stress occurs strongly in certain channels, resulting in early degradation of those channels. This can reduce the overall lifespan of the circuit, so it needs to be resolved [11], [12]. This paper proposes a method for resolving the problem of inter- channel imbalance by configuring eight channel Class E power amplifiers

II. PROPOSED CIRCUIT AND OPERATIONAL PRINCIPLE

The proposed circuit apply an inter-channel imbalance calibration method to 2kW class E power amplifier operating at 13.56 MHz, (Fig.1). Imbalance in the multichannel PA is caused by variations in the inductor (L) and capacitor (C) elements constituting the class E power amplifier and parasitic components in the layout (Fig. 2(a)). Although the same frequency was applied to ch1, ch2, and ch3, due to different effective LC values, the VDS waveforms for confirming zero voltage switching (ZVS) during switch-off times are different. That is, the voltage value becomes large in a specific channel (PA1), which causes continuous application of high voltage and results in reduced lifespan.

To solve this problem, this paper propose a method to detect the imbalance between channels and solve it through variable frequencies for each channel (Fig 2(b)). The VDS waveform of each channel in the circuit shown in Fig.1 is sensed and passed through a peak detector composed of RC. When this peak detector passes through an error amplifier composed of an opamp, an error with respect to the reference voltage Vref occurs. This error is converted into frequency information through a voltage-controlled oscillator (VCO). Then, the connection to the gate drivers of each channel generates the frequencies of each channel to be modified accordingly.

The control circuit uses the closed loop feedback. The difference between the peak value and a certain reference voltage changes the frequency. After changing the frequency and sensing the peak value again, the peak voltage becomes equal to the reference voltage. For example, in multiple channels, the VDS waveforms are different from each other, resulting in outputs of 1.01V, 1V, and 1.02V for PA1,2,3, respectively at the voltage sensor. Then, after passing through an error amplifier with a negative terminal reference

979-8-3315-1612-3/25 $31.00 © 2025 IEEE 2547

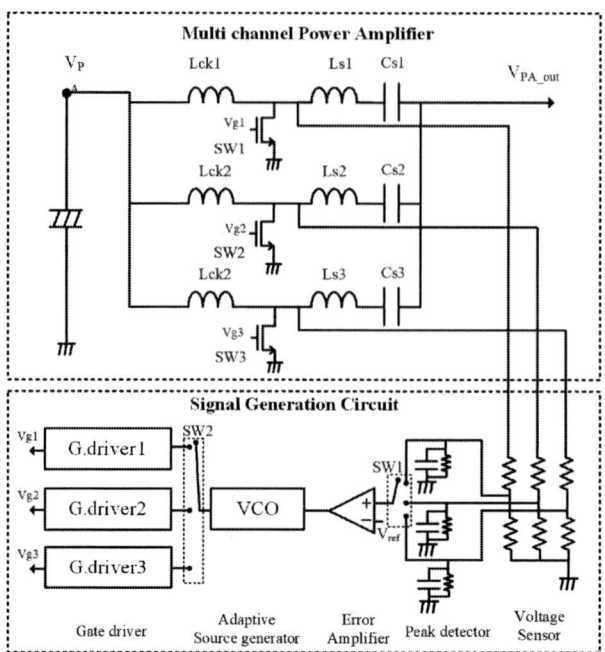

Fig. 1. Structure of the proposed circuit.

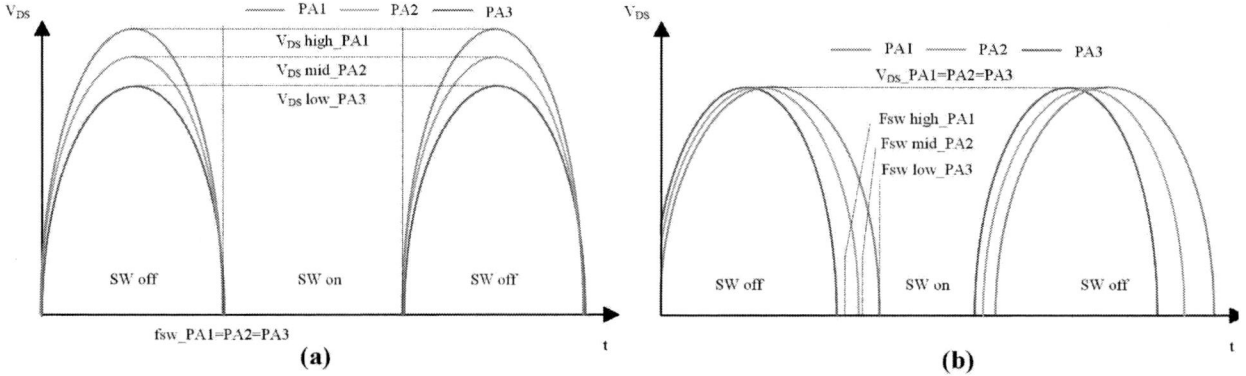

Fig. 2. Operational principle of the proposed methods; (a) unbalanced circuit, (b) balancing by the frequency variation.

of 1V, sequentially output values of 0.1V, 0V, and 0.2V are obtained for each channel. These values are applied to the VCO input, resulting in outputs of Fsw mid_PA2 (13.40 MHz), Fsw high_PA1 (13.56 MHz), and Fsw low_PA3 (13.30 MHz). (The output frequencies according to voltages can be set by adjusting the VCO.) The peak voltages of Channels 1 and 3 are decreased by frequency change. That is, all channels have the same peak voltage through several closed-loop controls.

III. EXPERIMENTAL RESULTS

A 2 kW class E power amplifier consisting of eight channels operating at 13.56 MHz was configured and tested (Fig.3(a)). This circuit was connected to a near field dryer (Fig. 3(b)) to verify its performance. A key point to

understanding the characteristics of class E power amplifiers is the drain voltage waveform of the main switch. Various information is indicated by the series resonant element and the switch parallel capacitor. The prototype was manufactured with a GaN switch and verified with a 50 ohm load.

The imbalance between channels was confirmed through the VDS waveform (Fig. 3(c)). The channel-to-channel imbalance phenomenon was identified by comparing the peak voltages of Vin= 100 V. As mentioned earlier, there are various causes for this imbalance. In general, a high frequency inductor has a manufacturing error of 20% and a radio frequency capacitor with pF units also has an error of 5%. Additionally, due to the presence of parasitic components in the tens of pF range on the PCB layout, such variations in waveforms are unavoidable.

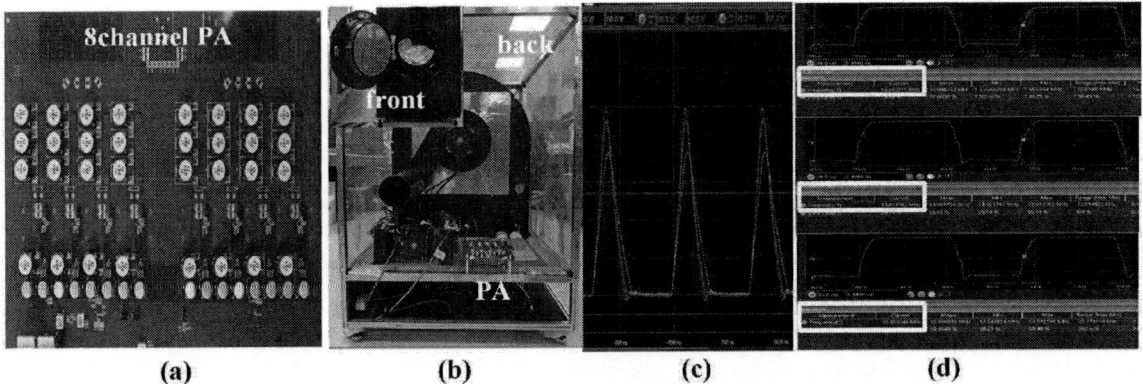

Fig. 3. Experimental results; (a) Prototype for verification of circuit, (b) near field dryer, (c) unbalanced drain voltage between multi-channels, and (d) the gate .signal of each PA.

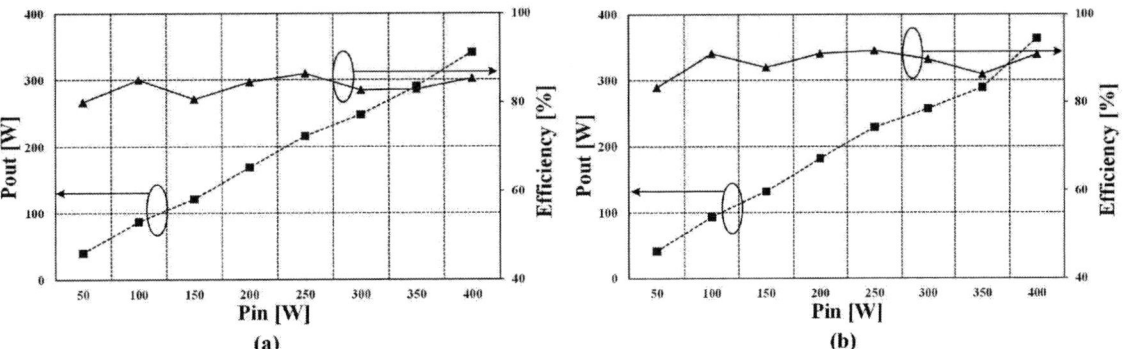

Fig. 4. Measured power conversion efficiency (a) conventional PA and (b) proposed PA.

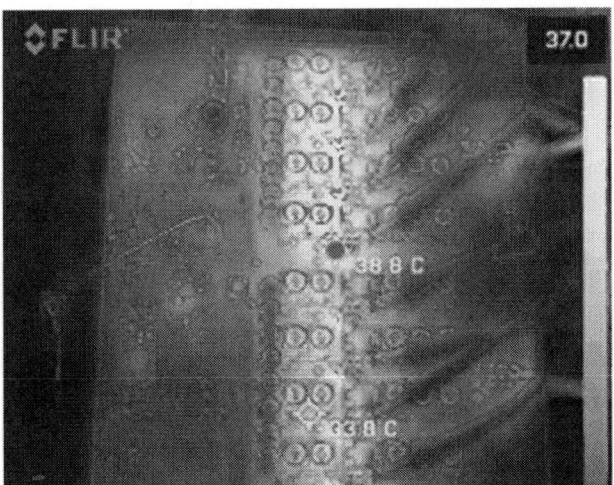

Fig. 5. The temperature of the prototype circuit.

To verify the proposed control circuit, the gate waveform of a class E power amplifier was measured (Fig. 3(d)). The gate signal frequencies for each of the three channels were measured to be 13.31 MHz, 13.41 MHz, and 13.56 MHz respectively. By varying the frequency in this way, the switch drain voltages of each channel are controlled while maintaining balance, which can extend the life of the circuit.

The power conversion efficiency of two channel power amplifier was measured. The conventional class E PA does not control the gate frequency, resulting in hard switching at

the power amp's switch and reduced efficiency as shown in Fig. 4(a). On the other hand, in this paper's frequency control, the efficiency is improved by zero voltage switching at the switch, as seen in Fig. 4(b). The maximum efficiency of the power amplifier was increased from 86% to 91%, a 5% improvement achieved by the proposed variable frequency control.

The balance between multi-channels can also be verified through the temperature of the circuit. The operating temperature of the prototype was measured to check the balance of the proposed circuit (Fig. 5). As a result of measurement, it was confirmed that there is no increase in temperature in any specific channel and all channels have uniform temperatures, thereby confirming the proper operation of the proposed method.

All of the above experiments were verified based on near field dryers rather than ideal resistors. In general, as the power increases, high-frequency noise occurs in the class E power amplifiers. To ensure the stability of circuits at high power, we conducted a strong verification of circuit reliability by verifying it based on real loads.

IV. CONCLUSION

This paper proposed a method to resolve the inter-channel imbalance that occurs in a parallel-driven 2kW class E power amplifier. The proposed circuit measures the switch drain voltage of individual channels and calibrates the voltage difference through frequency control. An error amplifier and a variable frequency VCO were used during this process. The imbalance was verified using a 3 channel 2kW class E power amplifier operating at 13.56 MHz.

REFERENCES

[1] J. Bohler, J. Huber, J. Wurz, M. Stransky, N. Uvaidov, S. srdic, and J. Kolar, "Ultra-High-Bandwidth Power Amplifiers: A Technology Overview and Future Prospects" *IEEE Access*, vol. 10, pp. 54613-54633, May. 2022.

[2] [2] A.Rawat, J. Rajendran, S. Mariappan, P. Shasidharan, N. Kumar, and B. Yarm, "A 919 MHz-923 MHz, 21 dBm CMOS PowerAmplifier With Bias Modulation Linearization Technique Achieving PAE of 29% for LoRa Application" *IEEE Access*, vol. 10, pp. 79365-79378, July. 2022.

[3] [3] X. Ding, J. Chen, Z. Zhao, and P. Luk, "A High-Precision and High-Efficiency PMSM Driver Based on Power Amplifiers and RTSPSs" *IEEE Transactions on Power Electronics*, vol. 36, no. 9, pp. 10470-10480, Mar. 2021.

[4] [4] X. Hao, J. Zou, K. Yin, X. Ma, and T. Dong, "Enhanced Power Conversion Capability of Class-E Power Amplifiers With GaN HEMT Based on Cross-Quadrant Operation" *IEEE Transactions on Power Electronics*, vol. 37, no. 11, pp. 13966-13977, Jun. 2022.

[5] [5] P. Niklaus, J. Kolar, and D. Bortis, "100 kHz Large-Signal Bandwidth GaN-Based 10 kVA Class-D Power Amplifier With 4.8 MHz Switching Frequency" *IEEE Transactions on Power Electronics*, vol. 38, no. 2, pp. 2307-2326, Oct. 2022.

[6] [6] S. Yerra, H. Krishnamoorthy, "Simplified Gate Driver Design Technique for Multi-MHz Switching GaN FETs in Three-Level Buck-Based 4G/5G Envelope Tracking" *IEEE Transactions on Industry Applications*, vol. 58, no. 5, pp. 6439-6447, Jul. 2022.

[7] [7] S. Yin, X. Xin, R. Wang, M. Dong, J. Lin, Y. Gu, and H. Li, "A 1-MHz GaN-Based LCLC Resonant Step-Up Converter With Air-Core Transformer for Satellite Electric Propulsion Application" *IEEE Transactions on Industrial Electronics*, vol. 69, no. 11, pp. 11035-11045, Oct. 2021.

[8] [8] J. Meng, X. Wu, T. Ye, J. Yu, L. Gu, Z. Zhang, and Y. Li, "Output Voltage Response Improvement and Ripple Reduction Control for Input-Parallel Output-Parallel High-Power DC Supply" *IEEE Transactions on Power Electronics*, vol. 38, no. 9, pp. 11102-11112, Jun. 2023.

[9] [9] S. Khan and D. Ahn, "Automatic Resonance Tuning With ON/OFF Soft Switching for Push–Pull Parallel-Resonant Inverter in Wireless Power Transfer" *IEEE Transactions on Power Electronics*, vol. 37, no. 9, pp. 10133-10138, Apr. 2022.

[10] [10] Y. Liu, X. Liu, F. Gao, D. Liu, and P. Wheeler, "Auxiliary Power Supply for Input-Series Output-Parallel Medium-Voltage Solid State Transformers" *IEEE Transactions on Power Electronics*, vol. 38, no. 6, pp. 7308-7321, Mar. 2023.

[11] [11] L. Zhou, J. Zhou, Q. Liu, J. Zhao, F. Huang, and Z. Zhu, "Dynamic Power Balancing Control Method for Energy Storage DC/DC Parallel Supply System With Low-Frequency Pulsed Load" *IEEE Transactions on Industrial Electronics*, vol. 71, no. 6, pp. 5766-5776, Jul. 2023.

[12] [12] C. Zhao, L. Wang, X. Yang, F. Zhang, and Y. Gan, "Comparative Investigation on Paralleling Suitability for SiC MOSFETs and SiC/Si Cascode Devices" *IEEE Transactions on Industrial Electronics*, vol. 69, no. 4, pp. 3503-3514, Apr. 2021.

High Temperature Operation of Digital Gate Driver Integrated into a Power Module

Kazuma Saiga[1], Shohei Zaizen[2], Satoshi Nakano[3], Shigeru Kusunoki[3], Kiyoto Watabe[3],
Katsuhiro Hata[4], Makoto Takamiya[5], Shin-ichi Nishizawa[3], and Wataru Saito[3*]

[1]Interdisciplinary Graduate School of Science and Engineering, Kyushu University, Fukuoka, Japan
[2]Department of Interdisciplinary Engineering, Kyushu University, Fukuoka, Japan
[3]Research Institute for Applied Mechanics, Kyushu University, Fukuoka, Japan
[4] College of Engineering, Shibaura Institute of Technology, Tokyo, Japan
[5]Institute of Industrial Science, The University of Tokyo, Tokyo, Japan
*email wataru3.saito@riam.kyushu-u.ac.jp

Abstract— This paper reports the experimental verification of high-temperature operation of a DGD-integrated power module, an area previously unexplored. Digital gate drivers (DGDs) have emerged as a promising technology to enhance the switching performance of power transistors, such as IGBTs and MOSFETs, compared to conventional gate drive circuits. DGDs provide significant improvements in energy efficiency and effectively mitigate voltage and current overshoots during turn-on and turn-off periods. In this study, a power module was fabricated with a 6-bit DGD, in which the DGD was mounted close to the IGBT to minimize gate voltage surge. The high-temperature operation was then evaluated to assess the influence of heat generation in the IGBT chip. Integrating the DGD reduced the gate voltage spike by approximately 97% while maintaining stable operation at ambient temperatures up to 130 °C. Despite the close proximity of the DGD and IGBT chips, heat transfer from the IGBT to the DGD was found to be negligible. With proper driver output stabilization and optimized thermal design, DGD-integrated power modules can achieve high-temperature performance comparable to that of conventional power modules.

Keywords—Power module, Digital gate driver, Module integration

I. INTRODUCTION

To reduce cost and increase the power density of power electronics systems, the power density of power modules has been continuously increased by reducing power device loss, increasing operating temperatures, and lowering the thermal resistance of power modules [1]-[3]. Reducing power device losses has been achieved mainly by improvements in device structure; however, in recent years, digital gate drivers (DGDs) have gained attention as a promising technology for improving the switching performance of power transistors. DGD can reduce switching losses while mitigating switching noise in power devices [4]-[5]. Additionally, configuring a digital interface enables the digitization of power modules, which could potentially enhance the power density and functionality of future power electronics systems [1], [6]-[9].

However, digital control circuits, composed of digital logic circuits similar to microcontrollers, present concerns regarding their lower noise immunity to high voltage and reduced maximum operating temperature compared to conventional gate drive ICs and power devices. Therefore, to integrate DGDs into power modules and achieve high power density and high temperature operation, it is necessary to address and verify these risks.

Previous studies have explored and demonstrated the integration of gate drivers into power modules, which consisted of Si-IGBTs or SiC-MOSFETs [10]-[15], as shown in Table I. Although high temperature operation was discussed, the integrated gate drivers have primally been based on analog circuits. As a result, risks specific to integrating DGDs into power modules have not been considered.

This paper reports on a prototype power module integrated with a DGD and presents the results of its first experimental verification, focusing on gate noise reduction and high-temperature operation to achieve highly reliable performance.

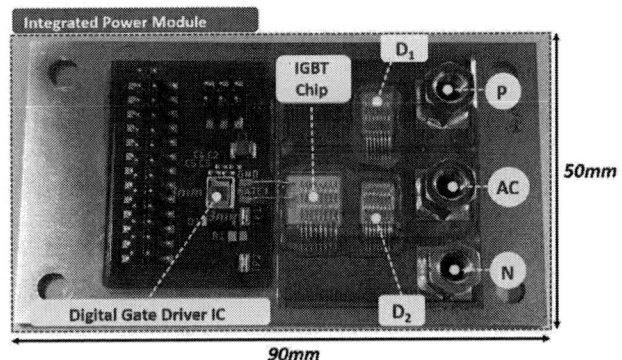

Fig. 1 Fabricated prototype of the DGD-integrated IGBT power module.

979-8-3315-1612-3/25 $31.00 © 2025 IEEE

Table I: A benchmark of gate driver integrated power modules.

Data Source	T-PE '15 [10]	CIPS '18 [11]	APEC '23 [12]	IECON '23 [13]	PEAS '23 [14]	ESTC '18 [15]	**This Work**
Main Switch	SiC-MOS	SiC-MOS	SiC-MOS	SiC-MOS	SiC-MOS	Si-IGBT	Si-IGBT
Rating Voltage	1200 V	1200 V	1200 V	1000 V	1200 V	650 V	1200 V
Rating Current	50 A	19 A	120 A	23 A	130 A	600 A	100 A
Gate Driver IC	Conventional (BCD-SOI)	Conventional (TTL+CMOS)	Conventional (Bipolar Based)	Conventional	Conventional (Bipolar Based)	Active Gate (Bipolar	**6-bit Digital (CMOS**
Driver IC Temperature	<225 °C	<120 °C	<230 °C	<70 °C	-	<120 °C	**<130 °C**
Gate Loop Inductance	-	4.6 nH	-	11 nH	11.5 nH	-	**9 nH**

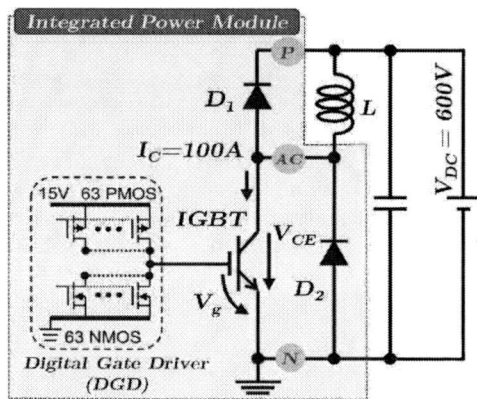

Fig. 2   Double pulse test schematic circuit diagram of the DGD-integrated IGBT power module in Fig. 1.

II. DESCRIPTION OF EXPERIMENTAL SETUP

A prototype power module was fabricated by integrating a DGD and 1200 V/100 A IGBT/FWD chips, as shown in Fig. 1. The DGD can modulate the gate drive current in 63 discrete steps using 6-bit control [4]. The switching characteristics were evaluated using a double pulse test, with the digital control signals generated by a LabVIEW program and a PXI digital pattern instrument, PXIe-6570. The test circuit is shown in Fig. 2.

Active gate current modulation via digital control effectively reduces switching losses and noise [4]. However, the gate drive current changes rapidly when the control signal varies, which causes an induced electromotive force due to gate loop inductance. This can result in significant gate voltage spikes, potentially leading to DGD failure due to overvoltage stress [16]. To mitigate this problem, the prototype integrated module minimized the gated loop inductance by placing the DGD close to the IGBT chips and connecting the driver outputs directly to the IGBT chips with bonding wire. Through simulation using COMSOL, the gated loop inductance of the fabricated module was estimated to be 9 nH.

The high-temperature operating characteristics of the DGD-integrated power module were evaluated to assess the trade-off characteristics between switching losses and noise, as well as to determine the maximum operating temperature. The trade-off in switching characteristics was evaluated in terms of turn-off loss and collector voltage surge. The maximum operating temperature was examined to ensure that the double-pulse test operated normally. All high operating temperatures were evaluated in an environment with an elevated ambient temperature using a hot plate.

Heat transfer from the IGBT chip to the DGD was also evaluated. The temperature rise of the DGD was investigated by self-heating the IGBT chip through continuous energization and measuring the temperature distribution in the module at that time with an infrared camera, FLIR T865. To analyze the thermal impedance of the module, the structure function was measured using T3STER in the MicReD® Power Tester 1500A. Heat transfer was also analyzed through the simulation using COMSOL.

III. RESULTS AND DISCUSSION

The switching waveforms at turn-off of the conventional module without an integrated DGD (non-integrated module) and the prototype DGD-integrated module fabricated for this study are shown in Fig. 3. As shown in Fig. 3(a), in the non-integrated module, the collector voltage change dV/dt is slowed down, and voltage surges are suppressed by switching the digital control signal from $N_1 = 63$ to $N_2 = 5$. However, a large spike of 8.5 V is generated in the gate voltage, and the peak value exceeds the drive voltage of 15 V [16]. This spike becomes a risk to lead the DGD failure. In contrast, as shown in Fig. 3(b), in the integrated module, the gate voltage spike was suppressed to 0.2 V, which is about 3% of the voltage spike in the non-integrated module, while maintaining the effect of dV/dt control through digital control. This is a result of the extremely small gate loop inductance of 9 nH in the integrated module compared to the gate loop inductance of 54 nH in the non-integrated module.

Figure 4 shows the switching trade-off characteristics between turn-off losses and surge voltage as well as turn-off switching waveforms as functions of the drive vector and temperature. The orange lines in Fig. 4(a) show switching characteristics controlled by conventional gate control. This conventional gate control is similar to the drive provided by conventional analog-based drivers. In DGD, the drive current is

(a)

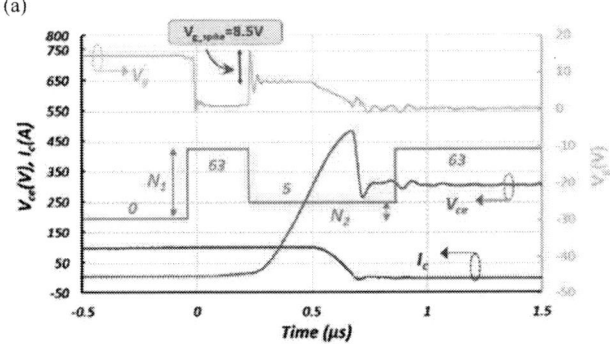

(b)

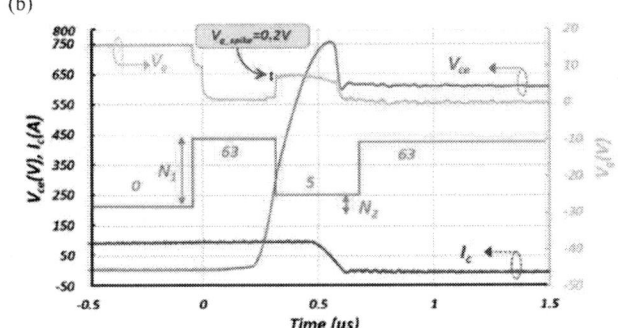

Fig. 3: Turn-off switching waveforms (a) DGD non-integrated module with $L_g = 54$ nH [9], (b) DGD integrated module with $L_g = 9$ nH.

adjusted according to the control vector values, which is similar to adjusting the drive current using an external gate resistance in analog-based drivers. Therefore, in conventional gate control, increasing N_1 reduces losses due to an increase in the gate drive current; however, it also increases the surge voltage, just as if the gate resistance were reduced. In contrast, with digital gate control, by setting an appropriate drive vector N_2, the collector voltage change dV/dt is suppressed, and the surge voltage is reduced. This improves the trade-off between turn-off loss and surge voltage.

The DGD-integrated power module operated normally even when the ambient temperature was increased to 130 °C. However, as the temperature increased, the trade-off curve shifts toward higher losses and lower surge voltage. As a result, the optimal drive vector N_2 at 25 °C was 5, whereas at 130 °C, N_2 changed to 8. This change occurs because the effective gate-collector capacitance C_{gc} increases as the carrier concentration in the IGBT increases with temperature [17]. A large C_{gc} results in slower turn-off. As shown in Fig. 4(b) and 4(c), the collector voltage change dV/dt during the N_2 period becomes smaller, and the mirror period becomes longer as the temperature rises. In other words, the effective C_{gc} is increasing. Therefore, to achieve the same dV/dt, the drive current must be increased. As shown in Fig. 4(d), increasing N_2 from 5 to 8 produces the same dV/dt as at 25 °C.

In the DGD used in this prototype module, it is necessary to manually set all drive vectors and control timings. As a result, when operating conditions such as temperature and current change, the optimal control conditions also vary. Recently, a function for automatic detection of control timing has been

(a)

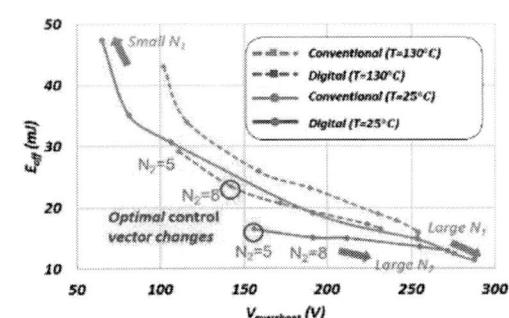

(b)

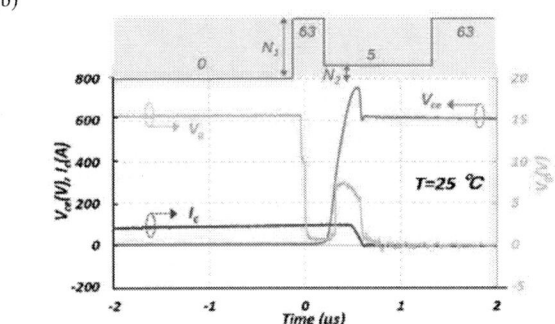

(c)

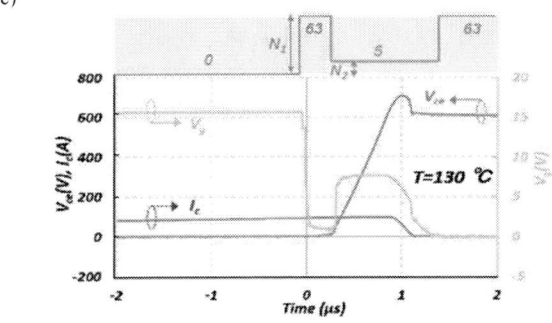

(d)

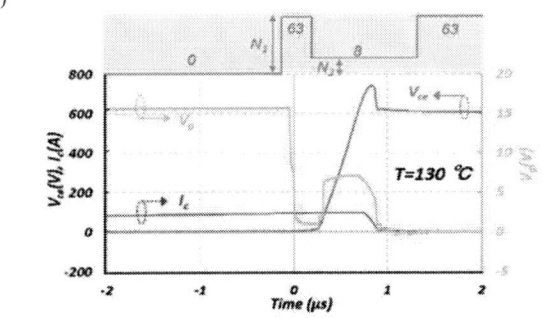

Fig. 4 (a) The switching trade-off characteristics between turn-off loss and surge voltage in case of conventional and digital control approaches under different operating temperatures, switching waveforms of DGD-integrated power module at (b) temperature of 25 °C and N_2 of 5, (c) temperature of 130 °C and N_2 of 5, and (d) temperature of 130 °C and N_2 of 8.

developed [5]. By integrating a DGD equipped with this function, it becomes possible to maintain improvements in trade-offs of switching characteristics, even when operating conditions change.

(a)

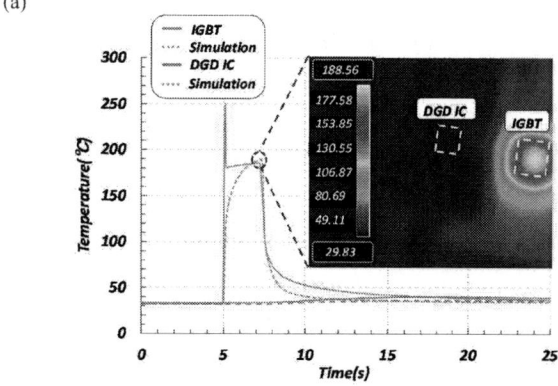

(b)

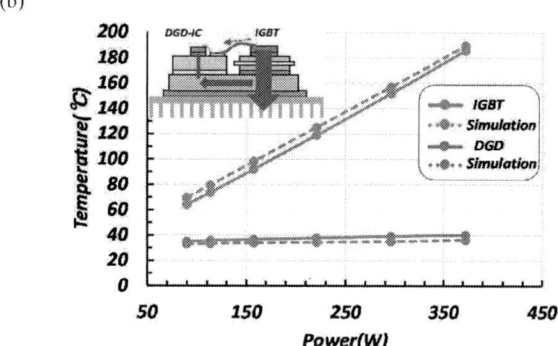

(c)

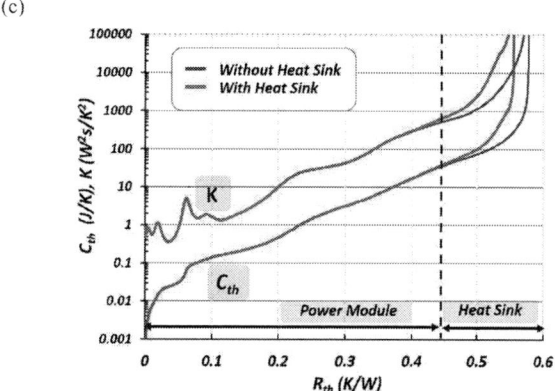

Fig. 5 (a) Temperature profiles of IGBT and DGD IC chips, (b) relation between junction temperature rising and input power, (c) structure function of DGD power module.

When the ambient temperature was increased to 140 °C, the DGD output became unstable. This instability may have been caused by an increase in the leakage current of MOSFETs in the driver output stage, resulting in a decrease in output impedance. This problem could be solved by a circuit that guarantees operation without a control signal.

The maximum operating temperature of analog-based gate drivers varies depending on the circuit configuration; therefore, a clear comparison with DGD has not been made at this point. In future work, we will clarify the circuits that determine the maximum operating temperature in DGD and conduct a comparison.

At high temperature, the effective C_{gc} of the IGBT increases; however, the gate charge does not change significantly. Therefore, the main gate drive current, set at $N = 63$, remains the same even at high temperatures and is equivalent to the conventional drive current. As a result, the drive loss of the DGD is small enough to be negligible.

Figure 5 shows the change in chip temperature when the IGBT chip was self-heated, along with the temperature distribution in the module at the maximum temperature and the simulation results. As shown in Fig. 5(a), the temperature of the IGBT chip rose to 186 °C due to self-heating during 2.4 seconds of continuous current flow. However, the temperature of the DGD chip rose by only 41 °C. The dotted lines represent the simulated junction temperatures of IGBT and DGD IC chips. When the same power was applied as in the experiment, the maximum temperature and the elapsed time were represented for both the IGBT and the DGD IC chips.

Fig. 5(b) shows the relationship between the input power due to self-heating of the IGBT chips and the junction temperature of each chip. Since both temperatures are proportional to the input power, it is expected that the temperature rise is determined by a simple thermal resistance model. From the input power and the temperature rise of the IGBT chip, the thermal resistance from the IGBT chip to the case is estimated to be 0.43 K/W. The dotted lines show simulation results of the relationship between the input power and the junction temperature of each chip. Since both chips results represent experimental results, the junction temperatures are determined by a simple thermal resistance model. Fig. 5(c) shows the structure function of the DGD-integrated power module measured by T3STER. The thermal resistance from the IGBT chip to the case, also estimated to be 0.43 K/W, aligns well with the value obtained from Fig. 5(b).

The rise in junction temperature of the DGD chip was one-twentieth that of the IGBT chip. The simulation results were in good agreement with the experimental results. These results indicate that, due to the cooling effect of the heatsink, the heat generated by the IGBT chip is primarily transferred downward, with minimal lateral heat transfer from the IGBT chip to the DGD chip via the bonding wires and the baseplate. As a result, even when the IGBT chip is mounted in close proximity to the DGD chip, the risk of malfunction in the DGD chip due to transient temperature rises is reduced.

To simplify control for the practical use of DGD-integrated modules, research will be conducted to realize a module with a capability to automatically control the timing of the control vector N_2. In addition, the reliability issue of long-term operation at high temperatures is not addressed. From the viewpoint of module reliability, it is necessary to evaluate not only high-temperature operation but also thermal cycle testing and power cycle testing. These reliability evaluations will be conducted in the future.

IV. CONCLUSIONS

A prototype Si-IGBT module, incorporating a 6-bit DGD, has been successfully developed and fabricated, demonstrating high-temperature operation of the DGD-integrated power module for the first time. The integration of the DGD effectively suppressed gate voltage spikes by minimizing the gate loop inductance, enabling stable operation at temperatures up to 130 °C. By enhancing driver output stability and optimizing the module design for efficient heat dissipation, the DGD-integrated power module can operate at high temperatures comparable to those of conventional power modules.

ACKNOWLEDGMENT

This work was supported by the New Energy and Industrial Technology Development Organization (NEDO) [Grant JPNP21009].

REFERENCES

[1] W. Saito, "A Future Outlook of Power Devices From the Viewpoint of Power Electronics Trends," IEEE Trans. Electron Devices, vol. 71, no. 3, pp. 1356-1364, 2024.

[2] G. Majumdar, "Advanced power semiconductor technologies for efficient energy conversion," in Proc. Slide Int. Workshop Junction Technol. (IWJT), 2013, pp. 1–26. [Online]. Available: https://www.researchgate.net/publication/261038917_Advanced_power_semiconductor_technologies_for_efficient_energy_conversion

[3] A. Beckmann, A. Christmann, I. Imperiale, C. Sandow, C. Ouvrard, and F. Wolter, "Design considerations for EDT3 750 V next generation IGBT technology for automotive drive applications," in Proc. PCIM Europe 2023, pp. 1–7.

[4] K. Miyazaki, S. Abe, M. Tsukuda, I. Omura, K. Wada, M. Takamiya, and T. Sakurai, "General-purpose clocked gate driver IC with programmable 63-level drivability to optimize overshoot and energy loss in switching by a simulated annealing algorithm," IEEE Trans. Ind. Appl., vol. 53, no. 3, pp. 2350–2357, May 2017.

[5] D. Zhang, K. Horii, K. Hata, and M. Takamiya, "Digital Gate Driver IC with Fully Integrated Automatic Timing Control Function in Stop-and-Go Gate Drive for IGBTs," 2023 IEEE Applied Power Electronics Conference and Exposition, 2023, pp. 1225-1231.

[6] J. Henn, C. Lüdecke, M. Laumen, S. Beushausen, S. Kalker, C. H. van der Broeck, "Intelligent Gate Drivers for Future Power Converters," IEEE Trans. Power Electronics, vol. 37, no. 3, pp. 3484-3503.

[7] K. Yamanokuchi, H. Watanabe, and J. Itoh, "Universal smart power module concept with high-speed controller for simplification of power conversion system design," in Proc. ECCE 2021, pp. 2484–2489.

[8] Y. Takahashi, Y. Ikeda, H. Watanabe, and J.-I. Itoh, "Universal smart power module (USPM) for carbon neutral society," in Proc. ECCE Asia (IPEC) 2022, pp. 1281–1287.

[9] F. Kahn, S. Islam, J. Major, A. Usman, G. Moreno, and S. Narumanchi, "A smart silicon carbide power module with pulse width modulation over Wi-Fi and wireless power transfer-enabled gate driver, featuring onboard state of health estimator and high-voltage scaling capabilities," in Proc. APEC. 2023, pp. 386–391.

[10] Z. Wang, X. Shi, L. M. Tolbert, F. Wang, Z. Liang, D. Costinett, and B. J. Blalock, "A High Temperature Silicon Carbide MOSFET Power Module With Integrated Silicon-On-Insulator-Based Gate Drive," IEEE Trans. Power Electronics, vol. 30, no. 3, pp. 1432-1445.

[11] A. B. Jorgensen, U. R. Nair, S. Munk-Nielsena, C. Uhrenfeldt, "A SiC MOSFET Power Module With Integrated Gate Drive for 2.5 MHz Class E Resonant Converters," in Proc. 10th International Conference on Integrated Power Electronics Systems (CIPS), 2018, pp. 128-133.

[12] S. Ahmed, P. Lai, S. Chinnaiyan, A. Mantooth, Z. Chen, "High-Temperature (250°C) SiC Power Module Integrated with LTCC-Based Isolated Gate Driver," in Proc. APEC, 2023, pp. 2588-2595.

[13] C. Cheung, Z. Gao, "Driver-Integrated Silicon Carbide Based Power Module with Self-Optimized Current-Sensorless Temperature-Driven Deadtime Control," in Proc. 49th Annual Conference of the IEEE Industrial Electronics Society, 2023.

[14] C. Zeng, C. Chen, Y. Kang, "A High Power Density Gate Driver Integrated SiC Multichip Power Module with Lower Parasitic Inductance," in Proc. IEEE 2nd International Power Electronics and Application Symposium, 2023, pp. 438-442.

[15] M. Jiao, Y. Li, J. Yu, J. Xie, P. Zeng, Z. Zhao, "Intelligent Power Module Featuring Optimised Active Gate Driver and IGBT Module Integration for Electric Vehicle Application," in Proc. 7th Electronic System-Integration Technology Conference, Sep. 2018.

[16] Z. Lou, M. Thatree, K. Hata, M. Takamiya, S. Nishizawa, W. Saito, "IGBT Power Module Design for Suppressing Gate Voltage Spike at Digital Gate Control," IEEE Access, vol. 11, p. 6632-6640, January 2023.

[17] V. Sundaramoorthy, E. Bianda, R. Bloch, I. Nistor, G. Knapp, A. Heinemann, "Online estimation of IGBT junction temperature (Tj) using gate-emitter voltage (Vge) at turn-off," 2013 15th European Conference on Power Electronics and Applications,2013.

Evaluation Index-Based Multiphysics Coupling Model and Analysis Methodology for High-Reliable Power Supply Module

Haoyu Wang
School of Integrated Circuits
Tsinghua University
Beijing, China
hy-wang24@mails.tsinghua.edu.cn

Xuliang Wang
School of Integrated Circuits
Tsinghua University
Beijing, China
xwangef@tsinghua.edu.cn

Yan Wang
School of Integrated Circuits
Tsinghua University
Beijing, China
wangy46@tsinghua.edu.cn

Xiaosen Liu
School of Integrated Circuits
Tsinghua University
Beijing, China
liuxiaosen@tsinghua.edu.cn

Abstract—The performance of a power supply is highly susceptible to environmental variations, which directly impacts the proper functioning of nonlinear loads such as lasers and LEDs. This paper proposes an evaluation index-based multiphysics coupling modeling (EIMCM) technique that enables comprehensive characterization of various physical properties for power supplies. By correlating heat, stress, electricity and magnetism, the technique explores the intrinsic connections between different physical fields and establishes a reliability evaluation model. For the large number of elements and indicators in the evaluation, a back-propagation neural network (BPNN) is employed to train the multi-objective parameters, thereby improving the performance of the module. To demonstrate the capabilities of the proposed methodology, a 600 W power supply for driving laser loads is designed and optimized for heat dissipation, strain, output power, and regulated voltage accuracy. The simulation and experimental results exhibit strong consistency as a systematic error less than 4%, validating the effectiveness of the proposed EIMCM methodology.

Keywords—Power Supply Module, High Reliability, Multiphysics Coupling, Evaluation Index.

I. INTRODUCTION

Nonlinear loads such as LED, laser, LiDAR applications require high-performance driving power supplies up to 500 W [1]-[4]. Its regulating accuracy and reliability are crucial to ensure the stable operation of these nonlinear loads [5][6]; however, they are highly susceptible to runtime variations in multiphysics parameters such as temperate, strain, electromagnetic interference, etc [7]-[10]. Most recent research on driving power supplies has focused on thermal management, protection circuits and modularity, while they explore only single indicator and lack a system-wide analysis of intrinsic physical mechanisms [11]-[15]. In addition to leveraging physical analytics, some researchers have utilized pre-existing data and applied machine learning techniques for prediction and processing [16][17].

To address the complex multiphysics design challenges of high-performance power supplies [18]-[20], this paper proposes a comprehensive methodology for modeling, analysis and optimization in three steps: (1) clarify the connections between different physical mechanisms and establish the EIMCM methodology; (2) calculate dependable evaluation metrics by correlating physical properties of power supplies; and (3) develop nonlinear equations and BPNN algorithms for quantitative analysis and provide optimized schemes. In this paper, a multi-physical field system is established to guide the design of the driving power supply. Both simulations and experiments are conducted to evaluate the accuracy and feasibility of the multiphysics coupling approach.

II. MULTIPHYSICS COUPLED MODEL

Currently there is a lack of the comprehensive and accurate methodologies for the design of highly reliable power module, leading to loads that often fail to operate as expected. Thus, a framework is developed as shown in Fig. 1, where the EIMCM is built based on the interconnections of design modules, heat, stress, electricity, and magnetism.

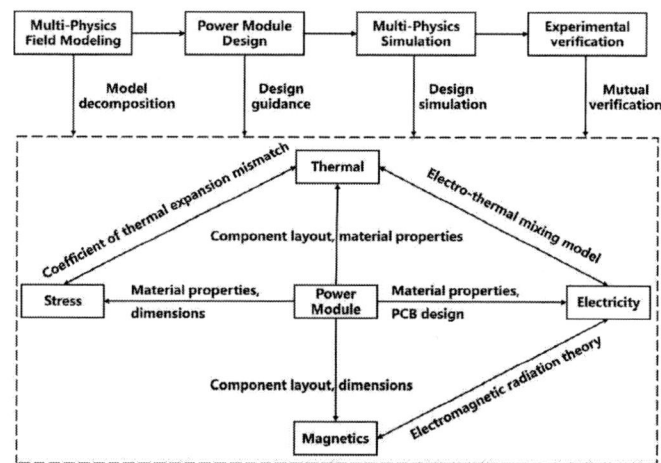

Fig. 1. The general framework used for EIMCM analysis.

Multiphysics phenomena have complex mechanism and are difficult to categorize under a single indicator. They influence the efficiency, ripple, speed and accuracy of the power supply, ultimately determining its overall reliability. To characterize and unify these phenomena, a parameter matrix R is proposed to incorporate heat (T), stress (S), electricity (E), magnetism (M) as follows,

$$R = f_R(T, S, E, M, P) \qquad (1)$$

979-8-3315-1612-3/25 $31.00 © 2025 IEEE

where T is thermal, S is the stress, E is electrical, M is magnetism, and P is the parameters of the designed power module including material properties, dimensions, spatial arrangement. Since the physical parameters interact with each other, a coupling equation is proposed as follows,

$$\begin{cases} T = f_T(T, S, E, M, P) \\ S = f_S(T, S, E, M, P) \\ E = f_E(T, S, E, M, P) \\ M = f_M(T, S, E, M, P) \end{cases} \quad (2)$$

However, the coupled model involves a large number of parameters, leading to significant computational complexities. To reduce the difficulty of the solution, the equation needs to be decoupled and simplified, allowing the T equation can be viewed as calculating the temperature distribution. Consequently, the current continuity, heat conduction, and thermal-electrical coupling equations are solved separately,

$$f_T(T, E, M, P) =$$
$$\begin{cases} \nabla[\bar{\sigma}(T)\nabla U(\bar{r}, t)] = -\dfrac{\partial q(t)}{\partial t} \\ \rho c \dfrac{\partial T(\bar{r}, t)}{\partial t} - \nabla[\kappa(T)\nabla T(\bar{r}, t)] = p(\bar{r}, t) \\ f_Q(\bar{r}, T, t) = p(\bar{r}, t) \end{cases}$$
$$(3)$$

where $\bar{\sigma}(T)$ is the conductivity, $U(\bar{r}, t)$ is the potential in space, ρ is the density, c is the heat capacity, $\kappa(T)$ is the thermal conductivity, $f_Q(\bar{r}, T, t)$ is the transient space heat source, $p(\bar{r}, t)$ is the Joule heat. Eq. (3) achieves the electro-thermal-modular coupling and solves for temperature based on the heat generation and transfer path. S can be converted into stress forces, thereby decoupling the electromagnetic effects and preserving the thermal-force coupling,

$$f_S(T, S, M, P) = \begin{cases} \varepsilon_{ij}^{Th} = \alpha \Delta T \\ \sigma_{ij} = \dfrac{1}{2}D_{ijkl}(u_{k,l}, u_{l,k}) - \alpha \Delta T D_{ijkl}\delta_{ij} \end{cases} \quad (4)$$

where ε_{ij}^{Th} is the thermal strain component, α is the coefficient of thermal expansion (CTE), σ_{ij} is the stress value, D_{ijkl} is the fourth-order strain coefficient, u is the displacement value, and δ_{ij} is the Dirac function. Eq. (4) establishes the relationship between thermal coupling and stress which is correlated by the CTE. For static analysis, E is excluded from other physical coupling and dominated by the effect of P. For example, parasitic inductance can be analyzed as,

$$f_E(P) = 0.0002L_t[\ln\left(\dfrac{2L_t}{W_t + T_c}\right) + 0.2235\left(\dfrac{W_t + T_c}{L_t}\right) + 0.5] \quad (5)$$

where L_t is the line length, W_t is the line width, and T_c is the copper thickness. M is calculated by the electromagnetic theory to illustrate the radiation amount and is decoupled from the thermal-stress effect,

$$f_M(E, M, P) = \begin{cases} E_Q = \dfrac{Z_0 I_c L_c \lambda_c}{8\pi^2 r^3} \\ M_Q = \dfrac{I_c L_c}{4\pi r^2} \end{cases} \quad (6)$$

where E_Q is the electric field strength, I_c is the common-mode current, L_c is the common-mode wire length, λ_c is the wavelength, r is the path length, and M_Q is the magnetic field strength.

Conventionally, we focus on interference from the electric near field. Once the objective functions of Eq. (2) are solved, Eq. (1) can be evaluated to consolidate systematic reliability as a figure of merit (FoM).

III. DESIGN AND ANALYSIS FOR POWER SUPPLY

High-power supplies typically demand greater stability, making the half-bridge drive topology a suitable choice for such scenarios. The half-bridge topology offers key advantages, including stable output voltage, high efficiency, and a simple structure, making it well-suited for high-power integration. As illustrated in Fig. 2, a 12 V isolated power supply is employed to provide the gate drive voltage and current boost necessary for driving the half-bridge MOSFETs.

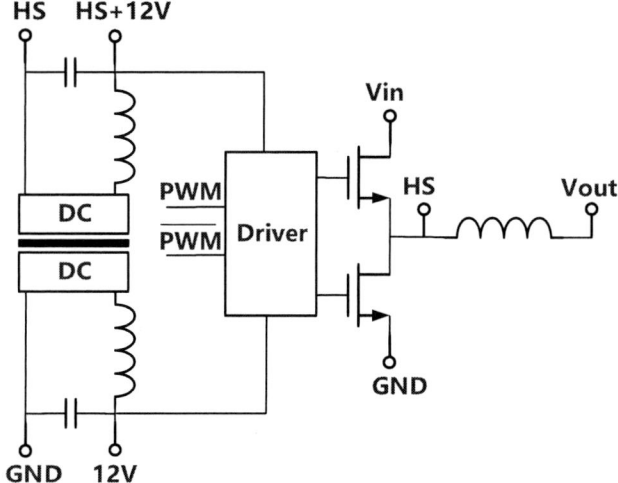

Fig. 2. Circuit structure of the driving power supply.

The analytical solutions for the evaluation criteria and physical parameters obtained in the previous section can be used to intuitively optimize the PCB layout. However, the highly nonlinear nature of the analytical formulas poses significant challenge to computational accuracy and efficiency when directly applied for optimization. To address this, a response surface method (RSM)-BPNN-genetic algorithm solution is proposed as illustrated in Fig. 3. The RSM simplifies complex physical formulas by approximating them as quadratic polynomials, which can then generate multiple sets of numerical solutions. Using this data, the back-propagation neural network (BPNN) mitigates overfitting and ensures accurate prediction of results. The BPNN outputs predicted combinations, which are then optimized using a genetic algorithm to derive the best optimization strategy.

Applying the optimization route shown in Fig. 3 to the PCB layout, the high-side and low-side MOSFETs were selected as key elements for analysis. Four optimization parameters were identified based on a 2D plane division. Using Eq. (2)-(5), 10-20 datasets were generated and fitted using RSM. A small noise evaluation index indicated that the fitting results were ideal, enabling the generation of numerous numerical solutions for

979-8-3315-1612-3/25 $31.00 © 2025 IEEE

input into BPNN. The BPNN model employs a double hidden-layer structure with 12 nodes per layer. The genetic algorithm was configured with a population size of 20 and executed for 300 iterations. The optimization process yielded the optimal layout coordinates for the high-side and low-side MOSFETs as (49.51, 8.51) and (64.41, 8.65), respectively, on a PCB with dimensions of 100×14.5 mm^2. Under this layout, the following performance metrics were achieved: 75.42 °C maximum temperature, 0.375 mm maximum deformation, < -35 dB return loss of S-parameter, 792.6 mV/(MHz) maximum electric field radiation.

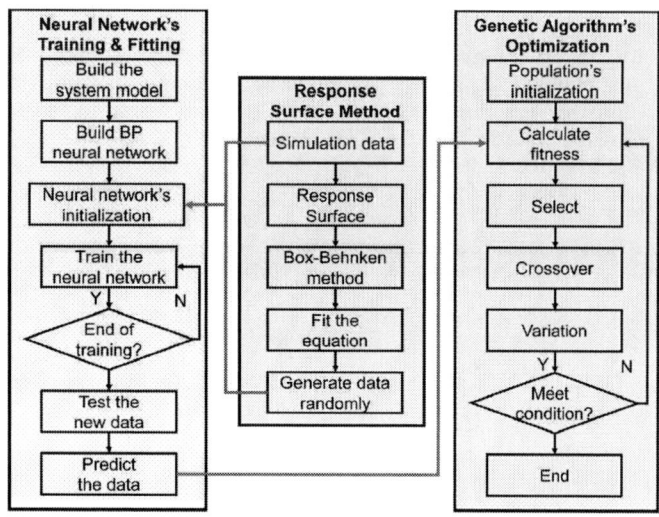

Fig. 3. RSM-BPNN-genetic algorithm solution routes

With full consideration of EIMCM methodology and BPNN intelligent algorithm, this paper designs the PCB layout as shown in Fig. 4. The whole board is a double-sided four-layer board, of which only the top layer for component placement. The bottom layer adds a blank heat dissipation area pad to facilitate efficient heat transfer with the water cooling equipment at the bottom.

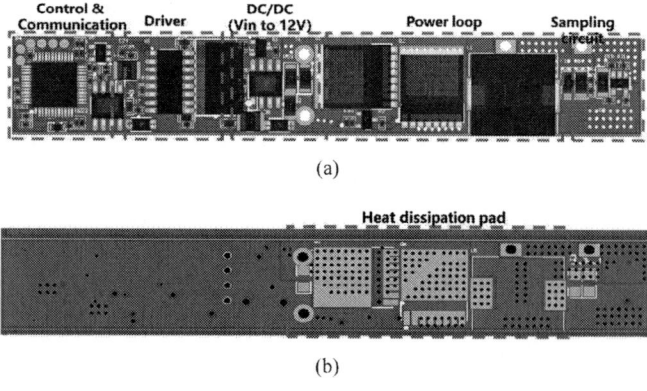

Fig. 4. (a) Top Layer PCB and (b) bottom Layer PCB.

After completing the circuit structure and PCB design, physical field simulation is performed to verify the feasibility of EIMCM methodology. Thermal simulation was first performed using Power DC with the target parameters of temperature and deformation, as shown in Fig. 5. The highest

temperature is 76.92 °C and the largest deformation is 0.369 mm, both occurring at the intersection of the high side and low side MOSFETs of the power loop.

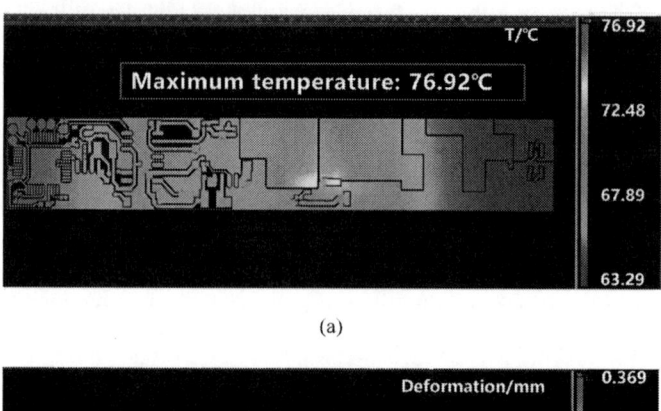

(a)

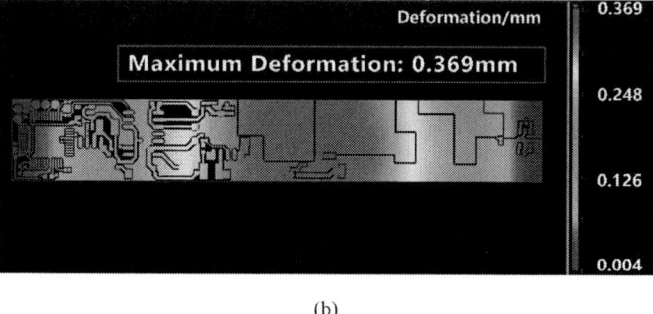

(b)

Fig. 5. (a) Temperature simulation and (b) deformation simulation results.

The electromagnetic simulation mainly measures the quality of the drive signals and the electromagnetic susceptibility, as shown in Fig. 6 for the simulation results of Power SI. The signal quality is evaluated in the form of S-parameters, and at a switching frequency of 500 kHz, the insertion loss is close to 0, and the return loss is less than -30 dB. In the EMI results, the maximum radiation is 856.1 mV/(MHz), which is almost unaffected by the surrounding electromagnetic influence.

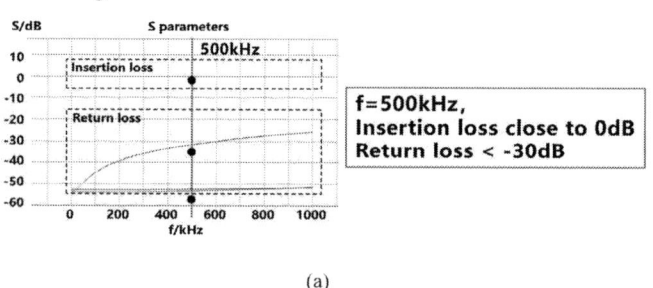

(a)

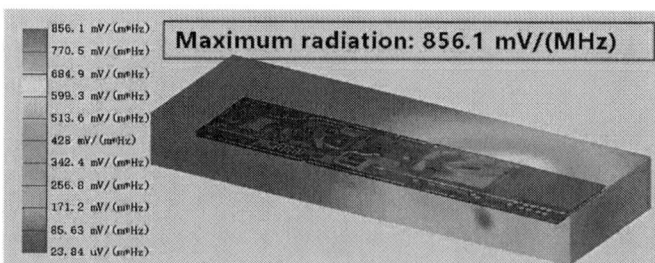

(b)

Fig. 6. (a) S-parameters simulation and (b) EMI simulation results.

The simulation data were compared with theoretical calculations, and results from unoptimized layout were also included as shown in Table I. After applying the RMS-BPNN-genetic algorithm optimization, significant improvements were observed across various parameters. Analytical solutions derived from the EIMCM demonstrated high consistency with the simulation results, maintaining an agreement level above 90%. It is worth noting that in some cases, the theoretically calculated values for unoptimized layouts appear to align more closely with simulation data. However, these comparisons are not directly valid because the overall layouts differ. The unoptimized layouts are included in this paper primarily to highlight the advantages of the optimized design. Comparisons based on consistent control variables, where theoretical calculations and simulation results correspond to the same layout, provide more meaningful insights into the effectiveness of the proposed optimization approach.

TABLE I. COMPARISON OF THEORETICAL AND SIMULATION RESULTS.

Parameters	Pre-optimization	After Optimization	Optimization effect	Simulation Results	EIMCM Accuracy
Temperature (℃)	78.61	75.42	4.06%	76.92	98.05%
Deformation (mm)	0.394	0.375	4.82%	0.369	98.40%
Parasitic S qualit (dB)	-37	-35	5.41%	-31.5	90.00%
Radiation (mV/MHz)	821.8	792.6	3.55%	856.1	92.58%

IV. EXPERIMENTAL VALIDATION OF MULTIPHYSICS ANALYSIS

The experimental platform was built as shown in Fig. 7. Considering the large amount of heat generated by the power supply and the laser during operation, a large water cooling plate was placed at the bottom.

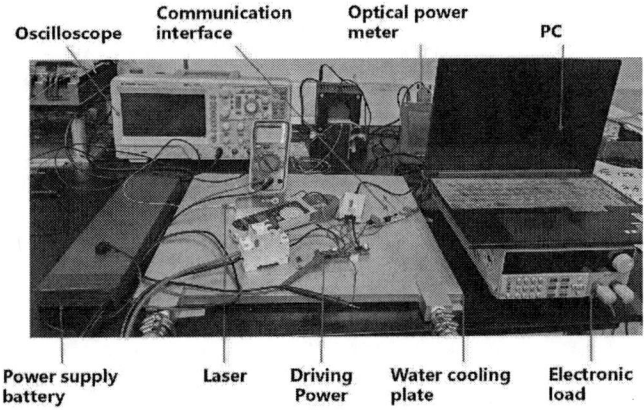

Fig. 7. Experimental platform for the laser measurement

The power supply was tested under maximum output conditions with the water-cooling system set to a temperature of 18 °C. Thermal imaging results, as shown in Fig. 8, indicate a maximum temperature of 72.8 °C. The water-cooling board effectively maximizes heat dissipation due to its well-designed thermal path, demonstrating excellent thermal management performance.

The deformation test was conducted using CHOTEST's VX9000 Scanning Image Measuring Machine, which utilizes optical principles and high-precision measurement algorithms. As shown in Fig. 9, the power supply was directly placed into the measuring machine, and the maximum PCB deformation was recorded as 0.41 mm, confirming the structural reliability of the design.

Fig. 8. Thermal performance with water cooling plate.

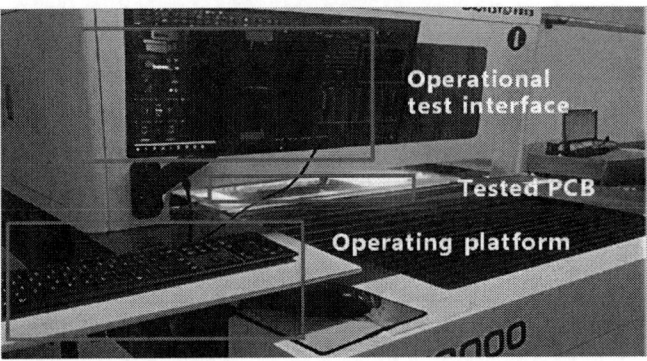

Fig. 9. PCB deformation test scenario.

Laser loads impose stringent requirements on the power supply's output current ripple, typically necessitating values within 1%. In this study, a current probe was employed to measure the current ripple under full power output conditions. As shown in Fig. 10, the current ripple was maintained at 0.8%, demonstrating compliance with laser load requirements and indirectly validating the effectiveness and rationality of the PCB layout.

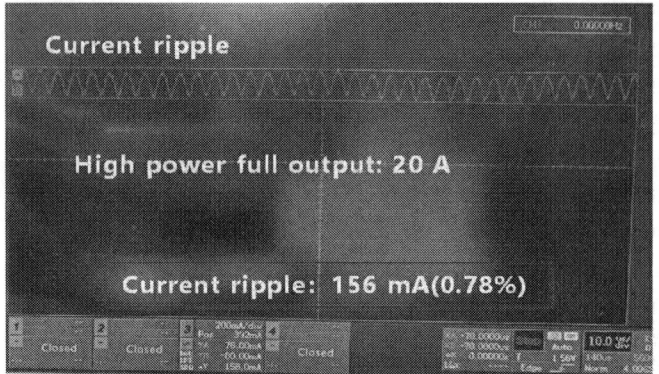

Fig. 10. Measured current ripple of 20 A.

The drive signal, inductive waveform, and response time provide insights into the effects of radiation and parasitics, as shown in the test results in Fig. 11. The initial drive signals generated by the MCU and gate driver include a certain dead time. The rising and falling edges of the drive voltage are steep, with no noticeable overshoot or ringing. The response time, defined as the duration from when the MCU receives the output command to when the output voltage stabilizes, was measured to be 40 ms, demonstrating efficient system response.

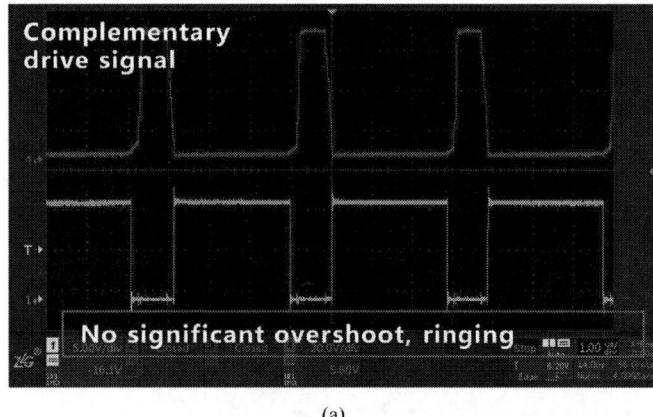

(a)

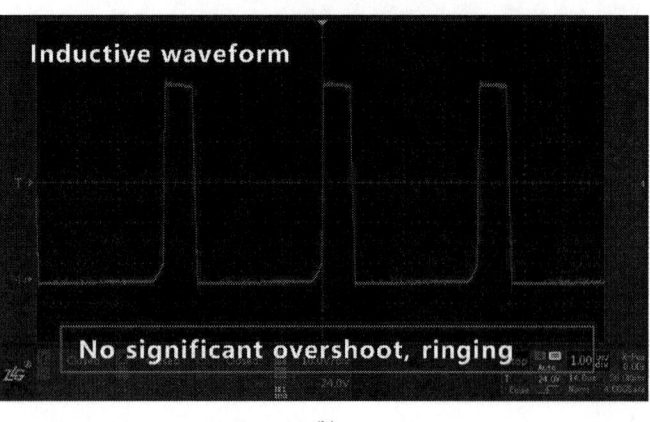

(b)

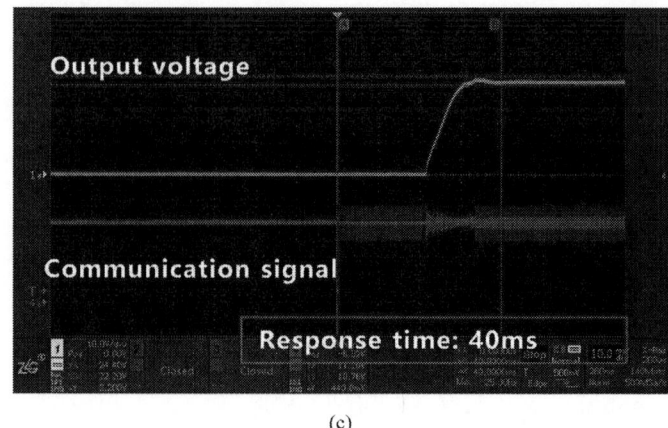

(c)

Fig. 11. Measured (a) drive signal, (b) inductive waveform, and (c) response time.

The laser load adjusts its light output by controlling the current. To ensure stable laser operation, the actual current value must closely align with the expected value. In this study, 11 groups of output currents with varying values were tested as summarized in Table II. The accuracy of each group exceeded 94%, with values surpassing 98% for high-power output, demonstrating precise current control and reliable performance.

TABLE II. OUTPUT CURRENT TEST RESULTS

Expected value (A)	Measured value (A)	Precision
2	2.121	94.30%
4	4.216	94.88%
6	6.213	96.57%
8	8.324	96.11%
10	10.196	98.08%
12	12.142	98.83%
14	14.208	98.54%
16	16.176	98.91%
18	18.243	98.67%
20	20.296	98.54%

As shown in Table III, the experimental and simulation results exhibit strong consistency with a 96% correlation factor.

TABLE III. COMPARISON OF EXPERIMENTAL AND SIMULATION RESULTS.

Parameters	Simulation	Experiment
Temperature (°C)	76.92	72.8
Deformation (mm)	0.365	0.410
Parasitic signal quality	S parameter up to standard	Low ripple, fast response time, rapid rising edge with no overshoots
Radiation (mV/MHz)	<856.1	Accuracy 98%, almost no radiation effect

V. CONCLUSION

The major contributions of our work are concluded as follows:

We proposed an evaluation index-based multiphysics coupling modeling (EIMCM) technique to guide and optimize the design of the driving power supplies. Compared to previous work, this paper establishes a comprehensive framework for power module design that includes modeling, evaluation, and optimization, making the evaluation index-based methodology more systematic. The coupling relationships are clarified at the physical theory level, and each parameter and system index are quantitatively analyzed with high accuracy and minimal computation cost. To validate the effectiveness of the proposed methodology, a high-power laser module experiment was conducted to demonstrate performance improvements under harsh environments. This power module not only drives 600 W laser module but also offers advantages such as low ripple, high efficiency and fast response. The experimental results show a maximum temperature of 72.8 °C, PCB deformation of 0.410 mm, high signal quality, low electromagnetic interference, and output accuracy exceeding 98%.

ACKNOWLEDGMENT

This work was supported by a grant from the National Key R&D Program of China (Project No. 2022YFB4401100) and National Natural Science Foundation of China (Grant No. 62374100).

REFERENCES

[1] N. Bode, Z. Holmes, S. Ng, B. Von Behren, D. Ottaway, and B. Willke, "Multiple beam coherent combination via an optical ring resonator," *Opt. Lett.*, vol. 48, no. 17, p. 4717, Sep. 2023.

[2] S. Xie *et al.*, "High-Power and High-Speed Multi-Junction VCSEL Arrays for Automotive LiDAR," in *2022 IEEE Photonics Conference (IPC)*, Vancouver, BC, Canada: IEEE, Nov. 2022, pp. 1–2.

[3] T. S. Bheemraj, V. Karthikeyan, and S. Pragaspathy, "A Hybrid Structured High Step-Up DC–DC Converter for Integration of Energy Storage Systems in Military Applications," *IEEE Trans. CIRCUITS Syst.*, vol. 70, no. 4, 2023.

[4] K. Hummler *et al.*, "High-power EUV light sources (>500w) for high throughput in next-generation EUV lithography tools," in *Optical and EUV Nanolithography XXXVII*, M. Burkhardt and C. Van Lare, Eds., San Jose, United States: SPIE, Apr. 2024, p. 43.

[5] H. Jiang, "Multi-output LED Drive Power System Design," in *2023 International Conference on Applied Physics and Computing (ICAPC)*, Ottawa, ON, Canada: IEEE, Dec. 2023, pp. 212–217.

[6] B. Feng, J. Zhao, H. Zhang, T. Li, and J. Mi, "Design of High-Performance Driving Power Supply for Semiconductor Laser," *Electronics*, vol. 12, no. 23, p. 4758, Nov. 2023.

[7] W. Guo, G. Xiao, K. Gao, and L. Wang, "A Compact Hybrid Sensor for Chip-level Online Current Sensing in Press-pack Power Module," in *2023*

IEEE Applied Power Electronics Conference and Exposition (APEC), Orlando, FL, USA: IEEE, Mar. 2023, pp. 2545–2549.

[8] S. Dulal, S. B. Sohid, H. Cui, G. Gu, D. J. Costinett, and L. M. Tolbert, "A Physics-Based Circuit Model for Nonlinear Magnetic Material Characteristics," in *2024 IEEE Applied Power Electronics Conference and Exposition (APEC)*, Long Beach, CA, USA: IEEE, Feb. 2024, pp. 396–401.

[9] C. Lu, W. Zhou, and K. Jin, "Electrothermal Modeling Based Digital Twin Method for Degradation Parameters Identification of DC-DC Converter," in *2023 IEEE Applied Power Electronics Conference and Exposition (APEC)*, Orlando, FL, USA: IEEE, Mar. 2023, pp. 1141–1144.

[10] A. Kundu, P. Korta, L. V. Iyer, and N. C. Kar, "Power Module Thermal Characterization Considering Aging Towards Online State-of-Health Monitoring," in *2023 IEEE Applied Power Electronics Conference and Exposition (APEC)*, Orlando, FL, USA: IEEE, Mar. 2023, pp. 1128–1134.

[11] G. Moreno, S. Narumanchi, X. Feng, P. Anschel, S. Myers, and P. Keller, "Electric-Drive Vehicle Power Electronics Thermal Management: Current Status, Challenges, and Future Directions," *J. Electron. Packag.*, vol. 144, no. 1, p. 011004, Mar. 2022.

[12] D. Bisi *et al.*, "Short-Circuit Protection for GaN Power Devices with Integrated Current Limiter and Commercial Gate Driver," in *2022 IEEE Applied Power Electronics Conference and Exposition (APEC)*, Houston, TX, USA: IEEE, Mar. 2022, pp. 181–185.

[13] R. Zhang, X. Li, J. Ding, S. Chen, H. Yang, and H. Guo, "Review of IGBT Intelligent Gate Drive and Protection Strategies," *IEEE Trans. Power Electron.*, vol. 39, no. 6, pp. 7392–7403, Jun. 2024.

[14] A. H. Mohamed, H. Vansompel, and P. Sergeant, "Polygon-Retrofitted Integrated Modular Motor Drive for Switched Reluctance Machines," *IEEE Trans. Ind. Electron.*, vol. 69, no. 12, pp. 12469–12479, Dec. 2022.

[15] A. H. Mohamed, H. Vansompel, and P. Sergeant, "An Integrated Modular Motor Drive With Shared Cooling for Axial Flux Motor Drives," *IEEE Trans. Ind. Electron.*, vol. 68, no. 11, pp. 10467–10476, Nov. 2021.

[16] E. Deleu *et al.*, "Multi-Material Power Magnetics Modeling with a Modular and Scalable Machine Learning Framework," in *2024 IEEE Applied Power Electronics Conference and Exposition (APEC)*, Long Beach, CA, USA: IEEE, Feb. 2024, pp. 370–377.

[17] Y. Cui, J. Hu, R. Tallam, R. Miklosovic, and N. Zargari, "Reliability Monitoring and Predictive Maintenance of Power Electronics with Physics and Data Driven Approach Based on Machine Learning," in *2023 IEEE Applied Power Electronics Conference and Exposition (APEC)*, Orlando, FL, USA: IEEE, Mar. 2023, pp.

[18] A. M. Muslu, V. Smet, and Y. Joshi, "Multi-physics Modeling of a Power Electronics Package with Integrated Cooling," in *2021 27th International Workshop on Thermal Investigations of ICs and Systems (THERMINIC)*, Berlin, Germany: IEEE, Sep. 2021, pp. 1–6.

[19] Electro-Thermal Co-design of a High-Density Power-Stage for a Reconfigurable-Battery-Assisted Electric-Vehicle Fast-Charger using Multi-Physics Co-simulation and Topology Optimization," in *2023 IEEE Applied Power Electronics Conference and Exposition (APEC)*, Orlando, FL, USA: IEEE, Mar. 2023, pp. 1808–1815.

[20] A. B. Jorgensen, S. Munk-Nielsen, and C. Uhrenfeldt, "Overview of Digital Design and Finite-Element Analysis in Modern Power Electronic Packaging," *IEEE Trans. Power Electron.*, vol. 35, no. 10, pp. 10892–10905, Oct. 2020.

Electrical Characterization of Modular 3D Packaging Assembled with Compressed Metal Foams

Paul Bruyere[1,2], Alexis Derbey[1], Betina Zynger-Capaverde[1], Yvan Avenas[1], Eric Vagnon[2], Jean-Luc Schanen[1], Jean-Michel Guichon[1], Omar Sanjakdar[3]

[1]Univ. Grenoble Alpes, CNRS, Grenoble INP, G2Elab, Grenoble, France
[2]Ecole Centrale de Lyon, INSA Lyon, Universite Claude Bernard Lyon 1, CNRS, Ampère, UMR5005, 69130 Ecully, France
[3]Univ. Grenoble Alpes, CEA, LITEN, INES, 73375 Le Bourget du Lac, France

Abstract—This paper proposes a novel modular packaging concept based on the 3D integration of 1.2kV SiC MOSFET with metal foams for electrical contacts with pressure assembly. These foams could allow improving the disassembly of switching cells by keeping low electrical resistance and inductance inside the package. Thus, static and dynamic experimental characterizations are carried out to evaluate their impact compared to a massive interconnection represented by copper interposers. Static measurements show that the electrical resistance of the package is 2.3mΩ which represents 7% of the total resistance. Dynamic characterizations demonstrate that the stray inductance of the switching cell is equal to 2.7nH. With results of the massive interconnection close to the ones of foams, the material used for the electrical interconnection have no impact on the electrical performances of the power module. The modification of pressure assembly of the power module shows however a global increase of the packaging resistance and stray inductance of the power module for both solutions. This increase is more important for the massive interconnection due the non-deformable mechanicals properties.

Keywords—Packaging, SiC devices, Modularity, Metal Foam

I. INTRODUCTION

One solution massively used in nowadays converters lies on Wide Band-Gap (WBG) semiconductors use [1] in replacement of classical Silicon (Si) material. The better intrinsic properties of WBG materials allow to enhance electrical and thermal characteristics of converters, but their implementation is very sensitive to parasitic elements, e.g. stray inductances and capacitances. The literature proposes some solutions to this issue. For instance, Huang et al. [2] and Morand et al. [3] present 1.2kV Silicon Carbide (SiC) power modules based on respectively Direct Bonded Copper (DBC) substrates and Printed Circuit Board (PCB) technology. The stray inductances of these very performant power modules range from 0.4nH to 2nH.

Despite their interesting electrical properties, these types of power modules, also known as Multi-Chip Modules (MCM), do not support modularity or disassembly criteria due to the strong integration of components and materials inside the packaging. Furthermore, their design being linked to a specific case (current rating, topology...), it is necessary to redesign the power module to meet a new set of specifications.

Another classical solution based on Single-Chip Modules (SCM) benefits much more from modularity aspects. Indeed, these discrete packages, e.g. TO-247, allow to select the most

interesting reference in regard to the application and to design the converter around several components to satisfy the specifications. This solution shows however worse electrical (parasitics, insulation) and thermal behaviors [4] than MCM. Also, their mechanical assembly being most of the time made with solders, their disassembly is possible but remains complicated.

A new type of integration is therefore proposed in this paper. Based on the SCM concept, one 1.2kV SiC MOSFET die is integrated in a prepackage allowing easy disassembly and with electrical and thermal behaviors close to the ones of MCM. In the following section, a switching cell assembly is described: the design of this new assembly entails to consider electrical, thermal and mechanical aspects in order to support modularity and disassembly criteria. Then, the developed assembly is electrically characterized in order to quantify and compare its electrical properties to literature integration concepts [2][3]. This paper focuses on the electrical characterization, thermal management issues will not be addressed.

II. MODULAR 3D PACKAGING CONCEPT PRESENTATION

A. Prepackage description

The concept developed in this paper is based on the integration of a 1.2kV - 134A - 17mΩ SiC baredie in a single prepackage. A synoptic pattern of this latter is represented in Fig. 1. The SiC baredie from *Wolfspeed* (ref. EP3M-1200-0017D1-R01) is firstly assembled between two Aluminum Nitride (AlN) DBC substrates which allow mechanical support of the die and double-side insulation and cooling. The first DBC is the simplest one with Drain potential connection, while the second one is patterned to report the others potentials, i.e. Gate, Source and Kelvin Source, thanks to copper bumps.

A simple power electronic structure, i.e. a switching cell, is obtained thanks to the assembly of diverse prepackages. The mechanical assembly of these prepackages is simplified by transferring the die potentials onto a larger area represented by a PCB. For that, a PCB with cavity (approximately baredie area) is inserted between both DBC and SAC305 solder paste is used to insure mechanical assembly of all the DBC and PCB substrates. Finally, to avoid any possible voltage withstand issues due to the proximity of the die terminals, a dielectric gel is injected in the cavity between both DBC. All materials of the different layers and associated thicknesses are depicted in Table 1.

979-8-3315-1612-3/25 $31.00 © 2025 IEEE

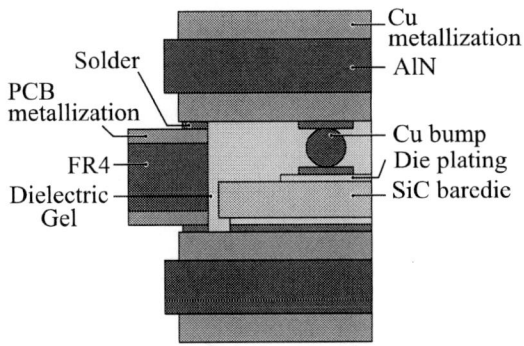

Fig. 1: Synoptic pattern of the developed prepackage.

Table 1: Different layers used for the prepackage assembly with associated materials and thicknesses.

Layer	Material	Thickness
DBC metallization	Cu	300μm
Substrate	AlN	1mm
Solder	SAC305	*not fixed*
Bump	Cu	1mm
Baredie	SiC	183μm
PCB metallization	NiAu	35μm
PCB	FR4	1.53mm

The final assembly of the developed prepackage is depicted in Fig. 2. With geometrical dimensions of 54x26x4.8mm, the baredie area (25mm²) represents only 1.8% of the total prepackage surface. This ratio can be explained by the transfer of die potentials from DBC to PCB in order to simplify the switching cell assembly developed in the next subsection.

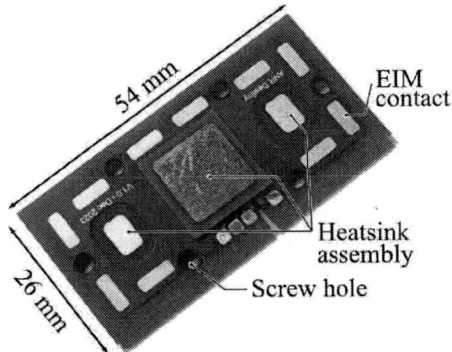

Fig. 2: Final assembly of the developed prepackage.

B. Power module description

An elementary switching cell built with two prepackages is described in Fig. 4. The classical phase-leg connection of both prepackages are made by two different PCB named as *motherboards*. Electrical contacts between the prepackages and the PCB motherboards are made through Electrical Interface Materials (EIM). These are made of compressible copper foams [5] with high porosity rate (>85%). By analogy with Thermal Interface Materials (TIM), the foam allows absorbing the possible thicknesses mismatch in the power module assembly and thus improving the current repartition in the structure. In this example, the switching cell assembly is done at low pressure with several screws to ensure alignment of the different layers of the power module and contact between each EIM and the PCB.

Each motherboard includes a gate driver circuit for one of the cell's MOSFET. The electrical connection between both MOSFET is made by the lower motherboard, while the upper one allows to connect the DC-link busbar to the switching cell. Finally, concerning the dynamic electrical behavior, the integration of ceramic decoupling capacitors as close as possible to the dies, i.e. on the upper motherboard, allows to minimize the stray inductance of the switching cell. This decoupling stage is here realized by ten ceramic capacitors (1kV - 6.8nF and *C0G* dielectric material) in parallel, leading therefore to a total decoupling stage of 68nF.

This assembly benefits from modular and easier disassembly possibilities thanks to respectively single chip prepackages and non-soldered electrical connections in the switching cell with low pressure and EIM. A final switching cell assembly is represented in Fig. 3. As this concept aims to investigate a new packaging technique, static and dynamic electrical characterizations are mandatory to assess the performances of this solution.

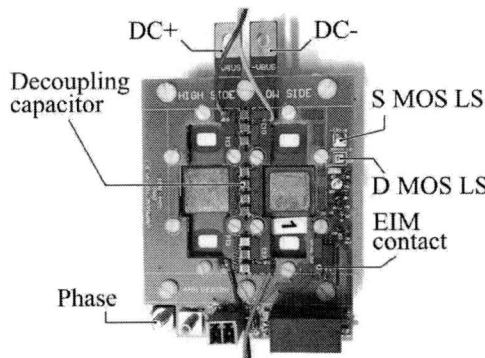

Fig. 3: Final assembly of the developed switching cell (top view).

III. ELECTRICAL CHARACTERIZATION METHODOLOGY

This section aims to present the overall methodology employed for the characterization of the developed power module. The first subsection presents the different EIM solutions implemented in this study. The second and third subsections present experimental tests benches and protocols used for the respectively static and dynamic characterizations. Finally, the last subsection proposes a parametric study related to the impact of pressure distribution on electrical performances of the power module.

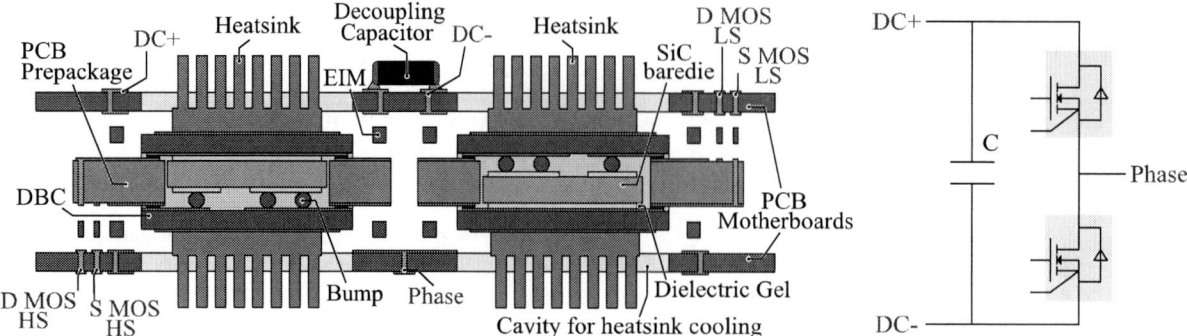

Fig. 4: Synoptic pattern of the developed switching cell with associated electrical schematic.

A. Copper foam EIM vs bare copper interposers

The main objective of the power module electrical characterization is to assess the benefit of electrical interconnection made with copper foams as EIM. For that, two different switching cells are characterized. The first switching cell is represented by an electrical interconnection with EIM made of copper foams represented in Fig. 5. In order to evaluate the interest of this material as power electrical interconnection, another more classical solution, where EIM are made of bare copper interposers, is developed. Keeping these two solutions for the comparison allows to quantify if a soft material, i.e. deformable, is more interesting in the case of pressure assembly power module compared to a massive interconnection represented by copper interposers. For these two solutions, the final thicknesses of the EIMs are 2mm when the switching cell is totally pressed with the different screws.

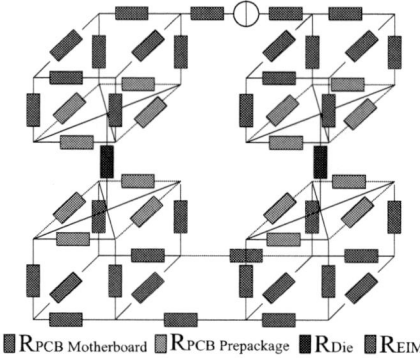

Fig. 6: Resistance network isometric view of the developed assembly with 4 EIMs represented for each interconnection.

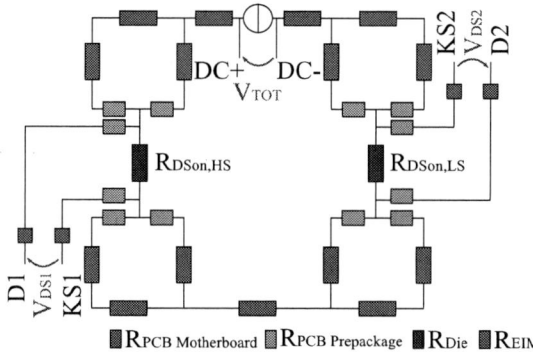

Fig. 7: 2D resistance network simplification used for power module resistance characterization and location of V_{DS} measurements.

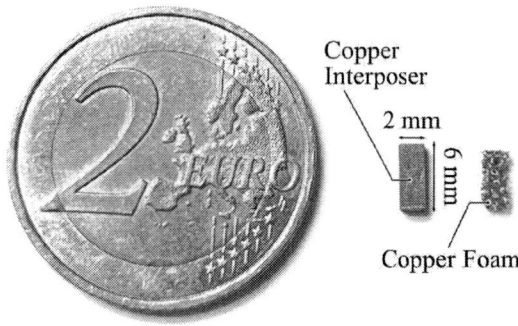

Fig. 5: Different EIM used for the switching cell assembly, each EIM final thickness is 2mm.

B. Characterization method for the power module resistance

The objective of the static characterization is to quantify the electrical resistance of the power module and to assess the impact of the electrical interconnection with EIM. The different resistances constituting the assembly, i.e. PCB motherboards, PCB prepackages, MOSFETs and EIMs, can be represented as a resistance network depicted in Fig. 6 with an isometric view. For simplicity, only four different EIMs are represented for each interconnection between PCB prepackages and motherboards. This resistance network is simplified in a two-dimensions (2D) representation depicted in Fig. 7 to represent the different measured resistances and the V_{DS} measurement locations.

The static characterization is done with each MOSFET in conduction mode (V_{GS}=15V), while a curve tracer (B1505A from *Agilent Technologies*) is used to inject pulsed current of 5A between DC+ and DC- terminals (see Fig. 7). Voltage measurements between different potentials of the switching cell represented in Fig. 7 allows to deduce several resistances of the power module assembly. By measuring V_{TOT}, V_{DS1} and V_{DS2}, resistances R_{TOT}, $R_{DSon,HS}$ and $R_{DSon,LS}$ can respectively be obtained. Considering these latter and applying (1), the packaging resistance R_{pack} can be deduced. This last resistance can be attributed to all electrical interconnections of the power module made of the PCB copper tracks (motherboards and prepackages) as well as the resistance of EIMs.

979-8-3315-1612-3/25 $31.00 © 2025 IEEE

$$R_{pack} = R_{TOT} - \left(R_{DSon,HS} + R_{DSon,LS}\right) \qquad (1)$$

C. Characterization method for the power module stray inductance

The main goal of this dynamic characterization is to quantify the power module parasitic stray inductance in order to compare with classical MCM concepts. Stray inductance is one of the most important electrical parameter to minimize in power modules. Indeed, WBG semiconductors being very sensitive to parasitic elements, a high stray inductance can lead to overvoltage and oscillations [6].

The Low-Voltage Oscillating Circuit (LVOC) method is used to characterize the stray inductance of the power module, experimental test bench of this method is depicted in Fig. 8. With Low-Side (LS) MOSFET always in conduction mode (V_{GS}=15V) and active control of the High-Side (HS) one, an oscillating circuit can be created between C and L_S when the decoupling stage is biased with low-voltage, in this case around 2V, and HS MOSFET is turned-on. The oscillating frequency F_{osc} measurement of the voltage across the decoupling stage allows to deduce the stray inductance of the power module by applying (2).

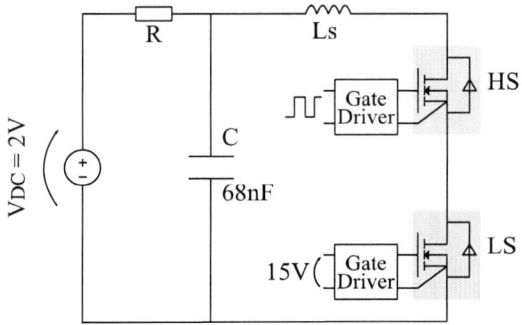

Fig. 8: Electrical schema of LVOC characterization method.

$$F_{osc} \approx \frac{1}{2\pi\sqrt{L_s C}} \qquad (2)$$

D. Impact of the pressure distribution on the electrical properties of the power module

In the case of pressure-based assembly power modules, the pressure distribution on the entire module area is an important aspect to control. Indeed, a non-regular clamping pressure can affect the contact resistance of electrical interconnections and lead to inhomogeneous current repartition, leading to a deterioration of electrical properties i.e. resistance and stray inductance [7]. This problematic can be pointed out in the case of EIM made of solid copper interposers due to their non-deformable structure with possible surface roughness and irregular planarity, resulting in high electrical contact resistance. On the contrary, EIM made of copper foams lie on more degrees of freedom in regards to mechanical properties due to their porous mechanical structure, making them deformable. In case of non-regular clamping pressure, the copper foam can accept mechanical deformation and thus bridge level differences or mechanical tilts. These two different

mechanisms resulting from non-regular pressure distribution are represented in Fig. 9.

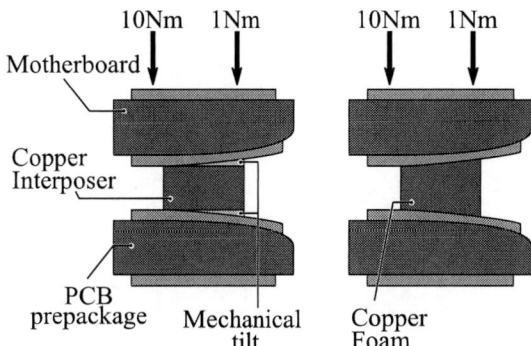

Fig. 9: Mechanical contact representation for each EIM in case of non-regular pressure distribution.

A top view of the power module with screws placement compared to the EIM and decoupling capacitors location is represented in Fig. 10. Due to the pressure assembly made by twelve different screws, the power module clamping pressure modification is simplified by tightening or not the screws of Fig. 10 containing a red star. The impact of pressure distribution on static and dynamic properties of foam and copper cells can therefore be evaluated. The screws concerned by this modification are chosen to have an impact on the high frequency current repartition, i.e. close to the decoupling stage. Additionally, external screws close to Gate and Kelvin Source of MOSFET, are kept tight to keep the controllability of both semiconductors.

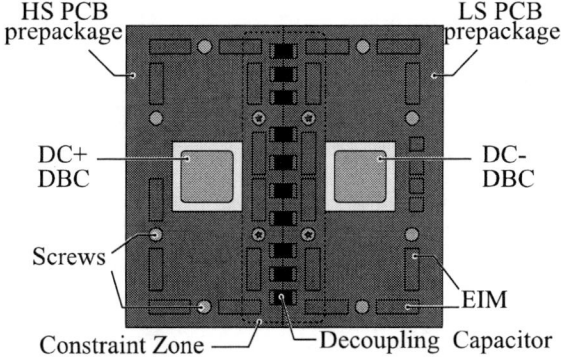

Fig. 10: Synoptic pattern of power module with screws location (top view). A star on a screw indicates that it can be untightened.

IV. EXPERIMENTAL RESULTS

The resistance and stray inductance results of the foam and copper switching cells are presented in a first subsection when considering a total clamping pressure of the power module with twelve screws. Then, the pressure distribution impact is quantified by tightening or not several screws.

A. Power module results for a total clamping pressure

Experimental results of both switching cell resistances are depicted in Table 2. Indicated resistances represent the average of eight different measurements on a single current pulse. All the results depicted in Table 2 represent therefore the average μ and associated standard deviation σ of each resistance.

Comparison of both switching cells results indicates very close static electrical performances. Indeed, with R_{TOT} of foam and copper cells respectively equal to 32.8mΩ and 32.9mΩ, no difference can be highlighted between both cells. This first result indicates that the total power module resistance is faintly sensitive to the EIM material. The same conclusion can be drawn regarding the on-state resistances of each MOSFET. Finally, packaging resistance of each cell can be computed thanks to (1). With R_{pack} of foam and copper cells respectively equal to 2.3mΩ and 2.4mΩ, a very low difference can be highlighted between both cells. Also, each R_{pack} represents 7% of the total power module resistance. This last conclusion indicates a non-negligible impact of the packaging resistance on power module static electrical performance, leading for example to higher conduction losses and possible lower conversion efficiency. This non-negligible impact of the packaging resistance can however be reduced by increasing the thickness of PCB copper tracks which is only 35 μm in this setup.

Table 2: Power modules resistances and stray inductances experimental results in the case of total clamping pressure.

	Foam Cell		Copper Cell	
	μ	σ	μ	σ
R_{TOT}	32.8mΩ	28.6μΩ	32.9mΩ	18.8μΩ
$R_{DSon,HS}$	15.0mΩ	36.3μΩ	15.0mΩ	31.5μΩ
$R_{DSon,LS}$	15.4mΩ	27.9μΩ	15.4mΩ	34.2μΩ
R_{pack}	2.3mΩ		2.4mΩ	
L_s	2.7nH	63.6pH	2.6nH	233pH

An example of LVOC experimental result of the copper foam switching cell is represented in Fig. 11. The oscillating frequency is clearly identified when the HS MOSFET is turned-on, i.e. when V_{GS} voltage is equal to 15V. With a frequency equals to 11.76MHz, a stray inductance of 2.7nH can be deduced with (2) for this test.

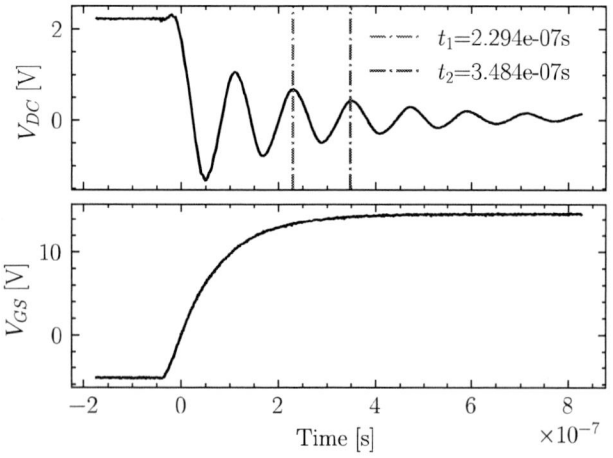

Fig. 11: Experimental result example of Cell A for stray inductance characterization with LVOC method.

All the results of this dynamic characterization are summarized in Table 2 for both switching cells. These results represent the average stray inductance with associated standard deviation on two different LVOC tests. Experimental results of respectively 2.7nH and 2.6nH can be obtained for foam and copper cells. This 5% difference allows to underline that the power module stray inductance is not sensitive to EIM solution in the present configuration. With stray inductance of the power module lower than 3nH, the proposed packaging concept can be compared to MCM concepts on dynamic behavior.

With these static and dynamic characterizations, it can be concluded that the both EIM solutions used for power electrical interconnections have not any influence on the power module electrical properties when 12 screws are tightened. It can be explained by the deformability of the PCBs under these pressure constraints.

The impact of the pressure distribution is investigated in the next subsection with the modification of the clamping pressure as described before in §III.C.

B. Modification of the clamping pressure

The same resistance and inductance characterizations are done for both switching cells with a modification of the pressure distribution. For that, different screws are tightened or not as described before. Table 3 summarized all the results. For each characterization, N represents the number of screws tightened. The different packaging resistance and stray inductance values represent the average result obtained on eight different measurements for the resistance and two different for the stray inductance.

Table 3: Power modules resistances and stray inductances experimental results in case of non-regular clamping pressure.

		N	12	11	10	9	8
Foam Cell	R_{pack} [mΩ]		2.4	2.4	2.5	2.8	2.7
	L_s [nH]		2.7	2.7	2.8	2.9	3.2
Copper Cell	R_{pack} [mΩ]		2.5	3.0	2.9	3.8	4.0
	L_s [nH]		2.6	3.1	3.3	4.2	5.2

Changes of the packaging resistance and the stray inductance as a function of N are depicted in respectively Fig. 12 and Fig. 13. For each switching cell, the packaging resistance increase as N decreases. Differences of 17% and 60% can be deduced between maximum and minimum packaging resistances for respectively foam and copper cells. The same behavior can be underlined for the stray inductance with differences of 19% and 100% between maximum and minimum results of each switching cells. With these results, the impact of pressure distribution is here clearly demonstrated.

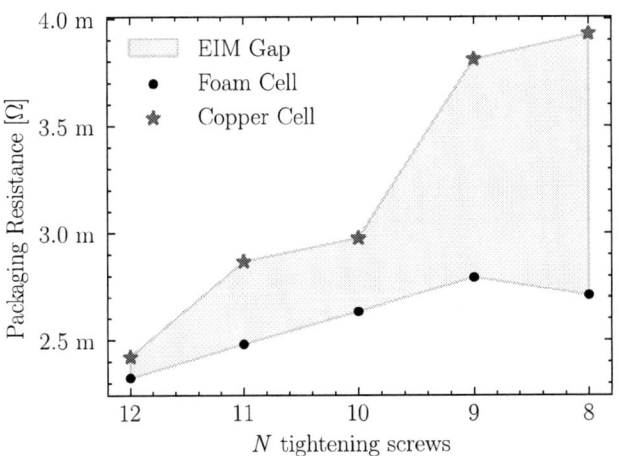

Fig. 12: Packaging resistance experimental result as function of pressure distribution.

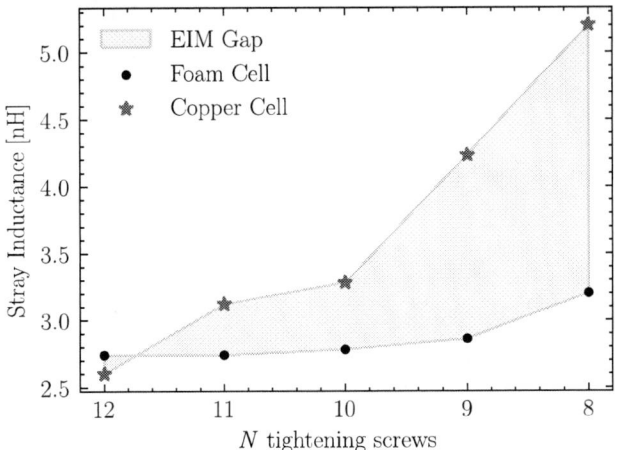

Fig. 13: Stray inductance experimental result as function of pressure distribution.

Results of Fig. 12 and Fig. 13 allow to underline a lower dependency to pressure distribution for foam cell. This latter being represented by EIM made of copper foam, better electrical properties can be considered for this type of material due to the possible mechanical tilt absorption thanks to deformable mechanical properties. Due to this lower increase of the packaging resistance and stray inductance, differences of respectively 48% and 63% can be pointed out for copper cell compared to foam cell when $N=8$. These differences are represented by EIM gap areas of each plots which represent the possible electrical properties improvement by changing EIM from massive to soft material interconnection.

V. CONCLUSION

This paper presents a modular 3D pressure-based power module developed with Single-Chip Module (SCM) packaging concept. This power module design holds on the assembly of two identical prepackages containing 1.2kV SiC baredie and manufactured thanks to DBC and PCB technologies. This prepackage allows a double-side insulated cooling solution of a single die compared to classical PCB-embedded switching cell integration. The mechanical and electrical assembly of both prepackages are respectively done by low pressure thanks to different screws, and combination of PCB motherboards and Electrical Interface Material (EIM) like copper foam. All these specifications are considered to support modularity and disassembly criteria of the power module by the development of a unique element, i.e. the prepackage, which can be extended to different power electronics topologies or set of specifications.

This new type of interconnection entails to consider the impact of EIM on static and dynamic properties of the power module. For that, two switching cells based on different EIM solutions, i.e. copper foams and massive copper interposers, are characterized. With differences between cells of 4% for packaging resistance and stray inductance, the power module electrical properties are not impacted by the EIM solution. The good results obtained for the massive interconnection is explained by the deformation of PCB when the pressure is applied regularly in the power module.

Pressure distribution is also investigated for these two types of interconnections. By tightening or not different screws of the power module assembly, the modification of resistance and stray inductance of the power module is quantified for different pressure distributions. For both cells, when screws are not tightened, the power module resistance and stray inductance increase due to worse electrical contact resistances on the EIM. This increase is however less pronounced in the case of copper foams due to their better mechanical deformability. This last conclusion entails therefore to consider soft materials for electrical interconnection in the case of pressure-based power modules. With these materials, electrical properties close to those of high-performance power modules which not support modular or disassembly criteria can be reached.

Other works aims to study a different pressure system compared to screws used here for simplicity. An external system could for example be investigated for pressure application, or the power module could be pressed between two cooling solutions. Finally, another electrical design could be proposed to increase the power density of the switching cell, and reduced packaging resistance and stray inductance to values closed to the ones of multi-chip optimized power modules.

ACKNOWLEDGMENT

The authors would like to thank Region Auvergne-Rhônes-Alpes (TAPIR project – Pack Ambition Recherche 2021) and ANR agency (ANR-21-CE05-0037) for the financial support of the project.

REFERENCES

[1] D. Cittanti, E. Vico, et I. R. Bojoi, « New FOM-Based Performance Evaluation of 600/650 V SiC and GaN Semiconductors for Next-Generation EV Drives », *IEEE Access*, vol. 10, p. 51693-51707, 2022, doi: 10.1109/ACCESS.2022.3174777.

[2] Z. Huang, C. Chen, Y. Xie, Y. Yan, Y. Kang, et F. Luo, « A High-Performance Embedded SiC Power Module Based on a DBC-Stacked Hybrid Packaging Structure », *IEEE Journal of Emerging and Selected Topics in Power Electronics*, vol. 8, n° 1, p. 351-366, mars 2020, doi: 10.1109/JESTPE.2019.2943635.

[3] J. Morand, R. Perrin, J. Le Lesle, et G. Lefevre, « Design and Evaluation of a Building Block for a 100kW DC/DC Converter Based on PCB Process », in *CIPS 2022; 12th International Conference on Integrated Power Electronics Systems*, mars 2022, p. 1-6.

[4] Z. Zhu, R. Yao, H. Li, J. Li, et Z. Chen, « Comparative Analysis on Switching Characteristics of Discrete SiC MOSFET in Press-pack Package and Wire-bonded Package », in *2022 IEEE International Power Electronics and Application Conference and Exposition (PEAC)*, nov. 2022, p. 66-71. doi: 10.1109/PEAC56338.2022.9959610.

[5] L. Wang, W. Wang, K. Zeng, J. Deng, G. Rietveld, et R. J. E. Hueting, « Opportunities and Challenges of Pressure Contact Packaging for Wide Bandgap Power Modules », *IEEE Transactions on Power Electronics*, vol. 39, nº 2, p. 2401-2419, févr. 2024, doi: 10.1109/TPEL.2023.3332050.

[6] M. Meisser, M. Schmenger, et T. Blank, « Parasitics in Power Electronic Modules: How parasitic inductance influences switching and how it can be minimized », in *Proceedings of PCIM Europe 2015; International Exhibition and Conference for Power Electronics, Intelligent Motion, Renewable Energy and Energy Management*, mai 2015, p. 1-8.

[7] Y. Yang, L. Dorn-Gomba, R. Rodriguez, C. Mak, et A. Emadi, « Automotive Power Module Packaging: Current Status and Future Trends », *IEEE Access*, vol. 8, p. 160126-160144, 2020, doi: 10.1109/ACCESS.2020.3019775.

Improvement in Short-Circuit Robustness of SiC-MOSFETs based Power Modules using Two-Level Turn-On (2LTO)

Muhammad Muneeb Alam*, Saad Khalid*, Nisar Ahmed Khan*, Ngoc Ho Tran* and Sebastian Strache[†]

*ME-PM/ENG3.3
Robert Bosch GmbH, Reutlingen, Germany
Email: muhammadmuneeb.alam@de.bosch.com
[†]ME-PM/PAH
Robert Bosch GmbH, Reutlingen, Germany

Abstract—One of the challenges in design of power modules is to have an increased robustness in failure modes and a superior performance during normal operation. The state-of-the-art methods to increase the short-circuit type 1 (SC1) robustness include changes in the design of a power semiconductor, a fixed decrease of on-state gate-source voltage and modifications in a power module layout. However, increasing the SC1 robustness usually results in a reduced conduction performance. This paper presents the use of a two-level turn-on (2LTO) method realized with an intelligent current-source gate driver (CSGD). It increases the robustness in SC1 using the first level and enhances the conduction performance using the second level of gate voltage. The double pulse (DP) and SC1 measurements validate that 2LTO method effectively increases SC1 robustness without significantly influencing the switching performance.

Index Terms—short-circuit type 1, two-level turn-on, current-source gate driver, constant-current, gate-shaping, switching loss optimization, blanking time, reaction time.

I. INTRODUCTION

The power modules in the automotive industry are designed with wide bandgap (WBG) semiconductors such as SiC-MOSFETs due to their faster switching speeds, lower conduction losses, better thermal conductivity and electron mobility compared to silicon (Si) devices [1], [2]. These attributes of WBG devices contribute to an enhanced performance during normal switching and conduction modes. However, the semiconductors in a power module are prone to certain failures over lifetime. Two such examples include short-circuit type 1 (SC1) and short-circuit type 2 (SC2) as mentioned in [3], [4].The SiC-MOSFETS have a shorter short-circuit withstand time (SCWT) compared to Si-IGBTs due to their better on-state conductivity. Therefore, a fast detection circuit is needed to clear a SC1 fault [5]. If the desaturation protection in the gate driver cannot detect a SC1 before SCWT, the stress during SC1 event can lead to a permanent breakdown of a semiconductor in a power module and loss of inverter [6]. The design of a gate-driver application to get a SCWT = 1.3 μs with a SiC-MOSFET automotive inverter and the influence of testing conditions on SC1 robustness is presented in [7]. The state-of-the-art methods focus on decreasing the SC1 current

by modifications in the design of the power semiconductors, power modules and gate drivers. An increase in the SCWT by 50% is achieved by adjusting the gate-oxide thickness and on-state gate voltage [8]. Other approaches include the addition of a clamping circuit in the gate driver and use of the common source inductance to decrease the peak SC current [9], [10]. However, these methods utilize voltage-source gate drivers (VSGDs), limiting the full potential of WBG semiconductors during normal switching operation as the current and voltage commutation slopes cannot be independently controlled. Compared to a VSGD, a CSGD in constant current (CC) and gate shaping (GS) modes with operation point based profiles reduces up to 22 % switching energy at peak performance (PP) operation point [11], [12].

In this paper, a 2LTO method using an intelligent CSGD is presented. This method aims to increase both the SC1 robustness and conduction performance without compromising the switching behavior.

II. METHODOLOGY

The proposed 2LTO method aims to enhance the SC1 robustness by minimizing stress on a power semiconductor during the SC1 event and subsequent turn-off event post SC1 detection. SC1 is more critical for power modules compared to SC2 due to the deactivation of desaturation protection during maximum blanking time ($t_{blanking,max}$), which is the sum of maximum turn-on time ($t_{on,max}$) during an active turn-on event and maximum comparator delays ($t_{comparator,max}$) as shown in Fig. 1. The $t_{on,max}$ is the duration between time t_0, when gate-source voltage (v_{GS}) starts charging from off-state negative voltage ($V_{GS,OFF}$) and time t_2, when v_{GS} reaches the end of plateau voltage ($V_{P,end}$).

Figure 1 illustrates the different gate-source voltages and the corresponding SC1 currents without and with 2LTO. Typically, a high on-state gate-source voltage ($V_{GS,ON2}$) is used in power modules to enhance the performance during conduction state. However, a high $V_{GS,ON2}$ leads to an elevated stress due to high peak SC1 current ($I_{SC1,wo,2LTO}$) during a SC1 event.

979-8-3315-1612-3/25 $31.00 © 2025 IEEE

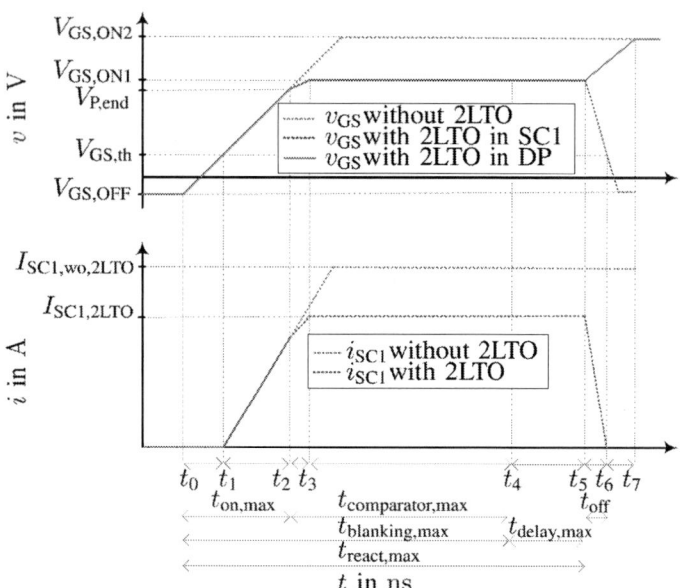

Fig. 1: Gate-source voltage signals without and with 2LTO and their influence on peak SC1 current.

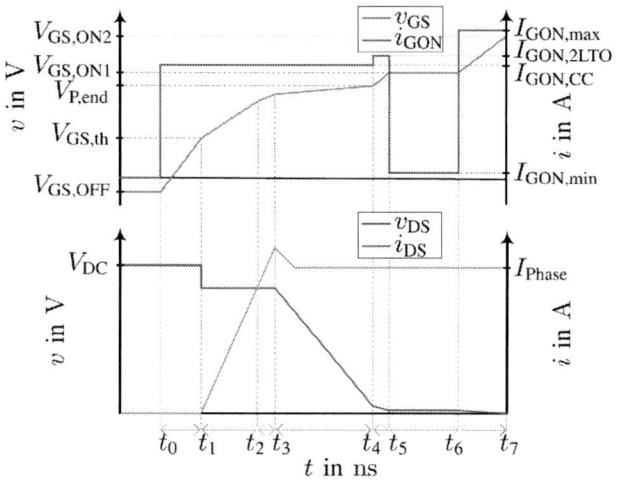

Fig. 2: CSGD in CC mode during an active turn-on event followed by 2LTO.

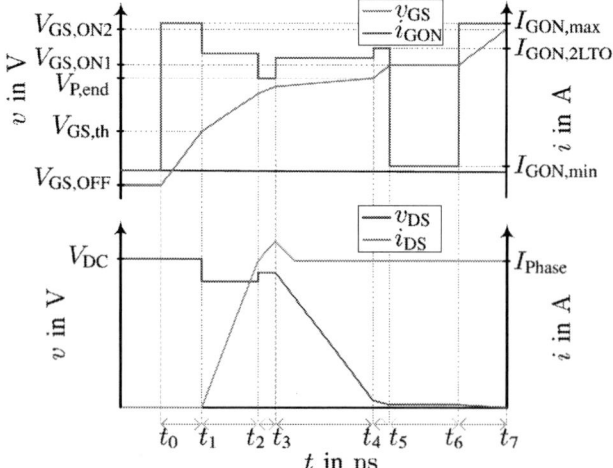

Fig. 3: CSGD in GS mode during an active turn-on event followed by 2LTO.

The 2LTO method maintains v_{GS} at the first level ($V_{GS,ON1}$) after the normal switching event and increases v_{GS} to the second level ($V_{GS,ON2}$) if a SC1 is not detected. $V_{GS,ON1}$ must be greater than the maximum $V_{P,end}$ to ensure the completion of the switching event and less than $V_{GS,ON2}$ to reduce stress during SC1. The SC1 current (i_{SC1}) is limited to a low saturation current ($I_{SC1,2LTO}$) with 2LTO due to a small gate-source voltage fixed at $V_{GS,ON1}$ as depicted by v_{GS} with 2LTO in SC1 signal. If a SC1 exists during normal hard switching operation, the desaturation protection detects it after $t_{blanking,max}$ and gate driver initiates a turn-off signal after the time delay ($t_{delay,max}$) without increasing v_{GS} further to $V_{GS,ON2}$ as shown by v_{GS} with 2LTO in SC1 signal. The reaction time ($t_{react,max}$) is defined from beginning of active turn-on event at time t_0 to start of the turn-off event at time t_5 post SC1 detection, representing the sum of $t_{on,max}$, $t_{comparator,max}$, and $t_{delay,max}$. It is used in SC1 measurements to quantify the influence of 2LTO. The $V_{GS,ON1}$ in 2LTO leads to slightly high conduction losses between the times t_3 and t_5. Therefore, this time should be kept as small as possible in an application. In case a SC1 is not detected during normal hard switching operation, the v_{GS} is increased further to $V_{GS,ON2}$ from time t_5 to time t_7 as shown by the v_{GS} with 2LTO in DP signal. The $V_{GS,ON2}$ with 2LTO can be set to a higher on-state gate-source voltage than a typical value without 2LTO to increase the conduction performance.

The 2LTO method can be implemented using both VSGD and CSGD. In this paper, the validation of 2LTO method is done with a CSGD. To optimize the switching losses, a CSGD can be used either in CC or GS mode during normal switching operation. The 2LTO method can be combined with both modes of CSGD as the two levels in 2LTO are introduced after the completion of switching event. Figure 2 shows the ideal control and power signals of CSGD in CC mode during an active turn-on event followed by 2LTO. It can be observed that the gate driver provides a constant gate current ($I_{GON,CC}$) during the active turn-on switching event from time t_1 to time t_4. Following the completion of the switching event, the gate current (i_{GON}) is set to 2LTO gate current ($I_{GON,2LTO}$) from time t_4 to time t_5. The $I_{GON,2LTO}$ can be calculated from the required 2LTO gate-charge ($Q_{GS,2LTO}$) to charge v_{GS} from $V_{P,end}$ to $V_{GS,ON1}$ and corresponding time as written in Eq. 1.

$$I_{GON,2LTO} = \frac{Q_{GS,2LTO}}{t_5 - t_4} \tag{1}$$

Once v_{GS} reaches the $V_{GS,ON1}$, the i_{GON} is set to the minimum gate current ($I_{GON,min}$) to maintain v_{GS} at $V_{GS,ON1}$ from time t_5 to time t_6. Subsequently, the i_{GON} is set to maximum gate current ($I_{GON,max}$) to charge the v_{GS} to $V_{GS,ON2}$ from time t_6 to time t_7. A six-step approach is presented to realize the 2LTO in an application with CSGD in CC mode followed by 2LTO.
Step 1: Selection of $I_{GON,CC}$ from DP measurements at PP operation point without 2LTO.
Step 2: Determination of maximum $V_{P,end}$ from DP measurements at different operating conditions without 2LTO.
Step 3: Selection of $V_{GS,ON1}$ from maximum value of $V_{P,end}$ and a safety margin of gate driver tolerances.

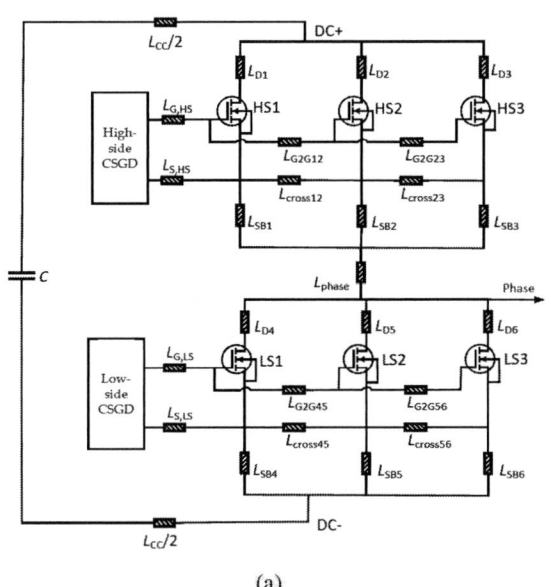

(b)

(a)

Fig. 4: (a) Circuit diagram of the power module with CSGD. (b) Hardware setup for DP and SC1 measurements.

Step 4: Determination of $I_{GON,2LTO}$, $I_{GON,min}$, and $I_{GON,max}$ for DP measurements with 2LTO at PP operation point and comparison of switching energies without and with 2LTO.
Step 5: Validation of 2LTO with DP measurements at different operating conditions.
Step 6: SC1 measurements without and with 2LTO at different operation points and $t_{react,max}$ to quantify influence of 2LTO. The control and power signals of 2LTO with CSGD in GS mode during an active turn-on event followed by 2LTO are depicted in Fig. 3. The six-step approach can be extended to CSGD in GS mode during active turn-on event.

III. MEASUREMENT SETUP

The power module for hardware testing comprises three parallel SiC-MOSFETs per logical switch in a common gate and common source configuration as depicted in Fig. 4a. The power MOSFETs have a rating of 1200 V, 130 A, and an internal gate resistance ($R_{G,int}$) of 5.5 Ω per chip. An AISC EG120 from Robert Bosch is used in CC mode followed by 2LTO, with one channel per high-side (HS) and low-side (LS) switch each. The LS switch serves as the device-under-test (DUT) and is actively controlled with a signal generator, while the HS switch is used passively in DP measurements. Parasitic drain and power source inductances per chip for HS and LS are denoted by L_D and L_{SB}, respectively. Additionally, L_{G2G}, and L_{cross} represent the parasitic gate-to-gate and source-to-source inductances connecting the chips within a logical switch. L_G and L_S denote the inductances in the gate-driver path of HS and LS chips. Lastly, the commutation cell inductance (L_{CC}) represents the inductance of filter capacitor (C) and the interface between the filter capacitor and power module. The hardware setup for DP and SC1 measurements is shown in Fig. 4b.

IV. MEASUREMENT RESULTS

This section details the comparison of DP and SC1 measurements on the power module employing typical SiC-MOSFETs using CSGD in CC mode followed by 2LTO. In step 1, the $I_{GON,CC}$ with CSGD during active turn-on event is defined to get the reference turn-on switching energy without 2LTO at PP operation point. This operation point encompasses battery voltage (V_{DC}) = 850 V, phase current (I_{phase}) = 300 A, virtual junction temperature (T_{vj}) = 150 °C, on-state gate-source voltage ($V_{GS,ON2}$) = 18 V, and off-state gate-source voltage (V_{OFF}) = −5 V. The $I_{GON,CC}$ is selected under the boundary condition that the maximum drain-source voltage on passive side ($V_{DS,passive,max}$) < 1200 V. Based on DP measurement results, an $I_{GON,CC}$ = 1.1 A is needed to fulfill the defined boundary condition at PP operation point and it is used in the following measurement steps.

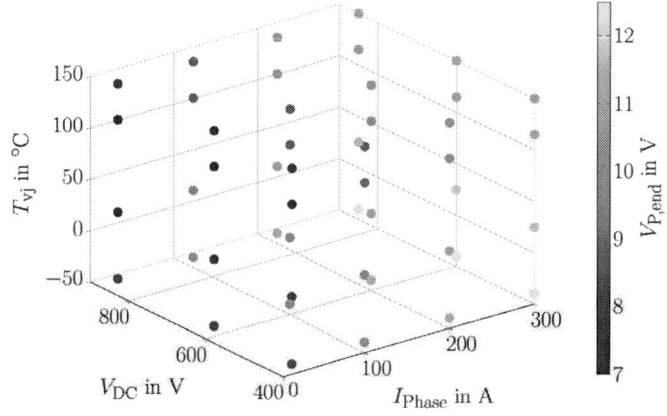

Fig. 5: Dependency of $V_{P,end}$ on V_{DC}, I_{phase}, and T_{vj}.

979-8-3315-1612-3/25 $31.00 © 2025 IEEE   2571

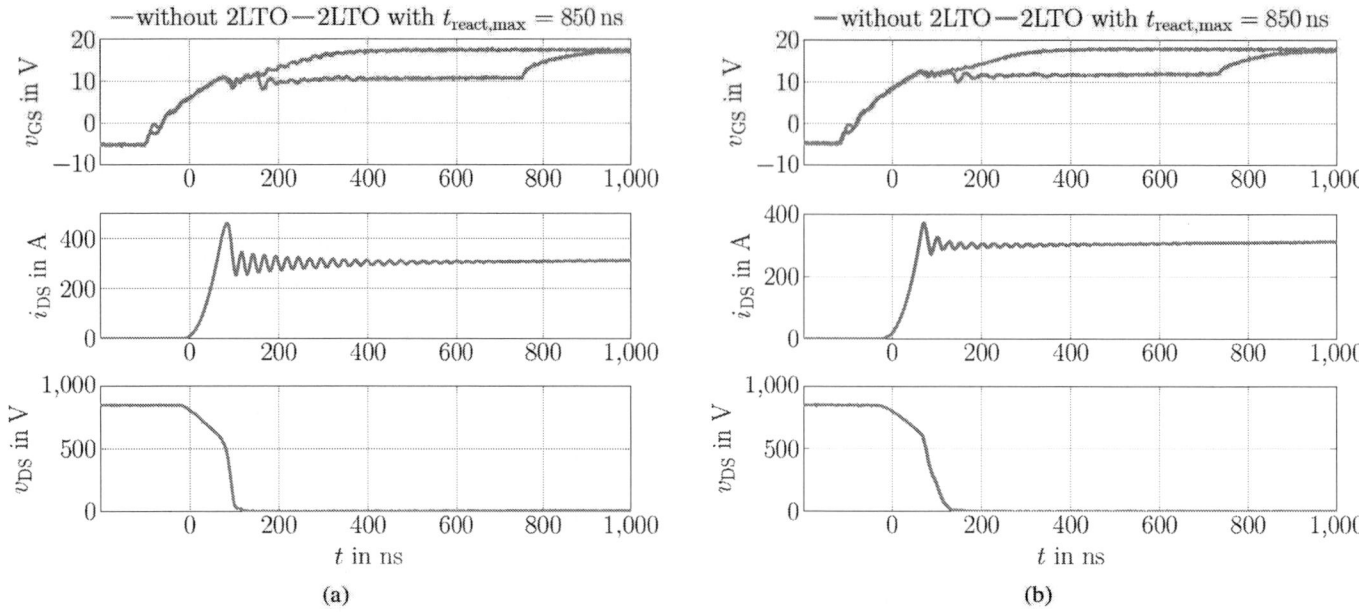

Fig. 6: DP measurements without and with 2LTO using $t_{react,max} = 850\,ns$ at 850 V, 300 A, and (a) 150 °C (b) −40 °C.

In step 2, the dependency of $V_{P,end}$ on different battery voltages, phase currents and virtual junction temperatures is examined. For this purpose, DP measurements are conducted with $i_{GON,CC} = 1.1\,A$ while sweeping $V_{DC} = 400\,V, 600\,V, 850\,V$; $I_{Phase} = 10\,A, 100\,A, 200\,A, 300\,A$; and $T_{vj} = -40\,°C, 25\,°C, 115\,°C, 150\,°C$. The $V_{P,end}$ is extracted from the DP measurements under all operating conditions. The dependency of $V_{P,end}$ on the V_{DC}, I_{Phase} and T_{vj} is shown in Fig 5. It can be observed that the $V_{P,end}$ has an inverse correlation with V_{DC} due to the drain-induced barrier lowering (DIBL) effect, an inverse correlation with T_{vj}, and a direct correlation with I_{Phase}. Therefore, the maximum $V_{P,end}$ of 12.5 V occurs at $V_{DC} = 400\,V$, $I_{phase} = 300\,V$, and $T_{vj} = -40\,°C$.

In step 3, the $V_{GS,ON1}$ is determined as the sum of the maximum value of $V_{P,end}$ with typical SiC-MOSFETs defined from step 2 and a safety margin of 0.5 V. This safety margin includes the tolerances in the gate driver. This leads to a $V_{GS,ON1}$ of 13 V.

Following the determination of $V_{GS,ON1}$, the DP measurements with 2LTO at PP operation point are performed in step 4. The gate currents to realize the 2LTO after the switching event are adjusted from times t_4 to t_7 as shown in Fig. 2. The $i_{GON,2LTO}$ from time t_4 to t_5 is set to 1.2 A to charge the v_{GS} from $V_{P,end}$ to $V_{GS,ON1}$. Subsequently, an $I_{GON,min} = 50\,mA$ is used to maintain the v_{GS} at $V_{GS,ON1}$ from time t_5 to time t_6. Once the selected $t_{react,max} = 850\,ns$ is reached at time t_6 in Fig. 2, the maximum turn-on gate current ($I_{GON,max}$) = 2.5 A is used to charge the v_{GS} from $V_{GS,ON1}$ to $V_{GS,ON2}$ from time t_6 to time t_7. The DP measurements without and with 2LTO with this $t_{react,max}$ at PP operation point are shown in Fig. 6a. The measured turn-on switching energy without 2LTO ($E_{ON,wo,2LTO}$) and turn-on switching energy with 2LTO ($E_{ON,2LTO}$) are 11.2 mJ and 11.3 mJ, respectively.

The switching energies are comparable, because the $I_{GON,CC}$ remains same during switching event leading to the same current and voltage commutation times in DP measurements without and with 2LTO. The v_{GS} signal with 2LTO shows a small reduction in gate-source voltage as i_{GON} is changed from $I_{GON,2LTO}$ to $I_{GON,min}$ due to a smaller voltage drop across $R_{G,int}$ with $I_{GON,min}$. The same gate profile with 2LTO is used to compare the switching behavior at 850 V, 300 A, and a minimum virtual junction temperature ($T_{vj,min}$) = −40 °C. Like DP measurements at PP operation point, there is no significant change in switching behavior without and with 2LTO at 850 V, 300 A, and −40 °C as shown in Fig. 6b. The measured $E_{ON,wo,2LTO}$ and $E_{ON,2LTO}$ at this operation point are 11.4 mJ and 11.8 mJ, respectively.

In step 5, the DP measurements are performed with 2LTO method at different operation points to compare the $E_{ON,wo,2LTO}$ and $E_{ON,2LTO}$ across a wide range of operating conditions. The same sweep of V_{DC}, I_{Phase} and T_{vj} is performed during DP measurements with 2LTO as described earlier for the $V_{P,end}$ without 2LTO in step 2. The gate profile with 2LTO at PP operation point from step 4 is used in DP measurements. Figure 7 shows the comparison of $E_{ON,wo,2LTO}$ and $E_{ON,2LTO}$ at different V_{DC}, I_{Phase}, and T_{vj}. It is evident that there is no significant difference in the switching energies without and with 2LTO at all operation points.

Hence, the DP measurements at different operating conditions confirm that the 2LTO method can be employed across the entire range of operating conditions without any reduction in the switching performance of the semiconductors in a power module. Similarly, the 2LTO can be combined with CSGD in GS mode for normal switching operation to minimize the switching losses compared to CSGD in CC mode with the ideal control and power signals illustrated in Fig. 3.

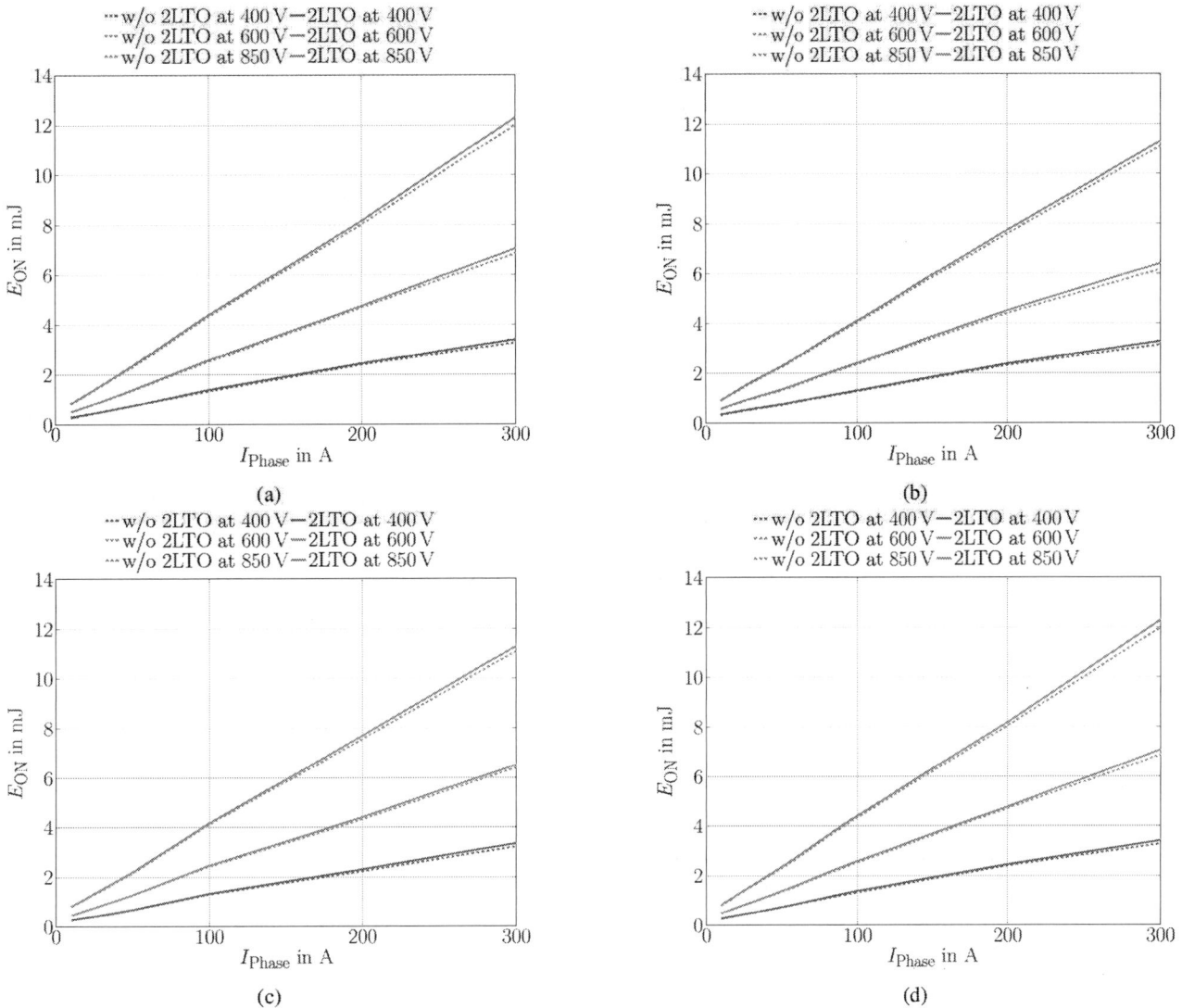

Fig. 7: Comparison of switching energies without and with 2LTO during active turn-on switching event using $t_{react,max} = 850\,ns$ at different V_{DC}, I_{Phase}, and (a) $-40\,°C$ (b) $25\,°C$ (c) $115\,°C$ (d) $150\,°C$.

In step 6, the SC1 measurements are performed on LS switch as DUT at different V_{DC} and T_{vj} without and with 2LTO. The gate profile from step 1 is applied to LS for SC1 measurement without 2LTO. For SC1 measurement with 2LTO, the gate profile from step 4 is used on LS switch. The $V_{GS,ON1} = 13\,V$ and $V_{GS,ON2} = 18\,V$ with this gate profile. The SC1 is artificially created by applying a positive gate voltage of $23\,V$ on HS switch to ensure that the peak SC1 current is not limited. The most critical operation point to test the SC1 robustness is the maximum battery voltage ($V_{DC,max}$) and $T_{vj,max}$. Therefore, this operation point is used to test the SC1 robustness of LS switch. The SC1 measurements are performed with the sweep of $t_{react,max}$ from $0.5\,μs$ to $1\,μs$ in $100\,ns$ steps. This variation in $t_{react,max}$ is aimed at determining the sensitivity of reduction in SC1 energy with 2LTO ($E_{SC1,2LTO}$) relative to SC1 energy without 2LTO ($E_{SC1,wo,2LTO}$). For turn-off after

SC1, a constant SC1 turn-off gate current ($I_{GOFF,SC1}$) = 0.7 A is used to keep the maximum drain-source voltage on LS ($V_{DS,LS,max}$) < 1200 V.

Figure 8a shows the SC1 measurement at $V_{DC,max} = 850\,V$, $T_{vj,max} = 150\,°C$, and $t_{react,max} = 1\,μs$. The peak SC1 current with 2LTO ($I_{SC1,2LTO}$) is reduced by 28 % to 2420 A compared to the peak SC1 current without 2LTO ($I_{SC1,wo,2LTO}$) = 3360 A. Additionally, an $I_{SC1,2LTO} < I_{SC1,wo,2LTO}$ at time instant $t = t_{react,max}$ and a shorter turn-off time with 2LTO ($t_{off,\ 2LTO}$) leads to a smaller SC1 energy during turn-off event. Therefore, $E_{SC1,2LTO}$ reduces by 24 % to 1.7 J in comparison to $E_{SC1,wo,2LTO} = 2.24\,J$. Table I shows that the relative reduction in $E_{SC1,2LTO}$ compared to $E_{SC1,wo,2LTO}$ decreases from 31 % to 24 % with the sweep of $t_{react,max}$ from $0.5\,μs$ to $1\,μs$. It means that the 2LTO method in SC1 decreases the stress by 24 % at most critical operation point with $t_{react,max} = 1\,μs$.

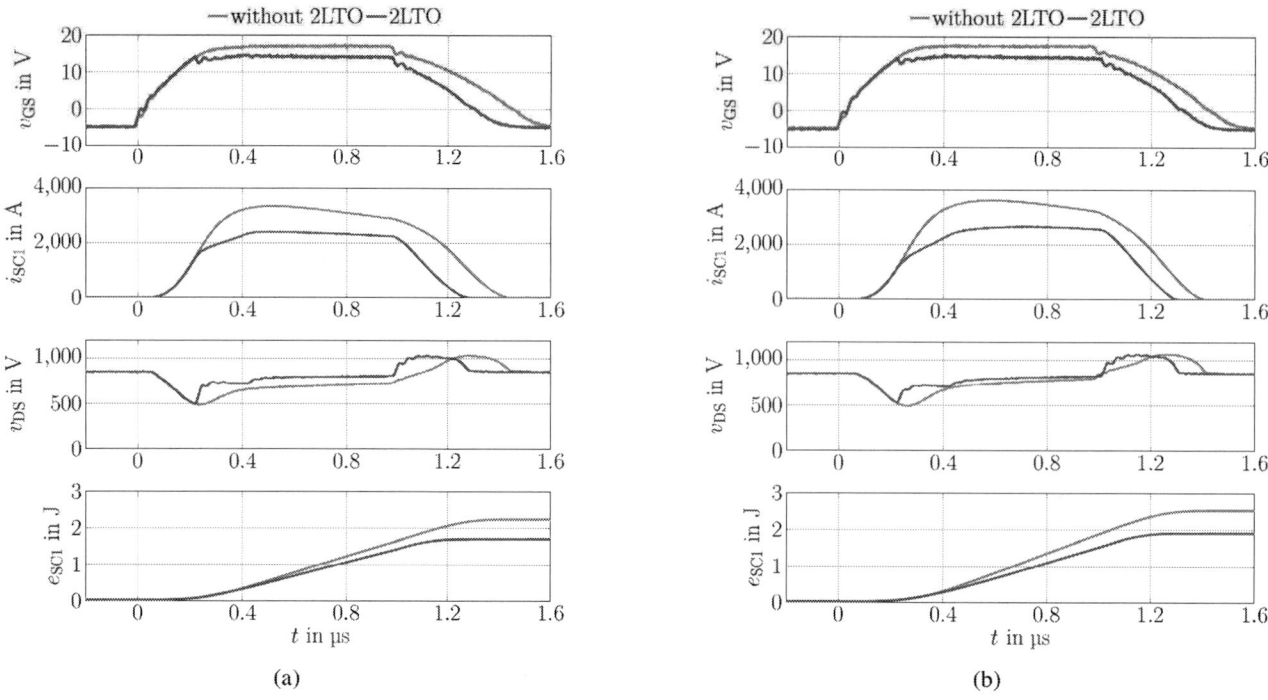

Fig. 8: SC1 measurements without and with 2LTO at 850 V, $t_{\text{react,max}} = 1\,\mu s$, and (a) 150 °C (b) −40 °C.

TABLE I: Comparison of $E_{\text{SC1,wo,2LTO}}$ and $E_{\text{SC1,2LTO}}$ at 850 V, 150 °C with the $t_{\text{react,max}}$ sweep from 0.5 μs to 1 μs.

$t_{\text{react,max}}$ in μs	$E_{\text{SC1,wo,2LTO}}$ in J	$E_{\text{SC1,2LTO}}$ in J	Relative Reduction
0.5	1.16	0.80	31 %
0.6	1.42	1.0	30 %
0.7	1.67	1.21	28 %
0.8	1.85	1.35	27 %
0.9	2.08	1.51	27 %
1	2.24	1.70	24 %

TABLE II: Comparison of $E_{\text{SC1,wo,2LTO}}$ and $E_{\text{SC1,2LTO}}$ at 850 V, −40 °C with the $t_{\text{react,max}}$ sweep from 0.5 μs to 1 μs.

$t_{\text{react,max}}$ in μs	$E_{\text{SC1,wo,2LTO}}$ in J	$E_{\text{SC1,2LTO}}$ in J	Relative Reduction
0.5	1.19	0.80	33 %
0.6	1.51	0.98	35 %
0.7	1.81	1.27	30 %
0.8	2.03	1.45	29 %
0.9	2.32	1.68	28 %
1	2.59	1.90	27 %

Finally, the 2LTO method is validated with SC1 measurements at $V_{\text{DC,max}} = 850\,V$, $T_{\text{vj,min}} = -40\,°C$, and sweep of $t_{\text{react,max}}$. Figure 8b shows the SC1 measurement at 850 V, −40 °C, and $t_{\text{react,max}} = 1\,\mu s$. There is a 27 % reduction in both $I_{\text{SC1,2LTO}}$ and $E_{\text{SC1,2LTO}}$ compared to SC1 measurements without 2LTO. The percentage reduction in $E_{\text{SC1,2LTO}}$ compared to $E_{\text{SC1,wo,2LTO}}$ with SC1 measurements at 850 V and −40 °C with the sweep of $t_{\text{react,max}}$ is shown in Tab. II. The relative reduction in $E_{\text{SC1,2LTO}}$ compared to $E_{\text{SC1,wo,2LTO}}$ decreases from 33 % to 27 % with an increase in $t_{\text{react,max}}$ from 0.5 μs to 1 μs.

The 2LTO method has also been validated at different V_{DC} and T_{vj} with the sweep of $t_{\text{react,max}}$. The percentage reduction in $E_{\text{SC1,2LTO}}$ relative to $E_{\text{SC1,wo,2LTO}}$ with different V_{DC} at $T_{\text{vj,min}}$ and $T_{\text{vj,max}}$ is comparable to the percentage reduction with $V_{\text{DC,max}}$ at $T_{\text{vj,min}}$ and $T_{\text{vj,max}}$. Based on SC1 measurements, it is validated that 2LTO method decreases the SC1 energy up to 24 % at $T_{\text{vj,max}} = 150\,°C$ and 27 % at $T_{\text{vj,min}} = -40\,°C$ with $V_{\text{DC,max}} = 850\,V$, and $t_{\text{react,max}} = 1\,\mu s$. This reduction in stress due to less SC1 energy decreases predamage of power semiconductors at end-of-line (EoL) tests and real SC1 events in field over lifetime. Moreover, 2LTO enables to use long t_{blanking} and t_{delay} in a gate driver application. The reliability of power semiconductor during the normal switching operation using 2LTO has not been investigated in this paper. Therefore, it is recommended to perform a dynamic reliability test with 2LTO before using this method in an application.

V. CONCLUSION

In this paper, the 2LTO method during the active turn-on event is presented and validated with DP and SC1 measurements using CSGD. The DP measurements without and with 2LTO show no significant difference in the active turn-on switching energies. The first level of $V_{\text{GS,ON1}}$ in proposed method decreases the saturation current and energy during SC1 event up to 24 % at most critical operation point. The second level of $V_{\text{GS,ON2}}$ increases the conduction performance by allowing a higher on-state gate voltage. A guideline is presented to select the $V_{\text{GS,ON1}}$ based on the DP measurements across a range of operating conditions. In order to maximize the advantages of 2LTO in an application, it is recommended to use the SiC-MOSFETs with low $V_{\text{P,end}}$ and $R_{\text{G,int}}$ values.

REFERENCES

[1] R. Adappa, K. Suryanarayana, H. Swathi Hatwar, and M. Ravikiran Rao, Review of sic based power semiconductor devices and their applications, in 2019 2nd International Conference on Intelligent Computing, Instrumentation and Control Technologies (ICICICT), vol. 1, 2019, pp. 11971202. DOI: 10.1109/ICICICT46008.2019.8993255.

[2] S. Schwaiger, K. Heyers, A. Martinez-Limia, K. Oberdieck, and C. Foerster, Advanced sic trench-mos technology for automotive application, in PCIM Europe 2023; International Exhibition and Conference for Power Electronics, Intelligent Motion, Renewable Energy and Energy Management, 2023, pp. 14. DOI: 10.30420/566091096.

[3] F. Scholl, J. Fuhrmann, J. da Cunha, and H.-G. Eckel, A new robust short-circuit protection gate-driver circuit for igbt with high desaturation current, in 2021 23rd European Conference on Power Electronics and Applications (EPE21 ECCE Europe), 2021, pp. 110. DOI: 10 . 23919 / EPE21ECCEEurope50061.2021.9570635.

[4] J. Schmitz, M. Meissner, and S. Bernet, Characterization of short circuit behavior of parallel con- nected gan hemt power semiconductors, in PCIM Europe 2023; International Exhibition and Confer- ence for Power Electronics, Intelligent Motion, Renewable Energy and Energy Management, 2023, pp. 110. DOI: 10.30420/566091362.

[5] M. Boesing and D. Schweiker, Design aspects in sic mosfet based high performance automotive and commercial vehicle inverters, in PCIM Europe 2023; International Exhibition and Conference for Power Electronics, Intelligent Motion, Renewable Energy and Energy Management, 2023, pp. 18. DOI: 10.30420/566091063.

[6] A. Maerz, T. Bertelshofer, R. Horff, and M.-M. Bakran, Requirements of short-circuit detection methods and turn-off for wide band gap semiconductors, in CIPS 2016; 9th International Conference on Integrated Power Electronics Systems, 2016, pp. 16.

[7] K. Oberdieck, S. B. Khelifa, M. Horvath, S. Strache and M. Boesing, "Short Circuit Robustness for Traction Inverters from an Application Point of View," PCIM Europe 2024; International Exhibition and Conference for Power Electronics, Intelligent Motion, Renewable Energy and Energy Management, Nrnberg, Germany, 2024, pp. 1823-1829, doi: 10.30420/566262251.

[8] A. Agarwal, A. Kanale, K. Han, and B. J. Baliga, Switching and short-circuit performance of 27 nm gate oxide, 650 v sic planar-gate mosfets with 10 to 15 v gate drive voltage, in 2020 32nd International Symposium on Power Semiconductor Devices and ICs (ISPSD), 2020, pp. 250253. DOI: 10.1109/ISPSD46842.2020.9170151.

[9] X. Liu, J. Kowalsky, and J. Lutz, A clamping circuit for short circuit ruggedness improvement of discrete igbt devices based on the di/dt feedback of emitter stray inductance, in 2019 20th International Symposium on Power Electronics (Ee), 2019, pp. 16. DOI: 10 . 1109 / PEE. 2019 . 8923349.

[10] D. A. Ruoff, Z. X. Sim and B. Ulrich, "Investigation of Common Source Feedback in SiC Power Modules Regarding Performance and Short Circuit Robustness," PCIM Europe 2024; International Exhibition and Conference for Power Electronics, Intelligent Motion, Renewable Energy and Energy Management, Nrnberg, Germany, 2024, pp. 1272-1277, doi: 10.30420/566262174.

[11] M. Riefer, J. Winkler, S. Strache and I. Kallfass, "Implementation of Current-Source Gate Driver with Open-Loop Slope Shaping for SiC-MOSFETs," PCIM Europe digital days 2021; International Exhibition and Conference for Power Electronics, Intelligent Motion, Renewable Energy and Energy Management, Online, 2021, pp. 1-8.

[12] M. M. Alam, S. Khalid and N. Ho, "Operation Point based Optimization of Switching Losses with Current-Source Gate Driver for SiC-based Power Modules," 2024 IEEE Applied Power Electronics Conference and Exposition (APEC), Long Beach, CA, USA, 2024, pp. 2502-2509, doi: 10.1109/APEC48139.2024.10509371.

GaN-Based Two Stage Point-of-Load (PoL) Converter with 2.5D Embedded Substrate Implementation

Samuel Defaz
Department of Electrical and Computer Engineering
Stony Brook Univeristy
Stony Brook, NY 11794-2350
samuel.defaz@stonybrook.edu

Yang Li
Department of Electrical and Computer Engineering
Stony Brook University
Stony Brook, NY 11794-2350
yang.li.9@stonybrook.edu

Fang Luo
Department of Electrical and Computer Engineering
Stony Brook University
Stony Brook, NY 11794-2350
fang.luo@stonybrook.edu

Abstract—**The high-density, high-step-down, high-current point-of-load (PoL) converter plays an important role in future computing and AI research. This paper presents a 48-1V POL target that reaches 92% with heterogeneous integration for AL/ML computing chips targeting a full load operation of 100A. This PoL converter will realize a two-stage cascade converter with the first stage being a hybrid switch capacitor converter at 200 kHz. The 2nd stage will be an interleaved 4-phase buck converter, tested at 1 MHz. The PoL package will implement a 2.5D implementation using embedded passive and actives to create an "active substrate" design by using an electrothermal co-design methodology with a goal to create a low parasitic inductance vertical power loop. An embedded prototype will be created using standard PCB manufacturing methods to create the smallest possible area. Different PCB substrates such as insulated metal core (IMS) PCBs will be explored to combine both the electrical and thermal design into a single package. Finally, a 3D implementation will be explored to improve modularity and to reach higher current demands.**

I. INTRODUCTION

Given the increase in data centers' power consumption, the processing power of AI processors, and innovations in super-computing research, there is a need for highly compact, highly efficient, highly modular VRMs that can be used to meet the ever-increasing demands for the computing industry. A typical VRM solution for CPU/XPU application involves a high step-down solution with a 48V voltage input, tight output voltage tolerances, and high load current draws. Currently, state-of-the-art VRMs implements a two stage solutions to distribute the overall step-down ratio. This implementation splits the voltage and current stress across the two converters, allowing for more freedom in optimizing the switching and conduction losses between each converter. Many techniques have been used to achieve smaller packages for VRMs. The main techniques has been to increase the operating frequency of the converter to take advantage of the overall reduction

in passive volumes. Typically in SMPS, the magnetics of a converter takes the majority of the size of each inductance. As a result, an investigation was done on which converter would minimize the magnetic and the operation of each converter [1]. In the end, a switch capacitor buck converter (SCBC) Fig. 1a for a 48V to 5V conversion was chosen for the first stage and an interleaved multi-phase buck converter (IMBC) Fig. 1b was selected for the 5V to 1V stage. The SCBC topology is seen as a major point of optimization to realize a soft charging implementation for a multi-converter 48V to 1V voltage regulator module (VRM) with a focus on maximizing current density. An updated hardware prototype was created to verify the operation at the 100 W operating point. The initial experiment tested the SCBC prototype without any load to measure the operation and slowly increase the current demand to the full 100 A operation. The IMBC topology is realized using a single board in parallel with each other. An FEA analysis will be done to find the optimal vertical loop design to ensure high-speed switching to reduce the values for the inductor so that can be properly miniaturized. Finally, the PoL Package will explore new PCB substrates such as insulated metal cores and diamond substrates which can manipulate the thermal path of modules by either coupling thermal loads together or by "insulating" them from each other. A prototype using 2 layers will be implemented using 3D printing methods. A fully embedded 2-layer active substrate will be realized and extended to a 4-layer 48V-to-1V converter using different substrates which will be optimized in terms of thermal resistance and current density

II. PROPOSED IMPLEMENTATION

The high current stage is achieved through a interleaved multi-phase buck converter topology which utilizes current splitting among multiple interleaved buck converters to reduce

979-8-3315-1612-3/25 $31.00 © 2025 IEEE

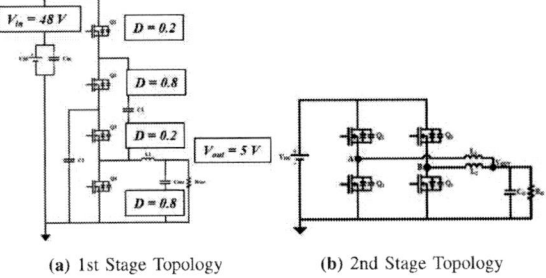

(a) 1st Stage Topology   **(b)** 2nd Stage Topology

Fig. 1: Two Stage Point-of-Load Converter

the individual current stress on the device, as seen in Fig. 8a. Interleaving nature of the converter allows for each phase to cancel out the ripple seen at the load. This allows for the number of output capacitors to be greatly reduced which allows for a more compact design. Each buck converter carries only part of the overall current, allowing for a reduced power rating for each switch and a smaller inductor value for each converter [2]. Given the low output capacitance and on-resistance of GaN devices, such as the EPC2067 [3], hard switching at 1 MHz is justified, as the high speed of Gallium devices allows for shorter time between the on and off transition where both current and voltage overlap. Reducing this overlap allow for lower hard switching losses. Initial SPICE simulations on a two-phase implementation reveal high conduction losses across the bottom side switch as seen in Fig 2a. In efforts to reduce this average current, Fig 2b compares the loss distribution between different phase configuration using SPICE models provided by the manufacturer. A 4-phase implementation gives the highest efficiency while keeping the average current between each branch low.

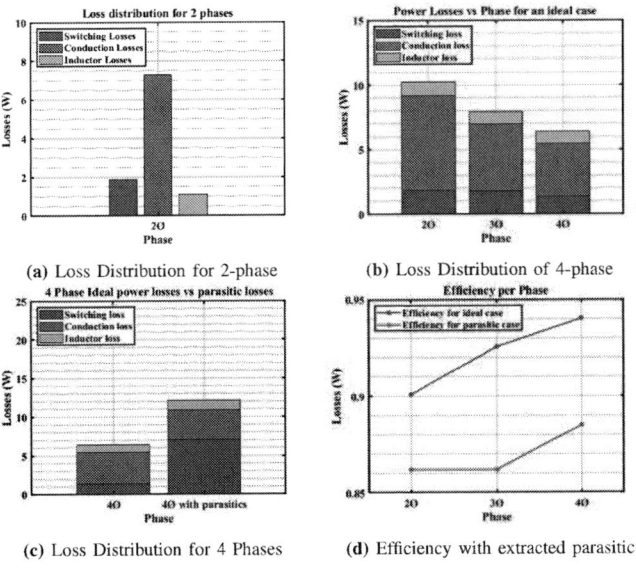

(a) Loss Distribution for 2-phase   **(b)** Loss Distribution of 4-phase

(c) Loss Distribution for 4 Phases   **(d)** Efficiency with extracted parasitic

Fig. 2: Estimated Efficiency for 5V to 1V Stage

The high voltage stage will be implemented through a hybrid switch capacitor buck converter. The benefit from this design is the soft charging capability which comes operating

the first stage at a lower switching frequency than the high current IMBC second stage. The SCBC topology also allows for a higher step-down conversion since the topology implements a flying capacitor stage which can be used to half the initial input voltage while maintaining a wide range duty cycle. This implementation also minimizes the semiconductor device count, the inductor count, and overall volume. The design methodology is to simulate the converter at the maximum operating point to approximate switching and conduction losses, to create a prototype using commercially off the shelf components, and to finally test the converter at different operating points. Doing a literature review on EPC devices, Table I shows the component used to realize the PoL converter. The focus was to minimize the area of the converter. The maximum surface area of the converter is the first-stage inductor which will be the limiting factor in minimizing surface area. Based on the RMS cycle distribution seen 3b, the device Q2 and Q4 will have the most conduction loss. The high RMS distribution is across the Q4 is due to this switching being the connecting path for the rectifying inductor for the buck converter stage. Overall, Q4 will experience the most thermal stress during operation. As opposed to the IMBC, switching losses dominate given the higher voltage step down and lower current demand.

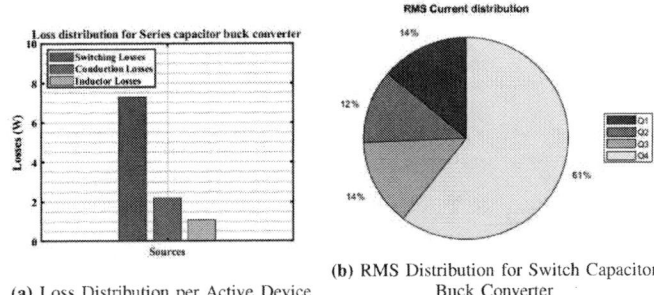

(a) Loss Distribution per Active Device   **(b)** RMS Distribution for Switch Capacitor Buck Converter

Fig. 3: Measured Efficiency for 48V to 5V Stage

A. Optimziation of Power Loop

In order to achieve high efficiency results for a hard switching implementation. It is imperative that the parasitic inductance of the power loop be minimized and isolated from the rest of the circuit. Based on the optimal layout suggested in [3], a proposed layout was designed where magnetic flux will be canceled both vertically and laterally by implementing through-hole vias and with a 4-layer PCB. Individual flux cancellation is achieved in Fig. 4a where interleaving vias will be routed in and out through the board before connected into the device. This implementation is proposed to try to minimize the common source inductance of the device and minimize any interaction between the gate driver and the power loop. A 4 layer PCB layout is initially proposed with 1 oz copper for inner layers and 2 oz copper for outer layers. Via sizes were limited to .4 mm to limit surface area across the entire converter. An CAD PCB model implementing this layout was designed, and the parasitic inductances/resistances will be extracted using Ansys Q3D software. The initial power

979-8-3315-1612-3/25 $31.00 © 2025 IEEE

loop path flows begins with the decoupling capacitors shown in Fig. 5b and Fig. 5a . The yellow arrows show the high frequency path flowing to the top device through an inner plane. Flux cancellation is achieved on the bottom device through the board while less cancellation is seen on the top device. The overall power loop inductance for this layout is 1.18 nH. Verification of this layout will be seen through implementation of hardware results for this PCB.

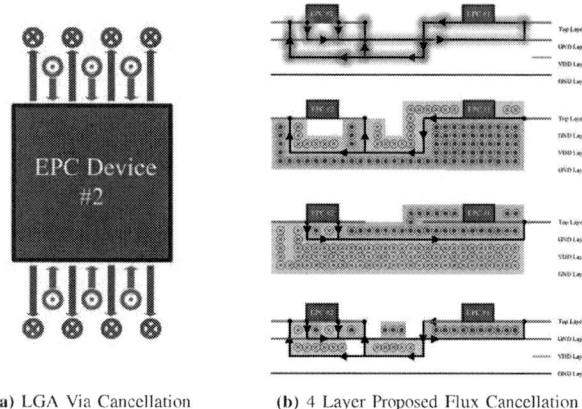

(a) LGA Via Cancellation   **(b)** 4 Layer Proposed Flux Cancellation

Fig. 4: Optimized Power Loop Layout

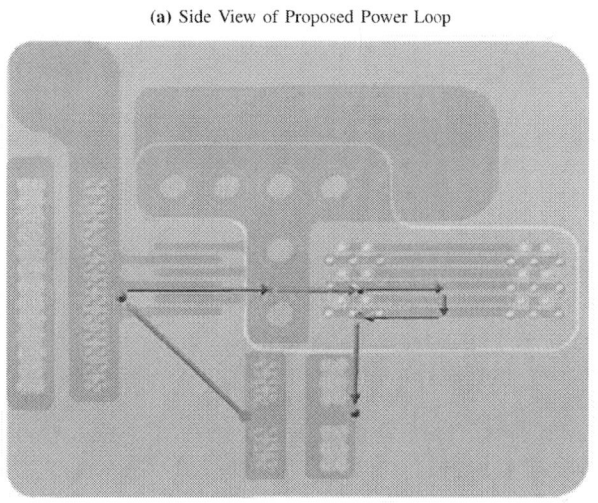

(a) Side View of Proposed Power Loop

(b) Top View of Proposed Power Loop

Fig. 5: ANSYS Q3D Model for Proposed Power Loop

III. Preliminary Hardware Results

The initial prototype for the IMBC is shown in Fig. 6a . The switch node ringing frequency is measured to be 62 MHz seen in Fig. 6b. Given the output capacitance of the EPC device, the measured parasitic inductance is seen as 1.8 nH. This result is a large deviation between the simulated result. Different operating points can led to this variation. This preliminary hardware results shows that the lateral and vertical

flux cancellation is able to achieve relatively low parasitic inductance of less than 2 nH. A 4 phase implementation using each board as a single phase was attempted but there was little canceling at the output of the converter. This design optimized for surface area at the expense of a heat sink. When trying to run multiple phases at the same time, most of the current will flow through the shortest impedance path. This was verified through thermal image where most of the current flow through a single phase rather than any other phase. This justifies an updated hardware design where all the phases are implemented on a single board where the impedance path between each of the phases are equal.

The intial prototype for the SCBC is shown in Fig. 7a. This implementation uses integrated gate driver and switches to minimize the gate loop and power loop distance by having the devices and the gate driver in a single package. The flying capacitor banks C1 and C2 were realized using multilayer ceramic capacitors (MLCCs) given the high power density when compared to other material. An initial size of 0805 case size for the capacitors were chosen as this size allows for high capacitance value with high blocking voltage and low ESL and ESR parameters. Initial testing and observation of the test node in Fig. 7b revealed that the 0805 case size and MLCCs in general have a high negative DC bias effect which results in low capacitance under high DC voltage bias. This leads to a switch node which is not a ideal square wave and a reduced output voltage. In the the updated implementation, high case size capacitors are needed which can handle a high DC bias and more capacitance needs to be paralleled to overcome this effect.

Updated design are designed and tested with the goal to fix the current issues seen in the preliminary converters. A single PCB 4-phase design to address the coupling issues the IMBC with a mountable heat-sink and an improved PCB layout with higher case size MLCCs for the SCBC. These updated hardware results are shown below.

IV. Updated Hardware Results

An updated design was created to fix the issues with the initial design. Fig. 8a is the updated version for the 4-phase IMBC. This version maintains the same power loop across the decoupling capacitors and the devices. Each phase is connected laterally and with each inductor connected to a single node layer where a large number of output capacitors are layout. A heat sink with can be adhered directly to the EPC devices. Fig. 8b shwos the operation of all 4 phase at a full load current of 80 A. Each phase is switching at a 1 MHz frequency. The ringing frequency for each phase is 92 MHz and the extracted power loop is 0.81 nH. This loop is smaller than the one designed with the preliminary design as this updated power loop moves the interleaving vias directly under the pads of the LGA socket for the EPC device. This shorten the distance of the power loop and accounts for the improved design. Despite the same board, the current sharing between each branch is maintain up to 50 A when PWM signals evenly spaced and with equal duration for a 4 phase converter. At 100

979-8-3315-1612-3/25 $31.00 © 2025 IEEE   2578

TABLE I: Optimization of Components in terms of Surface Area

Device	Name	Surface Area	# of Components	Name	Surface Area	# of Components
		48V - 5V Stage			**5V - 1V Stage**	
Transistor	EPC23102	$17.5\ mm^2$	4	EPC2067	$9.26\ mm^2$	4
Inductor	XAL1350	$156\ mm$	1	SLC7649S-101	$50\ mm^2$	4
Level Shifter	IL711-3E	$15.5\ mm^2$	1	-	-	-
Total Stage Area		$241.5\ mm^2$			$237.4\ mm^2$	

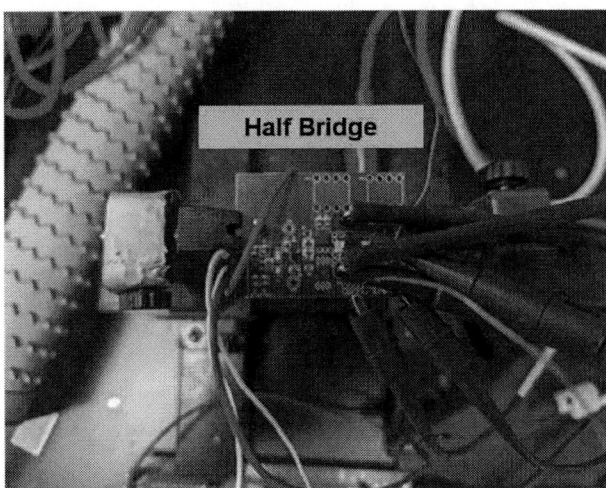

(a) Initial Half Bridge Implementation

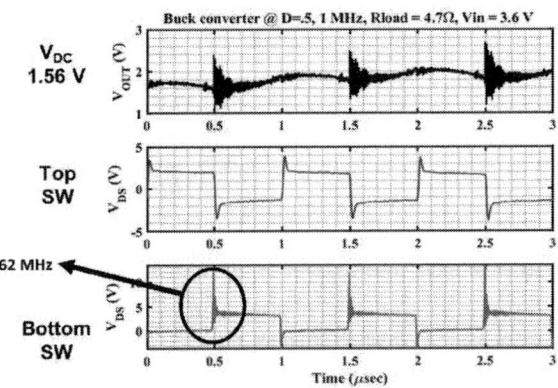

(b) Initial Power Loop Implementation

Fig. 6: Discrete Implementation of GaN Based Half Bridge

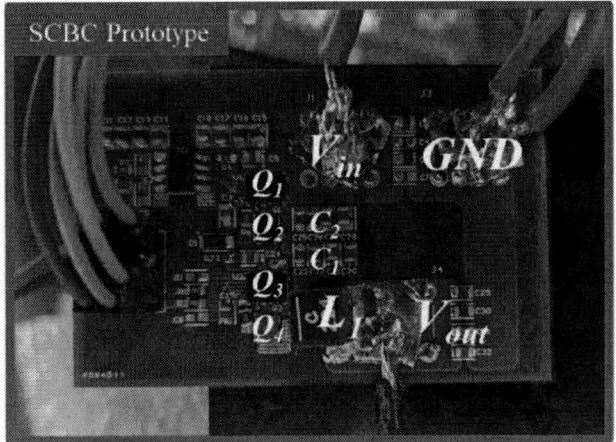

(a) Switch Capacitor Buck Converter Organic PCB implementation

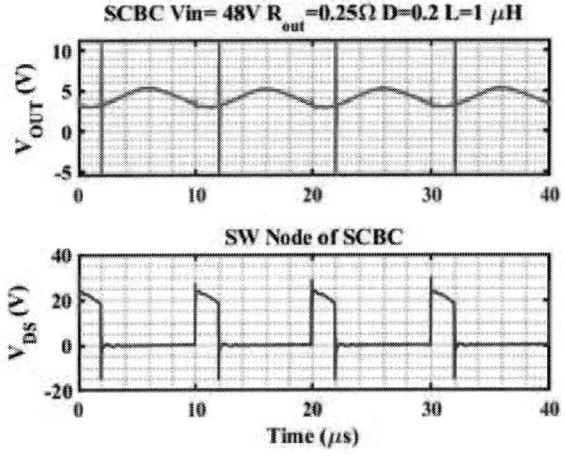

(b) Switch Capacitor Buck Converter at 25 A

Fig. 7: Discrete Implementation of Switch Capacitor Buck Converter

A, The phase which shares the shortest impedance still has the highest temperature of 125 Celsius. The phase farthest away from the source voltage has a temperature of 55 Celsius. This implies that the current sharing between the farthest branch and the closest branch to the load is not equal. Extraction of the parasitic resistance between these two phase are .5 mOhms and 1.7 mOhms respectively. Optimization of the resistance between each branch must also be taken into account in order to operate this converter without the uses of a control scheme. The peak efficiency reached for a 1-phase converter is 71% and 82% for a 4-phase implementation. This change in efficiency was attributed to improvements in dead-time optimization and a reduction in DCR of the proposed inductor in Table I.

Overshoot seen at the switch node of the converter is shown to reach a peak of 8.5 V. The undershoot reaches at peak voltage of -14 V. In efforts to minimize this results, the gate resistance of each phase was increased to 4.7 Ohms. This increase in gate resistance causes an increase in switching losses given the reduce speed of the switching device. The increased temperature of the device also causes an increases in the on-resistance of the EPC device. The reduction in both results in a lower peak efficiency which is seen in Fig. 9b. The proposed model will attempt to address the issues of thermal management and power loop inductance in the next section.

979-8-3315-1612-3/25 $31.00 © 2025 IEEE

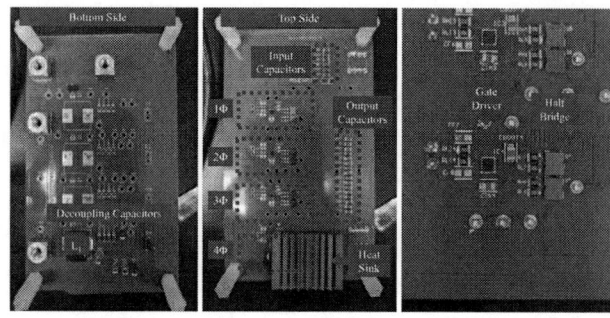

(a) Interleaved 4-phase Organic PCB Implementation

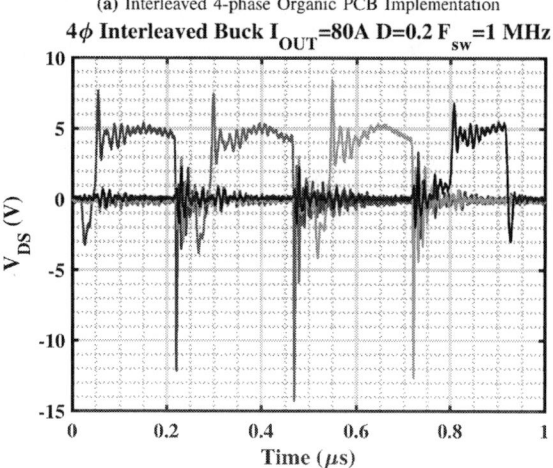

(b) Interleaved 4-phase Buck Converter operating at 80 A

Fig. 8: Discrete Implementation of 5V to 1V 4-phase converter

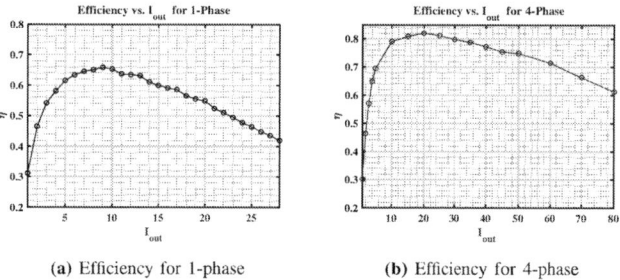

(a) Efficiency for 1-phase   **(b)** Efficiency for 4-phase

Fig. 9: Measured Efficiency for 5V to 1V Stage

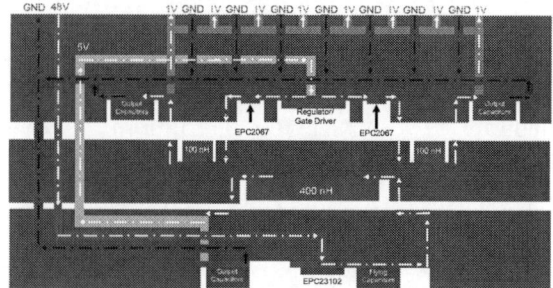

Fig. 10: Active Substrate Implementation

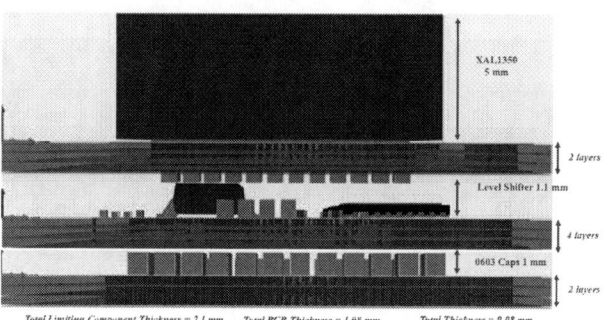

(a) 48V to 5V Layout

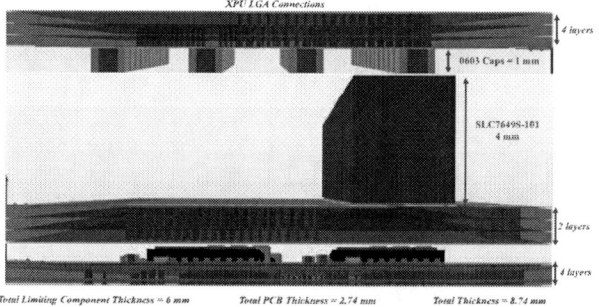

(b) 5V to 1V Layout

(c) Realization of 1-phase converter

Fig. 11: 2.5D Implementation of IMBC and SCBC

V. PROPOSED PACKAGE

Currently, power delivery for loads reaching 1000s of amps is moving from lateral delivery towards vertical power delivery. As a result, the packaging of each converter needs to have a vertical power path in mind. Fig . 10 shows a cross-section view of an active power socket model with the two cascaded converters which were tested in the previous section. Each converter stage will have the devices and the magnetics embedded in the PCB board to minimize the volume and the current losses of the path. The embedded nature of the design will allow for flat connections between each board which allows for easy connections between the board. The board is split into three parts where the embedded active

components will be separated into two boards and all the magnetic components will be contained on a separate board. Fig. 11a shows the placement where each of these components will sit in the 48V to 5V stage in Fig. 11b. The CPU/XPU will sit on top of the socket which will be connected with decoupling capacitors inside a cavity under the XPU. The thermal material between each active component will be FR4 material This package model will give insight on techniques that can be translated towards the system on wafer process for this same system. Fig. 11c shows a realized PCB sample for a single phase buck converter. The surface dimension for the converter is 20mm by 20 mm. Testing of this device is on-going to verify both the parasitic power loop and the overall thermal capabilities.

VI. CONCLUSION AND FUTURE WORK

The preliminary and updated prototypes verify the operation of both converter stages. Currently an embedded module will be implemented by testing the capabilities of a 3D PCB printer such as the Bot-Factory SV 2 Professional. Implementing this design will first be done with standard FR4 material and will be compared with different substrates such as IMS and Glass substrates. Electrical and thermal testing will be done to verify the effectiveness of the different substrates. This glass interposer will be positioned between the dies and the XPU and will act as a thermal "decouple" between the dies thermal load and the XPU thermal load. An insulated metal core substrate (IMS) could also be realized for this part of the module where a parametric study will be done to determine the dimensions of the IMS to optimize thermal resistance.

REFERENCES

[1] Y. Lu, J. Huang, Z. Tong, T. Hu, W.-L. Zeng, M. Huang, X. Mao, and G. Cai, "An overview of hybrid dc-dc converters: From seeds to leaves," *IEEE Open Journal of the Solid-State Circuits Society*, vol. 4, pp. 12–24, jan 2024. [Online]. Available: https://doi.org/10.1109/ojsscs.2023.3334228

[2] "ti.com," https://www.ti.com/lit/an/slva882b/slva882b.pdf, [Accessed 11-07-2024].

[3] A. Lidow, M. de Rooij, J. Strydom, D. Reusch, and J. Glaser, *GaN Transistors for Efficient Power Conversion Third Edition.* WILEY, 2019.

Near-Field Coupling Mitigation of the Noise from High Voltage DC-Link Decoupling Capacitors in Voltage Source Converters

Yuxuan Wu
Electrical and Computer Engineering
Stony Brook University
Stony Brook, NY, US
yuxuan.wu@stonybrook.edu

Kushan Choksi
Electrical and Computer Engineering
Stony Brook University
Stony Brook, NY, US
choksi.kushan@stonybrook.edu

Samuel Defaz
Electrical and Computer Engineering
Stony Brook University
Stony Brook, NY, US
samuel.defaz@stonybrook.edu

Dr. Fang Luo
Electrical and Computer Engineering
Stony Brook University
Stony Brook, NY, US
fang.luo@stonybrook.edu

Abstract—With the adoption of wide-bandgap (WBG) devices, high-voltage DC-link decoupling capacitors in high-density bidirectional converters (HDBCs) are increasingly prone to generating near-field (NF) coupling noise, which can affect surrounding circuits. This NF coupling can negatively impact the stable operation of voltage source converters, leading to issues such as mistriggering of switching devices and electromagnetic interference (EMI) in low-voltage systems. This paper investigates strategies to mitigate NF coupling noise from high-voltage DC-link decoupling capacitors (DCAPs) through a comprehensive workflow combining finite element analysis (FEA) and circuit simulation tools. The study provides design guidelines for optimizing DCAP placement to minimize NF radiation while achieving effective EMI suppression. This research aims to bridge the gap between the application of DCAPs for EMI differential mode (DM) noise control and NF coupling management in highly integrated power converter designs.

Index Terms—Near-field, DC-link design, Electromagnetic interference, Gate drives, Decoupling capacitor

I. INTRODUCTION

The deployment of high-voltage wide-bandgap (WBG) devices, such as Silicon Carbide (SiC) MOSFETs and Gallium Nitride (GaN) HEMTs, has significantly advanced the design of power converters by enabling higher switching frequencies. This improvement reduces thermal stress on switching devices and allows for the use of smaller filtering and cooling components [1]. Aligning with the U.S. Department of Energy's (DOE) goal of achieving a power density of 33 kW/L for a 100 kW traction drive by 2025 [2], manufacturers are transitioning from modular to highly integrated designs. These designs aim to streamline production and assembly processes, placing auxiliary power, control, and sensing circuits in close proximity to high-voltage and high-current power circuits, with minimal creepage and clearance.

While increasing power density, such close integration introduces new challenges, including near-field (NF) coupling effects that degrade system performance. Issues such as mistriggering of gate drives (GDs), electromagnetic interference (EMI), and signal interference stem from the high electric and magnetic fields generated by high-voltage, high-current circuits [3], [4]. Furthermore, as the switching frequencies increase, the intensity of these fields escalates, making mitigation more critical. These challenges can destabilize the system, impair efficiency, and compromise power quality, making NF coupling mitigation essential for stable operation in high-density voltage source converters (VSCs). Therefore, a comprehensive approach to modeling and optimizing these effects is required.

A. Near-Field Coupling in High-Voltage Converters

NF coupling, driven by high dV/dt and di/dt in power loops, introduces electromagnetic (EM) field radiation that impacts nearby GDs and control circuits. This coupling can cause mis-triggering of switching devices, EMI issues in low-voltage systems, and circulating common-mode (CM) currents between power and auxiliary loops, resulting in increased losses and instability. As shown in [5], NF coupling mechanisms can originate from parasitic CM paths in GD circuits or unintended noise propagation through signal traces. These interactions often intensify as power density and switching speeds increase, emphasizing the importance of NF coupling mitigation strategies in advanced designs.

Research has focused on EMI filter design to mitigate NF coupling by optimizing component layouts, magnetic winding configurations, and shielding techniques [6]–[8]. For instance, field modeling with finite element analysis (FEA) tools has been applied to reduce stray inductance in capacitors and

---

979-8-3315-1612-3/25 $31.00 © 2025 IEEE

optimize CM choke designs [9], [10]. However, in high-density converters, high-frequency noise from DC-link decoupling capacitors (DCAPs) requires more refined mitigation strategies. Addressing these challenges with precision can ensure cleaner signal transmission, reduced noise levels, and improved overall converter performance.

B. See-Through Simulation for NF Coupling Analysis

Advanced NF coupling analysis employs FEA tools integrated with circuit simulators for co-simulation of magnetic and electric field distributions. FEA-based modeling, grounded in Maxwell's equations, allows detailed evaluation of parasitic elements and field behavior [9], [11]. Coupling circuit simulations with FEA tools provides accurate excitation data, enabling the design of layouts and GD circuits that minimize NF coupling.

By visualizing near-magnetic field (NMF) distributions, see-through co-simulations help refine power loop layouts and optimize GD placements. These simulations are particularly useful for understanding the interaction of high-frequency fields in high-voltage DC-link systems, where DCAPs and GDs are closely integrated.

C. Organization of the Paper

This paper investigates NF coupling mitigation techniques in two DC-AC converters with power ratings of 25 kW and 200 kW, both operating at 800 V DC-link voltage and designed for low-inductance motor drives requiring high switching frequencies [12]. These converters represent distinct mechanical structures commonly used in HDBCs.

The 25 kW HDBC employs a PCB-based power loop integrating sensors, GDs, and high-voltage circuits (Fig. 1a), while the 200 kW HDBC uses copper busbars for the high-voltage power loop (Fig. 1b) [13], [14].

(a) 25 kW PCB-based. (b) 200 kW copper busbar-based.

Fig. 1: Evaluated high-density bidirectional converters.

This paper employs FEA tools and circuit simulation to address NF coupling challenges in DCAPs design and application. Section II has described the see-through simulation method that is used to identify the EM field distribution that can cause NF coupling. Section III has identified the cause of NF coupling induced by DC-link H-field leakage, and proposed design and optimization methods to reduce impact. Section IV discusses DC-link DCAPs optimization method to reduce NF coupling between phases. Section V focuses on DCAPs placement that is targeted to reduce H-field loop antenna effect NF coupling.

II. FEA MODELING INTEGRATED WITH CIRCUIT SIMULATION FOR NF COUPLING ANALYSIS

The physical geometry of a power converter significantly impacts NF coupling, and the parasitic effects caused by the layout are often difficult to predict with basic calculations. To address this, FEA tools such as Ansys Electronic Desktop (Ansys EDT) are essential for conducting precise NF coupling analyses. The workflow for leveraging integrated FEA and circuit simulation to optimize NF coupling is illustrated in Fig. 2.

For NF distribution analysis, High-Frequency Structure Simulator (HFSS) and Quasi-static 3D Extractor (Q3D) within Ansys EDT are employed. The design file of the converter is imported into EDT, where reference planes and ports for parameter extraction are configured. Ports corresponding to decoupling capacitors (DCAPs) and DC ripple current capacitors are also set up to facilitate comprehensive analysis.

A. Circuit Simulation Incorporating Switching Devices and Component Parasitics

The Nexxim Circuit Simulator within Ansys EDT enables seamless integration of frequency-domain results from HFSS or Q3D for transient simulations. To capture the effects of decoupling components, S-parameters or equivalent RLC models for passive elements, such as capacitors and inductors, are employed. While switching device SPICE models are commonly provided by manufacturers, compatibility issues can arise. For instance, the Wolfspeed SPICE model utilizes the unsupported *ddt* function in Nexxim Circuit [15]. To overcome this, an idealized switching model incorporating parasitics was developed to replace the SPICE models in simulations.

B. EM Field Visualization for Design Optimization

Transient simulation outputs are used as excitation sources for ports in the FEA models, enabling visualization of the resulting EM field distributions. These visualizations guide design refinements, such as adjustments to trace layouts, component orientations, and DCAP placements, to mitigate NF coupling effectively.

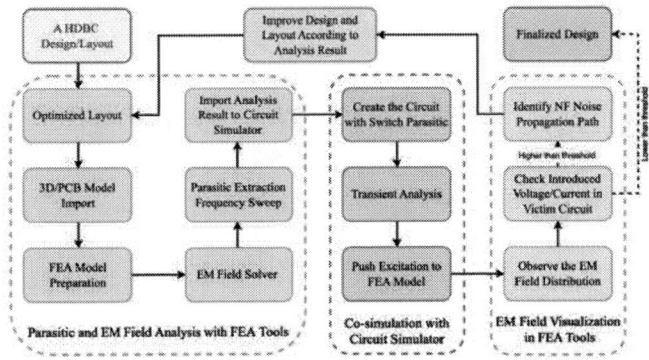

Fig. 2: Workflow for NF coupling optimization using FEA and circuit simulation.

C. Summary

This workflow extends the application of see-through analysis—commonly used in EMI filter design—to NF coupling challenges in critical components such as chokes, inductors, transformers, and capacitors. The integration of EM field analysis for GD layout optimization enables visualization-driven design improvements, helping to avoid high-coupling layouts and enhance converter performance.

III. IMPACT OF DC-LINK DESIGN ON NF COUPLING

High-frequency (HF) DC currents, arising from parasitic capacitance (C_{oss}) and DC-link parasitic inductance (L_{stray}), contribute to significant voltage overshoot and HF DM noise in converters. These currents act as noise sources, inducing NMF coupling to nearby circuitry. Therefore, optimizing the DC busbar design and carefully selecting the values and positioning of DCAPs is crucial for mitigating NF coupling effects.

As demonstrated in [16], [17], laminated DC-link busbars effectively reduce L_{stray} in the commutation loop, minimizing voltage overshoot and suppressing ringing amplitudes during transients. Additionally, laminated designs reduce current loop parasitics, DM noise, and CM noise by incorporating a grounding layer between the DC-link layers [18].

Beyond managing parasitics, practical considerations for busbar design include insulation, thermal performance, mechanical structure, and current-carrying capacity. For example, copper and aluminum busbar thicknesses must align with material constraints. In [19], three laminated busbar designs for a 200 kW system were evaluated. To accommodate gate driver positioning, a DC-link busbar surrounding the gate drivers was implemented. Due to the limited current capacity of PCB busbars, the design used 5 mm aluminum and 3 mm solid copper layers to meet the 200 A current requirement.

The 200 kW HDBC, equipped with a partially laminated ring-shaped DC busbar, showed no mistriggering during DPTs for switching validation (Fig. 3a). However, during continuous 3-phase operation (Fig. 3b), mistriggering occurred due to HF noise coupling from the nearby phase leg. This behavior, identified as HF noise-induced NF coupling, was mitigated by introducing a copper shield to cover the GD board, which effectively reduced mistriggering.

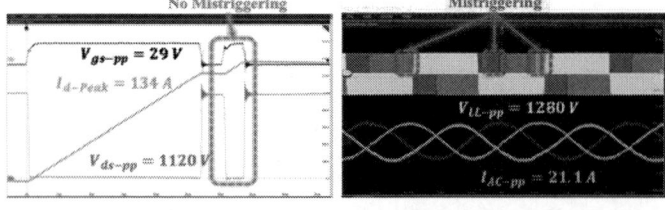

(a) DPT result (1 kV input) without mistriggering.

(b) 3-phase test (600 V input) with mistriggering.

Fig. 3: 200 kW HDBC testing results with partially laminated ring-shaped busbar.

A. Diagnosing Mistriggering and Noise Sources

To analyze GD mistriggering, differential probes measured voltages at the GS terminals (V_{gs}), GDIC output ($V_{gd-ICout}$), isolated DC-DC converter terminals (V_{DC-Pri} and V_{DC-Iso}), fiber optic output (V_{fiber}), and GDIC signal input ($V_{gd-ICin}$) for the affected phase. Although DC-link DCAPs were mounted atop the busbar to minimize L_{stray}, ripple currents on the partially laminated ring-shaped busbar (Fig. 4a) caused phase mistriggering in nearby GDs.

The overshoot voltage ringing frequency at the drain-source terminals during switching turn-off was measured at 28 MHz. Table I shows the ringing amplitudes at 28 MHz across various GD locations. While the V_{gs} threshold (4 V) and measured bias (-5 V) indicated that induced voltages (2.3 V) were insufficient for direct mistriggering [14], $V_{gd-ICin}$ entered an unstable region at 1.5 V due to its lower threshold (1 V for high, 2 V for low) [20], contributing to GD mistriggering.

TABLE I: Ringing Amplitude at Different GD Positions.

Terminal	Amp@28MHz Ring-shaped (V)	Amp@28MHz Fully Laminated (V)
V_{gs}	2.3	1.2
$V_{gd-ICout}$	0.3	0.2
V_{DC-Pri}	0.9	0.3
V_{DC-Iso}	1.2	0.8
V_{fiber}	0.3	0.2
$V_{gd-ICin}$	1.5	0.3

FEA tools were employed to investigate the induced voltage at GDs caused by NMF coupling. The H-field distribution analysis (Fig. 4a) revealed that flux leakage from the partially laminated busbar increased H-field intensity near GD1 by 1.8 times compared to GD2.

B. Fully Laminated Busbar Design for Mitigation

A fully laminated busbar design was introduced to mitigate NF coupling effects. This design replaced the partially laminated ring-shaped busbar, positioning GDs atop the fully laminated structure. The laminated copper layers generated eddy currents that suppressed magnetic field radiation from power loops (Fig. 4b). As shown in Table I, the fully laminated design significantly reduced HF ringing voltages across GD circuits.

C. Additional Considerations for Laminated PCB Busbars

Laminated PCB busbars feature multi-layer designs, providing larger surface areas for HF noise management and minimizing loop inductance. Thin insulation layers between laminates further reduce parasitic inductance while increasing parasitic capacitance on the DC link. However, laminated PCB busbars are generally limited to low-power applications due to constrained copper thickness, which balances cost and performance [19].

IV. DC-LINK DECOUPLING CAPACITORS OPTIMIZATION

As discussed in [19], optimizing DC link performance requires a strategic combination of DC link capacitors. The

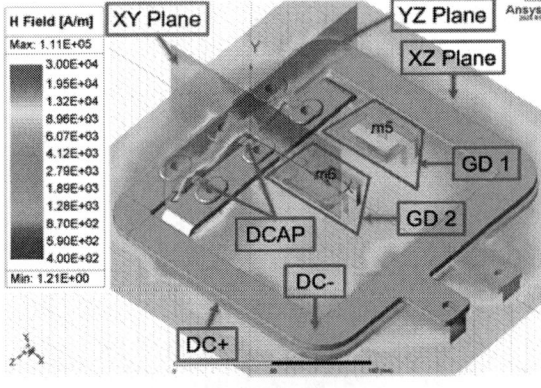

(a) Partially laminated ring-shaped busbar.

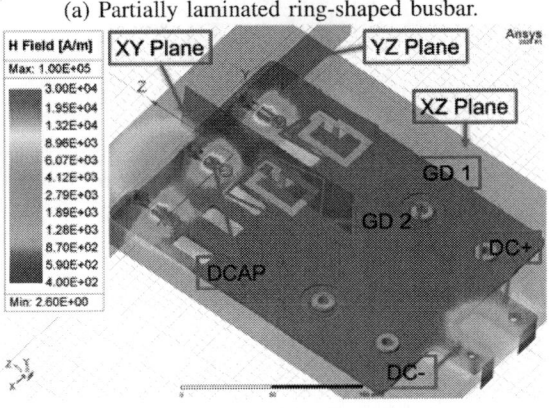

(b) Fully laminated busbar.

Fig. 4: H-field distribution for DC-link designs at resonance frequency.

DC link typically includes DC-link ripple current filter capacitors. DCAPs should be placed as close to the switching device as possible to mitigate the loop inductance. Multi-layer ceramic capacitors (MLCCs) with low equivalent series resistance (ESR) and low equivalent series inductance (ESL) are commonly used as DCAPs to mitigate HF noise. Film capacitors are employed to handle switching current ripples.

A. Impact of Decoupling Path and Path Impedance on NF Coupling

In [21], Adamczyk demonstrated that adding multiple capacitor values as DCAPs does not always reduce the impedance of the decoupling path. To use DCAPs effectively, the parasitic inductance and capacitance of switching devices and power loops need to be considered, and the DCAPs must have a resonance frequency (f_{res}) higher than the transient response.

To further validate the frequency inducing NF coupling, DPT tests were performed with a fully laminated busbar. As discussed in [22], when the number and package of DCAPs are constant, using DCAP combinations aligned with the power loop's resonance frequency can suppress voltage overshoot in the converter. Four different DCAP configurations were tested for performance validation (Fig. 5):

(a) $3 \times 3 \times 3.3\,\mathrm{nF}$ in 1825 package for three phases on a single board.

(b) $10 \times 82\,\mathrm{nF} + 10 \times 4.7\,\mathrm{nF}$ in 2220 package for three phases on a single board.

(c) $3 \times 3.3\,\mathrm{nF}$ in 1825 package for a single phase.

(d) $2 \times 56\,\mathrm{nF}$ in 2220 + $3 \times 3.3\,\mathrm{nF}$ in 1825 package for a single phase.

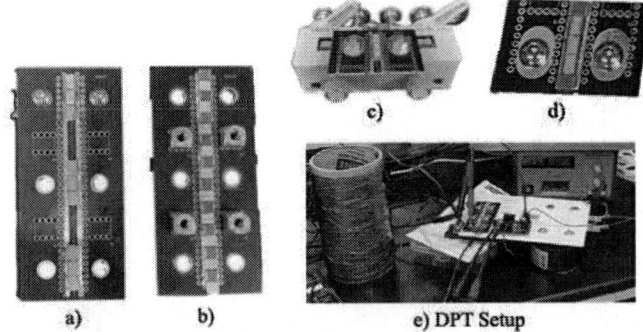

Fig. 5: (a) to (d): four different DCAP configurations for validation; (c): DCAP with fixture for impedance measurement; (e): DPT setup for DCAP performance validation.

The validation platform used a 200 kW power stage, and the DPT setup is presented in Fig. 5(e). Devices with integrated diodes exhibit higher C_{oss}, leading to a lower resonance frequency than modules with body diodes only. The f_{res} of the system was approximately 30 MHz, and the DCAPs were designed to mitigate overshoot ringing at this frequency. The impedance of the DCAPs at various frequencies for different configurations is shown in Fig. 6a.

During switching transients, the DC power loop experiences large current surges. The resulting H-field generated by high di/dt during device turn-on/turn-off can couple noise into the gate drivers. Fig. 4b illustrates the two primary H-field radiation points: (a) the power module's DC+ and DC- terminals and (b) the DCAPs.

As shown in Fig. 6a, configurations (a) and (c), where DCAPs for three phases are mounted on the same board, exhibit multiple resonance points. This causes performance inconsistency; for example, phase B performs better than phases A and C due to its symmetrical current path. Moreover, the shared DCAP board provides a low-impedance path between phases, allowing HF noise to circulate between them. Independent DCAP configurations for each phase present higher impedance to inter-phase HF noise coupling, reducing interference with nearby gate drivers.

By comparing the DPT results in Fig. 6, configurations (a) and (c), with $f_{res-High}$ matching the converter, reduce HF noise amplitude at 32 MHz, achieving modest reductions in V_{gs} ringing by 1.2 dB. However, at lower frequencies (6 MHz and 10 MHz), the induced voltage reduction is more pronounced. Configuration (d) demonstrated superior performance, achieving a 4.3 dB reduction in V_{gs} ringing at $\sim 6\,\mathrm{MHz}$.

B. Summary and Recommendations

NF coupling between the commutation loop and the GS loop is most significant in the 1–10 MHz frequency range. To

979-8-3315-1612-3/25 $31.00 © 2025 IEEE   2585

minimize this coupling:

- Use DCAPs with low ESL and small capacitance values, providing higher resonance frequencies.
- Introduce large-value DCAPs in similar packages to suppress low-frequency ringing.
- Separate DCAP boards/position for individual phases to break low-impedance paths for HF noise.

This comprehensive approach effectively reduces NF coupling from the power loop to the gate drivers, enhancing system reliability and noise immunity.

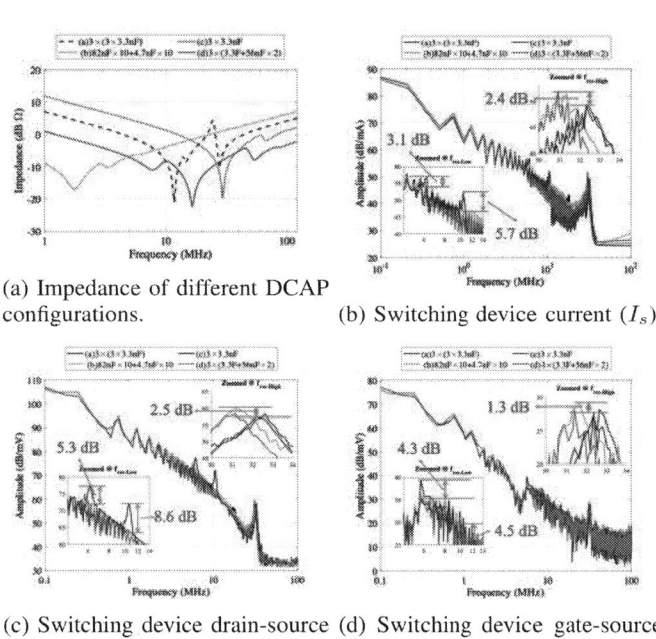

(a) Impedance of different DCAP configurations.

(b) Switching device current (I_s).

(c) Switching device drain-source voltage (V_{ds}).

(d) Switching device gate-source voltage (V_{gs}).

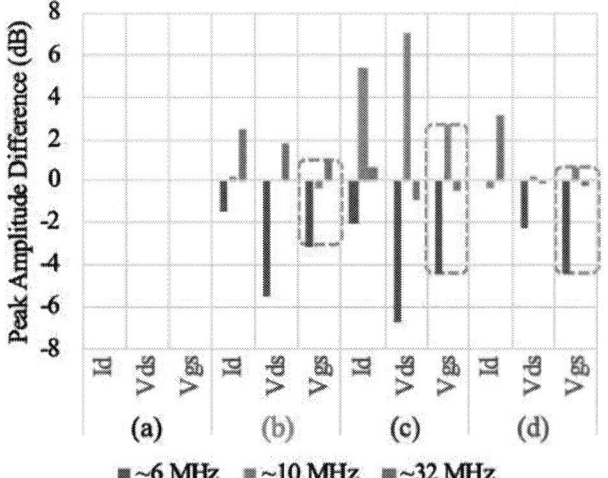

(e) Peak amplitude difference in the FFT result at three different frequencies. *Note: noise peak amplitude difference calculated with DCAPs (a) as baseline.*

Fig. 6: Frequency domain analysis for various DCAP configurations.

V. DECOUPLING CAPACITORS PLACEMENT

In [23], Adamczyk and Teune investigated the impact of varying DCAP placement relative to the noise source. Their results demonstrated that for a PCB with plane pairs spaced 30 mils (0.76 mm) apart, increasing the distance of DCAPs from the noise source from 1 inch to 2 inches resulted in a 17.5 dB reduction in impedance at 50 MHz. Therefore, placing the DCAPs as close to the noise soure is most effective. However, this section have identified that placing the DCAPs on different side of the PCB will also impact the NF coupling in HDBCs.

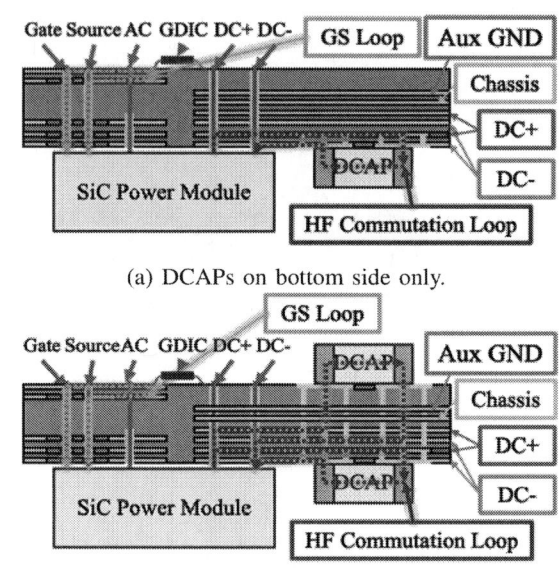

(a) DCAPs on bottom side only.

(b) DCAPs on top and bottom sides.

Fig. 7: Diagram illustrating varying DCAP positions.

A. H-Field and E-Field Effects from HF Loops

HF commutation loops create H-field antenna effects, as illustrated in Fig. 7. The geometry of these loops significantly influences HF radiation. For instance, Fig. 7a demonstrates that when DCAPs are placed only on the bottom layer, HF noise primarily propagates through the bottom two layers. The copper layers above act as effective shields, reducing NMF coupling between the GS loop and the HF commutation loop.

Conversely, Fig. 7b shows that when DCAPs are placed on both the top and bottom sides, they not only enhance H-field radiation but also enable HF currents to travel along the top surface of the PCB. Since the top surface lacks shielding copper layers, it allows for unimpeded field radiation.

Fig. 8a & 8b further illustrates that positioning DCAPs on both sides of the PCB results in strong H-field radiation across all DC link layers, especially on the top surface. In contrast, DCAPs placed solely on the bottom layer benefit from the upper copper layers shielding HF radiation. However, as presneted in Fig. 8c & 8d,for the DCAPs on the bottom surface, the H-field intensity is relative lower. That is because the power loop copper layers are on the bottom of the PCB, and reduce the H-field intensity.

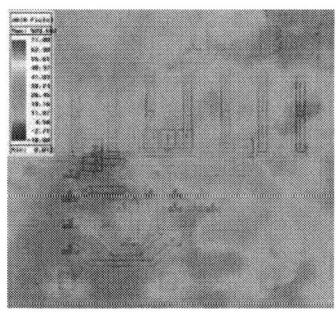 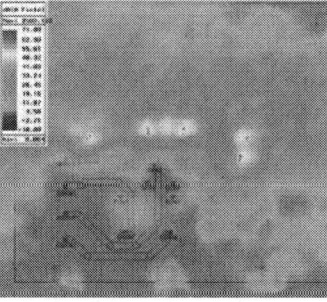

(a) Top field with DCAPs on bottom side only.

(b) Top field with DCAPs on top and bottom sides.

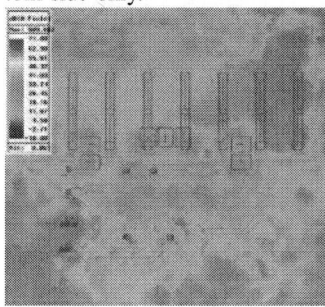 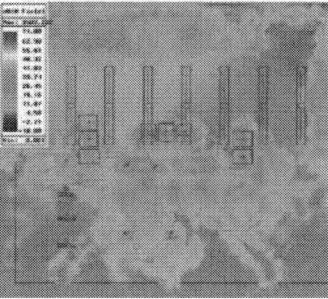

(c) Bottom field with DCAPs on bottom side only.

(d) Bottom field with DCAPs on top and bottom sides.

Fig. 8: H-field distribution for 25 kW HDBC with varying DCAP positions, analyzed using see-through simulation.

B. Recommendations for GD Circuit and DCAP Placement

The placement of GD circuits and DCAPs must avoid being on the same PCB side to minimize NF coupling. DCAPs are inherently noise sources, and shielding them is challenging due to the high voltage potential on their terminals. Positioning DCAPs on the opposite side of the PCB from GDs allows intervening copper layers or traces to act as effective shields, significantly reducing HF coupling from the DC link to the GD circuits.

VI. Conclusion

This paper presented a comprehensive analysis of near-field (NF) coupling effects in high-density voltage source converters (VSCs), focusing on the mitigation of noise from high-voltage DC-link decoupling capacitors (DCAPs). Through detailed modeling, field simulations, and layout optimization, the study highlighted the critical role of high switching frequencies and compact layouts in exacerbating NF coupling challenges. The integration of finite element analysis (FEA) tools and circuit simulators enabled accurate visualization of electromagnetic fields, facilitating the identification of design vulnerabilities and the development of robust solutions.

Key strategies were proposed to minimize NF coupling, including the optimization of DC-link layout and design, decoupling path and path impedance optimization, and DCAPs layouts. Hardware validation demonstrated the effectiveness of these methods in reducing EMI, mis-triggering, and system

instability. These findings underscore the importance of co-simulation and holistic design approaches in achieving reliable, efficient, and high-density VSCs, particularly as power electronics move towards increasingly compact and integrated designs.

References

[1] Y. Li, M. ul Hassan, A. B. Mirza, Y. Xie, S. Deng, S. S. Vala, F. Luo, X. Feng, S. Narumanchi, and J. D. Flicker, "State-of-the-art medium- and high-voltage silicon carbide power modules, challenges and mitigation techniques: A review," *IEEE Transactions on Components, Packaging and Manufacturing Technology*, pp. 1–1, 2024.

[2] S. Chowdhury, E. Gurpinar, G.-J. Su, T. Raminosoa, T. A. Burress, and B. Ozpineci, "Enabling technologies for compact integrated electric drives for automotive traction applications," in *2019 IEEE Transportation Electrification Conference and Expo (ITEC)*, 2019, pp. 1–8.

[3] F. Blaabjerg, H. Wang, I. Vernica, B. Liu, and P. Davari, "Reliability of power electronic systems for ev/hev applications," *Proceedings of the IEEE*, vol. 109, no. 6, pp. 1060–1076, 2021.

[4] S. Ozdemir, N. Altin, A. Nasiri, and R. Cuzner, "Review of standards on insulation coordination for medium voltage power converters," *IEEE Open Journal of Power Electronics*, vol. 2, pp. 236–249, 2021.

[5] H. Song and D. Boroyevich, "Investigation on noise caused by gate driver ic and near field coupling within gate driver pcbs for medium voltage sic-based converters," in *2024 IEEE Applied Power Electronics Conference and Exposition (APEC)*, 2024, pp. 1465–1471.

[6] Y. Li, S. Wang, H. Sheng, and S. Lakshmikanthan, "Reduction and cancellation techniques for the near field capacitive coupling and parasitic capacitance of inductors," in *2018 IEEE Symposium on Electromagnetic Compatibility, Signal Integrity and Power Integrity (EMC, SI & PI)*, 2018, pp. 432–437.

[7] Y. Lai, S. Wang, and B. Zhang, "Investigation of magnetic field immunity and near magnetic field reduction for the inductors in high power density design," *IEEE Transactions on Power Electronics*, vol. 34, no. 6, pp. 5340–5351, 2019.

[8] K. Takahashi, Y. Murata, Y. Tsubaki, T. Fujiwara, H. Maniwa, and N. Uehara, "Mechanism of near-field coupling between noise source and emi filter in power electronic converter and its required shielding," *IEEE Transactions on Electromagnetic Compatibility*, vol. 61, no. 5, pp. 1663–1672, 2019.

[9] I. F. Kovačević, T. Friedli, A. M. Müsing, and J. W. Kolar, "3-d electromagnetic modeling of parasitics and mutual coupling in emi filters," *IEEE Transactions on Power Electronics*, vol. 29, no. 1, pp. 135–149, 2014.

[10] P. Wang, B. N. Sanusi, C. Liu, J. Huang, T.-G. Zsurzsan, M. A. E. Andersen, and Z. Ouyang, "Utilizing mutual coupling to eliminate parasitic inductance in emi filter capacitors for noise reduction," *IEEE Transactions on Industry Applications*, vol. 60, no. 5, pp. 7206–7215, 2024.

[11] B. Touré, J.-L. Schanen, L. Gerbaud, T. Meynard, J. Roudet, and R. Ruelland, "Emc modeling of drives for aircraft applications: Modeling process, emi filter optimization, and technological choice," *IEEE Transactions on Power Electronics*, vol. 28, no. 3, pp. 1145–1156, 2013.

[12] H.-g. Choi, J.-I. Ha, and K. Lee, "Ultra-low inductance pmsm drive using a dc current sensor," *IEEE Transactions on Industrial Electronics*, vol. 71, no. 6, pp. 5597–5607, 2024.

[13] Wolfspeed, Inc., "CCB021M12FM3 Power Module Datasheet," Wolfspeed official datasheet, 2023, accessed: 2024-10-31. [Online]. Available: https://assets.wolfspeed.com

[14] Sansha Electric Manufacturing Co., Ltd., "FCA150AC120 Power Module Datasheet," Sansha Electric official datasheet, 2024, accessed: 2024-10-31. [Online]. Available: https://www.sansha.co.jp/eng/

[15] ANSYS, Inc., *ANSYS Electronics Desktop (EDT) Circuit Manual*, ANSYS, Inc., Canonsburg, PA, USA, 2024, version 2024 R1. [Online]. Available: https://ansyshelp.ansys.com

[16] C. Chen, X. Pei, Y. Chen, and Y. Kang, "Investigation, evaluation, and optimization of stray inductance in laminated busbar," *IEEE Transactions on Power Electronics*, vol. 29, no. 7, pp. 3679–3693, 2014.

979-8-3315-1612-3/25 $31.00 © 2025 IEEE

[17] A. D. Callegaro, J. Guo, M. Eull, B. Danen, J. Gibson, M. Preindl, B. Bilgin, and A. Emadi, "Bus bar design for high-power inverters," *IEEE Transactions on Power Electronics*, vol. 33, no. 3, pp. 2354–2367, 2018.

[18] D. O. Thomas, M. Sylvain, G. Jean-Michel, J.-L. Schanen, and A. Perregaux, "Reduction of conducted emc using busbar stray elements," in *2009 Twenty-Fourth Annual IEEE Applied Power Electronics Conference and Exposition*, 2009, pp. 2028–2033.

[19] Y. Wu, M. Ul-Hassan, and F. Luo, "Busbar design and optimization for high power three-phase inverter with wbg device," in *2022 IEEE 9th Workshop on Wide Bandgap Power Devices & Applications (WiPDA)*, 2022, pp. 132–137.

[20] Analog Devices, Inc., "ADuM4146 Isolated Gate Driver Datasheet," Analog Devices official datasheet, 2024, accessed: 2024-10-31. [Online]. Available: https://www.analog.com/media/en/technical-documentation/data-sheets/adum4146.pdf

[21] B. Adamczyk and J. Teune, "Impact of decoupling capacitors and embedded capacitance on impedance of power and ground planes: Part ii," *In Compliance Magazine*, vol. April 2020, 2020.

[22] Z. Chen, D. Boroyevich, P. Mattavelli, and K. Ngo, "A frequency-domain study on the effect of dc-link decoupling capacitors," in *2013 IEEE Energy Conversion Congress and Exposition*, 2013, pp. 1886–1893.

[23] B. Adamczyk and J. Teune, "Impact of decoupling capacitors and embedded capacitance on impedance of power and ground planes: Part i," *In Compliance Magazine*, vol. 16, no. 3, pp. 32–38, March 2020.

Advantages of Paralleling SiC MOSFETS in High-Performance Power Modules

Steffen Beushausen, Dominik Alexander Ruoff, Wenqi Zhou and Karl Oberdieck

Mobility Electronics
Robert Bosch GmbH
Reutlingen, Germany
Steffen.Beushausen@de.bosch.com

Abstract—Multi-chip power modules with bond-wire contacts at the source area of the die are sensitive to self-excited high-frequency chip-to-chip oscillations (SE-oscillations) between parallelized power semiconductor devices. The magnitude of occuring SE-oscillations can be damped with an internal gate resistance on chip level in the gate path of the semiconductor (e.g. Si/SiC MOSFET or Si IGBT). This additional internal gate resistance needs to be carefully balanced with limitations in the application, e.g. an unwanted increase of switching losses or the use of Miller clampers to stay in the gate safe-operating area of the semiconductor, or to avoid a parasitic turn-on. In this work, we analyze the performance of a multi-chip power module design with advanced assembly and interconnection technology (AIT) in regards to its susceptibility to SE-oscillations. A direct comparison to power modules with bond-wire source contacts shows a minimal necessary internal gate resistance to mitigate these oscillations. The lower internal gate resistance extends a high flexibility to the application engineer, who can select from a broader range of external gate resistors or utilize a current-source gate driver, matching the needs of the application. Furthermore, shifting the bulk gate resistance to outside of the semiconductor allows for a very stiff Miller clamping, guaranteeing a safe operation. From another perspective, our design is able to fully utilize the performance of all known SiC technologies today without suffering from SE-oscillations. This allows for easy integration of current and future high-performance SiC MOSFETs in this robust power module design.

I. INTRODUCTION

To increase the output power of semiconductor modules, multiple semiconductor dies (e.g. IGBTs or MOSFETs) are assembled inside a package in parallel for one logical switch. For today's generation power modules using bond-wires is a defacto assembly and interconnection technology (AIT) standard for the top bare-die attach. However, previous work shows that self-excited high-frequency oscillations (SE-oscillations) inbetween individual dies can occur in such power modules[1]–[7]. This effect is due to the amplifying nature of power semiconductors

and can lead to catastrophic failure. Without sufficient damping, there is a risk in every operating point of the power module that these oscillations spontaneously appear during transients. In order to increase damping and to avoid SE-oscillations, a common solution is to add an additional internal resistance into the semiconductors gate path on chip level. Unfortunately, the required internal gate resistance to avoid these type of oscillations depends on the combination of electrical parameters from the semiconductor and the package parasitics and must be balanced individually for each combination.

From an application point of view, this additional internal gate resistance reduces the degree of controllability via the gate terminal of the semiconductor. During active switching the internal gate resistance has additional tolerances and a temperature dependence which needs to be considered. Furthermore, an additional voltage drop on the internal gate resistance lowers the possible gate current range if a current source gate driver is used. Additionally, a parasitic turn-on (PTO[1]) of the device usually needs to be avoided while acting as a commutation partner. To ensure this, a separate low-resistive path on the gate driver board is created, which is called a Miller clamper. However, an additional internal gate resistance increases the minimum impedance in the Miller clampers path, and hence, lowers its effectiveness. This leads to an increase in PTO risk for a bridge-leg configuration with a hard-switched inductive load.

In the following sections, the trend in power SiC MOSFETs is discussed and a developed high-performance advanced AIT power semiconductor module for next-gen SiC MOSFETs is presented. To highlight its advantages, the developed power mod-

---

[1]Increase of the gate-source voltage of a power semiconductor over its respective threshold voltage.

979-8-3315-1612-3/25 $31.00 © 2025 IEEE

ule is compared to a standard AIT bond-wire power module. In this comparison, the focus lies on SE-oscillations as a key phenomenon limiting usability of power semiconductors in multi-chip packaging. It is shown that the high-performance package has major advantages compared to standard AIT bond-wire packages regarding choice of used power semiconductors.

II. NEXT-GEN SiC-MOSFETs

Due to a wider bandgap, higher thermal conductivity and superior electrical characteristics, SiC MOSFETs enable high-efficient, and more compact high-performance power electronic systems[8]. Figure 1 shows the small-signal electrical equivalent circuit (EEC) of a power MOSFET. In the following, it is briefly discussed which element of the EEC is optimized to achieve the design goals regarding a SiC MOSFET optimization.

The transconductance g_m, corresponding to the controlled current source in Fig. 1, is defined as (1).

$$g_\mathrm{m} = \frac{\partial I_\mathrm{D}}{\partial U_\mathrm{GS}} \quad (1)$$

Figure 2 shows the transfer characteristics of two generic MOSFETs. From the curves, it is evident that a higher g_m leads to a lower $R_\mathrm{DS,on}$. Thus, a high g_m is one optimization target to reduce conduction loss and increase MOSFET efficiency. A robust short-circuit performance requires a low MOSFET peak short-circuit current, which translates into a reduction of the drain-induced barrier lowering (DIBL) effect. Hence, the newer generations of SiC MOSFETs need to be designed with a reduced output conductance

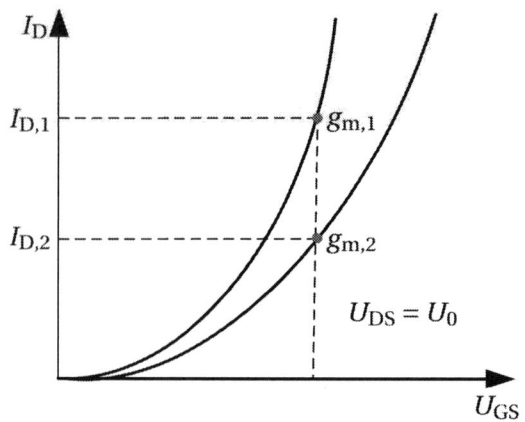

Fig. 2: Transfer curve.

g_o in saturation to achieve this target. Figure 3 shows the output characteristics of Bosch generation 1 and 2 SiC MOSFETs. The new generation features a reduced g_o in saturation, defined as (2).

$$g_\mathrm{o} = \frac{\partial I_\mathrm{D}}{\partial U_\mathrm{DS}} \quad (2)$$

This helps to significantly reduce the peak current during short-circuit events.

SiC technologies are currently in rapid development to comply with the need of automotive applications, as depicted in Fig. 4 for Bosch SiC MOSFET generations. To achieve the best performance in various applications, both conduction and switching losses are reduced in SiC MOSFETs[8]. Each new SiC MOSFET generation improves in performance, which is mainly

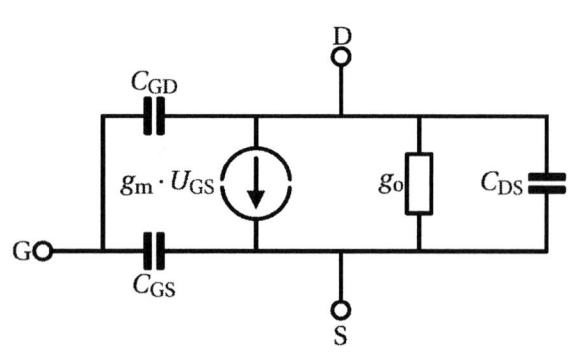

Fig. 1: Small signal electrical equivalent circuit of a power MOSFET.

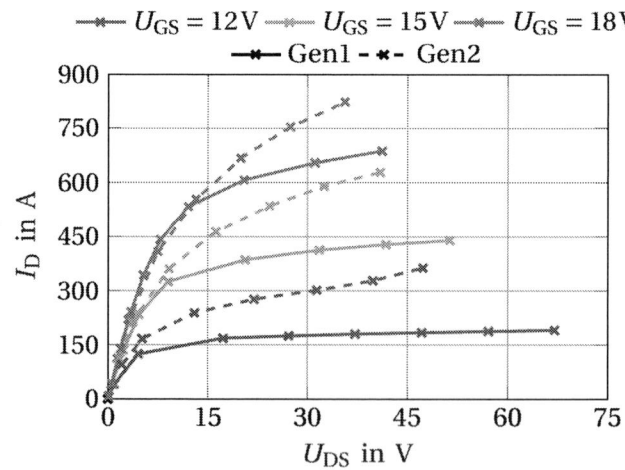

Fig. 3: Output characteristics Gen 1 and Gen 2 SiC MOSFETs.

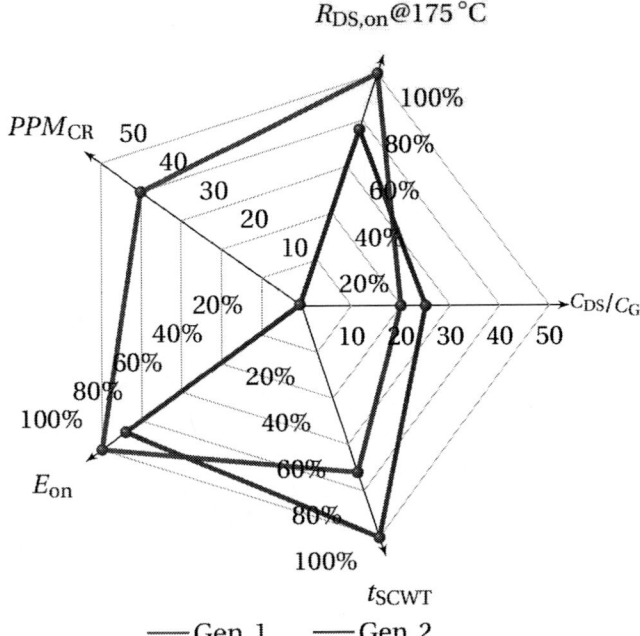

Fig. 4: SiC Technology for 1200 V.

$$—— Gen 1 \quad —— Gen 2$$

quantified in $R_{DS,on}@175\,°C$. For example the $R_{DS,on}$ of Gen 1 to Gen 2 was improved by over 25 %.

Beyond $R_{DS,on}@175\,°C$, further parameters are of significant interest. Four further parameters are briefly discussed in the following. A higher robustness for cosmic radiation enables lower failure rates. This was improved from Gen 1 to Gen 2 over a factor of 40[9]. Furthermore, the short-circuit robustness is an important parameter within a traction inverter. Even though a fast detection is possible, longer short-circuit times are more convenient from an application point of view[10]. This is why the short-circuit robustness increases over the generations. Besides the conduction losses, the switching losses are a main loss contributor which need to be reduced. For example the softness of the body diode at high temperature was significantly improved in Gen 2 compared to Gen 1[8]. This results in a reduction of overall turn-on losses (E_{on}), as gate resistances are chosen for the worst-case operating point. Finally, a trend in the capacitance ratio C_{DS}/C_{GD} is expected. The Miller capacitance C_{GD} is a key parameter in MOSFET switching behavior. To reduce the switching losses and also to prevent PTO, a smaller C_{GD} is beneficial. The total $R_{DS,on}$ can be further reduced with an optmized drift zone, which in a trade-off results in a higher drain-source capacitance C_{DS}. In general it can be said that over the generations a slight trend to higher values C_{DS}/C_{GD} is observed.

III. HIGH-PERFORMANCE POWER MODULE

A standard bond wire top bare-die attach for SiC MOSFETs is depicted in Fig. 5a. The individual dies in a parallel connection are sintered onto an active-metal brazed substrate (AMB) with their drain contact located on the bottom of the bare-die. Thus, all the semiconductors are contacted with a wide surface area on their drain side, resulting in a very low contact resistance and parasitic inductance between the paralleled dies. Contrary to the connection on the bottom of the bare dies, the source potential on the top of the bare-dies is contacted with individual power bond-wires towards the AMB. The electrical loop between the paralleled bare dies now consists of the AMB copper on the drain side, and bond-wires as well as AMB copper on the source side. Hence, the AIT resistance and parasitic inductance between the paralleled dies source potential depends significantly on the diameter, loop height, and amount of bond-wires contacting the top bare-die towards the AMB. Furthermore, static and dynamic current sharing between paralleled dies is difficult to achieve with standard AIT bond-wire modules. This is due to the lack of freedom in load current path design, as the bond wires only are able to transport a sufficient load current for a short length. The main current path is always located on the AMB in a 2D routing approach.

For the proposed high-performance power module the die attach is depicted schematically in Fig. 5b. Contrary to the standard AIT bond-wire module it

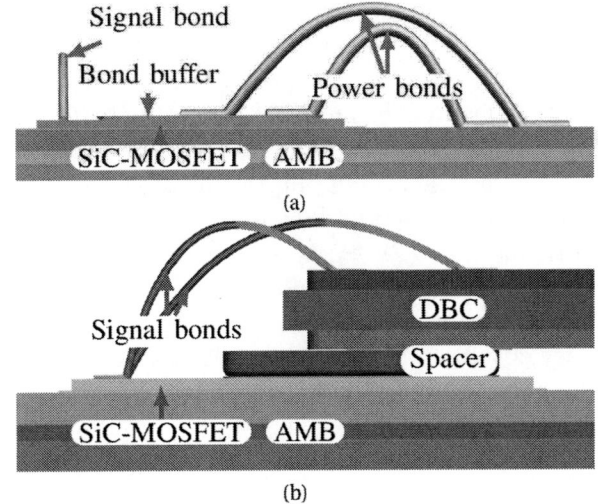

Fig. 5: Schematic of a power semiconductor module with (a) a standard bond wire top bare-die attach and (b) a sinter top bare-die attach.

Fig. 6: Bosch Gen 6 power module.

consists of a two sided sinter attach for the bare-die. This means that in addition to the bottom of the bare-die being sintered onto an AMB, the top of the bare-die is connected via a sinter process to a second substrate. Hence, in case of a SiC MOSFET the low-ohmic and low-inductive connection already established on the drain side of the semiconductor can now also be established on the source side of the bare-die.

The Bosch 6. generation power module (PM6), depicted in Fig. 6, is using this technology to provide a compact design and unique power density with a module dimension of 44.7 mm x 62.3 mm x 4.9 mm. A flexible scaling cabability of 4, 8 or 12 dies per module provides a wide RMS current range from 210 A to 600 A at up to 920 V DC-link voltage within one package. With appropriate structuring of the top-side substrate a very low module commutation inductance, AIT resistance, and parasitic inductance between individual paralleled dies is achieved. The basic AIT advantages of the PM6 are compared to a bond-wire module in table I. Furthermore, the PM6 is designed with a very high symmetry for both static and dynamic current sharing between paralleled semiconductors, as the second substrate for top bare-die attach introduces a new degree of freedom for current flow control.

The aforementioned improvements are also observed in double-side cooled modules using similar technology[11], [12]. Although, while the PM6 offers electrical performance and compact design advantages, there is no need for a complex double-side cooler design on system level. This provides cost benefits, particularly for applications with low to medium

TABLE I: Power module parasitics comparison.

Power Module	L_m	R_m	L_{SS}
Bond-wire module	6 nH	520 μΩ	6.2 nH
Bosch Gen6 power module	4 nH	280 μΩ	2.5 nH

phase current requirements.

IV. SELF-EXCITED OSCILLATIONS WITH NEXT-GEN SiC-MOSFETs

Package parasitics together with the electrical parameters of the power semiconductor form a *LCR* oscillation circuit for multi-chip power modules, see Figs. 1 and 7. Subsequently, this can lead to SE-oscillations due to the amplifying nature of power semiconductors, as long as the resulting *LCR* oscillation circuit lacks sufficient damping[2]–[4], [6], [13]–[16]. In general, it is well investigated in literature which parameters of the *LCR* circuit influence the risk towards SE-oscillations. Table II summarizes the impact of the *LCR* circuit parameters on both the SE-oscillations and the system loss[6] The relation between the parasitic inductances of the power module schematic in Fig. 7 and the relevant inductances in the table are given by (3) to (5).

$$L_{DD} = L_{D,i} + L_{D,j}, \qquad i,j = 1,2,\ldots,n; i \neq j \quad (3)$$
$$L_{GG} = L_{G,i} + L_{G,j}, \qquad i,j = 1,2,\ldots,n; i \neq j \quad (4)$$
$$L_{SS} = L_{S,i} + L_{S,j}, \qquad i,j = 1,2,\ldots,n; i \neq j \quad (5)$$

A major impact of the drain-source (C_{DS}) and gate-drain (C_{GD}) chip capacitances, more specifically the ratio C_{DS}/C_{GD}, on the oscillations has been derived in literature[2], [14], [15]. Furthermore, the source-sided inductance L_{SS} between paralleled dies and the drain-source chip capacitance are the only parameters where an optimization towards reduced SE-oscillations has a beneficial impact on the system loss. Hence, those three parameters are used as benchmarking parameters to compare the risk of SE-oscillations for different power semiconductor technologies and/or power modules[2], [14]. Additionally,

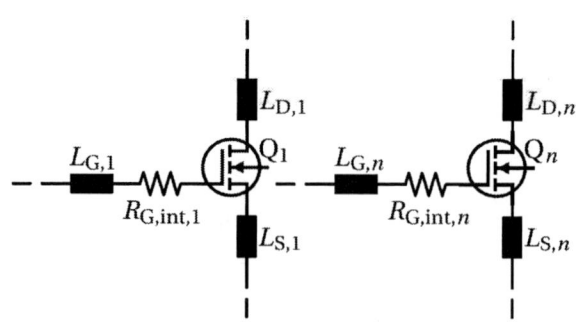

Fig. 7: Lumped parameter electrical equivalent circuit of paralleld dies in a power module.

TABLE II: System parameters and their impact on key characteristics[6].

Parameters		Change	SE-oscillations	Losses
MOSFET parameters	C_{DS}	↓	⇊	↓
	C_{GD}	↑	⇊	↑
	C_{GS}	↑	↓	↑
	g_m	↓	↓	↑
	g_o	↑	↓	↑
	$R_{G,int}$	↑	↓	↑
Power module chip-to-chip parasitics	L_{DD}	↑	↓	↑
	L_{GG}	↑	↓	↑
	L_{SS}	↓	↓	↓

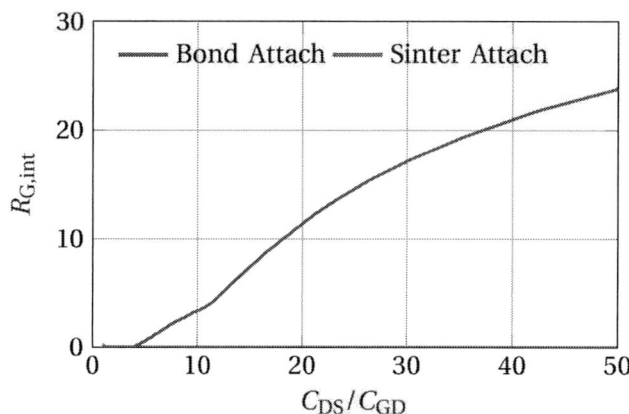

Fig. 8: Necessary damping resistance over capacitance ratio for the two compared power modules.

the symmetry in current sharing and the signal paths for the paralleled semiconductors plays a role in determining whether SE-oscillations will occur or not[16].

In the following, a bond-wire module and the previously introduced PM6, see section III and Figs. 5a and 5b, are compared regarding their respective SE-oscillation risk according to the three benchmark parameters C_{DS}, C_{GD}, and L_{SS}. This is done with a pole-zero (PZ) or eigenvalue analysis as is standard practice in literature[2]–[4], [6], [7], [13]–[16]. The two power modules herein represent a variation of L_{SS}. Different SiC MOSFET technologies are represented in this analysis with a variation of the ratio C_{DS}/C_{GD}, c.f. II.

There are commonly two approaches to calculate the stability of a power module as a system in respect to SE-oscillations. The first approach is the description of the system with a closed-loop transfer function $G(s)$ [7], [15], [16]. Subsequently, the poles of $G(s)$ are analyzed to gain an insight into the stability of the system. A different approach in literature and also used for this work is the use of a state-space representation, c.f. (6), of the system[3], [13]. The state-space representation is found according to the method described in [17], due to its simplicity and avoidance of co-dependent states.

$$\dot{x}(t) = A \cdot x(t) \qquad (6)$$

Knowledge about the stability of the system is obtained by calculating the eigenvalues of the system matrix A in (6). Avoidance of co-dependent states leads to a faster calculation time due to the significantly reduced order of the state matrix A.

To conduct the PZ-analysis, the AIT resistances and parasitic inductances are extracted for a CAD model of a standard bond-wire power module and the PM6 in Ansys Q3D, respectively. The bond-wire module

features a maximum L_{SS} of approximately 6.2 nH, whereas the PM6 only exhibits a maximum L_{SS} of around 2.5 nH inbetween any pair of paralleled dies, cf. table I. This amounts to a reduction of approximately 60 % for the worse case paralleled die pairing in case of the PM6. Furthermore, the difference between maximum and average L_{SS} is 2.5 nH and 0.5 nH for the bond-wire module and the PM6, respectively. Hence, the PM6 layout is very symmetrical, which can be explained by good bare-die placement for each logical switch on the AMB enabled by the additional degree of freedom in the design, due to the second substrate for top bare-die attach.

All necessary MOSFET parameters for the PZ-analysis are evaluated at 600 V drain-source voltage and nominal current density for a state-of-the-art example SiC MOSFET. A transconductance of 75 S, an output conductance of 64 mS and a gate-source capacitance of 3.6 nF are determined as the characteristic parameters of the 1.2 kV SiC MOSFET used for this comparison. For the drain-source capacitance a value of 212 pF is measured and, to vary the ratio the drain-gate capacitance is a sweep from 1 pF to 212 pF.

Figure 8 shows the minimum necessary internal gate resistance $R_{G,int}$ to stabilize the two different power modules, calculated for a varying C_{DS}/C_{GD} ratio as key parameter. It is evident that the PM6 is able to utilize power semiconductors with minimum necessary damping resistance for a wide range of gate-drain capacitances. This is due to the previously stated overall lower L_{SS} and other layouting benefits, e.g. current sharing symmetry, for this packaging type.

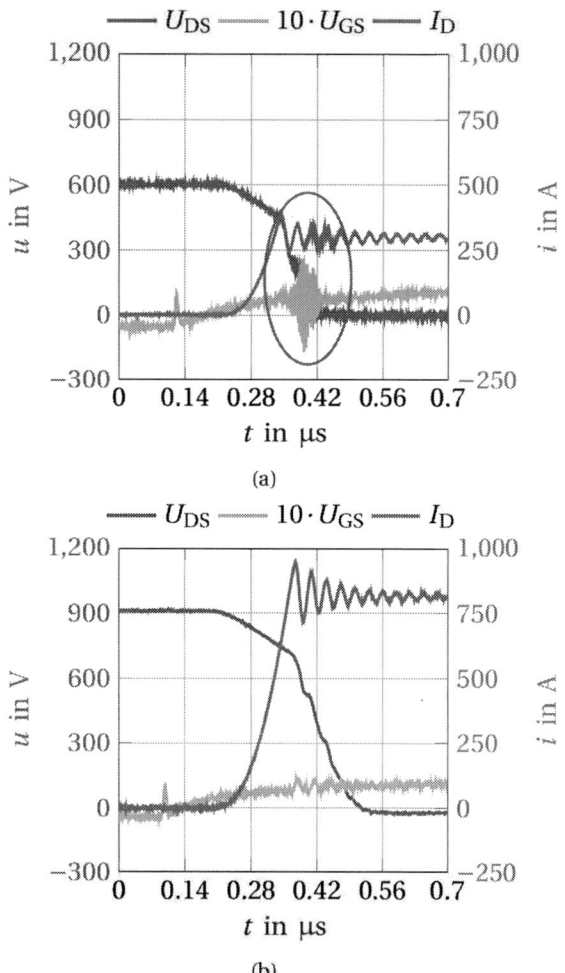

(a)

(b)

Fig. 9: Turn-on event for (a) the standard bond-wire module and (b) the advanced AIT power module.

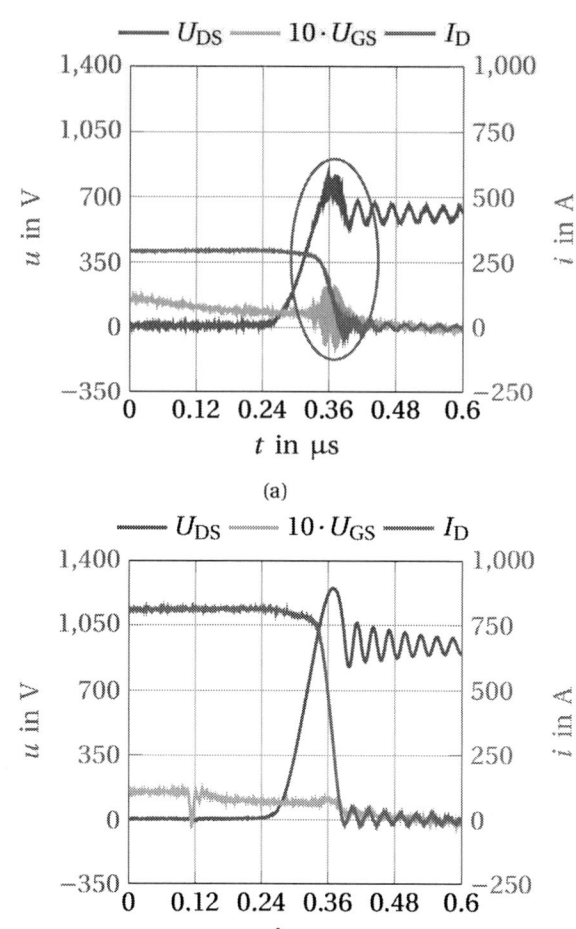

(a)

(b)

Fig. 10: Turn-off event for (a) the standard bond-wire module and (b) the advanced AIT power module.

V. MEASUREMENT RESULTS

To validate the previous statements and results regarding SE-oscillations, a standard bond-wire power module and a Bosch PM6 are populated with the SiC MOSFETs used for the theoretical analysis. The used SiC MOSFETs feature a peak c_{DS}/c_{DG} ratio of 25 during a switching event and a low $R_{G,int}$ in the range of 2 Ω. All the other key parameters are equal to the values stated in the previous section at a drain-source voltage of 600 V and nominal current density.

Both power modules are subject to a double-pulse test to evaluate the switching performance. Nominal DC-link voltage for both power modules is 920 V, whereas the switched current differs between the two due to the different overall SiC area utilized. Peak switched current for the investigated PM6 is 800 A, whereas the bond-wire module reaches its peak

switched current at around 700 A.

Figure 9 shows the turn-on event of the bond-wire and the PM6 power module, respectively. For the bond-wire module SE-oscillations are visible in the measured gate-source voltage, drain-source voltage, and drain current at a DC-link voltage of 600 V and for a switched current of 300 A, cf. Fig. 9a. This effect was noticed during the ramp-up of the double-pulse test and a further increase in DC-link voltage or switched current was halted, as this increased the magnitude and duration of the SE-oscillations. Hence, it was not possible to test the bond-wire module for nominal DC-link voltage and peak switched current, without risking catastrophic failure of the module due to SE-oscillations. Contrary to the bond-wire module, the PM6 was able to fully utilize the SiC MOSFETs up until the nominal DC-link voltage and peak switched current, cf. Fig. 9b. There are no SE-oscillations visible for the PM6, confirming the stability of the power module against inter-chip oscillations.

Similar results are obtained when investigating the turn-off event in the double-pulse test. Figure 10a shows SE-oscillations for the bond-wire module during turn-off, starting at a DC-link voltage of 600 V and a switched current of 300 A. A further increase of DC-link voltage or switched current was not possible for the bond-wire module with this SiC MOSFET, due to the risk of catastrophic failure of the module during testing. The PM6 is showing no signs of SE-oscillations whatsoever for the turn-off event, even for the nominal DC-link voltage and peak switched current, cf. Fig. 10b. Only the common commutation oscillation is visible in the waveforms for the PM6 turn-off event.

VI. CONCLUSION

As the SiC MOSFETs performance is increased in all domains, the trade-off is applicability in multi-chip power modules, c.f. sections II and IV. Together with unfavorable power module parasitics, e.g. high source-side inductance between parallelized dies, self-excited (SE) oscillations will occur. Hence, next-gen SiC MOSFETs may not be applicable for standard bond-wire power modules due to the increased source-side inductance between paralleled dies, c.f. sections III and IV.

The presented Bosch 6th generation power module is based on advanced AIT with double-side sintering of bare-dies. This approach leads to a favorable setup regarding AIT resistances and parasitic inductances for the power module, mitigating the risk for SE-oscillations even with next-gen SiC MOSFETs. A PZ-analysis with varying ratio of c_{DS}/c_{GD} and a subsequent experimental proof show the stability of the PM6 with minimal damping resistance, c.f. sections IV and V. For the same experimental setup, a standard bond-wire power module is not able to reach its nominal operating point without showing SE-oscillations, c.f. section V.

REFERENCES

[1] A. Nisch, C. Kloeffer, J. Weigold, W. Wondrak, C. Schweikert, and L. Beaurenaut, "Effects of a SiC TMOSFET Tractions Inverters on the Electric Vehicle Drivetrain," in *PCIM Europe 2018; International Exhibition and Conference for Power Electronics, Intelligent Motion, Renewable Energy and Energy Management*, 2018, pp. 1–8.

[2] I. Kasko, S. E. Berberich, M. Spang, and S. Oehling, "SiC MOS Power Module in Direct Pressed Die Technology and some Challenges for Implementation," in *2020 32nd International Symposium on Power Semiconductor Devices and ICs (ISPSD)*, Vienna, Austria: IEEE, Sep. 2020, pp. 364–367. DOI: 10.1109/ISPSD46842.2020.9170196.

[3] Z. Dong, R. Ren, W. Zhang, F. F. Wang, and L. M. Tolbert, "Instability Issue of Paralleled Dies in an SiC Power Module in Solid-State Circuit Breaker Applications," *IEEE Transactions on Power Electronics*, vol. 36, no. 10, pp. 11 763–11 773, Oct. 2021. DOI: 10.1109/TPEL.2021.3068608.

[4] M. M. Alam, S. Beushausen, and S. Khalid, *A Novel Approach to Suppress Self-Excited Oscillations in SiC-Based Power Modules*, eng. DE: VDE VERLAG GMBH, Jun. 2023.

[5] X. Wu, Y. Wu, Y. Wu, M. Rong, C. Tao, *et al.*, "Mitigating Gate Oscillations of Parallel SiC MOSFETs for Enhanced Performance in DC Solid-State Circuit Breaker," *IEEE Journal of Emerging and Selected Topics in Power Electronics*, pp. 1–1, 2024. DOI: 10.1109/JESTPE.2024.3351756.

[6] F. Sawallich and H.-G. Eckel, "Mitigating Inter-Chip Oscillation of paralleled SiC MOSFETs," in *2023 25th European Conference on Power Electronics and Applications (EPE'23 ECCE Europe)*, Aalborg, Denmark: IEEE, Sep. 2023, pp. 1–11. DOI: 10.23919/EPE23ECCEEurope58414.2023.10264236.

[7] Y. Shen, X. Dong, T. Schuetz, and R. Roesner, "Stability Modeling for Multichip SiC MOSFET Power Modules," in *CIPS 2022; 12th International Conference on Integrated Power Electronics Systems*, Berlin, Germany: VDE VERLAG GMBH, Mar. 2022, pp. 1–6.

[8] S. Schwaiger, A. Martinez-Limia, and K. Oberdieck, *Advanced SiC Trench-MOS Technology for Automotive Application*, eng. DE: VDE VERLAG GMBH, Jun. 2023.

[9] H. Syed, *Radiation Hardness of SiC Based Inverters Based on an EV Mission Profile*, eng. DE: VDE VERLAG GMBH, Jul. 2024.

[10] K. Oberdieck, *Short Circuit Robustness for Traction Inverters from an Application Point of View*, eng. DE: VDE VERLAG GMBH, Jul. 2024.

[11] A. P. Pai, M. Ebli, T. Simmet, A. Lis, and M. Beninger-Bina, "Characteristics of a SiC MOSFET-based Double Side Cooled High Performance Power Module for Automotive Traction Inverter Applications," in *2022 IEEE Transportation Electrification Conference & Expo (ITEC)*, Anaheim, CA, USA: IEEE, Jun. 2022, pp. 831–836. DOI: 10.1109/ITEC53557.2022.9813878.

[12] Y. Kang, P. Ning, X. Hui, T. Fan, D. Li, *et al.*, "Design of a Novel Double Sided Cooling SiC Power Module

979-8-3315-1612-3/25 $31.00 © 2025 IEEE

Based on Multiple DBC-Stacked," in *2023 IEEE 2nd International Power Electronics and Application Symposium (PEAS)*, Guangzhou, China: IEEE, Nov. 2023, pp. 418–422. DOI: 10.1109/PEAS58692.2023.10395598.

[13] M. Spang, S. Buetow, and G. Katzenberger, "Differential-mode oscillations between parallel IGBTs in power modules," in *2015 17th European Conference on Power Electronics and Applications (EPE'15 ECCE-Europe)*, Geneva: IEEE, Sep. 2015, pp. 1–10. DOI: 10.1109/EPE.2015.7309089.

[14] K. Saito, T. Miyoshi, D. Kawase, S. Hayakawa, T. Masuda, and Y. Sasajima, "Simplified Model Analysis of Self-Excited Oscillation and Its Suppression in a High-Voltage Common Package for Si-IGBT and SiC-MOS," *IEEE Transactions on Electron Devices*, vol. 65, no. 3, pp. 1063–1071, Mar. 2018. DOI: 10.1109/TED.2018.2796314.

[15] M. Nagel, I. Kovacevic-Badstuebner, R. Salvatore, D. Popescu, B. Popescu, *et al.*, "Stability Analysis of Parallel SiC power MOSFETs based on a Virtual Prototype," in *2023 35th International Symposium on Power Semiconductor Devices and ICs (ISPSD)*, Hong Kong: IEEE, May 2023, pp. 338–341. DOI: 10.1109/ISPSD57135.2023.10147660.

[16] F. Sawallich, *Inter-Chip Oscillation of Paralleled SiC MOSFETs*, eng. DE: VDE VERLAG GMBH, Jun. 2023.

[17] C. H. van der Broeck, D. Bura, and L. Camurca, "Simulating Switching of Power Electronic Devices via Generic State-Space Models," in *2023 IEEE Energy Conversion Congress and Exposition (ECCE)*, Nashville, TN, USA: IEEE, Oct. 2023, pp. 5447–5454. DOI: 10.1109/ECCE53617.2023.10362718.

A SiC Half-bridge Power Module Based on Liquid Metal Packaging for High Performance and Low Thermal Stress

Wei Mu
[1]*Department of Engineering*
University of Cambridge,
Cambridge, United Kingdom
wm338@cam.ac.uk

Ameer Janabi
[1]*Department of Engineering*
University of Cambridge
Cambridge, United Kingdom
ahaj2@cam.ac.uk

Luke Shillaber
[1]*Department of Engineering*
University of Cambridge
Cambridge, United Kingdom
ls669@cam.ac.uk

Borong Hu
[1]*Department of Engineering*
University of Cambridge
Cambridge, United Kingdom
bh529@cam.ac.uk

Teng Long
[1]*Department of Engineering*
University of Cambridge
Cambridge, United Kingdom
tl322@cam.ac.uk

Abstract— **Power electronics modules and converters are subjected to significant thermomechanical stress during operation, primarily due to the rigid bonding of multiple material layers with differing coefficients of thermal expansion (CTE). This stress arises when each material layer expands and contracts at different rates during thermal cycling, which induces mechanical strain. This issue is particularly critical for silicon carbide (SiC) power modules, which are known for their notably high Young's Modulus, making them more susceptible to stress-related failures such as cracking or delamination. To address this challenge, this paper demonstrates the integration of a liquid metal (LM)-based SiC packaging design into a half-bridge power module, featuring integrated gate drivers and water cooling. The LM-based package incorporates a floating die structure and LM fluidic connections, which act to decouple thermal strain, thereby allowing for significantly reduced thermal stress and minimized warpage during operation. This design suggests enhanced reliability, improved thermal management, and an extended operational lifetime for the module. Experimental testing conducted on half-bridge LM module prototypes validates that the proposed LM-based SiC package delivers uncompromised electrical performance under various operating conditions, emphasizing its suitability for next-generation power electronics applications.**

Keywords— *Thermomechanical stress, liquid metal, packaging, SiC power converter reliability*

I. INTRODUCTION

Silicon carbide (SiC) power devices offer significant advantages, including lower on-resistance per unit chip area, higher blocking voltage, and faster switching [1]. Traditional wire bond-based interconnection can introduce high parasitic inductance, which negatively impacts switching speed and overall module efficiency. In addition, the smaller chip area of SiC devices limits the number of wires that can be bonded to the device, which also potentially increases parasitic resistance and inductance, further degrading performance.

In order to unlock the full potential of SiC power devices, new packaging designs are attempting to eliminate bond wires entirely. For instance, copper clips are being soldered on top of the device to enhance current-carrying capacity for new-generation EV power modules [2]. These copper clips not only reduce parasitic inductance but also improve thermal dissipation, allowing for more efficient operation at high currents. Additionally, double-sided cooled (DSC) modules are becoming increasingly popular in automotive applications because they can provide enhanced thermal performance with an additional heat flow path, which significantly improves thermal management and module reliability [3].

PCB/DBC hybrid modules have the potential to combine the advantages of both PCB and DBC by using PCB for electrical integration and DBC for heat dissipation. In this configuration, both low inductance and low thermal resistance can be achieved [4]. Emon et al. [5] demonstrated this concept by sandwiching GaN devices between DBC and PCB with an integrated gate drive to achieve a low gate loop inductance of 1.24 nH, power loop inductance of 0.72 nH, and low thermal resistance of 0.26 K/W. They further modified the design to achieve a 3D hybrid module with double-sided cooling, which resulted in even better thermal performance [6]. Similar PCB/DBC hybrid GaN power modules with low thermal resistance and inductance are also reported in [7]-[9], demonstrating the feasibility of these advanced packaging approaches.

However, all these studies primarily focus on optimizing thermal and electrical performance, while the thermomechanical stress and its implications on reliability have not been thoroughly investigated. For hybrid power module designs, the mismatch in CTE between PCB and ceramic materials can cause

979-8-3315-1612-3/25 $31.00 © 2025 IEEE

significant stress in solder joints, potentially leading to fatigue and failure over time. This problem is even more pronounced for hybrid SiC modules due to the high Young's modulus of SiC, which makes it more susceptible to stress-induced damage. The small footprint of the SiC die can further increase heat flux density, resulting in greater thermal stress, especially in regions with steep temperature gradients [10]. These challenges contribute to why hybrid packaging solutions are less popular for SiC devices compared to GaN devices, despite their inherent electrical advantages.

In summary, these wire bond-less power modules inevitably employ rigid connections on both sides of the device. The mismatch in CTE among different material layers, coupled with the high Young's modulus of SiC devices, can generate considerable thermal stress, raising significant reliability concerns [11]. These concerns must be addressed to fully realize the benefits of SiC power devices in next-generation applications, particularly those that demand high efficiency and long-term durability.

To reduce thermal stress while maintaining the high performance of wire bond-less design, a liquid metal (LM) based package was proposed [12]. The LM-based SiC package comprises a flexible PCB (FPCB) and active metal brazed (AMB) substrate with a step-etched cavity to embed the die inside. Instead of using rigid connections such as solder joints and nano-silver sintering, a LM layer is adopted to drastically reduce the thermomechanical stress of the die attach layer and the joint between AMB and FPCB. The fluidic nature of LM can decouple the thermal strain at the interface of two different materials while maintaining thermal and electrical connections. Therefore, thermal stress can be significantly reduced without affecting the overall thermal and electrical conductivity. In this paper, a half-bridge power module with integrated gate drivers and a cooling plate is built using LM-based package. The highly integrated LM-based half-bridge module not only exhibits high thermal and electrical performance, but also significantly reduces thermal stress and minimizes warpage during operation, contributing to better overall converter reliability.

II. INTEGRATION OF LM PACKAGE INTO HALF-BRIDGE MODULE

Fig. 1 presents an exploded view of the integrated half-bridge power module, the module adopted a modular design with various components and layers involved in its construction. At the top, the gate driver circuit is shown, which is responsible for controlling SiC devices. Just below it is the main half-bridge power stage, The SiC package, based on the liquid metal (LM) package, sits beneath the main PCB. This package is sandwiched between the main power PCB and a liquid-cooled copper heatsink. The heatsink is made of copper instead of aluminum not only because copper can provide better cooling due to its higher thermal conductivity, but also to prevent corrosion of the heatsink caused by LM.

Beneath the PCB, two LM-based SiC packages are sandwiched between the board and a liquid-cooled heatsink, forming the core of the half-bridge. The exploded view at the bottom of the figure provides a detailed structure of the LM-based SiC package. The module utilizes Ag coated AMB substrate. The Ag coating can greatly enhance the wettability of

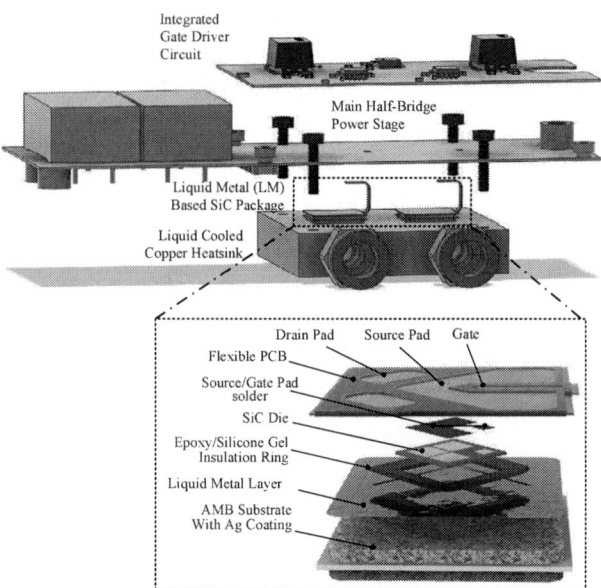

Fig. 1. The exploded view of the LM based half bridge module with detailed structure of LM based embedded SiC package

TABLE I. PROPERTIES OF SOLID MATERIALS USED IN THIS STUDY

Material	Properties				
	Thermal conductivity (W/m·K)	Densiy (kg/m3)	Elasticity modulus (Mpa)	Poisson's ratio	CTE (ppm/°C)
Chip/SIC	370	3210	501000	0.45	4.3
Copper layer	387.6	8933	110000	0.32	18
Si3N4	80	3200	310000	0.26	3.2
Polyimide	0.3	1900	2500	0.34	18
FR4	0.29	1850	24000	0.136	14
SAC 305 Solder	58.7	7370	43060	0.4	23.5

the LM with the substrate. This can make sure that the LM layer has good contact with the substrate, enhancing the overall thermal and electrical conductivity. The top copper of the substrate is etched down to have a cavity of 7 mm by 7mm to embed the SiC die. The SiC MOSFET die used in this study is the SCT116N120G3DXAG from ST microelectronics, which is rated at 1200V, 130A, with an on-resistance of 14.6 mΩ. The drain side of the die is floating on a liquid metal layer inside the cavity. This unique floating structure, coupled with the fluidic connections provided by the liquid metal, effectively decouples thermal strain while maintaining sufficient thermal and electrical conductivity. The AMB substrate cavity is filled with liquid metal, while an insulation ring surrounds the die to prevent any liquid metal leakage to the top side of the die that could cause short circuits.

To achieve precise gate and source connections, the SiC die is soldered to a flexible PCB (FPCB), which is further soldered to the main power stage PCB. "It is worth mentioning that these two solder layers won't cause high thermal stress due to the flexible PCB's properties. As shown in Table I, the core of

FPCB is made of polyimide which has a Young's modulus of 24 GPa. This value is only one-tenth of that of FR4, which is the core material of a normal PCB. Also, polyimide is characterized by its high dielectric breakdown voltage and high thermal stability. Previous studies have demonstrated that PI's dielectric strength remained high (>200 kV/mm) even above 225°C [13]. The polyimide layer of the flex-PCB has a thickness of 0.29 mm, which theoretically can withstand 58 kV. Therefore, FPCB is both thin and has a low Young's modulus, allowing it to effectively accommodate thermal strain. The copper layer of FPCB is 2 Oz, providing enough current-carrying capability.

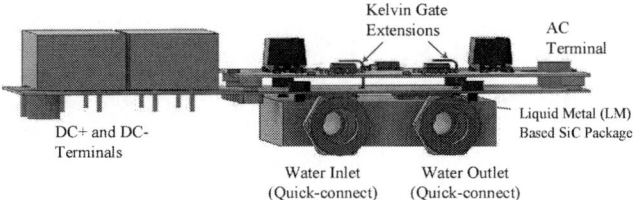

Fig. 2 The assembly of the proposed half-bridge power stack based on the LM package.

Fig. 2 presents an assembled view of the module, providing a side perspective of how the components are arranged. The DC+ and DC- terminals are clearly visible on the left side of the main PCB, while the AC terminal is on the right side, which allows for easier connections to external circuits. The gate drive PCB is positioned above the main power PCB with screws to connect them. The gate drive PCB is connected to the LM package using the FPCB to form Kelvin gate extensions. The top side of the FPCB extension is the gate pad and the bottom side of the Kelvin gate extension is the source connection. The thin FPCB can reduce the size of the gate loop to reduce parasitic inductance and enhance switching performance.

The LM-based SiC package is sandwiched between the main power stage PCB and the heatsink. Screws are tightened using a torque screwdriver to apply even pressure to press the package against the heatsink, guaranteeing good thermal contact. Because the FPCB and the die are connected to the AMB substrate only by a liquid metal fluidic layer, some pressure is necessary to maintain the package's integrity, particularly in a vibrating environment. In the meantime, this pressure is also enough for the LM module to achieve optimal thermal and electrical performance [12]. The copper heatsink is also equipped with quick-connect water inlet and outlet fittings, allowing easy connection with the water circulating system.

III. FINITE ELEMENTS ANALYSIS AND RESULTS

Thermo-mechanical coupled finite element analysis (FEA) is employed to evaluate the advantages of utilizing a liquid metal (LM)-based package within a half-bridge integrated module. To provide a meaningful comparison, a conventional soldered module—where both the SiC die and the FPCB are soldered to the AMB substrate—is used as the baseline. The study focuses on assessing the differences in thermal stress between the LM-based package and the conventional soldered package under identical operating conditions.

In order to simplify the model and provide a focused analysis, the simulation concentrates on the thermal stress

between the SiC package and the main power PCB, with each device dissipating a power of 100 W. The simplified model ensures that the key interactions between the components are clearly represented, especially the thermal stress caused by the main solder joint and package, so as to provide a comparison of the two packaging techniques. The results of this analysis are illustrated in Fig. 3

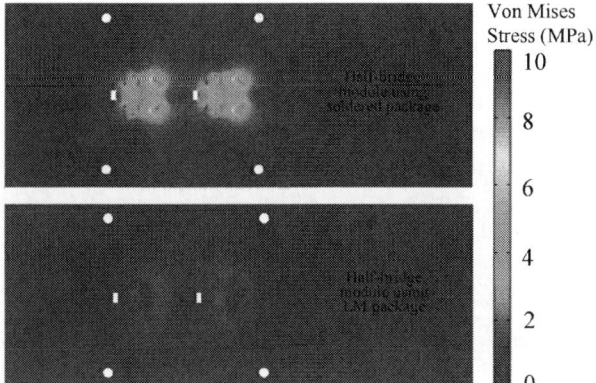

Fig. 3 Thermal stress comparison of main power stage PCB using conventional soldered package (top) and proposed LM-based package (bottom)

The results clearly demonstrate that the LM-based package exhibits significantly lower thermal stress in comparison to the conventional soldered counterpart. This substantial reduction in stress can be attributed to the unique characteristics of the LM layer, which effectively decouples the thermal strain originating from the AMB substrate. By acting as a compliant interface, the LM layer mitigates the transfer of thermal strain to the PCB, thereby substantially lowering the overall thermal stress experienced by both the PCB solder layer and the PCB itself.

This decoupling effect has important implications for the reliability and durability of the half-bridge integrated module. By minimizing the thermal stress on the main PCB, the LM-based package reduces the risk of thermal fatigue and material degradation, which are common failure modes in power electronics operating under high temperatures and thermal cycling conditions.

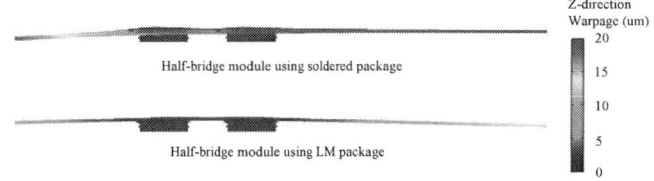

Fig. 4. Main Power stage PCB Warpage (z-direction) comparison

An additional and direct benefit of using the LM-based package in the half-bridge assembly is a significant reduction in warpage during operation, as shown in Figure 3. Warpage is a critical issue that often arises due to the mismatch in thermal expansion coefficients among the different materials in the module. In conventional soldered modules, this mismatch can lead to considerable warpage, causing mechanical stress that potentially degrades both electrical and thermal performance

979-8-3315-1612-3/25 $31.00 © 2025 IEEE   2599

over time. However, with the LM-based package, the decoupling effect of the LM layer ensures that the thermal expansion is better accommodated, resulting in significantly less warpage.

This reduced warpage directly correlates to a longer lifetime of the entire converter assembly, as indicated by experimental findings and discussed in reference [14]. The lower thermal stress and minimized warpage mean that the mechanical and thermal integrity of the components is preserved, even under harsh operational conditions.

IV. EXPERIMENT VERIFICATION RESULTS

A. Prototype Module and Test Setup

Prototypes of the LM-based half-bridge modules have been fabricated, and the main components of the module are shown in Fig. 5. The flexible printed circuit board, which integrates the die, is already soldered onto the backside of the main power PCB. The flexibility of the FPCB allows the Kelvin gate connection on the FPCB to be bent and pass through an opening in the corresponding location on the main power stage PCB and the gate driver PCB. This connection is finally achieved by soldering the pads on the top (gate) and bottom (source) layers of the FPCB to the gate driver PCB as illustrated in Fig. 6. Although the Kelvin gate connection extends approximately 1 cm from the PCB, the thin FPCB design ensures that the gate and source traces fully overlap in a small area. This overlap enables effective flux cancellation, resulting in a minimal gate loop. As a result, the parasitic inductance is significantly reduced, which facilitates faster switching speeds and improves overall performance.

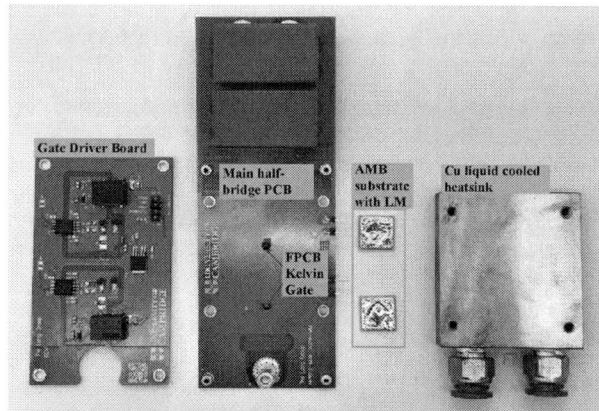

Fig. 5. The main parts of the LM-based half-bridge module.

Fig. 6. The assembled LM-based half-bridge module.

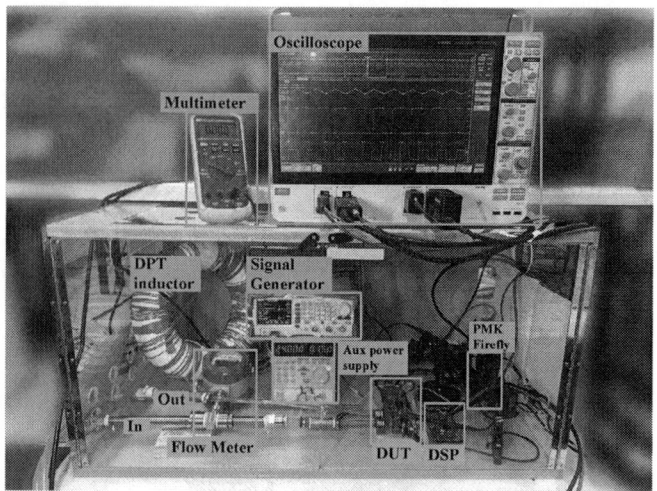

Fig. 7. Test setup for both DPT and Continuous Test.

Fig. 7 depicts the experimental setup designed for both Double Pulse Testing (DPT) and continuous operation testing, aimed at evaluating the performance and thermal behavior of the Device Under Test (DUT). At the top of the setup, a Tektronix 6 series oscilloscope is used to monitor real-time voltage and current waveforms. To its left, a multimeter is used to measure the DC link voltage. The DPT inductor, visible on the left side of the setup, is made of air core and has an inductance of 36 uH. In the center of the setup, the signal generator supplies gate drive signals to control the pulse duration during DPT. During the continuous test, the control signal is controlled by the DSP. The flow meter, positioned near the inlet, measures the coolant flow rate during continuous operation testing. The isolated probe PMK Firefly is utilized to measure the high-side voltages. The current during DPT is measured by a Rogowski coil which is inserted between the DC- and low-side switch. The inductor current during the continuous test is measured by a current probe.

B. Double Pulse Test

To evaluate the module's performance, a Double Pulse Test (DPT) is conducted to characterize its fast transient behavior under high-speed switching conditions. In this setup, the high-side LM package includes a 36 µH air-core inductor connected in parallel. Also, for the high-side switch, the gate terminal is shorted to the source to ensure it remains in the off state during the test.

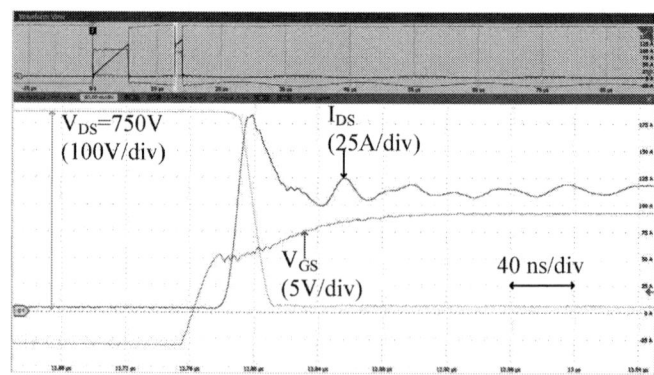

Fig. 8 The turn-on switching event of the double pulse test.

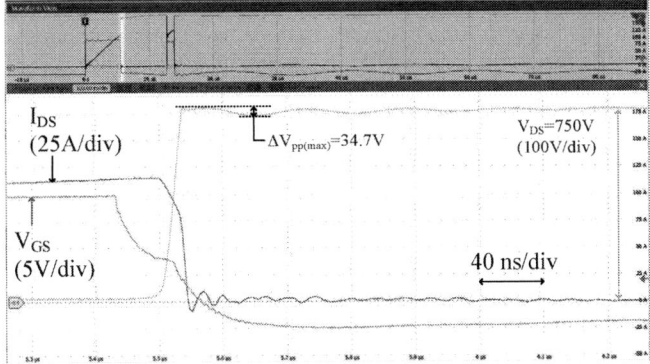

Fig. 9. The turn-off switching event of the double pulse test.

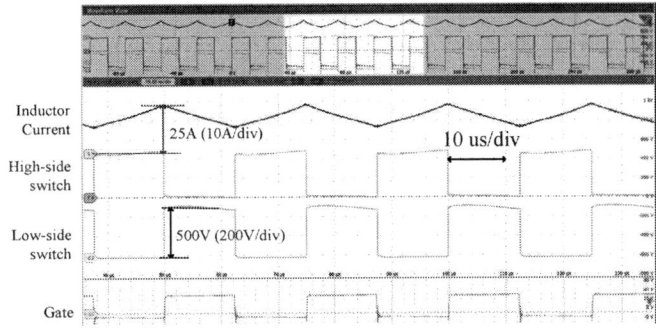

Fig. 11. Continuous test waveform of synchronous buck converter at 40 kHz.

The DPT is carried out with a 750 V DC link voltage, simulating the voltage stresses the module would encounter in practical electric vehicle applications. During the test, the current builds up to a peak of 132 A at the end of the second pulse, validating the module's capability to handle substantial pulsating current. The result presented in Fig. 8 reveals the module's fast switching characteristics during the turn-on transient. Specifically, the voltage slew rate reaches 40 V/ns, while the current slew rate peaks at 15 A/ns. These values highlight the rapid switching capability of the LM-based power module, a crucial factor in achieving high efficiency and reduced switching losses. Additionally, the turn-off transient performance shown in Fig. 9 is equally noteworthy. The voltage slew rate during turn-off is measured at 25 V/ns, with a maximum voltage overshoot limited to less than 18 V. The low overshoot is indicative of low power loop inductance that mitigates switching stress and improves reliability. These results validate the proposed LM-based module's ability to achieve high-speed, low-loss switching, making it well-suited for high-frequency power electronics applications.

C. Continuous Testing Using a Synchronous Buck Converter

A synchronous buck converter is constructed to conduct continuous testing of the LM-based half-bridge module, ensuring its performance is evaluated under realistic operating conditions. The converter is running at 50% duty cycle with a dead time of 100 ns. To validate the module's behavior under different scenarios, the converter is tested at different operation points. The waveforms of working under 400 V at 100 kHz is shown in Fig. 10. illustrating the inductor current and voltages of both switch position and gate. The peak inductor current is around 10 A at this operation point.

The LM-based half bridge is also tested under lower switching frequencies but higher inductor current. Fig. 11 presents the waveform of the buck converter operating at 40 kHz, where the peak inductor current reaches 25 A. The results demonstrate that the module performs reliably under these conditions, exhibiting very minimal ringing and overshoot, even at high current levels. However, the current waveform reveals that the inductor is approaching saturation. This is indicated by the slight distortion in the shape of the current waveform, which suggests a reduction in the inductor's effective inductance as the magnetic core nears its saturation limit. Despite this, the LM-based module maintains stable operation, highlighting its robustness in handling challenging operating points with high current demands.

V. CONCLUSION

This paper introduced a novel silicon carbide (SiC) half-bridge power module featuring a liquid metal (LM)-based packaging approach designed to address the thermal stress challenges commonly found in conventional wire-bond-less designs. The LM package effectively mitigates thermal stress by decoupling the thermal expansion mismatch between the components. Experimental validation under various operating conditions, including high voltage, high frequency, and high current scenarios, demonstrated that the LM-based half-bridge module maintained excellent electrical performance with minimal ringing and overshoot. The results confirm the effectiveness of the proposed design in achieving superior thermal and mechanical reliability while ensuring high-speed switching performance. This innovative packaging approach presents a promising solution for enhancing the reliability of power converters without affecting efficiency or adding excessive cost.

REFERENCES

[1] F. Yang *et al.*, "Compact-Interleaved Packaging Method of Power Module With Dynamic Characterization of 4H-SiC MOSFET and Development of Power Electronic Converter at Extremely High Junction Temperature," in *IEEE Transactions on Power Electronics*, vol. 38, no. 1, pp. 417-434, Jan. 2023.

[2] Kim, D., Lee, B., Lee, T. I., Noh, S., Choe, C., Park, S., & Kim, M. S. (2022). Power cycling tests under driving ΔTj= 125° C on the Cu clip bonded EV power module. *Microelectronics Reliability*, Volume 138, 114652, November 2022

[3] M. Liu, A. Coppola, M. Alvi and M. Anwar, "Comprehensive Review and State of Development of Double-Sided Cooled Package Technology for Automotive Power Modules," in *IEEE Open Journal of Power*

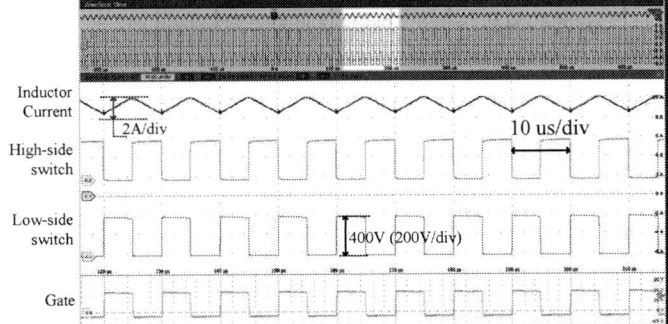

Fig. 10. Continuous test waveform of synchronous buck converter at 100kHz.

Electronics, vol. 3, pp. 271-289, 2022, doi: 10.1109/OJPEL.2022.3166684

[4] A. Janabi *et al.*, "Substrate Embedded Power Electronics Packaging for Silicon Carbide mosfets," in *IEEE Transactions on Power Electronics*, vol. 39, no. 8, pp. 9614-9628, Aug. 2024, doi: 10.1109/TPEL.2024.3396779.

[5] Emon et al., "A 650V/60A Gate Driver Integrated Wire-bondless Multichip GaN Module," 2021 IEEE 12th International Symposium on Power Electronics for Distributed Generation Systems (PEDG), 2021, pp. 1-6,

[6] A. I. Emon et al., "Design and Optimization of Gate Driver Integrated Multichip 3-D GaN Power Module," in IEEE Transactions on Transportation Electrification, vol. 8, no. 4, pp. 4391-4407, Dec. 2022

[7] A. B. Jørgensen, S. Bęczkowski, C. Uhrenfeldt, N. H. Petersen, S. Jørgensen and S. Munk-Nielsen, "A Fast-Switching Integrated Full-Bridge Power Module Based on GaN eHEMT Devices," in IEEE Transactions on Power Electronics March 2019

[8] S. Lu, T. Zhao, Z. Zhang, K. D. T. Ngo, R. Burgos and G. -Q. Lu, "Low Parasitic-Inductance Packaging of a 650 V/150 A Half-Bridge Module Using Enhancement-Mode Gallium-Nitride High Electron Mobility Transistors," in *IEEE Transactions on Industrial Electronics*, Jan. 2023

[9] X. Tian et al., "PCB-on-DBC GaN Power Module Design with High-Density Integration and Double-Sided Cooling," in IEEE Transactions on Power Electronics, doi: 10.1109/TPEL.2023.3311440.

[10] W. Mu et al., "Direct Integration of Optimized Phase-Change Heat Spreaders Into SiC Power Module for Thermal Performance Improvements Under High Heat Flux," in IEEE Transactions on Power Electronics, vol. 37, no. 5, pp. 5398-5410, May 2022

[11] B. Hu *et al.*, "Failure and Reliability Analysis of a SiC Power Module Based on Stress Comparison to a Si Device," in *IEEE Transactions on Device and Materials Reliability*, vol. 17, no. 4, pp. 727-737, Dec. 2017

[12] W. Mu, A. Janabi, B. Hu, L. Shillaber and T. Long, "Liquid Metal Fluidic Connection and Floating Die Structure for Ultralow Thermomechanical Stress of SiC Power Electronics Packaging," in *IEEE Transactions on Power Electronics*, vol. 39, no. 7, pp. 7808-7814, July 2024, doi: 10.1109/TPEL.2024.3379121.

[13] M. Nagao, G. Sawa, M. Fukui, and M. Ieda, "Dielectric Breakdown of Polyimide Film in High Temperature Region", *Japanese J. Appl. Phys.*, Vol. 15, pp. 1813-1814, 1976.

[14] Guo, Y.; Liu, M.; Yin, M.; Yan, Y. Reliability Sensibility Analysis of the PCB Assembly concerning Warpage during the Reflow Soldering Process. *Mathematics* **2022**, *10*, 3055. https://doi.org/10.3390/math10173055

Analysis and Modeling of Radiated EMI Considering Coupling between Power Converter and Power Cable with LC-Type EMI Filter

Qinghui Huang, Yingjie Zhang, Shuo Wang, Yirui Yang, Zhedong Ma, and Yanwen Lai
Power Electronics and Electrical Power Research Lab
University of Florida, Gainesville, FL, USA
Email: qinghuihuang@ufl.edu

Abstract- **This paper explores the use of an LC-type filter (CM choke and Y-capacitor) in a dual active bridge (DAB) converter to reduce radiated EMI from the antenna. However, a notable radiated EMI issue is identified between 100MHz and 400MHz, resulting from an overlooked antenna formed by the input/output cable and power converter. A new radiation model is proposed considering coupling between the power converter and input/output cables, with experimental verification. Additionally, an improved L and C arrangement is suggested for better EMI performance.**

Index Terms—Isolated converter, Radiated EMI, LC filter, CL filter

I. INTRODUCTION

Modern power electronics systems need to continuously improve the switching frequency to enhance the efficiency and power density of power converters [1]-[5] to meet the demands of datacenter, renewable energy, electrical vehicle applications[6][7]. WBG (wide band gap) devices have been widely used in power electronics systems to improve system efficiency and density[8]-[10]. WBG can also help to achieve the application of a planar transformer/inductor which makes the manufacturing process of magnetic components much easier[11][12]. However, high dI/dt and dV/dt would generate more EMI noise which would be very tricky problems for power electronics engineers[13].

Radiated EMI remains a challenging issue for power converters using cables acting as radiation antennas [15]. The existing method to reduce radiated EMI can be divided into two categories. The first category is to optimize the parameters of converters to reduce radiated EMI. Ground impedance is found to be a critical factor causing radiated EMI noise[16] Paper [17] proposed achieving a transformer balance technique to reduce radiated EMI noise.

Previous literature [19] investigates CM chokes should be inserted close to the power module to reduce radiated EMI noise. A radiated EMI model was developed to analyze the interaction between the cable antenna, power converter, and EMI filters [20]. However, the analyzed EMI filters were single-element solutions, either CM choke or Y-capacitor.

This paper introduces an LC-type filter, combining a CM choke and Y-capacitor, applied to a dual active bridge (DAB) converter. The LC-type filter effectively reduces radiated EMI noise from the antenna (input and output cables). A previously overlooked antenna comprising the input/output cable, and the power converter is found to be a dominant radiated EMI source. The main contribution of the digest includes: 1. a high-radiated EMI noise which is caused by a neglected antenna comprising the input/output cable and the power converter is identified. 2. A new radiation model considering the coupling between the power converter and input/output cable with LC filters is proposed and verified with experiments. 3. an improved L and C arrangement with better radiated EMI performance is proposed.

II. INVESTIGATION OF RADIATION OF DAB CONVERTER WITH LC-TYPE FILTER

A. Introduction of existing radiated EMI mode of power converters with cables

Power cables in power electronics systems work as an antenna generating high frequency radiated emissions. High dV/dt and dI/dt caused by switching are the excitations to the voltage across the antenna. The cable antenna can be modeled as a radiation impedance, as presented in Fig. 1 (b). The antenna impedance includes three parts: radiation resistance R_r representing the radiated power, loss resistance R_l representing ohmic power loss, and reactance jX_A representing the energy stored in the near field. I_A indicates the equivalent current flowing through the antenna. Power converters can be modeled as a voltage source in serials with an equivalent impedance $Z_s = R_s + jX_s$ based on Thevenin's theory. The final equivalent radiated EMI model for a DAB converter with cables is shown in Fig. 1(c). With this model, the current flowing through the antenna can be derived from (1). The maximum electric field can be calculated with (2) where r is the distance of observation. G_o is the gain of the antenna, which is determined by the antenna structure. η is the intrinsic impedance of free space. K_I and K_E are the current and electric field factors, which can be utilized to evaluate the CM current and electric field values.

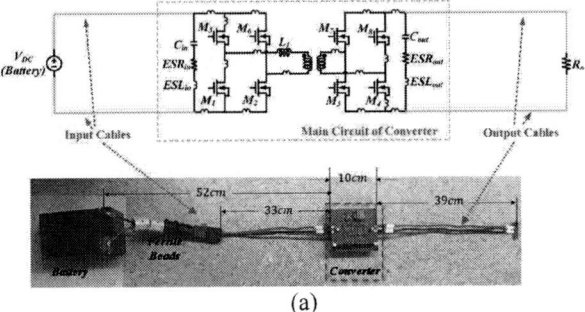

(a)

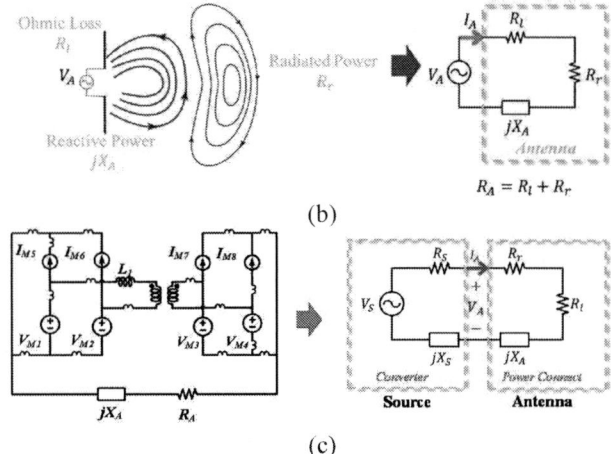

Fig. 1. (a) a DAB converter, (b) antenna model, (c) converter's radiated EMI model.

$$| I_A | = \frac{|V_S|}{\sqrt{(R_S + R_A)^2 + (X_S + X_A)^2}} \quad (1)$$

$$E_{max} = \sqrt{\frac{\eta G_o}{4\pi r^2}} \times |V_S| \times \frac{\sqrt{R_A}}{\sqrt{(R_S + R_A)^2 + (X_S + X_A)^2}} \quad (2)$$

$$K_I = \frac{1}{\sqrt{(R_S + R_A)^2 + (X_S + X_A)^2}} \quad (3)$$

$$K_E = \frac{\sqrt{R_A}}{\sqrt{(R_S + R_A)^2 + (X_S + X_A)^2}} \quad (4)$$

The DAB converter shown in Fig. 1 (a) is used for experiments.

B.   Applying LC-type filter to reduce radiated EMI

EMI filters could be an efficient way to reduce the radiated EMI of the converters with input and output cables. Impedance mismatch criteria are usually used to select the topology of the EMI filter [21]. The EMI noise source impedance and load impedance should be determined first. As for the radiated EMI model mentioned above, the source impedance is the converter impedance Z_S and the load impedance is the impedance of the antenna Z_A, as shown in Fig. 2.

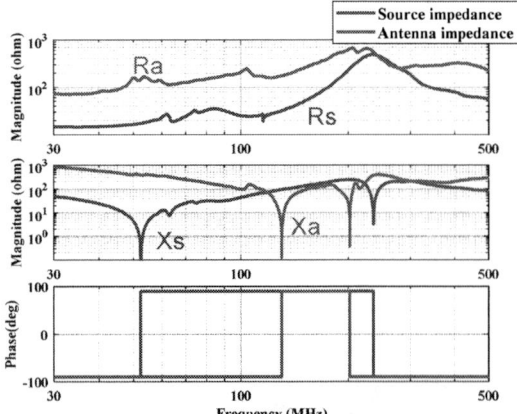

Fig. 2. Impedance curve of converter source impedance and antenna impedance.

To minimize the noise from the source to the load, the impedance of the inserted EMI filters should be mismatched with the source impedance and load impedance. For example, the source impedance is much

smaller compared to the load impedance in this case, a big impedance such as a CM choke can be added in serial with the source impedance to transfer more noise voltage across the CM choke, as shown in Fig. 3 (b). In other words, the CM choke helps increase the converter's total source impedance. A small impedance such as a capacitor can be added in parallel with the big load impedance. As a result, the noise current flowing through the load can be bypassed by the Y-cap. The specific radiated EMI filter implementation in a DAB converter is shown in Fig. 3 (b). A CM choke is inserted in serial with the output cable to increase the converter source impedance. A Y-capacitor is added between the negative terminal of the output cables and input cables. The radiated EMI model can be simplified as Fig. 3 (a).

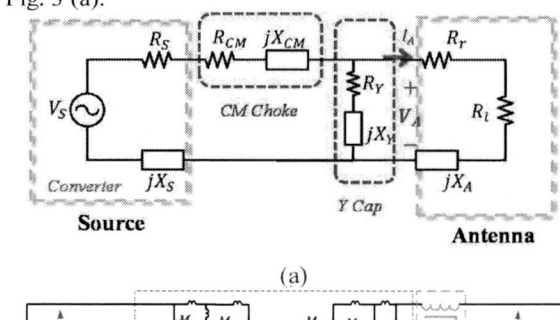

(a)

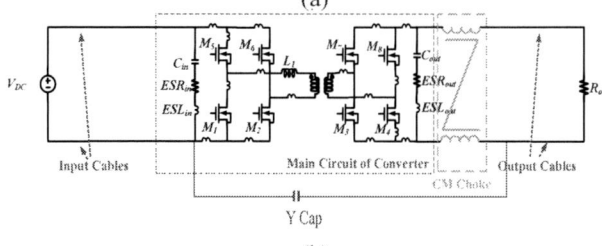

(b)

Fig. 3. LC-type filter applied to DAB converter.

With the radiated EMI model in Fig. 3 (b), the current factor K_I and electric field factor K_E should be rewritten as

$$K_I = \left| \frac{Z_Y \| Z_A}{Z_{CM} + Z_S + Z_Y \| Z_A} \right| \times \frac{1}{|Z_A|} \quad (5)$$

$$K_E = \left| \frac{Z_Y \| Z_A}{Z_{CM} + Z_S + Z_Y \| Z_A} \right| \times \frac{\sqrt{R_A}}{|Z_A|} \quad (6)$$

Z_Y is the impedance of the Y-capacitor which is in parallel with the antenna impedance Z_A. Z_{CM} is the impedance of the CM choke which is serial with the source impedance Z_S.

The calculated K_I and K_E with and without EMI filters are shown in Fig. 4 (a) and Fig. 4 (c). In case (a), the blue line is the baseline case without any EMI filters. The red line is the case with only the Y-capacitor (100pF). The yellow line is the case with the CM inductor and Y-capacitor (100pF). A similar EMI filter configuration is applied to case (c) but the Y-capacitor is 470pF. The measured radiated EMI results are shown in Fig. 4 (b) and (d). For cases with only a Y-cap filter, calculated K_E reduction (20dB for 100pF, 8dB for 470pF) matches measured radiated EMI (17dBuV/m for 100pF, 7dBuV/m for 470pF). However, for the LC-type filter case, there is a significant difference between theoretical K_E reduction (40dB for 100pF, 30dB for 470pF) and measured radiated EMI (11.5dB for 100pF, 11dB for 470pF). The LC filter cases show a larger radiated EMI from 100MHz to 400MHz compared to the cases with only Y-capacitors.

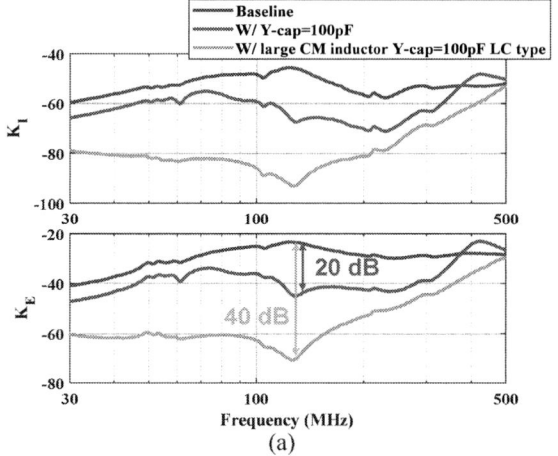

(a)

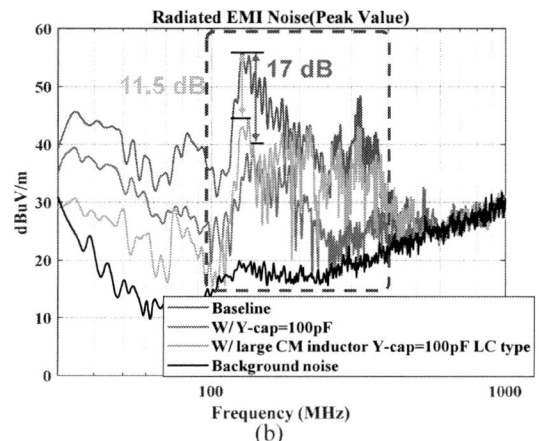

(b)

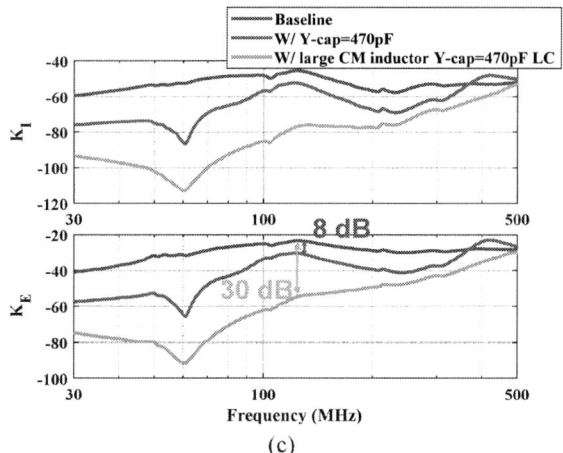

(c)

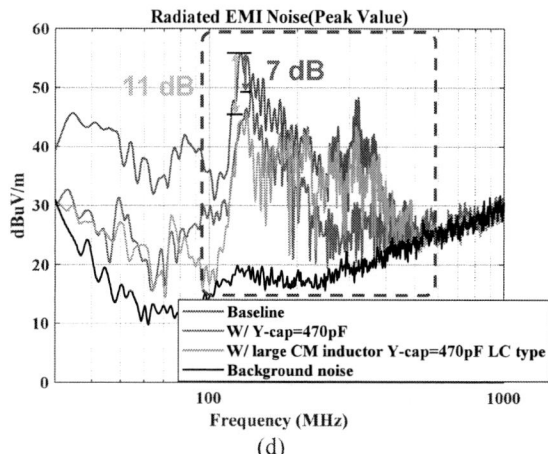

(d)

Fig. 4. Comparison of calculated K_I and K_E and measured radiated EMI.

C. Identifying dominant radiation antenna

The significant difference between theoretically calculated reduction K_E and measured radiated EMI reduction in the 100MHz to 400MHz range indicates that the previously radiated EMI antenna, comprising the input and output cables, is not dominant. The LC filter largely reduced the voltage across the input cable and output cable, but the radiated EMI noise raised within 100MHz to 400MHz range, which means the radiated EMI noise from the antenna made up of input and output cables is not dominant.

To identify the new dominant antenna of radiated EMI noise, three more experiments were conducted: short input cable (53 cm) with normal output cable (39cm), short output cable (53 cm) with normal input cable (34cm), and shortening both input and output cables. , as shown in Fig. 5. All of these cases are equipped with LC EMI filters.

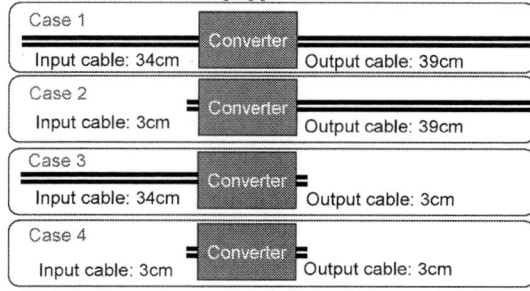

Fig. 5. Experiment setups to identify main radiation antenna.

The abbreviation of the input cable had minimal impact on the radiated EMI while altering the output cable greatly affected the radiated EMI in the 100MHz to 400MHz range. This suggests that the radiated EMI is mainly from the output cable, not the input cable. In addition, the original radiation antenna made up of input and output cables is not the dominant radiated EMI noise source.

979-8-3315-1612-3/25 $31.00 © 2025 IEEE

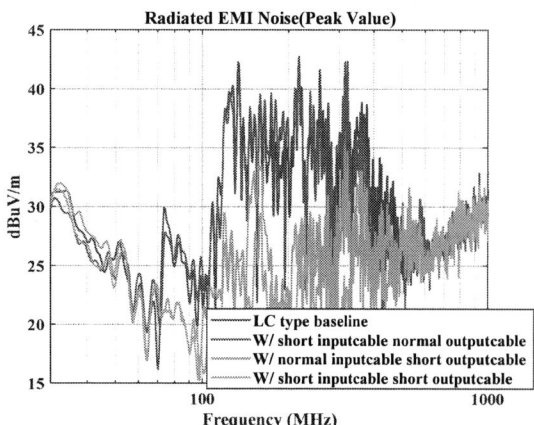

Fig. 6. Radiated EMI with input/output cable changing.

III. RADIATED EMI MODEL FOR DAB CONVERTER WITH LC FILTER

A. Analysis of radiation antennas in power converter systems

The antenna, which is comprised of input and cables, can easily become the dominant radiated emission due to its large electrical dimension, as shown in Fig. 7. The E-field between the input and output cables causes detached electromagnetic field emission to free space. The substitution theory is widely used to model the non-linear behavior of switches in building the model for the radiated EMI. The upper switches in the H-bridge are usually replaced with current sources, I_{Q1}, I_{Q3}, $I_{Q5,}$ and I_{Q6}. The lower switches in the H-bridge are usually replaced with voltage sources V_{Q2}, V_{Q4}, V_{Q7}, V_{Q8}. The current sources do not contribute to the voltage across the input and output antenna since the impedance of the DC-link capacitor is much smaller. Then only the contribution from voltage sources V_{Q2}, V_{Q4}, V_{Q7}, V_{Q8} are considered in Fig. 7.

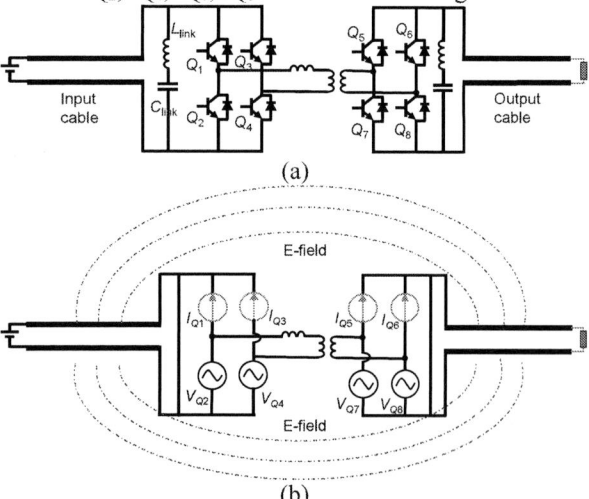

Fig. 7. Radiation antenna comprised of input and output cables.

To reduce the voltage across the input cable and output cable, the LC-type filter is introduced into the DAB inverter, as shown in Fig. 8(a). Based on the superposition theory, the contribution of voltage sources V_{Q2} and V_{Q4} can be analyzed first with V_{Q7} and V_{Q8} as short-circuits, as shown in Fig. 8 (b). The impedance of the Y-cap is much smaller than that of the CM choke. As a result, the Y-cap can be treated as a short circuit. Since the impedance of the CM choke is also much larger than the impedance of the power converter, most of the CM voltage would be applied across the CM choke. One side of the CM choke is connected to the power converter which consists of a transformer, an inductor, and traces. The other side of the CM choke is connected to input and output cables. As a result, there are strong electrical fields between the power converter copper area and the input/output cables copper area due to the voltage across the CM choke. This stray electrical field would generate radiated emission to the free space. Similarly, the V_{Q7} and V_{Q8} can generate large CM voltage across CM choke. The voltage of the CM choke can also induce electrical fields between the power converter and input/output cables.

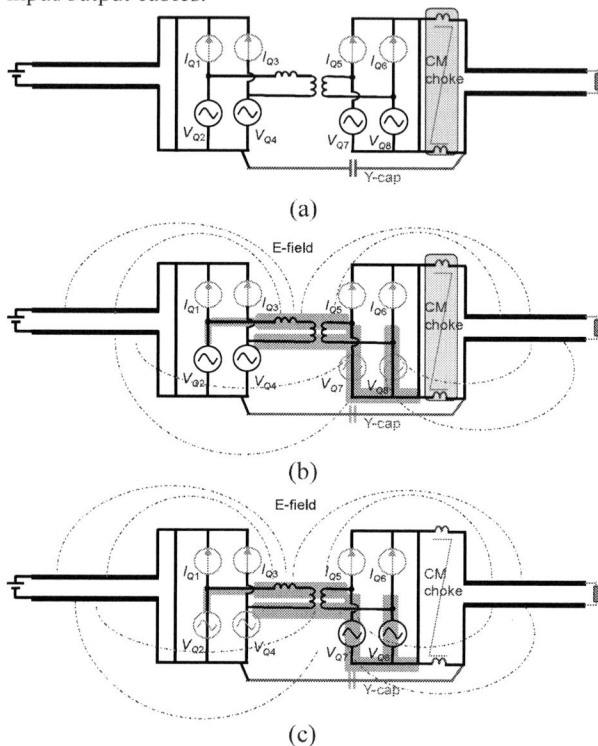

Fig. 8. Radiation antenna between power converter and input/output cables.

B. Experiments to verify the antenna between the power converter and input/output cables

To verify that the voltage across the CM choke is the radiated EMI excitation, a square wave generated by the signal generator is added to the CM choke, as shown in Fig. 9. The transfer gain from the voltage across the CM choke to radiated EMI can be extracted with the formula $Gain = \frac{E}{V}$. The transfer gain with the power converter working in normal conditions can also be extracted by measuring the voltage across the CM choke and radiated EMI simultaneously. The two transfer gain curves are shown in Fig. 10. The close match indicates that the radiated EMI primarily comes from the combination of the power converter, input/output cable, and CM choke.

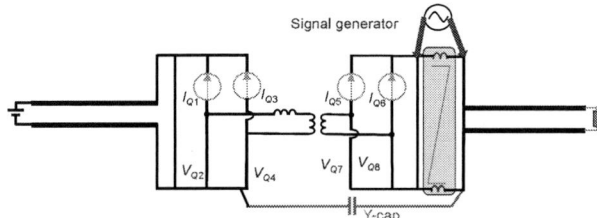

Fig. 9. Radiated EMI measurement with signal generator generation.

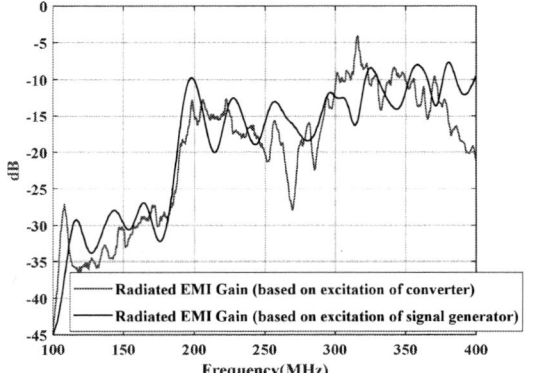

Fig. 10. Comparison of radiated EMI gain between excitations.

Adding shielding to boards to reduce the EMI noise

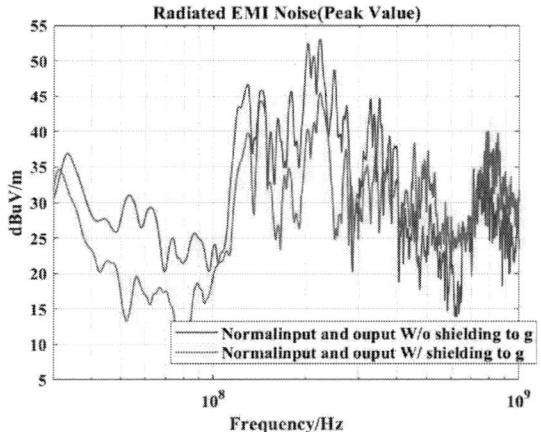

Fig. 11. Radiated EMI with and without shielding.

IV. TECHNIQUES TO REDUCE RADIATED EMI NOISE

The voltage across the CM choke is the excitation which causes the radiated EMI noise between 100MHz and 400MHz. To reduce the EMI noise between this range, the voltage across the CM choke should be reduced.

A. Applying capacitor to decrease voltage across CM choke

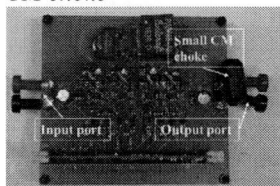 

Fig. 12. Picture of small CM choke and large CM choke.

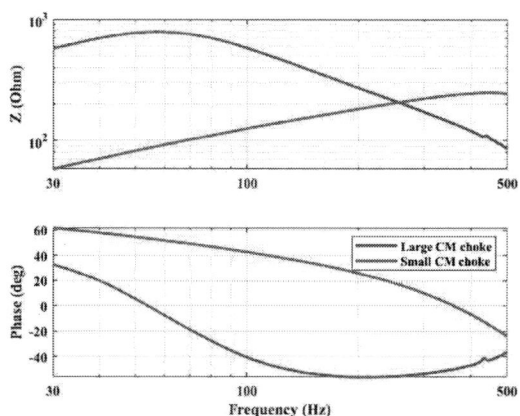

Fig. 13. Measured impedance curves of small CM choke and large CM choke.

Fig. 12 displays the physical pictures of large and small CM choke. The impedance curves of the two CM chokes are shown in Fig. 13. Fig. 14 (a) compares radiated EMI with different CM chokes. The radiated EMI noise between 100MHz and 400MHz is higher with a large CM choke, followed by a smaller one, and without any CM choke, which matches with the theoretical analysis before. Furthermore, two 330 pF capacitors are added in parallel with the CM choke. As a result, it can help to reduce high-frequency voltage and subsequently decrease radiated EMI noise within the 100MHz to 400MHz range, as shown in Fig. 14 (b).

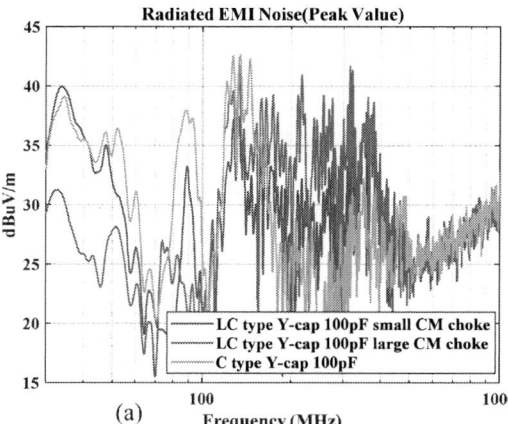

(a)

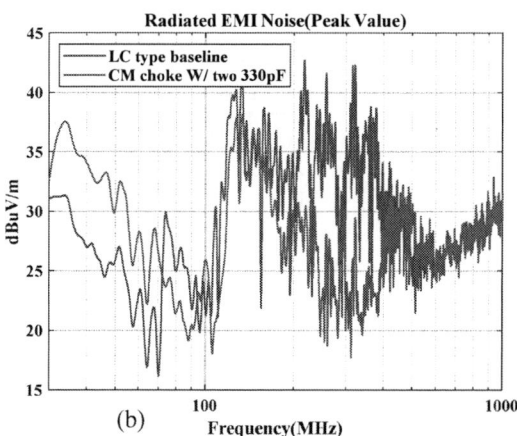

(b)

Fig. 14. (a) comparison of radiated EMI with large and small CM choke, (b) radiated EMI of CM choke with two 330pF capacitors.

B. Transfer LC-type filter to CL-type filter

The LC-type filter would naturally cause a larger voltage across the power converter and cables due to the large voltage across the CM choke. Then this LC-type filter can be changed to a CL-type filter which can help to reduce the voltage across the CM choke. In a CL-type filter, most of the CM voltage would be across the source impedance of the converter instead of the CM choke, which can also help to reduce the EMI noise between 100MHz to 400MHz. The comparison of radiated EMI results is shown in Fig. 17. The CL-type EMI filter shows lower EMI noise (15dB-20dB reduction) between 100MHz and 400MHz.

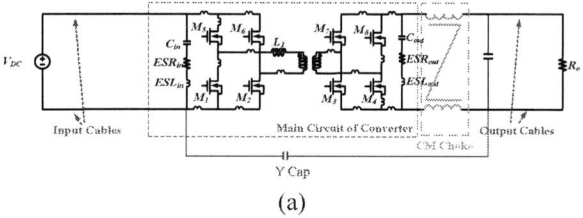

(a)

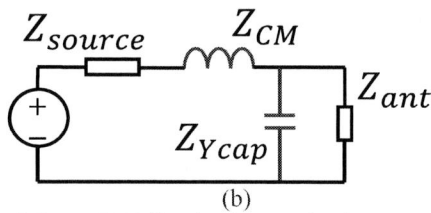

(b)

Fig. 15. LC-type EMI filter implementation in DAB converter.

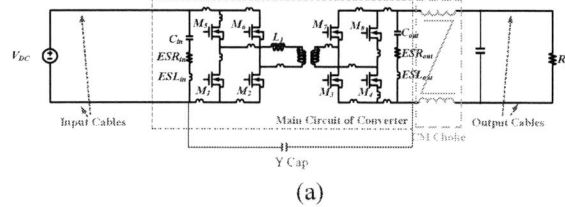

(a)

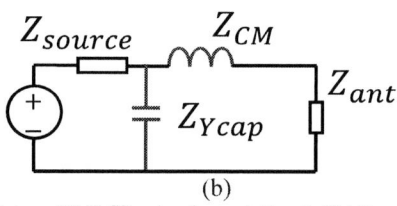

(b)

Fig. 16. CL-type EMI filter implementation in DAB converter.

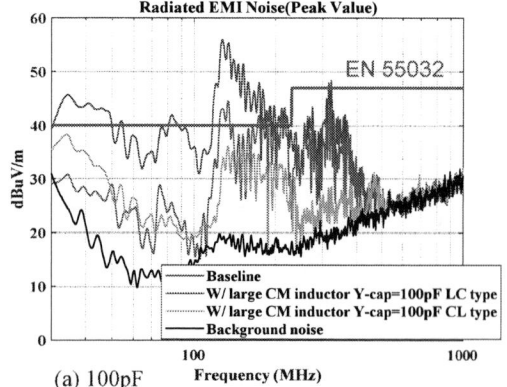

(a) 100pF

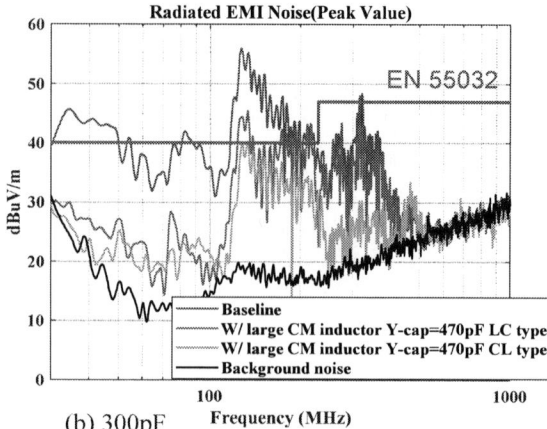

(b) 300pF

Fig. 17. Comparison of radiated EMI results between LC and CL type filters.

VI. CONCLUSION

The paper identifies a new dominant radiation antenna including the power converter and the input/output cable, after employing an LC-type filter. The antenna is mainly driven by the voltage across the CM choke. Furthermore, a radiated EMI model with an LC-type filter is proposed and validated with experiment results. Finally, two techniques are proposed to reduce the radiation EMI noise from the newly identified antenna including reducing the CM impedance of the CM choke and changing the EMI filter structure from LC-type to CL-type.

References

[1] H. Meng, M. Qiu, Z. Sun, X. Liu and D. Cao, "A 9x Matrix Autotransformer Switched-Capacitor DC-DC Converter for Datacenter Application," 2023 IEEE Energy Conversion Congress and Exposition (ECCE), Nashville, TN, USA, 2023, pp. 3305-3312, doi: 10.1109/ECCE53617.2023.10362716.

[2] X. Yang, X. Liu, Z. Zhang, C. Garcia and J. Rodríguez, "Two Effective Spectrum-Shaped FCS-MPC Approaches for Three-Level Neutral-Point-Clamped Power Converters," 2020 IEEE 9th International Power Electronics and Motion Control Conference (IPEMC2020-ECCE Asia), Nanjing, China, 2020, pp. 1011-1016, doi: 10.1109/IPEMC-ECCEAsia48364.2020.9367780.

[3] H. Gao et al., "Improved Short-Circuit Protection Method for SiC MOSFET Split-Output Power Module Achieving Ultra-Low Fault Dissipated Energy," in IEEE Transactions on Power Electronics, vol. 39, no. 2, pp. 2270-2280, Feb. 2024, doi: 10.1109/TPEL.2023.3338621.

[4] M. Qiu, M. Wei, X. Liu, H. Meng and D. Cao, "A Matrix Autotransformer Switched-Capacitor Converter for Data Center Application," in IEEE Transactions on Power Electronics, vol. 38, no. 12, pp. 14982-14999, Dec. 2023, doi: 10.1109/TPEL.2023.3308888.

[5] H. Cao, N. Lin, P. Darvish, Y. Yang, Z. Wang and Y. Zhao, "Enhanced Triple Phase Shift Modulation Strategy for ANPC-DAB Converter to Extend Soft Switching Range," 2024 IEEE Applied Power Electronics Conference and Exposition (APEC), Long Beach, CA, USA, 2024, pp. 445-452.

[6] S. Fan et al., "Inherent SM Voltage Balance for Multilevel Circulant Modulation in Modular Multilevel DC–DC Converters," in IEEE Transactions on Power Electronics, vol. 37, no. 2, pp. 1352-1368, Feb. 2022, doi: 10.1109/TPEL.2021.3105122.

[7] T. Shi et al., "Detecting Speed Improvement and System Stability Enhancement for DC Microgrids Islanding Detection Based on Impedance Characteristic Analysis," in IEEE Transactions on Power Electronics, vol. 38, no. 3, pp. 3785-3802, March 2023, doi: 10.1109/TPEL.2022.3221741.

[8] Q. Yang, A. Nabih, R. Zhang, Q. Li and Y. Zhang, "A Converter Based Switching Loss Measurement Method for WBG Device," 2023 IEEE Applied Power Electronics Conference and Exposition (APEC), Orlando, FL, USA, 2023, pp. 8-13, doi: 10.1109/APEC43580.2023.10131509.

[9] J. Zhou, H. Liu, T. Zhao, X. Xu and Y. Wang, "SiC Bidirectional Solid-State Circuit Breaker with Soft-Start Function for Motor Control Center," 2023 IEEE Applied Power Electronics Conference and Exposition (APEC), Orlando, FL, USA, 2023, pp. 2307-2312, doi: 10.1109/APEC43580.2023.10131339.

[10] Q. Yang, F. Jin and Q. Li, "An Accurate Temperature-Based Method for Fast Switching Loss Extraction of WBG Device," 2024 IEEE Applied Power Electronics Conference and Exposition (APEC), Long Beach, CA, USA, 2024, pp. 2183-2187, doi: 10.1109/APEC48139.2024.10509222.

[11] X. Liu, H. Peng and H. Xiao, "A Non-Uniform Planar Coil in Electromagnetic Vibration Energy Harvesting for Enhanced Output Power," in *IEEE Transactions on Magnetics*, vol. 58, no. 8, pp. 1-10, Aug. 2022, Art no. 8002610, doi: 10.1109/TMAG.2022.3177276.

[12] H. Cao, G. Zhu, F. Diao and Y. Zhao, "Novel Power Decoupling Methods for Three-Port Triple-Active-Bridge Converters," 2022 IEEE Applied Power Electronics Conference and Exposition (APEC), 2022, pp. 1833-1837.

[13] Q. Huang, Y. Yang, Y. Lai, Z. Ma and S. Wang, "A Survey of CM EMI Modeling and Reduction Technique of Transformer for Isolated Converters,s" 2024 IEEE Applied Power Electronics Conference and Exposition (APEC), Long Beach, CA, USA, 2024, pp. 1484-1490, doi: 10.1109/APEC48139.2024.10509196.

[14] A. Zhu *et al.*, "Accurate Multiport Network Model for Forced Oscillations Analysis and Suppression of Multichip IGBT Power Modules," in *IEEE Transactions on Power Electronics*, vol. 39, no. 1, pp. 552-569, Jan. 2024, doi: 10.1109/TPEL.2023.3325246.

[15] Z. Ma, S. Wang, Q. Huang and Y. Yang, "A Review of Radiated EMI Research in Power Electronics Systems," in IEEE Journal of Emerging and Selected Topics in Power Electronics, vol. 12, no. 1, pp. 675-694, Feb. 2024, doi: 10.1109/JESTPE.2023.3335972.

[16] J. Yao, S. Wang and Z. Luo, "Modeling, Analysis, and Reduction of Radiated EMI Due to the Voltage Across Input and Output Cables in an Automotive Non-Isolated Power Converter," in IEEE Transactions on Power Electronics, vol. 37, no. 5, pp. 5455-5465, May 2022, doi: 10.1109/TPEL.2021.3128628.

[17] Z. Ma, Y. Lai, Y. Yang, Q. Huang and S. Wang, "Review of Radiated EMI Modeling and Mitigation Techniques in Power Electronics Systems," 2023 IEEE Applied Power Electronics Conference and Exposition (APEC), Orlando, FL, USA, 2023, pp. 1776-1783, doi: 10.1109/APEC43580.2023.10131552.

[18] Q. Huang, Y. Li, Z. Ma, Y. Yang, Y. Lai and S. Wang, "RLC Balance Technique of Transformer to Reduce CM EMI for Isolated DC-DC Converters," *2023 IEEE Energy Conversion Congress and Exposition (ECCE)*, Nashville, TN, USA, 2023, pp. 2945-2952.

[19] N. Jia, X. Tian, L. Xue, H. Bai, L. M. Tolbert and H. Cui, "Integrated Common-Mode Filter for GaN Power Module With Improved High-Frequency EMI Performance," in *IEEE Transactions on Power Electronics*, vol. 38, no. 6, pp. 6897-6901, June 2023, doi: 10.1109/TPEL.2023.3248092.

[20] Y. Zhang and S. Wang, "Characterization and Design of Filter Inductors and Capacitors to Suppress the Radiated EMI in A Power Converter," 2022 International Power Electronics Conference (IPEC-Himeji 2022- ECCE Asia), Himeji, Japan, 2022, pp. 1082-1089, doi: 10.23919/IPEC-Himeji2022-ECCE53331.2022.9807145.

[21] Shuo Wang, F. C. Lee and W. G. Odendaal, "Using scattering parameters to characterize EMI filters," 2004 IEEE 35th Annual Power Electronics Specialists Conference (IEEE Cat. No.04CH37551), 2004, pp. 297-303 Vol.1, doi: 10.1109/PESC.2004.1355759.

Simple Prediction Method for Impacts of Switching Characteristics on EMI Noise of a Three-phase PWM Inverter

Shinobu Nagasawa[*], Toshiya Tadakuma[†] and Keita Takahashi[*]
Email: {Nagasawa.Shinobu, Tadakuma.Toshiya, Takahashi.Keita}@{da, bp, df}.MitsubishiElectric.co.jp
[*]Advanced Technology R&D Center, Mitsubishi Electric Corporation, Hyogo, Japan
[†]Power Device Works, Mitsubishi Electric Corporation, Fukuoka, Japan

Abstract— This article proposes a simple prediction method for impacts of switching characteristics on electromagnetic interference (EMI) noise of a three-phase pulse width modulated (PWM) inverter. In this method, target EMI noises are obtained by multiplying transfer functions of noise propagation paths and frequency spectra converted from time-domain switching voltage and current waveforms of a power device as noise sources. This method has two key features. First, it emulates the time domain scanning method in compliance with CISPR16-1-1, enabling predictions of peak, quasi-peak (QP) and average value of EMI noise. Second, it incorporates the dependence of the switching waveforms on output current into the EMI noise prediction. Finally, this article presents comparisons between measurement and prediction results of radiated EMI noise generated from a three-phase motor drive system with a three-phase PWM inverter, showing good agreement, respectively.

Keywords—electromagnetic interference (EMI), power electronics, S-parameter, transfer function, Fourier transform, time domain scanning, CISPR 16-1-1

I. INTRODUCTION

The switching speed of power devices has been getting faster with the transition from Si devices to wide band gap devices such as SiC and GaN, which contributes to higher energy efficiency but also makes EMI issues more severe [1], [2]. Therefore, it is important to quickly identify the preferred power devices that offer the optimal balance between high efficiency and low EMI noise without expending significant resources. To address this issue, methods for predicting the efficiency in the early stages of power device design and production process have been proposed [3], [4]. Regarding EMI noise, time-domain simulations with equivalent circuit models have often been proposed [5]. However, they often require the models for cables and parasitic couplings between power electronic components caused by high-density packaging for size reduction [6]. The models are constructed with a huge number of RLC circuit elements [5], [6], thereby taking long computation times. Therefore, we have focused on a full-wave 3-D electromagnetic simulation in frequency-domain, which does not require a huge number of RLC circuit elements, and proposed a dedicated noise-source model for a single-phased power circuit [7]. However, it requires expert skills for the 3-D modeling.

Therefore, this article proposes a new EMI noise prediction method without complicated simulations. In addition, this method covers more complicated circuit such as a three-phase PWM inverter. In this method, target EMI noises are obtained by multiplying frequency spectra converted from time-domain switching voltage and current waveforms of a power device as noise sources and transfer functions of noise propagation paths from the noise sources to an EMI noise measurement apparatus (e.g., EMI test receiver). Here, if the EMI noise measurement setup is fixed, it is enough to measure the transfer functions only once. In other words, by using the measured transfer functions repeatedly, EMI noises in various power devices can be evaluated without reconstructing the setup each time. Therefore, this method is useful for predicting and evaluating the impacts of power device characteristics on the EMI noise performance of the final product before the installation.

II. PROPOSED METHOD

First, in order to multiply the frequency-domain transfer functions, the time-domain switching waveforms must be converted into the frequency-domain spectra based on a half-bridge circuit where two pairs of switching and rectifying elements connected in parallel are arranged vertically. In this article, all elements having switching and rectifying functions in power devices are referred to as transistors and diodes, respectively. Note that the frequency-domain simulation can only represent systems as a no-varying topology [8]. Hence, the half-bridge circuit is represented as two stable states of a commutation cell composed of a transistor and a diode: diode ON, transistor OFF and transistor ON, diode OFF shown in Fig. 1. Here, the ON-state is represented by a voltage source and a series-connected ON-resistance of the power device because of its low impedance, whereas the OFF-state is represented by a current source and a parallel-connected junction capacitance of the power device because of its high impedance. Furthermore, when the direction of the output current of the commutation cell is forward, the transistor and the diode are respectively arranged in the upper and lower arm, while when the direction is reverse, they are arranged in opposite positions.

The waveforms to be injected into each voltage and current source are acquired by the double pulse test described later. Each acquired waveform is converted into the frequency spectrum by multiplying the window function that satisfies the overall selectivity specified in CISPR16-1-1 [9] and then applying complex Fourier transform. Kaiser window is used for the window function [10]. The method for converting the waveforms into the frequency spectra is described in [7, eq. (6)].

979-8-3315-1612-3/25 $31.00 © 2025 IEEE

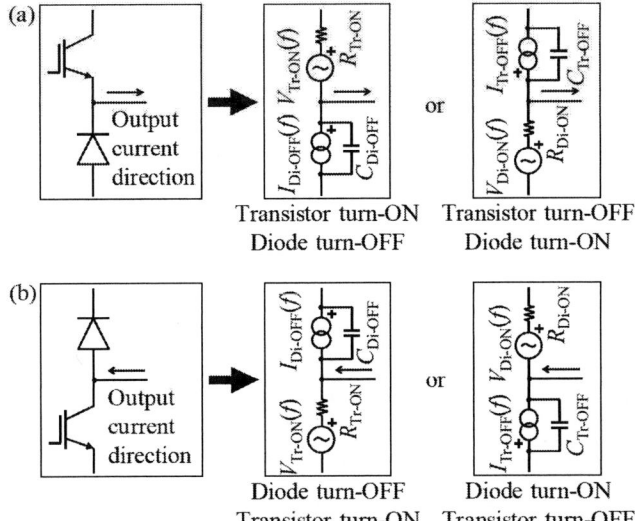

Fig. 1. Two stable states of a commutation cell when the direction of output current is (a) forward and (b) backward.

Note that the length of the Kaiser window T is not 22.3 μs in this article, different from [7]. It is determined to satisfy the following conditions. First, in [7, eq. (3)], −6 dB is used instead of −3 dB in order to satisfy the overall selectivity specified in CISPR16-1-1 [9]. Moreover, resolution bandwidth (RBW) f_{RBW} is set to 120 kHz for the Band C/D of CISPR16-1-1 (30 − 1000 MHz) [9]. As a result, T is derived as 26.25 μs.

Second, the transfer functions are described. The transfer functions are obtained by replacing the each noise source (i.e., the voltage and the current source described above) and the EMI noise measurement apparatus (e.g., EMI test receiver) with port i and j of a vector network analyzer (VNA), respectively. Then the S-parameter matrix is measured. Afterward, a reference impedance transformation is performed because the internal impedances of the VNA and the EMI test receiver are 50 Ω, whereas those of the noise sources are not 50 Ω. The internal impedances of the noise sources in the ON and OFF-state are determined by the ON-resistance and the junction capacitance of the power device, respectively. Therefore, the voltages across the EMI noise measurement apparatus located at the j-th port induced by the voltage source $V(f)$ and the current source $I(f)$ located at the i-th port, $V_{receiver(V-source)}$ and $V_{receiver(I-source)}$, are respectively expressed as:

$$V_{receiver(V-source)} = V(f) \cdot \frac{S'_{ji}}{2} = V(f) \cdot G_{ji(V-source)}(f) \quad (1)$$

$$
\begin{aligned}
V_{receiver(I-source)} &= I(f) \cdot \frac{1}{j2\pi fC} \cdot \frac{S'_{ji}}{2} \\
&= I(f) \cdot G_{ji(I-source)}(f)
\end{aligned}
\quad (2)
$$

where S'_{ji} is the S-parameter after the reference impedance transformation [11]. j in $1/(j2\pi fC)$ is imaginary unit. C is the junction capacitance of the power device. Injecting the diode and the transistor switching current spectrum $I_{Di-OFF}(f)$ and $I_{Tr-OFF}(f)$ into $I(f)$ results in injecting the diode and the transistor junction capacitance C_{Di-OFF} and C_{Tr-OFF} into C, respectively. $G_{ji(V-source)}(f)$ and $G_{ji(I-source)}(f)$ are the transfer functions from the port i where the voltage and the current sources are located to the port j where the EMI noise measurement apparatus is located, respectively.

Then $V_{receiver(V-source)}$ and $V_{receiver(I-source)}$ are summed as complex numbers in order to consider the time difference between the voltage and the current source. Note that when injecting the waveforms that include ringing noises into the voltage and current sources, the ON-resistances and the junction capacitances must be neglected as short and open, respectively. This is because the ringing noises are caused by the ON-resistances and the junction capacitances, and failing to neglect them would result in duplicating the ringing noises [7]. These relationships are summarized in Fig. 2.

Finally, the method for combining the voltages across the EMI noise measurement apparatus generated by turn-ON and OFF of the transistor derived above is described. Each of the voltages generated at the turn-ON and OFF of the transistor is multiplied by the window function value as a weighting factor and then summed as complex numbers in order to reflect the time difference between the turn-ON and OFF of the transistor on the EMI noise prediction result. Since the transistor turns ON and OFF periodically with the period T_g (e.g., the fundamental wave period in PWM) in the equipment-under-test (EUT) for EMI noise measurement, the window function needs to be swept over the period T_g as shown in Fig. 3. The window function is swept discretely with 93% overlap with the previous one, to ensure that the deviation from the EMI noise prediction result obtained when the window function is swept continuously is 0.4 dB or less [12]. The noise voltage spectrum obtained by the k-th window function, V_k, is expressed as:

$$
V_k = \sum_{j=1}^{\infty} \left\{
\begin{aligned}
&w\left(t_{ONj} - \frac{T}{2} - 0.07kT\right) \cdot \\
&V_{receiver(Tr-ON)}\left(i_o(t_{ONj})\right) e^{-j2\pi f t_{ONj}} \\
&+ w\left(t_{OFFj} - \frac{T}{2} - 0.07kT\right) \cdot \\
&V_{receiver(Tr-OFF)}\left(i_o(t_{OFFj})\right) e^{-j2\pi f t_{OFFj}}
\end{aligned}
\right\}
\quad (3)
$$

where $w(t)$ is a characteristic of the Kaiser window used as a window function, described in detail in [7, eq. (1)]. Note that w_0 is not multiplied in $w(t)$ in this article because the w_0 is multiplied in [7, eq. (6)], otherwise w_0 is duplicated. T is the length of the Kaiser window. t_{ONj} and t_{OFFj} are the times at the j-th turn-ON and OFF of the transistor, respectively. $V_{receiver(Tr-ON)}(i_o(t_{ONj}))$ and $V_{receiver(Tr-OFF)}(i_o(t_{OFFj}))$ are the voltages across the EMI noise measurement apparatus at the j-th turn-ON and OFF of the transistor, respectively. $i_o(t_{ONj})$ and $i_o(t_{OFFj})$ are the output current values of the EUT for EMI noise measurement at the j-th turn-ON and OFF of the transistor, respectively. By (3), it is possible to consider the switching waveforms causing the noise voltage spectra depend on the output current. $e^{-j2\pi ft}$ is a time delay term for considering the time difference between each turn-ON and OFF of the transistor. j in $e^{-j2\pi ft}$ is imaginary unit. Equation (3) is calculated using complex numbers.

Equation (3) is for the topology having one arm. For the topology having N arms, the (3) is summed for N gate signals whose phases are shifted by $2\pi/N$. For example, $N=3$ for a three-phase PWM inverter. Furthermore, for a three-phase PWM inverter, the voltage and current sources representing the noise

(a) When the direction of output current is forward

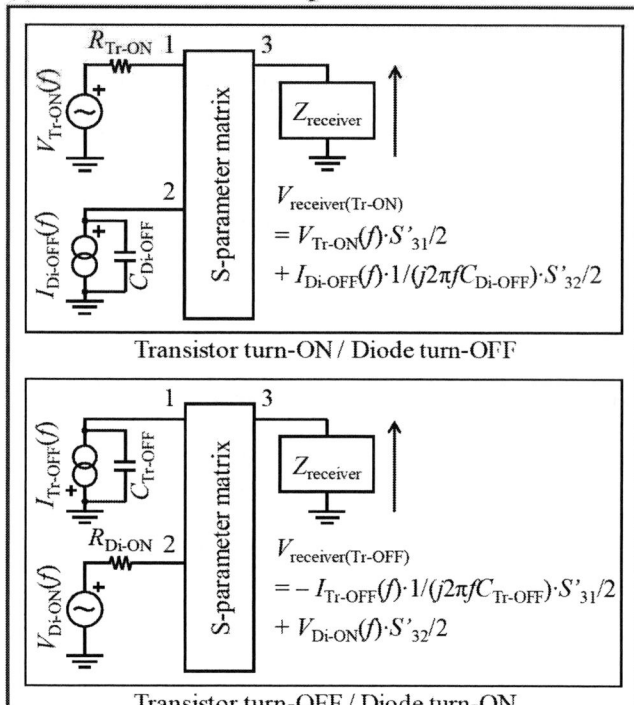

Transistor turn-ON / Diode turn-OFF

$$V_{\text{receiver(Tr-ON)}} = V_{\text{Tr-ON}}(f)\cdot S'_{31}/2 + I_{\text{Di-OFF}}(f)\cdot 1/(j2\pi f C_{\text{Di-OFF}})\cdot S'_{32}/2$$

Transistor turn-OFF / Diode turn-ON

$$V_{\text{receiver(Tr-OFF)}} = -I_{\text{Tr-OFF}}(f)\cdot 1/(j2\pi f C_{\text{Tr-OFF}})\cdot S'_{31}/2 + V_{\text{Di-ON}}(f)\cdot S'_{32}/2$$

(b) When the direction of output current is backward

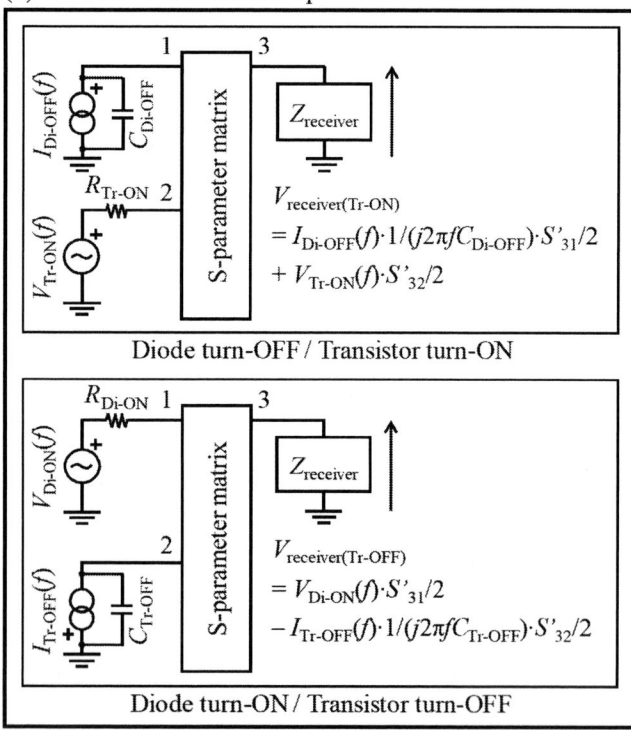

Diode turn-OFF / Transistor turn-ON

$$V_{\text{receiver(Tr-ON)}} = I_{\text{Di-OFF}}(f)\cdot 1/(j2\pi f C_{\text{Di-OFF}})\cdot S'_{31}/2 + V_{\text{Tr-ON}}(f)\cdot S'_{32}/2$$

Diode turn-ON / Transistor turn-OFF

$$V_{\text{receiver(Tr-OFF)}} = V_{\text{Di-ON}}(f)\cdot S'_{31}/2 - I_{\text{Tr-OFF}}(f)\cdot 1/(j2\pi f C_{\text{Tr-OFF}})\cdot S'_{32}/2$$

Fig. 2. Voltages across the EMI noise measurement apparatus expressed by S-parameter when the direction of output current is (a) forward and (b) backward. Z_{receiver} is the impedance of the EMI noise measurement apparatus.

sources are assigned to only one of the three arms, while the other two arms are modeled such that their diagonal switch pairs

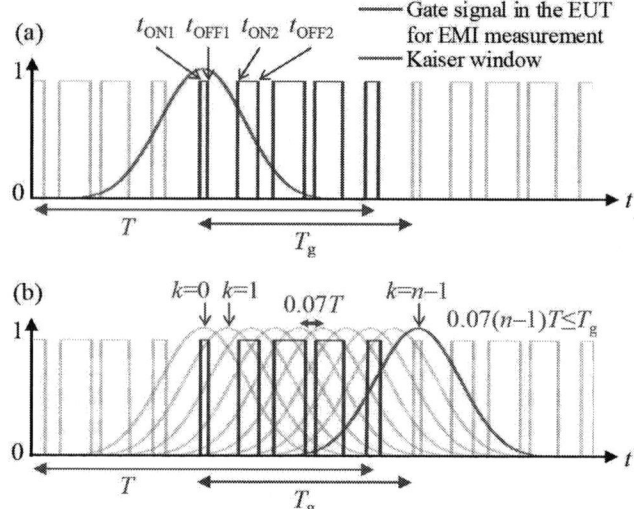

Fig. 3. Method for combining the voltages across the EMI noise measurement apparatus generated by turn-ON and OFF of the transistor. The window function is swept discretely with 93% overlap with the previous one over T_{g}.

turn-ON and OFF simultaneously, respectively. That is, it is assumed that the same EMI noise is generated by any arm switching. Although assigning the noise source to every arm brings the EMI noise prediction result closer to the actual measurement result, it is not preferable due to the increased computational complexity.

The values corresponding to the peak and the average value of the EMI test receiver are obtained by calculating the maximum and the average value of V_k in (3) where k is varied from 0 to $n-1$, respectively. That of the QP value is obtained by inputting V_k into the digitized QP detector where k is varied from 0 to $n-1$. A conventional analog QP detector composed of a combination of an RC charging circuit and a critically damped meter can be digitized with first-order infinite impulse response (IIR1) filters shown in Fig. 4 [13]. The filter coefficients b_0, b_1, a_1, b_{0m}, b_{1m} and a_{1m} are derived based on the bilinear transformation and the time constants of the conventional analog QP detector. The RC circuit is charged with the time constant τ_c if the input signal $u(k\Delta t)$ is above the output signal $u_2(k\Delta t)$.

Initial condition: $u(0)=u_2(0)=u_3(0)=u_4(0)=0$

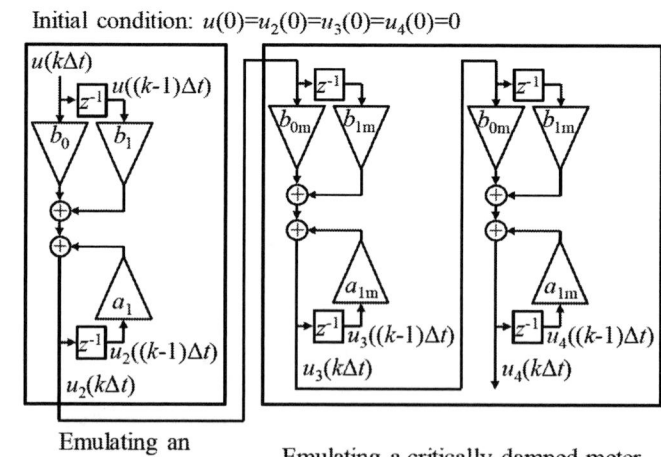

Emulating an RC charging circuit   Emulating a critically damped meter

Fig. 4. Digitized QP detector with IIR1 filters.

Meanwhile, it is discharged with the time constant τ_d if $u(k\Delta t)$ is lower than $u_2(k\Delta t)$. Here, $k\Delta t$ is the k-th discrete time. Δt corresponds to $0.07T$ in (3). The critically damped meter displays the amplitude of the output of the RC circuit with the time constant τ_m. Therefore, the filter coefficients are expressed as:

$$b_0 = b_1 = \begin{cases} \dfrac{\tan\left(\dfrac{\Delta t}{2\tau_c}\right)}{1 + \tan\left(\dfrac{\Delta t}{2\tau_c}\right)}, u(k\Delta t) \geq u_2(k\Delta t) \\ 0, \qquad u(k\Delta t) < u_2(k\Delta t) \end{cases} \quad (4)$$

$$a_1 = \begin{cases} \dfrac{1 - \tan\left(\dfrac{\Delta t}{2\tau_c}\right)}{1 + \tan\left(\dfrac{\Delta t}{2\tau_c}\right)}, u(k\Delta t) \geq u_2(k\Delta t) \\ \dfrac{1 - \tan\left(\dfrac{\Delta t}{2\tau_d}\right)}{1 + \tan\left(\dfrac{\Delta t}{2\tau_d}\right)}, u(k\Delta t) < u_2(k\Delta t) \end{cases} \quad (5)$$

$$b_{0m} = b_{1m} = \dfrac{\tan\left(\dfrac{\Delta t}{2\tau_m}\right)}{1 + \tan\left(\dfrac{\Delta t}{2\tau_m}\right)} \quad (6)$$

$$a_{1m} = \dfrac{1 - \tan\left(\dfrac{\Delta t}{2\tau_m}\right)}{1 + \tan\left(\dfrac{\Delta t}{2\tau_m}\right)} \quad (7)$$

where τ_c, τ_d and τ_m are 1, 550, 100 ms, respectively for the Band C/D of CISPR16-1-1 (30 − 1000 MHz) [9]. The maximum value of the output signal of the digitized QP detector $u_4(k\Delta t)$ corresponds to the QP value of the EMI test receiver. Note that the input signal $u(k\Delta t)$ must be provided up to 2 s to satisfy the requirement of CISPR16-1-1 [9].

III. VALIDATION

The proposed method is verified using a three-phase motor drive system with a three-phase PWM inverter, constructed in the anechoic chamber to measure the radiated EMI noise, as shown in Fig. 5. The inverter circuit consists of a six-in-one intelligent power module (IPM) that includes three Si-IGBTs for high side, three Si-IGBTs for low side and ICs that integrates gate drive function, which connect to each gate terminal of the IGBTs. The gate signals input to the IPM are generated by an external gate signal source, which is constituted by a micro controller unit (MCU). The DC link voltage is charged to 300 V using a 370 μF capacitor. Both the IPM and the MCU are activated by external DC power supplies. The radiated EMI noise is measured by a vertical Bilog antenna (Schwarzbeck, VULB9160) and an EMI test receiver (Rohde&Schwarz, ESR) with time domain scanning function to obtain peak, QP and average value. The antenna is positioned 3 m away from the EUT. The antenna and the EUT are placed 1 and 0.9 m above the floor, respectively. RBW of the EMI test receiver is set to 120 kHz. The measured frequency range is 10 to 100 MHz, considering the bandwidth of the current probes (Tektronix, TCP0030A, 120 MHz) used to obtain the switching current waveforms of the IPM as described below.

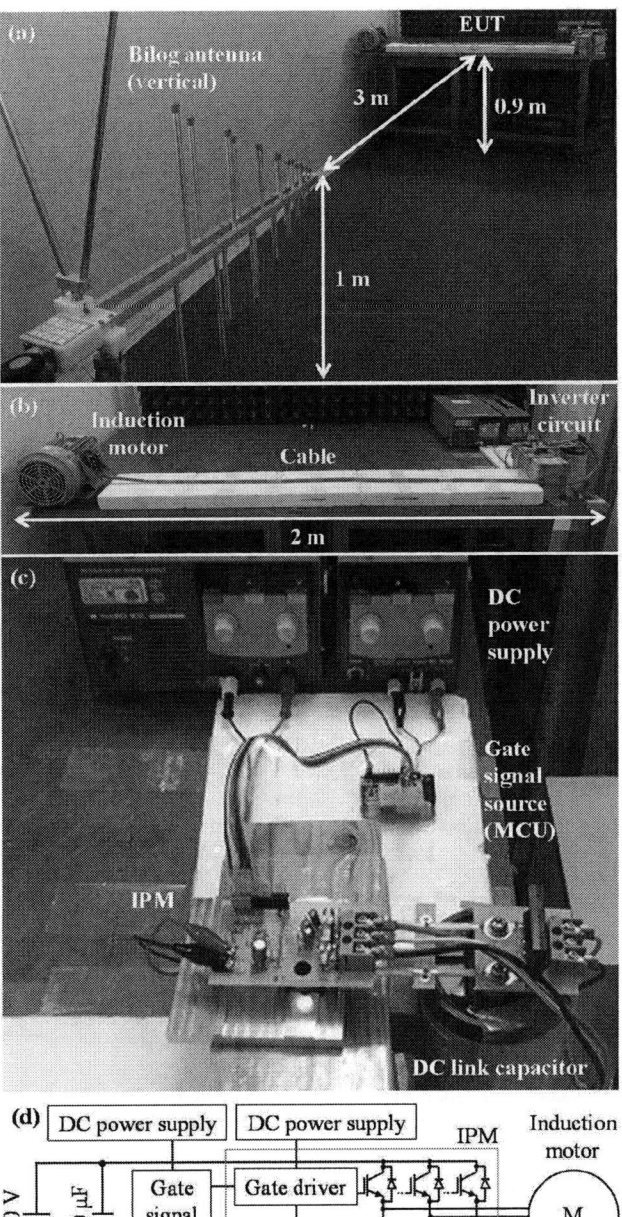

Fig. 5. Experimental setup for the validation.

The gate signals and the output current waveforms of the EUT, which are required to predict the radiated EMI noise, are also measured simultaneously using a logic probe (Tektronix, TLP058), current probes (Tektronix, TCP0030A) and an oscilloscope (Tektronix, MSO66B). The obtained output current waveforms i_o were sine waves with 1 A peak at 50 Hz as shown in Fig. 6. The obtained gate signals were sinusoidal PWM waveforms with 50 Hz of fundamental frequency, 5 kHz of carrier frequency, 0.93 of modulation index and 3 μs of dead time between upper and lower arm. These measured gate signals and the output current waveforms can be repeatedly used for the

EMI noise prediction.

The transfer functions must be measured in the same instrument layout as the EMI noise measurement. First, the upper and the lower arm of any one of the three arms of the IPM and the EMI test receiver are replaced with port 1, 2 and 3 of a VNA (Keysight, E5061B), respectively. Then the 3-port S-parameter matrix is measured. Note that each port connected to a VNA for measuring S-parameters shares common ground (GND) with the VNA. However, the upper and the lower arm of the IPM and the Bilog antenna do not share common GND while the radiated EMI noise measurement. Therefore, a pair of optical fiber link modules (SEIKOH GIKEN, ET-615 and ER-615) is used to measure S_{13}, S_{31}, S_{23} and S_{32}, which prevents them from sharing common GND. Another reason for using the optical fiber link modules is to prevent the creation of unintended sources of the radiated EMI noise. In other words, coaxial cables have been identified as the unintended sources. The ports not used during the S-parameter measurement are terminated with 50 Ω. Fig. 7 shows the S_{31} and S_{32} of the measured S-parameter matrix and the transfer functions converted from the measured S-parameter matrix. In this article, the transfer functions are calculated while neglecting the ON-resistances and the junction capacitances of the IPM as short and open, respectively. Instead, voltage and current waveforms including ringing noises are measured and injected into the voltage and the current sources, respectively, as described below. The obtained transfer functions can be repeatedly used for the EMI noise prediction unless instrument layouts are changed.

The switching voltage and current waveforms of the IPM can be measured either in the motor drive system for measuring radiated EMI noise described above or in a double pulse test. In this article, a double pulse test was selected. In the double pulse test, only one of the three arms of the IPM is used, while the remaining unused terminals are opened. The switching waveforms under various current conditions (0.2, 0.3, 0.5, 0.7, 0.9 and 1.1 A) were obtained by changing the first pulse width. Fig. 8 shows the measured waveforms and the frequency spectra converted from the measured waveforms under 0.2, 0.7 and 1.1 A current conditions. The waveforms were obtained by passive voltage probes (Tektronix, TPP1000), current probes (Tektronix, TCP0030A) and an oscilloscope (Tektronix, MSO66B).

As described above, in the proposed method, the switching voltage and current spectra under the current conditions corresponding to the output current of the EUT for the EMI noise measurement are called up to predict the EMI noise by (3). In this article, the output current can take an infinite number of values in the range of −1 to +1 A as shown in Fig. 6, but it is impossible to store the switching voltage and current spectra under an infinite number of current conditions corresponding to the output current. To address this issue, the switching voltage and current spectra for a finite number of current conditions are stored. Then, for conditions that do not match the stored current conditions, the switching voltage and current spectra are generated by linearly interpolating and extrapolating from the stored finite set of spectra, as illustrated in Fig. 9. Here, the current spectra are extrapolated to approach zero as the output current approaches zero.

Fig. 10 shows comparisons between the predicted and the

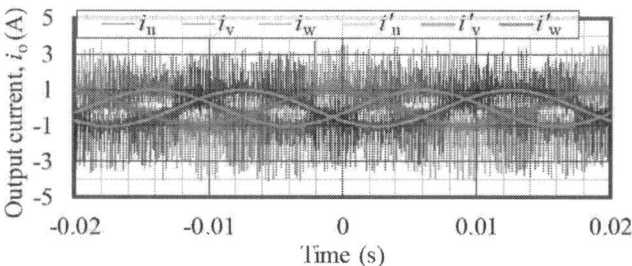

Fig. 6. Measured output current waveforms of the EUT, i_o, required to predict the radiated EMI noise. i_u, i_v and i_w are the measured output current waveforms in U, V and W phase of the EUT, respectively. i'_u, i'_v and i'_w are the waveforms after removing noises due to the PWM inverter switching from i_u, i_v and i_w, respectively.

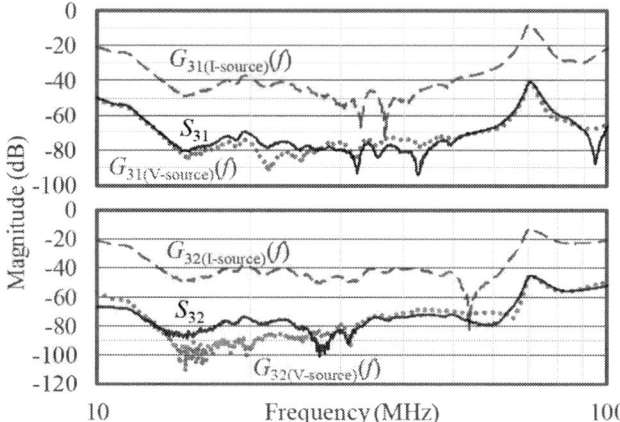

Fig. 7. Measured S-parameters and transfer functions converted from the measured S-parameter matrix.

measured results of antenna voltage as radiated EMI noise, demonstrating that the predicted peak, QP and average values agree with the measured values within approximately 10 dB, respectively. These deviations between the predicted and the measured results are believed to be caused by slight differences in the setup, such as the presence or absence of the optical fiber link module. In other words, the optical fiber link module and/or the coaxial connector between the EUT and it are considered to act as sources of the radiated EMI noise or shields.

IV. CONCLUSIONS

A simple and quick method was proposed to predict impacts of switching characteristics on EMI noise of a three-phase PWM inverter. Target EMI noises were obtained by multiplying transfer functions of noise propagation paths and frequency spectra converted from switching voltage and current waveforms of a IPM. This method enabled to predict peak, QP and average value of EMI noise by emulating the time domain scanning method in compliance with CISPR16-1-1 and by digitizing a conventional analog QP detector. The measurement and the prediction results of radiated EMI noise generated from a three-phase motor drive system using a three-phase PWM inverter agreed within approximately 10 dB in the range of 10 to 100 MHz, respectively. In future, it is possible to predict and evaluate the impacts of power device characteristics on the EMI noise performance of the final product before the installation.

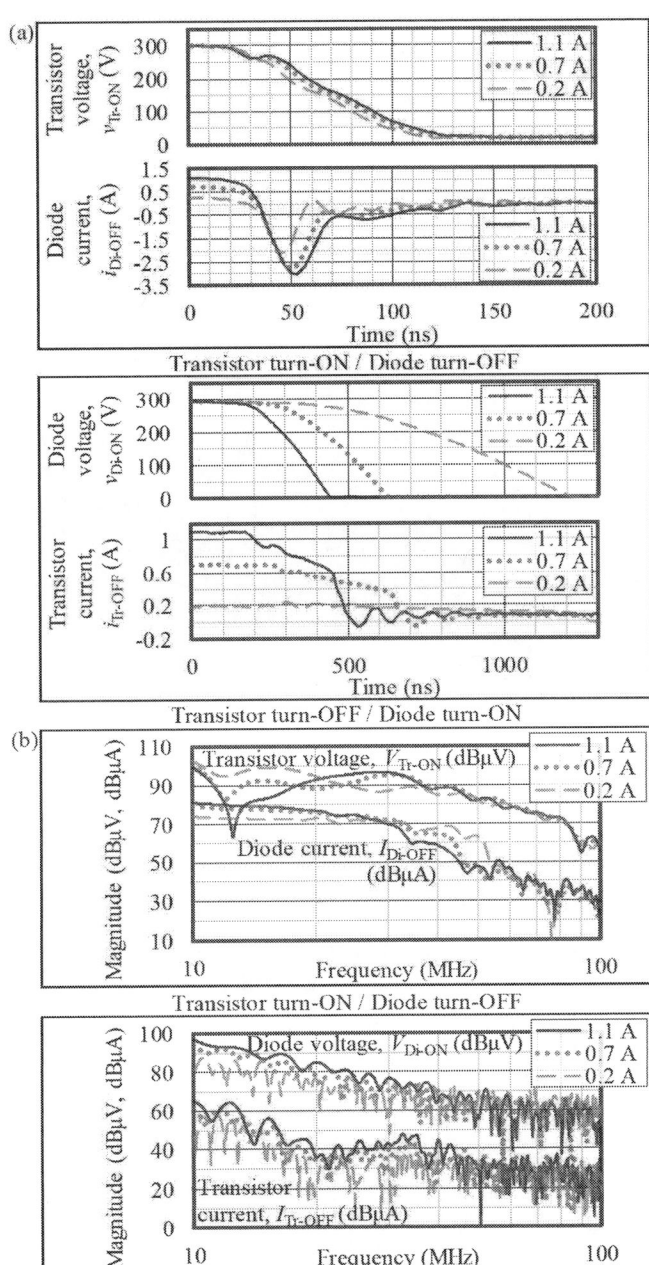

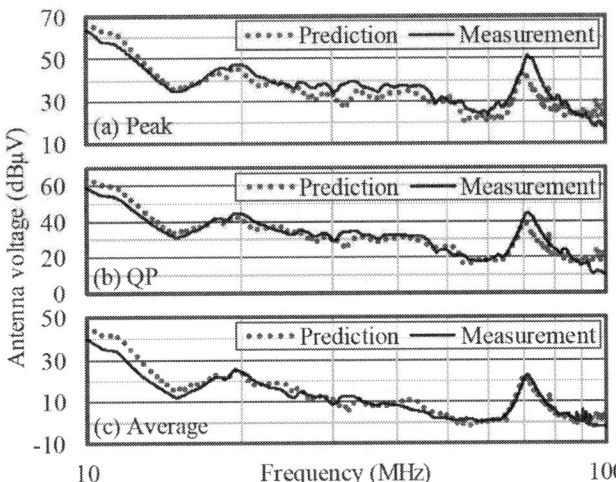

Fig. 10. Comparisons between the prediction and the measurement results of antenna voltage as radiated EMI noise.

Fig. 8. (a) Measured switching waveforms and (b) the frequency spectra converted from the measured waveforms of the IPM.

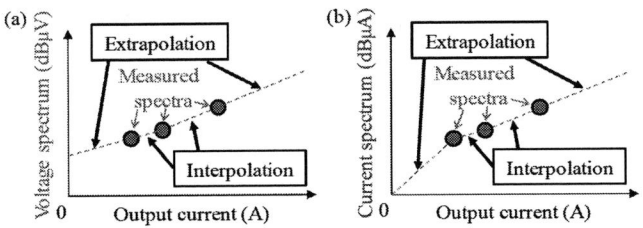

Fig. 9. How to interpolate and extrapolate to obtain the swtching (a) voltage and (b) current spectra under the output current conditions that do not match the stored current conditions.

REFERENCES

[1] N. Oswald, P. Anthony, N. McNeill, and B. H. Stark, "An experimental investigation of the tradeoff between switching losses and EMI generation with hard-switched all-Si, Si-SiC, and all-SiC device combinations," *IEEE Trans. Power Electron.*, vol. 29, no. 5, pp. 2393-2407, May 2014.

[2] D. Han, S. Li, W. Lee, W. Choi, and B. Sarlioglu, "Trade-off between switching loss and common mode EMI generation of GaN devices- analysis and solution", in *IEEE 2017 Appl. Power Electron. Conf. Expo.*, 2017, pp. 843-847.

[3] L. Zhang and X. Yuan, "Performance evaluation of high-power SiC MOSFET modules in comparison to Si IGBT modules," *IEEE Trans. Power Electron.*, vol. 34, no. 2, pp. 1181-1196, Feb. 2019.

[4] M. Amyotte, E. S. Glitz, C. G. Perez, and M. Ordonez, "GaN power switches: A comprehensive approach to power loss estimation," in *IEEE 2018 Energy Convers. Congr. Expo.*, 2018, pp. 1926-1931.

[5] M. Moreau, N. Idir, and P. L. Moigne, "Modeling of conducted EMI in adjustable speed drives," *IEEE Trans. Electromagn. Compat.*, vol. 51, no. 3, pp. 665-672, Aug. 2009.

[6] K. Takahashi, Y. Murata, Y. Tsubaki, T. Fujiwara, H. Maniwa, and N. Uehara, "Mechanism of near-field coupling between noise source and EMI filter in power electronic converter and its required shielding," *IEEE Trans. Electromagn. Compat.*, vol. 61, no. 5, pp. 1663–1672, Oct. 2019.

[7] K. Takahashi, T. Ibuchi, and T. Funaki, "Noise-source model for frequency-domain EMI simulation of a single-phased power circuit," *IEEE Trans. Electromagn. Compat.*, vol. 63, no. 3, pp. 772-782, Jun. 2021.

[8] C. Marlier, A. Videt, N. Idir, H. Moussa, and R. Meuret, "Modeling of switching transients for frequency-domain EMC analysis of power converters," in *15th Int. Power Electron. Motion Control Conf.*, pp. DS1e.1-1-DS1e.1-8, Sep. 2012.

[9] *Specification for radio disturbance and immunity measuring apparatus and methods—Part 1-1: Radio disturbance and immunity measuring apparatus—Measuring Apparatus*, CISPR 16-1-1 edition 5.0, May 2019.

[10] J. F. Kaiser and R.W. Schafer, "On the use of the I0-sinh window for spectrum analysis," *IEEE Trans. Acoust., Speech, Signal Process.*, vol. 28, no. 1, pp. 105-107, Feb. 1980.

[11] J. C. Tippet and R. A. Speciale, "A rigorous technique for measuring the scattering matrix of a multiport device with a 2-port network analyzer," *IEEE Trans. Microw. Theory Techn.*, vol. 30, no. 5, pp. 661-666, May 1982.

[12] M. Keller, "Comparison of time domain scans and stepped frequency scans in EMI test receivers," *Rohde&Schwarz White Paper*, Dec. 2013.

[13] F. Krug and P. Russer, "Quasi-peak detector model for a time-domain measurement system," *IEEE Trans. Electromagn. Compat.*, vol. 47, no. 2, pp. 320-326, May 2005.

Coaxially Nested 3.3 kV SiC MOSFET Packages with Uniform Interpackage Electric Field Distribution

Jack Knoll
Center for Power Electronic Systems
Virginia Polytechnic Institute and State University
Arlington, Virginia
knolljs@vt.edu

Mark Cairnie
Center for Power Electronic Systems
Virginia Polytechnic Institute and State University
Arlington, Virginia
mcairnie@vt.edu

Christina DiMarino
Center for Power Electronic Systems
Virginia Polytechnic Institute and State University
Arlington, Virginia
dimarino@vt.edu

Abstract—This work proposes a power semiconductor package architecture that scales well with voltage while simultaneously limiting commutation loop inductance and promoting good current balance between parallel devices. The target application is a coaxial power electronics conversion system which inherits the power density and voltage scaling benefits of a high voltage cable. This architecture is composed of a pair of coaxially nested single switch packages with similar internal structures and fabrication procedures. Aided by their coaxial commutation loop, the 3.3 kV SiC MOSFET single switch packages have a combined contribution to the commutation loop inductance of just 2.8 nH at 100 kHz in simulation. Static characteristics of the packages are comparable to bare die specifications from the datasheet and dynamic characterization demonstrates the impact of the low commutation loop inductance achieved with this structure. Using a prototype of the system's thermal management solution, the measured junction-to-case thermal resistance was found to be 0.073 °C/W for the input switch.

Keywords—*packaging, SiC MOSFET, electric field, coaxial, commutation loop inductance, current sharing, thermal resistance*

I. INTRODUCTION

Conventional planar power electronics solutions do not scale well with voltage due to their asymmetric electric fields, limiting their adoption in high voltage applications. A coaxial power converter system first proposed in [1] and seen in Fig. 1 could revolutionize distribution systems by mimicking the geometry of high voltage cables with the goal of seamless integration with the line while inheriting properties such as uniform electric fields. The unique constraints imposed by such a system disqualify many commercial packages and conventional packaging technologies. Power semiconductor packaging technology which utilizes ceramic and organic substrates for electrical isolation are limited in their ability to scale to high voltages, necessitating dramatic changes to the package architecture [2]. Discrete semiconductor packages such as TO-247 devices suffer from high internal parasitic inductances [3] and their rectangular form factor with leaded electrical interconnects curtails the benefits of the converter's coaxial structure.

Press-pack style packages are the closest to meeting the demands of such a system. They have a round form-factor and

The authors acknowledge the funding support from the U.S. Department of Energy, Advanced Research Project Agency – Energy (ARPA-E) through award: DE-AR0001568.

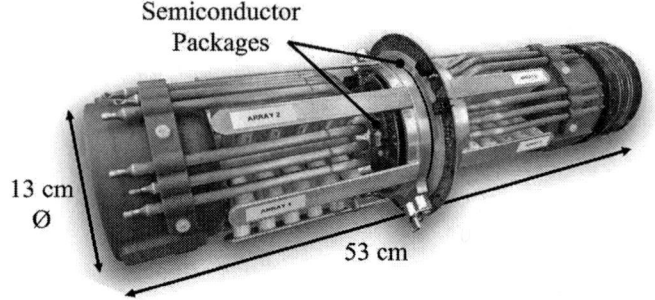

Fig. 1. 50 kW coaxial converter prototype [1].

circumvent the voltage scaling limitations of packages that contain substrates by omitting the substrate, leaving the electrical isolation to be dealt with elsewhere. However, press-packs rely on dry contacts that can increase thermal resistance, necessitate costly hermetic seals to contain the gas encapsulation, impose a strict pressure uniformity requirement, or limit current sharing among parallel devices depending on the structure chosen [4]. Press-packs also contain a single switch position which, along with their form factor, would complicate the formation of a tight commutation loop to limit parasitic inductance.

The set of coaxial 3.3 kV SiC MOSFET packages designed, fabricated, and tested in this work are, therefore, a critical component of the coaxial converter system. The proposed pair of coaxial single switch position power modules combine the voltage scaling benefit of substrate-less packaging with wetted contacts which enables the use of lower cost encapsulation materials and alleviates the uniform pressure requirement. In addition, the coaxial nesting of the modules allows for low commutation loop inductance and improved module-to-module voltage scaling. The axially symmetric distribution of parallel devices also promotes good current sharing with identical parasitics and an equivalent temperature distribution among parallel die.

II. PACKAGE ARCHITECTURE AND FABRICATION

The internal structure of the input switch package designed in this work is seen in Fig. 2 and the coaxially nested configuration of the input and output modules along with the circuit schematic of a single converter cell is seen in Fig. 3. The first subsection discussed the design decisions that led to these package structures and the benefits of axial symmetry. The next subsection describes the fabrication procedure in

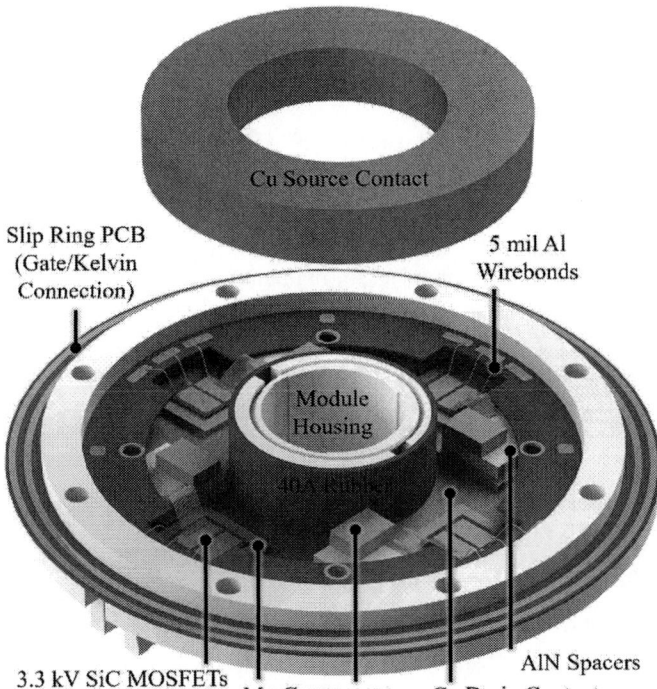

Fig. 2. A 3D render of the input package with the source contact exploded to reveal the internal structure with the critical components labeled.

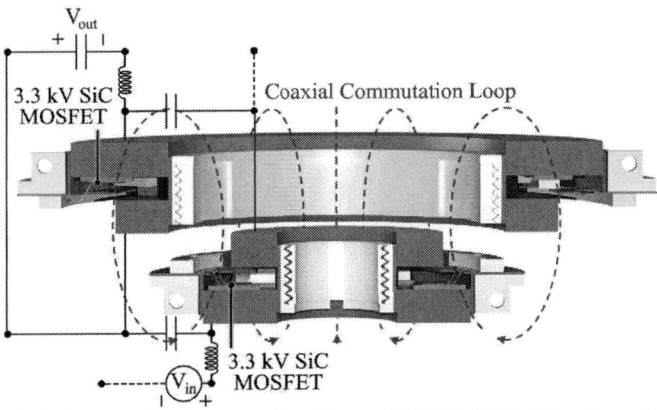

Fig. 3. A 3D rendered cross section of both 3.3 kV SiC MOSFET packages in their coaxial configuration overlaid with a schematic of a single converter cell.

detail. Finally, the benefits of the coaxially nested arrangement are discussed in the last subsection.

A. Package Design

The internal structures of both packages are intentionally similar. This strategy limits unique component counts as both packages share many of the same components. Any components unique to one package are scaled replicas of components from the other package. These include the drain and source contacts, the housing components, and the slip ring PCBs. After initially designing these components for the inner module, a simple change in radii is all that is needed to generate the design files for the outer module. The identical stack-up of the modules also allows for the use of similar fabrication procedures, limiting debugging time. Since it has

now been established that both packages have similar designs, the remaining discussion about the package design applies to both packages.

Starting with some of the more common materials and components seen in state-of-the-art power semiconductor packages, copper (Cu) was chosen for the drain and source contacts due to its high thermal conductivity (400 W/mK) and electrical conductivity (59.6×10^6 S/m). Less common place is the toroidal formfactor of these Cu components which create axially symmetric electrical and thermal contact to the coaxial converter prototype see in Fig. 1. Also uncommon is the lack of electrical isolation in the package's substrates. Here, the necessary electrical isolation is handled elsewhere in the system where the heat flux is lower to limit temperature drop at the low thermal conductivity electrical isolation barrier.

The die used in this work are 3.3 kV, 25 mΩ SiC MOSFETs from Microchip (MSC027SMA330D/S). Aluminum nitride (AlN) spacers that are 10x10x2 mm^3 and plated with 10 µm of Ag on the top and bottom are used to set the distance between the source and drain potentials within the package while providing electrical isolation between the two. The molybdenum (Mo) components, with a coefficient of thermal expansion (CTE) of 4.9 10^{-6}/°C, provide a thermal expansion buffer between the high CTE Cu contacts (17 10^{-6}/°C) and the low CTE SiC MOSFETs (3.7 10^{-6}/°C) and AlN components (4.5 10^{-6}/°C) [5]. This buffer limits the thermomechanical stress on the die, increasing package lifetime. There are two sizes of Mo components in these packages, the 12x12x1 mm^3 stages attached to the drain contact and the 5x7x2.11 mm^3 posts attached to the top of the AlN spacers.

These packages contain one component that the authors believe to be novel in the context of power semiconductor packaging: the slip ring PCBs which form the gate and kelvin electrical connections to the gate drivers. Slip ring PCBs have long been used in applications where electrical signals or power need to be passed between a rotating body and a stationary body [6]. Slip rings are commonly implemented in applications with continuous rotation such as motors or generators, so the lifetime of slip ring contacts and brushes remains a critical area of research due to their tendency to experience mechanical wear [7]. The slip rings used here are

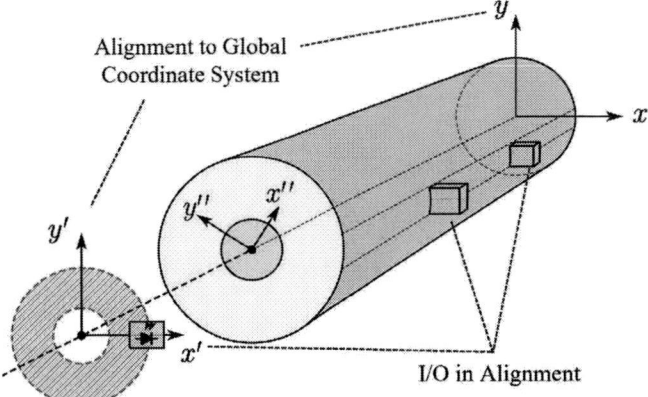

Fig. 4. Diagram demonstrating the need to align the inputs and outputs of the gate drivers to the global coordinate system of the converter.

979-8-3315-1612-3/25 $31.00 © 2025 IEEE

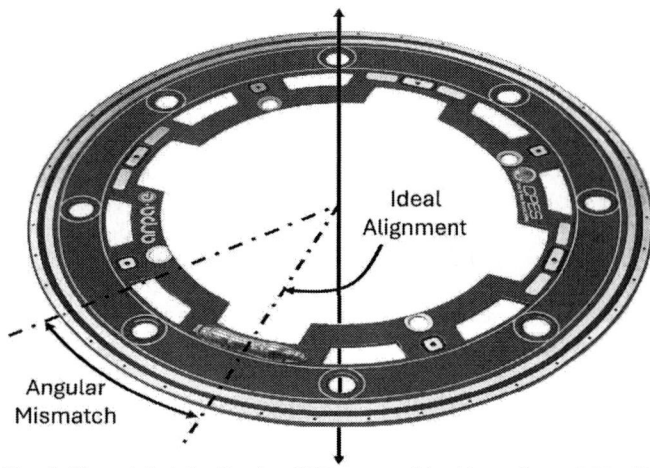

Fig. 5. The gate/kelvin slip ring PCB proposed in this work overlaid with graphics defining angular mismatch.

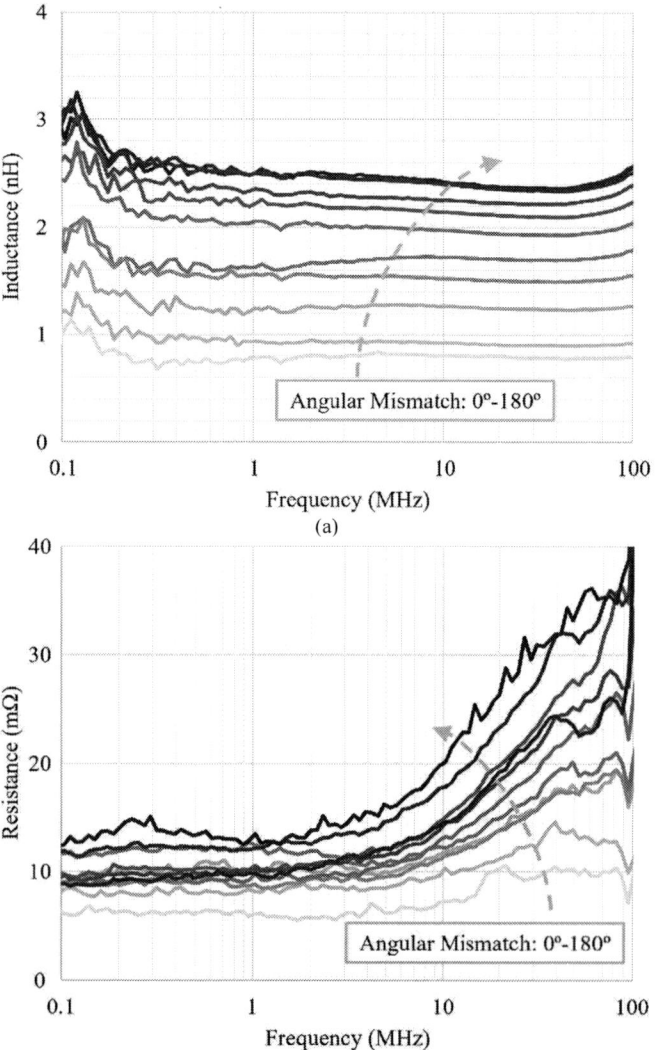

Fig. 6. The measured (a) inductance and (b) resistance contributions of the gate/kelvin slip ring PCB to the gate loop measured at different angular mismatches as a function of frequency. Note that measurements are taken at every other via along the gate trace seen in Fig. 5 or about every 18°.

not expected to experience continuous rotation, so there is no concern about their ability to withstand mechanical wear. In this case, the PCB slip rings provide modularity between the gate drivers and packages while allowing the packages to achieve axially symmetric electrical contact. The gate drivers must be clocked properly with respect to the global coordinate system of the converter to align the inputs and outputs appropriately as seen in Fig. 4. The modules and other components in the system have no such requirement. The slip ring allows the modules to maintain axial symmetry and modularity while interfacing to the gate driver with minimal impact on the drive signal.

To demonstrate the slip ring's ability to provide a dry electrical connection between the packages and gate drivers with minimal impact on the drive signal, impedance measurements were taken at various locations along the gate

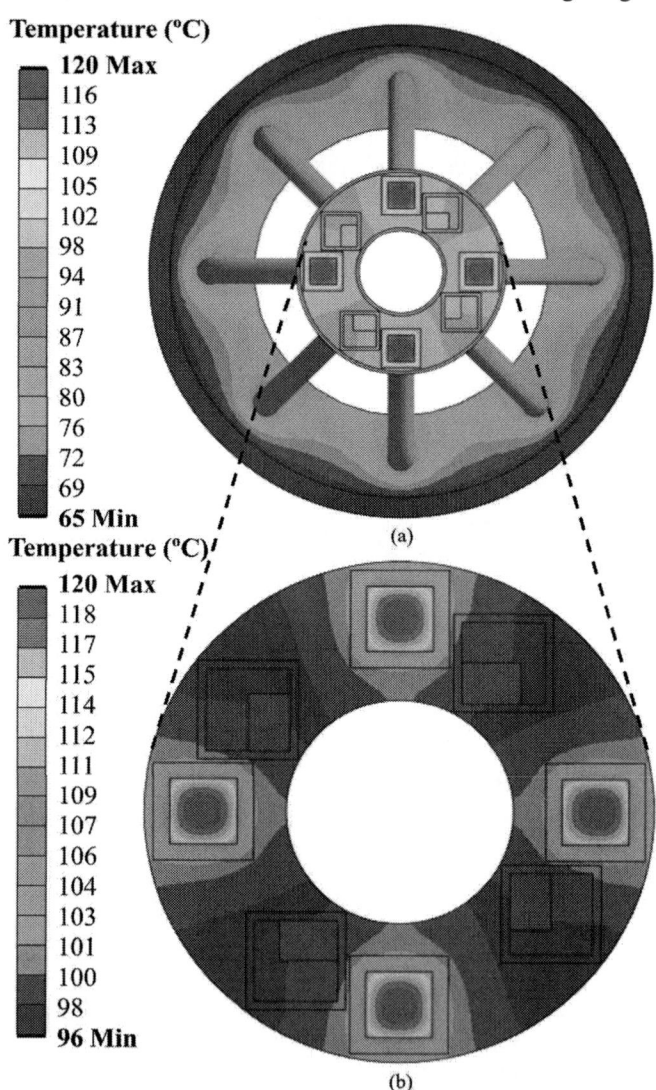

Fig. 7. Temperature contour plots of the (a) input module and thermal prototype and (b) input module generated in ANSYS Workbench steady state thermal. The ambient temperature is 21.7 °C, a loss of 79 W was applied to each die, and a convection coefficient of 73.5 W/m²°C was applied to the outside of the aluminum tube in place of the fins. Note that the source contact has been hidden in both plots to expose the die.

979-8-3315-1612-3/25 $31.00 © 2025 IEEE

and kelvin traces of the slip ring. As seen in Fig. 5, the gate and kelvin wirebond pads for one die in the slip ring were shorted using solder. The impedance as a function of angular mismatch between this location on the PCB and the connection to the gate driver is measured using an Agilent 4294A Precision Impedance Analyzer with a Keysight 42941A Impedance Probe. Here, 0° is considered ideal alignment with the die and 180° is considered the worst possible misalignment. Measurements were taken every 18° because the vias in the gate trace seen in Fig. 5 are spaced 9° apart and provide a good visual reference for aligning the probe tip. The measured inductance and resistance contributions of the gate/kelvin slip ring PCB at different angular mismatches as a function of frequency are plotted in Fig. 6. The close arrangement of the gate and kelvin traces in the slip ring allows for excellent mutual inductance cancellation, limiting the difference in inductance magnitude over the entire angular mismatch range to just 3 nH across the entire frequency range of interest as seen in Fig. 6(a). At the worst case, the difference in resistance is approximately 30 mΩ over the entire angular mismatch range at a frequency of 100 MHz as seen in Fig. 6(b).

There are a few ways to reduce the variation in parasitic resistance and inductance as a function of angular mismatch. The first of these is to reduce the diameters of the slip ring traces. By doing so, the arc length that the current is required to travel is reduced. Another way to address this issue is to distribute sets of current boosters and associated brushes that interface to the slip rings in an axisymmetric manner and in multiples of the number of die. This approach both limits the worst-case angular mismatch and ensures the parasitic inductances and resistances associated with each parallel die are equal. Here, for example, four sets of current boosters and associated brushes were distributed axially symmetrically about the gate drive PCB to limit the worst-case angular mismatch to 45°.

The circular shape of the modules comes with some inherent benefits independent of their coaxial configuration. The root of these benefits is symmetry. Asymmetrical electrical routing to parallel power semiconductors can lead to unequal electrical parasitics. The literature has found that unequal electrical parasitics can cause current imbalance among parallel devices, especially during fast switching transients [8]. Nonuniform current sharing can unevenly stress parallel die, reducing the packages overall lifetime. For this reason, the inherent axial symmetry of the packages is leveraged, and care is taken to distribute the die and their associated components in an axially symmetric manner to promote current sharing by ensuring that the electrical parasitics associated with the die are nearly identical to each other. This approach only addresses current sharing problems associated with the parasitics introduced by the package and device-related inconsistencies like variation in gate threshold voltage (V_{TH}) or on resistance ($R_{DS,ON}$) may still lead to current imbalance among parallel die [9].

Another benefit of the circular structure of the packages and axially symmetric die distribution is the uniform thermal coupling between parallel die. Even when care is taken to ensure parallel die have equal electrical parasitics in conventional rectangular packages, the die closest to each corner tend to enjoy a lower junction temperature than those located in the center [10]. This phenomenon is caused by nonuniform thermal coupling between co-packaged die. Die in the center of such a rectangular module may be thermally coupled to two or more other die, whereas die at the corner may be thermally coupled to as few as one other die [11]. As with current sharing, unequal heating of parallel die causes them to be stressed unevenly, reducing the reliability of the package. The axially symmetric structure of the packages designed in this work ensures the die's junction temperatures are all identical as seen in the temperature contour plot in Fig. 7. This plot was generated in an ANSYS Workbench steady

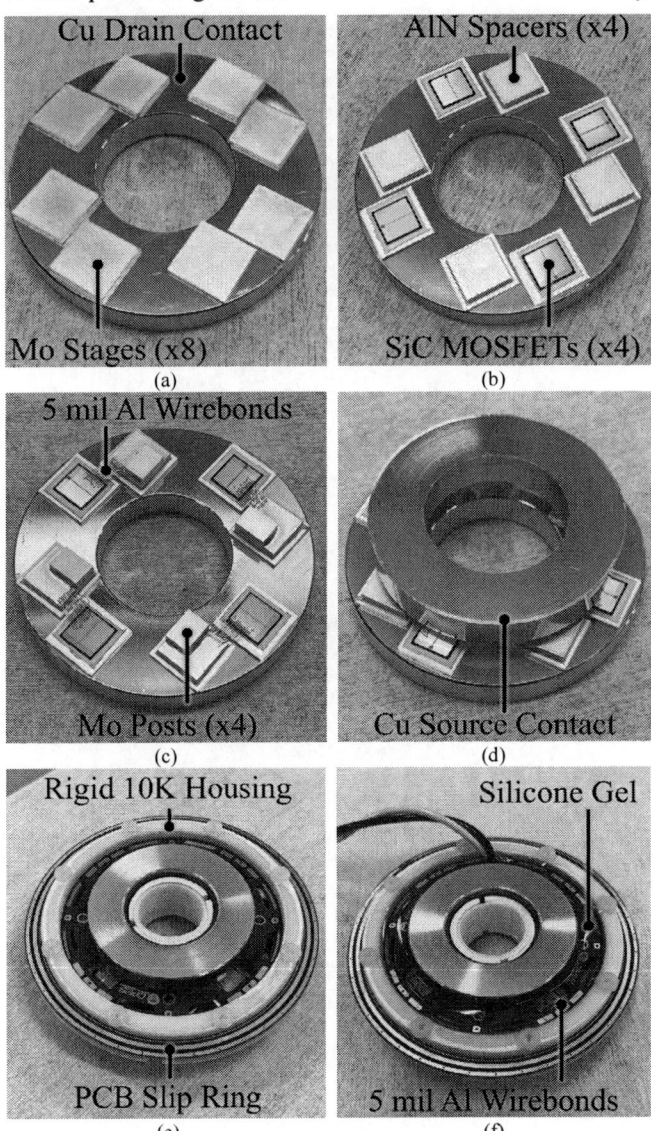

Fig. 8. Input module at various stages in the fabrication procedure. (a) Mo stages bonded to Cu drain contact with NBE Preform; (b) 3.3 kV SiC MOSFETs and AlN spacers bonded to the Mo stages with Indium Corp. silver paste; (c) 5 mil wirebonds made between the die and the top of the AlN spacers and Mo posts bonded to the top of the AlN spacers with Sn96.3/Ag3.7 solder; (d) Cu source contact soldered to the top of the Mo posts with Sn63/Pb37 solder; (e) rigid 10K housing and PCB slip ring attached to contacts; and (d) 5 mil Al wirebonds made between the die and PCB slip ring and encapsulated with Wacker SILGEL 612.

979-8-3315-1612-3/25 $31.00 © 2025 IEEE   2619

Ø 8.7 cm

(a)

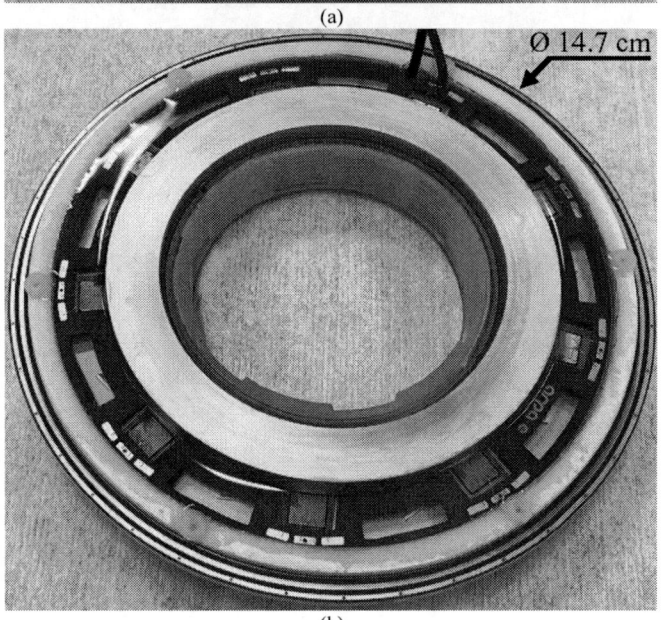

Ø 14.7 cm

(b)

Fig. 9. The 3.3 kV SiC MOSFET (a) input and (b) output packages.

state thermal simulation which aimed to provide a comparison point for the thermal testing discussed later in the paper. The temperature distribution of the module with the thermal prototype is seen in Fig. 7(a) and the temperature distribution of just the module is seen in Fig. 7(b). In this simulation, the ambient temperature was 21.7 °C, a loss of 79 W was applied to the active area of each die, and a convection coefficient of 73.5 W/m²K was applied to the outside of the aluminum tube in place of the fins to replicate the conditions of the experiment.

B. Package Fabrication

The fabrication procedure begins with sintering Mo components plated with 2-5 μm of Ag on 1-2 μm of Ni to the toroidal Cu drain contact plated with 1.25 μm of Au on 5 μm of Ni using NBE nanosilver preform as seen in Fig. 8(a). The AlN

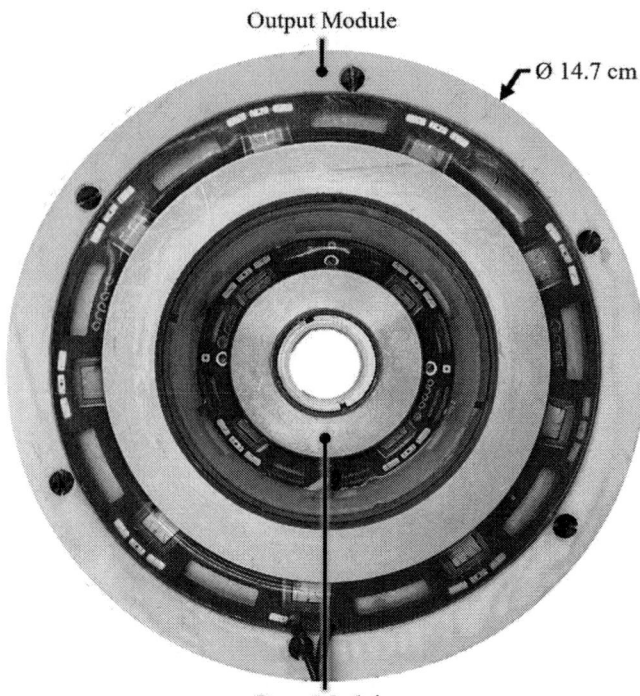

Output Module

Ø 14.7 cm

Input Module

Fig. 10. The 3.3 kV SiC MOSFET input and output switch packages in their coaxial configuration.

components and 3.3 kV, 25 mΩ SiC MOSFETs are then sintered to the Mo components using stencil-printed Ag paste from Indium Corp as seen in Fig. 8(b). The source pads of the SiC MOSFETs are then electrically connected to the top Ag surfaces of the AlN components using 5 mil Al wirebonds and another set of Mo components are soldered to the same Ag surface of the AlN components with Sn96.3/Ag3.7 solder as seen in Fig. 8(c).

The Cu source contact plated with 1.25 μm of Au on 5 μm of Ni is then soldered to the top surfaces of the Mo components with Sn63/Pb37 solder as seen in Fig. 8(d). 3D printed housing components made from Formlabs' Rigid 10K material are affixed to the contacts along with 40A silicone rubber to create a seal for the encapsulation step as seen in Fig. 8(e). Two housing components clamp the slip ring PCBs in place and 5 mil Al wirebonds made between the gate and source pads on the SiC MOSFETs and the associated pads on the slip ring PCBs form the gate and kelvin connections. Finally, Wacker SILGEL 612 encapsulates the modules as seen in Fig. 8(f). The resultant input module with four die and output module with nine die are seen in Fig. 9(a) and Fig. 9(b), respectively.

C. Coaxial Nesting

The two packages seen in Fig. 9 nest coaxially as seen in Fig. 10 to take advantage of the mutual inductance cancellation using the coaxial commutation loop seen in Fig. 3. As with coaxial conductors, this coaxial nesting promotes excellent mutual inductance cancellation. For this reason, the two packages contribute just 2.8 nH at 100 kHz to the commutation loop inductance according to Ansys Q3D despite their large size.

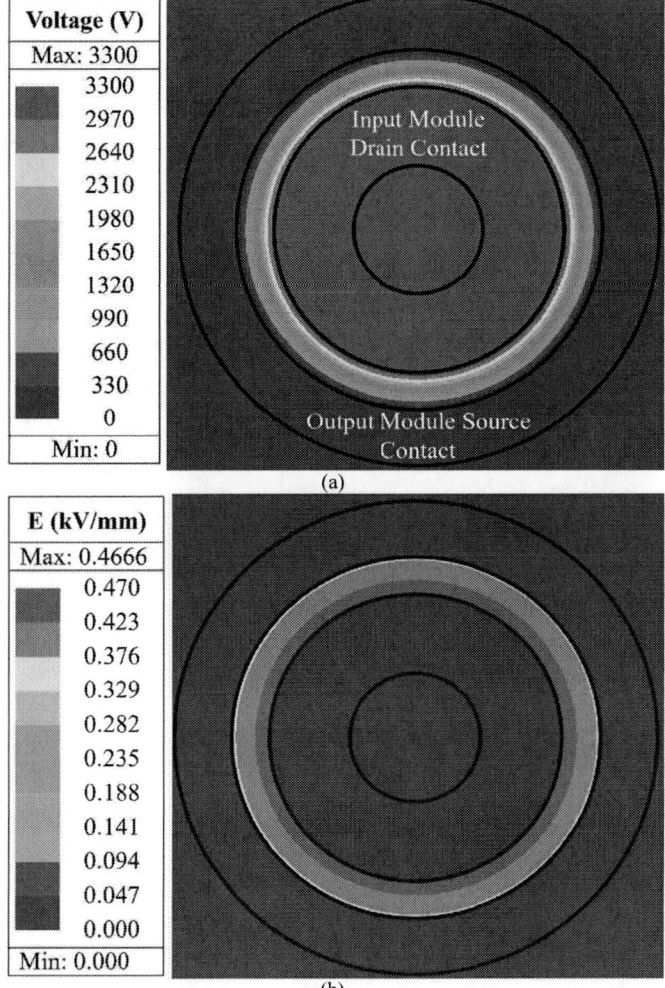

Fig. 11. The (a) voltage and (b) electric field distributions around the input module drain contact and the output module source contact.

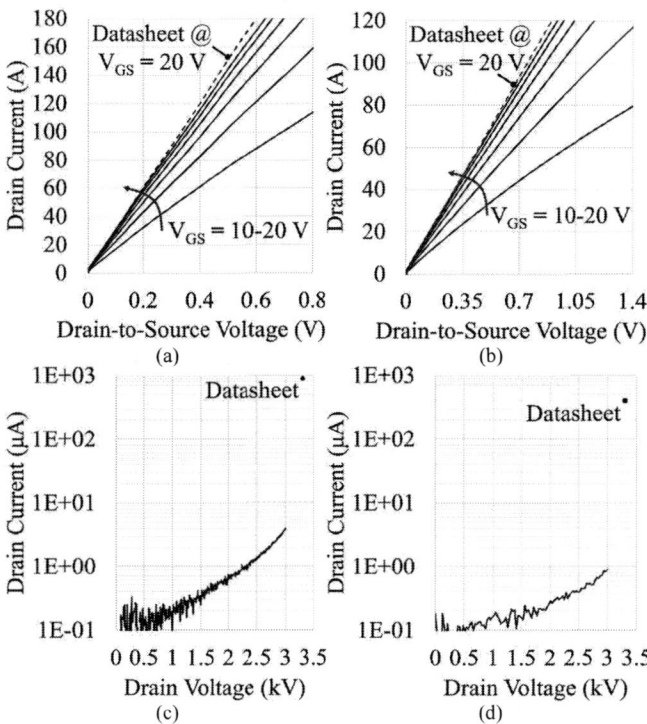

Fig. 12. The output characteristics for the (a) output module and (b) input module and the breakdown voltage for the (c) output module and (d) input module.

TABLE I. MEASURED AND DATASHEET STATIC CHARACTERISTICS

Parameter	Measured		Datasheet (1 Die)		
	Input (4 die)	Output (9 Die)	Min	Typ	Max
$R_{DS,ON}$ (mΩ) @ V_{DS} = 20 V, ID = 40 A/die	8.2	3.6	-	25	31
V_{TH} (V) @ I_D = 7 mA/die	3.1	3.1	1.9	2.7	-
V_F (V) @ I_{SD} = 40 A/die	4.0	4.1	-	3.65	-
I_{DSS} (µA) @ VDS =3 kV	0.88	4.0	-	-	100
I_{GSS} (nA) @ VGS = 20 V	2.3	3.7	-	-	100

The other benefit of coaxial nesting is the uniform voltage and electric field distributions between the input and output modules as seen in Fig. 11(a) and Fig. 11(b), respectively. In these simulations, the insulator between the two packages is assumed to be ambient air. The observed uniformity allows for improved module-to-module voltage scaling. If a change to the maximum input voltage rating of the system is required, the radial distance between the two modules is the only aspect that needs to be adjusted. It should be noted that the round form factor of the packages also provides a low peak electric field in addition to a uniform peak electric field offered by the coaxial configuration.

III. ELECTRICAL CHARACTERIZATION

The electrical characteristics of the packages are measured using a two-fold approach. First, the static characteristics are extracted and compared with the SiC MOSFET datasheet to determine what impact the package has on the device performance. The packages are then tested in the coaxial converter prototype seen in Fig. 1 to evaluate their dynamic electrical performance.

A. Static Characterization

A Keysight B1505A Curve Tracer was used to measure the static characteristics of both the input (Fig. 9(a)) and output (Fig. 9(b)) packages. The breakdown voltage and output characteristics of the input and output module are seen in Fig. 12. Additional static characteristics for both the input and output packages are listed in Table I along with the bare die characteristics from the datasheet. All static characteristics are within the datasheet specifications with some tolerance for package contributions and measurement error.

B. Dynamic Characterization

The dynamic characterization of the packages is carried out using the 50 kW prototype coaxial converter seen in Fig. 1 that was first presented in [1]. It should be noted that the converter was not allowed to reach thermal steady state during this testing as the goal was to demonstrate the electrical performance of the power semiconductor packages and other converter components. The experimental drain-to-source voltages (V_{DS}) and gate-to-source voltages of the input switch ($V_{DS,Input}$ and $V_{GS,Input}$) and output switch ($V_{DS,Output}$ and $V_{GS,Output}$) are seen in Fig. 13. Despite experiencing a peak dV/dt of 72 V/ns during the input switch turn-off, the switches

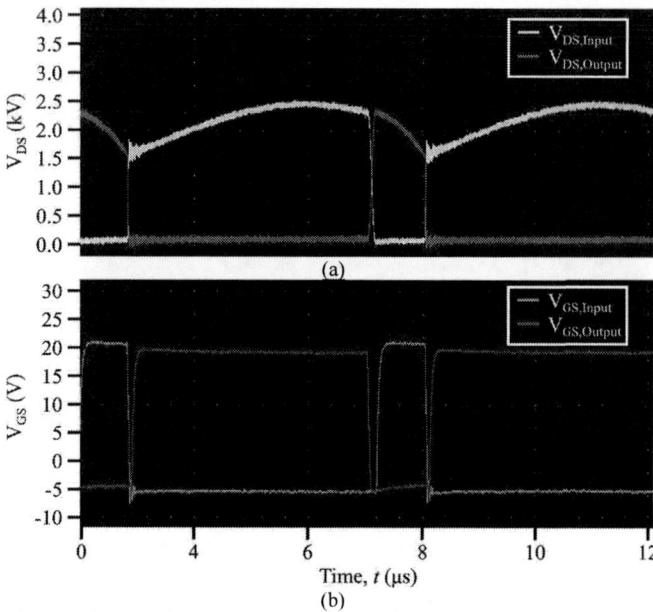

Fig. 13. The experimental V_{DS} waveforms for the input and output switches during a 1.9 kV DC to 380 V DC, 49 kW test.

experience negligible V_{DS} overshoot in part due to their low commutation loop inductances. The two packages contribute just 2.8 nH at 100 kHz to the commutation loop inductance according to Ansys Q3D due to their coaxial nesting.

IV. THERMAL CHARACTERIZATION

The junction-to-case thermal resistance ($R_{th,jc}$) of the inner module has been characterized using the Phase 12B Power Semiconductor Thermal Analyzer from Analysis Tech and a prototype of the system's thermal management system as seen in Fig. 14(a). The thermal management system is composed of Cu heat pipes to spread the heat from the module along the length of converter, an aluminum fin tube heat sink, and ParkerLord Cooltherm™ SC-324 high thermal conductivity (3 W/mK) silicone which electrically isolates the aluminum fin tube heat sink from the heat pipes. The $R_{th,jc}$ is found according to the JEDEC JESD51-14 transient dual interface method (TDIM) [12]. The TDIM calls for two measurements of the junction-to-ambient thermal impedance ($Z_{th,ja}$) with two different thermal interface materials (TIMs) applied between the case of the package and the thermal management system.

Here, the Heat Spring TIM from Indium Corp. was selected as this will be used in the system and ambient air was used for the other set of $Z_{th,ja}$ measurements. In this test, the maximum junction temperature was limited to 120 °C to provide a sufficiently large ΔT between junction and ambient without overstressing the package unnecessarily. The ambient temperature during these tests was 22 °C. The red die spring seen in Fig. 14(a) was used to apply a constant 60 PSI during all tests by compressing the spring to a set distance calculated using the spring constant.

The $R_{th,jc}$ is then found by extracting the two cumulative structure functions (CSFs) from the raw $Z_{th,ja}$ data and identifying where they diverge. An example of two of these CSFs overlayed on top of one another is seen in Fig. 14(b).

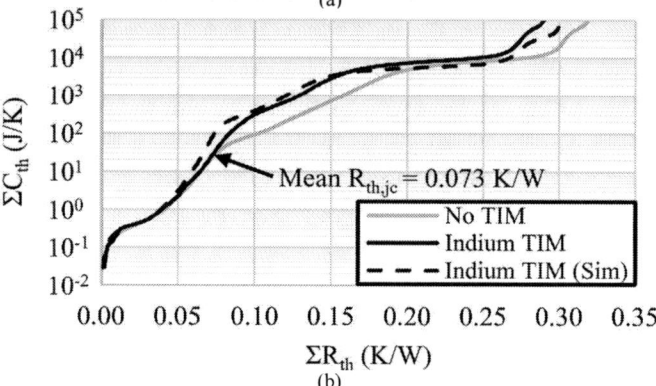

Fig. 14. The (a) experimental $R_{TH,JC}$ measurement setup and (b) measured and simulated CSFs.

TABLE II. SAMPLING RATES USED IN THE TRANSIENT THERMAL SIMULATION

Step End Time (s)	Sample Period (s)
1E-04	1E-06
0.01	1E-05
0.05	2E-04
0.1	0.002
1	0.01
10	0.1
300	1

Twelve $R_{th,jc}$ measurements were taken to account for measurement error and the mean was found to be 0.073 K/W with a sample standard deviation of 0.0032 K/W.

Also included in Fig. 14(b) is a simulated CSF of the module and thermal prototype with indium as the TIM. To obtain this CSF, an ANSYS Workbench transient thermal simulation was performed using the same model of the module and thermal prototype used to generate the steady state temperature contour plots seen in Fig. 7. As in the steady state thermal simulation, the ambient temperature was set to 21.7 °C, a loss of 79 W was applied to the active area of each die, and a convection coefficient of 73.5 W/m²°C was applied to the outside of the aluminum tube. In a transient thermal simulation, the sampling rates are critical to obtain sufficiently accurate results. The sampling rates used here are listed in Table II.

To generate a CSF using a transient thermal simulation, the output of interest is the maximum junction temperature as a function of time. This data is then converted to thermal impedance as a function of time by subtracting the ambient temperature and dividing by the total loss. The thermal impedance information is then reformatted and fed into the Phase 12B to generate the CSF. This process ensures that the measured and simulated CSFs are calculated in a similar manner.

The simulated and measured CSFs in Fig. 14(b) are in good agreement with each other. The region internal to the package, which occurs below the indicated divergence point, appears to be in particularly good agreement. Thermal impedance has been shown to be a good measure of bond quality [13]. Since the bonds in our simulation are modeled as ideal (full contact to both the die and substrate with no voids), one conclusion that can be made from the good agreement between the simulated and measured CSFs is that the bonds in the critical thermal path of the package are good quality.

V. CONCLUSION

This work proposes a pair of coaxially nested 3.3 kV SiC MOSFET packages. The packages have similar structures to reduce engineering time and cost. Their round form factors allow each parallel die to experience similar electrical parasitics and temperature distributions which aid in improving package lifetime. The coaxial arrangement of the modules helps them scale easily with system input voltage and limits commutation loop inductance. The commutation loop inductance is found to be 2.8 nH at 100 kHz in simulation because the coaxial nesting of the packages takes advantage of the mutual inductance cancellation benefits of coaxial conductors. This low commutation loop inductance contributed to the negligible V_{DS} overshoot achieved during converter testing. Opting to electrically isolate at the aluminum fin tube heat sink contributes to the low measured $R_{th,jc}$ of 0.073 K/W.

ACKNOWLEDGMENT

The authors greatly appreciate in-kind donations from NBE Tech LLC for the nano-silver sintering paste and preform as well as from Microchip Technology Inc. who provided the 3.3 kV SiC MOSFET die (PN: MSC027SMA330D/S). The authors would also like to thank the rest of the team from the ARPA-E SCALED project for their support. Special thanks to Aakash Kamalapur and Dr. James Bae for providing the dynamic characterization waveforms.

REFERENCES

[1] M. Cairnie et al., "Modular Coaxial Power Converter for High-Density Integration into Medium-Voltage Cables," in *PCIM Europe 2024; International Exhibition and Conference for Power Electronics, Intelligent Motion, Renewable Energy and Energy Management*, Nuremberg, Germany, 2024.

[2] C. M. DiMarino, B. Mouawad, C. M. Johnson, D. Boroyevich, and R. Burgos, "10-kV SiC MOSFET Power Module With Reduced Common-Mode Noise and Electric Field," *IEEE Transactions on Power Electronics*, vol. 35, no. 6, pp. 6050–6060, Jun. 2020, doi: 10.1109/TPEL.2019.2952633.

[3] M. Mandal, S. K. Roy, and K. Basu, "Analytical Switching Transient Model of TO-247-4 Packaged SiC

MOSFETs and Comparison with TO-247-3 Devices," in *2022 IEEE Energy Conversion Congress and Exposition (ECCE)*, Oct. 2022, pp. 1–8. doi: 10.1109/ECCE50734.2022.9947557.

[4] R. Simpson, A. Plumpton, M. Varley, C. Tonner, P. Taylor, and X. Dai, "Press-pack IGBTs for HVDC and FACTs," *CSEE Journal of Power and Energy Systems*, vol. 3, no. 3, pp. 302–310, Sep. 2017, doi: 10.17775/CSEEJPES.2016.01740.

[5] M. Wang et al., "Reliability Improvement of a Double-Sided IGBT Module by Lowering Stress Gradient Using Molybdenum Buffers," *IEEE Journal of Emerging and Selected Topics in Power Electronics*, vol. 7, no. 3, pp. 1637–1648, Sep. 2019, doi: 10.1109/JESTPE.2019.2920254.

[6] D. D. Phinney, "Slip ring experience in long duration space applications," May 1986. Accessed: Oct. 14, 2024. [Online]. Available: https://ntrs.nasa.gov/citations/19870006892

[7] X. Zuo, R. Zhu, and Y. Zhou, "Online tracking and prediction of slip ring degradation using chaos theory based on LSTM neural network," *Meas. Sci. Technol.*, vol. 34, no. 5, p. 055010, Feb. 2023, doi: 10.1088/1361-6501/acb5b6.

[8] R. Wu, L. Smirnova, H. Wang, F. Iannuzzo, and F. Blaabjerg, "Comprehensive investigation on current imbalance among parallel chips inside MW-scale IGBT power modules," in *2015 9th International Conference on Power Electronics and ECCE Asia (ICPE-ECCE Asia)*, Jun. 2015, pp. 850–856. doi: 10.1109/ICPE.2015.7167881.

[9] J. Ke, Z. Zhao, P. Sun, H. Huang, J. Abuogo, and X. Cui, "New Screening Method for Improving Transient Current sharing of Paralleled SiC MOSFETs," in *2018 International Power Electronics Conference (IPEC-Niigata 2018 -ECCE Asia)*, May 2018, pp. 1125–1130. doi: 10.23919/IPEC.2018.8507893.

[10] C. DiMarino et al., "Design of a novel, high-density, high-speed 10 kV SiC MOSFET module," in *2017 IEEE Energy Conversion Congress and Exposition (ECCE)*, Oct. 2017, pp. 4003–4010. doi: 10.1109/ECCE.2017.8096699.

[11] H. Li et al., "Improved thermal couple impedance model and thermal analysis of multi-chip paralleled IGBT module," in *2015 IEEE Energy Conversion Congress and Exposition (ECCE)*, Sep. 2015, pp. 3748–3753. doi: 10.1109/ECCE.2015.7310189.

[12] *Transient Dual Interface Test Method for the Measurement of the Thermal Resistance Junction-to-Case of Semiconductor Devices with Heat Flow Through a Single Path*. JESD51-14, 2010.

[13] H. Zheng, K. D. T. Ngo, and G.-Q. Lu, "Thermal characterization system for transient thermal impedance measurement and power cycling of IGBT modules," *Microelectronics Reliability*, vol. 55, no. 12, Part A, pp. 2575–2581, Dec. 2015, doi: 10.1016/j.microrel.2015.08.016.

Thermal Modeling and Performance of a Bare-Die Embedded PCB for High Power Density Converters Design

1st Shahid Aziz Khan †   2nd Feng Zhou †   3rd Mengqi Wang *   4th DucDung Le   5th Shivam Chaturvedi

University of Michigan
Michigan, USA
shahidkh@umich.edu
†Co-First Authors

Toyota Research Institute
of North America
Ann Arbor, MI, USA
feng.zhou@toyota.com

University of Michigan
Michigan, USA
mengqiw@umich.edu
* Corresponding Author

University of Michigan
Michigan, USA
dungle@umich.edu

University of Michigan
Michigan, USA
shivamc@umich.edu

Abstract—**The key goals in developing power electronics converters for the demanding electrified transportation industry are achieving high power density and efficiency with optimal thermal performance. This research focuses on the thermal aspect of a new bare-die embedding technique that incorporates the bare-die of the switching devices (such as MOSFETs) into the layers of the printed circuit board (PCB). This technique ensures high power density and minimal parasitics, improving efficiency. However, the thermal design of such an approach is challenging and needs special design considerations. A thermal design framework and the thermal model are proposed and validated, and the superior thermal performance of the bare-die embedded technique compared to the existing packaging techniques is proved through hardware and experimental results.**

Index Terms—**High Power Density, Bare-Die, Electric Vehicles, Power Converters**

I. INTRODUCTION

Electrified transportation is considered as a major breakthrough in tackling the global climate change and Carbon emissions. Electric Vehicles are the backbone of the transition towards clean energy and to meet the zero-emission targets [1]. The transportation sector is responsible for contributing 24% to the global carbon emissions [2], and vehicle transport directly emits 70% of the emissions [3]. To reduce these emissions and to slow the threats of world climate change, electrified transportation in general and electric vehicles (EVs) in particular have attained great attention due to their zero emissions operation, low noise, improved performance, and efficiency. The U.S. government mandates that the average fuel economy of a passenger car should rise to 54.5 miles/gallon by 2025 [4]. These targets cannot be achieved by using the internal combustion engine technology. Instead, electric machines driven by power electronics converters can provide such higher efficiencies. Therefore, electrification is the only way forward. Electrified transportation vision includes utilizing more electrical energy to power traction and non-traction loads in the vehicle. It's still a hot research and development area due to the various technical challenges the EV industry is facing

such as battery cost, charging infrastructure, power electronics reliability and converter's functionality in the drive train. Specifically, the challenges in the power converter's domain include a large number of components leading to low power density, high current stress, high switching loss, slow dynamic response, and computational complexity [5]. Moreover, Power Density and cost are the key design criteria for the modern power electronics unit in Electric Vehicles (EVs). However, the vertical space on the chassis for charging electronics is very limited. Moreover, with a non-compact package comes a high damaging probability due to unpredictable road conditions, higher costs, and vehicle integration issues.Markets for electric cars are expanding exponentially, with sales expected to reach 10 million by 2022. In 2022, 14% of all new automobiles sold were electric, an increase from less than 5% in 2020 and approximately 9% in 2021. Global sales were led by three markets. China led the way once more, making up over 60% of sales of electric vehicles worldwide. China currently has more electric vehicles on the road than any other country, and the nation has already surpassed its 2025 new energy vehicle sales target. Sales of electric cars rose by more than 15% in 2022 in Europe, the second-largest market, indicating that more than one in five vehicles sold there were electric. The third-largest market in the world, the United States, saw a 55% growth in electric vehicle sales in 2022, reaching an 8% market share. Sales of electric vehicles are predicted to rise significantly through 2023. In the first quarter, over 2.3 million electric automobiles were sold, which represents a 25% increase over the same period the previous year. By the end of 2023, we project sales of 14 million, a 35% annual growth, with new purchases picking up speed in the second part of the current year. Accordingly, during the course of the entire calendar year, sales of automobiles might be made up of 18% electric vehicles [6].

The thermal management of power electronics converters has recently gained significant attention and importance due to the developing needs of the electrified transportation industry, such as electric vehicles, to develop very compact power

979-8-3315-1612-3/25 $31.00 © 2025 IEEE

converters with high power density. The evolution of semi-conductor devices led to the possibility of high operating temperatures and thermal cycles [7]. The Department of Energy (DOE) plans to achieve 33KW/L power density for a 100KW traction drive system by 2025, which translates to 88% volume reduction. Similarly, the electrification industry is constantly trying to shrink the size of power train electronics and drive systems to make more space for batteries on the chassis and reduce production and material costs [8]. To realize such targets, apart from size reduction in magnetics through the usage of wide band gap devices that enable the high-frequency operation and planar magnetics design, low profile module usage, etc., innovation and improvement are required in the packaging and heat or thermal management domains [9], as these are the bottlenecks in realizing very high-power density converters. Researchers have explored several techniques to address the upcoming electrical and thermal challenges for semiconductor device packaging. In a direct bonded copper (DBC), the bare die is attached to the top side of the copper and offers good thermal conductivity [10]. Aluminium bond wires connect the die, substrate, and frame in the DBC package. These bond wires become the principal source of parasitic, especially in the case of wide band gap devices that operate from kilo to Mega Hertz frequency, which increases losses, poor EMI performance, and reduces reliability as the WBG devices, especially Gallium Nitride (GaN's), are very sensitive to parasitics [11]. Moreover, bond lifting due to high temperature is another common phenomenon causing power device failures [12]- [13]. To solve all these issues a bare-die embedded PCB approach is presented where the bare-die of the power device is embedded into the PCB layers and the electrical connections are made via laser drilled vias, resulting in minimal gate and power loop parasitics along with the improved reliability. This paper focuses on the thermal design considerations and thermal performance of this proposed abre-die technique via thermal modelling, thermal resistance equivalent circuit, and performance comparison with existing packaging technology. For the thermal performance comparison, a benchmarked motor drive from the market is compared with the one designed using the bare-die embedded technique, keeping the topology and every other component the same. The final full-length paper will present an in-depth literature review and comparative analysis.

II. THERMAL MODELING AND ESTIMATION OF THE BARE-DIE EMBEDDED APPROACH

Due to the growing power densities and the tendency toward electronic system downsizing, thermal performance and thermal design are crucial in chip-embedded PCBs. Heat dissipation gets more complicated when ICs are buried within PCB layers, and conventional cooling techniques like fans or heat sinks are frequently impracticable. Thermal hotspots, which impair performance and dependability, can result from heat buildup in the chip and surrounding materials. Elevated temperatures can cause thermal expansion mismatches, increase leakage currents, and speed up electromigration, all

of which can lead to an early device failure. Recent studies have suggested a number of thermal management techniques to overcome these problems.

Its construction and design must be well understood to understand the thermal modelling and performance of the bare-die approach. This research designed a four-layered PCB with the bare die embedded between the second and the third layer. Copper layers above and below the bare die are deposited, and the die is sandwiched between them. Laser vias are drilled to form the electrical connections for the gate, source, and drain. Copper-filled vias are used here to source and drain parts and enhance thermal performance. Moreover, instead of single via, multiple vias in the via stitching pattern are designed to divide the current stress and achieve better thermal performance. The bare die embedded PCB design process is shown in Fig 1a, the designed layer stack is shown in Fig 1b, and the designed motor drive PCB is depicted in Fig 1c.

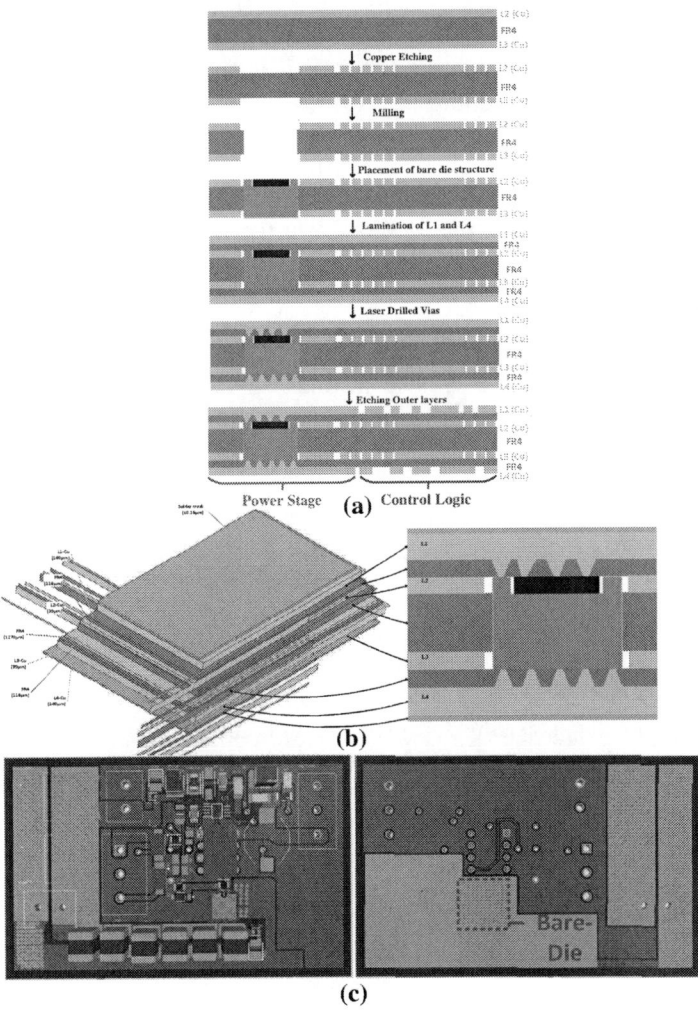

Fig. 1. a) Bare-Die Embedding PCB Process) Designed Layer Stack c) Designed Embedded Bare-Die PCB-Top and Bottom View View

As the structure of the bare-die embedded is distinct from the traditional package, the traditional thermal equivalent circuit, which estimates the thermal resistance, is not applicable. For this research, the developed thermal equivalent circuit is depicted in Fig 2. there is no junction to case $R_{th(J-C)}$ and case to ambient $R_{th(C-A)}$ resistance in the presented bare-die technique, which reduces the total thermal resistance of the packaging and implies better thermal performance. The bare-die equivalent thermal resistance is presented in equation (1).

$$\Theta_{RTH} = \frac{\Theta_{JA}(\Theta_{JBK}+\Theta_{RTBoard}+\Theta_{TIM}+\Theta_{HSA})}{\Theta_{JBK}+\Theta_{RTBoard}+\Theta_{TIM}+\Theta_{HSA}+\Theta_{JA}} \quad (1) \; ,$$

$$\Psi_{JB} = (T_J - T_{board})/P_D \quad (2)$$

where Θ_{JBK} is the thermal resistance from the bare-die junction to the back side. Θ_{JA} is the junction to ambient, and $\Theta_{RTBoard}$ is the total thermal resistance, including the board and drilled thermal vias, Θ_{TIM} is the thermal interface material resistance and Θ_{HSA} is the heatsink to ambient thermal resistance. Vertical heat transfer is realized by using copper-filled thermal vias, also avoiding the low thermal conductivity path of the FR4 material. The relationship of the junction of the bare die and the board can be formed through a junction-to-board characterization parameter Ψ_{JB} in Eq(2). If T_{board} is the temperature measured on the board surface above the bare-die location, then the junction temperature is presented in Eq(3). Where T_J and P_D represent the junction temperature and dissipated power, respectively. The designed thermal vias have three paths that includes via filler (Θ_{filler}), via barrel (Θ_{barrel}) and cooper and FR4 layers ($\Theta_{cu}+\Theta_{FR4}$) shown in Eq(4-6). The thermal conductivity $k(W/(m^2 K))$ of each of them can be presented as,

$$(T_J = T_{board}) + P_D \Psi_{JB} \quad (3) \; ,$$

$$\Theta_{filler} = \frac{t_{PCB}}{k_{filler}\pi(\phi/2-t_{PTH})^2} \quad (4),$$

$$\Theta_{barrel} = \frac{t_{PCB}}{k_{Cu}\pi t_{PTH}(\phi/2-t_{PTH})} \quad (5)$$

t_{PCB} is the thickness of the PCB, t_{PTH} is the plated hole thickness, t_{Cu} is the copper layer thickness, s is the inter via spacing and ϕ is the via diameter. The total vertical thermal resistance of the board is computed through Eq (7).

$$\Theta_{Cu} + \Theta_{FR4} = \left[\frac{t_{Cu}}{K_{Cu}} + \frac{t_{Cu}}{K_{FR4}}\right]\left[\frac{4}{(\pi+s)^2 * - \pi\phi^2/4}\right](6)$$

$$\Theta_{RTBoard} = \Theta_{filler}||\Theta_{barrel}||(\Theta_{Cu} + \Theta_{FR4})(7)$$

$$\Theta_{RTBoard} = \cfrac{1}{\cfrac{k_{filler}\pi(\phi/2-t_{PTH})^2}{t_{PCB}} + \cfrac{k_{Cu}\pi t_{PTH}(\phi/2-t_{PTH})}{t_{PCB}} + \cfrac{(\pi+s)^2 * - \pi\phi^2/4}{4t_{Cu}/K_{Cu}+t_{PCB}-4t_{Cu}/K_{FR4}}}$$

For the proposed thermal model validation, the approach shown in Fig2c is adopted where the model was first verified in Simulink, and the results obtained were compared with the experimental results that use the manufactured bare-die embedded PCB.

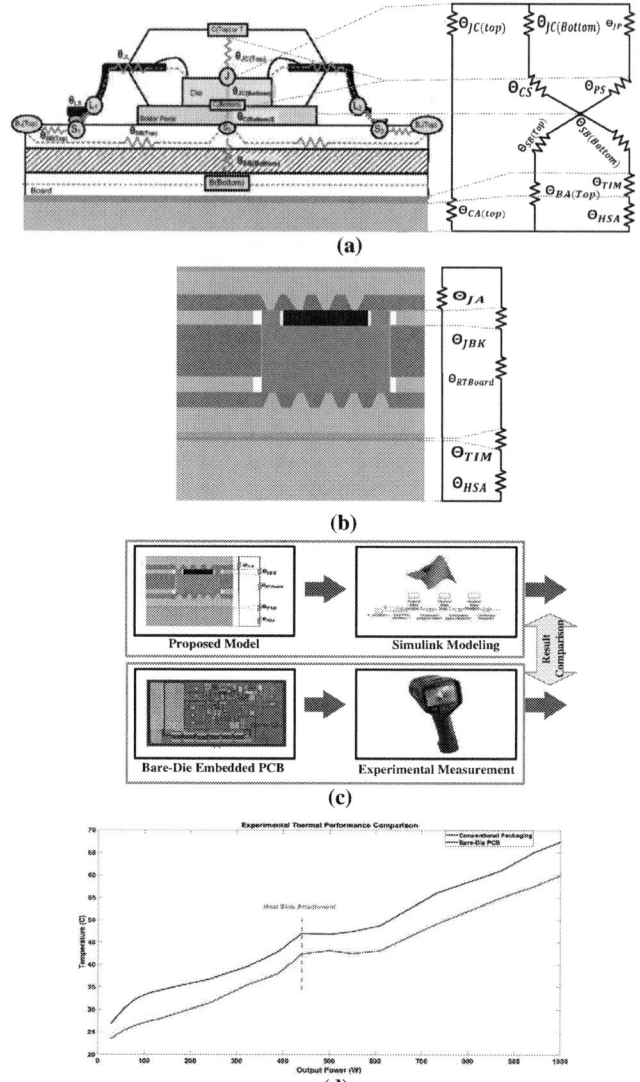

Fig. 2. (a) Conventional Package Thermal Modeling (b)Embedded Bare Die Package Thermal Modeling (c) Methodology of Thermal Validation (d) Experimental Thermal Performance Comparison

III. EXPERIMENTAL RESULTS AND ANALYSIS

To evaluate the thermal performance of the bare-die embedded PCB, the designed 1KW motor drive was tested under dynamic load conditions, and the performance was compared with the benchmark motor drive using the traditional packaging technique. The thermal performance data obtained is plotted in a graphical form and shown in Fig2d for a better illustration, where it can be observed that the proposed bare-die embedded packaging motor drive performed better. However, for a deeper understanding of the power changes and with respect to the temperature, Fig3a present the surface plot of the bare-die packaged PCB, which shows the variation and change in temperature corresponding to the changes in the load. Apart from the hardware experiments, the designed bare-die PCB was also simulated for thermal performance before manufacturing in the ANSYS-ICEPAK tool to predict the thermal performance and make necessary design adjustments. Fig3b and 3c show the simulated and the actual thermal performance,w, which match closely. Moreover, Fig3d depicts the thermal image of the benchmark motor drive, which can be compared to the thermal images of the designed bare-die motor drive to see the temperature difference. The experimental setup is shown in Fig 3e. The full-length paper will include detailed explanations, waveforms, and extended analysis.

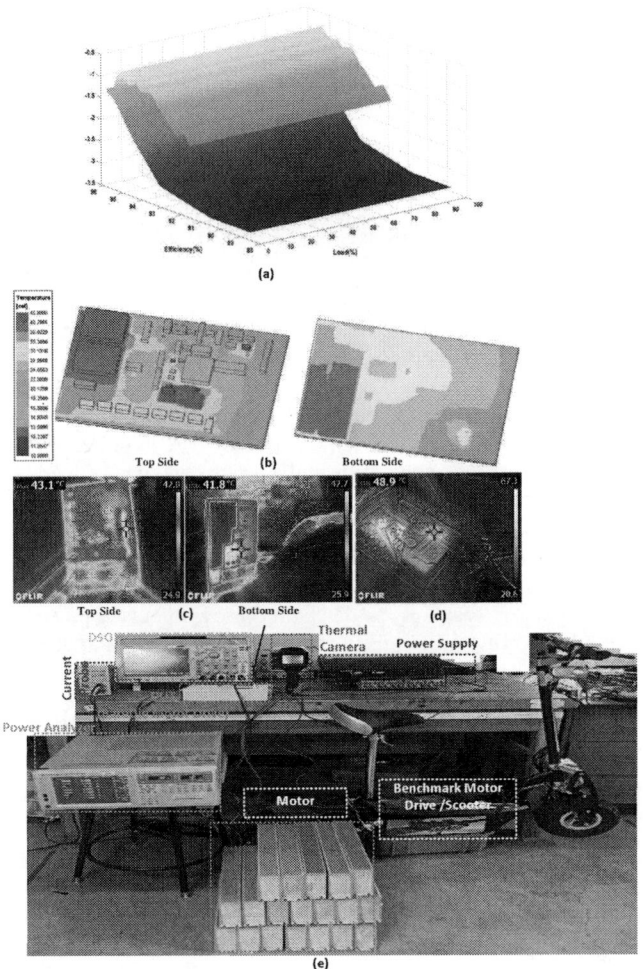

Fig. 3. (a) Surface Plot- Thermal Performance (b)ANSYS-ICEPAK Thermal Analysis of Bare-Die Embedded PCB at 600W (c) Experimental Thermal Image of Benchmark Motor Drive at 600W - Top and Bottom side (d) Benchmarked/Conventional Packaged Motor Drive PCB Thermal Image(e) Experimental Setup

IV. CONCLUSION

In conclusion, this research focuses on the thermal aspects of a novel bare-die embedding technique which can be utilised to increase the power density of power electronics units for small mobility applications. To illustrate the effectiveness of the proposed technique, a comparative analysis was done with a benchmark motor drive, which proved the superior

thermal performance of the designed bare-die embedded PCB. Moreover, the equivalent circuit modelling, mathematical expressions for the thermal vias design in such embedded PCBs and the design approach of such bare-die embedded PCBs were introduced. This work is the foundation to illustrate the potential of the newly proposed bare-die embedding technique to accelerate the electrification revolution.

REFERENCES

[1] title=Here's how EU legislation accelerates the EV revolution, journal=Retrieved Online on: 02 November 2024, vol. Retrieved From: HTTPs://www.virta.global/.,

[2] title=EA. Tracking Transport 2020 retrieved From:: https://www.iea.org/reports/tracking-transport-2020/electricvehicles , journal=Retrieved Online on: 12 January 2023,

[3] title=EEA. Greenhouse Gas Emissions from Transport in Europe—European Environment Agency. retrieved From::https://www.eea.europa.eu/data-and-maps/indicators/transport-emissions-of-greenhouse-gases/transportemissions-of-greenhouse-gases-12 , journal=Retrieved Online on: 14 January 2023,

[4] title=U.S. Environmental Protection Agency (EPA). Regulations and Standards: Light-Duty. retrieved From::http://www.epa.gov/ , journal=Retrieved Online on: 14 March 2024,

[5] author=Lipu, M.S.H.; Faisal, M.; Ansari, S.; Hannan, M.A.; Karim, T.F.; Ayob, A.; Hussain, A.; Miah, M.S.; Saad, M.H.M., title=Review of Electric Vehicle Converter Configurations, Control Schemes and Optimizations: Challenges and Suggestions., journal=Electronics 2021, 10, 477. https:// doi.org/10.3390/electronics10040477,

[6] title=IEA, Electric car sales, 2016-2023, IEA, Paris Light-Duty. Retrieved From::https://www.iea.org/data-and-statistics/charts/electric-car-sales-2016-2023, IEA. Licence: CC BY 4.0, journal=Retrieved Online on: 14 March 2024,

[7] S. M. I. Rahman et al., "Emerging Trends and Challenges in Thermal Management of Power Electronic Converters: A State of the Art Review," in IEEE Access, vol. 12, pp. 50633-50672, 2024, doi: 10.1109/ACCESS.2024.3385429.

[8] Electrical and Electronics Technical Team Roadmap, U.S. Drive Department of Energy, Oct. 2017.

[9] S. Akhtar, L. Kareem, A. Arif, M. Siddiqui, and A. Hakeem, "Development of a ceramic-based composite for direct bonded copper substrate," Ceramics International, vol. 43, no. 6, pp. 5236–5246, 2017. [Online]. Available: https://www.sciencedirect.com/science/article/pii/S0272884217300627

[10] H.-Y. Zhang, X. Li, H.-N. Jiang, Y.-H. Mei, and G.-Q. Lu, "Largearea substrate bonding with single-printing silver paste sintering for power modules," IEEE Transactions on Components, Packaging and Manufacturing Technology, vol. 11, no. 1, pp. 11–18, 2021.

[11] E. Gurpinar, B. Ozpineci, and S. Chowdhury, "Design, analysis, comparison and experimental validation of insulated metal substrates for high power wide-bandgap power modules," Journal of Electronic Packaging, vol. 142, 06 2020.

[12] B. Zhang and S. Wang, "Parasitic inductance modeling and reduction for wire-bonded half-bridge sic multichip power modules," IEEE Transactions on Power Electronics, vol. 36, no. 5, pp. 5892–5903, 2021.

[13] T. Moldaschl, S. Woetzel, R. Latella, M. Galvano, and A. Binder, "Comparison of the parasitic impedances from the drain-source path of power transistor packages at up to 2ghz," Engineering Reports, vol. 4, no. 5, p. e12489, 2022. [Online]. Available: https://onlinelibrary.wiley.com/doi/abs/10.1002/eng2.12489

Research on the Voltage Fluctuation Suppression Strategy in Weak Grid Under Pulsed Power Load Integration

Xi Chen
School of Electrical and Information Engineering
Tianjin University
Tianjin,China
2022234001@tju.edu.cn

Jiazheng Zhang
School of Electrical and Information Engineering
Tianjin University
Tianjin,China
2022234388@tju.edu.cn

Mingjun Bao
School of Electrical and Information Engineering
Tianjin University
Tianjin,China
baomingjun159@tju.edu.cn

Abstract—The power fluctuation phenomenon caused by pulsed power loads has a significant impact on the stability of weak grids. In this paper, a voltage fluctuation suppression strategy in weak grid under pulsed power load integration is proposed, utilizing the supercapacitor capacity to compensate for load power fluctuations. To enhance the stability margin and dynamic response to load power fluctuations, a combined control method of VSG control for the mean power and ripple current control for the ripple power is introduced. This paper thor-oughly analyzes the control strategy and validates its effectiveness through experiments.

Keywords—Pulsed power load,Voltage fluctuation, VSG control , Ripple current control.

I. Introduction

With the development of power electronics technology, pulsed power loads containing numerous power electronic switching devices are becoming increasingly prevalent[1]. Pulsed power loads, whose power exhibits a low-frequency pulsatile characteristic, are a significant factor contributing to power oscillations in electrical systems and can cause severe voltage fluctuations in weak grids, threatening the safe and stable operation of the grid. To eliminate the negative impacts of pulsed power load power fluctuations on the AC source side, it is necessary to introduce a voltage fluctuation suppression mechanism. Targeting the power fluctuation issue of pulsed power loads, an active secondary power pulsation suppression circuit and its control method have been proposed[2] [3], which essentially transfers the secondary power pulsations to capacitors, inductors, or other energy storage elements using a bidirectional circuit connected to the load. However, this control method has limited flexibility and fully utilize the supercapacitor's capacity to completely suppress power fluctuations of the pulsed power load. In this paper, a voltage fluctuation suppression strategy in weak grid under pulsed power load integration is proposed, utilizing a converter system based on supercapacitor energy storage paralleled at the weak grid node to suppress voltage fluctuations. By combining VSG

control for the mean component of power and ripple current control for the ripple component of power, the proposed approach increases the stability margin in weak grid conditions and provides rapid compensation for power fluctuations, thus suppressing voltage fluctuations.

II. The Proposed System

Fig.1 depicts the diagram of the converter system based on supercapacitor energy storage in Weak Grid. The DC/DC side connects to the DC bus and supercapacitor, while the DC/AC side is paralleled with the grid and connected to the load. Without the system enabled, the grid supplies the pulsed power load, generating pulsed currents.,which leads to voltage fluctuations at the point of common coupling (PCC) due to the higher imped-ance in weak grids.The proposed voltage fluctuation suppression strategy in weak grid under pulsed power load integration utilize the supercapacitor's capacity to detect and provide pulsed power to the load, mitigating pulsed currents flowing into the grid. Additionally, to address low stability margins in grid-following control and phase-locked loop(PLL) instability in weak grid scenarios, Virtual Synchronous Generator(VSG) control is applied to the mean component of the system's output power, enabling voltage stabilization and precise reactive power regulation via the VSG power controller.

A. The VSG control for the mean component of power with power controller

Figure 2 illustrates the proposed VSG control for the mean component of power with power controller. When voltage fluctuation suppression is enabled, it can be considered that the ripple component of the pulsed power load's power is entirely provided by the proposed system, leading to the following formula:

$$\begin{cases} P_{o,m} = P_o - P_{o,ripple} \approx P_o - P_{L,ripple} \\ Q_{o,m} = Q_o - Q_{o,ripple} \approx Q_o - Q_{L,ripple} \end{cases} \quad (1)$$

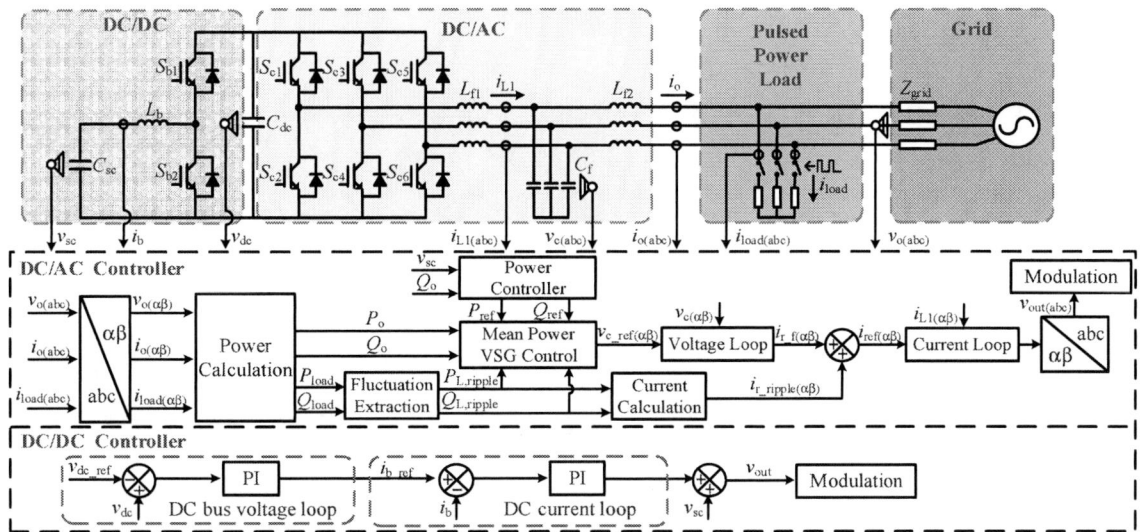

Figure 1: Diagram of the proposed approach for the converter system based on supercapacitor energy storage in Weak Grid.

Fig. 2(a) demonstrates the VSG control for the mean component of power, employing grid-forming control without using a PLL, which enhances the system's stability margin under voltage fluctuation scenarios in weak grids. Fig. 2(b) illustrates the power controller, where the Supercapacitor Voltage Controller adjusts the active power reference of the VSG to stabilize the mean component of the supercapacitor voltage. The Reactive Power Controller regulates the reactive power reference of the VSG to achieve zero-error adjustment of the system's output reactive power.

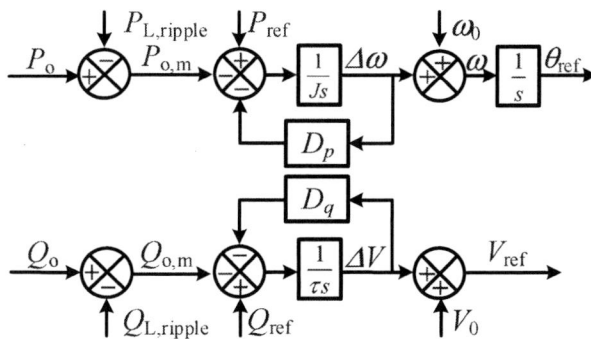

(a) Diagram of VSG control for the mean component of power

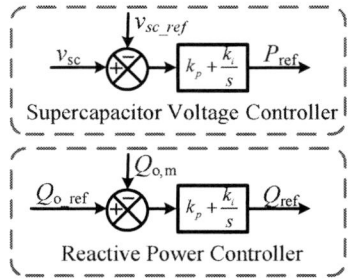

(b) Diagram of power controller

Figure 2: Diagram of VSG control for the mean component of power with power controller.

B. The voltage fluctuation suppression strategy

Fig. 3 presents the proposed voltage fluctuation suppression strategy. Due to the relatively slow response of

VSG control to changes in power references, VSG control is only applied to the mean component of power. For the ripple component of power, the proposed strategy employs current control for voltage fluctuation sup-pression. Initially, a filter is used for the fluctuation extraction, as shown in the following equation:

$$
\begin{cases}
P_{L,ripple} = P_{load} - P_{L,m} = \dfrac{s^2 + 2\omega_{c1}s}{s^2 + 2\omega_{c1}s + \omega_{c1}^2} P_{load} \\[2mm]
Q_{L,ripple} = Q_{load} - Q_{L,m} = \dfrac{s^2 + 2\omega_{c1}s}{s^2 + 2\omega_{c1}s + \omega_{c1}^2} Q_{load}
\end{cases}
\tag{2}
$$

where represents the cutoff frequency of the low-pass filter. Subsequently, based on the obtained ripple component of power, the reference of the ripple component of current is obtained as:

$$
\begin{cases}
i_{r\_ripple,\alpha} = (v_{o,\alpha}P_{L,ripple} + v_{o,\beta}Q_{L,ripple})/(v_{o,\alpha}^2 + v_{o,\beta}^2) \\[2mm]
i_{r\_ripple,\beta} = (v_{o,\alpha}Q_{L,ripple} - v_{o,\beta}P_{L,ripple})/(v_{o,\alpha}^2 + v_{o,\beta}^2)
\end{cases}
\tag{3}
$$

The reference of the ripple component of current is added to the reference of the fundamental component of current generated by the VSG control and outer voltage loop control, resulting in the inner loop reference. By directly controlling the ripple component through current control, the system compensates for the load power fluctuations, achieving rapid response to the ripple component of power of the pulsed

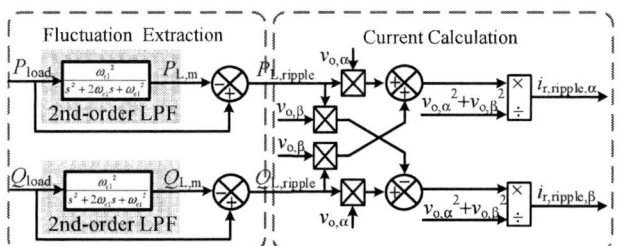

Figure 3: Diagram of the voltage fluctuation suppression strategy.

power load, thereby suppressing voltage fluctuations.

III. MODEL ESTABLISHMENT AND ANALYSIS

Due to the combination of average power VSG control and fluctuating current control, the base frequency current control and fluctuating current control are decoupled. Therefore, the control and circuit can be modeled separately based on the base frequency component and the fluctuating component, and the effectiveness of the control can be analyzed. The model of the fluctuating component primarily analyzes the system's effectiveness in suppressing voltage fluctuations; the model of the base frequency component primarily analyzes the system's stability under weak grid conditions.

A. Model of the fluctuating component

The transfer functions Gi(s) and Gv(s) in Fig. 4 are

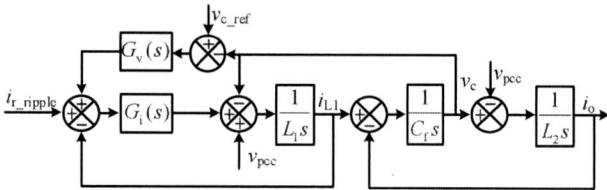

Figure 4: Diagram of the transfer function for the current control loop.

obtained as:

$$\begin{cases} G_i(s) = k_{pi} + \dfrac{2k_{ri}\omega_b s}{s^2 + 2\omega_b s + \omega_0^2} \\ G_v(s) = k_{pv} + \dfrac{2k_{rv}\omega_b s}{s^2 + 2\omega_b s + \omega_0^2} \end{cases} \quad (4)$$

where k_{pv}、 k_{pi} and k_{rv}、 k_{ri} is the proportional coefficient and resonant coefficient, ω_0 is the rated angular speed, ω_b is the controller bandwidth.

From Fig. 4, it can be seen that the proposed system controls the current. By simplifying the control block diagram, the closed-loop response of the output current is obtained as:

$$I_o = G_{ref,i}(s) \cdot i_{r\_ripple} + G_{dis,i}(s) \cdot v_{c\_ref} - Y_{eq}(s) \cdot I_o \quad (5)$$

where $G_{ref,i}(s)$ is the closed-loop gain of the current fluctuation reference, $G_{dis,i}(s)$ is the closed-loop gain of the disturbance, and $Y_{eq}(s)$ is the series closed-loop output impedance of the controller harmonic voltage source. The detailed expressions of these gains are obtained as:

$$\begin{cases} G_{ref,i}(s) = \dfrac{G_i(s)}{L_1 L_2 C_f s^3 + L_2 C_f G_i(s)s^2 + G_i(s)G_v(s)L_2 s + G_i(s)} \\ G_{dis,i}(s) = \dfrac{G_i(s)G_v(s)}{L_1 L_2 C_f s^3 + L_2 C_f G_i(s)s^2 + G_i(s)G_v(s)L_2 s + G_i(s)} \\ Y_{eq}(s) = \dfrac{L_1 C_f s^2 + C_f G_i(s)s + G_i(s)G_v(s)}{L_1 L_2 C_f s^3 + L_2 C_f G_i(s)s^2 + G_i(s)G_v(s)L_2 s + G_i(s)} \end{cases} \quad (6)$$

From this, the simplified equivalent circuit for the current response can be obtained, as shown in the figure.

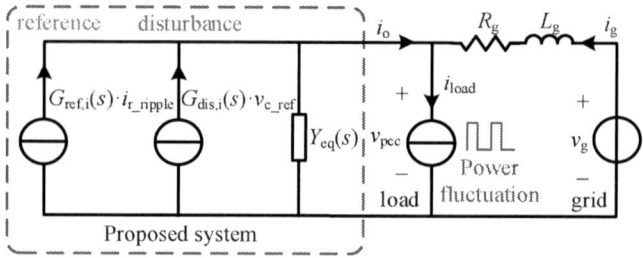

Figure 5: Simplified equivalent circuit for current response.

In this model, the focus is on the impact of the fluctuating components, so the fundamental control reference, which acts as a disturbance, is neglected. The AC voltage at the point of common coupling v_{pcc} is obtained as:

$$\begin{aligned} v_{pcc} = &\dfrac{1}{Z_g(s)Y_{eq}(s)+1}v_g + \dfrac{Z_g(s)G_{ref,i}(s)}{Z_g(s)Y_{eq}(s)+1}i_{r\_ripple} \\ &- \dfrac{Z_g(s)}{Z_g(s)Y_{eq}(s)+1}i_{load} \end{aligned} \quad (7)$$

The above equation describes the dynamic behavior of the AC voltage at the point of common coupling (PCC) of the proposed system. Considering that $\theta = (1/s)\cdot\omega_0$,the above equation can be linearized, and in practical systems, the pulsating fluctuations of the grid voltage are almost zero, the disturbance component of the grid voltage is neglected. A small-signal model of the PCC AC voltage amplitude V_o is established as follows:

$$\Delta V_o = \underbrace{H_p(s)\Delta i_{p,load}}_{\text{Active Power Fluctuation of Load}} + \underbrace{H_q(s)\Delta i_{q,load}}_{\text{Reactive Power Fluctuation of Load}} \quad (8)$$

In the equation, the operator Δ represents the small-signal perturbation around the system's equilibrium point. It can be noted that Eq. (10) is the sum of two independent terms, each representing the response of the PCC voltage amplitude to a specific disturbance source. Specifically, the first term and the second term represent the specific impacts of the active and reactive power components of the load power fluctuation on the PCC voltage amplitude, respectively. The transfer functions of the two independent terms are as follows:

$$\begin{cases} H_p(s) = G_R(s)G_H(s) - (L_g s + R_g) \\ H_q(s) = G_I(s)G_H(s) - \omega_0 L_g \end{cases} \quad (9)$$

In the equation, the transfer functions $G_R(s)$、 $G_I(s)$ mainly depend on the control parameters and circuit parameters of the proposed system. Their expressions are shown at the bottom of the next page.

From this, Bode plots of $H_p(s)$ and $H_q(s)$ can be made to analyze the effectiveness of the proposed control strategy in suppressing the fluctuations of the PCC voltage amplitude under different disturbance sources, as shown in the figure below. From the Bode plot in Fig. 6, it can be observed that under low-frequency load power fluctuation disturbances,

979-8-3315-1612-3/25 $31.00 © 2025 IEEE   2630

the proposed fluctuating current control strategy can effectively suppress the fluctuations of the PCC voltage

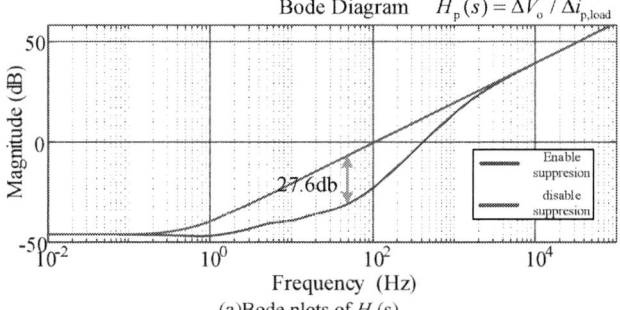

(a)Bode plots of $H_p(s)$

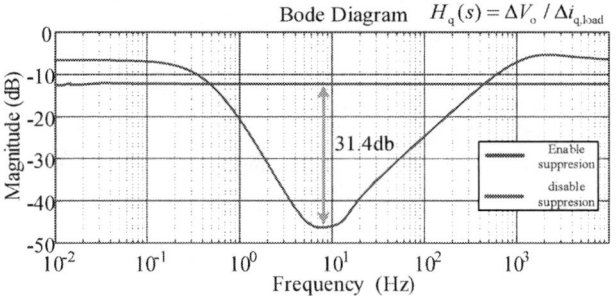

(b) Bode plots of $H_q(s)$

Figure 6: Bode plots of $H_p(s)$ and $H_q(s)$.

amplitude V_o caused by different disturbance sources, compared to when the control is not enabled..

B. Model of the base frequency component

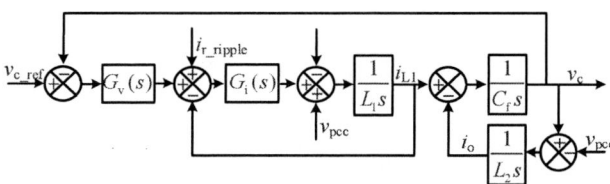

Figure 7: Diagram of the transfer function for the voltage control loop.

A diagram of the transfer function for the voltage control loop is made, primarily to analyze the effectiveness of the proposed system in controlling the fundamental voltage. Similar to the previous section, the simplified equivalent

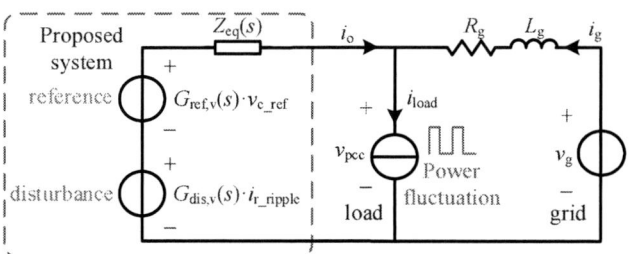

Figure 8: Simplified equivalent circuit for current response.

circuit for the voltage response can be obtained, as shown in the figure.

Combining the VSG control of the proposed system at the base frequency, as shown in Fig. 2, the small-signal expression for the base frequency voltage reference is obtained as:

$$
\begin{cases}
\Delta\omega_{ref} = \dfrac{\Delta P_{ref} - \Delta P_{o,m}}{Js + D_p} \\[3mm]
\Delta V_{ref} = \dfrac{\Delta Q_{ref} - \Delta Q_{o,m}}{\tau s + D_q}
\end{cases}
\tag{10}
$$

Neglecting the influence of the disturbance terms and the equivalent impedance $Z_{eq}(s)$, and knowing that the gain of the reference term $G_{ref,v}(s)$ is approximately 1 at the base frequency[4], the following can be obtained:

$$
\begin{cases}
\Delta\omega_{ref} \approx \Delta\omega_o \\
\Delta V_{ref} \approx \Delta V_o
\end{cases}
\tag{11}
$$

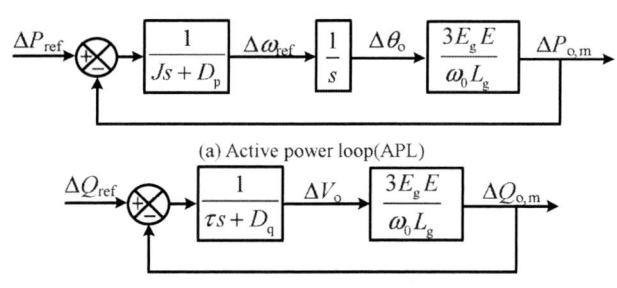

(a) Active power loop(APL)

(b) Reactive power loop(RPL)

Figure 9: Simplified equivalent circuit for current response.

$$
H_R(s) = k_{pi} + k_{ri} \frac{2\omega_b s\left(s^2 + 2\omega_b s\right) + 4\omega_b \omega_0^2\left(s + \omega_b\right)}{\left(s^2 + 2\omega_b s\right)^2 + 4\omega_0^2\left(s + \omega_b\right)^2}
\tag{12}
$$

$$
H_I(s) = k_{ri} \frac{-4\omega_b \omega_0 s\left(s + \omega_b\right) + 2\omega_b \omega_0\left(s^2 + 2\omega_b s\right)}{\left(s^2 + 2\omega_b s\right)^2 + 4\omega_0^2\left(s + \omega_b\right)^2}
\tag{13}
$$

$$
G_R(s) = \frac{\left(Z_g H_R(s) - \omega_0 L_g H_I(s)\right)\left[(L_1 + L_2)s + H_R(s)\right] + \left(Z_g H_I(s) + \omega_0 L_g H_R(s)\right)\left[(L_1 + L_2)\omega_0 + H_I(s)\right]}{\left[(L_1 + L_2)s + H_R(s)\right]^2 + \left[(L_1 + L_2)\omega_0 + H_I(s)\right]^2}
\tag{14}
$$

$$
G_I(s) = \frac{\left(Z_g H_I(s) + \omega_0 L_g H_R(s)\right)\left[(L_1 + L_2)s + H_R(s)\right] - \left(Z_g H_R(s) + \omega_0 L_g H_I(s)\right)\left[(L_1 + L_2)\omega_0 + H_I(s)\right]}{\left[(L_1 + L_2)s + H_R(s)\right]^2 + \left[(L_1 + L_2)\omega_0 + H_I(s)\right]^2}
\tag{15}
$$

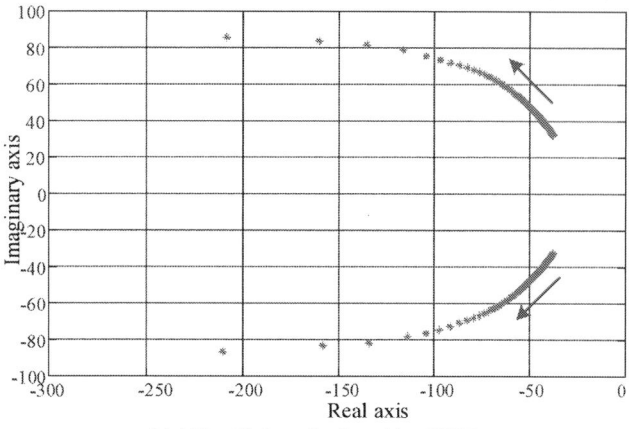

(a) APL with D_p varies from 50 to 10000

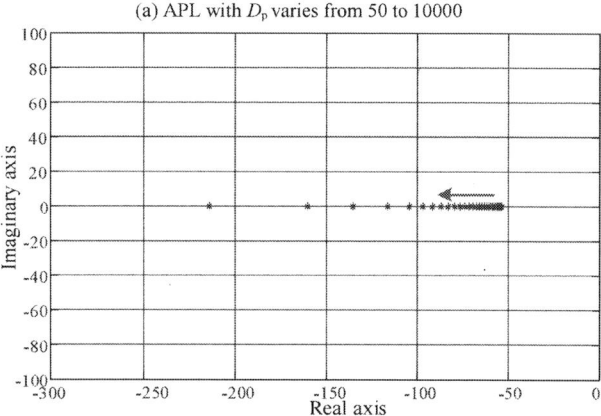

(b) RPL with D_q varies from 50 to 10000

Figure 10: Root locus plot of the APL and RPL.

Combining the above equations, the small-signal control block diagram of the VSG at the base frequency can be drawn.

From this, the root locus plots of the APL and RPL can be made, as shown in Fig. 9. From the root locus in Fig. 9, it can be seen that the dominant poles of the transfer function of the VSG control for the average power are located in the left half-plane. The proposed VSG control for the average power at the base frequency can maintain good stability under weak grid conditions, providing a large stability margin.

IV. EXPERIMENTAL VERIFICATION

This paper presents the experimental results obtained from the prototype of the proposed system with voltage fluctuation suppression. The experimental platform is shown in Fig. 11. Fig. 12 demonstrates the control effectiveness of the proposed system under different voltage fluctuation scenarios. The first row shows the point-of-common-coupling (PCC) voltage waveforms, and the second row displays the grid current waveforms. Fig.12(a) corresponds to the situation where the load generates a 5 Hz power fluctuation, Fig.12(b) shows the case of an 8 Hz power fluctuation, and Fig.12(c) represents the situation with a 10 Hz power fluctuation. It can be observed that before enabling the proposed control, the grid current exhibits significant fluctuations, and in weak grid scenarios, a large ripple component is superimposed on the PCC voltage via the grid impedance, causing PCC voltage fluctuations exceeding 10%. After activating the proposed control strategy, the

System parameters	
Rated Grid Voltage	V_N = 220VRMS/50Hz
Rated DC Bus Voltage	V_{dc}= 750V;
Load Active Power and Reactive Power	P_L = 16kW; Q_L = 1kvar
Circuit parameters	
Filter for Converter	L_f = 0.45mH; C_f= 60uF; L_s = 0.05mH L_h= 0.8mH;
Supercapacitor	C_{sc} = 15F;
DC Bus Capacitor	C_{dc} = 5000uF;
Switching Frequency	F_{sw} = 10kHz;

(a) Experimental parameters

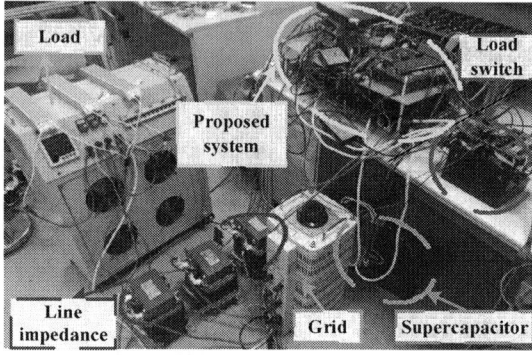

(b) Experimental platform

Figure 11: The experimental platform and its parameters.

system quickly utilizes the capacity of the supercapacitor to compensate for the grid current fluctuations. As shown in Fig. 5, under different load power fluctuation conditions, the proposed control rapidly suppresses the grid current fluctuations, thereby quickly reducing the PCC voltage fluctuations to below 3%.This ensures the system's stability under weak grid voltage fluctuations while quickly

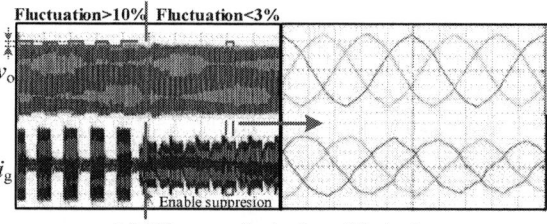

(a) 5 Hz power fluctuation of the load

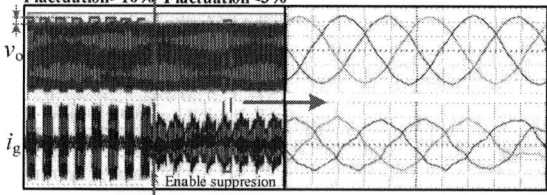

(b) 8 Hz power fluctuation of the load

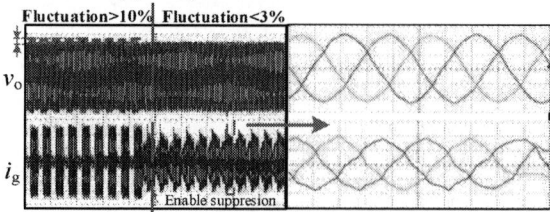

(c) 10 Hz power fluctuation of the load

Figure 12: Experimental waveforms.

suppressing voltage fluctuations caused by load power fluctuations.

CONCLUSION

In this paper, a Voltage Fluctuation Suppression Strategy is proposed to address voltage fluctuation issues in weak grids under pulsed power loads integration. By combining VSG control for the mean component of power and ripple current control for the ripple component of power, the proposed approach increases the sta-bility margin in weak grid conditions, avoids the impact of voltage fluctuations on PLL stability, and provides rapid compensation for power fluctuations, thus suppressing voltage fluctuations at the point of common cou-pling. The strategy offers both stability and fast response. Compared to previous methods, the proposed ap-proach need for external circuit solutions for each pulsed power load and can be more easily integrated into the grid.

Reference

[1] S. Qu, Q. Zhaohui, L. Zhaowei, S. MingMing, H. Yuchen and L. Zhenhua, "Energy Storage Active and Reactive Power Coordinated Control Considering DC Commutation Failue Voltage Recovery and Re-straining AC Line Active Power Fluctuation," 2022 7th Asia Conference on Power and Electrical En-gineering (ACPEE), Hangzhou, China, 2022, pp. 1163-1168.

[2] K. Wang, Y. Li, Z. Zheng and L. Xu, "Voltage Balancing and Fluctuation-Suppression Methods of Floating Capacitors in a New Modular Multilevel Converter," in IEEE Transactions on Industrial Electronics, vol. 60, no. 5, pp. 1943-1954, May 2013, doi: 10.1109/TIE.2012.2201433.

[3] Y. Sun, L. Jia, M. Hua and S. Lu, "Energy Storage Capacitor Optimization of Power Fluctuation Sup-pression System with Pulse Load," 2021 IEEE Sustainable Power and Energy Conference (iSPEC), Nanjing, China, 2021, pp. 3096-3101.

[4] H. Wu et al., "Small-Signal Modeling and Parameters Design for Virtual Synchronous Generators," in IEEE Transactions on Industrial Electronics, vol. 63, no. 7, pp. 4292-4303, July 2016, doi: 10.1109/TIE.2016.2543181.

An Optimized Firmware-Based Cycle-by-Cycle Current Limiting Method for Power Electronic Converters in UPS

Teng Wu, and Hong Liu
Secure Power Division
Schneider Electric (China) Co., Ltd. Shanghai Pudong Branch
Shanghai 200120, China
warren.wu@se.com, hong.liu@se.com

Abstract—**Cycle-by-cycle current limiting is a control strategy used to protect power electronic converters and loads by ensuring that the current does not exceed safe operating limits on a very fast, cycle-by-cycle basis, where a cycle typically refers to the switching period of a power electronic converter. However, Due to the current sampling delay caused by sampling circuit and analog-digital conversion rate, the current signal obtained by the firmware processor is lagging to the actual current, which causes delay in current limiting and protection. This paper proposes an optimized firmware-based cycle-by-cycle current limiting method, where the sampled current value is corrected on switching period scale to get a more accurate reconstruction of the actual current despite the sampling delay. The accuracy of this method is proved through experimental verification.**

Keywords—cycle-by-cycle current limiting, power electronic converter, protection

I. INTRODUCTION

Uninterruptible Power Supply (UPS) is a power protection equipment designed to continuously provide electricity to loads when the main power source fails. It protects critical load from the effects of power issues such as voltage fluctuations, power outages, and frequency changes [1].

Based on working principles and design structures, high-power UPS is generally divided into three types: standby, online, and online interactive [2-4]. Online UPS does not have switchover time issues, and its adaptability to the input mains power is stronger. It can effectively isolate distortion and interference from the input side. Thus, online UPS is more suitable for occasions that have strict requirements for power supply quality and continuity, especially data centers.

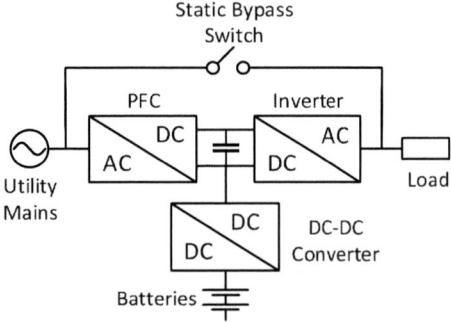

Figure 1. Block diagram of an online UPS

The power module of an online UPS includes a Power Factor Corrector (PFC), an Inverter, and a DC-DC converter that interfaces with the battery [5-7], as illustrated in Figure 1. These three converters are crucial for ensuring the continuity of load supply, so fast and accurate overcurrent protection functions are required for each of these converters.

For fast overcurrent protection of converters, many solutions are based on hardware protection circuits [8-10], which lead to increased costs and circuit size. However, in some cost-leading projects, a firmware-based solution is more attractive in terms of cost.

Cycle-by-cycle current limiting (CCL) is widely used in high-power converters to protect the power stage devices or loads in a fast way [11]. CCL can also be hardware-based [11-13]. Though for the firmware-based CCL, a fast and rapid current ripple prediction algorithm is required. Current ripple prediction is discussed in some literatures [14-16]. However, the current ripple is predicted based on a current prediction model, which is highly related to the parameters of the converter system. It is not easy to extend to other power electronic converters since it is not a universal model.

Generally, the firmware-based CCL is conducted as soon as the firmware processor gets the sampled value of current. If the sampled value exceeds a certain threshold, the relevant semi-conductor switches will be opened immediately until next switching cycle. Through this way, the processor can rapidly prevent current from rising beyond the physical current limitation of the power stage devices or loads.

CCL highly relies on the accuracy of the sampling of current. However, due to the current sampling delay caused by sampling circuit and analog-digital conversion (ADC) rate, the current signal obtained by firmware processor is lagging to the actual current. As shown in Figure 2, the envelope of sampled current is lagging to actual current by a certain delay which is caused by the sampling circuit, including the current sensor, the signal conditioning circuit, and the discrete ADC process. The total delay can be as long as tens of microseconds.

Consequently, due to the delay in current sampling, when over current occurs, the CCL would not be timely triggered. To achieve an effective current limiting and protection, this paper proposes an optimized firmware-based CCL method, where the sampled current value is corrected on switching period scale to get a more accurate reconstruction of the actual current despite the sampling delay. This guarantees a fast and accurate trigger of current limiting and protection. The proposed CCL method can be extended to various power converters. Simulation and experiments have also been conducted in different types of converter in UPS to verify the effectiveness and accuracy of this method.

Section II describes the principle of the proposed CCL method. Section III gives detailed design criteria for key parameters. Simulation and experimental results are shown in Section IV and V, respectively. Finally, the work is summarized in Section VI.

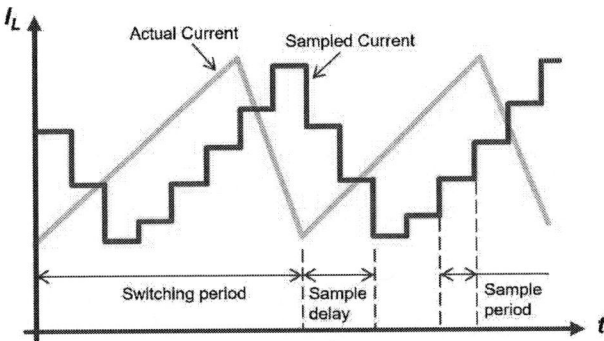

Figure 2. Sketch diagram of the actual current and sampled current

II. DESCRIPTION OF THE OPTIMIZED CCL METHOD

The basic formular of the optimized CCL method is described in (1):

$$I_{cor} = I_{samp} + m/8 \times [(K + n/4) \times \Delta_{max}] \tag{1}$$

where I_{cor} is the corrected current value, I_{samp} is the sampled current value, Δ_{max} is the maximum difference among adjacent samples, K is the correction factor, which equals to the proportion of the sampling delay time to the sampling period, m is the peak-shaving (PS) coefficient ranging from 0 to 8, and n is the over-sample coefficient ranging from 1 to 4.

For example, current is sampled every 5.2 μs. The difference between two sampled values is calculated and then recorded into a buffer of length 16. Δ_{max} is the maximum one of the 16 values in the buffer, indicating the current rising rate, which is updated every 1.3 μs.

In order to increase the prediction rate to 4 times per sampling, which is so-called over-sampling, K is added with $n/4$. The over-sampling within one sampling period is illustrated in the red circle in Figure 3. By adding $n/4$, current can be corrected every 1.3 μs.

Basically, the corrected current is the envelope of the sampled current plus a correction part. It is like displacing the envelope of the sampled curve upward as the dashed curve shown in Figure 4. Obviously, over correction exists if the peak of the dashed curve is not handled. In order to mitigate the peak, the correction part is multiplied with $m/8$, as described in (1), where the peak-shaving coefficient m decreases when current begins to fall and increases when current begins to rise.

In conclusion, by applying the formular described in (1), the corrected current shall be like the peak shaved solid curve shown in Figure 4, where the rising edge of the actual current can be accurately reconstructed.

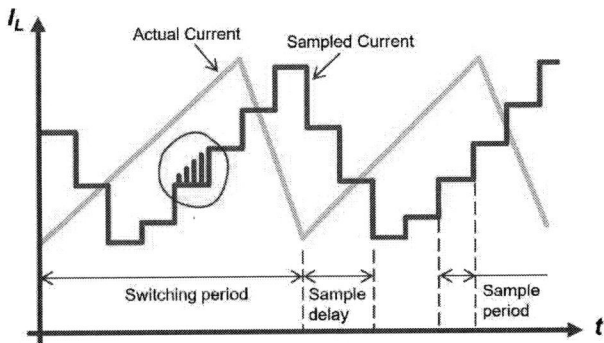

Figure 3. Illustration of over-sampling within one sampling period

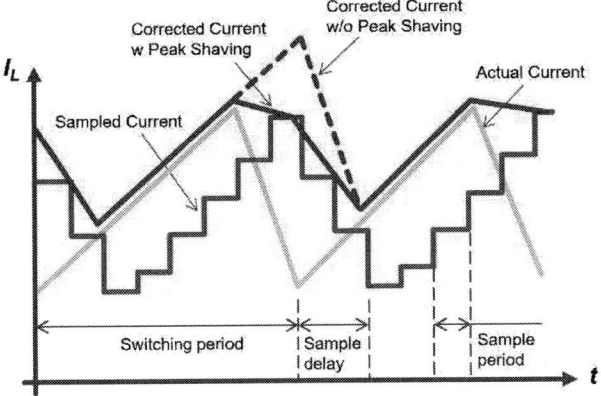

Figure 4. Sketch diagram of the corrected current based on the proposed CCL method

III. DESIGN OF KEY PARAMETERS

A. Correction factor K

The correction factor K is determined by how far back in time the data is expected to be corrected. Therefore, it is highly related to the total sampling delay. The value of K is derived from (2):

$$K = T_{delay}/T_{samp} \tag{2}$$

where T_{delay} and T_{samp} indicate the sampling delay and the sampling period, respectively.

For example, the total sampling delay caused by sampling circuit and ADC is 9.73 μs, and the sampling period of processor is 5.2 μs, which should have a K value of 1.87.

B. Peak-shaving coefficient m

As explained in Section II, the peak-shaving coefficient m is aimed to avoid over correction by adjusting the weight of correction. The peak-shaving coefficient m changes by one every correction period (1.3 μs) until it reaches the ceiling value (set to 8 for example) or the floor value (set to 0). The change of m is illustrated in Figure 5.

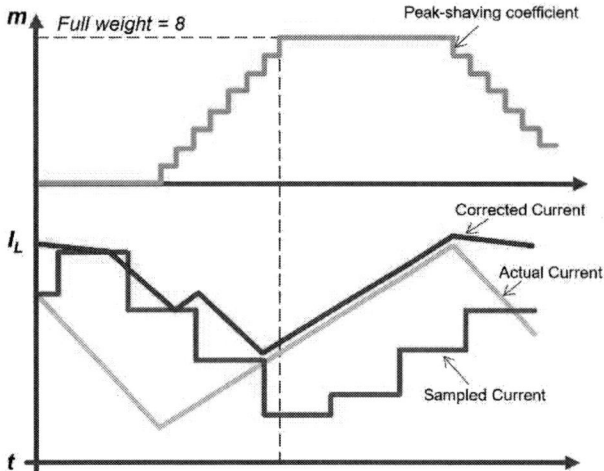

Figure 5. Illustration of the change of the peak-shaving coefficient

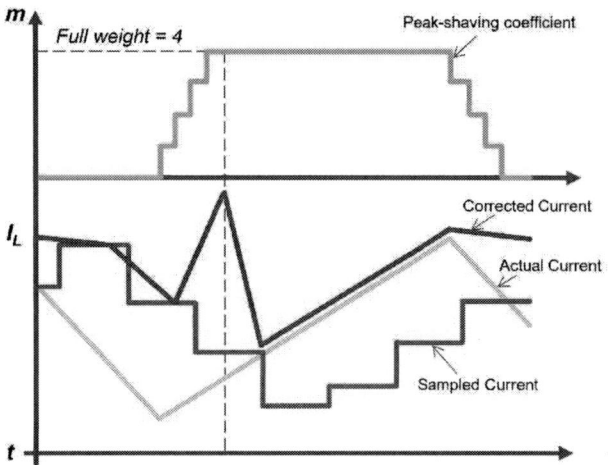

Figure 6. Over correction in case of a lower ceiling value of m

Whether is the ceiling value setting to 8 reasonable? If the ceiling value is higher than 8, the correction weight is adjusting too slowly to achieve a good correction. In opposite, the correction part should not take full effect before the sampled current ripple reaching the bottom. Figure 6 explains why over correction might happen if the ceiling value is smaller (for example 4). The over correction is produced because the sampled current is still high when the weight of the correction part already reaches 100%. Therefore, the sampling delay should be taken into consideration in the design of the ceiling value m_{ceil} of m, as described in (3):

$$m_{ceil} = round\left(T_{delay}/T_{cor}\right) \tag{3}$$

where T_{cor} indicates the correction period.

C. Limit of Δ_{max}

To calculate the rising rate of the current, the difference between two adjacent samples is recorded and stored into a buffer with length of 16. To avoid under correction, the maximum value in the buffer is symbolized as Δ_{max} and is used to calculate the rising rate of current. However, Δ_{max} should be well limited in order to avoid over correction caused by sampling error. The positive limit is analyzed as example in this section. The setting criteria of negative limit for negative current is the same as that of positive limit.

The upper limit of Δ_{max} should be set as the same as the current deviation within one sampling period, which is derived from (4):

$$\Delta i_{max} = V_{Lmax}/L_{min} \cdot T_{samp} \tag{4}$$

where Δi_{max} is the maximum current deviation within one sampling period, V_{Lmax} is the maximum voltage across the inductor, and L_{min} is the minimum inductance of the inductor.

Take the DC-DC converter in UPS as example. One of the task of this converter is to boost battery voltage to DC bus voltage. As shown in Figure 7, the maximum voltage across the inductor occurs when the boost switch closes, which is the upper limit of the battery voltage. Besides, the inductor will be saturated and inductance will be reduced when current flowing through it increases. Back to (4), the maximum current deviation Δi_{max}, which is also the upper limit of Δ_{max} can be derived by taking V_{Lmax} and L_{min} into calculation.

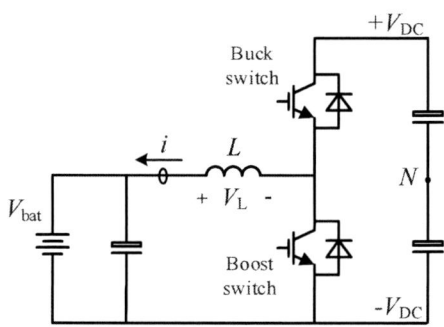

Figure 7. Topology of DC-DC converter in UPS

D. CCL threshold

Three parts should be taken into consideration in the setting of limiting threshold: the peak value of fundamental current, the peak value of inductor current ripple and the maximum possible correction error.

1) Inverter

Regarding a three-phase inverter, the peak value of fundamental current I_{peak} can be derived following (5):

$$I_{peak} = P_{rt}/3/\left(V_{rt}/\sqrt{3}\right) \cdot K_{crest} \tag{5}$$

where P_{rt} and V_{rt} indicate the rated power and rated phase-phase voltage, respectively, and K_{crest} is the crest factor to cover higher current peak in case of non-linear load.

The peak-peak value of current ripple is described in (6):

$$I_{rip} = V_L/L \cdot T_{on} \tag{6}$$

where I_{rip} is the peak-peak value of current ripple, V_L is the voltage across the inverter inductor, L is the inductance of the inverter inductor, and T_{on} is the time duration of the closed state of switch.

979-8-3315-1612-3/25 $31.00 © 2025 IEEE

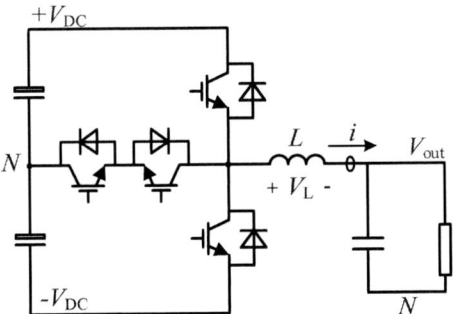

Figure 8. Single phase diagram of inverter topology in UPS

As illustrated in Figure 8, the voltage across the inverter-side inductor is the voltage difference between the DC bus voltage V_{DC} and the output voltage V_{out}. T_{on} is the product of duty ratio D and the switching period T_{sw}. Therefore, Equation (6) can be further transformed to (7) as below.

$$I_{rip} = (V_{DC} - V_{out})/L \cdot D \cdot T_{sw} \quad (7)$$

Furthermore, the circuit of inverter's each phase is a buck circuit. Therefore, (7) can be further transformed to (8) as below.

$$I_{rip} = (V_{DC} - V_{out})/L \cdot V_{out}/V_{DC} \cdot T_{sw} \quad (8)$$

It can be seen that I_{rip} is a quadratic function of V_{out}. When V_{out} is half V_{DC}, I_{rip} reaches the maximum value, which is

$$T_{sw} \cdot V_{DC}/(4 \cdot L)$$

The correction error is caused by the discrete processing of firmware. The maximum possible error I_{err} is the error caused by the execution delay of the algorithm, which equals to the correction period T_{cor}. The error is the current deviation within T_{cor}, which can be derived from (9) as below.

$$I_{err} = V_L/L \cdot T_{cor} \quad (9)$$

2) DC-DC converter

Regarding the DC-DC converter, the maximum DC current depends on whether it is operated in charging mode or discharging mode. In charging mode, the maximum DC current is usually limited to 10%-60% of the rated power of power factor corrector (PFC) stage. In discharging mode, the maximum DC current is usually limited to 150% of the rated power to cover temporary overload.

The topology of DC-DC converter under investigation is shown in Figure 7. The current ripple of the DC-DC converter also follows the formula described in (6). When the battery is discharging, it operates like a boost converter. Therefore, when the boost switch closes, the voltage across the battery-side inductor is the battery voltage. Equation (6) can be transformed to (10):

$$I_{rip} = V_{bat}/L \cdot (2V_{DC} - V_{bat})/2V_{DC} \cdot T_{sw} \quad (10)$$

where V_{bat} is the battery voltage. Note that VDC is multiplied by 2 because the inverter under investigation is a 3-level T-type inverter as illustrated in Figure 8.

When the battery is charging, it operates like a buck converter. Therefore, when the buck switch closes, the voltage across the battery-side inductor is the difference between double DC bus voltage and battery voltage. The current ripple can be also derived from (10).

It can be seen from (10) that I_{rip} is a quadratic function of V_{bat}. When V_{bat} is half V_{DC}, I_{rip} reaches the maximum value, which is

$$T_{sw} \cdot V_{DC}/(2 \cdot L)$$

Same as inverter current, the correction error of DC-DC current can also be derived from (9).

3) Power factor corrector (PFC)

The current ripple of PFC also follows the formula described in (6). As illustrated in Figure 9, each phase of PFC operates like a boost converter. the voltage across the PFC-side inductor is the input voltage. Therefore, equation (6) can be transformed to (11):

$$I_{rip} = V_{in}/L \cdot (V_{DC} - V_{in})/V_{DC} \cdot T_{sw} \quad (11)$$

where V_{in} is the input voltage of PFC.

It can be seen that I_{rip} is a quadratic function of V_{in}. When V_{in} is set to the lower limit of the input voltage and V_{DC} set to the upper limit of DC bus voltage, I_{rip} reaches the maximum value.

Same as inverter current, the correction error of PFC current can also be derived from (9).

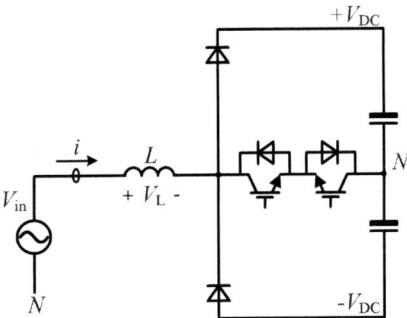

Figure 9. Single phase diagram of PFC topology in UPS

4) Summary

In conclusion, the CCL threshold I_{thrd} can be derived from (12) as below.

$$I_{thrd} = I_{peak} + I_{rip}/2 \times 1.08 + I_{err} + 10 \quad (12)$$

Note that the ripple current is multiplied by 1.08 to cover the deviation of inductance. Besides, a buffer of 10A is also taken into calculation.

IV. SIMULATION VERIFICATION

The current tracking performance of the proposed current correction method has been verified in Simulink.

Figure 10 shows the correction results of inverter current and DC-DC discharging current. It can be seen that during the current rising period, the corrected current tracks well the actual current, which lays the foundation for accurate CCL protection.

Note that the current is not corrected during falling period because current drop implies lower thermal risk.

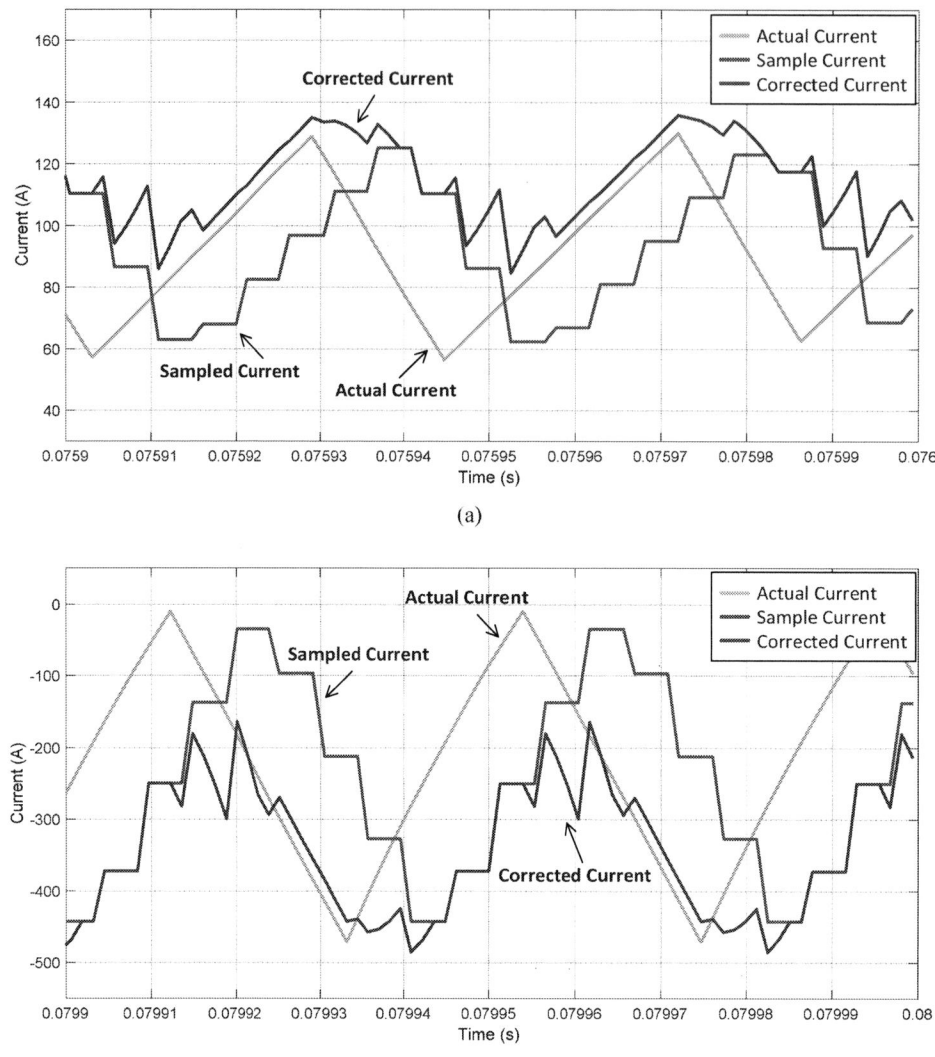

(a)

(b)

Figure 10. Simulation waveform of the proposed CCL method for (a) inverter, and (b) DC-DC converter

Also note that when battery is discharging to DC bus, the sampled current is negative because the positive direction of the current sensor is aligned with the charging current from DC bus to battery.

V. EXPERIMENTAL VERIFICATION

The experimental platform consists of a Schneider hardware-in-loop (HIL) UPS model. The proposed CCL method is verified both in the inverter and DC-DC converter of the HIL UPS. To test the accuracy of the proposed CCL method, 300% overload is applied in the HIL UPS and over-current protection accuracy (which is guaranteed by the CCL accuracy) is investigated.

The first test is conducted with UPS inverter. The proposed CCL method is implemented into the UPS inverter and the over-current protection threshold in the firmware is set to 365A. The load step is from zero to 300% rated power. The waveforms are shown in Figure 11(a). It can be seen when 300% load is applied in the inverter, the output current rises quickly beyond 365A within the first switching cycle. The over-current is timely detected by the proposed CCL method and over-current protection is triggered, which leads to the disable of IGBT driving and the falling of current. The actual current peak is 362A which is very closed to the over-current protection threshold setting value.

The second test is conducted with UPS DC-DC converter. The proposed CCL method is implemented into the UPS DC-DC converter and the over-current protection threshold in the firmware is set to -305A. The load step is from 100% to 300% rated power. The waveforms are shown in Figure 11(b). It can be seen when 300% load is applied in the DC-DC converter, the output current rises quickly beyond -305A (boost DC-DC current is sensed as negative current). The over-current is timely

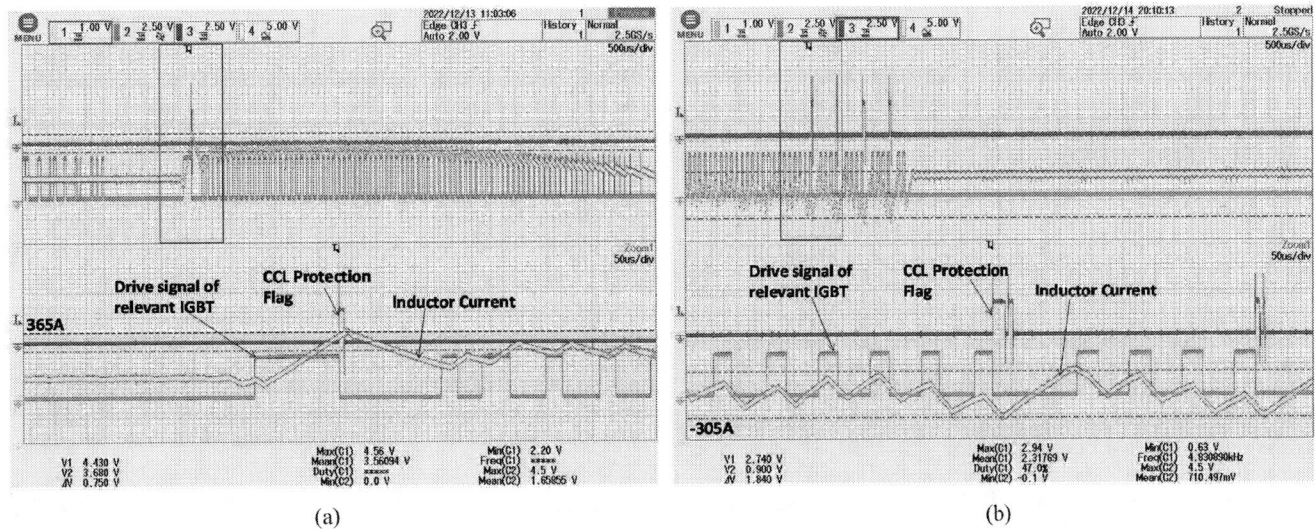

Figure 11. Experimental waveform of the proposed CCL method for (a) inverter, and (b) DC-DC converter

detected by the proposed CCL method and over-current protection is triggered, which leads to the disable of IGBT driving and the falling of current. The actual current peak is -300A which is very closed to the over-current protection threshold setting value.

VI. CONCLUSION

This paper proposes an optimized universal firmware-based cycle-by-cycle current limiting method to provide accurate over-current protection for power electronic converters. Through the proposed CCL method, the sampled current value is corrected in firmware to get a rapid and accurate reconstruction of the actual current despite the sampling delay, which lays a solid foundation for accurate over-current protection. In this paper, the design of key parameters is illustration and the effectiveness of the proposed method is verified in different types of power converter of UPS through both simulation and experiments. Lastly, the proposed CCL method is a universal method which is not only limited in UPS application but also can be extended to other power electronic applications.

REFERENCES

[1] Y. Chen, K. Shi, M. Chen, and D. Xu, "Data Center Power Supply Systems: From Grid Edge to Point-of-Load," IEEE Journal of Emerging and Selected Topics in Power Electronics, vol. 11, no. 3, pp. 2441–2456, 2023.

[2] W. Solter, "A new international UPS classi-fication by IEC 62040-3," In Proceedings of IEEE Telecommunications Energy Conference, Montreal, QC, Canada, 2002, pp. 541–545.

[3] S. Karve, "Three of a kind [UPS topologies, IEC standard]," IEE Review, vol. 46, no. 2, pp. 27–31, 2000.

[4] S. B. Bekiarov, and A. Emadi, "Uninterruptible power supplies: Classification, operation, dynamics, and control," In Proceedings of IEEE Applied Power Electronics Conference and Exposition, Dallas, TX, 2002, pp. 597–604.

[5] C. C. Yeh, and M. D. Manjrekar, "A Reconfigurable Uninterruptible Power Supply System for Multiple Power Quality Applications," IEEE Transactions on Power Electronics, vol. 22, no. 4, pp. 1361–1372, 2007.

[6] J. H. Choi, J. M. Kwon, J. H. Jung, and B. H. Kwon, "High-performance online UPS using three-leg type converter," IEEE Transactions on Industrial Electronics, vol. 52, no. 3, pp. 889–897, 2005.

[7] C. H. Lai, and Y. Y. Tzou, "DSP-embedded UPS controller for high-performance single-phase on-line UPS system," In Proceedings of IEEE Conference of the Industrial Electronics Society, Seville, Spain, 2002, pp. 268–273.

[8] F. F. Ma, W. Z. Chen, and J. C. Wu, "A Monolithic Current-Mode Buck Converter With Advanced Control and Protection Circuits," IEEE Transactions on Power Electronics, vol. 22, no. 5, pp. 1836–1846, 2007.

[9] M. E. Baran, and N. R. Mahajan, "Overcurrent Protection on Voltage-Source-Converter-Based Multiterminal DC Distribution Systems," IEEE Transactions on Power Delivery, vol. 22, no. 1, pp. 406–412, 2007.

[10] X. Xie, J. Zhang, C. Zhao, Z. Zhao, and Z. Qian, "Analysis and Optimization of LLC Resonant Converter With a Novel Over-Current Protection Circuit," IEEE Transactions on Power Electronics, vol. 22, no. 2, pp. 435–443, 2007.

[11] Y. Yang, J. Liu, and D. Zhou, "Pulse by pulse current limiting technique for SPWM inverters," in Proc. Int. Conf. Power Electron. Drive Syst. (PEDS), 1999, pp. 1021–1026.

[12] X. Gao, L. Zhang, J. Zhu, F. Wu, Y. Fang, and Y. Xing, "A Switch Partial Turned OFF Cycle by Cycle Current Limiting Method for 3L-NPC Half-Bridge Inverters," IEEE Access, vol. 8, pp. 15632–15639, 2020.

[13] LM5021 Datasheet, Texas Instruments, Dallas, TX, USA, 2014. [Online]. Available: http://www.ti.com/lit/ds/symlink/lm5021.pdf

[14] D. Jiang, and F. Wang, "Current-Ripple Prediction for Three-Phase PWM Converters," IEEE Transactions on Industry Applications, vol. 50, no. 1, pp. 531–538, 2014.

[15] L. Chang, and T. M. Jahns, "Prediction and Evaluation of PWM-Induced Current Ripple in IPM Machines Incorporating Slotting, Saturation, and Cross-Coupling Effects," IEEE Transactions on Industry Applications, vol. 54, no. 6, pp. 6015–6026, 2018.

[16] Z. Shen, and D. Jiang, "Dead-Time Effect Compensation Method Based on Current Ripple Prediction for Voltage-Source Inverters," IEEE Transactions on Power Electronics, vol. 34, no. 1, pp. 971–983, 2018.

Frequency Stop-Band Management System for DC-DC Converters

Alessandro Bertolini, Alberto Cattani, Claudio Luise and Alessandro Gasparini

STMicroelectronics, APMS R&D, Milan

Email: alessandro.bertolini@st.com

Abstract—**Ever complex electronic systems have pushed the requirements on power management to new limits, with applications needing to satisfy stringent operating frequency constraints while maintaining high efficiency. This article addresses the requirement to avoid any operation of a DC-DC converter in specific frequency bands when regulating light loads in discontinuous-conduction-mode (DCM) and operating in pulse-skip-mode (PSM). The presented stop-band-management-system (SBMS) is a versatile mixed-signal solution that has been designed for an inverting buck-boost (IBB) converter, which generates a programmable output voltage in the range of -1V to -9V from a Li-Ion battery. Targeting consumer Active-Matrix-Organic-Light-Emitting-Diode (AMOLED) display applications, a power-management-integrated-circuit (PMIC) has been fabricated in a 180nm bipolar-CMOS-DMOS (BCD) process, and the effectiveness of the proposed solution is shown with experimental results.**

Index Terms—**power management, PMIC, inverting buck-boost, DC-DC, switching regulator, BCD, power converter, DCM, stop-band, PSM, current-mode, EMI, PFM, AMOLED, SMPS.**

I. INTRODUCTION

IN many applications, the need for high-quality, ultra-low-noise power rails poses new challenges in providing such regulated rails. High-end Active-Matrix-Organic-Light-Emitting-Diode (AMOLED) panels are among these applications, requiring a clean and stable power supply to avoid flickering-on-screen (FOS) and not impact the user experience [1]. Any noise on such rails affects the voltages of each single pixel's thin-film transistor, manifesting as noise in the luminescence of the display [2], [3], [4], eventually causing eye fatigue. In addition, in densely packed consumer devices noise interference with other sub-systems is a concern, and thus both conducted and radiated electromagnetic interference (EMI) emissions in specific frequency bands must be avoided. De facto, in AMOLED display FOS can be also induced by any frequency beating with the panel operating frequencies. Concurrently, efficiency cannot be traded-off in providing these power rails, especially in battery-powered applications. For these reasons and considering the efficiency penalty posed by low-dropout (LDO) linear regulators (especially at high load), the design community has been recently pushed to investigate on switching-mode-power-supplies (SMPS) capable of generating high-quality regulated rails [5], [6]. Furthermore, to the best of the authors' knowledge, the request to seamlessly avoid any operation of clocked DC-DC converters in specific frequency bands has never been disclosed. Generally, the

involved DC-DC converters exploit the widely used peak-current-mode control (PCM) [7], [8], due to the inherent current limitation and simpler compensation compared to voltage-mode-control (VMC) [9]. At high load (e.g., up to 1 A), when the power converters are operated in continuous-conduction-mode (CCM), they typically operate with a switching frequency ($F_S = 1/T_S$) well above any stop-band SB (e.g., $F_S = 1.5\,\text{MHz}$). On the contrary, the focus is on light load conditions, where the converter is regulating in discontinuous-conduction-mode (DCM) and its switching activity is reduced (e.g., [10], [11], [12]). In fact, the aforementioned requirement is intended both in terms of i) switching frequency in CCM and/or DCM, and ii) pulse-skip-mode (PSM [13]) frequency or pulse-frequency-modulation (PFM [14]) activity. Typically, to allow a proper measurement a sampling-window SW up to a maximum of 1 ms is provided. The power converter is thus required to: 1) auto-detect its activity when steadily lying within an undesired band, and 2) "move" its operation outside forbidden bands without compromising the regulation performance. For the latter, in "moving" the converter operation outside forbidden bands the controller is required to minimize any output variation, guaranteeing a seamless correction, meaning ideally not visible on V_{OUT}. In this process, the output voltage ripple and the efficiency should not be degraded, making the forced-CCM operation a non-viable option. In Table I are reported some exemplary stop-band specifications, typically programmed from kHz up to hundreds of kHz and dynamically selected on-the-fly.

TABLE I
STOP-BAND SPECIFICATIONS

Stop-Band	F_{SB}^{MIN}	F_{SB}^{MAX}
SB1	15 kHz	35 kHz
SB2	40 kHz	60 kHz
SB3	90 kHz	110 kHz
SB4	135 kHz	155 kHz

The remainder of this article is organized as follows. The proposed solution and system architecture are described in Section II, with detailed implementation of the digital controller in Section II-B, circuit design insights are given in Section II-C, and an exemplary simulation showing the system behavior is presented in Section II-D. Experimental results are presented in Section III, while the conclusions are drawn in Section IV.

979-8-3315-1612-3/25 $31.00 © 2025 IEEE

II. PROPOSED SOLUTION & IMPLEMENTATION

A. System Architecture

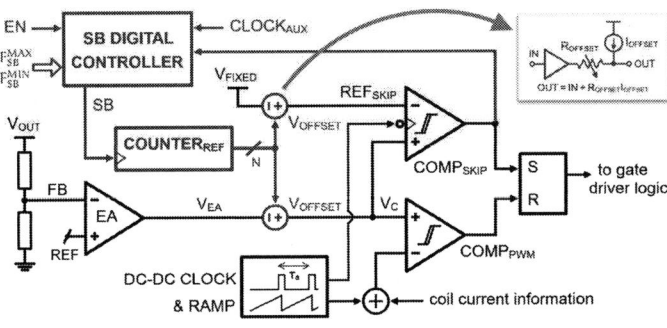

Fig. 1. Simplified schematic of the proposed solution (drawn in blue) applied to a generic peak-current-mode-controlled DC-DC converter (depicted in black) featuring PSM for light load operation.

Figure 1 shows the simplified schematic of the proposed solution (drawn in blue) applied to a generic DC-DC converter with peak-current-mode-control[1] (depicted in black) and is now briefly introduced. As a simple convention, the beginning of each switching cycle is defined by the DC-DC converter clock falling edge. In this precise moment, the skip comparator $COMP_{SKIP}$ is checked. By monitoring the control voltage V_C with respect to a fixed voltage REF_{SKIP}, this comparator decides whether to perform a new switching cycle (e.g., in CCM) or to halt the switching activity. As a normal PWM-based operation, the PWM comparator sets the converter duty-cycle for proper V_{OUT} regulation. At light load, when V_C goes below REF_{SKIP} (i.e., $COMP_{SKIP}$ does not trigger high), the converter activity is reduced to enhance efficiency [15]. Particularly, the DC-DC converter regulates in PSM (also called *burst-mode* [16]) performing K subsequent switching cycles followed by M succeeding cycles with no operation. In this case, the power converter operates with a coil peak current set by REF_{SKIP}, and regulates V_{OUT} by modulating K and/or M. The term F_{SKIP} is simply used to refer to the converter repetition rate in PSM.

Given the requirements described in Section I, it is necessary to 1) smartly measure F_{SKIP} to detect if/when the converter is operating within a SB, and 2) find a corrective action that robustly allows it to hop away from the SB in a seamless way (i.e., without generating any transient on V_{OUT}) with no performance degradation.

1) Agile Detection: First of all, the idea is to monitor $COMP_{SKIP}$ and compare its behavior with a time reference signal to understand if the converter is operating with a repetition frequency F_{SKIP} that is not allowed. $COMP_{SKIP}$ is oversampled by exploiting an auxiliary clock, $CLOCK_{AUX}$, with a fixed frequency F_{CK}^{AUX} sufficiently higher than the highest F_{SB}^{MAX}. Specifically, if both the following conditions

are satisfied, then it is possible to state that the DC-DC converter is operating within a SB (defined as the range of frequencies from F_{SB}^{MIN} to F_{SB}^{MAX}):

1) counting the number of auxiliary clock periods between two consecutive $COMP_{SKIP}$ rising edges

$$\left\lfloor \frac{F_{CK}^{AUX}}{F_{SB}^{MAX}} \right\rfloor \leq \text{number of } CLOCK_{AUX} \text{ periods} \leq \left\lceil \frac{F_{CK}^{AUX}}{F_{SB}^{MIN}} \right\rceil \quad (1)$$

2) in a sampling-window of duration t_{SW}

$$t_{SW} F_{SB}^{MIN} \leq \text{number of } COMP_{SKIP} \text{ rising edges} \leq t_{SW} F_{SB}^{MAX} \quad (2)$$

Condition (2) accounts for the average behavior of the converter within the sampling-window, but does not provide any information about the actual/instantaneous F_{SKIP}. Note that assessing only the second condition would be dangerous, since it does not provide any information about the density of $COMP_{SKIP}$ activity. In fact, condition (2) can be true for both the two following exemplary situations: 1) the converter is steadily performing in PSM with a given F_{SKIP1} that satisfies (2), and 2) the converter is initially performing with $F_{SKIP2} \gg F_{SKIP1}$ and then experiences a transient drastically lowering $F_{SKIP2} \ll F_{SKIP1}$. In the second case, in average F_{SKIP2} could still satisfy (2), but a corrective action would be required only in the first case, while in the latter the DC-DC converter should simply be allowed to settle, without the need of any action.

On the contrary, condition (1) provides insights about the present F_{SKIP}, but with a coarse resolution limited by F_{CK}^{AUX}. Moreover, not considering for a moment this limited resolution (i.e., ideally supposing to afford a super fast clock), relying only on condition (1) would still be deceptive, since it would not be possible to understand if the converter is in steady state or experiencing a transient (e.g., the DC-DC converter may "just be passing inside" the SB and eventually settling outside). Now it should be clear that checking only a single condition would be misleading and not sufficient to faithfully infer about the power converter's behavior, especially during transients. To increase the detection speed and maximize the performance during the converter transients, the sampling-window is continuously reset (i.e., shifted ahead in time) until condition (1) is not satisfied. When condition (1) is satisfied, it means that the present F_{SKIP} is coarsely within - or close to - the SB and thus a fine check is necessary. Here the sampling-window is started, and condition (2) is checked after t_{SW}. In other words, the precise evaluation is done only at the end of a sampling-window during which the converter has been operating with F_{SKIP} coarsely within or close to the forbidden band. Automatically, any transient is filtered out, avoiding misleading assessments.

The proposed algorithm is implemented[2] within the SB digital controller described in Section II-A.

2) Flexible Corrective Action: When the controller recognizes that the converter operation is within a SB, it initiates a

---

[1]The presented solution can be applied in the same way in valley-current-mode-control.

[2]For both the outlined conditions, a margin accounting for the non-idealities (e.g., limited clock frequency accuracy) has been applied.

corrective action by raising the flag SB in Figure 1 to increase $COUNTER_{REF}$ value. The idea is to slightly change the "programmed" coil peak current (i.e., REF_{SKIP}) and thus alter the skip behavior of the converter (i.e., change M and/or K). Iteratively modifying the added offset voltage V_{OFFSET} by means of a simple digital-to-analog converter (DAC), REF_{SKIP} is slightly changed until the converter finally operates outside of the SB. To minimize any transient on V_{OUT} when REF_{SKIP} is varied, the very same offset generator is also added at the error amplifier (EA) output (V_{EA}), which allows the system to automatically maintain V_C close to REF_{SKIP}. Proceeding by iteration allows minimal effects on the PSM performance, as the converter's working point is adjusted only as much as needed to move outside of the forbidden band. Such added offset gets iteratively increased until the converter is pushed out from the pre-selected SB. Besides, $COUNTER_{REF}$ is allowed to roll over, meaning that when REF_{SKIP} reaches its maximum it goes back to the minimum. This feature is key for the system to exit any future steady state condition in which the power converter may again operate within a SB[3]. The idea to proceed by iteration stems from the overwhelming complexity in understanding if a new SB violation is due to a new operative condition or a previous insufficient corrective action. For the sake of clarity, Table II provides an example of how the offsets are iteratively managed.

TABLE II
EXEMPLARY ITERATION VS OFFSET APPLIED

Iteration #	$COUNTER_{REF}$	V_{OFFSET}
0 (default)	0	0
1	1	+25 mV
2	2	+50 mV
3	3	+75 mV
4	4	+100 mV
5	5	-25 mV
6	6	-50 mV
7	7	-75 mV
8	back to iteration 1	

Practically, the offset steps are sized to exit from any SB in any operating condition, with a limited number of iterations (e.g., 7 in our case).
It is worth noting that as long as V_{EA} remains within the EA allowed output dynamic range, the voltage shift introduced by the programmable offset is not relevant for the DC–DC converter itself. In fact, the integral action present in the DC-DC converter loop forces the required compensator output V_C to maintain a specified regulation setpoint by changing the duty-cycle[4].

---

[3]In general, in PSM F_{SKIP} depends on the operative conditions (i.e., V_{IN}, V_{OUT}, I_{LOAD}) and the DC-DC converter parameters (e.g., L, C_{OUT}, etc.).

[4]This is always valid, no matter the DC-DC operating mode (i.e., CCM, DCM, PSM/PFM).

B. Digital Controller

The proposed algorithm has been implemented exploiting two asynchronous-finite-state-machines (A-FSM) [17], digital counters and combinatorial logic. The first A-FSM, labeled $AFSM_1$, is in charge of understanding if the actual burst is within the pre-selected SB (i.e., check the first condition) providing the feedback OK_1 to the second A-FSM ($AFSM_2$). This feedback gets reset when the present burst is outside the SB (i.e., condition (1) not satisfied). In Figure 2 the state-flow describing $AFSM_1$ and some of the additional logic controlled by it are depicted. In each state the corresponding outputs (colored numbers) are shown, while each arc is activated only as soon as the condition associated with the arc is true [18]. A counter ($COUNTER1$), is used to keep track of how many $CLOCK_{AUX}$ periods (i.e., rising edges) occur in a $COMP_{SKIP}$ period (i.e., between two consecutive $COMP_{SKIP}$ rising edges). This counter is initially enabled by $AFSM_1$ at the beginning of a switching cycle (i.e., $COMP_{SKIP}$ rising edge) and remains enabled and counting until: i) the next switching cycle, or ii) there is no need to wait for the next switching cycle since condition (1) cannot be satisfied[5]. Precisely, at the next switching cycle (i.e., $COMP_{SKIP}$ rising edge) condition (1) is evaluated[6], and the flag OK_1 is set high. Until F_{SKIP} of the actual burst remains within the SB, this flag remains set (i.e., $AFSM_1$ follows the states 1-2-3), while it gets reset as soon as a burst with F_{SKIP} outside the SB occurs.

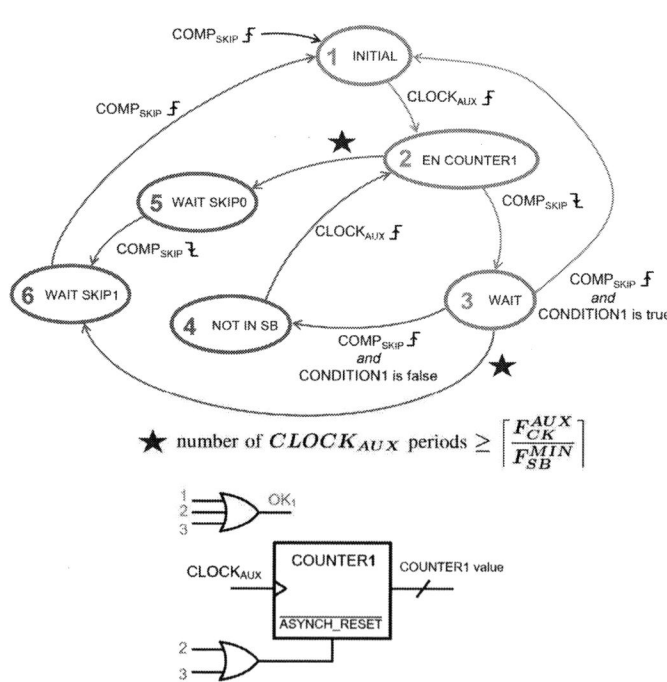

Fig. 2. State-flow describing $AFSM_1$ with some of the controlled logic.

---

[5]This is the case when the converter is operating with F_{SKIP} much below the SB.

[6]The combinatorial logic in charge of computing such evaluation, starting from the value of $COUNTER1$, is not shown.

The state-flow describing $AFSM_2$ is depicted in Figure 3. This state-machine is in charge of understanding: i) if within a sampling-window the number of $COMP_{SKIP}$ periods (i.e., $COMP_{SKIP}$ rising edges) falls within the SB (i.e., check the second condition), and ii) eventually decide if a corrective action is needed, outputting the flag SB = 1. To this end, $AFSM_2$ directly controls a burst counter ($COUNTER2$) that keeps track of how many $COMP_{SKIP}$ periods (i.e., rising edges) occur in a sampling-window lasting t_{SW}, while another counter clocked by $CLOCK_{AUX}$ is simply exploited as a timer for the sampling-window. Until $COMP_{SKIP}$ is 0, $AFSM_2$ remains in idle, looping in states 1 and 2. When $COMP_{SKIP}$ goes high (state 2), the timer and $COUNTER2$ are enabled to count. If the timer reaches its timeout, condition (2) is evaluated: if positive, the state-machine moves in state 3 setting SB = 1, otherwise, state 3 is reached and SB remains low. In both states 2 and 3 the timer and $COUNTER2$ are reset and the system is immediately ready for a new sampling-window. The presented behavior holds if the flag OK_1 remains set. In fact, in state 2 if the flag OK_1 goes to 0 there is no need to wait for the end of t_{SW} to understand if a corrective action is needed, since it is already acknowledged by $AFSM_1$ that the converter is performing outside the SB. The same applies if, before the timeout, the number of occurred $COMP_{SKIP}$ rising-edges goes above the maximum limit of $t_{SW}F_{SB}^{MAX}$.

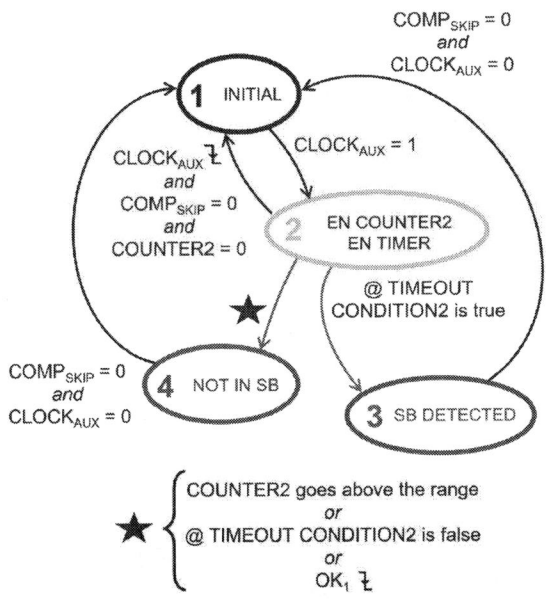

Fig. 3. State-flow describing $AFSM_2$.

To better understand the A-FSMs interactions, consider the following case in which the DC-DC converter is experiencing a transient (e.g., load, line, etc.). Depending on the perturbation's magnitude, F_{SKIP} changes typically over a wide range and finally settles at a given value. Even only a single burst that is recognized by $AFSM_1$ to be outside the SB is enough to promptly reset the sampling-window and the burst-counter (i.e., $COUNTER2$). In other words, $AFSM_1$ keeps shifting ahead the sampling-window until the actual F_{SKIP} is not

coarsely within - or close to - the forbidden band. On the contrary, when $AFSM_1$ acknowledges that the actual F_{SKIP} is coarsely within the SB, the timer and $COUNTER1$ are enabled, since only at this point it is necessary to finely check if the averaged F_{SKIP} is within the SB. A filter rejecting the converter transients has been obtained, since the need to update the corrective action is evaluated only at the end of a steady state SW. When the SB is changed on-the-fly, both the state-machines are reset (i.e., back to their initial state).

C. Circuit Design

The proposed stop-band-management-system (SBMS) has been designed for a dual-phase inverting buck-boost (IBB) converter, which provides the negative supply rail (V_{OUT}) for the AMOLED display, from the Li-Ion battery (V_{IN}). This power converter has been designed within a custom power-management-integrated-circuit (PMIC) fabricated in a $180\,\text{nm}$ BCD (bipolar-CMOS-DMOS) process. At light load, the IBB is operating only a single-phase (i.e., the second phase is disabled), and thus for the SBMS only the single-phase operation is relevant. An auxiliary clock with $F_{CK}^{AUX} = 3\,\text{MHz}$ has been exploited, since it is already available within the PMIC for other purposes, while the DC-DC converter is clocked at $F_S = 1.5\,\text{MHz}$ (i.e., $T_S = 666\,\text{ns}$). An open-loop-amplifier (OLA) Type-II compensator has been designed according to [19], while the clocked skip comparator is a P-type input StrongARM latch [20], [21]. As depicted in Figure 4, the digitally-controlled-offset-generators (DCOG) are implemented by means of a simple open-loop voltage buffer, made by a CMOS common drain stage [22] to decouple from the EA, followed by a programmable resistor (R_{OFFSET}) biased with a constant current I_{OFFSET}. Considering that this analog level-shifter retains the ability to only introduce programmable positive offsets, a baseline voltage is simply added to the values reported in Table II. As explained in Section II-A2, the absolute value of such baseline voltage is not relevant for the DC-DC converter, and neither for the SB corrective action.

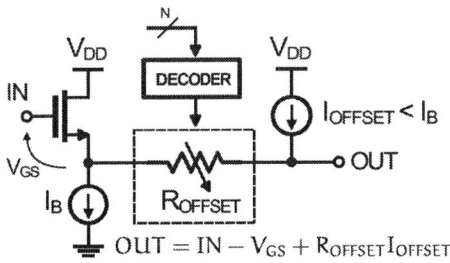

Fig. 4. Schematic of the digitally-controlled offset voltage generator.

D. Exemplary Simulation

In Figure 5 a simulation example shows the SBMS in action. The SB is programmed from $40\,\text{kHz}$ to $60\,\text{kHz}$ (top dashed black line boundaries) with a sampling-window of $t_{SW} = 0.5\,\text{ms}$. Initially, the SBMS is not enabled (EN = 0, second

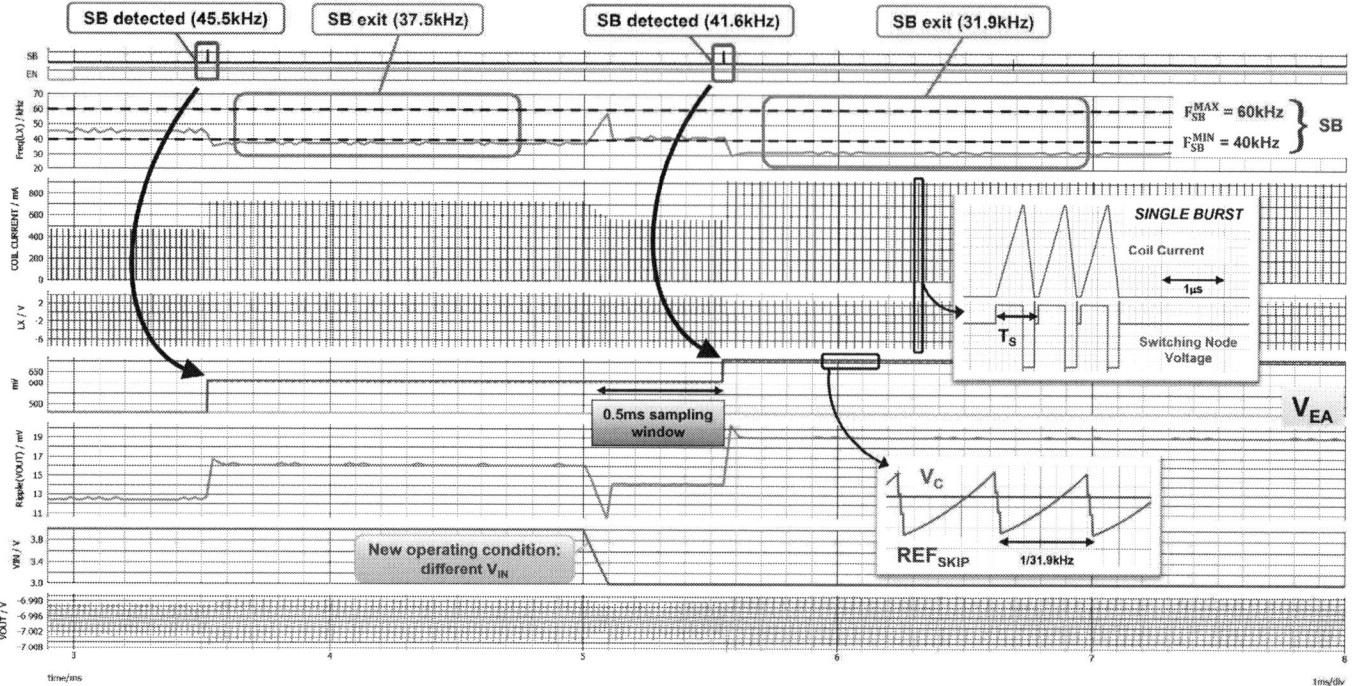

Fig. 5. Exemplary simulation of SBMS applied to an inverting buck-boost regulating $V_{OUT} = -7\,\text{V}$ in PSM with a load of $-6\,\text{mA}$: the SB is programmed from 40 kHz to 60 kHz ($t_{SW} = 0.5\,\text{ms}$).

waveform in yellow) and the IBB is steadily regulating V_{OUT} at $-7\,\text{V}$ (green bottom waveform) from $V_{IN} = 4\,\text{V}$ (red bottom waveform). With a load current of $-6\,\text{mA}$, the IBB is operating in PSM, as visible from the coil current (blue top waveform) and the switching node (magenta waveform in the middle), with F_{SKIP} around 45 kHz (magenta top waveform). At 3 ms, the proposed solution is enabled (EN = 1), and after about 0.5 ms (i.e., a sampling-window), it recognizes that a corrective action is needed (SB = 1 in the top black waveform). Consequently, REF_{SKIP} (blue waveform in the middle) is changed (increased in this case). This action proves effective in moving the IBB operation outside the SB, leading to $F_{SKIP} = 37.5\,\text{kHz}$. At 5 ms, a line transient occurs, which pushes back the IBB operation within the SB with $F_{SKIP} = 42.6\,\text{kHz}$. This is recognized by our solution (SB = 1) after 0.5 ms of steady operation, and REF_{SKIP} is thus changed again. With this corrective action the IBB finally settles outside of the forbidden band, with $F_{SKIP} = 31.9\,\text{kHz}$ and the output voltage ripple is minimally affected (green waveform in the middle). Thanks to the replica V_{OFFSET} on the EA output, both V_{EA} (yellow waveform in the middle) and V_{OUT} do not experience any perturbation, guaranteeing the seamless specification.

III. EXPERIMENTAL RESULTS

Figure 6 shows the die micrograph of the fabricated dual-phase IBB converter, with the SBMS key blocks highlighted in red. Considering the whole silicon area of the IBB controller (roughly $0.36\,\text{mm}^2$ for the single-phase), the SBMS area remains negligible (i.e., $\approx 6\,\%$). The dual-phase IBB converter

is part of a PMIC for AMOLED power supply and is designed to regulate a programmed output voltage V_{OUT} in the range of -1 V down to -9 V, providing in single-phase up to -500 mA from an input supply V_{IN} varying between 2.2 V and 4.9 V. At light load the IBB operates in DCM/PSM, as exhibited from the oscilloscope capture of Figure 7. In this case, with $V_{IN} = 4.8\,\text{V}$ and an output load of -40 mA, V_{OUT} is regulated at -3 V with $F_{SKIP} \approx 73\,\text{kHz}$, and each burst is composed of 3 consecutive pulses.

Figure 8 showcases oscilloscope acquisitions of the IBB regulating in PSM with a programmed SB from 40 kHz to 60 kHz (V_{OUT} = -3 V, V_{IN} = 4.3 V, I_{LOAD} = -6 mA). To

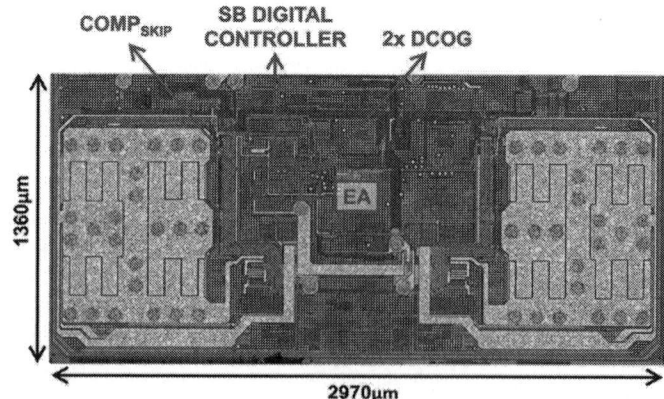

Fig. 6. Chip micrograph: dual-phase IBB with relevant SBMS blocks highlighted.

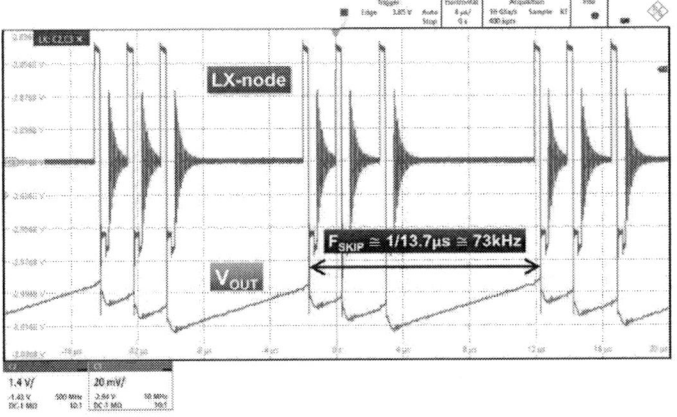

Fig. 7. Oscilloscope acquisition of the IBB operating in PSM at $F_{SKIP} \approx$ 73 kHz, with $V_{IN} = 4.8$ V, $I_{LOAD} = -40$ mA and $V_{OUT} = -3$ V (green trace). The trigger is on the DC-DC converter switching node (magenta trace).

check the repetition period of the IBB burst (i.e., F_{SKIP}), the oscilloscope is triggered on the IBB switching node and set with display persistence of multiple acquisitions. The upper acquisition in Figure 8 shows that without enabling the proposed SBMS feature (EN = 0) the IBB is operating within the SB. On the contrary, with SBMS enabled (EN = 1)

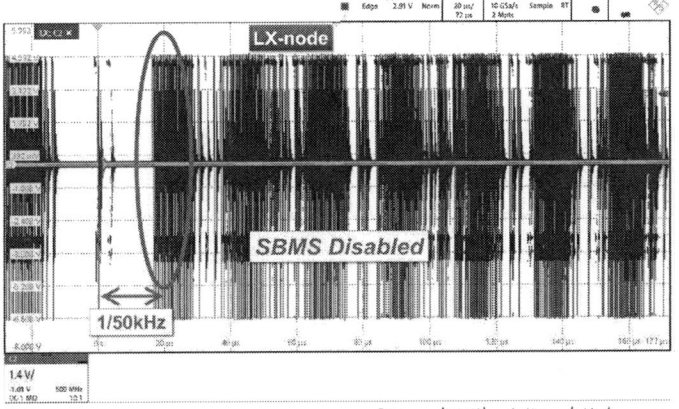

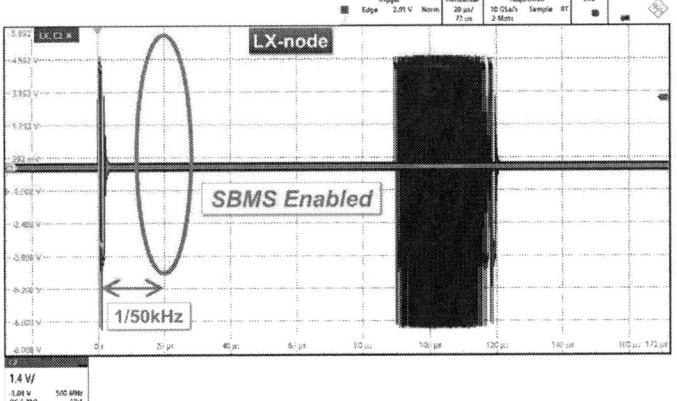

Fig. 8. Oscilloscope acquisitions of the IBB operating in PSM (trigger on the switching node): in the upper acquisition SBMS is disabled, while in the lower it is enabled with SB programmed from 40 kHz to 60 kHz.

under the same IBB conditions, the absence of bursts around 1/50 kHz proves the effectiveness of our solution (bottom acquisition in Figure 8).

To showcase how the proposed SBMS affects the power efficiency, in Figure 9 are presented two measured efficiency curves in the same operative conditions (i.e., $V_{IN} = 2.8$ V, $V_{OUT} = -7$ V, SBMS enabled and programmed SB from 40 kHz to 60 kHz). These measurements were performed on the very same sample by means of an automatic setup finely sweeping I_{LOAD}. During the first sweep-measurement (red trace) the SBMS applied a corrective action at about -26 mA, as visible in the zoomed inset (visibile discontinuity). Immediately without resetting the PMIC, a second sweep-measurement (blue line) was performed. Here the SBMS did not trigger and from -26 mA the efficiency overlaps the trace from the previous sweep. In this case, at light load the efficiency variation is limited to 6 %.

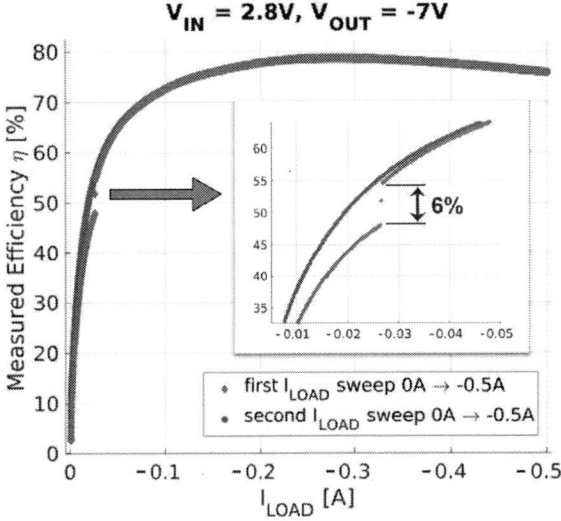

Fig. 9. Measured efficiency with $V_{IN} = 2.8$ V and $V_{OUT} = -7$ V when two automatic load-sweeps are performed consecutively: during the first sweep (red curve) at about -26 mA the SBMS applies a corrective action.

IV. CONCLUSION

The presented solution proves effective in preventing the DC-DC converter from operating within forbidden frequency bands in DCM at light load. The switching activity of the converter is first oversampled, followed by a fine averaging within a sampling-window, which allows to finally understand if a programmed SB is violated. This dual-step check allows the oversampling operation to not require any high frequency clock signal, easing the implementation. An offset voltage changing the set DCM peak-current is applied as corrective action, in turn altering the converter switching activity. The application of the same offset voltage on the EA output allows to seamlessly maintain V_{OUT} unaffected, without any visible transient when a corrective action is exerted. Working in closed-loop, the obtained frequency-band-exclusion system is robust against any process-voltage-temperature (PVT) variation, as well as operative conditions (e.g., V_{IN}, V_{OUT}, I_{LOAD}) and parameters (e.g., inductor, output capacitor, etc.). Thanks to its digital nature, there are no limitations on the number and range of SB, allowing maximum flexibility. The outlined algorithm properly filters out any transient that the power converter is experiencing, preventing any inappropriate and premature correction. Moreover, the iterative criteria allows to adapt the corrective action, minimally affecting the converter behavior and performance (e.g., output ripple, power efficiency). Furthermore, considering that the DCM peak-current value set by REF_{SKIP} is meaningless in CCM, the CCM stability and performance are not impacted by the SBMS. The presented SBMS is a digitally-assisted versatile add-on that can be applied to any DC-DC converter topology and light load operation mode (e.g., PFM and/or burst-mode). It has been designed and validated in a IBB converter providing a selectable output voltage in the range -1 V to -9 V, fed from Li-Ion battery. This DC-DC converter is part of a PMIC for AMOLED display power supply, fabricated in a $0.18 \mu m$ BCD technology.

REFERENCES

[1] F. Mao et al., "A hybrid single-inductor bipolar triple-output dc–dc converter with high-quality positive outputs for amoled displays," in *IEEE Trans. on Circ. and Sys. I*, vol. 70, no. 1, pp. 506–517, Jan. 2023.

[2] S. Forrset et al., "The dawn of organic electronics," in *IEEE Spectrum*, vol. 37, no. 8, pp. 2–34, 2000.

[3] M. K. Barwar et al., "Demystifying the Devices Behind the LED Light: LED Driver Circuits," in *IEEE Industrial Electronics Magazine*, vol. 17, no. 1, pp. 55–66, 2023.

[4] M. Kimura et al., "Low-temperature polysilicon thin-film transistor driving with integrated driver for high-resolution light emitting polymer display," in *IEEE Trans. on Elec. Dev.*, vol. 46, no. 12, pp. 2282–2288, 1999.

[5] Amir Besharati Rad et al., "An Ultra-Low-Noise Buck Converter for Noise-Sensitive Applications," in *IEEE Transactions on Power Electronics*, vol. 39, no. 2, pp. 2169–2178, February 2024.

[6] M. Alecci et al., "Noise Analysis in Voltage Mode Controlled DC-DC Converters," in *18th Conference on Ph.D Research in Microelectronics and Electronics (PRIME)*, June 2023.

[7] R. B. Ridley, "A New Small-Signal Model for Current-Mode Control," Ridley Engineering, Inc., 1999, https://ridleyengineering.com/education/books/books-current-mode-control.html.

[8] R. Erickson et al, Fundamentals of Power Electronics, 3nd ed., Springer, 2020.

[9] R. Sheehan et al., Switch-mode power converter compensation made easy, Sep. 2016.

[10] Wan-Rone Liou et al., "A High Efficiency Dual-Mode Buck Converter IC For Portable Applications," in *IEEE Trans. Power Electronics*, vol. 23, no. 2, pp. 667-677, March 2008.

[11] B. Sahue et al., "A high-efficiency, dual-mode, dynamic, buck-boost power supply IC for portable applications," in *18th International Conference on VLSI Design*, Jan. 2005.

[12] X. Zhou et al., "Improved lightload efficiency for synchronous rectifier voltage regulator module," in *IEEE Trans. Power Electronics*, vol. 15, no. 5, pp. 826-834, Sept. 2000.

[13] Alevoor et al, "A 95.2% Efficiency dc–dc Boost Converter Using Peak Current Fast Feedback Control (PFFC) for Improved Load Transient Response," in *IEEE Trans. Circuits Syst.—I*, vol. 70, no. 3, pp. 1097–1109, March 2023.

[14] Amir Besharati Rad et al., "A Wide-Input-/Output-Voltage-Range Buck Converter With Adaptive light load Efficiency Improvement and Seamless Mode Transition," in *IEEE Trans. on Power Elec.*, vol. 39, no. 2, pp. 2200–2212, February 2024.

[15] M. Zhao et al, "An Ultra-Low Quiescent Current Tri-Mode DC-DC Buck Converter With 92.1% Peak Efficiency for IoT Applications," in *IEEE Trans. Circuits Syst.—I*, vol. 69, no. 1, pp. 428–439, Jan. 2022.

[16] B. Yuan et al., "Hybrid Buck Converter With Constant Mode Changing Point and Smooth Mode Transition for High-Frequency Applications," in *IEEE Transactions on Industrial Electronics*, vol. 69, no. 1, pp. 1466–1474, February 2020.

[17] S. Dalle Feste et al., "A New Digital Flow for Automotive Robust Design of DC-DC Converters," in *AEIT International Conference of Electrical and Electronic Technologies for Automotive (AEIT AUTOMOTIVE)*, June 2019.

[18] R. Pellegrini et al., "A Seamless and Unified Flow for Robust Development of DC-DC Digital Controllers," in *18th International Conference on Synthesis, Modeling, Analysis and Simulation Methods and Applications to Circuit Design (SMACD)*, June 2022.

[19] A. Bertolini et al., "Practical aspects for analog compensators design in integrated dc–dc converters," in *IEEE Journal of Emerging and Selected Topics in Power Electronics*, vol. 12, no. 1, pp. 446–460, April 2024.

[20] B. Razavi, "The StrongARM latch [a circuit for all seasons]," in *IEEE Solid-State Circuits Magazine*, vol. 7, no. 2, pp. 12-17, 2015.

[21] Abdullah Alshehri et al., "A 19 fJ/op, Low-Offset StrongARM Latch Comparator for Low-Power High-Speed Applications," in *IEEE International Symposium on Circuits and Systems (ISCAS)*, May 2024.

[22] B. Razavi, Design of Analog CMOS Integrated Circuits, 1st ed. McGraw-Hill Education, 2001, ch. Basic MOS Device Physics, pp. 21–36.

Multi-Stage Model Predictive Control with Enhanced Discrete-Time Models for Multilevel Inverters

Hoang Le*, Apparao Dekka*, Deepak Ronanki†, and Abdul R. Beig‡

*Department of Electrical and Computer Engineering, Lakehead University, Thunder Bay, Ontario, Canada
†Department of Engineering Design, Indian Institute of Technology Madras, Chennai, India
‡Advanced Power and Energy Systems, Electrical Engineering Department, Khalifa University, Abu Dhabi, UAE
E-mail: hoangvle@ieee.org, dapparao@ieee.org, dronanki@ieee.org, and balanthi.beig@ku.ac.ae

Abstract—**Model accuracy, computational complexity, and weighting factor dependency are three primary challenges associated with multi-stage model predictive control (MS-MPC) of multilevel inverters (MLIs). Typically, existing MS-MPC employs the forward Euler integration method to discretize continuous-time models, due to the ease of implementation in real-time controllers. This approach faces a significant deterioration in computational accuracy as sampling period increases. To address this issue, this paper proposes a new formulation of MS-MPC for MLIs. Also, an enhanced discrete-time model of MLI by using Heun's integration method is proposed to implement the proposed MS-MPC, resulting in a more precise formulation and indirect minimization of common-mode voltage (CMV). Unlike the existing methods, the proposed method does not employ a cost function with weighting factor or offline selection of voltage vectors (VVs) to minimize the CMV. Furthermore, it does not increase computational complexity compared to the existing MS-MPCs. Simulation studies are conducted on a four-level MLI system to validate the efficacy of the proposed MS-MPC, and its performance is compared with the existing methods.**

Index Terms—**Discretization method, mathematical models, model accuracy, multilevel inverter, model predictive control.**

I. INTRODUCTION

Model predictive control (MPC) methods have gained prominency in achieving multiple objectives of multilevel inverters (MLIs) and their applications [1], including renewable energy [2], [3], electrified transportation [4], [5], medium-voltage drives [6], high-voltage direct transmission systems [7], and so on. Typically, the control objectives of MLIs are achieved in a single-stage optimization for all possible switching states, which is commonly referred to as a finite control-set MPC (FCS-MPC) in the literature [8]. This method has shown a superior reference tracking at steady-state and fast response during transients compared to the linear control methods [9]. However, the mathematical models accuracy, computational complexity, and weighting factor selection affect the real-time implementation and performance of FCS-MPC methods for MLIs [10].

Traditionally, the three-phase modeling philosophy has been adopted in the formulation of FCS-MPC, resulting in a higher computational complexity, and it drastically increases with the number of voltage levels of MLIs [11]. To address this

issue, the per-phase philosophy-based FCS-MPC formulation has been introduced for MLIs [12]. Irrespective of the formulation, the forward Euler integration method has been widely employed in the discretization of continuous-time (CT) models of MLIs, which affects the reference tracking performance at large sampling periods [13]. In [14], Heun's integration method has been proposed to formulate the discrete-time models for FCS-MPC, along with the per-phase implementation philosophy. This approach improves the harmonic performance and reduces the computational complexity. However, neither three-phase nor per-phase-based FCS-MPC methods have been able to escape the impact of weighting factor selection process [15]. To adjust the weighting factors dynamically, artificial neural network (ANN) [16] and adaptive approaches [17] have been proposed, but they further increases the real-time implementation complexity of FCS-MPC methods.

Alternatively, multi-stage model predictive control (MS-MPC) methods with an aim of computational complexity reduction and elimination of weighting factor dependency are well studied for MLIs [18], [19]. Depending on the number of control objectives, MS-MPC methods are designed with either two or three optimization stages by considering the redundancy of converter voltage vectors (VVs) and switching states [20], [21]. Typically, the exhaustive search is used to find the optimal VV and the corresponding switching state through multi-stage optimization process [22]. However, the exhaustive search becomes cumbersome with the increase in voltage levels of MLIs. In [23], [24], the VVs with the lowest common-mode voltage (CMV) are pre-selected offline and subsequently employed them in the optimization process. This process further reduces the computational complexity and eliminates the weighting factor dependency, but it affects the steady-state and transient performance due to the use of limited VVs in the optimization process [25]. On the other hand, the MS-MPC with indirect CMV minimization has been proposed for MLIs [26]. This approach is based on a per-phase philosophy, where the optimal voltage level for each phase is determined through an exhaustive search-based optimization process to fulfill the each phase' control objectives. It has shown a superior performance without compromising the transient response and computational complexity, but it leads to a high switching frequency operation.

To further simplify the MS-MPC methods, the predictive current models are reformulated in terms of predictive voltage models, thereby adapting the space vector philosophy to identify the suitable VVs for the optimization process [27]. In [28], the hysteresis comparator philosophy has been introduce to determine the optimal hexagon and sector, thereby identifying the nearest VVs and the corresponding switching states. In [29], the $g-h$ reference frame theory has been employed to identify the nearest VVs for optimization process. This philosophy has been further extended to incorporate the virtual VVs in the optimization process to address the capacitor voltage ripple reduction [30]. In [31], the sub-optimal VVs are identified based on the CMV variation and these resultant sub-optimal VVs are subsequently utilized in the cost function optimization to select the final optimal VV.

In [32], [33], the VVs are pre-selected based on the location of reference current tracking error rather than optimization process. Irrespective of the implementation philosophy, the existing MS-MPC methods are mainly targeted to reduce the computational complexity and eliminate the weighting factors, while fulfilling the control needs of MLIs and their applications. Moreover, these methods mainly use the forward Euler integration method-based models in their implementation, resulting in prediction inaccuracy as sampling period increases. To reduce impact of prediction error on the system performance, the multi-vector-based MS-MPC methods have been proposed for MLIs [34]–[36]. The use of more vector in each sampling time leads to higher device switching frequency compared to the single-vector approaches [37].

To improve the prediction accuracy, enhanced discrete-time (DT) models are proposed to implement the MS-MPC for MLIs in this article. The DT models are derived from CT models by using Heun's integration method. Moreover, the proposed MS-MPC is formulated with an indirect minimization of CMV objective to eliminate the weighting factor dependency, while maintaining a low computational burden. The enhanced DT models will also improve the harmonic performance and reduce the flying capacitor (FC) voltage ripple in MLIs. The efficacy of the proposed MS-MPC is evaluated through MATLAB simulations and its performance is further compared with the existing MS-MPC methods.

II. MODELING OF THE PROPOSED MS-MPC

The proposed MS-MPC is applied to a four-level MLI (4L-MLI), and its circuit configuration is shown in Fig. 1 [38]. The 4L-MLI is composed of eight switches (S_{w1}-S_{w8}) and two FCs (C_{w1}-C_{w2}) in each phase, where $w \in \{p, q, r\}$ denotes the ac output terminal. Table I shows the switching states corresponding to a 4L-MLI and their impact on FC voltages. Each FC is designed with a rated voltage of one-third of the dc-link voltage (V_{dc}) for a four-level operation of the inverter. The ac load is formed with a resistor (R) and an inductor (L) as shown in Fig. 1. In this article, the proposed MS-MPC is formulated to control the load currents and FC voltages

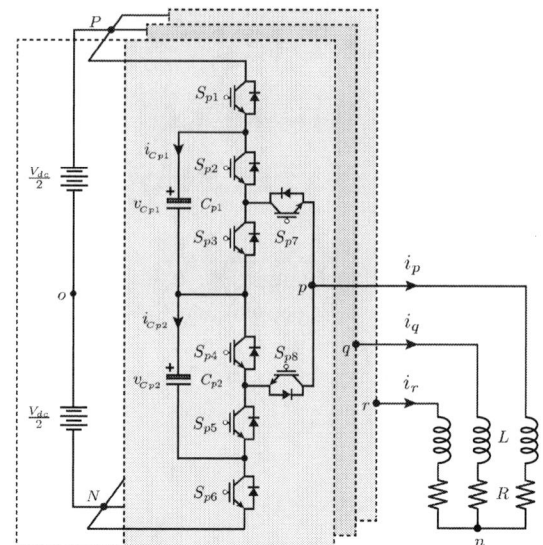

Figure 1.   4L-MLI circuit configuration [38].

of the 4L-MLI, while minimizing the CMV indirectly. The detailed formulation and mathematical models used in the implementation of the proposed MS-MPC are described as follows:

A. Formulation of Load Current Models

From Fig. 1, the CT model of the load current of phase-w can be written as [38],

$$\frac{di_w}{dt} = \frac{1}{L} v_{wn}(t) - \frac{R}{L} i_w(t) \tag{1}$$

where v_{wn} is the load voltage and i_w is the load current of phase-w.

By applying the KVL to the 4L-MLI shown in Fig. 1, the load voltage can be expressed as,

$$v_{wn}(t) = v_{wo}(t) - v_{no}(t) \tag{2}$$

where v_{wo} is the inverter voltage and v_{no} is the CMV.

To minimize the CMV indirectly in the proposed MS-MPC, the CMV objective is incorporated into load current control objective by postulating $v_{no}^*(t) = v_{no}(t) \approx 0$, leading to $v_{wo}(t) = v_{wn}(t)$ [26]. This assumption eliminates the CMV term in the load current formulation, allowing for a per-phase implementation philosophy to handle the inverter objectives.

Typically, the forward Euler method is widely used in the formulation of DT models for MS-MPC due to its simplicity and ease of implementation in real-time controllers. However, the accuracy of these models degrades as the sampling period increases and they are given as,

$$i_w^p(h+1) = \left(1 - \frac{T_s R}{L}\right) i_w(h) + \frac{T_s}{L} v_{wo}(h) \tag{3}$$

where T_s is the sampling period.

To enhance model accuracy, Heun's integration method is employed to discretize the CT model in (1) [13]. The resultant

Table I
SWITCHING STATES OF A FOUR-LEVEL MLI [38]

S_w	S_{w1}	S_{w2}	S_{w3}	S_{w4}	S_{w5}	S_{w6}	S_{w7}	S_{w8}	v_{Cw1}	v_{Cw2}	v_{wN}
0	0	0	0	0	1	1	0	1	No Change	No Change	0
1	1	0	0	0	1	0	0	1	Charge ($i_w>0$)	Charge ($i_w>0$)	$V_{dc}-v_{Cw1}-v_{Cw2}=\frac{V_{dc}}{3}$
	0	0	0	1	0	1	0	1	No Change	Discharge ($i_w>0$)	$v_{Cw2}=\frac{V_{dc}}{3}$
2	0	1	0	0	0	1	1	0	Discharge ($i_w>0$)	Discharge ($i_w>0$)	$v_{Cw1}+v_{Cw2}=\frac{2V_{dc}}{3}$
	1	0	1	0	0	0	1	0	Charge ($i_w>0$)	No Change	$V_{dc}-v_{Cw1}=\frac{2V_{dc}}{3}$
3	1	1	0	0	0	0	1	0	No Change	No Change	V_{dc}

DT model of the load current is given as,

$$i_w^p(h+1) = \left(\frac{1}{2}\left(\frac{T_s R}{L}\right)^2 - \frac{T_s R}{L} + 1\right) i_w(h) + \left(\frac{T_s}{L} - \frac{R}{2}\left(\frac{T_s}{L}\right)^2\right) v_{wo}(h) \tag{4}$$

Considering the switching states given in Table I, the 4L-MLI output voltage with respect to the dc-link mid-point (o) can be written as,

$$v_{wo}(h) = \mathbf{S}_w (\mathbf{S}_w - 1)(3 - \mathbf{S}_w)\frac{v_{Cw1}(h)}{2} + \mathbf{S}_w (3 - \mathbf{S}_w)\frac{v_{Cw2}(h)}{2} + \mathbf{S}_w (\mathbf{S}_w - 1)(\mathbf{S}_w - 2)\frac{V_{dc}}{6} - \frac{V_{dc}}{2} \tag{5}$$

where $\mathbf{S}_w$ is 4L-MLI phase voltage level.

The cost function for the load current optimization process is formulated as,

$$J_{w1} = \left(\widehat{i_w^*}(h+1) - i_w^p(h+1)\right)^2 \tag{6}$$

B. Formulation of FC Voltage and Current Models

According to circuit configuration in Fig. 1, the 4L-MLI is composed of 2 FCs in each phase. The CT model of the kth FC voltage in phase-w is given as,

$$\frac{d v_{Cwk}}{dt} = \frac{i_{Cwk}(t)}{C_{wk}} \tag{7}$$

where v_{Cwk} is the FC voltage, i_{Cwk} is the FC current, C_{wk} is the capacitance of FCs, and $k \in \{1,2\}$ is the FC index number.

The DT model of predictive FC voltage based on the forward Euler integration method is given as [26],

$$v_{Cwk}^p(h+1) = v_{Cwk}(h) + T_s \frac{i_{Cwk}(h)}{C_{wk}} \tag{8}$$

To obtain a more precise model, Heun's integration method is applied to the CT model in (7) [13], and resultant model is

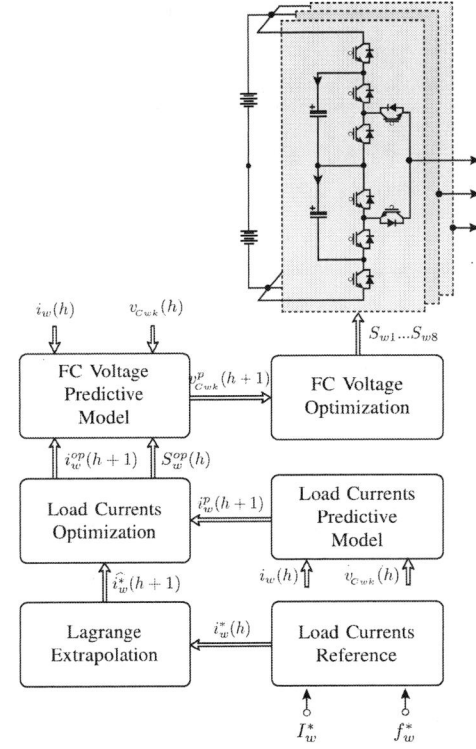

Figure 2. Block diagram of the proposed MS-MPC.

given as,

$$v_{Cwk}^p(h+1) = v_{Cwk}(h) + T_s \left(\frac{i_{Cwk}(h)}{2 C_{wk}} + \frac{i_{Cwk}(h+1)}{2 C_{wk}}\right) \tag{9}$$

The predictive FC currents at the (h)th sampling instant are given according to Table I as,

$$i_{Cw1}(h) = (S_{w1} - S_{w2}) i_w(h)$$
$$i_{Cw2}(h) = (S_{w5} - S_{w6}) i_w(h) \tag{10}$$

Similarly, the FC currents at the ($h+1$)th sampling instant can be written as,

$$i_{Cw1}(h+1) = (S_{w1} - S_{w2}) i_w^{op}(h+1)$$
$$i_{Cw2}(h+1) = (S_{w5} - S_{w6}) i_w^{op}(h+1) \tag{11}$$

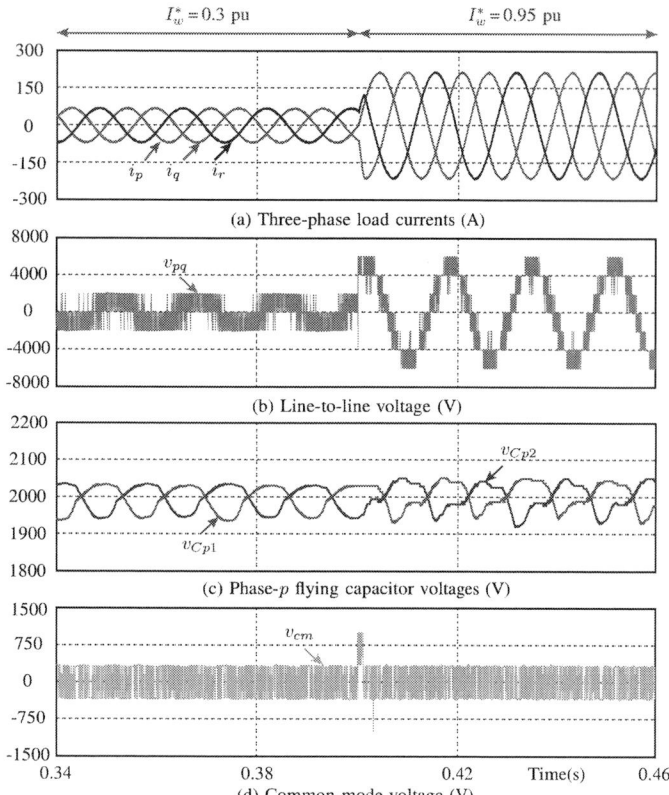

Figure 3. Proposed MS-MPC performance with current magnitude step-change at $f_w^* = 60$ Hz.

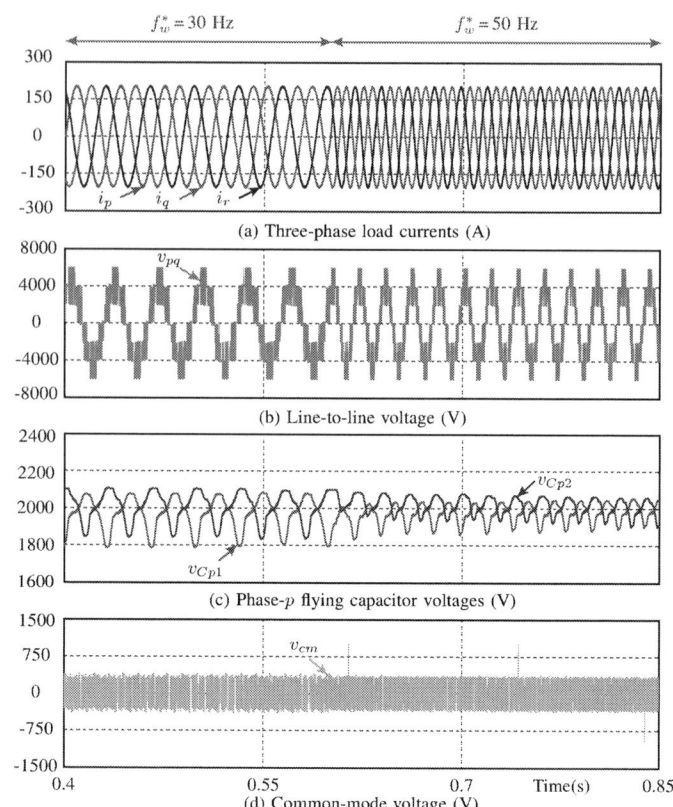

Figure 4. Proposed MS-MPC performance with frequency step-change at $I_w^* = 0.9$ pu.

Table II
ANALYSIS WITH I_w^* VARIATION AT $f_w^* = 60$ HZ

Performance index	$I_w^* = 0.3$ pu	$I_w^* = 0.95$ pu
%THD$_i$	2.41	0.79
%THD$_v$	101.88	31.06
Δv_C (V)	97	132

Table III
ANALYSIS WITH f_w^* VARIATION AT $I_w^* = 0.9$ PU

Performance index	$f_w^* = 30$ Hz	$f_w^* = 50$ Hz
%THD$_i$	0.87	0.86
%THD$_v$	38.30	34
Δv_C (V)	280	150

where $i_w^{op}(h+1)$ is the optimal value of phase-w current corresponding to the optimal voltage level of S_w^{op}.

The cost function (J_{w2}) is formulated with the predictive and reference FC voltages and it is given as

$$J_{w2} = \left(v_{Cwk}^*(h+1) - v_{Cwk}^p(h+1)\right)^2 \quad (12)$$

where v_{Cwk}^* represents reference FC voltage, which is set to $V_{dc}/3$ in a 4L-MLI.

III. IMPLEMENTATION OF THE PROPOSED MS-MPC

The proposed MS-MPC method for a 4L-MLI is implemented in two stages and their design steps are depicted in Fig. 2. In this method, the load current references (i_w^*) at the hth sampling instant are generated with a specified magnitude of I_w^* and a frequency of f_w^*, and they are given as,

$$i_w^*(h) = I_w^* \times sin(2\pi f_w^* + \theta_w) \quad (13)$$

where $\theta_w \in \left\{0, \frac{2\pi}{3}, \frac{4\pi}{3}\right\}$ denotes phase angles.

Through Lagrange extrapolation, the load current references are extrapolated to the $(h+1)$th sampling instant from (h)th sampling instant, and they are given as,

$$\begin{aligned}\widehat{i_w^*}(h+1) = &\, 4\, i_w^*(h) - 6\, i_w^*(h-1) \\ &+ 4\, i_w^*(h-2) - i_w^*(h-3)\end{aligned} \quad (14)$$

The load currents are predicted at the $(h+1)$th sampling instant by using (4). These currents are used in the cost function (6), together with the reference load currents. Each phase cost function is evaluated for a total of 4 voltage levels to determine the optimal voltage level (S_w^{op}) and the corresponding optimal load current value (i_w^{op}). These values together with the measured FC voltages are used in the prediction process of FC voltages given in (9)–(11). The predicted and reference FC voltages are used to form a cost function J_{w2} in (12). This cost function is evaluated for switching states corresponding to

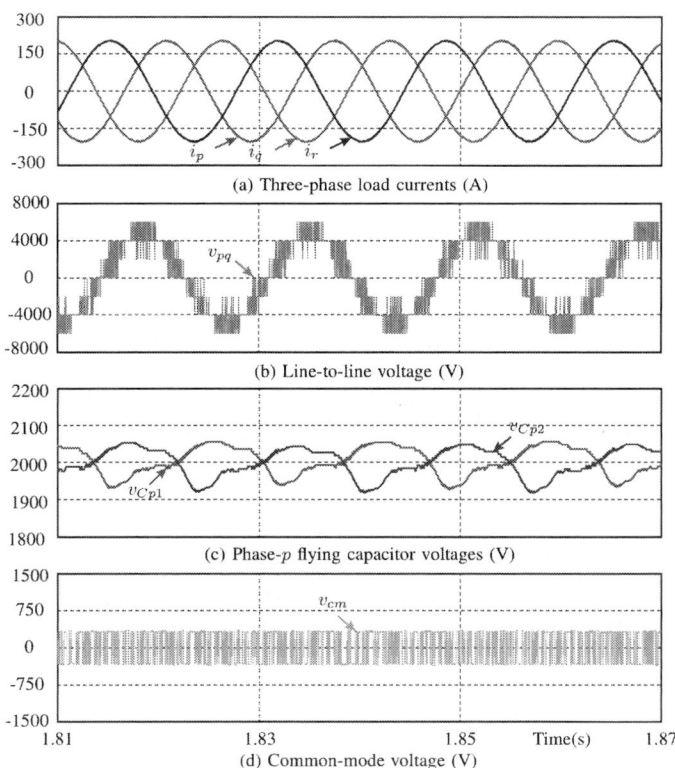

Figure 5. Existing MS-MPC performance at $I_w^* = 0.9$ pu and $f_w^* = 60$ Hz.

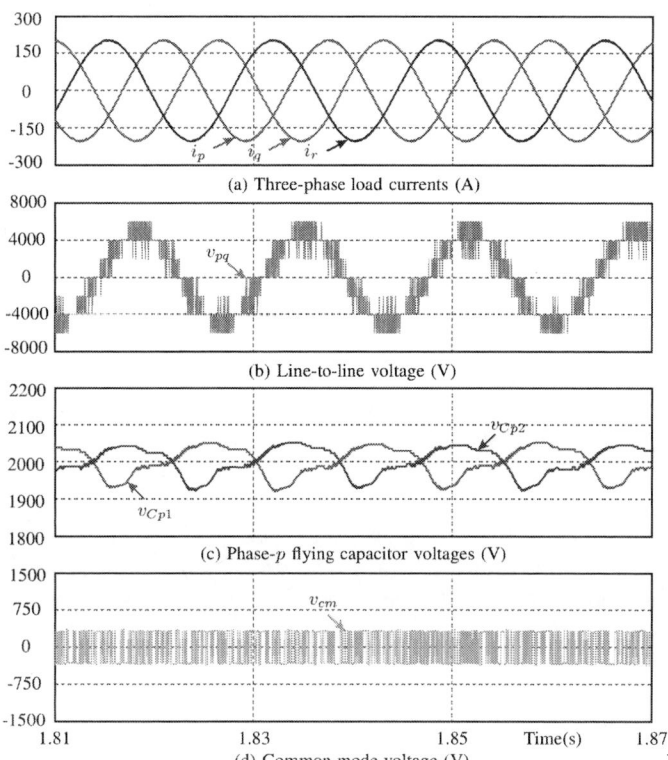

Figure 6. Proposed MS-MPC performance at $I_w^* = 0.9$ pu and $f_w^* = 60$ Hz.

Table IV
COMPARISON ANALYSIS OF EXISTING AND PROPOSED MS-MPC AT $I_w^* = 0.9$ PU AND $f_w^* = 60$ HZ

Performance index	Existing MS-MPC	Proposed MS-MPC
%THD$_i$	0.87	0.81
%THD$_v$	34.86	33.82
Δv_C (V)	138	129

IV. SIMULATION RESULTS AND ANALYSIS

The performance of the proposed MS-MPC under various working scenarios are validated on MATLAB simulation environment. The 4L-MLI is designed with the following specifications: rated power $(S_o) = 1.1$ MVA, rated voltage (line-to-line, rms) $(v_o) = 4000$ V, dc-link voltage $(V_{dc}) = 6000$ V, FC capacitance $(C_{wk}), = 1100 \mu$F, rated FC voltage $(v_{Cwk}^*) = 2000$ V, load resistance $(R) = 13 \Omega$, load inductance $(L) = 13$ mH, and sampling period $(T_s) = 40 \mu$s. Performance metrics used to evaluate the system performance include current harmonic distortion (%THD$_i$), voltage harmonic distortion (%THD$_v$), and FC voltage ripple (ΔV_c).

Fig. 3 shows the performance of the proposed MS-MPC during a step change in current magnitude from $I_w^* = 0.3$ pu to 0.95 pu, with an operating frequency (f_w^*) of 60 Hz. The results show that, regardless of working conditions, the three-phase load currents are perfectly sinusoidal with a %THD$_i$ of 2.41 and 0.79 at $I_w^* = 0.3$ pu and 0.95 pu, respectively, as shown in Fig. 3(a) and summarized in Table II. With increasing current magnitude, the line voltage rises from 3 steps to 7 steps (see Fig. 3(b)), resulting in a decrease in the %THD$_v$ from 101.88 to 31.06, as shown in Table II. Throughout the process, FC voltages are perfectly regulated at their rated values of 2000 V each as shown in Fig. 3(c). They have a ΔV_c of

97 V at $I_w^* = 0.3$ pu and 132 V at $I_w^* = 0.95$ pu. The results indicate that the higher current magnitude leads to a greater FC voltages ripple. Regardless of operating conditions, the 4L-MLI produces the least CMV at 358 V (peak) as shown in Fig. 3(d), and there is a spike in voltage at the moment of transient (at $t = 0.4$ s).

The performance of the proposed MS-MPC under the step change in frequency is shown in Fig. 4. In this study, the reference frequency (f_w^*) changes from 30 Hz to 50 Hz at $t = 0.6$ s while maintaining I_w^* at 0.9 pu. The results show that, regardless of operating conditions, the three-phase load currents have a %THD$_i$ of 0.87 at 30 Hz and 0.86 at 50 Hz, as shown in Fig. 4(a), and the same details are given in Table III. It is also observed that the change in frequency has no impact on the %THD$_i$ performance of the proposed MS-MPC. The line voltage remains 7 steps throughout the operation, and it has a %THD$_v$ of 38.30 and 34 at 30 Hz and 50 Hz, respectively, as shown in Fig. 4(b). Throughout the operation, each FC voltage is regulated at 2000 V as shown in

Fig. 4(c). These FCs have a ΔV_c of 280 V and 150 V at 30 Hz and 50 Hz, respectively. During the process, the proposed MS-MPC maintains the minimal CMV at 358 V (peak), as shown in Fig. 4(d).

To evaluate the effectiveness of the proposed MS-MPC, its performance is compared with the existing MS-MPC (forward Euler-based) [26], and the results are shown in Figs. 5 and 6. Irrespective of control methods, the three-phase load currents are perfectly sinusoidal as shown in Figs. 5(a) and 6(a). However, while the existing MS-MPC has a $\%THD_i$ of 0.87, the proposed MS-MPC produces a lower $\%THD_i$ of 0.81, as shown in Table IV. The line voltage has 7 steps with a $\%THD_v$ of 34.61 and 33.82 for existing MS-MPC and proposed MS-MPC, respectively, as shown in Figs. 5(b) and 6(b). Regardless of control techniques, FC voltages are consistently maintained at their reference values with a ΔV_c of 138 V for the existing method (see Fig. 5(c)) and a reduced ΔV_c of 129 V for the proposed MS-MPC (see Fig. 6(c)). During the process, both methods produce the same CMV with 358 V (peak), as shown in Figs. 5(d) and 6(d). It is interesting to note that both methods require the same execution time.

V. CONCLUSIONS

In this paper, an enhanced DT models using Heun's integration method for MS-MPC is proposed and applied to a 4L-MLI. The detailed formulation of DT models of the 4L-MLI with Heun's method and implementation steps of the proposed MS-MPC are also presented. The proposed MS-MPC is also studied on a 4L-MLI through MATLAB simulations under various operating conditions. The results show that the proposed method effectively reduces the current and voltage harmonic distortions and FC voltage ripples, while producing the lowest CMV. Furthermore, this approach does not increase computational complexity compared to existing MS-MPC. Overall, the proposed MS-MPC shows a superior performance with the enhanced DT models over the forward Euler method-based models, and it is highly suitable for MLI-fed electric motor drive applications.

REFERENCES

[1] P. Karamanakos, E. Liegmann, T. Geyer, and R. Kennel, "Model predictive control of power electronic systems: Methods, results, and challenges," *IEEE Open J. Ind. Appl.*, vol. 1, pp. 95–114, 2020.

[2] Y. Arias-Esquivel, R. Cárdenas, M. Diaz, and L. Tarisciotti, "Continuous control set model predictive control of a hybrid modular multilevel converter for wind energy applications," *IEEE Trans. Ind. Electron.*, vol. 71, no. 11, pp. 14 287–14 297, Nov 2024.

[3] T. Jin, X. Shen, T. Su, and R. C. C. Flesch, "Model predictive voltage control based on finite control set with computation time delay compensation for pv systems," *IEEE Trans. Energy Convers.*, vol. 34, no. 1, pp. 330–338, Mar 2019.

[4] D. Ronanki and H. Karneddi, "Electric vehicle charging infrastructure: Review, cyber security considerations, potential impacts, countermeasures, and future trends," *IEEE J. Emerg. Sel. Topics Power Electron.*, vol. 12, no. 1, pp. 242–256, Feb 2024.

[5] A. Andersson and T. Thiringer, "Assessment of an improved finite control set model predictive current controller for automotive propulsion applications," *IEEE Trans. Ind. Electron.*, vol. 67, no. 1, pp. 91–100, Jan 2020.

[6] H. Le, A. Dekka, and D. Ronanki, "Modeling and control of a new five-level converter for medium-voltage drive systems," *IEEE Trans. Transport. Electrific.*, vol. 10, no. 2, pp. 3782–3791, Jun 2024.

[7] J. Carmona Sánchez, O. Marjanovic, M. Barnes, and P. R. Green, "Secondary model predictive control architecture for vsc-hvdc networks interfacing wind power," *IEEE Trans. Power Del.*, vol. 35, no. 5, pp. 2329–2341, Oct 2020.

[8] P. Karamanakos, T. Geyer, and R. Kennel, "On the choice of norm in finite control set model predictive control," *IEEE Trans. Power Electron.*, vol. 33, no. 8, pp. 7105–7117, 2018.

[9] J. Rodriguez, C. Garcia, A. Mora, S. A. Davari, J. Rodas, D. F. Valencia, M. Elmorshedy, F. Wang, K. Zuo, L. Tarisciotti, F. Flores-Bahamonde, W. Xu, Z. Zhang, Y. Zhang, M. Norambuena, A. Emadi, T. Geyer, R. Kennel, T. Dragicevic, D. A. Khaburi, Z. Zhang, M. Abdelrahem, and N. Mijatovic, "Latest advances of model predictive control in electrical drives–part ii: Applications and benchmarking with classical control methods," *IEEE Trans. Power Electron.*, vol. 37, no. 5, pp. 5047–5061, May 2022.

[10] I. Harbi, J. Rodriguez, E. Liegmann, H. Makhamreh, M. L. Heldwein, M. Novak, M. Rossi, M. Abdelrahem, M. Trabelsi, M. Ahmed, P. Karamanakos, S. Xu, T. Dragičević, and R. Kennel, "Model-predictive control of multilevel inverters: Challenges, recent advances, and trends," *IEEE Trans. Power Electron.*, vol. 38, no. 9, pp. 10 845–10 868, Sept 2023.

[11] N. Chai, W. Tian, X. Gao, J. Rodriguez, M. L. Heldwein, and R. Kennel, "Three-phase model-based predictive control methods with reduced calculation burden for modular multilevel converters," *IEEE J. Emerg. Sel. Topics Power Electron.*, vol. 10, no. 6, pp. 7037–7048, Dec 2022.

[12] A. Bahrami, M. Norambuena, M. Narimani, and J. Rodriguez, "Model predictive current control of a seven-level inverter with reduced computational burden," *IEEE Trans. Power Electron.*, vol. 35, no. 6, pp. 5729–5740, Jun 2020.

[13] H. Le, A. Dekka, D. Ronanki, and J. Rodriguez, "A new predictive current control with reduced current tracking error and switching frequency for multilevel inverters," *IEEE Trans. Power Electron.*, vol. 38, no. 9, pp. 10 798–10 809, Sept 2023.

[14] D. Prajapati, A. Dekka, D. Ronanki, and J. Rodriguez, "Low-complexity heun's method-based fcs-mpc with reduced common-mode voltage for a five-level inverter," *IEEE Trans. Power Electron.*, vol. 39, no. 3, pp. 3329–3338, 2024.

[15] E. Zerdali, M. Rivera, and P. Wheeler, "A review on weighting factor design of finite control set model predictive control strategies for ac electric drives," *IEEE Trans. Power Electron.*, vol. 39, no. 8, pp. 9967–9981, Aug 2024.

[16] T. Dragičević and M. Novak, "Weighting factor design in model predictive control of power electronic converters: An artificial neural network approach," *IEEE Trans. Ind. Electron.*, vol. 66, no. 11, pp. 8870–8880, Nov 2019.

[17] C. Tang and T. Thiringer, "A model predictive control method with adaptive weighting factors for enhancing performance of modular multilevel converters," *IEEE J. Emerg. Sel. Topics Power Electron.*, vol. 12, no. 4, pp. 3887–3899, Aug 2024.

[18] Y. Yang, H. Wen, M. Fan, M. Xie, S. Peng, M. Norambuena, and J. Rodriguez, "Computation-efficient model predictive control with common-mode voltage elimination for five-level anpc converters," *IEEE Trans. Transport. Electrific.*, vol. 6, no. 3, pp. 970–984, Sept 2020.

[19] W. Wu, D. Wang, and L. Liu, "A multi-layer sequential model predictive control of three-phase two-leg seven-level t-type nested neutral point clamped converter without weighting factors," *IEEE Access*, vol. 7, pp. 162 735–162 746, 2019.

[20] Y. Yang, R. Chen, M. Fan, Y. Xiao, X. Zhang, M. Norambuena, and J. Rodriguez, "Improved model predictive current control for three-phase three-level converters with neutral-point voltage ripple and common mode voltage reduction," *IEEE Trans. Energy Convers.*, vol. 36, no. 4, pp. 3053–3062, Dec 2021.

[21] H. Lin, S. Niu, Y. Chen, W. Wang, X. Tang, P. Zhang, and Z. Yuan, "Three-stage duty cycle-based deadbeat predictive torque control for three-phase spmsms with cmv reduction," *IEEE Trans. Power Electron.*, vol. 38, no. 9, pp. 11 385–11 398, Sept 2023.

[22] Y. Zhang and W. Xie, "Low complexity model predictive control-single vector-based approach," *IEEE Trans. Power Electron.*, vol. 29, no. 10, pp. 5532–5541, Oct 2014.

[23] N. S. P. Musunuru and S. Srirama, "Cascaded predictive control of a single power supply-driven four-level open-end winding induction motor drive without weighting factors," *IEEE J. Emerg. Sel. Topics Power Electron.*, vol. 9, no. 3, pp. 2858–2867, Jun 2021.

[24] U. R. Muduli, A. R. Beig, R. K. Behera, K. A. Jaafari, and J. Y. Alsawalhi, "Predictive control with battery power sharing scheme for

dual open-end-winding induction motor based four-wheel drive electric vehicle," *IEEE Trans. Ind. Electron.*, vol. 69, no. 6, pp. 5557–5568, Jun 2022.

[25] D. Prajapati, A. Dekka, D. Ronanki, and J. Rodriguez, "High-performance sequential model predictive control of a four-level inverter for electric transportation applications," *IEEE J. Emerg. Sel. Topics Ind. Electron.*, vol. 5, no. 1, pp. 253–262, Jan 2024.

[26] H. Le, A. Dekka, and D. Ronanki, "Model predictive control with inherent cmv reduction capability for multilevel inverters," *IEEE Trans. Ind. Electron.*, vol. 71, no. 7, pp. 6730–6737, 2024.

[27] Y. Wang, Y. Yang, S. Chen, R. Chen, J. Hu, W. Wu, Y. Wang, H. Yang, L. Zhou, and J. Rodriguez, "An improved model predictive voltage control with reduced computational burden for t-type three-phase three-level inverters," *IEEE Trans. Power Electron.*, vol. 39, no. 2, pp. 2115–2127, Feb 2024.

[28] D. Zhou, L. Ding, and Y. Li, "Two-stage optimization-based model predictive control of 5l-anpc converter-fed pmsm drives," *IEEE Trans. Ind. Electron.*, vol. 68, no. 5, pp. 3739–3749, 2021.

[29] Y. Li, F. Diao, and Y. Zhao, "A generic two-vector model predictive control for hybrid multilevel converters," *IEEE J. Emerg. Sel. Topics Power Electron.*, vol. 9, no. 5, pp. 6008–6018, Oct 2021.

[30] Y. Li and Y. Zhao, "A virtual space vector model predictive control for a seven-level hybrid multilevel converter," *IEEE Trans. Power Electron.*, vol. 36, no. 3, pp. 3396–3407, March 2021.

[31] Z. Ni, A. H. Abuelnaga, Y. Pan, A. Elezab, O. Zayed, M. Narimani, and J. Rodriguez, "A new mpc formulation based on suboptimal voltage vectors for multilevel inverters," *IEEE J. Emerg. Sel. Topics Power Electron.*, vol. 10, no. 6, pp. 7261–7270, Dec 2022.

[32] M. S. R. Saeed, W. Song, B. Yu, A.-R. Youssef, E. E. M. Mohamed, and A. I. M. Ali, "Computationally efficient pre-selection-based model predictive control for five-phase three-level drives," *IEEE Trans. Transport. Electrific.*, pp. 1–1, 2024.

[33] M. L. Parvathy and V. K. Thippiripati, "A simplified voltage vector preselection-based multivector predictive current control for improved torque performance of pmsm drive," *IEEE Trans. Power Electron.*, vol. 38, no. 7, pp. 8775–8785, Jul 2023.

[34] Y. Yang, J. Pan, H. Wen, Z. Zhang, Z. Ke, and L. Xu, "Double-vector model predictive control for single-phase five-level actively clamped converters," *IEEE Trans. Transport. Electrific.*, vol. 5, no. 4, pp. 1202–1213, Dec 2019.

[35] X. Li, Z. Xue, L. Zhang, and W. Hua, "A low-complexity three-vector-based model predictive torque control for spmsm," *IEEE Trans. Power Electron.*, vol. 36, no. 11, pp. 13 002–13 012, Nov 2021.

[36] I. Harbi, M. Ahmed, C. M. Hackl, J. Rodriguez, R. Kennel, and M. Abdelrahem, "Low-complexity dual-vector model predictive control for single-phase nine-level anpc-based converter," *IEEE Trans. Power Electron.*, vol. 38, no. 3, pp. 2956–2971, Mar 2023.

[37] W. Hua, F. Chen, W. Huang, G. Zhang, W. Wang, and W. Xia, "Multivector-based model predictive control with geometric solution of a five-phase flux-switching permanent magnet motor," *IEEE Trans. Ind. Electron.*, vol. 67, no. 12, pp. 10 035–10 045, Dec 2020.

[38] H. Le, A. Dekka, D. Ronanki, and J. Rodriguez, "Predictive current control of a new four-level voltage source inverter," in *2022 IEEE International Conference on Power Electronics, Drives and Energy Systems (PEDES)*, 2022, pp. 1–5.

AUTHOR INDEX

Abarzadeh, Mostafa 1261
Abbas, Asad 2973
Abotaleb, Youssef 1850
Abrams, Kerry J. 1781
Abramson, Rose A. 291, 2805
Abu-Rub, Omar 3071
Abu-Zaher, Mustafa 2327
Acero, Jesús 2468
Addin, Ali Sharaf 2960
Adeli, Mohammad Hassan 1489
Ademane, Harsha 3133
Adisurja, Ananda Tjakra 1255
Adragna, Claudio 958
Afrasiabi, Seyedeh Nazanin 1279
Afridi, Khurram K. 1640, 1646
Agarwal, Anant 2986
Ahammed, Md Tanvir 2220
Ahmad, Faheem 175
Aider, Youssef 1026
Aiello, Natale 738
Ajmal, Aidha Muhammad 3024
Akamatsu, Keiji 1728
Akter, Tanzila 1844, 2407
Akuta, Hector 761
Alam, Md Didarul 1746
Alam, Muhammad Muneeb 1051, 2569
Alassi, Abdulrahman 3071
Alathamneh, Mohammad 1953
Al-Durra, Ahmed 622, 2871, 3064
Alenezi, Ali 1217
Alexander, Mark 2162
Aleyasin, Seyed Hossein 1408
Ali, Abdelrahman 429
Ali, Jana A. Sheikh 3071
Ali, Kawsar 1529
Alkhatib, Mohamed 1940
Allen, Mark G. 1791
Allgeier, Jan 1919
Allioua, Abdelmoumin 2125
Alou, Pedro 1197
Al-Smadi, Mohammad K. 2779, 2840
Altin, Necmi 1489
Álvarez, Ignacio 3109
Aly, Mokhtar 746, 895, 2290, 2327
Alzahrani, Ahmad 1230
Alzate, Cesar 401
Amano, Yoshiki 3096
Amarathunga, Supun 3030

Amirabadi, Mahshid 1465, 1983
Amitkumar, K. S. 1279
Amler, Adrian 1759, 1767
Amor, Yacine Ayachi 1781
An, Jongchan 3000
Anand, Aniket 1096, 3147
Anand, Sandeep 69
Anantha, Neeraj 1121
Andapally, Bharadwaj Reddy 3119
Andersen, Michael A. E. 246
Anderson, Blake 1850
Ando, Masato 2681
Anekal, Latha 1224
Anjum, Waseah 1148
Antoszczuk, Pablo Daniel 479
Anurag, Anup 9, 442, 1318
Ao, Chengkang 171
Arai, Takamasa 2821
Araki, Hideo 3077
Aravind, G. 1610, 2785
Arduini, Douglas 1159
Asadi, Peyman 2162
Asel, Thaddeus J. 2419
Ashikaga, Toru 2284
Asllani, Besar 2051
Atkinson, Joshua 401
Attanasio, Rosario 3133, 3304
Attukadavil, Jenson Joseph C. 1481
Atwimah, Samuel K. 185, 207
Aunsborg, Thore Stig 175
Avenas, Yvan 1396, 2562
Aygun, Deniz 195
Azzopardi, Stéphane 2718
Bader, Samuel 1681
Bae, Jung-Soo 2228
Bae, Youngmin 3000
Baek, Jaeil 491
Bagci, F. Selin 880
Bahrami-Fard, Milad 930
Bak, Yeongsu 1734
Bakhshai, Alireza 3083
Balakrishnan, Manu 1286
Balamurali, Aiswarya 1096
Balda, Juan C. 27
Balen, Gleisson 1935
Balutto, Mattia 479
Banaie, Amin 1184
Banerjee, Arijit 3089

Bansal, Divyanshu 1610, 2785
Bantemits, Georgios 479
Bao, Mingjun 2628, 2968
Bao, Xiaokun 1143
Barbosa, Peter M. 2002
Barbosa, Peter 9, 442, 1318, 2082, 2296
Barik, Tapas 1184
Barros, Stayner Nóbrega 689
Barzegarkhoo, Reza 90
Basu, Arka 3253
Basu, Shibaji 464
Batard, Christophe 1076
Bau, Plinio 195
Bauer, Pavol 609
Bavi, Danial 385
Bazzi, Ali 2332, 2510
Beckemeyer, Randy 2082
Beig, Abdul R. 2647
Beinarys, Rytis 2009
Belanger, Matthew 2833
Belikov, Juri 1622
Belkhode, Satish 164, 3334
Benson, Mikayla 2413
Bergveld, H.J. 1451
Bertolini, Alessandro 2640
Beura, Kalpana 1940
Beushausen, Steffen 2589
Bezerra, Pedro A.M. 2361
Bhagat, Chinmay 3285, 3291
Bhambay, Rajul 2920
Bhattacharya, Subhashish 370, 552, 1347, 1866
Bhuse, Tejas 1, 54
Biadene, D. 2014
Bian, Fengwei 3312
Bien, Franklin 1629
Blaabjerg, Frede 696, 912, 1501
Blanco, Cristian 1935
Blaquière, Jean-Marc 2718
Blij, Nils Hans Van Der 479
Boby, Mathews 1279
Boisseau, Sébastien 828
Boisson, Guillaume Piquet 1396
Bojoi, Radu 472, 1408
Bolaños, Robert E. 1190
Boles, Jessica D. 1012
Bonanno, Giovanni 1666
Borowy, Bogdan S. 1326
Boroyevich, Dushan 2228
Bosch, Michael 2387
Boutet, Jérôme 828
Bracken, Christopher 544, 3119
Bradford, Paul 2393

Brandão, Danilo I. 1355
Brandão, Dener A. de L. 1355
Briz, P. 147
Briz, Pablo 3109
Brown, Alyssa 231
Brown III, Buck F. 1153
Brückner, Thomas 2960
Bruyere, Paul 2562
Bu, Jiankang 854
Bugade, Vikas 821
Burdío, José-Miguel 2468
Burgos, Rolando 111, 409, 1495, 1551, 2992
Burnett, Hunter 401
Burt, Graeme 3167
Buttay, Cyril 2051
Cairnie, Mark 2228, 2616
Calabretta, Michele 1070
Cammarata, Federica 252
Campbell, Steven 2797
Cao, Hanqing 1810
Cao, Hui 27
Cao, Yue 3036, 3048
Carretero, Claudio 2468
Castro, Alejandro 1427
Catanoso, Matthew 1791
Cattani, Alberto 2640
Cazzaniga, Daniele 958
Cazzitti, Sacha J. 1512
Cerutti, Stefano 738
Cervera, Pedro Alou 788
Cervone, Andrea 1305
Chae, Jongyoon 1899
Chagas, Rafael Bogo Portal 2446
Chakkalakkal, Sreejith 3147
Chakraborty, Shiladri 69
Chambon, Clément 828
Chamorro, Luis Ruiz 788
Chandrasekhar, Nurani 2889
Chang, Che-Wei 1495, 1551, 1564
Chang, Chuan-En 664
Chang, Jun-Yang 16
Chang, Yi-Chun 2143
Chareyron, Mathilde 828
Chatterjee, Bhaskar 1919
Chatterjee, Kallol 821
Chaturvedi, Shivam 2624, 3155
Chaturvedi, V. 1451
Chaudhary, Jai Aditya 3304
Chavarria, Jose 491
Chavez, Fredo 385
Cheema, Muhammad Ali Masood 768
Chellamuthu, Anand 1286

Chen, Cai 2343, 2369, 2426
Chen, Ching-Jan 664, 2131, 2143, 2725, 2735, 2741
Chen, Chun-Yen 938
Chen, Eric 1274
Chen, Guozhu 2059
Chen, Hao 906
Chen, Hongyu 2968
Chen, Hua 518
Chen, Hung-Chi 2687
Chen, Jiahong 1114
Chen, Jiann-Fuh 2043
Chen, Kai-Hui 887
Chen, Kevin J. 1047
Chen, Minjie 139, 349, 510, 566, 1274, 1693, 1882, 2438
Chen, Qiling 2375
Chen, Shih-Gang 938
Chen, Tianxiao 2361
Chen, Ting 2846
Chen, Wanjun 192
Chen, Xi 2628, 2968
Chen, Xingyu 1537, 1741
Chen, Yilun 2535
Cheng, Eric Ka-Wai 3227
Cheng, Jinpeng 1832
Cheng, Kuang-Yao 2157
Cheng, Lin 966, 1687
Cheng, Qi 2236
Cheng, Tzu-Ping 2900
Cheng, Yan 1047
Cheng, Yun-Keng 887
Cheshire, Audrey 682
Chetri, Chandan 3114
Chiu, Huang-Jen 389
Chiu, Jui-Yang 16, 900
Cho, Jaeyong 3187
Choi, Beomseok 491
Choi, Dongho 2311
Choi, Dongmin 1899
Choi, Jinsoo 3187
Choi, Jungwon 1874
Choi, Seokwon 2268
Choi, Seungdeog 943, 951, 1026, 1420, 1858
Choi, Sunghyuk 1659
Choi, Sungjin 3006
Choksi, Kushan 2582
Choo, Vin Loong 1576
Choong, Yin Quen 505
Chowdhury, Vikram Roy 645, 761, 1465, 3059
Chuang, Cheng-Ta 664, 2725
Chung, Henry Shu-Hung 98, 1507, 1582
Ciabattoni, Matteo 1646
Ciardo, S. Yuri 252

Clark, Landon 919
Cobos, Álvaro 1427
Cobos, José A. 1427
Coday, Samantha 971, 2249
Collings, William M. 185
Cong, Yizhou 2986
Contreras-Barrios, René 629
Coomans, Bart 195
Corradini, Luca 1, 54, 334, 2764
Costa, Levy F. 1334, 1341
Costinett, Daniel 3253, 3267
Cox, James 538
Cronin, Jared 2865
Croston, José Andrés Aguilar 2051
Crovetti, Paolo Stefano 738
Cruz, Alfonso 860
Cruz, Mario F. 670
Cui, Hongchang 202
Cui, Wen Tao 1108
Cui, Yujia 2932
Curbow, Austin 1167
D'Amato, Davide 689
Da-Cunha-Alves, Wendell 429
Dai, Hang 3174
Dang, Yongliang 278, 2482
Dannehl, Kai 1774
Dardeer, Mostafa 906
Darvish, Peyman 2453
Das, Shuvangkar Chandra 1184
Datta, Kishalay 1715
Datta, Promit 586
Davari, S. Alireza 2290
De, Vivek 518, 1681
Deboi, Brian 1167
Deboy, Gerald 1444, 2260
Defaz, Samuel 2576, 2582
Dekka, Apparao 2647
Delmar, Aria 1242
Deneke, Niklas 848
Deng, Jianting 3312
Deniz, Erkan 1489
Deppe, Conner 2393
Derbey, Alexis 2562
Desai, Nachiket 1681
Descamps, Anne-Sophie 1076
Deshpande, Ankit Vivek 1459
Dev, Archit 2920
DeVoto, Douglas 1824
Diao, Naizhe 757
Dieckerhoff, Sibylle 2361
DiMarino, Christina 586, 1836, 2228, 2616
Ding, Peiyang 2375

Ding, Wenlong 2713
Divan, Deepak 164, 3334
Do, Huong 491, 1681
Dobakhshari, Sina Salehi 1673
Dominguez, Miguel Alvarez 1640
Dong, Minhai 2075
Dong, ... 111, 1495, 1551, 1564, 2992
Driesen, Johan 3124
Driussi, Francesco 479
Dryden, Daniel M. 2419
Du, Bangli 436, 2752
Duan, Bin 2713
Dujic, Drazen 266, 1063, 1305
Dutta, Soham 711
Dworakowski, Piotr 2051
Eguchi, Shinichiro 2828
Ekuewa, Oluwaseun Isaiah 2973
Elasser, Youssef 510, 566, 2438
Elezab, Ahmed 670
El-Fouly, Tarek H.M. 622
Ellis, Nathan M. 2276
Ellis, Philip 1781
El-Refaie, Ayman M. 1551
El-Refaie, Ayman 1230, 1495
El-Saadany, Ehab F. 622
Elsanabary, Ahmed 746
Elshaer, Mohamed 3155
Emadi, Ali 670, 3147
Endo, Shun 2681
Eni, Emanuel 2746
Enjeti, Prasad 727, 1217, 1459, 3054
Enomoto, Jun 3194
Enslin, Johan 1153
Espinar, Alberto 3100
Espinoza, Angel 214
Estrin, Julia 132
Etta, Dheeraj 1640, 1646
Evzelman, Michael 594
Expósito, Alberto Delgado 788, 1803
Fahimi, Babak 930, 3160
Fahmy, Youssef A. 272
Falkenberg, Niklas 2772
Fan, Junchong 1203
Fan, Yucheng 2981
Farantatos, Evangelos 1184
Farivar, Glen G. 1927
Fassi, Youssof 828
Fein, Martin 2348
Feng, Hao 1832
Feng, Kaiyuan 2894
Feng, Wenda 3174
Fernandes, Arnold 1311

Fernandes, Baylon G. 1481
Ferrari, Maximiliano 637
Figueroa, Alejandro 1427
Filho, Braz de J.C. 1355, 1615
Fiore, Michele 1070
Flannery, John 285
Flaten, Paul 682
Forouzesh, Mojtaba 1673, 1892
Forsyth, Andrew J. 1512
Foster, Geoffrey M. 207
Fox, Aidan P. 185
Fox, Matthew 1791
Francés, Airán 868, 3298
Francois, Thomas W. 1311
Frank, Simon 2348
Freeman, Andrew 1242
Fu, Minfan 809, 2846, 3206
Fu, Pengyu 1203, 2986
Fujisaki, Keisuke 1797
Fujita, Jun 1383
Funaki, Tsuyoshi 2813
Funatani, Kenji 2654
Funatsu, Shohei 1237
Furukawa, Akihiko 1383
Gaafar, Mahmoud A. 775, 906, 2327
Gajare, Siddhesh 214
Galamb, Andrew 2527
Gallage, Nirashi Polwaththa 874
Gangadhar, Pratheesh 2920
Gao, Alex 2149
Gao, Ju 171, 225
Gao, Mingze 2992
Gao, Xiang 2846
Gao, Xiaoguang 2070
Gao, Yuan 524, 1034
Gao, Yuntian 278, 2482
Garcia, Enrique 538
García, Pablo 1935
Garcia, Ricardo 214
García, Sofía 3298
García-Espinosa, Antoni 1774
Garza-Arias, Enrique 1459
Gasparini, Alessandro 2640
Gato, Jose 3119
Gautam, Sushanta 185
Gauthier, Jean-Yves 2051
Gauttam, Gaureej 3316
Geboers, Tim 436
Gennaro, Francesco 738
Georgescu, Sorin 180
Georgiev, Daniel G. 185, 207
Gessner, Joerg 1889

Ghanayem, Haneen	1953
Ghartemani, Masoud Karimi	943
Ghitelman, Kolman Puterman	2101
Ghosh, Mohendro Kumar	1326
Ghosh, Prosenjit	2541
Ghosh, Subarto Kumar	1420
Giardine, Francesca	151
Gil, Pablo M.	1701
Ginot, Nicolas	1076
Giuffrida, Simone	472
Gockel, Hendrik	2125
Goetz, Stefan M.	1754, 2846, 3206
Goicoechea, Javier	1427
Gomez-Rivera, Luis F.	1774
Gong, Jiakun	219
Gong, Minxiang	518
Gong, Taehyeon	3006
Gong, Xiaowu	1114
Gonzalez, Reynaldo S.	1190
Gonzalez-Castaño, Catalina	629
Goodrich, Dakota	719
Goto, Akiko	1569
Gouy, Louison	1076
Graber, Lukas	860, 3321
Grainger, Brandon	544, 1326
Green, Andrew J.	2419
Griepentrog, Gerd	2125
Grigoryan, Davit	566, 1882, 2438
Groon, Fabian	90
Guan, Quanxue	895
Guenther, Robert	1203
Guichon, Jean-Michel	2562
Guillod, Thomas	1816
Gunawardena, Pasan	3030
Guo, Heng	2713
Guo, Jiacheng	2375
Guo, Weisheng	3181
Guo, Xiaoqiang	757
Guo, Zhengchen	2703
Guo, Zhongyin	2070
Gurudiwan, Shubhangi	719, 2194
Guthrie, Travis	2162
Gutierrez, Harold	1159
Ha, Jung-Ik	457, 1659, 2268, 2937
Habibolahi, Zahra Sadat	2202
Haddadi, Aboutaleb	1184
Hajisadeghian, Hossein	1666
Halawa, Ali	1473
Hamani, Rachid	1889
Hameed, Aamna Nasir	1673
Hameed, Asad	1972
Han, Yi	103

Hanhart, Michael	2757
Hanna, Rachelle	1396
Hansen, Frederik Lillebæk	2380
Hanson, Alex J.	231, 1121, 2521
Hanson, Alex	77, 2857
Hao, Weijia	2109
Harbi, Ibrahim	895, 2290
Haryani, Nidhi	442
Hasan, Abu Shahir Md Khalid	1844, 2407
Hasan, Md Zakir	1026
Hasan, Syed Imam	1294, 2698
Hassan, Alaaeldien	2327
Hassan, Najam Ul	834
Hassan, Nazmul	1746
Hata, Katsuhiro	1084, 1102, 2284, 2551
Hayashi, Tetsuya	423
He, Bill	3129
He, Binghui	1673
He, JiangBiao	919, 1368
He, Jiayin	171, 225
He, Junlei	3129
He, Xinlong	2066
Heckel, Thomas	1759
Hedenik, Marina	1519
Hedeshi, Hamid Montazeri	2202
Hegde, Anantha	1728
Heinen, Stefan	2757
Heiries, Vincent	828
Heldwein, Marcelo Lobo	2446
Hemming, Samuel	670
Heo, Go Woon	1723
Herbert, Edward	2495
Hernandez, Arturo Sanchez	530
Herzer, Stefan	1286
Higashiyama, Koji	1728
Hiller, Marc	1919, 2348
Hiraki, Eiji	321, 2654
Hiraoka, Toshio	285
Hirase, Yuko	1946
Hisamochi, Hirofumi	1414
Hobart, Karl D.	185, 207
Hoene, Eckart	2361
Hokmabad, Hossein Nourollahi	1622
Hong, Kang	2096
Hontz, Micheal R.	207
Horibe, Masahiro	2821
Hornbuckle, Malachi	363, 2241
Horowitz, Logan	151, 2276
Hosani, Khalifa Al	1940, 2871, 3064
Hossain, Md Maksudul	2407
Hossain, Mohammad Safayet	1184
Hou, Ting	2375

Hou, Zhengming 1913, 2851
Houska, Brad 3334
Howell, Brandon... 2162
Hsieh, Chun-Yu.. 2735
Hsieh, Hsin-Che... 815
Hsu, Jun-Ming 938
Hu, Borong 1439, 2597
Hu, Changsheng... 2894
Hu, Changyu 3129
Hu, Jhih-Cheng.. 2692
Hu, Jiangang 2932
Hu, Shoudong 2764
Huang, Alex Q. 1786
Huang, Cheng 505, 1946
Huang, Hao-Ran 664, 2131
Huang, Ming-Shi 938, 2692
Huang, PengHao 1217
Huang, Peng-Hao 727
Huang, Qinghui.................................... 1173, 2603
Huber, J. 2014
Huber, Jonas 1318
Huber, Laszlo... 442
Hudgins, Jerry L. 2877
Hudgins, Jerry.. 1850
Huh, Kum-Kang 3174
Hui, Shu Yuen Ron 3275
Hung, Chien-Chih.................................... 16, 900
Hung, Yu-Ting 2735
Huo, Zhenguo 2660
Husain, Iqbal.. 1746
Husev, Oleksandr 1622, 2173
Hussain, Amir 1990
Hwang, Yun Seong.. 733
Iannuzzo, Francesco................................ 738, 1070
Ibáñez-Muñoz, Esteban 629
Ibrahim, Ahmed 775
Ibrahim, Eltaib Abdeen D. 775
Ibrahim, Hasan..................................... 727, 3054
Ibrahim, Mohamed 670
Ide, Tomoya 1946
Ikriannikov, Alexandr 2149
Iliæ, Milan 2764
Ilka, Reza.. 1368
Imaeda, Yuta 2431
Imaoka, Jun... 2431
Imperiali, Luc 1318
Inokuchi, Seiichiro.. 2356
Inoue, Shuntaro... 782
Irie, Yusuke... 2828
Ishido, Ryosuke 1797
Ishihara, Masataka 321, 2654
Ishikura, Yuki.................................... 3285, 3291

Ishizuka, Yoichi...2828
Ishraq, Naveed 34, 1135
Islam, Md Khurshedul.............................943, 951
Islam, Md Majharul......................................2407
Islam, Nasherul 2059
Islam, Sarwar 1824
Ismail, Ahmed H. 2453
Isobe, Takanori..1946
Ito, Yuki..3248
Itoh, Jun-Ichi 21, 48, 2913
Ivimey, Arjun 464
Iwabuchi, Akio 2828
Iwamoto, Motomitsu.....................................1108
Iyer, Rahul K. 157
Iyer, Vignesh 2764
Jacobs, Alan G 207
Jafarian, Yousefreza......................................3083
Jahns, Thomas 3174
Jain, Akshat 658
Jain, Praveen 464, 616, 2953, 3083
Jalakas, Tanel 1622
Jalalabadi, Esmaeil 416
Janabi, Ameer 2597
Jayalath, Sampath 3212
Jeong, Seogyong 834
Jeong, Won Hyo...2937
Jerez, Raiphy 2249
Jha, Kunal 1519
Ji, Shengchang 278, 2482
Ji, Shiqi 2857
Ji, Yichao 966, 1687
Ji, Yingfeng 2889
Jia, Xiaoting 1564
Jiang, C.Q..795, 3181
Jiang, W.L..1451
Jiang, Wei 2343, 2426
Jiang, Xi 1114
Jiang, Yang 978
Jiang, Yongbin...3220
Jiao, Dong 1913, 2851
Jiao, Yang 416
Jin, Feng 429
Jin, Liyang..1020, 1564
Jin, Sicong 3018
Jin, Zhiyang.. 860
Jing, Mengmeng...2713
Jo, Hyeonu...3200
Jo, Hyunkyeong 1629
Jochmans, Thomas 258
Johnson, Brian...711
Johnson, Ken...2510
Jørgensen, Asger Bjørn 357, 1034

Jørgensen, Jannick Kjær	175, 357
Joshi, Kishor	943
Juds, Mark A.	1326, 3119
Jung, Jee-Hoon	834
Jung, Jun-Hyung	689
Jurkov, Alexander	124, 132
Kabashima, Takamune	1728
Kachura, Avram	449, 1905
Kai, Toshihiro	423
Kalathy, Abirami	616, 2953
Kallfass, Ingmar	2387, 3241
Kamalapur, Aakash	2228
Kamran,	252
Kanakri, Haitham	2029
Kanathipan, Kajanan	768
Kandeel, Youssef	285
Kang, Byeong-Woo	2948
Kang, Doug	180
Kang, Eunjin	3012
Kang, Gyeong-Gu	566, 2438
Kang, Seung Hyun	733
Kang, Yong	2066, 2343, 2369, 2426
Kano, Yuko	782
Kanungo, Gautam Dey	821
Kar, Narayan C.	1096
Karanth, Shashank	2746
Karimi-Ghartemani, Masoud	1858
Kataoka, Soya	1237
Katsura, Kenshiro	1299
Kaufmann, Maik	1286
Kawahara, Chihiro	2356
Kawamoto, Keisuke	1569
Kawano, Akihiro	1977
Kelkar, Kapil	1519
Kennel, Ralph	895
Kerekes, Tamás	738, 3042
Khaburi, Davood Arab	895
Khadka, Purushottam	1040, 2400
Khalid, Saad	2569
Khalife, Khalil	479
Khan, Faisal	1824
Khan, N.	1451
Khan, Nisar Ahmed	2569
Khan, Shahid Aziz	2624, 3155
Khandelwal, Sourabh	385
Khandla, Dhaval	2920
Khanna, Mudit	854
Khanna, Raghav	185, 207
Khatua, Mausamjeet	1681
Khorasani, Ramin Rahimzadeh	2101
Kim, Byeong-Il	1734
Kim, Chae-Lyn	3200

Kim, Daehyun	3187
Kim, Dong Hwan	1723
Kim, Dongmin	1899
Kim, Han-Gyu	951
Kim, Hongrae	1746
Kim, Hyeon Soo	733
Kim, Jae-Seong	925
Kim, Jaewon	727, 3054
Kim, Jeonghun	761
Kim, Jonghoon	2973, 3000, 3006, 3012
Kim, Jong-Hun	834
Kim, Joon-Seok	1734
Kim, Jungho	1629
Kim, Katherine A.	880
Kim, Minhyeok	3012
Kim, Min-Sik	834
Kim, Myeong-Ho	834
Kim, Namwon	703
Kim, Sung-Oh	2943
Kim, Yura	3006
Kimball, Jonathan W.	1311
Kimpara, Renata	703
Kirtley, James L.	2474
Kisacikoglu, Mithat John	1602
Kishikawa, Ryoko	2821
Kishimoto, Sumiaki	285
Kitano, Junichi	3194
Klidbari, Mohammadreza Khodaparast	2202
Klymenko, Mariia	1590
Knapp, Jeffrey	854
Knappstein, Lukas	2772
Knoll, Jack	2228, 2616
Ko, Bomyeong	3006
Kobayashi, Takumi	3248
Koch, Dominik	2387
Koehler, Andrew D.	185, 207
Koga, Shunsaku	3194
Koga, Takahiro	2828
Kokkonda, Raj Kumar	1347
Kolar, J.W.	2014
Kolar, Johann W.	1318
Kolli, Nithin	1347
Komiyama, Yutaro	3248
Komo, Hideo	2356
Kondo, Hiroki	1102
Kondo, Ryota	2813
Kong, Jiaze	2167
Kong, Jie	2380
Kong, Rui	696
Konishi, Akihiro	3248
Koppolu, Manoj	2920
Korrani, Majid Ghasemi	930, 3160

Kosaka, Takashi 1237, 3096
Koseoglou, Sokratis 479
Kotani, Junichi 579
Kouro, Samir 775, 2327
Kozak, Joseph P. 1211
Kozielski, Kyle 3147
Kragl, Robert 1051
Krishnamoorthy, Harish S. 3316
Krishnamurthy, Harish 1681
Krishnamurthy, Karthik 3129
Krishnan, Sahana 151, 291, 2805
Kritprajun, Paychuda 1184
Ku, Han 900
Kubulus, Pawel Piotr 1034
Kularatna, Nihal 378, 874
Kularatna-Abeywardana, Dulsha 874
Kulasekaran, Siddharth 491
Kumar, Misha 2002, 2082
Kumar, Pavan 530, 3312
Kusaka, Keisuke 3261
Kusunoki, Shigeru 2551
Kutrolli, Uiliam 2332
Kwak, Jin Woong 2541
Kwon, Hyukjae 566, 2438
Kwon, Man Jae 733
Ladhar, Manraj Singh 2322
Laha, Arpan 616, 2953
Lahuerta, Óscar 2468
Lai, Jih-Sheng 815, 1058, 1913, 2851
Lai, Rixin 2138
Lai, Yanwen 1173, 2603
Laird, Ian 2088
Lam, John 768, 2022
Lamar, Diego G. 1701, 1959
Lawniczak, Celine 1129
Lawson, Wayne 1403, 1781
Lazzarin, Telles Brunelli 342
Le, Duc Dung 3155
Le, DucDung 2624
Le, Hoang 2647
Le, Thanh-Long 2718
Leary, Alex M. 2516
Lee, Bonyoung 1629
Lee, Byoung Kuk 733, 1723, 3200
Lee, Byunghun 834
Lee, Chen-Chan 1058
Lee, Dongcheol 3000
Lee, Dong-Choon 1267
Lee, Dongsu 457
Lee, Eun Woo 2311
Lee, Hoi 2236
Lee, Jaea 3012

Lee, Jaehyeong 3006
Lee, Ju-A 3200
Lee, Jun Young 2311
Lee, June-Seok 1734, 2311
Lee, Justin 2138
Lee, Juwon 457
Lee, Kahyun 2937
Lee, Kangbeen 2413
Lee, Kevin 327, 1261, 2907
Lee, Kyo-Beum 925, 2317, 2943, 2948
Lee, Kyungmin 2547, 3281
Lee, Miyoung 3000
Lee, Po-Chang 900
Lee, Seongkyu 3012
Lee, Seunghyun 3012
Lee, Sungjun 3006
Lee, Taewoo 1659
Lee, Ting-Lun 2143
Lee, Wen-Hsuan 2043
Lee, Woongkul 1473, 2413
Lee, Yun-Jin 2317
Lehman, Brad 761, 1465, 1983
Lehmeier, Thomas 1767
Lei, Weihao 1143
Lei, Yiming 3312
Leslie, Alec 401
Leyrer, Thomas 2920
Li, Bing 2932
Li, Chun-I 2741
Li, Duo 307
Li, Haoran 510, 566, 1882, 2438
Li, Heyuan 809, 2846, 3206
Li, Hui 1248, 2075
Li, Jiajun 1590
Li, Lingyun 524
Li, Peidong 3036, 3048
Li, Pengwei 2332
Li, Qiang 202, 299, 429, 498, 1433, 1537, 1557,
.... 1741, 2228, 2488
Li, Ruqi 1159
Li, Sichao 3129
Li, Tien-Sheng 111
Li, Xiang 3312
Li, Xiaoling 1824
Li, Xindong 3212
Li, Xinze 1143
Li, Xuewen 751
Li, Yang 2576
Li, Yanqiao 1590
Li, Yaohua 3220
Li, Yi 1153
Li, Yilei 2035

Li, Yiming 1173
Li, Yuan 761, 1465
Li, Yunwei 3030
Li, Zehui 485
Li, Zhenchao 1305
Lian, Zhina 1090
Liang, Gaowen 1927
Liang, Jingyuan 1108
Liang, Katherine 363, 2241
Liang, Tsorng-Juu 887
Liang, Yaogan 1084
Liao, Hong-Xuan 2692
Liao, Hsuan 2043
Liao, Kuo Fu 2043
Liao, Mian 139, 349, 1882
Libbos, Elie 3089
Lim, Gyu Cheol 2937
Lim, Je-Yeong 1723
Lim, Jong-Hun 1723
Lin, Fanfan 1143
Lin, Jesse 1211
Lin, Jinshu 2075
Lin, Lei 2535
Lin, Qing 409
Lin, River 1159
Lin, Wei-Ren 258
Linares, Daniel Ríos 1375
Liserre, Marco 90, 118, 689, 1148
Liske, Andreas 2348
Liu, Baihan 2343, 2426
Liu, Caifeng 2066
Liu, Chen 3042
Liu, Chien-Lung 2692
Liu, Ching-Yao 1058
Liu, Christopher 1403
Liu, Chun-Hung 1026
Liu, Gao 357, 1034
Liu, Hanbing 3232
Liu, Haoyang 2361
Liu, Hong 2634
Liu, Hongru 2675
Liu, Hualong 1363, 1597
Liu, Jia 751
Liu, Jiahong 3042
Liu, Jiaxin 2343, 2369, 2426
Liu, Jinjie 1114
Liu, Jinjun 751
Liu, Kevin 1274
Liu, Liming 1746
Liu, Ming 2521
Liu, Sijia 2369, 2426
Liu, Wen-Chin B. 315

Liu, Wentao 1090
Liu, Xiaosen 1544, 2556, 2675
Liu, Xiaoshan 429
Liu, Y. 1451
Liu, Yan-Fei 1673, 1892
Liu, Yang 2675, 3321
Liu, Yifu 3129
Liu, Yongjie 1501, 3042
Liu, Yu-Chen 2179
Liu, Yunting 1179
Liu, Zeguo 1687
Liu, Zengyang 2488
Liu, Zhan 2521
Liu, Zhanlei 278, 2482
Liu, Ziheng 171, 225
Locher, Fabrice 2495
Locke, William 3141
Lodge, Finlay 3167
Logi, Sean 880
Long, Haihong 651, 2981
Long, Teng 1439, 2597
Loparo, Kenneth A. 2698
Lope, Ignacio 2468
López, Abraham 1959
Lopez-Torres, Carlos 1774
Lu, Che-Yu 1967, 2900
Lu, Fengwang 98
Lu, Guo-Quan 586, 2228
Lu, Lucas 416
Lu, Mowei 1754, 2846, 3206
Lu, Wei 2117
Luan, Shaokang 1034
Lucía, O. 147
Lucía, Óscar 3109
Luckett, Benjamin 919
Luise, Claudio 2640
Lukic, Srdjan 2527
Lumod, Phen 1159
Luo, Fang 2576, 2582
Luo, Tianming 2035
Lv, Jianwei 2343, 2369, 2426
Ma, D. Brian 2541
Ma, Dingkun 2375
Ma, Guangji 2070
Ma, Hangxiao 978
Ma, Tianlu 3181
Ma, Zhedong 1173, 2603
Ma, Zhiyuan 1786
Ma, Zhuxuan 27
Maaz, Syed Mohammad 1267
Mabuchi, Yuuichi 2681
MacFadyen, Martin 3167

Madadi, Mehrnaz...370
Maddela, Avinash...1715
Maekawa, Sari...2924
Mahbub, S. Tahmid..157
Maheshwari, Anuj...3089
Maji, Sounak..1640
Major, Joshua...1824
Mak, Pui-In..978
Maksimoviæ, Dragan............... 1, 54, 334, 682, 2764
Malannino, Claudia..252
Mallik, Ayan 34, 1135
Mallik, Ranajay..658
Mandrile, Fabio..472
Manjrekar, Madhav..2883
Mannan, Tahmid Ibne 1420, 1858
Manos, Konstantinos 1274, 1693
Mansour, Mahmoud...719
Mantooth, H. Alan 1844, 2407
Manzoni, Stefano..958
Marcault, Emmanuel..1396
Marellapudi, Aniruddh 164, 3334
Marianne, Julien...828
Marin, Brandon...491
Marquardt, Rainer...2960
Martin, Alexander 1211
Martin, Sébastien...828
Martin, Trent..1, 54
Martinez, Wilmar................. 238, 258, 436, 2167, 2752, 3124
Martinez-Limia, Alberto 1051
Martins, João R.R.O. 1889
Martins, Rui P. 978
März, Martin...1759, 1767
Mather, Barry...645, 3059
Mathieu, Frédéric..2495
Mathúna, Cian Ó...285
Matiushkin, Oleksandr 1622, 2173
Matsumori, Hiroaki 1237, 3096
Matsumoto, Hirokazu 579
Matsumoto, Yohei..2681
Matsuo, Takayoshi...2932
Mattavelli, P. 2014
Mattavelli, Paolo..2667
Maureira-Riquelme, Ángel 629
Mauromicale, Giuseppe......................................1070
Mavencamp, Dan..2157
Mazariegos, Pablo..1427
Mazzer, Simone 1444, 2254, 2260
McDonald, Brent 42
McGrew, Tyler 1557
Mekhilef, Saad..746
Mendes, Arthur...2992
Mercier, Patrick P. 315

Metwly, Mohamed Y. 919
Meyer, Stefan...1034
Miao, Honglei...2059
Michelis, Stefano 479
Milivojeviæ, Nikola...1, 54
Min, Hao...1090
Min, Hyungki..1629
Min, Run...2109
Minato, Yuichiro 2913
Mirafzal, Behrooz...2461
Mirkoviæ, Nikola...788
Mishima, Taichi..3248
Mishra, Santanu K..2213
Mitcheson, Paul D..1653
Mitrovic, Vladimir 409
Mitsui, Koji...1299
Miyamae, Masaki...2681
Miyanjou, Kazuki..1977
Miyazaki, Tatsuya..1797
Mo, Liping 795
Mo, Xianghao...1375
Mohammad, Mostak..1635
Mohammadi, Sajjad..2474
Mohseni, Parham 2173
Moniruzzaman, Md.....................943, 951, 1420
Montejano, Misael 637
Monticone, Francesco 1646
Montoya-Acevedo, Diego 629
Moon, Gun-Woo..1899
Moon, Jinyeong..1473, 2413
Moorthy, Radha Sree Krishna 2797
Morris, Lauryn 1311
Moschopoulos, Gerry...1972
Motoori, Shuichiro...1977
Motto, Eric...2356
Mou, Di..2857
Mou, Shin...2419
Mounesi, Reza..1791
Moursi, Mohamed Shawky El...................2871, 3064
Mousavi, Mahdi S...2290
Mu, Qiang 1388, 2790, 3328
Mu, Wei..2597
Mu, Xuchu 978
Muduli, Utkal Ranjan...................1940, 2871, 3064
Mueller, Lukas 538
Muenz, Ulrich 1184
Mühlethaler, Jonas 2495
Mujica, Gabriel..868, 3298
Mukhopadhyay, Anwesha....................................3267
Mukunoki, Yasushige..2356
Müller, Kilian...3241
Mulumudi, Guru Abhilash....................................1135

Munk-Nielsen, Stig 175, 357, 1034
Murakami, Haruhiko 1569
Muravleva, Ekaterina 1850
Murillo-Yarce, Duberney 1959
Murray, Samantha K. 1905
Murukesan, Karthick 180
Muscat, Isaac 449
Musolino, Francesco 738
Mustakin, Zaheen 1388, 2790
Na, Woonki 2973, 3000, 3006, 3012
Nabila, Kashfia Tajmim 2877
Nabizadah, Ahmad 3160
Nag, Kumar Joy 990, 997
Nagahara, Teruaki 1569
Nagai, Yoshiyuki 423
Nagano, Masanori 285
Nagar, Anshul 2973
Nagasawa, Shinobu 2610
Nagayoshi, Kenichi 1102, 3096
Nakagaki, Akito 2654
Nakagawa, Shigeki 1797
Nakamura, Hirokazu 1728
Nakamura, Keisuke 1237
Nakano, Satoshi 2551
Nakashima, Junichi 2356
Nakata, Yosuke 1383
Nakata, Yuki 21, 2913
Nam, David 2992
Namadmalan, Alireza 2474
Namba, Akira 1797
Namburi, Krishna 1294
Naradhipa, Adhistira M. 498
Narasimhan, Sneha 1866
Narumanchi, Sreekant 1824
Nasiri, Adel 1489, 1791
Nassaji, Abolfazl 2290
Nassar, Rajaie 586, 2228
Nations, Mark 552
Naval, Sourav 1012
Navarro-Rodríguez, Ángel 1935
Neal, Adam T. 2419
Nelms, R.M. 1953, 2703
Nelson, Blake 395, 1167
Nelson, Tolen M. 207
Nelson, Tolen 185
Ng, Wai Tung 983, 1108
Ngo, Khai D. T. 2228
Ngo, Khai 586
Ngo, Minh 111
Nguyen, Allen T. 840
Nguyen, Calvin 1274
Nguyen, Duy T. 231, 1121

Nguyen, Hien 1, 54
Nguyen, Kien 3248
Nguyen, Tung-Tan 389
Ni, Chuan 2117
Nielsen, Morten Rahr 357
Nikmaram, Behnam 2290
Ning, Guangdong 809
Ning, Guangfu 2096
Ning, Shangxian 2660
Nishijima, Kimihiro 1977
Nishimura, Keigo 3096
Nishio, Haruhiko 1108
Nishizawa, Shin-Ichi 2551
Nitta, Honami 1797
Noesges, Brenton A. 2419
Noguchi, Koichiro 1569
Noh, Young-Seok 518
Norman, Patrick 3167
Notake, Koki 1299, 1414
Núñez, Guillermo 1197
Nuzzo, Jeremy 2387
O'Driscoll, Seamus 285, 2009
Oberdieck, Karl 1051, 2589
Oboreh-Snapps, Oroghene 1311
Ochiai, Yuki 3261
Ohi, Toshi 2821
Ohno, Takashi 21
Ohodnicki, Paul R. 544, 2516, 3119
Ohodnicki, Paul 370, 1326
Okamoto, Takahiro 321
Olalla, David 3100
Olimmah, Marshal 395
Onar, Omer C. 1635
Onishi, Hiroyuki 2431
Onuma, Naoto 2681
Opificius, Julian 401
Orabi, Mohamed 775, 906, 2327
Orlando, Tailan 342
Orr, Allison 1211
Oruganti, V.S.R.Varaprasad 801
Ota, Hiroaki 3248
Ou, Shuyu 1501, 3042
Ouyang, Ziwei 246, 252, 1810
Pahlevani, Majid 616, 2953
Pakala, Sriharsh 505
Palani, Praveenkumar 62
Pallantla, Manikanta 2708
Palmal, Manas 1874
Pan, Ci 192
Pan, Qishan 2207
Panja, Pijush Kanti 821
Paplham, Tyler W. 2516

Parashar, Sanket 1347
Paredes-Camacho, Alejandro 1774
Park, Junhyeong 3187
Park, Sung-Bum 3187
Parkhideh, Babak 1388, 2790
Parreiras, Thiago M. 1355, 1615
Pasupuleti, Sai Sushma 3316
Patle, Nagesh 2805
Paul, Sayan 1, 54, 334
Paulino, Glaucio H. 1274
Pavone, Mario Giuseppe 738
Peña-Alzola, Rafael 1935, 3167
Peng, Fang Z. 761
Peng, Hongjie 171, 225
Peng, Xiaochuan 1090
Penof, David 1519
Pereira, Joao 637
Pereira, Lucas 1388, 2790
Pereira, Thiago Antonio 118
Peretz, Mor Mordechai 594
Pérez, Fernando 868, 3298
Pérez, Sara 1197
Perez-Farre, Quirc 1774
Perreault, David J. 132
Perreault, David 124
Petriæ, Ivan Z. 157
Petriæ, Ivan 2764
Petrillo, Gaia 266
Petucco, Andrea 2667
Pfost, Martin 573, 1129, 1576, 2772
Philippe, Antoine 1396
Phukan, Ripun 2082, 2296
Phung, Thanh Hai 195
Picot-Digoix, Mathis 2718
Piel, Joshua J. 2419
Pietrini, Giorgio 670
Pigott, J. 1451
Pilawa-Podgurski, Robert C. N. 151, 157, 291, 558,
.... 2276, 2805
Pillonnet, Gaël 315
Pirson, Nicolas 258
Pizzuto, Matteo 1096
Plum, Thomas 1919
Pong, Man-Hay 389
Pool-Mazun, Erick 1459
Popoviæ, Zoya 682
Porras, David A. 27
Porter, Matthew 1020, 1564
Pou, Josep 1927
Pourjafar, Saeid 2173
Prabhakar, Siva 69
Pradhan, Rachit 670

Prakash, Surya 1940
Preindl, Matthias 272, 1255
Prodiæ, Aleksandar 307, 990, 997
Punjabi, Shobhana 1159
Qahouq, Jaber A. Abu 2779, 2840
Qi, Nianzun 357, 1034
Qian, Ting 2117
Qian, Yijie 524
Qiblawey, Yazan 3071
Qin, Yuan 1564
Qin, Zian 609
Qiu, Tian 1040, 2400
Queiroz, Samuel S. 1334, 1341
Quenette, Vincent 1889
Rabenold, Elizabeth 2249
Radhakrishnan, Kaladhar 491, 1681
Radici, Christian 1403, 1512, 1781
Rafiq, Aamir 395
Rahman, Md Rashedur 943, 951
Rahman, Mohammad Dehan 1844, 2407
Rahouma, Ahmed 27
Rajagopal, Narayanan 1836
Rajpurohit, Chirayu 2764
Raju, Soniya 378
Rallabandi, Vandana 1635, 3174
Ram, Achala 2920
Ramasubramanian, Deepak 1184
Ramirez, Juan 1211
Ramkumar, S. 2708
Ramos, Gabriel V. 1355, 1615
Ramos, Regina 1197, 1375
Ran, Li 1832
Rana, Dilip 1040
Rana, Mandeep S. 2213
Rao, Yifan 1274
Rashid, Syed Saeed 1640, 1646
Rathore, Vikas Kumar 594
Raval, Vishwam 727, 3054
Ravichandran, Krishnan 1681
Rawat, Shubham 1347
Raychowdhury, Arijit 518
Reddy, Narsimha 3054
Redondo, Alejandro 868, 3298
Reinotas, Jurgis 1754
Ren, Linhao 2343
Ren, Sheng 3181
Ren, Xufu 1439
Restrepo, Carlos 629
Rettner, Cornelius 1759
Richardeau, Frédéric 2718
Rikiishi, Yasuhiro 2284
Ripamonti, Giacomo 479

Ristic-Smith, Aleksandar 1529
Rivas-Davila, Juan 363, 2241
Rizkalla, Maher 2029
Rizzolatti, Roberto 1444, 2254, 2260
Roberts, Gianluca 307
Rodgers, Aidan 1242
Rodriguez, Ezequiel Ramos 1927
Rodriguez, Fernando 3100
Rodriguez, José 746, 895, 2290, 2327
Rodríguez, Juan 1701, 1959
Rogers, Daniel 1529
Rogers, Michael 1569, 2356
Ronanki, Deepak 2647
Rong, Mingzhe 3220
Rong, Zhenshuai 1439
Rosa, Bruno M.G. 1653
Round, Simon 1803
Roy, Soham 1121
Rubinic, Jaksa 416
Rueß, Manuel 3241
Ruiz, Juan M. 2002
Ruiz, Juan 442, 2082, 2296
Ruoff, Dominik Alexander 2589
Ruppert, Daniel 1759
Russo, Andrea 252
Ruszczyk, Adam 1803
Sa, Satyam 103, 449
Saberi, Sajad 2840
Sadasivan, Arya 2461
Sadilek, Tomas 401
Sado, Kerry 2833
Saeedifard, Maryam 2051, 3071, 3321
Saelens, Jonathan 1311
Saggini, Stefano 479, 1444, 2260
Saha, Subrata 1237
Saha, Tarak 3174
Sahoo, Subham 696, 912, 1501
Sai, Ranajit 285
Saiga, Kazuma 2551
Saito, Shoji 1569
Saito, Wataru 2551
Sakai, Hiroto 3077
Salari, Omid 3083
Salehi, Maryam 2883
Samanta, Akash 3141
Sambo, Haifah B. 291
Sandoval, Rolando 1459
Sangwongwanich, Ariya 738, 1501, 3042
Sanjakdar, Omar 1396, 2562
Santi, Enrico 2833, 2865
Santos, Ion Leandro Dos 342
Santos Jr., Euzeli Cipriano Dos 2029

Sanusi, Bima Nugraha 246, 1810, 2035
Saraf, Pushkar 77
Sarajian, Ali 895
Sarda, Radhika 62, 1927
Sarlioglu, Bulent 3174
Sarnago, H. 147
Sarnago, Héctor 3109
Sarofim, Seif 449
Sati, Shraf Eldin 622
Sato, Yuji 1383
Sato, Yuki 579
Satterlee, Ryan 3133
Satyamsetti, Vijayakrishna 1403
Sauter, Bailey 2764
Sayed-Ahmed, Ahmed 2932
Sba, Baher Abu 2453
Sbabo, Paolo 2667
Schanen, Jean-Luc 2562
Scheideler, William 1590
Scherer, Yohannes Amilcar Tekle 342
Sebastián, Javier 1959
Sebata, Kohei 1977
Sekiya, Hiroo 3248
Selvarasu, Uthandi 761, 1465
Sen, Paresh C. 1892
Sen, Tanuj 139, 349, 1882
Sengstock, Jonathan 1242
Sengupta, Arkadeb 90, 118, 1148
Seo, Gab-Su 602, 645, 3059
Seo, Seoktae 1629
Sethupandi, Abishek 62
Seugnet, Léo 2718
Shadmand, Mohammad B. 3071
Shafei, Ahmad El 1326
Shah, Shreyas B. 670
Shahbazi, Reza 1179
Shahsavar, Tala Hemmati 1622
Shang, Shuye 3227
Shao, Hang 2138
Shao, Linbo 1020, 1564
Sharma, Mohit 3141
Shen, Andy 3129
Shen, Xiaobing 436, 2167
Shi, Guannan 1564
Shillaber, Luke 2597
Shimada, Takae 2681
Shimosako, Shumei 3077
Shin, Se-Un 834
Shivdikar, Saumil 2400
Shoji, Tomokazu 2821
Shrestha, Niranjan 801
Shu, Wenze 524

Siddiquee, Ashraf 1294, 1602
Silveira, Hector Bessa 342
Sim, Dong Hyeon 3200
Sim, Si Yuan 505
Singh, Anurag 1, 54, 334
Singh, Prashant 1026
Singla, Rishabh 3054
Siraj, Ahmed 1040, 2400
Sitta, Alessandro 1070
Smith, John 637
Solecki, Alex 1242
Solomentsev, Michael 77
Son, Gibong 1741
Song, Chen 2075
Song, Keqi 1507
Song, Minwoo 3012
Song, Qihao 202
Song, Xiaoqing 1844, 2407
Song, Yubo 696
Song, Zhihao 327
Sönmez, Ertuðrul 1051
Soundararajan, Soundhariya G. 238
Souri, Naser 2202
Sowers, Elizabeth A. 2419
Sozer, Yilmaz 1294, 1602, 2698
Spiazzi, Giorgio 2667
Spieler, Matthias 1495, 1551
Sridhar, Sundaramoorthy 299
Sriram, Vaisambhayana B. 62, 1927
Srivastava, Shubham 2213
Starke, Michael 637, 703
Stauth, Jason T. 1590, 1715
Steiner, Mark 2356
Stella, Fausto 1408
Steyaert, Bernard 1255
Steyn-Ross, Alistair 378, 874
Stillwell, Andrew 1242
Stokowski, Nicole 1242
Strache, Sebastian 2569
Strathman, Sophia A. 1311
Streit, Jochen 1051
Strezelecki, Ryszard 2173
Stricula, Justin 401
Sturdivant, Maurice 544
Su, Gui-Jia 1635
Su, Mei 2096
Sugie, Hisashi 2730
Sui, Qingcheng 436, 2752
Sukita, Yohei 1102
Sullivan, Charles R. 840, 1816
Sun, Bosheng 1990
Sun, Kai 2488

Sun, Lingwei 1108
Sun, Peiyuan 2375
Sun, Ruize 192
Sun, Weifeng 524
Sun, Xiuhu 3328
Sun, Zhen 3275
Sun, Ziang 2981
Sund, Jade 971
Sune, Joseph Benzaquen 164, 3334
Suntharalingam, Piranavan 670
Suzuki, Asamira 1728
Swaminathan, Madhavan 2101
Sweet, Mark 3167
Swoboda, Philipp 2348
Syed, Hadiuzzaman 1051
Szczublewski, Austin M. 185
Tadakuma, Toshiya 2610
Taguchi, Koichi 2356
Taha, Wesam 3147
Tajima, Shin 782
Takahashi, Keita 2610
Takahashi, Yoshiaki 1414
Takamiya, Makoto 1084, 1102, 2551
Takamura, Yota 1797
Takayama, Naoki 2681
Takeda, Ryo 2821
Takeuchi, Kosuke 21
Takeuchi, Toshiro 2828
Takishima, Kenta 423
Takizawa, Sota 2924
Tan, Matthew 1882
Tan, Siew-Chong 3227
Tanaka, Kenichiro 579
Tanaka, Ryota 423
Tanaka, Shinsaku 2284
Tanaka, Toshiyuki 2828
Tang, Ho-Tin 1507, 1582
Tang, Wenyuan 1363, 1597
Tang, Yi 3220
Tant, Mike 1824
Tariquzzaman, Md. 3036, 3048
Tarutani, Masayoshi 1383
Tatetsu, Riku 1977
Tayebi, Milad 854
Teng, Fei 2527
Teng, Yiyina 757
Terauchi, Naoya 285
Terzija, Vladimir 757
Thacker, Thimothy 2992
Then, Han Wui 1681
Thevar, Madasamy Palavesha 62
Thike, Rajendra 1279

Thirumoorthi, Sathya Rupan 1866
Thurlbeck, Alastair P. 1602
Tian, Fanghao 2167, 3124
Tian, Jiachen 2327
Tian, Xiaoyang 1754
Tingbari, Vincent Masabiar 2973
Tomey, Hala 1211
Tomioka, Shohei 579
Tong, Junhong 1786
Tong, Qiaoling 2109
Torres, Javier 1197
Torres, Renato Amorim 1495
Touhami, Mustapha 1012
Tran, Ngoc Ho 2569
Trescases, O. 1451
Trescases, Olivier 103, 449, 676, 1905
Tripathi, Anshuman 62, 1927
Tsai, Chieh-Ju 664, 2131, 2143, 2725, 2735, 2741
Tschanz, James 1681
Tseng, Chien-Hao 2725
Tsou, Ming-Chang 887
Tsuchida, Takayuki 285
Tuzizila, Jeremie 401
Uchida, Yasuo 48
Uddarraju, Praneeth 1311
Uddin, Muhammad Fasih 2453
Uegaki, Shin 1383
Uematsu, Takeshi 3248
Ulrich, Burkhard 1707, 2303, 2502
Umanand, L. 1610, 2785
Umar, Jamil 2973
Umar, Muhammad F. 3071
Umetani, Kazuhiro 321, 2654
Ursino, Mario 1444, 2254, 2260
Uzum, Alper 1294, 2698
Vagnon, Eric 2562
Vanderwegen, Wout 238
Varadarajan, Kamal 180
Vasiæ, Miroslav 788, 1375, 1803
Vedula, Inder 1, 54
Vergès, Gaël 1905
Vico, Enrico 1408
Vines, Peter 1403, 1512, 1781
Vinnac, Sébastien 2718
Vitale, Gianni 3133, 3304
Vohl, Kenny 2757
Wagner, Tomas 3100
Walters, Andrew 951
Wang, Cheng Feng 103, 449
Wang, Daming 2369
Wang, Haiyan 3312
Wang, Haoyu 485, 983, 1544, 2207, 2556, 2675, 2857

Wang, Hongjie 719, 2393
Wang, Huai 912, 2380, 3042
Wang, Jin 1203, 2986
Wang, Jinyan 171, 225
Wang, Jun 538, 1850, 2088
Wang, Kaiyuan 3227
Wang, Kejia 505
Wang, Kun 2088
Wang, Kunrong 2162
Wang, Laili 2375
Wang, Lei 2162
Wang, Liang 983
Wang, Libing 3174
Wang, Lichong 757
Wang, Linguo 2070
Wang, Lisheng 1248
Wang, Liwei 1153
Wang, Maojun 225
Wang, Meng 3312
Wang, Mengqi 2624, 3155
Wang, Pinhe 246, 2035
Wang, Qiong 498
Wang, Rudy 9, 1318
Wang, Rui 1063
Wang, Shaozhe 1459
Wang, Shukai 566, 1882, 2438
Wang, Shumeng 761, 1465
Wang, Shuo 1173, 2603
Wang, Sunqing 3018
Wang, Wei 2692
Wang, Xiao 219
Wang, Xiaohua 3220
Wang, Xiaosheng 795
Wang, Xiaoting 3030
Wang, Xiaoyu 416
Wang, Xinlin 809
Wang, Xiongfei 1615
Wang, Xuan 1791
Wang, Xuliang 1544, 2556, 2675
Wang, Yan 1544, 2556, 2675
Wang, Yao 3275
Wang, Yibo 795, 3181
Wang, Yicheng 3147
Wang, Yiju 1368
Wang, Yulei 219
Wang, Yunxin 2675
Wang, Zijian 1537
Wang, Ziyao 485
Wang, Zuoshuai 3018
Watabe, Kiyoto 2551
Watanabe, Hiroki 21, 48
Watanabe, Kenichi 1102, 3096

Wehr, Erik	2757
Wei, Anran	1983
Wei, Bo	327
Wei, Jinxiao	1832
Wei, Xing	3042
Wei, Xuanjing	1403
Wei, Yuxin	2713
Weihs, Leon	2757
Weiser, Mathias C.J.	2387, 3241
Weng, Sheldon	1681
Wens, Mike	195
Wheeler, Patrick	895
Wicht, Bernhard	848
Wick, Lukas	2380
Williamson, Sheldon	801, 1224, 2322, 3114, 3141
Wilson, Marcus	378
Winkler, Joseph	848
Wipprecht, Lukas	2303
Wojewoda, Leigh	491
Wong, Andy	2833
Wouters, Hans	238, 258, 2167
Wright, Jason	401
Wu, Alan	307
Wu, Chih-Chiang	2687
Wu, Hsiang-Kai	2687
Wu, Shang-Syun	2179
Wu, Taotao	1090
Wu, Teng	2634, 2660
Wu, Tsai-Fu	16, 900
Wu, Xin	651, 2981
Wu, Xinke	1995
Wu, Yang	139, 349
Wu, Yanqing	1995
Wu, Yingzhe	1248
Wu, Yue	1114, 3220
Wu, Yuxuan	2582
Wunderlich, Andrew	2865
Wunderlich, Ralf	2757
Xi, Zichen	1020
Xia, Xiaoyi	2022
Xiang, Zhangwei	429
Xiao, Junjie	609
Xie, Biyun	919
Xu, Dehong	651, 2894, 2981
Xu, Guo	2096
Xu, Haoran	2109
Xu, Huangsheng	2207
Xu, Limei	2075
Xu, Shen	524
Xu, Wentao	1012
Xu, Wenzhe	2426
Xu, Xinmiao	1433

Xu, Yun	1114
Xu, Ziyang	2521
Xue, Hui	2117
Xue, Yuxiang	1248
Yabuta, Shigenori	2821
Yagielski, John	3174
Yamaguchi, Koji	1299, 1414
Yamamoto, Keisuke	3194
Yamamoto, Masayoshi	2431
Yamanaka, Kimito	1797
Yan, Decheng	3334
Yan, Yiyang	2343, 2426
Yan, Zhaoheng	1114
Yan, Zhixing	357, 1034
Yang, Bowen	602
Yang, Garam	3000
Yang, Hélène T.W. Ma	983
Yang, Juchen	1203
Yang, Liu	2369
Yang, Qichen	860
Yang, Qiuzhe	1537
Yang, Robert	180
Yang, Xin	1020, 1564
Yang, Xingyu	1953
Yang, Xinliang	409
Yang, Yirui	1173, 2603
Yang, Yongheng	3024
Yang, Yun	3227, 3275
Yang, Zineng	1020
Yao, Wenxi	327
Yao, Yuzhou	1203
Yasko, Mohamed	3124
Yato, Shinji	3077
Ye, Liang	285
Ye, Zhengyu	2117
Yeo, Howe Li	62
Yeo, Sungku	2547, 3281
Yi, Lifang	2413
Yi, Zheyuan	2488
Yin, Shan	1248, 2075
Yin, Tianxiang	2535
Yoneyama, Rei	2356
Yoshimoto, Kantaro	423
Yoshimura, Yuto	2654
You, Longxiang	3018
Youssef, Mohamed Z.	3083
Yu, Hao	2343, 2426
Yu, Jingshu	1681
Yu, Ruiyang	854
Yu, Sheng-Han	2131
Yu, Sheng-Yang	42, 1990
Yu, Wensong	2220

Yu, Xiang 1892
Yuan, Hao 2117
Yuan, Huan 3220
Yuan, Jiaqi 670
Yuan, Jingyi 966
Yuan, Song 1114
Yuan, Tianlong 429
Yuan, Tianshu 2375
Yun, Dam 1659, 2268
Zaabi, Omar Al 1940
Zade, Aditya 719, 1004, 2194
Zaitsu, Toshiyuki 1977, 2654
Zaizen, Shohei 2551
Zaman, Mohammad Shawkat 676
Zan, Xin 3232
Zane, Regan 719, 1004, 2194
Zeineldin, Hatem H. 622
Zekorn, Tobias 2757
Zeng, Hank 2138
Zeng, Jia-En 1967
Zeng, Wenliang 510
Zeng, Zheng 219
Zhan, Cao................................. 278, 2482
Zhang, Ben 3181
Zhang, Bing 2070
Zhang, Bo 192
Zhang, Bohua 573
Zhang, Boran 2675
Zhang, Boyi 1850, 2296
Zhang, Cheng 1512, 3212
Zhang, Chenghui 2713
Zhang, Chi 9
Zhang, Desheng 2109
Zhang, Fuxing 2059
Zhang, Haijin 3312
Zhang, Hely 1286
Zhang, Heng 2369
Zhang, Hengbin 1248
Zhang, Hong 2375
Zhang, Honglang 2075
Zhang, Jiazheng 2628, 2968
Zhang, Jincheng 978
Zhang, Jinfeng 1439
Zhang, Li 2535
Zhang, Qingzheng 2894
Zhang, Renjie 3220
Zhang, Shengke 214
Zhang, Shiqi 757
Zhang, Tianyi 2066
Zhang, Weihang 978
Zhang, Xiangrong 3275
Zhang, Xin 505, 1143, 3018

Zhang, Xiong 2343, 2426
Zhang, Yi 2380
Zhang, Yichi 2380
Zhang, Yifan 2343, 2369, 2426
Zhang, Yifu 2746
Zhang, Yingjie 2603
Zhang, Yuanxin 2857
Zhang, Yuhao 202, 1020, 1564
Zhang, Yuxin 2973
Zhang, Zhe 2907
Zhang, Zhenbin 2846, 3206
Zhang, Zheyu 1040, 1153, 2400
Zhang, Zhi Jin 3321
Zhang, Zhining 1203
Zhang, Zichen 1850
Zhao, Delin 3220
Zhao, Fangzhou 1615
Zhao, Hongbo 357, 1034
Zhao, Shuang 2369, 3227
Zhao, Shuofeng 1824
Zhao, Tiefu 1388, 2790, 3328
Zhao, Tuo 1274
Zhao, Wending 1995
Zhao, Yifan 2521, 2846, 3206
Zhao, Yue 27, 2453
Zheng, Zexiang 2343, 2369
Zhou, Daniel H. 1693
Zhou, Daniel 566, 1274, 2438
Zhou, Dao 2380
Zhou, Fei 2541
Zhou, Feng 2624
Zhou, Jiale 1388, 2790, 3328
Zhou, Kunxiao 809
Zhou, Lufan 1803
Zhou, Mingde 2207
Zhou, Wenqi 2589
Zhou, Xigen 2157
Zhou, Yan 1767
Zhou, Yi 651
Zhou, Yuan 2535
Zhou, Yuebin 2369
Zhou, Zongjie 1047
Zhu, Jiaqi 3312
Zhu, Jinli 761, 1465
Zhu, Junjie 2070
Zhu, Lingyu 278, 2482
Zhu, Liyan 1020
Zhu, Yicheng 558, 2276
Zhu, Zhenhai 1995
Zhuo, Fang 2327
Zolfi, Pouya 1230
Zou, Huanghaohe 1786

Zou, Jiaao 2066
Zou, Jiarui 558, 2276, 2805
Zou, Mingrui 219
Zou, Xudong 2066
Zou, Xuecheng 2109
Zufferli, Kevin 1444, 2260
Zuo, Yu 258, 436, 2167, 2752
Zuo, Zhiling 2070
Zynger-Capaverde, Betina 2562

—

IEEE
445 Hoes Lane
Piscataway, NJ 08854-4141

ISBN 979-8-3315-1612-3

9 798331 516123